D0619272

Industrial Engineering
Handbook

ALJIAN · *Purchasing Handbook*
ARKIN · *Handbook of Sampling for Auditing and Accounting*
BAUMEISTER AND MARKS · *Standard Handbook for Mechanical Engineers*
BRADY · *Materials Handbook*
BUELL AND HEYEL · *Handbook of Modern Marketing*
CONDON AND ODISHAW · *Handbook of Physics*
CONOVER · *Grounds Maintenance Handbook*
DAVIDSON · *Handbook of Modern Accounting*
DICHTER · *Handbook of Consumer Motivations*
DUNN · *International Handbook of Advertising*
FACTORY MUTUAL ENGINEERING DIVISION · *Handbook of Industrial Loss Prevention*
FINK AND CARROLL · *Standard Handbook for Electrical Engineers*
GREENE · *Production and Inventory Control Handbook*
GREENWALD · *McGraw-Hill Dictionary of Modern Economics*
HEYEL · *The Foreman's Handbook*
HUSKEY AND KORN · *Computer Handbook*
IRESON · *Reliability Handbook*
JURAN · *Quality Control Handbook*
KLERER AND KORN · *Digital Computer User's Handbook*
KORN AND KORN · *Mathematical Handbook for Scientists and Engineers*
LANGE · *Handbook of Chemistry*
LASSER · *Business Management Handbook*
LASSER · *Standard Handbook for Accountants*
MAGILL, HOLDEN, AND ACKLEY · *Air Pollution Handbook*
MANAS · *National Plumbing Code Handbook*
MARKUS · *Electronics and Nucleonics Dictionary*
MAYNARD · *Handbook of Business Administration*
MAYNARD · *Handbook of Modern Manufacturing Management*
MAYNARD · *Top Management Handbook*
MELCHER AND LARRICK · *Printing and Promotion Handbook*
MERRITT · *Building Construction Handbook*
MERRITT · *Standard Handbook for Civil Engineers*
MORROW · *Maintenance Engineering Handbook*
O'BRIEN · *Scheduling Handbook*
PERRY · *Engineering Manual*
ROSSNAGEL · *Handbook of Rigging*
SCHMIDT · *Construction Lending Guide*
SOCIETY OF MANUFACTURING ENGINEERS · *Manufacturing Planning and Estimating Handbook*
STANIAR · *Plant Engineering Handbook*
STANLEY · *Handbook of International Marketing*
STEPHENSON · *Handbook of Public Relations*
STETKA · *NFPA Handbook of the National Electrical Code*
YODER, HENEMAN, TURNBULL, AND STONE · *Handbook of Personnel Management and Labor Relations*

Industrial Engineering Handbook

H. B. MAYNARD *editor-in-chief*

President, Maynard Research Council Incorporated
Pittsburgh, Pennsylvania

STATE BOARD FOR TECHNICAL AND COMPREHENSIVE EDUCATION

Third Edition

McGRAW-HILL BOOK COMPANY

New York   St. Louis   San Francisco   Düsseldorf   Johannesburg
Kuala Lumpur   London   Mexico   Montreal   New Delhi
Panama   Rio de Janeiro   Singapore   Sydney   Toronto

INDUSTRIAL ENGINEERING HANDBOOK

07-041084-4

567890 COCO 75

This book was set in Caledonia by The Maple Press Company, and printed and bound by The Colonial Press Inc.
The editors were Harold B. Crawford, Daniel N. Fischel, and Stanley E. Redka. The designer was Naomi Auerbach.
Stephen J. Boldish supervised production.

To all industrial engineers who, by their
competent application of traditional industrial
engineering procedures and their readiness to
learn to use skillfully the newer techniques
and procedures which are constantly expanding
the effectiveness and fields of application of
industrial engineering, are contributing so
importantly to the ever-increasing usefulness
of their profession.

Contents

x   **Contents**

INDEX FOLLOWS SECTION 13.

Contributors

KIPLING ADAMS *Senior Consultant, H. B. Maynard and Company, Incorporated, Pittsburgh, Pennsylvania*

WILLIAM M. AIKEN *Senior Vice President, H. B. Maynard and Company, Incorporated, Pittsburgh, Pennsylvania*

ROY L. ALLEN *Manager, Management Systems, Columbus Division of North American Rockwell Corporation, Columbus, Ohio*

MRS. MARIA ALTSCHUL *Department of Operations Research, Case Western Reserve University, Cleveland, Ohio*

DR. CLINTON J. ANCKER, JR. *Chairman, Department of Industrial and Systems Engineering, University of Southern California, Los Angeles, California*

DR. CLIFTON A. ANDERSON *Head, Department of Industrial Engineering, North Carolina State University, Raleigh, North Carolina*

WILLIAM ANTIS *Technical Director, Maynard Research Council Incorporated, Pittsburgh, Pennsylvania*

PROFESSOR JAMES M. APPLE *Professor of Industrial and Systems Engineering, Georgia Institute of Technology, Atlanta, Georgia*

GUY J. BACCI *General Supervisor—Industrial Engineering, International Harvester Company, Hinsdale, Illinois*

A. J. BERGFELD *President, Case and Company, Inc., New York, New York*

PROFESSOR JOHN E. BIEGEL *Associate Professor, Department of Industrial Engineering, Syracuse University, Syracuse, New York*

EARL K. BOWMAN *U.S. Department of Agriculture, University of Florida, Gainesville, Florida*

DR. C. L. BRISLEY *Professor and Associate Chairman, Department of Engineering, University Extension, The University of Wisconsin, Milwaukee, Wisconsin*

PHILIP F. CANNON *Director, Coloney, Cannon, Main & Pursell, Inc., New York, New York*

MRS. RITA M. CARLSON *Executive Assistant, Maynard Research Council Incorporated, Pittsburgh, Pennsylvania*

FRANK J. CARR *Director, Informations Systems Laboratory, Westinghouse Electric Corporation, Pittsburgh, Pennsylvania*

DAVID J. CHESLER   *Program Director, Navy Training Research Laboratory, Naval Personnel Research Activity, San Diego, California*

JAMES J. CHILDS   *James J. Childs Associates, Alexandria, Virginia*

RICHARD R. CONARROE   *President, Walden Public Relations, Inc., New York, New York*

H. L. DAVIS   *Supervisor, Human Factors Group, Industrial Engineering Division, Kodak Park Division, Eastman Kodak Company, Rochester, New York*

DR. BURTON V. DEAN   *Chairman, Department of Operations Research, Case Western Reserve University, Cleveland, Ohio*

PROFESSOR J. WAYNE DEEGAN   *Chairman of Industrial Engineering, The University of Iowa, Iowa City, Iowa*

DR. MORRIS H. DeGROOT   *Head, Department of Statistics, Carnegie-Mellon University, Pittsburgh, Pennsylvania*

DR. RALPH L. DISNEY   *Professor, Department of Industrial Engineering, The University of Michigan, Ann Arbor, Michigan*

JAMES H. DUNCAN   *President, The Science Management Corporation, Moorestown, New Jersey; and Chairman of the Board, The Wofac Company, Moorestown, New Jersey*

PROFESSOR SALAH E. ELMAGHRABY   *Professor and Chairman, Operations Research Committee, North Carolina State University, Raleigh, North Carolina*

CARLOS FALLON   *Manager, Value Analysis, RCA Corporate Staff, Camden, New Jersey; and Vice President, Professional Development, Society of American Value Engineers*

MITCHELL FEIN   *Professional Engineer, Hillsdale, New Jersey*

DR. JOHN R. FREEMAN   *Director, Systems Development, The Medicus Corporation, Dallas, Texas*

DUANE C. GEITGEY   *Director of Training Program Development, Maynard Research Council Incorporated, Pittsburgh, Pennsylvania*

GEORGE H. GUSTAT   *Director, Industrial Engineering Division, Eastman Kodak Company, Rochester, New York*

ROSS W. HAMMOND   *Chief, Industrial Development Division, Georgia Institute of Technology, Atlanta, Georgia*

DR. WALTON M. HANCOCK   *Professor, Department of Industrial Engineering, The University of Michigan, Ann Arbor, Michigan*

JOHN W. HANNON   *Executive Vice President, Maynard Research Council Incorporated, Pittsburgh, Pennsylvania*

GÖRAN HEDBERG   *AB Svenska MEC, Gothenburg, Sweden*

JOHN H. HILDENBIDDLE, JR.   *Director, Industrial Engineering, Penn Central Company, Philadelphia, Pennsylvania*

PROFESSOR LAWRENCE S. HILL   *Professor of Science Management, California State College, Los Angeles, California*

WILLIAM K. HODSON   *President, H. B. Maynard and Company, Incorporated, Pittsburgh, Pennsylvania*

R. P. HOELSCHER   *Professor Emeritus, General Engineering Department, University of Illinois, Urbana, Illinois*

DR. A. G. HOLZMAN   *Professor and Chairman, Department of Industrial Engineering, Systems Management Engineering, and Operations Research, University of Pittsburgh, Pittsburgh, Pennsylvania*

WALKER T. HOWELL   *Corporate Manager—Industrial Engineering and Cost Control, The Bendix Corporation, Southfield, Michigan*

DR. DALE JONES   *President, Uniquest, Albuquerque, New Mexico*

DR. **LAWRENCE L. KAVANAU** *President, Systems Associates, Inc., Long Beach, California*

A. D. **KIDD** *Principal, Case & Company, Inc., New York, New York*

KERRY KILPATRICK *Research Assistant, Department of Industrial Engineering, The University of Michigan, Ann Arbor, Michigan*

WARREN D. KNAPP *Manager, Armaflex & Expanded Products, Millroom & Service Operations, Armstrong Cork Company, Braintree, Massachusetts*

KENNETH KNOTT *Managing Director, Maynard Training Centre, Birmingham, England*

JERRY KOVACH *Vice President—Management Operations, Research for Management Science, Inc., Appleton, Wisconsin*

RICHARD H. LEUKART *Secretary, National Screw and Manufacturing Company, Cleveland, Ohio*

EDMUND J. McCORMICK *Chief Executive Officer and Chairman, McCormick & Company, Yonkers, New York*

RUSSELL W. McDONALD *Director, Maynard Training Center, Cleveland, Ohio*

PROFESSOR PAUL E. MACHOVINA *Late Professor of Engineering Drawing, The Ohio State University, Columbus, Ohio*

LOWRIE W. McINTOSH *Weston, Connecticut (formerly Vice President, The Northern Trust Company, Chicago, Illinois)*

DR. **RALPH A. MAGGIO** *Associate Professor, Department of Industrial Engineering, Systems Management Engineering, and Operations Research, University of Pittsburgh, Pittsburgh, Pennsylvania*

DONALD G. MALCOLM *Director, Western Operations, Research Analysis Corporation, Los Angeles, California*

JAMES A. MALCOLM, JR. *Vice President and Director of Research, The Wofac Company, Moorestown, New Jersey*

JOHN C. MARTIN *Staff Assistant for Industrial Engineering, Headquarters Manufacturing Planning and Controls, Westinghouse Electric Corporation, Pittsburgh, Pennsylvania*

MORLEY H. MATHEWSON *Director, Long Range Strategic Planning, American Management Association, Inc., New York, New York*

WILLIAM J. MATTERN *Counselor, Maynard Research Council Incorporated, Pittsburgh, Pennsylvania*

WILLIAM S. MAXWELL *Administrative Manager, Maynard Research Council Incorporated, Pittsburgh, Pennsylvania*

DR. **HAROLD B. MAYNARD** *President, Maynard Research Council Incorporated, Pittsburgh, Pennsylvania*

DR. **C. I. MILLER** *Physician, Medical Consultant—Human Factors Group, Kodak Park Division, Eastman Kodak Company, Rochester, New York*

ALLAN H. MOGENSEN *Founder and Director, Work Simplification Conferences, Lake Placid, New York*

JACKSON E. MORRIS *Long Beach, California*

JOHN H. MORRIS *Management Consultant, Westerville, Columbus, Ohio*

BRUNO A. MOSKI *Assistant to General Manager, Yale Materials Handling Division, Eaton Yale & Towne, Inc., Philadelphia, Pennsylvania*

PROFESSOR WILLIAM ROBERT MULLEE *Professor, Industrial Engineering, Loyola University, Los Angeles, California*

DR. **MARVIN E. MUNDEL** *M. E. Mundel and Associates, Silver Spring, Maryland*

RICHARD MUTHER *Richard Muther & Associates, Inc., Kansas City, Missouri*

DR. **M. SCOTT MYERS** *Management Research Consultant, Texas Instruments In-*

corporated, Dallas, Texas; and Visiting Professor of Organizational Psychology and Management, Sloan School of Management, Massachusetts Institute of Technology, Cambridge, Massachusetts

JOSEPH J. NARESKY *Chief, Reliability and Compatibility Division, Rome Air Development Center, United States Air Force, Air Force Systems Command, Griffiss Air Force Base, Rome, New York*

R. H. NEWELL *Manager, Systems Research, Goodyear Tire & Rubber Company, Akron, Ohio*

DR. PETER V. NORDEN *Program Administrator, Management Sciences, Data Systems Division, International Business Machines Corporation, White Plains, New York; and Professor of Industrial and Management Engineering, Columbia University, New York, New York*

ARNOLD OCKENE *Vice President, Simulation Associates, Inc., White Plains, New York*

P. D. O'DONNELL *Director, Headquarters Manufacturing Planning and Controls, Westinghouse Electric Corporation, Pittsburgh, Pennsylvania*

PROFESSOR JAY L. OTIS *Professor of Psychology; and Director, Psychological Research Services, Case Western Reserve University, Cleveland, Ohio*

DR. RICHARD G. PEARSON *Professor of Industrial Engineering and Psychology, North Carolina State University, Raleigh, North Carolina*

DAVID B. PORTER *Professor Emeritus, Department of Industrial Engineering and Operations Research, New York University, Bronx, New York*

RALPH PRESGRAVE *Professor Emeritus, School of Business, University of Toronto, Toronto, Canada*

JOSEPH H. QUICK *Senior Vice President, The Science Management Corporation, Moorestown, New Jersey; and Senior Vice President, The Wofac Company, Moorestown, New Jersey*

JAMES A. RICHARDSON *Supervisor, Industrial Engineering Division, Kodak Park Division, Eastman Kodak Company, Rochester, New York*

THOMAS S. RIORDAN *Director of Manufacturing, Fluid Power Division, The Bendix Corporation, Utica, New York*

J. N. SALAPATAS *Florida Power and Light Company, Miami, Florida*

NORMAN F. SCHNEIDEWIND *Systems Consultant, Pacific Palisades, California*

JOHN L. SCHWAB *John L. Schwab Associates, Fairfield, Connecticut*

HARRY T. SCHWAN *Executive Vice President, Daniel D. Howard Associates, Inc., Chicago, Illinois*

D. W. SCHWEPPE *Principal, Case and Company, Inc., New York, New York*

JOSEPH E. SCOTT *Vice President, Meylan Stopwatch Corporation, New York, New York*

SIR WALTER SCOTT *Governing Director, W. D. Scott & Co., Pty. Ltd., North Sydney, Australia*

DR. STANLEY J. SEIMER *Professor of Organization and Management, College of Business Administration, Syracuse University, Syracuse, New York*

CLIFFORD SELLIE *President, Standards International Inc., Chicago, Illinois*

DORIAN SHAININ *Vice President and Director of Reliability and Quality Control, Rath & Strong, Inc., Lexington, Massachusetts*

MISS ANNE G. SHAW *Chairman, The Anne Shaw Organisation, Ltd., Brook Lane, Alderly Edge, Cheshire, England*

TULLY SHELLEY, JR. *Principal, McKinsey & Company, Inc., New York, New York*

GERALD J. SKERRETT *Technical Editor, Eastman Kodak Company, Rochester, New York*

KENNETH J. SLEYMAN *Director of Management Systems Development, H. B. Maynard and Company, Incorporated, Pittsburgh, Pennsylvania*

DR. HAROLD E. SMALLEY *Regents' Professor of Industrial Engineering; and Director, Health Systems Research Center, Georgia Institute of Technology, Atlanta, Georgia*

PROFESSOR WILBERT STEFFY *Department of Industrial Engineering, The University of Michigan, Ann Arbor, Michigan*

G. J. STEGEMERTEN *Executive Advisor, Maynard Research Council Incorporated, Pittsburgh, Pennsylvania*

C. F. STEPHENSON *President, The John B. Adt Co., York, Pennsylvania*

ULF SVENSÉN *AB Svenska MEC, Gothenburg, Sweden*

PROFESSOR RALPH O. SWALM *Professor, Department of Industrial Engineering, Syracuse University, Syracuse, New York*

ROLF TIEFENTHAL *H. B. Maynard and Company AS, Copenhagen, Denmark*

KENDALL C. WHITE *Vice President, Manufacturing Services, TRW Inc., Cleveland, Ohio*

DR. WILLIAM W. WHITE *New York Scientific Center, Data Processing Division, International Business Machines Corporation, New York, New York; and Assistant Professor of Mathematical Methods and Operations Research, Columbia University, New York, New York*

JOHN J. WILKINSON *Vice President, H. B. Maynard and Company, Incorporated, Pittsburgh, Pennsylvania*

LEONARD C. YASEEN *Chairman, The Fantus Company, New York, New York*

FREDERICK H. YOUNG *Manager, Corporate Industrial Engineering, Stop and Shop, Inc., Boston, Massachusetts*

Preface

The field of industrial engineering has continued to expand as predicted since the publication of the first and second editions of this Handbook in 1956 and 1963. Originally concerned largely with the design of manufacturing plants, methods improvement, work measurement, the design and administration of wage payment systems, cost control, quality control, production control, and the like, industrial engineering activities have progressed in many ways. For convenience, these may be classified into the following four categories.

1. Modification and improvement of existing techniques
2. Development and application of new techniques and procedures
3. Expanded interest in other disciplines having a close relationship to industrial engineering
4. Increased application of industrial engineering outside the manufacturing activity area

As a function begins to mature, it is customary to find that the techniques which originally comprised it are constantly being modified, refined, and improved. Standard data, for example, have been used since the time of Frederick W. Taylor, yet the standard data practices of today are a far cry from those proposed and used by Taylor. MTM (methods time measurement) basic data, originally published in the 1940s, have changed very little since that time, but modifications of those data better to meet specific application requirements have produced several useful data sets of second and even third generation data. A discussion of these and many other important modifications and improvements of existing techniques is included in this third edition of the Handbook.

A large number of new techniques and procedures have been developed since 1963. New mathematical, statistical, and programming procedures have been made available for those industrial engineers who are able to master and apply them. Computers, systems design procedures, numerical control devices, and various other new developments are also now available for industrial engineering use. Whether a given development is considered to belong within the field of industrial engineering or to a related discipline depends on how industrial engineering is defined. The following definition of industrial engineering, adopted by the American Institute of Industrial Engineers in 1955, is broad enough to permit the inclusion of most of the new techniques and procedures just mentioned, as well as others discussed in this Handbook.

> Industrial engineering is concerned with the design, improvement, and installation of integrated systems of men, materials, and equipment. It draws upon specialized knowledge and skill in the mathematical, physical, and social sciences, together with the principles and methods of engineering analysis and design, to specify, predict, and evaluate the results to be obtained from such systems.

At the same time, advances have been made in other disciplines which, although not generally considered to belong in the field of industrial engineering, are nevertheless of great interest to the industrial engineer. Behavioral scientists, for example, have long been critical of the philosophies and activities of the industrial engineer, yet many industrial engineers are ready and even eager to use any findings of the behavioral scientists which they judge to be practical and useful in achieving their own industrial engineering goals.

Although industrial engineering was originally oriented largely to manufacturing applications, it is now becoming increasingly difficult to identify any major activities to which industrial engineering has not been applied successfully to some degree. Some of the more important of these applications are discussed in Section 13.

In developing the contents of the third edition of this Handbook, it was evident from the start that to include all the new material of interest to industrial engineers while retaining all the time proven discussions which have made the previous editions so useful would require some careful planning. There is a top limit to the size of a single-volume handbook which cannot be exceeded, and it was clear that careful evaluation of the usefulness to the readers of any contemplated chapters would have to be made if the Handbook was to cover the field of industrial engineering with necessary thoroughness. Accordingly, an Editorial Advisory Committee composed of experts in the more important areas of industrial engineering was invited to assist the Editor-in-Chief in planning the contents of the Handbook.

The Committee was composed of the following members:

DANIEL N. FISCHEL   *General Manager, Professional and Reference Book Division, McGraw-Hill Book Company, New York, New York*

LEE S. HARDING   *Director, Productivity Engineering Directorate, Office of the Assistant Secretary of Defense, Washington, D.C.*

WILLIAM K. HODSON   *President, H. B. Maynard and Company, Incorporated, Pittsburgh, Pennsylvania*

DR. A. G. HOLZMAN   *Professor and Chairman, Department of Industrial Engineering, Systems Management Engineering, and Operations Research, University of Pittsburgh, Pittsburgh, Pennsylvania*

J. F. JERICHO   *Executive Director, American Institute of Industrial Engineers, Inc., New York, New York*

PROFESSOR BERT H. NOREM   *Chairman, Department of Industrial Engineering, Syracuse University, Syracuse, New York*

P. D. O'DONNELL   *Director, Headquarters Manufacturing Planning and Controls, Westinghouse Electric Corporation, Pittsburgh, Pennsylvania*

JOHN B. STOYA   *Assistant Controller, Manufacturers Hanover Trust Company, New York, New York*

Thus the committee discussions were influenced by the viewpoints of practicing industrial engineers, teachers of industrial engineering, consultants, and users of industrial engineering in manufacturing, government, and clerical activity areas.

The customary target for a handbook revision is one-third new material, one-third revised material, and one-third old material. Due to the rapid expansion of the field of industrial engineering in recent years, the planned contents of this third edition of the *Industrial Engineering Handbook,* as developed in the several meetings of the Editorial Advisory Committee, differed considerably from this target. Of the 87 chapters, 29 are completely new, and 22—which discuss subjects included in previous editions—were written by new authors. Thus, in effect, 51 chapters, or 59 percent, of the Handbook present new material. Of the remaining chapters, 15 have undergone major revisions, while only 21 chapters present material little changed from the second edition.

Due to space limitations, much of the material which appeared in the first and second editions could not be retained in the third edition. Most of this material is just as useful and up to date as ever. Those who possess copies of the first or second editions of the Handbook will find it advisable to keep them, for they will be useful for reference purposes for some time to come.

As one examines the pages of this edition of the Handbook, it becomes evident that all procedures, both old and new, have as their objective the improvement of existing or planned activities. Improvement in most cases means increased productivity and reduced costs. In the inflationary times in which we are living today, increased productivity is perhaps the most effective way of maintaining economic health that exists. Every company that makes effective use of industrial engineering thus contributes to the soundness of the economy of the country in which it is located.

Some companies—particularly when business is bad and profits are reduced or nonexistent—still feel that they cannot afford an adequate staff of industrial engineers. A recent survey of the industrial engineering profession shows how shortsighted this viewpoint is. Some 2,000 industrial engineers were questioned regarding the amount of savings they were able to achieve each year. The answers revealed that an industrial engineer saves each year a median amount equal to about five times his salary. Thus if the salary of a typical industrial engineer is $15,000 a year, he may be expected to produce about $75,000 a year in savings. Can any company afford to neglect this proven opportunity for profit improvement—in good times or bad?

A handbook is a reference book of practical, how-to-do-it information. An industrial engineer with a problem will be able to turn to this Handbook and find the help he needs to solve that problem. But in addition to this, there is another important use which can and should be made. The field of industrial engineering has been growing so rapidly that all industrial engineers are faced with the problem of keeping up to date. Industrial engineers, for example, working largely in the area of methods improvement and work measurement need not only be familiar with the latest and most advanced practices in these areas, but they should also be aware of what is going on in other areas of industrial engineering. To grow with their profession, they will need to extend their interests and skills, and even a casual reading of the material included in this Handbook will help them to move ahead in the right direction.

Similarly, the industrial engineers who are working in the newer areas often tend to feel that the older, traditional industrial engineering techniques are so simple that little besides clerical skill is needed to apply them satisfactorily. A study of the chapters dealing with these established techniques will result in a more intelligent understanding of them and will hasten the day when all the procedures discussed in the Handbook will be accepted by the majority of industrial engineers as belonging within their profession. The rivalry which once existed between time study men and motion study advocates may seem strange today to those who have long regarded time and motion study as integral parts of the work study procedure, but similar divisive rivalries are to be found today between practitioners of the older and the newer techniques. With greater understanding of the contributions of each group to the overall success of the enterprise they both serve, these rivalries should begin to disappear. It is to be hoped that this Handbook will contribute to the achievement of this desirable state of affairs.

In addition to serving the needs of industrial engineers, it is to be hoped that the Handbook will be used by other management men whose work is related to or affected by the industrial engineer. In particular, top management will find it profitable to scan these pages to refresh understanding of what industrial engineering can do. Profitable operation must always be the goal of top management, and in the welter of decisions involving marketing, product development, computerized information systems, mergers and acquisitions, and other potentially profitable actions, top management should never lose its interest in and appreciation for the steady, day-after-day profit improvement contributions brought about by properly applied industrial engineering effort. Industrial engineers need the understanding support of top managers, and given this support, they can do much to boost the all-important earnings-per-share figure that top management and stockholders alike watch so intently.

Finally, it should be pointed out that management and industrial engineering students will find the Handbook of value in more than one way. Its use as a reference source is obvious. But in addition, by studying the table of contents and sampling the key sentences and paragraphs of each chapter, the student can obtain—in a way not possible from the in-depth study of specific techniques—an understanding of the place of industrial engineers in human affairs and the vital role industrial engineering plays in our society in increasing the productivity of the people working in almost every kind of enterprise.

Inevitably, the student will be led to the conclusion that, although industrial engineering is concerned primarily with the application of engineering, mathematical, statistical, and systematic if empirical approaches to all areas where objectives are accomplished through the efforts of people, at no time is the human factor overlooked. He will see that nearly every phase of industrial engineering involves people—both the people who do the work and the people who are affected by it. He will see that the industrial engineer recognizes clearly that he cannot be fully effective without the full cooperation of those with whom he works and will see how important the work of behavioral and social scientists can be to him.

This Handbook is the result of the cooperative efforts of a large number of people. The initial planning effort of the Editorial Advisory Committee has already been acknowledged. In addition, the Editor-in-Chief wishes to thank the 108 authors, individually and collectively, who have contributed so importantly by their practical, how-to-do-it discussions of the techniques and procedures used in their particular areas of interest. Special thanks are extended to Professor A. G. Holzman, who as a member of the Editorial Advisory Committee, author, and editorial assistant is largely responsible for Section 10, Mathematical, Statistical, and Programming Procedures.

And finally, the outstanding contributions of Mrs. Rita Carlson are acknowledged with gratitude. Not only was she tireless in dealing with schedules, correspondence, and proofreading activities, but she also contributed a chapter to the Handbook and prepared the Handbook index.

H. B. MAYNARD
Editor-in-Chief

The Industrial Engineering Function

Chapter 1

The History and Development of Industrial Engineering

ROSS W. HAMMOND

**Chief, Industrial Development Division,
Georgia Institute of Technology, Atlanta, Georgia**

What industrial engineering is today and aspires to be in the future is determined by what has gone before. Industrial engineering had its roots in the Industrial Revolution; it was nourished by individuals who sought to advance organization and management principles at an early date; it emerged as a separate discipline and was formalized in the late nineteenth and early twentieth centuries; and it achieved maturity after World War II.

The Industrial Revolution. The Industrial Revolution, beginning in the eighteenth and nineteenth centuries, resulted from a number of factors. The advent of new inventions, especially in the textile industry, such as the spinning jenny, spinning frame, and the power loom, was important. The invention of the steam engine permitted factories to break away from water as the principal source of power. Advances were made in metal cutting and the production of machine tools. These and economic considerations led to the beginnings of factories with relatively large numbers of workers and the decline of the home craftsman and the "cottage system."

With the advent of the factory came the beginnings of management and management thinking. Sir Richard Arkwright, the inventor of the spinning frame, is credited with devising and administering a successful code of factory discipline. Matthew Boulton and his partner, James Watt, Jr., were progressive managers of a foundry enterprise in the late 1700s. They pioneered many manufacturing innovations and developed a remarkable cost accounting and control system. Charles Babbage is credited with developing an early calculating machine, writing an early definitive work, *The Economy of Machinery and Manufactures,* and various analyses of manufacturing operations.

The activities of these individuals were carried on independently in England, with no attempt to develop a formal body of knowledge about management. It is noteworthy, though, that these individual attempts to systematize industrial management thinking stemmed from the same application of the "scientific method" which was revolutionizing the industrial world. The application of the scientific method of analysis, experimentation, and practical demonstration had been extended to the production of machine tools, more complicated processes, and better products. Now it was being extended to man's thinking on organization and management principles and methods. "Scientific management" as a professional approach was yet to come, waiting on the works of pioneers in the field.

The Development of the Scientific Management Concept. At the end of the nineteenth century and beginning of the twentieth century, a body of management knowledge began to emerge as a result of the work of a number of individuals in several countries, but primarily in the United States. The latter included Frederick Taylor, Henry Gantt, Frank and Lillian Gilbreth, and Harrington Emerson. The contributions of these individuals are discussed below. Indeed, the contributions of the individuals can scarcely be separated from the fields of study in which they made their contribution. Thus Taylor's name is associated with methods studies, among other activities. Gantt's name is associated with developing management principles and procedures and a humanistic approach. Frank Gilbreth is identified with motion study, along with his wife, who went on to adaptations of industrial engineering procedures to the home and similar environments, as well as to the psychological aspects of human endeavor. Harrington Emerson wrote and expounded on efficient operation and developed a bonus pay plan.

Taylor and his contemporaries conceived of their contributions and attempts to formulate basic principles as a scientific approach to management, and this activity soon became known as "scientific management." It was from these beginnings in areas of scientific thought, now generally described as organization, methods, and work measurement, that industrial engineering sprang.

The work of these pioneers was aided and encouraged by the American Society of Mechanical Engineers. The Society provided a forum for a discussion of the work of Taylor and his peers at a time when no management society existed. Then in 1912, the Society to Promote the Science of Management was organized; in 1915 it became the Taylor Society, and in 1934 it merged with a Society of Industrial Engineers to become the Society for Advancement of Management. The latter, as well as the American Management Association, provided additional forums for advanced management thinking and the advocation and testing of new techniques.

Evolution of the Term "Industrial Engineering." Unfortunately, the term "scientific management," which seemed appropriate to the evolving principles, soon developed a measure of disrepute. Management became disenchanted with it because of its use in certain court cases, especially the Eastern Rate Case. Organized labor resisted the concept as being detrimental to the best interests of labor.

The early practitioners of scientific management had, for the most part, engineering backgrounds, and many of them came to consider themselves "industrial engineers." Others attracted to the management field came to look upon themselves as management consultants or members of "management." As curricula evolved in institutions of higher learning, business administration or industrial management programs became common in business schools, while the engineering college counterpart which developed was first centered in the mechanical engineering curricula (as perhaps a natural consequence of the interest of the American Society of Mechanical Engineers in the field).

As the principle and methodology of industrial engineering became focused, separate industrial engineering curricula and degree programs began to appear. Ultimately, university and college industrial engineering enrollments became quite large, and the graduation of large numbers of industrial engineers into business and industry further identified the emerging profession and its potential. At the same time, the field of industrial engineering was broadening greatly and encompassing many new activities and applications. The result has been acceptance by business and industry of the term "industrial engineering," and an awareness of the importance of this function.

The Modern Definition of Industrial Engineering. The generally used definition of industrial engineering was developed in 1955, although the profession had existed for many years prior to that. The definition evolved from the thinking of industrial engineering leaders in the business, industrial, academic, consulting, and government fields.

> Industrial engineering is concerned with the design, improvement, and installation of integrated systems of men, materials, and equipment. It draws upon specialized knowledge and skill in the mathematical, physical, and social sciences, together with the principles and methods of engineering analysis and design, to specify, predict, and evaluate the results to be obtained from such systems.

This definition of industrial engineering has wide acceptance and the endorsement of the American Institute of Industrial Engineers, Inc. The definition is broad enough to cover the diverse activities of the practitioners in the field. However, it does not describe the specific activities customarily categorized as being part of the discipline of industrial engineering. These primary activities are spelled out by the American Institute of Industrial Engineers as follows:[1]

1. Selection of processes and assembling methods
2. Selection and design of tools and equipment
3. Design of facilities, including layout of buildings, machines, and equipment; material handling equipment; raw materials and product storage facilities
4. Design and/or improvement of planning and control systems for distribution of goods and services, production, inventory, quality, plant maintenance and engineering, or any other function
5. Development of cost control systems such as budgetary controls, cost analysis, and standard cost systems
6. Product development
7. Design and installation of value engineering and analysis systems
8. Design and installation of management information systems
9. Development and installation of wage incentive systems
10. Development of performance measures and standards (including work measurement and evaluation systems)
11. Development and installation of job evaluation systems
12. Evaluation of reliability and performance
13. Operations research, including such items as mathematical analyses, systems simulation, linear programming, and decision theory
14. Design and installation of data processing systems
15. Office systems, procedures, and policies
16. Organizational planning
17. Plant location surveys which consider potential market for plant, raw material sources, labor supply, financing, taxes

THE EARLY PIONEERS AND THE TRADITIONAL FIELDS OF INDUSTRIAL ENGINEERING

The early development of what came to be known as industrial engineering is inextricably intertwined with the activities of a group of Americans who were

[1] American Institute of Industrial Engineers, Inc., *Membership Qualifications Manual*, sec. A, pp. 6–7.

active in the last decades of the nineteenth century. The impact that these individuals had on the profession endures and merits discussion of their contributions.

Frederick W. Taylor and the Beginning of Methods Analysis. The individual generally credited with being the father of scientific management and industrial engineering is Frederick W. Taylor (1856–1915). Taylor was a mechanical engineer who, during his early career in the steel industry, initiated investigations of better work methods and went on to become the first individual to develop an integrated theory of management principles and methodologies.

While a foreman at Midvale Steel Company, Taylor first began trying to solve the problems of "What is the best way to do this job?" and "What should constitute a day's work?"

A few years later, while employed at the Bethlehem Steel Works, he conducted his famous studies on shoveling and the handling of pig iron. Some 400 to 600 men were employed in shoveling iron ore, coal, and other materials. He conducted lengthy studies on the sizes of shovels and the materials which could be moved with the shovels. One of the results was the discovery that if the shoveler could move 21½ pounds of material on the shovel, the result would be movement of the maximum tonnage in a day's time. By providing different-size shovels to the men shoveling different materials, it was possible to reduce the size of the yard gang to 140 men in three years and still move the same volume of material. The cost of shoveling was reduced from 7 to 8 cents per ton to 3 to 4 cents per ton by these procedures.

Another of the large-scale operations at Bethlehem was the loading of pig iron on gondola cars. This involved the worker getting the pig iron from a pile on the ground, moving it up an inclined plank, and unloading it in the gondola car. Taylor studied this simple series of movements in depth and developed a standard time schedule for the job. Workers then were trained to do the job on a piece rate basis rather than getting paid by the day. Under the new method, a first-class worker moved 47 tons of pig iron a day and earned $1.85 per day. Under the original system, the worker averaged 12 tons of material a day and was paid $1.15. Initially, the workers resisted Taylor's approach to these jobs, but in the face of an incentive pay system soon were demanding to be put on a piece rate basis.

While Taylor was at Bethlehem, he effected many improvements in shop methods and standardized jobs. Henry L. Gantt joined him and worked on establishing piece rates for all production jobs. Carl Barth, working for Taylor, perfected the metal-cutting slide rule. Taylor and Maunsel White discovered "high speed" tool steel.

One of the organizational concepts of Taylor was functional foremanship. He conceived of a foreman's job as composed of eight major functions, each of which could be assigned to a specialized foreman who would concentrate on the one function. Each worker would then have eight specialized bosses. Although intriguing in theory, this concept has not been widely accepted.

Taylor was a strong-minded individual, and personal differences with Bethlehem Steel's management led to his dismissal in 1901. From that time until his death, Taylor became a management consultant, an advocate of the scientific method to solve business and industry problems, and a prolific writer and lecturer. He studied the organization of enterprises, wage payment plans, and many other aspects of management. His last book, *The Principles of Scientific Management,*[2] was the first attempt to delineate a whole philosophy of management. He was the first to see the interconnection between various elements of management and to attempt to evolve a unified concept. His formula for maximum production includes three elements—a definite task, a definite time, and a definite method—and is a fundamental concept of industrial engineering. One of his key concepts was that management and labor should have a unity of interests in the success of the company.

[2] Frederick W. Taylor, *The Principles of Scientific Management,* Harper & Brothers, New York, 1911.

One writer, James E. Chapman, summarizes the unique Taylor contribution as follows:[3]

1. Scientifically determined work standards
2. Differential piece rate system
3. Functional foremanship
4. The "mental revolution" that Taylor described as precedent to the establishment of "scientific management"

Of course, this simple listing does not purport to represent the broad range of interests of Frederick Taylor.

Henry L. Gantt and His Contributions. Gantt, an engineering contemporary of Taylor, had a profound impact on the development of management thinking. His numerous contributions, derived from long years of work with Frederick Taylor in various industries and as a consultant to industry, included the following:

1. Work in the motivation field and development of the task and bonus plan, a highly successful incentive plan
2. Greater consideration of the worker than was customarily accorded to him by management in Gantt's time
3. Advocation of training of workers by management
4. Recognition of the social responsibility of business and industry
5. Measurement of management results, through Gantt charts and other techniques
6. Extensive writing on management concepts, including three books: *Industrial Leadership; Work, Wages, and Profits;* and *Organizing for Work*

Gantt was very much influenced by Taylor, for whom he worked on several occasions. However, he approached scientific management with a much more humanistic slant than Taylor, who was interested primarily in the technological and scientific features of work in industry.

Frank and Lillian Gilbreth and the Advancement of Motion Studies. One of the great husband and wife teams of science and engineering, Frank Bunker Gilbreth and Lillian Moller Gilbreth, early in the 1900s collaborated on the development of motion study as an engineering and management technique. Frank Gilbreth was much concerned, until his death in 1924, with the relationship between human beings and human effort.

His well-known work in improving bricklaying in the construction industry is typical of his approach. He early observed that bricklayers had their own particular ways of working and that no two used the same method or motions. These observations led him to seek the best way to perform the task; in the course of finding the best method to use for bricklaying, he also developed many equipment improvements. He perfected an adjustable working platform so that the bricklayer would always be at the most convenient working level. The platform had a shelf for bricks and mortar, saving the workman from bending down and picking up each brick. He had bricks prestacked on wooden frames, with the best side of the brick always in the same position, eliminating the need to turn the brick over and over to find the best side. The workplace arrangement was such that the worker could pick up a brick with one hand and the mortar with the other. His system reduced the number of motions made in laying a brick from eighteen to five.

He and his wife continued their motion study and analysis in other fields and pioneered in the use of motion pictures for studying tasks and workers. Frank Gilbreth developed micromotion study, a breakdown of work into fundamental elements called "therbligs." Both Frank and Lillian Gilbreth were prolific writers and lecturers on management subjects and were dedicated to determining the "one best way" to perform tasks. They were contemporaries and friends of Henry Gantt and knew Frederick Taylor.

After Frank Gilbreth's death, Lillian Gilbreth continued actively consulting, writ-

[3] "Frederick Winslow Taylor, The Father of Scientific Management," *The Atlanta Economic Review*, Georgia State College, July, 1968, p. 9.

ing, and lecturing until the late 1960s. Her concern in the industrial psychology field is manifest in many of her writings, and her contributions have been great in the areas of assistance to the handicapped, studies of fatigue allowance, management in the home, and allied subjects.

Harrington Emerson, Efficiency and Organization. Harrington Emerson developed his managerial concepts simultaneously with Taylor, Gantt, and the Gilbreths. He applied his concepts while with the Santa Fe Railroad in the early 1900s. Among his contributions is the Emerson Efficiency Bonus Plan, an incentive plan which guarantees the base day rate and pays a graduated bonus. One of his books, *The Twelve Principles of Efficiency*,[4] defines the essentials of successful organization. Emerson's twelve principles of efficiency follow:

1. Clearly defined ideals
2. Common sense
3. Competent counsel
4. Discipline
5. Fair deal
6. Reliable, immediate, and adequate records
7. Dispatching
8. Standards and schedules
9. Standardized conditions
10. Standard operations
11. Written standard practice instructions
12. Efficiency reward

Emerson was a great exponent of the line and staff type of organization.

Other Pioneers. There were, of course, other individuals who made contributions to the development of scientific management and industrial engineering thinking. It would be difficult, and perhaps impossible, to try to list them all. But a few do deserve mention because of some special contribution.

Morris L. Cooke contributed, by the application of the Taylor principles, to his work as director of the Department of Public Works for the City of Philadelphia, utilizing Taylor's concept of functional organization. He applied the Taylor and Gilbreth beliefs of standardization of contract specification and performance to many phases of city government administration. His major contribution was the application of scientific management thinking to the area of government operations. His criticism of the operation of engineering societies stimulated changes leading to a dramatic broadening of their activities, increase of membership, and improvement of the public image of engineering.

Independently, but at the same time, Henri Fayol, a French mining engineer and manager, made great contributions to the field of management at the higher administrative levels. His book, *Administration Industrielle et Generale*,[5] describes his philosophy and approach. Fayol divided business and industrial operations into six groups: technical, commercial, financial, security, accounting, and administrative. He pointed out that these are interdependent, and management's job is to ensure the smooth working of all these groups. He recognized that there must be a single head for an enterprise and that each individual in the organization should report to only one individual. His work in defining the principles and elements of administration is well known and generally accepted.

Although the foregoing discussion concerned itself primarily with the thinking and achievements of individuals, it should be remembered that these persons were breaking new ground and developing a unified theory of scientific management. In part, this came about through cross-fertilization of ideas with others in the field. Although many of the precepts and methods which these individuals espoused have been modified by later practitioners, their thinking is still the basis for modern industrial engineering. The traditional fields of methods, motion and time study,

---

[4] Harrington Emerson, *The Twelve Principles of Efficiency*, John R. Dunlap, 1911 (subsequently copyrighted by *Engineering Magazine*, 1913).

[5] Henri Fayol, *Administration Industrielle et Generale*, International Management Institute, Geneva, 1925.

organization, and management, which these individuals pioneered, have been expanded and refined. New areas of activity have opened for industrial engineers.

Industrial Engineering in the 1920s. Scientific management and industrial engineering suffered a loss of popularity and acceptance in the years immediately before World War I. However, with the American participation in that conflict and the subsequent reconstruction, a period of heightened industrial activity ensued. This was a period of rapid standardization and mass production, providing an environment in which industrial engineering flourished. The techniques and principles of industrial engineering were applied to new fields of endeavor and on a much broader scale.

Organized labor, which had looked askance at early industrial engineering efforts, became interested in promoting productivity and raising the worker's standard of living. Interest was revived in industrial psychology, and increasing concern was expressed for the welfare of labor.

A deterrent to consistent growth of the profession during this period was the activities of unqualified practitioners in the field, the so-called "efficiency experts." These tended to give scientific management and industrial engineering a bad name by association. Ill-prepared and poorly disciplined, they were prone to recommend standard rate cutting, not because of the inherent characteristics of the task being studied, but to effect cost savings for management. This and other unscientific, if not unethical, practices cast a pall over the profession which took some time to dissipate. Suffice it to say that today's industrial engineer, with a rich background of technical and scientific knowledge, bears no resemblance to the "efficiency expert" of yesteryear.

One of the significant reports of the era, published in 1921, was *Waste in Industry*, produced under the auspices of the American Engineering Council of the Federated American Engineering Societies. This report was a study of inefficiency in six types of industry and involved information from more than 200 manufacturing plants in the selected fields of study. It indicated that little real progress had been made in the application of scientific management methods between the publication of Taylor's book, *Principles of Scientific Management*, in 1911 and the issuance of the report in 1921.

Industrial Engineering in the 1930s. The 1930s were a decade of economic hardship and social unrest. The great depression made management extremely cost conscious and created an environment in which industrial engineering principles and techniques were given serious consideration and fairly widespread application. Impressive research on the production and attitudes of workers was carried on over a twelve-year period in the Hawthorne Works of the Western Electric Company, terminating in the early Thirties. This was an early example of an industrial concern which was willing to finance an outside group in a study of its own plant workers.

At the close of the decade, the impact of World War II was beginning to be felt in terms of increased industrial production; this provided a further stimulus to industrial engineering precepts and practices. Trade unionism grew extensively in the 1930s, and workers felt less fear of rate cutting and more awareness of their ability to seek redress for incorrect wage rates. This resulted in less resistance to the industrial engineering movement. Concurrently, the methods study advocates and the time study advocates, who earlier had pictured themselves as being opposed in viewpoint, began to consider themselves in the same profession.

Also in the Thirties, a number of individuals wrote a notable series of books which delineated more fully the field of methods study. In 1932, Allan H. Mogensen published *Common Sense Applied to Motion and Time Study*, which related primarily to work simplification. S. M. Lowry, H. B. Maynard, and G. J. Stegemerten published the second edition of their book, *Time and Motion Study and Formulas for Wage Incentives;* Maynard and Stegemerten wrote *Operation Analysis;* and Ralph M. Barnes produced *Motion and Time Study.*

Early in the decade, H. B. Maynard and others associated with him developed methods engineering, a concept which embraced many aspects of methods work in one of the early systems approaches to industrial problem resolution.

These developments opened an era of intensive work in the methods and work simplification fields.

Other Industrial Engineering Activities. Although the emphasis in this review has been on the development of methods and work measurement activities, the pioneers and later practitioners of industrial engineering had wide-ranging interests and had done work in a number of other areas.

Wage and salary plans are closely tied to work measurement, methods work, motion and time study, and setting of standard rates; they have always played an important role in industrial engineering. Taylor's differential piece rates, Gantt's task and bonus system, the Emerson efficiency bonus plan, and the Bedaux system were developed early, and are indicative of some of the plans of compensation for workers which are tied to their productivity. Much industrial engineering effort is focused on incentive programs aimed at increasing worker productivity and at adequate compensation to the worker for that increase.

With such infinite pains being taken to study the elemental motions required in a task, it was inevitable that just as much emphasis should develop on accurate descriptions of job responsibilities and duties. Job analysis and job evaluation, as well as merit rating of individuals and wage administration, became activities in which industrial engineers naturally had a great deal of interest.

There has always been a continuing need in industrial plants to obtain and locate properly in the plant the necessary equipment, to maintain the equipment in good working order, to replace it with new and better equipment at the appropriate time, and to adapt to new space and industry techniques. These activities fall under the general name of plant facilities, a function in which industrial engineers have been very active. In this category should also be placed the functions of plant location and tool and gage design.

Another field for the industrial engineer has been the design and implementation of control procedures. This started out with bookkeeping and simple inventory controls as the factory system began developing. With the advent of larger organizations and multiplant companies, the need for expansion of these elemental systems into more complex systems grew as well. Interchangeable parts called for quality controls to ensure that replacement parts would work. Mass production and production line techniques called for planning and scheduling and better materials and inventory control systems. Tools for maintaining control, such as the Gantt chart and flow charts, were devised.

Morley H. Mathewson, in the second edition of the *Industrial Engineering Handbook*,[6] summarizes the traditional industrial engineering functions as a prelude to a discussion of some later fields of emphasis for industrial engineers. He included the following general headings:

1. Methods engineering—operations analysis, motion study, material handling, production planning, safety, and standardization
2. Work measurement—time study, predetermined elemental time standards
3. Control determination—production control, inventory control, quality control, cost control, and budgetary control
4. Wage and job evaluations—wage incentives, profit sharing, job evaluation, merit rating, wage and salary administration
5. Plant facilities and design—plant layout, equipment procurement and replacement, product design, tool and gage design

This listing covers the main activities of industrial engineering practiced widely in the period before World War II.

MODERN INDUSTRIAL ENGINEERING

The Development of Predetermined Time Standards. To this point in the development of industrial engineering, most of the effort in the methods field had been corrective. Now the thinking was to turn to prevention. This concept in-

---

[6] Morley H. Mathewson, in H. B. Maynard (ed.), *Industrial Engineering Handbook*, 2d ed., McGraw-Hill Book Company, New York, 1963, pp. 1–18.

volved studying methods before they were put into effect rather than after, in terms of design of work.

A number of groups of individuals developed systems of predetermined time standards independently at approximately the same time.

Gilbreth's concept of elementary motions led others to consider the possibility of combining these elements into normal tasks performed by industrial workers to arrive at a standard time for the task. A. B. Segur, prior to 1930, had utilized such elementary time standards, but his findings were not widely disseminated.

In 1940, Westinghouse Electric Corporation sponsored a study by H. B. Maynard and his associates. This was intensive research into the elemental motions required in drill press work. It led in 1948 to a book by Maynard, Stegemerten, and Schwab, *Methods-Time Measurement*,[7] which described the procedure developed to determine time standards in advance of the job performance. This concept immediately had wide national and international acceptance.

Meantime, three industrial engineers (J. H. Quick, W. J. Shea, and R. E. Koehler), working for RCA, developed a system which they called the Work-Factor system. It was described first in *Factory Management and Maintenance*, in 1945. This, too, became widely accepted as a means of predetermining time standards.

Industrial Engineering during World War II. As happens during most wars of any size or duration, a vast enlargement of industrial activities occurred in the United States as it sought to achieve military supremacy for itself and its allies. Not only were industrial capacities greatly enlarged, as in the case of production of aircraft, but greater conversions took place of peacetime products to materials needed for the war. Automobile plants converted from making civilian vehicles to the production of trucks, tanks, and other needed vehicles. Synthetic rubber production became necessary as the overseas sources of natural rubber disappeared. A vast effort was needed by the petroleum industry to produce various chemicals and 100-octane gasoline. Complete new industries were born, such as the atomic energy plants at Oak Ridge, Hanford, and elsewhere.

The result of all this was to put a great strain on the productive capacity of the nation, which of course was true of other nations as well, and to accelerate new developments in industrial engineering.

Many modern industrial engineering techniques had their genesis during the period 1940 to 1946. Predetermined time standards (such as MTM and Work-Factor), value engineering, and systems analysis are a few of these. They were expanded, refined, and applied in subsequent years, but they were developed and utilized initially in the environment of the war effort.

Operations Research during World War II. One of the fascinating products of World War II was the activity which came to be known as operations research. Operations research, basically, is the process of applying statistical and higher mathematical techniques to the solution of real-world problems. Developed as an adjunct to the war effort in Great Britain, it was quickly taken up by the United States. Its original military use has been adapted to the solution of business and industry problems all over the globe.

During World War II, the British used the operations research approach on many problems, including the early use of radar, determining the optimum size of merchant convoys to keep losses at a minimum, and the probability of interception of enemy aircraft.

Ideally, operations research utilizes a team approach to problem solving where individuals of differing backgrounds and knowledge pool their abilities. The team first studies and agrees on the formulation of the problem. It then applies common sense and various mathematical techniques to the variables in the problem and attempts to find optimum solutions.

The industrial engineer, with his broad knowledge of business and industry operations and his schooling in higher mathematics, fits admirably in many operations research team efforts.

[7] H. B. Maynard, G. J. Stegemerten, and J. L. Schwab, *Methods-Time Measurement*, McGraw-Hill Book Company, New York, 1948.

The Post-World War II Period. A highly significant era in the development of industrial engineering began after World War II. A great many new activities developed, and the application of principles and techniques was vastly broadened. The pace of technology development called for the increasing use of industrial engineering in many fields of endeavor, resulting in an unprecedented demand for people with training in this discipline.

In surveying the postwar period, many significant factors could be discussed, the impact of some of which is not yet fully apparent. Space limitations preclude a comprehensive discussion of all new developments; instead, six significant industrial engineering activities and techniques which have been widely used in the period will be reviewed. These are the following:

1. Industrial engineering and the computer
2. The development of systems analysis and design
3. The application of mathematical and statistical tools
4. Network planning techniques and their application
5. Value engineering
6. Behavioral science and human factors

Industrial Engineering and the Computer. The development of the first electronic digital computer (ENIAC) in 1946 ushered in a new era of sophisticated, high-speed calculation and information storage and retrieval. By 1948, a stored program computer was developed. The early uses of the computer were restricted primarily to high-speed manipulation of industry records, such as accounting records and billing. More sophisticated uses developed, and technology changes greatly increased the capability for storage, computation, and retrieval of information. The vacuum tube computers were followed by solid state computers, creating an entirely new generation of such devices—computers which were less bulky, used less power, cost less, and had greater capacity.

Continued technological advances in micro-solid state devices have resulted in further dramatic changes in the computers on the market which make them within economic reach of all but the smallest organizations. The latter also have the option of access to the computer through the computer service organizations which are available in most communities of any size.

Industrial engineers are involved with the computer in many ways. They are concerned with the design of computer installations from the management point of view, to make the information output most useful in the decision making process. Industrial engineers use the computer to solve complicated problems in industry, such as those involving the use of programming techniques. The computer can be used to simulate business and industry conditions and problems and to provide almost instantaneous answers for varying sets of conditions. Computers are used widely for training by simulation of various business conditions. The first computerized business game which was useful for training purposes was developed for the American Management Association by industrial engineers in 1957.

Industrial engineers also have been active in the design and installation of computers for the purpose of controlling and operating continuous flow processes. Originally used in chemical process plants and in the petroleum industry, these special-purpose computers are found in many other types of plants, and the potential for more widespread use is great wherever an automatic process can be installed.

In the 1950s, the application of computers to control entire machine tool processes came into being, and numerically controlled machine tool applications have been increasing dramatically since that time. Industrial engineers have played a large part in the development and design of these systems.

Computers are also used widely in the traditional fields of industrial engineering, such as work measurement, plant location, methods engineering, and control systems.

In the 1960s, perhaps the outstanding single endeavor of mankind was the space program centered primarily in the United States and the U.S.S.R. This program, leading to the exploration of the moon, and ultimately, of the solar system and beyond, would not have been feasible or possible without the development of the computer. In the United States, the NASA program relied heavily on the use of industrial engineers for many activities, not the least of which was in the use of computers and in the total systems engineering of the project.

The Development of Systems Analysis and Design. In the general area of planning and control procedures, much attention was focused in the 1950s and 1960s on overall systems analysis and design. This is based on the recognition that industrial activities (or for that matter, most of man's endeavors) are composed of systems of activities which are made up of subsystems. A great many variables affect the system and subsystems. Therefore, it is necessary to study the system or subsystems to alleviate or eliminate problems and to provide a corporate structure and mechanism which will optimize operations.

The systems analysis and design approach had its origins in World War II, and by the early 1950s, a number of companies had converted the wartime approach to their own use. A good example of this is the approach that United Air Lines took in 1955 to the design of an airline reservation system. The reservation function was a sizable subfunction of the company. The approach used was described by A. Weston in a presentation at a symposium sponsored by the American Institute of Industrial Engineers.[8]   In his conclusion, he said:

> The project dramatically highlighted the importance of taking an overall industrial engineering approach to the analysis and design of complex systems which involve both technical and human aspects. An important point in the approach taken is the development of system requirements which manufacturers are asked to meet as opposed to the current practice of trying to adopt and integrate existing hardware to meet business needs. The results, both in satisfying the company's requirements and in improving productivity, point up the value of allocating industrial engineering effort to the research and development of business systems.

The use of systems analysis and design is increasing as companies recognize the benefits to be obtained from the overall approach to problem solving.

The Application of Mathematical and Statistical Tools. Another major occurrence in postwar industrial engineering has been the acceptance of the fact that higher mathematical techniques can be successfully applied to the solution of business and industrial problems. Stimulated by the development of operations research, the use of these techniques has permeated most of the activity areas of industrial engineering.

Statistical theory has been applied to many problems. Statistical quality control is an extremely important aspect of many industrial operations. Directly related to it is reliability control, which has great importance in the military aircraft and space programs, although not restricted to these. Both these fields utilize probability theory, probabilistic parameters, density functions, and distribution functions.

The practical statistical methods found in probability concepts, sampling, variability, central measures, testing, regression, and correlation, all have become part of the tool kit of the industrial engineer. Waiting line theory has been extensively used in the design of facilities where queuing problems need to be solved.

The concept of developing mathematical models to represent the variables and situations found in real-life problems has been greatly advanced and is a factor in many of the industrial engineering activities which relate to complex processes.

Mathematical programming has been widely applied to the solution of complex industrial problems, where variables can be identified and placed in equations which can be solved by matrix algebra. Here again, the computer is frequently used because of the complexities of such variables and solutions.

Linear programming is based on work done by von Neumann in 1928, Hitchcock in 1941, Stigler in 1945, and Koopmans in 1947. However, it was a Department of the Air Force group led by George B. Dantzig, Marshall Wood, and their associates in 1947 which unified the theory and applications, while developing the widely used simplex method. It has a wide applicability to such functions

---

[8] A. Weston, "The Emerging Role of Industrial Engineering," *The Journal of Industrial Engineering*, March–April, 1961.

as multiproduct assembly lines, plant location, market research, and distribution studies.

In the late 1950s, an approach called systems simulation was tried with success by a number of companies. It is a process which developed from the early operations research activity. Basically, it consists of experimenting with the conditions which may exist in a system under varying circumstances. This is done by building a simulation model which reflects the variables in the system and which can be manipulated to measure the effect of changing one or more of the variables. The model may be a written statement, a flow diagram, a mathematical expression, or a financial statement. Where calculations are complex and numerous, the computer may be used to provide quick and accurate pictures of the effect of changing variables on the entire simulation. This technique has been used extensively in industrial processes, hospital management planning, public transportation problems, and various other activities of a complex nature. Systems simulation is listed in the general category of mathematical and statistical approaches because it generally utilizes probability, queuing, reliability, and other techniques, as the problem may demand.

Network Planning Techniques and Their Application. Another technique which has come into wide use is that of network planning. This received its original development from the United States missile program.

In the mid-1950s, the Navy Special Projects Office was involved with the development of the Polaris missile, a complex program which involved many millions of dollars and a critical need for speed. Part of the problem lay in the fact that the design time for the missile system was difficult to predict and many components of the missile had to be designed and produced simultaneously in some cases and sequentially in others. A team of individuals, led by Donald Malcolm, earlier associated with the development of computer business games, developed during 1958 and 1959 the scheduling and time compression network technique now called PERT. These initials originally stood for Program Evaluation Research Task, but it is now generally called Program Evaluation and Review Technique.

The original PERT network consisted of a complex simultaneous and sequential set of events relating to all the activities and events associated with the development of the Polaris missile system. The longest path through the network in terms of time became the critical path, and all efforts were made to reduce the time involved in the critical path so as to complete the program in the minimum time. Much of the credit for the speed and success of the Polaris development is given to the PERT technique.

A later variation of this technique is called the critical path method (CPM). The critical path method of network planning has subsequently been applied to many other activities of a planning nature. It is used effectively in the construction industry and many other industries where complex sequential situations exist. Network planning methods, such as PERT and CPM, are additional techniques where the use of the computer is frequently necessary to make the many calculations relating to probability and the numerous events in the network. Industrial engineers were involved in the development of this approach, and they have utilized the technique extensively.

Value Engineering. Value engineering, as a recognized technique, was another product of World War II. This was a period when critical materials were difficult to obtain and a great many substitutions had to be made. Harry Erlicher, a vice president of purchasing for General Electric Company, noted that frequently these substitutions resulted in lower costs and improvement of the final product. L. D. Miles was assigned the task of formally investigating substitute materials, and he developed an approach to the field which was initially called value analysis. In 1954, in an effort to reduce the cost of ships and equipment, the Bureau of Ships undertook a formal program based on the General Electric value analysis program. The success of this program, called value engineering, led to its general acceptance in the armed forces and among the subcontractors who supply material and weapons to the armed forces.

Value engineering is defined as a systematic, creative approach to ensure that the essential function of a product, process, or administrative procedure is provided at a minimum overall cost. It involves five basic phases: information, speculation, analysis, planning and decision, and summary and conclusion. Although none of these phases is new to the engineering profession, the systematic approach employed in value engineering has led to many startling cost improvements in industry. Industrial engineers are heavily involved in the value engineering function in industry.

Behavioral Science and Human Factors. The industrial engineer's emphasis on working with integrated systems of men, materials, and equipment has differentiated it somewhat from other fields of engineering. The inclusion of "men" as a principal responsibility recognizes the concern of the early practitioners of industrial engineering with the human element in the industrial scene.

Consideration of the behavioral sciences and human factors began with Gantt's appreciation of worker motivation and the Gilbreths' interest in the worker as a human being and in the field of applied psychology. Walter D. Scott, Hugo Munsterberg, and Anne Shaw were early contributors in industrial psychology. Much motivational and psychological study has been done since that time. This work has been applied to several fields of industrial engineering, particularly the approaches that industrial engineers must make in working with factory workers.

In addition, a great deal of study has been given to the physiological aspects of work, both in industry and in higher education. The original concern with fatigue has led to extensive studies of the body's reactions to work and of its physical capabilities. A great many devices have been invented to measure human effort, the adaptability of the human body to the workplace, and the human engineering of machinery and equipment to permit man to work with them most efficiently.

The sociological aspects of automation have been studied in depth. Biomechanics and bioengineering are promising fields relating to the human body and its biological reactions under work and other conditions. In all these activities, industrial engineers have played a large, often a leading, role.

The six major trends in the post-World War II period which have just been discussed are part of a dynamically changing discipline of engineering. These and other trends, which are discussed more fully in other chapters of this Handbook, have come about in an environment which anticipates and requires change. This environment has been aided by two other significant factors—education and professionalism.

INDUSTRIAL ENGINEERING EDUCATION AND THE GROWTH OF PROFESSIONALISM

In 1908, the first separate departments of industrial engineering were established at Pennsylvania State University and at Syracuse University. The department at Syracuse was discontinued shortly thereafter and not reestablished until 1925. In the 1920s, a number of schools initiated industrial engineering options in their departments of mechanical engineering. This trend continued in the 1930s and up until World War II. After the war, a great many returning veterans were interested in industrial engineering programs, and many other universities and colleges established departments of industrial engineering (with some variation of the title). At the same time, many of the programs which had existed as options in mechanical engineering became part of new industrial engineering departments.

By 1960, there were 74 schools offering curricula in industrial engineering, of which 48 were accredited in this discipline by the Engineering Council for Professional Development (ECDP). By 1968, the number of schools offering curricula had grown to 126, of which 57 were accredited.

This rapid growth in sources of industrial engineering education was matched by the enormous expansion in the numbers of students pursuing degree programs. In 1966, there were 9,800 students in undergraduate and graduate industrial engineering programs, which made it rank fifth in enrollment among the engineering disciplines, behind electrical, mechanical, civil, and chemical.

Correspondingly, in Canada, Europe, Latin America, and Australia, curricula had developed which involved many of the same courses taught in the United States, although the programs sometimes were called by names other than industrial engineering.

One of the characteristics of industrial engineering education in America has been its adaptability and willingness to change. New principles and new techniques have been introduced in curricula as they have become known and accepted. The emergence of skills based on higher mathematics and knowledge of computers has revolutionized academic approaches to industrial engineering. This schooling today graduates a highly sophisticated and technically capable individual, who has at his command tools and techniques for solving industrial and business problems which would have astounded the pioneers of the field.

Many two- and four-year junior colleges and technical institutes are now graduating industrial engineering technicians. The curricula of these institutions resemble those of senior colleges of an earlier decade.

The emergence of the separate industrial engineering curriculum has served to advance the professionalism so necessary to a separate discipline of engineering. A professional esprit de corps among practitioners has developed rapidly. A major group contributing to this evolutionary pattern has been the faculties of colleges and universities, which, though relatively small in number, have exercised an influence on the profession far out of proportion to their numbers. Through education and research, through technical publications and books, and in many other ways, they have enhanced the professional status of industrial engineers.

Another significant factor in the growth of professionalism has been the activity of professional societies. The role that the American Society of Mechanical Engineers, as well as other societies, such as the Society for Advancement of Management, played in the early and mid-1900s has been recounted. However, because the main focus of these societies was on other disciplines, they did not provide the industrial engineer with a society which he could feel was truly "his."

Such an organization came into being in 1948 with the establishment of the American Institute of Industrial Engineers (AIIE). This developed into a fast-growing, vigorous society with members from many countries of the world and many chapters in the United States and Canada. Dedicated to advancement of the profession and education of its members, its *Journal of Industrial Engineering* (now *Industrial Engineering*) since 1949 has provided a forum for much of the modern industrial engineering thinking and acted as a current awareness vehicle for the profession. Organized in geographical chapters and functional divisions, the AIIE satisfies the interests of industrial engineering practitioners, provides a sounding board for new developments, and serves as an effective means of advancing the status of the profession.

The widespread dissemination of industrial knowledge, the application of this knowledge to business and industrial problems, the increasing availability of graduate industrial engineers, and the continuing industrial expansion of the world's economy have all been factors in the development and acceptance of the profession of industrial engineering.

CONCLUSION

Significant trends are at work in the field of industrial engineering. There is a steadily increasing trend toward involvement of industrial engineers in other than industrial pursuits. Nonindustrial applications of industrial engineering are numerous in sales, distribution, banking, insurance, financial, service, and government activities. Industrial engineers are found in almost all companies of large size and many companies of medium and small size.

Companies such as Eastman Kodak, Procter and Gamble, United Air Lines, Union Camp Corporation, and many others have developed large staffs of industrial engineers with broad responsibilities in many functional activities of the companies. This trend will no doubt persist—the industrial engineering function will grow in size and responsibility, and larger staffs will lead to increased specialization in personnel. The principles and methodologies of industrial engineering are being applied in increasing measure to consideration of man's environmental problems—social, economic, and political—in line with the industrial engineer's awareness of the individual worker and his motivational requirements. The challenge to all industrial engineers is to learn to apply their skills and knowledge to the solution of problems in these exploding fields much as the earlier practitioners did in the traditional fields and industries in which the profession incubated.

BIBLIOGRAPHY

Barnes, Ralph M., *Motion and Time Study*, 6th ed., John Wiley & Sons, Inc., New York, 1968.

Chapman, James E., "Frederick Winslow Taylor, Father of Scientific Management," *Atlanta Economic Review*, July, 1968, pp. 9–11, 20–21.

Filipetti, George, *Industrial Management in Transition*, Richard D. Irwin, Inc., Homewood, Ill., 1953.

Maynard, H. B. (ed.), *Industrial Engineering Handbook*, 2d ed., McGraw-Hill Book Company, New York, 1963.

Spriegel, W. R., and R. H. Lansburgh, *Industrial Management*, John Wiley & Sons, Inc., New York, 1955.

Taylor, Frederick W., *The Principles of Scientific Management*, Harper & Brothers, New York, 1911.

Chapter **2**

Profile of
an Industrial Engineer

RICHARD R. CONARROE

President, Walden Public Relations, Inc., New York, New York

How does the "typical" industrial engineer see himself? In what light does he regard his work and how does he evaluate his value to his company? What, in his opinion, do others think of him?

To find the answers to these and similar questions, a survey was conducted among members of the industrial engineering profession in cooperation with the American Institute of Industrial Engineers.

The conclusions of the survey, which form the basis of this chapter, were based on the responses of more than 1,600 industrial engineers. Of that number, nearly two-thirds had more than ten years' experience in their field.

HIGHLIGHTS OF THE SURVEY FINDINGS

The "typical" industrial engineer, the survey indicates, holds the following views: (1) that the industrial engineering profession is based on certain standard techniques (time study, operation analysis, and the like), and that these techniques will never be outmoded by the newer, more sophisticated techniques that have recently been developed; (2) that the industrial engineer is not yet accepted as a fully professional engineer by top corporate management; (3) that many members of his profession limit themselves to a somewhat narrow and shortsighted interpretation of their work and functions; (4) that the human relations problems that stem from industrial engineering will be solved in the future, although they have not been handled

1-18

too successfully in the past; and significantly, (5) that the industrial engineering profession as a whole is due for some fresh new thinking.

A Composite Picture of the Industrial Engineer. "Industrial engineering" is an umbrella that covers a multitude of functions and encompasses nearly every phase of a company's operations.

How, then, do industrial engineers spend their working time?

Much of it, according to the survey, is devoted to two functions: methods studies and the development of performance standards and measurements. This much is clear from the fact that these functions are performed in fully 83 percent of the companies for which industrial engineers work.

In two-thirds of the companies, industrial engineers also work at the following: design of facilities and layout of buildings, machines, and equipment; and design and improvement of planning and control systems for distribution of goods, production, inventory, quality, and plant maintenance.

Other functions which occupy the industrial engineers' time include the development of management control systems for financial planning and cost analysis; the development, installation, and maintenance of wage incentive systems; the development of job evaluation systems; and the design and implementation of data processing systems.

In addition, industrial engineers conduct organization studies, design tools and equipment, make plant location studies, evaluate quality and reliability performance, administer wage and salary functions, solve complex business problems by the use of operations research, and develop new products and product applications.

THE TECHNIQUES OF INDUSTRIAL ENGINEERING

In recent years, a number of advanced, rather glamorous new industrial engineering techniques have come into vogue. Examples include linear and mathematical programming, queuing, matrix algebra, and others.

How widespread is the use of these techniques? Have they altered in any way the use of standard, time-honored techniques?

Some of the new methods have gained considerable acceptance, particularly in larger companies. In smaller companies, however, they do not seem to have as much applicability. And in both large and small companies, industrial engineers agree, the new techniques will never obsolete such traditional procedures as time study, process charting, operation analysis, and the like.

Work measurement techniques such as time study, work sampling, and standard data or time formulas are an excellent example of the endurance of traditional industrial engineering methods.

Industrial engineers report that the time study method is used by three-fourths of the companies they work for, while work sampling and standard data or time formulas are in use in nearly 60 percent of these companies. Furthermore, these traditional techniques have equal applicability in small and large companies.

TRAINING FOR INDUSTRIAL ENGINEERS

Nearly everyone agrees that it takes a long time for an industrial engineering graduate to become a full-fledged, contributing professional.

What are the reasons for the lengthy apprenticeship?

Many of the industrial engineers feel that the reason is that it takes a long time to get to know a company. In addition, a large number, particularly among the older industrial engineers, feel that colleges are not teaching the right basic skills, which must therefore be learned on the job at a considerable expenditure of time. Although the younger industrial engineers do not generally share this view, they do agree that no college can adequately do the entire training job.

The consensus among industrial engineers is that the best ways to train new graduates are through on-the-job work involving basic skills, through tutoring by more experienced industrial engineers, or through a combination of these two methods.

The industrial engineers also stress the importance of special training which can significantly reduce the time required to turn out a productive engineer. The training emphasis, the industrial engineers feel, should be in the following areas (listed in decreasing order of importance):

1. Human relations
2. Supervision and management
3. Methods improvement programs
4. Systems design
5. Statistical techniques
6. Operations research
7. Mathematical decision tools
8. Work measurement techniques

HOW INDUSTRIAL ENGINEERS SEE ONE ANOTHER

In the opinion of the responding industrial engineers, too many of their colleagues spend too much time merely cutting company costs. Cost cutting is certainly commendable, they agree, but only if it does not disturb the company's overall profit picture. In other words, cost reduction, a short-term goal, should not supersede profit optimization, a long-range and far more valuable goal.

There are probably a number of reasons why so many industrial engineers choose the cost reduction course. First, it is the quickest way to produce tangible financial results, and it can be instituted with much less difficulty than a profit optimization program.

Then, too, many industrial engineers feel that they are pressured by management to reduce costs. Although this is probably often true, the industrial engineer must nonetheless accept full responsibility for what he does on the job.

How Management Sees the Industrial Engineer. The industrial engineer, as he sees himself, has not yet gained full acceptance by management. There still exists a wide gulf of misunderstanding, and industrial engineers feel—almost to a man—that they are not yet thought of as professional engineers.

In the view of the industrial engineer, many managers do not understand what he does or why he is needed. By many, he is thought of as a "necessary evil." To others, he is still "just an efficiency expert."

But industrial engineers also accept most of the blame for the misunderstanding and feel they can do much to correct it.

How Industrial Engineers Can Improve Their Image. These are some of the steps industrial engineers feel would help improve their image:

1. Promote professionalism by acting more professionally, restricting the use of the industrial engineer's title to actual industrial engineers; eliminate mere technicians.
2. Educate management. Interpret to management the nature, aims, and potentialities of industrial engineering.
3. Improve performance and competence within the profession.
4. Broaden the industrial engineer's knowledge, outlook, activities, and responsibilities.
5. Work more closely with management.

As industrial engineers see it, then, the way to gain the professional recognition they want is to become more professional in their industrial engineering practices. As one leader in the profession put it, "The true function of the industrial engineer is to devise new systems for profit. To do this effectively, he must take the initiative in selling his concepts to higher management. Instead of merely responding to orders, he must prescribe specific courses of action."

THE HUMAN SIDE OF INDUSTRIAL ENGINEERING

Although most aspects of industrial engineering can be measured and predicted, there is one aspect that is both unmeasurable and unpredictable. This is the human element.

Most industrial engineers believe the human relations aspect of their work is quite important; nonetheless, nearly half are of the opinion that the profession has not made much headway in solving the human relations problems that often stem from industrial engineering work.

There are some signs of change, however. For example, it is generally the older industrial engineers who are more pessimistic about solving human relations problems. Among those with less than ten years' experience, there is increased optimism about the chances of dealing successfully with the human side of their work.

CONCLUSION

The industrial engineer of today bears little resemblance to the industrial engineer of thirty years ago. Most of the differences are due to the fact that industrial engineering, which is based on an ever-widening body of knowledge, is much changed from what it used to be.

Within the profession, there is widespread feeling that the time has come for a critical new look at industrial engineering. Industrial engineers point to the fact that they spend most of their efforts working toward short-range goals, when they should in fact be aiming at long-range profits. They realize that they must work toward greater professionalism in their field if they are ever to be fully accepted as professionals.

Increasingly, they are concerned about the human relations problems that are engendered by industrial engineering, although at the same time they are hopeful about being able to solve these problems, perhaps with the help of behavioral scientists. They are similarly concerned about finding ever-better methods for training new members of their profession.

Underlying these winds of change, however, is a solid foundation of fundamental industrial engineering knowledge. That foundation, the industrial engineers agree, will continue to serve them well.

BIBLIOGRAPHY

Haefele, H. A., "Planning a Dynamic Industrial Engineering Role in Organizations," *Proceedings*, American Institute of Industrial Engineers Conference, 1966, p. 41.

Hall, N. A., "What the Profession Wants of the IE Curriculum," *Proceedings*, American Institute of Industrial Engineers Conference, 1962, p. 207.

Hammond, Ross W., *Your Future in Industrial Engineering*, Richards Rosen Press, Inc., New York, 1965.

Hannon, John W., "A Fresh Look at Industrial Engineering," *The Journal of Industrial Engineering*, May, 1968.

Leimkuhl, F. F., "The Future of Operations Research and Industrial Engineering," *Proceedings*, American Institute of Industrial Engineers Conference, 1967, p. 240.

Scheid, P. N., "The Industrial Engineer: An Internal Consultant," *Proceedings*, American Institute of Industrial Engineers Conference, 1966, p. 49.

Chapter **3**

Organizing for Industrial Engineering

CLIFFORD SELLIE
President, Standards International Inc., Chicago, Illinois

TULLY SHELLEY, JR.
Principal, McKinsey & Company, Inc., New York, New York

Organizing for industrial engineering is no different from organizing for any company function; the same accepted principles of organization must be applied. Because this Handbook is not concerned with the theory of organization as such, any interested reader is referred to the many excellent texts on the subject.[1]

In simple outline, the organization of an industrial engineering department demands three steps:

1. The authority, responsibility, and accountability of the department must be clearly defined.
2. The department must be integrated in the company plan of organization.
3. Every provision must be made so that the department can effectively perform its assigned tasks.

In administering the department, the chief industrial engineer should, of course, strive to have a well-knit, smoothly functioning unit. His operating plan must also enable him to provide all the services expected of his group.

[1] To mention two good studies: John M. Pfiffner and Frank P. Sherwood, *Administrative Organization*, Prentice-Hall, Inc., Englewood Cliffs, N.J., 1960; J. G. March and H. A. Simon, *Organizations*, John Wiley & Sons, Inc., New York, 1958.

With this introduction to the general problem, the subject can now be examined more closely. To that end, this chapter is divided into the following main sections:

1. The role of industrial engineering in the business
2. Place of industrial engineering department in company organization structure
3. General forms of internal organization of industrial engineering department
4. Typical forms of internal organization
5. Special forms of organization of the industrial engineering department
6. Industrial engineering personnel
7. Departmental administration
8. Getting action on recommended programs

THE ROLE OF INDUSTRIAL ENGINEERING IN THE BUSINESS

A glance at the table of contents of this Handbook will show the variety of activities an industrial engineering department can perform Rarely, however, does a specific department do more than a part of those listed. The scope of its activities in any company is a deciding factor in determining where to fit the industrial engineering department in the company plan of organization and how to organize it internally. Equally important considerations are the nature of the services provided and the organization plan of the company.

The Industrial Engineering Approach. In carrying out its functions, the industrial engineering department uses what is sometimes referred to as the scientific approach. In other words, the industrial engineer gathers and analyzes facts, forms tentative conclusions, compares and tests the alternatives, and finally reaches and presents his findings, conclusions, and recommendations.

To ensure that the actions it recommends are beneficial to the company as a whole and have a good chance for adoption, the industrial engineering department must operate with as much objectivity as possible. Three suggestions here will illustrate how that objectivity may be put into practice:

1. In approaching a problem, the industrial engineering department must listen to and evaluate objectively the viewpoints of all the departments affected. In making recommendations, it should support its elected course of action by sound reasons proving that the proposal submitted offers the best possible solution.
2. The industrial engineering department must be prepared to meet prejudiced points of view and to treat them understandingly but firmly. If the department has succeeded in gaining the confidence of line management, it should be able to explain why the prejudice is ill-founded. It should never, of course, ridicule or expose the prejudiced man.
3. Although the industrial engineering department must never arbitrarily disregard the opinion of line management, it should not lose sight of the fact that its first concern is to strengthen the overall operations of the company. This may sometimes require the department to act contrary to the wishes of production supervisors.

Basic Objective of an Industrial Engineering Department. The basic objective of an industrial engineering department is usually twofold: (1) to establish methods for controlling production costs, and (2) to develop programs for reducing those costs. Both the methods and the programs are carried out by line management.

In most cases, the industrial engineering department exists primarily to provide specialized services to the production division. The services may be few or many and normally include such functions as time study, methods study, and the development of wage incentive programs. The industrial engineering department is ordinarily held fully responsible for the applicability and accuracy of the programs developed and recommended.

Other divisions may use its services infrequently or extensively. Should their demands consistently be at a high level, it is often advisable to set up service departments in the other divisions so that industrial engineering can concentrate on its primary responsibility to the production division. If this step is taken, formal organizational relationships between the service departments are seldom necessary because they will have very little in common beyond a uniform approach to problem solving. When industrial engineering is occasionally asked to take on work in addition to its continuing services, it customarily acts only as an expert advisor. Responsibility for the actions taken rests jointly with the department requesting assistance and the industrial engineering department.

In some cases, the industrial engineering department is assigned to head projects that will probably never recur.   In other cases, it may participate in a less important role.   The reputation of the department will undoubtedly have an important influence on the scope of its responsibilities in these studies. Whoever is in charge, however, should define the role and responsibility of every participant. Industrial engineering can then execute its task and assume its defined share of the total task.

The Department Must Weigh Company-wide Effects of Its Recommendations. Beyond the specific and obvious responsibility of the industrial engineering department to tie in its work in the production division with other divisions is its responsibility to recognize how the effects of its recommendations may reach beyond the area studied. The operating divisions are concerned with regular planning and execution of agreed-upon company programs; each division plays a well-understood part. The industrial engineering department, however, is for the most part an advocate of change. When it recommends changes in the production division, it must anticipate the probable consequences of those changes for other divisions. This point is emphasized for two reasons:

1. The action recommended in a production department may open up improvement possibilities in other divisions. Sometimes, of course, these secondary improvement possibilities turn out to be rather unimportant. Often, however, they can lead to dramatic changes that not only benefit the nonproduction department but reinforce the action taken in the production department.

2. Occasionally the value of an apparent improvement in the production divisions may be more than offset by the problems and additional costs if will create in another division. Sometimes the bad effects can be erased by modifying the recommendations; other times the recommendations should not be made.

Conduct of Department Members Important to Their Success. Industrial engineering can materially enhance its standing in the company by the conduct of its staff. Some hints on promoting good relationships with line personnel are:

Recognize the Service Role of the Department. Members of the industrial engineering department must recognize that they hold a unique position in the company. As troubleshooters with special training, their work usually calls for upsetting the status quo. Because that may be an unpopular assignment, the successful industrial engineers must build confidence, take care not to disparage production efforts, and observe department protocol if they wish to be accepted by line management.

Steer Clear of Company Politics. An industrial engineer is occasionally assigned problems loaded with political implications. In such a case, it is important for him to be completely objective both in attitude and in action. Any other course will jeopardize his future acceptability and effectiveness.

Respect Line Management's Prerogatives. In his service capacity, the industrial engineer must work through line supervision. Before undertaking a study in any department, he must see the supervisor or foreman, explain the purpose of his study, and solicit assistance. The department supervisor should be made a part of its staff. Some hints on promoting good relationships with line personnel are: give all orders to departmental personnel necessary for the conduct of the study.

Give Proper Credit for Contributions. The industrial engineer should be quick to give others credit for contributing to the development of sound recommendations. Such recognition does not detract from his own reputation as an idea man; on the contrary, it enhances his standing. It gives him a reputation as a catalyst. It will also make his work easier, for operating men will come to him more readily with good suggestions.

PLACE OF INDUSTRIAL ENGINEERING DEPARTMENT
IN COMPANY ORGANIZATION STRUCTURE

With an idea of the nature of the industrial engineering department, its responsibilities, and its working relationships within a company, where it belongs in the organization structure can now be considered. One guideline might be mentioned here, apparent though it may seem: the industrial engineering department should report to the executive that has line responsibility for the departments it regularly serves. Thus if a vice president for operations is responsible for coordinating research, engineering, production planning, sales, quality control, personnel, and so forth, he should probably have industrial engineering, too. If, however, there is a works manager to whom production activities report, the industrial engineering department also should report to him (see Figure 3-1).

This guideline admits many interpretations; for, practically speaking, there can be no hard and fast rule governing the place of industrial engineering in the organization structure of all companies and plants. To learn what forms have been effective, a number of companies were reviewed on this question: "To what executive does the industrial engineering department report, and what other executives are on the same level as that executive?" At the same time, an explanation was sought of any line or functional direction the industrial engineering department received from any other executive, from a headquarters industrial engineering group, and the like.

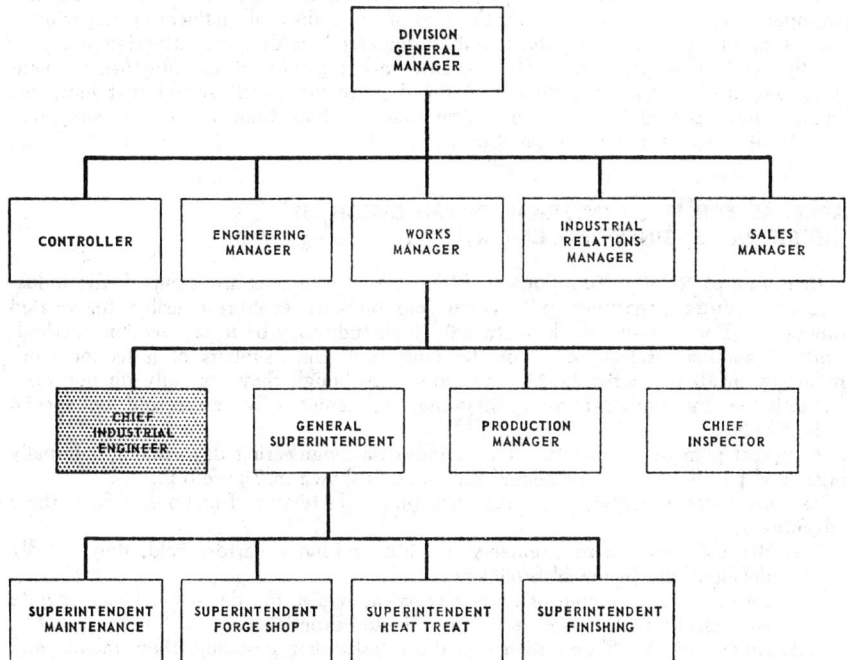

Fig. 3-1. Organization chart illustrating a typical location for the chief industrial engineer in a division of approximately 1,000 employees.

Although the review showed no single, typical position in the organization structure for the industrial engineering department, it did point up the factors that influence its position. These are: (1) the number of direct labor employees, (2) the scope of industrial engineering activities, and (3) the complexity of manufacturing operations.

Number of Direct Labor Employees. The formal recognition of industrial engineering as a full-time activity grows as the number of employees grows; at the same time, however, its organizational rank seems to become lower.

In the small companies, it was indicated that the top manufacturing man did all the industrial engineering, in addition to other duties, and that no formal department existed.

In companies of 300 to 500 employees, the men performing industrial engineering activities reported variously to executive vice presidents and production or operations vice presidents. In these cases, the chief industrial engineer was on the same organizational level as the superintendent and chief product engineer (but not necessarily considered as important).

In companies with 600 and over hourly paid employees, industrial engineering was formally recognized as a department and occupied a lower place in the organization plan. In these companies, the chief manufacturing executive did not have authority over engineering, quality control, and other activities, which are often grouped under one executive in smaller companies, and the industrial engineering department reported to a factory manager, works manager, or superintendent.

Scope of Industrial Engineering Activities. The activities of the industrial engineering department influence its place in the organization structure to this extent: it reports to the executive that is responsible for the departments in which it does the bulk of its work. In most cases, however, it provides some services to other departments.

Complexity of Manufacturing Operations. The review showed that as manufacturing operations grew more complex, the size of the industrial engineering department increased, but the scope of its activities decreased, with more attention going to strictly production problems. The increase in complexity of manufacturing operations and the corresponding growth of the department usually meant that industrial engineering reported at lower organizational echelons than it did in companies with less complex production operations.

GENERAL FORMS OF INTERNAL ORGANIZATION OF INDUSTRIAL ENGINEERING DEPARTMENT

Grouping of Related Functions. When related functions are grouped, the industrial engineering department is divided into sections, each responsible for related functions. For example, work relating to time study may be in one section, methods work in another, and so on. For the most part, the members of a section work regularly on all the activities in the section. Although they normally do not work on activities in another section, they may be temporarily transferred on special assignments.

A typical plan of organization for an industrial engineering department, internally organized by the grouping of related functions, is shown in Figure 3-2.

As an organization device, the grouping of related functions offers these advantages:

1. Because department engineers specialize within a narrow field, they rapidly develop high technical proficiency.
2. By limiting the range of its engineers' activities, the department can operate successfully with personnel of more limited experience.
3. Engineers can be easily assigned to tasks that best suit their talents and interests.
4. Line supervisors, assigned to the department for special training before advancement, can learn much about industrial engineering in a short time.

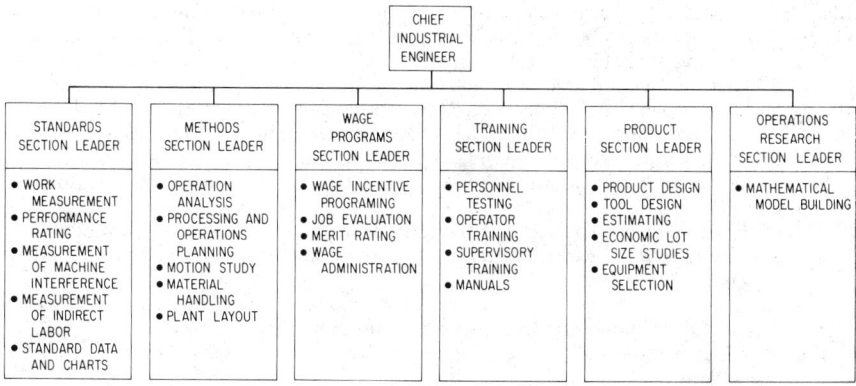

FIG. 3-2. Industrial engineering department organization by grouping of related functions.

The disadvantages of such an organization are:

1. Because the engineers become specialists and work with all departments, they are less likely to develop a close working relationship with line supervisors in any single department.
2. The specialist is less likely to be able to relate the effect of his work to the overall operation of the department in which he is working.
3. It is more difficult to transfer a man with specialized experience from one section of the industrial engineering department to another or from the department into a line supervisory position.
4. If there are insufficient specialists in one section of the industrial engineering department, the solution of high priority problems in some operating departments may be delayed until they become available.

Organization Paralleling That of Production Division. The other common pattern for the internal organization of the industrial engineering department is to parallel the organization of the production division. In a plant that has, for example, a heavy machine department, a light machine department, and an assembly department, industrial engineering may be organized in sections to provide all or almost all the industrial engineering services to each (see Figure 3-3). The engineers in each section work with flexibility on all the problems of their allied produc-

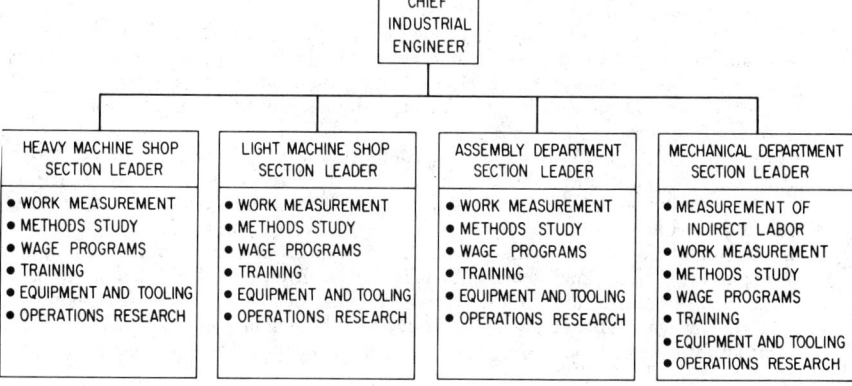

FIG. 3-3. Industrial engineering organization paralleling that of production division.

tion department.   They are also transferred much more freely among sections than are their counterparts in departments organized by a grouping of related functions.

It is interesting to note that all companies with this type of organization usually had one "floating" section that was not tied to any specific production department. It was ordinarily concerned with such activities as supervisory training, job evaluation, merit rating, and so forth.   The engineers working on those activities did so in all production departments served by the industrial engineering department.

Several advantages are ascribed to an organization patterned along production division lines:

1. This arrangement ensures that the highest priority problem in each production department gets earliest attention.   Rarely are all engineers in a section engaged on a special project.
2. An easy and friendly working relationship usually develops earlier between production department supervisors and the industrial engineers.
3. The supervisors gain a good perspective of the production department's operations.
4. This form of organization provides excellent specialized training for individuals being groomed as line supervisors of the department served.
5. When men are transferred between sections of the industrial engineering department, they bring a fresh viewpoint that facilitates problem solving.
6. Men working on all aspects of a department's problem receive very broad training and a good picture of how industrial engineering services complete and complement the production department.   This enlarges their growth opportunities and promotes a better industrial engineering department.

Some of the disadvantages of this form of organization are:

1. The department must have available personnel with broader interests and greater talent than those required under a system which permits individuals to specialize in a single industrial engineering technique.
2. An individual's proficiency in any one field is not likely to be obtained so quickly or in so high a degree as that of a specialist who works with only one kind of problem.

Organization on a Project Basis.   Regardless of how the industrial engineering department is organized, all companies assemble special teams from time to time to conduct certain projects.   Some companies regularly maintain a floating section for just that purpose.   In most cases, however, the special project team includes engineers temporarily recruited from the various sections of the industrial engineering department.   Project teams may be formed for an assignment in the production division—for example, to overhaul the wage incentive plan—or for special work in another division not ordinarily served by the department.

Factors Governing Choice of Organization Pattern.   Why a company chooses one form of industrial engineering department organization in preference to the other is due to some of these considerations:

1. Complexity of production operations may make it inadvisable to move an industrial engineer from one operation to another.   In such a case, it is best to have an industrial engineering organization which parallels that of the production division so that an engineer need learn only one phase of the total process in detail.
2. Where the industrial engineering department is the major service group in the company and regularly works in other divisions besides production, it is usually organized by a grouping of related functions so that specialists will be available for assignment anywhere in the company.
3. Multiplant companies with a home office industrial engineering group organized on a functional basis (that is, by a grouping of related activities) tend to organize the industrial engineering departments of the plants in a similar way. This is especially apparent where the plants produce different products, the reason being that it enables headquarters to coordinate the

plants effectively despite the dissimilarities in production. In multiplant companies with highly autonomous plants, the pattern is the reverse. There industrial engineering is organized to parallel the department organization of production divisions.

4. If the plant or production department is small, a large staff of industrial engineers may not be justified; so specialization by function may not be possible.

5. If there are natural subdivisions of the business such as fabricating, machinery, and welding, the industrial engineering organization should follow those lines. If the business is highly integrated, however, with activities in one area having a substantial impact on activities in another, a centralized department will ensure that the business is treated as a single unit.

The choice as to whether the industrial engineering organization should be centralized or decentralized depends on some of the following features:

1. A large company may find it necessary to decentralize its industrial engineering to promote a more detailed understanding of operations by the industrial engineer and to improve relations with operating personnel.

2. When profits are low, a company may not be able to afford a large decentralized industrial engineering staff, but must content itself with a small centralized staff which emphasizes profit improvement and cost reduction projects.

3. If the industrial engineering department is in its early growth stages, the training of large numbers of new personnel can often best be carried out with a centralized form of organization.

4. When labor costs represent a large proportion of sales price, the profit improvement potential of industrial engineering is highest. A large decentralized staff with a centralized core of specialists will often assure that profit improvement opportunities in all manufacturing areas are given proper attention.

TYPICAL FORMS OF INTERNAL ORGANIZATION

The number and variety of forms that the internal organization of an industrial engineering department may take are practically limitless. It will be useful to examine several organization patterns which are quite commonly used.

The Small Company (100 to 300 on Direct Labor). Small companies commonly operate with an industrial engineering department of one or two engineers. The industrial engineering activity is typically organized as shown in Figure 3-4. The senior industrial engineer should be a man with a wide variety of capabilities because the functions of industrial engineering that need to be performed in a small company are not limited in number, but only in degree. Typically, the emphasis will be on work standards, methods, and cost reduction. Plant layout, estimating, processing, material handling, and the like will receive lesser emphasis.

The Medium Company (300 to 600 on Direct Labor). In a medium-size company, the organization generally takes the form of individual sections for the various facets of industrial engineering. The need for sophistication in controlling flow through the more numerous and larger manufacturing departments, the need for greater control of labor hours, and the need for better methods, equipment, and training, all tend to create a demand for the industrial engineer who is an expert in one of the functions of this field. Hence, the organization takes on the appearance of Figure 3-2. Functions are well defined, and the title, "industrial engineer," generally used in the small company, is of lesser prominence. Such titles as work measurement engineer, methods analyst, process engineer, and material handling engineer are used to indicate the division of responsibility for industrial engineering service.

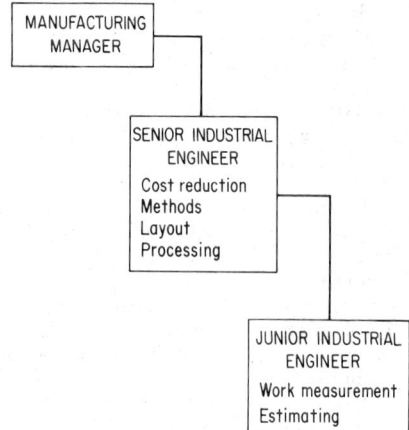

Fɪɢ. 3-4.   Typical organization for small companies.

The primary emphasis will not radically change from that of the small company. Work standards, methods, and cost reduction will still be primary, but will be carried out by individuals who are expert in each of these fields.   Processing becomes of greater importance and will receive more emphasis.

The Large Company (over 600 on Direct Labor).   The internal organization of a large company often also assumes the form of Figure 3-2.   In the large company, increased emphasis is placed on such functions as value engineering, operations research, training, and wage programs, none of which could feasibly be done in depth under the paralleling of production method.   The necessity for increasing the depth, intellectually and in numbers, and yet retaining control of the various activities, is still another reason for using the method of grouping by related functions.

The Corporate Level.   Perhaps the most important function of the corporate industrial engineering group is to coordinate the efforts of the industrial engineering

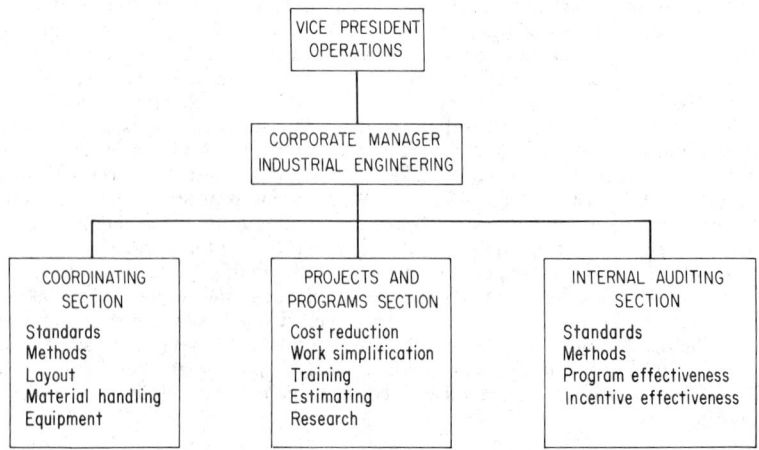

Fɪɢ. 3-5.   Organization at corporate level.

groups in all the corporation's subsidiaries, divisions, or plants. The corporate group should also be responsible for initiating special projects and programs as required. It should be in a position to report objectively to corporate management on conditions and progress of industrial engineering activities in the suborganizations.

These functions can be accomplished by organizing the corporate group into three basic subgroups, with the corporate manager of industrial engineering reporting to the vice president of operations, as shown in Figure 3-5. The three basic subgroups are designated as (1) the coordinating section, (2) the projects and programs section, and (3) the internal auditing section. The size of the corporate group will depend upon the corporation size, the number of outlying plants, financial return on industrial engineering effort, and the like. The basic functions remain, however, no matter what the size.

With a corporate industrial engineering group in place, the industrial engineering groups in the subsidiaries, divisions, or plants are often organized to parallel the production division organization, as shown in Figure 3-3.

SPECIAL FORMS OF ORGANIZATION OF THE INDUSTRIAL ENGINEERING DEPARTMENT

The organization form chosen by any company will be governed partly by the philosophies of its management group, partly by the nature of the work it performs, and partly by the size of the company. An examination of some special forms of organization will show this clearly.

Industrial Engineering Combined with Industrial Relations. A unique organization, shown by Figure 3-6, combines industrial engineering and industrial relations. The department is headed by an industrial relations manager who is responsible for the labor relations, personnel, employment, and industrial engineering functions of the division. He reports directly to the divisional general manager.

INDUSTRIAL RELATIONS DEPARTMENT

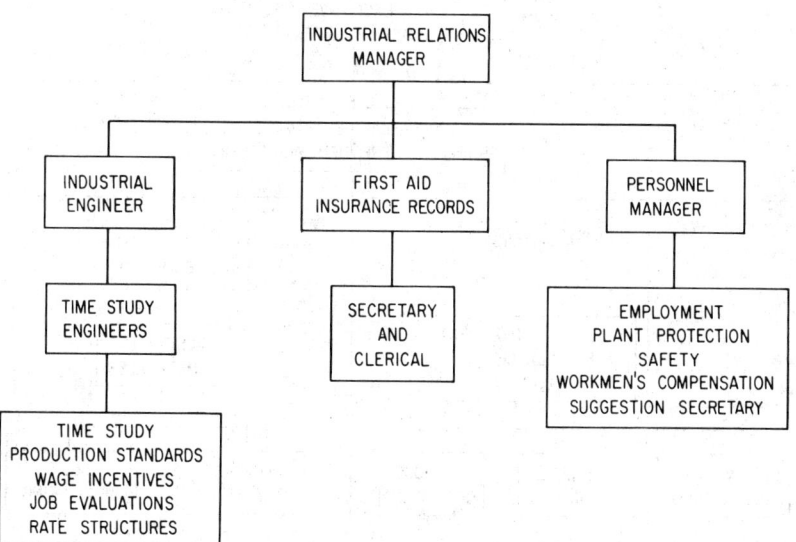

Fig. 3-6. Organization combining industrial engineering and industrial relations. (*Based on information available through the courtesy of J. M. Eikenberg, General Manager, Clinton Division, Revere Copper and Brass, Inc.*)

Although all members of the staff are at one point or another involved in the labor relations or grievance procedure function, the industrial relations manager normally reserves for himself the top level grievance and negotiating duties. The industrial engineering department, normally headed by an industrial engineer having time study engineers reporting to him, is responsible for time studies, production standards, wage incentives, job evaluations, rate structures, and cost reduction studies, plus any other assigned duties. The personnel department, normally headed by a personnel manager, is responsible for employment, protection, safety, suggestions, communications, and any of the normal personnel functions. Newsletters, house organs, written bulletins, and memos are handled either by the industrial relations manager personally or through the personnel department.

The separation of industrial engineering duties from the person heading production in the plant is felt to be a strong plus. It ensures two separate viewpoints and serves to provide checks and balances.

The combination of industrial engineering and personnel with labor relations fosters a more understanding approach to the common problems faced by each function. It also provides the head labor relations man with a somewhat different outlook from the person who functions only through the regular personnel department channels.

Banks. Figure 3-7 shows how a bank has organized its industrial engineering function into three divisions: technical planning, methods research, and operations analysis.

All three divisions are in the operating department, which is headed by a senior vice president and cashier, who reports to an executive vice president. The operating department has the largest number of employees within the bank and is responsible for the organization and execution of the bank's many services to customers, employees, and the general public. All physical facilities as well as supporting services are within the operating department's responsibilities.

Two divisions—technical planning and methods research—report to a second vice president, who in turn reports to a vice president in the operating department.

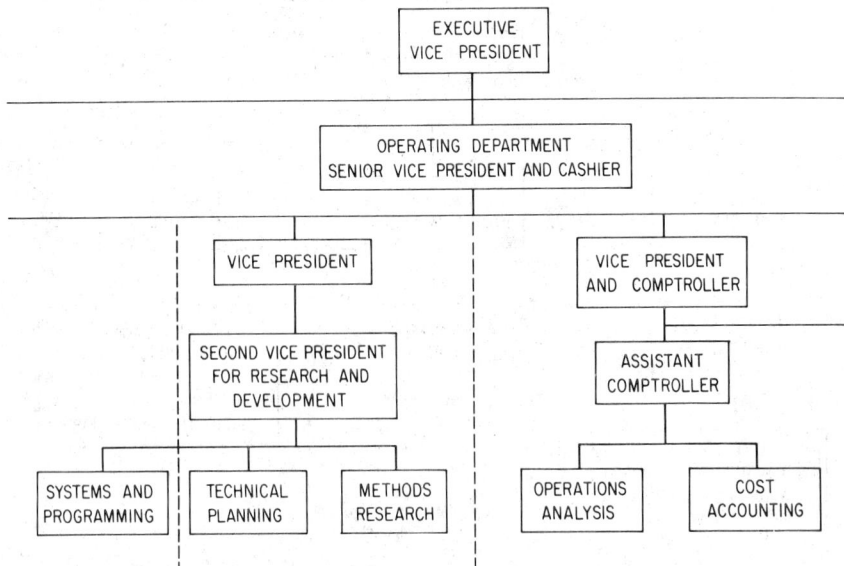

FIG. 3-7. Organization of the industrial engineering function in a bank. (*Courtesy of Alfred E. Clem, Public Affairs Officer, Continental Illinois National Bank and Trust Company of Chicago.*)

PRIME FUNCTION

TO ASSIST MANAGEMENT TO OBTAIN OPTIMUM OPERATIONAL COST THROUGH THE USE OF INDUSTRIAL ENGINEERING TECHNIQUES DESIGNED TOWARD ACHIEVING MAXIMUM UTILIZATION OF MANPOWER, EQUIPMENT, AND INVESTMENT DOLLAR.

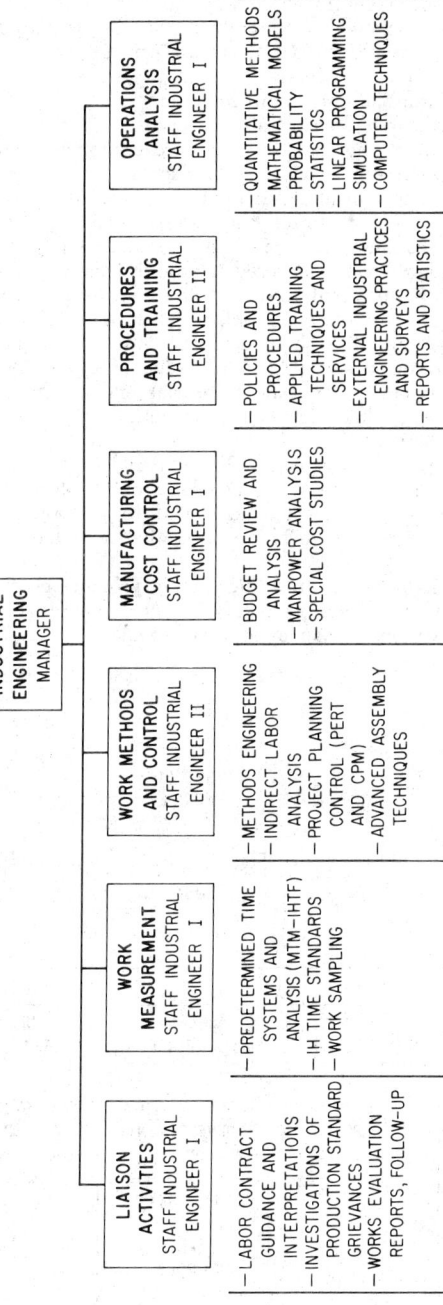

INDUSTRIAL ENGINEERING MANAGER

LIAISON ACTIVITIES STAFF INDUSTRIAL ENGINEER I
- LABOR CONTRACT GUIDANCE AND INTERPRETATIONS
- INVESTIGATIONS OF PRODUCTION STANDARD GRIEVANCES
- WORKS EVALUATION REPORTS, FOLLOW-UP

WORK MEASUREMENT STAFF INDUSTRIAL ENGINEER I
- PREDETERMINED TIME SYSTEMS AND ANALYSIS (MTM–IHTF)
- IH TIME STANDARDS
- WORK SAMPLING

WORK METHODS AND CONTROL STAFF INDUSTRIAL ENGINEER II
- METHODS ENGINEERING
- INDIRECT LABOR ANALYSIS
- PROJECT PLANNING CONTROL (PERT AND CPM)
- ADVANCED ASSEMBLY TECHNIQUES

MANUFACTURING COST CONTROL STAFF INDUSTRIAL ENGINEER I
- BUDGET REVIEW AND ANALYSIS
- MANPOWER ANALYSIS
- SPECIAL COST STUDIES

PROCEDURES AND TRAINING STAFF INDUSTRIAL ENGINEER II
- POLICIES AND PROCEDURES
- APPLIED TRAINING TECHNIQUES AND SERVICES
- EXTERNAL INDUSTRIAL ENGINEERING PRACTICES AND SURVEYS
- REPORTS AND STATISTICS

OPERATIONS ANALYSIS STAFF INDUSTRIAL ENGINEER I
- QUANTITATIVE METHODS
- MATHEMATICAL MODELS
- PROBABILITY
- STATISTICS
- LINEAR PROGRAMMING
- SIMULATION
- COMPUTER TECHNIQUES

FIG. 3-8. Industrial engineering organization—corporate level. (*Courtesy of H. W. Haupt, Manufacturing Research, International Harvester Company.*)

The third group—operating analysis—is part of the accounting division, reporting through an assistant comptroller to the vice president and comptroller, and thus to the senior vice president and cashier who heads the operating department.

The technical planning division has two primary functions: quantitative studies and computer hardware evaluation. "Quantitative studies" is the bank's term for what many other firms call operations research.

The methods research division also has two main jobs: evaluation of equipment (other than computers, but including tab card equipment) and management consulting, including the determination of costs associated with the operations under study.

As part of the accounting function, the operations analysis group examines the bank's manual systems, performs time study analyses, makes work simplification recommendations, conducts division-level operations analyses, and provides as a by-product time standards which are used to establish standard costs.

Corporate and Plant Combination.   The establishment of industrial engineering departments at both corporate and plant levels is quite common in large corporations. Figures 3-8 and 3-9 show how the industrial engineering organizations at the corporate and plant levels vary in design and primary functions in one large plant. At the corporate level, the industrial engineering organization is designed on a functional basis, while the plant organization is formulated according to specific production activities.

The prime purpose of the corporate industrial engineering section is to provide consulting services for the divisions and plants on all phases of industrial engineering and manufacturing analysis. It is designed to promote and develop the use of new industrial engineering techniques in the manufacturing areas of industrial operations.

The corporate industrial engineering section provides corporate and plant management with feasibility studies designed to reveal potential cost reductions through the utilization of various management science techniques, including computer applica-

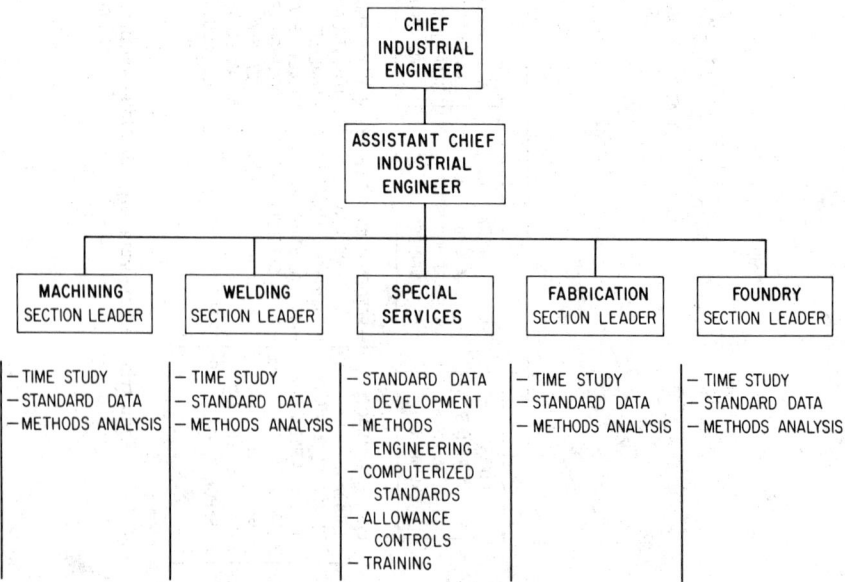

FIG. 3-9.   Industrial engineering organization—plant level. (*Courtesy of H. W. Haupt, Manufacturing Research, International Harvester Company.*)

tions. It further provides training and guidance in such areas as work measurement, work methods, manufacturing costs, and operations analysis activities. In labor relations, it develops and recommends labor contract language and renders counsel on grievances. These functions are divided into six major areas of industrial engineering activities, with a staff industrial engineer responsible for each.

The major function of the industrial engineering department at the plant level is to establish incentive production standards by means of stopwatch time studies and the application of predetermined (MTM) time standards.

Other areas of activity include the development of new standard data, methods analysis studies, manufacturing cost studies, and the administration and control of direct labor allowances.

Training courses are conducted periodically by designated instructors in methods time measurement, methods engineering, labor allowance control, and application of standard data. Performance rating sessions are also conducted on a monthly basis.

The section is organized by specific production activities within the manufacturing plant, such as machining, welding, fabrication, and foundry. Each production activity is supervised by a section head who is responsible for all industrial engineering activities within his assigned department.

INDUSTRIAL ENGINEERING PERSONNEL

The same principles of personnel administration that apply to other line and functional departments apply likewise to industrial engineering. The reader is undoubtedly familiar with them. What he may not realize are the special personnel considerations that set industrial engineering a little apart from the general run of departments. The major distinctions are discussed below.

Industrial Engineering as a Career. Often the industrial engineering department is staffed largely by recent college graduates whose major qualification is educational background. These men may lack the stature and maturity to influence action. In selling incentive piece rates to workers, in gaining the confidence of line supervisors, and in persuading executives to act, maturity and firsthand experience are beneficial. It therefore appears desirable that the well-staffed department have enough thoroughly experienced senior engineers to balance the younger group.

Training. An industrial engineering department can provide excellent training for men who will eventually assume line responsibility, but too often nothing is done to ensure that the training is complete and effective. This is as shortsighted as turning a man loose in a research laboratory and hoping that, with all the experiments going on, he will learn a lot. The experience may be interesting but not productive.

Training in industrial engineering should be formalized. The first obvious step is a good orientation program in which the new man learns what the department's mission is. If he is not familiar with operations in the departments which industrial engineering serves, he needs that orientation too. Then he must get specific training in the following areas:

1. Methods and procedures used by the industrial engineering department
2. The formal and informal organization structures of the company, the production division, and the industrial engineering department
3. The technique of operating in a service atmosphere and dealing with workers, supervisors, management, and union officials
4. Pertinent technical phases of production operations
5. Company policies on time study, wage administration, areas of participation by the union, craft observance, and the like

If he is a new employee, he should also have the regular company indoctrination.

Recognition. Recognition of both the department's and the individual's contributions is important. First, it stimulates further good work. Second, it builds the reputation of the department and thus encourages operating departments to make wider use of its services.

To ensure that its contributions and accomplishments are properly recognized, the industrial engineering department might take these steps:

1. Prepare annual programs of work and anticipated attainments in dollar savings, just as a line department does; prepare period progress control reports listing accomplishments to date.

2. Promote internal recognition of accomplishment and outside-the-company publicity on work done, and where applicable, give the names of the department members making major contributions. Technical society papers and magazine articles are excellent outlets. Internal recognition can be gained by reports which point up individual and group accomplishments, assignment to greater responsibilities or more complex problems, and pay increases.

DEPARTMENTAL ADMINISTRATION

Successful administration of a functional department such as industrial engineering is perhaps more difficult to achieve than in the case of a line department. One reason is that it is more difficult to develop and administer standards of performance and to measure attainment. That inherent difficulty may, however, be turned to the advantage of the department. Because department members receive less direction and control than operating personnel, they must be self-starters for work to get done. Their pride in doing a job well and quickly is a great asset to successful administration.

GETTING ACTION ON RECOMMENDED PROGRAMS

Because industrial engineering is a service department without line authority, it often has difficulty getting people to act on its recommended programs. This can be minimized by the manner in which the study is made, that is, whether or not the department sells as it goes along. It can also be minimized if the recommendation is in the form of an action program that makes it easy for management to act. Other steps the industrial engineering department can take to induce action include:

Emphasize Benefits to Entire Company. The broader the range of significant benefits, the more likely that action will result. First, more members of management will be interested in achieving the benefits; second, the action cannot die through the inertia of one executive.

Point up Compensating Benefits to Individuals. If a recommended action program will relieve executives of problems, the desirability of action will be enhanced. If the recommendations reduce the number of workers reporting to an executive or otherwise lessen his prestige or authority, the attendant benefits must be strongly emphasized to get him to act.

Spell out the Action Program. The action program should describe each step and designate each participant's responsibility and the date projected for completion of his assignment. Such a program will make it easy for management to take action.

Provide Progress Checkpoints in the Installation Plan for Use by Line Supervision. The installation plan should have a procedure for controlling the progress of the installation similar to that recommended for projects within the industrial engineering department. Where practicable, a member of line operating management, rather than a member of the industrial engineering department, should follow up the installation program.

CONCLUSION

This chapter covers the broad subject of organizing for industrial engineering. After reading it, the reader will, it is hoped, agree with the conclusion that the effectiveness with which an industrial engineering department operates depends

largely on how well the following factors are understood and weighted. The industrial engineering department must be considered in terms of its:
1. Role in the business
2. Working relationships with other departments
3. Place in the company plan of organization
4. Form of organization
5. Personnel
6. Administration
7. Getting action on recommended programs

Where these factors are properly related, a company can expect continuing improvements in control over production costs and steady reductions in actual production costs.

BIBLIOGRAPHY

Carrabino, Joseph D., "The Metamorphosis of Industrial Engineering into Management Engineering: Plea for Change in Name," *The Journal of Industrial Engineering,* January–February, 1961.
"Conference Board Reports," Studies in Business Policy, no. 78, National Industrial Conference Board, Inc., New York, 1956.
Forberg, Richard A., "Effective Control of the Industrial Engineering Function," *Proceedings 12th Annual SAM-ASME Management Engineering Conference,* April, 1957, pp. 204–223.
Gilbreth, L. M., "Organizing for Industrial Engineering," *Proceedings of the Seventh Annual Conference,* American Institute of Industrial Engineers, New York, 1956, pp. 14–24.
Kelly, Thomas, "How to Measure and Appraise Your Industrial Engineering Department Performance," *The Journal of Industrial Engineering,* March–April, 1959, p. 144.
Maynard, H. B. (ed.), *Handbook of Business Administration,* sec. 2, McGraw-Hill Book Company, New York, 1967.
Mundel, M. E., "Improving Organization and Performance of an Industrial Engineering Department," *Advanced Management,* July, 1957, pp. 8–12.
Rathe, Alex W., and Frank M. Gryna, *Applying Industrial Engineering to Management Problems,* Research Study 97, American Management Association, New York.
Villers, Raymond, "Organizing for Better Utilization of Industrial Engineers," *Proceedings of the Tenth Annual Conference,* American Institute of Industrial Engineers, New York, 1959.

Chapter 4

Source, Training, and Development of Industrial Engineers

MORLEY H. MATHEWSON

**Director, Long Range Strategic Planning,
American Management Association, Inc., New York, New York**

Americans living in the United States of America have enjoyed one of the highest living standards in the world because of high productivity and the continuing emphasis placed upon improving all phases of our society. Although everyone in society has contributed to these improvements, it is the industrial engineer who devotes all his efforts to finding, developing, and implementing better ways of effectively utilizing human and natural resources to benefit all mankind.

Beyond the United States, recognition of the contributions which industrial engineers can make to increase productivity has grown throughout the countries of the free world since World War II. Thus it is understandable that increasing attention and emphasis must of necessity be given to the education, training, and development of professional industrial engineers.

SOURCE OF PROFESSIONAL INDUSTRIAL ENGINEERS

There is a great deal of concern in education, government, and industry over the decline in the number of students who have been awarded engineering degrees during the years 1959 to 1968. The rate of decline is reflected in the number

TABLE 4-1. Undergraduate Industrial Engineering Degrees and Enrollment
ECPD Accredited Engineering Curriculum

Engineering curriculum	Year	No. first engineering degrees	No. increase or (decrease)	Percent increase or (decrease)	Rank	No. enrolled first engineering degrees	No. increase or (decrease)	Percent increase or (decrease)	Rank
Electrical.........	1959	9,837				53,849			
	1968	9,322	(515)	(5.2)	4	52,509	(1,340)	(2.5)	3
Mechanical.........	1959	8,300				40,051			
	1968	6,774	(1,526)	(18.4)	5	38,655	(1,396)	(3.5)	4
Civil.........	1959	4,939				25,280			
	1968	5,063	124	2.5	3	27,574	2,294	8.1	2
Chemical.........	1959	3,013				17,166			
	1968	3,177	164	5.4	2	16,166	(1,000)	(5.8)	5
Industrial.........	1959	1,904				6,632			
	1968	2,139	235	12.3	1	8,509	1,877	28.3	1
Total.........	1959	27,993							
	1968	26,475	(1,518)	(5.4)					

of first engineering degrees awarded to the five leading engineering disciplines in all United States schools with curricula accredited by the Engineers' Council for Professional Development. This downward trend in the first engineering degrees granted to each of the five leading engineering curricula shows a drop from 27,993 in 1959 to 26,475 in 1968—a total decrease of 5.4 percent in ten years. The significance of this drop is realized when compared with the unprecedented expansion in the number of college degrees awarded in fields other than engineering.

Industrial engineering is one of the youngest and fastest growing professions in the engineering field. At the turn of this century, the outstanding pioneers in industrial engineering began to be recognized for their important contributions. Industrial engineering's rapid growth and wide recognition in society have taken place since 1945. In this same period, an increasing number of schools have offered courses in this profession. One measure of its youth was the founding of the American Institute of Industrial Engineers, Inc. (AIIE) in 1948. From a position of relative obscurity at the turn of this century, industrial engineering is now ranked with the top five leading engineering professions.

The growing demand for professional industrial engineers is reflected by the fact that the number of professional industrial engineers graduating with first engineering degrees from institutions with curricula accredited by the Engineers' Council for Professional Development (ECPD) has been growing in the face of declining engineering enrollment.

Industrial engineering ranked fifth behind electrical, mechanical, civil, and chemical engineering in the number of first engineering degrees for the years 1959 and 1968. However, industrial engineering ranked first in terms of percentage growth during this period for degrees awarded and engineering enrollment (ECPD accredited curriculum). These trends are noted in Table 4-1.

Of equal interest was the increasing number of schools offering industrial engineering curricula. This trend for the five top engineering curricula was as follows:

Engineering curriculum	Year	No. of schools*	No. increase or (decrease)	Percent increase or (decrease)	Rank
Electrical...............	1959	145			1
	1968	209	64	44.1	
Mechanical...............	1959	141			3
	1968	198	57	40.4	
Civil....................	1959	141			4
	1968	185	44	31.2	
Chemical.................	1959	113			5
	1968	140	27	23.9	
Industrial...............	1959	71			2
	1968	102	31	43.7	

* This refers to the number of schools offering each curriculum. A particular curriculum in a given school may or may not be specifically approved by ECPD.

A major source of professional industrial engineers is the graduates from the following institutions with ECPD accredited curricula leading to first degrees in industrial engineering.

INDUSTRIAL ENGINEERING
ACCREDITED CURRICULA—1968

Alabama, University of
Arkansas, University of
Auburn, University of
Bradley University

California, University of (Berkeley)
Florida, University of
Georgia Institute of Technology
Houston, University of

Illinois Institute of Technology
Illinois, University of
Iowa State University
Iowa, University of
Johns Hopkins University
Kansas State University
Lafayette College
Lamar State College of Technology
Lehigh University
Louisiana State University
Massachusetts, University of
Miami, University of
Michigan, University of
Missouri, University of
Montana State University
Newark College of Engineering
New York at Buffalo, State University of
New York University
North Carolina State University at Raleigh
Northeastern University
Northwestern University

Ohio State University
Oklahoma State University
Oklahoma, University of
Oregon State University
Pennsylvania State University
Pittsburgh, University of
Purdue University
Rhode Island, University of
Rutgers—The State University
Saint Louis University
Southern California, University of
Southern Methodist, University of
Stanford University
Syracuse University
Tennessee, University of
Texas A & M University
Texas at Arlington, University of
Texas Technological College
Virginia Polytechnic Institute
Wayne State University
West Virginia University

OPTIONS AS PART OF OTHER
ACCREDITED CURRICULA

Akron, University of (M.E.)
Arizona State University (ENG.)

Cornell University (M.E.)
Toledo, University of (M.E.)

*Industrial and Management
Engineering*
Columbia University

*Industrial and Systems
Engineering*
Ohio University
San Jose State College

The quality and scope of the fifty-seven institutions with ECPD industrial engineering accredited curricula have varied widely. Some have been restricted to the more traditional technique courses. Some confine themselves almost entirely to manufacturing. However, an increasing number have adopted the "new look" of interweaving the newer developments in operations research and management sciences with the more traditional curriculum pattern.

Students and employers interested in obtaining more information relative to the general program content and objectives of the schools can write to the institutions for the course catalogs. These catalogs are an excellent reference source to learn about the industrial engineering philosophy of the schools approved by ECPD.

To name only a few, Georgia Institute of Technology, Ohio State University, Purdue, Northwestern, Stanford, and Johns Hopkins were considered to have outstanding, forward-thinking industrial engineering programs. Many other schools could be mentioned for their excellent work in specific areas of study.

Professional industrial engineers graduating from accredited institutions are in great demand. In fact, all engineering disciplines are in demand, as evidenced by the salaries being paid to beginning engineers. Starting engineering salaries continue to be among the highest of all categories of college graduates. The trend in salary offers for the top five engineering curricula is shown in the table on page 1-42.

Another source of industrial engineers is the industrial engineering technologist who may have a degree from outside the field of engineering. Some students who have earned degrees in such diverse fields as liberal arts, social sciences, mathematics, or physics make good industrial engineers once they have been properly trained. Personnel with multieducational backgrounds are sought by organizations moving toward the systems engineering concept. Industrial engineers are moving into the

Engineering curriculum	Average offers, dollars per month		Percent increase over last year
	1967-1968*	1966-1967	
Electrical............	774	728	6.3
Mechanical...........	768	720	6.7
Civil................	750	706	6.2
Chemical............	790	733	7.8
Industrial...........	757	707	7.1

* A study of *1967–68 Beginning Offers, Final Report, June, 1968,*
The College Placement Council, Bethlehem, Pa.

role of the systems engineer.   This is important because management is looking for engineers who can work on problems dealing with total systems as well as subsystems. For example, the design of a total management information computer system to serve the entire enterprise is the type of problem that requires the attention of systems engineers.   The industrial engineer is assuming the systems engineering role and working with teams with multieducational backgrounds to solve complex problems.

GRADUATE INDUSTRIAL ENGINEERING PROGRAMS FOR MASTER'S DEGREES AND DOCTORAL DEGREES

Many of the ECPD accredited schools offer graduate work. However, graduate programs are not inspected and accredited by ECPD. It is reasonable to assume that accreditation of an undergraduate program is a direct reflection of the capability to provide meaningful graduate work.

The graduate programs in industrial engineering frequently are not restricted to individuals with undergraduate degrees in industrial engineering. Usually, these programs require an engineering or science background, and many schools require the nonindustrial engineer to take noncredit courses in the undergraduate industrial engineering area as part of the graduate program. Almost all the programs are engineering oriented. As in the case of undergraduate studies, the graduate programs vary widely in quality and scope and also in level.

In the large-city areas, many of the graduate programs have been almost entirely evening programs. Columbia University, Stevens Institute of Technology, and Illinois Institute of Technology are good examples. These schools have both day and evening programs but their evening programs predominate. Other schools, such as Northwestern, Purdue University, and Cornell University, have had primarily day programs.

Some of the schools, such as Cornell, Johns Hopkins, Stanford, Ohio State, and Northwestern, have directed their efforts toward the doctoral program and tended to work toward a de-emphasis of a terminal master's program.

NONINDUSTRIAL ENGINEERING GRADUATE PROGRAMS

Some nonindustrial engineering graduate programs have provided a combination of industrial engineering and advanced business courses that allow the engineering graduate to obtain a good industrial engineering background. The schools in this category that have offered good opportunities for obtaining industrial engineering education are Carnegie-Mellon University, formerly Carnegie Institute of Technology (Graduate School of Industrial Administration), Yale University (Department of Industrial Administration), Massachusetts Institute of Technology (School of Management), University of Chicago (Graduate School of Business), and University of California at Los Angeles (Graduate School of Business Administration). These business schools are mentioned because they take advantage of the engineering background of engineering graduates. Such a famous school as Harvard University's Graduate School of Business places emphasis in other areas.

TRAINING PROFESSIONAL INDUSTRIAL ENGINEERS

Since the days of Taylor, Gantt, and the Gilbreths, industrial engineering has been a changing profession. In the early days, industrial engineering was largely time study oriented. By 1960, the dynamic nature of modern industrial engineering was reflected by the broadening scope of industrial engineering to serve all levels of management in problem solving.

Because of the tremendous changes that have taken place within sectors of industrial engineering, an educational problem of major magnitude, both academically and in industry, has developed. Both experienced industrial engineers and recent graduates find difficulty in keeping themselves up to date in their rapidly developing field.

Many companies have recognized the seriousness of the industrial engineering education problem and have sponsored educational activities of various sorts. Many pay tuition and fees and also release time for evening programs. A few grant an employee a leave of absence with at least partial pay for on-campus advanced industrial engineering education. A very few provide full support to their promising industrial engineers for pursuing a master's or doctoral degree program.

A number of companies provide special courses for their own personnel along the lines of industrial engineering and selected topics of a modern nature. Sometimes these programs are offered in conjunction with local universities, sometimes on the basis of having the services of a faculty member, sometimes by means of consultant services, sometimes by use of their own staff personnel, and frequently as a combination of the above.

ADVANCED TRAINING IN INDUSTRIAL ENGINEERING

Such companies as Eastman Kodak, United Air Lines, DuPont, Procter & Gamble, and the International Business Machines Corporation have developed excellent industrial engineering training programs. The Procter & Gamble program will be described briefly to illustrate the kind of advanced industrial engineering training that the more progressive companies consider necessary. Although Procter & Gamble is a large company, with one of the largest industrial engineering organizations in the country, this same type of program, scaled down, can be followed by medium and small companies.

Adaptation of industrial engineering to increasingly complex problems has been accomplished at Procter & Gamble, both by improved selection and by continuing development of personnel.

Training of personnel is accomplished in a variety of ways. Each individual is exposed to training at such times and frequencies as appear appropriate to his background, interests, and current assignments. The major methods of training are (1) formal training courses, (2) formal seminars, (3) informal seminars, and (4) professional society meetings.

Formal Training Courses. Formal training courses are used which are either internally organized or externally available. The internal training courses include the following:

Technique-oriented Courses
  1. Basic statistics
      a. Content—probability, discrete and continuous distributions, analysis of variance, special experimental designs
      b. Text—special P & G manual
      c. Schedule—five sessions of 8 hours each
      d. Frequency—two per year

    2. Advanced statistics

      a. Content—fractional factorials, response surface designs, multiple regression, multivariate analysis, advanced sampling techniques

      b. Text—special P & G manual

      c. Schedule—five sessions of 8 hours each

      d. Frequency—one per year

    3. Finite mathematics

      a. Content—symbolic logic, probability, matrix algebra, Markov chains, linear programming, game theory

      b. Text—selected

      c. Schedule—twenty sessions of 2 hours each

      d. Frequency—one per 2 years

    4. Decision theory

      a. Content—sets and functions, losses versus risks, admissible strategies, a priori versus a posteriori probabilities, Bayesian strategies

      b. Text—selected

      c. Schedule—twenty sessions of 2 hours each

      d. Frequency—one per 2 years

Application-oriented Courses

    1. Multiple regression applications

      a. Schedule—three sessions of 8 hours each

      b. Frequency—four per year or as needed

    2. Linear programming applications

    3. Special courses for specific systems applications

These courses supplement the regular courses in such areas as work simplification. Externally available training courses which have been found helpful include data processing courses provided by such companies as IBM.

Formal Seminars. Formal seminars are considered to be those in which communication is primarily from the speakers to the audience. These seminars, again, are either internally organized or externally available. The internally organized formal seminars generally use the services of outside management consultants. Typical topics for such seminars are linear programming, adaptive systems, feedback systems, and critical path scheduling. They range in duration from one to three days.

The formal seminars which are externally available are organized mainly by the management organizations and by the universities. Seminars such as those conducted by the American Management Association have proved helpful for general orientation in certain areas. The more technical seminars, conducted by such institutions as Massachusetts Institute of Technology, Case Institute of Technology, and Johns Hopkins University, have proved useful for more penetrating study. Typical subjects in both instances would include operations research, systems analysis, management control systems, management information systems, and organization theory.

Informal Seminars. Informal seminars are those in which communications go in both directions between the speakers and the audience. This is feasible only in a relatively small group. Some of these seminars are internally organized, often within a section or department. Subject matter covers a wide range suited to the specific interests of the group. Each individual in the organization serves as the resource person at some time for such seminars.

Other seminars are organized jointly with other companies, either singly or in groups. Mutual exchanges of views with such companies as DuPont, Westinghouse Electric, Eastman Kodak, and Eli Lilly are frequently arranged.

Professional Society Meetings. Attendance at and participation in meetings of various technical societies are encouraged. The societies with which close contact is maintained include the American Institute of Industrial Engineers, the Institute of Management Sciences, and the Operations Research Society of America.

All these techniques supplement regular on-the-job training. An extensive technical library is available. On-the-job training is more effective when the new man is apprenticed to an experienced analyst in the early stages. "Jam sessions" with co-workers, in which deliberate attempts are made to challenge the validity of new approaches, are regularly used to sharpen analytical ability. All these techniques occur naturally in the context of actual operational problem solving.

DEVELOPMENT OF PROFESSIONAL INDUSTRIAL ENGINEERS

Leading educational institutions, educators, and the professional society, the American Institute of Industrial Engineers, all work together to advance the development of professional industrial engineers.

The founding of AIIE in 1948 had a profound effect on the professional status of industrial engineers. AIIE in 1969 had over 17,000 members in 180 chapters in the United States and Canada. The growth and broadening scope of modern industrial engineering and the development of high professional standards attracted many scientists and scholars into the profession and the institute. It also increased management interest in the industrial engineer and his activities in business, government, service organizations, and industry.

Professional industrial engineers have constantly broadened their horizons to serve all levels of management. Using Frederick W. Taylor's scientific management concepts as a springboard, professional industrial engineers have used operations research and systems engineering tools and concepts in developing solutions to complex and difficult problems.

The professional industrial engineer has become concerned with the design, improvement, and installation of systems involving men, materials, and equipment. He therefore must have fundamental training in mathematics, physics, chemistry, and the engineering sciences. The treatment of the human component of the system requires knowledge of the behavior, motivation, capabilities, and limitations of people. Knowledge derived from training in the social sciences, such as psychology and sociology, is necessary to predict and describe the variations in the performance of men, equipment, and processes and the environmental factors affecting them.

CONCLUSION

The modern professional industrial engineer provides solutions to problems before the fact rather than after; he provides more creative assistance in support of dynamic modern management; and he provides factual and unbiased solutions based on the increasing ability to measure, understand, simulate, and manipulate existing and future systems. To meet the increasing demand for modern industrial engineers, greater emphasis and attention must of necessity be focused on the education, training, and development of professional industrial engineers.

BIBLIOGRAPHY

Churchman, C. West, Russell L. Ackoff, and E. Leonard Arnoff, *Introduction to Operations Research*, John Wiley & Sons, Inc., New York, 1957.

Cleland, David I., and William R. King, *Systems Analysis and Project Management*, McGraw-Hill Book Company, New York, 1968.

Forester, John, *Statistical Selection of Business Strategies*, Richard D. Irwin, Inc., Homewood, Ill., 1968.

Hertz, David B., and Roger T. Eddison, *Progress in Operations Research*, John Wiley & Sons, Inc., New York, 1964.

King, William R., *Quantitative Analysis for Marketing Management*, McGraw-Hill Book Company, New York, 1967.

Lehrer, Robert N., *The Management of Improvement; Concepts, Organization, and Strategy*, Reinhold Publishing Corp., New York, 1965.

Malcolm, Donald G., and Alan J. Rowe, *Management Control Systems*, John Wiley & Sons, Inc., New York, 1962.

Morris, William T., *Decentralization in Management Systems: An Introduction to Design*, Ohio State University Press, Columbus, 1968.

Optner, Stanford L., *Systems Analysis for Business Management*, Prentice-Hall, Inc., Englewood Cliffs, N.J., 1968.

Vaughn, Richard C., *Introduction to Industrial Engineering*, Iowa State University Press, Ames, 1967.

Chapter **5**

Winning Acceptance
for Industrial Engineering

KENDALL C. WHITE

Vice President—Manufacturing Services,
TRW Inc., Cleveland, Ohio

Unlike many other business functions, industrial engineering is, in some measure, an optional activity. Because it is undoubtedly heresy to make such a statement in an industrial engineering handbook, an explanation of the meaning follows.

THE NECESSITY FOR WINNING ACCEPTANCE

Every successful business—manufacturing, service, or other—has a recognized sales function to contact the customer, bring in the order, and service the order day in and day out. No manufacturing business will run for long without a recognized finance function or legal function or engineering product design function. These, and other activities, are considered essential parts of a business organization. Companies of any size accept as necessary the personnel function, the purchasing function, and others. But this recognition is not always accorded industrial engineering. Even though certain industrial engineering activities may be carried out in the operation of the business, they may not be on a recognized basis. Hence the statement that industrial engineering is, in some measure, an optional function. While just as necessary in a well-run business as, say, purchasing or personnel or engineering, it is not always recognized.

Why is this so of industrial engineering, and why, therefore, does it have to be concerned with "winning acceptance"? There are a number of reasons.

Industrial Engineering Not Understood. First, industrial engineering is not understood. It is a function which does not immediately conjure up in one's mind a concise concept of what is covered by the term "industrial engineering" as do the words sales, purchasing, finance, product design, law, and the like. The scope of industrial engineering is broad and impinges on many interests, as described in Chapter 2, Profile of an Industrial Engineer. It cannot be clearly and simply described in a sentence or two for quick understanding.

Scope of Responsibilities Differs. Second, even though the general purpose of industrial engineering may be understood, the scope of responsibilities differs from one company to another. It is not uncommon to hear managers of industrial engineering activities describing quite differently to one another what they do in their departments. Consequently, it is not to be wondered why others outside industrial engineering, including business and manufacturing managers, ask, "What is industrial engineering?" and "What do you do in industrial engineering?" These questions, of course, go beyond the concept of winning acceptance to the basic understanding of the function. But before acceptance can be gained, understanding is necessary.

Industrial Engineering Is Frequently Splintered. Third, the function of industrial engineering is frequently splintered. That is, while many of its interests exist in a business organization, they may not be organized into an integrated department titled "industrial engineering." Standards work may come under a comptroller or finance department, methods and layout under a manufacturing engineering department, and perhaps job evaluation under the personnel department. With this splintering of the function, a cohesive program cannot be planned to the best benefit of the organization because the function per se is not recognized.

Status Quo versus Change. Fourth, industrial engineering is not recognized in many places because there is a desire to live with status quo. Industrial engineering searches for improvement—the one best way which is never attained. Where a management is at least temporarily successful and complacent in its success, it does not desire change. This can be interpreted to mean it does not desire industrial engineering. In smaller businesses, the ingenious manager may himself establish the layouts, machines, tooling, fixturing, and material handling methods. If any effort is made at cost and labor performance control, rough rule-of-thumb standards can be picked from the air. If competitive conditions result in profitable operations under such circumstances, the manager may have no desire to do better.

But even though a lack of understanding or "head in the sand" attitude may not exist—and fortunately there are literally thousands of progressive companies where this is the case—promotional effort by the industrial engineer is still necessary. It is an essential ingredient of his function. There are two fundamental reasons why this constant selling effort is necessary to a successful industrial engineering program.

Dealings with Tomorrow's Problems. First, industrial engineering deals in a very large measure with tomorrow's operation. While the accountant is dealing with facts and numbers of today and yesterday and the sales department is negotiating for today's orders and promoting future known or potential orders, the fundamental characteristic of industrial engineering is to deal with change and how to do better tomorrow that which is being done today. It may involve procedures and forms design; it may involve manufacturing methods and machines or any of many practices in both direct production and overhead service functions. In this sense, industrial engineering is essentially a planning function—working on what is to be, or analyzing current situations to detect weaknesses such as high costs or inefficiencies which create the need for change. Good planning, by its very nature, particularly when done by a staff department such as industrial engineering, entails a need to convince others that sound thinking has been applied, that all factors have been considered and properly weighed, and that, in fact, with a minimum of risk, a change should be made. The very essence of the business of the industrial engineer is to make recommendations and to convince others that they should proceed with some new and frequently strange or relatively untried venture. This requires selling.

Resistance to Change. Second, even though it may be generally recognized by many that the recommended change which has resulted from good planning denotes progress for the company, there is still to be reckoned with the inherent characteristic of people—resistance to change. With industrial engineering dealing in such large measure with change, this inherent resistance creates a need for a much stronger "winning acceptance" effort than is necessary in most other departments.

Resistance to change may be of two types. It may be due to whim or conservatism or complacency, all of a quasi-unreal nature. In such instances, the objections to change may take on peculiar forms of unrelated reasoning as to why something should not be done. On the other hand, there are many situations where the resistance to change is very real and must be handled with understanding. A new machine or method may eliminate an operator's job or require effort on his part to learn new habit patterns. A foreman is handed a "headache" when he must shake down and bring into smooth production a new machine or process or hire and train new operators. Shakedown costs of a new process may adversely affect the earnings picture of a department for some period of time. This can cause a manager, under heavy cost pressure, to resist new methods—even though in the end they could be his salvation.

Both types of resistance to change must be recognized by the industrial engineer as being inherent in the nature of his work. By understanding such attitudes, real or unreal, he can take steps to minimize the hardships, overcome the unwarranted attitudes, and generally create an atmosphere for the easier adoption of changes.

LEGITIMATE PROMOTION VERSUS EMPIRE BUILDING

For reasons noted above, it is apparent that a fair share of the success of an industrial engineering program involves a conscious effort not only to do a good job but properly to promote acceptance. This is necessary all the way from the purely educational standpoint of "What is industrial engineering and what does it do?" to the very real problem of convincing a foreman that there will be a payoff from the new methods which are causing him problems both from his men and from excessive costs temporarily incurred.

This winning of acceptance by the industrial engineer, however, must not be based on empire building. It is not from any desire to operate a larger department that an industrial engineer promotes acceptance of his work; rather it is because of an honest desire to make understood the value of sound industrial engineering. Good industrial engineering, soundly applied, can be and often is the salvation of unprofitable organizations. A company which is in profit trouble, in an effort to economize, too often makes the mistake of cutting its creative thinking, much of which comes from industrial engineering. Where a company is in trouble because of high costs of production, it can very well be that it needs redoubled industrial engineering effort. Or it may be that industrial engineering has not been turned loose on the problem, because acceptance has not been properly promoted and won.

THE NATURE OF THE INDUSTRIAL ENGINEER

Generally speaking, the industrial engineer is not promotionally minded. He is more apt to be quiet and modest in manner, inclined to reflect on a situation rather than to act quickly. He is technically trained and accustomed to dealing with facts in his daily work. Having been trained as an engineer to gather and analyze data, to reflect on the data, and to design a solution to the problem based upon the data, he has a tendency to feel that facts speak for themselves. Consequently, the promotional effort necessary to win acceptance is something to which he must give conscious effort. Often, it does not come easily to him to promote the worth of his effort.

DEALINGS WITH PEOPLE

Few positions in an organization entail dealing with people and personalities more than the position filled by the industrial engineer. In a sense, he is dealing with the nervous system of a business in his work, which ranges from machines and methods and layout, through standards and systems for payment and control. And it is people who make up the nervous system of the organization. The industrial engineer must be able to get along with people and win their acceptance, even though he is not inherently sales-minded.

The relationships of the industrial engineer with people occur in five fundamental areas:

1. Relationships with his superior and those in higher levels in the organization
2. Relationships with positions on essentially the same reporting level as the industrial engineer—the cost accountant, manufacturing manager, design engineer, and the like
3. Relationships with frontline management or supervisory positions—the foreman, inventory control manager, toolroom superintendent, and the like
4. Relationships with direct labor, service, and clerical personnel
5. Relationships with union representatives

Each of these groups is influenced by the work of the industrial engineer. The degree of his success and the degree of constructive improvement he can bring to an organization are directly related to the degree to which he is able to work harmoniously with, and gain the confidence of, each group. Each group of persons represents different shades of interest, different knowledge and understanding, and in some measure, different purposes and outlooks on the company and their jobs. Hence the approaches to winning the acceptance of each of these groups must be somewhat different.

THE AUTHORITY OF THE INDUSTRIAL ENGINEER

But first, the authority of the industrial engineer and how he must work through formal and informal channels of communication toward the acceptance of his recommendations will be considered. Essentially, industrial engineering is a staff function, as has been developed more fully in Chapter 3, Organizing for Industrial Engineering. Industrial engineering serves in a staff capacity to a division manager, a manufacturing manager, or other reasonably high level position in the organization. As a staff position, it does not have authority to direct specific action. Rather, its duty is to study situations, gather and analyze facts, and prepare recommendations as a basis for others to take action. If the recommendations are sound and are accepted, it means that the industrial engineer does have considerable power to influence the course of events, even though he must work through the command authority of the line manager.

Obtaining Concurrences. Although there is always the technical line of command to resort to in bringing some recommendation into effect, complete staff work involves obtaining concurrences from all parties involved or statements of nonconcurrence where disagreements exist. When all parties concur in a recommendation, the position of the line manager in approving that recommendation is made far easier. Hence the industrial engineer may gain his superior's acceptance as much by his ability to obtain the concurrences of others as by the correctness of his analysis. When concurrence of all affected members of an organization is not obtained, with the courage of conviction the industrial engineer may still obtain an "OK" from his superior to go forward with some action. At this point, the technical responsibility for the decision rests with the line manager, but the industrial engineer in his staff capacity cannot shrink from a sense of responsibility. If he is to assume some credit if the decision turns out to be good, he must stand up to criticism if it proves to be poor.

Good Acceptance Minimizes Need for Line Authority. The industrial engineer should also recognize that it is not always necessary or advisable to resort to the line authority of his superior to accomplish his objectives. A great deal can be accomplished by persuading subordinate management levels to take direct action upon recommendations made to them. Care must be taken, however, that proper channels of contact have been closely adhered to and that there is a free flow of knowledge to superior positions when direct working arrangements exist with subordinate positions. Nothing can stir up quicker resentment than jumping over a channel in the line of command without either a formal or informal prior understanding. Direct action by a subordinate, where there is authority to act, can save much time for a superior and can also develop satisfaction of accomplishment on the part of the subordinate. It should be a cardinal rule of staff work not to resort to command decision whenever the same accomplishment can be attained (within established limits of authority and responsibility) through a proper working relationship with subordinate levels of management.

RELATIONSHIPS WITH SUPERIORS

In the industrial engineer's relationships with his line manager—a works manager, division manager, or manufacturing vice president—he must recognize that his superior will have many other diverse interests. He may not be oriented toward industrial engineering by background, because top managers often arrive at their positions through financial, sales, legal, or channels other than engineering or manufacturing. But even though oriented toward industrial engineering, it is probable that, as a top manager, he deals essentially in plans, programs, concepts, and policies. Hence it is not unlikely that he will wish to govern the industrial engineering activity in this manner.

Importance of a Program. A program for the industrial engineering activity together with a budget of manpower and expenses can be most useful in winning the acceptance of superiors. It may be a formal or an informal program, depending upon the size of the company and the customary reporting procedures. It is characteristic of all good industrial engineering effort, however, that a forward look is taken to identify the problem areas, that an assessment is made of the present status of these problem areas, and that a program is devised to bring about improvement.

To accept, cope with, and solve the day-to-day problems as they arise is not enough to win the full support of a top manager. He may accept the fact that a good job is being done because he does not receive complaints. But the good top manager will wish to set objectives and then budget and control toward those objectives. In this way, he becomes part of the activity, and industrial engineering can feel that it exists to do something more than merely keep the manager out of trouble.

In consequence, one of the first ways to win acceptance is to lay out a plan or program for the problems that exist, present it to the manager with the manpower and budget involved, and get the manager's concurrence in the program proposed.

Reporting Progress. After the program has been laid out, it is important to report accomplishments against the program, both formally and informally, to the superior. Whenever tangible measures of performance such as the results of cost reduction effort, direct labor hours on standards, or the like can be shown by chart or graph, they should be displayed routinely against the target of planned objectives. To keep such knowledge under cover can serve no purpose in building confidence in the manager that a satisfactory or superior job is being done. To display it provides the knowledge of the facts displayed and creates a favorable impression that control of the operation exists.

Gearing to Other Traits of Top Managers. The industrial engineer should recognize the likes and dislikes of his superior—whether he prefers tables or graphs, the degree and manner in which he explores details, and other personal traits. The top-level manager cannot explore all details. Nevertheless, he will wish to know that they have not been overlooked by others. He will wish to have a quick, concise presentation of fundamental facts bearing on a problem; some indication that the problem has been explored in sufficient detail; a statement of the alternatives

2. *Starting times* should be realistically set and rigidly adhered to. Wherever practical, ending times as well as starting times should be established to permit good planning of the day by all participants.

3. *Conclude* what can be concluded in the set time limit and carry that which cannot be concluded over to the next meeting. The opportunity to digest information between meetings will often make the succeeding meeting more productive.

4. *Objectives* should be set. In calling the meeting, the purpose should be briefly stated. It should be restated in opening the meeting and the interests of the individuals participating should be defined. Major issues and priorities should be identified, and the chairman should, at least mentally, prorate the time allotted for the meeting to the subjects to be covered.

5. *Interplay* of different interests must be recognized. Where opposing views exist, it is generally best to get these out on the table early so that the pros and cons can be established and the problem can be seen in its entirety. If controversial aspects develop after tentative decisions are reached, backtracking to reconsider earlier facts or information may be necessary.

6. *Different personalities* in a meeting must be handled adroitly. Draw out the quiet person who hesitates to state his views or contribute his knowledge. Control the aggressive person who tries to dominate the meeting. Recognize various interests, and address questions to various individuals to obtain a proper mixture of knowledge and attitudes. Avoid warping decisions or making wrong decisions simply because of personality traits or improper mixture of knowledge.

7. *Summarize* the discussion, the conclusions and decisions reached, the areas to be explored further, if any, and special assignments of responsibility which result from the meeting.

8. *Issue minutes.* People often do not absorb everything said or decided at meetings and do not interpret what is said in the same way. Minutes are the only protection.

RELATIONSHIPS WITH FRONTLINE MANAGEMENT POSITIONS

Much of the success of the industrial engineer depends upon the attitudes of foremen and other frontline supervisors. In fact gathering, they can cooperate willingly and provide not only what is asked for but voluntary valuable suggestions as well. On the other hand, they can easily confuse or fog information, intentionally or unintentionally.

The Approach to Frontline Management. The industrial engineer must recognize the true nature of his function when dealing with frontline management. The industrial engineer's job is to help analyze for improvements. He is trained to break a job into its elements—subdivide and conquer, so to speak. He obtains, analyzes, and organizes facts concerning each element, but he is not expected to be an expert in all things. It is his position to learn and ask why, much like a pupil, not to tell and state why, like a superior. But while adopting the attitude of a pupil, the industrial engineer has an opportunity for true leadership— leadership in developing creative thinking on the part of frontline management directed toward change and improvement.

Having gained a thorough understanding of the elements of a situation, the industrial engineer must view objectively what he has learned to assess alternatives and apply knowledge of past situations in the design of improvements. In doing so, however, he should draw on the contributions of frontline management, for they are the experts. The industrial engineer's duty is to furnish the occasion, the objectivity, the atmosphere, and the direction—and perhaps to correlate the interests of several departments. It is not his position to mastermind the improvement from behind closed doors. He must contribute to the proposed improvements from his background knowledge, but it should be in the sense of a contributor or advisor.

Keep Them Informed. It is at the frontline level of management that fear of and significant resistance to change may begin to become manifest. A foreman or supervisor is expected to run his department and to direct the efforts of his men. This is his job and he is likely to be rightfully proud of it. Hence it is easy for him to interpret a study of things taking place in his department as criticism of his own performance. He begins to wonder what it is all about unless he is told. And if he begins to wonder, it is safe to assume that fences will begin to rise and smog begin to appear. If he is informed, if the reasons for the study are explained, if the objects sought are clearly understood by him, his worries will usually disappear.

Create Their Desire to Ask for Aid. It is better for the foreman to request the help of the industrial engineer in solving a problem rather than for the industrial engineer to approach the foreman with a problem area to be studied, seeking his support. A foreman is expected to run his department. This means to manage it with respect to cost, quality, and delivery. If upper line management detects a weakness or area which needs to be analyzed for improvement, it can be discussed with the foreman with the suggestion that he, the foreman, go to industrial engineering and ask for assistance in the analysis. With a reputation for assistance built up, it is a good bet the foreman will be back for more.

Give Credit Where Due. Actually, it is usually not difficult to win the support at the frontline supervisory level. This is where much firsthand knowledge of what takes place in an operation resides. Many good suggestions can and will come from the supervisors for improving specific methods. This being the case, it is the perfect opportunity to give credit where credit is due. Appropriate recognition of good thinking—a letter, or something more tangible—may add to the personal pleasure derived from seeing one's ideas in actual production. It is the "seeing in production," though, that normally provides the real stimulus. This in essence is the approach of a work simplification training program. Once it catches fire, tremendous improvements in methods and general operations can result.

RELATIONSHIPS WITH DIRECT LABOR, SERVICE, AND CLERICAL PERSONNEL

Although it may not be generally recognized, it is frequently the direct labor, service, and clerical positions which stand to gain tremendously, both directly and indirectly, from industrial engineering efforts. A machine, tool, or fixture, well designed from the standpoint of motion economy, is much easier and safer to operate than one poorly designed. Long reaches, which extend sometimes to stretches, require much energy to perform and hence are more tiring than short reaches. Smooth, balanced rhythm of motion is less tiring physically and requires less concentration mentally than unbalanced, awkward motions. Well-designed forms and office equipment are easier to use than poorly designed forms and equipment. The fact that good methods are easier and less tiring makes those who use them more productive. Employees are more productive not as the result of greater effort but because easier actions are required. And it is productivity that promotes a healthy, prosperous firm which is capable of remaining in business to provide jobs.

Beyond this, work measurement is the foundation for fairness. Consistent standards promote fairness, in that operators who are performing well receive proper recognition for their good performance. It is these aspects of industrial engineering which must be understood by the direct operators if acceptance is to be won. Where the true purposes are explained clearly and where they are fairly pursued, fear of or resistance to standards and good methods work rarely is serious. But there must be understanding. The fairness and purposes of industrial engineering must be explained in such a manner that they can be appreciated. Under-the-table practices, which many years ago were employed by a few so-called "efficiency experts," developed an unfortunate attitude toward standards and methods work. But with prior discussion with employees of the whys and wherefores of modern industrial engineering practices, the support or at least the willing acceptance of industrial engineering practices can be obtained.

The problem here is—how do you get the story across to the people, how do you promote it so that it is sincerely felt and believed?

First, it goes without saying that management's attitude must be one of a desire to be fair in all respects.   This must be demonstrated whenever the occasion arises.   There must be a willingness to explore grievances under generally established grievance procedures, and if warranted, to change incorrect standards and to improve poor working conditions.   All such matters must be dealt with willingly, in the open, and with reasonableness.

Second, employees should be kept informed as much in advance as practical of significant changes brought about by industrial engineering activities—a new standards program, a new job evaluation system, or major methods changes.   Here is where much can be done to minimize the operator's resistance to change, to allay his fears, and to promote acceptance.   Benefits to the employees from such changes should be specifically emphasized in appropriate company literature—posters, letters, newspapers, and so on.   A close working relationship with the personnel department can be an invaluable aid to the industrial engineering department on such occasions.

Third, daily contacts among industrial engineers and the employees can be invaluable in promoting the acceptance of new methods.   If better working conditions and methods do truly benefit the worker, there will be many occasions where a few comments now and then will help develop an understanding of this fact.

Fourth, a good suggestion system, administered by industrial engineering, can bring the employee in closer contact with the thinking of the industrial engineer.   In this manner, a better understanding which leads to better acceptance will result.

These and similar actions can assist significantly toward winning acceptance by direct operators of the industrial engineering function.   The important thing again is to tell them, to keep them informed.

RELATIONSHIPS WITH UNION REPRESENTATIVES

An understanding by union representatives of the industrial engineering activity as it may influence their constituents is essential.   Here again, prior notification of significant changes affecting employees, with a full disclosure of the reasons, to the officers and stewards or other representatives of the employees is necessary.   Generally, officers and stewards of the union will have a broader understanding of basic company problems and objectives than do the individual employees.   Hence, if they understand and accept what is proposed, they can be of real assistance in helping to win the acceptance of employees of changes which will affect them.

If a new system of setting direct labor standards is to be used or if a radical change in machinery is to be made, it is desirable to discuss with the union representatives all the reasons and the benefits which are expected to result.   If fair and sound causes prompt the changes and if the reasons are well presented factually in frank terms which depict long-range benefits, the support of union representatives can usually be obtained, even though the changes may involve some short-range or local discomforts.   Where this is achieved, it significantly eases the work of the industrial engineer in bringing about changes.

APPROACHES AND AIDS TO WINNING ACCEPTANCE

From the foregoing, it is apparent that the industrial engineer must work for, with, and through people and interests at all echelons of the organization.   How he approaches his task and the role which he sets for himself can do much to spell success for his programs and for himself.   Although he is called upon to deal in facts typical of any engineer in the engineering profession, he is also called upon to deal with the human side of a situation in very large measure if acceptance is to be won.   The traditional approach of the engineer when designing a product, drawing up its specifications, and causing it to be made is not adequate for most industrial engineering programs and installations.   If a cost reduction program is to be installed, the industrial engineer can cast his role as designer of new methods and perhaps skim some cream for a short time on a few obvious cost saving items. Or he can cast his role as an organizer and perform a service with his specialized knowledge to line supervision in making studies at their request of methods, facilities, handling, standards, or other matters of interest to the supervision. In this latter role, the industrial engineer organizes the program, acts as consultant/advisor to the line, and administers the results. The cost reduction program, however, becomes one of participation. Line supervisors are held responsible for and are credited with the end result in their individual departments. The industrial engineer is credited with the overall influence and for the service given in attaining these end results. The industrial engineer who approaches his task in this manner, provided he has competence in his technical skills, will find his cost reduction work and other industrial engineering opportunities never-ending.

Vital Importance of Behavioral Sciences. The significant difference between these two approaches cannot be overemphasized. The importance of the human side of industrial engineering and the hazards of masterminding a solution have long been recognized. However, the vital significance of the human side has come into far sharper focus, and the means of dealing with it is far better understood through the development of knowledge in the behavioral sciences—group dynamics, team development, interpersonal relations, and the like—as discussed in Chapters 1 and 2 of Section 7. Knowledge and skills in the behavioral sciences—in consulting, communicating, understanding motivation, and group decision making—have become as much a vital part of winning acceptance for the industrial engineer as has been his knowledge of the traditional skills of the industrial engineer in work measurement, methods analysis, economic justification, and other tools of this profession.

Richard L. Betke sharply points up a successful industrial engineering program whose acceptance was won through this approach.[1] He sets forth the requirements shown in Figure 5–1 as essential in industrial engineering team building.

Two Kinds of Audience. There are essentially two kinds of audiences in an organization. First, there is the "public" of the organization in general—management, supervision, and employee alike—in whom a generalized understanding of and confidence in industrial engineering should be promoted. Where there is a general atmosphere of acceptance of the work of the industrial engineer, it is far easier to sell a specific proposal affecting some individual, department, or problem area when the occasion arises. Second, there are specific persons, such as department heads or top management, who are considering some specific problem. In the first instance, when promoting acceptance by the "public," the development of a climate or atmosphere which is right for acceptance of industrial engineering in general is dealt with. In the second instance, when making the specific proposal, the concentration is on a narrower audience with focus on a specific project.

Creating the Proper Climate or Atmosphere. The approach to creating the proper climate or atmosphere is essentially one of taking advantage of any good opportunity for effective publicity. Useful communications methods include:

1. News items in a company paper relative to cost reduction, new methods, and other newsworthy items
2. Separate brochure setting forth organization, functions, and responsibilities of the industrial engineering department
3. Bulletin board notice of significant programs under way or outstanding results accomplished
4. Participation in professional society meetings with due publicity accorded
5. Instructing in company-sponsored programs in standards, job evaluation, manufacturing methods, and other similar topics
6. Orientation programs especially prepared for management and supervisory levels to present and discuss either general responsibilities or specific proposed programs

[1] Richard L. Betke, "Application of Behavioral Science to the Practice of Industrial Engineering," *Journal of Industrial Engineering*, May, 1967.

An industrial engineer is most effective if he is	Which requires training in
1. First and foremost skilled in industrial engineering techniques	Latest industrial engineering techniques

2. Knowledgeable of group dynamics:
  a. Motivation
  b. Impact of change
  c. Organizational behavior
	Group dynamics

3. A good listener	Active listening (7) Communication skills
4. A catalyst in developing an understanding of needs	Consulting skills Communication skills
5. A change *agent*	Consulting skills
6. An interpreter of results	Industrial engineering skills Consulting skills Communication skills

7. Supportive of industrial engineering group objectives:
  a. Philosophy
  b. Policy
  c. Professional
	Sensitivity training
	Group problem solving
	Group decision making

8. An identifier/worker of peripheral/related problems	A developed understanding that industrial engineering does not solve *all* problems Group dynamics/motivation
9. Open with clients/others	Sensitivity training Active listening Relationship improvement program such as HDI programs (8)
10. A user of available resources	Knowledge of what resources are available and how to use them
11. Knowledgeable of company goals and plans, administration, and engineering systems	General business subjects

FIG. 5-1.   Industrial engineering team training requirements. (*Source:* Journal of Industrial Engineering, *May,* 1967, *p.* 297.)

    7. Personal contacts wherein conversation gives recognition to the good work of a member of the industrial engineering department, or the worth of a program

Conscious effort can do a great deal to establish a climate wherein acceptance of the industrial engineering function will grow. If a good industrial engineering program is in existence, there is ample material to use for this purpose.

Making the Specific Proposal.   In determining how best to present a specific proposal, a great deal depends upon the magnitude of the proposal and the *modus operandi* of the department head or heads to whom the proposal is to be presented. Due thought must be given to selecting, organizing, and presenting the pertinent points which will win approval. Most decisions of managers, while based to the extent practical on facts and knowledge, in some measure involve the exercise of judgment. The judgment exercised is of two sorts: (1) judgment relative to the facts at hand and (2) judgment relative to whether or not an adequate analysis has been made. If there is confidence in the industrial engineer, the manager whose approval is being sought may probe the adequacy of the analysis made as much as the final conclusion and recommendations reached. He may accept the fact that proper conclusions and recommendations were reached if a thorough study was conducted. A poor report or presentation will leave the impression that a superficial analysis was made, and the industrial engineer will have difficulty in getting approval. On the other hand, a well-conceived presentation generates confidence that a good study has been made. Attention becomes focused on the merits of the proposal itself rather than the demerits of the presentation.

If a presentation is made by written report, the following aspects will promote a favorable impression:

1. A neat cover and title page
2. The body of the report so organized as to present readily and quickly the scope, conclusions, and recommendations of the study, backed up by detail
3. A presentation of details which develops through more extensive discussion the approaches taken, the alternatives considered, the facts and data collected, the advantages and disadvantages of alternatives, and the general reasoning which led to conclusions and recommendations
4. A display of facts and data by pictures, charts, or graphs, well organized and scaled material

If a personal presentation is to be made, whether to a single person, to a small audience, or to a larger group, careful preparation with selection of material well organized is still necessary. Even with a relatively small audience, appropriate visual aids selected from the following list can be a tremendous aid to quick and effective understanding of the subject under consideration.

1. Chart board	9. Scale and cutaway models
2. Flip charts	10. Sketches
3. Flannel-board charts	11. Cartoons
4. Blackboard	12. Projected slides
5. Pictures	13. Viewgraph charts
6. Graphs	14. Movies
7. Drawings	15. Personal visit to observe
8. Samples	

The selection of the best visual aid for a particular audience depends very largely on the topic matter concerned, the size of the audience, and the room conditions. All visual aids, however, play on the premise that "seeing is believing." The nearer to realism the visual aids can be made the better. Hence a personal visit to see a display of actual circumstances—a floor layout situation, a new machine, or a mock-up of a new workplace layout—is perhaps the most effective visual aid. Naturally, this is not practical or justified in many circumstances, but other forms of visual aids, such as pictures, slides, movies, and models, can be employed to create the "see for yourself" atmosphere.

Generally speaking, it is better to display too little rather than too much on any given chart or slide because it is the mental image rather than detailed facts that are significant. Too much detail can detract from the mental image it is desired to create. Discussion can cover the detail.

As mentioned in the discussion on relationships with superiors, it is important to recognize the degree of understanding a manager may have of a subject and to pitch the presentation to that degree of understanding. Too much detail and painting of background can result in boredom or impatience, yet lack of it may result in a poor understanding of the matter by the manager. Therefore, it is essential in making a presentation to consider not only the subject itself but also what the audience knows about the subject.

THE IMPORTANCE OF THE "MAKE GOOD" REPORT

Much of the work of the industrial engineer involves changes which are major in scope and for which there is not an immediate or sharply measurable result. The benefits derived from a major piece of equipment or the rearrangement of layout may be difficult to display or may take a long time to become apparent. Too frequently, by the time the benefits do become available, other problems are pressing and an after-the-fact appraisal of results is not made. Or as soon as a new machine is working smoothly, it may become obvious that the new is better than the old, and the real story of the facts as to why is not told.

A good practice which will go a long way to confidence building is routinely to prepare "make good" reports after the lapse of a suitable time period succinctly and objectively to evaluate the worth of an installation or program. And the bad as well as the good should be told. Lessons are learned from mistakes or weaknesses and are either corrected in a current installation or prevented in the future. From the standpoint of winning acceptance for industrial engineering, such after-the-fact evaluation of results—cost improvements accomplished, lessons learned, benefits derived—is the payoff on which to demonstrate and bring into focus the true worth of industrial engineering. It is not enough to let the payoff from good industrial engineering effort reflect only in the ultimate profit and loss statements. There are too many other influences that also contribute to improved profit. With actual facts collected by evaluating the results of completed projects, there can be direct and effective placing of credit where credit is due.

THE SELLING FORMULA TO WIN ACCEPTANCE

In summary, the following are some of the fundamental considerations involved in selling or winning acceptance of any idea, plan, program, or function.

1. *Prepare Well.* Learn as much as practical about both the subject and the audience. Presumably the subject, whatever it may be, is well understood by the industrial engineer, but it must be well thought through to highlight the significant points and temper the less significant. Fit the presentation to the time available and prorate time for various points to obtain the best emphasis. Often the audience is known personally. If it is the boss and he prefers tables to charts, give him tables. If he likes to read a proposal, hand it to him to read. If he prefers to be told the content, be prepared to summarize and discuss the report verbally. Consider the traits of other audiences in the same manner. Do not pitch over their heads or too far beneath them.

2. *Inform Persons Affected, Explain Objectives.* Do this as early as practical. Start selling when you start a program. An informed group is an intelligent group. You cannot expect support from an ignorant group. Fear of the unknown begets resistance. No one who should know likes to be left out. Informed persons who know objectives may drop a real nugget of wisdom to help a program along. Use them wherever practical—see point 7 below.

3. *State and Illustrate Advantages Early.* Management will wish to see proposed changes reflected in profits; cost savings; better man and machine utilization; better and safer working conditions; and improved employee, customer, and public relations. And all employees like a prosperous company with good working conditions and relationships. These are the items which sell change most easily.

4. *Balance the Positive against the Negative.* Management knows it will not get something for nothing. Tell them how much it will cost and other drawbacks. Unfortunately, everyone can't win all the time. The question is how many lose and in what degree. Lay the facts on the table without prejudice. If only positive factors existed there would be no need for a decision. It would be automatic. So get both on top of the table.

5. *Emphasize Results, Not the Tool or Mechanics.* Sell the comfortable ride, the pleasing appearance, the fuel economy—not the wheels, the motor, the frame, and the steel in the body. These may be important means to an end, but they are not the end in itself. It is *results* from good standards or a good layout—and not the mechanics or procedures—that are most important in winning approval to proceed.

6. *Do Not Exaggerate the Benefits.* E. C. Hill, in his article, "When You Send an Idea Up the Line," says "one trap lies in our enthusiasm which can, unless we are stern with ourselves, lead us into disastrous overstatements. It is dangerously easy to overstate a case: to overestimate rewards and underestimate costs and practical obstacles. This may create so many doubts among the people considering

a proposal that the idea itself, with its lesser but certainly real advantages, is never even judged."

7. *Obtain Participation.* Participation leads to familiarity, perhaps to pride of authorship. The manager who has contributed some part to a proposal, the foreman who has suggested or contributed to a new layout, the operator who has proposed a simplified fixture, are supporters who need no winning. No one person has a corner on all the good ideas. Pooling of ideas is an essential ingredient, especially of good industrial engineering. The true experts are often those closest to the machines, the materials, the files, and the procedures. The industrial engineer is the catalyst to draw out, organize, and put detailed knowledge to work.

8. *Consider Alternatives.* The saying goes, "There are many ways to skin a cat." And so with most problems. Although the end result may be the same, the alternative approaches offer different shades of advantages and disadvantages. Weigh them all, one against the other. The boss who catches you off guard with a "Did you consider so and so?" has you on your way back to the drawing board.

9. *Anticipate and Forestall Objections.* Know the shortcomings or weaknesses of the proposal. If you do not, others will find them and will have you on the defensive. Do not sweep them under the rug or try to gloss over them. Deal with them in the open. If objections are fancied, have an answer to allay a fear but not ridicule the thought. If they are real, deal with them realistically and objectively. If they must be lived with, so state and why. If they can be minimized, indicate how and when. Face facts with facts, do not run away from them. They will catch you every time.

10. *If Practical, Use a Trial Run.* It is nice to drive a new car before you buy. And so with a new machine. If five are needed, propose installation and shakedown of one first. Let the proof be in the pudding. Begin modestly and expand. Foresight is seldom 100 percent accurate, so learn as you go whenever the opportunity is right to do so.

11. *Consider the Timing.* When the boss is known to be busy, do not barge through the door with a million dollar proposal. Stay away and then plow and till the soil for the big ones when the climate is right. Do not propose a major facilities expansion when markets are dull, profits meager. Tackle floor working conditions when new offices are being erected. Try to avoid the "It's a good plan, but the timing isn't right." Or if there is cause to go forward in spite of timing, recognize its impact and be prepared to answer "Why?"

12. *Be Brief but Tell the Story.* Enough said.

13. *Permit Full Discussion* (as long as pertinent). Keep the subject on the track, sidetrack the irrelevant without offense. If answers are not known but the question is insignificant, so indicate and tactfully state why; if significant, call for or get the answer at the earliest opportunity. When discussion drags, cut it off.

14. *Drive to a Close.* Provide a place and make it easy to agree or say "yes." Do not wind up with a "Well, what do you think?" atmosphere. State positive conviction and ask for concurrence. Show where initials or a signature should be placed to bind approval—and have a spare pen ready.

Chapter 6

Effective Management of Industrial Engineering Time

WILLIAM S. MAXWELL

**Administrative Manager, Maynard Research Council Incorporated,
Pittsburgh, Pennsylvania**

As the scope of industrial engineering activities broadens, the demands upon the available industrial engineering manpower increase. Industrial engineering management is faced with the problem of providing effective service in an ever-expanding area of operations while maintaining an acceptable industrial engineering cost relationship. People are the tools of the industrial engineering manager. These tools must be employed economically, effectively, and in such a way as to provide the maximum utility to the user. The manager should have a sound procedure for estimating, budgeting, and allocating his manpower resources. This chapter will discuss a practical way of effectively managing industrial engineering time.

THE BROADENING SCOPE OF INDUSTRIAL ENGINEERING ACTIVITIES

"Dynamic" is a word that well describes the industrial engineering profession. Webster[1] defines dynamic as ". . . marked by continuous usually productive activity or change. . . ." The profession has changed and is continually changing. From the days of Frederick W. Taylor to the modern space age, the technology of

---

[1] By permission. From *Webster's Seventh New Collegiate Dictionary* © 1970 by G. & C. Merriam Co., Publishers of the Merriam-Webster Dictionaries.

industrial engineering has kept abreast of industry demands and will continue to do so as new requirements are dictated by the industrial and business world.

Industrial engineering had its birth in the mills and shops of industrial America. Advances in industrial technology have led to the development of industrial engineering technology. Nonindustrial business has taken heed of past industrial engineering contributions and is employing industrial engineers in banks, stores, and municipalities, to mention only a few.

A new vocabulary is appearing in the profession. Terms such as biomechanics, simulation, management sciences, and stochastics are in everyday usage in the profession, and as time goes on, more and more will be added.

Industry, too, has found uses for industrial engineering outside of the manufacturing activities. Industrial engineers can be found engaged in assignments in areas such as systems design, operations research, finance, training, marketing and distribution, and many others. The accepted definition of industrial engineering states that:[2] "Industrial engineering is concerned with the design, improvement, and installation of integrated systems of men, materials, and equipment. It draws upon specialized knowledge and skill in the mathematical, physical, and social sciences, together with the principles and methods of engineering analysis and design, to specify, predict, and evaluate the results to be obtained from such systems." This definition, coupled with the functions of a staff organization to advise, assist, control, and coordinate, does not restrict the activities of industrial engineering to manufacturing.

"Optimization" is another word which assists in understanding the industrial engineering function. The term "dynamic optimization" results from the combination of "dynamic" and "optimization," causing the evolution of the partial definition ". . . to optimize the use of continually changing industrial resources." Not only must the industrial engineer concern himself with today, the current problem, but he must look toward tomorrow. Top management needs him to look into the future. Business decisions must be projected for their time effect on the business. Jay W. Forrester, in *Industrial Dynamics,* clearly illustrates the principle of business decision stimuli and their amplification through time.[3] Thus, the industrial engineering manager not only must manage an efficient organization, but he must utilize his manpower to optimize the use of the industrial resources of a many-faceted organization in a constant state of change.

THREE-PHASE APPROACH TO EFFECTIVE MANAGEMENT OF INDUSTRIAL ENGINEERING TIME

There is always an overabundance of potential projects before the industrial engineering manager. These projects all have one thing in common—they are important to the persons requesting their addition to the backlog of potential assignments.

The industrial engineering manager who never says "no" to project requests is in as much trouble as the manager who is too restrictive in the allocation of industrial engineering time. In the first instance, so many irons are in the fire that thorough investigations are not possible, results are not accurate, and the industrial engineers themselves become frustrated because of having to operate in an environment which is unorganized and consists of little more than fire fighting. On the other hand, the manager who seldom says "yes" is usually taking too narrow a view in his project selection process. If industrial engineers work only on the most immediately lucrative projects, the total organization soon refrains from requesting assistance from the industrial engineering department, and thus the total effectiveness of the function is impaired. The project selection process should consider the total integrated organization, not just one segment or one criterion.

[2] Anthony J. Brewer, "The Expanding Role of Industrial Engineering in Industry," *The Journal of Industrial Engineering,* May, 1966, pp. 233–238.

[3] Jay W. Forrester, *Industrial Dynamics,* M. I. T. Press and John Wiley & Sons, Inc., New York, 1961.

After project selection is accomplished, the industrial engineering manager must concern himself with the control of assigned projects. An industrial engineer's work is rarely restricted to one or two projects. It usually involves many and varied projects, all of which are important. The industrial engineering manager is responsible for seeing that assigned projects are completed on time in a professional and competent manner. The sooner a project is completed satisfactorily, the sooner another project can be assigned. As with any other product, industrial engineering assistance must be delivered by the industrial engineering manager on time, at the highest quality level, and at the optimum cost to his customer. To accomplish these goals, the industrial engineering manager must control the operations of his department properly.

After an assigned project is completed, a final analysis or review is essential to determine performance in relation to (1) scheduled time required, (2) results obtained, (3) costs incurred, and (4) information useful for other purposes. By accumulating this information, the industrial engineering manager can gain insight into the competence of his engineers, refine his estimating procedures, and develop budgetary data for his department. The better the estimating data are, the more efficiently can the manpower resource be allocated to the backlog and the more accurately can the manager determine his manpower budgets for future periods. Thus, better and more economical service can be rendered to the total organization.

The industrial engineering manager's job in effective utilization of industrial engineering time is threefold:

1. Selection
2. Control
3. Project review

Each of these phases is as important as the others and must be made a part of the manager's operating plan.

THE SELECTION PROCEDURE

The industrial engineering manager must establish a means whereby he can evaluate the backlog of potential projects and assign priorities to each. The total systems concept of industrial business thinking compounds his problems of project selection. The potential project, when considered by itself, may not appear to be deserving of a high priority; however, when viewed as a segment of a total integrated system, it may very well be the keystone to a structure which would warrant a high priority.

The increasing tendency is to broaden the scope of the system being investigated. There are several reasons for this: new methodologies evolving from research studies; new technologies which make broader investigations feasible; and a more critical look which is taken at the entire organization as an integrated whole.

Work assignments encompass the past, present, and future and may be classified as follows:

1. Maintenance of existing systems
2. Current improvement projects
3. Future planning projects

Each is equally important, and the system of priorities must consider all three areas.

Maintenance of Existing Systems. Past industrial engineering projects have resulted in systems which were deemed worthwhile and profitable, and until proved otherwise, must be maintained and serviced. Among these are work measurement systems, wage incentive plans, cost analysis and control, material and paperwork handling analysis, and coordination of continuing projects such as new product developments and machine economy studies. No system is effective unless it is used correctly and its performance maintained. The most complete, carefully planned scrap reporting system, for example, will be useless unless the information on it is gathered correctly, analyzed properly, and acted upon by those responsible for corrective action. A large share of the industrial engineering department's time is usually expended in the maintenance of existing systems, and sufficient time must be allocated for this important work.

Current Improvement Projects. Current improvement projects are those requiring investigation and study for the improvement of current operations. Typical examples include methods improvements, new equipment justification, material flow analysis, and new process studies.

Each of these current projects can be evaluated on the basis of expected return on investment, time required for the study, and total system effects.

Future Planning Projects. Future planning projects are those projects which do not deal with an immediate need but with some aspect of the future business. Examples of typical future projects are projected expansion plans of the physical plant, production requirements for future product line expansion, and research activities in the areas of future equipment and tooling developments. These projects require looking ahead to the future as opposed to what is happening today and what changes can be made today.

Included in the future planning category are also those projects which are so complex and involved that a great deal of time is required by individuals or task forces for study. They are immediate or current in that they will be implemented when complete, but the time involved is so great that they cannot be considered "current problems of today." Detailed planning of building expansion; specification and design of a transfer line machining center, including tooling and loading; and new plant location studies are examples of this type of project.

A future planning project has one other important characteristic. Because of the nature of the project and, usually, the magnitude of the investment required, top management will be actively involved in the decision to proceed with the project. This in itself tends to establish a high priority for the project and to place it in the industrial engineering work schedule at or near the "top of the pile."

Establishing Priorities. The order in which projects will be undertaken is determined at two levels: top management and industrial engineering management. Organizationally, the industrial engineering manager functions in a vertical chain of command, as shown by Figure 6-1. The kinds of projects his department will undertake is established between him and those to whom he reports. Once this has been done, the industrial engineering manager must establish priorities for specific projects within his department.

Projects in the future planning category are usually assigned by top management because of their very nature. The industrial engineering manager should advise top management of the impact of these projects on his departmental schedule and backlog. Alternative ways of providing the manpower required should be pre-

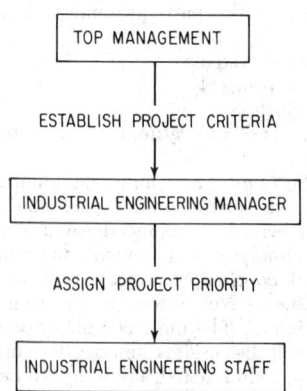

Fig. 6-1.   Industrial engineering function in a vertical chain of command.

sented, considering the amount and type of manpower available, the manpower required for the project, the backlog of projects in the department, and the projects under way at the present time.

This same kind of analysis will fix the position of the other two project classifications: maintenance of systems and current improvement projects. Existing systems which are considered to be worthwhile must be supported and maintained. Thus, when establishing priorities, the industrial engineering manager must allocate sufficient time for support and service. If sufficient manpower is not available to cover these activities adequately, (1) he has not accurately projected and budgeted his manpower requirements; (2) he has not made his superior aware of the impact of high priority future planning projects on his available manpower; or (3) he is not getting full effectiveness from his available manpower.

The broadest and most extensive project category remains—current improvement projects. In this category are those many projects that are expected to show cost savings returns, or that will improve product quality, reduce lead time, or generally improve today's operations. This is the category within which priorities and control are especially necessary.

There are several factors which the industrial engineering manager must consider while determining priorities of current improvement projects.

Purpose and Objective of Project. The purpose and objective should be considered as one factor in determining priority. The emphasis placed on a specific aspect of the project might well place it above another project which exhibits more obvious returns. For example, a quality improvement might be considered more important than, say, a direct cost reduction.

Anticipated Return on the Project. A number of measures can be used to estimate the anticipated return on the project. Return on investment (first-year or long-term), straight payback, and modified MAPI analysis (a form of first-year return on investment) are a few of the measures that can be used, depending on the policy of the company.

Man-hours Required. An estimate of man-hours required is used to (1) allocate man-hours in the departmental schedule and (2) fit the projects into the schedule.

Importance to User. The industrial engineering manager should determine from the department requesting the project how important the project is. This is particularly vital when a number of requests from one department are in the backlog. The industrial engineering manager must be prepared to find that all projects seem important to the department requesting assistance. He must use his own judgment, therefore, as well as the desires of the requesting department, in establishing realistic priorities.

Special Engineering Skills Involved. Certain projects will require the application of specific skills. The schedules of the industrial engineers possessing the needed skills or knowledge to accomplish the objectives of a project must be considered in establishing priorities.

Development Opportunities. If a certain project will expose an industrial engineer to a situation where a new skill may be learned or creativity exhibited, the priority may be adjusted to take advantage of this opportunity. The industrial engineering manager must consider the development of his manpower if his department is to continue to serve the organization effectively.

THE CONTROL PROCEDURE

After establishing a system of project selection criteria, the industrial engineering manager must establish a means of controlling the activities of his manpower on assigned projects. This requires establishing a project time schedule and seeing that the schedule is maintained.

Scheduling. When scheduling a project, manpower resources are allocated to the project on the basis of the estimated manpower requirement. This is the same as allocating an amount of money from a budget for the purchase of a capital good. The industrial engineering manpower is the money from the industrial engineering budget, and the project is the capital good which will produce a return on the investment either directly or indirectly.

Scheduling involves two elements: estimating the time requirement and assigning the project to available engineers. History is the basis for estimating time requirements. Performance on past similar projects will help the industrial engineering manager assess each project accurately and determine the hours, weeks, or months needed for completion. For this reason, the procedures outlined later under Project Review are essential.

After the total time required for completion of the project is estimated, the industrial engineering manager must determine the distribution of this time. Will an engineer be required full time until completion? Part time? Or a combination of the two? This information is essential for assigning the project to a specific engineer or group of engineers. Experience, manpower availability, and the requirements of the requesting department are factors affecting the distribution of the time requirement.

Engineers are creative and innovative people. Emergency problems can and do arise in day-to-day operations. For these two reasons, no manager can afford to schedule his engineers so closely that time is not available for individual creative innovation in unassigned areas or for coping with emergency situations which arise. If the industrial engineering manager oversystematizes his operation, his people become too confined, too specialized, and too narrow.

Technological obsolescence must also be considered. Time must be made available for the engineer to keep abreast of new developments in his profession. Providing work assignments with developmental potential has been mentioned, but time for study and retraining must also be made available.

One way of making it possible for the engineer to get the work done and also to develop so that he can continue to be effective and current is to use industrial engineering technologists to perform routine tasks and projects. They are accountable to the industrial engineer to whom the project is assigned in the same manner that the industrial engineer is accountable to the industrial engineering manager. This allows the industrial engineer to allocate time for self-development. Of course, the expense of the technologists must be justified by the industrial engineering manager.

Control. The control procedure must start with the individual engineer. When a project is assigned, the total time and the distribution of that time should be posted on the individual schedule of the engineer. At this same time, the industrial engineering manager should make a record of the project to use for a follow-up. The information required includes:

Project number
Project description
Area involved
Person requesting
Total time involved
Distribution of time
Engineer assigned
Date assigned
Date completed

The project number becomes a control and file number under which everything generated by or pertaining to the project is identified.

The record of projects assigned to each industrial engineer should be maintained by both the engineer and the manager. Through this device, the manager and the engineer can determine how much time is available and when it will be available. Figure 6-2 illustrates the record of projects assigned to one industrial engineer.

Control is implemented through the proper use of two tools: the daily or weekly time card and the project progress report. The time card is a record of time spent by the engineer on the various projects assigned to him. Figure 6-3 is an example of such a record.

The information collected on the time card is used as the basis for reporting the status of each project on the project progress report shown in Figure 6-4. Figure 6-3 shows the manager and the engineer where the time of each engineer was spent day by day; Figure 6-4 is in effect a weekly summary of time spent by project, plus a progress report for each project assigned to the engineer. Depending upon the degree of sophistication desired, the data for the project progress report

Name JOHN JONES			Year 19 –					
Control number and start date	Est. total hours required	Est. completion date	Time Distribution					Disposition and date
A 2121 6/1	160	9/12	WEEK 23 40 HRS	WEEK 24 20 HRS	WEEK 35 28 HRS	WEEK 36 32 HRS	WEEK 37 40 HRS	COMPLETED 9/12

FIG. 6-2.   Project assignment record.

Name JOHN JONES				Week Beginning 8/25/–35) (WEEK		
Control number	Monday	Tuesday	Wednesday	Thursday	Friday	Total
A2121	8 HRS	4	4	8	4	28 HRS

FIG. 6-3.   Weekly time reporting card.

Name JOHN JONES					Week Ending 8/29/– (WEEK 35)
Control number	Week no.	Weekly sched. hours	Total hours scheduled to date	Total actual hours to date	Comments
A2121	35	28	88	88	PROJECT ON SCHEDULE. EQUIPMENT DELIVERED 8/25, PROVE-OUT STARTED

FIG. 6-4.   Project progress report.

can be accumulated and displayed in departmental, sectional, or area groupings through the use of data processing techniques.

PROJECT REVIEW

No reporting or scheduling system, regardless of the time, effort, and expense involved in designing and using it, is worthwhile unless it is used regularly to assess the performance of the organization. An essential part of the system discussed in this chapter is the regular and continuous review of project progress by the industrial engineering manager.

At regular intervals, generally weekly, a project progress review should be made with each industrial engineer to whom a project has been assigned. The industrial enginer is in effect the project manager and must be prepared to report on the status of each project assigned to him. Management by exception does not apply here, because it is essential that the industrial engineering manager know not only what projects are lagging behind schedule, but which ones are ahead of schedule. Deviations from schedule, both ahead and behind, must be appraised for their effect on the total departmental work load, present and future.

At each review meeting, adjustments are made to the projected schedules of each engineer so that time can be accurately assigned or reassigned in the future. Reassignments can be made to supplement projects which are running behind schedule with engineering help from those ahead of schedule.

The capabilities of individual industrial engineers, from the standpoint of time budgets, technological competence, ability to direct others, and general project management and control, are brought to light by a systematic job review. As projects assigned become increasingly complex in nature, the management task usually becomes increasingly complex. The degree to which an engineer continues his self-education and creativeness while accomplishing his assigned project goals can also give an indication of the potential value of the individual. The opportunity for retraining, self-improvement, and creativeness must be provided by the industrial engineering manager—but the individual engineer must make use of the opportunity.

The industrial engineering manager's efficiency is also increased by the job review meetings. He keeps current on what his department is doing and does not have to seek information on a crash basis. His engineers, knowing that there are regular meetings, tend to hold minor questions until that time instead of constantly raising the questions as they come up. This is not to be confused with instituting a "closed door" policy.

VALUE OF SYSTEMATIC APPROACH TO THE MANAGEMENT OF INDUSTRIAL ENGINEERING TIME

The plan discussed here is a systematic approach to managing a many-faceted operation in a many-faceted organization. "Seat of the pants" management is not sufficient to control effectively an operation as broadly based as modern industrial engineering. The plan enables the industrial engineering manager to make continual improvements in the planning and scheduling of his activity. As time progresses, he will develop a data bank which can be used to determine times more accurately in scheduling future projects. (See Figure 6-5.) As projects of complexity sufficient to warrant the use of a control procedure similar to PERT (Program Evaluation and Review Technique) are encountered, accurate data are essential. As more refined data are obtained, the schedules developed from the data will become more realistic; and the industrial engineers, realizing this, will strive to attain scheduled completion dates more earnestly.

Through the use of the control and review procedures, a delay can be spotted early in the project, and corrective action can be taken to bring the project back on schedule before serious consequences result. The corrective action taken may be internal to the industrial engineering activity, or it may be in some other area through lateral or vertical reporting procedures. The important point is that manage-

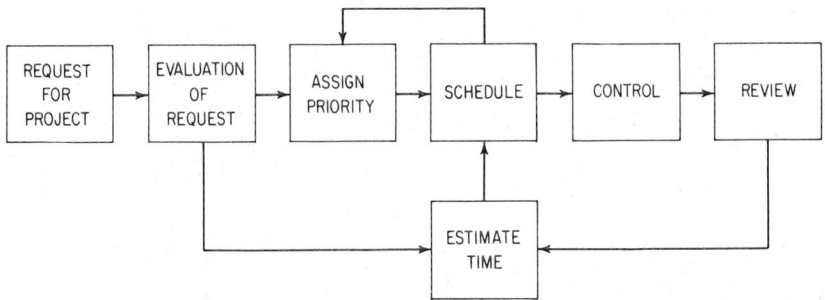

Fig. 6-5. Information flow in systematic approach to management of industrial engineering function.

ment will be aware of a delay early, so it can take action before the entire project is threatened.

The project review meetings are invaluable to the manager in appraising the managerial ability of his industrial engineers. As the work load for each engineer increases, his ability to manage becomes apparent. If technologists have been assigned to him, his ability to handle and coordinate people can be evaluated.

CONCLUSION

In summary, to manage industrial engineering time effectively, a systematic approach should be followed. The steps are:

    Evaluation of projects
    Establishing priorities
    Scheduling and allocating time
    Control of projects
    Project review

BIBLIOGRAPHY

Brewer, Anthony J., "The Expanding Role of Industrial Engineering in Industry," *The Journal of Industrial Engineering,* May, 1966, pp. 233–238.

Forrester, Jay W., *Industrial Dynamics,* M. I. T. Press and John Wiley & Sons, Inc., New York, 1961.

Gallagher, James D., *Management Information Systems and the Computer,* Research Study 51, American Management Association, New York, 1961.

Leavitt, H. L., and T. L. Whisler, "Management in the 1980s," *Harvard Business Review,* November–December, 1958.

Malcolm, Donald G., Alan J. Rowe, and Lorimer F. McConnell (eds.), *Management Control Systems,* John Wiley & Sons, Inc., New York, 1960.

Maynard, H. B. (ed.), *Handbook of Modern Manufacturing Management,* McGraw-Hill Book Company, New York, 1970.

Neuschel, Richard F., *Management by System,* 2d ed., McGraw-Hill Book Company, New York, 1960.

Peyton, Gilbert L., "Managing Technical Manpower in an Emerging Economy," *Advanced Management Journal,* April, 1969.

Methods

Chapter 1

Methods Engineering

DUANE C. GEITGEY

**Director of Training Program Development,
Maynard Research Council Incorporated, Pittsburgh, Pennsylvania**

Methods engineering enables the industrial engineer or line manager to subject each operation within the scope of his methods study to a very precise and systematic analysis. The objective of methods engineering is to eliminate every unnecessary element or operation and to achieve the quickest and best method of performing those elements or operations that are determined necessary.

The term "methods engineering" is used to describe a collection of analysis techniques which focus attention on improving the effectiveness of men and machines. Because increased efficiency must be the objective of any successful manager, the techniques of methods engineering should not be restricted to the industrial engineering department. In fact, these techniques can be used by any member of an organization with sufficient training. Methods engineering should not be confined to a single industry or business, nor should it be confined to any major functional area within an industry or business. Because of its great potential, it can be utilized by any function.

THE HISTORY OF METHODS ENGINEERING

The earliest recorded methods study was in 1760, when a Frenchman, M. P. Perronet, is said to have studied the manufacture of pins.[1] About 1830, Charles Babbage, an Englishman, also studied the manufacture of pins. These studies,

[1] R. L. Morrow, *Motion Economy and Work Measurement*, The Ronald Press Company, New York, 1957.

however, appear to have consisted of little more than timing the complete process of making pins.

The first systematic approach to methods improvement was taken by Frederick W. Taylor in 1883. Taylor divided a task into individual work elements and studied each element separately; the term "time study" first appears in the writings of Taylor. He was the first to apply the scientific approach to engineering better work methods.

A somewhat more searching analysis was being developed about the same time—by Frank B. and Lillian M. Gilbreth. Their attention was directed to subdividing a specific task into what they regarded as the most fundamental elements of movement, studying these elements separately and in relation to one another, and then rebuilding the task with the elimination of what they regarded as wasteful elements. This synthesis of the remaining elements was arranged to provide what was regarded as the best combination and sequence. The Gilbreths referred to their work as motion study. From the beginning, the Gilbreths showed an appreciation of rhythm and automaticity not possessed by Taylor. Taylor, it is true, had considered processes and motions roughly, but it was the Gilbreths who indicated their full significance.

In 1912 the Gilbreths presented a refinement of their original motion study technique before the American Society of Mechanical Engineers. They termed the new development micromotion study. Essentially this consisted of studying the fundamental elements of movement with the aid of motion pictures.

In the field of time study, Charles E. Bedaux had been busy since 1911 experimenting with the idea of measuring all human physical work in terms of a common unit. This unit, known as the Bedaux unit of human power measurement, or "B," was to consist of a combination of work and rest. The proportions between these two items were dependent upon the physical nature of the work and the subsequent rest required to compensate for it. As tasks varied, the ratio of work to rest within the unit was to vary, but the unit itself was to remain constant at a time value of one minute.

The rest value percentages applicable to various classes of work under varying conditions were said to have been compiled from studies made of the handling of different weights in varied positions, using different members of the body, attention being given to the relative sequence of motions and the frequency of their repetition.

The principle of measuring all physical work in terms of a common denominator so that it was possible to assess not merely individual but departmental performance provided a basis for control and comparison which appealed strongly to many managements.

The next development or modification in this subject was work simplification. Its originator, Allan H. Mogensen, was an industrial engineer from Cornell University. About 1930, he defined work simplification as the organized application of common sense to finding better and easier ways of doing every job, from eliminating waste motion in minor hand operations to complete rearrangement of plant layout. Every employee was to be taught to ask "Why?" about everything he did and encouraged to use initiative in furthering savings of time, energy, and material.

Although methods improvement and work simplification studies had been conducted by pioneers in the field of work methods, it was not until the 1930s that H. B. Maynard and his associate, G. J. Stegemerten, applied the name "methods engineering" to a coordinated and systematic approach to improving work methods. Maynard and Stegemerten recognized that the common goal of all workers in this field was to secure maximum labor effectiveness. They also recognized that this could not be secured by isolated studies of operators' movements, work standards, increased effort, or incentives, but by a combination of proven work study techniques, scientifically applied.

ANALYTICAL TECHNIQUES OF METHODS ENGINEERING

Methods engineering refers not only to the establishment of the method itself, but also to the standardization of all aspects of the job. The methods engineer has at his disposal a wide variety of analytical techniques which can be used either individually or in combination, depending upon the depth of study desired.

The key to the successful use of each methods engineering technique lies in the development of a questioning attitude; these techniques are tools with which the analyst can systematically question and analyze every aspect of a process.

By carefully selecting the methods engineering technique to be used, the methods engineer can regulate the depth and concentration of study; thus he can select a study technique which will provide a depth of analysis that is proportional to the potential cost savings.

Major methods engineering techniques include process charts, operation analysis, motion study, work sampling, work measurement, and value engineering.

PROCESS CHARTS

Process charts graphically present the events which occur during a series of actions or operations so that they can be easily visualized and analyzed. A process chart classifies the activities which occur during a process into five classifications: operations, transportations, inspections, delays, and storages. The classifications and their symbols have been defined by the American Society of Mechanical Engineers in their booklet number 101.

Operation Process Charts. An operation process chart (Figure 1-1) shows only the operations and inspections performed during a process. It is designed to give a quick understanding of the work which must be done to produce a given product. It makes possible a study of the operations and inspections so that the best sequence may be developed.

The greatest advantage of an operation process chart is its simplicity; it enables the methods engineer to visualize the relationship between operations or processes without showing the sometimes confusing material handling activities. For this reason, the operation process chart is an effective means of illustrating a process to persons who are unfamiliar with the sequence of operations and inspections.

The analyst, through the use of the operation process chart and the questioning attitude, can sometimes discover significant cost reductions. These cost reductions usually result from the combination or elimination of certain of the operations and inspections.

Flow Process Charts. Flow process charts are similar to operation process charts, but they include material handling and storage activities. Like operation process charts, flow process charts also aid in discovering means of combining or eliminating operations or inspections. They are particularly valuable because they graphically illustrate material handling operations which represent a major portion of most product costs. Because of the high cost of material handling, it may be desirable to analyze a job in detail and to study with a flow chart the events which occur between operations, as well as the operations themselves.

Flow process charts may include information such as the time required to complete an activity or the distance moved. They may be made in either the material type, which presents the process in terms of the events that occur to the material, or the man type, which presents the process in terms of the activities of the man. By systematically questioning the need for each activity recorded on the process chart and carefully analyzing the need for better material handling procedures, it is often possible to reduce substantially the cost of performing a process.

Flow process charts can be utilized effectively by production supervisors and foreman as well as by methods engineers. They have been used in factories, offices, banks, department stores, and hotels with excellent results.

Multiple Activity Charts. Multiple activity process charts graphically present the coordinated working and idle time of two or more men, two or more machines, or any combination of men and machines; for this reason, the multiple activity chart is sometimes called the "man and machine chart." A multiple activity chart consists of bars drawn against a time scale to represent the relationship between the working and idle time.

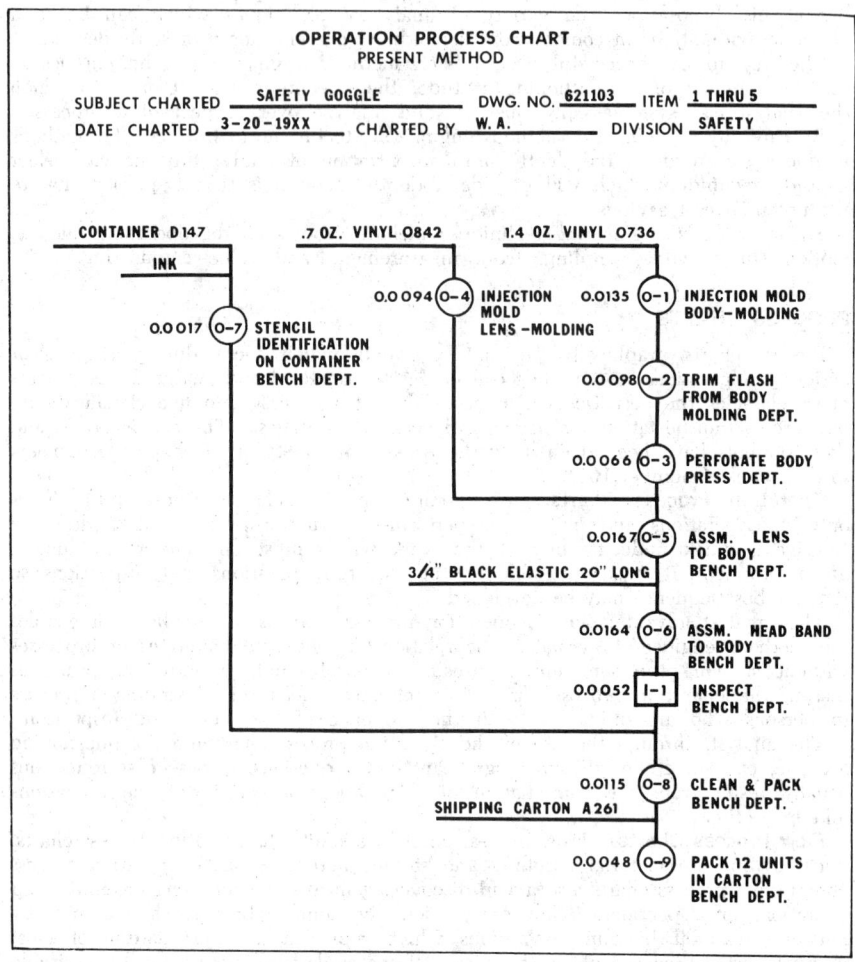

Fig. 1-1. Operation process chart.

By using a multiple activity chart, the analyst can rearrange the work cycle of either the man or the machine or both, and thus develop a more effective combination of activities. It is sometimes possible to include the performance of additional work during the machine cycle or to eliminate the additional labor time involved in an operation previously performed independently, outside of the machine cycle.

Operation Analysis. Operation analysis has been called "common sense, systematically applied." It is a procedure used to study the factors that affect the method of performing an operation to achieve maximum overall economy. The operation analysis technique consists of applying the questioning attitude separately to each part of an operation. The operation analysis form, shown in Chapter 4 of this section, guides the person making the methods improvement study through a systematic analysis of the operation. A thorough examination is made of each of the following ten points of primary analysis:

1. Purpose of operation
2. Design of part

3. Process analysis
4. Inspection requirements
5. Material
6. Material handling
7. Workplace layout, setup, and tool equipment
8. Common possibilities for job improvement
9. Working conditions
10. Method

The methods analyst applies the questioning attitude to each of the points of primary analysis to determine as much information as possible. He combines the information obtained through his application of the questioning attitude with his own knowledge of possible alternatives to develop suggestions for improvement. As the analyst conducts his study, he records the suggestions for improvement on the analysis form; these suggestions are the basis for further action. Experience has shown that a thorough application of the operation analysis procedure will almost always uncover opportunities for methods improvement. However, the potential methods improvements must be acted upon before cost savings can be realized.

Although the operation analysis form provides a means for recording potential methods improvements as they arise, the analysis actually takes place in the mind of the analyst. The operation analysis form itself ensures only that the points of primary analysis, which are the most important factors associated with an operation, are adequately considered.

The operation analysis procedure has received its greatest use in industry. However, the same systematic application of the questioning attitude can be utilized effectively in offices, retail stores, banks, and so forth. The clerical operation analysis form is illustrated in Chapter 4 of this section.

MOTION STUDY

Motion study consists of dividing work into the most basic elements possible and studying these elements separately and in relation to one another, both qualitatively and quantitatively. These elements are called Gilbreth Basic Elements, or therbligs. An understanding of the nature of the elements will provide the methods engineer with the key to improving work station layout and motion sequence. Elements can be classified into three groups: Group I, those which accomplish work; Group II, those which retard accomplishment; and Group III, those which do not accomplish. The elements in each group are shown by Figure 1-2.

As a result of motion study, the most efficient method of performing the operation can be synthesized. Although all the methods engineering procedures such as process charting and operation analysis can be utilized effectively to eliminate unnecessary operations, those operations which remain can be subjected to closer scrutiny by the use of motion study. If the operation being studied is highly repetitive, the elimination of what may seem to be inconsequential motions can result in significant savings.

Group I	Group II	Group III
Reach	Change direction	Hold
Move	Pre-position	Avoidable delay
Grasp	Search	Unavoidable delay
Position	Select	Rest to overcome fatigue
Disengage	Plan	
Release	Balancing delay	
Examine		
Do		

FIG. 1-2.   Gilbreth Basic Elements classified with respect to accomplishment.

When conducting a motion study, the analyst first reduces the job to its Gilbreth Basic Elements, such as Reach, Move, Grasp, Position, and so forth. After this detailed breakdown of the present method, individual motions are studied to develop an improved method. The improved method is also carefully analyzed to determine if further method improvements can be found. This procedure is continued until no further improvement appears possible.

The motion study analyst is governed and guided by the principles of motion economy, which have been established as a result of thorough study of the motion capabilities of the human body. The principles of motion economy are based on the fact that certain motion patterns can be performed more easily, more quickly, and with less fatigue than others. An operation which conforms to the principles of motion economy can be considered effective. An operation which does not, can be improved by subjecting it to motion study if justified by its repetitiveness.

Motion study may be performed in various levels of detail. An analyst who is thoroughly familiar with the principles of motion economy may, through observation only, develop a more effective motion pattern; however, motion study can be conducted in much greater detail with the use of motion picture analysis. Detailed motion studies without the use of motion pictures can also be made with the aid of a predetermined motion time system such as MTM.

WORK SAMPLING

One of the most effective and most frequently used methods engineering techniques is work sampling. The work sampling procedure is based on the fact that a small number of chance observations tend to follow the same distribution pattern that actually occurs in the situation being studied. In a work sampling study, observations are taken at random intervals. During the observations, the types of activities observed are recorded in predefined categories. From the distribution of random observations, the proportions of activity in each of the predefined categories can be predicted for the particular work situation.

Work sampling enables the observer to collect facts about an operation, process, or other activity inexpensively and accurately. These facts can later be used to improve the effectiveness of the operation and reduce costs. The advantages of work sampling are:

1. It is less expensive than continuous observation techniques.
2. It can be used by observers with little special training or skill.
3. The number of observations can be adjusted to meet desired levels of accuracy.
4. It is an effective means of collecting facts that would not normally be collected by other means.
5. It produces less anxiety and agitation in the person being observed.
6. It results in little interference with the operator's normal routine.

TIME STUDY

Time study is a procedure employed to measure the length of time required to perform a given task in accordance with a specified method, by an operator of average skill, working with average effort, under normal conditions. Time study is a specialized task which should be done only by trained time study engineers, but the time study procedure itself should be readily understood by all levels of management.

The time study procedure is as follows:

1. The method should be studied, improved, and standardized before the time study is begun.
2. The operator to be studied must be selected and informed that the study will be made.
3. The operation is subdivided into a number of smaller operations known as elements. Each element is exactly defined, so that one who is familiar

with the class of work can visualize every step of the operation merely by reading over the list.

4. The operation is timed with the aid of a stopwatch, allowing sufficient readings for each element to ensure accuracy.

5. The elapsed time for each element is secured by subtracting successive readings if the watch runs continuously, or directly from the watch itself if the watch is returned to zero at the end of each element.

6. Irregular or unusual elements are carefully recorded so that every moment of time during the study is accounted for.

7. The elapsed time for each element is determined and recorded on the time study form.   Abnormal values must be carefully examined to determine if they represent a typical situation.

8. Elemental times are adjusted by multiplying each one by a leveling factor which is determined by rating the skill and effort of the operator in the performance of the job.   Leveling is necessary to adjust the observed time to reflect the performance of an operator working at the average performance level.

9. Allowances for personal and unavoidable delays, fatigue, and any special conditions are added to the total leveled times for all elements.

10. Each elemental time is multiplied by the number of times it occurs per piece or cycle in the operation.

11. The resulting times are then added to determine the total time allowed for each piece.

To be satisfactory for business or industrial use, a time study must have the following characteristics:

1. It must be an accurate measure of the time required to perform the operation or process.

2. It must be clearly defined and readily understandable to anyone familiar with the time study procedure.

3. It must be salable, so that the operator can be convinced of the fairness and correctness of the time value.

METHODS TIME MEASUREMENT (MTM) ANALYSIS

Methods time measurement (MTM) is a work measurement procedure which analyzes the basic motions required to perform any manual operation or method and assigns to each motion a predetermined time standard based upon the nature of the motion and conditions under which it is made.   The MTM procedure is particularly effective because it enables the analyst to make a detailed study of method during the development of a time standard.   This feature of MTM assures that work standards will not be developed without prior consideration of method.

In the first step of the MTM procedure, the analyst makes a systematic analysis of the motions required to perform the operation.   This analysis can be performed at the work station; in a methods laboratory, prior to the establishment of the work station; or by visualization, if the analyst is thoroughly familiar with the class of work being studied.

After the analyst has recorded the basic motions, he obtains the time required for their performance from MTM data tables.   The data tables are based on extensive research into the normal time required to perform basic motions under various conditions.   The physical conditions involved in the performance of the motion and the conditions under which the motion is performed at the time of the analysis are considered.

Motion patterns are described by a system of motion symbols that are immediately recognizable by any person trained in the MTM procedure.   These recorded motion patterns can be quickly compared with the actual motion pattern to assure continued accuracy after the initial standard is set.   Each motion is assigned an MTM time value stated in time measurement units (TMU).   Each TMU represents 0.00001 hour.

MTM is particularly effective for the establishment of preproduction standards and methods. Using MTM, various methods of performing an operation can be compared, and the best method of performing the operation can be selected in advance of production.

VALUE ENGINEERING

Value engineering is a more recent addition to the general area of methods engineering. It has two unique characteristics that make it particularly valuable for reducing the costs of processes, procedures, systems, and services. These characteristics are (1) the emphasis on function and (2) the job plan, which is a systematic approach to value improvement.

The five phases of the value engineering job plan are as follows:

1. Information phase
2. Creative phase
3. Evaluation phase
4. Investigation phase
5. Implementation phase

The value engineering approach consists of selecting the project and accumulating complete cost data. The functions are exactly defined at this point. Using the brainstorming technique, alternative methods of accomplishing each function are developed.

The creative ideas obtained during the brainstorming phase are carefully evaluated and redefined until a workable, low-cost means of accomplishing the function is found.

During the investigation phase, alternatives developed during the creative and evaluation phases are examined and compared to achieve the best possible solution.

In the final phase, a recommended course of action is submitted for management approval.

The value engineering functional approach to cost reduction can be used to augment other methods engineering techniques which concentrate primarily on the processing of the product.

DETERMINING THE DEPTH OF STUDY

The depth to which the methods engineering study is conducted will depend upon a number of factors. To help him decide which of the analytical techniques he will employ in any particular case, the methods engineer must consider:

1. The repetitiveness of the activity
2. The hourly labor rate paid for the activity
3. The labor content of the activity
4. The anticipated life of the activity

The first two factors may be combined into an index figure giving the annual labor cost per 0.0001 hour: annual activity $\times$ hourly rate $\times$ 0.0001.

The significance of this index may be shown by the following example:

If an activity is performed three million times a year by a person whose hourly wage is $3.50, then the index shows an annual cost of $1,050 for 0.0001 hour. This means that it will be quite profitable to save even as small an amount of labor as that which may be done in 0.0001 hour; that is, it will be profitable to study the activity in great detail. The annual cost per 0.0001 hour is a valuable guide for other purposes; if it is low, only the simplest fixtures, tools, or equipment revisions are justified.

It is advisable to consider collectively the four factors mentioned above, for no one of them is sufficient to determine the amount of study warranted in a particular instance. The techniques of making a methods engineering study can be used in various combinations, and the methods engineer must determine which will be most effective in each particular case.

METHODS ENGINEERING TRAINING

The principles and techniques of methods engineering should be applied to some extent by every member of the organization. Therefore, methods engineering training is an essential part of any methods engineering program.

In-depth training should be provided to all methods analysts for whom methods improvements are a primary job responsibility. Other persons in the organization can be trained to varying levels of detail, depending upon their methods improvement responsibility. To achieve maximum results from a methods engineering program, all personnel should receive methods engineering training to at least an understanding level.

Analyst Training. Methods engineering analysts must make detailed methods studies and therefore should receive thorough training in methods engineering techniques. They should, for example, know the specifics of all types of process charts, their advantages, disadvantages, and uses. They should receive training not only in the uses of work sampling, but also in work sampling theory in such areas as accuracy, confidence, and number of observations. Their knowledge of work measurement techniques should be sufficient to enable them to make a work study utilizing the technique adopted by their particular company. In addition to the initial training, they should receive periodic refresher training to maintain a high level of capability.

Supervisory Training. One of the most important aspects of a sound methods engineering program is a cooperative attitude on the part of supervisors, inspectors, tool designers, and other key men who must participate in the methods improvement program. These men must understand the fundamentals of methods engineering to be able to cooperate fully with the methods engineer during the methods study phase, and later, during the installation of methods improvements. In addition, supervisors have a unique opportunity to recommend areas for methods study, because they are in constant contact with the company's operations and operators.

Supervisory training in methods engineering should be on a less technical level than the training received by the analyst. However, supervisors should be given sufficient training to enable them to understand the value of methods engineering techniques and the mechanics of the techniques. Generally, they need not receive the theoretical or technical training provided for the analyst.

Management Training. Because a sound methods engineering program must receive support from higher management, methods engineering training on an appreciation level for higher management personnel is essential.

The management training program in methods engineering can be presented in an overview format; it should present the potential advantages of each technique so that managers can recognize possible applications of the technique. Overview training will also enable managers to discuss and evaluate methods improvements generated by the technical or supervisory personnel.

Operator Training. Obviously, most potential methods improvements cannot be realized until the operator is actually using the new method. Thus a sound operator training program is an essential part of the methods engineering approach to work study. Operators who have been trained to follow a good method and to recognize the important part that methods engineering plays in reducing fatigue and increasing efficiency will benefit any methods engineering program.

ATTITUDES REQUIRED FOR EFFECTIVE METHODS ENGINEERING

Although methods engineering techniques provide the tools necessary to ensure the most effective means of accomplishing work, they cannot be successfully applied without positive attitudes on the part of both the methods engineer and management.

Every successful methods engineer must possess the following attitudes:

1. Belief that with sufficient study any method can be improved
2. The questioning attitude

Experience with the application of methods engineering techniques has shown that a methods engineer who earnestly and diligently applies the principles and techniques to any operation will ultimately find a better method of accomplishing that operation. Although the cost of making the methods improvement may not be justified by the nature of the operation itself, improvement nevertheless will be found. Despite the fact that a "perfect" method seems to have been achieved, experience has shown that further improvement can always be made, if only as a result of technological advance.

The questioning attitude is the basis for all methods engineering techniques. Anyone who seeks methods improvement must inquire into all aspects of an operation and ask the question "Why?" as he conducts his analysis. Every methods engineering technique provides a tool with which the methods engineer can question the factors surrounding an operation. Methods engineering itself represents a systematic means of applying the questioning attitude.

Successful application of methods engineering depends not only on the attitude of the methods engineer, but also on management's attitude toward methods improvement. Few things can be more frustrating than the realization that a potential methods improvement cannot be implemented because of the apathy or obstructionism of others in the organization. Management must promote an attitude of cooperation and understanding to realize significant methods improvement. Therefore, the importance of indoctrinating all levels of management in the value and use of methods engineering cannot be underestimated.

CONCLUSION

Methods engineering is an organized, systematic approach to increasing the efficiency of work. It represents a series of analytical techniques of proved effectiveness with which the analyst can study and improve methods.

In addition to having a thorough understanding of each technique, a methods engineer must believe that with sufficient study any method can be improved. He must be willing to apply the questioning attitude to every aspect of his study. He must also have the management support necessary to implement the methods improvements he develops.

A sound methods engineering program, with adequate training and management support, will significantly reduce costs in any business or industrial situation.

BIBLIOGRAPHY

Barnes, R. M., *Motion and Time Study*, 6th ed., John Wiley & Sons, Inc., New York, 1968.

Carroll, Phil, *How to Control Production Costs*, McGraw-Hill Book Company, New York, 1953.

Heyel, Carl (ed.), *The Encyclopedia of Management*, Reinhold Publishing Corporation, New York, 1963.

Krick, Edward V., *Methods Engineering*, John Wiley & Sons, Inc., New York, 1962.

Lowry, S. M., H. B. Maynard, and G. J. Stegemerten, *Time and Motion Study and Formulas for Wage Incentives*, 3d ed., McGraw-Hill Book Company, New York, 1940.

Maynard, H. B., G. J. Stegemerten, and J. L. Schwab, *Methods-Time Measurement*, McGraw-Hill Book Company, New York, 1948.

Niebel, B. W., *Motion and Time Study*, Richard D. Irwin, Inc., Homewood, Ill., 1967.

Chapter **2**

Processing and Operation Planning

THOMAS S. RIORDAN

Director of Manufacturing, Fluid Power Division,
The Bendix Corporation, Utica, New York

A product is depicted by a drawing or package of drawings and pertinent specifications outlining the material required, product configuration, and functional capabilities. The manufacturing activities necessary to accomplish the production of an end product must be processed or arranged in an orderly, workable sequence. Processing, in a broad sense, encompasses every phase of industrial engineering in establishing a manufacturing plan. The hub of this plan is the initial processing and operation planning.

Any task which involves adding value with manpower, material, or equipment may be improved with an adequately defined process. Indeed, it may be difficult to perform at all without this information. The degree of complexity of the process can vary from a list of simple operations handwritten on a sketch of a part to a detailed right- and left-hand description of every element of every operation with explicit descriptions of the tools, gages, and workplace layouts to be employed. Processing and operation planning is not limited to industrial-type activities, but is equally applicable to nonmanufacturing operations.

This chapter will restrict discussion to basic concepts of processing related to industrial applications. No attempt will be made to cover the full scope of establishing specific processes. The main objective is to present a guide to systems and techniques which will accomplish the processing segment of the manufacturing cycle effectively.

ORGANIZATIONAL SLOTTING

The function of preparing the manufacturing process can be accomplished in a number of ways, depending upon the size and type of industrial activity.

Small Job Shops. In a small machine shop, foundry, stamping shop, or tool shop, the shop foreman can effectively establish the process on a relatively informal basis. The skill of the production worker and his close relationship with supervision preclude the need for extensive detail.

Small- to Medium-size Production Plants. Where a product is manufactured repetitively and where there is expectancy of some reasonable stability in design, a separate support group is desirable. This may involve one or more technically competent manufacturing engineers. Their assignment will be to specify the process in sufficient detail to allow semiskilled production workers to produce a quality product at a reasonable cost. The manufacturing engineers may also be responsible for the design of tooling and establishing work standards.

Large, Single or Multiproduct Plants. In large plants producing one or more products repetitively, manufacturing specialists who specialize in only a segment of the operation may be desirable. An individual or group of individuals may be assigned to primary machining, foundry operations, sheet metal forming, or assembly. The process engineer normally would not be assigned to any function other than developing a workable process. The advantages of this approach are appreciable, provided there is adequate supervision to prevent a loss of continuity among all specialists involved. Also, there must be a close liaison with all other contributing manufacturing support functions.

Level in Organization. In establishing the location of the processing function in any organization, it should be recognized that the process is the primary basis for establishing manufacturing costs and the allocation of resources. A process which cannot be enforced or which is not maintained as an accurate workable definition of the task will present a false manufacturing cost and a false definition of resources required. Therefore, the processing function should be slotted within the manufacturing organization at a level of authority which can enforce adherence and control changes.

ESTABLISHING AN EFFECTIVE MANUFACTURING PROCESS

The sequence of events necessary to establish an effective manufacturing process usually includes:

1. Preparation of preproduction drawings
2. Review of manufacturing feasibility
3. Deciding to make or buy
4. Developing the process
5. Monitoring the pilot run
6. Making process changes

Preparation of Preproduction Drawings. The design draftsman seldom produces an initial drawing which is ready for production. His primary goal is to represent the functional requirements of a product by a series of detailed drawings of each component, subassembly, and assembly. The initial material selected may also consider only structural needs, with no thought given to machinability. The net result is that the first release of drawings must be considered as preliminary or preproduction.

Review of Manufacturing Feasibility. The process engineer should thoroughly review every preproduction drawing for the following:

1. Are dimensioning and datum surfaces compatible with accepted machining practices?
2. Are sufficient stock allowances provided on castings, forgings, and stampings to allow for anticipated mismatch or distortion in heat treatment?
3. Are sufficient clearance and access allowed for proper assembly of all components?

4. Are maximum allowable tolerances applied to nonfunctional characteristics?
5. Are tolerances on functional characteristics realistic, and is statistical tolerancing used where possible?
6. Are adequate clamping and locating surfaces needed for manufacturing provided?
7. Are value engineering suggestions for lower cost applicable?

The process engineer collects information on the above items in the form of comments and reviews them with the responsible design engineer. The acceptable suggestions or trade-offs are agreed upon and incorporated into each applicable drawing. The revised drawings become the production drawings.

Deciding to Make or Buy. The next step is to decide whether to make or buy. A team consisting of responsible individuals from purchasing, the estimating section, and the processing section should prepare an analysis for deciding whether to produce an item in-plant, to purchase it complete, or to purchase it semifinished and complete it in the plant. Proper decisions are based upon true cost comparisons, in-plant work loads, lead time comparisons, and in-plant versus vendor capability. The process engineer must be able to present a factual estimate of all parameters of producing each item in-plant.

Developing the Process. The method of developing a process for any given part or assembly will vary greatly, depending on the nature of the product, level of production, lead time allowed, and the like. A typical process will establish:

1. Material required
2. Tooling required
3. Machines required
4. Sequence of operations relating to what is to be done with the required tooling and machines
5. The necessary control counting points
6. The required inspection or quality audit points and the procedure of inspection or test

The process is not complete until the specific tooling, machines, and time standards are defined.

The process engineer will normally present his process to a tool design section for the design of tooling not available as standard. The process engineer preferably should approve each new tool design prior to its release for manufacture.

The section responsible for work measurement will either establish an estimate of the allowed time for each operation or establish an engineered standard time if predetermined data are available. The process engineer may find it necessary to revise his approach if he finds the cost of new tooling or direct labor excessive. This review is usually profitable prior to releasing a final approved process.

Monitoring the Pilot Run. The first production run using a new, untried process will normally require some adjustments. A practical approach is to assign a process engineer to monitor each operation of the first run. The production shop must be required to follow his process to the letter and to deviate only when he has had the opportunity to investigate and approve any need for a change. Too often, panic created on the first run will add sizable elements of cost which could be prevented by sound evaluation of each trouble spot. Another serious hazard on the first production run is design deficiencies. If the process is not followed and approved deviations are not documented, it can be very difficult to establish the difference between a product design problem and a manufacturing process problem.

Making Process Changes. A process change after the first production run must also consider:

1. Parts or assemblies in process and material on hand
2. Cost of the change, including effect on tooling, material, and delivery schedule
3. Anticipated savings or added cost

Although cost reduction is a constant philosophy of manufacturing, changes should be made with caution. A change to accomplish a minor saving may create unknown problems and losses. Each change requires another pilot run to prove the validity of the new process. There is much to be said for maintaining stability in manufacturing. Too often, in eagerness to show an improvement, consideration is not given to the cost of rebalancing machine loads and assembly lines, the cost of retraining, and potential scrap and rework losses.

SOURCE DATA REQUIRED FOR PROCESSING

The successful process engineer must be able to research existing and new techniques rapidly when the need arises. He must constantly look for new approaches specifically applicable to his type of manufacturing. A wealth of reference material is available in the form of textbooks, technical papers, technical data supplied by the machine tool industry, and technical journals.

In-plant Data.   Pertinent in-plant data are required to allow the process engineer to zero in on the specific details of producing in his plant. Much of this information can be provided and maintained inexpensively with electronic data processing. Examples of such data are:

1. *Machine inventory.*   An active listing of existing production equipment with specifications on size, capability, accessory tooling, feed and speed ranges, and tolerance capabilities.
2. *Machine load forecast.*   The forward sales forecast exploded by machine time required. This will guide the process engineer to where open time is available, or it may point out the need to request new equipment before a future production bottleneck occurs.
3. *Material specifications.*   Should include machinability ratings as well as acceptable thermal treatments.
4. *Standard tool inventory.*   Should identify quantity and application of existing standard tools.
5. *Gaging standards.*   Should include standard gages available, applications, and frequency of gaging required.
6. *Scrap and rework history.*   The causes of high scrap and rework may relate to marginal processing.
7. *Cutting fluid applications.*   The process engineer should develop or be provided with a standard application procedure for applying the proper cutting fluid.
8. *Speeds and feeds.*   Textbook speeds and feeds are intended to be conservative starting points. Each plant should develop accurate, proved feeds and speeds which will yield optimum production within an acceptable quality level.
9. *Work measurement data.*   If the process engineer is either trained in work measurement or provided with standard data, he can select the best approach to keep production time to a minimum.
10. *Abbreviations.*   An acceptable manual of abbreviations for all repetitive terms used in describing a process will save countless hours and space.

PROCESS FORMAT AND CONTENT

There is no set standard form for describing a manufacturing process. Like industries, however, will tend to follow similar patterns. The methods of duplication, distribution, and presentation to the production operator have considerable bearing on format.

Audiovisual equipment, using synchronized audio tape and slides or film strip, may preclude the need for a written process sheet. This technique, which is particularly successful in assembly, provides the operator with a visual picture of each step of the operation, accompanied by audio instruction about sequence and method. The process engineer develops the process with a recorder and a camera. Instant changes can be made by erasing or splicing tape and changing or adding film frames. This technique is also effective for a series of inspection points.

Many industries are using computer or data processing equipment to reproduce a process. This is usually accomplished by off-line storage, using discs, tape, or punched card master decks. The process engineer can have remote location access to all stored data. He may scan, change, add, or delete by keying his requests through a special keyboard, and he can check the results on a high-speed printout or a visual scope.

The standard approach is to develop a format and to preprint master reproducible forms. The details of each process are then printed, typed, or sketched onto a master reproducible form. Figures 2-1 to 2-4 are examples of typical process sheets for assembly and fabrication work.

Process Content. There are many theories on how much or how little should be included in a written process. The following items should at least be considered for inclusion.

Part Identity. A part number, pattern number, or catalog number with the latest drawing revision should be used along with the description of the part or assembly.

Stock Identification. The first sheet of the process should identify the material specification and size required if bar stock or sheet stock is used.

Operation Number. Each operation should be identified with a separate number. Care should be taken in the numbering system to allow room for added operations. Consideration should also be given to data processing acceptance of all numbers required.

Work Station Identity. A description of the machine or assembly station involved should be considered. A machine or work station code may be used to allow for proper machine loading and for collecting data on machine utilization.

Production Time Standard. Each operation should have a standard in hours per 100 pieces or allowed hours or minutes per piece. It is also a good practice to show the standard in pieces per hour and to include a standard for setup when applicable.

Operation Description. In machining operations, at least a basic description of the machining elements such as turn, face, drill, mill, or grind, together with the dimension and tolerance of each cut, is required. The sequence of performing the machining is also important, as well as the gaging procedure and frequency.

Assembly operations are normally described in more detail, even to the extent of specifying motion patterns (see Figure 2-2).

Tools and Gages Required. It is a good practice to identify all tools and gages required for each operation on the process sheet, using a standard system of identification, to permit the operator to draw out the necessary tooling with assurance that only the proper tooling will be used (see Figure 2-3).

Tool Layouts. A separate form may be developed for each machine using multiple tools, to define how the machine is to be set up. The majority of machine tool builders will provide such forms for their machines (see Figure 2-4).

Speeds and Feeds. The speed and feed for each machining element should be predetermined and included on the process sheet. It may even be advantageous to predict the number of cuts before a tool is to be changed because of wear.

Workplace Layout. The optimum workplace layout is normally shown for assembly operations. In some instances, this may also be desirable for machining or other operations.

Pictorial View of Operation. Particularly for machining operations, a sketch of the part showing what surfaces are to be machined is helpful. Such a sketch can also identify how to locate and grip the part as well as accurately pinpoint surfaces to be deburred (see Figure 2-3).

Special Handling and Protection. The process should define proper material handling, rust protection, and cleaning requirements.

PROCESS CONTROL PROCEDURES

A well-managed manufacturing organization must establish and enforce strict control procedures to cover necessary process changes and allowable deviations. The following controls should be considered.

PART NO. ⊗		MODEL NO.			PROD. CODE ⊗ 3		PIECES/HOUR ⊗ 91
PART NAME PUMP ASSEMBLY 34E66-3W		PROC. ENG. W.H.	JOB ORDER				SELECT TIME
							LINE BALANCE
DEPT. 117 OPER. 40 OF		USED ON ASSEMBLY NO.		STD. ENG. W.S	PAGE 10 OF	DATE 2/7/	TOTAL SELECT TIME
							FATIGUE %
							PERSONAL %
REFERENCE DRAWING AND SPEC. NO'S. P.S. 242 METHOD K 34E66-3W ASSEMBLY					CHANGE LETTER		DELAY %
							TOTAL ALLOWED TIME 6.59
							TOTAL HRS/C 10.935

T.E.C.R. CHANGES

DATE	CHK'D BY	SEQ.	CHANGE DESCRIPTION

TOOLING AND FIXTURES

SEQ.	MACHINE, TOOL, OR FIXTURE
	HOLDING FIXTURE CT-34E56-DN
	TORQUE WRENCH STD (30-40 IN lbs)
	BLAKESLEE FREON ULTRA SONIC DEGREASER

TOOLING AND FIXTURES

SEQ.	MACHINE, TOOL, OR FIXTURE

WORKPLACE LAYOUT

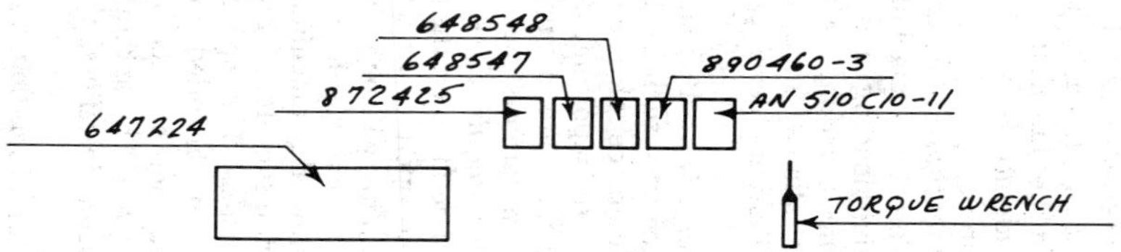

648548
648547
872425
647224
890460-3
AN 510 C10-11
TORQUE WRENCH

FIG. 2-1. Typical assembly operation first sheet.

2-18

PART NUMBER [X]	OPER. NO. 40	PART NAME	ASSEMBLY [X]	PAGE NO. //	MODEL NO. 34E [X] 66-3W
		PUMP			

SEQ. NO.	B.M. QTY NO.	PART NO.	PART DESCRIPTION	DESCRIPTION OF OPERATION
				PRECLEAN & ULTRASONIC CLEAN IN
		647224	GUIDE ASM.	FREON DEGREASER & PLACE IN WORK AREA
		472425	SIDE PLATE	
		648547	ROLLER BRG	
		648548	PIN	
	1	647224	GUIDE ASM.	PICK UP GUIDE ASM & PLACE IN WORK AREA
	4	872425	SIDE PLATE	PICK UP SIDE PLATES & ROLLER BRG & POSITION
	2	648547	ROLLER BRG	IN GUIDE ASM. & ASSEMBLE PIN
	1	648548	PIN	
	4	872425	SIDE PLATE	PICK UP SIDE PLATES & ROLLER BRG & POSITION
	2	648547	ROLLER BRG	IN GUIDE ASM.& ASSEMBLE PIN
	1	648548	PIN	
	2	890460-3	NUT	PICK UP NUTS & SCREWS & ASSEMBLE IN GUIDE ASM.
	2	AN5IOCID-11	SCREW	
				PLACE GUIDE ASM. ON HOLDING FIXTURE
				PICK UP TORQUE WRENCH & TORQUE SCREWS
				TO 30-40 IN LBS-LAY ASIDE TORQUE WRENCH
				REMOVE GUIDE ASM. FROM FIXTURE & LAY ASIDE

FIG. 2-2. Typical assembly operation detail sheet.

NAME: NUT SPLINE

BENDIX UTICA
PROCESS ROUTING SHEET

PART NO. 2493257

PER	DEPT	MACHINE	GRP NO	HRS/100		OPERATION
10	102	#4 W.B.S	206	18.867	S	SEE TOOL LAYOUT

SPEED	FEED	TOOLS & GAGES NAME	NO.
		1 1/8 COLLET PADS	STD-
		S.P. DRILL	T-155573.6-DN
		BORING TOOL	T-647497-PX
		GROOVE TOOL	T-121252-CF
		.010-.025 x 45°	VERNIER
		PIN GR. .178-.198	G-79645-F
		ADT. SN. GR. 1.510-1.515	MT-25833-4X
		ADT. SN. GR. 2.390-2.410	MT-25833-6
		PL. GR. "GO" 1.3101-Y	
		PL. GR. "NO GO" 1.314-Y	
		PL. GR. "GO" 1.27228-2	
		PL. GR. "NO GO" 1.276-2	
		GROOVE GAGE 1.305-1.315	G-121252-T
		PL. GR. 1.036-Y "GO"	
		PL. GR. 1.045-X "NO GO"	
		DEPTH MICS 1.021-1.041 1.204-1.209	

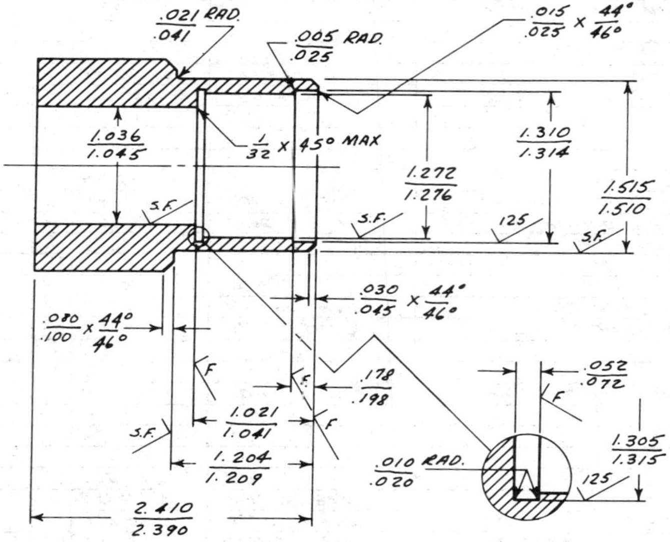

MACH. RATING - 60%

FIG. 2-3. Typical fabrication process with pictorial view of operation.

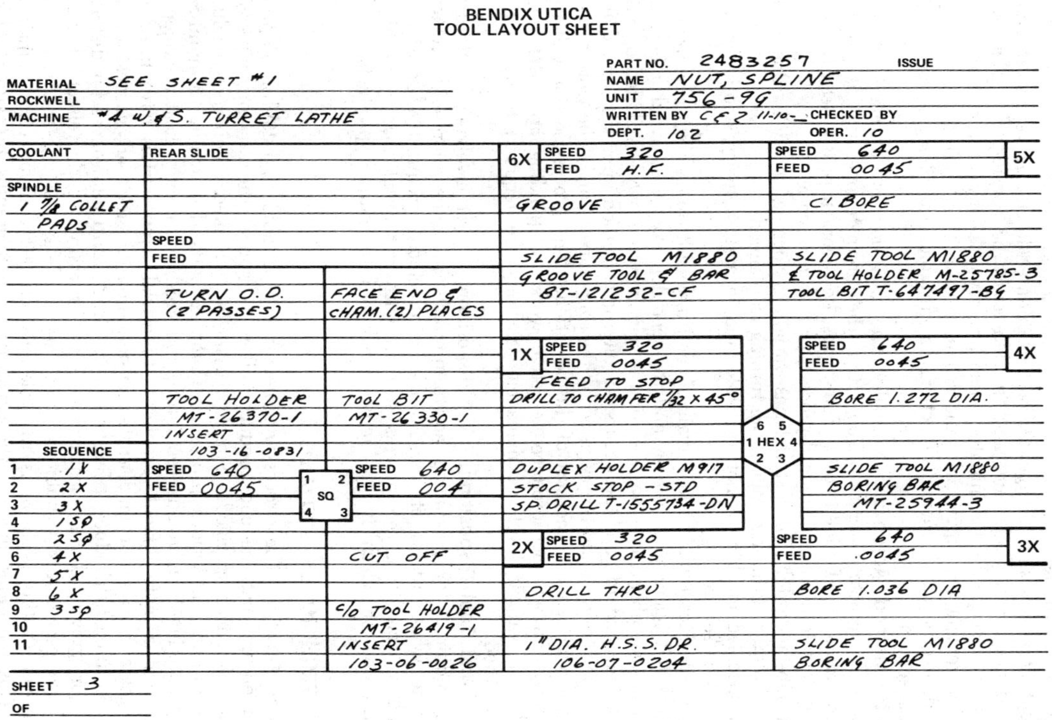

BENDIX UTICA
TOOL LAYOUT SHEET

PART NO. 2483257   ISSUE
NAME NUT, SPLINE
UNIT 756-96
WRITTEN BY C.E.? 11-10- CHECKED BY
DEPT. 102   OPER. 10

MATERIAL SEE SHEET #1
ROCKWELL
MACHINE #4 W&S TURRET LATHE

COOLANT

SPINDLE
1 7/8 COLLET PADS

REAR SLIDE

6X   SPEED 320 H.F.   SPEED 640 FEED .0045   5X
GROOVE   C'BORE

SPEED
FEED

TURN O.D.   FACE END &   SLIDE TOOL M1880   SLIDE TOOL M1880
(2 PASSES)   CHAM. (2) PLACES GROOVE TOOL & BAR & TOOL HOLDER M-25785-3
                               BT-121252-CF   TOOL BIT T-647497-B4

TOOL HOLDER   TOOL BIT   1X SPEED 320 FEED .0045   SPEED 640 FEED .0045   4X
MT-26370-1   MT-26330-1   FEED TO STOP   BORE 1.272 DIA.
INSERT   DRILL TO CHAMFER 1/32 x 45°
103-16-0031

SEQUENCE   SPEED 640   SPEED 640   DUPLEX HOLDER M917   SLIDE TOOL M1880
1 1X   FEED .0045   FEED .004   STOCK STOP - STD   BORING BAR
2 2X   SP. DRILL T-165574A-DN MT-25944-3
3 3X   1 2
4 1SP   SQ 3
5 2SP   4
6 4X   CUT OFF   2X SPEED 320 FEED .0045   SPEED 640 FEED .0045   3X
7 5X   DRILL THRU   BORE 1.036 DIA.
8 6X   C/O TOOL HOLDER
9 3SP   MT-26419-1
10   INSERT   1" DIA. H.S.S. DR.   SLIDE TOOL M1880
11   103-06-0026   106-07-0204   BORING BAR

SHEET 3
OF

FIG. 2-4. Typical tool layout sheet related to operation shown in Figure 2-3.

2-21

Change Request Procedure. Every process sheet should be considered as a required blueprint for manufacturing. It should be subject to change only when such change is properly reviewed for validity and approved by responsible supervision. Process change request forms should be provided to responsible shop supervision and technical support personnel. Any necessary (in the opinion of the requester) change should be documented, stating details of the process to be changed and the reason for the request.

In accepting a requested change, the cost to initiate the change versus the savings, effect on quality, tool lead time, and the like must be considered. If a change request is not accepted, the initiator should receive a copy of the request, with a positive statement explaining why the request was rejected. If a change request is accepted, a clear-cut sequence of actions to introduce the change properly must be established.

Alternate Method Approval. Normally, the inclusion of alternate operations in a process is a poor practice. Too often, the alternate operation is performed without consideration of added cost. Sometimes, however, there is a need to provide a means to deviate from the standard process for reasons such as:

1. A substandard lot of material has variations such as spotty hardness which may require grinding rather than turning as planned.
2. A specified machine is not available, and to meet delivery, another machine must be selected.
3. The specified tooling is not available and makeshift tooling must be used.
4. A skilled operator is absent, and to meet delivery or to avoid a bottleneck in an assembly line, a work station is split between two less skilled operators.

When a nonstandard conditon occurs, an alternate method approval form defining a requested deviation should be prepared by the shop supervisor. The work measurement section will estimate the cost of the deviation. The process section will verify that the deviation will not produce defective work. Depending upon the amount of added cost, various levels of management should be required to approve the alternate method.

Machine and Tool Disposition. Any disposition request involving facilities that might be required for producing active products must be thoroughly reviewed to determine what process, if any, would be affected. Frequently, decisions are made without consulting the process section, with the result that a machine or tool is disposed of which is still actively required for current products.

PROCESS TECHNIQUES AND AIDS

Tolerance Charts. The process engineer must be able to establish in-process control to ensure that the end product will in fact meet the drawing dimensions within the tolerances specified. Simple parts with a limited number of machining operations present no serious problem. Complex parts that require a number of machining cuts involving rough, semifinish, and finish machining; heat treatment; or plating present a more difficult problem. Industrial scrap barrels have been filled over the years because of inadequate consideration of stock allowances for each operation required to produce a finished part. Haphazard allowances for distortion, shrinkage, or growth during heat-treating, and guessing at the effect of plating buildup have cost industry millions of dollars in delays, scrap, and rework.

The answer to this problem is the tolerance chart technique of manufacturing control. This technique involves the construction of a graphic breakdown of each plane involved, showing every dimension machined and the locating surface from which it is machined. All tolerances are shown as equal bilateral tolerances. This allows for adding or subtracting dimensions as required, while the mean tolerances are always added. A properly constructed tolerance chart will trace the stock removal of every operation, define the stock allowed for each subsequent operation, and accumulate the tolerances of every dimension machined (see Figure 2-5).

The tolerance chart is a tool which can be used by value engineering to effect sizable cost reductions. It will allow fast determination of where process dimensions

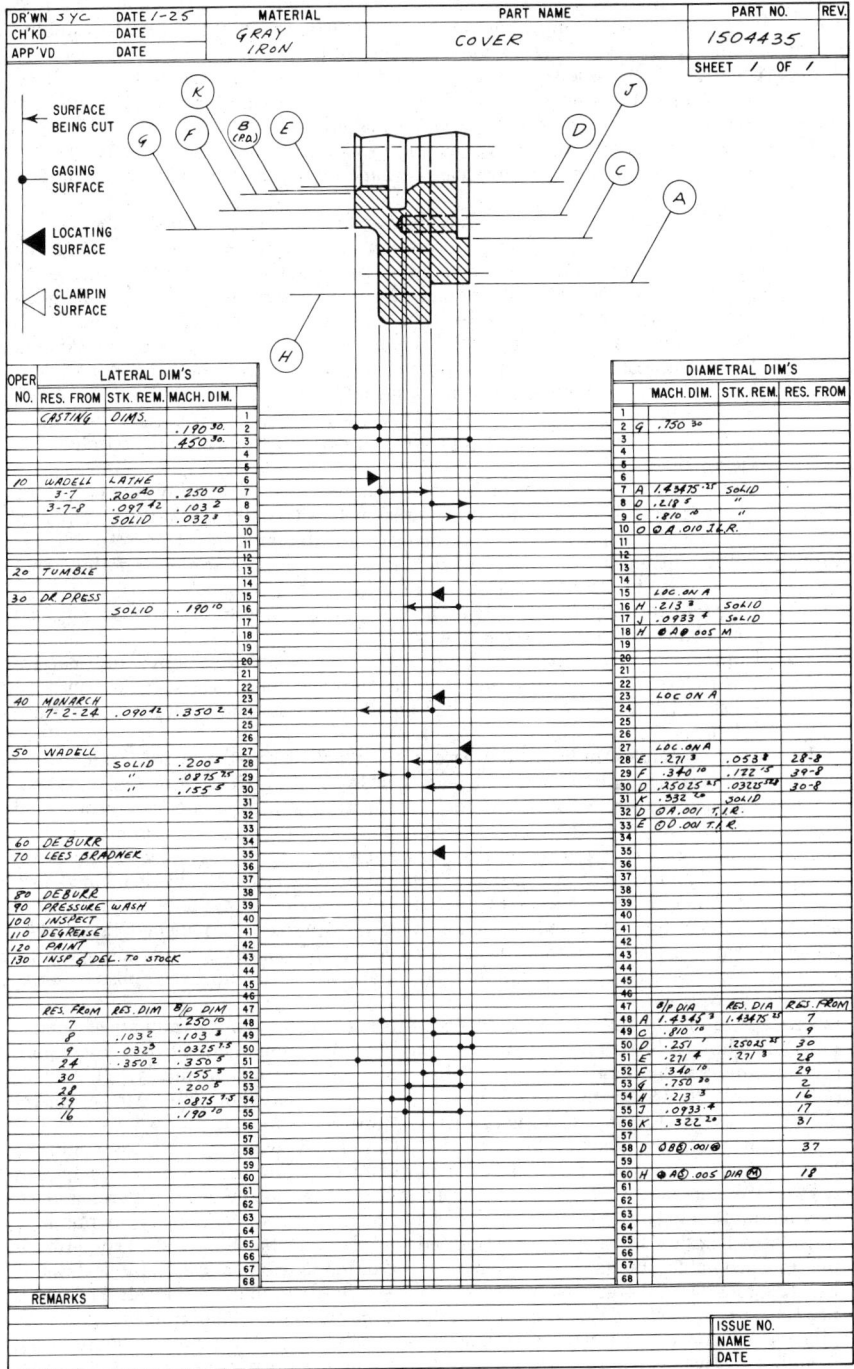

Fig. 2-5. Typical tolerance chart for a machined part.

are held tighter than is necessary. It will allow trade-off of tolerances so that the least expensive machining operations are applied. It can be used to establish statistical tolerancing, allowing tolerance relief with proper distribution control.

Line Balance. Any situation where a total manufacturing task is broken down into a series of operations which, by their nature, cannot be performed independently on a lot or batch basis, requires a line balance analysis.

The process engineer must arrive at the optimum approach considering:

1. What portion of the assembly can be handled as subassemblies on an off-line basis?
2. What portion of the assembly must be handled on a continuous-line basis?
3. Considering the total time involved on the continuous-line portion of the task, how many operations are required to meet the scheduled requirements?
4. Can the total task be split equally into operations with reasonable break-points? The problem of loading the early operations lightly to ensure that the units do not bottleneck in the beginning of the line must be considered. It is not unreasonable to have a differential of 10 percent between the allowed times of the first operation and the last operation.
5. At what point in the assembly sequence is it necessary to process inspection operations? This is largely dependent upon accessibility and expected frequency of rejects.

Computer Assists. The advent of the computer opened many doors to the processing function. There are numerous areas where the impossible tasks of yesterday can be handled quickly and accurately. In the area of data retrieval, by proper coding and input to punched cards, tape, discs, or core storage, the following typical information can be retrieved at will:

1. All product or part numbers which are processed to be run at a given work station or to use a given tool
2. The location and quantity available of all tools and gages, active and inactive
3. A listing of all parts produced from the same material specification
4. A listing of all parts which utilize an identical thermal treatment or surface treatment

The above examples indicate the ease with which across-the-board changes can be accomplished without the constant fear of missing an application or tool involved.

The process engineer responsible for manufacturing with numerically controlled equipment will require the service of a computer for programming all but the simple point-to-point operations. The approach used is to train the engineer in applying the technique of converting the physical dimensions to be machined into a machine language acceptable to the computer. The computer, in turn, using a post processor for the specific machine involved, prepares a punched tape. The punched tape, inserted in the control console of the machine, will automatically sequence the tooling, machine spindles, and part orientation to accomplish a series of machine cuts.

The advantage to the process engineer of using a computer for numerical control (NC) machining is beyond measure. Many complex profile surfaces would require several hundred thousand separate calculations if programmed manually. The computer is able to accept the equivalent of the equations defining the part geometry and then in seconds compute all necessary machine commands.

All major machine tool builders in the NC industry will train process engineers in the programming techniques for their equipment. For those smaller companies without an adequate computer, outside computer time can be readily purchased.

Outside Processing Assistance. It is common practice in industry to purchase outside tool design service. A well-trained tool designer has little difficulty performing to acceptable standards for a number of different companies. This is feasible because tool design practices are reasonably standard throughout industry. This is not the case, however, in processing and operation planning. The firm selling outside processing service must first learn the total capability of the plant to be serviced, including active machine inventory, standard tool inventory, conditions of machines and tools, and the like. All in-plant accepted practices and drawing interpretations, as well as the skill level of the production operators, must be fully understood. When and if a need arises to consider outside processing assistance, careful, thorough selection and indoctrination of the outside firm will be required. It must also be realized that the outside firm may not be available to troubleshoot or make changes at later dates. Many companies cross-train process engineers and tool designers. When a peak load occurs, the tool designers are shifted to processing, and tool design rather than processing is subcontracted.

BIBLIOGRAPHY

American Society of Tool and Manufacturing Engineers, *Tool Engineers Handbook,* 2d ed., McGraw-Hill Book Company, New York, 1959.
Cook, Nathan H., *Manufacturing Analysis,* Addison-Wesley Publishing Company, Inc., Reading, Mass., 1966.
Eary, Donald F., and Gerald E. Johnson, *Process Engineering for Manufacturing,* Prentice-Hall, Inc., Englewood Cliffs, N.J., 1962.
Machining Data Handbook, Metcut Research Associates, Cincinnati, Ohio, 1966.
Niebel, Benjamin W., "Process Analysis," sec. 7, chap. 3, in H. B. Maynard (ed.), *Handbook of Business Administration,* McGraw-Hill Book Company, New York, 1967.
Niebel, Benjamin W., "Process Engineering," sec. 3, chap. 3, in H. B. Maynard (ed.), *Handbook of Modern Manufacturing Management,* McGraw-Hill Book Company, New York, 1970.
Wade, Oliver R., *Tolerance Control in Design and Manufacturing,* Industrial Press, New York, 1967.

Chapter **3**

Process Chart Procedures

WILLIAM ROBERT MULLEE

**Professor, Industrial Engineering,
Loyola University, Los Angeles, California**

DAVID B. PORTER

**Professor Emeritus, Industrial Engineering,
New York University, New York, New York**

The term "process chart" refers to a family of charts, including operation process charts, flow process charts (single or multicolumn), multiple activity (man and machine or work planning) charts, workplace (right- and left-hand) charts, and simultaneous motion cycle (simo) charts. The first four are treated in this chapter. The last is covered by Section 2, Chapter 5, Motion Study.

OBJECTIVES OF PROCESS CHARTING

Process charts provide a systematic description of a process or work cycle, with sufficient detail for analysis to develop methods improvements. Each member of the process chart family is designed to help the analyst clearly visualize the present procedure. A standardized format provides a common language so that several people can visualize problems together. This stimulates an exchange or cross-pollination of ideas. Most charts combine written, graphic, and pictorial visualization which promotes full participation by everyone concerned. Finally, the charts are excellent tools for presentation of proposals for improved methods to all management levels.

PROCESS CHART ACTIVITIES

In accordance with the "Standard for Operation and Flow Process Charts" adopted by the American Society of Mechanical Engineers in 1947, process chart activities are classified under five headings: operations, transportations, inspections, delays, and storages. The following definitions cover these activities.

○

To change

Operation. An operation occurs when an object is intentionally changed in any of its physical or chemical characteristics, is assembled or dissassembled from another object, or is arranged for another operation, transportation, inspection, or storage. An operation also occurs when information is given or received or when planning or calculating takes place.

⇨

To move

Transportation. A transportation occurs when an object is moved from one place to another, except when such movements are a part of the operation or are caused by the operator at the work station during an operation or an inspection.

☐

To verify

Inspection. An inspection occurs when an object is examined for identification or is verified for quantity or quality in any of its characteristics.

D

To wait

Delay. A delay occurs to an object when conditions, except those which intentionally change the physical or chemical characteristics of the object, do not permit or require immediate performance of the next planned action.

▽

To protect

Storage. A storage occurs when an object is kept and protected against unauthorized removal.

⬯

Combined Activity. When two activities are performed concurrently, or at the same work station, the symbols may be combined. The example shown represents a combined operation and inspection.

ROLE OF THE PROCESS CHART IN PROBLEM SOLVING

The five-step pattern approach to solving production problems is:

Step 1. Select and define the problem.

Step 2. Break down and visualize in detail.

Step 3. Question with an open mind.

Step 4. Develop an improvement proposal.

Step 5. Install the proposal.

The process chart is used to aid in carrying out step 2. Most modern flow process charts have preprinted symbols and include the question part of step 3. Some provide space for the idea part of step 4. The five most commonly used types of process chart are:

Chart	*Designed to visualize*
Operation process..........	Entire process or assembly with all components
Flow process..............	One person or component through a process
Multiple activity..........	Combinations of men and machines
Workplace................	Motions of both hands and the arrangement of tools and materials at a workplace
Simo.....................	Workplace chart in micromotion detail

OPERATION PROCESS CHARTS

An operation process chart is a graphic representation of the points at which materials are introduced into the process and of the sequence of inspections and all operations except those involved in material handling. The chart may include a detailed description of the materials, the time required for each operation and inspection, and the location where it is performed.

Operation process charts are used by engineers, chemists, cost accountants, plant managers, and others who want an overall view of the entire process.

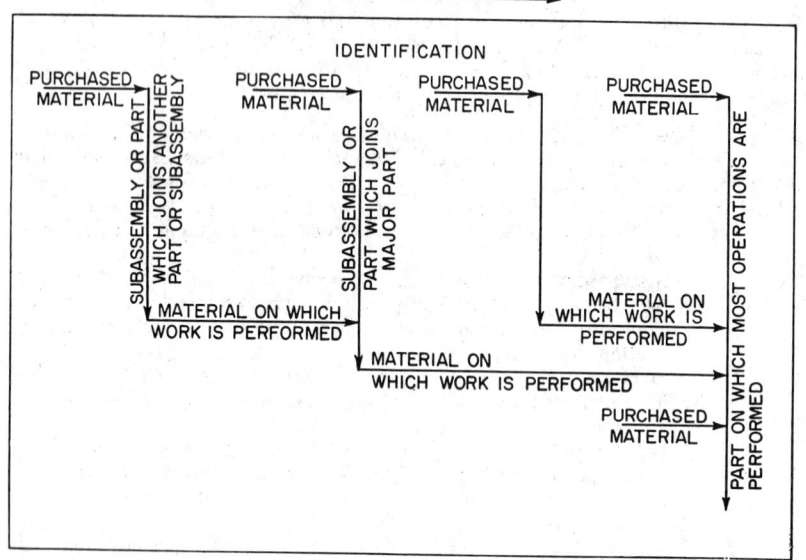

Fɪɢ. 3-1.   Graphic representation of principles of operation process chart construction.

Because of the wide range of applications, no preprinted form has been devised for general use.   A plain sheet of wide paper can be used.   All steps should be listed in proper sequence for each component, working vertically from top to bottom.   The major component or chassis is conventionally shown at the far right, and all other components are allotted space to the left of this component.   The image is like a conveyor line with components fed into the chassis in proper sequence. See the diagram for material feeding into a process, Figure 3-1.

To aid in a thorough analysis of the materials, all important information on alloy, finish, shape, and the like should be listed.   The descriptions of the operations or inspections should be brief.   Use a descriptive shop word (drill, tap, bore, and so on) with the name of the departmental location.   For inspections, indicate whether for quantity or quality, sampling or 100 percent, and for what characteristics.

The only symbols used in this chart are for operations and inspections.   The symbols are numbered in sequence, beginning with the first step on the major part or chassis, as indicated in Figure 3-2 for a mechanical assembly.   Note that the numbering starts with the first step on the chassis and continues up to the point where the first component is assembled.   The numbering then shifts to this component, continues up to the point of assembly, and then shifts back to the chassis.

Time values are usually expressed in decimal hours for both operations and inspections.   This helps evaluate the significance of each step in terms of potential savings.

Analyzing the Operation Process Chart.   With only four major considerations— materials, operations, inspections, and time—the subject of material is analyzed first.   All alternative materials, finishes, and tolerances are evaluated as to function, reliability, service, and cost.   Next, the operations are reviewed for possible alternative processing, fabrication, machining, or assembling methods and for changed tooling and equipment.   Can operations be eliminated, combined, changed, or simplified?   Inspections are reviewed for quality level, for replacement with in-process sampling techniques, or by job enlargement on related operations.   Time values are reviewed in terms of alternative methods, tooling, and of course, use of outside services or special-purpose equipment. For a more comprehensive discussion, see Section 2, Chapter 4, Operation Analysis.

FLOW PROCESS CHARTS

A flow process chart is a graphic representation of the sequence of all operations, transportations, inspections, delays, and storages occurring during a process or procedure. The material-type chart follows the steps performed on one component

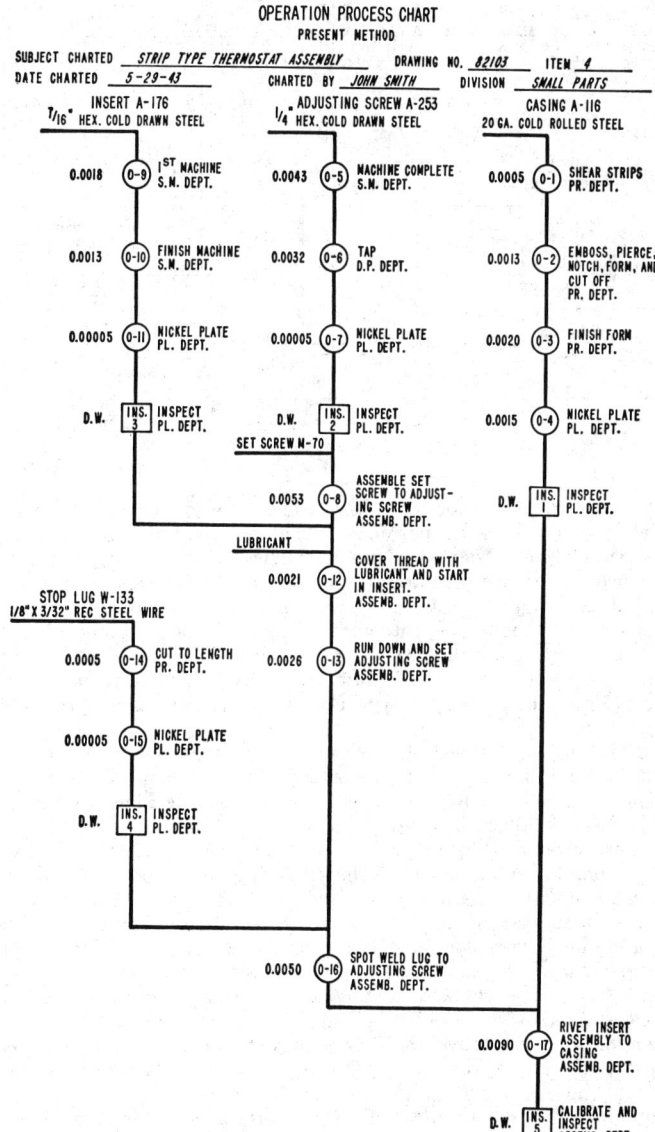

Fig. 3-2. Typical operation process chart.

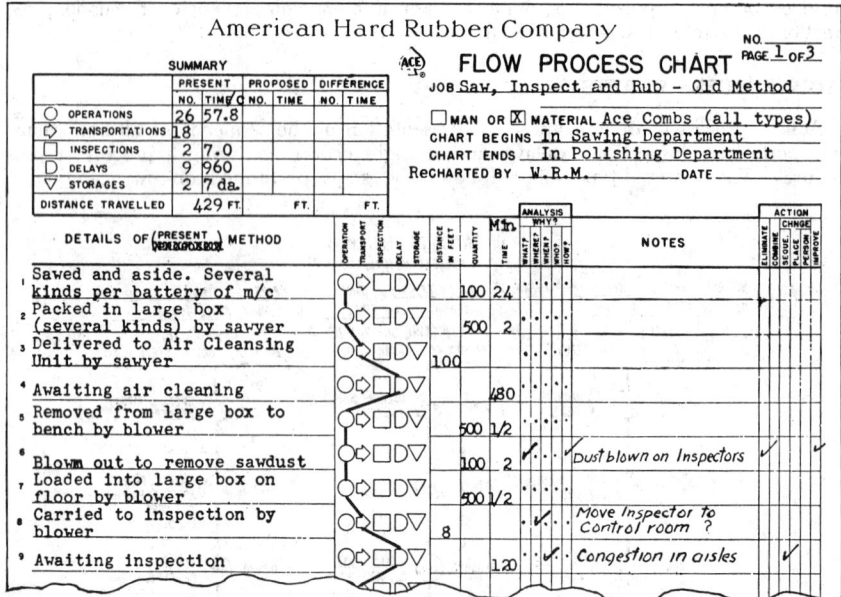

Fɪɢ. 3-3. Flow process chart (material type) using dot and check technique and preprinted symbols.

or material during the process or procedure. The man-type chart follows one person, indicating all the activities he performs. The material-type chart is most useful for a bird's-eye view of production operations, while the man-type chart is better for maintenance or service operations. These should be separate charts.

When using the preprinted form shown in Figure 3-3, the data required are obvious. Information should be gathered by actually following the object charted. No attempt should be made to chart from memory. Descriptions should be brief, as in a telegram. For a man-type chart, the active voice is used, as drills, taps, grinds, and so on. For a material-type chart, the passive voice is used, as drilled, tapped, ground, and the like.

Time and distance are shown for all important steps, but may be omitted for minor steps. Everything that happens at one work station during an operation or inspection is usually shown on one line. Avoid breaking an operation into minor details such as "place in basket," "remove from basket," and the like. These details are best examined in a workplace chart. Delays should be listed when important, but omitted when trivial. The chart should not be cluttered with minor details. Greater detail is permitted on the man-type chart, Figure 3-4.

The notes column can be used to continue the description when it cannot be condensed into the details column. Otherwise, this column is available to record ideas developed during analysis.

Each chart should be marked to indicate whether it portrays the present or the proposed method. The symbols selected for each item should be connected. This will emphasize the relative value of each step and aid in totaling the data for the summary at the top of the chart. Operations have the greatest customer value, with decreasing value for each symbol moving to the right.

Analyzing Flow Process Charts. To avoid resistance to change, it is desirable to use the six questions: why, what, where, when, who, and how. The questions, in proper sequence of use, and the actions expected are as follows.

Question	Followed by	Action expected
1. What is the purpose?	Why?	1. Eliminate unnecessary activity.
2. Where should this be done?	Why?	2. Combine or change place.
3. When should this be done?	Why?	3. Combine or change time or sequence.
4. Who should do this?	Why?	4. Combine or change person.
5. How should this be done?	Why?	5. Simplify or improve method.

A simple but effective way to apply the six questions to each item on a flow process chart has been developed for supervisory use. This method is called the "dot and check technique." The supervisor rests his pencil in each of the question

NO. 1
PAGE 1 OF 1
FLOW PROCESS CHART

JOB Receive air freight package and bring to outgoing freight area
☒ MAN OR ☐ MATERIAL Baggage handler
CHART BEGINS At receiving dock
CHART ENDS Outgoing freight area
CHARTED BY A.S.   DATE 9/26/—

SUMMARY

	PRESENT		PROPOSED		DIFFERENCE	
	NO.	TIME	NO.	TIME	NO.	TIME
○ OPERATIONS	50	6.6				
⇨ TRANSPORTATIONS	43	21.3				
☐ INSPECTIONS	17	21.9				
D DELAYS	1	5.5				
▽ STORAGES	-	-				
DISTANCE TRAVELED	1471	FT.		FT.		FT.

Fig. 3-4. Flow process chart (man type).

columns, leaving a pencil dot as he thinks through the implications of the question as applied to this particular item. If he gets an idea from this study, he places a pencil check mark opposite the proper question. In the proper action column, he checks "eliminate," "combine," or any of the indicated actions as shown in Figure 3-3 and places supplementary details in the notes column. This approach is particularly effective when discussion groups review the chart, because it fixes attention on one item at a time. Everyone may participate in developing the proposed method. Guided by the action checked plus the supplementary notes, a "proposed method" flow process chart may be constructed.

The Flow Diagram. The flow diagram is a sketch of the layout of floors and of buildings which shows the location of all activities appearing on a flow process chart. The path of movement of the material or man that has been flow process charted is traced on the flow diagram by lines or string. Each activity is located and identified on the flow diagram by symbol and number corresponding to that appearing on the flow process chart. The direction of movement is shown by placing arrows so that they point in the direction of flow.

If a movement backtracks over the same path or is repeated again in the same direction, separate lines should be drawn for each movement to give emphasis to this backtracking. If string is used, it may be wound around pins and laid up in layers to show repetitive movement. Figure 3-5 is an example of a flow diagram.

Where it is desirable to show the movement of more than one item or person on the same flow diagram, each may be identified by a different colored line or string. If one item or person is being followed, one color may be used for the present method and another color for the proposed method.

The flow diagram becomes a necessary adjunct to the flow process chart wherever movement is an important factor. It shows up backtracking, excessive travel, and points of traffic congestion and acts as a guide for an improved layout.

When a re-layout is contemplated, it is customary to use floor, building, or yard plans drawn to scale and templates of all machines and equipment made to the same scale. For the nontechnical supervisor or executive, it is better to use three-dimensional models. These permit greater participation in the development of a new layout. This may produce a better layout and create better acceptance of it, because more of those affected by it were able to take part in its development.

Charting Paper Work. Single-copy forms can be flow process charted (FPC) like a material.

Multicopy forms (usually a carbon-backed or NCR set) are first charted individually (FPC) and then assembled into a multicolumn procedure chart (MPC).

In preparation for the MPC, some modifications of symbols are necessary. The operation symbol will have three variations:

Create or originate	◎	Creation of a form, tape, card, or the like by the initial writing, typing, keypunching, or printout.
Add information	⊘	Addition of data to a paper, card, tape, or the like already in existence.
Handling operation	◯	Stapled, folded, sorted, collated, assembled, filed, retrieved, and so on. Omit trivia such as "place in basket," "remove from basket," "turn over," and similar minor operations.
☐ ⇨ D ▽		Same as previously defined.

Delays should be shown only when significant. Listing minor delays may clutter up the chart so that the procedure cannot be clearly seen because of the trivia. Remember that a storage of paperwork is always preceded by the operation "filed"

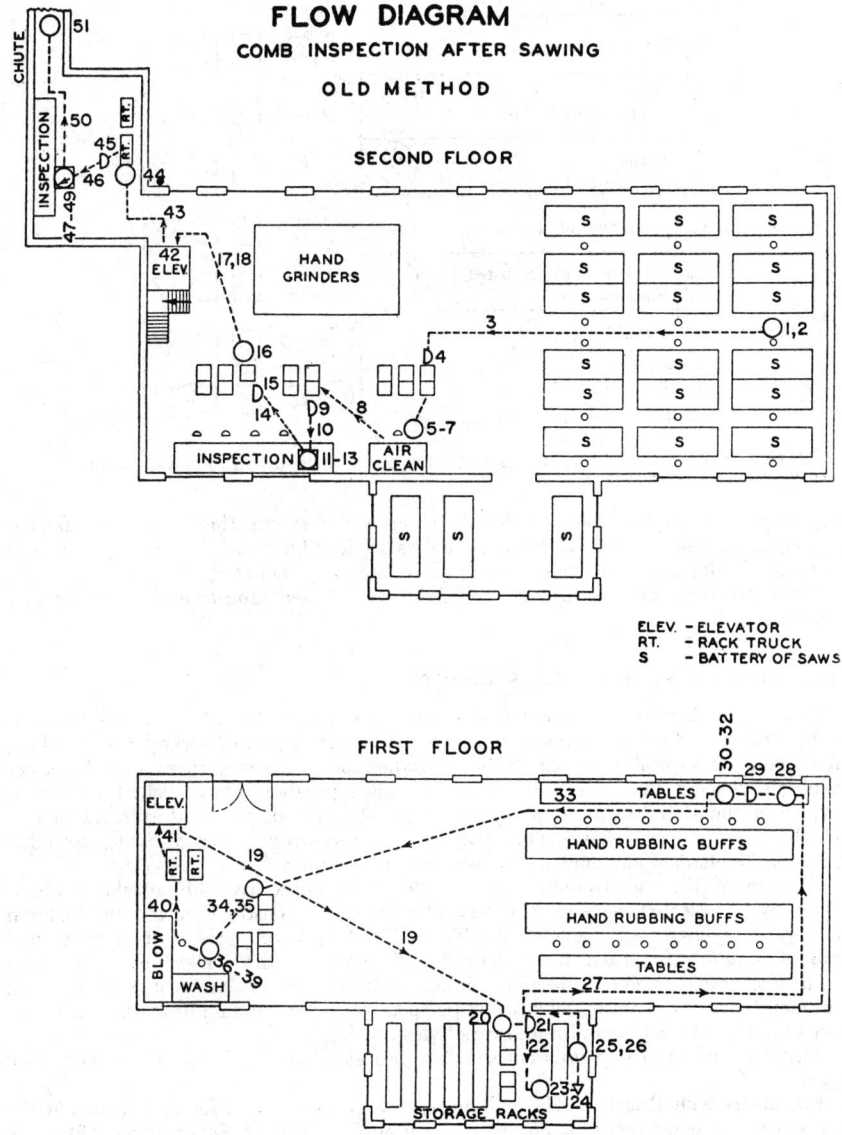

FIG. 3-5. Flow diagram for flow process chart shown in Figure 3-3.

and followed by the operation "retrieved" or "destroyed."   When a paper is destroyed, it can be indicated thus:

▽ (Circular file)   or   ⟩ (Destruction)

Flow Lines. FPC, in preparation for an MPC, will require modification of the flow line technique. For all steps that occur directly on the individual paper charted, the regular preprinted symbols are used. They will be modified by super-

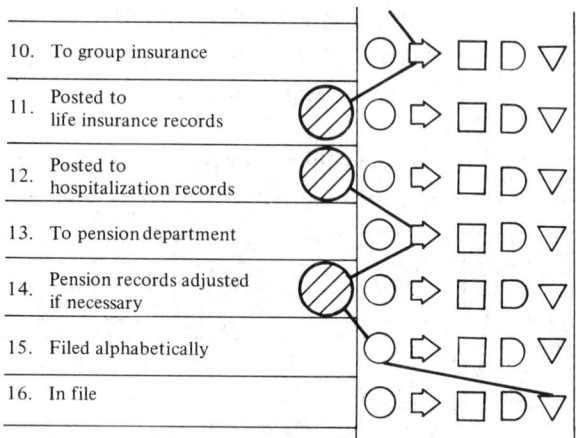

FIG. 3-6.   Convention for showing events that occur on affected papers.

imposing the double circle or cross-hatching.   For events that occur on affected papers, draw the symbol by hand as indicated in Figure 3-6.   Items 11, 12, and 14 occur on affected papers and are caused by the paper being charted.

For instructions on assembling a procedure chart, see Multicolumn Flow Process Charts.

MULTICOLUMN FLOW PROCESS CHARTS

The regular flow process chart is designed to visualize the action on one material or by one man during a process or a procedure.   If there are several men working in a gang or several components in a product or procedure, there may be great value in viewing the overall relationship.   The operation process chart does this, but lacks the detailed information on transportations, delays, and storages shown on the flow process chart.   One way to get the overall picture is to assemble a composite chart of all components, as shown in Figure 3-7.

Because of the inconvenience of handling a complex assembly of flow process charts by a vertical-type chart, and to standardize on the construction, the multicolumn flow process chart was developed.   This is a preprinted form with a horizontal line of symbols for each item charted.   Mounted on the wall, the entire chart is at eye level.   When explaining the chart or during analysis by a group, one can jump back and forth without losing perspective.   Viewed on a desk, the horizontal chart is again easier to scan for an overall view.

There are two types of multicolumn flow process charts: procedure type and gang type.

Procedure-type Charts.   The individual flow process charts for each paper in the procedure are constructed as outlined earlier under Charting Paper Work.   Because each detail will be brainstormed while in flow chart form, the multicolumn chart condenses the description of each step.   Its function is to show relationships.   See Figure 3-8 for a seven-part procedure for receiving returned goods.

For simplicity in visualizing the action and to avoid crossed lines where possible, arrange the individual charts on a desk so that those related to each other are positioned together.   For example, if copy 3 causes things to happen to copy 5, they should be on adjacent lines or tracks.

Now the flow lines and the symbols can be transferred from the FPC to the MPC.   Figuratively, the vertical line of symbols is picked up and placed horizontally on the MPC.   The preprinted symbols will fall on the track reserved for the individual paper.   The hand-drawn symbols will fall between the tracks.

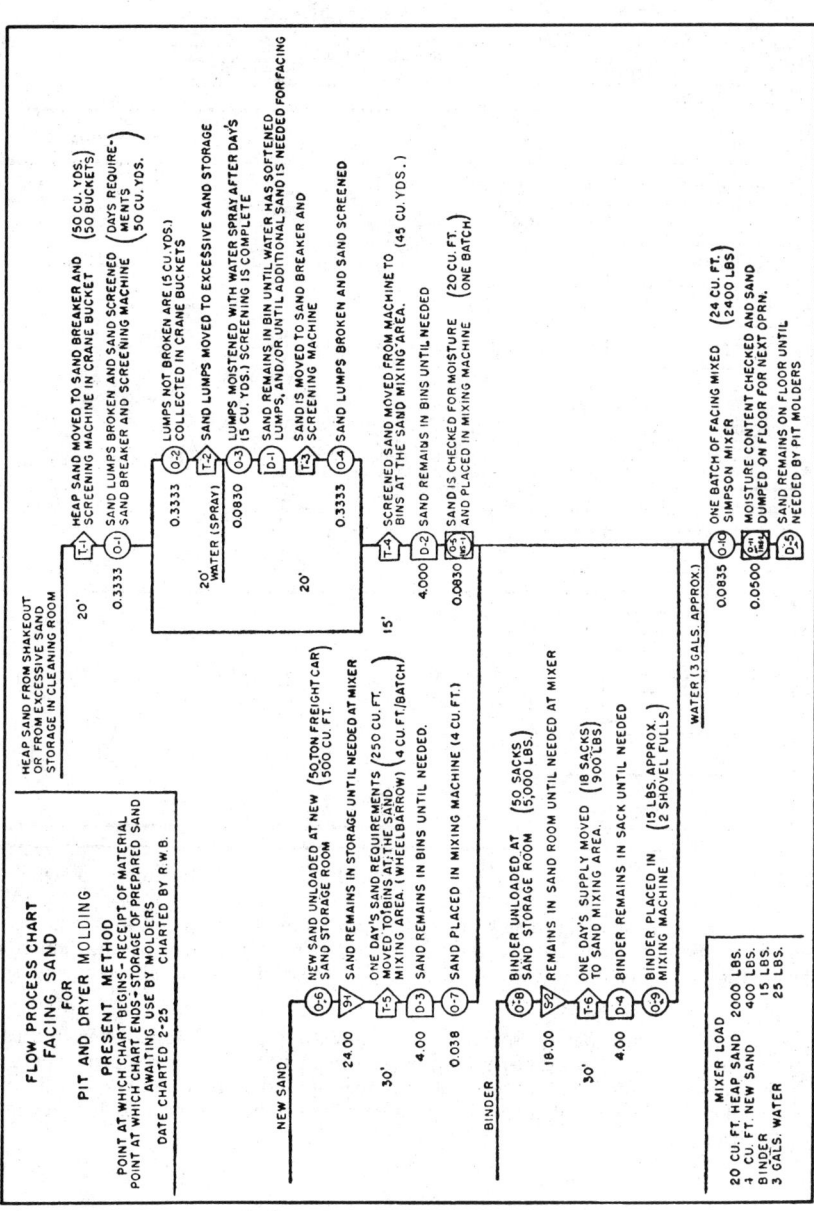

FIG. 3-7. Flow process chart of the material type showing manner in which several components are processed and brought together.

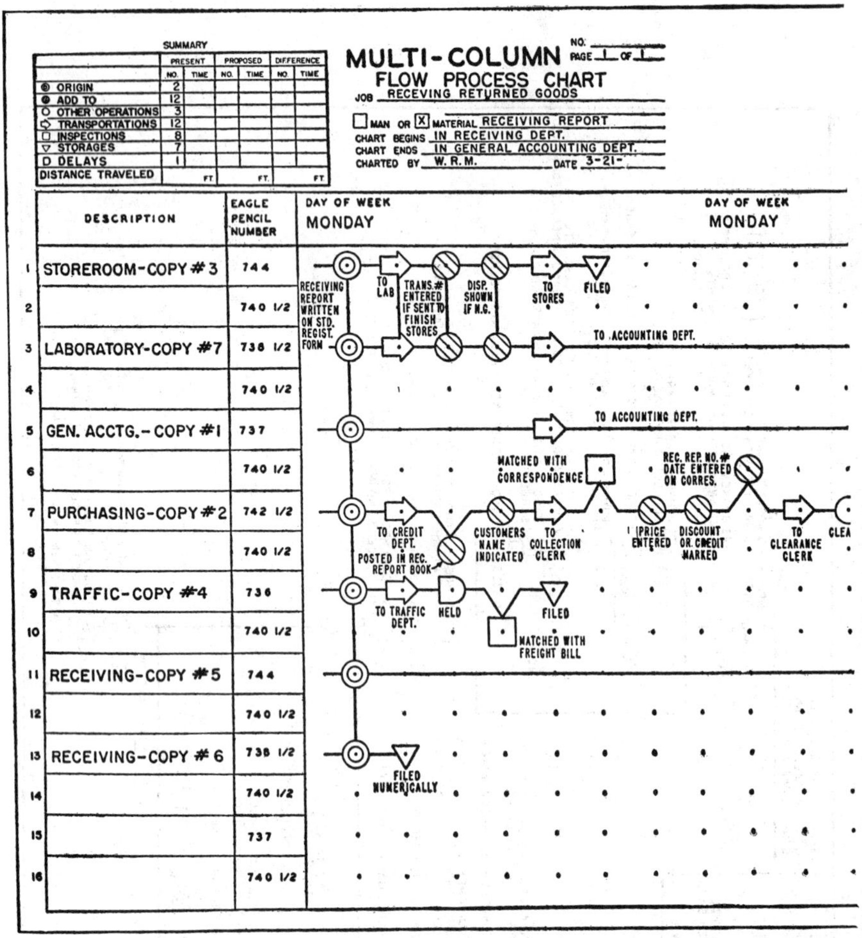

SUMMARY					
	PRESENT		PROPOSED		DIFFERENCE
	NO.	TIME	NO.	TIME	NO.	TIME
◉ ORIGIN	2					
● ADD TO	12					
○ OTHER OPERATIONS	3					
⇨ TRANSPORTATIONS	12					
□ INSPECTIONS	8					
▽ STORAGES	7					
D DELAYS	1					
DISTANCE TRAVELED		FT.		FT.		FT.

MULTI-COLUMN NO: _____ PAGE _1_ OF _1_
FLOW PROCESS CHART
JOB __RECEIVING RETURNED GOODS__

☐ MAN OR ☒ MATERIAL __RECEIVING REPORT__
CHART BEGINS __IN RECEIVING DEPT.__
CHART ENDS __IN GENERAL ACCOUNTING DEPT.__
CHARTED BY __W. R. M.__ DATE __3-21-__

	DESCRIPTION	EAGLE PENCIL NUMBER	DAY OF WEEK MONDAY	DAY OF WEEK MONDAY
1	STOREROOM-COPY #3	744		
2		740 1/2		
3	LABORATORY-COPY #7	738 1/2		
4		740 1/2		
5	GEN. ACCTG.-COPY #1	737		
6		740 1/2		
7	PURCHASING-COPY #2	742 1/2		
8		740 1/2		
9	TRAFFIC-COPY #4	736		
10		740 1/2		
11	RECEIVING-COPY #5	744		
12		740 1/2		
13	RECEIVING-COPY #6	738 1/2		
14		740 1/2		
15		737		
16		740 1/2		

FIG. 3-8.   Multicolumn flow process

In this manner, all steps that happen in a procedure are arranged in chronological sequence, reading from left to right, and there is a full view of the action on the multicopy set charted and a partial view of the action on all affected papers.

Charting Techniques

◎ ORIGIN

Origin of a Set of Forms. This symbol is used each time a new record is created. If, during the procedure, a three-part form is written, it would be shown as follows. (Note that each part is indicated and each has its own flow line upon which are shown the symbols indicating what is happening to that part.)

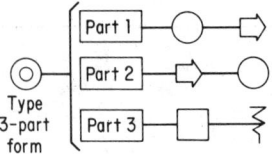

Designed, 1950, by William Robert Mullee
Published by Work Simplification Round Tables – New York University, N.Y.C.
A-PURPOSE: For a-bird's-eye view of a number of events and their chronological relationships. To develop a better product or procedure at a lower cost.
B-CONSTRUCTION: Use A.S.M.E. symbols ◯⯈☐▽ (MODIFICATIONS: ⊚ ORIGIN OF RECORD, ⊘ ADD TO RECORD)
  1-Material Type Chart (use passive voice, i.e., Typed, Data entered, Checked, etc.)
    a-Multicopy or Multiproduct: Chart each on a separate line. Use the 74-0½ line for posting to other papers, etc.
    b-Single Copy or product: Use a separate line for each station and indicate movement from station to station.
  2-Man Type (use active voice, i.e., Types, Enters data, Checks, etc.)
    a-Multi-Person: Use a separate line for each person, and chart like a chronological series of snap shots.
    b-Single Person: Chart from station to station to show travel.
C-IDENTIFICATION: To dramatize different items, fill out symbols with colors.
D-ANALYSIS: Steam shovel approach. Use 6 questions (why, what, where, when, who, how) to get the actions (eliminate, combine, change sequence, simplify). If detailed analysis is required, prepare a regular Flow Process Chart for each paper, material, or man.

DAY OF WEEK TUESDAY	DAY OF WEEK TUESDAY	WED.

(process chart with symbols and flow lines, labeled:)

FILED

MATCHED · POSTED TO ACCT. # · FILED

CREDIT INVOICE TYPED ⊚

POSTED TO DAILY SALES SHEET

TO ACCT. DEPT. · AUDITED · TO BILLING · CREDIT # ENTERED · PRICES CHECKED · WT. DOZ. ENTERED · EXTENDED · DEDUCTIONS OR ALLOWANCES CHECKED · CHECKED AND ATTACHED · INITIALED · TO SALES ACCT. · TO GENERAL ACCT.

ECKED WITH IN. TO RETURN

FILED WITH AUTH. TO RETURN

chart of the procedure type.

"ADD TO" OPERATIONS

Adding Information. This symbol is used to show anything added to a record. (It includes posting, stamping, signing, and the like.) When one of the papers being charted is responsible for an "add to" operation on another record, it is shown as follows.

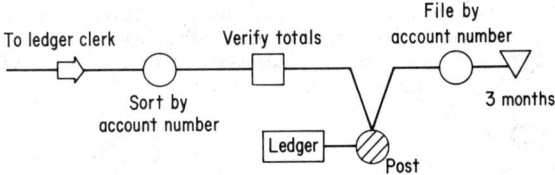

To ledger clerk   Verify totals   File by account number   3 months

Sort by account number

Ledger — Post

AFFECT LINE

One Form Affecting Another. This line, going from one flow line to another (or from a flow line to a separate record as shown above), indicates that one paper in the system has an effect upon another. The

"V" that the affect line creates may go up or down and may be as long as needed to reach the flow line of the paper affected. The following illustration shows a two-part form, with one part having an effect upon the other.

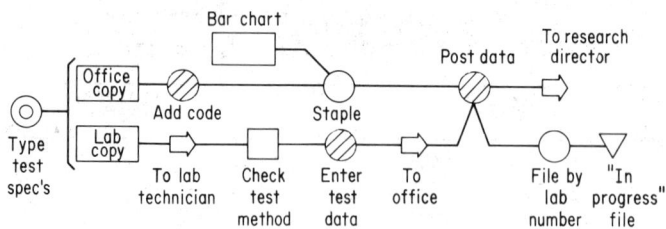

Simultaneous Action. When the same action is taken on two papers at the same time, this construction is used. In this case, it shows that the inspection was performed equally on both papers represented by the two flow lines.

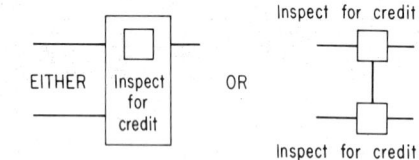

Alternate Action. This construction is used when a piece of paper may be processed in different ways, depending on circumstances. The normal or most common way is shown on the regular flow line. The exception is shown on the alternate flow line.

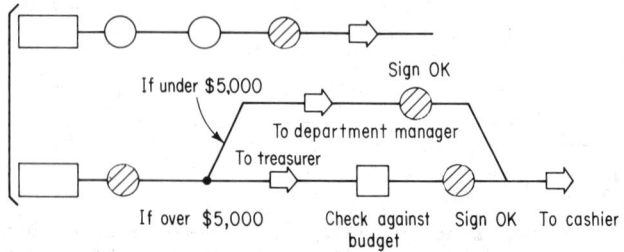

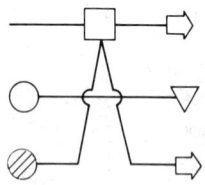

Skipping Over. The skip is used, usually with an affect line, as a charting convenience. It may not always be practical to have forms which affect each other on adjacent flow lines. It may be necessary to skip over a flow line to get to the line of the form affected.

Physically Attached Set of Forms. Often, a carbon-interleaved set of forms goes through a number of steps in a procedure as a set before being separated. Rather than draw many individual symbols or simultaneous action blocks, the flow lines can be brought together and the symbols shown on one flow line for the entire set. This technique may also be used when two or more forms are temporarily attached by staple or paper clip and go through several steps as a set.

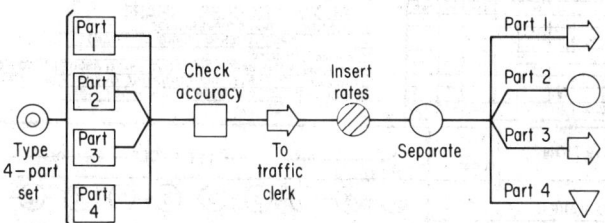

Chronological Time. As far as possible, multicolumn charts should be drawn to show the relative time sequence of the various steps. If one part of a form is idle for some time, its flow line should be blank for some distance.

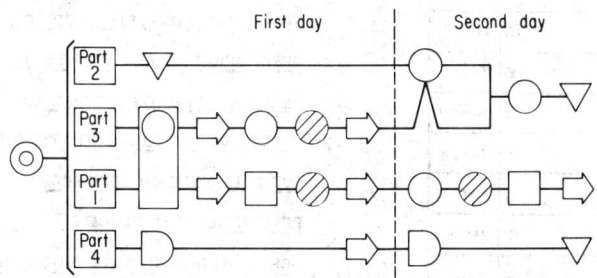

Gang Charts. For groups of men on warehousing, maintenance, rigging, or other material handling operations, the multicolumn form can be used as a gang chart as shown in Figure 3-9. The regular ASME symbols are used, and a line on the chart is assigned to each member of the gang. Plotted horizontally, the symbols are arranged like parallel flow process charts, one above the other. Moving from left to right, a vertical column of symbols represents simultaneous activity of all members of the gang, like a snapshot of the gang's activity. These snapshots can be taken at frequent intervals so that all changes in activity are reflected by changes of symbols. Gang operations are usually not planned to the same degree as individual jobs, and the chart may show many delay symbols. Substantial cost reduction may be had by rearrangement of symbols to cancel out delays. Eliminating delays permits transfer of excess personnel and shortens the production cycle. Combining operations or rearranging sequences saves makeready and put-away. Study of individual operations or inspections permits many simplifications, followed by further rearrangement of the gang operation to reduce the overall cycle.

Symbols are numbered to explain work details. The storage symbol is not used on gang charts.

PRINCIPLES AND PRACTICES FOR CONSTRUCTION OF MULTIPLE ACTIVITY CHARTS

A multiple activity chart, also called man and machine chart or work planning chart, is a graphic representation of the coordinated working and waiting time of two or more men or any combination of men and machines. The duration of the activities is represented by bars drawn to length against a time scale.

Activities Defined. It is helpful to distinguish between the work of an operator when working on a machine or with another operator and when working independently of a machine or another operator. Similarly, it is helpful to distinguish between the operating time of a machine when operating independently of an operator

MULTI-COLUMN FLOW PROCESS CHART

NO. ___ PAGE ___ OF ___

SUMMARY	PRESENT		PROPOSED		DIFFERENCE	
	NO.	TIME	NO.	TIME	NO.	TIME
◉ ORIGIN						
◉ ADD TO						
○ OTHER OPERATIONS	16					
⇨ TRANSPORTATIONS	2					
☐ INSPECTIONS						
▽ STORAGES						
D DELAYS	12					
DISTANCE TRAVELED	60	FT.		FT.		FT.

JOB MAKING LADLE STOPPER RODS
(ACTIVITIES OF GANG OF 3 MEN)

☑ MAN OR ☐ MATERIAL ___

CHART BEGINS WITH BARE ROD (ABOUT 2¼" ⊖ ABOUT 8'0" TO 9'0" LG.) STRAIGHTENED.

CHART ENDS COMPLETED-ROD-IN-DRYING OVEN

CHARTED BY INDUSTRIAL CREW(4) DATE 6-15-

DESCRIPTION	EAGLE PENCIL NUMBER	├──────1ST CYCLE──────┤
STOPPER ROD MAKER	1	744	①-③-③-③-③-③-⑨-⑩-⑫-D-①-③
FIRST HELPER	2	740 1/2	①-④-⑥-⑧-D-D-D-D-⑪-D-D-①-④
SECOND HELPER	3	738 1/2	⟦2⟧-⑤-⟨7⟩-D-D-D-D-D-D-⑬-⟦2⟧-⑤
	4	740 1/2	LEGEND ① MOVE STOPPER ROD TO WORK PLACE __ __
	5	737	__ ⟦2⟧ TRANSPORT COMPLETED ROD TO DRYING OVEN-A
	6	740 1/2	__ ③ APPLY MUD TO SLEEVE JOINTS __ __ __ •
	7	742 1/2	__ ④ PLACE SLEEVES ON STOPPER ROD, ONE AT A TIME
	8	740 1/2	__ ⑤ PLACE COMPLETED ROD IN DRYING OVEN __ •
	9	738	__ ⑥ PUT BULL NOSE ON NEXT ROD __ __ ▬
	10	740 1/2	__ ⟨7⟩ RETURN TO WORK PLACE __ __ ▬ ▬
	11	744	__ ⑧ SCREW RETAINING NUTS ONTO ROD __ ▬
	12	740 1/2	__ ⑨ PRELIMINARY TIGHTENING OF NUTS __ ▬
	13	738 1/2	__ ⑩ COMPLETE TIGHTENING OF NUTS __ ▬ __ •
	14	740 1/2	__ ⑪ VISE-HOLD ROD, BY MEANS OF CHAIN, AGAINST T
	15	737	__ ⑫ WIPE EXCESS MUD FROM ALL SLEEVE JOINTS •
	16	740 1/2	__ ⑬ PICK UP COMPLETED ROD BY MEANS OF MONO-F

FIG. 3-9. Multicolumn flow

and when being operated or serviced by an operator. It is also useful to distinguish between the time when a machine is waiting to be serviced and when it is being set up, loaded, or unloaded. To provide for this, the following classifications of working and waiting, with their graphic representations, are used.

Independent Work. For the operator, this classification means working independently of the machine or other operator, such as when getting and preparing material, inspecting finished product, and performing other work not connected with the operation of the machine.

For the machine, it includes the time it is actually doing its work without the services of the operator.

Combined Work. For the operator, this classification includes working with a machine or other operator while setting up, loading, and operating a machine with hand feed or working in cooperation with other operators.

For the machine, it includes the time it is operating and requiring the services of an operator and the time it is being set up, loaded, or unloaded.

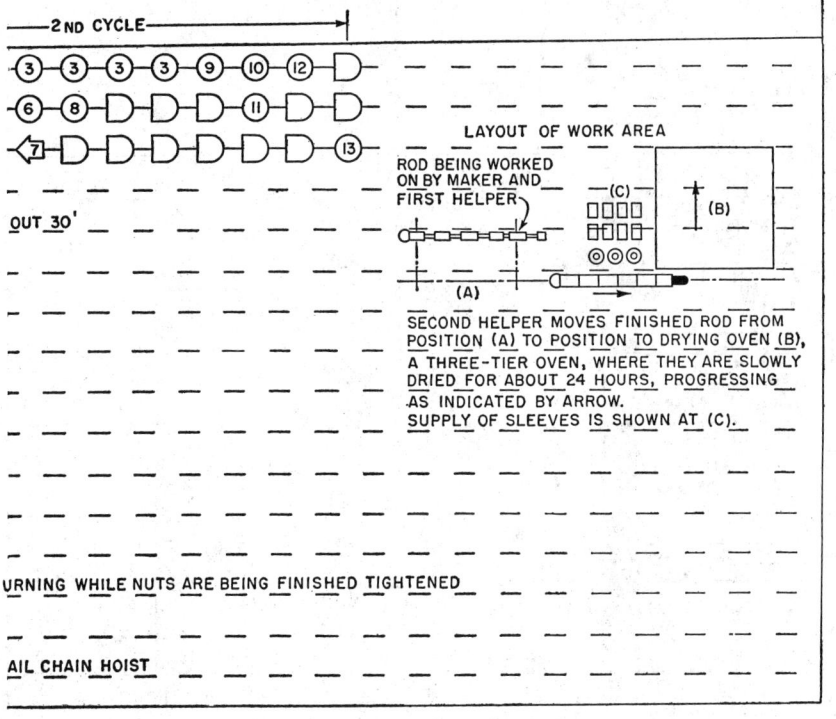

process chart of the gang type.

It is convenient, when analyzing a cycle of man and machine time, to differentiate between their times when working independently (independent work) and when one depends upon the other (combined work). The blocks of time representing independent work may be shifted around independently of each other, whereas the blocks representing combined work must not be shifted with reference to each other.

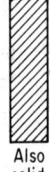
Also
solid
red

Waiting. This classification includes waiting on the part of either an operator or a machine. It occurs when one is waiting for the other. Work of an operator which prevents a machine from running, but which might be rearranged to allow the machine to operate, should be classed as independent work, and the corresponding time of the machine should be classed as waiting. This classification and graphic code place emphasis on real waiting of the machine and focus attention on work of the operator which may be rearranged to occur during machine operating time and thereby reduce machine waiting.

An example of a multiple activity chart covering a man and several machines is shown in Figure 3-10.

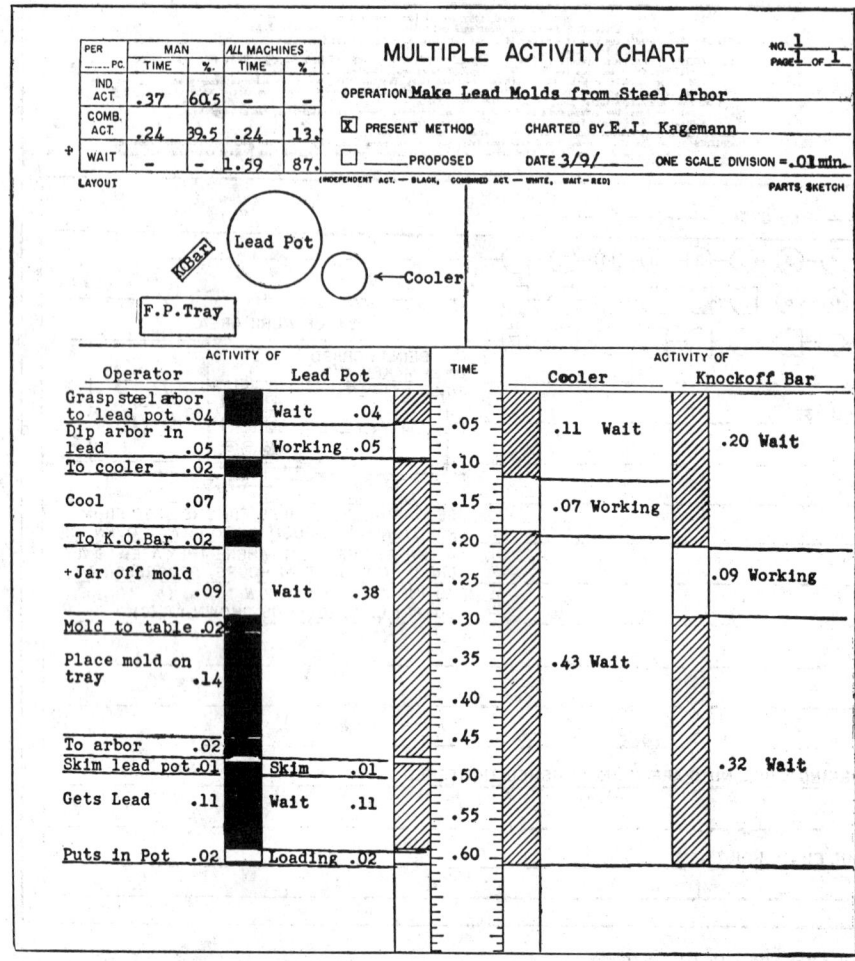

PER	MAN		ALL MACHINES		MULTIPLE ACTIVITY CHART	NO. 1 PAGE 1 OF 1
___PC.	TIME	%	TIME	%		
IND. ACT.	.37	60.5	–	–	OPERATION Make Lead Molds from Steel Arbor	
COMB. ACT.	.24	39.5	.24	13.	☒ PRESENT METHOD CHARTED BY E.J. Kagemann	
WAIT	–	–	1.59	87.	☐ ___PROPOSED DATE 3/9/ ONE SCALE DIVISION =.01 min.	

LAYOUT   (INDEPENDENT ACT. – BLACK, COMBINED ACT. – WHITE, WAIT – RED)   PARTS, SKETCH

ACTIVITY OF			TIME	ACTIVITY OF	
Operator	Lead Pot			Cooler	Knockoff Bar
Grasp steel arbor to lead pot .04	Wait	.04	.05	.11 Wait	.20 Wait
Dip arbor in lead .05	Working	.05	.10		
To cooler .02					
Cool .07			.15	.07 Working	
To K.O.Bar .02			.20		
+Jar off mold .09	Wait	.38	.25		.09 Working
Mold to table .02			.30		
Place mold on tray .14			.35	.43 Wait	
			.40		
To arbor .02			.45		
Skim lead pot .01	Skim	.01	.50		.32 Wait
Gets Lead .11	Wait	.11	.55		
Puts in Pot .02	Loading	.02	.60		

FIG. 3-10.   Multiple activity chart of a man and several machines.

The details into which the activities are divided might follow the pattern of a time study, using a stopwatch for measuring time. However, where several men or machines are being studied, it is usually more convenient to record the activities on a motion picture film. This also has the advantage of recording and measuring time for smaller elements.

A simpler type of multiple activity chart does not give the actual time for each element on a time scale. Instead, it uses the symbols for working time and delays, balancing these as well as possible by estimating relative times.

Another form without time values, which uses the flow process chart symbols progressing in a horizontal direction, has already been described and is shown by Figure 3-9 as the gang process chart.

Analyzing Multiple Activity Charts. The same procedure is followed in analyzing multiple activity charts as is used in analyzing flow process charts, namely, the questioning technique and group participation described previously. However, before challenging each step of the work items, substantial savings may often be found by eliminating waiting time for man and machine. This may frequently be done by a simple rearrangement of the work cycle or by giving the man other work to do.

PRINCIPLES AND PRACTICES FOR CONSTRUCTION OF WORKPLACE CHARTS

A workplace or right- and left-hand chart is a graphic representation of the coordinated activities of the right and left hands.

Where the job is sufficiently repetitive to warrant a detailed study of the right and left hands, a workplace study may be made. Moves, operations, holds, or delays performed by each hand may be charted, using flow process chart symbols.

Figure 3-11 shows a workplace chart, also known as an RH-LH chart or an operator process chart. In parallel vertical columns, the elements are charted to show the simultaneous activity of each hand. A brief description is provided alongside the regular ASME symbols. Whenever a change occurs to either or both hands, it is shown on the next line of the chart, regardless of how short or long the time element.

Counting the symbols and recording the totals in a summary at the top of the chart provides a comparison between the present and the proposed methods. Although the time for different elements will vary, the overall count of the symbols is a reasonable measure of the comparable time. The workplace chart may be considered as two detailed flow process charts, one for each hand. The symbols are later connected to form an activity pattern. This permits comparison of both hands as to similarity of work performed. A distance column on the form provides a record of the travel of each hand. Moving the work closer to the operator may improve the method and reduce the time.

The layout of the workplace is indicated by a grid of ¼-inch squares, against which the arrangement of bins, fixtures, and parts may be shown. The convenient work area is indicated by two semicircles described by the operator's forearms when seated at the workplace. Provision is made for parts sketches in the upper right-hand portion of the chart.

The position of each part on the layout is indicated by L_1, L_2, and so on, for items to the left of the operator, with L_1 nearest to center. The same convention is used for items to the right of the operator: R_1, R_2, and so on. Items directly in front are called C. If a second row or level is used, the designation is R_{1A}, L_{1A}, C_{1A}, and so on. The contents of each location are listed in the table at the upper left-hand side of the chart.

TWENTY PRINCIPLES OF MOTION ECONOMY

To help analyze the chart for improvements in method, the twenty principles of motion economy are shown in abbreviated form along the left-hand margin. This list of principles is a modification of the original Gilbreth list and has been found helpful in developing methods improvements for factory operations. The complete wording of these principles is as follows:

1. Begin each element simultaneously with both hands.
2. End each element simultaneously with both hands.
3. Use simultaneous arm motions, in opposite and symmetrical directions.
4. Use hand motions of lowest classification for satisfactory operations.
5. Keep motion path within normal working area.
6. Avoid sharp changes of direction. Plan a smoothly curved motion path.
7. Slide small objects. Avoid pickup and carry.
8. Locate tools and materials in proper sequence, at fixed work stations.
9. Use fewest elements to obtain shortest time.
10. Use rhythm and automaticity to increase output and lessen fatigue.
11. Relieve hands with foot pedals where possible.
12. Avoid holding. Use vise or fixture, freeing hands to move pieces.
13. Provide ejectors to remove finished pieces.

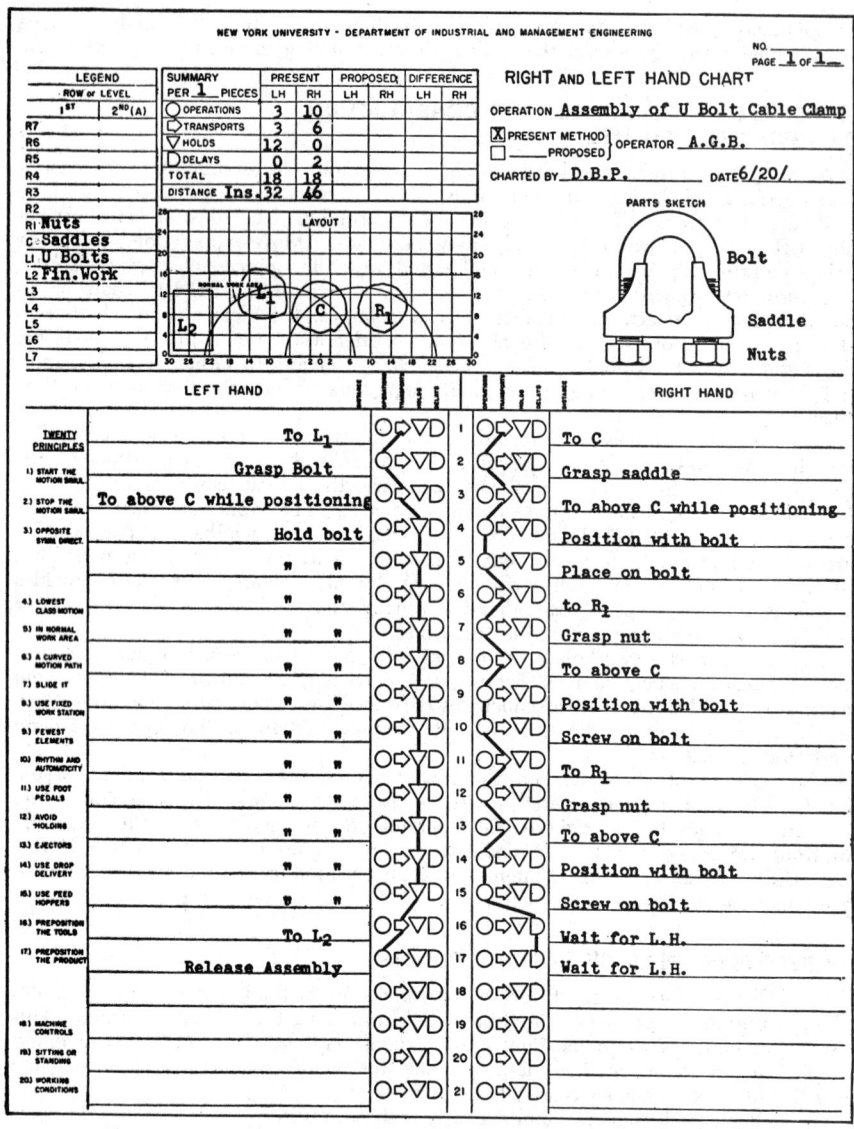

Fig. 3-11.   Workplace (RH-LH) chart.

14. Use drop delivery where possible.
15. Shorten transports by keeping materials nearby in gravity-feed hoppers.
16. Pre-position tools for quick grasp.
17. Pre-position product for next operation.
18. Locate machine controls nearby for ease of operation.
19. Design workplace height for sitting-standing arrangement, and provide proper height chair with comfortable seat and backrest for good posture.
20. Provide pleasant working conditions, considering illumination, temperature, humidity, dust, fumes, ventilation, noise level, color scheme, orderliness, and the like.

Chapter 4

Operation Analysis

G. J. STEGEMERTEN

Executive Advisor, Maynard Research Council Incorporated,
Pittsburgh, Pennsylvania

DUANE C. GEITGEY

Director of Training Program Development,
Maynard Research Council Incorporated, Pittsburgh, Pennsylvania

The factors that surround the simplest process or operation are many and varied. Accordingly, small progress will be made toward methods improvement and automation if the job is studied as a whole. The first step in any study that will produce results is to resolve the job into its component parts or elements. Each part may then be considered separately, and the study of the process or operation becomes a series of studies of fairly simple problems.

This kind of analytical work is covered by the term "operation analysis." The study of each process or operation really consists of two analyses. The primary analysis breaks the job down into such factors as material, inspection requirements, and material handling. Each of these factors is then examined critically to discover the broad possibilities for methods improvements and automation. The secondary analysis is essentially a more detailed examination of some of the same factors, with the emphasis on the manual motions, or on the motions made by automatic equipment, required to do the job.

The operation analysis procedure is the basis for all the manufacturing research work that is being done in industry today.

OPERATION ANALYSIS DEFINED

Operation analysis may be defined as "A systematic procedure, employed to study all the factors which affect the method of performing an operation, to achieve maximum overall economy. Through this study, the best available method of performing each necessary part of an operation is found, and new manufacturing and maintenance developments are incorporated as they become available, in the continuing effort to move every job one step closer to continuous automatic accomplishment."

APPLICATIONS AND LIMITATIONS OF OPERATION ANALYSIS

The feeling which is often prevalent in the mind of the manager who is acquainted only in a general way with methods engineering techniques is that, although operation analysis may be able to produce worthwhile accomplishments in some lines of work or in certain industries, his work is different and the techniques will be of little or no value to him.

Human nature is such that we all feel that "Our work is different." The best-intentioned manager with a cost problem consequently feels that the technique will work better in every other situation than it will in his own. However, an industrial engineer will have sufficient experience with the technique to know of the beneficial results that are always obtained as a result of applying it.

The principles of operation analysis are fundamental. They can be applied to any kind or class of work. It makes no difference if the manager's cost problem is in the maintenance area or in a partially mechanized high-volume production line.

This broad application is possible because all work may be resolved in terms that are more or less basic. Working methods used on widely varying jobs present points of remarkable similarity when closely analyzed. The motion made by a maintenance mechanic in reaching for a power drill is the same as that made by a cutting room operator in reaching for a pair of scissors. Similarly, the problems involved in lifting a ladle of metal with a crane in a high-production foundry are the same as the problems encountered by the maintenance crew in lifting a ladle of molten metal that will become a new bearing surface. Every manager can cite countless similar examples that can be found in the various jobs that are being done around him every day. A look at the steps of the operation analysis approach shown below emphasizes the fact that the technique may be applied to any job and that the principles of operation analysis are not limited in any way by the nature of the work being done.

Operation Analysis Approach to Improvement and Automation
1. Observe or visualize operation.
2. Ask questions.
3. Estimate degree of improvement or automation possible.
4. Investigate ten approaches to improvement and automation:
    a. Design of part or assembly
    b. Material specification
    c. Process of manufacture
    d. Purpose of operation
    e. Tolerances and inspection requirements
    f. Tools and speed, feed, and depth of cut
    g. Equipment analysis
    h. Workplace layout and motion analysis
    i. Material flow
    j. Plant layout
5. Compare old and new methods.

The repetitiveness of the operation is another factor that must be considered. If a large number of man-hours are being expended in a certain type of activity, a 1 percent saving may be important. On the other hand, a 10 percent saving on work that is infrequently performed may not offset the cost of making the analysis. It is obviously more profitable to study the work with the greatest activity; however, this does not mean that only mass production work can be studied. This is true because activity is measured on a type of work taken as a whole, rather than on individual jobs.

For example, the industrial engineer frequently finds operations that are repetitive from an operation analysis viewpoint even in maintenance work. In this instance, when a number of different jobs are reduced to their elements, he finds that several elements are common to many jobs. If such common elements can be shortened through the selection of better materials, equipment, or manual methods, a saving is obtained each time the elements are performed, regardless of the larger task that the maintenance man may be doing.

The rapid progress that is being made in all fields—materials, tools, and manufacturing processes—requires that every manager and every industrial engineer constantly search for continuing job improvement. They should never speak of the "best" method without using some qualifying clause which implies that some improvement is possible, even though economic reasons may make it impractical to make the improvement at the present time. This principle applies to all types of work. As a result, operation analysis is not limited to mass production work but may be applied to produce savings in any line of work in which a fair number of man-hours are expended. Conversely, it will probably not be profitable to study a varied line of work if only one man is engaged on it only part time.

TYPES OF METHODS STUDIES

The manager who is searching for the solution to a cost problem has many industrial engineering procedures available. The number of effective industrial engineering procedures or tools has multiplied over the years. Some of the tools enable him to achieve an accuracy that was impossible with older ones, with only a slight increase in cost. Others enable him to obtain needed information at a cost far less than previously possible, with only a slight reduction in accuracy. The industrial engineer, then, has a wide variety of tools to call on when he undertakes a methods improvement project. In selecting the proper tools, it can generally be said that the savings that the study will produce must equal or exceed the cost of making the study. The manager and the industrial engineer must determine which tools to use, and to what extent each should be used, on the basis of their estimate of the degree of methods improvement that is possible in the particular job to be studied.

The large number of available industrial engineering tools may be combined in many ways. However, for practical purposes these may be combined to yield six classes of methods study that, with only minor modification, may be used to cover all types of activity. These six types are:

1. Written analysis using process charts and operation analysis charts. Methods analysis employing methods time measurement (MTM). Detailed analysis of all available automation devices, tools, and equipment. Operator methods training utilizing audiovisual devices.

2. Written analysis using operation analysis charts. Methods analysis using MTM. Analysis of all tools and equipment. Operator methods training by supervisors or trainers familiar with the MTM procedure. Providing MTM-type written methods instructions to operators.

3. Mental analysis using the points described on the operation analysis chart as a guide. Methods analysis based on MTM-General Purpose Data. Operator training given by supervisors or trainers familiar with the MTM procedure.

4. Written job analysis of class of work, using process charts and operation analysis charts for representative jobs. Methods analysis of representative jobs, using MTM to determine best methods. Operator training on specific jobs by supervisors or trainers familiar with MTM.

5. Mental analysis during general survey of class of work, using points on operation analysis chart as a guide. Methods analysis, using second generation predetermined

time data. Operator training in use of standardized tools for the class of work by supervisors familiar with MTM.

6. Use of second generation predetermined time data as a guide to performance.

All the industrial engineering techniques mentioned in these six types of methods studies are described in detail in other chapters of this Handbook.

FACTORS DETERMINING FIELD OF APPLICATION OF THE SIX TYPES OF METHODS STUDIES

The kind and amount of study that can be justified on any job or class of work are determined by three principal factors. These are the repetitiveness of the job or class of work, the amount of human attention required, and the expected life of the job or class of work. These factors must be considered together in selecting the type of methods study to be used, because no one of them is sufficient for the determination.

Repetitiveness. For the purpose of determining the field of application of the various types of methods studies, the repetitiveness of the job or class of work may be divided into four classes: high, medium, low, and jobbing. The following general descriptions can serve as a guide in fixing the repetitiveness of the job or class of work being considered.

High. A job or class of work may be considered highly repetitive if it occurs at least 2,000 times a year and requires a total of not less than 1,000 hours to perform.

Medium. A job or class of work may be said to be mediumly repetitive if it occurs at least 500 times per year and covers an elapsed time of one to six months.

Low. A job or class of work may be said to be of low repetitiveness if it occurs at least 50 times per year and covers an elapsed time of two weeks to one month.

Jobbing. A job or class of work may be said to be jobbing if it occurs less than 50 times per year, lasts less than two weeks, and is not expected to be repeated in the foreseeable future.

Judgment must be used in applying these descriptions because in many instances the job or class of work will not fit any of them exactly. However, the descriptions will serve as a guide.

Human Attention. The portion of the job or class of work that requires human attention has an important bearing on the type of study that should be made. The term "human attention" includes any part of the job or class of work that is manually performed by human labor, and it also includes the time when the operator must be attentive to the equipment (watching or listening) to ensure its proper operation, even though he may have no specific manual motions to make.

The human attention required by a job or a class of work may be classified as high, medium, or low. The maximum condition, of course, is when all parts of the job or class of work are performed by the operator by hand or with simple, unpowered hand tools. The minimum condition is when the job is done entirely automatically by machinery, where the machine stops itself and signals the operator if there is a malfunction, so that the operator's attention does not need to be focused continuously on any one machine. The class into which any job falls with respect to human attention may be determined as follows:

High. Where human attention is required by the individual job or class of work more than 75 percent of the time.

Medium. Where human attention is required by the individual job or class of work between 25 and 75 percent of the time.

Low. Where human attention is required less than 25 percent of the time.

Life of Job. The life of the job or the class of work is another factor which must be considered along with repetitiveness and human attention. The more detailed types of methods study are expensive, and the manager and industrial engineer must determine whether or not the estimated life of the job will justify the expenditure.

The length of life of the job or class of work can be divided into three classes: over twelve months; from six to twelve months; under six months.

TABULATION OF FACTORS

Table 4-1 has been compiled to aid in the selection of the type of methods study which is economically justified under any given conditions. Remember that the manager or industrial engineer who must make the choice must use his judgment and be guided by the particular conditions surrounding the job or the class of work to be studied.

The most difficult factor to determine when using the table is the repetitiveness of the job. The degree of human attention and the expected life of the job or class of work may be quickly checked. Repetitiveness, in the sense in which it is used in connection with the table, is affected by the number of occurrences per year, the length of the particular cycle of the job or class of work being studied, and the total length of the job.

In determining the repetitiveness of a job or a class of work, the number of occurrences, the hours required to complete the work, and the time allowed per occurrence must be considered. By definition, a job or class of work is considered to be highly repetitive if it consists of not less than 2,000 occurrences and requires not less than 1,000 man-hours to complete. However, because the time allowed per occurrence should be considered, the three factors can only be related alge-

TABLE 4-1. Tabulation of Factors Which Determine the
Type of Methods Study to Be Employed

Repetitiveness of job or class of work	Human attention	Life of job, months	Type of study indicated
High............	High	Over 12	1
		6 to 12	1 or 2
		Under 6	2 or 3
	Medium	Over 12	1 or 2
		6 to 12	2 or 3
		Under 6	3
	Low	Over 12	2
		6 to 12	2 or 3
		Under 6	3
Medium..........	High	Over 12	2
		6 to 12	2 or 3
		Under 6	3
	Medium	Over 12	2 or 3
		6 to 12	3
		Under 6	3 or 4
	Low	Over 12	3 or 5
		6 to 12	3, 5, or 6
		Under 6	6
Low.............	High	Over 12	3 or 4
		6 to 12	3, 4, or 5
		Under 6	3 or 5
	Medium	Over 12	3, 4, or 5
		6 to 12	3 or 5
		Under 6	3, 5, or 6
	Low	Over 12	3 or 5
		6 to 12	3, 5, or 6
		Under 6	6
Jobbing..........	High	Under 6	5
	Medium	Under 6	5 or 6
	Low	Under 6	6

braically. Therefore, if the following formula is satisfied, the job may be classed as highly repetitive:

$$\frac{N \times T}{1,000} \geqq 1$$

where N = number of pieces (not less than 2,000)
    T = time allowed

By definition, a job is mediumly repetitive if it has not less than 500 pieces per year and lasts one to six months. To be considered mediumly repetitive, therefore, the following formula must be satisfied:

$$\frac{N_1 \times T}{167} \geqq 1$$

where N_1 = number of pieces (not less than 500)

A job of low repetitiveness consists of not less than 50 pieces per year and lasts two weeks to one month. The formula which must be satisfied is

$$\frac{N_2 \times T}{80} \geqq 1$$

where N_2 = number of pieces (not less than 50)

The use of the formulas and the table may be illustrated by the following example. In a production machine shop doing miscellaneous work, several representative jobs are selected to test the type of methods study that is economical to make. On the first job considered, the activity is estimated to be 5,000 pieces per year. The first operation is a lathe operation that requires 0.392 hour to perform. Substituting these figures in the formula for highly repetitive jobs, a value of 1.96 is obtained. The expression is thus satisfied, and the operation is classed as highly repetitive.

There are several long cuts involved in the operation during which the machine is in complete control. The human attention required during the whole operation is estimated at 45 percent and is classed as medium. There is every indication that this operation will continue to be performed for several years in the future; thus its life is over twelve months. Referring to the table, it can be seen that a highly repetitive job, requiring medium human attention, lasting over 12 months, calls for a type 1 or a type 2 study. In this particular case, the manager or the industrial engineer would consider the fact that the human attention is comparatively low, would estimate from experience that the possibilities for improvement through detailed motion study or the application of additional available mechanical devices appear limited, and would specify a type 2 study.

As pointed out in this example, the manager or the industrial engineer must recognize that judgment should be used when applying the formulas and using the table.

TO DO SUCCESSFUL ANALYSIS WORK, A SUITABLE MENTAL ATTITUDE MUST BE DEVELOPED

Human nature is such that the proper attitude toward analysis work does not develop naturally. Instead, people tend to get smug about their knowledge of a particular activity. They feel that they have reached the goal and do not need to strive any more. This attitude may be commendable as a means of securing peace of mind in everyday affairs, but it makes successful analysis impossible. If an analyst feels that he knows everything about a certain point and that he does not need to consider it further, he ensures that he will make no improvement on that point. To improve any process or operation, the analyst must approach it with a firm conviction that it can be improved.

As a result of many experiences with continuing job improvement, industrial engineers never speak of the "best" method. Rather, they refer to the "best available method" or the "best method yet devised." Carrying this thought to its logical conclusion, it might be stated: "Every time a man uses his hands, there is a continuing opportunity for methods improvement. This opportunity exists until the operation is mechanized to the extent that human attention is completely eliminated and the mechanical devices used are of ultimate simplicity."

This statement makes it clear that simple, automatic operation is the ultimate goal of any methods improvement program. The best method of doing an operation from an economy point of view is reached only when the human attention required has been reduced to zero and all complicated production equipment has been eliminated or simplified. Until this point has been reached, further improvement is always possible.

This principle furnishes the foundation for a sound approach to universal operation analysis for methods improvement and automation. If the analyst appreciates its logic, he will have an open mind. If he accepts it, he will not be bothered with such mental obstacles as "It won't work" and "We tried it before and it can't be done." Lack of success in improving or automating any job should not be interpreted to mean that the job cannot be improved. Such an occurrence is only an indication that the analyst is not aware of any developments that would improve the job or that available equipment is still too expensive to be economical. Acceptance of the continuous-opportunity-for-improvement principle will combat any tendency to feel content with things as they are, and it will inspire fresh attacks from new angles. It leads to progress.

An open mind paves the way for successful analytical work, but it is not sufficient in itself. One can be open-minded in the passive sense of being receptive to suggestions, but this type of open-mindedness will not lead to accomplishment. To get results, the analyst must take the initiative in originating suggestions.

In a world where it is often said that there is nothing new, the greatest amount of originality—or what passes for originality—comes from people who have an inquiring turn of mind. The man who constantly asks questions and takes nothing for granted disturbs the complacent members of the organization, but he originates new and better ways of doing things. Progress begins with doubt. Improvement begins with analyzing what is being done and then inquiring into what new techniques are available so that it may be done better.

Once this point is understood, the industrial engineer should conscientiously develop what is known as the "questioning attitude." The questioning attitude is a state of mind that prevents anything being taken for granted in the investigation of a job. It questions everything and determines answers on the basis of facts. It guards against the influence of emotions, likes or dislikes, and prejudices.

The man who is successful in bringing about improvements has only one deep-seated conviction: that the method can be improved. He accepts nothing as being right just because it exists. Instead, he asks questions and gathers answers. He evaluates the various possible answers in the light of his knowledge and experience. He questions everything. He investigates all phases of the job to the extent that time permits. He asks questions when the answers appear obvious, because the obvious things frequently hide valuable improvement opportunities.

The questions that the industrial engineer asks take the general form of what, why, how, who, where, and when. What is the operation? Why is it performed? How is it done? Who does it? Where is it done? When is it done in relation to other operations? These questions, in one form or another, should be asked about every factor connected with the job or class of work being analyzed.

When a job is examined in systematic detail and all factors related to it are questioned, possibilities for improvement are certain to be uncovered. The action that is taken on these possibilities will depend on the position of the person who uncovers them. If he has the authority to take action and approve expenditures, he will undoubtedly go ahead and make the improvement without delay. If he does not have that authority, he must present his ideas in the form of suggestions to the person or persons who do have that authority.

There are certain pitfalls to be avoided when making suggestions. In the first place, the real advantage of every suggested methods improvement should be carefully evaluated before it is offered. If an individual establishes a reputation for offering only meritorious suggestions, he will be assured of an attentive hearing of his ideas. If, on the other hand, he continually offers just any idea, he will find that those who receive his suggestions will soon stop taking the time to separate the good from the impractical, and all his offerings will be rejected.

The best way to prove the merit of any suggestion is to make an estimate of both the cost of making the improvement and the total yearly savings that the improvement will produce. If an improvement will cost $1,000 to adopt and will save $100 a year, it is not worth presenting—unless it would solve some pressing related problem such as operator safety. If, on the other hand, the expenditure will be returned in a reasonable length of time, the suggestion is worthy of careful consideration. Most companies have established criteria for determining what constitutes a reasonable length of time, and many have developed forms to standardize the manner in which the information on costs and savings is developed.

TEN PRIMARY POINTS OF ANALYSIS

When analyzing a job or an activity, there are so many questions which should be asked that, unless a systematic procedure is followed, it is quite possible that certain points may be forgotten. More than one analysis has proceeded to the point where elaborate suggestions for improvement have been presented only to have all the work discarded because some simple question like "Are all parts necessary to the function of the assembly?" was not previously asked, and the person to whom the suggestion was submitted recognized that the job should be eliminated rather than improved.

To avoid wasted effort and to make sure that all important points are considered, the analyst should keep clearly in mind the factors that should be examined in every operation. These factors should be considered in detail whether the analysis is mental or written. The ten main points or factors which should be considered in every operation—arranged in order of consideration—are as follows▸

1. Purpose of operation
2. Design of part
3. Process analysis
4. Inspection requirements
5. Material
6. Material handling
7. Workplace layout, setup, and tool equipment
8. Common possibilities for job improvement
9. Working conditions
10. Method

When actually making an analysis, it is seldom possible to complete the analysis of one of these factors at a time and then leave it for good. Almost all the factors are interdependent, and a change in one will cause a change in one or more of the others. The list, however, indicates in a general way the course along which the analysis will best proceed.

THE ANALYSIS SHEET

To simplify the work of making an analysis, a form known as the operation analysis sheet has been designed. Wherever the form is regularly used, the number of suggestions for improvement increase. The form, of course, does not accomplish this through any mystic property of its own, but its use ensures that none of the factors which should be considered will be neglected. A typical form of the operation analysis sheet is shown in Figure 4-1.

The form is equally useful whether analysis is to be mental or written. The mental analysis, made with the form as a guide, is quicker, but it is also less

Date started _____ Department_____

Dwg. or spec. _____ Item or part no. _____ Material _____

Description of part _____

Operation _____

Yearly activity _____ Expected life _____ Yearly labor cost per .0001 hr. _____

DETERMINE AND DESCRIBE	DETAILS OF ANALYSIS	ACTION
1. PURPOSE OF OPERATION _____	Is the operation necessary?	
	Does the operation accomplish the intended result?	
	Can the operation be eliminated by doing a better job on preceding operations?	
	Can the material supplier perform the operation more economically?	
	Can the operation accomplish additional results to simplify succeeding operations?	
2. DESIGN OF PART (suggest improvements, make sketches where necessary)		
	Are all parts necessary?	
	Could standard parts be substituted?	
	Does design permit least costly processing and assembly?	
	What design features do competitors use?	
	Will design allow eventual automation?	
3. PROCESS ANALYSIS (complete list of all operations performed on part)		
No. Description Work Sta. Dept.		
1. _____	Can operation being analyzed	
2. _____	be eliminated?	
3. _____	be combined with another?	
	be performed during idle period of another?	
4. _____	Is sequence of operations best possible?	
5. _____		
6. _____	Should operation be done in another dept. to save cost or handling?	
7. _____		
8. _____		
9. _____		
10. _____		

FIG. 4-1.   Operation analysis form.

4. INSPECTION REQUIREMENTS

Tolerances and specifications _____

_____

_____

_____

_____

Inspection procedures (suggest improvements) _____

_____

_____

_____

Are tolerance, allowance, finish, and other requirements
  necessary?
  too costly?
  suitable to purpose?

Should statistical quality control be used?

Is inspection procedure effective and efficient?

---

5. MATERIAL (suggest better material)

_____

_____

How can scrap costs be reduced? _____

_____

_____

Processing materials _____

_____

_____

Consider size, suitability, straightness, and condition.

Can cheaper material be substituted?

Will tool modifications permit use of lighter material or thinner sections?

Would a more expensive material lower machining and processing costs?

Is packaging suitable?

---

6. MATERIAL HANDLING (suggest improvements)

Brought by _____

_____

_____

Removed by _____

_____

_____

Handled at work stations by _____

_____

_____

_____

_____

_____

Can incoming materials be delivered directly to the work station?

Can signals such as lights or bells be used to notify material handlers that material is ready to be moved?

Should crane, gravity conveyors, tote pans, or special trucks be used?

Consider layout with respect to distance moved.

Are containers correctly sized?

FIG. 4-1 (*continued*).   Operation analysis form.

7. WORKPLACE LAYOUT, SETUP, AND TOOL EQUIPMENT
(suggest improvements, making sketches where necessary)

Arrangement of work area

Placement of tools, materials, supplies

How are dwgs. and tools secured?

Can setup be improved?

Trial pieces

Machine adjustments

TOOLS

Suitable?

Provided?

Ratchet tools

Power tools

Special purpose tools

Jigs, vises

Special clamps

Fixtures

Multiple

Duplicate

8. COMMON POSSIBILITIES FOR JOB IMPROVEMENT (consider the following)   RECOMMENDED ACTION

1. Install gravity delivery chutes.

2. Use drop delivery.

3. Compare methods if more than one operator is working on same job.

4. Provide correct chair for operator.

5. Improve jigs and fixtures by providing ejectors, quick-acting clamps, and the like.

6. Use foot-operated mechanisms.

7. Arrange for two-handed operation.

8. Arrange tools and parts within normal working area.

9. Change layout to eliminate backtracking and to permit coupling of machines.

10. Utilize all improvements developed for other jobs.

9. WORKING CONDITIONS (suggest improvements)

Light

Heat

Ventilation, fumes

Drinking fountains

Washrooms

Safety aspects

Design of part

Clerical work required (to fill out time cards and so on)

Probability of delays

Probable mfg. quantities

Fɪɢ. 4-1 (*continued*).   Operation analysis form.

10. **METHOD** (accompany with sketches or process charts if necessary)

    a. Before analysis and motion study _____

_____

_____

_____

    b. After analysis and motion study _____

_____

_____

_____

_____

Are hand motions symmetrical?

Are parts transferred between hands?

Is a more detailed motion study needed?

Has safety been considered?

Working posture

Does method follow Laws of Motion Economy?

Are lowest classes of movements used?

RECOMMENDATIONS FOR FURTHER IMPROVEMENTS IF THE ACTIVITY INCREASES:

RECORD OF ACTION TAKEN:

Proposal	Date	Referred To:	Action Taken

COMMENTS:_____

Date completed _____ Analyzed by _____

FIG. 4-1 (*continued*). Operation analysis form.

satisfactory than a written analysis. When a mental analysis is made, records are seldom kept, and if they are, they are usually not systematic or complete. This lack of records is a liability in the event that a change in the type of study required is later decided on, in which case the analysis will have to be repeated. However, mental analyses made systematically with the form as a guide will produce many good results on jobs where low activity or low human attention makes it uneconomical to undertake a more elaborate study.

A written analysis using the operation analysis sheet has several obvious advantages. The written analysis is more likely to be carefully made. The fact that the answer to each question must be committed to writing will ensure that proper consideration is given to each factor. The data that are usually collected in the preparation of a written analysis will support the suggestions made for the improvement of the job or class of work.

It should be unnecessary to stress the importance of identifying all the supporting paper work connected with an analysis, but experience shows that, unless this point is emphasized and reemphasized, the identification of the supporting papers is seldom complete.

USE OF THE ANALYSIS FORM

The analysis form acts as a guide to systematic operation analysis. It directs the person making the analysis through the factors to be considered and ensures that none of them will be overlooked.

The analysis itself actually takes place in the mind of the analyst. He questions each point as it is raised, gathers all the known facts, and combines these facts with his own knowledge of alternatives. In this way, he arrives at his suggestions for improvement. The nature and extent of these suggestions depend on the analyst's knowledge of what is taking place in the areas of new materials, tools, and manufacturing techniques. However, the systematic procedure outlined on the analysis form will help him achieve maximum results. As he makes the analysis, he records all facts and ideas for improvement at the time they occur. The form should include enough detail to provide a record of the conditions that exist at the time of the analysis and to suggest any improvements that come to mind. All descriptions should be recorded clearly and concisely.

Space is provided in the heading of the operation analysis form for recording all information necessary to identify the job or class of work. In the following paragraphs, each of the ten factors to be considered will be discussed in detail. Specific questions to be asked by the analyst are shown on the operation analysis form in Figure 4-1.

Purpose of Operation. Although most operations are properly set up the first time a job is performed, changes in design or material specification may cause an operation to become incorrect or even unnecessary. In industry and business, as in other phases of life, nothing remains constant for any great length of time. As a result, slight changes in preceding or subsequent processes may affect the efficiency of or the necessity for a particular operation. In fact, the application of the operation analysis procedure has uncovered a surprising number of operations found to be unnecessary after further study. Unfortunately, those most familiar with the unnecessary operations often fail to recognize that they are superfluous. The analyst, then, must be alert to the possibility that work that is being performed is no longer necessary. In some instances, the material supplier can perform the operation more economically, or the operation can be eliminated by doing a better job on either preceding or subsequent operations.

Design of the Part. Although the person making the analysis is seldom a design engineer, it is important that he consider the design before proceeding to the other points of analysis. Often, the design engineer does not have time to reconsider his design after the decision to manufacture is made. Therefore, the person making the operation analysis must ensure that the design is correct and desirable. Consideration of this point can ensure that expensive details originally designed into the part are still necessary. Many unnecessary design features have been eliminated, with resultant large dollar savings, because of attention paid to this factor of the analysis.

To analyze this factor adequately, one must take enough time to understand the essential functions of the part and the assembly being studied. He must ensure that the design of the part and the assembly is a least-cost design.

Process Analysis. No single operation can be studied by itself. It must be considered as a part of the total process. The effect of any changes that are suggested must be considered in the light of the process. Only in this way can the analyst be sure that the suggested improvement will produce results. By carefully reviewing all the operations performed on a part, the analyst can determine whether the operation under study can be eliminated, combined with another, or performed during the idle time of another operation.

Because of the rapid development of new processes and techniques, the analyst must keep abreast of the newest developments in the area of study. With this knowledge, he will be able to recommend changes which will simultaneously improve quality and reduce cost by improving or eliminating outmoded or unnecessary operations.

Inspection Requirements. Quality requirements established by the designer or originator of a process play an important part in the selection of operations and methods to be used. In fact, these quality requirements often force the selection of a specific process and method. On machined metal parts, for example, if a designer allows little or no variation from parallel for several holes through a part, the process engineer may be forced to specify the use of a precision boring mill rather than a drill press. Similarly, if the designer specifies too high a pressure test for welds on plant power piping, shop welding techniques may have to be substituted for simpler, less costly in-place welding.

The inspector, of course, also plays an important part in determining methods, because too literal an interpretation of the quality specifications can result in a more costly method.

Through application of the operation analysis procedure, the analyst will determine whether the quality requirements are consistent with the use to which the finished job will be put. After he has determined that the requirements are consistent with the use, he can then determine whether the operation under study will produce a result that will meet the requirement economically. In this way, he can ensure that the company is not paying for unnecessary requirements, and that properly established requirements are being uniformly enforced and met.

Material. Material costs are an important part of the total cost of any job or class of work. The kind of material from which parts are made is usually fixed by the nature of the part and the service conditions that it must withstand. However, materials that were originally specified by the designer or originator of a process may no longer be the most suitable. Unfortunately, design budgets seldom provide for periodic review of materials. Thus, the investigation of materials during the conduct of an operation analysis can sometimes result in significant savings. The analyst must be familiar with recent developments in new materials so that he may recognize when a currently specified material is no longer the best material available for the job. During the course of his study, the analyst must consider the size, suitability, and condition of existing materials and the possibility of substitute materials. He should also consider the use of supplies related to the operation.

Material Handling. The flow of material through a plant or business is usually accomplished by a number of separate transportations. These transportations may be into and out of storage locations or to and away from work stations. The analyst, by carefully studying the need for transporting the material and the nature of the material handling activity, can often significantly reduce this major cost.

Many devices have been developed through research to expedite the flow of materials and eliminate the problems connected with material handling. The bulkier the part, the more advantageous it is to think in terms of orderly, continuous flow rather than batch handling.   For example, storage conveyors are now used for almost every imaginable type of material to provide desired flow and to allow selection of items brought from storage areas without manual handling.

Workplace Layout, Setup, and Tool Equipment.   The workplace layout provided for an operator determines the motions that he must use while performing a job. Almost every industrial engineer is familiar with the attention given to manually performed bench-type operations.   He also recognizes that most machine tool manufacturers now acknowledge the importance of locating controls in the most effective manner.

Despite the emphasis on workplace layout, there are still many examples of unplanned work areas and lack of standardization.   Maintenance work, in particular, often suffers from poorly conceived workplace layout.   Although some people contend that this class of work does not lend itself to reasonable workplace layout, many companies have provided maintenance crews with methods training and suitable tool carts and other equipment which minimize manual motions.

The statement, "With sufficient study, any method can be improved," certainly applies to workplace layout.   In studying the workplace layout, the analyst must consider the placement and use of all materials and tools.   He must also consider such factors as the manner in which the job is assigned; how the operator receives job instruction; and how he obtains auxiliary equipment such as drawings, special tools, and measuring devices.   During operator instruction and learning time on certain kinds of repetitive work, it may be desirable to provide an audiovisual training device.   Although these devices take up valuable space, experience with them has proved their effectiveness.   Whether the workplace layout involves a machine tool, a clerk's desk, or a bench, the application of operation analysis will result in an improved arrangement of the work area.

Common Possibilities for Job Improvement.   During the application of operation analysis, there are certain factors to be considered that are particularly effective in improving almost any type of operation.   These factors, which are based on the principles of motion economy, are considered as common possibilities for job improvement.   They involve consideration of the use of such devices as delivery chutes, ejectors, quick-acting clamps, and foot-operated mechanisms.   They also guide the analyst in the consideration of operator comfort and the motion pattern employed during the performance of the operation.   Although these factors may be covered during the consideration of the other points of primary analysis, they have resulted in such significant improvements that they are listed separately as item 8 on the operation analysis form.

Working Conditions.   Although much attention is paid to the motions that a worker must perform and to the requirements for an effective process, the environment in which work is carried out also plays an important part in maintaining worker comfort and efficiency.   Extremes of heat or light, poor ventilation, or safety hazards may cause unnecessary operator fatigue or concern.   These factors have a direct bearing on output.   To be most effective, an operator should have optimum environmental conditions.   During the operation analysis procedure, the analyst must consider the effect of factors associated with operator comfort, safety, and well-being.

Method.   Although it may seem unusual to consider method last during the application of operation analysis, each of the preceding points of primary analysis directly affect the final step, which is establishing the best method.   When considering method, the analyst must first carefully examine the present method to find its weaknesses.   Each of the ten points of primary analysis help him in his examination.   After he has completed a thorough analysis of the present method, he is prepared to develop the improved method.   The operation analysis form provides space for description of both the original method and the improved method.   It also provides space for recommendations for further improvements, if the activity of the operation increases.   Periodic review of the operation analysis form and comparisons between levels of activity will help to pinpoint subsequent methods improvements.

OFFICE OPERATION ANALYSIS SHEET

Date _____ Dept. _____

Operation _____ Operator _____

DETERMINE AND DESCRIBE	DETAILS OF ANALYSIS
1. PURPOSE OF OPERATION	Why is the operation performed? Can purpose be accomplished better otherwise?

2. COMPLETE LIST OF ALL OPERATIONS PERFORMED IN PROCEDURE

No. Description Work station Dept.	
1. _____	Can operation being analyzed be eliminated? be combined with another?
2. _____	
3. _____	
4. _____	Is sequence of operations best possible?
5. _____	
6. _____	
7. _____	Should operation be done in another dept. to save time or handling?
8. _____	
9. _____	
10. _____	

3. QUALITY REQUIREMENTS	
a. Of previous operation	Are audit and other detail requirements necessary? too loose? too tight? suitable to purpose?
b. Of this operation	
c. Of next operation	
4. MATERIAL COSTS – FORMS AND SUPPLIES	Consider size, cost, condition, and advantage of standardization. Should form be prepared? Are forms best suited for insertion of data? ease in use?
5. TRANSMISSION OF INFORMATION	Consider means of transmitting data.
a. Brought by	Are means adequate? Is there excess backtracking? Is it done by "overqualified" personnel? Would conveyor system be practical?
b. Removed by	
6. WORKPLACE ARRANGEMENT (accompany description with sketches if necessary)	Can workplace arrangement be improved? Can distances be shortened?
A. Equipment	Equipment
Present	Should hand operation be mechanized? Is equipment suitable? in good condition? used to best advantage?
Suggestions	Typewriters Other office machines File drawers and cabinets Reproducing equipment

FIG. 4-2.   Office operation analysis sheet.

7. CONSIDER THE FOLLOWING POSSIBILITIES:	RECOMMENDED ACTION

1. Compare methods if more than one operator is working on same job.
2. Reassign duties
   a. To even work load?
   b. To handle peak loads?
   c. For better use of personal qualifications?
3. Use of snap-out forms.
4. Revision of existing form to serve additional purposes.
5. Arrange for two-handed operation.
6. Utilize all improvements developed for other jobs.
7. USE OF STAMPS, STAPLER, SIMPLE TOOLS.	

8. WORKING CONDITIONS a. Other conditions	Light Heat Ventilation Drinking fountains Washrooms Safety aspects Probability of delays

9. METHOD (accompany with sketches or process charts if necessary)
 	
   a. Before analysis and motion study.
 	
   b. After analysis and motion study.
	Arrangement of work area
	
	Placement of:
	  equipment
	  materials
	  supplies
	
	Working posture
	
	Does method follow Laws of
	  Motion Economy?
	
	Are lowest classes of
	  movements used?
	
	See supplementary report
	  entitled
	
	Date

Observer _____ Approved by _____

FIG. 4-2 (*continued*).   Office operation analysis sheet.

The operation analysis form also includes a record of action taken. This record assists management in evaluating the disposition of methods improvements recognized during the operation analysis.

Office Operation Analysis. Although operation analysis has most often been applied to manufacturing operations, its principles can also be effectively applied to office operations. Although the points of primary analysis differ somewhat from the operation analysis performed for manufacturing operations, the principles remain the same.   The office operation analysis form is illustrated in Figure 4-2.   It will provide a systematic, written analysis of clerical operations where the level of activity justifies such a written analysis.

OPERATION ANALYSIS CHECK SHEET[1]

The operation analysis sheet previously referred to is a guide to be used by the industrial engineer when analyzing any operation.   The sheet is in abbreviated outline form; to make a more thorough analysis of any operation, the analyst must expand on the list of questions that appears on the form.   With an abbreviated form, there is a danger that the analysis may be made too hurriedly and that proper consideration will not be given to each of the factors involved.   To overcome this, an operation analysis check sheet may be developed.

The check sheet is an expansion of the shorter analysis form.   On it are listed all the important questions which should be covered when considering the ten major points of the analysis.   Because a check sheet is quite detailed, a great deal of time is required to fill it in properly.   Accordingly, it should be used only when it can be economically justified.   Ordinarily, the check sheet would not be used unless a type 1 study, or occasionally a type 2 study, is indicated.

It must be remembered that the operation analysis form and the analysis check sheet are merely tools which can be employed when analyzing any operation. They are designed principally to guide the analyst and to keep clearly before him the points that he should study and seek to improve.   However, any improvements that are made will be the result of the ability and knowledge of the analyst himself rather than the tools which he uses, for no form will take the place of sound reasoning, constructive thinking, and creative ability based on knowledge.

The opportunities for improving industrial operations are unlimited.   The operation analysis technique is a powerful tool for accomplishing the methods improvement that every company constantly needs.

BIBLIOGRAPHY

Hammond, Ross W., "Industrial Engineering," sec. 10, chap. 2, in H. B. Maynard (ed.), *Handbook of Modern Manufacturing Management,* McGraw-Hill Book Company, New York, 1970.

Maynard, H. B. (ed.), *Handbook of Business Administration,* sec. 7, McGraw-Hill Book Company, New York, 1967.

Maynard, H. B., and G. J. Stegemerten, *Operation Analysis,* McGraw-Hill Book Company, New York, 1939.

O'Donnell, Paul D., and John C. Martin, "Developing Improved Methods," sec. 3, chap. 8, in H. B. Maynard (ed.), *Handbook of Modern Manufacturing Management,* McGraw-Hill Book Company, New York, 1970.

[1] A typical operation analysis check sheet may be found on pages 264-280 of *Operation Analysis,* by H. B. Maynard and G. J. Stegemerten, McGraw-Hill Book Company, New York, 1939.

Chapter **5**

Motion Study*

ANNE G. SHAW

Chairman, The Anne Shaw Organisation, Ltd.,
Brook Lane, Alderly Edge, Cheshire, England

Motion study is the investigation and measurement of the movements involved in the performance of any work, their subsequent improvement, and the application of easier and more productive methods. The study of the needs and problems of the operators is the starting point of any motion study investigation. Its final purpose is to enable them to work with minimum effort and maximum efficiency. With this end in view, the investigator studies not only the workers but also the conditions surrounding the work, including the movement of materials, tools, and equipment and the organization and layout of work. These factors directly influence the efficiency and well-being of the workers.

SCOPE OF MOTION STUDY

To many people, the term "motion study" means the study of operators' hand movements only. Its techniques, however, are in no way so restricted in scope. The limits of the application of motion study are the limits of movement itself. Because most of the processes in industry and commerce involve movement of some kind, the techniques of motion study have a very wide field of application. They can be used to study in greater or lesser detail such differing forms of movement as are involved in the flow of material through a plant, the activity of an operator at a bench, or the intricacies of paper work or office procedures.

* Adapted from *The Purpose and Practice of Motion Study*, 2d ed., by Anne G. Shaw, published by Columbine Press, Manchester and London, 1960.

Different kinds of problems may require some variation in the application of the motion study techniques to suit the nature of the work to be studied, but the fundamental approach to every problem and its analysis is always the same.

PRELIMINARY SURVEY

To make the most economical use of motion study, before any investigation is undertaken, a survey should be made to determine the limits of the investigation and to assess how long it will take, how much it is likely to save, and to what degree of detail it is profitable to go. Such a survey is the normal practice of a consultant, but it should be made in the same way for every internal investigation. In most cases, a survey will establish that part of the suggested subject should be studied broadly while certain other parts merit detailed study. A few hours spent on a survey at this stage may save a great deal of wasted time later. (See Chapter 4 of Section 2.)

The preliminary survey will set out not only to balance time spent on an investigation against possible savings, but also to assess the likely duration of the job or the product itself because this must affect the time schedule and the nature of the investigation. For example, when the product to be studied is seasonal or a fashion article or something liable to frequent changes in design, it may still be worth studying, but the investigation will have to be made to a tight schedule and limited to such alterations as can be made simply and quickly. Alternatively, the investigation may concentrate on those operations which are unlikely to be altered—for example, those parts of a garment which do not change fundamentally with the fashions—making provision for reasonable variation in other operations but not wasting time on uneconomic detailed study.

Another point that may be questioned in a preliminary survey is the scope for improvement where much of the work is connected with machines. Where machine efficiency is already high, increased output per machine is only partly in the operator's control. Increased efficiency of operator movement can give very little increase in output unless the machine can be modified to run faster. In such a case, it would be necessary to decide at this stage which of the three alternatives should be the object of the investigation: a slight increase in output obtained by making it possible for the operator to raise the machine efficiency the small amount that is available; a larger increase in output per operator obtained by allowing one operator to run two or more machines at lower efficiency; or finally, an investigation into the operations of the machine and the possibility of running it faster or more continuously.

PLANNING THE INVESTIGATION

It will be found that only about 25 percent of the time expended on the average motion study investigation is spent in developing improved methods. The remaining 75 percent is occupied in establishing them in the factory, convincing operators and supervisors of the need for change, and reassuring them that their personal interests will be safeguarded and that quality of production will not suffer. This large proportion of time can, however, be considerably reduced by anticipating some of the difficulties in advance, even before the first chart is drawn, and making provision either to meet or avoid them.

The first provision to be made will be some arrangement for adjusting the existing labor force to the likely requirements of the improved method. Here the preliminary survey should provide information as to what improvement is possible. It is then necessary to know whether the object of the investigation is to be increased production from the same labor force or the same production from fewer operators. The first alternative presents little difficulty from the point of view of the deployment of labor, but the investigator will be wise to check to see that material for the increased total production has been provided for at the time when it will be needed and that sales are fully ready to absorb the increase when it comes. The second alternative, a reduction in the labor force required, calls for even earlier action. Where it is essential to guarantee continuity of employment, the surplus labor can usually be absorbed by the normal labor turnover if plans are laid far enough in advance. This requires cooperation from those responsible for employment and placement in the plant as a whole and from the shop supervisor. They must be told in advance and kept informed of the progress of the investigation so that they are ready to absorb the surplus labor when it becomes available.

The next point to be considered concerns inspection standards. Much friction is often caused during the later experimental stages of an investigation by an increase in inspection standards, and therefore in rejects, to the discouragement of both operators and supervisors. It is natural that inspectors should take an increased interest when experiments are being made, and it is inevitable that any existing slackness should be tightened up without any specific instructions being issued to that effect. To avoid friction, it should be possible to fix a reference point for quality standards before any experimental work begins.

A further point to be covered before beginning an investigation is the question of wage payment. This is a general management responsibility, but it is advisable to check that a policy has been established, that it will be applied, and that it is clearly understood by both operators and supervisors.

Plans should be made at this stage for some sort of training course in simplified motion study of the work simplification type if those who will be concerned with the investigation, both supervisors and operators, have not already had such training. A broad knowledge of motion study will allow everyone to contribute something to the development of the final methods, making the investigation a team job for the whole group and acceptable as such, rather than a change imposed by an investigator from outside the group.

RECORDING PRESENT PRACTICE

So much space devoted to preliminaries may seem out of proportion in a comparatively short, factual account of motion study. Only careful planning, however, will prevent later difficulties and ensure an economical, well-directed investigation. It is equally essential to the proper application of motion study that the techniques for recording movement should be selected and used with discrimination.

Range of Techniques. As devised and used by Frank Gilbreth in the early years of the twentieth century, the techniques of motion study were extremely flexible, and properly applied today, that flexibility has in no way diminished. It has, in fact, been increased by modern developments in apparatus and experience. Unfortunately, however, there has been a tendency in recent years to attempt to standardize, and therefore limit, the use of particular motion study techniques. For example, it is often recommended that the process chart technique is sufficient in nine out of ten investigations and that micromotion study should be confined to mass production and large-scale repetitive operations, dismissing the study of the path of movement altogether, except for the chronocyclegraph which is regarded as a laboratory technique suitable only for research.

As a matter of fact, these three main techniques or groups of techniques are designed to record three different aspects of movement. They are not merely coarser or finer sizes of the same tool but three distinct tools or types of tool, all of which may be needed to complete a job. Each, when applied in its appropriate form, can be used to record as much or as little detail as is required, but it is a different kind of detail in each case. Process charts, for example, range from the flow process chart covering in broad outline the complete flow of material through a series of processes, to the most detailed two-handed chart recording small movements of each hand. Whatever the scale, the analysis is directed toward setting out the sequence of operations divided into categories according to the kind of work achieved. Micromotion study in the same way ranges from its coarsest application in memomotion, through therblig charts made from direct observation, to simo charts giving the fullest possible accuracy of detail, including the time relationship of the various sections of the work. But in each case the analysis is always on the basis of a consideration and classification of the purpose of each element. The third aspect, the path of movement, is studied by a range of related tools rather than by a set of identical tools in different sizes. The string diagram can be made on a scale as broad or broader than the flow process chart or as fine as the two-handed process chart. Movements of shorter duration may be recorded roughly by pencil sketches of the movement path or with the greatest accuracy by chronocyclegraphs. In many cases, a simple mental picture may be enough to examine this aspect of a series of movements making up a method of work.

Selection of Appropriate Techniques. The choice of techniques or of their degree of application will not always be decided by the aspect of movement needing analysis or by the degree of detail or accuracy required. Certain of the techniques have valuable secondary uses. For example, when a visual therblig analysis or a rough observation of the movement path would probably be sufficient, it is quite often worthwhile to take a micromotion film or a chronocyclegraph photograph because of its value in convincing operators and supervisors of the need for change and the logic of the changes suggested. This is likely to shorten, perhaps considerably, the time taken to get the improved method into operation. Later, supplemented by a similar record of the new method, it may again be invaluable for teaching detailed motion study to students of industrial engineering, or simplified motion study to executives, supervisors, and operators. The skill and time required for taking films or chronocyclegraphs is nearly always exaggerated. Both can be very simple operations. The cost of the film compared with the time likely to be saved at a later stage of the investigation often means a considerable economy.

TECHNIQUES FOR RECORDING PRESENT PRACTICE—PROCESS CHARTS

The process chart was the last of the three main groups of motion study techniques to be developed by Gilbreth. His work on process charts is set out in a paper read to the American Society of Mechanical Engineers in 1921.[1] In this paper, he defined his technique as follows:

> The process chart is a device for visualizing a process as a means of improving it. Every detail of a process is more or less affected by every other detail; therefore the entire process must be presented in such form that it can be visualized all at once before any changes are made in any of its subdivisions. In any subdivision of the process under examination, any changes made without due consideration of *all* the decisions and all the motions that precede and follow that subdivision will be found unsuited to the ultimate plan of operation. . . . It is not only the first step in visualizing the "one best way to do work" but is useful in every stage of deriving it.

The Gilbreth process chart symbols as published in the 1921 paper have been superseded by a revision which has been adopted as a standard by the American Society of Mechanical Engineers. These revised symbols and a complete description of process charting are contained in Chapter 3 of Section 2.

TECHNIQUES FOR RECORDING PRESENT PRACTICE— MICROMOTION STUDY

Gilbreth developed micromotion study among the first of his motion study techniques. His earliest published work on the subject was a paper read to the American Management Association in 1912, nine years before his first publication about process charts. He divided all movements into "elements" according to their purpose and called these "therbligs," his own name spelled backwards.

[1] F. B. and L. M. Gilbreth, "Process Charts: First Steps in Finding the One Best Way to Do Work," paper presented to the Annual Meeting of the American Society of Mechanical Engineers, New York, Dec. 5–9, 1921.

At first, micromotion analysis into therbligs was made from visual observation. When motion picture photography became practical, motion picture films were analyzed in the same way and with greater accuracy. The study of the micromotion film is now the central technique of a group of three micromotion study techniques. Visual therblig analysis is still used where less accurate information is sufficient, and memomotion study[2] is a technique developed to cover large-scale movements and multioperator analysis.

Therbligs.   In the same way that the process chart symbols are used as a basis for a variety of process charts from flow process charts to two-handed process charts, the therbligs are the basis of the various micromotion techniques. Gilbreth did not base his therbligs on a physiological analysis of movement, because it was recognized that the least observable movement might involve whole groups of muscles. His method of classification was based on an analysis of the purpose for which a movement or part of a movement was performed, not on the nature of the movement itself. For example, if the empty hand moves toward an object, its purpose is to reach that object. Having reached it, the purpose changes first to picking it up, and second to moving it from one place to another. Three different therbligs are needed to distinguish between these three distinct movements: "transport empty," "grasp," and "transport loaded." The operator's next intention may be to put it into a container—to "assemble" it. Before this "assemble," some rearrangement may be necessary which involves the therblig "position." If, on the other hand, the purpose of a movement is to separate two parts, "disassemble" will be the therblig used. If a tool is held, the therblig will be "use," and in putting down, "release load." If, while the hand is transporting it, the tool is arranged for easy use at a later stage in the work, the arranging will be described by the therblig "pre-position."

All these therbligs are productive or service elements. There are other parts of a work cycle which are only ancillary to the productive part. Sometimes, after the preliminary "transport empty" and before the other therbligs can begin, it may be necessary to "search" for, or "find," or "select" the object to be "used" or "assembled." Again, there may be a point in the work cycle where the purpose of a movement or the reason for an absence of movement is the need to "inspect" or test an article. Other pauses in the activity may have other causes, and they will be classified as "unavoidable delay" (outside the operator's control), "avoidable delay" (where the pause is deliberate), or even "rest for overcoming fatigue."

The original sixteen elements were used by Gilbreth to analyze movement. Thirteen were active elements and three accounted for absence of activity. A seventeenth, "plan," was added later to the second group. Those who have used the micromotion technique in recent years have added one more to the first group, "hold." This new therblig is sometimes useful, though Gilbreth would have called it a prolonged "grasp."

When Gilbreth devised the therbligs, he gave each the descriptive symbols shown in Figure 5-1. These symbols are pictorial and devised for easy memorizing.

The colors indicated in Figure 5-1 were added to the symbols later when film analysis became possible and the simultaneous motion cycle chart or simo chart was developed.

Therblig Charts Made from Direct Observation.   Therblig analysis was originally developed before motion picture photography was a manageable technique. Until recently, however, few people have made use of it in its simple form, as a paper-and-pencil technique for recording a direct visual observation of movement. Most textbooks and many exponents of motion study have treated therblig analysis exclusively as a means of analyzing a film. They have concentrated on the fact that the film analysis has the added advantage of giving information about the time relationship of the therbligs.

Those who have used therblig analysis from direct observation, as Gilbreth used it, have found it a very useful motion study technique. A two-handed therblig

[2] See Section 3, Chapter 7.

THERBLIGS

SYMBOL	NAME	COLOR	SYMBOL	NAME	COLOR
	SEARCH	BLACK		INSPECT	BURNT OCHRE
	FIND	GREY		PRE-POSITION	SKY BLUE
	SELECT	LIGHT GREY		RELEASE LOAD	CARMINE RED
	GRASP	LAKE RED		TRANSPORT EMPTY	OLIVE GREEN
	TRANSPORT LOADED	GREEN		HOLD	GOLD OCHRE
	POSITION	BLUE		REST FOR OVER-COMING FATIGUE	ORANGE
	ASSEMBLE	VIOLET		UNAVOIDABLE DELAY	YELLOW OCHRE
	USE	PURPLE		AVOIDABLE DELAY	LEMON YELLOW
	DISASSEMBLE	LIGHT VIOLET		PLAN	BROWN

FIG. 5-1.   Therblig symbols and colors.

chart can be made on the same pattern as a two-handed process chart. The result will be an analysis that is more accurate than the two-handed process chart. Being based on therblig elements differentiating the purpose of the movements rather than on process chart symbols differentiating what is happening to tools and materials, the movements will be seen in a different light.

The publication of such predetermined motion time systems as those referred to in Section 5 has focused attention on the practical possibilities of direct observation of this kind.

The Study of the Micromotion Film—Simo Charting. The simultaneous motion cycle chart or simo chart made from an analysis of a micromotion film is the most detailed technique in the micromotion group and gives the greatest degree of accuracy in recording movement. It is one of the most effective of all the motion study techniques in suggesting improvements. The film itself has a valuable secondary use as a means of showing operators and supervisors exactly where change is needed and convincing them of the value of proposed improvements.

In the past, much has been said about the time that is needed to make a simo chart analysis and about the cost of the film used. Actually, filming time can be almost negligible. Using modern, fast film and modern camera lenses, many operations can be filmed in the natural lighting of the workshop, and many more require only a very little additional lighting. The cost of film, unless used extravagantly, is only a few dollars. Both filming time and film costs are usually repaid many times over in time saved at the installation stage of the job and in the additional savings made possible by such an accurate analysis. Finally, the results of a simo chart analysis are so much more fruitful and reliable than the results of almost any other type of analysis using other techniques that it is worthy of much greater use on these grounds alone.

Micromotion Photography. Gilbreth's original purpose in taking motion pictures was to record the movement of an expert for examination and demonstration, but he soon discovered their use for analysis into therbligs. He was obliged to do his filming with a hand-cranked camera and to vary his camera speed still further because of deficiencies in lighting. He therefore introduced into his picture a chronometer so that he could check this speed when the film was projected. This chronometer or counter had a pointer making 20 revolutions per minute around a dial marked off into 100 divisions. Each of these divisions therefore represented 1/2,000 of a minute. Gilbreth called this division of time a "wink." Modern counters are electrically driven and a second or even a third pointer can be introduced to record the revolutions of the first and second pointers. The "wink," however, has not been changed. Having introduced the chronometer for this purpose of checking the camera speed, Gilbreth found that it gave him new opportunities in film analysis because it provided an accurate timing device which allowed an examination of the time relationship of the different therbligs. Although, as his work progressed, the technique of motion picture photography became much easier and more accurate, he maintained the chronometer or counter as essential for the measurement, not of total time, but of the time relationships between the different parts of the body. (Technical aspects of film making are discussed in Chapter 4 of Section 12 of this Handbook.)

Analyzing the Film. The main object of a micromotion study is its analysis and the subsequent development of a new method. The analysis is begun by projecting the film a number of times at normal speed to gain familiarity with the general sequence of movements. The form of the work cycle is next examined to decide at which point to begin the analysis. This point must be the same in each cycle. It will not necessarily be the moment when the operator begins work on an article unless that point is consistent in each cycle and is easily identified. It is best, wherever possible, to choose a place where both hands finish their therbligs together, but whatever point is selected, the therblig that begins that part of the work cycle is used to start the analysis, and the therblig preceding it becomes the finishing point of the last cycle.

The dividing point between the cycles having been decided, the whole film is then examined, taking chronometer readings at that point whenever it appears. This will give a list of cycles with their lengths in terms of "winks." It is then possible to select those which is most important to analyze first. They will probably be either the longest and the shortest cycles or cycles that are particularly outstanding for some other reason. Each consecutive frame of the chosen cycles is then examined for changes in movement of any part of the body. As one therblig changes to another, the chronometer reading is recorded and the appropriate therblig is entered on a film analysis sheet. All changes of movement are noted whenever they appear, right- and left-arm movements being observed and recorded together. Recording the movements of all parts of the body at the same time makes film analysis more difficult for beginners but has the great advantage of keeping a complete picture of the whole movement in view throughout the analysis. If the analysis is divided and all the movements of the right hand are considered before the left-hand movements are examined, the interrelationship of the two may be missed. If this happens, it will not be rediscovered until the simo chart is completed at a later stage.

Making a Simo Chart. When the film analysis into therbligs is complete, all relevant information has been extracted from the film, but it is not in a form that it can easily be visualized as a whole. It is necessary, then, before attempting to develop a new method, to put the facts into chart form. The Gilbreths developed a chart in which the various parts of the body were recorded along the top of the paper and the chronometer readings down the left-hand side, with the therblig symbols appearing against the correct counter reading and the vertical columns filled in with the appropriate therblig color. These charts were called simultaneous motion cycle charts or simo charts. They are usually drawn up on one-tenth-inch cross-section paper with the inch line heavily ruled.

The horizontal scale can be made to include any member of the body. Most industrial processes do not require the analysis of anything but finger and arm movements with an occasional foot movement. The typical chart (see Figure 5-2) uses only the following headings: upper arm; lower arm; wrist; thumb; first, second, third, and fourth fingers; and palm. Further columns are added for other body members where they are needed. Very few movements of the wrist or palm are observed, however, and the spaces allotted to them are therefore frequently left blank.

Having compiled the headings and decided upon the scale of the chart, the therbligs from the analysis sheet are recorded on the chart against their appropriate chronometer readings. In some particularly complex analyses, the work of each finger may have to be differentiated, but only such details as are needed are charted in each particular case. The space covered by each therblig is filled in in its appropriate color. The complete chart is an accurate, graphic record of the details of the various parts of the movements and of their time relationship to one another. To an experienced investigator, the distribution and extent of the different colors on the chart are of immediate significance. Attention is at once drawn to the form of the work cycle and to those parts which most need improvement.

Memomotion Study. Memomotion study is a variation of a micromotion technique developed by Dr. Marvin E. Mundel[3] at Purdue University. It is particularly useful for long-period studies and where a team of operators is at work. Its use is fully described in Chapter 7 of Section 3.

As its principal result, memomotion gives information about the time relationship of elements of movement. An adaptation of the simo chart is a useful way of presenting the information gathered during a memomotion study. A chart of this type is shown in Figure 5-3. Just as process chart symbols are used in different process charts to represent elements of very different sizes, so therbligs can be made to represent on a memomotion chart much larger and more comprehensive elements than they normally cover when the detailed movements of the limbs are studied. The therblig analysis of a memomotion film treats the operator as a whole, not separating the body members, but using therbligs to differentiate the purpose of wide groups of movement. To an investigator used to simo charts, the same coloring and symbols used on the micromotion chart but applied more generally are familiar and therefore helpful and suggestive of improvements. There seems little point in devising special symbols or colors.

TECHNIQUES FOR RECORDING PRESENT PRACTICE— THE STUDY OF THE PATH OF MOVEMENT

In the previous discussions on process charting and micromotion study, a work cycle has been considered as a series of individual movements or therblig elements rather than as a single pattern of movement. Neither technique, however, provides any means of recording a picture of the path of movement, its orbit, or its shape. In many operations, the path of the movement has a particular significance. It is often this aspect which presents the greatest opportunities for economy in fatigue. There are two main types of work where it is particularly important to study the movement path.

In work consisting of short, repetitive cycles of movement, the length of the path often matters less than its shape and its freedom from obstruction or changes in direction. In this type of work, most of the movements are of the hand or arm, and they must be studied in considerable detail. They can be recorded

[3] M. E. Mundel, *Motion and Time Study: Principles and Practice*, 4th ed., Prentice-Hall, Inc., Englewood Cliffs, N.J., 1970.

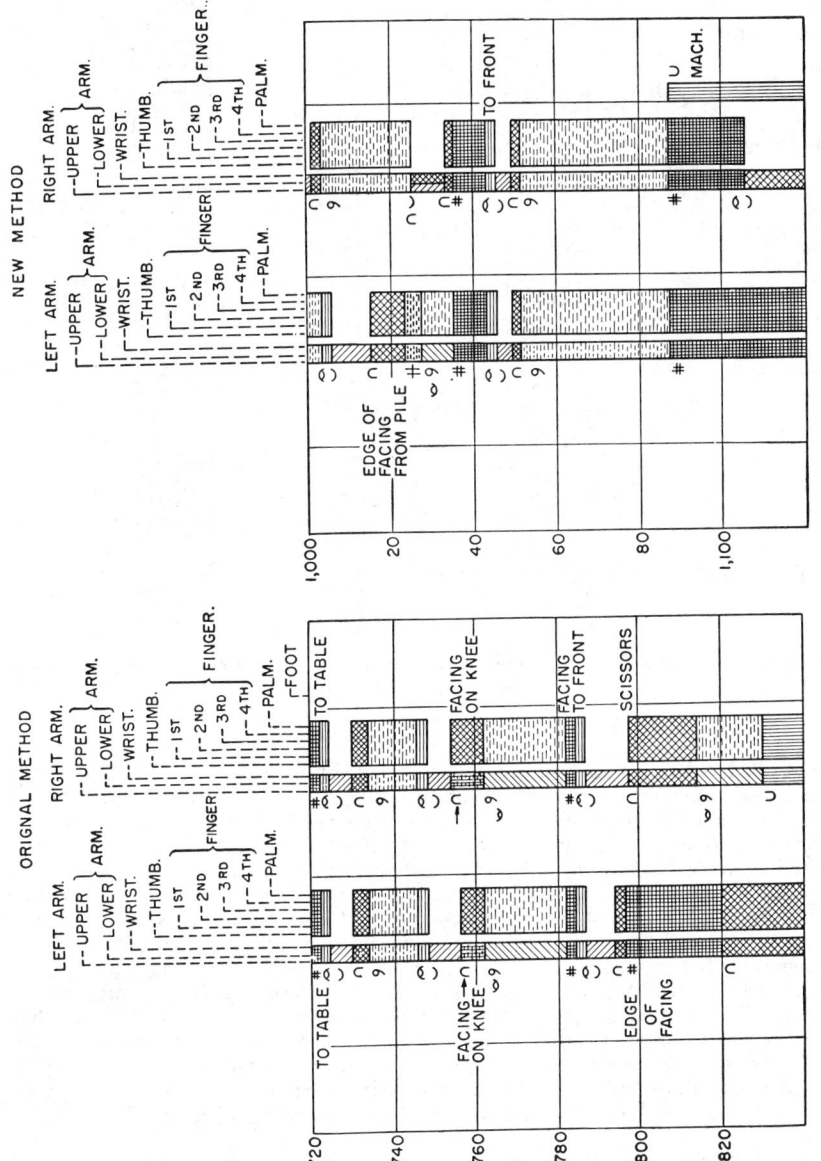

Fig. 5-2. Simo chart of original and new methods of sewing front facings to coat overalls.

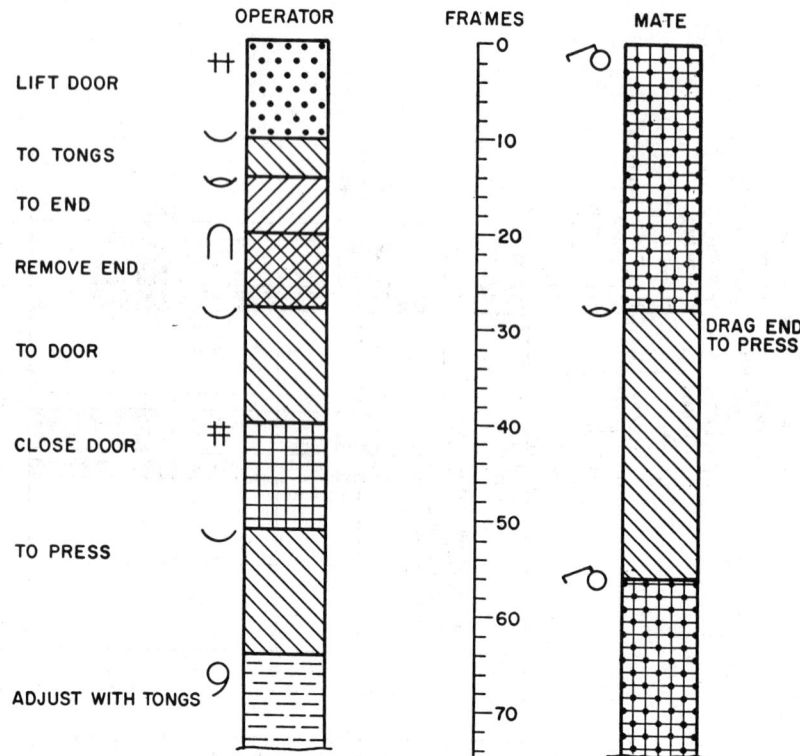

FIG. 5-3.   Fragment of a simo chart from a memomotion film of a furnace and press operated by two men.

and studied either by means of a chronocyclegraph, or if this is not practicable, by a pencil sketch of the movement path made from direct observation. If a three-dimensional study is needed, a model may be made using wire.

The second main type of movement path that must be studied is that of the general movements made by operators as they perform work which is spread over a wide area of floor space and which occupies a considerable period of time. Where such movements follow a regularly repeated cycle of no great complexity, flow diagrams from direct observation may be sufficient; where they follow an irregular sequence or are very complex, string diagrams are made from visual observations. Where greater accuracy or more detailed information is needed, a memomotion film can be analyzed to give a record of the path of movement which can then be expressed either as a flow diagram or as a string diagram. Where a number of operators work together as a team or gang, a multioperator string diagram can be made from the memomotion film. The string diagram can also be used to record finer hand or arm movements or the movement of material within the workplace.

Chronocyclegraphs.   The chronocyclegraph technique is the most accurate and detailed of the group of techniques used to study the path of movement. It is a photographic technique recording the path of movement as a figure composed of pear-shaped spots. This figure is obtained by taking a photograph, preferably stereoscopically, of the path of movement followed by an operator in a single cycle of work. Lights are attached to the operator's hands. A film is exposed for the whole length of the cycle. Because, in making it, the shutter of the camera is open for the whole length of a work cycle, a chronocyclegraph records as one picture on a single film the path of a movement that takes place over a period of time. The investigator is able to see the movement as a whole and also to compare and examine its different parts. The shape and spacing of the spots give him information about acceleration and deceleration which makes it possible for him to pick out hesitations that suggest obstacles in the way of the movement. He can judge the effect of sudden changes in the direction of the movement. By taking a series of chronocyclegraphs of different cycles, superimposed either on a single film or during viewing, he can discover whether rhythm is being built up by the variation, or absence of variation, in the movement path.

Uses of the Chronocyclegraph. The standardizing of the means of making chronocyclegraphs has made them so much easier to take that they can now be used freely to assist in the recording and measuring of movements—the first stage of any motion study investigation. Because of the very specialized nature of the technique, it has not as universal an application as many of the other motion study techniques, but chronocyclegraphs are extremely useful in the following ways:

1. *In recording an unrestricted movement.* A chronocyclegraph is of assistance in the development of a new method, particularly in cases where there is no definite workplace layout of fixtures to shape the movements as, for example, in the folding of a garment.

2. *In analyzing a complex single-purpose movement.* If the path of an extensive movement is of greater significance than its purpose, it cannot be properly recorded in accurate detail except by means of a chronocyclegraph. This technique should be used in addition to micromotion study to obtain a complete record. In the same way, a single operation on a process chart may be studied more fully by means of a chronocyclegraph. The chronocyclegraph of the folding of a towel, Figure 5-4, illustrates these two points. It shows a particularly extensive path of movement because the work cycle is made up of sweeping arm movements. There are no fixtures, and the workplace consists only of a flat table.

3. *In the experimental stage of a new method.* In the experimental stage of developing a new method by means of any of the other motion study techniques, chronocyclegraphs of the whole or any part of the work cycle may be of considerable assistance in obtaining a really satisfactory result. They are particularly useful in the following circumstances:

   a. It is often possible to show by means of a chronocyclegraph not only that existing machine controls are wrongly placed but also how the shape of handles or levers can be modified to allow an easier grasp.

   b. Such hand tools as the screwdriver used in assembly work or the scissors of a sewing machine operator can often be so placed at a point directly in the path of movement of the hands that they can be picked up without any change in the direction of the movement. A chronocyclegraph is useful here to check the position chosen. This chronocyclegraph will not only check the position of the tool in the work area but will also show whether the hand can pick it up without hesitating.

   If a tool is lying or hanging against a flat surface, a marked hesitation will be seen as the hand approaches it. Figure 5-5A shows a chronocyclegraph of a hand picking up a small screwdriver which is lying flat on a bench. There are eight spots at the point where the hand is in position to grasp the screwdriver and before it is ready to carry it away to position it in the slot of the screw. This shows that there is a hesitation in grasping the handle which is perhaps caused by an unconscious fear of striking the surface behind the tool at too high a speed. If the tool is arranged so that its handle is free on all sides, the hand will pick it up easily. Figure 5-5B shows the effect of placing the same screwdriver on a raised support with the handle projecting about an inch above the bench. In this chronocyclegraph, only three spots are seen at the same

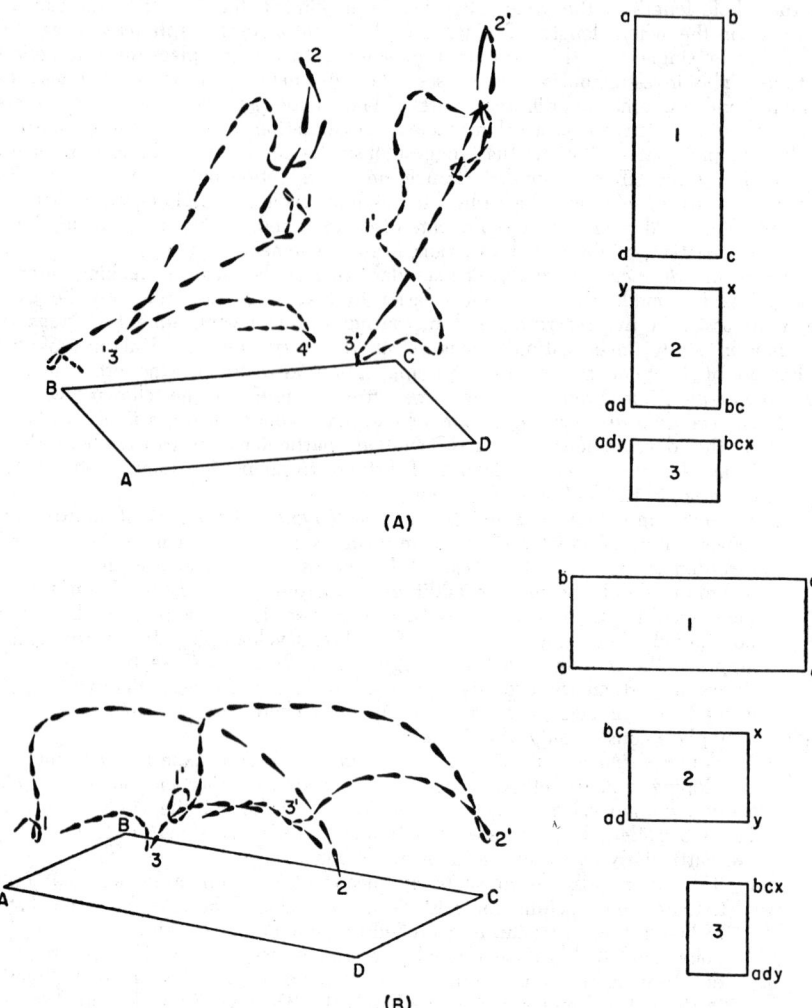

Fɪɢ. 5-4. Chronocyclegraph of folding a towel. Old method (*A*): making the first fold in the air; new method (*B*): turning the towel through a right angle and folding while on the table.

point where there were eight before. This particular tool rest was designed for small screwdrivers in instrument assembly work.

4. *As a check on the training of operators.* Chronocyclegraphs made at each stage in the training of operators will act as tests of progress. They can also be used to explain their movements to trainees to encourage them to aim at better and easier movements. It is possible, in the final stages of training operators, to discover when the movement path has become habitual. This can be achieved by superimposing a number of chronocyclegraphs of different work cycles and finding whether the paths of movement are uniform.

5. *As a teaching device.* Chronocyclegraphs can be used in training methods men to observe movements. They can also help those whose previous

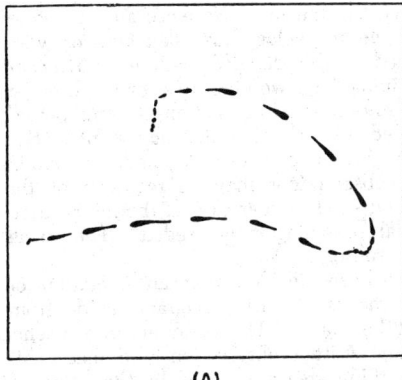
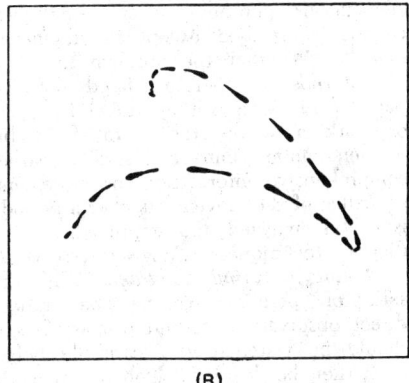

(A)   (B)

Fig. 5-5.   Chronocyclegraph of picking up a small screwdriver from (A) a flat
surface and (B) a raised tool rest.

experience has been limited to time study to develop the habit of seeing the
shape and character of a movement instead of concentrating largely on the
time taken to perform it.

As a means of demonstrating motion study to a wider circle, such as
supervisors, workers' representatives, and management, chronocyclegraphs
made of everyday jobs are helpful.

The chronocyclegraph presents information about the path of movement in the
most accurate form. Only the chronocyclegraph can put on paper the subtle difference between a good and a bad movement. This distinction can often be made
automatically by an experienced methods man, but it is very difficult to explain
to the inexperienced. The chronocyclegraph of a satisfactory movement presents
a picture of the shape of the movement, giving an impression of ease and rhythm.
Until an investigator has learned to see movement in this way, he will not be
capable of achieving the best results.

Sketching the Movement Path from Direct Observation. Where the movement
path to be studied is very simple and does not justify the taking of a chronocyclegraph or is too long or dispersed over too wide an area for a chronocyclegraph to be practicable, it is often useful to make a pencil sketch of the
operator's movements from direct observation. A sketch of this kind gives less
accurate information than a chronocyclegraph, but it is not just a substitute. It
can be made from an angle impossible to the camera. It may give a picture that
could never have been photographed, because materials or limbs blocking a view of
the path can be ignored. Chronocyclegraphs are often limited to a very restricted
angle of view stereoscopically. They are also relatively ineffective where movements
are very small.

Although the experienced observer may dispense with the sketch and rely on
visual observation alone, the less experienced will find that the making of the
sketch will bring out points not fully appreciated until the attempt has been made
to record them.

If a three-dimensional record is needed, a model of the movement path can
be made in wire.

The movement of materials or the sequence of an operator's broader movement
path can be recorded on a layout drawing of the room or working area as a
flow diagram with or without process chart symbols.

String Diagrams. The techniques for studying the path of movement so far
described are useful only where the cycles of movement are repeated time and
again in the same sequence and form. Many types of work—for example in textile
mills—show movement cycles which, while perhaps made up of constantly repeated

elements or operations, never present these twice in the same sequence and therefore cannot be studied except in fragments as chronocyclegraphs, sketches, or wire models. Whether the problem to be studied is the detailed path of movement of materials, the operator's hands within the immediate working area, or the broader path of his feet as they travel from work station to work station, a long period of work must be studied to get a balanced picture of such movement paths. A memomotion film as described earlier in this chapter can be made to obtain comprehensive information, or more simply, observations may be recorded of the sequence of the movements over a period of hours. In either case, if the information is to be analyzed and improved, some visual presentation is needed. The string diagram technique was devised to serve this purpose.

Making a String Diagram. When it has been decided that an operation or series of operations can best be studied by means of string diagrams made from direct observation, certain preparations must be made. The operator's movements should be watched very carefully before any written observations are made. It must then be decided which places in the working area are visited by the operator. Each of these points will be given a code number. This may be one of a simple series, or where the points are more numerous or complex, it may be built up from three or more digits, each representing a different class of object or place on the machine or in the workshop. At the same time, the various operations may be referred to by means of letters.

When a satisfactory code has been developed to provide for all foreseeable contingencies, it is used to record each movement of the operator. When the observations are complete, pins or nails are driven into all the numbered points of a scale diagram of the layout. Extra pins are located at the corners of the machines and anywhere else there may be an obstruction in the path of movement. String or cotton, previously measured or marked off into yards or feet, is wound around the pins following the operator's path of movement as recorded in the observations. When the diagram is complete, the remaining string length is subtracted from the original length to discover how much has been used.

The information of the observation sheet has other, secondary uses. For instance, it is often helpful to know how many times a day each machine is stopped, or what the proportion of the different operations is. The observations made of a new method will frequently show surprising contrasts with old method observations. These may be invaluable in planning further improvements.

Uses of the String Diagram. The completed diagram is used to examine the movements as a whole and to find where the greatest concentration of movement lies. The technique is used largely as a means of locating weak spots which need fuller analysis. The string follows the path of movement of either operator or material. The resulting pattern shows the investigator which parts of the job are causing the largest number of movements. These will be the parts which should receive further detailed investigation. At the same time, the string concentration among the various sections of the layout will show the closeness of the relationship among them and suggest changes in layout or routine to shorten the path of movement.

Typical instances of movement paths suitable for analysis by means of the string diagram technique are found where the work is bulky or heavy. In the assembling of the bases of large instruments, because of the size of the material and the difficulty of handling it, the heavy bases were laid out along a bench. The operator walked from one to another, adding one smaller component to each and returning to add the next until all the assemblies were completed at the same time. A string diagram resulted in this excessive walking being eliminated.

Another example is to be found where material is not particularly large or heavy but where there is something in the nature of the process, such as drying or baking, which causes a pause in the assembly process and makes it necessary to work on several pieces at the same time.

A second type of work particularly suitable for investigation by the string diagram technique is found in some forms of machine tending. Where one operator is tending several machines and performing a variety of different operations in an irregular sequence controlled only by the demands of the machines, a string diagram can be made to record his movements. This will reveal any irregularities or complexities in the movement path and will suggest where some improvement in the organization of the work can be made to reduce the length of the path and give the operator more opportunity to meet the demands of the machine.

String diagrams are of particular value in making department or plant layouts where the movement of both material and operators is of great significance. If an existing layout is to be modified, separate string diagrams should be made of all movements of operators and materials before anything is changed. The diagrams should then be compared and the changes that each suggests coordinated into a single, theoretical, modified layout drawn to the same scale. The observations used in making the original string diagrams can then be plotted on copies of the modified drawing and these in turn can be compared. It may be necessary to repeat the process several times until a layout is found to suit all circumstances. In the same way, two different suggested arrangements can be compared and assessed. In laying out a new department or plant, string diagrams can be useful, but the path of movement traced can only be theoretical. The diagrams must therefore be made from existing knowledge of the processes rather than from direct observation.

DEVELOPING AN IMPROVED METHOD FROM THE ANALYSIS OF A MOTION STUDY RECORD

Once a work method has been recorded on paper by means of any of the motion study techniques, it is ready to be analyzed. In most investigations, more than one technique will be used and more than one record made. The order in which the records are made is important. It is usual to begin with a broad type of record on the scale of a flow process chart, some form of memomotion record, or a large-scale string diagram to cover the foreground and background of the work which will be the main object of the study. This record should be analyzed before any more detailed record is attempted. Otherwise, time may be wasted recording and analyzing separately whole operations which might have been eliminated altogether or combined with another operation elsewhere in the sequence.

Until the first broad analysis has been made, it is unwise to decide what further records are required or in what order the different sections of the job should be studied.

The object of analyzing a motion study chart or diagram of any kind is to obtain ideas for improvement. The analysis of the first broad record will usually suggest indefinite ideas, the effect and validity of which have to be checked by making and analyzing further charts and diagrams. The subsequent records may be on the same scale but using a different technique to examine another aspect of the problem; for example, a process chart supplemented by a string diagram. But more frequently, a broad examination will be followed by a finer analysis of some sections. For example, an operator process chart might be followed by a two-handed therblig chart of one of the operations on the process chart, or a flow process chart might be followed by an operator or operation process chart of its central section.

General Principles of Analysis. Whatever the record and whatever its scale, it is analyzed basically in the same way.

The analysis is made by subjecting first the whole chart, sketch, or diagram, then each section of it, and finally each symbol or item to a critical examination under the following headings:

1. *Necessity.* Many movements and operations occur which are unnecessary or unproductive.
2. *Sequence.* A change in the order in which operations are performed may lead to a saving in movements.
3. *Combination.* An operation or a sequence of operations may be combined with others to give a reduction in the number of movements.

4. *Simplification.* In many instances, work may be made easier by a change of layout, the provision of special tools or fixtures, or a modification of the design of the product.

It is essential, in adopting these four headings as a basis of analysis, to use them in the right order. The idea of eliminating whole sections or items should obviously be explored before considering any combination or rearrangement in sequence. Most important of all, simplification—the provision or modification of tools, jigs, or other special equipment or the redesigning of layouts—must wait until all other possibilities have been explored. Any other order of procedure will result in a waste of time and effort on developing equipment which may later prove to be either totally unnecessary or in need of further expensive modification.

In every case, at the end of the analysis, a possibility chart or diagram should be drawn up embodying the suggestions that have been made. This should be compared with the original and in its turn analyzed once more in the same way, to make sure that nothing has been overlooked.

This four-point plan of analysis can be adapted to any motion study chart or record with only slight differences in application. Because the analysis of any record is closely related to the making of that record, the following examples, chosen as representing the three main groups of techniques, describe to some extent the making of each record as well as its analysis.

The Analysis of a Process Chart. In the analysis of a process chart, the routine of analysis described above is applied without modification. Each part and symbol is questioned under the four headings of necessity, sequence, combination, and simplification. The suggestions which arise out of this examination are incorporated into a possibility process chart drawn to the same scale. Symbols from the original chart with their original numbers are transferred to the possibility process chart to allow a comparison to be made.

Analysis of Detailed Process Chart—Inspection of Bomb Exploders. Figure 5-6 is an example of a detailed chart showing the work of two hands. The investigation was concerned only with the inspection of the finished article. Therefore, it was not necessary in the first instance to chart anything other than the inspection process. As this largely consisted of picking up and putting down exploders and gages, it was necessary to study in considerable detail the work performed by each hand.

A preliminary survey showed that the work cycle consisted of four inspection operations. These were performed in batches for each separate gaging. It was important that the chart should show this clearly in addition to showing the details of the work of each hand.

The chart was made in two columns, one for each hand. As it was intended to emphasize that the work was done in batches, the temporary storages between batches were brought in to a center line, as was the picking up and putting down of gages between batches. The double lines of symbols were therefore left to describe the work of the two hands in the sections which were repetitive.

The chart was analyzed under the usual headings:

Necessity. The need to inspect for the four points, "outer wrapper," "outer diameter," "bore," and "overall length," was accepted. The shape of the chart indicated clearly that, by inspecting the exploders in batches, each exploder was picked up and put down four times before it was completely inspected. This suggested that three of these handlings might be eliminated if all the inspections on each exploder could be completed at one time so that it was picked up and put down only once.

Sequence and Combination. As the method existed, it was not possible to inspect all four points at once without picking up and putting down the gage required for each inspection. This would have been no improvement, and a means was therefore sought of mounting the gages so that no handling of them was necessary. In considering this, the sequence of the inspections and the movement path of the hands holding the exploders were studied so that they could be performed as easily as possible. It was arranged that, wherever it was practicable, a movement

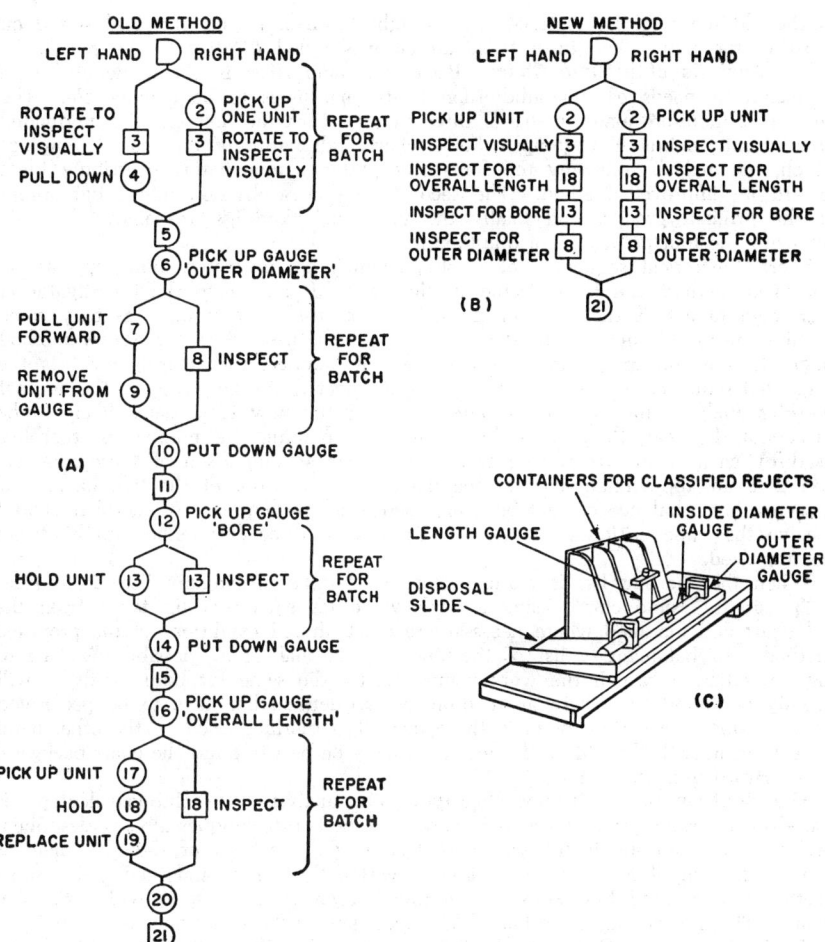

FIG. 5-6. Process charts of the old and new methods for the inspection of bomb exploders, together with a sketch of the workplace layout developed for the new method.

out of one gage should lead directly into the next. In addition, containers were provided to take the different types of rejects which had previously not been segregated.

Simplification. A new method was finally developed. Because the mounting of the gages freed one hand entirely, duplicate gages were provided so that both hands could introduce exploders simultaneously. Two exploders were picked up, one in each hand, visually inspected, and placed in the overall length gages which were fixed so that they gaged the exploders in a vertical position—the easiest method after the visual inspection. From these gages, they moved to the bore gages which were arranged in line with the ring gages for the outside diameter. The movement of removing them from the bore gages passed them through the ring gages, and if they were correct, they rolled down slides directly to the packer. If there was a reject, the inspector was instructed to place it in the appropriate container and to complete the inspection of the other exploder with one hand.

In addition to this, where the inspectors collected and removed their own material in the old method, the material was brought to them by laborers and stored on racks in the new. The inspected material was rolled directly to the packer.

The Analysis of a Simo Chart. Because a simo chart is a very detailed type of record, it needs careful subdivision if its analysis is to be manageable. The analysis is therefore made in two or three stages. First the simo chart is subdivided broadly into operational groups. Some charts fall naturally into such groupings which, on analysis following the four-point pattern, immediately contribute ideas toward an improved method. The second stage is a division into subgroupings of the original groups, and finally the individual therbligs are analyzed—again following the four-point plan of questioning.

Short, theoretical sequences of therbligs which are ultimately built up into an improved method can be collected in this way. It may help the investigator at this stage to act as the operator himself and to try out any ideas as they come to him during his analysis of the simo chart. As these ideas for a new method begin to fall into place and are tried out in practice, the investigator lists the suggested sequence of therbligs on an analysis sheet in the same way as the original therblig analysis, but without a time scale. If the new idea seems likely to be successful, he may then check his results by counting the number of therbligs used by each hand. It is frequently found that a comparison of these numbers will give an approximate idea of the balance of the work of the two hands. If it seems that a balance has not been obtained, the new sequence of therbligs should receive the same criticism as the old sequence to discover whether anything can be improved.

When the therblig list is reasonably complete, the investigator uses it to form a "possibility" simo chart, using a wink value for each therblig taken from the old chart and modified where necessary to meet altered conditions in the proposed method. In balancing a list of therbligs and in making a "possibility" chart of any operation in which the work is not exactly the same for both hands, it will usually be found that there are certain key sequences that have to be performed by one hand at a certain point in the cycle. The therbligs used by the other hand have to fit in with this. It will therefore usually be best to build the chart backward and forward from this point.

The Analysis of a Chronocyclegraph. The analysis of a chronocyclegraph is based on the same principles as the analysis of the two techniques already described, and the same routine is followed. First, the whole pattern of the movement is examined; then, if necessary, each section within the whole; and finally, if a satisfactory new method has not yet been found, each group of spots within the sections. The questioning again takes place according to the same four-point plan.

Application of a Common Analysis Routine to All Techniques. The same routine as described above is followed in analyzing any other type of chart or diagram, such as, for example, a two-handed therblig chart, a flow process chart, a string diagram, or a flow diagram. The same four points of analysis are used in the same order, and ideas or fragments of charts for a new method are written down at every stage. Finally, possibility charts or diagrams are made. It cannot be emphasized too much that possibility charts are essential. Without them, there is a very real danger of "inventing downward" as Frank Gilbreth would have said. An honest possibility chart often prevents mistakes, and great care should be taken to check its accuracy. It is useful only if it covers exactly the same ground as the original chart and is drawn to the same scale.

CRITERIA FOR DETERMINING AN EASY MOVEMENT

In using motion study to develop a new method of work, the improvement is made directly from a careful analysis of present practice. No problem is approached with a ready-made solution in mind, because no two problems are exactly alike. Each has its own peculiar background and circumstances. Nevertheless, there are certain similarities and common characteristics to be found in all good methods. These can be used to assist in the development of new motion study methods after an analysis of present practice has been made, or where, in a new job, there is no present practice to analyze.

Principles of Motion Economy. Motion study investigators, from the Gilbreths onward, have beeen conscious of this common background. They have listed various maxims which they have called "principles of motion economy." The Gilbreths listed twenty, and Professor Barnes,[4] at a later date, twenty-two.

Both Gilbreth and Barnes included in their lists points of varying significance. A briefer list called "characteristics of easy movement" has been developed by the author and found to be quite useful.[5] In revised form, it is as follows:

1. Movements should be simultaneous.
2. Movements should be symmetrical.
3. Movements should be natural.
4. The path of movement should be rhythmical.
5. The path of movement should be habitual.

These five characteristics are general, and they are intended to be interpreted broadly. The first three characteristics apply to detailed movements; the remaining two to the whole movement cycle. All five are interdependent and must not be isolated from one another. They should be interpreted on a broad basis. It will be found that the longer lists of Barnes and the Gilbreths do not contradict this list, but are either elaborations of it or recommendations for the necessary features in the arrangement of a workplace or the design of tools which the characteristics demand. The advantage of the shorter list is that the characteristics are few enough to be considered at once and sufficiently general to be applicable in all circumstances.

Simultaneous Movements. The movements of a motion study method should be so arranged that both hands and both arms work together. They should, if possible, be performing the same operation at the same time. One hand should not be idle while the other is working. One hand should never be used merely to hold something on which the other hand is working. At the same time, there is no advantage in working with both hands if they are allowed to move one at a time. Not only should both hands do useful work throughout the cycle, they should also begin and finish their movement sequences at the same instant. This is easiest to achieve when both hands are doing identical work. It may be possible to arrange for an operator to work on two units at once if the single unit does not allow both hands to do the same work at the same time. Where the single unit does not lend itself to identical movements of both hands and where two cannot be worked on at once, it may be necessary for the two hands to do different work. This is less easy because the movements must be balanced so that the left hand completes its movements at the same time as the right hand. If one hand is allowed to finish earlier than the other, it may tend to start on the next operation too soon. The whole cycle will then become progressively more unbalanced.

Symmetrical Movements. After planning the sequence of movements in a motion study method so that they occur simultaneously, they should be arranged, as far as possible, so that they can be performed symmetrically about an imaginary line through the center of the body. Because of the symmetrical structure of the body, arm movements can most easily be made when they are symmetrical, each arm moving in and out from the center. In most types of assembly work, the arms should move away from the body to pick up material and inward to the center to assemble it.

When movements are performed symmetrically and simultaneously, they achieve not only a time balance but also a balanced equilibrium of the whole body, which makes them easier to perform.

Natural Movements. A natural movement is easy and makes the best use of

[4] Ralph M. Barnes, *Motion and Time Study*, 6th ed., John Wiley & Sons, Inc., New York, 1968.
[5] A. G. Shaw, *An Introduction to the Theory and Application of Motion Study*, H. M. Stationery Office, London, 1945.

the shape and arrangement of all parts of the body. Just as symmetrical movements are easy because the body is symmetrical, so there are other types of easy movement depending on other features of the structure of the body. In developing a new series of movements, these features must be kept in mind.

Natural movements are curved, not straight. This is easily understood if the structure of the body is considered. The hand, for example, moves in an arc centered on the elbow or the shoulder; the foot swings from the knee or hip; the shoulder describes an arc as the body turns. Correctly designed workplaces, footrests, and foot pedals allow for these curved movements.

In heavy work, great care should be taken to use the mechanical advantages that the position and posture of the body can give. Obviously, as much heavy work as possible should be eliminated by the introduction of suitable mechanical devices or lifting tackle, but there will always be some work that must be done manually. It should be remembered, for example, that a weight lifted at a distance from the body, which is the center of gravity, not only feels heavier but actually puts more strain on the muscles.

In designing detailed movements, advantage should be taken of the natural shape and position of the hand and fingers. During assembly work or other fine work, individual fingers can hold small tools against the palm in a natural position for use at a later point in the work cycle. Again, the fingers move more strongly and accurately in toward the palm than in any other direction.

In most motion study manuals, arm movements are classified into five groups:

1. Movements of the fingers
2. Movements of the fingers and wrist
3. Movements of fingers, wrist, and forearm
4. Movements of fingers, wrist, forearm, and upper arm
5. Movements of fingers, wrist, forearm, upper arm, and shoulder

It is often stated that hand movements should be confined to the lowest category that is consistent with doing the job properly. This must not be taken to mean that, wherever possible, only finger movements should be used. Movements repeated at short intervals in cycles lasting from a few seconds to a minute or two should not involve the use of the whole stretch of the arm, but in arranging smaller movements, care must be taken to see that the operator does not feel cramped.

If any doubt arises as to the merits of different sequences of movement, chrono-cyclegraphs may be made of each suggested series and the results compared to see which looks easiest. Normally, the easiest movement will be the most natural, but there is one danger in assuming that a movement that is easy and comfortable for one operator is the best and most natural for others. Long practice may cause an operator to perform with ease and comfort movements which are neither natural nor easy to others. On the other hand, a very small amount of practice will often bring ease in the performance of movements which at the first attempt seem difficult and unnatural. Decisions should therefore be made only after a reasonable practice period and after taking all these factors into account.

Rhythmical Movements. Simultaneous, symmetrical, and natural movements are desirable characteristics of detailed sequences of movements that are part of a full work cycle, but no good motion study method is a mere addition of details. It is a whole pattern in itself. One of the main characteristics of a good movement pattern is the rhythm it develops when it is repeated. Neither is a cycle complete in itself. The last movements of one cycle should run easily into the first of the next just as the movement sequences within the cycle are linked together.

Any unnecessary change in the direction of movement tends to hinder a smooth rhythm. Changes of direction should be kept to a minimum in any case because slowing up, stopping, and changing direction waste time and energy.

Habitual Movements. When motion study techniques have been used to develop a new method, the movement cycle will probably contain simultaneous, symmetrical, and natural movements, and the complete cycle will be rhythmical. Because a method of this sort will be intended for frequent repetition, it must be planned so that each movement can be made in exactly the same way each time. In developing a rhythm, the operator will also develop movement habits. By making the movements habitual, they will become automatic and require no conscious direction. Much mental fatigue and strain will be eliminated in this way.

The formation of movement habits is a process which will always occur unless there is something in the layout that definitely prevents it. In a job which has not been studied, the movements may only be partly habitual because there will be certain sections in which material and tools come to hand in a different way in each cycle. These irregularities will be eliminated in a motion study method. The workplace will be so arranged that every opportunity is given to the operator to form habits. In designing a workplace for simultaneous, symmetrical, and natural movements, tools and materials will inevitably be given fixed locations as far as possible. The question of standardizing material will have been considered. Therefore it will be much easier for movements to become habitual in a motion study method than in a method which has not been studied.

LAYING OUT THE WORKPLACE

In developing a motion study method, the movements should be worked out before the design of the tools and workplace is considered. The workplace should then be built around the method of work and the operator who is to use that method. An improved method will demand an improved workplace, but the performance of the method should come first and the design of the workplace second. The layout of the workplace is part of the third stage of a motion study investigation. In the first stage the existing method is recorded, and in the second it is analyzed and the new method developed theoretically. The third stage covers the experimental work of turning theory into practice, and here the workplace must be designed before the first operator can be trained.

Position of the Operator. Before studying the details of a workplace layout, the investigator will check that the operator's position in it is as comfortable as possible. It will usually be found that he will become less fatigued if he is able to stand or sit at will. This can be arranged by making the working surface the correct height when he is standing and increasing the height of the chair to bring him into the normal sitting position at the higher working surface. A footrest is necessary if the chair is raised. As a general rule, the working surface, which is not necessarily the tabletop but possibly a fixture attached to it, will be at the correct height if the operator's elbows are level with it when his arms hang from the shoulders with the elbows bent at a right angle.

It is particularly important for comfort that the operator should be able to sit with his knees under the working surface. Anything in the design of the workbench that prevents this, such as a very deep edge or cupboards placed beneath it, should be cleared away.

In addition to the chair and its footrest, which should support the foot fully, the operator needs some place for personal possessions. Either a drawer in the workbench or a small locker built into the chair may be the solution. If nothing is provided in the working area, some part of the workplace which is intended for another purpose will inevitably be used to hold personal possessions.

When the seating and working heights have been adjusted, the lighting of the workplace should be considered. A good, diffused light is safer and better for most types of work than more concentrated, individual lighting, though there are always certain types of work that require special conditions. It is enough to state here that the best possible overall lighting should be installed so that there are as few dark corners as possible.

Extent of the Working Area. Having decided how the operator is to sit and having arranged for his comfort, it is necessary to decide upon the extent of the workplace; what area is available for the placing of tools, materials, fixtures, and machines; and what areas of the working surface are most easily reached by the hands. These areas are called "areas of easiest reach." They are found by describing arcs with each hand pivoting on the elbows at the level of the working surface.

Wider arcs are then described by the arms pivoting on the shoulders. In this way, areas are mapped out (Figure 5-7A) in which the hands can most easily reach tools and materials. The smaller arcs enclose areas in which the hands can work without stretching the arms. The larger arcs enclose the areas which can be reached by movements of the whole arm. Figure 5-7B shows the area inside which it is easiest to pick up material using both hands simultaneously, and 5-7C, the limited area where it is possible for the eyes to follow both hands working symmetrically.

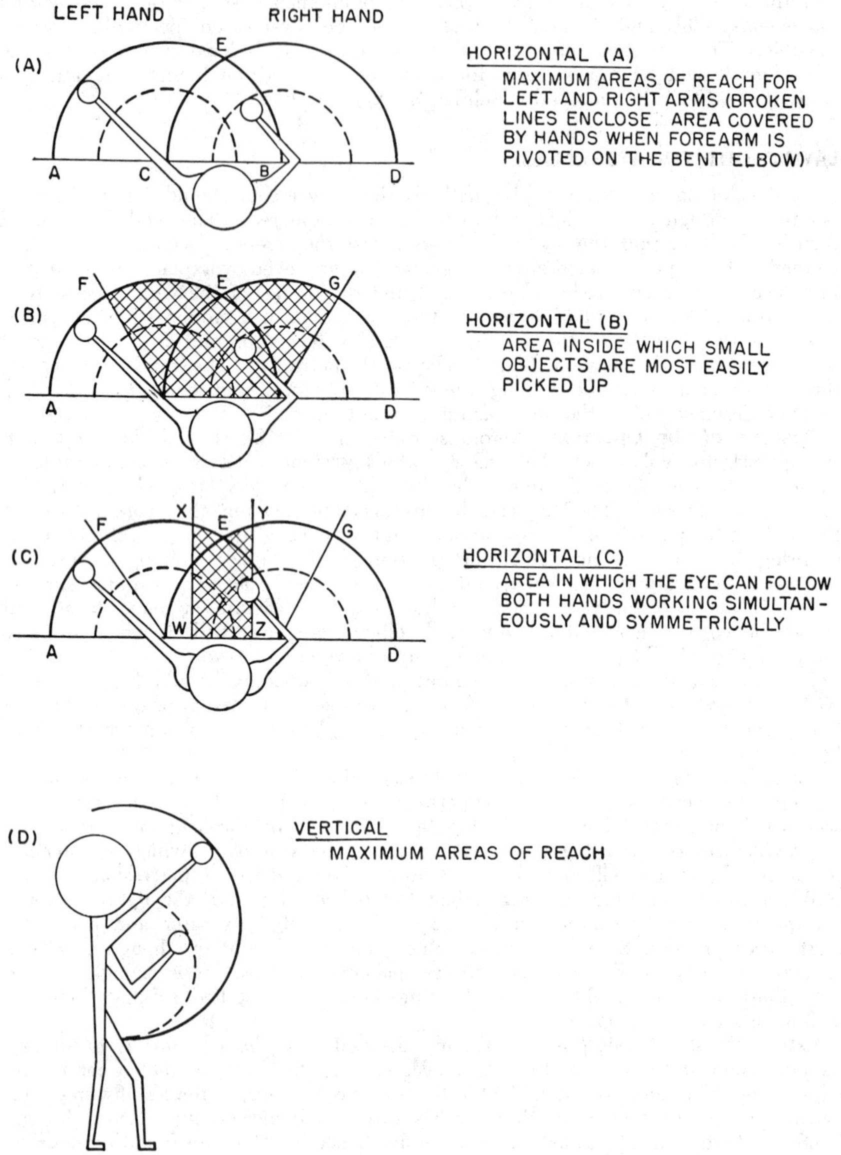

LEFT HAND   RIGHT HAND

(A)

HORIZONTAL (A)
MAXIMUM AREAS OF REACH FOR LEFT AND RIGHT ARMS (BROKEN LINES ENCLOSE AREA COVERED BY HANDS WHEN FOREARM IS PIVOTED ON THE BENT ELBOW)

(B)

HORIZONTAL (B)
AREA INSIDE WHICH SMALL OBJECTS ARE MOST EASILY PICKED UP

(C)

HORIZONTAL (C)
AREA IN WHICH THE EYE CAN FOLLOW BOTH HANDS WORKING SIMULTAN-EOUSLY AND SYMMETRICALLY

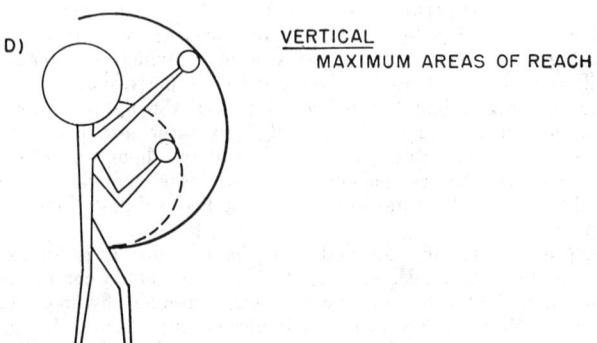

(D)

VERTICAL
MAXIMUM AREAS OF REACH

Fig. 5-7.   Areas of easiest reach.

So far, the working area has been considered as if it existed entirely on a horizontal plane. Actually it is three-dimensional and extends into the vertical plane in the same way (Figure 5-7D). The areas of easiest reach are not left behind by an operator when he gets up from his bench to perform a further operation in another place. They go with him and exist wherever work is to be done. For example, in a textile mill where an operator is responsible for three or four separate operations at various points on one or two machines or along a spinning frame, consideration must be given to keeping the work within the areas of easiest reach at all these points as well as to keeping within his areas of easiest reach the work he performs when standing in one place.

Contents of the Workplace. When the position of the operator in the working area and the extent of that area have been settled, attention turns toward the contents of the workplace—its equipment and furnishing. Before this equipment can be arranged to the best advantage, its design must be checked. In changing from an old method to a motion study method, some redesigning is inevitable. The new method will call for some alterations, and well-designed equipment will help to promote in a method of work the characteristics of easy movement previously outlined and to perpetuate the correct method.

The main types of equipment found in a workplace of any kind are material containers, tools, machines, and fixtures. Each type of equipment presents different technical problems of design, but the aims of the investigator are the same whether he is designing a tool, a box to hold material, or a fixture to hold the work and free the hands.

Arrangement of the Workplace. It is only possible to give general examples of the arrangement of the workplace. The details must be worked out to suit the requirements of each individual job and operator. However, certain types of workplaces have common features and common problems.

In workplaces designed for assembly or packing jobs, the contents of the workplace are primarily the materials that are to be assembled or packed and the finished work. There may also be small tools such as screwdrivers, hammers, scissors, pencils, paste brushes, or label-moistening devices, and there will be containers to hold the material. The material, in its containers, must be arranged somewhere within the areas of easiest reach so that it can be picked up easily by whichever hand the method requires and in the correct order. If both hands are to be used, a supply of material must be provided for each. If it is necessary to use the eyes to select material, the containers must be placed within the small area (see Figure 5-7C) where it is possible for the eyes to watch the hands without causing the operator to turn the head. In a case of this type, both hands can collect material from the same workbin, if its entrance is wide enough. Where all the material must be selected by sight and is too bulky or varied for it all to be placed within the normal area for visual selection, this area can be extended by sacrificing the symmetry of the movements. The material can then be arranged so that both hands move together around the workplace, selecting material from the same or adjacent containers. Before coming to the decision that symmetry must be sacrificed in this way, it should, however, be remembered that in many cases an operator may learn to select his material entirely by touch if he is allowed time to practice.

The primary consideration in deciding the position of hand tools must be that they should be picked up with the least disturbance to a smooth and rhythmical path of movement. As far as possible, an operator should be able to pick up a tool as his hand moves from one part of the work to the next, without making a special journey for it.

It is not enough to arrange tools so that they are easily picked up. It is equally important that they should be easy to replace in the same fixed position. If some form of automatic return is impracticable, the location of the next piece of material to be picked up should be arranged, if possible, to allow the operator to put the tool away as his hand goes toward it.

When an operation has been completed, the finished material must be removed. If it is robust enough to be dropped, it can be disposed of by releasing it through a hole in the bench or down a chute, to fall into a container beneath. The chute or hole should be placed so that the finished article can be made to fall into it as soon as the work is completed. If, for any reason, this immediate disposal cannot be arranged, the hole or chute should be placed so that the hand can drop the article on its way to pick up the first piece of material required in the next cycle of work, without using extra movements or changing its direction. Such arrangements for disposing of finished material are usually known as drop deliveries.

Where an article is easily damaged and the use of drop deliveries is not possible, or where a finished article is placed directly into its final packing, the hand movements involved in placing it in its container should be carefully considered so that the minimum of unproductive movement is used. For example, material placed on flat trays should be arranged so that the tray is filled from the back forward, each piece reaching its place with a single free movement without being lifted over the others.

In considering the arrangement of a workplace at a machine, many minor modifications to external machine features may be found necessary. Switches may be brought into the areas of easiest reach. Levers and foot pedals may be made easier to use or extended so that they can be operated by a worker standing or sitting in several different positions. Racks for material or finished articles can be redesigned and heavy controls or clumsy release mechanisms improved. Within the limits of mechanical efficiency, the comfort and ease of movement of the operator should be the first consideration.

In planning the layout of a group of workplaces for a sequence of operations, the arrangement of individual workplaces should be planned before they are considered as a group. Whether the work is line assembly, where parts are stacked between the different operations, or conveyor assembly, where parts are mechanically handled between operators, the disposal of material from one operation should be so organized that it reaches the next operator in the most convenient position for the next operation. Instead of using a conveyor, work can often be passed between operators by means of chutes or turntables. In this type of group work, it is often better to arrange the workplaces on both sides of a bench. This allows more flexibility than a straight-line arrangement, as it makes it possible for two operators to take material from one operator where necessary. Such an arrangement often makes it easier to balance the line. It may allow extra operators to be put on to deal with sudden demands for increased output or to supplement the work of a slow operator or of a temporary worker who is replacing an absentee.

Individual workplaces on conveyors present some of the most complicated problems of workplace layout. They are very difficult to arrange satisfactorily. Because of the structure of most conveyors, it is not easy to arrange a comfortable sitting position for the operator. There is usually insufficient knee room for the operators to sit facing the belt. If they sit at right angles to it, they are obliged to turn to pick up material. Where the work is very simple, such as the packing of small articles into boxes, and the material comes down the conveyor quickly, there is often a tendency to bend over the belt and therefore to sit all day in an unhealthy position.

The design of a conveyor and its use should be thoroughly investigated when an existing method of work is studied, and the obvious difficulties of conveyor workplaces should be considered when recommendations for an improved method are made. If, after due consideration, it is still felt that a belt conveyor is the most satisfactory form of transport between operations, careful arrangements must be made for the comfort of the operator. The disposal of the finished work must also be considered.

The height of the conveyor is very important, and where material is taken directly from the conveyor belt, it should be at the correct working height. Many mistakes are made here and many examples could be quoted. During World War II, a conveyor was designed to transport bomb fuses weighing 4½ pounds each. It was placed at such a height that each fuse had to be lifted down from the belt a distance of 1½ feet and put back again when the operation was finished. Merely by dropping the conveyor to bench level, an increase in output of 25 percent and a considerable reduction in fatigue were obtained.

The Wider Working Area. From the study of interrelated workplaces for a sequence of operations, it is a short step to the study of the layout of whole sections and departments and eventually of the whole plant. The subject of plant layout has been given a complete chapter in Section 11 of this Handbook, and there is no need to do more here than point out the influence exercised by plant layout on workplace layout and therefore on methods.

MAKING THE FULLEST USE OF MOTION STUDY

There are two main types of motion study application. The first is the full-scale investigation employing all available apparatus and techniques in the analysis of complex problems. The second is the much less thorough investigation of smaller and simpler problems, using only the simpler versions of each technique to make a much more superficial improvement.

It cannot be too strongly emphasized that any program planned to make the fullest use of motion study must consist of a combination of full-scale investigations made by experienced investigators working full time and using the full range of techniques, and the simpler improvements made by other members of the staff in the course of their daily work, using simplified techniques. The specialist should receive help from the supervisor and operator; they in their turn should receive help from the specialist.

Such a coordinated program implies the careful training of everyone in the plant to be what Frank Gilbreth called "motion-minded." Short courses of training in the use of the simpler techniques should be continuously available and should be directed at making the general principles of motion study understood by all. But the more difficult techniques should not be made a matter of mystery. Those who want more detail should be encouraged to learn as much as they like, though the emphasis should always be upon each employee using motion study in the way that best helps his regular work. Everyone has a part to play as a member of a team, but all have different parts.

BIBLIOGRAPHY

Barnes, Ralph M., *Motion and Time Study*, 6th ed., John Wiley & Sons, Inc., New York, 1968.

Gilbreth, Frank B., and Lillian M. Gilbreth, "Process Charts: First Steps in Finding the One Best Way to Do Work," paper presented to the Annual Meeting of the American Society of Mechanical Engineers, New York, Dec. 5–9, 1921.

Mundel, Marvin E., *Motion and Time Study: Principles and Practice*, 4th ed., Prentice-Hall, Inc., Englewood Cliffs, N.J., 1970.

Shaw, Anne G., *The Purpose and Practice of Motion Study*, 2d ed., Columbine Press, Ltd., Manchester and London, 1960.

Spriegel, William R., and Clark E. Meyers (eds.), *The Writings of the Gilbreths*, Richard D. Irwin, Inc., Homewood, Ill., 1953.

Chapter **6**

Manufacturing Research

BRUNO A. MOSKI

Assistant to General Manager, Yale Materials Handling Division,
Eaton Yale & Towne Inc., Philadelphia, Pennsylvania

Manufacturing research is one of the distinguishing characteristics of industrial engineering. This is illustrated by the testimony of Frederick W. Taylor in 1912 before the House of Representatives Special Committee to Investigate the Taylor and Other Systems of Shop Management. He said:

> The way to shovel refractory stuff is to press the forearm hard against the upper part of the right leg just below the thigh, like this, take the end of the shovel in your right hand, and when you push the shovel into the pile, instead of using the muscular effort of your arms, which is tiresome, throw the weight of your body on the shovel like this; that pushes your shovel in the pile with hardly any exertion, and without tiring the arms in the least.

He also said:

> Instead of allowing each shoveler to select and own his own shovel, it became necessary to provide some eight to ten different kinds of shovels, each one appropriate to handling a given type of material; not only so as to enable the men to handle an average load of twenty-one pounds, the ideal weight for least fatigue and greatest productivity, but also to adapt the shovel to several other requirements which become perfectly evident, when this work is studied as a science.

The three salient points of this testimony—(1) press the forearm hard against the upper part of the right leg just below the thigh, (2) eight to ten different kinds of shovels, and (3) average load of 21 pounds—were certainly the result of manufacturing research on the part of the father of scientific management.

OBJECTIVE OF MANUFACTURING RESEARCH

The clear-cut objective of manufacturing research is to introduce fresh concepts and to improve manufacturing and industrial engineering techniques within a given plant. In any specific situation, the technological level of manufacturing processes and the approaches to the establishment of detailed methods, the measurement of work, wage and salary administration, and other management controls are at a certain point in the scale of progress, ranging from primitive to sophisticated. Any step which is taken to advance the level of processes, methods, measurement, and controls which results in added value per dollar of cost constitutes practical manufacturing research.

ORGANIZATION FOR MANUFACTURING RESEARCH

The "shovel" testimony of Frederick W. Taylor provides a guide to basic organization for manufacturing research. Each industrial engineer, regardless of the specific area of industrial engineering to which he may currently be assigned, has a personal obligation, from the standpoint of his professional advancement, to devote some portion of his time and effort to manufacturing research. The direction of this research may be to expand the body of knowledge within his day-to-day work, or it may be pointed to other areas where apparent deficiencies exist.

Industrial engineering supervisors have a responsibility to stimulate and encourage all their subordinates to engage in manufacturing research as a part of their daily activities. When certain men demonstrate outstanding initiative and results beyond the targeted objectives of their current assignments, every effort should be made by supervision to formalize definite research projects to be conducted by them on a full-time basis.

In large companies operating on a multidivision basis, the divisional chief industrial engineers should meet regularly for research seminars. At these seminars, progress made at various divisions can be discussed and agreement can be reached on alternative approaches and new projects. Research seminars are particularly helpful where the divisions are diversified in character, for the participants benefit from the resultant "cross-fertilization" of ideas.

Where several divisions are similar in nature but are located in different geographical regions, it may be desirable to concentrate manufacturing research at a single research center or methods laboratory. The concentrated efforts of a few highly competent men can then result in contributions which will benefit a number of divisions. Particularly where substantial capital investments are necessary to conduct research affecting several divisions, a research center is essential.

SOURCES OF RESEARCH INFORMATION

The sources of manufacturing research information cover the full field of human knowledge, from the personal experience and creativity of the individual industrial engineer to the vast expanse of scientific, manufacturing, and human principles and practices which represent the accumulated knowledge of mankind. The following listing suggests some of the sources to be considered when approaching a practical manufacturing research problem.

Individual Personal Education. The industrial engineer is often a college graduate with an engineering degree. His education includes an excellent cross section of basic principles in scientific, engineering, and technical fields. It probably also includes considerable work associated with administration, management disciplines, and human behavior.

With this foundation, the young industrial engineer can look at established manufacturing practices from a fresh viewpoint and consider how the concepts he has studied at college can be applied to the realities of industry. He may well encounter the human trait of resisting change, but a persuasive and persistent approach can chip away at the status quo and demonstrate progress.

Individual Personal Experience. When becoming associated with a specific manufacturing plant, an industrial engineer may or may not have had previous industrial experience. If he has worked with another company, he has had the benefit of observing other manufacturing practices and the tangible results of the principles which he has applied.

With the advantage of previous industrial experience, whether related or unrelated to the present situation, an industrial engineer can bring a more analytical and mature viewpoint to specific manufacturing problems. By explaining time-tested principles, he can stimulate the organization he has joined to consider how these techniques may be applied to increase productivity, decrease costs, and enhance total return on capital investment.

Existing Human Resources. In any industrial organization, a relatively untapped wealth of practical talent exists at any point in time. For one reason or another, innumerable individuals who can contribute to progress are silent in the area of creativity, however articulate they may be in normal business relationships. Previous experience with blunt rejections of constructive ideas may have conditioned them to avoid taking the initiative in uncharted fields. In many cases, the potential danger of the elimination of their own jobs may influence them to remain silent.

The perceptive industrial engineer recognizes that the man on the job, whatever that job may be, is devoting at least one-third of his life to conscious or subconscious thought about his work. The probability is extremely high that at one time or another the man on the job has had one or more flashes of insight which might revolutionize his job. The research-minded industrial engineer will steadily draw out the innermost thoughts of both supervisory and nonsupervisory employees. By encouraging open discussions of alternative approaches to manufacturing practices, the concept of Russell Conwell's *Acres of Diamonds* may become a gratifying reality.

Manufacturing Handbooks. Supplementing his formal education and practical manufacturing experience, the professional industrial engineer reinforces his personal skills by consulting the accumulated know-how of the numerous authorities who have contributed to manufacturing handbooks. Such handbooks form the cornerstone of company and personal libraries of technical literature.

Management Literature. The progressive industrial engineer should regularly add to a company or personal library noteworthy books which are published in the area of management. These books will help broaden the field of interest of the industrial engineer while sharpening his mind in the conduct of manufacturing research.

Internal Market Research. Of serious concern to any company is the level of manufacturing technology of its major competitors. Unless a company maintains a rate of progress which approximates or exceeds that of its competitors, the inexorable forces of the free enterprise system will lead to inevitable attrition—and eventually lingering or even sudden death.

The minimum objective of any manufacturing research program is to keep pace with competition. Learning what competition is doing in manufacturing, however, poses a serious problem of "business intelligence." One feasible plan is to enlist the cooperation of all employees who may be in a position to have some pertinent information. Sales representatives frequently hear unrelated bits of data when competing with other companies for orders. They should be encouraged to report this information to the industrial engineering department. Similarly, buyers in the purchasing department, in their discussions with vendor representatives, are in an excellent position to gather information on competitive manufacturing practices. Occasionally, new employees who have been associated with competitive companies enter the organization. They should be queried in depth. By piecing together all the information which is thus obtained, the analytical industrial engineer can develop a good understanding of competitors' practices.

Business and Trade Publications.   A practical approach, which an industrial engineer can utilize to keep in touch with technological developments and to stimulate manufacturing research in specific areas, is to scan a selected list of business and trade publications. By regularly leafing through such periodicals, enough ideas can be generated to maintain a productive research program.

Supplier Literature and Representatives.   Among the most fruitful sources of information which can lead to tangible manufacturing research are vendor representatives and supplier literature.   Every purchasing department is constantly beseiged by sales representatives who are attempting to market new production equipment and improvements in manufacturing techniques.   A practical alliance between the purchasing and industrial engineering departments can result in screening literature and verbal claims to identify ideas which justify more detailed investigation.

Technical Associations and Seminars.   The professional character of industrial engineering is constantly enhanced by the strength and maturity of numerous technical and management associations which offer membership to individual industrial engineers.   Through their published journals and through dinner meetings, conferences, and seminars planned to promote group discussions, these associations open the doors to a host of manufacturing research projects.

Management Consulting Firms.   There are a number of reputable management consulting firms which extend beyond the scope of existing industrial engineering organization capabilities and offer services with a more immediate impact than the other sources of manufacturing research information which have been discussed. By their very nature, these firms attract highly motivated industrial engineers who are primarily interested in demonstrating economic results in an organized manner. These firms can render an invaluable service in pinpointing research projects which will yield a substantial return on invested capital and in following through with the completion of the projects.

Research Foundations.   When all other sources of manufacturing research information have been exhausted, and when it is apparent that there is a need for greater depth in research, the consideration of research foundations is recommended.   This approach is suggested when a company is prepared to make, over a period of two or more years, a substantial investment in the possibility of accomplishing a revolutionary breakthrough in manufacturing technology.

PRACTICAL AREAS OF RESEARCH

Individual industrial engineers and their supervisors and managers tend to be particularly capable in one or more areas of the industrial engineering field.   When planning a manufacturing research program, it is essential that careful attention be given to the complete field rather than concentrating upon areas of individual interest.   An objective survey is necessary to establish priorities, seeking maximum return on invested capital.

Total Systems Approach.   In planning a manufacturing research program, a total systems approach should be adopted.   Under this concept, all employees, land, buildings and office space, productive and nonproductive equipment, purchased materials, supplies, and operating procedures represent capital investment and elements of cost.   At any point in time, some or all of the elements of cost may be completely disproportionate to the value of their contribution to net financial profit.

Industrial engineering has a major responsibility to obtain a maximum return on invested capital by reducing the necessary capital to a minimum and increasing profitable turnover of the capital to a maximum.   To execute this responsibility, it is imperative that the fixed and variable nature of all costs be known and that all costs be measured to determine what inefficiencies exist.   Corrective action must be taken to eliminate the inefficiencies, and organized research must be conducted to improve upon the established cost relationships.

Electronics and Computer Technology.   For the first half of the twentieth century, the total systems approach depended largely upon the intuitive judgment and hunches of industrial engineers and managers. With the breakthroughs accomplished in the fields of electronics and computer technology, the total systems approach can become a tangible reality in numerous instances.

In the application of computers to industrial production, it is well to consider the lightning speed of electronic computers in locating specific information, making mathematical calculations, and recording the results. In measuring the speed of computers, time is expressed in microseconds and nanoseconds. A microsecond is one-millionth of a second; a nanosecond is one-billionth of a second. These tiny intervals of time are almost incomprehensible to the human mind.

Computers are available which specify 200 nanoseconds as their internal read, compute, and write cycle. In other words, in one-fifth of one-millionth of a second, the computer can locate a number which has been stored in its memory; then add, subtract, multiply, or divide that number with relation to another number; and record the answer in its memory, making it available for the next transaction to be performed. With such a capability at the disposal of the industrial engineer, the development of the total systems approach poses a genuine challenge to his ingenuity.

From a practical viewpoint, limiting factors which create an information gap between the genuine needs for computer services and computer output information which meets those needs consist of programming problems and providing organized computer input information. Industrial engineers must become sufficiently familiar with programming language to assist computer programming specialists in planning the computer programs required to meet specific needs. Care must be taken to train all employees who are responsible for providing source data to the computer to understand and conform to the precise procedures which are developed for specific programs.

Computerized Production Control. Production planning and control constitute a major phase of the total systems approach. Significant strides have been made by many companies in the development of computerized production control.

The various aspects of computerized production control include:

1. Translating product bills of material into mechanized information for utilization by the computer
2. Building forecast customer sales orders into the computer
3. Exploding forecast sales schedules into component part requirements related to calendar time intervals based on purchasing and manufacturing lead times
4. Maintaining an adequate inventory of raw materials, work in process, finished parts, and complete products
5. Issuing purchase requisitions and production orders to maintain a balanced inventory
6. Evaluating manufacturing machine load requirements generated by production orders to guide management decisions on manpower needs, overtime schedules, subcontracting requirements, and the need for additional manufacturing facilities.

Industrial engineers supplement the work of production control specialists by:

1. Properly measuring setup and processing time for all manufacturing operations
2. Defining homogeneous productive equipment in the establishment of a complete system of production stations
3. Determining alternative production stations which may be utilized when given production stations are overloaded
4. Determining groups of similar units of productive equipment which may be controlled by one or more direct labor employees
5. Establishing progressive machining lines to minimize manufacturing cycle times
6. Establishing and balancing progressive assembly lines to minimize assembly costs

Material Handling Systems. The high cost of material handling, which is rarely measured with the precision accorded to direct labor, offers excellent opportunities for the reduction of indirect labor costs through a fully integrated material handling system. Such a system, placing maximum emphasis upon the continuous flow of materials, reduces substantially the overall manufacturing cycle, with a corresponding decrease in manufacturing inventory and a substantial increase in the overall financial return on invested capital.

The ingenuity of many manufacturers has resulted in a broad diversification of conveyors, forklift trucks and attachments, overhead hoisting equipment, sophisticated stacker cranes, and supporting racks, bins, and pallets. Industrial engineers are in a position to consider all the building blocks which are available and to integrate the various elements, with the assistance of electronic controls and within the practical limits imposed by the products and processes of the manufacturing plant, into an approximation of the "push-button" factory.

NUMERICAL CONTROL EQUIPMENT

A workpiece machined under numerical control will be the same size and shape as one manufactured on a machine that is not numerically controlled. This being the case, what is the difference between the two systems?

The primary difference lies in the method of supplying input data and obtaining feedback signals. With numerical control, automatic operation is achieved by means of numerical instructions expressed in code. These instructions or programs are prepared in advance. Recorded on tape, these coded instructions control the sequence of machining operations, machine positions, spindle speed and rotational direction, distance and direction of movement of the tool or workpiece, flow of coolant, table indexing, and even the selection and changing of the cutting tool for each operation.

The coded tapes are placed on a control unit which consists of a system of electronic interpreting devices. When activated, the control unit guides the machine tool through the programmed operations and movements with little or no human intervention.

Numerical control, therefore, has been defined as "a system in which the direct insertion of programmed numerical values, stored on some form of input medium, are automatically read and decoded to cause a corresponding movement of the machine which it is controlling."

Advantages of Numerical Control. A number of substantial advantages have been demonstrated by numerical control equipment, justifying considerable research in this area.

Reduction of Tooling Costs. With conventional machining, complicated fixtures and drill jigs are necessary to hold the workpieces during the drilling, milling, boring, and other cutting operations. With NC, such tooling is essentially eliminated, and the simplest form of clamping fixtures is usually adequate. The cost of tool design and toolroom labor is thus substantially reduced.

Reduction of Setup Costs. With conventional machine tools, particularly of the automatic variety, a high degree of setup skill is required on the part of the employee, and the setup time is considerable. With NC, the setup skill is largely transferred to the programmer in the manufacturing engineering department, and the actual setup by the direct labor employee is relatively simple. The setup cost for each job lot is thus considerably decreased.

Reduction of Noncutting Time. With conventional machine tools, the employee devotes a large percentage of his time to the physical manipulation of the machine and its accessories, during which time the cutting tools are not functioning. With NC, the machine operates automatically with a high degree of acceleration and speed. The net result is a major reduction in the percentage of nonproductive time or noncutting time to the total cycle.

Standardization of Cutting Time. With conventional machine tools and competent industrial engineers and foremen, it should theoretically be possible to obtain optimum speeds and feeds for specific jobs. With NC, however, the rigid disciplines which are necessary for sound administration of numerical control provide increased assurance that speeds and feeds are properly selected. The standardization of cutting time generally results in some cost reduction.

Better Machine Utilization. The combination of the above factors of setup time, noncutting time, and cutting results in substantial improvements in overall utilization of the spindles controlling the cutting tools. The experience of many companies indicates that total actual time in cutting chips can be increased to more than 75 percent, as compared with frequent instances of 25 percent with conventional machines.

Reduced Scrap. Because of the consistently high accuracy of NC machines and because human errors are almost entirely eliminated, scrap is considerably reduced. As one example, a reduction of 84 percent in scrap was experienced when a manufacturer of business machines changed from conventional to NC machines.

Decreased Manufacturing Lead Time. Because the total time of tape preparation, setup, and production of the parts is reduced, the total manufacturing lead time is decreased. This is a substantial advantage both in the development of a new product and in meeting customer delivery schedules.

Inventory Reduction. The formula for economic lot quantities, which balances the cost of setup against the carrying charges of the inventory, results in substantial inventory quantities with conventional machines. With NC, the lower setup costs result in significantly lower economic lot quantities.

Disadvantages of Numerical Control. Partially offsetting the advantages are the disadvantages of NC equipment.

Increased Capital Investment. Numerical control involves a more substantial initial investment than is needed for conventional machines. NC machines require a more rigid frame and heavier lead screws, bearings, and other actuating mechanisms to permit the machine tool to achieve a high acceleration speed for efficient positioning times. To this cost of the basic NC machine must be added the cost of the tape preparation equipment and the data processing materials needed for preparing tapes for the NC machine. The net result is a capital investment ranging from approximately $20,000 for a small, two-axis, point-to-point drilling machine to $500,000 or more for a highly sophisticated five-axis machine capable of continuous path contouring in three dimensions.

Programmer Training. The success of NC is largely dependent upon the ability of the parts programmer to communicate effectively with the machine and the operator. Programmers must be selected who have a firsthand knowledge of machining operations and tooling and who also have the analytical ability to detail every step in the machining process.

Initial Operator Problems. Because of the radical differences between conventional and NC machines in actual operation, it is important to select skilled machinists or toolmakers in the initial phases of an installation. These men can be of great help in getting the new programs debugged and in designing simple and effective holding fixtures. There is frequently resistance to NC equipment on the part of older operators. After the programming and operating problems have been largely resolved, however, the younger operators frequently welcome the opportunity to work with NC machines.

Product Engineering Orientation. To a large extent, NC offers the product engineers greater flexibility in design, and engineering changes are less costly to put into effect. On the other hand, product engineers must become familiar with the requirements of NC and must dimension product drawings, together with tolerances, in a manner to facilitate the work of the programmer.

Controlled Tooling Environment. With NC, tooling environment must be carefully controlled with respect to:

1. Standardized cutting tools, tool holders, and tool drivers
2. Standardized fixtures, locaters, and work holders
3. Organized, detailed documentation for planning every aspect of the NC job
4. An airtight system for total tool control

Maintenance Technicians. The typical machine maintenance mechanic needs considerable retraining for NC equipment. The major maintenance problems in NC equipment require electronic maintenance technicians for troubleshooting the electronic control. Former radar operators and electronic technicians from the armed forces are excellent candidates for this job. Experienced shop electricians require less retraining than do typical maintenance mechanics.

Computers and NC for Integrated Manufacturing. Manufacturing research directed to the goal of integrated manufacturing can make substantial progress by harnessing the capabilities of computers to numerical control manufacturing technology.

In the ultimate integrated manufacturing plan, all detailed product engineering, manufacturing engineering, production control, and quality control information provides input to computers which then direct numerical control equipment to perform manufacturing operations and move materials, parts, and tools to and from storage.

Several computer systems are available which are designed to accomplish major segments of an integrated manufacturing plan.

As the first step toward an integrated manufacturing plan, a user of NC equipment may well consider the General Electric Computer Numerical Control Data Controller. The basic intent of the CNC/DC is to provide, at the relatively modest cost of $7,500 per NC machine, an opportunity to make performance and economic evaluations of computer-directed systems on a step-by-step basis before making the substantial investment that will be necessary to put an entire parts making operation "on-line."

ROHR INTEGRATED SYSTEM

Mass production plants lend themselves well to the application of integrated manufacturing principles. In a number of cases, however, the problems which must be solved by industrial engineers lie in the field of comparatively low volume production.

An excellent example which provides practical insight into the solution of integrated manufacturing problems under job lot conditions is the system installed at the Rohr Corporation, Chula Vista, California. The system is the result of the creative thinking of Burt F. Raynes, president of Rohr Corporation, and is described in detail in the February, 1969, issue of *Modern Materials Handling* under the direction of Miles J. Rowan, editor and assistant publisher.

Production Scope of Rohr Corporation. In any case study, it is desirable to have enough information about the scope and nature of the manufacturing problem to facilitate the application of the basic principles involved to another industry or firm. The relevant factors at Rohr may be summarized as follows:

1. Rohr is in the business of producing jet engine power plants and other major aircraft structural assemblies.
2. The manufacturing task consists of a high volume of small-size job lots in production at any time.
3. A total of 120,000 part numbers are on file, and the parts require as many as fifteen fabrication operations, using different tools.
4. At least 95 percent of the parts are made to order.
5. As many as 30,000 orders are in planning or processing stages at any one time.
6. The orders are for small lot quantities, generally from 50 to 400 pieces.
7. Hundreds of small lot jobs are competing for attention at fifteen manufacturing steps.
8. Finally, these parts involve as many as 600 to 800 engineering changes per month.

Concept of Rohr Integrated System. With a monumental flow control problem in manufacturing parts, the Rohr Integrated System has synchronized parts with associated tooling to be available when needed for closely scheduled production without any congestion in the plant.   Human effort has been reduced to a minimum in the scheduling, dispatching, material handling, and storage operations.

The fundamental problem of job lot flow control was solved with a blend of computer systems and material handling systems.   Computer-centered systems were developed for planning, scheduling, and follow-up of parts manufacturing operations. Computer-centered systems were established for the storage and delivery of parts and tools to support the manufacturing operations.

Mechanics of Rohr Integrated System.   To illustrate the manner of executing the basic concept of the Rohr Integrated System, the following outline of the mechanics of the system lists the physical facilities and personnel involved and the principal steps in the system.

Management Information System.   The basic management information system consists of a group of planners and schedulers equipped with voice input and output terminals connecting them with the central computer.   Production schedules are originated and revised at this point.

Central Computer.   The central computer, or data processing and storage system, is a multiple processor system using IBM 360 model 50 and model 60 equipment. The two central processing units share the random access disc files, the central data bank for flow control applications.   A separate magnetic drum contains a 79-word vocabulary, used to give voice replies over 136 voice input-output terminals distributed throughout two manufacturing plants 125 miles apart.

The central computer is linked with a satellite computer for controlling the storage and delivery of material, parts, and tools to each manufacturing operation. The central computer is also linked with all planners, schedulers, production clerks, and dispatchers within the two plants, through voice input-output terminals, for instant updating of all random access files to maintain an accurate record of the current status of all production orders.

The central computer works with the master production schedule and bill of material files to develop time-phased parts requirements, including precise quantities and the planned start and completion dates for production orders.

It also prints out a variable strip.   Using this document, an order clerk sets up detailed short-term schedule requirements—due dates on an "M day" basis—for each step on the process sheet.   Taken together, these documents become a production order for the shop.

Automated Storage of Parts and Tools.   The central computer maintains a record of the location of all parts in process and all manufacturing tools which are stored in 118,000 pigeonholes.   Instructions for moving the parts to the next manufacturing operation and making the necessary tools available at that operation are transmitted by the central computer to a DEC-PDP-8 satellite computer which controls physical movements.

Facilities for transportation of parts and tools include sixty stacker cranes and automatically guided tractors between conveyor stations in warehouse and production areas.

Voice Terminals.   One of the fascinating aspects of the Rohr system is that the computer, through voice terminals, actually "talks" to people as far away as 125 miles.

Normally, Touch Tone dial telephones can be used as voice terminals.   Because such dialing was not available on the West Coast at the time of the Rohr installation, the Rohr system uses regular telephones with attached Touch Tone pads.   Pacific Bell Telephone instruments and lines are used, with a rental charge of $11 per month per terminal and a terminal relocation charge, when necessary, of $6 per unit.

Planners, schedulers, production clerks, dispatchers, and others use the terminals to report all changes in job order status, including new orders, completed operations, shortages, and moves.   They also use the terminals to get up-to-the-moment facts about job orders.   The computer's "voice" answers their inquiries.

The terminals are used for twenty-three kinds of transactions.   In general, however, an individual employee is concerned with only three to five kinds.

For each transaction, an employee presses the proper buttons on the Touch Tone pad. This code connects the terminal with the appropriate voice recording at the central computer for the specific transaction required. For each transaction, the voice recording asks for details, step by step. The employee hears the voice on the telephone receiver and enters the requested data by the proper push buttons on the Touch Tone pad, step by step.

Return on Investment. As in any manufacturing project which requires capital investment or monthly rental charges, economic justification is essential. The Rohr Integrated System has met the acid test of economic justification.

The use of high-cube storage in stacker racks has reduced floor space requirements by 126,000 square feet. Making a comparison with conventional storage systems, a saving of approximately $250,000 in land and construction costs has been achieved.

In material handling costs alone, a reduction of $400,000 per year has been accomplished. It is estimated that the total cost of the complete system will pay for itself within only 3½ years of operation.

SUMMARY

The individual industrial engineer can make a major contribution to the development of a manufacturing research program through demonstrated results in a specific area. As progress is made on individual projects, with other members of the manufacturing organization participating in the advances, a more general spirit of enthusiasm may be generated. With top management understanding and support, together with tangible recognition of actual contributions which are made, continuous substantial results can be achieved through manufacturing research.

BIBLIOGRAPHY

Bright, James R., *Automation and Management,* Harvard Business School, Division of Research, Boston, 1958.
Brosheer, Ben C., and James C. De Sollar, "Variable Mission Machining," *American Machinist,* September, 1968.
Rowan, Miles J., "Doing the 'Impossible': Complete Control of Job Lot Flow," *Modern Materials Handling,* February, 1969.
Schaffer, George, "NC and the Computer," Special Report 626, *American Machinist,* March, 1969.
Taylor, Frederick W., "On Scientific Management, 1912," *An American Primer,* University of Chicago Press, Chicago, 1966.
Turnkey Systems for Moving Materials, Eaton Yale & Towne Inc., Automated Handling Systems Division, Washington, D.C., 1968.
Weir, Stanley H., *Order Selection: A Focal Point for Developing Warehouse Machine/Systems,* American Management Association, New York, 1968.
Wright, Herbert L., *Beginner's Course in Numerical Control,* Cincinnati Milling Machine Co., Cincinnati, Ohio, 1968.

Chapter 7

Value Engineering*

CARLOS FALLON

Manager, Value Analysis, RCA Corporate Staff, Camden, New Jersey;
and Vice President, Professional Development,
Society of American Value Engineers

Value engineering is a method for improving product value by improving the relationship between the function of a product and its cost. This chapter will explain how industrial engineers can use task groups to bring together the diverse specialists who contribute value, so that they can relate each element of product worth to its corresponding elements of product cost, in order to provide the function of the product at least cost in resources.

At any rate, that is the traditional goal of value engineering. But in a growing economy, least cost is not the customer's major consideration. As his discretionary income rises, the customer demands something better every year. Value engineering, therefore, is as much concerned with putting the money where it will do the most good as it is with reducing cost.

The term "value engineering" will be used in this chapter to include the parent discipline of "value analysis" and such variations as "value assurance," "value services," and the more descriptive "value improvement."

The word "product" is used here for products, services, and operations; "worth" for the monetary appraisal of the usefulness or pleasure provided by the product; and "cost" for the monetary measure of every expense that must be incurred to gain such usefulness or pleasure.

* Abstracted from a book by Carlos Fallon on value analysis and value engineering, scheduled for publication by John Wiley & Sons, Inc., in 1971.

THE INDUSTRIAL ENGINEER AND VALUE IMPROVEMENT

Can the industrial engineer improve product value? Of course he can! He has been doing it since the days of Taylor and Fayol. By dramatically reducing the cost of manufacturing, industrial engineering has provided the basis for the prosperity of Western Europe, the British Commonwealth, North America, and Japan. So much so that the cost of manufacturing—despite higher wages—continues to be an ever-smaller portion of total product cost.

Reducing the cost of manufacturing a given product, therefore, calls for ever-increasing effort which nevertheless yields diminishing returns. Some could say, "The job has been done too well!" But there are other worlds to conquer.

Improving the Product Itself. Instead of changing the factory to suit the product, how about changing the product to take advantage of improved manufacturing technology and modern industrial engineering?

"Changing the product . . . ? But that is not in our charter," is the usual wail. "We are supposed to make it like the drawing, applying our ingenuity and resourcefulness to manufacture something no matter how weirdly it has been designed, how horrible it looks, or how badly the materials have been procured." True or false?

Do industrial engineers ever change a product because it cannot be made like the drawing? Of course they do. As the result of a factory problem sheet, a methods man and a design engineer get together, make a product more producible, and often make it better all around. Their joint effort produces better results than the sum of their individual efforts, yet this successful teamwork did not happen on purpose. It was the result of an emergency. Why?

Because an industrial organization, although hopefully called a team, lacks some of the characteristics of a team. Like a team, it combines diverse skills and a pattern for their interaction. Unlike a team, the interaction is seldom left to the initiative of the players who are physically handling the ball.

IMPROVING THE INDUSTRIAL TEAM

Traditionally, engineering designs a product at the request, direction, or suggestion of marketing and styling; finance provides the resources and rides herd on them; purchasing buys materials and some specialized design services; and manufacturing makes the product.

But here is the rub. Engineering and manufacturing are expected to work mainly through the physical sciences, while buying, selling, styling, and financing are part of the social sciences, or so the universities have taught us. Faithful to the traditional classification of knowledge, we march our separate ways, forgetting that classification does nothing to the real relationship among the elements classified.

No matter how engineering and manufacturing are separated from finance, marketing, styling, and purchasing in the organization, the value of the products depends upon all of them together, upon their effective interaction as much as their departmental contribution. Two requisites for working together smoothly among departments are timely information and simple, straightforward communication.

Timely Information and Effective Communication. To provide a mechanism for face-to-face communication among the players who actually handle the ball, value engineering gathers the key specialists at the working level into task groups which can be truly called teams. Such teams dynamically exchange and digest in a few hours the information that normally takes weeks to circulate in written, drawn, or coded form.

How the value task groups are organized will be covered later. Why they are organized is a matter of information and communication, of bringing together the suppliers and the users of information.

Traditionally, each specialist gathers, digests, and generates the data related to his own specialty, thus spinning a single strand of information which will later form part of a network. Based on what he thinks they need to know, he sends his specialized strands of information to the other specialists.

The minute the various strands come in contact with each other, sparks fly, discrepancies show up, and much of the plant's energy goes into revisions, corrections, problems, and fixes.

The purpose of the modern value task group is to have the suppliers and users of information analyze, not their particular strands, but the network of information. All this information comes together for one chief purpose—to create or enhance the value of the company's products. An understanding of product value, therefore, can be of great use to the industrial engineer.

ENHANCING THE VALUE OF INDUSTRIAL PRODUCTS

The components of product value are:
1. Customers with money and unsatisfied wants—a market.
2. Utility to such customers—the product must suit the market.
3. Scarcity or difficulty in attainment—the product must be hard to get.
4. Total cost to the customer—an inverse component of value. Given difficulty in attainment, the customer wants to pay the least for overcoming that difficulty.
5. The customers' other options—competition.

The interaction of these components provides a definition: The value of industrial products is determined by the special relationship of utility to cost, which conforms to the customer's wants and resources in a given situation.

Kinds of Value. The human urge to classify—in order to understand—has led men to define various "kinds" of value. But classification in no way changes the nature of whatever is being classified. Classification itself, however, is governed by the interests of the classifier and the information available to him.

Use Value and Exchange Value. The first formal classification of values, which we owe to Aristotle, was colored by the classical Greeks' love of individuality and by their contempt for trade; hence the use of a sandal to be put on the foot was "more proper" than the use of a sandal to be given in exchange.

But if the sandal cannot be put on the foot, it has no exchange value. If it can be put on the foot but looks ugly and smells bad, it has very little exchange value. If it can be put on the foot, looks good, and smells like rich leather but the town is inundated with such sandals, it has exchange value only in another market.

Exchange value, therefore, is affected by use value, esteem value, and market value. All these kinds of value interact, as Alfred Marshall put it, "like a number of balls resting against one another in a basin."

From the value analysis standpoint, the aspects of economic value can be classified as:

Use value
Esteem value
Exchange value
Market value

But it should not be assumed that by thus classifying them we are separating them from each other in real life. Analysis takes things apart to understand them. To function, they have to go back together.

Isolating and condemning esteem value is perhaps the biggest trap in this respect. The right feel and appearance of military equipment contribute to the morale of the men and are a desirable form of esteem value.

Esteem Value. In the great stream of American industry, which is, after all, the backbone of defense supply, esteem value serves two purposes: it provides competitive advantage—a beautiful product is often less costly to make and easier to sell than an ugly product; and it guides the customer in the selection of functionally better products because, in good design as in nature, form follows function.

Market Value. Understanding market value helps avoid another trap—reducing cost at the expense of customer acceptance. The customers speak through the market. As James M. Roche, General Motor's chairman of the board, said at the 1968 stockholders' meeting, "In the dynamic and changing market for new cars, our customers through their purchases tell us what they want."

Use value and esteem value are related to the physical properties which make a product do what it is supposed to. Market and exchange value are related to its economic characteristics. Value engineering relates the physical to the economic characteristics of a product to give the customer more for his money.

Value and the Customer. Why the customer and not the manufacturer? In 1947, value to the manufacturer had already been thoroughly examined. It had been quantified as return on investment and profit, and qualified by appraisals of customer acceptance, share of the market, and growth. Then the professional customers—the purchasing specialists—within a large corporation decided to analyze value from their own point of view as customers.

Customer orientation posed the questions, "What does it do?" and "What is this worth?" and focused thinking on the dynamic task of customer satisfaction, on the function.

Value engineering aims at value to the customer, for the elementary reason that practically all aspects of value to the manufacturer, from profit and return on investment to share of the market and growth, are dependent upon value to the customer. But value to the customer, from the practical standpoint of the man in industry or the military buyer, has not been studied as closely or measured as carefully as value to the manufacturer; yet value to the customer is the source of all industrial value.

THE VALUE TASK GROUP

Although the value task group is not the only way to do value engineering, it has proved to be the fastest and most productive way. In planning such a task effort, a balanced combination of skills to match the requirements of the product is the most important consideration. Table 7-1 shows how this match is accomplished.

The groups listed were actual profit-oriented task groups working, not on demonstration or training exercises, but on real products. Each group happened to have five participants—which is the ideal number—but the groups can number anywhere from three to seven.

It will be noted that five major categories are represented in each group. These categories are related to the five components of product value, roughly as follows.

Task group element	*Effect on value*
Customer viewpoint............	To suit the market
Product design................	To create utility
Manufacturing................	To overcome difficulty in attainment
Finance......................	To allocate costs
Purchasing...................	Three roles: (1) to provide supplier information, (2) to appraise competitive products, and (3) to ask customer questions

The Supporting Group. A number of marketing and finance specialists, in addition to those assigned to the teams, may support a value workshop made up of five or six teams. The main reason is that nature has not produced enough cost estimators and that marketing people are often on the road. The teams therefore have to share the specialists available. Any necessary skill not represented in more than two teams out of five should be included in the supporting group.

Project Selection. The value task groups should be tailored to suit the projects, but the projects must be selected to meet the company's most pressing needs. Prime targets for a value task group are:

1. Jobs which have to advance the state of the art. The fiercest competition is in the race for new products—new ways of doing things. For this leap

TABLE 7-1.   Typical Value Task Groups

Product and industry	Task group members
Steam trap Process industry	Process control foreman Piping engineer Steam engineer Buyer Cost estimator
TV cabinet Consumer TV industry	Acoustic engineer Furniture designer Styling specialist Raw materials buyer Manufacturing engineer
TV chassis Consumer TV industry	Assembly line foreman Electronic engineer Mechanical engineer Methods engineer Mechanical buyer
Antibiotic Drug industry	Physician Biochemist Formulating chemist Chemical process engineer Buyer
Radar Defense industry	Circuit design engineer Applications engineer Manufacturing engineer Electrical buyer Quality engineer

into the unknown, the responsible product manager needs fresh information on new materials, new suppliers, and new manufacturing methods.

2. Jobs which must be delivered ahead of schedule—getting there first. When there is no time for the step-by-step approach characteristic of yesterday's gentler competition, a value task group can do concurrently much of the work that is usually done in sequence.

3. Jobs which cost more than they should, either because the price would exceed what the customer can pay or because the gross margin does not yield enough profit.

Figure 7-1 shows one way of comparing projects for selection. In this case, the "skills" column refers to the personnel available for assignment to the value engineering workshop. The same chart is used later by each task group to determine what area of their project to work on.

Matching Skills to Projects. Once the projects have been selected, participants are assigned, not arbitrarily, but with the specific approval and support of their department heads. These department heads have already given their general approval to the workshop. What is required now is approval for the participation of a given individual on specific dates. Figure 7-2 shows a handy form for negotiating such approval and for achieving a balanced combination of skills.

Team or Committee. Committees were originally developed, within the parliamentary process, for the purpose of reconciling conflicting interests. Committee meetings are not, therefore, the most pleasant of chores. Compare the lamentation, "I've just come from a miserable committee meeting," with the cheerful, "I had a lot of fun," often heard after a value analysis session.

While a committee meets to reconcile departmental interests, a task group meets for the interaction of departmental skills independently of departmental interests. It has a predetermined goal, participating leadership, and subordination of professional and departmental interests to the task at hand.

After the value task group has determined the best combination of product

Selection criteria	Potential	Feasibility	Information	Time	Skills	Probability of implementation
	What is the proposed improvement worth?	What are the chances it will work as planned?	Do we have or can we get all the information required?	Do we have the time we need to develop this proposal?	How suitable are our skills for developing this proposal?	
	Estimated in money	100 % = certainty	100 % = full information	100 % = full time	100 % = suitable	%
Choices	Multiply the potential gains by percentages expressed in decimals, as in the following example.					Yield forecast
Example: magnetic head	$1,000	90 % × 0.90 = 900	80 % × 0.80 = 720	100 % × 1.00 = 720	90 % × 0.90 = 648	80 % × 0.80 = $518

FIG. 7-1.   One way of comparing projects for selection.

Workshop organization	C	P	M	F	B
Teams 1 to 3	Customer viewpoint	Product design	Manufacturing	Finance	Buying
Team no. 1 Project:_____					
Team no. 2 Project:_____					
Team no. 3 Project:_____					

Activities from which workshop participants are usually selected.

C—Marketing, styling, quality assurance, and so on
P—Engineering or new-product development
M—Industrial engineering, manufacturing engineering, time and methods
F—Cost estimating, cost accounting, auditing
B—Purchasing and related fields such as production control

FIG. 7-2.   Selecting a balanced combination of skills.

benefits for a given level of resources, it may well reconvene as a committee to make sure that the interests of the various departments are safeguarded. But that must come later, for without a good product, all departmental jockeying for advantage is fruitless.

Riddle of Group Dynamics. We know that in certain instances group dynamics works, and works well, but we do not really know what it is. Behavioral scientists, of course, have observed the advantages of face-to-face communication, the buildup of creative potential, and the improved personal relations, but all these are results, not explanations. No one has really explained the magic that turns a collection of individuals into a task group which becomes a fountainhead of imagination, productivity, and common sense.

Without pretending to know what group dynamics cannot do, value engineers do know that it is a source of energy and enthusiasm in a value task group.

We did not start out with groups for the sake of groups, much less with the intention of using group dynamics. Much early value engineering was practiced by loners. Most designers and buyers do some value engineering on that basis, but product value is not contributed by designers and buyers independently of each other and of the other specialists involved.

It is worth repeating that one thread of information about a product is not truly meaningful unless it is viewed in relation to the other threads of the network. To come up with a good product, it is necessary to look at the network as a net.

Lacking a Leonardo da Vinci, a Leibnitz, or a Benjamin Franklin in the average industrial plant, we have to make do with a collection of specialists, each of whom understands one of the threads, and all of whom can discuss the net as a net.

This collection is usually made up of people varying widely in authority, responsibility, education, and experience, but three factors tend to draw them together:
1. Each has information the others need.
2. Each has been assigned a task jointly with the others.
3. Each will benefit personally if the task is successful.

Because the dynamics of human groups existed long before Dr. Kurt Lewin brilliantly formalized their study, this collection of individuals begins to interact as a group right away.

Efficiency of the Task Group. Reference has been made to the value task group as a team. It *is* a team, a special kind of team.

The Interdiscipline Approach. Hopefully, this team includes the principal skills that contribute value to the product under analysis. The mixture of suitable skills is a major advantage, never to be sacrificed for the sake of getting a workshop going. Such a combination of specialists, communicating face-to-face, can see at once the effect that their actions have upon the activities of others. They benefit right away from the interchange of firsthand information. What is even more important, they jointly appraise each thread of information in relation to the whole network.

Commitment to the Group and Its Task. Having cast aside such ploys as, "We are all working for the same company, so you have to do thus and so—which will make you look bad—but it is for the good of the company and of my operation," the members go to work on improving the product so that they all look good. They also take care that nobody suffers either in the group or among the people who have entrusted their product to them. Rather than Pollyanna sweetness and light, this regard for the professional reputation of the people concerned is a matter of cold business logic. Too much time is wasted in building defenses against criticism. The simplest way to reduce this defensiveness is to eliminate the attack. All the effort that previously went into sharpening knives, procuring Band-Aids, writing "memos to file," and protecting vulnerable areas now goes into improving the product.

Exploiting the Advantages. To get the most out of their interaction, team members learn to achieve a measure of objectivity, to communicate without jargon, to listen, to inquire, and to work well together. This makes for a powerful tool indeed. Now for the way to use this tool.

THE VALUE ANALYSIS JOB PLAN

The task of analyzing value follows the general pattern of the scientific method, incorporating problem solving and innovation techniques with the teamwork characteristic of group dynamics. The sequence follows this outline:

I. Information Phase
  A. What is it? The initial value engineering question.
    1. Identify the project.
    2. Define scope of the study.
    3. Determine quantities and product life.
    4. Learn marketing requirements.
    5. Review cost data.
    6. Search for the additional information required.
    7. Explore for unexpected information.
    8. Pinpoint the most significant facts.

II. Analytic Phase
  A. What does it do? The key value engineering question.
    1. Define the function (see Figure 7-3).
    2. Identify the benefits.
      a. Determine their relative importance.
  B. What does it cost? (See Table 7-2.)
    1. Cost to find and select the product.
    2. Cost of acquisition.
    3. Cost of operation.
    4. Cost of maintenance.
    5. Cost of repair and overhaul.
    6. Cost of downtime.
    7. Cost of disposal.
  C. What should it cost?
    1. Evaluate by comparison. Determine the basic or lowest cost to perform the function.
      a. Cost for a single operation.
      b. Cost for multiple operations.

System, assembly, part, or process	What does it do?		Degree				Estimated cost
	Verb	Noun	1st	2nd	3rd	4th	

FIG. 7-3. Defining the function. A definition has different purposes in varying situations. The purpose here is to pinpoint the most important aspect of the functions by boiling down the concept into a verb and a noun. The primary function is checked off in the first column under "Degree"; less important functions, in the other columns. The "Cost" column will serve later to determine if cost is proportional to degree of importance.

TABLE 7-2. What Does It Cost?
Actual ( ) or estimated ( ) cost of:
Present approach ( ), choice A ( ), choice B ( ), choice C ( ), choice D ( )

Name or description	Part or drawing no.	Quantity over 12 months

Nonrecurring costs per unit	*Recurring costs per unit*
Tools................. $	Raw materials......... $
Facilities.............. $	Purchased parts........ $
MHX* (%)........ $	Subcontracts.......... $
Total materials............... $	Other materials........ $
Engineering........... $	Subtotal............ $
Drafting.............. $	MHX* (%)....... $
Model shop........... $	Total materials.............. $
Testing............... $	Direct labor hours...... $
Qualification.......... $	Average rate........... $
Setup................ $	Direct labor cost....... $
Documentation........ $	ESE† (%).......... $
Subtotal............ $	Total labor................. $
ESE† (%)........ $	
Total labor................. $	Other...................... $
Other...................... $	Total recurring costs.......... $
Total nonrecurring costs.. $	Total unit cost.............. $

* MHX—material handling expense.
† ESE—employee service expense.
With these two exceptions, all the above are "unloaded" costs.

       c. Cost per year.
       d. Cost per unit of output.
       e. Cost per unit of measure.
       f. Cost per unit of service.
   III. Creative Phase
     A. What will do it better?
       1. Providing a welcome for new ideas.
       2. Freeing the inventive personality.
       3. Environment for discovery.
       4. Fostering inventive traits.
       5. The process of invention and discovery.
     B. Innovation.
       1. Identify barriers to innovation.
       2. Surmount barriers to innovation.
       3. Discuss the mechanism of innovation.
       4. Adapt to change.
       5. Create change.
       6. Stand back from the product.
       7. Identify with the product.
       8. Look for analogies.
       9. Look for "way out" ideas.
       10. Screen the choices developed.
     C. Optimization.
     D. Simplification.
   IV. Evaluation Phase
     A. The Combinex method.

1. Optimization and outright improvement.
2. The art and science of combination.
3. Meaningful numbers.
4. Effectiveness and utility.
5. A measure of customer satisfaction.
6. Setting upper and lower bounds.
7. Measures of relative importance.
8. Effectiveness of the various choices.
9. Use of submatrices.
10. Racking up the results.
B. Selecting a manufacturing method.
1. Identifying the expected benefits.
2. Assigning weights.
3. Setting up the scoreboard.
C. Comparing value analysis options.
1. Benefits from whose point of view?
2. Upper and lower bounds.
3. Weighting techniques.
4. Adjusting for utility and effectiveness.
5. Using the scoreboard.
V. Verification
A. What will this buy us?
1. Improved schedules.
2. Increased customer acceptance.
3. Dollar savings.
B. What will it cost us?
1. Schedule delay.
2. Disruption.
3. Nonrecurring costs.
4. Recurring costs.
C. What are the risks?
1. Definition of risk.
2. Performance problems.
3. Procurement problems.
4. Manufacturing problems.
5. Market hazards.
6. Other hazards.
7. Preparing the written report.
VI. Recommendations
A. Summarizing the proposed course of action.
B. Preparing the plan of implementation.
1. Who will authorize the action?
2. Who will implement it?
3. Who should see the draft?
4. Who will verify the figures?
VII. Implementation
A. Factors governing acceptance.
1. Choice—which opportunity to grasp.
2. Unexpected effort.
3. Unexpected costs.
B. Requirements for acceptance.
1. Well-substantiated information.
a. Gains, costs, risk.
2. An honest forecast.
a. Effect on overall goals.
b. Effect on departmental schedules.
C. Mechanism of implementation.
1. Determine who will do what.

a. Who will see that it gets done?
b. Who will do the work?
c. Who will monitor progress?
d. Who will provide the funds?
e. Who will control costs?
f. Who will report progress?
g. Who will set priorities and cut-in points?
2. Determine what areas will be affected.
   a. Where will the work be done?
   b. Where will the funds be charged?
3. Determine timing and sequence.
   a. When will the effort begin?
   b. How long will it take?
   c. How will it fit in the general schedule?
   d. How long to pay off cost of implementation?
   e. How many units before break-even point?
4. Establish follow-up procedures.
   a. Milestones.
   b. Progress reports.
   c. Contingency plans to bring the program back on track.
5. Measure results.
   a. Audit all gains, both dollars and elapsed time.
   b. Audit all costs, both dollars and time lost.
   c. Compare net gains with the full cost of the value effort.

HOW THE JOB PLAN WORKS

Two thousand years of military history continuously hammer home the importance of information. Lest we forget its economic importance, consider the delightful adventures of those professional treasure hunters who search the bottom of the sea.

Competing with many well-financed and lavishly equipped search vessels, Robert Sténuit, a Belgian professional diver, and Marc Jasinski, his photographer friend, discovered and salvaged a dream haul of priceless relics of the Spanish Armada.[1] Their equipment? A rubber raft, an outboard motor, rented diving gear, and a vast store of scientifically gathered information. These young men had spent years in the archives and libraries of Europe studying every letter, every report, every legend referring to the Spanish ships *Duquesa Santa Ana, Sancta Maria Encoronada,* and the galleass *Girona.* How they found the treasure of all three ships in one spot is a fine example of the value of information when it comes from the best sources.

The creative phase of the job plan is laconically described in the outline, but the analytic and evaluation phases deserve fuller treatment for two reasons: (1) because the very foundation stone of value analysis is the function of the product, which takes up a major part of the analytic phase; and (2) because the evaluation phase, in this particular job plan, includes the author's Combinex method, which may prove useful to industrial engineers in many other applications.

THE FUNCTION

"What does it do?" is the first question that the captain of a value engineering team must ask when he looks at an industrial product. "What is its function?" he may continue. "Is this the primary function? What about secondary functions?" These questions free the mind from the static limitations of structure, leading it to consider the dynamics of usefulness.

In the sense of performance toward a given end, the word "function" is tied to the concepts of direction and end use, and it should be borne in mind that the end use is entirely within the province of the customer. He is interested

---

[1] "Priceless Relics of the Spanish Armada," *National Geographic,* June, 1969.

in what the product will do for him according to his own standards. Satisfying a customer is the end purpose of all industrial products in the free world. He may be satisfied in terms of end use, esteem, or both.

If the customer wants a little beauty in his life, and he is willing to pay for it, and he wants it in living color, that is what we must give him! Conversely, if the god of war wants more military equipment for his money, even if the equipment looks horrible, we simply have to comply. A word of warning, however—the god of war is less austere than he pretends to be. If he secretly wants the equipment to look good and if it does not, he may turn it down for "quality."

The principle of usefulness calls for two tasks:
1. Finding out what the customer really wants the product to do for him
2. Designing, making, or buying a product that will meet his needs and desires in proportion to the relative importance he gives each one of them

Whoever the customer is, the product must be worth more to him than he pays for it; otherwise he would have no reason to make the exchange. In addition to giving him his money's worth, the supplier therefore has to compensate the customer for the risk and effort of buying. A good reputation, opportune delivery, and good service usually provide such compensation over and above the sale price. The sale price then remains as the minimum measure or lower limit of what the function itself is worth to the customer. We have to give him that much utility at the very least.

But breaking even on utility and compensating the customer for the risk and effort of buying are not enough. To be competitive, we have to improve the function, reduce the sale price, or improve the way the function is performed.

Define the Function. If we take a complex idea and condense it into two words, say a verb and noun, we lose information. If we do it on purpose, we have to decide what information we can afford to lose. Then we are left with the most significant aspect of the product's purpose—the primary function (see Figure 7-3).

It is important to bear in mind that we are not attempting a definition for the benefit of all mankind, but are merely looking for one or two words that will pinpoint the primary function of the product.

This simple definition, however, should be at the highest level of abstraction compatible with the company's business. Take the function of the bent hose which transfers waste water from a clothes washer to the laundry tubs. If we are in the hose manufacturing business, we can call the function "transfer water." Raising it to a higher level of abstraction we could say "eliminate water" and recommend that the clothes washer be connected directly to the sewer. But we make hoses; so our ceiling of abstraction is still "transfer water."

To raise the level of abstraction we ask "Why?"
To lower the level of abstraction we ask "How?"

THE COMBINEX METHOD

The Combinex method consists of (1) analysis of the objective to identify its requirements or expected benefits; (2) a weighting technique based on the principle of limited resources; (3) a bounded interval scale which excludes both the inadequate and the excessive in order to measure variations within a practical range of choice; (4) a set of commensurable rating factors representing the contribution made by each of the available choices toward each of the requirements or benefits; and (5) the Combinex scoreboard, which serves a different purpose at each stage of the work. First, it helps analyze the function, task, or objective of a product by breaking it down into a number of requirements. Once these requirements have been determined, the scoreboard aligns them with their measure of relative importance or weighting factors.

The same scoreboard can be used as a submatrix for combining the various benefits that make up a complex requirement and for noting their relative contribution toward meeting this requirement. Finally, the scoreboard provides a framework for evaluating various courses of action to select: the best combination of desirable characteristics for a given level of resources, or a given combination of desirable characteristics for the least cost in resources.

More important, putting numbers into this framework makes people think. The matrix serves as a two-dimensional checklist which forces them to face, one at a time, the major factors leading to a decision. For instance, at this point they are obliged to study the choice between the best combination for the given resources, or the least resources for a given combination. Trying for both the best combination and the least resources is like trying to get both ends of a seesaw to go up at the same time. No technique of mathematical optimization can achieve this.

Optimization and Outright Improvement. Mathematics can only work with what we have to begin with, but other sciences can be used to improve what we have. Applying physics, we can put a jack under the seesaw, and by adding some effort, we can then raise both ends of the seesaw at the same time. The Combinex method can point out those areas where the available effort can be spent to greatest advantage.

The scoreboard is set up with the requirements along the top and the various choices in a column down the left. Here is a hypothetical example: A transportable battlefield radar is to be designed for sale to one of the larger South American nations. Their needs include high mobility; low initial cost, which we will call initial economy; low field cost, or field economy; good performance; and a property which the customer calls combat endurance. This latter property includes reliability, low repair time, negligible maintenance, the capacity to withstand the shock of near misses, and the capacity to withstand rugged transportation conditions.

The Art and Science of Combination. Once the customer's requirements have been defined, they must be studied systematically to generate a number of satisfactory combinations. The key concept here is that of a balanced combination rather than the most of the best of everything. This is the difference between well-allocated effort and maximum effort in all directions.

The most promising combinations of requirements are then compared by expressing, in numerical measures, the extent to which each element of each combination contributes to the particular requirement it is supposed to meet.

When numerical measures are suggested, people frown. They look at each other and someone says, "You can't compare peaches and pears!" But you can. It is no problem if you take one property at a time. A pound of peaches weighs exactly as much as a pound of pears. A carload of melons occupies the same number of cubic feet as a carload of potatoes. A dozen elephants is just one dozen, and it has the same number of units as a dozen Ping-Pong balls. At this point, someone will say, "Ha! Just what property do endurance, mobility, and cost have in common?"

Meaningful Numbers. Endurance, mobility, and cost all contribute to the mission of the equipment. How does cost contribute to the mission? It contributes numerically. Cost is the inverse measure of the number of equipments obtainable from the resources committed. Cost, size, and weight determine how many equipments can be provided in a given place at a given time. This contribution, this satisfaction of the customer's needs, is the common property that must be measured for all major characteristics.

In school, "reading, 'riting, and 'rithmetic" are measured numerically on a standard scale in which 70 is passing, 90 is very good, and 100 is perfect. This scale has the advantage that it is immediately meaningful to everyone—engineer, shop foreman, buyer.

We can use this standard scale, but we may not want to pay the full price for perfection implied by the grade 100. Absolute perfection could cost too much and take too long. So the real end points of the scale are 70 and 90. Figure 7-4 shows how system weight can be transformed into a standard scale. This figure is based on the assumption that the correlation is linear. Often, however, the relationship is not linear.

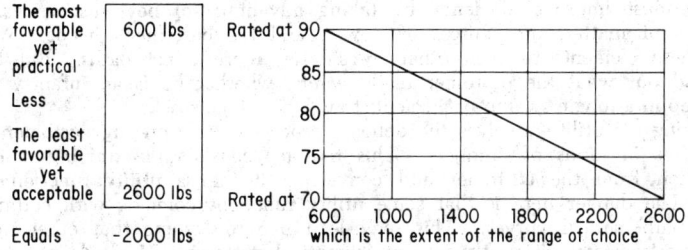

The most
favorable   600 lbs
yet
practical

Less

The least
favorable
yet
acceptable   2600 lbs

Equals   -2000 lbs

FIG. 7-4.   Rating the effect of system weight: a linear normalization.

Effectiveness and Utility. As any marine engineer knows, the amount of fuel consumed is not directly proportional to shaft horsepower delivered. In the same manner, weight reduced is not directly proportional to improved mobility. In Figure 7-5, one can see that as certain objects get lighter and lighter they do not get uniformly more mobile. At first, you cannot budge them, then you can barely budge them, and finally they begin to move. This nonlinear relationship in physical interaction is a counterpart of the utility function in economics. Such nonlinearity should be suspected, searched for, and taken into account.

It seems easy enough to understand that an article can be worth more to one person than to another or that it can be worth more for one particular application than for another, yet the concept is hard to express in numbers. Economic utility, or worth to a particular customer for a particular application, becomes increasingly important as the number of free, individual customers increases. Daniel Bernoulli formulated a good mathematical approximation of utility as early as 1738. He was followed by Herman Gossen, William Jevons, Maffeo Pantaleoni, and Alfred Marshall.

John von Neumann and Oskar Morgenstern combined the previous groundwork with their own innovations to synthesize a theory of utility applicable to present-day economics and to the modern analysis of value. The customer's concept of value starts out with his personal utility. The product must do something for him. He has to want it before he even considers the cost. Then he compares what the product is worth to him and what it will cost him. A favorable relation between product worth and product cost is the sort of value sought for in value analysis.

A Measurement of Customer Satisfaction. We try to give the customer either a little more for his money or something every bit as good for less money. The techniques of innovation, simplification, updating, and better use of people sometimes make it possible to give the customer something better for less money. To do

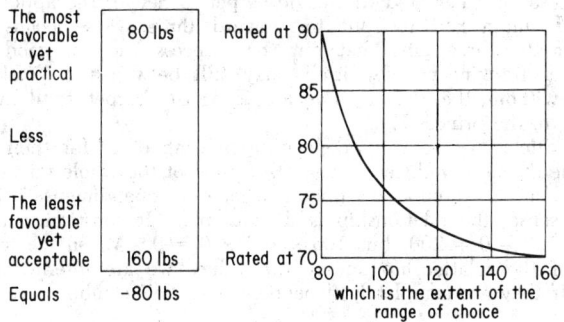

The most
favorable   80 lbs
yet
practical

Less

The least
favorable
yet
acceptable   160 lbs

Equals   -80 lbs

FIG. 7-5.   Rating the effect of weight of the heaviest case: utility function.

this, we must improve efficiency by taking advantage of new discoveries in the properties of matter, the sources of power, and the customer's needs. We then relate these elements to each other by shorter, more direct paths. But first we must find out what the customer really wants, whether he is an infantry captain knee-deep in a jungle swamp or a scientist analyzing lunar rocks.

Adjusting for utility involves the delicate process of interpreting customers' needs and desires in terms of numbers. This requires considerable training in eliciting information from the customer and converting it into a utility function on the graph. The danger here is that some utility functions coincide with certain well-known mathematical curves. This may lead one to assume that other unknown utility functions will fit neatly into mathematical patterns. Many do not. A most useful explanation of utility for our purposes can be found in Chernoff and Moses.[2]

Setting Upper and Lower Bounds. The objective more often than not combines several requirements. Each of these requirements offers a range of choice between a lower and an upper bound of demand. The threshold or lower bound is the "least favorable but adequate condition." This level must coincide with the equivalent condition of the other requirements and with the "passing grade" of the standard scale, usually 70. The upper bound is the "best practical condition." It must coincide with the equivalent condition of the other requirements and with the "very good" grade of the standard scale, usually 90. If 90 is "best practical," then 91 is "excessive" and 100 indicates the absolute maximum. These extra benefits can be accepted when they are not achieved at the expense of actual requirements within the working range.

The Additivity Assumption. Performance, safety, comfort, and style add up to a good car. Antipasto with anchovies, followed by chicken cacciatore, and topped off with spumoni ice cream makes up a good Italian meal. The first example is truly additive. The second is additive only in sequence—adding anchovies to the spumoni ice cream may not improve it.

Four factors govern additivity among the benefits: compatibility, balance and proportion, sequencing and timing, and interaction. The first two are self-explanatory; the other two call for some thought.

Sequencing and Timing. Though anchovies and spumoni ice cream may be incompatible, proper sequencing can create compatibility. Anchovies as hors d'oeuvres and spumoni as dessert, with a main course in between, do add up to a good meal.

Interaction. Interaction among the benefits is the one aspect of the additivity assumption that calls for the most precautions to avoid error.

In the Christmas story, the three gifts brought by the Wise Men of the East provide an example of additive values. Say they were worth 100 talents each: 100 talents of gold, 100 talents of frankincense, and 100 talents of myrrh. The newborn Babe then received 300 talents worth of this world's goods. Compare these gifts with the magic mirror to find the sick princess—in the Arabian Nights' tale—the apple to cure her, and the magic carpet to deliver the apple. The happy but bewildered sultan had to split the reward three ways because it was the combination of the three gifts that put the princess back in good health. 100 talents worth of frankincense, by itself, may still be worth 100 talents, but the magic apple, without the mirror to locate it or the carpet to deliver it, is not worth anything to the princess.

The magic articles of the story depend upon each other for their performance. When the value of one is reduced to zero, the value of the whole mission disappears, just as the volume of a cube disappears when any one dimension is reduced to zero. In this sense, the relationship is dimensional. It constitutes a product, not a sum: $100 + 100 + 0 = 200$, but $100 \times 100 \times 0 = 0$. When we understand the exact nature of the relationship among the values we are intermingling, there is no problem. If they are completely dependent and their combined value constitutes

---

[2] Herman Chernoff and Lincoln E. Moses, *Elementary Decision Theory*, (section on "Utility"), John Wiley & Sons, Inc., New York, 1959.

a true product, we merely multiply, as in Figure 7-1. Say the three magic articles are each 90 percent efficient: $0.90 \times 0.90 \times 0.90 = 0.729$, or a 73 percent probability that the princess will get well. But in many cases, the relationship among requirements or benefits is neither a true product nor a true sum. In most cases it approaches a sum, the interdependence being slight or uncertain. Figure 7-1 is an example of interdependence.

It is possible to bring ponderous mathematical tools to bear on this problem, provided the nature and degree of independence in each case is fully understood. A more robust approach is to assume additivity and safeguard the work from serious error through the use of procedures which minimize or exclude the nonadditive portion of mixed values.

The most important safeguard is setting lower bounds of customer acceptance. These bounds have a threshold nature—below them, the customer will not buy. This situation can be created by introducing into the process of optimization one or more benefits below the threshold level of adequacy, or as happens more frequently, permitting a supposedly unimportant benefit to be driven below its threshold level by the human tendency to maximize the more glamorous ones. This error is guarded against by excluding from the Combinex scoreboard any input that falls below a previously established level of adequacy.

Another safeguard consists of preventing the introduction of an excessive measure of any one benefit into the Combinex scoreboard. Such an excess would displace an equivalent amount of desirable or necessary measures of other benefits. When excessive quality, excessive economy, and excessive reliability masquerade as desirable benefits, they introduce an error which leads to sheer waste.

To exclude this error, there must be an upper bound to the desired benefits. It is not for the manufacturer to tell the customer how much is too much or what he needs or does not need. The customer himself sets these bounds through his buying practices. The manufacturer therefore must learn the upper bounds of customer satisfaction from his own marketing, styling, and quality control people.

The reason for this extraordinary care in the measurement of what the customer wants is well stated in the U.S. Department of Commerce booklet, *Profits and the American Economy:* "Producing too much or too little, or at a price that is too high or too low, or at a level of quality uncalled for or inadequate, can wipe out profits."

Weighting Factors. Assigning measures of relative importance to the requirements or benefits calls for the greatest possible information on the intended use of the equipment and on the objectives of the decision maker. It also calls upon judgment in the highest degree.

The total importance of the requirements or benefits is set equal to unity, which is then divided by the number of requirements, under the initial assumption that they are all equally important. This is seldom the case, but it is a good way to start. If there are five requirements, each weighted at 0.2, some will have to be raised and others lowered.

These measures of relative importance or weighting factors are placed under their corresponding requirements (see Figure 7-7). The decision is really made when this is settled, but there is much backing and filling before it is settled. Once the relative importance of the objectives has been determined, all that remains is to find out which of the available choices best satisfies that decision.

Effectiveness of the Various Choices. Each choice or course of action contributes to each of the requirements or desired benefits in varying degrees of effectiveness. For example, system weight and weight of largest case make a negative contribution to mobility. We want a simple additive matrix using nonnegative numbers; so we must convert this contribution into its complement of lightness or positive contribution to mobility, as shown in Figures 7-4 and 7-5.

Use of Submatrices. The various elements of mobility are combined in a submatrix (see Figure 7-6) and weighted for relative importance. The measures of effectiveness introduced from the graphs in Figures 7-4 and 7-5 are entered in the body of this submatrix in the upper left-hand corner of each cell, as are

Benefits	TOTAL WEIGHT	HEAVIEST CASE	LARGEST CASE	FEWER UNITS	SETUP TIME	
Weights	0.30	0.20	0.15	0.10	0.25	
Old design	70 ... 21	70 ... 14	70 ... 10	80 ... 8	75 ... 19	MERIT 72
High performance design	73 ... 22	80 ... 16	74 ... 11	82 ... 8	75 ... 19	76
High endurance design	80 ... 24	74 ... 15	70 ... 10	74 ... 7	71 ... 18	74
High mobility design	90 ... 27	89 ... 18	89 ... 13	80 ... 8	90 ... 22	88

FIG. 7-6.   Example of a submatrix: elements of mobility.   The last column, showing the relative merit of each design's mobility, is entered in the "Mobility" column of the Combinex scoreboard (Figure 7-7).

the other numbers developed in the same manner.   These numbers are multiplied by the weighting factor at the top of their respective columns, and the results entered in the lower right-hand corner of each cell, and then added across to yield the right-hand column of relative merit, representing the effectiveness with which each choice contributes to mobility in the final matrix.   Note that the numbers now appear on the upper left-hand corner of the cells in column 3 of Figure 7-7.

Racking Up the Results.   The measures of effectiveness in the final matrix (Figure 7-7) are multiplied by their weighting factors as before, and then added across to yield the relative merit of each of the choices.

Suppose that instead of selecting the best out of four alternatives, we were allowed to make changes resulting in different and perhaps better combinations.   As a Persian mathematician once put it, "Could we but shatter this scheme of things entire, and make it nearer to the heart's desire!"

Sometimes we can!   Not as a committee investigating someone else's work, but as the personal staff of whoever is responsible for a given product or service.

If we provide him with specialists in the relevant disciplines, other than his own, the man responsible can make new decisions involving conflicting requirements, intricate and unsuspected rates of exchange, and the leverage exercised by each element upon the value of the whole system.

It is important to bear in mind, however, that scientific aids to decision making only align the information for the exercise of judgment by the responsible risk taker.   These aids provide neither wisdom nor insight, but they do reveal previously unperceived magnitudes and relations which bring into play the wisdom and insight of the responsible managers.   For this reason, the Combinex and similar methods are as useful in presenting recommendations as they are in arriving at them.

Benefits	PERFORM-ANCE	EN-DURANCE	MOBILITY	INITIAL ECONOMY	FIELD ECONOMY	
Weights	0.20	0.18	0.22	0.23	0.17	
Old design	70 ... 14	73 ... 13	72 ... 16	70 ... 16	75 ... 13	MERIT 72
High performance design	89 ... 18	79 ... 14	76 ... 17	88 ... 20	81 ... 14	83
High endurance design	76 ... 15	89 ... 16	74 ... 16	71 ... 16	87 ... 15	78
High mobility design	70 ... 14	72 ... 13	88 ... 19	75 ... 17	70 ... 12	75

FIG. 7-7.   The Combinex scoreboard as used in selecting a field radar.

SELECTING A MANUFACTURING METHOD

The task of converting an engineering design or a purchasing specification into reality can present perplexing choices. In such a case, the Combinex method may be helpful.

Identifying the Expected Benefits. Assume that we are ordering a complex wave guide structure requiring a smooth interior surface finish between the bounds of 8 and 64 microinches, where 8 is the smoothest we can practically use and 64 is the roughest we can accept. Dimensional precision is another critical requirement which must fall between an upper bound of ±0.0005 inch and a lower bound of ±0.0003 inch. Weight is a penalty because the equipment must be transported rapidly over rough terrain. Cost can be decisive to the contractor because the contract will be awarded on competitive bids, and decisive to the user because cost determines the number of equipments he will receive for the resources available.

Upper and lower bounds, in dollars and pounds, respectively, have been established for economy of production and for lightness of weight, the lightest weight practical being 150 pounds and the heaviest acceptable being 190 pounds. In the same way, low initial cost is normalized into initial economy, between the bounds of lowest cost practical, $1,500, and highest cost acceptable, $5,500.

Rating the Benefits. Having identified the benefits and defined their upper and lower bounds, there is a range of choice within each benefit. Such a range of choice permits the free play among the benefits which is necessary to arrive at the best combination.

This altering of proportions is what artists do when they mix paints, what poets do when they combine words, and what cooks do when they calculate the effects of moisture, temperature, time, and seasoning on the flavor and texture of well-prepared food.

In this case, we will have to combine microinches of surface finish with tolerances in decimals of an inch, with pounds of weight, and with dollars of cost.

All these peaches and pears, apples and oranges, contribute to the function of the equipment, and we can measure that. This example was chosen because the raw measures are negative—something the reader should learn how to handle. The less surface roughness, the smaller the tolerances, the lighter the weight, and the lower the cost, the better. Not only must these "baddies" be made to fit the same measure, but they must be turned into "goodies," for we want to measure their positive contribution to the function of the equipment. Two examples will be enough to show how the transformation is accomplished:

1. Smoothness of surface finish, transformed from the raw data which were measured in microinches of surface roughness (see Figure 7-8)
2. Dimensional precision, transformed from tolerances measured in decimals of an inch (see Figure 7-9)

The straight line sloping down from the left in Figure 7-8 not only inverts the relationship so that low numbers get a high rating, but also transforms the raw data into the 70-to-90 rating scale.

The curved line sloping down to the left in Figure 7-9 also inverts (downward slope) and transforms (from one scale into another), but in the course of transformation it takes into account the difference between tolerances allowed and the effect of dimensional precision on the function of the equipment.

Assigning Weights. There is a balance and proportion among the benefits which will yield the best performance in a product and the greatest effectiveness in a system. One way of describing such a balance, when we know what it should be, is to assign measures of relative importance, or weighting factors, to the benefits, as outlined in the description of the Combinex method. Even when we do not know what the balance should be, preliminary weighting factors serve as a basis for experimentation, calculation, or simulation in the search for the best balance.

Experience with the performance of electronic wave guides which form part of transportable military equipment reveals that dimensional precision is by far

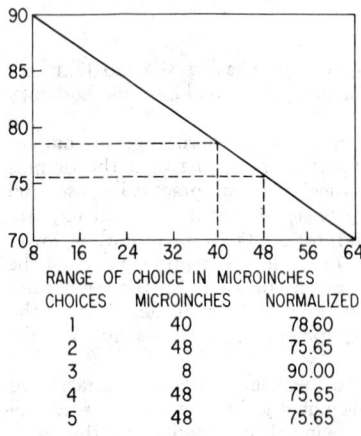

CHOICES	MICROINCHES	NORMALIZED
1	40	78.60
2	48	75.65
3	8	90.00
4	48	75.65
5	48	75.65

RANGE OF CHOICE IN MICROINCHES

FIG. 7-8. Microinches of surface roughness into smoothness of surface finish. The column of normalized ratings is entered under B_1 in the Combinex scoreboard (Figure 7-10).

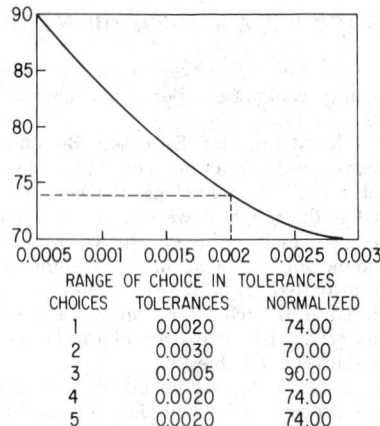

CHOICES	TOLERANCES	NORMALIZED
1	0.0020	74.00
2	0.0030	70.00
3	0.0005	90.00
4	0.0020	74.00
5	0.0020	74.00

RANGE OF CHOICE IN TOLERANCES

FIG. 7-9. Tolerances into dimensional precision. The column of normalized ratings is entered under B_3 in the Combinex scoreboard (Figure 7-10).

the most important consideration, being as important as the other three benefits put together. It is therefore assigned half the total weight, or 0.5 (see Figure 7-10). Smoothness of surface finish and economy of production are each considered twice as important as lightness of weight; so the remaining 0.5 weight is distributed on a 0.2, 0.2, 0.1 ratio, the total adding up to unity.

Setting Up the Scoreboard. Figure 7-10 shows the Combinex scoreboard with all the work completed. The benefits B_1 through B_4 are listed along the top, with each of the weights beneath its corresponding benefit, and the choices or options C_1 through C_5 are listed in the column at the left.

BENEFITS WEIGHTINGS / CHOICES	Smooth B_1 finish		Production B_2 economy		Dimensional B_3 precision		Light weight B_4		Relative merit
	0.2		0.2		0.5		0.1		
C_1 STANDARD WAVE GUIDE COMPONENTS SPECIAL JOINTS	78.60	15.72	70.00	14.00	74.00	37.00	72.70	7.27	73.99
C_2 FORMED AND PUNCHED ALUM. SHEET DIP OR OVEN BRAZED	75.65	15.13	90.00	18.00	70.00	35.00	90.00	9.00	77.13
C_3 ELECTROFORMED ON FIXTURED MANDRELS	90.00	18.00	78.80	15.76	90.00	45.00	88.80	8.88	87.64
C_4 FORMED AND PUNCHED ALUM. SHEET, ELECTRON BEAM WELDED	75.65	15.13	81.20	16.24	74.00	37.00	90.00	9.00	77.37
C_5 INVESTMENT AND DIE-CAST COMPONENTS BRAZED TOGETHER	75.65	15.13	71.90	14.38	74.00	37.00	70.00	7.00	73.51

FIG. 7-10. The Combinex scoreboard: selecting a manufacturing method.

Now look at Figure 7-8. Choice 1 has a surface roughness of 40 microinches. The dotted line rising from the 40 along the horizontal axis shows how this number is transformed to its rating of 78.60 on the vertical axis. Choices 2, 4, and 5 all have a surface roughness of 48, which becomes 75.65; and choice 3, 8 microinches, the best finish, is rated at 90.00. The ratings for the five choices then appear in the column under "Smoothness of finish" in Figure 7-10. They are written in the upper left-hand corner of each box.

The same procedure is followed for the other benefits whose ratings are written in the upper left-hand corner of the boxes under them.

These ratings are then multiplied by the weights above them, and the results—written in the lower right-hand corner of each box—are added across to give a figure of relative merit for each choice.

Looking at the "Relative merit" column in Figure 7-10, choice 3, electroforming, rates highest for this particular application.

COMPARING VALUE ANALYSIS OPTIONS

Among the options developed in a value analysis study, one or two may stand out as superior to the others. In that case, a simple appraisal of performance and cost reveals the best one.

The Combinex method should be used only when the choice is difficult or when the reasons for the choice must be explained to people who have not participated in the study.

Up to this point, we have been concerned with providing a better product or service for the money, or providing the same product or service for less money. In any case, we have been studying the product or service itself, or more accurately, what it does for the customer.

Now we have to provide information on the merits of our proposals. **Benefits from Whose Point of View?** Before value analysis proposals can do anything for the customer, they must do something for the company; and before they do anything for the company, they must do something for the people who approve them and implement them—the risk takers.

Who must approve our proposals? What are his needs? What are his problems?

Who will allocate the funds? From what sources?

Who will do the work? When? Where?

What department will gain the most? Are they represented in the task group? Do they understand the benefits they will derive?

What department will suffer? Loss of work load? Overload? Schedule delay? Loss of control?

Assessing the Chances of Acceptance. Answering these questions, and many more like them, will reveal many advantages and disadvantages associated with a given proposal. Earlier comparisons helped in selecting those options most beneficial to the customer. The present comparison will help in selecting from options which are known to be good for the customer, the ones more beneficial to management.

In appraising advantages and disadvantages, the first pitfall to avoid is counting them. Their number is not what matters. One single advantage, such as being a pretty girl, often outweighs many disadvantages in a secretary. "Oh well, you can always teach them to read and write," is often given as justification for the obvious choice.

A single disadvantage, too, can outweigh all the advantages, such as the speed, endurance, and spirit of the horse Borysthenes, who has the disadvantage of having been dead for 1,900 years.

The second pitfall to avoid is canceling out disadvantages against equally important but unrelated advantages, as was done by the merchant of Ahmadabad who claimed full price for a slave girl, saying, "She is missing two fingers, but she has twelve toes."

Some Bounds Are Still Required. Although the mechanics of the Combinex method seldom need to be applied to the advantages and disadvantages of accepting a proposal, the ground rules do establish certain safeguards such as setting bounds. In the foregoing example, judiciously set bounds would have excluded illiterate secretaries and dead horses.

Disadvantages cannot be inverted too meaningfully into positive benefits. By definition, disadvantages are incorrigible "baddies." The disadvantage of schedule delay would have to masquerade as somewhat reduced delay. The phrase has a hollow ring. Risk would have to become safety, security, or even invulnerability. Yet we had no trouble inverting surface roughness into smoothness of surface finish, or tolerances into dimensional precision. Why the difference?

Two Different Kinds of Animals. Design or service requirements, often in competition for their share of resources, must have both an upper and a lower bound because accepting the excessive in any one benefit will detract from the useful in another.

Once a satisfactory mix has been achieved, its advantages are usually open-ended at the top. If being pretty is an advantage in a secretary, we do not mind if she is very pretty or quite beautiful. If dollar gains are an advantage, we do not mind how much more money we make.

We must set lower bounds, however, to the advantages. If the appearance of secretarial applicants ranges from "her face stops clocks" through "her appearance is bearable on a rainy Monday morning" all the way to "she is divinely beautiful," the lower bound should be set no lower than "her appearance cheers you up on a rainy Monday morning."

The concepts of utility and effectiveness keep reappearing in this chapter when there is a difference between an input, such as the optical image of the girl, and the effect of that input. Let us say that our clock-stopper has a very pleasant manner, likes her work, and is happy to see her boss. He could well rate her as a girl who "creates sunshine on a rainy Monday morning," cheerfully adding, "Who needs clocks!"

If advantages are open-ended at the top, disadvantages are open-ended at the bottom. We do not care how small they get and we are happy if they disappear, but we do not want them growing beyond unacceptable limits. We may rate risk simply as low, moderate, and high, but then we must set a bound at unwarrantable.

Importance of Time, Money, and Risk. At this stage, we have a list of advantages and disadvantages, all open-ended at the favorable end and all bounded by the least favorable condition acceptable. We know that some are more important than others and that we cannot match disadvantages against unrelated advantages.

Matching related advantages and disadvantages is another story. We can match them when they differ only in polarity or direction. Dollar losses can be matched with dollar gains within a given time frame. Time lost can be matched with time gained within the same cycle, provided the change does not alter necessary sequence.

One definition of risk is the probability of an unfavorable event multiplied by the cost of that event. An even chance of losing $1,000 yields an expected loss of $500.

Suppose we estimate $1,000 as the cost of implementing a proposal provided we do not run into problems. If we do run into problems, it may cost twice as much, and there is an even chance that we will run into problems.

Running into problems would raise the cost of implementation from $1,000 to $2,000, that is, $1,000 more than anticipated. Chances of that happening are 50 percent; so we must add $500, as risk capital, to the estimated cost of implementation.

The Three Baskets. Most advantages and disadvantages can be put into three baskets: time, money, and risk. The relative importance of these major categories in the eyes of management varies in each situation. In selecting those proposals most likely to be accepted, the value task group must ascertain management's present aims in the areas of gross margin, cash flow, product turnover, dependable delivery, and sales. Cost savings affect the first two most directly. Saving time affects the next two. Sales are affected by both.

In addition to saving time and money, a task group truly concerned with value will identify areas where additional money invested or additional time spent will yield as good or better a return than routine investment opportunities.

In summary, proposals should be screened for acceptability, bearing in mind that acceptance may depend not only on savings but also on improvement.

COMPANY BILLINGS AND EXCHANGE VALUE

The money a retailer receives in his cash register and a company receives in payment for invoices is an exact measure of the exchange value placed by the customers on what they buy from the company.

Why did the customers make the exchange? Why does anybody make an exchange? Because what he receives is worth more to him that what he gives. The relation of worth to cost in the eyes of the customer can therefore serve as a basis for measuring exchange value. Figure 7-11 shows the importance of the various relations between worth and cost. This chart helps us to understand the difference among value, gain, and return.

The index of value is a dimensionless number analogous to efficiency in mechanics. Both are ratios of output over input, but the analogy stops there. In mechanics, the output never exceeds the input; in profitable trade, the return must always exceed the outlay.

The measure of gain or loss is the difference between worth and cost expressed in a common dimension. It is analogous to profit in business. It tells, in numbers, how much the customer came out ahead or how much he lost, but it tells nothing about the amount of resources invested. To find out whether the gain was worth committing those resources, the measure of gain or loss must be related to the index of value—neither is fully meaningful without the other.

The rate of return, of course, is the gain or loss expressed as a percentage of the resources invested.

A study of the relationships illustrated in the chart reveals that there is much more to improving value than merely minimizing cost. The task is to optimize value—to provide the required effectiveness of a mission or product for the least

A customer appraises a product:				
Calculating that:	He calls the ratio:		And he calls the difference:	Which yields:
The function is worth—$20 and the product costs—$10	A good value	2.0	A $10 gain	100 % return
The function is worth—$10 and the product costs—$10	An even exchange	1.0	No gain, no loss	No return
The function is worth—$ 5 and the product costs—$10	A poor value	0.5	A $5 loss	50 % loss
Ratio of worth to cost	Qualitative rating	Index of value	Measure of gain or loss	Rate of return

Fig. 7-11. How customers quantify the utility of a product's function in terms of what it is worth to them.

cost, or often more important, to provide the greatest mission effectiveness or the highest product utility for a given cost.

A QUANTITATIVE FORMULA FOR VALUE

How can the index of value be made more meaningful? It is a dimensionless number because the dollar dimensions canceled out. By reintroducing the cost dimension, but relating it, not to what the function is worth in dollars, but to what the function accomplishes in terms of performance, we arrive at a quantitative formula for the value of a product or service:

$$\frac{\text{Function}}{\text{Cost}} = \text{value}$$

where function is entered in units of performance, cost is entered in monetary units corresponding to the expenditure of resources, and value is a figure of merit expressing the relation between what a product or service accomplishes and what it costs.

Feet per second, miles per hour, revolutions per minute, cubic yards per hour, cubic feet per minute, gallons per minute, passenger miles, ton miles, pounds of payload, horsepower, and kilowatt-hours are units frequently used to measure performance.

Deutsche marks (DM), francs (FR), and dollars ($) are typical units for measuring cost. Although money offers an excellent system of units for comparing the expenditure of resources, cost need not always be expressed in money.

When the expenditure being considered is primarily that of a single resource, such as coal, the units of cost may be entered directly in units of that resource. The concept of resource availability can be extended to such limiting factors as weight, bulk, elapsed time—whatever penalty must be incurred to obtain the required result. The cost then may be entered in units of the limiting factor, such as the weight of fuel that a steamship must carry to complete a given voyage. The weight of fuel displaces payload and is an element of cost. Cubic feet of space occupied by the life-support equipment in a space capsule is an element of cost because it limits the payload.

By and large, however, cost enters the formula for value in units of currency. The term for value, then, is expressed in units of performance per unit of currency. Say the function is to travel from Philadelphia to Chicago, first class, by air:

$$\frac{\text{Function}}{\text{Cost}} = \text{value} \qquad \frac{666 \text{ air miles}}{\$54 \text{ air fare}} = 12.33 \text{ miles}/\$1$$

The purpose of this formula is to tell us what we get for our dollar, so that we can compare relative value. Traveling from Philadelphia to Los Angeles gives us more mileage per dollar, 14.42 miles/$1, while from Philadelphia to New York we get only 5.8 miles/$1. We were already aware that long-range air travel is more economical than short hops. Now we know how much more.

The formula follows all the laws of transposition:

$$(\text{Value})(\text{cost}) = \text{function} \qquad \frac{\text{Function}}{\text{Value}} = \text{cost}$$

$$(12.33 \text{ miles}/\$1)(\$54) = 666 \text{ miles traveled}$$

$$\frac{666 \text{ miles traveled}}{12.33 \text{ miles}/\$1} = \$54$$

The unit of value, miles per dollar, is of necessity a compound unit made up of two dimensions: performance and cost. Sometimes it may include more dimensions, as when performance itself is given in compound units, such as miles per hour. In that case, the unit of value would be miles per hour per dollar, having three dimensions: distance, time, and cost.

The following example from the process industry illustrates how such compound units must be handled.

A continuous process treats 600,000 gallons of product in a 24-hour day at a daily cost of $12,000. Both the industrial efficiency and the economic value of the process are subject to a measure which answers the question, "What do we get for our dollar?"

We enter the measure of performance (600,000 gallons/day), and the measure of cost ($12,000) in the formula for value:

$$\frac{\text{The function is 600,000 gallons/day}}{\text{The cost is \$12,000}} = \text{value}$$

A simple equation? Not at all. It is an equality of ratios made up of two statements of proportionality, each using a different combination of units.

The first relates gallons to dollars:

$$\frac{600,000 \text{ gallons}}{\$12,000} = 50 \text{ gallons/\$1}$$

Units of product per dollar	Units of time per dollar	Units of cost in dollars
$A = 50$ gallons	$B = 0.002$ hour	$C = \$1$
To find time and cost for a given quantity, divide by A]]] and multiply by B and C:		
1,000 gal/$A = 20$ $\longrightarrow$	$20 \times B = 0.04$ hr $\longrightarrow$	$20 \times C = \$20$ $\longrightarrow$
= 1,000 gal each 0.04 hr at a cost of $20		
25,000 gal/$A = 500$ $\longrightarrow$	$500 \times B = 1.0$ hr $\longrightarrow$	$500 \times C = \$500$ $\longrightarrow$
= 25,000 gal each hour at a cost of $500		
To find quantity and cost for a given time, divide by B and multiply by A and C:		
$8.3 \times A = 415$ gal ↑└	1 min = 0.0166 hr 0.0166 hr/$B = 8.3$ $\longleftarrow$ $\longrightarrow$	$8.3 \times C = \$8.30$ $\longrightarrow$
= 415 gal each minute at a cost of $8.30		
← $4,000 \times A = 200,000$ gal	← 8 hr/$B = 4,000$ $\longrightarrow$	$4,000 \times C = \$4,000$ $\longrightarrow$
= 200,000 gal each 8-hr shift at a cost of $4,000		

Fig. 7-12.   Computation with compound units.

that is, we process 50 units of product per dollar.

But when we pay for the processing effort, we also pay for processing time; so we have a second statement which relates dollars to units of time.

$$\frac{24 \text{ hours}}{\$12,000} = 0.002 \text{ hour}/\$1$$

The time and effort we get for our dollar, that is, a measure of the industrial efficiency and economic value of the process is then

$$50 \text{ gallons}/0.002 \text{ hour}/\$1$$

So long as we keep track of the three types of units—units of product, units of time, and units of cost—we can use this single expression to give us time and cost for any given quantity, or quantity and cost for any given time interval, as shown in Figure 7-12.

SUMMARY

Fundamental characteristics of value engineering, which have not changed since Lawrence D. Miles founded the discipline in the 1940s, are:

Value. Deliberately stepping from matters of fact to matters of value, from what is to what should be.

The Function. Concentration on the function of a product or service, rather than the structure of the product or the form of the service—a results-oriented approach.

The Method. A proved sequence, called the job plan, which systematically applies information search patterns and creative techniques to the analysis and improvement of value.

BIBLIOGRAPHY

American Society of Tool and Manufacturing Engineers, *Value Engineering in Manufacturing,* Prentice-Hall, Inc., Englewoods Cliffs, N.J., 1967.

Chernoff, Herman, and Lincoln E. Moses, *Elementary Decision Theory,* John Wiley & Sons, Inc., New York, 1959.

Churchman, C. W., *Prediction and Optimal Decision,* (chapter on "Additivity of Values"), Prentice-Hall, Inc., Englewood Cliffs, N.J., 1961.

Falcon, William D., *Value Analysis/Value Engineering,* American Management Association, New York, 1964.

Fishburn, Peter C., *Decision and Value Theory,* John Wiley & Sons, Inc., New York, 1964.

Hicks, John R., *Value and Capital,* Oxford University Press, London, England, 1946.

Miles, Lawrence D., *Techniques of Value Analysis and Engineering,* McGraw-Hill Book Company, New York, 1961.

Value Engineering Digest, published twice a month by Sci/Tech Digests, Washington, D.C. (This is a must for the professional.)

von Neumann, J., and O. Morgenstern, *Theory of Games and Economic Behavior,* rev. ed., (chapter on "Utility"), John Wiley & Sons, Inc., New York, 1953.

Note: The two books below, in German and Swedish, respectively, are worth getting if for no other reason than the charts, forms, and diagrams. They are so good that a value analyst or value engineer will end up learning the language.

Kourim, Gunther, *Wertanalyse,* R. Oldenbourg, Verlag, Munich, 1968.

Ollner, Jan, et al., *Vardeanalysis,* Sveriges Mekanforbung, Stockholm, 1967.

Section **3**

Work Measurement Techniques

Establishing and Maintaining Sound Standards

RALPH PRESGRAVE

Professor Emeritus, School of Business, University of Toronto, Toronto, Canada

It should hardly be necessary to point out that the term "standards" in the chapter title refers to time standards, that is, to the times established for the performance of designated tasks, taking due account of standards of quality, dimension, and so on, that may also be required. The all-embracing term is "work standard," but the concern here is almost entirely with the element of time, whether or not the standard is expressed in terms of time per quantity, of output quotas, or even of money.

The numerous technical procedures for establishing standard times are discussed in subsequent chapters and will barely be touched on in this one. Rather, the emphasis will be on the time standard itself—its nature and environment, and above all, the crucial need for establishing and maintaining standards that are sound and viable. The soundness is a matter of technical correctness by whatever criteria are set up. The viability relates mainly to acceptability by management and employees.

Soundness and Viability. It is no exaggeration to say that the problems of maintaining sound standards far overshadow those of establishing sound standards in the first place. Combined, they form what is almost certainly the most difficult and continuous problem with which the industrial engineer must cope. Indeed, it must also rate high among the most frustrating problems of industry.

This is not surprising when one considers that it deals directly with what a man is required to do during his working hours and what he will be paid for doing it. That is to say, it moves head on into the inexorability of economic laws as they clash with the present near-inexorability of human demands.

It will be seen from the foregoing that work standards are considered here only as they are used in incentive plans or in enforced quota systems, that is, where a man's pay and what he does to earn it are directly affected. If time study is used merely to schedule production or to facilitate cost estimating, then the standards are never truly put to the test and the problems raised are minor.

WORK STUDY IS THE CORE OF INDUSTRIAL ENGINEERING

The problems of costs and standards and human relations should be apparent to all; but increasingly there has been a tendency, among those who are not directly involved in the problem, to be carried away by the mystique of automation and cybernation and to conclude that work study is an obsolescent technique.

Such a conclusion obscures the hard fact that a time standard—whether of machine time or of man time and whether precise or not—is essential in the control of production, the synthesis of cost, the payment of wages, and the projection of manufacturing budgets. Time study is the medium whereby technological gains are quantified and consolidated.

This is not to deny that there are areas in which time study is of little value, or that there are many instances where the timing of the man is of minor consequence as against the timing of the machine or the mechanized system. However, the reverse may also be true, and in other areas, automation is indirectly intensifying the need for the most advanced work study. There is still heavy employment in those labor-intensive industries where full automation seems unlikely to be achieved soon.

Wages naturally tend to be higher in the highly automated industries with minimal labor content. This tends also to force up wages in the nonautomated industries whose most serious competition often comes from countries with lower wages and lower standards of living. The result is a steady demand for the highest analytical and measurement skills in the development of waste-free work standards.

Recognition of Constraints and Tolerances. It is a further anomaly that not least among the required skills are the judgment, imagination, and flexibility that will permit the practitioner to understand the constraints of the situation and the nature and extent of the tolerances he may admit in the standards he sets.

For instance, a highly repetitive operation performed by a large group of people over long periods will perforce be analyzed and measured with scrupulous attention to minute detail to prevent embodying potential waste in the final standard. The latter will be precise and its tolerances negligible—possibly of the order of ±2 percent. Many clerical operations (for example, card punching) will eventually fall into this class. The studies that result in the standard will be long and costly, but they will be justified. At the other extreme, a lengthy, nonrepetitive maintenance job performed on occasion by one person might not be worthwhile studying at all, except perhaps in a brief way by reference to past records or to crude standard data. The tolerances might be ±25 percent, and the standard might bear little relationship to what it would be if the operation were a repetitive production job. The economics might even be such that no specific standard would be set, especially in the smaller plants.

Most work lies between these two extremes and in general provides the area to which this chapter relates. In setting standards throughout the range, the perfectionist may find it difficult to depart from his meticulous ways, and his standards may cost more to produce than they are worth. The more impatient type will produce standards that incorporate waste of time, and while cheap to produce, will in the end be costly.

Both practitioners may be classed ignominiously as rate setters. The golden mean is the intelligent and perceptive industrial engineer with keen judgment and a sense of values.   It is, of course, an absolute requisite that he have high technical skills.

Consistency Is the Basic Essential.   Referring again to waste of time, it is well to note that in the last analysis all waste is the waste of time unless its recovery or prevention would entail a still greater expenditure of time.

This is oversimplification of a fairly complex process, but before looking at some of the problems, it should be borne in mind that, in spite of what has been said about precision, there can be no such thing as an intrinsically correct time standard.   All time standards—in fact, all standards—are peculiarly the result of arbitrary decisions, not necessarily random or irrational, but arbitrary nevertheless.

For instance, in setting time standards it is necessary to select a base.   This may be the statistical concept of the average operator, the slow operator, the fast operator, the third quartile operator, or any other operator.   Once this base is chosen, the test of any standard is the degree of consistency with which it relates to all other standards embodying the same basic concept.   This is not the place to explore the implications and ramifications of this, but a careful reading of subsequent chapters in this Handbook should help bring it into perspective, particularly the chapters discussing such matters as the rationale of conventional stopwatch time study, which perforce uses the selection and rating of elemental times in default of statistical information; predetermined motion time systems, which are or should be based entirely on the compilation of valid statistical numbers; or work sampling, which, although itself a statistical device, operates without recording the actual time taken to perform any part of the operation that is being studied.

PROBLEM AREAS IN PRACTICAL APPLICATION

The difficulties encountered in the continuing application of sound work standards are divided into four main areas:
1. The initial establishment and application of work standards
2. The subsequent and continuing introduction of work standards on new operations
3. The correction of existing work standards when the work content has been reduced
4. The correction of existing work standards when the work content has been increased

Initial Application.   The elements of technical skill and consistency exist equally in all four of the above situations.   The problems of acceptance exist in varying degree and kind throughout.   The initial application should present little other than a straight technical challenge once the principle of direct incentives or of mandatory quotas has been accepted.   For this reason, the initial application will not be discussed here.   Rather, the reader is referred to subsequent chapters dealing with the different types of work measurement and the manner of their application.

Subsequent New Standards.   The second area, the continuing introduction of work standards as new jobs come into being, differs from the first area only when and to the extent that area 3 (the correction of loose standards) becomes a problem.   It appears to be almost a law of nature that standards that are not constantly monitored tend to become loose.   If this looseness is permitted to continue to the point of significantly inflating actual or potential earnings, it will become increasingly difficult to introduce standards on entirely new operations, when such standards conform with the original basic criteria.

Correction for Reduced Work Content.   Area 3, the correction of standards that have become loose, is one of the touchiest problem areas of modern manufacturing.   There are many pressures that tend toward excessive standards, and squatter's rights set in early if corrections are not made even earlier.   The problem of loose standards is discussed in some detail below.

Correction for Increased Work Content.   Area 4 mainly refers to standards that have become tight—a situation that occurs far less frequently than looseness and creates few serious difficulties with employees.   Once the error is discovered, the standard is corrected, usually with alacrity. Neither management nor industrial engineering is averse to seizing the opportunity to demonstrate the fairness of the system and of themselves.

Apart from this, the investigation of once correct standards that now appear to be too low presents a unique opportunity to the time study engineer. This will be dealt with at the end of this chapter.

Creeping Change. The most serious manifestation of the problem of loose standards is the so-called creeping change, which may be defined as the gradual loosening of a time standard to an embarrassing extent, without there being any clearly defined point at which revision appeared practicable. Also included must be those instances where there undoubtedly was a juncture at which a change in standard would have been warranted, but the moment passed either unnoticed, unreported, or deliberately ignored. There are also the changes in method introduced by the operator and skillfully concealed to forestall a reduction in a piece rate or a time standard. Again, it may be that a standard was incorrect in the first place and was not taken overt advantage of by the operators. The looseness may have been a straight error, or the standard may well have included certain elements of care or inspection that had been challenged by the observer, defended by the foreman and the operator and possibly by the merchandising department, and subsequently abandoned without the fact being reported. In such cases, the time study man unwittingly sows the seeds of his own subsequent problems.

The Need for Correction. These developments can lead eventually to the discrediting and even the collapse of the entire work standard and wage incentive system. The more extensive the looseness and the longer it is permitted to exist, the more difficult it is to restore consistency except by some sweeping capitulation such as the revision of basic criteria to compensate for the looseness. There is strong and understandable resistance by operators to corrections that might result in lower earnings. This is so even when the operator is aware that his current earnings are excessive in comparison with other operators and possibly with what his own earnings were a short time earlier.

It is natural to assume that if the work study had been correct in the first place, the problem would not have arisen. To some extent this is true; but the fact that the dilemma created by loose and inconsistent standards is an almost universal experience suggests that any doctrine of perfection will not eliminate it.

It is, of course, imperative that all waste should be taken out of time standards before they are introduced, and that the standards should be carefully policed after they have been introduced. In the case of the former, the analytical and descriptive possibilities of predetermined motion time systems should be exploited to whatever extent is feasible.

The Policing of Standards. In reference to the policing of standards, constant surveillance of individual performance must become a matter of routine and all deviations from the normal range of output for the individual must be reported at once by the payroll department. Only in this way can the deviations be accounted for and dealt with. It is also necessary that all who could have knowledge of changes in methods, materials, equipment, and service be required to report these changes to the industrial engineering department as promptly as possible.

As noted, squatter's rights set in early in respect to loose standards and become more positively asserted as time passes. It is in this area that the problems become more acute and more complex, respective rights come into conflict, and serious difficulties arise in industrial and personnel relations. Because of this, this phase of the general problem is discussed more extensively than any of the other three.

SIX PHASES OF REDUCTIONS IN JOB CONTENT

Entirely apart from initial errors in the work study, there are at least six fairly distinct phases of reductions in job content that may cause a work standard to permit unduly high earnings. They are listed below, more or less in ascending order of the difficulty of correcting the time standard.

1. Specific change introduced by the employer and immediately studied
2. Change resulting from an abrupt improvement in working conditions
3. Change resulting from a gradual improvement in working conditions
4. Change in method unknowingly introduced by operator
5. Change in method consciously introduced by operator
6. Change inherited by operator

Original Error and Delay. In addition to these six points, there is the less important matter of original error in the standard that has gone unnoticed and uncorrected for an unduly long time.

There is no assurance that a loose standard caused by such error will show up in disparities in earnings or performance. It may be obscured by the mix of other standards; it may come into the shop so intermittently that the operator will not get into his stride; or the operator may recognize the looseness and gear his pace to his own earnings target.

Whether or not the standard can be corrected will depend on how long it has been permitted to run, on the terms of the union agreement, and on the nature of the error. Inevitably, the longer an error persists and the longer a higher level of earnings opportunity is maintained, the greater will be the resistance to change.

A manifest error in arithmetic may well be covered in the agreement, and in any event is easier to correct than an error in recording the job content. The latter in turn is easier to correct than an error in the rating of operator performance. Somewhat akin to error in its effects is undue delay in issuing a revision in standard after a change in job content. The remedy is obvious.

1. *Specific Company Change.* There should be little difficulty when a change is introduced by the company. Most union agreements provide for reductions in standards to the same extent that the change reduces the work content. This provision is not as permissive as it first seems, however, for the work standard may have been loose before the change was made. Much depends on the manner of subtraction and whether it is possible to take entire elements out of the original study or whether only the correct value of the subtracted elements may be used. Also, some question may arise as to whether all the original elements were equally loose.

Some union agreements appear to permit the elimination of overall looseness when a legitimate change is made, but in general, the intent of the clause appears to be that earnings opportunity shall not be affected. This will depend on how long a relatively excessive level of earnings opportunity has been allowed to persist. Thus it is often the case that only the net amount of the change can be subtracted and a loose standard remains loose—or even looser.

Variations may develop if elements in the original study were not equally loose. All in all, this type of change is the least vexing of the problems of maintenance, and may cause no difficulty at all if the original standards have been well monitored.

2. *Sudden Improvement in Conditions.* If a sudden improvement is positive and apparent, it should present no more difficulty than the preceding if the standard is issued promptly. If this is not done, the difficulty may be greater. The reference is to such improvements as a transfer to materials with fewer flaws or to a system of production control that ensures a steadier flow of work. Even a marked improvement in comfort (for example, air conditioning) might justify small reductions in the fatigue allowance. It would have to be shown that the original fatigue allowance had been excessive by comparison with others.

3. *Gradual Improvement in Conditions.* Gradual improvement might be of the same kind as the preceding (as for example, a step-by-step improvement in materials). The gains might not be noticed or might not be considered worth taking advantage of until earnings had risen significantly. However, it should not be difficult to effect a change in a standard if it is introduced after prior and repeated

warning. One factor that is usually annoying to the industrial engineer and to management, but which might be useful here, is the frequently encountered clause forbidding changes in standard of less than 5 percent but permitting lesser changes to be cumulative.

Occasionally, gradual changes in conditions may be so late in being noticed that it is difficult if not impossible to identify them specifically. In this instance, the problem will be little different from that outlined below in changes 4, 5, and 6.

4. *Unwitting Change by Operator.* A more difficult situation is the creeping change, where serious resistance by employees and by the union is likely to be encountered when moves are made to correct the standard.

It is by no means unusual for operators to be unaware that they are falling gradually into shorter methods by such things as the elimination of pauses for regrasping, inspecting, and so on. Some discover that they have exceptional innate dexterity, reaction time, or speed of movement. If the operator proves to be in that small minority classed as "superskilled," there may be no problem and the standard will not be loose for the general run. However, it is all too easy to assume this to be so and therefore not to make the effort to find out whether the skill may be teachable. If the revised method proves to be generally transferrable, then the creeping change has come into play and must be dealt with.

This appears to be an appropriate place to introduce two related matters. The first is that when an operator is found to be using a shortcut, it is sometimes the practice to insist on the job being done in exactly the same way that it was done when initially studied.

If the new method endangers quality, this may be necessary. However, the chances are better than even that quality will not be affected, in which case the insistence on reverting to the original motion pattern becomes ridiculous, even though it does appear to maintain a principle. If it continues to be insisted on, potential savings will be lost, and it would be reasonable to assume that the operator will continue to use the new method but will restrict his output. It would be more sensible to face the problem by first correcting the standard and developing some form of compensation. This is discussed later.

The second relevant matter is that in a normal group there is a tendency for average earnings to rise gradually, in parallel with a tendency for better than average operators to be attracted to and to remain in the group while those whose abilities confine them to lower output tend to drift away. There is a risk of attributing this narrowing of the normal distribution curve to the existence of loose standards rather than to a shift in the mode. In practice, the two will probably be mixed and will require high industrial engineering judgment to effect a separation.

5. *Change Consciously Introduced by Operator.* It is probable that change consciously introduced by the operator will be more troublesome than change which is inadvertent. This is because of the specific awareness on the part of the operator who, having knowingly developed a shortcut or a new method, is then obliged to decide how he will exploit it in the absence, say, of a company award for such things, or perhaps even if there is an award. He must decide whether he will use it to increase his earnings or whether he will take it easy on the job. Probably there will be a combination of the two, skillfully concealed.

The improvement may go unnoticed for years, even by other operators, although it is likely that some will be aware of it and will be under some compulsion to follow the same course as the originator. In the end, it will no doubt be discovered, but so late that any change in the standard will be strongly resisted.

6. *Change Inherited by Operator.* The inherited change, the most difficult situation, can be exemplified by the typical instance of a new operator who is taught by one who has adopted a shorter method and has kept it secret but reveals it to the new operator because he, the old operator, is about to leave. The new operator, unaware of the total situation, may soon proceed to earn at an exceptionally high level. If any attempt is made to bring the standard into line, the operator will naturally and justly protest that the standard was posted when he started,

that he has followed instructions to the letter, and that no change in method has taken place since he started on the job. The personal, ethical, and possibly legal aspects are more apparent than is the method of resolution.

THE CORRECTION OF INCONSISTENT STANDARDS

The manifestations of the creeping change come into focus better when one considers that because a once acceptable standard still remains, progressive looseness does not mean a rise in cost. Rather, it unlatches the door to cost reduction, although the full opening of the door may be difficult to achieve. On the other hand, developing tightness almost always means an increase in cost.

However, because loose standards are invariably inconsistent, they are also discriminatory. Their proliferation stultifies time study and the incentive principle and leads to deterioration in employee-management relationships, ragged morale, and dubious motivation. Thus it is imperative that they be corrected as soon as possible, no matter how or by whom they are introduced or discovered.

The correction of the standards themselves is routine industrial engineering. The real problem is what to do about the resulting reduction in earnings opportunity. The problem is touchy and complex and deserves more detailed treatment than can be given here. Some of the broad areas of approach are:

1. Install without compensating operators.
2. Place a ceiling on earnings.
3. Leave earnings potential unchanged.
4. Apply "red-circling."
5. Compensate in cash—"buy-back."
6. Introduce gradually.
7. Effect compromise.

Fallacies and Remedies. It is unfair and is rarely practicable to reduce current earnings opportunity without warning or compensation. The argument is used that the operators have enjoyed illicit "gravy" and should be satisfied with that windfall, which has now ended by management prerogative according to established criteria. This approach can be rationalized, but it is likely to bring on bitter opposition. If it can be imposed, it is preferable in the long run to the fairly common practice of leaving part of the looseness in the standards. The latter does not solve the problem of inconsistency. However, some would go so far as to leave all the looseness in, on the grounds that, having made a change in the face of a guaranteed standard and a presumably irreducible job content, the operator should not have to share the profit on his skill and ingenuity. This untenable point of view embodies several fallacies. For instance, it cannot be extended to its logical conclusion or else an operator could demand full wages for doing nothing in the quite possible event that he had found how to eliminate the job entirely. In any case, it is often difficult or impossible to distinguish among changes introduced by the operator, changes inherited by the operator, and changes that develop out of altered circumstances.

Some actually impose a ceiling on earnings, which, on the face of it, seems to be a sort of head-in-the-sand solution. Operators will be quick to get the message and to restrict output accordingly, thus producing an ostensibly satisfactory situation which obscures the problem. Any marked consistency in the output index of a large group of operators can be taken as evidence of loose standards.

Red-circling. More realistic are several plans which offer the operator full or partial separate compensation for the reduction. Red-circling is one of these. It permits the standard to be corrected but pays the difference in an overriding and separately recorded supplement to the standard. Presumably those hired after the standard has been corrected do not receive the supplement. The anomaly of people working side by side with different pay for the same output creates obvious difficulties and is not likely to be favored by a union.

Across-the-board "Add-on." Another partial remedy is referred to as the "add-on." When an hourly wage increase is negotiated, it is sometimes added uniformly to all earnings after the incentive calculations have been made. This has the effect of dampening the inflationary aspect of loose standards, but it eventually destroys the incentive element and reduces the system to the equivalent of daywork.

Increase in Base Rates. Another plan which may have considerable merit is one in which the base rates are increased by a percentage equivalent to the established average looseness of all standards. At the same time, correct standards are installed. The plan requires able selling, and it may be possible for the base rate increase to be something less than equivalent to the looseness. It means, of course, that those who have had excessive earnings will tend to lose part of them, while those who happened to have been on more correct standards or who have not cashed in on looseness will tend to gain.

Buy-back. If there is too much opposition from those whose earnings are threatened, it may be possible to supplement the plan with another that has had some currency. This is sometimes called the "buy-back" and is a plan for paying a lump sum equivalent to the vested looseness for, say, one year.

As a sole plan, this presents difficulties, not the least of which is the return to lower earnings. It is likely to be strongly opposed by unions. Some employers are averse to the inevitable operator who accepts a substantial buy-back and immediately quits. The problem may be minor, however, because he will be replaced by a new operator who will not receive extra compensation. The cost will be the same whether or not the old operator leaves. The buy-back is much the same as treating the looseness as if it were a suggestion under an unusually generous suggestion plan. The difference is that the suggestion plan would pay to one individual, while the buy-back could go to many, regardless of whether any had actually contributed to effecting the change.

Need for Discussion. The preceding methods of correcting loose standards do not exhaust the possibilities. They are intended to be suggestive only. None is especially recommended. There are innumerable variations, all presenting difficulties to one side or the other. The problem calls for open discussion between management and employees, less in the sense of bargaining than in the sense of finding a rational solution to a joint dilemma. Whatever the solution, the fundamental requirement is the restoration of correct standards—correct, that is, by specific criteria, and therefore consistent. This must, of course, be supported by continuing rigid and complete monitoring of all standards.

THE TIGHT STANDARD AND ITS USES

To this point, the problems of the loose standard have been discussed. Some comments follow on the standard that has become or appears to have become too tight. This is usually made evident by an operator failing to do as well as could reasonably be expected, as revealed by payroll or incentive records or by complaint by the operator. If time study uncovers the unreported addition of some bona fide work element, the standard will be corrected, possibly retroactively—a procedure that is rarely feasible in the case of loose standards.

If, as may well be the case, the original standard appears to be correct, the unimaginative rate setter announces the fact and moves on to other studies. But the imaginative observer begins to seek causes and speculates as to the innumerable possibilities of the situation.

A few of these are mentioned to illustrate the constructive uses of time study as an aid to administration. The work standard is more than the one indispensable quantification in industrial engineering. Its intelligent maintenance is the key to possible error in management procedures at all levels.

Operator Failure. The first point of speculation would probably be the operator himself. Was his reduction in output intentional, for reasons of his own or for union purposes? Has he the necessary skill, stamina, and temperament for the job? Should he be moved to a more suitable job? Has he been properly instructed? Was he chosen by test and interview? Are the hiring methods adequate? Did induction get him off to a poor start?

Personnel Administration. Moving away from the man to the administration of personnel, because personnel policies are extensions of general policy, such questions may be raised as: Is employment of suitable people hampered by a niggardly wage policy? Do the company's actions and reputation tend to sift applicants down to the less desirable, as well as to create low morale?

Deterrents to Full Performance. Turning to the immediate surroundings of the job as being more likely to cause low earnings, the observer will explore the possibility of interruptions and other deterrents to rhythm and steady output. For instance, machine delays due to breakdowns in excess of those provided for in the standards may come from indifferent maintenance, which in turn may be related to inadequate, poorly trained, or poorly directed mechanical and electrical staff. On the other hand, the controller's office may be at fault in failing to analyze fully the relationship between maintenance costs and replacement costs.

Another cause of low productivity is uneven flow of work to the operator. The cause may be lagging service by the floor help or intermittent flow from the preceding department. Foremanship may be at fault, or the erratic flow may result from unorganized production releases. These last may or may not be the fault of production control. They could stem from undue delays in issuing sales projections, with the inevitable rush releases and sudden changes.

Production Control. Somewhat similar in effect is the overall drop in production caused by small batches. Apart from the loss caused by excessive setup time, there is almost certain to be a loss of rhythm and speed by the operator, especially in manually controlled jobs. The sources of the unduly small batch are obvious. There may be an overzealous attempt to hold inventory to an irreducible minimum—a common end-of-season or pre-financial-statement phenomenon. There may be delay in anticipating customer requirements, with the result that, in order to enter the shipping season with a full range, small lots will have to be scheduled. Once started, this may well prevail through the season. Again, end-of-the-season inventory control may necessitate small batches to balance out and run out the stock.

Variation in Materials. Faulty or substandard material may also make it necessary to supplement standards with special allowances. Again questions arise. Has the purchasing department tried to save money without warning manufacturing? Were the savings in material cost counterbalanced by a rise in labor cost? Did the laboratory fail in its testing? Has the supplier erred, or inched on quality? Has a purchasing officer been got at?

Status of Industrial Engineer. Enough! The possibilities are virtually limitless. It is then essential that those who must isolate, and if possible eliminate, the causes of the excess cost inherent in tight standards shall be intelligent and imaginative persons with sufficient status for their findings to be fully considered by officials with power to act.

Chapter **2**

Stopwatch Time Study

WILLIAM ANTIS

Technical Director, Maynard Research Council Incorporated,
Pittsburgh, Pennsylvania

Stopwatch time study is a tool of work measurement which has been used throughout industry to determine the time required to do work since it was first developed by Frederick W. Taylor before the turn of the century. A number of time study procedures, all of which are quite similar, have been developed and used with success in all types of industry and business. One of the widely used procedures is discussed in this chapter.

DEFINITION OF TIME STUDY

Time study may be defined as follows: Time study is a procedure used to measure the time required by a qualified operator working at the normal performance level to perform a given task in accordance with a specified method.

In practice, it is difficult to separate methods study and time study completely. The industrial engineer, often called a time study engineer or technician, cannot help but study methods while he is making a detailed time study. The definition of time study states that the task measured is performed with a specified method. This chapter discusses the skills required to use the time study procedure to establish allowed times. Section 2 of this Handbook is devoted to a detailed discussion of methods; so they will not be discussed here.

Figure 2-1 presents a graphic analysis of the steps involved in establishing a time standard. To emphasize the fact that the method should always be studied, improved, and standardized before the time study is begun, these steps appear before time study on the chart. Time study itself begins with "Selection of Operator."

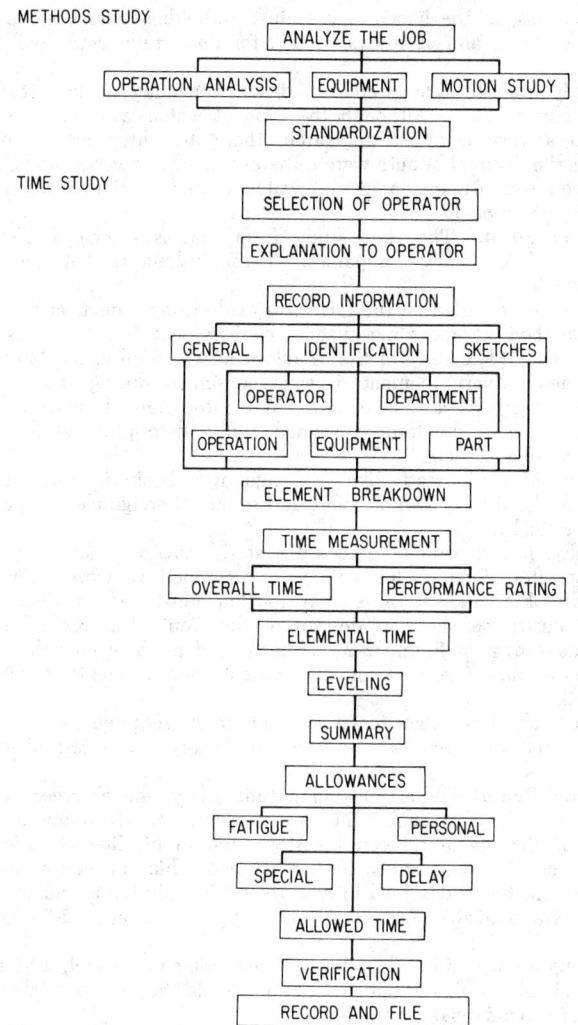

FIG. 2-1.   Graphic analysis of the steps involved in establishing a time standard.

TIME STUDY TOOLS

Although it has often been said, and sometimes proved, that a good time study technician can make a usable time study with only the back of an envelope, his wristwatch, and a stubby pencil, there are a few essential tools that are needed by the time study engineer to make a good time study.   They are:

1. Time study watch
2. Time study form
3. Observation board
4. Pencil
5. Ten-foot tape
6. Speed indicator
7. Slide rule

These few tools, in the hands of a trained individual who has the temperament, intelligence, abilities, and aptitudes needed for time study work, will produce good results.

Time Study Watch. The watch used when making a time study is perhaps the most important tool. Although the type of watch will vary from an ordinary timepiece to a very complex stopwatch, there are three main types which are widely used: the decimal minute watch, the decimal hour watch, and the split-second watch. A complete discussion of the watches used in time study may be found in Chapter 3 of Section 12.

Time Study Form. The time study form that is shown in Figures 2-2 and 2-3 is one used by many companies. Through long use, it has proved to be entirely adequate.

The time study recorded on the form was made using a decimal hour watch. The continuous method of recording watch readings was followed. As a result, the study tells the complete story of the sequence in which all elements of the operation were performed. Every moment of time consumed during the period in which the study was made is accounted for. When the normal sequence of operations was interrupted, both the time consumed and a description of the nature of the interruption were recorded.

The completed study reads like the page of a book; it starts at the top left, reads across to the right, down one line, and so on. Foreign elements or interruptions appear as marginal notes.

Occasionally, it is desirable to record a study vertically rather than horizontally. This is most often done on low-volume or nonrepetitive work. For this purpose, a time study observation form similar to that illustrated in Figure 2-4 may be used. The illustration shows a portion of the front of a vertical form. Part of the same data shown in Figure 2-2 has been used to show how they would appear on a vertical form. The back of the vertical form is identical with that shown in Figure 2-3.

Any time study form that meets the minimum requirement of permitting the technician to tell the complete story is satisfactory. The test of the form is in its use.

Observation Board. When making a time study, the observer generally stands and moves about as he makes his observations. He is required to watch the movements of the operator, keep his stopwatch in his line of vision, and record his readings on the time study form. To assist him in performing his job, he needs a thin light board designed to hold the time study forms and watch. Observation boards are available from different suppliers. Figure 2-5 shows a typical board.

Other Equipment. Little need be said regarding the pencil, 10-foot tape, speed indicator, and slide rule except that they should be of acceptable quality. All these items are commercially available.

TIME STUDY PROCEDURE

When the method has been established, conditions standardized, and the operators trained to follow the standard method, the job is ready for study.

Selection of Operator. A qualified time study engineer can study any operator he wishes so long as that operator is using the accepted method. By applying the performance rating procedure correctly, he will arrive at the same final time allowance within practical limits, regardless of whether he studies the fastest or the slowest operator. From the engineer's viewpoint, however, his work is made somewhat easier if he studies an intelligent, cooperative worker performing at an acceptable performance level. If there is only one operator doing the job, there is, of course, no choice.

Explanation to Operator. The manner in which the operator is approached at the beginning of the study is important. In his approach, the time study engineer tries to be courteous and unassuming and shows a recognition of and respect

FIG. 2-2. Time study form—front.

STUDY NO. __2__ DATE __7-20-__

OPERATION __PACK TERMINAL BLOCK__

DEPARTMENT __M12__

OPERATOR	NAME	M. SMITH	No. 167
~~Men~~ Woman			

MOULD __TB 9/3__   PATTERN _____

PART DESCRIPTION __TERMINAL BLOCK 463 207__

DIE _____   INS. SPEC. __22A 207__

DWG. __B 7194__   STYLE __463 207__   SUB __9__   ITEM __1-4__

L. SPEC. _____

MATERIAL __ASSM.__

EQUIPMENT

MACHINE TOOL NO.

SPECIAL TOOLS, JIGS, FIXTURES, ETC.

CONDITIONS

OBSERVER _____   APPROVED BY _____

SKETCH

No.	ELEMENTS	SMALL TOOL NOS FEED SPEED, DEPTH OF CUT. ETC	ELEMENTAL TIME ALLOWED (BOTTOM LINE OTHER SIDE) OR CYCLE	OCCURRENCES PER PIECE	TOTAL TIME ALLOWED
1	GET BOX D412, GET NUMBER STAMP & STAMP SERIAL # ON FLAP AT TOP OF BOX	INK PAD	.00164	1	.00164
2	CREASE & CLOSE BOTTOM OF BOX		.00173	1	.00173
3	GET, VISUAL INSPECT, & PUT BLOCK IN BOX		.00159	1	.00159
4	GET BAG #1714 & 2 SCREWS, PUT SCREWS IN BAG		.00117	1	.00117
5	FOLD OVER BAG		.00262	1	.00262
6	PUT BAG IN BOX		.00099	1	.00099
7	CREASE & CLOSE TOP OF BOX		.00174	1	.00174
8	SET BOX IN TOTE PAN		.00045	1	.00045
				TOTAL	.01193

TIME ALLOWED, SET UP _____   EACH PIECE __.0119 HR__

REMARKS: MATERIAL HANDLER KEEPS BINS REPLENISHED, MOVES FULL TOTE PANS TO SHIPPING, & SUPPLIES EMPTY TOTE PANS.

CHECK WITH "METHODS" TO IMPROVE ELEMENT 5.

OBSERVATION SHEET

Sketch labels: BAG 1714, SCREWS, TERMINAL BLOCKS, SELF-INKING NUMBER STAMP, TOTE PAN, 26" HIGH RACK UNDER, OPERATOR SEATED, BOX D 412 (FLAT), 16", 12", OPERATOR

FIG. 2-3. Time study form—back.

NO	ELEMENTS	T	R	T	R	T	R	T	R	T	R	T	R	T	R
1	Get box. Stamp serial no. on flap	11	11	12	105	13	93	13	87	15	403	12	97	14	601
2	Close bottom of box	15	26	16	21	14	207	11	98	13	16	16	513	12	13
3	Get, check, & put block in box	12	38	14	35	10	17	A12	334	14	30	11	24	14	27
4	Get bag & 2 screws. Put in bag.	09	47	08	43	10	27	11	45	10	40	09	33	11	38
5	Fold bag	22	69	16	59	18	45	21	66	20	60	25	58	19	57
6	Put bag in box	06	75	06	65	08	53	06	72	08	68	09	67	07	64
7	Close top of box	14	89	12	77	17	70	13	85	14	82	16	83	17	81
8	Lay aside in tote pan	04	93	03	80	04	74	03	88	03	85	04	87	03	84

Fig. 2-4.   Vertical time study form.

for the problems of the worker. He is frank in dealing with the man and is willing to explain what he is doing and how he does it at any time in clear, nontechnical language. The time study engineer is dealing with facts, and he has nothing to hide. His work would be much easier if all the men he studied were familiar with the details of the time study technique that is being employed.

There is nothing in the time study procedure which could be considered anything but fair by anyone who understands it. When time study work is properly handled, in cases where the correctness of a time value is questioned, the operator himself is the first to request a check time study, knowing that the time study will settle the question with fairness to all concerned.

Record Information. A time study, to be of value for future use, must tell the entire story of a job in such a way that it will be understood by anyone familiar with the time study procedure. A good test of a time study is to use it to reconstruct a job as it was originally performed.

Provision is made on the back of the time study form to record all identifying and other pertinent information. These data should be recorded at the time the study is made. Records should be made to show complete identification of the operator; the part upon which the operation is being performed; the machines, tools, and equipment being used; the operation; and the department in which the operation is performed. Sketches, for which space is provided, are generally a desirable and necessary adjunct to verbal descriptions. These sketches preferably

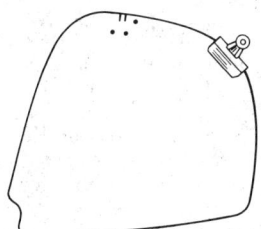

Fig. 2-5.   Observation board.

should show the workplace layout, or they can illustrate the part upon which the operation is performed. Some companies attach a photograph of the workplace or piece to the time study form to illustrate the conditions at the time the study was made.

The recording of complete information is of the utmost importance and cannot be too highly stressed. Figure 2-3 shows typical information recorded on the time study form. Note that under "Remarks" the method of bringing material to the workplace and removing it from the workplace has been described. Such data, unless recorded at the time the study is made, become lost in limbo when later trying to reconstruct the conditions to which the standard applies.

Most of the headings of the spaces used to record information are self-explanatory. The "Ins. Spec." heading is used to record the number of the inspection specification, if such data are available in the plant. The "L. Spec." heading identifies the electrical specifications number where it is pertinent.

Element Breakdown. The first step in making observations is to subdivide the operation into a number of smaller operations which will be studied and timed separately. These subdivisions are known as elements or elemental operations. An element is a subdivision of an operation that is distinct and measurable; it contains a logical portion of the work.

An element, to be usable, must meet all the qualifications stated in the definition. For example, the element description "Move piece to vise" is neither distinct nor does it contain a logical portion of the work. In this example, the end point of the element is indefinite. At what point over the vise does the element end? This point can vary in the eyes of the individual observer, as well as being indefinite in the eyes of other observers. A more acceptable element would be "Move piece and place in vise." In this case, the end point of the element is definite. The element ends when the piece is in the vise.

Many companies develop a list of standard elements that are completely described. In such companies, the standard description of the element "Move piece and place in vise" might be "The element begins as the operator grasps the part to be moved to the vise with one or both hands, depending on size, shape, and weight. It includes the total time required to move the part to the vise and insert it between the vise jaws, and ends as the part is located between the vise jaws."

With standard element descriptions of this type available to all the time study observers in the company, there is a consistency obtained in all time studies that contain the standard elements. Data collected in such a manner are readily usable in developing standard data and making comparisons and in checking the variables.

When the element breakdown has been completed for a study, a short description of each element is recorded on the front of the time study form in the space provided. A more complete description of the elements is recorded on the back of the time study form in the "Elements" column. Figure 2-6 illustrates the method of recording the element description on both the front and back of the observation sheet.

Time Measurement. The operation is timed with the aid of a stopwatch. The two most popular types used in time study are the decimal minute and the decimal hour watches. The major difference between the two watches is that the decimal hour watch is slightly more accurate because it runs faster and thus measures smaller increments of time.

There are two principal ways of reading the watch when making a time study: the snapback method and the continuous method.

Using the snapback method, at the termination of each element the observer reads the watch and as nearly as possible at the same instant snaps the hand of the watch back to zero. The advantage of this method is that the clerical work of making subtractions, required when the continuous method is used, is saved. It also facilitates the recording of readings for elements which are performed out of the normal sequence.

Because only elapsed times are recorded, a study made with the snapback method does not present a clear picture of the sequence in which the elements were per-

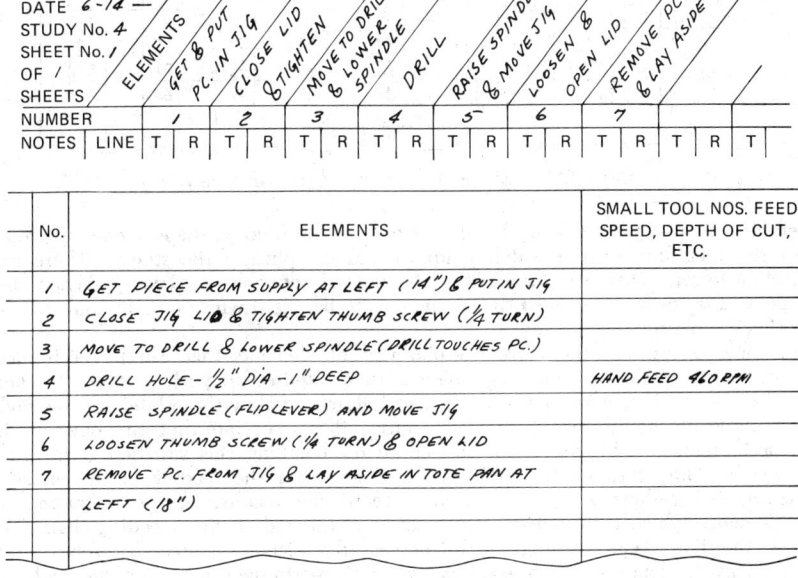

| | SMALL TOOL NOS. FEED |
No.	ELEMENTS	SPEED, DEPTH OF CUT, ETC.
1	GET PIECE FROM SUPPLY AT LEFT (14")& PUT IN JIG	
2	CLOSE JIG LID & TIGHTEN THUMB SCREW (¼ TURN)	
3	MOVE TO DRILL & LOWER SPINDLE (DRILL TOUCHES PC.)	
4	DRILL HOLE - ½" DIA - 1" DEEP	HAND FEED 460 RPM
5	RAISE SPINDLE (FLIP LEVER) AND MOVE JIG	
6	LOOSEN THUMB SCREW (¼ TURN) & OPEN LID	
7	REMOVE PC. FROM JIG & LAY ASIDE IN TOTE PAN AT	
	LEFT (18")	

FIG. 2-6.   Element descriptions.

formed.   The extent and nature of foreign elements as they occur are not always recorded.   This causes some studies to be incomplete and increases the selling job required to gain acceptance of the resulting time value.

Another disadvantage of the snapback method of recording watch readings is the inherent error occasioned by the time required to snap the watch back to zero.   Laboratory studies made with the aid of slow-motion pictures show that an error of 3 percent to over 9 percent will occur on each element of 0.0010-hour duration.   The amount of error is greater on shorter elements and less on longer ones.   The objection to lost time using the snapback method can be overcome by using multiple watches.

When using the continuous method of watch reading in making a time study, the watch is allowed to run continuously from the beginning of the study to the end.   At the end of each element, the time is noted and recorded on the time study form.   With this method, it is possible to record every event that occurs during the actual study and to trace back and determine what was being done at each instant of the study.   Additional clerical work is required to develop the elemental times using the continuous method, but this can be delegated to a clerk if desired.

Occasionally, variations from the regular sequence of elemental operations occur which the observer must be prepared to handle and record without confusion.   Variations may be divided into four general classes:

1. Elements performed out of regular order
2. Elements missed by the observer
3. Elements omitted by the operator
4. Foreign elements

By the continuous method of watch reading, the recording of all these variations can be handled quite simply by the observer.   By the snapback method, it is difficult to reconstruct the study to show when and how these variations occurred.

Variations in Sequence.   When the time study engineer has established the sequence of elements, he should insist upon this sequence being followed by the

NUMBER		/		2		3		4		5		6		7		8	
NOTES	LINE	T	R	T	R	T	R	T	R	T	R	T	R	T	R	T	R	T
	1		08		19		28		43		58		71		79		89	
	2		98		107		18		$\frac{48}{32}$		$\frac{32}{18}$		$\frac{66}{47}$		73		85	
	3		93		202		13		28		41		57		65		75	
	4																	
	5																	

FIG. 2-7.   Elements performed out of order.

operator.   Any departure from the regular order of performing the elements interferes with the recording of the watch readings and complicates the study.   There are times, however, when variations from the normal sequence do occur.   When this happens, the technician must be prepared to handle them properly and record the times correctly.

In some operations, there are combinations of elements that can be performed in different sequence without affecting the time required to perform them.   When this does occur, if it is not an indication of failure to follow the established method, the technician should record the variations rather than interrupt the operator in the performance of his job.   The procedure for handling this situation is to draw a horizontal line through the middle of the space for the reading of the element which is being done out of order, and record the reading at the beginning of the element—which is the same as the reading at the end of the preceding element— below the line.   The reading at the end of the element is recorded above the line.   This is done for each element which is performed out of order and for the first element after the regular sequence is resumed.   This convention is illustrated in Figure 2-7.

There are times when an element will be omitted by the operator.   This does not happen too often, and when it does, it probably means that the element was not necessary on that particular piece.   If the same element is omitted frequently during the course of the study, further methods study is indicated.   To record information on an element omitted by the operator, it is necessary only to draw a horizontal line across the column in the space where the reading would have been entered, omit recording any reading, and then continue on to the next element.   Figure 2-8 shows the method by which an element omitted by the operator is recorded.

There are times when the observer might miss the end point of an element and fail to make the reading.   When this occurs, he does not attempt to guess at the reading and record the guess, but records it honestly as a missed reading.   This is done by placing the letter "M" in the space where the reading should have gone, as shown in Figure 2-9.   When this occurs, the time for the following element cannot be determined.

Foreign elements generally occur as interruptions to the normal sequence of elements.   They may be a necessary part of the work or they may be completely unnecessary.   For example, if the operator is required to replenish a supply of

NUMBER		/		2		3		4		5	
NOTES	LINE	T	R	T	R	T	R	T	R	T	R	T
	1		40		102		64		210		49	
	2		92		349		418		72		508	
	3		52		620		90		735			
	4		73		839		911		65		1006	
	5											

FIG. 2-8.   Element omitted by operator.

1		2		3		4		5		
T	R	T	R	T	R	T	R	T	R	T
32		40		52		68		83		
91		M		113		31		45		

FIG. 2-9.   Element missed by observer.

material, this would be recorded as a foreign element and would be necessary for the performance of the job. However, if he were interrupted by a supervisor, clerk, or other employee, the interruption might or might not be necessary to the performance of his job. All interruptions are recorded as foreign elements. The occurrence of an interruption is indicated on the time study form by inserting a letter symbol in the time space of the element where the interruption occurred. The time for the foreign element is recorded below and above the line of the corresponding letter space under the heading "Foreign elements" on the right side of the time study form. This is illustrated in Figure 2-2.

Number of Observations. Nomographs and equations have been developed to aid in determining the number of observations to be made.[1] There are some businesses that either through company policy or labor contracts have established the number of cycles to be studied or have specified a minimum elapsed time for the study.

Although mathematical methods for determining the number of observations which must be made to determine time at a desired confidence level may have an appeal to statistically oriented people, they may not take into account all the factors that affect a study.

Generally speaking, the number of pieces to be studied can safely be left largely to the judgment of the observer. There are many factors which will influence this judgment. They include length of cycle, number of repetitive elements in the cycle, skill of operator, variations in the workpiece, and the number of interruptions. The observer should study enough cycles to be sure that his study is representative of normal conditions. Studies that extend beyond this point suffer from the law of diminishing returns. The advantages do not increase in proportion to the time and effort expended.

TABLE 2-1. Average Elemental Time

Element no.	After 10 cycles	After 16 cycles
1	0.00133	0.00132
2	0.00138	0.00139
3	0.00127	0.00128
4	0.00095	0.00094
5	0.00207	0.00211
6	0.00080	0.00080
7	0.00140	0.00140
8	0.00034	0.00036
Total..........	0.00954	0.00960

[1] Gerald Nadler, "How Many Time Study Readings to Take?" *Factory*, February, 1955; and B. L. Hanson, "A Graphic Method for Finding Required Number of Time Study Readings," *The Journal of Industrial Engineering*, May–June, 1957.

An analysis of the elemental times in the time study shown in Figure 2-2 indicates that the average elemental times, after ten cycles, were not materially affected by extending the study for six additional cycles. These data are shown in Table 2-1.

Overall Time. Every time study should have recorded on it the time of day that the study started and the time of day at which it stopped. From this, the overall elapsed time for the time study may be determined. It should be approximately equal to the sum of all the detail times and the foreign elements. When this time is divided by the number of pieces completed, it will show the average overall time for each piece.

By recording the overall time of a study, it is possible for the observer to explain to the operator in detail exactly what occurred during the course of the study, how many pieces were completed, and the overall time of the study (which the operator himself can check if he wishes). This has a tendency to make the time value established from the study more salable to the operator.

Performance Rating. Performance rating is discussed in the next chapter. The rating system used for the illustrations in this chapter has as a base the following concept of average performance level.

The characteristics of average performance level are as follows:
1. Equivalent to a fair day's work.
2. The point at which incentive pay begins under many wage payment plans.
3. Usually expressed as 100 percent performance.
4. May seem somewhat slow when observed.
5. Can be accelerated rather easily.
6. Easily maintained over long periods by the physically normal operator without requiring him to draw upon his reserves of energy.
7. The operators perform without undue hesitation, planning, or errors.

The prerequisites of average performance level are:
1. Working conditions are those usually prevailing.
2. The operator is qualified for the job.
3. The operator has been given sufficient training and practice to enable him to gain acceptable proficiency at the job.
4. Adequate allowances are provided.

It is vital for the observer to record the performance of the operator when the study is being made and before he leaves the work station. At this point, his judgment must be at its best, for the final allowed time will be based upon his evaluation of the performance of the operator. It is important to remember that regardless of the performance rating technique employed, all time study engineers should be soundly trained in its application.

Elemental Time. When the observations and the information have been recorded, the remainder of the work necessary to complete a time study and to determine the allowed time for the job consists almost entirely of calculations. Much of this work is of such a routine nature that it may be assigned to a clerk. It must be remembered, however, that accuracy is essential.

If the snapback method of recording watch readings has been used, it is not necessary to make the subtractions required by the continuous method of watch recordings. However, all other procedures in working up the study are identical. Figure 2-10 illustrates the entries that are made in the summary section of the time study form.

Subtractions. When the continuous method of recording watch readings is used, the first step of working up the study is to determine the elemental elapsed times. This is done by subtracting successive watch readings and recording the resulting time value in the T column in the space between the watch readings which determine it.

Referring to Figure 2-2, the first elapsed time reading for element 1 is 11. Because the watch was started at 0, the termination of the element is also the elapsed time for that element. In the T column for element 2, the elapsed time is recorded

TOTALS "T"	0299	0218	0045	0167	0189
NO. OBSERVATIONS	14	12	14	14	12
AVERAGE "T"	00214	00182	00032	00119	00158
MINIMUM "T"	0018	0014	0002	0011	0011
MAXIMUM "T"	0028	0020	0004	0015	0018
RATING (SK - EF)	C/C/	DD	C/C/	C/C/	C/C/
LEVELING FACTOR	1.11	1.00	1.11	1.11	1.11
L.F. X AVE. "T."	00238	00182	00036	00132	00175
% ALLOWANCE	15	15	15	15	15
TIME ALLOWED	00274	00209	00041	00152	00201

Fig. 2-10. Summary section of time study form.

as 15. This was obtained by subtracting 11, the first reading, from 26, the second reading. This procedure is followed throughout the entire study. The values 11 and 15 are in actuality 0.0011 hour and 0.0015 hour. The study was made using a decimal hour watch.

Abnormal Values. At this point in working up the study, the engineer who made the time study should carefully examine the elapsed times for abnormal values. If any are found, they should be indicated so that they can be excluded from the element summary. The careful, competent observer should never have an abnormal value that he has not recognized. He will have recorded in the "Notes" column anything that would have caused the value to be extremely high or low. If several abnormal situations occurred, he would have continued his study to cover enough cycles to eliminate the effect of the abnormally high or low values.

There are times when a value will be extremely high because a small foreign element occurred during the performance of the element. This should have been noted by the competent observer, and his line note should indicate the cause of the excessive time.

Total Elemental Time. When the abnormal values, if any, have been excluded from the summary, the next step is to add the elapsed times for each element. The totals are recorded on the line headed "Totals *T*" of the summary. Because of its routine nature, the actual addition can be performed on a calculator by a competent clerk.

Number of Occurrences. The number of occurrences of each element should next be recorded on the line headed "No. observations." If more than one column has been used for an element, the values appearing in all of them should be summarized in the first column devoted to the element. This applies to the number of occurrences also. No summary would appear, then, in any of the other columns that have been used for the element.

Average Time. The next step is to divide the total *T* by the number of observations. The result will be the average time for the element. It is recorded on the "Average *T*" line.

Minimum and Maximum Times. Looking over the elapsed time values for each element, the minimum time and the maximum time for each element are easily recognized. These values should be recorded on the fourth and fifth lines of the summary. These time values are not used to determine the final allowed time, but are used for analysis purposes. Too great a difference in maximum and minimum times may indicate that some abnormal or unstandardized condition existed and that further study of the element is required.

Rating. At the time the study is made, the observer records the performance rating of the operator. If the operator gives the same performance on all elements of the operation, only one rating is used. If, however, the operator performed at different levels for various elements, this will be noted and recorded on the time study form.

With the ratings thus established, it is necessary that they be converted to a multiplier. This multiplier is recorded on the line headed "Leveling factor." If all elements have been rated the same, the figure that will appear on this line will be a constant. However, if there were variations in the performance of the individual elements, then different rating factors will be entered on this line.

Standard Time. The leveling factor multiplied by the average time for the element gives the standard time for the element. This standard time represents the time required by a qualified operator working at the normal performance level to perform the element. It is not influenced by geographical location, physical makeup of the operator, or other outside factors. It is the time required by a qualified operator to perform the element using the prescribed method.

Allowances. If the operator were able to work continuously without interruption, the standard time would be the allowed time for the operation. Constant application to the job is something almost impossible to attain. In the course of the day, there are some interruptions for which allowances must be made to establish the allowed time.

Determining and making correct allowances is a very important step in making time studies. The time study engineer must make a thorough study of the conditions surrounding the job to determine the kind and amount of allowances that will be included in the study.

Allowances cannot be calculated for each individual study as it is made. For this reason, a percentage allowance is calculated for the various types of interruptions that occur during the normal workday, and this allowance is included in every time standard.

The four major classifications of interruptions that the average operator will experience, which must be covered by the allowance, are personal, fatigue, delay, and special.

Personal Allowances. As the name implies, personal allowances cover the time required by the average operator to take care of personal needs. It includes such items as getting a drink of water, going to the rest room, washing hands, and so on. It does not include personal time included in rest periods that are paid for as a separate item. Neither does it include time for eating lunch, if regular lunch periods are designated. Obviously, the personal needs of individuals will vary with the physical makeup of these individuals. For this reason, the time required by the average person should be determined as an allowance for all personnel.

Fatigue Allowances. One of the most perplexing problems faced by the time study engineer is the measurement of fatigue. The measurement methods which have thus far been developed are at best only partially satisfactory, and an intelligent approach to the subject cannot be made unless the limitations of the procedures are clearly understood.

In the first place, although the word "fatigue" is used quite freely in industrial circles, and although almost everyone who has not studied the matter carefully feels that he knows what fatigue is, not even the nature of fatigue is clearly understood. One school of thought believes that there is no such thing as fatigue, but rather that what is called fatigue is a manifestation of monotony susceptibility, boredom, and the like. Another theory accepts the existence of fatigue but recognizes different types of fatigue, such as those caused by excessive heat, prolonged exertion, nerve strain, or monotony.

The engineer finds these various theories interesting but conflicting, and to arrive at a practical starting point, he comes to the conclusion, based on analysis and personal experience, that even under the best conditions some workers experience an exhaustion of strength, either physical or mental or both, as the working day progresses, which for want of a better name may be called fatigue. His problem then is to measure the effects of fatigue as manifested by a loss of production as fatigue is experienced.

Before he can do this, he must first understand just what effect fatigue may be expected to have on output. Analysis indicates that output builds up during

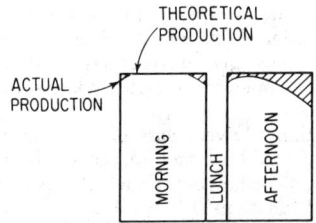

FIG. 2-11.   Graphic effect of fatigue.

the first half hour or so of the working day as the worker gets into his stride.   At the very start, there is quite likely to be a muscular or mental stiffness resulting from the fatigues of the day before, and this stiffness must be worked out before the worker can reach the peak of his productive ability.

When the operator has once reached his peak, he should be able to continue to produce at his maximum rate until he begins to experience exhaustion of his bodily and mental resources.   Then his output should begin to decrease until, at the end of the working period, it has fallen off appreciably.

In an 8-hour working day divided into two 4-hour periods with an interval for lunch, the output of this theoretical worker would be as shown graphically by Figure 2-11.   The curve of afternoon production would be similar in shape to the curve of morning production, but somewhat lower because of the fatigue accumulated during the morning, from which it was not possible to recuperate entirely during the lunch interval.

If this analysis is correct, it should be relatively easy to measure the retarding effect of fatigue.   Graphically, the shaded area shows the amount of output that is lost because of the effects of fatigue.   It is this area that the engineer attempts to measure quantitatively.

Realizing the variability of human beings, it is well to be suspicious of the results of fatigue studies that are too nearly perfect.   Fatigue studies are often made with the full cooperation of the worker.   The purpose of the study is explained to him, and the expected results are suggested.   If the worker is truly cooperative, it is quite likely that the expected results will be obtained.   The worker may not consciously vary his pace to suit the expectations of the observer, but the power of suggestion is strong, and if the worker is asked from time to time if he is beginning to feel tired, it is most likely that he will answer "yes" in all good faith and work more slowly.

It has been pointed out how individuals vary in their characteristics and in their reactions to fatigue.   Similarly, they differ in their working habits.   Some like to work in short bursts of speed.   Others like to work uninterruptedly for longer periods.   Variations in work habits will materially affect the shape of the theoretical curves of Figure 2-11.   The most satisfactory way devised up to the present time for handling the problem of fatigue allowances seems to be to add a percentage allowance to all time values in recognition of the fatigue element.

The amount that should be allowed to cover fatigue will vary with the nature of the job and the conditions surrounding it.   The most satisfactory way of arriving at the proper allowances is to make a number of all-day studies, average all usable results, and finally establish a fixed percentage for that class of work, based upon the results of the studies plus a liberal amount of judgment.

Delays.   During the course of a day, an operator is interrupted from time to time either to perform work necessary for the completion of the operation or to do things outside the scope of the operation.   Where delays occur and are necessary, a percentage allowance is determined for those delays.

Some of the delays which may be recognized as being necessary are replenishing material, rejecting substandard parts, and making minor repairs to tools and equipment. Other delays which are necessary but not connected with any particular type of work are those occasioned by interruptions by the foreman or other supervisory personnel to ask questions, give instructions, or otherwise communicate. The lost time experienced by the operator for these kinds of delays must be covered by the allowances.

Special Allowances. Within a given class of work or within an individual plant, there may be certain functions that are not covered by the time study. These are classified as special delays. They cover such things as clean and oil the machine, if this is a regularly assigned function.

The kind and amount of allowances should never be estimated or arbitrarily decided upon, but should be the result of careful study and analysis. In general, allowances are established for a type of activity, for a department, or in some instances, for a plant.

With allowances established, the percentage by which the standard time is to be increased should be entered on the line headed "Percent allowance."

Allowed Time. When the allowance percentage has been recorded, the final step is to determine the allowed time for each element. This is done by multiplying the standard time by 1 plus the percent allowance. These times are then entered on the line headed "Time allowed." When these calculations have been completed on the front of the form, the allowed elemental times are transferred to the back of the time study form, as illustrated by Figure 2-3.

When all entries have been completed, the allowed elemental times, multiplied by the frequency with which they occur, are added together, and a total allowed time per piece is determined. This is the time value for the operation. It covers only the repetitive part of the operation and does not include makeready or setup.

Verification. With the allowed time established, it is possible to verify, with some degree of accuracy, the correctness of the time standard. Unless there have been extremely long foreign elements occurring during the course of the study, one way is to multiply the allowed time by the number of pieces completed. This figure, divided by the elapsed time of the study, will give the performance of the operator, unless there were many long foreign elements included in the study. If this closely parallels the performance rating that was used, a good verification of the standard is obtained.

Record and File. The allowed time, as determined by the time study, is recorded on the routing sheet, manufacturing information sheet, or other permanent papers related to the operation. The time study itself is placed in a permanent file and is available for reference at all times.

CONCLUSION

This chapter has discussed a time study procedure that is fair and equitable. The three main steps involved—recording information, timing the operation, and working up the study—have been presented in detail.

Every time study department should have a time study manual which sets forth the policies, procedures, and rules that have been adopted by the company for the use of time study. Advantage should be taken of every opportunity to explain the contents of the manual to all concerned. Time study has nothing to hide, and the better this is understood, the more harmonious will be the relationships among time study engineers, supervisors, and the work force.

BIBLIOGRAPHY

Barnes, R. M., *Motion and Time Study*, 6th ed., John Wiley & Sons, Inc., New York, 1968.

Chaffin, D. B., "Physical Fatigue %: What It Is—How It Is Predicted," *MTM Journal*, July–August, 1969.

Hoag, L., "Prediction of Physiological Strain and Performance under Conditions of High Physiological Stress," Ph.D. thesis, University of Michigan, Ann Arbor, Mich., 1969.

Lowry, S. M., H. B. Maynard, and G. J. Stegemerten, *Time and Motion Study and Formulas for Wage Incentives*, 3d ed., McGraw-Hill Book Company, New York, 1940.

Nadler, G., *Work Design*, Richard D. Irwin, Inc., Homewood, Ill., 1963.

Niebel, B. W., *Motion and Time Study*, rev. ed., Richard D. Irwin, Inc., Homewood, Ill., 1958.

Chapter **3**

Performance Rating

CLIFTON A. ANDERSON

Head, Department of Industrial Engineering, North Carolina State University, Raleigh, North Carolina

During the recording of the elemental readings in the course of a time study, attention should be especially directed toward the kind of performance the operator is displaying. Thus, it may be asked, "Is the work being performed rapidly?" or "Is the worker deliberately taking more time than he needs to do his work?" Seldom is a time study recorded that is truly representative of the output of a group of several men who may be employed on the same job. It is necessary to insert a step in the time study procedure which evaluates this variation in output and adjust the results to those of a normal performance. It is this phase of the procedure which will be examined in some detail in this chapter.

A DEFINITION OF PERFORMANCE RATING

The term "performance rating" is used throughout this chapter in a comprehensive sense. It includes all procedures which have as their purpose the adjustment of observed time values to correspond more closely to the time which is deemed to be reasonable and fair for doing the work in question. No single solution to the rating problem is universally accepted at present. In general it is recognized that judgment must be employed in the process. One school of thought bases its approach on the comparison of pace or speed of work evidenced by the operator with that of a defined standard of normal. Another approach is to define the factors which affect rate of production and then to evaluate the operator's performance in terms of these definitions. Still another method for the solution of the problem is found in the accumulation of standard data which are then employed without the need for continuing application of judgment.

Later in this chapter, more space will be given to examining several of the techniques which have become known by certain names. Thus, speed rating, effort rating, leveling, and other procedures will be examined in turn. Each has essentially the same objective but reaches it by somewhat different means.

Most authors of time study texts have included in their writings a definition of what constitutes performance rating. The Society for Advancement of Management's National Committee on the Rating of Time Studies has operated with this definition:[1] "Rating is that process during which the time study engineer compares the performance of the operator under observation with the observer's own concept of proper performance." Phil Carroll says:[2] "Rating is the gaging of the operator's pace during the time study in terms of a normal pace." Lowry, Maynard, and Stegemerten state:[3] "In order that the time standard established from a time study for any degree of skill or effort may be a standard representing average performance, it is necessary to use some method of adjustment of the recorded elemental times if the operator studied gave other than an average performance."

Each of these statements recognizes that the particular operator under observation may work at a performance level which does not coincide exactly with what is expected of an operator on the job.

The time study departments in some plants attempt to avoid the need for including the rating procedure by requiring the operator to perform at a normal or average pace while the study is in progress. Obviously, the necessity for assessing operator performance is still present, and someone must say when the proper (expected) performance level is achieved. It is not always possible to get the operator to perform at this level.

The process of performance rating should under no circumstances be confused with allowances. Performance rating is confined to the actual assessment of operator performance in relation to some standard level of performance. The elemental times actually consumed doing the necessary work are then adjusted to the times that are judged normal for performing these work elements. Once the adjustment to normal is made, additions to the normal time to cover personal needs, unavoidable delays of a minor character, and fatigue may be made. The two steps should be handled separately and ordinarily are.

MEANING OF STANDARD PERFORMANCE AND A FAIR DAY'S WORK

Later in the chapter, the variables that affect output or its evaluation will be examined. It is desirable that the preconceived standard of performance which is to be designated as normal be defined in such terms that it may be recognized, duplicated, and measured. Different terms have been used to describe the standard or task that must be met to entitle the operator to his base pay. Terms such as normal performance, standard performance, average performance, expected performance, and in some cases, a fair day's work are typical of those which name the concept. Words are used to define these terms which must themselves be defined to impart objective meaning to the whole concept. This type of definition permits a rather broad interpretation and a corresponding variation in the concept of a standard performance. Research such as carried out by the Society for Advancement of Management's Committee on the Rating of Time Studies has indicated the extent of the variation in the interpretation of what is actually expected of the operator to earn his base pay.

Another approach to the solution of the problem of defining the concept of

[1] *Advanced Management*, vol. 6, July–September, 1941, p. 110.
[2] Phil Carroll, *Time Study Fundamentals for Foremen*, 2d ed., McGraw-Hill Book Company, New York, 1951, p. 81.
[3] S. M. Lowry, H. B. Maynard, and G. J. Stegemerten, *Time and Motion Study and Formulas for Wage Incentives*, 3d ed., McGraw-Hill Book Company, New York, 1940.

standard performance is to select representative jobs which are readily reproduced or duplicated. The definition is then expressed in terms of output on these jobs. It is necessary to define method and the work requirements very carefully in the case of such bench mark jobs to avoid uncertainty and possible error in the interpretation of normal. It is advisable to establish these measures of performance by careful research rather than to select arbitrarily a group of jobs and establish certain production rates as normal. It is important that the work content of the several jobs be consistent so that the standards are truly comparable in every case. However, the objectivity which such definitions present is a very desirable attribute.

In the definitions that have been written, the emphasis is on what should be done and on what is expected of the operator in order that he conform to the concept of standard. The operator is expected to perform at a reasonable rate, to be qualified for the job, and to use the abilities required to good advantage. He should work at a natural pace that can be maintained all day without undue or cumulative fatigue. There are many ways of wording these definitions. All have the objective of stating the concept of standard which will serve as the reference for all comparisons with other jobs.

A fair day's work is usually defined in terms of the normal performance. It is considered to be the amount of work that will be performed by a qualified operator, working at a normal pace and properly utilizing his time. This definition carries with it the idea of sustained performance over a longer period of time. The time standard includes allowances for delays, personal needs, and fatigue. Both the concept of normal performance and a fair day's work are directly related to the base pay which the operator is entitled to as a result of meeting the standard.

WHY PERFORMANCE RATING IS NECESSARY

When several operators are performing the same job, their output will seldom be the same. There is usually one operator who consistently produces more than others in the group. His superiority may be due in part to a better method of doing the work, but when all men are supposedly following the same method, these differences still persist. On the other hand, there may be one or two operators who are decidedly slower than the others and achieve less production for this reason. Obviously, it would not be fair to the workers to study the fast man and submit, as a standard for all the group, the results of such a study. On the other hand, the study taken on the low-production worker may result in a loose standard which would be reflected in excessively high earnings for some of the group, and consequently, an excessively high labor cost for the product.

An example is spelled out to make clearer the need for a device to assist in arriving at a fair standard for any and all operators and jobs studied. Assume that there is a group of three men who are employed on the same work. The method has been standardized and taught to all three men. One man is new on the job and has not practiced the method long enough to be up to expected performance. The second operator has been on the job for several months, is considered to be a steady worker, and comes very close to our concept of a normal operator. The third man is clearly outstanding, consistently outproduces the other two men, maintains high quality, and generally excels. A careful time study is made on each of these operators. In order to compare the results on a given element of the study better, the readings on the several cycles are arranged in order from the smallest to the largest for each of the operators, that is, a frequency distribution of the readings has been prepared. Figure 3-1 shows the number of occurrences, N, of each elemental time, T.

If it may be assumed that operator A conforms to an acceptable performance standard, then all the elemental times for operator B, with the exception of two, were shorter than the average time of operator A. Because the examples have been selected so that the arithmetic mean, the median, and the modal time are very nearly the same values, the argument holds for comparisons on any of the

	OPERATOR A			OPERATOR B			OPERATOR C	
	N	T		N	T		N	T
				3	.13 Min.			
				9	.14			
	1	.15 Min.		16	.15			
	5	.16		3	.16			
	6	.17		3	.17		1	.17 Min.
	12	.18		2	.18		1	.18
	7	.19					2	.19
	4	.20					5	.20
	0	.21					6	.21
	1	.22					8	.22
							5	.23
							2	.24
							3	.25
							2	.26
							1	.27
Total	36	6.48		36	5.40		36	7.92
Average		.18			.15			.22

FIG. 3-1.   Frequency distributions of elapsed times for three operators.

three bases. Operator C, on the other hand, had only two readings that were as short as the average of operator A. Unless some device were used to adjust the results of the findings on operators B and C to compare with the findings on operator A, inconsistencies in the final answer would result. Had operator A not been present at the time it was necessary to study this job, the time study man would have been compelled to choose between the two remaining operators. Regardless of which was chosen, an injustice would have been done. A tight standard or a loose standard would have resulted, depending on his choice of operators.

To carry the example a step farther, it may be pointed out that in many plants all the operators on a given job may be of the caliber of operator B. In this case, any of the operators studied may have yielded the low elemental times and a resulting tight standard. While all the operators could meet the standard, there would be small opportunity for added incentive earnings. In comparison with other jobs, the standards would not be fair to the operators.

The time study man can avoid this difficulty if he is able to observe the performance of operator B and assess his performance level as equivalent to 20 percent above that expected. If this is assumed to be the case, then operator B must have taken a correspondingly shorter time to do each piece than is expected. By increasing his average time by 20 percent ($0.15 \times 1.20 = 0.18$), the result coincides with the average time of the normal operator. Similarly, by observing the performance of operator C and concluding that his performance is approximately 20 percent below what is expected, the average of his elemental time should be reduced by this amount ($0.22 \times 0.80 = 0.18$).

It is this process of assessing the performance of the operator and interpreting it in terms of a factor which may be applied to the observed elemental times of the time study that is commonly referred to as performance rating.

The above discussion implies that one standard is to be set for all persons. This assumes that equipment, machines, methods, and other variables ordinarily assigned as the responsibility of the management function are controlled within close limits. The extent of variation of these factors allowable within the standards established should be defined so that further departures may be compensated for in adjusted time values. Differences in performance as a result of the qualifications, aptitudes, and motivations of the operator are reflected in different levels of output. Hence different levels of earnings result where a monetary incentive plan is present.

EXTENT OF INDIVIDUAL DIFFERENCES

Attention is first drawn to the possible reasons for and the extent of the differences which were found in the studies of the three operators referred to in the illustration cited above. Differences among individuals do exist. The differences may be in physical capacities, training or practice, aptitudes for a given type of work, or in a number of other respects. Good personnel management practices attempt to minimize these differences within the work group by careful selection and training of the employee and otherwise taking steps to assure his suitability for the job.

The extent of these individual differences in terms of productive output has been variously estimated. An average ratio of 1 to 2.25 is most often quoted. Obviously, this large a range is more likely to be found in large groups than in small groups. It is not intended to imply that the best operator is expected to be 2¼ times as productive as the poorest operator in small groups. In a large random sample not confined to a factory, the ratio is found to be higher than 1 to 2.25 for many types of activities. For a large group of equally motivated persons selected at random, the plot of the frequency distribution of the group output is expected to approximate the form of the normal curve. See Figure 3-2.

In terms of operator performance, the variability will be determined by selection, training, practice, and other differences previously mentioned. The term most frequently used to designate the amount of dispersion or variability is the small Greek letter sigma (σ). The average or arithmetic mean is commonly designated by $\overline{X}$. In the normal distribution curve, the percent of the total area beneath the curve is as follows:

Limits	Percent of total area within limits
$\overline{X} \pm \sigma$	68.26
$\overline{X} \pm 2\sigma$	95.46
$\overline{X} \pm 3\sigma$	99.75

Thus it can be predicted that fewer than three cases in 1,000 would be expected to fall outside the 3 σ limits.

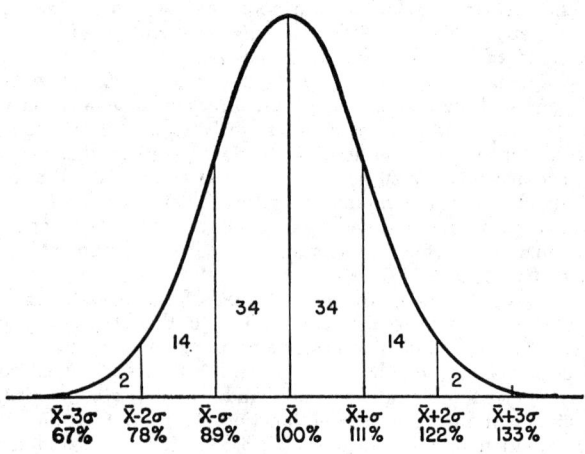

$\overline{X}-3\sigma$	$\overline{X}-2\sigma$	$\overline{X}-\sigma$	$\overline{X}$	$\overline{X}+\sigma$	$\overline{X}+2\sigma$	$\overline{X}+3\sigma$
67%	78%	89%	100%	111%	122%	133%

FIG. 3-2. Normal distribution curve.

Assume as an example that the ratio of the slowest $(\overline{X} - 3\sigma)$ to the fastest operator $(\overline{X} + 3\sigma)$ is 1 to 2 on a given operation. The limits may be drawn on the normal curve, together with values for sigma, and the number of persons predicted to be found in each of the areas. With the mean $(\overline{X})$ taken as 100 percent, and sigma taken as 11 percent, $\overline{X} - 3\sigma$ is 67 percent, and $\overline{X} + 3\sigma$ is 133 percent. The $\overline{X} - 3\sigma$ to $\overline{X} + 3\sigma$ ratio is 1 to 2. For a group of 100 workers, approximately 68 would be expected to perform between 89 and 111 percent. Approximately 14 would fall into the bracket 78 to 89 percent, and the same number would perform between 111 and 122 percent. Two would fall below 78 percent and two above 122 percent. The number of workers in each section of the diagram corresponds to the area under the curve.

With a group of workers who are well motivated through an incentive plan, the average performance is expected to be above 100 percent. Because the employees are usually selected for the job and a weeding-out process operates to eliminate the misfit or the very poor performer, both the mean and the dispersion will be altered. Obviously, in cases where the operator wishes to restrict his output, his production may fall below the lower limits. Experience has shown that evaluation of performance is more difficult at the extremes than at performances near to normal.

REQUIREMENTS OF GOOD RATING SYSTEMS

The all-inclusive criterion which determines the superiority of one rating system over another is accuracy of results. Thus, if a plan could be found which would always yield the correct answer within very close limits, say plus or minus 1 percent, other features making it difficult to administer or explain can be rationalized. Obviously, some error is certain to be present in every application of a rating factor. As long as the error is minor (normally assumed as within plus or minus 5 percent) the plan is deemed satisfactory. There are a number of characteristics of rating systems that contribute to the overall accuracy and make one more acceptable than another.

Perhaps the most significant feature that a system may possess is the attainment of consistent results. Consistency does not necessarily imply absolute accuracy, but means rather that the error should be in one direction and very nearly the same on all applications. Consistency of results may be thought of in terms of the single individual, a department consisting of several time study men, or a plant with several locations of its various departments. Consistency in all is desirable. Thus, the individual time study man should be able to make his time studies over the entire range of his work and set a standard in each case that will result in nearly the same average efficiency for any normal operator working at the same performance level. Should this time study man tend to be on the low side, a general correction can easily be made for such an error. Difficulties arise when the time study man using the plan cannot achieve consistency in his results. The tendency toward inconsistency may be inherent in the man, or the plan may not be rendering the assistance to him that it should.

Consistency should be achieved between time study men in the same department to avoid a condition where one man gets a reputation for setting tight standards and another loose standards. When this happens, the operator will soon show a preference for the more generous judgment. It should be expected that standards set for exactly the same job in different departments should be very nearly the same. Unless this is true, inequities are present which do an injustice to the operators involved. Although time study men may differ significantly in their abilities to do consistent rating of operator performance, it has been found that the qualified time study man possessing the necessary aptitudes will improve his rating consistency and accuracy when he has been carefully trained and fully understands what he is trying to do.

A good rating system is simple. Its operation can be explained in simple terms so that the operator can understand how it operates and how his performance is reflected in the rating assigned him for a particular study. Involved computations of deviation ratios, averages of averages, and the like do not necessarily impress the operator with the soundness of the basis on which the plan is conceived. Relating the performance of the operator directly to a factor by which his elapsed time is adjusted is more satisfactory.

Perhaps one of the greatest needs of the time study man who does performance rating is a basis for justifying his judgment. Thus, the provision of definitions which will clearly describe a given level of performance so that others can agree on the soundness of the judgment is needed. Devices which will make these definitions as objective as possible enhance the value of the system.

Objectivity in definitions is very difficult to attain in word definitions. When a normal operator is talked about, other descriptive terms must be employed to make clear what is meant by normal. Thus, the operator may be qualified for the job and have enough experience on the job to entitle him to be considered a normal operator. However, the term "qualified for the job" must be defined in more specific terms. This difficulty in interpretation of exactly what is meant by the subjective term has given rise to a number of attempts to state normal output in specific terms. Several standards for comparison have been suggested, each of which is simple and readily duplicated. This practice can be extended to filming operators possessing varying degrees of dexterity and using these films as a comparison against other operators in the shop. By means of frame-by-frame analysis of the motion picture, the actual reduction in time due to the improved manipulative ability of one operator over another may be quantified. This analysis thus serves as an aid in judging similar operations and operators.

This discussion cannot very well be concluded without referring to accuracy considerations. The qualified and trained time study practitioner who understands his system of rating and its application is capable of rating within the limits of plus or minus 5 percent. Variations beyond these limits are likely with a person not as well qualified nor as carefully trained. The plus or minus 5 percent variation must be determined on the basis of the individual readings and their respective departures from the normal. This means that none of several persons' results will differ from the mean by more than 5 percent. For such comparisons, normal is usually taken as the average result of several persons making the same rating. It must be stressed that only through careful selection and training will it be possible to attain such results.

WHO SHOULD DO THE RATING?

Perhaps the first consideration is one of qualifications and training. It is not easy to rate performance and get consistent results.

Arguments are presented from time to time by unions and others that rating is subject to negotiation and bargaining. One suggested means of accomplishing this is to have a union man accompany the time study man, make simultaneous studies, rate their respective studies, and work up standards. Then, prior to issuing the standard to the operator, the results are compared and any differences resolved. While there can be a definite advantage in pooling judgment, this practice would seem to encourage bias and bargaining on each side because both men would not have the same objective. Each of the persons making the study would be encouraged to rate high or low, depending on which way he knows he may be forced to compromise. By starting with a margin of safety, he can bargain and still come out with a satisfactory deal. So the question is asked, "Why shouldn't one person be taught to do the job as nearly correctly as possible, putting the emphasis on setting a rate that is equitable for both parties?"

The position of the union time study man is a most difficult one. When he accepts a standard as being fair, then he has the job of selling the standard to the operator. To avoid disagreements with his constituents, he must, in effect, find out what the operator considers to be a fair standard and get his result to conform with the predetermined answer. It is recommended, therefore, that these questions be handled through the usual grievance procedures with the union committees and the operator directly involved.

Placing the performance rating procedure in the area of bargainable procedure is admitting its lack of objectivity and accepting one person's judgment as being as good as another's. Trained judgment exercised by competent persons offers the best chance of success and should be employed by plant management to handle this problem successfully in its work measurement program.

RATING MUST BE DONE AT THE TIME THE STUDY IS MADE

The whole purpose of rating is defeated when a time study man works backward from a preconceived notion of the proper standard to the actual time used to do the work, to get a factor which will give him the desired answer. The answer is honestly found when the performance is assessed in terms of clearly defined standards of performance and the correction applied to the actual time observed. Obviously, the performance cycles to be rated are those which produce the readings which comprise the study. Therefore the rating must be done simultaneously with the observation of the elemental times.

Some factories require that the foreman of the department check the performance at the time of the study and that agreement, or the lack of it, on the level of the observed performance be noted on the time study. Any difference between the time study man and the foreman must be resolved at once or appealed for decision to a higher authority. The matter of foreman approval may extend considerably beyond performance rating and be required for method, quality of finished pieces, and other standards.

Certainly there can be no argument that, if the time study is to be in fact what it pretends to be, then the rating must be done as a separate act, independent of the final answer or any bias that it may invite. The actual leveling must be in terms of some independently established standard, uniform for all of the organization.

BASIC TECHNIQUES EMPLOYED IN RATING PERFORMANCE

There are three general approaches to the problem of rating, with respect to the manner in which the rating is to be accomplished. These are:

1. Statistical or mathematical treatment of time study data
2. Assessment of observed performance level on the basis of trained judgment
3. Comparison of actual time with an accepted standard time

The first approach is a strictly statistical procedure. By a series of mathematical or arithmetical manipulations, a factor is derived which is expected to reveal the extent of the departure of the operator's performance from the standard sought. Different statistical methods have been followed, all of which are presented as devices that make it unnecessary to exercise judgment on the part of the time study man. Perhaps the best known of these was proposed by Merrick. The idea of variation, or the lack of it as evidenced by consistency, was taken as the predominant characteristic of good performance. Perhaps this is a logical conclusion if the operator is doing his very best to give a top performance. But it has been found that the technique falls down badly when the operator slows down his pace and concentrates on obtaining consistent time values for each repetition of the cycle. With such a bias imposed on the data, the factor obtained could very well indicate a high level of performance although it was slow and deliberate, with the production falling far below the expected level. Mainly for these reasons, the technique has been abandoned in favor of other methods of assessing operator performance.

The second approach—now generally accepted—requires the time study man to apply trained judgment to establish a numerical factor that will adjust observed time to standard or normal performance. All the variations of this technique call for evaluating observed performance in terms of a preconceived standard. The performance as a whole is compared with the standard, or the performance is assessed in terms of component factors, each of which may be evaluated by itself. The combined effect is then used as the factor by which the time is adjusted. Later in the chapter, several of the techniques are looked at in detail.

The third approach to performance rating is called synthetic leveling. In this procedure, the time study man accurately times certain easily recognized movements or elements with a stopwatch, and then compares these results with accepted time values or previous values that are well substantiated through long use. Thus, if the observed elements take 10 percent more time to perform than the standard values for these same movements, it is assumed that the entire study is performed at this same level. The accuracy of results of such an evaluation depends upon the accuracy of the stopwatch readings, the extent to which the observed elements actually duplicate the standard elements with which they are compared, and the accuracy of the standards themselves.

In addition to the three general procedures cited is the idea of standard data and its use for completely developing a time standard for a job without the use of the stopwatch or the necessity of applying the judgment factor in rating performance. The use of standard basic motion times for each of the movements required of the operator eliminates the need for evaluating the speed of the performance of the operator because the time study itself is no longer needed. The determination of method takes on added importance because it must be detailed within close limits. The types of motions required must be identified and the time value corresponding to these motions assigned.

In the assembly of the standard data, it was necessary, of course, actually to rate the many performances which provided the original data. These data are based on broader coverage and represent a composite of opinion as expressed in the averaging of many cases. Further, the data have had ample trial study to assure their acceptance as suitable standards. However, the discussion in this section will be confined to consideration of the second of the procedures cited. Here the time study man will be called upon to observe operator performance, and through the application of trained judgment, evaluate the performance in terms of a normal or standard.

OTHER FACTORS FOR DISCUSSION AND CLARIFICATION

There are several minor questions that inevitably appear in connection with the process of rating performance. For example, what should be the relationship between base pay and the operator's efficiency when he is putting out a fair day's work? Should he be receiving just his base rate when he is achieving a normal performance? Obviously, the answer is also related to various incentive plans that may be employed. Nonetheless, it is important to understand clearly just what the relationship is. In some plants, the operator is expected to perform at 130 or 135 percent of standard, especially where the process is dependent upon this operator's functioning. Obviously, the same interpretation of the several systems for assessing performance cannot be applied if the operator's earnings must bear a relationship to his base pay of 100 percent, 135 percent, or some other ratio when the normal performance is achieved. Such incentive features are best introduced as an incentive factor and kept independent of the rating procedure, but frequently they are combined. In the discussion which follows, the performance rating is assumed to be such that normal performance will be reflected as 100 percent output and earnings.

In the installation of any time study program in a plant, a definite procedure must be included which tells just how to derive or select each elemental time value to which the performance rating factor is to be applied. The arithmetic mean or average value for the several cycles studied is often used. Statistically, it can be proved that this provides the best estimate of the actual time which the operator has required to do the work.

Recognizing that a judgment factor is to be applied next, the selection of the value is sometimes done by less time-consuming methods. The modal time is the reading which occurs most frequently in the series for a given element. It tends to disregard the extremes and the skew in the distribution. The median value is the one which has an equal number of readings above and below when the readings for an element are arranged in ascending order. In most distributions found on time studies, there is usually little difference in the actual value of the three. There are ways, of course, of arriving more quickly at the time to be used and thereby avoiding averaging, or arranging readings in some order. The arithmetic mean is recommended because it is easily understood and is accepted as being fair.

Why is a straightforward statistical procedure which determines the average or some other statistic not sufficient without the use of the judgment factor? The best single word answer seems to be bias. No time study man needs to be told that an operator can deliberately slow down his movements and hence take longer than necessary to perform the job. Nor does he need to be reminded that there are also operators who find satisfaction in demonstrating their superiority. Most statistical procedures are dependent upon an unbiased sample to predict the nature of the parent population. If, as a result of improved motivation, the performance level is substantially raised after the study is completed, then the sample information is of small use in predicting the outcome in the new situation. It is necessary for valid results that the two situations remain essentially the same. Hence, the hope that the straight statistical result will yield a satisfactory solution is not founded on sound assumptions. A further deterrent lies in the need for large samples to assure reliability of the results. Bias is introduced not only by the deliberate intent of the operator but also by the status of operator proficiency as gained through practice. There is pressure by both management and the operator to set standards as quickly as possible. The performance on the tenth piece is by no means going to duplicate the performance on the 10,000th piece, when learning is much more advanced. This is actually manifested in what might be called improved motion patterns on the part of the operator. For example, in early stages the operator cannot work with both hands simultaneously when trying to do certain things. With practice, this is not only possible but entirely to be expected.

Rating the Study versus Elemental Rating. There is some difference of opinion about whether to rate the study as a whole, to rate the elements, or to rate the individual reading. When elemental times are as short as 0.05 minute, it is virtually impossible for the time study man to exercise a careful judgment for a succession of such elements. As the elemental times increase in length, the time available to the time study man to observe and record his readings will permit this detailed analysis to be carried out. In many cases, an operator may be called upon to perform a cycle in which there are certain elements which are new to him while other elements have been performed often. These familiar elements are performed proportionately faster than the new elements in the cycle. After observing the operator for several cycles, and through the study, it is quite obvious to the observer that this difference exists. As a result, he assesses the performance of the elements separately. When such a situation pertains, it is possible and desirable to follow such a procedure. In short cycles consisting of only a few elements, all of which are familiar to the operator, it is usually satisfactory to assign one rating to the entire study. It is recommended on any study where variations are not distinguishable.

Rating Machine-controlled Elements in the Cycle. Where the machine controls a portion of the cycle through a power cut or a process time of some sort, there is nothing that the operator can do to change the time required to perform that function. This is implied in the definition that it is machine controlled. If the operator has nothing to do during this period, the time taken should be allowed 100 percent. For additional allowances and for incentive earnings, special arrangements are sometimes made. If the operator has internal elements to perform, that is, if he is doing manual work while the machine is making its power cut, then his work should be observed both for the time required and the level of his performance. The normal working time for the internal work elements and the computed waiting time are determined for the machine-controlled portion of the cycle. This information is important if consideration is ever given to machine coupling. Eventually it may be possible to arrange a production line or to couple machines in such a way that the operator has more work to do in the period in which the machine is occupied than he can do at a normal pace. As a consequence, his incentive opportunity approaches the expected incentive earnings level. The incentive plan may be designed so that the incentive opportunity available to the operator for this part of his cycle is proportional to the percent of time that he is occupied. His potential earnings opportunity increases as more work is provided in the machine-controlled portion of his cycle. It is hard to see why an operator would ever voluntarily participate in machine coupling if he is presently paid 130 or 135 percent earnings during a part of his cycle when the machine is occupied and he has nothing to do. From the standpoint of performance rating, it is necessary to measure the manual work done during the machine-controlled part of the cycle and convert the values to normal working time. This does not preclude the assignment of delay allowances and personal allowances.

THE MORE IMPORTANT TECHNIQUES IN GENERAL USAGE

Speed Rating. In the rating technique termed "speed rating," the speed of the movements of the operator is given as the only factor which lends itself to measurement and hence to rating procedure. The extent of method and skill present are revealed in the time study by the actual elements that appear on the study. Differences in the speed of the operators are reflected in the actual time required to perform like elements.

The rating procedure thus consists of judging the pace or speed of the operator's movements in relation to a normal pace and noting the relationship as a factor. The rating makes no attempt to correct for deviations from the standard method. This rating is applied to each element and in some cases is applied to each individual reading on the study. The observed time for each element is multiplied by the ratio of observed speed to expected speed to arrive at the expected or normal time for the element. Deviations from the standard method must be handled independently of the rating factor.

The factor is sometimes expressed on the basis of a standard hour equal to 60 and deviations in speed indicated as a fraction of 60. Therefore, $\frac{50}{60}$ would indicate a speed of movements $\frac{5}{6}$ or $83\frac{1}{3}$ percent of that expected. A factor of $\frac{75}{60}$ would indicate a performance 25 percent above normal. This factor is applied directly to the actual time selected to adjust it to normal. This expression of the factor was derived from the idea that normal is the amount of work that should be done in one unit of time, which is conveniently taken as 1 minute. Thus, a rating of $\frac{50}{60}$ indicates that 50 minutes of work are being performed in the hour.

The factor may also be expressed as a percentage, with 100 percent being considered as normal. In this case, a rating of 90 percent indicates a speed equal to 90 percent of that expected. A rating of 125 percent indicates a speed 25 percent above normal. This notation seems to have the advantage of already being in the form in which it will be used when the actual time is extended to normal. It is also the form most often used to make ratio comparisons.

In speed rating, the process of rating is confined to the comparison of speed of movements with a concept of normal speed. On the basis of this assumption, the rating process is made simpler and, with training in developing the concept of normal pace, the observer may become quite proficient in his judgment.

Effort Rating. The technique for assessing operator performance, which is termed "effort rating," closely approximates speed rating. Although effort rating may be used by many to name the procedure they employ, this term is associated with Ralph Presgrave, and the following comments are based on his book,[4] *The Dynamics*

[4] Ralph Presgrave, *The Dynamics of Time Study*, 2d ed., McGraw-Hill Book Company, New York, 1945.

of Time Study, wherein it is discussed.   He explains that the term has been selected because of its wide acceptability, and when time study men speak of effort, they have in mind relative production rates.   However, the meaning of effort is confined to the concept of speed of movements.

For an excellent analysis of the problems incident to the evaluation of operator performance, the reader is referred to Presgrave's treatment of the subject.   The case for rating speed of movement as the only measurable factor in the process is very well stated.   Although skill in the broad sense is recognized as contributing to both method and speed of movement, it is not segregated.   Method must be determined and established as a function outside of the rating procedure and the rating procedure limited to judgment of the speed of movements.

In this and the speed rating technique, the concept of a normal speed or effort is built around one speed or tempo.   A problem results when trying to evaluate the performance of operators on work which may be differentiated as light or heavy.   The tempo of light bench operations is not the same as that evidenced by the operator doing heavy lifting, pushing, or pulling.   It is necessary, therefore, to make an additional judgment to distinguish between the two situations if proper time standards are to be derived in all cases.

Pace Rating.   The term "pace rating" is employed in some companies, notably the U.S. Steel Corporation, to describe the system of performance evaluation in use.   Although the technique incorporates most of the ideas of speed rating and effort rating, two other devices are used to assist the person doing the rating and to extend the scope of the application.   Thus it is recognized that all jobs are not performed at the same tempo, so that the pace or speed observed must be related to a concept of normal for the type of work involved.   The time study man uses a number of concepts of normal, depending on the type of work being observed.   Where his work is limited to one type or a few, the standards or normals would be correspondingly limited.

To assist the time study man in the acquisition of a set of concepts that is uniform for all time study men, bench marks have been provided in different types of work.   These have been quantified in terms of specific rates of production.   Thus, walking on a smooth level surface, without load, at X miles per hour is one standard.   This and other standards can be duplicated or viewed on a motion picture screen and thereby provide an objective interpretation of the pace described.   Rating is expressed as a performance percentage above, below, or at normal, and the ratio or factor is applied to the selected time for the element.

An attempt is made to minimize the effects of other variables by studying those operators who are judged to be adequately qualified and trained to do the job in question.

Objective Rating.   A procedure originated by Mundel[5] is perhaps the next logical step following the idea of speed rating.   Mundel does rating in two steps, namely, (1) pace rating and (2) a secondary adjustment to compensate for job difficulty.

A normal concept of pace is established against which all jobs are compared.   Because no attention is given to job difficulty, a single standard is given as adequate for this comparison rather than standards for each type of work.   To assist in judgment of deviations from a normal pace, several step films, each representing some known departure from normal, are recommended for training and practice purposes.   Following the judgment of pace or speed, the job is evaluated in terms of its difficulty.   The factors or categories for which secondary adjustments are added include: (1) amount of body used, (2) foot pedals, (3) bimanualness, (4) eye-hand coordination, (5) handling or sensory requirements, and (6) weight handled or resistance encountered.

From experiments and other sources, numerical values were assigned to different degrees of these factors.   The summation of the percentage adjustments for each of these factors comprises the secondary adjustment.

[5] M. E. Mundel, *Motion and Time Study*, 2d ed., Prentice-Hall, Inc., Englewood Cliffs, N.J., 1955.

Leveling. The method of assessing performance which is referred to as leveling is described in detail in Lowry, Maynard, and Stegemerten's book on time and motion study.[6] The technique is usually described in any collection of information about performance rating. It is in use in many plants and has gained quite wide acceptance. Close reading of the authors indicates their stress on the need for full understanding and adequate training in the use of the technique in order to get consistent and accurate results.

Four elements are given as constituting the important factors which determine the rate of production that an operator achieves. These four factors are skill, effort, conditions, and consistency. The first two are by far the most important. Each of the four elements carries a somewhat special or limited meaning. It is important that these meanings be understood prior to the application of the technique.

Skill is defined as "proficiency at following a given method." Method is thus excluded from the concept of skill in the definition. The time study man judges the level of skill by observing such things as hesitations, precision of movements, interruptions to the normal cycle by improper performance, and the general coordination and rhythm of working pace manifested by the operator. Departures from average, or the definition of normal skill, are indicated by grades. Each of these grades is in turn defined and indicated in chart form as poor, fair, average, good, excellent, or superskill. Thus, the skill manifested is judged in terms of definitions and compared with a concept of normal or departures therefrom.

Effort is defined simply as "the will to work." Effort is considered to be within the control of the operator at all times. It is not measured in terms of foot-pounds of work done but rather is judged in terms of the spirit in which the operator attacks his job. It may range all the way from idleness to excess. As will be pointed out later, it is necessary to confine the range of effort observed to narrower limits if the technique is to be entirely successful. Six gradations of effort level are defined, including poor, fair, average, good, excellent, and excessive. Thus, poor effort is manifested by a slow-motion style of working which is very obvious to the observer. The introduction of unnecessary work is also indicative of poor effort. This obviously has an effect on method also. It is generally taken to be a result of poor attitude on the part of the workman. Average effort is defined as that manifested by the operator who works steadily and with fairly good system. Lost motions are reduced and the man takes some interest in his work. Excellent effort is exhibited by the operator who plans ahead, reduces lost motions to a minimum, uses the best method available, and takes a keen interest in his work. He is anxious to demonstrate his superiority.

Conditions are narrowly defined as those conditions which affect the operator rather than the operation. Light, heat, and ventilation, or rather the variation of these conditions from what is normally provided for the given operation, are included in consideration for leveling purposes. Corrections for this factor cover only minor departures from standard. Major items which may affect the method should be corrected before work measurement is begun.

Consistency is established primarily as a factor to call attention to the extent of consistency or lack of it. The recommendation is made that the cause of inconsistency should be determined and corrected rather than graded. The correction for perfect consistency or poor consistency is a minor factor.

In general, the study is rated as a unit. However, where the operator may be doing familiar work in one part of the cycle and have elements included in another part of the cycle in which he is not practiced, the recommended procedure is to grade those elements individually which are not performed at the same level as the major part of the study. For long studies, it is recommended that performance evaluations be made periodically to take account of any changes that may develop over the longer period of time.

Numerical equivalents have been provided for each of the grades or levels of

[6] *Op. cit.*

Skill			Effort		
+0.15 +0.13	A1 A2	Superskill	+0.13 +0.12	A1 A2	Excessive
+0.11 +0.08	B1 B2	Excellent	+0.10 +0.08	B1 B2	Excellent
+0.06 +0.03	C1 C2	Good	+0.05 +0.02	C1 C2	Good
0.00	D	Average	0.00	D	Average
−0.05 −0.10	E1 E2	Fair	−0.04 −0.08	E1 E2	Fair
−0.16 −0.22	F1 F2	Poor	−0.12 −0.17	F1 F2	Poor

Conditions			Consistency		
+0.06	A	Ideal	+0.04	A	Perfect
+0.04	B	Excellent	+0.03	B	Excellent
+0.02	C	Good	+0.01	C	Good
0.00	D	Average	0.00	D	Average
−0.03	E	Fair	−0.02	E	Fair
−0.07	F	Poor	−0.04	F	Poor

Fig. 3-3. Performance rating table for leveling. (*From S. M. Lowry, H. B. Maynard, and G. J. Stegemerten, Time and Motion Study and Formulas for Wage Incentives, 3d ed., McGraw-Hill Book Company, New York, 1940, p. 233.*)

the factors. These equivalents are shown in Figure 3-3. To determine the correction factor or leveling factor, the assigned ratings for each of the four factors are noted and their respective numerical equivalents are added algebraically. The result is added to numeral 1. As an example, a rating of B2 skill, C1 effort, E conditions, and D consistency would provide correction factors of +0.08, +0.05, −0.03, and 0.00. The algebraic sum is +0.10, and added to 1, the resultant leveling factor is 1.10. This factor is applied to the average of the observed time for the element or elements to derive the average (normal) time.

It should be noted that this technique limits the variation that can be compensated for. When an operator slows down to half speed, it is impossible to make adequate adjustment through the leveling factor to correct the actual time to normal time. Within limits of about plus or minus 25 percent of normal, the trained observer can get consistent results utilizing the technique.

It is helpful to utilize bench mark performances as a training and checking device, just as for the other methods of performance rating. The definitions lack objectivity in themselves, and unless the various levels of performance can be demonstrated, there is a tendency toward inconsistency in interpretation of the various gradations.

The important difference between leveling and speed rating lies in the attempt in leveling to relate the performance displayed by the operator to the causes which result in the various levels of performance. When the operator works in a particular manner, as indicated by the definition which describes his performance, the resulting productivity will differ from normal by the amount indicated by the numerical values assigned to each gradation. Somewhat wider scope is thus assigned to the leveling procedure than in speed rating or effort rating where the judgment is limited to speed of movements only.

Synthetic Leveling. This procedure for determining the performance rate of the operator is presented by R. L. Morrow[1] as a means of taking much of the need for judgment out of the rating procedure, thereby attaining more accurate and more consistent time values. The procedure consists of comparing the times for as many of the elements as possible to known standards. There are usually a number of elements in a time study which are common to many time studies, and these at least may be compared. The relation of the elemental standard times to the observed elemental times indicates the level of performance on this study for these elements. The rating is then extended to the entire study. It is important in the interests of accuracy that the elements compared contain the same work requirements. End points, method, and actual work requirements enter into the accuracy of the technique. The assumption that the entire study should be graded alike is necessary also.

Use of Standard Data as a Substitute for Time Studies and Leveling. This is not the proper section to enter into a full discussion of the use of some form of standard data as a substitute for time studies. Much has been done in this field. The Methods Time Measurement technique has achieved considerable prominence. Work-Factor is another technique which has the same objective but achieves it in a somewhat different way. Section 5 of this Handbook gives a fuller discussion of these procedures. Basically, each of these systems has established a body of data from studies of past operations which are applied to new work situations. It should be pointed out that conventional standard data (Section 3, Chapter 8) do essentially the same thing, but with more limited application for a given body of data. Whereas time study standard data are of necessity applied to the class of work from which they were derived, the more fundamental data have much wider application.

PROFICIENCY IN RATING

Proficiency in rating is measured in terms of how closely the time study man is able to approximate the correct normal time after the rating factor has been applied. Thus, theoretically, every performance may be transformed into a normal performance. Because of errors in judgment on the part of the observer, this does not always happen. However, if the resulting time can be adjusted to fall within the limits of plus or minus 5 percent, it is generally accepted as a satisfactory standard.

The question naturally arises, "How is the correct standard or normal derived so that the judgment of the observer may be assessed in terms of a true standard?" Obviously, there will be no such standard on the individual job which is being studied. About the only way to derive such a standard for comparison is to pool the judgments of several people and assess the results of the individual in terms of the group judgment. Another way is to prove the standard over a long period of time with a number of operators working on incentive. If the earnings of these operators fall within the limits expected, the standard is considered satisfactory. At the time of the study, and when the standard is presented to the operator, no such information is available. Therefore, the time study man must develop a record for setting standards correctly in order to have them accepted without too much difficulty. His standards must prove to be consistent and very near what is expected as the interpretation of a fair day's work.

Where several operators may be studied on the same job, it is of course possible to make studies on each. With proper rating, the resulting time standard from

---

[1] Robert Lee Morrow, *Time Study and Motion Economy*, The Ronald Press Company, New York, 1946.

each individual time study will be within plus or minus 5 percent of the mean of the several studies. Placing all the results within this narrow range does not prohibit the entire set of judgments from being biased on the high or low side. Consistency is perhaps the major quality to be achieved. Whether or not the judgment of the group of time study men in a plant conforms with the national interpretation of a fair day's work is not as important as that they agree with each other.

In achieving consistency of rating and conformance to the opinion of the group, there is no substitute for frequent training sessions in which the individual may continually check his judgment with the group and with some known standard. The observer should have opportunity to make judgments against known answers. If there are errors in judgment occurring, then the reasons for the failure to judge correctly should be examined with a view toward refreshing understanding of the components of performance or the concept of normality.

There are no validated psychological tests for selecting individuals who will be proficient in performance rating. However, it is likely that, unless the individual achieves some proficiency in rating in a short time following proper training, he may never master the technique. Consistency is the most important trait which the individual may possess. Bias on either the high or low side can be corrected with a factor. Inconsistencies are unpredictable and hence conducive to large error.

Proper training is necessary if a new time study man is to become adept at performance rating. It is doubtful if a man will ever become proficient at rating if left to his own resources to master it.

TRAINING DEVICES

Several means are employed to help the time study observer improve his accuracy and consistency in rating. Spot rating practice is designed to provide a frequent check of the observer's judgment. In using spot rating, a supervisor may accompany one or more of his time study men about the plant and have each independently rate various operators. Ratings are recorded and then compared with the supervisor and within the group to reveal the degree of consistency among the raters. Where deviations too great to be acceptable are occurring, the actual operation may be looked at in more detail while the various individuals advance the reasons for their ratings. This contributes to a better understanding of what to look for and also helps to build the concept of a normal performance.

Specific instructions can be given in the form of exactly what to look for in observing the performance. For example, the observer may be misled in his judgment when an operator performs a long hand movement very rapidly but takes a longer time than normal to perform the following positioning operation because of the previous rapid movement. Finger dexterity, certainty of movements, and a blending of movements are indications of superior performance. Pointing out these differences between individuals and assessing their effect on productivity is helpful training to the practicing time study man.

Films depicting different operations performed at various levels are receiving considerable attention. Several films have been prepared which provide standards by which the individual may assess and thereby compare his judgment of the performances shown with the pooled judgment of experienced time study men who have rated the film. Practice rating or leveling in real-life situations is desirable because that is where the work will eventually be done. It is found that practice in rating filmed operations is helpful to the time study man. This is particularly true where the operation is reviewed following the original rating. In the review, the small differences between operator movements may be examined in detail and the reasons for the differences in output pointed out so that they will be recognized in the future. Regardless of the device used, it is well established that it is necessary to train the beginner carefully and to continue to train him as long as he is asked to do performance rating. Such a planned program will yield good results in the form of more accurate and more consistent time standards.

RECORDING PROGRESS

It is advisable to keep records of the results of any performance rating training. Where films are used with known levels of performance provided for the instructor, the extent of each individual's deviation from the given rating can be expressed as a percent of the correct rating. Successive ratings may be plotted on a chart along the X axis and the percent deviation, plus or minus, plotted as the ordinate as in Figure 3-4. As a result of such a plot, the subject doing the rating can see graphically the extent of his variation, whether he tends to rate high or low, and whether he improves with additional practice. The same information is provided the supervisor for each member of his group so that he may assess the relative competence of his men and the competence of the department as a whole.

Inasmuch as the measure of accuracy of rating is determined by the individual reading, the practice of averaging several ratings for plotting purposes is misleading. As an example, three ratings which were 10, 15, and 20 percent above the known levels of performance may be offset entirely by three other ratings which are 10, 15, and 20 percent below the given ratings. Although the six ratings averaged together would apparently indicate perfect conformity, the variation within the set of readings when considered individually would be quite unsatisfactory. Each time value, and hence each rating, must meet the standard of accuracy. The statement that ratings are accurate to within plus or minus 5 percent is correct only when the individual ratings are being considered and not the average of several. Statistically, it may be shown that averaging four readings will result in a plot with only one-half the variation evidenced by the individual readings. The relationship of variation between unaveraged and averaged readings is inversely proportional to the square root of the number of readings averaged.

Another way of plotting results of rating is to construct a graph in which known ratings are scaled on the X axis and the subject's evaluation is scaled on the ordinate. If all ratings are judged correctly, the coordinates of each rating fall on a line which has a slope of 1 or makes an angle of 45 degrees with the axes when the same scale is used on each axis. Loose ratings will show up as points above the line and tight ratings will fall below the line. As plots are made of performance judgments which depart farther and farther from normal performance, a tendency develops which is sometimes referred to as conservatism.

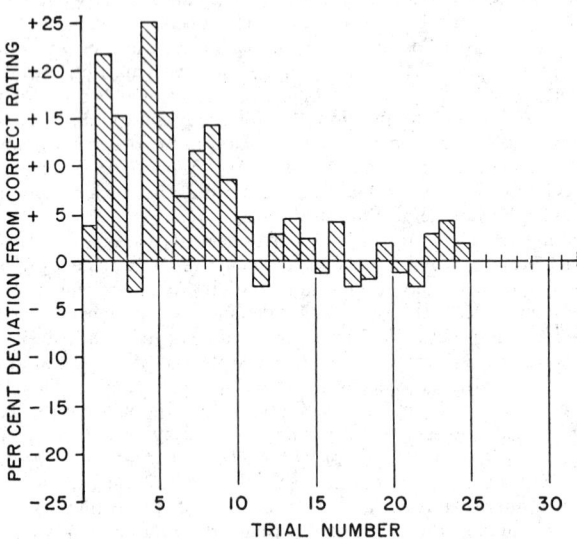

Fig. 3-4.   Progress chart of variation in practice rating.

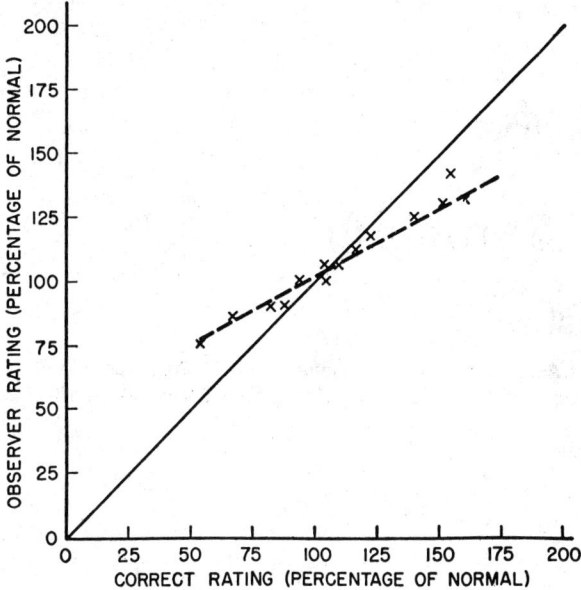

Fig. 3-5.   Chart showing tendency to conservative rating.

This term describes the result of rating low performances too high and high performances too low. The graph, Figure 3-5, shows the correct rating line and a line representing the actual results when conservatism was present. This tendency seems to be present with all the rating techniques which depend on judgment for their accuracy. This is primarily a failing on the part of the observer to recognize extremes and to compensate sufficiently for them.

CONCLUSION

Time study men will find it necessary to continue making judgments of operator performance. A substitute procedure that will eliminate this requirement is not yet available when stopwatch time study is used. Because the problem cannot be eliminated, the logical approach is to do the best possible job with the tools at hand. Further, the way these tools are used must be continually examined to assure that they are being employed to the very best advantage. Thorough indoctrination and training followed with a continuing program of checking progress is a "must" so that the performance rating technique may do the job that is required of it.

Chapter **4**

Work Sampling

C. L. BRISLEY

Professor and Associate Chairman, Department of Engineering,
University Extension, The University of Wisconsin, Milwaukee, Wisconsin

Private firms, hospitals, educational institutions, and government have realized the necessity for improving individual and general efficiency. Among other areas, research has been focused upon chance occurrence and behavioral patterns of engineers, managers, nurses, and clerical and factory employees.

Work sampling is used for work measurement and analysis. No stopwatch is used. The technique tends to be somewhat less personal than stopwatch studies because usually it is applied to groups of people or machines. Work sampling, as a technique, was introduced in England by L. H. C. Tippett in 1934.[1] It has been increasingly applied in many areas that were not formerly measured. Because this original paper is not available in most libraries and because many people interested in work sampling desire to study it, Ralph M. Barnes reproduced it with the permission of the author in his book on work sampling.[2]

Work sampling consists of random observations to determine the ratio of those observations of various delays and elements of work to the total number of observations in the process. The ratio or percentage of observations recorded in a given state tends to measure the average percentage of time it is in that state. The number of observations depends on how accurate the answers need to be. A larger number of observations provides a greater accuracy.

[1] L. H. C. Tippett, "Use of the Binomial and Poisson Distribution: A Snap Reading Method of Making Time Studies of Machines and Operations in Factory Surveys," *Shirley Institute Memoirs*, vol. 13, November, 1934, pp. 35–93.
[2] Ralph M. Barnes, *Work Sampling*, John Wiley & Sons, Inc., New York, 1957.

WHY DOES WORK SAMPLING WORK?

Work sampling is based on the law of probability. It works because a smaller number of chance occurrences tends to follow the same distribution pattern that a larger number produces. The simplest way to explain the phenomenon is by reference to tossing a coin or throwing dice.

Examples of Probability. When a coin is tossed, the result is one of two possibilities, heads or tails. The law of chance says there should be 50 heads and 50 tails in 100 tosses of a coin. That is the ratio of the average possibility. It does not mean it will come out 50–50 on the button every 100 tosses. The score may be 60–40, 45–55, or some other ratio. But it has been proved that the law becomes increasingly accurate as the number of tosses increases. The percentage of possible error decreases.

Another example of how the law of probability works is dice throwing. It is a little more complex than coin tossing because one throw of two dice has 36 possible results, instead of two (see Figure 4-1). It is an idea that will help sell work sampling to others by helping them understand it.

Thirty-six throws will tend to produce one 2, two 3s, three 4s, four 5s, five 6s, six 7s, five 8s, and so on. Note that it *tends* to produce. Each series of 36 throws will not duplicate the pattern exactly. But the results get closer to the pattern (probability curve) as the number of series of 36 throws is increased. The percentage of error decreases.

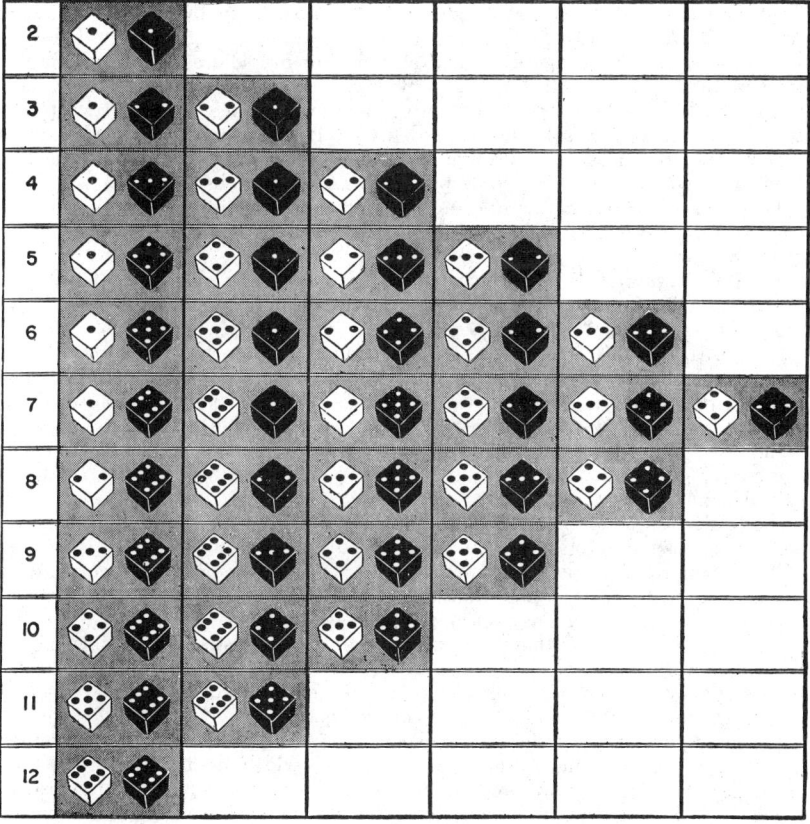

Fig. 4-1.   Probability curve for throwing a pair of dice.

The law of probability may be proved by making a number of series of 36 throws of a pair of dice. By keeping a tab on the number of throws in order not to lose count and by charting the scores, the results should tend to equal the theoretical results as the number of series of 36 throws is increased.

HOW TO PREPARE FOR A WORK SAMPLING STUDY

First, Sell Work Sampling. Although work sampling may seem simple enough, it will be found that many people will not believe it—in a five-minute introduction. Operators will not buy a brief explanation either. The very natural reaction is that one cannot possibly get a true picture if he does not watch the operation continuously.

Communications Important in Initiating Work Sampling. In nearly every organization, the failure has been in communications.

Harold Smiddy, former vice president of General Electric, said, "A great deal could be accomplished if we would consider communications in the light of this simple, four-word formula, 'Talk to the guy.' "

So often, in installing work sampling programs, management takes the general attitude that this is its prerogative and that it is not necessary to relate to the union, or to those to be observed, just what this technique is all about. Although much progress has developed with regard to refinements of work sampling, one of the greatest needs today is in this matter of gaining acceptance of the programs from the personnel involved. In short, it is necessary to "talk to the guys" more.

Two grievances that were received in one company on the subject of work sampling follow.

First Grievance. "The aggrieved and all our Unit people are aroused and highly perturbed by the actions of the Company in regard to the so-called survey now being conducted in the shops. The answer to the grievance in the first step was that only the facility was being checked. We say this is false. The checker in our shop on one occasion was running around going 'nuts,' not for the facility, but for the operator, when for quite a spell he was nowhere to be seen.

"No matter what heading or title tags it—'industrial engineering,' 'work samples,' 'spying,' 'brain washing,' or whatever it is—it is a violation of our Agreement. We request this survey be stopped at once."

Second Grievance. "It is the contention of the aggrieved and of the Union that the way the Company is making this survey or work sample is not conducive to good employee-company relations.

"To help clarify this, a member of the Industrial Engineering Department walks along a prearranged route to this department and others and checks predetermined facilities (and operators) approximately every 6–10 minutes for the course of the entire day. We have been informed that this is to continue for three weeks.

"The way this type of survey is being done, it is undemocratic, un-American in principle, costly, detrimental to the Union people, and will not be condoned by the Union.

"The Union requests that this new method, survey, work sample, or whatever the Company may wish to apply to it, cease immediately.

"The Union requests the Company use the accepted methods of time studying to determine work loads, crane waits, etc., instead of these police state methods instituted by the Industrial Engineering Department."

Company Answer. "In these grievances, the Union is strenuously objecting to the 'random work sampling' program instituted by Management of various machine facilities. They contend that Unit employees are being subjected to undue pressure, harassment, intimidation, etc., by this type of observation.

"As thoroughly explained during discussion, 'work sampling is a method of gathering data pertinent to manufacturing operations and is widely used by industries.' Information gathered during this program will be used for the sole purpose of improving operating efficiencies and in no way can this program be construed as a means of harassment, intimidation, or coercing of our employees.

"We do not expect employees to conduct themselves during the course of this 'work sampling program' in any other manner than is normally expected of them. Accordingly, there should be no cause for alarm. We do not agree that the program places any undue pressure on our employees. Accordingly, the request that this study be discontinued is not granted, and any alleged violation of the Agreement is denied."

The reaction of employees reflects a great need to do a better job with respect to the sociological and psychological aspects of the work of industrial engineering. Just what happened here? Having had some unfavorable experience in the past of applying work sampling, the chief industrial engineer attempted to do a good job in this case. But his attempt failed. The approach that he used was to talk to the general superintendents about the work sampling study. He then asked them to convey to the superintendents, the general foremen, and the foremen the approach that the observers expected to use in making the work sampling study. The chief industrial engineer asked them to relate to the employees through the line of command that "a work sampling was to be made for the purpose of evaluating both the equipment and the personnel to determine how they expended their time."

On all work sampling studies since this one, the chief industrial engineer now asks for an opportunity to explain this information to each level of the personnel below them. He "talks to the guys" directly, with the line organization members present.

There is no need to emphasize further how important it is to have the complete understanding and confidence of the people who are concerned with the results of a work sampling study. The best way to sell work sampling is in the explanation to "Why does work sampling work?" By all means, include the dice and coin demonstrations. They are more effective than a long talk in getting the idea across.

Define the Problem. Determine exactly what information is required. It is usually well to make a preliminary survey—observe the operation for a day—to get a list of the operation elements.

For instance, if the causes and amount of downtime on a machine are sought, it will be necessary to define all the possible causes. It is much the same as the preparation for a time study.

Make an Observation Recording Form for the Job. The form used for recording the observations made during the course of a work sampling study must be individually designed in each case. Its design will depend upon the number of work stations or people to be observed and the classification of the activities upon which it is desired to obtain data. Figure 4-2 shows a typical observation record form which was designed for a study of draw-bench activities.

Select the Frequency of Observation

Nature of the Operation. If it is a short-cycle, repetitive operation in which all the desired elements occur frequently, the observations can be spread out over a period of time. If it is a nonrepetitive operation or one in which some elements occur infrequently, it is better to make more observations in a day. This improves the chance of getting all the details.

Physical Limits. If there is just one observer and a long route is necessary to make one round of observations, he will be able to make relatively few observations in a day. For instance, a study of maintenance crews by one observer would probably require a long route.

Total Number of Observations Required and Time Limit. If 1,600 observations are needed for desired accuracy, and there are only ten working days to make them, it is evident that an average of 160 observations a day will be needed.

Determine Time of Trips

On a Random Basis. Making 20 random samplings which follow no set pattern is quite often difficult. The safest way is to use a table of random numbers, because the human mind has a tendency to follow a set pattern.

Sampling can be randomized by the day, within an hour, or within any other period of time—90 minutes, 2 hours, or the like. For example, 20 random samplings per hour could be observed (called stratified random sampling), or 160 random

STUDY	DATE		OBSERVER		
ITEM	BENCH NO. A-24	BENCH NO. A-23	BENCH NO. A-22	BENCH NO. A-21	TOTAL
CYCLE	UH UH IIII	UH UH UH UH I	UH UH UH UH	UH UH IIII	71
SET UP	I	III		I	5
NOT OPERATING	IUH UH IIII	III		UH UH III	30
OPERATOR ABSENT			I		1
OPERATOR IDLE			II		2
POINT — HEAVY					
SMALL					
BENT		I	I		2
JAWS — NOT GRABBING		I			1
NOT RELEASING					
STOCK — TANGLED			I		1
BREAKING			I		1
HANDLING	II	I	II	II	7
WAITING — CRANE	I	I			2
STOCK	I	I		I	3
TAIL IN DIE					
THREADING MANUALLY			I		1
HOOK WON'T ENGAGE					
TUBE RELEASED TOO SOON					
GUIDE TUBE INTO DIE					
MAINTENANCE		I		I	2
ADJUSTING RODS				I	1
CHECKING PINS & DIES				I	1
CLEAN UP	I	I	I		3
REC. INSTRUCTIONS			II		2
					136

FIG. 4-2.   Observation record form designed for study of draw-bench activities. Each mark represents one random observation.

samplings per day could be made, with certain hours having more observations than others. Randomness of observation is stressed to reduce sampling errors.

On a Fixed Interval Basis. Haines[3] and many others, including the author, have employed work sampling on a continuous basis whereby the same operation or operator may be observed every one minute, five minutes, or some other small fixed interval of time. Flowerdew and Malin,[4] through simulation studies, concluded that systematic activity (fixed interval) sampling gives more accurate results than random sampling when a sampling interval shorter than the shortest elements can be used. However, it is necessary to make a short preliminary timing to determine the frequency of the observations because the cycle and elemental times will not be known at the start.

[3] I. Landis Haines, "Work Sampling by Fixed Interval Studies," The Journal of Industrial Engineering, July–August, 1958, pp. 266–268.
[4] A. D. J. Flowerdew and P. W. Malin, "Systematic Activity Sampling," The Journal of Industrial Engineering, July–August, 1963, pp. 201–207.

This sampling, as noted by Davidson,[5] is fundamentally the same process as measuring with a stopwatch where the intervals of sampling are shorter than the activity duration. Also, this method approaches the group timing technique (see Chapter 5 of this section).

Estimate the Number of Observations That Will Be Needed. This information is needed to plan the frequency, number of observers, and length of the study. The number depends on how accurate the answers need to be. A larger number of observations provides greater accuracy. Experience with work sampling and a knowledge of the operation will enable the analyst to make a fair off-the-cuff estimate. As the study progresses, he can check the results to see when he has enough observations. Later in this chapter, a simple chart that signals the end is explained.

There is a mathematical method of preestimating the number of observations needed to give the practical accuracy desired, which is described later on in this chapter. The theory on which it is based is the same as that used in statistical quality control.

Table 4-1 shows the required number of observations for some common combinations of percentage occurrence and relative accuracy. Absolute or "precision interval" accuracy is shown on the alignment chart for determining sample size, Figure 4-3.

TABLE 4-1. How Number of Required Observations Varies in Relation to Percentage Occurrence of Element and Relative Accuracy Desired in Work Sampling Results

Percentage occurrence of element	Number of observations required		
	1 % accuracy	5 % accuracy	10 % accuracy
1	3,960,000	158,400	39,600
5	760,000	30,400	7,600
10	360,000	14,400	3,600
20	160,000	6,400	1,600
30	93,330	3,730	930
40	60,000	2,400	600
50	40,000	1,600	400

Evaluate Methods by Which Biased Readings May Be Reduced. It should be pointed out that the inefficient motions or elements of an operation will not necessarily show up through a work sampling study. In some instances, the operator may be working during the downtime of the machine when this work could be done during the machine operating cycle. The work sampling observation would indicate that the operator was working, but a more refined motion study would show that this work could be performed during the machine cycle time. It should also be pointed out that a work sampling study will not show if the operators are limiting production or pacing themselves. If the observers are qualified to do so, some indication of effort level can be obtained by rating the operator on those observations when he is recorded as performing the working elements of his job. It should be stressed to the observers that it is very important to make their observations at the same spot each time so that their readings will not be biased by making their observations at some distance from the designated spot.

Quite often, certain elements of work occur only once a year and that may

[5] H. O. Davidson, W. W. Hines, and T. L. Newberry, "The Error of Estimate in Systematic Activity Sampling," *The Journal of Industrial Engineering*, July–August, 1960, pp. 290–292.

ALIGNMENT CHART FOR DETERMINING SAMPLE SIZE

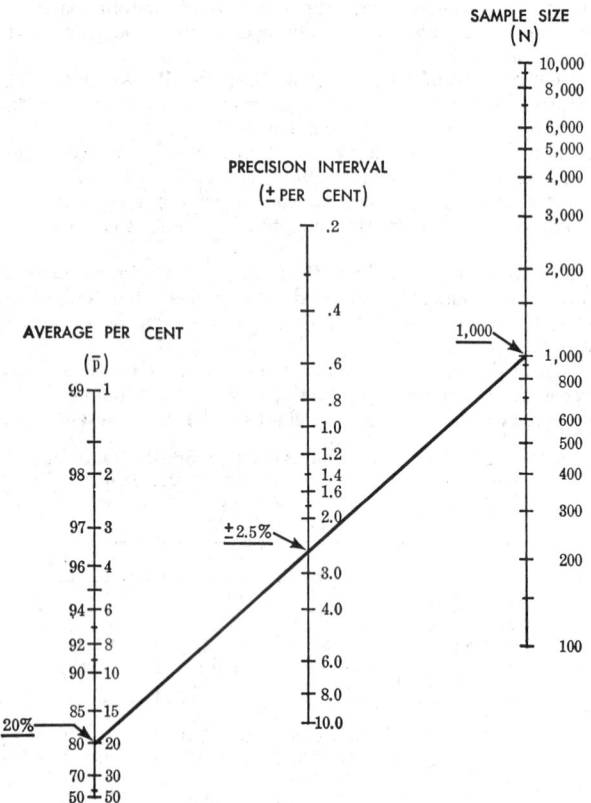

FIG. 4-3.   Alignment chart for determining number of observations required to obtain percent occurrences within given absolute limits of error and confidence level of 95 percent.

be during the period of the work sampling study, for example, machines down for a major overhaul.   This abnormal maintenance might be weighted accordingly if it is known, for example, that the machine is usually down for abnormal maintenance because of overhaul for 5 days out of a year, and that this overhaul occurred during the 15-day study.   Assuming that there are 252 working days to the year, the weighted percentage of downtime because of maintenance would be 2 percent instead of the recorded 33.3 percent.   Obviously, it is necessary to rely on fairly accurate maintenance records to weight these factors with any degree of accuracy.

The question has often been raised with respect to the effect of an operator seeing an observer coming toward him and then getting to work.   It is true that a study is thrown off at the beginning because of this situation.   However, it has been found through experience that as observers pass the operators day in and day out over a sufficiently long period of time, this influence on the readings levels off.   It is strongly recommended that no work sampling studies be taken without the knowledge of the person being studied.   The supervision directly over the operators should explain to them what the study is and what the purpose of it is.   After the studies have been completed, the supervision and those studied should be enlisted in improving the conditions that are pinpointed by the study.

Prior to the Start of a Study, Have a Session with the Observer or Observers. Clearly define and discuss each element to be observed and recorded. This step is very important where two or more observers study the same operation. Without it, they may not be consistent in how they designate what they see.

EXAMPLES OF WORK SAMPLING

As an example of work sampling, assume it is desired to find out how much time a selected machine operator spends on operations, setup, maintenance, and delay.

Using the work sampling technique, the machine is visited a predetermined number of times a day or hour. Assume that 10 random samplings that follow no set pattern are wanted during the day. The safest way to accomplish this is to use a table of random numbers such as Table 4-2, assigning times to these numbers. The numbers must be arranged in sequence.

Assuming that the period in which the study will be made is from 8:00 A.M. until 5:00 P.M., not including the lunch period, the 8 hours or 480 minutes may be divided into forty-eight 10-minute periods. Time intervals of 10 minutes each, beginning with 8:00 A.M. and going to 8:10, 8:20, and so on, will be numbered consecutively from 1 to 48. The analyst then makes use of the random numbers by choosing as many as the number of observations he wishes to make during the day. In Table 4-2, column 3, for example, ignoring numbers over 48, the first 10 numbers would be 43, 24, 17, 12, 07, 38, 40, 28, 17, and 18.

The numbers are then arranged in sequence and used to determine the observation times. The intervals between the random numbers determine the number of 10-minute periods between each observation as shown.

Sequence	Time
07	9:00
12	9:50
15	10:20
17	10:40
18	10:50
24	11:50
28	1:30
38	3:10
40	3:30
43	4:00

There are two ways of randomizing observations. Numbers from 1 to the number of periods into which the day has been divided can be written on slips of paper. By drawing them from a convenient container, random numbers will be selected. In the above example, these slips would be numbered from 1 to 48. Another method of assuring a random selection of observation times would be to utilize the tumbler and numbered disks so common to Bingo games.

The element which is occurring at the instant of each visit is recorded.

At the end of 10 days, the record may read:

Element	Observations	Percent of total
Operation.......	60	60
Setup...........	18	18
Maintenance.....	10	10
Delay..........	12	12
Total........	100	100

The percentage of distribution of the various elements, as they occurred during

TABLE 4-2.   A Table of Random Numbers

```
03 47  43  73 86   36 96 47 36 61   46 98 63 71 62   33 26 16 80 45   60 11 14 10 95
97 74  24  67 62   42 81 14 57 20   42 53 32 37 32   27 07 36 07 51   24 51 79 89 73
16 76  62  27 66   56 50 26 71 07   32 90 79 78 53   13 55 38 58 59   88 97 54 14 10
12 56  85  99 26   96 96 68 27 31   05 03 72 93 15   57 12 10 14 21   88 26 49 81 76
55 59  56  35 64   38 54 82 46 22   31 62 43 09 90   06 18 44 32 53   23 83 01 30 30

16 22  77  94 39   49 54 43 54 82   17 37 93 23 78   87 35 20 96 43   84 26 34 91 64
84 42  17  53 31   57 24 55 06 88   77 04 74 47 67   21 76 33 50 25   83 92 12 06 76
63 01  63  78 59   16 95 55 67 19   98 10 50 71 75   12 86 73 58 07   44 39 52 38 79
33 21  12  34 29   78 64 56 07 82   52 42 07 44 38   15 51 00 13 42   99 66 02 79 54
57 60  86  32 44   09 47 27 96 54   49 17 46 09 62   90 52 84 77 27   08 02 73 43 28

18 18  07  92 46   44 17 16 58 09   79 83 86 19 62   06 76 50 03 10   55 23 64 05 05
26 62  38  97 75   84 16 07 44 99   83 11 46 32 24   20 14 85 88 45   10 93 72 88 71
23 42  40  64 74   82 97 77 77 81   07 45 32 14 08   32 98 94 07 72   93 85 79 10 75
52 36  28  19 95   50 92 26 11 97   00 56 76 31 38   80 22 02 53 53   86 60 42 04 53
37 85  94  35 12   83 39 50 08 30   42 34 07 96 88   54 42 06 87 98   35 85 29 48 39

70 29  17  12 13   40 33 20 38 26   13 89 51 03 74   17 76 37 13 04   07 74 21 19 30
56 62  18  37 35   96 83 50 87 75   97 12 25 93 47   70 33 24 03 54   97 77 46 44 80
99 49  57  22 77   88 42 95 45 72   16 64 36 16 00   04 43 18 66 79   94 77 24 21 90
16 08  15  04 72   33 27 14 34 09   45 59 34 68 49   12 72 07 34 45   99 27 72 95 14
31 16  93  32 43   50 27 89 87 19   20 15 37 00 49   52 85 66 60 44   38 68 88 11 80

68 34  30  13 70   55 74 30 77 40   44 22 78 84 26   04 33 46 09 52   68 07 97 06 57
74 57  25  65 76   59 29 97 68 60   71 91 38 67 54   13 58 18 24 76   15 54 55 95 52
27 42  37  86 53   48 55 90 65 72   96 57 69 36 10   96 46 92 42 45   97 60 49 04 91
00 39  68  29 61   66 37 32 20 30   77 84 57 03 29   10 45 65 04 26   11 04 96 67 24
29 94  98  94 24   68 49 69 10 82   53 75 91 93 30   34 25 20 57 27   40 48 73 51 92

16 90  82  66 59   83 62 64 11 12   67 19 00 71 74   60 47 21 29 68   02 02 37 03 31
11 27  94  75 06   06 09 19 74 66   02 94 37 34 02   76 70 90 30 86   38 45 94 30 38
35 24  10  16 20   33 32 51 26 38   79 78 45 04 91   16 92 53 56 16   02 75 50 95 98
38 23  16  86 38   42 38 97 01 50   87 75 66 81 41   40 01 74 91 62   48 51 84 08 32
31 96  25  91 47   96 44 33 49 13   34 86 82 53 91   00 52 43 48 85   27 55 26 89 62

66 67  40  67 14   64 05 71 95 86   11 05 65 09 68   76 83 20 37 90   57 16 00 11 66
14 90  84  45 11   75 73 88 05 90   52 27 41 14 86   22 98 12 22 08   07 52 74 95 80
68 05  51  18 00   33 96 02 75 19   07 60 62 93 55   59 33 82 43 90   49 37 38 44 59
20 46  78  73 90   97 51 40 14 02   04 02 33 31 08   39 54 16 49 36   47 95 93 13 30
64 19  58  97 79   15 06 15 93 20   01 90 10 75 06   40 78 78 89 62   02 67 74 17 33

05 26  93  70 60   22 35 85 15 13   92 03 51 59 77   59 56 78 06 83   52 91 05 70 74
07 97  10  88 23   09 98 42 99 64   61 71 62 99 15   06 51 29 16 93   58 05 77 09 51
68 71  86  85 85   54 87 66 47 54   73 32 08 11 12   44 95 92 63 16   29 56 24 29 48
26 99  61  65 53   58 37 78 80 70   42 10 50 67 42   32 17 55 85 74   94 44 67 16 94
14 65  52  68 75   87 59 36 22 41   26 78 63 06 55   13 08 27 01 50   15 29 39 39 43

17 53  77  58 71   71 41 61 50 72   12 41 94 96 26   44 95 27 36 99   02 96 74 30 83
90 26  59  21 19   23 52 23 33 12   96 93 02 18 39   07 02 18 36 07   25 99 32 70 23
41 23  52  55 99   31 04 49 69 96   10 47 48 45 88   13 41 43 89 20   97 17 14 49 17
60 20  50  81 69   31 99 73 68 68   35 81 33 03 76   24 30 12 48 60   18 99 10 72 34
91 25  38  05 90   94 58 28 41 36   45 37 59 03 09   90 35 57 29 12   82 62 54 65 60
```

the random observations, tends to equal the percentage of the time spent on these activities that would be found by continuous observation.

The key to the accuracy of the work sampling study is in the number of observations. A greater number of observations provides a higher degree of accuracy, provided the study is designed to reduce bias. But nearly all plant or business problems have a point beyond which greater accuracy of data is not worthwhile. In the example, 100 observations may or may not be enough, depending upon the accuracy required.

Over a sufficiently long study, the number of times a man or machine is observed—idle, working, or in any other condition—tends to equal the percentage of time in that state. This is true whether the occurrences are very short or extremely long, regular or irregular, or many or few. It should be emphasized that the study can be as detailed as one cares to make it; but the more detailed it is, the greater are the number of observations necessary to obtain the degree of accuracy that might be desired for all the elements.

ACCURACY AND PRECISION OF WORK SAMPLING

Work sampling recognizes the variability inherent in work measurement. However, R. W. Conway[6] stresses that there is lacking in much of the literature on work sampling an explicit recognition of the difference between accuracy and precision of an estimate. He defines these terms as follows.

"Accuracy is the measure of the degree of bias in measuring. Bias is the amount by which the long-run observed mean value of a set of measurements differs from the 'true' value of the quantity.

"Precision is a measure of the reproducibility of the measured value of a given quantity without regard to the 'true' value of that quantity."

Bias can be prevented only by the proper design and execution of the sampling process. Possible sources of bias are in:

1. The precise definition of the population to be sampled
2. The ambiguity of the definition of various states of activity
3. The latitude on the part of the observer in choosing the moment of observation
4. The method of selecting the observation times
5. The extent that the worker can anticipate the time of observation and is able to alter the state of activity that will be observed

During the design stage of the study, a period should be selected that will avoid some unusual circumstance. The period of study should be at least as long as the longest period of any cyclical behavior of the characteristic being studied. Likewise, the population upon which the estimate is based must be similar to and representative of the period to which the estimate is to be applied.

The amount that the observer contributes to bias can be investigated by having two, three, or more observers perform simultaneous studies on the same operations. Likewise, through multiple studies, individual workers can be evaluated. Some studies have been made using continuous time study along with the application of work sampling to determine the degree of bias.

The purpose of work sampling is to establish the value of $\bar{p}$ in the binomial distribution. The normal distribution describes the probabilities of the various values of $\bar{p}$ that might occur. It has a mean average of p. p is the percentage occurrence of the element being observed, expressed as a decimal. The parameters p and N (number of observations) are used in the binomial distribution, and p and σ (sigma) in the normal distribution. However, the binomial values p and N are used to measure the parameter σ. The relation between p, N, and σ is given as the equation

$$\sigma = \sqrt{\frac{\bar{p}(1 - \bar{p})}{N}}$$

Sigma (σ) is the standard deviation. The statistical derivation which shows that 68 percent of the time an observation can be expected not to deviate from the mean in the normal distribution by more than $\pm$ sigma can also show the probability, a, associated with more than one sigma.

[6] R. W. Conway, "Some Statistical Considerations in Work Sampling," *The Journal of Industrial Engineering*, vol. 8, no. 2, March–April, 1957, pp. 107–111.

The percentage of the area under the normal curve, a, or the selected level of confidence, a, between a perpendicular erected at the arithmetic mean and a perpendicular erected at specific points to the left and right of the mean may be determined from Table 4-3. The distance between the arithmetic mean and the selected point is expressed in terms of standard deviations as C.

TABLE 4-3.   Selected
Level of Confidence

C	
$(+ \text{ and } -)$	a
1.000	0.68
1.645	0.90
1.960	0.95
2.576	0.99

There is a mathematical method of preestimating the number of observations needed to give the practical accuracy that is desired. It will be necessary to watch the progress of observations too. Sometimes it will be possible to tell with great accuracy that it is not necessary to take as many observations as originally

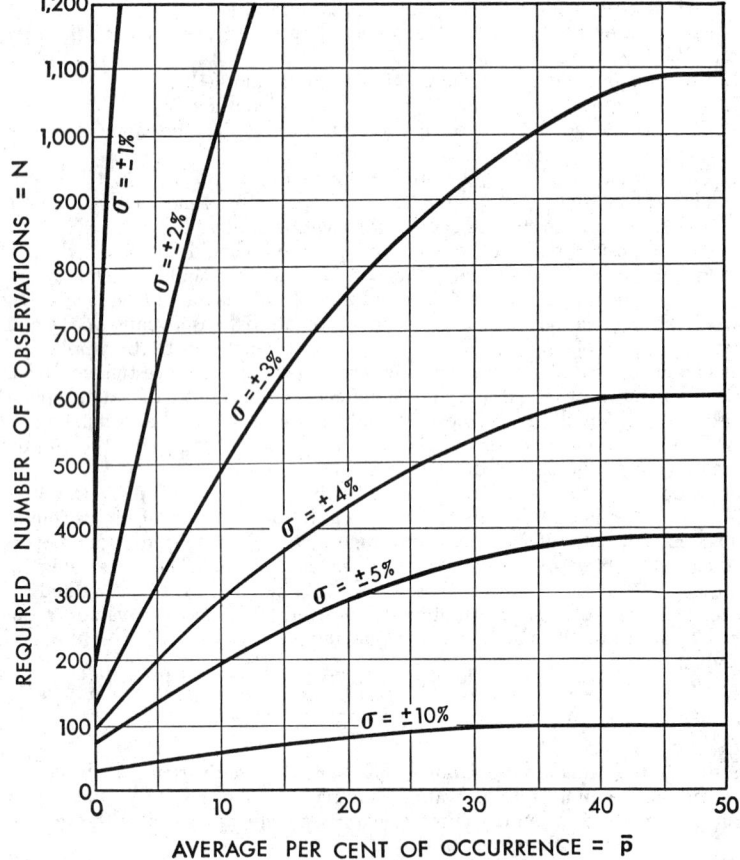

FIG. 4-4. Number of observations required to obtain work sampling percent occurrences within given absolute limits of error and confidence level of 95 percent.

planned. Or perhaps, to be safe, more must be taken than was originally figured. The formula to be used to solve for N, the number of observations required, is as follows:

$$N = \frac{C^2 \bar{p}(1 - \bar{p})}{\sigma^2}$$

To facilitate these computations, Allderige[7] recommends the use of alignment charts to determine the number of observations to meet the needed precision requirements (Figure 4-3).

An alternative to the alignment chart in determining the required sample size, N, given a predetermined standard deviation (Figure 4-4) is suggested by Sammet and Malcolm.[8] Also, this diagram shows that, with the sample size constant, the standard error decreases as $\bar{p}$ becomes smaller and increases as $\bar{p}$ becomes larger.

CONTROL CHART

The control chart, similar to those used in quality control, is advocated by many who apply work sampling to ascertain that the daily percent plots are within one-, two-, or three-sigma limits. This chart (Figure 4-5) enables the work sampling observers to know that the data are in a state of statistical control and are homogeneous and consistent. The control chart is considered pertinent in determining equitable delay allowances. It has an additional and important advantage in that the effect of a change in operating conditions can be checked to determine whether it produces a significant change in the delay percentage. The observer should be alert to strive to bring about greater control as a result of the use of a control chart.

Control charts may be determined on a daily basis, using constant upper and lower control limits for $\bar{p}$. The first few random trips of observations produce percentage $p_1, p_2, p_3, \ldots, p_n$, or the average of $\bar{p}$. This $\bar{p}$ is set up as the center line of the control chart. The upper and lower control limits are the $\bar{p} \pm \sigma$, 2σ, or 3σ, depending upon the confidence level desired.

[7] John M. Allderige, "Work Sampling without Formulas," *Factory Management and Maintenance*, vol. 112, no. 3, March, 1954, pp. 136–138.

[8] L. L. Sammet and D. G. Malcolm, "Work Sampling Studies: Guides to Analysis and Accuracy Criteria," *The Journal of Industrial Engineering*, July, 1954, pp. 9–12ff.

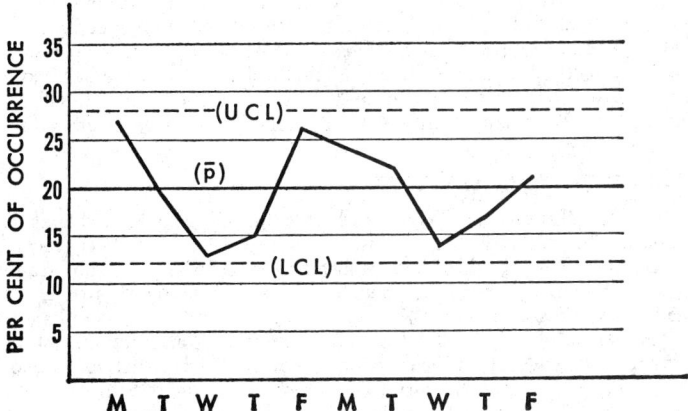

FIG. 4-5. Control chart on a daily basis using constant upper and lower control limits.

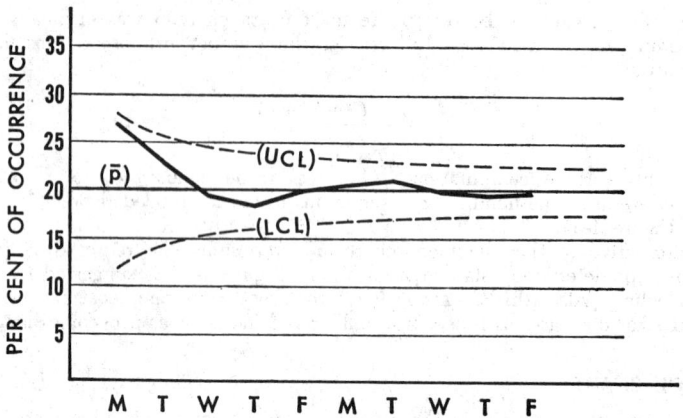

Fɪɢ. 4-6.   Control chart based on an accumulative number of daily observations.

Example. If $\bar{p} = 20$ percent and $N = 100$ observations per day,

$$\sigma = \sqrt{\frac{\bar{p}(1 - \bar{p})}{N}}$$

$$= \sqrt{\frac{0.20(0.80)}{100}}$$

$$= \sqrt{0.0016}$$
$$= 0.04$$

Therefore, $2\sigma = 0.08$
                                            $3\sigma = 0.12$
Upper control limit at $+2\sigma = 0.20 + 0.08$
                                  $= 0.28$
Lower control limit at $-2\sigma = 0.20 - 0.08$
                                  $= 0.12$

Another approach to emphasize that the percentage of error reduces as the number of observations increases is to compute 2σ deviations on an accumulative number of daily observations (Figure 4-6). Assuming 100 observations per day, σ is determined for 100, 200, 300, and so on, observations. This chart may also be used as an empirical method of determining the length of the study. When the variation from day to day is reduced to the desired level for the element chosen to control the study, it can be assumed that enough observations have been gathered.

APPLICATIONS OF WORK SAMPLING

Setting Delay Allowances. When work sampling is used for setting allowances, it is important to standardize methods. This precaution has been recommended for half a century with regard to the establishment of standards for time studies. Yet this care is not emphasized to the degree that it should be regarding allowances established by work sampling. This is probably because work sampling has been used to pinpoint problem areas. In such cases, the work sampling study shows what is happening without any correction. However, when the technique is used for setting allowances, it is mandatory that methods be standardized.

As a part of the process of standardizing the methods, work sampling should be used in conjunction with work simplification. People. like to participate. They are anxious to be a part of the refinement process. Goals can be set to reduce the indicated allowable activities to a certain percentage below where they were first observed, if this goal is desirable. There may be some activities that it is desired to increase, such as inspection time, which may bring about higher quality. After these efforts, another work sampling study should be made to assure that the goals have been met. Some delays, such as personal time, must be established arbitrarily as a policy matter.

Delay Elements Important Part of Standard Data. At the 12th Annual National Conference of the AIIE in 1961, Jack M. Waite[9] of AC Spark Plug, Division of General Motors, expressed very well a point of disagreement in work measurement.

We believe the outstanding advantage of standard data is the determination of all delay elements affecting a job. An act-breakdown analysis and stopwatch will readily give us the cycle time, but disagreement seldom arises over cycle time. Invariably, if a question arises, it concerns the time to get a machine going, clean chips, stock-handling, inspection, tool trouble, salvage, reworking, etc. Many of these items will not be observed during a half-hour to one-hour study, but if studies are taken over a long period of time at all hours of the shift, then specific times can be established for all noncycle and irregular elements.

Work Measurement Sampling (WMS). Schmid[10] shows how it is now possible under many common conditions in industry to establish job standards of substantially the same accuracy obtainable with conventional time study for less than 10 percent of the usual cost. For certain broad categories of jobs, this is possible by the use of a statistical sampling technique that he elected to call work measurement sampling.

This technique is best adapted to situations where few if any standards exist and reasonably accurate standards are desired quickly. Although it is theoretically possible to use work measurement sampling on short-cycle and highly repetitive jobs, it is often not economically feasible to do so because in these cases traditional methods can usually develop job standards less expensively. However, certain short-cycle studies have been run successfully using WMS. For example, WMS was used to develop standards in a hospital clinical laboratory where cycle times were from 3 to 20 minutes.

The essential features of the technique are:

1. Collection of data on a sampling basis from a number (10 to 50) of job positions simultaneously
2. The reduction of these observations to punched cards, one card per observation
3. The processing of these data on a computer to give job standards directly for all the jobs processed at each of the positions during a study

Statistical Basis for Work Management Sampling. The statistical procedure on which work measurement sampling is based requires that each time a job is observed being worked on, four points in time be identified. These four points are referred to here as O_1, O_2, O_3, and O_4, that is, the four observations that straddle the beginning and end of job $ABCD$.

Consider the time axis in Figure 4-7, with a series of random observations (X points) on it. Assume that at each point in time (represented by the X points on the time axis), an observation is made of a man working at a given position to determine what he is working on. From the viewpoint of job $ABCD$, consider only that either he is working on job $ABCD$ or he is not working on job $ABCD$. If then it is observed that (1) at point O_1 he is not working on job $ABCD$, (2) at point O_2 and each subsequent point to and including O_3, he is working on job $ABCD$, and (3) at O_4 he is not working on job $ABCD$, it is logical to conclude that job $ABCD$ is at least $O_3 - O_2$ long and is not longer than $O_4 - O_1$.

Now consider Figure 4-8. Here, using the smallest unit of time (the length of the time axis representing a single point) as 0.01 hour, select the arbitrary

[9] Jack M. Waite, "Development and Application of Standard Time Data," *Proceedings*, 12th Annual National Conference and Convention, American Institute of Industrial Engineers, May 11–13, 1961, pp. 221–228.

[10] Merle D. Schmid, *Work Measurement Sampling*, University of Dayton, Dayton, Ohio, 1965.

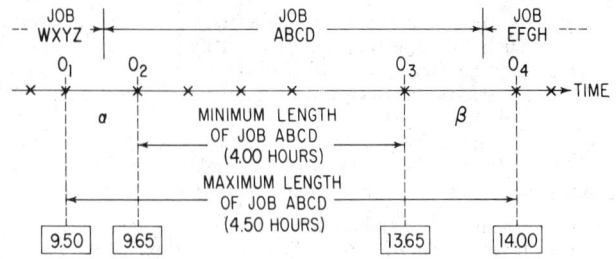

FIG. 4-7.   Schematic length of job *ABCD*.

time values posted on Figure 4-7. It can be seen that in this case, $\alpha = 0.15$ hour and $\beta = 0.35$ hour.   Because the values O_1, O_2, O_3, and O_4 are known (Figure 4-7), a good estimate of the length of job *ABCD* would be the median of all possible lengths of job *ABCD* that could have occurred under the restrictions of the known values of O_1, O_2, O_3, and O_4.   This, of course, would be[11]

$$\frac{(O_3 - O_2) + (O_4 - O_1)}{2} = \frac{4.00 + 4.50}{2} = 4.25 \text{ hours}$$

Work Standards by Work Sampling.   At the 1966 Industrial Management Society Clinic, Forsythe[12] explained the establishment of labor standards through work sampling on the more variable, less frequent jobs at Trans World Airlines, Inc.
His explanation of just how standards were set follows:

> We were already receiving a monthly report that provided us with the actual hours spent on each operation in our shops, so all we really had to do was develop the work sampling procedure.   We decided to establish the cost center as the unit to be sampled since in our shops the cost center is normally the responsibility of one foreman and usually consists of from twenty to forty mechanics.   We decided we could obtain the accuracy we wanted by making approximately twenty trips per week through each cost center at random times.   Twenty trips per week in a cost center of twenty people gave us an accuracy of ±5% on an element that occurred 80% of the time as our working element normally does.

[11] *Ibid.*
[12] Arthur J. Forsythe, "Work Measurement by Work Sampling," *IMS Clinic Proceedings 1966*, Industrial Management Society, Chicago, pp. 42–44.

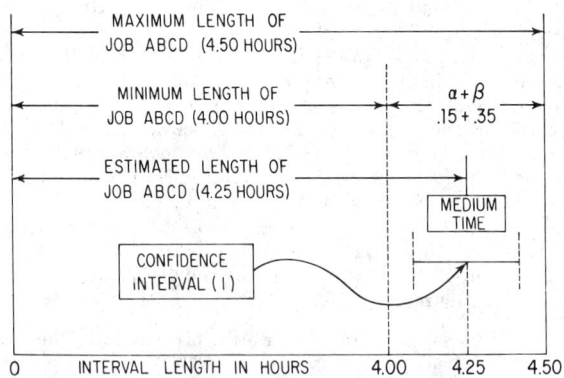

FIG. 4-8.   Confidence interval for job *ABCD*.

Our work sampling studies serve two purposes:
1. The reports themselves are used by the foremen to control lost and idle time.
2. We use the data to establish job standards.

We have been using this sampling technique for the past two years and find that the accuracy of our standards is improving as we gain additional data. We occasionally take time studies of jobs with standards that were established by work sampling and find that there is an acceptable degree of consistency.

In most organizations where work sampling is employed to measure work, the personnel being sampled are performance rated and a work count is maintained. Sometimes, one of the difficult tasks is determining what the work count should be. This requires some innovative approaches on the part of the industrial engineer. But usually, he can key the work load to work orders completed, shipments, or some common kind of units to which the department devotes its major attention. The standard is then derived for the rated elements by the following formula:

$$\text{Standard} = \frac{\text{actual hours in study} \times \% \text{ utilized} \times \text{average rating}}{\text{number of units produced}}$$

To these elements are added the unrated elements or allowances, which may be adjusted according to policy decisions relating to personal and idle time, allowable instruction time, and the like.

Utilization of Engineers' Time—Drafting. A number of companies have taken steps to evaluate and measure engineering and scientific employees. In one firm, a work sampling study was conducted in the product engineering drafting section, the purpose of which was to analyze methods, work load, and employee efficiency.

Errors Reduced. The most important gains resulting from this study were made by reducing a high percentage of discussion time between the draftsmen and the checkers. The discussions were centered on errors that had to be corrected on drawings made by draftsmen. A program was initiated to determine the types of errors made and the people making them. Each time errors were made on a drawing, the checker marked the number and the nature of the errors. At the end of the week, the supervisor charted each individual's amount of errors, the nature of the errors, and the weekly productivity. Each employee was reviewed individually, periodically, by the supervisor, and his errors were discussed and corrective measures recommended. This procedure reduced the errors by 65 percent and resulted in the elimination of one of the three checkers.

First Things First. The work sampling approach provided a tool for the supervisor that enabled him effectively to determine with exactness the inefficient areas in his department. Consequently, this enabled him to increase efficiency by working on the major problems first.

Work Sampling of Engineers Combined with an Interview. Another company, on a random basis, combined the work sampling study with an interview with the engineer concerning the considerable amount of detailed information desired. Because of the length of the interviews, which amounted to approximately five minutes per engineer, the random sampling took place fewer times per day and over a fairly long period of time. The purpose of the interviews was to determine why, when, where, and how much communication is done by the engineers and research workers.

Communication. It was determined that a minimum of 50 percent of an engineer's or research man's day was spent in communicating in one form or another. Out of a 40-hour week, 29 hours per week, on the average, were expended in communication as follows:

Category	Hours per week
Telephone	3.0
Reading	3.6
Writing	8.0
Meetings	14.4

Telephone. An analysis was made of the distribution of telephone communication and the frequency of telephone calls against the time of day. Likewise, communications relating to reading and writing were evaluated.

Reading. Type of reading was noted as follows:
1. Pleasure (little concentration)
2. Scanning for general knowledge of content
3. Deep concentration

Writing. Communication relating to writing was analyzed with regard to:
1. Dictation (to secretary)
2. Dictation (to dictating machine)
3. Outline to final form
4. Final form prepared immediately

The number of pages read or written was noted, together with the amount of time required to read or write a document.

Meetings. A part of this study concerned daily meetings with respect to:
1. Preparation time
2. Actual meeting time
3. Postmortem time
4. How and why meetings take place
5. Size of meeting versus time in minutes per meeting

As a result of the analysis of the data collected by this work sampling interview approach, it was recommended that the following steps be taken:
1. Analyze all meetings to improve their effectiveness.
2. Determine who should be in the audience at each meeting so as to reduce the number of meeting man-hours.
3. Assist engineers in preparation for meetings and in properly recording the minutes and results of a meeting.
4. Encourage the use of visual aids, where applicable.

It was interesting to note that, as the number of people attending the meeting increased, so did the total time of the meeting. The average meeting increased by six minutes for each person added over the minimum of two people. Most of this time was due to the fact that personnel were not familiar with the subject matter, and time had to be spent in orienting them. Also, a need for conference leadership training was apparent.

WORK SAMPLING APPLIED TO MAINTENANCE WORK

Work sampling has been utilized more and more to analyze plant maintenance work. Comparisons have been made of all-day logs of typical crews and work sampling of these crews.

Study of Maintenance Electricians. In an aircraft plant, an effort was made to concentrate on industrial engineering functions as they related to maintenance. To concentrate efforts, it was decided to locate first of all where good cooperation might be expected and a considerable improvement effected. The area selected was the maintenance electricians' operation.

To carry out this study, one of the first things done was to talk to the foremen about the types of work that were performed in the unit. Some of these activities were "trouble-call" maintenance, preventive maintenance, and the like.

After determining what the maintenance electricians' work elements were, the advantages of running a work sampling study to find out the percentage of time that each man expended in these activities were presented to supervision. As the technique was explained, supervision seemed well satisfied with this work sampling approach. Therefore, arrangements were made to orient the men on how the work sampling study was set up and would be carried out.

To begin with, a few questions were asked. The industrial engineer who was responsible for the meeting and the work sampling study emphasized that the purpose of the study was to make a good operation even better. He stressed that the object of the work sampling study was to bring about improvement. This seemed to interest the men and to trigger them into making suggestions on how the operation might be improved. They were really anxious that management learn what they were thinking and suggesting. Before the study started, some of categories at first and then summarize them later if it seemed desirable. The effect before the study started.

Setting Up Categories of Activities. In setting up the activities for the work sampling study, it was decided that it would be better to use a large number of categories at first and then summarize them later if it seemed desirable. The purpose was to break down the elements as fine as possible, knowing that it would be much easier to consolidate elements than to separate them later on.

The list of categories initially decided upon for this study were as follows:

1. Not at job site
2. Get assignment, or telephone the maintenance crib
3. Handle trouble tickets and other paper work
4. Make ready for the job and clean up
5. Trade work (the normal electrical repair work including analysis of the trouble)
6. Personal and idle time
7. Crew unbalanced (the worker busy while the helper was idle or vice versa)
8. Delays by other trades
9. Delays by production using a machine
10. Other delays
11. Talk with the supervisor or the lead man
12. Other serious talk time (with engineers)
13. Miscellaneous delays

During the course of the study, these categories of work seemed quite sufficient. However, some other activities were added as the study continued. For instance, there was a category added for "looking at machine manuals." This involved studying to find out the details of the machine in order to repair it. Another element was added for "looking at blueprints."

Locating the Electricians. Because the maintenance electricians served not only the main manufacturing area but the complete facility, it was a problem to locate them. In any study of this type, the building and equipment must be marked so that the worker can be located easily. In this plant, the buildings were numbered and letters and numbers were on the columns. This aided greatly. It was ascertained from the supervisor where the workers were scheduled to be located by building number and column numbers.

In a few cases, there was some difficulty in locating the worker. In "shooting trouble," he might have to move along a conduit or to some other area. The electrical trouble may not have occurred at the exact location of the machine or equipment to which the worker was sent. Because of the very large area covered by this work sampling study, it was necessary for the observer to use a bicycle to gather information expeditiously.

Also, in some instances, the worker may have started on a job at a particular location but may have completed or performed some work on the job back at the maintenance crib. If in the course of the round of observations the man could not be located at the machine, he was first checked "not at job site," after which the maintenance crib was checked; or if it was known that the man might be in another area, the area was checked first. Then the corrected observation check was indicated.

Work Behavior. During the course of this study, one could not help but get an impression of some of the habits and attitudes of the worker. Even though during the orientation it was mentioned that this was a study of the unit, that is, the whole group, and not of the worker himself, each employee still seemed anxious at times for the observer to know just what he was doing. Sometimes, an electrician would contact the observer and explain to him what he was doing and why he was not exactly at the machine where he should be. The cooperation of the maintenance workers was excellent. Only one worker seemed to show any real signs of nervousness in being observed. However, he was one of the most conscientious of the workers.

Summary of Data. After the study was summarized, the results were presented to supervision first, and then later discussed with the men.

After hearing the results of the study in another meeting, the men again made a great many additional suggestions for improvement. Action was taken to investigate these suggestions to put them into effect, if feasible. If they were not applicable, each man was told why his particular suggestion could not be put into effect.

The morale of the electricians was high because they had the opportunity of making suggestions and their accepted suggestions were put into effect.

Measuring Different Elements of Maintenance Work. Allard[13] differentiates three separate elements of the total maintenance program: direct work, indirect work, and travel.

Direct work is the actual work done with the tools of the trade: an electrician making a connection or pulling wires through a conduit; a pipe fitter threading pipe or attaching a fitting.

Indirect work includes planning, which is usually the largest item, and getting and preparing tools and materials.

Travel involves going to the job site and back, as well as additional trips for more tools, materials, or information.

It must be recognized that these three separate elements require different kinds of measurement. Performance ratings, such as those used in time study, can be applied to sampling observations of direct work. But evaluating indirect and travel observations requires a different technique.

A more sophisticated approach for determining normal or standard indirect and travel time for a given amount of direct work may be set up as follows. A series of point values, comparable to job evaluation, are used for a relative measurement of the four characteristics of indirect work: tools, material, planning, and auxiliary or extra work.

Seven degrees are established for each of these characteristics. The first degree is the least complex; in the case of tools, this means that the only tools required are those the man carries with him normally (such as those an electrician carries on his belt). The seventh degree would represent a complex situation which might involve an expensive piece of equipment locked up in storage. The man has to get permission to use it, obtain the key, and so on. This involves considerable time before he actually obtains the tool.

The same breakdown is used for materials. The first degree would be "no material involved" or a limited amount that the man could carry with him. The seventh degree would be "reusing salvage material on a replacement job or some other complex situation involving considerable time in getting material to the job and ready to use."

In planning, the first degree would be something requiring practically no planning, such as running an errand or changing a light bulb. The in-between degrees would include situations in which a good part of the job is laid out so that the man knows what is required before he gets there. The seventh degree would be "a completed job where there has been no advance planning."

To analyze conditions surrounding a typical job, the work sampling analyst establishes a code number (1 to 7) for the indirect part. It is based on the degree of complexity of a particular job and the extent to which tools, materials, and planning have been previously lined up.

Then the indirect allowance that would be normal or standard should be added to the direct work measured by work sampling. This takes care of one of the big variables. On the scale of 1 to 7, indirect work can vary from 15 to 40 percent of direct work. In the typical plant, it will average about 25 percent.

[13] Harold F. Allard, "Work Sampling: Valuable Maintenance Aid," *Plant Engineer,* September 19, 1968, pp. 84–86.

Work sampling can tell the maintenance manager not only how his workers are performing, but also how much it is costing him for excess travel, excess indirect activities, and waiting time at the storeroom and indicate other weak spots.

Many companies with widespread operations at the manufacturing site have discovered through work sampling the extensive and unnecessary travel time involved; they have helped solve this problem by installing better mobile maintenance equipment with two-way radios for communicating with the central shop.

WORK SAMPLING EXECUTIVES

A work sampling study made of executives showed that the inefficient executive:

Never finds himself alone
Allows chance and interruption to govern his day
Works "too hard"
Lingers on long distance calls while duties wait
Cannot delegate responsibility
Overconsults, overdirects, overrepeats
Does not have time to think or plan
Continues doing the job from which he was promoted
Does not define his job
Lets personalities influence decisions

One of the striking revelations of this study is the need for greater efficiency in oral communication, because from 80 to 85 percent of a manager's time is consumed in talking. There is evidence that more "plan ahead" time is required to bring about greater effectiveness in meetings, discussions, conversations, telephone calls, and dictation. Management personnel do not spend enough time planning what they are going to say to people. Likewise, work sampling accentuates the need for management to allow work measurement to be applied to themselves. By such application, coupled with a determination to improve management performance, better decision making and policy formulation practices will result. This objective can and will materialize when management is willing to subject itself to the same scientific management techniques that it believes should be applied to subordinates.

Work sampling analysis enables the executive to judge if his daily activities really are directed with the fullest possible efficiency toward the realization of his goals. Examination of the old timetable generally leads to an entirely new one for the future, incorporating the answers to questions like: Why did I do it? Was I the right man for the job? Did I do it at the right moment? Couldn't I have eliminated that phone call? Should I have projected what was likely to happen at the meeting? Why must I meet with Joe twice a day instead of once a week?

WORK SAMPLING IN A HOSPITAL

One of the first work sampling studies in a hospital situation took place at Harper Hospital in Detroit, Michigan, in 1950. This study[14] involved analyzing the activities of head nurses, registered nurses, practical nurses, nurses aides, orderlies, and housekeeping maids on a 24-hour-a-day basis for 7 days. Each of these groups of personnel participated in analyzing the data collected to make improvements in their jobs. In this study, the participation theory was used. It was assumed that people would be interested in improving their own jobs if they reviewed the work sampling data themselves and made suggestions on how various activities might be shortened or increased, depending upon the need to bring about greater efficiency. The result was that many tasks being handled by higher skilled people were passed on to lower skilled personnel, thus enhancing the work of each level.

[14] Marion J. Wright, *Improvement of Patient Care*, G. P. Putnam's Sons, New York, 1954, pp. 109–125.

As a result of the work at Harper Hospital, the American Hospital Association became keenly interested, and many similar projects throughout the country were developed.

Rising and Millen[15] applied work sampling to the care of the elderly in a hospital rehabilitation unit. Three separate work sampling studies were made on staff activities. The sampling of staff activities was based upon the work schedule of the various classifications of skill and the staff.

The following list briefly presents the nineteen subcategories organized into the five major categories:

I. Rehabilitative activities
    A. General ADL (activities of daily living)
    B. Therapy
    C. Transportation for rehabilitative purposes
    D. Other rehabilitative activities
II. Custodial activities
    A. General ADL
    B. Cleaning activities
        1. Patient category—after soiling bed, before retiring, and so on
        2. Staff category—care of patient environment
    C. Dietary
    D. Transportation for custodial purposes
    E. Other custodial activities
III. Rehabilitative and custodial activities
    A. General ADL
    B. Dietary
    C. Other rehabilitative and custodial activities
IV. Nursing activities
    A. Medication
    B. Dressings and supports
        1. Patient category—application, removal, and change
        2. Staff category—administrative and clerical
    C. Patient care
    D. Other nursing activities
V. Other activities
    A. In bed
        1. Patient category—resting or sleeping in bed
        2. Staff category—rest period and breaks
    B. Out of bed
        1. Patient category—resting or sleeping out of bed
        2. Staff category—personal and idle time
    C. Miscellaneous
        1. Patient category—miscellaneous: talking to each other, nurses, and visitors; reading; TV; radio; and so on
        2. Staff category—miscellaneous

A careful view of the above definitions reveals an immediate possibility for confusion between categories I and III, and II and III. The concept of an activity that is at once both custodial and rehabilitative is not hard to develop at a conceptual level: an activity is custodial in the sense that it is necessary to sustain the patient, for example, feeding; but if the activity is deliberately performed in such a way that it becomes instruction for the patient in one of the activities of daily living (ADL), it is also a rehabilitative activity. This dual concept could have proved difficult to interpret because it also was necessary to recognize both "pure" rehabilitative activities and "pure" custodial activities. Special care had to be taken to develop the concepts of these three categories for those who collected the data,

[15] Edward J. Rising and Roger N. Millen, "Work Sampling in a Hospital Rehabilitation Unit," *Proceedings,* 17th Annual National Conference and Convention, American Institute of Industrial Engineers, May 26–28, 1966, pp. 176–182.

and special attention needed to be given to the definition of categories in operational terms.

To avoid any kind of bias in this study, it was decided to randomize both the times of observation and the person to be observed at a particular time. An interval of at least five minutes between observations had to be provided because of the nature of the activities under observation. In the preliminary stages, 120 readings per day (60 per shift) were considered satisfactory. A total of 1,268 staff observations and 1,247 patient observations were obtained during 10 two-shift days.

Results and Conclusions. The most important information that was revealed from these studies of staff time can be induced from Tables 4-4 and 4-5.

TABLE 4-4. Degree of Agreement among Three Studies, in Percent

Category	Study 1	Study 2	Study 3
	Patient load		
	14–17	10–12	11–15
Rehabilitative...............	3.4	3.4	3.1
Custodial...................	43.8	46.7	43.9
Rehabilitative and custodial......	2.5	2.3	2.1
Nursing.....................	25.0	21.2	28.4
Other......................	25.3	26.4	22.5
Total..................	100.0	100.0	100.0

TABLE 4-5. Rehabilitative, Custodial, and Nursing Activities, in Percent

Category	Nurse		Aide		Housekeeper	
	Shift 1	Shift 2	Shift 1	Shift 2	Shift 1	Shift 2
Rehabilitative.........	7.2	0.4	6.2	0.4	0.3	0.0
Custodial.............	8.4	24.0	42.5	35.2	85.0	83.5
Rehabilitative and cus-todial.............	2.1	0.8	4.9	0.8	0.4	0.0
Nursing..............	67.1	50.0	20.5	31.4	0.0	0.0
Other................	15.2	24.8	25.9	32.2	14.3	16.5
Total..............	100.0	100.0	100.0	100.0	100.0	100.0

1. The rehabilitative objectives of the unit were met with very small percentage of staff time devoted to purely "rehabilitative" and "rehabilitative and custodial" activities.
2. A high degree of agreement was obtained among the three studies performed by different (although similarly trained) observers. This tends to substantiate these figures.
3. The results appear to be stable over time, because they were performed at different times of the year over a two-year period.
4. The patient load varied over wide limits and did not materially change the percent allocation of staff time to the various categories (although it is obvious that the per patient allocation of time must change with the size patient load if the staff idle time remains the same).

WORK SAMPLING IN THE OFFICE

Work sampling originated in the plant. However, its utilization in the office has also been great. This technique is predicated upon observation and analysis of the behavior of clerical workers.

Some programs are designed to measure the percentage of time expended in various activities, after which suggestions are sought for improvement in efficiency. In one study, the first step was to have each employee make up a detailed observation sheet listing all the important elements of his job. These were reviewed by the immediate supervisor in the department to make certain that every function of each job was listed on the observation sheet. Included were such activities as:

Receiving instructions
Discussions in department
Calculating, extending, and totaling
Preparing mailings
Revising card system
Typing invoices
Sorting invoices
Answering telephone
Personal
Absences

Two bells—one placed at each end of the office to make certain everyone would hear them—were set up to go off at random intervals during the day. The bells were operated through a punched tape on the master time clock. When the bells sounded, each employee marked down on his observation sheet whatever he was doing at the moment.

Because the bells rang thirty times a day, workers with less than thirty check marks on their observation sheets knew how many absent periods to record.

When the five weeks of work sampling were over, each employee was asked to make suggestions on how his job could be made more efficient. These recommendations were discussed with the supervisor.

In making work sampling studies among office workers, the effectiveness of the employees' suggestions depends greatly on the interest and enthusiasm the supervisor shows in the sampling studies made on his own job.

Also, with certain jobs, such as that of supervisor, it is better to use an observer rather than have the individual jot down what he is doing every time the bells ring. In this way, the supervisor is not interrupted continually while in meetings or talking to visitors.

No system is perfect, of course, and there are some bugs in this particular method of work sampling. The main drawbacks are the frequent work interruptions and the fact that much of the information obtained depends on how honestly the workers fill in their observation sheets.

The overall purpose of establishing a work sampling program among office workers in this example was to reduce costs. This work sampling was utilized in conjunction with work simplification.

Usually, it is well to have an observer who is well acquainted with the kind of work being performed. An observer skilled in office work simplification can stimulate many suggestions for improvements in office methods and procedures. The elements of work should be carefully defined, and if a standard is to be set, those elements on which a performance rating is to be established should be indicated.

Work sampling will help office supervision to:
1. Identify productive and nonproductive activities
2. Improve supervision
3. Identify peaks and valleys
4. Determine if no peaks exist
5. Support requests for additional personnel

6. Identify job content
7. Set up performance standards
8. Study equipment needs
9. Make cost allocations

CONCLUSION

It has been found that work sampling can be applied in areas in which many variables are involved. This statistical technique has been found to be advantageous in measuring all aspects of work. Work sampling is a useful technique for gathering facts inexpensively and accurately about an operation, process, or any other activity—facts which make it possible to reach decisions to reduce costs, to develop controls, and to improve manpower effectiveness.

The work sampling technique has these advantages:

Gets the facts at one-third to one-sixth the cost of continuous observation.

Does not require observers with special skill and training.

Requires little observer training. However, the observer should be familiar with the type of work to be observed.

Running totals may be kept to keep track of period-to-period progress.

Provides the accuracy required. (It is likely to produce even more accurate results than continuous observation.)

Makes it practical to get facts one would not otherwise try to collect.

Produces fewer complaints from individuals under study than continuous observation.

Produces less distortion in the individual's normal work route than continuous observation.

Chapter **5**

Group Timing Technique*

ROLF TIEFENTHAL

H. B. Maynard and Company AS, Copenhagen, Denmark

Group timing technique (GTT) is an efficient and versatile measuring procedure easily applicable to work measurement and quantitative surveying tasks in industry.

Its one major drawback as a work measurement tool is that no direct connection between method and time is provided. Nevertheless, GTT can still compete favorably with alternative procedures in many practical situations. GTT can often replace traditional stopwatch time study, providing equal or better data at a lower cost.

This chapter will present the basic method of GTT, identify with examples its most important industrial applications, and describe application procedures, manual as well as computerized. The mathematical theory of GTT will be briefly presented, as well as some useful formulas for control of the statistical accuracy of time values derived from GTT application.

WHAT IS GTT?

Group timing technique is a work measurement procedure for multiple activities that enables one observer using a stopwatch to make a detailed elemental time study on from two to fifteen men or machines at the same time. Continuous elemental observations are made at predetermined fixed intervals and are recorded as tallies on a form listing the elements of the job. Elements that will vary in

* Much of the material in this chapter was taken from *Industrial Engineering Handbook*, 2d ed., sec. 3, chap. 7, by George Dew, one of the originators of the group timing technique.

time because of operator performance may be leveled. Either or both elemental time values and percentage allowances, including fatigue, may be established from a single group timing technique study or series of studies.

Basic Procedure. Assume that a certain production operation is performed separately by four operators. It consists of three work categories, the duration of which is to be ascertained. These are:

A. Preparation and finishing time

B. Process time

C. Allowances

After this element selection, each of the operators is observed at constant or fixed time intervals of each second minute. At each observation and for each operator, a record is made of which of the three activities is taking place.

The number of observations necessary to obtain a result that is acceptably close to the result that would be obtained through a continuous study can be computed. The statistical accuracy needed for the purpose of the study can be chosen. The length of the study will then be determined by the number of observations needed and the observation interval, modified if necessary to allow for the inherent variability of the process itself.

After completion of the study, the tallies indicating the number of observations for each individual work category, activity, or element are summarized. The elapsed time or duration of one activity as a percent of the total time of the study is equal to the number of observations for the activity as a percent of total observations.

If the total number of observations is 1,000 and activity B accounts for 600 of these, it may be concluded that process time amounts to 60 percent of the total time. Figure 5-1 shows how the measurements are made.

Absolute measurements can also be made. In the example, the total duration of activity B should be close to 600 registrations of approximately 2 minutes duration each, that is, 1,200 minutes, or 20 hours. If the number of units processed during the study was 40 units, the average process time per unit will be close to 0.5 hour.

During the study, the observer can performance-rate operators and elements so that normalized time standards can be established. The total GTT procedure can be symbolized as shown in Figure 5-2.

Comparison with Other Work Measurement Tools. GTT studies look so much like work sampling studies that they could easily but mistakenly pass as such. In

Activity	Operator				Total
	1	2	3	4	
A	83	45	29	63	220
B	124	155	160	161	600
C	43	50	61	26	180
Total	250	250	250	250	1,000

Process time (B) for operator 2:

$$\frac{155}{250} \times 100 = 68\%$$

Total process time (B):

$$\frac{600}{1,000} \times 100 = 60\%$$

FIG. 5-1.   Computation of percent process time by GTT.

both cases, samples are taken that permit drawing certain conclusions about the nature of the population from which they are taken (the actual process being studied). These conclusions are of course subject to some statistical error.

The apparently slight difference in the sampling plan, however—fixed versus random observation intervals—relates the two methods to entirely different mathematical models. The computations that relate the number of observations to the statistical error are different for the two models. It can be shown that, for the same level of accuracy and all other features being equal, GTT will normally call for a considerably smaller number of total observations than work sampling.

Since its origin in the United States in 1955 under the name "group timing technique," the procedure described has been discussed and used under other names such as "systematic activity sampling," "fixed interval study," and the like.

Advantages and Disadvantages of GTT. The following list gives the principal advantages and disadvantages of GTT.

Advantages

1. One observer can time as many operations or objects as are active within a reasonable observation range. The maximum practical number of operators or objects per observer is about fifteen.
2. GTT has all the advantages (except for some sacrifice in accuracy on short elements) of a conventional time study using continuous watch readings. One hundred percent of the total elapsed time is accounted for. If a little time is lost on one element, it is picked up on another.
3. The working up of final results usually takes considerably less time than for a stopwatch time study. Computation of an eight-hour study will take a maximum of two hours.
4. For a given level of accuracy, a GTT study requires less time and fewer observations than a work sampling study.
5. The interrelationships among operator, machine, and material are easier to identify and observe with GTT than with work sampling.
6. GTT permits the full utilization of performance rating procedures.
7. The same GTT study can provide both leveled elemental time values and allowances, including fatigue.
8. GTT observers need no special skill or training except familiarity with the process under study. Intelligent individuals with one hour of instruction have made good GTT studies. They require special training, of course, if performance rating is required.
9. Psychologically, a group study may be more acceptable to the operators than a series of studies of individual workers.

Disadvantages

1. GTT is essentially a passive technique based on recording what an observer observes. It lacks an aggressive attitude toward methods and does not enforce constructive methods study, as do predetermined time systems such as MTM.
2. In situations where the observer must make a number of moves between observations, the study may be broken up into several substudies, made at different geographical locations of activity. The study bases, either locations or work gangs, are picked at random.

    If it is not possible to handle the location spread in this way, it is recommended that other means of measuring, such as work sampling, be used. In these special cases, there is a limitation to the applicability of GTT.
3. In situations where the process being studied is repetitive, with very long cycles showing instability (such as seasonal, monthly, or weekly variation), caution must be used to ensure that the study periods selected are truly representative of the process one wishes to measure.

    If in doubt, it is preferable to make periodic substudies and evaluate through statistical examination whether differences are significant. In these cases, full use of the limited period of observation in comparison with work sampling cannot be realized.

WHERE AND WHEN GTT CAN BE USED

GTT can be used for many different tasks and purposes. It has proved to be particularly useful for investigating or measuring easily observable activities taking place within limited localities where several operators, machines, or other activity centers are to be studied.

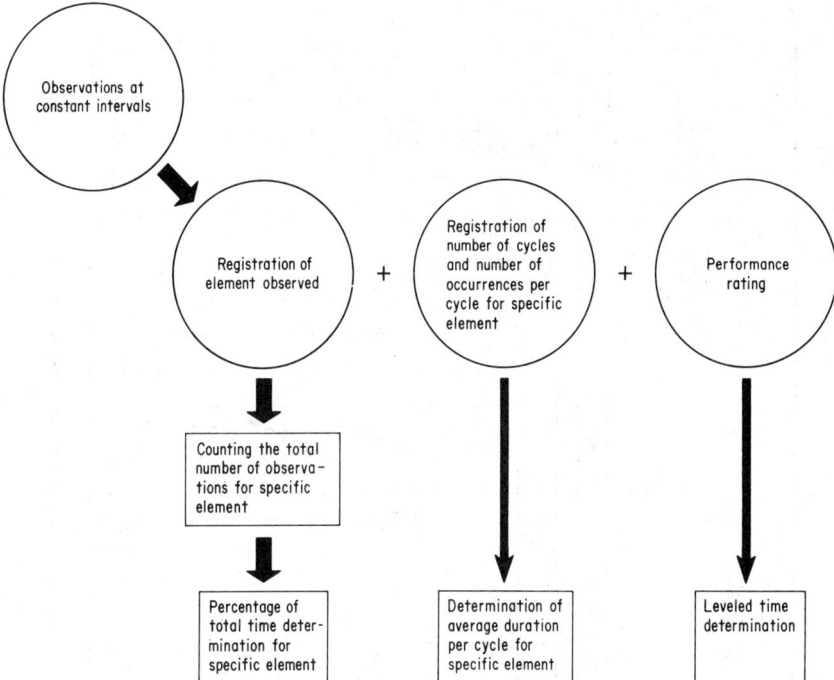

FIG. 5-2.   Chart summarizing the GTT procedure.   (*Source: Rolf Mattson, "GTT: Work Measurement through Fixed Interval Observations,"* Svenska Arbetsgivarföreningen, 1968.)

GTT can be used in production, maintenance, and offices for the following purposes:

1. Determination of leveled each-piece times for setting work standards, production norms, and incentives
2. Determination of allowances, including fatigue, also in combination with a simultaneous check on standards previously set through other means (control study)
3. Determination of the load on groups of operators or machines
4. Surveying or fact-finding studies on organization and processes

Some typical examples of GTT application are described in the following paragraphs.

Use of GTT to Determine Each-piece Time.   Figure 5-3 shows the front of a study made of two men shaking out sand molds in a steel foundry. If desired, tallies could have been recorded each 0.0050 hour instead of 0.0100 hour, and the tallies for each operator could have been recorded in separate columns. The 20 percent allowance had been established by earlier eight-hour GTT studies in the foundry. Most of the idle avoidable delay time occurred soon after 8:00 P.M. when these men started their work shift. These men were dayworkers when the study was taken, as suggested by their performance ratings and foreign elements C and E. The job was completed at 9:45 P.M.; so the study was limited to 1.75 hours in length.

Figure 5-4 is the back of Figure 5-3 and follows the design of a widely used time study form. Because 12 flasks were shaken out, the occurrences per flask on each element were $\frac{1}{12}$, except on element 11, where the operator carried aside the 12 bottom boards from these flasks plus 8 bottom boards remaining from an earlier job, making a frequency of $\frac{1}{20}$.

GROUP TIMING TECHNIQUE STUDY

Tallies recorded every .0100 elapsed hour
Operation: Shake Out with Jib Crane

Day: Wednesday
Date: February 22
GTT Study No.: S-64
Sheet No.: 1 of 1 Sheets

Observer: George Dew

Operators: G. Appleby, T. Zutic

	Skill	Effort	L.F.
	C1 +.06	C2 +.02	1.08
	D .00	E1 -.04	.96
		Ave. L.F.	1.02

Study Finished 9:45 PM
Study Started 8:00 PM
Over-all time 1.7500 hours
2 x 1.7500 = 3.500 man-hours

175 Tallies x .0100 = 1.7500 Hours

SUMMARY

Line	Elements	Total Tallies	Total Man-hrs.	Leveling Factor	Leveled Hours	% Allowance	Allowed Hours	Line
1	Remove clamps and wedges	8	.0800	1.02	.0816	20%	.0979	1
2	Wedges, gaggers, etc. aside	6	.0600	"	.0612	"	.0734	2
3	Move empty crane to flask	5	.0500	"	.0510	"	.0612	3
4	Hook flask with chain	12	.1200	"	.1224	"	.1469	4
5	Move flask by crane to shake out area	20	.2000	"	.2040	"	.2448	5
6	Mallet flask	33	.3300	"	.3366	"	.4039	6
7	Operate crane during shake out	4	.0400	"	.0408	"	.0490	7
8	Set down empty flask on 2nd flask	8	.0800	"	.0816	"	.0979	8
9	Stack empty flasks by crane	22	.2200	"	.2244	"	.2693	9
10	Unhook flask	7	.0700	"	.0714	"	.0857	10
11	Bottom board aside by hand	19	.1900	"	.1938	"	.2326	11
12	Move crane to casting	3	.0300	"	.0306	"	.0367	12
13	Hook casting with chain	14	.1400	"	.1428	"	.1714	13
14	Move casting aside by crane	2	.0200	"	.0204	"	.0245	14
15	Unhook casting	4	.0400	"	.0408	"	.0490	15
16	Set down moulds brought by cab crane	4	.0400	"	.0408	"	.0490	16
17	Wood flasks aside by hand	32	.3200	"	.3264	"	.3917	17
	Foreign Elements							
A	Get tools	2	.0200	—	—	—	—	A
B	Clean work area	2	.0200	—	—	—	—	B
C	Idle A.D.	109	1.0900	—	—	—	—	C
D	Idle U.D.	8	.0800	—	—	—	—	D
E	Personal delays	26	.2600	—	—	—	—	E
	TOTALS	350	3.5000					

Fig. 5-3. Front of GTT study form used to establish an incentive time standard.

Study No.	S-64	Date February 22					Dwg. Style _____		Sub Item _____

Operation: Shake out with Jib Crane

Department Steel Fdry.	Operator			Mould					Sub.
	Man	Name Appleby Zulic		3 fabricated steel flasks 24 x 36 x 20"					
				5 fabricated steel flasks 20 x 32 x 29"					
				4 fabricated steel flasks 18 x 24 x 16"					

Equipment: Jib crane with chain slings

Pattern _____ Ins. Spec. _____ Die _____ L. Spec. _____

Part Description: Moulds made on J & J Roll Over Machine

Material Manganese Steel Castings

No.	Elements	Elem. Time Allowed	Occurrences/pc	Sub.	Total Time Allowed
1	Remove clamps and wedges	.0979	1/12		.0082
2	Wedges, gaggers, etc. aside	.0734	1/12		.0061
3	Move empty crane to flask	.0612	1/12		.0051
4	Hook flask with chain	.1469	1/12		.0122
5	Move flask by crane to shake out area	.2448	1/12		.0204
6	Mallet flask	.4039	1/12		.0337
7	Operate crane during shake out	.0490	1/12		.0041
8	Set down empty flask on 2nd flask	.0979	1/12		.0082
9	Stock empty flasks by crane	.2693	1/12		.0224
10	Unhook flask	.0857	1/12		.0071
11	Bottom board aside by hand	.2326	1/20		.0016
	TOTAL ALLOWED HOURS for Magnetic Castings				.1391
12	Move crane to casting	.0367	1/12		.0031
13	Hook casting with chain	.1714	1/12		.0143
14	Move casting aside by crane	.0245	1/12		.0020
15	Unhook casting	.0490	1/12		.0041
	TOTAL ALLOWED HOURS for Non-Magnetic Castings				.1626

Special tools, jigs, fixtures, etc.: Wood mallets   Mach. Tool No.

Conditions: Average

Observer: Dew   Sketch   Approved by

Time All., Set Up | Each Piece | Total

Remarks: Elements 16 and 17 are foreign to this study. Magnetic castings are lifted out of the sand by a magnet later on. Non-magnetic castings, such as some manganese steels are hooked with a chain by the shake-out men and set aside to be hooked with a chain by the cab crane later.

FIG. 5-4. Back of GTT study form shown in Figure 5-3.

Use of GTT to Measure Allowances and Fatigue. As has already been pointed out, the GTT procedure can be used to measure the allowances which should be made for fatigue and special, unavoidable, and personal delays.

Figure 5-5 (study 314) shows one sheet of an eight-hour study taken on eight sewing machine operators. The allowances can readily be calculated from this study. First, the results of the all-day study are summarized as shown in the following table.

Element no.	Description	Total, 8 girls	Average, 1 girl	Classification of time
		Elapsed hours		
1	Sew	51.94	6.4925	Working element
2	Handle bundle	2.22	0.2775	Working element
3	Arrange work	0.86	0.1075	Working element
4	Change thread	0.24	0.0300	Special delay
5	Change tubing or elastic	0.16	0.0200	Special delay
6	Thread breakage	1.16	0.1450	Special delay
7	Down—machine trouble	0.10	0.0125	Special delay
8	Talk to supervisor	0.14	0.0175	Unavoidable delay
9	Idle—unavoidable	0.46	0.0575	Unavoidable delay
10	Start late	0.02	0.0025	Avoidable delay
11	Quit early	0.20	0.0250	Avoidable delay
12	Rest period	1.68	0.2100	⎰ Consider all as personal
13	Work during rest period	0.88	0.1100	⎱ delays
14	Additional personal time	0.22	0.0275	Personal delay
15	Idle—avoidable	0.16	0.0200	Avoidable delay
16	Clip parts apart	2.62	0.3275	Working element
17	Clean up machine	0.68	0.0850	Special delay
18	Fill in P. W. report	0.26	0.0325	Special delay
	Total.............	64.00	8.0000	

Next, the allowances for special, unavoidable, and personal delays are calculated:

Working elements		Special delays		Unavoidable delays		Personal delays	
(1)	6.4925	(4)	0.0300	(8)	0.0175	(12)	0.2100
(2)	0.2775	(5)	0.0200	(9)	0.0575	(13)	0.1100
(3)	0.1075	(6)	0.1450			(14)	0.0275
(16)	0.3275	(7)	0.0125				
		(17)	0.0850				
		(18)	0.0325				
Total.......	7.2050		0.3250		0.0750		0.3475
% allowances.........		$\frac{0.3250}{7.2050} \times 100 = 4.51\%$		$\frac{0.0750}{7.2050} \times 100 = 1.04\%$		$\frac{0.3475}{7.2050} \times 100 = 4.82\%$	

Finally, a reasonable allowance for fatigue is determined, using the method developed by Lowry, Maynard, and Stegemerten for use with all-day allowance time studies.[1]

$$\% \text{ fatigue} = \left(\frac{OL}{NS} - 1\right) 100$$

[1] S. M. Lowry, H. B. Maynard, and G. J. Stegemerten, *Time and Motion Study and Formulas for Wage Incentives,* McGraw-Hill Book Company, New York, 1940, p. 259.

where O = overall working element time
$\quad L$ = leveling factor at the point of maximum performance during the day
$\quad N$ = number of pieces produced during the day
$\quad S$ = leveled time per piece
OL for study 314 is calculated as follows:

Operator	Elapsed working element hours	Maximum leveling factor for day	Leveled hours at maximum
1	7.32	1.19	8.71
2	6.96	1.08	7.52
3	7.30	1.11	8.10
4	7.04	1.065	7.50
5	7.34	1.11	8.15
6	7.30	0.95	6.94
7	7.14	1.13	8.07
8	7.24	1.095	7.93
		Total for 8 operators....	62.92
		OL (average for 1 operator)....	7.8650

NS is calculated as follows:

The 8 operators worked on 15 operations during the all-day study. Their piece counts, N, were obtained from the verified production reports submitted to the payroll department. The leveled time for each operation was taken from previously established time standards.

NS is obtained by multiplying the number of pieces produced on each operation by the leveled time for the operation. The total NS for the 8 operators is 60.24 hours.

$$\frac{60.24}{8} = 7.5300 \ (NS \text{ for 1 operator})$$

Substituting in the equation,

$$\% \text{ fatigue} = \left(\frac{OL}{NS} - 1\right) 100 = \left(\frac{7.8650}{7.5300} - 1\right) 100$$

$$\% \text{ fatigue} = 4.45\%$$

Total allowances for fatigue and special, unavoidable, and personal delays are therefore, in percent,

Special delays.............	4.51
Unavoidable delays........	1.04
Personal delays...........	4.82
Fatigue..................	4.45
	14.82
Call..................	15

Use of GTT as a Survey Tool. GTT is a gross measuring tool compared with time study, MTM, or other predetermined motion time systems. By the same token, it should often be used first to uncover major inefficiencies. On jobs that have not been carefully studied and improved or where methods that formerly were good have become obsolete, GTT will frequently give a greater return for the time and effort invested in the study than any return that may be subsequently realized from more precise techniques. It will usually be highly desirable to follow GTT with MTM, but the law of diminishing returns will be functioning.

The following cases are representative of the kinds of improvements that result from GTT studies. The first five resulted from studies made by several inexperienced

		1 Grace	2 Sarah	3 Verna	4 Lizzie	5 Esther	6 Margaret	7 Bessie	8 Dornelda	
Study on 8 Sewing Machine Operators					Sheet Started 10:30 A.M.			Finished 11:30 A.M.	GTT Study No. 314	
Sheet 4 of 8 Sheets					Each Tally = .0200 Hour		Wednesday, March 26		Observer George Dew	
Over-all Tallies		†††† †††† ††††	†††† †††† ††††	†††† †††† ††††	†††† †††† ††††					
Operator		1 Grace	2 Sarah	3 Verna	4 Lizzie	5 Esther	6 Margaret	7 Bessie	8 Dornelda	No.
Machine		Flatlock	U. Special Sew Tubing	Button Machine	Flatlock	U. Special Sew Elastic	Flatlock	Flatlock	Flatlock	
Skill-Effort L.F.		B1B2 1.19	C2C 1.065	C1C1 1.11	C2C 1.065	C1C1 1.11	E2C .935	B2C1 1.13	C1C1 1.11	
No.										
1	Sew	†††† †††† †††† / †††† †††† †††	†††† †††† †††† / †††† †††† †††	†††† †††† †††† / †††† †††† ††††	†††† †††† †††† / †††† †††† ††††	†††† †††† †††† / †††† †††† ††††	†††† †††† †††† / †††† †††† ††††	†††† †††† †††† / †††† †††† †††	†††† †††† †††† / †††† †††† †††	
1	Sew	†††† †††† ††††	†††† †††† †††	†††† †††† ††††	††††	†††† †††† ††††	†††† †††† ††††	†††† †††† †††	††††	
1	Sew	46	43	46	34	44	45	43	35	35
2	Handle bundle	1	IIII	1	I II	2 I	1	I	1	1
3	Arrange work	1	1	1	1	III	1	1	1	1
4	Change thread									
5	Change tubing or elastic						IIII			
6	Thread breakage	II	I	II	2 †††† III III	13	4		II	2
7	Down-machine trouble									
8	Talk to supervisor									
9	Idle - Unavoidable	I	I	I	I	I				
10	Start late									
11	Quit early									
12	Rest period									
13	Work during rest period									
14	Personal time	I	I	I	I	2 II				
15	Idle - Avoidable							†††† I	†††† †††† I	
16	Clip parts apart							6	11	11
17	Clean up machine									
18	Fill in PW Report									
	Totals	50	50	50	50	50	50	50	50	50

FIG. 5-5. GTT study form used to measure percent allowances and fatigue.

GTT observers working under the direction of an experienced industrial engineer during a fourteen-day period in two small refractory plants. Their studies pinpointed inefficiencies that led to annual savings of $27,375.

Case A. An observer made a sixteen-hour study, containing 29 elements, of a four-man crew setting ware (firebricks) in a periodic kiln. The crew consisted of one setter and three wheelers. The study showed that 25 percent of the total man-hours was wasted. This was corrected at an annual saving of $5,000. Most of the saving resulted from better planning and supervision and instructing the wheelers to set ware when they were otherwise idle.

Case B. An observer made studies on a three-man crew loading ware in regular railroad boxcars and compared this method with two men loading DF loader-type boxcars. He found the saving to be as follows:

Item saved	Saving per car
Boxmaker's wages in cutting blocking and making end gates....	$10.00
Lumber for blocking and end gates.........................	12.00
Steel strapping and nails.................................	8.00
Reducing loading crew from 3 to 2 men.....................	3.75
Total..	$33.75

The company had never requested DF loaders in the past because no one appreciated the savings. By changing to DF loaders, the company saved $7,000 per year.

Case C. Studies on refractory molders and nearby tunnel kiln setters showed that one laborer could service both groups at an annual saving of $2,500.

Case D. A tractor driver and three laborers unloaded Indian kyanite and bauxite from a periodic kiln. The tractor driver placed one box outside one door of the kiln (the other door, although available, was not used) and then stood idle while the laborers loosened the material with picks and carried one lump at a time by hand or one small shovelful to the box. After the box was loaded, the three laborers stood idle while the tractor driver dumped and returned the empty box. A new method was established that required:

1. Half the kiln to be unloaded through one door and half through the other to reduce walking time
2. Two boxes at each kiln door so the men always had one to load while the other was being emptied
3. The use of wheelbarrows after the men worked a few feet in from a door, which eliminated about 95 percent of their walking
4. The tractor driver to shovel and wheel in the kiln when he was not driving his tractor

These changes saved $9,000 per year at no cost, because plenty of wheelbarrows and boxes were available.

Case E. A crew of three men used cardboard and steel strapping to pack pallets of bricks for shipment. An 18-element GTT study showed one man was idle much of the time and that two men were often attempting to hold cardboard or straps for the third man who was working. Only one assistant was required. The crew was reduced to two men with no reduction in output. The saving was $5,000 per year.

Case F. Studies in a steel foundry showed that the average time to wait for cab crane was 0.1780 hour per wait. To collect more data, a GTT study around the clock for 5 days (5×24 hours = 120 consecutive hours) was made. This study, taken on 50 molders and cranemen, required 10 GTT observers (400 observer-hours). It measured 2,000 worker-hours (1 man-year) in a calendar week.

The study showed that the addition of one cab craneman and one crane follower (16 man-hours) during the night would save 51 molder-hours per day for a net saving of 35 man-hours per day. At $3 per hour, this saved $105 per day, or $27,000 per year.

Case G. A group of ten men used air hammers to chip steel castings. They were dissatisfied because they were averaging only a 20 percent bonus under an old piecework plan applied to each individual. An observer took two eight-hour GTT studies covering all ten men in each study.

These studies showed the chippers to be spending 48.5 percent of their total clock hours on avoidable delays (start late, avoidable idleness, and quit early). This was in addition to personal delays, unavoidable delays, and crane waits. Another 14.7 percent of the total clock hours was spent doing unnecessary work.

The conclusion was that the performance of the chippers was unsatisfactory because of loose piecework prices, overmanning, poor methods, poor supervision, and poor worker application to the job.

Case H. An industrial engineer wanted to study one operator running two spring-crimping machines. He prepared a GTT form listing 25 man-elements in a column on the left and ruled off four columns for his tallies headed as follows:

1. Both machines operating
2. Machine A operating, machine B down
3. Machine A down, machine B operating
4. Both machines down

This form enabled the observer to make a GTT study that showed what each machine was doing while the operator was performing each of the 25 elements. Several GTT studies gave information that enabled the engineer to plan a more effective sequence of elements for the operator. Then the engineer took additional studies to establish time standards for operating the two machines. Time standards for one man operating two or more machines are difficult to set by time study. With GTT, the problem is more easily solved.

Use of GTT to Measure Scattered or Roving Workers. An observer making a typical GTT study stays in one location where he can see all the operators. GTT engineers in one plant wished to make a study on nine operators working fairly close together. However, two large machines obstructed the view so no one location could be found where the observer could see all nine operators. He could observe four men from one point, three from another, and two from a third point. The observer lengthened the time between observations to 0.0300 hour and walked to each of the three observation points in succession. This type of study is known as a walking GTT study.

Groups of roving maintenance men in large plants are difficult to study, but the following procedure gives excellent results. Using GTT, an observer stations himself at the maintenance shop, attaches himself to the next group dispatched on a job, and stays with the group until it returns to the shop. The group may consist of from two to fifteen men and may increase or decrease in size several times during the trip. The trip may last ten minutes or eight hours. These factors create no special problems. The observer stays with his group, records his observations and performance ratings, and records necessary production counts and other information.

Many walking and other GTT studies have been made using a maximum interval between observations of 0.0500 hour, or three minutes. When a longer interval is required, a work sampling study is made. The three-minute maximum interval used on GTT studies is an arbitrary limit that represents the current thinking of industrial engineers having considerable experience with GTT and time study. In their opinion, each worker is under observation a sufficient amount of time to justify rating his performance in the same manner as used on a continuous time study.

Use of GTT to Develop Time Formulas. GTT may be used to develop time formulas either alone or with predetermined motion time standards, time study, or other forms of work measurement. In such instances, GTT is most advantageous where time standards must cover a wide range of variations in methods when it is impossible or undesirable to attempt to standardize on a single good method.

For example, in a certain bronze foundry, the jolt-squeeze molders pushed their completed molds aside on gravity roller conveyors. A group of five men poured the molds and shook them out by dumping the sand through gratings in the floor onto a belt conveyor. When a shake-out man completed other duties, he dumped molds alone, using method A. Later he was joined by a second shake-out man, and they split the operation, using method B. Later a third man joined them, and the operation was split three ways, using method C. When a fourth man joined them, the men usually split into two crews on separate roller conveyors, using method B, but if no other work was available, all four men worked on one conveyor, using method D. Occasionally five men worked on one conveyor, using method E. An occasional bent or damaged flask or other irregularity caused other methods to be used at times. A total of at least a dozen acceptable methods occurred with unpredictable frequencies during a single day.

A single GTT time formula was made, covering all fifteen hourly workers in the bronze foundry other than the molders who were already working on wage incentives. The operations were as follows: weigh and mix metal, charge furnaces, tend pot heaters, melt metal, skim ladles, pour metal, shake out molds, move castings to cleaning area with tractor, clean molding machines, operate sand system, tend pattern storage, replace pots in Stroman furnaces, and sweep floor.

The time formula was based on ten GTT studies and four time studies (used when an individual man was studied). Twelve of these studies each covered a full nine-hour day. The same studies were used to measure both allowances and each-piece time. The studies required a total of 113.5 observer-hours. The overall project, including a preliminary methods study and the preparation of a time formula report, required six man-weeks.

Final time standards were as follows:

Description	Unit measured	Allowed man-hours
Setup time:		
Clean molding machine and sweep.....	1 machine-day	0.4129
Replace pot in Stroman furnace.......	1 replacement	4.0013
Operating time:		
Operate furnaces....................	100 lb melted	0.1108
Mix metal, charge, tend pot heaters...	100 lb melted	0.0921
Operate sand system................	1 mold	0.0072
Pour and shake out................	1 mold	0.0458
Tend pattern storage...............	1 pattern set used	0.0772

All the indirect men were placed in one wage incentive group. The installation operated well, and the saving was relatively great in proportion to the cost of making the study.

STEPS IN MAKING A GTT STUDY

GTT has many similarities to both stopwatch time study and work sampling; so an industrial engineer experienced in these procedures will have no difficulty in making a GTT study. The steps employed in making a GTT study are as follows:

1. *Decide on the purpose of the study.* It might be one of the following:
   a. To provide quick information on idle time, proper crew size, waiting time, minor work performed, and similar data
   b. To measure percent delays
   c. To measure percent fatigue
   d. To measure leveled time or allowed time

2. *Select the operation, group of operators, and time periods to be studied.*

3. *Decide whether or not it is necessary to separate the time for each operator.* If not, the GTT form will be easier to design and the observations can be recorded in less time, thus permitting the use of a shorter time interval between observations if desired.

4. *Decide on the recording method.* One of the following procedures should be chosen: by observer(s), by operator(s), or by memomotion filming. Using observers is frequently the most suitable procedure. They register tallies on a study form, prepare punched cards, or employ a portable data input terminal for subsequent data processing.

Recordings can be made by operators if the purpose of the study is suitable and will not interfere too much with the work. In this case, the element breakdown must be held to a minimum. The use of fully preprinted registration forms is recommended (see Figure 5-6).

Memomotion study, discussed in Chapter 7 of Section 3, employs a motion picture camera to record observations at intervals of 0.5 to 10 minutes when used in connection with GTT.

5. *Select and define elements.* If unfamiliar with the operation, preliminary observations should be made to determine what elements to include. At this time, it should also be decided how detailed the study needs to be. The number of elements should be minimized without jeopardizing the purpose of the study. Some open space should be left on the study form for later addition of unexpected elements. The element breakdown should be logical and with natural limits easily definable and observable. For large studies, particularly when using several observers, it may be desirable to work out a brief written element definition. In many cases, element definitions can be standardized for repeated use within a company.

6. *Design and reproduce study form.* GTT is used in so many varying situations that no standard form suitable for all purposes exists (compare Figures 5-3, 5-5, and 5-6). A suitable form should be designed for each study, trying to list all elements on one 8½ by 11 sheet if possible. If not, two sheets may be used, but this will require turning the sheets back and forth while making the study. The forms should be reproduced in sufficient number to allow for spoilage and for

Punch code	Time		Activity code	Remark	Main activity	Sub-activity code	Sub-activity
43		00	31			9	
44		10	31	①		2	
45	8	20	12		Drafting	2	
46		30				2	
47		40				8	
48		50				8	
49		00				8	
50		10					
51	9	20			Computing		
52		30					
53		40					
54		50					
55		00			Technical literature		
56		10					
57	10	20					
		30					

① = Remark concerning activity 31

Fig. 5-6.   Example of preprinted study form to be filled out by draftsmen in a design office.

summarizing the final results.   In making the study, each observer will usually start a new form each hour.

7. *Select interval size.*   The proper interval size is primarily dependent on the number of operators or objects and their location.   The observer should not be so pressured that he does not have time to observe thoughtfully what is going on.   The following table suggests intervals that an observer can use comfortably during an eight-hour study.   The interval, however, must be shorter than the smallest

Number of workers in group	Interval measured on stopwatch	
	Decimal hour watch	Decimal minute watch
1	Use time study or MTM	
2	0.0050 hour	0.5 minute
3– 6	0.0100 hour	1.0 minute
7–10	0.0200 hour	2.0 minutes
11–15	$\begin{cases} 0.0300 \text{ hour} \\ 0.0400 \text{ hour} \\ 0.0500 \text{ hour} \end{cases}$	3.0 minutes
Over 15	Use 2 or more observers	

element to be measured.   Also, any rigid cycle time inherent in the process should not coincide with multiples of the observation interval, to prevent possible over-representation of some activities.

8. *Determine study periods and duration.*   Several factors influence the length of the study.   The minimum number of total observations or observations for the critical activity should be computed as described later under the subheading, Statistical Accuracy of GTT.   The minimum duration of the study is the minimum number of total observations needed to ensure statistical accuracy, multiplied by the observation interval.

To ensure that the data obtained are truly representative of normal conditions, the duration of the study must be sufficient to include and level out the natural, long-cycle variations that may exist in the process.   Where such variations exist, the total number of observations can be divided into a suitable number of substudies and spread out to cover a longer period of time or a wider range of conditions.

Daily plotting of diagrams such as shown in Figure 5-7 should be considered. This continuous check on the daily and accumulated values of one or more critical activities may be helpful in determining the stability of conditions and evaluating the necessary duration of the study.

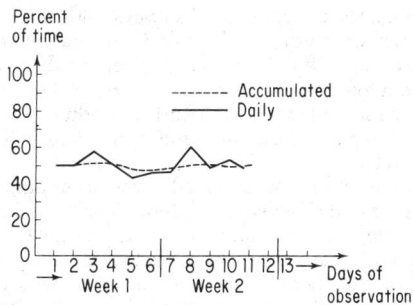

FIG. 5-7.   Control curves used to determine stability of observed conditions.

9. *If operators and their supervisors are not familiar with GTT, inform them of its scope and purpose and how it is performed.*

10. *Consider a trial study.* A brief trial study will provide a good test of the soundness of the preparatory decisions such as element selection, element definition, interval size, and the like.

11. *Make the study.* Record carefully the names or numbers of the operations, operators, interval size, and other identifying information.

Start the stopwatch and allow it to run continuously. Record the starting time from a regular watch or clock to the closest half minute.

Make distinct observations and mark tallies at the predetermined interval. If five men are being timed each 0.01 hour, first record a tally in the overall tallies box when the decimal hour stopwatch reads 0.01. Then record a tally for each of the five men. Repeat this action in the same sequences each 0.01 hour for the duration of the study.

If the number of operators is large or if the observer has to move between different locations, it is possible to establish fixed time delays between groups of observations to allow for comfortable observation.

Change sheet at fixed intervals, such as every hour or second hour. Record the time for each switch on both the old and the new form. If desired, the performance rating for each operator is recorded once on each sheet.

On a half-day or all-day GTT study, account for the full four hours or eight hours of each man by means of tallies. This may require elements such as start late or quit early, coffee break, safety meeting, and the like. This will make the study more understandable, reduce errors, and make the study easier to summarize.

An exception may be made to this rule in the case of a "floating" worker who is sometimes a member of the group and at other times is outside the group or the observation area.

The element "work on foreign operation" is used if an operator temporarily performs some task that is not part of his job.

When more than two or three men are studied, an element such as "absent—reason unknown" is often required. Every effort should be made to hold the number of tallies recorded for this element to a minimum by discovering its reason.

On an each-piece study, the piece count and other pertinent information should be recorded.

At the completion of the study, the time finished and the skill and effort ratings for the last sheet or sheets should be recorded.

The study should be worked up the same as a time study or a work sampling study, depending on the type of GTT study taken.

COMPUTERIZED GTT

Making a simple GTT study is an easy routine task and not very time consuming, as previously indicated. In certain cases, the data volume and transactions involved grow rapidly. Individual, large-scope GTT surveys with many interrelated variables and possibly a need for structuring the result in several dimensions—such as "per foreman area," "per work shift," "per craft," "per product," and so on—become rather laborious if computed manually.

Also, if GTT is established as a standard procedure for standards setting in a large enterprise, the resulting load on engineering time will soon make computer applications quite feasible.

Observations can be registered on punched cards according to the "Port-a-punch" or "mark-sensing" methods developed by International Business Machines Corporation. The former employs a manually operated, portable card punching device. The second uses automatic scanning of manually pencil-marked cards to produce regular punched cards. In both cases, the observer will produce one punched card for each operation for later computer processing according to a suitable program, resulting in a regular alphanumeric printout of the results or plotting of diagrams such as shown in Figure 5-8.

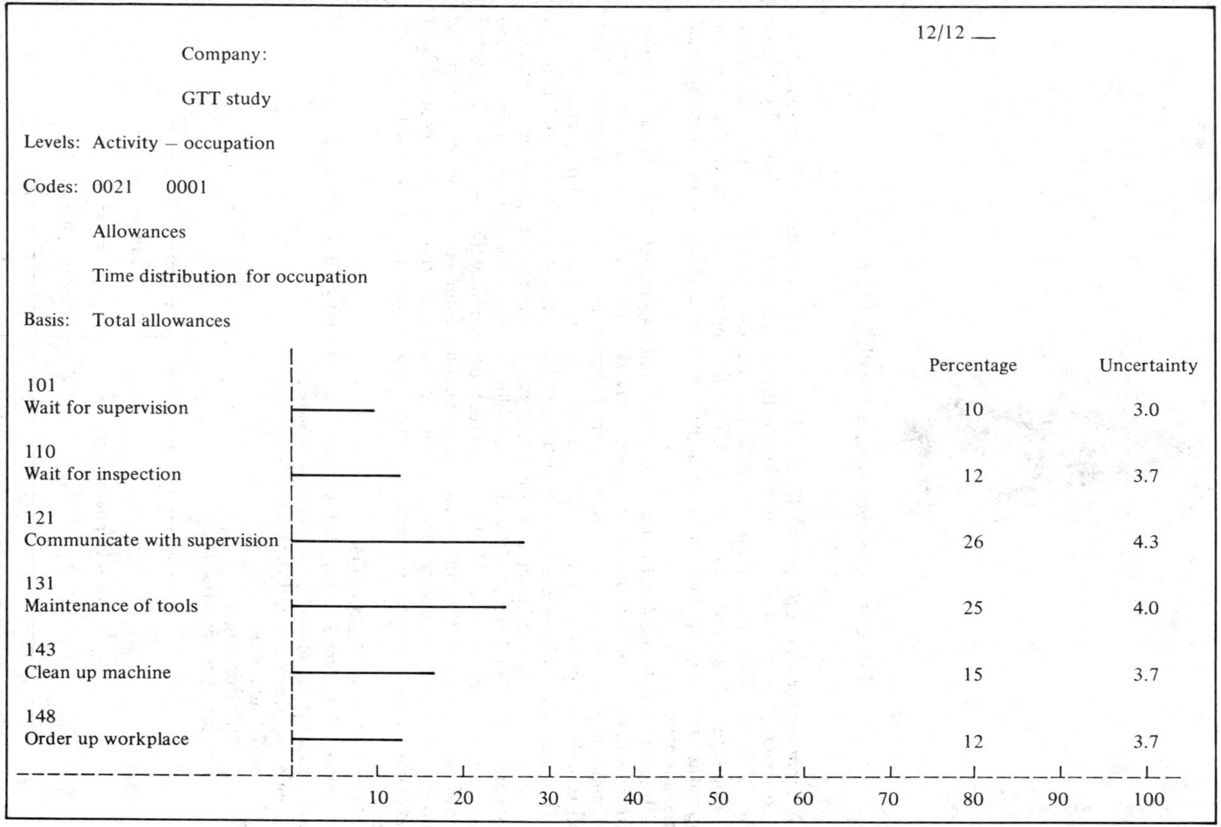

Company:

GTT study

Levels: Activity – occupation

Codes: 0021 0001

Allowances

Time distribution for occupation

Basis: Total allowances

Code	Activity	Percentage	Uncertainty
101	Wait for supervision	10	3.0
110	Wait for inspection	12	3.7
121	Communicate with supervision	26	4.3
131	Maintenance of tools	25	4.0
143	Clean up machine	15	3.7
148	Order up workplace	12	3.7

10 20 30 40 50 60 70 80 90 100

Fig. 5-8. EDP plotting chart summarizing GTT study.

3-85

FIG. 5-9.   Observer equipped with ETM recorder.

A more recent development of H. B. Maynard and Company is "electronic time measurement," employing the portable electronic data recorder shown by Figure 5-9. This portable data input terminal can be used for GTT, work sampling, or stopwatch time study. If set for GTT, the internal electronic timer of the apparatus will also produce acoustic signals to initiate observation at preselected time intervals.

Observation data are fed through a keyboard onto a magnetic tape in the recorder. After completion of the study, data are converted onto punched tape or ½-inch magnetic tape in a separate process before being read into a computer which will process and print out according to one of a number of standard programs.

The recorder weighs less than 5 pounds. The input capacity is 30,000 marks per cassette and a minimum of eight hours continuous recording before recharging the batteries. Ten interval sizes are available between 0.25 and 6 minutes. Time and cost savings of from 30 to 50 percent compared with manually conducted studies are reported, the higher savings stemming from larger and more complex studies. Two typical time comparisons between electronic time measurement (ETM) and conventional procedures indicate how and where the time reductions are achieved.

Case A.   Area: Design office
Purpose: Combined study for organization planning and the establishment of time standards for project planning and control
Scope: 25 designers and draftsmen
   70 activities
   400 drawings
   12,500 observations

| | Man-weeks | |
Study phase	Conventional	ETM
Planning, preparation, and information...	1	1
Observation and registration.............	4	4
Computation and reporting.............	5	1
Total.................................	10	6

Case B.   Area: Shipyard, plateshop
Purpose: To establish allowances for welding operations
Scope: 35 workers in 15 crews
        4 cranes
        2 work shifts for 2 weeks
        25,000 observations

Study phase	Man-weeks	
	Conventional	ETM
Planning, preparation, and information...	1	1
Observation and registration.............	4	4
Computation and reporting.............	9	0.5
Total.............................	14	5.5

STATISTICAL ACCURACY OF GTT

Being a sampling procedure, GTT produces time estimates containing some statistical error in relation to the true time values of a pattern of activity. These errors can be determined.

The formulas discussed below will provide an answer to questions such as: What accuracy did we get by this study? How many observations do we need in order to measure safely within a given accuracy level? They will also give the GTT practitioner an increased understanding of what he is doing, so that he can determine, for example, if some modification of the study procedure that he is considering for practical reasons is permissible within the basic procedure.

Explanations. The following symbols and abbreviations will be used:

T total length (time) of study
T_a total time for activity a
t time for one work cycle
t_a time for one occurrence of activity a
N total number of observations
N_a total number of observations for activity a
K_a total number of occurrences for activity a
C_a number of occurrences during one work cycle for activity a
i interval size
r_{T_a} relative error for T_a at the 95 percent confidence level
r_T relative error for T_a in percent of total time T at the 95 percent confidence

$$\text{level} \left(r_T = r_{T_a} \times \frac{T_a}{T} \right)$$

s_{T_a} standard deviation for T_a
n_a number of observations for one occurrence of activity a
a_a time between the previous observation to the actual start of activity a
b_a time between the actual end of activity a and the following observation

Formulas. The following formulas (all valid at the 95 percent confidence level) are used in the practical planning and evaluation of GTT studies.

Determination of Total Time

$$T_a = N_a \times i \tag{1}$$
$$T = N \times i \tag{2}$$

Determination of Statistical Error

$$r_{T_a} = \pm \frac{80 \sqrt{K_a}}{N_a} \quad \% \tag{3}$$

$$r_T = \pm \frac{80 \sqrt{K_a}}{N} \quad \% \tag{4}$$

Determination of Number of Observations

$$N = \frac{6{,}400}{r_T{}^2} \times \frac{i \times C_a}{t} \tag{5}$$

$$N = \frac{6{,}400 \times i \times t \times C_a}{r_{T_a}{}^2 \times t_a{}^2} \tag{6}$$

$$N_a = \frac{6{,}400 \times i \times C_a}{r_{T_a}{}^2 \times t_a} \tag{7}$$

Theory. Three different cases of observation exist for a specific occurrence j of activity a according to Figure 5-10. For all three cases

$$t_{aj} = (n_{aj} + 1)i - (a_{aj} + b_{aj})$$

Because, according to definition,

$$T_a = \sum_{j=1}^{K_a} t_{aj}$$

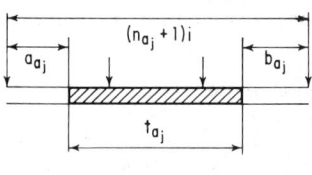

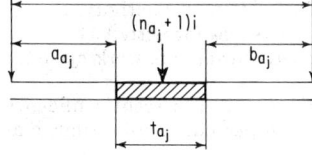

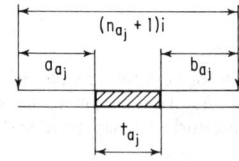

Fig. 5-10. Three different cases of observation for a specific occurrence of activity a.

we get

$$T_a = \sum_{j=1}^{K_a} [(n_{a_j} + 1)i - (a_{a_j} + b_{a_j})]$$

$$= N_a \times i + K_a \times i - \sum_{j=1}^{K_a} (a_{a_j} + b_{a_j})$$

a_a and b_a, however, are rectangularly distributed between the limit values 0 and i (equal probability for a_{a_j} or b_{a_j} to assume any value between 0 and i). Consequently, with increasing K_a,

$$\sum_{j=1}^{K_a} a_{a_j} \quad \text{and} \quad \sum_{j=1}^{K_a} b_{a_j} \to 0,5 \times i \times K_a$$

and

$$T_a = N_a \times i + (K_a \times i) - (K_a \times i) \qquad (1)$$
$$= N_a \times i$$

Similarly,

$$T = N \times i \qquad (2)$$

To establish the error of T_a in formula (1), we study the standard deviation, s, for the rectangular distribution between the limit values 0 and i.

$s = i/\sqrt{12}$ according to statistical reference literature.

For t_{a_j}, the standard deviation $s_{t_{a_j}}$ is the square root of the square sum of two standard deviations corresponding to the rectangular distributions in the beginning and in the end of t_{a_j}:

$$s_{t_{a_j}} = \sqrt{\left(\frac{i}{\sqrt{12}}\right)^2 + \left(\frac{i}{\sqrt{12}}\right)^2} = \frac{i}{\sqrt{6}}$$

Further,

$$s_{T_a} = \sqrt{\sum_{j=1}^{K_a} s_{t_{a_j}}{}^2} = \sqrt{K_a \times \frac{i^2}{6}} = \frac{i \times \sqrt{K_a}}{\sqrt{6}}$$

At the 95 percent confidence level,

$$r_{T_a} = \frac{1.96 \times s_{T_a}}{T_a} \times 100 = \frac{1.96 \times \dfrac{i \sqrt{K_a}}{\sqrt{6}}}{N_a \times i} \times 100$$

$$= \frac{80 \sqrt{K_a}}{N_a} \qquad (3)$$

Similarly, according to definition,

$$r_T = \frac{1.96 \times s_{T_a}}{T} \times 100$$

$$= \frac{80 \sqrt{K_a}}{N} \qquad (4)$$

If, for cyclic work, the number of occurrences during one work cycle for activity a is C_a, the total number of occurrences

$$K_a = \frac{T \times C_a}{t} = \frac{N \times i \times C_a}{t}$$

This will turn formula (4) into

$$r_T = 80 \sqrt{\frac{i \times C_a}{N \times t}}$$

or

$$N = \frac{6,400 \times i \times C_a}{r_T^2 \times t} \tag{5}$$

which can be transformed into

$$N = \frac{6,400 \times i \times t \times C_a}{r_{T_a}^2 \times t_a^2} \tag{6}$$

and

$$N_a = \frac{6,400 \times i \times C_a}{r_{T_a}^2 \times t_a} \tag{7}$$

Application of the Theory. According to formula (3), the largest statistical error will occur on an activity showing a high value for K_a and a low one for N_a. The critical activity showing these properties is typically one with frequent occurrences, each occurrence being of short duration.

Formulas (3) to (7) are used for two practical purposes:

1. To estimate before the GTT study is begun the necessary number of observations needed to reach a certain accuracy level of measurement
2. To determine the statistical accuracy of measurement after the study has been made

For practical reasons, the control of accuracy will be based either on the selection of one or a few critical activities or elements or on the basis of an average activity or element.

The computations of statistical accuracy give control of only the methodical error in the GTT itself. The actual result may be influenced by a number of other and more intricate sources of error, such as an unrepresentative period of study, errors made by the observer in classification and performance rating, and the like. This reservation should call attention to these matters, but it in no way detracts from the obvious value of being able to forecast and control statistical deviation in GTT measurements in the simple manner illustrated by the following examples.

Case A. Forecasting the Accuracy of a Delay Study. Repair work. The study is planned to take 27 hours, with an observation interval of 2 minutes. The shortest delay time is estimated to be 6 minutes, and the average to be 9 minutes (according to available statistics or a limited pilot study). Total delay time is expected to be 15 percent of available time. Determine the relative error in the value for total delay time which will be determined by this study.

$$T = 27 \text{ hr} = 1,620 \text{ min}$$
$$i = 2 \text{ min}$$
$$t_{\text{average}} = 9 \text{ min}$$
$$N_a = \frac{T_a}{i} = \frac{15 \times 1,620}{100 \times 2} = 122 \qquad \text{formula (1)}$$
$$K_a = \frac{T_a}{t_{\text{average}}} = \frac{15 \times 1,620}{100 \times 9} = 27$$
$$r_{T_a} = \frac{80 \times \sqrt{K_a}}{N_a} = \frac{80 \sqrt{27}}{122} = \pm 3.4\% \qquad \text{formula (3)}$$

Case B. Planning a Study to Establish Piecework Standards. Assembly work by two operators and one assistant. The smallest suboperation (element) occurring once per cycle was estimated to be 3 minutes, and the total cycle time to be 30 minutes. A relative error of no less than 5 percent is desired for the smallest element. Observation interval: 0.5 minute.

Determine the number of observations, the duration of the study, and the number of work cycles.

$$i = 0.5 \text{ min}$$
$$t = 30 \text{ min}$$
$$r_{T_a} = \pm 5\%$$
$$t_a = 3 \text{ min}$$
$$C_a = 1$$

$$N = \frac{6,400 \times i \times t \times C_a}{r_{T_a}{}^2 \times t_a{}^2} = \frac{6,400 \times 0.5 \times 30 \times 1}{25 \times 9} = 427 \qquad \text{formula (6)}$$

$$T = N \times i = 427 \times 0.5 = 214 \text{ min}$$

$$K_a = \frac{T}{t} = \frac{214}{30} = 7.1, \text{ or } 8$$

CONCLUSION

GTT is a convenient method of studying man and machine processes in industry and elsewhere. The procedure is simple and easy to learn and can be partly automated if used extensively. In cases where systematic process improvement through detailed methods, tool, equipment, and workplace design is sought, alternative techniques which take more account of methods may be preferred.

BIBLIOGRAPHY

Conway, Richard W., "Some Statistical Considerations in Work Sampling," *The Journal of Industrial Engineering*, March–April, 1957.
Davidson, H. O., W. W. Hines, and T. L. Newberry, "The Error of Estimate in Systematic Activity Sampling," *The Journal of Industrial Engineering*, July–August, 1960.
Flowerdew, A. D. J., and P. W. Malin, "Systematic Activity Sampling," *The Journal of Industrial Engineering*, July–August, 1963.
Haines, I. Landis, "Work Sampling by Fixed Interval Study," *The Journal of Industrial Engineering*, July–August, 1958.
Jones, N. G., and P. M. Ghare, "Confidence Intervals for Systematic Activity Sampling," *The Journal of Industrial Engineering*, May–June, 1964.
Ljungqvist, Nils-Olof, and Rolf Tiefenthal, "GTT: A Quick and Reliable Work Measurement Technique" (Swedish), *Affärsekonomi*, 1966.
Mattson, Rolf, "GTT: Work Measurement through Fixed Interval Observations" (Swedish), *Svenska Arbetsgivarföreningen*, 1968.

Chapter **6**

Work Measurement
of Multimachine Assignments

DALE JONES

President, Uniquest, Albuquerque, New Mexico

A multimachine assignment is one where more than one machine is operated (or tended) by a single operator (or operator-helper team working together). The primary purpose of this chapter is to give information that will provide help in determining how many machines to assign to an operator and how to establish equitable production standards on these multimachine assignments.

The solution to many multimachine assignments involves the calculation of machine interference idleness and operator idleness which are caused by assigning more than one machine to a single operator. Machine interference idleness (or machine interference) is the time that a machine is idle because the operator is servicing another machine in the group. Operator idleness is the time that the operator is idle because all the machines in the group are running automatically.

Two general types of solutions to multimachine assignment problems are presented here, depending on whether machines are randomly serviced or systematically serviced by the operator.

Looms, spinning machines, and cartoning machines are examples of machines that have random servicing demands. In the case of weaving with looms, the occurrence of yarn breakage, and therefore loom stoppage, is entirely random. Solutions based on the laws of probability are used for these assignments.

Other machines, such as plastic-mold presses and gear-hobbing machines, have regular (systematic) servicing demands. The loading and automatic processing times of these machines are predictable as to order of occurrence and elapsed time required for both operating and servicing. Solutions to problems involving these machines are determined by systematic analysis.

The efficiency of a machine individually tended by one operator is apparently unity minus the portion of time the machine is nonproductive while being unloaded, loaded, and the like. However, the efficiency of a machine tended in multiple is unity minus the portion of nonproductive time required for unloading, loading, and so on, minus the portion of time the machine experiences interference. Thus, refined work load planning and incentive payment of multimachine activities require preestimation of the degrees of machine interference idleness to be experienced by the assigned machines. The same can be said in behalf of refined cost estimation and scheduling.

DIRECT MEASUREMENT OF RANDOM MACHINE INTERFERENCE

Considerable effort has been spent in direct measurement of random machine interference, mainly via stopwatch study. However, because of difficulties encountered in timing simultaneous events, rating the operator, and the like, and because of the necessity of developing interference curves or tables covering the whole range of possible assignments entailing different numbers of machines and work loads, these direct timing efforts have generally proved fruitless. Thus statistical methods of preestimating machine interference, such as those to be described, have replaced the costly, empirical, directly timed interference allowances in achieving refined work load planning and incentive payment of randomly serviced assignments.

The discussion to follow will first treat random machine interference of multimachine assignments having random servicing demands. Then the more regular, repeating machine interference resulting from synchronized multimachine assignments will be considered. The discussion will be predicated on multimachine assignments tended by individual operators or tended by operator and helper combinations working together as teams. However, the principles of measurement apply to all multiactivities where individual machines, processes, work stations, or operators have simultaneous servicing demands resulting in interference.

APPLICATION OF THE RANDOM MACHINE INTERFERENCE TABLE

The use of the random machine interference table (Table 6-1) and the mathematics on which it is based can be understood by considering how four looms are assigned to one operator. Each loom is weaving the same type and size of cloth. Time study and frequency study show that this product would require, on an average, 1 minute of servicing for each 6 minutes of elapsed time if the operator were to tend only one loom from a point of average walking distance when four such looms are tended by a single operator. Thus the operator's work load (on an individual attention basis) for each of the four looms is $\frac{1}{6}$ or 16.67 percent, and the total work load is $4(\frac{1}{6})$, or 66.67 percent. On rare occasions, all four or three of the four looms chance to become idle at the same time. And on more frequent occasions, two, one, and none of the looms will be idle. When two or more looms are idle together, all but the one being serviced must incur interference idleness. And, as the operator subsequently services these looms, still others may chance to demand servicing and therefore incur machine interference idleness.

The average interference per loom for the assumed assignment can be estimated from Table 6-1 by locating four machines in the left vertical "No. machs." column and reading to the 65 and 70 percent columns for "Operator's total work load on individual attention basis." Direct interpolation of the italicized 5.8 and 6.9 percent values (to estimate the value for 66.67 percent load) gives 6.2 average percent interference per machine. In other words, on an average, for 100 minutes of elapsed assignment time, each of the four looms will be idle 6.2 minutes because of machine interference.

The operator percent unavoidable idleness inherent in the assumed assignment

TABLE 6-1. Random Machine Interference Table*

This table shows average percent (of elapsed time) interference per machine (in italic type) and percent (of elapsed time) operator idle time (in nonitalic type), in randomly serviced multimachine assignments tended by one operator, when servicing demands are random and approximately the same for each assigned machine. Use the accompanying machine interference adjustment chart (Figure 6-1) to obtain correction factors to be applied to these tabulated machine interference values when the work loads of the assigned machine are significantly different.

No. machs.	Operator's total percent interference-causing work load† on individual attention basis														
	50	55	60	65	70	75	80	85	90	95	100	105	110	115	120
2	*3.9*	*4.9*	*5.9*	*7.0*	*8.2*	*9.5*	*10.7*	*12.4*	*13.9*	*15.5*	*17.3*	*18.8*	*20.8*	*22.6*	*24.4*
	52	48	44	40	36	32	29	25	23	18	17	15	13	11	9
3	*3.7*	*4.7*	*5.7*	*6.7*	*7.8*	*9.2*	*10.5*	*12.2*	*13.8*	*15.4*	*17.2*	*18.8*	*20.8*	*22.6*	*24.4*
	52	48	43	39	35	32	28	25	22	18	17	15	13	11	9
4	*3.2*	*3.9*	*4.9*	*5.8*	*6.9*	*8.2*	*9.4*	*10.9*	*12.6*	*14.2*	*15.9*	*17.7*	*19.6*	*21.6*	*23.5*
	52	47	43	39	35	31	28	24	21	18	16	14	12	10	8
5	*2.7*	*3.5*	*4.3*	*5.2*	*6.2*	*7.3*	*8.6*	*9.9*	*11.4*	*13.0*	*14.7*	*16.5*	*18.5*	*20.6*	*22.6*
	52	47	43	39	34	31	27	23	20	17	15	12	11	8	7
6	*2.3*	*3.1*	*3.7*	*4.5*	*5.5*	*6.6*	*7.8*	*9.1*	*10.6*	*12.2*	*13.8*	*15.7*	*17.6*	*19.6*	*21.7*
	51	47	42	38	34	30	26	23	20	17	14	11	9	7	6
7	*2.1*	*2.7*	*3.3*	*4.0*	*5.0*	*6.0*	*7.1*	*8.4*	*9.8*	*11.3*	*13.0*	*14.9*	*16.8*	*18.8*	*20.9*
	51	46	42	38	34	30	26	22	19	16	13	11	8	7	5
8	*1.9*	*2.4*	*3.0*	*3.7*	*4.6*	*5.4*	*6.6*	*7.7*	*9.0*	*10.6*	*12.3*	*14.1*	*16.0*	*18.2*	*20.2*
	51	46	42	37	33	29	25	22	18	15	12	10	8	6	4
9	*1.7*	*2.2*	*2.8*	*3.3*	*4.3*	*5.0*	*6.0*	*7.1*	*8.5*	*9.9*	*11.6*	*13.4*	*15.4*	*17.5*	*19.6*
	51	46	42	37	33	29	25	21	17	15	12	9	7	5	4
10	*1.5*	*2.0*	*2.6*	*3.1*	*4.0*	*4.6*	*5.6*	*6.6*	*8.0*	*9.4*	*11.0*	*12.8*	*14.8*	*17.0*	*19.0*
	51	46	42	37	33	28	25	21	17	14	11	8	6	4	3
11	*1.4*	*1.9*	*2.4*	*2.9*	*3.7*	*4.3*	*5.3*	*6.2*	*7.5*	*8.9*	*10.5*	*12.4*	*14.4*	*16.5*	*18.5*
	51	46	41	37	33	28	24	20	17	13	11	8	6	4	3
12	*1.3*	*1.8*	*2.2*	*2.7*	*3.4*	*4.1*	*5.0*	*5.9*	*7.1*	*8.5*	*10.1*	*12.0*	*14.0*	*16.1*	*18.2*
	51	46	41	37	32	28	24	20	16	13	10	7	5	4	2
13	*1.3*	*1.7*	*2.1*	*2.5*	*3.2*	*3.9*	*4.7*	*5.6*	*6.8*	*8.1*	*9.7*	*11.6*	*13.6*	*15.7*	*18.0*
	51	46	41	37	32	28	24	20	16	13	9	7	5	3	2
14	*1.2*	*1.6*	*2.0*	*2.4*	*3.0*	*3.7*	*4.4*	*5.3*	*6.4*	*7.7*	*9.3*	*11.2*	*13.3*	*15.4*	*17.8*
	51	46	41	37	32	28	24	20	16	12	9	7	5	3	2
15	*1.1*	*1.5*	*1.9*	*2.2*	*2.8*	*3.5*	*4.2*	*5.1*	*6.1*	*7.4*	*8.9*	*10.8*	*13.0*	*15.1*	*17.6*
	51	46	41	36	32	28	23	19	15	12	9	6	4	3	1
16	*1.1*	*1.4*	*1.8*	*2.1*	*2.6*	*3.2*	*4.0*	*4.8*	*5.9*	*7.0*	*8.5*	*10.5*	*12.6*	*14.8*	*17.4*
	51	46	41	36	32	27	23	19	15	12	9	6	4	2	1
17	*1.1*	*1.3*	*1.7*	*2.0*	*2.5*	*3.1*	*3.8*	*4.6*	*5.6*	*6.7*	*8.2*	*10.2*	*12.3*	*14.5*	*17.2*
	50	46	41	36	32	27	23	19	15	11	8	6	3	2	1
18	*1.0*	*1.2*	*1.6*	*1.9*	*2.4*	*3.0*	*3.6*	*4.4*	*5.4*	*6.5*	*7.9*	*9.9*	*12.0*	*14.4*	*17.1*
	50	46	41	36	32	27	23	19	15	11	8	5	3	2	1
19	*1.0*	*1.1*	*1.5*	*1.9*	*2.3*	*2.9*	*3.5*	*4.2*	*5.2*	*6.2*	*7.7*	*9.6*	*11.8*	*14.3*	*17.0*
	50	46	41	36	32	27	23	19	15	11	8	5	3	1	—
20	*0.9*	*1.1*	*1.4*	*1.8*	*2.2*	*2.8*	*3.4*	*4.1*	*5.0*	*6.0*	*7.5*	*9.3*	*11.5*	*14.1*	*16.9*
	50	46	41	36	32	27	23	18	15	11	8	5	3	1	—
25	*0.8*	*0.9*	*1.2*	*1.5*	*1.8*	*2.3*	*2.8*	*3.4*	*4.2*	*5.2*	*6.6*	*8.4*	*10.8*	*13.7*	*16.8*
	50	45	41	36	31	27	22	18	14	10	7	4	2	—	—
30	*0.7*	*0.8*	*1.0*	*1.2*	*1.6*	*2.0*	*2.4*	*2.9*	*3.7*	*4.6*	*5.9*	*7.7*	*10.2*	*13.4*	*16.7*
	50	45	41	36	31	26	22	17	13	9	6	3	1	—	—
40	*0.5*	*0.6*	*0.8*	*1.0*	*1.2*	*1.5*	*1.8*	*2.3*	*2.9*	*3.7*	*5.0*	*6.8*	*9.4*	*13.2*	*16.6*
	50	45	40	36	31	26	21	17	13	9	5	2	—	—	—
50	*0.3*	*0.5*	*0.6*	*0.8*	*0.9*	*1.2*	*1.5*	*2.0*	*2.5*	*3.2*	*4.3*	*6.1*	*9.3*	*13.1*	*16.6*
	50	45	40	36	31	26	21	17	12	8	4	1	—	—	—
75	*0.2*	*0.3*	*0.4*	*0.5*	*0.6*	*0.8*	*1.0*	*1.3*	*1.7*	*2.1*	*3.3*	*5.2*	*9.1*	*13.0*	*16.5*
	50	45	40	35	30	26	21	16	12	7	3	—	—	—	—
100	*0.2*	*0.2*	*0.3*	*0.4*	*0.5*	*0.7*	*0.8*	*1.0*	*1.2*	*1.6*	*2.8*	*5.1*	*9.0*	*13.0*	*16.5*
	50	45	40	35	30	26	21	16	11	7	3	—	—	—	—

* Prepared by Dale Jones, Michigan State University, East Lansing, Mich.

† Figure the work load at expected operator servicing pace and exclude deferrable "internal" duties (those which can be performed when all machines are producing): this gives the correct estimate of machine interference when there is little variation of the work loads of assigned products. Then, to estimate percent operator idle time, deduct from the respective estimated percent operator idle time the estimated percent of time to be spent on the deferrable internal duties. When there is great variation of work loads of assigned products, adjust the interference and operator idleness values as explained in the accompanying example.

can be estimated from Table 6-1 by direct interpolation between the nonitalicized values of 39 and 35 percent shown below the italicized interference values for 65 and 70 percent loads, the value being 37.7 percent for the 66.67 percent work load. Actually, this value would be 37.5 percent had the nonitalicized values been expressed to the nearest one one-hundredth percent. In other words, on an average, all four looms would be running together and the operator would therefore be idle 37.5 minutes per elapsed 100 minutes of the assignment.

MATHEMATICAL EVALUATION OF RANDOM MACHINE INTERFERENCE

As will be seen in the following example, it is necessary to estimate values of interference successively until a calculated value equals the respective estimated value. The principal advantage of the random machine interference table is that it eliminates these laborious calculations.

When the running and servicing of the assigned machines are independent and when each machine has the same random attention demand value, A, the probabilities of the various combinations of A demands can be predicted through use of the binomial distribution theorem. And, by multiplying the various probabilities of simultaneous A demands by the respective numbers of interference waits, then totaling and dividing by the number of assigned machines, n, the average percent interference per machine, i (decimally expressed), is established for the assumed A value.

Now let us calculate the average percent interference per machine for the assumed assignment of four machines, each having an individual attention work load, S, of $\frac{1}{6}$. The percent attention time, A, for machines tended in multiple, decimally expressed, is as follows:

$$A = S(1 - i) + i$$

where S and i are decimally expressed percentages as previously defined. In the assignment under consideration,

$$A = 0.1667(1 - 0.062) + 0.062 = 0.2184$$

As previously mentioned, calculation of i requires successive estimation of i for substitution in $A = S(1 - i) + i$, required for the calculations to be described, until a calculated i equals a respective assumed i. In this case, however, having already obtained $i = 6.2$ percent from Table 6-1, the value $i = 0.062$ is being assumed in solving for A. Thus the calculated i should equal this assumed $i = 0.062$.

Given $A =$ probability of a machine's requiring attention at any given moment $= 0.2184$, let $B =$ probability of a machine's not requiring attention at any given moment; $B = 1 - A = 0.7816$. Now, expanding $(A + B)^n$ or $(0.2184 + 0.7816)^4$ and factoring by interference waits, the average percent interference per machine can be evaluated as follows:

Possibility	Probability	Inter-ference waits	Inter-ference
4 machines requiring attention	1 $(0.2184)^4$	3	0.0068
3 machines requiring attention	4 $(0.2184)^3 (0.7816)$	2	0.0652
2 machines requiring attention	6 $(0.2184)^2 (0.7816)^2$	1	0.1750
1 machine requiring attention	4 $(0.2184) (0.7816)^3$	0	
0 machine requiring attention	1 $(0.7816)^4$	0	
		Total	0.247

Average interference per machine $= 0.247 \div 4 = 0.062$

Note the calculated 0.062 interference per machine agrees with the assumed value of 0.062 used in its calculation, which was obtained from Table 6-1. *No other value of i, when used in A, could give a like calculated value of i for the assumed assignment.* Thus the average percent of elapsed assignment time each of the four machines would experience interference has been established as being 6.2 percent.

THE RANDOM MACHINE INTERFERENCE EQUATION

The above calculated 6.2 percent average interference per machine could have been evaluated by a shorter method as follows (refer to the above calculations):

1. The probability that no machines will require attention at any given moment $= (1 - A)^n$.
2. The probability that one or more machines will require attention at any given moment $= 1 - (1 - A)^n$. This is the probability that the operator will work at any given moment.
3. The portion of elapsed time the operator will work on (service) each machine when the n machines are tended in multiple is therefore $[1 - (1 - A)^n]/n$; however, as previously mentioned, this also equals $S(1 - i)$. Thus,

$$S(1 - i) = \frac{1 - (1 - A)^n}{n}$$

but since $A = S(1 - i) + i$ as previously developed, and $S(1 - i)$ therefore equals $A - i$,

$$i = A - \frac{1 - (1 - A)^n}{n}$$

Substituting the previously established A and n values in this equation, the previously calculated $i = 6.2$ percent is verified.

$$i = 0.2184 - \frac{1 - (1 - 0.2184)^4}{4} = 0.062, \text{ or } 6.2 \text{ percent}$$

This equation was used in developing the interference versus work load curves required to prepare the italicized interference values of Table 6-1.

OPERATOR IDLENESS IN RANDOMLY SERVICED ASSIGNMENTS

If the total work load and average interference per machine in the assignment are known, it is easy to calculate the operator idleness inherent in randomly serviced assignments. The method used to establish the nonitalicized operator idleness values in Table 6-1 is shown below:

$$OIT = 100 - \Sigma S(1 - i)$$

where OIT = percent operator idle time, S = a machine's percent work load on an individual attention basis, and i = average percent interference per machine, decimally expressed. Thus, for the assumed assignment,

$$OIT = 100 - 4(16.67)(1.00 - 0.062) = 37.5 \text{ percent}$$

as previously interpolated from Table 6-1.

In reference to machine work load (percent operator servicing requirement), it is to be noted that it would be S if the machine were individually tended, in which case the machine would not incur interference idleness; but it would be $S(1 - i)$ if the machine were tended in multiple and the machine therefore incurred interference idleness. Of course, the operator's total work load when tending n machines would be $\Sigma S(1 - i)$. This explains why the operator can incur idle time in assignments having work loads ΣS greater than 100 percent on an individual attention basis.

RANDOM ASSIGNMENTS WHERE WORK LOADS DIFFER

The values in Table 6-1 assume each machine in any assignment under consideration has the same work load (S value). Frequently, however, work loads of assigned machines are significantly different, posing the question of validity of the Table 6-1 values when applied to such assignments. While employed at Georgia Tech, the writer developed a machine interference computer, permitting simulation of assignments of two to ten machines randomly tended by one operator, and electronic evaluation of resultant machine interference and operation idle time.[1] Results obtained from 100 simulated assignments involving various degrees of nonuniformity of machine servicing demand revealed that average interference per machine reduces, for any given total assigned work load for any given number of machines, as the degree of difference of the individual work loads of the assigned machines increases. Subsequent research has yielded adjustment factors which, when applied to the Table 6-1 values, provide close estimates of average interference per machine when different machines within any given assignment have significantly different work loads (S values).

From 100 experiments utilizing random number tables, the following equation was developed:

$$F = 1 - \frac{\sigma}{\sigma_m}$$

where F is a decimally expressed adjustment factor to be applied to the interference value obtained from Table 6-1, σ is the standard deviation of the work loads (S values) of the machines in the assignment under consideration, and σ_m is the maximum possible standard deviation of any assignment involving the number of machines and total work load under consideration.

Insight into and appreciation of the validity of this equation can be had by considering a hypothetical extreme assignment situation. Assume one operator is to service randomly four machines having a total of 66.7 percent work load as the previously assumed assignment had, but assume the individual S values are 0.1 percent for each of three of the machines and 66.4 percent for the fourth machine. Obviously there would be virtually no machine interference in this assignment, which entails virtually the maximum possible standard deviation of work loads for an assignment of four machines having 66.7 percent total work load. Thus, for these assumed conditions, σ is virtually the same as σ_m, and when substituted in $F = 1 - (\sigma/\sigma_m)$, gives $F =$ virtually 0, meaning the Table 6-1 value of 6.2 percent interference for four machines having 66.7 percent total work load should be factored by virtually 0 to give virtually 0 interference per machine.

In the previously assumed assignment, each of the four machines has the *same* work load, 16.67 percent. Thus the σ of the assigned work loads is 0. Accordingly, substituting $\sigma = 0$ in $F = 1 - (\sigma/\sigma_m)$ gives $F = 1$, which calls for no adjustment of the 6.2 percent table value.

THE MACHINE INTERFERENCE ADJUSTMENT CHART

The machine interference adjustment chart shown in Figure 6-1 permits rapid determination of F factors to be applied to the Table 6-1 values. Let us now use this chart for an assignment where the machines have significantly different work loads (S values) when, as before, the servicing demands of the machines are random.

A six-machine assignment will be used as the case example. The time values, which have been slightly adjusted to simplify the calculations, are based on stopwatch time studies and frequency studies. External servicing duties are those which must be performed while the machine is nonproductive. Internal servicing duties are those which should be performed while the machine is producing. The normal

---

[1] Dale Jones, "Mathematical and Experimental Calculation of Machine Interference Time," *The Research Engineer*, January, 1949.

time values reflect the conventional low-task 100 percent or "fair day's work" speed, and the expected times assume the operator will work at the 125 percent brisk pace when performing the servicing duties. Estimated average walking requirements have been included in establishing the servicing times. All the time values represent time (minutes) per 1,000 pieces. The assignment is as follows:

Product	Normal servicing time		Expected servicing time		Automatic time (R)	Expected work load,* $S = \dfrac{Sd + Sr}{Sd + R}$
	External	Internal	External (Sd)	Internal (Sr)		
A	6.0	3.0	4.8	2.4	34.8	18.2
A	6.0	3.0	4.8	2.4	34.8	18.2
B	7.0	2.0	5.6	1.6	41.4	15.3
C	5.0	3.0	4.0	2.4	31.0	18.3
D	6.0	2.0	4.8	1.6	27.2	20.0
E	4.0	3.0	3.2	2.4	24.8	20.0
					Total load =	110.0

* Each of these values represents the percent of elapsed assignment time the operator would be engaged in servicing each of the respective machines if such machines were individually tended, that is, if there were no machine interference and if the operator worked at 125 percent pace.

To estimate the average percent of elapsed time the machines will be idle because of machine interference, and the operator idle because of work load limitations:

1. Determine the σ of the work loads (S values): $\sigma = \sqrt{\Sigma(\overline{S} - S)^2/n}$. For this assignment, $\sigma = 1.6$ percent.
2. Using the machine interference adjustment chart (Figure 6-1), locate 110 percent on the left vertical work load scale. Project horizontally to the right to intersect the six-machine curve. At the point of intersection, project vertically downward to the σ_m scale. Align this point with the calculated $\sigma = 1.6$ on the lower horizontal σ scale. Read $F = 0.97$ at the point of intersection of the F scale.
3. Using Table 6-1, obtain 17.6 percent average interference per machine for the six-machine, 110 percent work load assignment.
4. Adjust the 17.6 percent table value by multiplying by the obtained $F = 0.97$ to obtain an adjusted 0.97 (17.6), or 17.1 percent average interference per machine.
5. The operator idle time values in Figure 6-1 assume an F factor of unity. In this case, however, $F = 0.97$, which gives an adjusted interference value of 17.1 percent. Thus, for these conditions, operator idle time $OIT = 100 - 110(1 - 0.171) = 9$ percent, which, because of the large F factor, does not differ from the 9 percent operator idle time value in Table 6-1 which assumes $F = 1.0$.

Internal duties which are deferrable (no deferrable internal duties were considered in the above example) and which may therefore be performed when all machines are producing, at no expense of machine interference, are not to be considered in deriving the operator's expected work load, for purposes of estimating machine interference (italicized values in Table 6-1). However, when such duties exist, the tabulated (nonitalicized) percent operator idle time should be adjusted as follows:

1. When the factor from Figure 6-1 is 0.9 to unity, estimate operator idle time by subtracting the estimated percent of time to be spent on deferrable

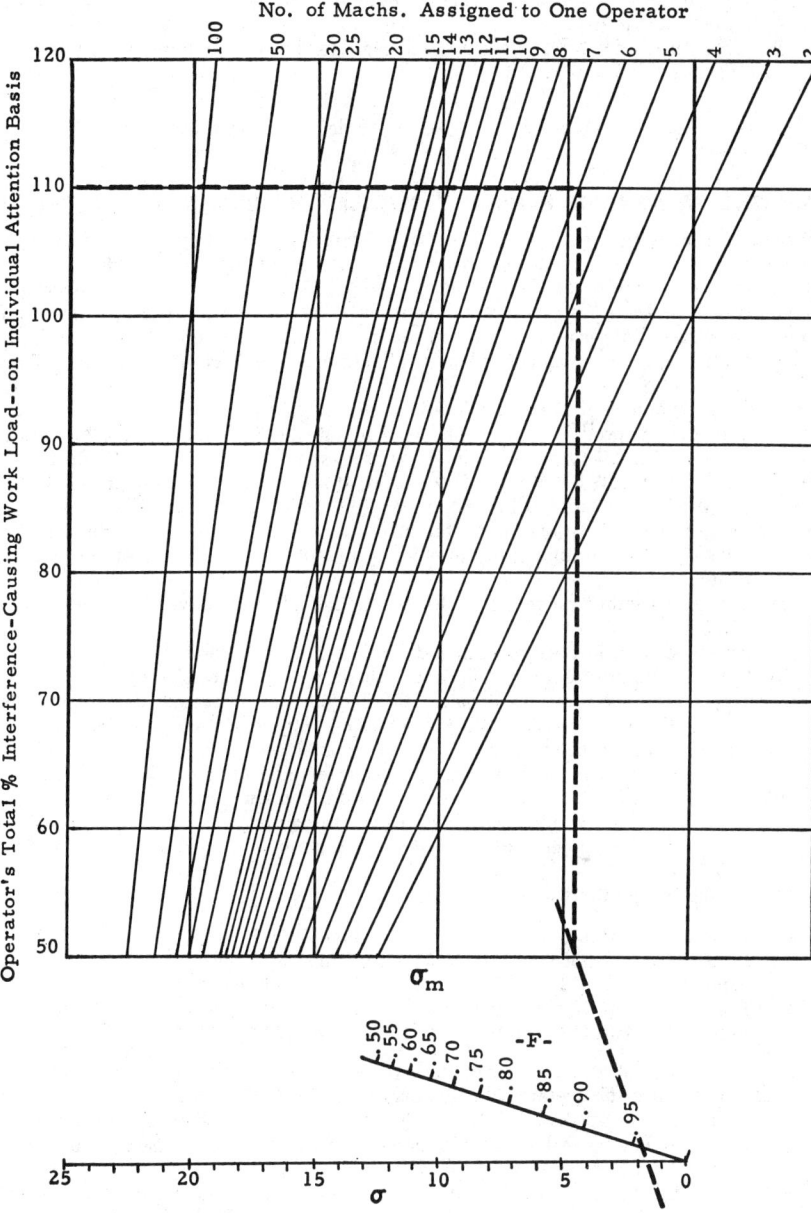

FIG. 6-1. Machine interference adjustment chart.

internal duties from the tabulated nonitalicized percent operator idle time value in Table 6-1.

2. When the factor is less than 0.9, calculate percent operator idle time (percent OIT) as follows:

$$\% \ OIT = 100 - \Sigma \ S \ \frac{100 - \%i}{100} - \% \ \text{deferrable internal duties}$$

INCENTIVE PAYMENT OF RANDOMLY SERVICED ASSIGNMENTS

When the operator total work necessarily varies considerably from assignment to assignment, incentive payment based on production standards only, with separate predetermined operator idleness allowance payments for specific assignments, is recommended. When, however, assignments can be planned so that operator total work load varies little from assignment to assignment, production time standards with included operator average idleness allowance should be used, as described below.

For the assignment under consideration, the servicing time standard in hours per 1,000 pieces for product A would be (6.0 minutes + 3.0 minutes) ÷ 60 minutes, or 0.150 hour per 1,000 pieces. Similarly, the standards for products B, C, D, and E would be 0.150 hour per 1,000 pieces, 0.133 hour per 1,000 pieces, 0.133 hour per 1,000 pieces, and 0.117 hour per 1,000 pieces, respectively. Now, the method of adjusting these standards to include expected average operator idleness is as follows (personal, fatigue, and minor unavoidable delay allowances are ignored at this point):

1. Expected average percent operator productive effort on servicing duties = 125 percent.
2. Expected percent operator unavoidable idle time = 9 percent.
3. Correct operator earning efficiency when working at 125 percent for 91 percent of the time and paid 100 percent for expected 9 percent idle time = 0.91(125) + 0.09(100) = 113.75 + 9.00 = 122.75 percent.
4. Operator earning efficiency if incentive standards are based on servicing time only and if no payment is made for idle time = 113.75 percent.
5. Factor which should be applied to normal servicing times to provide time standards which would yield 122.75 percent earning efficiency when operator works at 125 percent and is *not* paid separately for idle time = 122.75/113.75 = 1.079.
6. Correct single time standards:

Product A = 0.150 × 1.079 = 0.1619 hour/1,000 pieces
Product B = 0.150 × 1.079 = 0.1619 hour/1,000 pieces
Product C = 0.133 × 1.079 = 0.1435 hour/1,000 pieces
Product D = 0.133 × 1.079 = 0.1435 hour/1,000 pieces
Product E = 0.117 × 1.079 = 0.1262 hour/1,000 pieces

The correctness of the foregoing method of deriving production time standards including allowance for operator idleness may be verified as follows:

1. Regardless of the combination of products assigned, when the operator has a 91 percent work load at 125 percent working pace, his earning efficiency should be 122.8 percent.
2. Assuming the operator produces only product A in the six-machine assignment, making 18.2 percent × 6, or 109.2 percent work load on an individual attention basis, giving 91 percent true work load when tended in multiple because the operator would have 9 percent idle time (see Table 6-1), his earning efficiency would be his paid time per 1,000 pieces divided by his actual work time per 1,000 pieces, or

$$\frac{0.1619 \ \text{hour}/1,000 \ \text{pieces} \times 91\% \ \text{of time worked}}{0.150 \ \text{hour}/1,000 \ \text{pieces at normal } 125\% \ \text{work pace}} = 122.8\%$$

The operator idleness allowance factors which should be applied to normal servicing times in deriving production time standards, for various expected operator productive efforts and various total assigned work loads, to yield 100 percent pay rate for operator idleness, are given in Table 6-2. Similar tables may be developed to yield 105, 110 percent, and so on, pay rates for operator idleness by substituting 105, 110, and so on, for 100 in making the calculations shown in step 3 above. However, it is to be kept in mind that payment of more than 100 percent base pay rate for operator idleness causes overpayment and increases the earnings difference between incentive paid and daywork paid operators. Also remember that payment of different percents of base pay for operator idleness to machine operators and manual operators causes pay inequities between these two groups.

TABLE 6-2.   Operator Idleness Allowance Factors Yielding 100 Percent Pay
Rate for Operator Idle Time

Expected average operator production effort (rating)	Expected percent operator idle time (see Fig. 6-1 and Table 6-1)									
	5	10	15	20	25	30	35	40	45	50
110	1.048	1.101	1.161	1.227	1.303	1.390	1.490	1.606	1.745	1.909
115	1.046	1.097	1.153	1.217	1.290	1.373	1.468	1.579	1.711	1.870
120	1.044	1.093	1.147	1.208	1.278	1.357	1.449	1.556	1.682	1.833
125	1.042	1.089	1.142	1.200	1.267	1.343	1.432	1.533	1.655	1.800
130	1.041	1.085	1.136	1.193	1.257	1.330	1.414	1.513	1.629	1.770
135	1.039	1.082	1.131	1.185	1.247	1.317	1.399	1.494	1.607	1.741

DEFERRABLE INTERNAL DUTIES

As stated in connection with the six-machine assignment under consideration, internal duties which are deferrable and which may therefore be performed when all assigned machines are producing, at no expense of machine interference, should not be considered a part of the operator's work load in estimating machine interference. However, the estimated percent of time to be spent on such duties should be subtracted from the tabulated (nonitalicized) percent operator idle time values in Table 6-1, in evaluating operator idleness for incentive payment purposes. The normal time required for performance of these duties should be added to the regular interference-causing normal external and internal servicing times when setting a product's time standard. To illustrate, for the assignment under consideration, assume deferrable internal duties on product A require 1.0 minute normal time/1,000 pieces (these duties being in addition to those requiring the stated 6.0-minute nondeferrable external servicing and 3.0-minute nondeferrable internal servicing) and assume the operators work at approximately 125 percent incentive pace, giving 0.8 minute expected time/1,000 pieces of product A for deferrable internal duties. Assume none of the other assigned products requires deferrable internal duties. The percent work load (on an individual attention basis) for these deferrable internal duties would be $0.8 \div (S_d + R)$ or $0.8 \div (39.6)$, or 2 percent on two of the six assigned machines, totaling 4 percent. When the machines are tended in multiple, this 4 percent reduces to 4 percent [(100 percent − 17.1 percent interference)/100] or 3.3 percent. Subtracting this value from the previously estimated 9 percent operator idleness gives 5.7 percent operator idleness upon which the operator idleness allowance factor (see Table 6-2) should be based. Also, the previously established 0.150 hour/1,000 pieces of A (which did not consider the deferrable internal duties requiring 1.0 minute normal time/1,000 pieces of A) should be modified as follows: normal minutes/1,000 = 6.0 external + 3.0 internal

$+ 1.0$ deferrable internal $= 10.0$ minutes. Thus, standard hours/1,000 $= {}^{10}\!/_{60} = 0.167$.

PERSONAL, FATIGUE, AND DELAY ALLOWANCES

The purpose of personal, fatigue, and delay (PFD) allowance is to compensate the operator for nonproductive time caused by personal requirements, on-the-job relax, and minor unavoidable delays so small in duration that it is not economical to account and pay for them separately from the incentive standards. If it is estimated that 10 percent of elapsed shift time should be taken for PFD on a manual operation having 1.0 minute estimated normal time per piece, the time standard should be $[60 - 10$ percent $(60)] \div 1.0$, or 54 pieces per hour. And if the operator experiences the 10 percent PFD and works at 125 percent efficiency on correct time standards, he will produce 540 pieces during the 432 working minutes of a 480-minute shift, giving 125 percent earning efficiency for the shift.

In mechanized work such as multimachine assignments, little assignment time is lost to PFD because of the automaticity of the machines and the use of relief operators which are justified by the high overhead expense of the assignment. Under such conditions, little if any PFD allowance should be applied when setting the time standards. On the other hand, if multimachine assignment operators are expected to leave the assignment, without being relieved, for personal and fatigue recuperation requirements, the percent of shift time warranted for this should be estimated and represented in the time standards. To illustrate, in reference to the assignment under consideration, assume the operator is allowed to leave the assignment for 12 minutes recess during the first and the second half of the 8-hour shift (5 percent of the shift). The correct time standard for product A would be 0.150 hour/(100 percent $-$ 5 percent), or 0.158 hour/1,000 pieces, the same reasoning applying to the other products produced in multiple. And as in the case of manual operations, when PFD allowances are thus applied in setting the time standards, the operator will be fairly paid in proportion to his productive effort, assuming, of course, that he actually experiences the nonproductive time reflected in the applied PFD allowance.

If, as assumed for product A above, a significant degree of PFD allowance is applied when setting the time standard, the allowance should be taken into consideration when establishing the product's work load, S. Again, assuming the 5 percent PFD in behalf of product A, its expected true work load S (previously established at 18.2 percent, ignoring PFD) would be developed as follows:

$$S = \frac{S_d/0.95 + S_r/0.95}{S_d/0.95 + R} = \frac{4.8/0.95 + 2.4/0.95}{4.8/0.95 + 34.8} = 19.0\%$$

Of course the product's work load including the PFD allowance (19.0 percent for product A) should be used in planning the assignment and incentive-paying the operator.

MACHINE EFFICIENCY IN RANDOMLY SERVICED ASSIGNMENTS

Production planning and scheduling of multimachine assignments require prediction of machine efficiencies during expected 8-hour shift assignments, with 100 percent efficiency regarded as complete automatic production by the machine. With this criterion in mind, a given machine's percent efficiency, E, in any given randomly serviced assignment may be stated as follows:

$$\%E = 100 - \frac{S_d}{S_d + R}\left(\frac{100 - i}{100}\right) - i$$

where S_d, S_r, and R are the previously defined servicing and automatic times per 1,000 pieces, 100 pieces, and so on, for the product under consideration (see previous case example), and i is the percent of elapsed assignment time that machine interference occurs when the machines are tended in multiple.

In evaluating i, remember the necessity of considering interference-causing servicing duties only, and the necessity of factoring the Table 6-1 values by F factors derived from Figure 6-1, in assignments where the work loads (S values) of individual machines are significantly different. Also remember that the obtained i value represents the average interference per machine in the assignment; the interferences of individual machines in the assignment may differ significantly from the average interference per assigned machine, the magnitude being dependent on differences of machine work loads (S values) and the duration of the assignment (the sample size). Of course, the i values are predicated, fundamentally, on the expected servicing and automatic times of the assigned machines (products). Accordingly, the validity of the i values depends on the degree to which the servicing and automatic times fulfill the expectancy upon which they are predicated.

SYNCHRONIZED MULTIMACHINE ASSIGNMENTS

The production cycles of most machines are regular rather than random. Typically, such machines are idle while being unloaded, loaded, and started. Then they produce automatically for predetermined fixed periods of time during which the operator's attention may or may not be required. When the operator's attention is not required during a machine's automatic producing time and when such time is proportionately great in relation to its required unloading and loading time, it is common practice to assign the operator several machines. These machines may be alike or different, producing the same or different products. When, however, the cycles of the machines are regular (not random) in regard to required operator servicing duties and when respective time requirements are of little variation from cycle to cycle, it is possible to arrange for a synchronized battery cycle involving several machines. The durations of operator servicing requirements and automatic running times can be quite different among machines in the battery, yet a synchronized operator servicing pattern is possible if the individual machine cycles are regular.

Given a means of predetermining machine interference and operator idleness in synchronized multimachine assignment alternatives, the engineer can determine which of several assignment possibilities is optimum. Then, having established the assignment, he can arrange for equitable incentive payment of the operator. This, of course, assumes that the most economical cycle has been established and specified for each product under consideration and that the respective standard cycle times are known.

The material to be presented will first treat the problems of predetermining machine interference and operator idleness in synchronized assignments. Then the recommended work load planning procedure will be discussed. Finally the recommended incentive payment method will be presented. An understanding of the following material will be facilitated by using some examples of compression molding assignments. First, assignments in which each press experiences one cycle per battery cycle will be considered. Then, assignments in which two or more presses experience two or more cycles per battery cycle will be considered. The principles treated in these examples apply to all multimachine assignments warranting synchronized servicing.

CONVENIENT SYMBOLOGY

The machine cycle of a compression mold press having automatically controlled close (or molding) time is illustrated in Figure 6-2. The width of the diagram and its parts should conform to a conveniently selected horizontal time scale. The crosshatched parts of the diagram represent operator work time. The space between the right-angle bends at each end of the heavy line denotes automatic machining (or processing) time; the duration of this automatic time is generally fixed by machine speed, feed, settings, and the like, and is usually not controllable by the operator of the machine. The crosshatched work time not lying between the right-angle bends represents work performed by the machine operator while the

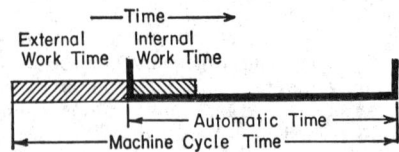

Fig. 6-2.   Graphic description of a machine cycle.

machine is nonproductive (it may be running but not producing). This type of work is conventionally called "external work," that is, work performed external to the automatic producing time. The crosshatched work time lying between the right-angle bends of any given heavy line (automatic time) represents work performed by the machine operator during the machine automatic producing time. This type of work is conventionally referred to as "internal work," that is, work performed in (during) automatic time. It is desirable, although not always possible, for the operator to perform a machine's internal work requirements immediately after the beginning of that machine's automatic production time. Frequently, however, some internal duties must be performed just prior to the completion of the machine's automatic production time.

It may or may not be necessary for the operator to observe closely the cutting, winding, or other processing requiring the automatic time. When attention is necessary, time spent on it should be classified as regular internal work. Also, work performed internally during a given machine's automatic time need not necessarily be performed at that machine; during the automatic time the operator may perform some assigned operation or inspection involving another product. Of course, the same could apply to external work time. However, because of the importance of achieving maximum machine efficiency, it is seldom that foreign operations or inspections which could be performed only as external work would be assigned to the machine operator.

SIMPLIFICATION OF MACHINE CYCLES

As for manual operations, the most efficient cycles of mechanized operations should be determined before setting the time standards. The following four ordered steps should be taken to minimize the machine cycle time for each product to be produced in single or synchronized multimachine assignments:
1. Reduce the automatic time.
2. Eliminate unnecessary external servicing duties.
3. Internalize external duties as much as possible.
4. Simplify remaining external duties.

Note that nothing is said about internal duties in the four steps above, because they do not affect the machine cycle time. However, the product's operator work load can be minimized by eliminating unnecessary internal duties and simplifying the necessary ones. Because the product's machine cycle time as well as its operator work load governs the economics of its production in synchronized multimachine assignments, both determinants should be minimized. In the examples to follow, it is assumed this recommended work simplification has been carried out for each product under consideration.

BATTERY CYCLE TIME WHEN EACH MACHINE IS SERVICED ONCE PER BATTERY CYCLE

A typical three-press compression molding assignment is graphically illustrated in Figure 6-3. During the external work time of any given mold press, the operator unloads, cleans, loads, and closes the mold. Then the mold remains closed for a predetermined automatically controlled period of time, after which it automatically opens. During the mold-close time of any given mold press, the operator may

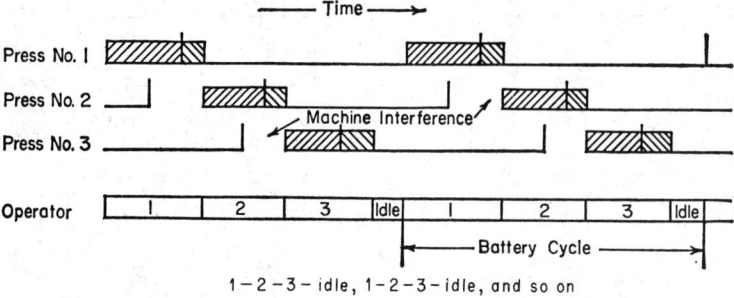

FIG. 6-3. Typical three-machine assignment where the operator is underassigned and where each machine is serviced once per battery cycle.

| Machine (press) | Graphed time data | |
	Expected machine cycle time	Expected service time/machine cycle
1	2.4	1.2
2	0.9	0.5
2	0.9	0.5
Total service time/battery cycle.......		2.2 minutes

perform various required internal duties such as breaking away flash, unloading and loading the cooling form, and placing molded pieces in the box. After this, he starts unloading and cleaning the next open mold in accordance with a planned servicing pattern. In Figure 6-3, the operator first services press No. 1. He then services press No. 2 and press No. 3, after which he must wait for completion of the automatic molding time of press No. 1. He then repeats the battery cycle, which may be described as 1-2-3-idle, 1-2-3-idle, and so on. The battery cycle time of this assignment is apparently 2.8 minutes, the longest of the individual machine cycle times (that of press No. 1).

Assume, now, that a new mold is set up in press No. 3. The machine cycle time of press No. 3 is now 2.6 minutes. The new assignment is graphically illustrated in Figure 6-4.

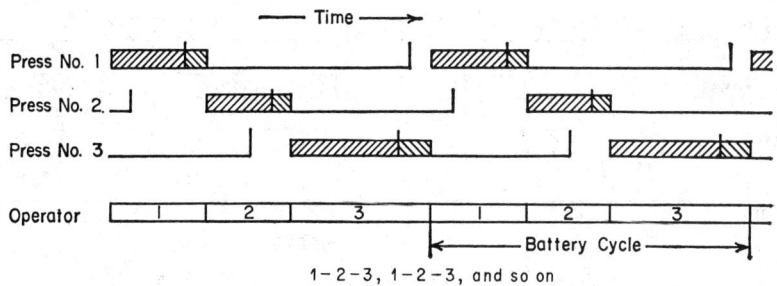

FIG. 6-4. Typical three-machine assignment where the operator is overassigned and where each machine is serviced once per battery cycle.

Machine (press)	Graphed time data	
	Expected machine cycle time	Expected service time/machine cycle
1	2.8	0.9
2	2.3	0.8
3	2.6	1.3
Total service time/battery cycle........		3.0 minutes

Note in Figure 6-4 that the operator services the mold presses in the 1-2-3 pattern as before. However, unlike the assignment represented by Figure 6-3, the battery cycle time of this assignment is not the longest machine cycle time but is the sum of the servicing times of the three assigned mold presses (0.9 + 0.8 + 1.3, or 3.0 minutes).

In reference to Figures 6-3 and 6-4, battery cycle time in synchronized assignments where each machine is serviced but once per battery cycle is the longest of the individual machine cycle times, or the total of the required operator servicing times per battery cycle, whichever is the greater.

BATTERY CYCLE TIME WHEN ONE OR MORE MACHINES ARE SERVICED MORE THAN ONCE PER BATTERY CYCLE

Figure 6-5 illustrates an assignment of three mold presses in which one press (No. 2) is serviced twice per battery cycle. The battery cycle time of this assignment, however, is 2.4 minutes, which is greater than the total of the operator servicing times per battery cycle (2.2 minutes) and greater than the greatest machine cycle time per battery cycle (2.2 minutes). It follows, therefore, that the rule for determining battery cycle time of assignments where each machine is serviced once per battery cycle should not be used in evaluating battery cycle time of assignments in which one or more machines are serviced more than once per battery cycle.

Battery cycle time for assignments where one or more machines are serviced more than once per battery cycle is determined by adding total operator servicing time per battery cycle to total operator idle time per battery cycle.

Total operator servicing time is easy to determine because it is simply the total of the individual machine servicing times in one battery cycle, including each

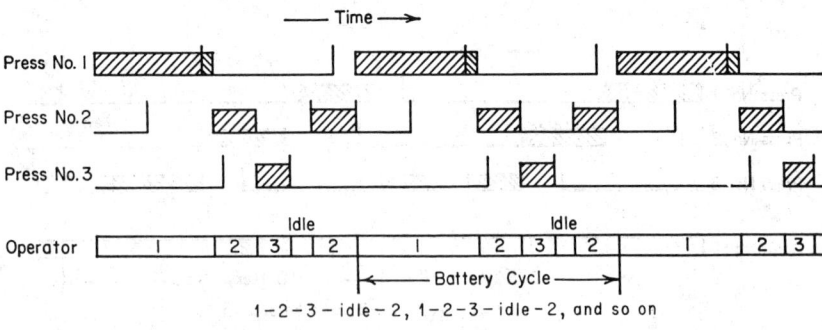

Fig. 6-5. Typical three-machine assignment where one machine (press No. 2) is serviced twice per battery cycle.

	Graphed time data	
Machine (press)	Expected machine cycle time	Expected service time/machine cycle
1	2.2	1.1
2	0.9	0.4
3	2.1	0.3
2	0.9	0.4
Total service time/battery cycle........		2.2 minutes

machine servicing time once for each time the individual machine is used during the battery cycle.

Total operator idle time is somewhat more difficult to determine but can be calculated as follows:

Step 1. Make a list of the machines included in the battery showing machine identification, expected machine cycle time, and expected service time per machine cycle. (Remember, "machine cycle time" is external service time plus machine running time, and "service time" is both external and internal service time.) Leave room at the right of the list for recording operator idle time. The list is tabulated as shown at the top of this page.

Step 2. Cross out all machine cycle times of machines serviced *once* per battery cycle which are equal to or less than the total operator servicing time per battery cycle, because these machines cannot cause any inherent operator idle time. See illustration below.

Step 3. Determine operator idle time for each remaining machine cycle occurrence in the battery cycle. Record this idle time in the "Operator idle time" column as shown below. When the machine cycle time for an individual machine is greater than the total service time for all machines, the idle time recorded in the "Operator idle time" column is the difference between these two figures (machine cycle time minus total service time). For the remaining machine cycle occurrences (which are for machines serviced more than once per battery cycle), the operator idle time to be recorded is the difference between the machine cycle time for a specific machine and the sum of the service times for that machine and subsequent machines before that machine is serviced again.

If only one operator idle time is found in this step, this idle time is the total operator idle time per battery cycle. (If more than one idle time is found in this step, step 4 must be performed to find total operator idle time.) The following tabulation derived from Figure 6-5 shows an operator idle time calculation that is complete at the end of step 3.

Machine	Cycle time	Service time	Operator idle time	Calculated operator idle time
1	~~2.2~~	1.1		
2	0.9	0.4	0.2	$0.9 - (0.4 + 0.3) = 0.2$
3	~~2.1~~	0.3		
2	0.9	0.4		$0.9 - (0.4 + 1.1) = -0.6$, or 0
Total........	...	2.2	0.2	

Battery cycle time $= 2.2 + 0.2 = 2.4$ minutes.

Step 4. Evaluate operator idle times determined in step 3 to see how they should be combined to give total operator idle time. The operator idle times are evaluated to determine whether they are concurrent or not. Any overlapping idle time should, of course, be allowed only once when determining total operator idle time. The following tabulation is similar to the one above, but it requires evaluation of two idle times. See Figure 6-6.

Machine	Cycle time	Service time	Operator idle time	Adjusted operator idle time	Calculated operator idle time
1	2.4	1.2	0.2	0	2.4 − 2.2 = 0.2
2	0.9	0.5	0.4	0.4	0.9 − 0.5 = 0.4
2	0.9	0.5			0.9 − (0.5 + 1.2) = −0.8, or 0
Total	...	2.2	...	0.4	

Battery cycle time = 2.2 + 0.4 = 2.6 minutes.

In determining the adjusted operator idle time for the above example, the 0.2-minute idle time originally calculated for machine No. 1 was canceled out because it was concurrent with the idle time for machine No. 2. In other words, when the 0.2-minute operator idleness due to machine No. 1 was calculated, the fact that the operator would be idle 0.4 minute because of machine No. 2 was not taken into consideration. Thus the calculated 0.2-minute operator idleness due to machine No. 1 would not actually exist and should therefore be canceled out.

MACHINE INTERFERENCE IN SYNCHRONIZED ASSIGNMENTS

Knowing the battery cycle time of an assignment, it is a simple matter to evaluate the interference any given machine experiences per battery cycle. This is done by subtracting the machine's total cycle time per battery cycle from the battery

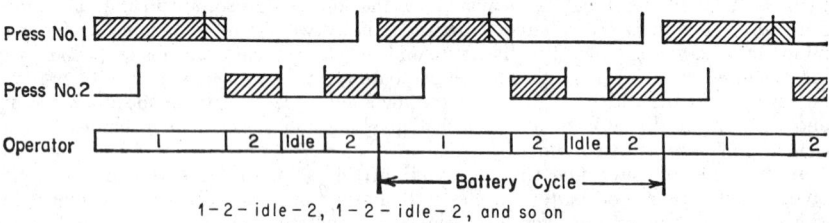

1 − 2 − idle − 2, 1 − 2 − idle − 2, and so on

Fig. 6-6. Typical two-machine assignment where one machine (press No. 2) is serviced twice per battery cycle.

Machine (press)	Graphed time data	
	Expected machine cycle time	Expected service time/machine cycle
1	2.4	1.2
2	0.9	0.5
2	0.9	0.5
Total service time/battery cycle........	2.2 minutes	

cycle time. Then the portion of elapsed assignment time the machine incurs interference can be determined by dividing its interference time per battery cycle by
the battery cycle time. To illustrate, referring to Figure 6-3, the battery cycle
time is 2.8 minutes, and press No. 2, which has 2.3 minutes machine cycle time
per battery cycle, therefore experiences 0.5 minute interference per battery cycle.
This is 0.5 ÷ 2.8 or 17.9 percent of elapsed assignment time. In reference to
Figure 6-5, having 2.4 minutes battery cycle time, press No. 2, which has 0.9 + 0.9,
or 1.8 minute total cycle time per battery cycle, experiences 0.6 minute interference
per battery cycle. This is 0.6 ÷ 2.4, or 25.0 percent of elapsed assignment time.

OPERATOR IDLENESS IN SYNCHRONIZED ASSIGNMENTS

As in the case of machine interference, after the battery cycle time of a possible
assignment is determined, it is a simple matter to evaluate the degree of inherent
operator idle time. This is done by subtracting from the battery cycle time the
total operator servicing time therein. Then the portion of elapsed assignment time
the operator is unavoidably idle is determined by dividing the operator idle time
per battery cycle by the battery cycle time. To illustrate, referring again to Figure
6-3, the battery cycle time is 2.8 minutes and the total operator servicing time
per battery cycle is 2.5 minutes. Thus the operator is unavoidably idle 0.3 minute
per battery cycle, or 0.3 ÷ 2.8, or 10.7 percent of elapsed assignment time. In
reference to Figure 6-5, having 2.4 minutes battery cycle time, the total operator
servicing time per battery cycle is 2.2 minutes. Thus the operator is unavoidably
idle 0.2 minute per battery cycle. This is 0.2 ÷ 2.4, or 8.3 percent of elapsed
assignment time.

MACHINE EFFICIENCY IN SYNCHRONIZED ASSIGNMENTS

Machine efficiency is apparently unity minus the portion of elapsed time the
machine is nonproductive. Machine efficiency can be expressed in many ways. As
previously stated, production planning and scheduling of multimachine assignments
require prediction of machine efficiencies during expected 8-hour shift assignments,
with 100 percent efficiency regarded as complete automatic production by the machines. With this criterion of efficiency in mind, a given machine's percent efficiency,
E, in a synchronized assignment may be stated as follows, where R is its total
automatic producing time per battery cycle (remember, in some assignments a
machine experiences more than one machine cycle per battery cycle) and where
B is the battery cycle time:

$$\%E = \frac{100R}{B}$$

PLANNING WORK LOADS OF SYNCHRONIZED ASSIGNMENTS

As in the case of randomly serviced multimachine assignments, the objective
of work load planning in synchronized assignments is to minimize machine interference and operator idleness expenses. Unlike randomly serviced assignments, however, it is possible under ideal conditions to arrange for synchronized assignments
having zero machine interference and operator idleness. This should be the goal
of the work load planner.

Zero machine interference and operator idleness would result in a synchronized
assignment where each product has the same machine cycle time per battery cycle
(some machines might be serviced more than once per battery cycle) and where
the total operator servicing time per battery cycle is equal to this common machine
cycle time per battery cycle. For example, products A, B, and C might each
have 4.0 minutes machine cycle time and a total of 4.0 minutes operator servicing
time per battery cycle. Assuming each machine is serviced once per battery cycle,
there would be no machine interference or operator idleness in this assignment.

For assignments where one or more machines are serviced more than once per battery cycle, it is also possible to have no machine interference or operator idleness. In practice, the planner of synchronized multimachine assignments has the problem of determining what products to add to assignments where production requirements on one or more machines in previously planned assignments have been (or soon are to be) completed. For example, in reference to Figure 6-3, assume the production has just been completed on press No. 1, leaving a remaining assignment of only press No. 2 and press No. 3, with machine cycle times of 2.3 and 2.4 minutes, respectively, and a total of 1.6 minutes expected servicing time per battery cycle for these two presses. Here it can be seen that the ideal product to assign to the now idle press No. 1 is one having 2.4 minutes machine cycle time and 2.4 − 1.6, or 0.8 minute, expected servicing time per machine cycle. This would result in a 2.4-minute battery cycle for the three-press assignment, giving zero operator idleness because the expected servicing time per battery cycle would also total 2.4 minutes, zero interference for press No. 1 and press No. 3, and only 0.1 minute interference per battery cycle for press No. 2.

Of course, the ideal, most compatible products, as just discussed, are seldom immediately available for assignment on available machines. Also, delivery promises, lengths of runs, locations of machines, and so on, are practical limitations to the achievement of the theoretical ideal. Thus the planner of synchronized assignments has the job of optimizing assignments, in the light of both the theoretical considerations of machine interference and operator idleness expense, and practical limitations opposed to the theoretical ideal. As in the case of randomly serviced assignments, however, it is economically very important that the planner constantly work toward the theoretical ideal, giving due consideration to the less tangible, practical considerations—rather than ignoring the theoretical in favor of unscientific rule-of-thumb planning governed only by the intangibles. Moreover, as in the case of randomly serviced assignments, refined and effective planning of synchronized assignments requires that the total work force on such assignments be varied in accordance with the resultant total effects of the work load optimizing procedure discussed. And, as previously mentioned, this does not mean that excess operators should be laid off as required or that needed operators should be hired as required; there should be provision for transfer of operators among different activities, planned production of deferrable standard inventoried items, and the like, as a means of stabilizing the total work force.

EVALUATING OPERATOR IDLENESS IN SYNCHRONIZED ASSIGNMENTS

The production time standards of three products to be produced in a typical three-press phenolic compression molding assignment, in a 1-2-3, 1-2-3 sequence, will now be derived, after which the procedure for estimating operator idleness will be presented.

Product	Standard servicing time/machine cycle			No. pieces/ cycle N	Standard service hours/ 100 pieces = 100 (total service) ÷ N ÷ 60
	External	Internal	Total service		
Toaster base.....	0.66	0.25	0.91	1	1.517
Desk pen holder..	0.50	0	0.50	1	0.833
Radio case.......	0.44	0.25	0.69	1	1.150
			Total = 2.10		

The average productive effort of incentive-paid manual operators in the plant under consideration is 125 percent. On the assumption that servicing duties of the assignment will be performed with this degree of productive effort, the expected servicing and machine cycle times for the three products under consideration are:

Product	Expected service time/machine cycle			Automatic time per machine cycle R	Expected cycle time = (external ÷ 1.25) + R
	External ÷ 1.25	Internal ÷ 1.25	Total		
Toaster base.......	0.53	0.20	0.73	1.47	2.00
Desk pen holder....	0.40	0	0.40	1.20	1.60
Radio case.........	0.35	0.20	0.55	1.40	1.75

As previously mentioned, these three products are to be molded via a 1-2-3 servicing sequence. Accordingly, the operator idle time per battery cycle is evaluated as follows:

Machine	Product	Expected cycle time	Expected servicing time
1	Toaster base	2.00	0.73
2	Desk pen holder	1.60	0.40
3	Radio case	1.75	0.55
			Total = 1.68

The expected operator idle time per battery cycle is the 1.68-minute total operator servicing time per battery cycle subtracted from the 2.00-minute total battery cycle time, or 0.32 minute.

SIGNIFICANCE OF KEY CYCLES

Because of the influence of key cycles on incentive-earning opportunity, it is important that the person responsible for developing and maintaining the assignments knows which products entail key cycles in existing and pending assignments. Then, in the interest of production economy and incentive-earning equity, he and the supervisor should strive to minimize key cycle times through job design as previously discussed. This does not mean that job design should be deferred until assignments are planned; such practice would not only hinder the planner in his efforts to minimize machine interference by combining into assignments products with compatible machine cycle times but would also rush, at an expense of quality, the job design function which must precede the setting of sound incentive standards. However, it does mean that the planner and/or the foreman should closely check the automatic time and external work content of key cycle products as they go into assignments, for the purpose of:

1. Ensuring that the shortest possible automatic time is incurred by the key cycle product
2. Ensuring that the operator is performing the key cycle external work in the most efficient manner
3. Ensuring that the operator is not externally performing key cycle internal duties at the expense of battery cycle time

PERSONAL, FATIGUE, AND DELAY ALLOWANCES IN SYNCHRONIZED ASSIGNMENTS

Like products produced in randomly serviced assignments, those produced in synchronized assignments sometimes warrant a significant degree of PFD allowance. For the sake of brevity, the reader is referred to the earlier discussion, under Personal, Fatigue, and Delay Allowances, when studying equitable methods of handling PFD allowances for developing time standards on products to be produced in synchronized assignments.

CONCLUSION

This chapter has described methods of estimating machine interference and operator idleness in random and synchronized multimachine assignments. Procedures for optimizing multimachine assignment operator work loads, and incentive paying the operators were also presented.

The procedures described reflect the most refined and effective applications of the principles presented, based on the experiences of several plants. Although the case examples treated multimachine assignments, the principles presented apply to all forms of multiactivity entailing the interference determinant conditions upon which the principles are predicated.

BIBLIOGRAPHY

Ashcroft, H., "The Productivity of Several Machines under the Care of One Operator," *Journal of the Royal Statistical Society*, series B, vol. 12, no. 1, 1950.

Bowman, E., and R. Fetter, *Analyses of Industrial Operations*, Richard D. Irwin, Inc., Homewood, Ill., 1959.

Feller, W., *An Introduction to Probability Theory and Its Applications*, vol. 1, John Wiley & Sons, Inc., New York, 1950.

Freeman, H., W. Wright, and W. Duvall, "Machine Interference," *Mechanical Engineering*, August, 1932.

Jones, D., "Mathematical and Experimental Calculation of Machine Interference Time," *The Research Engineer*, January, 1949.

Jones, D., "A Simple Way to Figure Machine Downtime," *Factory Management and Maintenance*, October, 1946.

Palm, C., "Arbetskraftens fordelning vid betjaning av automatmaskiner," *Industritidningen Norden*, vol. 75, 1947.

Chapter 7

Memomotion Study[*]

MARVIN E. MUNDEL

M. E. Mundel and Associates, Silver Spring, Maryland

The human eye, ear, and hand, when used in a real-time data recording system, impose a severe restriction with respect to the number of bits of information which may be recorded per unit of time. Further, they limit the number of aspects of a situation which may be maintained under simultaneous observation. In addition, the eye-ear-hand data recording system, when extended to its limit, is subject to errors of both observation and recording. There are many work measurement study situations wherein some additional data recording aid is needed to meet the requirements of effective data collection. Such situations are encountered when analysis requires the study, in detail, of the relationships among a large and complex crew, the simultaneous study of multiple-cause events carried on in a restricted time frame, or the study of events in a situation "hostile" to the eye, such as with unanticipatable arc welding or in outer space.

In situations such as have been listed, a variety of photographic and electronic data recording aids have been used. These aids assist by recording data from real-time events and holding the information in a form permitting its transcription into a form serving the purpose of analysis. These aids serve as a time buffer; they record within the time frame set by the events under scrutiny but may be transcribed, in a time frame imposed by human limitations, into a form suitable for analysis.

[*] Adapted from *Motion and Time Study: Principles and Practice*, 4th ed., by M. E. Mundel, published by Prentice-Hall, Inc., Englewood Cliffs, N.J., 1970, chaps. 14, 15, and 18.

MEMOMOTION STUDY

Memomotion study is the name given to the analysis of the special forms of film or video tape study in which pictures are taken at unusually slow speeds. (With tape, the shift to the equivalent of slow speed is obtained by manipulation of the analysis set.) Sixty frames per minute (one per second) and one hundred frames per minute are the speeds most commonly used. Like all film or tape study, memomotion study requires three phases: filming, film analysis, and data presentation. Memomotion study may also be used to study the time and activity relationships among the flow of material or the use of material handling equipment in an area, or to study simultaneously the man-work, equipment usage, and flow of material. In such cases, if a time-shortened visual presentation is desired, film must be used or a special film made from the tape record. In this way, one hour of activity can be viewed in about four minutes. The information contained on the film may be analyzed in numerous ways, and alternative presentations of the data are possible, depending on the objectives of the study. The development of an adequate basis of analysis will be discussed later in this chapter.

With the advent of relatively low priced, portable video tape recorders, the motion picture camera and the video tape recorder may be considered as almost interchangeable devices. They both employ a camera. The motion picture camera records on photographic film; the video tape recorder on magnetic tape. They both can record visual and auditory information simultaneously. The speed of the motion picture camera and of the video tape recorder can be held constant with respect to a time datum such as the frequency of the alternating current in use. The video record can be converted to a film record, and vice versa. There are, however, some small differences which affect the desirability of employing one or the other of these devices in certain situations. In most cases, that which is given as a characteristic of motion picture photography should be understood as applying equally to the use of the video tape recorder.

AREAS OF APPPLICATION

Memomotion study has been applied with advantageous results to the study of activities such as the following:[1]

1. Gas company street work
2. Twenty-four-man steel casting mold line
3. Prefabricated house section manufacture
4. Railroad car humping in a classification yard
5. Aircraft service on the ramp at a commercial airport
6. Dry-salt meat-packing line
7. Stripping at the delivery end of a cutting press
8. Package handling at a packinghouse sorting center
9. Two-man welding crew on water heater assembly line
10. Municipal garbage handling
11. Dental activity
12. Household activities[2]
13. Department store clerks
14. Fifty-man papermaking-machine repair crew
15. Icehouse crew
16. Railroad carloading crew
17. Auto and passenger pattern at airport passenger terminal entrance

Memomotion study is particularly advantageous for studying situations which show any one or any combination of the following:

1. Long cycles
2. Crew activities
3. Irregular cycles
4. Long-period studies of changing events

[1] The list cited is not exhaustive, but merely indicates the wide range of activities to which the technique may be usefully applied.
[2] See "Easier Homemaking," *Life*, Sept. 9, 1946.

SPECIAL ADVANTAGES OF MEMOMOTION STUDY

Memomotion is frequently chosen because of the desirability of one or more of the six following unique advantages of photographic or electronic aids for data recording:

1. *Permits greater detailing than eye observation.* Of course, the analyst should obtain from the record only that degree of detail best suited to his analysis requirements. However, with crew work or work with irregular cycles, the simultaneity of actions of the crew or the sequence, as it occurs, may be recorded with complete details by the camera or tape far beyond the capacity of any observer to observe and record without aids. For instance, a film study record taken at 1 frame per second of a pharmacist revealed that his prescription-filling time was consumed as follows:[3]

Description of activity	Rank importance	Percent of time used
Work on labels or prescription blanks.	1	23.3
Work wrapping.	2	10.5
Work putting material into prescription containers or with containers.	3	10.1
Inspection of prescription blanks.	4	7.4
Work applying labels.	5	7.2
Travel to and from register.	6	5.4
Work counting items.	7	5.2
Travel to shelves or cupboards for material.	8	5.0
Work with balance and accessories.	9	4.3
Work getting down items.	10	4.1
Work compounding.	11	3.9
Work on drugs.	12	3.3
Talking to customers.	13	2.8
Work at cash register.	14	2.3
Inspection of shelves.	15	2.0
Inspection of drug containers or contents.	16	1.3
Travel to shelves or cupboards to put away.	17	1.1
Work putting up items.	18	0.6
Work with liquid measures.	19	0.2
Total.	...	100.0

Such data could have been gathered without film only with difficulty, because the work pattern was highly irregular. This analysis almost automatically suggests methods improvement steps that might be taken to facilitate the work. For instance, the time expended on the most time-consuming part of the work could be reduced with a special attachment for labels on the platen of the typewriter; the next most time-consuming task could be facilitated through the use of a prescription bag (introduced later); and so forth.

2. *Provides greater accuracy than pencil, paper, and watch techniques.* The time from picture to picture is a function of camera speed. Slow speeds of 100 frames per minute (0.01 minute per frame), 60 frames per minute (1 second per frame) and 50 frames per minute (0.02 minute per frame) are often extremely useful and economical. Of course, even larger time units may be used if desired. Speeds of 60 frames per minute appear to be just about right for maximum film economy and adequate detail, despite the disadvantage of using seconds instead of decimal minutes. In some industrial applications, however, the advantages of

[3] Study made for "The Pharmaceutical Survey" of the American Council on Education.

decimal minutes may outweigh the film economics, making 100 frames per minute preferable. In any case, the time value used should be selected by considering all the requirements of the problem. Video tape may be used in a similar fashion.

3. *Provides greater convenience.* The operation may be studied after a short run or after an experimental run. The film or tape can be stopped at will when being studied, so that each phase of the operation can be studied without bothering the operators. Even with such irregular jobs as building custom truck bodies, a single unit could be usefully studied by means of 1-per-second films (or a video tape used to produce equivalent data), developing data for better methods and better cost estimates on any other order that even partially resembled the one studied. Films or tapes also permit group observation without interrupting factory routine.

4. *Is more adaptable to hostile environments.* Intermittent arcing of welders, low temperature or pressure in space, underwater environment, and so forth, do not interfere with the full recording of motions, actions, or events and related time data. The slow filming speeds of micromotion provide an exposure opportunity that allows extremely faint illumination to suffice.

5. *Reduces the cost of a study.* Memomotion study reduces film cost to about 6 percent of the cost with normal film speeds, and consequently reduces the amount of film to be analyzed without reducing the period covered.

6. *Compressed time scale viewing is possible.* When film is used, it permits rapid visual review of an extended period of performance. When a film taken at 1 frame per second is projected at the normal speed of 16 frames per second, it permits viewing a film of an hour of operation in four minutes. In addition to saving time, viewing with a compressed time scale frequently brings to light novel aspects of the subject being studied which are often instrumental in developing new ideas for better methods.

MAKING MEMOMOTION FILMS OR TAPES

Equipment

The 16-mm Camera. A 16-mm camera (super 8-mm film may be used in many cases) with an *f*1.5, 1-inch lens or better or an *f*1.2, wide-angle (12-mm) lens. The wide-angle lens is used most frequently. The camera should have a shaft to which the motor drive can be attached.

The ideal setup includes a synchronous motor drive for the camera which provides speeds of 60 and 100 frames per minute, with a gear shift for rapid change from speed to speed, as well as a speed of 1,000 frames per minute for running leader or for normal speed pictures.

Tripod for Camera. A tripod helps get better pictures for little extra money. It does away with jerkiness often caused by unsteady hands. If a tripod is used, it should have a pan-and-tilt head.

Exposure Meter. Any of the electronic exposure meters presently available are suitable for assisting in determining the proper exposure. A regular exposure meter, rather than one specially set up for ciné work, is to be preferred because of the odd exposure intervals used with memomotion speeds.

Synchronous Motor Drive. A synchronous motor drive is a necessity with memomotion filming to assure the accuracy of the time intervals from frame to frame.

Lighting. Since the introduction of modern high-speed lenses and films, lights are not required for micromotion filming. Indeed, modern lenses and films permit the making of good-quality motion pictures in almost any location where there is enough light for the worker to see what he is doing. The use of high-intensity photographic lights in the factory is a source of disturbance. The glare is not only annoying to the operator, but it makes the operator the focus of attention. Certainly, the conditions filmed are different from normal. The employment of available-light photography also reduces the setup time for picture taking. Equipment without lights greatly reduces the bulk which must be brought to the job, making it possible for one man to carry, set up, and employ the equipment.

Filming. Filming for data recording involves only six simple rules:

1. Obtain the cooperation of the operator and his foreman. It is often worthwhile to notify them a day ahead, so that the operator may dress accordingly, if so inclined. On the other hand, with available-light photography, the use of motion pictures may become so routine that this procedure may be abandoned as attaching undue importance to the films. In a union shop it may be worthwhile obtaining the cooperation of the steward. No one should have anything to fear from pictures, for they are merely a more reliable means of studying what would otherwise be visually studied with less accuracy. When filming large crews, subsequent analysis of the film may be much easier if the workers are issued inexpensive T-shirts with large numbers on the front and back. If so, adequate explanations for the shirts should be made well beforehand and all questions answered.

2. Place the camera as close to the action of the job as possible, but be sure the view of the camera includes all the activities wanted for analysis. Set the lens focus and check the view through the finder. It should be borne in mind that the films are to gather data and not to represent an artistic effort; they should be taken with as little fuss as possible.

3. Select the speed of picture taking and set the camera speed control or motor drive accordingly.

4. Get the best exposure possible. Quality films are not a requisite, but better exposures make the analysis easier. With modern high-speed film, surprisingly little light is required. Exposure instructions accompany both camera and exposure meters and are also published in book form. Basically, the length of exposure is controlled by the number of frames per second in a manner explained in the instruction book accompanying the camera. The exposure meter has a calculator attached to it that takes into account:

   a. The exposure interval.

   b. The sensitivity of the film, which is given either on the film carton or in a folder in the film carton.

   c. The amount of light on the subject (pictures for data recording should usually be exposed for the darkest part of the picture in which detail is wanted). The use of available light is usually accompanied by more uniform lighting than with photographic lighting; this simplifies the exposure problem.

The calculator on the exposure meter is used to determine the correct lens aperture for any condition of these three variables.

5. Make a record of the speed at which the camera is run so the time interval will be known.

6. Make a record of exposure data and pertinent job information. This record permits discovery of the reasons for poor exposures, if they occur, and the application of corrective measures on future films. It also provides a place, identifiable against the film, to record important job information that does not appear on the film.

Video Tape Recording. Video tape recording differs little from film recording. A special camera is used, and this camera is connected to a special tape recorder.[4] The equipment is handled in a manner similar to that used with film. A tripod, exposure meter, and synchronous motor drive are all necessary, and all the human problems remain the same.

Film Analysis. The films must be processed after they are taken. Inasmuch as motion picture film processing is available on a one-day or same-day basis in most major cities, the film is rapidly available for analysis. Plants located in cities without processing service can usually arrange for mail service from the nearest processor; a three-day turnaround time is usually the maximum. (When time was of the essence, forty-five minute service has been obtained.)

A projector used for film analysis must have specific features somewhat different from the conventional projector. The optical system must have heat filtering suffi-

---

[4] The make or model numbers of suggested equipment are not given because of the extreme rate of change and improvement taking place in commercially available equipment.

cient to permit prolonged examination of single frames. A frame-by-frame advance of the film, forward or backward, must be convenient; the projector should also have a built-in frame counter.

Obtaining the exact relationship between the members of a crew or any other phenomena recorded on the film with respect to time is much easier with film than with actual observation, because the film can be stopped and the action held still from step to step during the analysis. Each frame may be individually examined, and notes made of the method and the time for each step. The analysis is usually made with a portable shadow box.

USING MEMOMOTION STUDY TO SET TIME STANDARDS

Although memomotion study is extensively used as a methods analysis technique, it is also used for setting time standards. Two basic procedures are available. Both procedures offer the advantage of providing an objective record which may be reviewed later by anyone, including the worker, and which is not "interpreted data" as with a stopwatch record. Different element breakdowns may be made at any time, as the need arises. In addition, more smaller elements may be timed. If standard data are being developed, this is often a decided advantage. Also, a whole crew may be timed simultaneously. The necessary data may be recorded on the film or tape more rapidly and more completely than with stopwatches. Indeed, most of the advantages given for photographic or electronic aids over visual techniques apply here also.

It is particularly suggested that disputed rates subject to a grievance or arbitration procedure may be most ideally studied with the full-camera technique as a more accurate and reliable record, because the objectiveness of the process safeguards the interests of both management and labor, if they are both concerned with accuracy, as they should be. Furthermore, rates set with these data may be discussed more factually and many grievances may be avoided. The two procedures are:

1. Partial use of camera (motion picture or video recorder)
2. Full use of camera

With either method, as with stopwatch studies, the worker should be aware that a time study is being made. Also, all the usual data taken prior to a stopwatch time study should be recorded except the motion pattern, which may later be taken from the film or tape. No microchronometer need be used, but the worker should be fully aware that time is being recorded.

Partial Use of Camera. In partial use of camera, the camera is used only at the 1-per-second speed or at the 100-frames-per-minute speed; hence the camera is used only to record method and time.[5] However, once set up, it does this automatically, leaving the observer free to devote his attention fully to the task, and provides a much better opportunity to formulate a rating than when using a stopwatch. This method, however, only begins to make use of the full potentialities of the camera.

The film, when developed and ready for analysis, is put into a hand-cranked or automatic indexing projector with a frame counter, and using a time study sheet, the data (which would have been recorded while the job was observed if a stopwatch were used) are entered in the blanks in a similar fashion, counter readings instead of stopwatch values being used. Otherwise, the procedures are identical. However, the record may be made at the analyst's own pace, small elements down to one or two seconds identified and recorded, and the film later reanalyzed and the elements changed if this is found desirable. The data of the film are permanent and accurate. All calculations are made in a fashion identical with stopwatch calculations.

The memomotion procedure is particularly advantageous when an answer to the problem of determining the elements to separate is not readily apparent from

---

[5] As was noted earlier, a video tape record may be used in an equivalent fashion.

a visual observation of the activity, or as noted earlier, when the situation overwhelms the human recording system.

Full Use of Camera. Full use of camera is the same as the partial procedure except that, either during the study or immediately after it, the gear shift on the drive is used to obtain some footage at 1,000 frames per minute so as to record a reviewable version of the pace exhibited by the operator during this study.[6] This will be used for rating later. Provided care is taken to see that what is filmed at 1,000 frames per minute is similar in pace to what was filmed at 1 frame per second, this procedure should aid time study work as follows:

1. Allow others to check the rating
2. Provide a record of the pace
3. Facilitate actual rating
4. Provide material for discussion rather than grievances

These features all improve what is the most difficult phase of time study.

STUDYING LONG-CYCLE CREW ACTIVITIES

A twenty-five-man molding crew at the American Steel Foundries plant at Granite City, Illinois, working on large molds, was time studied with the aid of memomotion camera timing. Because of the extreme area involved, the men were given numbered jerseys to wear to increase the ease of identifying them on the films. The gang was filmed in six sections. The films were taken at 100 frames per minute to obtain values similar to those obtained by stopwatch methods. The advantages of using films rather than direct observations have been summarized by Mr. C. H. Walcher, works manager, American Steel Foundries, Granite City, Illinois, as follows:

> The pictures were taken as a result of a grievance. The pictures gave us a very adequate coverage and enabled us to show all concerned parties exactly how the rate was determined. Because the time study men were not involved in recording elemental breakdowns, much more attention was given to pace determination, which resulted in better leveling factors. The pictures proved that the present rate on this job was not tight. All men on the unit were covered in one day, which represents a tremendous reduction of in-shop time to obtain the necessary information.

Mr. L. Randolph, the chief industrial engineer at the plant, reports that the pictures taken have been used continually as reference points for methods descriptions. They have also, he reports, been used for method instruction for the old as well as the new employees.

STUDYING IRREGULAR CYCLES

From its records, a public utility found that the time to install a gas service renewal averaged 64 man-hours, although some crews consistently took only 24 man-hours. A nine-man crew was used on the job, although only three or four were employed on a single site at any one time. A memomotion study was made of one of the superior crews to determine the method that made this low-cost performance possible.

The analysis categories were created by compounding the phases of the work (preparing the site, opening the excavations, making the connections, closing the excavations) with the basic man activities (do, move, inspect, wait) and with the tools and equipment used (shovel, air drill, torch, and the like). Typical categories were:

WPB   work preparing barricades for traffic
WOD   work opening excavation with air drill
MPA   move air compressor to location

[6] When many time studies are being taken on film, some plants have found it economical to employ a second camera to do the 1,000-frame-per-minute sampling.

The need to develop extensive categories such as these is typical of studies of long, irregular tasks.

The analysis revealed that the quick performance was due primarily to the organization of the work, and not to the speed of working. In addition, numerous ways of making the task easier and less costly were found. In this respect, particular attention was paid to the most time-consuming activity categories. A summary of the analysis categories and a multiman chart (with only the major steps) were produced to use both as training aids and as a guide to supervision. These led to a considerable reduction of costs.

STUDYING LONG PERIODS

Long periods may be studied by memomotion study when the situation is such that:

1. The frequency of random events, such as would be detected with work sampling, is desired.

2. Queue-type data for simulation are needed.

Studies of the second type have been made in railroad classification yards, airport ramp and passenger loading facilities, and so on.

The method of analysis of long-period studies is almost always peculiar to the study and depends upon the events for which time data are desired.

SUMMARY

Memomotion analysis refers to obtaining work activity and work time data from motion picture films taken at unusually slow speeds, such as 50, 60, or 100 frames per minute. Video tape may also be used.

Memomotion study is particularly advantageous when visual observations are not readily feasible because of the duration or complexity of the situation to be observed. Many types of activities and time analyses are feasible with this approach.

BIBLIOGRAPHY

Barnes, R. M., *Motion and Time Study*, 6th ed., John Wiley & Sons, Inc., New York, 1968.

Coggan, B. F., "Why Not Try Area-wide Camera Studies?" *Modern Materials Handling*, December, 1953, pp. 74–76.

Mundel, M. E., "Memomotion," *Time and Motion Study*, March, 1958, pp. 32–43.

Mundel, M. E., "Memomotion Study Technique Simplifies Work Analysis," *Factory*, June, 1949.

Mundel, M. E., *Motion and Time Study: Principles and Practice*, 4th ed., Prentice-Hall, Inc., Englewood Cliffs, N.J., 1970.

Niebel, B. W., *Motion and Time Study*, 4th ed., Richard D. Irwin, Inc., Homewood, Ill., 1967.

Richardson, W. J., "Memomotion and Fork Truck Time Standards," *Modern Materials Handling*, April, 1953, pp. 67–70.

Chapter **8**

Standard Data Concepts*

HAROLD B. MAYNARD

President, Maynard Research Council Incorporated, Pittsburgh, Pennsylvania

WILLIAM K. HODSON

President, H. B. Maynard and Company, Incorporated, Pittsburgh, Pennsylvania

The time required to establish time standards by the detailed study of individual jobs is considerable. This fact has led industrial engineers to seek ways of reducing standards setting time. Their aim has been to find ways of establishing standards quickly without unduly affecting the accuracy of the results. A number of different procedures have been developed for accomplishing this. The more successful and widely used of these techniques will be discussed in this chapter under the major topics of:

1. Standard data building blocks
2. Elemental standard data
3. Task data and time formulas
4. Standard work groups and benchwork standards
5. Application procedures

During the discussion, predetermined motion time systems, and especially methods time measurement, or MTM, will be referred to frequently. Those who are not acquainted with these procedures will find it helpful to study Section 5 of this Handbook before reading this chapter.

* The material in this chapter has been drawn largely from the book, *Design and Application of Industrial Time Measurement Systems*, by H. B. Maynard and W. K. Hodson, not yet released for publication.

ORIGIN OF STANDARD DATA CONCEPTS

When Frederick W. Taylor was developing his time and motion study procedure before the turn of the century, he early recognized the importance of subdividing an operation into a series of small elements so that a clear insight into the method used to do the work could be gained. When unnecessary elements were recognized, they were eliminated, which resulted in the development of an improved method for doing the work.

Frank and Lillian Gilbreth in their micromotion study procedure carried the subdividing step much further, breaking the work down into the fundamental motions, or "therbligs," used by the operator in doing the task.

When time values were developed for each element of a time study, it became obvious that the time required to perform the operation by any method which used some combination of those same elements could be computed by simple arithmetic. Carrying this a step further, Taylor saw that if the time required to perform every element in every trade could be measured, it would no longer be necessary to make time studies of each new job that came along. The time values could be determined synthetically from tables of elemental times. He predicted a handbook which would contain such tables.

Frank Gilbreth also recognized that if fair time standards could be developed for each therblig, the time for performing any task by any conceivable method could be calculated.

Taylor never developed his handbook of elemental times, nor has anyone else done it since his time. The number of different methods that can be used to accomplish any given element are many, so that an elemental time without a statement of the exact method to which it applies is not at all precise.

Gilbreth could describe a method quite clearly in terms of therbligs, but he never completed the research necessary to establish time values for each of them.

It was not until predetermined motion time systems were developed that techniques became available for describing methods with exactitude and for computing corresponding standards. But it was the pioneering work of Taylor and the Gilbreths in subdividing an operation into smaller elements or fundamental motions that provided the basis for all standard data concepts.

STANDARD DATA BUILDING BLOCKS

The concept of standard data building blocks likens the procedure for developing a standard from a number of basic or detailed time values to the construction of a large building from a number of smaller components.

Assume, for example, that it is desired to determine the time (and hence the labor cost) for making an end bracket for an electric motor. The total time is represented in Figure 8-1 by the large rectangle, A. The first step will be to determine the operations required to produce the end bracket, a step often called processing or process engineering (see Section 2, Chapter 2). Necessary operations such as make cores; make mold; pour metal; clean casting; turn and bore; mill slots; and paint will be specified. These operations are represented in Figure 8-1 as divisions of the large rectangle, A.

The next step will be to determine the time for performing each of the operations. The first operation is represented by the shaded area of the large rectangle, A. The time required for performing it can be built up from various kinds of building blocks or detailed time values. These building blocks come in various sizes. Some of them are very small, such as those represented by B in Figure 8-1. These are the individual motion times provided by a predetermined motion time system such as basic MTM. Other types of blocks are larger, ranging from the still fairly small blocks shown by C up to the very large blocks represented by F.

It is evident that the time required to establish a standard varies inversely with the size of the blocks used. The smaller the blocks, the longer it takes to build

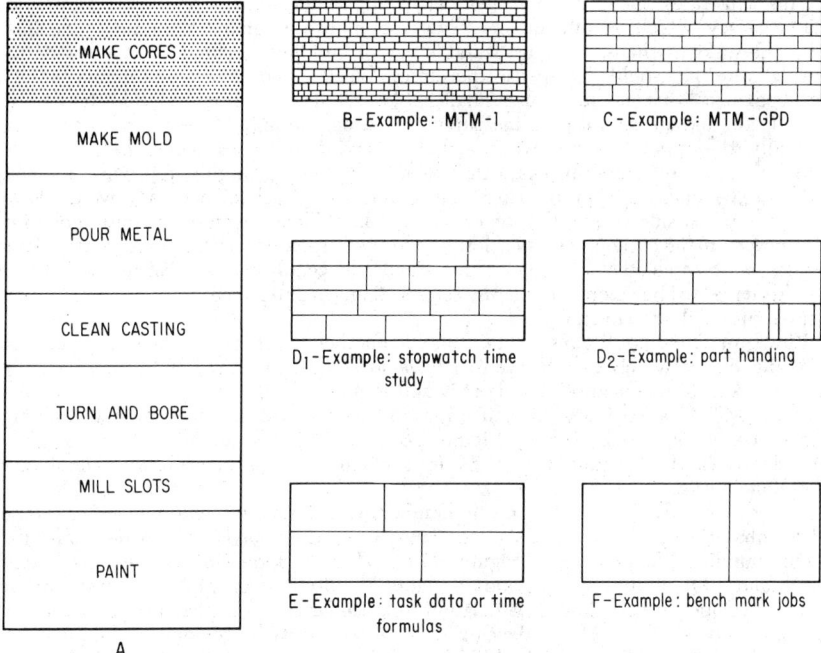

B-Example: MTM-1

C-Example: MTM-GPD

D_1-Example: stopwatch time study

D_2-Example: part handing

E-Example: task data or time formulas

F-Example: bench mark jobs

A

FIG. 8-1.   Standard data building block concept.

the standard. This fact has caused industrial engineers to develop larger and larger blocks.

It does not follow, however, that the larger blocks are always desirable. If only two sizes of blocks are available for use, the size of the building that can be built with them is quite limited. In Figure 8-2, two sizes of blocks, A and B, are shown. Using one A block and one B block, the structure X can be built. Using one A block and two B blocks, the structure Y results. If the size of the structure which should be built (the correct standard) is that represented by the dashed rectangle, it is evident that structure X will be too small and structure Y will be too large. To get a better fit, it will be necessary to use smaller building blocks.

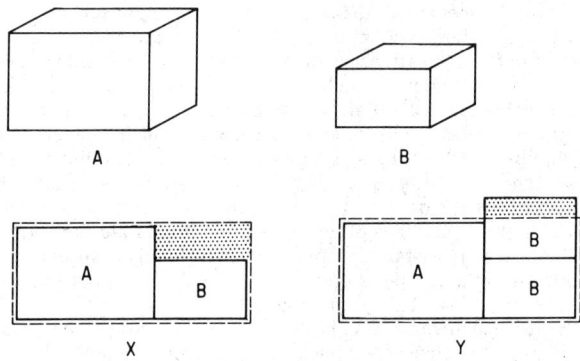

A

B

X

Y

FIG. 8-2.   Effect of large building blocks on accuracy.

This illustrates the fact that, generally, the smaller the blocks, the greater is the accuracy which is attainable. Thus the industrial engineer faces a dilemma when he seeks to choose the most suitable size of building blocks. To save standards setting time, he would like to use the largest blocks. To achieve greater accuracy, he recognizes that he must use smaller blocks. Thus, when he seeks to design a workable and successful time measurement system, he finds it necessary to consider carefully the nature of the work and the conditions under which his standards will be applied. Large blocks will not be suitable for highly repetitive manual work on which a number of people are employed. Lack of accuracy would have serious repercussions. On the other hand, small blocks would be unsuitable for the measurement of nonrepetitive, long-cycle work such as maintenance work. Jobs would be completed long before the standards could be established, and there would have to be more standards setters than maintenance men to obtain any degree of standards coverage.

Kinds of Building Blocks. There are a number of different kinds and sizes of building blocks which can be used to develop standards. Five of the most widely used blocks will be discussed briefly now and in more detail later.

Basic MTM, sometimes referred to as MTM-1, provides the smallest building blocks commonly used (see B, Figure 8-1). It requires the most time to apply but results in the greatest accuracy. It is much used for developing other, larger building blocks.

To save standards setting time, a number of different systems which make use of combinations of basic MTM data have been developed to provide somewhat larger building blocks (see C, Figure 8-1). These include USD, or universal standard data; MSD, or master standard data; MTM-GPD, or MTM general-purpose data, developed by the MTM Association for Standards and Research after surveying a number of similar systems developed by American management consulting firms and industrial companies; and MTM-2, developed by the Applied Research Committee of the International MTM Directorate after a study of the systems in use in the several member countries. All these systems are quite similar in that they reduce the time required to set standards at some sacrifice of accuracy.

The next largest building blocks consist of elemental data of two different types. The first, D_1 of Figure 8-1, is the elemental data derived from stopwatch time study, which was the first of the building blocks to be used. Time studies of individual jobs, such as the one shown by Figure 8-21a and b, produce time values for each of the elements into which the job is divided. Lists of elemental times obtained from a number of time studies made on a given class of work provide building blocks from which time standards for other similar jobs can be developed.

The second class of elemental data, D_2 of Figure 8-1, is designed to cover elements common to a number of classes of work. Parts handling, tool use, and gaging are examples of this type of elemental data. The elements may be larger or smaller than the first type of elemental data derived from stopwatch time study.

Elemental data of either type may be built up from time study, MTM-1, or any of the systems of combined basic data. They can be developed for a process such as welding, for an industry such as railroads, and for work which includes mental activity such as clerical, drafting, or design work, as well as for individual operations or product lines. The relative ease with which elemental data can be compiled and applied makes them a much used type of building block.

Task data or time formulas, E of Figure 8-1, can be developed from any of the building blocks thus far discussed. They often save considerable time in setting standards with little if any sacrifice of accuracy. Task data are derived by combining elemental data to develop a standard time for performing a specific piece of work which is not necessarily a complete operation or job. Task data constitute building blocks of various sizes, many of them quite large.

Finally, there are bench mark standards, F of Figure 8-1. These are the largest building blocks of all. They are quickly used, and although they may not be altogether accurate for any one job, when applied to a number of jobs over a period of time such as a workweek, they give a statistically acceptable measure of performance for the time period.

Figure 8-3 shows graphically the five types of building blocks discussed above and indicates how they are used to develop standards or other, still larger building blocks.

Building Block Development. Figure 8-4 shows an example of how smaller building blocks are used to develop ever-larger building blocks until a bench mark standard is reached. The operation covered is "Mount and connect medium-size electric junction box."

Development starts with basic MTM. Basic motion times are combined to produce elemental data. The building blocks used in this step range from 2 to 70 TMU[1] in size, or $\frac{1}{14}$ second to $2\frac{1}{2}$ seconds.

In the case illustrated, the use of combinations of basic MTM data is not shown. This is something that is done only once, when the system is first developed. Thereafter, the combined data may be used as the starting point instead of the basic data if desired. Often, however, it is felt that the greater accuracy of basic MTM is more important than the small amount of application time saved when the object is to develop not a single standard, but elemental data, time formula, and bench mark building blocks which will be used over and over again.

In Figure 8-4, a dotted line is shown dividing the elemental data column in two. This is done to distinguish the size and the method of developing the two types of elemental data building blocks discussed above. The first type is developed directly from basic MTM or time study and generally ranges in size from 30 to 3,000 TMU, or 1 second to 2 minutes. The other type, which covers elements common to a number of different classes of work, comes partly from basic MTM and partly from other previously developed data and ranges in size from 40 to 10,000 TMU, or 1⅓ seconds to 6 minutes.

The next data level, task data, is shown as being derived from elemental data with the addition of certain other basic MTM data to fill the gaps not covered

---

[1] A TMU (time measurement unit) is 0.00001 hour.

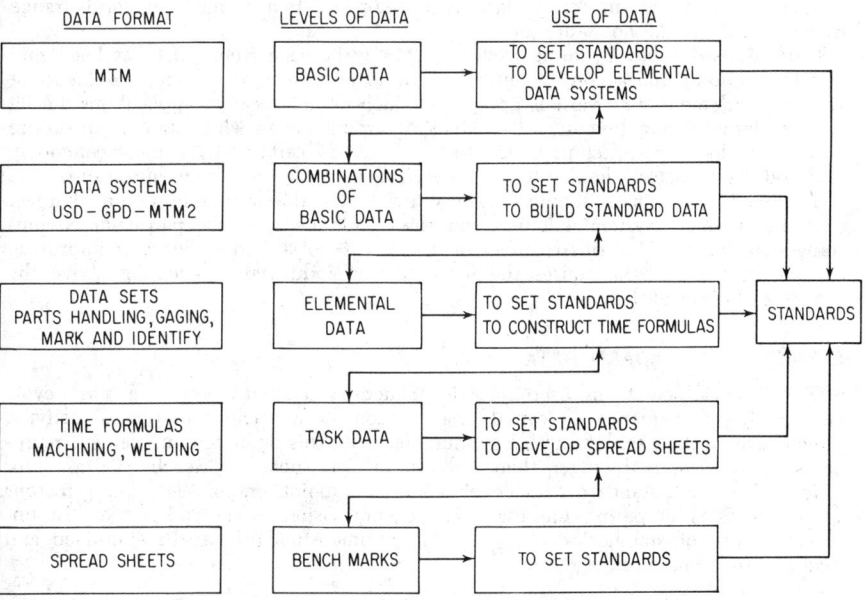

Fig. 8-3. Types of building blocks and their uses.

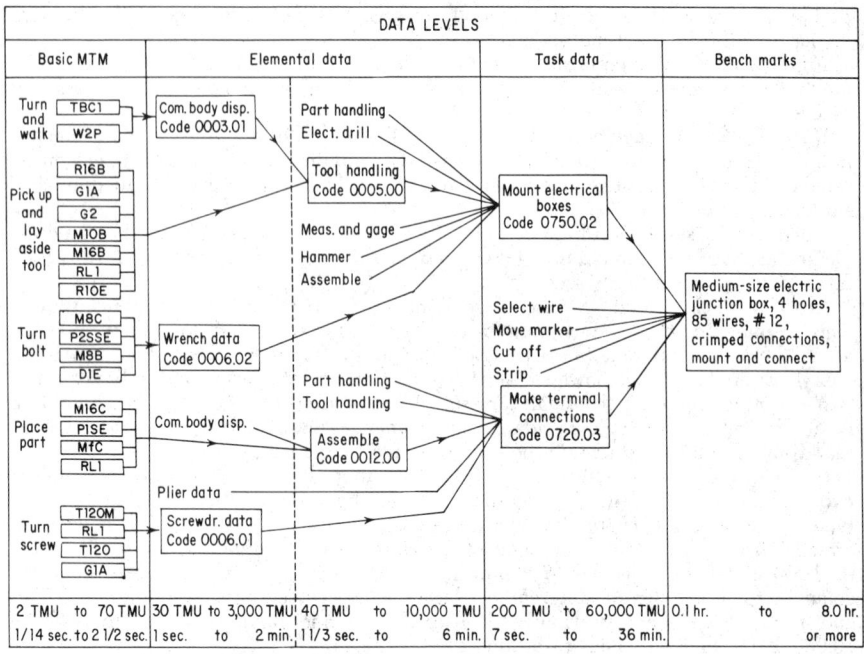

DATA LEVELS			
Basic MTM	Elemental data	Task data	Bench marks

2 TMU to 70 TMU	30 TMU to 3,000 TMU	40 TMU to 10,000 TMU	200 TMU to 60,000 TMU	0.1 hr. to 8.0 hr.
1/14 sec. to 2 1/2 sec.	1 sec. to 2 min.	1 1/3 sec. to 6 min.	7 sec. to 36 min.	or more

FIG. 8-4.   Building block development.

by the elemental data. The size of the task data building blocks usually ranges from 200 to 60,000 TMU, or 7 seconds to 36 minutes.

Finally, the bench mark standard is developed. Bench mark standards range from about 0.1 to 8.0 hours or more.

Time Measurement Systems Design. It may be seen from what has been said that the building block concept underlies the arrangement of time measurements of motions or elements into various groupings which permit speedier applications during the standards setting process. The kinds of arrangements which can be made are many, and the above attempt to identify and classify certain of the more commonly used building blocks should not be regarded as excluding other approaches.

The industrial engineer should recognize that he has the latitude to design whatever time measurement system will best suit the requirements of the particular circumstances he faces. It is this freedom to design and to seek fresh ways of improving existing approaches that raises the level of true industrial engineering above the merely technical level.

ELEMENTAL STANDARD DATA

The term "element" as used here is defined as a subdivision of a work cycle composed of a sequence of several basic motions or a machine or process activity, which is distinct, describable, and measurable.[2] By this definition, an element represents a time interval longer than a basic motion such as "Reach," "Grasp," or "Move." From a standard data development standpoint, one of the primary reasons for elements is to permit the use of large subdivisions of a work cycle. In this way, the use of standard data to develop a time standard can be simplified and made more economical.

[2] Adapted from *Industrial Engineering Terminology*, ASME, New York, 1955.

METHODS ANALYSIS CHART					REFERENCE No._____

PART_____ DATE_____ STUDY No. _____

OPERATION_____ ANALYST_____ SHEET No. __OF__ SHEETS

DESCRIPTION – LEFT HAND	No.	LH	TMU	RH	No.	DESCRIPTION – RIGHT HAND
M1 Sit down on chair or stool at workbench						
			29.0	B		Bend over
		R-B	- -	R-B		Reach to side of chair
		G1A	2.0	G1A		Grasp chair
			34.7	SIT		Lower body to chair
		LM12	14.3	LM12		Move legs forward
		M10B	12.2	M10B		Move chair forward
		RL1	2.0	RL1		Release chair
		R22E	18.0	R22E		Place arms on bench
		LM12	- -	LM12		Move feet under bench
			112.2			

Fig. 8-5.   Typical MTM motion pattern for using a chair or stool.

Example of Elemental Data.   An example may help to clarify this point.   One of the basic motions of MTM is "Sit." This motion may be described as ". . . the seating of oneself *after* the body has fully positioned itself to the chair.   It consists of bending the knees and lowering the body until contact is made with the chair or bench." From this description, it can readily be seen that, in almost every case, additional motions before and after the "Sit" motion must be performed by a person who is going to use a chair or a stool.

The MTM analysis, Figure 8-5, shows a typical motion pattern for the element "Sit down on chair or stool at workbench." This occurs when one stands in front of a desk or bench with a chair behind him.   To sit at the desk or bench, it is necessary to bend, reach down, grasp the chair, and lower the body to sit.   This requires four basic motions: B; R__B; G1A; and Sit.   The R__B is combined with the Bend.   Before work can be done at the desk or bench, it is necessary to pull the chair forward.   This is done by first moving both feet forward and then pulling the chair toward the desk.   The arms are then placed in position on top of the desk as the feet shift to a comfortable position away from the chair.   All this takes five additional motions after the body is seated.

A study of a number of jobs will show that this motion pattern, with slight variations, will occur again and again.   Analyzing this pattern with basic MTM each time it occurs requires considerable time.   A better approach is to record the motion pattern carefully once only.   This then becomes a standard data element.   The element can be described and coded and the value rounded off to the nearest TMU—112.   It can then be used whenever "Sit" occurs.   It is never necessary to go through the detailed MTM analysis of this particular motion pattern again.

Combining Related Elements.   This elemental data approach can be simplified still further.   It is obvious that if a person performs the element of sitting, he must inevitably perform the reverse, "Arise from chair or stool at workbench." The MTM analysis for this element is shown by Figure 8-6.

Because every "Sit" must be followed by a "Stand," the problem of classifying, coding, and retrieving elements and time values may be simplified still further by combining these two elements and using the average time of the two.   The

DESCRIPTION – LEFT HAND	No.	L H	TMU	R H	No.	DESCRIPTION – RIGHT HAND
N - 1 Arise from chair or stool at work bench						
		R22A	14.0	R22A		Reach to chair
		G1A	2.0	G1A		Grasp chair
		LM12	- -	LM12		Place feet under chair
		M12B	13.4	M12B		Push chair out
		LM12	14.3	LM12		Move legs back under chair
			43.4	STD		Stand up
			87.1			

METHODS ANALYSIS CHART   REFERENCE No. _____

PART _____ DATE _____ STUDY No. _____

OPERATION _____ ANALYST _____ SHEET No. __ OF ____ SHEETS

Fig. 8-6.   Element "Arise from chair or stool at workbench."

element will then be called "Use chair or stool at bench or desk—Sit or Arise"—100 TMU.   The value of 100 TMU is the average of the two elements:

$$\frac{112 + 87}{2} = 100 \text{ TMU}$$

Thus, two fairly complex motion patterns are reduced to one simple element and one time value.

Elemental Data from Time Studies. From this one example, the important saving in application time that can be gained by using elemental data instead of basic data can be seen.

In this example, basic MTM was used to:

1. Describe or define the element
2. Specify the beginning and end points
3. Establish the leveled time

The element could also have been established by means of time study, but the procedure would be somewhat more complex.   To be certain that different time study men time exactly the same elements, it is important that each analyst have a clear definition of the elements to be studied.   In the case example, the element of "Sit down on chair or stool at workbench" might be described as follows:

Begins: When operator is in a position to start lowering his body to the chair or stool
Includes: Pulling the chair into the bench
Ends: When the operator's hands are on the bench ready to start the next element

Because there are several possibilities for errors in obtaining a correct leveled time for this element, it is desirable to obtain a series of values from a number of time studies made by several different men.   This will tend to minimize errors due to rating the operator, reading the watch, and inconsistent start and stop points.   With training and experience, these errors can be minimized.

There is little question but that predetermined motion time systems are superior to time study for the development of elemental standard data.   They are, however, limited to data on purely manual motions. Consequently, time study or other measurement techniques must be used to develop elemental data on mechanically controlled elements or process times.

Types of Elemental Data. There are two types of elements: constant elements and variable elements. If an element is of such a nature that a constant or nearly constant time is required to perform it, it is classified as a constant. If the time required to perform the element varies with one or more variable factors or parameters, it is classified as a variable.

Constant Elements. A good example of a constant elemental time value is the element "Sit down on chair or stool at workbench." This is an example of an element that will require an almost constant time whenever it is performed—"almost" because the distances reached or the lengths of the leg motions can vary slightly. For practical purposes, however, the element can be considered as a constant. There are many cases in actual practice where an elemental time value will not be truly a constant time value.

Fortunately, the accuracy requirements of most time standards permit treating many "almost" constant elements as constant values. When an elemental time value is constant, it greatly simplifies the subsequent use of the element in developing standard data.

Variable Elements. There are some elements of work whose time values will vary substantially in relation to one or more parameters or variable factors. Many machine time or process time elements are of this nature. For example, the time required to drill a hole in metal will vary in proportion to a number of variable factors such as:

1. Size of the hole	6. Pressure on drill spindle
2. Depth of the hole	7. Sharpness of drill
3. Type of metal	8. Angle of drill tip
4. Type of drill	9. Relief angle of drill
5. Speed of drill	10. Lubricant used

It is obvious that the determination of elemental machine time values can become quite complex because of the great number of variables that must be considered. A similar problem exists in developing elemental time values for process times. Hand-spray painting is a good example. Some of the variables that must be considered are the area to be covered, the shape of the part, whether the coat is a base coat or a finish coat, the finish quality requirements, the viscosity of the paint, and the pattern generated by the nozzle.

There are also manually controlled elements that are variable. The factor causing the variability may be the size of the part, the nature of the material, the shape or weight of the part, or the like. From a work measurement standpoint, the problem is first to identify the variable and then to determine the relationship between the variable and the time required to perform the element.

The solution to this problem varies considerably between time study and predetermined motion time system techniques. These differences can be illustrated by an example using the element "Pick up part and place in chuck," an element which occurs frequently in engine lathe work.

Analysis of Variables by Time Study. To determine the time required to perform a variable element, a number of time studies, preferably by more than one observer, should be made of the full range of parts. This requirement is sometimes difficult to fulfill, because the small parts and the large parts may not always be going through the shop when the studies are being made. If the studies are not well planned, there are likely to be a large number of studies on parts in the middle of the size range and very few studies of parts at each extreme. It sometimes takes several months of normal flow of work through the shop before a representative sample of parts has been studied. This is one of the practical difficulties encountered in developing elemental data by the time study approach.

When sufficient studies have been obtained, they are posted on a time study spread sheet or master table as shown by Figure 8-7. The vertical columns of this form provide space for recording the essential information for each time study.

MASTER TABLE OF DETAIL TIME STUDIES

FORMULA Course C #1
DATE Oct. 14, 19 —
PART Brass Clamps for Type X Regulators
OPERATION Mill Slot
PERFORMED ON Horizontal Milling Machines
COMPILED BY M.E.C.

JOB CHARACTERISTICS

STUDY	S-1	S-2	S-3	S-4	S-5	S-6	S-7	S-8	S-9	S-10
OPERATOR	#1 6/1/	#2 6/1/	#1 6/4/	#2 6/4/	#3 6/4/	#1 6/5/	#2 6/5/	#1 6/6/	#2 6/6/	#1 6/7/
	GROSS	WILLIAMS	SMITH	GROSS	SMITH	WILLIAMS	WILLIAMS	SMITH	WILLIAMS	GROSS
SKILL	D	D	D	D	C	D	D	C	D	C-1
EFFORT	C-1	D	C	C-1	C	D	D	C-2	D	D
MACHINE	#3 Le Blond	#2 Cincinnati	#2 Milwaukee	#3 Le Blond	#2 Milwaukee	#2 Cincinnati	#2 Cincinnati	#2 Milwaukee	#2 Cincinnati	#3 Le Blond
DWG. NO.	22289	326907	89210	61918	99201	33213	82112	63800	92678	55210
ITEM	1	9	4	5	16	3	11	6	4	10
FEED	6"/Min.	4.75"/Min.	5.5"/Min.	5.5"/Min.	6.00"/Min.	5"/Min.				
SPEED	140 R.P.M.	140 R.P.M.	140 R.P.M.	140 R.P.M.	140 R.P.M.	140 R.P.M.				
CUTTER DIA.	6"	6"	5 1/2"	5 1/2"	5"	6"	6"	5 1/2"	6"	5"
MACHINE NO.	3589	863	248	3589	248	863	863	248	863	3589
CLAMP VOLUME	2.23	1.20	11.8	30.9	6.1	63.9				
SLOT DIMENSIONS	7/8"–1 3/8"	4"–1 1/2"	2 1/4"–1"	3"–1 1/8"	4 1/2"–7/8"	3 1/2"–1 1/4"				

Symbol	Operation Description	Time Allowed (Hours)	Reference	Operation Class	S-1	S-2	S-3	S-4	S-5	S-6	S-7	S-8	S-9	S-10
A	Pick up part from table	CURVE A		V	.0007	.0015	.0010	.0012	.0009	.0014				
B	Place in vise	CURVE B		V	.0009	.0013	.0009	.0010	.0009	.0011				
C	Tighten vise	.0024	S-3,5	C	.0021	.0028	.0024	.0026	.0024	.0023				
D	Start machine	.0003	S-1,2,3,4,6	C	.0003	.0003	.0003	.0003	.0004	.0003				
E	Run table forward - per inch	.0007	S-2,4,5	C	.0003	.0007	.0006	.0007	.0007	.0004				
F	Engage feed	.0003	S-1,2,4,5,6	C	.0003	.0003	.0004	.0003	.0003	.0003				
G	Mill slot	SEE MACHINING TABLE		V	.0080	.0178	.0108	.0129	.0082	.0158				
H	Stop machine	.0014	S-3,4,6	C	.0015	.0012	.0014	.0014	.0013	.0014				
I	Return table - per inch	.0005	S-2,3,6	C	.0003	.0005	.0005	.0006	.0004	.0005				
J	Release vise	.0009	S-3,6	C	.0013	.0008	.0009	.0008	.0010	.0009				
K	Lay aside part in tote pan	CURVE C		V	.0009	.0020	.0011	.0014	.0010	.0017				
L	Brush vise	.0010	S-3,6	C	.0009	.0012	.0010	.0011	.0009	.0010				
	SET UP													
M	Get time card	.0406	S-9								.0398	.0450	.0406	.0400
N	Get job and drawing	.0500	S-7								.0500	.0483	.0513	.0605
O	Clock time on card	.0178	S-8								.0136	.0178	.0201	.0179
P	Get tools from tool room	.1303	S-10								.1435	.1181	.1250	.1303
Q	Check operation with drawing	.0122	S-9								.0125	.0098	.0122	.0135
R	Clean vise and table	.0105	S-10								.0115	.0094	.0101	.0105
S	Put vise on table	.0084	S-9								.0075	.0091	.0084	.0087
T	Get 2 bolts and set in table slot	.0180	S-9								.0199	.0171	.0180	.0176
U	Tighten vise to table (2 bolts)	.0156	S-7,9								.0156	.0124	.0156	.0182
V	Put on cutter and collars	.0087	S-7,10								.0091	.0091	.0085	.0087
W	Loosen vise from table	.0078	S-8								.0045	.0078	.0072	.0074
X	Remove 2 bolts to locker	.0052	S-9								.0045	.0060	.0052	.0047
Y	Remove vise from table	.0117	S-9								.0118	.0105	.0125	.0117
Z	Remove cutter and collars	.0077	S-10								.0078	.0080	.0074	.0077

FIG. 8-7. Time study spread sheet.

The horizontal lines provide space for posting the elemental times determined by the various time studies.

The next step is to analyze the variable elements to seek a relationship between the elemental time values and some characteristic of the part. A good approach is to plot the elemental times for each study against some possible variable such as weight. When this is done for the element "Pick up part and place in chuck," the result is shown by Figure 8-8. The only conclusion that can be reached from this plot is that there is probably a general relationship between time and weight. The spread, however, from any least squares line that might be drawn through these points is too great to be usable. The reason for this is that there is evidently more than one variable influencing the time required to perform the element.

Further analysis and observation reveal that there is a variation in the method of handling the parts. Very small parts are usually dumped onto the bed of the lathe. The small size permits the operator to pick up a new part during the cutting time of the part in the machine. When the finished part is removed from the chuck, the new part is immediately placed in the chuck. The finished part is then set aside after the machine is started again. The variable here is more the location of the part rather than its weight. Further analysis shows that this method is limited to parts that can be easily held in one hand.

The next larger size parts are usually placed on a workbench next to the lathe. Because of the bulk or weight of these parts, the operator first removes and sets aside the finished part and then picks up and places the new part in the chuck.

For still larger parts, the practice is to deliver them on skids which are placed behind the operator. The picking up of the part now involves a number of body motions, and the placing of the larger part in the chuck becomes more cumbersome.

When the data are plotted to distinguish among these three different methods, the result is as shown by Figure 8-9. From this plot, it is evident that the really significant variable is not the weight of the part, but rather the method employed in picking it up and placing it in the chuck. Within certain limits, the weight is a factor, because a 20-pound part would be too large to be placed in the bed of the lathe. On the other hand, a 6-pound piece of bar stock might be small enough to be placed in the bed, but a 6-pound piece of thin-wall tubing might be so bulky that it would be placed on the bench. This accounts for the overlap of the data when plotted by weight. The conclusion in this case is that, instead of treating the element as a variable element, it should be treated as three different constant elements.

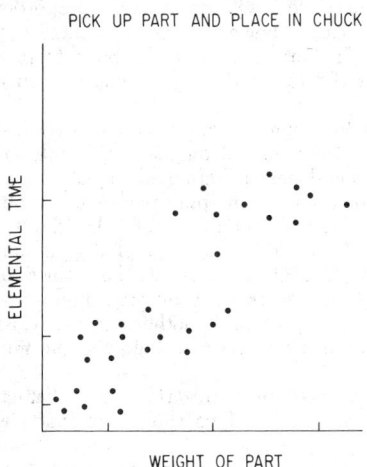

PICK UP PART AND PLACE IN CHUCK

ELEMENTAL TIME

WEIGHT OF PART

FIG. 8-8. Plot of elemental time values and the weight of the parts.

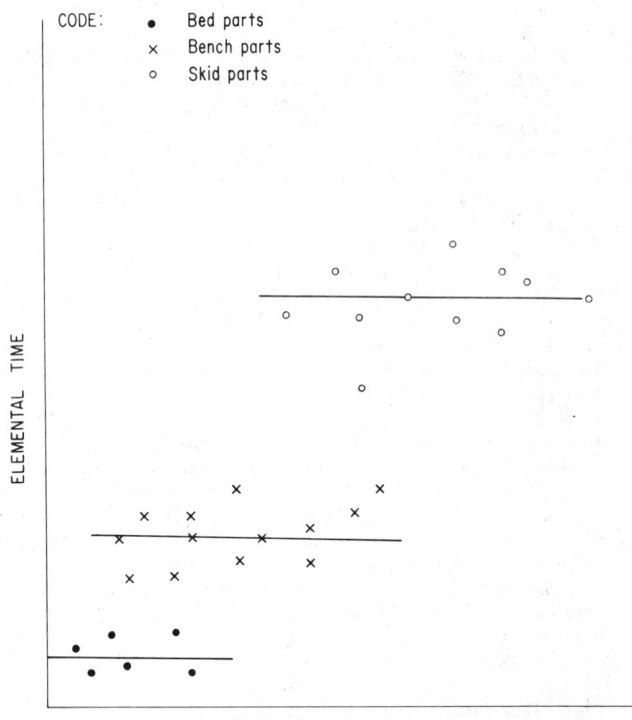

WEIGHT OF PART

FIG. 8-9.   Points plotted by method employed in performing element.

There are many other situations where the elemental time does vary in direct proportion to a single variable.   In foundry work, for example, the time for the element "Shovel sand in mold" will be directly proportional to the net volume of the mold.   In other cases, however, the elemental time may be affected by two or more variables.   In fact, it is often desirable to proceed on the assumption that this is so to avoid the temptation of forcing the curve for a single variable to fit the data.

The best approach is to prepare a chart with the vertical axis representing time and the most obvious variable plotted on the horizontal axis.   If a second variable is suspected, the data should next be grouped in batches, with the second variable as a constant value.   For example, in spray painting, the obvious variable is surface area to be covered.   A second variable might be the type of paint or quality of finish.   In this case, it would be well to plot separately the values for prime coat, first coat, and finish coat, all against the values of time and area.   The result will be a family of curves as shown in Figure 8-10.   Here, the second variable is fairly easy to ascertain.   Another approach is to use a modification of the statistical quality control chart techniques, plotting the $\bar{X}$ and R values for the data.[3]

When an element is affected by two variable job characteristics and the method of treating one of them as a series of constants does not yield satisfactory accuracy,

---

[3] For further details, refer to Gerald Nadler, *Motion and Time Study,* McGraw-Hill Book Company, New York, 1955.

HAND-SPRAY PAINTING TIMES

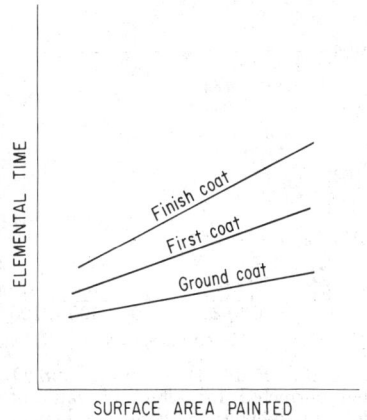

SURFACE AREA PAINTED

FIG. 8-10. Family of curves for spray painting.

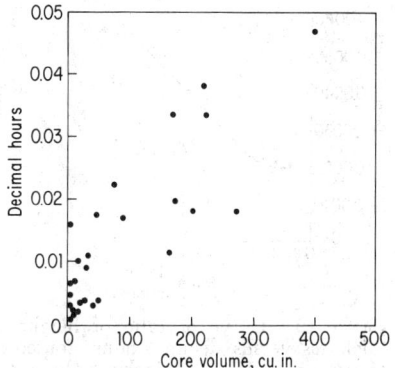

FIG. 8-11. Points plotted from time study data of time required to fill core box with sand and peen.

another way of treating the data must be found. For example, when standard data were being developed for the element "Fill core box with sand and peen," it was anticipated that filling and peening time should vary with the volume of sand handled. Accordingly, points were plotted of volume against time, with the result shown by Figure 8-11. Inspection showed that some other factor undoubtedly affected the data.

Further analysis indicated that the relationship between the height and the thickness of the core might also affect the time. It was reasoned that where the thickness of the core is great in comparison with the height, all sand may be put in the box and peened at one time. Where the thickness of the core is small as compared with the height, sand must be put in a little at a time and peened frequently. Thus it seemed evident that, in addition to core volume, the ratio of height to thickness would affect filling and peening time.

An attempt was next made to classify the work according to this ratio, plotting one curve for ratios up to one, another for ratios between one and two, and so on. The results were unsatisfactory, however, and this approach was abandoned.

It was then decided to use two curves in conjunction with each other. A curve of time against ratio of height to thickness was plotted for cores having an approximately constant volume. Because volume was constant, the curve, Figure 8-12, showed a true relationship unaffected by volume. Note that the curve intercepts the Y axis at 0.0020 hour. The time scale was changed by calling this point of interception "1"; the point at which the time doubled, or 0.0040 hour, "2"; and so on. This produced the factor X scale shown on the right of Figure 8-12.

Next, to get a curve of volume against time unaffected by the height and thickness of the core, each time value taken from the original data was divided by a factor as determined from the curve by the ratio of height to thickness. These values plotted against the corresponding volumes yielded a time-volume curve unaffected by the height and thickness of the core. The points now lined up into two sets as shown by Figure 8-13.

Additional analysis of the data showed that the higher values were all obtained from studies of cores which were difficult to make. A further classification of jobs as simple or complex permitted the plotting of the final curves shown by Figure 8-14.

With the curves now available, to arrive at filling and peening time for any core, a factor X is found corresponding to the ratio of the height and the thickness

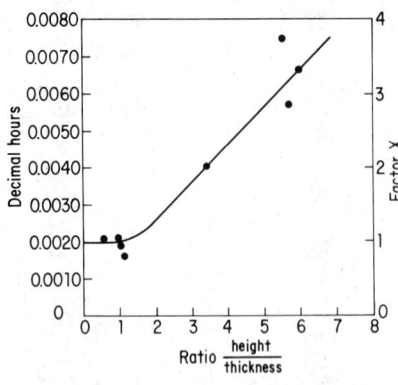

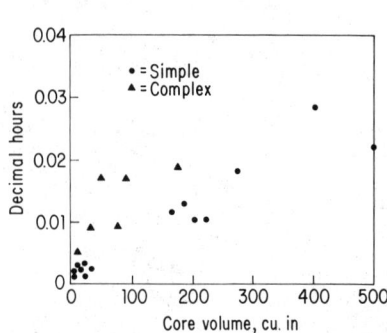

FIG. 8-12. Curve of ratio of height to thickness versus decimal hours for cores of from 5 to 10 cubic inches volume for element "Fill core box with sand and peen."

FIG. 8-13. Points shown in Fig. 8-11 when corrected for effect of ratio of height to thickness of core for element "Fill core box with sand and peen."

of the core, and a base time for the volume is found from the time-volume curve. These two values multiplied together give the true time required for filling and peening. To illustrate, a simple core having a ratio of height to thickness of 4.2 and a volume of 130 cubic inches would have a standard time for filling and peening of $2.4 \times 0.0092 = 0.0221$ hour.

A more sophisticated solution can be achieved by the use of multivariate analysis. For example, the time to pick parts from a bin might be represented as

$$Y = a_0 + a_1X_1 + a_2X_2 + a_3X_3$$

where Y = normal time to obtain one part
X_1 = number of parts grasped at one time
X_2 = location of bin
X_3 = weight of part

To calculate or estimate the coefficients (a_0, a_1, a_2), it is necessary to collect ele-

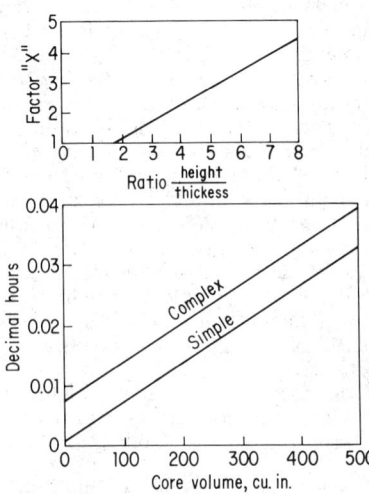

FIG. 8-14. Final curve sheet for element "Fill core box with sand and peen."

mental data for a variety of conditions or variables. In this case, a minimum of four data sets are needed to provide a solution. The greater the number of data sets collected, the higher the reliability of the estimates. Either manual or computer solutions may be used to solve the equation.

Although the above equation is for a linear relationship, nonlinear relationships may also be solved by the same model. For example, for the equation

$$Y = a_0 + a_1X_1 + a_2X_2 + a_3X_1{}^2 + a_4X_1X_2 + a_5X_2{}^2$$

one can set $X_1 = Z_1$; $X_2 = Z_2$; $X_1{}^2 = Z_3$; $X_1X_2 = Z_4$; and in this way, proceed to estimate the linear regression equation described above.[4]

Analysis of Variables by Predetermined Motion Time Standards. When analyzing a variable element by using predetermined motion time standards to supply the elemental time data, the first step is to record the motion pattern required over the full range of part sizes or other variable. This can be done either by observation or by visualizing the motion pattern. If parts at the extremes of the size range are not being worked on in the shop, recording the motion pattern for the extremes by visualization will frequently save many months of waiting time. If the analyst himself is not sufficiently familiar with the work to visualize the motion pattern for different size parts, he can usually have someone else familiar with the operation go through the motions required to perform the element being studied. By recording the motion pattern for different size ranges or other variable, the variations in method, and therefore the variations in time, become quite evident.

For example, when analyzing the element "Pick up part and place in chuck," the analyst will first study the small parts and determine the motion pattern used. He will note that the part is picked up from the bed of the lathe during the cutting time of the previous part. He will see that when the finished part is removed from the chuck by the left hand, the new part is placed in the chuck by the right hand, using the following simple pattern:

10.3	M6C	Move part to chuck
5.6	P1SE	Position part to chuck
3.4	M2C	Insert in chuck
19.3	TMU	

The additional motions required to tighten the chuck jaws and start the lathe are covered by other elements.

As soon as the part becomes too large to be easily held in one hand, the analyst will note that the operator uses a different method. The parts are set on a workbench adjacent to the lathe during the cutting time, when the operator would otherwise be idle. When the part in the lathe is finished, it is removed and set on one end of the bench. The operator then picks up an unfinished part and places it in the chuck, with the following pattern:

Reach to unfinished part	R20B	18.6	R20B	Reach to unfinished part
Grasp part	G1A	2.0	G1A	Grasp part
Move to chuck	M24C10	32.0	M24C10	Move to chuck
Position to chuck	P1SD	11.2	P1SD	Position to chuck
Insert in chuck	M2C	3.4	M2C	Insert in chuck
		67.2	TMU	

If the parts are too large to fit on the bench next to the lathe, they are placed on a skid. The skid is located on the floor behind the operator. The finished part is placed on one side of the skid, and an unfinished part is picked up and placed in the chuck. This motion pattern is similar to the last one, but involves body motions and heavier weights.

[4] For further details, see R. C. Jelinek and W. Steffy, "Use of Multivariate Techniques for the Analysis of Work Measurement Data," *AIIE Journal of Industrial Engineering*, February, 1966.

Reach to unfinished part	R20B	18.6	R20B	Reach to unfinished part
Grasp part	G1A	2.0	G1A	Grasp part
Move for better grasp	M3B30	18.4	M3B30	Move for better grasp
		5.6	G2	Regrasp for better grip
Regrasp for better grip	G2	5.6		
		31.9	AS	Arise from stoop
		18.6	TBC1	Turn to machine and
		34.0	W2PO	Walk to machine
Move to chuck	M6C30	13.7	M6C30	Move to chuck
Position to chuck	P1SD	11.2	P1SD	Position to chuck
Insert in chuck	M2C30	15.3	M2C30	Insert in chuck
		174.9	TMU	

From these analyses, it becomes evident that there are three different methods employed, with a nearly constant elemental time for each method. The time for the element will vary slightly because of the effect of weight on the "Move" time. So there is a relationship of a sort between the weight and the method used. The heavier the part, the bulkier it is. The following table can be set up to take this into account.

Small parts up to 5 pounds.......... 0.0002 hour
Medium parts up to 20 pounds....... 0.0007 hour
Large parts up to 60 pounds.......... 0.0017 hour

When predetermined motion time standards are used to analyze elements, it will be found that most variable elements break down into a series or family of different methods and motion patterns. Consequently, most elements can be handled with a group of constants. Where an apparent variable does exist, as in the case of filling a mold with sand, it is found that even here the time is made up of a varying number of repetitions of a constant element consisting of the time required to shovel one shovelful of sand. The variable is not the elemental time, but the frequency with which the element is performed. In effect, then, predetermined motion time standards permit the conversion of nearly all variable elements into constants. This greatly simplifies the development and use of elemental time values and is one of the major advantages for the use of predetermined motion time standards.

A good example of how MTM can be used to handle a variable element is shown by Figure 8-15. This is a chart giving the time required for turning or cranking the handwheels which are found on almost every machine tool. The variables affecting the element are the diameter of the crank; the distance reached to the crank to grasp it; the resistance to the cranking motion; and of course, the number of revolutions. These variables and the influence they have on time are quite easily determined by an MTM analysis. The table is built up by adding one constant to another. To develop this same table by time study methods would be a much more difficult and time-consuming job.

Applications of Elemental Standard Data. Elemental standard data have been established by many companies on many types of work. The most common are those covering machine shop operations and assembly work. Figure 8-16 is an example of elemental data covering the time required to clamp parts into position prior to machining. Similar charts can be prepared for all elements involved in machine shop work.

In assembly operations, an important portion of the assembly time is devoted to part handling. A set of elemental data covering this work is shown by Figure 8-17. The time unit used in this table is one ten-thousandth of an hour—TTH—or 0.0001 hour.

Data can also be developed rather conveniently for the use of hand tools of all types. A typical set of data for the use of screwdrivers is shown by Figure 8-18.

Data of this type need not be limited to industrial or shop-type operations. They can also be developed for such work as clerical, drafting, and laboratory operations.

CHART 6-10
CRANKING

Diam. of Crank	Dist. Reached to Crank	Number of Revolutions											
		1 Resistance			2 Resistance			3 Resistance			4 Resistance		
		To 2.0#	2.1 to 10#	10.1 to 35#	To 2.0#	2.1 to 10#	10.1 to 35#	To 2.0#	2.1 to 10#	10.1 to 35#	To 2.0#	2.1 to 10#	10.1 to 35#
3	6"	28.4	31.6	39.2	39.0	42.8	52.1	49.6	54.0	65.0	60.2	65.2	77.9
	10"	31.3	34.5	42.1	41.9	45.7	55.0	52.5	56.9	67.9	63.1	68.1	80.8
	14"	34.2	37.4	47.0	44.8	48.6	57.9	55.4	59.8	70.8	66.0	71.0	83.7
4	6"	29.2	32.4	40.2	40.6	44.6	54.1	52.0	56.8	68.0	63.4	69.0	81.9
	10"	32.1	35.3	43.1	43.5	47.5	57.0	54.9	59.7	70.9	66.3	71.9	84.8
	14"	35.0	38.2	46.0	46.4	50.4	59.9	57.8	62.6	73.8	69.2	74.8	87.7
5	6"	29.9	33.2	41.1	42.0	46.0	55.9	54.1	58.8	70.7	66.2	71.6	85.5
	12"	34.2	37.5	45.4	46.3	50.3	60.2	58.4	63.1	75.0	70.5	75.9	89.8
	16"	37.1	40.4	48.3	49.2	53.2	63.1	61.3	66.0	77.9	73.4	78.8	92.7
6	6"	30.5	33.8	41.8	43.2	47.3	57.3	55.9	60.8	72.8	68.6	74.3	88.3
	12"	34.8	38.1	46.1	47.5	51.6	61.6	60.2	65.1	77.1	72.9	78.6	92.6
	16"	37.7	41.0	49.0	50.4	54.5	64.5	63.1	68.0	80.0	75.8	81.5	95.5
7	6"	31.0	34.3	42.4	44.2	48.3	58.5	57.4	62.3	74.6	70.6	76.3	90.7
	12"	35.3	38.6	46.7	48.5	92.6	62.8	61.7	66.6	78.9	74.9	80.6	95.0
	16"	38.9	41.5	47.6	51.4	95.5	65.7	64.6	69.5	81.8	77.8	83.5	97.9
8	8"	32.9	36.2	44.4	46.5	50.6	61.0	60.1	65.0	77.6	73.7	79.4	94.2
	12"	35.7	40.5	48.7	49.3	54.9	65.3	62.9	69.3	81.9	76.5	83.7	98.5
	18"	40.0	44.8	53.0	53.6	59.2	69.6	67.2	73.6	86.2	80.8	88.0	102.8
9	8"	33.3	36.7	44.9	47.3	51.6	62.0	61.3	66.5	79.1	75.3	81.4	96.2
	12"	36.1	39.5	47.7	50.1	54.4	64.8	64.1	69.3	81.9	78.1	84.2	99.0
	18"	40.4	43.8	52.0	54.4	58.7	69.1	68.4	73.6	86.2	82.4	88.5	103.3
10	8"	33.7	37.1	45.4	48.1	52.4	63.0	62.5	67.7	80.6	76.9	83.0	98.2
	14"	38.0	41.4	49.7	52.4	56.7	67.3	66.8	72.0	84.9	81.2	87.3	102.5
	20"	42.2	45.6	53.9	56.6	60.9	71.5	71.0	76.2	89.1	85.4	91.5	106.7
12	8"	34.3	37.7	46.1	49.3	53.6	64.4	64.3	69.5	82.7	79.3	85.4	101.0
	14"	38.6	42.0	50.4	53.6	57.9	68.7	68.6	73.8	87.0	83.6	89.7	105.3
	20"	42.8	44.2 7.5#	54.6 22.5#	57.8	62.1	72.9	72.8	78.0	91.2	87.8	93.9	109.5

FIG. 8-15.   Chart covering the element of cranking handwheels.

CHART 3-4
CLAMP PART IN POSITION FOR MACHINING

Number of clamps required	TYPE I CLAMPS		TYPE II CLAMPS		TYPE III CLAMPS		TYPE IV	ALL TYPES
	Nut and clamp remain on stud KN2	Nut and clamp removed from stud KR2	Nut and clamp remain on stud KP2	Nut and clamp removed from stud KS2	Nut and clamp remain on stud KQ2	Nut and clamp removed from stud KT2	Nut or bolt remains – clamp removed KU2	Relieve strain of clamp KV2
1	.006	.010	.006	.010	.007	.011	.006	.003
2	.012	.019	.011	.019	.014	.021	.011	.006
3	.017	.027	.016	.027	.020	.031	.016	.009
4	.022	.036	.021	.036	.026	.041	.020	.012
5	.028	.045	.026	.045	.032	.051	.025	.015
6	.033	.053	.031	.053	.039	.061	.030	.018
7	.038	.062	.036	.062	.045	.071	.035	.021
8	.043	.071	.041	.071	.051	.081	.040	.024
9	.049	.080	.046	.080	.057	.091	.045	.027
10	.054	.088	.051	.088	.064	.101	.050	.030
Condition	A	B	A	B	A	B	C	
Brief descriptive sketch	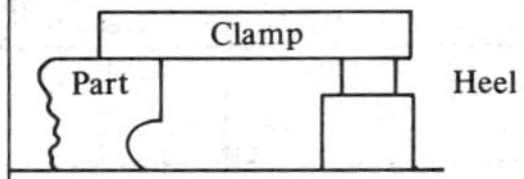		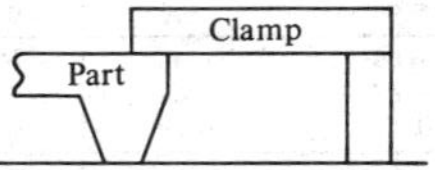		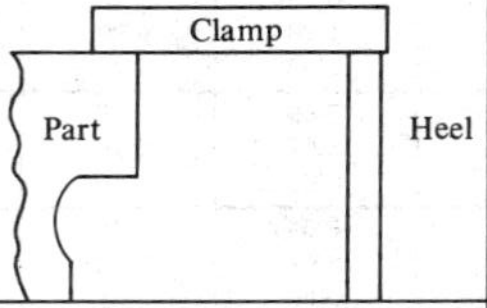		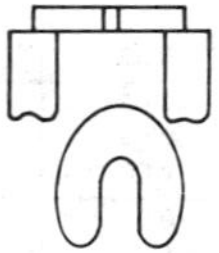	
Basic characteristic	Several blocks stacked to build up heel		Clamp and heel one unit		Clamp considerably above table level – heel will in most cases not stand by itself		Slotted washer	

The chart includes time to pick up wrench, loosen clamp, lay aside wrench, run nut off stud or provide more clearance, reverse clamp, replace clamp, run nut down or back on stud, pick up wrench, tighten nut, lay aside wrench.

Fig. 8-16. Elemental data covering the time required to clamp parts into position prior to machining.

		PART HANDLING				

Operation		Variable or Type			Symbol	TTH
Get and place or Get and dispose	Easy pickup (G1A)	Variable location	6″	EV	06	2
			12″		12	3
			20″		20	4
		Loose position (P21)	6″	EL	06	3
			12″		12	4
			20″		20	5
		Close position (P22, P23)	6″	EC	06	4
			12″		12	5
			20″		20	6
	Jumbled or complex pickup (G4B)	Variable location	6″	JV	06	3
			12″		12	4
			20″		20	6
		Loose position (P21)	6″	JL	06	4
			12″		12	5
			20″		20	6
		Close position (P22, P23)	6″	JC	06	5
			12″		12	6
			20″		20	7
Miscellaneous Additives	Secondary position	Loose (P21)		SP	01	1
		Close (P22, P23)			02	2
	Difficult handling			DH	01	1
	Apply pressure			AP	01	1
	Disengage – recoil			DE	01	1
	New hold			NH	01	1
	Weight factor (ENW)	10 –25 lbs.		WF	01	1
		over 25 lbs.			02	2

FIG. 8-17. Elemental standard data covering part handling time during assembly.

Although a tremendous amount of time has been devoted to the development of elemental time standards by thousands of companies, very little of this work has ever been published.[5] There are two basic reasons for this. One reason is that the data are frequently considered proprietary. They, in effect, indicate the time required for a company to perform an operation or job, and this type of information is considered confidential. Another reason is the difficulty involved in fully describing the data and their application or in transferring them from one set of working conditions to another. When time study is used as the basis for the development of the elemental standard data, it is difficult to describe fully and clearly the beginning and end points of the elements and the methods employed. The use of predetermined motion time standards helps overcome this problem as well as the problem of adjusting the time study performance ratings from one plant to those of another.

Even with predetermined motion time data, however, it is often difficult to transfer the data from one plant to another. Although the data were designed from the beginning to be universal in nature rather than for a specific and limited work application, the elemental times developed from the data must be carefully documented with detailed analyses and write-ups covering their application and limitations

[5] Exceptions are H. B. Maynard, W. M. Aiken, and J. F. Lewis, *Practical Control of Office Costs,* Maynard Research Council Incorporated, Pittsburgh, 1960; and Arthur A. Hadden and Victor K. Genger, *Handbook of Standard Time Data,* The Ronald Press Company, New York, 1954.

SCREWDRIVER

Operation				Turn screw											
Type of screwdriver				Single move	Number of threads										
					1	2	3	4	5	6	7	8	9	10	
Name	Variable	Degrees turned per cycle	Symbol	A	B	C	D	E	F	G	H	J	L	M	
Conventional slot or star	Finger turns — 1/4" hdl.	360	CV — A	1	1	2	3	4	5	5	6	7	8	9	
	Finger turns — 1/2" hdl.	180	B	1	2	4	5	7	9	11	13	14	16	18	
	turns — 1" hdl.	120	C	1	3	5	8	11	14	16	19	22	24	27	
	turns — 1 1/2" hdl.	120	D	1	4	8	12	16	20	23	28	31	35	39	
	Wrist turns	120	E	2	6	13	19	25	32	38	44	50	57	63	
Ratchet	Wrist turns	120	RA A	2	5	10	15	20	26	31	36	41	46	51	
Spiral	Stroke length — 2" 1 rev.		SP A	---	1	2	3	4	5	6	7	8	9	10	
	Stroke length — 4" 2 rev.		B	---	1	2	3	3	4	5	6	6	7	8	
	length — 6" 3 rev.		C	---	1	2	2	3	3	4	5	5	6	7	
Tighten or loosen	Normal		TL 01	2											
	Heavy		02	4											
Engage and disengage	Regular		01	3											
	Magnetic		ED 02	5											
	Split wedge or equivalent		03	7											

Data do not include tool or part handling.

Fig. 8-18.   Typical set of data for the use of screwdrivers.

if they are to be applied successfully in plants other than the one in which they were developed.

TASK DATA AND TIME FORMULAS

The discussion up to this point has shown how, under the building block concept, basic motion time data can be put together into larger combined data units and how either basic motion time data or combined data can be put together into still larger elemental data units. This enlarging process can be carried a step further by putting together basic motion time data, combined data, or elemental data—including elemental data developed by time study—into still larger units called task data.

Task data may be defined as the standard time for performing a specific piece of work. A task may be the same as what has been called an operation up to this point, such as "Assemble type X junction box complete," or it may be a part of many different operations, such as part handling, tool handling, or gaging. A task must be a unit of work which is repeated frequently enough to justify the cost of developing task data, but it need not consist of repetitive direct labor operations. Indeed, the concept of task data sprang from the problems encountered in attempting to establish standards on maintenance work. Maintenance work is nonrepetitive in the sense that very few maintenance jobs are done in exactly the same way very often. It is, however, composed of a number of smaller tasks which are performed over and over again in various combinations and sequences. Figure 8-4 shows that the maintenance job of "Mount and connect medium-size electric junction box" is composed of the tasks of "Mount electrical boxes" and "Make terminal connections," combined with the smaller elemental operations of "Select wire," "Move marker," "Cut off," and "Strip."

The procedure used to develop task data is a modification of the procedure which has long been used to compile time formulas from elemental standard data. The concept of time formulas came from the work of Taylor and Gantt. Although time formulas antedated task data, they now are looked upon as a special case of task data. They are developed largely for direct labor operations where the accuracy of individual standards is a matter of primary concern. They are therefore developed carefully from extensive elemental data and are described in comprehensive reports which set forth and justify every step of the derivation procedure.

Task data which are applied to a large number of different jobs do not require this same degree of accuracy because errors tend to offset one another in accordance with well-defined statistical principles. Hence the development of task data is a somewhat less exacting procedure, although carelessness or avoidable inaccuracies are not condoned.

An understanding of task data development can best be gained by a study of the more exacting time formula derivation procedure. Therefore, an explanation of this procedure will be given first in some detail.

Time Formulas Defined. A formula may be defined as the expression of a general fact, rule, or principle by algebraic symbols. It is a convenient way of expressing the manner of variation between two or more interdependent variables. When all but one of the variable quantities are known in a given situation, it is possible to find the unknown quantity by substitution in and solution of an algebraic formula.

A time formula is a collection of standard time data arranged in the form of an algebraic expression for determining the standard time for an operation. The industrial engineer, when developing a formula for a given class of work, seeks to arrange pertinent time data into such a simple form that anyone can determine the time required to perform a given piece of work on the part by determining quantitatively certain characteristics of the part and substituting these data in the time formula expression.

From Elemental Standard Data to Time Formulas. Various kinds of elemental data have been discussed above. These data can be compiled into a list of all the different elements that are used for doing a given class of work with representative or standard time values for each element shown. Every element that differs even slightly from any other element will have its own time value.

When a job comes up on which no time value has previously been established, it is analyzed either mentally or, preferably, by direct observation, and the elements required to perform it are determined. Time values are then selected for each element from the standard data list. Their sum gives the time standard for the job.

This method of establishing time standards is a great improvement from a time, cost, and consistency standpoint over individual detailed time study, but it is capable of further refinement and improvement. A moment's thought will show that on a given class of work certain elements will be performed on every piece produced, while other elements will be performed only when a piece has certain characteristics. Furthermore, there are cases where the performance of a certain element will always require the subsequent performance of another element. In addition, it can be seen that the time for performing certain elements will be the same, regardless of the characteristics of the part being worked upon, while the time for performing certain other elements will vary with the nature of the part.

For example, to machine a piece on an engine lathe, the element "Pick up part" must always be performed. On the other hand, the element "Place in chuck" will be performed only when the characteristics of the part are such that a chuck is required to hold it. Whenever the element "Start machine" is performed, the subsequent performance of the element "Stop machine" will always be required. An element such as "Engage feed" will require the same time to perform whether the part in the machine is large or small, but the time for performing the element "Lay part aside" will be affected by the size and shape of the part.

Recognition of these facts leads the industrial engineer to the idea of simplifying the application of standard elemental time data by developing time formulas. A time formula is merely a convenient arrangement of elemental data which facilitates their correct application. Much of the analysis which is necessary each time elemental data are applied is done once and for all when the formula is derived. The job characteristics which make the performance of certain elements or groups of elements necessary are determined, and the formula is expressed in terms of these characteristics.

For example, every time a hole is drilled using a sensitive drill press, it is necessary to perform the elements "Lower spindle" and "Raise spindle." If the time for lowering the spindle is 0.0004 hour and for raising it is 0.0003 hour, in developing the formula for this work, these two times would be added together. The following term would then be included in the formula expression

$$0.0007N$$

where N = number of holes drilled

Later, when applying the formula to establish a standard, instead of having to visualize and record every performance of the element "Lower spindle" and every performance of the element "Raise spindle," it will be necessary merely to determine from the drawing the number of holes to be drilled and multiply this number by 0.0007. This same procedure is followed for all other elements or groups of elements.

The great amount of time which the use of formulas saves those charged with the task of setting standards is readily apparent. The time required to make and work up an individual time study or MTM study will be as much as one to four hours when the length of the cycle is quite short. It will be much longer on larger work where 100 or more hours may be required to perform the operation once. Elemental time data will save a great deal of this time, but a time formula will usually save even more. The time required to establish a time standard from a formula will, in the majority of cases, range from one to fifteen minutes, depending upon the complexity of the formula and the amount of time required to determine the variable characteristics of the job. Where all necessary information may be obtained from a drawing or the part, the time standard may generally be computed in less than five minutes.

From Time Formulas to Task Data. The same comments apply to task data. They save a great deal of time in establishing standards for portions of large jobs or operations. Task data may be expressed in algebraic form, but are often presented in tabular form in one or more tables. The same is true of time formulas. Once a formula has been developed, it is often transformed into a chart. The term "time formula" thus becomes something of a misnomer, but because it has been used for so many years to describe combined elemental standard data, it will continue to be used in the present discussion.

Task data stemmed logically from the time formula approach. Originally, if time formulas were being developed in a certain plant, say, for drill presses, milling machines, and slotters, the derivation of each formula was approached as a unique project. It became evident, however, that each formula had certain elements in common. Part handling and clamping parts to the machine table are examples. Thus task data were used to speed up time formula derivations before they were so named and before their wider application was recognized.

To explain more fully how time formulas and task data are developed, the step-by-step procedure for deriving a time formula will now be described. This will be followed by an example of typical task data which shows how elemental data derived as discussed above are organized in a manner which permits their application to a wide variety of tasks.

Steps of Time Formula Development. Time formulas may be divided into two classes, job formulas and operation formulas. Job formulas cover a given operation performed upon a number of different but similar parts. The turning of shafts for small electric motors up to 1 horsepower would be covered by a job formula. The same machine or machines are used to do the work, and the shafts are similar in shape and material although varying in dimensions.

Operation formulas are of more general application and cover all the work which can be done at a given work station or on a single class of machines such as engine lathes, planers, steam hammers, or molding machines. Operation formulas are more complex and require more work to derive than job formulas, but they have a much wider application.

Figure 8-19 shows graphically and in chronological order the steps required to develop either kind of formula. The first step is to make a general analysis and survey of the work to be covered. This analysis is similar to that made when making an individual study, but it covers a class of work as a whole and not merely the individual job. The analysis and improvement made will affect a large number of jobs.

After a clear idea has been formed of what the formula will cover and what must be done to develop it, the next step is the actual collection of time data. Time data may be obtained from previously established elemental times or from time studies or MTM studies of a number of representative jobs. These data are then posted on a spread sheet or master table.

Next comes the step of classifying each elemental operation as either a constant or a variable. The class in which an element belongs is not always readily apparent, and it requires considerable analytical ability to make the classification correctly.

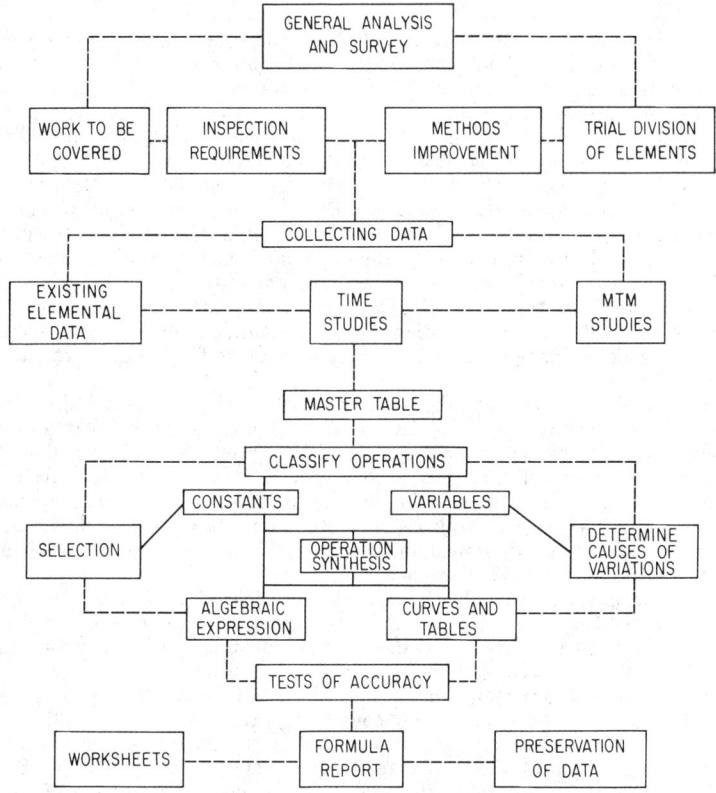

Fig. 8-19.   Steps of time formula development.

When the classifications have been made, it is then necessary to select a definite time for each elemental constant and to make a further analysis of each elemental variable to determine just how and with what it varies.

When all constants and variables have been determined, if they are short and numerous they are often combined into larger elements by a step called operation synthesis. The elements are then further combined and assembled into a formula expression. The resulting formula is tested for accuracy, statistically or otherwise. The final step is to write a report which explains clearly the way the formula was derived and how it is to be applied.

General Analysis and Survey. Before starting to collect data for the compilation of a formula, a general analysis is made to determine the field to which the formula is to apply. The operation to be covered is identified and defined. The number of operators engaged in doing the work and the machines and tools being used are surveyed. The range in size of parts to be covered is decided upon. If the industrial engineer is unfamiliar with the technology of the process, he investigates it thoroughly.

He also investigates the inspection requirements to determine what work must be performed to meet them and what standards of accuracy, finish, and the like must be maintained. He observes the work as it is being done and makes an overall study of methods. If the industrial engineer has previously established standards on the work by individual studies, he will already have a good understanding of the operation. Nevertheless, he will find it profitable to spend some further time considering the work as a class. It is quite possible that he will recognize opportunities for improving methods which he overlooked before when concentrating on single jobs.

Trial Division of Elements and Selection of Jobs for Study. When elemental times are to be determined by time study, it will be wise to make a trial division of the operation into its elements before beginning to collect data. All data are later to be posted on a master table. By comparing the time taken for an element on one study with the time taken for the same element on other studies, it is possible to determine if the element is a constant, and if not, how it varies. This comparison is meaningful only if the operation has been divided into its elements in exactly the same way on all studies. For example, the act of picking up a part, placing it in a box jig, and closing the cover of the jig can be subdivided in several ways. It would not do to divide it into "Get part from table," "Place in jig," and "Close cover" on one study; "Get part from table and place in jig" and "Close cover" on another; and "Get part from table" and "Place in jig and close cover" on a third. Such variations in divisions of elements will more than double the work of compiling a formula and will seriously affect the accuracy of the results.

When data are to be collected by more than one person, it is especially important to establish in advance not only the elements into which the operations are to be divided, but also the exact starting and ending points of each element. Unless this is done, the data collected will probably be useless for time formula development. Of course, if the elements are developed from predetermined motion time standards, the initial way in which the operation is divided into elements is not nearly as critical. Any adjustment of beginning and ending points which is later found to be necessary can easily be made.

A division of elements should first be made on a job selected at random, dividing the operation into as small elements as is consistent with accuracy. Constant and variable elements, to the extent that they can be identified by analysis only, should be kept separated.

Next, the jobs that are going to be studied should be selected. These should include the full range of jobs to be covered by the formula, from the smallest to the largest and the simplest to the most complex. The trial division of elements should be applied to all these jobs to see if it applies adequately to the class of work as a whole. If it does, the industrial engineer is ready to proceed with the actual collection of data.

Collecting Data. If elemental data already exist for the elements as they have been defined for formula derivation purposes, these data may be used and supplemented by additional studies to obtain any elemental data which may be missing. This, however, is seldom the case.

Often, there are existing studies which were previously made to establish standards for individual jobs. If these are stopwatch time studies, it is unlikely that the elements will have been established exactly as needed for formula derivation. If they are MTM studies, the data can be rearranged into the desired elemental groupings, regardless of how the elements may have been originally designated.

Usually, however, the quickest procedure is to make all new studies. As a result of his preliminary analysis and survey, the industrial engineer knows exactly the jobs he needs to study and the elements into which they should be subdivided. All that he needs to do is to collect the required data just as though he were making individual studies—with one exception. A complete record of all identifying information is even more important on jobs which are to be used for formula development than on studies which are used to set only a single standard. The studies that are to be used for formula compilation may be made several weeks or months before they are so used. They will often be referred to several years after the formula has been put into use to find out if conditions or methods have changed. Thus, the value of the studies used for formula development is in direct proportion to the completeness of the recorded descriptive information.

Arraying Data on Spread Sheet. The industrial engineer uses a spread sheet, often called a master table, to tabulate the rather extensive data he has collected and put them in a form which is convenient for analysis. The master table illustrated by Figure 8-7 is 22 by 17 inches in size. When folded once in each direction, it may be conveniently filed with 8½ by 11 inch papers.

All usable data are posted on the master table neatly and preferably in ink. Values which are suspected of being incorrect and elements which are obviously unnecessary need not be posted. Under "Operation description" is recorded the name of each elemental operation. In the space for "Job characteristics" is recorded the date when the study was made, the name of the operator studied, and a complete record of job identification and variable job characteristics. On some classes of work, it will be desirable to make a small, neat sketch of the part at the bottom of the "Job characteristics" space.

In the columns for elapsed time, the leveled time value is posted with or without allowances as may seem most desirable. This is done for each element of each study. Thus, a vertical column is a list of the elemental times which occurred on one particular study. A horizontal line is a list of the time values which occurred on every study for one particular elemental operation.

It is not necessary to list the elements in the order of their occurrence. When the first study is posted, the elements will be tabulated as they occurred. For the succeeding studies, whenever an element is encountered that occurred on the first study, the time value is posted on the line reserved for that element. Otherwise, the name of the element must be recorded on a separate line. Each elemental operation is given an alphabetical symbol, which is recorded in the first column of the master table. When all data have been posted, they are ready for further analysis.

Constant and Variable Elements. Analysis alone will generally be sufficient to tell an experienced industrial engineer if the time for performing an element should be constant on all jobs, but he should also be guided by the data. If analysis shows that an element should be constant, and if the time values posted on the master table appear to be of the same magnitude with only a slight variation between the maximum and the minimum, then it is safe to classify that element as a constant. Similarly, if there is a wide range between the maximum and the minimum time values of an element which analysis shows should be a variable, it is again proper to classify the element accordingly. Only in cases where analysis and the data do not lead to the same conclusion is it necessary to make a further investigation.

After the elements have been classified, a time value must be selected for each element that has been recognized as a constant. The selected value should be the time in which qualified men engaged on the line of work can perform the element, working with an average performance under normal conditions when using the prescribed method.

The value chosen should be, where possible, one which occurred on several different studies and which is about midway between the minimum and maximum values. If sufficient data have been carefully collected, such values will be found in the majority of cases. Where there is an unexplainably large variation and where no value is repeated on two or more studies, all values may be averaged, and the actual value which is closest to this average value may be used.

The treatment of variables requires judgment and good analytical ability coupled with some knowledge of algebra and curve plotting. The simplest way of handling variables is to divide the work into several classes—such as small, medium, and large; or simple, average, and complex—and to select a constant value for each, as was explained on page 3-136. Often, however, it is preferable to express variables in curve or regression line form. Time is used as the ordinate and the other variable as the abscissa. When the regression line is drawn, the way the points line up will indicate whether or not a correct analysis of the variable element has been made.

There are also a number of statistical treatments that may be accorded both constants and variables. The industrial engineer who is familiar with them may find that their application gives him greater confidence in the accuracy of his final results. He will also be able to define and defend whatever degree of accuracy he has achieved to others who have a good grounding in statistical concepts. Statistical treatments, however, often tend to confuse less mathematically sophisticated people such as workers and foremen and may even cast doubt on the practical value of the results. The advisability of using statistical jargon in explaining to them the way the formula was derived is something which should be considered carefully.

Operation Synthesis. When time studies are made, the operation is generally subdivided into as small elements as can be accurately timed with a stopwatch. When MTM is used, the elements often become even smaller. The time for the individual elements is thus determined quite accurately, but on complex work, the analyzing or breaking down of the operation into very short elements results in a large mass of data which may become unwieldy and difficult to manipulate. It is often desirable to synthesize or add together groups of these small elements into fewer larger elements before the derivation of the formula expression is attempted.

Figure 8-20 shows an operation synthesis page taken from a rather complicated turret lathe formula. The larger elements which result from combining smaller elements are shown by K numbers. K32, for example, is the element "Set stop—hex turret." The time allowed for performing it is given in decimal hours. It is 0.0130 hour.

The smaller elements from which K32 was developed are shown directly below. The reference shows the source of the elemental data. A given element may have been taken from the spread sheet or from another similar formula, or it may have been developed by MTM analysis. Each of the elements required to "Set stop—hex turret" is listed, with the time required to perform it shown in TMU. The sum of the times for all the elements is 1,345.2 TMU, which, rounded off, becomes the time for K32, or 0.013 hour.

The formula from which Figure 8-20 was taken contained 92 K elements. It required sixteen pages in the formula report to show how they were derived. Although 92 is still a goodly number of elements, it is evident that there would have been between 400 and 500 much smaller elements to handle had the synthesis step not been employed.

Element Analysis. A complete record of the reasoning and deductions made in classifying elements, choosing the times for the constants, and establishing curves

SYM	REF	OPERATION OR ELEMENT DESCRIPTION	TMU	FREQ.	TOTAL
		OPERATION SYNTHESIS			
		CODE 2222.01			
K32		Set stop – hex turret			.013
	KW2	Loosen and tighten lock screw			483.1
	SKM3	Get and return tool – toolbox			685.6
	Y1	Pick up wrench			35.0
	A2	Lay aside wrench			17.8
	Chart 6-11	Turn screw in or out, 6/12/8/2			123.7
					1345.2
K33		Set tool to cut – hex turret (additional cuts)			.002
	Chart 6-1	Start and stop spindle, 14/2			23.0
	Chart 6-2	Engage and disengage feed, 14/2 + AP	43.6	2	87.2
	Chart 6-2	Move to job, 14/4	38.5	2	77.0
					187.2
K34		Set tool to cut – hex turret, first cut			.003
	K33	Set tool to cut			187.2
	Chart 6-2	Index turret, 16/14	39.9	2	79.8
					267.0
K35		Set up sliding head – hex turret			.013
	SKL3	Get and return tool holder			664.6
	Y1	Pick up wrench	35.0	2	70.0
	A2	Lay aside wrench	17.8	2	35.6
	KW2	Loosen and tighten bolt (1)			483.1
	U11	Place holder in head			47.1
	X11	Remove holder			27.4
					1327.8
K36		Change tool – Jacobs chuck			.011
	SKM3	Get and return tool			756.6
	J4	Loosen chuck			125.2
	K4	Tighten chuck			133.8
	Y1	Pick up tool			35.0
	A2	Lay aside tool			17.8
	U11	Position key in chuck			33.0
					1101.4
K37		Place sleeve – hex turret			.007
	Chart 2-1	Place and remove sleeve, Class III/A			113.5
	SKL3	Get and return sleeve			593.6
					707.1

FIG. 8-20.   Typical operation synthesis.

and tables is of considerable value. This record is known as the element analysis. In the element analysis, the prescribed method for performing each element is described carefully and completely. The reasons for treating it as a constant or a variable are described. The selected values for all constants are explained and justified. The derivation of the curve or table for each variable element is given and its basis justified, either statistically or otherwise. Every step and thought which led to the final result is recorded.

Should it become necessary to investigate or revise the formula at any time in the future, a review of the element analysis will give the industrial engineer who made the formula, or any other industrial engineer, a complete understanding of the methods on which the formula was based, the way in which it was derived, and the reasons for each elemental value.

Deriving the Formula Expression. The formula expression is developed by combining constant and variable elemental time values wherever possible. It should not be the aim to make an imposing or complicated expression, but rather one that can be applied in the shortest possible time with the minimum chance of error.

It requires a good deal of ingenuity to find all possible ways of simplifying a formula expression. More than one novice in formula derivation work has brought a formula expression to his supervisor for approval only to have it pointed out to him how it could be reduced to half the number of terms.

Certain kinds of data should be compiled into tables and referred to in the formula by a single symbol. Curves can often be added or combined and then put into tabular form. Tables are generally easier to use accurately than curves, and are especially desirable when those who will apply the formula have little understanding of graphics.

When all constants have been combined and all charts and tables developed, there remains only the task of placing them in the most convenient form for reference. The algebraic expression is usually entirely satisfactory. It shows clearly all factors which must be considered and precludes the possibility of any omissions.

The constant which applies to every job, if there is one, should come first. Then follow other constants with the symbols, which, together with the symbol explanation, show when and how often each constant should be used. Lastly should follow reference to tables and curves.

The final formula expression should be in the simplest terms to which the data may be reduced. All possible combinations and contractions should have been made. In most cases, even for rather complex work, the final formula expression will be rather short and easy to solve.

To illustrate, Figures 8-21a and b show a detailed time study made to establish a standard on a simple milling machine operation. The same standard can be derived much more quickly from the formula

$$\text{Table 1} + \text{Table 2} = \text{each-piece time}$$

Table 1 combines the times for the variable elements "Pick up part from table," "Place in vise," and "Lay aside part in tote pan" with the times for the constant elements "Tighten vise," "Start machine," "Run table forward 3 inches," "Engage feed," "Stop machine," "Release vise," and "Brush vise." Table 2 combines the times for "Mill slot" and "Return table." The standard time for milling a slot in a brass clamp of any size is computed by determining the variable characteristics from the drawing—in this case, the volume of the part and the perimeter of the cut—and adding together the times read from Table 1 and Table 2.

Testing the Formula. When the formula has been put into its final shape, the industrial engineer must satisfy himself of its accuracy. Before he can attempt to sell the accuracy of the formula to anyone else, he must have complete confidence in it himself.

The industrial engineer may test his formula by checking time values derived from the formula against existing time values if they are known to be accurate, by checking against the studies used to develop the formula, by checking against detailed studies or overall checks made expressly for the purpose, or by a combination of all these.

If the formula values check closely with the values with which they are being compared and if at least one of each type of job which will come within the range covered by the formula is included in the sample, the industrial engineer may decide by inspection that his formula is sufficiently accurate for its intended

DATE 2-1 — STUDY NO. 1 SHEET NO. 1 OF 1 SHEETS

S-1

COLLECTING AND TABULATING DATA — MILL SLOT FORMULA

NUMBER		1		2		3		4		5		6		7		8		9		10		11		12	
ELEMENTS		PICK UP PART FROM TABLE		PLACE IN VISE		TIGHTEN VISE		START MACHINE		RUN TABLE FORWARD 3"		ENGAGE FEED		MILL SLOT		STOP MACHINE		RETURN TABLE 5.5"		RELEASE VISE		LAY ASIDE PART IN TOTE PAN		BRUSH VISE	
NOTES	LINE	T	R	T	R	T	R	T	R	T	R	T	R	T	R	T	R	T	R	T	R	T	R	T	R
	1	06	06	06	12	23	35	03	38	08	46	02	48	72	120	16	36	13	49	12	61	09	70	11	81
	2	05	86	08	94	21	215	03	18	09	27	03	30	76	306	13	19	12	31	14	45	07	52	06	58
	3	07	65	07	72	21	93	02	95	10	405	03	08	70	78	12	90	-	M	-	M	-	513	08	21
	4	05	26	A 07	46	14	60	03	63	08	71	03	74	73	647	13	60	16	76	12	88	08	96	08	704
	5	07	11	07	18	19	37	04	41	09	50	02	52	71	823	11	34	14	48	09	57	07	64	08	72
	6	07	79	08	87	17	904	03	07	10	17	02	19	72	91	11	1002	13	15	11	26	09	35	—	
	7	07	42	08	50	22	72	03	75	12	87	03	90	71	1161	16	77	14	91	09	1200	06	06	07	13
	8	05	18	06	24	15	39	03	42	B 10	64	02	66	76	1342	12	54	16	70	10	80	06	86	06	92
	9	06	98	08	1406	12	18	02	20	13	33	03	36	74	1510	14	24	15	39	10	49	06	55	06	61
	10	06	67	08	75	14	89	03	92	12	1604	04	08	71	79	12	91	16	1707	11	18	07	25	08	33
	11	05	38	06	44	16	60	02	62	09	71	03	74	75	1849	13	62	12	74	09	83	08	91	07	98
	12	C 05	1918	08	26	14	45	03	48	12	60	04	64	71	2035	15	50	16	66	13	79	07	86	07	99/92
	13	06	92/86	07	2106/99	14	20	04	24	13	37	02	39	75	2214	14	28	15	43	11	54	08	62	09	71
	14	(15)	86	07	93	13	2306	03	09	12	21	04	25	71	96	13	2409	15	24	10	34	09	43	07	50
	15	06	56	08	64	17	81	03	84	13	97	03	2500	D 72	2694	16	2710	16	26	10	36	08	44	08	52
	16																								
	17																								
	18																								
	19																								
	20																								

SUMMARY

	1	2	3	4	5	6	7	8	9	10	11	12
	.0083	.0109	.0257	.0044	.0160	.0043	.1090	.0201	.0203	.0151	.0105	.0095
NO. OBSERVATIONS	14	15	15	15	15	15	15	15	14	14	14	13
AVERAGE "T"	.00059	.00073	.00171	.00029	.00107	.00029	.00727	.00134	.00145	.00108	.00075	.00073
MINIMUM "T"	.0005	.0006	.0012	.0002	.0008	.0002	.0070	.0011	.0012	.0009	.0006	.0006
MAXIMUM "T"	.0007	.0008	.0023	.0004	.0013	.0004	.0076	.0016	.0016	.0014	.0009	.0009
RATING (S.E.C. & CY.)	D C1											
LEVELING FACTOR	1.05											
L.F. x AVE. "T"	.00062	.00077	.00180	.00030	.00112	.00030	.00727	.00141	.00152	.00113	.00079	.00077
% ALLOWANCE	15	15	15	10	10	10	10	10	10	15	15	15
TIME ALLOWED	.0007	.0009	.0021	.0003	.0012	.0003	.0080	.0015	.0017	.0013	.0009	.0009

FOREIGN ELEMENTS

SYM	R	T	DESCRIPTION
A	539 / 526	13	PIECE IN VISE WRONG
B	1256 / 1244	12	MACHINE STOPPED
C	1913 / 1898	15	CLEAN HANDS
D	2656 / 2534	122	PIECE PULLED OUT OF VISE
E			
F			
G			
H			
I			
J			
K			
L			
M			
N			
O			

SKILL			EFFORT		
	A1, A2	SUPER		A1, A2	EXCESSIVE
	B1, B2	EXCELLENT		B1, B2	EXCELLENT
	C1, C2	GOOD	✓ C1	C1, C2	GOOD
✓	D	AVERAGE		D	AVERAGE
	E1, E2	FAIR		E1, E2	FAIR
	F1, F2	POOR		F1, F2	POOR

CONDITIONS		CONSISTENCY	
A	IDEAL	A	PERFECT
B	EXCELLENT	B	EXCELLENT
C	GOOD	C	GOOD
D	AVERAGE	D	AVERAGE
E	FAIR	E	FAIR
F	POOR	F	POOR

	SKILL	EFFORT	COND	CONSY
GENERAL RATING FOR STUDY	.00	+.05		

STUDY STARTED	STUDY FINISHED	OVERALL TIME
2:54 P.M.	3:10½ P.M.	.28 HRS.

FIG. 8-21*a*. Detailed time study of a simple milling machine operation—front.

STUDY /   DATE 2-1-   OPERATION /

MILL SLOT

DWG. 22289   SUB. /   STYLE   ITEM /   L.SPEC.   SUB.

MATERIAL COMMON BRASS

DEPARTMENT 10	OPERATOR MAN/WOMAN NAME GROSS NO. 33	
MOULD	PATTERN 9341-R	DIE
	PART DESCRIPTION CLAMP FOR REGULATOR - TYPE X-4	INS. SPEC.

EQUIPMENT #3 LE BLOND HORIZONTAL MILLING MACHINE   MACHINE TOOL NO. 35P9

SPECIAL TOOLS, JIGS, FIXTURES, ETC.: 6" DIA SPL SIDE CUTTER

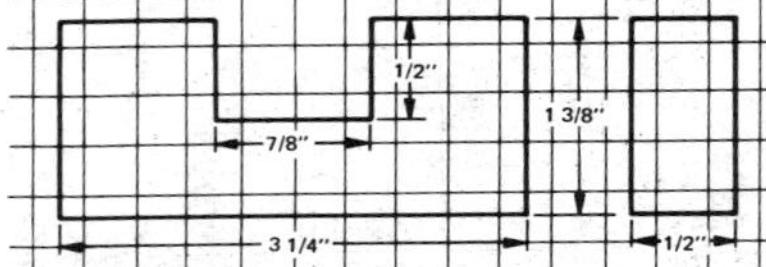

NO.	ELEMENTS	SMALL TOOL NOS. FEED SPEED, DEPTH OF CUT, ETC.	ELEMENTAL TIME ALLOWED (BOTTOM LINE OTHER SIDE)	OCCURRENCES PER PIECE OR CYCLE	TOTAL TIME ALLOWED
1.	PICK UP PART FROM TABLE				.0007
2.	PLACE IN VISE				.0009
3.	TIGHTEN VISE				.0021
4.	START MACHINE				.0003
5.	RUN TABLE FORWARD 3"				.0012
6.	ENGAGE FEED				.0003
7.	MILL SLOT	6" DIA SPL SIDE CUTTER 190 R.P.M. 6"/MIN.			.0080
8.	STOP MACHINE				.0015
9.	RETURN TABLE 5.5"				.0017
10.	RELEASE VISE				.0013
11.	LAY ASIDE PART IN TOTE PAN				.0009
12.	BRUSH VISE				.0009

EACH PIECE   TOTAL .0198

TIME ALLOWED, SET UP

REMARKS: OPERATOR REMOVES PARTS FROM TOTE PAN AND PLACES THEM ON TABLE WHILE MACHINE IS MAKING CUT. HE ALSO CLEANS CUTTINGS FROM TABLE AT THIS TIME. CUTTING SPEED FOR THIS LINE OF WORK IS HELD CONSTANT AT 190 R.P.M. FEED VARIES WITH WIDTH AND DEPTH OF CUT. ON THIS JOB, FEED IS 6" PER MINUTE

OBSERVER   APPROVED BY   SKETCH

3/4"   7/8"   1/2"   1 3/8"   1/2"

OBSERVATION SHEET

FIG. 8-21b. Detailed time study of a simple milling machine operation—back.

3-150

purpose. Or he may go further and compute a coefficient of correlation to determine more surely just how accurate his formula is.

The Formula Report. When the formula has been compiled and tested, there remains only the writing of the formula report. The formula report tells in full the details of the construction of the formula, and it also explains just how the formula is to be applied. It should enable anyone to check back at any future period to see how and where each value was obtained. It should also make it possible for anyone who is familiar with formulas in general to apply this formula even though he has never seen it before.

The formula report should be written up in standard form, both to ensure uniformity of practice throughout the plant, and more important, to minimize the chance for the omission of any pertinent facts. The standard subdivisions generally found in all reports are as follows.

Formula number
Date

Part:
Operation:
Work station:
Allowed time:
Application:
Analysis:
Procedure:
List of studies:
Element times:
Synthesis:
Inspection:
Payment:
Approvals:

The element analysis, which is often quite detailed and lengthy, is generally included as an appendix to the report.

In addition to these standard subdivisions, any other set of facts which it is desirable to treat separately may be placed in an additional subdivision with an appropriate subheading. At the end of the report, space should be provided for the signature of the industrial engineer who made the formula and for the approval of his immediate supervisor.

Example of a Time Formula

The following (including "Approvals" on page 3-153) is an example of a time formula. Because of space limitations, the list of studies, elemental times, and synthesis are not shown. A study of the example will serve to clarify the foregoing discussion of time formula reports.

Formula no. 2222.01
August 15, 19—
MACHINING TIME FORMULA
TURRET LATHES—RAM

Part: Steel, cast iron, and nonferrous metals, bars, plates, castings, forgings, and so forth.

Operation: Turning, boring, facing, forming, tapping, die head chasing, threading chasing (follower type), pipe tapping, drilling, straight reaming, taper reaming, special taper reaming, burning, and polishing.

Work station: Ram-type bar and turret lathes

Turret lathe Machine reference numbers	Bar lathe Machine reference numbers
110	97
	119
	113
	1399
	1400
	1401
	1455

Allowed time: Calculate time from worksheets and charts.

Application: This formula applies to all the operations mentioned above performed on a variety of metals. It also applies to all setup and manual operations necessary to perform the machining operations.

Analysis: The operations are performed under the conditions described below. As long as the conditions remain as described in this formula, the time values are also applicable to parts machined on the saddle-type turret lathe and bar machines. If there is a change in conditions, methods, material, or equipment, the time values must be revised.

Raw material, tools, and drawings are brought to the lathe by material handlers. The operator punches in at the time recorder and commences to make the new setup. Part handling is done by the machine operator with the assistance of a bridge crane or jib hoist. Machine time is calculated on feed, speed, and machinability of material based on surface speed.

The following is a list of tools associated with this formula:

Dial indicator	Universal surface gage
Open end wrenches	Thickness gages
Files	Screw pitch gages
File cleaning brushes	Center gage
Oil cans	Radius gage
Grease guns	Outside calipers
Babbitt hammer	Inside calipers
C clamps, bolts, nuts,	Hermaphrodite calipers
and the like	Divider calipers
Allen wrenches	Bevel protractor
Special wrenches	Combination square with
Inside micrometers 2″ to	center head
largest	Depth gage
Outside micrometers 2″ to 23″	Depth micrometer 0″ to 3″
Thread gages—plug and ring	Inside micrometers 1½″ to 6″
Torque wrenches	Outside micrometers 0″ to 3″
Vernier calipers	2″ scale
Large-radius gages	6″ hook scale
Taper gages	12″ scale
Thread micrometer	18″ scale
Precision squares	24″ scale
Vernier protractor	Adjustable wrenches
Brushes and rags	Tape measure
Box wrenches 1⅛″ and up	Rail indicator
All cutting tools	Radius gages up to 3″
All special gages	Ball peen hammer
Storage boxes and cupboards	Toolbox

The machines and tools must be in good working condition; the shop well lighted with fluorescent lamps and incandescent bulbs; the aisles well laid out and clear of all work in process; and benches and toolboxes conveniently located near the machine. A drinking fountain and a rest room are located in the general shop area.

The foreman and inspector should be in or near the department area at all times. The foreman, along with his other functions, plans the work for each operator and sees that all the tools, gages, and prints are on hand for a given job before it starts. In the event that everything necessary to perform the next job has not been provided by the foreman or others at the machine, the operator will check out on a charge account while he gathers the tools, gages, and prints.

The capabilities of a typical machine were studied and listed so that near-maximum performance could be secured from each machine, considering:

1. Continuity of cut
2. Cutting tool size

3. Clamping strength
4. Distortion of workpiece
5. Material specifications
6. Overhang of tool or workpiece
7. Finish requirements

Cutting charts were prepared to facilitate the calculation of the cutting time necessary to machine the work to the required specifications.

Included in the standard time is a percentage allowance providing time for personal needs, unavoidable delays, rest periods, and minor work elements.

Procedure: The material, tools, and drawings are brought to and removed from the work station by material handlers. This movement of material is the responsibility of the foreman. The foreman gives a new job to an operator, preferably before the operator finishes the previous job. When the operator finishes a job, he runs the job card and clock card through the time recorder and commences with the next setup and machine operation. The unmachined parts and finished parts (pallets) are located as near as possible to the machine. The operator sets up the machine, fixtures, clamps, tools, gages, and so on, for the job. The part may be loaded manually or by jib crane or bridge crane, depending on the weight and bulk of the object. The part is machined as required and gaged when necessary to make sure that it meets drawing or engineering specifications. The finished part is unclamped or unlocked and removed to an area adjacent to the machine. When the work order is completed, the setup of the machine is dismantled and set aside. The machinist cleans the machine and tools when necessary, especially at the end of the shift.

Inspection: There is no allowance for delay due to inspection performed by the inspection department. The part is generally inspected after removal from the machine. In rare cases, where there may be delay due to inspection, the operator should clock out on a charge account. Gaging that is performed by the operator is included in the each-piece time.

Payment: The formula is used to develop standards for use in checking performance under the company's measured daywork plan.

Approvals:

J. G. Merten	S. S. Essex
Chief Ind. Engineer	Senior Ind. Engineer

(*End of time formula*)

Steps of Task Data Development. A time formula of the type just discussed is designed for application to a specific class of work in a specific environmental situation. The accuracy of each standard established by applying the formula is considered important. Thus, time formulas developed in the manner described are used largely for setting standards on direct labor operations where there is a good deal of repetition.

Task data, on the other hand, are designed for universal application. Although the data are developed carefully and painstakingly, it is not expected that they will yield the exact time required to perform an element or a task under every conceivable situation. Rather, they are designed to give the average time which will be required to perform the task under the conditions commonly encountered in industry. Individual time standards which are higher or lower than the actual times required to perform the tasks are expected to average out. Performance is measured by comparing the time taken to perform a number of jobs with the total time allowed by the task data for these jobs.

The steps of task data development are shown graphically by Figure 8-22. Task data which are sufficiently comprehensive to cover all types of work performed in all types of industry involve many thousands of elemental and task standards. If it is to be possible to retrieve these data easily when needed, a coding system must be designed and used consistently from the time that data collection starts.

The collection of data is a lengthy task. Generally, data are developed as the need for them arises. Elemental data are developed from time study or predeter-

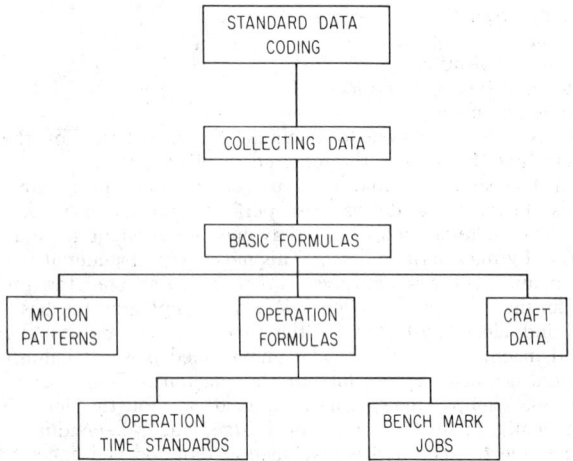

FIG. 8-22.   Steps of task data development.

mined motion time standards as described above. They are combined into groups to make task data as task standards are found necessary. In due course, sufficient elemental data are assembled to make it possible to develop most new task data needed.

Task data are then written up as basic formulas covering frequently used motion patterns; operation formulas (similar in many ways to the time formulas just discussed) covering whole classes of work such as automotive repair; or craft data covering tool handling, gage use, and the like.

Finally, task data are used to establish standards for specific operations or to develop the bench mark job building blocks discussed below.

Example of Task Data

Figure 8-23 shows the analysis sheet used to develop a bench mark standard for testing the compression of a six-cylinder gasoline engine. The standard is the sum of twenty-seven different operation times. These twenty-seven times are obtained from previously established task data. The code numbers in the "Reference symbol" column show the source of the task times.

For example, the first operation performed by the operator once he is in the general work area is to move to the location where the work is to be done. The reference symbol shown is 03.0104. This means that the time for moving to location was obtained from basic formula code 0003.01 and that the value from this formula identified as 04 was selected. The following is basic formula code 0003.01.

BASIC FORMULA CODE 0003.01
COMMON BODY DISPLACEMENTS

Operations covered:
1. Small body displacements (B, SS, TB)
2. Walk displacements
3. Climbing steps
4. Crawling
5. Using a chair

Application: The data in this basic formula (whose method of development is shown in Figure 8-24) apply to the common body displacements that are necessary to do a job at the job site. The body displacements are for a movement in

BENCH MARK ANALYSIS SHEET

CODE 1490

Description	Test compression:		Date: 7/20/—	BM # 19
	6-cylinder gas engine		Craft: Auto repair	
			Dwgs:	
		No. of men 1	Analyst: I.B.	Sh. 1 of 1

Line	Men	Operation description	Reference symbol	Unit time	Freq.	Total time
1	1	Move to location	03.0104			.0015
2	1	Open hood	04.0108	.0016	2	.0032
3	1	Body displacements	03.0103	.0009	3	.0027
4	1	Remove and replace spark plugs	1420.0403	.0297	6	.1782
5	1	Get test kit	04.0103			.0017
6	1	Open box	12.0006			.0017
7	1	Get adapter	04.0102			.0011
8	1	Get and lay aside wrench	05.0002	.0015	2 X 6	.0180
9	1	Install adapter − later remove	06.0202	.0085	2 X 6	.1020
10	1	Use jumper wire	1420.0416	.0174	2	.0348
11	1	Crank engine	13.0003	.0050	6	.0300
12	1	Get and lay aside tester	04.0102	.0011	2 X 6	.0132
13	1	Read tester	13.0001	.0012	6	.0072
14	1	Assemble and disassemble to adapter	12.0006	.0017	2 X 6	.0204
15	1	Get and lay aside pencil	05.0006	.0012	6	.0072
16	1	Body displacements	03.0102	.0005	2 X 6	.0060
17	1	Record gage reading	08.0113	.0007	6	.0042
18	1	Misc. body displacements	03.0103	.0009	15	.0135
19	1	Check and inspect	13.0002	.0025	4	.0100
20	1	Get and lay aside oil can	04.0103	.0017	2	.0034
21	1	Use oil can	14.0021	.0016	6	.0096
22	1	Repeat lines 11−17				.0882
23	1	Put tester in box	04.0102	.0011	4	.0044
24	1	Close box	12.0006			.0017
25	1	Start and stop engine	02.0020	.0100	2	.0200
26	1	Body displacements	03.0103	.0009	6	.0054
27	1	Close hood	04.0108	.0016	2	.0032
28						
29						
30						
Notes:			Bench mark time			.5925
			Standard work group			D

FIG. 8-23. Bench mark analysis sheet.

one direction only—they are not "round trip" values. They are used in the development of other basic and craft formulas and in the analysis and evaluation of bench mark jobs. These data are not to be confused with the basic formula for area travel, code 0002.00.

When an unusual body displacement occurs, it is not necessary to develop another table value. The body displacement in question will be the equivalent of a displacement in the table, a combination of the displacements, or a case of applying the correct frequency of occurrence. The type of operation and how it should be performed will determine what is required.

Analysis: The body displacements that are most frequently used at the job site are the small body and walk displacements. Therefore, they appear in the first part of the table and are presented in convenient ranges for ease of application.

The small body displacement is an average of five MTM body motions. This is representative of the condition that is often encountered when a worker must shift his body to a new location to perform the next element of work. This is set up in the table to cover a chargeable displacement up to about two feet.

COMMON BODY DISPLACEMENTS
DATA TABLE   CODE 0003.01

Type of Displacement	Range of Displacement	Symbol	Hours
Small body displacement	Up to 2 feet	01	.0003
	2.1 to 5 feet	02	.0005
	5.1 to 15 feet	03	.0009
*Walk displacement	15.1 to 25 feet	04	.0015
	Over 25 feet - per each 10 feet	05	.0006
	1 to 5 steps	06	.0008
Climb or descend steps	6 to 15 steps	07	.0018
	16 to 25 steps	08	.0033
Crawl on hands and knees about 5 feet		09	.0023
Use chair - - complete act of sitting or standing		10	.0016

*When loads of greater than 35 pounds are encountered, or when the load is bulky, apply a factor of 1 1/2.

OPERATION SYNTHESIS   CODE 0003.01

SYM	REF	OPERATION OR ELEMENT DESCRIPTION	TMU	FREQ.	TOTAL
03	C	Walk displacement of about 10 feet in one direction (5.1 to 15 feet)			.0009 93.6
04	D	Walk displacement of about 20 feet in one direction (15.1 to 25 feet)			.0015 153.6
05	E	Walk displacement for each 10-foot increment in direction - - for distances greater than 25 feet			.0006 60.0

ELEMENT SUMMARY   CODE 0003.01

SYM	ELEMENT DESCRIPTION	TMU
C	Walk displacement of about 10 feet in one direction (5.1 to 15 feet)	93.6
D	Walk displacement of about 20 feet in one direction (15.1 to 25 feet)	153.6
E	Walk displacement for each 10-foot increment in one direction - - for distances greater than 25 feet	60.0

ELEMENT ANALYSIS CHART   CODE 0003.01

DESCRIPTION – LEFT HAND	NO.	LH	TMU	RH	NO.	DESCRIPTION – RIGHT HAND
C Walk displacement of about 10 feet (5.1 - 15)						C
			18.6	TBC1		
			75.0	W5P		
			93.6			
D Walk displacement of about 20 feet (15.1 - 25)						D
			18.6	TBC1		
			135.0	W9P		
			153.6			
E Walk displacement for each 10-foot increment for distances over 25 feet						E
			60.0	W4P		

FIG. 8-24.   Illustration of standard data development procedure.

The walk displacements are analyzed on the basis of a pace equaling 15 TMU. The number of paces required for the ranges established takes into consideration the short paces that are required for starting and stopping. The walk displacement for the 2.1 to 5 feet range is frequently encountered when a worker is making a repair in the production shop. The larger walk displacements of 5.1 to 15 feet and 15.1 to 25 feet are often required on installation projects. Occasionally, a walk displacement of greater than 25 feet is required. This can be encountered when a worker is making a series of machine inspections. A table value is set up for this, and it is per each 10 feet of displacement.

The latter part of the table covers the step and crawl displacements and the use of a chair. Normally, these are not used as frequently as the small body and walk displacements. Therefore, the three ranges set up in the table for climbing or descending steps have a much greater time spread. Only single values are set up for the crawl displacement and the use of a chair.

When loads of greater than 35 pounds are encountered or when the load is bulky, factor the walk displacements by 1½. This factor is approximately equal to the ratio of the length of pace and the TMU value per pace for walking with loads of 5 to 35 pounds and greater than 35 pounds, or $\left(\dfrac{30''}{24''} \text{ pace} \times \dfrac{17}{15} \text{ TMU} \right)$ = approximately 1.5.

The time value identified by the symbol 04 will be found in the common body displacements data table in Figure 8-24. This shows that the analyst allowed for a walk displacement of from 15.1 to 25 feet.

Operation 4 in Figure 8-23 is described as "Remove and replace spark plugs." The reference symbol is 1420.0403. This refers to formula 1420.04. Excerpts from this report follow, ending with Figure 8-25.

TABLE 8-1.   Automotive Electrical Table

Code: 1420.04

Operation		Symbol	Hours
Spark plug	Remove and replace	03	.0297
	Install new	04	.0339
	Gap	01	.0211
	Clean	02	.0190
Check and fill battery	12 volt	05	.0927
	6 volt	06	.0503
Distributor	Remove and replace dist.	10	.1099
	Remove and replace points	09	.1303
	Check points and oil cam.	12	.0338
Adjust engine timing		08	.1056
Check voltage regulator (A.V.R. tester)		14	.0733
Replace taillight bulb		11	.0433
Replace sealed beam unit		13	.0853
Check lights and instruments		07	.0520
Replace section of electrical wire		15	.0553
Disconnect wire term. and tag 2 ends		18	.0299
Remove and replace light bulb		17	.0288
Use jumper wire		16	.0174

Analyst: I. Bergsson   Formula: 1420.04

Approved: J. H. Brooks   Date: July 23, 19___

Approved: E. O. Helland   Page: 1 of 11

FORMULA REPORT

Operation: Automotive electrical. Replace and adjust various automotive electrical components.

Work station: Automotive repair shops in diesel building and assigned repair areas.

Leveled time: Times shown are in leveled hours. Area travel and allowances are not included.

Application: This formula applies to all automotive equipment in the automotive repair shops.

Analysis: This formula consists of various operations pertaining to automotive electrical systems as they are performed in the automotive repair shops.

SYM	REF	OPERATION OR ELEMENT DESCRIPTION	TMU	FREQ.	TOTAL
02		Clean spark plug			0.0190
	04.0101	Get and lay aside plug	30	2	60
	12.0002	Position plug in cleaner	70	2	140
	B	Sand blast	500	2	1000
	12.0002	Remove from cleaner	70	2	140
	13.0002	Inspect plug	250	2	500
	03.0101	Body displacement	30	2	60
					1900
03		Remove and replace spark plug			0.0297
	05.0002	Get and lay aside wrench			150
	03.0101	Bend to engine	30	2	60
	12.0003	Remove and replace wire	100	2	200
	06.0212	Remove and replace plug	1030	2	2060
	04.0102	Get and lay aside plug	110	2	220
	03.0101	Arise from engine	30	2	60
	04.0102	Get and lay aside wire	110	2	220
					2970
04		Install new spark plug			0.0339
	04.0102	Get and lay aside box	110	2	220
	12.0003	Open box			100
	12.0003	Remove plug from box			100
	03	Remove and replace plug			2970
					3390

FIG. 8-25. Automotive electrical operation synthesis, code 1420.04.

The automotive electrical table, Table 8-1, shows the times for four operations related to spark plugs. The operation identified by symbol 03 is "Remove and replace." The time for this per spark plug is 0.0297 hour. The derivation of this time value is shown in the operation synthesis, Figure 8-25. It shows that the operation of "Remove and replace spark plug" consists of the elements "Get and lay aside wrench," "Bend to engine," "Remove and replace wire," "Remove and replace plug," "Get and lay aside plug," "Arise from engine," and "Get and lay aside wire." Each element but the first is performed twice per spark plug.

Application of Task Data. The use of task data building blocks to build up still larger building blocks is quite simple. Once the coding system is understood, any desired value can be quickly found.

In developing standards from task data, the analyst can concentrate on establishing the proper method for doing the task. He does not need to concern himself with the time that should be allowed for each element, for this has already been determined for normal or representative conditions. At the same time, if he wishes to check the motion pattern on which any elemental or operation time was originally based, he need only refer to the basic formula reports to see exactly what was allowed.

STANDARD WORK GROUPS AND BENCH MARK STANDARDS

Industrial engineers for many years have sought ways of economically applying work measurement techniques to nonrepetitive operations. Many of these operations are in job shops and indirect labor areas. They include such activities as receiving, inspection, transportation and material handling, shipping, toolroom operations, model shops, sanitation work, plant maintenance and repair, and job shop operations of all types. By the use of one or more of the standards setting techniques discussed so far in this chapter, it is possible to apply work measurement to most of these areas.

Perhaps the most difficult of all areas is that of plant maintenance and repair. For a number of years, engineers have looked for some way of developing a satisfactory solution to this problem. The first efforts in this field started in the early 1940s. The approach basically was to use a combination of time formula and worksheet techniques. Because of the wide variety of maintenance work, the number of different crafts involved, and the low degree of job repetition, the ratio of standards applicators to craftsmen was very high—in the range of one applicator to ten or fifteen craftsmen. In addition, the coverage of jobs performed on standards to the total work performed was rather low—60 to 75 percent.

In spite of the high cost of administering a work measurement program of this sort, impressive results were obtained. It was often possible to double the productivity of maintenance operations by the use of standards for planning, scheduling, and controlling the maintenance activities. But even with these excellent results, many managers were unwilling to spend 8 or 10 percent of their maintenance labor dollars on a work measurement and control activity.

At length, a new approach to the problem was developed. It stemmed from a recognition of the fact that approaches which worked well for the measurement of repetitive work were not really applicable to maintenance work, because the nature of maintenance work is quite different. One of the questions frequently asked by practical maintenance managers is, "How can you predetermine the time required, say, to remove and replace a valve in a pipeline without knowing in advance how badly the threads are corroded?" The truth of the matter is that a standard cannot be accurately predetermined in situations of this kind. Accuracy can be approximated by having an analyst go to each job to observe any obstructions or difficult working conditions that may exist, but this takes time and increases the cost of setting standards.

This thinking led to the idea of using the variable nature of maintenance work as an asset rather than looking upon it as a problem. Industrial engineers had thought that they had to set accurate standards for each maintenance job because of their experience with production work. On production work, a worker may perform the same job day after day, or at least hour after hour. On a job where the operator is turning out 300 pieces an hour, an error in the standard of as little as one second could result in an error of nearly 10 percent. If the man were working on incentive, his pay would be affected by this amount. Thus, industrial engineers had been strongly impressed over the years with the importance of accuracy in setting standards.

On maintenance work, however, it is not necessary to set job standards that are accurate to the nearest second, but in many cases, it is only necessary to keep them accurate to a quarter of an hour. The very variability of maintenance work does away with the need for a high degree of accuracy. If engineered standards are developed for a large number of different jobs requiring from 1½ to 2½ hours to complete, a rectangular type of distribution curve can be expected. There will be just as many jobs in the normal mix of maintenance work that require 1½ hours to perform as there are jobs requiring 1¾, 2, 2¼, and 2½ hours to perform. If a hundred jobs are selected at random, it can safely be predicted with a high degree of accuracy that they will require 2 hours per job—the average of the time range—or a total of 200 standard hours to complete. For maintenance planning and control purposes, it is not essential to say that job A will take exactly 1.763 hours and job F will take 2.094 hours. This is important only if one man is going to perform job A or job F over and over again. Thus, for the great bulk of maintenance work, it is not necessary for standards to be set with pinpoint accuracy. It is perfectly adequate merely to say that a given job will be performed within a given range of time. For planning and scheduling purposes, it is awkward to work with time ranges for each job. This problem is easily solved by using the average of the range as the standard for the job.

As a result, the question of the practical maintenance manager can be answered by saying, "No, we cannot tell you with any high degree of accuracy how long it will take to change a specific valve; but we can tell you that 95 times out of 100, it will take between 1½ and 2½ hours to do the job, and you can plan on scheduling 2 hours of work for such a job." This concept leads to the establishing of what are called "standard work groups."

Bench Mark Jobs. Once the principle of standard work groups has been accepted, it becomes necessary to develop another concept called "bench mark jobs." This concept is based on the assumption that all jobs for any craft can be related or compared to a limited number of bench mark jobs. Engineered time standards are developed for selected bench mark jobs by the use of conventional work measurement techniques. The bench mark jobs are then placed in the appropriate standard work groups. By comparing jobs of unknown work content with the bench mark jobs, the standard work groups in which they belong are quickly determined. This is the bench mark concept of setting standards.

When a standard work group has been established and a series of bench mark jobs has been developed, the data are arranged in a format called a spread sheet. Figure 8-26 shows a spread sheet for the electrical craft. This spread sheet covers only one specific task area of electrical maintenance work—general installation. Other spread sheets will cover other task areas such as motor repair, replacement wiring, and the like.

The spread sheet shown by Figure 8-26 covers only the standard work groups for ½ to 3½ hours. Other spread sheets cover work groups above and below this range. For convenience, each standard work group is assigned a letter. Work group E, for example, covers the time range from 0.9 to 1.5 hours. The average time for this range is 1.2 hours.

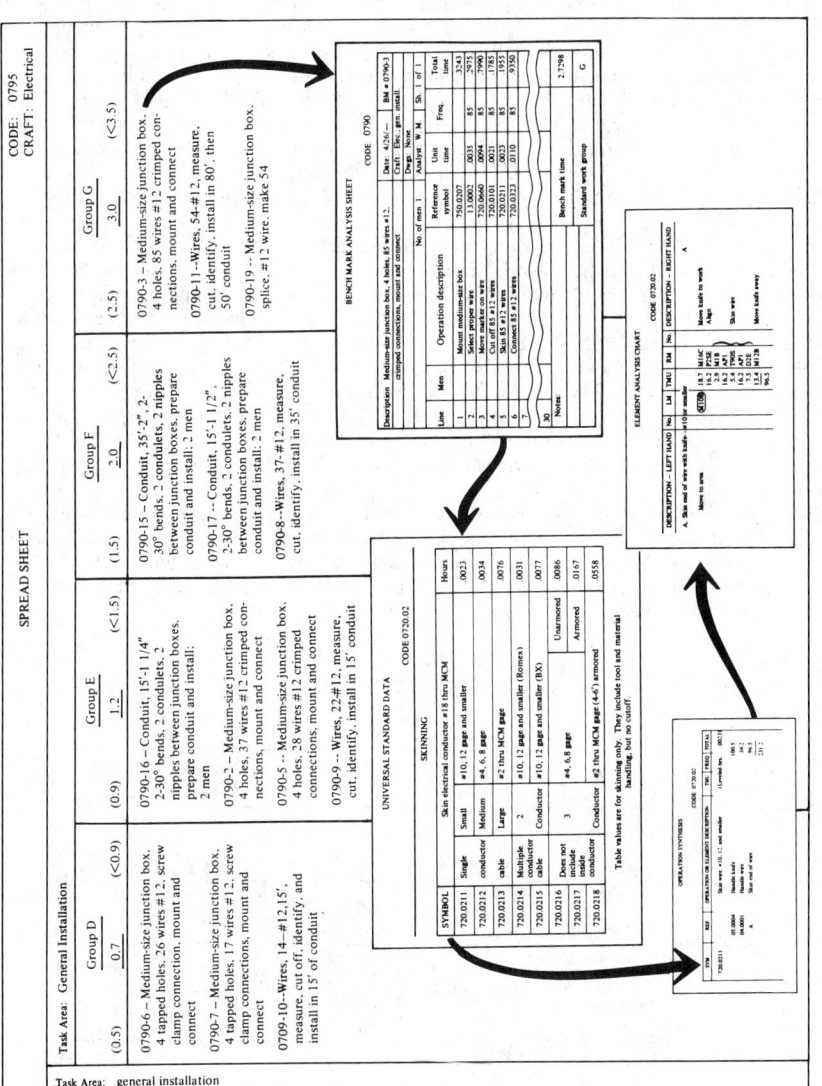

FIG. 8-26. Spread sheet for general electrical installation work.

General installation work includes the work of installing conduits; pulling wires; mounting junction boxes, outlets, and switches; and connecting the wiring to terminals. This work will vary somewhat in size and scope from one plant to another; consequently, it is desirable to tailor spread sheets of this sort to a specific plant and type of work.

The spread sheet is backed up by data sheets showing how the bench mark time standards were developed. For example, the first bench mark job listed under standard work group G is described as "Medium-size junction box, 4 holes, 85 wires #12 crimped connections, mount and connect." It has a code number, 0790-3, for reference purposes and for use in data retrieval. This job was studied in detail, using a combination of time formulas and elemental data to establish a relatively accurate standard time. The bench mark analysis sheet developed for the job is shown by Figure 8-27. The actual standard time for this job was 2.7298 hours. Work group G covers the range 2.5 to 3.5 hours, so the job was placed in this time group.

Figure 8-26 shows a number of other bench mark jobs for other time ranges as well. Assume that a new work order is received for a job described as "Mount and connect medium-size junction box with 34 #12 wires, crimped connections." What time will be allowed for this job? Inspection of the spread sheet shows that in standard work group E there are two very similar types of jobs. Job number 0790-2 is to mount a similar box with 37 wires, and job number 0790-5 a box with 28 wires. They both fall in the same time range, although the first job will obviously take a little longer. In this case, it does not require much experience or analysis to determine quickly that the new job will fall in the same time range. Consequently, it will have a standard of 1.2 hours, the average of the E range.

Now assume that the work order received is for a job similar in all respects except that it requires connecting 60 wires. Because this job appears to be about halfway between the 37-wire job (group E) and the 85-wire job (group G), group F will be a logical choice. There are no similar jobs listed in group F, however; so the decision cannot be made with absolute certainty.

BENCH MARK ANALYSIS SHEET

CODE 0790

Description Medium-size junction box, 4 holes, 85 wires #12, crimped connections, mount and connect			Date: 4/26/—		BM # 0790-3
			Craft: Elec.; gen. install.		
			Dwgs: None		
		No. of men 1	Analyst: W. M.	Sh. 1 of 1	
Line	Men	Operation description	Reference symbol	Unit time	Freq.	Total time
1		Mount medium-size box	750.0207			.3243
2		Select proper wire	13.0002	.0035	85	.2975
3		Move marker on wire	720.0660	.0094	85	.7990
4		Cut off 85 #12 wires	720.0101	.0021	85	.1785
5		Skin 85 #12 wires	720.0211	.0023	85	.1955
6		Connect 85 #12 wires	720.0323	.0110	85	.9350
7						
30						
Notes:			Bench mark time		2.7298
			Standard work group		G

FIG. 8-27. Typical bench mark analysis sheet.

This example raises several questions. One question might be: Why not study a job of about 60 wires in order to provide a bench mark job of this type for future use as a guide in group F? This could be done. There is a tendency, however, for analysts to include too many jobs in each of the time ranges in an effort to reduce their areas of indecision. The more jobs there are in each range, the longer it takes to look through them in an attempt to find a comparable job. Experience has shown that it is best to split the craft into as many logical task areas as possible to simplify proper referencing to the appropriate spread sheet, but to limit the number of bench mark jobs in any one task area to from three to five jobs.

When analysts first start using spread sheets and come to the problem just described, they are frequently tempted to go back to the files to dig out the bench mark analysis sheets and calculate the time for the job under study. For example, one could quickly calculate the time for a 60-wire job from the information given on the bench mark analysis sheet, Figure 8-27. The unit times for all elements of work related to 1 wire add up to 0.0283 hour. Because there are 25 fewer wires between the bench mark job and the 60-wire job, the time for the new job will be 25×0.0283 hour, or 0.7075 hour less. If this is subtracted from the bench mark time of 2.7298 hours, a standard for the new job of 2.0223 hours is obtained. As can be seen, however, this happens to fit into the exact middle of the group F range of 1.5 to 2.5 hours; so the original slotting was correct.

In practice, once analysts are trained, they are strongly discouraged from going through computations of the sort just described in an attempt to be sure that they are always right. It is too time consuming and contributes very little to the overall accuracy of the total measurement process. As a matter of fact, it is recognized that the analyst will not always slot the job in the proper work group. If, for example, there is another job with 75 wires, the true time required will be on the border of the F and G ranges of 2.5 hours. Without going through any calculations, the job can obviously be mis-slotted by one standard work group. This degree of error is expected, and it is unimportant. Experience has shown that if several hundred jobs are classified over a period of a week, the errors resulting from occasionally placing a job in the wrong work group will largely be offsetting, and the effect on the cumulative standard times is negligible.

Figure 8-26 shows the details behind the development of a bench mark job, and a study of these details will help provide a better understanding of the building block concept of standard data. Shown in Figure 8-26 is a complete range of building blocks, from the smallest to the largest. The largest block, of course, is the bench mark. In maintenance work, some of the bench mark jobs will cover work of 16 hours or more. Of course, some maintenance work orders may require as many as several hundred man-hours to complete. The time for such work orders is calculated by adding a number of different bench mark jobs from a number of different spread sheets and different crafts.

The bench mark analysis sheet, illustrated more clearly in Figure 8-27, shows the individual operations involved in the 85-wire job. The operation of mounting a medium-size box was studied only once. It was then assigned a reference symbol. Once the standard has been established, it is not necessary to study this task over again for any other bench mark job containing this same operation. The other elements of the job are taken from various tables of data contained in a set of electrical maintenance formulas. Figure 8-28 is an example of a data sheet used for obtaining the time for item 5 of the bench mark job "Skin 85 #12 wires." It shows the time for skinning wires of all types and gages. The time for skinning 12-gage wire is 0.0023 hour. If further information is desired on how this time was developed or on the method involved, it is only necessary to refer to the code number for this item (720.0211) and pull the information from the file. This is shown by Figure 8-29.

The operation of skinning wire was synthesized from three elements of work: "Handle knife," "Handle wire," and "Skin wire." The detailed MTM element analysis for the element of "Skin wire" is shown by Figure 8-30. By carefully

UNIVERSAL STANDARD DATA				
			CODE 0720.02	
SKINNING				

SYMBOL	Skin electrical conductor #18 thru MCM			Hours
720.0211	Single	Small	#10, 12 gage and smaller	.0023
720.0212	conductor	Medium	#4, 6, 8 gage	.0034
720.0213	cable	Large	#2 thru MCM gage	.0076
720.0214	Multiple conductor	2	#10, 12 gage and smaller (Romex)	.0031
720.0215	cable	Conductor	#10, 12 gage and smaller (BX)	.0077
720.0216	Does not include inside conductor	3	#4, 6, 8 gauge — Unarmored	.0086
720.0217			#4, 6, 8 gauge — Armored	.0167
720.0218		Conductor	#2 thru MCM gage (4-6') armored	.0558

Table values are for skinning only. They include tool and material handling, but no cutoff.

FIG. 8-28.   Table of standard data for skinning wire.

coding and classifying data in this fashion, the details behind any bench mark job are quickly available. More importantly, the work involved in building up the data in this fashion is greatly simplified, because any one element or job is never studied more than once.

Figure 8-31 shows a typical set of standard time intervals developed for use with maintenance standards. The time ranges have been selected to minimize errors in slotting. The interval between the ranges in quite small at the beginning and increases up to an interval of 2 hours for standard work group L, 8 to 10 hours.

Other Applications. The bench mark concept was developed to solve the specific problem of developing time standards for maintenance work. Maintenance work

OPERATION SYNTHESIS					
			CODE 0720.02		
SYM	REF	OPERATION OR ELEMENT DESCRIPTION	TMU	FREQ	TOTAL
720.0211		Skin wire; #10, 12, and smaller	(Leveled hrs.		.0023)
	05.0004	Handle knife			100.5
	04.0001	Handle wire			34.2
	A	Skin end of wire			96.5
					231.2

FIG. 8-29.   Operation synthesis—one of the steps in the building block process.

ELEMENT ANALYSIS CHART

CODE 0720.02

DESCRIPTION – LEFT HAND	No.	LM	TMU	RM	No.	DESCRIPTION – RIGHT HAND
A. Skin end of wire with knife- -	#10	or smaller				A
Move to area		(M10B)	18.7	M16C		Move knife to work
			16.2	P2SE		Align
			2.9	M1B		
			16.2	AP1		
			5.4	T90S		Skin wire
			16.2	AP1		
			7.5	D2E		
			13.4	M12B		Move knife away
			96.5			

FIG. 8-30.   A detailed MTM element analysis.

Work group	Standard	Time range (hours)	
		From	Up to
A	0.1	0.00	0.15
B	0.2	0.15	0.25
C	0.4	0.25	0.50
D	0.7	0.5	0.9
E	1.2	0.9	1.5
F	2.0	1.5	2.5
G	3.0	2.5	3.5
H	4.0	3.5	4.5
I	5.0	4.5	5.5
J	6.0	5.5	6.5
K	7.3	6.5	8.0
L	9.0	8.0	10.0
M	11.0	10.0	12.0
N	13.0	12.0	14.0
O	15.0	14.0	16.0
P	17.0	16.0	18.0
Q	19.0	18.0	20.0
R	22.0	20.0	24.0
S	26.0	24.0	28.0
T	30.0	28.0	32.0

FIG. 8-31.   Typical set of standard work groups for maintenance work.

is typically nonrepetitive to a high degree and fairly long cycle. As more experience was gained with bench mark standards and standard work groups, the possibility of applying the same approach to other types of nonrepetitive work became evident. The same approach was at length applied to a large job machine shop. This shop was basically involved in manufacturing special-purpose equipment used in the plant's own manufacturing operations. The installation proved to be so successful that the standards were subsequently used as the basis for the installation of a wage incentive program.

After this first application to work outside the maintenance area, a number of other installations were made to such operations as plate layout, fabrication and erection, assembly of large machines and equipment, fit and weld, and small assembly operations. Figure 8-32 shows a spread sheet for the operations involved in small assembly: assemble, drill, pin, peen, rivet, and pressfit. Other spread sheets for this same company cover other task areas such as screw, bolt, fit, solder, electrical, and so on.

APPLICATION PROCEDURES

The procedures used for establishing standards from time formulas, task data, and standard work groups should be carefully designed to ensure both accuracy

.03 Group D .04 / .05	.05 Group E .06 / .07	.07 Group F .10 / .12	.12 Group G .15 / .19
24401 05 4/60 (.034)	25120 05 4/60 (.053)	37988 05 4/60 (.091)	36090 05 3/60 (.133)

SOCKET
Assemble socket, post, washer, sleeve & PIN; assem screw knob & peen. PEEN THREAD; screw knobs onto "T" handle; SCREW into post
  (6)S   (3)M
  (4) Thd. con.
	LATERAL CASSETTE HOLDER
	UNWRAP tray & tube; DRILL & PIN tube to support; remove paper around screw holes & screw tray to support
	  (2)S (1)M
	  (1) Drill & pin
	  (3) Screws
	  Drill press
		WEINBERGER HAND TRACTION
		UNWRAP angle & plate; RIVET angle to plate; bolt (5) rubber wedges to plate & PEEN ends
		  (7)S
		  (4) Rivets
		  (5) Screws, washers, nuts
		  (5) Peen
			SANITIZER RACK
			Assemble (8) sets (shoulder pins, spacers, washers, wheels) and STAKE to rack
			  (40)S (1)M
			  (8) Stake

22669 05 5/59 (.039)	12855 05 4/62 (.058)	B7201 05 6/60 (.099)	53207 05 8/63 (.151)

WASTE VALVE STEM
Assem DRILL, PIN handle to stem; assem nut, gland packing, cover, washer, disc; DRIVE PIN thru stem
  (5)S   (3)M
  (1) Groove pin
  (1) Drill & pin
	SWITCH ASSEMBLY
	DRILL kick fork; RIVET kick fork to shaft; assem roller & pin to kick fork & PEEN ends; DRIVE PIN in shaft; DRIVE weight lever thru shaft & PEEN; RIVET spring to frame w/clip
	  (10)S (2)M
	  (5) Rivets
	  (3) Peen
	  (1) Pin
	  (1) Fit
		ADJUSTABLE FLASK HOLDER
		UNWRAP flask handle & knob; SCREW thrust nut to handle; assem plunger, plunger screw, knob & guide screw to holder; DRILL & PIN knob to plunger screw; assem (2) rubber bumpers; TEST & WRAP
		  (6)S (1)M
		  (1) Screw
		  (1) Drill & pin
		  (2) Thd. con.
		  Drill press
			VACAMATIC SHAFT
			ASSEMBLE & ALIGN (6) cams to shaft; DRILL & PIN (6) cams set to shaft
			  (6)S (1)M
			  (6) Set screws
			  (6) Drill & pins
			  Drill press
			  Fixture

33191 05 12/62 (.044)	18422 05 4/61 (.053)	22036 05 12/59 (.110)	53153 05 12/62 (.157)

HANDLE & PINION ASSEMBLY
Assem DRILL & PIN pinion to shaft; assem plate to shaft; assem DRILL, PIN "T" hand to shaft
  (3)S   (1)M
  (2) Drill pins
	OIL CHECK AEROFLUSH
	UNWRAP cap & tube; assem spring, gasket & cap to plunger; PRESS FIT sleeve to plunger; DRILL, TAPER, REAM & PIN connector to plunger; FILL TUBE with oil, insert plunger, screw up cap tight
	  (1) Drill taper pin no. 2
	  (1) Press fit
	  Drill
	  Arbor press
		COMPER KNEE & FOOT REST
		UNWRAP foot rest & knee crutch; (2) DRILL & PIN foot rest to rod; REAM & assem foot rest, wing nut assem & stop screw to knee crutch; assem stud, set screw, post assem & wing nut to knee crutch; CLEAN & prepare for pack
		  (3)S (4)M
		  (2) Screws
		  (2) Drill & pin
		  (1) Thd. con.
		  Drill press
			HEAD REST
			Assem, DRILL & PIN (2) socket assemblies to weldment; assem coupling, washers, bolt & wing nut to weldment; DRILL & PIN stud to weldment; CLEAN and assem 3 rubber pads
			  (11)S (1)M
			  (3) Drill & pins
			  (1) Speed nut & pin
			  (1) Thd. con.
			  Drill press

59123 05 11/61 (.050)		26186 05 3/60 (.096)	30142 05 8/60 (.160)

NEUROSURGICAL ATTACHMENT
Screw support assembly in vertical support; DRILL & PIN; CLEAN
  (2)M
  (1) Groove pin
		SUPPORT ASSEMBLY
		UNWRAP clamp & socket assembly, screw in stud, PIN knob to stud; assemble ball socket, set screw & wing nut
		  (8)S (1)M
		  (1) Screw
		  (1) Drive pin
		  (3) Thd. con.
			UNIVERSAL SOCKET
			ASSEM "T" handle to socket & SCREW on (2) knob ends, insert pad in socket; ASSEM & DRILL & PIN bolt to sleeve; screw swivel on bracket & screw in set screw; assem bolt assy. & washer to swivel & screw on "T" handle assy: assem stop pin & set screw; CLEAN UNIT
			  (8)S (2)M
			  (3) Screws
			  (1) Roll pin
			  (4) Thd. con.
			  Fixture

Dept. ASSEMBLY	Task Area SUBASSEMBLY	Activity ASSEMBLE, DRILL, PIN, PEEN, RIVET, PRESS FIT	Code No. 59400T10

FIG. 8-32.   Spread sheet for short-cycle light assembly work.

and low application time. Procedures which have been used successfully in applying time formulas and standard work groups are discussed in some detail below. Much of what is said applies also to task data application.

Studying the Formula Report. The final step of time formula development is the writing of the formula report. The industrial engineer describes in the report how the formula was developed and how it is to be applied for establishing standards. He tries to do this clearly enough so that anyone who reads the report in the future will find answers to any questions that may arise about the derivation or application of the total formula or any element of the work which it covers.

It follows, therefore, that the first thing anyone must do who hopes to use the formula properly for standards setting purposes is to study the formula report carefully and in detail. Some reports are clearer than others because of variations in the writing skills of the developers, but the applicator should never feel satisfied until he has mastered the meaning of every statement made in the report and sees clearly exactly how the final formula expression was derived. In some cases, he may find it advisable to get from the files the individual detailed studies from which the formula was developed, to gain a clearer understanding of what jobs were studied, the methods used, the conditions which existed at the time, and the like.

All this is necessary, not only to be able to apply the formula correctly, but to defend the correctness of the standards the applicator sets should their accuracy ever be questioned. In addition, by clearly understanding the methods and conditions on which the formula was based, he will be in a position to call attention to the necessity for formula revision whenever methods or conditions change.

Constraints and Limitations. Once a formula is developed, it saves so much time in setting standards that there is an ever-present temptation to "stretch" its field of application and to use it for jobs it was never intended to cover. For example, if the formula includes a curve of handling time for parts from 1 to 50 pounds, and it becomes necessary to establish a standard for a part weighing 55 pounds, there is a temptation to extrapolate. Yielding to this temptation can only lead to difficulties. At some point unknown to the applicator, as weight increases, handling methods must be changed; so the value obtained from the extrapolated curve will be incorrect. It is important to recognize the limitations of the formula and to act accordingly.

In machining work, it is necessary to recognize the constraints imposed by the machine itself. Figure 8-33 shows a chart of recommended cutting speeds for single-point carbide and high-speed tools. This chart is from the formula shown in part beginning on page 3-149. This chart might show that, to machine a certain part, a speed of 200 revolutions per minute and a cross-feed of 0.0085 should be used.

Figure 8-34 shows the speeds and feeds available on machine no. 97, a Warner and Swasey no. 4 bar lathe—ram type. It shows that speeds of 177 revolutions per minute or 232 revolutions per minute are available, but not 200 revolutions per minute. Similarly, it has carriage cross-feeds of 0.0065 and 0.010, but not 0.0085. The applicator must recognize these constraints and must base his standards on available feeds and speeds if he wishes his standard to be accurate for a specific machine.

Charts. When a formula is being developed, every effort should be made to save subsequent computation time. Charts offer one means of doing this. If an involved calculation must be made each time a standard is set, it may be better to make the calculations all at one time for all conditions likely to be met and arrange the resulting data in chart form, rather than to require the applicator to make a calculation each time the formula is applied. Time will be saved and the chances of errors in calculation will be greatly reduced.

For example, Figure 8-35 shows the equation for calculating threading time from the turret lathe formula. It is evident that it will require considerable time to solve this equation when establishing a time standard for threading and that there is liability of making mistakes in arithmetic.

LATHES

CHART B1

RECOMMENDED CUTTING SPEED (SFPM)

SINGLE-POINT CARBIDE AND HIGH-SPEED TOOLS

Cut depth Cut feed	.005 to .031 .002 to .008 .00009 sq. in.		.032 to .060 .008 to .018 .00038 sq. in.		.061 to .156 .010 - .020 .00176 sq. in.		.157 to .375 .015 - .030 .00554 sq. in.		.316 to .750 .020 to .035 .0141 sq. in.		.750 to 1.00 .035 to .060 .042 sq. in.		Mach. index	Unit H.P.
Material class	HSS	Carbide	HSS	Carbide	HSS	Carbide	HSS	Carbide	HSS	Carbide	HSS	Carbide		
#1-Metal #1-Cast iron	90 70	275 200	85 60	225 125	60 45	175 100	40 35	120 90	25 25	75 75	20	75	10 to 25	1.30
#2-Metal #2-Cast iron	125 90	400 250	120 80	325 200	85 60	250 175	50 45	175 150	35 35	150 100	25	90	25 to 40	1.10
#3-Metal #3-Cast iron	175 120	550 370	160 100	450 300	115 80	350 250	70 60	250 190	45 45	175 125	30	110	40 to 50	.90
#4-Metal #4-Cast iron	225 140	650 450	200 125	520 350	140 90	380 300	90 70	300 250	55 55	200 190	35	130	50 to 65	.75
#5-Metal #5-Cast iron	250 160	875 575	225 140	660 480	170 110	525 350	110 80	400 300	65 65	250 225	40	135	65 to 85	.60
#6	290	950	270	750	200	625	125	500	70	325	40	140	85 to 110	.50
#7	325	1000	275	850	240	720	200	600	100	400	45	145	110 to 135	.45
#8	375	1500	325	1200	265	900	220	775	120	480	50	150	135 to 190	.40
#9	500	1800	400	1500	300	1100	230	900	150	580	60	155	190 to 300	.35
#10	700	2500	575	2200	475	1500	325	1250	200	700	70	180	300 to 500	.30
#11	1200	4000	900	3000	750	2400	600	1800	400	1000	80	200	500 to 2000	.20

Fig. 8-33.   Chart of recommended cutting speeds.

Figure 8-36 shows a chart from which threading time per inch may be read directly for most of the diameters, threads per inch, and materials likely to be encountered in the course of formula application. It is evident how much more quickly and accurately threading time can be obtained from this chart than from the equation shown in Figure 8-35.

Worksheets. Formula application time can also often be saved and accuracy increased by the preparation of carefully designed worksheets. The worksheet design depends upon the work it covers, but in general it lists all elements or detailed operations which are likely to occur in the total class of work covered by the formula; shows the time allowed for each; and provides space for recording the

AVAILABLE SPEEDS AND FEEDS
BAR LATHE—RAM TYPE
WARNER AND SWASEY NO. 4
No. 97

Spindle revolutions per minute:
    30, 39, 54, 72, 94, 130, 177, 232, 320, 766, 554, 423
Carriage feeds:
    Cross-feed: 0.0025, 0.0065, 0.004, 0.010, 0.026, 0.017
    Longitudinal feed: 12 percent of cross-feed
Hexagon turret feed:
    Longitudinal feed: 0.0045, 0.0075, 0.003, 0.018, 0.030, 0.012
    Cross feed: None

Fig. 8-34.   Available speeds and feeds on machine no. 97.

TURRET LATHES
CHART WA
SINGLE-POINT THREADING
(Equation for calculating threading time when it is not covered by Chart W)

Equation: $T = \left\{ L \left[N \left(0.0167 \times \dfrac{F}{R} \right) \right] \right\} + 0.0025N$

where T = hours per part, including handling time
    L = total length of threads in inches
    F = threads per inch
    R = RPM, based on recommended SFPM from section I below
    0.0167 = one minute in decimal hours
    0.0025 = handling constant, K92, in hours
    N = number of cuts, based on recommendations from section II below

Section I: For materials listed, use given surface feet up to a maximum of 175 RPM.

Mat'l. class	1	2	3	4	5	6	7	8	9	10	11
SFPM	7	9	14	19	28	38	48	57	77	95	134

Section II: For threads per inch listed, use given number of cuts.

TPI	Cuts	TPI	Cuts	TPI	Cuts	TPI	Cuts	TPI	Cuts	TPI	Cuts	TPI	Cuts
80	5	48	5	32	5	18	7	12	9	9	11	5	15
72	5	44	5	28	5	16	7	$11\frac{1}{2}$	9	8	11	$4\frac{1}{2}$	17
64	5	40	7	14	7	14	7	11	9	7	12	4	18
56	5	36	5	13	9	10	10	10	6	6	14		

Example: 1.927″ dia. − 18 thds. × 518″ long
    Class 4 material
    $R = 39$ $N = 7$

    $T = \left\{ 0.625 \left[7 \left(0.0167 \times \dfrac{18}{29} \right) \right] \right\} + 0.0025 \times 7$

Fig. 8-35.   Equation for calculating single-point threading time.

number of occurrences and making all necessary calculations. Figure 8-37 shows such a worksheet and illustrates the way it would be filled in by an applicator when establishing a time standard.

A different type of worksheet is shown by Figure 8-38. This one is used to assist in calculating machining time in an orderly manner. Figure 8-39 shows still a different kind of worksheet, used for calculating gaging time. The three worksheets together are used when establishing a standard from the turret lathe formula.

Application Instructions. It is important, of course, that a formula should be applied correctly. Therefore, the industrial engineer who develops it often prepares a set of instructions designed to guide the applicator in the correct use of the formula or the worksheets. Figure 8-40 is the first page of a six-page set of instructions for using the worksheet, Figure 8-37.

Application Shortcuts. When there are a large number of standards to be set from a time formula, it will be worthwhile to view the standards setting task as a repetitive operation and to seek to improve the methods used for performing

TURRET LATHES
CHART W
APPLICATION CHART — SINGLE-POINT THREADING

Diameter		Threads	Threading Time Per Inch Based on Material Class											Handling Time
			1	2	3	4	5	6	7	8	9	10	11	
#8	(.164)	32	.0155	.0155	.0155	.0155	.0155	.0155	.0155	.0155	.0155	.0155	.0155	.0125
		36	.0170	.0170	.0170	.0170	.0170	.0170	.0170	.0170	.0170	.0170	.0170	.0125
#10	(.190)	24	.0189	.0161	.0161	.0161	.0161	.0161	.0161	.0161	.0161	.0161	.0161	.0175
		32	.0175	.0155	.0155	.0155	.0155	.0155	.0155	.0155	.0155	.0155	.0155	.0125
#12	(.216)	24	.0210	.0161	.0161	.0161	.0161	.0161	.0161	.0161	.0161	.0161	.0161	.0175
		28	.0175	.0135	.0135	.0135	.0135	.0135	.0135	.0135	.0135	.0135	.0135	.0125
$\frac{1}{4}$	(.250)	20	.0203	.0154	.0133	.0133	.0133	.0133	.0133	.0133	.0133	.0133	.0133	.0175
		28	.0205	.0155	.0135	.0135	.0135	.0135	.0135	.0135	.0135	.0135	.0135	.0125
$\frac{5}{16}$	(.313)	18	.0224	.0175	.0119	.0119	.0119	.0119	.0119	.0119	.0119	.0119	.0119	.0175
		24	.0308	.0231	.0161	.0161	.0161	.0161	.0161	.0161	.0161	.0161	.0161	.0175
$\frac{3}{8}$	(.375)	16	.0245	.0189	.0119	.0105	.0105	.0105	.0105	.0105	.0105	.0105	.0105	.0175
		24	.0370	.0273	.0182	.0161	.0161	.0161	.0161	.0161	.0161	.0161	.0161	.0175
$\frac{7}{16}$	(.438)	14	.0252	.0189	.0126	.0091	.0091	.0091	.0091	.0091	.0091	.0091	.0091	.0175
		20	.0357	.0266	.0182	.0133	.0133	.0133	.0133	.0133	.0133	.0133	.0133	.0175
$\frac{1}{2}$	(.500)	13	.0342	.0261	.0171	.0126	.0108	.0108	.0108	.0108	.0108	.0108	.0108	.0225
		20	.0413	.0308	.0203	.0154	.0133	.0133	.0133	.0133	.0133	.0133	.0133	.0175
$\frac{9}{16}$	(.563)	12	.0351	.0261	.0171	.0135	.0099	.0099	.0099	.0099	.0099	.0099	.0099	.0225
		18	.0413	.0308	.0203	.0154	.0119	.0119	.0119	.0119	.0119	.0119	.0119	.0175
$\frac{5}{8}$	(.625)	11	.0360	.0270	.0180	.0135	.0090	.0090	.0090	.0090	.0090	.0090	.0090	.0225
		18	.0455	.0343	.0231	.0175	.0119	.0119	.0119	.0119	.0119	.0119	.0119	.0175
$\frac{3}{4}$	(.750)	10	.0440	.0330	.0220	.0160	.0110	.0100	.0100	.0100	.0100	.0100	.0100	.0250
		16	.0490	.0364	.0245	.0182	.0119	.0105	.0105	.0105	.0105	.0105	.0105	.0175
$\frac{7}{8}$	(.875)	9	.0506	.0374	.0253	.0187	.0121	.0099	.0099	.0099	.0099	.0099	.0099	.0275
		14	.0497	.0745	.0252	.0189	.0126	.0091	.0091	.0091	.0091	.0091	.0091	.0175
1	(1.000)	8	.0506	.0385	.0253	.0198	.0132	.0099	.0088	.0088	.0088	.0088	.0088	.0275
		14	.0567	.0434	.0287	.0217	.0140	.0105	.0091	.0091	.0091	.0091	.0091	.0175
$1\frac{1}{8}$	(1.125)	7	.0564	.0408	.0276	.0204	.0132	.0108	.0084	.0084	.0084	.0084	.0084	.0300
		12	.0720	.0531	.0351	.0261	.0180	.0135	.0108	.0099	.0099	.0099	.0099	.0225
$1\frac{1}{4}$	(1.250)	7	.0612	.0456	.0300	.0228	.0156	.0120	.0096	.0084	.0084	.0084	.0084	.0300
		12	.0783	.0585	.0396	.0297	.0198	.0144	.0099	.0099	.0099	.0099	.0099	.0225
$1\frac{3}{8}$	(1.375)	6	.0672	.0504	.0336	.0252	.0168	.0126	.0098	.0084	.0084	.0084	.0084	.0350
		12	.0855	.0648	.0432	.0324	.0216	.0162	.0126	.0108	.0099	.0099	.0099	.0225
$1\frac{1}{2}$	(1.500)	6	.0742	.0560	.0364	.0280	.0182	.0140	.0112	.0098	.0084	.0084	.0084	.0350
		12	.0954	.0720	.0477	.0333	.0234	.0180	.0144	.0117	.0099	.0099	.0099	.0225
$1\frac{3}{4}$	(1.750)	5	.0780	.0570	.0375	.0285	;0195	.0150	.0120	.0090	.0075	.0075	.0075	.0375
2	(2.000)	$4\frac{1}{2}$	.0918	.0680	.0442	.0340	.0221	.0170	.0136	.0119	.0085	.0068	.0068	.0425
$2\frac{1}{4}$	(2.250)	$4\frac{1}{2}$	.0986	.0748	.0510	.0374	.0255	.0187	.0153	.0136	.0102	.0068	.0068	.0425
$2\frac{1}{2}$	(2.500)	4	.1098	.0810	.0522	.0396	.0270	.0198	.0162	.0126	.0090	.0072	.0072	.0450
$2\frac{3}{4}$	(2.750)	4	.1206	.0864	.0576	.0432	.0288	.0216	.0180	.0144	.0108	.0090	.0072	.0450
3	(3.000)	4	.1206	.0918	.0630	.0486	.0324	.0234	.0180	.0162	.0126	.0090	.0072	.0450
$3\frac{1}{4}$	(3.250)	4	.1332	.1008	.0666	.0504	.0342	.0252	.0198	.0180	.0126	.0108	.0072	.0450
$3\frac{1}{2}$	(3.500)	4	.1512	.1098	.0756	.0540	.0360	.0270	.0216	.0180	.0144	.0108	.0072	.0450
$3\frac{3}{4}$	(3.750)	4	.1512	.1206	.0810	.0594	.0396	.0288	.0234	.0198	.0144	.0126	.0090	.0450
4	(4.000)	4	.1926	.1206	.0864	.0630	.0414	.0324	.0252	.0216	.0162	.0126	.0090	.0450

Note 1. To determine total threading time, multiply threading time per inch by the length to be threaded and add (1) handling time.
2. For threading not listed above, use Chart WA (Figure 8-35).

FIG. 8-36.   Chart for determining single-point threading time.

Work Center	/5 20		LATHES – RAM TYPE		Part No.	390620
Part Name	SHAFT				Oper. No.	30
Apparatus	PUMP		WORKSHEET "A"		Material	CL.2
Date	10-2- —				Applicator	W.B

OPERATION DESCRIPTION							Occ.	Setup	Occ.	Each Piece
Shop setup:				K1	.108		/	.108		

LOAD AND UNLOAD CHUCK

		Up to 5#	5.1 to 20#	20.1 to 60#	Over 60# Jib Hoist	Over 60# Chain Hoist	Occ.	Setup	Occ.	Each Piece
A	Load, unloaded, and 3-Jaw	K62 .006	K62 .009	K63 .012	K65 .042	K66 .069			/	.0120
B	Secure in chuck 4-Jaw	K67 .010	K68 .013	K69 .016	K70 .046	K71 .073				
E	Turn part end for end 3-Jaw	K72 .006	K73 .008	K74 .009	K75 .024	K76 .051				
	and secure in chuck 4-Jaw	K77 .010	K78 .012	K79 .013	K80 .028	K81 .055				
	Load and unload fixture	K2 .003	K4 .006	K6 .009	K9 .039	K11 .066				

C	Tighten and loosen bolts K61	1	2	3	4	5	6	7	8
		.005	.009	.012	.016	.020	.024	.027	.031

OPERATION DESCRIPTION		Occ.	Setup	Occ.	Each Piece
Set up plug fixture	K44 .090				
Use plug fixture	K45 .005				

D	Advance bar stock and lay aside piece	K82 .005	K83 .007	K84 .010	-	-				
	Load bar stock	K85 .026	K86 .029	K87 .032	K88 .062	K89 .089				
	Reset bar feed per 32" of stock used				K58	.006				
	Change collet pads				K59	.028	/	.028		

Description		3-jaw		4-jaw	
Adjust chuck jaws		3-jaw	K13 .027	4-jaw	K19 .052
Change jaws		3-jaw	K16 .071	4-jaw	K22 .060
Reverse jaws		3-jaw	K17 .055	4-jaw	K23 .041
Remove and replace chuck		3-jaw	K15 .141	4-jaw	K21 .149
Cut soft jaws		3-jaw	K18 .144	-	-
Use copper shims		-	-	4-jaw	K24 .019
Relieve strain		3-jaw	K55 .002	4-jaw	K56 .004

Description		K-code	Occ.	Setup	Occ.	Each Piece
Set up tool	square turret	K25 .028	3	.084		
Set up formed tool	square turret	K26 .033	/	.033		
Set stop	square turret	K27 .014	5	.070		
Index turret – 1st cut	square turret	K28 .006			4	.0240
Each additional cut	square turret	K29 .003			2	.0060
Set up tool holder	hexagon turret	K30 .023				
Place tool in holder	hexagon turret	K31 .017				
Set stop	hexagon turret	K32 .013				
Set up sliding head	hexagon turret	K35 .013				
Place sleeve on holder	hexagon turret	K37 .007				
Change tool – Jacobs chuck	hexagon turret	K36 .011				
Set up die head	hexagon turret	K52 .204	/	.204		
Use die head or roller turner	hexagon turret	K53 .003			/	.0030
Place and remove drill	hexagon turret	K47 .017				
Use tap in holder	hexagon turret	K46 .001				
Set up roller turner	hexagon turret	K91 .022				
Remove flanged tool holder or roller turner	hexagon turret	K92 .039				
Index turret – 1st cut	hexagon turret	K34 .003			/	.0030
Each additional cut	hexagon turret	K33 .002				
Change speed		K39 .0003			6	.0018
Change feed		K40 .0008			5	.0040
Set dial clip		K48 .001				
Adjust coolant		K50 .0007	5	.005		
Set up indicator		K42 .015				
Use indicator per spot		K43 .035				
Use rail indicator		K51 .005				
Align splits per split (surface gage)		K60 .051				
Polish, file, or deburr per 1" length X 1" diameter		K54 .0004				
Index or change carbide insert		K90 .011			/	.0004

GAGE PART – WORKSHEET C		.062	.0460

COMMENTS				
	Total net time		.594	.1002
	Allowances		15%	15%
	Allowed standard time		.683	.1152
	Machine time worksheet B			.0383
	Machine time allowed			.0440
	Total standard time		.683	.159
	Total standard time – per piece			.159

FIG. 8-37. Worksheet "A" for setup and handling elements of turret lathe formula.

it. With a little ingenuity, it is often possible to reduce standards setting costs rather substantially.

For example, a formula was developed covering the making of cores in a large brass foundry. The foundry worked on about fifty different jobs each day. It would have been possible to set standards daily on all the new jobs worked on, but it was estimated that this would require the full time of an applicator for the first few months until jobs began to repeat and the standards setting load began to lessen, and at least half time thereafter for a considerable period of time.

All the information necessary to establish a standard could be obtained by

Work Center __1520__
Part Name __SHAFT__ __PUMP__
Apparatus _____
Date __10-2-__

Part No. __390620__
Operation No. __30__
Material __CL. 2__
Applicator __W.B.__

TURRET LATHES – RAM AND SADDLE
WORKSHEET "B" – MACHINING TIME

Description of Cut Sequence No.	Tool Descpt.	Matl.	Dia. of Cut	Depth of Cut	Speed S.F.P.M.	Speed R.P.M.	Feed	Length of Cut	Tool Lead	Total Length	Hours Per Inch	Hours Per Cut	No. of Cuts	Tool Life	Rigidity Factor	Total Hours
F	C		1½	⅛	250	645	.012	0.9	.1	1.0	.0020	.0020	1	N	S	.0020
T			1⅜	1/16	325	1110	.012	0.3		0.4	.0020	.0008	1			.0008
RT			⅞	¼	175	720	.020	1.2		1.3	.0010	.0013	1			.0013
FT			⅞	1/16	325	1370	.008	1.2		1.3	.0020	.0026	1			.0026
F.U'CUT	HSS		0.8	3/16	30	142	.003	0.4		0.5	.0390	.0195	1			.0195
THD	C		⅞	14	—	65	—	1.0		1.1	.0079 .0028	.0115	1			.0115
CHFR. ⅛	C				DIRECT FROM CHART M					—	—	.0006	1			.0006
																.0393

FIG. 8-38. Worksheet "B" for calculating machining time.

WORKSHEET "C" – GAGING

Description			Setup				Each Piece			
	Lgth.	Dia.	Sym.	Hours	Occ.	Total hours	Sym.	Hours	Occ.	Total hours
Outside micrometers	3"	to 4"	SKR	.007	/	.007	KQ	.003	//	.006
		4.1–12"	SKD1	.010			↑	.006		
		>12"	SKE1	.025				.008		
	3.1 to 10"	to 4"	SKR	.007				.006		
		4.1–12"	SKD1	.010				.011		
		>12"	SKE1	.025				.014		
	>10"	to 4"	SKR	.007				.009		
		4.1–12"	SKD1	.010			↓	.015		
		>12"	SKE1	.025			KQ	.020		
Inside micrometers	3"	to 4"	SKH1	.032			KR	.005		
		4.1–12"	SKJ1	.035			↑	.006		
		>12"	SKL1	.052				.007		
	3.1 to 10"	to 4"	SKH1	.032				.010		
		4.1–12"	SKJ1	.035				.012		
		>12"	SKL1	.052				.014		
	>10"	to 4"	SKH1	.032				.015		
		4.1–12"	SKJ1	.035			↓	.017		
		>12"	SKL1	.052			KR	.021		
Depth micrometers	to 3"		SKU	.014		.014	KU	.005	///	.015
	over 3"			.014				.006		
Telescope gage			SKT	.008			KT	.009		
Calipers — Spring joint	O.D. to 8"		SKX	.017			KA1	.006		
	I.D. to 8"			.017				.008		
Calipers — Firm joint	O.D. 8.1–14"		SKY	.010			KB1	.009		
	>14"		SKZ	.010				.010		
	8.1–14"		SKY	.010			KC1	.011		
	>14"		SKZ	.010				.012		
Calipers — Vernier	to 12"		SKA1	.009			KD1	.011		
	>12"			.009				.014		
Plug gages	to 2"		SLF1	.031			KY	.009		
	>2"		SKF1	.031			KY	.015		
Ring gages	tapered plain tapered		SKF1	.031			KZ	.005		
Threaded plug gages	to 1" 1.1–4" 4.1–8"		SKF1	.031	/	.031	KJ1	.010	/	.010
			↑	↑			↑	.012		
								.014		
	>8"		SKF1	.031			KJ1	.015		
Threaded ring gages	to 1" 1.1–4" 4.1–8" 8.1–12"		SKF1	.031			KH1	.011		
			↑	↑			↑	.011		
								.012		
								.018		
	>12"		SKF1	.031			KH1	.022		
Sq.	Combination – hand		SKW	.010			KW	.002		
	Solid – machine		SKQ	.005				.003		
Protractor	Comb. bevel – hand		SKS	.022				.002		
	Assembled	Hand		.008			KW	.002		
		Machine		.008				.003		
Profile gage			SKQ	.005			KV	.003		
Vernier height gage			SKB1	.011			KE1	.013		
Feeler gage			SKC1	.007			KF1	.001		
Scale	6"			.005	/	.005		.002	///	.006
	12" to 18"			.005	/	.005		.003	///	.009
Total			Setup			.062	Each piece			.046

FIG. 8-39. Worksheet "C" for calculating gaging time.

TURRET LATHES—RAM TYPE
Formula No. 2222.01
INSTRUCTIONS FOR USING WORKSHEET "A"

Ref.	Includes time for	How and when to apply
K1	*Shop setup*—consists of elements needed to perform setup such as get instructions, get drawing, record time, clean machine, and turn coolant off and on.	Apply each time a new job is set up. Apply to setup.
K2–K7 K3–K8 K4–K9 K5–K10 K6–K11 K12	*Get and lay aside part*—consists of time to place part in fixture. Element selection depends on weight of piece, size, length, how handled, and distance moved.	Included in K62–K89.
K13	*Adjust 3-jaw chuck*—consists of time to move jaws in closer to chucking diameter.	Apply this element when it is necessary to adjust jaws to hold piece. Apply to setup.
K14	*Hold part in 3-jaw chuck*—consists of time to tighten and loosen jaws.	Included in K85–K89, K62–K66, and K72–K76.
K15	*Place and remove 3-jaw chuck*—consists of time needed to change chucks.	Apply each time the chuck must be replaced. Apply to setup.
K16	*Change jaws, 3-jaw chuck*—consists of time needed to remove and replace jaws.	Apply each time jaws must be removed to place soft jaws in chuck or to place another set on chuck. Apply to setup.
K17	*Reverse jaws, 3-jaw chuck*—consists of time to remove and replace jaws in reverse.	Apply each time jaws are reversed, determined by piece being placed in chuck. Apply to setup.
K18	*Bore soft jaws*—consists of time to adjust, hold plug, cut jaws, set tool, and change feed and speed.	Apply each time soft jaws are required to hold a part. Apply to setup.
K19	*Adjust 4-jaw chuck*—consists of time to adjust jaws to hold new part by turning in or out.	Apply each time jaws must be moved or adjusted to hold part. Apply to setup.

FIG. 8-40.   First page of instructions for using worksheet "A."

measuring the core box and determining by looking at it the kind and amount of core reinforcing (wires, nails, rods) that would be required. There were about 20,000 core boxes in the core box storage racks.

After careful analysis, it was decided that considerable economies would be realized if the necessary information were gathered from all the core boxes at one time instead of doing it over a long period of time as each core box was used. A worksheet—in this case, a 3 by 5 inch card—was developed for recording the needed information for each core box. A portable desk was designed which could be easily moved about the core storage area.

A team of two men then proceeded to gather the needed data for each core box. One man was a standards applicator thoroughly familiar with the time formula. The other was an experienced core maker who was qualified to determine from inspection of the core box what reinforcing would be required in the finished core.

As they moved through the core box storage area, the first man would pick up each core box in turn, call off the core box number, and make the necessary measurements, calling off the figures as he obtained them. The second man, sitting at the portable desk, would record on the 3 by 5 inch cards the information furnished by the first man. He would look at the core box, mentally determine the reinforcing necessary, if any, and add this information to the worksheet. The whole procedure took less than a minute per core box, and they were able to cover the entire core storage area in less than 2½ months.

The time standards then had to be computed. This could have been done quite rapidly with the aid of a computer. However, it was recognized that many of the core boxes in storage would probably never be used again. After analyzing the situation carefully, it was decided that it would be least costly to compute the standards each day as new jobs came along. The data cards were taken from the file as the core boxes were ordered out of storage. A clerk performed the necessary calculations and recorded the resulting standards on the work orders which were sent to the foundry.

A somewhat different standards setting problem occurred in a large jobbing machine shop employing about 300 men. Both management and the workers wished to have a wage incentive plan installed. The workers expressed a strong preference for having the plan installed at one time throughout the shop rather than piecemeal a group at a time. Management instructed the industrial engineering department to do this.

Time formulas were first developed for every machine in the shop. Because the average lot size was three pieces and because jobs repeated on the average only once every six months, it was recognized that it would be a major problem for some months to establish standards for every operation on every job and to have them ready before the jobs reached the shop.

The problem was solved by setting up the equivalent of a production line. At the head of the line was the desk of the process engineer. He studied the drawings of the jobs about to be put through the shop and developed the sequence of operations to be followed in producing each job, which he recorded on a master operation card.

The drawing and the master operation card were then passed on to the standards men. Each man was assigned the task of establishing standards on from one to five kinds of operations, depending upon the complexity of the operation and the time formula covering it. Horizontal boring mill work, for example, required the full time of one standards man, while the man who handled the slotters had time for several other simple operations as well.

The drawings and master operation card for each job passed over the desks of the standards men approximately in the order in which the operations were listed on the master card. Each standards man filled out worksheets for the operations he handled but performed no calculations. The drawing with the master card and the worksheets attached eventually arrived at the desk of a clerk. The clerk then calculated all standard times from the worksheets and typed up a finished master operation card. Economies of specialization were realized from this arrangement, and the standards setting task was accomplished with a minimum of lost motion and delay.

Use of Formulas for Estimating. Time formulas, with or without modification, can be used to good advantage for estimating purposes. When the exact nature of the product or job is known, the time for making it can be determined exactly by using the applicable time formulas. The standards which are developed for the estimate will be the same as those which will be used in the shop if the work is subsequently done. Hence, it may be seen that time formulas provide the means of making highly accurate estimates.

It takes a certain amount of time, of course, to develop standards from time formulas, and in certain cases where a large number of estimates are to be made or where only preliminary "ball park" figures are desired or where the finished design work has not been completed, it may not seem necessary or desirable to take the time to go through the formula application process.

In such cases, it may prove desirable to modify the time formulas so that they may be applied more quickly at the sacrifice of some accuracy. For example, by plotting accurate times developed from formulas—as for representative jobs in a given class of work—against some major job parameter, a regression line may sometimes be developed which will permit the quick determination of a time estimate accurate enough for the purposes for which it will be used.

Use of Formulas for Building Task Data and Bench Mark Jobs. There are situations involving low-quantity or nonrepetitive work where the time required to set a standard with a time formula is not justified. The use of task data and bench mark jobs in such cases has been discussed.

It should be noted at this point that time formulas sometimes provide a convenient means of developing task data and the standards for bench mark jobs. Figure 8-22 illustrates the steps which are followed. The point to be emphasized here is that the industrial engineer should be alert to the possibilities of using his time formulas for more than just establishing standards on the classes of work for which they were developed. A formula developed for job shop engine lathe work, for example, may be quite useful in the development of maintenance bench mark jobs where an engine lathe is used to accomplish all or part of the work.

Determining Allowed Time under the Standard Work Group Concept. The time values shown on the spread sheets like Figure 8-26 cover only the leveled time for the job itself. To arrive at an allowed time for a work order, it is necessary to include allowances and values for delays, travel time, and preparing for the job. In maintenance work, it is especially desirable to determine separately the time required for travel and the time for job preparation.

Travel time should be handled separately, because it can vary considerably. Frequently, the travel time may exceed the craft time. A substantial error may be introduced if travel time is merely averaged into the job time. At the U.S. Naval Maintenance Center at Pearl Harbor, a job requiring travel time to Ford Island involves a ferry boat trip and takes an hour and a half. Although this is an extreme case, travel time can be a substantial factor in a large plant. A simple way of handling this problem is to measure the time required for travel from any one building of the plant to any other building. The data can then be set up in the format of a mileage chart, showing the time required for any given trip. The format can be modified slightly to show also the time required from one point to two or more points in the plant to cover cases where a man is assigned two or more jobs at one time to reduce travel time. An example of such a format is shown by Figure 8-41.

Job preparation time should also be handled separately, because it frequently is independent of the job or craft time. Going to the stockroom to get an electric light bulb and then filling out a requisition and a job card may require more time than the job of removing and replacing the bulb. On the other hand, a long job like excavating a deep trench with a pick and shovel may have very little job preparation time but a long job time. For these reasons, it is good practice to add job preparation time as a separate factor rather than attempting to apply it as a percentage of the job time. Job preparation for most maintenance crafts can be divided into three categories: simple, average, and complex. A typical description of these three categories of job preparation follows.

> *Simple.* Job paper work, minimum instructions (up to approximately 3 minutes), use of regular tools, and safety precautions. No time allowed for special tools or stores materials or parts.
>
> *Average.* Includes simple preparation plus additional time for more detailed instructions (up to approximately 5 minutes), for obtaining 3 or 4 tools from personal or nearby tool area, for obtaining as many as 5 different parts from stores, and for reading a simple sketch.
>
> *Complex.* Includes average preparation plus additional time for further instructions (up to approximately 10 minutes), for obtaining special tools and materials, and for reading more complicated sketches or prints.

The time required to perform these operations will depend to some extent upon the procedures and the paperwork requirements of each individual company. In

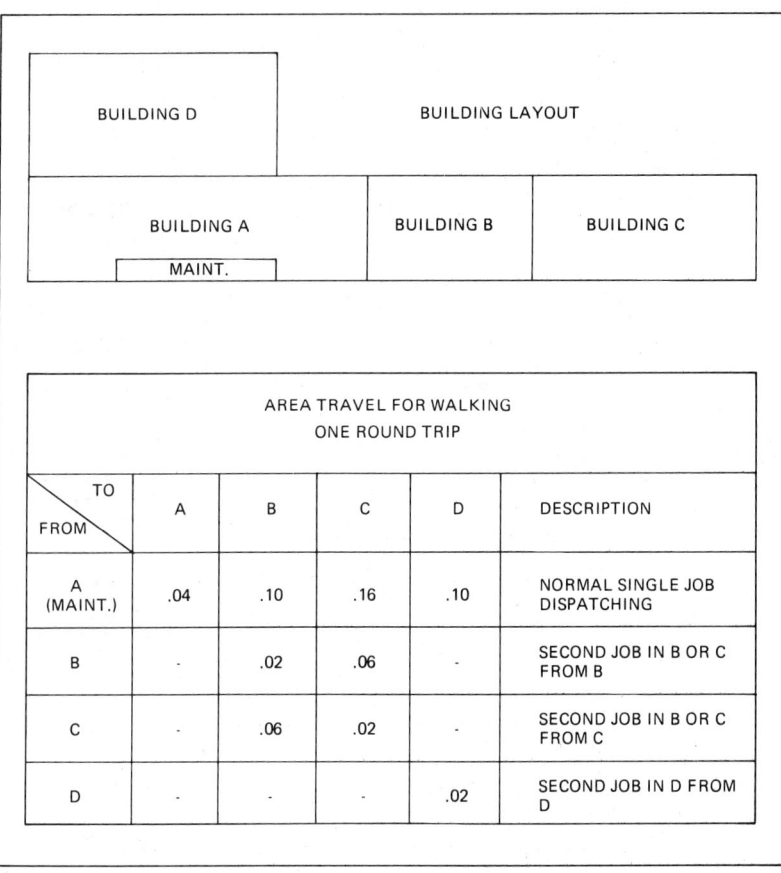

	TO A	B	C	D	DESCRIPTION
A (MAINT.)	.04	.10	.16	.10	NORMAL SINGLE JOB DISPATCHING
B	-	.02	.06	-	SECOND JOB IN B OR C FROM B
C	-	.06	.02	-	SECOND JOB IN B OR C FROM C
D	-	-	-	.02	SECOND JOB IN D FROM D

Fig. 8-41.   Building layout and area travel time values in hours.

practice, however, these three classifications seem to work quite well in a wide variety of companies and maintenance activities.

To arrive at the total allowed time for a work order, it is necessary to add the standard work group time, the travel time, and the job preparation time. This total is then multiplied by an allowance factor. The allowance factor covers the time lost due to unavoidable delays, personal time, fatigue, and the like. Various types of nomographs have been devised to perform this calculation. After a great deal of experimentation, a direct reading table appears to be the best answer. Figure 8-42 illustrates such a table. The table provides the allowed time in hours for various combinations of standard work groups, travel time, job preparation, and allowances. Only groups A, B, and C are covered on this sheet. Other sheets cover the larger work groups. Figure 8-43 shows another version of this chart, which provides for a wider variation in travel times. It also provides direct readings for two-man jobs. In this case, a separate sheet is prepared for each standard work group. The sheet shown is for group A, which has a leveled time of 0.1 hour.

Application to Plate Shop Work.   The standard work group concept was developed originally as a solution to the problem of setting time standards on nonrepetitive work of reasonably long cycle times.   When this concept had been used successfully

Standard work group time	Job preparation	One-man jobs, areas or zones						Two-man jobs, areas or zones					
		A		B, D		C		A		B, D		C	
		Round trips						Round trips					
		1	2	1	2	1	2	1	2	1	2	1	2
(A) 0.1	Simple	0.3	0.3	0.3	0.5	0.4	0.6	0.4	0.5	0.6	0.8	0.7	1.1
	Average	0.4	0.4	0.4	0.6	0.5	0.7	0.6	0.7	0.8	1.0	0.9	1.3
	Complex	0.6	0.6	0.6	0.8	0.7	0.9	1.0	1.1	1.2	1.4	1.3	1.7
(B) 0.2	Simple	0.4	0.4	0.5	0.6	0.5	0.7	0.5	0.6	0.7	0.9	0.8	1.2
	Average	0.5	0.5	0.6	0.7	0.6	0.8	0.7	0.8	0.9	1.1	1.0	1.4
	Complex	0.7	0.7	0.8	0.9	0.8	1.0	1.1	1.2	1.3	1.5	1.4	1.8
(C) 0.4	Simple	0.6	0.7	0.7	0.8	0.8	1.0	0.8	0.9	0.9	1.2	1.1	1.4
	Average	0.7	0.8	0.8	0.9	0.9	1.1	1.0	1.1	1.1	1.4	1.3	1.7
	Complex	0.9	1.0	1.0	1.1	1.1	1.3	1.4	1.5	1.5	1.8	1.7	2.0

1. Allowed hours for one-man jobs =
(standard work group time + job preparation + area travel) $\times$ 120 %
2. Allowed hours for two-man jobs =
[standard work group time + 2(job preparation) + 2(area travel)] $\times$ 120 %

FIG. 8-42.   A direct reading table to convert leveled craft time into allowed time.

on maintenance work for several years, it was quite logical to extend its use to production work of a job shop nature. A successful application of this technique mentioned earlier was in a plate shop engaged in fabricating steel tanks.[6] The first approach to the problem involved the development of time formulas and the application of standard data. Forty-three formulas were developed for such operations as machining, fitting, welding, burning, rolling, and assembly.

The time required by the standards analysts to set one standard is shown by Figure 8-44. The first standards were set in September of the first year of the installation. At the start of the installation, every standard required an average time of 1.4 hours to develop. This high time was largely due to the inexperience of the analysts and partly due to the fact that they were concentrating on setting standards on complicated machining operations. Within one year's time, the standards setting time had been reduced fourfold to 0.35 hour. For the next several months, little improvement was made, and the standards coverage in the plant hit a plateau. With the objective of speeding up the standards application and increasing coverage, a decision was made to try the standard work group approach. Bench mark jobs were selected from jobs that had previously been analyzed using the time formulas and standard data procedure.   Spread sheets were then constructed for the various areas of work.

[6] R. E. Duvall, "Accurate Time Standards in Less Time," *Management Services,* July–August, 1967.

STD. WORK GROUP	AREA TRAVEL	ONE-MAN JOBS			TWO-MAN JOBS		
		JOB PREPARATION					
		SIM.	AVE.	COM.	SIM.	AVE.	COM.
	.02		0.3	0.5	0.4	0.6	1.0
		0.3					
	.10		0.4	0.6	0.5	0.7	1.1
(A)					0.6	0.8	1.2
0.1		0.4	0.5	0.7		0.9	1.3
	.18				0.7		
					0.8	1.0	1.4
		0.5	0.6	0.8		1.1	1.5
	.26				0.9	1.2	
					1.0		1.6
		0.6	0.7	0.9		1.3	
					1.1		1.7

1. ALLOWED HOURS FOR ONE-MAN JOBS —
   [STD. WORK GROUP TIME + JOB PREP + AREA TRAVEL] X 120%
2. ALLOWED HOURS FOR TWO-MAN JOBS —
   [STD. WORK GROUP TIME + 2 (JOB PREP.) + 2 (AREA TRAVEL)]
   X 120%

FIG. 8-43. Another table for adding travel, job preparation, and allowances to obtain allowed times.

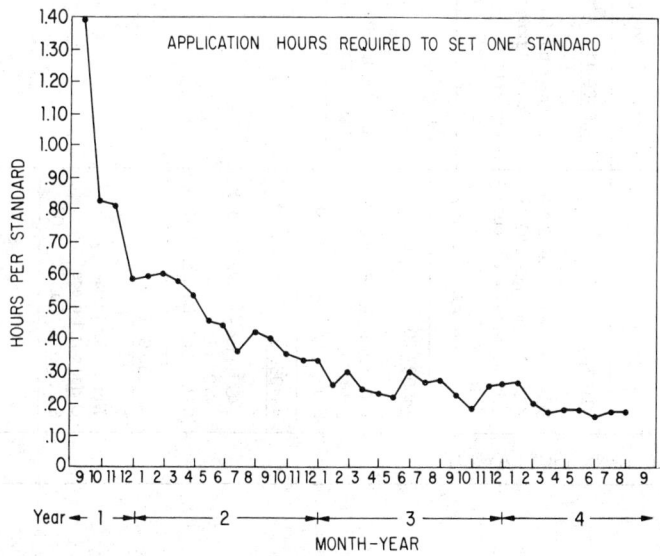

FIG. 8-44. A graph showing how the time required to establish a single standard was reduced by standard work groups.

Task area: TANK SHOP LAYOUT

Group A			Group B			Group C			Group D		
(0)	0.10	(0.15)	(.151)	0.20	(.25)	(.251)	0.4	(.50)	(.51)	0.7	(.90)
831970-3 Spec. shipping brace—2 1/2 X 2 1/2L—56" lg.—L/O for ∠ shearing plus, order out 2 parts S.U. = .17 Ea. pc. = .137			636669-41 L/O gage bar for oil level indicator 1 hole plus 1 bend line S.U. = .17 Ea. pc. = .19			411260-11 L/O flange—32" O.D. X 24" I.D. S.U. = .17 Ea. pc. = .27			411260-SA01 L/O nozzle development—All strt. lines 154" X 38" S.U. = .17 Ea. pc. = .84		
448507-20 L/O flange—9 7/8" O.D. X 3 7/8" I.D. S.U. = .17 Ea. pc. = .10			831971-6 L/O 2 holes in lugs on 30" dia. X 15/16" plate S.U. = .17 Ea. pc. = .19			411260-4 L/O flange— 59 1/2" O.D. X 48" I.D. S.U. = .17 Ea. pc. = .47			690812-5 L/O plate & location of 2 hdls. + flng. for oil tank cover S.U. = .17 Ea. pc. = .84		
434055-2 L/O flange—11 7/8" O.D. X 1 1/2" I.D. S.U. = .17 Ea. pc. = .13						670739 L/O form template for oil tank bottom S.U. = .17 Ea. pc. = .43			815784-4 19 blade imp. 20" O.D. X 7 1/2" I.D. S.U. = .35 Ea. pc. = .76		
411260-9,10 Water inlet flange Order out 1 piece S.U. = .17 Ea. pc. = .06						843100-1 Soleplate—3" X 28" X 24" L/O 4 strt. lines S.U. = 17 Ea. pc. = .40			833748-4 15" O.D. flanged brg. hous'g. end cover. 1 temp., 1 circle, 2 strt. lines, 1 order out S.U. = .242 Ea. pc. = .521		

Task area: TANK SHOP LAYOUT

Fig. 8-45. Spread sheet for plate layout operation in a tank shop.

The first standards set by work groups were installed around the first of year 3. By August of the next year, the time required to set one standard had been cut in half to 0.16 hour. During this same period, the standards coverage increased from a plateau of about 35 percent to almost 80 percent.

Other experiences have demonstrated that the number of standards set by the use of standard work groups can be increased two- to threefold over the time required by a good system of time formulas and standard data. A typical spread sheet for the operation of layout in the tank shop is shown by Figure 8-45.

To test the validity of the standard work group concept, a number of time standard comparisons were made. First, a group of analysts slotted selected jobs according to the standard work groups. Another group of analysts used the time formula data to calculate detailed standards for the same jobs. A comparison was then made of the two sets of standards for the operations required to fit, tack, and weld a tank. The results of this comparison are shown by Figure 8-46. The slotted standards for the first operation were 4 percent higher than the calculated standards. On the last operation, the slotted standards were 4 percent lower than the calculated standards. For a combination of both operations, the difference in time was 0.4 hour out of a total time of 50 hours. Individual job comparisons vary by 30 percent. This is to be expected for this type of application. It is not designed to provide accurate standards for individual jobs. It will, however, provide reasonably accurate standards over a period of time for a series or group of jobs.

Application to Short-cycle Assembly Work. Although the standard work group concept was originally developed to cover long-cycle nonrepetitive jobs, it can also be used to excellent advantage on short-cycle jobs involving order quantities of up to 25 or 50 pieces.

	Calculation of bench marks by work content comparison								
Derivation of data								
Fit, tack, and weld—rectangular oil tanks								
Part no.	First operation				Last operation				
	Op. no.	Calc. std.	Weld lgth., in.	Slotted std.	Op. no.	Calc. std.	Weld lgth., in.	Slotted std.
670702-20	70 and 80	1.70	70	1.2	110	2.15	164	2.0
670739-17	60	1.77	88	2.0	90	2.55	160	2.0
670604-20	70	1.70	91	2.0	100	3.40	136	2.0
670605-20	70	1.80	92	2.0				
670633-20	70 and 80	1.75	92	2.0	130	3.65	272	4.0
670737-20	80	2.07	92	2.0	...	3.92	260	4.0
825271-20	70 and 80	2.14	122	2.0	130	2.97	206	3.0
670607-20	90 and 100	2.47	126	2.0	130	3.08	316	4.0
825269-24	80 and 90	2.72	173	3.0	140	4.69	321	4.0
670843-28	80	3.21	296	4.0	100	3.85	392	*
815487-39	100 and 110	4.92	764	*	140	7.52	544	*
670822-17					100	2.90	196	3.0
Total		21.33		22.2		29.31		28.0

$$\frac{22.2}{21.3} = 1.04 \qquad \frac{28.0}{29.3} = 0.96$$

$$\frac{50.2}{50.6} = 0.999$$

* Not covered by graph—outside validity limit.

Fig. 8-46.   Time comparison for a series of jobs between time formula standards and slotted standards.

Standard work groups cannot be used, however, where the repetitiveness of an operation requires a high degree of accuracy for individual standards. It is best applied where the total hours of work being measured is large enough to permit statistical equalization. For this reason, it is better to use the standards to calculate performances over a period of a week or more, rather than on a job-by-job or day-by-day basis. If the nature of the operation is such that the performance of two or more people can be used for control purposes, the standard work groups will provide good results even on the more repetitive work. The more people in a group, the greater the number of different jobs performed. This increases the sample size and permits statistical equalization to be reached on a group basis where it could not be reached on an individual basis.

Selecting Analysts. Too much emphasis cannot be placed upon the importance of selecting qualified analysts. The single most important qualification is that they have actual work experience as craftsmen. It is very difficult for anyone to do an effective job of work content comparison if he is not very familiar with the work. An individual who has had experience as a general-purpose mechanic can usually do a good job of slotting a variety of other craft work such as carpentry, sheet metal, painting, and the like. Certain special crafts, like electrical and toolmaking, require specific experience in the crafts to slot jobs effectively.

There are a number of other characteristics that a good standards applicator must possess. He should have the ability to visualize spatial relationships. He should be able to read drawings. Some knowledge of shop mathematics is also essential. Many craftsmen do not take well to the job of standards analyst because they lack an aptitude for clerical work. This aptitude is essential.

The temperament of the standards applicator is as important as his skill and experience. If he does not get along well with his fellow workers, he will have great difficulty in this type of work. Equally important, he must also be respected as a skilled craftsman. If he is not, the standards that he establishes will be constantly subjected to question. Finally, he must have the ability to learn the technical aspects of the standard data system, the application of the standard data to a job, and the fundamentals of work measurement and methods engineering. This requires a level of intelligence in the 100 to 110 IQ range. A high school education is desirable but not essential if the IQ requirements are met.

The use of IQ, temperament, and aptitude tests has proved to be quite effective as an aid in making selections from shop personnel. The tests also tend to minimize charges of favoritism or bias in the selection of the analyst. They tend to place the selection procedure on a merit and ability basis.

COMPUTER APPLICATIONS FOR STANDARDS SETTING

The development of a time standard for a typical manufactured part is a time-consuming and expensive procedure. It consists of two basic steps. First, the part to be manufactured must be analyzed and a proper sequence of operations established. Then the time standard must be developed for each operation. This requires considerable engineering time, particularly if the process is at all complex.

With the advent of computers, it was only natural that they were applied to the problem of developing time standards. The first attempts in this direction were quite elementary. They involved nothing more than merely keypunching all the information from a time study form and having the computer make subtractions, add the elemental readings, determine the average elemental time, apply leveling and allowance factors to the raw elemental times, and compute the standard. In short, the computer was used to replace a comptometer operator. Because of the additional keypunching operation, the economics of such applications are highly suspect.

A more sophisticated computer application to standards setting is a system called Autorate, promoted by the Service Bureau Corporation. Essentially, the system provides for the storage of extensive files of standard data properly coded with an alpha mnemonic system that was designed to facilitate easy recall on the part of the analyst. The system requires the analyst to study the operation in sufficient detail so that he can record all the basic motions or tasks needed to complete the operation. The computer program then retrieves and prints out the appropriate element descriptions and time values. It also totals the element values to obtain operation times. The basic problem with this approach to computerized standards setting is that it does very little to relieve the standards applicator of any significant amount of detailed study and analysis. It merely provides him with a rapid table look-up device and calculating machine capabilities. Further, it does little or nothing to simplify the processing work that precedes the standards setting procedure.

To make a significant contribution to the work of developing and applying time standards, it is desirable that any computerized application simplify the amount of input needed in both the processing and the standards setting steps. There have been some significant developments in this direction in a few very limited or specialized areas of work. These applications provide an indication of the important benefits that can be provided by computer applications. A good example is a computerized program covering both the processing and standards setting operations for large boring mills producing a limited family of parts. The parts are illustrated by Figure 8-47. Part A is a rectangular ring which is machined from

RECTANGULAR RING–PART A

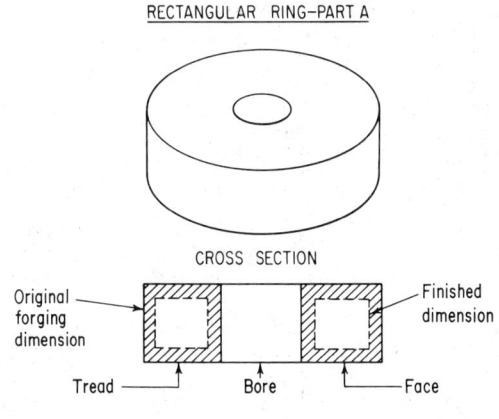

PIPE FLANGE–PART B

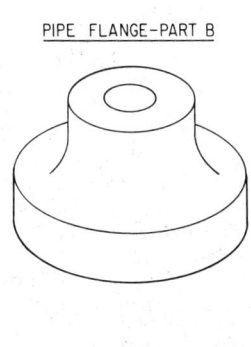

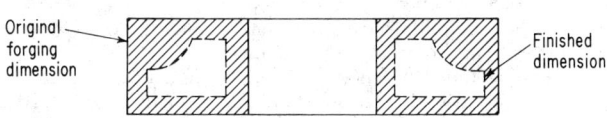

FIG. 8-47.   Parts to be produced on a boring mill.

a rough forging. In some cases, two or more rings may be machined from a single forging. Part B is also machined from a forging, but it has a pipe flange on one side. For either basic part, there are endless variations that may be due to differences in size, surface finish, tolerances, initial forging dimensions, and the like.

The input information for this program was simplified and limited to items that can easily be determined by a technical clerk. The input includes:

1. Type of part—rectangular or flanged
2. Rough dimensions of forging
3. Finished dimensions of part
4. Surface finish and tolerance

With this limited information, the computer then uses the logic built into the program to calculate and print out the following information:

1. Which kind of tool type (carbide or high-speed steel) to use for each cut (ten types to select from).
2. Number of tools required (one or two cutting heads may be used).
3. Sequence of operations—rough, semifinish, finish.
4. Cut description—face, bore, turn, and chamfer.
5. Start and finish dimensions for each cut.
6. Amount of stock to be left for next operation.
7. Optimum depth of cut for minimum machining time.
8. Surface speed and RPM.
9. Rate of feed.
10. Length of each cut.
11. Machining hours per inch of cut.
12. Hours per cut.
13. Number of cuts.
14. Total machine time.
15. Number of heads to be used for each cut. When possible, two heads are used simultaneously to produce the least machining time.
16. Whether part should be chucked on outside diameter or inside diameter of ring.
17. The limiting manual elements of the work. These include time for get and lay aside part; clamp and true part; set up cutting tools in holders; gage; feed and speed changes; and the like.

Allowances are then applied to both the machine time and the manual time. Finally, the time standard is calculated for the complete operation, and the sequence of cuts is tabulated.

The computer printout for a typical part is shown by Figures 8-48 and 8-49. Figure 8-48 shows all the basic input information, a listing of all the cuts, and the time required to complete the machining. Figure 8-49 shows a listing of the manual operations required as well as the standard data reference code and time for each operation. The bottom of the form shows the application of allowances and the total time standard, 5.161 hours. The two printouts provide complete instructions to the operator on how the job is to be done.

For purposes of comparison, the worksheets for the manual method of developing time standards used before the introduction of the computer program are shown in Figures 8-50 and 8-51. Figure 8-50 shows the development of the machining time, and Figure 8-51 shows the development of the manual time. The manual standards application method produced a standard of 5.45 hours, or about 0.3 hour more than the computer-developed standard. Most of the difference is in the machining time; so the operator does not have to work any harder. Rather, he gets better instructions, because the computer is programmed to optimize all machine times. The computer-developed time standards were consistently about 5 to 10 percent lower than those developed manually.

The average running time on a 360/30 computer for a typical part is 12 seconds. The average time required to establish a standard manually is about 20 minutes. In checking the computer standards against the manual standards, it was discovered

SSOD 72.875   ID 62.250   W 6.750   FIN   BP 52762 / OD 72.125 / TOL. OD 72.125   SER / ID 63.000   MAT SP 181-1 / W 6.000 FIN CODE   BHN / F 500   MAT 5 / T 500 / B 250   MACH 5

MAT	TOOL NO.	CUT DESCRIPTION		DIAMETER START	FINISH	STOCK	DEPTH OF CUT	SPEED SFM	RPM	FEED	FIN HOURS /INCH	LENGTH OF CUT	HOURS /CUT	NO. CUTS	TOTAL HOURS	HEADS I	HEADS II
CA	2	RF	OD	72.875	62.250	0.375	0.250	180.	9.4	0.044	0.039	5.312	0.211	2	0.422	−0.211	0.211
CA	3	RT	OD	72.875	72.125	0.375	0.250	180.	9.4	0.044	0.039	3.750	0.149	2	0.298	0.000	0.298
CA	4	RB	OD	62.250	62.750	0.250	0.250	153.	9.4	0.052	0.033	6.750	0.229	1	0.229	−0.229	0.000
CA	5	SB	OD	62.750	62.999	0.125	0.125	164.	10.0	0.083	0.020	6.375	0.128	1	0.128	0.128	0.000
HS	1	KF	OD	72.125	62.999			75.	3.9			4.562	0.083	1	0.083	0.083	
HS	2	KT	OD	72.125	72.125			75.	3.9			2.750	0.046	1	0.046	0.046	
HS	3	KB	OD	72.999	63.000			75.	3.9			6.375	0.087	1	0.087	0.087	
CA	6	RF	OD	72.875	62.999	0.375	0.250	180.	9.5	0.044	0.039	4.937	0.194	2	0.388	0.000	0.388
CA	7	RT	OD	72.875	72.125	0.375	0.250	180.	9.5	0.044	0.039	3.000	0.118	2	0.236	−0.236	0.000
CA	9	RH	OD	72.125	64.250	3.250	0.375	155.	8.2	0.059	0.059	3.937	0.232	9	2.091	1.161	−0.929
CA	10	RH	OD	64.250	66.250	0.000	0.375	155.	8.2	0.034	0.059	3.250	0.191	1	0.191	0.191	
HS	4	KF	OD	64.250	63.000			75.	3.9			0.625	0.016	1	0.016	0.016	
HS	5	KF	OD	72.125	66.250			75.	3.9			2.937	0.047	1	0.047	0.047	
CA	6	KH	OD	64.250	66.250			75.	4.5			3.250	0.047	1.5	0.071	0.071	
		RADIUS		64.250		0.250								1	0.027	0.027	
		RADIUS		66.250		0.500								1	0.066	0.066	

TOTAL HOURS 2.802

Fig. 8-48.   Printout of cutting data and time allowed.

Operation	Code	Time	No.	Total	Code	Time	No.	Total
SHOP SETUP—FIRST-CHUCKING	K216	0.139	1	0.139				
SUB-CHUCKING	K217	0.116	1	0.116				
GET AND LAY ASIDE PART OD OVER 72 HT TO 18					K 10	0.070	2	0.140
PUT ASIDE FIN PART					K144	0.031	2	0.062
CLAMP AND TRUE SINGLE STOP — FIRST-CHUCKING	K222	0.328	1	0.328	K218	0.152	1	0.152
SUB-CHUCKING	K222	0.328	0	0.000	K220	0.105	1	0.105
TOOLING								
SET UP TURNING TOOL					K210	0.019	16-3	0.246
SET UP FORMING OR SLICING TOOL					K242	0.029	2	0.058
SET TOOL TO FIRST CUT					K201	0.022	10	0.220
SET TOOL TO ADDITIONAL CUT					K202	0.013	10	0.129
SET KNIFE TOOL TO FIRST CUT					K198	0.020	6	0.120
GAGING								
TRAMMEL OR ROD	K 98	0.025	4	0.100	K 70	0.015	4	0.060
TEMPLATE OR SQUARE					K138	0.012	2	0.024
CALIPERS—WIDTH	K187	0.010	2	0.020	K134	0.015	3	0.045
MISC								
CHANGE SPEED					K 85	0.002	6	0.012
CHANGE FEED					K 84	0.001	5	0.004
TOTAL LEVEL MANUAL TIME				0.702				1.379
TOTAL MACHINE TIME								2.802
TOTAL LEVEL TIME				0.702				4.182
PERCENT ALLOWANCE				0.070				0.418
PERCENT ALLOWANCE				0.000				0.560
TOTAL ALLOWED TIME				0.773				5.161

Fig. 3-46 District of manual operations

DATE:
APPLICATOR:
APPROVED:

VERTICAL BORING MILL MACHINING TIME
WORKSHEET "B"
MACHINE GROUP – V

VBM-I – VI
B.P. NO. 5276-2-RH
OPER. NO. 10

LINE (a)	TOOL MAT. (b)	No.	CUT DESCRIPTION (c)	DIAMETER START (d)	FINISH (e)	STOCK (f)	DEPTH OF CUT (g)	SPEED SFM (h)	RPM (l)	FEED (m)	HOURS PER INCH (n)	LENGTH OF CUT (p)	HOURS PER CUT (q)	NO. OF CUTS (r)	TOTAL HOURS (s)	HEAD I (t)	HEAD II (u)	MAX. STOCK (v)
1	C	1	RF (OD)	72 7/8	62 1/4	3/8	1/4	180	9	.041	.045	5 1/4	.24	2	.48	(44)	(44)	7
2	'	2	RT TO 3"BUTT	72 7/8	72 1/8	"	"	'	'	'	'	3 3/4	.17	2	.34	34		73 1/8
3	'	3	RB	62 1/4	62 3/4	1/4	"	'	'	'	'	6 3/4		1	.30	30	—	62 1/4
4	'	3	SFB	62 3/4	63	1/8	"	155	10	.083	.020	6 1/2		1	.13	13		
5	HSS	4	KF	72 1/8	63	—	—	75	3.8	—	(	4 1/2		1	.09	09	(	
6	'	5	KT BELOW COLLAR		72 1/8	—	|	'	'	—	(	3		1	.06	06	(	
7	'	6	KB		63	—	|	'	4.6	—	|	6 1/2		1	.09	09	(	
8	C	7	RF (OD)	72 7/8	63	3/8	1/4	180	9	.041	.045	5	.23	2	.46	46	(22)	
9	'	8	RT BUTT	72 1/8	72 1/8	"	"	'	'	'	'	3	.14	2	.28	18		
10	'	7-9	RS HUB	72 1/8	64 1/4	3/4	3/8	180	10	.031	.056	4	.22	9	1.98	110	(68)	
11	'	10	" "	64 1/4	66 1/4	—	"	"	"	"	"	3 1/4		1	.18	18	(	
12	HSS	11	KF	64 1/4	63	—	(	75	4.6	—	(	3/4		1	.02	02	(	
13	'	11	KF COLLAR	72 1/8	66 1/4	—	(	'	3.8	—	(	3		1	.06	06	—	
14	'	12	K HUB	64 1/4	66 1/4	—	(	'	4.2	(	—	3 1/4		1 1/2	.08	08	—	
15	C	7-8	FS 1/2 R(1) F3 WELD BEVEL		66 1/4 / 62 1/4									1	.05 / .03	.05 / .03	1.1 –	
											TOTAL HOURS	w				2.99		

MATERIAL DETAILS		SPECIAL REQUIREMENTS	
SPEC.	121-I	LINE	
HEAT TREAT	–	TOL.	
BRINELL	–	RMS	
GROUP	5	REMARKS	

LIMITING MACHINE TIME: y

LEVEL MACHINE TIME: z

Fig. 8-50. Manually developed cutting data and time allowed.

DATE: 10/5/—
APPLICATOR: a73
APPROVED:

VERTICAL BORING MILLS
WORKSHEET "A"
60″ and 80″ NILES

BP NO. 52762-RH
OPER. NO. 10
CLASS 2

LINE		ELEMENT DESCRIPTION	SETUP Sym.	SETUP Unit hrs.	SETUP Occ	SETUP Total hours	EACH PIECE Sym.	EACH PIECE Unit hrs.	EACH PIECE Occ	EACH PIECE Total hours
1		Shop setup - 1st chucking	K216	.139	1	.139				
2		Shop setup - sub. chucking	K217	.116	1	.116				
	GET AND LAY ASIDE PART	O.D. / I.D. / HEIGHT								
3		To 36″ / -- / To 18″					K5	.042		
4		24″ - 36″ / 18″ to 48″					K8	.082		
5		36″ to 72″ / -- / To 18″					K7	.061		
6		To 72″ / Over 36″ / 18″ to 48″					K14	.082		
7		Over 72″ / -- / To 18″					K10	.070	2	.140
8		Over 72″ / -- / 18″ to 48″					K16	.091		
9		Put aside fin. mach. part					K144	.031	2	.062
10	CLAMP AND TRUE	Sgl. stop - 1st chucking	K222	.328	1	.328	K218	.152	1	.152
11		Sgl. stop - sub. chucking	K222	.328			K220	.105	1	.105
12		Stop change	K223	.158						
13		Dbl. stop - (thin section)	K224	.486			K227	.140		
14		Forgings - 1st chucking								
15		sub. chucking								
16	TOOLING	Setup tool-rgh. or knife tool	K197	.028			K210	.019	12	.228
17		Setup tool-form or slice	K221	.038			K242	.029	2	.058
18		Set tool to 1st cut on surface					K201	.022	9	.198
19		Set tool to add. cut on surface					K202	.013	13	.169
20		Set tool to 1st cut-knife tool only					K198	.020	6	.120
21		Set tool to add. cut - knife tool only					K215	.005		
22		Trial cuts-tol. > ± 1/64					K225	.027		
23		Trial cuts-tol. < ± 1/64					K226	.055		
24		Set head to angle - normal	K234	.046			K234	.046		
25	GAGING	Trammel or rod	K98	.025	4	.100	K70	.015	4	.060
26		I.D. micrometer-to 36″ dia.	K181	.052			K136	.034		
27		I.D. micrometer-over 36″ dia.	K181	.052			K137	.051		
28		Template or square					K138	.012	2	.024
29		Calipers	K187	.010	2	.020	K134	.015	3	.045
30	MISC.	Change speed					K207	.005	9	.045
31		Change feed					K208	.003	4	.012
32		TOTAL LEVEL MANUAL TIME				.703				1.418
33		MACHINE TIME - WORKSHEET "B"				—				2.99
34		TOTAL LEVELED TIME				.70				4.41
35		% ALLOWANCE				07				.44 .60
36		TOTAL ALLOWED TIME	SETUP			.77	EACH PIECE			5.45

FIG. 8-51. Conventional development of data for manual elements.

that about one manual worksheet out of four contained a significant error which resulted in either a tight or a loose standard.

Although the results of this particular example are quite dramatic, it must be remembered that the systems and programming time required to develop this program was quite substantial. It was economically worthwhile to make the investment in this case, because four men were required to handle the volume of work under the manual system, whereas the computer reduced this number to one.

The example indicates the tremendous potential that exists for the use of computers for processing and standards setting. It further suggests that the best approach may be the development of a series of programs for a wide variety of standard machines and a diverse family of parts. Because these programs will have universal application, they could perhaps be developed as software packages and sold separately to speed the recovery of the development cost. The other possibility is the establishment of a real-time computer center with access to the center by remote terminals in each shop. In this way, the capacity of a large computer could be made available to even the smallest shop, and the systems and development cost of the programming could be spread over a large number of users.

CONCLUSION

It may be seen from this discussion of standard data concepts that the possibilities of reducing standards setting costs and improving the accuracy and consistency of the standards are many and varied. A number of techniques have been developed and successfully applied which eliminate the necessity of using individual studies to develop time standards. The best technique to use in any given situation will be governed by the nature of the work, degree of repetition, length of cycle, and many other factors. There is opportunity for considerable ingenuity and even creativeness in designing the time measurement system which will best meet the requirements of a given standards setting application. For this reason, standard data development can be one of the most interesting and challenging tasks of the industrial engineer.

BIBLIOGRAPHY

Barnes, Ralph M., *Motion and Time Study*, 6th ed., John Wiley & Sons, Inc., New York, 1968.

"Computerized Time Study Gaining Momentum," *Steel*, Oct. 21, 1963.

Haddad, J. A., "Change in Engineering (Contribution of Computers)," *Dun's Review and Modern Industry*, September, 1965.

Honeycutt, John M., Jr., William Antis, and Edward N. Koch, *The Basic Motions of MTM*, The Maynard Foundation, Pittsburgh, Pa., 1968.

Karger, Delmar W., and Franklin H. Bayha, *Engineered Work Measurement*, Industrial Press, New York, 1966.

Krick, Edward V., *Methods Engineering, Design and Measurement of Work Methods*, John Wiley & Sons, Inc., New York, 1962.

Maynard, H. B., G. J. Stegemerten, and J. L. Schwab, *Methods-Time Measurement*, McGraw-Hill Book Company, New York, 1948.

Merrihew, Hawley B., and Joseph H. Redding, "Computer-aided Industrial Engineering," *IAG Quarterly Journal of the International Federation for Information Processors*, vol. 1, no. 1, Amsterdam, 1968, pp. 83–92.

Pappas, Frank G., and Robert A. Dimberg, *Practical Work Standards*, McGraw-Hill Book Company, New York, 1962.

Quick, Joseph H., James H. Duncan, and James A. Malcolm, Jr., *Work-Factor Time Standards*, McGraw-Hill Book Company, New York, 1962.

Wilkinson, John J., "How to Manage Maintenance," *Harvard Business Review*, March–April, 1968.

Applied Work Measurement

Measurement of Repetitive Work

HAROLD B. MAYNARD

President, Maynard Research Council Incorporated,
Pittsburgh, Pennsylvania

The measurement of repetitive work is fairly common throughout industry. The benefits which stem from work measurement are widely recognized, and the majority of managers look to work measurement, with or without the use of incentive wage payment, as a practical way of increasing labor productivity and reducing labor costs. Indeed, the benefits which have resulted from the measurement of repetitive work have been so apparent that over the years the trend has been to extend work measurement to areas where the work is not repetitive, using techniques and procedures similar to those described in Chapter 8 of Section 3.

In spite of the wide acceptance of the value of measuring repetitive work and in spite of the long and varied experience which management has had in this area, there have been too many applications of measurement which have been something less than satisfactory. This chapter discusses some of the problems that exist in the measurement of repetitive work and points out some of the pitfalls that should be avoided.

DECEPTIVE SIMPLICITY OF MEASURING REPETITIVE WORK

Repetitive work is apparently very easy to measure. There are still companies who feel that it is so easy that formal work measurement is unnecessary, and they rely on the foreman to establish production standards or piece rates on the basis of past performance records, some overall checks of floor-to-floor time, or perhaps just "experience." These methods sometimes seem quite satisfactory. If the workers peg their production, their earnings on incentive can be made to appear consistent and within the range where management expects them to be. More refined work measurement, however, will usually show that productivity is well below what is readily achievable.

A somewhat more sophisticated approach is to use time study for work measurement purposes. This is certainly better than no objective measurement at all, but once again the measurement task is likely to appear deceptively simple. Management often feels that it can be accomplished by clerks after a brief period of training in the techniques of stopwatch time study. Even industrial engineers themselves often feel that work measurement is a low-level task which should be delegated to technicians, while they concentrate on the more exotic aspects of industrial engineering.

It is true that the time study procedure itself can be mastered by one of average or better intelligence in a comparatively short time. But the proper application of the time study procedure requires an understanding in depth of many, many factors. Contrary to common belief, the more repetitive an operation is, the more difficult it is to measure it satisfactorily.

ACCURACY OF MEASUREMENT

The typical pattern of a work measurement installation on repetitive work, particularly where incentive payment is used, is somewhat as follows. The operators who are doing the work are time studied and time standards are established. When the standards are first announced, there is usually a feeling—shared by foremen and workers alike—that they are too tight. After a period of argument and discussion, sometimes accompanied by the adjustment of a few standards, the standards are accepted. The operators then go to work and for a while everything seems quite satisfactory. Earnings climb to the anticipated level, the operators exert a good incentive effort, and production rises and costs decrease.

Presently, however, particularly if they feel that they will be supported by a strong union, the earnings of some of the workers will begin to climb. Some workers will earn more than others, perhaps with seemingly less effort, and dissatisfaction will begin to be expressed with the tighter standards. As earnings levels continue to rise, management will begin to feel that it is paying too much for the value of the work being done. If standards are guaranteed, as they usually are in a modern installation unless methods or conditions change, there will be a continuing effort on the part of management to prove that there have been changes in methods and a continuing countereffort on the part of the workers to prove that there have not. If the time studies do not show clearly the methods and conditions which were in effect when the study was made—and they seldom do—it is difficult to reach agreement on the propriety of changing the standards. The situation continues to deteriorate until remedial action can no longer be postponed. Incentive payment may be discontinued, management may seek a new location to make a fresh start, or management and the workers may agree to a program of complete restudy which will result in a return to more realistic standards and earnings.

There are a number of factors which cause the pattern just described to develop, many of which are discussed in Chapter 1 of Section 3. In most cases, however, there is tendency to blame the accuracy of the original work measurement. The solution, in part at least, therefore appears to lie in the development of more accurate measurement procedures. Performance rating, for example, is a notorious area of uncertainty; so performance rating procedures are overhauled and training in using them is given to all time study personnel. The adequacy of the sample studied is often questioned, and mathematical formulas are developed to show how many cycles must be studied to achieve a given degree of accuracy. Statistical tests are introduced to ascertain the limits of accuracy and confidence level of a given study.

Certainly it is desirable to be as accurate as is reasonably practical when making time studies, but the fundamental problem goes much deeper than that. A time study measures what is being done at the time the study is made. There is no assurance that what is being done at that time will continue to be done one month, six months, or one year later. Indeed, experience shows that there will almost inevitably be change.

METHODS CHANGE

The fact that methods will change with the passing of time is increasingly recognized. In the early days of time study, this was not so. Although the procedure was called time and motion study almost from the start, many practitioners concentrated largely on time and accepted whatever motions the worker happened to be using. The result was that when the standard was set the worker could usually improve the method rather easily and could thus increase his earnings substantially without increasing his effort.

With the advent of the methods engineering concept which preceded the study of time by a study of methods, much of the potential for subsequent methods improvement was engineered out of the job in advance. This was certainly an important step in the right direction and helped reduce the magnitude of swing from satisfactory to unsatisfactory described above. It did not fully solve the problem, however, especially in the case of highly repetitive work. When the improved method was standardized and taught to the worker, it was still necessary to study what he was doing at the time. Again there was no assurance that this is what he would be doing at any given time in the future.

THE LEARNING PROBLEM

Research has shown that methods seldom if ever remain the same over a period of time. Even if the method appears to the untrained observer to be the same from one time to another, refined motion study will usually show that changes have occurred. As an operation is performed over and over again, the operator, either consciously or unconsciously, constantly learns how to perform it a little better each time. The result is that performance time continues to decrease with repetition as shown by the learning curve in Figure 1-1.

The learning curve is different quantitatively for different jobs or classes of work. For any individual, it is seldom the smooth curve portrayed by the illustration. There may be several plateaus where little or no learning occurs for a period of time. There may be periods of regression caused, for example, by poor health, where time even increases for a while. In general, however, learning continues, probably for the life of all but the longest jobs.

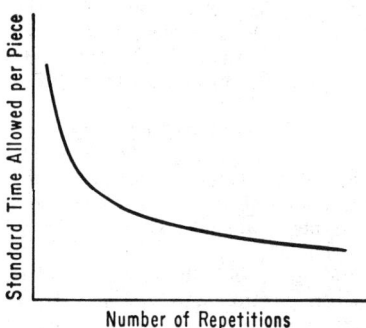

Fig. 1-1. Learning curve showing how time per piece decreases as the number of times a task is performed increases.

The methods changes which ordinarily occur with practice may be classed for discussion purposes as:
1. Easily observable methods changes
2. Subtle methods changes

Continuing research, particularly into the second classification of methods change, is giving increased understanding of what happens in specific cases to make the learning curve behave the way it does. The following generalized discussion is by no means exhaustive, but it may help the industrial engineer to understand more clearly the problems he faces when he measures repetitive work. Chapter 5 of Section 7 sheds additional light on the subject.

Easily Observable Methods Changes. The term "easily observable methods changes" is something of a misnomer. This type of change is easily recognized by the trained observer who is looking for it, but it may not be recognized by the untrained observer or the trained observer who is not consciously looking closely for method changes. When the change is discovered and pointed out, however, it is easily observable to almost anyone.

This kind of a change occurs when an operator learns to omit one or more motions from the cycle. This may be possible because the operator finds that the motions are unnecessary, or he may introduce changes which make the motions unnecessary. For example, an assembler may learn to seat two parts with two taps of his hammer instead of three or to clean a part with four wiping motions instead of six. Sometimes these methods changes are teachable to other operators and sometimes they can be used only by those with superior coordination.

In a plant making first-aid products, an industrial engineer changed the method of applying sealing wax to the end of a cardboard tube containing an iodine swab from a one-handed method to a two-handed method. The operator had been picking up a tube with her right hand, dipping it in hot sealing wax, tapping it on the edge of the container to break the thread that trailed along from the sealing wax, turning the tube to an upright position, twirling it in her fingers to form a smooth sealing wax button, and setting it aside.

The industrial engineer recognized that this could be done by the right and the left hands simultaneously and he so instructed the operator. He then took motion pictures of the one- and two-handed methods so that he could demonstrate the improvement which had been made. Over the next few months, he showed the pictures to others several times and thus became thoroughly familiar with the method the operator had used. One day he happened to pass the operator as he walked through the department. Because of his familiarity with it, he immediately recognized a methods change. The operator was no longer tapping the tubes on the side of the container to break off the trailing sealing-wax thread. In fact there was no thread.

Investigation showed that the operator had discovered that, if she kept the sealing wax a few degrees hotter than before, no thread would form. Thus she was able to introduce an easily recognizable methods change once it was spotted. The change also permitted another change which, while easily observable when one knew what to look for, was not quite so obvious. With the sealing wax at a higher temperature, she was able to form a smooth button with two twirling motions instead of three.

Changes of this kind are teachable once they are known. If they had not been recognized, there would have been an apparent loose standard and unexplainably high earnings. Recognition of the change made it possible to adjust the standard and teach the method to others. The problems involved in compensating the operator who discovered the better method are discussed in Chapter 1 of Section 3 and need not be repeated here.

Sometimes, operators find it possible to change whole motion sequences. For example, the assembly of an electronic tube required putting together a molded base and a glass bulb from which four fine wires projected. The wires had to be inserted in four tiny holes in the base before the base and the bulb would seat together. When the operation was first established, the operators would insert each wire separately in each of the four holes in the base. It was an awkward, tedious task. Presently, one girl learned that, if she straightened the wires first, she could insert them in the base much more easily. With practice, she learned to assemble the base to the bulb with virtually a single, rapid motion. The other assemblers copied her method and soon most of them were able to use it.

Subtle Methods Changes. What are here called subtle methods changes arise chiefly from three sources:

1. Skill development through practice
2. Muscular development
3. Unique physical characteristics

Anyone who has attempted to master a repetitive manual operation—even a home handicraft operation like knitting—or who has watched someone else learning it will recognize how greatly skill can develop with practice. At first, each motion is performed separately with hesitations and obvious mental activity between motions. The motions are slow, awkward, and poorly controlled. After a period of practice, the hesitations begin to disappear and the motions begin to blend into one another. Speed is gained and awkwardness disappears. With further practice, additional improvement is observable. Not only are the motions blended into one seemingly continuous easy pattern, but they can be performed for the most part without visual guidance.

The development of predetermined motion time systems has helped gain an understanding of what happens. In the MTM system, for example, a case B Reach is a reach to an object in a location which may vary slightly from cycle to cycle. It is a motion requiring visual control. A case A Reach, on the other hand, is a reach to an object in a fixed location which requires no visual control. The case A Reach is faster than the case B Reach.

When an operator first begins to learn an operation, he will use case B Reaches even to objects in fixed locations. Being unfamiliar with the workplace, he will have to look toward every object reached for. Shortly, however, as he becomes oriented to the workplace, he will begin to make his Reaches to objects in fixed locations without looking toward them. He has learned to replace case B Reaches with case A Reaches and thus has learned to perform the work in a shorter time.

As time goes on, he may eventually be able, without looking, to make Reaches to objects in a location which may vary slightly from cycle to cycle and thus may be able to employ case A Reaches. During the original MTM research, for example, the motions used by a highly skilled machine molder in a foundry were studied. The molder, after using his bag of parting sand, would lay it down on the yoke of his machine. This would put it in a location which would normally be expected to vary slightly from cycle to cycle. When he reached for the parting sand again, however, film analysis showed that he was able to do this in the time required for a case A Reach. After years of repetition, he had learned to place the bag of parting sand so nearly in the same location each time as he laid it aside that he could reach for it again as though it actually were in a fixed location.

Position is another basic motion which is affected by long practice. Operators often acquire the ability to eliminate the Position motion altogether as they repeatedly bring the same parts together over a long period of time.

MTM recognizes the fact that three different types of motions are employed when making Reaches and Moves. A type 1 motion is one where the hand is not in motion at either the beginning or the end of the Reach or Move. A type 2 motion occurs when the hand is in motion at either the beginning or the end of the motion. In type 3 motions, the hand is moving at both the beginning and the end of the motion. With practice, operators are often able to replace type 1 motions with type 2 motions which, of course, require less time to perform. Occasionally they are able to develop sufficient skill to use type 3 motions. These are subtle differences which are difficult to detect without concentrated study. The untrained observer, while recognizing the smooth blending of motions, would probably say that the method which used type 1 motions and the method which used type 2 and type 3 motions were the same. Actually they are not. They would be described differently by an MTM analysis and the time required would be different.

A further area for improvement through subtle methods changes lies in the overlapping or combining of motions. The inexperienced operator will tend to work first with one hand and then the other. With practice, he will begin to learn to do two things simultaneously with both hands. He will learn first to perform simultaneously the motions which are the same for each hand, as reaching for and picking up parts with the right and left hands simultaneously instead of first with one hand and then the other. With further practice, he will become able to do different things with each hand at the same time. Improvement through overlapping or combining motions can continue for a long time, months or even years if the operation lasts that long.

Frank and Lillian Gilbreth in their original motion study research found that operators would use different motion patterns when working quickly than when working slowly. More recent research, when developing predetermined motion times, has confirmed this finding. Therefore, in addition to practice opportunity, the pace at which the operators work will have an influence on method, usually in the area of the subtle changes just discussed.

This has an important bearing on the standards setting problem. When an operation is new, the working pace is quite likely to be slow while the operators are developing their skills. After practice, when the motions are coming more easily, the slow pace may continue for a while. Eventually, however, under the stimulus of incentive payment or when some change occurs which arouses an interest in better performance, the pace will pick up. When this occurs, production will increase not only because the operators are working faster, but because at the faster pace they will use a superior motion pattern, usually unconsciously. The result is a two-way gain, fewer and often better motions performed with a higher degree of effort.

Improvements of this sort can be made more than once. Indeed, over a long period of time, they may occur several times as the operators move from one performance plateau to another. The famous Hawthorne experiment[1] may well have been such a case. The descriptions of the study appear to assume that the method used by the operators remained constant throughout. In view of the findings of recent research, this seems extremely unlikely. The learning curve in Figure 1-1 was undoubtedly operative during that experiment as in any other repetitive work situation and would help to account for the constant improvement in productivity which the researchers observed as they varied some of the other conditions under which the work was performed.

Subtle methods improvements can result from muscular development which occurs as practice continues. Heavy jobs become lighter as operators develop new muscles and learn to apply their strength properly without waste or strain. Working postures which may have seemed tiring at first become comfortable after a period of time. Skin hardens, calluses form, and other physical changes take place which enable the operator to perform the work in an improved manner. For example, when an operator is giving the final twist to seat a screw being driven with a hand screwdriver, he will at first usually take a fresh grip on the screwdriver before applying the tightening pressure. This is done for a variety of reasons, one of which is to get the flesh of the hand and fingers distributed in a way which will avoid painful pinching as the final tightening pressure is applied. As the hand becomes toughened with the constant repetition of this motion, the regrasp of the screwdriver handle becomes less necessary. Eventually it may be discontinued altogether, with an accompanying increase in productivity.

The third factor mentioned as leading to subtle methods changes was the physical characteristics of the operator. It is quite evident that a tall operator can reach

[1] E. Mayo, *The Human Problems of an Industrial Civilization*, The Macmillan Company, New York, 1933.

to high locations more easily than a short operator. An operator with long arms can reach farther without body motion than an operator with short arms. There are also other more subtle differences. For example, one large overweight operator constantly outperformed other operators of slighter build. The work was done in a seated position. Close study showed that the heavy operator was able to reduce the length of his Reach and Move arm motions by shifting his body simultaneously on the heavy layer of flesh on which he was seated.

FACTORS AFFECTING OUTPUT

From the foregoing discussion, which is admittedly incomplete, it may be seen that there are many factors which affect output and which therefore complicate the problem of work measurement.

In the first place, there is the original method established for doing the work. The industrial engineer must study this carefully and develop the best method he can design at the time. He must watch carefully thereafter for methods changes introduced by others if he wishes to keep his standards in line with the method currently being used.

Even so, if he establishes his standard based on what the operator is actually doing toward the start of a long-run job, he can expect to find that the standard will seem more and more loose as time goes on if he judges it on the basis of the performance efficiency of the operator.

The reasons for this should now be evident. With practice opportunity, the operators will develop constantly improved methods. Some of these are recognizable and teachable and present no particular difficulties if there is an established procedure for revising standards when methods change.

The more subtle improvements are harder to handle. They are often developed unconsciously by the operator. He will have difficulty recognizing these subtle improvements as methods changes but will feel that they are a function of his own increasing skill which should be rewarded by higher earnings.

This problem is not as acute on measured daywork as it is when incentive payment is used. On measured daywork, earnings do not vary with output. When the operator has learned to produce the quota specified by the standard, he has done what is required of him by management. The incentive to improve beyond that point, therefore, is not strong. Any subtle methods improvements which he may introduce will mean merely that he can get his day's work done more easily. The learning curve typified by Figure 1-1 is not likely to be so steep or to continue for so long as in the case of a well-administered incentive plan. This is one of the reasons why some companies favor measured daywork. The administrative problem caused by inconsistent earnings due to methods changes does not occur. At the same time, productivity is not likely to increase so much, because of the lack of incentive to keep on improving.

When incentives are used, the desire of the operator to improve is strong. He will attempt from time to time to increase his working pace, which as has been seen will result in additional subtle methods changes. He will take advantage of any exceptional physical capability he may have to increase output still further.

As a result of all this, the earnings of people on highly repetitive work are likely to become higher than management originally contemplated when the incentive plan was first established. In addition, they are likely to be less consistent from operator to operator because all operators are not endowed with the same skills, attitudes, and physical characteristics.

The longer a given job lasts, the greater the earnings and the inconsistencies are likely to be. Management can perhaps accept the high earnings because it receives high output in return. The inconsistencies, however, cause greater problems. No operator likes to see another earning much more than he himself earns for equivalent work. It is the human problems caused by an inconsistent earnings picture which cause management the greatest difficulty. These same problems cause difficulties for union leadership as well, for dissatisfactions which are not resolved by management are quickly referred to the union.

STANDARDS BASED ON PREDETERMINED ELEMENTAL TIMES

It has been shown how difficult it is to establish by time study a standard for a repetitive job which will be satisfactory over a period of time. The time study man must measure what is being done at the time he makes his study. To try to adjust the data by performance rating to give a standard which will not quickly become loose requires such a low rating as to be unrealistic when compared with the performance actually given and hence unacceptable to supervision and workers alike. Even if the low performance rating were accepted because the reasons for it were understood, it would at best be an estimate unsupported by any factual data.

Standards based on predetermined elemental times can help solve some but not all of the problems which have been discussed. With a procedure like MTM, the industrial engineer can establish a standard for any method he believes will be used at any time in the future. If he has a thorough knowledge of MTM and a thorough understanding of the class of work he is studying, he should be able to predict fairly closely the method that will be used at any point on the learning curve and the standard which would be proper at that point. He will have a factual and defensible basis for his standard which he cannot have if time study is his only tool of work measurement.

In predicting what the method will be at some future time, the industrial engineer must be careful to distinguish between teachable methods and methods which can be followed only by exceptionally endowed operators. The earnings of an operator who obviously has greater skills and capabilities because of unusual coordination, exceptional size or strength, or whatever are seldom resented by those of lesser ability.

MAINTAINING CONSISTENT PERFORMANCE

A number of different approaches have been used to maintain a consistent earnings situation under incentive payment or a consistent effort level under measured daywork. The objective in all cases has been largely to reduce the human problems which are always present when inconsistencies exist, rather than to seek cost reductions by tightening standards, although this can be important in a difficult competitive situation.

Typical approaches to maintaining consistency are as follows:
1. Revise standards as methods change.
2. Establish different standards for different amounts of repetition.
3. Establish different "normal" performance levels for different amounts of repetition.
4. Establish a single standard and a decreasing learning factor.
5. Use the equitable bonus plan or the equivalent.

Revise Standards as Methods Change. There is generally a provision in the administrative policies governing a standards installation that if the method is changed on any job the standard will be changed. Sometimes this provision is accompanied by restrictions such as that the change in method must result in a change in time of 5 percent or more before the standard will be changed or that only the elements affected by the changed method will be revised.

The provision that standards may be changed makes it possible for management to keep standards consistent and in line. It requires continuing watchfulness and firm administration, however. Some companies have established the position of standards auditor. The auditor may be an employee of the company or he may be an impartial outsider, but in any case he has no standards setting responsibility himself. His task is to audit standards periodically to determine if the methods on which they were based are still in effect. He may do this on a continuing basis at the rate of so many per day or per week or he may do it on a sampling basis every three or six months. A sound policy followed by some companies is that no standard may be in effect for more than a year without audit. If, as the result of the audit, methods are found to have changed, the standard must be revised immediately. This is the only way of maintaining consistency.

Obvious as this is, many managers may hesitate to take action for what may appear at the time to be good reasons. Available industrial engineering time may be needed for developing new standards, and this may seem more important than changing existing standards. The foreman may rationalize the situation by telling himself that, because there are other standards which are tight, it is only right to have a few loose ones. Management may justify inaction by pointing out that contract negotiations with the union will begin shortly and that it would be unwise to "stir things up" by seeking to revise standards "at this time." With resistance to change almost certain on the part of the affected operators, it is all too easy to find good reasons for not carrying out the established administrative policy of changing standards when methods change.

Hence, the maintenance of consistent standards and earnings, where there is provision for revising standards as methods change, is largely a function of the skill and aggressiveness of management in carrying out its own policies. If a satisfactory installation is allowed to deteriorate and get out of hand, placing the blame on the industrial engineer or the union is seldom justified.

When standards and methods are allowed to get out of line and when no corrective action is taken for a long period of time, it becomes increasingly difficult to start enforcing the policy of revising standards. The operators will point out forcefully that management accepted the standards as being satisfactory for months or perhaps years and will usually sincerely feel that the belated attempt to change them is nothing but a thinly disguised rate cut. Usually in a situation of this kind, it is better to have a complete change through what is often called a "rationalization program" rather than to attempt to bring standards in line one at a time.

Establish Different Standards for Different Amounts of Repetition. Some companies have approached the problem of maintaining consistent standards and earnings opportunity by establishing different standards to be applied under different amounts of repetition. One way of administering this is to establish a standard and limit the number of pieces to which it is applicable. A set of standard data designed for a jobbing machine shop, for example, might carry a limiting note such as the following:

Application. The standard data for small engine lathes apply to engine lathe work on nonferrous parts as performed in department 0-4 on orders of not more than 100 pieces.

If orders of more than 100 pieces are received, a new (and lower) standard is developed and applied.

Another approach is to establish and publish a list of standards which decrease as the number of pieces produced increases. A typical series of standards would appear as follows:

Piece no.	*Standard*
1–10	.0100
11–100	.0090
101–500	.0081
501–2,000	.0073
2,001–10,000	.0065
10,000 up	.0063

This approach is an attempt to establish standards which will approximate the learning curve in Figure 1-1. It is technically feasible to do this, particularly if predetermined time standards are used. The approach has been acceptable to the operators in situations where it has been used, but the reasoning underlying it must be clearly explained and thoroughly understood. Because of obvious administrative difficulties, the use of this approach has been limited.

Establish Different "Normal" Performance Levels for Different Amounts of Repetition. When measured daywork is used (see Chapter 3 of Section 6), a standard is established for each operation which represents the amount of production that is considered normal. The standard in effect indicates what management considers to be a "fair day's work."

The performance of the operator is determined by dividing the standard hours allowed by the hours worked. When the standard hours allowed are equal to the hours worked, the performance of the operator is said to be 100 percent. This is the goal which all operators but the learners are expected to attain under many measured daywork plans.

The 100 percent performance level recognized by a measured daywork plan is not necessarily the same as the average skill and effort level used for performance rating purposes during time study. (See Chapters 2 and 3 of Section 3.) The latter average or normal performance level is a fixed level established by definition. It is not influenced by the present or anticipated future repetitiveness of the work.

When work is not very repetitive, the average or 100 percent performance level of performance rating and the normal or 100 percent performance level expected from the operator on measured daywork may be the same. As the work grows more repetitive, however, the output of the operator may be expected to increase, for the reasons already discussed. Therefore, the amount of production which will be considered to represent a fair day's work changes. For a certain degree of repetition, 100 percent performance on the part of the measured dayworker—the normal expected quota—might be rated by a time study analyst as 105 percent in terms of skill and effort. In other words, greater production on highly repetitive work is the normal expectancy.

The concept is virtually the same as that discussed under the heading, Establish Different Standards for Different Amounts of Repetition. It may not be quite as easy to grasp or explain to the operators, however, because the several different normal performance levels are all considered to represent 100 percent performance.

Establish a Single Standard and a Decreasing Learning Factor. Another approach to coping with the realities of the learning problem is to establish a standard which will represent the time in which experienced operators may be expected to do it toward the end of the life of the job, and then apply a learning factor which is high at the start and decreases toward unity as time goes on. The hours earned at any time are determined from the formula

$$N \times L \times S$$

where N is the number of pieces produced, L is the learning factor, and S is the standard time allowed.

If a standard which will be correct several months later is established at the start of the job, the learning factors will usually have to be rather high at first. An example of a table of learning factors might be as follows:

Period	Learning factor
1st week	2.50
2nd week	2.00
3rd week	1.60
4th week	1.35
5th to 8th week	1.20
9th to 12th week	1.10
13th to 18th week	1.06
19th to 30th week	1.03
31st week on	1.00

It should be understood that these are illustrative figures only and that they are not to be applied to any specific job. The correct learning factors for any job can be determined only by a study of that job. Again, the objective of the approach is to approximate the realities of the learning curve as closely as possible. It is an approach which has been used frequently when a class of work is first changed from unmeasured daywork to incentive. It gives the operators a chance to make incentive earnings during the period when they are learning to come up to the incentive level of performance and avoids the discouragement which would exist if only their true performance against standard were reported.

This approach is equally applicable when standards are to be established for highly repetitive work. Some years ago, a time study man said, "I have found that no standard is any good if the operators can meet it in under three months." He recognized, without knowing fully the reasons for it, that output and hence earnings tend to increase continuously as time goes on. Because he felt himself responsible for maintaining earnings levels within a range acceptable to management, his solution was to establish standards which were impossible to meet at first.

This approach was acceptable, or at least accepted, in the 1920s. The operators were willing to keep plugging away at their minimum guaranteed rates, trying hard to meet the standard for a long period of time until they at last succeeded. It is doubtful if they would do so in a modern environment where immediate results are expected by almost everyone. If, after trying for a few days to meet the standard, the operators would find that they could not come close to it, the normal reaction would be to enter a grievance against the standard and then work at a daywork pace until something was done about it.

Where an attitude such as this is likely to prevail, the approach of establishing a tight standard and providing a decreasing learning factor has much to recommend it.

Use the Equitable Bonus Plan or the Equivalent. Another approach to the problem of maintaining a consistent earnings picture has been through the design of the wage payment plan. Many years ago, the Rowan plan was developed to prevent runaway earnings. The bonus earned on any job was computed from the formula

$$\frac{\text{Time allowed} - \text{time taken}}{\text{Time allowed}} \times 100 = \text{percent bonus}$$

If the time taken were reduced to zero, the bonus would become a maximum of 100 percent. Actually it was much lower even when standards were very loose.

A plan of this type puts a ceiling on earnings. It also puts a ceiling on production, for the operators quickly learn that it does not pay them to produce more than a certain amount. The same result can be accomplished under any incentive payment formula by putting an arbitrary ceiling on earnings.

To avoid the limitation of production which approaches of this sort produce, a plan known as the equitable bonus plan has been developed. It has several interesting features. First, management and the union take joint responsibility for making the plan operate satisfactorily. Management assumes the responsibility for training the operators and for maintaining conditions which will permit the operators to meet standards. The union assumes the responsibility for seeing that an acceptable level of working effort is maintained.

The operators work in small groups whose performance is computed for a pay period, usually one week. Any group that fails to achieve 100 percent performance is paid at its guaranteed rate. The performance of all groups who make out—that is, exceed 100 percent performance—is averaged, and a single department-wide or plant-wide performance figure is computed. All groups that make out are then paid on the basis of this average performance.

If any group exceeds 135 percent performance, this is taken as evidence that a methods change has occurred. The work is restudied and new standards are established. However, the net saving resulting from the methods improvement for the next six months is paid to the operator or operators who developed the improved method. Thus there is a strong incentive to do as well as possible at all times, for only good can come from it.

There are any number of variations to this plan which can be introduced to meet specific situations and conditions. The approach, however, offers yet another method of avoiding the difficult human problems which result when serious inconsistencies in earnings develop.

CONCLUSION

A few conclusions may be drawn from this discussion of the measurement of repetitive work.

First, if repetitive work is to be measured successfully, the industrial engineer must understand thoroughly all the problems which are inherent in the situation. In particular, he must recognize the inevitability of continuing learning and the reasons which underlie it.

Next, he must have a thorough understanding of the work he is to measure. If he is to be able to develop standards which will be somewhat correct at any point of the learning curve, he must be able to recognize the subtle methods changes which may be expected to occur as time goes on. A knowledge of a predetermined elemental time system like MTM will be essential, but he will further need to know the elemental motions used for the specific class of work he is studying, worker attitudes and motivations, general effort levels, and any other factors which may influence the learning curve quantitatively.

Armed with this knowledge, the industrial engineer must be able to develop the application policies and procedures which best fit the requirements of the specific situations with which he is dealing. Here he runs into management policies and attitudes, the educational level of the supervisors and the work force, union attitudes, and many other things.

It can be seen, therefore, that the measurement of highly repetitive work is not a simple task which can be handled by a partly trained technician, but rather that it is something which calls upon the skills and the judgment to be found only in the competent industrial engineer. Any extra expense that may be incurred in employing the more qualified man will quickly be recovered several times over by the superior work which he will do.

BIBLIOGRAPHY

Cochran, E. B., "Learning: New Dimensions in Labor Standards," *Journal of Industrial Engineering*, January, 1969.

Cochran, E. B., "New Concepts of the Learning Curve," *Journal of Industrial Engineering*, July–August, 1960.

Hancock, Walton M., *Learning Curve Research on Manual Operations*, Research Report 113A, The MTM Association for Standards and Research, Fair Lawn, N.J., 1969.

Hancock, Walton M., *Learning Curve Research on Short Cycle Operations: Phase I, Laboratory Experiments*, Research Report 112, The MTM Association for Standards and Research, Fair Lawn, N.J., 1963.

Hirschmann, Winfred B., "Profit from Learning Curve," *Harvard Business Review*, January-February, 1964.

Powers, F. J., "Costs Strike Out with Learning Curve Incentive," *Factory*, October, 1961.

Turban, Efraim, "Incentives during Learning: An Application of the Learning Curve Theory and a Survey of Other Methods," *Journal of Industrial Engineering*, December, 1968.

Young, S. L., "Misapplication of the Learning Curve Concept," *Journal of Industrial Engineering*, August, 1966.

Chapter 2

Measurement of Low-quantity Work

JOHN J. WILKINSON

Vice President, H. B. Maynard and Company, Incorporated,
Pittsburgh, Pennsylvania

This chapter covers the measurement of low-quantity work as it occurs in toolrooms and in job shops and in development and experimental work. It is necessary to show how this work measurement can be economically carried out and how to overcome data development and application problems for both measured daywork and incentive payment.

HISTORICAL ASPECTS

From its inception, work measurement centered mainly on repetitive work which was seemingly easy to measure and control. Methods improvements similarly were concentrated in the same areas. The ability to achieve results appeared to rely on lot sizes large enough to justify detailed methods analysis and time study. Management and industrial engineers began to use the same procedures and techniques for less repetitive work but achieved mixed results. Failure or economically unjustifiable installations were the rule rather than the exception. The few brave men who investigated really low quantity work admitted defeat and categorically stated that methods improvement and work measurement were not economically justified because of excessive time required for development of data and for standards application. As can be seen in Section 5 of this Handbook, the development of predetermined motion time systems overcame some of these difficulties for mediumly repetitive work. But there still was the question of economic justification for both development of data and application of standards to the really low quantity work.

The next development, special forms of standard data and new concepts in work measurement and application, made methods improvement and work measurement for low-quantity work worthwhile.

ECONOMIC JUSTIFICATION

These advances in work measurement procedures made it possible to justify economically the measurement of low-quantity work. The basic philosophies in two main areas are given below.

Development of Data and Procedures. In the development of work measurement data for low-quantity work, the most important concept is that pinpoint accuracy and detailed individual standards are not necessary. They are not an essential factor in realizing really worthwhile cost reductions on low-quantity work.

Acceptance of overall methods improvement and an "average condition" concept for work measurement, combined with less detailed methods and control procedures, have made possible economically justifiable measurement of low-quantity work.

Application. The application procedures for this type of work measurement installation must be the simplest possible. A simple work card with only a brief outline of the work to be done and a good suggested method, along with a standard time which is within a chosen range of accuracy something less than exact, achieves the desired results for the low-quantity production items. The calculation of several hours of standard time becomes a matter of minutes instead of hours.

These concepts, however, at first appear somewhat radical and bring forth many questions regarding accuracy and validity. Many problems arise, and careful explanation of the principles involved is necessary.

PROBLEMS, SOLUTIONS, AND APPROACHES USED

Even with these newer procedures, problems have arisen and have been overcome.

Methods. The question of how to specify methods improvement for low-quantity work invariably arises. How can you achieve cost reductions if you do not improve methods; and how can you make changes if you are not able to analyze, improve, and teach better methods for individual operations? The answer is to concentrate on the optimum use of the best method for both equipment and hand tools. This does not necessarily mean the most expensive and the most complex tools. The realistically justifiable items must be available and must be used in the proper manner.

For metal-cutting machine tools, it is neither justified nor necessary to specify the optimum speed, feed, and depth of cut for every individual operation. But it is necessary to prepare and issue complete data for each type of machine, showing the rate of stock removal used as a basis for standard-time calculation for each class of material to be machined. The experienced machinist will probably make his own decision as to whether this cut should be a heavy feed and light depth of cut, or vice versa; but he must know what rate of stock removal is demanded to meet the issued standard time and to earn incentive pay. The operator should have available the correct standard tools and should be given instructions for their best use under different representative conditions. Figure 2-1 shows a typical cutting data table—in this case for high-speed steel single-point cutting tools used on shaping, turning, and similar operations. With a given depth of stock to be removed, the depth of cut, feed, area, and cutting surface feet per minute can be quickly determined for any material.

For assembly work, teaching the correct use of ring spanners, ratchet wrenches, and adjustable wrenches frequently achieves better results with less capital expense than installing power tools and semiautomatic equipment for this low-quantity work.

SPEED IN FEET PER MINUTE FOR SINGLE POINT TOOL CUTTING (H.S.S.)
RELATED TO FEED AND DEPTH OF CUT

DEPTH OF CUT	FEED	AREA OF CUT	MATERIAL CLASS						
			I	II	III	IV	V	VI	VII
			Aluminum	Brass	Soft C.I.	Hard C.I.	Soft Steel	Med. Steel	Hard Steel
.020 .024 .028	.005 .006 .007	.00010 .00014 .00020	Max.	Max.	211	97	314	218	127
.032 .036	.008 .009	.00026 .00032	Max.	946	201	93	298	203	120
.060	.0150	.00090	Max.	598	154	72	208	135	84
.066	.0165	.00109	Max.	522	146	69	192	126	78
.069	.0172	.00119	Max.	510	143	67	184	122	75
.072	.0182	.00131	Max.	493	140	66	178	118	73
.480 .510	.048 .051	.0230 .0260	193	73	61	28	53	35	21
.540 .560	.054 .056	.0292 .0314	168	65	58	27	49	32	20

FIG. 2-1.   Typical cutting data table.

A simple "universal" tool or handling device can save many man-hours yet costs very little money to install. As an example, Figure 2-2 shows a simple lifting device for heavy flat die sets, stripper plates, backup plates, and raw stock.

This is the "ice tongs" principle applied to minimize the use of slings or chains and eyebolts in a toolroom. In one company, appreciable handling time was saved for bench work and tryout in the die flipper, and the safety hazard was greatly reduced.

Specification of methods for the best use of standard tools and for repetitive elements of work is essential. Specification plus training for best methods on typical assembly or fitting operations also can be successfully developed to the point where all associated work is improved likewise.

Simplified Data. When the use of simplified work measurement procedures and simplified data is initiated, it is found that many new "gimmicks" and titles for the new procedures have been used; but the basic principle is common to most of them and must be kept in mind at all times:

For successful measurement of low-quantity work, data must be accurate enough for the class of work they are intended to cover but simple enough to apply easily and quickly to the wide variety of work found within that class.

Any data used must have a firm foundation of accurate measurement and synthesis—especially the representative elements and basic data.

Problems have arisen from acceptance of the word "universal" applied to simplified work measurement data. There is a risk that data may be used without adequate validation, with the result that good data are applied to the wrong work. Such bad results tend to undermine confidence in the data and may develop strongly negative reactions. Even the best data must be carefully checked and validated

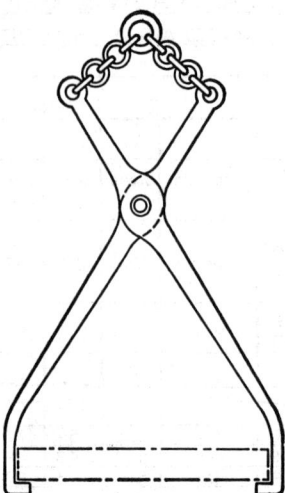

FIG. 2-2.   Lifting tool for heavy die plates.

against the type of work which they are intended to cover—they cannot accurately be used universally for all types of conditions and work areas.

For example, the element of time "remove sharp edges" can vary from just touching the corner with a file to very extensive radiusing and polishing for certain assembly work. The difference in time required is large, and the standard data must reflect the difference and must be adequately described to avoid errors in application.

Standardization. The wrong use of good data has been discussed and the careless use of "universal data" has been warned against. However, in addition, it is necessary to stress that successful work measurement of low-quantity work relies also on the standardization of general procedures and overall methods. Carefully analyzed averages and typical conditions must be developed and used as a basis for standard data which apply to low-quantity work. Yet they cannot be too detailed or too specialized, or the development and application again become completely uneconomical.

Averages and Range of Time Concept. For successful measurement of low-quantity work, the fact must be accepted that a given element of work or a given job may be done in several different ways, each of which is reasonably

"Get" Elements: Time in TMU (1/100,000 Hour)

Distance, inches	Weight, pounds				
	3–10	11–20	21–30	31–40	41–50
3–5	16	20	24	28	33
6–14	25	30	35	40	44
15–22	38	44	49	54	60
23–30	49	55	61	67	72

FIG. 2-3.   Typical use of ranges of distance and ranges of weight to simplify application of standard data.

acceptable. Similarly, conditions may change from time to time, and frequently the changes are not of sufficient magnitude or permanent enough to justify revision of the standard time for that portion of work. In many forms of simplified standard data, an average distance and an average weight are used—thus eliminating the need for revision of the standard time for minor distance or weight variations. Figure 2-3 shows such a table for "get" elements.

The principle used in most forms of standards application to low-quantity work is that an average time, based on good measured data, will be issued for the work. It is anticipated that the work will be completed within a given range of time—sometimes a given percentage above or below the average time allocated. For instance, in the Universal Maintenance Standards (UMS) procedure, the standard times issued for the jobs are understood to represent ranges of time, as illustrated by the letters and time values shown in Figure 2-4.

Similarly, for toolroom fitting work, certain operations which really have a very large number of slightly different values difficult to isolate for each occurrence may be represented only by three classifications—simple, average, and complex—each with its own time value. This can be seen on the calculation data card shown in Figure 2-5.

Maintaining Required Validity and Accuracy. Another major problem, which cannot be overcome by the initial development of data or by the original application procedures, is the maintenance of the required validity and accuracy of the standard times. This is a management and industrial engineering responsibility and is absolutely necessary on a continuing basis. There is real need for regular, systematic audit of the data and the application procedure.

Regular spot checks are necessary of the actual time taken and time recording. Standard times and performance have been severely criticized sometimes, yet investi-

Work group	Standard	Time range, hours
A	0.1	0.00–0.15
B	0.2	0.15–0.25
C	0.4	0.25–0.50
D	0.7	0.5–0.9
E	1.2	0.9–1.5
F	2.0	1.5–2.5
G	3.0	2.5–3.5
H	4.0	3.5–4.5
I	5.0	4.5–5.5
J	6.0	5.5–6.5
K	7.3	6.5–8.0
L	9.0	8.0–10.0
M	11.0	10.0–12.0
N	13.0	12.0–14.0
O	15.0	14.0–16.0
P	17.0	16.0–18.0
Q	19.0	18.0–20.0
R	22.0	20.0–24.0
S	26.0	24.0–28.0
T	30.0	28.0–32.0

FIG. 2-4.   Typical Universal Maintenance Standards work groupings.

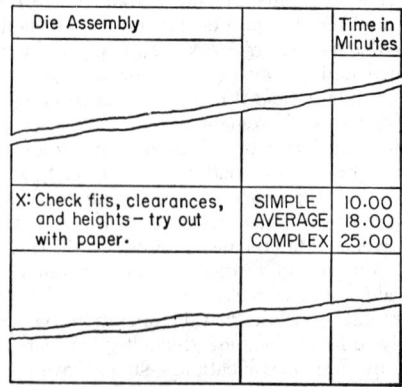

Die Assembly		Time in Minutes
X: Check fits, clearances, and heights— try out with paper.	SIMPLE AVERAGE COMPLEX	10.00 18.00 25.00

Fig. 2-5.   Example of assembly time averages on degree of complexity.

gation has proved that there was incorrect time recording against the job. Off-standard hours and waiting time had been given to increase reported performance in some cases. Extra work was being done with no standard hours credit given, resulting in lower than actual performance being reported in other cases.

As for any work measurement installation, the effects of equipment changes on the standard times must be considered. For low-quantity work, minor methods changes and minor quality requirement changes can be ignored, provided they do not change the standard job times outside the time ranges or beyond the desired percent variation in accuracy. However, major methods changes must be carefully considered; necessary changes must be made in the standard times, as for any other standards application.

For the economic application of standards to low-quantity work, it is sometimes necessary to use judgment and to try to make objective decisions between certain alternative data. This decision making must be strictly policed and carefully checked for possible bias even though the overall net effect of the bias is usually small. For instance, in the aligning of a shaft when assembling a piece of new equipment, the calculator has to make a choice among "simple," "average," or "complex" alignment. Despite the fact that he has specific guide rules for making the choice, he may gradually, almost subconsciously, start to give a higher or a lower value than is justified. Spot checks, retraining, and occasional reanalysis of a complete job time independently by a second calculator are excellent means of policing the judgment being used in standards application.

Technical Development. Technically, the development of data for low-quantity work is appreciably different from work measurement of highly repetitive work. The analysis and synthesis of data must definitely take into consideration the degree of repetition of jobs and also of specific motions—there is little or no opportunity for practice of specific jobs by repetition with this type of work. Frequently, also, planning or thinking time must be allowed in addition to actual working time.

During MTM analysis of low-quantity work, for example, simultaneous motions do not occur as often as on repetitive work—the bolt and washer will usually be picked up and inserted separately before reaching for the wrench when there is little repetition and opportunity for practice. The Reach for the chuck key will take place after the machine has stopped and the time will not be limited out by machine time when the analysis is made for application to low-quantity work. In the specific case of an element "open and close the part-holding vise on a milling machine," the allowed hours are 1½ times greater for low-quantity application than for repetitive work. The difference is mainly because the Reach to the lever plus Grasp, and the Release plus hand away, are limited out by other motions or machine time when doing repetitive work. For low-quantity work, where frequently only one or two pieces are involved, checking time and decision making time must be allowed either by eye focuses and blueprint reading time or by measured allowances added to the physical activity time.

These ranges of time, averages, and different analyses may bring forward accusations of deliberately increasing the allowed time by an indefinite amount with no real foundation for change. In actual fact, careful study of the detailed analysis and application procedures completely repudiates these accusations; a factual basis for the differences can be seen by anyone experienced in work measurement.

REMAINING PROBLEMS

By using the new approaches to development of data, and with a good understanding of the pitfalls, many of the problems of measurement of low-quantity work can be overcome. It becomes relatively routine to apply good work measurement and control procedures to low-quantity work successfully and economically. In many cases, companies pay incentives for performance exceeding these standards.

However, there remain several problems which must be overcome, either by careful handling of each situation as it arises or by the long-term development of more and more proof by demonstration of the practical success of this phase of industrial engineering work.

Technical Aspects. From a technical viewpoint, the data and procedures developed by industry and the major consulting companies are statistically and theoretically sound when they are correctly applied and have been proved by successful practical installations. But individuals, who criticize and detract from this approach to the measurement of low-quantity work, find alleged technical faults. This happens because, although a procedure may be completely valid and be completely justified by a "bell curve" distribution of the averages or ranges of time used, the individual examples when taken separately from the whole may appear to contradict these claims of accuracy and statistical validity.

For example, when the range of time concept is used in UMS (Universal Maintenance Standards), the possible deviations from the group time on the very short elements of work in the lower work groupings appear very large if considered individually. Yet a series of such jobs considered together shows very good overall accuracy compared with other data and with the working time taken. There may be some individual jobs where the performance is unrealistically high or unrealistically low, but several jobs must be considered together if the "averaging concept" is to be applicable and the real accuracy and validity are to be correctly assessed. For example, groups of several weeks' completed maintenance jobs were plotted to show a distribution curve of the performance against man-hours and against number of men. The bell curve peak came almost exactly at the average performance reported each week for that maintenance department. However, as can be seen from this distribution, shown in Figure 2-6, there were a few hours when a performance of 130 percent was recorded and a few hours when a performance of less than 50 percent was recorded. Taken separately, both of these may be called unrealistic when compared with the 84 percent overall recorded performance. Statistically, this spread is to be expected, and in no way does it detract from the overall accuracy of the standards application procedure, or from its value for control purposes or incentive payment.

Remember also that the time recording on the individual jobs is frequently not so accurate as we might desire. In the example used for this distribution, incorrect clocking in and out at the beginning and end of various jobs contributed as much to the apparent high and low performance as did the "inaccuracy" of the individual standard times.

Human Problems. In addition to technical problems, of course, there are human problems when management, individual workers, and union representatives are in-

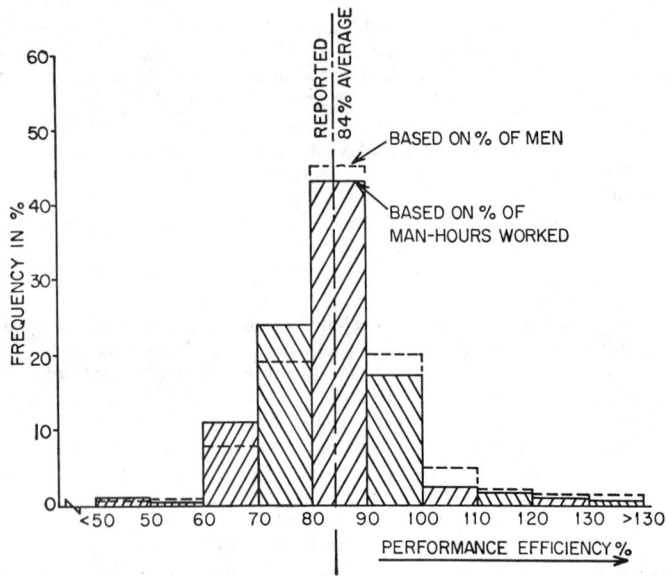

FIG. 2-6.   Distribution of UMS performance—efficiency by man-hours and number of men.

volved. It is always difficult to explain work measurement procedures fully in writing, and it is frequently difficult to do so verbally—especially when such discussions fringe on criticism of the individual or group of men.

One way of overcoming these problems is by very careful preparation and technical education of the people involved. It has been found that practical demonstrations and a completely "open book" approach have been most successful. For example, the random choice approach of the tombola wheel has been used to explain the use of the standard work groupings and ranges of time used in the UMS procedure. In just the same way as a bell-shaped distribution curve is often explained most successfully by the random falling of balls into slots, the equalizing effect of the time ranges and even the neutralizing effects of errors have been proved by this practical demonstration. Demonstration of theories and practical results has proved successful in answering doubts and overcoming some of the human problems which arise from any new work measurement and application procedures.

A more unusual human problem concerns the men who develop and apply the data for low-quantity work. Industrial engineers have been trained for many years that accuracy and preciseness are essential to good work measurement. MTM and other predetermined time systems have been developed, and men have been trained in their use, specifically for detailed methods analysis and application of time for repetitive work. This experience and training forces engineers to think clearly and accurately for all work measurement data. Then they are asked to apply their knowledge to the measurement of low-quantity work. They have problems!

The changeover to a more gross or overall approach is sometimes very difficult for them. To realize that simultaneous motions may not be so important, may not even occur; to measure the work done by only an average method, not the best they can devise; to speak in terms of units of several minutes or several hours instead of fractions of a second—these are changes not easy to make. Training of new men, and even retraining of experienced men, for measurement of low-quantity work must lead carefully away from the detailed accurate analysis which is the necessary foundation, into the wider overall approach to analyzing, synthesizing, and applying work measurement data.   These men must be taught to appreciate and to see in practice that a 20 percent variance in an element or operation which is 20 percent of the whole results in only a 4 percent overall error; and this can be a plus or minus error which will average out over several elements, provided there is no intentional bias present.   Similarly, if a work element occurs only one time in twenty, the accuracy of analysis can be appreciably reduced—or differences in this element from occurrence to occurrence may even be ignored completely—without appreciable overall loss of desired accuracy.

Coincident with overcoming technical and human problems with the development of data, there is the real need for achieving understanding and acceptance by the workers to whom the work measurement applies.   It appears that worker acceptance is usually a combination of personal confidence instilled by full explanation and demonstration and by proved results from practical application over a reasonably long period.   For example, it has been found that a program to measure low-quantity work has started with very low performance against standards.   After several weeks, very little improvement is realized; then suddenly both supervisor and worker confidence in the data and application procedures appears to develop, and performance rapidly increases to acceptable levels.   After careful explanation and demonstration, the men have gained understanding of the validity and fairness of the measurement used as a basis for their reported performance and for their incentive pay.

Management Support.   Despite these new approaches to work measurement and to reducing human problems, the measurement of low-quantity work cannot be successfully accomplished without the strong support of top management.   Top management support and close liaison with the industrial engineer are absolutely essential during the development and installation phases of this work.   Open, outspoken and authoritative management support is necessary to develop confidence and acceptance from everyone concerned.   Negative or even neutral top management has offset all technical success on more than one occasion.

Management usually accepts the need for its strong support and control in the production areas.   They may not be aware of the greater need for the same thinking and even stronger action in the less repetitive or low-quantity work areas.   Management must be represented at all major discussions and must be available for decision making and action taking whenever necessary.

After installation, as for any work measurement program, management must continue its active support and insist on systematic periodic reporting and checking of progress and of the results being achieved.   An active industrial engineering team, with management support, must continue and maintain the program after installation or the investment will be wasted.   An installation of measurement for low-quantity work can deteriorate and lose its control or incentive effect even more easily than a similar program in repetitive work areas.

CASE HISTORIES

Some of the technical and human problems to be overcome when measuring low-quantity work and applying standard times either for incentive pay or for control purposes have been discussed.   Now some examples of the standards application procedures for specific cases can be looked at.

Low-quantity Machining.   In toolrooms, development shops, and true jobbing machine shops, usually only one, two, or perhaps five pieces from any one drawing are made.   It is not possible, or not worthwhile, to set stops and gages or to make special tools and cutters.   Frequently, the "each piece" handling and gaging cover most of the machine setup time which is associated with repetitive work. Using the most simple machine to illustrate the calculation procedure (the end result of the time formula development described in Chapter 8 of Section 3), Figure 2-7 shows the worksheet A used to calculate the machine time for any drilling, reaming, or tapping operation.   This involves conventional use of standard

WORKSHEET (A) – MACHINING TIME   Sensitive drill presses

Apparatus _____ Part name _____ Oper. no. _____ Part no. _____

Material _____ Date _____ Applicator _____

D-Drill	R-Straight ream	T-Tap	SF-Spotface	CB-Counterbore	C-Chamfer
RD-Redrill	<R-Taper ream	PT-Pipe tap	BF-Backface	CS-Countersink	DB-Deburr
FBD-Flat bottom drill					

Operation			Machine settings			Time per hole (hrs.)					Total time (hrs.)		Comments
Tool sym.	Dia.	Chart no.	RPM	Feed	Time per/in.	Hole depth	Time per hole	Lead time	Relief time	Total time per hole	No. of holes	Total time	
												Total Machining Time

Fig. 2-7. Typical calculation sheet for machining time.

data on surface speed, feed, and depth of cut for different types of material to calculate actual time for machine cutting.

In Figure 2-8, worksheet B shows the calculation procedure for the handling or manual time for doing the drilling operations. It is necessary to decide what holding devices will be used, how the part will be turned, and any special tools used. The handling time can then be calculated easily by multiplying the element standard data by the frequency of occurrence for each item and totaling.

On this sheet, the machine time, each-piece time, setup time, and gaging time (calculated on another worksheet) for all operations done on the drill press are brought together. Basic standard time is calculated, and then allowances for personal, fatigue, and minor delays are added—in a typical toolroom they amount to 20 percent. Because the machine time is based on *optimum* speeds, feeds, and depth of cut, a "machine time allowance" is usually added separately to this pure machine cutting time if management decides a bonus should be earned when these optimum cutting rates are met. Similar worksheets are available for all machining operations, and standards are calculated in a similar manner for turning, milling, grinding, shaping, drilling, and other operations.

In practice, it is sometimes found that certain classes of items occur frequently—for instance, small pins, shafts, bushings, or clamp plates. A further step of simplified standards application can then be taken if desired; tables are developed showing ranges of sizes and shapes for each class of item along with the work group time for each typical item. To develop these tables, the actual standard time for producing items in each range is calculated, using worksheets A and B. Figure 2-9 shows an extract from such a table for sleeves and bushings produced on a lathe. By using a "work content comparison" approach, the standard time for almost any size or shape or tolerance of bushing can be read from these tables

SENSITIVE DRILL PRESSES
WORKSHEET (B) – HANDLING TIME

Part no. _____

Apparatus _____ Part name _____ Oper. no. _____

Material _____ Date _____ Applicator _____

Operation description and symbol					Setup		Each Piece	
					Occ.	Total hrs.	Occ.	Total hrs.
Shop Setup (K1-.091)						-	-	
Load and unload	To 5#	To 20#	To 60#	Over 60#	-	-	-	-
Get and remove part	K2-.003	K3-.006	K4-.009	K5-.039	-	-		
Turn part over	K6-.002	K7-.004	K8-.006	K9-.045	-	-		
Type of holding device		Setup	Each piece		-	-	-	-
Angle plate		K11-.023				-	-	
V-blocks (2)		K14-.023				-	-	
Vise		K10-.027	K13-.004					
Bushing plate		SC- -.001	K15-.006					
Parallel bars (2)		SKG2-.009	K16-.004					
Fixture		K12-.031	Chart			-	-	

Type of fixture fastener		Number of fasteners							
	Sym.	1	2	3	4	-	-	-	-
Bolt or capscrew	KW2	.005	.009	.012	.016	-	-		
Thumbscrew	KX2	.002	.005	.007	.009	-	-		
Handwheel	KY2	.002	.005	.007	.010	-	-		
Cam or eccentric clamp	KZ2	.001	.003	.004	.005	-	-		
Plug or pin	KP	.002	.004	.005	.007	-	-		

Type of clamp				Number of clamps								
	Sym.		1		2		3		4	-	-	-	-
	S.U.	Pc.	S.U.	Pc.	S.U.	Pc.	S.U.	Pc.	S.U.	Pc.	-	-
C clamp	SKN3	KM	.009	.006	.009	.012	.009	.018	.012	.024	-	-
B, N, & W	SK32	KW2	.021	.005	.036	.009	.051	.012	.066	.016		
B, N, W, & C	SKJ2	KU2	.037	.006	.053	.011	.069	.016	.087	.021		
B, N, W, C, & H	SKY3	KN2	.051	.006	.073	.012	.095	.017	.119	.022		

B - bolt   N- nut   W - washer   C - clamp   H - heel

Set up and align part		Type of device	Setup	Each piece	-	-	-	-
		Scale	SKQ-.005	KG1-.002				
		Level or square	SKQ-.005	KW -.003				
		Protractor	SKS-.022	KW -.003				
		Feeler gage	SKG1-.007	KF1-.001				
		Surface gage	K33-.001	A36-.002				
Gage part		WORKSHEET C						

Change tool –	Keyed chuck (Jacobs)	K18	.004				
	Tapered tool in spindle	K19	.006				
	Install sleeve or adapter to tool	K20	.004				
Manipulation –	Position part to tool (same spindle) (fixture, predrilled)	K21	.001	-	-	-	-
	Position part to tool (same spindle) (to layout hole)	K22	.002	-	-		
	Additional-position part from spindle to spindle	S38	.001	-	-		
	Place and remove slip bushing in fixture	K32	.002	-	-		
	Brush on lubricant	M3	.001	-	-		
	Blow out chips from hole drilled to depth	K33	.001	-	-		
	Raise or lower spindle head	K34	.021	-	-		
	Raise or lower table (Avey only)	K23	.010				
	Set feed (Avey only)	K27	.018				
	Set speed (Avey only)	K30	.001				
	Set speed (Delta only)	K31	.006				
	Set depth stop	K35	.007				
	Deburr drilled hole			-	-		

Comments: _____	Total net time	-	
_____	Allowances	-	
_____	Allowed std. time	-	
_____			
_____	Machining time (Worksheet A) _____	-	-	-
_____	Mach. time allow. _____	-	-	-
_____	Total std. time _____	-	-	
	Total standard time - per piece	-	-	

FIG. 2-8. Typical calculation sheet for handling time.

4-25

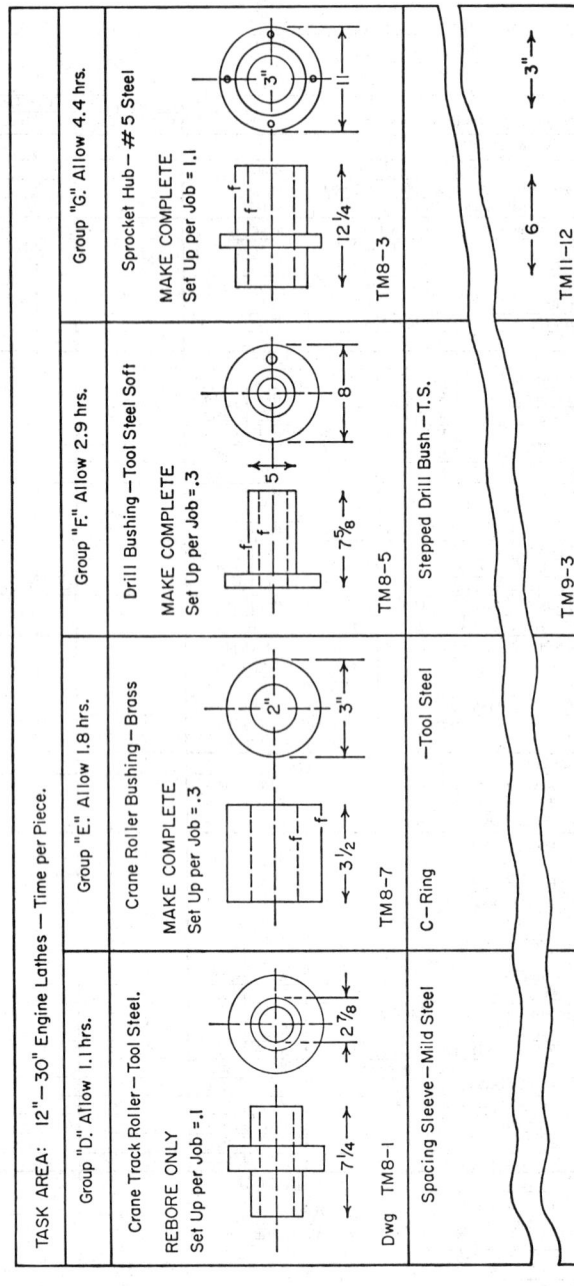

Fɪɢ. 2-9. Extract from table or spread sheet for similar lathe operations, used for work content comparison standard setting.

or spread sheets. If difficulties arise because of special factors for a certain item, the specific item can be calculated from the worksheets as before.

The main feature of these two approaches is speed of standards application while still maintaining the necessary degree of accuracy. Difficulties arise if lot sizes change, and precautions must be taken to avoid misuse of such data. For example, in one toolroom, data for the lathe operations were giving excellent results until a change in policy required lot sizes of 100 for standard punches, bushings, and C rings to be put into stock. The data were still correct, but simplified application procedures set up for lot sizes from 1 to 5 were not accurate for lot sizes of 100 where multiple cuts, machine stops, and gages were used. All simplified data and application procedures must specify limitations for their use and must be checked periodically for any necessary revisions.

Maintenance and Field Assembly. For the manual work associated with maintenance and field assembly, there are major problems relating to the variety of the work, finding out what actually has to be done, what differences occur, and then economically applying the standard times. Elsewhere in this Handbook, solutions to these problems are discussed in detail. Sufficient here to say that acceptance of the various special criteria for the measurement of low-quantity work has made economically possible the development and application of standards to this difficult maintenance and field assembly area. To illustrate the results from application of UMS standards and controls for one actual installation, Figure 2-10 shows the first eight periods' reporting of performance against standard, standards coverage, man-hours used per two weeks, and backlog man-hours for one maintenance department in a large steel plant. The benefit of performance increase from 68 to 90 percent (wih no incentive pay) is confirmed by the 30 percent reduction in man-hours per week used for the same maintenance work load, and the drastic reduction in backlog hours.

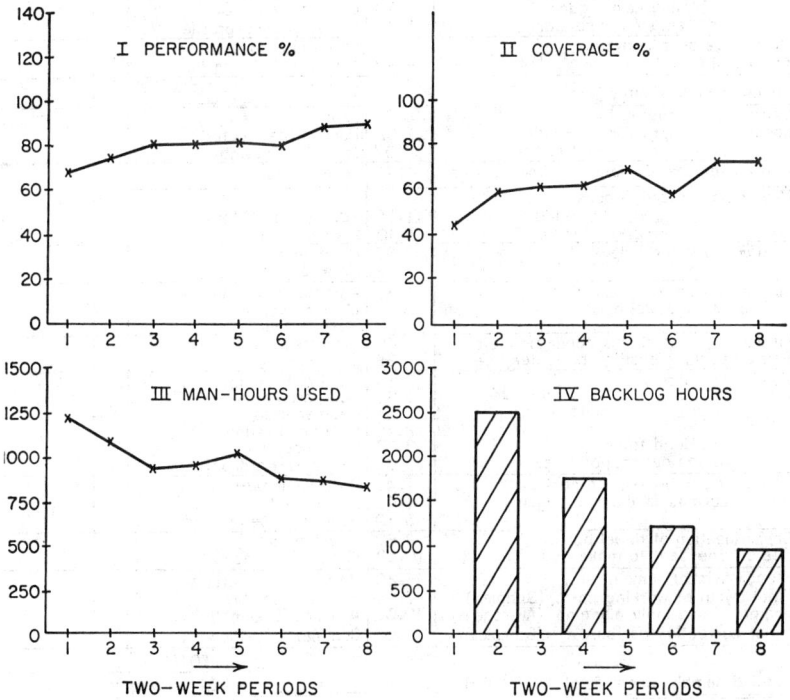

Fɪɢ. 2-10.   UMS installation results—first eight periods' reporting—steel mill.

Toolroom Fitting. Procedures for applying standards to the machining of various components for tools and dies have been described. The fitting, assembly, and repair of tools and dies is a further case of measurement of low-quantity work where special data and application procedures must be used. Methods time measurement has been especially applicable for the development of data for this toolroom fitting work. Time formulas or standard data are developed for all representative work, and in many cases can be subdivided as:

Use hand and power tools
Layout
Fit mating parts
Assemble and disassemble ⎫
Try out   ⎬ subdivided according to type of die or tool
Repairs   ⎭

TEMPLATES, JIGS & FIXTURES
BENCH WORKSHEET VIII

Symbol_____Order No._____Quantity_____Date_____

	Time	Allow for each	Use	Lev. Mins.
A. Prepare Parts.				
1. Job preparation I-54-60-20	34.50	Job		
2. Move large parts to work area.	5.42	Part over 50 lbs		
3. Check with B.M. for parts.	1.00	Item on B M		
4. Check completion of mach. parts.	1.00	Machined part		
5. Deburr parts. Small (under 50 lbs)	.74	Small part		
Large (over 50 lbs)	2.86	Large part		
6. Trial assm. of mating parts.	.68	Trial assm. crit. fit		
7. File parts to make a fit or match profile.	8.63	Part to be fit or profile		
PLUS	1.25	Inch to be filed		
8. Layout (a) Using comb. square	.86	Pair crosslines		
(b) Using angle plate	.11	Use of angle plate		
(c) Using height gage	.50	Line scribed/H.G.		
(d) Temp. or comp. part	3.81	Scribe around profile		
9. Move large part to layout.	.32	Large part to layout		
10. Apply bluing.	6.62	Area over 360 sq. in.		
B. Fitting and Assembly		TOTAL "A"		
1. Extra trips to tool crib.	8.72	Job		
2. Locate part approx. Hand	2.62	⎫ Parts to be located		
Hoist	6.97	⎭ for drill & assmb.		
3. Apply & remove "C" clamp or similar device.	1.40	Located part		
4. Move part to mach. & back.				
Hand	1.14	Part under 50 lbs		
Hoist	9.10	Part over 50 lbs		
5. Drill & tap fixing hole.	3.69	Machine set-up		
	2.22	Hole tapped		
6. Locate parts accurately.	7.33	Part accurately located		
7. Drill & ream dowel hole.	3.69	Machine set-up		
	2.53	Hole reamed		
8. Fit dowels, locating pins, taper pins, etc.	.40	Pin fitted		
9. Bolt all parts in position & tighten.	.29	Bolt used		
10. Hand operations (ave.)	⎧ 2.84	Hand drill setup		
Hand drill	1.65	Hole drill. by hand		
Hand ream	.51	Reamer used		
	2.55	Hole hand reamed		
Hand tap hole	3.37	Hole tapped		
Clean out tapped hole	⎩ 2.00	Tap. hole-Hard. pc.		
11. Fit bushing & pressed in parts.	1.47	Pressed-in part		
12. Ream bearings (1" diam. x 2" lg.)	1.02	Reamer set-up		
	5.10	Bearing to be reamed		
13. Check location of dowel parts.	2.62	Doweled part		
14. File scrape-seat to make square or flat.	10.57	10 sq." of surface		
15. Stencil symbol number, etc.	7.20	Unit up to 75 letters		
16. Final tryout of working parts, (Simple)	1.60	Simple assembly		
gaging of assembly, planning (Average)	10.00	Average assembly		
minor corrective action. (Complex)	40.00	Complex assembly		
		TOTAL "B"		

Total assembly time = Total A + Total B =_____+_____=_____x 1.24 allowance

FIG. 2-11.   Calculation sheet for toolroom benchwork standards.

Figure 2-11 shows one of several worksheets used in one toolroom standards installation for calculation of the complete man-hours required for all fitting and benchwork done in that shop. This example shows the subdivision of the work—"templates, jigs, and fixtures." Other worksheets cover various fit and assemble operations on all types of dies. The standard data, developed by MTM and tabu-

---

S.B.,F.P.,D.D.,S.O. Dies   Bench Work Order — 3 —

Date _____   Work Order No. B_____

Die No._____   Die Class (S)

Die Description _____

Regular Service ◯ Replacement ◯ Change ◯ Rebuild ◯

		Tool Room	Die Vault	Press Room	Clean
Complete	)	◯ .125	◯ .245	◯ .275	◯ .096
Top or Btm. Shoe	) From	◯ .076		◯ .226	◯ .036
Other Sub-assem.	)	◯ .008		◯ .060	◯ .012
Complete	)	◯ .125	◯ .245	◯ .275	
Top or Btm. Shoe	) To	◯ .076		◯ .226	
Other Sub-assem.	)	◯ .008		◯ .060	

BOTTOM SHOE Cradle	x .066=	Change: From	
Die Sections	x .061 =	To	
Bumper Blocks	x .042 =	Replace New Parts:	
Full Stripper	x .086=	Piercing Punch	x .098=
Stock Equalizer	x .078=	P.O. Slot Punch	x .098=
Other Sections	x .066=	K.O. Punch	x .098=
TOP SHOE Punch Holders	x .104=	(Ex. C.L.)	
Other Sections	x .066=	Contour Punch	x .098=
PUNCHES Piercing.	x .039=	Trimming Punch	x .270=
		Die Section	x .303=
Knock-out	x .039=	Bushings	x .098=
Stencil	x .049=	Pilot Pin	x .098=
Trimming	x .068=	Round Insert	x .098=
Lancing	x .059=	Dowel Pin	x .095=
Dimpling	x .039=	Quill	x .098=
Extruding	x .039=		
Embossing	x .039=	Restore Radius – Hard	
		Clearance – Inches	x .020 =
BUSHINGS, INSERTS & MISC. Stencil Bumper	x .039=	Draw or Form – Hand	
Forming Insert	x .074=	Stock | x 3 =	
Cut-off Insert	x .069=	Plus .020	
Embossing Insert	x .049=	Inches |x| =	
Bushing	x .045=	BROKEN BOLTS:	
Pilot	x .039=	Large Small	
Spring Pad	x .104=	Proj. .300 .250	
Lifter Pin	x .056=	Flush	
Push Back	x .056=	Blind 1.000 .750	
Quill	x .039=	Flush	
Spring Stripper	x .152=	Open .750 .500	
Guard	x .043=	BROKEN SET SCREWS:	
Method of Mounting:		1.000	
Bolted Only	x .043=	Get Parts Welded ◯ .250 =	
Ditto w/springs		Use Die Flipper ◯ .747 =	
or shims	x .068 =	Re-dowel each part \[\] x .366=	
OTHER PARTS Bolted & Dowelled	x .063=	Ear Form (Octagon ◯ .458 =	
Ditto w/springs		Adjust (Utility ◯ .308 =	
or shims	x .078=		
Press Fit	x .045=		
Slide or Tap Fit	x .039=		
		Job Constant .120	
		Total	
		Times 1.40 = Allowed Hrs.	

FIG. 2-12. Calculation sheet and work card for toolroom die repair standards.

lated on this worksheet, must be used with a frequency decided by standard procedures and the die drawings; but once the operations and frequency are analyzed by an experienced engineer or toolmaker, the calculation of the standard time is very simple. Figure 2-12 shows one of six work cards used in another toolroom incentive installation for calculation of the complete man-hours required to carry out any type of repair to six types of punch, trim, blank, and draw dies. Again, MTM-developed standard data are used as a basis, and the standard time can be quickly calculated for any frequency of individual operation. In this case, these work cards are issued to the worker by the foreman, with apparently necessary repair elements checked off. Additional elements of work are added as required, subject to foreman and industrial engineering approval, since these standards are being used as the basis for incentive payment.

In both these toolroom installations, the simplified application procedures enabled one calculator to issue work standards for incentive payment to an average of twenty-five toolroom machinists and benchworkers. Both installations led to appreciable reductions in man-hours and to increased incentive earning for the machinists and toolmakers.

SUMMARY

Measurement of low-quantity work is completely practical—both technically and economically—as far as development of data and the application of standards are concerned. Special difficulties must be overcome, and a more overall or average approach must be used. The benefits to be realized are great. Good labor savings can be gained if industrial engineers and managers work together. Low-quantity work standards can be used for control purposes and for incentive payment. Increased money can be earned by men in toolrooms, job shops, and maintenance departments for work done against good standards based on accurate work measurement. Management can really control the labor and machine costs for this low-quantity work and realize major cost reductions.

In this chapter, some of the problems, some of the solutions, and some of the technical procedures used have been outlined. Often, only a relatively small proportion of low-quantity work is covered with standards, because management frequently considers development and installation problems to be too difficult. This feeling is unjustified, as shown by this presentation of solutions and examples of results achieved.

BIBLIOGRAPHY

Wilkinson, J. J., "How to Manage Maintenance," *Harvard Business Review*, March–April, 1968.
Wilkinson, J. J., "Maintenance Cost Control," sec. 6, chap. 3, in H. B. Maynard (ed.), *Handbook of Modern Manufacturing Management*, McGraw-Hill Book Company, New York, 1970.
Wilkinson, J. J., "Maintenance Management," four-article series in *Plant Engineering* beginning Jan. 9, 1969.

Chapter 3

Fundamentals of Indirect Labor Measurement

WILLIAM K. HODSON

President, H. B. Maynard and Company, Incorporated,
Pittsburgh, Pennsylvania

Historically, work measurement has been concerned with the measurement of direct labor. One reason is that some direct labor measurement is required in most cases for product costing and pricing. Another major reason is that direct labor historically accounted for the bulk of the hourly labor in a plant and was the most significant cost to measure and control. From an industrial engineering viewpoint, the measurement of direct labor operations is a great deal easier and consequently less costly than indirect labor. For these reasons, the emphasis in the past has been on the measurement of direct labor operations.

Both evolutionary and revolutionary changes have been taking place in our economy. These changes tend to place much greater emphasis on the measurement of indirect labor. Because of the impact of automation, the traditional relationships between direct and indirect labor are often reversed. An example is the automated machining transfer line, where the direct workers are completely replaced by setup and maintenance men. It is no longer unusual to find a factory where the indirect workers outnumber the direct workers. This condition is further compounded as clerical, technical, and administrative activities become a larger part of the economy. These activities are central to most service industries, which are growing at a rate greater than the economy as a whole. Undoubtedly, the computer and advanced technological improvements such as numerically controlled machine tools are accelerating the growing importance of indirect labor operations. All these developments have created a greater need for the measurement and control of indirect labor.

DISTINCTION BETWEEN DIRECT AND INDIRECT LABOR

The manufacturing cost of a product is usually divided into three major components: labor, materials, and overhead. Labor cost normally is composed of only the direct labor cost. This is the cost of labor directly involved in the physical manufacturing of the part. Each direct labor operation has some effect on the part or assembly. The amount of labor required for each part or assembly can be measured directly.

A major part of overhead costs is the cost of indirect labor. A typical indirect labor cost is janitorial service. Although the cost of janitorial labor can be easily determined, it is difficult to relate the cost of this labor directly to a specific part or product. As a consequence, the cost of indirect labor operations is usually allocated to specific products on a percentage or prorated basis. A typical approach is to allocate indirect labor costs to products in proportion to their direct labor hours. This approach assumes that there is a direct relationship between the indirect labor hours and the direct labor hours. Typical indirect labor costs include material handling, shipping, receiving, warehousing, tool cribs, toolmaking, maintenance, janitorial service, inspect and test, factory service and repair, field installation and servicing, clerical workers, chemical and physical laboratory workers, and draftsmen.

Some of these activities—for example, "inspect and test"—may sometimes be classified as direct labor if the work content of the operation can be directly related to the job. In industries, such as aerospace, which are oriented to government contracts, a strong attempt is made to classify as many manual operations as possible as direct labor. This classification tends to reduce the indirect labor or overhead costs and to increase the direct labor hours. Because overhead is frequently expressed as a percentage of direct labor costs, this greatly reduces the overhead percentage. The magnitude of overhead costs is frequently a factor in the government's evaluation of the efficiency of a contractor. In many cases, however, the apparent efficiency of a contractor is established in the accounting department, and not on the shop floor. Low overhead rates may be the result of accounting treatment, and not truly the reflection of low costs.

FACTORS INFLUENCING INDIRECT LABOR MEASUREMENT

Although the basic work measurement techniques used to measure indirect work are the same as those used to measure direct labor, there is a significant difference in emphasis and in the method of application. Other, more serious differences exist in related areas such as methods, work counts, standard hour calculations, group application of standards, and the like. An analysis of these factors will provide a better understanding of some of the reasons for the differences between direct and indirect labor measurement.

Methods. Both direct and indirect labor measurement require that the method employed in performing the operation be standardized before the standard time is established. The application of this rule varies substantially between direct and indirect work. On very repetitive direct labor work, it is not at all unusual to spend from fifty to one hundred man-hours on methods work for every one hour spent on work measurement. Furthermore, great emphasis is placed upon training the operator to follow the established method with a high degree of precision. In some cases, each individual motion is analyzed and subjected to the operation analysis and motion study approach. The number of motions required to perform the work is reduced to a minimum, and the motions that remain are reduced to the simplest form of motion. Unless an operator is specifically trained to follow the method, motion by motion, it will be difficult for him to perform the operation in the standard time allowed.

The other extreme of methods analysis exists in most maintenance activities. Although methods study can contribute substantially to improving maintenance work, the emphasis is different. Instead of concentrating on the methods employed to perform a specific job or operation, attention is focused on the overall methods employed by the craftsmen. The overall methods are applicable to all jobs. For example, considerable time should be devoted to studying the various tools used on the job and the methods for transporting and storing these tools. Frequency studies can quickly disclose which tools are most often used. It can then be objectively determined which tools should be attached to a tool belt where they are quickly available, and which tools should be kept in a tool box, or in some cases, in a specially designed tool cart.

If the maintenance men have come up through the shop with little or no formal craft training, the methods work may have to be supplemented by basic training in the proper use of hand tools. The principle to follow is to concentrate on methods studies that will have universal application and to minimize methods studies for specific operations.

Maintenance work represents an extreme example of nonrepetitiveness in indirect labor work. Other forms of indirect labor, such as janitorial work, tend to be more repetitive, and the methods study can be directed more toward the methods to be used in performing a specific job. Often, very little thought is given to janitorial work from a methods viewpoint. It seems to be regarded by all concerned as a simple job—one that anyone can do. Consequently, most janitorial people are given very little, if any, instruction on how to do the job, how frequently jobs should be performed, and the like. Therefore, methods work in this area can produce substantial returns.

This point illustrates one reason why the application of work measurement to indirect labor produces such worthwhile returns. Practically all direct labor work is subject to some form of methods study and improvement. This is particularly true in those cases where work measurement is applied. As a consequence, the work methods used by direct labor are usually fairly effective. On the other hand, it is unusual to find any application of methods study to indirect labor operations. In fact, the complexity and variety of work performed by the indirect workers require a much higher order of methods study than do repetitive production operations performed under a standard set of conditions, but the return can be greater than the return on the typical direct labor job.

Units of Work. Another characteristic of indirect labor that makes it difficult to measure is the lack of a single unit of work. The unit of work for most direct labor operations is the "piece." A time standard can be expressed in terms of minutes or standard hours per piece. The number of pieces produced by the operator can then be counted. The pieces produced multiplied by the standard hours allowed per piece quickly gives the number of standard hours earned for a particular job.

Consider, however, the problems involved in attempting to establish a unit of work for an order picker. An order picker is commonly employed in a finished goods warehouse. An order for a number of different items in differing quantities is received by the picker. His job is to fill the order by walking to the different bins or locations where the items are stored, selecting the correct quantity of each item, assembling the items in a box or truck, and returning the completed order to a packing station.

Obviously, the unit of work cannot be "an order." The number of different items, usually called "line items," can vary from one to twenty or more. The standard time per order will vary accordingly. It is not possible to use the line item as the unit of work because there are certain elements of work associated with each order that are somewhat fixed regardless of the number of line items. Further, the time for a line item will vary substantially, depending on the shape of the item, its location, whether it is ordered in standard package quantities or less than package quantities, whether the item has to be weighed, and so on. In most indirect labor applications, it is necessary to use a number of units of work and to establish a routine for recording the units of work performed. The standard hours earned for each of the units can then be calculated. The problem of establishing work counts is generally a great deal more complicated and expensive than for direct labor. In many cases, however, some ingenuity and imagination on the part of the industrial engineer can simplify this problem. In order picking, for example, the job of calculating the hours earned for each order can frequently be combined with the pricing and price extension of the order. If this is computerized, the data base for the time standards can be incorporated in the pricing program. In this way, the standards calculation can be performed as a by-product of the pricing program.

With some ingenuity, the job of collecting work counts and applying time standards can be greatly simplified despite the fact that many more variables are involved than in most direct labor operations. This does, however, tend to increase the administrative costs of indirect labor programs.

Measurement of the Work. The standard techniques for measuring direct labor are time study, predetermined motion times, standard data, and work sampling. All these are also used for measuring indirect labor, although their application is quite different.

One of the most common studies on direct labor is an individual time study of a single operation. The operation is normally divided into a number of elements and each element is repeated in a standard sequence. A time study is then made of the operation, usually covering ten or twenty pieces. The same sequence of elements is followed for each piece.

This type of study rarely occurs on indirect work. This may be quite frustrating to the industrial engineer whose previous work measurement experience has been limited to direct labor. A main characteristic of indirect labor is that it is nonrepetitive by nature. If a series of time studies is made of the order picking operation, one study for each order picked, it is likely that there will be twenty different studies for twenty different orders. The number of line items, the location of the items, and the quantity of the items picked will vary considerably from one study to the next.

Because of the nonrepetitive nature of indirect labor operations, it is important that the work be carefully analyzed and planned before any attempt is made to measure it. One of the primary purposes of this preplanning is to develop methods improvements and to standardize the methods used for performing the work. The standardization of the work is important because it enables the predetermination and synthesis of what the operator should do to perform a given job. A secondary benefit of this analysis is a greater insight into the nature of the work and its characteristics, the major elements of the work, the variable factors that exist in these elements, the sequence of the elements, and the like. All this information is essential to the design of the work measurement system.

Because of the variable sequence of the work elements and the variables that may exist within a given work element, it is essential to rely heavily on the standard data and the building block concept of work measurement. This concept is explained in detail in Chapter 8 of Section 3. Either time study or predetermined motion times can be used to develop the detailed elements needed to construct the standard data, but predetermined motion times are usually better suited to this task than time study elemental times, and their use is recommended.

Multiple Regression Analysis. The procedures discussed in this chapter for measuring indirect labor emphasize for the most part the traditional approach to this problem. This approach involves studying each activity of the indirect labor operation and establishing time standards for them by the use of time study, predetermined motion times, standard data, or work sampling techniques. Although the returns from most indirect labor measurement programs are well worthwhile, the measurement procedure is often tedious, time consuming, and relatively expensive. A number of attempts have been made to simplify the work measurement aspect of the problem by the use of mathematical techniques, the most promising of which is multiple regression analysis combined with work sampling.

Two or more units of measure must usually be used to measure the output of indirect labor operations accurately. The multiple regression analysis technique provides a means of correlating several units of measures, called "predictors" in multiple regression terminology, with the time required to perform the job.

The multiple regression model is an equation that relates the predictors (units of measure or variables) to the response (standard time to perform the work). The standard format is

$$Y = b_0 + b_1X_1 + b_2X_2 + b_3X_3 + b_4X_4 + \cdots + b_iX_i$$

Assume that a shipping operation consists of filling orders by the use of a fork truck. The truck operator picks up an empty pallet and proceeds to the various stock locations where he loads the necessary cases for the orders on the pallet until the orders are filled or the pallet is loaded. He then transports the pallet to an order assembly area. The objective is to determine a method for measuring the time required per pallet load. A preliminary analysis of the operation indicates that the predictors, in this case, may be any of the following:

X_1 = number of cases packed per pallet load
X_2 = number of orders per pallet load
X_3 = weight of packed material per pallet load
X_4 = volume of cases per pallet load

A work sampling study is then made of the operation. During the course of the study, observations are made of the work and nonwork elements for each of the operators. The data are adjusted by performance rating factors to provide a leveled observation percentage of the working time. Records are also maintained of the actual hours worked, the number of cases packed, the number of orders, the weight of the packed material, and the volume of the cases for each pallet loaded. All the data for each pallet load are then placed in a computer program for multiple regression analysis. Because of the complexity of the calculations, a computer is the only economical way of using multiple regression for problems of this magnitude.

The computer solution provides values for the coefficients b_0, b_1, b_2, b_3, and b_4. The program also provides correlation coefficients for each of the predictors.

By examining the various correlation coefficients and the correlation between the predictors, an individual firmly grounded in the mathematics of regression analysis can determine which of the predictors are significant and which are not. After such an analysis, it may be decided that the predictors of weight and space provide very little additional predictive capability to the predictors of number of cases and number of orders, and the final formula might look like this:

$$Y = 5.02 + 3.37X_1 + 0.45X_2$$

where Y = standard minutes per pallet loaded
X_1 = number of cases packed per pallet
X_2 = number of orders per pallet load
The values for the coefficients b_0, b_1, and b_2 are, respectively, 5.02, 3.37, and 0.45 minutes. The standard time for a pallet made up of six cases and two orders can be calculated by inserting the predictor quantities into the formula and solving for the standard time:

$$Y = 5.02 + 3.37 \times 6 + 0.45 \times 2$$
$$= 26.14 \text{ minutes}$$

Although regression analysis is a powerful tool and shows great promise in the field of indirect labor measurement, it has the potential of being grossly misapplied by industrial engineers who use the programs on a "cookbook" basis. It should not be used without the counsel of an individual who is completely familiar with the mathematics of regression analysis. Further information on this technique may be found in Section 10 and in the bibliography.

Individual or Group Application. Once the units of work have been established and time standards have been determined for each unit of work, the question arises of how to apply the standards. Should they be applied to individual operators or to groups of operators? There are many points of view on this question, and the subject is treated more thoroughly in Chapter 4 of Section 6. There is a definite tendency to apply standards to indirect operations on a group basis. One reason is the difficulty of establishing units of measure that are directly related to individual performance. This occurs whenever the work is performed by a team of people and not by individuals.

Another reason arises from the manner in which the standards are established. Because of variations in methods and conditions of performing indirect jobs, it is frequently necessary to rely heavily on averages. If standards make extensive use of averaging conditions or methods, then group application is desirable. If standards are applied on an individual basis, the operators may be inclined to pick and choose from available jobs with variations on the favorable side of the average and to let the more difficult jobs wait until later. If the performance of the group as a whole is measured, they will be less inclined to do this.

In other cases, the work of one operator may have a significant influence on the work of the next operator. When this is the case, group application is essential. An example of this arises in a toolroom. A part for a die is first milled to approximate size, heat-treated, and then ground to the exact tolerance. If the toolmaker performing the milling operation leaves too much stock on the part, the grinding time may be far in excess of the time allowed by the standard toolroom practice manual. Under a group application, it is to the interest of all members of the group to simplify the work to be performed by any other member of the group. This is not true in individual applications.

There are some cases, however, where the individual application of standards is quite logical for indirect work—janitorial work, for example. In most janitorial installations, a specific schedule of work is prepared for each man for each day of the week. If a worker completes every item on the schedule, he will be working at the desired performance level. This level may be 80 to 100 percent for daywork or 120 to 130 percent for incentive applications. In either event, it is entirely practical to apply the standards on an individual basis.

The selection of the unit of work will also have an effect upon the decision whether to apply the standards on a group or an individual basis. The order picking operation can be used to illustrate this point. The reverse of order picking is replenishing items in the bins or storage areas. Sometimes, the order picking job is combined with the stock replenishment job, and they are both performed by the same person. In other cases, they are performed by different people. Girls can sometimes be used for order picking, but stock replenishment involves heavy lifting and is performed by a man. In this example, assume that the latter is the case. The time standards for order picking and stock replenishment are established separately. If the standards are to be applied on an individual basis, it will be necessary to keep track of two sets of work counts. One will be the orders and items picked; the other, the quantity and type of items replenished. If, however, the standards are applied on a group basis, only one set of work counts will be maintained. This is the count for the pickers. The time standard applied for the items picked will also include the time for replenishing that same item. This approach will tend to simplify both the work count procedure and the standards application procedure. Wherever possible, time standards for two or more operations should be combined in this fashion and then applied on a group basis to simplify the administrative cost of maintaining the system.

Incentive or Measured Daywork. It is often asked if the design of the time standards needs to be altered in any way, depending upon whether the standards are to be used for incentives or for measured daywork controls. Although there may be specific exceptions, the general answer is no. There should be no difference in the design of the time standards. It may be desirable to have some variation in the work count procedures if incentives are used, but the time standards should be equally good for either application. If the standards for a measured daywork installation are any less accurate than those established for an incentive installation, their weaknesses will quickly become apparent to both the workers and their supervisors, and the program will not be effective.

Quality Control. It is just as important to install some form of quality control over the work produced by indirect labor as it is to maintain quality control on direct labor operations. Because many indirect labor operations are more in the nature of a service rather than the producing of a product, however, the job of establishing quality controls is more difficult.

This may be illustrated by the problems involved in setting quality standards and controls on janitorial services. When is a window satisfactorily clean? When is a floor well waxed? Because most janitorial work is set up on a scheduled basis, it is also necessary to verify that the items on the schedule have, in fact, been performed. Figure 3-1 illustrates a form used for quality control and frequency checks by the janitorial supervisor. The frequency of items to be performed is indicated by "D" for daily, "W" for weekly, and "2W" for twice weekly. The number of items to be cleaned is also shown.

The quality of other indirect labor operations can sometimes best be measured by the end result of the job. In most maintenance work, the quality of the job is usually determined by whether or not the machine has been put back in operation or the leak has been stopped. The quality of a jig or fixture produced in the toolroom is perhaps best measured by whether or not it performs, to the proper tolerances, the function it was designed to perform.

In still other cases, it may be necessary to rely upon the number of complaints received from internal managers or external customers as a measure of the quality of the work performed. Order filling operations, material handling, maintenance, and a number of other service-oriented activities are in this category. If the source

Formula: ___1090.01_____

Date ___7-28-_____   Sheet ___1___ of ___3___ Sheets

Building ___Main office___   Floor ___1st floor___   Room ___Men's washroom___   Sq. ft.___200___

REFERENCE	ELEMENT DESCRIPTION	NO. ITEMS	FREQ.	HOURS	TIME		
					DAILY	OTHER	STD. HRS.
20.0601	Clean drinking fountain			.0083			
20.0401	Clean dispenser	4	D 1.0	.0071	.0284		.0284
70.0101	Service dispenser	4	D 1.0	.0079	.0316		.0316
70.0201	Service eye glass cleaning station			.0098			
20.0416	Clean paper cup dispenser			.0098			
70.0104	Service paper cup dispenser			.0038			
70.0103	Service toilet tissue dispenser	2	2W .4	.0021		.0042	.0017
20.0412	Preflush urinal or commode	2	D 1.0	.0008	.0016		.0016
50.0103	Empty and wipe wall-hung ashtray			.0086			
50.0104	Empty and wipe wall-hung ashtray, water type			.011			
50.0106	Empty and refill sand urn			.0177			
50.0105	Remove butts from sand urn			.0089			
40.0003	Pick up heavy debris	1	D 1.0	.0006	.0006		.0006
50.0202	Empty wastebasket			.0045			
20.0502	Clean wastebasket			.0097			
50.0201	Empty oily rag receptacle			.0014			
50.0307	Empty waste receptacle liner - metal or cloth	1	D 1.0	.0092	.0092		.0092
20.0501	Clean exterior of waste receptacle	1	W .2	.0144		.0144	.0029

Fɪɢ. 3-1.   Evaluation report and standards calculation sheet for janitorial work.

of complaint is a sensitive source, such as a customer, a large number of complaints may require a more formal inspection or spot checking. This is particularly true of factory repair and service operations. The work of such a group should be subjected to an inspection procedure at least as rigorous as the inspection of the original product.

WAREHOUSE OPERATION

A typical application of indirect labor measurement may be illustrated by a warehouse operation. The warehouse was a large, single-story building of 410,000 square feet. It was serviced by an 800-foot railroad siding and twelve truck docks. Prior to the work measurement program, the warehouse was operated by thirty-eight warehousemen.

The products stored in this warehouse consisted of a variety of large industrial equipment. Each piece of equipment was of a standardized shape and configuration but of different size. The various kinds of equipment were divided into seven general classes.

The equipment was manufactured in adjacent assembly buildings and transported to the warehouse by trailer trains; it was palletized after assembly and stacked and handled within the warehouse by the use of fork trucks.

As a first step in the design of a work measurement plan, the warehouse operations were broken down into five basic activities:

1. Transport equipment from assembly areas to warehouse.
2. Receive and store incoming material.
3. Pick items to fill orders and accumulate in holding areas.
4. Prepare items for shipment by crating, labeling, banding, stenciling, and the like.
5. Load trucks and railroad cars, including cribbing and bracing cars.

Methods Analysis. The first step in this installation was to carry out an intensive methods analysis program to standardize the work and develop improvements.

After a preliminary study, it quickly became apparent that there were substantial variations in the loads transported by identical fork trucks. One operator picked up and transported one unit at a time; another, a stack of three units at a time. This problem was assigned to a committee composed of the warehouse foreman, the safety engineer, the shipping clerk, group leaders, and an industrial engineer. They reviewed each commodity stored in the warehouse and then established a standard unit load in terms of the number of pieces that could safely be transported at one time. These unit loads were then used as the standard in the development of standard times.

A significant improvement was made by combining the jobs of the operators accumulating the equipment and the operators involved in labeling the equipment in the accumulation area. Previously, the labelers lost a great deal of time in locating the material. Further, it was difficult to balance the work loads so that some idle time did not exist. Combining the jobs solved both problems.

A study of the railroad car cribbing and blocking operation resulted in a simplified method that not only saved on labor but resulted in a significant saving in lumber blocking materials as well.

The location of material within the warehouse and the methods of stacking the equipment were changed. This accomplished two things. It made much better use of the cubic capacity of the warehouse, which in turn eliminated a great deal of the restacking work formerly required. Changing item locations permitted the storage of fast-moving items to be closer to the assembly areas, and this reduced the average travel time of the fork trucks.

A simple signal light was installed in the warehouse, with a switch in the assembly area. The light signaled the trailer train operator when a load was ready to be moved. This saved a substantial amount of waiting time on the part of the driver and reduced delays at the end of the assembly line caused by waiting for trains to be moved. A number of other miscellaneous improvements were made. Each improvement was designed either to standardize the method of performing the job or to simplify the work to be done. Frequently, standardization resulted in simplification as well.

Work Measurement. Only after standardization and methods improvement were completed was a start made on the actual measurement of the work. The work was not measured operation by operation. For example, studies were not made of unloading the trailer trains and placing the equipment into storage areas. Nor were studies made of removing equipment from storage and placing it in accumulating areas, or of moving equipment from material handling areas to trucks or railroad cars. Instead, sets of standard data were established for each of the basic operations or pieces of equipment. The data developed included:

1. Tow truck operating data
2. Fork truck operating data
3. Steel banding
4. Railroad car cribbing
5. Stenciling
6. Crating and the like

By developing standard data first, a great deal of flexibility was obtained, so that standards could be developed to fit the units of work used to measure the output of the warehouse.

Also, this approach resulted in a library of standard data that simplified the development of standards in other areas of indirect labor measurement. In fact, the powered truck data were used extensively later in the development of intra- and interdepartment material handling standards in the manufacturing departments. An example of the tow truck data is shown in Figure 3-2.

Units of Work. The cost of having a clerk check each load shipped to determine the number of units handled, the distance traveled, and the like was clearly prohibitive. The idea of having each warehouseman report the unit loads handled each day was also rejected. An independent source of information on the quantity of work completed was desired.

There were two established reports that seemed to cover most of the warehouse activities. One was a report on the number of trailer loads of equipment shipped from the final assembly area to the warehouse. The other was the shipping order

TOW TRUCK				CODE 3002.04	
Operation	Variable or type	Range	Sym.	.0001 hours	
Trip constant	Start and stop - empty or loaded	–	01	22	
	†Hook and unhook - tow truck to lead trailer		02	95	
Travel time	Open level floor or roadway and little or no congestion	Ea. 10 ft.	11	3	
Open and close door	Free swinging and/or bumper type	–	21	82	
	†Sliding, overhead or equiv., and manually operated †Power contr. door - actuating switch at door		22	185	
	Power contr. door - act. sw. on approach path to door		23	59	
Miscellaneous	Mount and dism. tow truck	–	31	24	

†Includes mount and dismount of tow truck

FIG. 3-2. Tow truck data.

prepared by the order entry group in the sales department and completed by the shipping department.

Using these two documents to measure the units of work, standards were developed for each item on the shipping orders and the trailer train reports. Every operation performed in the warehouse was related in some way to these two documents. Because the standard data were already available, the time standards for these two units of work were developed merely by taking frequency studies to determine what elements of work were required and how frequently they were performed for each item handled.

Performance Measurement. A procedure for measuring the units of work and time standards for each unit of work was now available. The time standards multiplied by the units of work gave the standard hours of work produced, which were then compared with the actual hours spent in carrying out the work. A ratio of the actual hours to the standard hours provided a performance index or a measure of the effectiveness of the operation.

At first, these measures of performance were applied to the work performed in previous months to establish a past performance level. This past performance level provided a bench mark by which future progress and savings could be measured. Next, a dry run period of four weeks was carried out to test completely the operation of the procedures and the adequacy of the standards. As bugs in the procedures developed, they were eliminated during this four-week trial period. In addition, the warehousemen had a firsthand opportunity to evaluate the system and the standards and to judge their ability to perform to the standards established. Finally, the system was installed and used as the basis for a wage incentive program. All the preinstallation work required a great deal of explanatory discussions with both the warehouse managers and the warehousemen. These information sessions were essential to showing both supervisors and workers the fairness of the system and the advantages it offered each group—better management controls on the one hand and improved earnings on the other.

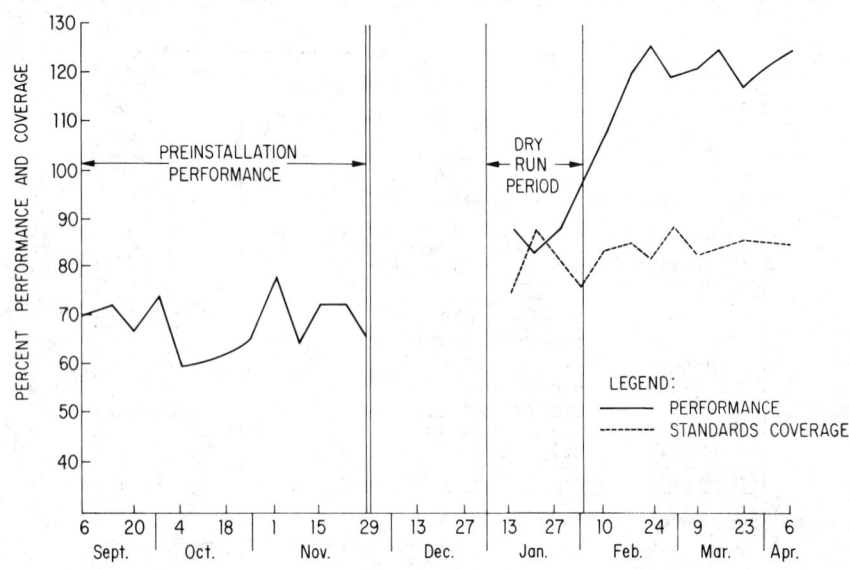

Fig. 3-3.   Record of increase in performance under wage incentive plan.

Figure 3-3 graphically illustrates the results that were achieved by this program. The period, September, October, and November, shows the performance of the group in the months preceding the dry run. Work on the project was suspended during December. The dry run started in January, and the incentive program started immediately after the dry run.

The performance of the group prior to the dry run was about 70 percent. This jumped substantially during the dry run period to around 100 percent at the end of the dry run. Performance then climbed to a level of 120 to 125 percent. Also shown on this same graph is the percentage of the work covered by standards. It varied between 80 and 90 percent, which is about typical for most indirect labor applications. There are always some items of work that are not covered by the standards and hence are not measured. In addition, nonstandard conditions may prevent the performance of the work according to the standard method, and consequently the time standards cannot be applied.

Results Obtained. As a result of this program, the number of warehousemen required to perform the work was reduced from thirty-eight to twenty-two. The surplus men were transferred to other departments. The earnings of the group were increased by 25 percent, and the overall cost of the operation was reduced by about one-third. Unexpectedly, a substantial saving resulted from a reduction in the number of fork trucks used in the warehouse. The twenty-six original trucks were reduced to seventeen, due to the increased productivity of the warehousemen. The saving on these nine trucks more than paid for the entire development costs of the program.

The total savings were in excess of $150,000 per year. The entire project required a total of fifty-five man-weeks to develop and install. Because of the simplified procedures used in developing the work counts and applying them to the standards, the application costs of the system were less than 1 percent of the labor costs.

TOOLROOM

One of the characteristics of indirect labor is the wide variety of work performed. This necessitates flexibility in the design of the work measurement program which will be best suited to the type of work being measured. Fundamental concepts remain the same, but their application varies, and it requires considerable ingenuity on the part of the industrial engineer to devise modifications for each different situation. A discussion of some of the problems involved in designing a work measurement program for a toolroom will illustrate some of the basic problems involved.

Importance of Planning. One feature common to almost all indirect labor work is that the work is performed with little or no formalized planning. In a typical production machine shop, it is normal practice to develop a route sheet for each part manufactured, showing the sequence of operations to be performed. In a toolroom, on the other hand, it is fairly common practice to do little or no planning of the work. Typically, one toolmaker may be given the entire job of making a fixture or die from beginning to end. He personally performs each operation on each part and does all the assembly and benchwork as well. About the only thing that might be done by another operator is some specialized operation such as heat-treating or plating.

The major virtue of this approach is that it simplifies the foreman's work and provides the advantages of the job enlargement concept to the toolmaker. On the other hand, it has a number of serious disadvantages. Contrary to popular belief, it does not produce the best-quality tool. The details on the drawing will frequently be revised by the toolmaker as a means of compensating for mistakes. If a shaft is ground undersized, the bearing can be made oversized to fit. This fitting and trying procedure is time consuming. Furthermore, the problems involved in replacing broken or worn parts are multiplied if the original parts were not made to dimension. Parts made separately by a group of operators, but made exactly to print, will produce a far better tool.

By requiring each operator to use every piece of equipment in a toolroom, he develops competence in many operations and expertise in none. The use of a fully qualified toolmaker to produce a simple shaft on a lathe is a waste of high-priced talent.

Another myth is that it is more economical to have the toolmaker do the complete job because this does not require the overhead cost of shop planners. Obviously, the planning involved in making a part or tool has to be performed by someone. Even though the entire job is turned over to the toolmaker, he still has to do some planning. Many toolmakers have never mastered shop arithmetic and geometry. They struggle through it somehow; but it is a struggle, and it is tedious, and it is slow. Arithmetic mistakes frequently result in high scrap costs or extra labor costs. The average toolmaker performing planning work is certainly no match for an expert specially trained for the job, performing it 100 percent of his time, day in and day out, and equipped with the reference material, electric calculators, and workplace required to perform the job effectively. The fact that planning is not done by a planner does not mean that the planning function is eliminated. In fact, the planning time spent by the toolmaker is probably two or three times greater.

Operators assigned to specific machines or equipment soon develop a proficiency and speed of operation that any toolmaker, no matter how well qualified he may be, will find difficult to match. Furthermore, the skill requirements and wage rates of machine tool operators are substantially less than those of a toolmaker.

A program designed to standardize toolroom operations and improve methods frequently involves a major reorganization of the toolroom and its conventional methods of making tools. However, a toolroom based upon the concept of detailed planning of work and the routing of work to specialized machine operators is far more effective than the traditional toolroom, which assigns a complete job to one man and expects him to perform all the required operations. This reorganization of the toolroom is often difficult to sell, but it is essential to the establishment of effective standard practices.

Machine Work. The bulk of the work performed in most toolrooms can be classified under one of two categories: machine work and benchwork. The machine work can be broken down into classes or types of machine tools. Time standards, usually in the form of a worksheet, can then be developed for each type of tool. Another type of worksheet or series of worksheets can be developed for the various types of benchwork performed. A few examples will illustrate this point.

Figure 3-4 shows a generalized worksheet which can be used for a variety of machine tools. It is designed primarily to be used when calculating the machining time for the operation. Supplementary sheets must be used to calculate the setup time, gaging time, part handling time, and the like. All these elements of time are totaled to get the time per piece. Tables of standard data are also needed to provide information on surface speeds, feeds, and RPM for a variety of materials and cutting tools. Some typical charts of standard data for drilling operations are shown in Figure 3-5. A worksheet is prepared to develop a time standard for each operation. The worksheets are completed by the planner, who must also lay out and plan each operation of each part of the tool to be manufactured. A route sheet is then prepared to route each piece through the shop.

Drawings. Many toolrooms attempt to work with a minimum of drawings. They rely on sketches and on the toolmaker's experience to translate the sketch into a finished part, complete with all the proper tolerances and finishes. Although a competent toolmaker has the ability to perform these functions, he is in effect acting as a draftsman. The idea that this approach eliminates the cost of drawings is another myth of toolroom tradition. The use of a toolmaker as a substitute for a draftsman results in high-priced drafting time, high frequency of errors, scrap material, and scrap labor dollars.

On the other hand, elaborate and overly detailed drawings can be expensive and are undesirable. Simplified drafting techniques can be used to excellent advantage in toolroom work. This approach not only speeds up drafting time, but

DRILL PRESS		WORKSHEET				MACHINE TIME			

M.W.O. _____   S.O. _____ Job description _____

Tag no. _____ Mat. class _____ Dwg. no. _____ Item no. _____ Date _____

1	2	3	4	5	6	7	8 (6X7)	9	10 (8X9)
Operation description	Diameter of cut	SFM	RPM	Feed	Actual time/ inch	Length of cut	Time /cut	No. of cuts	Total time/ piece

Total machine time [   ]

Operation description code:
D = drill, R = ream, T = tap, CS = countersink, RD = redrill, RLF = relief, BT = blind tap, CB = counterbore

A. (Setup time □ + setup gaging time □) X 1.21 allow.   =   [   ]

B. Total mach. time/pc. □ X 1.25 allow. X no. pcs. O   =   [   ]

C. Total const. time/pc. □
   + Total gaging time/pc. □ X 1.25 allow. X no. pcs. O   =   [   ]

Total std. hours/job   =   [   ]

Fig. 3-4.   Generalized worksheet for calculating machine tool time.

it can have a beneficial effect in the shop as well. Figure 3-6a shows a die stripper part drawn in a conventional manner, while Figure 3-6b shows a simplified drawing of the same part. The use of coordinates in lieu of dimension lines greatly reduces the calculation time of the jig borer operator. All the coordinates are already calculated for him. The lack of clutter also reduces the time required for an operator to read and interpret the drawing. Simplified drawings have a double-barrel effect. They simplify both the drafting work and the shop work.

Benchwork. Because of its variations, benchwork in a toolroom is by far the most difficult to measure and control. The variety of benchwork can be simplified somewhat by separating the work into several different classifications and then developing worksheets for each classification. An obvious classification is by the type of tools. Dies, jigs, fixtures, and special tools might be appropriate classifications for one shop but may not be at all applicable to another shop. The work classification therefore must be tailored to the toolroom in question. Another classification of work within a given class of tools, such as dies, can be by type of work performed. For example, benchwork on dies might be classified into new, rebuild, and periodic service work. The primary purpose of establishing classes of work is to establish parameters, which in turn limit variations and tend to standardize and simplify the problem of establishing time standards for the work performed.

Figure 3-7 shows a worksheet designed to be used when developing standards on a very difficult type of toolroom benchwork, namely, the job of repairing and servicing dies. In making a new die, the work to be performed is clearly specified. In die repair work, however, the work to be performed is frequently unknown. If a die is not functioning properly in a press, it is pulled from the press and sent to the die repair section for repair. A brief description of the die trouble might

CHART I
SF/M RECOMMENDATIONS

Material	Drill or redrill	Tap	Counterbore	Ream
Class I	300	70	200	200
Class II	200	70	130	130
Class III	90	50	60	60
Class IV	80	40	50	50
Class V	40	25	30	30
Transite*	160	40	100	

* Use feed equal to twice drill feed of soft metal:
Class I   —aluminum
Class II —bronze and brass
Class III—cast iron
Class IV—soft steel
Class V—hard steel

NOTE: To counterbore—(a) flat bottom drill bit—use standard drill to depth desired minus length of drill point, then flat bottom drill at counterbore SFM and drill feed for distance equal to the length of the point for the diameter of the drill used in the redrilling; (b) "fly-cutting"—use counterbore SFM and .010 feed.

CHART II
FEEDS—DRILLING AND REAMING: DRILL POINT LENGTH

Drill dia.	Drill pt.	Drill feed/ rev.	Reamer feed	Redrill feed*
1/8	.04	.002	.004	
5/32	.05	.002	.004	
3/16	.06	.004	.008	
7/32	.06	.004	.008	
1/4	.08	.006	.012	
5/16	.10	.006	.012	
3/8	.11	.008	.016	
7/16	.13	.008	.016	
1/2	.15	.010	.020	.012
9/16	.17	.010	.020	.012
5/8	.19	.010	.020	.012
11/16	.21	.012	.024	.015
3/4	.23	.012	.024	.015
13/16	.24	.012	.024	.015
7/8	.26	.012	.024	.015
15/16	.28	.014	.028	.017
1	.30	.014	.028	.017
1 1/16	.32	.014	.028	.017
1 1/8	.34	.014	.028	.017
1 1/4	.38	.017	.034	.021
1 3/8	.41	.017	.034	.021
1 1/2	.45	.017	.034	.021
1 5/8	.49	.017	.034	.021
1 3/4	.53	.017	.034	.021
1 7/8	.57	.017	.034	.021
2	.60	.017	.034	.021
2 1/8	.64	.017	.034	.021
2 1/4	.68	.017	.034	.021
2 3/8	.71	.017	.034	.021
2 1/2	.75	.017	.034	.021
2 5/8	.79	.017	.034	.021
2 3/4	.83	.017	.034	.021
2 7/8	.86	.017	.034	.021
3	.90	.017	.034	.021

* 120 % of drill feed.

CHART III
DRILL PRESS—COUNTERSINK HOLE

Flat head screw size	Countersink diameter	Countersink depth	.010 feed RPM	Decimal hours
#4 (.112)	.25	.092	686	.0002
#12 (.216)	.50	.190	506	.0006
3/8	.78	.270	393	.0011
1/2	1.00	.333	280	.0020
5/8	1.25	.417	166	.0042
3/4	1.50	.500	140	.0060

FIG. 3-5. Typical standard data for drilling operations.

be recorded on the work order, and this is all the die repairman has to go on. How are time standards set under these conditions? It really is not difficult. Incidentally, this is true of many indirect labor situations. After a first look at a problem, it may be thought that the work cannot be standardized or controlled. Further analysis will invariably lead to a solution.

In this particular case, the various dies were first broken down into a number of families. A worksheet similar to that shown by Figure 3-7 was then developed for each family. This was done by observing the various types of repairs and faults that could occur in a die. These repairs were then grouped by the major sections of the die. The sections were the bottom shoe, top shoe, punches, bushing, inserts, pins, and the like. Further, the work was generally limited to disassembling the die, inspecting the parts, and reassembling the components. All the broken parts were either replaced from stock if they were standard parts like punches or bushings, or if special, an order was written and the parts were made by other toolroom personnel.

By limiting the work in this way, it was not difficult to develop unit time values for disassembling the die, removing and replacing defective parts, and then reassembling it. These are the time values shown next to each part or item on the worksheet, Figure 3-7.

The major problem in this case was to establish a procedure for determining the frequency of work units. After study, it was decided to permit the operator to record the frequency of work directly on the worksheet. The worksheet was then extended by a clerk. Because this work was performed under a wage incentive program, there was some fear that the operators might pad the report on unit frequencies to produce higher pay. To guard against this and to eliminate a temptation that might otherwise exist, a system of spot inspections was instituted. These inspections checked for both the quality of the work performed and the frequency of work units. It was quite simple for a qualified person to determine how many new parts had been replaced. Interestingly enough, this system of inspection disclosed the exact opposite of what everyone had expected. Invariably, the die repairmen overlooked or forgot to record items more frequently than they reported an item in error. This was so consistent that a percentage was finally added to the allowance factor to compensate the operators for items they forgot to record. Although the percentage adjustment was small, the favorable psychological effect on the die repairmen more than offset any negative reactions to the spot checks.

MAINTENANCE AND REPAIR ACTIVITIES

From a work measurement and control viewpoint, one of the most difficult indirect labor activities is plant maintenance and repair. A unique approach to the measurement of maintenance work and similar nonrepetitive activities has been developed. This approach involves the use of bench mark jobs and a range of time concept. Instead of attempting to measure a job to an accuracy of a hundredth of a minute, jobs are slotted into standardized time ranges. The average of the time range is used as the standard time for the job. This concept is explained in some detail in Chapter 8 of Section 3, Standard Data Concepts.

USES OF INDIRECT LABOR MEASUREMENT

It has been said that management is based on measurement plus control. The primary essential to the control of any process is to provide some method of measurement. On direct labor operations, the simplest method of measurement is to count the number of pieces or quantity of product produced. This measurement, although it is only relative, does provide a basic means of control. With the advent of the scientific management movement, Taylor and others developed more sophisticated means of measuring labor in relation to some established norms of performance. Thus, the use of piece counts as a relative measure of work gave way to the concept of the standard hour of work as an absolute measure of performance. Because

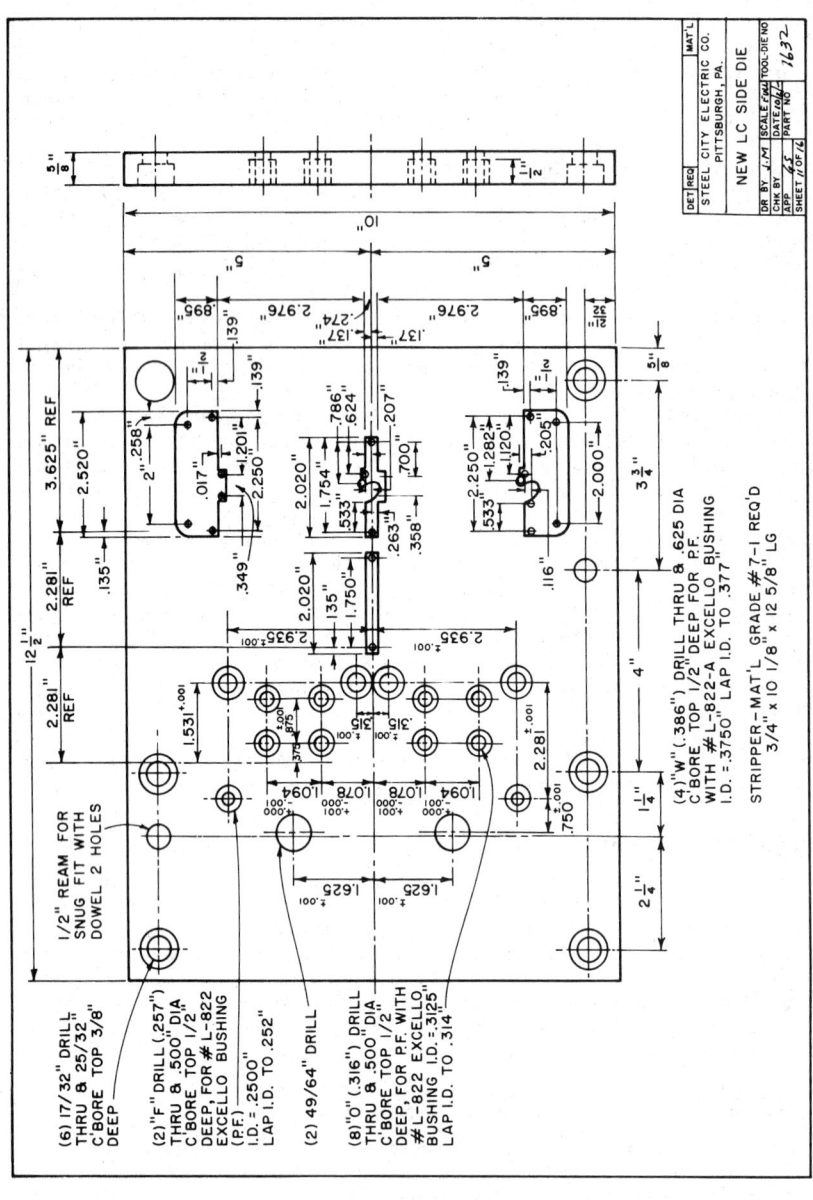

Fig. 3-6a. A conventional drawing for a die stripper part.

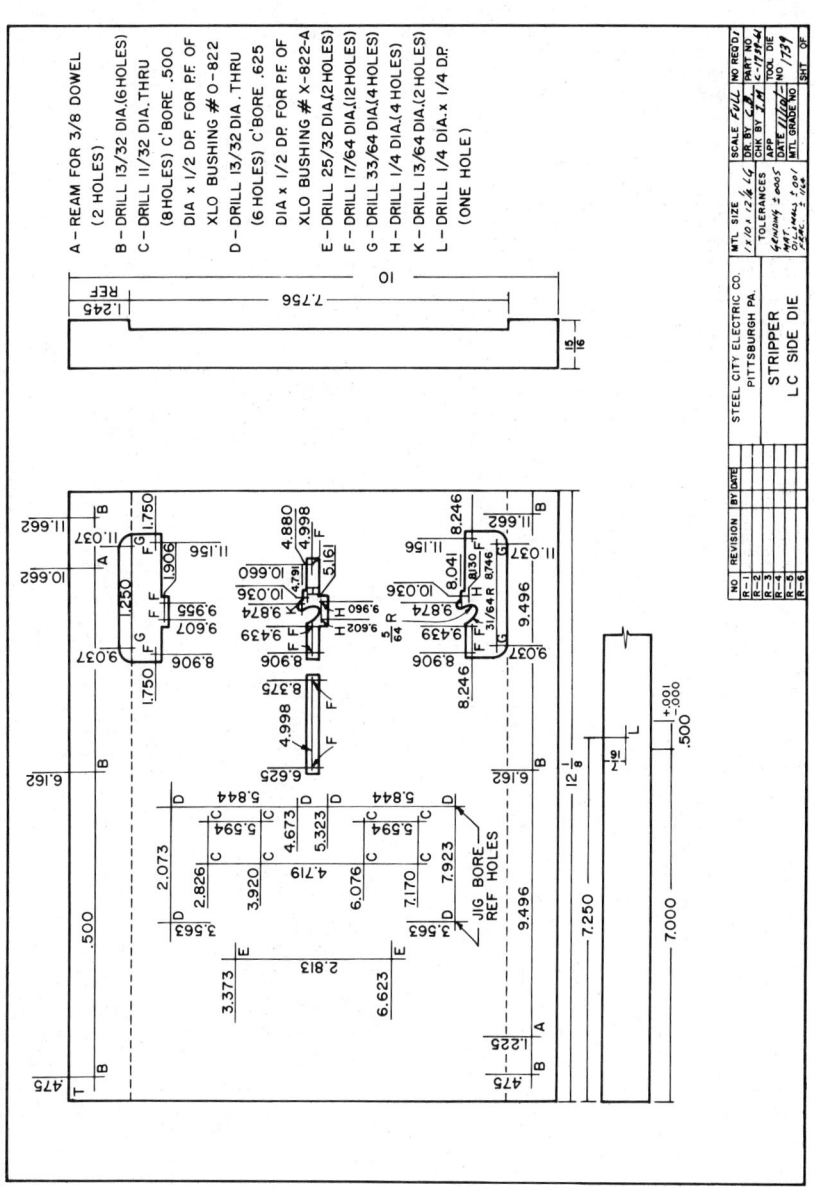

FIG. 3-6b. A simplified version of Figure 3-6a.

S.B., F.P., D.D., S.O. dies
Date _____
Die no. _____
Die description _____
Regular service◯ replacement◯ change◯ rebuild◯

Benchwork order-3-
Work order No. B _____
Die class S _____

		Toolroom	Die vault	Press room	Clean
Complete	⎫	◯.125	◯.245	◯.275	◯.096
Top or btm. shoe	⎬ from	◯.076		◯.226	◯.036
Other subassem.	⎭	◯.008		◯.060	◯.012
Complete	⎫	◯.125	◯.245	◯.275	
Top or btm. shoe	⎬ to	◯.076		◯.226	
Other subassem.	⎭	◯.008		◯.060	

	Item		X=		Item		X=
BOTTOM SHOE	Cradle		X.066 =		Change: from		
	Die sections		X.061 =		to		
	Bumper blocks		X.042 =		Replace new parts:		
	Full stripper		X.086 =		Piercing punch		X.098 =
	Stock equalizer		X.078 =		P.O. slot punch		X.098 =
	Other sections		X.066 =		K.O. punch (Ex. C.L.)		X.098 =
TOP SHOE	Punch holders		X.104 =				
	Other sections		X.066 =		Contour punch		X.098 =
	Piercing		X.039 =		Trimming punch		X.270 =
	Knockout		X.039 =		Die section		X.303 =
	Stencil		X.049 =		Bushings		X.098 =
PUNCHES	Trimming		X.068 =		Pilot pin		X.098 =
	Lancing		X.059 =		Round insert		X.098 =
	Dimpling		X.039 =		Dowel pin		X.095 =
	Extruding		X.039 =		Quill		X.098 =
	Embossing		X.039 =				
					Restore radius—hard		
					Clearance—inches		X.020 =
BUSHINGS, INSERTS, & MISC.	Stencil bumper		X.039 =		Draw or form—hand		
	Forming insert		X.074 =		Stock | | X3 =		
	Cutoff insert		X.069 =		plus .020		
	Embossing insert		X.049 =		inches | X | =		
	Bushing		X.045 =		Broken bolts:	Large	Small
	Pilot		X.039 =		Proj.	.300	.250
	Spring pad		X.104 =		Flush		
	Lifter pin		X.056 =		blind	1.000	.750
	Push back		X.056 =		Flush		
	Quill		X.039 =		open	.750	.500
	Spring stripper		X.152 =		Broken set screws:		
	Guard		X.043 =			|1.000|	
	Method of mounting:						
	Bolted only		X.043 =		Get parts welded	◯.250 =	
					Use die flipper	◯.747 =	
	Ditto w/springs or shims		X.068 =		Re-dowel each part	☐ X.366 =	
	Bolted & dowelled		X.063 =		Ear form ⎰ octagon	◯.458 =	
	Ditto w/springs or shims		X.078 =		Adjust ⎱ utility	◯.308 =	
OTHER PARTS	Press fit		X.045 =				
	Slide or tap fit		X.039 =				

Job constant		.120
Total		
Times 1.40 = allowed hrs.		

FIG. 3-7. Worksheet for benchwork operations on repairing and servicing dies.

piece counts have little application to the measurement of indirect labor operations, the standard hour is the only practical way of measuring indirect labor operations. The primary use of indirect labor measurement is in the management of the indirect labor work force. Measurement provides not only a means for measuring the performance of individuals or groups of workers, but also the basic tool needed to measure the value of improved methods. It also provides a means for planning and scheduling the work to be performed.

MEASURED DAYWORK AND INCENTIVES

Indirect labor measurement is used in two basic ways as a means of controlling the performance of the work force. One method is to use the standards as the basis for measured daywork controls. The other is to use the standards as the basis for incentive wage payment. There are many pros and cons to the use of these two basic methods of control which are discussed in Chapters 2 and 3 of Section 6.

An example illustrating the results to be expected from indirect labor measurement will show the use of both methods of control. This is a particularly interesting case. The management of a company decided to measure as many of the indirect labor operations as was practical from a work measurement standpoint and to let each group of employees determine by a vote whether they wanted an incentive basis of payment or not. If they elected not to go for incentives, management would use the work measurement results as the basis for a measured daywork program.

At the time the standards were developed, the industrial engineers had no way of knowing whether the group involved was going to decide for or against incentives. As a consequence, all the time standards and the work measurement procedures were identical for incentive and measured daywork programs. A comparison of the two applications could therefore be made with the assurance that there were no basic differences in the time standards or in the performance level concept. In some cases, the employees first elected not to go on incentive and later changed their minds and elected to accept incentives. In these cases, the same time standards were used without change or adjustment.

Results of Indirect Labor Measurement. At the start of the indirect labor measurement program, the plant work force was made up of 1,400 direct workers and 1,050 indirect workers. Thus, the indirect force was 75 percent of the direct force. Company records showed that the percentage of indirect to direct workers had been increasing steadily for the past ten years. Management began to realize that, if something was not done to change this trend, the cost of indirect labor would soon exceed the cost of direct labor. Although all sorts of measurements and controls were available for the direct work, very little control, other than budgets, was applied to the indirect work. Despite an excellent system of budgetary controls, the cost of the indirect operations continued to grow each year, and management finally decided that a more effective control was needed to halt the steady increase in costs.

Eight years after management made a decision to undertake a program of indirect measurement, about 60 percent of all indirect workers had been covered. During this period, the overall volume of the company had grown slightly; it now employed about 1,500 direct workers. If management had been successful in holding the ratio of indirect workers at 75 percent, about 1,100 indirect workers would have been employed. The number was actually 800. This represented a saving of 300 people, or about 27 percent. Most managers felt that they probably would have employed even more than 1,100 people, based upon past history and the established trend of increasing indirect ratios between direct and indirect.

Figure 3-8 shows the number of employees covered by incentives and measured daywork each year and the net saving that accrued from the program each year. The cost of installing the program was deducted from the gross savings to arrive at the net saving.

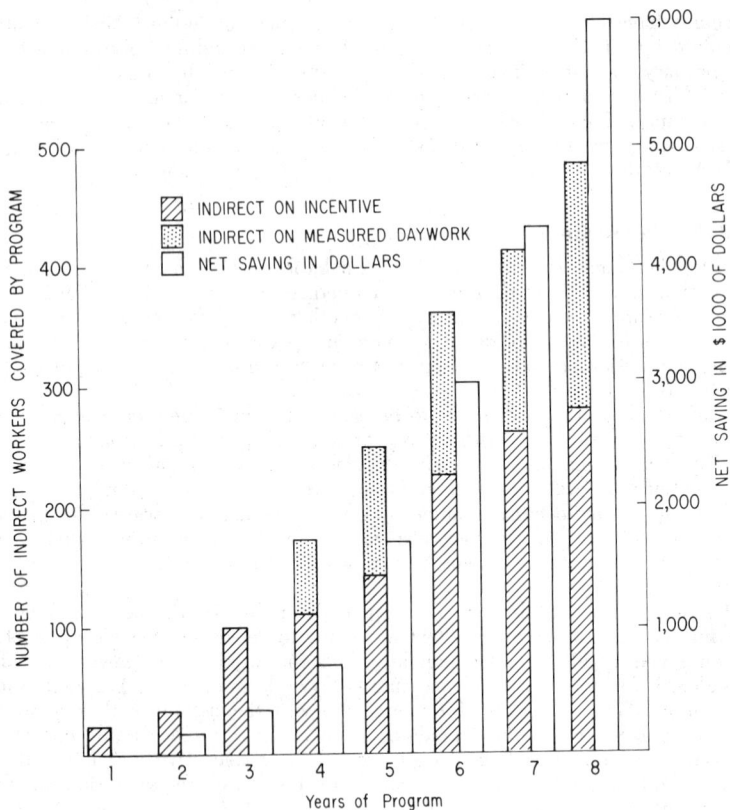

Fig. 3-8.   Results obtained in a typical indirect labor standards program.

By the end of the first year, the program was just breaking even—the cost of the installation equaled the gross savings. By the end of the second year, the program showed a net saving of about $100,000. By the end of eight years, the program covered a total of 480 people and was producing savings at an annual rate of $1,700,000 per year. The cumulative savings for the eight-year period amounted to $6,000,000. These savings would have been almost twice this amount if all employees had elected to accept the incentive program. Employee take-home pay was about 25 percent greater for incentive personnel.

In this example, the company elected to install most of the program with their own technical personnel. The program could have been accelerated substantially by greater use of outside assistance. Although the installation cost would have been greater, the savings would have accrued at a faster rate, and the cumulative net saving at the end of the eight-year period would have been significantly greater. Working under identical time standards, the employees under the measured daywork plans worked consistently at a performance level of 80 percent; those working under the incentive program consistently worked at a level of 125 percent.

IMPORTANCE OF MAINTAINING STANDARDS

The importance of maintaining standards to reflect changes in working methods or conditions cannot be overemphasized. If wage incentives are employed, management will be subjected to considerable pressure from time to time to relax the

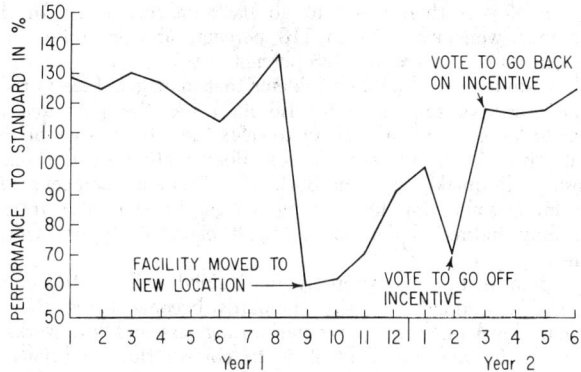

FIG. 3-9.   Effect of change in conditions on incentive earnings.

standards in one way or another to permit higher earnings for the same effort or equal earnings for less effort. Despite good intentions upon embarking on an incentive program, managers sometimes find it extremely difficult to resist this pressure.

A typical example of the pressure that can be exerted on management is illustrated by Figure 3-9. This chart shows the performance of a group of repairmen working under a group incentive plan. The work involved the overhaul and repair of machinery requiring factory rebuild and servicing. The group had been on incentive for a year before the beginning of the chart, and performance consistently averaged about 125 percent of standard.

Due to a re-layout of the plant, the rebuild and service operations were moved to a new location. At the same time, a number of new methods, tools, and equipment were introduced to simplify the work and reduce costs. The industrial engineers completely reworked all the old standards to reflect the changes in methods, tooling, equipment, and working conditions.

Just before the move, in month 9, the group was allowed credit for a number of miscellaneous jobs that had accumulated over several months, but which were cleared out during the move. This accounted for the somewhat higher than normal performance in month 8.

Immediately following the move, the new standards were applied to the work, and the employees promptly filed a grievance claiming the standards to be too tight. Production dropped off, and performance sank to 60 percent.

Management then had the industrial engineering group review all the new standards to be certain that they truly reflected actual working conditions. This took about one month, and a few changes were made. Management appealed to the employees to give the new standards a fair trial. This trial took place over months 11, 12, and 1. During this period, performance gradually climbed toward 100 percent. In the meantime, the backlog of work had built up and customers were complaining about the delays in shipping dates. The employees were also complaining vigorously about their loss in earnings. The industrial engineers firmly maintained their position that the standards were correct and that the group could equal their previous performance if they really went to work.

Management was clearly on the spot. What should be done? Should management instruct the industrial engineers to loosen the standards, or should it "stick to its guns" and put up with the pressure from customers and the union? While management was wrestling with this problem, the employees voted to discontinue the incentive program.

Performance immediately dropped to 70 percent in month 2. After two weeks of being off incentive, one employee asked to be transferred to another department.

The remaining employees then voted to go back on incentive. In the next two weeks, performance went from 70 to 118 percent, and in a few months' time, was once again consistently averaging 125 percent.

This example is quite typical of the problems that managers face in the administration of a wage incentive program. It also illustrates the pressures that will be brought to bear to loosen the standards or to relax the maintenance of the standards. If management gives in to these pressures, degeneration of the plan will soon follow and costs will quickly get out of hand. From a manager's viewpoint, an incentive program is something like having a tiger by the tail. It is a powerful tool for controlling indirect labor costs, but it can quickly eat you up once it gets out of hand.

An incentive plan that is not properly maintained will, in the course of time, become loose and outmoded. As the standards become loose, the workers can maintain the same level of earnings by performing less and less work. The looser the plan becomes, the more difficult it is to correct the conditions causing the looseness. The plan must be strictly maintained from the first day of installation. It cannot be used as a subject for bargaining or negotiation. If it is, it will collapse.

COST OF MAINTAINING AND OPERATING A STANDARDS PROGRAM

The cost of operating and maintaining a work measurement program on indirect labor operations is a factor to be considered. Although the savings which accrue from such a program can be substantial, a portion of the savings must be devoted to the operation and maintenance of the system. These costs will vary somewhat and will depend on the units of work used in applying the standards, the degree of stability of the physical conditions and the product line, and many other factors. A conservative estimate of these costs for typical indirect labor operations is:

Day-to-day application costs.........................	3 %
Maintenance of the standard data base................	2 %
Timekeeping, unit work counts, and payroll costs.......	1 %
Total...	6 %

The total cost of operating and maintaining an incentive program on indirect labor operations will thus be about 6 percent of the indirect labor cost payroll. About half the cost is spent on developing and applying standards for the day-to-day operation of the system. About 30 percent is spent on the maintenance of the system to reflect changes in equipment, methods, tooling, new factory locations, and the like. Twenty percent is spent on keeping track of the hours worked and the units of work produced and calculating incentive earnings on performance.

Frequently, these administrative and maintenance costs will start out at a higher rate and gradually reduce. A definite learning curve effect seems to apply. Figure 3-10 shows the learning curve effect on standards application and maintenance costs for a typical installation.

LONG-RANGE RESULTS

One of the major advantages of indirect labor controls based upon sound work measurement is that they provide a sound basis for long-term cost reduction and continuous control. Many managers have been tempted to use a quick and easy approach to controlling indirect labor costs. This approach frequently involves the establishment of approximate or estimated standards and concentrates on the scheduling of work in small batches with close supervision and follow-up. This has produced quick results in amazingly short periods of time. The results have generally proved to be short-lived, however, because the standards are not soundly established or clearly based upon a given set of methods, equipment, and working conditions. As is true with so many management problems, the quick and easy approach in the short range turns out to be the least effective and the most expensive in the long range. Although the quick and easy approach is tempting when seen

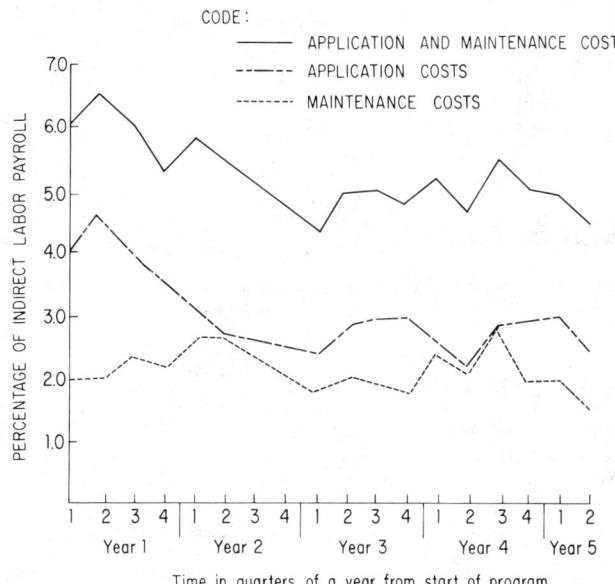

CODE:
——————— APPLICATION AND MAINTENANCE COSTS
– – – – – APPLICATION COSTS
– – – – – – MAINTENANCE COSTS

FIG. 3-10.   Effect of learning on cost of applying and maintaining time standards.

from a short-range profit viewpoint, a more soundly conceived program, based upon soundly developed work standards, is a much superior long-range solution to the problem of controlling indirect labor costs.

BIBLIOGRAPHY

"Be Practical? or Precise? in Measuring Indirect Work," *Factory*, December, 1964.
Bennett, K. W., "Putting Incentive Back into Wage Incentives," *Iron Age*, Nov. 25, 1965.
Crockwell, D. C., "Intercorrelation and Multiple Regression in Industrial Engineering," *Journal of Industrial Engineering*, January, 1967.
Devaney, R. J., and R. G. Lee, "Your Incentive Plan: Does It Need Revising?" *Administrative Management*, July, 1966.
Grove, V. A., and R. Reul, "Wage Incentives for Maintenance—Do They Pay Off?" *Factory*, February, 1964.
Ladd, G. W., "Regression Method—Visual Inspection Times," *Journal of Industrial Engineering*, September–October, 1960, p. 411.
Louden, J. D., and J. W. Deegan, *Wage Incentives*, 2d ed., John Wiley & Sons, Inc., New York, 1959.
Mangum, G. L., *Wage Incentive Systems*, University of California, Institute of Industrial Relations, Berkeley, 1964.
Parks, G. M., "Multiple Criteria Sequential Work Sampling," *Journal of Industrial Engineering*, July–August, 1964, p. 221.
Salem, M. D., "Multiple Linear Regression Analysis for Work Measurement," *Journal of Industrial Engineering*, May, 1967, p. 314.
Secor, H. W., "Regression Analysis—Standards Efficiency," *Journal of Industrial Engineering*, January, 1966, p. 33.
Thelwel, D. R., "Linear Programming and Multiple Regression Analysis in Estimating Manpower Needs," *Journal of Industrial Engineering*, March, 1967, p. 227.
Wein, W. H., and others, "Sensitivity Analysis for a Wage Incentive System," *Business Topics*, Summer, 1966.
Wilkinson, John J., "How to Manage Maintenance," *Harvard Business Review*, March–April, 1968.
Williams, J. A., "Master Plan for Office Incentives," *Administrative Management*, April, 1965.

Chapter **4**

Measurement of Automated Processes

GEORGE H. GUSTAT

Director, Industrial Engineering Division,
Eastman Kodak Company,
Rochester, New York

Measurement of automated processes may be more appropriately viewed as measurement of highly mechanized processes, for true automation is seldom if ever achieved. Industrial engineering activities revolve around automated processes in terms of labor and quality controls, interruptions of production, interferences among both the human and the material elements involved, and the supportive organization needed to prevent problems.

The degree of mechanization may vary from very little to very high, depending on the economics of the situation. The reasons for and the principles of measuring automated systems are much the same whatever the degree of mechanization.

Automated systems make it possible to transfer the burden of physical work from men to machines. They can also reduce the unit cost of product if they are well designed and properly utilized. Other factors which may affect the degree of mechanization justified are uniformity of quality, material utilization, safety, availability of qualified workers, availability of capital, and probable life of product being made.

IMPORTANCE OF MEASURING PROCESSES

There are wide divergences of thought and attitudes of management on the importance of detailed measurement of highly mechanized work and the need to

4-54

motivate operators working on such jobs. A popular belief is that the more automatic a production system is, the less important it is to measure it carefully, if at all, or to spend much time and money on motivational efforts for the operators involved. Reasons for this belief are based on the facts that the process is machine paced or automatically controlled and that the operators have very little or no influence on the productivity of the equipment.

If this reasoning is followed very far, it is almost certain to lead to a low utilization of the equipment; low enough often to ruin the high expectations and predicted economic justification for the equipment. Without adequate measures of performance, the predicted performance of the new process or equipment will be a far cry from what is actually realized.

There are several ways to evaluate the benefits that can accrue from a well-organized automated system. One is the comparison of the labor cost and overhead costs—such as depreciation, space, and the like—of the old manual system with those of the new one that replaces it. These costs alone, however, do not truly represent the entire effect on the company's profitability.

Perhaps a more realistic evaluation can be made in terms of the loss of profit on unproduced units caused by downtime. It is not unusual for a mechanized system to replace forty or fifty manual jobs. If one of the manual operators loses an hour in a day, the cost is only one hour's worth of labor. In addition, there are a certain number of units of product that are not produced. The profit on these units is likely to amount to something more than the labor cost.

If the new automated system goes down for an hour, it may cost one or two hours' worth of labor, but the number of units not produced may be as high as, say, 25,000. The loss of profit on this many units can be serious and of course is much greater than the cost of labor and overhead for that one hour.

There are often explanations in financial statements that a lower profit in this quarter in comparison with the same quarter last year is due to the start-up of new mechanized systems in new plants. This is another way of expressing the effects of downtime on profit.

Thus, an important objective of careful measurement of automated systems is to have current knowledge of how well they are doing against their optimum profit potential.

CHARACTERISTICS OF HIGHLY MECHANIZED SYSTEMS

A highly mechanized system has certain characteristics which must be taken into account before deciding how its performance can best be measured.

1. Reasons for mechanization include the reduction of cost per unit of output, greater production uniformity, and quality improvement. High construction costs place an increasing emphasis upon better space utilization by consolidation of many independent machines or hand operations into a mechanized process. Behavioral considerations include automating out of the job those elements that are of a highly repetitive nature and have a high volume.

2. Capital investment is usually very high in comparison with manual or partially mechanized systems. In fact, it can extend from tens of thousands of dollars per worker to a 25- to 50-million-dollar installation which is monitored and serviced by a small work crew.

3. Variability in production is inversely influenced by the number of producing units. When many units are producing, variability tends to be offset from one unit to another. Thus, a certain level of production can be closely predicted and relied upon. Conversely, a single but larger production unit is either on run time or downtime on an "on-off" basis. When on, it may or may not be producing 100 percent product, but when off, the system is at a dead loss.

4. Probability of failure or downtime is much greater in a single, highly complex producing unit. If a simple machine has a 95 percent chance of being

in "run" mode, then 95 percent of a number of these units will be pretty sure to be producing. If a single, much more complex machine has several stages arranged in series, each with a 95 percent probability of running, the overall probability of run time is the product of the independent probabilities. Thus, a four-stage machine would have a probability of running of only 0.95 × 0.95 × 0.95 × 0.95, or 81.5 percent.

5. Investment justification requires a high degree of utilization of expensive equipment, to the extent of two- or three-shift operation, sometimes for seven days a week. Where overtime formerly could compensate for production breakdowns on one- or two-shift operations, the new system compresses all variability into one unit running around the clock, and time lost is much more difficult to replace. Fluctuating production schedules may compound the problem of lost production time and capacity, even resulting in lost customer sales. These facts often necessitate backup equipment, requiring ever-higher capital investment.

6. Maintenance costs are sure to run higher in automated systems than in less mechanized operations, because of higher mechanical skills needed, longer shutdowns during production runs, and the requirement of around-the-clock troubleshooting service. The maintenance service usually reports to other than the production foreman, and this is a source of functional and jurisdictional difficulties.

7. Labor content in highly mechanized processes is low, relative to depreciation or amortization costs, which continue even during a shutdown.

   Operators who run highly mechanized equipment usually have a high sense of responsibility and involvement in the task (equipment, process, output). They have a strong feeling of accomplishment within the process. This must never be neglected, because it is always a real, though sometimes overlooked, asset.

8. Know-how must, of necessity and by definition, be substituted for physical effort. Because the machine is supposedly indifferent to the operator's pace or intensity of effort, it follows that the operators address themselves to the following:

   a. Monitoring and control of quality on-stream to preserve the level of quality engineered into the product.

   b. Maintaining supply, inventory, and disposal conditions on-stream to extend run time.

   c. Exercising preventive maintenance on-stream to avoid downtime through minor servicing and adjustments that do not require shutdowns.

   d. Troubleshooting to minimize downtime as quickly as possible whenever breakdowns occur. This requires maximum knowledge and communication among operators and maintenance or service people.

   e. Changing products or machine setups as quickly as possible to avoid excessive downtime. This requires maximum communication and cooperation among members of a work crew or among several crews.

   f. Rendering effective assistance to co-workers or other specialized service staff during scheduled maintenance periods.

THE MEASUREMENT SYSTEM

Primarily, a measurement system provides ways of measuring units of performance and establishing the goals and objectives to be attained. It is an information system used to collect and report key critical data on performance for feedback to the concerned work crews, to supervision, and to management.

The measurement system should include the organization's philosophy on the particulars of payment to the people involved—from individuals, to crews, to whole departments or divisions.[1] It should be a dynamic process of information generation whose primary goal is to establish direction for continuing improvement in productivity, lower costs, and better products.

MEASURES OF PERFORMANCE

There are literally dozens of ways to count things or quantities of work being done.[2]

1. Quantitative measures—such as units per hour; pounds, pieces, gallons, or footage; or distinctive yields relative to the process being measured—are the most obvious ones and generally the easiest to establish.

2. Qualitative measures may be more meaningful to the process but more difficult to collect automatically. They include data such as run time, downtime, service times, delays or wait times, and quality ratings and yield or reject statistics; and data on housekeeping, preventive maintenance, and troubleshooting.

3. Multiple production lines may demand various labor combinations, according to machine setups or combinations of products being run. The labor need may or may not be relevant to traditional output data. Thus, specific machine setups or combinations may require labor measurements independent of output, just to keep them running as needed.

4. Labor crew requirements may be the best way to define the needs of the process, either because output is at a fixed rate, or if variable, because a certain crew size is needed both to turn out the work and to make necessary on-stream adjustments. In case of trouble or shutdown, extra people may be needed temporarily and may represent the system's maximum limit of manpower need.

It should be apparent that counts are meaningful only when related to goals and objectives, and occasionally to the functional operating restrictions of both the production systems involved and the people concerned.

GOALS AND OBJECTIVES

Highly mechanized processes are designed to deliver consistent production rather than to keep operators busy. Measures of performance or of how well they are doing tend to be based primarily on the output of the process. However, the control of process costs may be more appropriately expressed in terms of proper crew size, as determined by measurement studies. Operational needs and safety may be overriding considerations, with physical effort being a much lesser requirement. Organizational goals such as meeting production schedules or maintaining labor crew sizes must also be made meaningful to the operators or crews. In addition to specific quantities, a typical general description of a process might include such goals and objectives as:

1. Producing high-quality product free of contaminants
2. Producing sufficient quantities of material to meet production requirements
3. Minimizing total labor and material costs, consistent with production and quality requirements

The individual within the crew can help attain the crew goals by:

1. Contributing his efforts in such a manner as to meet production requirements

[1] G. H. Gustat and J. A. Richardson, "Motivation Principles," sec. 9, chap. 1, in H. B. Maynard (ed.), Handbook of Modern Manufacturing Management, McGraw-Hill Book Company, New York, 1970.

[2] Ralph M. Barnes, Motion and Time Study, 4th ed., John Wiley & Sons, Inc., New York, 1958, chap. 20.

2. Performing his duties in a correct and precise manner to assure high-quality products
3. Recognizing the need for continuing improvement of his knowledge of all phases of the operation to ensure better self-utilization, versatility, and flexibility in crew cooperation
4. Being effective in all assignments
5. Maintaining a safe and clean work area
6. Assisting in performing good preventive maintenance within the system and its equipment

Note the general nature of these descriptions. Although it is always useful and desirable to assemble complete job descriptions for training purposes, it is even more critical to promote a basis for *esprit de corps* within the working crew, which will provide an active dynamic apprenticeship for the newcomer into the crew. Job descriptions and operating manuals can be maintained by the men themselves. The emphasis should always be on encouraging the working crew to prove, for themselves and to others, that they can meet reasonable goals, even exceed them, and to develop ways of continually improving upon and surpassing past goals.

OTHER USES OF MEASUREMENT DATA

Automated processes need to be measured for many other reasons than labor control, production output, and pay. Areas in which measurement can be helpful include:

1. Debugging new equipment for improved performance characteristics
2. Preventive maintenance procedures
3. Redesign of troublesome components
4. Job conditions and training descriptions of standard operating procedures
5. Cooperation and effectiveness of operators and maintenance men
6. Interference patterns within and between man-equipment groupings, from simple to very complex
7. Labor/equipment/capacity scheduling capabilities
8. Work assignment combinations
9. Establishment of standard time values for estimating and predicting more effective match-ups of manpower and equipment
10. Methods analyses and comparisons of alternatives applied to small or large pieces of the total system
11. Gathering of information for machine control purposes

MEASUREMENT TECHNIQUES

There are many ways of measuring and arriving at the numerical goals and objectives that should be continuously monitored for the best utilization of complex production processes. These techniques, many of which are discussed elsewhere in this Handbook, include:

1. Production and operation history
2. Detailed machine logs
3. Statistical—rolling averages, moving annual totals, measures of dispersion, control limit charting, and the like
4. Stopwatch time study
5. Predetermined elemental time data
6. Work sampling
7. Standard data
8. Factoring of data—various forms of exponential smoothing
9. Regression analysis—simple linear, multiple linear, nonlinear
10. Predictive models such as queuing and simulation techniques

Often, the designers' preliminary forecasts of percent run time for new equipment will be used for assumptions of overall productivity, but these are seldom accurate.

Hence, there is a need to use other data gathering and evaluative techniques. Management must be able to predict the effective run time of the process in advance, in order to meet overall goals and objectives of production, equipment, and labor scheduling.

INFORMATION FEEDBACK

Feedback of useful information is vital to the successful operation of the process. The required data are unique to each process. They should be easily and accurately collectable and should be at least grossly monitored for credibility. The characteristic terminology of the process reflects the types of necessary data which may be reported, such as:

1. Quantities of standard quality product, through counters or other production stations downstream.
2. Nonstandard product, through deviation and waste reports.
3. Downtime of the various components of the process. (This information is used for concentrated debugging efforts or as the basis for redesign of ineffective components.)
4. Standard downtime, from expected setups or maintenance pauses.
5. Nonstandard downtime, machine or operator responsible, from deviation reports.
6. Product run time.
7. Experimental run time.
8. Standard or unit costs resulting from production history.
9. Indices of performance, actual versus planned or expected.

CONCLUSION

Measurement data should be reported on forms designed for utilization within the total system's normal information reporting procedures. It is vitally important that these data be transmitted first and most directly to the production operators concerned in the work crew. Second, levels of supervision and management should also be informed, in summary if not in detail, but by the time the information reaches them, the necessary corrective actions should already have been accomplished or be in progress. The information should be fed, first and soonest, to those who will be making the response and who have the best and most intimate knowledge of what needs to be done.

BIBLIOGRAPHY

Barnes, Ralph M., *Motion and Time Study*, 6th ed., John Wiley & Sons, Inc., New York, 1968.
Bekker, John A., "Modern Manufacturing Environment Is Creating New Work Concepts," *Automation*, October, 1964.
Bright, James R., *Automation and Management*, Harvard University Press, Cambridge, Mass., 1958.
Bright, James R., and Yale Material Handling Division, "Management Guide to Production," Eaton Yale & Towne Manufacturing Company, Philadelphia, Pa.
Drucker, Peter F., "The Promise of Automation," *Harper's*, April, 1955.
Einzig, Paul, *The Economic Consequences of Automation*, W. W. Norton & Company, Inc., New York, 1956.
Gustat, G. H., and J. A. Richardson, "Motivation Principles," sec. 9, chap. 1, in H. B. Maynard (ed.), *Handbook of Modern Manufacturing Management*, McGraw-Hill Book Company, New York, 1970.

Chapter **5**

Uses of Time Standards

J. WAYNE DEEGAN

Chairman of Industrial Engineering,
University of Iowa, Iowa City, Iowa

Time standards are one of the most important fundamental units of information which make scientific management possible. It is essential to understand the uses which may be made of these data to provide for their maximum utilization. It is sobering to realize that in all operating systems planning and control activities, even those utilizing the most sophisticated mathematical and computer techniques, the ensuing decisions are irrevocably dependent upon the bits of performance information (time standards) that are utilized as input. This chapter will be devoted to a discussion of the applications of standards followed by a presentation of procedures for standards development which will assist in their application.

For a time standard to have significance and utility, it must have sufficient descriptive detail as to operation and development. Because the amount of detail required varies with its intended use, time standards are usually developed in a series of steps which may be considered as separate parts. An outline of the different parts of time standards may help in understanding the detailed discussion of their use.

THE COMPOSITION OF TIME STANDARDS

Time standards consist of three parts: (1) a work standard, (2) an itemized operating method, and (3) the details of standards development.

    1. The work standard comprises:

        a. A general statement of the work to be performed, the location, the part number, and the like

        b. The time standard in units of time or money

    The work standard, therefore, is the bare information needed to identify the job itself together with the time required to perform the operation. It may take the form of an entry on an operation routing sheet or a job ticket and is normally stated as so many production units per hour, or its converse, so many hours or parts of an hour per production unit.

2. The itemized operating method includes:
   a. A detailed description of the working procedure
   b. The elemental standard times used in combination to determine the operation time (including standard data tables and/or formulas)

This itemized operating method may be recorded in a variety of ways. In the case of standards developed from individual time studies, it might appear only as a part of the time study itself. In most cases, however, it will take the form of a standard job instruction sheet. It may also appear as a part of a bulletin or report describing the incentive application for the particular operation.

3. Details of standards development include:
   a. Alternate work methods
   b. Elemental analyses of right- and left-hand motions and similar data
   c. Summary or recap sheets of time study data
   d. Time studies themselves

These data include the graphs, charts, calculations, approximations, procedures, methods, descriptions, and measurements upon which the standard is based.

USES OF TIME STANDARDS

In the discussion which follows, it is assumed that in all cases time standards will be available or can be developed.[1] In the case of existing operations, if a standard is not available it can be obtained by observing the work and making a suitable form of direct time study. On new work which is not yet being performed, the standards can either be established by estimate—which usually results in doubtful accuracy—or they can be established through the use of one of the predetermined elemental time standard procedures described in Section 5 of this Handbook.

Use of Time Standards for Design of Product. The functional perfection desired by the mechanical designer, the styling desired by the sales department for greater customer appeal, the trouble-free performance desired by the service department, and the low cost of product desired by everyone should be balanced against a factual standard for evaluation. Such evaluation without a yardstick is extremely difficult.

The time standard can provide the key to the evaluation of these diverse interests by providing a basis for estimating their cost.

The information which the standards must give is that which will help determine:

1. The effect of alternate methods of manufacture on the cost of the product (particularly if different quality is involved)
2. The effect of alternate materials on the cost of manufacturing the product
3. The effect of different designs on the method of manufacture, and consequently, on the cost of manufacturing the product

When a new product is to be produced, the sales representative, the process engineer, the product engineer, the operating superintendent and foreman, and the industrial engineer are usually involved. If materials and quality are important considerations, the purchasing agent and the quality control manager may be involved.

The amount of detail required and the frequency with which time standards are used are highly variable. For example, if a new design represents only a slight increase in wall thickness of a thermosetting plastic part requiring a slightly longer curing time, all that might be necessary is a standard data table giving

---

[1] This is not to deny the necessity for new and extended techniques to cope with new problems created by the extension of the industrial engineer's concern with ever more areas of human endeavor. These extensions range from manual work with microminiature parts at the one extreme to work of a creative nature and so-called "global" systems at the other.

the curing times for various wall thicknesses. If, on the other hand, the increase in curing time were great enough, it might call for a change in operating method to make maximum use of the operator time available during the curing cycle.

Use of Time Standards for Design of Productive Equipment. Time standards are useful in the design of equipment to determine:

1. The ultimate cost of manufacture of the equipment
2. The effect of alternate methods of manufacture upon the cost of manufacturing the equipment; for example, welding versus casting versus plastic molding
3. The effect of alternate designs upon the method of manufacturing, and consequently, upon the cost of making the equipment
4. The effect of alternate designs upon the method of operation, and consequently, upon the cost of using the equipment
5. The economic utility of the equipment

Whether the equipment in question is to be used as productive equipment in the manufacturer's own shop or that of another manufacturer, the possibility for use of time standards remains essentially the same. In the latter case, however, more people, including the sales department, become vitally interested in their utility.

Although the design of new or altered equipment might conceivably involve almost everyone in a plant, those usually most immediately concerned are the product and process engineer, the operating personnel such as the department foremen and superintendent, and the industrial engineer.

Because any revised designs may change the operating methods considerably, it is most desirable to have available the maximum amount of detailed information about exact operating methods.

It is likewise essential in work of this type that the standard times for setup, piece handling, machine handling, and machine or process operating times be separately available and that the work which may be performed concurrently be clearly designated. In this connection, it is frequently helpful to utilize the details of standards development because of the finer breakdown of job methods and alternate operations which might be included.

Use of Time Standards for Selection of Equipment. When consideration is being given to the choice of new equipment either for expansion or for replacement of existing facilities, performance standards are used in determining the economic utility of the equipment, including:

1. Cost of operation
2. Cost of setup
3. Quantity of equipment required (machine capacity)

Generally speaking, these comparisons for alternate choices are made on a direct cost basis so that little modification in the standards is required other than to provide a conversion into money units.

Again, the maximum detail as to operating method may prove most helpful to the operating personnel and to the product, process, and industrial engineers who are usually most concerned with equipment selection. In addition, the details of standards development may be used in the consideration of the balance of machine handling, machining or processing, and the piece handling times, to develop proper relative cost figures.

Use of Time Standards for Processing and Operation Planning. Standards are essential to operation planning to determine:

1. The total time for each operation
2. The most economic method of scheduling by evaluating the possibilities for:
   a. Multiple cuts
   b. Concurrent operations
   c. Multiple machine operations
3. The excess cost of alternate methods which may be necessary during periods requiring maximum productivity

The responsibility for processing or operation planning frequently falls upon the process engineer, or the industrial engineer, or the operating department. It is possible to do an adequate job of process planning through the use of condensed tables of standard data and formulas. However, the more familiar those responsible for this phase of planning are with the operating characteristics of the equipment available and with the method of developing the data, the better the job can be performed.

The complete details of operating methods should be available for immediate reference by those people even though they may not require their use too frequently. If additional details of alternate methods can be gleaned from the details of standard development, it is highly possible that a special summary of estimating data should be prepared for the use of those responsible for this function.

Use of Time Standards for Design of Tools, Jigs, and Fixtures. The uses of time standards in the design of tools, jigs, and fixtures are essentially the same as those covered in the section for design of equipment. Such standards are used for determining:

1. The cost of the tool, jig, or fixture
2. The effect of alternate materials or designs on the cost of manufacture of the tool, jig, or fixture
3. The effect of alternate designs or materials on the method of operating, and consequently, upon the operating costs
4. The economic utility of the tool, jig, or fixture

The tool designers, the process engineers, the operating foremen, and frequently the industrial engineers are most involved in this application of the standards.

As does the designer of productive equipment, the tool designer needs the breakdown of time values by the setup, machine handling, piece handling, and machining or processing time values. He also needs the breakdown in operating detail to the finest elements for which standards have been developed.

Use of Time Standards for Production Scheduling. Time standards are used for production scheduling to determine:

1. The work load for a man or machine
2. The machine and man load that a department is to carry
3. The effect of alternate methods on 1 and 2, above
4. Possible shipping dates (days in process)

In the first three of these applications, the operating superintendent and foremen are vitally interested and may be working with the production planning manager or his representative. The sales department becomes vitally interested in shipping dates because of its concern with the customer promises. If the standard is for performance of an office task, the office manager is most involved.

To do this scheduling properly, the standards used by the scheduler must take into consideration the effect of variances in production not usually considered in the original performance standard of production. Some of these variances might be:

1. Machine breakdowns greater than those allowed as miscellaneous delay
2. Machinery overhaul
3. Below-standard operator performance
4. Above-standard operator performance
5. Absenteeism
6. Lack of materials

In other words, the standards used by the scheduler should be modified by actual operating experience. This modification need not be applied against every job, but can be applied against the overall load for the processes in question.

It is not usually necessary for the production scheduler to have detailed information about either the operating method or the development of the standard. Consequently, the amount of descriptive detail required for this use is not great, and the work standard itself is usually in enough detail for this purpose. In case alternate methods of scheduling are listed, these should be available for use in arriving at the production schedules, and the mechanism should be provided to indicate any excess cost which will be incurred through the use of the alternate method.

Use of Time Standards for Plant Layout and Material Handling. Time standards are essential for evaluating the economics of the many alternatives which are available to the plant layout engineer. The factors which assume major importance are the transportation times for the materials and the operators.

The aspects of plant layout which cause variations in these transportation factors are:

1. Location of equipment
2. Number of machines operated by an individual
3. Number of floors
4. Choice of intradepartmental material handling methods
5. Choice of interdepartmental material handling methods
6. Product- versus process-type layout

The placement of equipment is closely allied with the choice of equipment itself. Because the factors which affect transportation also affect operating procedures, all engineering departments and all levels of production supervision are vitally concerned with problems in layout. This is true whether the problem is one of a completely new layout or a problem of revising an existing layout. The frequency of use of standards for layout planning in this situation will vary markedly from one plant to another.

The need for detail will also vary greatly from plant to plant. If there is considerable layout work to be done, it may be desirable to combine basic standards for walking, trucking, package handling, conveyor loading, and other material handling data into a table especially for use of the layout engineer. This provides him with a rapid means of calculating the approximate worth of various ideas.

It will also be advisable to provide him with all the details of operating method, including the machine and piece handling and processing times for the specific equipment to be utilized in the layout.

Use of Time Standards for Budgeting and Cost Control. The development of sound operating budgets must take into consideration all the predictable cost factors involved in the operation of the business. In this respect, time standards may be used to determine:

1. Standard costs for direct labor operations
2. Indirect labor costs at various operating volumes
3. Redistribution of duties required at various operating volumes of activity
4. The man and machine activities at various operating volumes
5. A basis for the distribution of overhead items

As a cost control tool, the time standard is fundamental. It provides the basis for measuring the effectiveness with which direct and indirect labor costs and machine utilization have been controlled.

The control of costs, of course, does not come from the performance indicators but from the intelligent action of the departmental supervisor relative to the utilization of personnel and equipment. Such action implies detailed knowledge of the operating methods upon which the performance standard is based.

Although it is usually thought that the control of costs is primarily a manufacturing function, many companies have obtained excellent results from cost control programs extended into other phases of their operations. In fact, it is logical that every supervisor has an element of cost control in his job for which he can utilize the standards information available.

The bulk of the work in calculating the actual performance against the standards usually becomes a cost department function. It is important, therefore, to furnish this department with the performance standards for all operations in such a way that the minimum amount of work is necessary to modify them and make them ready for application.

Use of Time Standards for Cost Estimating—Setting Sales Prices. In many of the functions previously discussed, the role of standards in cost estimating for economic calculations is mentioned. The results of these activities may frequently have a bearing on the sales price of an article. Although price schedules may frequently not be set on the basis of cost alone, it is extremely helpful to those establishing sales prices to know the actual cost of production as well as the variations in direct costs for the various products in their line.

In this connection, time standards may be used to help determine:

1. The cost of the product now being or to be made
2. The effect of size, shape, color, material, or tolerances upon the cost of the product
3. The effect of lot size (that is, setup cost versus storage cost) upon the cost of the product

The cost so obtained can then further be used to relate the effect of lot size upon sales price, which may be reflected in the use of quantity discounts.

Before the standards information can be effectively utilized for this purpose by the sales department, it usually requires modification by the cost department. It is necessary to make a conversion of the performance standard from pieces per hour or hours per unit to a dollars-per-unit basis. This may also include the addition of burden costs which are usually provided by some section of the accounting or controller department.

Very little detail is necessary for the sales department, because it is usually concerned only with the total cost per unit sold. Usually, then, the performance standard itself will provide all the information which can be effectively utilized by the sales department.

Use of Time Standards for Manpower Planning. In the area of manpower planning, properly developed standards may be used to help determine:

1. Number of people required
2. Skills involved
3. Training schedules
4. Training and labor turnover costs

A detailed description of the motion pattern required to perform an operation will give the skilled job trainer a good start on the development of the training procedure. It is then usually advisable that the individual who is responsible for training have the details of operating method for each job as well as the time standards.

Actual manpower planning should be performed by everyone who is in a supervisory capacity as well as those in the personnel department. In many departments, this work is performed by the supervision with or without help from the personnel department or from the industrial engineering department.

Use of Time Standards for Employee Relations. In the day-to-day working relationships between the supervisor and the individual worker, time standards play a very important part because they may be used to determine:

1. Job assignments
2. Work loads to be handled
3. Guides for job instruction
4. Methods of operation
5. Identification of nonstandard methods

Because these relationships are fundamental to the industrial climate of operation, it is apparent that whenever the performance standards enter into a discussion, all information relative to their utilization should be immediately available. This implies the need for standards and for details of operating methods by each supervisor for those jobs under his supervision.

Use of Time Standards for Job Evaluation. The task of determining the relative worth of various jobs can be considerably speeded through the use of production standards that are properly prepared. These standards will be of particular help in:

1. Obtaining the basic job description
2. Obtaining the relative amounts of time spent on various elements of the job—this may be essential in a consideration of the amount of physical or mental effort involved in a job

From the nature of the above, it is apparent that the information needed for this use would include the detailed operating procedure.

Those usually most immediately concerned with job evaluation are the operating personnel and the job analyst. Because job evaluation is performed for almost all occupations in many companies, however, the standards may be used in this connection by all supervisors having the direction of personnel for whom standards have been developed.

Use of Time Standards for Wage Incentives. In wage incentive plans, the time standard is fundamental, as it determines the productive rate which can be maintained. Because the operating department is responsible for the administration of its wage incentive program, it is apparent that detailed information about the operating method must be available in the department. For further discussion of wage incentives, refer to Chapters 1, 2, and 4 of Section 6.

Use of Time Standards for Collective Bargaining. Because the scope of collective bargaining has generally been interpreted to apply to all matters relating to the wages, hours, and other terms and conditions of employment, it becomes apparent that management may be requested to submit to such bargaining some of the requirements indicated through the use of time standards. The exact matters to be bargained about will be subject to the individual working agreement (contract) that the company holds with the union involved.

Likewise, the amount of information relating to such standards which is to be submitted to the union is likely to be subject to similar stipulation within the bargaining agreement.

Representing the company in collective bargaining procedures are usually the production supervision and the industrial relations department. This type of discussion will usually cover the detailed operating procedure and, in some cases, the details of standards development. When the details of standards development are to be discussed, the industrial engineering department should be represented to assure that the details are being correctly interpreted.

Use of Time Standards for Methods Improvement. Standards are especially helpful in methods improvement work to determine:

1. The wisdom of investigation
2. The comparative costs of alternative methods

If the methods improvement function is a part of the duties of the standards section, all the standards information which is needed will be at hand.

Those attempting to do the most effective job of methods improvement will be aided by all the detail, including:

1. The work standard itself
2. Details of operating method
3. Details of standards development

Nonindustrial Areas of Applications for Time Standards. Industrial engineering and management science techniques are applied to many activities outside the areas of the production and distribution of goods. Worthy of note are the applications in the health professions, especially hospital and nursing administration; government agencies such as the post office and the New York Port Authority; the entire transportation industry; and research and development activities.

On the surface, the applications of standards in these areas may seem significantly different from those previously discussed. In these applications, quite obviously the major and sometimes only applications of standards may be in the universal areas of manpower planning, cost estimating, and budgeting. Many opportunities exist, however, for their use in the design and selection of equipment, operations planning and scheduling, and layout and materials in ways roughly analogous to those described on the previous pages.

PREPARATION OF STANDARDS INFORMATION

The logical objective in the preparation of standards information is to make the information available in a usable form at a low cost.

It is apparent that almost everyone performing a management function has some use for time standards. It might, then, erroneously be assumed that the simplest procedure would be to plan to provide copies of the complete standards to all those with management responsibilities. It should be recognized, however, that whenever the work standard only is required, the complete standard will be unnecessary.

Summary of Uses of Data. To plan an economic procedure for the preparation and presentation of time standards, it is necessary to consider for each part of the standard:
1. Who needs the data?
2. How frequently are they needed?
3. How much detail is required?
4. Does the basic time standard require modification?
5. Who has information for modification?
6. Does the individual needing the data have sufficient knowledge of the specific job to utilize the data effectively? (Is interpretation needed?)

After detailed consideration of the individual uses for time standards, a tabular summary of requirements can be prepared. One such tabulation for a typical situation is shown by Table 5-1. It should be emphasized that although the example used here was developed for a manufacturing concern, the general method can be used for visualizing and designing any standards development and dissemination system.

Flow of Standards Information. With the information contained in Table 5-1, it becomes somewhat simpler to select a suitable method for preparing and reproducing time standards information. By assigning rough frequencies and costs to the instances of use, a quick appraisal can be made of the extent of a systems design project which can be justified. The final methods may vary from a highly sophisticated computerized information system to the use of the most simple duplicator and filing processes. Seldom, however, will one method be best for all parts of the standard. Consequently, the following discussion presents a typical systems analysis essential for a viable design and is developed for each part of the standard.

To aid in the development of this discussion and in following the flow of the parts of a time standard, a flow chart (Figure 5-1) has been prepared showing

TABLE 5-1. Specimen Summary of Individual Use of Various Parts of a Standard

	Work standards	Itemized operating methods	Details of standards development		
			Alternate work methods	Elemental analysis	Time studies and summaries
Factory manager........	Occasional	Occasional			
Operating superintendent...............	Frequent	Frequent	Occasional	Occasional	
Department foreman...	Continual	Continual	Occasional	Occasional	
Production planning manager...........	Continual*				
Personnel manager.....	Frequent	Frequent	Occasional		
Product engineer.......	Occasional	Frequent	Occasional	Occasional	
Process engineer.......	Occasional	Frequent	Occasional	Occasional	
Industrial engineer.....	Continual	Frequent	Frequent	Frequent	Occasional
Tool engineer.........	Occasional	Occasional	Frequent	Frequent	
Plant controller.	Continual	Occasional			
Sales executive........	Frequent*				
Industrial relations manager...........	Frequent	Frequent		Occasional	
Office manager........	Frequent	Frequent	Occasional		

* Work standard as issued by standards department requires modification for use by these individuals.

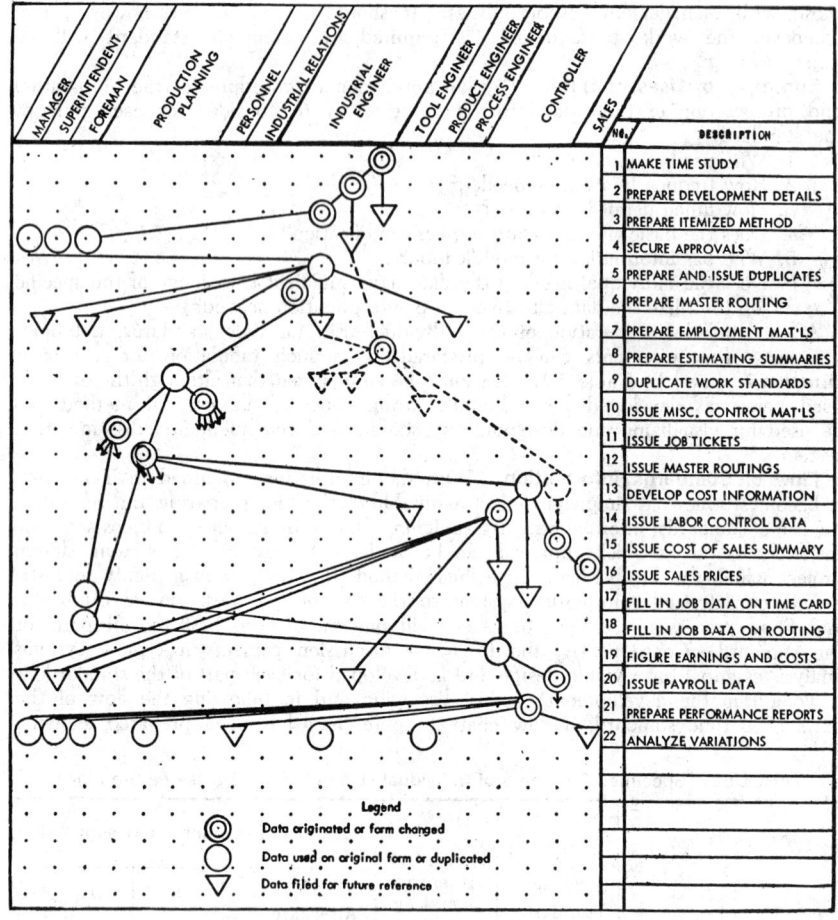

FIG. 5-1.   Specimen flow of standards information through an industrial organization.

the path of development and use of this material. The steps referred to in the following discussion refer to the steps depicted in this flow chart. Steps 1 to 8 depict the work done by the industrial engineering department in the preparation of the time standards. Steps 9 to 22 depict the logical flow of the standards material through the rest of the organization.

Step 1. Make time studies.

Step 2. Prepare details of standards development.

All recap sheets, charts, graphs, tables, elemental summaries, and right- and left-hand analyses are prepared. Copies of these are used by the industrial engineer and tool engineer.

Step 3. Prepare the itemized operating method and standard.

The summary in Table 5-1 indicates that the details of the operating method are used at least frequently by eight different groups. However, product and process engineers are usually working either on different jobs or as a team on one certain job. These individuals are part of a central engineering organization and are located in adjacent sections. Consequently, one copy of the operating method data is usually sufficient for these two groups. Likewise, the industrial engineers and tool engineers are able to utilize the industrial engineering copies, and the industrial relations and personnel people can utilize a single set of data. This requires five copies to be reproduced ordinarily. Occasionally a conflict in use requires an additional copy. If the itemized operating methods are posted in the shop as standard procedures or are used for operator training, then additional copies will be necessary.

Step 4. Secure approvals of the itemized operating method and standards.

The original copy of the itemized operating method and standards is routed to the foreman, superintendent, and plant manager for approval.

Step 5. Issue duplicates of itemized operating methods.

Step 6. Prepare master routing of work standards.

After consideration of the data shown in Table 5-1, it can be seen that nine groups or individuals have at least frequent use for the work standards. Closer analysis shows, however, that the standards included with the itemized operating method are usually sufficient for the operating superintendent, the personnel department, and the engineering department. It is evident that six copies of the work standards might be sufficient to allow the departments other than industrial engineering to proceed with their work.

The work standard information for the individual part is therefore placed on a master routing together with the work standards for the other operations on that part. This master is then sent to the production planning department.

Step 7. Prepare employment and training materials.

The personnel department develops procedures and time tables for training from the itemized operating method information.

Step 8. Prepare summaries for estimating.

Using the information from groups of standards and groups of operations, the industrial engineering department prepares summaries for rapid estimating for new and changed methods, materials, and parts.

Step 9. Duplicate copies of work standards from master routing.

Copies of many different forms are prepared from the master routing. These forms are described in steps 10, 11, and 12.

Step 10. Issue miscellaneous production planning and control materials.

From the information contained on the heading of the master routing, it is possible to preprint shipping notices, material requisitions, move tickets, tool cards, inspection tickets, shipping labels, and storage tickets for each order. The order number and other information pertaining to an individual order is supplied by the insertion of a variable section into the master.

It is to be noted that before the time data can be effectively used for scheduling by the production control department, they often require modification for actual performance against standard. It is obvious that this modification can most easily be handled by the production scheduler on the basis of average operation performance rather than by performance on each job individually. For example, if the actual production for a group of operations shows consistent daily performances 20 percent above standard, the schedules are based on 9.6 operating hours instead of 8. If hourly schedules are required, they are based on 1.2 hours.

Step 11. Issue job tickets.

Individual job tickets are preprinted from the master routing and are sent to the various departments where the job is to be processed.

Step 12. Issue copies of master routing.

Additional copies of the master routing are duplicated for various uses.

One copy is used as a tracer for tracing the job through the shop.

One copy is used by the production scheduler to plan the production schedule.

One copy is used by the controller department for cost analysis.

Step 13. Develop cost information.

A copy of the master routing printed on a special folded form is provided to the cost department. When opened, this provides a cost analysis sheet, thus permitting direct extension and accumulation of additional items of cost without copy work.

Step 14. Issue performance indicator control data.

From the time and cost data obtained from the analysis sheets and other sources, the standard cost and budget information is compiled and issued to department heads.

Step 15. Issue cost of sales summary.

The cost information is summarized and transmitted to the sales department for pricing purposes.

Step 16. Issue sales prices.

The sales department, using the cost department's calculations together with information from the economists, competitors, and the like, determines the sales price to be charged for the items.

Step 17. Fill in job ticket.

As the job progresses through the shop, the department clerk and operator fill in such pertinent information as time taken, operator, and date. This ticket is then sent to the controller's department for cost analysis and payroll calculation.

Step 18. Fill in master routing.

Information relative to the process as a whole is accumulated and reported by operation on the departmental copy of the master routing. When a job cost is desired, this copy is then sent to the controller's department for cost summarizing.

Step 19. Calculate earnings and costs.

The production and time information contained on the job ticket is used to determine the earnings and the performance of the operator.

Step 20. Issue payroll data.

The payroll data are entered on each individual's prepunched tabulator clock card for the day. This clock card is sent to the tabulation department where the amount is punched on the card. The total for the week is obtained and the corresponding check written automatically by the tabulator equipment.

Step 21. Prepare and issue performance reports.

The performance data calculated in step 19 are next summarized into a daily performance report for each individual and each department. These reports show both daily and to-date totals for each individual department and are issued before noon of the workday following the day on which the work is performed.

Step 22. Analyze performance data.

When the summarized information is back in the hands of the foreman, causes for variance are noted and plans for future operations are made.

BIBLIOGRAPHY

Amrine, Harold T., John A. Ritchey, and Oliver S. Hulley, *Manufacturing Organization and Management*, Prentice-Hall, Inc., Englewood Cliffs, N.J., 1966.

Apple, James M., *Plant Layout and Materials Handling*, The Ronald Press Company, New York, 1963.

Barnes, Ralph M., *Motion and Time Study: Design and Measurement of Work*, 6th ed., John Wiley & Sons, Inc., New York, 1968.

Bennett, Edward, James Degan, and Joseph Spiegel, *Human Factors in Technology*, McGraw-Hill Book Company, New York, 1963.

Buffa, Elwood S., *Operations Management: Problems and Models*, John Wiley & Sons, Inc., New York, 1968.

Cleland, David I., and William R. King, *Systems, Organizations, Analysis, Management: A Book of Readings*, McGraw-Hill Book Company, New York, 1969.

Kronenberg, M., *Machining Science and Application*, Pergamon Press, New York, 1966.

McCormick, Ernest J., *Human Factors Engineering*, 2d ed., McGraw-Hill Book Company, New York, 1964.

Mundel, Marvin E., "A Philosophy of Work Measurement," *Systems & Procedures Journal*, January–February, 1965, pp. 14–19.

Reed, Ruddell, Jr., *Plant Layout: Factors, Principles, and Techniques*, Richard D. Irwin, Inc., Homewood, Ill., 1961.

Smalley, Harold E., and John R. Freeman, *Hospital Industrial Engineering*, Reinhold Publishing Corp., New York, 1966.

Starr, Martin Kenneth, *Production Management Systems and Synthesis*, Prentice-Hall, Inc., Englewood Cliffs, N.J., 1964.

Chapter 6

Administrative and Control Procedures

WILLIAM K. HODSON

President, H. B. Maynard and Company, Incorporated,
Pittsburgh, Pennsylvania

Practically every text on the subject of industrial engineering deals with specific technical procedures and techniques. Process charts, operation analysis, motion study, time study, predetermined motion times, wage incentives—all are covered by many texts. On the other hand, there is little specific information available on the ordinary, everyday procedures and routines required for the effective operation of any industrial engineering department. The purpose of this chapter is to provide this type of information. The details of administrative procedures and clerical routines will, of course, vary considerably from plant to plant. They will vary in accordance with the size of the plant, the nature of the operations, the organizational setup, whether or not wage incentives are employed, and a host of other factors. There are, however, certain principles and procedures of administration that apply to all industrial engineering departments. In the pages that follow, these principles and procedures are discussed from the standpoint of how they apply to the typical industrial engineering department.

ESTABLISHMENT AND MAINTENANCE OF TIME STANDARDS

Establishing and maintaining time standards is one of the major responsibilities of any industrial engineering department. It is also one of the industrial engineer's most troublesome and complex tasks. Because time standards provide the basic data required for manpower controls, standard costs, production control, machine utilization, manning requirements, budgets, and wage incentives, it is essential that they be established accurately and consistently.

If time standards are to be accurate and consistent, they must be based on proven techniques of work measurement. Most inconsistencies in standards are the result of poor maintenance. If standards are to remain effective, they must be continually revised to reflect changes brought about by changes in methods, processing, and specifications.

The basic administrative procedures required to establish and maintain a system of time standards are illustrated graphically by Figure 6-1.

A time standard must be established whenever a new operation is introduced or an existing operation is changed. New operations are normally introduced by the addition of new parts or products. Existing operations are normally changed as the result of a change in:

1. Method
2. Processing
3. Design

New Operations

Process Sheet Prepared. New operations are created by the introduction of a new part or product. Normally, the new part is originated by the engineering, technical service, or similar department through the medium of drawings or technical specifications.

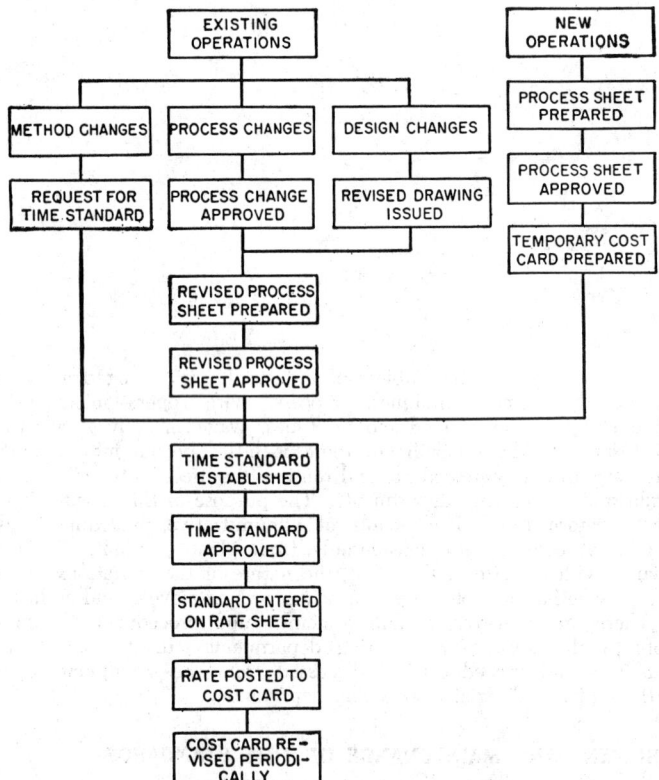

Fig. 6-1.   A graphic analysis of the steps required to establish and maintain time standards.

OPERATION RECORD						MATERIAL	PC. NO.	DRAWING NO.
NAME			NET WT.		PATTERN NO.			
			GROSS WT.					
ENG. SIZE	CONTD. CARD	CARD	WRITTEN BY	CARD NO.	RECORDED	RECORDED		
	NO.		DATE	OF CARDS	ROUTING DIV.	COST DIV.		

DEPARTMENT AND MACHINE	JIG NO. OR FLASK SIZE	OPERATION DESCRIPTION	DATE SET-UP TIME VALUE	DATE EACH PIECE TIME VALUE	HOW SET	CLASS

Fig. 6-2.   A typical process sheet form.

Before the new item can be manufactured, however, it is necessary to prepare a process sheet. The information included on this sheet will vary considerably. In general, it includes a sequence of the operations required to manufacture the part, along with pertinent technical information on each process or operation. A typical process sheet is shown by Figure 6-2. The sheet is prepared by a processing group working under either the mechanical engineering or the industrial engineering department. Frequently, the process sheet is combined with the cost card and all the information required by both the manufacturing and the cost accounting departments is included on one form. For further details on processing and process sheets, see Chapter 2 of Section 2.

Process Sheet Approved. If the process sheet is prepared by other than the industrial engineering group, it is forwarded to industrial engineering for approval. At this time, the manufacturing processes and the sequence of operations are reviewed for possible improvements and cost reductions.

If satisfactory, the process sheet is approved and distributed. Its basic purpose is to provide the manufacturing departments with the information required to process and manufacture the item.

When standard data or time formulas are used to establish time standards prior to production, the standards are entered on the process sheet or cost card at this time. If standards cannot be established until the operations are performed, then the standards are either left off the process sheet and cost card, or temporary estimated standards are used. In this case, a temporary cost card is prepared and issued. When permanent standards are developed, a permanent cost card is issued.

Temporary Cost Card. The only purpose of a temporary cost card is to provide a temporary means of charging labor and material costs to the proper account until such time as permanent standards are established. The cost card provides the necessary authorization to perform the operations specified.

Whenever it is practical to do so, standard data and time formulas provide an ideal means of establishing time standards. They not only provide the advantage of more consistent standards, but they also permit the establishment of standards prior to production. This eliminates the necessity of establishing temporary or estimated standards. This advantage is particularly important where the standards are used as the basis for a wage incentive plan.

When standard data or formulas are not available, it is desirable to issue temporary standards rather than no standards at all. When temporary standards are used, they are normally limited to a specific time period, usually thirty days. At the end of this period, they are replaced by permanent standards. Thirty days usually provide sufficient time to iron out the production problems normally encountered in manufacturing a new item and to establish permanent time standards. During this period, the temporary standards provide a basis for incentive wage payment, production control, cost control, and the like.

At this point, a permanent time standard should be established (see Figure 6-1). Because the procedure followed in establishing a time standard is basically the same whether the operation is new or existing, this procedure will be discussed after the steps required to establish a standard on an existing operation have been analyzed.

Existing Operations. The importance of providing procedures that will ensure the proper maintenance of time standards has already been stressed. Time and time again, a successful system of time standards and wage incentives has been installed only to fall into disrepute for lack of adequate maintenance.

As shown by Figure 6-1, there are three common changes that occur continuously. These changes occur in methods, processing, or design.

Methods Changes. Normally, methods changes refer to all changes that occur on an operation that do not affect the basic process and are not caused by changes in materials or specifications. They include changes in the workplace, changes in the motion pattern employed, minor modification of jigs and fixtures, and the like. They can usually be made without requiring a change in the process sheet, drawing, or specifications.

Because these changes are, by nature, subtle changes, they are the most difficult type to control. On the other hand, they may result in sizable changes in the standard itself.

One of the basic problems encountered in this regard is determining when a method has changed. On highly repetitive, short-cycle operations, the elimination or combination of a few motions will frequently have a major effect on the time standard. For this reason, the importance of recording a complete description of the method on which the standard is based cannot be stressed too strongly. A procedure employing predetermined motion times such as methods time measurement is particularly well suited for this purpose.

Normally, the foreman is held responsible for reporting to the industrial engineering department any methods changes originating in his department. When a request for a new time standard is received by the industrial engineering department, the operation is analyzed to determine whether or not a change in standard is required. If it is, then the standard is established in the conventional manner.

Because methods changes normally do not affect the process sheet, it is not necessary to revise this form except to change the standard for the operation concerned. This step is shown in Figure 6-1.

Process Changes. Process changes are those changes that do not require a change in either the drawings or material specifications, but do require a change in the process sheet itself. They normally are basic changes in the manufacturing process. Typical examples are changes in equipment or tooling. An operation may be transferred from an engine lathe to a turret lathe; a drill fixture may be substituted for layout and center punching; a form tool may be substituted for a single-point tool. The basic rule is that the change requires a revision of the process sheet but not the engineering drawings or material specifications.

A request for a change in the process sheet will originate with either supervision, industrial engineering, or mechanical engineering. Changes originated by the processing group are prepared in the form of a revised process sheet. Requests for changes from other departments are first investigated, and if the change is justified or desirable, a revised process sheet is prepared to incorporate the changes.

Design Engineering Changes. The third major reason for changes in time standards is a design change. Design changes are those changes in the part or material that require a change in the engineering drawings or material specifications.

While these changes normally originate in the engineering department, they are sometimes originated by the foreman, processing group, or industrial engineering. Final approval for the change rests with the engineering department, although it may consult with processing or industrial engineering before the change is made.

Because changes in drawings or specifications normally require changes in processing, Figure 6-1 shows this type of change resulting in a revised process sheet being prepared to accommodate the change. Revised process sheets resulting from design changes follow the same routine as any other process change, as shown in Figure 6-1.

Setting Standards. All the steps shown in Figure 6-1, for both new and existing operations up to the point of actually establishing the time standard, have now been covered. The mechanics of establishing the time standard are essentially the same whether it is a new operation or a change in an existing operation. The choice of techniques that may be used, either separately or combined, is discussed in detail in other chapters of the Handbook.

Time Standard Authorization. Normally, a daily log is maintained by industrial engineering of all the standards established each day (see Figure 6-3). This rate authorization sheet serves two purposes. First, it acts as an authorization, by the industrial engineering department, for the accounting department to enter or adjust a standard on the cost card and process sheet. A second and extremely important use of this form is to provide a record and summary of time standard revisions. Columns are provided for comparing the new standard to the existing standard. The saving or loss resulting from the revision and any pertinent remarks are entered in the appropriate columns. A summary of the net saving or loss is made on the bottom of the form. The daily figures are usually summarized by weekly and monthly control reports, which will be discussed further later in this chapter.

Upon receipt of the rate authorization sheet, the accounting department enters the new or revised standards on the appropriate cost cards. Normally, it is good accounting practice to revise cost cards only once a year, usually upon completion of a physical inventory. For this reason, the revisions in standards are merely entered in pencil or other temporary fashion. The revised standard is used for wage incentive or control purposes but not for standard cost purposes.

If both a process sheet and cost card are used and if both have provisions for time standards, then the revision of the process sheet is made by the industrial engineering department. This is done at the same time that the standard is posted to the rate authorization sheet.

Time Standard Controls. Any system of standards is continuously subject to pressures that tend to loosen them. The pressure is exerted primarily by labor, but in a number of cases, by management as well. The pressure from labor takes a number of forms. It may range anywhere from a friendly jibe during a time study to slowdowns and strikes. Pressure from management may come from any level—from the foreman, interested in meeting his budget by means of more lenient standards, to the president, concerned about an impending strike resulting from a grievance on time standards.

When standards are established on the basis of individual time studies, these pressures are exerted on every standard established. Either consciously or unconsciously, the engineer is constantly aware of the pressures that encourage him

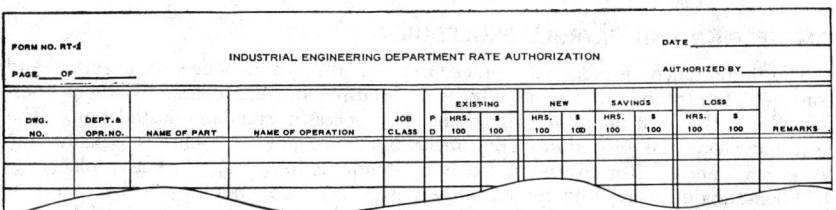

Fig. 6-3.   A form used to provide a daily log of all changes in time standards.

to rate the performance of the operator liberally, or to overlook obviously poor methods, or to allow time for unnecessary elements of work. By continuously setting standards on the liberal side, he avoids the discord that may result from a standard set on the tight side. It is not long, however, before management is faced with either going out of business or bringing inflated standards back to normal.

It is desirable, from the standpoint of control, to establish individual standards by means of standard data or time formulas rather than by individual time studies. Their use relieves the engineer from the pressures exerted on individual time studies. Once the data are established and proved, they are less subject to dispute and grievances. By their very nature, they ensure consistency and the maintenance of a uniform level of performance.

If a system of standards is to be controlled, it is essential to employ the time standard controls mentioned previously. These controls are designed for permanent methods changes. Occasionally, however, a situation is encountered where it is necessary to make a temporary change in method. Such a change may result from an unusually hard lot of castings that require additional machine time, or an imperfect lot of parts that require a slight additional operation before they can be assembled, or any of many different reasons. The only characteristic that these changes have in common is that they are normally limited to a specific lot of parts. If this change is not to become a permanent one, procedures must be established to handle it.

When the need for a special standard arises, the foreman notifies the standards department of the change and requests a special allowance. A new standard is established on the operation and the change is entered on the special allowance form. One copy is sent to payroll or accounting, and the other is retained by the standards department. It is important that the special allowance be issued only for a specific lot size or quantity. Once the special lot is completed, the original standard will automatically become effective. Normally, the chart of accounts is so designed that the extra time required can be charged to a special account. In this way, a control is established to prevent the abuse of the special allowance authorization.

Because time standards form the heart of labor cost controls, it is essential that all time standards be authorized only by responsible personnel, usually the chief industrial engineer. The form used for this formal authorization, Figure 6-3, can also be used to good advantage as a means of controlling time standards. In addition to providing an authorization for a change in a standard, it also provides a comparison of the new and the existing standards to indicate whether the change resulted in a saving or loss. This form is usually compiled on a daily basis, but the totals can be compiled in the form of a monthly report. Under normal conditions, any monthly summary of changes in standards should show a net saving. Any changes in standards that result in a loss can be carefully examined before they are approved.

In addition to providing the chief industrial engineer with a method for controlling changes in standards, this report can also be used by top management as a means of estimating the effectiveness of a general methods improvement program. If desired, the report can also be summarized by departments to provide additional information and control.

TIMEKEEPING AND PAYROLL PROCEDURES

The importance of adequate procedures for the maintenance of accurate and consistent time standards has already been discussed in some detail. Without question, this is the most important ingredient of a successful standards installation. The next most essential procedures are timekeeping and payroll. This is equally true for either standards for cost control only or standards for a wage incentive plan.

When standards are used for incentive pay, everyone is more apt to be cognizant of the importance of good timekeeping and payroll procedures. This is so because they have an important bearing on both wages and labor costs. When standards are used for cost control purposes only, timekeeping and payroll procedures are not normally installed and administered with the same care. In either case, however, they are seldom given the attention they deserve. Experiences of several companies that have come close to bankruptcy because of poor timekeeping and payroll procedures emphasize the importance of these two procedures.

The industrial engineering and accounting departments are normally responsible for developing and installing timekeeping and payroll procedures. Responsibility for the accuracy of the time and piece count records, however, belongs to the line supervisors. The periodic audits and controls required to ensure that this responsibility is properly carried out are again the responsibility of the industrial engineering and accounting departments. These procedures can never be considered as satisfactory unless provisions are made for specific controls and periodic audits.

It is almost universal practice to have all accounting records audited by a firm of professional accountants. More and more companies, however, are now following the practice of having a similar audit made of time standards and associated procedures by a firm of professional management engineers. This has a great deal of merit. Poor practices are usually introduced over a long period of time—a minor change here, a slight deviation from procedures there. Each change, in itself, is relatively unimportant. Over a period of years, these creeping changes become established practice. Management, being so close to the scene, frequently does not realize what has happened until the changes have become firmly embedded. By this time, it is extremely difficult to change these practices. As a result, more and more organizations are following the policy of having periodic standards audits made to uncover changes, deviations, and poor practices before they become firmly rooted.

Basic Requirements of Timekeeping and Payroll Systems. There are a great many basic systems of timekeeping and payroll procedures. For each of these basic procedures, there are innumerable variations and modifications. Looked at in this light, the subject is exceedingly complex and beyond the scope of a chapter of this sort. On the other hand, when the fundamental requirements of a good timekeeping and payroll procedure are examined, the problem becomes quite simple. Therefore, the basic requirements of any system of timekeeping and payroll procedures will be analyzed first, and then a few specific systems to determine how well they satisfy these requirements will be examined.

For purposes of this discussion, timekeeping can be thought of as the routines and procedures connected with originating source records on the hours worked and the work performed. Payroll procedures are the clerical routines required to translate source records into payrolls, reports, posting to accounts, distribution of labor costs, and the like. With this distinction in mind, the following can be listed as the major requirements for good timekeeping and payroll procedures:

1. Provide a record of hours worked
2. Provide a record of work performed
3. Provide a means for distributing labor costs
4. Be capable of providing information for reports, analyses, and controls
5. Be accurate
6. Be economical to administer
7. Satisfy requirements of Federal and state laws on deductions, wages, and hours

The importance of any one factor will depend on the particular case at hand. As far as payroll systems are concerned, a record of the work performed, for example, is more important to a plant with an incentive wage payment plan in effect, and less important to a shop working on straight daywork. A detailed discussion of each of the requirements listed above is desirable for two reasons. First, it will provide a better understanding of the importance of each requirement for a specific case. Second, it will provide a means for explaining how these requirements may be best satisfied, that is, the mechanics of the procedures and routines.

Record of Hours Worked. The one requirement essential to any system of timekeeping is a record of the hours worked. This is true regardless of the size of the organization, the method of payment, or the type of accounting system employed. In all cases, it is necessary to have some record of attendance.

Although a few companies have established the policy of banning any type of time clock, most attempts along this line have been based more on dramatics and psychology than reality. There is no Federal law requiring the use of a time clock to record attendance, but the Federal Wage and Hour Law does require a record of attendance. The only legal proof of attendance is a time clock. Several states have passed laws making the use of a time clock compulsory.

It is common practice to provide time clocks at some point convenient to the employees' entrance. In all cases, it is necessary for each employee to ring in his arrival time and quitting time. In addition, many companies require employees to ring in and out for lunch periods. If this practice is required, it is necessary to locate attendance clock cards at a point more convenient to the shop floor. Normally, one time clock is sufficient for 250 employees.

There is a wide variety of clock card forms available. They all provide space for recording a full week's work. Time clocks are available recording in units of hours and minutes, and recording on a decimal hour basis. The reading 16.75, for example, is 4.75 hours past 12 noon, or 4:45 P.M. The decimal hour clock simplifies payroll calculations considerably, because it is not necessary to convert minutes to fractions of an hour when computing pay.

Accurate Distribution of Labor Costs. For practically all systems of accounting, it is necessary to distribute an employee's working time to various jobs or accounts. This is done by the use of a job clock card and job time recorders. These cards and clocks are subsidiary to the attendance cards and time clocks. In a few cases, however, they are combined and only one time card and time clock are used. When they are used, the job clocks are invariably designed to record time to a tenth of an hour rather than hours and minutes.

There are three basic types of job cards or tickets:

1. Individual job ticket
2. Man or group ticket
3. Continuous job ticket

As the name implies, the individual job ticket is prepared for each job, that is, each operation on each part. The ticket may be used for only one day or part of a day, or it may be allowed to run until the end of the pay period if the job runs for more than one day. This ticket is used for either daywork or incentive pay. It simplifies the distribution of labor to jobs or accounts, but makes it difficult to summarize standard hours for an individual.

The man or group ticket is normally used when incentive pay is involved. It serves as a record for all jobs performed by one individual or a group of operators when they are paid on a group basis. The ticket is designed to be used on either a day or a pay period basis. Although this type of ticket simplifies incentive payroll computations, it is difficult to use for distributing labor to jobs or accounts.

The continuous job type of ticket (Figure 6-4) is used for daywork pay where costs are distributed by jobs. In effect, it automatically distributes labor costs by jobs because only one card is used for a job for the pay period. Any operator working on the job records his time on the job ticket. In this way, all labor costs for that job are accumulated automatically. The primary difficulty encountered in the use of this type of card is in balancing the time worked on various jobs with the attendance clock cards.

A preprinted job ticket is merely a variation of any of the three basic tickets. Its use has become increasingly popular, however, due to the time saved by the operators in filling out the card and the reduction of errors. In preprinting the cards, two alternatives are available. In one case, all the information that is constant per job can be printed in advance. Such things as the job number, the operation description, time standard, feed, speed, or other job specifications are preprinted. In this case, the timekeeper fills in by hand the information pertaining to the operator working on the job.

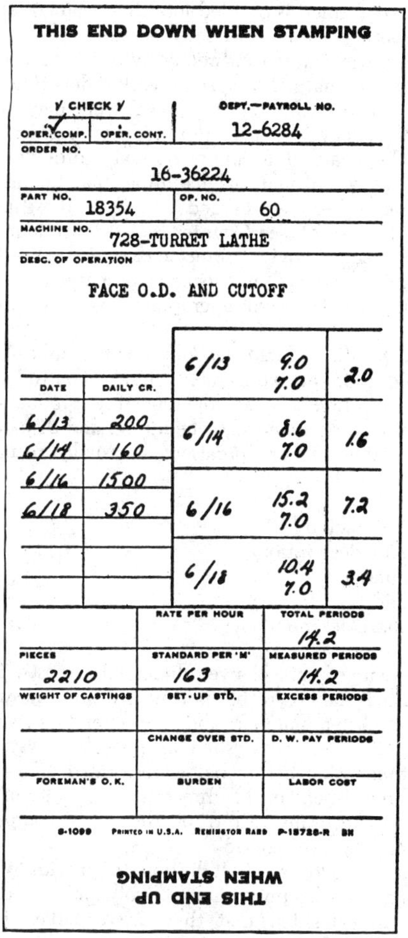
THIS END DOWN WHEN STAMPING

✓ CHECK ✓ | OPER. COMP. ✓ | OPER. CONT. | DEPT.—PAYROLL NO. 12-6284

ORDER NO. 16-36224

PART NO. 18354 | OP. NO. 60

MACHINE NO. 728-TURRET LATHE

DESC. OF OPERATION FACE O.D. AND CUTOFF

DATE	DAILY CR.
6/13	200
6/14	160
6/16	1500
6/18	350

6/13	9.0 7.0	2.0
6/14	8.6 7.0	1.6
6/16	15.2 7.0	7.2
6/18	10.4 7.0	3.4

	RATE PER HOUR	TOTAL PERIODS 14.2
PIECES 2210	STANDARD PER 'M' 163	MEASURED PERIODS 14.2
WEIGHT OF CASTINGS	SET-UP STD.	EXCESS PERIODS
	CHANGE OVER STD.	D. W. PAY PERIODS
FOREMAN'S O. K.	BURDEN	LABOR COST

8-1099 PRINTED IN U.S.A. REMINGTON RAND P-18728-R BH

THIS END UP WHEN STAMPING

Fig. 6-4. A form used to accumulate hours worked by jobs.

The second alternative is to preprint the job ticket for each operator. The operator's name, clock number, department, shift, and rate are all printed in advance. In this case, the timekeeper completes by hand the information pertaining to the job.

Both alternatives have their advantages and disadvantages. The preprinted ticket with job information as a constant eliminates more of the handwritten information. The difficulty encountered here is in determining how many tickets to preprint for each job or operation. By preprinting the job ticket with the man information as a constant, this disadvantage is overcome because the tickets can be used at any time. It has the disadvantage of requiring more handwritten information by the timekeeper or operator. Either type, however, possesses advantages not encountered in the more commonly used blank job ticket.

Record of Work Performed. A record of the work performed is second in importance only to the record of hours worked. When incentive wage payment is used, the work performed is perhaps even more important than the hours worked. In this case, the work performed will determine the amount of a man's paycheck. Therefore it is extremely important that the work performed be accurately recorded.

In general, the work performed is divided into three categories:
1. Direct labor—work performed directly on the product itself
2. Indirect labor—labor charged indirectly by overhead accounts
3. Special labor—work charged to special work orders or accounts

Reporting work performed on indirect or special work may or may not be a simple procedure. The indirect labor is usually reported by classification or category rather than by quantities. The chart of accounts normally provides several classifications of indirect labor, such as setup, material handling, maintenance, and the like. The hours worked on indirect labor are charged to one of these accounts. For a more detailed discussion of this subject, see Chapter 3 of this section.

Most work order systems provide some means for reporting work performed on special orders. Special orders are usually prepared for products that are to be used either for the factory's own operations, experimental work, or something of a similar nature.

The work performed by direct labor accounts for the bulk of most factory payrolls. Consequently it is the type of work most closely controlled. In addition, it is necessary to have an accurate system for reporting production work to maintain an effective system of production control, scheduling, and dispatching.

Work performed on direct labor operations is usually recorded in the following terms:

> Number of units completed
> Operation number or description
> Part or drawing number
> Work order number
> Operator's name and clock number

To provide some system of control over direct labor costs, it is common practice not to permit any direct labor to be performed on any work unless the work is covered by a work order number and the operation is specifically listed on the operation sheet. All other work, such as rework or extra operations, requires special work order authorization before being performed. In this way, it is possible to control closely the hours spent on all direct labor operations.

In order to keep an accurate record of the units completed, it is extremely important to provide some type of control.

Reports, Analyses, and Controls. In addition to providing basic information for payroll and accounting, timekeeping and payroll procedures must provide time and work data in such a fashion that they are capable of also being used for reports, analyses, and controls. Because this aspect of timekeeping and payroll procedures is so important to the effective operation of any plant, this subject will be discussed separately shortly.

Accuracy Requirements. Obviously, timekeeping and payroll procedures are no better than the accuracy of the original records. Good procedures originate in the shop. It does little good to have an automatic checkwriting machine that will prepare 2,000 payroll checks an hour when the checks may be written for fraudulent amounts. There are three items of payroll information that must be absolutely accurate if the system is to operate satisfactorily:
1. Hours worked
2. Work performed
3. Distribution of work

Both hours worked and work performed may be deliberately falsified because they affect payroll computations when wage incentives are employed.

Time clocks, both attendance and job recorders, are used to ensure the accuracy of hours worked. Time clocks in themselves, however, will not ensure correct records unless they are used in accordance with adequate regulations governing their use. It is also essential that those regulations be properly enforced. The best rule to follow in this regard is to establish facilities and regulations in such a manner that all temptation to juggle hours worked is eliminated. While it is never possible to develop procedures and facilities that will eliminate fraud, they should be so established that it is necessary to commit a deliberate fraud in order to turn in an inaccurate record. If procedures are so lax that errors can occur by honest mistakes, the company is unduly tempting employees to turn in inaccurate records.

The accuracy of records on work produced is not so easily controlled as the hours worked.

Errors in distributing work are more the result of misunderstandings than deliberate intent. Nevertheless, inaccuracies in this regard will result in inaccurate accounting records, reports, and controls which may seriously affect top management plans and decisions. Undoubtedly, many of the misunderstandings are due to overzealous accountants. Some charts of accounts are so complicated that they require a CPA to interpret them. If the distribution of labor to accounts is to be made by the operators themselves or by their supervisor, they must be reasonably simple. If, for accounting reasons, they cannot be simplified, then it should be recognized that this complexity will probably require the use of a full-time timekeeper in each of the various departments.

A large number of the errors that occur in distributing labor costs can be avoided by the use of preprinted time cards. This is particularly true where the time cards are preprinted with all the information that is a variable of the job. This requires the operator to complete only the information on his name, clock number, and hours worked. For this reason, more and more companies are turning to preprinted job and time tickets.

Economy of Administration. There are a number of timekeeping and payroll procedures in use that do an excellent job of satisfying all of the requirements discussed so far. Quite frequently, however, these procedures are so complex that the cost of administration becomes excessive.

The automatic business machine manufacturers have done an excellent job of reducing the cost of administering timekeeping and payroll procedures. Complete equipment is available that will prepare payrolls and payroll checks, distribute labor costs to accounts, and prepare labor cost reports and controls. A word of caution is in order at this point, however. A number of companies, in a desire to reduce administration costs, have installed business machines only to discover that their costs have increased rather than decreased. These machines do have their limitations. The most frequent mistake is to mechanize an installation that does not have sufficient volume to utilize a machine effectively. There are two aspects to this particular point. First, business machines are expensive, and like any other item of equipment, they must be utilized to pay for themselves. Buying a five-spindle automatic turret lathe for a machine shop is not considered unless the volume is such that it can be utilized effectively. This point is so obvious that it seems somewhat elementary even to mention it. In the clerical field, however, a number of companies have installed punched card equipment, the automatics of the business machine field, when the volume was so small that the equipment was idle most of the time. The second aspect to this point is the fact that the complexity of payrolls, distribution, reports, and the like are directly proportional to volume. For example, when volume is small, time tickets can be sorted into accounts manually with little or no difficulty. As the volume increases, it becomes almost impossible to sort time tickets manually, and the time required per ticket increases quite rapidly. So volume has a twofold effect. It not only increases the total number of computations and sortings, but it increases the time per computation or sorting when it is done manually. On the other hand, business machines become more efficient as volume increases and setups are reduced. It must be certain, therefore, that there is sufficient volume before going to business machine operations.

For companies that have too large a volume to handle manually but not quite enough volume to justify fully automatic equipment, there is a happy alternative—the key sort procedure. For in-between volumes, this procedure has most of the advantages of mechanization and few of the disadvantages of manual operation.

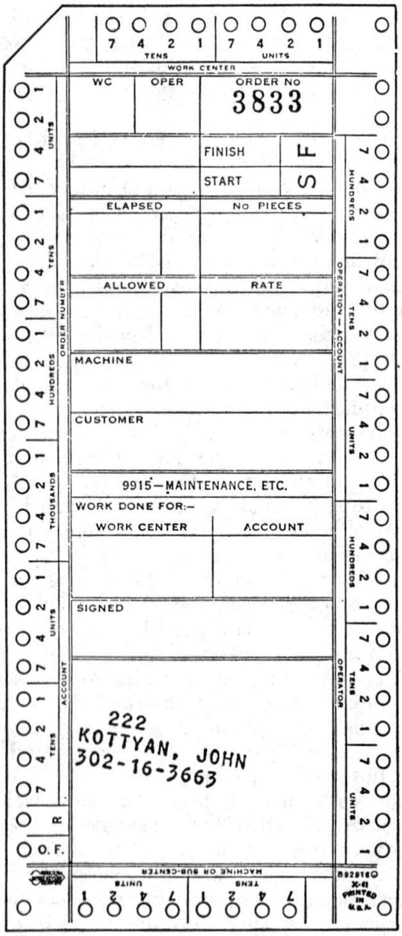

Fig. 6-5.   A typical key sort job card preprinted with employee's name, clock number, and social security number.

Furthermore, the initial investment is small.   The procedure makes use of a punched card but substitutes manual sorting, tabulating, and computing for more elaborate automatic equipment.   It can be used for preparing payrolls, distributing labor, and preparing reports as well as for inventory and production control procedures. Figure 6-5 shows a typical key sort job card.   In this example, the card is preprinted with the operator's name, clock number, and social security number by the use of an addressograph machine.

The second mistake most frequently made in installing automatic business machine equipment is that the installation is not integrated into all phases of the clerical system.   The use of punched card tabulating equipment just to prepare a payroll and payroll checks, for example, is a poor application regardless of volume.   The time required to punch the card containing the basic data is the most expensive operation.   Once the card is punched, the more applications that can be made of it, the lower the cost per application.   A good rule of thumb to follow in this respect is that one card must be used for a minimum of three different applications to be economically justified.   For example, a punched card time ticket may be used for preparing payrolls, preparing payroll checks, distributing labor costs to accounts, preparing performance reports, direct and indirect labor cost reports by departments, and the like. Furthermore, these applications must be essential to the effective operation of the business. A very common mistake in using punched card tabulating equipment is to justify its use by preparing a number of reports of doubtful value merely to increase the number of applications.

Mention has been made of a preprinted job ticket that is used as one means of increasing the accuracy of source records. The use of preprinted job tickets is particularly advantageous when punched cards are used. The preprinting is combined with a prepunching. This not only reduces errors and increases accuracy, but equally important, it reduces the cost of keypunching considerably because the cards can be gang-punched.

In other cases, all the job tickets for a particular item or assembly can be prepunched automatically by the use of a master deck of punched cards, one deck for each item or assembly. This greatly reduces the rather expensive operation of keypunching cards individually. It is more suited to mass-production operations, however, than it is to jobbing operations.

Federal and State Laws Affecting Timekeeping and Payroll Procedures. Obviously, an essential requirement of any timekeeping and payroll procedure is that it satisfy Federal and state laws.

One part of this problem has already been discussed in connection with a record of the hours worked. Although the Federal Wage and Hour Law does not specifically require the use of time clocks, it does require a record of the hours worked. The most practical way of maintaining this record is by the use of time clocks and time cards. Other provisions of this law that pertain to the keeping of records also influence the work of the timekeeping and payroll department.

One section of this law establishes the minimum wage that can be legally paid to an employee. Another section states that a worker must be paid at the rate of at least one and a half times his regular rate of pay for all hours worked over forty in a given week. When an incentive plan is in effect, his regular rate of pay is his average earned rate, that is, all earnings before overtime premium is calculated, divided by the hours worked. This act does not stipulate the payment of overtime for work in excess of eight hours for any one day.

Numerous other provisions of this law also dictate to a certain extent the timekeeping and payroll procedures that must be followed.

The Walsh-Healy Act is applicable to certain government contracts and specifies that overtime premiums must be paid on an eight-hour day or forty-hour week, whichever is the greatest. In other words, a comparison must be made between the overtime hours on a daily basis and the overtime hours on a weekly basis. The act does not specify the payment of overtime for Saturday and Sunday, as such, but only when hours worked previous to those days exceed forty.

State laws must also be considered. For example, Pennsylvania has a Women's Labor Law that governs such things as the maximum number of hours per week and the periods during a day when a woman can be legally employed.

Copies of Federal laws and interpretive bulletins can be obtained through any of the offices of the U.S. Department of Labor. Information concerning the laws of specific states must, however, be obtained from the department of the state under consideration that handles laws pertaining to labor and industry.

In addition to satisfying Federal and state laws, it is also necessary that timekeeping and payroll procedures comply with the provisions of collective bargaining agreements and company policies.

Types of Forms. An individual job card that is particularly well suited to job cost accounting procedures is illustrated in Figure 6-6. This is a two-part form. The first part is illustrated. It has a carbon backing which imprints on the second part. The second part of this form is identical in makeup to the first part, but it is printed on cardboard stock which is perforated at each double line separating the information pertaining to each job.

The first part of the form is a man-type job ticket, that is, it summarizes all

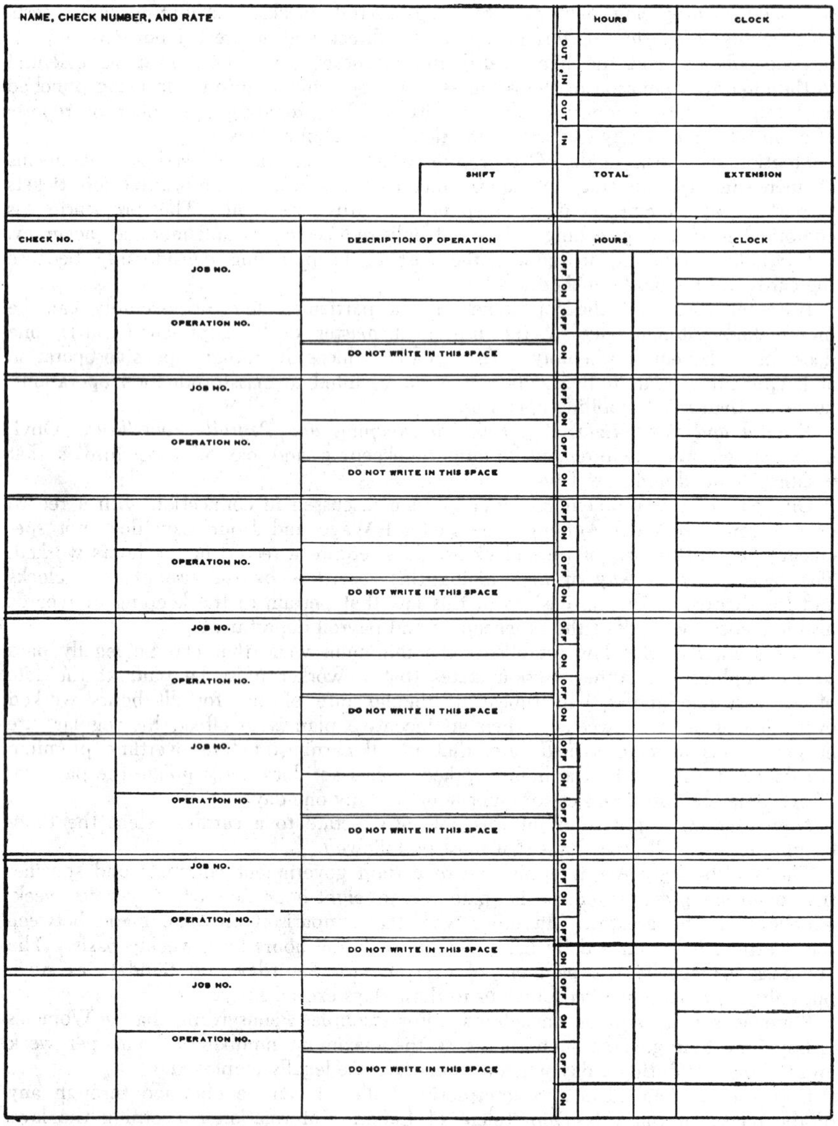

FIG. 6-6.   A job card form designed for manual clerical procedures that is particularly well designed for easy labor distribution.

the work performed by one man for a day. This part of the form is used by the payroll department to prepare the payroll. All the information pertaining to hours worked is thus consolidated on the form. Part one is used as a permanent record of hours worked.

The second part of the form is separated from the first by the payroll department after all information has been checked for completeness. The second part is then forwarded to the accounting department where it is used to distribute labor to the various job orders. This labor distribution is performed quite simply merely by tearing the card along the perforated lines. This separates the card into a number of subjob cards, each of which contains all the information pertaining to a single job. The subjob cards are then sorted manually by job numbers and posted to a job order summary sheet.

This card, in effect, combines the advantages of the man and the individual job cards. For ease in payroll computations, the first part summarizes all information pertaining to the man. The second part permits the separation of the man card into individual job cards to facilitate the distribution of labor to job orders.

Also notice that this card combines the attendance clock card and the job card. The upper right-hand corner of the card is the attendance clock card. It provides space for punching in and out at the beginning and end of each shift and again during the lunch hour. The upper left-hand portion has been designed to be used with an addressograph plate. This permits the card to be preprinted with all the information pertaining to the man.

A typical job card for incentive wage payment is illustrated by Figure 6-7. Notice that this card not only provides space for all the information required for mechanical systems, but it also provides space for payroll extensions and earnings calculations. The upper portion of the card is used for operations performed on incentive. The lower portion is used for daywork operations. This type of card can be used to good advantage when a standard cost system of accounting is used, because it eliminates the necessity of distributing labor costs to individual job orders. If a job order cost system is used and if a large number of job orders are continuously in process, this type of job card is not practical.

Whenever a group is formed for the purpose of wage incentive payment or cost control, it is usually desirable to provide a special job card. A typical group job card (Figure 6-8) provides space for the names and clock numbers of all members of the group and their particular job in the group. Because a separate record is maintained for the type and number of pieces produced by the group, no space is provided for this information on this particular card. If separate records are not maintained on piece counts, then a form similar to that shown in Figure

DAILY TIME SLIP

Dept.________ Shift________ Date________ Employee No.________

TIME	PRODUCTION ORDERS (INCENTIVE)													
	HRS. WKD.	ORDER NO.	DRAWING NO.	ITEM NO	OPR. NO.	OPERATION	TOTAL PIECES	SET UP	EA. PC. TIME	HOURS EARNED	EXTRA SET UP	EXTRA EA. PC.	CHG.	EX. HRS. EARNED

DAY WORK				EARNINGS					
	HRS. WKD.	CHARGE	OPERATION	HRS. WORKED	HRS. EARNED	HRS. EARN. + 10%	% EFF.	RATE	EARNINGS
OPERATOR'S NAME			MALE FEMALE	TOTAL HRS. DAY WORK					

APPROVED FOREMAN________ EXTENSION O K. PAYROLL________ TOTAL EARNINGS

FIG. 6-7. Form combining the conventional job card with payroll calculation sheet.

GROUP SHEET

DATE_____ SHIFT _____ FURN._____ DEP'T._____

GROUP NO._____ MACH. NO._____OPERATION_____

NO.	NAME	JOB	HRS. IN GROUP	HRS. EARNED	RATE	EARNINGS

NO. OF CARDS		HRS. ON LEARN		X	=		
		LEARN. JOB		LEARN. TIME		GROUP	
D. W.		TOT. HRS.		LEARN FACTOR		EFFICIENCY	
EARNED		JOB TIME		OP. LEARN.			
		NET		X	=	_____%	
TOTAL				TOTAL			

MISCELLANEOUS EXPENSE

CHGE.	OPERATION	NO.	PCS.	PER PC.	TOTAL	RATE	AMOUNT

FOREMAN _____TIME STUDY_____PAY ROLL_____

FIG. 6-8.   A job card form specially designed for employees working on group operations.

6-9 is used. This provides adequate space for the item or part number, the operation performed, and the quantity produced.

Although all the job cards shown here are designed to be used on a daily basis, that is, one card for each day of the week, they can also be designed to be used on a weekly basis. If the same individual or group works on the same part and operation for several days, a weekly card is practicable. This assumes, of course, that if an incentive system is in use, it is based on a weekly pay period. If the day or job is the period of payment, it is essential that a daily or individual job card be used.

A form used for preparing a payroll manually is shown in Figure 6-10. This particular form is designed for a payroll period of two weeks. Information pertaining to the employee's name, clock number, and the like is entered on the form by means of an addressograph plate. This same plate is also used for preprinting job tickets. The several horizontal lines provided for each employee can be used to summarize the hours worked when more than one job ticket is used for any one day. One horizontal line is also used for tabulating overtime hours. The

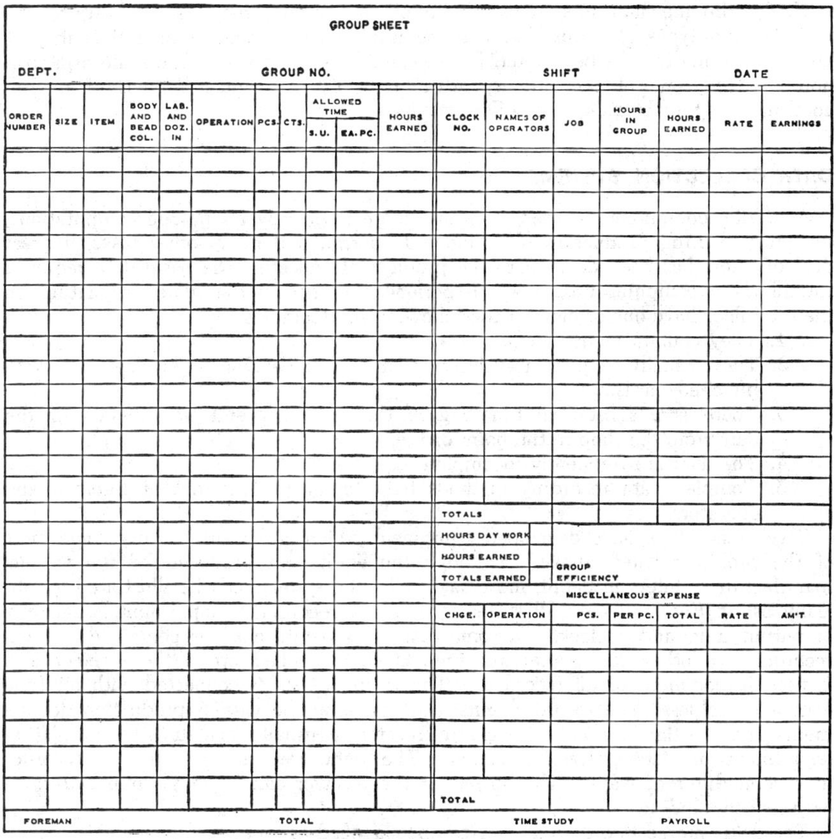

GROUP SHEET																	
DEPT.					GROUP NO.						SHIFT						DATE

Fig. 6-9.   A group job card designed for recording piece counts as well as operator's time and earnings.

deduction columns on the right-hand side of the payroll sheet are so designed that they match the deduction stub on the payroll check and the form for the employee's earnings record. By the use of a mechanical alignment board, it is possible to produce the information required on the payroll sheets, the deduction stub of the payroll check, and the employee's earnings record form in one writing.

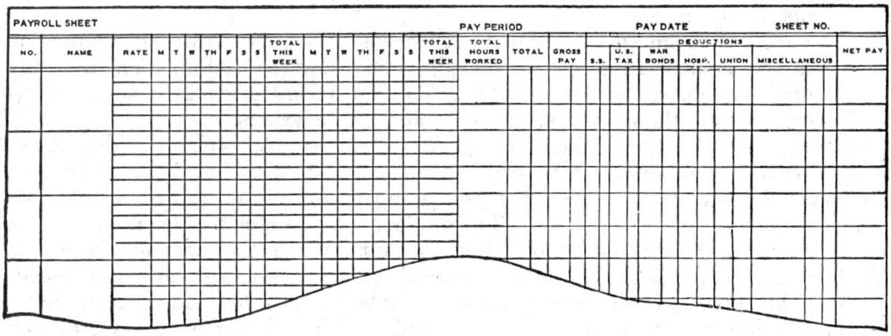

Fig. 6-10.   A payroll summary sheet designed for manual clerical systems.

Unlike the mechanical systems of payroll accounting, there is no limit to the varieties and types of forms that may be used where a manual payroll is in use. Usually, no one form is best suited for all conditions, even within the same organization. Indeed, one of the primary advantages of the manual systems of payroll accounting is the versatility they offer.

DATA COLLECTION SYSTEMS

As digital computers are used more and more frequently for payroll computations, inventory control, production planning and control, and many other tasks, a great deal of effort has been devoted to the problem of collecting the basic data required for all these computer programs. Experience has shown that data originating on the shop floor have one or more of the following shortcomings:
1. They contain many errors.
2. They usually require transcription from handwritten information to keypunched cards.
3. There is a substantial time lag in the recording and transmission of the data from the shop to the point of use.
4. The data are frequently incomplete.
5. Complex data recording methods tend to make clerks out of foremen and operators.

Several suppliers have developed equipment which is designed to overcome most of the problems listed above. Although the equipment produced by the various manufacturers differs in detail, most data collection systems can be illustrated graphically as by Figure 6-11. All systems contain a recorder or input unit, a receiver or output unit, and a clock. In some cases, a separate clock is contained in each recorder. In other cases, a master time clock is hooked up to the receiver and is used in common for all recorders. The entire system is connected with multiple conductor cables. The recorder units are located in the various production departments just like time clocks. The central receiver is usually located in the tabulating department or data processing center. The data are entered into the recorder in several different ways. The output of the receiver can be either punched cards or punched tape.

The data entered into the recorder can be classified as follows:
1. Fixed station identification. This is a station identification number or cost center code. It is recorded automatically with each transmission to identify the transmitting station.
2. Man information. This is usually the man's clock number, but it might also include other information such as wage rate classification. In many units, this information is placed on a coded identification badge which records automatically when inserted into the recorder.
3. Fixed job information. This includes such things as the job order number, the operation number, and standard time. This information is entered either manually or by means of a tabulating job card which is inserted into a slot in the recorder.
4. Variable job information and miscellaneous. This includes such information as the number of pieces produced, scrap or rework information, and any other variable data desired. These data are recorded manually by the operator through the use of dials or other numeric indicators.

In a typical installation, when the operator receives a job from the dispatcher, he also receives a tabulating card which contains all the fixed job information—the job order, the operation number, the time standard for the operation, the number of pieces on the order, and so on. In addition, the operator has in his possession an identification badge with tabulating punches. When the operator finishes a job, he counts the number of pieces produced and the number of pieces scrapped; walks to his departmental or cost center recorder; inserts his badge; then places the job tabulating card in the proper slot; and sets the number of pieces produced

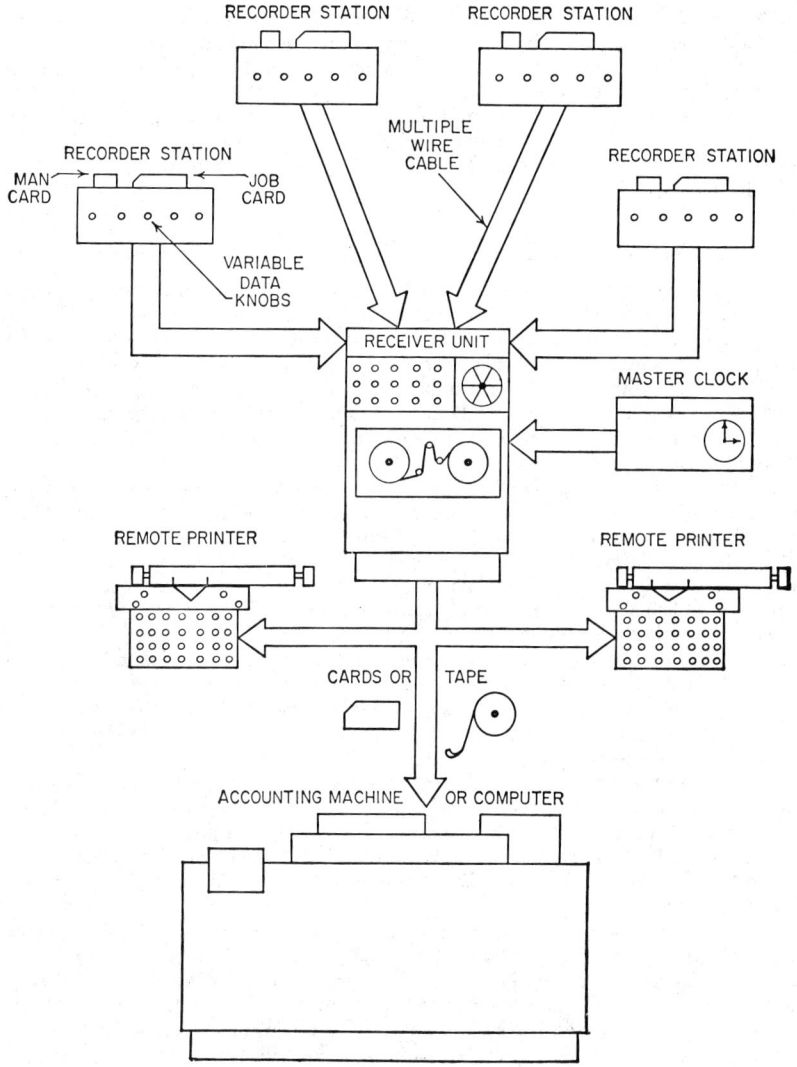

Fig. 6-11.   Graphic representation of an electronic data collection system.

and the number of pieces scrapped on the appropriately labeled dials. When this has been done, he presses the transmit button which automatically transmits all this information, plus the department or cost center code, to the central receiver. The central receiver also adds the time at which the transmission is received.

All this information is automatically punched on either cards or tape. This output is then used as the input data to either a tabulating system or a computer program. In the case of multiple plants, the data may be transmitted by leased lines or wide-area telephone service to a central data processing center at the home office.

In some systems, one or more remote printers may be coupled to the receiver to record all or part of the transmitted data. For example, one printer may be located in the production control office where it prints out only that portion of the data concerned with job order number, operation number, and pieces produced. In this way, the production control department can be informed of the exact status of every operation on every job as soon as the operation is completed. Other modifications of this basic system can be used to report the movement, receipt, or issue of materials or special tooling. Maintenance and other indirect labor personnel can make use of the same system to report the completion of repair orders and the like.

PERFORMANCE CONTROLS

Scientific management has frequently been defined as being based on measurement plus control. Although this is probably the shortest definition ever made of the complex thing generally referred to as "management," it is probably the most definitive. Before control can take place, it is necessary to have some system of measurement. On the other hand, measurement in itself is of little value unless it is used as the basis for control. One of the primary functions of any industrial engineering group is to provide management with the means for measuring and controlling labor costs.

A large portion of this Handbook is concerned with the techniques and procedures employed in measuring the effectiveness of labor. Reading the various chapters will disclose the immense amount of knowledge that has been accumulated on this subject. Although all the problems of work measurement techniques have not been solved, progress has been made to the point where production standards can be established with sufficient accuracy for most types of work. Work measurement is becoming less and less of a problem to the qualified engineer. The industrial engineer is now becoming more concerned with the problem of maintaining accurate standards once they are established, and how to use these standards to provide a maximum of control over labor costs. The purpose of this portion of the chapter is to discuss techniques employed in maintaining accurate standards and the use of standards for the control of labor costs.

Count Control Procedures. In any manufacturing operation, it is essential to have an accurate count of the work performed in order to obtain any measure of control. A record of the work performed is essential to any system of production control as well as to the control of labor costs. Effective count control procedures become even more important where such counts are used as the basis for wage incentive payment.

With the addition of complex cost accounting procedures, payroll accounting methods, government requirements on social security and income tax deductions, plus hospitalization and other types of deduction accounts, it is all too easy to lose track of certain fundamentals. The simplest system of payroll accounting and count control existed before the addition of the complexities noted above. For example: in a small needle trade plant in a loft in New York City, the owner employed an effective system of controls. He assembled all the pieces of cloth required to make a dozen dresses into a bundle. To the bundle he attached a ticket which gave the style and size information plus the piece rate for the dress. Upon completing their operations, the operators turned in the finished bundles with the ticket to the owner. Without further ado, he rang up the cash register, paid the operator, and placed the ticket in the cash register as his receipt for the work performed.

It is quite easy to understand why, in this establishment, a ticket was looked upon with the same respect as a payroll check, for in effect it was. An operator would no more think of altering a ticket than altering a payroll check. Because of the complexities of current accounting procedures and practices, however, count control forms are, in the minds of an operator, far removed from a payroll check. There is little reluctance, in some cases, to report incorrect or at least inaccurate information. For these reasons, it is particularly important that count control procedures be carefully developed.

In establishing a system of count control, there is something of a dilemma. On the one hand, accurate records are wanted, and on the other, they are wanted with a minimum of cost. Accurate records are not impossible to obtain. The banks do a very effective job. They do so, however, at a considerable cost. Furthermore, the things they are interested in controlling—cash, checks, notes, and the like—are easily counted. In industrial operations, however, everything under the sun from cotter pins to automobiles must be counted. Obviously, this problem is one that requires a great deal of planning and ingenuity if it is to be solved satisfactorily.

Counting Mechanisms and Procedures. The most common type of counting mechanism is the ordinary machine counter. It normally is mounted on a machine and actuated by a moving part of the machine. It is generally actuated once per machine cycle. On a punch press, for example, it is actuated by the ram of the press and records the number of press strokes. The counter dial is very much like a speedometer trip recorder in appearance. It is usually equipped with a ratchet-type knob that permits the dials to be turned backward to zero. The dial can only be advanced by tripping the trigger mechanism. With some ingenuity, and a variety of mounting brackets and adapters, this counter can be adapted to fit most machines. It has the additional advantage of being comparatively inexpensive. The counter is also used on bench or assembly operations where the operator trips the counter by hand or with a foot attachment upon the completion of a standard cycle.

Although the machine counter is inexpensive and simple to operate, it is far from being foolproof. On a punch press, for example, it will be actuated on every stroke of the press regardless of whether the press is producing parts or merely cycling. Consequently, it will record the number of press strokes rather than the number of pieces completed. A counter mounted on a machine can usually be tripped quite readily by hand without the machine cycling at all.

More elaborate variations of this basic machine counter are available for special types of installations. If more than one type of part is being counted by one operator, the machine counters can be arranged in rows or banks and operated with a typewriter-like keyboard. This same arrangement can be used to good advantage when the same part is to be counted and sorted into a number of different classifications such as size, weight, color, or the like.

Another variation of the mechanical counter is an electrically operated counter. It may, for example, be connected to the motor of a punch press. It is usually designed so that it will only be actuated by a full-current load. In this way, the counter does not operate if the press is merely cycling. Consequently, it records only productive strokes of the press. A further advantage is that the counter cannot be tripped manually. If desired, it may be located some distance from the machine itself—in the production control office, for example. This type of equipment is, of course, considerably more expensive than the simple mechanical counters.

A further modification of the electrical counter is the graph-type recorder. In addition to showing cycle times, this type of recorder can also be used to show variations in the speed of equipment or downtime against a time scale. It is frequently used on large papermaking machines, printing presses, and the like as a means of recording production and as a control over downtime and other nonproductive periods.

Another frequently used counting mechanism is the proportional scale. This mechanism is best suited to the counting of small parts, such things as nuts, bolts, rivets, and washers, that are fairly uniform in weight. The parts to be counted are placed on the platform of the scale. The scale arm is then balanced to zero by placing the necessary number of parts in a balancing pan on the scale arm. The proportion between the weight of the parts on the platform and the weight of the parts in the balancing pan can be adjusted to any desired ratio—25 to 1, 50 to 1, 100 to 1, or 1,000 to 1. The number of parts in the balancing pan multiplied by the appropriate ratio gives the number of parts on the scale platform.

This technique is not only used for counting parts between production operations, but it is also used extensively in receiving departments as a means for checking purchased parts.

Quite frequently it is more convenient to establish count controls on a unit of measure other than pieces. For example, on an automatic screw machine it may be simpler to count the number of bars of stock used for a uniform size part than it is to count the number of parts themselves. The number of bars used times the number of pieces obtained from a standard bar will give the number of parts produced. For other operations, such units of measure as coils, drums, barrels, kegs, and boxes may be used to eliminate the necessity of counting individual parts or pieces produced from these standard quantities.

Another approach to simplified counting procedures is the use of standardized containers, racks, trays, tote pans, pallet loads, and the like. By placing standard quantities in a standard container, only the containers themselves need be counted. To be accurate, however, the parts and their position in the container must be such that a quick visual inspection will reveal any partially filled containers. Where possible, the containers should be designed to hold standard quantities.

Whenever possible, counts should be recorded by someone other than the operator producing the work. They may be recorded by a move man, timekeeper, dispatcher, inspector, group leader, or the like. If practical, all parts should remain in a standardized lot. This practice permits the count of each subsequent operation to act as a check on the previous operation.

In setting up any system of count control, a good rule to follow is to design the system so that it does not become necessary to rely upon or to question the honesty of anyone who works under it. By doing so, incidents which may well become serious industrial relations problems are avoided altogether.

Efficiency Reports. Figure 6-12 shows a report prepared on individual employees' incentive rates. This report may be prepared with any type of payroll accounting procedures but is most easily compiled by punched card systems. It is not limited to an analysis of incentive rates only, but may be used with equally good results in analyzing any system of time standards. This report is a summary by operation and part number of all work performed during any given period (five months in this case).

The first group of entries covers the operation, "turn, operation no. 1," for part "no. 14936, gear stud." This report shows that this operation was performed on January 13 and 14 by employee number 1560. On January 13, he worked 4.2 hours and produced 5.0 standard hours for an efficiency of 119 percent. On January 14, he worked at an efficiency of 106 percent on the same job. On March 5, employee number 206 worked on the same job. He worked 8 hours and produced only 7.3 hours for an efficiency of 91 percent. In a similar fashion, all operations for all parts can be summarized and recorded.

The key column is the one labeled "percent efficiency." Consistently low or high percent efficiency will tend to indicate tight or loose standards. In either case, corrective action should be taken by first determining the reason for the abnormal performance. An analysis of this kind will frequently uncover unreported methods changes. It is also helpful in uncovering any general trend toward loose or tight standards on particular operations such as turning, drilling, tapping, and the like.

The two control reports illustrated by Figure 6-13 are typical of reports used for controlling performance. The top report is a daily summary of the work performed by each operator. The key column in this report is also the one labeled "percent efficiency." This figure is obtained by dividing the standard hours by the actual hours. This report provides a day-to-day analysis of the performance of each operator on each job worked.

If this type of report is to be effective, it must be prepared daily. The foreman or operator can hardly be expected to remember all the reasons for a poor performance that occurred on a job a week or two back. For this reason, individual performance reports are usually prepared today for the work performed yesterday.

ANALYSIS OF INCENTIVE RATES

DATE MO. DAY	EMPLOYEE NO.	ORDER NO.	PART NUMBER	PART DESCRIPTION	OPERATION DESCRIPTION	DEPT. NO.	MACH. NO.	PIECES FINISHED	STANDARD PER C	HOURS ACTUAL	HOURS STANDARD	EARNED	BONUS EARNINGS	% EFFICIENCY

FIG. 6-12. Typical payroll analysis report prepared by punched card accounting machines.

4-93

DAILY PERFORMANCE RECORD

14 MACHINE SHOP

PART NUMBER	PART DESCRIPTION	OPERATION DESCRIPTION	MACH. NO.	PIECES NO.	PIECES FINISHED	HOURS ACTUAL	O.T. PREM.	STANDARD	EARNED	% EFFICIENCY	DEPT.	EMPLOYEE NUMBER
14396	GEAR STUD	TURN	1	6	7 5	2:8		3:0	:2	107	14	206
14940	NOZZLE HEAD	TURN	3	6	2 6 0	5:2		6:4	1:2	123	14	206
9325	COLLAR	GRIND O DIAM	2	1 0	9 0	3:0		4:0	1:0	133	14	342
10190	BUSHING	GRIND O DIAM	4	1 0	1 4 0	3:5		3:8	:3	109	14	342
4672	VALVE SOCKET	ROUGH BORE	1 0	4	9 4	8:0		9:4	1:4	118	14	518
36125	SUPPORT ROD	ROUGH GRIND	7	2 0	5 4	4:0		3:6	:4 -	90	14	615
36125	SUPPORT ROD	FINISH GRIND	8	2 0	5 0	6:0	1:0	6:2	:2	103	14	615
	CARING	DRIL		9	6 6	2:0			:2	110	14	703
15130	MOTOR		UM	2	1 2 2	8:0		1 0:2	2:2	1 2 8	14	20
17160	TOP COVER	DRILL	2	1 8	9 2	5:0		4:6	:4 -	92	14	2183
18310	PUSH ROD	DRILL	3	1 8	7 2	3:0		2:4	:6 -	80	14	2183
5895	ROLLER	ROUGH GRIND	8	1	5 0	4:6		5:0	:4	109	14	2470
16721	ROLLER	FINISH GRIND	9	1	4 8	3:4		4:0	:6	118	14	2470
8673	SUPPORT	DRILL	1 4		1 8 6	8:0		9:3	1:3	116	14	2492
7346	BUSHING	DRILL	3	1 3	1 3 2	8:0		8:8	:8	110	14	2896
9462	CLEVIS PIN	DRILL	5	8	1 8 2	8:0		9:1	1:1	114	14	3720
						13 1:7	2:0	14 5:9	14:2			
		WAITING WORK	9 8	1 0		1:5					14	342
		MACHINE DOWN	9 9	9		3:0					14	703
		MACHINE DOWN	9 9	3		3:8					14	1304
		MACHINE DOWN	9 9	1 2		1:7					14	6123
						1 0:0						

WEEKLY STATEMENT OF BONUS EARNINGS

DEPT. NO.	EMPLOYEE NO.	NAME OF EMPLOYEE		HOURLY RATE REGULAR	BONUS
1 4	2 0 6	W V ASTUR		1,200	.600

PART NUMBER	OPER. NO.	PIECES FINISHED	DATE MO.	DATE DAY	REGULAR HOURS	REGULAR EARNINGS	O.T. PREM HOURS	STANDARD HOURS	EARNED HOURS	% EFFICIENCY	BONUS EARNINGS
14396	1	7 5	5	2 5	2:8	3:3 6		3:0	:2	107	:1 2
14940	3	2 6 0	5	2 5	5:2	6:2 4		6:4	1:2	123	:7 2
14396	1	1 0 0	5	2 6	3:0	3:6 0		4:0	1:0	133	:6 0
		6 2									
28832		3	5	2 8	5:0	6:0 0		1:6	1:2	1 3 2	:9 0
17755	3	7 8	5	2 9	4:0	4:8 0		5:2	1:2	130	:7 2
36362	2	5 5	5	2 9	4:0	4:8 0		5:5	1:5	138	:9 0
					4 4:0 ✪	5 2:8 0 ✪	2:0 ✪	5 4:7 ✪	1 1:7 ✪		7:0 2 ✪

Fig. 6-13. Control reports on departmental and individual performance records prepared by punched card accounting machines.

This particular report will be of little or no value unless it is used as the basis for corrective action. This action is usually initiated by the use of a form similar to the one shown in Figure 6-14. It is made up by the payroll department for all individuals that failed to meet standard performance the previous day.

It is then forwarded to the foreman who, in consultation with the operator concerned, fills out the reasons for the falldown. It is then forwarded through the industrial engineering department to the appropriate supervisor. The "Remarks" space is reserved for the use of the industrial engineering department.

The primary purpose of the falldown report is to uncover conditions or situations that are causing substandard performance and to initiate corrective action designed to correct the substandard condition. If the condition happens to be poor effort on the part of the operator, then it should result in corrective action along this line. Another report for controlling individual performance, however, is illustrated in the lower portion of Figure 6-13. This weekly report tends to iron out the variations that occur in performance from one job to the next. It provides a good picture of individual performance on a weekly or pay period basis and can be used to initiate corrective action whenever an operator shows consistently poor performance.

The reports and controls required for the effective administration of any standards program need not require a great deal of clerical effort if they are properly integrated into the payroll procedures. Most time standard programs include such a system of reports and controls when they are first installed. One of two things frequently happens, however, after the system has been in effect for a year or two. Either the controls and reports are eliminated as an unnecessary clerical operation, or they are continued but not used for control purposes. In either event, the normal result is

```
┌─────────────────────────────────────────────────────────────────┐
│                      FALLDOWN REPORT                              │
├─────────────────────────────────────────────────────────────────┤
│  DEPT.              OPERATION                        DATE          │
│  OPERATOR OR GROUP                                                │
│  HOURS EARNED                                                     │
│  HOURS WORKED                                                     │
│  REASON                                                           │
│                                                                   │
│                                                                   │
│                                                                   │
│                                              FOREMAN              │
│  REMARKS                                                          │
│                                                                   │
│                                                                   │
└─────────────────────────────────────────────────────────────────┘
```

FIG. 6-14.   A form used to notify a supervisor of substandard individual performance.

that the standards are so poorly maintained that they are worse than no standards at all.   The company is then faced with the problem of overcoming high labor costs and completely rebuilding their standards program.   This is certainly one place where an ounce of prevention is worth a pound of cure.

FILING SYSTEMS

An industrial engineering department, regardless of size, must maintain certain files and records.   They may vary in complexity from a pocket notebook and a desk drawer to a complete battery of file cabinets for correspondence, index files, special blueprint files, and the like.   Because there are several excellent texts on the subject, the purpose of the following pages is not to discuss filing systems from a comprehensive standpoint, but to discuss them as they apply to an industrial engineering department.

Once again the problem arises of discussing procedures that will vary widely from one organization to the next.   This can be minimized, however, by first discussing the types of filing equipment commonly used in practically all organizations.   Then the types of materials that may be filed by an industrial engineering group will be listed, followed by the various methods and procedures employed in filing them.

Types of Files.   The wide variety of forms, records, and reports maintained by an industrial engineering department makes it extremely difficult to develop a uniform system of files and indexing procedures.   After looking at the types of things that must be filed, better understanding is gained of how to file each item and how to index it for easy reference.   The following items are generally maintained in the files of an industrial engineering department:

1. Individual time studies or predetermined elemental motion time studies
2. Production or time standards on individual operations
3. Job instruction cards
4. Process or route sheets
5. Cost cards
6. Request for time studies
7. Rate authorization sheets
8. Time formula worksheets

   9. Special allowance authorizations
   10. Blueprints
   11. Time formula reports
   12. Falldown reports
   13. Cost reduction reports
   14. Physical data on plant equipment
   15. Reports on special projects
   16. Standard daily, weekly, and monthly reports
   17. Occupational job rates and classifications
   18. General correspondence and memoranda
   19. Plant layout drawings
   20. Miscellaneous files

The above list is by no means an all-inclusive list. There are a host of other forms such as flow process charts, operation process charts, operation analysis forms, and the like which must also be filed. Generally, however, these forms are in the nature of worksheets and are normally filed with the material listed under one of the headings above rather than in separate files. For example, the process charts and operation analysis forms that are prepared in connection with a cost reduction project are filed with all the other papers relating to that project. All these things are covered in item 13, cost reduction reports. These same forms may also be prepared in connection with a plant layout. In this case, they are filed with all the other worksheets relating to the layout involved.

The principal indexing classification for industrial engineering files is the part number, drawing number, or catalog number. This is the number by which a part or assembly is referred to by the shop. Filing materials that can be identified to a specific piece or part are filed in the appropriate part number folder. All industrial time studies on specific parts or assemblies are filed by part number. Consequently, any one folder may have several time studies filed in it—one or more for each operation on the part or assembly. In a similar fashion, all other forms pertaining to a single part or assembly are filed in the same folder. Such forms as instruction cards, process sheets, cost cards, requests for time studies, rate authorization sheets, time formula worksheets, special allowance authorizations, and blueprints are filed in a single part number folder. In this way, any single item of information is easily located and all information pertaining to any one part is available in one folder.

When forms or papers cannot be specifically identified to a single part number, then they should not be filed in the part number file. A typical example is a time formula report. The formula may cover all operations on a No. 3 Warner and Swasey turret lathe; it cannot be specifically identified to any single part. There are several ways in which information of this sort may be indexed. One method is to assign a code number to each formula, such as 15-6. In this case, the "15" indicates the department number to which the formula applies. The "6" indicates that the formula is the sixth one prepared for department 15. Another option is to index the formulas by type and size and arrange the file alphabetically with this classification. In this case, the index may read something like this:

   Lathe, engine, Lodge and Shipley 12 inches
   Lathe, engine, South Bend 4 inches
   Lathe, turret, Warner and Swasey No. 2
   Lathe, turret, Warner and Swasey No. 3
   Lathe, turret, Warner and Swasey No. 3A
   Milling machine, horizontal, Cincinnati No. 5
   Milling machine, horizontal, Kearney and Trecker No. 3
   Milling machine, vertical, Bridgeport No. 3

If the formula file is voluminous, it can be further subdivided into machine shop, foundry, plate shop, assembly shop, and other classifications. This same classification procedure can also be used for other types of filing material such as cost reduction reports, reports on special projects, and the like.

Another convenient filing classification is department number. A great deal of filing material can be specifically identified by department numbers. Plant layout worksheets, sketches, and drawings are good examples.

Clock card numbers provide another means of indexing for any material that is identified by individual operators. Falldown reports by individual operators are a good example. In cases such as this, it may be desirable to file more than one copy of a record or form for cross-reference purposes. The purpose of a falldown report is to keep a record of the operators that failed to meet standard performance for the operations involved. The falldown may be due to poor performance on the part of the operator or to a tight standard for the operation. By filing two copies of the same form—one by clock number and the other by part number—the true reason for the falldown will soon become apparent. A large accumulation of reports in one file or the other will indicate the true reason for the falldown.

Occasionally, special files or folders are required to file material properly. Physical data on plant equipment are an example. Data of this sort require a great deal of time to collect and organize. Consequently, they should not be subject to the loss or misuse that may occur if they are merely filed away in a common file folder. These data are usually assembled and bound in separate hard-cover post or ring binders with appropriate tab indexes.

Periodic reports—daily, weekly, and monthly—are usually filed separately, one folder for each report. The reports should be arranged chronologically within the folder with the latest report on top.

General correspondence and memoranda are best filed alphabetically by company name for outside correspondence and by individual name for internal correspondence. If desired, new file folders can be prepared each year with a distinctive color tab for each year. In this way, the folders do not become cluttered up with old correspondence. At the same time, any correspondence can be readily located if the approximate date is known.

At one time or another, everyone has had the exasperating experience of hunting for an important piece of paper that has been misfiled or misplaced. Only one experience of this sort is sufficient to be convincing of the benefits to be offered by an efficient filing system. Unimportant papers become extremely important when they cannot be found. To be effective, a filing system must be carefully planned and strictly adhered to. Wherever possible, it is desirable to have only one individual responsible for all files—not only for the filing but for the removal of papers as well. If everyone is free to remove material from the files without some system of accounting for it, it will be extremely difficult to maintain a truly effective filing system.

ADMINISTRATION

A major problem faced by any chief industrial engineer is, "What can I do to ensure the successful performance of my group?" Throughout the pages of this Handbook, there are various techniques and procedures that can be employed to good advantage in improving the technical competence of the industrial engineering group. In the final analysis, however, the success of any group will depend primarily on the people in it. The surest road to success, therefore, is to ensure that the group is composed of an adequate number of properly qualified people. The first question that arises is "What is an adequate staff?" The second then is, "How can I obtain properly qualified personnel?" These questions will be discussed one at a time.

Size of Staff. What is an adequate staff to carry on the routine work of the industrial engineering department? The answer to this depends upon three factors:

1. The responsibilities of the industrial engineering group
2. The size of the organization
3. The nature of the organization

Although there are a number of other factors that have some influence on the size of an industrial engineering department, the three listed above are the essential ones. Before going any further, it will be necessary first to define the responsibilities of the industrial engineering group. Even though these responsibilities vary considerably from one organization to another, most industrial engineering groups are charged with the following major responsibilities:

1. Time standards
2. Methods
3. Plant layout
4. Labor controls
5. Wage payment

The importance of the last factor will depend on whether or not wage incentives are employed. For purposes of discussion, it is assumed that they are.

Although these five areas of responsibility are those that most industrial engineering departments are assigned, there are other closely related areas in which industrial engineers must have a working knowledge. In the small plant, for instance—where a high degree of specialization is not possible or even desirable—safety, operator training, budgetary and cost control, inventory control, and so on may actually be the responsibility of the industrial engineering department. In many instances, industrial engineering departments are called upon to set up the techniques and procedures that will be followed by those who do work in these areas.

The size of the organization, for the purpose here, is measured best in terms of the number of hourly paid operators, both direct and indirect. The work of the industrial engineering group is primarily concerned with operating personnel. Therefore this figure provides a convenient index.

The complexity of the industrial engineering responsibilities is determined by the nature of the organization's operations. For this purpose, all operations can be classified into one of three categories:

1. Highly repetitive
2. Fairly repetitive
3. Job lots

The highly repetitive operations are typified by the mass-production industries: automobiles, refrigerators, sewing machines, textiles, and the like. These classifications, however, are not as descriptive as the ones listed above, because custom-built automobiles, for example, fall in the job lot classification.

"Fairly repetitive" is the description that generally fits most manufacturing operations. Kitchen cabinets are a good example. Although they are mass-produced by the larger manufacturers, there are so many style and size variations to contend with that the operation is best described as fairly repetitive.

Job lot operations are typified by the jobbing machine shop. Seldom is work produced for stock. It is normally produced to meet specific customer specifications and frequently jobs do not repeat. Lot sizes are comparatively small—from 1 to 100 pieces.

The volume of methods work that must be carried on in a company varies almost directly with the number of different production jobs that must be performed during a given period. This is also true of the number of standards that must be set. With this in mind, it is logical to expect that the amount of industrial engineering work to be done is much greater in a job shop than it is in a company with highly repetitive operations, even though the two companies may have the same number of employees.

Now the question, "What is an adequate staff for carrying on routine activities?" can be answered. The following table shows the number of hourly employees normally handled by one industrial engineer for each of the above classifications.

Classification	Number of hourly employees per industrial engineer
Highly repetitive...............	200–300
Fairly repetitive...............	100–200
Job lots.....................	50–100

The figures listed above are approximations only and should be used merely as rules of thumb. They have been developed from personal observations of many organizations in various parts of the country and are provided here merely to supply some tangible rule of thumb answer to the question, "What is an adequate staff for carrying on routine activities?"

If the industrial engineeering department has duties other than the routine maintenance of methods, standards, and wage incentives, it may well be considerably larger than the above figures indicate. If it becomes necessary to develop standards data or time formulas, it will require more manpower during the development stage than when the standards data are available for application. If the company initiates an intensive cost reduction program, the size of the industrial engineeering staff should probably be expanded; or if a program of retooling or extensive methods research is introduced, it will require additional manpower.

In determining the size of an adequate total staff, therefore, the work which is to be done by the staff should be considered. In one plant of 500 people, a staff of 5 men might be too large; in another of the same size, a staff of 15 might be inadequate—it depends upon the work to be done. As long as an industrial engineer can return a satisfactory profit on the money invested in his salary, it is good business to employ him. Thus, management's concern should be with the effective utilization of its industrial engineers rather than with the total number employed.

Qualifications of Personnel. "How can I obtain qualified personnel?" is the second question that was raised in connection with the problem of ensuring the successful performance of an industrial engineering group. Before this question can be answered, it will be necessary to determine what is meant by the adjective "qualified."

The term "qualified" usually refers to two entirely different things at the same time. First, it refers to the individual's technical qualifications—his training, education, experience, and so forth. Second, it refers to the individual's personal qualifications—his temperament, intelligence, interests, and aptitudes. To obtain qualified personnel, it is important to keep these two different sets of qualifications clearly in mind. Generally speaking, the first group of qualifications are all things that can be acquired. The second group of qualifications are an inherent part of the individual. There is very little that can be done to change them. The second group of qualifications will largely determine such things as: will the individual get along with people; can he sell his ideas to operators, foremen, or management; will he be able to see the forest for the trees; does he have future growth potential; does he have rigidly fixed ideas; and a host of other questions.

In selecting industrial engineers, it is essential to understand fully the importance of this second group of qualifications. By the very nature of their work, industrial engineers must work very closely with other people at all levels of the organization. Unless they have the basic personal qualifications required to get along with other people and to sell their ideas or courses of action, they will be of little value, regardless of their technical qualifications. Industrial engineering is at least 50 percent human relations. Qualified people therefore must satisfy the second group of qualifications. The first group of qualifications can always be acquired by training, experience, and the like, which can be provided. However, a man cannot be provided with intelligence, temperament, or aptitude.

The first step, therefore, is to obtain individuals with satisfactory personal qualifications. If these individuals also have the necessary technical qualifications, so much the better. If not, they can always be provided by means of training and experience.

Selection of Industrial Engineering Personnel. Although personnel testing has not yet reached the state of development where it can be relied upon 100 percent, it is sheer folly not to take advantage of this technique in selecting personnel. The results of psychological tests should be looked at in the light of impressions obtained from interviews, checks on previous employment, and other nontest data.

Training. It is not always possible or even desirable to employ fully experienced industrial engineers as members of an industrial engineering department. First, some natural promotional sequence must be established within the department to provide an incentive for advancement. To do this, some junior or trainee engineers must be employed. Second, the industrial engineering department is frequently used as a training ground for potential supervisory personnel. This practice should be encouraged because it provides the department with the opportunity of thoroughly indoctrinating future supervisors with the various functions of the department. In addition, it provides the future supervisor with experience that will be valuable to him in supervisory positions. Finally, the industrial engineering department is frequently called upon to train or at least indoctrinate union personnel in the elements of industrial engineering, particularly those pertaining to the establishment of production standards. For these reasons, training is an important function of any industrial engineering department.

The question now arises as to how to go about training a trainee or junior industrial engineer. Fortunately, there are a number of alternatives.

A number of universities and management consulting firms offer courses in the basic principles and fundamentals of industrial engineering. The new men can be sent to one of these courses to receive training in the fundamentals of the subject away from the day-to-day pressures and interruptions encountered when training a man on the job. This option has the additional advantage of reducing the work load on the normal departmental personnel. A typical course outline of a program of this sort consists of the following subjects:

1. Introduction to the principles of industrial engineering
2. Aims, fundamentals, and development of industrial engineering
3. Use and construction of process charts
4. Operation analysis
5. Principles and application of motion study
6. Principles and application of time study and/or predetermined elemental motion times
7. Work sampling studies
8. Determining allowances
9. Principles and application of time formula construction
10. Wage payment
11. Plant layout

A second option is to conduct the training at the plant. If this course is selected, several alternatives are available. A number of home study courses are available in the field of industrial engineering. Because home study courses are designed to require little or no supervision, this option considerably reduces the burden of training new men. A second alternative is to prepare a list of recommended reading materials and to supervise closely the progress and retentive powers of the trainee. The reading course can be supplemented by appropriate practical assignments in the shop under the supervision of an experienced engineer. A third alternative is to assign the responsibility of training all new men to one experienced engineer. In larger organizations, his responsibility can be expanded to include such things as the preparation of standard course outlines, company manuals, standardized exercises, examinations, and the like.

A very effective method of training industrial engineers is through the use of multimedia training programs. These programs employ such techniques as programmed instruction, audiovisual presentations, motion pictures, and learn-by-doing sessions. The technique that will best teach a particular subject is used throughout. As a result, training to the professional level of competence is achieved. Multimedia programs are available on a lease basis for in-plant presentation, or they may be obtained from training centers located in various parts of the country.

This Handbook provides a wealth of information on all phases of industrial engineering. It can be used to excellent advantage as a master textbook for any industrial engineering training program. Because it covers all subjects pertaining to industrial engineering, both basic techniques and more advanced or related techniques and procedures, it may be used as the basis for training new engineers in fundamentals, as well as older engineers in more advanced techniques.

If the man to be trained is a recent college graduate without practical industrial experience, he will need some shop experience before starting his training. He should be able to use shop terms and talk a common language with the operators to gain their acceptance. For this reason, it is better to give a recent college man a few months on the shop floor where he can associate with operators, as one of them, before starting his training as an industrial engineer.

Regardless of which option is selected, all training should receive the close attention of the chief industrial engineer. The effectiveness of the training program will eventually determine the effectiveness of the department as a whole.

Work Assignments. The assignment of work presents few problems to the small department. Where two or three engineers constitute the entire department, it is almost essential that they be completely familiar with all the functions of the department and the various departments of the organization. In a word, they must be interchangeable.

As the size of an industrial engineering department increases, however, a certain amount of specialization among engineers is almost inevitable. Some engineers, because of their background, training, and experience, will specialize in machine shop work. Others, for the same reasons, will specialize in foundry operations, and so on. In addition, there is also the tendency for engineers to specialize by function. Some will become expert on material handling, others on work measurement problems, still others on plant layout or methods, and so on. Although perhaps there is some justification in the very large organizations for specialization of this sort, it generally should be discouraged. Engineers should be encouraged to be generalists rather than specialists. There are a number of reasons supporting this viewpoint.

Having a number of specialists in an industrial engineering department is comparable to having a number of special-purpose tools in a jobbing machine shop. It severely limits the flexibility of operation.

If an organization is planning a new plant construction, a number of engineers will be required with experience in plant layout. If emphasis is suddenly shifted to the establishment of standards on indirect labor operations, a group of men with experience in work measurement techniques will be needed. An overly strong reliance on specialists for these particular functions will cause narrowness and inflexibility. Finally, transfers and terminations present difficult replacement problems if the use of specialists is leaned upon too strongly.

Besides the disadvantage of inflexibility, there is a question as to the efficiency of specialists in the industrial engineering field. Very few assignments can be handled by a single specialist. A plant layout assignment, for example, cannot be performed completely by a specialist in layout alone. It is impossible to plan any layout adequately without giving serious consideration to such factors as material handling equipment, methods, workplace layouts, production standards, and the like. By relying too strongly on specialists, four or five specialists may have to be assigned to a project that could be better coordinated and completed by one generalist.

Another approach to the assignment of work is to assign one man to one or more departments and to hold him responsible for all industrial engineering work required by these departments. This approach has a number of advantages and is quite generally employed. The engineer can become better acquainted with the department's operations, its problems, and most of all, the people in the department—both operators and supervisors. From an organization standpoint, it has the advantage of providing clean-cut responsibilities. One man is clearly responsible for all industrial engineering activities in the departments assigned to him. There is the additional advantage of providing the engineer with the opportunity of applying all industrial engineering techniques rather than specializing in only one or two areas.

This approach to the assignment of work and the organization of a department is not free of disadvantages. Some of the disadvantages common to specialization by function are also present in this approach of specializing by departments or areas. Replacement of personnel becomes a major problem. When only one man is familiar with a department or area, he tends to become indispensable. Specializing by department also results in narrowing an engineer's perspective. It tends to encourage the "our work is different" attitude. This, in turn, acts as a sedative on the "questioning attitude" which is so essential to good industrial engineering work.

One way of reducing these disadvantages and at the same time retaining most of the advantages, is to assign two or three men to several departments or areas. One individual is assigned the responsibility of the group's performance. All individuals work in all the departments assigned to the group. This arrangement provides an additional advantage of flexibility. All individuals can be assigned to work in one particular department, if necessary, to complete special projects or assignments. Very little lost motion is encountered when such a situation does arise because all men are familiar with all departments. This also permits the removal of one or more men from any one group for special assignments outside of their normal area with a minimum of disruption to everyday routines.

In the final analysis, of course, the assignment of work to the staff must be tailored to meet the specific conditions faced by a specific organization. The primary purpose of this discussion is to present the various possibilities and their advantages and disadvantages, so that the approach which best suits particular requirements can be decided.

Assignment of Projects. A major portion of the work assignments in any industrial engineering department is rather routine in nature and consists of such things as establishing time standards on new operations or revising standards on existing operations, preparing periodic reports, developing improved working methods, and the like. All these things are handled in a standardized, routine fashion. They require no particular follow-up but merely periodic checks to ensure that work is not piling up.

On the other hand, a good deal of the work performed by industrial engineering departments is of a project nature. This work includes such things as: developing a new layout for a shipping department; setting up a production line for an old, new, or revised product; preparing a survey of material handling equipment and methods; placing a toolroom on an incentive basis of wage payment; and the like. These assignments are out of the normal routine. To be performed satisfactorily within a reasonable time, they usually require considerable attention and close follow-up.

If the project is of such magnitude that it places an unduly heavy burden on the regular staff or if it requires the use of specialized talents or experience, the employment of a management consulting firm to handle all or part of the project should be considered. If the project is to be handled internally, one or more individuals should be assigned to the project on a full-time basis. Assigning a special project to an individual "in addition to his normal duties" is the least desirable approach. Either his normal duties or the project will suffer as a result.

The first step in the preparation of any special project is a clean-cut statement of objectives. Frequently, projects are started with only a vague idea of what is to be accomplished. If the work is to be started properly, the goals to be obtained should be clearly set down in writing before the project is started. The typical statement of objectives shown in Figure 6-15 includes an estimate of the time required to complete the project as well as an estimate of the saving to result.

The next step is to plan the method of approach and to list the major steps required to complete the project. If this is done on a Gantt chart similar to the one illustrated in Figure 6-16, it will help considerably in following the progress of the assignment. At the beginning of the assignment, a schedule is made of the time required to complete each individual step. This is shown graphically on the progress chart. As the assignment progresses, a record is maintained of the time actually spent on each step and the percentage of completion of each step. In this way, the form shows at a glance the status of the assignment. Actual progress against estimated progress is immediately apparent. This permits

STATEMENT OF OBJECTIVES

DATE ___November 7_____   ASSIGNED TO ___W. C. Kling____

WRITTEN BY ___K. J. James_____   STARTING DATE ____June 14_____

ESTIMATED SAVINGS $75,000 per year COMPLETION DATE January 2

Improved methods and set standards on all office and clerical procedures including accounting, purchasing, storekeeping, printing, mailing, personnel testing, and the like.

The manner of accomplishing this work will be, in general, to analyze the methods and procedures now in use, to develop improved methods and procedures, and to have them accepted and put into use. After they have become familiar to those using them, establish time standards for their performance. The standards should be in a form that will promote their use in determining manpower requirements, scheduling work, and developing an incentive wage payment plan.

Finally, an office procedures manual will be prepared for use in controlling methods and training new employees. In addition, a system for controlling costs and measuring performance is to be established.

Fig. 6-15. A statement of objectives.

corrective action to be taken before the assignment becomes hopelessly behind schedule. Steps that were overlooked at the start of the project or added to it after it started can be added to the form as the job progresses. The additional time required for these additional steps can be estimated as they arise. In this fashion, both the engineer and his supervisor are provided with a fairly accurate timetable on the progress of the assignment.

The final step on all major projects is to prepare a final report. Essentially, this report should summarize the results obtained in accomplishing the original statement of objectives.

Reports to Management. In addition to reports on special projects, a number of industrial engineering departments find it desirable to prepare a periodic report of their activities either weekly or monthly.

Because cost reductions are a major responsibility of most industrial engineering groups and because dollar savings make interesting and agreeable reading, they are commonly used as the basis for periodic reports. Reports of this type can cause a great deal of dissension, however, if they are not properly handled.

It takes more than the efforts of any one department to effect most cost reductions. If the industrial engineering department is inclined to take sole credit for cost reduction work by means of periodic reports to management, they are first of all unfair. Second, and more important, they will risk the possibility of developing an uncooperative attitude on future cost reduction programs in other departments. Although cost reductions do form a major portion of the industrial engineering

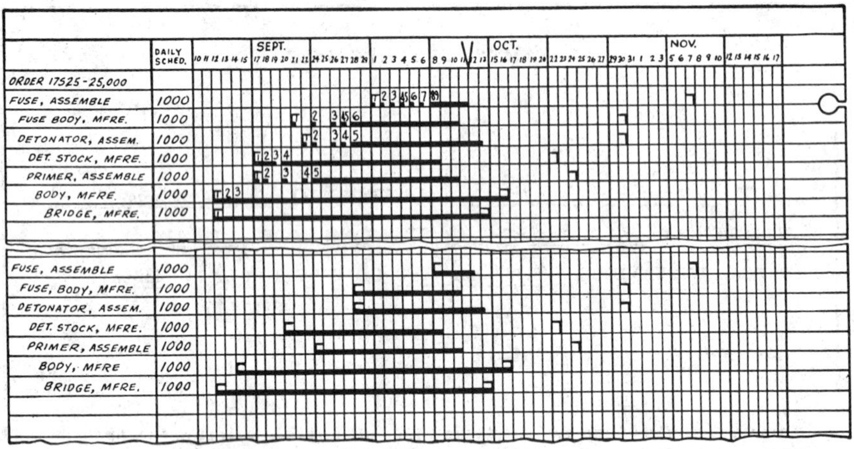

FIG. 6-16. Progress-type Gantt chart used to plan the manufacture of a detonating fuse.

activity, they should only be presented with due credit to the other departments concerned.

A typical periodic industrial engineering report form is illustrated by Figure 6-17. This particular form is designed for a weekly report. The left-hand portion of the form provides space for a daily listing of the most important activities in daily form. The right-hand portion of the form provides space for reporting on special projects and routine activities. Special projects are reported on by means of a project progress chart. Below this, space is provided for reporting on cost reduction and work measurement activities. This report provides a simple yet effective means of keeping management informed of the department's activities.

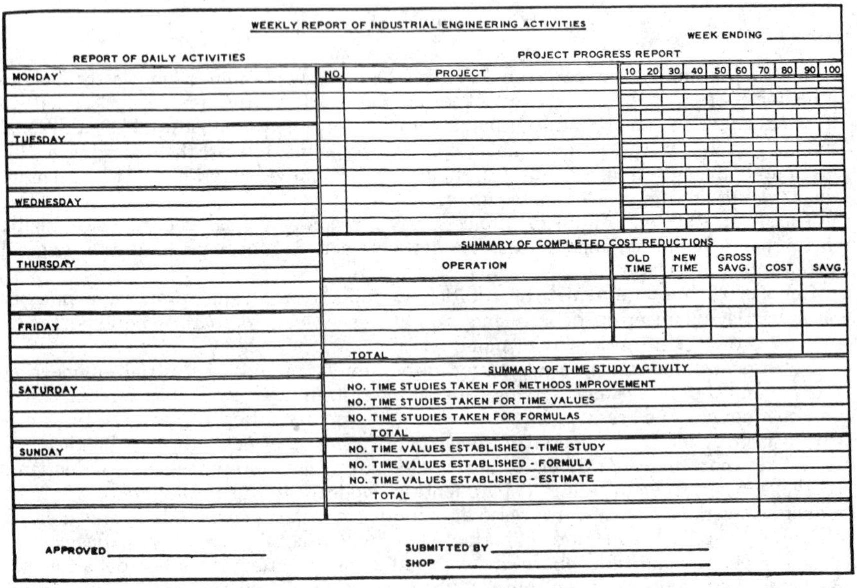
WEEKLY REPORT OF INDUSTRIAL ENGINEERING ACTIVITIES

WEEK ENDING ______

REPORT OF DAILY ACTIVITIES

MONDAY

TUESDAY

WEDNESDAY

THURSDAY

FRIDAY

SATURDAY

SUNDAY

PROJECT PROGRESS REPORT

NO.	PROJECT	10	20	30	40	50	60	70	80	90	100

SUMMARY OF COMPLETED COST REDUCTIONS

OPERATION	OLD TIME	NEW TIME	GROSS SAVG.	COST	SAVG.
TOTAL					

SUMMARY OF TIME STUDY ACTIVITY

NO. TIME STUDIES TAKEN FOR METHODS IMPROVEMENT	
NO. TIME STUDIES TAKEN FOR TIME VALUES	
NO. TIME STUDIES TAKEN FOR FORMULAS	
TOTAL	
NO. TIME VALUES ESTABLISHED - TIME STUDY	
NO. TIME VALUES ESTABLISHED - FORMULA	
NO. TIME VALUES ESTABLISHED - ESTIMATE	
TOTAL	

APPROVED ______

SUBMITTED BY ______
SHOP ______

FIG. 6-17. A form used for a weekly report to management of an industrial engineering department's activities.

Predetermined
Time Standards

Chapter 1

Uses of Predetermined Time Standards

WILBERT STEFFY

**Professor of Industrial Engineering, The University of Michigan,
Ann Arbor, Michigan**

The development of a time standard to perform a task or operation has been and still is controversial. Frederick W. Taylor, the father of time study, wrote in the latter part of the nineteenth century that to establish an operation standard, it is necessary to subdivide the operation into elements of work, write a description of each element of work, time each element of work with a stopwatch, and add certain allowances to cover unavoidable delays and fatigue.

Several years after Taylor began his work with time study, Frank B. Gilbreth developed his technique of using the motion picture camera to study the motions required to perform certain tasks. Gilbreth subdivided the Taylor elements into basic motions which he called therbligs. These therbligs were used to build up a time standard in much the same way that Taylor used elements of work for this same purpose.

The Taylor and Gilbreth methods have resulted in a number of techniques that are used to develop time standards. Developing and administering time standards is often considered to be expensive. However, much of this cost can be reduced through the use of effective methods for setting time standards.

Traditionally, there are several methods for determining time standards, of which the most important are:

1. Predetermined time systems
2. Time study

3. Work sampling
4. Standard data development

Predetermined time standards, the primary concern of this chapter, have become increasingly important to industrial companies and financial establishments in setting time standards for various work tasks. Well-known systems using this approach are Motion Time Analysis (MTA), Work-Factor (WF), Basic Motion Timestudy (BMT), and Methods Time Measurement (MTM). Other systems of a somewhat proprietary nature are those used by Western Electric, called the Standard 400 System; and the General Electric systems: Engstrom, Motion Time Standards (MTS), and Dimension Motion Time (DMT).

TYPES OF PREDETERMINED TIME SYSTEMS

It took almost a quarter of a century to attach basic times to the original Gilbreth motion therbligs. It is obvious that the motion study technique will permit a more refined analysis of method than can be obtained with a watch and the naked eye. Then, too, the observer who is trained to perceive in terms of motions instead of combinations of them, such as the time study elements, may frequently be able to improve a method without elaborate laboratory investigation. It was quite natural that Gilbreth, using the micromotion process intensively in 1912, should emphasize the paramount importance of methods rather than time. The development of basic times for the basic motions, however, has provided an even more effective tool for methods improvement than motion study alone.

Time study elements can be quickly analyzed through the micromotion process used by nearly all predetermined time systems. The following shows these relationships:

Time study element	*Micromotion*
1. Get part from tote pan............	Reach
	Search
	Select
	Grasp
2. Place part in fixture..............	Move
	Pre-position
	Assemble
	Release

In chronological order, the best known predetermined time systems are listed below.

System	Originator	Approximate date
1. Methods Time Analysis (MTA)........	A. B. Segur	1925
2. Work-Factor (WF).................	Joseph H. Quick, James H. Duncan, and James A. Malcolm, Jr.	1938
3. Engstrom.......................	Harold Engstrom and General Electric	1940
4. 400 System......................	Western Electric	1944
5. Methods Time Measurement (MTM)..	H. B. Maynard, G. J. Stegemerten, and J. L. Schwab	1948
6. Methods Time Standards (MTS)......	General Electric	1950
7. Basic Motion Timestudy (BMT)......	J. D. Woods and Gordon Ltd.	1951
8. Dimension Motion Time (DMT)..	General Electric	1954

Some of these systems were developed through independent research, while others were derived from or influenced by the independent systems. Several of the inde-

pendent systems spend significant sums of money on continuing research to explore new applications and develop new areas of use. Brief descriptions of several predetermined time systems follow.
Methods Time Analysis (MTA). A. B. Segur of Oak Park, Illinois, was one of the first to establish the relationship between the time element and the motion itself. His ambition to integrate time with motions led to the development of his MTA system. From his research, he discovered the law of fundamental times which finally made Taylor's dream of universally applicable standards in industry a working possibility. The law was stated as follows: "Within reasonable limits, the time required by experts to perform a fundamental motion is a constant."

This discovery enabled Segur to develop an analytical method which could be applied to a great variety of manual or manual/machine operations. He emphasized that the time required to accomplish an act depends upon how the work is performed or the method used by the operator. To determine the time for an act, one must know precisely how the act is performed.

When an operation is studied, it is generally discovered that the operation consists of getting something, moving it to some location, processing or assembling it, and then releasing it. For example, the operation of writing with a pen might be motion analyzed as follows:

Description—right hand	*Motion*
1. Move hand to penholder................	Transport Empty
2. Grasp pen in penholder..................	Grasp
3. Move pen to paper......................	Transport Loaded
4. Write on paper.........................	Use
5. Move pen to penholder..................	Transport Loaded
6. Pre-position pen to penholder.............	Pre-position
7. Assemble pen to penholder...............	Assemble
8. Release pen............................	Release
9. Move hand back to paper................	Transport Empty

This example is simple, but it illustrates the number of motions involved in a proper motion analysis. In addition to analyzing the motions involved, certain additional information must be collected, such as the distances moved, the type of grasp, the body members required, the type of release, and the like. This additional information allows the analyst to focus on the proper time to allow for each motion and helps answer the question "How?" How was the Transport Empty, Pre-position, or Grasp performed? The answer to this question allows one to define the method precisely, thereby enabling the engineer to choose the proper time for performing the motions.

Segur stated that the method must be well defined before an attempt is made to time-analyze the motions involved. He developed a table of improvement principles involving many of his basic motions, such as Hold, Grasp, Pre-position, Position, Avoidable Delay, and Balance Delay. The following chart shows the motion breakdown for the operation of starting a stud into the tapped hole of a single bracket.

Left hand	Motion	symbol	Right hand
1. Move hand to tote pan........	TE	TE	1. Move hand to tote pan
2. Grasp bracket...............	G	G	2. Grasp stud
3. Move bracket to work area....	TL	TL	3. Move stud above hole in bracket
4. Hold bracket in work area.....	H	PP	4. Pre-position stud
5. Hold bracket in work area.....	H	A	5. Turn stud in hole
6. Dispose assembly to tote pan..	TL	BD	6. Balancing delay

The improvement principle involved here is in the elimination of the left hand as a holding device. Segur suggests the possibility of a simple mechanical fixture where two brackets could be assembled simultaneously, thus eliminating the hand hold.

In the MTA system, motion values are carried out to the fifth decimal—0.00150 minute, for example. These base times include no allowance for fatigue and delay, which must be provided for as a separate addition to the base times. All base times are based on maximum practical performance speed.

Table 1-1 shows the motions used by the MTA system.

The Work-Factor System (WF).[1] During the years 1935 and 1936, a group of time study engineers were working on the development of a "second operation" punch press formula to be used in establishing rates for piercing, forming, and other types of punch press operations following the original blanking. Complete and detailed information was recorded for each work motion involved in the operations. These data consisted of such information as the distance moved; the body member used; the weight or resistance involved; size and type of tools, jigs, and fixtures required; and the like. After several months of work on the second operation formula, it became evident that the type of data collected could be applied to many operations. This led to a broadening of the project into other types of factory operations. Hundreds of different types of work motions were studied

[1] Work-Factor is the trademark of the Work-Factor Company, now known as the Science Management Corporation, which identified its services as consultants to industry and its system as a predetermined fundamental motion time system.

TABLE 1-1. Motion Time Analysis—Table of Motions

Abbreviation	Therblig	Definition
TL	Transport Loaded	The act of moving a Transportation Means with a load or against a resistance
TE	Transport Empty	The act of moving a Transportation Means without a load or to a point from which it can be moved against a resistance
D	Direct	The act of guiding actions with sensory movements
G	Grasp	The act of gaining complete managing control
H	Hold	The act of maintaining complete managing control
RL	Release Load	The act of completely relinquishing managing control
UD	Unavoidable Delay	The delay in the operation which is beyond the control of the operator
AD	Avoidable Delay	The delay in the operation which is under the control of the operator
BD	Balance Delay	The delay in the operation caused by the nervous limitations of the human body
R	Rest	The delay in the operation which permits elimination of fatigue
PP	Pre-position	The act of rearranging Transportation Means, the part being transported, or any other part to have them in readiness for continuing the main operation
P	Position	The act of bringing two parts to an exact and predetermined relationship with each other after the transportation is complete
SE	Select	The act of making a choice between two or more pieces which are in a known location
S	Search	The act of determining the location of anything
I	Inspect	The act of examining the characteristics of anything
PL	Plan	The act of determining a method for accomplishing anything
U	Use	The act of performing a mechanical or chemical operation

and recorded. The format used in collecting the original data for establishing standard moving times was as follows:

1. Each operation was subdivided into its many basic motions in such a way that the variables affecting time would be clearly distinguishable. Thus each motion was classified according to the distance moved, body member used, the type and degree of manual control involved, and the weight or resistance encountered.

2. Each time study was taken by two experienced time study engineers using two 6-second stopwatches mounted in a brace. A bar was passed over the crowns of both watches and was so designed that watch *A* could be started at the beginning of the motion, watch *B* at the end of the motion, and subsequently the two watches could be stopped simultaneously. By subtracting the reading on watch *B* from the reading on watch *A*, the elapsed time for the motion could be derived. From ten to twenty readings were taken on each motion, and overall readings were taken on the entire operation. Each motion was leveled independently by the two time study engineers, after which the results were compared. After mutual agreement on the leveling factor for each motion, the overall cycle time was leveled.

3. The time values for each element of every study were tabulated according to the variable factors described above, and by means of conventional statistical methods were analyzed and correlated to determine moving times.

This approach resulted in the eventual development of the Work-Factor system described in some detail in Chapter 3 of this section.

Methods Time Measurement (MTM). The Methods Time Measurement procedure may be defined as follows: "Methods Time Measurement is a procedure which analyzes any manual operation or method into the basic motions required to perform it and assigns to each motion a predetermined time standard which is determined by the nature of the motion and the conditions under which it was made."[2]

The primary object of MTM is to improve methods of operation. Methods work is often primarily correction of some previous method established by a worker, foreman, or engineer. But MTM establishes methods accurately, before production starts, by determining correct times and motions for operations. Because most operators object to changes, it is of inestimable value to establish correct methods at the start.

In 1940, while seeking a way of establishing good working methods in advance of putting new jobs into production, the developers of MTM took motion pictures of a large number of experienced workers working in a number of plants in widely separated geographical areas. The films were analyzed in detail, using essentially the micromotion study techniques developed by the Gilbreths. The number of motion picture frames between the beginning and the end of each basic motion or therblig were counted and recorded. Because the motion picture film had been exposed at the rate of 16 frames per second, this procedure gave the time required to perform each motion in sixteenths of a second. The data thus secured were carefully analyzed to determine the variables which affected the time for performing each basic motion. Eventually, the methods time measurement application data illustrated by Figure 2-1 in the next chapter were developed. They were found after an extensive period of testing to have universal application and have been used throughout the world virtually without change since they were placed in the public domain in 1948.

The MTM procedure can be used to establish production standards without the use of a stopwatch. By using the MTM time data, the rating of the performance of the operator is eliminated. Because of the difficulty of rating operators correctly during stopwatch studies, standards often tend to vary on the same job. MTM overcomes this problem.

MTM can be applied by using two general procedures:

1. By observation

[2] H. B. Maynard, G. J. Stegemerten, and J. L. Schwab, *Methods-Time Measurement*, McGraw-Hill Book Company, New York, 1948, p. 12.

2. By visualization

The first is the analysis of actual jobs in production; the second is the analysis of jobs to be put into production. The MTM procedure and the way it is applied are described in further detail in Chapter 2 of this section.

The MTM Association for Standards and Research sponsors continuing research in the predetermined time system area. The Industrial Engineering Department of the University of Michigan has been conducting research for the Association since 1953. This research will be discussed in more detail later.

Basic Motion Timestudy (BMT). Basic Motion Timestudy was developed and is taught by J. P. Woods and Gordon, Limited, Toronto, Canada. Like other predetermined motion time systems, all manual activity has been divided into basic motions.

A basic motion, according to Woods and Gordon, is defined as "Any motion which starts from rest, moves through space, and ends at rest." Hand activity is divided into three classifications:

(Type 1) *Reach:* The basic element employed when the predominant purpose is to move the hand or finger from one position to another.

(Type 2) *Move:* The basic element employed when the predominant purpose is to move an object from one position to another.

(Type 3) *Turn:* The motion employed to turn the hand, either empty or full by a movement that rotates the hand, wrist, and forearm about the long axis of the forearm.

The body motion and symbols are very similar to the body motions employed by MTM. The only difference lies in the Side Step, where the distance measured is the distance the foot travels.

In use, the basic Reaches, Moves, and Turns are classified and recorded, and additional allowances are added for any Precision, Simultaneous Allowance, or Force (weight) that is involved.

The variables which affect the time to perform a Reach or Move are:
1. Distance
2. Control
3. Precision
4. Force
5. Change Direction

The distance is the arc distance transversed by the large knuckle of the forefinger. If finger motions are involved, the measurement is made from the fingertips. In all measurement of distance, any assist of the body, arm, or wrist is ignored and the gross arc distance is used.

The control variable is divided into five classes. These five classifications include Turns as well as Reaches and Moves.

Class A. No muscular or visual control is necessary to stop the motion. The motion is terminated entirely by contact.

R___A   "Movement of empty hand requiring no visual control, and terminated abruptly by resistance offered by a rigid object."

M___A   "Move object against a stop."

Class B. Muscular but no visual control is necessary for the termination of this type of motion.

R___B   "Movement of empty hand requiring no visual control and terminated in space by muscular control."

M___B   "Toss object aside, where visual control is not required to direct the object to its destination."

Class BV. Both muscular and visual control are necessary to terminate the motion. To warrant the "V" classification, the eyes are required to direct the motion to its destination. An additional requirement is that the eyes of necessity must be focused at some point other than the terminating point when the motion in question began.

R___BV "Movement of the empty hand requiring visual control and terminated in space by muscular control."

M___BV "Toss object aside where visual control is required to direct the object to its destination."

Class C. The reach is terminated by a simple "Pick Up" or "Grasp," and the move is terminated by a "Put Down" or "Release."

R___C "Movement of the empty hand requiring no visual control and terminated by grasping an object. Usually, the object is in a fixed location or in the other hand."

M___C "Set object aside, where visual control is not required to direct the object to its destination."

Class CV. Motions which are similar to the Class C motions, but which also require visual control, are designated Class CV.

R___CV "Movement of the empty hand requiring visual control and terminated by grasping an object. Usually applies if the object is not in a fixed location and not in the other hand."

M___CV "Set object aside, where visual control is required to direct the object to its destination."

Precision is defined as the distance limits within which the center of the fingertips must be located to achieve a satisfactory grasp or to complete a Move satisfactorily. To determine the proper Precision, it is necessary to consider the distance limits in three directions (visualize grasping an object from a flat surface):
1. The direction which the fingers travel in grasping an object
2. The horizontal line perpendicular to this direction of travel
3. The vertical direction (except that the vertical Precision can be ignored "provided the height of the object above the surface on which it rests is one-eighth of an inch or more")

Force is generally associated with the pressure of four factors:
1. Apply Pressure
2. Start
3. Stop
4. Weight

Each of these, if present, will require an addition to the basic Move time. In connection with Moves, Apply Pressure is the time required to gain control of the weight, that is, to get it on the verge of moving—to bring it into equilibrium. Start is the time required to set a weight in motion and accelerate it to its maximum speed. Stop is the time required to decelerate and halt a weight already in motion. Perhaps an example will make the treatment of weight clear.

If a 10# weight is picked up, moved 24 inches, placed down, and no visual control is necessary, the treatment would be:

Time
(ten-thousandths of a minute)

M24C......................................	111
10AP24"..................................	16
10ST24"..................................	16
10SP24"..................................	16
Total......................................	159

The actual recording would be:

```
M24C.....................................  111
F10#.....................................   48
    Total.................................  159
```

If the weight need not be picked up but is already held and then placed on the table, only the Start and Stop allowances for weight are used. If a weight is held and then tossed aside, only the Start time is allowed, and so on. When an object is slid and friction is involved, the treatment is in the same manner, except that spring scales are used to determine the magnitude of the friction. The Apply Pressure portion is equal to the force of friction; the force to Start is the deadweight plus the friction; and the force to Stop is the deadweight minus the friction.

Apply Pressure is also used independently when force is needed, but no motion results. Determining the amount of force is not too difficult if a pair of spring scales is used for its determination. It should be noted that the force table applies to the movement of weight by one hand only. In general, it was found that if two hands were used, the effect was to reduce the weight allowance to 80 percent of its one-hand value.

Change Direction is defined as a marked change in the path followed by the hand between the terminal points of the motion. But the majority of motions will follow a path that shows little or no noticeable curvature.

Change Direction can occur in an infinite number of degrees. Its effect on time is small unless the amount of change is quite marked.

The procedure for recognizing the effect on time for Change Direction is summarized as follows:

1. Small CD: Use one motion and compensate by measuring the arc distance.
2. Very marked CD: Use two motions, "B" up, followed by whatever is called for on the way down.
3. In between: Use an average of the two above, if the motion is an important part of the cycle.

Exactly the same treatment applies to type 2 and 3 motions:

1. If the slowdown is almost a stop, use two motions.
2. If no slowdown is noticeable (between a Move-Reach combination), use the total distance and the classification of the last occurring motion.
3. In between: Use average of the two above, if the motion is an important part of the cycle.

Simultaneous Motions. The two variables which affect the time in performing arm motions simultaneously are:

1. The precision with which the motions must be completed
2. The distance between the completion points of the motions (separate distance)

The cause for the difference in time is that the eye has to direct both motions to their destinations. No Simultaneous Allowance is given unless both motions must be so directed (they may or may not be classified as "V"). As the tolerance is reduced, more eye control is needed, and as the separation distance is increased, more time is required for the eyes to travel between the two points; hence the two variables and the increase in time.

The effect of Precision, Simultaneous Motions, and Force was found to be additive; there was no multiplying effect. If all three allowances were involved, the time for the Move can be determined, to which are added:

1. The allowance for Precision
2. The allowance for Simultaneous Motions
3. The allowance for Force

Body Motions. A Body Motion is defined as a motion that is generated by body members other than the fingers, hands, or arms. The following symbols and descriptions are indicative of the body members covered in the BMT system.

Motion term	Symbol	Description
Foot Motion............	FM	Lateral or vertical motions hinged at the ankle.
Leg Motion.............	LM	Forward or backward motions hinged either at the hip or the knee.
Side Step..............	SS	Leg motions in a lateral direction.
Bend..................	B	Trunk bend from waist.
Stoop.................	S	A squatting action done by bending the knees.
Kneel.................	K 1	Kneeling when one knee is placed on floor.
	K 2	Kneeling when both knees are placed on floor.
Arise.................	A	Applies to the bend, stoop, and kneel, and is symbolized by prefixing the letter A to the appropriate symbol. Thus, Arise from Bend would be AB.
Sit....................	SIT	Sitting down on a chair.
Stand.................	STAND	Standing from a position of sitting on a chair.
Turn Body............	TB	Turning the body to a new position accomplished by moving the feet.
Walk.................	W	Walking in a normal manner, that is, opposed to a Side Step or a leg motion to operate a foot control on a machine.

Eye Time. In certain cases, one will find an operation in which a series of hand and body motions is interrupted by eye activity, which is distinct from the manual motions that precede and follow.

Usually, this eye activity is a matter of eye travel from one point to another, followed by focusing of the eyes on the second point.

It may be remembered that the eye function is taken into account both in the use of "V" where it is necessary, and somewhat in the Simultaneous Motions allowance.

Recoil is defined as the presence of a motion that may be termed involuntary. In connection with disengages or other situations where Recoil is involved, the treatment would be generally to allow any loosening motions, force (Apply Pressure), and finally two "B" class Moves for the actual Recoil.

It is beyond the scope of this chapter to cover all the detail and unique features of the BMT system. The system is completely workable, as are the other predetermined time systems summarized previously. A table of basic motion times used in this system is shown by Figure 1-1.

Some Proprietary Systems

Western Electric System. The Western Electric basic "400" system made its first appearance in 1949. Because this system is proprietary in nature, only a general presentation will be given. This system presents all time values in units of 0.0001 minute.

Time tables are provided for the following motions:
1. Transport (TE or TL)
2. Transport (sliding, dragging, or pushing)
3. Apply or Release Pressure
4. Obstacle in transport path
5. Pre-position
6. Grasp
7. Specialized Grasp
8. In Movement or Out Movement
9. Simultaneous Motions
10. Projection Position
11. Flat Surface Position
12. Bar Alignment Position

REACH OR MOVE

Inches	½	1	2	3	4	5	6	7	8	9	10
A	27	30	36	39	42	45	47	50	52	54	56
B	32	36	42	46	49	52	55	58	60	62	64
BV	36	42	48	53	57	60	63	66	68	70	73
C	41	48	55	60	64	68	71	74	77	79	81
CV	45	54	62	67	72	76	79	82	85	87	90

Inches	12	14	16	18	20	22	24	26	28	30
A	60	64	68	72	76	80	84	88	92	96
B	68	72	76	80	84	88	92	96	100	104
BV	77	81	85	89	93	97	101	105	109	113
C	86	90	94	98	102	107	111	115	119	123
CV	95	99	104	108	112	116	120	124	128	132

PRECISION

Inches	1	2	3	4	5	6	7	8	9	10
½″ tol.	3	4	6	7	8	9	10	11	12	13
¼″ tol.	13	16	18	21	23	25	27	29	31	32
⅛″ tol.	33	37	41	45	48	52	55	58	60	62
1/16″ tol.	60	65	69	73	76	80	83	87	90	93
1/32″ tol.	90	97	102	106	110	114	117	120	123	126

Inches	12	14	16	18	20	22	24	26	28	30
½″ tol.	14	16	17	18	19	20	21	22	23	24
¼″ tol.	36	39	42	45	48	51	53	55	57	59
⅛″ tol.	67	72	76	80	83	87	91	94	98	101
1/16″ tol.	98	103	107	112	115	119	123	127	131	135
1/32″ tol.	131	135	139	143	147	150	153	157	161	165

SIMULTANEOUS MOTIONS

Separation distance	0	2	4	6	8	10	12	14	16	18	20	22	24
¼″ tolerance and over	0	10	18	27	34	41	47	54	59	65	69	74	78
⅛″ tolerance and over	0	12	21	30	37	44	51	57	63	68	73	78	82
1/16″ tolerance and over	0	15	27	37	45	53	61	68	75	80	86	91	96
1/32″ tolerance and over	0	19	34	47	58	68	77	84	90	97	103	107	111

FORCE

Apply Pressure, Start, or Stop

Inches	6	12	24
2 pounds	2	3	3
4........	6	6	7
6........	8	9	10
8........	10	11	13
10........	13	14	16
15........	18	20	22
20........	23	26	28
30........	31	35	38
40........	38	43	47
50........	45	50	55

TURN

Degrees	30	45	60	75	90	120	150	180
A	26	29	32	34	37	43	49	54
B	33	36	40	43	47	54	60	67
BV	40	44	48	52	56	65	72	80
C	56	60	64	68	72	81	88	96
CV	73	77	81	85	89	98	105	113

EYE TIME

80

BODY MOTIONS

LM (up to 6″).........	50	Leg Motion	TB₁................ 110	Turn Body
Add per inch..........	2		TB₂................ 220	Turn Body
			B................ 180	Bend
FM....................	55	Foot Motion	S.................. 180	Stoop
W....................	100	Walk One Pace	K₁................ 180	Kneel on One Knee
SS₁ (up to 6″)..........	60	Side Step	AB, etc............. 200	Arise
Add per inch..........	2		K₂................ 440	Kneel on Knees
SS₂ (up to 6″)..........	120	Side Step	AK₂................ 480	Arise from Knees
Add per inch..........	4		SIT............... 220	Sit
			STAND........... 270	Stand

FIG. 1-1.   Predetermined times used in Basic Motion Timestudy system (expressed in ten-thousands of a minute).

13. Two side by side
14. Use of tweezers
15. Blind Position
16. Release

Transport (TE or TL) is the action of moving the hand, with or without load, through a measurable distance. The factors influencing the time to perform these motions are:
1. The distance of hand movement
2. The weight or force exerted during the motion
3. The work area in which the motion is performed

The distance is measured along the arc path of the motion. If obstacles are encountered in this curved line path, additional times are allowed to overcome the obstruction.

Where weight and force exerted are encountered, for the purpose of practicability, the time allowances are grouped into two categories as follows:
1. Objects handled with one hand
2. Objects handled with two hands

Each of these two major categories is subdivided into three groups, with pounds being used as the basis for the divisions.

The work area in which the motions are confined has been divided into areas that can be covered by:
1. Shoulder, arm, hand, and finger motions
2. Trunk and body movements

Pre-position is defined as a required supplementary movement of the fingers, hand, and arm performed before a Grasp or a Position.

Grasp is defined as the action of gaining control of an object by the hand, starting when the fingers first start to close on the object and ending when the opposing fingers hold the object under control so that it can be raised.

Factors influencing the time for the Grasp motion are:
1. The number of objects grasped at one time
2. The type of surface on which the object rests
3. The arrangement of the objects in the workplace
4. The size of the object

Many different classes of Grasp are used in the system.

Position is defined as the action of locating an object in a predetermined relationship with another object. Factors influencing the time of Position are:
1. The clearance between the object and the clearance location
2. The size of the object
3. The type of Position location

The "400" system developed clearance dimensions with appropriate times for many conditions of the Position motion. These conditions include those for straight plugs, chamfered points of entry, tapered plugs, nonsymmetrical cross sections, symmetrical cross sections, flexible plugs, and the like. Many size variations were included for the proper analysis of the Position motion under the many conditions which are encountered.

Simultaneous motions where both hands are working at the same time may require adjustments to the elemental time values, because the activity of one hand may not entirely compensate for the activity of the other. The following work conditions may be encountered where simultaneous additions to time may be required:
1. Both hands get and place separate objects simultaneously.
2. One hand gets and other hand places separate objects.
3. Transfer object from one hand to the other.

The In Movement and Out Movement motions as defined in the manual may require some explanation.

The In Movement is the distance the object is moved from the entering point to the point of release by the operator. An example would be placing a plug into a tube for assembling the two objects together. Time values for In Movement are dependent upon clearances and length of movement and therefore require an extensive series of values for the large amount of variables involved.

The Out Movement involves the removal of a plug, for example, from an object. When the same direction of movement is continued after removal toward the ultimate disposal area, transport distance is measured at the point of grasp and regular transport time provides for the removal of the object. When the direction path of the removal is different from the direction path of disposal, the initial removal is not part of the transport distance and additional time must be provided (the initial removal is called an Out Movement).

Many of the unique time applications, especially in the important Position area, cannot be discussed because of space limitations. The "400" system is well designed and is one that can be applied with a great amount of objectivity.

General Electric Systems. The first system of moving times developed by General Electric was initiated by Harold Engstrom, working in their Bridgeport, Connecticut, plant in 1940. The system derived by Engstrom was successfully applied to operations dominated by Get and Place work acts such as punch press and stamping operations. This system, at least initially, did not divide the work elements into basic motions; the work elements consisted of more than one basic motion. For example, the Get act consists of at least two basic motions—Reach and Grasp.

The Engstrom system influenced subsequent applications of predetermined times at General Electric. For example, one of their systems, identified as the MTS Data System (Motion Time Standard), was developed at the General Electric Schenectady plant in 1950. This system provides time tables for the following motion activities:

1. Transport
2. Turn
3. Walk
4. Get
5. Place
6. Precision
7. Miscellaneous

The times in the tables are presented in minutes per hundred occurrences. The Transport tables are tabulated in inches of distance moved; the Turn tables in degrees turned; and Walk in time per 100 paces, where one pace is defined as 30 inches. The Get table is defined in terms of Contact, Grip, Sliding Grip, Down Grip, Lift, and Untangle. Place is divided into Drop Release, Full Release, Down and Release, and Pre-position. The Precision table includes time for one-hand and two-hand operations working in restricted areas. The areas are defined in terms of inches, using the descriptive words of under ———" to and including ———".

The Miscellaneous table provides for such motions as:

1. Get up or sit down in chair
2. Press solenoid button
3. Step on pedal with foot
4. Cut wire with pliers
5. Cut strip with scissors

Another system developed by General Electric in their Bridgeport, Connecticut, plant about 1953 is called Dimension Motion Time (DMT) and is currently used in several General Electric plants.

USES OF PREDETERMINED TIME STANDARDS

The systems for operation time development summarily described above have made significant contributions to the setting of reasonable time standards for many work tasks encountered in the industrial and financial environments. In addition, many other uses for predetermined time systems have been developed since their introduction. The list of uses is an ever-expanding one. Some of the most important applications are in the following areas:

1. Manual learning curves

2. Manual skill training
3. Developing effective methods in advance of beginning production
4. Establishing time standards
5. Developing standard data and time formulas
6. Training supervision to become methods conscious
7. Settling grievances
8. Research

Manual Learning Curves. Extensive research effort has been directed toward the understanding and prediction of learning in manual operations. Many job evaluation plans include factors for learning a particular job. For example, one job may have a three-week learning factor while another may have a factor of one week. In many of these plans, the learning times are estimates that have not been subjected to close scientific scrutiny.

Before learning times can be predicted accurately, the establishing of good methods is mandatory. Therefore, the purpose of training operators in the best method is to ensure that they learn the correct method as quickly as possible. Correct methods can be established in the preproduction stage by the use of predetermined time systems, thus avoiding the expensive trial and error approach to correct methods.

Research performed at the University of Michigan Industrial Engineering Department under the sponsorship of the MTM Association for Standards and Research (see Chapter 5 of Section 7) indicated that a number of different factors affect the time for an operator to learn a prescribed method. The following factors seemed to be the most important:

1. *Kinesthetic ability.* The ability to position a body member in a relatively precise location while keeping the other sensory factors at a minimum.
2. *Decision ability.* The ability to make discretionary judgment between two or more alternatives.
3. *Recording ability.* The ability to reduce sensory information time. The amount of cues necessary to perform a whole series of events is the important factor in this category.
4. *Transfer learning.* Learning rates are definitely affected by past performance on various manual operations.
5. *Operation environment.* The environment is a factor in learning. Is the environment noisy, dusty, poorly lighted, or the like?
6. *The training method.* Is the training accomplished by trial and error, written instructions, verbal instructions, or otherwise?
7. *The number of skill elements in the operation.* The relative frequency of occurrence of the high-skill elements in an operation.

Some work done by Professors Ralph M. Barnes and Harold C. Amrine on learning times for repetitive operations (screwdriver operations) resulted in the following statements:

1. The total cycle time for the 3,400th cycle decreased 43 percent.
2. Move Motion times decreased 15 percent.
3. Turning Motion times decreased approximately 50 percent.
4. The decrease in time, that is, the rate of learning, was highest for about the first 900 cycles. However, there was no indication that learning could not have been improved after 3,400 cycles, but at a decreased rate.

It seems to be evident that important labor savings can be obtained by using the insight gained from learning research based on the elemental approach expressed in many predetermined time systems.

Manual Skill Training. The ability to learn to perform a specific task varies. There are certain contributing factors that enable individuals to perform a specific task. These can be categorized as follows:

1. Mental capacities
2. Physical capacities
3. Motivation
4. Transfer learning level

The mental and physical capacities are most important to the learning ability of an individual.   These so-called human capacities can be divided into three basic categories:

1. Sensory capacities—threshold levels of the eyes, ears, nose, mouth, skin, and limb position
2. Psychomotor capacities—decision process and coordination limitations
3. Motor capacities—physical limitations on exertion of force, speed, and endurance

Thus, the important factors in manual skill training are many and varied and difficult to measure.   No simple relationship exists between variations in performance of two individuals working on different jobs.   However, jobs and trainees can be carefully analyzed for common factors which can aid a training program to achieve a desired performance level in the shortest possible time.

Developing Effective Methods in Advance of Beginning Production.   The problem of developing the best method in advance of performing an operation has been perplexing and time consuming.   However, techniques in the field of predetermined time standards have given tremendous aid to engineers in deciding what method should be used to attain the least cost and the highest production per unit of time.

The time study groups of Taylor and the motion study groups of Gilbreth both emphasized the importance of using correct methods in the performance of a task. Job descriptions were given to operators to aid them in following the prescribed methods.   But the one item lacking in these early job descriptions was that of time.

It was not easy to compare one method with another from a time standpoint until after the job was placed in production.   The procedures of Taylor and Gilbreth considered method and time separately, but it is evident that the method used determines the time required to perform a task.   With predetermined times established for the various motions necessary to perform a task, comparing one method with another becomes relatively easy, even before the job is placed in production.

Preproduction methods will not eliminate all changes in methods, because a certain amount of change is inevitable in industry.   Also, it is often necessary to see the part in production before a change may suggest itself.   These production changes will, in the long run, be beneficial.   But a great many of these changes can be anticipated by proper methods study before production starts.

Before a study of a particular production method is begun, management services such as schedule control, tools and equipment, working conditions, and supervision should be considered, so that the worker will not be delayed even though the correct production method is installed.   From this preliminary survey, it will become evident if distances between motions are longer than they should be, if mechanical devices can be used to replace manual motions, and if some motions can be eliminated.

When the industrial engineer approaches the problem of engineering effective methods in advance of beginning production, he is confronted immediately with a number of different ways of performing the same job.   By visualizing the movements necessary to perform the operation, he can develop the best workplace layout, determine the type and position of tools, and prepare an instruction sheet for training the operator in the best method.   This is true because the use of predetermined times for each required motion allows the industrial engineer to determine the cycle time for the best method.   Where machine time is necessary, it is added to the motion time to get the cycle time per unit.

Establishing Time Standards.   Establishing correct time standards for industrial operations is important to several necessary phases of successful manufacturing. Different phases in which time standards can be used to advantage include the following:

1. Basic management record of time to perform operations
2. Cost estimate to get business
3. Cost check on measured productive labor
4. Line balance of operations

  5. Calculation of number of machines an operator can use effectively
  6. Calculation of load in plant for scheduling purposes
  7. Basis for incentive pay
  8. Calculation of percent efficiency of labor operations
  9. Determination of correct method
10. Time formula derivation

Thus, when one considers all the varied uses of time standards, one can see that they are essential to the day-to-day operation of a successful business. The technique of using predetermined times for establishing time standards can be summarized as follows:

  1. Securing necessary information
  2. Dividing operation into elements
  3. Dividing elements into motions
  4. Applying predetermined times to each motion
  5. Determining allowances (personal, fatigue, and unavoidable delays)
  6. Calculating the standard time

Recording the necessary information involves a description of the method of work, material characteristics, a sketch of the piece and workplace layout, and machine characteristics.

Elements should begin and end at well-defined points in the cycle and should not include more than twenty motions. Ten are preferable.

Manual handling should be separated from machine time. Where power feeds are used, cutting formulas can be applied as a check against stopwatch times for the machining part of the cycle. The separation of manual from machine time is essential for the construction of formulas.

Constants should be separated from variables. A constant is independent of the size and weight of a part, while a variable depends upon part characteristics.

After the elements have been divided properly, the motions for each element must be determined. These motions must be broken down into various Reaches, Grasps, Moves, Positions, Releases, and other basic motions.

When the predetermined times are applied to each motion, the elemental times for the operation can be established. By adding the elemental times together, the normal time for the operation is determined. Then the allowances necessary for fatigue, personal, and unavoidable delays must be added to the normal time to get the standard time for the operation.

Developing Standard Data and Time Formulas. Standard data are designed to cover not one but many operations belonging to common groups called classes of work or families. Significant mathematical relationships are developed for the variable elements, and these are presented in the form of an equation, with the dependent variable representing the time standard. (See Chapter 8 of Section 3.)

A recent development for determining time standards is the use of "set theory" and computer regression techniques. Set theory is used for part family determination, and computers are used to determine the time-predicting equation. To discover which variables are closely correlated with time presents some challenging problems. To develop a predicting equation that is meaningful, useful, and accurate, all the factors which may influence the standard time should be used and integrated into the basic formula. Also, some means should be found to determine the effect of interaction between these variables on the dependent variable of time.

Trying to generate a regression equation with only five independent variables can cost more than a thousand dollars in labor and provide no assurance of significant results. This has motivated research in computer applications for this expensive problem of equation development. This research has developed the use of a stepwise, simple computer program designed to compute regression equations that may contain as many as 100 significant variables. The use of computers to develop standard data equations is a new approach but an ever-expanding one. The marriage between computers and predetermined time systems is inevitable and will aid materially in advancing the art of developing better and more accurate equations for operation time determination.

Training Supervisors to Become Methods Conscious. A common definition of methods study is the analysis of operations so that the maximum output of the desired quality can be produced in the shortest time.

The purposes of methods study are generally considered to include the following:
1. Elimination of unnecessary motions
2. Reduction of effort and fatigue
3. Improvement of working conditions
4. Training of supervisors and operators in correct methods
5. Reduction of time by better methods
6. Design of necessary controls so levers are in normal work areas
7. Reduction of long motions by keeping activity within easy reach

The efficient supervisor must be able to teach the correct method to the worker, help solve mechanical difficulties, and give and encourage cooperation. To teach correct methods, the supervisor should be able to break a job down into the motions required, arrange the workplace to correspond to these motions, select the best worker for the job, explain the operation to the worker in terms of motions, and demonstrate the method.

It has been stated by competent engineers that of the saving in time from methods analysis, 20 percent comes from the workplace layout and 80 percent comes from operator training in the use of the correct method. Proper training of the operator by the supervisor is essential, because each operator will otherwise perform differently. Some are well coordinated, others are not. Bad work habits should be avoided by setting up correct methods at the start. Correct methods usually result from a study of the motions used by the operator. Faculty research project No. 420 at the University of Michigan showed some of the reasons for the differences in time between fast and slow workers. It was observed, through film analysis of several operators, that the slow worker used:
1. Distinct and abrupt changes in direction
2. No eye anticipation
3. No merging of movements such as Grasp and Move
4. No circularity of movements
5. High arcs for Reach and Move above the working plane

The fast operator used the reverse in each of the five points. To teach correct methods, the supervisor must recognize and correct faulty adjustment to an operation, and then establish precise motion habits.

The supervisor can be trained to observe the practice of faulty methods through the use of predetermined elemental times. After the supervisor becomes familiar with the times for various motions, he will be able to reduce motions to a minimum and attain the best method for the job. Inefficient motions such as Transfer Grasps can often be eliminated by a change in workplace layout. The training of supervisors in elemental times can be extended to tool designers, time study engineers, foremen, and other groups interested in getting better methods and least cost.

Settling Grievances. Time standards, in many cases, have been difficult to sell to workers. Some workers fear that to attain the production standard would:
1. Require a killing pace
2. Emphasize disabilities of old and slow workers
3. Establish a basis for cutting rates

These fears can be minimized by informing the workers of company policies and training supervisors in the technique of establishing correct methods. A troublesome spot has been suspicion of the operators toward the performance rating procedures used to establish normal times for an operation in connection with stopwatch time study. However, by using predetermined elemental times, the engineer is relieved of trying to explain the fairness of the performance rating factor used in time study. The operator can see the motions used and the times applied for each motion. By use of this approach, a great many grievances can be settled at the foreman level, especially where the foreman has been trained in methods study. After the operator is shown the precise motions of the Grasps, Reaches, Moves, Positions, and so on required to attain the standard time for the operation, the reasons for not meeting the standards can usually be isolated. When this has been done, an amicable solution to the problem generally results. A casual observation of elements is, in most cases, not sufficient to reveal minor changes in method which the operator may have introduced, intentionally or inadvertently. "Are you using the right method?" should be one of the first questions asked when the grievance occurs.

Research. Predetermined elemental time systems open the door to a number of interesting and important research projects. Some of these projects include a study of methods changes in operations where the operation is performed by different operators, the learning time required for an operator to become proficient in performance, and a study of performance rating when slight methods changes occur because of differences in operator effort, skill, and speed.

These variations come through minute changes in the motion pattern and the use of overlaps in the motion sequence. These minute changes in method can result in important differences in production. See Chapter 1 of Section 4. Operators who can overlap motions, such as Grasp or Move, and who can change the motion pattern, for example, by lowering the arc of a Reach or Move, will in all probability be classified as fast operators. Perhaps it may be possible to teach all operators to overlap motions and to lower the arcs of movement, thus increasing the production of the slow operators and thereby getting a higher average performance from the group.

BIBLIOGRAPHY

Antis, William, John M. Honeycutt, Jr., and Edward N. Koch, *The Basic Motions of MTM,* The Maynard Foundation, Pittsburgh, 1968.

Barnes, R. M., *Motion and Time Study,* 6th ed., John Wiley & Sons, Inc., New York, 1968.

Basic Motion Timestudy Manual, J. D. Woods and Gordon, Limited, Toronto, Canada, 1951.

Buffa, Elwood S., *Operations Management,* 2d ed., John Wiley & Sons, Inc., New York, 1964.

Chaffin, D. B., et al, *Factors in Manual Skill Training,* Research Report 114, The MTM Association for Standards and Research, Fair Lawn, N.J., 1966.

Dockx, S., and P. Bernays, *Information and Prediction in Science,* Academic Press, Inc., New York, 1965.

Foulke, J. A., et al, "An Orientation to MTM Manual Learning Curve Research," *MTM Journal,* September–October, 1966.

Hancock, Walton M., *Learning Curve Research on Manual Operations,* Research Report 113, The MTM Association for Standards and Research, Fair Lawn, N.J., 1965.

Hancock, Walton M., et al, "Computation of Learning Curves," *MTM Journal,* July–August, 1966.

Hancock, Walton M., et al, "Effects of Learning Curves on Short Cycle Operations," Research Information Paper 2, The MTM Association for Standards and Research, Fair Lawn, N.J., 1965.

Hancock, Walton M., et al, *A Study of Positioning Movements,* Research Reports 109 and 110, The MTM Association for Standards and Research, Fair Lawn, N.J., 1965.

Maynard, H. B., G. J. Stegemerten, and J. L. Schwab, *Methods-Time Measurement,* McGraw-Hill Book Company, New York, 1948.

Quick, J. H., J. H. Duncan, and J. A. Malcolm, Jr., *Work-Factor Time Standards,* McGraw-Hill Book Company, New York, 1962.

Segur, A. B., *Motion Time Analysis Instruction Manual,* A. B. Segur Company, Oak Park, Ill., 1946.

Steffy, Wilbert, et al, *Computer Generated Time Standards,* The University of Michigan, Institute of Science and Technology and Industrial Engineering Department, Ann Arbor, Mich., 1968.

Steffy, Wilbert, et al, *Motion and Time Study Notes,* The University of Michigan, Industrial Engineering Department, Ann Arbor, Mich., 1952.

Western Electric Company, *Elemental Time Standards for Basic Manual Work,* New York, 1949.

Chapter **2**

Methods Time Measurement

JOHN L. SCHWAB

John L. Schwab Associates,
Fairfield, Connecticut

The original and continuing objective of MTM is the establishment of tangible, understandable, and acceptable data for the scientific measurement of human effort. Once this objective is achieved, a major step in the progress of management to its ultimate goal of becoming a true science will be realized.

MTM, like other measurement devices which underlie every true science, has been designed to enable its users to:

1. Plan for and predetermine an end result
2. Evaluate these results
3. Analyze causes of deviations
4. Transmit knowledge

Since its publication in 1948, the wide acceptance of MTM throughout the world tends to indicate that the original objectives are within reach, as proved by the results and the achievements of methods men, work simplification engineers, time study analysts, and many others in related fields of management. Because of its refinement, understandability, ease of application, and consistency, it has enabled those interested in the industrial engineering field to develop answers to pressing problems quickly, consistently, and with a greater understanding of the underlying causes surrounding them.

USES OF METHODS TIME MEASUREMENT

The uses to which this tool has been put are almost infinite in scope. In general, they may be summarized in the following twelve categories:
1. Developing effective methods and plans in advance of beginning production
2. Improving existing methods
3. Establishing time standards
4. Developing time formulas for standard data
5. Cost estimating
6. Aiding product designs
7. Developing effective tool designs
8. Selecting effective equipment
9. Training supervisors to become methods-conscious
10. Settling time study and wage rate grievances
11. Operator training
12. Research in such subjects as operating methods, training, and performance rating

It must be emphasized that this chapter of the *Industrial Engineering Handbook* devoted to methods time measurement is a condensed description of the whole procedure. As such, much that is important to the user is left unsaid. Caution is therefore urged. The reader should delve much more deeply into the subject before attempting to apply MTM to his problems. Otherwise, errors through misunderstanding are bound to be created, and a true picture of the value of this important tool of management cannot be obtained.

THE METHODS TIME MEASUREMENT PROCEDURE

Methods time measurement is defined as follows: "Methods time measurement is a procedure which analyzes any manual operation or method into the basic motions required to perform it and assigns to each motion a predetermined time standard which is determined by the nature of the motion and the conditions under which it is made."[1]

From this definition, it can be seen that the procedure provides established basic work motions and the time required for their performance to serve as a basis for the measurement of any manual operation. In addition, it establishes the laws and concepts of how and why motion patterns are made by persons of normal mental and physical qualifications.

The methods time measurement procedure not only consists of data tables which establish the normal times for certain basic motions under varying conditions, but it also establishes the laws about the sequences these motions will follow in much the same manner as the laws of chemistry or physics explain mathematically the expected material results which will be encountered under varying physical conditions.

THE METHODS TIME MEASUREMENT DATA

The methods time measurement procedure recognizes eight manual movements, nine pedal and trunk movements, and two ocular movements. Thus, there are nineteen fundamental motions to be considered in the establishment of any motion pattern. The time for each of these motions is determined not only by the physical conditions involved in the motion's performance, but also by the nature of the conditions under which it is made. Thus, the time for a given motion is affected by a combination of physical and mental conditions.

Since their original development, substantial research has been conducted to refine and expand the MTM data to increase the extent of their application. This chapter includes data developed since the original work was published.

[1] H. B. Maynard, G. J. Stegemerten, and J. L. Schwab, *Methods-Time Measurement*, McGraw-Hill Book Company, New York, 1948, p. 12.

A substantial portion of the new data has been validated and authorized for use by the MTM Association for Standards and Research for general acceptance. In addition, this chapter includes data not yet validated or authorized by the MTM Association, which have, however, been proved accurate and applicable in practice. They have been included as a further guide for future refinement and expansion.

PRINCIPLE OF THE LIMITING MOTION

The law governing the usage of the motions in the methods time measurement technique (their sequences and combinations) has been called the "principle of the limiting motion." By a detailed observation and analysis of a large number of people in the industrial and business world, it has been determined that a physically and mentally qualified operator can perform certain motions simultaneously, while others cannot be performed at the same time. This information has been carefully established and tested in wide usage.

Additional research has been conducted on this phase of the methods time measurement technique because the principle of the limiting motion was developed from a study of effects rather than from a knowledge of causes. The rules and laws which were used in the early applications of MTM proved practical in application but were not validated as scientifically as were the data for the motion times. Subsequent research clarified the understanding of simultaneous motions considerably but did not materially alter application practices.

It should be stressed that these laws are equally as important as the time data tables, for regardless of the accuracy of the standard times established for individual motions, the job standard which is determined from their application will be inaccurate if the principle of the limiting motion is not taken into consideration.

APPLICATION TECHNIQUE OF METHODS TIME MEASUREMENT

The use of the methods time measurement procedure for determining a standard follows exactly the same pattern as that used in the solution of any engineering problem. The steps taken are:
  1. A systematic analysis is made of ideas for performing a given job.
  2. Applications of measurement are made to the ideas.
  3. Computations are made to determine the end result which will be achieved if the idea is physically created.

To explain this approach further, consider the steps taken by a mechanical engineer in designing a new type of equipment. Beginning with a knowledge of the result which he wishes to achieve, he starts visualizing the form which the equipment will take. Then, by using the known and established measurements of mechanical engineering and physics, he develops one or more designs of the contemplated equipment which will produce the desired result. When his ideas have fully jelled (guided by the measurements which he has applied during the machine's development), he knows in advance the result which the machine will produce when constructed.

With the methods time measurement procedure, the industrial engineer follows exactly the same approach. He visualizes one or more ways of performing a given operation. By application of the measurement data and the principle of the limiting motion, he is able to determine in advance the quantity to be produced by a qualified person before the method is installed. Thus, the industrial engineer has truly engineered the job, because productivity, the end result with which he is concerned, is known in advance.

RESULTS WHICH CAN BE EXPECTED

Successful application of MTM will inevitably result in many benefits, such as:
  1. By truly engineering jobs, methods and costs are known in advance and later costly changes are greatly reduced.

2. Industrial relations problems, resulting from disputes over standards established by less objective techniques, are reduced.
3. Consistency is attained in production standards. Thus, one of the major sources of industrial relations problems is eliminated. It is generally recognized that those who complain about piece rates and time standards are concerned about their earnings in relation to the earnings of others working in similar occupations.
4. The objectivity of the approach and its understandability will inevitably result in greater acceptance of all scientific management principles which are related to the field of work measurement.

It should be emphasized that the methods time measurement procedure is not a panacea nor is it the entire answer to the problem of work measurement. It has several limitations which should be mentioned at least briefly.

MTM measures human effort. Consequently, it cannot be used to measure work elements outside the control of the employee, such as chemical process times, cutting times determined by machine feeds and speeds, and the like. However, it has been successfully used in measuring unusual types of partially machine-controlled work such as precision hand polishing, metal spinning by hand, hand feeding tools such as drill presses and lathes, decorative paint spraying, and the like. It has also been used successfully for establishing time standards for the primarily mental activities of visual inspection, reading, blueprint interpretations, and so on. Its successful use in conjunction with frequency and work sampling studies has also been significant, for it frequently indicates the causes of intermittent and unpredictable occurrences, which in turn enables the analyst to correct them and to gain a new insight into the problem under study and develop predictable patterns of occurrences.

DEVELOPMENT OF MTM

MTM was developed originally from an analysis of many hundred feet of motion picture films. The films were taken on industrial operations performed by qualified operators in many geographical areas.

The camera used was driven by a constant-speed motor. Thus, with the film exposed at a constant speed, the use of timing devices and other distracting accessories was eliminated, enabling the operator to work under normal operating conditions.

The film was exposed at 16 frames per second. Exposure at a slower speed made it more difficult to analyze the film in the early research phase of MTM development because element starting and stopping points were not easily identifiable.

Body, leg, and foot motions were derived later by detailed time study, assisted by already developed MTM data. Simultaneous equations and statistical methods were employed to determine the published time standards. Constant research since the development of the original data has led to refinements and additions to the original data, but the original data have been proved by the authors and the many independent agencies and schools who have conducted research on this subject to be reproducible to a highly satisfactory degree.

Continuing research has indicated the existence of additional and hitherto unmeasured MTM elements of accomplishment, especially in the areas of purely mental actions and reactions. Additional refinements to the original published data have been developed, definitions have been made more precise, and the objectivity and accuracy of the original work have been improved. Many of these new findings which are useful and needed by the MTM practitioner are included in this chapter. Although some of them have not yet been universally adopted, sufficient tests and proofs exist to indicate their validity. The new data will prove valuable by providing information heretofore unavailable.

Methods Time Data. The methods time measurement time data and definitions are summarized in Figure 2-1. A quick glance will indicate the simplicity of the procedure.

METHODS-TIME MEASUREMENT
APPLICATION DATA IN T M U

1 TMU = .00001 hour
      = .0006   minute
      = .036   second

Do not attempt to use this chart or apply Methods-Time Measurement in any way unless you understand the proper application of the data. This statement is included as a word of caution to prevent difficulties resulting from mis-application of the data.

MTM ASSOCIATION FOR STANDARDS AND RESEARCH
9-10 Saddle River Road
FAIR LAWN, NEW JERSEY 07410

MTMA 101
PRINTED IN U.S.A.   APRIL 1965

FIG. 2-1.   Methods time measurement application data.

TABLE I—REACH—R

Distance Moved Inches	Time TMU				Hand In Motion		CASE AND DESCRIPTION
	A	B	C or D	E	A	B	
¾ or less	2.0	2.0	2.0	2.0	1.6	1.6	A Reach to object in fixed location, or to object in other hand or on which other hand rests.
1	2.5	2.5	3.6	2.4	2.3	2.3	
2	4.0	4.0	5.9	3.8	3.5	2.7	
3	5.3	5.3	7.3	5.3	4.5	3.6	
4	6.1	6.4	8.4	6.8	4.9	4.3	B Reach to single object in location which may vary slightly from cycle to cycle.
5	6.5	7.8	9.4	7.4	5.3	5.0	
6	7.0	8.6	10.1	8.0	5.7	5.7	
7	7.4	9.3	10.8	8.7	6.1	6.5	
8	7.9	10.1	11.5	9.3	6.5	7.2	C Reach to object jumbled with other objects in a group so that search and select occur.
9	8.3	10.8	12.2	9.9	6.9	7.9	
10	8.7	11.5	12.9	10.5	7.3	8.6	
12	9.6	12.9	14.2	11.8	8.1	10.1	
14	10.5	14.4	15.6	13.0	8.9	11.5	D Reach to a very small object or where accurate grasp is required.
16	11.4	15.8	17.0	14.2	9.7	12.9	
18	12.3	17.2	18.4	15.5	10.5	14.4	
20	13.1	18.6	19.8	16.7	11.3	15.8	
22	14.0	20.1	21.2	18.0	12.1	17.3	E Reach to indefinite location to get hand in position for body balance or next motion or out of way.
24	14.9	21.5	22.5	19.2	12.9	18.8	
26	15.8	22.9	23.9	20.4	13.7	20.2	
28	16.7	24.4	25.3	21.7	14.5	21.7	
30	17.5	25.8	26.7	22.9	15.3	23.2	

TABLE II—MOVE—M

Distance Moved Inches	Time TMU			Hand In Motion B	Wt. Allowance			CASE AND DESCRIPTION
	A	B	C		Wt. (lb.) Up to	Factor	Constant TMU	
¾ or less	2.0	2.0	2.0	1.7	2.5	1.00	0	
1	2.5	2.9	3.4	2.3				
2	3.6	4.6	5.2	2.9	7.5	1.06	2.2	A Move object to other hand or against stop.
3	4.9	5.7	6.7	3.6				
4	6.1	6.9	8.0	4.3				
5	7.3	8.0	9.2	5.0	12.5	1.11	3.9	
6	8.1	8.9	10.3	5.7				
7	8.9	9.7	11.1	6.5	17.5	1.17	5.6	
8	9.7	10.6	11.8	7.2				
9	10.5	11.5	12.7	7.9	22.5	1.22	7.4	B Move object to approximate or indefinite location.
10	11.3	12.2	13.5	8.6				
12	12.9	13.4	15.2	10.0	27.5	1.28	9.1	
14	14.4	14.6	16.9	11.4				
16	16.0	15.8	18.7	12.8	32.5	1.33	10.8	
18	17.6	17.0	20.4	14.2				
20	19.2	18.2	22.1	15.6				
22	20.8	19.4	23.8	17.0	37.5	1.39	12.5	
24	22.4	20.6	25.5	18.4				C Move object to exact location.
26	24.0	21.8	27.3	19.8	42.5	1.44	14.3	
28	25.5	23.1	29.0	21.2				
30	27.1	24.3	30.7	22.7	47.5	1.50	16.0	

TABLE III—TURN AND APPLY PRESSURE—T AND AP

Weight	Time TMU for Degrees Turned											
	30°	45°	60°	75°	90°	105°	120°	135°	150°	165°	180°	
Small— 0 to 2 Pounds	2.8	3.5	4.1	4.8	5.4	6.1	6.8	7.4	8.1	8.7	9.4	
Medium—2.1 to 10 Pounds	4.4	5.5	6.5	7.5	8.5	9.6	10.6	11.6	12.7	13.7	14.8	
Large— 10.1 to 35 Pounds	8.4	10.5	12.3	14.4	16.2	18.3	20.4	22.2	24.3	26.1	28.2	

APPLY PRESSURE CASE 1—16.2 TMU.   APPLY PRESSURE CASE 2—10.6 TMU

FIG. 2-1 (continued). Methods time measurement application data.

TABLE IV—GRASP—G

Case	Time TMU	DESCRIPTION
1A	2.0	Pick Up Grasp—Small, medium or large object by itself, easily grasped.
1B	3.5	Very small object or object lying close against a flat surface.
1C1	7.3	Interference with grasp on bottom and one side of nearly cylindrical object. Diameter larger than ½".
1C2	8.7	Interference with grasp on bottom and one side of nearly cylindrical object. Diameter ¼" to ½".
1C3	10.8	Interference with grasp on bottom and one side of nearly cylindrical object. Diameter less than ¼".
2	5.6	Regrasp.
3	5.6	Transfer Grasp.
4A	7.3	Object jumbled with other objects so search and select occur. Larger than 1" x 1" x 1".
4B	9.1	Object jumbled with other objects so search and select occur. ¼" x ¼" x ⅛" to 1" x 1" x 1".
4C	12.9	Object jumbled with other objects so search and select occur. Smaller than ¼" x ¼" x ⅛".
5	0	Contact, sliding or hook grasp.

TABLE V—POSITION*—P

CLASS OF FIT		Symmetry	Easy To Handle	Difficult To Handle
1—Loose	No pressure required	S	5.6	11.2
		SS	9.1	14.7
		NS	10.4	16.0
2—Close	Light pressure required	S	16.2	21.8
		SS	19.7	25.3
		NS	21.0	26.6
3—Exact	Heavy pressure required.	S	43.0	48.6
		SS	46.5	52.1
		NS	47.8	53.4

*Distance moved to engage—1" or less.

TABLE VI—RELEASE—RL

Case	Time TMU	DESCRIPTION
1	2.0	Normal release performed by opening fingers as independent motion.
2	0	Contact Release.

TABLE VII—DISENGAGE—D

CLASS OF FIT	Easy to Handle	Difficult to Handle
1—Loose—Very slight effort, blends with subsequent move.	4.0	5.7
2—Close — Normal effort, slight recoil.	7.5	11.8
3—Tight — Considerable effort, hand recoils markedly.	22.9	34.7

TABLE VIII—EYE TRAVEL TIME AND EYE FOCUS—ET AND EF

Eye Travel Time $=15.2 \times \frac{T}{D}$ TMU, with a maximum value of 20 TMU.

where T = the distance between points from and to which the eye travels.
D = the perpendicular distance from the eye to the line of travel T.

Eye Focus Time = 7.3 TMU.

FIG. 2-1 (*continued*). Methods time measurement application data.

TABLE IX—BODY, LEG AND FOOT MOTIONS

DESCRIPTION	SYMBOL	DISTANCE	TIME TMU
Foot Motion—Hinged at Ankle.	FM	Up to 4"	8.5
With heavy pressure.	FMP		19.1
Leg or Foreleg Motion.	LM —	Up to 6"	7.1
		Each add'l. inch	1.2
Sidestep—Case 1—Complete when leading leg contacts floor.	SS-C1	Less than 12" 12" Each add'l. inch	Use REACH or MOVE Time 17.0 .6
Case 2—Lagging leg must contact floor before next motion can be made.	SS-C2	12" Each add'l. inch	34.1 1.1
Bend, Stoop, or Kneel on One Knee.	B,S,KOK		29.0
Arise.	AB,AS,AKOK		31.9
Kneel on Floor—Both Knees.	KBK		69.4
Arise.	AKBK		76.7
Sit.	SIT		34.7
Stand from Sitting Position.	STD		43.4
Turn Body 45 to 90 degrees— Case 1—Complete when leading leg contacts floor.	TBC1		18.6
Case 2—Lagging leg must contact floor before next motion can be made.	TBC2		37.2
Walk.	W-FT.	Per Foot	5.3
Walk.	W-P	Per Pace	15.0

TABLE X—SIMULTANEOUS MOTIONS

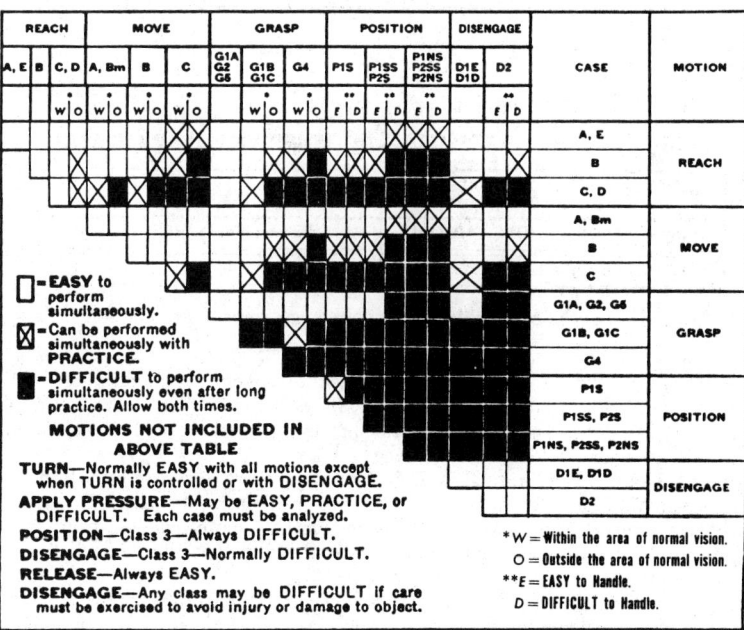

FIG. 2-1 (*continued*). Methods time measurement application data.

SUPPLEMENTARY MTM DATA

> Tables 1 and 2 are supplementary data. For proper explanation and usage, refer to MTM Application Training Supplements No. 8 and No. 9.

TABLE 1—POSITION—P (SUPPLEMENTARY DATA)

Class of Fit and Clearance	Case of† Symmetry	Align Only	Depth of Insertion (per ¼")			
			0	2	4	6
21 .150'—.350'	S	3.0	3.4	6.6	7.7	8.8
	SS	3.0	10.3	13.5	14.6	15.7
	NS	4.8	15.5	18.7	19.8	20.9
22 .025'—.149'	S	7.2	7.2	11.9	13.0	14.2
	SS	8.0	14.9	19.6	20.7	21.9
	NS	9.5	20.2	24.9	26.0	27.2
23* .005'—.024'	S	9.5	9.5	16.3	18.7	21.0
	SS	10.4	17.3	24.1	26.5	28.8
	NS	12.2	22.9	29.7	32.1	34.4

*BINDING—Add observed number of Apply Pressure
DIFFICULT HANDLING—Add observed number of G2.

†Determine symmetry by geometric properties, except use S case when object is oriented prior to preceding Move.

TABLE 2—APPLY PRESSURE—AP (SUPPLEMENTARY DATA)

Apply Force (AF) = 1.0+(0.3×lbs.). TMU for up to 10 lb. = 4.0 TMU max. for 10 lb. and over	
Dwell, Minimum (DM) = 4.2 TMU	Release Force (RLF) = 3.0 TMU
AP = AF+Dwell+RLF	APB = AP+G2

Fig. 2-1 (*continued*). Methods time measurement application data.

The definitions are, of necessity, generalized to cover the majority of conditions. However, a detailed understanding of MTM is necessary in order to apply the technique to all situations. Too liberal an interpretation of the definitions can lead to serious errors.

Units of Time. The original MTM time data were expressed in terms of the number of motion picture frames required for each basic element. It was deemed advisable to convert this unit of $\frac{1}{16}$ second into a more widely recognized time unit. Because of its widespread usage, it was decided that the decimal hour would be most desirable.

One-sixteenth of a second is the equivalent of 0.00001735 hour. To overcome the awkwardness encountered in using such extended decimal fractions, the unit 0.00001 hour was arbitrarily chosen. This, expressed for simplicity as unity or

Motion	Symbol	Motion detail	Convention
Reach..........	R	Reach, 8 inches, Case A, hand in motion at start	mR8A
Move..........	M	Move, 10 inches, Case B, weight 15 pounds	M10B15
		Move, 16 inches, Case A, hand in motion at end	M16Am
Turn...........	T	Turn, 30 degree arc, small load	T30S
Apply Pressure...	AP	Apply Pressure, Case 2	AP2
Grasp..........	G	Grasp, Case 1B	G1B
Release.........	RL	Contact release	RL2
Position........	P	Position, close fit, nonsymmetrical fit, part easy to handle	P2NSE
Disengage.......	D	Disengage, close fit, part easy to handle	D2E

FIG. 2-2. Typical conventions for identifying basic hand, arm, and finger motions.

1, was given the name, time measurement unit or TMU. Thus, one motion picture frame, expressed in TMU and carried out to only one decimal place, becomes 1.7 TMU.

The time data tables are expressed in leveled or normal TMU. They contain no allowances for personal needs, fatigue, or delays. To convert these units to other units of time, the following conversion factors are used:

$$1 \text{ TMU} = 0.00001 \text{ hour}$$
$$1 \text{ TMU} = 0.0006 \text{ minute}$$
$$1 \text{ TMU} = 0.036 \text{ second}$$

Conventions for Recording Methods Time Data. To simplify recording the individual MTM motions, a system of MTM conventions has been developed. By using this system, every detail of a motion can be easily recorded. Consequently, anyone who is familiar with the conventions can determine exactly how a motion was performed.

Figure 2-2 shows all the motion symbols other than those for the body motions, the variable factors which affect the motion, the manner and sequence of their use, and the convention which fully describes each motion.

REACH

Reach is the basic element employed when the predominant purpose is to move the hand or finger to a destination or general location. The time for making a Reach varies with the following factors:

1. The conditions under which the motion is made
2. The length of the motion
3. The absence of acceleration or deceleration in the motion

Classes of Reach. There are five different classes of Reach. The difference is found in the nature of the object reached for and the surroundings of the object, which in turn influence the amount of mental concentration necessary to complete the motion.

Case A Reach (Symbol: R Distance A). This Reach takes little or no mental concentration and does not usually require sight. It may be (1) to an object in a fixed location, such as the control lever of a machine, (2) to an object held in the other hand near the point of contact, or (3) to an object on which the other hand rests. It is a motion of habit or reflex and is the fastest case of Reach.

Case B Reach (Symbol: R Distance B). A Case B Reach occurs when the hand moves to a single object whose general location is known. It requires sight to ascertain the exact location, which may vary somewhat from cycle to cycle.

Case C Reach (Symbol: R Distance C). This Reach is to an object jumbled with other objects in a group. It is the most difficult Reach to perform and requires the utmost concentration. The operator must decide during the motion which object will be selected and control his reach in order to perform the succeeding motion.

Case D Reach (Symbol: R Distance D). This is a careful Reach to a single object where an accurate grasp is required. The Reach is made with caution because of the precision required for the next motion or possible hazard to the hand or object. The motion requires both sight and concentration and is identical in time to the Case C Reach.

Case E Reach (Symbol: R Distance E). A Case E Reach is performed to get the hand into position for body balance, for the next motion, or out of the way.

Measuring the Length of a Motion. Care must be taken in measuring the lengths of motions, because this factor has a major influence on the time taken. There are three rules to observe:

1. Measure the actual path taken by the hand or finger, rather than the straight-line distance between the points.
2. Be consistent in choosing a measuring point. For a finger motion, use the fingertip. For an arm motion, the knuckle of the forefinger should be used. If the full arm is brought into play, however, body assistance to the Reach may exist.
3. Consider the amount of assistance which the body may lend the arm in performing long or arduous motions. To determine the effect of body movement on the Reach, observe the position of the shoulder before and after the motion; then measure the distance of the shoulder motion. The effective Reach distance can be determined by subtracting the distance moved because of assistance by the shoulder from the total Reach distance.

Types of Reach. There are three general types of Reaches. They are:

1. Standard—the hand first accelerates, travels at a rapid rate of speed, and then decelerates.
2. Hand in motion at beginning or end—where the hand is moving at a rapid rate of speed when the Reach begins and then decelerates, or where the hand is moving at a rapid rate of speed when the Reach ends.
3. The hand is moving at a rapid rate of speed at both start and end of the Reach. Types 1 and 2 Reaches are those most usually encountered in industrial operations.

MOVE

Move is the basic element employed when the predominant purpose is to transport an object to a destination.

Move time is varied, like Reach, by (1) the conditions present, (2) the distance moved, and (3) whether the hand is in motion at the beginning or end of the Move. In addition, the factor of weight or resistance has an effect on the Move time.

Classes of Move. There are three distinct classes of Move. Each Move, however, is not comparable with the Reach designated by the same letter.

Case A Move (Symbol: M Distance A). Move Case A is employed to move an object against a stop or to the other hand. It requires less sight and concentration than the other Moves. Care must be taken to prevent damage to the object when moving it against a stop.

Case B Move (Symbol: M Distance B). Move Case B is employed to move an object to a general or an indefinite location. It requires a reasonable amount of sight and concentration to move the object to its destination.

Case C Move (Symbol: M Distance C). Move Case C is required to move an object to an exact location. It is a careful, precise motion which demands sight and a maximum of physical control and mental concentration. It is the most accurate Move. However, if it is necessary to have greater precision, the Case C Move will be followed by a Position element for final adjustment.

Effect of Distance on Move. The length of a Move will affect the time for performing it in the same manner that it affects Reach. Care must be taken to maintain a single measuring point and to follow the actual path of the motion. The weight of an object sometimes makes the length of the Move greater than that of a Move with little weight.

Effect of Weight or Resistance on Move. The weight of an object has a dual effect on Move time. Not only does the hand carrying a heavy object travel more slowly than when carrying a light object, but it also sometimes hesitates prior to motion. This hesitation is the time required for the body to adjust grasping pressure in anticipation of the weight to be overcome in the move.

A table of factors and constants has been developed and is included in the Move table. To determine the time allowed for a Move with weight, it is necessary to multiply the basic Move time by the appropriate weight factor, and to this product, when required, add the appropriate weight constant.

Increased weight may also influence the motion pattern. The distance moved will be greater as the hand, wrist, and arm cannot be readily flexed when carrying a heavy load.

In analyzing the effect of weight on Move, the care so necessary for the successful application of MTM must be observed by considering the entire motion pattern as well as each individual motion.

TURN

Turn is the motion employed to turn the hand, either empty or loaded, by a movement that rotates the hand, wrist, and forearm about the long axis of the forearm. The length of turn is measured in terms of degrees turned (see Figure 2-3). The length of the Turn motion is considered to vary from 30 to 180 degrees in increments of 15 degrees. The weight factor is handled by three classifications as follows:

1. Small—loads up to 2 pounds in weight
2. Medium—loads from 2.1 to 10 pounds
3. Large—loads from 10.1 to 35 pounds

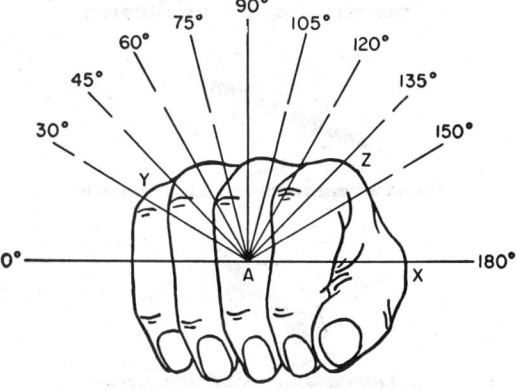

Fig. 2-3.   Concept employed in determining length of Turn motions.

APPLY PRESSURE

Apply Pressure is the basic element used to overcome resistance or to exert precise control. It appears as a distinct pause or hesitation and is required to overcome an amount of pressure or precision which is abnormal for the body member used to perform the action. The two types of Apply Pressure are:

AP1—Regrasp or squeeze and application of pressure. An AP1 is caused by the need for force or precision greater than normally required. It contains an element of surprise; hence it requires readjustment of the fingers to avoid injury or to secure additional control before action is taken.

AP2—Application of pressure only. Positive AP2 is recognized by the need for pressure (mental), tensing of muscles (mental), physical movement or lack of movement (physical and mental), and release of tension.

GRASP

Grasp is defined as the basic element employed when the predominant purpose is to secure sufficient control of one or more objects with the fingers or the hand to permit the performance of the next required basic element. It begins at the end of the preceding basic element and it ends when the next basic element begins. Thus, mental elements such as Search and Select are included in the Grasp values.

The Grasps described here are those most commonly encountered in industry. However, special complex Grasps are frequently found under special circumstances. The time for these Grasps should be developed by using combinations of the other basic MTM elements.

G1A—Simple closing of fingers to secure control of a single object.

G2—Regrasp or shifting of object in hand to gain better control (see Figure 2-4).

G3—Changing control of an object from one hand to the other.

G5—Sliding, Hook, or Contact Grasp. Operator has sufficient control when hand contacts object.

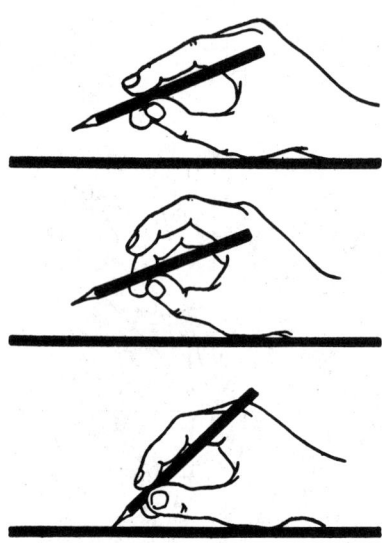

Fig. 2-4.   Regrasp.

FIG. 2-5. GIB Grasp used when cloth or paper is stacked in layers.

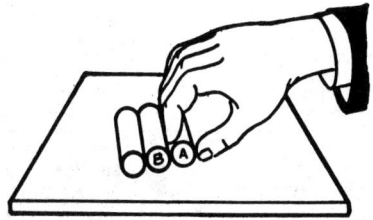

FIG. 2-6. G1C Grasp used when cylindrical objects are in contact with one another.

G1B—Securing control of a single object lying close against a flat surface (see Figure 2-5).

G1C—Interference with Grasp on bottom and one side of object (see Figure 2-6).

G4 Series—All G4 Grasps secure control of an object jumbled with other objects in a group. The difficulty of the grasp increases as the size of the object decreases (see Figure 2-7).

POSITION

The MTM element Position is defined as "the basic element employed to align, orient, and engage an object with another object, where the motions used are so minor that they do not justify classification as other basic elements." A Position usually follows a Case C Move. In some instances, where a body member such as the finger is being positioned to an object, it may follow a Case C Reach.

There are three major variables affecting Position. They are:

1. Class of fit
2. Symmetry
3. Ease of handling

Each of these must be carefully considered in detail and in order by even the most experienced analyst, so that Position can be accurately classified. Because

FIG. 2-7. G4 Grasp used when object is jumbled with other objects in a group.

Position Table—Approximate Tolerances		
Class of fit	Pressure required	Precision of placement
Class 1, loose Class 2, close Class 3, exact	Gravity sufficient to seat part Light pressure required Heavy pressure required	Align to tolerances $\frac{1}{32}$ to $\frac{1}{2}$ inch Align to tolerances $\frac{1}{32}$ inch or less

FIG. 2-8. Bench marks for identifying class of fit used for basic element, Position.

the time standard for the element Position is large, improper classification can result in serious errors in MTM analyses.

Class of Fit. Three classes of fit corresponding to terminology used to identify screw-thread fits have been used in MTM. They are:

Class 1—Loose
Class 2—Close
Class 3—Exact

Class of fit describes two conditions: pressure to overcome resistance and precision of placement. The table shown in Figure 2-8 gives a good rule of thumb, explanation, and guide to the determination of the proper class of fit.

To use the table, determine the definitions which are applicable. The more stringent of the two determines the class of fit.

Symmetry. The symmetry of the parts affects the time for orienting one or both around the point of engagement. There are three cases of symmetry (see Figure 2-9), divided as follows:

Symmetrical. Object can be positioned in an infinite number of ways about the axis that coincides with the direction of travel. Example: a cylindrical pin in a round hole.

Semisymmetrical. Object can be positioned in several ways about the axis that coincides with the direction of travel. Example: hexagonal pin in a hexagonal hole.

Nonsymmetrical. Object can be positioned in only one way about the axis that coincides with the direction of travel. Example: irregularly shaped object in a hole of the same shape.

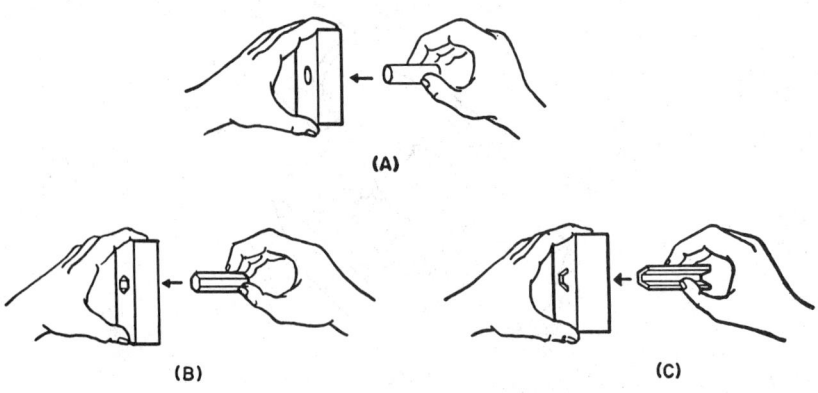

(A)

(B)   (C)

FIG. 2-9. Classes of symmetry illustrated: (*a*) symmetrical, (*b*) semisymmetrical, and (*c*) nonsymmetrical.

Ease of Handling. The third variable which must be considered to classify Position is Ease of Handling. The part may be considered easy to handle when the location of Grasp of the object being positioned is such that the fingers need not be shifted to complete the engagement. An Easy to Handle condition also occurs when the part is held close to the point of engagement, enabling the object to be aligned through its entire length.

If a part is grasped several inches from the point of initial engagement, alignment at both the front and rear of the part must be made before a complete engagement can be accomplished. The position is therefore considered as Difficult to Handle. It is impossible to define this condition in terms of distances because a rigid part may be held as much as six inches from the point of engagement, while a thin, flexible part, such as a thread, must be aligned at front and rear even though it is held only an inch or two from point of engagement. In general, however, flexible parts can be classed as Difficult to Handle except where there is a very loose or Class 1 fit. Also, very small parts such as flat, thin pieces are considered Difficult to Handle in most cases because of the necessity to Regrasp to complete the Position.

DISENGAGE

Disengage is the basic element employed to break the contact between one object and another. It is characterized by an involuntary movement caused by the sudden ending of resistance.

There are three variables that have been found to affect the time for Disengage:
1. Class of fit
2. Ease of handling
3. Care of handling

Class of Fit. Three classes of fit (see Figure 2-10) have been established as follows:
1. Loose—very slight effort—recoil blends with subsequent move.
2. Close—normal effort—slight recoil. Up to 5 inches in length.
3. Tight—considerable effort—hand recoils considerably. Up to 10 inches in length.

Ease of Handling. Parts which can be readily grasped and which can be disengaged without binding are considered Easy to Handle.

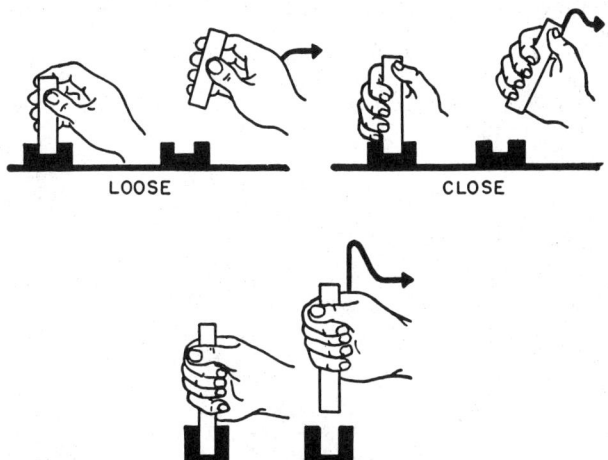

LOOSE   CLOSE

TIGHT

FIG. 2-10.   Classes of Disengage.

Difficult to Handle parts are those which cannot be grasped readily. Additional grasping motions over and above the initial Grasp must be employed. The time for these motions is included in the Difficult to Handle classification.

Care of Handling. In certain cases, extra care must be exercised in performing Disengage. This may be necessary to prevent damage to the objects being separated or injury to the hand from an uncontrolled recoil. Where the fit would normally be classed as Class 1, it is satisfactory to use the Class 2 time to compensate for the extra time required to exert care in handling. Where the fit would normally be Class 2, use the Class 3 time.

If extra care must be used in performing Disengage where the fit would normally be considered Class 3, the method will usually be changed to avoid the final recoil. Situations of this kind must be observed to determine the method that is used in each case.

RELEASE LOAD

Release Load is the basic element employed to relinquish control of an object by the fingers or the hand. The two classifications of release are as follows:

Case RL1—characterized by a simple opening of the fingers.

Case RL2—contact release—the release begins and is completed at the instant the following Reach occurs. No time is allowed for this motion.

EYE TIMES

Although the eyes are in constant use during nearly every work cycle, the time for moving and focusing them is usually not limiting. In the large majority of operations, hand, arm, and body motions are limiting, and eye time need not be considered.

Occasionally, however, operations are encountered where eye time is limiting. Therefore, the industrial engineer must be prepared with eye time data and a knowledge of how to use it when such cases occur.

There are two types of eye time, Eye Focus time and Eye Travel time.

Eye Focus Time. Eye Focus time is the time required to focus the eyes on an object and look at it long enough to determine certain readily distinguishable characteristics within the area which may be seen without shifting the eyes. The time for an Eye Focus as shown on the data card is 7.3 TMU.

Eye Travel Time. Eye Travel is somewhat more complicated than Eye Focus because the variable of distance enters in (see Figure 2-11).

The eyes must move from object 1 to object 2. In doing so, they move along the line T. This line T is a distance, D, from the eyes. If the values for T and D are known, Eye Travel time can be determined from the following formula given on the data card:

$$\text{Eye Travel time} = 15.2 \frac{T}{D}$$

SUPPLEMENTARY MTM DATA

In April, 1965, the MTM Association for Standards and Research released a new MTM data card that contains supplementary MTM data for Position and

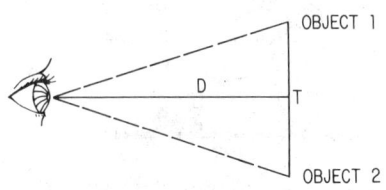

FIG. 2-11. Eye Travel.

Apply Pressure. This is the first change in the MTM data card since 1955, when short Reaches and Moves, the G1A Grasp, and the RL1 Release were adjusted and when the new Weight Allowance table was provided that included a static factor and a dynamic component. These earlier changes came about as a result of extensive research conducted by the MTM Association on *An Analysis of Short Reaches and Moves* and *A Study of Arm Movements Involving Weights*, Report Studies 106 and 108, respectively.

Following this research in 1955, additional research was conducted on Position and Apply Pressure. As a result of this new research, the Board of Directors of the U.S.-Canada MTM Association reviewed and approved the addition of Supplementary Position and Apply Pressure data to the official MTM data card. The supplementary data were then released in Application Training Supplements[2] 8 and 9 in January, 1965. In April, 1965, the new data were added to the official MTM card and titled "Supplementary MTM Data" (Figure 2-1).

Position Movements. Position movements are those motions necessary to transport an object to a predetermined destination and to seat it in or on this destination in a precise manner. The positioning movements can be grouped into two classifications: transporting motions and adjustive motions.

Transporting Motions. Transporting motions include a Move component which brings the object to be positioned to the vicinity of the positioning destination, and an engage component which seats the object in or on the positioning destination. This engage component is further broken down into a primary engage which seats the object in the destination.

Adjustive Motions. Adjustive motions are concerned with placing the object in a proper relation with the destination so that the transporting motions can be completed successfully. They include an orient component which adjusts the object through rotational movements and an align component which adjusts the object through linear movements.

Total Position. Total Position includes the hand and arm motions which occur from the moment an object is grasped until it is released after being positioned. The total positioning movement (Figure 2-12), which includes all the positioning components just mentioned, is affected by the Move variables of travel distance and weight and by the three internal variables of fit or clearance, distance of secondary engagement, and maximum possible orientation of the object. Total positioning movement time increases with decreasing clearance, with increasing distance of secondary engagement, and with increasing maximum possible orientation of the object.

Position Proper (Position). Position Proper (Figure 2-12) includes the motions align, orient, primary engage, and secondary engage, which may be required in addition to the basic transporting Move motion. Except for the orient component, these components of Position Proper must be performed following the Move.

Primary Engage Component. This is the motion in Position Proper which brings the object to the destination surface. Its performance time is relatively constant for all types of positioning movements.

Secondary Engage Component. This is the motion in Position Proper which seats the object in the destination. It is the last component motion of a given positioning and is affected by the variables of distance of secondary engagement and fit or clearance.

---

[2] Published by The MTM Association for Standards and Research.

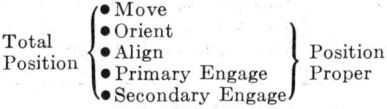

Fig. 2-12. Total Position and Position Proper.

Class of fit	Clearance (radial)	
	Inches	Centimeters
Class 21.........	0.150 to 0.350 inclusive	0.381 to 0.889 inclusive
Class 22.........	0.025 to 0.149 inclusive	0.064 to 0.380 inclusive
Class 23.........	0.005 to 0.024 inclusive	0.013 to 0.063 inclusive

FIG. 2-13.   Classes of fit and clearance.

Align Component. Align is the linear adjustment of the object or the tilting motion required to make the axis of the object coincide with the axis of the positioning destination. The align component is affected mostly by the variable of fit or clearance and to a lesser extent by the variable of maximum possible orientation.

Orient Component. Orient is the rotational adjustment of the object or the turning motion required to match the cross-sectional shapes of the object and the positioning destination. This component is affected solely by the maximum possible orientation of the positional object.

Internal Variables of Position. The internal variables of Position are fit or clearance, depth of insertion of secondary engagement, and symmetry.

Fit or Clearance. Fit or clearance is the minimum distance between the positional object and the positioning destination when the object is centrally located at the plane of initial engagement, which is the surface of the destination. There are three recognized classes of fit, and these are illustrated in Figure 2-13.

Depth of Insertion. The depth of insertion for secondary engagement is the travel distance from the plane of initial engagement to the end of the insertion, with a maximum value of 1¾ inches. Travel beyond the 1¾-inch limit should be carefully analyzed and the proper motion noted. The depth of insertion is illustrated in Figure 2-14.

Symmetry. Symmetry is a measure of the amount of orientation required by the positional object necessary to permit insertion into the positioning destination. The cases of symmetry are judged directly by the geometric properties of the cross sections of the object and the destination, when projected on a plane perpendicular to the axis of insertion. These are illustrated in Figure 2-15.

Performance Times for the Position Components. The performance times for primary engage, secondary engage, align, and orient can be expressed independently of one another.

Primary Engage. Primary engage is similar to an M1C to the surface of the positioning destination, both as to type of movement and as to the time. It is not affected by the class of fit, the case of symmetry, or the depth of insertion. It has a constant performance time of 3.4 TMU.

Depth of insertion, inches	Research range, inches	Symbol
0	0(0 to ⅛ inclusive in application)	0
½	Over ⅛ to ¾ inclusive	2
1	Over ¾ to 1¼ inclusive	4
1½	Over 1¼ to 1¾ inclusive	6

FIG. 2-14.   Depth of insertion.

Case of symmetry	Symbol	Maximum possible orientation	Insertion description
Symmetrical	S	0	Insertion is possible in any orientation
Semisymmetrical	SS	Between 0 and 90	Insertion is possible in two or several orientations.
Nonsymmetrical	NS	180	Insertion is possible in only one orientation.

FIG. 2-15.   Symmetry cases.

Secondary Engage.   Secondary engage, when performed, is the last motion in a positioning sequence. It is affected by the class of fit and by the depth of insertion but is not influenced by the case of symmetry. The performance times in TMU, with the depth of insertion expressed in quarter-inch increments, are illustrated in Figure 2-16.

Align.   Align is sensitive to changes in class of fit and case of symmetry but is unaffected by the depth of insertion. Normally, it limits the primary engage time.   Performance times in TMU are illustrated in Figure 2-17.

Orient.   Orient performance time changes only with the case of symmetry. The performance times in TMU are shown in Figure 2-18.

The supplementary Position data table, illustrated in Figure 2-1, has been con-

Class of fit	Depth of insertion symbols, per $\frac{1}{4}''$ increments			
	0	2	4	6
Class 21	0	3.2	4.3	5.4
Class 22	0	4.7	5.8	7.0
Class 23	0	6.8	9.2	11.5

FIG. 2-16.   Secondary engage performance times.

Class of fit	Case of symmetry		
	S	SS	NS
Class 21	3.0	3.0	4.8
Class 22	7.2	8.0	9.5
Class 23	9.5	10.4	12.2

FIG. 2-17.   Align performance times.

Case of symmetry		
S	SS	NS
0.0	6.9	10.7

FIG. 2-18.   Orient performance times.

structed by combining the time increments for the four Position components in accordance with the research results.

Supplementary Apply Pressure. Apply Pressure is an application of controlled muscular force to overcome object resistance, accompanied by little or no motion. The total cycle of Apply Pressure includes three components performed sequentially in a manner suggested by the force-time diagram in Figure 2-19.

Apply Force (AF). Apply force is the period of time during which no movement occurs while an increasing controlled muscular force is being applied to an object. The force application builds up steadily, but the performance time required levels off at a constant value of 4.0 TMU after 10 pounds of force has been applied.

Minimum Dwell (DM). Minimum dwell is the period during which reaction occurs for the reversal of force, where the force is held relatively constant during the reaction interval. A constant reaction time of 4.2 TMU is allowed for this component of Apply Pressure.

Release Force (RLF). Release force is the period of time required for an operator to release muscular force. The mean value for this component is 3.0 TMU.

A total cycle of Apply Pressure occurs each time force is applied and then released with minimum dwell period intervening. If a portion of the Apply Pressure is limited by combined or simultaneous motions, the proper time required may still be determined by using the independent components.

WALKING

Walking is the basic element employed to move the body from one point to another by the use of the legs. Walking time is based on an average time per pace for the average distance the body moves during one pace.

To arrive at an average distance and time for one pace, a considerable number of men and women were studied. In this group, ages ranged from seventeen to sixty-five years. Heights ranged from 5 feet to 6 feet 4 inches. From this representative group, an average time of 15.0 TMU per pace was determined.

The factors that influence walking time at an average performance level are age and weight. Although the length of the step taken by the operator varies, the walking time per pace at average performance is a constant. The average pace was found to be 34 inches. A pace of 34 inches taking 15.0 TMU results in a time per foot of 5.3 TMU.

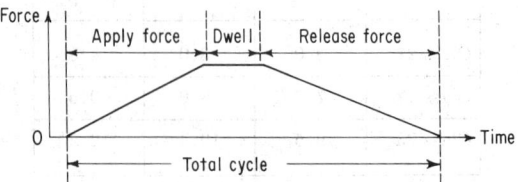

FIG. 2-19.   Apply Pressure force-time diagram.

Normally the value of 15.0 TMU per pace is used. When it is difficult or impractical to determine the actual number of paces required, the value of 5.3 TMU per foot is used. Both these values are for walking on level surfaces without load.

When walking under load, a shorter pace is used. As the load increases over 50 pounds, the time per pace is also increased. Following are the values for walking under load.

Load, pounds	Length of pace, inches	Time per pace, TMU
5–35	30	15.0
35–50	24	15.0
Over 50	24	17.0

A higher time per pace must be used for walking through obstructed areas, on sand, on railroad ties where the pace is restricted, and the like. The value is 17 TMU per pace. The number of paces required should be determined by observation.

FOOT MOTION

Foot Motions are those where the foot is moved with the ankle serving as a hinge; or the instep serving as a fulcrum of the motion. It is usually made in a vertical direction (Figure 2-20). The range of movements of this type is limited by the nature of the body member involved so that an average time value of 8.5 TMU may be used. When heavy pressure is required, it becomes necessary to add an Apply Pressure, Case 2, and the convention is FMP — 19.1 TMU.

LEG MOTION

Leg Motions are movements of the foreleg or the entire leg where either the knee, as in Figure 2-21, or the hip, as in Figure 2-22, serves as a pivot. There is no appreciable difference in time between a foreleg motion and one requiring a full leg movement.

FIG. 2-20.   Typical Foot Motion.

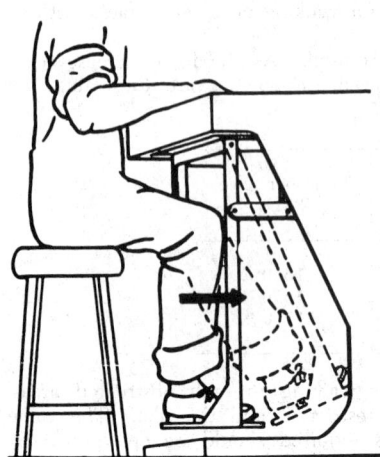

Fig. 2-21.   Leg Motion pivoted about the knee.

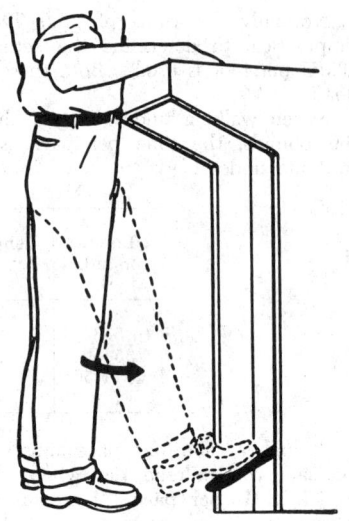

Fig. 2-22.   Leg Motion pivoted about the hip.

SIDE STEP

A Side Step occurs when the body must be displaced sideways from one location to another in the immediate area without turning or taking more than one step. There are two cases of Side Step (see Figure 2-23) as follows:

> Case 1—Side Step is completed when the leading leg makes contact with the floor.
> Case 2—Side Step occurs when the lagging leg must be brought into position beside the leading leg before the next motion can be made.

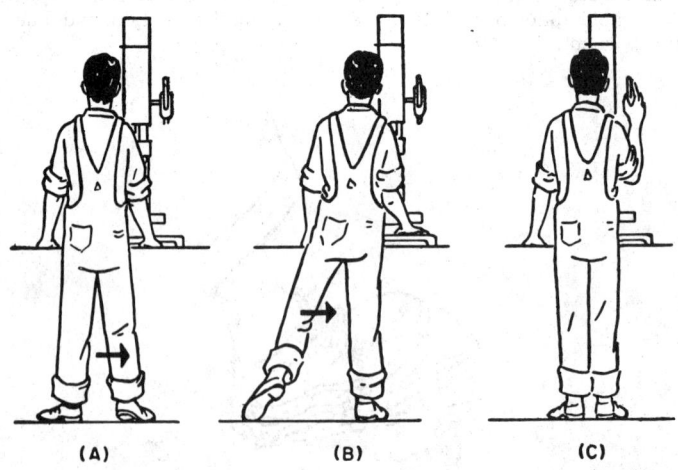

(A)　　　　　**(B)**　　　　　**(C)**

Fig. 2-23.   (A) Side Step, Case 1; (B) Side Step, Case 2; and (C) normal body position.

Fɪɢ. 2-24. Bend.

TURN BODY

Turn Body, a variation of Side Step, occurs in cases of turning the body to a new location while stepping from a work station. There are also two cases, as in Side Step, which are:

Case 1—occurs when the hands take over as the leading leg makes contact with the floor.

Case 2—occurs when the lagging leg must be brought alongside the leading leg before work can begin.

Turn Body also applies to turning the upper torso with the body seated. Such a Turn Body is always Case 1.

BEND

Bend occurs when the body is bowed at the waist with the upper portion of the torso lowered to bring the hands to within reach of an object that cannot be obtained when the body is held erect, as in Figure 2-24.

A Bend begins with the downward motion of the shoulders and ends when the hands are at the level of the knee or slightly below, with the arms fully extended. A Bend seldom blends smoothly with the preceding motion.

STOOP

When the body is lowered by bending the knees as in Figure 2-25, a Stoop occurs. A Stoop begins with the downward movement of the body and ends when the hands touch the floor or object close to the floor.

KNEEL ON ONE KNEE

A Kneel on One Knee begins with the downward movement of the body and ends as the knee just touches the floor. Figure 2-26 illustrates this.

The motions of arising from the positions assumed in performing Bend, Stoop, or Kneel on One Knee are easy to recognize.

Fig. 2-25.   Stoop.

Fig. 2-26.   Kneel on One Knee.

KNEEL ON BOTH KNEES

When the body is lowered to kneel on both knees, a Kneel on One Knee is performed first, the weight of the body is applied to the knee carefully, and the remaining leg is then lowered to a kneeling position.

SIT AND STAND FROM A SITTING POSITION

Sit, or the seating of oneself, begins when the body is positioned at the location from which the body can be lowered to a chair or bench, without further movement of the body or feet, and ends when the body is seated. Stand begins when the feet are in position on the floor and ends when the body has assumed an erect stance.

STEP[3]

Step is the basic element required to change the body location by a movement of the legs alone without the assistance of body momentum. In accordance with

[3] These data have not yet been validated or authorized by the MTM Association for Standards and Research.

this definition, a Step involves:
1. Shifting of body weight from one foot to the other
2. A forward, lateral, or backward leg motion
These data are illustrated as follows:

Symbol	Distance	TMU	Description
SB SS SF	Up to 12 in. Each additional in.	17.0 0.6	Step forward, to side, or back Complete when leg contacts floor

Variables Affecting Step. The time for performing the element Step is affected by two variables—the distance the body must be moved and the nature of the motion which succeeds it in the methods pattern.

The distance of the Step is measured at the waist, with the belt buckle as an excellent reference point. Thus, the distance of the step is the distance moved by the body rather than the feet.

The effect of distance moved and the time for making a Step is minor, comparatively speaking. The greater portion of the time is consumed in overcoming inertia or stopping momentum, with the "in motion" time of the body attaining an average speed of 0.6 TMU per inch. The effect of distance on the time for Step has been impossible to detect with existing measuring devices. Hence, the constant value, or 17.0 TMU per Step, has been established for any distance up to 12 inches.

Occurrence of Step. The element Step is required for a great number of reasons, indicative of a need for improvement of the job method. Among the most frequently encountered conditions which necessitate a Step in the motion pattern are:
1. To start and to stop walking
2. Long Reaches and Moves whose termination is beyond the limits of the hands and arms
3. Restricted walking where momentum cannot be achieved, such as walking on slippery floors, sandy beaches, up ramps, or on a congested or dangerous floor area
4. Awkward and consequently dangerous locations from which heavy or bulky material must be moved
5. Overcoming a weight greater than that which can be normally handled by the hand and arm
6. Avoiding an obstruction, such as sidestepping between a chair and desk after standing
7. Walking with very heavy weights

The principal purpose of Step is to place the torso in a location from which the hands and arms can work properly and effectively. An example is where a part must be secured which is beyond the reach of the hand. To bring the hand within range of the part, the body must be shifted as the hand performs a Reach. Thus, the Step makes it physically possible to accomplish the task.

Many cases exist where a Step is made for convenience rather than necessity. If a part is to be moved which weighs more than the amount which can be comfortably handled by the hand and arm in its present location, the legs will shift the body by a Step to a position which permits the back muscles to assist the arm in overcoming the resistance.

Accuracy of the succeeding motions can likewise necessitate a Step to bring the body to a better working position. It is obvious that careful Alignment and Position can be made more easily if the hand and arm are in a comfortable rather than an extended position.

Symbol	Displacement						Description
	2″	4″	8″	12″	18″	24″	
B	13.2	14.7	17.2	18.9	21.4	24.1	Bend—forward or backward motion of the body, measured in inches moved at the neck
	Degrees turned						
	10°	20°	30°	45°	60°	90°	
TB1	12.8	15.8	17.7	20.1	22.6	26.7	Turn body Case 1—Standing—complete when leading leg contacts floor Sitting —complete when body motion stops
TB2	....	33.3	35.0	36.7	37.7	38.4	Case 2—Complete when lagging leg contacts floor

Fig. 2-27.   Partial Body motions.

PARTIAL BODY MOTIONS[4]

The Partial Body motion time data (Figure 2-27) may be used when one Body motion is frequently and repetitively used to accomplish a given task. It is usually required when the hands and arms do not have sufficient strength or are not in a satisfactory position to accomplish the operation without body assistance.

Partial Bend Body Motions. Examples of Partial Bend Body motions occur in hand sawing heavy timbers or pipes, hand scraping the beds of machine tools, moving heavy, bulky objects short distances on a table (see Figure 2-28), and extending the arms beyond their normal limits at an extended workplace area.

[4] *Ibid.*

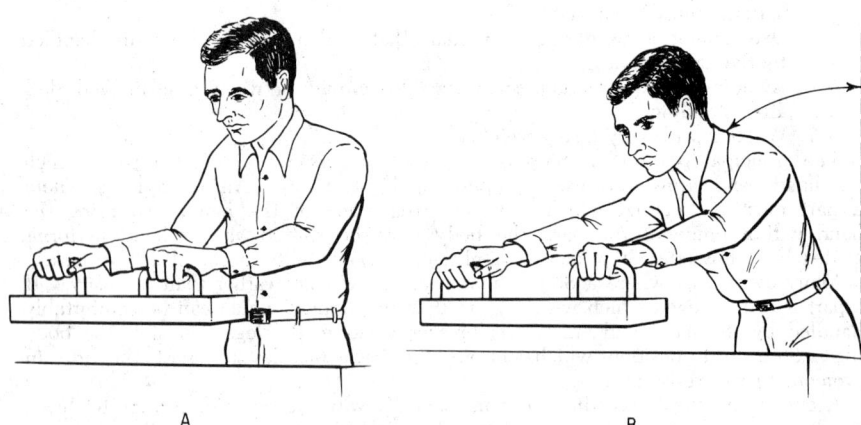

A   B

Fig. 2-28.   Partial Bend Body.

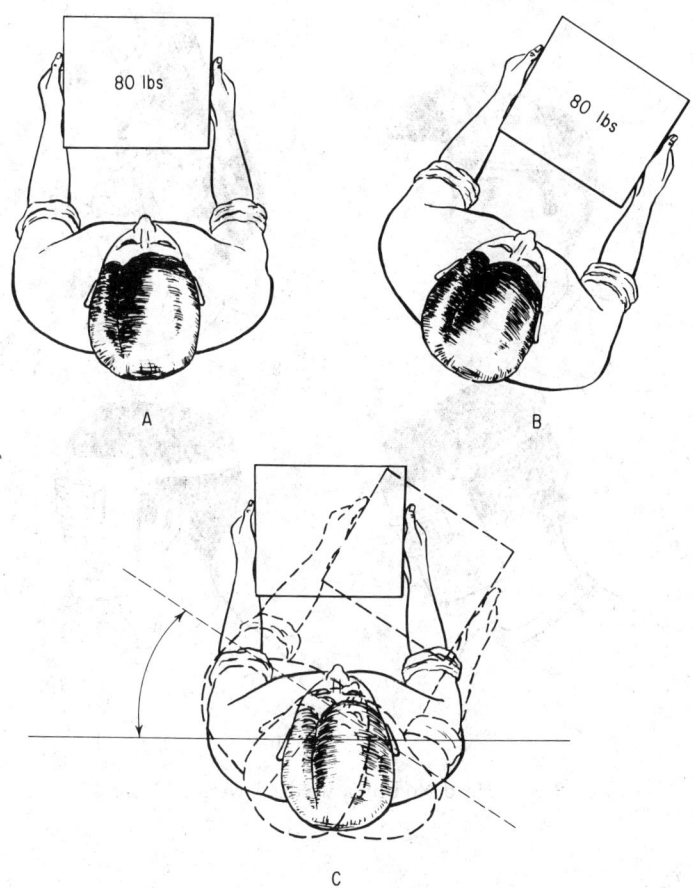

FIG. 2-29.   Partial Turn Body.

Another frequent cause of Partial Bend Body motions is the need to move the head and the body sufficiently close to an inspection point to enable the eyes to see properly.

Partial Turn Body Motions. Partial Turn Body motions are torsional body movements with or without the movement of the feet. The degrees turned are measured by the angle between the plane of the shoulders and the plane of the hips. Such motions are frequently encountered in yard labor operations such as shoveling, in repetitive material handling work such as moving heavy containers (Figure 2-29), in loading and unloading conveyors with heavy or bulky items, and in similar situations where the arms have insufficient strength to do the job alone.

HEAD MOTIONS[5]

A Head Motion is the vertical travel of the head up or down or the horizontal travel of the head from side to side using the neck as a pivot. A vertical Head Motion or Nod is exemplified by the actions of a welder who lowers his mask

[5] *Ibid.*

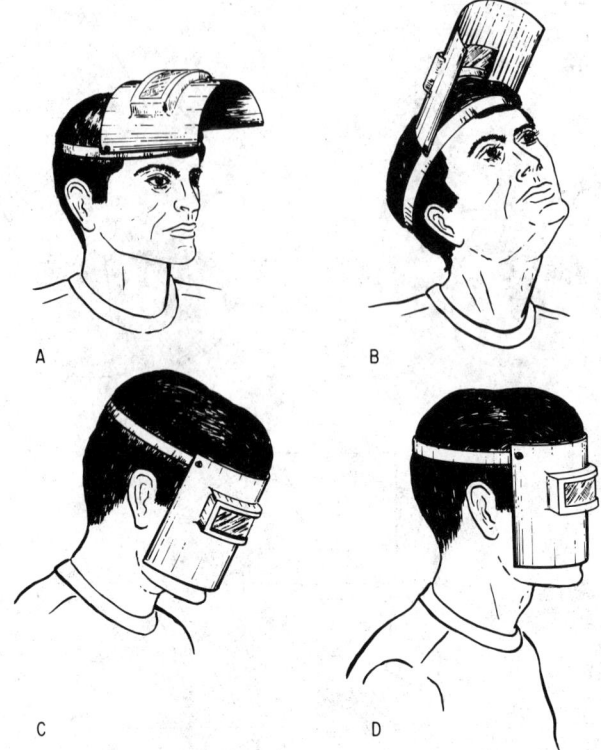

FIG. 2-30. Head Nod by a welder lowering his mask.

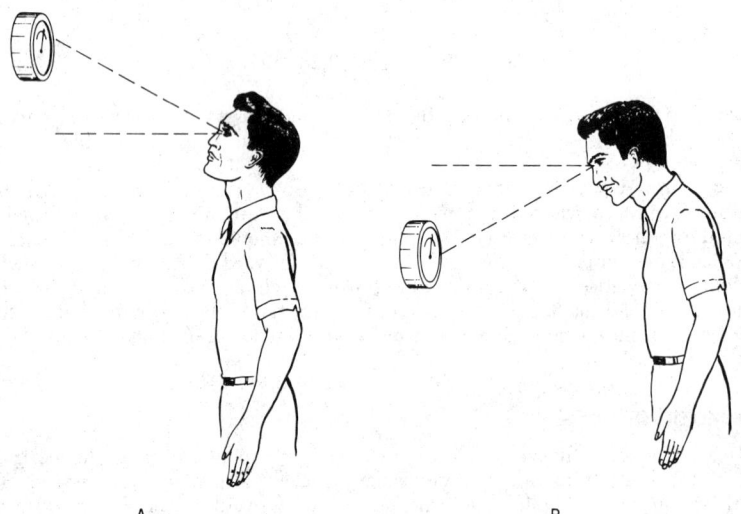

FIG. 2-31. Head Nod to read a meter.

Symbol	30°	45°	60°	90°	Description
HN	10.1	12.5	14.8	17.1	Head Nod—vertical travel up or down
HT	10.1	11.9	13.6	15.4	Head Turn—horizontal travel side to side

FIG. 2-32. Head motions.

by tilting his head back, snapping it forward, and returning his head to the normal position as the mask falls in place, as illustrated in Figure 2-30. Other examples include the lowering of the head to bring the eye to a microscope lens while seated; raising the head to read a meter or dial well above the normal head position during an assembly operation, as in Figure 2-31; and the like. The time for Head Motions is shown in Figure 2-32.

One of the most frequent causes for Head Turns is the inability of the eyes to turn sufficiently to perform an inspection operation without turning the head. If two widely separated meters are to be read, the head must be turned before the eyes can be brought into focus. This motion is illustrated in Figure 2-33.

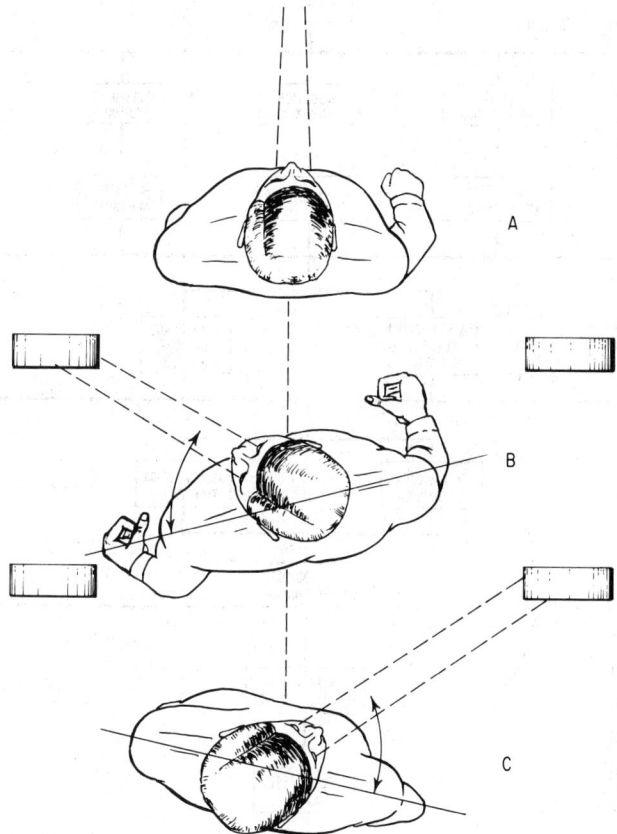

FIG. 2-33. Head Turn between meter readings.

MTM APPLICATION PROCEDURE

The methods time measurement procedure may be applied in two ways:
1. By visualizing an operation not yet existent
2. By observing an already established operation

The approach to either method is similar, except that the application of MTM to a visualized operation requires more attention to detail to avoid error. The procedure for applying MTM by either method can be divided into certain basic steps as illustrated by Figure 2-34. These are:

Visualization	*Observation*
Create·the operation	Observe the operation
Visualize and organize information	Broadly analyze and record existing information
Plan the operation method	Record the operation method
Analyze operation details and establish time	Analyze the method and establish the time

Step 1. Establish Basic Method. The first step in either procedure is to establish the basic method. Whether it is visualized or observed by the analyst, a broad description of the purpose of the operation and the general details should be concretely established.

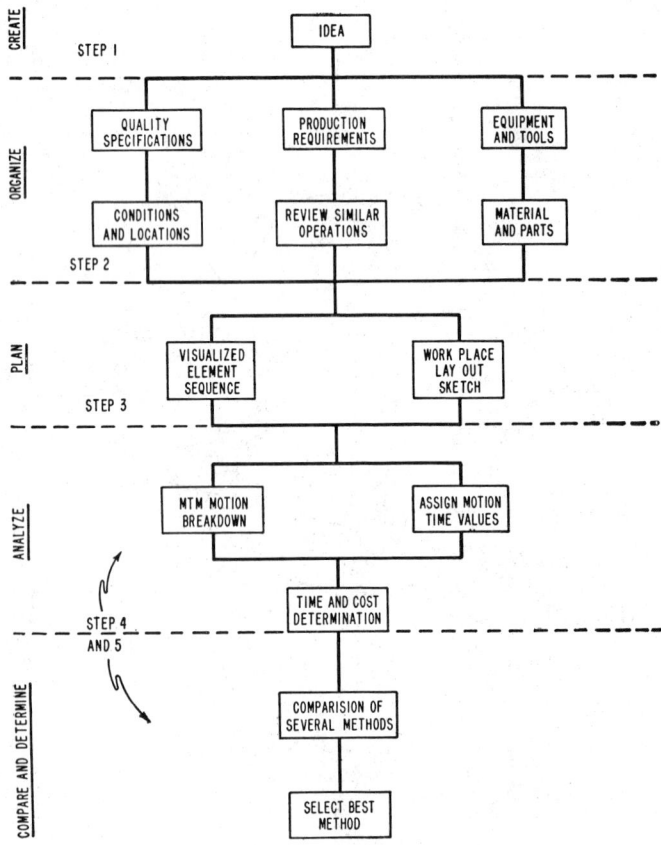

Fɪɢ. 2-34. Graphic analysis of MTM application procedure.

Step 2. Organize Information. The second step is to organize all tangible information on a more detailed scale. This will include determination of the following:
1. Quality specifications
2. Production requirements
3. Equipment and tools
4. Location and conditions
5. Materials and parts
6. Pertinent information which can be supplied by a review of similar operations already in existence

Once this information is gathered, it should be reported on the back of the MTM analysis sheet as shown in Figure 2-35 and should cover such specific information as the following:
1. Operation—the operation should be clearly and concisely defined. Operation numbers should be used where possible.
2. Location—the location should be described sufficiently well to enable any interested person to determine where the job is to be performed. Machine numbers, tool numbers, and the like should be carefully inserted in the form to aid in identification of the location.
3. Parts—every means for identifying the part, such as part number, name, and the like, should be used. This should be supplemented by sketches of the part on the section of the form so designated. A freehand sketch is usually satisfactory.
4. Material—weight, flexibility, size, color, or other material specifications which may have an effect on the motion pattern used should be recorded.
5. Equipment—all equipment should be clearly identified by names, numbers, or sketches. Auxiliary devices which have a bearing on the method should also be fully described and sketched if necessary.
6. Quality requirements—quality has a major bearing on the motion pattern to be used. Therefore, all quality requirements such as tolerance, care of handling, finish, and the like must be completely described. A write-up may be made and attached to the MTM analysis sheet when the quality requirements are of sufficient importance.
7. Tool and part sketches—freehand sketches of tools and parts must be included on the section of the form devoted to sketches. Such sketches will aid in determining the hand to be used in performing any motion, the type and degree of positions required, grasps which must be utilized, and the like.

Step 3. Prepare Detailed Elemental Breakdown. A detailed elemental breakdown of the method for performing the operation can now be established. The elements should be written in detail similar to that used to describe those in an accurately developed time study.

The workplace layout should then be accurately drawn, using the element sequence as a guide. The sketch should be drawn to scale with all parts identified and located in their approximate locations. Containers, tools, and work simplification devices which are visualized should be roughly sketched and placed in their proper locations on the workplace layout plan.

The analyst should then follow the regular work simplification approach to methods development. Elements should be studied to be certain that they are in the proper sequence, to determine whether or not manual elements may be overlapped with machine or process time, that the rules of motion economy are being followed, and the like. Once this detailed analysis has been made, the correct element sequence and revised workplace layout may be prepared in sketch form.

Step 4. Make MTM Analysis. The fourth step to be followed in applying MTM is as follows:
1. Determine a finely drawn elemental breakdown.
   a. Separate constant and variable elements.

PART NAME 1/2" PIPE UNION – KNURLED ENDS – HEX NUT   PART NO. 652   DEPT. 20

OPERATION NAME ASSEMBLE   APPROVED: JAMES B. MAINWARING   OPER. NO. 5   DATE 12-1

ANALYST: WALTER SKERPAN

ACTIVITY 5,000 PER WK.   MTL. HANDLING: DELIVERY BY   RELATED ANALYSES: OTHER TYPES AND

EQUIPMENT: CHAIR AND BENCH   MATERIAL HANDLER   SIZES OF PIPE – UNIONS
      5-STACK BINS   OPERATOR FILLS OWN BINS –
      1-REJECT-BOX   REMOVAL BY MATERIAL HANDLER

TOOLS: 2-HOLE FIXTURE – U123

SAFETY:

REMARKS: SEE IMPROVED METHOD

MATERIALS: NUT – BUSHING – SWIVEL   QUALITY: REJECT OBVIOUS DEFECTS
      (STEEL CADMIUM PLATED)   SEATED AND FINGER-TIGHT ASS'Y.

SKETCH OF WORK PLACE, WITH DIMENSIONS

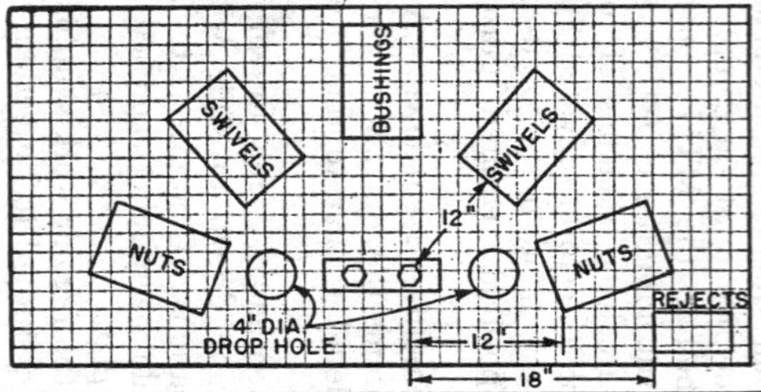

SKETCH OF PARTS, TOOLS, ETC., WITH DIMENSIONS

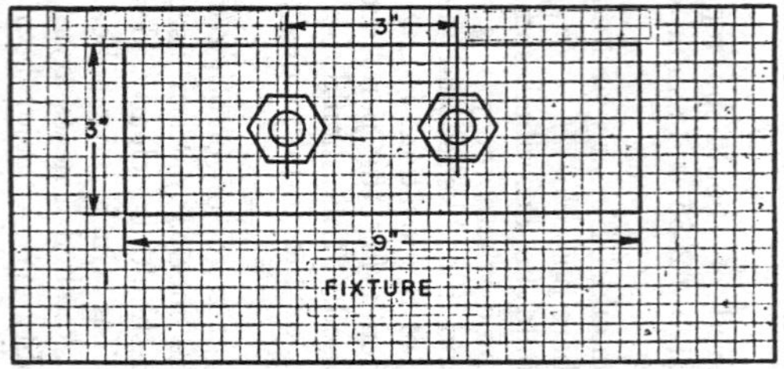

BUSHINGS   SWIVELS   NUTS   REJECTS   NUTS   SWIVELS   4" DIA. DROP HOLE   12"   18"

FIXTURE   3"   9"   3"

ELEMENT	DESCRIPTION	TMU REQUIRED	LEVELED TIME	% ALLOWANCE	TIME PLUS ALLOWANCE	OCC. PER CYCLE	TOTAL TIME REQ.
A	LAY ASIDE ASSEMBLIES AND GET NUTS	50.1	.000501	15	.000576	1/2	.000288
B	NUTS INTO FIXTURE	16.3	.000163	15	.000187	1/2	.000094
C	SWIVELS PLACED INTO NUTS	51.6	.000516	15	.000593	1/2	.000297
D	GET BUSHINGS INTO NUTS	69.1	.000691	15	.000794	1/2	.000397
E	SCREW BUSHINGS INTO NUTS – FINGER-TIGHT	92.1	.000921	15	.001059	1/2	.000530
F	GET EMPTY BUSHING BIN AND TAKE TO TOTE-BOX	259.2	.002592	15	.002986	1/250	.000012
G	FILL BIN WITH PARTS AND DELIVER TO BENCH – SIT DOWN	1040.3	.010403	15	.011964	1/250	.000050
H	GET EMPTY SWIVEL BIN AND TAKE TO TOTE-BOX	295.2	.002952	15	.002980	2/500	.000012
I	REPEAT G – FOR SWIVELS (OR NUTS)	658.8	.006588	15	.007576	4/500	.000061
J	REPEAT G – FOR NUTS (OR SWIVELS)	381.5	.003815	15	.004387	2/500	.000018
K	FILLED NUT BIN TO BENCH – SIT DOWN	215.5	.002155	15	.002478	2/500	.000010
L	NUT BIN TO TOTE-BOX	259.2	.002592	15	.002980	2/500	.000012
	ALLOWED TIME						.001781
	TOTALS						

SHEET ____ OF ____ SHEETS

FORM NO. MTM-4

FIG. 2-35. Back of MTM observation sheet with identifying information and summary of time data recorded.

b. Make the elements short. Attempt, if possible, to have no more than 10 individual motions, for each hand, to an element.

c. List all elements in proper sequence with reference to the broad elements previously developed.

2. Describe each element in detail and record its description at the top of an individual MTM analysis sheet (Figure 2-36).

3. Visualize or observe and then record on the MTM analysis sheet the motions required by both the left and the right hands.

    a. Record all motions.

    b. Circle nonlimiting simultaneous motions and cross out nonlimiting combined motions.

    c. Record foot, leg, or body motions under the column headed "Right Hand."

    d. Analyze and complete only one element at a time.

    e. Be complete in the written description of each motion required. Descriptions should be adequate to explain fully the reason for the assignment of each MTM motion time.

4. Check the motion sequence for errors in visualization, observation, and recording.

5. Sign and date the study.

With the elemental breakdown determined and the motions described, the study can be completed by assigning the motion time values from the data card. The motions should be recorded by using the conventions previously described.

Step 5. Complete Study. To complete the study, the following steps, illustrated in Figure 2-35, should be taken:

1. Identify and describe the elements in sequence and assign the TMU required as developed on the analysis sheet.

2. Convert the TMU to equivalent time values. To convert, use the conversion factors given in this chapter under Units of Time.

3. Insert the allowance factor to compensate for personal needs, unavoidable delays, fatigue, and the like in the "Percent allowance" column.

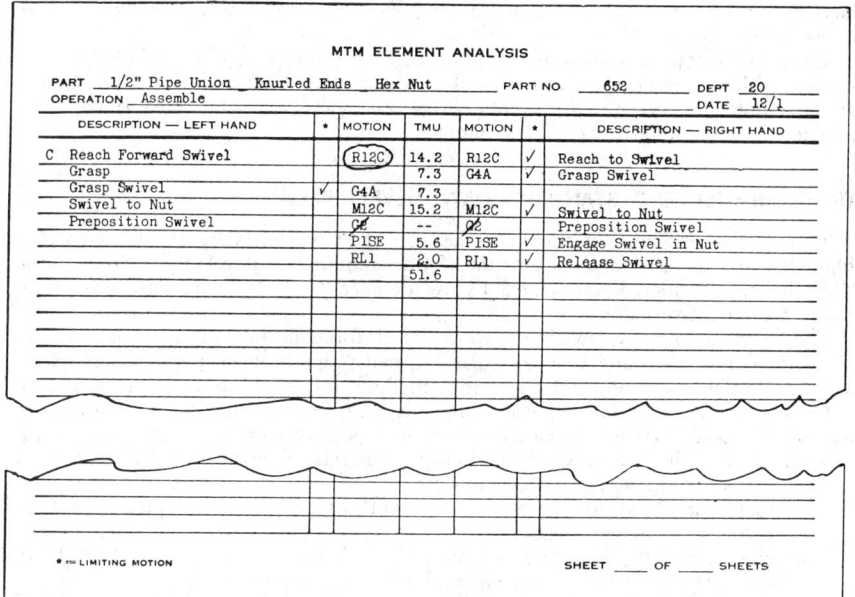

FIG. 2-36. MTM analysis of element *C* of the operation shown by Figure 2-35.

4. Compute and record the allowed element time under the column "Time plus allowance."
5. Insert the frequency of occurrence of each element under "Occurrence per cycle."
6. Determine and record the allowed time per element per cycle in the "Total time required" column.

The allowed time per unit may be computed by totaling the last column. It is then recorded on the bottom line as allowed time.

The study should always be carefully checked for mathematical errors as well as errors in applying the MTM procedure. It may finally be signed to indicate completion.

LIMITING MOTIONS

In performing most industrial operations, it is desirable to have more than one body member in motion at a time. Usually, by having two or more body members in motion at one time, the most effective method of performing an operation is approached. If two or more motions are combined or overlapped, all can be performed in the time required to perform the one demanding the greatest amount of time, or the limiting motion.

Combination motions are those which occur when two or more motions are performed by the same body member at the same time. For example, in turning a part in the hand while moving it to a destination, the Turn is performed in combination with the Move. Also, while moving a part to a destination, the hand may Regrasp or Reposition the part. Combination motions in most cases have a Move as one of the motions. When such motions are encountered, the one requiring the greatest amount of time is considered the limiting motion. The standard value for the limiting motion is the value for the combined motions.

Simultaneous motions are those which are performed simultaneously by two or more body members. When the time values of the simultaneous motions are unequal, the natural rhythm of the body tends to make the members begin and end their motions at the same time. Consequently, the one with the higher time value is the limiting one.

The simultaneous motions table, a basic part of Figure 2-1, is a guide to limiting motions although it does not apply in every case. It is broken down as to whether an operation is nonrepetitive, semirepetitive, or repetitive. The table is compact, easy to follow, and will cover most situations.

TIME FORMULA AND STANDARD DATA DEVELOPMENT

Two of the original objections to the use of stopwatch time studies were the high cost and the considerable amount of time required to develop the time values. Thus it was considered impractical to use detailed time study on jobs other than those of a repetitive nature.

The answers to these objections were found through the development and use of standard time data and time formulas. When properly developed, both provided a quick, consistent, and understandable method for establishing time standards.

Further refinements in operating methods and mass-production techniques, as well as the need for time measurement of nonrepetitive work such as maintenance, toolmaking, and the like, created previously unforeseen objections to data developed by time studies. Among these objections are:

1. Such data tend to be inconsistent and inaccurate when applied to high or low production extremes.
2. They are costly to develop because of the length of time required to obtain sufficient studies to measure all conditions and cases accurately.

MTM has provided a solution to these objections. Standard data developed through MTM or a combination of MTM and time study enable one to:

1. Develop standard data in much less than the time required to construct similar data by use of conventional time study techniques.
2. Obtain far greater accuracy and scope of coverage.
3. Minimize and isolate variables; hence increase understandability.
4. Record methods and time in far greater detail. Hence, questions which may arise at some future date can be more easily answered.
5. Achieve universal application.

Standard data have been successfully developed by the use of MTM for nearly every type of industrial and clerical work, ranging from clerical and business equipment operations through maintenance and building construction. These applications have shown that the application of MTM to this field is almost limitless. The development of time formulas and standard data is thoroughly discussed in Chapter 8 of Section 3.

MTM AND PRACTICAL OFFICE COST CONTROLS

Both business and government agencies are faced with an ever-increasing burden of clerical work. As the task of managing becomes more complex, managers need more reports, information, and other operating documents at a faster pace than ever before. This need has caused administrative costs to soar. At the same time, it has led to demands for more clerical help, supervisors, data processing technicians, and other administrative manpower. To complicate the problem further, labor costs have simultaneously increased and shortages of skilled clerical help have developed. Improvements in methods and procedures and equipment help reduce costs. But in addition, some form of cost control is needed.

Clerical work measurement using the MTM procedure has been widely and successfully used to determine how long it should take to perform all kinds of clerical work. With the work load evaluated, a practical balance between work load and manpower can be obtained.

Clerical Standard Data. The MTM procedure may be applied to the complete range of office and clerical work, ranging from the repetitive use of calculators and typewriters to reading, writing, and other similar semicreative, nonstandardized types of clerical jobs. The advantages of MTM clerical data are similar to those found in the application of MTM to manufacturing operations. Among the major uses are:

1. Determining manning schedules
2. Equipment selection
3. Forms design
4. Developing effective work methods
5. Employee evaluations
6. Wage incentives
7. Operator training
8. Systems comparisons, improvements, and design

MTM can be most effectively used for clerical work measurement in the standard data format. The flexibility of MTM clerical standard data provides clerical work standards which can be consistent and precise. The standard data can be constructed, tested, and proved in a fraction of the time normally required to develop data of equal accuracy by conventional work measurement techniques.

The Clerical Control Procedure. To obtain the maximum benefit from work measurement, it is important that the work be analyzed in the proper manner. Generally, the routine steps the analyst follows are:

1. Obtain all pertinent information about personnel, forms, reports, equipment, and schedules.
2. Improve methods by analyzing the workplaces, forms, reports, equipment used, and operator methods.
3. Select and measure the tasks. After improvements have been made or at least decided upon, list the duties of the worker and develop the time required for the work.

4. Establish the necessary controls. This involves the counting of production and the necessary control reports.

A common mistake is to start an office cost control project with step 3. The collection of all the information related to the operation is often neglected so as to set standards quickly. Methods improvement and standardization possibilities are ignored. Shortcutting the procedure in this manner will result only in incorrect and inconsistent clerical standards which will give little or no control. Figure 2-37 illustrates the steps which should be followed in applying the clerical control procedure.

Forms, Reports, and Equipment. A list of all forms used, all reports prepared, and all office equipment used should first be compiled. A typical example of a list of forms is shown in Figure 2-38. The information includes number and title of the form, number of copies per set, annual use, where used, and any other data considered pertinent. The report list is usually shorter than the forms list and will include reports made up by the supervisor as well as his employees. The information normally collected includes such things as identification, frequency, preparer, and the estimated hours used to prepare the report. In preparing a list of office equipment, the analyst is concerned with the kind and amount of

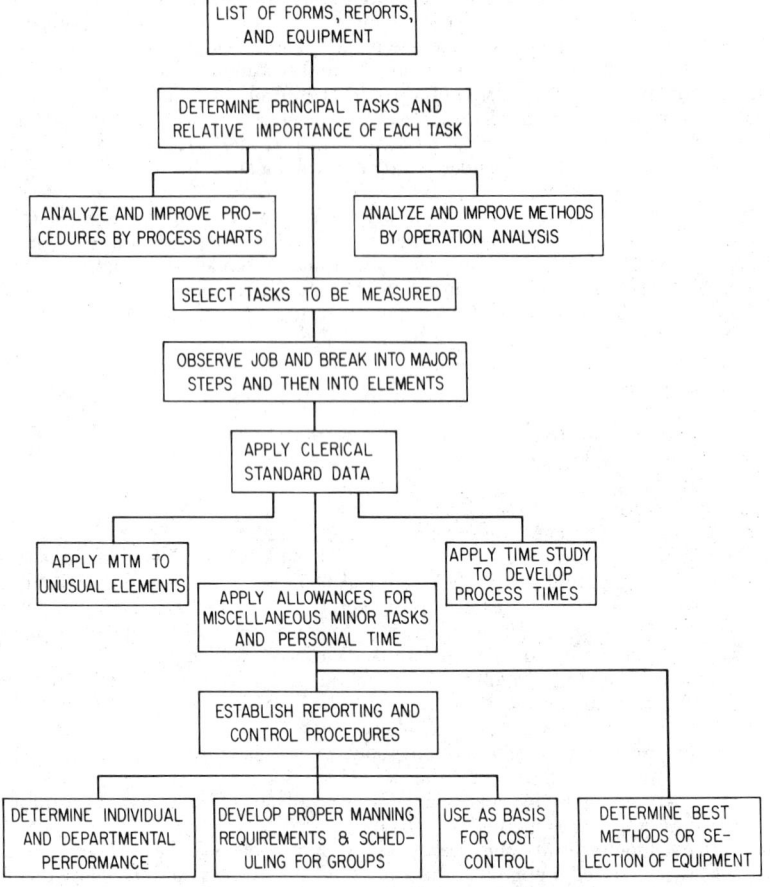

FIG. 2-37. Clerical cost control procedure.

List of Forms

Unit: Central Order Section					Date: 6/14/—
No.	Form no.	Title	No. copies	Annual use	Used by	Notes
1	6 MC 52	Previous Purchase Record	1	8000	C.O.S.	Posted to and filed
2	6 MC 518	(Postal card)	1	5000	C.O.S.	Typed and mailed
3	6 MC 384	Bidder's List Record	1	1000	C.O.S.	Posted to and filed
4						

Fig. 2-38. Typical list of forms.

equipment and the hours that it is operated daily. Once the analyst knows the various forms, reports, and equipment, he has a good general idea of the kind of work as well as how work counts can be obtained.

Determine Principal Tasks. For each employee in the area covered, a task list should be prepared, including the estimated time spent on each task. When the lists are completed, they should be reviewed and approved by the immediate supervisor to obtain agreement on the relative importance of each task. Figure 2-39 illustrates a task list with five duties, including such daily tasks as "mail invoices" and such intermittent work as "maintain record cards of customers' names and addresses." Task lists for all positions are then consolidated into a single form known as the Work Distribution Chart for the purpose of getting an overall picture of the work in the department. As can be observed from Figure 2-40, the Work Distribution Chart

Name: Mary Smith	Occupation or title: file clerk		Classification:
Department billing	Section	Supervisor: Brown	Date: 7/31/—

Task number	DESCRIPTION	Performed	Quantity	Hours per week
1	Check customer's name and address for billing	Daily		15
2	Mail invoices – last 2 hours each day	Daily	200	10
3	File invoices – when time permits, about 2 hours a day	Daily		10
4	Maintain record cards of customer's name and address			
	– approximately 1 1/2 hours, 2 times a week	Weekly		3
5	Weekly supply order	Weekly		2
			Total	40

Fig. 2-39. Task list.

Billing Department

No.	Task	Total hours	Carey	Bailey	Smith	Jones	Evans	Anderson	Harris	Roberts	Murphy
	Process invoices:										
1	Domestic	48	8	40							
2	Export	12	12								
3	Check customer name	25			15						10
4	Maintain customer name records	3			3						
5	Price invoices	25				25					
6	Maintain price records	5				5					
7	Extend invoices	25				5	20				
8	Check extensions	20				5	15				
	Type invoices:										
9	Domestic	60						30	20	5	5
10	Export	15						10	5		
11	Type letters	10							10		
12	Type reports	4								4	
	Proofread invoices:										
13	Domestic	35								15	20
14	Export	10								10	
15	Control invoices	15	15								
16	Mail invoices	10			10						
17	File invoices	10			10						
18	Special projects	12	5						3	4	
19	Miscellaneous	16			2		5		2	2	5
	Total hours	360	40	40	40	40	40	40	40	40	40

FIG. 2-40. Work Distribution Chart.

clearly brings out that a number of different people work at the same tasks, and further, that the hours spent by each employee on the same task may differ considerably.   The distribution also shows what tasks require the most hours.

Analyze and Improve Methods and Procedures.   Before any time standards are developed, it is important that major improvements be made in methods and that the work methods of the operators be standardized.   The time standards, when they are developed, will then reflect the improved procedures.

In offices where little or no methods and procedure work has been performed, certain techniques are very useful in analyzing operations for improvement.   Two of the most important and widely used techniques for work improvement are process charts, such as the one shown by Figure 2-41, and the operation analysis procedure discussed in Chapter 4 of Section 2.

Select Tasks to Be Measured. In selecting the tasks to be measured, the Work Distribution Chart, Figure 2-40, should be utilized. The objective is to measure as large a percentage of the tasks as is economically justifiable. If an office employee works one day a month on a scrap report, this task might not be practical to study, nor would the ten to twenty minutes a day she might spend obtaining supplies. The tasks selected for clerical work measurement should require significant time, they should have a relatively large volume of output, and they should be easy to count.

Although some of the work will not be measured and thus will be cataloged as nonstandard, this need not cause an appreciable loss of control as long as the nonstandard work does not represent a large percentage of the total hours worked by the employees.

Operation number			Inv./ week	Est. time* per week
		Mail received		
1		Open, stamp, and sort by group	3,500	14
2		Check extension – selected invoices	1,500	15
		Carry invoices to vendor code clerk		
3		Alphabetize invoices, apply vendor code	3,500	70
		To checker		
4		Check vendor code, discount and separate	3,500	45
		To process clerk		
5		Check accounting code, batch invoices for IBM control	3,500	100
		To tabulating room for card punching		
		Cards punched		
		Invoices returned from tabulating room		
6		Alphabetize and interfile invoices	3,500	45
		Hold for weekly payment		

*From task lists

Fig. 2-41.   Typical process chart for preparing invoices for payment on tabulating equipment.

Observe the Operation. When studying a specific operation, the analyst will first observe it directly and break it into major steps to simplify the analysis. The steps are arbitrary subdivisions and have no bearing on the accuracy of the work standard. For example, when the analyst observes tasks 16 and 17, "mail invoices" and "file invoices" (Figure 2-40), he finds that the two tasks are performed in succession by the same person. When the clerk finishes mailing the day's invoices, she normally proceeds to file copies of them. Because both tasks are measured by the same count and are performed by the same person, the two tasks can be combined into one task for measurement and control. The major steps of the operation "mail and file invoices" might be identified as follows:

1. Prepare invoices for mailing.
2. Put in envelope and deliver.
3. Log the order number and sort for file.
4. File.
5. Make new file folder.

The analyst will then break each step into elements for the purpose of evaluating the operation. An element is a series of motions which accomplishes some part of the operation. Thus, step 3 had a number of elements such as open desk drawer, get invoices, and copy order number in register. This is illustrated in Figure 2-42 by line entries 1, 3, and 7.

No.	Element	Element		Batch		Invoice	
		Ref.	TMU	No.	Time	No.	Time
1	*Make ready* Open/close desk drawer	2A	74		74		
2	Get/replace register	1A	54	2	108		
3	Get invoices	1A	65	3	195		
4	Get and use pencil	1A	52		52		
5	*Post order number* Note number on top invoice	5A	45	3	135		
6	Find page in register	5D	126	3	378		
7	Copy order number on register	4B	120				120
8	Turn over invoice	1A	32				32
9	*Sort invoices* Get and jog invoices	1C	107	2	214		
10	Turn sheets to jog ends	1C	33	2	66		
11	Sort invoice by class	1E	45				45
12	Pick up second sheet	1A	30			⅓	10
13	Staple sheets	2C	47			⅓	16
14	Arrange invoices in sequence	1E	121				121
	Total				1222		344

FIG. 2-42.   Element 3: log order number and sort for file.

2A
DESK DRAWER
OPEN AND/OR CLOSE DESK DRAWER
Start: hand over desk   End: hand on drawer

Drawer	Distance opened or closed	Open and close	Open drawer only	Close drawer only: location of hand when "close" starts	
				In or at drawer	Over desk
Center...................	Part 5″	39	25	14	17
	Full 12″	104	55	49	53
Side-top..................	Part 9″	54	33	21	31
	Full 18″	71	38	33	41
Side-middle...............	Part 9″	57	36	21	32
	Full 18″	74	41	33	43
File drawer...............	Part 9″	71	43	28	39
	Full 18″	90	49	41	51

FIG. 2-43. Standard data for desk drawer.

Apply Office Standard Data and MTM. In the example, Figure 2-42, the analyst determines at the outset that subelements fall into two main categories. There are subelements that occur once for each batch of invoices handled and subelements that occur for each invoice. As the analyst makes his study, he keeps the time for these two categories separate.

The clerk begins by getting the sales register which is kept in a desk drawer. The first subelement then is to open the desk drawer. This is covered by standard data table 2A (Figure 2-43). Table 2A shows that the analyst must know which drawer is used and how it is used. In this case, it is the side-middle drawer, and it must be opened all the way or approximately 18 inches.

Under these conditions, the time to open the drawer is 41 TMU. This value could be used for the first subelement. However, the analyst realizes that when the clerk has finished this batch of invoices, she must replace the register in the drawer. She does not need to close the drawer until the end of the job. For convenience, the analyst combines the two subelements into a single element, "open and close desk drawer." This saves the writing of one subelement and eliminates the possibility that the analyst might overlook the closing of the drawer at the end of the analysis. The time for "open and close desk drawer" is 74 TMU, as shown in table 2A (Figure 2-43). Therefore, 74 TMU and the reference 2A are recorded after the first subelement (see Figure 2-42). The time is also recorded in the column for each "batch." All succeeding subelements are analyzed in the same manner.

The derivation of the time in table 2A for "open and close desk drawer" is illustrated in Figure 2-44. The MTM analysis shows that a motion pattern for "open drawer only" is reach, grasp, pull drawer out, and release. The time for this motion pattern is 41 TMU. For "close drawer," the reach and close motions require 33 TMU. The time for "open and close desk drawer" is then 74 TMU.

Apply Allowances. Allowances must be added to cover the many interruptions the employees have throughout the day, the incidental duties too frequent or too small to be measured directly, and the personal time that is needed by the employees. To obtain the total allowed time for the task, the necessary allowances must be added to the basic time.

Figure 2-45 illustrates the application of allowances and the computation of

Open desk drawer, side-middle, and 18″:		
Description	Symbol	TMU
Reach 22″....................	R22B	20.1
Grasp.......................	G1A	2.0
Pull drawer out..............	M18B	17.0
Release.....................	RL1	2.0
Total.....................		41.1, or 41
Close drawer:		
Reach 16″....................	R16B	15.8
Close drawer................	M18A	17.6
Total.....................		33.4, or 33
Open and close drawer, total........................ 74		

Fig. 2-44. MTM analysis of "open and close desk drawer."

the final standard for mailing and filing invoices. The basic time per invoice is 953 TMU. The basic time per batch is 4,617 TMU. This, divided by the average batch size of 200, gives a time per invoice of 23 TMU. The total time per invoice is thus 976 TMU. The allowances of 8 percent for official talking, related duties, and variability and 10 percent for personal time are then added to give a total standard time of 1,159 TMU, or 0.0116 hour. This means that a clerk handling this job should process 100 invoices in 1.16 hours. She can actually do the job in less time by steady application to the work.

Establish Reporting and Control Procedures. The final and one of the most

Element no.	Description	Time per batch	Time per invoice		
			Unit time	Freq.	Invoice time
1	Prepare invoices for mailing	1,025	194	1	194
2	Put in envelope and deliver	1,610	165	1	165
3	Log order no. and sort for file	1,222	344	1	344
4	File	760	225	1	225
5	Make a new file folder		615	1/25	25
	Basic time	4,617			953
	Average batch	200			
	Batch time per invoice	4,617/200			23
	Total basic time				976
	Miscellaneous minor tasks:				
	Official talking	5%			
	Related duties	2			
	Variability	1			
	Total	8%			78
	Subtotal				1,054
	Personal time	10%			105
	Total standard time, TMU				1,159
	Total time, hours				0.0116
	Number invoices per hour				86

Fig. 2-45. Computation of final standard in mailing and filing invoices.

ADMINISTRATIVE MANPOWER REPORT

Department	Actual hours used			Output utilization		Budget cost		
	Holi-day vaca-tion	Ex-cused absent	Avail-able worked	Hours earned*	Utili-zation %	Budget $	Actual $	Gain (loss)
Order service.......	21	100	1,000	900	90	2,700	2,730	(30)
Accounts payable......	20	25	485	390	80	1,170	1,300	(130)
Accounts receivable....	10	19	300	250	83	750	700	50
Accounting pool.........	8	8	34	20	60	60	120	(60)
Monthly total..	59	152	1,819	1,560	85	4,680	4,850	(170)
Year to date...	1,318	684	18,198	14,558	80	44,200	47,000	(2,800)
Percentage.....	7	4.0	100	80	...	100	106	−6

* Volume times allowed hours.

FIG. 2-46.   Illustration of manpower report.

important steps of the clerical control procedure is to institute controls. Standards by themselves, without good control procedures, have limited value in reducing office costs. Both the supervisor and his employee may agree that the employee can handle twenty documents an hour, but if no record is kept of how many documents are actually handled, passing time will soon erase the importance of twenty documents per hour as a yardstick of performance. Target times influence action by management, but to bring about real and lasting results, the final step—control—must be carefully installed and maintained.

The reports may show performances, both individual and departmental; the percent utilization of personnel; scheduling of the various jobs; and manning requirements, both present and future. Manpower reports such as the one shown in Figure 2-46 are based on key operations or key indicators of activity level, such as sales orders prepared, invoices sent to customers, or payroll checks issued.

In summary, extensive applications have proved the MTM procedure to be a most effective tool for the development of clerical controls which are applicable to the wide range of conditions that exist in office work. Equally important is its use in employee training. With clerical methods accurately described as an inherent requirement of the MTM procedure, the effectiveness of clerical training techniques is substantially improved.

MTM AND DETAILED MOTION STUDIES WITHOUT MOTION PICTURES

The MTM procedure is a most effective tool for making detailed motion studies. Refined motion studies are often made with the aid of motion pictures and the micromotion study technique. The drawback here is that there first must be a qualified operator doing the job. Expensive camera equipment and an experienced camera operator are required. Other deterrents are setting up the camera equipment, preparing the operator and workplace for filming, and later, the delay in processing the film before the detailed motion analysis can even begin. All this is eliminated when motion studies are made with MTM.

Figure 2-36 illustrates how the MTM procedure can be used to make detailed motion studies without resorting to motion pictures.   The MTM conventions and symbols that denote the basic motions and their related variables, along with brief descriptions of the motions, present a quick understanding of the motion pattern and its characteristics.

BIBLIOGRAPHY

Annual Conference Proceedings, The MTM Association for Standards and Research, Fair Lawn, N.J., annually since 1952.

Antis, William, John M. Honeycutt, Jr., and Edward N. Koch, *The Basic Motions of MTM*, The Maynard Foundation, Pittsburgh, 1968.

Hancock, Walton M., and Ulf Åberg, *Design Criteria of Predetermined Time Systems, with Special Reference to the MTM System*, International MTM Directorate, Stockholm, Sweden, 1968.

Hasselqvist, Olle, Per Söderström, and Alf Wiklund, *MTM:s Grundrörelser*, Utgivare Svenska MTM-gruppen AB, Stockholm, Sweden, 1962.

Karger, Delmar W., and Franklin H. Bayha, *Engineered Work Measurement*, Industrial Press, New York, 1966.

Maynard, H. B., W. M. Aiken, and J. F. Lewis, *Practical Control of Office Costs*, Maynard Research Council Incorporated, Pittsburgh (formerly published by Management Publishing Corporation), 1960.

Maynard, H. B., G. J. Stegemerten, and J. L. Schwab, *Methods-Time Measurement*, McGraw-Hill Book Company, New York, 1948.

Methods Time Measurement Application Training Supplements: *Position*, no. 8; *Apply Pressure*, no. 9, The MTM Association for Standards and Research, Fair Lawn, N.J., 1965.

MTM Journal, The MTM Association for Standards and Research, Fair Lawn, N.J.

Chapter 3

The Work-Factor System

JOSEPH H. QUICK

Senior Vice President, The Science Management Corporation, Moorestown, New Jersey; and Senior Vice President, The Wofac Company, Moorestown, New Jersey

JAMES H. DUNCAN

President, The Science Management Corporation, Moorestown, New Jersey; and Chairman of the Board, The Wofac Company, Moorestown, New Jersey

JAMES A. MALCOLM, JR.

Vice President and Director of Research, The Wofac Company, Moorestown, New Jersey

Work-Factor[1] is an elemental time system for compiling time standards to establish the expected productivity of the human when performing useful manual and mental work. Stopwatches and other timing devices are not used with Work-Factor except for machine or process times. Procedures for specifying motion and work element

[1] "Work-Factor" is the registered service mark (trademark) of the Science Management Corporation.

difficulties do not use comparative terms or verbal classification. In the Work-Factor system, these procedures are specified numerically and in terms which do not require interpretation by the analyst. Hence, consistency is achieved worldwide.

DEVELOPMENT AND TEST OF THE WORK-FACTOR SYSTEM

The idea for the system was conceived in Philadelphia, Pennsylvania, between 1930 and 1934, because rate setting with stopwatch time study was unacceptable to union employees working under tightly controlled incentive pay systems. The original motion time data were developed between 1934 and 1938 by a group of experienced time study engineers directed by Joseph H. Quick, assisted by Samuel F. Benner, William J. Shea, and Robert E. Koehler. The purpose of the 1934 research was to develop an objective work measurement technique to eliminate the stopwatch and the observer's performance rating judgments in establishing work standards and setting rates.

The Motion Times were first published in May, 1945.[2] In 1947, the Work-Factor system was introduced to industry and business at large. James H. Duncan and James A. Malcolm, working with Quick, developed techniques to simplify its application. Slow-motion picture enlargements, analytical procedures, and test timings made it possible to identify and count the number of sequential motions in complex work elements such as Grasp part from pile, Pre-position part for assemble, Assemble parts together, and the like. From these studies, the Work-Factor Tables of Standard Element Times were established, thereby eliminating the need for much of the detailed motion analysis in standards setting. In 1962, the Work-Factor textbook was published.[3]

In 1949, research and investigation were begun by Quick, Duncan, and Malcolm on the problems of measuring the human mental processes occurring in useful work. These studies led to the compilation of the Mento-Factor Manual of Detailed Work-Factor Mental Process Times in 1965. The Mento-Factor system is now applied without the stopwatch, to measure all useful mental work activity except the creative mental functions.

Four Levels of Application Detail. The requirements of work measurement vary according to the factors of operation, uniformity, frequency of repetition, purpose of measurement, and the like. The Work-Factor system consists of four integrated systems with levels of detail to accommodate the measurement situation.

Detailed Work-Factor—for highly repetitive short-cycle operations. The Work-Factor Time Unit equals 0.0001 minute.

Ready Work-Factor—for longer cycle and less repetitive operations. The Ready Work-Factor Time Unit equals 0.0010 minute.

Abbreviated Work-Factor—for custom operations, with very little repetition such as short-order work, maintenance, or toolroom. The Abbreviated Work-Factor Time Unit equals 0.0050 minute.

Work-Factor Standard Data—for rapid standards and rate setting in the measurement of all human work situations.

Although no changes in motion time values have been required since their publication in 1945, development of the Work-Factor System continues for refinement of its Rules of Application, extension of its conventions to cover technological changes in work methods and environments, and adaptation to high-speed computer procedures.

    [2] J. H. Quick, W. J. Shea, and R. E. Koehler, "Motion-Time Standards," *Factory Management and Maintenance* (now *Modern Manufacturing*), May, 1945. As a result of this publication, the Work-Factor system became known in Europe for a short time as the QSK system, after the authors' initials.
    [3] Joseph H. Quick, James H. Duncan, and James A. Malcolm, Jr., *Work-Factor Time Standards*, McGraw-Hill Book Company, New York, 1962.

DEFINITIONS OF WORK-FACTOR TERMS

Work-Factor Time. Work-Factor Time[4] is defined as that time required for the Average Experienced Operator working with good skill and good effort (commensurate with physical and mental well-being) and under standard working conditions to perform one work cycle or operation, on one unit or piece, according to prescribed method and specified quality. Work-Factor Time includes no allowance for personal needs, fatigue, environmental unavoidable delays, or incentive payment.

Work-Factor Time is not comparable with times referred to as normal, daywork performance, sixty-minute-hour performance, or other terms used to indicate the work pace expected of the average worker who performs without incentive or at a level of productivity commensurate with base rate output.

The Work-Factor Motion Time Tables provide time values in terms of Work-Factor Time. Allowances for fatigue, environmental unavoidable delays, and personal needs are added to the Work-Factor Time in accordance with the working conditions involved. Once equitably established, they need not be varied except for special situations involving unusual working conditions. For abnormal cases, it is good practice to make work sampling studies to determine appropriate changes to the established allowances.

Adequate Task Physique. Adequate Task Physique is the Work-Factor term for the physical capability of workers to perform a given operation or task, at rates commensurate with Work-Factor Time. When the worker has an Adequate Task Physique, he works without undue fatigue, hazard, or physical strain, and maintains physical well-being throughout the working day.

Adequate Task Intelligence. Adequate Task Intelligence embraces the concept that the Average Experienced Operator has sufficient intelligence to perform the task involved at a rate commensurate with Work-Factor Time. It is expected that the subject, intent, procedures, tools, reading material, and all other items related to the task are within the operator's understanding and experience.

Adequate Task Vision. Adequate Task Vision is the Work-Factor term for visual capability adequate to perform properly the Visual Task involved at a rate commensurate with Work-Factor Time. Visual Tasks with fine detail, poor illumination, fine hue distinctions, or weak contrasts may require better than so-called normal vision. Visual Tasks involving large shapes and masses, bright hues and contrasts, and the like often can be adequately performed with poor or so-called subnormal vision.

Work-Factor Analysis. Work-Factor is a method of determining the time for a given motion pattern by (1) making a Detailed Analysis of each motion based on the identification of the Four Major Variables of work and the use of Work-Factors as a unit of measurement and (2) applying to each motion the proper Work-Factor Time Value contained in the Motion Time Table.

Basic Motion. A Basic Motion is any motion the performance of which involves the least amount of difficulty or precision for any given distance and body member combination, for example, tossing a small, nonfragile object into tote pan.

Work-Factor. Work-Factor is a unit used as the index of additional time required over and above the Basic Time when motions are performed involving the following Work-Factor Variables:
1. Manual Control
2. Weight or Resistance

[4] The times for manual movements or thought processes established by Work-Factor times are not comparable with the maximum or possible speed which can be attained by humans. Emergency reactions, such as in the application of automobile foot brakes, striking in self-defense, or other sudden mental or manual actions, represent a special study not contemplated in these data. The term "Work-Factor Time" as used here replaces the term "Work-Factor Select Time" formerly used. There is no change in concept.

Work-Factor Standard Elements of Work. Work-Factor Standard Element is the term applied to the basic divisions of work, such as Transport, Grasp, and Assemble.   All manual work consists of one or more Work-Factor Standard Elements, which are themselves composed of one or more motions.
Motion Analysis. The Motion Analysis is the combination of Work-Factor symbols used to identify a motion for direct reference to the Motion Time Table in selecting the proper Work-Factor Time Value.

THE FOUR MAJOR VARIABLES

The Work-Factor system is based on the principle that the Four Major Variables which affect the time to perform manual motions are:
1. Body Member used (identified by exact definition)
2. Distance Moved (measured in inches)
3. Manual Control required (measured in Work-Factors, defined or dimensional)
4. Weight or Resistance involved (measured in pounds, converted to Work-Factors)

Body Member. The Body Member used is identified by observation through the application of an exact definition.   Work-Factor Motion Times have been compiled for each significant Body Member.   The distinguishing characteristics of each of these members is shown below.

Finger—Hand. Includes any movement of the fingers and thumb above the knuckles, or any finger joint and movements of the hand about the wrist joint.   May also involve simultaneous movements of the fingers and thumb.

Arm. Includes all movements of the lower arm around the elbow (except swivels) and all movements of the whole arm hinged at the shoulder (except swivels).   Movements of the hand, lower arm, and fingers may occur simultaneously.

Forearm Swivel. Lower arm swivels at the elbow as in turning a doorknob (rotates about the axis of the forearm), or full arm is extended and swivel point is at the shoulder (rotates about the axis of the extended arm).

Trunk. Includes any movement of the trunk of the body in a forward, backward, sidewise, or swiveling motion.

Foot. The foot pivots at the ankle.   The upper and lower leg normally remain in practically a fixed position.

Leg. Includes movements of the lower leg about the knee, movements of the knee when the foot remains in a fixed position, movements of the leg about the hip joint, and movements in which the knees bend and the feet remain fixed, such as coming to or arising from a sitting position.

Head. Head Motions involve moving in relation to the trunk.   Head Motions frequently occur simultaneously with other Body Member Motions requiring greater time.

Distance. Distance is perhaps the easiest of the variables to recognize and determine because it is readily measured with a scale.   All Distances, except those with a change in direction, are measured as a straight line between the starting and stopping points of the motion arc described by the Body Member.   The actual motion path is measured only when a change in direction is involved.   Below is a list of the points at which Distance should be measured for the various Body Members:

Body Member	Point of measurement	Body Member	Point of measurement
Finger or hand......	Fingertip	Trunk.........	Shoulder
Arm..............	Knuckles (use knuckle having greatest travel)	Foot...........	Toe
		Leg............	Ankle, knee, or hip
Forearm swivel......	Knuckles	Head Turn......	Nose

Manual Control. Control is the most complex of the variables influencing manual motions. Its evaluation is made more difficult because there is no physical measuring device, such as the foot rule or balance, with which its effects can be checked. To provide definite units of measurement, as inches or pounds, the Manual Control Work-Factor has been established and given specific dimensional limitations and positive Rules of Application for identification as a means of accurately evaluating the types and degrees of Manual Control, and hence difficulty. Although a great many control types could be established, the following four, singly or in combination, are most commonly encountered in manual work. (Classifications based on research data show that 95 percent of all work motions can be so identified.)

Definite Stop Work-Factor. Manual Control required to terminate a motion with a Definite Stop Work-Factor is limited to movements terminated at the will of the operator and does not include movements arrested by a physical obstruction.

Steer Work-Factor. Manual Control required to direct or steer a motion through a limited clearance or toward a small target area.

Precaution Work-Factor. Manual Control required, or precaution exercised, to prevent damage or injury, or to maintain Manual Control as a necessary function of the motion (other than directional).

Direction Change Work-Factor. Manual Control required to change the direction of motion, such as that required in moving around an obstruction.

Weight or Resistance. Weight or Resistance involved in manual motions is measured in pounds for all Body Members with the exception of the Forearm Swivel. Because Forearm Swivel Motions are rotary, they are measured in pound-inches of torque.

An extensive study was made of the effect of weight and resistance on motion time. It was found that the effect of weight on time varies with several factors, most important of which are:

1. Body Member used
2. Sex of operator

As in the case of Manual Control, the Work-Factor for Weight (Resistance) is used as an index of the effect of weight on the time required to perform manual motions.

The factor of Resistance is commonly found in movements such as bending heavy wire, pushing against a spring, assembling tight-fitting parts, and rubbing and sanding. The research data indicate that the same principle governs the time effect of Resistance and that of Weight.

THE WORK-FACTOR

All Four Major Variables actually affect the time required to make manual motions. The Work-Factor is a unit devised for identifying the effect of only two of the four variables, Weight and Control. The other two Variables, Distance and Body Member, are measured in terms of the inches and the Body Member used, respectively. The Work-Factor can be considered as merely a means of describing the motion according to the amount of Control or Weight (Resistance) involved in its performance. Because the value of a Work-Factor in terms of time has been established in tabular form, only specific dimensions and rules are needed to determine the number of Work-Factors involved in a given motion.

There is an unlimited number of Basic, 1 Work-Factor, 2 Work-Factor, or 3 Work-Factor motions, all varying in function, yet requiring equal time to perform as long as an equal number of Work-Factors is involved. The complexity of the motion is dependent on the amount of Manual Control, Weight, or Resistance involved, which in turn determines the number of Work-Factors.

The simplest or Basic Motion involves no Work-Factors. As complexities are introduced to a motion, they add Work-Factors, and consequently, time. It makes no difference whether a given motion involves 2 Work-Factors of Control and 1 of Weight, or 2 of Weight and 1 of Control—the motion is valued at 3 Work-Factors, and can be identified as such on the Motion Time Table. The nature of the individual Work-Factors does not affect time; it is the total number of Work-Factors that determines the time required.

The use of the Work-Factor makes it possible to construct for any Body Member a simple and concise motion time table with only two Variables, Distance Moved and number of Work-Factors.

THE WORK-FACTOR MOTION TIME TABLE

The Work-Factor Motion Time Table (see Table 3-1) includes all Work-Factor Motion Time Values in tabular form. When a motion has been identified according

TABLE 3-1. Work-Factor Motion Time Table for Detailed Analysis Time in Work-Factor Units

Transport											
Motion Distance, in.	Basic	Work-Factors				Motion Distance, in.	Basic	Work-Factors			
		1	2	3	4			1	2	3	4
Arm (A): measured at knuckles						Leg (L): measured at ankle					
1	18	26	34	40	46	1	21	30	39	46	53
2	20	29	37	44	50	2	23	33	42	51	58
3	22	32	41	50	57	3	26	37	48	57	65
4	26	38	48	58	66	4	30	43	55	66	76
5	29	43	55	65	75	5	34	49	63	75	86
6	32	47	60	72	83	6	37	54	69	83	95
7	35	51	65	78	90	7	40	59	75	90	103
8	38	54	70	84	96	8	43	63	80	96	110
9	40	58	74	89	102	9	46	66	85	102	117
10	42	61	78	93	107	10	48	70	89	107	123
11	44	63	81	98	112	11	50	72	94	112	129
12	46	65	85	102	117	12	52	75	97	117	134
13	47	67	88	105	121	13	54	77	101	121	139
14	49	69	90	109	125	14	56	80	103	125	144
15	51	71	92	113	129	15	58	82	106	130	149
16	52	73	94	115	133	16	60	84	108	133	153
17	54	75	96	118	137	17	62	86	111	135	158
18	55	76	98	120	140	18	63	88	113	137	161
19	56	78	100	122	142	19	65	90	115	140	164
20	58	80	102	124	144	20	67	92	117	142	166
22	61	83	106	128	148	22	70	96	121	147	171
24	63	86	109	131	152	24	73	99	126	151	175
26	66	90	113	135	156	26	75	103	130	155	179
28	68	93	116	139	159	28	78	107	134	159	183
30	70	96	119	142	163	30	81	110	137	163	187
35	76	103	128	151	171	35	87	118	147	173	197
40	81	109	135	159	179	40	93	126	155	182	206
Weight, lb:						Weight, lb:					
Male....	−2	−7	−13	−20	>20	Male......	−8	−42	>42		
Female..	−1	−3½	−6½	−10	>10	Female....	−4	−21	>21		

TABLE 3-1. (Continued)

Trunk (T): measured at shoulder

1	26	38	49	58	67
2	29	42	53	64	73
3	32	47	60	72	82
4	38	55	70	84	96
5	43	62	79	95	109
6	47	68	87	105	120
7	51	74	95	114	130
8	54	79	101	121	139
9	58	84	107	128	147
10	61	88	113	135	155
12	66	94	123	147	169
14	71	100	130	158	182
16	75	105	136	167	193
18	80	111	142	173	203
20	84	116	148	179	209
22	88	121	153	185	215
24	92	125	158	190	220
26	95	130	163	196	226
28	99	134	168	201	231
30	102	139	173	206	236

Weight, lb:

Male....	−11	−58	>58
Female..	−5½	−29	>29

Walk

Type	30-in. paces		
	1	2	Over 2
General....	Analyze from table	260	120 + 80/pace
Restricted.		300	120 + 100/pace

Add 100 for >120° − 180° Turn at Start or Finish of Walk

Up steps (8-in. rise, 10-in. flat).. 126/step
Down steps................. 100/step

Finger—Hand (F, H): measured at fingertip

1	16	23	29	35	40
2	17	25	32	38	44
3	19	28	36	43	49
4	23	33	42	50	58

Weight, lb:

Male....	−⅔	−2½	−4	>4
Female..	−⅓	−1¼	−2	>2

Foot (Ft): measured at toe

1	20	29	37	44	51
2	22	32	40	48	55
3	24	35	45	55	63
4	29	41	53	64	73

Weight, lb:

Male....	−5	−22	>22
Female..	−2½	−11	>11

Forearm Swivel (FS): measured at knuckle

45°	17	22	28	32	37
90°	23	30	37	43	49
135°	28	36	44	52	58
180°	31	40	49	57	65

Torque, lb-in.:

Male....	−3	−13	>13
Female..	−1½	−6½	>6½

Head Turn (HT): measured at nose tip

Degrees turn	Dist-ance, in.	Number of Work-Factors		
		Basic or 1	2	3	4
>22½−45	>2−4	40	51	58	66
>45−90	>4−8	60	76	86	99

Mental Process (MP)—Simple

Focus (Fo)........................ 20
React (Rn)....................... 20
Inspect (I)....................... 30
Mento (Mt).................... 10

to the Four Major Variables, the correct time value can be selected quickly from the following arrangement of the Tables.

Body Member. There is a separate table for each Body Member.

Distance. Each table contains values for various Distances measured in inches or in degrees of rotation for Forearm Swivel.

Weight. At the bottom of each Body Member Table is a Weight section with weight classes in pounds for both male and female operators. These indicate the number of Work-Factors to be used for various Weights or Resistances which may be encountered. Forearm Swivel Motions are measured in pound-inches of torque.

Manual Control. Each table contains five columns of time values. The first, headed Basic, applies to all motions which, according to the Work-Factor Rules of Application and Definitions, have no Work-Factors. The other four columns —headed by 1, 2, 3, and 4 Work-Factors—apply to motions involving the respective number of Work-Factors.

All time values on the Work-Factor Motion Time Table (Table 3-1) are in Work-Factor Time Units of 0.0001 minute.

Selecting Time Values from the Motion Time Table

Example 1. The arm moves 10 inches to toss small object aside (Basic Motion). Because this is an Arm Motion, refer to Arm Table. In the first or Distance Moved column, find 10. Opposite this 10, in the column headed Basic, find the value 42. This is the number of Work-Factor Time Units, or 0.0042 minute for the motion.

Example 2. A man moves a building brick, weighing 4 pounds, 30 inches from a pile to place it on a worktable (Definite Stop and Weight Motion).

First refer to the bottom of the Arm Table to determine how many Work-Factors are involved because of the weight of the brick. Because a male is involved in this case, refer to the line marked Male. The brick weighs 4 pounds and is between the Classes 2 and 7 (pounds). Therefore, the Weight involved calls for one Weight Work-Factor. The motion also requires one Work-Factor to compensate for the Definite Stop involved at the end of the motion in placing the brick on the table. Two Work-Factors are therefore involved. In the Distance Moved column, find 30 inches. Opposite the 30, under the column headed 2 Work-Factors, find the value 119 Work-Factor Time Units, or 0.0119 minute.

WORK-FACTOR NOTATION

Simple notations are used when applying the Work-Factor technique. The motions and conditions involved are described completely so that direct reference can be made to the tables. Each motion is identified in terms of the Body Member used, the Distance Moved in inches, and the number and type of Work-Factors involved. Symbols are employed for the Body Members and Work-Factors. The use of these symbols in combination to identify motions is called the Motion Analysis.

Body members	Symbol	Work-Factors (written in this sequence)	Symbol
Finger.........................	F	Weight (Resistance)...........	W
Hand.........................	H	Steer.......................	S
Arm.........................	A	Precaution..................	P
Forearm Swivel.................	FS	Direction Change.............	U
Trunk.......................	T	Definite Stop................	D
Foot........................	Ft		
Leg.........................	L		
Head Turn...................	HT		

In recording Distances, fractional inches are not used. Motions of 1 inch or less are recorded as exactly 1 inch. Motions more than 1 inch in length are recorded as the nearest integral number. A motion of 10¼ inches is recorded as 10 inches, and a motion of 10¾ inches is recorded as 11 inches.

Recording the Analysis. In recording a Motion Analysis, the Body Member is indicated first; the Distance Moved, second; and the Work-Factors, third. Referring to the examples given previously, the correct recording of these motions is as follows:

Description of motion	Motion analysis
1. Toss small part aside 10 in. (Basic Motion)	A10
2. Move 4-lb brick 30 in. from pile to place on worktable (Weight, Definite Stop Motion)	A30WD

Recurring Motions. When the same motion occurs more than once in sequence, the number of occurrences is written before the Body Member symbol in the Motion Analysis.

Examples

1. If the end-for-end Pre-positioning of a small shaft requires three Basic 1-inch Finger Motions, this series of motions is analyzed and written as 3F1. In operations where certain motions do not occur in every cycle, but do occur a percentage of the time in accordance with the laws of probability, the motions are recorded as described above, and the Analysis is followed by the Percentage of Occurrence.

2. It may be necessary to Pre-position the shafts 50 percent of the times they are picked up, because in 50 percent of the cases they may be grasped in the correct position for use. The Motion Analysis is then written 3F1-50%.

In certain operations, the Distances Moved vary from cycle to cycle. Reaching to grasp bushings from a pinboard is a situation in which the Distances might, for example, vary from 12 to 18 inches, depending on their location on the pinboard. To simplify the Analysis, an average Distance of 15 inches can be used. This is recorded as VA15D. The symbol V is written preceding the Motion Analysis to indicate that the distance is an average Distance. In situations where the number of motions varies, as in the aligning of a shaft to a hole, the average number of motions may be recorded in a similar manner, for example, V2A1S. This signifies an average of two Align Motions.

WORK-FACTOR ABBREVIATIONS AND SYMBOLS

Align	Aln
Apply Pressure	AP
Arm	A
Assemble	Asy
Average	V
Balancing Delay	BD
Blind	B
Blind Distance	bd
Both Hands	BH
Change of Direction	U
Closed Target—Mechanical	CT
Closed Target—Surface	CTS
Contact Grasp	Gr-C
Contact Release	Rl-C
Cylinder	Cyl
Definite Stop	D
Disassemble	Dsy
Dislodge	Dsl
Distance Between Targets	DB
Entangled	e

WORK-FACTOR ABBREVIATIONS AND SYMBOLS (Continued)

Finger... F
Focus... Fo
Foot... Ft
Forearm Swivel................................... FS

Greater than...................................... >
Grasp... Gr
Gripping Distance................................. gd

Hand... H
Head Turn.. HT

Index... Ind
Insert.. Ins
Inspect Interval.................................. I

Left Hand.. LH
Leg... L
Less than... <

Mental Process................................... MP
Mento.. Mt
Move... M

Negligible.. Neg
Nested... n

Open Target—Mechanical.......................... OT
Open Target—Surface............................. OTS

Pace—General.................................... PaG
Pace—Restricted................................. PaR
Permanent Blind.................................. PB
Pick Up.. PU
Plug Dimension................................... p
Precaution....................................... P
Pre-position...................................... PP
Pre-position—Large Objects...................... PP-L
Pre-position—Medium One Hand................... PP-M
Pre-position—Medium Two Hands.................. PP-M$_2$
Pre-position—Optimum Size Object................ PP-O
Pre-position—Very Small Object.................. PP-V

Ratio—Plug to Target............................ r
Reach.. R
React.. Rn
React after Inspect............................... RnI
Reciprocating Motion............................. RM
Regrasp.. ReGr
Relax Pressure................................... RP
Release.. Rl
Right and Left Hands............................. R & LH
Right Hand....................................... RH

Seat... St
Separate... Sep
Short Motion..................................... SM
Simo... s
Simo Factor...................................... SF
Simultaneous..................................... Simo
Slippery.. slp
Solid... Sol
Steering.. S

Target Dimension................................. t
Temporary Blind.................................. TB
Thin Flat... TF
Tolerance.. Tol
Transfer Grasp—Pinch........................... Tr-P

WORK-FACTOR ABBREVIATIONS AND SYMBOLS (Continued)

Transfer Grasp—Wraparound........................ Tr-W
Trunk... T
Turn.. Tu
Tweezers.. tw

Unwrap Release.................................. Rl-U
Up to and including............................. —
Upright... Up
Use... Use

Visual.. v

Weight.. W
Work Area....................................... WA

WORK-FACTOR STANDARD ELEMENTS OF WORK

Work-Factor Standard Elements of Work have been set up to represent the basic divisions of work. They may be composed of a single motion or a series of motions. There are eight Work-Factor Standard Elements of Work.

1. Transport (Reach and Move) (R, M)
2. Grasp (Gr)
3. Pre-position (PP)
4. Assemble (Asy)
5. Use (Manual, Process, or Machine Time) (Use)
6. Disassemble (Dsy)
7. Mental Process (MP)
8. Release (Rl)

During the development of the Work-Factor system, it became apparent that, for ease and simplicity of description, the eight divisions listed above are adequate and practical. The Rules of Application of the Work-Factor system are set up according to the following description of the Standard Elements.

Transport. The Standard Element Transport normally involves a single motion by a given body member. If the body member is moved for the purpose of reaching to a destination, location, or object, the motion is called a Reach. If the body member is moved for the purpose of transporting an object to a new location, the motion is called a Move. Other Standard Elements, such as Grasp and Assemble, may contain a sequence of Transport Motions. The Work-Factor Times for all Reaches and Moves are read as single values directly from the Work-Factor Motion Time Table.

Grasp. The Standard Element Grasp contains all the movements and manipulations which are necessary to obtain Manual Control of an object. It normally begins after the Element Reach, and normally ends as soon as Manual Control of the object or objects has been obtained. Grasp may consist of a single motion or may involve a series of motions.

Pre-position. The Standard Element Pre-position consists of one or a series of motions which are required to orient an object in preparation for performing subsequent work elements with it. Pre-position frequently occurs on a percentage basis, particularly when an object is grasped from a Random Pile. In such instances, the object will sometimes be grasped initially in a usable position and sometimes in an unusable position which will require Pre-positioning.

Assemble. The Standard Element Assemble is the act of joining objects together. It may involve the actual insertion or linking of two or more objects (Mechanical Assemble), or it may involve the simple placing of one object against another in a specific relationship (Surface Assemble).

The Element begins after a move to a point of assembly and ends when the objects have been satisfactorily joined so that the next Element of work can begin.

Use. The Standard Element Use may consist solely of machine time, may be a combination of machine time and manual motions, or may include only manual motions. The manual motions which are involved may be either machine-controlled or operator-controlled. Such Elements as pouring liquids, painting, heating, and cooling are also considered under the Use classification.

Disassemble. The Standard Element Disassemble involves the separating of objects. Although essentially the reverse of the Standard Element Assemble, it is usually simpler in nature.

Mental Process. The Standard Element Mental Process includes all time intervals which depend entirely or partially upon mental reactions or nerve impulses such as Visual Inspect, Decide Time, Compute Time, and Read Time.

Release. The Standard Element Release is the reverse of Grasp. It includes all the motions and manipulations necessary before a body member can move away from an object which has been grasped.

THE WORK-FACTOR STANDARD ELEMENT TRANSPORT (R and M)

As previously explained, the time for a Transport Motion depends upon the Body Member, the Distance, the Weight or Resistance involved, and the Manual Control required to make the motion. The Body Member, Distance Moved, and Weight or Resistance are readily determined by observation or measurement.

The Distance of travel for a Transport Motion is the straight-line Distance from the beginning to the end of the motion, unless it is necessary for the Body Member to bypass an obstruction. Figure 3-1 shows how the Distance Moved is measured in the situation illustrated.

The number of Work-Factors required for Manual Control is determined by applying the Work-Factor Rules of Application.

Basic. A Basic Transport Motion is a simple motion which involves negligible weight or resistance and requires no Manual Control. Work-Factors are therefore not involved. A motion can be Basic only when it falls in one of the following categories:

    1. A tossing motion
    2. A motion stopped by a rigid object or requiring no control by the worker
    3. A motion made to an indefinite location
    4. A motion to return a body member to a normal, relaxed position

Weight or Resistance Work-Factor (W). The number of Work-Factors involved in moving a weight or overcoming a resistance is shown in the lower portion of each Body Member Table. The Weight Classes apply to single body members. When the weight or resistance is borne equally by two or more body members, the total weight or resistance is divided by the number of members involved.

Definite Stop Work-Factor (D). All Reaches preceding Grasp of an object are classified as Definite Stop Motions. All Moves which are made to place an object or to Assemble an object, or which precede the Inspect of an object, involve

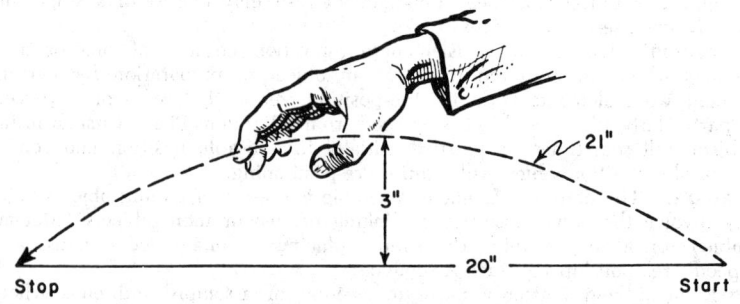

Use 20 inch distance

Fig. 3-1. Determination of Distance Moved.

the Definite Stop Work-Factor. Motions involving only the Definite Stop Work-Factor can be terminated in a space greater than a sphere 2 inches in diameter. Not more than one Definite Stop Work-Factor can be involved in one motion.

Steer Work-Factor (S). When a motion must be terminated in a space equal to or less than a sphere 2 inches in diameter, the Steer Work-Factor is applied in addition to the Definite Stop (D) Work-Factor. All motions except Align, Upright, and Index (occurring in the Standard Element Assemble) requiring the S Work-Factor must also include the D Work-Factor.

When a motion is terminated within the space of a sphere equal to or less than 0.625 inch, it is always followed by additional motions of the Standard Element Assemble. All Moves immediately preceding Assemble Aligns require both the Steer and Definite Stop Work-Factors. Only one Steer Work-Factor is used in one motion.

Precaution Work-Factor (P). The Precaution Work-Factor is involved:

1. If the Body Member must pass within 1 inch of an object which could cause injury to the Body Member or damage to the objects or equipment involved

2. In the movement of a container of fluid which is filled to within 1½ inches of the brim, or other opening

3. When an object must be moved between two real or imaginary parallel lines 1 inch or less apart

Only one Precaution Work-Factor is assigned to one motion.

Direction Change Work-Factor (U). When a motion is caused to curve more sharply than its normal arc by an obstruction between start and stop points on the same plane, additional manual control and therefore more time is required. The additional time is determined by the following Work-Factor Analytic Procedure, illustrated by Figure 3-2.

ADC is the plane of a motion in either direction between *A* and *C*.

BD is the dimension of the obstruction.

The arc *ABC* is the actual motion path required by the obstruction. When $BD \leq \frac{1}{3}AD$, the Motion Distance is the line *AC*, and no U Work-Factor is required.

When $BD > \frac{1}{3}AD$ but $\leq AD$, the Motion Distance is the total of the dotted lines *AB* and *BC* and no U Work-Factor is required.

When $BD > AD$ but $\leq 3AD$, the Motion Distance is the total of the dotted lines *AB* and *BC*, and the U Work-Factor is applied to the motion.

When $BD > 3AD$, two separate motions *AB* and *BC* are required.

The U Work-Factor is not applied to either motion. Not more than one Direction Change Work-Factor is used in a single motion, except in Work-Factor Circular Motions for turning cranks or winding thread, yarn, wire, and the like.

THE WORK-FACTOR STANDARD ELEMENT GRASP (Gr)

The term Grasp refers to the act of obtaining Manual Control of an object or objects for the purpose of transporting, using, or holding. In Work-Factor terminology, there are four general classifications of Grasps:

1. Simple Grasps

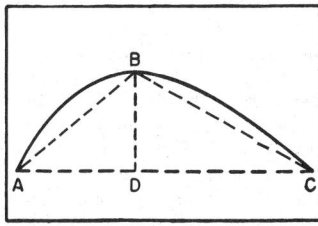

FIG. 3-2.   Determination of Direction Change.

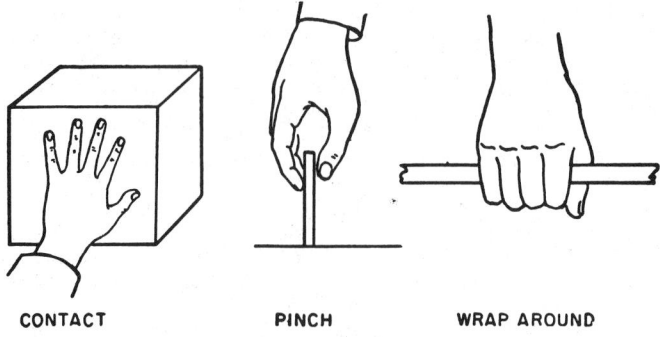

CONTACT   PINCH   WRAP AROUND

FIG. 3-3.   Three types of Simple Grasps.

2. Manipulative Grasps
3. Complex Grasps
4. Special Grasps

Simple Grasps.   The Simple Grasp is defined as one requiring not more than a single motion of the fingers.   It normally occurs when grasping isolated objects. There are three classes of Simple Grasps:

1. Contact Grasp
2. Pinch Grasp
3. Wraparound Grasp

Figure 3-3 illustrates each of these classes.

Contact Grasp.   This occurs when it is necessary only to press the body member (usually the fingers or hand) against the object.   No time is involved in a Contact Grasp because this contact is actually made during the final portion of a preceding Reach and no additional grasping movements are required.   The Contact Grasp occurs most frequently when an object is to be pushed or slid.   Examples of Contact Grasp are:

1. Placing finger on coin prior to sliding it to edge of table
2. Placing knee against control lever to start sewing machine

Pinch Grasp.   When an object can be grasped between the thumb and fingers without wrapping them around the object, the grasp is called a Pinch Grasp.   The Analysis of a simple Pinch Grasp occurring immediately after the hand has been at rest is F1.   This involves 16 Work-Factor Time Units, or 0.0016 minute.

When a Pinch Grasp immediately follows a Reach to an object, the Grasp Motion normally begins before the Reach is completed.   Under this condition, the Grasp is described as ½F1, which is equivalent to 8 Work-Factor Time Units, or 0.0008 minute.   When a heavy object is grasped using a Pinch Grasp so that the surfaces in contact with fingers are vertical, gravity tends to slide the object out of the . fingers.   In this situation, additional time is required for the Grasp.   This occurs because the fingers not only move to the object, but also must apply sufficient pressure to keep it from slipping out of the fingers.   This involves Weight Work-Factors, which are determined by the Work-Factor Rules of Application in the Work-Factor Manual.

Wraparound Grasp.   The Wraparound Grasp is similar to the Pinch Grasp except that the fingers wrap around the object.   Although the Wraparound Grasp is also a Single-motion Grasp, it differs from the Pinch Grasp in two important ways:

1. The distance moved by the fingertips may exceed 1 inch.
2. The tendency of the object to slip away from a Wraparound Grasp due to gravity or resistance is normally much less.   However, if the object grasped is in such a position that it has the same tendency to slip away as was discussed under Pinch Grasp, then the Weight Work-Factors are involved to the same extent as in the Pinch Grasp.

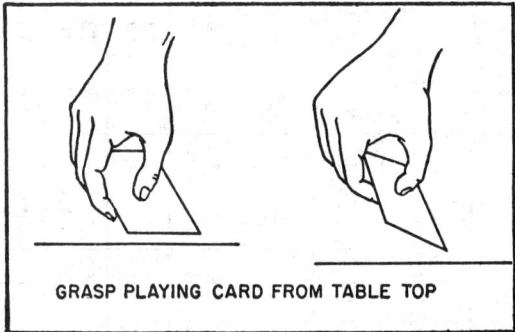

GRASP PLAYING CARD FROM TABLE TOP

FIG. 3-4.   Example of Manipulative Grasp.

Reach Motions Preceding Simple Grasps.   Reach Motions preceding Simple (one-motion) Grasps normally involve only the Definite Stop Work-Factor.   An S Work-Factor for Steer is involved only when the object must be grasped at a specific point in a specific manner, or is small enough to slip through the interstices formed by adjacent fingertips, thumbs, and a flat surface.

The Manipulative Grasp.   All Grasps of isolated or orderly stacked objects which require more than one motion of the fingers are classified as Manipulative Grasps.

A Manipulative Grasp may involve one or more Arm Motions in addition to the Finger Motions required.

The number of combinations of Manipulative Grasp motions which are possible is virtually unlimited because of the variations which occur in the objects being grasped and in the surrounding conditions.   Figure 3-4 shows an example of a Manipulative Grasp.

To evaluate the time for a Manipulative Grasp, it is necessary to list the motions which are involved and to assign the proper time values to them in accordance with the Work-Factor Rules of Application for Transport Motions.   The time for the Manipulative Grasp is the total time for these motions.

The Complex Grasp.   The Complex Grasp is defined as the Grasp of an object from a Random Pile.   Complex Grasps involve more than one motion of the fingers and frequently include arm movements.

The Complex Grasp is composed of the following motions.

Search, which involves one or more motions when it is necessary to find the object prior to First Grasp.

A *First Grasp*, which is an F1 motion unless Precaution or Weight is present, in which case Work-Factors are involved.   Determination of these Work-Factors is made in accordance with applicable Work-Factor Rules of Application.

A *Second Grasp* to compensate for fumbles (unsuccessful First Grasps).   This requires 2 F1 Motions unless Precaution or Weight is present, in which case Work-Factors are involved.   Because a Second Grasp does not occur in every cycle, the frequency is determined on a percentage basis.

Separate Motions to roll or slide surplus objects out of the fingers.   Usually one or two Finger Motions are required.   These motions also occur on a percentage basis.

In general, as objects become larger, the number of Second Grasps increases and the number of Separates decreases.   If an object is very small, it becomes almost impossible to complete a First Grasp without obtaining several parts.

As objects become larger than 1 inch in three dimensions, the likelihood of grasping more than the desired object decreases rapidly.

To reduce the need for Detailed Analysis of Grasp Times, tables such as Table 3-2 have been prepared, composed of values covering the kinds of Complex Grasps most frequently encountered.   These Tables provide values for about 90 percent of the Complex Grasp situations.   The values are based on the accumulated results of 12

TABLE 3-2. A Portion of the Work-Factor Complex Grasp Time Table

Size (Major Dimension or Length), inches	Solids and Brackets, thickness in inches >0.047 (>3/64) Blind		Visual		Cylinders and Regular Cross-sectioned Solids, diameter in inches −0.063 (−1/16) Blind		−0.125 (−1/8) Blind		−0.188 (−3/16) Blind		−0.500 (−1/2) Blind		−0.500 (−1/2) Visual		>0.500 (>1/2) Blind		>0.500 (>1/2) Visual		Add for Entangled, Nested, or Slippery Objects	
	n	s	n	s	n	s	n	s	n	s	n	s	n	s	n	s	n	s	n	s
−0.063 / −1/16	120	172	B	B	S	S	S	S	S	S	S	S	S	S	S	S	S	S	17	26
−0.125 / −1/8	79	111	B	B	85	120	S	S	S	S	S	S	S	S	S	S	S	S	12	18
−0.188 / −3/16	64	88	B	B	79	111	74	103	S	S	S	S	S	S	S	S	S	S	12	18
−0.250 / −1/4	48	64	B	B	79	111	68	94	64	88	S	S	S	S	S	S	S	S	8	12
−0.500 / −1/2	40	52	B	B	62	85	56	76	56	76	44	58	B	B	S	S	S	S	8	12
−1.000 / −1	40	52	32	40	62	85	56	76	48	64	48	64	44	58	40	52	32	40	8	12
−4.000 / −4	37	48	20	22	56	76	48	64	40	52	40	52	36	46	37	48	20	22	8	12
>4.000 / >4	46	61	20	22	56	76	48	64	40	52	40	52	36	46	37	48	20	22	9	14

n = non-Simo   s = Simo   S = use Solid Table
B = use Blind column since Visual offers no advantage

years of Work-Factor use and are derived from several thousand detailed Grasp Analyses. The values have been tested and revised where necessary, before being included in the Table. They are shown as single values for the sake of simplicity.

Objects for which Grasp values have been provided are classified into three basic types in the Tables:
1. Solids and Brackets
2. Thin Flat Objects
3. Cylinders and Regular Cross-sectioned Objects

Time values are included for various-size objects in each of these types, for both Visual and Blind Grasp conditions, and for Simo (two hands working together) and non-Simo operation. There are also time values in the Tables which have been established to compensate for Entangled, Nested, or Slippery Objects. As in the case of the Motion Time Table, the fundamental factors affecting Grasp Time have been the basis of the tabular arrangement. As a result, 220 tabulated time values provide times for more than 800 individual Complex Grasp situations.

Example of the Use of the Grasp Table (Table 3-2). Visual Grasp of wooden dowel pin 2 inches long and ¼ inch in diameter from Random Pile—non-Simo.

Since the object is a Cylinder, refer to the section of the Table marked Diameter of Cylinders and Regular Cross-sectioned Objects. Since its diameter is ¼ inch, refer to column headed − 0.500 (−½). In this column under Visual and non-Simo and opposite the line representing the object's Major Dimension, −4, will be found the total Grasp Time, which is 36 Work-Factor Time Units, or 0.0036 minute.

Special Grasps. There are three classes of Special Grasps:
1. Transfer Grasp
2. Exact Quantity Grasp
3. Group Grasp

Transfer Grasp. This Grasp occurs when an object is transferred directly from one hand to another. It usually involves a Pinch or Wraparound Grasp with one hand and a Gravity or Unwrap Release with the opposite hand.

Exact Quantity Grasp. This Grasp occurs when it is necessary to Grasp an exact number of objects. Such Grasps require a careful and specific Detailed Analysis which may involve Inspect and React Time in addition to motions of the fingers, hand, and arm.

Group Grasp. This Grasp occurs when it is necessary to Grasp an indefinite quantity of objects. Either one or two hands may be used.

When the hand can be easily inserted into a pile of objects, a Scoop-type Group Grasp is employed. This usually involves an Arm Motion for insertion, a Finger Motion to close the fingers, and another Arm Motion to withdraw from the pile.

Because of the great number of variations in Group Grasps, each case must be analyzed separately.

THE WORK-FACTOR STANDARD ELEMENT PRE-POSITION (PP)

Pre-position is the act of turning or orienting an object to a correct position for a subsequent Standard Element of work.

Simple Pre-positions requiring not more than a single Finger Motion or wrist turn are made simultaneously with a Move without a significant effect on the time for that motion. Consequently, no time value is applied to Pre-positions of this type. Complex Pre-positions made by several motions of one or more body members require separate analyses and time values. Complex Pre-positions are usually analyzed as occurring immediately following Grasp. A time table of thirty of the most common values for Complex Pre-position has been prepared based on Detailed Analysis of the motions required by the variables—shape, size, weight of object—and Body Members involved in the Pre-position, and the percentage of occurrence of Pre-position that will be required when the object is grasped from a random arrangement or a stack.

For example, a bolt 2 inches long weighing 3 ounces can be pre-positioned for assembly into a tapped hole by turning it end-for-end with one hand. An average of three Finger Motions is required to complete the Pre-position. When the bolt is grasped from a Random Pile, it will be in the correct position for the assembly in half of the occurrences, and will need pre-positioning in the other half of the occurrences. Therefore, the analysis for this Pre-position is V3F1-50%.

THE WORK-FACTOR STANDARD ELEMENT ASSEMBLE (Asy)

Assemble is the act of joining objects together or placing them in useful locations. Assemble begins immediately after the Move bringing the objects to a position where Assemble can start. Assemble ends when the objects are so placed that manual control can be released or another Standard Element can begin.

All Work-Factor Assembles utilize the concept of Plug and Target, where a Plug is inserted into or placed on a Target. Work-Factor Assembles are further classified according to the following characteristics:

Classes of Assemble

1. *Simple Mechanical.* A mechanical aid to the joining is present, and at least either the Plug or Target is round, permitting the total time to be read directly from the Work-Factor Assemble Time Table.

2. *Complex Mechanical.* A mechanical aid to the joining is also present, but many more variables influence the time for Assemble, and the total time is determined by analysis of all the subelements present in the Assemble.

3. *Surface.* The fitting of the Plug onto the Target without mechanical aid of any kind, where the completion of the Assemble depends entirely on the muscular and visual control of the operator.

Types of Target. Both Mechanical and Surface Assemble Targets are classified by the number of quadrants on their periphery to which the plug must be aligned.

1. *Closed,* requiring alignment in four directions: left, right, back, and front. (See Figure 3-5.)

2. *Open,* requiring alignment in only two opposite directions. (See Figure 3-6.)

Target Dimension. The effective mating (with the Plug) dimension allowing the longest assembly time, when there are several different dimensions on a nonround Target opening. The Target Dimension of circles is the diameter; of rectangles,

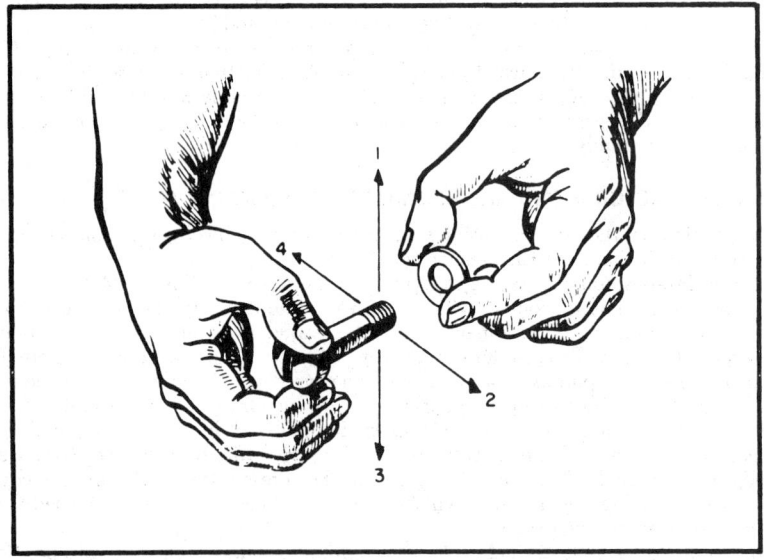

Fig. 3-5.   Closed Target showing Alignment in 4 Directions.

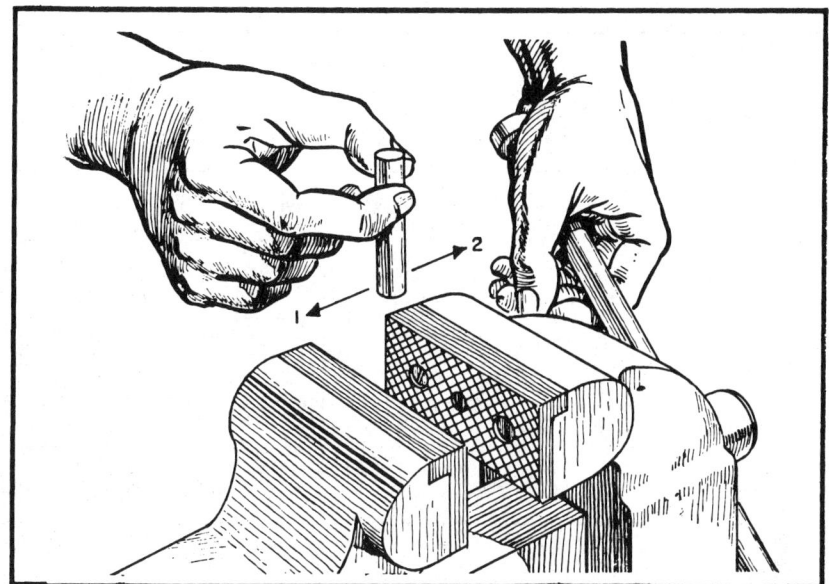

FIG. 3-6.   Open Target showing Alignment in 2 Directions.

the shorter side; and of complex, irregular shapes, generally the smallest nondiagonal dimension. Beveled or chamfered Targets are measured at the extreme outside of the bevel.

Plug Dimension. The dimension of the Plug at the Mating Target Dimension. Flat-ended Plugs are measured across the flat.   Pointed, rounded-end, or bullet-nose Plug Dimensions are taken as one-third of the dimension between parallel surfaces.

Plug-Target Ratio. The Plug Dimension divided by the Target Dimension.   The ratio will always be less than unity.

Simple Mechanical Assembles. Simple Mechanical Assembles contain at most only three subelements.

1. *Align Motions* are short locating motions made after the Move to the Target, but before the actual Insert of the Plug into the Target. Align motions are always analyzed as A1S, unless W or P Work-Factors are also required.

2. *Upright Motions* are used to bring the plug axis to a coinciding position when the Plug-Target Ratio is greater than 0.900. The Plug is literally raised upright to avoid jamming against the sides of the Target Well, as might happen during a "cocked" Insert. Upright is normally an A1S motion, but may involve a longer distance, and may also contain W or P Work-Factors.

3. *Insert* is the actual joining of the two objects. Usually, Insert is a Basic Motion of variable length. When the Plug-Target Ratio is >0.935, the Insert includes a P Work-Factor to ensure a smooth and continuous motion. The D Work-Factor is also used when the depth of the Insert must be controlled by the operator.

Tables incorporating these three subelements have been prepared for Open and Closed Target Simple Mechanical Assembles. The Closed Target Table is shown by Table 3-3.

The numbers in parentheses represent the average number of Aligns for Assembles in each table class. Aligns are variable frequency Motions. For example, in the Assemble of a 1$\frac{1}{16}$- by 2-inch-long round pin into a 0.0650-inch hole, the Ratio is 0.0625/0.0650 = 0.96.   The Aligns are V3A1S, and the total time including Upright and Insert is 130 Work-Factor Time Units.

TABLE 3-3. Work-Factor Assemble Time Table

Target dimension, inches	Closed Targets											
	Ratio of plug diameter to target dimension											
	−.225		−.290		−.415		−.900		−.935*		>.935†	
>.875	(D‡)	18	(D‡)	18	(D‡)	18	(¼)	25	(¼)	51	(¼)	59
−.875	(D‡)	18	(D‡)	18	(SD‡)	18	(¼)	25	(¼)	51	(¼)	59
−.625	(SD‡)	18	(SD‡)	18	(¼)	25	(½)	31	(½)	57	(½)	65
−.375	(½)	31	(1)	44	(1)	44	(1½)	57	(1½)	83	(1½)	91
−.225	(1)	44	(1)	44	(1)	44	(1½)	57	(1½)	83	(1½)	91
−.175	(1)	44	(1¼)	51	(1½)	57	(1½)	57	(1½)	83	(1½)	91
−.124	(2½)	83	(2½)	83	(2½)	83	(2½)	83	(2½)	109	(2½)	117
>.025 −.074	(3)	96	(3)	96	(3)	96	(3)	96	(3)	122	(3)	130

* Requires A(X)S Upright for all ratios >.900 (Table value includes A1S Upright and A1 Insert).

† Requires A(Y)S Upright and A(Z)P Insert for all ratios >.935 (Table value includes A1S Upright and A1P Insert).

‡ Letters indicate Work-Factors in Move preceding Assemble.

Complex Mechanical Assembles. Complex Mechanical Assembles include additional variables and subelements besides those contained in Simple Mechanical Assemble.

1. *Distance Between Targets* is applied when two disconnected Plugs, one in each hand, are assembled to two Targets at the same time, or when a single object with two connected Plugs is assembled to two Targets by one or by both hands. The increased difficulty caused by this condition is compensated for by percentage increases to the number of Align Motions according to a Work-Factor table of varied distance classes.

2. *Gripping Distance* is also allowed as percentage increases to the number of Align Motions when the hand grips an object at a significant distance from the point of alignment.

3. *Blind Assembles* occur when the Plug, the Target, or both are obscured from view. This condition is compensated for by percentages added to the Align Motions, tabulated according to the Blind Distance. In the Work-Factor Rules of Application, two separate and distinct Blind Conditions are recognized: (1) Temporary Blind and (2) Permanent Blind, the more difficult condition.

The "add-ons" to Align required by any or all of these three conditions are calculated separately in the analysis of a Complex Mechanical Assemble, but are not compounded or pyramided.

4. *Index* is a subelement to turn or rotate nonround Plugs and nonround Targets so that their cross-sectional outlines will mate. The Assemble of a shaft with an inserted key into a collar with a keyway is an example of the requirement for Index. The Work-Factor Rules of Application define the conditions where Index is required, the frequency with which it is allowed, and the Body Members and Distances used. F1S and A1S are the most common types of Index.

5. *Seat* is the last of the five subelements of Assemble. It is required when the first Insert does not complete the Assemble of the objects. There are two general types of Seat: (1) Dislodge from a shoulder on the Plug, or a step inside the Target—in some cases, when the shoulder or step width is sufficiently wide,

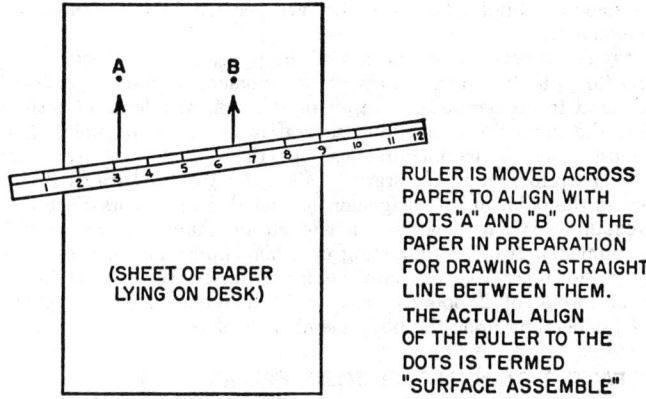

FIG. 3-7.   Example of Surface Assemble.

a complete second Assemble may be required—and (2) Engage and Lock manipulations which lock the Plug and Target together after Insert—the actual motions required must be analyzed.

Because of the number of different variable percentages and subelements involved, it is not practical to set up tables for the total time values of Complex Mechanical Assembles. Each case must be analyzed in detail from the various applicable Work-Factor Assemble Tables.

Surface Assemble. A Surface Assemble is defined as an Assemble in which the Plug and the Target merely rest one on the other and there is no fitting of one into the other. A typical example of a Surface Assemble is given in Figure 3-7.

Assembles of this nature provide no mechanical stops to aid Align. Analysis of Surface Assembles is accomplished by applying the basic Work-Factor Principles and Rules of Application relating to Mechanical Assemble plus the use of appropriate Work-Factors and Mental Process Time applicable to the Surface Assemble.

THE WORK-FACTOR STANDARD ELEMENT USE (Use)

Time values for manual motions which occur during the Element Use are taken from the Motion Time Table if they are controlled by the operator. If the motions are controlled by the process or machine, the time is obtained by mathematical formulas, empirical curves, standard tables, or stopwatch or other timing devices.

Typical elements of work classed as Use are:

Use Elements Wherein the Machine or Process Is the Controlling Factor
1. Machine tool cutting time
2. Time for liquids to pour
3. Plating time in electroplating
4. Time for objects to slide in chute

Use Elements Normally Controlled by the Operator
1. Rubbing to finish furniture
2. Driving screws with hand driver
3. Moving levers and foot pedals
4. Driving nails with a hammer

The number and variety of Use Elements are, of course, unlimited.

THE WORK-FACTOR STANDARD ELEMENT DISASSEMBLE (Dsy)

Because the Element Disassemble is the reverse of Assemble and consists primarily of Transport Motions, the majority of Disassemble Motions can be analyzed following the instructions already outlined.

There are some conditions, however, that are peculiar to Disassemble and require additional comment.

Post-disengage Travel. If, in the act of disengaging two objects, it is necessary to exert force to pull them apart, then at the moment of disengagement, the body member will tend to accelerate as a result of the sudden release of resistance. Because of this, the motion will continue several inches past the point of disengagement. In most cases, it is not practical for the operator to arrest this motion immediately. Therefore, it is necessary to allow for the total travel distance. This distance beyond the point of disengagement is included in a Work-Factor Table.

Relax Pressure. When removing a wrench or similar device from a nut or other object immediately after tightening or other application of pressure or torque, it is necessary to relax the pressure before Disassemble can begin. Relaxation of pressure or torque of 2 pounds or more is analyzed as either ½A1W, ½F1W, or ½FS45°W, depending upon the body member involved.

THE WORK-FACTOR STANDARD ELEMENT RELEASE (R1)

The Standard Element Release is the act of relinquishing control of an object. It is one of the least complex of the Elements of Work. There are three Work-Factor classes of Releases.

Contact Release. The Contact Release occurs when the fingers (or other body member) move away from an object against which they have been resting. It is the opposite of the Contact Grasp.

Contact Releases involve no time unless they follow application of pressure or movements which involve pressures in excess of 2 pounds. If pressure of more than 2 pounds immediately precedes the Contact Release, the Release is assigned the value ½(X)1W, where X represents the Body Member involved. These values are Time Equivalents for Relax Pressure.

Gravity Release. Gravity Release is the opposite of the Pinch Grasp and occurs when an object can drop from the fingers without the necessity of unwrapping them or when the hand can be pulled away from an object as soon as the muscles are relaxed. The Gravity Release is ½F1 when not preceded by pressure of more than 2 pounds, and ½F1W when it is preceded by pressures of more than 2 pounds.

The Unwrap Release. Unwrap Release is the opposite of Wraparound Grasp. It involves the opening of the fingers which have been wrapped around an object. The Unwrap Release is normally F1, except when immediately following pressure of more than 2 pounds, in which case it is F1W. In situations where the fingers must move more than 1 inch before the hand can move away from the object, the actual distance which must be moved by the fingertips is used.

No time value is assigned to Releases which occur in connection with tossing motions, because the fingers performing the Release do so simultaneously with the tossing Arm Motion.

THE WORK-FACTOR STANDARD ELEMENT MENTAL PROCESS (MP)

Mental Process is the act of using the senses, brain, and connecting nervous system to perform useful mental work. Following extensive research, the Wofac Company has developed and published a complete set of Fundamental Mental Process Times for thirteen functions, including Eye Shift, Nerve Conduct, Discriminate, Span, Identify, Decide, and Transfer Attention.[5] The use of the Tables contained therein is beyond the scope of this chapter. The technique presented here is termed Simple Mental Process. It is generally adequate for measuring office and factory operations where Sequential Mental Process Time is only a relatively small portion of total cycle time. The Work-Factor system divides Mental Process into two major, mutually exclusive classes by their effect on work cycle times.

    1. Concurrent Mental Processes. Concurrent Mental Processes are concurrent

---

[5] *Mento-Factor Manual of Detailed Work-Factor Mental Process Times,* 5th ed., Wofac Company, Moorestown, N.J., 1968.

with the Body Member Motions they control. They have no observable or measurable time effect independent of the motion time. Concurrent Mental Processes are considered to produce a time effect in two ways: (1) as a uniform constant in all work cycles, regulating the worker's speed so that he can maintain a productive level of output during the workday and workyear and (2) as the variable degree of required Manual Control identified by the Work-Factors.

2. Sequential Mental Processes. Sequential Mental Processes are not concurrent with Body Member Motions and have an observable, measurable time consumption.

There are four types or subelements of Sequential Mental Processes. The designations, symbols, and time values are shown in Table 3-4.

Focus. A Focus (20 Work-Factor Time Units) is used each time the eyes must shift to a new location. If the head must turn before the Focus begins, time is required for Head Turn in addition to Focus. At Normal Fix Distances (12 to 18 inches, averaging 15 inches), one Focus is used for each Inspection Unit (each 3-inch square) which requires inspection. When it is necessary for the eyes to adjust to a new location, a second Focus is required.

Inspect. At completion of Focus, the operator is not considered to have seen the object or area being inspected in terms of ability to evaluate or make a useful decision. Focus time must therefore be accompanied by time for at least one Inspect Interval (30 Work-Factor Time Units).

Inspect is the act of seeing, evaluating, and decision making. The Inspect Interval provides the time required for these functions.

At a specified viewing distance and under specified lighting conditions, one Inspect Interval occurs for:

1. Determining the presence and identity of or the absence of one or any Group of visible and distinguishable characters, objects, symbols, or characteristics, hereinafter referred to as Work-Factor Inspection Characters. A visible and distinguishable Character subtends a Visual Angle of not less than one minute. At Normal Fix Distance, a Character Dimension subtending a one-minute Visual Angle is 0.004 inch.
2. Determining the presence or absence of exactly one, exactly two, exactly three, or exactly four Inspection Characters.
3. Recognizing up to and including three digits of a number.
4. Recognizing words up to and including 6 letters, provided the word is common to the language of the reader. Two Inspect Intervals are applied for familiar words of 7 or 8 letters. For words of 9 or more letters, and for unfamiliar words, one Inspect Interval is applied for each 3 letters.

One Inspect Interval is used for each Group of numbers, letters, or symbols, even though there are fewer than 3 Items in the Group.

When an Inspection Unit is inspected for the presence or absence of a specified Inspection Character, one Inspect Interval is used for the initial Inspect of the Inspection Unit and one Inspect Interval is used for each Character, such as a scratch, spot, stain, or scar, which could be mistaken for the specified Inspection Character.

React. This process may occur as an integral part of an Inspect Interval, immediately following the Inspect of 30 Work-Factor Time Units. React also occurs

TABLE 3-4. Sequential Mental Processes

Type	Symbol	Work-Factor Time Units
Focus...	Fo	20
Inspect..	I	30
React..	Rn	20
Complex Mental Processes (Mento).........	Mt	10

completely separate from Inspect. The React following an Inspect is always a "Choice" type, where the operator must make a selection from several Alternate Actions already known to him.

One React Time of 20 Work-Factor Time Units is used for more than two and including five Alternate Actions following the Inspect. Two React Times are allowed for more than five Alternate Actions. If, for example, an operator places parts in bins following an Inspect, the React Times applied to three different levels of sorting are:

Type of sort	Analysis	Work-Factor Time Units
Accept + Reject........................	—	—
Accept + Reject + Repairable............	Rn	20
(5) Acceptable Grades + Reject...........	2 Rn	40

React occurring separately from an Inspect is either Simple or Choice type. Simple React is the response to a single stimulus, called an Action Signal, such as a light, bell, or presence or absence of some other signal. If the Action Signal is anticipated, no React Time is used. A light flashing at the end of each machine cycle is an example of an Anticipated Action Signal. One React Time of 20 Work-Factor Time Units is allowed when the Action Signal is Unanticipated; that is, the operator knows what the stimulus will be, but not when it will occur.

When Choice React occurs independently of Inspect, it is always considered as Unanticipated, because there are a number of Action Signals from which to select, and the operator cannot know which one will appear. In Choice React, a React Time of 20 Work-Factor Time Units is used for each two Alternate Actions, not exceeding eight such Actions, which could result from the stimulus or Action Signal. When the number of Alternate Action Signals exceeds eight, additional time is not required.

Complex Mental Processes. The Work-Factor analysis of operation cycles may require additional Mental Process times besides Focus, Inspect, and React. Three other Mental Processes are measured in Mentos of 10 Work-Factor Time Units each. These are briefly described as follows.

"Memorize" is the Work-Factor term for the act of fixing a number, word, or idea in the mind so that it can be used by the operator a few seconds or minutes later. This Memorize is not comparable to memorizing a speech, poem, or table of numbers. The time for Memorize is 2 Mt = 20 Work-Factor Time Units.

"Recall" is the act of recalling information. When at least one Motion or Mental Process other than an Inspect Interval intervenes between Memorize and the need to use the information memorized, a Recall of two Mentos is required.

"Compute" is the term for adding, subtracting, multiplying, or dividing two one-digit numbers. Compute of two or more digit numbers is analyzed as a series of steps involving one digit of each number. The time for Compute is two Mentos.

TRUNK, LEG, AND FOOT MOTIONS—BODY TURNS AND WALK

Trunk Motions. The Work-Factor Rules of Application discussed previously apply to all body members, including the trunk. However, there are several conventions which should be observed to ensure consistent application of Work-Factor Time Values.

Combination of Trunk and Arm Motions. When both the trunk and arm move simultaneously, it is necessary to analyze both to ascertain which requires the greater time. If the trunk is the Controlling Motion when reaching to Grasp an object or moving an object to a location, it must be considered to contain a Definite Stop Work-Factor, but not one for Steer.

Standing Erect. When returning the trunk to its normally erect or forward position, no Definite Stop is involved. When arising from a bent or stooped position, the trunk is normally allowed to return to the erect position before the worker is expected to start to Walk or make a 180-degree Turn. Exceptions are made only when careful Analysis reveals the specific reasons. These are recorded on the Analysis.

Leg Motions. The Work-Factor Rules of Application for Transports also apply to Leg Motions. In addition, several Special Applications are made using the Leg Table. These are defined in the Work-Factor Manual.

Foot Motions. Foot Motions that must be analyzed are not common. They usually occur when the heel remains in a relatively fixed position and the toes move about a pivot at the ankle. Some types of machine pedals require Foot Motions, as for example, accelerators on automobiles. However, many controls require that the entire foot be lifted or lowered. The motion then becomes a Leg Motion. All such movements should be carefully analyzed to determine whether they are Foot or Leg Motions. The Definite Stop Work-Factor is rare in Foot Motions.

Head Turns. The movement of the head is an essential element in some types of operations, especially where inspection of two widely separate points is required. A common occurrence, other than in Inspect, is nodding or shaking the head when lowering the visor on a welder's visor-type helmet.

Body Turns and Single Steps with One Foot (not followed or preceded by Walk). When an operator is standing and moves one foot forward, backward, or to the side, or moves the foot to the side and back so as to turn the body, the Analysis is a Leg Motion with two Weight Work-Factors. In this case, the Weight Work-Factor compensates for the extra time to obtain body balance and is not for body weight in the normal sense.

Body Turns and Single Steps with Both Feet (not followed or preceded by Walk). In these situations, both feet move so that at the end of the sequence, the body has taken a new position. The motion of the first leg is analyzed identically with that of a single step involving one foot, that is, as a Leg Motion with two Weight Work-Factors. The movement of the second leg is analyzed as a Basic Leg Motion. In most cases, actual work can start as soon as the first motion is completed; consequently the second motion is entirely Simo.

Walk. (See portion of Table 3-1 on Walk Time.) Walk is analyzed in terms of 30-inch paces (paces which advance the body 30 inches with the movement of each leg). In using the Walk Time Table, it should be noted that taking a single pace is not classified as Walk but as a Single Step. Turns exceeding 120 degrees at the start or end of Walk require an addition of 100 Work-Factor Time Units for either General or Restricted Walk.

General Walk applies to Walk without heavy loads, on surfaces which present good footing, and through areas where no obstructions hamper normal movements.

Restricted Walk involves cases where loads are carried, pushed, or pulled or where walk conditions are difficult.

THE SIMPLE ANALYSIS—DETAILED WORK-FACTOR

Operations involving primarily one hand or two hands working in sequence, without the simultaneous occurrence of Complex Grasps, Pre-positions, or Assembles, can be analyzed simply and without need for a Right- and Left-hand Analysis. The simple Analysis should contain:

1. An accurate and complete identification and description of all physical and other conditions which have a significant effect on the time required to perform the work being studied. This includes part names, part numbers, or other identification of objects; types of tools, machines, or equipment; machine or process speeds and feeds; company department numbers or other identification of the work station; date of study; name of worker or workers; and the name of the observer.

2. A concise yet complete description of each Standard Element involved in the performance of the work being studied.
3. The Work-Factor Motion Analysis for each Element.
4. The Work-Factor Time Units for each Element.
5. The total Work-Factor Time for the operation.
6. Conversion of the total Time into standard time, pieces per hour, hours per hundred, or other units, depending upon company policy or the end use to which the time value will be put.

Work-Factor terminology, abbreviations, and symbols should be used throughout. Weights, Target and Plug Dimensions, and other pertinent dimension information

WORK-FACTOR ONE-HAND ANALYSIS FORM

Part Name COMPRESSOR HEAD	Sheet No. 1 of 1	Department 10	Part No. A96-521		Sub. 0	Oper. No. 18
Operation Name & Description TIGHTEN 9/16" HEX HEAD BOLT WITH BOX-END WRENCH (BOLT PREVIOUSLY HAND TIGHTENED)					Company JOHN DOE AND SON	

No	Elemental Description	Hand L	Hand R	Analysis	Time Units Elemental	Time Units Cumulative	No
1	R to wrench on bench top		X	A15D	71	71	1
2	Gr wrench		X	F1	16	87	2
3	M wrench to bolt		X	A15SD	92	179	3
4	Asy wrench to bolt (t = 21/32; p = 9/16"; r = 86)		X	TABLE	25	204	4
5	Gr Distance 2 3/4"		X	10% ½A1S	1	205	5
6	Index to hex of bolt		X	A1S	26	231	6
7	Seat wrench over bolt		X	A1	18	249	7
8	M wrench for 1st twist (Average resistance under						8
9	1 pound)		X	VA5	29	278	9
10	Dsy wrench from bolt		X	A1	18	296	10
11	M wrench back to bolt at starting position		X	A5SD	55	351	11
12	Re-Asy wrench to bolt		X	#4 thru #7	70	421	12
13	M wrench for 2nd twist (Average resistance 5 lbs.)		X	VA5W	43	464	13
14	Rl pressure		X	½A1W	13	477	14
15	Dsy, M back, and Re-Asy wrench to bolt		X	#9 thru #11	143	620	15
16	M wrench for 3rd twist (Average resistance 12 lbs.		X	VA5WW	55	675	16
17	Rl pressure, Dsy, M back and Re-Asy wrench to bolt		X	#13 & #14	156	831	17
18	M wrench for final tightening (Average resistance						18
19	15 lbs.)		X	A2WWW	44	875	19
20	Rl pressure		X	½A1W	13	888	20
21	Dsy wrench from bolt		X	A1	18	906	21
22	M wrench back to bench top		X	A15D	71	977	22
23	Rl wrench		X	½F1	8	985	23
24							24
25							25
26							26
27							27
28							28
29							29
30							30
31							31
32							32
33							33
34							34

wofac	Date 1-20-	Analyst J. SMITH	Total 985	Time in Minutes .0985	Multiplier 2.45	13.5/1 (69/1)

Fig. 3-8. Example of a simple Analysis prepared using Detailed Work-Factor.

which is required in computing the correct time values must be included. Figure 3-8 is a typical example of a simple Analysis prepared using Detailed Work-Factor.

EFFECT OF SIMULTANEOUS ELEMENTS ON TIME (SIMO-FACTOR)

In the performance of manual work, the use of two or more body members simultaneously and independently to make simple motions does not significantly retard the worker. However, when Complex Manipulative Elements of work are performed simultaneously (by each of the worker's hands independently, for example), the elapsed time required to complete the two elements together is greater than when only one of the same elements of work is performed using one hand only.

A Complex Manipulative Element of work involves a series of deft motions (usually varying in number from cycle to cycle) performed in rapid succession. When two such elements are performed simultaneously, the variation in the number of motions required creates differences in the amount of work and the time required. This difference is not constant but changes from cycle to cycle, according to the irregularities encountered by each hand. As a result, one hand may complete its element in advance of its mate. Because the Average Experienced Operator normally moves his hands in phase, the first hand which completes its element will normally and unconsciously be retarded until the slower hand has completed its element.

The net effect during a series of cycles is to increase the overall time required to perform an average cycle when the hands are working simultaneously. When elements of work are performed in this manner, they are termed Simo Elements.

On an average basis, a 50 percent increase in time for the occurrence of the variable elements on Simo operations is both adequate and equitable for most situations. This increase in time is termed Simo-Factor, and is applied to Work-Factor Analyses as follows:

1. All Simo Elements of Grasp, Pre-position, and Assemble involve the Simo-Factor if they contain motions which occur on a Percentage or Average (Variable) basis.
2. The Simo-Factor is 50 percent of the Variable portion of each Simo Element.
3. The Simo-Factor is applied if any portion of any Element containing Variable Motions occurs in the work cycle simultaneously with any portion of another Element containing Variable Motions.

THE RIGHT- AND LEFT-HAND ANALYSIS

Operations involving the use of both hands are normally analyzed with the Right- and Left-hand Analysis Form shown by Figure 3-9. On the left side of the Form (reading from outer edge toward center), spaces are provided for the Elemental Description, the Motion Analysis, the Elemental Time, and the Cumulative Time for the elements of work performed by the left hand. Similar spaces are provided on the right side of the Form for the elements performed by the right hand.

Necessary elements of work in the operation under study are listed according to the hand which performs them. In recording the Right- and Left-hand Analysis, it is important that a suitable starting point in the cycle be selected. Preferably, the first elements recorded on the form should be at a point in the work cycle wherein the two hands are beginning or ending a work element at the same instant, or a point at which it is possible to establish some other type of exact time relationship between the two hands. Elements in which both hands move the same object jointly or in which both hands start to move away from an object at the same instant are usually suitable starting points. The Right- and Left-hand Analysis includes necessary elements of work performed and Delays, Holds, and Waits experienced by each hand.

Balancing Delay. A Balancing Delay is the difference in working times between the right and left hands when working together and when, at some point or points in the cycle, the two hands must be in exact coordination.

WORK-FACTOR TWO-HAND ANALYSIS FORM

Part Name: HOUSING AND PLUG	Sheet No. 1 of 1	Company: JOHN DOE AND SONS		Department 11	Part No. 65-427	Sub. 0	Oper. No. 25

Operation Name & Description: ASSEMBLE HOUSING AND PLUG TO CARTON ON CONVEYOR

No	LEFT HAND — Elemental Description	Analysis	Time Units	Cumulative Time	Time Units	RIGHT HAND — Analysis	Elemental Description	No
1		O R I G I N A L	S E Q U E N C E					1
2	R to carton	A6D	47	47	96	96	A30D	R to housing	2
3	Gr carton	F1	16	63	125	29	F1WW	Gr housing	3
4	Slide carton to WA	A6D	47	110					4
5	Balancing Delay	BD	157	267	267	142	A30WWD	M housing to carton	5
6					322	55	A5WW	Ins housing in carton	6
7	Hold carton upright	BD	78	345	345	23	F1W	R1 housing	7
8	R1 carton	F1	16	361	399	54	A8D	R to plug	8
9					415	16	F1	Gr plug	9
10					469	54	A8D	M plug to carton	10
11	Wait	BD	116	477	477	8	4f1	R1 plug to carton	11
12									12
13		I M P R O V E D	S E Q U E N C E					13
14	R1 carton	F1	16	16	96	96	A30D	R to housing	14
15	R to plug	A8D	54	79	125	29	F1WW	Gr housing	15
16	Gr plug	F1	16	86					16
17	M plug to carton	A8D	54	140					17
18	R1 plug to carton	4F1	8	148					18
19	R to next carton	A6D	47	195					19
20	Gr carton	F1	16	211					20
21	Slide carton to WA	A6D	47	258					21
22	Wait	BD	9	267	267	142	A30WWD	M housing to WA	22
23					322	55	A5WW	Ins housing in carton	23
24	Hold carton upright	BD	78	345	345	23	F1W	R1 housing	24

wofac — Date 1-20-	Analyst J. SMITH	Total 345	Time in Minutes .0345	Multiplier (Improved Sequence)	115/2/ (69/1)

FIG. 3-9.   Right- and Left-hand Analysis highlighting poor motion sequences.

Hold.   The term Hold refers to that portion of a cycle during which the hand holds an object as a function in the operation.   Obviously, Hold is an undesirable element and violates the principle of motion economy which states that the hand should not be used as a holding device.

Wait.   The term Wait refers to a period of time during which a hand is idle, is not holding an object, and is not waiting for the opposite hand to come back into balance because of unequal Motion Distances or manipulations.   Wait occurs during a period when no work is assigned to the hand over some portion of the cycle.

All work, Balancing Delays, Holds, and Waits must be recorded for each hand, and corresponding time values assigned.   Values for these elements are totaled in the Cumulative Column in the same manner as all other elements.   Thus all activities and times for each hand are accounted for.

Because the Analysis for each hand begins at zero (0) time, the study will end with identical Cumulative Totals for each hand.   If the totals are not identical, an error has been made.

The inclusion of Balancing Delays, Holds, and Waits not only is necessary for an accurate cycle time, but also provides an essential tool for making improvements in the motion pattern and the method.   The appearance of any of these elements should immediately signal the engineer to review the operation and to arrange for economies which will eliminate such costs as far as possible.

Figure 3-9 shows how the Right- and Left-hand Analysis highlights poor motion sequences.   In the original Analysis, the left hand had the following nonproductive elements:

> Balancing Delay .... 157 units
> Hold .... 78 units
> Wait .... 116 units
>
> Total .... 351 units

By transferring work from the right to the left hand, this was reduced in the improved sequence to:

Balancing Delay	9 units
Hold	78 units
Wait	None
Total	87 units

READY WORK-FACTOR

Ready Work-Factor is based on Detailed Work-Factor, and is entirely consistent in all respects with the other Work-Factor techniques. Although Ready Work-Factor was originally developed for persons not engaged in work study as an occupation, a wide range of applications over several years has demonstrated that time values established with Ready Work-Factor are entirely suitable for establishing both day-work and incentive standards. When correctly applied, time standards developed by Ready Work-Factor for operations with cycle times greater than 0.15 minute may be expected to be very close to corresponding standards established by the Detailed technique. As a result of the relatively short training period required to achieve the work time concepts of Ready Work-Factor, many companies find it suited to their need for a modern technique based on proved time standards.

Time Values in Ready Work-Factor are expressed in Ready Work-Factor Time Units (RU). One Ready Unit equals ten Detailed Work-Factor Time Units (0.0010 minute). Where practical, the terms and definitions used in Ready Work-Factor are the same as those used in Detailed and Abbreviated Work-Factor. One Motion Time Table applies for all Body Member Motions (Finger, Hand, Arm, Foot, Leg, and Trunk Motions). Standard Elements are the same as in the other Work-Factor Techniques.

Dimensions used in conjunction with Rules of Application are few. Where possible, Work-Factors are associated with the presence or absence of physical characteristics readily seen and identified by the analyst. The Ready Work-Factor Time Table, Table 3-5, provides values for all the Standard Elements.

In Ready Work-Factor Analysis notation, the numbers 0, 1, 2, 3, and 4 are used as symbols of the number of Work-Factors involved. The same degrees of difficulty are used for other Standard Elements.

Referring to the Transport section of Table 3-5, a 15-inch Arm Motion with two Work-Factors requires 9 Ready Units, or 0.009 minute.

Measurements to determine Motion Distances are for the most part unnecessary, because the practical analyst can usually determine them within the accuracy of the Ready Work-Factor System. If the Distance is measured, the nearest whole number of inches is used. Transport values are easily memorized.

The upper half of the Transport section of the Table provides Weight ranges in pounds for each Body Member corresponding to the various Weight classes. For example, if the arm moves a 5-pound weight, the Weight class is listed on the line with Arm in the middle column as "—6" (4- to 6-pound class, and indicates that the motion requires two Weight Work-Factors for Weight. Similarly, a Leg Motion involving 10 pounds requires one Weight Work-Factor, because it falls in the second column on the line with Leg.

Time values for Pick Up and the Work-Factor Standard Elements Grasp, Pre-position, and Release are also listed on the Ready Work-Factor Time Table in five columns according to difficulty. In the Assemble section of the Table, time values for total Assemble Time and Align Time are listed according to type. This is consistent with other Work-Factor Tables. Add-on values are listed according to Distance involved. The remaining times for Body Turns, Mental Process, Release Pressure, Walk, and Circular Motions are presented as individual values.

An Analysis using the Ready Work-Factor Two-hand Graphic Analysis Form is shown in Figure 3-10.

TABLE 3-5. Ready Work-Factor Time Table

All times (bold face numbers) are expressed as Ready Work-Factor Time Units. One Ready Work-Factor Time Unit = 0.0010 minute.

Symbols: — indicates "up to and including." > indicates "greater than."

Work-Factors

	0 Very Easy	1 Easy	2 Average	3 Difficult	4 Very Difficult
TRANSPORT		Weight limits, lb			
Finger, Hand	-1	-2	-3	-5	>5
Arm	-2	-4	-6	-10	>10
Foot	-3	-8	>8	....	....
Leg	-5	-16	>16	....	....
Trunk	-7	-32	>32	....	....

Motion Distance,[1] in.

	0	1	2	3	4
Very Short -4 A[2]	2	3	4	5	6
Short -10 B	4	5	6	7	8
Medium -20 C	5	7	9	11	13
Long -30 D	7	9	11	13	15
Very Long -40 E	9	11	13	15	17

GRASP

Simple	Pinch[3]	Wrap and Transfer[3]	Average No. Motions		
Manipulative	...	...	-2	-3	-4

Complex[7]			
Major Dimension, in.	>¼		-¼
Diameter, in.	>¼	-¼	
Thickness, in.	>³⁄₆₄	-³⁄₆₄	All

Visual	1	2	3	5	8
Blind			4	6	

PICK UP[4]

Distance, in.					
-4 A	8	9	10	12	15
-10 B	12	13	14	16	19
-20 C	17	18	19	21	24
-30 D	21	22	23	25	28
-40 E	25	26	27	29	32

Distance, in.	Add-on for Weight				
-10	...	1	2	3	4
>10	...	2	4	6	8

Add-on for:			
Blind	...	1	...
Simo	...	2 5	

PRE-POSITION

	One hand		Two hands		
Major Dimension, in.	>³⁄₈ -4	-10	-³⁄₈	-10	>10
% Occurrence					
25	1		2		
50	2		3		4
75	3	4	5	5	6
100	4	5	6	7	8
Simo	Add 50% to time		...	...	

RELEASE

	Gravity	Unwrap
	1	2

ASSEMBLE Mechanical[6]

Target, in.	Open (I) -.4	-.9	>.9	Closed (X) -.4	-.9	>.9
		Ratios			Ratios	
>³⁄₈	2	3^1	7^1	2	3^1	7^1
-³⁄₈	3^1	4^2	8^2	5^3	6^4	10^4
-⅛	6^4	6^4	10^4	9^7	9^7	13^7

Tolerance, in.	Surface[6]	
>³⁄₈	3	3
-³⁄₈	5^2	6^3
-⅛	8^5	12^9

% Add-ons for Assemble

Distance, in.	-1	-2	-3	-5	-7	-15	>15
Gripping Distance	...	...	10	20	30	50	70
Distance Between	...	20	30	50	70	2 Asy	2 Asy +5
Temporary Blind						150	...
Permanent Blind	30	50	70	150	250	500	...

Miscellaneous

Index: Mechanical 3 Surface 4
Simo: 50% of (Aligns + Add-ons)
Seat: Turn; Use Transport Rules
Dislodge; Use Assemble Rules

Weight, lb	...	-2	-4	-6	-10	>10
% Add-on to total Asy time	0		30	50	70	100

MENTAL PROCESS Focus 2, Inspect 3, React 2

BODY TURN	Head Turns: -45° 4, -90° 6 Body Turns: -90° (1 Foot) 10 -90° (2 Feet) 20, -180° 26
WALK (in terms of 30-inch Paces)	General 12 + 8 per Pace Restricted 12 + 10 per Pace Up and Down Steps 10 per Step Stand Up 13 Sit Down 9

RELAX PRESSURE	1	APPLY PRESSURE	See note 2

CIRCULAR MOTIONS Diameter, in.	No. Work-Factors	
	Single circle	Multiple circles
-1	0	1
>2	0	0

[1] For Trunk Motion multiply Distance by 2.
[2] Use Transport class A for Apply Pressure and for Forearm Swivel.
[3] For weight over 3 pounds multiply time by 2.
[4] Times based on Visual Grasp. Work-Factors in Pick Up are considered to be identical with those in Grasp. Times include Reach, Grasp, and Move, not Release.
[5] Add Blind value also.
[6] Small numbers are Align times, large numbers are total Assemble times.
[7] For Complex Grasp: add 2 Units for Simo and 1 Unit each for Entangled, Nested, and Slippery.

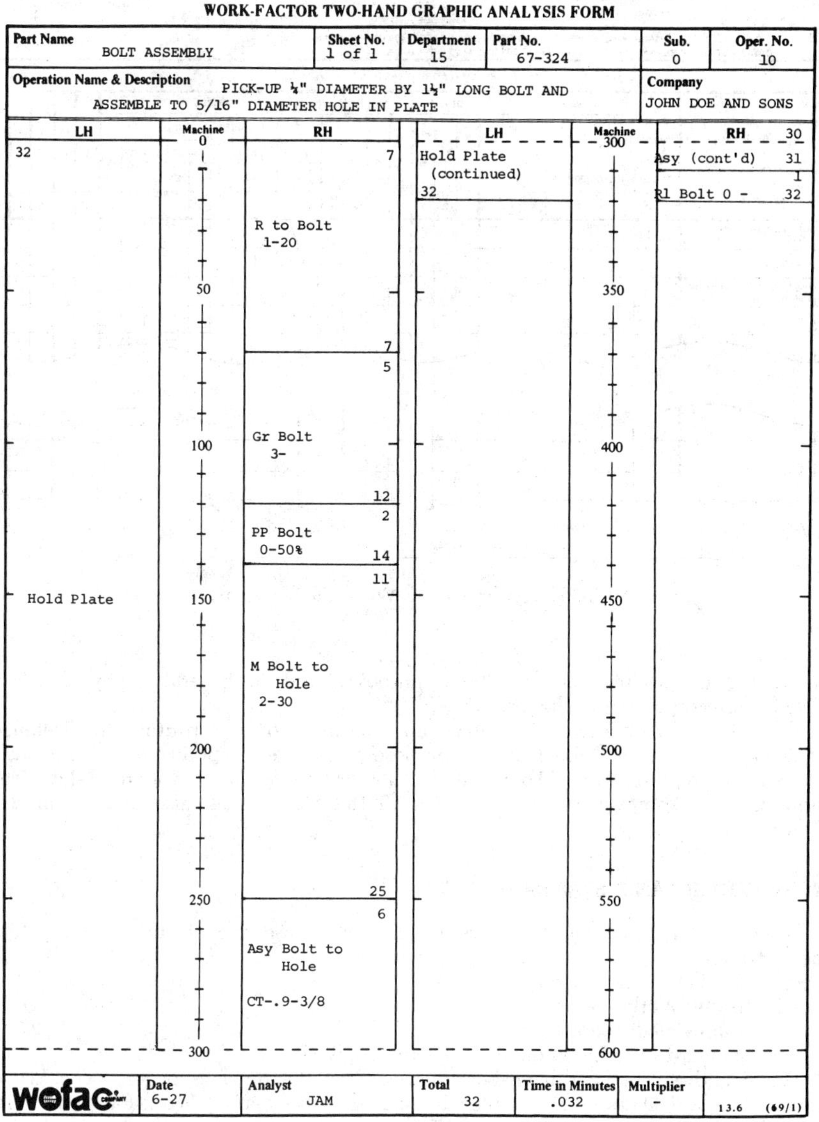

WORK-FACTOR TWO-HAND GRAPHIC ANALYSIS FORM

Part Name BOLT ASSEMBLY	Sheet No. 1 of 1	Department 15	Part No. 67-324	Sub. 0	Oper. No. 10
Operation Name & Description PICK-UP ¼" DIAMETER BY 1½" LONG BOLT AND ASSEMBLE TO 5/16" DIAMETER HOLE IN PLATE				Company JOHN DOE AND SONS	

LH	Machine 0	RH		LH	Machine 300	RH 30
32		7		Hold Plate (continued) 32		Asy (cont'd) 31 Rl Bolt 0 – 32
		R to Bolt 1-20				
	50				350	
		7 5				
	100	Gr Bolt 3-			400	
		12 2 PP Bolt 0-50% 14 11				
Hold Plate	150				450	
		M Bolt to Hole 2-30				
	200				500	
	250	25 6			550	
		Asy Bolt to Hole CT-.9-3/8				
	300				600	

WOFAC	Date 6-27	Analyst JAM	Total 32	Time in Minutes .032	Multiplier –	13.6 (69/1)

Fig. 3-10.   Two-hand Graphic Analysis of Bolt Assembly.

ABBREVIATED WORK-FACTOR TIME STANDARDS

Abbreviated Work-Factor provides a measurement procedure where rapidity is a requisite to the success of the measurement or rate-setting program. The mechanics of applying Abbreviated Work-Factor are quite different than for either Detailed or Ready Work-Factor. All time values in Abbreviated are a part of the Analysis form, thus eliminating the need for reference to either Manual or Tables. The Abbreviated Time Unit (AU) is 0.005 minute. A minimum amount

First Name		Sheet No.		COMPANY		Section No.	Part No.		Sub.	Oper. No.
CASE		1 of 1		JOHN DOE AND SON		37	48-719		0	7

Operation Name & Description   MACH. #1031 BLISS DOUBLE ACTION DRAW PRESS 240 TONS — 72.5 RPM
1st DRAW — 2 PAN DIE BLANK — CR STEEL THK .050 — .003 DIA. 19.25 WT. 4.04 LBS.

ABBREVIATED WORK — FACTOR * ANALYSIS

No.	DESCRIPTION OF ELEMENTS OF WORK	TRANSPORT							GRASP		PP				ASSEMBLE												T O T A L
1	PU blank, place in die, pull lever	2/2		2	3		1																				10
2	Machine time																									28	28
3	Remove part from press, toss aside, turn to WA	1			3	1/1																		1	2		9

			TOTAL	Time in Minutes	Multiplier		
wofac	S. BROWN		47 X .005				47
	ENGINEER DATE		1-20-	.24			

Copyright 1951   AIOO - 2-5-51

FIG. 3-11. Abbreviated Work-Factor Analysis.

of writing is required for descriptive purposes, and in general, many shortcuts have been incorporated in the procedure.

Abbreviated Work-Factor is neither an alternate nor a substitute for Detailed or Ready Work-Factor. Each has its purpose and range of application as previously defined. A representative Abbreviated Analysis is shown in Figure 3-11. The form used in Abbreviated, together with all the time values, also appears in this illustration.

WORK-FACTOR AND STANDARD DATA

Work studies based on the Work-Factor System use one of the following four methods:

1. Detailed Work-Factor
2. Ready Work-Factor
3. Abbreviated Work-Factor
4. Standard Data based on Work-Factor Analyses

The first three methods have been discussed in the previous pages. The Work-Factor Standard Data method, which is used extensively in all Work-Factor installations, is a very important part of the Work-Factor system. Because of space limitations, the following discussion of this method is brief.

The Work-Factor method which is the most suitable for a given situation is readily determined by briefly considering the following factors:

1. Quantity of parts to be produced
2. Length of cycle time required to complete one piece or operation
3. Similarity of work content
4. Complexity of work involved
5. Labor content

When consideration of these factors indicates that individual Work Segments of Standard Data would be used repeatedly if developed, it is usually found that the Standard Data method is the most practical method of measurement for the particular work involved. Experience indicates that the Standard Data method is the most useful and desirable measurement technique for approximately 90 percent of the measurement situations encountered in general industrial and office work.

In the broad sense, Work-Factor Elemental Time Standards are in themselves, classed as General Standard Data as opposed to Specific Standard Data. As generally understood, however, Standard Data are time values of a larger magnitude used to determine the time required to perform a specific type of work such as punch press, machine shop, mechanical assembly, foundry, typing, filing, or any other class or type of operation. Because the industrial engineer must set his labor standards promptly, accurately, and inexpensively, Work-Factor Standard Data are tabulated and reduced to the simplest terms and flexibility required for the specific type of work for which they are to be used. With proper organization of data, often the industrial engineer can set forty to fifty production standards daily, many of them from drawings only and in advance of production. This type of data organization is termed Specific Standard Data.

Developing Work-Factor Specific Standard Data. The Detailed and Ready Work-Factor systems are normally used for developing Specific Standard Data.

One of the chief problems in compiling Standard Data with a timing device is the need for great numbers of time studies to accumulate sufficient data for correctly establishing time-variable relationships. This is not only time consuming but also difficult from the standpoint that many of the desired conditions, such as variations in parts, sizes, materials, and the like, are not always available for study. This difficulty is largely eliminated when developing Standard Data by Work-Factor because the actual performance of all elements under all conditions is not required. The physical availability of all sizes and shapes of parts which are to be covered by the standards is also not essential. As a result, it is usually possible to establish Work-Factor Specific Standard Data in much less time than would be required to establish Specific Standard Data of comparable detail and accuracy using conventional stopwatch time study. This not only reduces the cost of Standard Data programs but also substantially shortens the elapsed time required for completion.

The establishment of Work-Factor Specific Standard Data normally involves the following steps:

1. Determination of the work to be analyzed and the scope the Standard Data will cover.
2. Observation to make certain that the proper methods are employed. Usually, it will be obvious at the start of the job that certain improvements can be put into effect. Other potential improvements will become apparent during the Work-Factor Analyses. All those improvements that are practical after consideration of quantities on order, tool costs, and the like should be put into effect and become the basis upon which the Standard Data are built. In no case should Standard Data be established where methods and other factors influencing the cost of the operation require improvement.
3. Division of the operation into elements which, because of their nature, size, and other similar characteristics, are suitable for the type of Standard Data being developed.
4. Analysis of each element of the operation using Work-Factor.
5. Summarization of the resulting time values on either Work-Factor Specific Standard Data Sheets or Work-Factor Pre-standard Forms.
6. Filing all Detailed Analyses and associated data in suitable form for future reference.
7. Checking of final results to ensure accuracy, completeness, and ease of application.
8. Training of industrial engineers to make certain that the Standard Data are applied correctly.

It frequently is desirable to establish production rates in advance of production.

This is practical where sufficient detailed information concerning tooling, production methods, product characteristics, and the like is made available prior to production. Without carefully and accurately prepared time standards, satisfactory pre-rating cannot be accomplished. For this reason, the time detail provided by the Work-Factor System is well suited to compiling Standard Data for establishing standards in advance of production (Pre-standards). A desirable feature of a good Standard Data system is the Pre-standard Form. The purpose of this Form is to so simplify the standards setting procedure that the tables of time values and the actual work study form can be one and the same. When the Pre-standard Form has been filled in by the industrial engineer, it establishes a complete, accurate,

WORK-FACTOR STANDARD DATA	Analyzed by Riser & Dombrowski Approved by J. B. Taggart Dist. Code A Ref. No. 03	Effective 11-30- Supersedes None Sheet No. 4

DESCRIPTION OF STANDARDS Riveting and machine operations	INDEX OR REFERENCE NUMBER

Analysis B – Pick up and assemble additional parts

This analysis covers the pickup and assembly of additional parts after the first two. As in analysis A, small parts are picked up using one hand, while larger parts (Class 3, 4 & 5) are picked up and assembled to the anvil using two hands. After the first two parts are assembled to the anvil, one hand moves the part to the next point of assembly, while the free hand picks up and pre-positions the next part.

Part	Elemental Description	Analysis	8"	12"	18"	24"	30"	40"
1A	R to part	A (X) D	54	65	76	86	96	
	Gr part (ave.)	Table (V)	51	51	51	51	51	
	PP part	3F1-100%	48	48	48	48	48	
	M to anvil	A (X) SD	70	85	98	109	119	
	Asy part to anvil	03-E1	64	64	64	64	64	
	R1 part	1/2 F1	8	8	8	8	8	
	Clear hand	A1	18	18	18	18	18	
			313	339	363	384	404	
1	R to Part	A (X) D	54	65	76	86	96	
	Gr part (ave.)	Table (V)	39	39	39	39	39	
	PP part	3F1-50%	24	24	24	24	24	
	M to anvil	A(X) SD	70	85	98	109	119	
	Asy part to anvil	03-E1	64	64	64	64	64	
	R1 (simo insert)		–	–	–	–	–	
			251	277	301	322	342	
2	R to part	A (X) D	54	65	76	86	96	
	Gr part (ave. 1" –4")	Table (V)	20	20	20	20	20	
	PP part	Simo move	–	–	–	–	–	
	M to anvil	A (X) SD	70	85	98	109	119	
	Asy part to anvil	03-E1	64	64	64	64	64	
			208	234	258	279	299	
3	R to part	A (X) D		65	76	86	96	109
	Gr part	F1		16	16	16	16	16
	M part to machine	A (X) D		65	76	86	96	109
	Slide into machine	VA2SPD		44	44	44	44	44
	Asy part to anvil	03-E1		64	64	64	64	64
				254	276	296	316	342
4	R1 first part	F1			16	16	16	16
	R 2nd part	A (X) D			76	86	96	109
	Gr part	F1W			23	23	23	23
	M part to machine	A (X) WD			98	109	119	135
	Slide into machine	VA4WSPD			66	66	66	66
	Asy part to anvil	03-E2			99	99	99	99
					378	399	419	448

FIG. 3-12. Representative sample of a Detailed Work-Factor Analysis used in compiling Specific Standard Data.

WORK-FACTOR RIVETING PRE-RATE SHEET

Part No. __345-12__   Part Name __Lid Assembly__   Date__9-21-__
Analyst ___I.D.N.___   Total Select Time __.208__   Pcs/Hr._____
A - Pick Up First Two Parts and Position on Anvil

CLASS OF PART	MOVING DISTANCE					
	8"	12"	18"	24"	30"	40"
1A	.044	.047	.049	.052	.054	
1	.039	.042	.044	.046	.048	
2	.031	.034	.036	.038	.040	
3		.026	.028	.030	.032	.035
4			.038	.040	(.042)	.045
5			.040	.042	.044	.047

Pick up is simo for class 1A, 1 and Class 2 parts

Class 3, 4 & 5 Parts, P/U is Separate. For 2nd Part use Table B.

.042
_____

B - Pick Up and Assemble Additional Parts to Anvil

CLASS OF PART	MOVING DISTANCE					
	8"	12"	18"	24"	30"	40"
1A	.031	.034	.036	.038	.040	
1	.025	(.028)	.030	.032	.034	
2	.021	.023	.026	.028	.030	
3		.025	.028	.030	.032	.034
4			.038	.040	.042	.045

.028
_____

C - Shift Assembly for Added Parts

CLASS	SMALL PART FIRST	LARGE PART FIRST
1A	.012	Simo P/U
1	.010	Simo P/U
2	.010	Simo P/U
3	.014	.014
4	.023	.019
5	.024	.019

D - Shift to Drive 2nd Rivet in the Same Two Parts

PART CLASS	2ND PART		
1ST PART	1A	1	2
1A	.012		
1	.012	.012	
2	.012	.012	.012
3	.012	.013	.013
4	.017	(.018)	.018
5	.017	.018	.019

x 3

.054
_____

Fig. 3-13. Typical Work-Factor Pre-standard Form.

and impartial record of both the method and the Analysis of the time required to perform the operation. It replaces the conventional time study form entirely.

As previously listed, the individual elements of Standard Data resulting from the Work-Factor Analyses are summarized in the form of Work-Factor Pre-standard Forms, or where practical, in the form of Standard Data Summary Sheets.

An example of a Detailed Work-Factor Analysis sheet which was compiled during an actual Standard Data program is shown by Figure 3-12.

The time standards developed in Figure 3-12, plus many others developed in a similar manner covering other elements of riveting work, are incorporated in Part B of the representative Pre-standard Form shown by Figure 3-13 under the heading "Pick Up and Assemble Additional Parts to Anvil."

The circled time values on the Pre-standard Form, together with the handwritten

E - Remove Assembly and Aside

PART CLASS	MOVING DISTANCE					
	8"	11"	18"	24"	30"	40"
1 Toss	.006	.006	.007	.008	.009	.010
2 Bench	.008	.009	.010	.011	.012	.014
Stack	.011	.013	.014	.015	.016	.017
3 Bench		.014	.015	.016	.017	.018
Stack		.018	.019	.020	.021	.022
4 Bench			.021	.022	.023	.025
Stack			.025	.026	(.027)	.029
5 Bench			.023	.024	.025	.027
Stack			.027	.028	.029	.031

.027

F - Machine Time   .008 x _4_ Strokes . . . .   *.032*

G - Additional Allowances

    1. Hand feed rivets   .028 per rivet   _____
    2. Hand P/U then tweezer feed rivet   .044 per rivet   _____
    3. Tweezer feed rivets from tray   .030 per rivet   _____
    4. Test part for looseness - One hand shake   .008 per part   _____
                            - Two hand twist or slide   .011 per part
    5. Flanged parts   **2** x .008 per rivet in or out   *.016*
    6. Blind asy. - Up to 3'' to visible area   .001 per hole   _____
              '' '' 5'' ''   ''   ''   .002 per hole   _____
    7. Fixture allowance - Nest part in fixture   .005 per part   _____
                    - Slide fixture in and out   .005 per hit   _____
    8. Rotate part - 90° Class 3 or 4   .016 per time   _____
               - 180° Class 3 or 4   .023 per time   _____
    9. Align 2nd hole to added part   **1** x .009 per occurence   *.009*
   10. Inspect part for mark, etc.   .010 per inspection   _____
   11. Invert chassis - Class 3   .015 . . . .   _____
               - Class 4   .021 . . . .
   12. Unload wire basket or truck - small parts Class 2   .018 per part   _____
                   Medium parts Class 3 & 4   .010 per part
                   Class 3 - 4 2/time   .020 per part
   13. Other _____   _____

                                                TOTAL   *.208 MiN.*

Parts Classification

1A. Washers, terminals, bushings, studs and parts which tend to cling together.

1. Brackets, switches, sockets, terminal strips and similar parts.

2. Large brackets, covers, shields, transformers, relays, etc., and small panels approximately 5" x 8".

3. Average panels and bases, approximately 8" x 12".

4. Transformers and parts over 3#, large panels and bases, approx. 12" x 18".

5. Very large panels and bases, approximately 12" x 24".

FIG. 3-13 (*continued*). Typical Work-Factor Pre-standard Form.

descriptive information and time values, illustrate how the Pre-standard Form is actually used in rating an operation and how it reduces a complex measurement requirement to its simplest form.

The operation which is shown on the Pre-standard Form in Figure 3-13 involves the riveting of a flat nameplate (2 by 4 inches in size) to a large metal lid (11 by 20 inches in size). Four rivets are used. The lid has a 1-inch flange.

USE OF DETAILED WORK-FACTOR IN MAINTAINING PRODUCTION STANDARDS

The Work-Factor System provides a means of effectively maintaining standards when proper procedures and administration are in force. Because individual work motions or any combination of work motions can be measured by Work-Factor, it is possible to determine the exact effect of methods improvements. Major or minor changes in the work content can therefore be rapidly evaluated and incorporated in the Standard Data or in the specific production standard.

Through continuous time adjustments made as work contents significantly change, it is possible to avoid the rate revision program and labor conflict which arise from a company's sudden and urgent need to correct several years of neglected maintenance of its time standards. It has been the repeated experience of the authors that accurate adjustment of production standards (up or down) plays as significant a part in confident employee-management relations as does accuracy in setting the original standards.

BIBLIOGRAPHY

"How C. & N. W. Has Cut Brake Overhaul Costs," *Railway Locomotives and Cars,* February, 1969.

Kattan, Appu, and Gerald Nadler, "Equations of Hand Motion Path for Work Space Design," *Human Factors,* vol. 11, no. 2, 1969.

Mento-Factor Manual of Detailed Work-Factor Mental Process Times, 5th ed., Wofac Company, Moorestown, N.J., 1968.

Nanda, Ravinder, "The Additivity of Elemental Times," *Journal of Industrial Engineering,* May, 1968.

Quick, Joseph H., James H. Duncan, and James A. Malcolm, Jr., *Work-Factor Time Standards,* McGraw-Hill Book Company, New York, 1962.

Taggart, John B., "Comments on 'An Experimental Evaluation of Predetermined Elemental Time Systems,'" *Journal of Industrial Engineering,* November–December, 1961.

Whitmore, Dennis A., *Work Study and Related Management Services,* William Heinemann Ltd., London, 1968.

Chapter **4**

MTM-GPD and Other
Second Generation Data

WILLIAM J. MATTERN
Counselor, Maynard Research Council Incorporated, Pittsburgh, Pennsylvania

KENNETH KNOTT
Managing Director, Maynard Training Centre, Birmingham, England

RUSSELL W. McDONALD
Director, Maynard Training Center, Cleveland, Ohio

From the time that the basic predetermined time systems described in Chapters 2 and 3 of this section were first developed and used, there has been an ongoing effort to develop modifications of these systems which could be applied more quickly and with less analysis effort without sacrificing too many of the advantages inherent in the basic systems. The modified systems are generally known as second generation data, implying that they are descended from the original basic data, as indeed they are.

Second generation data developed by the Wofac Company have already been discussed in Chapter 3. They are comprised of three systems known as Ready Work-Factor, Abbreviated Work-Factor, and Work-Factor Standard Data. This chapter will discuss certain of the second generation data developed from basic MTM.

METHODS TIME MEASUREMENT

SIMPLIFIED DATA

(All times on this Simplified Data Table include 15% allowance)

HAND AND ARM MOTIONS	BODY, LEG, AND EYE MOTIONS
REACH or MOVE TMU 1″ .... 2 2″ .... 4 3″ to 12″ 4 + length of motion over 12″ 3 + length of motion (For TYPE 2 REACHES AND MOVES use length of motion only)	TMU Simple foot motion....... 10 Foot motion with pressure 20 Leg motion .... 10 Side step case 1......... 20 Side step case 2......... 40 Turn body case 1........ 20 Turn body case 2........ 45
POSITION	
Fit Symmetrical Other Loose 10 15 Close 20 25 Exact 50 55	Eye time.............. 10
TURN—APPLY PRESSURE TURN.............. 6 APPLY PRESSURE.. 20	Bend, stoop or kneel on one knee............ 35 Arise.................. 35
GRASP Simple.............. 2 Regrasp or Transfer... 6 Complex............ 10	Kneel on both knees..... 80 Arise.................. 90 Sit..................... 40 Stand................. 50 Walk per pace.......... 17
DISENGAGE Loose.............. 5 Close.............. 10 Exact............... 30	1 TMU = .00001 hour = .0006 minute = .036 second

Fig. 4-1. Methods time measurement simplified data card.

DEVELOPMENT OF SECOND GENERATION MTM DATA

The first attempt to simplify the basic MTM data was made by the developers of the methods time measurement procedure, H. B. Maynard, G. J. Stegemerten, and J. L. Schwab. They devised a simplified data card (Figure 4-1) which reduced the number of motion classifications that had to be considered, and averaged certain of the motion times. To these times was added a 15 percent allowance factor, considered suitable in many industries for personal, fatigue, and unavoidable delay allowances. When the resulting values were rounded off, they became a set of a relatively few, easily remembered whole numbers. The simplified data were published in 1948,[1] and for a time, were quite widely used.

[1] H. B. Maynard, G. J. Stegemerten, and J. L. Schwab, *Methods-Time Measurement*, McGraw-Hill Book Company, New York, 1948.

Because MTM was placed in the public domain as soon as its development was completed, it was only natural that a number of different people and groups, both in the United States and abroad, would seek to develop modified or second generation data designed to meet a particular problem which they faced. This resulted in a considerable number of systems, each of which presumably was considered satisfactory by its developers.

Two of these systems were published and are described briefly below. The first was Universal Standard Data, or USD.[2] The procedure was developed in 1954 by a group of consulting management engineers of Methods Engineering Council, who faced the problem of establishing a large number of standards in a Swedish plant assembling a number of different models of farm tractors on a common progressive assembly line. To do this with basic MTM would be time consuming and expensive. They therefore developed a procedure which was much quicker to apply, although not quite as accurate as basic MTM. Later, the procedure was modified further to give it universal application. The second procedure was Master Standard Data, or MSD.[3] It was developed by a group of engineers from Serge Birn Associates, again in an effort to devise a system which was quicker and easier to apply than MTM and for use where extreme accuracy was not important. Both systems have been widely used.

A number of similar systems were developed from 1954 to 1962, all with the same general objectives. At length, it was recognized that it would be desirable if this multiplicity of systems could be combined into a single system which all MTM practitioners could use. The MTM Association for Standards and Research undertook to do this task. Data developed over the years by industrial and professional members, including USD and MSD, were donated to the Association. After thorough research, a set of data known as Methods Time Measurement General Purpose Data, or MTM-GPD, was developed and published by the Association in 1962.[4] The procedure is described briefly below.

During this same period, second generation data were also being developed in other countries as well. Again, the multiplicity of systems was felt to be undesirable, and the International MTM Directorate[5] undertook the development of still another system, which it was hoped would embrace the best features of all systems yet developed, including MTM-GPD. The result of their work is known as MTM-2. It also is described below.

NEED FOR A QUICKLY APPLIED PREDETERMINED TIME SYSTEM

The burgeoning development of second generation data described above was sparked by the desire of industrial engineers to set standards on low-quantity and long-cycle work without resorting to stopwatch time study. Basic MTM proved to be ideal for short-cycle, highly repetitive work. It was more accurate and consistent than stopwatch time study, it eliminated the always controversial procedure of performance rating, and it gave an insight into methods superior in some respects (because the element of time is included) to that resulting from motion picture film analysis. The procedure required time and effort to apply, but the methods improvements which almost always accompanied its application made its use economically justified.

As applications of basic MTM were made to less and less repetitive work, however, it soon became evident that there was a point at which the resulting economies

[2] W. K. Hodson and W. J. Mattern, "Universal Standard Data," sec. 3, chap. 12, in H. B. Maynard (ed.), *Industrial Engineering Handbook*, 2d ed., McGraw-Hill Book Company, New York, 1963.

[3] R. M. Crossan and H. W. Nance, *Master Standard Data*, McGraw-Hill Book Company, New York, 1962.

[4] *MTM General Purpose Data (MTM-GPD)*, The MTM Association for Standards and Research, Fair Lawn, N.J., 1962.

[5] Member countries include Finland, France, Germany, Japan, The Netherlands, Norway, Sweden, Switzerland, United Kingdom, and United States/Canada.

would not pay for the cost of making the studies. Nevertheless, the desire to use the procedure or something like it remained strong, and the development of second generation data began. The objective was to develop a procedure which could be applied more quickly than basic MTM while retaining as many of its advantages as possible. The objective was accomplished by a trade-off of a certain amount of accuracy and a lessened intensity of methods study for a reduction in application time.

CHARACTERISTICS OF SECOND GENERATION DATA

Second generation data seek to reduce the number of decisions that must be made when developing a standard. Variables such as class of motion, distance traveled, and weight are averaged to reflect the most commonly encountered work conditions. Certain of the basic motions are combined with others. As a result, fewer motions have to be recognized, identified, and recorded. It therefore requires' less time to make a study, but because the averages do not always correspond to the actuals, there is a loss of accuracy. The longer the cycle is, the less serious this loss of accuracy is, because the highs and the lows tend to average out.

Figure 4-2 shows a comparison of studies made with basic MTM and MTM-GPD. Each study divided the operation into ten elements. To record the method using basic MTM required 168 lines of motion pattern descriptions. Using MTM-GPD, only 93 lines were required. The number of time values that had to be looked up and recorded were reduced in the same proportion. The number of decisions that were required to recognize and identify individual motions were reduced even more. As a result, the standard was set more quickly by MTM-GPD.

The difference in MTM-GPD element times from those established with basic MTM ranged from −9 to +16 percent. This would be an unacceptable amount of inaccuracy if individual standards were to be established for each element. When the element times are added, however, the MTM-GPD standard differs from the

COMPARISON OF ANALYSIS USING MTM-GENERAL PURPOSE DATA

Operation: Film analysis no. 112—from the MTM film library
Title: Slide projector—Package slide projector together with accessory items in carton ready for shipping.

Element no.	Description	MTM		MTM-GPD		TMU difference, %
		No. of lines	Total TMU	No. of lines	Total TMU	
1	Obtain carton and open	13	79.8	5	81	+ 1
2	Get projector and accessories	16	105.7	9	114	+ 8
3	Fold cardboard spacer	16	108.1	12	113	+ 5
4	Get slide frame	13	70.1	8	64	− 9
5	Place literature	15	58.7	8	57	− 3
6	Place slide magazine	9	51.5	4	51	− 1
7	Place and fasten cover	27	129.0	14	134	+ 4
8	Place pulp inserts	12	121.3	6	119	− 2
9	Pack carton	25	148.0	14	172	+16
10	Staple and palletize	22	158.7	13	149	− 6
	Total	168	1,030.9	93	1,054	+ 2

FIG. 4-2. Reliability comparison. (*Source: MTM-General Purpose Data Manual. Copyright 1962, The MTM Association for Standards and Research. No reprint permission is granted without express written consent of the MTM Association, 9–10 Saddle River Road, Fair Lawn, N.J.*)

basic MTM by only 2 percent. This would be considered acceptable by most industrial engineers. If MTM-GPD standards are established for a class of work and if the operator works on a number of different jobs, a further averaging occurs, and total performance is measured about as accurately by one system as the other.

Thus, second generation data permit establishing standards more quickly, with a loss of accuracy which ranges from small to negligible where cycles are long and the jobs worked on are many. The major disadvantage is the loss of detailed methods analysis which occurs. The more simplified the second generation data are, the less valuable they are as a tool of methods analysis. This has caused a system like MTM-2 to be referred to as a "paper stopwatch."

UNIVERSAL STANDARD DATA

As mentioned above, the basic concept of Universal Standard Data (USD) was first formulated in 1954. At that time, the Methods Engineering Council was working on the development of a common progressive assembly line for the assembly of a wide range of farm tractors. The cycle time at each work station was rather long, and there were a number of variations in the assembly procedures for each of the many different types of tractors involved. Thus the engineers were faced with the task of developing a large number of standards.[6] They recognized that a less time-consuming approach than basic MTM was needed, and they developed the basic concept of USD.

The procedure worked well on this initial application, and the originators were encouraged to try the same approach on other types of work. This they did in both Europe and the United States. USD was soon used on a wide variety of assembly work ranging from the assembly of electronic digital computers to the construction of concrete batching and mixing plants for field construction work. With each new application, the data were refined and expanded to cover a variety of conditions and types of work. New procedures and concepts for application of the basic data were also developed, and USD soon came to be used to construct time formulas and higher levels of standard data with great speed and consistency. This led to the extension of the data to cover all types of machining operations, fabricating operations, production bench assembly, material handling, and a wide variety of indirect labor operations. The end result of all this work was a set of basic data that were not limited to any particular operation or process. Consequently, the name chosen for the data was Universal Standard Data.

The Data Card. USD consists of seven brief tables of data:

> Table I Get Object—a combination of Reach and Grasp motions
> Table II Place Object, Nominal Weight—a combination of Move, Position, and Release motions for weights up to 2½ pounds
> Table III Place Object, Significant Weight—a combination of Move, Position, and Release for weights over 2½ pounds
> Table IV Get Turn and Place Turn—a combination of Turn and Grasp, and Turn and Release
> Table V Walk Displacement—a combination of Turn Body and Walk
> Table VI Miscellaneous Body Motions—Body, Leg, and Foot combinations
> Table VII Cranking—motions employed in using a crank or handwheel

Portions of the USD data card are shown in Figure 4-3. A quick review of these two tables illustrates the combination of MTM motions, the grouping of distances, and coding. For example, the symbol G12S in Table I denotes a reach for an easily grasped object where the distance reached is in the 12-inch range, or from 10 to 14 inches.

Condensed basic MTM tables (not illustrated here) are also a part of the USD card. The purpose of these tables is to supplement the USD whenever the nature of the operation or its degree of repetition justifies its use. Tables are provided for Reach, Move, Grasp, Release, Position, Turn, Apply Pressure, Disengage, Eye

---

[6] "Quick Way to Measure Long Jobs," *Factory Management*, October, 1958.

TABLE I. GET OBJECT

Symbol	Distance reached, inches							Type of grasp	Description
	f	2	5	8	12	18	26		
G__S	4	6	10	12	15	19	25	Simple pickup	Easily grasped object (G1A, G1B, G5)
G__E	10	14	18	20	22	27	32	Jumbled or Inter- ference	Easily jumbled or some interference (G4A, G4B, G1C1, G1C2)
G__A	15	19	22	24	27	31	37		Average jumbled or medium interference (G4C, G1C3)
G__D	21	24	28	30	33	37	42		Difficult jumbled, separa- tion problem (G4C+G2+G4A+2G2)
G__N	6	8	12	14	17	21	27	New hold	Get new Grasp (RL1+R__B+G1A)
G__T	8	9	13	15	19	23	30	Transfer	One hand to other hand (M__A+G3)

TABLE VI. MISCELLANEOUS BODY

Symbol	TMU	Description	Symbol	TMU	Description
BD1	18	SSC1, TBC1	FL	13	FM, FMP, LM10
BD2	32	SSC2, TBC2, B, S, KOK, A. . . , W2P			
BD3	73	KBK, AKBK	ST	39	SIT, STD

FIG. 4-3. Examples of USD data. (*Copyright* 1962, *H. B. Maynard and Company, Incorporated.*)

Travel, Eye Focus, and Body, Leg, and Foot Motions. All USD table values are expressed in time measurement units (TMU) rounded off to the nearest whole number. These USD data are applied in the same manner as the MTM-GPD data illustrated in Figure 4-7.

Accuracy. One of the primary questions in the development of USD was the effect that the grouping of distances and motion patterns would have on the final time standards. The users of the USD procedure recognized that the complexities of the problem did not permit a mathematical solution; so they utilized a simulation approach[7] as a means of providing answers on the accuracy of the procedure when applied to a variety of jobs under various conditions. For the purpose of this simulation, a set of ground rules was established to include the many variables and their respective probabilities. A program was then developed for an IBM

---

[7] Hodson and Mattern, *op. cit.*

704 to use all these factors in a simulation program. By generating a series of random numbers, the computer developed a large number of simulated USD studies for a variety of jobs within the limitations of the program.

The computer readout pointed out several significant relationships. First, it became readily apparent that as the job length increases, the job error decreases. Second, the data obtained from the simulated program indicated that basic MTM data, rather than USD data, should be used for jobs 0 to 0.002 hour long. In this range, approximately 50 percent of all jobs had an error in excess of 5 percent. In the job length interval 0.003 to 0.005 hour, the error for 85 to 90 percent of the jobs was close to 5 percent. And in job length interval 0.010 to 0.015 hour, the error for all the jobs was less than 5 percent. For long jobs, the indications were that basic USD provides greater accuracy than is actually needed. In these cases, it is desirable to develop larger units of standard data and thus reduce the time required to establish standards. The general rule in this regard is that as the job length increases, standard data made up of larger and larger building blocks may be employed without exceeding a given level of accuracy.

MASTER STANDARD DATA

Serge Birn Associates originated Master Standard Data (MSD). The system was published by McGraw-Hill Book Company in 1962.[8] MSD is similar to USD in approach; that is, both are simplified approaches in which basic MTM motions are combined.

Long an MTM user, the Birn engineering group had developed a standards approach for work in furniture and woodworking industries, a standards approach for material handling, and a standards approach for office and clerical work. With this work behind them, they conceived the idea that it should be possible to improve work measurement, standards setting, and methods improvement procedures, and decided to start some new research on MTM. The idea was not to change basic MTM in any way, but to improve and simplify its application procedure.

At this time, an informal survey was made of a cross section of the companies using MTM. It showed that most of those trained in MTM were not using it to the full extent of its potential application possibilities. For example, one company would use it for improving methods but not for setting standards. Another would employ it for setting standards and forget that the predetermined times could be better used through the media of standard data and formulas. One organization would use MTM only for highly repetitive work, another only for indirect work, and so on, but seldom was there any definite pattern. They also found that to supplement basic MTM, some of its users developed offshoots of the basic technique, which applied only to certain specific areas such as needle trades, machine shops, or maintenance work. The question then was, would it be necessary to tailor MTM for everything? With this question in mind, the engineers researched the approach to work measurement application based on MTM and developed a procedure that they felt was truly basic. What was really convincing was when a large computer manufacturer, after huge expenditures for standard data using conventional measurement techniques, switched to the new concept. Using the new procedure, they were able to develop corporate-wide standard data for direct and indirect work, nationally and internationally, and for very much less than the original investment. Installation time was reduced to less than one-half of the time previously planned. The technique was named Master Standard Data. It is applicable to any type of work and can be used quickly and economically to measure work never before considered measurable from an economic standpoint.

The Data Card. The complete MSD data card is shown in Figure 4-4. This card, and nothing more, is all that is required to measure and describe the elements of work that are contained in an industrial situation. This of course assumes that the analyst has been adequately trained. MSD consists of six tables of data:

[8] Crossan and Nance, *op. cit.*

Table 1 Obtain—a combination of Reach, Grasp, and Release motions
Table 2 Place—a combination of Move and Position motions
Table 3 Rotate—motion patterns employed in turning an object about its axis using the fingers or the wrist, or the cranking pattern
Table 4 Use—motions employed to move the fingers, hand, or hand and arm, back and forth as characterized by such patterns as sawing, filing, hammering, and the like
Table 5 Finger Shift and Exert Force—motions employed to Regrasp and Apply Pressure
Table 6 Body Motions—a combination of the horizontal and vertical body displacement motions and the Foot motion

Figure 4-4 illustrates the groupings of distances in Obtain and Place and coding. For example, the code RHF in Table 3 denotes a rotational pattern with the fingers; and the code RCL, a cranking pattern of about 8 inches in diameter. The RHF is employed, for example, every time an operator runs a stud down with his fingers where there is little resistance to turning. An example of the RCL is turning a handwheel, which is over 6 inches in diameter, one revolution.

As in the case of USD and MTM-GPD, the MSD table values are also expressed in time measurement units (TMU) and are rounded off to the nearest whole number. When applying these data, the procedure follows the same general pattern as the MTM-GPD procedure illustrated in Figure 4-7; that is, the principle of application is the same.

Table I

OBTAIN-O

Distance in inches	Degree of control			
	Some-S		High-H	
	1	2	1	2
2	8	8	17	30
6	13	13	21	34
12	17	17	25	38
18	21	21	30	42

Table 2

Place-P

Distance in inches	Location					
	Other hand	General	Exact			
			Loose-L		Close-C	
	O	G	1	2	1	2
2	7	5	11	26	21	47
6	11	9	16	31	27	52
12	15	13	21	36	31	57
18	19	17	26	41	37	62

Table 3

ROTATE		
R		
H	F	9
	W	15
C	S	17
	L	19

Table 4

USE	
U	
V	4
L	8
M	13
H	17

Table 5

FINGER SHIFT	
FS	6
EXERT FORCE	
EF	11

Table 6

BODY MOTIONS		
B		
A	Arise-sit	108
F	Foot	9
V	Vertical	61
W	Walk	17

FIG. 4-4. The MSD data card. (*Source:* Master Standard Data *by Richard M. Crossan and Harold W. Nance. Copyright* 1962, *McGraw-Hill Book Company, New York. Used by permission.*)

Accuracy. How accurate is MSD? In answering this question, the originators always arrived at the same answer: "It depends on what you are comparing it with." MSD can be compared for accuracy in three different ways, namely, with leveled time study, with conventional standard data based on MTM, and with detailed MTM analyses. Regardless of which is compared:

1. Comparisons must be based on the same method.
2. Both systems must use the same performance levels.
3. The same allowances, where applicable, must be used.
4. The conditions must be the same.

When these four factors are taken into consideration so that like things are being compared, MSD measures up well, regardless of what it is compared with. In thousands of instances, it has proved accurate, well within the accepted work measurement limits of plus or minus 5 percent.

When MSD was compared with time study, the results were favorable. In one comparison with a large manufacturer of washing machines, the comparison was made on work cycles of varying lengths and complexity; it covered a wide range of operator performances, from a rated low performance of 60 percent to a rated high of 125 percent; and it included quite a wide variety of work, from placing batts of insulation to the simple driving of several screws. The MSD times, when compared with the rated time study times, showed a maximum negative variation on an individual study of minus 10 percent and a maximum positive variation on an individual study of plus 10 percent. However, when comparing the total operation times, MSD and time study checked closely when the individual inconsistencies in ratings and so forth were allowed to compensate. The overall comparison was minus 3 percent.

MSD was also compared with detailed MTM studies and with conventional standard data that had been developed with MTM over a period of years. This comparison was made by the staff of one of the world's largest corporations and was not limited to any single plant or operation. Comparisons were made in five different locations, and the operations covered everything from heavy milling machines to precision electrical assembly. A summary of this comparison is shown in Figure 4-5.

The MSD-based standards varied from the detailed MTM studies by only a plus or minus 5 percent, whereas the existing data standards varied from a high of plus 46 percent to a low of minus 12 percent. The greatest variation for the existing data occurred on the very short cycle operations.

METHODS TIME MEASUREMENT—GENERAL PURPOSE DATA

In 1955, at the Fourth Annual MTM Conference, forward-thinking members of the MTM Association for Standards and Research suggested a design of standard data that could be used in many types of manufacturing processes. In 1958, several systems were presented at the Seventh Annual International MTM Conference. Each design or development offered a different approach to the compilation of standard data. Again, at the Eighth Annual Conference in 1959, still other systems of standard data development were presented.

At length, it was recognized that it would be highly desirable if one system could be developed for all MTM practitioners. In a paper presented for the Long Range Planning Committee of the MTM Association at the Board of Directors meeting of December, 1961, Dr. Maynard said in part:

> Finally, although MTM provides basic motion time data which are the same for every company in every industry, industrial engineers in each company use it independently to solve afresh the same problems over and over again. Every company develops its own time formulas or standard data, something which could easily be done just once for many different machines and processes if there were a central agency working on this problem. The cost of all this duplication of effort represents a tremendous waste which could easily be avoided.

Plant no.	Part no.	Oprn. no.	Detailed MTM	Existing data	New std. data	Compared to MTM New std. data, %	Compared to MTM Existing data, %
	6404	...	1237	1365	1293	+4.5	+10.3
1	3600	227	1960	...	1883	−4.0	...
	6308	10	695	778	670	−3.0	+12.0
	1102	201	4994	4882	4961	−0.7	− 2.3
	201	20	91,599	98,326	91,933	−0.4	+ 7.3
2	1062	20	8226	...	8279	+0.6	...
	1250	30	4721	...	4686	−0.7	...
2 & 3	Rockers	135	991	...	1034	+3.8	...
	510	Boring 30	439	633	432	−2.0	+44.0
	401	Heavy Mill 10	1385	1906	1420	+3.0	+34.0
3	314	Vert. Mill 29	1146	1799	1157	+1.0	...
	873	10	1202	1100	1264	+5.0	− 8.5
	280	37	994	1071	967	−3.0	+ 8.0
	280	17	1099	...	1059	−4.0	...
	Blocks	11	7907	8241	7845	−0.8	+ 4.2
4	3707	21	9747	9879	9871	+1.3	+ 1.3
	6262	11	3486	3061	3416	−2.0	−12.5
	6217	20	...	102,045	100,596	−1.4	...
	5331	430	394.6	...	387.0	−2.0	...
	5332	430	434.4	6650	423.0	−0.9	+46.0
5	5335	430	733.8	6650	733.0	−0.1	+ 8.0
	4845	170	900	...	862	+4.4	...
		60	17,911	20,896	18,850	+5.0	−10.0

FIG. 4-5. How accurate is MSD? (*Source:* Master Standard Data *by Richard M. Crossan and Harold W. Nance. Copyright 1962, McGraw-Hill Book Company, New York. Used by permission.*)

As a result of this report, the Board of Directors of the MTM Association established the Applied Research Implementation Committee early in 1962. The Committee first made a survey of the needs of industry. Second, they reviewed the contributions of standard data as submitted by a number of the Association members. Finally, their efforts led to the publication of the first volume of the general purpose data (GPD). This first volume included ten categories of basic data: Get, Place, Elemental, Body Motions, Read, Threaded Fastener, Dip, Lubricate, Write, and Actuate.

After the release of the first volume of data, the Applied Research Committee, which was made up of industrial, professional, and academic members who donated large amounts of time to the development of the data, foresaw the need for expanding the system.

In August, 1963, the MTM Association released three new categories of basic data to complete the first level of data. The three new categories were Clean, Inspect and Test, and Tool Use.

All elements in the first level of MTM-GPD are identified as "B," basic. They comprise the Volume 1 data and consist of the frequently used MTM patterns as well as certain individual MTM conventions.

Multipurpose is another level of data. It is a combination of basic and other multipurpose elements which depict a higher level of work occurring in a wide variety of operations. This level of data was published in Volume 2. At the present time, Volume 2 consists of Multi-Get, Multi-Place, Multi-Body Motions, Multi-Clamping, and Multi-Vising. These data were released in September, 1964.

The Data Card. MTM-GPD consists of two volumes of data. These data are presented on three data cards: Volume I, Basic Data Card #1; Volume I, Basic Data Card #2; and Volume II, Multipurpose Data.

The basic data on Card #1 (see Figure 4-6) consists of the following four data tables:

1. Get—a combination of Reach, Grasp, and Release motions
2. Place—a combination of Move and Position motions
3. Elemental—a listing of certain individual MTM motions, a combination of the Turn motions, and weight factors
4. Body Motions—a combination of Foot, Leg, and Body motions

These data tables are comparable with the USD data partially illustrated in Figure 4-3 and the MSD data illustrated in Figure 4-4.

The remaining data of MTM-GPD in the basic level (not illustrated here) contain the following nine sets of data:

1. Clean—various cleaning patterns, with and without resistance
2. Dip—various combinations of dipping
3. Lubricate—various methods and patterns of applying lubrication
4. Threaded Fastener—various patterns for "running" threaded fasteners with the fingers and the hand
5. Read—for reading individual words or a sequence of words
6. Write—for various types of writing and punctuation
7. Inspect and Test—various patterns for a limited number of gages
8. Actuate—numerous patterns for turning cranks, knobs, dials, and switches; and moving levers, wheels, and the like
9. Tool Use—a comprehensive set of data for using the most common type of hand tools

Volume 2, multipurpose data (not illustrated here), consists of five data tables:

1. Multi-Get—a combination of the basic Get and basic Place patterns
2. Multi-Place—a combination of basic Get and basic Place as characterized by tool handling, that is, Get, Place to use, and Place to lay aside
3. Multi-Body Motions—a combination of the basic body motions for displacing the body in horizontal and vertical directions
4. Multi-Clamping—various patterns for using several common clamps
5. Multi-Vising—patterns for using a vise

All GPD table values are expressed in time measurement units (TMU), and like the other data presented, are rounded off to the nearest whole number. These data are applied in the format illustrated in Figure 4-7. This is an example of a GPD worksheet that shows several elements of the reliability comparison, Figure 4-2. The format of the GPD worksheet is simple. The work element is described first. This is followed by the GPD code. The TMU value and frequency are recorded next, and finally the total TMU are listed in the right-hand column of the worksheet.

Data Coding. MTM-GPD are coded using a seven-position alpha mnemonic code. The first position of the code denotes the level of data, such as "B" for basic and "M" for multi. Positions 2 and 3 denote a major variable or a verb. Examples of this are Get (GT) and Vising (VS). Positions 4 and 5 denote a sub-variable or a modifying adverb. Examples of this are Contact Fixed (CF) and Tighten or Loosen (TL). The last two positions of the code, positions 6 and 7, denote a further variable, such as distance or the length of the vise handle. A distance of 12 inches is coded 12. The code for a vise handle up to 9 inches in

BGT - BASIC GET			Distance	1"	1-3"	3-9"	9-15"	15-21"	21-27"
			Code	01	02	06	12	18	24
CF	CONTACT FIXED			2	4	7	10	12	15
CO	CONTACT OBJECT — PLAIN REACH			2	4	9	13	17	22
E	EASILY GRASPED	F	Fixed Location	6	8	11	14	16	19
		V	Variable Location	6	8	13	17	21	26
		A	Additional Object	17	19	—	—	—	—
J	JUMBLED	O	One Hand	13	17	21	25	30	34
		S	Simo	24	28	32	36	41	45
		A	Additional Object	24	28	—	—	—	—
		H	Handful	33	35	39	44	48	52

BPL — BASIC PLACE			Distance	0	1"	1-3	3-9	9-15	15-21	21-27
			Code	00	01	02	06	12	18	24
AL	APPROXIMATE LOCATION			—	2	5	9	13	17	21
L	LOOSE	S	Symmetrical	6	8	11	16	21	26	31
		N	Not Symmetrical	9	11	14	19	24	30	35
C	CLOSE	S	Symmetrical	16	18	21	27	31	37	42
		N	Not Symmetrical	20	22	25	30	35	40	45
E	EXACT	S	Symmetrical	43	45	48	53	58	63	69
		N	Not Symmetrical	47	49	52	57	62	67	72
OH	OTHER HAND			6	8	9	14	19	23	28
S	START THREADED FASTENER	V	Visible	—	26	29	34	39	44	49
		B	Blind	—	60	63	68	73	78	83

BEL — BASIC ELEMENTAL				Code	TMU
AP	APPLY PRESSURE	Case 1		01	16
		Case 2		02	11
DE	DISENGAGE	Loose		01	4
		Close		02	8
		Tight		03	23
EF	EYE FOCUS			01	7
ET	EYE TRAVEL	Per Inch At 15" Distance		01	1
		Per Foot At 30" Distance		02	6
		Maximum 70°		03	20
RG	REGRASP			01	6
T	TURN WRIST	W Wrist Turn Only	Up to 90°	01	4
			90° to 180°	02	7
		S Shift Grasp And Turn	Up to 90°	01	12
			90° to 180°	02	19
W	WEIGHT FACTORS	F First Static And Dynamic	2.5 to 10 E.N.W.	10	3
			10 to 20 E.N.W.	20	8
			20 to 30 E.N.W.	30	12
			30 to 40 E.N.W.	40	17
			40 to 50 E.N.W.	50	22
		A Additional Dynamic Only	2.5 to 10 E.N.W.	10	1
			10 to 20 E.N.W.	20	2
			20 to 30 E.N.W.	30	3
			30 to 40 E.N.W.	40	4
			40 to 50 E.N.W.	50	6

BBM — BASIC BODY MOTIONS				Code	TMU
FM	FOOT MOTION			01	9
LM	LEG MOTION	Up to 9" Length		06	7
		9" to 15" Length		12	14
		15" to 21" Length		18	22
HC	HORIZONTAL CHANGE — TURN BODY OR SIDE STEP			01	19
VC	VERTICAL CHANGE	Bend, Stoop, Or Kneel & Arise		01	61
		Kneel On Both Knees & Arise		02	146
SS	SIT AND STAND	Chair Stationary		01	108
		Chair Moved		02	172
W	WALK	O Obstructed — Per Pace		01	17
		U Unobstructed	Per Pace	01	15
			Per Ten Feet	02	53

FIG. 4-6. MTM-GPD, volume 1, basic data card #1. (*Copyright* 1965, *The MTM Association for Standards and Research. No reprint permission is granted without express written consent of the MTM Association, 9–10 Saddle River Road, Fair Lawn, N.J.*)

GPD DATA WORKSHEET

Study No. Film Analysis #112   Date 9/19/—
By_____   Page 1 of 3

Symbol	Description	GPD Motion	TMU	No.	Total TMU
I.	**OBTAIN CARTON AND OPEN**				
	Sidestep from pallet to cartons	BBM-HC-01	19	2	38
	Hand to carton	BGT-CO-12	13	Int.	–
	Guide carton to table	BPL-AL-24			21
	Contact inside flaps	BGT-CO-12			13
	Spread flaps open	BPL-AL-06			9
					81
II.	**GET & PLACE PROJECTOR, RUBBERBAND, & ELECTRIC CORD**				
	Get cord and top	BGT-JO-12			25
	Top out of carton to cord box	BPL-AL-18			17
	Get cord from left hand	BGT-EV-12			17
	Get rubberband	BGT-JO-18			30
	Get projector	BGT-EV-18	21	Int.	
	Move to jig	BPL-AL-18			17
	Add for weight	BEL-WF-10			3
	Get projector to guide W/RH	BGT-CO-16	17	Int.	
	Place cord and rubberband	BPL-AL-02			5
					114
III.	**FOLD CARDBOARD SPACER**				
	Get cardboard	BGT-EV-12			17
	Transfer to RH	BPL-OH-12			19
	Get edge "B"	BGT-EV-12			17
	Relocate cardboard	BPL-AL-06			9
	Break seams	BPL-AL-12			13
	Bend seams 2, 3, 4	BPL-AL-06			9
	Regrasp	BEL-RG-01			6
	Fold	BPL-AL-02			5
	"	BEL-TW-01			4
	Reach clear from under CB	BGT-CO-02			4
	Align flap with slot	BPL-AL-01			2
	Engage and transfer to LH	BPL-OH-01			8
					113
IV.	**GET SLIDE FRAME & PLACE W/CARDBOARD SPACER**				
	Get slide frame	BGT-EV-12			17
	To cardboard pocket	BPL-AL-18			17
	Into place	BPL-AL-01			2
	Transfer to RH	BPL-OH-02			9
	Spacer to projector	BPL-AL-12			13
	Contact side of projector	BGT-CO-06	9	Int.	–
	Cardboard forward	BPL-AL-01			2
	Thumb to cardboard	BGT-CO-02			4
					64

Fig. 4-7. GPD data worksheet. (*Source: MTM-General Purpose Data Manual. Copyright 1962, The MTM Association for Standards and Research. No reprint permission is granted without express written consent of the MTM Association, 9–10 Saddle River Road, Fair Lawn, N.J.*)

length is 01. Thus the complete codes for these examples are BGT-CF-12 for a value of 10 TMU, and MVS-TL-01 for a value of 31 TMU.

Accuracy. Figure 4-2, which shows a comparison between basic MTM and MTM-GPD, was discussed briefly on page 5-105. It was pointed out that although MTM-GPD did not measure individual elements accurately in comparison with MTM, when several elements were taken together, the overall accuracy was quite good. Thus, the question that arises is: At what point of the time spectrum does MTM-GPD become an acceptable tool? This is a fairly clear-cut question, but it is difficult to provide a clear-cut answer. To shed more light on the problem, the simulation approach was utilized.[9] For the purpose of this simulation, a set of input rules was established which included the many variables and their related probabilities.

One of the computer readouts that resulted from this simulation study is shown by Figure 4-8. Upon reviewing these data, it is readily apparent that as the job time interval increases, the job error decreases. The data obtained from the simulation program, and for the conditions specified in this particular simulation, indicate that basic MTM data should be used for jobs in the job length interval of 0 to 0.002 hour. In this interval, approximately 35 percent of the jobs had an error that was in excess of 5 percent. In job length interval 0.003 to 0.005 hour, approximately 15 percent of the jobs had an error in excess of 5 percent. In job length interval 0.010 to 0.015 hour, only 1 percent of the jobs was in excess of 5 percent. Thus, 99 percent of the jobs had an error of less than 5 percent.

MTM-2, INTERNATIONAL COMBINED MTM DATA

The international situation on combined data systems was similar to that which existed in the United States prior to the acceptance of MTM-GPD. A number of combined data systems were in use throughout the member national associations. The International MTM Directorate (IMD) recognized that, by adopting one system, the duplication of cost and effort by the many groups involved in developing data systems could be eliminated. As a result, an International Applied Research Committee (IARC) was created, with representatives from each of the national associations. At a meeting in Stockholm in October, 1964, the IARC was directed to make a survey of the needs of industry for standard data in various manufacturing operations, and to recommend one system of synthesized first-level MTM data, and to suggest the structure of other levels. The committee was also instructed to assure the IMD that the synthesized data were logically constructed and capable of being transferred internationally. After months of hard work and committee meetings in Germany, Sweden, and The Netherlands, MTM-2 was presented to the International MTM Directorate in Munich, Germany, in June, 1965. The Directorate approved the data card values, the card format, and the coding concept for each of the values.

The official definition of MTM-2 describes it as a system of synthesized first-level MTM data. It is based exclusively on basic MTM motions. It is adapted to the operator and is not limited by the workplace or the equipment used.

The Data Card. MTM-2 was constructed by combining and averaging basic MTM motion times. MTM analyses from different industries and work areas with different degrees of mechanization were collected and checked by the technicians of the IARC. These analyses were then summarized by a computer which was programmed to yield information about motion sequences and motion frequencies. The number of motions summarized totaled over 14,000, and the analyses represented 30,000 man-hours of work.

The MTM-2 data card, Figure 4-9, has eleven categories, two of which—Get and Put—have three subcategories. The subcategories each have five distance classes. Seven other categories of body motions have one value each, and there are two categories for weight factors. The data card has a total of thirty-nine values. A brief description of each MTM-2 category follows.

1. Get—a combination of Reach, Grasp, and Release motions—coded GA, GB, and GC
2. Put—a combination of Move and Position motions—coded PA, PB, and PC

[9] W. K. Hodson, "Accuracy of MTM-GPD," *The Journal of Methods-Time Measurement,* September–October, 1963, p. 9.

Conditions: Experienced applicator
Table value repeat probability: .40 – high
Job value repeat probability: .90

Job time intervals	Job error frequencies										Total number of jobs
	0–1 %	1–3 %	3–5 %	5–8 %	8–10 %	10–15 %	15–20 %	20–25 %	25–35 %	35–99 %	
.000–.002	36	40	47	24	9	13	1	19	0	0	189
.002–.003	30	32	122	9	18	22	0	0	0	0	233
.003–.005	122	217	148	58	13	0	0	0	0	0	558
.005–.007	120	128	28	15	0	12	0	0	0	0	303
.007–.010	246	322	84	46	0	0	0	0	0	0	698
.010–.015	408	525	34	8	0	0	0	0	0	0	975
.015–.020	391	318	20	0	0	0	0	0	0	0	729
.020–.025	252	85	30	0	0	0	0	0	0	0	367
.025–.030	257	179	0	0	0	0	0	0	0	0	436
.030–.040	141	135	8	0	0	0	0	0	0	0	284
.040–.050	326	253	0	0	0	0	0	0	0	0	579
.050–.070	381	310	0	0	0	0	0	0	0	0	691
.070–.100	94	100	0	0	0	0	0	0	0	0	194
.100–.150	309	209	0	0	0	0	0	0	0	0	518
.150–.200	104	106	0	0	0	0	0	0	0	0	210
.200–.250	77	42	0	0	0	0	0	0	0	0	119
.250–.300	16	19	0	0	0	0	0	0	0	0	35
.300–.400	79	70	0	0	0	0	0	0	0	0	149
.400–.500	45	27	0	0	0	0	0	0	0	0	72
.500–.750	13	22	0	0	0	0	0	0	0	0	35
Error frequency by week	6	4	0	0	0	0	0	0	0	0	10

Note: Figures to the left of the heavy black line represent approximately 95 % of the jobs, or 2 sigma limits.

FIG. 4-8.   Computer simulation error analysis.   (*Source: William K. Hodson, "Accuracy of MTM-GPD," Journal of Methods-Time Measurement, September–October, 1963. No reprint permission is granted without express written consent of the MTM Association, 9–10 Saddle River Road, Fair Lawn, N.J.*)

MTM-2

Code	GA	GB	GC	PA	PB	PC
− 5	3	7	14	3	10	21
− 15	6	10	19	6	15	26
− 30	9	14	23	11	19	30
− 45	13	18	27	15	24	36
− 80	17	23	32	20	30	41

GW 1-1 Kg.   PW 1-5 Kg.

A	R	E	C	S	F	B
14	6	7	15	18	9	61

Warning: Do not attempt to use these data unless you have been trained and qualified under a scheme approved by the International MTM Directorate.

FIG. 4-9.   The MTM-2 data card. (*Source: MTM-Association of the United Kingdom.*)

3. Get Weight—the time necessary for the muscles and arm to take up a weight prior to moving it—coded GW
4. Put Weight—the additional time necessary to perform a Put with weight—coded PW
5. Apply Pressure—the force required to overcome resistance—coded A
6. Regrasp—shifting an object in the hand—coded R
7. Eye Action—Eye Focus and Eye Travel Time—coded E
8. Crank—the cranking action—coded C
9. Step—the motion of the leg or one in which the trunk is moved—coded S
10. Foot Motion—a motion of the leg and one in which the trunk is not moved—coded F
11. Bend and Arise—lowering the trunk and arising—coded B

The predominant categories in the MTM-2 data are Get and Put. An example of Get is code GB15, which means a simple pickup of an object with a motion path of over 5 centimeters (2 inches) to 15 centimeters (6 inches). From the data table, 10 TMU are allowed for a GB15. All table values are expressed in time measurement units (TMU) and are rounded off to the nearest whole number. The MTM-2 data can be applied in the same manner as the MTM-GPD data illustrated in Figure 4-7, or they can be recorded on a left-hand and right-hand Methods Analysis Chart form, as illustrated in Figure 4-10.

The system error[10] when MTM-2 is compared with basic MTM is illustrated in Figure 4-11. In nineteen cases out of twenty, the system error of MTM-2 at cycle lengths of 2,500 TMU will be less than 4 percent. However, exclusive concern for system error is incorrect because total error is also affected by error of application. The evidence is that the applicator error of basic MTM is bigger than the system error. Because MTM-2 has a fewer number of variables to consider and has carefully constructed training decision models, codes, and application rules, the error in application appears to be smaller than in the case of basic MTM. The true situation, then, is that parity with basic MTM is reached at approximately

[10] *Manual of MTM-2*, MTM Association of the United Kingdom, 1965.

METHODS ANALYSIS CHART   REFERENCE No. _AGS-65_

PART _ROCKING CHAIR_   DATE_____ STUDY No. _4728_

OPERATION ___ASSEMBLY___   ANALYST _BFM_____ SHEET No._11_OF_12_SHEETS

DESCRIPTION – LEFT HAND	No.	LH	TMU	RH	No.	DESCRIPTION – RIGHT HAND
Pick up assembly and turn		GB15	10	GB15		Pick up assembly and turn
to bench		GW4	4	GW4		to bench
		PA80	20	PA80		
		R		R		
		PW5	1	PW5		
			256	GC80	8	Rails to seat
			328	PC80	8	
				R	8	
			23	GB80		
Hold rail		GB-	41	PC80		Pick up back
				R		
	7	GB15	70			
Position rails			147	PC5	7	Headrest located
	7	PA5	21			
			32	GC80		Pick up mallet
			20	PA80		
			48	PA15	8	Hammer down
			20	PA80		Hammer aside
		Total	1051	TMU		

FIG. 4-10.   An MTM-2 analysis.

1,600 TMU, or 0.0160-hour cycle length. Figure 4-11 illustrates the total error, and although analyst and system errors are not delineated, the limitation imposed by the cycle time is emphasized.

When the cycle time is less than 1,600 TMU (approximately one minute), when the cycle is highly repetitive, and when the cycle contains a high proportion of complex simultaneous finger motions, basic MTM should be used. As the job length increases, MTM-2 provides a greater accuracy than is actually needed. In these cases, it is desirable to develop larger units of standard data and further reduce the time required to develop standards.

AREAS OF APPLICATION

Second generation MTM data provide the industrial engineer with an additional tool for work measurement. It is a versatile tool, but it is not the proper tool to use in every case. A clear understanding of its characteristics and limitations is essential if misapplications are to be avoided.

The principal uses which have been made of second generation data include:
1. Establishing standards on work of low repetitiveness
2. Establishing standards on long-cycle operations
3. Developing elemental data
4. Estimating and preproduction planning
5. Training

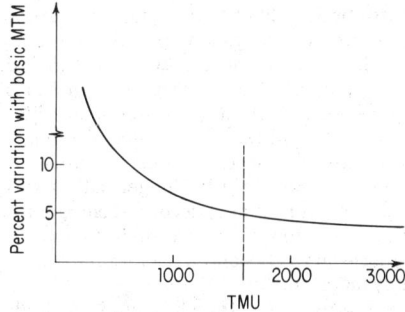

FIG. 4-11.   Comparison of MTM-2 with basic MTM.

Establishing Standards on Work of Low Repetitiveness. One of the most compelling reasons for developing second generation MTM data was the desire for a procedure similar to basic MTM which would be quick enough to apply to make its use economically justifiable on low-quantity work. Therefore, with the tool available, it is only to be expected that it has been widely used for establishing standards where work is not highly repetitive.

Just where the dividing line is between the area where basic MTM should be used and the area where second generation data are more properly applicable is not always clearly defined. An industrial engineer who is thoroughly familiar with basic MTM and believes in its effectiveness as a tool of methods improvement will be inclined to apply it to considerably less than highly repetitive work. Conversely, an industrial engineer who likes the quickness and ease of application of second generation data will be found using it where the more thorough analysis required by basic MTM will be desirable. The industrial engineering manager should be aware of these possibilities and should define the areas where each procedure is to be applied in his plant.

Establishing Standards on Long-cycle Operations. Long-cycle operations, even though they may be repeated day after day and year after year, are suitable for second generation data application. They are tedious to analyze with basic MTM. They are long enough so that any inaccuracies in elemental times caused by second generation data application will tend to average out.

Some elements or groups of elements of long-cycle operations may be repeated so frequently that they become highly repetitive and the time for performing them becomes an important percentage of the total cycle time. It will then be profitable to establish the times for these repetitive elements with basic MTM while using second generation data to establish the times for the other, less repetitive elements. Combining the application of basic MTM and second generation data is perfectly feasible, for the systems are compatible.

Developing Elemental Data. Chapter 8 of Section 3 discusses the building block concept of standard data development in some detail. Second generation data are larger building blocks built up from the smaller building blocks of basic MTM. In the same way, second generation data can be used to develop still larger building blocks. These blocks are usually elements which in turn are used for developing the even larger building blocks of task data, time formulas, and bench mark jobs.

Whether second generation data or basic MTM should be used for developing elemental data depends on the accuracy requirements of the elemental times and the amount of attention that should be devoted to methods study before the standard is established. These are a function of repetitiveness. Thus, second generation data will be entirely satisfactory for developing elemental data which will eventually be used for setting maintenance standards, whereas basic MTM will be preferable for developing elemental data which will be used for deriving a time formula for a class of work of medium to high repetitiveness.

Estimating and Preproduction Planning. During preproduction planning, time estimates of reasonable accuracy are often required for costing, production planning and scheduling, determining manpower requirements, tool design evaluation, preliminary methods analysis, and the like. The precise methods which will eventually be established are not yet known, but various preliminary ideas are being considered and there is a need for at least "ball park" estimates of time. As planning progresses and ideas begin to crystallize, the need for more accurate estimates develops. At some point, general overall estimates are no longer satisfactory, and a better estimation of time is required. At this point, second generation data often supply the need. Basic MTM would be too time consuming, and the degree of accuracy it provides is not yet necessary. Second generation data can be used to provide reasonably good estimates more quickly.

Training. Management often feels that it would be highly desirable if all key people in the organization, including manufacturing managers, supervisors, tool designers, production planners, quality control men, union officials, and many others, were trained in basic MTM so that they will clearly understand the principles of methods improvement, work measurement, and time and cost relationships. Usually, however, it is felt that thorough training in basic MTM for such a large group would be too time consuming and costly to be practical.

Although this is not necessarily the case, as companies who have given the broad training have demonstrated, training in second generation data application, even if only to the appreciation level, will accomplish at least part of management's objectives quickly and inexpensively.

IMPORTANCE OF TRAINING OF INDUSTRIAL ENGINEERS

Second generation data systems are generally described in detail in manuals published by the sponsoring organizations. The MTM Association for Standards and Research, for example, tries to control the use of the manual on MTM-GPD by specifying the conditions under which it is sold, and monitors the use and further development of the system through its Applied Research Committee. It emphatically stresses that no one should attempt to apply MTM-GPD without proper training.

Nevertheless, it is inevitable that once a manual has been published and distributed, it will be read by many people. On paper, most second generation data systems seem quite easy to understand. Their application appears to be deceptively simple. Therefore, there is a tendency for people to attempt to apply the system without further training and guidance.

Such attempts generally lead only to disappointing results. The untrained applicator will usually create more problems than he solves, and confidence in the value and soundness of the system will be destroyed. Any of the systems discussed in this chapter will give good results if competently applied by people who thoroughly understand them, in areas where they are applicable. But thorough training is a prerequisite for success, and its importance cannot be too highly stressed.

It is often asked if training in basic MTM is necessary as a preliminary to training in the use of a second generation data system. An applicator can undoubtedly be trained in the mechanics of applying a second generation data system without basic MTM training. The second generation data system is only a tool, however, and a clear understanding of its characteristics and limitations is necessary for its successful use. If one knows only the second generation data system, he has only one tool to use and therefore no choice. Misapplications are thus inevitable. Thorough training in basic MTM is necessary if the applicator is to have the competence necessary for professional-level work.

CONCLUSION

A number of second generation predetermined time systems have been developed to meet the requirements of certain work measurement situations. They provide an additional tool in the kit of the industrial engineer, which in certain situations is the best one to use. The industrial engineer who does not already possess this tool should investigate its applicability to the work he handles. If he finds that it is applicable and that the benefits it offers are worthwhile, he should insist on receiving proper training in the system. Only then should he begin to put it to work.

BIBLIOGRAPHY

Crossan, R. M., and H. W. Nance, *Master Standard Data,* McGraw-Hill Book Company, New York, 1962.

Hodson, W. K., "Accuracy of MTM-GPD," *The Journal of Methods-Time Measurement,* September–October, 1963, pp. 9–17.

Mabry, J. E., "MTM-GPD General Purpose Data Original Development and Future Expansion," *The Journal of Methods-Time Measurement,* September–October, 1963, pp. 18–22.

Mabry, J. E., "MTM-2: International Combined MTM Data," *The Journal of Methods-Time Measurement,* March–April, 1966.

Manual of MTM-2, MTM Association of United Kingdom, 1965.

Methods-Time Measurement General Purpose Data (MTM-GPD) Instruction Manual, MTM Association for Standards and Research, Fair Lawn, N.J., 1962.

Wage and Salary Administration

Chapter 1

Wage Administration

GUY J. BACCI

General Supervisor—Industrial Engineering,
International Harvester Company,
Hinsdale, Illinois

The primary purpose of wage administration is to assure management and employees of equitable compensation for services rendered. This purpose, although simply stated, is difficult to translate into a practical program for implementation. Equitable compensation is of utmost importance to all. From the new employee, who must make his decision based on outside knowledge of the company, to the long-service worker, who has intimate knowledge of the company's policies and procedures, the wage plan offers the opportunities to raise his standard of living and that of his family. This is the motivation necessary for successful business ventures, and wage administration can play an important role in promoting this motivation. This was discussed by Smyth and Murphy in their coverage of the subject.[1] They said:

> Monetary income is the most important phase of the employee-employer relationship. As prerequisites to sound industrial relations, the individual employee (1) should receive an absolute amount of income sufficient to sustain him and his dependents adequately, and (2) should feel generally satisfied with the relationship between his income and the income of other persons performing the same class of work in the concern and in the community or industry.

Although the wage administration plan contributes greatly to the public image of the company or business concern, it is not entirely an internal company matter.

[1] Richard C. Smyth and Matthew J. Murphy, *Job Evaluation and Employee Rating*, McGraw-Hill Book Company, New York, 1946, p. 3.

The design and function of the wage plan is influenced greatly by outside forces such as the government, employee associations, and labor unions. The government's role is executed at local, state, and national levels through various means of legislation such as the Walsh-Healey Act and the Fair Labor Standards Act. Employee associations and labor unions are primarily concerned with the wages of the employees they represent and therefore influence the wage plan through collective bargaining and labor contract administration.

The objective of this chapter is to present the subject of wage administration in outline form and to cover the important parameters that are inherent, integral parts of a successful wage administration program. These parameters are:

1. Purpose and objectives of the wage program
2. Organizational status
3. Mechanics of a sound wage program
4. Control and appraisal of the program

The general principles outlined in this chapter offer an excellent starting point for the industrial engineer or manager to launch a sound wage administration program. Other aspects of wage matters that are closely aligned with a successful wage program will be found in other chapters of Section 6.

PURPOSE AND OBJECTIVES OF THE WAGE PROGRAM

The skillful management of wage compensation is essential to assure the attaining of cost or profit objectives. The achieving of these cost or profit objectives is accomplished through maintaining proper balance between the input costs and the resultant income from output. This balance begins with good organization of the wage program, continuing surveillance and maintenance of the program, and proper controls adjusted to react to certain feedback indices. An immediate conflict of objectives arises, because the wage administrator's aim is to satisfy differing goals of employer, owner, or stockholder; employees; and outside interests which include suppliers and consumers.

Labor Relations Aspect. A key ingredient to solving this conflict rests on the understanding and acceptance of the wage program by the employee.

Whether the industry contains a large degree of labor content or a relatively small degree, the importance of good management of labor is apparent. Even with the ultimate sophistication of complete automation, some labor is essential to maintain efficient operations. A well-engineered and smooth-functioning wage program contributes greatly to harmonious labor relations and understanding between management and labor. This point is stressed by Lawrence A. Appley, former president of the American Management Association, in the foreword to the AMA handbook on this subject:[2] "A fair day's pay for every job well done, from the least skilled to the most complex, is a fundamental goal and responsibility of good management and a foundation stone of industrial peace."

Employer (Owner or Stockholder) Aspects. To assure the proper balance between expenses and income, an understanding of the inherent factors of operational cost is a must for good management. A sound and equitable wage program is influenced to a great degree by many factors, the most important of which are the following:

1. General economic climate of the industry
2. Anticipated rate of return on investment as compared with other companies in the industry
3. Governmental influence
   a. Legislation
   b. Taxation
4. Effects of collective bargaining

[2] M. Joseph Dooher and Vivienne Marquis (eds.), *The AMA Handbook of Wage and Salary Administration*, American Management Association, New York, 1950.

With these factors clearly in mind, the objective of a sound wage administration program can be subdivided into specific subgoals.

These goals, which represent the milestones along the path to successful and equitable wage compensation, are:

1. Equitable wage payment in proportion to relative worth to the organization
2. Consistency of wages between comparable occupations
3. Adjustment of wages in relation to changes in the labor market
4. Recognition of individual capability and proficiency
5. Comprehension of the plan by supervision and management
6. Procedures to solve wage problems rationally

ORGANIZATIONAL STATUS

The organizational form of wage administration varies, depending on the industry and the emphasis placed on the function by management. From a corporate standpoint, it is normally an integral part of the industrial relations department and plays an important role in the total policy making of this department. As mentioned earlier, sound wage administration enables the achieving of important industrial relations objectives. In a corporate or divisional organization, the principal duties of wage administration are the establishment of policies and procedures, the researching of new and changing concepts in the field of compensation methods, and the auditing of the operational levels of wage administration to assure adherence to policies and procedures.

Operational Level Hierarchy. The manufacturing manager's job can normally be separated into a number of principal duties, as shown by Figure 1-1. It consists of guiding and directing the work connected with the following functions:

Materials Management—consisting of procuring, scheduling, and delivering materials, both direct and indirect.

Financial Management—consisting of securing the necessary funds for operation and accounting for their expenditure.

Personnel Management—consisting of establishing employment policies and procedures for operation.

Technical Services Management—consisting of the technical staff services necessary to support and sustain the production effort.

Production Management—consisting of the fabrication and shipment of quality products in accordance with specifications and production schedules.

Personnel Management—the personnel management or industrial relations organization is typically further subdivided into at least six major functions, as shown in Figure 1-2. These comprise the basic elements inherent in handling personnel relations.

Wage and Salary Administration. At the operational level, wage and salary administration has basic responsibilities for assuring fair and equitable compensation. Meeting these responsibilities requires concentration in the major areas which comprise this function. These areas can be depicted as shown in Figure 1-3.

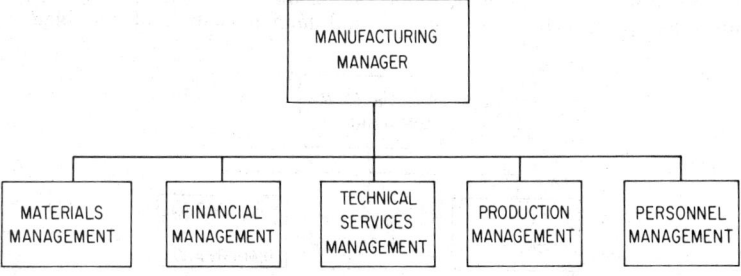

Fig. 1-1.   Functions of a typical manufacturing organization.

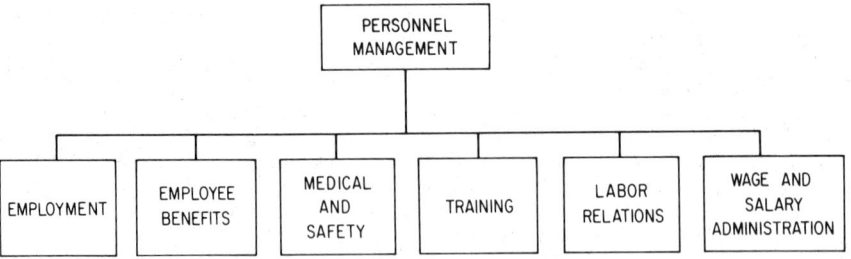

Fig. 1-2.   Functions of a typical personnel management organization.

Wage Administration. The importance of using different approaches to solve wage problems as compared with salary problems cannot be overemphasized. However, the overall general policy of equitable compensation in relation to relative worth to the company and in comparison with other companies must be maintained. This relationship between the blue-collar worker and the white-collar worker can only be maintained by a proper coordination between the wage function and the salary function, as depicted organizationally in Figure 1-3.

MECHANICS OF A SOUND WAGE PROGRAM

A sound wage program can result only when the wage plan is well thought out, thoroughly understood by employees at all levels, and properly executed by management, and when it remains flexible to demands of the personnel involved. The typical wage plan is complex when viewed in its entirety, but the key to comprehension for the manager or industrial engineer is knowledge and understanding of the components that comprise it. These components, along with the important core of the program, namely, the necessary planning, the fundamental policies, and the operational procedures, are illustrated in Figure 1-4.

Planning. The planning for implementation of a new or revised wage plan must be carefully executed. The success or failure of the program depends heavily on this foundation. A number of questions immediately require answering, such as:

1. Who will develop the program?
2. What jobs will be included?
3. Which specific plan will be selected?

The answers to these questions will take many forms, depending on the character of the company and the environment in which it operates. In most organizations, the wage administration department will be responsible for developing the program. To achieve the objective of equitable compensation, it is imperative to include all hourly paid jobs in the wage program, whether direct or indirect labor. The form of the wage plan will vary considerably, depending on type of work performed, skills required, availability of labor, and the like.

Policies—Procedures. The development of the plan requires competent wage administration personnel. The approval and implementation of the plan require

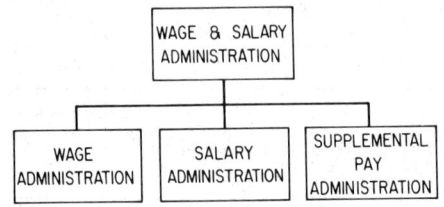

Fig. 1-3.   Wage and salary administration functions.

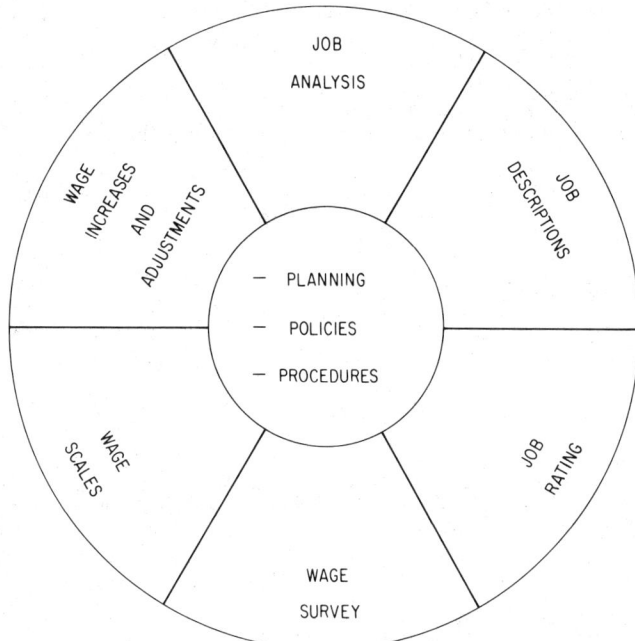

Fɪɢ. 1-4.   Components of a wage program.

equally competent and dedicated top management and supervision. The importance of comprehension and effective decision making in wage matters cannot be over-emphasized. This point was well covered by Charles W. Brennan in his writings on the subject:[3]

> The formulation of policies and the establishment of procedures is a difficult and complex undertaking, requiring careful study, analysis, and judgment. They should be worked out by the person in charge of the program, who recommends them to top management for review and authorization. However, while the policies are formulated by the person in charge of the program, the communication of the major decisions should originate in top management.

The policies and procedures are normally reviewed and approved by a wage committee consisting of the wage administrator and other members of management. In some cases where specific coverage is included in the labor contracts, union officials also serve in this capacity. The policies and procedures must be clear and understandable and must permit quick reaction to questions and problems by supervision and workers. To assist in the accomplishment of this objective, all new employees are normally given an orientation session that includes the subject of wage administration. To augment this, training classes on wage administration policies and procedures are conducted for supervision.

Job Analysis. Job analysis is the important and beginning step that follows planning and initiates the mechanics of the wage plan. During this process, all pertinent factors are collected and analyzed by the job analyst. To assist the analyst in procuring the necessary information, a questionnaire or worksheet is normally utilized. This worksheet is designed specifically to stimulate the analyst's

[3] Charles W. Brennan, *Wage Administration*, Richard D. Irwin, Inc., Homewood, Ill., 1963, p. 73.

approach by investigating the four basic aspects of the job, namely:
1. What work is performed?
2. How is this work performed?
3. Why is it performed?
4. What are the skills required to perform it?

An illustration of a typical job analysis worksheet is shown by Figure 1-5.

JOB ANALYSIS SHEET

Job Title   Set up automatic screw machine

Dept. No.   40   Job. No.   201   Date   Sept. 20, 19

Code No.   Sex   Male

Dept. Name   Miscellaneous Machining

A. Job Duties:

Works from simple drawing. Sets up single and multiple spindle automatics. Uses precision measuring instruments. Selects feeds and speeds. Makes daily inspection of equipment, minor adjustments. Assists operator in maintaining production standards. Repetitive work.

B. Working Conditions:

Analyst's Value

1. Job condition   Ordinary shop conditions.

2. Safety for others   None.

3. Hazards   Possibility of infection from screw cutting oil.

4. Responsibility of   Possibility of machine damage due to setting.
   equipment and material   Loss of approx. $250.

C. Skilled Job Requirements:

1. Knowledge:
   Equipment and tools   Expert knowledge of average to complex tools.
   Method   Knowledge of average method of doing work.
   Material   Working knowledge of wide variety of materials (ferrous).

2. Education: 8 years of school. Read simple drawings. Knowledge about speeds and feeds. Writing simple reports.

3. Experience: 3 years of diversified experience as a screw machine operator.

4. Trust and Confidence: Minor.

5. Supervision: Indirect supervision.

6. Judgment and Initiative: Requires judgment of average nature, some degree of initiative.

7. Mental Capability: Decisions to maintain standard average initiative.

8. Physical Skill: Reaction time of the worker is not a factor within reasonable limits.

FIG. 1-5. Job analysis sheet.

The job analysis function was well summarized by Lawrence C. Lovejoy in his definition of the subject:[4]

> For the purpose of this text, job analysis is defined as the process of ascertaining and recording information about a specific job. This information will include the nature of, and time spent on, job duties; equipment and materials used; the responsibilities, authorities, and relationships involved; the qualifications of a suitable employee; and the conditions under which the work is performed.

Job Descriptions. The end result of the job analysis function is a clear and concise job description. This description must accurately portray the basic and necessary information inherent in performing the job. This information is found in many forms in industry, but whatever the form, there are basic items that must be covered. These are:

Job Title. The title should briefly describe the job and be definitive in nature. A typical example of a job title is "drill press operator—multiple spindle."

Summary of the Job. The summary should be in the form of a brief statement covering the purpose and function of the job. A summary of the job titled above might be: "Set up and operate a multiple spindle drill press to machine gray iron casting parts, including drilling, reaming, boring, tapping, and chamfering to close tolerances."

Job Duties and Work Performed. This includes an accurate description of the specific tasks of work performed and the occurrence factors associated with them.

Equipment, Tools, and Material. A good description of equipment, tools, and material is essential to indicate the nature and complexity of the job.

Working Conditions. The physical surroundings and job-related hazards should be carefully described.

Job Requirements. The general requirements of education and experience necessary for job performance should be determined and recorded. A typical job description is illustrated by Figure 1-6.

Job Rating. The process of job rating, the third step in the overall procedure of wage administration, is extremely important, because it establishes the relative worth of the job. This relative worth is a beginning point for the eventual determination of rates of pay. The criteria for determination of this relative worth can be classified as either quantitative or nonquantitative.

Quantitative Methods. Typical of the quantitative types of job rating is the job evaluation technique. This technique subdivides each job into its essential job factors, which are then weighted in accordance with a rating scale. Other types of evaluation plans include the comparison of these factors of each job and the ranking of these factors between jobs. More detailed coverage of this subject will be found in Chapter 6 of this section.

Nonquantitative Methods. The main difference between nonquantitative methods of job rating and the quantitative approach lies in the fact that the latter determines the essential factors which constitute the job, while the former compares one job with another on the basis of total job content. Each job is ranked by the overall evaluation of all its elements in comparison with the other jobs. The chief advantage of this method of job rating is its simplicity and ease of application.

Wage Survey. Once the internal relative worth of the job has been established, it becomes necessary to compare this worth with the prevailing wage for comparable jobs in the industry or community. This can be accomplished by a survey of the labor market to determine current compensation rates and practices.

Methods and Format. The formats used are as numerous as the industries they survey, but some basic information is inherent in each form. Because this review is only a survey, it becomes extremely important to select key jobs for comparison. In addition, a truly representative sample of companies must be selected. This sample must include companies having similar wage policies and objectives. The

[4] Lawrence C. Lovejoy, *Wage and Salary Administration*, The Ronald Press Company, New York, 1959, p. 122.

JOB DESCRIPTION SHEET

JOB TITLE: _Power truck operator_   JOB SYMBOL:_____

DEPARTMENT: _Plant engineering_   DATE OF ANALYSIS:_____

CODE NO.: _____

SUMMARY: Under immediate supervision, to operate industrial power trucks such as lift trucks, trailer and tractor type, in moving, hauling, and transporting material inside and outside the building.

DUTIES: Load and unload materials, transporting them by power truck to and from various locations. Operate lifting and elevating mechanisms of the truck in storing and tiering materials according to approved practices.

Perform minor adjustments and repairs to power trucks, fueling and checking operations periodically, reporting major failures when such become apparent. Perform related work as assigned.

EDUCATION: Graduation from high school or equivalent educational training and study.

TRAINING AND EXPERIENCE: Previous satisfactory experience in the operation of industrial power trucking equipment.

EQUIPMENT, TOOLS, AND MATERIAL: Lift trucks, trailers or a tractor type, up to a maximum of 2½ tons rated capacity.

A well-lighted and ventilated plant. Job subjected to traffic hazards and noise; requires special clothes.

PROMOTION TO: No formal line of promotion.

SUPERVISED BY: Foreman of the department.

Fig. 1-6.   Typical job description.

information contained in the survey can be obtained by personal contact or by mail. Other important sources for this information are the various organizations that periodically conduct compensation surveys. Typical of these are:

1. Personnel associations
2. Chambers of commerce
3. Trade and employee associations
4. Various governmental organizations such as the Department of Labor

An example of a wage survey form is shown in Figure 1-7.

The primary use of the information obtained in the wage survey is to establish a comparison of the wages paid for similar jobs in different companies. However, there are other uses for this information. Some of these are:

1. The establishment of minimum hiring rates
2. The current assessment of pay rate trends and their effect on product cost
3. The determination of nonwage benefits offered in industry

Wage Scales. The final step in the determination of the relative worth of a job requires a review of some of the important parameters inherent in the company's

WAGE SURVEY DATA SHEET

Name of participating company _____

Address _____

Nature of business _____

Survey no. _____

. Data furnished by _____ Title _____ Date _____

	Hourly	Incentive	Total
1. No. of employees in your company:			
2. Minimum hiring rate:			
3. Average no. of hours worked			
per week:			
per year:			
4. Do you use single rate for each job?			
Do you use rate range for each job?			
5. Method of progression			
within the range:			
automatic increase:			
merit increase:			
part automatic:			
6. Are you granting rest periods?			
7. Method of overtime payment:			
8. Percentage of base rate paid as supplemental wages:			
9. Do you use a wage incentive plan?			
If so, describe:			
10. What are the average incentive earnings as percentage of base rate?			
11. Do you guarantee employees an annual income?			
12. Do you have night shifts?			
13. Is clean-up time granted at the end of each shift?			
14. If sick leave is granted with pay, how is it paid?			
full pay			
% base rate			

15. Holidays observed:
    Jan. 1, Feb. 12, Feb. 22, May 30, July 4, Labor Day, Thanksgiving, Dec. 25

    Hourly
    Incentive

16. Do you pay supplemental wages (fringe benefits)?
    Name the fringe benefits:

17. If you have employee benefit plans, excluding social security
    and workman's compensation, how are they contributed:

	Yes	No	Company only	Employee only	Both Company %	Employee %
Death						
Accident						
Sickness						
Hospitalization						
Pension						
Savings						
Other						

18. Do you have a suggestion system?
    If yes, how are employees rewarded?

19. Do you pay a separation allowance at the time of termination?
    If yes, explain the basis:

20. Do you supply work clothes and laundry?

FIG. 1-7. Wage survey data sheet.

business picture, such as the following:
1. The current anticipated profit potential
2. The availability of an adequate labor supply
3. The effects of collective bargaining
4. The benefits of paying a leading wage
5. The important effects of nonwage benefits

After careful review and study, the results of all the preceding steps can be combined to result in an equitable wage scale designed to accomplish the overall objectives of the company.

Graphical Method of Wage Scale Development. A graphical method of developing the wage curve starts with the selection of key jobs that cover the entire range of jobs from nonskilled laboring jobs to highly skilled apprenticeable trades. The midpoints of these key jobs are plotted as shown in Figure 1-8. In this manner, the range of labor grades is determined graphically. The smoothness of the wage curve is assured by using mathematical correlation methods.

Wage Increases and Adjustments. The culmination of all the preceding steps in the wage determination process is the equitable distribution of wages. The vehicle for achieving this result is a thorough understanding of the process by supervision and employees. This must be augmented by fair and equitable compensation of all individuals. Although some jobs may be paid on a single rate basis and therefore leave no room for merit consideration, performance on these jobs is given definite consideration toward future promotion.

There are a number of causes for wage adjustments, such as change in internal relative worth, change in worth in outside companies, and so on. The most common type of adjustment, however, results from changes in the economy. As technological advances are achieved, the results of productivity gains are usually shared with the wage earners in the form of annual wage increases. In addition to this, wage increases or decreases have also been tied to cost-of-living changes. These usually are determined by a formula based on the Cost of Living Index published by the Department of Labor.

Automatic Wage Increases. In many of the jobs assigned to rate ranges, the beginning portion of the rate range is used for automatic wage increases. These

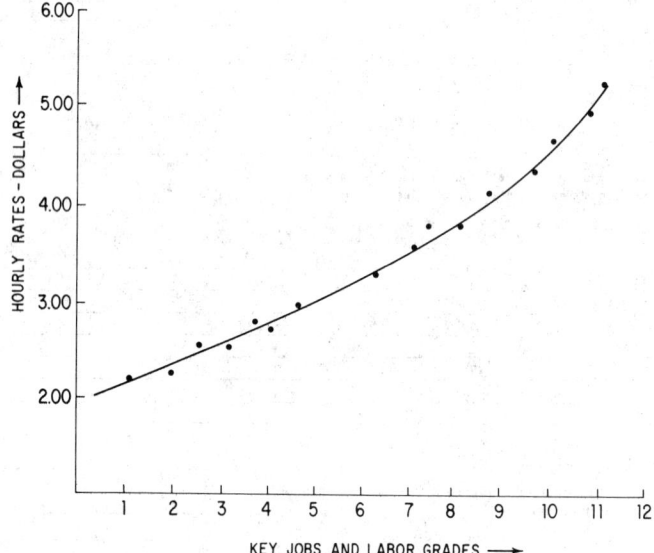

Fɪɢ. 1-8.   Determination of line of best fit to wage data of key jobs.

are based on the time spent on the job and have no relationship to skill or ability. This concept began with many of the lower rated occupations but has spread even to apprenticeable trades.

Supplementary Compensation Methods. In addition to the real wage received by employees, the effect of supplemental pay policies has become increasingly important. Payment for the fringe benefits of paid vacation, holidays, life insurance, hospitalization, pensions, and the like have a definite effect on product costs and the resultant profit. Therefore, the wage administrator must remain cognizant of this effect and take this in consideration when instituting or altering the wage plan. Statistics showing the cost effects of fringe benefits are published periodically by the U.S. Chamber of Commerce. This subject is thoroughly discussed by Adolph Langsner and Herbert G. Zollitsch.[5]

CONTROL AND APPRAISAL OF THE PROGRAM

The wage administration program, like many other programs, must be constantly surveyed and appraised. The conditions that affect its important factors are dynamic and change almost continuously. New models are introduced and processes change, which result in changed work requirements, materials, specifications, and methods. These produce shifting values of job skills and job conditions which influence job descriptions and standards.

Internal Company Implications. A constant appraisal is mandatory to maintain the goal of equitable wage compensation. This can be accomplished by an active program of internal company wage surveys. These surveys must be continuous and made in accordance with a schedule established to review all job assignments at least once a year. The survey takes the form of a continuous job analysis, with the check sheet serving as the means of assessing the success of the program. As changes in job content are discovered, immediate reaction should be forthcoming. This aspect of rapid feedback on changed situations goes a long way toward assuring employee understanding and cooperation in the wage plan.

Legal and Contractual Implications. In addition to constant knowledge of changing internal conditions, the wage administrator must keep fully informed of government regulations and other legal aspects. As new legislation becomes law, the wage administrator must immediately react to implement the necessary changes. At the same time, the cost effects of these changes must be evaluated and reported to executive management.

[5] Adolph Langsner and Herbert G. Zollitsch, *Wage and Salary Administration,* South-Western Publishing Company, Incorporated, Cincinnati, 1961, pp. 593–632.

ARTICLE XII—SECTION 1(b)
JOB CLASSIFICATION

If an appropriate classification does not exist in the appropriate Works Occupational Rating Book of the Works involved for new daywork or piecework, and in the case of changed daywork or piecework if the change has made the former classification inappropriate under the principles of the arbitration awards on classification under the prior Contract, the Company shall initially determine the classification, job description and wage group, including those cases under Article VII of this Contract in which an Arbitrator has decided that no appropriate classification exists. Such determination by the Company shall become final unless challenged by the Union within a thirty (30) day period after the Company informs the Local Union and the International Union of such determination. If challenged by the Union within such period, the issue shall become the subject of collective bargaining between the Company and the Local Union without undue delay. However, at the request of either the International Union or the Labor Relations Department of the Company, negotiations will be conducted on a central level. The Company's determination shall continue to be applied unless changed as the result of such collective bargaining.

Fig. 1-9.   Typical wage contractual language found in many labor contracts.

Labor Contract Implications. Wage matters are more and more being included in labor contracts. Following national labor contract negotiations and wage pattern setting, the results of the collective bargaining process are made an integral part of the labor contract. Due to this inclusion, the procedural aspects of wage matters are also covered. An example of this coverage is illustrated by Figure 1-9.

As this type of coverage is expanded, more and more collective bargaining sessions will deal with details of the wage administration plan. This will necessitate the inclusion of the wage administrator in many of the face-to-face negotiation sessions.

CONCLUSION

Wage administration plays an important role in assisting management to achieve competitive costs. A sound wage program can also assure management of the ability to hire and retain a competent work force. In the changing world of automation and job complexities, this can be a great asset toward achieving overall company goals.

BIBLIOGRAPHY

Belcher, David W., *Wage and Salary Administration*, Prentice-Hall, Inc., Englewood Cliffs, N.J., 1962.

Bradfield, E. G., *Wage and Salary Administration: Meeting the Challenge of Changing Conditions*, Personnel Series 157, American Management Association, New York, 1954.

Brennan, Charles W., *Wage Administration*, Richard D. Irwin, Inc., Homewood, Ill., 1963.

Dooher, Joseph M., and Vivienne Marquis (eds.), *The AMA Handbook of Wage and Salary Administration*, American Management Association, New York, 1950.

Dunlop, John T., *The Theory of Wage Determination*, St. Martin's Press, Inc., New York, 1957.

Ells, Ralph W., *Salary and Wage Administration*, McGraw-Hill Book Company, New York, 1945.

Langsner, Adolph, and Herbert G. Zollitsch, *Wage and Salary Administration*, South-Western Publishing Company, Incorporated, Cincinnati, 1961.

Lanham, Elizabeth, *Administration of Wages and Salaries*, Harper & Row, Publishers, Incorporated, New York, 1963.

Lovejoy, Lawrence C., *Wage and Salary Administration*, The Ronald Press Company, New York, 1959.

Maynard, H. B. (ed.), *Handbook of Business Administration*, McGraw-Hill Book Company, New York, 1967, sec. 11.

Chapter 2

Wage Incentive Plans

MITCHELL FEIN

Professional Engineer, Hillsdale, New Jersey

Managers agree that high employee productivity depends on two essentials: effective managing and the will of the employees to work. But managers are not agreed on how to motivate employees. Psychologists who have delved into productivity questions have proposed approaches which gain the intellectual acceptance of managers, but managing styles in plants have not changed appreciably. Managers still largely use the authoritarian approach.[1]

This chapter analyzes financial incentive principles and practices so that managers will better understand when incentives should be used and why some incentive plans succeed and others fail. It is not an objective comparison of incentive and daywork systems, nor is it designed to extoll the virtues of incentives over daywork, although the entire discussion is devoted to incentives. Determining which system is best for a particular plant depends upon the total environment of the plant rather than the efficacy of the system of employee compensation.

The material is organized to present broad questions and problems which affect incentives. The reader will then be better prepared to understand and work with the design and installation of incentive plans. This presentation discusses wage incentives largely for hourly rated employees. Supervisors and other groups of salaried employees are not covered.

VIEWS ON INCENTIVES

Managing Concepts. The approaches of managers to managing center in large measure around their concepts of how best to motivate employees to produce.

[1] Mason Haire, Edwin E. Ghiselli, and Lyman W. Porter, *Managerial Thinking: An International Study*, John Wiley & Sons, Inc., New York, 1966.

The main views of managers fall into broad categories, varying combinations of which may be found in the same plant:

Some managers use financial incentives to motivate workers; they believe managing with incentives raises productivity.

Measured daywork advocates believe management of the plant and maintenance of productivity are management's responsibility and that incentives are not necessary to attain higher productivity. These managers rely on close supervision, work measurement, and labor controls.

Some traditional managers rely solely on supervisors to spur production, with no work measurement or labor controls. Most smaller plants fall into this category.

Some managers run their plants along the lines proposed by human behaviorists, relying on the motivational theories of Maslow, Likert, McGregor, Meyers, et al.

Although management literature often extolls the advantages of a particular approach, managers find that theory must be weighed against all factors in the plant. Some managers use combinations of approaches to obtain most effective overall results.

Measured Daywork versus Incentives. Methods of compensation are closely tied to management's policies on managing. There are two methods: payment by time and payment by performance. Payment by time is based on the principle that workers' wages are established periodically by agreement and that increased worker productivity is largely effected by management capability and should not be reflected in wages between the wage settlements. Payment by performance is based on the principle that workers are entitled to higher pay than was bargained with the union or established as equitable by management in the absence of a union, when their productivity is increased through their own efforts.

Measured daywork encompasses payment by time and labor control through work measurement. It compares closely with wage incentives; the main difference is that, on incentive, workers earn extra pay when their production exceeds standard, while on measured daywork the pay does not change.

Comparison of measured daywork to incentives is meaningful only in the context of a given plant, although often the advantages of one system or the other are cited. The administration costs of an incentive plan are essentially the same as the costs of maintaining the labor controls of measured daywork; the procedures for both are similar. When measured daywork procedures are loosely administered with an inadequate staff, comparing measured daywork with a well-run incentive plan is not valid.

Effect of Incentives on Management Policies. As incentives are introduced into a plant, significant changes evolve in management's policies toward managing and in relations with employees. When operating on daywork, management has the burden of raising productivity through various management and motivational approaches. Incentive-oriented managers believe the burden is eased when employees on incentive accept the principle of extra pay for extra work.

How management administers the incentive plan personifies to the employees management's attitude toward them. Because policies are readily reflected in the workers' pay and in how hard they have to work to earn it, workers become extra sensitive to management's actions. Production decisions which involve the incentive plan become instantly quantified; the impact is readily sensed and measured. Many management decisions outside of incentives also affect pay, job security, and working conditions, but none as consistently and sharply as incentives.

Effect of Incentives on Job Security and Morale. Managers sometimes erroneously believe that the desire of employees to increase their earnings will motivate them to greater productivity. However, employees and especially unions will consider the effect that increased productivity will have on their job security. The response of employees to incentives will in some measure reflect their outlook on the way in which the incentives may affect their employment.

Numerous problems arise under incentives when employees attempt to protect their job security, as when employees try to maintain larger work crews than management deems necessary. Self-imposed production limitations are sometimes instituted as a means of ensuring against displacement of employees. Successful incentive administration requires that the security apprehensions of the employees be relieved by sound policies which protect them and do not require that they work against their own interests. If increased productivity will cause some employees to be laid off, the increased pay to those remaining will not necessarily be sufficient to gain acceptance of an incentive plan.

Prevalence of Incentives. Large segments of American industry operate with incentives. In several studies covering the period from 1945 to 1968,[2] the Bureau of Labor Statistics (BLS) found that about 26% of the nation's production work force was on incentive. The coverage differed by industry, ranging from 80% in men's garments, 66% in basic steel, and 31% in automotive parts, meat packing, and textiles, to practically none in the chemical and process industries. Although there were changes within industries over the period due to mechanization and other reasons, the national level remained fairly constant over the entire period. Table 2-1 shows data by industry.

Studies of the prevalence of incentives in England for all industries show that in 1938 it was 25%, rising to 32% in 1945, and then remaining fairly level until 1961, when it was 33%. These are the last overall figures available. Studies of the engineering industries—which include iron and steel, chemical, shipbuilding, aircraft, motor vehicles, and electrical and mechanical industries—show an overall of 48.5% in 1964 and 47.2% in 1967. These figures are not comparable with the overall for all industries. The data for England indicate a fairly constant usage of incentives in industry.[3]

Although incentives are generally thought of as tools of capitalism, the Soviet Union has found them powerful motivators to increased productivity. The Soviet Union makes the most extensive use of and obtains the greatest gains from financial incentives of any country. Reports are that about 60% of Russian workers are covered by incentives.[4] The strong convictions of the Soviet managers on a subject which is so basic to the capitalist system warrant serious study by American managers. American industry may be missing substantial advantages by not making greater use of incentives. Crawford H. Greenwalt, formerly president of E. I. du Pont de Nemours & Company, was troubled on this point when he stated: "It seems quite clear that, if financial rewards in the two countries are examined quantitatively, the Russians are not far behind us. What troubles me is to note that, at the very time the Soviets are embracing our principle of incentives and show signs of benefiting thereby, we ourselves seem intent upon abandoning it."[5]

Articles in the management literature suggesting that large numbers of managers are becoming disenchanted with incentives for workers are not supported by industry statistics. Incentives have major application where workers exercise control over their output. As mechanization increases, there is reduced need for traditional money incentives. Where major problems have been encountered in incentive plants, some have been converted to measured daywork. It is probable that, although some widely publicized incentive plans have been discontinued, there has been an expansion of incentives in other plants. Industry statistics do not show a decline in the use of incentives.

[2] Earl L. Lewis, "Extent of Incentive Pay in Manufacturing," *Monthly Labor Review,* May, 1960; *Wages and Related Benefits,* part II, Bulletin 1345-83, U.S. Department of Labor, Bureau of Labor Statistics, June, 1964; George L. Stelluto, "Report on Incentive Pay in Manufacturing Industries," *Monthly Labor Review,* July, 1969; and personal correspondence with Bureau of Labor Statistics.

[3] *Payment by Results Systems,* Report 65, National Board for Prices and Incomes, Her Majesty's Stationery Office, May, 1968, pp. 76–78.

[4] Abram Bergson, *The Economics of Soviet Planning,* Yale University Press, New Haven, Conn., 1964.

[5] Crawford H. Greenwalt, *The Uncommon Man,* McGraw-Hill Book Company, New York, 1959.

TABLE 2-1.   Number of Production and Related Workers and Percent Paid under Incentive Wage Plans in Selected Manufacturing Industries,* 1963–68

Industry and payroll reference	Total number of production and related workers	Percent paid on an incentive basis	Industry and payroll reference	Total number of production and related workers	Percent paid on an incentive basis
Food and kindred products			Paints and varnishes (Nov. 1965).	31,147	1
			Fertilizer (Mar.–Apr. 1966).......	25,484	1
Meat products (Nov. 1963):			*Petroleum and coal products*		
Meatpacking.................	131,965	30	Petroleum refining (Dec. 1965)....	73,318	†
Prepared meat products.......	39,071	8			
Flour and other grain mill products (Feb. 1967)..........	12,565	2	*Rubber and plastics products*		
Candy and confectionery products (Sept. 1965).................	49,736	25	Miscellaneous plastics products (June 1964).................	109,482	13
Tobacco manufacturers			*Leather and leather products*		
Cigarette (July–Aug. 1965).......	31,507	†	Leather tanning and finishing (Jan. 1968).................	23,712	53
Cigars (Mar. 1967).............	16,552	57	Footwear (Mar. 1968)...........	172,381	70
Textile mill products			*Stone, clay, and glass products*		
Cotton textiles (Sept. 1965).......	240,996	34	Pressed or blown glass and glassware (May 1964):		
Synthetic textiles (Sept. 1965).....	104,136	26	Glass containers..............	51,848	38
Woolen textiles (Nov. 1966):			Other pressed or blown glass....	29,900	36
Yarn and broadwoven fabric....	41,765	27	Structural clay products (July–Aug. 1964).............	51,324	28
Dyeing and finishing..........	3,559	9			
Scouring and combing.........	4,041	6	*Primary metal industries*		
Hosiery (Sept. 1967):			Basic iron and steel (Sept. 1967)..	452,977	66
Women's....................	44,545	70	Iron and steel foundries (Nov. 1967):		
Men's.....................	20,078	65	Gray iron, except pipe and fittings.................	90,317	21
Children's.................	15,255	70	Gray iron pipe and fittings......	20,991	23
Textile dyeing and finishing (Winter 1965–1966)...........	54,774	11	Malleable iron................	21,934	33
			Steel........................	51,994	26
Apparel and other textile products			Nonferrous foundries (June–July 1965)..............	57,507	18
Men's and boys' suits and coats (Apr. 1967)...................	98,354	74	*Fabricated metal products*		
Men's and boys' shirts, except work shirts, and nightwear (Apr.–June 1964).............	96,935	81	Fabricated structural steel (Oct.–Nov. 1964).............	55,429	8
Work clothing (Feb. 1968).......	62,775	82	*Machinery, except electrical*		
Furniture and fixtures			Machinery manufacturing (Mid-1966).................	1,171,278	17
Wood household furniture, except upholstered (May–June 1965)...	120,000	18	Engines and turbines..........	66,017	22
			Farm machinery..............	100,259	34
Paper and allied products			Construction and related machinery..................	193,725	13
Pulp, paper, and paperboard mills (Oct. 1967)..................	197,919	3	Construction machinery.......	98,645	11
Paperboard containers and boxes (Nov. 1964):			Metalworking machinery.......	225,091	13
Folding paperboard boxes.......	29,201	6	Special dies, tools, jigs, and fixtures	79,803	1
Setup paperboard boxes.......	16,545	18	Machine-tool accessories......	40,408	14
Corrugated and solid fiber boxes.	57,132	36	Special industrial machinery....	125,728	13
Sanitary food containers.......	18,625	19	General industrial machinery ...	180,281	16
Fiber cans, tubes, drums, and similar products............	9,484	37	Office and computing machines.	95,366	24
			Service industry machines......	82,060	18
Chemicals and allied products			*Transportation equipment*		
Industrial chemicals (Nov. 1965)..	168,515	5	Motor vehicle parts (Apr. 1963)...	186,684	31
Synthetic fibers (Feb.–Apr. 1966):			Motor vehicles (Apr. 1963).......	460,798	2
Cellulose...................	26,712	15			
Noncellulose...............	35,695	1			

* Data are based on BLS nationwide occupational wage surveys in selected manufacturing industries, conducted between 1963 and the spring of 1968.   The industry studies nearly always have a minimum establishment size cutoff; establishments under the cutoff usually account for less than a tenth of the industries' work force and if included would not substantially affect the percentages provided above.   The cutoff was 20 workers for all industries except the following: Prepared meat products (10), cigars (8), men's and boys' suits and coats (5), pulp, paper, and paperboard mills (50), fertilizers (8), industrial chemicals (50), paints and varnishes (8), petroleum refining (100), footware (50), basic iron and steel (250), nonferrous foundries (8), special dies, tools, jigs, and fixtures (8), machine-tool accessories (8), and motor vehicles parts (50).   There was no minimum size cutoff for cigarettes, synthetic fibers, and motor vehicles.

† Less than 0.5 percent.

Source: George L. Stelluto, "Report on Incentive Pay in Manufacturing Industries," *Monthly Labor Review*, July, 1969.

That management has, in general, a healthy respect for financial incentives as a way of motivating people other than production personnel is borne out by National Industrial Conference Board studies, which show that over one-half of industry's top executives[6] and three-fourths of the sales forces[7] are on some form of incentive. The studies indicate that incentive coverage of these groups will continue to rise; financial incentives are high in management's thinking on how to motivate people.

Union Attitudes toward Wage Incentives. The official position of many in the leadership of the AFL-CIO and in a number of the large international unions is opposition to wage incentive plans. However, the practices in individual plants and the attitudes of local unions vary considerably from that of their parent international unions. Local unions have considerable autonomy and the right to make their own decisions based on local conditions.

The attitude of unions as expressed by practice can be seen from the most recent study by the Bureau of Labor Statistics regarding the prevalence of wage incentives in industry, summarized in Table 2-1. Incentive coverage varies with industry. Some industries are traditionally oriented toward incentives, with management and employees dependent upon incentives to promote productivity. Highly mechanized or process industries do not feel as strong a need for incentives.

The United Automobile Workers (UAW) are officially opposed to incentives. The BLS study shows that 2% of the workers in automotive assembly are on incentive and 31% of the auto parts plants are on incentive. The low coverage in automotive assembly primarily stems from management's decision to operate without incentives because of the pacing afforded by conveyorized assembly. The parts plants are well suited to incentives, and many plants are on incentive.

The steel industry traditionally has been heavily oriented toward incentives, with coverage at about 66%. The trend in the steel industry, in response to demands from the union, is for additional coverage.

The policy that emerges from a study of local union attitudes toward incentives is one of accommodation with management to local conditions in ways to increase employees' earnings where the employees' interests can be safeguarded. Management's attitude is a prime factor which unions consider in deciding whether to permit the installation or removal of an incentive plan. Where relations are fairly harmonious and there is mutual trust between the union and the employees, management's desire to install an incentive plan is often granted.

When management wishes to establish an incentive plan, the matter must be discussed and negotiated with the union involved, if there is one. Where there is no union, the matter must nevertheless be brought to the employees in a manner which will elicit their cooperation and support for the program. In the final analysis, with a union or not, the success of an incentive plan depends completely upon the attitude of the employees working under it. Where they perceive that their interests are enhanced by the plan, they will support it. Should they feel that their interests are undermined by the plan and that in the long run they will suffer by it, they will oppose it and ultimately cause the plan to be removed.

Views of Psychologists. Psychologists generally agree that, as expressed by Gellerman: "There is no doubt that money has an important effect on the thinking and behavior of production workers; but this effect is neither as simple nor as strong as management has often assumed."[8] Yet there is a divergence of opinion among psychologists concerning the efficacy of wage incentives as motivators to increased productivity.

Some psychologists believe that incentives will motivate people. Leavitt says: "Money incentives have come to occupy a central place because money is a common

[6] Harland Fox, *Top Executive Compensation,* Personnel Policy Study 204, National Industrial Conference Board, New York, 1966.

[7] David A. Weeks, *Compensating Field Sales Representatives,* Personnel Policy Study 202, National Industrial Conference Board, New York, 1966.

[8] Saul W. Gellerman, *Motivation and Productivity,* American Management Association, New York, 1963.

means for satisfying all sorts of diverse needs in our society and because money may be handled and measured. Money is 'real'; it is communicable. Many other means to need satisfaction are abstract and ephemeral. Moreover, money incentives fit with our culture's conception of what work means. . . ."[9] Vroom supports this approach: "Much of the evidence that is available concerning the effect of wages on performance is consistent with the assumption that people strive to maximize the amount of their wages."[10]

Some psychologists see incentive pay as more than just money; it carries recognition with it which is a source of satisfaction to the employee. A high productivity level raises earnings and the standing of the employee on the recognition scale. Where incentives do not operate freely but are controlled by the workers, this obviously does not occur.

Whyte takes a more restrained attitude toward incentives: "Systems of financial incentive in industry yield a net gain in productivity, but most of them will fail to release more than a small portion of the energy and intelligence workers have to give to their jobs."[11] There are others who hold that money is not as strong a motivator as is commonly believed; that workers have nonfinancial needs which, if satisfied, will motivate workers to increase their productivity.

Numerous studies of the relation between job satisfaction and performance indicate that a high level of satisfaction does not necessarily motivate workers to expend increased energies directed toward improved performance. Simply put, happy workers may just continue to be happy workers, not necessarily high producers. Job satisfaction will improve the climate for increased productivity; it will reduce turnover and secure other desirable attitudes. There is no assurance, however, that job satisfaction without motivation will improve productivity.[12]

Reporting on the role of financial compensation in industrial motivation, Opsahl and Dunnette, in a broad study of current research, state that: "Strangely, in spite of the large amounts of money spent and the obvious relevance of behavioral theory for industrial compensation practices, there is probably less solid research in this area than in any other field related to worker performance. We know amazingly little about how money either interacts with other factors or how it acts individually to affect job behavior. Although the relevant literature is voluminous, much more has been written about the subject than is actually known. Speculation, accompanied by compensation fads and fashions, abounds; research studies designed to answer fundamental questions about the role of money in human motivation are all too rare."[13]

A number of psychologists have reported on plants with wage incentives. Gellerman states that "their firsthand accounts of what actually happens when incentives are introduced form the basis for the most comprehensive analysis yet of how the dollar affects the worker attitudes."[14] This aspect of the behavioral scientists' findings is often missed and the writings are used to judge incentive plans, whereas the writings are not critiques of incentive plans but only report workers' reactions to incentives. These must be viewed for what they are: case histories reporting primarily the opinions of workers who obviously reacted to their environment. In many of the reported cases, the incentives appeared to be deficient. Inferences cannot be drawn in such plants of how workers react to incentives in general, but only to deficient incentives. The writings of the psychologists do not seem

[9] Harold J. Leavitt, *Managerial Psychology*, 2d ed., The University of Chicago Press, Chicago, 1964.

[10] Victor H. Vroom, *Motivation in Management*, American Foundation for Management Research, New York, 1965.

[11] William F. Whyte, "Economic Incentives in Human Relations," *Harvard Business Review*, March–April, 1952.

[12] Vroom, *op. cit.*, pp. 187, 262.

[13] Robert L. Opsahl and Marvin D. Dunnette, "The Role of Financial Compensation in Industrial Motivation," *Psychological Bulletin*, vol. 66, no. 2, 1966.

[14] Gellerman, *op. cit.*

to take this into account, perhaps because they could not accurately identify deteriorating incentives.

To those not expert in the operation of incentive plans, the anecdotes and case histories of the incentive games played between workers and management reported by the behavioral scientists make interesting reading. Much of the behavioral literature depicts management as almost continually in trouble with incentives. To someone not experienced in incentives, reading this literature is comparable to a layman's reading medical case histories of sick people. After a while, it appears as though the world were filled only with sick people. Although a few successful plans are reported, many of the cases describe problem-ridden incentives. The behavioral writings on incentives can prove misleading where these are drawn from limited case histories.

Although some psychologists suggest that increased pay will not necessarily motivate workers because there are some needs which money cannot satisfy, it is generally believed by managers that money is the prime reason that people work and that people in general endeavor to maximize their earnings.

Managers looking for clear recommendations from the psychologists will be disappointed; support can be found for and against incentives in their writings. But there is much to be gleaned from the behaviorists' findings by managers who understand the temper of their plants and who put together the right combinations of people relationships and motivation.

Profit Sharing as an Incentive Plan. Profit sharing is not a form of incentive as discussed in this chapter. The Profit Sharing Research Foundation defines profit sharing as: "A plan in which the company contribution to employees is based upon business profits, regardless of whether the benefit payments are made in cash, or deferred, or a combination of the two." Profit sharing varies with profits and not necessarily with employee productivity. More often, profits come from management's efforts and sound decisions. Under incentives, employees can see their efforts translated into daily or weekly earnings. Profit sharing is long-term, deferred compensation, often contributed to a pension fund. Profit sharing and wage incentives are complementary and compatible. One does not displace the other.

MANAGING INCENTIVE PROBLEMS

Symptoms of Incentive Problems. Superficially, it may seem that when incentives operate successfully, it is because the employees are motivated to increase their earnings. Actually, the incentive plan and all the forces operating around it form a system which is quite complex. Because incentives become an important part of employees' earnings, pressures are brought to bear on the operation of the plan to increase earnings. Despite these pressures from all sides, those in management entrusted with responsibility to maintain the system must make sure that the system operates equitably. The success of an incentive plan is in a large way the measure of management's capabilities.

Managers often erroneously blame their employees when difficulties arise with incentive operations, claiming that it is the employees' desire to increase earnings which encourages excesses and frictions with management. For a clear perspective on incentive operations, managers must differentiate between symptoms and causes. Excessively high earnings and runaway standards, withholding of production, excessive grievances, and numerous other difficulties are all symptoms of deteriorating incentives. Frequently, attempts are made to eliminate these symptoms which management labels as its problems. But attempts to remove symptoms of deteriorating incentives are no more successful than using aspirin to eliminate a bodily disorder. Unless the root cause is eliminated, the problem will continue. When the causes of problems are pinpointed and diagnosed, those responsible for operation of incentive plans will achieve more effective operations.

What Goes Wrong with Incentives? The greatest problem with incentives is that they have a tendency to deteriorate over a period of years, sometimes to an extent which seriously interferes with management's ability to operate a plan efficiently. Some managers would rather forego the benefits of incentives than chance the possibility of problems.

What Causes Incentives to Deteriorate? Incorrectly pinpointing the causes of incentive degeneration prevents managers from eliminating their problems. In each plant where incentives have deteriorated, the story is usually the same. The disillusioned manager blames the downfall of the incentives on the workers' insistent pounding away at the standards over the years to obtain concessions from management. Managers attribute the workers' pressures to their desire to earn as much as they can with as little effort as they can get away with. The harassed manager who has tough production schedules to meet firmly believes the workers are ganged up against him and that it is their unfair pressures which are destroying the incentive plan and raising costs. This superficial reasoning gets nowhere.

An approach to understanding incentive problems requires, first, a delineation of the essential principles of a sound incentive plan:

1. The plan and its administration must be fair to the workers and to management.
2. Extra pay is earned for increased productivity as measured by time standards.
3. The incentive potential should be specified for manual work when working at an incentive pace, and for machine or process controlled operations.
4. A time standard represents not pieces produced but work requirements in terms of materials, equipment, process, methods, and work conditions.
5. Any change in the work requirements must be reflected in a changed time standard.

Where these principles are maintained in practice, the incentive plan will not deteriorate. There may be rough spots, grievances, and bargaining, but the plan will continue to benefit the workers and management. It is management's responsibility to safeguard the principles and uphold the integrity of the plan.

The major causes of incentive deterioration are proscribed practices which negate these principles and are clearly outside the incentive agreement. These practices occur when:

1. The employees find that there are other ways to increase their earnings than through increased productivity, the only method provided by the plan
2. Management does not fulfill its responsibilities to administer a sound plan

Practices which most frequently contribute to deterioration of incentives involve:

1. Changes in work requirements which are not reflected in the time standards
2. Establishment of time standards which vary in incentive earnings potential from the agreement
3. Cheating by employees

Deterioration rarely arises from negotiated changes in the incentive plan. More frequently, it is caused by shortsighted managers who permit degenerative practices to creep in. Others are introduced surreptitiously by workers, but almost always with some knowledge of plant supervision.

Probably more incentive plans are wrecked by erosion of the relationship between the time standard and the work requirements, through what is commonly called "creeping changes," than all other causes combined. Creeping changes are abetted on all sides by careless managers, by supervisors who think the changes are inconsequential, and by setup men and operators who stand to benefit from the changes. Despite universal acceptance of the principle that a change in work requirements must be reflected in an appropriate change in time standard, this is often flagrantly overlooked in practice.

Another serious degenerative practice is management consciously setting time standards which exceed the agreed-upon potential incentive earnings. For instance, many steel industry contracts provide that machine or process controlled operations are to permit 25% incentive earnings. Invariably, the actual earnings are closer to from 60 to over 100%. The argument by union representatives, when employees' earnings rise abnormally high on manual work, is that the workers have gained high skills. There is no such justification for the earnings level of machine paced work. It is impossible for employees to produce at 190% productivity on machine controlled elements if the standard is properly set at 125%. The answer to the discrepancy is that the rolling mill speed was increased and the standard was not reset by management, or the standard was set to maintain past practice. In either event, the incentive agreement is violated and a degenerative practice obtains.

No one condones cheating, and anyone caught can be fired on the spot; the union will rarely intercede. But "making money with the pencil" is widely practiced in various forms. Management can eliminate this evil if it wants to, but often too many are involved. Foremen permit the practice as a way of buying cooperation or paying off individuals. Cheating sometimes exists because of inept management personnel who do not detect it. Where cheating is permitted, it undermines the workers' respect for management and the incentive plan.

The Effect of Morale on Incentives. The question of employees' morale and their confidence in management is often overlooked in assessing why incentive plans fail. Sound administration of incentives requires that management continually support fair play, regardless of the workers' pressure in their own interests. Once relations break down, the plan is in for difficulties.

The administration of an incentive plan involves the most difficult aspects of labor relations and therefore requires skillful handling. Many incentive plans founder when employees lose confidence in management because grievances are not given prompt attention, management promises are not kept, obvious inequities are not corrected, or because of numerous other nonmonetary reasons. Not only is substantial take-home pay involved, but when earnings vary between operations, serious questions of relative equity are raised. A worker may be fully satisfied with his earnings-effort relationship, but when he sees someone across the aisle making more money with obviously less effort, he becomes dissatisfied. Sometimes he may not ask that his standard be loosened, but he will express complaints about foreman favoritism. Skilled handling of labor relations is critically needed to ensure confidence and support by the workers.

How to Prevent Deterioration. Deterioration of incentive plans can be prevented by taking the following precautions:

1. Do not permit dilution of the five essential incentive principles through proscribed practices.
2. Develop a fair plan; clearly delineate all aspects.
3. Administer the plan equitably; do not attempt to gain unfair advantage.
4. Attend to grievances and complaints promptly; keep workers informed.
5. Administer plan with capable staff; employ sound work measurement techniques.
6. Establish an audit program to monitor incentive operations continuously.

The most important precaution to prevent deterioration is point 1 above—not to permit dilution of the essential incentive principles through proscribed practices. This requires that managers comprehend the difference between proscribed practices and legitimate bargaining.

Some aspects of incentives are proper subjects of bargaining between workers and management. Concessions in such areas no more dilute management's control over incentives than do wage increases. For example, the incentive earnings opportunity when working at an incentive pace, or for machine paced operations, is clearly bargainable. Unless time standards are specifically excluded from bargaining in its labor-management contract, management cannot take the position that its standards are "correct" and therefore not subject to collective bargaining. Despite the belief of industrial engineers that the bargaining process loosens standards, there is no way to avoid it.

Bargaining and making concessions on individual standards need not erode the plan, although it will increase costs. Suppose the plan provides for 30% incentive pay potential and management maintains its position on this point. The setting of individual loose standards by compromise will permit excessive earnings on these jobs, but this should not affect other standards which are still governed by the 30% agreement. With a sound work measurement program, management should be prepared to go to arbitration on such issues and prevail.

Some aspects of incentives must not be bargained; otherwise the plan will be seriously impaired. The principle of not permitting changes in work content without changing the standard is not a bargainable issue. This principle is a main source of protection for workers against abuses of increased work loads, and they are vigilant in protecting their interests. Although management well understands the importance of this principle, practice often finds management negligent in these matters. This principle is management's main protection to assure that improvements in methods and process are recovered.

Preventing Creeping Changes. When changes are made by management personnel which are not reflected in the standards, the workers quietly adjust to the changes and do not "kill the rates," which would alert the industrial engineers. Changes innovated by workers are put into effect slowly to avoid detection by supervision.

Why do workers try to outsmart management? Some managers reason that it is the workers' way of getting back at the company for past unsettled scores. In some cases this may be so, but it is not the main reason. Workers oppose management on this issue because they have every incentive to withhold production and hide their innovations. Turning their ideas over to management yields them very little and frequently earns the scorn of the other workers. Few plants have bona fide suggestion plans which offer sufficient bonus to encourage the workers to present their ideas to management. A good idea which escapes detection benefits all the workers on the job, and if in effect long enough, becomes permanent. An extra benefit to the workers is that the loose standard becomes leverage to loosen other standards.

Management must protect its position by clearly providing in the incentive agreement that a change is a change, even if innovated by the employees. But when management "captures" a change created by a worker, it is understandable that the worker is antagonistic and adopts the attitude that management is cutting rates and acting unfairly. A worker who improves on a method and in effect reduces costs is entitled to consideration. Take the following: if a manager decides to acquire a piece of equipment or make a change because it will reduce costs, he is prepared to pay for it. Should he not be equally prepared to pay for an improvement in a method introduced by a worker? The change is easily measured: it is the difference between the original and the new time standard. If management were to pay for workers' innovations, this would reduce hostile attitudes and even encourage workers to try different methods. Where managers pursue participative management theories to encourage workers' ingenuity and involvement in their work, compensation for creative ideas becomes mandatory.

There is only one way to prevent creeping changes: institute an audit program to detect changes. Some managers spend endless energy attempting to prevent unauthorized changes from going into effect without notification to the industrial engineers, but this is usually not effective. Changes and improvement should be welcomed by management from whatever source. It is not change which causes looseness, but undetected change. Such changes can be detected by monitoring operations, following suggestions for an audit program described later.

Rules versus Practices. Successful operation of an incentive plan over a long period of time requires that the original plan be sound, eminently fair to both parties, and equitably administered. A plan which has unfair features for either management or the employees soon sets up pressures to correct the shortcomings, which will weaken the plan.

Because success of the plan depends on the free exercise of employee effort, the employees must see the plan as fair to them and they must have confidence in management's integrity. Where they find that management attempts to obtain unfair advantages, they will lose confidence. The reaction to this is usually an increase in proscribed practices, which now become sanctified in the eyes of the workers.

Many incentive problems arise through carelessness and sometimes from lack of attention. When such problems go to arbitration, it is found that, although arbitrators are not bound by hard and fast rules in areas of incentive differences, there is fairly close agreement on basic precepts. For example, arbitrators will not rescue managers who had remedies available to correct standards loosened by creeping changes, but who allowed several years to pass without taking action. Similarly, arbitrators will give greater weight to incentive practices than to the written agreement. The point is that managers must be alert to all occurrences and take timely actions. This is one area where lack of action is ruinous.

Practices which are clearly against management's interests and which would never have been agreed to in negotiations sometimes come into existence almost imperceptibly over a period of time, mainly through management shortsightedness. Examples include unreasonable restrictions on how management's engineers are to take time studies and calculate time standards. Through some employees' objections to specific events, or because of an overzealous shop steward, minor agreements blossom into full-fledged practices. These may have no value to the employees, but if they irritate management or tie its hands, the employees may persist in continuing them.

The subject of equitable earnings opportunity often finds management on the short end of the bargain after a number of years. An incentive plan starts out with the proposition that extra pay is earned for extra effort. Employees complain, with justification, that when a machine breaks down they are denied, through no fault of theirs, the opportunity to earn extra pay. When management agrees to pay some form of average hourly earnings, it is usually difficult to restrict the practice thereafter. Many incentive plans suffer when average hourly earnings payments go to extremes. But there are ample remedies for these situations which management can apply as, for example, reducing all downtime to a minimum and maintaining high incentive coverage. This will reduce nonincentive work and minimize the effect of average hourly earnings. But this requires alert managers; too frequently, managers do not pay sufficient attention to these matters, and so costs rise. It is not fair to deny employees their legitimate earnings because inept management cannot cope with its problems. To cover its incompetence, management frequently blames average hourly earnings for the degeneration of the incentive plan.

Where management agrees to increased incentive earnings potentials, this should be incorporated into the incentive agreement and adhered to in practice. Even if management is saddled with high incentive earnings through past practice, this should be recorded in writing so that it can be maintained. If this is not done, the earnings level will surely rise still further. It is the clinging to the fantasy of one set of standards for the agreement and another for plant practices which destroys the sanctity of the agreement.

Responsibilities of Management. When management undertakes the establishment of an incentive plan, it must fully comprehend the magnitude of the undertaking and the skill and quality of the personnel who will be needed for its administration. Many incentive difficulties arise because the program is not adequately staffed, particularly with competent industrial engineers. Although seemingly simple, incentive plans become complex in operation, and considerable expertise is needed in the proper handling of problems which arise.

Supervisors must be trained so they fully understand the plan and are able to handle the problems which arise in their departments. Too often, supervisors are not involved in the incentive plan, placing a tremendous load on the industrial engineers.

Administration of the incentive plan is a high-order management responsibility. Incentives cannot be operated as a part-time activity. Because labor relations are always deeply affected, critical decisions must be made on a high level. Personnel responsible for the day-to-day administration of the plan must not be permitted to make or change policy, and especially not to initiate deals. A responsible manager must continuously oversee the operation of the plan.

Problems with Time Standards. Many managers have an oversimplified concept of work measurement, believing that there are fairly simple, universally accepted techniques which can be used to establish valid time standards. They often think the problems with time standards come primarily from employees' general opposition to all things which favor management. If this were true, the problems could be solved with ease.

Though work measurement techniques may seem simple, there are difficulties built into all work measurement systems. It is vital that management fully understand the nature of work measurement limitations so that the systems in their plants operate as effectively as possible.

Because measurements are made in decimal units of time, there is the notion that time standards are precise in the sense that physical measurements are precise. Because the subjective judgment of the time study analyst is always necessary for performance rating—an essential step in stopwatch time study—the time values cannot be labeled as completely objective. Predetermined time systems eliminate the necessity for performance rating and hence produce more consistent standards. The determination of the work pattern, however, also requires subjective judgments.

Despite the technical difficulties of work measurement, numerous plants set time standards which are reliable and acceptable to workers and management. Recognizing the problems which can arise, managers in these plants safeguard their measurement systems by taking a fundamental step: they make sure the program is directed by a fully qualified person, and they fully support the activities with budget and personnel.

Management Attitudes under Incentives.   There are often significant differences in management attitudes under incentives from those that exist in measured daywork plants. In the incentive plant, management sometimes takes advantage of the workers' desire to earn extra pay, counting on this to keep the workers going in the face of difficulties. In a daywork plant, it is more likely that workers will stop the operation if they have difficulty. Under measured daywork, management relies primarily on supervision to obtain higher productivity. Greater priority is given to the training and upgrading of supervisors. In the incentive plant, management expects that the motivation of the workers will obviate the need for close supervision.

This shifting of management responsibility to the worker not only affects the operation of the incentive plan, but it also undermines the relationship between worker and manager, demeaning the image of management formed by the workers. The workers recognize their role in the organization, and they readily sense when management shifts its burden to them.

When managers complain that they lose control over their operations under incentives, they miss the fact that much of this is due to their attitude toward supervision; they encourage employees to exercise greater initiative. After years of working in this sort of environment, it is little wonder that employees on incentive are more independent and vocal regarding production problems than are employees in measured daywork plants.

Nothing can take the place of good supervision, especially not incentives. Where managers establish incentives to make up for deficiencies in supervision, the incentives will surely deteriorate and further add to management's difficulties. Sound operation of an incentive plan over a long period of time requires competent supervision. The same plant on measured daywork may require greater supervisory attention to maintain high productivity. Although some managers of incentive plants cite the pressures of employees to earn more for less effort as the major cause of problems, the primary causes of incentive problems are more often found in deficiencies in managing.

EXERCISING CONTROL OVER INCENTIVES

The difficulty in controlling incentive operations arises primarily because the stability of the plan is based on the maintenance of given conditions and practices, while the elements involved in the incentive plan are constantly changing; operations change, tooling is revised, quality requirements are altered, pressures are exerted by the employees, and concessions are made by management. Changes which conform with the plan do not cause problems. But when changes violate the plan, erosion sets in.

Controlling Change. Control of the incentive plan in large measure depends upon the control of change. Because many of the changes are not readily detected when they occur, or because the long-term effect may not be discernible, appropriate procedures must be established to detect and evaluate change. The most effective way to detect changes involving the incentive plan is through techniques similar to the audit and control procedures established for the fiscal activities of the business. Every chief executive requires that cash, inventory, receivables, payables, assets, and salient operation details be verified by independent auditors. No one questions these practices.

The operation of an incentive plan is less commonly audited, although it is the basis for the payment of much of the production payroll. Not only are substantial sums involved, but the productivity level of the enterprise is at stake because a degenerating plan lowers productivity.

Control over an incentive plan is most effectively established by a control system incorporating independent audit.

Responsibility for Audit. The accounting and fiscal activities of an enterprise are the responsibility of the comptroller and the treasurer who administer the various systems and procedures, process the data, and prepare the needed reports. Management uses their internal interim reports for guidance and decision making; but it is the final, verified reports prepared by independent auditors which check and validate management's data. Of perhaps greater worth are the suggestions of the auditors on ways to improve the performance of the areas audited.

The same logic must be followed with wage incentive plans. The industrial engineering manager is responsible for administration of the plan. The error frequently made is that he is also held responsible for audit. For some reason, management does not see clearly the need for independent auditing of this critical activity of the business. Audit is a top management responsibility for incentive plans as well as for fiscal matters.

Purpose of Audit. Specifically applied to incentives, audit is a continuous review and appraisal of work measurement and incentive operations to determine if these are carried out in conformity with management policy and objectives. The primary purposes of audit are:

1. Detection of change so that it can be examined
2. Evaluation of change
3. Provision of meaningful data upon which to formulate decisions for acting on change

Some of the more important detailed reasons for audit are:

1. To assure that incentives operate as designed
2. To detect changes; to determine how and why these came into practice and the effect they have on the incentive plan, on costs, and on employee earnings; and to recommend remedial action (note that remedial action is a management responsibility and not a part of the audit function)
3. To evaluate the effectiveness of the work measurement system and wage incentive plan; to form judgments on the capabilities of the industrial engineering and other management personnel working in this area; and to recommend improvements

By preventing erosion of the incentive plan, audit will assist in maintaining the productivity level and incentive earnings in line with the incentive agreement and management's objectives. It is vital that it does not become the purpose of audit to limit or depress employee earnings. Audit will, in the long run, hold earnings in line—the line being the incentive agreement between labor and management—and management acts properly when it upholds the agreement.

Management's Right to Audit. Audit is a basic management right indispensable to the monitoring of operations and the gathering of information and to the evaluation of activities in order to make decisions. However, when incentives are involved, workers will not naively accept management's prerogative without assurances that it will not be misused.

Management must meet with its workers to explain clearly the purpose of the program, how it will work, and the actions that management expects to take. The employees' rights must be clearly established so that they are fully protected. Because this is a sensitive issue which can easily be misinterpreted, management must be completely candid with the workers so that there is a full understanding by employees and management personnel of the audit procedures.

The reception given to management's audit program will invariably depend on the relationship between employees and management and the trust employees have in management's concepts of fair play.

Who Should Audit.   The auditor must have a sound background in the operations being audited.   This is not a fact-finding mission in the ordinary sense, for considerable judgment is involved at various stages.   A high order of skill and capability is requisite for the auditor.

There are two aspects to audit which can be separated in purpose and in execution, although both are conducted in essentially the same manner.

1. Routine audit of operations to detect changes and variance from standard procedures
2. Management audit of the overall operation of the incentive program and of the techniques and approaches used by the industrial engineering department in the performance of its work

The routine audit should be undertaken by the members of the industrial engineering department as a regular part of its activities, and it should be performed continuously.

The management audit should be conducted by an engineer not connected with the department, and preferably from outside the company.   The high standards established for auditing the fiscal activities of the company should be followed for the management audit to assure that management obtains the advice and information it needs.

Criteria for Audit.   The essential criteria for audit are:

1. The wage incentive agreement, including the relevant sections of the labor-management contract, if there is one
2. Definitions of how to measure, particularly the bench marks for "normal"
3. The work requirements and conditions of the operation when the time standard was originally established

The criteria must be clearly delineated so that these are not lost sight of or changed with the passage of time.   Most important of the criteria is the incentive agreement which establishes the basis for the incentive program.

Practices are sometimes introduced which are not included in the agreement. Where these violate the written agreement, management should clarify and remove discrepancies to avoid administration problems and to delineate the criteria for auditing.   For example, the agreement may specify that, when working at an incentive pace, the employees have the opportunity to earn 30% extra pay.   But when the plant averages 50% and the employees on the average are not "superskilled," management must examine its definitions.   The engineers who establish such time standards violate the agreement.   If management has in effect agreed to a higher incentive opportunity through past practice, then this should be incorporated into the agreement so that the engineers always follow the agreement.

Management benefits when it updates the agreement to reflect current practice. Where most of the employees earn 50% incentive pay and the agreement provides for 30%, management has as much chance of reducing the incentive earnings as it has of convincing the employees to take a 20% cut in their hourly pay.   If these are the facts of life, the written agreement, providing for 30% incentive pay, has no meaning; there is in fact no agreement.   When incentive earnings later inch upward above 50%, management will have no agreed basis for setting standards.   This type of situation works against management's interests.

Where the incentive agreement does not quantitatively specify the relationship between incentive earnings and incentive pace, management must expect a continual rise in earnings year after year. Many incentive agreements which are deficient in this regard can be rectified only by bargaining with the employees and establishing a figure. Any figure, no matter how high, is preferable to no figure.

Concepts of "Normal." The bench marks for "normal" must be specified without ambiguity. Commonly used terms which cannot be quantified should be avoided. For example, in referring to the level of incentive earnings when working at incentive pace, words such as "adequate" and "fair" have no meaning.

The lack of a clear definition of normal often causes problems in many plants in setting time standards and in auditing. Normal can be expressed in relation to incentive pace, where incentive pace is defined as:

> *Motivated Productivity Level (MPL).* The work pace of a motivated worker, possessing sufficient skill to do the job properly; physically fit for the job, after adjustment to it; and working at an incentive pace that can be maintained day after day without harmful effect.

Having formulated incentive pace, normal is expressed as:

> *Acceptable Productivity Level (APL).* The work pace which is established by management, or jointly by management and labor, at a level which is considered satisfactory; it is established at a given relationship to the motivated productivity level.

Many of the difficulties in work measurement and in audit arise from the subjective basis of the entire measurement process, which cannot be avoided in measuring human work performance.[15] One way of minimizing such problems is provided by the definition of normal, or acceptable productivity level, where normal is related quantitatively to incentive pace. This requires that management establish the percent earnings to be paid when working at an incentive pace. Because the incentive earnings potential is a bargainable figure, it can be set at any level.

Normal (APL) varies in relation to incentive pace (MPL) by the agreed-upon spread between normal and incentive pace. Normal is not fixed under this concept; it varies. It is incentive pace which is fixed.

Stress is placed on the definition of normal because, in auditing, the question arises of how much a worker should earn for a given level of application of effort. The auditor must have clear guides against which to evaluate current productivity.

How to Audit. The techniques of audit must be organized to achieve the main purpose: detection of change and of deviations from standard practice. The changes which management seeks are those which are not readily detected by observation or by examination of productivity reports.

Audit studies are primarily concerned with operator performance on specific operations, to ascertain whether changes have occurred in the work requirements of the operation as it was established when the time standard was set. In auditing a time standard, the criteria are the original work methods and the conditions of the operation. The level of earnings of the worker at the time of the audit is not germane to the audit, although the earnings may be symptomatic of what has occurred. Sound audit requires that the data supporting the time standards be clearly set forth and available. Experience in numerous plants has shown serious deficiencies in record keeping, often preventing management from supporting its case for changes made in operations. Auditing requires clear-cut criteria. Management must ensure that records are well kept and fully substantiable.

Effective audit is based on the following essentials:

1. Checks or studies should be made on randomly selected operations.
2. No person other than the auditor is to have advance notice of audit.
3. The auditor is not to consult any files or background material concerning the operation before making the study; no bias is permitted.

[15] Mitchell Fein, "A Rational Approach to Normal in Work Measurement," *Journal of Industrial Engineering*, June, 1967.

4. After selecting an operation for study, the auditor in effect samples the operation to determine how it is then performed; all pertinent information is obtained. A time study is made of the operation in the conventional manner; and a time standard is calculated, reflecting the existing conditions. If a predetermined time standard system is used, the study will conform to plant practice.

5. The data are then compared with the original data upon which the existing time standard was calculated. The work content is compared element by element, and any changes which may have occurred are noted.

6. Differences in any work conditions and requirements are investigated to determine how and when these were made. The investigation is an important part of audit, because it will turn up practices and looseness which undoubtedly affect other standards.

7. All the facts, together with recommendations, should then be submitted to management. The audit is primarily a fact-finding function. It does not include taking corrective action, which is the responsibility of management.

The search for undetected changes will also uncover management practices which are deficient. Ineffective management practices and personnel will also be disclosed. The entire process will permit evaluation of the work measurement and incentive program and make data available upon which to form judgments and recommendations for improvement.

REMEDIES FOR A DETERIORATED PLAN

Appraising the Situation. An attempt to salvage a deteriorated plan is a positive step forward and in no way an indication of defeat. Too frequently, managers are fearful that attempts to rectify former errors will reflect on their ability as managers. Or they are fearful of exorbitant demands by the employees as the price of making the changes.

The first step in planning a course of action is to make a complete and constructive analysis of the plan from its origin. The findings should primarily show where and how various practices arose which eroded the plan. The various proscribed practices should be detailed, with an analysis showing the effect each has on manufacturing costs, productivity, and the incentive plan.

No conclusion should be made at the start that the plan should be retained or discontinued. This final step will come after much analysis and bargaining. With clear facts before it, management must firmly face the problems. Because the incentive plan is probably deeply entrenched and because the incentive earnings are a substantial part of take-home pay, serious problems lie ahead if management decides to eliminate the plan. Alternatives must be examined not only for immediate cost but also for their effects upon productivity, management's control of the operations, and other essential factors.

Suppose the incentives are discontinued. Will this permit management to control its operations? Managers sometimes jump from the frying pan into the fire, thinking that under measured daywork many of the productivity problems will be eliminated and management can then proceed unhampered. But grievances over standards arise just as before. Although incentive pay is not involved, many of the problems under incentives also occur under measured daywork.

Under measured daywork, management depends primarily on supervision to obtain desired levels of productivity; the worker does not have as much motivation to assist in the process as he has under incentives. Before discontinuing incentives, managers should consider upgrading the capability of supervisors, as they would have to do under measured daywork, and determine whether this would help remedy some of the problems under incentives.

Buy-out Costs to Eliminate Incentives. The buy-out cost to eliminate incentives is often very high. It usually requires that the incentive workers receive about the average take-home pay previously earned under incentive. In a study conducted by

Ralph M. Barnes of 72 companies,[16] 96% of which had incentives, it was the opinion of those responsible for incentives in 67 plants that, if they abandoned their incentives on direct labor and instead paid average earnings as their hourly wages, unit labor costs would increase from 10 to 200%; only one reported that there would be no change.[17]

Eliminating incentives will probably also raise nonproductive costs. Under incentives, the nonincentive workers are told that the production workers earn extra pay for extra effort; this justifies their increased pay. When the incentive worker is placed back on time work, however, this argument no longer holds. Setting new base rates for the incentive workers will also require that the pay of the nonincentive workers be increased; otherwise, serious inequities will be created between the production workers and those in the indirect categories. Raising the pay of these workers to the new levels of the production workers will substantially raise costs. Raising plant wage rates may also affect office jobs.

The decision to eliminate incentives is not easily made; there are numerous difficult cost and labor relations problems to be overcome. In most circumstances, managers would do well to discuss their problems and plans with the union or the workers. Such talks may lead to alternatives which management might not think of from its own position. Despite the seeming opposition of workers to management's interests, they are concerned with the company's welfare. Where serious problems are encountered which will affect the company and the workers, management will usually find that the workers will be sympathetic and will bargain in good faith.

There are various ways to eliminate incentive plans. Some companies have taken firm positions with their employees, which often result in protracted strikes and long periods of strife. Other companies have developed buy-out formulas. There are no pat approaches to recommend for these situations. Management in each company must make its own decisions. Much, though, can be gained from meeting with the workers and arriving at decisions which are mutually advantageous.

Too frequently, managers think in terms of black and white: if an incentive plan is troublesome, discontinue it. Most of the time this should not be done. Where shortcomings and problems are faced squarely, there are ways to remedy deficiencies and reestablish incentives on a sound basis. Incentives are powerful motivators which offer substantial benefits over operating under measured daywork conditions.

Salvaging a Deteriorated Plan. Where incentives have seriously deteriorated and the decision is to rectify the incentives and establish a new plan, this must be undertaken in collective bargaining where a union is involved. The most successful approaches have been where management has undertaken open and frank discussions with the leadership of the workers, laying all issues squarely on the table. Such discussions are often expedited by a third party with an understanding of the problems, who has the confidence of management and the employees and joins the discussions as a neutral. Because important issues are involved, the discussions should not be rushed. Resolution of problems may take a year or more before mutual agreement can be had.

Before engaging in discussions, management must evaluate the effect of the runaway incentives and establish the magnitude of the extra costs which are incurred. For example, where incentive earnings are 20 percentage points higher than agreed upon or warranted under the circumstances, if the average hourly base rate is $3 per hour, the unwarranted incentive earnings are $24 for a forty-hour week, equal to about $1,250 per year per employee. One approach might be to offer the employees a lump cash settlement of six or twelve months' extra incentive

---

[16] The plants ranged from small up to 45,000 employees, with an average of 4,217 per plant. All plants had organized industrial engineering departments, with a total of 3,740 industrial engineers.

[17] Ralph M. Barnes, "Industrial Engineering Survey—1967," *Journal of Industrial Engineering,* December, 1967.

earnings, with management then getting the right to reset all time standards to an agreed-upon level of incentive earnings.

Another approach is to red-circle the earnings of all employees who produce above a given productivity level and guarantee these employees an hourly add-on equal to the red-circle difference. Assume the plant average incentive earnings equal 55%, and management and the employees agree that 35% incentive pay would be equitable for the output, which is measured at 155%. Time standards would be reset to the agreed level, but each employee then in the plant would receive a personal guarantee, based on his past average hourly earnings for six months to a year, as a red-circle add-on computed as the difference between his average incentive earnings on the original standards and his earnings on the new standards. For example, the red-circle for an employee with a past average of 160% would be calculated as $160 - (135/155 \times 160) = 21$ percentage points. An employee at 140% would be calculated as $140 - (135/155 \times 140) = 18$ percentage points. Each employee would then receive his new earnings plus the percentage points or the money equivalent of his red-circle add-on for all hours worked on incentive. New employees would not have a red-circle add-on, and they would receive their regular incentive pay based on the new time standards. The advantage to management under this arrangement is that cost savings are made on new employees, even though present employees continue with their past average hourly earnings. As discussed later, hourly add-ons have the effect of changing a one-for-one incentive plan to less than one for one. This type of add-on is not to be confused with contractual annual wage increases, which are sometimes not added to base rates for calculating incentive pay.

These two illustrations can be expanded and developed into varied combinations, depending on the needs of the situation and the attitudes of the employees and management. There may be substantial cost savings possible even with seemingly high buy-out costs. In some cases, the time standards may be so loose that the employees hold back production for fear that their excessive earnings will cause problems. With an agreed buy-out and the establishment of new time standards, the employees can increase their output to higher productive levels and further increase their earnings, yet at the same time permit management to reduce its costs. These types of arrangements offer benefits to the employees and to management, without the need to destroy an incentive plan which has the potential of benefiting both parties.

DESIGN OF INCENTIVE PLANS

Many industrial engineers use the approach that 100% normal is a self-supporting measurement base from which all measurements are made. Although this approach is valid when used within a given plant, it cannot always be used to compare time standards or incentive plans of different plants. The work measurement concepts of one plant using 100% normal may be different from the concepts used in other plants. The fact that a number of plants all use the figure of 100% as normal does not mean that each plant's measurements are comparable with all other plants.

When comparing time standards or incentive plans between two plants, it is essential that the measurement systems in the two plants be set to the same measurement base. The following discussion is based on the proposition that the common reference base between plants is incentive pace. Normal should be established for each plant, based on specific criteria for that plant, in relation to incentive pace. After normal is defined and established for a plant, it can thereafter be used as a reference base for that plant only.

Incentive Plan Principle. Incentive plans are designed on the following principle: Employees are compensated for increased productivity above an acceptable productivity level, at a predetermined participation ratio, in proportion to the increase in productivity or in accordance with an established plan. The various factors involved in this principle require discussion and delineation.

Major Factors in Design of Incentive Plans. All incentive plans require (1) the establishment of reference points from which to evaluate productivity and (2) specifications on how increased productivity is to be shared between management and the employees. The major factors involved are:
1. Criteria to establish time standards and to evaluate performance, based on the relationship of normal to incentive pace
2. The ratio of labor participation in the increased productivity above normal
3. The level of productivity prevailing before the introduction of incentives, to evaluate the efficacy of the plan and as a guide in determining the ratio of labor participation

The following definitions of terms are essential to a full understanding of the discussion.

Measurement Criteria

Measurement Reference Base. A reference base which is definable, reproducible, and stable; it is common to the phenomena being measured. The only base in work measurement between plants is incentive pace.

Incentive Pace. The work pace of a worker motivated by incentive pay to produce at a high productivity level which can be maintained day after day. This pace is a subjective judgment which may be supported by bench marks or previously agreed-upon time standards for specifically defined operations or fundamental elements of work. Incentive pace concepts are translated through work measurement techniques into time standards.

The commonly used terms to describe normal and low task, and incentive pace and high task, may be replaced by the following two more meaningful terms.[18]

Motivated Productivity Level (MPL). The work pace of a motivated worker, possessing sufficient skill to do the job properly; physically fit for the job, after adjustment to it; and working at an incentive pace that can be maintained day after day without harmful effect.[19]

Normal or Acceptable Productivity Level (APL). The work pace which is established by management, or jointly by management and labor, at a level which is considered satisfactory; it is established at a given relationship to the motivated productivity level.

APL is the productivity level at which incentive pay starts. MPL is a function of human work ability. APL is established, by agreement, in relation to MPL.

Standard. An established measurement base within a given plant, usually stated as time per unit of production or per task, from which productivity measurements are made; may also be stated as units per interval of time. Standard is usually set at either normal (APL) = 100% or incentive pace (MPL) = 100%. Technically, it can be any base which can be defined and reproduced. Standard need not coincide with either normal (APL) or incentive pace (MPL). The uses of standard are discussed later.

Productivity. The ratio of actual production to standard expressed as a percentage.

Incentive Plan Criteria

Labor Participation Ratio. The percent of the increased productivity above APL which employees receive in the form of incentive pay as their share of the increased productivity.

Incentive Expectancy. The increase in pay which is earned by an employee, either as a percent or in money value, when working at incentive pace (MPL), over the pay earned at APL. Incentive expectancy practices are discussed later under Relation of Incentive Potential to Effective Motivation.

Clarification of Normal (APL) and Incentive Pace (MPL) as Measurement Bases. In establishing a new incentive plan or revising a deteriorated plan, the

---

[18] Fein, *op. cit.*
[19] This definition is derived from Marvin E. Mundel's formulation of a time standard in *Motion and Time Study*, 3d ed., Prentice-Hall, Inc., Englewood Cliffs, N.J., 1960, pp. 333–339.

definition of the measurement base is critical. Improper or loosely conceived concepts will permit earnings levels not envisioned for the plan and cause numerous operating problems.

An understanding of APL and MPL as measurement bases can be obtained from the following illustration. Suppose management wants to establish an incentive plan under which the employees will receive full credit for all production above normal (APL). This is called a one-for-one or 100% participation plan and is described in detail later. The question then is how to establish normal (APL), above which employees will receive incentive pay. Assume a well-defined manual assembly operation on daywork with production at 62.5 pieces per hour. Through work measurement, it is established that incentive pace (MPL) for this operation is 100 pieces per hour. To establish normal, management and the employees must first agree on the incentive expectancy of the incentive plan, because the incentive expectancy then establishes the relation of normal to incentive pace. Table 2-2 shows the relationship between normal (APL) and incentive pace (MPL) at different levels of incentive expectancy for an operation which has an incentive pace standard of 100 pieces per hour, under a 100% participation plan.

When a time standard is established at incentive pace, or MPL, this standard will not change as the incentive expectancy of the plan is changed, as shown in Table 2-2. As the incentive expectancy is increased, the normal (APL) pieces per hour are reduced. The vital point is that normal (APL) is set in relation to incentive pace (MPL) by the incentive expectancy of the incentive plan.

Having established that normal (APL) is set in relation to incentive pace (MPL), normal can now be stated in two ways:

1. Normal as 100%, with incentive pace shown in relation to normal
2. Incentive pace as 100%, with normal shown in relation to incentive pace

This relationship is shown in Table 2-3 for incentive plans of varying incentive expectancies.

Note in Table 2-3, whether normal is expressed as 100% (column 2) or incentive pace as 100% (column 8), the pieces per hour at incentive pace (column 5 or 9) remain at 100 pieces per hour. Also note that for each level of incentive expectancy percent, the pieces per hour for normal (column 3 or 7) are the same whether normal is 100% (column 2) or the equivalent lower percent (column 6) when incentive pace is 100%. For example, at 25% incentive expectancy, normal is 80 pieces per hour (column 3 or 7), whether normal is called 100% (column 2) or 80% (column 6). At 30% expectancy, normal is 77 pieces per hour (column 3 or 7), whether normal is called 100% (column 2) or 77% (column 6).

Suppose that each horizontal line in Table 2-3 represented a different plant and that management in each plant had established as the incentive expectancy percent for its plant the figure shown on that line. Note that if normal is expressed as 100% (column 2), then each plant designates its normal as 100%. Yet the

TABLE 2-2.   Relation of Normal to Incentive Pace at Varying Levels of Incentive Expectancy for an Operation Which Has an Incentive Pace Standard of 100 Pieces per Hour, under a 100% Participation Plan

Percent incentive pay expectancy of incentive plan	Normal (APL) as percent of incentive pace (MPL)	Normal as pieces/hr with MPL at 100 pieces/hr	Incentive pace (MPL), pieces/hr
20	83.3	83.3	100
25	80.0	80.0	100
30	77.0	77.0	100
40	71.4	71.4	100
50	66.7	66.7	100

TABLE 2-3. Comparison of Normal as 100% or Incentive Pace as 100%, with Different Levels of Incentive Expectancy, under a 100% Participation Plan

Incentive expectancy percent	Normal as 100 %				Incentive pace as 100 %			
	Normal (APL)		Incentive pace (MPL)		Normal (APL)		Incentive pace (MPL)	
	Per-cent	Pieces/hr	Per-cent	Pieces/hr	Per-cent	Pieces/hr	Per-cent	Pieces/hr
(1)	(2)	(3)	(4)	(5)	(6)	(7)	(8)	(9)
20	100	83.3	120	100	83.3	83.3	100	100
25	100	80.0	125	100	80.0	80.0	100	100
30	100	77.0	130	100	77.0	77.0	100	100
40	100	71.4	140	100	71.4	71.4	100	100
50	100	66.7	150	100	66.7	66.7	100	100

pieces per hour (column 3) for each plant's normal are different. Such normals as measurement bases cannot be compared between companies. But when normal is expressed in relation to incentive pace as 100% (column 6), the pieces per hour at incentive pace (column 9) for each plant do not change. Now these normals for each plant fully represent the relative productivity level of each plant with respect to the others, assuming that the incentive pace concepts are the same, which they properly must be.

Most managers and industrial engineers prefer to use 100% as the base for normal (APL) for incentive plans of different expectancies, because 100% in everyday common usage denotes an easily recognizable base. If incentive pace is denoted as 100%, normal (APL) becomes 77% under the 30% expectancy plan. This is an odd-sounding percent and appears to be more difficult to use than normal = 100%. Actually, the measurement base of normal = 100% and incentive pace = 130% is exactly the same as normal = 77% and incentive pace = 100%.

Standard as a Measurement Base. Standard is the measurement base within a given plant for the establishment of time standards. The preceding discussion of normal (APL) and incentive pace (MPL) as measurement bases shows how time standards are treated when either is used as the basis for measurement. In these situations, either APL or MPL is synonymous with standard.

Some plants use standard as the basis for measurement different from APL or MPL. For example, the following are possible relationships within a plant among standard, APL, and MPL.

Case	Standard	APL	MPL
A	100 %	100 %	130 %
B	100 %	80 %	100 %
C	100 %	80 %	125 %

Case A shows standard and APL as the same. Case B shows standard and MPL as the same. Case C shows standard at 100%, with APL at 80%. This could be a sharing plan. Time standards in this plant would be set to a base of standard = 100%, with incentive payment starting at 80% productivity.

Standard can be established at any level which management finds appropriate for its plant. Once set, time standards are set to the defined base.

Labor Participation Ratio. Every plan must specify the ratio at which increased productivity above normal (APL) will be shared. The most favorable approach for management is 100% participation, under which the employees receive 100% credit for all production in excess of normal. Such plans are called one-for-one plans because employees receive 1% increase in pay for each 1% increase in productivity above normal. Employees also usually prefer 100% participation.

Once an incentive plan has been established and stabilized, the starting point for incentives is accepted. Thereafter employees look upon production exceeding standard as that which should accrue solely to them because it is produced through their efforts. Employees in general and unions in particular are opposed to incentive plans which pay less than one for one, unless there are circumstances which the employees agree require different ratios of incentive payment. Less than 100% participation plans are generally used for groups or plant-wide incentive plans, where it may be desirable to start incentive payments at a lower APL, but with an agreed-upon incentive expectancy at incentive pace.

In comparing a sharing plan with a one-for-one plan, if both plans pay the same incentive expectancy at incentive pace, unit labor costs under a 100% participation plan are lower at every level of productivity below incentive pace than the sharing plan. The reasons become evident when the underlying factors are examined in the following illustration. Assume the conditions stated for the previous example, in which preincentive production is 62.5 pieces per hour and 100 pieces per hour at incentive pace. Assume also that management decides the plan is to offer 30% incentive expectancy at incentive pace. Table 2-3 shows that normal (APL) under a 30% plan is set at 77 pieces per hour when incentive pace is equal to 100 pieces per hour. Under a 100% participation plan, the employees would receive no incentive pay until their production rises from 62.5 to 77.0. Above 77, they receive 1% increase in pay for each 1% increase in productivity. When production reaches normal, or 77, labor costs have been reduced by 18.83% compared with the preincentive costs at 62.5 per hour. The employees at 77 per hour receive no additional pay, although they have increased production from 62.5 to 77.

If the employees refuse to increase output from 62.5 to 77 with no increase in pay, there is an alternative approach: reduced participation. Keeping the 30% incentive expectancy at incentive pace equal to 100 pieces per hour, a new normal (APL) can be set at 62.5 pieces per hour, which is the present level of productivity. But instead of 100% participation, it can be set at 50%. (*Note:* The reduced APL and the ratio of participation can be any levels desired or agreed upon. The figures used are examples only.) The increase from 62.5 pieces per hour to 100 is 60%. Under the 50% participation plan, the employees receive 50% of the increased productivity, which is 50% of 60%, or 30%. This approach satisfies the employees in that they receive incentive pay starting from present productivity and they receive 30% at the incentive pace level. But they receive credit for only 50% of the increase above normal for the 50% plan.

Under the 100% plan, management receives the full benefit of the increase in production from 62.5 to 77. Above 77, the employees receive full credit. Unit costs are stabilized at 77 pieces per hour. Under the 50% plan, management and the employees each receive 50% credit for the increase in production above 62.5. At incentive pace, the labor costs are identical for both plans. Above incentive pace of 100 pieces per hour, labor costs are lower for the 50% plan. Unit labor costs under the sharing plan, including the incentive earnings paid to the employees, reduce as production is increased. The relationship of costs under both plans can be calculated from the formulas provided in the discussion of incentive plans which follows.

Note that the normal (APL) for both plans is different: normal for the 100% plan is 77 per hour, and for the 50% plan it is 62.5. Both plans yield 30% expectancy at incentive pace, which in this example is 100 pieces per hour. For the sake of emphasis, note that incentive pace is always the same regardless of incentive expectancy or participation ratio. Only the normal (APL) changes according to the incentive expectancy and the participation ratio of the plan.

Analysis of Wage Incentive Plans. The basic incentive plans in use in industry are analyzed to show the relationship between productivity, earnings, and costs under the various plans. The definitions of normal (APL) and incentive pace (MPL) employed in this material make possible considerably simplified analysis of plans, and especially comparisons between plans.

The following notations are used in the incentive plan formulas:

x = ratio of any given productivity level to the measurement base of the incentive plan, which can be either normal (APL) = 100%, or incentive pace = 100%, or any other definable base. In comparing plans, only incentive pace = 100% should be used.

y_w = ratio of wages at any productivity level to the base rate. The base rate will be assumed to be the wage paid at the productivity at which incentive begins. At this point, $y_w = 1.00$.

y_c = ratio of labor cost at any point to the labor cost at normal (APL).

p = ratio of labor participation in incentives. It is calculated: (% incentive pay)/(% increase in productivity above normal at point measured).

s = ratio of normal (APL) to incentive pace (MPL) for the plan.

General Comments about Incentive Plans. The cases which follow are all based on normal (APL) = 100% as the start of incentive payment. Therefore only two cases, 100% participation and other than 100% participation, are needed to describe all plans, except plans which combine both cases. When several plans are compared, the base is changed to incentive pace = 100%. Each incentive plan described guarantees the employee's base rate should productivity fall below normal (APL).

Case 1. Daywork. Daywork is a method of wage payment, not an incentive plan. It is discussed here because aspects of daywork apply to incentive operations when productivity is below normal and base rates are guaranteed. The normal (APL) for daywork can be established at any relation to incentive pace which management deems appropriate for its plant, even though incentive pace is not expected under the circumstances. The level designated as 1.00 in Figure 2-1 is the APL. To compare daywork to incentive, it is first necessary to determine the true daywork APL against the scale MPL = 1. The APL for daywork is generally lower with respect to MPL than the APL for incentive plans.

Under daywork, the employee is paid the same hourly rate regardless of productivity. The relations are therefore

Earnings: $$y_w = 1 \tag{1}$$

Costs: $$y_c = \frac{1}{x} \tag{2}$$

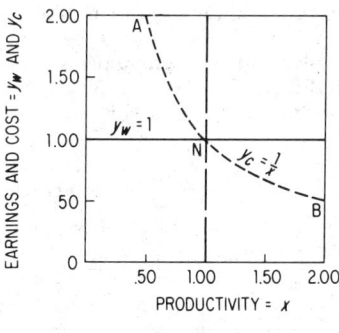

FIG. 2-1. Relationship between costs, earnings, and productivity in the daywork wage payment plan.

Case 1A. *Measured Daywork (as Daywork).* The commonly used definition of measured daywork (MDW) is the payment of a fixed hourly wage for work which is measured by time standards. This is shown above.

Case 1B. *Measured Daywork (as Incentive).* Some plants use measured daywork as a form of incentive under which employees receive an hourly wage rate which is adjusted periodically—generally monthly or quarterly—based on measured productivity for the past period. Base rates are guaranteed, with additional increments earned based on measured increased productivity.

Case 2. *Incentive Beginning at Normal (APL) = 100% with 100% Participation.* This plan increases pay 1% for each 1% in productivity above normal, shown by line *NE* in Figure 2-2.

	Below normal		*Above normal*	
Earnings are:	$y_w = 1$	(1)	$y_w = x$	(3)
Costs are:	$y_c = \dfrac{1}{x}$	(2)	$y_c = 1$	(4)

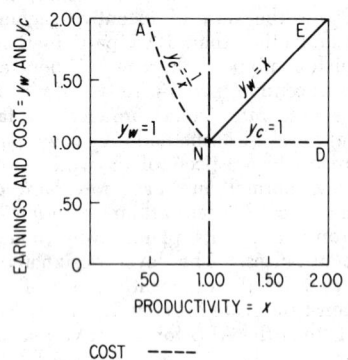

COST — — — —
EARNINGS ————

Fig. 2-2. Relationship between costs, earnings, and productivity in the incentive plan with incentive beginning at normal (APL) = 100%, with 100% participation.

Case 3. *Incentive Beginning at Normal = 100% with Other than 100% Participation.* These plans are practically always less than 100% participation and are called sharing plans. Some of the earlier plans called Bedaux, Haynes, and others were of this type. The Scanlon plan and other group plans are also of this type.

Earnings are shown in Figure 2-3 on line *NF*. Line *NE* shows earnings for case 2, as a comparison with case 3. When normal (APL) = 100%, the following applies:

	Below normal		*Above normal*	
Earnings are:	$y_w = 1$	(1)	$y_w = 1 + p(x - 1)$	(5)
Costs are:	$y_c = \dfrac{1}{x}$	(2)	$y_c = \dfrac{y_w}{x}$	
			$y_c = \dfrac{1 + p(x - 1)}{x}$	(6)

When incentive pace = 100%, with normal expressed in relation to incentive pace, equations (1), (2), (5), and (6) change to:

	Below normal		*Above normal*	
Earnings are:	$y_w = 1$	(1)	$y_w = 1 + p\left(\dfrac{x}{s} - 1\right)$	(7)
Costs are:	$y_c = \dfrac{s}{x}$	(9)	$y_c = \dfrac{s[1 + p(x/s - 1)]}{x}$	(8)

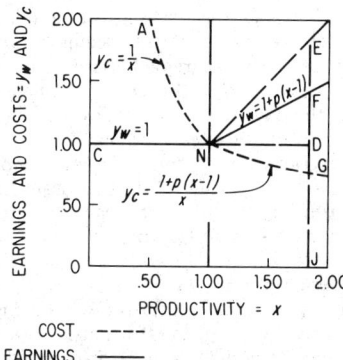

FIG. 2-3. Relationship between costs, earnings, and productivity in the incentive plan with incentive beginning at normal (APL) = 100%, with less than 100% participation.

The participation ratio can be increased to over 100% for special cases where automatic machines are involved and the objective of the plan is to secure maximum operation of the machines. For example, a plan could offer 3% increase in pay for 1% increase in productivity, or some other such ratio. If the APL were set at 85%, with maximum output of the machine at 100%, $s = 0.85$. The participation ratio = 3. At 100% productivity, the earnings potential of the plan, following equation (7), would be

$$y_w = 1 + p\left(\frac{x}{s} - 1\right)$$
$$= 1 + 3(1/0.85 - 1)$$
$$= 1.529 = 52.9\%$$

Case 4. Incentive Beginning at Lower than 100% Productivity but with 100% Participation. This plan in reality is the plan described in case 2 with a reduced normal (APL). Such situations arise when management decides that the normal (APL) established at a given incentive expectancy is too high with respect to preincentive productivity. The various levels shown in Table 2-3 fit this case. Suppose management decided initially that 25% incentive expectancy was appropriate for its plant. Normal (APL) = 100% would then be set, using Table 2-3 as an example, at 80 pieces per hour. With preincentive production at 62.5, the employees would have to increase their output by 28% to reach normal.

If the productivity gap is too wide to be bridged under the circumstances, management can reduce the gap by reducing the normal (APL) to any other lower level shown in Table 2-3. This increases the incentive expectancy of the plan and preserves meaningful guidelines. Some engineers suggest that the time standards not be changed but that the incentive plan start paying off at 90%, 80%, or the like, rather than at 100%. But if this done, standards set at 100% are misleading with respect to incentive potential when incentive payments start at 80%. What is the incentive expectancy of the incentive plan: the spread from 100% to incentive pace or from 80% to incentive pace? For example, assume the plan provides that standard = 100%, incentive expectancy = 30%, with incentive payments to start at 80% of standard with 100% participation. In effect, normal (APL) has been set at 80%. Although the plan is stated as providing 30% incentive expectancy from standard to incentive pace, the true expectancy is $(1.3 - 0.80)/0.80 = 62.5\%$. This use of standard = 100% as the measurement base for time standards serves no useful purpose. When an employee in such a plant works at 85%, he is above normal. But 85% sounds substandard. When the employee is at 100%, he has exceeded normal by $0.20/0.80 = 25\%$, but 100% seems to denote normal. Establishing normal (APL) = 100% as the measurement base avoids these misleading indices and permits the incentive expectancy to be clearly expressed. As discussed earlier in this chapter, preserving the incentive expectancy will help avoid runaway incentive earnings and deterioration of the incentive plan.

Case 5. Combinations of Incentive Plans. Varied combinations can be devised, using cases 2 and 3 as the basis. Case 2 is the basic one-for-one plan. Varied incentive expectancies can be built into the plan, as shown in Table 2-3. The slope of the wage payment line can also be decreased or increased to establish sharing plans, as shown in case 3.

Where preincentive productivity is low in relation to a one-for-one plan with a 30% expectancy, a sharing plan can be established starting at present productivity as normal (APL). Selecting an appropriate participation ratio, a sharing plan can be developed which will turn into a one-for-one plan at a productivity level where the earnings of both plans are equal. This technique is described in detail below.

Comparison of Different Incentive Plans. To compare incentive plans which have different incentive expectancies, use incentive pace (MPL) = 100% as the measurement base. The earnings above normal for such plans are obtained from equation (7):

$$y_w = 1 + p \left(\frac{x}{s} - 1 \right)$$

To determine the productivity level at which two plans, A and B, have equal earnings, equate the above equation for each plan:

$$1 + p_a \left(\frac{x}{s_a} - 1 \right) = 1 + p_b \left(\frac{x}{s_b} - 1 \right)$$

This reduces to

$$x = \frac{p_a - p_b}{p_a/s_a - p_b/s_b} \tag{10}$$

Compare two plans with the following specifications; note that both plans are set to incentive pace (MPL) = 100% as the common point.

	Symbol	Plan A, one for one	Plan B, sharing
Incentive pace................	...	1.00	1.00
Normal (APL)................	s	.80	0.667
Incentive expectancy..........	...	25 %	25 %
Participation ratio............	p	1.00	0.50
Base rates...................	...	Same under both plans	

Comparison of plans A and B: to determine the incentive earnings level at which plans A and B are the same, use equation (10) above.

$$x = \frac{1.00 - 0.50}{1.00/0.80 - 0.50/0.667} = \frac{0.50}{0.50} = 1.00 = 100\%$$

Plans A and B will yield the same incentive earnings at 1.00, which is incentive pace = 100%.

Plans A and B are very different. Plan A is a relatively tight one-for-one plan, starting incentive payment at 80% of incentive pace with 25% expectancy. Plan B pays the same incentive pay at incentive pace as plan A, but it starts to pay at 66.7% of incentive pace and at 50% participation.

Suppose management wants to design a plan C to fit in between A and B, with the following features: incentive expectancy = 30%, so $s = 0.714$;[20] participation ratio = 75%. At what productivity level will plans A and C break even in earnings? Using equation (10),

$$x = \frac{1.00 - 0.75}{1.00/0.80 - 0.75/0.714} = \frac{0.25}{0.20} = 1.25 = 125\%$$

The basic characteristics of the three plans compare as follows:

	A	B	C
1. With MPL = 100%, plans start payment at [s]	80%	66.7%	71.4%
2. Incentive expectancy	25%	25%	30%
3. Participation ratio [p]	100%	50%	75%

4. Break-even in earnings	
A with B	At 100% MPL
A with C	At 125% MPL

To compare costs at any level, use the appropriate cost equation for the plan. Inspection of the wage lines on Figure 2-4 shows that below 100% MPL plan B is more liberal to the employees than plan A. Above 100%, the reverse occurs. Plan C is more liberal than plan A for all levels below 125% MPL. To compare costs at any level, use the appropriate cost equation.

Combination Plans. Plans A and B or A and C can be combined if desired to provide a sharing plan at lower levels of productivity to bridge wide differences in productivity between preincentive levels and incentive standards. Then, after reaching the break-even point between the plans, all productivity above break-even can be paid at one for one, which is plan A. Figure 2-4 shows a graphic comparison of plans A, B, and C in relation to each other.

Relation of Incentive Potential to Effective Motivation. The motivating influence of incentives varies with the individual employee and the circumstances. Employees who are a secondary income source for the family generally will not be motivated to as high a degree as employees who are the sole income source. Other factors will also affect the degree to which workers extend their efforts, as for example, their confidence in management's integrity not to cut standards should earnings rise abnormally high, their regard for job security matters, and other such things.

There are no hard and fast rules as to the ideal relation of the incentive potential

[20] s is calculated as follows. Using equation (7),

$$y_w = 1 + p\left(\frac{x}{s} - 1\right)$$

$$s = \frac{px}{y_w - 1 + p}$$

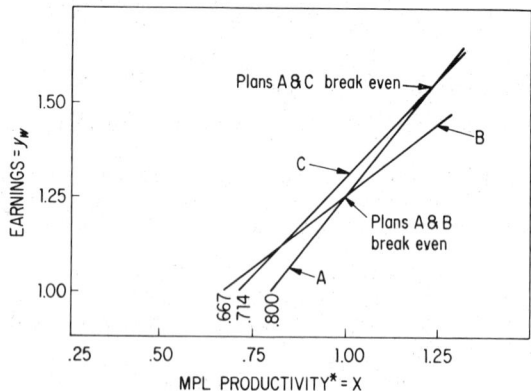

*Note: Because plans are compared at MPL-100%, x in this graph
is not at the same base as Figures 2-1, 2-2, and 2-3,
which are at APL = 100%.

Fig. 2-4.   Comparison of incentive plans A, B, and C, showing earnings at varying
levels of productivity.

in percent over base pay. There are, however, numerous experiences in plants around the world on this question. During the 1930s, an incentive potential of from 10 to 15% was often considered adequate by management. With the growth of unions and recovery from depression days, at the beginning of World War II the level in industry was generally from 20 to 25%. During World War II, all new wage incentive plans and changes had to be approved by the War Labor Board, which ruled that 30% incentive potential was fair and equitable. Union and management representatives now generally agree that the amount of the incentive potential should be established through collective bargaining, in the same manner as other working conditions are decided. Thirty percent incentive potential is probably the most widely used figure in American industry.

Experiences in other countries parallel those in this country. Reports of the operation of wage incentives in the Soviet Union show that 30% is frequently the potential incorporated into its incentive plans. Similar figures appear in reports from the other communist countries. When Communist China established incentives, they were patterned after the Russian incentive programs, with 25% as the incentive goal.[21] The recent cultural revolution policies in China have opposed financial incentives in favor of nonmaterial rewards.

From numerous experiences in this country and abroad, it is found that as the incentive potential diminishes, greater numbers of employees lose interest in the plan. But the converse holds true only to about 30%. Increasing the potential above 30% does not appreciably motivate employees to increase their physical efforts over what would have been exerted under a 30% plan.

The Pros and Cons of a Ceiling on Incentive Earnings. Practically all incentive plans in this country and abroad operate on the principle of no limit on incentive earnings for two reasons: to encourage high output and to ensure employees against rate cutting. In the early development of incentive plans, particularly before the advent of trade unions, it was not uncommon for management to reduce piecework rates when in its opinion employee earnings had risen beyond what it deemed reasonable. Employee opposition to this practice led to the principle that rates would be changed only in proportion to changes in methods and that there would be no ceiling on earnings.

[21] Charles Hoffman, *Work Incentive Practices and Policies in the People's Republic of China, 1953 to 1965*, State University of New York Press, New York, 1967.

The assurances given to employees against incentive abuses were warranted and necessary. However, the guarantees are often widely misunderstood by managers and employees. When employees are told that "the sky is the limit" on incentive earnings and that their earnings are limited solely by their skill and application, they take this literally. But this proposition is patently not true by definition or in fact. When a time standard is established with a definite relationship between normal and incentive pace, this sets a limit range. For example, if the standard is set to yield 30% incentive pay at incentive pace, it is expected that a normally proficient employee working with incentive effort on a manual operation will earn 30%. With extra effort he may earn 40%, 50%, and perhaps more. If machine paced elements are involved in the task, the limit will probably be lower. In every situation, there is a limit based on an employee's capabilities. Many wage incentive agreements provide guides to incentive earnings opportunities which also serve as limits. Even where the guides are not clearly specified, management and employees usually have an understanding on incentive potential. There are limiting factors which completely negate the literal interpretation of no limit on earnings.

The concept of no limit on earnings has caused far more difficulty in the administration of incentive plans than is generally recognized. An employee on incentive is entitled to increased earnings resulting from increased productivity. Employees often interpret increased productivity as increased physical output and believe that they are entitled to the fruits not only of their physical efforts but also of their ingenuity in developing better methods. Employees righteously point to the extra pieces produced as resulting from their efforts and at no extra cost to management. However, productivity in the work measurement view is a measure of output in relation to the accomplishment of a given task employing specific methods, procedures, materials, equipment, and the like. Because this definition is contrary to employees' views, or to management's stipulation that there is no limit to earnings, employees will surreptitiously employ ingenuity to change methods to increase their incentive pay further.

When employees institute methods changes and thereby increase their output, this destroys the relationship between earnings on comparable jobs and causes serious differences between management and the employees. The earnings level achieved on an operation where methods changes have been made will then serve as the bench mark for judging the equity of earnings on other operations. Where comparable earnings are then not possible, because an employee may not be as ingenious or the potential for changes does not exist, the result will be employee dissatisfaction. Pressure is then placed on management to increase the earnings potential of other jobs, thereby eroding the incentive agreement.

The concept of no limit on incentive earnings also provides employees with the incentive to employ proscribed practices and various tactics labeled on the plant floor as "making money with a pencil." This adds to cost, as differentiated from increased earnings resulting from changes in methods which permit increased output.

The game relationship in the plant between management and the employees is further intensified by the no ceiling principle. Employees can increase their earnings and reduce their work loads by beating the game. Management is then encouraged to employ methods to reduce excessive earnings. Inevitably, this results in a struggle between management and the employees for control of the incentive plan.

There is much to be gained when employees and management establish a ceiling on earnings. The elimination of the no limit principle will remove the incentive for surreptitious earnings and reduce some of the tensions brought on by the incentive plan. When a ceiling is established, the work measurement and wage incentive principles and practices must be fully detailed.

Some managers and engineers may argue that the ceiling on earnings may become a cover-up for poor work measurement and management practices. This may occur, but where unsound practices are employed, management will suffer under either policy regarding earnings. The important point is that the ceiling removes the evils and pressures brought on by the "sky is the limit" concept. It then permits management and the employees to develop and maintain equitable earnings pursuant to their agreement.

Individual versus Group Incentives. Where operations can be performed on an individual basis, that is, where one operator's productivity does not depend upon the adjoining operations, it is usually found that the overall productivity of a group will be higher when they are on individual rather than on group incentive. In the operation of a group, there is a leveling effect in which the best operators may not work as hard for the group as they would for themselves. On the other hand, the group will raise the level of the lowest workers through the effect of the group's persuasion, the morale of the group, attention by better workers to lower skilled workers, and the like.

There are, however, compensating advantages to the group over individual incentives. With the group, reporting becomes simplified, because one report takes care of the entire group. There is less need to determine which operator worked on which operation, because the total output of the group is used for measurement against total hours input.

Some forms of work can be measured only by the group, as for example, the output of an assembly line or a crew working on a single piece of equipment. In such cases, individual measurement has no meaning. However, in computing the group standard, studies should be made of each individual operation so that an effective balance can be attained between the operators on the line or in the group and to permit setting standards based on the group bottleneck operations.

The group system of wage payment is discussed in more detail in Chapter 4 of this section.

Plant-wide Sharing Plans: Scanlon, Kaiser Steel, Eddy-Rucker-Nickels. Judging from incentive practices in industry in this country, plant-wide plans are not as desirable as conventional incentives in the view of managers and employee representatives. Although plant-wide plans present potentials for major breakthroughs in productivity increases and in improved labor-management relations, such plans are seldom adopted.

Plant-wide plans should not be casually disregarded, because in some instances the plant-wide approach offers substantial advantages over conventional individual or small-group incentives. A major advantage is that the entire work force is involved as against about half in many conventional plans. With increased mechanization, machine operators become machine tenders, and productivity is in the hands of setup and maintenance employees, who are often omitted from conventional incentives.

When plant-wide plans are established, the first several years usually show substantial gains. But then invariably, productivity sharing declines, and both employees and management lose interest. A major obstacle to using the plant-wide plans is the problem of how to measure productivity and especially how to adjust for technological changes and the introduction of new products. These problems often cause demise of the plan.

The concept of union-management cooperation toward increased productivity and the sharing of the benefits was promoted by Joseph N. Scanlon, formerly research and engineering director of the United Steelworkers of America and later professor at Massachusetts Institute of Technology. At the end of World War II, some companies saw merit in the cooperative approach and established Scanlon-type sharing plans. After Scanlon's death in 1956, Frederick G. Lesieur carried on his work in promoting and developing Scanlon plans.

The Scanlon plan is described in a book of essays by people with experience with the plan. Douglas McGregor's "The Scanlon Plan through a Psychologist's Eyes" captures the potentials of the plan.[22]

Lesieur reports that when the above book was published there were about

---

[22] Frederick G. Lesieur, *The Scanlon Plan: A Frontier in Labor-Management Cooperation*, The M.I.T. Press, Cambridge, Mass., 1958.

50 to 60 plants using the plan. In 1968, ten years later, there were about 130 plants using a Scanlon plan. In addition, there were from 300 to 500 plants using some version of the plan.[23] More recent versions of the Scanlon plan and several case histories are presented in an article.[24]

The Kaiser Steel Corporation and the United Steelworkers established a plant-wide sharing plan in 1962 to replace the deteriorated and outdated conventional incentives then in effect for 6,000 workers at the Fontana, California, plant. Through an agreement containing a substantial cash buy-out, employees were given a choice of whether to remain on their old incentives or to change over to the plant-wide plan. Reports of the first two years showed $20.9 million shared, with $7.6 million going to the employees. Total average bonuses each year per employee were $1,213, ranging from 8 to 28% of base pay by month.[25] Declining bonuses in subsequent years forced management and the union to negotiate changes. In January, 1968, the plan was extended for an additional four years. Reports on operation of the plan have been scanty. Neither industry nor unions have apparently been impressed by the results obtained by the plan.

The Eddy-Rucker-Nickels plan[26] is a plant-wide plan measuring productivity on the concept of "value added," defined as the sales value of production less purchased materials. The merit of this concept derives from cooperation by employees and management to reduce the cost of converting materials to finished product and so share in the savings. But as with other plant-wide plans, after these plans have been in effect, stresses develop from changes in production technology, product mix, plant additions, and the like, and the effectiveness of the plans declines.

Name Plans. During the 1920s and 1930s, a number of incentive plans were established and identified by the innovator's name. All the plans are some variation of case 3 described above.

Bedaux. The usual installation has 75% participation to the employees and 25% to supervision. Standards are expressed as "B's," with 100% set at a 60B hour. 80B represents 133% productivity.

Halsey. Typically, the Halsey plans are sharing plans with normal at 100% with less than 100% participation.

Haynes. The participation is the same as Bedaux. The standards are expressed as "manits," abbreviated from man-minute. Standard is 60, corresponding to 100% on the conventional scale.

Gantt. Incentive starts at normal = 100% with less than 100% participation. But a step increase in base rate is established at 100% to provide extra motivation for employees to reach and exceed 100%. Below 100%, the employee's base rate is guaranteed.

Incentives for Increased Yield. Where yield is controllable by the employees, it can serve as the basis for incentives, often with excellent results. Simply stated, yield is the ratio of units of output related to units of material input. For example, in the production of steel, yield is the ratio of pounds of finished production related to pounds of input materials, compared with standards established for base periods. In wood manufacturing, yield is units of finished product related to board feet input. In process industries, yield is the ratio of units of output to units of input, based on material costs or other criteria.

Despite the diligence of management and supervisors in operating a plant, when employees have a personal stake in increasing yield, somehow the yield manages to increase. When the value of the increased yield is sufficiently high, it can

[23] Frederick G. Lesieur and Elbridge S. Puckett, "The Scanlon Plan: Past, Present, and Future," a paper presented at the Annual Meetings of the Applied Science Associations, Chicago, December, 1968.

[24] Frederick G. Lesieur and Elbridge S. Puckett, "The Scanlon Plan Has Proved Itself," *Harvard Business Review,* September–October, 1969.

[25] "The Kaiser Sharing Plan's Second Year," *The Conference Board Record,* National Industrial Conference Board, New York, July, 1965.

[26] Allen W. Rucker, *Gearing Wages to Productivity,* The Eddy-Rucker-Nickels Company, Inc., Cambridge, Mass., 1962.

serve as the basis for a sound incentive plan. As with wage incentive plans based on productivity, incentive plans for increased yield must be based on soundly conceived standards which reflect increased yield due to efforts of the employees.

Incentives for Nonproductive Employees. The term "nonproductive" is often a misnomer in many plants, particularly with increased mechanization where maintenance mechanics and other skilled crafts are more important to continuing production than employees tending the machines. However, because the operators are closely related to the equipment, and the so-called nonproductive workers service a number of machines, nonproductive employees are often omitted from consideration in the incentive plan. When some employees earn extra pay, even where increased efforts are required, employees who do not have an opportunity for increased pay become dissatisfied and inequities arise.

Some managers take the easy way out and pay nonproductive employees such as material handlers, setup men, and other such classifications the average of their department or group. This simple expedient may satisfy the employees involved, but it may not be equitable to management under varying conditions. Some companies consider receiving, storekeeping, shipping, and material handling as measurable and establish incentives as they do for production operations. A growing number of companies establish incentives for difficult-to-measure areas such as maintenance.

The sound approach is taken by managers who follow the principle that increased pay is earned for increased output when measured by suitable criteria. Tying nonproductive to a department may not fully measure their contribution. Sounder control may be obtained by establishing standards consisting of ratios of nonproductive to productive hours, or nonproductive hours per unit of product, or other such variations. The fundamentals of indirect labor measurement are discussed in some detail in Chapter 3 of Section 4.

Payment for Employee Improvements. Management broadly accepts the principle that employee suggestions should be rewarded. But employee suggestion plans often tap only a small portion of the potential which exists in every plant. Employees often do better by not turning in a suggestion to management because they can use the improvement to their own gain under an incentive plan or utilize the improvement to gain additional idle time during the day.

The proposal contained in the discussion, Preventing Creeping Changes, is an approach which can easily be put into effect in any plant. If an incentive plan is in operation, the value of an employee's suggestions can readily be calculated as the difference between the old standard and the newly established standard, multiplied by the total pieces to be produced over a given period of time. Several approaches can be taken to paying for such improvements. The employee suggesting the change can receive 50% or any other percentage of the value of the savings for a year or more. When other employees are also affected by the change but are not involved in innovating it, one approach might be to give the innovator of the idea 50% of the reward, with the balance to be spread equally in his department, machine group, or other natural work unit. Or some portion might go into a plant-wide pool to be distributed semiannually or annually.

This approach of passing along substantial cash rewards for new ideas will motivate employees to exercise their ingenuity primarily to benefit themselves and at the same time reduce costs. Where management attempts to recoup such changes to its sole advantage, changing the time standards to obtain increased output, it will create considerable antagonism among the employees.

INSTALLING AN INCENTIVE PLAN

An incentive plan should be established only when management is convinced that the plan is warranted and will accomplish desired objectives. The plan should be carefully evaluated for the effects on all aspects of the company's operations and should include consideration of the following:

1. Can productivity be effectively measured?
2. What is the value of the increased productivity which the incentives will

encourage, considering the value of labor separately from the costs of operating the facilities generally included as overhead?

3. How effectively does management control its operations?
4. To what extent can the employees affect productivity?
5. Can management increase productivity by improved supervision without incentives? What is the difference between this increase and that which will be obtained through incentives?
6. What extra costs will be incurred in operating the incentive plan, including the need for clerical and engineering staff? (Costs should not be charged to operating the incentives that would normally be involved under measured daywork or that are presently needed.)
7. What is the value to management of the increased take-home pay which the employees will earn through their own efforts? For example, 30% increased earnings on a $3 base rate is equal to $0.90 per hour. How much better off is management when its employees take home $3.90 per hour instead of $3, particularly in a tight labor market?
8. To what extent will there be less pressure by the employees for wage increases when they earn extra pay under incentives, as compared with measured daywork?
9. What effect will incentives have on labor-management relations? In some ways, incentives cause stresses. In other ways, especially because employees willingly increase their productivity and management is relieved of having to apply pressure, incentives may improve relations.

Managers should carefully examine all factors which will affect operations. All too often, increased output and reduced costs are the major considerations in installing an incentive plan, without weighing the overall effect of the incentives. Incentive plans are more easily established than discontinued. If an incentive plan fails because it has deteriorated while the employees' earnings have increased substantially, the cost of discontinuing the plan may be very high, as discussed previously. This factor should not be minimized.

Installation Steps. The most important initial step in installing an incentive plan is obtaining competent personnel who will design and install the incentive plan to suit the conditions at the company. Most of the difficulties encountered with incentives arise through faulty management decisions during the design and installation of the incentive plan. Very rarely do difficulties originate from pressures by the employees.

The following are the principal steps involved in the installation of an incentive plan.

Preliminary Discussions with Employees. Apprise the employees of management's plans and solicit their opinions; communications are extremely important, particularly at the beginning.

Draw Up Incentive Agreement. Prepare a detailed agreement specifying protection for employees and management; spell out incentive potential, definitions of standards, how standards will be changed, basic guarantees, and other necessary provisions.

Eliminate Wage Rate Inequities. Incentives must never be used to eliminate inequities; where inequities exist, these will be further aggravated by the incentives.

Timing of Installation. The installation should preferably be timed at the beginning of the season, to take advantage of low employment levels in the plant. Where incentives cause labor displacement, use attrition to prevent layoffs. Unions will generally oppose layoffs, but will go along with attrition.

Installation of Plan. Preferably, start with a department at the beginning of the production process, to avoid running out of work. Balance employees against actual needs as productivity rises. The engineers must work closely with the areas in which incentives are initially installed, especially to take care of employee problems.

Schedule Closely. Production scheduling is critical, particularly in the initial stages of the incentive installation, when the productivity balance of the plant will be disrupted as part of the plant is on incentives and other parts are on daywork. It is vital that departments going on incentive do not run out of work.

Monitor Installation. The engineers must closely watch productivity of all employees, day by day. Attention must be paid to areas of low productivity to determine causes. Remedial action and employee training must be used to raise low productivity.

Handle Grievances Promptly. Employee complaints and questions should be answered without delay, particularly at the beginning. Employees' grievances regarding time standards should be looked into promptly so that employees do not become discouraged. Disciplinary measures should be avoided, even where warranted. The overall effort should be to raise productivity of all employees in a positive and constructive manner.

Supervisor Training. Train supervisors thoroughly in the design and operation of incentives, how time standards are established, and how to handle employees' grievances and other problems that will arise. This is a very important part of the installation. Completely spell out the role of the supervisor in the operation of the incentives.

Prepare Wage Incentive Manual. Detail the entire plan, how it works, work measurement practices, and other necessary instructions. Make copies available to the employees.

Need for Disclosure of Data. There is a twofold need to make available to the employees all data and information regarding the incentive plan and time standards: (1) the National Labor Relations Board and the courts have clearly established the principle of disclosure; (2) secrecy breeds suspicion and doubts concerning management's objectives.

The NLRB and the courts have broadly upheld the employees' right to all material relative to the establishment of hourly rates of pay; the data and studies used to establish time standards; and manuals, instructions, and procedures used in the development of time standards and the administration of incentive plans.

Experience shows that full disclosure of data engenders an understanding attitude by the employees which reduces grievances. Some managers fear that disclosure provides the employees with information to counter management and so to loosen standards. Where managers play games with the employees, problems are certain to arise. Management usually comes out second best in such contests.

Incentive Payment Period. The period for the measurement of productivity should be as short as possible. It should be based on the cycle time of the operations performed. For example, if a large assembly takes three or four days to complete, then measurement by the week may be meaningless. In this case, measurement should be based on two- or three-week periods. Where the cycle times are short, as for example, several hundred pieces per hour, then measurement by the day is most desirable.

When measuring productivity by the day and paying incentive earnings by the week, it is often advisable that earnings should be guaranteed by the day and not averaged over the week. What happens is this: when an employee has one or more low productivity days at the beginning of the week and productivity is averaged by the week, the employee will sometimes not try to make up the loss during the rest of the week, and the entire week will suffer. Both the employee and management lose. Where the employee has several good days and then suffers low productivity toward the end of the week, wiping out gains made on previous days, it will affect employee morale. It is found that employees will attempt to maintain a fairly level output which they deem within their capabilities and desire for increased income. Barring problems associated with the work or for personal reasons, once employees are on incentive, they attempt to maintain their output levels. From the viewpoint of the employee and management, when an employee has low productivity, it should not count against incentive pay earned on other days, regardless of the cause of the drop in productivity.

Where the operations have long cycle times, weekly measurement with a moving average will level out peaks and valleys due to the nature of the work performed.

The moving average is calculated as follows. Assume the following are the weekly productivities for five successive weeks: 130%, 85%, 120%, 105%, 90%. If a three-week moving average is used, the average of the first three weeks is 335/3 = 112%. On the fourth week, drop the first week of 130% and add the fourth week of 105%, which gives a total of 310/3 = 103%. On the next week, drop the second week of 85% and add the fifth week of 90%, which produces an average of 105%. Though the weekly averages fluctuate widely, the three-week moving averages dampen the swings. In using the moving average, the employees have to make up for productivity which is below 100% by the higher productivity in other weeks. But this may be equitable if the cycle time is greater than one week, because the weekly productivity may not truly represent the week.

Group incentive plans covering entire departments or plants usually work best by calculating productivity on a weekly or monthly basis. The moving average may also be used to help dampen wide swings in productivity.

Incentive Payments for Overtime. The Fair Labor Standards Act establishes requirements which must be followed in the payment of overtime. The provisions applying to incentive pay are explained in Interpretative Bulletin Title 29, Part 778, of the Code of Federal Regulations.

A major point in the regulations is that time and a half must be paid for all hours exceeding forty hours during the week, based on average hourly earnings, which include incentive earnings. These regulations are rigidly enforced.

Hourly Add-ons. The increases in base rate with standard hour plans are automatically calculated into employee incentive earnings because incentive payments are calculated on hours earned. Should a wage increase be granted which is not to be included in incentive payments, this is treated as an hourly add-on. In incentive payment with hourly add-ons, the calculations are made the same as with a standard hour plan but the hourly add-on is multiplied only by the total hours worked. The hourly add-on will effectively reduce a one-for-one standard hour plan to less than one for one. As the hourly add-on becomes larger, the employee receives proportionately less for each piece produced. This weakens the motivating power of the incentive plan. For example: the base rates of a plant twenty years ago may have been $1.50 per hour, with a one-for-one incentive plan. As wage increases were given over the years, these were not added to the incentive base rate. Twenty years later, incentives are still one for one on the original $1.50 per hour, but with $1.50 added per hour worked for increases during the twenty years. This effectively converts the one-for-one plan to 50% participation. The employees do not have incentive earnings opportunity on the $1.50 add-on. Though management appears to save the incentive earnings not paid on the hourly add-on, this may be illusory. The employees have their own concepts of equity, and they usually are sufficiently resourceful to find ways of increasing their incentive earnings to make up for the incentive earnings loss on the add-on.

Adding the add-on to the base rates in the above example will substantially increase costs. Before attempting a solution, management should thoroughly study the overall situation, including the soundness of the time standards and the incentive plan, to determine the extent to which looseness has developed over the years in all areas. Correction of the add-on inequity should then be made part of an overall program of correcting inequities.

CONCLUSION

To achieve high productivity, management must manage effectively and enhance the will of the employees to work. The will to work is enhanced by removing restraints by assuring employees' income and providing motivation.

The motivation of workers to raise their productivity can be approached in several ways. It is not so much the selection of a technique or an approach which determines the outcome, but the attitude of employees and management toward each other. Attitudes in the plant originate from and are nurtured by management-employee relations over a long period of time. The sum total of job experiences, income security, economics, and numerous other factors merge and are expressed in the employees' will to work.

There is general agreement that employees must be motivated to achieve higher levels of productivity, but there is a divergence of views on how to motivate. Some managers rely primarily on nonfinancial motivators, while others favor financial incentives. Everyone, worker or manager, strives to increase the material returns for his efforts. It is largely a question of what he gives up for what he receives. The workers' response to incentives is a combination of many factors, of which money is only one, but it is pointless to argue whether money is first or last in the workers' view.

Wage incentives are not as simple as implied in the proposition "extra pay for extra work." Successful operation of incentives requires that managers fully understand their responsibilities and apply the efforts and resources necessary for the task.

Deficiencies in managing are the primary cause of incentive degeneration. Often, the pressures of workers to earn more for less effort are erroneously cited as the major cause of management's problems. This failure by managers to recognize their shortcomings clearly blinds some managers to the solutions of their problems. Nothing can take the place of good supervision. Increased motivation of workers cannot make up for ineffective management.

BIBLIOGRAPHY

Barnes, Ralph M., *Motion and Time Study: Design and Measurement of Work,* 6th ed., John Wiley & Sons, Inc., New York, 1968.
Carroll, Phil, *Better Wage Incentives,* McGraw-Hill Book Company, New York, 1957.
Louden, J. Keith, and J. Wayne Deegan, *Wage Incentives,* 2d ed., John Wiley & Sons, Inc., New York, 1959.
Mundel, Marvin E., *Motion and Time Study: Principles and Practice,* 4th ed., Prentice-Hall, Inc., Englewood Cliffs, N.J., 1970.

Chapter 3

Measured Daywork

P. D. O'DONNELL

Director, Headquarters Manufacturing Planning and Controls,
Westinghouse Electric Corporation, Pittsburgh, Pennsylvania

The word "daywork" carries with it the clear implication of an absence of wage incentives. The term "measured daywork" was applied in previous years to a form of wage incentives in which the base rate was raised or lowered in steps at specified intervals, depending on performance during the preceding interval. "Step incentive applied quarterly" (or monthly or whatever the period may be) is suggested as a more descriptive and acceptable name for such wage incentive systems. The term "measured daywork" can only be confusing when applied in such cases, partly because it is incorrect in implication and partly because the term has broader usage in correctly categorizing the method of control which this chapter describes.

Following a definition of measured daywork, this chapter will touch on some of the concepts that are considered important in making it an effective management tool. The broad scope of this method of control will be emphasized. Some of the techniques normally involved in its application will be discussed. Its potential advantages will be listed, followed by a summary of the operating principles which should be observed if a successful measured daywork plan is to be installed and properly administered.

DEFINITION

Measured daywork is a form of managerial control involving (1) the development of time standards for the performance of productive operations; (2) regular comparisons between elapsed hours and standard hours; and (3) an hourly rate of

6-51

pay that is not raised or lowered by the level of performance. Although the principle of "a fair day's work for a fair day's pay" should be recognized and strongly encouraged by management, it is fundamental that the measured daywork system does not include incentive features based on either wage bonuses or elapsed work time.

The Time Standards. Time standards are derived through the application of work measurement procedures, including the use of time study or a predetermined motion time system such as MTM and the use of standard data or formulas. The target standards should represent what is considered a fair day's work under good working conditions.

Importance of Line and Staff Management Functions. The manner in which unit supervisors discharge their responsibilities becomes of prime importance in achieving desired results with measured daywork. Of equal importance is the provision of adequate staff assistance so that the supervisors will be free to devote their time largely to direct supervision. It is the function of plant or departmental management to see that shop conditions are maintained that will make it possible for unit supervisors to achieve their objectives. Performance reports at regular intervals are relied on to show where attention and follow-up are required.

GENERAL CONCEPTS

An understanding of the measured daywork concepts can be obtained by considering how worker earnings and management controls under this system compare with earnings and controls under (1) unmeasured daywork and (2) typical wage incentive plans.

Characteristics of Unmeasured Daywork. In the daywork form of wage payment—measured or unmeasured—hourly worker income is predetermined rather than fluctuating. This eliminates a source of potential labor problems concerning wages that is inherent in wage incentives.

A characteristic of unmeasured daywork is that the level of worker performance in a shop section is not determined factually. The costs of specific jobs may also be unknown. Even with a system that accumulates job costs, it is possible only to compare costs on one job with the costs of others that may be similar, and such comparisons are normally possible only after the work has been done and the product shipped. If a new job must be handled for which past performance history is lacking, there is little basis for planning the completion date or for scheduling the work in the shop.

Daywork without performance standards leaves much to be desired because of this lack of a basis for planning and control. This is true not only for direct operations, but for indirect work as well. The need for better controls is evident from the efforts of many companies to apply work standards or estimates in the planning and control of traditionally unmeasured operations such as maintenance and office jobs.

Characteristics of Incentives. A wage incentive system differs from a daywork system in the areas of operator earnings and management control capabilities. From the operators' standpoint, the essential feature is that earnings depend on productive output as determined by time standards that have been established by representatives of management. Thus, there tend to be frequent "discussions" among the operators and supervisors and industrial engineers about how standards are established and when they should be changed.

From management's standpoint, the periodic reports of performance by operators and by shop sections can serve as an effective way to control labor costs. The time standards in wage incentives provide a reliable basis for planning and auxiliary controls.

Characteristics of Measured Daywork. Measured daywork is identical with unmeasured daywork in that worker income is based on a fixed hourly rate. It is like an incentive system in that time standards are available for planning and cost control. It is unlike a carefully applied wage incentive plan in that operators do not tend to have a high level of interest in keeping productively occupied or in delivering a fair day's work. It is essential in the administration of measured daywork that line and staff management take a direct and active part in controlling the availability of material and shop equipment so that operators can be expected to attain the established performance standards.

DUTIES AND RESPONSIBILITIES OF PRODUCTIVE EMPLOYEES, UNIT SUPERVISORS, AND INDUSTRIAL ENGINEERS

An understanding of the characteristics of measured daywork may be gained by considering the duties and responsibilities of those who are affected by it.

Productive Employees. Through selection and training, employees should be capable of performing their assigned work in a proficient manner. Except for the usual differences in rates of pay to account for job classifications, shift and merit differentials, or overtime, there are no variations in hourly rates. The rate of pay will be fair for the job performed, relative to other jobs in the plant, when a proper job evaluation system is in effect. If the rates established are at least equal to the average area rates paid for comparable jobs, there will be a basis for consistently obtaining qualified operators.

In return for this rate of pay, employees are expected to remain at work during the full shift (less allowed personal time), use prescribed methods and facilities, and meet quality requirements. The time standards are based on a pace which is ordinarily expected of a qualified workman under capable supervision. If conveyor lines are utilized, their established rate of movement may serve to indicate the standard level of production. In other cases, standards are generally made known to operators so that they will understand what is considered a fair day's work.

The performance of new operators in relation to standard can properly be checked during a probationary period as a means of evaluating their qualifications. Thereafter, when operators are not performing satisfactorily, they should be informed of specific areas where improvement is required.

Unit Supervisors. The number of employees one supervisor can handle adequately under measured daywork will vary according to area covered, type of work, and other factors. This number commonly falls in the range of about fifteen to forty people. It is desirable that supervision by a management representative be on a direct basis, without group leaders, leadmen, or assistant foremen acting as barriers to effective communication. A unit supervisor's section may include setup men, production control expediters, material handlers, and other indirect personnel as required. There may be staff personnel assigned to the section, such as industrial engineers. The unit supervisor, however, is expected to manage his section in all respects. This includes the following partial list of duties:

> Final scheduling of work to be done
> Screening of new operators
> Giving work assignments
> Checking the level of performance
> Controlling quality
> Ensuring that operators are trained to follow correct methods
> Preventing delays from lack of material or tools
> Controlling indirect expenses
> Ensuring good housekeeping
> Handling employee relations

The unit supervisor is responsible for correcting situations where an employee consistently fails to meet time standards. To know where such action is required for conveyor operations may be merely a matter of regular observation. On individual repetitive operations, it may be sufficient to observe whether operators remain on the job and follow the correct method at a suitable work pace. For longer cycle jobs, however, it may be desirable to establish systematic data collection in some form. This may be through electronic input stations, with provision for supervisors within sections to obtain job performance data as required. Or, simple shop records may be established on which stop and start times are entered.

If action is needed to correct poor performance, the supervisor, as a first step, is expected to determine the reasons for not meeting standards. Such reasons often point to delays beyond the control of the operator, such as material not available when needed, material abnormally defective, tools or equipment in poor condition, or lack of balance between operations. The supervisor is responsible for correcting such conditions, using staff assistance as required.

At other times, however, the operator may be at fault. Improper handling or operating methods, including wrong machine cutting speeds or feeds, will cause substandard performance. Training and closer supervision are called for in such cases. When delays or incorrect methods are ruled out as causes of poor performance, the supervisor may determine through observation that specific operators are failing to work effectively. The supervisor will be expected to follow up such situations until they are corrected. In measured daywork, all should understand that a fair day's work will be a condition of continued employment.

Industrial Engineers. The first major function of industrial engineers in a measured daywork system is to establish how a job is to be performed. The usual routine is for drawings and basic manufacturing information to be transmitted to the industrial engineer for the preparation of routings. This may properly include the designation of machines; shop sections; operation sequence; speeds and feeds; tool, jig, or fixture requirements; labor grade required; and detailed methods descriptions. For short-cycle, high-volume operations, a methods description for use at each work station is desirable.

The next major function is to establish time standards and administer the reporting procedures, as described later.

The final major area of responsibility is to improve current work methods. Unless a sound job is done along this line, much of the potential advantage of measured daywork will not be realized. It is a good practice for industrial engineers to be assigned by specific plant areas—working closely with the supervisors involved to see that the layout includes provisions for sound workplace methods and that the prescribed methods are being followed.

ESTABLISHING STANDARDS AND REPORTING PERFORMANCE

A measured daywork program requires sound standards. They must be sound when first applied, and they must be revised as methods change.

Basis for Standard Level of Performance. Although a collection of words can never serve to convey a precise definition of a performance level, the following definition brings out an important point.

> A measured daywork performance standard is the time, *under good shop conditions,* which is determined to be necessary for a qualified workman, working at a pace which is ordinarily expected under capable supervision, to do a defined amount of work of specified quality when following the prescribed method.

The italicized words are intended to emphasize management's responsibilities for maintaining good shop conditions. The standards are used to evaluate sectional performance and should be designed accordingly. The working pace attainable by an operator will differ according to the amount of repetition which the work involves. This is because learning time is needed for an operator to develop rhythm and sureness of motion.[1]

The target standard in measured daywork has no real relationship to the 100 percent point in any wage incentive system. Because various wage incentive plans differ widely in their concept of the 100 percent pace, there is nothing to gain

---

[1] For a more complete discussion of this, see Chapter 1 of Section 4.

from trying to define expected measured daywork pace in terms of a wage incentive "normal." It is a function of plant management—through the industrial engineering staff—to develop a concept of 100 percent that correctly applies to the work at hand. Films of plant operations will be helpful in doing this. The following summary may also be helpful.

The expected measured daywork pace is not:

1. An easy-going, effortless work level
2. An excessively tiring work level
3. The top level at which a qualified operator can work
4. The same for low-volume work as for high-volume jobs
5. The same as 100 percent or "normal" for a specific wage incentive system

The expected measured daywork pace is:

1. A brisk performance level that shows reasonable effort
2. A level that will not involve undue physical fatigue while working a full shift, provided the operator is fitted to the job
3. The pace a qualified operator should be expected to average during a full work shift (less personal time), while meeting other requirements
4. A level that will vary according to the degree of repetition inherent in the work involved
5. A level that can be recognized, through training and experience

Any valid system of time study or predetermined motion times may be applied to establish measured daywork standards. Standard data should of course be used where applicable. There are several points to be considered, however: (1) performance rating methods used for time study should be consistent with the concept of 100 percent pace that is developed to fit work within the plant; (2) general allowances should be applied only for identifiable and expected conditions; and (3) predetermined time systems that define a wage incentive normal may require an adjustment factor for measured daywork target standards.

Possible Additional Uses of Standards. The performance standards that are essential in a measured daywork plan have many secondary uses for control purposes. Some of these follow.

Production Planning and Control Procedures. Reliable performance standards permit developing reliable production schedules. This helps ensure an uninterrupted flow of work from one plant section to another and helps meet delivery schedules with a minimum of lead time.

Manpower Planning and Control. Regulating manpower requirements to fit the anticipated load is a necessary preliminary step to securing a satisfactory level of performance. This is an integral part of the measured daywork program and can be done best when performance standards and other data are dependable.

Determination of Equipment and Plant Capacity. Capacity estimates are essential for advance production planning and for decisions about the purchase of specific equipment. The production standards provide a sound basis for making such estimates.

Standard Cost System of Accounting. Time standards provide the foundation for a standard cost system. They are essential for the kind of cost control discussed in Chapter 7 of Section 8.

Evaluation of Cost Reduction Possibilities. In many cases, the justification for new equipment, layout changes, and methods changes of various kinds involves making comparisons with existing standards. Systematic improvements in product design for more profitable production are facilitated by cost estimation through the use of standard data.

Developing Expense Budgets. Calculation of total standard hours planned for future work provides a reliable basis for flexible budgets covering the manufacturing overhead expense requirements.

Performance Reports. A major element of control in a measured daywork plant is obtained through the use of standards to measure the unit supervisor's effectiveness. The same data provide an indication of the effectiveness of employees in carrying out their assignments. Thus the performance records are used by the supervisors to help them manage the group they supervise. Periodic reports to management, either daily or weekly as justified, show the level of productivity for each department. These reports permit the supervisor to chart his own progress, determine where improvement should be made, and know how he stands with respect to established goals. The performance reports provide higher management with the same information for a larger section of the plant and provide a means of evaluating individual supervisors. They provide a basis for questioning substandard performance so that causes can be determined and corrected. The reports also provide staff departments, such as industrial engineering, with information to show where increased activity on their part is needed to assist the unit supervisors.

The performance report illustrated in Figure 3-1 stresses the performance ratios for specific units of the plant. The ratio is defined as standard hours for planned work divided by the difference between total elapsed hours and any hours for planned work not covered by standards. In practice, this is approximately the ratio of standard hours which can be sold to the customer divided by total hours worked. The illustrated report format shows the performance ratio goal which each unit supervisor is expected to achieve.

ADVANTAGES OF MEASURED DAYWORK

Promotes Increased Productivity. Measured daywork allows productivity to be increased more rapidly than is likely to be the case with other forms of wage payment. To understand the reasons for this, consider the ways in which plant productivity may be improved. This can be through increased worker effort or through the introduction of new methods and equipment. Manufacturing management cannot expect employees who are already exceeding standard to work harder each year than they did the last. Industrial engineers stress working "smarter, not harder." Productivity improvements that continue to occur year after year must come largely from improved equipment and methods rather than from increased worker effort.

In the first half of this century, there were opportunities within industry for operators to increase their output through extra effort. With increased mechanization, however, many of these opportunities have disappeared. Much of the equipment in modern manufacturing plants is either semiautomatic or conveyorized, with the cycle times largely controlled by the equipment rather than by the operators. Finding ways to improve methods and equipment still further is the sure road to continued productivity improvement. The wage plan should encourage this, not hinder it. Measured daywork can assist in creating a favorable climate for improving methods and equipment.

To retain this condition, those who are concerned with the administration of job classifications should follow closely the methods changes that are made. If there is a change in job content, a revision of the job classification may be justified. New classifications may be higher or lower. In the latter case, employee resistance should be anticipated and dealt with. Under measured daywork, there is increased pressure concerning job classifications, but this is far less than the pressure on time values under an incentive system.

Reasons for a more rapid increase of productivity with measured daywork than with other wage payment plans may be summarized as follows:

    1. Methods changes are easier to put into operation, resulting in a proportional increase in productivity.

UNIT SUPERVISOR AND UNIT NO.	NO. OF OPERATORS ON PROD. WORK	ELAPSED HOURS				STANDARD HOURS PRODUCED		PERFORMANCE RATIOS		
		TOTAL (B+C+D)	WITHOUT STANDARDS		WITH STANDARDS	PLANNED	UNPLANNED	OPERATORS (E+F)/D	MANAGEMENT	
			PLANNED	UNPLANNED					E/(A−B)	GOAL
		A	B	C	D	E	F	G	H	J
M1 − T. A. Roberts	25	980	45	36	899	785	60	94%	84%	88%
M2 − John Jones	29	1170	30	50	1090	1005	40	96%	88%	90%

FIG. 3-1. Weekly performance report format.

2. Industrial engineers can devote more time to methods instead of being burdened with the administration and grievances in an incentive system.
3. Constructive ideas from the workers themselves are more likely to be obtained through the plant suggestion system.

Aids Technological Progress. In an age of massive engineering, with frequent product innovations, manufacturing technology is increasing at a rapid rate. Processes used in many of the newer industries were unknown a few years ago, and similar changes in technology may be expected to continue in the future. Frequent changes in methods and equipment are required to permit building an acceptable, reliable product. Measured daywork aids this technological progress.

Provides Sound Basis for Supervision. The emphasis on direct control by supervisors, along with attention to the operator/supervisor ratio, places the supervisors in a position to manage their sections more effectively. Not only are fewer questions likely to occur on the troublesome subjects of standards, bonus earnings, and related topics, but supervisors are in a better position to resolve the inevitable questions that do occur on other subjects. With the line of communication from top management to operators clearly defined, management policies can be explained properly and questions are more likely to be settled by the supervisors without staff assistance.

Emphasis on Overall Productivity. The attention of management at all levels in measured daywork plants is directed toward evaluating and improving productivity. This covers more than the efficiency of the operators. Management attempts to minimize the repair work, machine downtime, and various other factors that also have a major influence on productivity.

More Practical Basis for Control of Indirect Work. With measured daywork, there is less difference between the cost control of direct work and approaches that may be taken by management to control indirect or service costs. In neither case are operators paid on an incentive basis. This similarity of approach provides a better climate for effective management of indirect and service personnel. The indirect work can more easily be covered by some form of performance evaluation.

Improves Labor Relations. The performance standards in measured daywork do not serve as a basis for incentive earnings. The failure to meet standards is not normally a direct cause for disciplinary action, because the supervisor is expected to look for the causes of poor performance when taking action. For these reasons, operators do not have impelling motives to question standards or argue about them. This tends to remove one of the handicaps to good labor relations.

Increases Flexibility of Operator Assignments. The fixed daywork rate prevents any lowering of income due to transfers within the same job classification. This provides a climate that is more conducive to the assignment of operators where they are needed. Better scheduling and equipment utilization can occur as a result.

Promotes Revision of Standards as Methods Change. The maintenance of time standards to keep the standards in line with methods is important. This is often difficult to accomplish, however, if the operators resist the changes that should be made. This does not tend to occur with measured daywork. Although changes of standards should be made concurrently with methods changes whenever possible, being able to revise them at any time is a definite advantage.

Simplifies Paper Work. Although adequate records should be maintained with measured daywork, as with any system involving comparisons between standards and actual times, payroll accounting becomes simpler and less costly when employees are all paid a daywork rate. There can therefore be some reduction in record keeping on the part of supervisors, time clerks, and industrial engineers.

OPERATING PRINCIPLES

To obtain the advantages that can occur with measured daywork, installation of the plan must be based on sound principles. Many of the detailed procedures can properly differ between plants because of management policies or inherent plant differences. The development of details, however, should be based on funda-mental concepts such as those listed below.

1. Qualified Unit Supervisors. Supervisors must be able to apply sound manage-ment techniques. A supervisor of incentive workers may "get by" while ignoring some of his responsibilities, but a measured daywork supervisor must function as the manager of his section in all respects.

2. Direct Supervisory Contact with Operators. The operators should report directly to the unit supervisor, without pseudo managers such as group leaders, leadmen, or assistants. The operators should be able to look to the supervisor for guidance and for the solution of work-related problems.

3. Limited Scope of Supervisory Coverage. In measured daywork, the super-visor must motivate his people to produce a fair day's work. He must also have the necessary time to follow up any production problems that occur and handle other required duties. A good job cannot be done if the supervisor is expected to handle an unreasonably large number of operators. Although the proper number will vary, depending on the nature of the work, size of the area, and other factors, a range of fifteen to forty employees has already been suggested.

4. Continuous Supervisory Coverage. To supervise properly and obtain a fair day's work, the unit supervisor should remain in his section during the normal working hours. His desk should be within the area. If he is required to leave for occasional meetings or for other reasons, a qualified substitute, such as an industrial engineer, should be assigned to manage during his absence.

5. Unit Supervisor Responsible for Work Habits of Operators. Acceptable work habits include applying a reasonable effort level, staying on the job during the shift interval (less allowed personal time), using prescribed methods and facilities, and meeting quality requirements.

6. Use of Time Standards by Unit Supervisor. Actual performance in rela-tion to the time standards should be used by the supervisor as a first step in evaluating performance. When specific individuals are not meeting the standard, the supervisor should determine the causes and take corrective action. Often the cause is a management-controlled problem. If the problem is with the individual worker, the supervisor should provide the necessary assistance, training, and instruc-tion. If disciplinary action becomes necessary, it should be based on the specific faults identified.

7. Management Evaluation of Sectional Productivity. Regular performance reports should provide an index of productivity within each section. This will reflect the influence of various factors under the supervisor's control, including the efficiency of the operators. A properly established goal should be indicated, which may be less than 100 percent if there are conditions such as equipment maintenance problems not subject to full control by the supervisor. When super-visors do not attain such goals, management should investigate the reasons carefully and take action as required.

8. Maintenance of Time Standards. Sound industrial engineering practices should be followed to establish reliable standards and revise them promptly when methods changes occur for any reason. An auditing procedure by industrial engi-neering is desirable to guard against the effect of creeping methods changes.

9. Content of Time Standards. The standards should include proper personal allowances but not allowances for items correctable by the supervisor. For example, if operators are forced to lose about 5 percent of their time because of substandard machine performance, there should be no allowance to cover this in the time stan-dards, provided the conditions can be corrected. It is the responsibility of the unit supervisor to see that the necessary repairs are made.

10. Review of Standards by Unit Supervisors. The supervisor must be con-vinced of the fairness of the work measurement data. Normally, this will include a review and approval of new standards developed by the industrial engineer. If disagreements occur, they should be resolved by higher management.

11. Making Standards Known to Operators. The operators should know what performance is expected of them. The time standards are usually made a part of the routine manufacturing information. In the case of paced operations, the line speed usually determines the standards, and there may be no need to publish them.

12. Follow-up of Detracting Conditions. Conditions not under the control of operators are often the primary detractors from satisfactory productivity. As manager of his unit, the supervisor is responsible for getting such conditions corrected, utilizing the available assistance of staff personnel. This does not exempt the industrial engineers, however, from actively following up such conditions on a staff basis.

13. Prompt Performance Reports. If the unit supervisor is to improve conditions where the need to do so exists, he must receive performance reports promptly. Records of performance while long-cycle jobs are in process may be essential. In the case of daily performance reports, these should be available at the start of work the next day.

SUMMARY

Measured daywork is a form of wage payment based on the application of time standards to productive operations and with the wages of the operators independent of their output. The primary advantage of this method of control is that it establishes a favorable climate for increasing productivity through the introduction of new production methods and equipment. The installation of a measured daywork plan should be based on the thirteen principles discussed above.

Chapter **4**

Group System of Wage Payment

HAROLD B. MAYNARD

President, Maynard Research Council Incorporated, Pittsburgh, Pennsylvania

The results obtained from the kind of individual wage incentive plans described in Chapter 2 of this section have been quite satisfactory in a large number of cases. There have been instances, however, where experience has shown that there are certain disadvantages which occur from motivating a number of workers to concentrate only on their own individual production. If each worker is paid only for what he produces, it is natural that he will strive to increase his own output and that he will not be much interested in the problems of anyone else, either supervision or fellow workers.

To correct such conditions, the group system of wage payment has been devised. It has proved successful in many situations in many different industries where conditions were such that the group system was applicable. It is not a panacea which will solve all wage payment problems, however, and attempts to use it indiscriminately as a cure-all have usually ended in disappointment. This chapter will describe the fundamental principles upon which the group system is based and will discuss its advantages and disadvantages.

CHARACTERISTICS OF THE GROUP SYSTEM

A group is made up of a number of workers who pool their entire output. The method of computing and distributing the earnings of the group among the individuals in the group is known as the group system of wage payment.

It is important to understand this definition. The group system is not per se another wage incentive formula like those described in Chapter 2. It is a method of computing and distributing earnings among a group of workers who pool their accomplishments.

Payment Plans Used with the Group System. The group system may be used with any good incentive plan. Indeed, many of the advantages of the group system are realized even when the daywork plan is used. Piecework is the least suitable form of wage incentive plan for group payment unless all workers in the group are doing the same class of work. This is so because of the computational difficulties involved in making an equitable distribution of earnings. At the same time, it must be recognized that there have been many successful applications of group piecework, so that the company that wishes to adhere to piecework does not necessarily have to forego the use of the group system.

The wage incentive plans most commonly used for groups are those which have standards expressed in units of time. The wage payment formula may be any of those covered in Chapter 2 of Section 6 or any variation of those formulas. From the standpoint of the workers themselves, the type of plan which pays a 1 percent increase in earnings for every 1 percent increase in output is generally the most acceptable.

Period for Checking Performance. As in the case of individual incentive work, the performance of a group may be computed on a job, day, or pay period basis. The job basis is ordinarily used only when the job is of fairly long duration and when all members of the group can begin and finish their work at approximately the same time.

The day basis is used most commonly on mass production work where it is possible to get an accurate count of the work completed each day. Assembly lines, for example, are usually handled on the day basis. Where the work is more varied and is complex or of a jobbing nature, the pay period is generally the most satisfactory basis for checking performance.

FIELD OF APPLICATION OF THE GROUP SYSTEM

From the standpoint of stimulating individual effort, the individual wage payment system is undoubtedly the best. If the nature of the work is such that the worker is unable to help or be helped by others, it is usually a mistake to force him to pool his accomplishments with others. If it is done, his tendency will be to limit his production to the average of the group, for he cannot see enough benefit to himself from producing more when he has to share his increased output with a number of others.

The group system applies best when:
1. There is a community of interest among the members of the group
2. The work is such that it is impossible to measure the contribution of the individual member accurately

By community of interest is meant anything which makes one worker interested in what the other workers do. This may be purely a psychological interest, as when each worker wants the earnings of all the other workers to be the same as his. This, of course, can easily be accomplished by the group system. More often, community of interest is inspired by the work itself. If one worker can help another by advice, giving him help in lifting a heavy object, doing his own work more carefully, taking his turn at doing certain undesirable work, or in any other way, a community of interest exists which may make the application of the group system advisable.

Probably the most common example of work where there is a community of interest is the assembly line. There a definite number of operations are always performed in the same sequence. The work can be divided so that each operator performs a small and approximately equal part of the total job. He thus becomes highly specialized and proportionately more efficient. By performing his part of the work well, he makes it possible for the other members of the group to do their work well, and the output of the whole group rises.

Other work involving a community of interest on which the group system has been highly successful is foundry molding; casting cleaning; assembly and fitting work, both standard and miscellaneous; coil winding; painting; transformer building; and machine work. The common denominator in these and the many other classes of work to which the group system has been successfully applied is "community of interest."

There are certain classes of work where it is impractical to measure the work which the individual does sufficiently accurately to permit individual incentive payment. Examples are storeroom crews, material handling crews, maintenance and construction workers, furnace crews in foundries, and the like. The only practical method of measuring performance is on a group basis. Fortunately, these classes of work involve a community of interest also; so the group system works successfully.

On some kinds of work such as maintenance, it is practically impossible to determine accurately the time required to do a given task each time it is done. The time required to replace a section of leaky pipe, for example, will depend in part on how badly the pipe joints are rusted. This cannot easily be determined in advance of doing the work. The most successful approach for handling this is to establish the range of time in which the job is likely to be done, using the midpoint of the range as the allowed time for any job. Over a period of time, say a pay period, the easy situations and the difficult situations will tend to average out, and a reasonably accurate measurement of true performance is obtained. When a number of maintenance workers work as a group, the averaging out process occurs over a large number of jobs, and the accuracy of performance measurement is correspondingly improved.

ADVANTAGES OF THE GROUP SYSTEM

The following advantages have been observed sufficiently frequently where the group system is used to justify mentioning them in this discussion. It should be emphasized, however, that all these advantages do not occur automatically every time a new group is formed. For example, people with a community of interest usually cooperate better when they work as a group rather than as individuals. A number of machine molders were formed into a group in a certain foundry. During pouring, they could help each other by planning the work so that certain molders got certain classes of work in which they were skilled, by using certain unavoidable delay periods to help others, and so on. There was a definite community of interest as far as the work was concerned.

But the group was a dismal failure. The group members cordially hated one another for a variety of reasons. They preferred to annoy one another by not helping rather than make more money. The group leader was a conscientious man but he was not strong enough to overcome the personal animosities. So the group failed to produce and continued to work far below standard until the group system was abandoned and individual incentives were installed. The molders then made informal arrangements to get help when needed from individuals with whom they were friendly, and everyone was soon turning out satisfactory production. Interestingly enough, while all this was going on, a similar group was organized on the other side of the foundry which was successful from the start. The members of this group liked—or at least respected—one another, and they quickly developed a high degree of cooperation.

Thus it may be seen that the human factor is very important in the success or failure of the group system. Establishing a successful group is more than a mechanical procedure. It is something of an art in which the ability of the organizer to sense and solve human problems is extremely important. In reviewing both the advantages and disadvantages of the group system, this should be kept in mind.

Better Cooperation. Ordinarily, if personal dislikes are not present, there is greater cooperation among men in a group than when they are being paid on an individual basis. Because everything that will aid in the completion of the work will mean more money for the group and hence for each group member, the individual is willing to help his fellow workers whenever necessary. This willingness to cooperate usually increases with the length of time the group has been working together, until a group spirit is built up which influences every man in the group.

Lost time caused by waiting for a move man to bring more material, waiting for the toolroom to grind tools, and other small delays is practically eliminated under the group system. If the operator is forced to stop his own work for any reason, he will help another man in the group on another operation or he will do some odd job which will help the group as a whole.

Again, a man will be more careful in performing his operation if he knows it will aid another group member in performing the following operation. For example, one man milling a casting which is next to be drilled will try harder to remove all burrs so that the piece will fit easily into the drill jig of the next man.

There are always some jobs in a given class of varied work which are not so desirable as others from the worker's viewpoint. The work may be heavier or more complicated than average, it may be more disagreeable, or it may be a short order which will not allow the worker to get into the swing of the work and thus work more efficiently and make higher earnings. No matter what the cause, if the job is undesirable, there is a tendency under the individual system for the operator to shun it in the hope that eventually someone else will do it. This makes it difficult for the production control department to get such jobs through the shop and often these orders are seriously delayed. A group, on the other hand, realizes that it will eventually have to do the job and that there is no particular advantage in setting it aside and favoring other work. Thus, the schedule clerk has merely to inform the group leader when the job is wanted to be reasonably sure of getting it. This materially lessens the work of the production control department, and it tends to reduce the number of overdue orders.

Reduction of Supervision. When a group is organized, one of the members is appointed group leader. It is his function to take on certain duties which would ordinarily be performed by each individual, such as getting drawings and tools, planning the method, counting the finished work, making out time and work reports, and so on. By having all these things done by one member of the group—the group leader—time can be saved for the group. This is the legitimate function of the group leader.

When one hundred men work as individuals, the foreman must deal with each man separately. Similarly, the production control department must assign jobs to one hundred different work centers. If the hundred men are organized into ten groups, both the foreman and the production clerk need to deal with only ten group leaders. This greatly simplifies their work and makes it possible to get the work done with fewer foremen and production clerks.

There is a danger inherent in this situation which must be guarded against, however. As long as the group leader's duties are confined to doing more efficiently the things that the group members would have to do anyway were he not there, the group system is being operated as it should be. But it is all too easy to pass on to the group leader functions that properly belong to a foreman or a salaried assistant foreman. In effect, the group leader who is not a part of management is asked to perform some of management's functions.

Thus, while the group system undoubtedly does lead to reduced supervision and overhead, management must be very careful not to abuse the system by requiring the group leader to perform functions that are not rightly his. And if it does err in this direction, it is equally unwise to try to correct the situation by going to the other extreme and giving a member of the management group functions that should be performed by the group leader.

Reduced Operator Training Time. Under the group system, the learning time of new people is usually reduced. Unless the group leader or another group member is allowed time to do training work and is paid for it as such, the initial training should be done by an instructor appointed by management (see Section 12, Chapter 8). But in the period which follows, when the new man is trying to develop speed through practice, the group leader and the other men in the group all help the new man to correct his mistakes and to go about his work more efficiently. The new man tends to learn more from the members of the group than he would from the instructor during this period, because they are in more constant contact with him.

The older men will often go out of their way to be helpful, for they realize that the sooner the new man is broken in, the sooner they will get the full advantage of his efforts. The new man receives an important psychological lift from the fact that the others show an interest in him. He becomes eager to show the group that he is capable of working with them, and he tends to try more earnestly to learn than he would under an instructor. The method of paying a new man so that the group does not have to bear the burden of the unproductive learning period is explained later.

Indirect Labor May Be Included in and Controlled by the Group. The group system offers some very real benefits to indirect labor. For example, it is often possible to include in the group material handlers and other indirect service people so that they may share in the efforts and earnings of the group. If one man is needed to bring materials to and remove finished work from a group of ten men, by increasing all time standards or piece rates applicable to the group by 10 percent, it will be possible to include the move man in the group. His earnings are affected by the group's performance and hence he will be quick to help them as much as possible. The group leader, on his part, will find simple jobs for the move man to do when he has no material handling work to occupy him. In time, this man will be able to learn to do the more difficult jobs. Eventually, he may be able to take his place in the group as a full-fledged artisan, and of course, will be able to increase his earnings materially.

When this procedure is followed consistently, the indirect labor people, as a whole, may come to realize that they have a chance to improve themselves and get ahead. It will then be possible to get a better class of men to accept and keep such jobs. Labor turnover will be reduced, and at the same time, new skilled workers will constantly be developed.

The method of including indirect labor in a group as just described is the one which should be followed whenever possible. Sometimes the mistake is made of paying the indirect labor the same percentage that is earned by the group, but he or they are not included in the group. When this is done, control of indirect labor is lost. In the example given above, one move man was added to a group of ten producers and the standards were increased by 10 percent. If work falls off so that only five producers are needed, the time standards would automatically allow for only half the time of one move man. If work increases so that twenty producers are needed, the time standards will allow for two full-time move men. Thus the indirect labor in the group is under the same control as the direct labor.

On the other hand, if the move man had been told that he would be paid the same percentage that the group of ten producers earned, he would have a proper incentive to be helpful as long as the size of the group remained the same. But if it fell off to five producers, he would still be paid whatever percentage they earned and would in effect maintain his previous earnings while doing half as much work. Theoretically, the foreman should find extra work for him to do. Actually, the conditions which prevailed when a setup of this kind is first made are quickly forgotten, and before long, unless he digs through a lot of records, the foreman does not know how many indirect workers he is supposed to have at various volume levels. The experience is that, as volume increases, he is asked to add them by the producers themselves who need the extra help, but that no one reminds him to take them off as volume decreases. Therefore, the only safe procedure is to include the time for the indirect workers in the group standards.

Timekeeping Simplified. In most cases, timekeeping and work reporting are somewhat simpler where the group system is used. On the individual payment basis, it is necessary to keep a record of the work produced by each individual. With the group system, it is necessary only to record the production of the group. For example, if ten people work as individuals on a product that has ten operations, it will be necessary to have a minimum of ten reports of work accomplished if each operator works on only one operation. If the operators switch about so that each one works on two or more operations during the course of the reporting period, the recording problem is much more complicated.

When the group system is used, it is customary to report completed units only. Thus one report takes the place of ten or more. Work on partially completed units is usually carried over from period to period without reporting it, as it tends to balance out.

There are two ways of reporting time worked and work accomplished under the group system. The time worked may be turned in on separate slips for each member of the group, and the work accomplished can be turned in on a single sheet like a shipping report. Or the time worked may be turned in on a single group report and a separate slip may be turned in for each job completed. There are also a number of variations that lie between these two procedures. In Section 4, Chapter 6, several time and work reporting forms are shown.

Quality of Product Is Improved. Where incentive systems are in effect, it is often stated that quality is sacrificed for the sake of production. Whether it is or not is largely dependent upon the quality standards which are established and the strictness with which they are maintained by the inspection department. For any given set of conditions, however, experience shows that quality is improved by the installation of the group system.

There are several reasons for this. In a group, there is little tendency to slight work either with the hope that it will not be traced to the responsible person or with the idea that someone else will make up the deficiencies later. The group as a whole is directly responsible for the quality of work produced. If one man in the group slights part of the work, another must make it up, and thus no one gains anything.

Furthermore, in the interests of higher production, the group leader will try to divide the work so that each man will always be doing the same kind of work. This tends to develop specialists within the group. Obviously, specialists will, because of their superior skill, turn out a better quality of work.

Frequently, it will be found that a group doing assembly work or performing a number of operations on the same part takes a pride in its work which the individual does not. The individual does one or two operations on a piece and then loses sight of it. The group sees the products from the time they start work on them until the time the products are ready to ship out. It is only natural that the group would take more interest in turning out a good job.

Checking Is Simplified. Checking the production of the individual operator to see that he has actually done all the work that he reports is no small task. If quality as well as piece count must be checked after each operation, the task is large indeed. Under the group system, it is not necessary to check each operation except in the case of groups that perform only one operation. The completed product as it is shipped from the group is all that must be checked. The number of checkers and inspectors neeeded is thus reduced, and overhead expense is lowered as a result.

Absenteeism Is Reduced. Each member of a group knows that the others expect him to work conscientiously and he expects them to do the same. He realizes the justice of this, for he knows how he would feel if he saw another man in the group idling while he was working.

Where work flows through a group—each worker performing an operation upon it and then passing it on the next—each man tends to feel that he is a link in the chain of production. He realizes that if he is absent it will retard the flow of production on that particular day. Not only will he lose what he might have earned that day, but also if performance is checked on the basis of the pay period, his bonus for the pay will be reduced because the group on that day was less efficient. For this reason and out of consideration for their fellow workers, some men, at least, working in this kind of a group tend to be more regular in attendance.

Wages Are Fairly Distributed. The earnings of a group are pooled for a given pay period and are then distributed among the members of the group. The share which each individual receives is dependent upon two factors: first, the number of hours he worked in the group, and second, his base wage rate. If the latter is a true reflection of his relative value in terms of such factors as skill, length of service, attitude, intelligence, and general makeup, he receives payment in proportion to his contribution to the company.

Other Advantages. There are several other advantages which have been experienced with the group system which should be mentioned briefly. Where actual costs are determined for every job that goes through the shop, costing is simplified by the group system. Instead of having to deal with the individual base rates of each operator, a single group rate may be used for costing purposes.

Wages from one pay period to the next do not fluctuate greatly under the group system. When a man is working as an individual, a difficult job or an unusual series of minor delays and mishaps may retard him so much that his wages for the pay period are seriously affected. In a group, these things tend to average out. When one operator is having trouble, another will be going along unusually well. Thus group earnings tend to be more nearly constant.

Most people like to do the things they can do well. When a man is working in a group, he is assured of getting the jobs he can do best, for his efficiency affects the group's efficiency. The group leader plans this very carefully, and because he is close to both the man and the work, he is able to give out jobs to the best advantage.

The working environment is often found to be more pleasant where the group system is used. A group, as soon as it is organized, tends to develop a team spirit. Each member realizes that he must cooperate with every other member if he is to benefit to the fullest extent. As the group continues in existence, this perfunctory cooperation often develops into a real group spirit. Each member has a friendly feeling toward the others that is often carried outside the working hours. Such an atmosphere cannot help but make the working hours more pleasant.

DISADVANTAGES OF THE GROUP SYSTEM

Against all these advantages, there are some disadvantages in the group system. These should be understood before deciding whether or not the group system is applicable in any given situation.

The Problem of Incomplete Jobs. Theoretically, a group should so plan its work that, at the end of the pay period, there are no incompleted jobs upon which some work has been done. Actually this is seldom possible. On work like an assembly line, this presents no real problem. The incomplete work may be ignored, for it is about the same at the end of each pay period, and therefore it does not affect earnings.

On more miscellaneous work, such as the building of specially designed electrical apparatus, the problem is more serious. It can easily happen that during a pay period the group will start a number of jobs but will complete none of them, perhaps because some part common to all of them is missing. During the next pay, the part comes in and all the jobs are completed. If the group is credited only with completed work, it would show 0 percent efficiency the first pay and perhaps 250 percent efficiency the second pay. This would be unsatisfactory to the group and unfair to the company, because for the first pay the group would receive its guaranteed rate and thus the company would in effect be paying twice for some of the hours earned.

This problem is solved by making an inventory of the work completed during the first pay period. If a job requiring 800 hours to assemble is found to be half finished at the end of the pay period, the group is credited with 400 allowed hours. The same is done for all other jobs. As a result, enough allowed hours are usually credited to allow the group to earn its normal bonus. When the jobs are finally completed, the number of allowed hours previously paid for is deducted from the total number of hours allowed by the standard and the remainder again represents about the normal number of allowed hours for the pay period.

This procedure is similar to that often followed when individuals work on incomplete jobs and theoretically it should present no difficulties. It will prove satisfactory if it is carefully controlled and if the inventory is taken carefully. Difficulties have occurred, however, when management has permitted the group leader to take the inventory without any check or control. There is a natural human tendency to estimate in one's own favor. At special periods of the year, such as just before Christmas when a big pay is much to be desired, there is a very real temptation to feel that a job which is actually 40 percent complete may be with a little imagination considered to be 60 percent complete. If the group leader continues little by little to borrow on the future in this way, the same thing happens as when a family constantly overspends its budget. Eventually the group goes broke. It cannot scrape together enough hours for a make-out, and it falls down badly.

Because the company must pay guaranteed base rates in the case of falldowns, it takes an understandably dim view of such situations. The result can be quite unhappy for all concerned, particularly for the group leader who may have drifted into the state of group bankruptcy without in any way meaning to. To prevent such a situation from occurring, management should take the responsibility for controlling the inventorying of incompleted work.

No Check on Standards on Individual Jobs. Under the individual system, it is comparatively easy to get a check on any time standard. If there has been no time juggling, both the number of hours worked and the number of hours earned will be shown on the time card for the job.

Under the group system, there is no way of checking an individual job other than timing the worker while he is doing it. Fortunately, the necessity for such checks is rare, particularly when a predetermined elemental time system is used. If it is really necessary, the actual observation will give a much closer check than the time card.

No Check on Individual Efficiency. It is not difficult to check the efficiency of an operator working on an individual basis. His earned bonuses will be an exact index of his working efficiency. When the same man is working in a group, his individual output is absorbed in the output of the group and is lost sight of. If it is desired to know how the individual is doing, it is necessary to make personal observations and checks.

This can be a very real problem when a group is first organized. Quite often at the start, before the group members really understand the new setup, there is a certain amount of conscious or unconscious holding back to see how things are going to work out. If everyone or nearly everyone holds back, things do not work out and the group fails to earn a bonus. If this continues for any length of time, the group will become discouraged and will cease to try to produce at an incentive pace. When that happens, it is very difficult to reestablish their interest and to build up their confidence that they can, after all, make out.

During those critical first days, it is impossible to tell from the group records which group members are not doing their share. Therefore management does not know where corrective efforts are needed. To get a group started successfully, therefore, the wise industrial engineer will make it a point to spend all his time with the group during the first few days of its existence. He will observe how everyone is working, will make frequent checks of individual performance, and will be on the alert to note anything which causes the group to lose time. By taking corrective action immediately, he can help the group to earn a bonus on the very first pay. Once the group is convinced that it can meet the standards and make out as a group, there is usually very little trouble.

Sometimes Hard to Find Right Man for Group Leader. When organizing a new group, it is not always possible to find the type of man who makes a good group leader. Although this is not a disadvantage of the group system itself, it is a practical difficulty that is sometimes encountered. And because the group leader is a key to the success of the group, the difficulty must be solved.

In an organization of any size, it usually is possible, after some searching and a consideration of the alternatives, to find someone who is capable of handling the group, at least on a trial basis. It is often found that a man who was put on the job only temporarily because of doubt of his ability to handle it actually turns out to be a satisfactory group leader. The chance to show what he can do will often spur a man to efforts of which he was not thought capable. This performance, coupled with the change of attitude which greater responsibility so often produces, may enable the man to become a first-class group leader.

Exceptional Ability Is Penalized. The criticism most often leveled against the group system by management and workers alike is that the exceptional ability of the highly skilled worker is penalized when he must work in a group, and that he is working hard and producing a lot for the other, less skilled men. Where all men in the group are paid the same base rates and earn the same bonus, this is undoubtedly true. If, on the other hand, the base rates are in proportion to the value of each individual, then the highest producer has the highest base rate and makes the highest earnings.

In actual practice, the situation is usually in between these two extremes. The exceptional worker does have a higher base rate than the others, but it may not be enough higher to compensate him fully for his superior performance.

Good Management Is Required. As has so often been said, an incentive plan is not a substitute for good management. This is true if an individual incentive payment plan is used and it is perhaps even more true under the group system. Because of the large number of advantages which result from the group system under favorable conditions, management's work is often very much simplified. As a result, there may be a tendency for management to relax its attention on plant management problems and to devote its attention to other things of seemingly greater importance.

This can often be disastrous. Human nature being what it is, it is only prudent to check frequently to see that work done is properly reported, that timekeeping is being done accurately, that material use is efficient and controlled, that machines and equipment are maintained in good working condition, that operators are properly instructed, and so on. If management neglects these things, conditions may soon get out of hand. It may start with the group leader incorrectly reporting the time taken on a job to compensate for a machine in poor operating condition. He may reason that this is perfectly justified, but if the incident passes unnoticed, it will be easier to do the same thing again, perhaps with less justification. This sort of thing can build up, until circumventing the system for the benefit of the group may seem to be an exciting battle of wits between men and management, and the question of what is right or wrong is lost sight of. Most men wish to be honest, but deceiving a company and deceiving a fellow man are not always looked upon in the same light.

This being so, the management which allows looseness to creep into its administrative practices must accept a share of the blame when things get out of hand.

The Group Leader Affects Industrial Relations. Industrial relations under a properly managed group system are usually quite good. Because management deals more with the group leader than the individual workers, however, there may be a tendency for the supervisor to grow apart from his men.

Indeed, the group leader may encourage in a desire to enhance his own importance. If unauthorized practices are going on within the group, the group leader may seek to act as a barrier to communications between the supervisor and the workers. This can lead to an unhealthy industrial relations situation and a feeling of frustration and loss of control on the part of the foreman.

SIZE OF THE GROUP

The success of a group is in part dependent upon its size. In general, the smaller the group, the better is the fellowship and cooperation among the members of the group, but this may be carried too far. Experience has shown that, when the work is not scattered over large areas and when it is not primarily a "line" operation, groups composed of as many as fifteen members work very smoothly and efficiently. Thus, as a general rule, it may be said that the size of a group should not exceed fifteen men.

There are, however, many exceptions to this rule. The most notable occurs on line work, particularly on work which is mechanically paced. Here each operator performs a small part of the total operation at a fixed work station following a carefully prescribed method. There is a definite community of interest, for if one operator falls behind or starts to produce poor quality work, the output of the whole line suffers. In this case, the whole line may be included in one group even if the number of operators is greatly in excess of fifteen. Groups of considerably more than a hundred operators have operated successfully on a mechanically paced production line.

On fairly simple work, such as casting cleaning in a foundry, groups of thirty or forty members have worked successfully. Here it is not practicable to divide work into smaller units. The work is simple and one group leader can see that all men are supplied with work and are using the proper methods.

On the other hand, a group of twenty-five men, widely separated physically and working on fairly complex work, was quite unsuccessful. The group leader had difficulty in getting around to each man, and each man could only see a very few other members of the group at any one time. Theoretically, there was a community of interest involved in the work, but it was so scattered physically that it was difficult for the group members to feel it. The result was that there was a general falling off of effort, each man feeling that he might be doing more than his share and being ready to blame the other members for the poor performance of the group. The problem was solved by regrouping the work physically into two areas and dividing the big group into two smaller groups. The performance of the smaller groups immediately improved and soon the situation was entirely satisfactory.

The minimum size of a group is, of course, two, and in some cases, it is desirable to form a group of this size. Where only two men are needed to do a particular class of work and where by working together they can work more efficiently than by working alone, the formation of a group is advisable.

ORGANIZING THE GROUP

Recommendations for the formation of a group are ordinarily originated by the industrial engineering department as part of their work in connection with the installation of incentives. The recommendation is usually made because it is felt that some or all of the advantages previously discussed will be realized.

The first and probably the hardest task in organizing is the selection of a suitable man for group leader. This is usually done after joint consultation of the superintendent, foreman, and chief industrial engineer. In some cases, the head of the union is also brought into the discussion. It is sometimes necessary to go outside the department to find a man who has the necessary qualifications for group leader. If it is at all possible, however, the group leader should be chosen from the men in the group. Not only is he sure to be more familiar with the details of the work, but also the psychological effect on the other men is better.

In some cases, a group leader may not be necessary. On line work, for example, where there is no work to be given out, where there is no instructing done after the simple task is once learned, and where there is no planning, the need for a group leader is eliminated. But on groups handling a more miscellaneous and complicated class of work, the group leader is the key to success and he should be carefully chosen.

The desirable qualifications for a group leader are many and it is probably impossible to find one man in whom they are all combined. First, the group leader should be reasonably familiar with the work the group is to do. He should be something of an instructor, for he guides all new men during their learning and practice periods while they are coming up to standard performance. He should also be able to give help and information to the older men on any difficulty that may occur.

The group leader should be able to recognize what each individual is best fitted for so that he can place him accordingly. He should be able to interpret drawings and manufacturing information correctly. It is desirable that he have planning ability, for he orders all supplies and tools and assigns all jobs to the individuals in the group. He must keep work ahead of each man and be certain that it is work that the man can do.

The ability to write legibly and to do simple arithmetic is also important. The group leader turns in all time and work reports for the group and generally, if at all interested in his work, keeps a record of group output as a check against possible mistakes in the payroll division. The group leader should have a sense of responsibility, for he is held responsible by both the company and the members of the group for the quality and quantity of the output produced. Last, and perhaps most important, he should be a leader of men, one who can inspire confidence and gain cooperation. He should have a good understanding of human nature so that he may know what to expect from his men. He should never take advantage of his position as group leader but should devote all his time not otherwise occupied to actual production.

After the group leader has been selected, the next step, that of picking the members of the group, is comparatively easy. Usually, the foreman and the industrial engineering department decide what operations are to be done by the group. Those operators who have previously been doing this work are usually those who will compose the group in the future.

The last step, that of completing the paper work and securing the approvals necessary for the formation of the group, will vary somewhat with the individual company. The key document for this step is the group organization sheet. A typical form for this purpose is shown by Figure 4-1. The sheet shows the members of the group, their base rates, occupation and job classifications, and any extra rates paid for shift differentials or group leading. The signatures of all who must approve the formation of the group are also shown. The group organization sheet will be made up with a sufficient number of copies to satisfy the requirements of the routines established within the company.

THE NEW EMPLOYEE

It is very important to handle the new man in a group in such a way that his introduction does not affect the group efficiency. It is also very important that each member understands how the new man is taken care of, to avoid having the older men feel that they are working for the new man while he is learning.

It will be readily recognized that a newly hired man will not be so efficient at first as a man who has been working in the group for some time, even though he may have had previous experience on a similar line of work elsewhere. For example, a man who has been running a milling machine in another plant will be less efficient than the milling machine operators in the plant to which he has just come, because it will be necessary for him to become familiar with such things as the location of the toolrooms and washrooms, the method the engineering department follows in specifying tolerances and finishes, the materials used, and the like. It will take him some time to get acclimated to his new surroundings and to settle to the routine of the work. During this time, the new man will have to ask many questions and receive many instructions. Not only will he produce

GROUP ORGANIZATION SHEET

ORGANIZATION ☐   REVISION ☐   DISCONTINUANCE ☐   DEPARTMENT_____

WAGE PAYMENT PLAN: GROUP NO. _____

DESCRIPTION OF WORK: DATE _____

CHECK NO.	NAME	BASE RATE	SHIFT DIF.	SUPV. RATE	OCCUPATION	CLASS	JOB NO.
	GRP. LDR.						

GROUP PARTICIPATION PERCENTAGES IN PAY PERIODS															
OCCUPATION NAME	CLASS	1	2	3	4	5	6	7	8	9	10	11	12	13	14

CHIEF IND. ENGINEER	GENERAL FOREMAN	WAGE ADMINISTRATION	SUPERINTENDENT
INDUSTRIAL ENGINEER	PAYROLL DIVISION	JOB EVALUATION SECTION	

FIG. 4-1.   Group organization sheet.

less than an experienced man, but he will also require considerable attention from the group leader and other members of the group and will slow them up to some extent.

The length of time it will take the new man to attain average group efficiency and the length of time it will take the group leader and other members of the group to train him to this point will depend upon the complexity of the work he is performing and whether or not it is of a repetitive nature. It will take longer for a new man to come up to average group efficiency if he is working on a difficult fitting job, where the quantities are small and the jobs repeat themselves only once every six months on the average, than it will for a man performing a simple drill press operation where the parts come through regularly in quantities of 5,000.

With this in mind, a chart similar to the one shown by Figure 4-2 may be developed. This chart shows job activity divided into five classes: highly repetitive, repetitive, partly repetitive, variable, and highly variable. Each of these classifications must, of course, be defined for the particular industry where the chart is to be used. A job that might be considered as variable or even exceedingly variable for an automobile plant might be considered repetitive in a shop doing strictly jobbing work.

The vertical columns on the chart are headed by the classes of work, determined as described in Chapter 6 of this section, Job Evaluation. With the activity and the class of work known, it is a simple matter to determine from the chart the amount of time expressed in pay periods that it will take the new man to reach average group efficiency.

During this learning time, the new man will, of course, be producing a certain amount of finished product, small at first, but in increasing quantities as the learning period approaches the end. To prevent the earnings of the group from being affected adversely by the introduction of the new man, all the output which he

ACTIVITY	LABOR CLASSIFICATION								
	9	8	7	6	5	4	3	2	1
Highly Repetitive	6	6	5	5	4	3	2	1	0
Repetitive	8	8	7	6	5	4	3	2	1
Partly Repetitive	10	10	9	8	7	6	5	3	1
Variable	12	12	11	10	9	8	7	5	2
Exceedingly Variable	14	14	13	12	11	10	8	6	3

FIG. 4-2.   Time in pay periods (two weeks) for a new man to reach average group efficiency.

produces should be credited to the group, but only a certain percentage of the hours he works, determined as shown below, should be charged against the hours worked by the group. The hours charged against the group are a fixed amount for a given week or pay period, but the output of the new worker is a variable depending on how well he is able to perform. It is thus to the advantage of the members of the group to help the new man learn the job as quickly as he can, so that his production will be as high as possible, thus adding to the earnings of the group.

For example, assume that a group customarily performs at 125 percent efficiency. Assume further that it has been found from experience that a new man, during the first pay period he works with the group, will produce five hours of finished product each day. To maintain the group earnings at their average level, only four of the hours the man works daily should be charged against the group. The rest of the time should be charged to the overhead account for learning. The new man's computed efficiency will be five hours earned divided by four hours worked, or 125 percent. Under these circumstances, the group earnings will be unaffected by the introduction of the new man.

Assume now that the group leader is alert for opportunities for increasing the group earnings. He spends a certain amount of time helping the new man to master his job more quickly and encourages the other group members to do likewise. As a result, the new man is able to produce six hours of finished product each day. His computed efficiency will then be six hours earned divided by four hours worked, or 150 percent. Obviously, the group benefits by the extra attention it gives the new man.

The new man, of course, is not actually meeting standard in either case because he is taking eight hours to turn out five or six standard hours of work. Some companies feel that he should receive none of the bonus earned by the group because his contribution is an artificial one, and that he should be paid at his base rate until he becomes a full-fledged member of the group and is doing his share to hold up the group efficiency without accounting aids. This policy, however, overlooks the effect that incentive payment will have on the motivation of the new man to learn the job as quickly as possible. By permitting him to share in the group's earnings for the hours of his time which are charged against the group, he has the motivation to exert his best efforts, which is what an incentive plan is designed to provide. He will not earn as much as a full-fledged member of the group, for he is paid bonus on only part of the hours he works, the part charged against the group. He does, however, have an incentive to contribute to raising the group's efficiency as much as he can. When both the group members and the new man have an incentive for the new man to produce as much as he can, the situation is most likely to result in higher output.

PAY PERIOD	TIME IN PAY PERIODS (2 WEEKS) TO REACH AVERAGE GROUP EFFICIENCY													
	1	2	3	4	5	6	7	8	9	10	11	12	13	14
1	75	70	70	65	60	55	50	45	45	45	40	40	35	30
2	100	90	85	75	70	65	60	55	55	55	50	45	40	35
3		100	95	85	80	75	70	65	65	60	55	50	45	40
4			100	95	90	85	80	75	70	65	60	55	50	45
5				100	95	90	85	80	75	70	65	60	55	50
6					100	95	90	85	80	75	70	65	60	55
7						100	95	90	85	80	75	70	65	60
8							100	95	90	85	80	75	70	65
9								100	95	90	85	80	75	70
10									100	95	90	85	80	75
11										100	95	90	85	80
12											100	95	90	85
13												100	95	90
14													100	95
15														100

Fig. 4-3.   Anticipated efficiency progress of new employees in terms of percentage.

Experience has shown that it is most satisfactory to establish a standard table of anticipated efficiency of a new worker throughout the learning period, similar to the one shown by Figure 4-3. This table was established empirically, based on past records of new-worker efficiencies. Where learning curves similar to the one shown by Figure 1-1 in Chapter 1 of Section 4 have been developed by more careful study, they may be used for developing a similar but hopefully more accurate table.

A chart of this sort is superior to the method sometimes used where the participation percentages are established each pay period for each new man on the basis of the combined judgment of the foreman and the industrial engineer. It gives more uniform and consistent results with a great deal less effort. Once the accuracy of the chart has been established, the proper introduction of a new man into a group may be handled in a routine manner by the payroll division.

When a fixed chart is used, the group leader and the other members of the group know that, at the end of each pay period, the time of the new man charged against them will be increased. They realize that it is to their advantage to see that the new man is able to do his share of the work during each period. The new man, in turn, will be anxious to show that he will be able to hold up his end so that he may help increase the earnings of the group. This mutual desire toward the same end serves to keep the learning period at a reasonable minimum. Thus the group system, with this feature of administration, permits the breaking in of new men in the minimum time at minimum expense with little or no interference with the other employees.

THE ELDERLY EMPLOYEE

In cases where groups have been long established, one or more of the older men may eventually become unable to meet the established standards because of old age or other physical handicaps. The other members of the group will recognize this and may ask that something be done so that they do not have to carry the older man along. Under these circumstances, charging the full time worked by the old employee against the group and granting him full participation in the group earnings would obviously be as unfair to the group as would be the extension of this privilege to new group members, and provision should be made for handling situations of this kind.

Unlike the case of new men entering the group, there is no general basis upon which to judge the value of these older men, and hence no definite charts can be established. Such factors as length and character of service, attitude, dependability, previous performance efficiency, and physical limitations must be taken into consideration. In some cases, it will be found best to transfer the older individual to some less strenuous occupation. In many other cases, however, it will be found that the man, because of his long experience on the line of work that he is doing, is of great value to the group. His advice and instructions on certain parts of the work will be most helpful and he will be of more value to the company and to the group if left where he is.

When such a decision is reached, the industrial engineer, the foreman, and the group leader should jointly determine the probable efficiency of the man from the standpoint of output. This will be expressed in terms of a participating percentage, as in the case of new men. It is to be expected that as time goes on the man will become less and less efficient. The tapering-off tendency will be so slow, however, that the participation percentage should be considered as a constant until further falling off of the man's output is detected at some later period.

Whether or not the elderly man should participate to any extent in the bonus earnings of the group is a matter of company policy. Technically, the man is no longer meeting the standard and therefore should receive only his guaranteed rate. On the other hand, he is contributing a certain amount to the general efficiency of the group by his advice and experience. Therefore, it would seem to be fair that he should receive at least part of the benefit of extra group earnings, probably in proportion to his participation percentage.

This method of handling elderly men makes it possible to reward in proportion to their true worth those loyal and worthy men who have been long in the employ of the company. The value of their output will be greatly in excess of that which could be obtained by downgrading them to such jobs as watchmen, elevator operators, or janitors.

SUPERVISION RATES FOR GROUP LEADERS

It has been shown that the group leader, in addition to his duties as a regular workman, must do a certain amount of supervisory work while directing the activities of his group. For this work, he must have special qualifications and abilities. Because these abilities are necessary for the success of the group, it is but fair to compensate the group leader for possessing and using them.

The amount of supervision which the group leader must do depends upon the number of members in his group and the nature of the work which the group is doing. Obviously, a leader of a group of but two men doing a simple machine operation has to exercise less supervision than a group leader with a group of fifteen men doing high-grade assembly work of a variable nature.

The chart shown by Figure 4-4 is based upon this concept. The actual money values given vary, of course, in different communities and industries. Because the nature of the work determines the length of the instruction period, the chart

NUMBER IN GROUP	LENGTH OF INSTRUCTION PERIOD IN PAY PERIODS (2 WEEKS)													
	1	2	3	4	5	6	7	8	9	10	11	12	13	14
2 to 4 inclusive	0.06	0.06	0.09	0.09	0.09	0.09	0.09	0.12	0.12	0.12	0.12	0.12	0.12	0.15
5 to 9 inclusive	0.09	0.09	0.12	0.12	0.12	0.12	0.12	0.15	0.15	0.15	0.15	0.18	0.18	0.18
10 to 15 inclusive	0.12	0.12	0.15	0.15	0.15	0.15	0.15	0.18	0.18	0.18	0.18	0.21	0.21	0.21
Over 15	0.15	0.15	0.18	0.18	0.18	0.18	0.18	0.21	0.21	0.21	0.21	0.24	0.24	0.24

Fig. 4-4. Supervision rates in dollars per hour for group leaders.

has been compiled using this latter factor. Furthermore, the chart of time in pay periods to reach average group efficiency (Figure 4-2) makes the use of this factor very convenient. Where several classes of work are done by the same group, the class calling for the longest instruction period should be used with the chart (Figure 4-4) to arrive at the supervision rate for the group leader, because he must have and use the ability to supervise the highest class of work done by the group.

EQUITABLE BONUS PLAN

One of the difficulties that is sometimes encountered with the type of group system that has been discussed thus far is that, for various reasons, different groups working with the same degree of effort make greatly different earnings. The reasons for this situation are largely connected with inaccurate work measurement practices and the improper maintenance of standards. This is not so likely to occur under the group system as under an individual payment plan. Nevertheless, there have been extreme cases where inconsistent earnings caused serious industrial relations problems that wage incentives had to be eliminated to restore harmony.

The equitable bonus plan was developed to satisfy the viewpoints of some managements and unions that it would be best, from the standpoint of industrial relations, if all operators would earn the same percentage bonus. It retains as many of the desirable features of the group system as possible and avoids some of the objections to the overall type of incentive plan where everyone in the plant is considered to be a member of one big group. It is, in effect, what might be called a group group system.

Its basic plan of operation is as follows. The operators working on a given class of work are considered as a group. One of their number is appointed as leader and is held responsible for the performance of the group. The amount of work accomplished by the group during the pay period in terms of allowed hours is totaled. The number of hours worked by the group during the same period is also totaled. The hours earned are then divided by the hours worked to determine the group efficiency.

The plan thus far is the regular group incentive plan. At this point, however, an additional computation is made. The groups that were not at least 100 percent efficient are dropped from the the the calculations and are paid only their guaranteed base rates. Then the hours earned and the hours worked by all groups that did make out are totaled. An average bonus percentage is computed by dividing the total hours earned by the total hours worked, and payment is made on this basis to everyone in the groups with an efficiency of 100 percent or better. Because the percentage bonus earned by all workers in all groups making out is the same, unsettling differentials are avoided.

The equitable bonus plan may be administered under various philosophies and policies, and various features may be added or omitted as the conditions of a specific installation may warrant. One type of arrangement developed in cooperation with the union is as follows.

The groups which do not make out receive no bonus. Therefore, there is a very strong incentive for them to try to make out. These groups and the groups which make out at less than the plant average—but which are paid the plant average—are considered to be a joint responsibility of management and labor. Management considers itself responsible for the accuracy of standards, for the maintenance of proper conditions, and for giving the operator adequate instructions. Labor, on the other hand, accepts the responsibility for the effort level. Thus, when a group that earns only 105 percent is paid the plant average of, say, 120 percent, management checks to make sure that it has not been at fault. If and when all conditions are proper, it expects 120 percent performance for 120 percent pay, and labor assists in seeing that this is attained.

Under the rules of administration of this particular equitable bonus plan installation, the maximum performance recognized for any group is 135 percent. It is agreed that this is the maximum that can be earned on a properly set standard by good performance alone, the factor which a wage incentive plan is designed to reward. Earnings higher than this are attributed to methods improvements made by either the operators or management. Jobs which have become too liberal because of management-sponsored methods improvements are restudied, as is done nearly everywhere else. Jobs too liberal because of methods improvements made by the operators are treated as follows. On jobs of this sort, if a group shows the ability to make more than 135 percent for more than one month, the work is restudied. An amount of money equal to the resulting savings for one year is immediately distributed among the members of the group. If the work will not continue for a year, the savings on the balance of the work to be done are distributed.

The equitable bonus plan possesses several distinct advantages. It offers the uniform earnings of an overall plant-wide incentive plan. At the same time, it retains all the advantages of requiring the setting of accurate standards and the checking of the efficiencies of small groups. It permits the inclusion for incentive payment of difficult-to-measure work such as maintenance work and material handling. This should be measured as accurately as possible for reasons of control, but errors in measurement will not cause earnings to fluctuate wildly. The plan requires management and labor to accept joint responsibility for its successful operation. They cannot consider the plan as something installed and hence to be forgotten, leaving the operators to bear the brunt of any substandard conditions which cause earnings to fall. At the same time, runaway earnings are prevented, although adequate rewards for methods improvements are permitted.

OVERALL PLANT-WIDE INCENTIVES

The idea of an overall plant-wide incentive plan based on some easily determined yardstick has been attractive to management for many years. The work of developing accurate standards for each operation in the plant is costly and time consuming, and the rate at which the incentive installation proceeds often seems undesirably slow.

For these and other reasons, managements have experimented with overall plant-wide incentives. Particularly during World War II, when the need to increase production quickly was so vital, plans of this sort were installed almost overnight. A yardstick, such as tons of material produced, number of carloads shipped per dollar of payroll, or some other readily obtainable figure, was chosen. The employees of the plant were then informed that a bonus would be paid to all of them in relation to the improvement shown beyond the established standard.

Some of these installations enjoyed temporary success and actually did increase production, reduce costs, and increase earnings. Others failed from the start. It is significant that even the successful plans were for the most part abandoned after the war as production switched to peacetime goods and that few of them were used thereafter.

It is generally conceded that overall plans work best where performance is quite poor at the time the installation is made. If 100 percent represents the performance which would be required to meet a measured standard, and if a plant is producing at a performance level of, say, 30 percent, then the installation of an overall plant-wide incentive plan may raise the performance level to 60 percent. This results in doubling output, which certainly is a very desirable result. The fact remains, however, that the plant is still producing far below its capacity.

One of the difficulties with plans of this sort is that the yardstick chosen may not bear a close relation to performance. The most popular single yardstick in the steel fabrication industry is tonnage. The effort required to produce a ton of fence posts, however, is very different from the effort required to produce a ton of fine-gage steel wire. Where the product varies greatly from pay period to pay period, earnings are likely to bear little relation to effort.

The other major difficulty lies in the size of the group. The larger the group, the more difficult it is for the operator to see the relation between his individual effort and his earnings. When the group gets too large, the effort level is bound to suffer.

BIBLIOGRAPHY

Aiken, William M., "Labor Performance Control," sec. 4, chap. 6, in H. B. Maynard (ed.), *Handbook of Modern Manufacturing Management*, McGraw-Hill Book Company, New York, 1970.

Chaffin, Don B., and Walton M. Hancock, *Factors in Manual Skill Training*, MTM Research Report 114, MTM Association for Standards and Research, Fair Lawn, N.J., 1966.

Hancock, Walton M., and James A. Foulke, *Effects of Learning on Short-cycle Operations*, Research Information Paper 2, MTM Association for Standards and Research, Fair Lawn, N.J., 1961.

Hancock, Walton M., et al., *Learning Curve Research on Manual Operations*, MTM Research Report 113A, MTM Association for Standards and Research, Fair Lawn, N.J., 1965.

Turban, Efraim, "Incentives during Learning—An Application of the Learning Curve Theory and a Survey of Other Methods," *Journal of Industrial Engineering*, December, 1968.

Chapter 5

Supervisory Incentive Plans

PHILIP F. CANNON

**Director, Coloney, Cannon, Main & Pursell, Inc.,
New York, New York**

Incentive pay systems for hourly paid workers at one end of the organizational scale and for members of top management at the other end of the scale have long been accepted methods of compensation. Frequently overlooked are the opportunity for and the value derived from the development and application of incentive pay plans to middle and lower levels of management. Nevertheless, under appropriate conditions, well-engineered incentive pay plans for supervisory personnel have significant value for the typical industrial company. The purpose of this chapter is to describe the development and application of such plans.

TYPE OF SUPERVISOR TO BE CONSIDERED

A discussion of the many different types of plans which might be applicable to various classes of supervisors in all types of commercial and industrial activity would be too long for these pages. However, the basic principles and techniques of incentive development have wide applicability, and it may be possible to apply these fundamentals to many more situations than will be discussed here.

The incentive plan discussed in this chapter is designed primarily for the industrial supervisor, usually at the level of department foreman, general foreman, or supervisor. For ease of reference, the term "foreman" will be used to cover this general stratum of management personnel. To be properly included under an incentive pay system, the foreman should:

1. Have the responsibility for costs properly assigned to him as part of his overall job responsibility. If this is assigned to him through the medium of a formal position analysis, so much the better.
2. Be able to exercise significant influence on the costs of his department's operation by virtue of increasing his skill and effort in managing.
3. Have a sufficient amount of controllable costs in his area to make the application of an incentive plan worthwhile to the company and to him.
4. Be assigned to an area of responsibility which corresponds to an accounting reporting area, so that his performance can be measured by the regular system of cost reporting.

Too often, the foreman's job is not considered to include cost responsibility, or the general understanding of his responsibility is limited and imperfect. The following extract from the organization manual of a well-known steel specialty manufacturer leaves no doubt of the foreman's role, in this particular company, in controlling costs:

> Assume direct responsibility for control of labor hours, material usage, and machine output.
> Operate within limits of expense budget for repairs, supplies, small tools, and other expenses assigned to his control under budget plan.
> Furnish information for and assistance in preparation of expense budgets as required.
> Recommend changes in methods of operation to reduce costs or increase production.

OBJECTIVES OF INCENTIVE PLAN

The supervisory incentive plan must be designed to accomplish these objectives:

1. Establish cost standards for the foreman to meet.
2. Measure the foreman's performance against cost standards; help him to do a better job by improving his performance against these standards.
3. Reduce manufacturing costs; maintain control thereafter to keep costs within predetermined limits.
4. Tap the skills and energies of the foreman; get him to realize that improvement in performance must stem from his doing something better in the future than he has done in the past.
5. Create an opportunity for the more capable foreman to increase his earnings as his value to his company increases.
6. Provide a useful training medium for development of more competent foremen.
7. Provide a commonly understood means of communication between layers of manufacturing management.
8. Give the foreman a greater sense of participation in the affairs of the company, particularly in the areas where he is directly concerned.

ELEMENTS OF A SOUND INCENTIVE PLAN

The elements of a sound incentive plan for foremen are quite similar to those of a good wage incentive plan. A plan must be:

1. Spelled out in the form of a policy and procedures manual.
2. Easy to understand on the part of the foreman; simple in design.
3. Easy to administer; it should avoid high clerical expenses in calculation of performance and pay.
4. Designed to reflect directly in higher incentive pay the extra effort and accomplishment by the foreman; it should contain enough incentive pay possibility to evoke active interest.

5. Able to reflect a reasonable relationship between effort (cause) and cost improvement (effect); the foreman must be able to see the results of his endeavor and to evaluate the results in terms of his improved performance.

6. Designed to promote teamwork among the supervisory group; this does not necessarily require a group incentive plan, but the atmosphere in which the plan operates should avoid competition among the foremen at the expense of the overall good of the company.

The plan should promote the "consultative" approach to management and should evoke the attitude of "how can I help you to do a better job?" from the foreman's immediate superior.

OTHER AREAS CONCERNED WITH THE PLAN

The major relationships on the part of the foremen with other areas of the organization in an incentive plan include close contact with manufacturing staff departments, the controller's division, and personnel relations.

Industrial engineering is deeply concerned with the development and administration of the incentive plan, inasmuch as a sound plan requires engineering of cost standards. The periodic revision of performance standards is part of industrial engineering's responsibility, and a close working relationship with the foremen is required to perform this function formally and effectively.

The budget department of the controller's division is likewise concerned with the establishment and revision of performance standards, in cooperation with members of industrial engineering. Explanation to the foremen of variations from standard and analysis of the means of improving performance are a continuing responsibility of the budget department if the plan is to promote understanding and acceptance by foremen of the company's system of reporting and controlling costs. Budgeting must be regarded as an aide to better manufacturing and not as the critic or the natural enemy of supervisors. Members of a budgeting department, who work with foremen to develop and administer standards, acquire a thoughtful understanding of the significance of the data with which they are constantly working.

Any phase of operations which revises basic relationships, involves new forms of communication, imposes new requirements for training, and highlights possible inadequacies in personnel is of deep concern to an alert and effective personnel relations department. Properly used, an incentive pay plan for foremen is a useful means of uncovering weaknesses or shortcomings in the capabilities and training of incumbents. Personnel relations has a major stake in seeing that the administration of the incentive plan is oriented to effect a maximum response from people who are learning new responsibilities and new techniques. Where necessary, personnel relations may make recommendations to replace, transfer, or upgrade supervisors according to the results of their performance against standards.

ORGANIZING TO SET BUDGETS

Because budgets should form the basis on which supervisory incentive standards are set, it is usual for the budget director to be responsible for directing the work of budget setting. However, the chief industrial engineer may be assigned to this task, as part of the general responsibility for cost control and incentive administration.

To assure success of the budget and incentive plans, the top manufacturing executive and his key management personnel will be directly concerned in the preparation of cost budgets and standards. Using sales forecasts and planned profit levels as guidelines, they must develop the associated manufacturing program required to meet these volume and profit objectives.

The assignment of responsibility for establishing budgets to a team composed of representatives of industrial engineering and of the budget department is a useful approach to organizing a budget team. The industrial engineer should be able to clarify historical costs and should emphasize the "engineered" aspects of setting cost standards. Further technical assistance in setting standards may be required of product or plant engineers, key quality control administrators, and other scientific personnel.

If foremen have little or no confidence in the result of the budgeting program, or if they are able to profess ignorance of the work done in setting the budgets in their departments, a major objective of the budgeting program is frustrated. The foremen must feel that they have had an important role in the budgeting program, and they must be aware of their responsibilities in making it successful.

REQUIRED SUPPORTING ELEMENTS FOR PLAN

Just as in the development and application of a wage incentive plan for hourly paid workers, the plan for supervisors must be strongly supported by adequate management techniques in related fields. The incentive plan cannot be a substitute for good production planning and scheduling, efficient factory layouts and operating methods, and proper selection and training of foremen; above all, an incentive plan cannot substitute for a sound salary administration plan; foremen's earnings from incentive performance cannot be regarded as a substitute for a proper level of base salary compensation.

It is not necessary, however, to wait for the final stages of perfection in plant operation before considering installation of a supervisory incentive plan. The actual installation of a properly engineered incentive plan designed to require basic supporting elements can be used to correct a number of organizational or procedural shortcomings as it is installed. This potential for correction of unsatisfactory related conditions is in itself one of the major benefits from an application of this sort.

INCENTIVE PLAN AND MANAGEMENT CONTROL SYSTEM

In the usual supervisory incentive plan, the cost standards which are used should be taken directly from the budget; the budget itself is the standard. The remainder of this discussion will use budget interchangeably with standards; to set the budget is to set the standards for the incentive system.

It is not possible to have a well-engineered incentive plan without some form of budget covering major manufacturing costs and without some planned financial goals for manufacturing operations. The incentive plan must be tied in to the regular reporting system of costs, using the same figures and data reporting plan. The budget used in the incentive plan should be the same one on which the company's profit and financial planning system is based.

One of the benefits of an incentive plan is to breathe life into a budget by getting the foreman actively interested in the budget and the related cost structures. Too often, the budget is a set of meaningless figures not directed to the foreman's knowledge and use and representing a vague threat to the foreman's security.

The traditional attitude of factory supervision to accounts and to budgets is one of vague fear and suspicion based on ignorance of the true function of the management control system. Through educating and making the foreman a party to actual administration of the budget, it is possible to break down these traditional attitudes. The accounting and budget functions should be regarded by the foreman as sources of help to him in his efforts to turn in a better job. The function is to help him manage, not to hold him up to criticism and penalty.

The starting point for the design of the plan is in the building or improving of a budget system which will distinguish between fixed and variable costs and will segregate controllable from noncontrollable costs.

COSTS COVERED BY PLAN

The major controllable costs of manufacture include direct labor, materials, and certain elements of overhead expense. Each of these major costs must be covered by standards to get maximum results from the supervisory incentive plan.

Factory Overhead. Another chapter of this Handbook deals with the types and techniques of setting budgets to control factory expenses. It should be studied in detail to accompany the development of the supervisory incentive plan.

For purposes of this discussion, there are three types of overhead costs to be considered in setting standards for either budgeting or incentive purposes. These costs are fixed, semifixed, and variable. Semifixed costs are sometimes known as programmed fixed costs because they can be planned ahead according to a manufacturing program to meet a forecasted level of sales activity. Once the decision to incur programmed costs has been made for a given fiscal period, these costs usually remain fixed unless significant revisions are made in schedules during the period covered by the programming.

In setting up the incentive plan for supervisors, only those costs which are controllable by the supervisory level to which the incentive applies will be discussed here. It is necessary to exclude fixed or noncontrollable costs in the plan to avoid the variances caused by changes in volume, and to keep the focus of supervisory attention on costs which they can influence.

Fixed Costs. Fixed costs are those costs which continue to be incurred regardless of the differences in levels of productive activity in the plant. The decision to build a plant in a given community under a certain tax system thereafter makes the costs of carrying the depreciation and taxes on that plant uncontrollable. Only the removal or closing down of the plant can significantly affect these continuing charges against the property.

Semifixed Costs. Semifixed costs are those which fluctuate according to various levels of productive activity within a plant but which do not vary in a direct proportion to changes in production, or at the same time that production schedules are altered.

In the noncontrollable category of semifixed costs are included such things as workmen's compensation, social security taxes, unemployment taxes, major building or process alterations, and employee training. Some of these costs are fixed by law, and others are the responsibility of top management. They must be included in the corporate budget; however, they should not be included in the section of the budget which is to be controlled by the supervisory group.

Within the semifixed category of costs, there are a number of controllable items such as maintenance, repair, overtime premiums, supervisory, factory clerical, and power. All these costs will fluctuate with levels of production, but not with the same sensitivity found in "pure" variable costs. In certain types of companies where sales forecasting can be developed to a high degree of accuracy because of the predictable nature of the business, it is possible to select the level of production activity required to meet sales forecasts. Knowing sales and profit goals, it is then possible to set up budgeted provision for semifixed controllable costs which are to be incurred according to anticipated levels of production.

In other types of industries, particularly in capital goods, it may not be practical to develop the budget plan for a period of as much as one year on the basis of sales forecasts. Unforeseen and substantial fluctuation in sales volume can create significant favorable or unfavorable volume variances in the budget plan to the extent that the cost standards for semifixed controllable items can be distorted beyond useful meaning.

In a large plant producing a product line associated with the food industry, the plant manager and supervisors are required to spend their planned allotments for maintenance of machines and equipment to ensure the expected and useful life of the fixed assets. Without this requirement to spend maintenance allotments, it would be possible to make short-term windfall budget gains by deferring vital maintenance programs. In time, however, this false short-term economy would be highly detrimental to the company.

Where the nature of a company's business limits the use of sales forecasts in programming semifixed costs, it is necessary to study the company's semifixed controllable expenses over a period of years to determine the extent of the fluctuation of these costs against production activities.

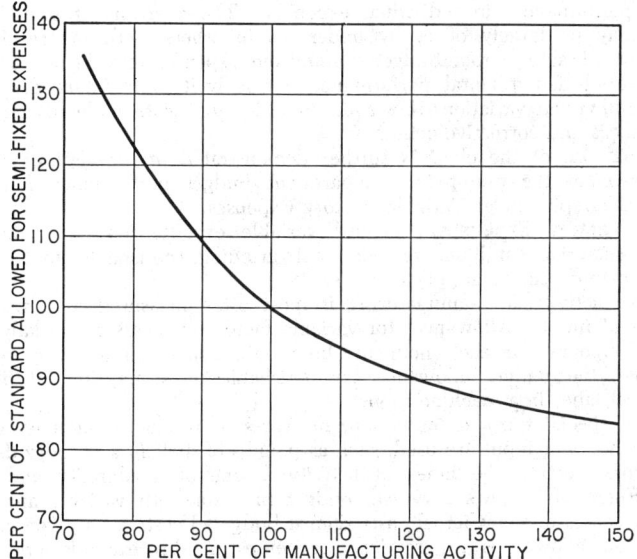

Fɪɢ. 5-1. Curve for determining allowance for semifixed expenses on the basis of manufacturing activity.

In studying the experience of a large manufacturer of capital equipment, it was found that the sum of all expenses included in the semifixed category rose neither so fast nor so far as did production activity in periods of high activity. In slower periods, these costs failed to decline at the same rate as production activity or "pure" variable costs. The extent and reliability of the correlation between semifixed costs and production were sufficient to be plotted on a chart (Figure 5-1), which was then used to determine allowances for semifixed expenses for a current budget period.

The chart is used in the following manner:

1. Determine the level of manufacturing activity in a production department by dividing the month's actual direct labor hours spent in the department by the normal, or 100 percent, direct labor hours in the department. The result of this calculation is the determination of the department's activity level.

2. Enter the chart at the calculated activity level for the department, and read up to the curve.

3. Read left from the point on the curve to the left-hand scale of the chart. The percentage reading on the scale is the correction factor to be applied against the allowance for semifixed expenses, based on the level of activity in the department.

4. Refer to a table of expenses to determine the "percent of direct labor" that is allowed for semifixed expenses at the 100 percent level of activity.

5. Multiply the allowance from item 4 by the department's actual direct labor dollars for the month. This product is the uncorrected budget dollar allowance for the month for semifixed expenses in the department.

6. Determine the corrected allowance for semifixed expenses by multiplying the uncorrected budget dollars from item 5 by the correction factor determined in item 3.

This somewhat empirical method of calculating allowances for semifixed expenses based on a curve such as is shown here may produce some distorted variations or erratic performances in individual expenses. The overruns of some categories, however, may be largely offset by underruns in others, with the result that the net variation of actual from budgeted semifixed expenses may be a true reflection of total overall departmental performance. It is well to maintain the routine of compiling individual variations, however, in order to isolate wide discrepancies for further analysis and corrective action.

The actual use of the chart is further demonstrated in developing some of the figures shown in the example of department budget performance found in the following paragraphs under Variable Factory Expenses.

Variable Factory Expenses. Typical variable expense items are departmental inspection, material handling, rework, tool grinding, perishable tools, and miscellaneous manufacturing supplies.

Productive activity in manufacturing is most often measured in terms of direct labor hours of input. Allowances for variable factory expenses are usually expressed as so much money for each hour of direct labor spent in a given department; for example, the budget allowance for perishable tools may be $0.15 for each hour of direct labor in a machine shop.

In certain special types of operation, measures of productive activity other than direct labor hours of input are used, such as pounds of melt in a foundry department. The test for selecting the index of activity is one of availability and reliability of data. Figure 5-2 shows a typical budget plan for both variable and semifixed overhead expenses to show actual costs against budgeted costs for a given month.

The budget allowances for semifixed expenses in this example are calculated with the help of Figure 5-1. At 113.6 percent of manufacturing activity the standard allowance for each of the five semivariable allowances is multiplied by 92.5 percent, which is the volume correction factor obtained from the Y axis of the curve.

Direct Labor Costs. The foreman may have the assistance of a well-engineered incentive plan to control direct labor costs in his department. Under a sound plan, the major controllable aspect of direct labor costs with which he is concerned is the amount of substandard labor costs that must be absorbed by the company through failure of operators to meet base standard production requirements. The presence of an undue amount of direct labor makeup allowance generally means

DEPARTMENTAL BUDGET PERFORMANCE

| ACTUAL DIR. LABOR COST: $19,688 |
| ACTUAL DIR. LABOR HRS: 8,750 |
| D.L. HRS. @ 100% ACTIVITY: 7,700   DEPT. NUMBER: 27 |
| % LEVEL OF ACTIVITY: 113.6%   DEPT. NAME: Genl. Machining |
| MONTH OF: June, 197 – |

VARIABLE EXPENSES ACCT. NO.			% OF DIRECT LABOR $ ALLOWED	STD. ALLOWANCE FOR MONTH	ACTUAL EXPENSES FOR MONTH	VARIANCE MONTHLY	STD. ALLOW. FOR QTR. TO DATE	ACT. EXP. FOR QTR. TO DATE	VARIANCE FOR QTR. TO DATE
10			18.2	$ 3583	$ 3394	$(189)	$11824	$10742	$(1082)
13			.2	39	65	26	129	205	76
14			1.3	256	400	144	845	1178	333
15			1.5	295	321	26	974	862	(112)
23			4.9	965	1116	151	3185	3005	(180)
26			3.5	689	650	(39)	2274	2105	(169)
31			.1	20	15	(5)	66	47	(19)
33			10.7	2107	1927	(180)	6953	6017	(936)
35			.3	59	50	(9)	195	340	145
36			.1	20	–	(20)	66	17	(49)
39			1.3	256	162	(94)	845	1100	255
44			.8	158	98	(60)	521	407	(114)
55-65			10.7	2107	1755	(352)	6953	5816	(1137)
75			.1	20	42	22	66	50	(16)
			53.7	$10574	$9995	$(579)	$34896	$31891	$(3005)

SEMIFIXED EXPENSES ACCT. NO.	% OF D.L. $ ALLOWED @ 100%	113.6% ACTIVITY CORRECT. FACT.*	% OF DIRECT LABOR $ ALLOWED	STD. ALLOWANCE FOR MONTH	ACTUAL EXPENSES FOR MONTH	VARIANCE MONTHLY	STD. ALLOW. FOR QTR. TO DATE	ACT. EXP. FOR QTR. TO DATE	VARIANCE FOR QTR. TO DATE
05	5.4	92.5	5.0	$ 984	$ 807	$(177)	$ 3050	$ 3117	$ 67
08	3.9	"	3.6	709	1114	405	2198	2400	202
12	15.4	"	14.2	2796	2361	(435)	8668	9162	494
16	2.6	"	2.4	473	712	239	1465	1312	(153)
86	5.9	"	5.5	1083	906	(177)	3355	3162	(193)
			30.7	$ 6045	$ 5900	$(145)	$18736	$19153	$ 417
			84.4	$16619	$15895	$(724)	$53632	$51044	$(2588)

*Taken from chart

FIG. 5-2.   Departmental budget performance sheet.

that operators are insufficiently trained, or that there are delays in the flow of materials, or that the quality of material on which the men are working is off standard to the point where standards cannot be met, or that the condition of tools and equipment is such as to slow down production. These reasons for low production are generally within the orbit of a foreman's control, and he should be held responsible for costs arising from nonstandard conditions.

Makeup hours, however, are not charged as direct labor to work in process but usually appear in the variable expense category of departmental overhead.

A company must have some form of measurement to determine whether its direct labor is being used effectively to meet required cost goals. A measured daywork plan may be established within a department, under which the departmental labor performance must equal or exceed, for instance, 85 percent of the predetermined standard. Unfavorable variances from this performance will usually be charged to the foreman as a manufacturing variance just as makeup allowances in an incentive situation, and such charge will also be made to an overhead account. The use of a standard cost plan, under which standards for direct labor operations are based on actual past average costs, will also require that unfavorable variances be charged as part of plant overhead.

Inability of the foreman to do his part in maintaining the proper physical environment and general tone of his department will be reflected in low output results. Where the investment in plant property per individual is high, it is essential that top production be maintained to get an economic return on plant investment. Higher input will also benefit the overhead cost structure by prorating fixed and semifixed costs to more units of output. It may thus be desirable to tailor the incentive plan to meet this high-productivity need by including added bonus incentives for greater departmental output.

Another area of direct labor cost control for which the foreman is responsible is in the proper assignment of operators to appropriate tasks according to their job grades. If, for instance, a first-class machinist is required to do simple drill press work for which the hourly rate is $0.40 an hour less than the machinist's base rate, the cost of this excess paid to get the drill press job done should be thrown into an unfavorable variance account and charged against the foreman.

It is perhaps conceivable that a budget plan could be established in a department where there are neither wage incentives, standard costs, nor measured daywork available as direct labor control instruments. There must be, however, some norm or standard established for each department to associate its through-put with some predetermined financial or cost objective; otherwise the company will be so lacking in basic management control systems as to make the installation of a supervisory incentive plan an entirely premature undertaking.

Raw Material Costs. The importance of control of raw material costs in many types of industries completely overshadows the significance of direct labor and controllable factory overhead costs. Such an example is provided by the soap industry, where the yield of finished product from raw material input holds a position of primary cost importance. Process industries can generally be identified where the manufacturing processes change the basic chemistry or nature of raw materials; it is in the conversion processes that major losses can occur. Standards of material usage or net yield from raw material input are essential not only to measure performance at periodic intervals but also to maintain continuing control over processes during the course of manufacture.

The development of material usage standards usually requires substantial engineering or scientific research, and highly sophisticated quality control procedures may be needed to assist supervision to maintain their operation within acceptable limits of material yield. The design of incentive standards and payment plans must reflect the importance to a company of securing maximum material yield.

In ordinary manufacturing, where only the dimensions or shape of material is changed, standards for material usage are also required. However, the nature of the control mechanisms is usually much simpler, and periodic reporting may be sufficient to meet control requirements. The control requirements extend beyond those of detecting nonstandard incoming purchased materials, responsibility for which may be primarily placed on receiving inspection personnel.

Excessive use of material and costs of scrapping partly processed parts for improper workmanship or bad tooling are usually the major sources of material loss in ordinary manufacturing. Such loss is usually reflected in the overhead account as excess material costs or as the cost of scrap, including material value and applied labor and overhead. Costs of rework and salvage are also reflected in factory overhead costs. Carefully engineered standards covering allowances, scrap, rework, and salvage are part of a budget's requirements.

The responsibility for purchasing production material according to predetermined standard costs is usually assigned to the purchasing agent or a high-ranking officer of the company, depending on the nature of the business. The cost control program discussed here is not concerned with questions of the price of purchased material but only with the yield secured from it or the costs of scrap and rework.

ACCOUNTING FOR BUDGET VARIANCES

The point to which a company's accounting system charges unfavorable budget variances of labor, material, and overhead is covered in another chapter of this Handbook. Whether variances are gathered into inventory values or charged directly to current manufacturing activity, the measurement of these variances for incentive determination should be the same.

SETTING BUDGET STANDARDS

The setting of budget standards for direct labor and production material is usually a straightforward project because of the direct measurement techniques available in these two areas. But in the setting of budgets for controllable overhead, the problem of arriving at valid and consistent standards is more difficult. Many types of expenses cannot be quantitatively linked to manufacturing activity through any set of fixed values or relationships. The soundest approach is based on "engineering" cost standards as far as possible—that is, analyzing and measuring each element of cost to be sure that allowances are based on tangible reasons.

A major weakness in using unalloyed historical costs as the basis for budget and incentive standards is that the foreman who has spent money carelessly in his department has created for himself an overstuffed standard—in contrast to the foreman who has operated his department prudently with a minimum of expense. Improvement in the latter department may be harder for the conscientious foreman because there is less water to squeeze out than there is in the department where there has been less restraint and care in incurring expenses. As a matter of practical experience, however, it has been observed that foremen who have been considered as poor operators before installation of the budget and incentive plans continue to have trouble in meeting their standards after installation even though it is recognized that their standards could be tighter. And it seems that the foreman who has earned the reputation of running a tight ship before the budget can continue to show cost reduction against standards which are undoubtedly tighter.

The role of the industrial engineer is important in straightening out as far as is possible the base for cost standards in each department. If he is knowledgeable in the operation of the department and if he will become conversant with its operating history, he should be able to extract from the cost records over the past years the out-of-line expenditures associated with that department's history. Engineering cost elements through a review of the operating history of the department is essential if a consistent basis for the budget in each department is to be attained.

One large chemical company solved its problem a number of years ago simply by allowing in each forthcoming year only 75 percent of the previous year's expenditure in each department on the assumption that the water would continue to flow out of these expense accounts as long as permitted. After a number of years of this jackscrew approach, the padding was taken out of all budgets, so that final standards represented hard-to-meet realistic and profitable financial goals for each department.

The argument can be advanced that an incentive plan should be deferred until the budget plan has had the benefit of two or three years of carefully controlled and supervised operation. The merit in this argument is obvious in that unrealistic allowances can be reduced to solid standards before tying in budget performance to the supervisor's pay. On the other hand, there is logic in the argument that tying the supervisory incentive plan to the budget as soon as standards have been installed is the more effective way to get the budget in a fully operative and effective condition. This latter point can be accepted if there are rules and procedures governing the continuing revision of budgets and standards and if there are qualified administrative personnel available to do the job of revision.

REVISION OF STANDARDS

The usual incentive pay system for factory workers contains restrictions on revision of standards unless associated changes in methods justify such revision. This is a sound restriction, inasmuch as a company's industrial engineering department has usually had a chance to engineer the method and the standard so that they represent the best possible means of doing the job.

Because the same quantitative engineering steps cannot be followed in establishing overhead cost standards, it is necessary that the budget be subject to revision. The need for revision is a continuing one due to the fact that the structure of factory costs in most companies undergoes constant change. Increased investment in machines and other laborsaving equipment tends to increase the ratio of overhead to direct labor—a condition which has been accelerating rapidly throughout American industry.

Continuing change occurs in manufacturing methods, requiring associated changes in direct labor standards. And the great progress in materials of recent years must be reflected in revisions of material costs.

The importance of making changes in the budget, however, is second only to the importance of making the changes in a manner which is understandable and acceptable to the foremen who are involved in such changes. A close working relationship between the budgeting group and the line supervisory organization will naturally facilitate budget revision. Revision should be a continuing matter wherever practical, and sweeping year-end changes that bear the stamp of an arbitrary decision from higher echelons must be avoided.

MAKING SPECIAL ALLOWANCES

It may be that some categories of controllable expenses may be beyond the control of department foremen under certain nontypical circumstances. One such example is furnished by a change from high-speed steel to carbide cutters in a general machining department. The initial investment in carbides will run to many thousands of dollars to get the program under way and to replace the existing inventory of high-speed steel cutters. Certainly, the regular budget allowance for perishable tooling will not cover a major expenditure of this sort. Not only will a special budget allowance be required to cover the initial purchases, but new budget standards for perishable tools and tool grinding expenses must be established.

In the early years of working with budgets, there is particular need for flexibility in their administration New refinements born of experience and understanding will dictate some change, and some special allowances may be necessary. Arbitrary or unreasonably fixed ideas on the part of top manufacturing management that the budget cannot be changed may breed a lack of confidence on the part of supervision in the fairness of the plan. The system should be the tool of management, not the master, and should be used with intelligence to accomplish its major objectives.

Tight control over budget changes and incentive standard allowances must be maintained by top manufacturing and financial management, upon recommendation of key budgeting or manufacturing line and staff people.

RESTRICTING ELEMENTS OF PLAN

Control over the three major costs of manufacture—direct labor, production material, and overhead—has been discussed so far in this treatment of a supervisory incentive plan. Unless unusual circumstances dictate otherwise, these three elements of cost control should be the only factors considered. For the incentive plan is an economic instrument, designed to measure and reward directly the performance of foremen in their economic role. Serious danger attaches to attempting to include in the plan the measurement of subjective factors such as loyalty, initiative, and ability to work with other people. Use of these factors may raise the suspicion of favoritism or the belief that tangible results on cost standards can be erased by inaccurate judgment of the intangibles.

If subjective rating of a supervisor is to be done in such a way as to affect his pay, the salary administration plan covering his base pay should be used. Merit rating can be used alongside and in addition to the incentive plan, in such a manner to cover evaluation of factors deliberately omitted from the incentive plan.

INDIVIDUAL VERSUS GROUP PLANS

The classic array of arguments concerning the merits of group and individual incentive plans can be brought to bear on the selection of one type over another—or a mixture of both.

When there is but little interdependence among foremen and where it is possible to measure accurately each one's performance, the individual plan is usually and properly preferred. Where the reverse is true, the group plan should be used—but only to cover those situations where there is true linkage and dependency in the group.

If the plan will foster a spirit of competition, there is nothing intrinsically wrong unless competition between individuals develops to the detriment of the company's operation. If the action of one foreman can directly and adversely affect the operation of another, the probability is that they should be grouped so that their combined effort can produce the greatest overall benefit.

THREE TYPICAL PLANS

Each individual company will have its own particular characteristics and problems to which its plan must be tailored, as shown in the three following examples:

Machinery Manufacturer. A large builder of production machinery focuses primary attention in its plan on the control of factory overhead. Material losses are noted and charged into overhead accounts covering scrap, rework, and salvage costs. Direct labor makeup hours are also charged to an overhead account.

Productivity of direct labor above the 100 percent required of standard labor performance does not have a primary effect on this company's overhead structure. And a major revision of the hourly wage incentive plan shortly after the development of the plan for supervisors has made productivity data difficult to obtain on a consistent basis. Substandard direct labor hours are charged to "direct labor makeup hours" as part of variable overhead.

The incentive scale permits a maximum incentive premium of 25 percent. No premium is earned when the budget is met, but any improvement of performance beyond meeting the budget is paid according to the curve shown by Figure 5-3.

To calculate "percent performance on standards," the total budgeted expenses are divided by the total actual expenses for the month.

The pay scale is plotted on an arithmetic scale in this example and pays a higher rate for each 1 percent of improvement as performance moves up the scale.

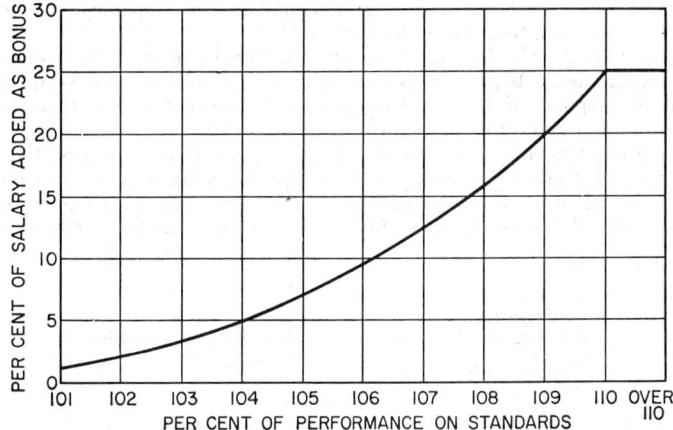

Fɪɢ. 5-3.   Projection of incentive scale for machine manufacturing company.

This is done on the theory that the last 1 percent of improvement is harder to attain than the first 1 percent over standard. If this incentive scale were plotted on a semilogarithmic scale, it would be a straight line connecting minimum and maximum earnings.

The plan is an individual system of payment; a few group bases exist, but only where individual performances cannot be sorted out accurately.

Assistant foremen who do not have direct responsibility for their department's controllable expenses are nevertheless of considerable help to the foremen or general foremen who are in charge. Assistant foremen in areas where the incentive is applied are given an incentive reward on their base salaries equal to one-half of the percentage of base pay earned as a premium by their associated foremen or general foremen.

This company operates its plan on the principle that only foremen who are directly connected with areas of incentive application will be paid incentive premium. Also, there are no "overrides" for the assistant superintendent or superintendent level.

In many instances, inclusion of these higher supervisory levels in the plan on a basis of the average performance of all their assigned departments can be done without detriment to the plan. However, each company's situation should be studied carefully to determine the feasibility of such inclusions.

Incentive earnings are paid quarterly, although each foreman's performance is reported and analyzed monthly. The bonus check is a separate one, and every effort is made to disassociate it from regular monthly base salary.

Package Manufacturer.   Material usage, spoilage, supplies, indirect labor, service and clerical labor, and direct labor are the budget headings. Actual dollars of costs are compared against budgeted costs, and performance is determined by dividing budgeted controllable costs by actual costs.

Under this plan, all eligible members of plant supervision and staff share equally on a plant-wide basis. The group system is used because it is thought that a higher degree of cooperation and plant efficiency is attained in this way. The scale showing the percent of salary earned for plant cost control performance is given in the table on page 6-90.

The limit of salary added for performance against plant cost budgets is 30 percent. Interim payments are made during the year, and the balance remaining is paid at year end. All payments are made in separate checks, distinct from salary earnings.

Chemical Manufacturer.   Four elements are measured in a large chemical company: raw material yield, direct labor performance in terms of output attained by equipment and machines, controllable manufacturing expense, and an overall rating of the individual foreman on qualitative factors.

The bonus scale for the foreman starts after 80 percent of machine or departmental efficiency is attained; 1 percent premium is added to his base salary for each 1 percent of increase of labor efficiency over 80 percent of the theoretical limit of output as established by process and machine capacity.

In calculating incentive earnings on material yield and scrap loss, the monthly totals of finished production and daily scrap and degrading reports are summarized, and raw material and work-in-process inventories are verified to determine performance on material standards. The following scale is used to determine incentive earnings:

Performance on material standards, %	Percent of salary added
95	1
96	3
97	5
98	7
99	9
100	11

Plant cost control performance expressed as a percentage of budgeted controllable costs		Percent of salary added to participants
From	To	
−2.50	−2.39	1
−2.38	−2.11	2
−2.10	−1.83	3
−1.82	−1.55	4
−1.54	−1.27	5
−1.26	−0.99	6
−0.98	−0.71	7
−0.70	−0.43	8
−0.42	−0.15	9
−0.14	+0.14	10
+0.15	0.42	11
0.43	0.70	12
0.71	0.98	13
0.99	1.26	14
1.27	1.54	15
1.55	1.82	16
1.83	2.10	17
2.11	2.38	18
2.39	2.66	19
2.67	2.94	20
2.95	3.24	21
3.25	3.52	22
3.53	3.80	23
3.81	4.08	24
4.09	4.36	25
4.37	4.64	26
4.65	4.93	27
4.94	5.21	28
5.22	5.49	29
5.50 and up	......	30

On budgeted overhead expenses, the following scale applies:

Performance on expense standards, %	Percent of salary added
95–96.9	1
97–98.9	2
99–100.9	3
101–102.9	4
103–104.9	5
105 and over	6

The fourth category includes qualitative factors such as departmental cleanliness, general level of product quality, relationships with departmental personnel, relationships with other supervisory and management personnel, attendance, and general attitude. The rating is done monthly by the plant manager and the foreman's superior, and the results are reviewed in a series of monthly progress meetings, when each department's performance is analyzed and reviewed. Additions to base salary range from 1 to 5 percent.

Performance below the established incentive base in one category will not deprive the foreman of a bonus earned in others. A theoretical top performance in each category would add 42 percent to a foreman's salary. In practice, such a result would be so difficult and unusual to attain as to be highly improbable. The actual range of incentive earnings is usually in the range of 10 to 30 percent.

Incentive for top plant management is based on the overall performance of the plant against the foregoing yardsticks, as well as further measurement against cost and performance standards tailored to these higher levels of control and responsibility.

PUTTING CEILINGS ON BONUS EARNINGS

Most plans have either an explicit or implied limit on the percent of salary added through incentive performance. The justifications for such limits include these valid arguments:

1. The foreman should be doing his job well enough on a straight salary basis as to make excessively high performances the product of windfall, rather than extra effort and skill in management.
2. Relationships of the foreman's take-home pay to the salary of nonincentive salaried personnel in other departments must be preserved on a reasonable basis.
3. It is not always possible to maintain strict accuracy and up-to-date revisions of standards; the company should be protected against excessive bonus payments which are not really earned.

ESSENTIAL INGREDIENTS OF SUCCESS

In each of the foregoing examples, there is the added ingredient of solid top management backing of the plan as one of the major management techniques for cost control. Each of the plans has the continuing problem of maintaining standards in an up-to-date condition, and each plan requires intelligent and consistent administration. Without these added factors, no plan can be assured of success.

BIBLIOGRAPHY

Boyden, Arthur C., "Supervisory Motivation and Compensation," sec. 9, chap. 5, in H. B. Maynard (ed.), *Handbook of Modern Manufacturing Management,* McGraw-Hill Book Company, New York, 1970.
Sibson, Robert E., *Wages and Salaries,* American Management Association, New York, 1960.
Supervisory Management Compensation Report, 11th ed., American Management Association, New York, 1966.

Chapter **6**

Job Evaluation

DAVID J. CHESLER

**Program Director, Navy Training Research Laboratory,
Naval Personnel Research Activity, San Diego, California**

RICHARD H. LEUKART

Secretary, National Screw and Manufacturing Company, Cleveland, Ohio

JAY L. OTIS

**Professor of Psychology; and Director, Psychological Research Services,
Case Western Reserve University, Cleveland, Ohio**

A textbook on job evaluation has for its subtitle, A Basis for Sound Wage Administration.[1] This phrase, the authors of this chapter believe, expresses most succinctly the purpose of job evaluation. Job evaluation will not automatically eliminate conflict between employer and employee, but in its absence, it is difficult to see how a basis for resolution of differences can be established. Constant vigilance in examining the content of jobs, the worker characteristics required to perform them, and employee remuneration is essential in industry today. This vigilance must be systematic rather than haphazard and must be based on something more than good and honest intentions.

[1] Jay L. Otis and Richard H. Leukart, *Job Evaluation*, 2d ed., Prentice-Hall, Inc., Englewood Cliffs, N.J., 1954.

6-92

This is not to say that high-minded intentions are not important. Job evaluation is not a precise method of measurement. In its essence, it is a judgment or rating process. The engineer accustomed to fine tolerances and measurement by devices that minimize errors to tiny fractions may be impatient with those of his brethren who, together with the industrial psychologist and personnel administrator, have attempted to apply the scientific method to wage administration. However, the challenge of industrial conflict cannot be sidestepped, and the best attack is one which emulates scientific method, coupled with integrity. Integrity is all the more important in evaluating jobs because job evaluation procedures have not, and never will, reach the degree of demonstrable exactness of many of the techniques to which the engineer is accustomed.

Another important point should be made. The final product of job evaluation is the pricing of jobs. However, more meets the eye here than mere dollars and cents. There are standard of living, the welfare of family, security for the present and the future, job satisfaction or dissatisfaction, job prestige, and acceptance or rejection of the proffered fruit of toil. One should add the factor of morale, the morale not only of the employee, but also of the employer. These are emotional factors that play an important part in the job evaluation process.

The most difficult part of job evaluation undoubtedly comes after the formal rating of job elements and worker specifications has been completed, the wage curve determined, and the wage classes superimposed on the curve. Whether the plan and the results can be imposed on the parties vitally affected is another question. This problem can, of course, be solved, as many instances in industry testify. More will be said of this later. For the moment, it is sufficient to note that a job evaluator must understand human relations. He must educate; he must be able to work with groups and to compromise differences; he must elicit confidence; and he must be patient.

WHAT IS JOB EVALUATION?

According to the Occupational Analysis and Industrial Services Division of the United States Employment Service, job evaluation is defined as:[2]

> The complete operation of determining the value of an individual job in an organization in relation to the other jobs in the organization. It begins with job analysis to obtain job descriptions and includes relating the descriptions by some system designed to determine the relative value of the jobs or groups of jobs. It also involves the pricing of these values by establishing minimum and maximum salaries for each group of jobs based on their relative value. The operation ends with the final checking of the resulting salary system.

The salient features of this definition are (1) job evaluation starts with job analysis; (2) relative rather than absolute values of jobs are obtained; (3) jobs are grouped into classes for which minimum and maximum wages or salaries are established.

Job evaluation has functions other than the pricing of jobs. Because it classifies jobs in terms of specific abilities and human characteristics, it aids in the selection of new employees. It can be a guide in the transfer and promotion of personnel. And of course, it should serve to improve employee-management relations and to increase job satisfaction among individual workers.

WHAT IS JOB ANALYSIS?

Of the three significant characteristics in the definition of job evaluation cited above, job analysis is probably the most important. This is true because it is

[2] *Industrial Job Evaluation Systems*, Department of Labor, United States Employment Service, Occupational Analysis Branch, Washington, D.C., 1947, p. 19. See also Robert D. Gray, *Systematic Wage Administration in the Southern California Aircraft Industry*, Industrial Relations Counselors, Inc., New York, 1943, p. 89.

the foundation for the whole job evaluation structure. Job analysis is used for many purposes other than job evaluation. The most important of these are selection, training, promotion, and transfer. In each case, the purpose determines to a large degree the form that the job analysis will take.

Job analysis has been formally defined as:[3]

> . . . the process of determining and reporting pertinent information relating to the nature of a specific job. It is the determination of the tasks which comprise the job and of the skills, knowledges, abilities, and responsibilities required of the worker for successful job performance.

WHAT IS A JOB?

For purposes of job analysis and job evaluation, the term "job" must have a precise meaning. To give it this exactness, two other terms must be defined: "task" and "position." The War Manpower Commission in its *Training and Reference Manual for Job Analysis* defines task and position as follows:[4]

> A task is created whenever human effort must be exerted for a specific purpose. The purpose may be physical, as pulling and lifting, or mental, as planning and explaining. The effort may be exerted to change a material or merely to maintain the status quo of a material. The material may be tangible, as boards and nails, or intangible, as numbers and words.
> When enough tasks accumulate to justify the employment of a worker, a *position* has been created. A position, therefore, is an aggregation of duties, tasks, and responsibilities requiring the services of one individual.

According to this definition, the number of positions in an organization is the same as the number of workers.

Analysis on the basis of positions is undesirable because two or more positions might have exactly the same or very similar descriptions. Hence a more basic term is needed, namely, job. A job has been defined as:[4]

> . . . a group of positions which are identical with respect to their major or significant tasks. Therefore, a job may be considered as a group of positions which are sufficiently alike to justify their being covered by a single analysis. Despite this, the analyst should always treat a job as being performed by a single worker even though his analysis is a composite of several positions.

These three definitions, for task, position, and job, respectively, are the ones upon which further discussion in this chapter will be based.

KINDS OF JOB EVALUATION SYSTEMS

Various classifications have been used to describe job evaluation systems. The most common one is that presented in *Industrial Job Evaluation Systems* published by the Department of Labor. This is a fourfold plan, as follows:

1. Nonquantitative
   a. Ranking system
   b. Grade description system
2. Quantitative
   a. Point system
   b. Factor comparison system

Variations of these four systems are in use, as are systems that are combinations of two of the four main types. For example, Johnson, Boise, and Pratt advocate a method that utilizes features of both the point system and the factor comparison system.[5]

---

[3] War Manpower Commission, Division of Occupational Analysis, *Training and Reference Manual for Job Analysis*, Government Printing Office, Washington, D.C., 1944, p. 7.

[4] *Ibid.*

[5] Forrest H. Johnson, Robert W. Boise, Jr., and Dudley Pratt, *Job Evaluation*, John Wiley & Sons, Inc., New York, 1946.

The authors of this chapter favor a quantitative rather than a nonquantitative system. It is believed that in the long run a quantitative system will be more acceptable to employees and will better serve additional purposes such as community wage surveys and the comparison of jobs in one organization with those in another.

From another point of view, all four systems may be considered "quantitative" in that they place jobs in hierarchical classes. The difference is one of degree. The ranking and grade description systems may be considered rough or gross techniques as compared with the point and factor comparison systems which are more precise and which utilize finer units of measurement.

Choosing a system is the first important decision that faces an organization planning a job study. Each system has its advantages and disadvantages in terms of complexity, time, cost, and ease of comprehension.

WHAT JOBS ARE INCLUDED IN A JOB STUDY PLAN?

All the jobs in an organization may be subjected to job evaluation, but it is poor practice to study them under one plan. Clerical jobs should not be evaluated with factory jobs. Jobs should be separated for study into large natural divisions or major groupings. In a manufacturing concern, it is best to evaluate hourly or factory jobs separately from the office jobs (clerical, supervisory, and administrative). Jobs of officers and executives are rarely included in a job study.

Thus, although an organization will use one of the four systems for evaluating its jobs, the number of specific plans will vary with the number of major groupings of jobs. A separate wage curve will be derived for each group, although these wage curves must, of course, be equitable with respect to each other. It is possible to construct a plan that would be applicable to the universe of jobs, but such a plan would be extremely difficult to apply in the everyday task of maintaining the wage administration system. Experience would seem to indicate that such a plan would lead to conflict and inequities rather than the solution of differences.

THE RANKING SYSTEM

The job ranking method is easy to understand and easy to explain to others. It usually takes less time than the other methods and requires fewer forms and less paper work. However, it is generally agreed that it is less accurate than either of the two quantitative methods, although probably more accurate than the grade description system. It is far superior to arbitrary job pricing, and where time is a major consideration, it is particularly useful. Its greatest limitation stems from the fact that ranking, as a method of evaluating jobs, persons, or anything else, merely places objects in a hierarchical order from high to low. It is as though one had three pieces of string of varying lengths. It is easy to determine which is longest, shortest, and intermediate in length. But it is not known *how much* longer one piece is than another. A frame of reference is needed such as a foot rule or yardstick. Given a foot rule, two or more persons could "evaluate" entirely different sets of strings and exchange information so that all the strings could be accurately compared with each other. This is impossible with a straight ranking method, although one can resort to various statistical procedures to obtain estimates of the comparative lengths.

Another disadvantage is that the judges must be thoroughly familiar with all the jobs to be included in the plan. In a large organization, this is a difficult condition to achieve. In addition, the judges are likely to be unduly biased by knowledge of the current wages for the jobs being ranked and by the quality of the incumbents of the jobs. These criticisms apply, of course, in some degree to all job evaluation systems.

Evaluating jobs by the ranking system consists of six major steps:
1. Job analysis
2. Selecting the jobs
3. Choosing the rankers

---

1. Job Title ___LARGE HEADER SETUP OPERATOR___   2. Department_____

3. No. on Job_____ 4. No. in Dept. _____ 5. Date _____

Statement of Job

The LARGE HEADER SETUP OPERATOR sets up and operates a battery of cold bolt headers using wire 3/8" diameter and over. Header operations include open and solid die work on one, two, or three blow headers. The LARGE HEADER SETUP OPERATOR must set up and adjust his machines and be able to handle a variety of materials and work to close tolerances. Occasionally tolerances of one one-thousandth of an inch must be held. Works from simple blueprints and usually has supervision on call when needed.

Duties

1. Sets up and operates a battery of cold bolt headers, using one, two, or three blows working on wire 3/8" diameter and over. Places and adjusts placement of hammers, solid or spring upset punches or hammers, and dies, solid or open. Occasionally must hold tolerances as small as one one-thousandth of an inch. Usual tolerance is two one-thousandths.

2. May have to touch up dies or hammers or change their shape or size slightly by polishing or scraping to achieve desired results. Frequently has to fit dies and hammers and adjust for wear of machine and errors in tool-

10. Contacts his supervisor for approval of the job and for further instructions on special and experimental products requiring more than usual supervision and instructions.

11. Checks production of previous shift to determine amount still to be run.

Worker Specifications

Normally two and one-half years of experience required to become familiar with the functions of the machine and to operate it competently. Tenth grade education or its equivalent is desirable. Must be able to understand simple drawings of product, use measuring instruments, and be able to understand all written instructions. Previous experience as a HEADUP JUNIOR SETUP OPERATOR is usually mandatory.

FIG. 6-1.   Portion of a job description for a factory job.

4. Ranking the jobs by departments
5. Statistical treatment of rankings
6. Integrating department rankings

Job Analysis. As already indicated, job analysis is the first step not only in the ranking system but in all systems. Figure 6-1 shows a portion of a job analysis for the job of large header setup operator.[6] This analysis contains certain identifying information (items 1 to 5, inclusive), a summary statement of the job, a listing of the tasks comprising the job, and the worker specifications. The length and amount of detail in job descriptions varies considerably in industrial and business organizations. Excellent examples of job analyses are those presented in the *Training and Reference Manual for Job Analysis*.[7] These are probably too detailed and lengthy for job evaluation purposes in most organizations, but they may well serve as a standard for format, clarity, and thoroughness.

---

[6] From *Handbook of Job Evaluation for Factory Jobs*, p. 24 (modified), courtesy of Industrial Fasteners Institute, Cleveland, Ohio (formerly American Institute of Bolt, Nut and Rivet Manufacturers), privately printed, not for distribution.

[7] War Manpower Commission, Division of Occupational Analysis, *op. cit.*, pp. 79–100.

In the ranking system, the job evaluator must study the complete description for each job. He can then write on a card, of a size convenient for sorting, the title and a brief description of the job. The summary statement is usually sufficient for the brief description.

Selecting the Jobs. As already indicated above, all the jobs in an organization are rarely, if ever, included in one job study plan. Factory jobs and office jobs are evaluated separately. If time is a consideration, the step of selecting the jobs may precede the writing of job descriptions, but only if the lines of demarcation among the major job groupings are clear-cut and unquestioned. In practice, however, there are usually some jobs which cannot unequivocally be placed in one major grouping or another. For example, certain foreman jobs may be more accurately compared with clerical, administrative, and supervisory jobs than with maintenance or factory jobs. It is good insurance against later conflict, therefore, to have job analyses completed for most, if not all, of the jobs in an organization as the first step in the job evaluation process.

Choosing the Rankers. In the ranking system, the first ranking of jobs is done by departments. This is necessary because few individuals know all the jobs in the company. Therefore, the first rankings are usually made by department supervisors such as foremen, assistant foremen, shop stewards, or other unit or assistant unit chiefs. Other persons who are well acquainted with the jobs in one or more departments may also perform initial rankings; such persons are industrial engineers, job analysts, personnel workers, and plant and office managers. Persons who are selected to rank jobs should not be expected to evaluate jobs they do not know and should limit their judgments to jobs about which they are well informed.

Ranking the Jobs by Departments. This is the process of arranging the jobs in each department in a hierarchy from highest worth to lowest worth.

If abbreviated job descriptions have been typed on small cards, these cards can be arranged in the rank order desired. The most difficult job is given the rank of 1, the next most difficult the rank of 2, and so on until all the jobs have been ranked. If two jobs are given the same rank, each job is given the average of that rank and the following rank. For example, if two jobs are ranked 11, they both receive the rank of 11.5, that is, the average of 11 and 12. The job ranking immediately below these two tied jobs is given the rank of 13. If three or more jobs tie, the same principle is followed. Evaluators should avoid giving the same rank to more than one job because this defeats the purpose of job evaluation, which is to differentiate jobs on the basis of difficulty. If the jobs are examined closely enough, a basis for ranking one higher than another can almost always be found.

Statistical Treatment of Rankings. One method which has been used to facilitate ranking is the method of paired comparison. This method has resulted in greater reliability and consistency of the judgments made. In this method, each job is compared separately with each of the other jobs, so that only two jobs are compared at a time. For example, suppose the following jobs were to be ranked:

Assembler
Welder, gas
Milling machine operator
Shaper operator, spindle
Machinist, general
Tool and die maker

Each of these jobs would be paired with each of the others as shown in Figure 6-2, taking care to randomize the pairs so that there is no systematic presentation of one job in the list, and also to place each job in the second position as often as in the first. The results of each rater's comparisons may be summarized as shown in Figure 6-3.

The paired comparison method has the advantage of simplifying the decisions

Instructions: In each of the pairs of jobs listed
below, underline the job which you believe is more
difficult and should receive the higher wage. Please
be sure to make a choice for each pair even though it
may be hard to distinguish between the difficulty
level of the two jobs.

Assembler	Milling Machine Operator
Milling Machine Operator	Tool and Die Maker
Machinist, General	Shaper Operator, Spindle
Welder, Gas	Assembler
Machinist, General	Milling Machine Operator
Tool and Die Maker	Machinist, General
Assembler	Machinist, General
Shaper Operator, Spindle	Tool and Die Maker
Assembler	Shaper Operator, Spindle
Welder, Gas	Shaper Operator, Spindle
Milling Machine Operator	Welder, Gas
Welder, Gas	Machinist, General
Shaper Operator, Spindle	Milling Machine Operator
Tool and Die Maker	Welder, Gas
Tool and Die Maker	Assembler

Fig. 6-2.   Method of paired comparisons.

that have to be made because only two jobs are compared at a time. It is very practical when the number of jobs involved is small. However, the number of comparisons to be made increases very rapidly as the number of jobs increases. The formula for the number of comparisons to be made for N jobs is $[N(N-1)/2]$. For 10 jobs, the number of comparisons is 45, and for 40 jobs it is 780.

Integrating Department Rankings. When all the jobs have been ranked separately by departments, the next step is to obtain a single set of rankings for all the jobs included in the job plan. This step is usually accomplished by a committee representing the various departments and including persons whose knowledge of the jobs cuts across the various departments.

JOB	NUMBER OF TIMES JUDGED MORE DIFFICULT	RANK
Assembler	0	6
Welder, Gas	3	3
Milling Machine Operator	2	4
Shaper Operator, Spindle	1	5
Machinist, General	4	2
Tool and Die Maker	5	1

Fig. 6-3.   Summary of paired comparison judgments.

In this step, the jobs in one department are compared with the jobs in another. Then the jobs in a third department are compared with those in the first two departments, and so on until all the jobs have been considered. The first result is one similar to that shown in Figure 6-4. Committee members should be provided with worksheets similar to Figure 6-4. Entries for each department are completed after the committee has arrived at a final placement of all the jobs.

THE GRADE DESCRIPTION SYSTEM

As its name implies, the grade description or job classification method consists essentially of a series of descriptions of job levels or job grades. Each job description is compared with the various grade descriptions and assigned to the one which most nearly defines the complexity and responsibilities of the job. Minimum and maximum pay rates are established for each grade level.

An example of a grade description plan for factory jobs is illustrated in Figure 6-5. This plan contains eight grade levels.

The crux of the grade description method lies in the definitions of the various grade levels. Such descriptions are not easy to write. Each grade level description must be general and at the same time specific enough to cover a variety of tasks and duties. This condition is often difficult to achieve because many jobs have some tasks which are at a low level of difficulty and other tasks at a high level. In addition, each grade level must be distinct from the grade levels adjacent to it and at the same time represent a logical step in a continuum, which must be a continuum in fact and not a scale with discernible gaps. In brief, it must be possible to assign each job under consideration to one of the grade levels without difficulty. If such difficulty is experienced with too many jobs, then the definitions of the grade levels must be revised or new levels must be added. It is less likely, although quite possible, that a grade level would be eliminated. This might occur if very few jobs were assigned to a particular job level, and if such jobs approximated very closely the worth of jobs in adjacent levels.

The process of establishing a grade description plan may be considered as fivefold:

1. Determining the type of position to be evaluated
2. Writing the job descriptions
3. Determining the number of grades
4. Writing the grade level descriptions
5. Assigning the jobs to grades

Determining Types of Positions to Be Evaluated. It was stated early in this chapter that a job evaluation plan is not usually designed to apply to all jobs in an organization. The first step, therefore, is to determine the types of positions for which it is to be used.

Writing the Job Descriptions. This step has already been discussed above as one of the steps in the ranking method.

Determining the Number of Grades. In most organizations there is a non-formalized, gross grade structure in the minds of both management and workers. Initial attempts at setting up the grade levels usually result in a formalization of this hitherto nonformalized structure. It may be expected that the broader the classification of jobs according to type, the greater the number of grade levels. The greater the range of remuneration, the greater the probability of additional grade levels. Similarly, it may be expected that the number of grade levels required will increase as the range of skills covered in the jobs increases. In an organization where upgrading is the method of advancing meritorious workers, the number of grade levels will probably be greater than in an organization where wage increase within grade level is preferred.

The number of grade levels ranges from a minimum of six to a maximum of about twenty. In small organizations with few jobs, less than six may occasionally be feasible. The trend is generally toward a small rather than a large number of levels, and this is understandable from the viewpoint of ease of administration of the plan. In an organization as large as the United States Civil Service, where

Total rank	Accounting	Clerical	Engineering	Administrative	Sales
1				President (1)	
2				Vice president (2)	
3					
4				General manager (3)	
5					
6	Controller (1)				
7					
8				Assistant general manager (4)	
9					
10	General auditor (2)		Senior mechanical engineer (1)		
11					
12			Mechanical engineer (2)		Sales manager (1)
13					
14	Auditor (3)			Executive assistant (5)	
15					
16	Senior accountant (4)		Chief designer (3)		Jobbing salesman (2)
17					
18	Accountant (5)		Senior designer (4)		
19		Chief clerk A (1)			
20					
21	Junior accountant (6)	Chief clerk B (2)	Designer (5)		Senior salesman (3)
22		Senior clerk A (3)			
23	Bookkeeper (7)	Senior clerk B (4)	Assistant designer (6)		Salesman (4)
24					
25	Assistant bookkeeper (8)	Clerk A (5)	Junior designer (7)		Junior salesman (5)
26			Senior draftsman (8)		
27		Clerk B (6)	Draftsman (9)		
28			Junior draftsman (10)		
29		Assistant clerk (7)	Tracer (11)		
30					
31		Junior clerk (8)	Junior tracer (12)		

FIG. 6-4.   Integrating rankings of jobs in different departments. (Numbers in parentheses indicate departmental ranking.)

GRADE 1

Jobs included in Grade 1 are very simple. None requires over 1 month of experience and most can be learned satisfactorily in 1 week. Light laboring jobs such as janitor or sweeper and other light unskilled jobs are in this grade. Many jobs having to do with packing the product come in this category. Often the most difficult part of the job is that the worker must be on his feet nearly all the time.

GRADE 2

This grade includes more jobs and more employees than any other. Most of these are concentrated on the numerous semiautomatic machine-feeding, sorting and inspecting jobs. Most of the laboring jobs in the plant also fall within this bracket. Helpers and servicemen as well as learners on machine jobs are for the most part included here, too. So also are operators of relatively simple equipment. As a rule the experience requirements for jobs in Grade 2 run between 1 week and 3 months. Responsibilities on these jobs are usually very small although they often rate high on effort.

GRADE 3

Almost as many jobs are included in Grade 3 as in Grade 2. Operators of machines of medium difficulty are included here. Inspection jobs involving responsibility and discretion are in this grade. Most of the jobs involving learning or helping to set up and operate complex machines are Grade 3 jobs, as are some maintenance jobs of semi-skilled variety. This grade is definitely one covering semiskilled jobs.

GRADE 4

The setting up and operating of most of the plant machines are included in this grade. Grade 4 also covers many maintenance jobs and a variety of individual jobs involving considerable skill. Few of these jobs can be learned in less than 1 year and most require from 1 to 3 years of experience. Responsibilities on these jobs are usually substantial.

GRADE 5

Jobs in this grade are all of a high degree of skill and as a rule take up to 5 years to learn. Most of them also involve substantial responsibilities for products and materials and frequently considerable responsibility for the work of others. Skilled maintenance jobs, setting up complex machines, and some floormen's jobs are included. These latter jobs include certain minor supervisory activities.

GRADE 6

Only jobs requiring a high degree of skill are in this grade. Most of the small number of jobs are floormen. On all these, both the experience and responsibility demands of the jobs are high. The most highly skilled maintenance department jobs are also in Grade 6, as are several machinist, toolmaker, and diemaker jobs. As a general rule, from 5 to 8 years of experience is required on these jobs. An ability to work independently with only a small amount of supervision is characteristic of most of these jobs.

GRADE 7

This grade covers jobs similar to those in Grade 6 except that these are a little more exacting. There are only three jobs in this grade. These are the pattern-maker, toolmaker, Grade B, and the most difficult floorman's job in the plant. Eight to ten years of experience is required.

GRADE 8

Jobs in this grade are the most difficult and require the most skill of any jobs in the plant. The men on these jobs are expected to be able to plan and carry out their work with little supervision. Only the top toolmaker and machinist jobs merit inclusion in this bracket, and these rank here largely because the men on this work must almost be machine designers to carry out their work. Much of the equipment in the plant is specially built, or is very old, and parts must be designed and made by these men. These jobs, calling for from 8 to 10 years of experience, are the top jobs in the plant.

FIG. 6-5. Grade description plan for factory jobs. (*From Jay L. Otis and Richard H. Leukart,* Job Evaluation *2d ed., Prentice-Hall, Inc., Englewood Cliffs, N.J., 1954, pp. 98–99. Reprinted by permission.*)

there are as many different jobs as one could reasonably expect to find anywhere, the grade description plan in use for "General Schedule" (GS) jobs contains 18 levels.[8]

Writing the Grade Level Descriptions. Grade level descriptions may be written after the jobs have been tentatively separated into groups according to difficulty, or they may be written without such an initial classification of jobs. Writing grade level descriptions is usually a task that is shared by the members of a committee rather than one individual. When the descriptions are written in advance of an initial classification of jobs, it is best to write the description for the lowest and highest levels first. This defines the limits of the grade description scale. At this point, it is advisable to select some of the most simple and most difficult jobs and see if they are satisfactorily classified by the two extreme definitions. Modifications of or additions to the two definitions may be in order.

Descriptions for the second highest grade and the second lowest grade are then prepared, and again appropriate jobs are compared with the descriptions to see if they are suitably classified. This process is continued until there are sufficient grade levels to encompass all the jobs in the study. Throughout the process, constant reference should be made to the job descriptions to make sure that important job elements have been included for each level.

If the jobs have been first sorted into groups ranging from low to high, it is easier to write the description for either the highest or lowest grade level and then those for the remaining grades in order. Here again, it is advisable to make frequent reference to the job descriptions.

In summary, it may be pointed out that constructing a grade description plan requires a knowledge not only of all the jobs involved but also of the principles and rules of rating scale construction, because in essence the grade description method is a graphic or descriptive rating scale designed to apply to a specified sample of jobs.

THE POINT SYSTEM

The point rating system is generally conceded to be the most commonly used system in industry. It utilizes rating scales to measure specific job characteristics or factors which are common to many jobs—for example, education, amount of supervision exercised, on-the-job training time, and the like. Each factor is assigned a certain number of points on the basis of its judged relative worth, as compared with the other factors. On the basis of this point value, each level or degree in the rating scale for each factor is assigned a point value. Each job is then considered separately for each factor and evaluated against the appropriate factor scale. For example, a particular job may receive 42 points for education, 16 points for physical effort, 60 points for experience, and so on. The points obtained by each job on the various factor rating scales are then summed, and the total represents the relative worth or difficulty of the job.

The point system is more complicated than either the ranking or grade description system. It is more difficult and requires more time to construct. Considerably more paper work is involved. However, the fact that it is the most popular method would seem to indicate that it has advantages which outweigh the disadvantages. Once in use, both management and workers can understand it with ease, and what is most important, independently reach similar point values for a job. Rater agreement is high, and this is important if the concept of rater agreement may be accepted as a criterion for validity of job ratings.

The process of constructing a point system and putting it into use may be described in nine steps:

    1. Determining type of position to be evaluated

    2. Selecting the factors

    3. Defining the factors

[8] Copies of civil service grade descriptions may be obtained from the U.S. Government Printing Office, Washington, D. C.

4. Defining degrees for each factor
5. Determining relative values of job factors
6. Assigning point values to degrees
7. Constructing the job evaluation manual
8. Preparing the job descriptions and job specifications
9. Rating the jobs and obtaining final point values

Determining Type of Position to Be Evaluated. The same general considerations exist here as in the ranking method. This first step is especially important in the point system because the factors selected for inclusion in the plan (step 2 above) will vary with the type of position to be evaluated.

Selecting the Factors. An examination of the job factors that have been used in existing job evaluation plans reveals that there are many of them. However, it is also clear that certain ones occur again and again, although under somewhat different names. A survey of seventeen point systems for factory jobs and twelve systems for clerical and supervisory jobs to determine what factors were used most frequently yielded the results shown in Figures 6-6 and 6-7. The factors were classified, as shown in these figures, under the headings of skill, effort, responsibility, and working conditions. The authors of this survey also report that some of the systems used these headings as factors, without any further breakdown. Figures 6-6 and 6-7 are interesting because they show what management and workers feel should be the main bases for job remuneration. It should be remembered that these lists represent the most frequent factors used and that, for certain types of jobs or situations, other factors may be very suitable. These lists can be helpful in determining the factors to be included in a point rating plan.

In the practical situation, the selection of factors is best done by a committee. Each member should independently compile a list of what he considers suitable factors. A final agreement can be obtained later by the full committee.

Certain principles should be kept in mind in the selection of factors:

1. The factors should be pertinent to the type of position included in the job study. If no supervisory positions are included, there is no point in having the factor, "supervision of others." However, "supervision received" might be very appropriate. If the physical surroundings are universally excellent, such a factor should be omitted.

2. Only important factors should be selected. Otherwise the final list will be unmanageable. It is suggested that fifteen factors be the maximum number and that preferably ten to twelve be used. It should be possible to cover the more important factors and some of the less important job characteristics within these limits.

3. The factors selected should not overlap in meaning. For example, "physical skill" and "dexterity" are practically the same in meaning. As much as possible, therefore, an effort should be made to select factors which are unique with respect to each other.

4. The factors chosen must be ones which lend themselves to differentiation in terms of "amount" of the job characteristics that they represent. This means that they must be ratable and that they can be described in terms of varying degrees. They must be quantifiable, at least in terms of brief verbal descriptions which show a clear hierarchy from high to low. To take an easy example, "education" can be described in precise terms, such as "eight years of school," "high school graduate," or "Ph.D. in economics." A factor such as pulchritude would be much more difficult to describe on a scale of varying values, although fortunately such a factor would rarely be included in a job study plan.

5. Factors should not be included on which all or most of the jobs are given the same rating. This is equivalent to adding a constant to the value of every job and does not contribute to differentiating among the jobs. For example, if all the jobs are performed under the same physical conditions, a factor for this job characteristic is not required.

6. The factors selected must be acceptable to both workers and management.

SKILL

Education	Job knowledge	Resourcefulness
Education or mental development	Knowledge of machinery and dexterity with tools	Versatility
Trade knowledge		Job skill
Schooling	Knowledge of materials and processes	Manual dexterity
Experience		Manual accuracy and quickness
Previous experience	Mentality	Dexterity
Experience and training	Mental capability	Degree of skill and accuracy
Training time	Accuracy	
Training required	Ingenuity	Physical skill
Time required to learn trade	Initiative and ingenuity	Ability to do detailed work
	Judgment and initiative	
Time required to adapt skill	Intelligence	Social skill

EFFORT

Mental effort	Fatigue due to eye strain	Muscular or nerve strain
Mental application	Physical effort	Fatigue
Mental or visual demand	Physical application	Monotony of work
Concentration	Physical demand	Monotony and comfort
Visual application	Physical or mental fatigue	

RESPONSIBILITY (FOR)

Safety of others	Supervision of others	Protection of materials
Material or product	Supervision exercised	Physical property
Material and equipment	Cost of errors	Plant and services
Equipment or process	Necessary accuracy in checking, counting, and weighing	Cooperation and personality
Equipment		
Product		Coordination
Machinery and equipment	Effect on other operations	Details to master
		Quality
Work of others	Spoilage of materials	

WORKING CONDITIONS

Unavoidable hazards	Occupational hazard disease	Surroundings
Hazards involved		Dirtiness of working conditions
Exposure to health hazard	Danger—accident from machinery or equipment	Environment
		Job conditions
Exposure to accident hazard	Danger—from lifting	Disagreeableness

FIG. 6-6. Factors selected for factory point rating systems. (*From Jay L. Otis and Richard H. Leukart, Job Evaluation, 2d ed., Prentice-Hall, Inc., Englewood Cliffs, N.J., 1954, pp. 118–119. Reprinted by permission.*)

If a sufficient number of jobs are extremely repetitive, omission of a "monotony" factor would lead to dissatisfaction, even though it could be demonstrated statistically that this factor contributes little to the final point ratings of the job, or that it could easily be covered under another factor.

Selection of the factors to be included in the plan is not a simple matter. The six principles enumerated above show some conflict among themselves, and careful judgment and common sense will have to be exercised in making the final selection. Although all or most of the jobs may show such little variation with respect to a "monotony" factor that it might well be omitted, nevertheless if the workers

SKILL

Mental requirement	Managerial techniques	Capacity for getting
Mentality	Difficulty of work	along with others
Mental application	Education	Capacity for self expres-
Creative ability	Preparation for the job	sion
Judgment	Essential education and	Social skill
Analytical ability	knowledge	Ability to do detailed
Initiative	Basic knowledge and	work
Resourcefulness	experience	Ability to do routine
Versatility	Experience, knowledge,	work
Skill requirement	and training necessary	Manual or motor skill
Complexity of duties	Previous experience	Office machine operation
Personal requirements	Training time	Manual dexterity
Ability to make decisions	Experience and training	

EFFORT

Physical requirement	Physical or mental	Mental effort
Physical application	fatigue	Volume of work
Physical effort	Manual effort	Attention demand
Physical demand	Pressure of work	

RESPONSIBILITY (FOR)

Executive responsibility	Dependability and	Methods
Personnel	accuracy	Determining company
Supervision of others	Accuracy	policy
Character of supervision	Details	Market
given	Quality	Contact with others
Work of others	Effect of errors	Contact with public, cus-
Monetary responsibility	Material	tomers, and personnel
Commitments, property,	Equipment	Good will and public
money, or records	Records	relations
Company cash	Confidential data	Cooperation and person-
		ality

WORKING CONDITIONS

Job conditions	Working conditions	Attention to details
Tangible surroundings	Personal hazard	Out-of-town travel
Intangible conditions	Monotony	

FIG. 6-7. Factors selected for clerical point rating systems. (*From Jay L. Otis and Richard H. Leukart, Job Evaluation, 2d ed., Prentice-Hall, Inc., Englewood Cliffs, N.J., 1954, pp. 118–119. Reprinted by permission.*)

feel strongly that this job characteristic is an important element in their work, then it had best be included.

Defining the Factors. By the time the persons responsible for selecting the factors have made their final decision, enough discussion should have taken place in completing that step to assist greatly in defining the factors. The job factors should be clearly defined and should mean the same thing to all persons involved in constructing the job study plan. Here again, it may be worthwhile for several persons to write definitions independently and then to compare and integrate them into a final definition. Varying interpretations on the part of others of ideas, concepts, words, and phrases should be anticipated as much as possible, and appropriate revisions made. For example, if "training" is a factor, it should be carefully stated whether this includes formal schooling only or, in addition, a period of apprenticeship or vestibule training. Some factors which are very specific, such as "responsibility for funds," will be relatively easy to define. Other more general factors, such as "complexity of duties," will be more difficult. An example of a definition for a specific factor is the following.[9]

> Experience and Training. Use this factor to record the time it usually takes an individual to acquire the ability needed for normal production and effective performance of the job's other duties. Give points for the experience factor *over and above* those given for education. In rating this factor, remember that experience is of two kinds: (*a*) previous experience on related work, either within or without the organization, or on lesser jobs, directly related to the productive attainment of this job; and (*b*) the breaking-in time, including special training courses, or period of adjustment, required to reach normal production. This factor does not include time spent in jobs owing to lack of turnover ahead. Use it to weigh only the actual learning time.

An example of a definition for a more general factor is the following:[10]

> Complexity of Duties. Use this factor to appraise the job's requirements for independent action, exercise of judgment, and creative effort in devising new methods or new products. Rate a job high in this factor if it requires a great deal of judgment, and ability to resolve complex data or problems into units that can be evaluated and compared. Rate the job low in this factor if it is circumscribed by standard practice.

Defining Degrees for Each Factor. This step consists essentially of constructing a rating scale for each factor. Each level or category of the scale is commonly referred to as a degree. For example, the factor "experience and training" cited above may contain six degrees, as shown in Figure 6-8.

[9] Johnson, Boise, and Pratt, *op. cit.*, p. 76.
[10] *Ibid.*, p. 77.

DEGREE	DEFINITION	TYPICAL JOBS
1	Up to three months	Key punch operator
2	Three to twelve months	Jr. engineer, aircraft design
3	One to three years	Engineer, layout, aircraft design; detail engineer, aircraft design
4	Three to five years	Methods engineer B; major estimator A; major estimator B
5	Five to seven years	Assistant group engineer, aircraft design; lead engineer, aircraft design
6	Seven to ten years	Lead technician A, motion-picture laboratory

FIG. 6-8.   Degree definitions for "experience and training."

Degree	Definition	Typical jobs
1	Simple routine duties, requiring the use of only a few definite procedures and little individual judgment, the work either being performed under immediate supervision or involving little choice as to methods of performance.	Keypunch operator, messenger
2	Duties are clearly prescribed by standard practice but require the use of several procedures and the making of minor decisions requiring some judgment.	Clerk C, stenographer A, typist A.
3	Duties involve an intensive knowledge of a restricted field and require the use of a wide range of procedures and the analysis of facts to determine what action, within the limits of standard practice, should be taken.	Detail engineer, aircraft design; junior engineer, aircraft design; methods engineer B; major estimator A; major estimator B; executive secretary; placement man, personnel; librarian A.
4	Duties involve general knowledge of company policies and procedures and their application to cases not previously covered. Duties require working independently toward general results, devising new methods, and modifying or adapting standard procedures to meet new conditions. Decisions, however, are based on precedent and company policy.	Major engineer, layout, aircraft design; engineer, layout, aircraft design; lead technician A, motion picture laboratory; senior auditor A, internal auditing.
5	Difficult work on highly technical or involved projects, presenting new or constantly changing problems. Duties require outstanding ability to deal with complex factors not easily evaluated, or the making of decisions based on conclusions for which there is little precedent.	Assistant group engineer, aircraft design; lead engineer, aircraft design.

FIG. 6-9.   Degree definitions for "complexity of duties." (*Modified from Forrest H. Johnson, Robert W. Boise, Jr., and Dudley Pratt,* Job Evaluation, *John Wiley & Sons, Inc., New York,* 1946, *pp.* 77–78.)

Another example of degree definitions is illustrated in Figure 6-9. These are the degree definitions for the general factor "complexity of duties" cited above.

It is apparent that degree definitions for a general factor, like the definition of the factor itself, are more difficult to write, primarily because such factors do not lend themselves easily to quantitative description.

The number of degrees should not be greater than are reasonably required to differentiate the jobs. There may be as few as two or three if all the jobs can be grouped into that many distinct categories with respect to the factor concerned.

FACTOR	PLAN			
	NEMA AND NMTA	GENERAL ELECTRIC	WESTINGHOUSE	U.S. STEEL
Skill	50%	62½%	60½%	45%
Effort	15%	12½%	22½%	16%
Responsibility	20%	12½%	13½%	24%
Job Conditions	15%	12½%	3½%	15%
Total	100%	100 %	100 %	100%

FIG. 6-10. Relative point values in four job evaluation plans.

This implies that all the factors included need not contain the same number of degrees.

Degree definitions should be stated in terms that are understandable to the workers. The language of the trade and the industry elicit better acceptance than words which are not customary. Whenever possible, objective terms should be used rather than subjective terms. It is better to say "machining operations involving tolerances below 0.001 inch" than "machining operations involving precise accuracy." Examples should be used to illustrate elements within the degree definitions and the degree itself.

Determining Relative Values of Job Factors. Some factors are more important than others in determining the values of jobs. Hence they should not all have the same weight. The problem of determining the relative weights of the factors is one of the crucial steps in the point system. However, this step may be made easier by considering the experience of other organizations. Figure 6-10 summarizes the experiences of four well-known plans—National Electrical Manufacturers Association and National Metal Trades Association, General Electric Company, Westinghouse Electric Corporation, and United States Steel Corporation. The major factor, "skill," receives the highest weight in all four plans.

There are enough differences among the four plans in Figure 6-10 to caution against adopting the weights used in any one of them for use in one's own organization. This merely emphasizes the fact that a point system should be tailor-made for one's own organization, although it is helpful to consult the plans of other organizations.

Relative values of factors are usually the result of the consensus of one or more committees. The method may be summarized as follows:[11]

1. One or more juries are selected to judge the relative values of the factors.
2. These juries are instructed to study the job evaluation manual carefully, especially the factor definitions and degree definitions.
3. Each member of the jury is asked to rank the factors in order from the one which contributes most to the total value of the jobs to the one which contributes least.
4. Instructions are then issued to members of the jury: Assuming that the relative values when totaled should equal 100 percent, distribute this 100 percent among the factors according to your judgment of their relative values. Make sure that the values assigned total 100 percent.
5. The relative values so obtained are averaged.

Some persons who are unfamiliar with rating methodology may have difficulty with the concepts involved in steps 3 and 4 above. The important thing is to think of the entire universe of jobs and not of one particular job or group of jobs. Factors are being ranked and then weighted—as they apply to all jobs. One should ask himself such questions as, "Should jobs be paid more for experience than for hazards?" or "Should education receive more consideration in pricing jobs than monotony?" and so on. The fact that a particular job which one may have

---

[11] Otis and Leukart, *op. cit.,* pp. 137, 138.

in mind is very monotonous and requires little education—as compared with one which requires a great deal of formal training and is not monotonous—does not mean that the monotony factor should receive more weight than education. When two such jobs are rated for the monotony factor, one will be rated high and the other low, and the converse will occur when the jobs are rated for the education factor.

Assigning Point Values to Degrees. After the relative weights of each factor have been determined, it is relatively simple to assign points to the degrees in each factor. The first step is to decide upon the theoretical range of total points that could be obtained on all the factors. This is the same as saying that a decision must be reached upon the lowest score that any job could possibly achieve and also the highest score. The former would be achieved by a job that was rated on the lowest degree for each factor, and the latter instance would occur if a job was rated on the highest degree for each factor. Actually, two such jobs are quite hypothetical, but the concept is useful in determining the theoretical range of points for the plan.

Because the theoretical range is an arbitrary matter, any two numbers, sufficiently far apart to permit differentiation among jobs, can be selected. A simple and convenient method is to allow the percentage weights for the factors to represent also the point values for the lowest degree in each factor. This is illustrated in Figure 6-11, where the entries for degree 1 are the same as the factor weights that were determined for this particular job evaluation plan. Theoretically, therefore, a job could receive a minimum of 100 points under the plan.

The theoretical maximum was determined by multiplying 100 by 5 = 500. The theoretical range for this plan therefore is 100 to 500. This allows a possible range of 400 points, which is sufficient to differentiate among jobs. The point value for the highest degree of each factor is obtained by multiplying the value of the lowest degree by 5. For example, the value of degree 6 of factor 1, "work experience," is $21 \times 5 = 105$.

The values for the intermediate degrees in Figure 6-11 have been assigned on

Factor	Degree					
	1	2	3	4	5	6
1. Work experience........................	21	38	55	71	88	105
2. Essential knowledge and training...........	17	31	44	58	71	85
3. Dexterity...............................	4	9	15	20		
4. Character of supervision...................	10	20	30	40	50	
5. Character of supervision given..............	11	22	33	44	55	
6. Number supervised.......................	7	13	18	24	29	35
7. Responsibilities for funds, securities, and other valuables...............................	6	11	16	20	25	30
8. Responsibility for confidential matters.......	6	18	30			
9. Responsibility for getting along with others...	6	14	22	30		
10. Responsibility for accuracy—effect of errors...	6	12	18	24	30	
11. Pressure of work.........................	4	9	15	20		
12. Unusual working conditions.................	2	6	10			

FIG. 6-11. Assignment of point values according to arithmetic progresssion. (*From Job Evaluation Manual for Clerical, Supervisory, and Administrative Positions, courtesy of Psychological Research Services, Case Western Reserve University, Cleveland, Ohio, 1968.*)

the basis of an arithmetic progression. Some plans utilize a geometric progression to obtain the intermediate degree values.

Constructing the Job Evaluation Manual. When the six steps described above have been accomplished, the job evaluation point system is complete, and if job descriptions and job specifications are available, one may proceed to rate the jobs. The job evaluation system is usually embodied in a manual for convenience of use.

Preparing the Job Descriptions and Job Specifications. Before the jobs can be accurately rated, job descriptions and job specifications must be prepared. In the point system, the job specifications are keyed to the factors of the job evaluation plan. The job specification consists essentially of a series of descriptions, one for each factor. Each description contains information showing the extent to which a factor is present in the job. This information must be complete and factual enough to permit an accurate assignment of the job to one of the degree levels of the factor scale. Care must be taken to avoid writing this information in the terminology of the degree definitions, because this amounts to prejudging the job with respect to that factor. A portion of a complete job description and job specification is presented in Figure 6-12.

Rating the Jobs and Obtaining Final Point Values. Given the job evaluation manual and the job descriptions and specifications, one is prepared to rate the jobs and to obtain their final point values. Needless to say, more than one person will be assigned this responsibility, and their ratings should be compared and differences of opinion resolved. The process consists essentially of reading each job description and specification carefully and deciding what degree level for each factor rating scale best describes the information that has been obtained under that factor heading in the job specification. In instances of doubt, the job analyst may desire to obtain additional information relating to the job description or to some particular factor in the specification. This should, by all means, be done. Having determined what degree level in each factor rating scale best describes a job, the job analyst then records the number of points assigned to the degree levels selected and totals these points to obtain the final point value for the job.

THE FACTOR COMPARISON SYSTEM

The factor comparison system is the most recent of the four basic systems. Its leading advocate has been Eugene J. Benge.[12] Benge objected to the point system on the grounds that (1) it assumes all jobs are composed of the factors that are selected; (2) point values are assigned to degree levels of the factors in an arbitrary manner, especially the upper limits of the factors; (3) the point system contains "seeming refinements" which are unjustified; (4) the unit of measurement, namely, the "point," is undefined; (5) factors are frequently not defined; (6) the final value of a job is based on a job analysis rather than a comparison of the job with other jobs.

Advocates of the point system reply to objections 1, 2, 3, 5, and 6 as applying to the factor comparison system also. Objection 4 is claimed to be an advantage rather than a disadvantage. Many instances are cited to show that factor comparison plans also use a point unit in spite of the alleged monetary base.

It would seem that the proponents of each of the two quantitative systems are basing their criticisms not on the system per se but on the manner in which it has been constructed or applied in some particular organization. Benge's objection 5 above is certainly in this category. The point system and the factor comparison system both have advantages and disadvantages. These are probably not nearly so important as the thoroughness and accuracy with which either method is established and maintained in any company.

One definite advantage of the factor comparison method is that it is tailor-made for each company, because the primary basis for evaluating jobs is to compare

[12] Eugene J. Benge, *Job Evaluation and Merit Rating*, National Foremen's Institute, Inc., New York, 1943.

---

1. Job Title ___BOLTMAKER SETUP OPERATOR___   2. Department _____

3. No. on Job _____   4. No. in Dept. _____   5. Date _____

Statement of Job

The BOLTMAKER SETUP OPERATOR sets up and operates two National Boltmakers on all types of jobs. Usually sets up and operates the more difficult jobs requiring very close tolerances with a minimum amount of supervision. Is able to select, grind, and adjust all types of dies and tools. Assists in the stocking and re-stocking of machines. Supervises one BOLTMAKER HELPER.

Duties

1. Makes complete setup and adjustments, and operates two boltmakers.

2. Sharpens on a pedestal grinder pointing tools, trimmer dies, grip fingers, cutter, and quilt. Grinds kicker pins to proper clearance.

Experience: Degree_____Points_____
   What is the length of time usually required by a worker to obtain sufficient work experience to perform the job duties effectively? Does the job require one or both of the following?
   Previous experience on related work, or lesser positions, within or outside the organization?

   A period of adjustment and/or "breaking in" on the job itself?
   The "Rating" is the total required experience before being placed on the job plus the period of adjustment or "breaking in" on the job itself.

   • Requires five years to attain full manual and mental proficiency on the job.

Education: Degree_____Points_____
   Refers to formal school training or its equivalent in general knowledge which requires some instruction. This factor measures the requirements for the use of shop mathematics, drawings, measuring instruments, and general educational background.

   • Ten years. Must be able to speak, read, and write the English language. Understand micrometers, measuring gauges, simple drawings, and all written instructions.

FIG. 6-12.   Portion of a job description and job specification for use in a point system job evaluation plan. (*From* Handbook of Job Evaluation for Factory Jobs, *pp.* 26, 39–41, *courtesy of Industrial Fasteners Institute, Cleveland, Ohio, privately printed, not for distribution.*)

them with selected "key jobs" in the company. The relative positions of these key jobs constitute the steps of the rating scales, and because key jobs in one company are rarely, if ever, the same as those in another, each company is forced to construct its own factor comparison plan. On the other hand, key jobs, like any job, are subject to change, and as soon as this happens, the scale of measurement also changes.

Probably the strongest criticism of the factor comparison method is the establishment of scales on the basis of monetary units. It is felt by many that this introduces a biasing or contaminating feature in that job analysts will be influenced by the current wages being paid for the jobs, and this in turn may prolong existing inequities, contrary to the purpose of a job evaluation plan.

As might be expected, therefore, there have been modifications of the factor comparison system, which in essence are combinations of the point system and factor comparison system.

The steps in the factor comparison system which follow are based primarily upon the description by Benge, Burk, and Hay.[13]

The method consists of eight basic steps:

1. Selecting the factors
2. Preparing the job descriptions and job specifications
3. Selecting the key jobs
4. Ranking of key jobs by factors
5. Apportioning pay rates among the factors
6. Setting up the job comparison scale
7. Adding supplementary key jobs to the job comparison scale
8. Evaluating the remaining jobs

Selecting the Factors. Benge suggests five basic factors to be used in a job comparison scale: (1) mental requirements, (2) skill, (3) physical requirements, (4) responsibilities, (5) working conditions. The definitions of these factors are presented in *Job Evaluation and Merit Rating*.[14]

Preparing the Job Descriptions and Job Specifications. This step is similar to the procedures followed in the point system. The job specifications will, of course, be keyed to the particular factors selected for the plan.

Selecting the Key Jobs. According to Benge, from fifteen to twenty "key jobs" should be selected by a committee. These jobs must be subject to no dispute as to duties and current pay rates. A further requirement is that they sample adequately the range of the pay scale; that is, they must vary in pay from very low to very high.

Although Benge and his colleagues have not emphasized the limitation of a job study plan to one type of position, their examples imply that this should be done. However, the same five factors have been used for all types of jobs.

Ranking of Key Jobs by Factors. When the key jobs have been selected, copies of their descriptions and specifications are distributed to the committee. The members independently rank the key jobs separately for each factor. The key job possessing the lowest amount of one of the factors is assigned the rank of 1. Each key job is therefore ranked five times—once for each factor.

The rankings by the committee members are given to the chairman or a chief job analyst. Benge suggests that the committee members complete three sets of rankings at intervals of two weeks. This provides the chief job analyst with a larger amount of ranking data upon which to base an average final rank for each key job. He can also study individual rater consistency, and if necessary, discuss with any rater any discrepancies which might indicate carelessness in the rankings or misunderstanding of the content of any particular job.

The results of the ranking and reranking activity are then presented to the entire committee, which decides upon a final ranking for each key job. Where unanimity or a clear majority cannot be achieved, the particular job involved may be deleted from the list of key jobs.

Apportioning Pay Rates among the Factors. The next step is to apportion the current wage or salary of each key job among the five factors. This step is also performed independently by each of the committee members. A convenient form to use for this step is illustrated in Figure 6-13. The columns headed Rank show the final rankings assigned to the key jobs for each of the factors. The columns headed Points show the proportion of the current hourly rate that has been assigned to the five factors. For each job, the sum of the points must equal the current pay rate.

Apportioning the pay rates among the factors may be difficult. Not only must the sum of the points for each job equal the current pay rate, but the entries columnwise must be consistent with the relative final rankings of all the key jobs. In

[13] Eugene J. Benge, Samuel L. H. Burk, and Edward N. Hay, *Manual of Job Evaluation*, Harper & Brothers, New York, 1941.
[14] Benge, *op. cit.*

Rater_____   Date_____

Key job	Current pay rate	Mental		Skill		Physical		Responsibility		Working conditions	
		Rank	Pts	Rank	Pts	Rank	Pts	Rank	Pts	Rank	Pts
Assembler.........	2.43	4	51	4	45	3	78	3.5	42	3	27
Automatic screw-machine operator.	4.59	13	102	14	120	4.5	84	13	78	15	75
Brake operator.....	3.21	7.5	57	9	72	6	87	5.5	48	13	57
Carpenter..........	3.96	10	78	12	96	10	108	11	63	10.5	51
Expediter..........	3.54	15	117	6	57	1.5	72	14	84	1.5	24
Janitor...........	2.40	2	36	2	30	11	(90)	2	24	5	36
Machinist..........	4.20	11	87	13	105	13	123	12	66	6.5	39
Material mover.....	2.49	1	27	1	24	15	150	1	18	4	30
Millwright.........	3.75	5.5	54	10	75	14	141	8	54	10.5	51
Painter...........	3.09	5.5	54	7	(78)	4.5	84	3.5	42	14	69
Pipe fitter........	3.30	7.5	57	8	69	7.5	99	7	51	12	54
Timekeeper........	2.85	12	96	3	33	1.5	(54)	9.5	60	1.5	24
Tool and die maker.	4.98	14	111	15	135	12	120	15	93	6.5	39
Truck driver.......	2.94	3	42	5	51	9	105	5.6	48	8.5	48
Turret lathe operator.........	3.72	9	72	11	93	7.5	99	9.5	60	8.5	48

Fig. 6-13. Key job data sheet for factor comparison system. (*Adapted from R. C. Smyth and M. J. Murphy,* Job Evaluation and Employee Rating, McGraw-Hill Book Company, New York, 1946, p. 21.)

other words, horizontal relationships must not violate vertical relationships. In actual practice, there are almost always a few jobs for which vertical-horizontal consistency cannot be achieved. One job may not have a large enough current pay rate to permit the rank desired in one or more factors. Another may have too large a pay rate to apportion among the factors without changing the final ranks. In such instances, a circle is drawn around those point entries which are inconsistent. In Figure 6-13, the jobs of janitor, painter, and timekeeper presented such difficulties, as shown by the circles around the skill factor entry for painter, and the physical factor entries for janitor and timekeeper.

When all the committee members have completed this step, their data sheets are handed in to the chairman. The committee members repeat the step of apportioning rates two more times at two-week intervals. At the completion of the third evaluation, the chairman averages the apportioned points and reports to the committee. Discrepancies are discussed and, if possible, they are resolved. Those jobs for which rank and apportioned points cannot be resolved are discarded as key jobs.

Setting up the Job Comparison Scale. The remaining key jobs are set up as a job comparison scale, as shown in Figure 6-14. This consists of five scales or "measuring sticks," one for each factor. It will be noted that the jobs of painter, janitor, and timekeeper do not appear because they were eliminated as key jobs. Each measuring stick has a number of points from low to high. Various points along each scale are defined in terms of one of the key jobs.

Adding Supplementary Key Jobs to the Job Comparison Scale. It is necessary to add additional key jobs to the job comparison scale for three reasons: (1) to fill the wide spread that may exist on any of the measuring sticks between two adjacent jobs, (2) to make up for the loss of original key jobs, and (3) to validate the selection of the remaining key jobs. Additional jobs are chosen as before, and all six steps described above are repeated. From thirty to fifty

Cents	Mental requirements	Skill requirements	Physical requirements	Responsibility	Working conditions
150			Material mover		
147					
144					
141			Millwright		
138					
135		Tool and die maker			
132					
129					
126					
123			Machinist		
120		Automatic screw-machine operator	Tool and die maker		
117	Expediter				
114					
111	Tool and die maker				
108			Carpenter		
105		Machinist	Truck driver		
102	Automatic screw-machine operator				
99			Turret lathe operator		
96		Carpenter	Pipe fitter		
93		Turret lathe operator		Tool and die maker	
90					
87	Machinist		Brake operator		
84			Automatic screw-machine operator	Expediter	
81					
78	Carpenter		Assembler	Automatic screw-machine operator	
75		Millwright			Automatic screw-machine operator
72	Turret lathe operator	Brake operator	Expediter		
69		Pipe fitter			
66				Machinist	
63				Carpenter	
60				Turret lathe operator	
57	Pipe fitter / Brake operator	Expediter			Brake operator
54	Millwright			Millwright	Pipe fitter
51	Assembler	Truck driver		Pipe fitter	Millwright
48				Truck driver / Brake operator	Carpenter / Turret lathe operator
45		Assembler			
42	Truck driver			Assembler	
39					Tool and die maker / Machinist
36					
33					
30					Material mover / Assembler / Expediter
27	Material mover				
24		Material mover			
21					
18				Material mover	
15					
12					
9					
6					
3					

Fig. 6-14. Job comparison scale for factor comparison plan. (*Adapted from R. C. Smyth and M. J. Murphy,* Job Evaluation and Employee Rating, *McGraw-Hill Book Company, New York, 1946, pp. 24–25.*)

supplementary key jobs should be sufficient to construct the final form of the job comparison scale. Evaluation of the supplementary key jobs should be accompanied by continual reexamination of the original key jobs.

The construction of a factor comparison plan, although simple and logical in design, is a time-consuming and sometimes difficult process. The individual job analyst or committee member is required to make many important decisions: selecting key jobs, ranking key jobs, rating key jobs and making ratings consistent with rankings, integrating his judgments with those of his fellow committee members and resolving differences of opinion, and selecting and evaluating additional key jobs. However, the fact that the method is widely used in industry and business would seem to indicate that its usefulness compensates for the difficulties involved in constructing the plan.

Evaluating the Remaining Jobs. Using the five job scales containing the key jobs and supplementary key jobs as guide points, the committee decides where on each scale the remaining jobs should be placed. This completes the job evaluation per se.

THE COMBINATION POINT–FACTOR COMPARISON SYSTEM

It has been stated above that some concerns use a system that is a combination of the point system and the factor comparison system. Such a plan is apparently an answer to the criticisms that have been made of both systems, at the same time taking advantage of the greater flexibility and ease of use of the point system. In a combination system, the same steps are followed as in the factor comparison system through step 6, Setting Up the Job Comparison Scale. One essential difference, however, is that the apportioned rates for each key job need not equal the current average pay rate for that job. If the entire committee is in agreement, the current rate may be disregarded. In addition, the rates are doubled or tripled and then rounded off to multiples of 5 or 10. The purpose of this is to decrease the biasing effect of knowledge of current rates on the various judgments that have to be made. In brief, an attempt is made to disguise the monetary aspect of the measuring sticks.

Once the job comparison scale is set up, the factors are weighted primarily on the basis of the range of points utilized for each. The remaining steps follow the point system. Definitions are written for the factors, degree levels for each factor are defined and assigned point values, and a job evaluation manual is constructed and used for the remaining jobs and for new jobs in the same manner as the point system.

REVIEW OF JOB EVALUATION RATINGS

After all the jobs in an organization have been evaluated, a procedure should be established for reviewing or checking the final ratings before the plan is formally installed. The process of review or verification consists essentially of evaluating and comparing the jobs by some method other than that used in the original evaluation. If the ranking system has been used, some sort of labor grade or job classification plan will have been established. One way of reviewing the final rankings is to write definitions for these labor grades in a manner similar to the grade description method. Job analysts or other qualified persons may then be given the task of classifying the jobs according to these labor grades. It is essential, of course, that these persons not be informed of the final rankings assigned by the original job evaluation committee. Jobs that are placed in grades other than those of the original evaluation should be carefully checked with respect to their job descriptions and specifications.

If the grade description method has been used, one may resort to the ranking technique for purposes of review and verification. Reviewers may be given the jobs that have been assigned to two or three adjacent grades and asked to rank them. Again, the reviewers should not be informed of the grades to which the jobs were originally assigned.

Evaluations obtained by the quantitative job evaluation systems lend themselves to verification much better than do the nonquantitative systems. If the point system has been used, the reviewers may be asked to rank the jobs according to overall value, or separately according to each factor. The former utilizes the ranking technique and the latter the job comparison technique as a reviewing procedure.

If the factor comparison method has been used for the basic evaluation, degree level definitions may be written for each of the factors, using as a guide the characteristics of selected key jobs. The reviewers may then assign the jobs to degree levels, as in the point system. Or levels may be defined as in the grade description method, and the jobs assigned to them.

A review of job evaluation ratings is important because it represents the last opportunity, so to speak, to change the point values of any jobs before these values are used as a criterion for the fairness of the wages paid the jobs.

CLASSIFICATION OF EMPLOYEES

After all the jobs have been described, identified, and evaluated in terms of rank, grade level, or points, it is important that each employee be identified with the title of the job which he performs. If the job descriptions are accurate, the task of classifying the employees will be all the easier, because accuracy of job descriptions implies that the job analysts have observed the employees at work and have obtained much of the job information by interviewing workers and supervisors.

Misclassification of employees inevitably leads to worker dissatisfaction on the part of either the misclassified individual or his fellow workers. The former will feel disgruntled if he is underpaid and perhaps uneasy if he is overpaid. The latter will attribute the existing injustice to the job evaluation system itself rather than to an error in the manner in which the system was applied.

Another serious result of employee misclassification is that the wage curve may be inaccurate. If the number of misclassifications is large enough, the slope of the curve may be affected, as will the final classification structure which is super-imposed on the curve.

The persons usually considered responsible for the classification of employees are their first-line supervisors. This means that supervisors must have the job descriptions for the employees in their departments. One of the best ways of obtaining a check on the classifications is to permit each employee to see his job description and to discuss any points of disagreement with his supervisor. Many concerns extend this privilege to union officers and shop stewards. The services of one of the job analysts should be available to both supervisors and employees. The job analyst may find it necessary to explain the principles underlying the job descriptions and the specifications.

THE WAGE CURVE AND JOB CLASSIFICATIONS

The next two major steps in achieving the final wage structure are (1) deriving the company wage curve, and (2) establishing job classes or labor grades. These two steps are closely related, and the problem of which should be accomplished first is largely a function of the type of job evaluation system used. A preliminary definition of these two steps is necessary in order to understand what is involved.

Job classification is the process of grouping the jobs into various categories on the basis of job difficulty value. In the ranking method, the job difficulty data consist of raw rankings or converted scores (for example, linear scale score). In the point system and the factor comparison system, the job difficulty data consist of point values. In these three methods, the wage curve is helpful in determining the grouping of jobs into similar grades. For this reason, and because job analysts will inevitably compute the wage curve as soon as they possess the necessary data, it is treated first in this discussion.

The Wage Curve. The wage curve is a graphic representation of the relationship between job difficulty values (points, ranks, or grade levels) and the wages paid the jobs. When this relationship is a straight line, it is often referred to as a wage line.

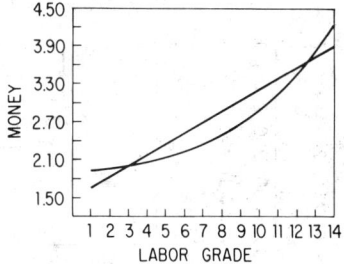

FIG. 6-15.   Straight-line and second-degree wage curves computed from the same data.

The wage curve is obtained by setting up a scattergram in which the abscissa represents point values or labor grades and the ordinate represents wages. Each employee's job is plotted on the chart, and a line of best fit is either drawn freehand through the plottings or computed by statistical methods. Statistical methods are definitely preferred to freehand or French curve drawings. The usual, or at least preliminary, procedure is to determine the straight-line relationship. However, if the plottings appear to follow a curve, it is advisable to compute a second-degree curve for the data. The computational procedures for obtaining the straight-line trend and the second-degree curve may be found in most advanced statistics texts.[15]

Figure 6-15 shows both the straight-line and the second-degree curves computed from the same wage data. It is readily apparent that a wage structure based on the second-degree curve pays higher pay rates for the jobs in the very high and very low labor grades, and lower pay rates for the intermediate labor grades, as compared with a wage structure based on the straight line.

It was stated above that the job of each employee should be plotted on the scattergram. In this case, the number of plottings is equal to the number of employees. Very often, wage curves are derived in which each job is plotted only once for all the employees on that job. When this is done, the median wage paid these employees is the ordinate value for the plotting, and the number of plottings is equal to the number of jobs. The two methods should yield practically the same curve. Plotting each employee has the advantage of showing to what extent individual employees deviate markedly from the general trend for the entire organization. This information will be needed to resolve individual wage inequities. Using median values means a saving of time to derive the curve. When several wage curves have to be computed, as in a community wage survey (see below), using median values has its advantages. The wage curve based on plottings of individuals will be more accurate if only because it is based on a greater amount of data and because the "pulling" effect of each employee on the wage curve is taken into account.

Job Classifications. Point values must ultimately be converted to monetary values. For ease of administration of the wage system, it is customary to group the jobs into classes and to establish a pay rate for each class. Otherwise, a pay rate would have to be established for each job, and a wage structure based on such a system would be cumbersome to operate.

Basic Principles. There are many ways of setting up job classes. Each method has its advantages and disadvantages. The method selected depends to a large extent upon the slope of the company's wage curve and the needs of the company. The various plans may best be discussed by referring to Figure 6-16.

[15] For a description of the method of computing a straight-line trend, see Otis and Leukart, *op. cit.*, pp. 418–427. For a description of the methods of computing a second-degree curve, see Frederick E. Croxton and Dudley J. Cowden, *Applied General Statistics,* Prentice-Hall, Inc., Englewood Cliffs, N.J., 1955, pp. 486–524.

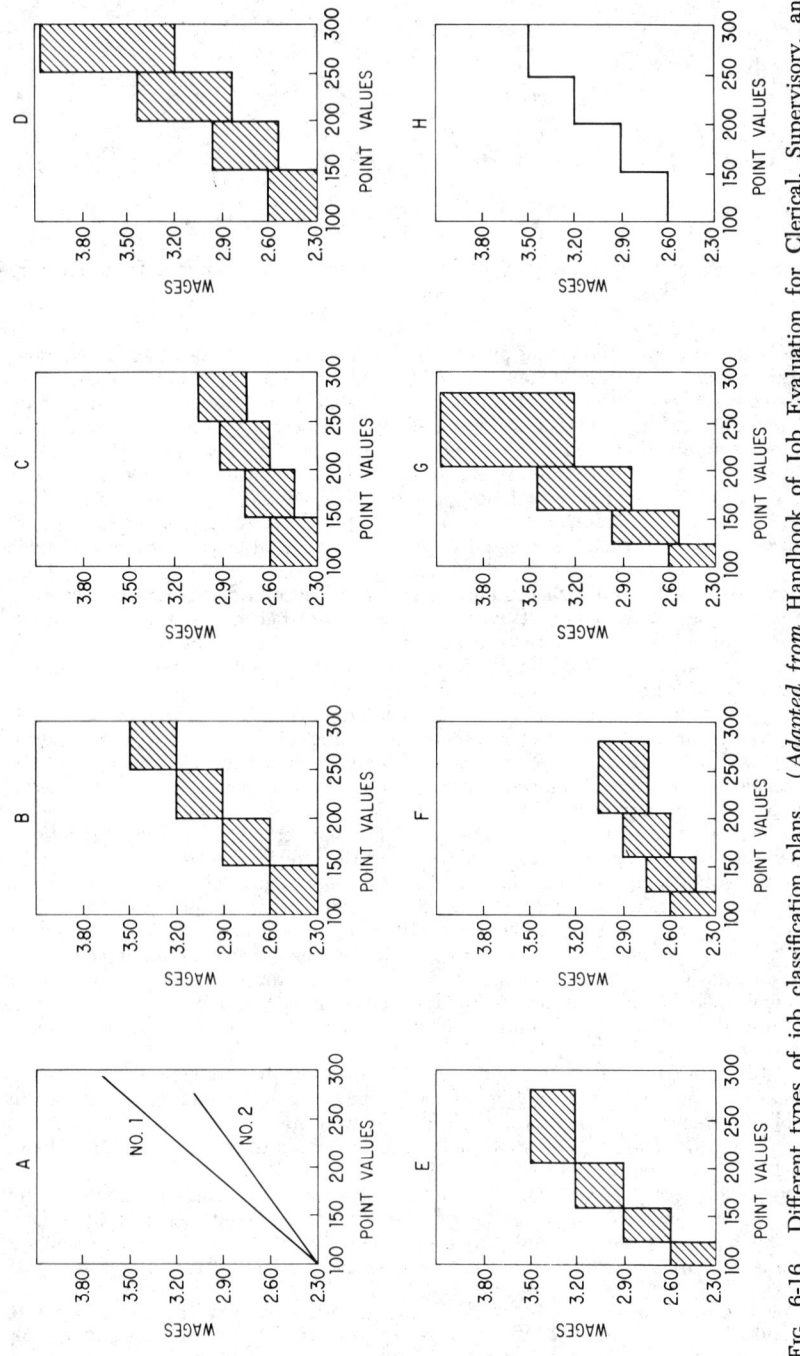

Fig. 6-16. Different types of job classification plans. (*Adapted from Handbook of Job Evaluation for Clerical, Supervisory, and Administrative Jobs, courtesy of Industrial Fasteners Institute, Cleveland, Ohio, privately printed, not for distribution.*)

Chart A in Figure 6-16 shows two wage curves. Curve 1 has the greater slope; curve 2 is "flat." Wages or salaries are obviously higher with curve 1. These two curves could have been obtained with the same set of job difficulty values. They are presented to illustrate the importance of the wage curve in setting up job classes. The steeper the curve, the less need for overlapping of wage ranges for the job classes. For example, compare Chart B with Chart C, or Chart E with Chart F. Charts B and E are based on the steeper curve.

Chart B shows job classes with equal point ranges, equal money spread, and no overlap. The range of point values for all the jobs has been divided into classes of 50 points each. Each class has a 30-cent wage range. The wage maximum of each class is the minimum for the next class.

Chart C is similar to Chart B except that it shows a money-spread overlap. This is due to the fact that Chart C is based on the flat curve in Chart A. The maximum wage for each class is no longer the minimum for the next class. Overlap of this sort has an advantage in that an older, more experienced employee on a low-rated job may receive more wages than a beginner on a more difficult job. However, the potential wage ceiling for the higher rated job is higher, which is the way it should be.

Chart D illustrates equal points, percent money spread, and overlap between job classes. Each job class still has a range of 50 points, as in Charts B and C. However, the wage range for each job class is larger than the preceding one. The wage range for the lowest class is 30 cents, or 13 percent of the minimum for the class. Each wage range is equal to the preceding wage range plus 40 percent of the preceding wage range. The use of a percent money spread makes it possible to have a wider wage range for the more difficult positions.

In Chart E, money spread is equal but the point classes are unequal. There is no overlap between job classes. Each job class is 40 percent larger than the one preceding it. The philosophy of this type of job classification is that it is easier to differentiate between job classes in the lower point values than in the higher point values; therefore a wider range of points should be used as the jobs become more difficult.

Chart F illustrates unequal point spread, equal money spread, and overlap. The overlap is due primarily to the fact that the wage structure is superimposed on a flat wage line.

Chart G illustrates unequal point spread, percent money spread, and overlap. This is probably the most common structure used in business and industry.

Chart H illustrates the use of a single wage for each class with equal point values for each class. According to the chart, all jobs in Class 1 would receive $2.60; all jobs in Class 2, $2.90; all jobs in Class 3, $3.20; and all jobs in Class 4, $3.50.

The problem of which of the eight methods to use is a matter of judgment and experience. There is no scientific solution to this problem, because the unit of job difficulty is abstract and merely a convenient way of quantifying a human judgment. When the wage curve is flat, overlapping is practically mandatory if very narrow wage ranges are to be avoided. If it is felt that the point unit represents the same amount of job difficulty along the entire abscissa, equal job classes are justified. However, many investigators feel that a point unit at the upper end of the scale represents more job difficulty than one at the lower end and that a correction or compensation is in order; hence they advocate successively greater job classes.

Number of Job Classes. The number of job classes depends on several factors: desired wage ranges, tradition in the company, the type of jobs included in the job study, and the range of job difficulty values. For clerical jobs alone, from six to ten classes are usually used. If supervisory and professional jobs are included, anywhere from three to eight more classes will be necessary. For factory jobs the number varies between eight and twelve for the most part, with an average of ten. The tendency is toward fewer rather than many job classes. Wide wage ranges make it possible to grant more merit or in-grade wage increases, thus increasing employee morale. When there is wide overlap, employees in one job class frequently receive lower wages than some employees in the next lower job class. The trend in wage administration is to reduce overlap as much as possible.

Resolving Wage Inequities. After the job classes and their wage ranges have been established, it will invariably be found that some employees fall above or below the wage maximum and wage minimum, respectively, of their job classes. The best way to show this is to plot each employee on the chart showing the job classes. Each job class with its established wage minimum and wage maximum will "box in" most of the workers, leaving a small number above and below each job class. Those above the maximum wage for each job class may be considered overpaid and those below the minimum are underpaid. Many wage administrators establish job class ranges and wage spreads in an empirical manner by boxing in as many employees as possible while at the same time attempting to follow some systematic scheme for setting minimums and maximums.

It may be expected that some adjustments and compromises will be made that involve a greater or smaller departure from the wage structure that has been established by formula.

Those employees falling below should be brought up to at least the minimum. This results in an increase in the payroll. Ideally, the workers who are above the maximum should have their wages reduced. In practice, however, this is very difficult to do without jeopardizing the success of the job study, and it illustrates a cardinal point. A job study almost always means initial increased payroll costs to management, and this increase is likely to continue for a long time. Workers who are above the maximum should be transferred wherever possible to more difficult jobs in higher job classes. If this cannot be done, they should be carried at their old pay rates so long as they perform their work efficiently. When normal turnover creates vacancies for their jobs, the new incumbents can be paid at a rate consistent with the established wage structure.

COMMUNITY AND INDUSTRY WAGE SURVEYS

It may be expected that an organization will derive its own wage curve or wage line as soon as it has compiled the wage data and job difficulty data required for the computation. This requires what seems like a great deal of information about the jobs and the employees in a company, and in truth it does represent a tremendous number of facts, most of which were unknown or only suspected before the job study began. However, the most important question concerning the current wage structure has not been answered. In its simplest form, this question is, "Is the general wage structure of the company good or bad?" This question is answered by means of a community or industry wage survey.

Major Wage Survey Methods. Wage surveys have been made on the basis of (1) job title, (2) job description, and (3) job difficulty. Of these, the job title method is the most inaccurate, the job description method is the most frequent, and the job difficulty method is probably the most accurate and defensible. To carry these comparisons further, one may say that the job title method is the easiest, and the job description method is the most difficult.

Wage Survey by Job Title. By this method, an employer who wished to know how the wages of his employees compared with those in the community would send other companies a list of job titles such as "receptionist," "machinist," or "stenographer." The cooperating companies would return the list with the pay rates entered after the job titles. The wage information for each job title would then be consolidated and returned to the cooperating organizations so that each could compare its pay rates with those of the others.

However, although a job title may be common to many companies, the content of the jobs it represents may vary considerably. A foreman in one company may be quite different from one in another. A tool and die maker may work under much poorer conditions or receive more supervision in one company than in another. For reasons such as these, wage surveys conducted on the basis of job title may be very inaccurate and misleading.

Wage Survey by Job Description. Wage surveys on the basis of job descriptions represent a decided improvement over the job title method. In this method, the cooperating organizations are furnished brief job descriptions which can be matched with the content of jobs in their own organizations.

Many variables are involved in a wage survey, and a great deal of planning and effort is required to ensure that the information obtained will be accurate. This is true regardless of the wage survey method used. In the job description method, careful attention must be given to the jobs that will be included in the survey. Smyth and Murphy give five criteria for selecting jobs:[16]

1. The jobs should be distributed over the whole range of evaluated jobs.
2. The jobs should have remained relatively stable in recent years.
3. The jobs should exist in nearby or competing companies.
4. The jobs should be filled by as large a number of workers as possible.
5. There should not be an unusual shortage or surplus of workers qualified to fill the jobs.

The number of jobs to survey in each participating company need not be large. It is now generally agreed that 25 to 40 carefully selected jobs will yield practically the same wage curve as 300 to 400 jobs.

The companies to be included in the survey should be those with jobs which the sponsoring organization has selected, and which are competing in the labor market for persons to fill these jobs. Participating organizations will therefore often be members of the same industry as the sponsoring organization, unless the job study is based on jobs common to several industries—for example, clerical and office jobs. It is generally agreed that twenty-five to thirty carefully selected companies represent the maximum number that need ever be included. It is more important to obtain a good sampling of companies than to strive for a large number. Many excellent surveys have included as few as five to ten companies.

The information to be obtained from each company requires very careful planning. The wages paid similar jobs in two different companies may be equal, but further analysis may reveal that the worker in one company is actually the richer because of items usually described as supplemental income, for example, paid holidays, hospitalization, and bonuses. For this reason, it is essential to differentiate base pay, earned pay, supplemental income, and hours worked. The necessary information is usually obtained by experienced job analysts who visit each company armed with appropriate questionnaires and forms. The objective is to obtain sufficient information about the same job in each of the companies so that a fair comparison may be made of the wages paid that job. To that end, it is necessary to take into account the differences among the companies with respect to hours and supplemental income for the particular job. The usual practice is to obtain the base hourly rate or base weekly salary of each employee on the job; this is done separately for each participating company, and the results are later consolidated for all the companies.

Wage Survey by Job Evaluation. A wage survey by job description leaves much to be desired. The job descriptions are usually brief—"thumbnail sketches"—with the result that one can never be sure that the key jobs have been matched contentwise in each of the participating organizations. The answer to this difficulty is to extend the methods of job evaluation into the wage survey itself.

The essential difference between the job description and job evaluation methods of conducting a wage survey is that in the latter, both wage data and job difficulty data are obtained for the jobs surveyed in each participating company. The job analyst goes into the participating company armed not only with his questionnaires and forms for recording wages and number of employees on each key job, but also with a job evaluation manual so that he can determine point values for the jobs. This means that in each company he must obtain the specification for each job in terms of the factors used in the job evaluation manual. Although this seems like additional work, it is believed that in the long run a saving of time

[16] C. R. Smyth and M. J. Murphy, *Job Evaluation and Employee Rating*, McGraw-Hill Book Company, New York, 1946, pp. 97–98.

will ensue because it is not so essential to match the key jobs in each company. The primary concern for the job analyst is to obtain a representative sampling of key jobs in each company. He need not be worried lest the job of "receptionist" in one company is less or more difficult than in another company, because such differences will be reflected in the point values that are obtained.

As in the job description method, a wage survey based on job evaluation should sample from twenty-five to forty jobs in each company. The primary consideration is that, in both the sponsoring company and the participating companies, the jobs be of the same type. In each company, they should be a representative sample of that type of job.

THE FINAL WAGE STRUCTURE

Upon completion of the wage survey, the wage analyst is ready to ask himself the sixty-four dollar question: "Where should my company's wages fall in the community?" Smyth and Murphy give an excellent answer which is quoted in full:[17]

> Those firms which secure the necessary facts by surveying the community or industry for information as to their wage scales have, upon analyzing the results of their surveys, three possible courses of action:
> 1. They may deliberately adopt the policy of being leaders (or may be forced into it). Such firms maintain a wage scale that is on the average 5 to 10 percent (or occasionally even more) above the average of either the concerns in their own industry or the better firms in the same geographic area. Since this policy decidedly facilitates employment, it is often encountered in periods of labor scarcity, particularly when the scarcity coincides with need for rapid expansion on the part of the firm or firms involved. Although it is easy enough to decide to become a leader in comparison with the other local or industry wage scales, it is not always so easy to withdraw from this sometimes unenviable position. Accordingly, it is very important that the long-term possibilities of such a policy be most carefully considered in advance. It is also wise to examine critically the claims that are likely to be encountered to the effect that a wage scale definitely above the industry or area average will attract employees who will be more productive per man-hour. Although such claims sound logical, it will be found that their substantiation is extremely difficult.
> 2. They may deliberately adopt the policy of being tailenders (or may be forced into it). Such firms normally maintain a wage scale that is appreciably lower than the average for the industry or area. There are a number of possible reasons for this:
>   *a.* The concern may be part of an industry that is forced to operate on a bare marginal profit, sometimes due to gradual displacement of the industry by technological change and sometimes due to other competitive problems peculiar to the industry.
>   *b.* The concern may be the victim of poor management, which has resulted in the dissipation of profits and depletion of capital.
>   *c.* The management of the concern may mistakenly feel that a minimal wage scale tends to increase an already acceptable margin of profit.
>   *d.* The concern may be in its formative stages, struggling for existence on small capital.
>   There is little doubt that in the long run the policy of maintaining a wage scale that is *appreciably* lower than the industry or area average makes the securing of an adequate labor supply a difficult task. As a general rule, a low scale also tends to increase labor turnover because employees readily leave to secure higher paid jobs elsewhere. Operating expense and overhead are increased as a consequence, since the training of new employees is expensive and they function at low efficiency during their break-in period.
> 3. They may adopt the policy of maintaining their wage scale at or near the average of the industry or area. In most instances, this policy will probably be found to be the most logical and satisfactory one for the majority of concerns.

[17] *Ibid.*, pp. 131–133.

The following four-point wage program may be regarded as basic to this approach. An average wage structure should be characterized by

a. A base rate structure equal to the average of the industry or area both as to the absolute amounts of money paid and the relations of those amounts to one another (i.e., the slope of the base rate curve).

b. Average earned rates that are comparable to the averages of the industry and area both as to the absolute amounts of money paid and the relation of those amounts to one another (slope of curve).

c. A minimum hiring rate that is equal to the average of the industry or area.

d. An aggregation of miscellaneous additions to income or industrial relations policies (such as paid vacations, subsidized insurance coverage, holidays, or night-shift premium) that are reasonably comparable to the practices of a majority of the concerns in the industry or area.

To the above advice, it might be added that a company must continually study the relationship of its wages to community rates. When wage rates in the community change, the company should attempt to maintain the same relative position.

The Finished System. After the wage analyst has modified his company's current wage structure in the light of community and industry rates, he is ready to present the finished system in a form that will be easily understood, especially by the workers and applicants for employment. Figure 6-17 is an example of a job evaluation system for salaried jobs based on the point method.

KEEPING THE WAGE STRUCTURE UP TO DATE

It is not uncommon for an excellent wage structure to be set up and then to become unsatisfactory in a short time because it has been inadequately policed and maintained. The content of jobs changes and sometimes in a subtle manner, so that job descriptions and specifications are outmoded. The level of wages in the community or industry also changes so that the entire wage curve of a company may be out of line. Policing and maintaining a wage structure is a big job and a full-time job, unless the company is so small that the employer knows all his employees and their jobs fairly well.

Administration of wage structures, being such a major responsibility, has usually been delegated to a committee who are assisted by one or more full-time wage and job analysts. The committee usually consists of the chief analyst and such

JOB CLASS, WAGE RANGE, & POINT VALUE RANGE		
CLASS	WAGE RANGE	POINT VALUE RANGE
1	$ 275 - 460	100 - 139
2	375 - 615	140 - 179
3	460 - 770	180 - 219
4	560 - 935	220 - 259
5	650 - 1090	260 - 299
6	750 - 1250	300 - 339
7	840 - 1400	340 - 379
8	940 - 1555	380 - 419
9	1030 - 1700	420 - 459

FIG. 6-17. Completed wage structure—production positions.

persons as the comptroller and department heads. Employee representatives are sometimes included.

MERIT INCREASES

In some organizations, department heads initiate recommendations for wage increases. In other organizations, this may be determined largely by union contract. In still others, for example, United States Civil Service positions, wage increases are automatic at specified intervals provided the employee's work record has been satisfactory. Whatever the method of granting wage increases for merit, it should be the responsibility of the committee to see that they are granted when due and when deserved.

With respect to wage administration, the following problems involving merit increases are mentioned at this time:

1. Wages must be kept within established ranges. They should not go above the maximum. This may be a very difficult principle to follow, especially when the labor market is tight. Making one exception inevitably results in pressures to except other employees.
2. When a merit increase is granted to one employee, the records of other employees should be examined to determine if they, too, are deserving of a merit increase.
3. The wage range of a job class should be utilized fully and in an intelligent manner. Workers who are barely satisfactory should be paid the minimum wage of the job class. Exceptional workers should receive the maximum. Workers of intermediate ability should receive an appropriate sum between the two extremes.
4. Increases granted solely on the basis of tenure on a job should be clearly identified as such and should be handled separately from merit increases.
5. Because department heads will not always treat their subordinates alike in recommending merit increases, the committee should review periodically the merit ratings and requests for wage increases of all employees.
6. Policies with respect to merit increases should be enforced in a consistent manner. If the maximum for one job class is raised, the maximums of other job classes should be adjusted accordingly. If workers with a certain amount of tenure are rewarded, other workers with the same amount of tenure should be rewarded.

ADMINISTRATIVE PROBLEMS

Administrative problems which arise are primarily those concerned with making adjustments for a tight labor market, transferring employees from one position to another, promoting individuals to higher positions, demoting employees, and increasing wages.

During periods of employee scarcity, it may be difficult to hire employees at the established minimums if other companies are hiring at increased rates. Under the circumstances, a company may have to modify its policy of hiring at the minimum by setting a new "hiring-on wage" for the lowest or the two lowest job classes.

When an employee is transferred from one position to another within a job class, his current wage or salary continues or may be increased. Occasionally, however, an employee must be transferred who is already receiving the maximum of his job class. If he is unable to perform as efficiently as other workers in the new position, they may resent his earning as much as they are. The only solution to this problem is to publicize the policy that persons who are transferred will not be reduced in pay, so that when such a transfer occurs, all employees concerned will know what to expect.

When an employee is promoted from one job class to the next higher class, there is always the possibility that he may not make good on the new job. Demoting him back to his old job is always embarrassing, especially if he has received a wage increase with the promotion. One of the best solutions to this problem is to have overlapping wage ranges and to promote those individuals who have attained a wage in their job class which is at least at the minimum of the next job class. This makes possible a tryout period on the new job without a pay increase. If the employee makes good, he may be granted a pay raise within his new job class; if not, he may be placed back on his former job without a decrease in pay.

Demotions from one job class to another are often necessary because of inefficiency, reduction in force, and slack periods. If the individual is above the maximum of the class to which he must be demoted, the problem arises as to whether he will continue to receive his old wage or the maximum wage of the lower job class. It is generally agreed that the new salary should be within the wage range of the new position. Some companies have adopted the policy of setting the new wage at the same point in the new wage range as it was in the old. For example, if an employee was at the midpoint of the range, upon demotion he would be placed at the midpoint of the lower wage range.

THE MANAGEMENT DECISION

This chapter, which has of necessity limited itself to the essentials, is sufficient to indicate that a job evaluation program is a major undertaking that requires a great deal of planning from beginning to end.

A major decision that must be made is whether a job study should be undertaken at all. This decision is important not only because of the expense involved, but because it will affect the personnel and production operations of the company for a long time. For this reason, it is practically a maxim that the decision to embark on a job study must be made by the top executive of the organization, and once the decision is made, it must receive his full support. The announcement of the study should therefore come from the chief executive officer rather than from the personnel department, the industrial relations department, or some other department.

WHO SHOULD DO THE JOB STUDY?

Once the decision has been made to do a job study, it is necessary to determine who will do it. Three ways are available. The company may have a firm of management experts do the entire job. It may do the job itself by utilizing qualified company employees. Or it may use a combination of these two arrangements by utilizing an outside firm on a consulting basis.

Hiring an outside firm is the easiest way and probably the most common. An outside firm is more likely to be fair and impartial because it has no ax to grind. An outside firm is likely to complete the job study as quickly as possible without taking company employees away from their regular assignments. It may be expected to provide trained and experienced professionals to do the work.

The argument for using company employees is that they will identify themselves with the project and will feel responsible for the progress of the work. Including company employees makes the later task of acquainting them with the results easier. In addition, company employees are acquainted with the traditions of the firm and are more sensitive to the feelings of employees about matters concerning job content, wages, and lines of authority. Using company employees also provides training for those who will be expected to police and maintain the wage structure after it has been established. Too often, when an outside firm has done the study, management accepts the results as a finished job which should not be modified with the passage of time. However, as emphasized earlier in this chapter, the job evaluation process is a continuous one and the wage structure must be reviewed periodically.

The use of an outside organization of experts on a consulting basis seems to combine the advantages and minimize the disadvantages of using an outside firm or company employees exclusively. The consulting firm should be retained to make periodic checks on the system.

BIBLIOGRAPHY

Belcher, David W., *Wage and Salary Administration*, 2d ed., Prentice-Hall, Inc., Englewood Cliffs, N. J., 1962.

Brennan, Charles N., *Wage Administration*, Richard D. Irwin, Inc., Homewood, Ill., 1963.

Langsner, Adolph, and Herbert G. Zollitsch, *Wage and Salary Administration*, South-Western Publishing Company, Incorporated, Cincinnati, 1961.

Lanham, Elizabeth, *Administration of Wages and Salaries*, Harper & Row, Publishers, Incorporated, New York, 1963.

Maynard, H. B. (ed.), *Handbook of Business Administration*, McGraw-Hill Book Company, New York, 1967, sec. 10.

Maynard, H. B. (ed.), *Handbook of Modern Manufacturing Management*, McGraw-Hill Book Company, New York, 1970, sec. 9.

Otis, Jay L., "A Psychologist Looks at Salary Administration," *Management of Personnel Quarterly*, Summer, 1963, pp. 22–27.

Patton, John A., C. L. Littlefield, and Stanley A. Self, *Job Evaluation Text and Cases*, 3d ed., Richard D. Irwin, Inc., Homewood, Ill., 1964.

Sibson, Robert E., *Wages and Salaries*, American Management Association, New York, 1960.

Walker, Robert L., "Maintaining Adequate Wage Differentials between Skilled and Unskilled Jobs," *Personnel Administrator*, vol. 6, no. 2, 1961, pp. 21–23.

Chapter 7

Salary Evaluation

HARRY T. SCHWAN

**Executive Vice President, Daniel D. Howard Associates, Inc.,
Chicago, Illinois**

This chapter presents current concepts and practices involved in the establishment of salaries for clerical, technical (or staff), and managerial positions. It avoids as much as possible duplication of materials contained in the preceding chapter on job evaluation. The objectives and mechanics of salary evaluation are similar and in many cases identical to those used in job evaluation. It is therefore recommended that the reader have a sound knowledge of job evaluation procedures. This can be obtained by reading the job evaluation chapter (Section 6, Chapter 6) and some or all of the suggested references.

WHAT JOB EVALUATION IS

Without going into details, the term "job evaluation" applies to a group of procedures whose basic objective is to determine the relative worth of the various jobs within a given company. By precedent and habit, more than for other specific reasons, job evaluation procedures are usually applied to hourly paid, nonsupervisory factory jobs. Typical of such jobs are those handled by janitors, material handlers, machine operators, assemblers, welders, electricians, pipe fitters, and toolmakers.

WHAT SALARY EVALUATION IS

The term "salary evaluation" also applies to a group of procedures whose basic objective is to determine relative worth of the various jobs within a given company. But in this case, the jobs are of a different type. Salary evaluation is the term

6-127

usually used when evaluating office and clerical, supervisory, staff, and managerial jobs. The jobs are often called positions and are usually paid on a weekly or monthly salary basis. Examples of people paid this way are file clerks, typists, cost clerks, foremen, engineers, accountants, designers, and executives.

WHY SALARY EVALUATION?

Many businesses of all types use job evaluation procedures. Fewer use salary evaluation procedures, however, especially for technical, staff, and managerial jobs. Perhaps the words "formal" or "systematic" should be inserted here. In actuality, every company evaluates salaries in some way or other. The question is whether it is done in an orderly and systematic way such that the entire salary structure is both reasonable and internally consistent.

Several benefits accrue to a company from a properly installed and operated salary evaluation program. But one benefit alone—internal consistency—makes the procedures not only worthwhile but a definite necessity in companies of even moderate size. It is much easier to explain and gain acceptance for an internally consistent salary structure than it is to handle the problems of a structure that averages somewhat higher but is not consistent. The notion that higher level technical or management people are not concerned about inconsistencies (real or fancied) between their salaries and the salaries of others doing similar work is unwarranted. These inconsistencies invariably come to light and cause problems of morale, turnover, and performance.

Another important benefit from a well set up salary evaluation plan is that it aids in establishing and maintaining a smoothly operating organization. This benefit applies particularly in the middle and higher levels of management rather than in clerical and lower levels of supervision.

Other benefits accrue because the work done in preparing descriptions and in the evaluating process is useful in considering selection, promotion, transfer, performance evaluation, and the like.

EXTENT OF SALARY EVALUATION

An increasing number of firms have been installing salary evaluation plans. A survey by the Bureau of National Affairs, Inc., Washington, D.C., showed how widely salary evaluation plans are used. The results of this survey are tabulated in Figure 7-1. The survey covered a random sample of 132 companies of all

Large companies are those with more than 1,000 employees. Small companies have 1,000 or fewer employees.

Coverage by Type of Plan

	Large	Small
Formal Plans	86%	57%
Informal Plans	14%	43%

Coverage by Type of Job

	Large	Small
Factory	78%	86%
Office	80%	71%
Supervisors	51%	54%
Executives	22%	26%
Sales Employees	28%	17%

FIG. 7-1.   Extent of job and salary evaluation.

sizes distributed throughout the United States. Other surveys have produced results indicating that as high as 55 percent of the companies covered have formal salary evaluation plans for management and staff personnel.

Obviously, salary evaluation is not a universal procedure, especially for higher levels of management. On the other hand, when from 25 to 50 percent of the companies are using these procedures, it is a good indication that they are worth major consideration. There has been a tendency for some companies to feel that stock option, stock purchase, pension, profit sharing, and various deferred-payment plans would remove any real need for salary evaluations plans. Many of those who have tried this approach have learned that they still have the same basic and fundamental problems. There is just no escape from the fact that basic salary is the foundation of all compensation structures. When this foundation is unsound, no amount of additional superstructure will correct it.

PREPARATION OF POSITION DESCRIPTIONS

There are no generally accepted procedures or formats for the preparation of position descriptions. The probable reason for this, outside of the personal preferences of various writers, is the fact that various companies have different needs as well as different policies and objectives in describing their positions. Some companies, the larger ones especially, need to have many details clearly stated. Others, where smaller numbers of people are involved and where a single position covers a wider range of duties and responsibilities, may obtain satisfactory results with less detail.

The preparation of position descriptions for routine clerical jobs is a comparatively simple procedure. The main problem is to get the basic nature of the job on record in terms that will readily identify the proper place to rate the job in the evaluation plan being used. The most important thing to avoid is the writing of the description as if it were a procedure or set of instructions covering how the various tasks are to be performed.

Because more problems and more care are involved in the preparation of managerial job descriptions, the following additional material will present some typical approaches in use at present. The reader will find it useful also to refer to "Dimensions of Executive Positions," by John K. Hemphill.[1]

The only common major divisions that are found in the format of practically all position descriptions are:

1. Identification
2. Basic nature of the position
3. Duties and responsibilities

These three headings are about the minimum and are found in nearly all cases. Identification has to do with merely listing such things as title, department, division, section, and code number. Usually this is followed by a short general statement covering the basic nature of the position. These statements are seldom more than a few short sentences in length. The section on duties and responsibilities may range from short to long and from few to many depending on the position and the objectives the company may have in mind for using the description.

Here are the outlines used by two companies in preparing their descriptions:

Example A

1. General responsibilities
2. Reports to
3. Primary duties
4. Key organizational relationships

---

[1] John K. Hemphill, "Dimensions of Executive Positions," Ohio State University, Bureau of Business Research, Research Monograph 98, Columbus, Ohio, 1960.

Example B

1. Function
2. Responsibilities and authorities
   a. General statement
   b. Objectives, policies, programs
   c. Organization and personnel

d. Money, facilities, materials
e. Functional activities
f. Appraisal of results
3. Relationships
4. Limits of authority

The best position descriptions are clear, concise, and complete and present all the distinguishing characteristics of the job. Brevity is desirable, but too little detail will not serve properly to differentiate one job from another and will not provide sufficient data to make equitable evaluation possible.

One company with considerable experience in this field has prepared a thorough set of instructions which are useful in preparing descriptions. The set of instructions which follows is presented in the same format in which the descriptions are written.

Broad Function. This is a short paragraph—or reader's digest—summarizing the tasks performed by the employee. The lead sentence should begin with the infinitive form of the verb.

Principal Responsibilities. Each specific task performed should be written out. This is the most important phase of the position description and should be carefully prepared.

The most common difficulty encountered in the preparation of such descriptions is the omission of pertinent information. However, descriptions occasionally contain entirely too much minute detail. The clarity of the description will be improved if the following four principles are generally adhered to:

1. Each sentence should begin with a functional verb.
2. The present tense should be used throughout.
3. A terse, direct style should be employed, omitting unnecessary words.
4. The description of the principal responsibilities should be grouped as follows:
   a. *Planning.* To develop and prepare a proposed method of action or procedure.
   b. *Directing.* To control or guide the activities of other employees, or a course of action. "Direction" is usually due to a superior position and emphasizes the idea of immediate supervision.
   c. *Executing.* To perform or carry out personally.
   d. *Servicing.* To supply data, information, or assistance for the use of others. The services furnished have been previously executed or prepared.
   e. *Advising.* To counsel or recommend action. The advisee may not necessarily follow the recommendations.
   f. *Miscellaneous.*

Relationships. Because meeting, dealing with, or influencing other persons is a responsibility that goes with certain positions, these intra (within), inter (between), and outside contacts should be considered. Consideration should be given to how the contacts are made, the level or rank of the contacts, and how often. This material should not duplicate the information described under Principal Responsibilities.

Authority and Reservations of Authority for Decision Making. This part of the position description should make clear the responsibilities, the general scope of authority, and matters on decision making authority which are reserved. In preparing this part, reservations of authority should refer only to nondelegated responsibilities and should be keyed to specific duties and subject matter. The general principle to be followed is that the authority for any particular position is complete, with respect to that needed to fulfill designated responsibility, except as specifically reserved. The reservations, therefore, call for careful particularization.

Measures of Accountability. This part of the position description should describe the criteria and standards to be used in appraising the performance of the incumbent. The aim should be to set the foundation for a personalized rating sheet—one that is designed to match the work of the position. A checklist of generalized "quality" items should be included, such as quality of climate, morale, and spirit in the organization. This should be augmented by suggestions for specific yardsticks, bogies, or budgets.

Word understanding is one of the most difficult problems to solve in preparing position descriptions. The following is a list of functional verbs used by a management consulting firm in describing positions, with definitions and examples of usage. This type of material is always useful, but it is a necessity when several persons are involved in preparing or using descriptions.

Functional Verbs Arranged by Categories

Planning (to develop)	Directing (by position)	Executing (to do personally)	Servicing (to supply something)	Advising (to counsel or recommend)	Miscellaneous
Create	Administer	Audit	Expedite	Advise	Assist
Develop	Conduct	Classify	Inform	Appraise	Cooperate
Establish	Control	Collect	Provide	Confer	Coordinate
Forecast	Decide	Compile	Report	Consult	Issue
Formulate	Delegate	Establish	Service	Contribute	Promote
Initiate	Determine	Evaluate		Counsel	Represent
Institute	Direct	Handle		Inform	
Organize	Guide	Interview		Interpret	
Plan	Instruct	Maintain		Recommend	
Write	Manage	Operate		Suggest	
	Order	Order			
	Prescribe	Perform			
	Supervise	Prepare			
	Train	Produce			

The following are some typical definitions and examples of these functional verbs which are used frequently in describing management jobs:

> *Administer.* To direct or manage the execution, application, or conduct of plans and personnel. May require negotiation and arbitration to bring individual and divergent interests into harmonious arrangement.
>
> *Control.* To exercise directing, guiding, or restraining power over plans or personnel.
>
> *Determine.* To come to a decision as a result of investigation or reasoning. Certainty and reliability of the decision are implied.
>
> *Evaluate.* To determine the value or amount of; to appraise.
>
> *Formulate.* To compose into or put in a systematized statement.
>
> *Initiate.* To introduce by a first act. Initiate emphasizes the act of being.
>
> *Review.* To go over or examine deliberately. Review emphasizes the idea of critical examination.
>
> *Plan.* To prepare a proposed method or action or procedure.

The definitions should follow standard dictionary usage, but where various usages are included in the dictionary, a selection should be made of one usage to be followed within a given company. The example should use terminology from the specific company.

Following is a position description for a research economist. It is perhaps slightly shorter than the typical description.

Research Economist[2]

Reports to. Assistant, economics research.

Supervises. Analyst, economics research.

Basic Function. Responsible for aiding the assistant, economics research, in

[2] SOURCE: National Industrial Conference Board, *Management Record*, vol. 23, no. 9, September, 1961.

the development, processing, and evaluation of economic data for special studies and for the preparation of long- and short-range forecasts.

Primary Responsibilities

1. Assists in analyzing economic and business trends and interpreting their effects on company operations.
2. Contributes to the development of short-range forecasts of economic and business trends as required (*a*) for operational guidance and (*b*) for profit planning activities.
3. Develops long-range forecasts of selected economic series as assigned (*a*) for individual line and staff studies and (*b*) for coordinated long-range planning activities.
4. Assists in the development of appropriate reports, charts, and other means of presentation which will assure proper dissemination of economic data affecting the operations or planning activities of the company.
5. Programs and conducts special research studies, develops market information, or supplies appropriate data.
6. Assists in reviewing the flow of economic publications from industrial, educational, consulting, and government sources.
    a. Examines and evaluates individual sources and renders opinions as to accuracy and completeness.
    b. Suggests additions to departmental library source material as required.
7. Reviews and/or develops economic data disseminated by the company as required by assignment.
    a. Economic material for publications and speeches.
    b. Economic data supplied to government agencies.

Internal Relationships. Line and staff division and department managers, as required in the derivation and dissemination of economic data for studies and guidance of planning activities.

Manager, profit planning, to provide short-term economic data as required.

External Relationships

Civic research and discussion groups.

Research organizations.

Industry, trade, and industrial management associations.

Local, state, and government fact-finding agencies, to attend meetings, seminars, conferences, and otherwise keep informed of economic trends, research activities, and new data development. To receive long-range forecasts and exchange appropriate information.

SALARY EVALUATION PLANS

Any salary evaluation plan must fit the situation in which it is being used. This seems to be obvious, and yet there are those who attempt to take a plan from some shelf and apply it blindly in their own company. Others have taken plans that have worked well on factory workers or low levels of factory supervision and have attempted to stretch the plans to cover research engineers, staff people, and executives. These approaches are almost certain to result in failure.

This does not imply that the whole field of compensation must be re-researched from the beginning in each new situation. What is needed is enough tailoring to make certain that the final plan used fits the situation and that the people involved think it fits the situation.

In the middle and higher levels of management and in a wide range of technical and administrative staff work, there is much similarity between companies in what people do in general areas such as planning, supervising, and problem solving. On the other hand, there is frequently wide variation in the importance of these general categories between companies. Further, every group of people is at least a little different and has been working in a different environment, with a different history and a different set of precedents and concepts.

A study made by W. R. Jacobson of the General Electric Company resulted in the preparation of a "Guide for Developing Factor Complexes." The definition of a "factor complex" is hard to express exactly, but it is built around the idea that a factor suitable for use in evaluating managerial positions represents an area of activity and is multidimensional in nature. In contrast, factors suitable for factory jobs are narrower in scope, are not so much interrelated with other factors, and are usually unidimensional in nature. This guide, which is presented below, is a useful aid in considering the usability of a plan or in developing or modifying a plan.

Guides for Developing Factor Complexes[3]

I. Supervision and Technical Leadership
  1. Kinds of work supervised or technically led—scope, complexity, and diversity.
  2. Numbers and types of those supervised or led.
  3. Highest extent of technical leadership; limitations—supervisory responsibilities reserved for superior.
      Consider where pertinent:
      a. Advice and aid available from others.
      b. Results or products required from group led or supervised—effect on work of others.
      c. Special skills required in technical areas other than those supervised.
II. Planning
  1. Time span of planning.
  2. Complexity and diversity of elements dealt with—known and unknown variables.
  3. Scope and organizational level planned for.
  4. Highest level of planning work; limitations—planning responsibilities reserved for superior.
      Consider where pertinent:
      Effect on men, money, time, and so on.
III. Analysis and Innovation
  1. Subject matter, complexity, and obscurity of problems and data dealt with.
  2. State of development of problems when assigned and at conclusion of individual's work on them. Are problems presented for solution or are they developed and defined by the man?
  3. Scope of analysis and innovation—characteristic breadth and nature. Does work involve development of new concepts, plans, and the like, or modification of existing ones; is work concerned with entire programs, a phase, a subphase, and so on?
  4. Highest level of this activity; limitations—responsibilities reserved for superior.
      Consider where pertinent:
      a. State of the art.
      b. Impact of success or failure.
      c. Unusual pressures of time, volume, and so on.
      d. Requirement for rare combinations of skills and knowledges—unusual "job mixes."
IV. Recommending and Decision Making
  1. Subject matter of recommendations (if this is not sufficiently brought out under III).
  2. Recommendations made: to whom—how—in what form?
  3. Highest level of this activity; limitations—recommending or decision making reserved for superior.
      Consider where pertinent:
      a. Precedent, guidance, and review available.
      b. How and by whom recommendations are ultimately used.
      c. Impact of right or wrong recommendations.
V. Relationships and Communications

[3] Source: Stanley E. Herman, "Compensation: Wrong Roads and Questionable Directions," *Personnel*, vol. 36, no. 5, September–October, 1959.

1. Purpose of contact: to instruct, provide information, persuade, sell?
2. Type and level of contact or communications (e.g., briefings to department manager)—nature of subject matter—expository, complex, controversial.
3. Highest level of this activity; limitations—relationships and communications reserved for superior.
    Consider where pertinent:
    a. Lone-wolf contacts, or are contemporaries available for support?
    b. Unusual pressures of time, volume, and so on.
VI. Man Considerations
(Answer this only if information is *not* brought out fully in preceding sections.)
    1. Unique nature of actions taken or subjects acted upon, necessitating special prestige, training, or education.
        Consider where pertinent:
        a. Situation or context in which prestige or recognition is required.
        b. Essential special preemployment training or education other than B.S. or M.S. in standard field. (In beginning-level position, distinction between B.S. and M.S. may be considered, if pertinent.)

SPECIFIC PLANS

Space will not permit the publication, in detail, of several typical salary evaluation plans. A presentation can be made, however, of the outline and basic idea of several plans which have had extensive use. Reference will be made to the source of more information about each plan.

Plan A. This plan was developed and put into use at the Revere Copper and Brass Company. It was used there on clerical jobs such as those found under the direction of the treasurer, office manager, and accounting department supervisors. The factors and their explanation are typical of many purely clerical plans.[4]

Elemental Factor Value. The elemental factor value in the clerical evaluation covers the minimum requirements of general availability and willingness to work; a certain modicum of ambition, neatness, personality, honesty, and dependability; and certain minimum physical attributes. These are static imponderables which are constant in all jobs. An attempt to measure them is usually impractical. Instead of assigning any variation in valuation among jobs for these characteristics, an initial constant value is placed on all jobs.

Education. This factor is given value according to the fundamental minimum education necessary for an understanding of the position. The error of placing a value on the education of the man rather than on the requirement for education which is an intrinsic part of the job must not be permitted. For instance, a master's degree in medieval history would be of no great benefit to a file clerk.

Practical Experience. This is the minimum experience necessary to fill the job. It is conceivable that there will be positions in which no experience is imperative. From that point on, as far as the factor extends, the measurement can be placed in terms of the time necessary to obtain the requisite amount of experience.

Analytical Requirement. Analytical requirement and complexity of work take into consideration the difficulties and complications of the assigned task, the convolutions necessary to make the necessary analysis, the ability to initiate, and the ingenuity to carry through ideas supported by data and to come to a final definite conclusion. All these elements are combined in this factor, which might be called "native ingenuity." It is the only factor which might make the man on the job much more valuable than the job itself.

Accuracy. Accuracy applies to the routine aspect of any job, grading the possibility of error and the importance that such an error would have. There are actually three considerations implied here:
    1. The possibility or frequency of error

[4] SOURCE: John H. Eikenberg, "Case History of an Office Job Evaluation Program," *AMA Handbook of Wage and Salary Administration*, American Management Association, 1950.

2. The possibility that the error will be detected and corrected before having its ultimate result

3. The importance of the error in time and money

Memory. Memory is applied only to the facility for memory of fact directly associated with the position at hand. A retentive memory is bound to add to the efficiency of the holder of most clerical jobs, but in its lowest application it has relatively little value. Certain positions may place a large dependence on the facility with which the holder can call to mind countless nonrepetitive and complicated facts, while others hold only the possibilities for associated memory allied to current issues.

Manual Dexterity. Manual dexterity has been based on the degree of manual ability necessary to operate various types of standard office machinery. It is a highly factual element which can be aligned in keeping with the complexity of the assigned machines. Once the degree of complexity is decided upon, the graduation can be based entirely on the type of machine operated.

Supervision. Supervisional requirements are also largely a factual element but take into consideration the nature of the work as well as the number of employees.

Working Conditions. Conditions of the work consider the physical effort expended and the conditions surrounding the place of work. It is common knowledge that it is a much less difficult proposition to complete a task in an atmosphere of solitude than to do so in the midst of hubbub.

Continuity of Work. Continuity of work is also essentially derived from the physical strain and fatigue imposed by the job but is based primarily on the opportunity for rest intervals. The monotony of the job in its entirety is pictured, then the opportunities for breaks in this monotony as a whole are considered. Unquestionably the job with variety, with frequent opportunities to stop for discussion or any other purpose, carries with it a lower fatigue element than the position which entails constant attendance at a routine task.

Physical Strain. Physical strain on the senses considers the tension at which the occupant of the job must fulfill his duties. In a clerical position, a large part of this tension is apt to be the result of eyestrain. Distraction from the task at hand and the consequent difficulty in returning to the interrupted continuity is also worthy of remuneration.

Public Relations. Public relations or contacts could easily be termed "the voice with a smile." Essentially, however, this element is a measurement of the extent and relative difficulty and importance of the work which brings the position holder into contact with other departments or companies. A consideration of the possible effect these contacts may have on plant or public relations is implied.

Plan B. This plan was developed especially for a firm which makes and sells a wide variety of electrical and mechanical equipment to industrial and utility companies. The plan covers sales engineers and all levels of management above them to the vice president. It also includes engineers of various types and other administrative management positions. The plan was put together only after an extensive study of the nature of the jobs and the feelings of the people involved concerning what was important, what the company really should pay for, and what was identifiable and measurable. This company, by tradition, places a considerable amount of emphasis on the long-range aspect of all its work and relationships.

The factors used in this plan are as follows:

Factor 1. Education. This factor measures the formal education required by a position. The evaluation of this factor plus experience (factor 2) measures the total know-how required by a position.

8 degrees   Point range 0–140

Factor 2. Experience. Experience measures the know-how gained from performance.

45 degrees   Point range 13–585
(Each degree represents 1 year of usable experience)

Factor 3. Judgment. Judgment is the ability to apply sound reasoning. It is common sense. It is the ability to realize that enough facts have been accumulated to support a decision, plus the ability to weigh the relative importance of each fact, plus the ability to combine and synthesize the evidence and produce the proper answer or decision. Judgment can be considered the ability to apply the "know-how" evaluated in the first two factors, education and experience. It is closely allied to analytical ability and problem solving ability. Judgment usually tends to grow with maturity and age. However, some men acquire it rapidly; others are immature and unreliable at the age of sixty.

15 degrees   Point range 0–167

Factor 4. Dealing with People. This factor measures the position requirement "ability to get along with people." The term, as used in this factor, implies the ability to get along in a positive, result-producing manner.

15 degrees   Point range 0–198

Factor 5. Persuasive Skill. Persuasive skill measures salesmanship. It is the ability to recognize the needs and viewpoints of others and to present benefits and facts about a proposed course of action in such a manner that the proposal is accepted and acted upon by others. It includes the sale of ideas and plans as well as the sale of products.

15 degrees   Point range 0–192

Factor 6. Counseling Skill. Counseling skill measures the ability to give advice. It includes the ability to understand the personal problem of another, the possession of sympathy and empathy, and the ability to listen and to advise a sensible course of action. This is essentially a managerial skill. It does not include counseling in subjects such as engineering, sales, or financial matters. Credit for the latter is given in other factors—in education, experience, judgment, informal speaking, and persuasive ability.

15 degrees   Point range 0–50

Factor 7. Informal Speaking Skill. Informal speaking skill is the ability to communicate through conversation. The ability to communicate and to be understood is required in varying degrees by all sales positions. No zero degree is listed.

15 degrees   Point range 1–91

Factor 8. Writing Skill. This factor measures the ability to communicate ideas effectively through the medium of clear and concise written material.

18 degrees   Point range 0–126

Factor 9. Creativeness. Creativeness measures the ability to produce new ideas. The ideas need not be entirely sound and practical. To achieve practicability, one must temper creativeness with factor 3, judgment. Factor 9 measures only the requirement for originality—the presence of the creative spark. The possession of patents and suggestion awards is unmistakable evidence of creativeness.

18 degrees   Point range 0–96

Factor 10. Initiative. Initiative measures self-starting ability.

15 degrees   Point range 0–10

Factor 11. Industriousness. Industriousness is the measure of the intensity and duration of application to work. It is the equivalent of the "effort" factor often found in job evaluation plans.

15 degrees   Point range 0–27

Factor 12. Responsibility. In determining the level of responsibility, consider dollar volume, dollar expenditure, dollar profit, number of employees, total dollar payroll, capital investments, investments of funds, confidential matters, sharing or total responsibility.

19 degrees (levels)   Point range 2–840

The handling of factor 12, responsibility, in this company is done in a rather unique way. The central personnel group prepares a number of position descriptions which include spelled-out statements of responsibility. Each of these descriptions is classified as to degree or level of responsibility. Each new position is compared with these examples and by comparison is placed into the appropriate degree or level. The bench mark descriptions are of course modified from time to time so that they reflect the company's current situation.

The way this company uses its plan is also quite interesting. It evaluates each position in two ways: first, in accordance with the minimum requirements for an acceptable performance; and second, in accordance with the maximum degree which the company could fairly be asked to include in a man's pay in the position. Then, each year, the incumbent of each position is rated in terms of the degree which he used in the job and in terms of the degree he possesses. This approach helps the company to maintain overall control of salary costs, but particularly it is useful in pointing out places where positions are being overfilled and talent is being wasted.

Plan C. This basic plan has been widely applied on all sorts of supervisory, technical, administrative, staff, and executive positions. This plan, like all others of a similar nature, requires some tailoring to bring it into line with the specific needs of a particular company. The plan, with proper tailoring, has been used in general manufacturing industries, process industries, and service organizations of several types.

It is normally composed of nine factors, although some factors have been added or subtracted in special situations. The usual factors and a general description of each follow.

1. Responsibility for Programs, Projects, or Operations. This factor covers the individual responsibility, effort, knowledge, and attention required by the position to organize and carry out a particular program or assignment according to approved company or division policies or procedures. The subfactors take into consideration the scope of the programs, the authority vested in the position, and the complexity of the field involved. As an example, the question could be asked, "Is this program or project something that involves the entire company, or is it merely something that involves the production operations within one division of the company?" Concerning the matter of authority, the question could be asked, "Does the person who occupies this position have full and complete authority to make a final decision and to direct the taking of action, or does he have authority merely to investigate the situation and make recommendations to someone else?" From the standpoint of complexity, such things could be asked as, "Does this program involve a field of activity in which very little complicated or advanced knowledge is needed, in which facts are easy to find and easy to assemble, and in which techniques and procedures are well known and tested?" or "Does it involve a highly complicated and nebulous field in which there is a dearth of knowledge available, in which facts and statistics are hard to come by, and in which many unknowns must be estimated or forecast?"

This factor studiously stays away from using the words "education" or "experience." Naturally, these two items are important, but it is quite easy to make them an end in themselves rather than a means to an end. The way these two items are reflected as part of relative worth of management positions is in the fact that the scope of the programs or the complexity of the field may be so demanding that the person who fills the job will have to have a great amount of experience or an advanced education.

2. Responsibility for Supervision of Personnel. This factor covers responsibility for supervision of personnel. This includes both the direct supervision of personnel and the supervision of them through various subordinates. This indicator is basically concerned with those elements of supervision required to get the work done at the right time, in the right amount, and of the right quality. It covers the type and level of jobs supervised, the variety of work supervised, and the size of employee group.

3. Responsibility for Human Relations. This factor covers responsibility for human relations; it has to do with administering to people's needs while the preceding indicator has to do with keeping them busy. This indicator covers such things as involvement in selection, hiring, and orientation of new employees; handling of wage and salary problems; adjustment of grievances; arranging for promotions, transfers, and dismissals; handling of union negotiations; and matters of a similar nature. It covers the nature of the activities, authority, relative difficulty, and size of group.

4. Responsibility for Contacts outside the Company. This factor has to do with responsibility for contacts outside the company. It is a fairly self-explanatory indicator. Naturally, it is important to assess more worth to the man who makes commercial agreements with the president of an outside company than to the man who discusses technical or manufacturing problems with a technician in an outside company. And, of course, a man who spends most of his time in outside contacts is of more worth than a man who has only infrequent outside contacts, if it is assumed that the other elements of the contact are the same.

5. *Responsibility for Internal Contacts.* This factor covers responsibility for internal contacts. It is very much the same as the fourth indicator except that it covers contacts within the company. It is a very important indicator, for experience has shown that keeping smooth internal relationships does a great deal toward improving the effective operation of a company. This factor considers type and level of contacts, relative difficulty and importance, and frequency.

6. *Responsibility for Fact Finding or Investigation.* This factor has to do with responsibility for fact finding or investigation. It deals with starting and conducting studies, surveys, tests, and other similar activities for the purpose of finding facts; developing ideas, designs, or processes; or determining the course of action to take in some particular field of endeavor. It considers scope or breadth of activity, authority to take action, and complexity of the field.

7. *Responsibility for Planning, Scheduling, or Forecasting.* This factor covers responsibility for planning, scheduling, or forecasting. It covers what might be called the "looking ahead" aspects of a manager's work. For the president of the company, the complexity of the plans in which he becomes involved is, of course, of the highest degree. His participation includes the matter of final decision, and the variety and importance are, of course, at a maximum value for his company. On the other hand, a man such as the person in charge of manufacturing engineering would probably get into some fairly complex plans, and his participation would probably include some minor decision making. From the variety and importance standpoint, he would be limited to those things which had to do with manufacturing planning rather than expanding his scope into such things as sales or personnel policies.

8. *Responsibility for Establishing Objectives and Procedures.* This factor covers the responsibility for establishing objectives and procedures. It involves the responsibility for formulating, testing, and putting into effect such things as policies, regulations, general orders, practices, procedures, engineering designs, and the like which are in the nature of being a standard. It would include a standard which was established as a continuing guide, such as a purchasing procedure, or standards for a specific situation, such as the design of a building. It reflects scope, authority, and complexity.

9. *Responsibility for Financial Benefit or Injury to the Company.* This factor covers the responsibility for financial benefit or injury to the company. This indicator gets into the matter of gain or loss as a result of the authority vested in the position and the decisions which come as a result of that authority. In this indicator, such things are considered as the gain or loss that may be realized or prevented through effecting economies, operating efficiently, utilizing creative opportunities, making appropriate contractual obligations and commitments, selecting proper courses of action, and maintaining proper standards. This factor considers scope, importance, and degree of control.

The way this plan is used illustrates the considerable amount of flexibility that can be obtained from it. Figure 7-2 shows the working chart for factor 1. Assume an evaluation was being made of the chief of the general accounting section of the controller's division of a typical single-plant manufacturing company. The probability is that his duties would be classed under item C as far as scope is concerned. The maximum extent of his authority, typically, would fall in column J. Where column J and line C intersect, there is an index number 6 (for vertical components) that is used in determining the final point value for this factor. The nature of the field would then be considered. Referring again to Figure 7-2, at the lower part of the chart, another index number that would probably be appropriate is

1. RESPONSIBILITY FOR PROGRAMS, PROJECTS, OR OPERATIONS   Factor No. 1

This factor covers the individual responsibility, effort, knowledge, and attention required by the position to organize and carry out a particular program or assignment according to approved company or division policies or procedures. It takes into consideration the level of responsibility and scope, authority vested in the position, and the nature of the field or the difficulties and problems involved in carrying out the duties and responsibilities of the position.

LEVEL OF RESPONSIBILITY FOR PROGRAMS (VERTICAL COMPONENT)

SCOPE OF PROGRAMS, PROJECTS, OR OPERATIONS AND LIMIT OF AUTHORITY	AUTHORITY VESTED IN THE POSITION						
	Carry out programs, follow regulations, and report unusual action taken. May recommend changes -- or assist in directing programs.		Direct programs, recommend changes and report results -- or assist in managing and co-ordinating major programs.		Manage and co-ordinate major programs -- take necessary action and report over-all results -- or assist in determining or administering major programs.		Determine and administer major programs. Responsible for over-all results.
	G Limited Authority	H General Authority	I Limited Authority	J General Authority	K Limited Authority	L General Authority	M Full Authority
F. Broad company-wide programs involving line or functional direction of major segments of the company's operations -- limited by general company policies and objectives.	6	7	8	9	10	11	12
E. Specialized company programs involving a number of large scale departments or several functions -- limited by company or functional policies and objectives.	5	6	7	8	9	10	
D. Broad divisional programs or specific programs involving a group of sections or activities -- limited by divisional policies and decisions.	4	5	6	7	8	9	
C. Specialized divisional programs involving several sections or specialized activities -- limited by divisional general instructions and orders.	3	4	5	6	7	8	
B. Specific programs or specialized activities at specific locations involving one or a small number of sections or activities -- limited by specific orders and regulations.	2	3	4	5	6		
A. Specific activities of programs or operations -- limited by specific instructions.	1	2	3	4			

NATURE OF FIELD (HORIZONTAL COMPONENT)

Standardized supervisory or staff activities involving well-established procedures in a limited area of knowledge or activity.		Supervisory, staff, or technical activities involving standardized methods in a specialized area of knowledge or activity.		Managerial, staff, or technical activities involving unstandardized methods in an advanced field of knowledge or activity.		Administrative, staff, or highly-technical activities involving undeveloped unrelated and intangible elements in a broad field of knowledge or activity.	
Generally Applicable	Completely Applicable	Generally Applicable	Completely Applicable	Generally Applicable	Completely Applicable	Generally Applicable	Completely Applicable
13	14	15	16	17	18	19	20

FIG. 7-2. Factor 1: Responsibility for programs, projects, or operations.

Horizontal Degrees

	13	14	15	16	17	18	19	20
12	132	135	140	146	156	171	193	225
11	99	102	106	112	121	134	154	183
10	75	78	82	87	95	107	125	151
9	57.	59	63	67	75	86	102	125
8	43	45	49	53	60	70	84	104
7	33	35	38	42	48	57	69	87
6	25	27	30	33	39	47	58	74
5	19	21	23	26	31	38	48	63
4	15	16	18	21	25	31	40	54
3	12	13	15	17	21	26	34	46
2	9	10	12	14	17	22	30	40
1	7	8	9	11	14	19	26	35

Vertical Degrees or Levels

FIG. 7-3.   Point rating chart for factor 1.

16 (for horizontal components). These two index numbers would then be used to read the point value from the chart shown in Figure 7-3. In this example, the point value would be 33 for this factor. This is obtained from reading the chart to the right of vertical component index 6 and below horizontal component index 16.

One very interesting and practical aspect of this plan can be seen by inspection of Figure 7-3. This chart shows that point values increase in greater than direct proportion as the scope, authority, complexity, and other elements of managerial jobs increase. This reflects the realistic aspects of organization structure and the modern ways of assigning responsibility to managerial people. Straight-line relationships are, of course, practical for factory jobs and very low levels of supervision.

THE SALARY STRUCTURE

Figure 7-4 is indicative of the appearance of a modern salary structure for staff and managerial positions. Typically, the structure is composed of a number of position classes or grades as indicated by the rectangles labeled A, B, C, D, E, and F. Each class or grade covers all positions whose point values fall within the limits of the range of points established for that grade. Also, the height of the rectangle indicates the range of salaries for all positions within the point range.

It is becoming increasingly a reality, consciously or otherwise, to reflect the effect of income taxes on salaries. Figure 7-4 illustrates how this has been done in several cases. First, of course, a salary survey is made in order to determine the going rates in the community and industry. It is then assumed that these rates are being paid to a man with a wife and two dependent children, and this amount of Federal income tax is deducted. The remaining values are charted and a typical line of best fit (or least squares) is drawn. This line is indicated by line A-A on Figure 7-4. By judgment or policy decision, this line may be accepted or altered to become the base line for starting the salary structure. Once a decision is made, the income tax is added back on and results in the development of line B-B. The concept on which lines A-A and B-B are built is important. In the illustration, they represent what the company is willing to pay for minimum acceptable performance on the job. These two lines could, of course, be constructed on the concept of average or even maximum for the job. Line C-C is also constructed on the basis of judgment and policy. In practice, line C-C may be set

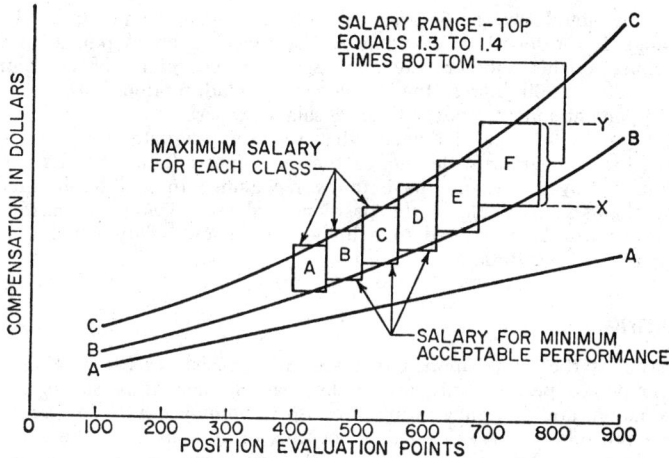

FIG. 7-4.   Modern salary structure for staff and managerial positions.

at a level ranging from 20 to 60 percent above line B-B. In the bulk of the cases, it falls between 30 and 40 percent.

NUMBER OF CLASSES OR GRADES

The number of classes or grades found in a salary structure is determined largely by judgment and policy. Obviously, the whole process of salary evaluation, while being systematic and reasonably objective, is not mathematically precise. It is doubtful that the process could hold tolerances any less than plus or minus 5 percent, even if these tolerances could be measured.

In actual practice, the number of points representing the top of the range for a given salary grade will be between 10 and 15 percent higher than the number of points representing the lower limit of the range. Naturally, the wider the percentage spread, the lower the number of salary grades that will result. Using both a 10 and a 15 percent factor and starting at 100 points, the range would appear as follows:

Grade	10 percent spread	15 percent spread
1	100–110	100–115
2	111–122	116–134
3	123–135	135–155
4	136–150	156–180
5	151–166	181–208
6	167–184	209–240
7	185–204	241–278
8	205–226	279–320
9	227–250	321–370
10	251–276	371–427
11	277–305	428–493
12	306–337	494–568
13	338–372	569–655
14	373–410	656–755
15	411–453	756–871

It should be noted that an increase of only 5 percentage points in the spread of each range has resulted in covering almost twice as many points on the total scale. Looking at this another way and assuming the spread of the entire scale to be from 100 to 300 points, the 10 percent spread produced 11 salary grades while the 15 percent spread produced only 8 salary grades.

The only guides to good judgment that seem appropriate are: first, keep the point spread for each grade wide enough that an impossible accuracy is not implied and that an employee does not have to be reclassified to a different grade when only minor changes are made in the position; and second, keep it narrow enough so that persons can be promoted to a higher grade and salary when a significant change is made in their duties.

MERIT RATING

Merit rating is the name applied to systematic procedures for judging how well a specific employee performs his job. Other names, such as personnel rating, employee evaluation, and efficiency rating, are also commonly used. In some instances, merit rating procedures are expanded to include an assessment of the overall worth of the employee to the company. They also may include an assessment of the employee's promotability to a job of higher responsibility.

This chapter will not attempt to give full treatment to the subject of merit rating. Instead, it will discuss briefly the basic procedure, cover some things to do that will increase the likelihood of successful operation, and in the bibliography, provide the reader with references from which detailed procedures may be obtained.

General Nature of Merit Rating. Merit rating procedures are usually developed and administered by the personnel department in a company. The actual ratings, however, are made by the employee's immediate supervisor. In most cases, the next higher supervisor reviews the rating. Typically, the employee is rated on from six to twelve or more personal attributes or measures of performance, which are called factors. The factors used almost always include quantity of work and quality of work. It is quite usual to find factors such as attendance, cooperation, versatility, safety habits, initiative, getting along with people, and similar more or less definable traits or characteristics. In some plans, the factors include such items as judgment, creativeness, integrity, honesty, loyalty, friendliness, attitude, intelligence, and so forth.

In a typical plan, the rater decides where the employee fits with regard to each factor and then marks the rating on either a numerical scale or a scale with degrees such as poor, fair, average, good, and excellent. Naturally, the rater must have some good reasons, backed up by evidence, for his rating.

Finally, most systems call for the rater to conduct a private interview with the employee, during which the ratings and the reasons therefor are explained and suggestions are made for improvement. The time interval between ratings is seldom less than six months or more than one year.

Who Uses Merit Rating. Every day, every hour, and in every business, some supervisor is evaluating the performance of some employee. So the question is not "Will we or will we not have merit rating?" but rather "To what extent and how will we have merit rating?" Very few small companies have any formal or organized procedures. About 60 percent of the medium-size companies have a procedure which may range in adequacy from very poor to excellent. Most, perhaps 80 percent, of the large companies have a reasonably well spelled out procedure.

Assuming the foregoing estimates are approximately correct, it is obvious that merit rating procedures are not overwhelmingly in style. Further, experience indicates that the skill and resoluteness of the efforts of those companies who do make an attempt is adequate in only about half of the cases.

Why is this? The answer is simply that many companies and individual managers fail, or perform inadequately, on those things that can make merit rating systems deliver worthwhile and identifiable results. These are discussed in the next paragraph.

Keys to Successful Merit Rating. Good merit rating procedures can make a worthwhile contribution to the achievement of company goals. Certainly, they are only one tool of many available and needed by managers to accomplish these goals. But merit rating procedures are a key part of the area of human relations and motivation, which, in turn, is a dominant factor in achieving company goals of profit and productivity. Recent studies in the behavioral science area have shown conclusively what could have been done in the past had the essentials of human motivation been known and understood. More important, however, these studies are pointing the way toward greater achievement in the future, if there is the wisdom to give attention and to act.

The first key to successful merit rating is in the hands of top management. They must believe that good procedures will be worth the effort. They must communicate their belief and desire for such procedures to their line and staff subordinates. And they must follow up to see that good procedures are installed and operated.

The second key to successful merit rating is a combination of skill and desire in the management group at all levels in making the ratings and in using the ratings in merit reviews. This skill and desire can be obtained only through training of a better grade and greater quantity than almost everyone has done in the past. This is the area in which a company has the greatest likelihood of assuring success or failure. Managers need to feel that they are skilled and adequate in using a procedure to have the confidence that motivates them actually to use it rather than avoid it. They must also feel that their time invested in using the procedure will pay off in identifiable and worthwhile results that contribute to the achievement of their own goals.

The third key to successful merit rating concerns the administrative support provided by the personnel department in handling routine operating procedures and in aiding line and staff supervisors to know about and use the information that the system can produce.

The fourth key is the plan or system itself. It should be as simple as possible. Even the best merit rating system has a large amount of subjectivity in its makeup. Constructors of such plans, in an effort to obtain maximum objectivity, have tended to add to the complications of using and understanding the plan. Obviously, there is much interdependence among all these keys to success.

Recommended Action. The use of merit rating procedures is recommended, even in small companies, if the company is willing to abide by the four keys to success mentioned above. Setting these procedures up requires some time and skill. Lacking these, professional help should be obtained.

BIBLIOGRAPHY

Benton, L. R., *A Guide to Creative Personnel Management,* Prentice-Hall, Inc., Englewood Cliffs, N.J., 1962, chap. 12.

Likert, Rensis, "Motivational Approach to Management Development," *Harvard Business Review,* July–August, 1959.

Maynard, H. B. (ed.), *Handbook of Business Administration,* McGraw-Hill Book Company, New York, 1967, sec. 11.

Maynard, H. B. (ed.), *Handbook of Modern Manufacturing Management,* McGraw-Hill Book Company, New York, 1970, sec. 9.

Sibson, Robert E., *Wages and Salaries,* American Management Association, New York, 1960.

Yoder, Dale, *Personnel Management and Industrial Relations,* 5th ed., Prentice-Hall, Inc., Englewood Cliffs, N.J., 1962, chaps. 15 and 17.

Chapter **8**

Grievance Handling

JOHN H. MORRIS
Management Consultant, Westerville, Columbus, Ohio

Industrial engineers work with facts and people. The engineer who cannot combine the two into one harmonious package is headed for trouble. A distorted fact may glare back from a time study report, a job evaluation form, or a man and machine chart; but a disgruntled employee who believes his earnings are too low or his work effort too high will make his opinions known loud and clear. When this occurs, the industrial engineer must recognize that he has obligations to the employee as well as to work measurement principles. He must weigh the possible consequences of violating one obligation at the expense of the other.

In this chapter, it will be assumed that the industrial engineer is technically competent. Technical competence, however, is secondary in a plant that is plagued by strikes, slowdowns, and employee discontent. Every labor relations man knows that no subject produces employee discontent so spontaneously as some action which workers believe will have an adverse effect on their earnings or work efforts. This chapter will discuss the importance of achieving and maintaining good employee relations, the causes of employee grievances, the manner in which grievances are resolved, and the industrial engineer's legal as well as moral obligation to promote industrial harmony.

RIGHTS OF MANAGEMENT AND EMPLOYEES

Employee complaints are natural. In fact, experienced labor relations men will worry when complaints are few or nonexistent. They know that the absence of employee complaints might indicate some underlying fear or pressure that could erupt unexpectedly into discontent, slowdowns, or strikes. Complaints, however, should never be ignored, especially if they involve such personal subjects as time standards, earnings, and work methods. To the contrary, each complaint should be examined and resolved as quickly as possible. This is not only good personnel relations, but it is also a moral and sometimes even a legal obligation.

For example, no employee or group of employees should be punished for the simple act of protesting conditions that affect their wages, hours, or working conditions. This is true even when a protest is presented with considerable heat and vigor. The right to protest is a legal right. In fact, Federal labor law, which includes the Labor Management Relations Act (Taft-Hartley Act) and the National Labor Relations Act, as amended, says employees have the right to engage in concerted activities for their mutual aid and protection. Most companies, regardless of size and importance, are covered by this law[1] or by similar state laws, even though their employees may not be represented by a union. Furthermore, the Taft-Hartley Act says the obligation to bargain collectively includes conferring in good faith over questions involving wages, hours, and other terms and conditions of employment,[2] not only during contract negotiations, but also during the term of a labor agreement. The law also says that any employee or group of employees has the right at any time to present grievances to their employer and to have the grievances adjusted with or without the intervention of a union representative. The union representative, however, referred to in the act as the "bargaining representative," must be given an opportunity to be present at the adjustment.[3]

An employee's right to question situations involving wages or working conditions and to file grievances concerning them does not obligate management to alter its position. Management, of course, should be sure that its position is correct, make changes if necessary, and discuss the situation with the individuals involved. But management, too, has rights, although these rights are not always clearly established. The decisions of courts, arbitrators, and government agencies have left management's rights in a state of uncertainty. This is especially true in plants where unions are established. If a union contract exists, the industrial engineer should become thoroughly familiar with its contents. His major interest, however, will be the contract's management rights clause, which, in most instances, will support his company's right to make economically motivated changes. A short but typical management rights clause might read as follows:

> Subject to the provisions of this Agreement, the management of the works and the direction of the working forces—including the right to hire, suspend,

---

[1] The Taft-Hartley Act applies to industries whose operations affect interstate commerce as well as those actually engaged in interstate commerce. In some situations, the dollar value of outflow or inflow determines whether or not a company is covered by the act. If coverage is in doubt, an attorney should be consulted.

[2] For the purposes of this Section, to bargain collectively is the performance of the mutual obligation of the employer and the representative of the employees to meet at reasonable times and confer in good faith with respect to wages, hours, and other terms and conditions of employment; or the negotiation of an agreement, or any question arising thereunder; or the execution of a written contract incorporating any agreement reached, if requested by either party; but such obligation does not compel either party to agree to a proposal or require the making of a concession. [Section 8(d) of the National Labor Relations Act, as amended, in part.]

[3] Representatives designated or selected for the purpose of collective bargaining by the majority of the employees in a unit appropriate for such purposes shall be the exclusive representatives of all the employees in such unit for the purposes of collective bargaining in respect to rates of pay, wages, hours of employment, or other conditions of employment: Provided, that any individual employee or a group of employees shall have the right at any time to present grievances to their employer and to have such grievances adjusted, without the intervention of the bargaining representative, as long as the adjustment is not inconsistent with the terms of a collective bargaining contract or agreement then in effect. Provided further, that the bargaining representative has been given an opportunity to be present at such adjustment. [Section 9(a) of the National Labor Relations Act, as amended.]

or discharge for proper cause; transfer; or change assignments, together with the right to relieve employees from duty because of lack of work or for other legitimate reasons—is vested exclusively in the Company; provided that this will not be used for purposes of discrimination against any member of the Union.

Notice that this clause contains a statement making it subject to other provisions of the labor agreement. Even without this statement, a management rights clause is sometimes referred to as a "general clause" and is frequently held to be subordinate to "specific" clauses that deal with individual subjects. For example, if a labor contract contains clauses, frequently referred to as articles, on "seniority," "rates of pay," or "new or changed jobs," the provisions of any of these clauses or articles could supersede the provisions of the management rights clause.

GRIEVANCE HANDLING

A grievance may be loosely defined as a violation of a worker's rights. Quite often, a worker only thinks his rights have been violated, and his grievance is nothing more than a "bum beef" or a gripe. Every complaint, however, is real to the man who entered it. It should be investigated and processed or a reason given explaining why it is not legitimate. This is a job for the company's labor relations representative, but the industrial engineer may play a key role in deciding if a wage or work method grievance is justified.

Studying the labor contract, where one exists, will give the industrial engineer a better understanding of his responsibilities and his limitations. It will also answer a most important question: What happens when grievances arise? This question, of course, can also be asked in plants where no union exists. In fact, the industrial engineer in a nonunion plant has an even greater obligation to help prevent as well as adjust grievances. More than most management representatives, he can create conditions which might result in union activity or even involve his company in legal technicalities.

A typical labor contract grievance procedure begins with a discussion of a grievance between an employee and his foreman. If they fail to settle the matter, the grievance proceeds through a series of steps, with each step bringing a higher level of management and union representatives into the action. The steps are frequently accompanied by time limits, and at some point in the procedure, the grievance is usually submitted in writing. A simple grievance procedure might read as follows:

> The Company and the Union encourage the highest possible degree of friendship and cooperative relationship between all employees. It is the intent and purpose of this agreement to promote and improve the relationship between the Union and the Company and to set forth the basic rules that have been agreed upon between the parties regarding wages, hours, and working conditions affecting employees covered by this contract. However, should any grievance arise between the Company and the Union, or its members employed by the Company, as to the meaning and application of the provisions of this agreement, earnest effort shall be made to settle such differences in the following manner:
> 1. The employee shall discuss the grievance with his foreman and may have his union committeeman present if he so desires.
> 2. In the event an agreement is not reached, the matter shall be referred to the Local Union Committee. The Local Union Committee and department head concerned shall meet to consider the matter within three (3) days.
> 3. If no settlement is reached in step 2, the Union shall present the grievance in writing to the Vice President of Industrial Relations, or his designated representative, requesting a meeting which shall be held within seven (7) days, unless a later date is mutually agreed to.
> 4. If no agreement is reached after step 3, the matter may be referred to the International President of the Union or his designated representative and the Vice President of Industrial Relations of the Company or his designated representative who shall meet with the parties concerned within ten (10) days. Their decision, if unanimous, shall be binding on both parties.

Some contracts contain special provisions for processing or expediting grievances over wages and work methods.

Grievance Handling Guidelines. When grievances involve activities of the industrial engineer, he may be required to be present and testify at any step in the grievance procedure. He may in fact be required to produce and explain his records. Because of this possibility, the industrial engineer should not only know what is expected of him at grievance hearings, but he should also have a general knowledge of grievance handling guidelines recommended for all management representatives to follow.

Turning to the guidelines first, management representatives should:

1. *Understand the Grievant's Position.* This is not always easy. As grievance hearings progress from step to step, it frequently becomes apparent that the grievant or the union wants something entirely different from what the union presented as the issue at the first hearing. This can happen even when the grievance is presented in writing. One means of avoiding a misunderstanding is to restate the grievance as management understands it and then ask if the restatement is an accurate account of the grievant's position.

2. *Ask for Time to Investigate or to Consider the Facts.* Nearly every grievant has had time to think through and prepare his position, but the management representative to whom he presents his complaint will be caught cold. If the management representative jumps to a conclusion without getting an understanding of the facts, he may establish an undesirable precedent or reach a decision that conflicts with company policy or perhaps with some practice that is well established in another department. Should there be any doubt about how to proceed, take time to investigate all possibilities and reply to the grievance at a mutually agreeable later date.

3. *Never Refuse to Listen.* This guideline is very important. Even the simplest grievance may go all the way through the grievance procedure, and management must gather all ideas, arguments, facts, and reasons to help in the ultimate decision. Two other reasons for giving a grievant every opportunity to state his views are:

   a. If the issue should go to arbitration—a subject to be discussed later—the arbitrator will be favorably influenced by the company's effort to understand the grievant's position.

   b. If the grievant discusses his problem at length, he may realize that he had no just cause for complaint in the first place.

4. *Keep a Complete Written Record.* The record may be only handwritten notes, but it should include all dates, names, subjects discussed, references to contract articles involved, reasons for actions taken, and recommendations for future action.

5. *Settle at the Lowest Possible Step in the Grievance Procedure.* All grievances should be settled at the lowest possible step in which settlement can be reached by mutual agreement after all facts are known and without horse trading or sacrificing principles.

6. *Reply in Detail.* This is another guideline that can be important if a case goes to arbitration. In general, all evidence should be considered during grievance hearings, and additional evidence that was known at the time of the grievance meetings should not be introduced during arbitration proceedings. In any event, give complete answers to all charges and try to convince the grievant that the company's position is correct. When the case is settled, make sure the grievant is told what happened and why.

Conduct at Grievance Hearings. When grievances involve earnings or work methods, the industrial engineer may be the most important management representative at the hearing. He is, of course, the management specialist in this field, but he must never forget that he is not the management specialist in charge of the hearing. Depending on how far the grievance has progressed through the grievance handling procedure, the management representative who presents and defends the company's position may be a foreman, a department head, the director of labor relations, or someone else in a position of authority. This person, not the industrial engineer, is responsible for management's strategy. He decides who will testify for management and what evidence should or should not be introduced.

The industrial engineer should make sure before the hearing that the management representative in charge is aware of all pertinent facts and evidence that he can contribute. He should also have available, but not necessarily in the hearing room, all essential supporting charts and other exhibits. Once he has done this, he should follow instructions of the management representative in charge.

As the hearing progresses, the industrial engineer should keep the management representative in charge of strategy informed on technical and other details that become important. He can do this by writing notes or by requesting brief private meetings or caucuses. Other than this, he should leave the conduct of the meeting to the management representative, especially during the final steps of the grievance procedure, when the management representative is more apt to be a man experienced in company policy and employee relations. It is this representative's responsibility to determine which company personnel should speak, when they should speak, and what they should discuss.

An industrial engineer who dominates a grievance meeting or who introduces exhibits merely because they are on hand will serve no purpose except to satisfy his own ego. In fact, he may damage his company's position by revealing too much too soon or by creating new and unnecessary problems. All too many simple misunderstandings have erupted into full-fledged labor disputes because industrial engineers or other management witnesses volunteered opinions or introduced unwanted evidence.

In all cases, the best time to settle a grievance is before it develops. This places an obligation on the industrial engineer to look at all problems from the worker's point of view, and where union engineers are present, to work as closely as practical with them.

ARBITRATION

Most grievances are withdrawn, forgotten, settled, or compromised as they progress through the grievance procedure. But what happens when a grievance remains unresolved after all procedural steps have been explored? The answer lies in the wording of the labor contract. In fact, some contracts say the union is free to strike at this point. The great majority of labor contracts, however, provide for arbitration as the ultimate means of achieving a firm and binding settlement of labor disputes after all other procedures have been tried and no agreement has been reached.

Arbitration is a semijudicial means of settling disputes in which both sides agree in advance to be bound by the decision of a neutral arbitrator or a panel of arbitrators. A neutral arbitrator can be anyone acceptable to both the company and the union. Customarily, he is an attorney, an educator, or some other competent and respected individual. He may be selected in advance by the company and the union, as a judge would be, to sit in judgment of all disputes that cannot be settled by the normal process of grievance handling. More commonly, he will be chosen on a case-by-case or ad hoc basis from a list of five to nine potential arbitrators which the company and the union can obtain from the Federal Mediation and Conciliation Service, a governmental agency, or from the American Arbitration Association, a private organization. Each side in turn strikes a name from the list until only one name, that of the neutral arbitrator, remains.

When a panel of arbitrators is used, the union customarily selects one member of the panel, and this member, in effect, serves as a union advocate. The company selects one panel member who acts as a company advocate. These two then select a third, or neutral, member, and all three try the case, with a majority decision being binding on all parties. If the union and company arbitration panel members cannot agree on a third man, the ad hoc method of selecting a neutral arbitrator is usually followed. Obviously, the selection of a suitable neutral is extremely important. A man considered competent to settle a job evaluation case, for example, might be entirely unacceptable to the company or the union in a discharge or discipline case. Proper selection of a neutral arbitrator is a time-consuming process involving an investigation into the background and attitudes of all potential candidates, together with studies of their prior decisions. It is a job for an experienced labor relations man.

Like all labor agreement provisions, arbitration procedures vary from contract to contract. A typical article calling for a tripartite panel of arbitrators and a typical article specifying ad hoc arbitration follow.

TRIPARTITE ARBITRATION

In the event that any grievance or dispute arising out of the interpretation or application of any clause of this contract remains unsettled after the steps provided by the grievance procedure have been taken, either party, within two weeks, may refer the matter to a tripartite arbitration panel consisting of one representative chosen by and from Management, one representative chosen by and from the Union, and a third mutually acceptable impartial arbitrator.

If the Union and Company representatives do not agree on an arbitrator, he shall be chosen from a list of five arbitrators proposed by the Federal Mediation and Conciliation Service. The Union and Company shall alternately strike one name from the list until only one name remains. The right to strike the first name shall be determined by lot. Final selection shall be made within thirty days.

The arbitrator shall have no power to add to, modify, or subtract from any of the terms of this agreement or any supplemental written agreement of the parties.

The expense of arbitration shall be borne one half by the Company and one half by the Union.

The decision of the arbitrator shall be made within thirty days and shall be binding upon the Company, the Union, and the employees involved.

AD HOC ARBITRATION

Section 1. Any grievance which has not been settled within two weeks in Step —— of Article —— may be referred to arbitration by either party within an additional two weeks.

Section 2. If, during the last two weeks mentioned in Section 1 above, the Union and the Company do not agree on an arbitrator to settle the grievance, he shall be chosen from a list of nine arbitrators proposed by the American Arbitration Association. The Director of Labor Relations of the Company, or his designated representative, and the International President of the Union, or his designated representative, shall alternately strike one name from the list of nine until only one name remains. The right to strike the first name shall be determined by lot.

Section 3. Any matter involving the effect, interpretation, application, or violation of the terms of this Agreement may be submitted to arbitration. The arbitrator shall have no power to modify any of the terms of this Agreement, nor shall he substitute his discretion for that of the parties thereto. The arbitrator shall have no power to establish or change wage rates for jobs nor shall he have power to set or change standards of production. This shall, however, in no way restrict his right to determine whether the principles of the job evaluation and incentive plans, as they affect rates and standards, have been fairly and properly applied.

Section 4. The decision of the arbitrator, which shall be rendered within thirty (30) days after the completion of the hearing, shall be binding on both parties to this Agreement and must be complied with within five (5) working days after it is announced.

Section 5. The fees and living and traveling expenses, if any, of the arbitrator shall be divided equally between the parties.

Arbitration is an extremely serious business. Every effort should be made to avoid arbitration by convincing the union of the correctness of the company's position. Of course, if there is equity in the union's claims, it should be acknowledged and adjustments should be made. If agreement is impossible and the dispute goes to arbitration, however, the company in effect bestows its right to manage on the arbitrator and agrees to abide by his decision. This means that an arbitrator who knows little or nothing about a company, its programs, or its principles can render a decision based on his own philosophy and principles, or lack of them, and thus alter or destroy carefully conceived plans and labor contract meanings. Furthermore, arbitration occupies a favored place in the opinions of Federal judges. It is very difficult to have an arbitrator's ruling set aside unless some serious charge, such as fraud, can be proved. In spite of all its dangers and faults, however, arbitration is the best method yet devised for the final settlement of industrial disputes. It is, in fact, the final barrier between labor peace and crippling strikes.

Because of the importance and finality of the arbitration process, every industrial engineer should know how arbitration is conducted and how he can help his company prepare and present an arbitration case. Here are checklists designed for the company representative who will present the case, together with guidelines intended for industrial engineers.

Preparing for Arbitration

1. *Study All Articles of the Contract.* Guideline: Remember what was said about the management rights article and "specific" contract clauses? Quite often, some forgotten contract wording will have a reference, however remote, to the grievance. If the reference is favorable, it can be exploited. If it is unfavorable, a defense should be built against it in the event the other side tries to exploit it.

2. *Get All the Facts; Check All Records.* Guideline: Let the company representative conduct the case, but be sure he has advance knowledge of all pertinent information, unfavorable as well as favorable. There is nothing more fatal at arbitration than to have the other side introduce evidence that should have been known but was not.

3. *Become Thoroughly Familiar with:*
   a. Bargaining history
   b. Past practices
   c. Settlement of related grievances
   d. Decisions in prior arbitration cases

Guideline: What has been done before will show intent, and intent is important if contract language is obscure. A knowledge of settlements reached in related grievances and in prior arbitration cases is equally pertinent. Grievance handling and arbitration are looked upon as extensions of the collective bargaining process, and grievance and arbitration decisions are frequently given as much weight as negotiated agreements. Present arbitrators are not obligated to reach the same decisions as prior arbitrators who have decided similar issues, but they customarily hesitate to reverse a prior arbitration award unless the award was clearly wrong. Even awards by other arbitrators under other contracts and involving other unions and companies are often cited as precedents. It is not the duty of the industrial engineer to research this type of information, but he should know how similar situations have been resolved in his company.

4. *Outline the Case and Review It with Others.* Guideline: The labor relations representative should perform this function. If he neglects to do so, the industrial engineer should prepare and review a case outline for his own information. When arbitration is involved, he should never rely on his own convictions without having someone else try to find flaws in them.

5. *Prepare the Case from the Union's Viewpoint.* Guideline: This is important. The industrial engineer will be expected to reply to the union's position, and he will be better prepared to reply to that position if he can anticipate what the position will be.

6. *Prepare the Witnesses.* Guideline: This is a job for the company's labor relations representative. Advice to witnesses that is applicable in grievance meetings and arbitration hearings alike is outlined on page 6-152.

Presenting the Company's Arbitration Case. The company's labor relations representative should be solely responsible for presenting the company's position to the arbitrator. He has a two-pronged duty:

1. Prove to the arbitrator that the company's position is correct and that the award should be made in its favor.

2. Supply the arbitrator with convincing reasons why the decision should favor the company so that the arbitrator can apply these reasons in supporting a favorable award.

The industrial engineer can make an important contribution in arbitration cases involving earnings and methods if he knows in advance the type of information the labor relations representative will need. The labor relations representative should:

1. *Have All Necessary Exhibits Available in Triplicate.* Guideline: The labor relations representative should assemble well in advance of the hearing all exhibits he wishes to introduce. All the exhibits should be in triplicate:

  a. One copy for the arbitrator
  b. One copy for the union
  c. One copy for the company's file

Copies of the labor contract are almost always submitted as exhibit no. 1, and are commonly introduced as a joint company and union exhibit. Other exhibits may include such items as copies of the grievance forms, plant rules, notices, charts, and diagrams. Of course, the union as well as the company may enter exhibits.

2. *Prepare a Good Opening Statement.* Guideline: When arbitration involves discharge or discipline, the company normally presents its case to the arbitrator before the union's side is heard. In other situations, including work assignment, wage, and methods cases, the union usually speaks first and tells why it brought the charge against the company.

When the company's turn to speak arrives, its labor relations representative has his first chance to gain the arbitrator's support for his cause. The best way to do this is through a carefully prepared opening statement which summarizes the company's position and leaves nothing to chance. In fact, in some forms of arbitration, union and management exchange prehearing statements and send copies to the arbitrator.

In drafting an opening statement, the labor relations representative should remember that the arbitrator probably knows little or nothing about the company, its product, its processes, or its problems. His statement should introduce the arbitrator to the company, and should make the company "real" to the arbitrator. The industrial engineer can contribute to it by providing charts, graphs, and pictures as needed.

3. *Have Only Essential People and Papers Present.* Guideline: The arbitrator is in complete charge of the hearing. Most arbitration hearings are orderly, but some arbitrators believe that arbitration has a therapeutic effect on employees when they are allowed to speak freely. As a result, some hearings may sound like an undisciplined debating society, with each participant free to question any other participant at any time. For this reason, it is better to bring only essential people and papers into the room. This presents a smaller target and lessens the chance of someone making a damaging remark.

The labor relations representative should decide who will speak for the company at the hearing and what data they should provide. He should also tell his witness not to speak extemporaneously nor to ask questions without his knowledge and consent, no matter how union participants conduct themselves.

4. *Emphasize These Points:*

  a. The responsibilities of the employee mentioned in the grievance
  b. The importance of the action involved
  c. How losing the grievance can adversely affect the company, the employees, and personnel relations throughout the organization
  d. The original intent of the parties
  e. How the company complied with the terms and intent of the labor agreement

Guideline: The points outlined above are intended to convince the arbitrator of the justice of the company's position. They should be brought out by the labor relations representative through the questioning of witnesses, if possible.

Advice to Witnesses. In cases involving industrial engineering, the union usually presents its case and questions its witnesses first. When questioning its own witnesses, neither the union nor the company should ask leading questions such as "You cannot make your quota, can you?" This type of question tells a friendly witness what answer is desired. If either side asks leading questions of its own witnesses, the other side may ask the arbitrator to rule whether the questions are proper.

When the union has finished questioning its witnesses, the company can cross-examine them if it wishes. Leading questions may be asked during cross-examinations. The arbitrator, too, may intervene at any time and ask questions of his own.

Once the union has completed its testimony, the company introduces its witnesses and develops its case by questioning them. The union then takes its turn at cross-examining.

If you are called as a witness, keep the following points in mind:

1. *Tell the Truth.*
2. *Stick to Facts.*
3. *Do Not Try to Justify Your Action.* Guideline: The original grievance may have been entered as a result of some action of yours. When testifying, however, do not try to justify that action unless the company's labor relations representative tells you to. Winning the case is his responsibility, not yours. If he knows his business, he will have a good defense prepared; and if you speak out of turn, you may weaken his arguments. Of course, he should go over the defense with you before the hearing starts so that you will be familiar with his plan of action.
4. *Be Polite.*
5. *Do Not Play Detective.* Guideline: Remember the advice given under Grievance Handling? If you think of something the labor relations director should know, send him a note or speak to him on the side, but do not try to mastermind his defense.
6. *Do Not Be too Precise.* Guideline: You should, of course, know in advance what questions the labor relations representative will ask, and you should try to anticipate what questions the union will ask on cross-examination. Your answers to questions, however, should not sound "canned" or coached. You cannot, for example, remember what happened at some exact moment in time several months or years ago.
7. *Never Volunteer Information.* Guideline: Answer questions briefly. A simple "yes" or "no" is sufficient if the question can be answered in one word. Remember that the more you speak, the more you tell the other side. Unnecessary testimony gives the other side a greater opportunity to confuse issues, to claim that there are flaws in your statement, and to discredit your testimony. It is the labor relations representative's job to ask more questions and bring out more information if he so desires. Of course, if he asks you to explain a point or describe a situation you should do so.

Arbitration Do's and Don'ts

1. *Hold Your Temper.* Guideline: Avoid getting in an argument with the union. The time to argue is during the grievance hearings. When a case reaches arbitration, your only responsibility is to convince the arbitrator. If you find yourself growing angry, try not to talk. Make an excuse to leave the room or ask the labor relations representative to call a private meeting or caucus until you regain complete control.
2. *Take Notes.* Guideline: Even if a court stenographer is present, keep a brief but accurate record of what is said, who said it, and why. It will be helpful when you testify, and it will be useful to the labor relations representative when he prepares his summation.
3. *Do Not Talk too Fast.* Guideline: When you are testifying, watch the arbitrator and speak slowly enough for him to write down what you say.
4. *Do Not Enter Unnecessary Evidence or Exhibits.* Guideline: Cases are won by good evidence, but too much evidence gives the other side a greater chance to cross-examine, divert attention to side issues, and attempt to discredit

all testimony. Prepare to defend the case from all angles, but do not be disappointed if most of your material is not actually used.

After all the testimony is in, each side will summarize its position. At this time, the labor relations director may ask for a caucus to organize his presentation. The notes you took and your basic knowledge should make this task easier for him. The same is true if a posthearing brief or follow-up statement is prepared and sent to the arbitrator after the arbitration hearing closes.

In general, the arbitrator should render a decision within thirty days of the hearing. This time will be extended, however, if posthearing briefs are filed. Some decisions are entered in less than thirty days, but many are delayed for additional weeks or months, depending on how complicated the case is and on the work load of the arbitrator.

The arbitrator's written decision may be either long or brief, but it will usually outline the grievance, the position of the company, the position of the union, the reasoning of the arbitrator, and a statement of award. The statement of award will grant the grievance, deny the grievance, or outline some other remedy. In any event, it will be final and binding in nearly all cases.

COLLECTIVE BARGAINING

The industrial engineer may never take part in negotiating a labor contract, but he should know what takes place there. This knowledge will give him a better understanding of labor agreements, their meaning, and their application.

According to the Labor Management Relations Act (previously identified as the Taft-Hartley Act):

> Sec. 7. Employees shall have the right to self-organization, to form, join, or assist labor organizations, to bargain collectively through representatives of their own choosing, and to engage in other concerted activities for the purpose of collective bargaining or other mutual aid or protection, and shall also have the right to refrain from any or all such activities except to the extent that such right may be affected by an agreement requiring membership in a labor organization as a condition of employment as authorized in section 8(a)(3).

Neither the rights of employees section, the duty to bargain section, nor any other section of the Taft-Hartley Act mentions the word "union." "Labor organization" is the expression used, and it is possible to create an illegal labor organization unintentionally by the simple process of bargaining with an informal group of employees. For the purposes of this discussion, it will be assumed that a legally recognized union exists, that management and union are about to negotiate a new labor contract, and that certain legal preliminaries concerning notification have been complied with. This is the time for proposals, or "demands," to be compiled. Demands are proposed changes in or additions to contract wording. They are compiled individually by both union and management, and they reflect experiences under the then existing labor agreement. In the case of the union, demands may also include all the expressed wishes of each and every union member. Many of the union's demands may be unreasonable or even dangerous to company and employee alike, but they are often presented and advocated vigorously by union leaders who know that the demands will eventually be rejected by the company. The union may support such demands for many reasons. For example, the demands may have trading value, they may allow union members to get things "off their chests," or union negotiators may lack the courage or the political ability to refuse to present them to management.

Demands may be made or exchanged in various ways, but usually they are presented during the opening bargaining sessions. The chronology of these events may be:

1. Each side in turn may present a demand which will be discussed, rejected temporarily or permanently accepted, or deferred for later consideration.

2. One side may present all its demands, and then the other side will reply to them and present demands of its own.

3. Rejected or dropped demands may reappear as bargaining progresses, and new demands may be added.

Each demand will be advocated or opposed with an intensity reflecting its personal appeal to the negotiators. This intensity may also be an indication of the political repercussions the negotiators fear will result if they fail to press certain demands or if they abandon them too soon.

From this point on, there will be a series of proposal exchanges, speeches, threats, pressure, concessions, and compromises. Ultimately, with or without a work stoppage, an agreement will be reached and hopefully will be a workable tool even though it may not be what either side actually desires.

All this may sound archaic to anyone who has never witnessed the bargaining process. Those who negotiated the contract would probably agree; but here again, no one has ever developed a satisfactory substitute. The industrial engineer who understands the process can help make it less objectionable by bringing needed contract changes to the attention of the company's labor relations representative and by pointing out hidden dangers in union proposals.

LABOR DISPUTES AND THE INDUSTRIAL ENGINEER

The amount of space devoted to grievance handling, arbitration, and collective bargaining indicates the importance of these subjects to the industrial engineer. More than those of other management representatives, his actions affect the income and work habits of employees, and these are the items most likely to produce worker discontent. They are also the items union representatives are most likely to pursue to the bitter end during grievance meetings and contract negotiations. Even when the union representative is himself well informed on industrial engineering principles, he will customarily defend a worker's complaints for political or other reasons.

Because the actions of the industrial engineer are so intimately related to worker sentiment and to labor relations in general, he should make a determined effort to ascertain the scope of his authority. He should know the limitations on his ability to introduce changes, and he should recognize the support or lack of support he can expect to receive from top management. This subject is vital. Degree of support may depend on company size, progressiveness, or type of ownership. It may depend also on the previous experiences, knowledge, aims, or philosophy of management. It is an exceptional management, however, that will support its industrial engineering department indefinitely against determined worker opposition.

The industrial engineer should also realize that a favorable decision through the grievance-arbitration procedure may not mean a final victory for engineering principles. The union may bide its time until the labor contract expires and then attempt to force acceptance of its demands as a price for signing a new contract. This is quite proper, for an unpopular position forced by management on labor or by labor on management is an inborn breeder of discontent. Discontent, in turn, encourages retaliation, disruption, and inefficiency. Therefore, an industrial engineer who is professionally competent must, for his own security as well as for the benefit of his company, be more than demonstrably accurate. He must also be able to sell his programs to workers, union, and management alike, and to recognize and adopt alternatives when necessary. A little "give" may well be the sign of cooperation that demonstrates understanding and results in acceptance.

USE THE KNOWLEDGE OF OTHERS

The U.S. Department of Labor edits a paperback book called *Federal Labor Laws and Programs.*[4] The book itself may be of little interest to industrial engineers

[4] *Federal Labor Laws and Programs,* U.S. Department of Labor Standards, Bulletin 262. For sale by the Superintendent of Documents, U.S. Government Printing Office, Washington, D.C., 20402.

except to note that the table of contents requires two pages of fine print just to list laws and agencies that affect industry and industrial relations. Even so, the table omits many current additions, and it makes no effort to include thousands of state labor regulations.

Obviously, no man can be acquainted with all the legal ramifications that could apply. There are, for example, regulations on wages, formulas for computing overtime, restrictions on work and work methods that vary with the type of industry and worker, and limitations on hours of work that apply in certain situations. There are also obligations to review certain contemplated changes with union representives, where plants are organized, even before decisions on the changes are adopted. In addition, there may be labor contract limitations on work content changes and the methods by which the changes are made. There are also personnel and human relations factors to consider whether unions are involved or not.

In short, the industrial engineer should take no new idea or innovation for granted. He is part of the management team, and he should review all contemplated programs with the production department, the wage administration department, and other groups whom his decisions may affect. A review of this type has many advantages. It keeps these groups informed in advance of coming changes; it gives specialists in many fields a chance to preview ideas; and it helps identify possible danger points, errors, or oversights before plans are put into practice.

No members of the management team should receive greater consideration from the industrial engineer than foremen and other members of frontline supervision. These are the ones with the most intimate and practical knowledge of the problems any worker-related program will encounter. They are also the ones who should promote the program. The industrial engineer must remember this and must also remember that frontline supervisors normally have deep pride in their work. Work-related programs introduced without their knowledge, understanding, or support can place them and their subordinates in doubt of their authority, deprive them of control, and destroy their morale. This is a high and unnecessary price to pay. Fortunately, it can be avoided. Work-related programs should be introduced through or in cooperation with frontline supervision after proper preparation, explanation, and if necessary, industrial training.

OUTSIDE INFLUENCES

An industrial engineer's success or failure may depend on conditions not related to standards or methods. There is, for example, the grapevine. Employee grapevines thrive in most plants and offices. Consider a case where new time standards are planned for direct workers and job evaluation for clerical employees. A rumor can be expected to start about a pending speedup and wage cut. Management and the industrial engineering department know better, but what about the employees directly involved? Workers at machines and typewriters who never heard of a decimal minute stopwatch or a job evaluation plan will be sure that a reduction in personnel is imminent. Thus the employee grapevine is outside influence number one. It can be controlled only by liberal doses of DDT: Disclose Details and Truth.

Outside influences, even employee grapevines, are not always considered harmful. At certain periods in industrial history, some companies depended upon an occasional scare to increase production. During the depression of the 1930s, for example, a crowd of unemployed men standing outside a hiring gate was enough to encourage workers inside to produce. Management attitudes such as this have long since changed. The costs of recruiting, hiring, testing, training, and retaining workers have contributed to a management desire, partly social and partly economic, to keep up with industrial neighbors in matters involving wages, fringe benefits, and good working conditions. The industrial engineer who can help management achieve these ends by introducing methods changes and improvements with a minimum of disruption and opposition will add to his stature and to his value to his employer.

Changes, however, usually produce uncertainty, and uncertainty induces fear. Fear can become one of the most unpredictable and unreasonable of all outside influences. People fear what they do not understand. Workers will frequently resist changing to modern methods even though they know that their old and obsolete procedures are resulting in high costs and lost orders. The industrial engineer has a moral as well as a practical duty to reduce workers' fears by making sure that every program is correct and is thoroughly understood. If he cannot do this, there is something wrong with his approach or his program.

The industrial engineer may be able to reduce if not eliminate employee fear and still meet resistance from union officials and self-appointed employee spokesmen. Overcoming this type of resistance will be difficult at best, but it will be infinitely harder if the company has defaulted on its obligation to communicate effectively with its employees. A neglected or nonexistent company communications program usually means that union or other influences have stepped into the void. Should this be the case, an announcement by high company officials of a new and beneficial methods program will be ineffective if the self-appointed "spokesmen" have other ideas. The industrial engineer who finds himself in the middle of this power play should pay special attention to the opinions of the industrial relations representative and production department officials. These men will probably recommend that the program be introduced through foremen who are closest to the workers, after an announcement has been made and after the foremen have been thoroughly briefed.

There are many reasons why a union representative might object to methods changes, job evaluation, or time studies; and all the reasons are classed as outside influences. The union representative may distrust industrial engineering innovations because of some unfortunate past experience, or the innovations may run contrary to the policy of his international union. He may belong to the old school of unionism that believes unions should resist anything management advocates. More likely, however, the workers themselves resist the change, and the union representative merely acts as their spokesman to secure his own political position. On the other hand, union objections may be proper and legitimate. This could happen if the labor contract forbids changes or specifies how changes should be introduced. If contract language is involved, the company's industrial relations representative becomes all-important. He may find it necessary to bargain with the union about the changes, make concessions, or even resort to such additional outside influences as utilizing the grievance procedure and arbitration.

A final set of important outside influences stems from the intervention of government agencies into labor and management relations. Some of these influences were discussed in detail under grievance handling, arbitration, and collective bargaining. One such potential influence whose extent is not yet fully defined is compliance with equal employment opportunity regulations. When the meanings of these regulations are finally understood, they might require the redesign of methods and machines to make them more adaptable to female employees and the disadvantaged.

One fact is certain, however: it is a rare management that will support industrial engineering changes indefinitely in the face of continued worker opposition, harassment, and strife. For his own protection, the industrial engineer should realize that he is part of the company team, coordinate his efforts with those of the labor relations department and other management groups, and make his proper contribution to employee communications, harmony, and training.

SUMMARY

Any activity that affects employees' earnings or work habits will also affect worker morale, labor relations, and productivity. The most technically correct work method is useless if it ignores human factors and encounters persistent employee opposition. The inevitable result will be an unstable work force, lower quality, and resistance to maximum production.

More than most management representatives, an industrial engineer can influence employee earnings and working conditions. If he is to succeed in increasing production and in reducing costs, the engineer must be more than a competent technician. He must also be a salesman capable of convincing management, workers, and union officials alike that his program contains advantages for them. This selling function is particularly important because workers have a legal, and in many cases a contractual, right to question matters that affect their income and working conditions. The legal right is derived from the Taft-Hartley Act, which permits workers, whether unionized or not, to confer with management over "wages, hours, and other terms and conditions of employment" and to "present grievances to their employer and to have such grievances adjusted." The contractual right is based on labor contracts, where they exist, and many include restrictions on methods changes or the manner in which they are introduced.

Industrial engineers employed by companies whose workers are represented by unions should be thoroughly familiar with all applicable labor contracts. These contracts have the force of law, and they may contain provisions limiting the engineer's activities or means of implementing wage or methods changes. For example, they will almost invariably prevent unilateral wage decreases or increases. They will also outline formal procedures through which workers can air their grievances and have them adjusted, possibly through arbitration. The industrial engineer must also know what is expected of him and how to conduct himself before and during grievance and arbitration hearings. A misstep at any of these points could not only cause his company to lose the current grievance but might also produce unfavorable precedents or rulings that could limit future policy.

The industrial engineer shares one condition with all management representatives from foreman to president. He is part of a team. His efforts, like the efforts of all team members, must be directed toward efficient operations. This requires team effort, the ability to recognize management problems other than his own, and willingness to make adjustments in the interest of overall objectives.

Finally, the industrial engineer should be able to recognize and adapt to the personality quirks of the people he encounters. For example, he should know and respect the ambitions and drives of other members of the management team. In addition, he should be aware of the aspirations and fears of employees and union officials. Among these groups, he will encounter distrust of anything that changes the status quo, a desire for greater control over conditions that affect work environment, and the existence of politically motivated decisions and compromises. The industrial engineer who can combine competence with personal considerations will make a real contribution to his company and his profession.

BIBLIOGRAPHY

Bittel, Lester R., *What Every Supervisor Should Know*, 2d ed., McGraw-Hill Book Company, New York, 1968.

Elkouri, Frank, and Edna Asper Elkouri, *How Arbitration Works*, rev. ed., Bureau of National Affairs, Washington, D.C., 1967.

Labor Law Course, Commerce Clearing House, Chicago, current edition.

Statistical Abstract of the United States, U.S. Department of Commerce, U.S. Government Printing Office, Washington, D.C.

Stone, Morris, *Labor Grievances and Decisions*, Harper & Row, Publishers, Incorporated, New York, 1965.

Chapter **9**

Standards and Wage Systems Audits

JOHN C. MARTIN

Staff Assistant for Industrial Engineering, Headquarters Manufacturing Planning and Controls, Westinghouse Electric Corporation, Pittsburgh, Pennsylvania

This chapter discusses ways of preventing deterioration in the productivity control plans that are applied throughout industry. Such plans include the various forms of wage incentives plus the nonincentive applications of work measurement that are referred to as measured daywork.

These systems are a necessary part of an effective manufacturing process, yet they are fragile in design, with many pitfalls. The operators concerned tend to seek loose standards. As representatives of management attempt to follow established rules, they must contend with frequent methods changes that are the order of the day. It is easy for inequities to creep into the picture, remaining unidentified while affecting overall results adversely. Ways of finding when this has occurred are needed so that corrective measures can be taken.

Those who have experienced at first hand the problems that can occur through misapplication of the system's components will need no reminder of the importance of adhering to sound principles. For others, who may be starting new control plans or considering such steps, the following discussion of the principles that apply may be helpful. Next, some of the ways are listed in which the framework of control plans and the time standards involved tend to be shifted away from a sound basis, even though properly applied at first. Methods of auditing and review that have proved successful, both for standards and for other wage system components, are then presented. Finally, a list of typical questions is given which

6-158

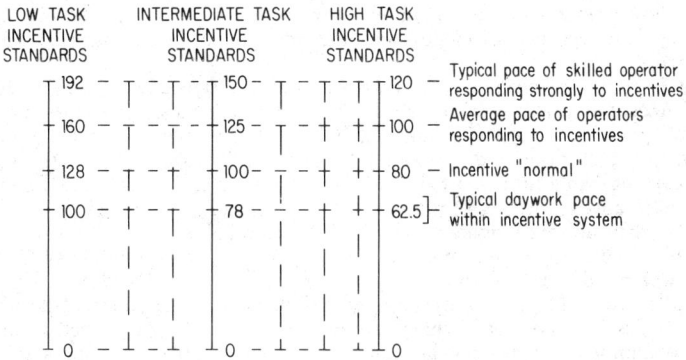

FIG. 9-1. The various task levels of wage incentive standards.

can be used to aid in the evaluation of systems in which work measurement is the basis for control.

VALID TIME STANDARDS

Successful applications of either wage incentives or measured daywork require reliable standards for the evaluation of the operator's performance. The comments that follow are applicable in either system.

Level of Performance Involved. Figure 9-1 illustrates the variations in work content represented by performance standards when they are established correctly for various incentive systems. There are many more than the three task levels specifically identified, as indicated by the vertical dotted lines. Developing standards that represent the appropriate task level is a function of the industrial engineering analysts, who are expected to follow the policy set by management when the incentive framework is first defined or later revised.

The measured daywork task level is not shown on the chart except as it may be represented by one of the vertical dotted lines. Although there is no inherent relation between the 100 percent point in measured daywork and that of any specific incentive system, there is evidence that the performance level represented by typical measured daywork target standards usually falls within the medium to high task range.

It is a function of the industrial engineering manager to see that the performance level represented by 100 percent in a plant's incentive or measured daywork control plan is adequately defined. This may be through a careful evaluation of performance rating films, followed by the use of such films to guide analysts who make time studies. It may also be through the selection of a predetermined time system having an appropriate performance level which is either built in or obtained through the use of multiplying factors.

Allowances Applied. A distinction should be made between irregular elements connected with specific jobs and general allowances that are applied. It will simplify later discussions to classify the usual allowances as follows:

1. Authorized personal time during work hours, when relief operators are not used
2. Short delay intervals that usually occur daily and that are considered unavoidable (examples: receiving instructions from supervisors, assigned machine lubrication, cleanup at the work station)
3. Short delay intervals that occur which are unavoidable by operators, but which could to a large extent be controlled by supervisory planning; or delays similar to category 2 but not likely to occur daily (examples: waiting on material availability, waiting for next job assignment, waiting for maintenance repairs)

4. Allowances for fatigue or undefined reasons which have the effect of increasing the recorded efficiency and incentive bonus when operators remain at work

Wage Incentive Allowances. Incentive allowances typically include all four categories. Accepted practice involves (1) a realistic allowance for category 1, often in the 5 percent range; (2) an unavoidable delay allowance determined from work sampling studies; and (3) an allowance for category 4, which is often in the 5 percent range for most jobs. Excessively tiring work calls for increased fatigue allowances, although performance rating can be applied so as to allow for greater than normal fatigue without added general allowances.

The allowances apply to manual elements. A strict interpretation of bonus earnings as a reward for the application of extra effort would call for dropping the fatigue allowances for elements controlled by machines or processes. Another point of view is possible, however, which recognizes the value to be gained from keeping expensive equipment operating at capacity. In such cases, allowances for process-controlled time may equal or exceed those for manual elements.

Measured Daywork Allowances. For measured daywork standards, a somewhat different concept of general allowances is customary. These standards are used to define a fair day's work, with emphasis on good supervisory control. Thus, the allowances described by category 3 would generally be omitted. When unavoidable delays occur, the reported performance for a supervisor's section will be reduced. The target standards which the supervisors expect operators to attain do not apply during these delay intervals. In addition, the category 4 allowances for measured daywork are not called for in cases where work measurement integrates average conditions of fatigue with the leveled performance time.

Selection of Ways to Establish Standards. There are a number of values to be gained from the application of work measurement. These include (1) the methods reviews that are involved, (2) improved performance through wage incentive or measured daywork follow-up, (3) uses of standards for production planning and control, and (4) possible applications of standard cost accounting methods. Despite these multiple uses, the cost of establishing standards is an important consideration which deserves attention.

Standard Data Applications. For operations that comprise a family of related jobs, work measurement can be accomplished through the use of standard data or formulas to develop a broad range of standards. In certain cases, it is practical to develop standard data through computer programming methods. Calculations by computer may also be practical as a basis for developing individual standards. The derivation of standard data or formulas often can be simplified through the use of predetermined time systems, which are themselves a form of standard data. Applications of standard data or formulas can serve to improve the reliability of specific standards as well as to reduce their development costs.

Deriving Standards through Analysis. Both in the development of standard data and where operations are not covered by such data, specific time standards must be established through analysis. In selecting an applicable technique, a compromise may be required between application costs on the one hand and precision plus methods improvement capabilities on the other. The following list of work measurement techniques is arranged in the approximate order of increased application cost, increased precision, and increased methods improvement capabilities.

1. Estimates based on records of past performance
2. Comparisons with standards on comparable jobs
3. Work measurement sampling, where applicable
4. Stopwatch time study plus performance rating
5. Condensed forms of predetermined times
6. Detailed predetermined times

There will be some valid exceptions to the listed order; so specific plants should develop policies to guide their own practice. Figure 9-2 illustrates how one plant has established in chart form a policy concerning the methods to be used for work measurement when standard data are not available.

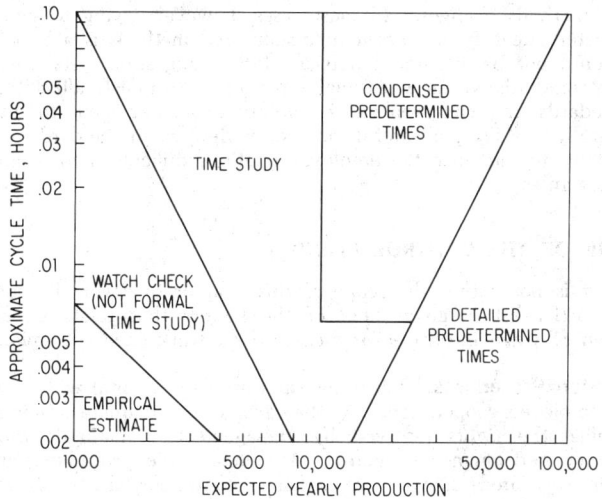

Fig. 9-2. Method used at one plant to designate a technique for establishing specific standards.

This chart suggests 10,000 cycles per year as a good point to switch from stopwatch time study to predetermined times, in order to concentrate on methods in more detail. First, the condensed forms of predetermined times are used except for cycles shorter than about 0.006 hour, where precision with this technique becomes doubtful. Detailed predetermined time systems are prescribed for considerably greater production quantities where finely developed motion patterns are important. For combinations of cycle time and yearly production that produce lower total work times, the expense of time study or predetermined times precludes such applications in setting individual standards. The chart illustrates the concepts involved, with the ranges prescribed being based on a mixture of application costs plus opinions on the results attainable. The technique usage ranges will vary for specific plants.

Maintenance of Standards. Manufacturing methods are frequently revised. As this occurs, the time standards require corresponding revisions. This fact is so obvious as to be almost a truism. Yet the failure to keep standards in line with methods is a major cause of deterioration in both wage incentive and measured daywork systems. Particularly with wage incentive systems is this a serious problem, as there may be contractual limitations on management's right to change standards unless methods revisions can be proved.

Knowing That Changes Have Occurred. Thorough documentation of the basis for both individual studies and standard data is essential for wage plan administration. This documentation normally occurs routinely with applications of predetermined times. When time study or other work measurement techniques are applied, adequate elemental descriptions are called for. This is an important area for checking in the audits which will be discussed.

In neither wage incentives nor measured daywork is there likely to be a reliable routine way to report the minor or even some of the major methods changes. An alert industrial engineering group can help do this. But periodic audits of standard data and long-term individual standards are a safeguard to assure adequate control.

Changes Originated by Operators. Questions may arise on whether methods improvements originated by operators should be accompanied by revised time standards in wage incentive plans. There is the natural feeling that the employees' contributions to methods are their own property. Plant suggestion systems give some weight to this feeling, because they often pay sizable awards to employees who initiate methods changes. In most cases, however, trying to separate all employee-originated ideas from management-sponsored methods is a hopeless task and one that should not be attempted outside of the suggestion system itself. Failure to revise the standards within a reasonable period can seriously affect the consistency of time standards and can weaken the ability of management to administer the plan successfully. Making a careful methods review when standards are first established will serve to minimize the employee relations difficulties that can result from enforcing such rules.

FRAMEWORK OF THE CONTROL PLANS

Although it is not within the scope of this chapter to cover all aspects of wage systems, a partial list is given here of the factors that relate closely to sound administration of wage incentives or measured daywork once such plans have been established.

Clear, Written Procedures. Clear-cut rules are fundamental with wage incentives to promote employee cooperation. At the same time, adequate administrative rules serve to define the rights reserved by management to maintain the consistency and validity of performance standards. Understandable procedures are also a requirement in measured daywork to define the responsibilities of line and staff personnel.

The Concept of Pace in Relation to Bonus Earnings. In wage incentive plans, thorny questions are inevitable regarding the basis for bonus earnings. Should such earnings be directly related to extra manual productivity above some defined normal point? Or, when semiautomated equipment limits manual productivity, is the opportunity for bonus payment to be considered a guaranteed right? As an alternative statement of the latter concept, is continued manning of very expensive equipment, such as numerically controlled machine tools, to be considered reason for bonus payment?

Historically, wage incentives have been applied for extra effort in job performance. As modern manufacturing methods have reduced the effort content of most manual jobs, the purpose of incentives has shifted to some extent toward getting operators to remain at work. Yet such a policy tends to conflict with the primary incentive concept. Operators who must be manually productive to earn a bonus will tend to raise questions when others receive similar payments for merely staying at their machines. Although a thorough discussion of this problem is outside the scope of this chapter, it is clear that answers are required which will withstand labor relations pressures.

In measured daywork applications, the problem is somewhat less severe. Job evaluation can be applied to adjust wage rates where heavy lifting or similar conditions make effort requirements high. It is equally important in measured daywork, however, to uphold the concept of fair work output for fair wages. When standards are established, they should be enforced. This enforcement can come through supervisory follow-up, as discussed in Chapter 3 of this section.

Consistency of Application. It is not sound for the incentive earnings of one department of a plant to be unusually high or unusually low if conditions are otherwise approximately the same. This leads to resentment and deterioration in overall results. The reasons for such variations may lie in standards application, standards maintenance, operator performance, detracting conditions that occur, or other factors. When inconsistencies are apparent, analysis is justified to determine the reasons so that corrective action can be initiated.

The same is true for operators' performance under measured daywork. The follow-up by direct supervisors to obtain compliance with standards should be consistent throughout the plant. On the other hand, there may properly be some inconsistencies in sectional productivity performance and goals due to existing conditions.

For individual wage incentive applications in which the operators respond as intended, consistently applied standards cannot be expected to produce uniform

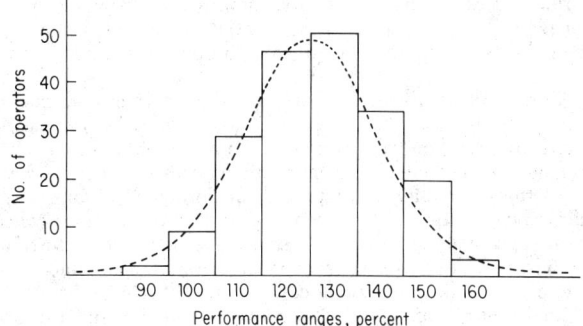

FIG. 9-3. Typical distribution of performances for operators responding adequately to intermediate task incentives that are applied on an individual basis.

performance results. Instead, there is evidence that a normal distribution of operator performance values will occur, as in Figure 9-3. If the plot of individual incentive performance data is considerably at variance from the normal distribution, as in Figure 9-4, this may indicate that something is wrong, perhaps in the application of standards.

Reliance on Tested Concepts. Those who decide to experiment with untested aspects of wage systems application should do so with full realization of the pitfalls that exist in the employee relations area. Over the years, certain concepts have been tested and found workable. They can be upheld as being fair and reasonable from the standpoints of both management and the operators. The operating fundamentals for wage incentive and measured daywork control plans are discussed in more detail in Chapters 2 and 3 of this section. Some of the fundamentals that should be studied and evaluated relate to conditions such as (1) the amount of wage incentive bonus earned at incentive pace and at other work levels, (2) conditions under wage incentives that justify "average earned rate" payments rather than the daywork rate, (3) the periods for averaging performance to calculate bonus earnings, and (4) job evaluation principles. Wage system audits should be expected to check the validity of existing practices in these areas. Other tested concepts to be followed are noted below.

Evaluation of Individual Performances Where Practical. Some work inherently involves group action, and the measurement of performance on a group basis is

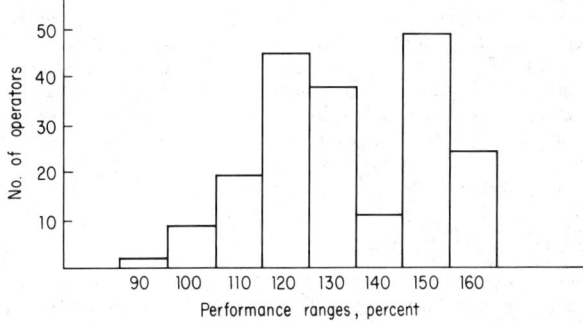

FIG. 9-4. A distribution of performances for operators on individual incentive which would indicate the probability of loose standards or other deficiencies.

necessary in such cases. Advantages from improved teamwork may result. It is generally conceded, however, that the potential effects of wage incentives or of measured daywork standards will be greater through applications on an individual basis.

Control of Extra Allowances. In both wage incentive and measured daywork plans, it is common practice to make allowances for delaying conditions which are not covered by the general allowances and which are not the fault of the operators. When not controlled, the extra allowances can be the basis for marked deterioration in the reliability of reported performances. One approach to the control of allowance claims is to require clock card or data collection records, with authorizing approval of the supervisor concerned, the industrial engineer, or both. Although other, less precise approaches may be suitable, periodic reviews should be made of extra allowance practices.

Adequate Supervisory Coverage. The requirements for supervision within plant sections will of course vary. For example, wage incentive applications may not call for the same ratio of foremen to operators that is justified under measured daywork. As another example, thirty assembly line operators working in the same area may require no more supervisory attention than half that number on individual, unrelated jobs. It is not possible to generalize on this subject except to emphasize the need for adequate direct supervision by a representative of management. Otherwise, there will be difficulties in making the wage system function satisfactorily.

Staff Follow-up. The control of methods and standards is not likely to be effective unless industrial engineers have time for the required follow-up. If the functions of operation routing, standards setting, tooling coordination, and problem solving take up all the time of an understaffed industrial engineering group, methods studies are likely to be neglected. A wage system review should not fail to consider this aspect.

Issuing Standards to Operators. Under incentives, the need for advance notice of time standards is self-evident. With measured daywork, informing operators of the standards is a matter of fairness, assuming that compliance with standards is a management goal. Everyone wants to know what is expected of him, and measured daywork operators are no exception.

Planning for a Full Day's Work. Production schedules on a short-term basis are highly desirable so that job assignments to operators can be made in advance of completing prior work. If a fair day's work is specified for each man in advance, there is a better chance of obtaining it.

Providing Reliable Status Reports. For wage incentives, status reports can properly be expected to show overall productivity ratios for each section, as well as the usual statistics related to bonus earnings. For measured daywork, the performance reports are particularly important. Prepared by accounting, they should be on a daily basis where applicable. Prompt report issuance is essential. The performance trends within each line supervisor's section should be evaluated with respect to objectives.

Adequate Record Systems. The preparation and routing of manufacturing information deserves careful attention. Clerical work may be simplified through data processing methods. For example, computer input may be obtained directly from tape output following typing of the route sheets. Repeat jobs can usually be routed to the shop without the need for handling by industrial engineering. The information transmitted to the shop should be concise but sufficiently detailed to minimize reference errors. Shop records to indicate the performance on specific operations should be dependable and simple in format.

REASONS FOR REGULAR REVIEWS

The preceding discussion of principles implies that periodic reviews by management's representatives may be required if standards and wage systems are to remain effective. This need not discourage those who are considering such methods of productivity control. It is easier to manage in this way than to struggle against competition without the aid that effective controls provide.   The following comments give further evidence of the need for periodic reviews.

Guarding against Deterioration in Standards.   Some of the tendencies that exist and which serve as reasons for standards audits are:

Methods Changes without Corresponding Revisions in Standards.   Operators cannot realistically be forced to give notice of methods changes which they originate, and these may be too subtle for ready detection by supervisors or others concerned. If foremen have been struggling to get operators' performances into the bonus column of a wage incentive system, there are temptations not to report minor changes in methods.   This may occur through oversight rather than intent.   Or, if measured daywork supervisors are closely evaluated according to their sectional productivity, the methods revisions that increase "paper" productivity will tend to be overlooked.

Therefore, it becomes the function of industrial engineering representatives to check existing shop methods against those on which the standards were based.   Yet this is sometimes not recognized as a high priority job.   Analysts assigned to an area may be occupied on other projects that leave little or no time for shop methods review.   The inevitable tendency of methods to change, however, is a compelling reason for periodic standards audits.

Loosening of Standards through Application Errors.   An engineering design can be viewed by a supervisor and checked to a large extent.   The format of a time study can be checked in the same way, but the performance rating on which the study is based must be accepted at face value unless the entire study is repeated. Similarly, the result from a predetermined time analysis is difficult to check closely without repeating much of what the analyst has done.

There are several influences that can lead to errors from time to time.   Prior experience of an analyst can lead to high or low ratings.   Performance rating is an art, and some are less adept at it than others.   Some line supervisors, because of their personalities and their close association with industrial engineering analysts, may have an effect on the way some standards are set.   Methods established may be less than ideal, as has been discussed.   Standards analysts are perhaps no more or less prone to error-free performance than those in other professions.

If the errors were randomly of a plus or minus nature, no great harm might result, because in many cases an operator works on a number of different jobs during the interval of performance evaluation.   However, errors in establishing methods and errors from other causes generally have a loosening effect.   This is because most operators and some supervisors are quick to call for a recheck when they find that new standards are on the tight side.   They tend to remain silent, however, when easy targets have been established.   Such tendencies need to be counterbalanced by periodic standards audits.

Although measured daywork target standards do not affect bonus earnings, they tend to act as production ceilings.   Thus, the standards have a monetary influence that is less direct but just as definite as when wage incentives are used.

Possible Deterioration of Wage Incentive Plans.   The following points indicate that wage incentive reviews are also justified from a broader standpoint than the analysis of individual standards.

Gradual Changes in Framework of the Plan.   Periodic negotiations may involve written or implied agreements that affect the incentive principles listed previously. Examples are (1) negotiated allowances, (2) changes in the concept that incentive earnings come largely through extra effort, (3) wage increases on a basis that hinders sound incentive administration, and (4) gradual changes in the application of average earned rate as a basis for payment when incentive work is not available. As such changes become cumulative, their net effect should be carefully reviewed as a basis for future management policy.

Effects of Increased Mechanization.   Equipment installations that occur within a few years' time can affect shop methods enough to produce marked changes in the value of an incentive plan that is aimed at promoting extra manual effort.   As semiautomatic equipment or paced lines go into effect, there may be valid reasons for considering revisions in the existing plan—individual to group incentives, extra effort incentives to machine usage incentives, incentives to measured daywork, and the like.

Changing Significance of Indirect Work. Where wage incentives have been applied, they have primarily covered the direct manufacturing operations. This is because of the relative ease and reliability of work measurement in the direct area as contrasted with work measurement in less routine activities such as receiving, inspection, and other indirect jobs that vary in work content. In an incentive shop, not providing the chance to earn a bonus on indirect work may cause the operators concerned to conclude that they are "not supposed to work effectively." Correspondingly, the performance of dayworkers within an incentive system is likely to be very low.

As the proportion of indirect employees to direct employees increases—which has been the trend throughout industry—it becomes more important to avoid the negative incentive which tends to exist where wage incentives cover only part of the work in a plant. The sampling of productivity in daywork sections is a valid phase of an incentive review. Such information can help guide future policies.

Lack of Uniform Application. The periodic incentive earnings reports can show clearly the individuals or groups with unusually high or low performances. Extreme variations are an indication that something is wrong in administering the program. There may be cases, however, where earnings are in the expected range but performance is pegged at a low level to conceal loose standards. Work sampling studies or careful observations during an incentive review can help bring such cases to light for correction.

A wide variation in the bonus earnings of individual employees can occur. Reviews are called for in such cases to determine whether this reflects variations in performance or whether there are inconsistencies in the way that time standards have been established.

Miscellaneous Practices. An incentive plan can be loosened by going along with employee claims for extra allowances because of unusual conditions such as defective purchased material, waiting for maintenance repairs, or extra operations. A supervisor who wants to avoid arguments may fail to monitor such claims. Other examples of practices that need periodic review are unresolved grievances, use of group leaders' time, group leader allowances in the standards, and control of job classifications.

Possible Deterioration of Measured Daywork Control Plans. Because measured daywork standards tend to establish performance ceilings, their validity needs to be checked. And because the management and staff functions under measured daywork are keys to the attainment of standards, these functions also need periodic reviews.

Supervisory and Staff Coverage. The direct supervisors, aided by industrial engineers and others, must be depended upon to apply measured daywork controls successfully. As business conditions change from time to time, there are temptations for the direct supervisors to be spread thin in coverage or to be given too few operators to manage. Similarly, industrial engineers may be assigned to special projects, leaving little or no time for shop methods follow-up or the development of sound standards. Because the effects of such neglect may not be immediately apparent, the conditions may be allowed to continue. Periodic reviews will assist in bringing them to light, with supporting evidence, so that sound operating conditions can be reestablished.

Shop Performance Records. Daily or weekly performance reports may be issued late. They may be inaccurate or misleading for various reasons. They may not show clearly how each section's performance relates to specific goals. There may be emphasis on sectional performance but a lack of records concerning the performance of individual operators to guide the supervisors in taking constructive action. An adequate review will identify such practices so that management can take action as justified.

Miscellaneous Practices. Most of the reasons noted for reviewing wage incentive plans apply also to measured daywork installations. General allowances should be checked. The level of performance on which standards are based is a very important factor. Consistency of application, allowances for detracting conditions, and clerical procedures involved are other aspects that should be checked to avoid deterioration.

METHODS OF REVIEW

The preceding discussion has pointed out some of the principles of effective wage incentive or measured daywork control plans. Many ways have been noted in which these principles can be compromised during the rush of day-to-day activity. The need for periodic auditing of standards and review of the control plan framework has been emphasized. A discussion of auditing and review techniques follows.

Auditing Time Standards. Time standards can best be audited through a low-key, continuing function rather than an intensive, short-term effort. The auditing, however, can be accomplished in various ways that range from sampling to 100 percent review. Some allowable tolerance such as plus or minus 5 percent should be established. A sampling of time standards in an area will indicate the probable reliability of the total number. If a periodic random sample indicates that the values are sound, it may be unnecessary to check further. Conversely, a random sample full of discrepancies signals the need for a more extensive review.

Establishing a continuous method for sampling the reliability of time standards is a desirable approach. This involves selecting the number of time standards that should be checked periodically to cover all outstanding values within a given interval. Consider the following:

$$n = \frac{N}{50T}$$

where n = number to be checked weekly within the section covered by one industrial engineering analyst

N = total number of values established within the section

T = time interval in which a 100 percent review is to be completed, such as two to three years

For a plant with approximately 6,000 time values, checking forty standards each week will cover them all in a three-year period. The jobs to be reviewed may be selected at random from those being performed during the week. An alternative approach is to give greater weight to checking the standards for long-run jobs. For example, jobs running in the range of 100 man-hours per year could be checked every three years, with yearly checks for jobs requiring approximately 1,000 man-hours per year. The methods for checking specific standards can be expected to vary according to existing conditions, as follows.

Values Set from Time Study. The original study can be reviewed for completeness, and the methods in effect compared with those noted on the study. Further checking might be considered unnecessary unless there are discrepancies.

Values Set from Predetermined Times. With the original data in hand, a brief observation of the job cycle will serve to point out where methods variations have occurred.

Comparing Time Study with Predetermined Times. If standards have been set by time study, occasional checks by applying predetermined times or vice versa are highly desirable. Performance rating can be evaluated in this way, or the analyst's skill in developing predetermined time aggregates can be verified.

Taking Action When Discrepancies Occur. With measured daywork applied as described in Chapter 3 of this section, the required changes in time standards can be made at once. The reasons should be identified for explanation to supervisors. In the case of wage incentives, there are usually restrictions that require methods changes before standards are revised. Discrepancies found through the auditing procedure will probably be due to such changes. It is then a matter of following through with the necessary explanations. When mistakes are discovered that are unrelated to methods revisions, a policy for taking action should be worked out.

The Cost of Auditing Standards. The added cost of auditing standards may be regarded as insurance against eventual deterioration of the wage system. It can be kept to a small percentage of the total methods and standards budget, say, 5 percent. This cost is negligible if it prevents major labor relations difficulties which usually result from deterioration of the wage system. In addition to this, however, there are factors that may apply to offset the auditing cost entirely. One is a probable reduction in the tangible and intangible costs of grievances caused by inequities. Another is the saving from correcting a larger proportion of standards where methods improvements have been made.

Review of Control Plans. The review of controls can be made a continuing function of the industrial engineering manager. In a single-division organization, no further formalities may be required to be sure that management is made aware of existing problems.

In multidivision organizations, there may be a diversity of opinions concerning some of the wage systems policies that should be followed. It may be necessary, even under decentralized management policies, to coordinate centrally those aspects which could affect labor negotiations and long-term corporate policies. Accordingly, the wage systems reviews, other than standards control, may be handled as a corporate staff function. Labor relations and industrial engineering specialists may be involved.

One large company has successfully utilized a three-man team for plant reviews. The team is comprised of the local industrial engineering manager, a man with similar functions from another division, and a corporate representative. Through this approach, plant and divisional management is assured of an appraisal that considers the problems involved from a broad standpoint.

TYPICAL QUESTIONS TO BE ASKED

An adequate review of the control plan in use involves a close look at industrial engineering and management functions. The principles that have been discussed apply. Particular attention should be given to the areas most likely to cause problems. To do this, a questioning approach is needed. This involves interviews with those concerned and frequent observations in the shop. The observations should be aimed at obtaining information that will furnish answers to questions such as the following, which have been phrased so that a "yes" answer is the eventual objective.

Basic Manufacturing Information

Completeness. Is clear-cut information being prepared for the shop, listing required operations and pertinent details? Are time standards indicated for each operation? Are feeds and speeds specified for machining operations? Are equipment and tooling indicated?

NC Equipment. For numerically controlled tools, are the needed programs developed by specialists and in time for required use? Is full advantage being taken of available techniques for aiding in program development?

Establishing Standards; Line Balancing

Basis for Standards. Are standards analysts trained adequately, both in performance rating for time studies and in predetermined time applications? Is the level of 100 percent performance, represented by the predetermined time system or by performance rating methods, applicable to the existing system of wage payment?

Standard Data. Are adequate standard time data available? Are they in easy-to-use format? Have they been carefully developed and checked? Is the scope of their coverage clearly defined? Are the data revised regularly to account for methods changes?

Paced Lines. Are line balancing techniques applied where a group works together on paced conveyors? Is the correct line speed indicated? Are operator assignments clearly specified?

Computer Applications. Are computer techniques applied to expedite and improve such functions as the routing of repeat jobs, production scheduling of a full day's work, standard data development, standard data application, and line balancing?

Validity of Standards. Are time standards audited on a continuing basis? Are clear-cut procedures followed to change time standards whenever methods are revised?

Methods Analysis and Improvement

Attention Received. Do industrial engineers spend an appreciable portion of their time on methods improvement? Are shop methods in good shape? Is there regular emphasis by management on cost improvement, with full participation by all within manufacturing line and staff management plus others such as design engineering?

Instructions. Do route sheets list the operational details as justified? Have lengthy specifications been condensed and simplified for reference by operators? Are operator instruction sheets issued for manual operators? Are they in effective use?

Administration of Wage Incentives

Framework. Is the incentive framework established in a sound manner? Are there clear policies for administering the system on both a short-range and long-range basis? Are supervisors and industrial engineers complying with existing procedures? Are incentives applied on an individual basis where practical?

Records Involved. Do the performance reports indicate sectional productivity ratios as well as operator efficiency data? Is there evidence that production quantities are correctly reported?

Results. Are the performance figures generally within the range that indicates effective response to wage incentives? Are they reasonably consistent? Are special allowances for detracting conditions held to nominal amounts? Do observations within the shop lead to concurrence with reported results? Do operators start and stop work on time?

Effects on Indirect Work. Does coverage of the wage incentive system extend as far as practical? Do indirect operators not covered by incentives work effectively? If not, is their number so small that this does not cause serious problems?

Daywork. Do the line supervisors recognize the need for close follow-up of dayworkers' assignments and working effectiveness? Is the work sampling technique applied regularly within such areas to assist in performance evaluation and provide a basis for control?

Administration of Measured Daywork

Framework. Do time standards represent the desired 100 percent target level? Are the standards made known to operators? Is performance reported reliably and promptly? Is individual performance determined wherever practical? Is sectional performance summarized? Does accounting handle the record keeping involved? For long-cycle jobs, do shop records indicate the performance of operators in relation to standards as the work progresses?

Follow-up. Do supervisors actively follow up cases of low operator performance to correct the deficiencies involved? Does industrial engineering actively assist with this follow-up? Are the detracting conditions that affect sectional performance reported and followed up as justified?

Results. Are the reported sectional performance values near the goals that have been established? Do operators start and stop work on time? Do other shop observations lead to concurrence with reported performances?

Manpower Control

Classifications. Are plant supervisors given data at regular intervals concerning the mix of labor classifications that should be carried? Are the recommendations followed closely?

Total Required. Are plant supervisors given data concerning total manpower requirements to handle planned work load for current intervals? Are reliable forecasts prepared for management showing manpower loads expected during future months and years?

Methods of Assigning Industrial Engineers. Are classifications established to recognize the need for technicians on some work and engineers on other work? Have the duties of each man been reviewed to determine the optimum grouping of duties? Are goals of accomplishment regularly established for each man with follow-up to check results?

Follow-up of Audits and Reviews. The standards audits are likely to involve no major questions in their follow-up, once it has been determined that specific corrections should be made. Of course, there may be problems in convincing operators that methods have changed, particularly after some time has passed. The responsibilities for handling such situations are usually clear, however.

Concerning the control system framework, some of the problems that may be uncovered through questioning may not have easy solutions. Convincing everyone concerned that an upsetting revision of procedures is necessary to preserve the value of a productivity control system may at times be difficult. But this should be done when called for.

By far the best approach is to administer the control system carefully over the years without disregarding fundamental principles. The next best approach is to initiate reviews, as this chapter describes, to find where slippage has occurred; then to take action as required.

BIBLIOGRAPHY

Barnes, Ralph M., *Motion and Time Study*, 6th ed., John Wiley & Sons, Inc., New York, 1968.

A Fair Day's Work, Report of Research Division, Society for Advancement of Management, New York, 1954.

Fein, Mitchell, "A Rational Basis for Normal in Work Measurement," *Journal of Industrial Engineering*, June, 1966.

Forberg, Richard A., "Effective Control of the Industrial Engineering Function," *Proceedings Twelfth Management Engineering Conference*, SAM-ASME, New York, April, 1957.

Hutchinson, John G., *Managing a Fair Day's Work*, University of Michigan, Bureau of Industrial Relations, Ann Arbor, 1963.

Krick, E. V., "Maintenance of Time Standards," *Journal of Industrial Engineering*, March–April, 1956.

Mundel, Marvin E., *Motion and Time Study*, 3d ed., Prentice-Hall, Inc., Englewood Cliffs, N.J., 1960.

Niebel, Benjamin W., *Motion and Time Study*, 4th ed., Richard D. Irwin, Inc., Homewood, Ill., 1967.

Patten, Thomas H., Jr., *The Foreman, Forgotten Man of Management*, American Management Association, New York, 1968.

Behavioral Science and Human Factors

Chapter 1

Behavioral Science and the Industrial Engineer

JAMES A. RICHARDSON

Supervisor, Industrial Engineering Division, Kodak Park Division, Eastman Kodak Company, Rochester, New York

An up-to-date industrial engineering handbook would be incomplete without significant reference to behavioral science. In this edition, there will be two chapters dealing with the subject, this one and the chapter entitled "Human Effectiveness Principles for the Industrial Engineer" by Dr. M. Scott Myers.

This chapter will make the case for the relevance of behavioral science to industrial engineering practice. This will be done by showing the ideas and theories of behavioral science which have become available and how these relate to the mission and technology of industrial engineering. At the same time, the special nature of behavioral science knowledge will be pointed out—"special" as it might be seen through the eyes of an engineer or one whose training has been oriented to the physical sciences.

Industrial engineering has been defined by the American Institute of Industrial Engineers as being concerned with

> the design, improvement, and installation of integrated systems of men, materials, and equipment. It draws upon specialized knowledge and skill in the mathematical, physical, and social sciences together with the principles and methods of engineering analysis and design to specify, predict, and evaluate the results to be obtained from such systems.

Industrial engineering is therefore depicted as a field of practice which is concerned with systems including people (social systems), whose body of theoretical knowledge includes systematic knowledge about human behavior (behavioral science), and whose practitioners have the skill to meld mathematical, physical, and behavioral sciences to design, specify, predict, and evaluate such systems.

This is a tall order, and when the definition was written about 1955, it was (and still is) certainly more of a hope than a statement of the existing state of knowledge or condition of the industrial engineering profession. Behavioral science courses are just beginning to find their way into industrial engineering curricula. The extent of behavioral science knowledge in the possession of practicing industrial engineers has mainly been obtained from short courses, personal outside reading, seminars, or behavioral science courses especially designed for and taught in business schools or in industry.

INDUSTRIAL ENGINEERING'S PAST RELATIONSHIP WITH BEHAVIORAL SCIENCE

As a point of departure for relating behavioral science and industrial engineering, it is useful to look back briefly to examine the earlier historical relationship of industrial engineering and behavioral science.

It is common to mark the turn of the century as the time industrial engineering as an entity emerged from mechanical engineering. Frederick W. Taylor, Frank and Lillian Gilbreth, and Henry L. Gantt were the prominent figures whose works gave industrial engineering its early formulation and direction.

Many behavioral science researchers of the past thirty years have criticized these industrial engineering pioneers for having had narrow views of people. Such criticism has overlooked that these people were, for their times, highly enlightened in their views of and attitudes toward working people. Any careful student of the early writings of these pioneers will note frequent references to which can be inferred motives that could only be classed as enlightened. These people were concerned with increasing productivity and at the same time lightening the load of the working man, all to the end of improving the human condition. That the technology they spawned led to numerous abuses and alleged exploitation of workers was not due to their motives. The ideas and technology they developed were perfectly consistent with if not in advance of the conventional wisdom about human behavior prevalent in those days. In fact, it is doubtful if the behavioral scientists of that era had any more insight into or scientific knowledge about human behavior than the industrial engineering pioneers.

Regardless of how enlightened or unenlightened the industrial engineering pioneers might have been, the industrial engineering practitioners have been viewed as villains on the scene by most behavioral science researchers. It has only been in the past few years that significant cooperative relationships have emerged between industrial engineering and behavioral science.

BEHAVIORAL SCIENCE

The term behavioral science is a relatively new one which has come into being to refer to that area of scientific inquiry relating to human behavior. To a considerable extent, it has supplanted the term social sciences as the collective descriptor.

Behavioral science is presently mainly taken to include psychology, sociology, and anthropology. Persons trained in these classical fields think of themselves as behavioral scientists. However, it is also possible to receive a graduate degree from some modern business schools having concentrated heavily on the behavioral aspects of people in organizational settings. This field is often called organizational behavior, and those trained in it may consider themselves behavioral scientists.

The older term, social sciences, was considered to include political science, economics, and even history. In some respects, these fields could be considered behavioral science, and the works of political scientists and economists particularly may be as relevant to industrial engineering as the works of a psychologist or sociologist.

Considering the imprecise boundaries of the field, for the balance of this chapter, behavioral science will be taken to mean any science which relates to the behavior of people. For our particular purposes, those whose research concentration has been on organizational behavior are of the most interest.

BEHAVIORAL SCIENCE "KNOWLEDGE"

To those educated in the "hard" or physical sciences or engineering, the behavioral sciences may hardly seem to qualify as sciences. Engineers, in particular, are educated to respect such relationships as Boyle's law, Ohm's law, $F = MA$, and $E = MC^2$. There are no such neat relationships in the behavioral sciences.

It has been said that the maturity of a science is in proportion to its ability to predict. In this regard, the behavioral sciences are considerably less mature than the physical sciences. At best, the predictive powers of the behavioral sciences are probabilistic in nature. It may be noted that behavioral scientists, on the whole, are much more sophisticated in their ability to use probability and statistics than are engineers. Their advanced state of statistical training is likely due to the sheer necessity for behavioral scientists to abstract their phenomena statistically.

Some feel for the nature of the behavioral sciences may be gained from the following quotes from eminent behavioral scientists. Everett Hughes remarks, "A considerable part of sociology consists of cleaning up the language in which common people talk of social and moral problems." In his classical book, *The Human Group*,[1] George C. Homans asks, "What single general proposition about human behavior have we established?" He answers, "None."

Bernard DeVoto observes: "Sociology has repeatedly tried to make usable generalizations about large organizations of people. . . . Such generalizations frequently yield valuable insights and may roughly describe parts, perhaps even large parts, of what actually happens. But they do not hold good throughout. . . . They do not correspond to all the facts nor can they be used throughout to serve the basic scientific functions of criticism, verification, and prediction."

It is not the intent of these cautionary remarks to downgrade the behavioral sciences or to imply that they are inferior to the physical sciences. The intent is to convey to the reader, who will probably have been educated in engineering, the nature of behavioral science. Throughout the balance of this chapter, the works of a number of behavioral scientists will be examined. Each has his own perspective and emphasizes different variables in the organizational environment. Each contributes to the yet incomplete mosaic of organizational behavior. None unfolds the whole truth. The concepts are intriguing and insightful. They shed light on perplexing dilemmas of long standing like the motivational power of money, but will yield no "law of compensation" or the like.

There is an additional property of some behavioral research. It is that once a new concept has been described or demonstrated, it seems to appear obvious; and one may wonder why he "didn't think of that." The reader will note in this chapter that, once stated, some behavioral science concepts and theories are reasonably verifiable in one's own experience and have the ring of common sense. But they were not so perceived prior to research.

There may be a number of ways to relate behavioral science and industrial engineering. This chapter will attempt to do so by examining:

1. The state of industrial engineering about 1930
2. The unfolding of key behavioral science "findings" between 1930 and 1970
3. The probable implications of forty years of behavioral science research for current industrial engineering education and practice

INDUSTRIAL ENGINEERING, 1930

1930 is chosen as a point of departure because by that time traditional industrial engineering technology had reached a plateau. The basic ideas of the pioneers

[1] George C. Homans, *The Human Group*, Harcourt, Brace & World, Inc., New York, 1950.

had been essentially formulated by 1920, and the decade 1920-1930 had seen the emergence of many consultants whose principal products were various proprietary kinds of incentive systems. By 1930, many companies across the United States had adopted systems which were rooted in the theory and mechanics of time and motion study.

Viewed in perspective, these approaches assumed a simple model of maximizing productivity which can be reconstructed as follows:

1. Division of operations into simple jobs which can be studied elementally by time and motion study approaches to establish "the one best method"
2. Establishment of performance standards based on stopwatch time study and rating
3. Wage payment proportionate to output through the use of any one of a variety of essentially piecework payment plans

Great productivity gains were apparently made—at least for the short run—but what is of particular interest to us are the behavioral assumptions which supported these approaches. They may be inferred as follows:

1. People come to work only to earn a living as individuals.
2. Working within their physical and mental capacities, the only thing which is important is how much money they can make on the job.

There were enlightened management people who saw the matter as more complex than this by 1930. A few companies had by this time adopted practices which viewed the working man as more than just an hourly producer and wage earner. However, the notions as stated had widespread acceptance and had become the basis of a primary productivity thrust of American industry.

BEHAVIORAL SCIENCE, 1930–1970

Elton Mayo—The Hawthorne Studies. There had been very little systematic behavioral research in industry prior to this time; but by 1930, the now famous Hawthorne studies were in progress. These studies, conducted between 1927 and 1932 by Elton Mayo and his colleagues at the Harvard Business School, were done at the Hawthorne Works of the Western Electric Company. The tale of the Hawthorne studies is a complex and fascinating one, upon which we can only touch briefly.

The key findings were:

1. The industrial work group is a human social group, and group behavior phenomena have a powerful influence upon individual members. It is not sufficient to consider man-at-work as an autonomous economic individual.
2. The work group, as a human social group, appears to fill legitimate human needs on the job—needs which heretofore, if they were considered at all, were felt to be the legitimate concern of the family, the church, or fraternal associations.
3. The work group can be a powerful force for or against productivity, but because of inept management, it usually engages in practices such as banking, restriction of output, and the substitution of informal leaders for management leadership—all to protect group members from external threat and promote group solidarity.

The Hawthorne studies were halted by the depression in 1932, and it was seven years later, in 1939, that the first definitive reports were published by Roethlesberger and Dixon as *Management and the Worker*. (Note: References not footnoted are found in the appended bibliography.)

The Hawthorne studies became widely publicized in a disjointed way. Most laymen remember them as being the study in which illumination levels were increased to increase production, and later when the levels were lowered, production continued to increase. This is often loosely referred to as the "Hawthorne effect" and popularly interpreted as another example of the perversity of human nature. A more insightful interpretation is that so long as people are treated as human beings whose needs require consideration and attention, they may cooperate in increasing productivity.

The Hawthorne studies were "openers" as examples of behavioral science research in industry. They stimulated many questions but did not seem to have had any major effect on changing industrial practices in the short run.

For the longer run, the Hawthorne studies stimulated an era of human relations, during which training courses of many kinds under this heading were instituted in many companies. These were honest attempts to bring a heightened awareness of people's needs to the industrial setting. They did not appear, however, to have had any lasting or discernible effects. The era of human relations training has been cynically referred to as the "be nice to people" era.

Perhaps the most important consequence is that the Hawthorne studies attracted the attention of behavioral scientists to the industrial environment as a fruitful field of research.

Abraham Maslow—Hierarchy of Human Needs. Through World War II, industrial interest in behavioral science continued to concentrate on the question of individual motivation, particularly the motivation to work.

During World War II, Dr. Abraham Maslow first published his theory of human needs and incorporated it into leadership training in the U.S. Air Force. For a number of years thereafter, this theory attracted much interest as a basic structure for at least broadly understanding human motivation.

Briefly, Maslow's theory posited a hierarchy of human needs as follows:

	Lower	(Ordering)		Higher	
	Survival or basic needs	Security or safety needs	Love or belonging needs	Status or esteem needs	Self-realization or self-actualization
Satisfied by things like	Air Food Water Shelter	Banking Storing Saving Insurance	Group acceptance and membership	Position Rank "Badges" Self-esteem	Being what one can become

The theory holds that individuals are motivated by lower order needs until these are relatively satisfied and higher order needs are then evoked.

Without discussing the many psychological interpretations and nuances of the theory, it can be said that it has a certain commonsense appeal and squares with reality. From it can be inferred several interpretations which explain organizational behavior patterns. It would, for instance, help to account for the Hawthorne findings. It would also account for the apparent waning power of money as a direct source of motivation in an already affluent society. It suggests that management, whose apparent concern has focused on lower order needs, must be more concerned with higher order needs. It does not suggest how management is to do this.

Douglas McGregor—Theory X and Theory Y. Probably no single concept has had as much widespread influence as the late Douglas McGregor's now famous Theory X and Theory Y.

In the middle 1950s, he had begun to introduce the ideas to his students at the Sloan School of Industrial Management at the Massachusetts Institute of Technology and had spoken and written on the subject. The complete rationale was finally published in *The Human Side of Enterprise* in 1960.

McGregor contended that traditional management practices had grown, rooted in assumptions about human nature, which were held by most managers and others as well. These assumptions, essentially pessimistic, were said by McGregor to be implicit in management thinking and rarely stated openly in any organized way. They had been absorbed from our culture and taken for granted. The assumptions, inferred from observing management practices, were labeled collectively Theory X and stated by McGregor to be as follows:

1. "The average human being has an inherent dislike of work and will avoid it if he can."
2. "Because of this human characteristic of dislike of work, most people must be coerced, controlled, directed, threatened with punishment to get them to put forth adequate effort toward achievement of organizational goals."
3. "The average human being prefers to be directed, wishes to avoid responsibility, has relatively little ambition, wants security above all."

On the whole, McGregor held that these assumptions led to a management strategy of direction and control. That is to say, if the assumptions are correct, management's job is to tell people as explicitly as possible what to do and how to do it and establish the necessary controls to see that they do it.

It is not very difficult to relate traditional industrial engineering practices such as the microdivision of work, tightly defined methods, and piecework incentives to such notions about people.

McGregor continued to say that such ideas would not have persisted had not managers been able to observe considerable behavioral evidence to support them. He also observed that Theory X ideas had the pernicious quality of self-confirming prophecies. For example, one observes Theory X behavior and accordingly devises management practices which reinforce continuation of such behavior. One's ideas are confirmed by continued observation of this behavior, and the circle is closed.

This circle can hopefully be broken and more productive and enlightened practices devised if a newer set of assumptions are used. These new assumptions, which McGregor contended were supported by modern behavioral science research, he called Theory Y and stated as follows:

1. "The expenditure of physical effort in work is as natural as play or rest. The average human being does not inherently dislike work. Depending upon controllable conditions work may be a source of satisfaction (and will be voluntarily performed) or a source of punishment (and will be avoided if possible)."
2. "External control and the threat of punishment are not the only means for bringing about effort toward organizational objectives. Man will exercise self-direction and self-control in the service of objectives to which he is committed."
3. "Commitment to objectives is a function of the rewards associated with their achievement. The most significant of such rewards, e.g., the satisfaction of ego and self-actualization needs, can be direct products of effort directed toward organizational objectives."
4. "The average human being learns under proper conditions not only to accept but to seek responsibility. Avoidance of responsibility, lack of ambition, and emphasis on security are generally consequences of experience, not inherent human characteristics."
5. "The capacity to exercise a relatively high degree of imagination, ingenuity, and creativity in the solution of organizational problems is widely, not narrowly, distributed in the population."
6. "Under conditions of modern industrial life, the intellectual potentials of the average human being are only partially utilized."

McGregor's proposed management strategy, which he held followed from Theory Y, was the principle of integration of goals and self-control. That is, he saw management's job as being the creating of conditions within which the individual could integrate his goals with those of the organization so that satisfaction of the former were coincident with achievement of the latter.

McGregor's ideas received widespread attention in the early 1960s. They undoubtedly led to much management soul searching and at least a beginning of the examination of many well-established practices.

It is very difficult to relate traditional industrial engineering approaches to the ideas inherent in Theory Y.

Frederick W. Herzberg—The Motivation to Work. In the late 1950s, Frederick W. Herzberg and his colleagues at Western Reserve University were researching the motivation to work.

It had been a part of conventional wisdom for a long time that productivity was positively related to happiness or satisfaction. Common sense indicated that happy workers were productive workers, and other generally accepted clichés.

To test this widely held theory, Herzberg initially surveyed every study which could be found that had tried in one way or another to relate productivity to happiness and satisfaction. Over 150 reasonably scientific studies had been made of the subject. The results of this survey indicated that no consistent relationship could be found. Some studies showed positive correlations and some negative. Many were inconclusive.

An obvious problem with studies relating productivity and satisfaction or happiness had been the difficulty of operationally defining either in any quantitative way.

Herzberg and his colleagues, Mausner and Snyderman, developed a theory and designed a study to verify it. The study, originally conducted using engineers and accountants in the Pittsburgh area as subjects, utilized depth interviews and a "critical incident" technique. The theory and findings hold as follows.

Instead of just one continuum of happiness or satisfaction, there are at least two major categories of factors affecting the motivation to work:

Dissatisfiers or hygiene factors	Satisfiers or motivators
Pay and benefits	Recognition
Company practices	Achievement (sense of)
Boss relationships	Responsibility (sense of)
Working conditions	Work itself
	Advancement
(Extrinsic to the job)	(Intrinsic to the job)

The dissatisfiers are those things Herzberg defines as extrinsic to or surrounding the job. They contain the tangible, after-the-fact payoffs which have always been considered by management as the proper rewards. As sources of motivation, they operate in an essentially negative fashion. That is, if satisfied, they do not demotivate, but merely produce a neutral situation. On the other hand, if not satisfied up to some point the recipient considers equitable, they are a source of negative motivation and even conflict. Herzberg uses the medical term "hygiene" to refer to these factors collectively, due to the analogy with a hygienic situation. In a hygienic environment, cause for disease is absent but persons in the environment are not necessarily positively healthy.

The factors labeled satisfiers or motivators are defined as intrinsic to the job or work situation itself. They can be found only in the job. These are the factors which Herzberg's theory holds are the positive sources of the motivation to work.

It is important to stress that in this theory the dissatisfiers or hygiene factors need to be satisfied as a basis of reaching the satisfiers or motivators. Most important is the corollary that oversatisfaction of the hygiene factors will yield no positive results in the form of productivity.

As mentioned, the original studies were conducted using engineers and accountants as subjects. Subsequently, this study has been replicated a number of times and reasonably verified over a wide range of job types. It must be mentioned, however, that Herzberg's theory is not without its detractors, who have attacked the research methodology in particular.

To the extent that it is valid, this theory has great impact on traditional industrial engineering approaches, emphasizing as they do the primary payment-by-results reward mechanism. In fact, many would argue that the traditional industrial engineering approach which emphasizes fine division of labor and job overspecialization actually works to thwart motivation because, inadvertently, the satisfiers are eliminated from jobs.

Arthur N. Turner and Paul R. Lawrence—Industrial Jobs and the Worker.
The four examples of research and theory given to this point have all received wide publicity. The study to be reported upon now has received considerably less attention—much less than it deserves, in the opinion of this author. It is included because it raises a fundamental question and also because it demonstrates how a research study, attempting to verify a theory, can through serendipity yield unanticipated findings.

In the early 1960s, Arthur N. Turner and Paul R. Lawrence of the Harvard Business School launched a study to examine the response of workers to the technologically determined attributes of their work. Technologically determined attributes are such things as variety, autonomy, responsibility, and opportunities for personal interaction which are present or absent as determined by required job duties. In some respects, these are similar to Herzberg's satisfiers in that they are intrinsic to or "in" the job situation.

The research was relatively complex and cleverly designed, including a way of scoring the job attributes and measuring responses. It provided for the study of nearly 500 people in forty-seven jobs in ten different industries ranging from transportation to manufacturing and consumed three years. The attempt was to relate the job attributes as independent variables to workers' response as dependent variables. The major hypothesis was that the high scoring jobs would be the high response jobs.

The hypothesis, in considerably more complex form than reported here, was generally validated but with certain unexplained ambiguities. In further searching the data to explain the ambiguity, the authors found a supplementary or intervening variable. This they called a "subcultural influence" or "subcultural predisposition" relating to work.

They discovered that workers from rural and small-town surroundings and workers from urban and city environments demonstrated measurably different responses. Briefly, town or rural workers tended to react positively to more complex work requiring greater skill and acceptance of responsibility. On the other hand, city workers tended to react negatively to this kind of work, but positively to simpler tasks requiring less skill and the acceptance of less responsibility. It must be noted that the simple classifications of town and city are oversimplified shorthand descriptions of groups which included ethnic and even religious orientations.

This study and its "accidental" findings raise at least two broad questions:

1. How universal and culture-free are the theories of Maslow, Herzberg, and McGregor?
2. As the country becomes more urbanized, are the cultural predispositions toward work likely to change for the worse?

At the very least, this study reminds us of previous findings, both scientific and intuitive, which have indicated that the attitude toward work or achievement may be more culturally determined than just being a matter of inherent individual psychological motivation.

Accordingly, it is dangerous to apply theories about work to differing cultures which may lack achievement orientation.

All the researches mentioned here have had as a central concern the matter of human work, achievement, or performance in an organizational environment. This is of primary concern to the practice of industrial engineering. There is, however, another focus of research which is highly relevant.

INNOVATION, PLANNED CHANGE, IMPROVEMENT

Concerned as he is with the fundamental matter of improving the performance of organizations, the industrial engineer has an understandable interest in the phenomenon of change. All improvement efforts involve changing from some existing state to some better or improved state. "Better" or "improved" are defined in terms of some hopefully measurable criteria which span the organization's goals.

The past decade has seen considerable behavioral science research on the phenomena of innovation/planned change/improvement. Early work emanated from W. Bennis, K. Benne, and R. Chin, who published *The Planning of Change* in 1961.

Essentially, the students of the change process are systematically examining the factors which stimulate or inhibit change. They examine relationships of tasks, leadership behavior, and the change process itself. Included are studies of management tactics and strategy most effective in promoting planned change, as well as the behavior of people labeled change agents—those whose behavior appears to catalyze and facilitate change.

Under the general heading of the "management of improvement," Dr. R. N. Lehrer of the Georgia Institute of Technology has sponsored annual management of improvement seminars, and his book, *The Management of Improvement,* was the first to focus this subject sharply in an industrial engineering context.

Professor Herbert Goodwin of the Massachusetts Institute of Technology has researched, spoken, and written on the subject of planned change.

Many additional researchers, particularly in graduate business schools, have taken up the subject, and any current literature survey of creativity, innovation, planned change, or improvement will yield many and varied studies.

One part of this research focuses on the matter of interpersonal relations. This is the basic stuff of which the influence process—how people impact upon and influence each other's behavior—is made.

Significant research has been done on the manner in which the level of trust between people in an organization can be raised, with the objective of improving the authenticity of interpersonal relations and reducing the game playing in organizations.

A major significance of this area of research to industrial engineering is that it can help clarify the role the industrial engineer has always been trying to play. It highlights the necessity of being concerned with the *process* of using technology of whatever kind as distinct from the *content* of the technology.

OTHER RESEARCH

The author has chosen in this chapter to emphasize the work of five major research efforts having as a central focus the matter of human motivation/human performance in organizations. In addition, the research on planned change has been examined very briefly. These choices were made as being among the most influential that have occurred over the past forty years.

In addition, there have been a number of other studies of relevance covering differing subject areas. A few, listed in terms of the authors' and researchers' names and prime subjects, are:

Kurt Lewin—Group dynamics
Rensis Likert—Leadership style
        Behavioral measures
W. F. Whyte—Work group behavior
        Money and motivation
A. Zaleznik, et al—Satisfaction and productivity
Eric Trist, E. Jacques, and A. Rice—Sociotechnical systems
L. Davis and R. Cantor—Job design
C. Argyris—The individual and the organization

SUMMARY

The researches of Mayo (Roethlesberger and Dixon), Maslow, McGregor, and Herzberg are remarkably consistent. They all paint a picture of the human being in an organizational setting who is a considerably more complex fellow than management practices have formally recognized. Clearly, he works for more than money and other benefits. The intrinsic meaningfulness of his job is of prime importance, provided he is already well compensated. Paradoxically, if he is well compensated and well treated, this is no guarantee of high performance in the absence of a job which makes sense to him.

He does not behave entirely as an autonomous individual, and his work group has a powerful influence over his on-the-job behavior.

There are suggestions from the research that his cultural (subcultural) origins have a significant impact on his predisposition to commit himself to work or not work.

IMPLICATIONS OF BEHAVIORAL SCIENCE RESEARCH FOR INDUSTRIAL ENGINEERING EDUCATION AND PRACTICE

At the beginning of this chapter, the formal AIIE definition of industrial engineering was quoted. This definition specifically commits the industrial engineering profession and its practitioners to "draw upon specialized knowledge and skill in the . . . social sciences. . . ."

Clearly, a first order of business should be to include in industrial engineering curricula some significant training in the substance of behavioral science research. Incomplete as it is, this body of the past forty years of research knowledge would seem to be indispensable to the graduating industrial engineer. This has barely begun in industrial engineering schools but has become a significant portion of a graduate business school education.

Beyond education in the substance of behavioral science research, there is a clear need for training in behavioral-science-based interpersonal skills if the practicing industrial engineer is to operate effectively as an agent of change in organizations. It is not clear that this can or should be a part of formal college training. At least, significant efforts in the business environment are in order to develop and refine personal skills.

To a considerable extent, the industrial engineering practitioners are left to synthesize for themselves the probable changes in their practice which should follow from behavioral science research. Most of what has been published relative to reduction-to-practice of behavioral science theory suffers from a lack of comprehensive treatment.

There are numerous and varied examples in print of approaches to job design, job enlargement, and job enrichment. These range from simple addition of job duties, to substitution of total jobs for assembly lines, to the assignment of total production and quality to small work groups. What has been reported has been successful. But one does not hear about the failures or inconclusive efforts.

Clearly, the job enrichment thrust is indicated from the research. However, up to this point there has emerged no job enrichment technology which can be used as a basis for designing improved jobs. What has been done seems to have been the result of determined efforts by engineers which have been more artistic than rational, and highly situational in nature.

Another area of probable impact lies in the broad area of communications, involving goals, goal setting, measures of performance, and information systems.

A packaged approach embodying these has come into being under the title of "management by objectives." More properly, it should be called "management by objectives and self-control," which, it will be remembered, is the strategy which Douglas McGregor said followed Theory Y beliefs.

MBO, as it is coming to be called, has mainly taken the form of a top-down attempt to align the goals of the organization and those of the individual. This is done mainly by stressing mutual goal setting and achievement review sessions between superiors and subordinates. These sessions supplant the former character- and trait-oriented merit rating approaches. Primarily, MBO has been limited to use for and by management and professional people in organizations.

More closely related to the industrial engineering role have been several reported attempts to improve performance by reviewing the entire communications process surrounding the goal setting, measurement, and feedback process.

Such efforts have stressed close examination of goals as to their validity and reality; the selection of measures of performance which truly reflect real organizational goals; and the process by which the information system transmits and displays performance and achievement to the entire organization.

All such efforts in the general area of communications relate to that portion of behavioral science research which indicates that if people are to work toward achievement of goals, they must be fully informed and be a part of the achievement process.

Industrial engineering has, in the past, had a strong association with the matter of wage payment and compensation. This has diminished in relative importance to industrial engineers as their involvement with incentive systems has diminished. Many, however, are still concerned with compensation responsibilities.

The major implication of behavioral science research on this activity is paradoxical. On the one hand, it is indicated that money and benefits are not the be-all and end-all of motivation they were once assumed to be. On the other hand, it is indicated that pay and related hygiene factors are primarily important in establishing a climate of equity. If this is not done, conflict and dissatisfaction will likely ensue. But if such a climate is achieved, it is apparently just a starting point for more positive motivational efforts through the intrinsic satisfiers. It would seem that no less diligence and vigilance is required to maintain a sound compensation system, and those industrial engineers engaged in this activity cannot be less instrumental despite the noise about other motivational factors. The question is not one of money versus other factors but money plus other factors.

CONCLUSION

The user of a handbook may be inclined to look for how-to-do-it information. Unfortunately, the state of development of behavioral science knowledge in industrial engineering practice has not reached the how-to-do-it stage.

This chapter has attempted to touch upon the major relationships between behavioral science and industrial engineering and to give some idea of how behavioral science research has affected and will probably continue to affect industrial engineering education and practice.

To the industrial engineer who is "turned on" by behavioral science is left the job of learning all that he can about such research and the difficult job of incorporating it into his practice. He must be ever aware that not all research results are valid and must guard against hasty adoption of ideas. He must, on the one hand, remain open-minded and searching, but on the other, careful to incorporate these kinds of changes into his practice in such a way that he does not damage the progress of the art.

BIBLIOGRAPHY

Argyris, C., *Personality and Organization,* Harper & Row, Publishers, Incorporated, New York, 1957.

Bennis, W., K. Benne, and R. Chin, *The Planning of Change,* Holt, Rinehart & Winston, Inc., New York, 1961.

Davis, L. E., and R. R. Cantor, "Job Design," *Journal of Industrial Engineering,* January–February, 1955.

Ford, R. N., *Motivation through the Work Itself,* American Management Association, New York, 1969.

Gellerman, S. W., *Management by Motivation,* American Management Association, New York, 1969.

Gellerman, S. W., *Motivation and Productivity,* American Management Association, New York, 1963.

Goodwin, H. F., "Improvement Must Be Managed," *AIIE Proceedings of 19th Annual Institute Conference,* American Institute of Industrial Engineers, New York, 1968.

Herzberg, F., B. Mausner, and B. Snyderman, *The Motivation to Work*, 2d ed., John Wiley & Sons, Inc., New York, 1959.

Lawrence, P., and A. Turner, *Industrial Jobs and the Worker*, Harvard Business School, Boston, Mass., 1964.

Lehrer, R. N., *The Management of Improvement*, Reinhold Publishing Corporation, New York, 1965.

Likert, R., *New Patterns of Management*, McGraw-Hill Book Company, New York, 1961.

Maslow, A. H., "A Theory of Human Motivation," *Psychological Review*, vol. 50, 1943, pp. 370–396.

McGregor, D., *The Human Side of Enterprise*, McGraw-Hill Book Company, New York, 1960.

McGregor, D. (W. Bennis and Caroline McGregor eds.), *The Professional Manager*, McGraw-Hill Book Company, New York, 1967.

Merrill, H. F., *Classics in Management*, American Management Association, New York, 1960.

Rice, A. K., *Productivity and Social Organization: The Ahmedabad Experiment*, Tavistock, London, and Barnes & Noble, Inc., New York, 1958.

Roethlesberger, F. I., and W. I. Dickson, *Management and the Worker*, Harvard University Press, Cambridge, Mass., 1939.

Smith, Henry Clay, *Sensitivity to People*, McGraw-Hill Book Company, New York, 1966.

Trist, E. L., and K. W. Bamforth, "Some Social and Psychological Consequences of the Longwall Method of Coal-getting." *Human Relations*, vol. 4, no. 1, 1951.

Whyte, W. F., *Money and Motivation*, Harper & Row, Publishers, Incorporated, New York, 1955.

Zaleznik, A., C. Christensen, and F. J. Roethlesberger, *The Motivation, Productivity, and Satisfaction of Workers: A Prediction Study*, Harvard Business School, Boston, Mass., 1958.

Zaleznik, A., and D. Moment, *The Dynamics of Interpersonal Behavior*, John Wiley & Sons, Inc., New York, 1964.

Chapter 2

Human Effectiveness Principles for the Industrial Engineer[*]

M. SCOTT MYERS

Management Research Consultant, Texas Instruments Incorporated, Dallas, Texas;
and Visiting Professor of Organizational Psychology and Management,
Sloan School of Management, Massachusetts Institute of Technology,
Cambridge, Massachusetts

Industrial engineering as a science and profession has come a long way since the turn of the century when Frederick W. Taylor's scientific management was making its initial impact on factory operations. Believing that the nature of work had gradually evolved from an art to a science, Taylor recommended that each job should be fractionated, analyzed for efficiency techniques, and given to the highest aptitude employees trained for one specific task. To maximize efficiency, Taylor further recommended that employees be motivated through piecework incentive systems of pay by which the most productive would earn the highest wages. Taylor's research, coupled with that of Frank B. Gilbreth, is now often known as time and motion study.

The discovery that incentives other than wages, hours, and working conditions motivate employees came as a surprise to many managers when Elton Mayo uncovered the importance of the attitude of the worker toward his job at Western Electric's Hawthorne plant in 1927. Moreover, he called attention to the effects of groups on productivity, noting that cohesive groups had the power to raise

[*] Excerpted from M. Scott Myers, *Every Employee a Manager*, McGraw-Hill Book Company, New York, 1970.

or lower production according to their attitudes toward their jobs and toward the company.

The shift in emphasis away from improving individual efficiency to human relations and improved group processes was a natural consequence of trends toward mass production and automation. While machines and processes have been made increasingly complex, the workers who monitor the machines are experiencing diminishing demands on their intellect, initiative, and creativity. Recognizing that automation is making man an appendage of machines, Charles Walker[1] calls attention to the need to return the machine to its proper role as an appendage of man. Attempts to remedy the stultifying relationship between man and machines initiated the concepts of human engineering and job enrichment, whereby machines are designed to meet the abilities and limitations of man and humans are taught to amplify the efficiency of machines.

Realizing that the dimensions of a job exceed the conventional formula of wages, hours, and working conditions, Walker cites several other work dimensions useful as analytic tools in determining both productivity and satisfaction on the job:

1. Knowledge and skill requirement
2. Pacing or rate of performance
3. Degree of repetitiveness or variety
4. Relation to the total product or process
5. Relationship with people
6. Style of supervision and of managerial controls
7. Degree of worker's autonomy in determining work methods
8. Relation of work to personal development

To the extent these dimensions are known about a job, and improved in terms of both technological and psychological factors, there is potential for putting meaning back into work. As employers recognize the need to design machines to fit man, they also see the importance of designing jobs to meet man's needs. Thus, there is a new dimension in the industrial engineer's job.

In answer to the question of how the design of a job affects the meaningfulness of a job, Peter Vaill[2] concluded, on the basis of research on the working lives of fifty factory workers, that jobs are more meaningful when (1) they offer the worker continuous opportunity to learn on his job, (2) they encourage quality workmanship, (3) they allow the worker to set his own standards and goals, (4) they are experienced by the worker as psychologically "whole," and (5) they show the relationship between the goals of a particular job holder and company goals.

Regarding the relationship between the design of a job and the working environment, Vaill concluded that there is an inverse relationship between the degree of concern with wages, hours, and working conditions and job challenge and complexity. Vaill found that the effect of improved job design resulted in greater willingness on the part of the workers to take an active rather than a passive role in the organization, thus leading to their increased commitment and self-confidence.

MEANINGFUL WORK

Meaning is given or returned to work through processes known as job enrichment or job enlargement. Job enrichment may result from horizontal or vertical job enlargement, or a combination of both. Horizontal job enlargement is characterized by increasing the variety of functions performed at a given level. As an intermediate step, it serves to reduce boredom and broaden the employee's perspective, thereby preparing him for vertical job enlargement.

[1] Charles R. Walker, "Changing Character of Human Work under the Impact of Technological Change," undated multilith.

[2] Peter B. Vaill, "Industrial Engineering and Sociotechnical Systems," paper presented before the AIIE, San Francisco, May 26, 1966, pp. 13–15.

For example, assemblers on a transformer assembly line each performed a single operation as the assembly moved by on the conveyor belt. Jobs were enlarged horizontally by setting up work stations to permit each operator to assemble the entire unit. Thus the operations performed by each operator included cabling, upending, winding, soldering, laminating, and symbolizing.

Vertically enlarged jobs enable employees to take part in the planning and control functions previously restricted to persons in supervisory and staff functions. In one plant, female electronic assemblers were given training in methods improvement and encouraged to make suggestions for improving manufacturing processes. Natural work groups of five to twenty assemblers each elected a "team captain" for a term of six months. In addition to performing her regular operations, a team captain collects work improvement ideas from her team, describes them on a standard form, credits the suggesters, presents the recommendations to their supervisor and superintendent at the end of the week, and gives the team feedback on the utilization of their ideas. Although most job operations remained the same, vertical job enlargement was achieved by providing increased opportunity for planning, reorganizing, and controlling their work.

Sometimes jobs can be enlarged both horizontally and vertically. In a clad metal rolling mill, jobs were enlarged horizontally by qualifying operators to work interchangeably on breakdown rolling; finishing rolling; and slitter, pickler, and abrader operations. After giving the operators training in methods improvement and basic metallurgy, jobs were enlarged vertically by involving the operators with engineering and supervisory personnel in problem solving and goal setting sessions for increasing production yields.

Management Functions. The functions of management are commonly defined in business school terminology as:

1. *Planning:* objectives, goals, strategies, programs, systems, policies, forecasts
2. *Organizing:* manpower, money, machines, materials, methods
3. *Leading:* communicating, motivating, instructing, delegating, mediating
4. *Controlling:* auditing, measuring, evaluating, correcting

These management functions generally refer to the job of a manager, but not to the job of the worker. For example, a manager in a typical auto assembly plant might describe his own job in terms of planning, organizing, leading, and controlling, and would see his fifty foremen concerned primarily with leading and controlling. Their main responsibility would be seen as supervising the 2,000 workers on the assembly line who are doing the work. This concept is reflected in Figure 2-1.

The Management-Labor Dichotomy. This typical point of view excludes employees from the realm of management and creates, subconsciously if not deliberately, a dichotomy of people at work—unintelligent, uninformed, uncreative, irresponsible, and immature workers who need the direction and control of intelligent, informed, creative, responsible, and mature managers. The consequences of this viewpoint are widely evident in industry and are reflected in Figure 2-2, which shows the cleavage between management and labor in terms of social distance and alienation.

Though the gap between the employer and the employed has a long heritage and in some respects seems inescapably inherent in the relationship, it has become more formalized and widened through the efforts of labor unions, whose charters depend on their success in convincing labor that management is their natural enemy. The union, while pressuring management to share its affluence and relinquish its prerogatives, has at the same time clearly defined the laboring man's charter as being separate from, and indeed in conflict with, that of management. Managers typically and naturally align themselves with the goals of the company, but workers divide their allegiance between the union and the company, often with a feeling of closer identification with the union than with the company.

Although reductive management, tradition, and labor union strategy all tend to perpetuate the two-class concept, two forces in America have the potential for narrowing or obliterating the gap between labor and management. One is the improving socioeconomic status and the consequent rising aspirations of the less

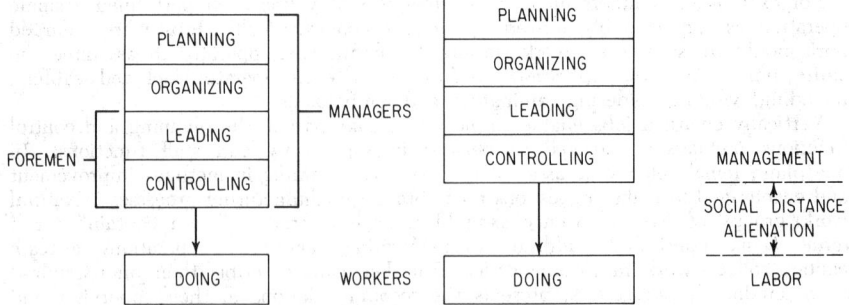

FIG. 2-1. The manager's traditional perception of his job.

FIG. 2-2. The management-labor dichotomy.

privileged, accelerated by legislated equality and increasing enlightenment in a democratic and increasingly affluent society. The other is a growing awareness on the part of managers of the inevitability of democracy as the pattern for successful competition in an entrepreneurial society, and their acceptance of their role in initiating and supporting it.

The Changing Needs of Man. Maslow's[3] hierarchy of needs theory is useful in understanding the consequences of the increasing affluence of man. Primeval man's efforts were directed primarily toward survival needs—safety, food, and shelter—leaving little time or energy for preoccupation with his latent higher order needs. As his survival needs were satisfied, he became sensitized to social and status needs. Finally, in the affluence of recent decades, these lower order or maintenance needs are being satisfied to the point that he is ready to realize his potential; to experience self-actualization in terms of intellectual, emotional, and aesthetic growth, that is, to satisfy his motivation needs.

Management and the union both have contributed to the worker's readiness for self-actualization. Industrial engineers, under the concepts of scientific management, simplified tasks and created mass production technology. Jobs were fractionated for efficiency in training (and to escape management's dependency on prima donna journeymen) and to satisfy the implicit assumption that workers would be happy and efficient doing easy work for high pay. And although mass production technology made man an appendage of tools and destroyed his journeyman's pride and autonomy, it helped get the cost of automobiles, washing machines, refrigerators, and other consumer products within his reach. These and other effects of the mass production economy accelerated the satisfaction of man's lower order needs and readied him to become aware of his dormant and unfulfilled self-actualization needs.

The union's role was just as vital in readying the worker for self-actualization, for it forced managers to share company success with the worker, thereby narrowing the economic gap between the manager and the worker, further enabling him to buy the products of mass production. However, the union, for reasons of self-preservation, sharpened the worker's identity as a member of labor rather than as a member of management—preserving the social gap that might otherwise have been reduced through economic trends.

When Work Is Meaningless. Although jobs may satisfy certain personal needs of people at work, work itself is a form of punishment in the eyes of many workers. It is uninteresting, demeaning oppressive, and generally unrelated to or in conflict with their personal goals. But it is an activity which they take in stride, or an

---

[3] Abraham H. Maslow, *Toward a Psychology of Being*, 2d ed., D. Van Nostrand Company, Inc., Princeton, N.J., 1968.

unpleasantness they are willing to endure, to get the money needed to buy goods and services which are related to personal goals. But the income itself is not the sole motive for working.

Apart from the needs satisfied through income earned on the job, work itself, however dull and menial, satisfies a wide variety of motives:

1. It reduces role ambiguity. It establishes the worker's identity, and although the self-image may not be an attractive one, for most it is better than an undefined role. Sometimes it is a form of escape from responsibility sought by people who are culturally conditioned to associate security with roles prescribed by authority.

2. It offers socializing opportunity. Close and sustained association with others having similar job goals, socioeconomic backgrounds, and interests are natural conditions for social compatibility. However, social relationships, in the absence of a unifying achievement mission, can result in group pressures disruptive to productivity. Broad-scale group cohesiveness and social interaction sometime occur among the members of a work force who can find no better basis for uniting than to defy the management Goliath.

3. It increases solidarity. The performance of similar tasks, however routine, is a shared ritual which provides a basis for equality and role acceptance. "Misery loves company" only because of the solidarity created by shared misery. The individual who is promoted or transferred from the unifying circumscribed role becomes an outsider whose solidarity needs must be satisfied elsewhere. The saying, "God must have loved common man because he made so many of them," finds grateful acceptance by people who need solace for their inescapable commonness.

4. It bolsters security feelings. Apart from the security related to economics, feelings of security for many persons require continuous affirmation from authority figures. Authority-oriented people, particularly when deprived of meaningful work roles, have unusually high requirements for feedback from the supervisor to satisfy their security and achievement needs. Dependency relationships to authority figures are also manifested by achievement-oriented people with thwarted achievement needs.

5. It is a substitute for unrealized potential. Keeping busy channels energy or thwarted intellectual capability and helps obscure the reality of unfulfilled potential. Although it is an escape mechanism, at least it is less punishing than alcoholism or other means of escape, and it helps buy freedom and the opportunity off the job which gives better expression to talent. Furthermore, fatigue from an honest day's work evokes social approval.

6. It is an escape from the home environment. Particularly for women whose homemaking roles are unbearable or completed, a job provides an opportunity for getting away from the home. Other reasons for wanting to get away from home include domestic conflict, neighborhood friction, unattractive home facilities, and loneliness.

7. It reduces feelings of guilt and anxiety. In an achieving society where dignity and pride are earned through the traits of ambition, initiative, industriousness, and perseverance, idleness violates deep-seated values, and work for work's sake is virtuous. By Horatio Alger or Protestant Ethic standards, idleness is the equivalent of stealing, and a strong conscience is a key motive for staying on the job.

Although the roles of meaningless work defined above relate to the personal needs of individuals, they are not constructively aligned with company goals. Moreover, these roles thwart long-range personal goals as they increase dependency relationships and discourage the development of talent. However, work itself, properly designed, can satisfy other needs which are related to the achievement of long-range personal and organization goals.

For example, the manager's job is usually found to be challenging, related to company goals, and generally aligned with his long-range personal goals. The difference in job attitude between manager and worker is usually ascribed to immaturity of the worker, overlooking the fact that maturity is developed or impaired as a function of opportunity to manage one's job.

Managers manage their jobs, while workers are managed by their jobs. Workers

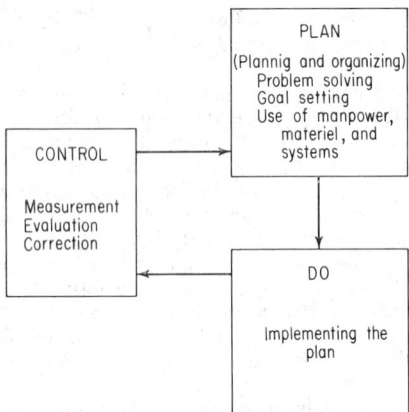

Fig. 2-3.   Meaningful work model.

are frequently only appendages of tools or links between them—doing what is necessary to satisfy the requirements of inflexible, inanimate monsters.

The Dimensions of Meaningful Work. Work itself, to be meaningful, must
make tools the appendage of man and place man in a role not restricted to obedient
doing. It must include planning and controlling, as well as doing, as illustrated
in Figure 2-3.

The plan phase includes the planning and organizing functions of work and
consists of problem solving, goal setting, and planning the use of manpower, materiel,
and systems. Planning is the ingredient of work which gives it meaning by aligning
it with goals. The do phase is the implementation of the plan, ideally involving
the coordinated expenditure of physical and mental effort, utilizing aptitudes and
special skills. Control includes measurement, evaluation, and correction—the feedback process for assessing achievements against goals. Feedback, even to a greater
extent than planning, gives work its meaning, and its absence is a common cause
of job dissatisfaction. The control phase is the basis for recycling planning, doing,
and controlling. People who work for themselves generally have meaningful work
in terms of a complete cycle of plan, do, and control.

The Meaningful Work of Managers. Managers in industry, though seldom having as much autonomy as self-employed entrepreneurs, typically have jobs rich
in plan, do, and control phases, particularly at the higher levels. Three typical
management jobs out of a seven-level hierarchy in a manufacturing organization
are identified below for analysis in terms of their usual planning, doing, and controlling phases.

President
● Operating vice president
Department manager
● Manufacturing manager
Superintendent
● Foreman
Operator

Figure 2-4 shows that the operating vice president, as division manager, plans
in the realm of economic and technological trends, facilities expansion, manpower
and management systems, and policy formulation. The doing aspect of his job
involves him routinely with key customers, public relations roles, visits to various
operating sites, and the exchange of business information. His control functions
include the measurement, evaluation, and correction of factors associated with cus-

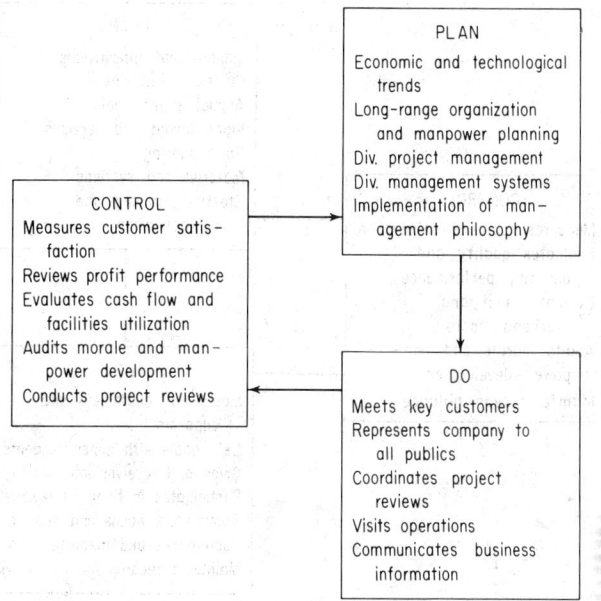

FIG. 2-4.   Meaningful work—operating vice president.

tomer satisfaction, net sales, profits, cash flow, facilities utilization, return on investment, morale, and manpower development. Hence, the division director's job is
rich in plan, do, and control, much like the self-employed individual.

Similarly, Figure 2-5 shows the manufacturing manager's job to be relatively rich
in the meaningful aspects of work. Although his job is narrower in scope and
two levels below the division manager's position, it is nonetheless rich in plan,
do, and control. A company is rarely plagued with the lack of commitment of
a manufacturing manager or the people above him.

Where Meaningful Work Usually Stops. Even the foreman's job, two levels
below the manufacturing manager, may be rich in terms of the ingredients of
meaningful work. Figure 2-6 indicates that the foreman's job, although narrower
in scope than the manufacturing manager's, offers him considerable latitude in
managing his work. This example depicts a traditional authority-oriented supervisor.

Although this foreman's job portrays a complete plan-do-control cycle, it is nonetheless unsatisfying to the incumbent because its authority orientation prevents
the delegation of a complete plan-do-control cycle of responsibility to the operator.
Under this foreman, the operator lives in a world circumscribed by conformity pressures to follow instructions, work harder, obey rules, get along with people, and be
loyal to the supervisor and the company, quashing any pleasure that work itself
might otherwise offer. His role puts him in a category with materiel, to be manipulated by managers exercising their management prerogatives (as kings once exercised
their "divine rights") in pursuit of "their" organizational goals. Conformity-oriented
workers tend to behave like adolescent children responding to punishments and rewards of authoritarian parents; and their prerogatives, which are generally expressed
in terms of rights wrested from management, are only incidentally aligned with
company goals.

The Impact of Supervisory Style. When meaningful work extends down to
the operator, the supervisor's role is revised to provide opportunity for people
to manage their own work, as portrayed in Figure 2-7. In contrast to the authority-
conformity–oriented roles of the supervisor and operator shown in Figure 2-6, each

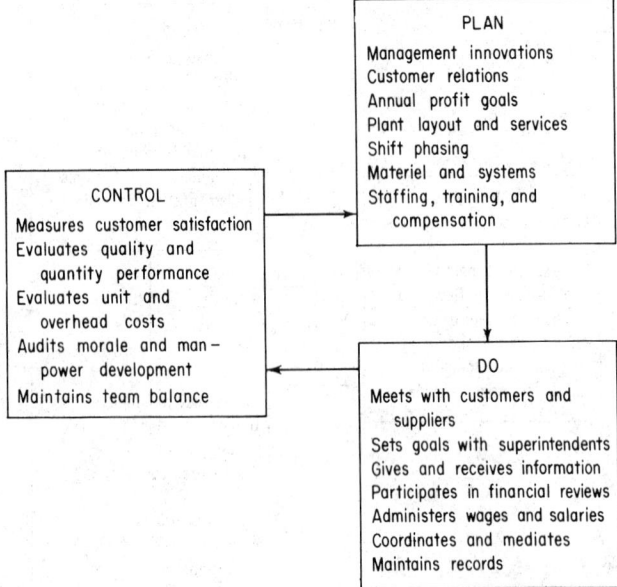

Fig. 2-5.   Meaningful work—manufacturing manager.

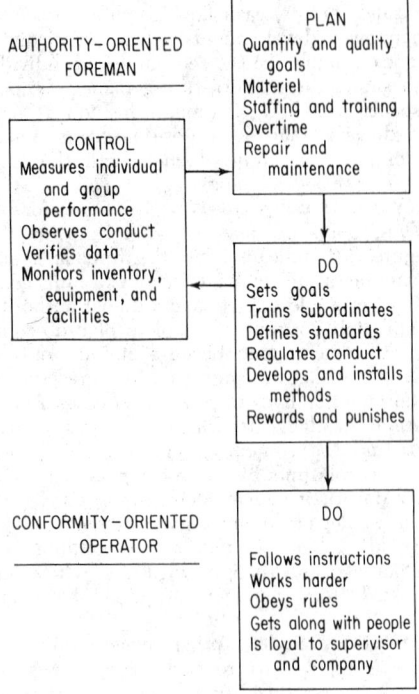

Fig. 2-6.   Authority-oriented relationship between foreman and operator.

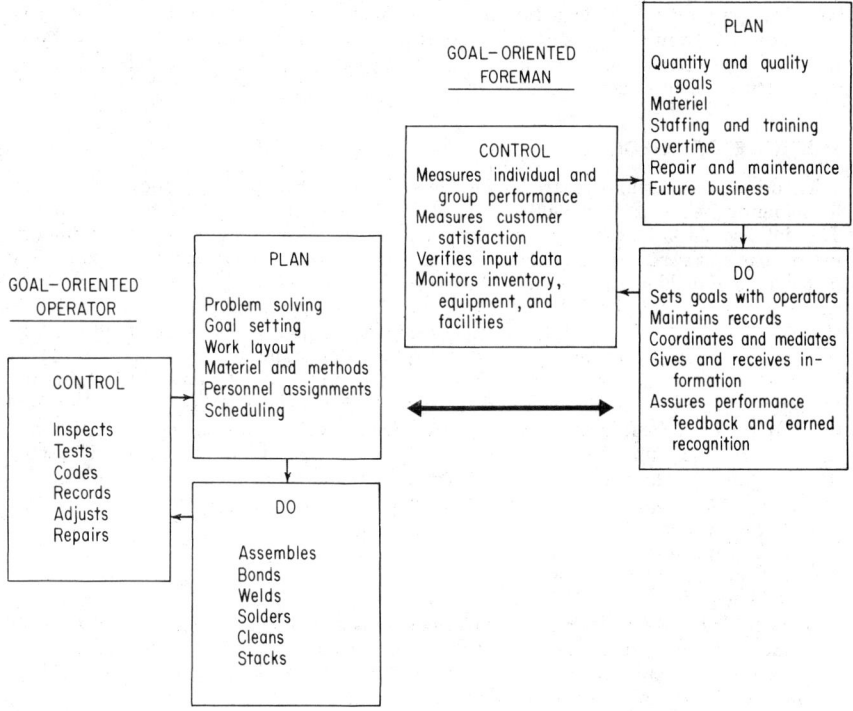

FIG. 2-7.   Goal-oriented relationship between foreman and operator.

now has a goal-oriented role in which the revised do phase of the supervisor and plan phase of the operator comprise the realm of interface between them. Figure 2-7 shows the goal-oriented supervisor to be a resource person whose involvement is invoked primarily at the initiative of the operator.

This role of the leader, evolved initially to illustrate the relationship between the foreman and operators, is a model representing ideal supervisory relationships at any level. Furthermore, enriching the operator's job has changed higher level jobs, in some cases making it possible to reduce the number of levels in the management hierarchy. The foreman, now freed of many detailed maintenance and control functions, has more time to be involved with his supervisor in higher level planning functions and is also more available to meet his responsibility as a mediator and resource person when needed by the natural work group under his supervision.

Job enrichment sometimes results naturally from the intuitive practices of goal-oriented, emotionally mature managers who evoke commitment through a "language of action" which grants freedom and reflects respect, confidence, and high expectations. Unfortunately, many managers still see job enlargement as a form of benevolent autocracy, and their unguided attempts to enlarge jobs fall more within the realm of manipulation than job enrichment. When job enrichment is attempted by reductive, authority-oriented managers, they usually fail to inspire the level of involvement and commitment achieved by goal-oriented managers. Their motives are suspect, and their language of action comes through as manipulation and exploitation rather than as acts of trust, confidence, and respect.

Hence, job enrichment depends on style of supervision as well as job requirements and is not merely a matter of duplicating patterns of work and relationships found to be successful elsewhere. Moreover, job enrichment is not a process that can be engineered by following a formula or detailed systems and procedures. Rather, it must stem from an understanding of the needs of people, and the conditions that must be satisfied in the workplace if people are to direct their talents toward the achievement of organizational goals.

THE ROLE OF THEORY

Theories are springboards to action and change. However, theories rarely lead to changing behavior until deliberate and intensive efforts are made to apply them. The intellectual understanding of management theory has about the same impact on a manager's supervisory style that the intellectual study of snow skiing has in teaching him how to ski. In either case, his competence is developed primarily through application—through the actual practice of supervision or by actually skiing. If he is satisfied with his style of supervising or skiing, he will expend little effort in learning and applying theories for the purpose of changing his style.

However, if he is dissatisfied with his performance to the point that his desire to improve exceeds his reluctance to accept assistance, he may approach the study of theory with a readiness to change. A theory will be useful to him if he can translate it into remedial action that will reward his efforts. A theory is useful to an industrial engineer when its application leads to desired changes in managerial behavior. The application of theory generally requires a four-step process:

1. Awareness
2. Understanding
3. Commitment to change
4. New habits

Step 1: Awareness may result from a convincing speech, reading a book, viewing an educational film, attending a public seminar, or simply through shop talk. This first step may occur for a manager when he gains at least a superficial insight into a new theory and the implied deficiency in his present style of managing.

Step 2: Understanding may result from activity precipitated by his awareness of the possible need to change. He may read numerous books and articles on the theory and selectively choose training programs and attend lectures on the subject. This step may be thought of as an intellectual conditioning process. He may become an articulate spokesman for his newly acquired insight, but his managerial style may continue to follow old habit patterns.

Step 3: Commitment to change occurs when he becomes aware of the discrepancy between his newly adopted theory and his everyday behavior, but only if he believes he will benefit personally through changing his style of management. Initial attempts are often discouraging, and if not reinforced by some type of rewarding feedback, may gradually be discontinued. Commitment and reinforcement must be strong and continuous to overcome established habit patterns. Moreover, his changed behavior is often viewed with suspicion by persons whose opinions about him have been crystallized by his previous style.

Step 4: New habits are established when sustained deliberate applications of the new theory finally result in attitude changes and automatic and natural expressions of the desired changes in style of management. Attainment of the new habit formation stage is a long and difficult process requiring perhaps five to ten years of sustained reinforcement from steps 2 and 3. Some individuals never progress beyond step 2, particularly when others in the organization upon whom he depends for continuing opportunity do not encourage him through their language of action and words.

The application of management theory in a business organization generally begins with a step 2 intellectual conditioning experience to prepare managers for the step 3 translation process. A motivation seminar at Texas Instruments, based on an amalgamation of several theories, illustrates the nature and role of an intellectual conditioning process. Dubbed the "motivation-maintenance theory," it defines media in the typical industrial organization through which the needs of people at work are satisfied, as illustrated in Figure 2-8.

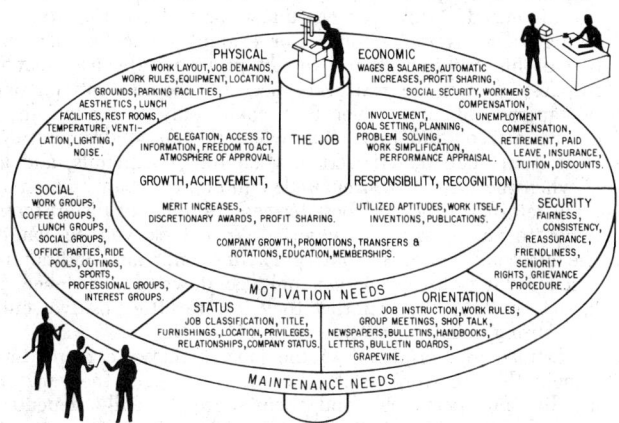

Fig. 2-8.   Employee needs—maintenance and motivational.

The Motivation-Maintenance Theory.   Maintenance needs are synonymous with Maslow's lower order needs, and the term "maintenance" is used to denote the fact that people, like buildings and machines, must be maintained. Motivation needs, synonymous with Maslow's higher order self-actualization needs, are satisfied when man is developing his potential through the pursuit of meaningful goals.

Maintenance Needs.   The maintenance needs of people at work are quite similar whether the individual is a machine operator—as illustrated in the diagram—or the president, the vice president, the middle manager, the foreman, the technician, the secretary, or the floor sweeper. Although the maintenance of people is not the key to motivating them in the business organization, it is usually a prerequisite for motivation. The same needs apply to people outside the industrial organization, such as housewives, policemen, clergymen, schoolteachers, and students. All require the satisfaction of their maintenance needs, defined here in terms of economic, security, orientation, status, social, and physical factors.

1. Economic maintenance needs involve wages, salaries, and supplemental benefits received almost automatically by virtue of being on the job. Economic maintenance needs do not include forms of compensation stemming from meritorious performance, mentioned later as reinforcements of motivation.

2. Security maintenance needs refer to feelings of people arising primarily from their perception of their supervisor as an impartial, consistent, reassuring, friendly type of person, and from the knowledge that they are protected by a just system in the job situation.

3. Orientation maintenance needs require knowledge of the company and the job. This information is supplied by the supervisor, by printed media such as newspapers and bulletins, or through the informal grapevine which exists in every organization.

4. Status maintenance needs are generally satisfied through job classifications, titles, furnishings, privileges, relationships, the company image, or the product image. The process of acquiring status relates to motivation factors of growth and achievement discussed later, but the possession of status or symbols of it is largely maintenance.

5. Social maintenance needs are satisfied through formal or informal group activities in work groups, luncheon groups, coffee groups, ride pools, or after-hours recreational activities.

6. Physical maintenance needs are satisfied by the work layout, parking facilities, air conditioning, lighting, rest rooms, eating facilities, noise levels, and other physical factors.

When maintained at adequate levels, dissatisfactions stemming from these lower order needs are minimized. However, the maintenance factors in these six categories have only fleeting value as motivators. For example, when a plant manager air-conditioned one of his buildings that had not previously been air-conditioned, the enthusiastic response during the first week seemed like motivation. However, enthusiasm soon tapered off to a level that could only be called the absence of dissatisfaction with air conditioning. But, when the air-conditioning system failed, the response was immediate dissatisfaction, vociferous complaints, and lowered production. And when the system was repaired, building occupants were not motivated; they were merely returned to a level best termed "absence of dissatisfaction." Their typical comments to the repairmen were: "What took you so long?" or "Why did you let the air conditioning go out?" Having built air conditioning into their expectations, their feelings toward it can only go downward. Hence, maintenance factors are characterized by the fact that they inspire little positive sentiment when added, but incite strong negative reaction when removed.

Maintenance factors are peripheral to the job, as they are more directly related to the environment than to work itself. For the most part, they are group administered, usually by staff personnel, and their success usually depends upon their being applied uniformly and equitably. Managers often fail to understand the ingratitude of employees toward maintenance factors such as the Christmas turkey, free coffee, and other expressions of well-intentioned paternalism. They are particularly disillusioned when these "gifts" become the subject of collective bargaining and become perpetuated as "rights of labor."

Since the turn of the century, supplemental benefits have increased in cost from less than 10 percent of the payroll budget to more than 30 percent. The more supplemental benefits added as maintenance factors, the higher their potential for dissatisfaction. However, it would be an oversimplification to conclude that maintenance factors are only increasing dissatisfaction, or that they are the primary sources of dissatisfaction.

Maintenance factors serve a necessary function that can be appreciated only in historical perspective. Looking back to the early decades of the century, the working man lived in a world of management prerogatives of arbitrary "hire and fire." He lived and worked in substandard conditions and received substandard wages. But over the years, two primary influences have brought about the affluent society:[4] (1) the intervention of labor unions, which forced the sharing of company wealth, and (2) the mass production technology which prices automobiles, washing machines, refrigerators, and other consumer products within the reach of increasingly higher percentages of the population. Maintenance factors embrace the wages, hours, and working conditions which have been the focus of collective bargaining for many years. The emphasis on maintenance factors by unions raises a question regarding the future rule of labor unions and their ability to survive through a continuing strategy based on improving wages, hours, and working conditions. Life in an affluent society where maintenance needs are satisfied would seem to preclude the need for unions. Because people's needs change, moving upward as lower needs are satisfied, employees in increasing numbers are failing to experience the satisfaction previously derived from improved maintenance factors and are aimlessly seeking, with mingled hope and despair, something more meaningful than comfortable working conditions and routinized work. The union's role has gradually, subtly, and inadvertently shifted from "people's defender" to a medium for displacing the aggression stemming from frustrations which it helped create.

Motivation Needs. The only constructive outlet for these frustrations in the business organization is upward through Maslow's hierarchy of needs to opportunities for satisfying self-actualization or motivation needs in terms of such factors as growth, achievement, responsibility, and recognition, illustrated in the inner circle of Figure 2-8.

1. Growth, in this context, refers to mental growth. Although physical growth

---

[4] J. K. Galbraith, *The Affluent Society,* Houghton Mifflin Company, Boston, 1958.

generally levels off before age twenty, mental growth may continue throughout the life span of the individual. A challenging job is one of the most effective antidotes to mental stagnation and vocational obsolescence.

2. Achievement refers to the need for achievement (nAch) that McClelland has shown to be a key motive when it can find expression. Individuals differ from each other in terms of their need for achievement, and a given individual's level of achievement motivation will vary with his opportunity to find expression for it. When jobs offer little opportunity for satisfying achievement needs, high nAch people seek outlets for their achievement needs within or outside the organization. Jobs rich in opportunity for growth and achievement attract and retain high achievers. Low nAch people in such an environment may develop nAch through the multiple influences of a challenging role, peer pressure, and image emulation. By the same token, jobs lacking in challenge tend to attract and retain low achievers whose needs are satisfied largely through the maintenance factors such as security, benefits, affiliation, and comfortable surroundings.

3. The term "responsibility" refers to a sense of commitment to a worthwhile job. It has long been recognized that a sense of responsibility is a function of level in the organization—people high in management having a proportionally higher sense of responsibility than people at lower levels. A study of factors relating to motivation of managers at Texas Instruments[5] demonstrated this relationship, but it also revealed that level of motivation was more strongly related to style of supervision than it was to level in the organization. It was found that Theory Y supervision, as discussed later on pages 7-32 to 7-36, evoked a greater sense of commitment than Theory X supervision.

4. Recognition, as a motivation need, refers to earned recognition stemming from meritorious performance. Unearned recognition or friendliness, defined as a condition of security in the outer circle, is needed for keeping communication channels open so that when vital issues arise, they may be surfaced and dealt with. But, within the inner circle, recognition as positive feedback for a job well done is a reinforcement of motivated behavior. Recognition at its best does not depend on the value judgments of an authority figure to translate achievements into praise, advancements, respect, awards, pay increases, and other rewards. Recognition dispensed by value judgment places unjustified faith in the objectivity, reliability, sensitivity, attentiveness, and competence of the judge, and tends to foster dependency relationships.

Ideally, recognition should not depend on an intermediary, but should be a natural expression of feedback from achievement itself. When the astronauts landed on the moon, or when Jonas Salk discovered polio vaccine, or when Babe Ruth hit a home run (or when Casey struck out!), they did not need a supervisor to give them recognition. They received feedback naturally and spontaneously, and the quality of this feedback was not distorted by the interpretation of an intermediary. Hence, recognition at its best is primarily an expression of direct feedback.

Although achievement and responsibility may be their own rewards, they too are reinforced when someone upon whom the individual depends for continuing opportunity recognizes him for achievements. So long as the supervisor's authority is the basis for continuing opportunity and his judgment the basis for pay and status changes, his feelings are a necessary part of the feedback.

The Need for Equilibrium. The relative importance of motivation and maintenance needs is situational, as is the relative importance of the six maintenance and the four motivation needs. A given maintenance need, such as physical, may loom as the most important if it is the main source of current dissatisfaction. But once the air conditioning, noise level problem, or other focus of concern is solved, other factors may assume greater importance. If product obsolescence and organizational stagnation stymie growth opportunity, growth needs assume greater importance.

[5] M. Scott Myers, "Conditions for Manager Motivation," *Harvard Business Review,* January–February, 1966.

However, thwarted growth needs are often misleadingly displaced and expressed as amplified concern for maintenance factors. Increasingly, the real problems of people at work are lack of inner circle opportunities, though outer circle factors linger as the issue of conflict. Both maintenance and motivational needs must be satisfied, not one to replace the other, but rather to provide better balance between the two.

People in upper levels of management have relatively more opportunity to satisfy inner circle needs, but they devote much of their effort to coping with labor problems stemming from the inability of people at lower levels to get into the inner circle. Management's mission, then, must be to get people into the inner circle, not for reasons of altruism or as missionaries for participation, but rather as a sound business strategy to provide outlets for human talent in the pursuit of organizational goals.

Other Useful Theories. The motivation-maintenance theory above is presented as only one of many theories which, when elaborated and translated, can serve as foundations for developing systems and programs for improving human effectiveness. Figure 2-9 portrays twelve additional theories of human effectiveness, selected more to reflect variety than comprehensiveness. Each of these theories is presented on a linear scale, the left end representing conditions conducive to ineffectiveness, and the right end conditions for greater effectiveness. These scales do not reflect the full complexity of these theories or their application, nor are they intended to define the scope or primary focus of their developers' professional competence. Rather, they are displayed as an aid in comprehending the commonality, as well as the uniqueness, of what might otherwise appear as a confusing and contradictory proliferation of theories. The top four theories place the focus on managerial styles or assumptions, the middle four are generally described as combinations of managerial styles and management systems, and the lower four are predominantly descriptors of the impact of managerial styles and systems. The first four theories have as much impact as the middle four on systems, as managerial styles and values inevitably find expression in system design and administration.

A discussion of these various theories is beyond the scope of this chapter. The reader is referred to the bibliography at the end of the chapter for further information.

The several theories implicitly show that human effectiveness is a function of three causal conditions:

1. Interpersonal competence
2. Meaningful goals
3. Helpful systems

The satisfaction of these three conditions leads to profitable organizational growth and self-actualization of its members. The circular self-perpetuating characteristics of these conditions are shown in Figure 2-10, and each of the three causal conditions are defined in the following discussion.

INTERPERSONAL COMPETENCE

Interpersonal competence in industry refers to ideal relationships among people at work, united in the pursuit of compatible personal and organizational goals. Interpersonal competence is generally a function of two factors—managerial style and the systems framework of the organization. Generally speaking, both managerial style and management systems reflect the values of managers and hence the philosophy of the organization. Interpersonal competence is discussed here in terms of the informal relationships among members of the organization and the influence of managerial assumptions, particularly those assumptions that find expression in the reductive or constructive use of authority.

The formal systems in business and industry reflect the values of the people who developed the systems and who administer them. The organization culture which arises in support of, in spite of, or in reaction to the formal structure of industry may be termed the informal organization of industry. It is not reflected on organization charts, nor is it always acknowledged by management. But none-

Fig. 2-9. Theories of human effectiveness.

	INEFFECTIVENESS				EFFECTIVENESS
MANAGERIAL STYLES					
ROBERT BLAKE	1, 1 Neutrality and indecision	MANAGERIAL GRID: 1, 9 Inadequate concern for production; 5, 5 Compromise, middle-of-the-road; 9, 1 Inadequate concern for people			9, 9 Integration of resources
JAY HALL	1, 1 Decisions by default and precedent	DECISION MAKING GRID: 1, 9 Inadequate concern for quality decision; 5, 5 Decision through bargaining; 9, 1 Inadequate concern for commitment			9, 9 Adequate concern for commitment and quality decisions
RENSIS LIKERT	SYSTEM 1 Exploitive authoritative	SYSTEM 2 Benevolent authoritative	SYSTEM 3 Consultative		SYSTEM 4 Participative group
DOUGLAS McGREGOR	THEORY X Reductive assumptions				THEORY Y Developmental assumptions
MANAGERIAL STYLES & SYSTEMS					
CHRIS ARGYRIS	AUTOCRATIC RELATIONSHIPS Conflict and conformity, alienation				AUTHENTIC RELATIONSHIPS Interpersonal and technical competence, commitment
WARREN BENNIS	BUREAUCRACY Authoritarian, restrictive management structure				DEMOCRACY Goal-oriented, adaptive management structure
FREDERICK HERZBERG	ENVIRONMENTAL COMFORT — — Hygiene seeking — —				MEANINGFUL WORK — — Motivation seeking — —
CONSEQUENCES OF MANAGERIAL STYLES & SYSTEMS					
JOHN PARÉ	BOSS POWER—Direction and control by authority; SYSTEM POWER—Bureaucratic controls; PEER POWER—Social pressure of group				GOAL POWER—Self-alignment with organizational goals
ERICH FROMM	ESCAPE FROM FREEDOM Conformity, domination, destructiveness				FREEDOM Self-reliance, spontaneity, responsible behavior
WILLIAM GLASSER	AVOIDANCE OF REALITY Maladjustment				COPING WITH REALITY Responsible behavior
ABRAHAM MASLOW	LOWER NEED FIXATION Halted growth				SELF-ACTUALIZATION Realizing potential
DAVID McCLELLAND	LOW nACH More interested in things like affiliation, security, money, possessions				HIGH nACH Achievement its own primary reward, high challenges, moderate risks, independence

7-29

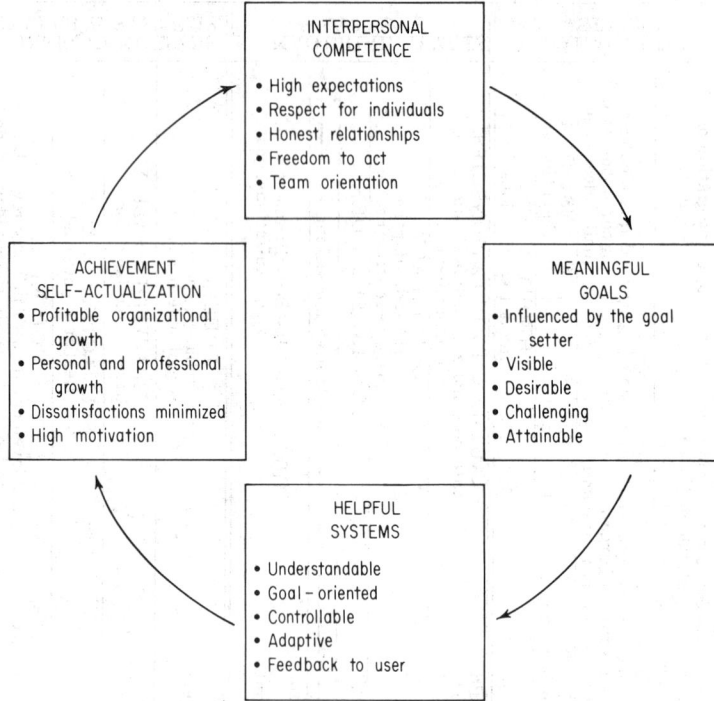

FIG. 2-10.   A system for human effectiveness.

theless it is the medium through which interpersonal competence or conflict finds expression.

The Informal Organization of Industry.   Organizational relationships are usually defined formally through organization charts and job descriptions.   Although formal relationships are the basis for many informal relationships, many social and technical (sociotechnical) relationships form quite independently of the formal organization. People may form social or "primary"[6] groups because they do the same kind of work, are of similar ethnic or regional origin, have similar interests, or are the same age, sex, or seniority in the firm.   More often, however, they come together merely because they are near each other in the work area.   The structure of a primary group is not stable, but changes in a fluid process as membership, job relationships, and work assignments change and events occur that alter the roles of individuals within the group.   Primary groups based on work station relationships may be preempted by other primary group memberships in other role relationships, such as in ride pools, the coffee bar or lunchroom, and bowling teams.   However, these other groups do not detract from the work group membership and in fact may even be a source of its enrichment.

Large organizations are comprised of small informal groups held together by the process of face-to-face communication.   A primary group will sometimes develop its own jargon, which tends to create solidarity and serves as a means of identifying group membership.   These small primary groups vary in size, but average between six to ten people.   Because problems of communication increase with the size of the group, a group tends to break up or subdivide after it has reached a certain critical size.

[6] J. A. C. Brown, *The Social Psychology of Industry*, Penguin Books, Inc., Baltimore, 1965, pp. 124–256.

When a person enters a job with the intention of permanent employment, he will strive naturally to succeed and develop primary group relationships. Interpersonal competence through primary groups is most likely to exist in skilled trades where turnover is low, where the plant is located in a relatively small and stable community, and in a stable work force not subject to fluctuations of seasonal employment, temporary help, layoffs, or high turnover for whatever reason.

The primary group is the medium through which individuals acquire their attitudes, values, and goals. It is also a fundamental source of discipline and social control. Members of primary or natural work groups expect a fair share of the group's work from each other and will rally against the member who benefits at the expense of another. Because the informal working group is the main source of social control, attempts to change human behavior should be made through the medium of the group rather than through the individual. The supervisor should try to exercise legitimate influence through such groups and should avoid breaking them up.

John Paul Jones[7] has enumerated five conditions, summarized below, which exist in an effective primary group or team:

1. Mutual Trust. Mutual trust takes a long time to build and can be destroyed quickly. It is established in a team when every member feels free to express his opinion, say how he feels about issues, ask questions which may display his ignorance, and disagree with any position, without concern for retaliation, ridicule, or negative consequences.

2. Mutual Support. Mutual support results from group members having genuine concern for each other's job welfare, growth, and personal success. If mutual support is established in a team, a member need not waste time and energy protecting himself or his function from anyone else. All will give and receive help to and from each other in accomplishing whatever object the team is working on.

3. Genuine Communication. Communication has two dimensions: the quality of openness and authenticity of the member who is speaking, and the quality of nonevaluative listening by other members. Open authentic communication takes place when mutual trust and support are so well established that no member feels he has to be guarded or cautious about what he says. It also means that members of a good team will not "play games" with each other, such as by asking "trap" questions or suggesting wrong answers to test another member's integrity. Nonevaluative listening means listening with "bias filters" removed to what the other person is trying to communicate. Most persons listen through an evaluative screen and tend to hear only those aspects of a communication which do not threaten status, roles, and convictions.

4. Accepting Conflicts as Normal and Working Them Through. Individuals differ uniquely from one another and will not agree on many things. An unproductive heritage left by the old school of "human relations" is the notion that people should strive for harmony at all costs. A good team (where mutual trust, mutual support, and genuine communication are well established) accepts conflict as normal, natural, and an asset, because it is from conflict that most growth and innovation are derived. It is also worth noting that conflict resolution is a group process, and the notion that a manager can resolve a conflict between or among subordinates is a myth.

5. Mutual Respect for Individual Differences. There are decisions which, in a goal-oriented team, must be team decisions because they require the commitment of most or all of the resources of the team and cannot be implemented without this commitment. However, a good team will not demand unnecessary conformity of its members. It is easy for a group to drift into the practice of making decisions for or forcing decisions on an individual where clearly, for his own growth and for the good of the organization, he should make the decision. The individual member should be free to ask advice from other members, who in turn will recognize that he is not obligated to take the advice. A good team delegates within itself. In

[7] John Paul Jones, *The Ties That Bind,* National Association of Manufacturers, New York, pp. 21–23.

a well-established team with high mutual trust and support, the leader, or any member, will be able to make a decision which commits the team. In such a team, only important issues need to be "worked through," and there is much delegation from leader to members, from members to member, and even from members to leader.

The leader of a primary group is not only a member of his own working unit, he also joins with other primary group leaders to form a higher echelon group whose members act as "linking pins"[8] in maintaining contact between the primary groups. This overlapping membership is necessary to make an organic whole of a larger complex organization, sometimes referred to as the secondary group. Attitudes of individuals toward the secondary group are usually determined by the extent to which its goals coincide or conflict with the goals of their own primary group. When all the primary groups within a secondary group direct their efforts toward a common goal, organizational cohesiveness is maximized.

A crowd or mob is different from the primary or secondary group. Members of a mob are usually not acting through primary or secondary group memberships, typically have no established interpersonal relationships, and do not subordinate their impulses to a functional task. Each member remains virtually anonymous and lost in the mass. The effect of the crowd on its members is largely contagious, and normally quiet and thoughtful citizens may cast aside inhibitions connected with their family, their neighbors, the law, and their work and participate in thoughtless violence.

Supervisory behavior is a key factor at all levels of the organization in permitting the natural formation of primary groups out of what might otherwise be a mob of anonymous individuals lost in the mass. Moreover, leadership style determines whether primary groups are united in support of or in rebellion against the formal organization.

A supervisor's behavior at any level is a function of his own values and his fulfillment of role expectations dictated by systems and people above him in the organization. Hence, the values of top management are the key determinant of organizational climate as created by supervisory behavior and management systems. Douglas McGregor has explained managerial style as a function of the manager's assumptions about people.[9] He defines assumptions in terms of their position on

[8] Rensis Likert, *The Human Organization*, McGraw-Hill Book Company, New York, 1967, p. 50.
[9] Douglas McGregor, *The Human Side of Enterprise*, McGraw-Hill Book Company, New York, 1960.

Theory X assumptions	Theory Y assumptions
People by nature:	People by nature:
1. Lack integrity	1. Have integrity
2. Are fundamentally lazy and desire to work as little as possible	2. Work hard toward objectives to which they are committed
3. Avoid responsibility	3. Assume responsibility within these commitments
4. Are not interested in achievement	4. Desire to achieve
5. Are incapable of directing their own behavior	5. Are capable of directing their own behavior
6. Are indifferent to organizational needs	6. Want their organization to succeed
7. Prefer to be directed by others	7. Are not passive and submissive
8. Avoid making decisions whenever possible	8. Will make decisions within their commitments
9. Are not very bright	9. Are not stupid

FIG. 2-11.   Supervisory assumptions.

the subordinate is free to contact those at his own level or below. The authority-oriented person tends to choose persons at his own level or above for luncheon dates or social activities, although he would find it acceptable to accompany a group of his subordinates to lunch. For him, social stratification within and even outside the organization is determined largely by position in the official hierarchy.

If one of his subordinates loses favor with a higher-up, for whatever reason, his own perception of the subordinate is altered, and he readily turns and "takes the second bite" at the heels of the subordinate. He may salve his conscience by uttering feeble demurrals, damning the subordinate with faint praise, but ultimately acquiescing and obediently performing the painful ceremony of admonishment, transfer, demotion, or termination with the courage displayed by all official ax men.

Occasionally, goal-oriented work groups form within the authoritarian's organization, made up of individuals whose professional competence immunizes them against his arbitrary use of power. He holds them in awe because he respects power of any kind. As long as they incur top management favor and support, he does not stand in their way. However, his relationship to them is not usually one of goal-oriented reciprocity. For the authoritarian, solidarity or equality is an uncomfortable experience. Hence he tends to abandon them or disengage himself from involvement in their efforts. If the group finally disbands and leaves his organization in reaction to his style of leadership and he is faced with the mission of restaffing the vacated positions, he vows he will never again let another "power base" develop within his organization!

Slightly more subtle are the cat-and-mouse tactics employed by the boss as if to remind underlings that their security and freedom exist only through his magnanimity. The arbitrary withholding and dispensing of information according to whim, the last-moment scheduling or cancellation of meetings, delaying and extending staff meetings, the extraction of "reasons" for personal leave, the selective distribution of homemade Christmas cakes to his favorites, and the dog-in-the-manger authorization to use company-financed club facilities are examples of more subtle misuse of authority.

When the authoritarian supervisor chairs a staff meeting—for example, to solve a problem or to evaluate a proposal—his subordinates develop an infallible method for taking the "right" position on any issue—they learn to read his facial expression! They sit in silence until the chief speaks and then converge on an elaboration of his viewpoint. So firmly established become these cue-reading patterns that subordinates, as well as the boss, are often deceived into interpreting conformity as consensus.

The boss's authority is often adopted by his secretary through a process of authority by association. His secretary's "request" for information frequently comes through as orders to subordinates and their secretaries. Exempted from timekeeping regulations, rotational relief assignments, and many standard ground rules, she gradually activates a conditioning process which increases subordinate acquiescence and alienation and her secretarial imperiousness. Attempts to give feedback to the boss regarding his secretary's "little dictator" syndrome characteristically evoke a defensiveness that effectively shuts off future feedback attempts.

Perhaps the most discouraging aspect of the misuse of authority is the supervisor's insensitivity to his syndrome. It should be noted that authority orientation is not usually an expression of intentional or deliberate malice. It is usually a form of pathology, perhaps a result of years of adapting to authoritarianism in the home, schools, the church, the armed forces, and previous jobs. The intelligent authoritarian typically experiences occasional flashes of insight and brief periods of remorse. In fact, a true authoritarian is also a masochist who pathologically enjoys punishment or criticism—as long as it comes from a respected authority and represents an opportunity for atonement from which he can recover. When he becomes aware of alienation in his group, he may attempt to win goodwill through the paternalistic generosity of an office party, a home barbecue, a Christmas turkey, or other irrelevant tactics. Paternalism, of course, only increases social distance. His attempts to change are sincere, but most such attempts are superficial veneers which fail to conceal his real personality, which he exposes through his day-by-day language of action.

Finally, it must be noted that the authority-oriented supervisor should not bear the full brunt of his ineptness. The managers who appointed him and reinforced his behavior through rewards and authoritarian systems and leadership styles above him are the primary problem. If he *can* change, he certainly will not until he gets different cues from above, as his behavior largely reflects his attempts, usually subconscious, to build himself in their image.

MEANINGFUL GOALS

Everyone has goals. Some goals are set by the individuals pursuing them, some are set with the participation of others, some are set exclusively by others. Generally speaking, individuals most actively pursue the goals they set themselves. When too many of a person's goals are set by others, he reacts individually or collectively to set goals to circumvent, violate, or change these goals. These goals of avoidance and rebellion then become personal goals. Therein lies the crux of the problem of goal setting in industry.

The higher a person's position in the organization, the more degrees of freedom he has to set goals. If the man at the top defines his goals with the genuine involvement of the people below him, his goals are also their goals.

But if goal setting is a top management function to be sold downward through the use of persuasion, authority, bribery, and manipulation, his people respond with words and actions that say, "Those are not my goals, they are management's goals." If management is seen as his enemy, the individual fights back in subtle or overt ways, often ingeniously, sometimes subconsciously, to thwart management goals. But if management is perceived as benevolent and friendly, he may curb his inner frustrations, turn out a fair day's work, and appreciate their well-intended praise and rewards. But most of the time he thinks and talks about his goals—which he finds off the job.

Some Dynamics of Goal Setting. Goals are related to satisfaction according to this equation:

$$\text{Satisfaction} = \frac{\text{achievements}}{\text{goals}}$$

Goals nearly always exceed achievements, and hence satisfaction increases as achievements approach goals. But in reality, this satisfaction is illusory and fleeting because, as an individual's achievements approach a particular goal, he begins raising his goals or directing them elsewhere. For example, a person's goal to limit his smoking to three cigarettes per day, when achieved, may be adjusted to stop smoking altogether. Or when hungry, his immediate goal is to eat. Having satisfied that goal, he redirects his aspiration, say, to reading a book or completing some office homework.

The equation also has long-range application. For example, a high school graduate's goal may be a bachelor's degree in electrical engineering. When he gains the satisfaction of attaining this goal (usually before graduation day), he sets a new goal—perhaps to get a job as an engineer in a certain company or to pursue an advanced degree. While working toward these long-range goals, he has, of course, set and achieved (or set, failed, and readjusted) many short-range goals, such as pledging a particular fraternity, earning an "A" in calculus, getting a passing "D" in history, taking a particular girl to the Spring dance, winning a tennis match, and learning to parallel on snow skis.

Victor Vroom[11] defines the attractiveness of a particular goal as a function of the net desirability of any number of consequences of its attainment. Further, the level of motivation with which an individual pursues a goal is a function

[11] V. H. Vroom, *Work and Motivation*, John Wiley & Sons, Inc., New York, 1964.

of the net value of the anticipated consequences of having achieved the goal.   Thus, a high school graduate may volunteer for an undesirable military assignment to earn educational assistance to get a college education which he values more than he dislikes the military assignment.   While performing his uninspiring military duties, he may unexpectedly discover an opportunity to apply for a military-sponsored educational assignment which is relevant to his professional interests and which will also grant him college credits.   He pursues the assignment with newly kindled enthusiasm, for now the presumed negative value of the military assignment itself has the potential of leading to a positive outcome to be coupled with the positive value of his long-range college plan.

Goal setting problems often arise from supervisory assumptions that people have little interest in organizational goals.   This view stems largely from the fact that many supervisors, not understanding the characteristics of meaningful goals, assign only tasks or duties to people.   Tasks and duties must be performed, of course, to achieve the supervisor's goals, but the supervisor often has little success in getting others to perform tasks and duties with the enthusiasm he feels toward his goals.   The corrective mission, in this case, becomes one of helping the supervisor understand the need for a framework or hierarchy of goals within the organization that is meaningful at any level.   Tasks and duties take on meaning when those who perform them can relate them to a meaningful chunk of a hierarchy of goals and see the relationship between their efforts and the attainment of these goals.   Job efforts have maximum meaning when they include the planning and measurement of achievements in accordance with the plan, do, and control concept described on page 7-20.

Finding a Goal Setting Arena.   Most people ultimately find the arena in life in which they can achieve goals, usually by trial-and-error processes.   The new college graduate entering industry usually finds ever-blossoming opportunities for setting and achieving increasingly higher goals.   Because his college degree opens most doors, his success in the company is largely a function of his ambition and talent.   His economic maintenance needs are routinely met through his expanding compensation package and are not his primary concern as long as his broadening professional role in the organization provides growth, responsibility, and recognition for the achievement of challenging goals.   Indeed, he may become so engrossed in self-actualizing experiences on the job that he gradually, voluntarily, and sometimes unconsciously disengages himself from outside interests to devote ever-increasing amounts of time and energy to the job which is *his* arena for goal setting and achievement.

But his former high school classmate who entered industry without a college degree usually finds the industrial workplace only temporarily rewarding.   Having escaped parental control and satisfied his immediate maintenance needs, he casts about impatiently for new opportunities, only to find them reserved for the degreed newcomer who is often younger and less experienced than he.   His alternatives are few, difficult, and not often satisfying.   He can do double duty and acquire the requisite academic credentials by attending classes after working hours; he can earn advancement through sheer talent, initiative, and perseverance; or he can abandon the organization—physically or mentally.

Most who enter industry without benefit of a college degree stay on the work force physically; job hopping occasionally; preoccupied during duty hours with wages, hours, and working conditions; finding and gradually accepting their identity through their work roles and memberships in peer groups.   But their compliant performance of simplified tasks is undemanding of their talents, and their interests and energies are channeled outward to *their* arena for self-actualization—off the job.

Off the job, the individual may satisfy growth needs through travel, reading, Toastmasters, stamp collecting, bird watching, ham radio, technical group memberships, and miscellaneous intellectual pursuits.   Achievement opportunities abound in a variety of activities such as bowling, fishing, hunting, skiing, flying, sailing, painting, stock speculation, photography, mountain climbing, linguistics, ceramics, and home workshop projects. Vicarious achievement is experienced through spectator sports, movies, television, and reading. He may experience a sense of responsibility as a scoutmaster or Sunday school teacher, and through school board membership, public office, PTA leadership, and participation in social action groups. Recognition needs may be satisfied through many of the foregoing, plus activities such as little theater, ballroom dancing, public speaking, competitive sports, and social group membership. While his degreed contemporary is gradually channeling more of his energy into the pursuit of organizational goals, the nondegree employee is more often disengaging himself from organizational commitment and finding expression for his talents in the pursuit of meaningful goals off the job.

Hence, the manager and worker go separate ways in pursuit of goals. The problem is circular and self-perpetuating. The manager finds he must do extra duty to make up for the lack of commitment to goals at the lower levels. But people at the lower level pursue goals outside the organization because managers have reserved the more interesting aspects of their jobs for themselves.

The Consequences of Overcommitment. It is not uncommon to encounter overcommitment to the job at the higher management levels—overcommitment in the sense that it deprives the individual of a well-rounded life of responsible citizenship. The avid corporate goal setter rarely has enough energy and time left over from his company duties to attend to his personal and professional growth, his family, and his community responsibilities. The more engaged he becomes in the pursuit of meaningful goals on the job, the more tunnel-visioned he becomes and the more disengaged he becomes from involvement with the members of his family, and hence the less opportunity he has to experience goal setting within family and community units. Moreover, his family's familiarity with his vocational role is usually so fragmentary that it offers little opportunity for them to experience his achievements vicariously.

Members of the corporate goal setter's family often have life roles similar in many respects to the work roles of lower level workers in his organization. Like the traditional hourly-paid worker, they do not share his higher order corporate goals as a foundation for their goal setting. It would be unrealistic, of course, for them to expect to share his job goals unless they were also members of his organization. Each family as an organization does, of course, have unique goals which need the involvement, support, and commitment of all members of the family unit, including the person who earns the income to pay the bills. But checkbook benevolence is not an adequate substitute for personal participation. However, many business organizations seem to thrive, at least temporarily, at the expense of a community whose wives and children display symptoms of ennui and neuroses as a function of absentee husbands and fathers overcommitted to corporate goals.

The Consequences of Undercommitment. The other population, comprised primarily of the hourly, nonexempt, may have just as much imbalance in their lives. Their goal-setting efforts within the organization, because of their alienation, are often unofficial and aimed at counteracting limitations imposed by corporate goal setters. Because they have little opportunity to apply their talents in influencing company goals or in managing challenging jobs which support them, their talents find expression at work in pursuit of goals associated with wages, hours, and working conditions, which tradition and labor legislation have placed within their realm of jurisdiction. Higher wages, paid leave, broadened insurance, liberal retirement benefits, and shorter hours are only intermediate goals, of course, as they provide the means to achieve the goals which are attractive to them off the job. In addition, the attainment of economic goals affords the satisfaction and excitement of thwarting management authority and thus gaining temporary respite from a monotonous existence.

In the absence of challenging jobs, workers' goals become associated with a wide spectrum of maintenance factors extrinsic to work itself. They seem to have capricious and vacillating interests in issues peripheral to the job, such as an improved grievance procedure, revising work rules, changing the content of the company newspaper, use of the bulletin boards, more prestigious job titles, changing the cafeteria menu, getting the new typewriter or the chair at the end of the assembly line, avoiding the noisy work area, being located near the lunch facilities and rest rooms, and having convenient parking facilities. Goals often relate to social needs such as gaining acceptance and status within work groups, meeting friends at the coffee bar, joining a particular group at lunch time, organizing office parties, finding a congenial ride pool, planning recreational outings, and participating in special-interest group activities.

Most nonproductive or unofficial workplace preoccupations serve primarily to make time at work more bearable or to reinforce off-the-job pursuits. As noted earlier, their after-hours goals involve them in sports and outings, professional societies, community projects, civic undertakings, social affairs, and family activities. Efforts directed toward these off-the-job activities sometimes seem wasteful to the corporate goal setter. But off-the-job activities have greater potential than company activities for involving the family and hence for contributing to family cohesiveness and community stability. Thus wage earners, because of their freedom from the organization, as well as their greater numbers, have a disproportionately greater influence on the values and behavior patterns of a culture.

Unfortunately, ability to meet responsible citizenship roles is inversely related to the time available to do so. When the growth and responsibility needs of people are thwarted on the job, as is often the case with wage earners, they become culturally conditioned to irresponsible maintenance-seeking habits and attitudes which handicap them for effective leadership roles in their families and communities. In contrast, the corporate goal setter, whose leadership talents are often being challenged and developed through his involvement in responsible roles on the job, has the least time to utilize this competence in the family and community. Ideally then, a balance should be sought in which more of the corporate goal setter's leadership can be devoted to the community and family, and more of the wage earner's leadership talents can be developed through responsible goal-oriented job activities.

Requirements for Meaningful Goals. Meaningful goals can give meaning to almost any type of activity, on or off the job. Ideally, of course, work itself is intrinsically interesting. However, even distasteful, enervating, and humdrum activities are usually tolerated as long as they lead to meaningful goals. Otherwise, diapers would not be changed, dishes would not be washed, and lawns would not be mowed. Factors which give meaning to goals and thus inspire people to achieve them may be defined in terms of the characteristics of goals themselves and the impact that the pursuit or attainment of goals has on the goal setter.

Goals which have maximum motivational value are

1. Influenced by the goal setter
2. Visible
3. Desirable
4. Challenging
5. Attainable

and they lead to the satisfaction of needs for

6. Growth
7. Achievement
8. Responsibility
9. Recognition
10. Affiliation
11. Security

Company success can be a motivational goal in satisfying the above conditions, but only in terms of criteria meaningful to each jobholder. To the president, it might be return on investment, share of the served available market, or profit before taxes. To a brand manager, company success may be capturing a greater share of the market from competitive brands. An engineer's goal might be a technological breakthrough needed to solve a product performance problem. Members of an assembly line contribute to company success when they are producing units to meet quantity and quality goals which they themselves have set. To the extent that each of these goal setters identifies his goal with company success and his goal meets the criteria enumerated above, then it can be said that company success is a meaningful goal.

Consider, for contrast, goal setting through the task force approach versus traditional supervisory goal setting. In the task force process, the supervisor convenes the operators in a conference room, shares cost information with them, explains the company commitment to meet a goal established by the competitive bidding process, and asks for their ideas. Suggestions obtained from the operators through this conference approach lead to process improvements, greater cooperation and commitment, and the attainment of goals which typically surpass the competitive bid requirements.

Prior to the group problem solving, goal setting process, the operators had been assigned to work stations on a line balanced by engineers and were given job instruction by supervisory and engineering personnel. Attempts to improve the line through better engineering and to motivate the girls by persuasion and enforcement of standards did not evoke the desired performance ultimately achieved through

Characteristics of meaningful goals	Task force goal setting	Supervisory goal setting
1. Influenced by goal setter	Operators participate with supervisor and others to help set goals based on analysis of problems.	Operators receive goals from supervisor, usually in terms of engineered standards.
2. Visible	Operators see goals as customer goals in terms of quantity, quality, and delivery dates.	Operators see performance goals in terms of standards established by management.
3. Desirable	Achievement of goals desirable for meeting personal commitments and to earn merit pay.	Achievement of goals desirable to earn merit pay and to avoid punishment.
4. Challenging	Both mental and physical challenges to raise and achieve self-established goals.	Physical challenge to meet quantity and quality goals, and sometimes mental challenge to beat the system.
5. Attainable	Attainability determined by group problem solving, consensus, and cooperation.	Goals usually established at levels where a majority can meet standard.
6. Growth	Operators broaden perspective, develop problem solving skills and mature attitudes.	Little on-the-job learning opportunity beyond immediate job skills.
7. Achievement	Achievement motive recurrently stimulated and satisfied by goal setting.	Achievement motive satisfied by attaining and exceeding standards or by thwarting system.
8. Responsibility	Responsibility for the project results naturally from voluntary commitment to goals.	Responsible for following instructions and being loyal to the supervisor and the company.
9. Recognition	Recognition from within and outside the group for attainment of goals, and from prestige of group membership.	Praise from supervision for high performance, and acceptance from peers for supporting unofficial goals.
10. Affiliation	Joint stake effort increases interpersonal competence and group cohesiveness.	Social needs satisfied through informal cliques.
11. Security	Feelings of self-confidence fostered by knowledge and freedom.	Feelings of insecurity fostered by unpredictability of the job situation.

FIG. 2-12.   A comparison of goal setting techniques on the assembly line.

the group process. Analysis of these two processes in terms of eleven criteria of meaningful goals is presented in Figure 2-12 and shows that the task force excels the traditional approach on every point.

The eleven criteria of meaningful goals are not presented as an exhaustive list of factors which can give meaning to goals, but rather as ideal characteristics and consequences of work itself. The relative importance of factors varies among individuals and may fluctuate for any given individual. Moreover, even meaningless work satisfies needs for money and other goals which may be extrinsic to the job. Most of the characteristics in Figure 2-12 can be and often are satisfied by off-the-job goals. In addition, it must be recognized that some goals are desirable because they represent stepping-stones to other goals.

HELPFUL SYSTEMS[12]

A management system is a process of people interacting to apply resources to achieve goals. Management system designers tend to place major emphasis on hardware and software (technology), but the primary emphasis must be on the human factor. Computers, machines, buildings, materials, and money lie idle and lifeless in the absence of humans. Hence all management is the management of human effort. The materiel with which people interact may be organized to facilitate their efforts and to inspire their commitment, or it may be organized in a way that impedes their efforts and evokes their opposition.

When people encounter difficulty in the pursuit of goals, there is a tendency to blame the system. For example, problems encountered in terminating a book-of-the-month club membership, in changing a mailing address, in getting a charge account error corrected, in clearing an expense account, or in getting an accounts receivable balanced are usually attributed to the system. Physical aspects of systems are convenient scapegoats at the "customer-complaint desk," and system administrators (users) often attribute their own limitations to restrictions imposed by the system. Attempts to remedy a system, of necessity, lead back to the system designer.

Role of the Management System Designer. The system designer can often defend his applications of software and hardware, and then assert that the system failed only because people misused it. Within that point of view lies the crux of most management systems problems.

The system designer is correct in diagnosing system failures as human failures. But he usually fails to recognize that his responsibility embraces the human factor—that the system designer's role is one facilitating human processes, and that helpful systems function as extensions of man, not man as an appendage of systems. Furthermore, the system designer often overlooks his responsibility for seeing that the system user is adequately trained to administer the system. System failures sometimes result from designer permissiveness in allowing the user to divest himself of the responsibility for helping design the system. Because of sheer job pressure, the user may welcome the staff man's take-over.

Sometimes, systems users try to participate in the design of their systems, and the systems designers may even urge them to do so. But the system user frequently encounters the same problem in talking to the system designer that the foreman sometimes encounters when he tries to talk to the personnel psychologist—in neither case is the staff man's jargon fully understood. The system user may be as confused with the designer's use of terms like SYSGEN, time sharing, syntax-directed, bombout, and real-time as the foreman is with the psychologist's use of emotional stability, exophoria, IQ, manic-depressive, and ego drive. In both cases, it is the staff man's dereliction in failing to adapt his terminology to the boundaries of his customer's language.

A system user can no more divest himself of responsibility for system design

[12] Much of this discussion is abstracted from documents coauthored with Charles L. Kettler, Coordinator of Texas Instruments' Management Systems Development Committee, and A. Graham Sterling, Manager of Control and Administration in TI's Materials Group.

than the foreman can delegate the handling of grievances and job instruction to the personnel department. When systems designers and personnel managers permit this type of disengagement, the results almost always are ineffective systems and inept foremen.

Informal Systems. Systems may be formal or informal, simple or complex, but all come into existence because of needs of individuals or groups at any level of the organization. For example, the office check pool (highest poker hand in paycheck serial number) is an informal and unofficial system that forms almost spontaneously and is perpetuated by a combination of social, financial, diversion, and risk taking needs of the members. Management may perceive the check pool as a violation of company rules on gambling and may try to develop a system for stopping it. Attempts to quash the check pool may be implemented through a system of posted notices, newspaper inserts, public pronouncements, and supervisory instruction, all reinforced by specific or veiled threats of punishment. But if the need to perpetuate the check pool is strong enough, or if the joy of circumventing authority is great enough, the pool system goes underground, thereby satisfying rebellion needs provoked by management edict, and perhaps increasing the system's value in satisfying social (group cohesiveness), diversion, and risk taking needs.

All systems are circular and give feedback to the user. Check pool members contribute their dollars and obtain feedback in terms of observed payoff to the highest poker hand. Attempts by management to intervene merely activate a countersystem for evading detection, which gives them feedback in terms of not getting caught; or if their system fails, getting caught. Similarly, management's control system for preventing participation in the check pool gives feedback to management in terms of official reports of conformity or violation. Feedback may or may not be valid.

Formal Systems Development. More formal and complex management systems are developed and refined through a continuous circular process that may be described in terms of the seven phases defined below. Although any phase, and particularly phase 5, may lead directly back to any preceding phase, a system's development generally follows a circular evolutionary process, as diagramed in Figure 2-13.

Phase 1. Goal Setting. Goal setting is initiated in response to a felt need to create a new system or modify an old one. Goals are expressed quantitatively and qualitatively in terms of end results desired and resources available.

Phase 2. Concept Study. The concept study is a systematic consideration of how to achieve the stated goals and what the long- and short-range impact of the system is on the user, on other systems, on uninvolved bystanders, and on the community in terms of social, economic, and legal considerations. The concept study ends with the selection of one of several alternative concepts to achieve the established goals.

Phase 3. Specifications. Specifications define the details of how the system will implement the concepts to achieve the goals. They are established in terms of costs, time schedules, personnel, machines, equipment, materials, facilities, responsibility, and evaluation criteria.

Phase 4. Implementation. The system is implemented by committing hardware, software, manpower, services, space, and budget; designing trial application; and defining error signals.

Phase 5. Trial Applications. Trial applications are made with real or simulated data on representative samples of users; errors are corrected; the system is refined; and management and user commitment is confirmed. Phase 5 may lead directly back to any previous phase.

Phase 6. Installation. The system is installed by instructing users, transferring system management to users, informing affected publics, establishing review schedules, and monitoring initial applications. Systems with potential for broad organizational application should be "fanned out" immediately to other operations to maximize system payout.

Phase 7. Evaluation. The system is evaluated by measuring performance against goals, impact on other systems, deviations from design, and goal adjustments required.

Characteristics of Effective and Ineffective Systems. This evolutionary process,

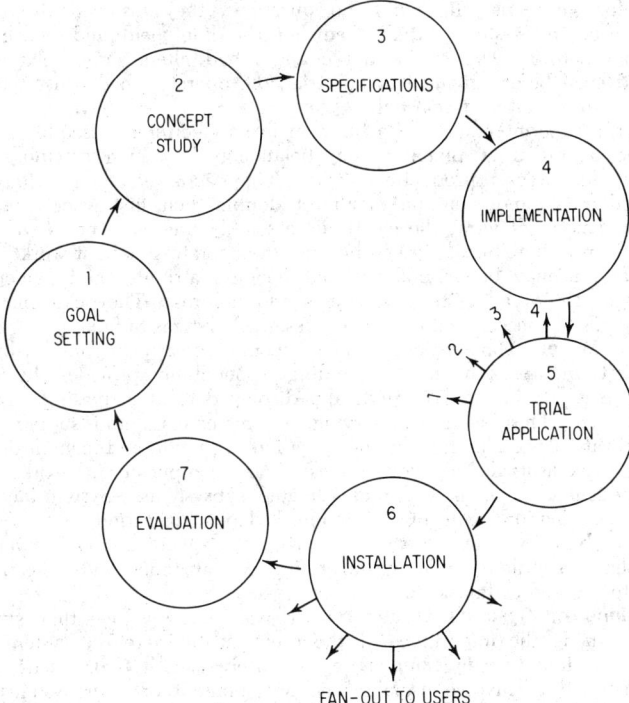

Fig. 2-13. Seven phases of system development.

perpetuated by feedback from the users, represents an ideal model of an effective, self-correcting management system. A management system is considered effective when the people whose job performance is influenced by it:

1. Understand its purpose
2. Agree with its purpose
3. Know how to use it
4. Are in control of it
5. Can influence its revision
6. Receive timely feedback from it

Stated negatively, as a basis for understanding system failure, it may be generalized that a management system is not effective when the people whose job performance is influenced by it:

1. Do not understand its purpose
2. Disagree with its purpose
3. Do not know how to use it
4. Feel they are unnecessarily restricted by it
5. Feel it is hopeless to try to change it
6. Receive inadequate feedback from it

The Key Role of the User. It is noteworthy that the conditions for effective and ineffective systems noted above are almost exclusively functions of systems users' attitudes and perceptions. This observation is illustrated by comparing the behavior of two groups of machine operators toward their assembly lines.

One group, in a paper carton factory, was idled for a few hours, with pay, while industrial engineers introduced improvements into their line. The operators clustered near the Coke machine, laughing, drinking Cokes, and smoking. When the engineers completed the installation, they briefed the operators on the changes and asked for questions. Receiving no questions, they assumed the installation completed. But the system actually reduced the line yield and hence was less effective than before. The engineered changes had altered role relationships on the line, and even before giving it a fair trial, the operators had conspired, perhaps unconsciously, to make the system fail.

In contrast, a superintendent and a foreman in an electronics assembly department involved the operators in planning and balancing their own assembly line and setting their first week's production goals. They achieved their Friday evening goal on Wednesday and went on to almost double their first week's goal. From an engineering point of view, the electronic assembly line was not as well designed as the paper carton line, but it worked because the operators made it work!

In summary, it may be generalized that people's attitudes and perceptions are the primary causes of all systems successes and failures. They can enable poorly designed systems to succeed and cause well-designed systems to fail.

Lost in the Maze. Management systems become increasingly complex by a spiraling process. Computer technology, expanding exponentially, provides physical ability to store, retrieve, manipulate, transmit, and display data at increasingly faster rates and lower costs. This capability, serving the mainstream and support functions of an organization, accelerates the organization's growth and complexity, at the same time increasing its dependency on massive and complicated networks of systems and their meticulously detailed and coordinated subsystems. System-imposed conformity, in combination with bigness-induced depersonalization, evokes alienation and apathy or hostility. Thus systems themselves, depending upon how they were developed, have a primary role in generating the attitudes and perceptions that cause them to succeed or fail.

Qualifications of Systems Designers. It was noted earlier that systems designers, recognizing the importance of the users' attitudes to the system's success, have taken the initiative in familiarizing themselves with their users' operations or have attempted to involve users in the development of their systems. It was also noted that these cooperative efforts were often discouraged by the job demands of the user or the system designer's jargon. Moreover, it was pointed out that sensitivity to the human factor—the causes of commitment and alienation—is an essential ingredient for developing and managing effective systems. Thus it is apparent that systems designers must apply three types of competence in developing workable systems:

1. Technical knowledge of electronic data processing and related technology
2. Knowledge of the functions or operations to be served by the system, and the proposed system's relation to and potential impact on other systems
3. Sensitivity to the factors evoking human commitment and alienation in the development and adminstration of management systems

Management's Role. The hardware and software available for facilitating modern management systems have almost unlimited potential, to be exploited or limited by the people who interact with them. People who man the work force also have untapped potential, to be utilized or limited by the systems with which they interact. If synergistic relationships are to exist between people and their systems, human development and systems development must be guided by persons who understand both people and systems.

Thus, formalized and effective management systems cannot be established as the warp and the woof of the organization if systems development is relegated to staff functions. Many organizations try it, just as they try to delegate the planning function to staff personnel. Neither will succeed, of course, as planning and control functions are mainstream processes which must be managed by the persons who are to implement them.

CONCLUSION

A manager is responsible for managing materiel, manpower, and technology to achieve organizational goals. The industrial engineer's responsibility in managing management systems is to see that systems development is not an isolated, unco-ordinated, or unilateral process but rather a joint or task force effort appropriately balanced with systems technology, mainstream user participation, and human effec-tiveness expertise.

BIBLIOGRAPHY

Argyris, Chris, *Integrating the Individual and the Organization*, John Wiley & Sons, Inc., New York, 1964.
Argyris, Chris, *Organization and Innovation*, Richard D. Irwin, Inc., Homewood, Ill., 1965.
Bennis, Warren, *Changing Organizations*, McGraw-Hill Book Company, New York, 1966.
Blake, Robert, and Jane Mouton, *Corporate Excellence through Grid Organizational Development*, Gulf Publishing Company, Houston, Tex., 1968.
Fromm, Erich, *Escape from Freedom*, Holt, Rinehart & Winston, Inc., New York, 1941.
Glasser, William, *Reality Therapy*, Harper & Row, Publishers, Incorporated, New York, 1965.
Hall, Jay, Vincent O'Leary, and Martha Williams, "The Decision Making Grid: A Model of Decision Making Styles," *California Management Review*, Winter, 1964.
Herzberg, Frederick, *Work and the Nature of Man*, World Publishing Company, Cleveland, 1966.
Likert, Rensis, *The Human Organization*, McGraw-Hill Book Company, New York, 1967.
Maslow, Abraham H., *Toward a Psychology of Being*, 2d ed., D. Van Nostrand Company, Inc., Princeton, N.J., 1968.
McClelland, David, *The Achieving Society*, D. Van Nostrand Company, Inc., Princeton, N.J., 1961.
McGregor, Douglas, *The Professional Manager*, McGraw-Hill Book Company, New York, 1967.
Myers, M. Scott, *Every Employee a Manager*, McGraw-Hill Book Company, New York, 1970.
Paré, John, "What's Your Power Structure?" *Canadian Business*, April, 1968.

Chapter **3**

Human Factors Engineering

RICHARD G. PEARSON

**Professor of Industrial Engineering and Psychology,
North Carolina State University, Raleigh, North Carolina**

Everything designed is ultimately for the use of or by man. To ensure safe and efficient use of equipment, engineering design and layout should take into account man's capabilities and limitations. Human factors engineering embraces a broad approach, with focus upon man's sensory input, motor response output, and information processing characteristics and their interaction with environmental conditions and system function requirements.

This chapter discusses a man-machine systems design approach in which man is given early attention in systems development and production plant design. System functions must be allocated effectively to man and machine. The human factors engineer's task is to define human performance requirements and ensure that these are not compromised by improper equipment design and layout. Thus, specific attention is given to design "from the man out," with focus on traditional human factors engineering topics such as (1) display and control design, (2) human physical characteristics, (3) environmental factors, and (4) design for maintenance. Also discussed are contemporary approaches to injury control and product design, as well as some misconceptions regarding fatigue and environmental stress.

THE HUMAN FACTORS ENGINEERING APPROACH

Industrial engineering has been concerned with the relations between men, machines, and materials, and accordingly has evolved effective methodology for dealing with problems of operator efficiency and workplace layout. A field born in the late 1940s, human factors engineering has potential for complementing traditional industrial engineering approaches.

Characteristics of the Field. The human factors engineering field is characterized by concern with the efficient and safe utilization of man in man-machine systems, with emphasis on the selection, design, and arrangement of system components so as to take into account both man's capabilities and his limitations. In his approach to man-machine systems problems, the human factors specialist tries to develop a "global view" of system requirements so that system efficiency is optimized at the interface of man, the equipment with which he works, and the work environment. In achieving such an "impedance match," the specialist may recognize accident probability and thus will incorporate design safety and injury control criteria into his overall approach to the problem.

Contrasts with Industrial Engineering. A distinctive aspect of the field involves the vast interdisciplinary body of knowledge and methodologies considered by the human factors specialist in his approach to man-machine systems problems. The human factors approach also places considerable focus on such skills as psychomotor coordination, complex information processing, and decision making. Interests here have long been identified with the domain of the experimental psychologist. Thus, as systems became more complex and demanding, thereby provoking an ever-increasing requirement for precise and timely responses on the part of man, it was a natural outcome that experimental psychologists should step forward with their interest in and knowledge of human skilled performance. Indeed, for many years, the great majority of the members in the Human Factors Society identified principally with the discipline of psychology. In 1968, society membership (over 1,600) was comprised of over 60 percent psychologists, as contrasted with about 12 percent engineers (only a little over 2 percent called themselves "industrial engineers"). However, a survey of activities in human factors[1] indicates that university departments of industrial engineering are a principal source of training in human factors, second only to psychology departments.

Education and the Communication Gap. Although psychologists have been a dominant force in human factors, the challenges of the field demand a departure from traditional training in psychology as well as in industrial engineering. In both industry and government, the human factors specialist commonly works with other engineers, often under the supervision of an engineer. In his position, he needs to know something about human sensory capabilities, learning processes, man's ability to process information and solve problems, characteristics of his motor responses, and performance under stress. Such subject matter is, with rare exception, taught and written about by people with training in experimental psychology. Yet the specialist working on a man-machine systems design problem must have some "hardware" sophistication; he is working in the world of the engineer and must be able to communicate. It is an unfortunate fact that much of the human performance data in the psychological literature are not readily translatable or interpretable for engineering design purposes. Although there are human engineering guides available to consult, it is often difficult to get a "quick and dirty" answer to a specific design question involving human factors considerations.

History of the Field. History provides many examples where little attention has been given during systems development to human functions and requirements. World War II brought the problem into the limelight when considerable and critical human error became commonplace in complex weapons systems, despite efforts to select and train effective personnel. Typically, human error was traced to (1) failure on the part of design engineers to consider man's limitations, (2) placing too much of a demand upon the operator, or (3) poor or improper equipment design and layout, which in effect set man up to make errors.

[1] Jack A. Kraft, *Human Factors and Biotechnology: A Status Survey for 1968–69*, Lockheed Missiles and Space Company, Report 687154, Sunnyvale, Calif., April, 1969.

Taking note of the early studies of human error, some corporations, engaged in the development of complex weapon systems that would demand skilled operator performance, recognized that human factors had best be considered from the start of systems development. Those who did not adopt this philosophy often were rudely reminded of human factors design deficiencies by the pains of retrofit necessary to meet contract requirements. Accordingly, there arose a demand for personnel with a broad outlook who could give attention to man's biological, psychological, biomechanical, and anthropometric characteristics. The emergence of a human factors team approach involving design engineers, experimental psychologists, applied physiologists, and physical anthropologists reflects the broad range of problems to which the field has application.

The Systems Concept. When a human factors engineer talks of systems, he generally means man-machine systems. Such systems can vary in size and complexity. An automobile and its driver are a man-machine system. Highways, automobiles, and drivers are also a system, and are in turn part of a much larger, national ground transportation system. The machinist at his lathe is a system; so is the housewife at her stove. Manufacturing plants and production lines are systems.

Roles of the Human Factors Engineer. The specialty of the human factors engineer is his approach to man-machine systems design, development, and evaluation. This approach puts prime emphasis on analysis and allocation of system functions to ensure the effective use of man in meeting system requirements. As Bennett notes[2] the human factors engineer's responsibility is not to develop efficient systems per se, but to fight for human efficiency, because he may be the only person who will. The adversary role which Bennett identifies is one that often has to be accepted, but can be difficult to maintain over an extended period. The project engineer will frequently be applying pressures to compromise optimal human requirements to meet systems specifications involving criteria of size, weight, reliability, costs, and the like.

Apart from playing the adversary, other roles are open to the human factors specialist. He may, for example, conduct laboratory research to ascertain the effects of certain system parameters or design concepts upon human performance, when such effects cannot be readily predicted. As system prototypes are developed, the specialist may conduct special evaluations involving field tests or laboratory simulation to verify that the design, under operational conditions, performs in accord with system requirements. In unique situations, he may design new tools or job aids required by operator tasks. Finally, the specialist can serve as an expert consultant when problems of work inefficiency emerge in existing systems.

Man as a System Element. To consider human performance in proper perspective, it is first necessary to understand exactly how man relates to and interacts with other system elements. A traditional way of representing man's role is that of an operator in a closed-loop system, as shown in Figure 3-1. Outputs of the machine part of the system are presented to man, generally in the form of visual or auditory displays. These inputs to man are then processed and responded to, the resulting outputs constituting control operations by man, which in turn affect the operation of the machine and ultimately, of course, the resultant display back to the operator. The figure portrays a simple system, but it is important to recognize that in larger systems it is possible to talk of several interacting control systems and display systems. The output from one control system may, for example, be the input to a second control system, and so on. In some systems, then, the operator can be quite remote from the machine parameters he is required to control, and feedback (knowledge of system response) can accordingly be delayed. This is important to recognize, because man should have prompt knowledge about the results of his actions.

Not to be overlooked in Figure 3-1 is the effect of stress on man. System

[2] Corwin A. Bennett, "The Adversary Role," *Human Factors Society Bulletin,* May, 1969.

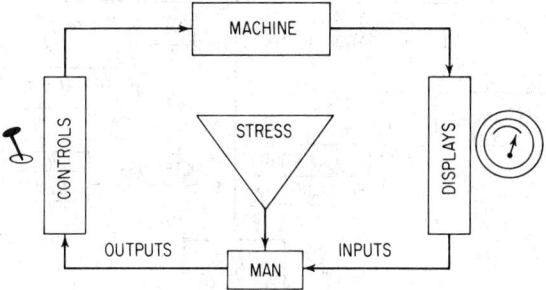

Fɪɢ. 3-1.   Man in a closed-loop system.

performance is not only a function of man and machine but also a function of the environmental context in which the system must operate.   An expanded categorization of factors that determine man-machine system efficiency appears in Table 3-1.   The items listed can serve as a guide to factors that must be considered in the development of a new system.   They also can be used as a checklist in evaluating an existing system which is functioning inefficiently.   Focusing on man, the analyst might ask whether operators have been effectively selected and trained for the job.   Satisfied here, he may then take a close look at system design and operation.   Has equipment been designed and arranged in accord with good human factors practice?   Are adequate job aids and communication channels available?   What about the work environment—illumination, noise, ventilation, and the like?   Assuming that the man-machine system design meets these criteria, then what could be wrong?   The answer could lie in the third category, where certain transient or undesirable factors creep into the picture to compromise systems efficiency.   We all know how a new manager can shake up the system!

The "black box" concept of man conveyed by Figure 3-1 implies that man can be treated as a simple input-output subsystem—but this is far from true. Depending upon how inputs are presented to man, there are several stages of processing that can be involved within man's sensory, nervous, and motor (muscular)

TABLE 3-1.   Determinants of Systems Performance

A. Personnel capability
    1. Selection
        a. Ability, experience
        b. Personality, attitude, interest
        c. Physical characteristics
    2. Training
B. The system
    1. Design
        a. Equipment design, displays, and controls
        b. Workplace layout
        c. Job aids, tool design
        d. Environmental considerations
    2. Operation
        a. Work load
        b. Communication channels
        c. Maintainability
C. Modifiers of context
    1. Supervision
    2. Administrative procedures, policy; rules, laws
    3. Health and morale
        a. Drugs, diet, sleep
        b. Family problems as source of mental distraction
    4. Work schedule and rotation
    5. Unusual environmental stress

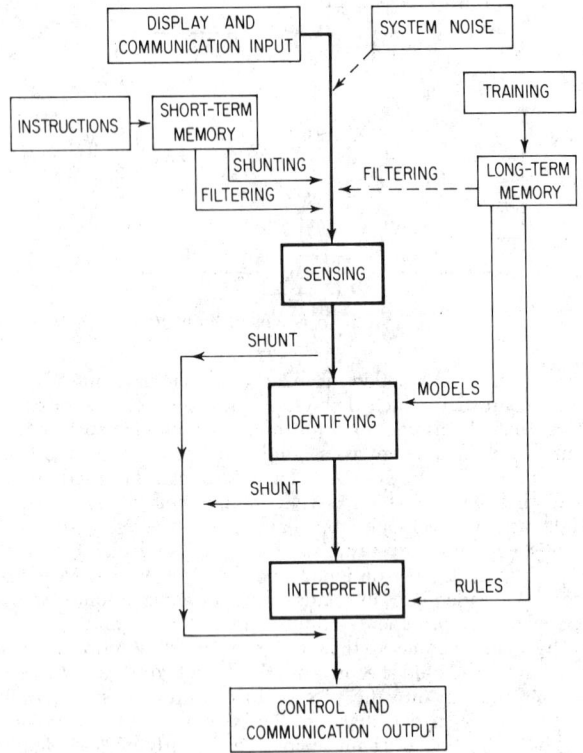

Fig. 3-2.   Model of man as an information processing system.

systems before a response (output) is made. Figure 3-2 portrays a model with three levels of human information processing. Each level has certain unique characteristics and varies in complexity. A brief summary of these characteristics is offered here, and a more complete discussion of the subject can be found in Gagné. (*Note:* References not footnoted are found in the appended bibliography.)

Sensing, the most elementary function, refers to the attention given to the existence of or to change in physical energy. It is associated with binary choice or go–no-go situations in which the operator is required to sense (hear, see) a particular signal or event. Filtering conditions determine the attention of the operator and can relate to whether an event is detected or missed. These may be derived from instructions (from the manager) stored in short-term memory, from attentional cues, or from scanning patterns stored in long-term memory as a product of training. It is hoped, of course, that system noise (electronic interference, distractions) does not preclude there being information to receive.

The more complex levels require greater amounts of brain processing time. In *identifying,* the operator must discriminate among different input qualities or distinguish the presence or absence of one input feature against a background of additional inputs. Thus, training of the operator in the task may be required so that models stored in long-term memory can provide standards against which inputs are compared.

Shunting results from instructions stored in short-term memory and determines the level of functioning at which an individual operates. *Interpreting,* the highest level of functioning, requires exercise of the two preceding functions and puts emphasis on the meaning or effects of inputs rather than on their appearance (as in identification); further, it offers courses of action based on mental activity (thinking). Again, training is needed to ensure that rules, stored in long-term memory, can dictate alternative responses. When a number of interpretations are made in sequence, in accord with feedback from successive outputs, the process is termed "decision making."

Characteristics of the three levels relate to other considerations in systems performance in addition to training. One purpose of a selection program is to screen out operators who fall below a given standard applicable to a particular sense mode; another is to identify individuals who have the capacity to acquire and store models and rules needed in higher level processing. Clearly written manuals and checklists come under scrutiny insofar as they establish sets for filtering and shunting or serve as an aid in maintaining retention of models and routines. Imagine how difficult it would be to analyze vacuum tubes with a tube tester without operational instructions and a list of settings for each tube.

The concepts of man as an element in a closed loop and as an information processing system constitute a useful base from which the human factors engineer can attack systems design and evaluation problems. Certain system functions are best performed by man, while others are best performed by machines. A clear conceptualization of man's role in systems is thus a necessary starting point.

THE SYSTEMS DEVELOPMENT CYCLE

Understanding of the concept of the systems development cycle is a core requirement for the human factors engineer. Although it is recognized that few industrial engineers commonly play the role of human factors protagonist in systems development (in the prevailing aerospace industry context), nonetheless, when a new manufacturing plant or production system is regarded as a man-machine system, the concept has direct relevance to the engineer involved in plant design and layout.

It is important to note that, as systems development progresses through a series of stages and sequential processes, different activities on the part of the human factors engineer are required. A simplified conceptualization of this is shown in Figure 3-3. There are iterative characteristics to the cycle in that performance requirements undergo refinement as the design freeze approaches and as equipment designs are modified at various points following analysis or evaluation of equipment drawings, block diagrams, models, mockups, and prototypes. The basic parallelogram character of the figure is intended to emphasize the relative importance and

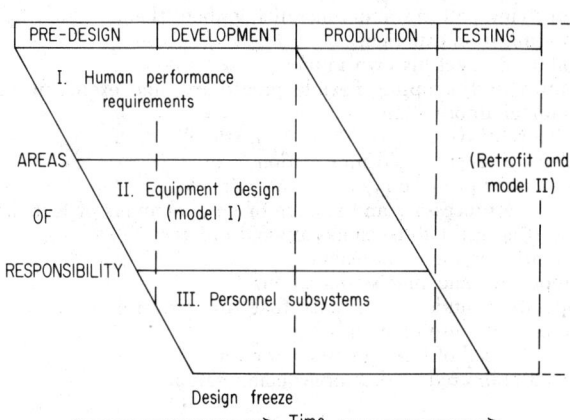

FIG. 3-3. Human factors activity in the systems development cycle.

sequential nature of the three major areas of work during the various stages of the cycle. The same parallelogram (cycle) can be merged with the basic figure to indicate activities associated with the development of a similar system (model II).

With large and complex systems, the cycle may take several years, and the concomitant variety of challenges the human factors specialist must face can require a considerable and versatile expertise. A most lucid and definitive treatise on human factors in systems development is found in a monograph edited by Folley.[3] Gagné is also a reasonable source. Alternative sources with relevant discussions, but less comprehensive on this specific topic, include McCormick, Meister and Rabideau, Morgan et al, and Woodson and Conover.

Human Performance Requirements. The identification and analysis of human performance requirements begin in the predesign stage of the systems development cycle and continue into the development stage. Three phases of application are associated with the analysis activity: (1) function analysis, (2) task derivation, and (3) task and skill analysis.

Function Analysis. The function analysis phase begins with a critical assessment of the functions to be performed by the system. The goal is to allocate functions to man, machine, or man-machine combinations so as to optimize overall systems efficiency. As a first step in function allocation, information on system operational requirements is sought. This may be obtained from a variety of sources, including requests for bids, requests for proposals, military handbooks and specifications (MILSPECS), market analyses, and evaluations of existing (or competitive) systems. Next, operational requirements are broken down into system functions, that is, capabilities for converting inputs to outputs. Such functions are often identifiable in system block diagrams and usually take the form of a verb, such as sense, receive, amplify, filter, integrate, monitor, transmit, process, store, record, measure, program, display, print, regulate, activate, sample, compare, compute, inspect, or verify. The final step involves assignment of functions to men and equipment. This requires collaboration between the design engineer and the human factors specialist in evolving tentative allocations, in evaluating them in terms of capabilities basic to system performance criteria, and in choosing among alternative man-machine combinations so as to maximize overall system effectiveness.

A natural question that arises at this point relates to which functions are performed better by man and by machine. The following "generalizations" can serve as a guide.

Man is better or unique:
1. At discriminating relevant from irrelevant signals
2. At innovation in problem solving
3. In reasoning inductively
4. In perceiving patterns and generalizing about them
5. In profiting from experience
6. In ability to select his own inputs
7. In improvising, adopting flexible procedures, and exercising judgment based on minimal information
8. In being sensitive to a wide variety of stimuli
9. In selective recall of old information

Machines are better or are unique:
1. For routine processing and storage of large amounts of facts and details
2. For exerting great force, smoothly and precisely
3. For monitoring men and machines
4. For repetitive and precise operations
5. In operating under conditions that are stressful or intolerable to man
6. For rapid response to signals
7. For rapid recall of large amounts of data
8. For accurate, rapid, and complex computations

[3] John D. Folley (ed.), *Human Factors Methods for System Design*, American Institute for Research, Report 290-60-FR-225, Pittsburgh, Pa., 1960.

9. For originating and sustaining many actions concurrently

10. In sensing stimuli beyond the range of human sensitivity

Certain physical characteristics of man obviously are also limiting factors in design. Direction and extent of limb movement is one example. Man can exert but limited amounts of force, with limited precision, for limited amounts of time. There also are limitations of attention and discrete time delays in response to signals which have a basis in man's sensory, muscular, and nervous systems.

The human factors specialist must, of course, be the defender of man's limitations at the time of function allocation, for once equipment design is specified, the character of human tasks is predetermined. Too often, the task has been to fit man into the system, rather than to design from the man out in accordance with his capabilities and physical characteristics.

Refinements of function allocation can be brought about by further analyses. These are discussed in Folley[4] as mission analysis and contingency analysis. The emphasis in mission analysis is on the sequence of tasks performed in a given environmental context over a work cycle of the system. Time line analyses of operator functions, which should be familiar to the industrial engineer, are used here. These may identify periods of operator overload that may require new or revised allocations of function. Mission analysis may also identify environmental stresses which need to be considered if operator efficiency or health is not to be impaired. Contingency analysis takes this last consideration a step further by appraising the effects on system performance of exposure to nonroutine operating conditions, atypical environments, or equipment and personnel malfunctions (accidents, failures).

During the process of function allocation, the human factors specialist should also evaluate systems design for good maintainability and safety design practice. These considerations should not be treated as an ancillary part of the process, because their neglect early in the cycle may eventually lead to overall system deficiencies and accidents, as well as the possibility of costly retrofits.

Task Derivation. Task derivation begins where the function allocation phase ends. Functions to be performed by man are considered in relation to the equipments with which he will work, the goal here being the identification of specific elements of behavior (tasks) required in the operational context. In early stages of the systems development cycle, the definition of tasks can hardly be specific; hence the process is continuous as the system evolves and design data become available. The job of task derivation is easier, of course, when off-the-shelf equipments are used in the design or when an existing, comparable system is available for comparison. As a system evolves, the industrial engineer who follows good human factors practice will also want to derive tasks that are specific to maintenance functions.

Task and Skill Analysis. The final phase of activity in determining human performance requirements is task and skill analysis. The goal here is a comprehensive delineation of the physical and psychological characteristics required by all system tasks, with accompanying specifications of the levels of skill and knowledge required. In the early work of task analysis, again the industrial engineer will find use for traditional tools, such as time line analysis, as functional and temporal relations among tasks are explored. The human factors specialist will be taking note of such things as the number and kinds of sensory inputs to the operator, the amount of time sharing of attention involved, the precision and force of actions required, the amount and level of mental activity, and the like. The work environment, equipments, and job aids for each task should be stated, for these elements will determine knowledge and skill requirements. Whether the task can be broken down into segments or has procedural characteristics is also important information. The time needed for performance and the frequency of occurrence of a task should be estimated, because these relate to the importance or weight assigned to skill and knowledge requirements. Finally, it will be necessary to identify task areas

---

[4] *Ibid.*

where probability of human error is significant. This may prompt redesign of equipment or reallocation of functions, if not requiring higher skill levels (reliability) of the operators.

Skill analysis, comparable to job analysis, focuses on the ability and knowledge requirements of each task derived. Personnel specialists may be valuable here as consultants. Each activity or response involved in a task or subtask should be identified and the skill level for each estimated. Skill rating can be an arbitrary process, but it is generally desirable to develop special category scales to rate task behaviors along such dimensions as difficulty, precision, dexterity, muscular effort or coordination, and sensory discrimination. In a like manner, the analyst may also rate knowledge requirements along dimensions of training and education levels, technical background, and professional specialization.

The specification of tasks and skills relates importantly to the later involvement of the human factors specialist with the systems development cycle. It says something about how equipment should be designed and arranged, what job aids will be required, and how personnel are to be selected and trained.

Equipment Design. Collaboration with design engineers in the design and layout of equipment is the second major area of human factors responsibility. Whether the system is a space vehicle, a stove, an automobile, or a production line, attention must be given to the interface between man and the equipment with which he interacts in performing his tasks. Having allocated certain functions to man, the specialist must next participate in equipment design planning to ensure the match necessary to efficient and safe task performance. Considerations may include display of information needed for decision and response, selection of controls for effective action, provision of necessary job aids, efficient layout of the workplace, and design for maintainability.

Design Trade-offs. As tentative designs are proposed, there will be inevitable conflicts in trying to satisfy both human factors and engineering design criteria. For example, the specialist may object to locations of displays or controls which are not optimal for human response; on the other hand, he may desire larger dial indicators than panel space will allow. Human factors design principles can conflict at times too, as, for example, when displays are grouped according to similarity of function and this precludes layout of displays on the basis of frequency of use or of criticality of attention. Criteria for safety, habitability, and survivability can also pose a challenge. In certain situations, if ideal comfort and safety criteria were met, it might mean that an operator's visual inputs and range of movements would be severely limited. If, for example, sufficient space for protection in crash deceleration were provided air passengers, then aircraft would contain only a few, widely separated seats.

Undesirable Design Characteristics. Analysis of accident data, case studies, and laboratory research has identified certain design practices which increase the probability of human error or inefficiency. These practices can be categorized and are briefly discussed here; specific design practices which attempt to avoid these are discussed later, beginning with the discussion of Display Design and Selection.

Designs that violate population stereotypes are a common source of error. A population stereotype is the way one expects a characteristic group of users to respond in a particular situation. Red and green are merely colors, but man has learned to expect that red means stop and green means go. Clockwise turning of a control wheel or moving a lever to the right is typically associated with vehicle movement to the right or an increase in the reading of a dial. Similarly, forward movement of a control or clockwise turning of a rotary control knob generally results in increases of power or current flow. Note, however, certain exceptions, such as in lathe operation or in plumbing practice, where a control (faucet) turned clockwise shuts down fluid flow. As a design comes into widespread use by consumers or operators, its operational characteristics can generate stereotyped behavior. Think of the confusion if either the direction of rotation or ordering of numbers were to be reversed in the design of the common circular telephone dial.

Industry-wide standardization of certain design practices, sometimes prompted

tors specialists, they are regarded here as beyond the scope of general interest to industrial engineers; their relation to the overall cycle of systems development should not, however, pass unnoticed. Both Folley[5] and Gagné are good sources for detailed discussion of the subject.

DISPLAY DESIGN AND SELECTION

Displays are necessary extensions to man's senses and provide both supplemental and prime information needed by human operators in making decisions and in effecting control responses. Human factors textbooks and guides commonly treat display systems before control systems insofar as it seems natural to regard input (to man) before output, or stimulus before response. So far as systems development is concerned, the problems of display and control design and selection should be treated within an overall workplace layout and display-control compatibility framework. There will be situations where system functions will predetermine the choice of certain control systems, which in turn will require specific types of information display for their effective operation. Conversely, there will be situations where system states (as displayed to the operator) will predetermine choices of response mode.

General Guidelines. A basic rule in display design is to present only that information required by the operator to make a decision or response—and no more. There are a number of types of displays from which to choose in terms of operator informational requirements and system functions. Overall workplace considerations also relate importantly to choice of design.

Conditions of Use. The distance and position of the operator relative to the display are important considerations. Obviously, as viewing distance is increased, the size of numerals and scale markings on dials will have to be increased. The Woodson and Conover guide provides useful data on selection of numeral or letter heights and stroke widths for viewing at distances up to 240 inches and under different levels of illumination; it also provides a useful table that relates dial diameter as a function of the number of scale indices required at a given viewing distance. For example, 150 graduation marks can be viewed effectively on a 2-inch-diameter display at 20 inches; at 6 feet, however, dial diameter must be increased to 8 inches for the same number of indices.

Optimal viewing distance for printed materials is typically considered to be about 16 inches, while for cathode ray tubes a range of 14 to 18 inches is acceptable. In workplace layout, consideration is generally given to operation of controls and viewing of several displays, so that common practice has led to specification of 28 inches as an optimal display panel distance. Accordingly, many of the handbook specifications for numeral size and scale markings are based on this viewing distance. The preferred angle of view is 90 degrees to the plane of the display. For the seated operator, panel surface is typically inclined 60 degrees from the horizontal; if the operator is standing, the recommended angle is 30 degrees; if he alternates between sitting and standing, the angle can be 45 degrees. Depending on the angle of view and on installation practices, glass covers on displays can be a source of problems involving glare, shadows, and parallax. Design can eliminate or minimize these problems.

Illumination is a specialty in itself, and appropriate guides should be consulted. Lighting should be sufficient and uniform for the task at hand without producing glare, and there should be suitable brightness contrast between task elements and the surround. Light sources should be directed at the task, not at the operator. Indirect lighting has advantages in providing evenly distributed light without glare and shadows. Distribution of light can also be effectively controlled through choice of colors on painted surfaces, such as walls or ceilings, and through use of barriers, such as "egg crates." Certain tasks, for example, radar air traffic control, require special illumination and special displays. Various kinds of colored lighting and

[5] *Ibid.*

indicator lighting (flood, indirect, edge, and rear) are available for special situations; their characteristics are discussed in detail in the guide edited by Morgan et al.

Types of Displays. There are various ways to categorize displays in terms of their purpose and use. Signs and labels are, in effect, displays of a static variety. Generally, displays are thought of in the dynamic sense of being subject to change over time. Most displays are symbolic in that they are an abstract representation of system conditions (rpm, pressure, temperature) or parameters related to performance (altitude, distance). Pictorial displays resemble the real world (a television display) or otherwise bear a schematic or geometric resemblance (a radar display). There are, of course, degrees of pictorial realism, so that some displays are, in effect, a combination of symbolic and pictorial, as in the case of many aircraft attitude displays. Pictorial displays typically are used for judging spatial relationships or for tracking a particular course or target; hence they convey information as well as suggest modes of response. Symbolic indicators are commonly of the so-called mechanical indicator variety and are used in presenting quantitative or qualitative information, or for check reading to determine that system function is in normal operating mode. Some displays are an integral part of a control, such as in some contemporary potentiometers and thumb-wheel selector switches, and are used for setting or entering values into a system. Sometimes several parameters can be integrated or combined for presentation in one display, ostensibly for human efficiency and savings of panel space. Higher cost, decreased reliability, and more difficult maintenance of integrated displays may offset advantages, however.

Visual Display Design Factors. Related to each of the uses and types of displays just mentioned are various design considerations that determine efficient information transfer.

Mechanical Indicators. Symbolic indicators are of three basic types: the direct-reading counter, the moving pointer-fixed scale, and the fixed pointer-moving scale. The counter is the display of choice for quantitative information, while the moving pointer display is generally preferred for qualitative information display, tracking, and check reading. Both are also useful for setting functions. About the only time a moving scale is preferred, in contrast to a moving pointer, is when a numerical value must be quickly read and this appears in a window segment of the display. In such an application, numbers should appear vertically and always increase in a clockwise direction.

The size of a dial-type indicator limits the amount of information that can be effectively presented. To extend the utility of a basic dial, one can go to multiple pointers or various types of subdials, but these approaches have their limitations. Generally preferred is a vertical-type instrument display which may take up less panel space but also prove more costly. Vertical or horizontal straight-scale, moving-pointer designs (so-called edgewise panel meters) are popular because of their space saving feature. They are good for check reading, especially when a series is aligned in a horizontal row; however, they have limited value in displaying any significant amount of quantitative information.

Considerable data exist in the human engineering handbooks regarding design of scale markings, letters, numerals, and pointers for the scale-pointer type of indicators. Only a brief overview of general considerations is presented here.

Capital letters without flourishes are preferred for display purposes. For black numbers and letters on a white background, stroke width should be about one-sixth to one-eighth of character height; height-to-width ratio of the character should be about 3 to 2. In selecting a scale, the designer will have to judge the reading precision required by the operator. He will then want to choose numbered interval values that will appear on the display, as well as the number of graduated marks to appear between major indices. The magnitude of indicated values should be arranged to increase in a clockwise fashion. Generally, there should be at least ½ inch between numbered scale points, with no more than 9 intervening graduation marks. Graduation intervals of 1 or 5 are preferred, as are major scale intervals of 1, 10, 100, and so on.

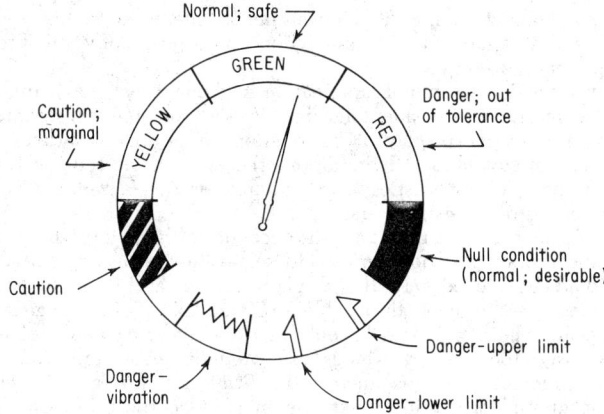

FIG. 3-4.   Examples of visual display zone marking.

Scales should be designed so as to avoid interpolation between graduation marks. Dimensions for the height, width, and spacing for major, intermediate, and minor graduation marks are found in standard human factors design guides. Whenever possible, numbers should be arranged on the perimeter or adjacent to the major scale points so that the pointer tip does not interfere with their reading. The pointer tip should be the same width as the smallest index, and have a minimum gap of ¹⁄₁₆ inch with the index. Pointers should be installed to avoid parallax and should be the same color as numbers and indices. Distortion of scale layouts for styling purposes (commonly noted on automobile speedometers) should be avoided.

Coding of zones or ranges on a dial can be a helpful adjunct, especially when certain critical conditions should be recognized quickly. Figure 3-4 is a composite of several zone markings commonly used to indicate various system conditions. Standard color codes are shown in the upper half of the figure; shape codes, shown in the bottom half, are recommended when ambient illumination does not permit color perception. Color coding is also useful in conjunction with multiple range meters; here, each scale can be color coded to permit association with discrete settings of selector switches where these are similarly coded.

Counter wheels avoid the problem of interpreting pointer-scale relationships and are the best choice for accurate, rapid display of quantitative information. Recommended height-to-width ratio of numerals for this indicator is 1 to 1. Black characters on a white background should be used for conditions of normal illumination; with low levels of illumination (darkrooms, radar rooms), a reverse combination should obtain. Counter numbers should move upward to reflect increasing values and should preferably snap into place rather than advance continuously. Window design should reveal only one number per wheel and avoid shadow effects.

Digital Readouts. Various kinds of digital readouts are commonly found in electronic equipment, such as digital voltmeters and high-speed counters. These include decimal arrays, matrix displays, register tubes (the so-called "Nixie" readout), edge-lighted plates, and rear-projector devices. The rear-projection readout is generally considered to be superior in terms of visual requirements, but does have the limitation of requiring occasional bulb replacement; the other displays either suffer in terms of character legibility or pose special viewing and illumination problems.

Pictorial and Cathode Ray Tube Displays. For industrial applications, care should be exercised in selecting those portions of a pictorial display that constitute reference and those that represent movement, so that the operator can comprehend natural spatial relationships. When the operator is interested in motions about himself, he and his environment should be represented as fixed elements, with objects represented as moving about him. If, however, he is interested in his own (vehicle) movement in space, again the space environment (geography) is fixed, but his position is represented to move.

Cathode ray tubes have common application in radar, sonar, navigation, electronic testing, and monitoring. Considerable data exist from military-oriented research that relate to the effective display of information on cathode ray tubes and can be summarized only briefly here.[6] In addition, there are advanced electronic display techniques that have characteristics worth noting for specialized applications, such as output devices for digital computers.

Accurate judgment of position and distance on cathode ray tube displays can be aided by incorporating various grids and scales on the display surface. Ideally, such aids should be etched into the display surface, as in the internal graticule found on oscilloscopes, to avoid the problem of parallax. Grids, range rings, coordinate markings, and the like should be only as elaborate as accuracy of interpolation requires; obviously, too many markings can produce some confusion in the task. Alternatively, scope size can be increased. Brightness contrast between signals, noise, and background is a critical variable in cathode ray tube displays. Special filters can be used to enhance signal visibility as well as to reduce problems of glare and reflection. Hoods also can be useful when ambient illumination and reflected light cannot be controlled properly. Finally, attention should be given to choice of phosphors. Some can produce considerable visual discomfort; some are better for viewing under low levels of illumination.

Indicator Lights, Signals, and Warning Devices. Lights and mechanical "flag" signals are useful to indicate status of a system or to provide cautions and warnings. The principal design requirements here involve (1) getting the attention of the operator and (2) telling him what is wrong or what he should do. Bright or large signals are generally better than flags, unless the latter are large and fast moving. Flags commonly are preferred for status (on-off) signals. Indicator lights can be used as status signals, but the problem often is that, with many signals, the operator may "tune out" their presence. An alternative here might be to use one conspicuous master warning light that indicates trouble, and then a remote, supplemental panel with signals, such as annunciator displays, to identify the locus of malfunction or to suggest a course of action. To attract attention, a warning light should be double the brightness of its background and be located within 30 degrees of the operator's normal line of sight. Lights flashing 3 to 10 times per second are good alerting signals but are sometimes found irritating by operators. Through appropriate grouping of lights or flags, it is possible to create patterns so that an alerting signal will stand out from the group. A final recommendation with indicator lamps is to ensure reliability with dual-lamp assemblies.

Labels. Good labels are a necessary adjunct for display identification and interpretation. Labels should be placed horizontal to the line of sight, close to the component being identified, and indicate parameters measured rather than the type of instrument. Clarity and brevity of wording are essential. Manufacturer's and trade names should be avoided on instrument faces. Labels for a number of grouped displays should be arranged consistently (preferably above) and in an orderly, logical pattern—as should be the displays themselves. Other general recommendations include: (1) use all capital letters; (2) avoid abbreviations; (3) letter width-to-height ratio should be 3 to 5; (4) stroke width should be about ⅛ character height; (5) for 28-inch panel distances or less, letter height should be at least 0.20 inch; and (6) print should be black on white matte background, except for situations where low levels of illumination obtain, and then the relations should be reversed.

Auditory Signals. Consideration is often given to the use of auditory displays (1) when the visual channel is overloaded, (2) when illumination or vibration levels

[6] A good discussion of display systems generally, as well as specialized applications, appears in H. R. Luxenberg and R. L. Kuehn (eds.), *Display Systems Engineering*, McGraw-Hill Book Company, New York, 1968.

are such as to preclude optimal use of a visual display, or (3) as an alerting cue or warning device. A unique characteristic of auditory displays is that they are omnidirectional; the operator need not face the display source. Obviously, the use of auditory signals, especially as warning devices, is limited when too many different auditory signals must be received and discriminated or when background noise is high, or masks signals, or otherwise injects confusion into the picture. Generally, auditory warnings should be at least 10 decibels above the ambient noise level. Auditory signals or cues often are preferred to speech if the message is simple and the operator has been trained in the code. For example, auditory codes varying in frequency, duration, and intensity can be used to direct operator attention to various aspects of a large panel containing large numbers of displays. They also are useful to indicate off-course or out-of-tolerance conditions in operator tracking tasks.

Buzzers, bells, chimes, sirens, and horns all have specific uses as warning devices. Characteristics to consider include the following: (1) for distances over 1,000 feet, use high-intensity signals less than 1,000 cycles per second; (2) if obstacles or partitions intervene, keep frequencies below 500 cycles per second; (3) to demand attention, use intermittent beeps or warbling sounds varying in frequency; and (4) choose signal frequencies to optimize differences from possible masking noises.

CONTROL DESIGN AND SELECTION

The selection of controls should be considered with regard to the functional requirements of the system. The amount of informational detail presented in a display commonly predetermines the choice of a control. Hence, for binary information, a simple activation control such as a push button will suffice. For control responses to qualitative information, discrete setting with selector switches or with levers having distinct (detent) positions is suggested. For control response to quantitative information, continuous controls such as knobs, wheels, or cranks can be chosen. Additional functional requirements that add refinement to control specifications include (1) the amount of force required by the response, (2) the speed and precision of operation required, and (3) desirable feedback to the operator concerning the results of his actions. Consideration should also be given in selection to the overall workplace layout and the total load on the operator. Available panel space may preclude certain choices. No one limb should be overburdened. Principals of motion economy should be considered. As a general rule, frequently used controls should be located between elbow and shoulder height whenever possible.

A wide variety of controls are available to meet these various specifications. Specific human factors considerations which relate to effective control operation are discussed next. Note should be made of a book by Kelley[7] which treats man's relation to automatic control systems. It further surveys in depth the relations between control and displays for manual operation, and includes material on unique control systems for specialized applications.

Design Factors. The following discussion applies to controls generally, although specific controls will receive some mention for purposes of illustration.

Control-Display Compatibility. A basic rule in panel or workplace layout is to make the relations between displays and their associated controls logical and clear-cut. In part, this rule is intended to capitalize on a concept discussed earlier—that of the population stereotype. For example, it is both logical and natural to expect that forward movement of a lever control such as a throttle leads to an increase in power and forward movement of a vehicle.

The problem of habit interference is brought about by (1) violations of population stereotypes or (2) lack of standardization in tasks which operators encounter in more than one situation. Figure 3-5 exemplifies a typical problem. Having been

[7] Charles R. Kelley, *Manual and Automatic Control,* John Wiley & Sons, Inc., New York, 1968.

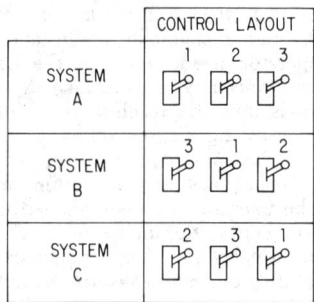

FIG. 3-5.   Habit interference in control operation.

trained in system *A*, one has to unlearn specific habits and acquire new ones if he later operates system *B*. Transferred still later to system *C*, the problem is further compounded. Under stress, it is not uncommon to operate the wrong control when one is frequently exposed to different control layouts as indicated. Persons who drive two different automobiles are frequently faced with a similar problem.

System response, system purpose, or display response can all be related to choice of control movement. Movement of lever controls up, forward, or to the right should result in increase in power output or temperature, or in activation or start as opposed to off or stop. Related displays should accordingly reflect increases with clockwise movement of pointers on circular dials, with upward movement on vertical displays, and with movement to the right for horizontal displays. Similarly, rotary controls turned clockwise should typically increase readings on displays in clockwise, upward, or rightward relations. One exception here is when a rotary control or wheel is located on the left side of a console or black box; then forward movement, although counterclockwise, should lead to an increase in display value. Push-pull controls often pose confusion in usage because forward motion is here associated with off or decrease rather than with application of power.

Correct direction-of-motion relationships can relate effectively to control operations by reducing reaction time, by reducing incorrect responses, and by facilitating speed and precision of adjustments.

Control-Display Ratios. The ratio of the distance of control movement to that of the moving element of the display or to movement of a vehicle under control is called the control-display ratio. It is comparable to the concept of gain in electronics. With a low *C/D* ratio, a small control action produces a large display movement and allows quick but gross adjustments. A high *C/D* ratio requires large control movements to effect a small display response and thus permits more precise adjustments. Both high and low ratios have their special applications, but generally, an intermediate, optimum ratio is desired, so that movement toward the desired setting is somewhat rapid yet permits final, precise adjustment. The small pocket transistor radio typically has a small thumb-wheel switch to select stations; inevitably, locking in on a station is a frustrating task. In contrast, the large dial for frequency selection found on many audio oscillators represents intelligent choice. For vehicle steering, again, the middle ground is preferred. A high ratio that requires two complete rotations of the wheel to turn the vehicle 90 degrees demands too much physical effort; at the other extreme, a low ratio with considerable vehicle response to wheel rotation through a small arc might place an inordinate demand upon operator attention and psychomotor coordination.

Control Resistance and Feedback. Resistance in control operation is a critical variable because it can affect a number of factors, including (1) the feedback to the operator regarding control position; (2) the speed, precision, and smoothness of control operation; and (3) the susceptibility of the control to accidental activation and vibration. Various types of resistance are an inherent part of or can be built into control operation. One type may be used to counteract adverse aspects of another, for example, friction versus excessive inertia. Controls should be selected with the types of resistance in mind that yield effective performance.

Spring-loading provides an elastic resistance that is proportional to displacement of the control, returns the control to and aids identifications of a null position, can minimize accidental activation, and provides good feedback of control position. Deadman switches and joysticks use spring-loading to good effect. With sufficient resistance, the operator can rest his hand or foot on the control. Frictional resistance, common to rotary controls such as on potentiometers, permits smooth movement and stable setting of a control in a desired position with little concern for accidental movement or displacement by vibration. Viscous damping is desirable when quick, gross movements are to be avoided; conversely, it aids smooth, fine adjustments and effectively damps out vibration effects. Inertial resistance is useful in effecting smooth changes in velocity but requires considerable operator effort; an overshoot tendency precludes precise adjustments. For certain controls, such as joysticks and levers, that operate over a distance, amplitude of movement may provide better feedback of control position than can force cues due to resistance. For controls with limited travel, however, it is important to have good pressure (control resistance) cues.

Accidental Activation. Controls can be bumped when an operator reaches to operate another control or can be activated by being caught in the cuff of a shirt. For panels, then, it may be desirable to recess controls if space permits or to put barriers around them. Critical controls can be isolated or arranged in a sequence that tends to preclude accidental operation. For levers and toggle switches mounted on display consoles, a horizontal mode of operation tends to avoid the shirt-cuff problem; in corridors, such controls should operate in a vertical mode so that operators do not activate them as they walk by. Critical controls that are not used frequently can be protected with hinged covers or locked in position, as is done with potentiometer settings on analog computers. Sometimes the operation of one control can be used as an effective precondition for operation of a second. Control resistance, discussed previously, can prevent accidental activation in many situations.

Control Coding. Operation of the wrong control at times can be frustrating, lead to inefficiency, or cause accidents. These problems can be reduced and performance can be enhanced by improving control recognition through one or more of the several approaches to coding. Choice of codes can be affected by such variables as the number of controls to be coded, available panel space, level and quality of illumination, operator work load, the criticality of control function, and whether coding has been used for other purposes (safety codes or display or target identification).

Shape and texture coding can be effective where illumination is low or where controls cannot be directly viewed. Recognition by touch alone is good so long as shapes are standardized and operators are well trained in the code. Standard shape-coded knobs have been adopted for various military usages as well as by the electronics industry. Shapes which reflect control function (for example, flap control in the form of an airfoil) are generally desirable. Texture coding is accomplished by fluting or knurling, and is commonly used for cylindrical and thumb-wheel knobs. Limitations of both shape and texture coding include (1) the small number of codes which can be effectively discriminated and (2) the reduction or loss of discrimination if gloves are worn. Size coding is not generally useful and is uncommon except for one situation where it is inherent in the design. This is in the case of ganged knobs mounted on concentric shafts, and is not generally recommended anyway because of space requirements and a problem of inadvertent operation. Location coding has some of the advantages of shape coding, the same disadvantage regarding limited discriminations, and a further disadvantage in requiring considerable space to permit effective coding. The mode in which a control operates can aid identification of controls; this approach has limited utility because the control must be activated for identification to occur. Finally, color coding and label coding can be useful, provided adequate illumination is assured. Labels permit a large number of controls to be identified, but can require space themselves. Color coding of controls can be quite useful when coordinated with display color codes, especially when large numbers of displays and controls are involved and associations are not readily apparent.

Specific Control Recommendations.   Various functional requirements and design features are summarized here for specific controls as follows:

1. Push buttons should have positive, snap (click) action for operator feedback and concave or rough surface tops to aid in fingering. Optimum diameter is ½ inch, with 1-inch separation from other controls or ½-inch separation between push buttons operated in a sequence. Separators between buttons are useful to minimize inadvertent operation.

2. Rocker keys provide desirable visual feedback of their state, but in horizontal layouts are prone to inadvertent operation.

3. Toggle switches provide a quick mode of response, require little space, and can be operated simultaneously with others in a bank. They also give both a visual and a tactual indication of their state. Space permitting, bat handles afford easier operation of toggles.

4. Rotary selector switches require a medium amount of space and can give good visual and tactual feedback of position, but require a trade-off in speed and accuracy if many settings are required. They should preferably be of the moving-pointer, fixed-scale type. Detents should be provided at each position. No more than twenty-four positions should be used. Skirts are desirable in preventing scratches of the panel surface and provide a place for pointers and labels. The knob portion should have indentations to facilitate grasp. For limited settings, a bar-type knob with tapered tip may be preferable.

5. Slide switches and thumb-wheels have limited use; for ease of operation, serrations are recommended.

6. Knobs for continuous control can be shape and color coded effectively, and with appropriate gearing, provide a considerably flexible choice of adjustments. Small crank handles, which fold out from the knob surface, can aid in rapid slewing. Surface knurling or serrating is generally desirable.

7. Cranks are used for high rates of adjustment over long distances. Position feedback is poor. They are not generally amenable to coding. The turning axis should be parallel to the frontal plane of the body when extreme torque must be applied but perpendicular when speed of movement is essential. Small cranks (up to 2-inch radius) are optimal for quick positioning, while larger cranks (5- to 7-inch radius) handle torque load better.

8. Levers are commonly used as gear shifts, throttles, or joysticks in positioning or tracking tasks. Spring-loaded levers have desirable feedback characteristics, discussed previously. Joystick lengths depend on the precision of movement required. To facilitate precision of control, it is desirable to provide support for the limb. If the control is operated with finger movements only, the control pivot point should be located below the surface which supports the hand.

9. Steering (hand) wheels are useful where large rotary forces must be applied, but they do demand a space penalty. Optimal displacement should not exceed a 120-degree turn. Recommended diameter is 12 to 18 inches, with a rim cross-sectional diameter 0.75 to 1.50 inches, the smaller figure being better for female operators. The rim should be contoured for good gripping.

10. Pedals have characteristics similar to handwheels—good for large force applications, poor in terms of space utility. Additionally, they do not permit the precision and speed of application of handwheels, but they are useful in distributing work load. Pedals operated by ankle operation should require no greater motion than about 30 degrees. The angle formed by the foot and lower leg as placed on the pedal surface should optimally be between 90 and 130 degrees. If the pedal surface is slanted more than 20 degrees above the horizontal, a heel rest should be provided. Pedals should be at least 3 inches wide and have a displacement of between 2 and 4 inches.

WORKSPACE LAYOUT

The effective layout of workspace is based on many considerations, including the nature of the tasks (speed and accuracy requirements, load, sequence, effort, criticality), allowable system tolerances, information inputs, response outputs, environmental conditions, equipment specifications, and maintenance requirements. Related to these are a number of human factors design considerations involving anthropometric characteristics of operators, equipment arrangement, and panel layout.

Principles of Arrangement. In the context of systems development discussed earlier, attention normally will have been given early to definition of overall work areas and to the delimitation of individual work stations. Similarly, in the context of industrial plant layout, the industrial engineer will have evaluated space allocations, utilizing flow charts of material, equipment, people, and operations.

Link Analysis. Link analysis is a valuable technique for determining the frequency and importance of interactions between men and machines so as to develop efficient workspace arrangements. A typical procedure is:

1. Construct a rough diagram using circles labeled with numbers to identify operator functions, and squares labeled with letters to identify equipments.
2. Determine and identify separately, through coded lines, the various control, visual, and talk links between all combinations of elements, that is, man-man, machine-machine, and man-machine.
3. Using a three-point scale, have experts weight each link in terms of both frequency and importance of use.
4. Multiply the two weightings for each link, and then list all components in rank order according to the sum of their composite link ratings.
5. Redraw the original diagram so that links with higher composite ratings are shorter than those with lower ratings, and try to reduce crossing links.
6. Modify the drawing in accordance with system constraints, including physical access to controls, field of vision, communication efficiency, and indices of walking and crowding.

Obviously, the importance given a particular constraint or criterion can have a considerable bearing on the final layout. Figure 3-6 depicts a final drawing in link analysis involving four operators and three consoles; operator 4, as viewed here, could be a monitor, coordinator, or supervisor.

Workplace Arrangement Criteria. Attention is next focused on the individual operator stations. Several principles, or criteria, to take into account are as follows:

1. Group together components of similar function.
2. Place important or critical controls in the best locations for rapid and easy use.
3. Arrange components operated or viewed in sequence in close and logical physical relationships.
4. Place components used most frequently in central locations, and those infrequently used in peripheral areas.
5. Recognize that certain controls require optimum locations in terms of reaction time, accuracy, or force that must be applied.

With large numbers of displays and controls, one cannot hope to meet all these criteria; inevitably, one criterion will conflict with another, for example, functional grouping versus sequence of use. As a general rule, frequency and sequence of use should receive major emphasis in layout problems.

Analytical Technique. Task analyses previously performed can provide useful information about information display, control response, and communication requirements at individual work stations. The data from motion and time study can further provide useful guidance. In systems that appear to be faulty, it may be desirable to interview or otherwise observe operators. Analysis of films taken at slow rates (memomotion—one frame per second) is one technique. Activity analyses can be made using a checklist and timing signal to indicate when to observe and record behavior. When responses are complex and quickly follow in succession, a stenotype may be used by trained observers who employ codes to identify specific

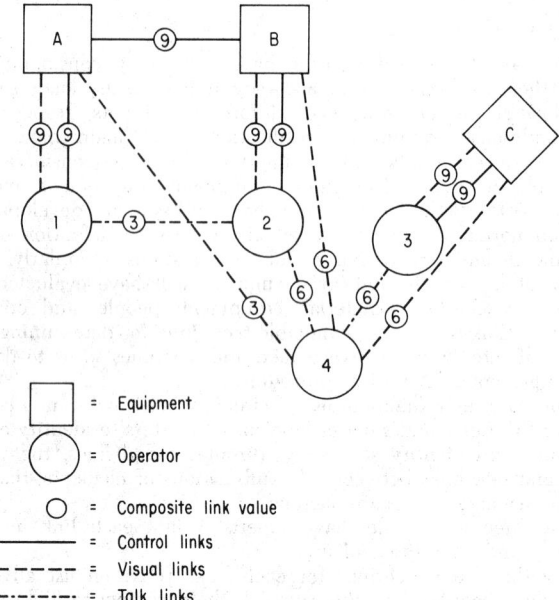

= Equipment

= Operator

= Composite link value

——————— = Control links

— — — — — = Visual links

—·—·—·—·— = Talk links

Fig. 3-6.   Arrangement of men and machines on the basis of link analysis.

task behaviors. In critical visual tasks, eye movement studies may be desirable, but good equipment for this purpose is expensive.

Displays, Controls, and Panel Layout. The problems of habit interference and compatibility discussed earlier are accentuated when controls and displays must be integrated within the larger context of panel layout. Direction of motion relationships are still critical, but now major consideration must also be given to relationships among groups of controls and displays. Figure 3-7 points up some of the alternative, desirable approaches to this problem. As shown in A, whenever practicable, controls should be placed close to the displays which they affect. Direction of motion compatibility should be observed. It should be consistent from one combination to another. Layout should follow an orderly sequence of operation, that is, scan or operate from left to right or top to bottom. If displays and controls must be separated, as in B, a similar geometric arrangement is recommended. Lines may be used to segment the panel to aid in identification of subsections of large display panels, and color coding may further be used as a supplemental cue to relate controls with displays. When conditions prohibit such orderliness, as in large operations displays, the approach shown in C, involving sensor lines, possibly coupled with color coding, is recommended. For rapid check reading of instruments, it is desirable to emphasize bands on display faces that represent null or normal operating conditions and orient these in a standard position—the 9 o'clock position is frequently best. As depicted in D, this approach permits rapid detection of an errant reading in contrast to one in which bands are arranged in an irregular fashion.

Other Design Considerations. There are countless other design problems and choices relating to workplace layout which cannot be properly discussed in detail in this chapter. Standard human factors guides include data on the design of aisles and corridors, doorways, hatches, ladders, stairs, ramps, work platforms, and special chairs and tables. Related to each of these applications is the need to consider human body size, structure, and physical movements. As a final step in workplace arrangement, it is important to mock up the panel design in cardboard and evaluate operator activities under simulated task conditions. This can be done

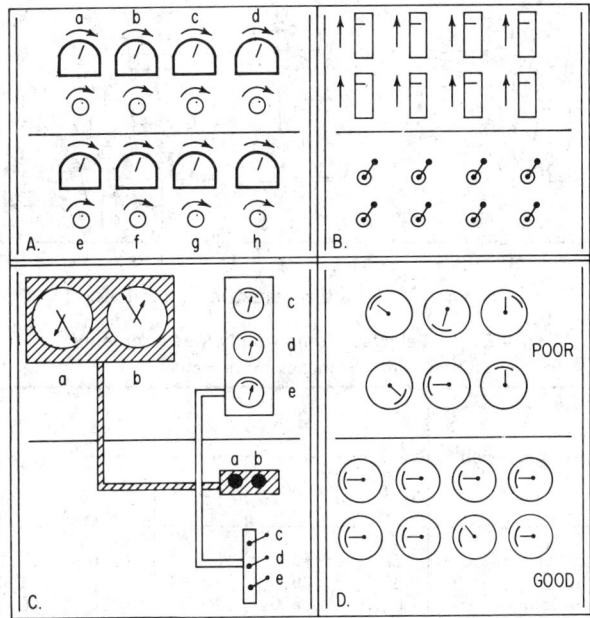

FIG. 3-7. Accepted practice in display and control layout.

full-scale with live operators, or at ⅛ scale with cardboard manikins. In either case, the designer should give attention to the physical characteristics of intended operators.

Anthropometric Factors. Failure to take into account human body structure and dynamic characteristics in workspace layout can have adverse effects on operator efficiency, health, and well-being. Both static and dynamic anthropometric data are available in detail in standard guides; the text by Damon et al is specific and comprehensive on the subject. The designer should be familiar with the kinds of data available and be briefed in their applications, their implications, and intelligent use.

Static Anthropometry. In a workspace layout problem, such human body measurements as indicated in Figure 3-8 may be of concern. Typical practice is to use the 5th and 95th percentiles for a given work population as a guide. Data given in Table 3-2 are based on U.S. Air Force personnel, and as representative of a large population of workers age 18 to 45, can be useful as "ball park" figures for an initial, rough layout or mock-up of workspace. It must be recognized that there is no such thing as an average man. A person average in one dimension may be far from average in another, and thus it is virtually impossible to design for all extremes of a workspace layout problem. Only in a few situations, such as workbench surface height, is design for the average man reasonable. In considering control installations, the 5th percentile for arm or leg reach should be used to ensure that operators can reach controls. Seat heights should be adjustable over the 5th to 95th percentile range. Clearances should be based on 95th percentile dimensions. In the design of auditorium seating, assuming staggered seat arrangement, the floor slope ideally should permit a female with a sitting eye height at the 5th percentile to see over a male with a sitting shoulder height at the 95th percentile.

Anthropometric data are further useful in the design of personal equipment (helmets, goggles, masks, earphones, gloves), escape hatches, worktables and desks, vehicles, prosthetic devices, furniture, appliances, office equipment, and passenger seats. Determination of the useful visual field is another application. It should

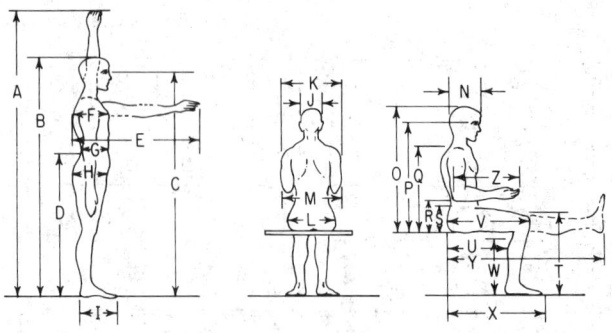

FIG. 3-8. Basic anthropometric dimensions.

TABLE 3-2. Selected Human Body Measurements

Dimensions in inches by percentiles

Key to Fig. 3-8	Body dimension	Male			Female		
		5	50	95	5	50	95
A	Vertical reach	77.0	82.5	89.0	69.0	....	81.0
B	Stature	65.0	69.0	73.0	60.0	64.0	68.0
C	Eye-to-floor	61.0	65.0	69.0	56.0	60.0	64.0
D	Elbow-to-floor	41.0	43.5	46.0	37.0	40.0	43.0
E	Arm reach from wall	31.9	34.6	37.3	29.0	31.5	34.0
F	Chest circumference	35.0	39.0	43.0	30.0	32.5	35.0
G	Waist circumference	28.0	32.0	38.0	24.0	26.0	29.0
H	Hip circumference	34.0	38.0	42.0	34.0	37.0	40.0
I	Foot length	9.8	10.5	11.3	8.7	9.5	10.2
J	Head breadth	5.7	6.1	6.4	5.4	5.7	6.1
K	Shoulder breadth	16.5	17.9	19.4	14.0	16.0	18.0
L	Hip breadth	12.7	13.9	15.4	13.0	15.0	17.0
M	Elbow-to-elbow	15.0	17.0	20.0	13.0	15.0	17.0
N	Head length	7.3	7.7	8.2	6.4	6.9	7.3
O	Seated height	34.0	36.0	38.0	32.0	34.0	36.0
P	Eye-to-seat	29.5	31.5	33.5	28.0	30.0	32.0
Q	Shoulder-to-seat	21.0	23.0	25.0	21.0	23.0	25.0
R	Elbow rest height	7.0	9.0	11.0	7.0	9.0	11.0
S	Thigh clearance height	4.8	5.6	6.5	4.0	5.5	7.0
T	Knee clearance to floor	20.1	21.7	23.3	18.0	20.0	22.0
U	Seat length	17.0	19.5	22.0	17.0	19.0	21.0
V	Buttock-knee length	21.9	23.6	25.4	20.0	22.5	25.0
W	Lower leg height	15.7	17.0	18.2	14.0	16.0	18.0
X	Buttock-toe clearance	32.0	....	37.0	27.0	....	37.0
Y	Buttock-leg length	39.0	43.0	46.0	34.0	....	49.0
Z	Forearm-hand length	17.6	18.9	20.2	14.0	....	18.0

NOTE: Measures are on nude subjects and thus make no allowance for clothing; data were extracted from USAF surveys summarized in WADC Technical Report 56-30; certain dimensions have been rounded to nearest inch or half inch; some female dimensions are approximated by subtraction of a constant from corresponding male data.

be noted that anthropometric data vary with a number of factors: age, sex, race, and nationality are examples. Less obvious are differences found among occupations. Perhaps not unexpectedly, research workers and truck drivers differ in several body dimensions. Clothing and personal equipment may also demand special consideration, especially with regard to movement, seating, and in the case of gloves, spatial separation among controls.

Dynamic Anthropometry. Dynamic anthropometry is concerned with the functional range and the pattern of body movements and with the operations that can be made by the limbs in various positions. Dynamic data for kneeling, crawling, and prone positions have unique application where work is spatially restricted, as is often the case with mechanics, plumbers, or repairmen. Range of limb function, for example, grasping and operating a control, and torso movement can be studied through analysis of slow motion photography. Contours of body element motion can then be plotted in each of three dimensions, and inferences drawn concerning optimum control location, distribution of effort, and the like. A similar approach can be used to study the kinematic behavior of the body under rapid deceleration so that protective devices and controls can be properly located for injury control. Specific data are available on arm and leg reach envelopes, on the forces that limbs in various positions can exert on controls, and on the volumetric requirements for effective use of hand tools, especially where access is limited.

Maintainability Design. Importantly related to the overall workspace layout problem are the design and arrangement of components for ease and reliability of maintenance. Poor maintainability design can contribute to maintenance errors, inefficiency, and accidents. Further, it is not uncommon in industry for the maintenance activity to be given an undue amount of authority that can cripple production. Hence, it is desirable to design equipment so that maintenance does not interfere with or unnecessarily disrupt its operation. Access is a necessary precondition, while interchangeable units and modularized components with spares can reduce downtime. For major systems, maintenance might be provided from a floor underneath the principal area of operation. Test points can be located at remote locations for ease of access. The guide by Morgan et al provides good coverage of the subject.

A brief survey of major design principles is as follows:

1. Locate parts, assemblies, and fuses so that their removal does not require removal of other components.
2. Provide sufficient access for test probes, soldering irons, and other tools.
3. Plug-in assemblies are generally preferred to solder connections.
4. Use a minimum of mounting hardware, consistent with vibration and stress.
5. For frequent access, use hinges, or pullouts with stops, and capitalize on access from both sides.
6. Use hand-operated fasteners or, alternatively, slotted, hex-head screws that permit choice of tools.
7. U-lugs are generally preferred to O-lugs for wire connections.
8. Label all components, and use color codes or labels to identify relations between wires and terminals and between cables and connectors.
9. Cables should be routed so they are not caught in doors, pinched, or tripped over by operators.
10. Aligning pins on plugs should have an asymmetrical arrangement to avoid incorrect insertions into receptacles.
11. Provide sufficient access doors or windows on components that require internal adjustments by technicians.
12. Consider the anthropometric characteristics of technicians, both for ease of access as well as for use of tools in close quarters.

BIOMECHANICS, SAFETY, AND PRODUCT DESIGN

Biomechanics is concerned with the relation between physical force and the human body. Many people are injured or subjected to discomfort by equipment or products that, from a biomechanics' viewpoint, are poorly designed. Research in biomechanics, a field involving collaboration among engineers and biomedical scientists, is focused on the reduction of such trauma. Good human factors practice should recognize this activity and incorporate appropriate safety and injury control criteria into its approach to equipment design.

Systems Safety Engineering. Systems safety engineering is a subspecialty that gives recognition to the above view by acknowledging that many accidents are inevitable. Safety and efficiency are recognized to be intrinsically related, not separate entities. Finally, it is much concerned with the problem of product liability.

Product Liability. Pew[8] describes a case involving finger amputation resulting from inadvertent activation of a power press. The operator brought suit against the equipment manufacturer, claiming negligence in design; the jury found the manufacturer liable. Another suit against a manufacturer involved a claim for back injury due to vibration from improper vehicle seat design and installation. Such suits are giving emphasis to the point that poor engineering design can cause or contribute to accidents and injury.

Industrial Biomechanics. Tichauer[9] has been a pioneer in studying the cumulative stress to body members caused by improper equipment design. He has used X rays, electromyography, tissue rheology, oxygen consumption analysis, and infrared photography, as well as conventional observational techniques to pinpoint stress effects. His findings show how poor seat designs can reduce lower leg circulation and create postural discomforts. Several hand tools were found to be discomforting, inefficient, and productive of pressure on blood vessels and nerves in the hand or of undesirable friction between bones in the wrist. New designs for pliers, paint scrapers, and ratchet screwdrivers have emerged from this work.

The discomforting effects of vibration upon the users of gasoline-powered chain saws have been studied, and designs developed to attentuate energy transmission to the hands.

Murphy's Law. Figure 3-9 is used to illustrate Murphy's law that, "if anything can go wrong, it will go wrong." The prime user application requires supply from a manufacturer of plate A with an arm bracket welded on as shown. In application, this is positioned over three studs and secured. Plate B, machined by the same supplier, is used in a different application. A cost-conscious engineer discovers that machining essentially the same plate to produce A is more costly than the savings in scrap; so he delivers B with the arm attached. Note that the assembly can now be installed over the three studs with the arm to the right—and it will be 50 percent of the time! One-way installation through intelligent design is the answer to this problem, one that is common in maintenance circles, and the cause of many accidents, especially in aviation.[10]

Injury-control Concepts. Much can be done through design to reduce the trauma associated with accidents. One key concept here is energy absorption. The idea is to soak up energy before it is applied to the body, or provide material that dissipates the energy when contact is made. The former can be accomplished through attenuation devices or collapsible mounts built into equipment such as seats; the latter is accomplished through use of materials that absorb energy gradually, such as dashboard padding, or otherwise spread the impact energy over a larger area. Thin, flat (frangible) surfaces are, for example, better in this respect than heavy, rigid tubing. A related concept is termed, often unpopularly, "delethalization," but it is adequately descriptive—reduce the injury potential of the workplace by removing or recessing all rigid and sharp objects.

System Error Data. Accident data can be a fertile source of information concern-

---

[8] Richard W. Pew, "Witness for the Human Factor," *Human Factors Society Bulletin*, March, 1968.

[9] E. R. Tichauer, "Ergonomics: The State of the Art," *American Industrial Hygiene Association Journal*, March–April, 1967.

[10] Such design efficiencies are documented effectively in a series of "design notes" published and distributed by the Flight Safety Foundation, Inc., Arlington, Va.

PLATE A

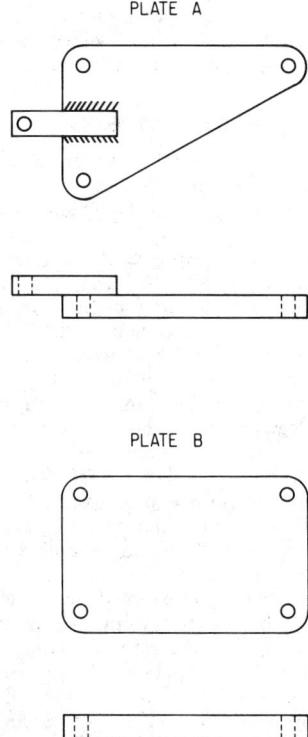

PLATE B

FIG. 3-9.   Example of design-induced installation error (Murphy's law).

ing failures in man-machine system design and operation. However, there are virtues in giving attention to near accidents or incident data. Typically, incident analyses provide a larger pool of data which, through evaluation, can identify major sources of system error. Further, such data can be collected from the outset of operations, possibly before a major system error (accident) occurs. Incidents can yield clues early; one has to wait for an accident to occur. A single incident analysis may identify a critical "quick fix" that may prevent later accidents. The implementation of a system error program should be treated as a major management effort and not as a "brush fire" response to a rash of errors. Specific system error forms for investigator use should be developed and a team of analysts trained. Special time line chronologies and worker report forms are useful in detailing the events surrounding an incident.

Product Design.   Human factors engineers are interested in product design because of their concern for efficiency and safety. Inevitably, there is conflict between the functional criteria of the engineer and the aesthetic criteria of the product designer or stylist. Consumer preference and market analyses can be overriding factors. Articles on product designs appearing in scientific journals emphasize adoption of human engineering criteria. But the human factors engineer is often frustrated by the absence of evaluative data; he wants to ask, "Is the product better?" and receive an answer based on comparisons or operational use.

But designs can be both well human engineered and aesthetically pleasing. Antiqued copper doorknobs are more functional than glass ones and equally attractive. Cosmetic bottles can be made that are attractive, with a surface that is not conducive to slipping out of the hands and in a shape that occupies minimal space. The simple ball-point pen can be designed for more efficient gripping. Actually, much can be done from a human factors viewpoint to improve home products and fixtures without sacrificing style and consumer appeal.

PERFORMANCE AND ENVIRONMENTAL STRESS

The human factors engineer is ultimately and principally concerned with skilled human performance. Beyond the equipment design and layout factors already discussed, there are stress and fatigue effects which should be considered in the total picture.

Environmental Stress. Detailed treatment of the effects of environmental stress upon human physiology can be found elsewhere.[11] It is important, however, to refute the common notion that noxious environmental factors typically do adversely affect performance. Man has learned to adapt to and accept a number of stresses in our complex urban-industrial world. A distinction can and must be made between the discomforts and unpleasantries of a job, for which the worker may be compensated, and the extreme stresses which can cause physical injury or impairment. True, high levels of noise can damage the auditory mechanism, and should be controlled. Moderate levels of noise, however, are an inescapable part of industry and a common source of worker complaint. Yet research findings typically fail to disclose an adverse effect of noise upon work efficiency. In fact, in many situations, performance appears to be enhanced by the presence of noise. This poses an interesting question for job evaluation—should work in a noise environment be given a higher weighting?

Fatigue and Work Decrement. Although it was suggested nearly fifty years ago that the word "fatigue" be banned from use, this has not happened, and the subject is still a source of perplexity to performance researchers. Countless studies have demonstrated that complaints of fatigue are generally unrelated to and not predictive of the course of performance. Physiological or muscular fatigue brought on by tasks requiring expenditure of considerable physical energy is another matter. The fatigue at issue here can be defined in terms of feelings of tiredness or boredom. Better yet, it can be taken as an index of task aversion—the worker is tired of performing a given task. But this does not mean (1) that he cannot perform the task or (2) that he is performing or about to perform the task poorly. Note how the worker, fatigued at 4:30 P.M., can run to his car at the 5 o'clock whistle and head for a round of eighteen holes of golf! Prime attention thus should be on work decrement; subjective aspects can raise hypotheses concerning improper task design or morale problems but often are indicative of nothing but a desire for change in routine.

[11] J. A. Gillies (ed.), A *Textbook of Aviation Physiology*, Pergamon Press, Oxford, England, 1965.

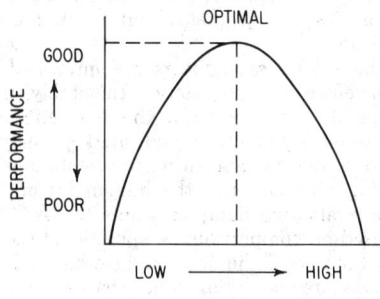

FIG. 3-10.   The relation of task inputs to operator performance.

Research does suggest that certain task parameters or elements can be manipulated to affect proficiency. This can best be viewed in the context of a monitoring or inspection task. As Figure 3-10 depicts, if too much information input per unit of time prevails upon man, he can be overloaded and performance will necessarily be degraded. On the other hand, if information input is low, the task may lack challenge, be uninteresting, and man's efficiency will drop off. If a moderate level or optimal variety of information input can be provided, then operator performance at a reasonable level of proficiency should be expected.

An interesting implication of this concept is that in some industrial situations it may be desirable to add to man's work load or introduce artificial signals into the task to keep the operator alert and operating efficiently. The same "inverted U" relationship indicated in the figure may apply to the element of task variety. It appears that many operators crave some reasonable amount of task variety, and when confined to a demanding and prolonged task, their efficiency suffers. This may be the basis for inefficiency observed in so-called "clean room" environments and similar tasks. In any event, much can be done using good human factors practice to optimize the variety and challenge of industrial tasks.

CONCLUSION

Human factors engineering, in method and concept, has as its principal concern the design of efficient and safe man-machine systems. This concern is compatible with and supplements traditional industrial engineering approaches to problems of performance and workspace layout. A considerable body of human factors knowledge and practice exists and should receive application.

BIBLIOGRAPHY

Bennett, E., J. Degan, and J. Spiegel (eds.), *Human Factors in Technology*, McGraw-Hill Book Company, New York, 1963.
Burns, N. M., R. M. Chambers, and E. Hendler (eds.), *Unusual Environments and Human Behavior*, The Free Press of Glencoe, Inc., New York, 1963.
Chapanis, A., *Research Techniques in Human Engineering*, The Johns Hopkins Press, Baltimore, 1959.
Damon, A., H. W. Stoudt, and R. A. McFarland, *The Human Body in Equipment Design*, Harvard University Press, Cambridge, Mass., 1966.
Gagné, R. M., *Psychological Principles in System Development*, Holt, Rinehart, and Winston, Inc., New York, 1962.
Haddon, W., Jr., E. A. Suchman, and D. Klein, *Accident Research: Methods and Approaches*, Harper & Row, Publishers, Incorporated, New York, 1964.
McCormick, E. J., *Human Factors Engineering*, 3d ed., McGraw-Hill Book Company, New York, 1970.
Meister, D., and G. F. Rabideau, *Human Factors Evaluation in System Development*, John Wiley & Sons, Inc., New York, 1965.
Morgan, C. T., J. S. Cook, III, A. Chapanis, and M. W. Lund, *Human Engineering Guide to Equipment Design*, McGraw-Hill Book Company, New York, 1963.
Murrell, K. F. H., *Human Performance in Industry*, Reinhold Publishing Corporation, New York, 1965.
Webb, P. (ed.), *Bioastronautics Data Book*, National Aeronautics and Space Administration, Publication NASA-SP-3006, Washington, D.C., 1964.
Woodson, W. E., and D. W. Conover, *Human Engineering Guide for Equipment Designers*, University of California Press, Berkeley, 1964.

Chapter **4**

Human Productivity
and Work Design

H. L. DAVIS

Supervisor, Human Factors Group, Industrial Engineering Division,
Kodak Park Division, Eastman Kodak Company, Rochester, New York

C. I. MILLER

Physician, Medical Consultant—Human Factors Group,
Kodak Park Division, Eastman Kodak Company, Rochester, New York

Technological changes in a production system often produce changes in the work design which may well produce effects on the operator that were not expected. Unfortunately, only rarely do these chance effects increase the output of the system. Failure to match the demands of the job with the capacity of the operator is much more likely to reduce output. Stress on some physiological or biomechanical system may go even farther and produce temporary operator breakdown. This kind of poor design may go unrecognized and reduce system outputs for long periods, because the human operator has such adaptive capabilities. The fact that a man can do a certain job does not indicate by any means that the job was properly designed. Man's short-term adaptability and positive motivation for the present situation may mask fundamental design defects to which other persons may not be able to adapt or which they may not care to tolerate.

If the industrial engineer is to design work for maximum productivity, he must learn the principles of ergonomics and how they can be applied.

ERGONOMICS IN INDUSTRY

It is quite necessary that we understand human capacity and capability in the same manner that we would wish to know system or machine capacity and capability. When designing a machine, we call upon engineering for precise information about materials, structures, power, capacities, and the like. When selecting the men to operate this machine, all too often we rely on chance placement or that elusive, ever-changing phenomenon called common sense. (It was once common sense to think of the world as being flat.) This is a mistake which is frequently made and which causes the costly breakdown of people and systems. Information about human physical and mental capacities is available or can be generated by a conglomerate of science called human factors ergonomics. This ever-growing subgroup of the scientific community is made up of physiologists, psychologists, anthropologists, physicians, and engineers. The body of knowledge which they are generating concerning man at work can be utilized in the design of machines, machine systems, workplaces, production methods, and the physical environment. To achieve greater efficiency and effectiveness of man-machine systems with minimum stress upon the individual, this knowledge of man, coupled with the knowledge of the designer, can lead to a more efficient man-machine-environment system with minimum stress upon the individual.

It was during World War II, when weapons systems became more complex and were capable of operation at much greater speeds, that the need for more scientific understanding of man in his working environment became more apparent. With a faster pace of working, the need to perceive information quickly, make rapid judgments and decisions, and perform accurate functions is becoming even greater. Unfortunately, the speed of basic human responses is approximately the same as it was centuries ago, and it appears that there will be no noticeable increase in this speed in the foreseeable future. It is obvious, therefore, that people with a knowledge of man's capabilities must team up with those who are designing man-machine systems to realize effective systems with minimal stress upon people. The term "machine" is used here to denote any component or combination of components which assist a human being in performing some function, or which make up the environment in which he is operating. Therefore, an automobile is a machine, but so also is a wrench, a machine tool, a room air-conditioner, or a radar set.

The physical environment as well as equipment design lends itself to improvement through ergonomic considerations. Such factors as general working conditions, lighting, temperature, humidity, and noise can be optimized to yield more efficient and less stressful conditions. In industry, standards for these factors are generally maintained to ensure health and safety, but very often they are not optimized to enhance efficiency. As an example, a healthy person doing hard physical work in a $72°F$ and 40 percent relative humidity environment would have his capacity for this work severely reduced if he were to do it in a $95°F$ and 65 percent relative humidity environment. To obtain the same degree of efficiency in the second situation, either the environment would have to be cooled and conditioned or the work design and rest pauses would have to be altered.

Man is a very adaptive animal, but if we are interested in efficiency, we must design man-machine systems that enhance, not hinder, man's capacities. The fact that man is able to operate poorly designed equipment in a poorly designed physical environment speaks well of his adaptive abilities, but does not mean that the most effective use is being made of man. A system should be designed to make the operator's role not only possible, but reasonable. Physical effort and mental strain should be carefully controlled so that he is free to devote his attention to those factors in his work where his judgment and flexibility, which he alone possesses, can be used to the greatest advantage.

PHYSICAL EFFORT TASKS

Physical work performed by man requires the use of groups of muscles. Some muscles are used to maintain the body posture for the particular task, while others are used to perform the work. Muscles, like other body tissues, obtain their energy for work from the food we eat and the oxygen we breathe. Muscle cells have an immediate need for oxygen in order to perform work. To meet this demand, certain physiological changes must take place as soon as we begin to do physical work. For example, the breathing rate increases to supply more oxygen to the blood, and the heart rate increases to circulate more of this oxygenated blood to the muscles. Thus, there is a direct relationship between the amount of physical work being performed and some basic physiological functions.

There are three main types of physical effort tasks:

1. Full body dynamic work
2. Localized muscular work
3. Static muscular work

Muscular work is divided into these groups because they must be measured by different techniques.

Full Body Work. Full body work is characterized by the utilization of the large muscle groups usually involving two-thirds or three-fourths of the body's total muscles. Walking, climbing, carrying, lifting, and full body pushing and pulling are examples of full body muscular work. The total energy expended in this type of activity ranges from moderate to very high.

Localized Muscular Work. Localized muscular work requires less energy expenditure than full body work because fewer muscle groups are used to perform the work. For example, in a bench-type job or an assembly operation, motion of one or both arms may be all that is required, while the legs are quiet and muscles of the trunk may only support the body in a sitting or standing position. This type of work may still be very difficult to perform, however, because of the weight involved, the speed of the operation, or the particular reaches involved in the task. Although this type of work is frequently considered to be very light in nature, it may actually be very demanding.

Static Muscular Work. In static muscular work, a force is exerted by the muscles but no mechanical work results. For example, holding a power sander against the ceiling is static work. Static work often requires a very forceful muscular contraction and therefore may also be very demanding. This type of work is also misleading, because it may appear to be quite light in nature.

THE MEASUREMENT OF PHYSICAL WORK

It was mentioned above that there is a direct relationship between the amount of work being done and certain physiological phenomena such as oxygen consumption and heart rate. A similar relationship also exists between other physiological functions such as blood pressure, ventilation, body temperature, and the sweating rate. Measurements of these can indeed give information about a man at work. Some of these, however, such as heart rate and oxygen consumption, have been found by Brouha and others to be directly related to the amount of energy being expended. In addition, these two functions can be measured easily enough to make them practical for on-the-job or in-the-laboratory use. For these reasons, heart rate, oxygen consumption, and total ventilation are the most common measurements made to determine the true physiological demands of certain tasks.

These particular physiological functions are useful because, when an individual is at rest, he is operating at a certain rather steady level which can be measured. This level has been found to be characteristic of the resting state and varies only slightly from one healthy individual to another. However, if an individual changes from the resting state to a new level of physical activity, these physiological functions also change to a new level. This new level can then be measured. The difference

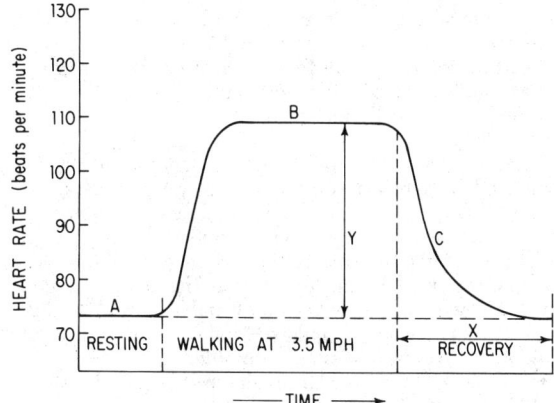

Fɪɢ. 4-1.   Change in heart rate with physical activity.

between it and the previous resting level has been found to be proportional to the change in the physical activity that has taken place.

In full body dynamic activity, the demands placed upon the cardiovascular system and the respiratory system are the greatest. The respiratory system must supply an adequate amount of oxygen to the circulatory system, which in turn must transport this oxygen to the active muscles. The result is a higher level of oxygen consumption and a higher heart rate.

For example, if the heart rate of an individual at rest is measured, it may be about 75 beats per minute. This is represented by that section of the curve in Figure 4-1 designated as A. Then, if this individual begins to walk at a steady rate of 3.5 miles per hour, his heart rate will increase rapidly and then level off after about 2 or 3 minutes at a rate of about 110 beats per minute. This is shown by section B on the curve. Section C shows how the heart rate returns to the original level when the resting state is again resumed. Thus, the difference between the heart rate at work and at rest is represented by the distance Y. As stated above, this distance is proportional to the amount of energy being expended. The line X represents the time required for the heart rate to return to the original resting level following a period of increased activity. This has also been found to be proportional to the energy expended during the previous period of activity. And so, from measuring heart rate alone during work and immediately following work, some notion of the energy expenditure during that period of work can be gained.

Oxygen consumption is that amount of oxygen, usually expressed in liters per minute, which an individual extracts from the air he is breathing. The curve for the change in oxygen consumption that occurs when a person changes his level of physical activity will look very similar to the one for heart rate. And again, the difference between the level at rest and at work will represent the amount of energy above the resting level which is being utilized to perform the task.

For purposes of convenience, energy expenditure is usually expressed in terms of calories. Oxygen consumption can be converted into calories as follows: 1 liter of oxygen = 5 calories. The actual energy expenditure measured during a certain activity is called the gross energy expenditure. This figure, minus the resting energy expenditure, is referred to as the net energy expenditure.

Measurement Techniques. These physiological principles can be rather easily used to help determine the level of activity of certain work situations. Of the two physiological functions, heart rate is somewhat easier to measure. The general principle of the electrocardiogram is utilized. Small electrodes are placed on the chest wall of the individual to be studied, and the tiny electrical current generated each time the heart beats is transmitted to the recording instrument by means of either a cable or a radio wave. The individual heartbeats can then be counted directly or converted electronically into beats per minute and recorded as a continuous curve of heart rate.

To measure the oxygen consumption, the individual being studied must wear a face mask connected by 1-inch flexible tubing to a small gas meter (respirometer), which he wears on his back. This weighs about 7 pounds and measures the total volume of air expired and at the same time collects a sample of the expired air. The percent of oxygen in this sample is measured by means of an oxygen analyzer. A comparison is then made between the oxygen content of the sample and of room air. Knowing these percentages and the total volume of air breathed, the oxygen consumption can be calculated and converted into calories of energy expended per minute by means of a simple formula.

Heart rate is ordinarily measured throughout the entire task or perhaps for an entire day or shift. Oxygen consumption, however, is usually sampled for a few minutes at a time when the heart rate indicates there is a prolonged change in physical activity.

When evaluating the total physiological stress of a certain task, the physiological reactions both during the work and during the recovery period must be considered. When mechanical work stops, physiological work continues above the resting rate until recovery is complete. This is an important concept for the engineer to keep in mind. From the point of view of the human operator, a work cycle must include the physiological cost of the work plus the physiological cost of the recovery period.

Factors Influencing Measurement Results. Physical fitness, sex, and age all affect the physiological response to muscular work.

Studies of the effect of age show that maximum capacity for physical work occurs at about 25 or 30 years, and from that point on, capacity steadily decreases. This is associated with a decreasing capacity of the cardiovascular system and the respiratory system. For example, at about age 20, the maximum heart rate is about 200 beats a minute. This steadily decreases, and at about age 70, the maximum heart rate is down to about 150.

Physical Fitness. Physical fitness has a marked effect on the capacity to do work, especially as we get older. For example, if a man in a poor state of physical fitness performs a severe muscular task for about 20 minutes, he might well attain a rate of about 170 beats per minute. After undergoing a training period of two months, this same man might perform the task with a heart rate that reaches only 140 beats per minute. The training required to produce this increase in physical fitness might consist of jogging 1 or 1½ miles about five days per week.

Female Capacity. Females have about 20 percent lower capacity for work than males. For a given level of heart rate, females are able to do about 20 percent less. This holds true for all but the very lightest work.

Environmental Factors. Besides the changes in the individual physiological response, there are also elements in the environment or in the job itself which may change the physiological response by changing the actual demands made upon the worker. Pace of the job, weight of the load being carried or lifted, body weight of the operator, and the temperature and humidity of the environment are all capable of changing the actual demands placed upon the operator and therefore the physiological response.

If the pace of the job is increased, the energy required to do the job will also be increased. As an example, if a person lifts a 40-pound weight from the floor to a height of 20 inches off the floor, his caloric expenditure will vary as the frequency of the lift varies: 6.7 calories per minute for a frequency of 6 times per minute, 7.9 calories per minute for a frequency of 8 times per minute, and 9.8 calories per minute for a frequency of 10 times per minute. Of course, increasing the weight of the load lifted or carried also increases the demands of the job. The fact that the energy required to do a certain job will be increased if the operator is heavier is not quite so obvious. However, this can be a very important factor when the task requires lifting a load from the ground or similar level which requires the operator to either bend or lower his own weight. When lifting the load, he also has to lift his own weight. Therefore, a 200-pound operator lifting a 20-pound box from the floor does considerably more work than a 150-pound operator lifting a 20-pound box from the floor. When this is multiplied by hundreds of times per day, it can become a very important factor. The exact method used in performing a task can also be an important factor. For instance, in lifting a box, an operator was found to consume fewer calories when he used the hand holds than when he grasped the corners of the box. Such small differences in methods are not easy to detect by simple observation of an operator performing the task.

Because all these various factors affect the total demand that a certain job places upon an operator, it is impossible to assess this man without performing an in-plant work physiology study.

Criteria for Full Body Muscular Work. It is not a simple matter to set limits for the amount of work that can be done by a human operator. No two people are just alike physiologically, even if they are the same age, height, and weight. Neither do they perform their jobs in the same way, even if they are doing the same job. Also, the "same" job may not be the same in another part of the same plant.

Because work intensity in most industrial jobs varies considerably over the workday, the maximum work can perhaps best be thought of in terms of a maximum average energy expenditure for the entire day. There is one condition where this is reliable—the job which has peak loads of very high intensity. This situation is quite rare and can be identified quite easily. A maximum physical activity level of this type which is based on the physiological responses of the man at work is the most reliable measure we have. It is based on the actual physiological capacity of man as determined by well-known physiologists in various parts of the world.

Criteria for an average male worker are:
1. A maximum average energy expenditure of 5.0 calories per minute over an 8-hour day
2. A maximum average heart rate of 115 to 120 beats per minute over an 8-hour day

Criteria for an average young female worker are:
1. A maximum average energy expenditure of 4.0 calories per minute for an 8-hour day
2. A maximum average heart rate of 115 to 120 beats per minute over an 8-hour day

Several restrictions must be taken into consideration when applying these criteria:
1. The worker who is being studied on the job must first be measured while performing standard tasks in a laboratory setting. It must be demonstrated that his oxygen consumption and heart rate responses fall within the normal range. While studying the demands of the job at the workplace, the man is the measuring tool, and we must be sure he is a "standard" man. This will prevent errors in interpretation of the data which individual variations might produce.
2. Again, the above criteria for oxygen consumption apply to jobs which demand essentially full body dynamic activity. Lifting, carrying, pushing, pulling, walking, climbing, and shoveling, all fall into this category.

Where work involves only part of the body—such as standing at a bench handling 15-pound loads repetitively—the heart rate criteria may be used. Limits for oxygen consumption for this type of work should be less, but the exact amount has not been determined.
3. The oxygen consumption criteria do not apply for hot environments. The oxygen consumed will not change appreciably as the environmental temperature rises. However, the heart rate will rise with the temperature and reflect the increased load placed upon the worker. Thus, an average for the day of 120 beats

per minute is a good indication that a job in a hot environment is well within the capacity of an average man.

4. The capacity of older workers is less, especially beyond the 40 to 50 age groups, and is a function of physiological, not chronological, age. Definite limits for various age groups have not been determined.

Work and Rest Cycles. Although the total work for the day should not average more than 5 calories per minute, a man can work at considerably higher levels for shorter periods provided an adequate rest period follows. Where jobs have elements of high energy demand scattered throughout the day, Müller[1] has given a practical way of determining adequate recovery periods. He recommends that the average man be considered as having an energy reserve of 25K calories. As long as work intensity remains below the level of 5K calories per minute, this reserve energy will not be utilized. However, when the job demands exceed this level, the reserve energy will be utilized at a rate equal to the amount by which the energy demands exceed 5K calories. For example, a man working at a level of 8K calories per minute would use his reserve at a rate of 3K calories per minute; thus it would be exhausted in about 8 minutes. This point then becomes the optimum time to begin a recovery period.

The length of the recovery period can be determined by monitoring the heart rate. The next work cycle should not be begun until the resting level is reached. The recovery time can also be estimated by the following formula:

$$\frac{\text{net K cal/min}}{4} - 1 \times 100 = \text{recovery time as a percent of length of work}$$

This will give recovery times for some common levels of work as follows:

Gross caloric expenditure, K cal/min	Rest as percent of work time
6	12
7	40
8	65
9	90

Although these figures may not be precise, they are close enough to be a good guide when actual recovery measurements cannot be made. When considering rest breaks or recovery periods following heavy work, it should be kept in mind that these do not necessarily need to be formal rest breaks taken away from the job. A man who has been working at 6.5K calories per minute can recover very well while doing a light standing job, such as ticket work, checking, or inspecting his finished work. A low level of energy expenditure is what is needed, and this can often be found as a useful part of the job content.

Examples. The amount of work represented by an energy expenditure of 5K calories per minute can be understood by walking on the level at a rate of 3.2 miles per hour. Also, walking 25 miles in an 8-hour shift will average out at about 5K calories per minute. This is a considerable amount of work. Table 4-1 gives the energy expenditure for a variety of industrial jobs. The figures shown are weighted working averages. They reflect all components of the individual's job except for rest breaks. The men performing job 1, for example, work exceedingly hard for short periods, giving them peak heart rates that are close to their maximum. However, it is not unusual to have 40 percent of their day spent in a standby situation, giving an acceptable average for the whole day. In this type of job, where the peaks of energy are short but very high, it is important to employ young, physically fit workers, even though the total caloric expenditure for the day does not average over 5 calories per minute.

[1] E. A. Müller, "The Physiological Basis of Rest Pauses in Heavy Work," *Quarterly Journal of Experimental Physiology,* vol. 38, 1953, p. 205.

TABLE 4-1.   Work Physiology Plant Studies

Jobs	Weighted working average	
	Energy expenditure, cal/min	Heart rate, beats/min
1. Unloading coal cars in power plants	8.0	150
2. Handling 38-lb cans of chemicals	6.5	123
3. Stitch and dispose cases of 8mm film	6.1	147
4. Mixing powdered chemicals	6.0	135
5. Loading corrugated cartons of product into boxcars		
a. Handling only	6.0	117
b. Handling and checking	4.8	97
6. Pulling 1,900 lb in cart		
a. Soft wheels on cart	5.8	110
b. Hard wheels on cart	3.7	101
c. Air pallet	4.9	103
7. Unloading wood cases from boxcar	5.8	107
8. Stitch and dispose cases of 16mm film	5.3	115
9. Tending cartoning machine	5.2	112
10. Stitch and dispose of cases of 8mm film		
a. Stitch and dispose	5.0	123
b. Miscellaneous handling	4.1	108
c. Combined (job rotation)	4.6	116
11. Tending cartoning machine 2		
a. One-man job: load and handle	5.0	112
b. Two-man job: (1) loading	4.0	106
(2) handling	4.0	101
c. Combined (job rotation)	4.0	104
12. Cleaning production dept. (C shift)	4.9	114
13. Handling boxes of sheet film products	4.5	116
14. Cleaning floors and tables	4.5	112
15. Handling 46-lb cans of chemicals	4.4	112
16. Operating square cutter (paper products)	4.4	110
17. Operating square cutter (various products) (with levelator)		
a. Simulation	4.3	115
b. Plant	4.2	116
18. Packing and handling sheet film products	4.2	108
19. Ash removal in power plant (94°F) (mechanized)	4.0	140
20. Packing on conveyor	3.7	113
21. Scrap sorting	3.6	100
22. Unloading rolls of film at slitter windup	3.5	108
23. Handling 1,300-lb boxes (on cart)	3.5	105
24. Heavy-duty packaging	3.5	95
25. Boiler cleanout in power plant (87°F) using power cleanout tools	3.2	113
26. Operating die-cutting press	3.1	98
27. Field machinist, fan deck (103°F)	2.9	135
28. Perforating movie film	2.7	101
29. Bag and pack paper rolls (women)	2.5	113
30. Coal displacer operator	2.1	102

The table gives a good indication of the energy levels that might be found in some typical industrial situations.

Biomechanical Factors.   So far, we have been discussing that capacity of man for full body muscular work in reference to repetitive dynamic work where the continuous muscular load places severe demands on the heart, lungs, and circulatory system. However, some jobs are designed so that there are only infrequent severe loads of very short duration, for example, a single lift of a very heavy load. This activity may have a total duration of 4 or 5 seconds. In this case, the task is over before the heart, lungs, or any portion of the circulatory system has had a chance to respond to the load. Thus, they are essentially not involved. There are, however, other anatomical structures that may be the weak link in the chain and limit this kind of activity. Generally speaking, the structures subject to failure under conditions of short-duration, severe loading belong to the framework and motor elements of the body called the musculoskeletal system. Occasionally, muscles and tendons may rupture under sudden heavy loading. More often, ligaments, joint capsules, or other coupling or supporting structures, such as the intervertebral discs between the vertebrae, are stretched or partially torn, and we say strain has occurred. Usually, these strains occur in tissues that are undergoing or have undergone degenerative changes. The most common condition, frequently attributed to industrial strain, is low-back pain.

Experience of Dr. M. L. Rowe of the Eastman Kodak Company with the evaluation and long-term observation of more than 500 low-back patients in industry has resulted in some tentative conclusions regarding back disability, which have application in work design. It appears that only about 10 percent of future low-back disability problems can be identified at the usual age of employment by present medical diagnostic methods. The incidence of low-back pain is about the same in white-collar as in blue-collar workers, although the blue-collar man is likely to have more difficulty working with a backache than is his white-collar contemporary and will therefore lose more time from the job. The vast majority of industrial low-back disability does not follow definable injury or unusual physical activity, but rather explodes suddenly without mechanical reason or while the man is performing his usual and customary activity in the ordinary and oft-repeated way. Structural failure of one of the two lowermost intervertebral discs seems to account for about 70 percent of the total low-back problems among men in industry. In most cases, failure of the disc appears to result from inborn tissue inadequacy rather than external physical stress. These observations tend to force the conclusion that low-back disability is unpreventable as well as unpredictable, and to a considerable extent this is probably true. Certain new ideas of back mechanics have, however, derived from the study, and these appear worthy of consideration in job design, with the aim of minimizing if not preventing disability.

The human spine can quite properly be viewed as a multijointed derrick boom capable of motion in all directions necessary to bring the working hands into proper position with the job. In vertical balance, very little work is necessary to maintain position. When off balance, relatively huge amounts of energy by short lever arm motors are required to counterbalance the long boom. One has only to recall the attempt to move a painter's ladder in a heavy breeze to realize the painful truth of this principle. A worker can exert tremendous and unnecessary strain upon his low-back area merely by an off-balance working position, even though he does not have to lift anything heavier than a pencil. An obvious and significant case in point is the draftsman who lifts and carries nothing but may have severe low-back pain and disability because he must maintain the weight of his torso, head, shoulders, and arms with his spinal boom inclined forward a few degrees from the vertical. Countless jobs in industry which do not involve the handling of significant loads produce backache and periodic disability because the designer has forgotten the principle of the painter's ladder.

The principle carries over into load-lifting jobs, and the work designer would do far better to worry about the job-dictated man position for lift than to lie awake nights figuring how to cut the load from 55 to 40 pounds.

With respect to necessary lifting work, some other principles are critical to job design. Because of the design of the human low back and pelvis, the major share of the weight of the boom is borne on the back third of the two lowermost discs, and this is the point of usual failure. A man who has to bend backward with a load, as in setting it on a shelf above the waist level, is superstressing the vulnerable areas and may hasten or precipitate a failure at this level. His chances of trouble are multiplied if he has to twist his trunk before setting the load down. How many jobs involve just this mechanism when a man lifts a load from a conveyor belt at or below waist level and twists 90 degrees to set the load down atop a stack on a skid alongside the production line? As a general rule, no load of any significant weight should be lifted above waist level. If possible, no workplace should demand simultaneous lifting and twisting.

Stress on the back third of the lower discs can be significantly decreased if the man can bend his knees, tilt his pelvis forward, and get under the load, thus making the lower discs horizontally bearing across their entire surface. Yet, how often are conveyor lines, benches, shelves, and storage facilities designed so that there is no room for a man to get under his load, and he is forced to stand back and lift? The weight of the load is important, of course, but not nearly so important as the conditions and directions of lifting imposed by faulty job and workplace design.

Weight as the sole criterion of liftability, and therefore suitability, for the human element in the job situation plays us false in other ways. Often, the load has a poor handle. A 75-pound ingot of silver can be handled with much less expenditure of energy and with much greater safety than a 3-foot-square box of parts weighing half that amount; or a 24-inch film roll, smooth, shiny, and without any hand hold; or a jar of liquid which slops about and must not be spilled. Job designers, then, must be at least as concerned about the handle of the load as about weight.

An observation with regard to the scheduling of job assignments may have application at this point. In general, the man who does only infrequent heavy or clumsy lifting is more apt to get into trouble than the regular lifter. As a corollary to this, the regular lifter is more apt to get into trouble with his back on resumption of work after a vacation or a period of illness. Quite obviously, these are examples of conditioning for lifting. This conditioning, oddly enough, does not involve the back muscles as much as it does the abdominal muscles. Men who lift successfully create, by contraction of abdominal muscles, a positive pressure air cushion with their abdominal and chest cavities with which they support the derrick boom from in front and thus diminish the stress on the boom base in the back. The occasional lifter never develops this vital auxiliary support mechanism, and the vacationing regular lifter may lose it temporarily while he loafs on the holiday. Both may be regarded as poorly conditioned athletes and therefore prone to injury if the matter of conditioning is overlooked. It may be better job planning so far as back disability is concerned if lifters are kept lifting rather than trying to spread out the lifting chores among a large number of other workers on a rotating basis.

In summary of these brief observations, it would seem that, just as doctors have altered their approach to one of minimizing periods of disability rather than seeking to prevent backache, so job designers might concentrate more on the conditions of lifting than on the weight lifted, with profit in terms of human comfort and productivity.

THE EFFECTS OF ENVIRONMENT ON PERFORMANCE: HEAT AND COLD

Heat and cold are extremely important factors which alter the effectiveness of a man at work. Extremes of temperature in either direction can drastically slow down and inhibit man's ability to perform any work. Extremes of heat, especially those which occur in industrial situations, are capable of producing drastic changes and even death. However, it is important to remember that we do not need extremes in temperature to make substantial changes in human performance. It is not uncommon for a few degrees of temperature change to make a significant performance change.

Most people work best at a temperature that does not vary more than a few degrees from 70°F. However, before discussing more details, it is important to understand some of the physiology of heat regulation. The human body attempts to maintain a relative constant internal temperature. Heat is being continuously generated inside the body by its own chemical process which is called body metabolism. This includes a tremendous number of chemical reactions involved in the process of oxidation of the food we eat by the oxygen from the air we breathe. The rate at which we produce this metabolic heat while we are at rest is called our basal metabolic rate. Everything we do above the resting level produces more heat, and the harder we work the more heat we produce. Unfortunately, during physical work, the body converts only about 20 percent of its chemical energy into mechanical power. The rest appears as heat. Thus, we can see that the body has a problem to regulate heat produced within itself as well as the heat produced by the outside environment. The human body, of course, like any other object, receives heat from its environment through convection, conduction, or radiation. On the other hand, if the environmental temperature is low, the body loses heat by the same mechanisms.

Body Heat Regulation. The body handles heat in the same manner whether it is generated from within or obtained from the surrounding environment. An increase in blood flow occurs along with a dilation of blood vessels which are close to the skin surface. If this mechanism should be inadequate to dissipate heat rapidly enough to maintain body temperature, then the sweat glands are stimulated and the skin's surface becomes wet. As this evaporates, heat is extracted from the skin's surface. This probably is the most effective cooling measure which the body has to call upon. If the environmental temperature is very hot, as much as one quart of water can be lost in sweat in a single hour. If this kind of situation persists, this fluid must be replaced, of course, along with some of the salt which is also lost in the sweat.

Fortunately, if a man is repeatedly exposed to hot environmental conditions, his mechanisms for coping with this become more efficient. This increase in efficiency will continue over a period of several weeks of daily exposure. When maximum efficiency is reached, a man is said to be acclimatized.

Thus, the human body can either gain or lose heat by means of convection, conduction, or radiation. It also loses heat by evaporation of water from the skin.

Reduction of Heat Stress. In the warm, moist type of environment, the greatest reduction in heat stress will be obtained by removing the excess water vapor from the air. This is a specialized job which should be done by the air-conditioning engineer.

There is another approach to this problem, which avoids changing the entire environment.[2] This involves providing a small air-conditioned room in which the worker can take rest breaks. This small room can be maintained at a reasonable temperature and humidity at a relatively small cost. Because workers in a hot and humid environment will undoubtedly have high heart rates, they will recover much more quickly in an air-conditioned environment. As a matter of fact, complete recovery will not occur even at complete rest in a hot and humid environment. If working conditions are particularly severe, a divided rest room will be even better. The first portion should be at temperature and humidity about halfway between the conditions of the work area and the second part of the rest room. The first portion should also have cool liquids available for the workers to drink. Thus, they can be replenishing their fluids while they adjust to the temperature change for a few minutes.

In the hot, dry environment, the most likely cause of problems is radiant heat. In this situation, air conditioning can be of help only in the mildest cases of heat stress. However, it may well be possible to erect simple inexpensive baffles which can be very effective, as illustrated by Figure 4-2. Aluminum foil on inexpensive, heat-resistant backing material can do a very good job. Sheet aluminum is also very good, provided the aluminum is always kept clean and polished. If the source

[2] L. Brouha, *Physiology in Industry*, Pergamon Press, Inc., New York, 1960, p. 120.

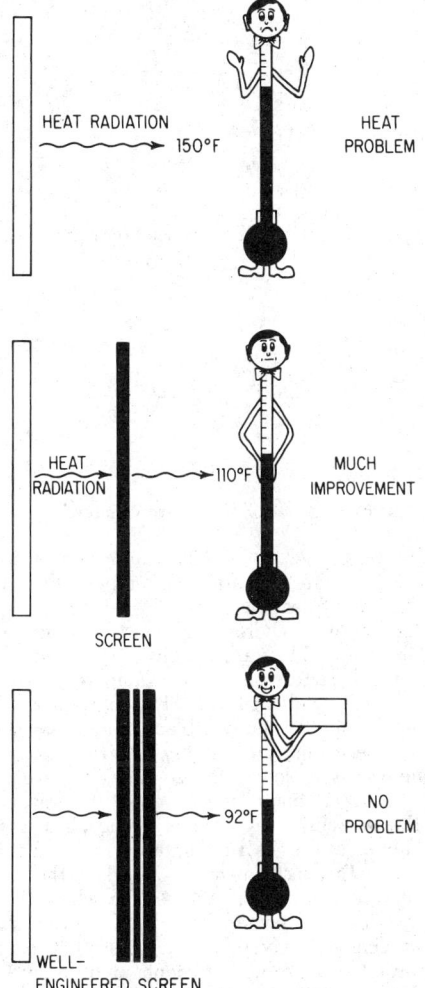

HEAT RADIATION
───────► 150°F

HEAT
PROBLEM

HEAT
RADIATION ─► 〰〰►110°F

MUCH
IMPROVEMENT

SCREEN

〰〰►   〰〰►92°F

NO
PROBLEM

WELL-
ENGINEERED SCREEN

FIG. 4-2.   Screens used to reduce heat radiation.

of radiant heat is relatively small and from only one direction, some type of protective clothing can solve the problem—perhaps a fabric coated with aluminum.[3] In this connection, one should remember that protective clothing should not cover the whole body unless absolutely necessary, because it will keep in heat and moisture generated by the body and thus defeat its very purpose. Remember, also, that the legs and arms dissipate the great portion of the body heat. Therefore, they should be covered as lightly as possible if not directly exposed to radiant heat. When severe conditions warrant covering the entire body, a special suit cooled with air may be the best answer.

[3] C. E. Lewis, R. F. Scherberger, and F. A. Miller, "A Study of Heat Stress in Extremely Hot Environments, and the Infrared Reflectance of Some Potential Shielding Materials," *British Journal of Industrial Medicine,* vol. 17, 1960, pp. 52–59.

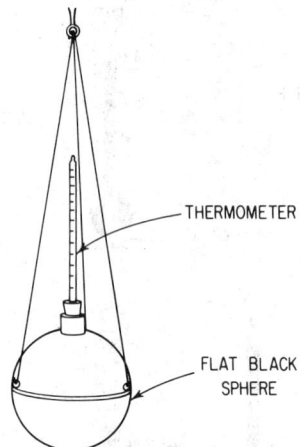

THERMOMETER

FLAT BLACK
SPHERE

FIG. 4-3. Globe thermometer.

Measurement of Environmental Heat. The important factors which determine whether we gain or lose heat are (1) ambient temperature, (2) relative humidity, (3) air velocity, and (4) radiant heat.

Any assessment of the thermal environment must include measurements of these factors. None can be omitted. They are closely interrelated. Methods and devices for measuring the first three factors should be familiar, but very few people have occasion to measure the fourth—radiant heat. This is measured by a globe thermometer, shown in Figure 4-3. This is usually a black sphere, several inches in diameter, with a standard mercury thermometer extending into the center of the sphere. This type of globe thermometer has a slow response time, however, and smaller versions of it have been made using a black sphere about the size of a Ping-Pong ball with a thermistor in the center. This type of device has a much faster response time. These devices can measure the radiant heat impinging upon the body from all directions. Thus, when this instrument is placed at the workplace, it will give a fairly accurate representation of the radiant heat to which the operator is exposed. Measurements of the working area should be made at least throughout one working day, because significant changes may take place during a single shift. In certain climates, it is also important to take measurements at different times of the year. A job that has almost no heat stress during the winter may possibly be almost intolerable during hot summer weather.

Estimation of Heat Load. How long can a man do a specific job in a particular thermal environment? This is often the question for which we would like a clear-cut, simple answer. Unfortunately, it is a very complex problem, and it is important to understand this. It is unwise to attempt simple shortcuts in this area. Many competent investigators have attempted to reduce the physical measurements of the hot environment to a simple physiological index which would predict the effect of this environment upon man. None has found this a simple task. One of the most useful of these indices is the "heat stress index" developed by Belding and Hatch.[4] This utilizes air velocity, wet bulb, dry bulb, and globe temperatures plus the physical work load placed on the man, expressed in British thermal units (Btu) per hour. By the use of a nomograph, a single number can be determined that will represent a particular thermal environment. However, as Lewis et al[5] point

[4] H. S. Belding and J. F. Hatch, "Index for Evaluating Heat Stress in Terms of Resulting Physiologic Strains," *Journal of American Society of Heating and Air-Conditioning Engineers*, August, 1955, pp. 129–136.
[5] *Op. cit.*

out, one must still be careful, because it is possible to have two different environments give the same heat stress index, but have maximum exposure times which differ by a factor of 4.

A detailed explanation of this index is beyond the scope of this chapter. However, the true basis for the index is that a safe tolerance limit can be considered as the time required for the heart rate to increase 45 beats per minute and the body temperature to rise 2°F.

PSYCHOMOTOR TASKS

Webster's Seventh New Collegiate Dictionary defines the adjective "psychomotor" as follows: "Of or relating to muscular action believed to ensue from prior esp. mental activity."[6] This definition may be used as a convenient point of departure for establishing a broad, general categorization of industrial tasks.

It is helpful to think of industrial tasks as forming a continuum. At one end of this continuum are high-energy tasks, in which the demands on the individual are predominantly physical. Emptying coal cars, handling containers in and out of storage, and loading over-the-road trailers with products are obvious examples. The human factors problems attached to high-energy tasks have been discussed earlier in this chapter.

At the other end of the continuum are the sensory perception tasks in which the demands are primarily on the individual's senses,[7] his ability to process internally the information thus acquired from the task environment, and in the making of appropriate but usually quite low energy motor responses. Examples of sensory perception tasks are various forms of inspection, particularly conveyorized and moving-web inspection, and the control of industrial processes through instrument monitoring.

Between the extremes of the continuum are the psychomotor tasks. An important characteristic of this task category is that the mix of perceptual, information processing, and motor demands on the individual may vary widely between tasks. The state of the art does not yet permit either an unequivocal definition of the term psychomotor or the drawing of a sharp line of demarcation along the continuum between high-energy and psychomotor tasks, on one hand, and psychomotor and sensory perception tasks, on the other.

Numerically, psychomotor tasks perhaps constitute the largest category of industrial tasks. Examples can be found in every industry. An everyday example of a psychomotor task is automobile driving. Because of the sheer number of these tasks, any significant contribution to existing procedures for evaluating them is of great value.

Task Evaluation. The problem of task evaluation in the psychomotor area is felt to be a matter of assessing the job's stress content. More specifically, we are concerned with evaluating the magnitude of the stress imposed by the job on the individual's sensory-neuromuscular chain. This is in contrast to the evaluation of high-energy tasks, where the significant stresses are imposed on the oxygen-transport function, as carried on by the pneumocardiovascular system. Because of this, it can readily be seen that radically different techniques of task evaluation are required in these two areas.

The underlying method used in the evaluation of psychomotor tasks involves the use of a simultaneous loading or preempting task.[8] The function of this loading

[6] By permission. From *Webster's Seventh New Collegiate Dictionary* © 1970 by G. & C. Merriam Co., Publishers of the Merriam-Webster Dictionaries.

[7] The principal senses involved, of course, are the visual, auditory, and tactile. Due to the need for working in darkness, the kinesthetic sense may also prove important in the photographic industry. The kinesthetic sense has to do with "sensing" the position of body members in the absence of information from the other senses.

[8] I. D. Brown and E. C. Poulton, "Measuring the Spare Mental Capacity of Car Drivers by a Subsidiary Task," *Ergonomics,* January, 1961, p. 35; I. D. Brown, "Measuring the Spare Mental Capacity of Car Drivers by a Subsidiary Auditory Task," *Ergonomics,* January, 1962, pp. 247–250.

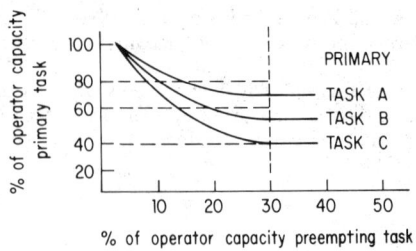

Fig. 4-4.   Method of evaluating psychomotor tasks.

task is to absorb part of the individual's capacity, thus forcing output decrements in the task being evaluated (primary task). These decrements in the primary task output, obtained under controlled reductions of operator capacity, are a measure of the demand imposed on the individual by the primary task.

In actual practice, a statistically meaningful sample of individuals performs the primary task while simultaneously performing a preempting task such as mental arithmetic. The output data for both tasks are normalized. They may be expressed as percentages of each individual's capacity and may be plotted as in Figure 4-4, thereby illustrating the task-ranking ability of this procedure. As an aside, it can readily be seen that the variance between people can be used as a predictor of individual tolerance to job stress. Task C is seen to be substantially more stressful to the individual than either task B or A, as shown by the fact that a statistically adequate sample of individuals produces at a lower average percentage (approximately 40 percent) of their individual capacities at this task when a given percentage (say, 30 percent) of their capacities is absorbed by the preempting task.

Measuring psychomotor stress as indicated is just beginning to be used experimentally in industry, and much research work needs to be done to develop acceptable tests and procedures. However, from this example, it can be seen that the simultaneous task approach is also potentially useful in evaluating proposed methods changes to ensure that increased output has not been obtained at the cost of increased job-stress content.

VISUAL INSPECTION TASKS

As machine speeds increase and production processes become more automated, the importance of inspection and control of quality becomes critical. Lags in reporting machine-produced defects result in larger amounts of waste. Machine-induced hypnosis sharply reduces inspector efficiency. Paced, on-line inspection workplaces approach and exceed human capabilities to scan for and identify defects. All this occurs while increasingly critical use of products demands even more critical standards for inspectors. Thus, inspection departments find themselves in a squeeze between increasing quality standards and decreasing time per unit of product in which to attain them. This is not, however, a problem which can be solved merely by more manpower. Technical innovations in inspection and quality control—based on knowledge of human characteristics, job methods design, and new inspection equipment—can in many cases allow this stage of the production process to keep pace with and contribute to the increasing efficiency and productivity of modern production processes.

Industrial inspection jobs fall into two categories: acceptance inspection and process control inspection. Acceptance inspection requires a worker to look at products to determine whether they pass an acceptable level of quality for sale or further use in the manufacturing process. The inspector usually sorts products into two categories: good and defective. In some cases, good product may be graded as to its relative goodness, and defective product may be divided into that which may be corrected by an additional operation or that which must be scrapped. Acceptance inspectors usually are expected to weigh an aggregate of defects and decide whether their combination amounts to a good or a defective product.

Process control inspection requires a worker to inspect raw or finished products to detect *any* defect and perhaps note its trend with time in order to make specific corrections in the manufacturing process. Process control inspection is the more difficult of the two because inspectors are usually expected to detect irregularities of almost any magnitude so as to make corrections before they develop into ruinous proportions. Acceptance inspection jobs, on the other hand, are often deliberately designed so the inspector will not detect inconsequential defects, the ideal system being one in which every detected irregularity is a basis for product rejection. This simplification purportedly reduces the information processing and decision making required and allows the worker to inspect more product per unit of time. It is important to note here that regardless of how such a situation is set up, there is usually a gradation of defects from the easily to the not-so-easily perceptible. If the proportion of the latter is large, there is evidence that allowing an inspector to grade product into as many as five categories of defect severity reduces decision time substantially.

The problem areas involved in the design of inspection jobs may be divided into the following categories:

1. Vision and illumination
2. Sensory memory
3. Psychological and social factors

Vision and Illumination. The detection of a visual signal is dependent both on properties of the signal and on human peculiarities. The signal characteristics include its size, brightness, hue, saturation, length of exposure, shape, contrast with background, edge quality, simultaneous presence of glare, type and degree of movement, and position in space. For each of these parameters, there are human limits in detection and discrimination and often optimal points facilitating detection. Also, on the purely human side, characteristic individual differences will exist due to age, state of dark adaptation, and inherent abilities.

Visibility and Acuity. A primary question which is often asked is: How small an object can the unaided eye see? How small a defect can an inspector be expected to see? Surprisingly enough, there is no definitive answer to this question. Despite a huge amount of research on the resolving power of the eye, data may not exist for the specific defect in question. To be more specific, the data which do exist usually regard the resolving power of the eye or acuity for dark objects on lighter backgrounds at a 50 percent chance of detection level. Thus, if one were interested in the smallest bright object an inspector could detect on a dark background 99 percent of the time, research data would probably not be readily available.

If we were to complicate the matter further by adding the condition that this must be at low light levels under narrow-band filtered light in $\frac{1}{25}$ of a second, almost certainly no data exist.

To begin with, the case of small bright objects against darker backgrounds has not been studied because it is recognized that the size of the object is not important for detection, but rather how much light it emits or reflects. Theoretically, a single photon of light can stimulate the retina and betray the presence or visibility of an object, though not its shape. In setting up inspection systems, then, small defects which are made to appear as bright spots against darker backgrounds will be detected more readily than if they are made to appear dark against lighter backgrounds. When this is not possible, acuity data on dark objects may be useful to predict detection and identification.

An estimate can be made of how small a dark object an inspector can see 50 percent of the time under optimal conditions, that is, at a close distance, with good vision, with a lighter background, and under maximum contrast. Data indicate that at a close viewing distance of 4 inches, objects as small as 15 microns can be seen. If the critical defects are smaller than this, magnification of some sort is required.

Other data from various investigators suggest that a white spot can be seen when it subtends an angle of about 10 minutes, but a black spot on a white background will have to subtend an angle of about 30 minutes to be seen. Now, if this black spot is extended so that it becomes a fine line, the line is visible when it subtends a visual angle of between 0.5 and 1 minute.

As a general rule, the image of an object will be sharpened when the aperture of the iris is very small, so that minimum separable acuity will be increased by increasing the brightness of the objects viewed. It then follows that to see fine details, the contrast must be as high as possible and the brightness of the object must be as intense as possible. Both of these are variables which can be manipulated by the designer for the benefit of the human operator.

Type and Degree of Movement. A question that is often asked is from what position an inspector should observe a moving product to detect defects best. Should he sit at the side and observe it passing him, or in front or in back of a moving conveyor or web and observe it approaching or receding from him, or perhaps be perched above and watch it passing from the top to bottom or bottom to top of his visual field? Some general principles can be of help.

1. Horizontally moving objects are more easily discriminated than vertically moving objects.[9] There appears to be some evidence that movements to the right are better than movements to the left, and movements up are better than movements down.
2. The effect of movement per se on acuity for small objects has been studied for the left-right direction. Results are unanimous in indicating a degrading of acuity at progressively faster speeds.

Conversely, there is some evidence that slow movement may actually increase acuity significantly.

Effects of Glare. Glare is important to a person engaged in a day-long task of looking for small critical details in the general visual environment. Distracting glare spots can occur either within or outside the immediate task area. This glare can be characterized as the presence of bright spots or reflections which contrast sharply with the visual task. Glare has the effect of creating fatigue because it continually attracts the eye away from the task.

The elimination of glare sources requires a careful survey of the visual field of the inspector in his work position. It may be necessary to provide dark backdrops, glare-free table tops, partial light shields, or perhaps even visors to eliminate glare.

Use of Polarized Light to Reduce Glare. Polarization of light can sometimes be used, provided the critical details depolarize light enough so that they do not become invisible under polarized light. Some important points regarding polarized light are:

1. Polarizing cuts the light intensity by one-half to one-third, and therefore requires brighter light sources.
2. The material inspected must not substantially depolarize light which strikes it.
3. The defect must polarize or depolarize light differently than the nondefective surround. They usually do not.
4. Polarized light should strike the inspected surface.

Dark Adaptation. In some industries, notably the photographic industry, inspectors work in dim light areas or darkrooms with safelights. A review of what is known about dark adaptation will be helpful to the person designing jobs in this area.

1. Exposures to white light of increasing brightness lengthen the period of adaptation to the dark.

---

[9] Leon G. Williams and Marion S. Borow, "The Effect of Rate and Direction of Display Movement upon Visual Search," *Human Factors*, April, 1963, pp. 139–146.

2. The longest period of adaptation is the first one of the day. Adaptation time is more rapid after subsequent exposures to bright light areas.
3. Time to readapt after leaving a brighter light area is more rapid if the visit has been less than one minute and the brightness less than 44 milli-lumens. Beyond that, a point of diminishing returns is reached.
4. Adaptation time is longer, the longer we have been exposed to brighter light. However, this is not a smooth one-to-one relationship.
5. Smoking lengthens the period of adaptation.

In general, for lights of equal brightness, those near the red end of the spectrum will allow faster dark adaptation than white light or filtered light near the blue end. The explanation for this is rather simple in that the retina appears to be composed of two types of receptors: rods and cones. Although the cones are sensitive to all wavelengths, the rods are not sensitive to the long wavelengths, the reds. It is also the rods which are most sensitive to dim lights. Therefore, the rods are literally dark adapting if one is exposed to red light or wearing red goggles.

Age and Vision. Predictable events occur with age to impair vision, which make it possible to redesign jobs to accommodate for these events or allow supervision at the proper time to transfer older inspectors out of critical jobs that only young eyes can be expected to perform.

1. Acuity. Advancing age brings poorer acuity for almost everyone. Although only one out of four may need glasses at age 19, better than four out of five need them at age 60. Unfortunately, even the best-known prescription cannot always correct visual defects caused by aging.

There is evidence that standard acuity tests do not foretell the whole story. Studies of untimed situations have shown that although older inspectors may achieve the same level of accuracy as younger inspectors, the increase in time needed is in the order of 2 to 1 between ages 26 and 50. The interpretation given these data is that there is a general slowing of processes concerned with vision and per-ception, dependent on the degradation of the optical system of the eye.

2. Low contrast. Evidence indicates that older people are particularly less able to perceive low-contrast objects, a condition not reflected by acuity tests.

3. Dark adaptation. Older people cannot dark-adapt as quickly and thoroughly as younger people. This is due to the pupil becoming smaller with age and therefore allowing less light to enter the eye, and to the yellowing of the lens. As an example, if a darkroom were illuminated to a brightness of 4.5 millilumens, it would take a 40- to 59-year-old 11 minutes on the average to adapt to it, but 15 minutes for a 60- to 69-year-old. The difference would be even greater for 20- versus 65-year-olds.

4. Presbyopia. Recession of the near point. One of the most easily and uni-versally noted phenomena occurring in vision with age is the inability to focus on near objects. The nearest point which can be brought into sharp focus gradually recedes from the eye from a matter of a few inches in the teens to beyond arm length in middle age. This has been attributed to the hardening of the lens. This condition can normally be corrected quite well with glasses. With older inspectors, an additional problem becomes the proper placement of the bifocal lens. The normal placement at the lower portion of the field may be poor for an inspector who must work close distances eight hours a day. Consequently, a recommendation often made is either for the bifocal lens to be placed in the middle portion of the field or for the entire lens to be prescribed for near use only, with separate glasses for street wear.

Sensory Memory. When a signal is detected, it must be interpreted. Part of the interpretation process involves comparison between aspects of the presently experienced stimulus and memories of previous stimuli. The discriminations or judgments of stimuli may be made on a relative or on an absolute basis. A relative judgment is one which is made when there is an opportunity to compare two or more stimuli; thus, one might compare two or more spots for size or in terms of density. In absolute judgments, there is no opportunity to make comparisons;

rather, the individual must identify a single stimulus as one of several or many possible stimuli of the same class (dimension). This would be true, for example, if one were to identify a given note on the piano, say, middle C, without being able to compare it with any others, or to identify a given color out of several possible colors when it is presented by itself.

As one might expect, people generally are able to make fewer discriminations on an absolute basis than on a relative basis. In fact, in some dimensions, the difference in the number of relative versus absolute discriminations is tremendous. It has been estimated that most people can differentiate as many as 100,000 to 300,000 different colors on a relative basis, for example, when comparing two at a time, taking into account variations in hue, brightness, and saturation. On the other hand, the number of colors that can be identified on an absolute basis is very limited. In one study, for example, it was found that only 11 to 15 different combinations of hue and saturation could be identified with reasonable accuracy.

The obvious lesson to be learned from this is to use comparative judgments whenever possible. For example, presenting visual examples of defects of various sizes, densities, and the like so they can be sorted into severity categories is a recommended practice.

A frequently used comparative system is with projection microscopes. Two fields rather than one are projected on a rear projection screen. One field contains the projected microphotograph of a standard passable item, and the other field contains the microprojection of the object being inspected. An easy comparison can be made between the two adjacent images.

Psychological and Social Factors. The psychological processes which control the behavior of an inspector are among the most important factors to be dealt with. Although the limits of acuity, influences of age, effects of object movement, and equipment characteristics set an upper limit on the effectiveness of an inspector, the great range of effectiveness below those limits is determined by such psychological factors as training, alertness, anticipation of most likely defective areas, subjective standards of severity, influences from supervision and other inspectors, motivation to do the job, and general methods of carrying out the task.

Training. Standardized training is recommended for all inspectors rather than individual on-the-job instruction. The latter type perpetuates false rules and poor methods which arise over time. On-the-job instruction also relies on the teaching ability and motivation to instruct of each instructor. A standardized program with a single instructor enables better control over the training process and of eventual standards and inspection methods used by inspectors.

The initiation of a training program begins with a job analysis. Unless this is done carefully and with some sensitivity for the whole job of the inspector and his learning and perceptual processes, it is likely to be ineffective. It is unlikely that a quality control inspector is only looking for defects, for instance. It is likely that he has various channels of information which he must use to gain knowledge of his day-to-day tasks. He must also learn to make decisions and communicate them back through the same channels or others. The task of learning to use and give information about products may be as important as locating and identifying defects.

The task of locating defects is not merely one of identifying some particular form, but rather of differentiating defects from a nonuniform background, commonly called noise. This is particularly true when inspecting a continuous web of a homogeneous product. It is important to be able to recognize noise as well as defects. In fact, by far the greatest number of decisions an experienced inspector makes are unrecorded ones regarding the presence and identity of noise, such as inconsequential defects.

Another important aspect of training is channeling the techniques by which inspectors will naturally attempt to simplify and speed up their jobs later on through search patterns and attention centering. Inspectors tend to spend more time inspecting certain areas of a product because past experience has led them to expect a high proportion of defects in that area. They will also tend to look for certain types of defects first because of their frequency or importance. Left on his own, the areas or defects that an inspector chooses may or may not correspond in importance to reality. This is due to his unique experiences, particularly reprimands he may have received for having missed particular defects, or false information from other inspectors. It is important that the expectancies of an inspector be properly channeled by providing him with all available manufacturing information and structuring the information he receives, such as reprimands or rewards, to make his behavior correspond with the demands of reality. The knowledge that a particular machine or operator has been involved in an item's manufacture has the effect of focusing the inspector's attention first on particular defects or areas where these defects usually occur. Much of this knowledge can be incorporated in a training program. The reprimands or rewards an inspector receives should be controlled during training and afterward to reflect the overall importance of the defect. A reprimand for missing a single, infrequently occurring defect would not accomplish anything constructive if it caused the inspector to revise an otherwise efficient search and attention centering pattern and thereby start missing even more defects.

The importance of search and attention centering patterns in the overall expertise of an inspector cannot be overemphasized. They evolve apparently due to the natural strategy on the part of the inspector to reduce the information processing load placed upon him and thereby decrease inspection time per item. It is known that some inspectors become unable to see certain characteristics of a product that an inexperienced observer is overwhelmed by at first glance. In one sense, the evolution of these patterns is good in that it results in more efficient, faster inspections. Studies have shown that people who are able to maintain a consistent search and inspection pattern indeed make better inspectors and detect more of the important defects than inconsistent inspectors. There is a penalty paid here, however, in that such patterns are likely to result in the missing of rare or new defects. The best known method of avoiding this situation is through retraining. This may take the form of a periodically changing display in the inspection area of defective material containing these rare defects. In this manner, some small expectancy is kept alive as to their possible occurrence without disrupting an otherwise good inspection routine. The recommendation here is that an inspector should always be made aware of all pertinent manufacturing information concerning a product. This will allow him to pick his most efficient routine for that item, with maximum time left over to do a random perusal for rare or new defects.

Alertness. Even when ideal viewing conditions and training are provided, other problems may have a significant bearing on efficiency, especially during routine inspection. It seems that where constant attention is required, a considerable strain is imposed on the inspector. For example, recent studies have shown that larger rest allowances are needed to maintain performance at a high level on continuous inspection work than on almost all other classes of industrial operations. The rate of decline in efficiency is associated with a number of unfavorable conditions such as:

1. Where a speed of working is imposed mechanically and is outside the inspector's control
2. Where faults occur either very rarely or less frequently than the inspector expects
3. Where no information is given to the inspector on how accurate he is in rejecting faults
4. Where the inspector has to check large batches of work without a break
5. Where viewing conditions are poor

Motivation. There are many facets to the problem of designing a job so that an inspector continually puts forth his maximum effort. Surprisingly, this may have little to do with pay or other extrinsic fringe benefits associated with company employment.

Some individuals prefer work in which they act much like automatons. They prefer to have everything precisely defined and organized. Under these conditions, they work well and hard. Other individuals become easily bored with this routine and prefer problem solving situations in which they enjoy challenge and competition in finding solutions to problems. An inspection job can be set up to accommodate either type of individual, but usually not both simultaneously. The former type appears to be in greater abundance and is better matched to many reject-type industrial inspection situations. The job can be set up to accommodate this type of individual by a thorough analysis of the nature of the inspection process, a training program based on this analysis, and satisfactory viewing conditions, ancillary equipment, and standards.

When the exact nature of the job is not analyzed or understood, however, the trainee is required to discover its essential features for himself by trial and error. It may take him months or even years to become fully efficient. He has to find out what is important by progressively trying out different ways of selecting from and responding to the information. He learns how successful he has been from the reactions of his direct supervisor and of other people in the organization, and so develops a way of doing his task which he considers to be in line with what is required. For those who are temperamentally inclined, this is a fine way to structure a job, although they may not continue in the job long once they have learned it completely and it has finally become routine.

Unless the department has need to use the inspection job to discover who its problem solvers and thinkers are for promotion to higher jobs, the former method of structuring the job is recommended for allowing the maximum number of people to reach a satisfactory performance level at the earliest possible time. An investment in a thorough job analysis and training program for the rapid training of a larger proportion of potential inspectors will show its greatest advantages in periods of accelerated labor turnover or department expansion.

Perceptual Organization and Learning. From a number of studies, it is clear that the quality that distinguishes a good inspector from an inefficient one is the degree to which he has developed a "perceptual reorganization," or the way he looks at and sees things as a result of knowledge and experience with the object inspected. This has also been termed "perceptual learning."

The perceptual learning of an inspector is, of course, complex and may take many months to become fully established. Although formal training plays some part in determining what the inspector learns, the development and maintenance of skill in selecting signals from the mass of incoming information and organizing them into a meaningful pattern are functions of the total situation within which he works. Some of the ways in which the experienced inspector uses his past experience to reduce the complexity of his task include the following:

1. Ways of viewing the product. This is partly a function of the type of illumination the product receives, partly the angle at which it is viewed, and partly a function of what is observed and which parts are ignored.
2. Knowledge of production. Knowledge of which machines, products, or operators were involved in the manufacturing process provides clues to existence of certain defects.
3. Knowledge of the types of defects. Inspectors know that certain defects are random; others occur in runs, and if one is found, others like it will probably follow. The likelihood of a certain position of a defect on a product is also used in searching procedures.

Summary. The following conditions can lead to success in the design of inspection jobs:

1. Ambiguity should not exist. Defects should be defined and documented and understood by all.
2. The work should be performed in small lots and should be interspersed with other work such as clerical work, obtaining and disposing of work, and the like.
3. Standards of acceptance and rejection should be explicitly defined.
4. A standardized training program should be given to all inspectors and supervision.
5. Adequate time should be allowed for the job, and inspectors should be able to decide their own work/break pattern.

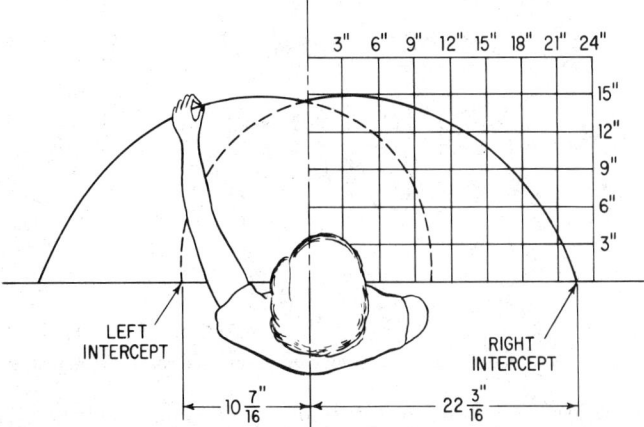

FIG. 4-5. Maximum reach at working surface (female). These data have been estimated from the female population.

6. Feedback is necessary. An inspector's work (passed and rejected) should be monitored, and knowledge of his performance should be given him. This must be done in a constructive fashion if he is to be positively motivated.

WORKPLACE DESIGN

The ideal design of any workplace should begin with the operator in mind. The design should ensure that the operator will have adequate and comfortable posture, that he can see what he must, and that he can operate his controls in an effective manner and without risk of error. If the workplace is not properly adapted to his dimensions and to his other typically human characteristics, he will not be able to perform his work with maximum efficiency. Despite the universal recognition of the importance of workplace design, poorly designed workplaces are still commonplace.

Dimensions of the Working Surface. Figures 4-5 and 4-6 show the recommended dimensions of the working surface for female and male operators.[10] The circumscribed line defines the maximum reach area for the operator. Any control or object that is to be grasped must be located within this area. It is the greatest distance from which small objects can be procured. Large and heavy objects

[10] These data are taken from a study by the Human Factors Group, Eastman Kodak Company, Rochester, N.Y.

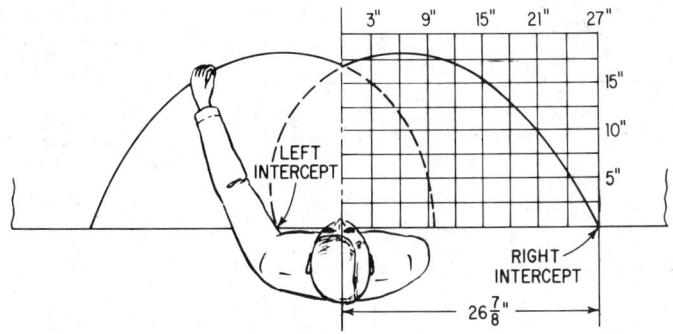

FIG. 4-6. Maximum reach at working surface (male).

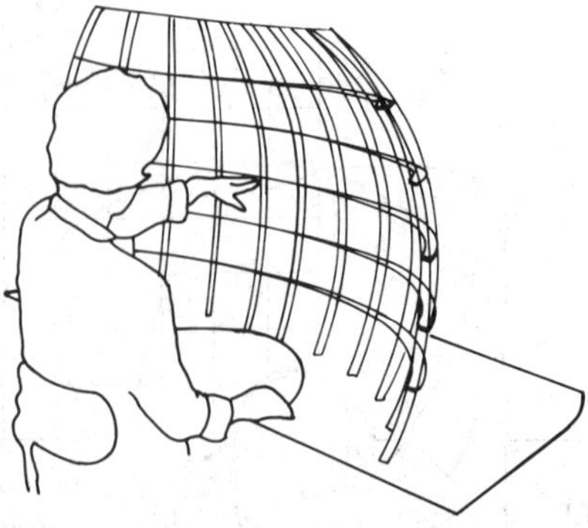

FIG. 4-7. Model of maximum working envelope for female workers.

will have to be located even closer to the body. All fine manipulations that require the participation of both hands should be performed in the area directly in front of the operator's body.

Dimensions of the Working Envelope. It is not adequate to think of an operator as performing a task only in a horizontal plane. Work is often done in the space above the horizontal working plane. This is particularly true in man-machine situations where the manipulation of controls is involved. To determine where parts or controls should be located, it is necessary to visualize a complex three-dimensional envelope of the space in front of the operator. Figure 4-7 is a sketch of a model of this working envelope. The dimensions for the right-hand side of the maximum working envelope for female and male workers are shown in Tables 4-2 and 4-3. The left-hand side can be treated as a mirror image of the right-hand side. The dimensions in the tables are expressed in terms of two coordinates. Each column represents a height above the working surface. The first column describes the maximum reach in a plane 1 inch above and parallel to the working surface. The rows are measured along the front edge of the workplace. Thus, Table 4-3 shows that at a point 9 inches to the right of the center line of the operator's body and 1 inch above the worktable surface, he can reach 18⅙ inches forward of the front edge of the workplace. The row labeled "Intercept" shows the point at which the curve actually crosses the front edge of the workplace. A physical model of the maximum reach envelope for females and males can be built and used in workplace design problems.

TABLE 4-2. Maximum Reach of Right Hand (Female)

Distance along front edge of workplace	Height above worktable surface						
	1″	6″	11″	16″	21″	26″	31″
0″ (center line)	14½″	15⅝″	15⅜″	15 7/16″	13½″	10⅞″	5 7/16″
3″ right	15	15⅞	16⅛	15⅞	13 15/16	11⅛	5⅞
6″ right	14¾	15¾	16	15¾	14	10 15/16	5¼
9″ right	14	15¼	15 9/16	15 5/16	13½	10¼	3⅝
12″ right	12⅞	14¼	14⅝	14¼	12⅛	8½	⅞
15″ right	11 3/16	12½	13	12⅝	10⅛	6	
18″ right	8½	9⅞	10½	9 13/16	7	¾	
21″ right	3⅞	6⅛	6⅞	5½	1½		
Intercept	22 3/16	23 9/16	23⅞	23¼	21 9/16	18¼	12⅝

TABLE 4-3.   Maximum Reach of Right Hand (Male)

Distance along front edge of workplace	Height above worktable surface						
	$1''$	$6''$	$11''$	$16''$	$21''$	$26''$	$31''$
0'' (center line) ...	$18''$	$19\frac{1}{8}''$	$19\frac{3}{8}''$	$18\frac{15}{16}''$	$16\frac{5}{16}''$	$14\frac{3}{8}''$	$8\frac{7}{8}''$
3'' right...........	$18\frac{1}{2}$	$19\frac{1}{2}$	$19\frac{3}{4}$	$19\frac{3}{8}$	$17\frac{1}{2}$	$14\frac{7}{8}$	$9\frac{1}{2}$
6'' right..........	$18\frac{9}{16}$	$19\frac{1}{2}$	$19\frac{3}{4}$	$19\frac{1}{2}$	$17\frac{9}{16}$	$14\frac{13}{16}$	$9\frac{3}{8}$
9'' right..........	$18\frac{3}{16}$	$19\frac{1}{8}$	$19\frac{7}{8}$	$19\frac{1}{4}$	$17\frac{5}{16}$	$14\frac{5}{16}$	$8\frac{3}{4}$
12'' right.........	$17\frac{1}{4}$	$18\frac{1}{8}$	$18\frac{13}{16}$	$18\frac{1}{2}$	$16\frac{9}{16}$	$13\frac{1}{8}$	$6\frac{5}{8}$
15'' right..........	$15\frac{15}{16}$	$17\frac{3}{8}$	$17\frac{5}{8}$	$17\frac{1}{4}$	$15\frac{1}{16}$	$11\frac{1}{2}$	4
18'' right..........	$14\frac{1}{2}$	$15\frac{1}{2}$	$15\frac{7}{8}$	$15\frac{3}{8}$	13	$8\frac{11}{16}$	
21'' right..........	$11\frac{1}{2}$	13	$13\frac{3}{8}$	13	$10\frac{5}{16}$	$5\frac{1}{4}$	
24'' right..........	$7\frac{1}{2}$	$9\frac{5}{8}$	$10\frac{1}{8}$	$9\frac{1}{8}$	6		
27'' right...........	.....	$3\frac{3}{4}$	$5\frac{1}{8}$	$3\frac{7}{8}$			
Intercept.........	$26\frac{7}{8}$	$28\frac{1}{8}$	$28\frac{15}{16}$	$28\frac{3}{8}$	$26\frac{3}{16}$	23	$17\frac{1}{2}$

Workplace Height. The correct working height depends on the nature of the task being performed. Most manual tasks are easily performed when the work is at elbow height. The recommendations given below assume that ordinary manual tasks are being performed. If the job requires the perception of fine visual detail, it will be necessary to raise the work above elbow height and bring it closer to the eye.

Sit-Stand and Standing Workplace. The dimensions of sit-stand workplaces are shown in Figures 4-8 and 4-9. In general, a sit-stand workplace is more desirable than either a sit or a stand workplace. If a sit-stand workplace is to be suitable for use by all operators, it must be provided with an adjustable-height chair and an adjustable footrest. Standing workplaces can be built as shown in Figures 4-8 and 4-9, but with the footrests omitted. However, the workplace for a standing job will be significantly improved if it is made adjustable in height. When this is done, the distance from the floor to the top of the working surface should be variable, from 36 to 42 inches for females and from 40 to 46 inches for males.

Sitting Workplaces. Workplaces where the operator will always be seated can be built with a lower working surface. These workplaces should also be provided

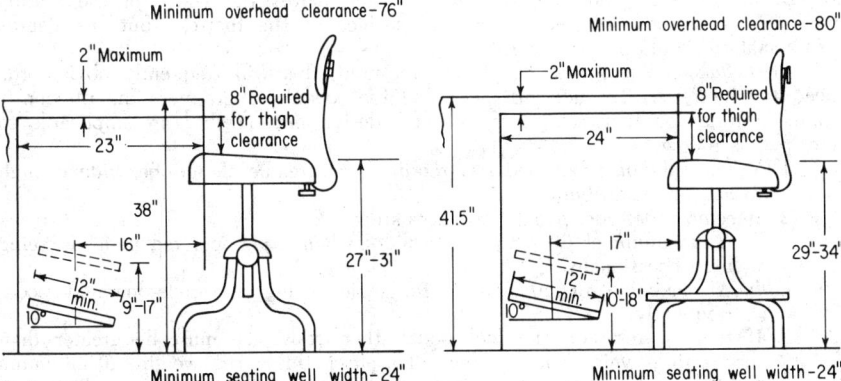

FIG. 4-8.   Sit-stand workplace for females.   FIG. 4-9.   Sit-stand workplace for males.

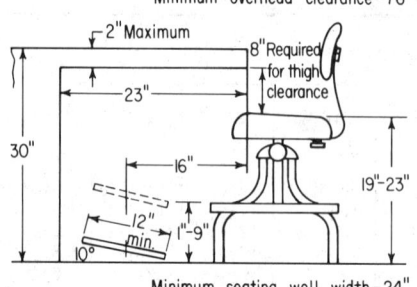

FIG. 4-10.   Sitting workplace for females.

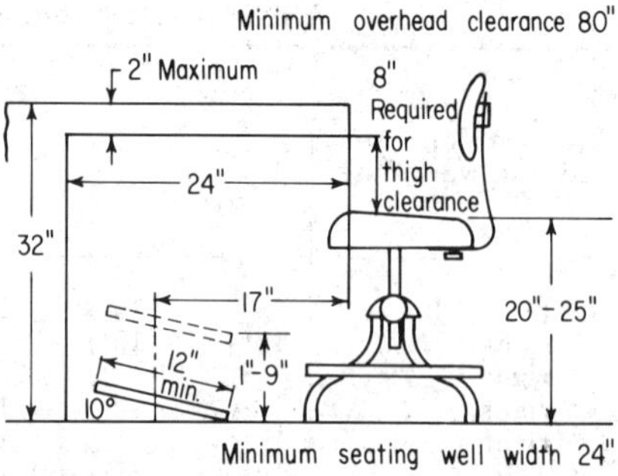

FIG. 4-11.   Sitting workplace for males.

with adjustable footrests.   The dimensions recommended for female and male operators are shown in Figures 4-10 and 4-11.

If it is not possible to provide an adjustable footrest for a sitting workplace, the height of the workplace must be lower than the dimensions shown in Figures 4-10 and 4-11.   Under these circumstances, the workplace should be 26 inches high for males and 24 inches high for females.   The chair for males should be adjustable from 15 to 21 inches, and the chair for females should be adjustable from 13 to 19 inches.   The depth of the seat well must be increased to 29 inches for males and 28 inches for females.   This type of workplace is, however, less satisfactory than one having an adjustable footrest.

The Design of Footrests and Foot Pedals

Footrests.   The importance of providing adequate footrests cannot be overstressed.   In 1919, Frank Gilbreth said, "If every manager were made to sit for a certain number of hours today with his feet hanging, there would be an enormous increase in the number of footrests in our industrial plants tomorrow morning." Unfortunately, this has not happened.   Many managers are still not aware of the importance of providing adequate foot support for workers.

Most workplaces require an adjustable footrest if they are to accommodate all sizes of operators.   An adjustable chair, by itself, is not sufficient.   In addition to the chair, either a footrest or the workplace must be adjustable in height.   It is usually less expensive and more convenient to vary the footrest.

Footrests must be large enough to support the soles of both feet.   A surface of 12 by 16 inches is adequate.   Footrests that provide line support, such as pipe or dowling, are unsatisfactory.   If the footrest is built into the workplace, it should be 12 inches wide and long enough to reach across the width of the seating well.   It is desirable to incline the top surface of the footrest, but the degree of inclination should not exceed 15°.

Foot Pedals.   Where foot pedals are required, this will frequently obviate the need for footrests.   In such instances, the foot pedal should meet the previously stated criteria for footrests.   Some general design criteria that are applicable to foot pedals follow:

1. When only one foot pedal is required, its treadle should be wide enough to accommodate both feet.
2. Recommended foot pedal resistances are:
   a. A maximum of 60 pounds for males when a leg-operated pedal is being used occasionally.
   b. A maximum of 10 pounds for males using an ankle-operated pedal frequently.
3. If the leg rests on the foot pedal, then resistance must be greater than the resting weight of the leg.   To offset this dead weight, a minimum resistance of 10 pounds is recommended for leg-operated pedals and 6 pounds for ankle-operated pedals.

4. The displacement for leg-operated pedals should be 2 or 4 inches. Ankle-operated pedals should move about 2 inches through an angle of 10° to 12°.

5. At seated workplaces, the foot pedal should be located so that the angle between the upper and lower leg is approximately 115°. The angle between the foot and the lower leg should be between 85° and 115°, with 90° the optimum.

Selection of Chairs. Many production workers spend their entire day sitting at a workplace. The chair they are sitting in, along with the footrest, is one of the most important elements of the workplace design. However, many factory chairs can actually interfere with the work cycle. For example, this can be caused by the backrest being so wide that the operator's elbows strike it unless the arms are raised. For most seated operations, the weight of the operator's arms is much greater than the weight of the product, the average female arm weight being slightly more than 10 pounds. Even if the operator is required to lift her arms only 1 inch to clear the backrest, a considerable amount of effort may have been added to the work cycle. This is especially true of highly repetitive operations.

The use of a fixed chair rather than a swivel chair can lead to poor performance in those jobs where the operator is required to work through a wide arc. In one situation, a group of operators had been supplied with fixed chairs and were working through an arc of almost 180°. They were working at a rate of 900 cycles per day. To work through this arc, the operators had to twist in their chairs, and on occasion, they actually picked up the chair and turned it. Because of the incorrect chair being used for this operation, there was an increase in both time and effort.

It is also possible for a perfectly well-designed chair to be used in a way that will force the operator into a poor posture. This frequently happens when insufficient attention is paid to the relationship between the chair and the footrest. There are far too many workplaces where the shorter operators cannot reach their footrests (if they have them at all). This leaves the operator with two alternatives that are equally poor. She can slide to the front edge of the seat and increase her effective leg reach, or she can slide back in the seat so that she can use her backrest. If she slides forward, she will not be able to use her backrest, and if she sits back, her feet will be dangling in the air. The point to observe here is that adequate back and leg support is necessary in a sitting position to reduce fatigue.

The importance of good seating cannot be overstressed. Many studies have been published and numerous articles written on the subject. Recommendations have been made based upon anthropometric and orthopedic considerations. The authorities tend to disagree on some of the recommended chair dimensions, but the amount of disagreement is small when compared with the actual variation in dimensions of chairs that are commercially available. The following dimensions and criteria, illustrated by Figure 4-12, have been found to be very acceptable for a large industrial population.[11]

1. The width of the seat, A, should be at least 17 inches. A seat that is narrower will be uncomfortable for heavier individuals. No harm is done if the seat is wider than 17 inches.

2. The depth of the seat, B, should be approximately 15½ inches. If the seat is not this deep, it will not provide adequate support for the larger individuals. If a seat is deeper than this, it will be uncomfortable for the smaller individuals. If the backrest can be adjusted "in-out," this depth of 15½ inches becomes less critical. The adjustable backrest provides a means of changing the effective depth of the seat (see dimension E).

3. The backrest, dimension C, should be between 6 and 9 inches high.

4. The backrest, dimension D, should be 12 to 14 inches wide. Backrests much wider than 14 inches may interfere with the elbows in some operations. The

---

[11] From an unpublished study on industrial seating by T. Faulkner et al, Human Factors Group, Eastman Kodak Company, Rochester, N.Y.

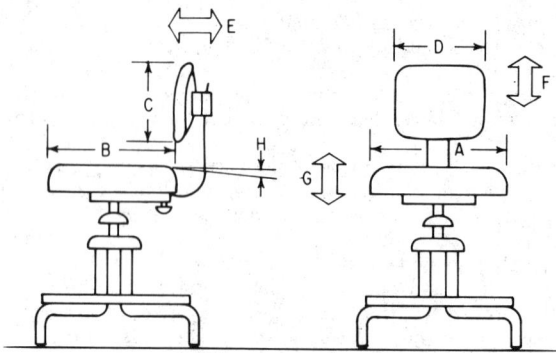

FIG. 4-12.   Work chair design.

backrest should be curved to conform to the shape of the back. The shape of the backrest is more important than the actual dimensions.

5. The backrest should be adjustable in an "in-out" direction, E, so that the effective depth of the seat can be varied. It should be possible to vary the distance between the front edge of the seat and the front surface of the backrest from 12 to 17 inches.

6. The backrest should also be adjustable in an "up-down" direction, F, to account for natural variations in back size and posture. A range of 6 inches will suffice for most situations.

7. The seat must be adjustable in an "up-down" direction to provide for workers of different heights. The greater the range of adjustment, the more flexible the chair will be. As a minimum, the range of adjustment, G, should be 6 inches. This assumes that the midpoint of the range of adjustment corresponds to the 50th percentile seat height for the workplace being considered.

8. The seat should slope to the rear. An inclination, H, of 3° to 5° will help in positioning the worker against the backrest.

It should be possible for the operator to make all the adjustments discussed in items 5 to 7.

Because the ideal chair probably does not exist, some compromise must be made in selecting a chair. The chair that will be selected will depend on the nature of the operation being considered.

CONCLUSION

Well-designed work methods and good workplace designs which take into consideration the capabilities and limitations of the operators who must use them can markedly improve performance and output and reduce operator fatigue and errors. This chapter has discussed some of the more important areas which should be considered by the industrial engineer who has the responsibility for evaluating job demands, work methods, and workplace design. The working environment designed with the human operator in mind goes a long way toward optimum use of industry's most valuable asset—man.

BIBLIOGRAPHY

Brouha, L., *Physiology in Industry,* Pergamon Press, Inc., New York, 1960, p. 120.
Brown, I. D., "Measuring the Spare Mental Capacity of Car Drivers by a Subsidiary Auditory Task," *Ergonomics,* January, 1962, pp. 247–250.
Brown, I. D., and E. C. Poulton, "Measuring the Spare Mental Capacity of Car Drivers by a Subsidiary Task," *Ergonomics,* January, 1961, p. 35.
Damon, Albert, Howard W. Stoudt, and Ross A. McFarland, *The Human Body in Equipment Design,* Harvard University Press, Cambridge, Mass., 1966.

Lewis, C. E., R. F. Scherberger, and F. A. Miller, "A Study of Heat Stress in Extremely Hot Environments, and the Infrared Reflectance of Some Potential Shielding Materials," *British Journal of Industrial Medicine,* vol. 17, 1960, pp. 52–59.

McCormick, E. J., *Human Factors Engineering,* 3d ed., McGraw-Hill Book Company, New York, 1970.

Müller, E. A., "The Physiological Basis of Rest Pauses in Heavy Work," *Quarterly Journal of Experimental Physiology,* vol. 38, 1953, p. 205.

Murrell, K. F. H., *Human Performance in Industry,* Reinhold Publishing Corporation, New York, 1965.

Morgan, C. T., et al, *Human Engineering Guide to Equipment Design,* McGraw-Hill Book Company, New York, 1963.

Chapter **5**

The Learning Curve

WALTON M. HANCOCK

**Professor, Department of Industrial Engineering,
The University of Michigan, Ann Arbor, Michigan**

The prediction of the rate at which a person can learn a manual task is a rather complex process. This is because humans differ in their ability to learn, and this ability interacts with the varying characteristics of the jobs that they are asked to learn. The situation is further complicated by the very significant effect of the method of training on the rate at which a person learns. What then can be done to obtain a good prediction? According to the findings of research studies described at some length in the references listed in the bibliography at the end of the chapter, the learning process can be broken into two segments: the threshold segment, where the person is learning the sequence of events that he has to do, and the reinforcement segment, where the individual knows the sequence and by repetition improves his ability to perform the task. This chapter will emphasize the prediction of the reinforcement phase, because it is in this area that more reliable predictions can be made. Greater consistency is obtained because the operator has already learned the sequence of events, so that the improvement in performance results primarily from the repeated practice of known sequences.

THRESHOLD LEARNING

In the threshold area, it has been found that the rate at which a person learns is highly dependent upon the quality of instruction. The literature on learning by parts, programmed instruction, and audiovisual techniques contains a wealth

7-102

of information on the most appropriate manner to train people in the threshold area. Unfortunately, much of this information is not being actively used to train operators. Part of the reason for this is that the benefits in terms of cost savings have not received proper emphasis. Also, the new training techniques are often considered to be too complicated to use in a routine manner. The available evidence of people versed in contemporary training techniques, however, indicates that the potential is great and that the savings are considerably more than the expenditures. Briefly, the results achieved are the following:

1. The total time to learn the sequences of events, that is, to complete the threshold area of training, is reduced by a factor of 10–20 to 1. This means that a typical learning period of several weeks can be reduced to a matter of days.
2. The error rates are reduced to as much as one-half of their previous values.
3. The attitudes of the employees toward their supervisors and their work are considerably improved.

It should be emphasized that great improvements are possible in this area because of the relatively poor techniques presently in wide use. The typical way people learn manual skills is by the trial and error method. This method consists of letting the employee try to do the operation with very little instruction. The assumption is that if he is allowed to try the operation, he will determine the proper procedure for himself. Unfortunately, this method of instruction has been repeatedly found to take the longest time and to result in the largest variation in methods of any of the techniques mentioned. Also, because little attention is paid to the differences among individuals, the training time allowed is usually the same for all people and is usually based on the learning rate of the poorest operator.

REINFORCEMENT LEARNING

Perhaps the best way to discuss reinforcement learning and at the same time provide a methodology which can be used to predict the learning rate is to give an example. The one that has been chosen is the assembly of four studs in a plate. This is used because it contains most of the factors that should be taken into consideration. Also, by giving an example, a method of solution can be presented which, as long as a computer is not available, simplifies the computations to the maximum extent possible.

Prediction of Number of Cycles to Standard. Figure 5-1 is the MTM analysis of the operation. Note that two columns headed "Motion type" have been added to the analysis. Codes are used to designate the appropriate classification of each motion. The meaning of these codes is as follows:

    O—the motion is performed while the other hand is inactive, that is, single-handed motion.

    S—the motion is performed simultaneously with a motion of the other hand.

    L—the motion is limited out. It is performed by the hand indicated in the same or less time than the motion being performed by the other hand. In the case of eye travels or eye focus, these are accomplished during the same time that the hands are active.

Figure 5-2 is a summary sheet of the MTM analysis shown in Figure 5-1. The motions are categorized by the motion classes as listed across the top of the figure and then further subdivided into whether or not the motion was performed either single-handed or as a simultaneous motion. The operation is also broken down into the frequency of occurrence of subcycles within the main cycle. The reason for this breakdown is that each of the factors has been found to affect the learning rate. The specific effect will become clear when the next step in the prediction methodology is examined.

Manual Learning Curve Computation. Figures 5-3 and 5-4 are calculation sheets used to determine how many cycles are needed to attain the standard time developed by the MTM analysis.

Description— left hand	No.	LH	Motion type	TMU	Motion type	RH	No.	Description— right hand
Get one plate		R12B	O	12.9				
		G1A	O	2.0				
Place on assembly stand		M10B	O	12.2				
		ET						
		EF						
		T90°S						
Release		RL1	O	2.0				
To jumbled studs	2	R12C	S	28.4	L	R12C	2	To jumbled studs
	2	ET	L					
	2	EF	L					
Grasp stud	2	G4A	O	14.6				
				14.6	O	G4A	2	Grasp one stud
To plate	2	M10C	O	27.0				
				27.0	O	M10C	2	To plate
		2ET	L					
		2EF	L					
Position into hole	2	P2SE	O	32.4				
					L	ET	2	
					L	EF	2	
				32.4	O	P2SE	2	
Seat stud	2	M4A	S	12.2	L	M4A	2	Seat stud
Release stud	2	RL1	S	4.0	L	RL1	2	Release stud
Total cycle time for operation				221.7				

FIG. 5-1.   Modified MTM analysis of assembly of studs into plate.

Two calculation sheets are necessary in this case because of the existence of the subcycle. The transporting and positioning of two bolts at a time is done twice within the main cycle. The reason for including this in the methodology is that there are an important number of cases, particularly in clerical operations, where the same sequence is done several times within the cycle. It has been found that this situation results in substantially faster learning rates.

The equations listed under "Pattern" in Figures 5-3 and 5-4 are normalized equations for each TMU of motions where the equation applies. For example, in the case of the M___A simultaneous motion, there are 12.2 TMU in the example. The procedure is to multiply the intercept and slope of the unit basis equation to obtain the learning equation for this motion category as follows:

$$12.2[1.115 - 0.000058(N)] = 13.603 - 0.000708(N) \qquad (1)$$

For each motion class, there are two sets of unit basis equations. One set is for 0–500 cycles, and the other for 500–1,000 cycles.

Because the unit basis equations are linear, the intercepts and slopes can be added to obtain a composite equation. The composite equations for the 0–500 and 500–1,000 cycles of the frequency 1 elements are

$$24.112 - 0.003368(N) \qquad (2)$$
$$22.436 - 0.004013(M) \qquad (3)$$

where N = number of cycles
$M + 500$ = number of cycles
Likewise, the composite equations of the frequency 2 equations are

$$188.097 - 0.038578(N) \qquad (4)$$
$$178.497 - 0.016922(M) \qquad (5)$$

Note, however, that one additional step is necessary where a subcycle exists. In this case, the slope of the composite equation must be increased to reflect the

Fig. 5-2. MTM element summary sheet.

Frequency	R_A		R_B, R_C, R_D, R_E		M_A		M_B, M_C		G1A		G1B, G1C, G2, G3		G4_		P1SE		P2SS_		P1SSE		P1NSE		Remainder	
	One	Simo	One	Simo	One	Simo	One	Simo	One	Simo	One	Simo	One	Simo	One	Simo	One	Simo	One	Simo	One	Simo	One	Simo
1			12.9				12.2		2.0														2.0	
2			12.9	28.4		12.2	12.2		2.0				14.6				32.4						2.0	4.0
							27.0						14.6				32.4							
							27.0																	
Total TMU				28.4		12.2	54.0						29.2				64.8							4.0

MTM motion	Type	TMU	0–500 cycle equation		500–1,000 cycle equation	
			Unit basis	Pattern	Unit basis	Pattern
R_A	One		$1.142 - 0.000065$ (N)*		$1.109 - 0.000193$ (M)	
	Simo		$1.403 - 0.000144$ (N)		$1.331 - 0.000193$ (M)	
R_B, R_C, R_D, R_E	One	12.9	$0.776 - 0.000075$ (N)	$10.010 - 0.000968$ (N)	$0.739 - 0.000094$ (M)	$9.533 - 0.001213$ (M)
	Simo		$0.964 - 0.000155$ (N)		$0.887 - 0.000094$ (M)	
M_A	One		$0.916 - 0.000022$ (N)		$0.905 + 0.000058$ (M)	
	Simo		$1.115 - 0.000058$ (N)		$1.086 + 0.000058$ (M)	
M_B, M_C	One	12.2	$0.580 + 0.000068$ (N)	$7.076 + 0.000830$ (N)	$0.614 - 0.000128$ (M)	$7.491 - 0.001562$ (M)
	Simo		$1.090 - 0.000336$ (N)		$0.922 + 0.000093$ (M)	
G1A	One	2.0	$2.365 - 0.001270$ (N)	$4.730 - 0.002540$ (N)	$1.730 - 0.000430$ (M)	$3.460 - 0.000860$ (M)
	Simo		$2.927 - 0.001702$ (N)		$2.076 - 0.000430$ (M)	
G1B, G1C_, G2, G3	One		$0.765 + 0.000048$ (N)		$0.789 - 0.000149$ (M)	
	Simo		$1.010 - 0.000126$ (N)		$0.947 - 0.000149$ (M)	
G4	One		$1.663 - 0.000318$ (N)		$1.504 - 0.000001$ (M)	
	Simo		$2.247 - 0.000412$ (N)		$2.041 - 0.000086$ (M)	
P1SE	One		$1.995 - 0.000182$ (N)		$1.904 + 0.000029$ (M)	
	Simo		$2.568 - 0.000608$ (N)		$2.264 - 0.000018$ (M)	
P1SSE	One		$1.309 + 0.000574$ (N)		$1.596 - 0.000346$ (M)	
	Simo		$2.208 - 0.000583$ (N)		$1.917 - 0.000209$ (M)	
P1NSE	One		$1.929 - 0.000737$ (N)		$1.561 + 0.000155$ (M)	
	Simo		$2.208 - 0.000583$ (N)		$1.917 - 0.000209$ (M)	
Remainder motions	One	2.0	$1.148 - 0.000345$ (N)	$2.296 - 0.000690$ (N)	$0.976 - 0.000189$ (M)	$1.952 - 0.000378$ (M)
	Simo		$1.581 - 0.000667$ (N)		$1.247 - 0.000100$ (M)	
P21S_, P22S_	One		$0.940 - 0.000091$ (N)		$0.895 + 0.000013$ (M)	
	Simo		$1.208 - 0.000284$ (N)		$1.066 - 0.000008$ (M)	
Total		29.1		$24.112 - 0.003368$ (N)		$22.436 - 0.004013$ (M)

Fig. 5-3. Cycles to standard calculation sheet, frequency = 1.

* N and M are the number of cycles.

Fig. 5-4. Cycles to standard calculation sheet, frequency = 2.

MTM motion	Type	TMU	0–500 cycle equation — Unit basis	0–500 cycle equation — Pattern	500–1,000 cycle equation — Unit basis	500–1,000 cycle equation — Pattern
R_A	One		$1.142 - 0.000065\ (N)$*		$1.109 - 0.000193\ (M)$	
	Simo		$1.403 - 0.000144\ (N)$		$1.331 - 0.000193\ (M)$	
R_B, R_C, R_D, R_E	One	28.4	$0.776 - 0.000075\ (N)$	$27.378 - 0.004402\ (N)$	$0.739 - 0.000094\ (M)$	$25.191 - 0.002670\ (M)$
	Simo		$0.964 - 0.000155\ (N)$		$0.887 - 0.000094\ (M)$	
M_A	One	12.2	$0.916 - 0.000022\ (N)$		$0.905 + 0.000058\ (M)$	
	Simo		$1.115 - 0.000058\ (N)$		$1.086 + 0.000058\ (M)$	
M_B, M_C	One	54.0	$0.580 + 0.000068\ (N)$	$13.603 - 0.000708\ (N)$	$0.614 + 0.000128\ (M)$	$13.249 + 0.000708\ (M)$
	Simo		$1.090 - 0.000336\ (N)$	$31.320 + 0.003672\ (N)$	$0.922 + 0.000093\ (M)$	$33.156 - 0.006912\ (M)$
G1A	One		$2.365 - 0.001270\ (N)$		$1.730 - 0.000430\ (M)$	
	Simo		$2.927 - 0.001702\ (N)$		$2.076 - 0.000430\ (M)$	
G1B, G1C__, G2, G3	One		$0.765 + 0.000048\ (N)$		$0.789 - 0.000149\ (M)$	
	Simo		$1.010 - 0.000126\ (N)$		$0.947 - 0.000149\ (M)$	
G4	One	29.2	$1.663 - 0.000318\ (N)$	$48.560 - 0.009286\ (N)$	$1.504 - 0.000001\ (M)$	$43.917 - 0.000029\ (M)$
	Simo		$2.247 - 0.000412\ (N)$		$2.041 - 0.000086\ (M)$	
P1SE	One		$1.995 - 0.000182\ (N)$		$1.904 + 0.000029\ (M)$	
	Simo		$2.568 - 0.000608\ (N)$		$2.264 - 0.000018\ (M)$	
P1SSE	One		$1.309 + 0.000574\ (N)$		$1.596 - 0.000346\ (M)$	
	Simo		$2.208 - 0.000583\ (N)$		$1.917 - 0.000209\ (M)$	
P1NSE	One		$1.929 - 0.000737\ (N)$		$1.561 + 0.000155\ (M)$	
	Simo		$2.208 - 0.000583\ (N)$		$1.917 - 0.000209\ (M)$	
Remainder motions	One	4.0	$1.148 - 0.000345\ (N)$		$0.976 - 0.000189\ (M)$	
	Simo		$1.581 - 0.000667\ (N)$	$6.324 - 0.002668\ (N)$	$1.247 - 0.000100\ (M)$	$4.988 - 0.000400\ (M)$
P21S__, P22S__	One	64.8	$0.940 - 0.000091\ (N)$		$0.895 + 0.000013\ (M)$	$57.996 + 0.000842\ (M)$
	Simo		$1.208 - 0.000284\ (N)$	$60.912 - 0.005897\ (N)$	$1.066 - 0.000008\ (M)$	
Total				$188.097 - 0.019289\ (N)$		$178.497 - 0.008461\ (M)$
Adjusted total				$188.097 - 0.038578\ (N)$		$178.497 - 0.016922\ (M)$

* N and M are the number of cycles.

accelerated learning rate. This is done by multiplying the slope by the subcycle frequency, which in this case is frequency 2.

Equations (2) and (3) are then added to equations (4) and (5), respectively, to obtain an overall learning expression. The resultant equations are

$$212.209 - 0.041946(N) \tag{6}$$
$$200.933 - 0.020935(M) \tag{7}$$

These are for a person who is performing at a level of approximately 120 percent. Therefore, a correction is necessary for a normal performance level. This is done by multiplying the intercept value by 1.2 as follows:

$$1.2(212.209) - 0.041946(N) = 254.651 - 0.041946(N) \tag{8}$$
$$1.2(200.933) - 0.020935(M) = 241.120 - 0.020935(M) \tag{9}$$

These equations are the normal manual learning equations for the example. Because the standard time for the operation is known (221.7 TMU), the number of cycles necessary to attain this value, N, can be computed.

$$254.651 - 0.041946(N) = 221.7$$
$$N = 785.55 = 786 \text{ cycles} \tag{10}$$

Because N is greater than 500 cycles, the 500–1,000 cycle equation must be used:

$$241.120 - 0.020935(M) = 221.7$$
$$M = 927.63$$
$$500 + M = 1{,}428 \text{ cycles}$$

Thus, it is predicted that 1,428 cycles are needed to attain the MTM standard time of 221.7 TMU.

Industrial validation of the manual learning expression on machine, clerical, and inspection operations has indicated that the above procedure gives a good prediction of the number of cycles that must be performed before standard is reached. However, the fits of the derived equations, (8) and (9), are not good fits to the learning curves that were obtained during research investigations. This is because the laboratory studies, where the unit basis equations were developed, were on relatively short-cycle operations where the interaction between the information processing and motor systems was relatively stable. In the industrial validation studies, it was found that there is a substantial change in the number of times the eyes are used while the operators are learning. The number of eye fixations at the beginning of the reinforcement phase is, on the average, 2.5 times the number which occurs after the person has attained the MTM standard times. It was also found that in the initial stages of reinforcement learning, the eye travels and focuses are not limited out, thereby causing substantial increases in the cycle times. As learning progresses, the use of the eyes decreases as the person uses his eyes to gain information at times when his hands are in motion.

The general form of the actual learning curves was

$$y_i = K x_i^{-a} \tag{11}$$

where K and a = constants
y_i = cycle time of the ith cycle
x_i = cycle number of the reinforcement stage

A more appropriate way to write this equation for graphical solution is

$$T_N = K \times T_s \times N^{-a} \tag{12}$$

where T_N = time for the Nth cycle
T_s = standard time, in TMU
N = cycle number
a = learning curve exponent

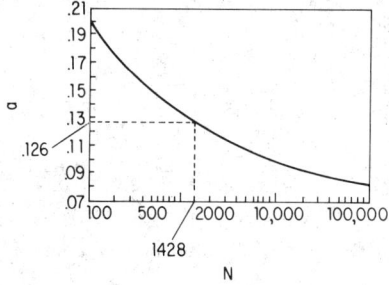

FIG. 5-5.   Exponent values of a for learning curves where $K = 2.5$.

In this expression, K will vary with the use of the eyes.   The most appropriate value for K has been found to be 2.5.   The equation then becomes

$$T_N = 2.5 \times T_s \times N^{-a} \tag{13}$$

The value of T_s is known from the MTM analysis, and it is only necessary to determine the value of a.   This can be done graphically by use of Figure 5-5 as follows:

1. Let N_s = the number of cycles to standard, which in the example equals 1,428 cycles.   Locate this value along the N scale.
2. Proceed vertically until the curve is intersected.
3. Proceed horizontally until the a scale is intersected.

The dashed lines indicate the procedure for this example.   The resultant value of a is 0.126.   The learning curve then is

$$T_N = 2.5 \times 221.7 \times N^{-0.126} \tag{14}$$

Computation of the Time at a Given Cycle Number.   Frequently, it is desirable to be able to check to see if an operator is performing according to prediction.   One way of doing this is to plot equation (9) on graph paper and then plot the operator's progress against the prediction.   Figure 5-6 is a plot of equation (9) using ordinary graph paper.'   However, to get sufficient data points to make a reasonable plot, one must do extensive calculations.   Because of this, the usual procedure is to plot equation (11) on log-log paper as in Figure 5-7.   Because log-log paper is used, a straight line is obtained which requires only two computations of T_N.

To reduce the computational load further, T_N may be obtained by graphical means.   To do this, use the following equation:

$$\text{Cycle time} = R_A \times T_s \tag{15}$$

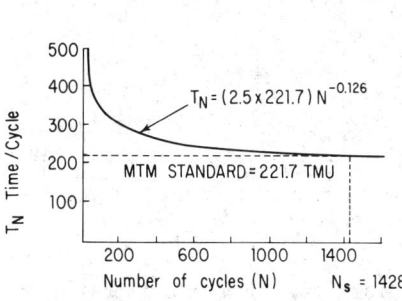

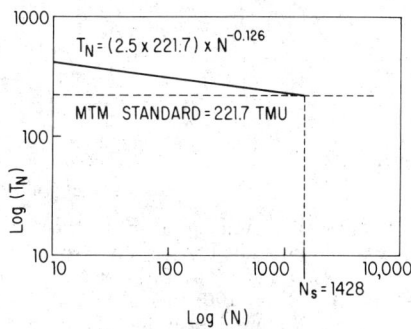

FIG. 5-6.   Plot of example learning curve.   FIG. 5-7.   Plot of example learning curve on log-log scale.

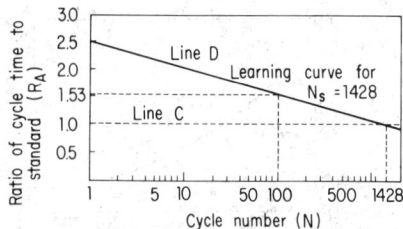

F‌ig. 5-8. Graph to determine cycle time at any point on the learning curve.

R_A may be obtained by use of Figure 5-8.   The procedure is as follows:
1. Draw a straight line from the value of K that will intersect line C at the value of N corresponding to N_s.   In the example, $K = 2.5$ and $N_s = 1,428$.
2. The cycle time at a given cycle number can then be obtained by drawing a vertical line from the value of N_s of interest until it intersects line D.   A horizontal line is then drawn to determine the value of R_A.

In the example, assume that the cycle time at 100 cycles is desired. R_A is 1.53. Therefore, using equation (15),

$$\text{Cycle time} = 1.53 \times 221.7 = 339.2 \text{ TMU}$$

Determining the Cumulative Time for a Given Number of Cycles. In job lot manufacturing or in cases where learning curves are used in conjunction with incentive systems, it is frequently necessary to determine the cumulative time required to perform a certain number of cycles.   This can be determined as follows:

$$\text{Cumulative cycle time} = \int_{N_1-0.5}^{N_2+0.5} KT_s N^{-a} \, dN \tag{16}$$

This equation can be simplified to the following:

$$\text{Cumulative cycle time} = \int_{N_1}^{N_2} KT_s N^{-a} \, dN \tag{17}$$

Performing the integral,

$$\text{Cumulative cycle time} = \frac{-KT_s(N_2)^{-a+1}}{a-1} + \frac{KT_s(N_1)^{-a+1}}{a-1} \tag{18}$$

In evaluating this integral from the first to the Nth cycle, N_1 would be equal to 0. Equation (18) then reduces to

$$\text{Cumulative cycle time} = \frac{-KT_s}{a-1} N_2^{-a+1} = \frac{KT_s}{1-a} N_2^{1-a} \tag{19}$$

Let

$$Z = \frac{N_2^{1-a}}{1-a} \tag{20}$$

Then

$$\text{Cumulative cycle time} = K \times T_s \times Z \tag{21}$$

Graphically, cumulative cycle times can be determined by using Figure 5-9 to determine the value of Z at any desired value of N as follows:
1. Using equation (20), let $N_2 = 1$; then $Z = 1/(1-a) = 1.144$.
2. Let $N_2 = 100$; then $Z = 100^{1-a}/(1-a) = 64.1$.
3. On Figure 5-9, draw a straight line between the values of Z determined for $N_2 = 1$ and $N_2 = 100$.
4. To determine the cumulative time, draw a vertical line from the value of N

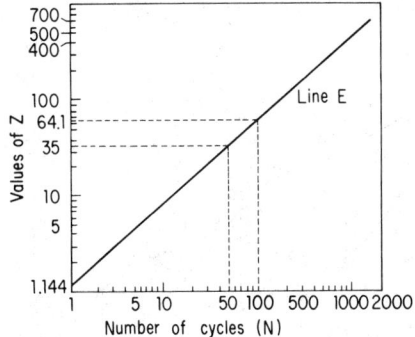

Fig. 5-9.   Chart for determining the cumulative time.

of interest to the line drawn in step 3, and draw a horizontal line to obtain the value of Z.

5. Use equation (21) to obtain the cumulative cycle time in TMU.

Line E is the line drawn to determine the value of Z for the example.   The values of Z were found to be 1.144 and 64.1 for values of $N_2 = 1$ and $N_2 = 100$, respectively.   The cumulative time for, say, 50 cycles may then be determined by using line E.   Z is found to be 35.   Therefore

Cumulative cycle time = 2.5 × 221.7 × 35 = 19,399 TMU

Calculation of the Average Time for a Given Number of Cycles.   The average time for a given number of cycles is also of interest, because it can be used to provide a correction factor for standard times that have been established under the assumption that the people are fully trained.   The following equation applies:

$$\text{Average time} = \frac{\text{cumulative time for } N \text{ cycles}}{N} \tag{22}$$

In the example, where $N = 50$ cycles

$$\text{Average time} = \frac{19,399}{50} = 388 \text{ TMU}$$

A comparison of this value with the MTM standard time value of 221.7 shows the effect of learning on performance.

Use of Computers in Learning Curve Predictions.   The procedure that has just been described requires a substantial amount of computations even when graphical techniques are employed.   An alternative method, which results in a further reduction of the computational time, is to do all the computations on a computer.   Sufficient details have been included here to enable an experienced programmer to accomplish this readily.   Such a program is under development by members of the MTM Association for Standards and Research.

The Effect of Breaks on Operator Learning.   A situation that frequently occurs is where an operator is working with small lot sizes.   In these cases, predictions of his learning rates have to be changed.   Figure 5-10 is an example of performance under these conditions.   The available data indicate that very short breaks, such as for coffee, have no adverse effect on the learning rate.   However, breaks of from one to ten days produce the result shown in Figure 5-10.   Present data provide the following information concerning the effect of breaks.

1. The amount of regression is a function of where the learning curve is broken. For example, at point T_A in Figure 5-10, the difference between the last cycle before stopping and the first cycle after the break is much greater than a break for the same length of time taken at points T_B and T_C.   The extent of the regression can be esti-

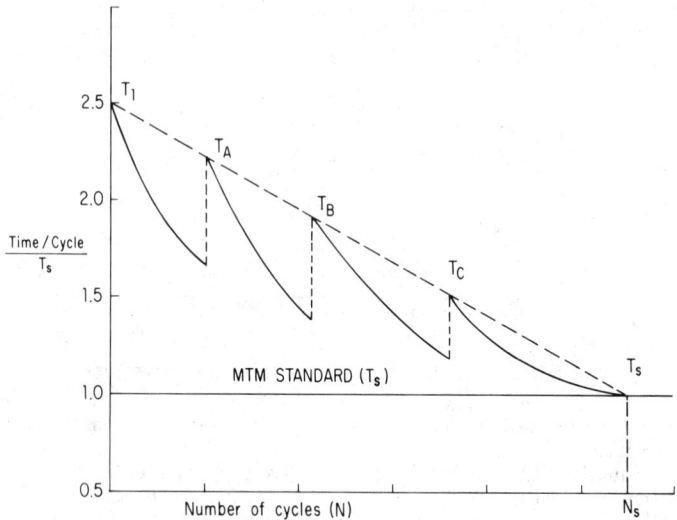

Fig. 5-10.   A learning curve indicating the effect of breaks.

mated by drawing a straight line between points T_1 and T_s in Figure 5-10.   Points T_A, T_B, and T_C will generally be along this line.   One way to predict the learning rate after a break, then, is to shift the origin of N so that the first cycle after the break is considered to be the first cycle.   The value of K, of course, has to be changed because the time per cycle will be less than 2.5.   The appropriate value of K can be found as follows:

$$K_N = K - \frac{K - 1.0}{N_s} N \tag{23}$$

In the example, assume a break at 99 cycles.   The value of K_N can be predicted, using equation (23), as follows:

$$
\begin{aligned}
K_N &= 2.5 - \frac{2.5 - 1.0}{1{,}428} \times 100 \\
&= 2.5 - 150/1{,}428 \\
&= 2.5 - 0.105 \\
&= 2.395
\end{aligned}
\tag{24}
$$

Because breaks affect the time per cycle and not the number of cycles to standard, the original predictions can be revised by making appropriate modifications in the equations given in this chapter to compute the values of a, T_N, and cumulative and average cycle times.

2. The average effect of a break is to increase the total learning time between the breaks by 10 percent.

In cases where more precision is needed, the magnitude of the break time can be determined by using the following expression:

$$BT = \sum_{i=1}^{m} \left(\int_{N_i}^{N_{i+1}} K_{N_i} T_s N^{-a}\, dN - \int_{N_i}^{N_{i+1}} K T_s N^{-a}\, dN \right) \tag{25}$$

where N_1, N_2, . . . , and so on indicate the cycles at break points.

An Example of the Use of Learning Curves to Predict Training and Turnover Costs.   Turnover and training costs can be broken down into two types: external and internal. External turnover costs are caused by the normal attrition of the work force and other factors such as competition and managerial policy. Internal turnover costs are the costs associated with the new jobs and repeated assignments.

Many companies keep elaborate records of their external labor turnover, and comparisons are frequently made. However, there is little information on the cost of separation of a given employee and the hiring and introduction to an organization of a new one. The costs that are quoted for the leaving of one person and the hiring of a new one are from \$600 to \$2,000. These figures usually do not consider the cost of teaching a person his first job.

A method to determine the relative magnitude of these costs is

$$ETC = \text{external turnover cost per year}$$
$$= 12 \times T \times S[C_1 + C_2(LT)] \tag{26}$$

where LT = learning time (excess time in addition to the standard time)
T = external turnover rate per month
S = size of the work force
C_1 = cost of separation, hiring an employee, and initial indoctrination
C_2 = cost per unit of time for the learning time

Internal turnover cost is much harder to determine, primarily because very little data have been kept by industrial firms. Different types of industries may have substantially different costs because of the flexibility required for their work, as well as the length of job assignments. The following is an expression which is a gross estimate of these costs:

$$ITC = \text{internal turnover costs}$$
$$= C_2S[(NJ \times LT) + OJ(0.10LT)] \tag{27}$$

where NJ = average number of new jobs assigned per employee per year
$0.10LT$ = approximate increase in learning time (LT) every time the job is assigned
OJ = number of old jobs repeated per employee per year where full learning has not occurred

Also, one of the assumptions of equation (27) is that the number of cycles per year is equal to or greater than N_s.

The total cost of training, therefore, is the sum of the internal and external costs. To get some idea of the relative costs, the following data from the example can be used:

N_s = average number of cycles to standard = 1,428
RL = average run length per job = 100
T_s = average cycle length = 221.7 TMU = 0.133 min

The equation for LT is

$$LT = \frac{KT_s}{1 - a} N_2{}^{1-a} - T_sN_s \tag{28}$$

Using the data from the example,

$$LT = \frac{2.5 \times 0.133}{0.874} \times 1,428^{(1-0.126)} - (0.133 \times 1,428)$$

$$= 217.5 - 189.9$$
$$= 28 \text{ min} = 0.47 \text{ hr}$$

Let C_1 = \$600, the lower value for separation and rehiring
C_2 = \$5 per hour, which is assumed for this example to be the total hourly cost of an employee
T = 2 percent per month, which is considered to be an average turnover rate

Using equation (26), the external turnover costs are

$$ETC = 12 \times 0.02 \times S[\$600 + (\$5 \times 0.47)]$$
$$= 0.24(S)(\$602.35)$$
$$= \$144.56(S)$$

Similarly, internal turnover costs can be computed.

Let $NJ = 12$, which is equivalent to being assigned one new job per month
OJ = number of repeated job assignments during the year, 24, which is equivalent to two job reassignments per month per employee
Using equation (27),

$$ITC = \$5(S)[(12 \times 0.47) + 24(0.10 \times 0.47)]$$
$$= \$5S(5.64 + 1.128)$$
$$= \$33.84(S)$$

Therefore, the total approximate cost of training and turnover for this example is

$$\text{Total cost per year} = \text{external cost} + \text{internal cost}$$
$$= \$(144.56 + 33.84)(S)$$
$$= \$(178.40)(S) = \$178.40(S) \quad (29)$$

where S is the size of the work force. It is important to point out that the external costs are frequently less than the internal costs, especially where job reassignments and T_s are higher. For example, in the case where T_s, the cycle time, equals 5 minutes, and N_s, the number of cycles to standard, equals 500, if a equals 0.147, LT equals 7 hours. Using the same cost parameters,

$$\text{Total cost per year} = \$(153 + 504)(S)$$

Here the internal costs far exceeded the external costs. The conclusion is that more effort should be devoted to the improvement of the internal training programs than to the reduction of turnover.

BIBLIOGRAPHY

Barnes, Ralph M., and Harold T. Amrine, "The Effect of Practice on Various Elements Used in Screwdriver Work," *Journal of Applied Psychology,* April, 1942, pp. 197–209.

Caplan, Stanley H., and Walton M. Hancock, "The Effect of Simultaneous Motions on Learning," *Journal of Methods-Time Measurement,* September–October, 1963, pp. 23–31.

Chaffin, Don B., and Walton M. Hancock, *Factors in Manual Skill Training,* MTM Association for Standards and Research, Research Report 114, Fair Lawn, N.J., 1966.

Clifford, Robert R., and Walton M. Hancock, "An Industrial Study of Learning," *Journal of Methods-Time Measurement,* January–February, 1964, pp. 12–27.

DeJong, J. R., "The Effects of Increasing Skill on Cycle Time and Its Consequences for Time Standards," *Ergonomics,* November, 1957, pp. 51–60.

Hancock, Walton M., and James A. Foulke, *Learning Curve Research on Short Cycle Operations: Phase I, Laboratory Experiments,* MTM Association for Standards and Research, Research Report 112, Fair Lawn, N.J., 1963.

Hancock, Walton M., and Prakash Sathé, *Learning Curve Research on Manual Operations: Phase II, Industrial Studies,* rev. ed., MTM Association for Standards and Research, Research Report 113A, Fair Lawn, N.J., 1969.

Paine, Russell R., "Experiment in Operator Learning at TRW," *Journal of Methods-Time Measurement,* May–June, 1964, pp. 6–15.

Schwartz, Shula, *The Learning Curve and Its Applications,* Chance Vought Aircraft, Inc., Library Bibliography 20, Dallas, Tex.

Smith, Patricia Ann, and Karl U. Smith, "Effects of Sustained Performance on Human Motion," *Perceptual and Motor Skills,* March, 1955, pp. 23–29.

Welford, A. T., *Ageing and Human Skill,* Oxford University Press, London, 1958.

Planning and Control
Procedures

Chapter 1

Systems Analysis, Design, and Operation Procedures[*]

LAWRENCE L. KAVANAU

President, Systems Associates, Inc., Long Beach, California

CLINTON J. ANCKER, JR.

Chairman, Department of Industrial and Systems Engineering,
University of Southern California, Los Angeles, California

NORMAN F. SCHNEIDEWIND

Systems Consultant, Pacific Palesades, California

The systems approach, consisting of systems analysis, systems engineering, and systems management, is very useful for solving large, complex problems involving men and machines. It is possible to model and anticipate the effects of change by conducting a detailed analysis of system elements or components. These include both hardware—with its associated construction—and operating procedures. The relationships between elements are analyzed and their individual functions are observed through the study of postulated scenarios. Proposed solutions can be evaluated for the system as a whole against various effectiveness criteria, incorporating practical constraints. Engineering development and management actions can then be taken to implement the selected solution.

[*] The authors are indebted to Russell L. Strom for his helpful contributions.

8-3

SYSTEMS APPROACH

The application of the scientific method to analyze systems is not new nor is it radical. In the past, however, when finite, man-made machines and their uses were studied, only the principal elements of such systems were then analyzed. The current level of science and technology enables the analyst to study these components to virtually any desired level of detail.

For example, to analyze one kind of transportation system, such as automobiles, it is necessary to look at the overall system requirements which encompass not only the automobile, but also trains, airplanes, and waterways. In other words, the systems approach to sociotechnical problems and socioeconomic problems requires a modification of the traditional scientific method. In effect, the system must be built up in order to analyze it, not broken down as in the traditional approach. These two diverse approaches toward applying the scientific method are comparable to looking through opposite ends of a telescope.

A key feature of the systems approach is its critical emphasis on analyzing the interrelationships among system elements. Rather than dividing the system into smaller and more manageable components prior to analysis, the elements are brought together and kept together for analysis and management. Only in this way can total system performance be estimated and measured. The systems approach thus provides the only means of determining the influence of each element on the characteristics of the total system as well as the effects that any set of decisions or operating conditions will have on characteristics of the system.

Definitions. A variety of meanings has been assigned to terms that describe the areas of complex systems, but no attempt will be made here to suggest definitions of universal application. Instead, the following definitions are in accordance with the meaning of the terms as they are used in this chapter.

Systems Approach. The term "systems approach" is used by different people in different contexts. Even within a single discipline, semantic confusion often occurs. Three basic categories of activities exist which, individually or collectively, may be considered as parts of the systems approach—systems analysis, systems engineering, and systems management. They are three-dimensional in that they operate in different domains. These three orthogonal vectors of the systems approach can be considered as forming a system space, as illustrated by Figure 1-1.

Systems Analysis. Systems analysis includes investigation of system objectives, selection of criteria for evaluating alternative solutions, conceptualization of alternative solutions, examination of the feasibility of proposed solutions, evaluations of

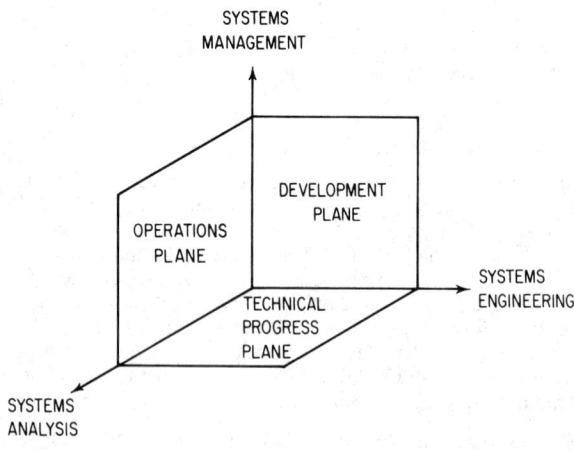

Fig. 1-1. System space.

feasible solutions, selection of the preferred solution, and development of functional specifications for the preferred solution.

Systems Engineering. Systems engineering is the top-level engineering process usually associated with the development or major modification of a complex system. It is a creative development process not unlike basic engineering, but it does not involve the detailed design and development of equipment or system elements. It has its origins in the total systems plan or concept which is developed in systems analysis. Systems engineering provides for continual refinement of the plan until a real system exists. It identifies and specifies the subsystem elements that will be assembled, engineered, procured, developed, tested, and evaluated in accordance with a development plan. The development plan is an action plan of tasks, activities, and milestones. It involves time and resources so that existing and new elements can be integrated into a new system able to meet predetermined system requirements.

Systems Management. Systems management includes the development of the procedures and the organizational structure for planning, directing, and controlling systems engineering activities and operations throughout the life cycle of the system.

Systems Approach Activities and Interrelationships. Figure 1-1 shows the vectors of system space, thus illustrating the systems approach in its three-dimensional set of activities. The first aspect is the systems analysis of a complex network of elements, either existing or planned. The second is a systems engineering approach that is used creatively to develop an entirely new system or to modify an existing system in a planned and orderly way. The third activity, systems management, includes the functions and procedures necessary to operate an existing system or to accomplish the development of a new system. The planes formed between these vectors are the technological, operations, and development planes. Most important for discussion in this Handbook are those activities related to the systems engineering process that is related to the development plane.

The systems analysis activity involves diagnosis of a real or hypothetical system within defined boundaries. Operations research is usually defined as systems analysis for real systems with components that already exist. Systems analysis evaluates the dynamics of the system and its sensitivity, not only to its boundary conditions, feedback, and environment, but also to the variable characteristics of each of its elements.

The only changes within an analytical system occur in the characteristics of its elements. Such a system might be a water distribution system which currently exists in a community, or it could be a representation of a system that is planned for the future. Because the ability to represent an element analytically is sometimes limited, simulation models or empirical data must sometimes be utilized.

The use of systems analysis and simulation not only enhances the ability to specify the important variables of a system, but also enables the analyst to measure or predict its performance with different environmental boundary conditions. This permits identification of those critical parameters and alternative solutions most relevant to practical constraints. Such capability facilitates decisions which can result in the greatest benefit or least cost to the ultimate beneficiary. For example, in the case of urban problems, this would be the city dweller.

Systems management is concerned with tasks and activities that involve both men and machines. It can be applied to the singular use or continuous operation of an existing system such as a military system, a manufacturing production facility, or the waste management system of an urban community. This is termed systems operations management. Another form of systems management is systems development management. Systems development management is required for the organization and scheduling of all activities and functions necessary during the engineering development of a system.

SYSTEMS ENGINEERING PROCESS

For the purpose of this discussion, four functions—systems analysis, design, operation planning, and management—are considered the elements of the systems engineer-

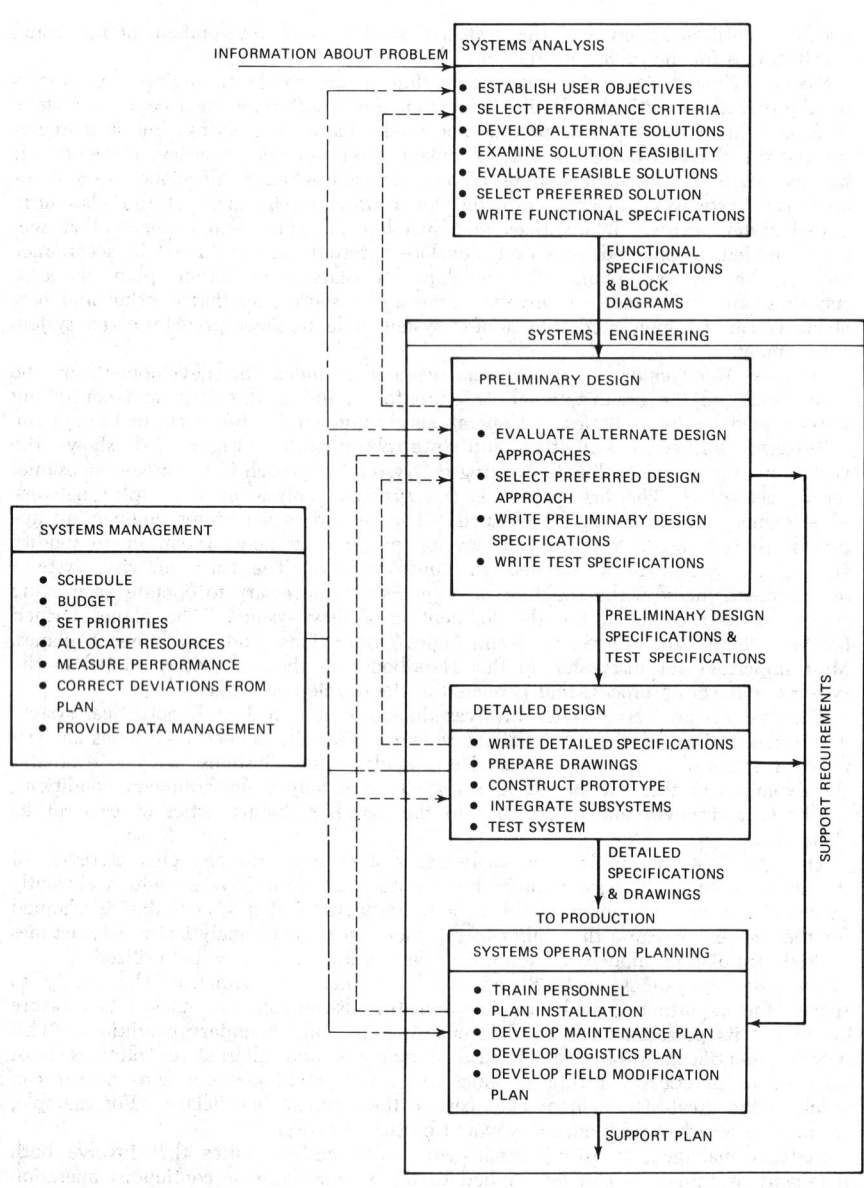

INFORMATION ABOUT PROBLEM

SYSTEMS ANALYSIS
- ESTABLISH USER OBJECTIVES
- SELECT PERFORMANCE CRITERIA
- DEVELOP ALTERNATE SOLUTIONS
- EXAMINE SOLUTION FEASIBILITY
- EVALUATE FEASIBLE SOLUTIONS
- SELECT PREFERRED SOLUTION
- WRITE FUNCTIONAL SPECIFICATIONS

FUNCTIONAL SPECIFICATIONS & BLOCK DIAGRAMS

SYSTEMS ENGINEERING

PRELIMINARY DESIGN
- EVALUATE ALTERNATE DESIGN APPROACHES
- SELECT PREFERRED DESIGN APPROACH
- WRITE PRELIMINARY DESIGN SPECIFICATIONS
- WRITE TEST SPECIFICATIONS

PRELIMINARY DESIGN SPECIFICATIONS & TEST SPECIFICATIONS

SYSTEMS MANAGEMENT
- SCHEDULE
- BUDGET
- SET PRIORITIES
- ALLOCATE RESOURCES
- MEASURE PERFORMANCE
- CORRECT DEVIATIONS FROM PLAN
- PROVIDE DATA MANAGEMENT

DETAILED DESIGN
- WRITE DETAILED SPECIFICATIONS
- PREPARE DRAWINGS
- CONSTRUCT PROTOTYPE
- INTEGRATE SUBSYSTEMS
- TEST SYSTEM

DETAILED SPECIFICATIONS & DRAWINGS
TO PRODUCTION

SYSTEMS OPERATION PLANNING
- TRAIN PERSONNEL
- PLAN INSTALLATION
- DEVELOP MAINTENANCE PLAN
- DEVELOP LOGISTICS PLAN
- DEVELOP FIELD MODIFICATION PLAN

SUPPORT PLAN

SUPPORT REQUIREMENTS

- - - - - - INDICATES ITERATIVE NATURE OF SYSTEMS ENGINEERING
— — - — INDICATES MANAGEMENT CONTROL FUNCTIONS

FIG. 1-2.   Systems engineering process.

ing process.   This process encompasses existing and future operations and places emphasis on the system life cycle—from the conception of a need to the operation, maintenance, and service of an installed system.   The relationship of these four functions of the systems engineering process is shown in Figure 1-2.

Systems Design.   Systems design is divided into two parts, preliminary design and detailed design. In preliminary design, the approaches for implementing the functional specifications developed during the systems analysis phase are evaluated and a design approach is selected. Detailed design involves translating the design approach developed during preliminary design into detailed hardware and software specifications, preparation of engineering drawings, construction or simulation of a prototype, integration of subsystems and components, and systems testing.

Systems Operation Planning. Planning includes the preinstallation activities required to support the system after delivery to the customer. These activities include training of operations and maintenance personnel; preparing technical manuals; developing plans for installation, maintenance, logistics, and failure reporting; and field modification of equipment.

Example: Urban Transit System. The following example illustrates the systems engineering process of Figure 1-2. The problem is to improve the passenger transportation system in an urban area, and the systems analysis phase will be concerned with evaluating the alternatives of building additional freeways, installing a rail transit system, improving the traffic control system, expanding the bus system, using minicars, staggering of work shifts, and similar possibilities. Only some of these alternatives may be feasible, because of constraints such as available funds, desired project completion date, state of technological development, and social and institutional factors (8 A.M. to 5 P.M. work shift).

As indicated in Figure 1-2, the systems analysis phase provides selection of an alternative or combination of alternatives and generates functional specifications and block diagrams. Functional specifications stipulate the requirements that must be satisfied by the system, independent of the particular hardware and software which will ultimately be employed.

In the example, if rail transit had been selected as the feasible solution, functional specifications would describe passenger carrying capacity, train routes, number and location of platforms, passenger loading at each platform, passenger wait times, and travel time. Type of propulsion, number of rails, number of cars, track gage, horsepower, speed, computer control system, and the like would not be included.

Based upon the functional specifications, several design approaches would be conceived and evaluated during the preliminary design phase. These might consist of trade-off studies of monorail versus dual rail, electric versus diesel propulsion, and manual control versus automated control. Preliminary specifications, which support the selected design approach or approaches, are the output of the preliminary design phase.

In the detailed design phase, detailed hardware and software specifications are developed to satisfy the requirements of the selected design approach. For example, detailed specifications would involve dimensions, weight, and materials of railway cars; number and size of seats and windows; track gage; station dwell time; headway; door-open time; speed between stations; speed control; and the like. The detailed design phase produces detailed specifications for the manufacture of railway equipment and the construction of railroads and stations. In addition, a prototype railway car might be built and tested, and a computer simulation of the entire train operation might be constructed.

Early in the preliminary design phase and continuing until system start-up, requirements are developed to support the system after it is operational. These include train and track maintenance programs, training of train operators and station controllers, number and location of service yards, spare parts supplies, and provisions for reporting equipment troubles and making equipment modifications after delivery.

Systems Management. Systems management is not a phase of systems engineering. It is the controlling function that operates throughout the life cycle of the system. Its purposes are to plan, monitor, and control. Among the myriad systems management activities in the example are obtaining right-of-way and street closure, rerouting traffic, procuring materials and equipment, hiring, identifying key milestones and the schedule critical path, approving specifications, and evaluating and selecting contractors.

Project documentation is one of the key control elements of systems management.

Placing documentation requirements on contractors allows the project status to be monitored and measured. Some of the required documents are design and test specifications, schedules, identification of schedules, identification of schedule bottlenecks, test results, engineering drawings, and fund expenditure reports.

Value of the Systems Engineering Process. The chief value of the systems engineering process is that it provides a logical procedure for identifying and evaluating objectively a range of possible solutions.

The methodology of systems engineering forces the analyst to define objectives, develop performance criteria, specify constraints, and weigh each alternative against all criteria. A common pitfall in problem analysis is to become involved in considerations of hardware and methods of implementation before all objectives have been clearly stated and before all alternatives have been identified and evaluated. Thus the solution becomes locked in at an early stage to a particular set of hardware specifications. The use of the systems engineering process will help the industrial engineer avoid this type of pitfall.

The systems engineering process is particularly applicable to the development of complex man-machine systems utilized in the industrial and business fields. Examples are:

1. Process and production control
2. Communication networks
3. Environmental management and control
4. Management control
5. Stock market analysis and reporting
6. Product distribution
7. Transportation
8. Information networks
9. Natural resource recovery and distribution

SYSTEMS ANALYSIS

Systems analysis consists of the following:
1. Problem definition
2. System objectives
3. System boundaries
4. User requirements analysis
5. System effectiveness measures
6. Functional analysis
7. Constraints evaluation
8. Delineation of feasible alternatives
9. Evaluation of feasible alternatives

Problem Definition. The first step in systems analysis is to define the problem. It is important to examine critically the statement of the problem by the user or customer. Does the statement express the real problem to be solved?

An interesting example of this principle is given by Quade:[1]

> For fiscal year 1952, Congress authorized approximately $3.5 billion for air base construction, about half to be spent overseas. The RAND Corporation was asked to suggest ways to acquire, construct, and maintain air bases in foreign countries at minimum cost. The analyst who reluctantly took on this problem regarded it at first as essentially one of logistics. He spent a long time—several months, in fact—thinking about it before he organized a study team.
>
> Although he had little of the information needed to make recommendations, he was able to see the problem in relation to the Air Force as a whole. He came to the conclusion that the real problem was not one of the logistics of foreign air bases, but the much broader one of where and how to operate

---

[1] E. S. Quade and W. I. Boucher (eds.), *Systems Analysis and Policy Planning: Applications in Defense*, RAND Corporation, June, 1968, pp. 36–37.

them in conjunction with the base system chosen. He argued that base choice would critically affect the composition, destructive power, and cost of the entire strategic force; therefore, it was not wise to rest a decision about base structure and location merely on economy in base cost alone. His views prevailed and he led the broader study. The results contributed to an Air Force decision to base SAC bombers in the continental United States and to use overseas installations only for refueling and restaging. An Air Force committee later estimated that the study recommendations saved over $1 billion in construction costs alone. In addition, it sparked a tremendous improvement in strategic capability, particularly with regard to survival, and stimulated a good deal of additional research on related questions.

System Objectives. It is also important to examine statements of objectives carefully for possible inconsistencies. An example of an inconsistent objective is the frequently expressed one of "maximizing effectiveness for the least cost." Because it is highly unlikely to maximize effectiveness and minimize cost simultaneously, the objective should be stated as the "maximization of effectiveness for a given cost" or the "minimization of cost for a given effectiveness." Another example of a poor statement of objective for a wholesale merchandising operation is "maintain inventory at a specified level." This statement of objective predetermines the solution to one that is probably not optimal in effectiveness or cost. A better way of stating the objective is either (1) the maximization of service to customers at a given total sales and inventory cost or (2) the minimization of total cost for a specified level of service.

Using the first criterion, it would be necessary to consider the requirements of the sales force in satisfying customer demands for the company's products, in addition to the problems and cost of handling inventory in the warehouse. This would involve making the optimum allocation of resources—personnel, equipment, and time—across both the sales and inventory functions for a fixed expenditure. Using the second criterion, the problem is to minimize the total cost of sales and warehousing operations for given quantities of product and the delivery times that must be achieved.

The choice of whether to hold cost or effectiveness fixed is largely a matter of the nature of the problem. The decision maker may have a fixed budget that he wishes to allocate optimally, or he may be committed to providing a specified level of service or effectiveness and desires to minimize his cost. Of course, the problem could be solved both ways.

System Boundaries. An impediment to system optimization occasionally arises because it is impractical or not permissible to analyze the entire problem. This situation may result from lack of resources to do the total analysis, the extreme complexity of the entire analysis, the organizational structure, or political factors that prevent the analyst from investigating the total problem. When an analysis of the total problem is not possible, optimization of each subsystem may be achieved but the total system may be suboptimal. System studies are usually restricted in some fashion, and most solutions are suboptimal. For example, optimal purchasing systems may be developed independently for each plant of a multiplant corporation. However, these individual systems may not provide an optimal solution to the total purchasing requirements of the corporation as a whole. A different situation is where it is possible to consider the whole problem, but the solution is restricted by political, economic, legal, or other factors so that optimization applies to only a limited set.

Because the scope of the problem limits the extent of system optimization, agreement on the boundaries of the problem should be reached between the sponsor of the study and the study team. The analyst should describe to the sponsor the solution he can expect to a problem specifically limited by the boundaries selected. The sponsor should specify problem boundaries for which he has obtained concurrence in his organization. Sometimes, a sponsor may specify system boundaries based on his desire or philosophy rather than on officially authorized constraints. A study team that accepts an assignment under this condition will be in a precarious position because of the difficulty in selling solutions to a problem when incomplete authority exists for solving the problem.

User Requirements Analysis.   User requirements analysis consists of the identification and evaluation of user needs.   Using a data system problem as an example, this analysis would provide answers to the following questions about user characteristics:

> What are the mission requirements?
> Who collects data?
> For what purpose (present and future) are data collected?
> Where are data collected?
> When are (were) data collected?
> Who uses the data?
> Why are the data used?
> Where are the data used?
> When are (were) data used?
> What are the additional user data requirements?
> What improvements are desired in the collection, processing, and display of data?

Additional information concerning the data that would have to be gathered includes:

> Type of data
> Type of media and nature of recording
> Relationships among data variables
> Accuracy and reliability
> Structure: content, hierarchy, and format of data
> Inhomogeneity: the degree to which the same or similar data elements are incompatible for multiple uses because of variations in format, accuracy, range, and so on

System Effectiveness Measures.   Before evaluating proposed solutions, it is necessary to establish a set of measures, or the criteria, by which the effectiveness of the system in satisfying user requirements can be judged.   Using the data system example again, the following effectiveness measures could be formulated.

Data collection:

> Usefulness of data to purpose of collection
> Usefulness of data to entire organization
> Volume of data collected per unit time
> Quality of data collected
> Rate of data obsolescence
> Time between data collection and availability to user
> Efficiency of data flow
> Provisions for privacy of information
> Extent of data compatibility among users
> Degree to which system can respond to changes in technology and user requirements

Data storage and processing:

> Size, variety, and quality of data bases
> Nature and percentage of requests that can be answered to the satisfaction of the requester
> Response time for answering requests
> Capability for absorbing increases in data collection volume, data analysis requirements, and requests for information
> Capability for rapid dissemination of data to user

Capability for providing correlations of related data, which may exist in diverse formats, codes, and data bases

Capability of providing various levels of data storage and response times compatible with user needs

Functional Analysis. Before quantitative values can be assigned to measure the effectiveness of each proposed solution, an analysis must be made of current and future functions that the system is to perform in satisfying user requirements. This is called functional analysis. As described by Hall.[2]

> This analysis starts with a statement of boundary conditions and desired inputs and outputs and proceeds to a detailed list of functions or operations that must be performed. Each function in a system possesses inputs and outputs. Inputs and outputs of functions are matched to determine the required sequence of operations or information flow. The problems that exist at the interface between functions are some of the most important to be resolved in system analysis. The problem frequently occurs because individual subsystems are independently designed, or the division of labor among subsystem activities results in some incompatibilities due to different approaches or lack of communication among the activities.

Block diagraming is important to functional analysis. It shows inputs, outputs, time relationships, information flow, and the functions to be performed at each stage of the system. The diagrams also indicate how inputs are transformed at each stage into outputs, which become the inputs to the next stage.

The elements to be identified in all block diagrams for each input and output are:

What is the source of the input?
When does the input arrive?
How frequently does the input arrive?
What is the delay between input and output?
What is done to the inputs to create an output?
When does the output occur?
How much output occurs?
How frequently is there an output?
What is the destination of the output?

The major characteristic of a block diagram is that it depicts flow—whether the flow be men, materials, money, or information. It also gives the time and sequence dependencies of the elements of the system. A portion of a system block diagram and a method of labeling system variables are shown in Figure 1-3. An input-output table is also important to functional analysis. Table 1-1 identifies for each system function the inputs and outputs, input and output characteristics, and operations performed by the function.

The use of block diagrams and flow analysis is illustrated by this information system example. Data collection activities are defined and forecast in terms of the total collection requirements as a function of time, and flow charts are developed for evaluating the nature and effectiveness of data flow. These charts show specific functions and organizations involved in data collection, storage, processing, and dissemination. The charts also identify time delays in the flow of data. Two types of data flow are involved in this analysis, interorganization and intraorganization.

Interorganization Flow Analysis. For interorganization flow analysis, it is necessary to identify the type, volume, frequency, and criticality of data flowing between organizations. Interorganization flow analysis identifies the following:

Importance of a given organization as a data provider to other organizations
Importance of an organization as a receiver of data from other organizations

---

[2] Arthur D. Hall, *A Methodology for Systems Engineering,* D. Van Nostrand Company, Inc., Princeton, N.J., 1962, pp. 111–112.

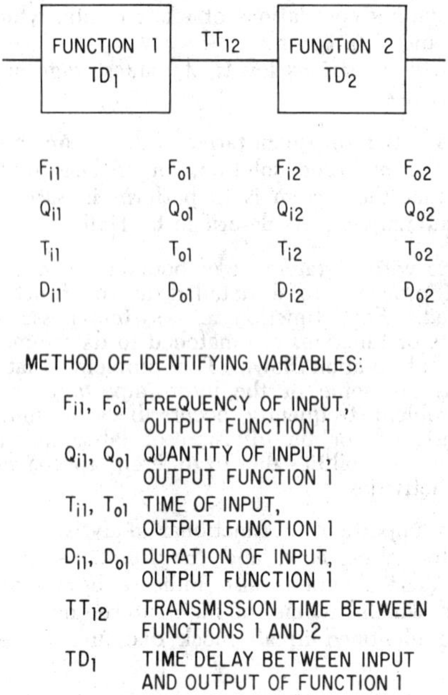

```
        ┌──────────────┐  TT₁₂  ┌──────────────┐
    ────┤  FUNCTION 1  │────────┤  FUNCTION 2  ├────
        │     TD₁      │        │     TD₂      │
        └──────────────┘        └──────────────┘
```

$$F_{i1} \qquad F_{o1} \qquad F_{i2} \qquad F_{o2}$$

$$Q_{i1} \qquad Q_{o1} \qquad Q_{i2} \qquad Q_{o2}$$

$$T_{i1} \qquad T_{o1} \qquad T_{i2} \qquad T_{o2}$$

$$D_{i1} \qquad D_{o1} \qquad D_{i2} \qquad D_{o2}$$

METHOD OF IDENTIFYING VARIABLES:

F_{i1}, F_{o1} FREQUENCY OF INPUT,
OUTPUT FUNCTION 1

Q_{i1}, Q_{o1} QUANTITY OF INPUT,
OUTPUT FUNCTION 1

T_{i1}, T_{o1} TIME OF INPUT,
OUTPUT FUNCTION 1

D_{i1}, D_{o1} DURATION OF INPUT,
OUTPUT FUNCTION 1

TT_{12} TRANSMISSION TIME BETWEEN
FUNCTIONS 1 AND 2

TD_1 TIME DELAY BETWEEN INPUT
AND OUTPUT OF FUNCTION 1

FIG. 1-3.   Example of block diagram used to describe system characteristics.

Amount of overlap which exists in providing data
Probable interorganization flows of data as new collection, processing, and
  dissemination requirements evolve
Diversity of data requirements
Data of critical importance
Areas where data deficiencies exist

Intraorganization Flow Analysis.   Intraorganization flow analysis is concerned with
describing and identifying the internal data requirements of organizations.   However,
it also describes the requirements at the input-output interface.   Intraorganization
flow analysis identifies the following:

Variety and magnitude of data inflows
Variety and magnitude of data outflows

TABLE 1-1.   Input-Output Table Example
Function: Inventory Control—New Stock Received

Characteristic	Input: stock received transaction	Output: excess inventory report
Frequency.........	Twice per month	As required
Quantity..........	200	1
Time.............	1st and 15th	When condition detected
Duration..........	1 day	1 per week

Processing performed: Add received stock to quantity on hand and compare
with upper stock limit.   Report if new stock level exceeds upper limit.

Variety and magnitude of data stored and processed

Current data handling capability, based on variety and magnitude of inflows/outflows, storage, and processing

Note that no mention has been made of hardware for carrying out systems functions. This is characteristic of functional analysis. As Hall cautions, discussion of hardware in the early stages of planning confuses the distinction between the job to be performed and the method of implementation. In the first stages of systems analysis, solutions that are constrained by equipment characteristics should be avoided. Specific hardware and software are considered during the systems design phase.

Other items important to functional analysts, such as simulation and queuing theory, are described in Sections 9 and 10.

Constraints Evaluation. A major part of the systems analysis task is the definition of the boundary between the system and the environment. As described previously, this task involves the clarification and establishment of the parameters of the problem to be solved and definition of the specific areas within the general problem to be studied. The steps involve consideration of general political, physical, sociological, economic, legal, technological, and temporal factors. In addition to these boundary conditions, there are some additional boundaries called constraints. These include all other boundaries that limit the area of feasible, acceptable, or permissible solution, and that fix many of the external and internal properties of the system. The identification of constraints and their effects on overall systems design cannot proceed independently of the other steps in the process; instead, the constraints are recognized and defined during the other steps. As a list of applicable constraints is derived, it is possible to explore the interrelation of their effects on the overall system and the sensitivity of system objectives to variations in the constraints.

The identification of constraints, together with their impact on system effectiveness, is an extremely important yet often overlooked facet of problem analysis. By acting as a filter through which idealistic systems needs are passed, constraints analysis provides a realistic perspective of what can be practically achieved.

Constraints may be classified according to their spheres of influence—political, legal, economic, technological, physical, environmental, organizational, attitudinal, and social. Specific instances of constraints related to the data system problem include:

Low quality data and inadequate recording devices
Lack of standardization in data recording and formats
Inadequate data collection and handling procedures
Gaps in data comprehensiveness and quality
Incompatibilities of hardware and software
Scarcity of appropriately trained personnel
Privacy of information considerations
Budgetary limitations
Conflicting objectives in system development
Vested interests
Lack of incentives and motivation
Lack of coordination in system development and operation

The evaluation of constraints includes:

Impact of constraint on the achievement of system objectives
Effect of constraint on the system operation
Effect on system effectiveness of reducing or eliminating the constraint and the cost of doing so

Delineation of Feasible Alternatives. At this point in systems analysis, alternative solutions that satisfy system constraints are developed. Each solution consists of a series of functions which are described by block diagrams with values of effectiveness measures appended. Effectiveness measures are quantified for each solution. In the context of the data system example, effectiveness measures—such as input and output rates, response time, storage capacity, or computational speed— are quantified. The objective of this phase is to develop alternative solutions without serious consideration of costs. Later, during the evaluation of feasible system alternatives, both effectiveness and costs are evaluated.

System flexibility is an important consideration in the development of alternatives. In an information system, for example, it is necessary to determine the degree of hardware and software expandability and modularity necessary to accommodate future data base size and data input and output rates. Data rates and volumes must be projected several years in advance. A modular design concept is evolved so that added increments of hardware capacity and additional and expanded computer programs can be implemented without major revisions in the systems or programs design. This is accomplished by predicting future hardware and software requirements and by designing programs that will be compatible with the peripheral equipment and memory sizes required in the future.

Evaluation of Feasible Alternatives.   After alternative system configurations have been synthesized and personnel and other requirements have been established for each alternative, it is necessary to compare the alternative systems. A typical trade study matrix used to conduct such a comparison is illustrated by Figure 1-4. It is desirable to analyze cost and effectiveness in terms of dollars. In many studies, the analysis is restricted to an evaluation of cost and to some physical attributes of the system, such as response time. An adequate analysis cannot be performed unless both parts of the relationship are evaluated in commensurate terms. Therefore, physical levels of effectiveness are translated into dollar values. Some qualitative factors, such as better and more up-to-date information, are occasionally difficult to translate into dollar terms. For an information system, however, this qualitative statement could be translated into quantitative measures of reduction in response time (for example, five days to five minutes) and greater data currency (current to within one week rather than one month). These improvements in physical attributes can then be translated into dollar values in terms of the labor time saved by the user in obtaining information and the cost of lost productivity in waiting for the information to arrive.

The cost effectiveness analysis, in terms of an information system example, consists of two phases:

1. An analysis is made of the total demand for services to be provided by the system for several years, so as to compute the total dollar benefits that will accrue to the users of the system over a five- to ten-year period.

2. To the extent possible, user benefits are translated into dollar amounts. This calculation is made by summing the savings accruing to all users from using the information system as compared with present methods. This procedure involves considerable cost analysis. A sampling procedure may be employed to estimate the total costs of using present data collection and analysis methods. Solutions are ranked on the basis of the summation of the difference in benefits and costs over the anticipated life of the system. A discount factor, representing the opportunity cost of alternative investment possibilities, is used in the calculations.

The synthesis of alternative systems configurations and the cost effectiveness analysis constitute an iterative process. Trade-offs between effectiveness and costs become apparent once the feasible alternatives are delineated. Some alternatives may be eliminated from further consideration; other alternative systems may be modified as a result of the cost effectiveness analysis.

The criteria for selecting the optimum solution are based on the principle of maximization of net benefits over the anticipated life of the system.

Finally, the effectiveness measures of the selected solution are documented in the form of functional specifications. These specifications are the major input to the next phase of systems engineering—preliminary systems design. These specifica-

NOMENCLATURE
Trade-Off Jettison Stage 1

COMPARISON MATRIX OF DESIGN APPROACHES

FUNCTIONAL AND TECHNICAL DESIGN REQUIREMENTS*	1 ELECTRICAL APPROACH	2 HYDRAULIC APPROACH	3 PNEUMATIC APPROACH	SELECTION
Power Use common airborne pneumatics (helium). Physical Locate helium pressure flask in Stage 1 assembly. Interface Use available 28 VDC airborne electrical system. Procurability Use commercially available parts where possible.	A signal from guidance through the autopilot programmer energizes a solenoid which physically disconnects the fastening device. Energy Source: Battery DISCUSSION* Pro 1. Simple, straightforward logic. Self-contained power source. 2. No problem to design effective manual override display and control. Con 1. The solenoid requires too much power, drain on battery. 2. Surges cause voltage spikes and interference with other electronic systems. 3. A relay is needed to translate the guidance signal into solenoid power control. Relays and solenoids have a poor reliability history.	A signal from guidance through the autopilot programmer controls a hydraulic valve. Energy to actuate the unfastening device is drawn from the anticipated airborne hydraulic power supply. DISCUSSION* Pro 1. The electrical energy required to operate a hydraulic valve is less than under Solution 1. 2. Hydraulic power will be available in both Stage 1 and Stage 2. 3. No problem to design effective manual override display and control. Con 1. This method would involve tapping the stability control system, causing frequency response problems. 2. Possible hydraulic leakage could cause a contamination problem.	A signal from guidance through the autopilot programmer activates a pneumatic valve (for trade-off selection) which releases pneumatic pressure into a cylinder/piston. A push-rod actuates the unfastening mechanism. Energy will be supplied by the airborne pneumatic system. DISCUSSION* Pro 1. The electrical energy required to operate the pneumatic valve is less than under Solution 1. 2. Leaks would not cause catastrophic failure. 3. Tapping the existing airborne pneumatic system does not cause interference with other consumers of pneumatic energy. 4. No problem to design effective manual override display and control. Con 1. Helium lines are difficult to seal.	Performance 3, 2, 1 Maintainability 1, 3, 2 Reliability 3, 2, 1 Safety 3, 1, 2 Procurability 1, 3, 2 SELECTION Solution 3 Note: See Section 4 of trade study report for reasons of selection.

Means shall be provided to cause either automatic or manual (override by astronaut) separation of the Stage 1 engine section from the launch vehicle within 800 milliseconds of a staging command. (Reference function 1.1.2 and RAS Document 2-00002 14 Feb. 1964.)

Jettisoning Stage 1 shall not cause damage to the launch vehicle. (Reference function 1.1.2 and RAS Document 2-00002 14 Feb. 1964.)

Staging status data shall be monitored and relayed to ground control and spacecraft. (Reference function 1.1.2 and RAS Document 2-00002 14 Feb. 1964.)

* For example purposes, only representative requirements and partial discussion are listed.

FIG. 1-4. Trade study matrix. (*Source:* Systems Management, Systems Engineering Management Procedures, *Air Force Systems Command Manual* 375-5, *March 10, 1966.*)

TABLE 1-2.   Example of a Contract
End Item Detail Specification

Section 3.	Requirements.
3.1	Performance.
3.1.1	Functional Characteristics.
3.1.1.1	Primary Performance Characteristics.
3.1.1.2	Secondary Performance Characteristics.
3.1.2	Operability.
3.1.2.1	Reliability.
3.1.2.2	Maintainability.
3.1.2.2.1	Maintenance and Repair Cycles.
3.1.2.2.2	Service and Access.
3.1.2.3	Useful Life.
3.1.2.4	Environmental.
3.1.2.5	Transportability.
3.1.2.6	Human Performance.
3.1.2.7	Safety.
3.1.2.7.1	Flight Safety.
3.1.2.7.2	Ground Safety.
3.1.2.7.3	Nuclear Safety.
3.1.2.7.4	Personnel Safety.
3.1.2.7.5	Explosive and/or Ordnance Safety.
3.2	CEI Definition.
3.2.1	Interface Requirements.
3.2.1.1	Schematic Arrangement.
3.2.1.2	Detailed Interface Definition.
3.2.2	Component Identification.
3.2.2.1	Government-furnished Property List.
3.2.2.2	Engineering Critical Components List.
3.2.2.3	Logistics Critical Components List.
3.3	Design and Construction.
3.3.1	General Design Features.
3.3.2	Selection of Specifications and Standards.
3.3.3	Materials, Parts, and Processes.
3.3.4	Standard and Commercial Parts.
3.3.5	Moisture and Fungus Resistance.
3.3.6	Corrosion of Metal Parts.
3.3.7	Interchangeability and Replaceability.
3.3.8	Workmanship.
3.3.9	Electromagnetic Interference.
3.3.10	Identification and Marking.
3.3.11	Storage.
Section 4.	Quality Assurance Provisions.
4.1	Category I Test.
4.1.1	Engineering Test and Evaluation.
4.1.2	Preliminary Qualification Tests.
4.1.3	Formal Qualification Tests.
4.1.3.1	Inspections.
4.1.3.2	Analyses.
4.1.3.3	Demonstrations.
4.1.3.4	Tests.
4.1.4	Reliability Tests and Analysis.
4.1.5	Engineering Critical Component Qualification.
4.2	Category II System Test Program.

SOURCE: Systems Management, Configuration Management during Definition and Acquisition Phases, *Air Force Systems Command Manual* 375-1, *June* 1, 1964.

tions are concerned only with the functions that the system must perform and not with hardware. Table 1-2 is condensed from a contract end item specification. For example, functional specifications for the data system would stipulate data storage capacity and access time and not the hardware such as core, drum, disc, or tape.

SYSTEMS DESIGN

Systems design may be divided into preliminary and detailed design phases.
Preliminary Design. As described by Asimow,[3] the methodology for preliminary design consists of the following:

Screening of design approaches
Development of mathematical model
Sensitivity analysis
Compatibility analysis
Stability analysis
Optimization
Projections into the future
Testing the design concept
Simplification of design

Screening of Design Approaches. Several feasible design approaches are identified by evaluating the most promising approaches with respect to the system effectiveness measures developed during systems analysis. Feasible approaches are those which equal or exceed the values of system effectiveness stipulated in the functional specifications, and cost no more than the available funds.

Development of Mathematical Model. A mathematical model of the physical system is developed to analyze the behavior of the system.

Sensitivity Analysis. By manipulating the mathematical model, a determination is made of which parameters critically affect system effectiveness and which do not.

Compatibility Analysis. The degree of compatibility between subsystems is determined. Two subsystems are compatible when the output characteristics of one subsystem are or can be converted to the input characteristics required by the succeeding subsystem. The subsystems having the least effect on system output, as determined from the sensitivity analysis, can receive the major adjustments required to achieve compatibility.

Stability Analysis. The system is analyzed for its ability to withstand perturbations. Damping must be such that the system will return to equilibrium within a reasonable time after the occurrence of the perturbation.

Optimization. The best of the feasible design approaches is identified by evaluating these approaches with respect to a criterion function. The criterion function is formed by weighting the various effectiveness measures according to their importance in achieving system objectives.

Projections into the Future. The adequacy of the proposed design must be tested against the demands of a changing environment and the influences of evolving technology. Examples of environmental factors are population growth, economic growth, and shifts in consumer needs and preferences. Changes in these factors may radically influence the demand for the system in the future. Future technological developments may cause a design approach to be obsolete once the system reaches the marketplace. The selected design must not become prematurely obsolete due to changes in the environment and technology. It is necessary to predict the behavior of the system under changing conditions.

Testing the Design Concept. Laboratory or field testing of a prototype model is used to prove or disprove the selected design approach.

Simplification of Design. The selected design approach is examined to reduce system complexity by eliminating requirements that contribute little to effectiveness and substantially increase the complexity and cost of the system.

Iteration. As in all facets of systems analysis and design, this sequence of steps is highly iterative and does not proceed in a single, sequential fashion. Many

[3] Morris Asimow, *Introduction to Design,* Prentice-Hall, Inc., Englewood Cliffs, N.J., 1962, pp. 24–33.

of the steps may take place concurrently, and there may be much doubling back to previous steps as improvements in the design approach are identified. Figure 1-5 shows this iterative process. In addition, much similarity exists between some of the steps and the procedures discussed under systems analysis. For example, in both systems analysis and systems design, identification of feasible alternatives and sensitivity analysis are major considerations. The primary difference is the level of detail and the concern of systems design with hardware and software— matters of little interest during the systems analysis phase.

Selection of a Design Approach. The core of preliminary design, the selection of a design approach, may be illustrated by considering the several design approaches that might be employed for the design of an information system:

1. Real-time system
2. Batch processing system
3. Time sharing system

The selection of a particular design approach will be governed by the user requirements identified during the systems analysis phase. Each design approach has associated with it a particular response time and mode of operation. Mode of operation is the manner in which data are acquired and transmitted to the processing facility, such as on-line or batched and transmitted at a convenient time. In some cases, there will be little latitude in the selection of a design approach, because the requirements of the problem may dictate that a particular approach be employed. This would occur, for example, in the design of a process control system in which the process must be controlled in real time. In other

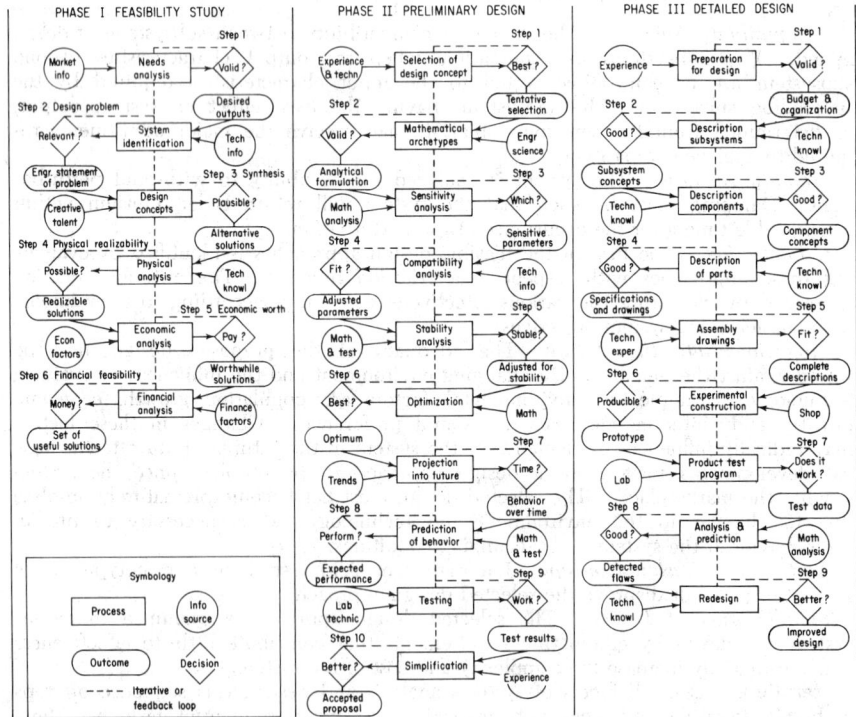

Fig. 1-5. Morphology of design. (*Morris Asimow,* Introduction to Design, © 1962. *Reprinted by permission of the author and Prentice-Hall, Inc., Englewood Cliffs, N.J.*)

cases, considerable leeway may be afforded in choosing among design alternatives. Such would be the case in the design of an engineering computational facility in which several methods could be employed to satisfy user requirements—batch processing, time sharing, or individual small computers.

Design Categories. One way to categorize design is to identify it as custom design (special purpose) or generalized design (general purpose). In the former, the objective is to satisfy the requirements of a single user or unique body of users, or to design a system for a specialized application. Examples of this type of design are the custom automobile and the airline reservation systems. These systems are characterized by lack of interchangeability of parts and subsystems with other systems, a limited market, and high performance and cost requirements. Generalized design strikes a balance among needs of a great body of users. Design requirements are set at average performance and cost levels, rather than at the maximum as in the case of custom design. An example is the mass production of automobiles and homes.

Another method of categorizing design is by centralization versus decentralization. In the former, system control resides in a single element of the system. In a decentralized system, control is distributed throughout the system. The advantage of centralization is the ability to maintain greater control over system operation and the reduction of duplication of services. Its disadvantage is less responsiveness, relative to a decentralized system, in meeting the needs of a particular geographic region or body of users. An example of a centralized system is one large central library which can provide services superior to those of many branch libraries but which also requires some users to travel long distances to utilize its services. The greatest application of decentralized systems occurs in situations in which the need to serve local markets outweighs the inefficiencies of duplication of services, such as in branch banking.

Although the choice between custom and generalized design and between centralized and decentralized system operation may be influenced by hardware considerations, it should be based primarily on system performance requirements. Hardware characteristics should be divorced from this analysis, at least during preliminary design. The reason for this is to prevent hardware characteristics from dictating the form of the design approach at the expense of performance.

Block Diagrams. The development of block diagrams depicting the design approach is an important element of preliminary design. These diagrams should indicate desired outputs, sources and characteristics of required inputs, and the transformations that must be performed on the inputs to convert them to useful outputs. These diagrams have the format of Figure 1-3. At this stage of design, specific types of hardware are not considered. Hardware details are added to the block diagrams during the detailed design phase.

Example of Preliminary Design Procedure. Consider the design of an airline passenger reservation system. Assume that the systems analysis phase has been completed and that functional specifications have been developed for the system. The next phase is the preliminary design, and the first step is the selection of the design approach. In this example, the selection of a design approach is constrained by the need to provide rapid service to customers who request reservations either over the telephone or in person at the agent's desk. Although an off-line system would result in some improvements in managing the seat inventory, it could neither satisfy the customer's desire for rapid information nor respond to rapidly changing seat bookings, cancellations, standbys, reconfirmations, and the like. Therefore, an on-line system is indicated for this application. However, within the context of an on-line system, a number of response time (time between passenger request and receipt of reply) versus system cost evaluations would be made to establish an acceptable response time which is not prohibitively costly. The evaluation of response time is one example of the use of effectiveness measures for systems design. Other evaluations suitable for effectiveness measures might be passenger volume and number and capacity of aircraft that can be accommodated in the system.

The primary use of mathematical modeling as a part of the preliminary design procedure is for queuing models, described in Section 10. In this manner, mathematical modeling aids in determining the number of agent terminals, the number of input-output channels, input-output channel speeds, main storage size and auxiliary storage size, and the access time required to satisfy specified response time and cost requirements. The design may also be simulated to study the behavior of the system over a long period in real time under a variety of passenger arrival and aircraft availability conditions. Simulation is also useful for analyzing the operation of parts of the system which are difficult to model mathematically, such as the computer operating (software) system.

Sensitivity analysis is employed to determine the effect on response time of changes in the arrival rate of messages, time distribution of message input, number of passenger reservation terminals, data transmission network configuration, main storage size and auxiliary storage size, and access time. The impact of these influences on response time must be determined for each reservation location, because the volume of passenger traffic and number of input terminals are different for each location. One of the objectives of this analysis is to determine the effect on response time of overload and slack conditions. If the response time is longer than desired for only a small percentage of the day, it may be tolerated if the cost of further reducing the response time is significant.

Compatibility analysis would be involved in achieving compatibility among passenger agent sets, data communication modes, common carrier communication facilities, and data communication control equipment. Differences in signal levels, data formats and codes, and data input and output rates would have to be resolved to have a properly functioning system when components are connected. The extent of incompatibility of commercially available hardware would be a strong factor in determining whether to make or to buy certain equipment. In addition to achieving hardware compatibility, the problem of "marrying" the human operator and the equipment requires solution. In this example, it would be necessary to identify human operator physical and psychological limitations which may be involved in the use of passenger agent sets under pressure and to design the agent sets to be in harmony with the requirements of the operator. This identification involves such considerations as type and size of displays, keyboard arrangement, number and types of agent controls, response time, and provisions for error recovery.

Stability analysis would be employed to determine the reaction of the system to unusual conditions such as a deluge of reservation cancellations, repeated reservations in the name of the same party (made by himself, his wife, and his secretary), or sudden cancellation of several flights as a result of mechanical difficulties or bad weather. Most systems can respond satisfactorily to average conditions. It is the response to unusual conditions that makes or breaks the system. Simulation has great advantages for stability analysis because the behavior of the system under unusual conditions must be ascertained before building hardware and designing computer programs and because unusual occurrences may be difficult to provide for in a mathematical model.

Although optimization can rarely be achieved in the sense that one system is clearly superior to all other systems with respect to a number of attributes, optimum seeking methods are used as an aid in selecting one or several systems from the set of feasible systems. In the example, several designs may provide the desired response time and be within the upper limit on cost. The optimum system could be defined as the least costly system of this set. It may be desirable to give further consideration to more than one system, if several are close in cost. In these situations, such other factors as the risk involved in achieving the system design may govern the selection.

For an airline reservation system, the performance of the selected design should be further evaluated in terms of projections of future environmental conditions. Such factors as future number of passengers and the resulting message traffic, number and capacity of aircraft, airline schedules, and number and capacity of airports would be considered. In addition to expected performance under future conditions, this analysis would also indicate the life of the initial investment. Therefore, equipment and software expansion capability and modularity are important. These attributes would be included in the set of design criteria referred to earlier.

In the case of an airline reservation system, a major test of the design approach occurs during mathematical modeling and simulation. While that phase of the preliminary design is in process, no actual piece of the system is available for test. For a system as large as an airline reservation system, it is impractical to build an experimental prototype of the entire system. However, it would be feasible to develop some of the key computer programs and to test these programs on a drastically scaled-down version of the ultimate equipment complex. This might be done for a single airport or a single passenger reservation terminal. Although this experiment could not be used to validate the total systems design, it might indicate deficiencies in design approach that could be corrected before committing funds and manpower to the total systems design.

After the design concept has been fully evaluated, a final examination is performed to identify possible simplification in design. Certain features of systems design may contribute an inordinate share of the total cost. An example of this in an airline reservation system is overdesign with respect to reliability; another is use of many systems which employ fully duplexed computer systems. In lieu of an overly costly design feature, moderate degradation of system operation may be considered, that is, longer response time for a short period, rather than full redundancy.

Detailed Design. In the detailed design activity, substance is given to the design concept by designing or selecting hardware and software systems that will achieve the objectives of the design approach. The following steps describe the procedure for the detailed design phase:

Preparation for design
Establishing hardware and software performance requirements
Preparation of design specifications
Design of subsystems and components
Design of parts
Preparation of assembly drawings
Experimental construction
System integration
Testing
Redesign

Preparation for Design. Technical and supporting manpower is assembled and organized. Budgets and schedules are established.

Establishing Hardware and Software Performance Requirements. During this portion of detailed system design, the design approach of the preliminary design is translated into detailed hardware and software specifications, the "build to" specifications. It is worthwhile to review the procedure used to determine hardware specifications. Functional "design to" specifications are developed during the systems analysis phase. These specifications are written in terms of system effectiveness requirements, system output characteristics, response time, and information flow rates. In the systems design phase, "build to" specifications must be derived, which translate the language of system effectiveness into the terminology of hardware and software characteristics. This can be achieved conveniently by the use of matrices in which the various hardware and software characteristics are related to system effectiveness measures. Table 1-3 shows an information system matrix application.

Preparation of Design Specifications. Specifications state the requirements that a system must meet. After hardware and software requirements are defined, a set of specifications is developed for such items as equipment, software, and personnel. An equipment specification format follows.

Equipment Specification
    System identification
    Subsystem identification
    Performance requirements

TABLE 1-3.   Effectiveness Measure—
Hardware/Software Characteristics Matrix

System effectiveness measure	Hardware/software characteristics
Data input rate*..........	Number and data rates of input unit
Request response time*....	Storage access time
	Input/output channel capacity and data rates
	Storage size
	Number of input terminals
Data base size*..........	Number and capacity of storage units
Data dissemination rate*...	Number and speed of output unit
Data compatibility†......	Hardware data representation
	Data conversion software
Generalized data management†..........	Availability of data management software

* Quantitative measure.
† Qualitative factor.

Reliability
Size
Weight
Materials
Power requirements
Air conditioning requirements
Environmental constraints
    Temperature
    Humidity
    Pressure
    Vibration
    Electrical
        Signal   Inputs   Outputs
    Name
    Source/destination
    Shape
    Amplitude
    Frequency
    Duration
    Time of occurrence
Other governing specifications
Equipment drawing reference
Assembly drawing reference
System block diagram reference
Test provisions
    Input signals
    Required outputs
    Duration of test
    Number of tests
    Temperature range
    Vibration range
    Humidity range
    Pressure range
    Criteria for acceptance

    Specifications are used by several activities—design engineering, production, quality control, and shipping.   Specifications should be tailored to the needs of each group.

For example, a production specification would place emphasis on manufacturing methods, whereas design specifications would emphasize performance requirements. The preceding example of an equipment specification format is suitable for engineering purposes, but not for production.

As more systems are designed for man-machine interaction as opposed to equipment acting alone, it is increasingly important to provide personnel specifications for some systems. The information required for a personnel specification includes:

1. Position
2. Job tasks
3. Environmental conditions
4. Requirements
    a. Attributes
        (1) Physical
        (2) Psychological
    b. Experience
    c. Age
    d. Education
5. Salary
6. Training program
7. Number required
8. Dates required

It is not necessary to define all systems design details before writing the first version of the specification. The specification can serve as a means of communication among systems and design engineers. Initial concepts can be documented in specification format and continually updated as changes occur. This procedure, however, should not be used to communicate with the production function because the design should be firm before any action is taken, such as ordering materials and establishing production procedures. Production planners and foremen should be invited to comment on the practicality of producing the proposed design, for this is a valuable source of design simplification and production cost reduction.

Design of Subsystems and Components. Each subsystem and component is treated as an entity and subjected to the same approach used during preliminary design.

Design of Parts. All design questions must be resolved by this stage. No ambiguities must remain regarding size, shape, or material. Purchase of parts is an alternative to in-house design and manufacture. Consideration of tolerances and related costs is important.

Preparation of Assembly Drawings. Assembly drawings are produced at various levels:

Subsystem—system
Component—subsystem
Part—component

During the preparation of these drawings, incompatibilities among subsystems, components, or parts may be discovered. Material, part, component, and subassembly lists are prepared.

Experimental Construction. A full-scale prototype is constructed. Production problems are considered at this point, and an attempt is made to simplify production procedures and to reduce the cost of production.

System Integration. A most important part of systems design involves the integration of all subsystems and components. This is normally a far more complex problem than the testing of individual subsystems and components, because all subsystems must operate as an entity. The accomplishment of this objective may be very difficult, due to several factors:

1. Conflicting objectives among subsystem designs
2. Differences in design approach as a result of separate teams assigned to each subsystem design
3. Incompatibilities in hardware and software among subsystems

4. Incompatibilities between the output of a subsystem and the inputs of the succeeding subsystem stage

To minimize these difficulties during system integration, it is important during preliminary design to give a great deal of attention to the problem of interfacing subsystems. Interfacing is the mating of two connecting subsystems so that the subsystems will perform according to specification when connected. The interfaces of concern are the input/output interfaces between subsystems. The outputs of one subsystem must be compatible with the input requirements of the succeeding stage. Inputs and outputs must be compatible in the following respects:

Signal level
Signal shape
Code
Format
Frequency
Duration
Time of occurrence

It is necessary to provide electrical compatibility by analog-to-digital converters, amplifiers, attenuators, and pulse shapers. Mechanical compatibility must also be achieved by choosing appropriate tolerances for mating parts and subsystems.

It is important to provide extensive coordination among groups that are designing the several subsystems. Complete documentation is required; specifications should be written for each interface. A system integration group should be established to deal specifically with the problem of ensuring the proper functioning of all subsystems as a system. This group should be independent of all other groups. It should be responsible for writing interface and system test specifications and for testing and accepting the complete system.

The system integration activity is initiated at the beginning of the preliminary design phase and continues until the prototype system has been successfully tested.

Testing. Test specifications are developed and tests are conducted on the various levels: component, subsystem, and system. Test specifications are derived from the systems design specifications. It is the responsibility of the systems integration group to perform or supervise testing of the entire system and to certify that the system is operating according to design specifications. If the system fails any tests, the systems integration group documents the nature of the failure, recommends corrective measures, and submits the problem to the design engineers for correction. After design modification, the system is retested by the systems integration group.

Redesign. If redesign is necessary, a useful criterion for selecting alternatives is to choose the alternative that will minimize the number of changes in other parts of the system. At this stage, the design concept is reasonably fixed, and major commitments have been made to the original concept. It is undesirable to introduce major changes at this point.

SYSTEMS OPERATION PLANNING

A relatively neglected facet of systems engineering is systems operation planning. Frequently, the systems engineering function is considered completed when the system is delivered to the customer. This view of systems engineering can contribute as much to the failure of a system as poor engineering design, because the user normally requires considerable assistance from the manufacturer in order to use the system effectively. This assistance includes maintenance, training, installation planning, spare parts supply, failure reporting, and field modification of equipment.

Systems operation has been neglected because designers are naturally more interested in solving the technical problem of equipment design than in problems related to the physical and psychological limitations of operating and maintenance personnel. In addition, the former approach of adapting the man to the machine consisted of selecting operations and maintenance personnel whose physical and psychological makeup was considered compatible with the characteristics of the equipment. This approach was employed during World War II for the selection of aviators. Later, it was recognized that man's limitations should be considered in systems design. The problems of systems operation must be resolved early in the design of the system rather than after the equipment is delivered. This is accomplished best by treating support functions such as equipment maintenance as an integral part of the systems design rather than as independent functions. Using this approach, repair policy is as important a design consideration as equipment performance.

As an illustration of the methodology of systems operation planning, maintainability (a major element of systems operation) is described in the following paragraphs.

Maintainability. As defined by the Department of Defense, "Maintainability is a characteristic of design and installation which is expressed as the probability that an item will conform to specified conditions within a given period of time when maintenance action is performed in accordance with prescribed procedures and resources."[4] Thus, maintainability is one of the two design factors associated with equipment availability; the other factor is reliability and is discussed in Chapter 6 of Section 8.

As system complexity has increased, more attention has been given to maintainability in systems design because of the high cost of system maintenance. Eleven percent of the Defense budget is spent on maintenance.[5]

Some complex weapons systems have been phased out because of excessive support costs. These excessive costs arose in part from inadequate support planning, due to contracts being awarded on the basis of initial hardware procurement costs rather than on total system life cycle costs—hardware costs plus support costs over the life of the system. This meant that little consideration was given in the design stage to the requirements and costs of supporting a system after it became operational. Maintainability, the largest support item, includes maintenance, preparation of technical manuals, spare parts provisioning, failure reporting, field modifications, and selection and training of maintenance personnel. To provide effective maintainability, the following factors must be considered early in the design and must be continually refined as the design progresses:

1. Repair policies
2. Analysis and prediction of maintainability
3. Design factors affecting maintainability
4. Human factors affecting maintainability

Repair Policies. Repair policies may be adopted which require the design of any of the following:

1. Fully repairable systems
2. Partially repairable systems
3. Nonrepairable systems

Each of these policies has a great influence on systems design and subsequent maintenance and field support activities. In general, the more time required for a system to be repaired to correct a malfunction, the greater the need for high reliability. In a nonrepairable system, maintenance is by replacement only; therefore, units are discarded rather than repaired. In this type of system, lower reliability can be tolerated for a given system availability because repair time is drastically reduced. An advantage of nonrepairable systems is that the caliber of user maintenance personnel does not have to be high, thus reducing personnel costs, training time, and training costs. Also, the amount and cost of maintenance manuals and quantity of spare parts can be drastically reduced. In nonrepairable systems, units can be encapsulated or sealed, thus providing greater protection against the environment and resulting in increased reliability. It has been estimated that nonrepairable design can reduce failure rate by 40 percent.[6] The disadvantage is the high cost of discarding units. A nonrepairable maintenance policy is particularly advantageous

[4] *Research and Development Material Maintainability Engineering,* Department of the Army Pamphlet 705-1, June, 1966.
[5] *Ibid.,* p. 6.
[6] *Ibid.,* p. 29.

categories of human factors considerations:

> Capabilities and limitations of personnel
> Equipment design
> Environmental effects

1. Capabilities and limitations of personnel. Consideration should be given to the requirements of the maintenance job for working space, weight lifting, and weight carrying. The capabilities and limitations of personnel to satisfy these requirements should be assessed. In addition, intelligence, aptitude, and skill also should be considered when developing maintenance policy.

One approach to organizing personnel for maintenance is to use the least capable personnel at user installations in conjunction with a no repair or minimum repair policy at the user level and to place higher skill personnel in central maintenance facilities.

2. Equipment design. Downtime can be reduced by providing free access for the inspection, service, repair, or replacement of equipment. The least reliable units should be the most accessible.

Frequently, the operation of equipment is made unnecessarily complicated by including maintenance controls with operator controls. The operator console should not be cluttered with maintenance controls. A good practice is to provide a separate maintenance panel for those and the displays required for maintenance and to place this panel well out of the visual range and working area of the operator. If feasible, the maintenance panel should be designed to be operated independently of the operator's console, and it should be possible to perform a limited amount of maintenance checking from this panel without interfering with the operation of the equipment. Thus, a limited amount of scheduled maintenance may be performed without halting equipment operation. Additionally, maintenance personnel are able to observe system performance during malfunctions and thus obtain better information concerning the source of the malfunction than is possible when the system is not in the operating mode.

Because it has been found that fault isolation accounts for 75 percent of repair time,[7] procedures for reducing this time are an important design objective. One procedure for achieving this objective is the application of a nonrepair policy at the user level, combined with centralized maintenance. By shifting the problem of fault location to the central repair facility, it can be accomplished under less demanding time constraints and with the aid of sophisticated diagnostic equipment.

3. Environmental effects. Considerable attention is usually given during equipment design to the environmental conditions under which equipment must operate. Unfortunately, the environmental conditions under which humans must operate are not always accorded equal attention. If a system is to operate in an adverse environment, this is another condition favorable to the application of a nonrepair policy at the user level, because the exposure time for performing maintenance can be reduced and units can be encapsulated for protection against the environment. This approach is highly attractive, because in certain types of field operations it may not be possible to control the environmental conditions under which maintenance must be performed.

Maintenance environmental conditions that should be considered in systems design include:

> *a.* Ambient air conditions: temperature, relative humidity, air circulation, and air purity
> *b.* Illumination
> *c.* Noise
> *d.* Vibration

SYSTEMS MANAGEMENT

Activities, Inputs, and Outputs. Systems management is comprised of the organization, activities, and procedures that are used to govern the systems engineering

[7] *Ibid.,* p. 105.

process. To understand the functions of systems management better, a review of the inputs and outputs of the four systems engineering functions previously described is presented below.

Systems engineering activity	Inputs	Outputs
Systems analysis........	Problem description User requirements Constraints Available funds, manpower, and time	Functional ("operate to") specifications of preferred solution System block diagrams
Preliminary design......	Functional ("operate to") specifications	Preliminary ("design to") specifications of selected design approach Test specifications Preliminary support requirements
Detailed design.........	Preliminary ("design to") specifications	Detailed hardware and software ("build to") specifications Drawings Prototype (constructed or simulated) Detailed support requirements
Systems operation planning	Support ("operate to") requirements	Support plan Staffing Training Installation Maintenance Logistics Field modification

Milestones. Probably the greatest contribution to effective systems management is the identification of checkpoints which can be used to monitor schedule, budget, and technical performance with sufficient lead time to correct deficiencies in project performance. These milestones, in addition to representing convenient points for monitoring project status, should also be chosen to provide decision points at which future courses of action can be selected without severely jeopardizing schedule or budget. A list of milestones, categorized by activity, follows:

Systems Analysis
  Approve systems analysis study plan
  Approve performance criteria
  Approve functional specifications
  Evaluate high risk areas (solutions which highly depend on the state of technological development)
  Prepare initial system cost and schedule estimates
  Prepare initial project management network (PERT)
  Establish project documentation requirements
  Recruit design personnel
Preliminary Design
  Approve design approach
  Approve preliminary design specifications
  Prepare subsystem cost estimates
  Prepare subsystem engineering and production schedules
  Update project management network
  Approve initial support requirements
  Prepare system hardware and software procurement plan (for items not produced internally)

Prepare plan for the delivery of systems to customers
Approve system test plan
Allocate personnel and other resources to design
Detailed Design
Approve detailed hardware and software specifications
Release specifications and drawings to production
Prepare component cost estimates
Prepare component engineering and production schedules
Update project management network
Approve detailed support requirements
Systems Operation Planning
Approve support plan
Staffing
Training
Installation (facilities planning)
Maintenance
Logistics
Field modification
Prepare spare parts production or procurement plan
Recruit maintenance and operations personnel (if not provided by user)
Prepare support plan cost estimates
Prepare support plan schedule
Update project management network

Project Monitoring. In addition to the management control afforded by the major milestones, it is also desirable to provide continuous monitoring for adherence to schedule, budget, event sequence, documentation requirements, and technical performance. A continuous monitoring capability is more easily accomplished with an automated data processing system than with manual procedures. This is due to the capability of an automated system to scan large files of schedules, costs, and technical data; to compare actual project status with desired results; and to produce reports of out-of-tolerance conditions. The management control system can be made much more effective if a predictive capability is provided in addition to the standard accountability functions. By predictive capability is meant procedures and reports that warn the program manager of projected cost overruns, slippage in schedule, failure to meet technical specifications, and the like, if the project continues according to present plans. An early warning system allows the program manager to take corrective action to prevent such problems from arising. This type of system also allows the program manager to forecast the outcome of proposed project changes by processing simulated instead of real inputs.

The basis of such a system is the continual updating of a PERT network with project activity and expenditure information so that remaining activity completion times, project completion time, critical path, and expenditure forecasts can be reforecast. Project control networks are described in Chapter 3 of Section 8.

In addition, it is desirable to schedule periodic design reviews to compare actual status with scheduled standards of performance, schedule, and cost. Such a review should be conducted in conformance with a previously established specific list of review parameters and their established target values and tolerances.

System Integration and Configuration Control. System integration and configuration control are two management activities that pervade every phase of systems engineering. Considering system integration first, systems management must ensure the following:

1. Compatibility, both mechanical and electrical, between components and between subsystems. Compatibility is achieved when a given component or subsystem delivers the voltage, current, torque, and so on required by the following stage in the system.

2. Compatibility of user organizations which must communicate if the system is to function. Compatibility is achieved when information from the generating organization is usable by the recipient organization. This involves

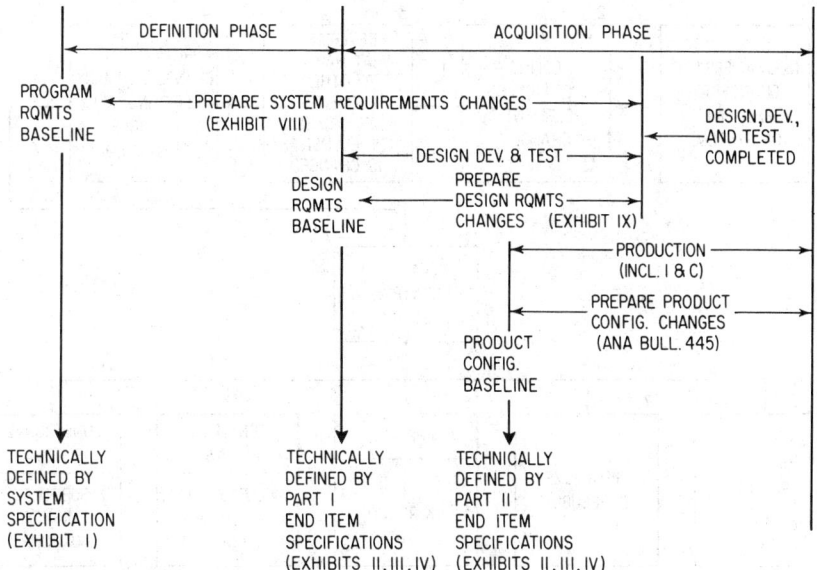

NOTE: EACH ENGINEERING CHANGE PROPOSAL (ECP) PREPARED USING EXHIBITS VIII & IX MUST
INCLUDE A SPECIFICATION CHANGE NOTICE (EXHIBIT VII).

FIG. 1-6. Baseline management.

consideration of format, quantity, content, frequency, timeliness, and routing of information.

A major mechanism for achieving system integration is to document the characteristics of all points in the system where hardware subsystems meet and where organizations must cooperate for successful operation of the system. This involves describing physical, organizational, and personnel characteristics at each side of an equipment or organization interface. It is the function of systems design to resolve incompatibilities. It is the function of systems management to verify that interface problems have been resolved. This involves review of specifications for conformance to design requirements, which have been specified to achieve compatibility, and the review and approval of test procedures and system test results.

Configuration control refers to procedures that are employed to control and document changes made to hardware and software at various stages in the system life cycle. Specifications are the authority and reference for making changes. Specifically, a proposed change must not degrade the system performance which is documented in the specifications that are controlling at the time of the proposed change. Also, the controlling specifications must be revised to reflect changes that are made. The controlling specifications and the applicable time period for configuration control are as follows:

Time period	*Controlling specifications*
During systems analysis and preliminary design...............................	Functional specifications and block diagrams
During detailed design..................	Preliminary design specifications and engineering drawings
During production......................	Detailed design specifications, engineering drawings, and manufacturing specifications
During system test.....................	Test specifications
During systems operation...............	Functional specifications

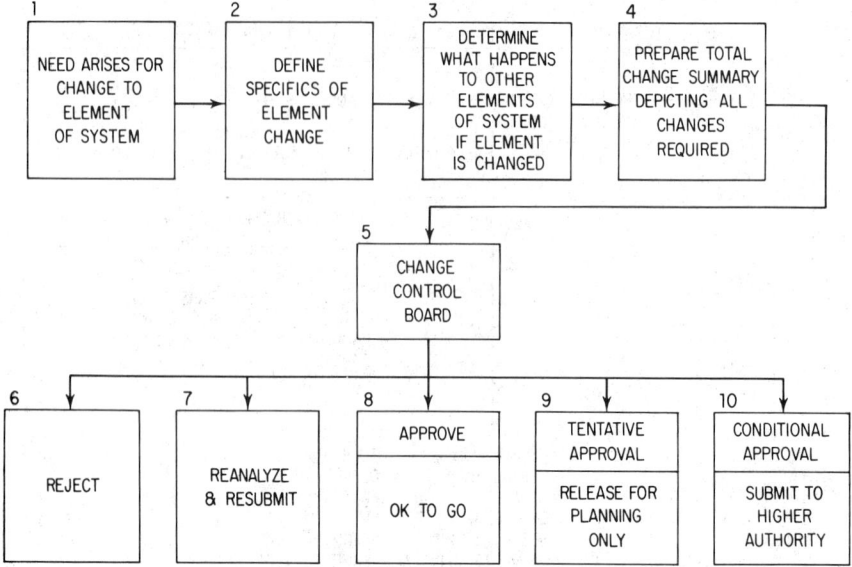

Fig. 1-7.   Coordinated change control system.

Because configuration change control is an extremely important process, an independent change control board is usually established to review and approve proposed changes that affect performance or cost as defined by the controlling specifications, and fully documented base lines are established sequentially for performance, design, and production, as illustrated by Figure 1-6.   The committee will approve changes consistent with overall system objectives which satisfy one or more of the following:

1. Increases performance
2. Reduces cost
3. Shortens system delivery time
4. Simplifies manufacturing processes
5. Results in faster procurement of components and parts
6. Results in obtaining better components and parts

Figure 1-7 shows some of the actions necessary to prepare for the change control board to carry out its functions properly, as well as some of the action categories it can use.

CONCLUSIONS

The systems approach to development or major modification of complexes composed of many singular components has been shown to be a totally integrated examination of the complete system as well as a part of the total environmental, logistic, maintenance, and operational atmosphere in which it must function.   This is a marked departure from the historical method of dealing with a system merely through individually segmented and understandable elements.

The strength of the systems approach is obvious in that all factors exerting an influence on the system from its initial inception throughout its operational life are considered.   The weakness of the approach is that the individuals involved in systems analysis, design, and operation tend to possess parochial viewpoints emphasizing their own specialty or technological area, which may be in the regimes of systems analysis, design, operational planning, or management.   In reducing the systems approach to fundamentals, it cannot be said that its objectives, tools, or techniques are new. What is new is that all these capabilities are utilized with a single, ultimate objective in mind; and all phases of analysis, development, and operational plans are pursued in this context, considering at the same time factors such as available technology, performance requirements, operational limitations, and budgetary constraints. Thus, it is the management aspect of the systems approach that provides the consolidating methodology to merge the available or projected techniques and capabilities into a unified treatment.

BIBLIOGRAPHY

Asimow, Morris, *Introduction to Design,* Prentice-Hall, Inc., Englewood Cliffs, N.J., 1962.

Chestnut, Harold, *Systems Engineering Methods,* John Wiley & Sons, Inc., New York, 1967.

Hall, Arthur D., *A Methodology for Systems Engineering,* D. Van Nostrand Company, Princeton, N.J., 1962.

Hare, Van Court, Jr., *Systems Analysis: A Diagnostic Approach,* Harcourt, Brace, & World, Inc., New York, 1967.

Krick, Edward V., *An Introduction to Engineering and Engineering Design,* 2d ed., John Wiley & Sons, Inc., New York, 1969.

Quade, E. S. (ed.), *Analysis for Military Decisions,* RAND Corporation, R-387-PR, November, 1964.

Quade, E. S., and W. I. Boucher (eds.), *Systems Analysis and Policy Planning: Applications in Defense,* RAND Corporation, R-439-PR, June, 1968.

Research and Development Material Maintainability Engineering, Department of the Army Pamphlet 705-1, June, 1966.

Specification Practices, Department of Defense, MIL-STD-490, October, 1968.

Systems Engineering Management, Department of Defense, MIL-STD-499, July, 1969.

Systems Management, Configuration Management during Definition and Acquisition Phases, U.S. Air Force, Air Force Systems Command, AFSCM 375-1, June, 1964.

Systems Management, Systems Engineering Management Procedures, U.S. Air Force, Air Force Systems Command, AFSCM 375-5, March, 1966.

Systems Management, Systems Program Management Procedures, U.S. Air Force, Air Force Systems Command, AFSCM 375-4, May, 1966.

Walton, Thomas F., *Technical Data Requirements for Systems Engineering and Support,* Prentice-Hall, Inc., Englewood Cliffs, N.J., 1965.

Wilson, Ira G., and Marthann E. Wilson, *Information, Computers, and Systems Design,* John Wiley & Sons, Inc., New York, 1965.

Chapter **2**

Operations Planning, Scheduling, and Control

JOHN E. BIEGEL

Associate Professor, Department of Industrial Engineering, Syracuse University, Syracuse, New York

The function of operations planning, scheduling, and control may or may not be closely related to the industrial engineering function in any given enterprise. If the function operates largely on an empirical or cut-and-try basis—delivery dates established by "guesstimates," with expediters chasing overdue orders, for example— there is generally a minimum of contact with the industrial engineering function. As planning, scheduling, and control procedures become more sophisticated, however, and as greater reliance is placed on quantitative methods, the industrial engineer is likely to be called on more and more for assistance. His mathematical background and objective, problem solving approaches fit him to render a helpful service to the planning, scheduling, and control people, who may not be as well grounded in basic quantitative methodologies.

This chapter will discuss a number of quantitative techniques that have been used successfully to solve production problems which the industrial engineer will find it useful to understand.

OPERATIONS PLANNING

Modern society requires that business and industry supply its needs and desires at reasonable cost, in adequate quantities, and with goods and services of satisfactory quality. Operations planning's objective is to make a "best" effort to provide the desired goods and services at the right time. The term "operations" is used in the general sense of a productive activity. Operations scheduling puts together the details of the plan and specifies the equipment, manpower, and materials necessary for the accomplishment of the objectives of the plan. The control aspect arises in specification of the reaction of the operation to departures from planned output.

Operations cover a wide variety of activities. An operation might be a transportation system, an appliance manufacturing line, a consulting firm, a dry cleaning establishment, or a tool and die shop. All these enterprises have the common problem of planning, scheduling, and controlling inputs and processes to be able to meet the demands of the customer. The output must be available at a time specified by the customer.

Operations planning will be broken down into two major areas: forecasting and planning. Generally, in a competitive economy, forecasts of demand for a service or product are necessary. These forecasts are for the output of the operation rather than for input. The objective of planning is to translate the projected output into requirements for inputs of raw materials, manpower, equipment, facilities, management, and the like. Both forecasting and planning must be done for the long range as well as the intermediate and short range. The length of the forecasting or planning period is highly dependent upon the industry or the particular operation. The same techniques may generally be used, but the expected accuracy decreases as the time span increases.

Forecasting. A forecast is a prediction of the future. One hardly expects it to be exact, yet it must be as accurate as possible. Therefore it must be monitored and revised when necessary.

There are many techniques for forecasting; some are quantitative and others qualitative. The computer makes it possible to use sophisticated techniques. It removes the requirement for long manual computations.

One mathematical-statistical approach to forecasting is the use of regression applied to past data. This approach presumes that the past is the best representation of what might occur in the future. A general regression form that provides for the inclusion of trends as well as cyclic components is

$$d' = a + bt + U \cos \frac{2\pi}{N} t + V \sin \frac{2\pi}{N} t \qquad (1)$$

where d' = forecast demand
a = intercept
b = slope
t = time period, $t = 1, 2, 3, \ldots$
N = number of periods per cycle
U, V = constants

The constants a, b, U, and V are parameters which must be determined by regression. The method of finding these constants can best be shown in the form of a determinant:

$$\begin{vmatrix} d' & 1 & t & \cos \theta & \sin \theta \\ \Sigma d & n & \Sigma t & 0 & 0 \\ \Sigma dt & \Sigma t & \Sigma t^2 & n/2 & \Sigma t \sin \theta \\ \Sigma d \cos \theta & 0 & n/2 & n/2 & 0 \\ \Sigma d \sin \theta & 0 & \Sigma t \sin \theta & 0 & n/2 \end{vmatrix} = 0 \qquad (2)$$

where d = actual demand
$\theta = (2\pi/N)t$
Σ = sum from 1 to n $\left(\sum\limits_{1}^{n} \right)$
n = number of periods used in the regression

Forecasts should be made based upon data from an integral number of cycles and starting at a point where the seasonal effect is minimal. If there are no cyclic components, the constants U and V will be zero. If no trend exists, the constant b will be zero.

Another approach is to try several regression functions and use the one with the minimum least squares $[\Sigma(d' - d)^2 = \text{minimum}]$. Some possible candidates are functions in which different combinations of constants in equation (1) are chosen to be zero. Some possible choices are

$$d' = a = \bar{d}$$
$$d' = a + bt \tag{3}$$
$$d' = a + U \cos \frac{2\pi}{n} t + V \sin \frac{2\pi}{n} t$$

Determining a Forecast from Past Data. The data in Table 2-1 represent the actual demand for a seasonal product. The forecasting function from equations (3) and (1) and the standard errors of estimate are shown in Table 2-2. The minimum standard error (best fit) occurs with

$$d' = 451.1 + 12.6t + 194.4 \cos \frac{\pi}{6} t + 47.3 \sin \frac{\pi}{6} t \tag{4}$$

Forecasts for the next twelve periods using equation (4) are presented in Table 2-3. These forecasts become the basis for production planning for the next twelve-month period.

A comparison of the actual demand for the base period with the regression estimates is shown in Table 2-4 and presented in Figure 2-1. Inspection of Figure 2-1 reveals that there was a slight upsurge in demand in the fifth and sixth months. The fifth month was December. The data are adapted from a seasonal service which would be expected to exhibit increased demand in the holiday season. If an additional correction were made for this, the fit would be still better. In other words, the standard error of 57.45 could be reduced to a smaller value.

TABLE 2-1. Monthly Demand for Seasonal Product

Month	1	2	3	4	5	6	7	8	9	10	11	12
Demand	704.4	578.8	524.6	361.2	451.0	407.8	254.8	450.8	471.6	652.8	719	818.6

TABLE 2-2. Forecasting Functions and Their Standard Errors

Function	Standard error
$d' = 533.0$	165.67
$d' = 446.3 + 13.3t$	158.5
$d' = 533 + 207.0 \cos \frac{\pi}{6} t + 0.3 \sin \frac{\pi}{6} t$	63.80
$d' = 451.1 + 12.6t + 194.4 \cos \frac{\pi}{6} t + 47.3 \sin \frac{\pi}{6} t$	57.45

TABLE 2-3. Forecasts for Next Twelve Months

Month	13	14	15	16	17	18	19	20	21	22	23	24
Forecast	806.8	765.6	687.3	596.3	520.5	483.3	498.3	564.7	668.2	784.3	885.4	947.7

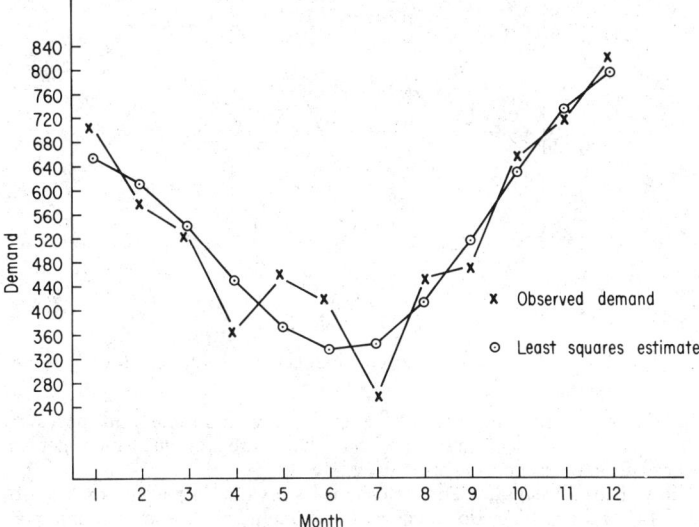

Fig. 2-1. Comparison of actual demand with regression estimates for base period.

TABLE 2-4. Comparison of Actual Data with Regression Estimates

Month.....	1	2	3	4	5	6	7	8	9	10	11	12
Demand ...	704.4	578.8	524.6	361.2	451.0	407.8	254.8	450.8	471.6	652.8	719.0	818.6
Estimate...	655.7	614.5	536.2	445.3	369.4	332.2	347.2	413.6	517.1	633.2	734.3	796.6
Difference..	+48.7	−35.7	−11.6	−84.1	+81.6	+75.6	−92.4	+37.2	−45.5	+19.6	−15.3	+22.0

Some Other Forecasting Techniques. Other techniques for forecasting vary from qualitative to quantitative. In general, it appears that a quantitative technique will be better than a qualitative technique, because the former lends itself to an analysis of its errors while qualitative procedures do not.

A method of forecasting that has gotten considerable attention is exponential smoothing. Exponential smoothing assumes that demand for the next period is some weighted average of the demands for the past periods. This is expressed as

$$d_1' = (1 - k)d_0 + (1 - k)kd_{-1} + (1 - k)k^2 d_{-2} + \cdots + (1 - k)k^n d_{-n} + \cdots \quad (5)$$

where k = weighting factor

d_{-i} = ith prior period

With equation (5), the forecasts are made period by period. By the inclusion of trend and cyclic components, it is possible to obtain forecasts farther into the future. Once the first forecast has been made, following forecasts can be made by using

$$d_1' = (1 - k)d_0 + k\bar{d}_0 \quad (6)$$

where $\bar{d}_0$ = forecast for the current period

The use of this method reduces the data storage in the computer, because only the current usage (d_0), the weighting factor (k), and the current forecast $(\bar{d}_0)$ are required for the next forecast.

To establish trend and cyclic components, it is necessary to have several cycles

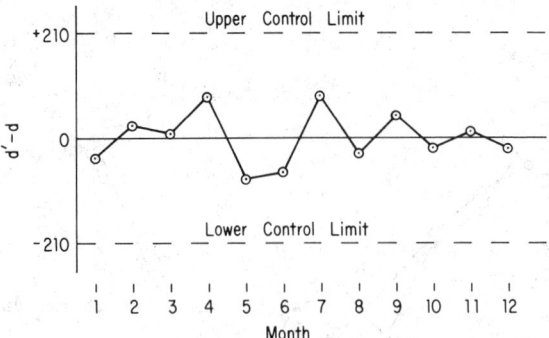

FIG. 2-2.   Moving range chart for seasonal product.

of data. It appears that at least three cycles are desirable and necessary. The use of this quantity of data presumes that the cause system which generated the demand was stable over the time span used.

Controlling the Forecast. The criterion for evaluating a forecast is its ability to predict the future. To do this, it must include all the components of the nonrandom variation of demand from period to period. It must exclude the random variations. One way of testing this conformity is by a control chart such as is used in quality control. A version of the control chart that is useful for this purpose is the moving range chart. The moving range is defined as the absolute value of the change of the "error" of the forecast for successive periods. Using the prior notation, we get

$$MR_i = |(d_i' - d_i) - (d_{i-1}' - d_{i-1})| \qquad i = 1, 2, \ldots, n \qquad (7)$$

The control limits for the chart are at $\pm 2.66\overline{MR}$, where $\overline{MR}$ is the average moving range. The center line of the chart is at zero, and variables to be plotted are the $(d_i' - d_i)$'s. The chart is constructed on the basis of the n observations used for determining the forecast. The moving range chart for the seasonal product has control limits of ± 210.4. The first moving range is $|(765.6 - 578.8) - (806.8 - 704.4)| = 84.4$. The average moving range is 7.91. The moving range chart is shown in Figure 2-2. The chart shows a stable pattern of variation. Thus, the function $d' = 451.1 + 12.6t + 194.4 \cos (\pi/6)t + 47.3 \sin (\pi/6)t$ has been determined to be a good forecasting function, and planning should proceed on the basis of the forecasts of Table 2-3.

The moving range chart can also be used to determine when a new forecasting function needs to be developed. To do this, the chart is extended in time as data become available. For the thirteenth month, the forecast is 806.8, while actual demand was found to be 827. The value plotted on the moving range chart is $806.8 - 827.0 = -20.2$. The actual demand was 20.2 units above the forecast, or the forecast underestimated the actual demand by 20.2 units. The actual demands generated for the next twelve months are shown in Table 2-5. The data in the last row of Table 2-5 are added to the moving range chart of Figure

TABLE 2-5.   Comparison of Forecast and Actual Demand for Seasonal Product

Month...	13	14	15	16	17	18	19	20	21	22	23	24
Forecast .	806.8	765.6	687.3	596.3	520.5	483.3	498.3	564.7	668.2	784.3	885.4	947.7
Actual ..	827.0	743.0	678.8	504.6	603.6	525.6	391.0	668.2	679.0	849.0	971.0	1112.2
Difference	−20.2	+22.6	+8.5	+91.7	−83.1	−42.3	+107.3	−103.5	−10.8	−64.7	−85.6	−164.5

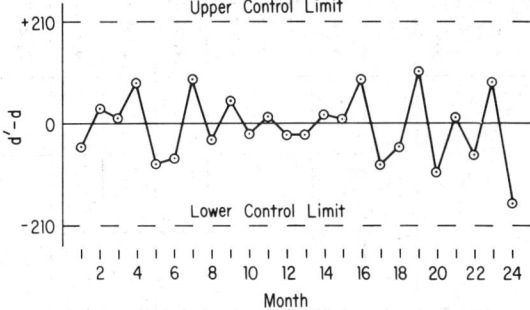

Fig. 2-3.   Moving range chart for seasonal product.

2-2 to obtain Figure 2-3. The chart exhibits statistical control; so it should be assumed that the same forecasting function can be used for more forecasts. When an out-of-control condition is observed on the control chart, a new forecasting function should be determined and used. (*Note:* In this case, the moving range chart shows control for an additional month. Thus, a forecast based on twelve months' data is good for thirteen months.)

OPERATIONS PLANNING METHODS

Mathematical programming and the electronic computer have had a tremendous impact on operations planning. If costs and incomes are assumed linear with the quantity of product manufactured or sold, or with the amount of service sold, then linear programming techniques are applicable. One simple yet effective planning technique uses a tabular solution of a linear programming problem. Assume each unit of seasonal product requires 4 man-hours of direct labor. Labor costs $5 per man-hour and overtime is paid at 1½ times regular time. Each man-hour of product carried in inventory costs $0.30 per man-hour per month. In the next year, there are 242 working days, or 1,936 man-hours per man; the total requirements are 8,208.4 units times 4 man-hours per unit, or 32,834 man-hours required. This means that 16 or 17 men are required. A tabular comparison of man-hours available with 17 men and the man-hours required is made in Table 2-6. The "Total"

TABLE 2-6.   Comparison of Man-hours Available and Man-hours Required

Month	Monthly man-hours available	Monthly man-hours required	Monthly difference	Cumulative monthly man-hours available	Cumulative monthly man-hours required	Cumulative difference
13	2,856	3,227.2	−371.2	2,856	3,227.2	−371.2
14	2,584	3,062.4	−478.4	5,440	6,289.6	−849.6
15	2,992	2,749.2	242.8	8,432	9,038.8	606.8
16	2,992	2,385.2	606.8	11,424	11,424.0	0.0
17	2,720	2,082.0	638.0	14,144	13,506.0	638.0
18	2,992	1,933.2	1,058.8	17,136	15,439.2	1,696.8
19	2,992	1,993.2	998.8	20,128	17,432.4	2,695.6
20	1,496	2,258.8	−762.8	21,624	19,691.2	1,932.8
21	2,856	2,672.8	183.2	24,480	22,364.0	2,116.0
22	2,856	3,137.2	−281.2	27,336	25,501.2	1,834.8
23	2,584	3,541.6	−957.6	29,920	29,042.8	877.2
24	2,992	3,790.8	−798.8	32,912	32,833.6	78.4
Total	32,912	32,833.6	78.4	32,912	32,833.6	78.4

TABLE 2-7. Production Plan for Seasonal Product

Month in which required	Amount required		13 RT	14 RT	14 OT	15 RT	16 RT	17 RT	18 RT	19 RT	20 RT	21 RT	22 RT	23 RT	24 RT	24 OT	Total man-hours required
(capacity)			2,856	2,584	646	2,992	2,992	2,720	2,992	2,992	1,496	2,856	2,856	2,584	2,992	748	
13	2,727.2	Avail.	2,856.0														2,727.2
		Cost	0.00														
		Plan	2,727.2														
14	3,062.4	Avail.	128.8	2,584.0	646.0												3,062.4
		Cost	0.30	0.00	2.50												
		Plan	128.8	2,584.0	349.6												
15	2,749.2	Avail.			296.4	2,992.0											2,749.2
		Cost			2.80	0.00											
		Plan				2,749.2											
16	2,385.2	Avail.			296.4	242.8	2,992.0										2,385.2
		Cost			3.10	0.30	0.00										
		Plan					2,385.2										
17	2,082.0	Avail.			296.4	242.8	606.8	2,720.0									2,082.0
		Cost			3.40	0.60	0.30	0.00									
		Plan						2,082.0									
18	1,933.2	Avail.			296.4	242.8	606.8	638.0	2,992.0								1,933.2
		Cost			3.70	0.90	0.60	0.30	0.00								
		Plan							1,933.2								
19	1,993.2	Avail.			296.4	242.8	606.8	638.0	1,058.8	2,992.0							1,993.2
		Cost			4.00	1.20	0.90	0.60	0.30	0.00							
		Plan								1,993.2							
20	2,258.8	Avail.			296.4	242.8	606.8	638.0	1,058.8	998.8	1,496.0						2,258.8
		Cost			4.30	1.50	1.20	0.90	0.60	0.30	0.00						
		Plan								762.8	1,496.0						
21	2,672.8	Avail.			296.4	242.8	606.8	638.0	1,058.8	236.0		2,856.0					2,672.8
		Cost			4.60	1.80	1.50	1.20	0.90	0.60		0.00					
		Plan										2,672.8					
22	3,137.2	Avail.			296.4	242.8	606.8	638.0	1,058.8	236.0		183.2	2,856.0				3,137.2
		Cost			4.90	2.10	1.80	1.50	1.20	0.90		0.30	0.00				
		Plan								98.0		183.2	2,856.0				
23	3,541.6	Avail.			296.4	242.8	606.8	638.0	1,058.8	138.0				2,584.0			3,541.6
		Cost			5.20	2.40	2.10	1.80	1.50	1.20				0.00			
		Plan							819.6	138.0				2,584.0			
24	4,790.8	Avail.			296.4	242.8	606.8	638.0	239.2						2,992.0	748.0	4,790.8
		Cost			5.50	2.70	2.40	2.10	1.80						0.00	2.50	
		Plan				242.8	606.8	638.0	239.2						2,992.0	72.0	
Total man-hours planned		RT	2,856.0	2,584.0		2,992.0	2,992.0	2,720.0	2,992.0	2,992.0	1,496.0	2,856.0	2,856.0	2,584.0	2,992.0		32,912.0
		OT			349.6											72.0	421.6

8-40

row shows that there are more man-hours available than required during the year. Examination of the "Monthly difference" column shows that months 13, 14, 20, 22, 23, and 24 have requirements exceeding the time available. The "Cumulative difference" column shows that the cumulative man-hours required exceed the cumulative man-hours available in months 13 and 14 only. Thus, if there are 371.2 overtime man-hours available in month 13 and 849.6 overtime man-hours available in months 13 and 14, all demands can be met when they occur. The "Total" row shows that there are 78.4 more hours available than required during the year. So, if 849.6 overtime hours are used and all regular time man-hours are used productively, there will be 928.0 man-hours in inventory at the end of the year. The above has assumed that all demands materialize as forecast and that production is at the assumed rate.

The production plan that follows assumes that all demands must be met when they occur. Also, it assumes that overtime is available up to one-quarter of the regular time available per month if required. Further, it assumes that 125 units, or 500 man-hours, were available in inventory at the beginning of month 13. The planned inventory for the beginning of month 25 is 1,000 man-hours, or 250 units. The production plan is shown in Table 2-7. The "Total man-hours planned" row shows both regular time planned and overtime planned. The last entry is the totals of RT and OT and should equal the sum of the "Total man-hours required" column. It should be noted that planning has been on the basis of having to pay for regular time hours not worked, in which case the incremental cost for using overtime in month 14 is $9.50 and month 24 is $6.50, because any overtime used there in place of regular time not used does cost those amounts per hour. Such decisions are necessary only at the end of the planning horizon.

The inventories planned are shown in Table 2-8. The inventory carrying costs are $5,987.88 [(128.8 + 242.8 + · · · + 1,000.0) × 0.30]. The overtime costs are $1,054.00 [(349.6 + 72.0) × 2.50]. The total incremental costs are $7,041.88, which is $2,638.12 less than the lower bound to using 18 men (18 × 1,936 × $5.00). To use 16 men would require an additional 1,936 man-hours of overtime, which would add $4,840.00 (1,936 × $2.50) to the total cost. Thus, if the requirement is to have a level work force, the minimum cost size of that work force is 17 men.

Operations Planning for Seasonal Service. The techniques for forecasting and forecast control are the same for the service industries as for the manufacturing industries. The techniques for operations planning are necessarily different, because it is impossible to carry inventories of service, while it is possible and frequently desirable to carry inventories of product. In the service industry, it may be possible to assume that services can be backlogged. If so, the planning must differ because only current or future capacity can be used to meet current demand. Recall, in the case where inventory can be carried, that it is past plus current capacity which is used to meet current demand. If the data for seasonal product are interpreted as the demand for a seasonal service, the plan of hours to be used would be made in much the same way. For example, if the backlog is 500 man-hours at the beginning of month 13 and is not to exceed 1,000 man-hours at the end of month 24 with a minimum use of overtime, the plan is shown in Table 2-9. An alternative to the use of 1,617.2 man-hours of overtime in the last three months would be to add another man during the latter part of the year. This seems justifiable because in months 25 through 36, the forecast indicates that the requirements for service will increase by 12 × 12 × 12.6, or 1,814.4 man-hours. This is almost the regular time for one additional man. If this man were

TABLE 2-8.   Planned Ending Inventories for Seasonal Product

Month......	13	14	15	16	17	18	19	20	21	22	23	24
Planned inventory (man-hours)	128.8	0	242.8	849.6	1,487.6	2,546.4	3,545.2	2,782.4	2,965.6	2,684.4	1,726.8	1,000

TABLE 2-9. Plan for Seasonal Service

Month	13	14	15	16	17	18	19	20	21	22	23	24	Total
Planned man-hours:													
RT	2,856.0	2,584.0	2,992.0	2,992.0	2,582.0	1,933.2	1,993.2	1,496.0	2,856.0	2,856.0	2,584.0	2,992.0	30,716.4
OT	0	0	0	0	0	0	0	0	0	223.2	646.0	748.0	1,617.2
Excess RT	0	0	0	0	138.0	1,058.8	998.8	0	0	0	0	0	2,195.6
Backlog man-hours	871.2	1,349.6	1,106.8	500.0	0	0	0	762.8	579.6	637.6	949.2	1,000.0	7,756.8

TABLE 2-10. Monthly Backlogs and Overtime with One Additional Man

Month	20	21	22	23	24	Total
Overtime man-hours....	0	0	0	73.2	792.0	865.2
Backlog man-hours.....	674.8	323.6	436.8	1,169.2	1,000.0	3,604.4

added in month 20, the backlogs would be reduced and the overtime would also be reduced. The new figures for backlogs and overtime for months 20 through 24 are shown in Table 2-10. The overtime has been reduced from 1,617.2 to 865.2 man-hours, and the total month-end backlogs from 7,756.8 to 7,432.0. The reduction in overtime (752.0 man-hours) results in a direct labor saving of $1,880.00, and there is an additional saving of a cost, such as loss of future business, which can be associated with the backlogs.

Operations Planning Where There Are No Seasonal Effects. If there are no seasonal effects, the same techniques can be used. The primary difference is that the planning will be simplified because not as many considerations need be made. In cases where there is neither a trend nor a seasonal pattern, there should be little need for overtime because the forecast demand is the same for all periods.

Operations Planning When the Work Force Is Variable. The prior discussions have assumed a constant work force or at most a discrete change in work force level. It is possible to treat the somewhat more complex but perhaps more realistic problem of a variable work force by use of mathematical programming. The model that follows assumes that the relationships are linear, but such a restriction can be removed if nonlinear programming techniques are used.

In the model to be used, the following definitions are made:

W_i = work force size in period i
I_i = inventory at end of period i
D_i = demand forecast for period i
h = cost of carrying one unit in inventory for one period
k = cost of having one unit of shortage at the end of a period
m = cost of adding one man
n = cost of laying off one man
p = number of man-hours per unit of product
r_i = number of man-hours per man in period i (not a decision variable)
s = cost of one man-hour of regular time

The resulting linear program to be solved is:

Minimize

$$Z = \sum_{i=1}^{N} [h(I_i^+) + k(I_i^-) + m(W_i - W_{i-1})^+ + n(W_i - W_{i-1})^-]$$

subject to

$$I_i^+ - I_i^- = I_{i-1}^+ - I_{i-1}^- + \frac{r_i}{p} W_i - D_i$$

$$W_i = \frac{p}{r_i} D_i - \left[\left(\frac{p}{r_i} D_i - W_i\right)^+ - \left(\frac{p}{r_i} D_i - W_i\right)^-\right]$$

$$(D_{i-1} - D_i) = [(I_i^+ - I_i^-) - 2(I_{i-1}^+ - I_{i-1}^-) + (I_{i-2}^+ - I_{i-2}^-)] - \left[\left(D_i - \frac{r_i}{p} W_i\right)^+\right.$$

$$\left. - \left(D_i - \frac{r_i}{p} W_i\right)^-\right] + \left[\left(D_{i-1} - \frac{r_{i-1}}{p} W_{i-1}\right)^+ - \left(D_{i-1} - \frac{r_{i-1}}{p} W_{i-1}\right)^-\right]$$

$$- \frac{r_i}{p} [(W_i - W_{i-1})^+ - (W_i - W_{i-1})^-]$$

$$W_i \geq 0$$

$$W_i - W_{i-1} = (W_i - W_{i-1})^+ - (W_i - W_{i-1})^-$$

$$D_i \geq 0$$

where
$$X^+ = \begin{cases} X & \text{if } X \geq 0 \\ 0 & \text{if } X < 0 \end{cases}$$

$$X^- = \begin{cases} 0 & \text{if } X > 0 \\ |X| & \text{if } X \leq 0 \end{cases}$$

The above formulation assumes that
1. All regular-time hours will be used to build product to meet current demand or to build product for inventory (there is no regular time paid for but not worked).
2. No overtime will be used.

Although the preceding appears to be a formidable problem (and it is, if attacked by manual methods), it is relatively easy to solve by computer. There are many routines available for the computer solution of linear programming problems.

ADJUSTING OPERATIONS PLANS

It is expected that there will be variations between actual demand and forecast demand and actual production and planned production. So long as these variations are random, and production variations can be tested in a manner similar to that suggested for demand variations, no corrections or changes in plans are required. However, most managers would feel more secure if adjustments were made.

In making adjustments to a production plan, it is necessary to decide if the adjustments should be linear or some weighted function of the variations seen. Further, it is necessary to decide on a time interval over which the adjustment will be spread.

Two systems which accomplish the above results are (1) leveling, which is a linear adjustment over N periods, and (2) smoothing by use of a weighting factor. In the weighted smoothing techniques, a larger correction is made in the first adjustment period and successively smaller ones in future periods. In leveling, the adjustment is spread equally over the N periods.

It must be realized that adjustments usually cannot be made in the period that immediately follows. This is true because (1) actual demand data may not be immediately available and (2) it takes time to change production plans, production schedules, procurement and delivery schedules, and the like. Therefore, adjustments are usually projected some time into the future.

An Illustration of Leveling. The technique of leveling will be illustrated by an application to seasonal product. The production plan for seasonal product is shown in Table 2-7. The actual demands are shown in Table 2-5. These data, along with the adjustments, are shown in Table 2-11. The adjustments in Table 2-11 have been made over a five-month period, with a two-month lag prior to making the adjustment effective. For example, after data for month 13 were available, it can be seen that actual demand had exceeded forecast demand by 20 units. The actual inventory was 20 units less than planned. The decision at this point was to increase production in months 16 through 20 by 4 units each, or a total of 20 units. The adjusted planned production for month 16 was $748 + 4 = 752$ units. When data for month 14 were available, actual inventory was 2 units above planned inventory, which seemed to indicate that no adjustment was necessary. However, 20 units of production were added for the correction made after month 13. This had to be removed; so the adjustment to months 17 through 21 was -4 units. The remainder of the table follows that pattern. The same result is more easily obtained by examining the differences between actual demand and forecast demand. (*Note:* The effects of rounding show up in some entries.)

TABLE 2-11. Adjustment by Leveling*

Month	Demand			Production			Inventory†		
	Actual	Forecast	Difference	Planned	Adjustments	Adjusted plan	Planned	Actual	Difference
13	827	807	20	714	0	714	32	12	20
14	743	766	−23	733	0	733	0	2	−2
15	679	687	−8	748	0	748	61	71	−10
16	505	596	−91	748	+4	752	212	318	−106
17	604	521	83	680	+4−4	680	372	394	−22
18	526	483	43	748	+4−4−2	746	637	614	23
19	391	498	−107	748	+4−4−2−19	727	886	950	−64
20	668	565	103	374	+4−4−2−19+16	369	696	651	45
21	679	668	11	714	−4−2−19+16+9	714	741	686	55
22	849	784	65	714	−2−19+16+9−22	696	671	533	138
23	971	885	86	646	−19+16+9−22+21	651	432	213	219
24	1112	948	164	766	+16+9−22+21+2	792	250	−107	357

* All quantities rounded to nearest unit of product.
† Opening stock was 125 units.

The net effect of any adjustment method is to arrive at a compromise between carrying excessive inventories and permitting excessive production variations. If production plans are assumed fixed, then any variations in demand must be met from inventories. This may require excessive inventory. The other extreme is to absorb all demand variations by production variations. In such a case, inventory can be held fixed but production must vary. The choice of the leveling factor (number of periods over which variations are spread) reflects the compromise that is being made between variations in production and variations in inventory. As the number of periods increases, the smaller the expected production variations become and the larger the expected inventory variations. As the number of periods over which variations are leveled is decreased, the larger the expected variations in production become but the smaller the expected variations in inventory.

Adjusting Operations Plans in the Service Industries. In service industries where inventories cannot be established but demand can be backlogged, any excess of demand over time available will be carried to the next period. Consequently, an adjustment of planned output during the next periods should be made in a manner similar to the case where product is involved. The period over which an adjustment should be made is dependent upon the resources available and the delay the customer will tolerate.

If the planning is for a service industry in which demand cannot be backlogged—that is, demands not met at the time they occur are lost—any adjustment must be on the basis of a change in anticipated or forecast demand. It is possible to increase the resources available to meet a demand and thus increase the probability that all demands will be met, but this must be done at the risk of incurring added costs due to idle time, which must be treated as added operating expense. This amounts to planning for a demand in excess of the demand that has been forecast. The economic justification for such planning lies in the realization that the loss of profit per unit of demand not served is greater than the expense of excess capacity not used.

OPERATIONS PLANS AS COMPANY GOALS

A significant result of operations planning is the translation of the operations plan into a company goal. Such a translation occurs because the establishment of the plan triggers actions and decisions that commit the company to the expenditure of funds and time which cannot be recovered. This goal establishment aspect of operations planning is inherent in the planning process. It is best illustrated by the purchasing of production tooling and equipment to meet anticipated increases in demand, by the training of new personnel, by the retraining of personnel currently on the payroll, by the commitment for additional raw materials, and the like. All these decisions are triggered by an operations plan which calls for the expansion of operations. Likewise, plans which indicate a reduction in demand will trigger decisions to release personnel, to sell equipment, to cut back on material commitments, and so on.

The response to changes indicated by the operations plan might be one of several. If the plan calls for a higher output, the decision might be (1) to try to meet the higher demand, in which case the actions would be of the nature indicated in the prior paragraph, or (2) to reduce sales and advertising effort and not expand output. If the forecast and plan call for a lower output, the decision might be (1) to reduce output or (2) to increase sales and advertising effort to regain or expand the market for the product.

The import of the above discussion is that operations plans can and do have a very significant effect on company operations. Therefore, they must be as accurate as possible. The accuracy of the operations plan is a direct result of the accuracy of the forecast.

PLANNING OPERATIONS FOR ECONOMIC INVENTORY LEVELS

The discussion of operations plans was based on the implicit assumption that the product would be manufactured on a production line and only that product would be manufactured on that line. In such a case, planned inventories of that product would be as shown in the production plan of Table 2-7.

The component parts that are used in the product may be purchased or they may be manufactured internally. In either case, the operations should be conducted in an economic manner, which implies purchasing and manufacturing in economic lot sizes. If several products are to be made on the same equipment, it is again necessary to schedule in an economic manner, but possibly suboptimally.

The economic lot size in purchasing will minimize the sum of two costs: the cost of purchasing and the cost of holding the average inventory. This can be expressed as:

Minimize $$C_p = \frac{Ad}{q} + i\frac{q}{2} \tag{8}$$

where C_p = purchase cost
 A = cost of placing an order
 d = expected demand
 q = quantity to be purchased
 i = per unit cost of carrying inventory
The solution to equation (8) is

$$q_p = \sqrt{\frac{2Ad}{i}} \tag{9}$$

If the order cost (A) is \$100, the annual usage (d) is 100,000 units, and the cost of carrying a unit in inventory (i) is \$5, then the economic quantity to purchase as a lot is q_p = 2,000 units. This means 50 orders per year.

The economic lot size in manufacturing minimizes the sum of the same two costs, but the average inventory is not $q/2$ but $[q(1-d/p)]/2$, where p is production rate (interpret A as setup cost). The total variable cost to be minimized is

$$C_m = \frac{Ad}{q} + i\frac{q}{2}\left(1 - \frac{d}{p}\right) \tag{10}$$

The total variable cost C_m will be minimum when

$$q_m = \sqrt{\frac{2Ad}{i(1 - d/p)}} \tag{11}$$

When $p \rightarrow \infty$, $q_m \rightarrow q_p$ for the same costs. Assume the prior example with p = 400,000 per year. In this case, q_m = 2,310, or q_m is greater than q_p.

When several products are manufactured on the same equipment, it may not be possible to manufacture all of them in economic quantities. Scheduling may require a compromise, in which case it is possible to suboptimize. The economic fraction of yearly demand to be manufactured at each run can be shown to be

$$f = \sqrt{\frac{2\Sigma A_k}{\Sigma d_k i_k(1 - d_k/p_k)}} \tag{12}$$

where the subscript k represents the kth product. The necessary data and calculations are shown in Table 2-12. The economic fraction is 0.0286, or there will be 35 lots of each product per year. This will keep the process busy $\frac{1}{4} + \frac{8}{25} + \frac{1}{3}$, or 90.3 percent of the time. The economic manufactured lot sizes are 2,310, 2,304, and 1,795, respectively. This implies that 43, 35, and 28 lots of each product would be manufactured each year. The schedule required to accomplish this would be quite complex.

Scheduling In-process Inventories. In-process inventories, or those existing between machines, operations, or departments, serve to decouple operations and to permit manufacturing in economic lot sizes. Such inventories should be minimal so that the investment is kept as low as possible. When operations are serially connected, there is no economy in having one operation or process manufacture

TABLE 2-12. Calculation of Economic Fraction of Yearly Demand

Product (k)	Demand (d_k)	Setup cost (A_k)	Production rate (p_k)	Inventory cost (i_k)	$1 - \dfrac{d_k}{p_k}$	$d_k i_k$	$d_k i_k \left(1 - \dfrac{d_k}{p_k}\right)$
1	100,000	100	400,000	5.00	0.750	500,000	375,000
2	80,000	90	250,000	4.00	0.680	320,000	217,600
3	50,000	150	150,000	7.00	0.667	350,000	233,333
Total	....	340	....	....	.....	....	825,933

$$f = \sqrt{\frac{680}{825,933}} = 0.0286$$

in quantities that are different from another operation or process that either succeeds or follows. For example, suppose machine A and machine B are serially connected, with all material being processed through machine A prior to machine B. If the lot size on machine A is one-half that of machine B, machine A must make two lots before machine B can make one lot. Thus, the average inventory following machine A is not $q_A/2$ but q_A, and this would be reflected in a higher inventory carrying cost than was presumed in the lot size determination. Actually, the setup cost is doubled also. If the lot size on the following machine is the smaller, then the same kind of analogy can be made. The economic approach is the same as for several products on one machine.

The inventories between work stations required to decouple one from another economically can be determined if certain data are available. Assume two machines are serially connected in what amounts to a production line. The output of machine A is fed into machine B. However, when machine A breaks down, it is desirable to have machine B remain operational. Therefore, some inventory is required between machines A and B. Assume that the following are known:

k = mean time between breakdowns on machine A
l = mean duration of a breakdown on machine A
$f(t)$ = probability density of t
c = cost of idle time on machine B
d = demand of machine B
i = cost of carrying an item in inventory per unit time

It is desired to determine I_o, the optimal inventory between machines A and B. The optimal inventory level between the machines is determined by determining the value of I_o which satisfies

$$F\left(\frac{I_o}{d}\right) = 1 - \frac{dik}{c} \tag{13}$$

where $F(t)$ is the cumulative distribution of t, $(F(t) = \int_o^t f(u)\,du)$

Suppose the following are true:

k = 16 hours
l = ½ hour
$f(t)$ = exponential $[\,2\exp(-2t)\,]$
c = \$50
d = 30 units/hour
i = \$0.001 per unit/hour

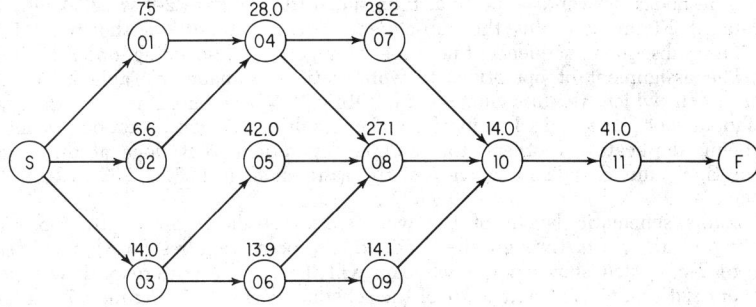

FIG. 2-4. A sequence of operations and times.

Then
$$F\left(\frac{I_o}{30}\right) = 1 - \frac{30(0.001)(16)}{50} = 0.9904$$

or
$$\int_o^{I_o/30} 2e^{-2t}\, dt = 0.9904$$

Then $I_o = -15 \times \ln 0.0096 = 70$ units. It is worthwhile to carry 70 units in inventory to avoid frequent shutdowns of machine B due to breakdowns on machine A.

SCHEDULING OPERATIONS ON A PRODUCTION LINE

The scheduling or sequencing of operations on a production line is generally accomplished heuristically. The first step is to determine the logical breakdown of the manufacturing process into its basic operations or elements and to specify their precedence relationships.

A second step is to determine either the number of work stations or the scheduled output. It is impossible to specify all three, because they are not independent. The usual first step toward an assignment, once the above data are known, is to determine a weighting factor for each operation. The weighting factor can be the total amount of time required by all succeeding operations, called the subsequent sequence time. The operations are then sorted and assigned in the order of subsequent sequence times, starting with the largest. A sequence of eleven operations and their times (in minutes) are given in Figure 2-4. Figure 2-4 is drawn as an

TABLE 2-13. Precedence Table

Operation	Immediately preceding operation(s)	Subsequent sequence time*	Duration
01	. . . .	110.1	7.5
02	. . . .	110.1	6.6
03	. . . .	124.1	14.0
04	01, 02	82.1	28.0
05	03	82.1	42.0
06	03	82.1	13.9
07	04	55.0	28.2
08	04, 05, 06	55.0	27.1
09	06	55.0	14.1
10	07, 08, 09	41.0	14.0
11	10	0.0	41.0

* By longest path.

activities-on-nodes network. The total time specified in Figure 2-4 is 236.4 minutes. The data of Figure 2-4 plus the subsequent sequence times are shown in Table 2-13. The subsequent sequence times are arranged in descending order in Table 2-14. The assignment of operations to work stations is made in Table 2-15. The decision upon which the assignments in Table 2-15 are based is that the cycle time should not exceed 14.1 minutes. The result is 17 work stations, some of these being duplicates of others; for example, operation 05 is done at three work stations: E, F, and G. The efficiency of the assignment is (236.4/239.7) × 100 = 98.6 percent.

A possible schematic layout of the work spaces is shown in Figure 2-5. The scheduling of the parts through the production process requires an approach such as Figure 2-6, which shows how each part will flow. The pattern of flow repeats itself every sixth part. A Gantt chart of the schedule is shown in Figure 2-7. Figure

TABLE 2-14. Operations in Order of
Descending Subsequent Sequence Times

Operation	Immediately preceding operation(s)	Subsequent sequence time*	Duration
03	....	124.1	14.0
01	....	110.1	7.5
02	....	110.1	6.6
04	01, 02	82.1	28.0
05	03	82.1	42.0
06	03	82.1	13.9
07	04	55.0	28.2
08	04, 05, 06	55.0	27.1
09	06	55.0	14.1
10	07, 08, 09	41.0	14.0
11	10	0.0	41.0

* By longest path.

TABLE 2-15. Assignment of
Operations to Work Stations

Work station	Operation(s) assigned	Duration
A	03	14.0
B	01, 02	14.1
C	04	28.0
D	04	28.0
E	05	42.0
F	05	42.0
G	05	42.0
H	06	13.9
J	07	28.2
K	07	28.2
L	08	27.1
M	08	27.1
N	09	14.1
P	10	14.0
Q	11	41.0
R	11	41.0
S	11	41.0

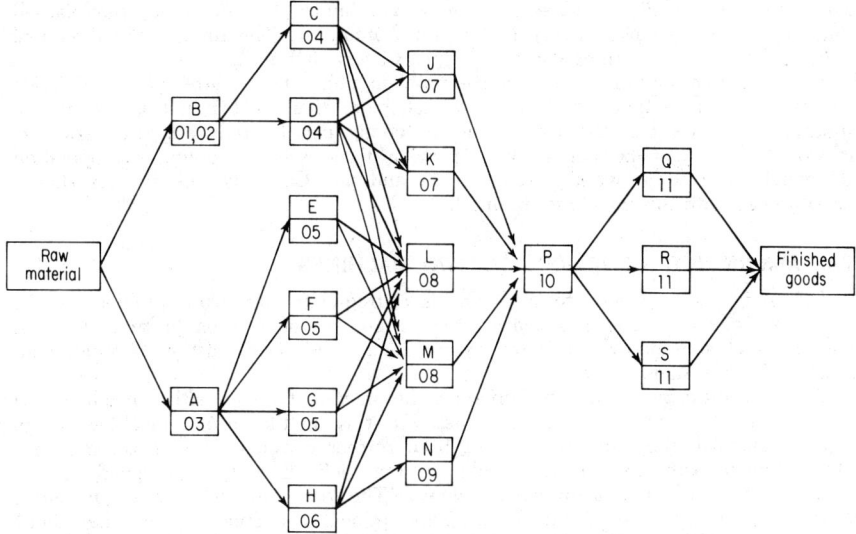

Fig. 2-5.   A schematic layout of the work spaces.

WORK STATION

Unit number	A	B	C	D	E	F	G	H	J	K	L	M	N	P	Q	R	S
1	X	X	X		X			X	X		X		X	X	X		
2	X	X		X		X		X		X	X	X		X		X	
3	X	X	X				X	X	X		X		X				X
4	X	X		X	X			X		X	X	X	X	X			
5	X	X	X			X		X	X		X		X	X		X	
6	X	X		X			X	X		X	X	X	X				X

X – Unit is processed through this work station.

Fig. 2-6.   Arrangement of flow through work stations.

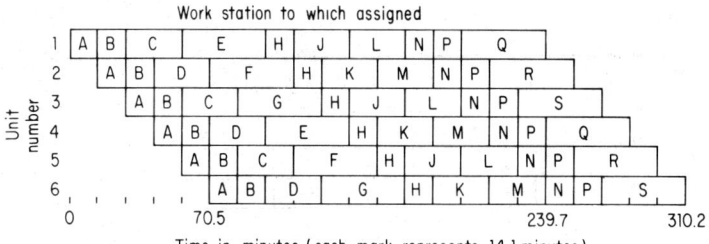

Fig. 2-7.   Gantt chart of assignment of units to work stations.

2-7 shows that while a cycle time of 14.1 minutes is maintained, the elapsed time from start to finish of any unit is 239.7 minutes. The routine of subsequent units follows the same arrangement and repeats every sixth unit.

An alternative solution to the above problem might be an attempt to subdivide all operations into time units that are not more than 14.1 minutes. If such a subdivision is possible, the investment in tooling can be reduced. For example, it would no longer be necessary to provide three sets of tooling for operation 05 which is done at work stations E, F, and G. Certainly, the process should be examined with this possibility in mind.

THE ASSIGNMENT OF N JOBS TO TWO MACHINES

The assignment of jobs to machines is a frequently occurring problem in the job shop industry. The problem of the assignment of N jobs to two machines has an exact solution which is readily obtained. (All jobs must go through both machines in the same order.)

The procedure is to find the job with the shortest time on either machine. If it is on the first machine, assign it first. If it is on the second machine, assign it last. Remove the job just assigned from further consideration. Proceed in the same manner with the remaining jobs until all jobs have been assigned.

An example of such a problem follows. The times for six jobs on machines 1 and 2 are shown in Table 2-16. The optional assignment (least idle time) is shown in Table 2-17. Figure 2-8 is a Gantt chart showing how the actual schedule is put together. Note that the objective of the assignment is to start machine 2 as soon as possible and to keep it operating during the maximum fraction of the time and have it finish as soon after machine 1 as possible. If the operation times on machine 2 had been shorter, Figure 2-8 would have shown some idle time. For example, if job F had required only 7 minutes instead of 9 minutes on machine 2, Figure 2-8 would show one unit of idle time for machine 2—from

TABLE 2-16. Operation Times for Six Jobs on Two Machines

Job...............	A	B	C	D	E	F
Machine 1...........	10	9	14	12	8	4
Machine 2..........	8	11	12	5	17	9

TABLE 2-17. Optimal Assignment of Six Jobs to Two Machines

Order of assignment	First	Second	Third	Fourth	Fifth	Sixth
Job..........	F	E	B	C	A	D

Fig. 2-8. Schedule of six jobs on two machines.

time 11 to time 12—and all later jobs on machine 2 would have been started and completed one unit sooner.

THE ASSIGNMENT OF N JOBS TO M (>2) MACHINES

When the preceding problem is expanded, the optimal assignment becomes more complex. An exact solution is obtained if all combinations of assignments are made and the elapsed time determined. The minimum of that set is optimal. However, this can become a large problem very rapidly. For six jobs, there are 6!, or 720 different assignments. As the number of jobs becomes larger, the number of possible assignments becomes astronomical. There are heuristic routines for computer usage which will provide optimal or near optimal solutions within reasonable computation times.

THE OPTIMAL ASSIGNMENT OF MEN TO JOBS

In a job shop or in a service industry, it is necessary to make a decision on the assignment of men to jobs for which they are not equally qualified. The objective of the assignment is to minimize the total time or the total cost of doing the given tasks. Each man must be assigned a unique job. If there are m jobs and m men and there is a measure of the efficiency of each man on each job, the optimal assignment can be determined by the "Hungarian" method. This is best illustrated by an example. In Table 2-18, the time for each man to do each job is given. The first step of the procedure is to subtract the minimum in each row from every element of that row. This result is shown in Table 2-19. Then subtract the minimum of each column from every element of that column. Draw a minimum number of lines through all zeros. These two steps are shown in Table 2-20. If the minimum number of lines is less than the number of rows or columns, an optimal solution has not been found. The lines in Table 2-20 are through rows 1, 3, 5, and 6 and column D. If it had required six lines to intersect all zeros, an optimal assignment would have been found. Because Table 2-20 does not contain an optimal solution, it is necessary to proceed with the algorithm. The next step is to find the smallest element not covered by a line. This element, 1, is subtracted from each uncovered element and added to all elements at the intersection of lines. The result of this step is shown in Table 2-21. The optimal assignment is also shown in Table 2-21. (The elements are enclosed in squares.) The cost of the assignment chosen is $9 + 10 + 13 + 6 + 7 + 8 = 53$. Only one man is assigned the job at which he is expected to take the least time. Another optimal assignment (there are several for this problem) is 1—E, 2—D, 3—C, 4—A, 5—B, and 6—F.

TABLE 2-18. Data for the Assignment Problem

Man	Job					
	A	B	C	D	E	F
1	10	8	7	8	6	9
2	13	13	12	10	12	15
3	13	15	11	12	14	18
4	8	9	7	5	6	10
5	10	7	9	8	6	12
6	11	13	8	10	8	11

TABLE 2-19. First Step of the Solution
to the Assignment Problem

Man	Job						Number subtracted
	A	B	C	D	E	F	
1	4	2	1	2	0	3	6
2	3	3	2	0	2	5	10
3	2	4	0	1	3	7	11
4	3	4	2	0	1	5	5
5	4	1	3	2	0	6	6
6	3	5	0	2	0	3	8

TABLE 2-20. First Evaluation of Assignment Solution

Man	Job					
	A	B	C	D	E	F
1	~~2~~	~~1~~	~~1~~	~~2~~	~~0~~	~~0~~
2	1	2	2	0	2	2
3	~~0~~	~~3~~	~~0~~	~~1~~	~~3~~	~~4~~
4	1	3	2	0	1	2
5	~~2~~	~~0~~	~~3~~	~~2~~	~~0~~	~~3~~
6	~~1~~	~~4~~	~~0~~	~~2~~	~~0~~	~~0~~
Number subtracted	2	1	0	0	0	3

TABLE 2-21. An Optimal Solution
to the Assignment Problem

Man	Job					
	A	B	C	D	E	F
1	2	1	1	3	0	[0]
2	0	1	1	[0]	1	1
3	[0]	3	0	2	3	4
4	0	2	1	0	[0]	1
5	2	[0]	3	3	0	3
6	1	4	[0]	3	0	0

THE ASSIGNMENT OF N ORDERS TO M MACHINES

In a job shop, it may be necessary to assign a set of N orders to M machines where each order may require time on only one machine. The assignments will be those that optimize (maximize) profit. To make such an assignment, it is necessary to know the order size, the selling price per unit, the time required to make one unit of each order on each machine, the cost of machine time for each machine, and the time available on each machine. This is a linear problem and can be solved exactly by the simplex method of linear programming and less exactly by the transportation or MODI methods. To illustrate the formulation:

Let a_{ij} = time to manufacture 1 unit of product i on machine j
  h_j = time available on machine j
  b_i = quantity of product i
  x_{ij} = number of units of product i to be made on machine j
  c_{ij} = profit per unit of product i made on machine j (= unit selling price of product i minus $a_{ij} \times$ cost of time on machine j)

The problem is to:

Maximize
$$\sum_{ij} c_{ij} x_{ij}$$

subject to
$$\sum_j x_{ij} = b_i$$

$$\sum_i a_{ij} x_{ij} \leq h_j$$

$$x_{ij}, b_i, h_j \geq 0$$

This problem can now be solved by the simplex method.

SCHEDULING WHEN THERE ARE TIME CONSTRAINTS

Frequently, it is necessary to schedule production or service operations when there are time constraints. In such cases, it is generally necessary to suboptimize. The best procedure appears to be to find first an optimum solution, and if it does not satisfy the time constraints, to back off to a less than optimum solution which does satisfy the time constraints. Suppose it had been necessary in the assignment of the six jobs to two machines to have job B completed by time 30 (see Figure 2-8). The logical alteration of the schedule would have been to interchange jobs B and E. The new schedule has an elapsed time of 66; so it does not affect the efficiency in this case. This will not always be true.

CONCLUSION

There are many problems connected with operations planning, scheduling, and control for which quantitative solutions can be developed by means of available mathematical procedures. The industrial engineer who has a good background in applied mathematics will find the solving of such problems an interesting challenge.

BIBLIOGRAPHY

Biegel, John E., *Production Control: A Quantitative Approach*, 2d ed., Prentice-Hall, Inc., Englewood Cliffs, N.J., 1971.
Buffa, Elwood S., *Production-Inventory Systems: Planning and Control*, Richard D. Irwin, Inc., Homewood, Ill., 1967.
Conway, Richard W., William L. Maxwell, and Louis W. Miller, *Theory of Scheduling*, Addison-Wesley Publishing Company, Inc., Reading, Mass., 1967.
Moore, Franklin G., and Ronald Jablonski, *Production Control*, 3d ed., McGraw-Hill Book Company, New York, 1969.
Plossl, G. W., and O. L. Wright, *Production and Inventory Control*, Prentice-Hall, Inc., Englewood Cliffs, N.J., 1967.

Chapter **3**

Graphical and Network Planning Techniques

DONALD G. MALCOLM

Director, Western Operations,
Research Analysis Corporation, Los Angeles, California

LAWRENCE S. HILL

Professor of Science Management,
California State College, Los Angeles, California

The functions of planning, organizing, directing, and controlling business operations of all varieties have become increasingly complex. It is hardly surprising, therefore, that management now requires improved techniques of planning and control that are capable of coping with this complexity and that are, at the same time, more accurate, less costly, and more effective than prior methods. Techniques are needed which allow management to forecast problems early enough to take corrective action; which provide means for stimulating alternative plans; and which integrate time, budget, and performance considerations. The procedural approach to and operating characteristics of four major graphical and network planning techniques developed to meet the needs of increasingly complex operating environments are presented in this chapter. Brief notation of some scheduling and control methods that predate network techniques is included below, but the emphasis is on the use of the Program Evaluation and Review Technique (PERT), the PERT Cost approach, the Critical Path Method (CPM), and the Line of Balance technique (LOB).

GANTT AND MILESTONE CHARTS

The first formal scheduling model used by management was the Gantt chart.[1] This technique provided a powerful tool to management for planning and controlling industrial operations. The Gantt chart has been most successfully applied to highly repetitive production operations.

Generally, a time scale is placed horizontally along the top of a Gantt chart. The rows represent machines, personnel, department, or whatever resources may be required to accomplish a job. The time scale may be subdivided into calendar time or selected temporal units. Charts may be prepared for various managerial levels and responsibilities, so that performance may be monitored and responsibility traced throughout the organization.

An example of a Gantt chart for machine loading is shown in Figure 3-1. In charts of this type, the machine number is listed in the left-hand column. A horizontal line is drawn on which the orders scheduled for processing on the corresponding machine are shown. A second line, depicting progress against this schedule, is then drawn.

The Gantt or bar chart is an effective planning and scheduling tool for production operations involving a minimum of interrelationships. As such, it does not make provision for treatment of uncertainty nor does it provide a means for treatment and forecasting when interrelationships between the various activities create constraints on performance. Thus, large-scale research and development projects, characteristic of many military and industrial programs, require additional planning

[1] Henry L. Gantt, a disciple of Frederick W. Taylor, made numerous contributions to the scientific management at the turn of the twentieth century. He is the inventor of the Gantt chart, a technique devised to display data required for scheduling purposes in manufacturing operations.

		JUNE 13 MON.	14 TUES.	15 WED.	16 THURS.	17 FRI.	18 SAT.	20 MON.	21 TUES.	22 WED.	23 THURS.	24 FRI.	25 SAT.
TOTAL OPERATING	TIME												
OF MACHINES IN	DEPT.												
DRILL PRESSES	TOTAL												
	1401												
	1344												
	623												
	869												
	373												
	858												
	1343												
	1333												
	1325												
	1336												
	1071												
MILLING MACHS.	TOTAL												
	1259												
	528												
	477												
	273												
BROACHING MACH.	TOTAL												
	1436												
	1198												
	1378												

Fig. 3-1. The machine record type of Gantt chart. (*Source: Wallace Clark*, The Gantt Chart, *The Ronald Press Company, New York*, 1922; *Sir Isaac Pitman & Sons, Ltd., London*, 1934.)

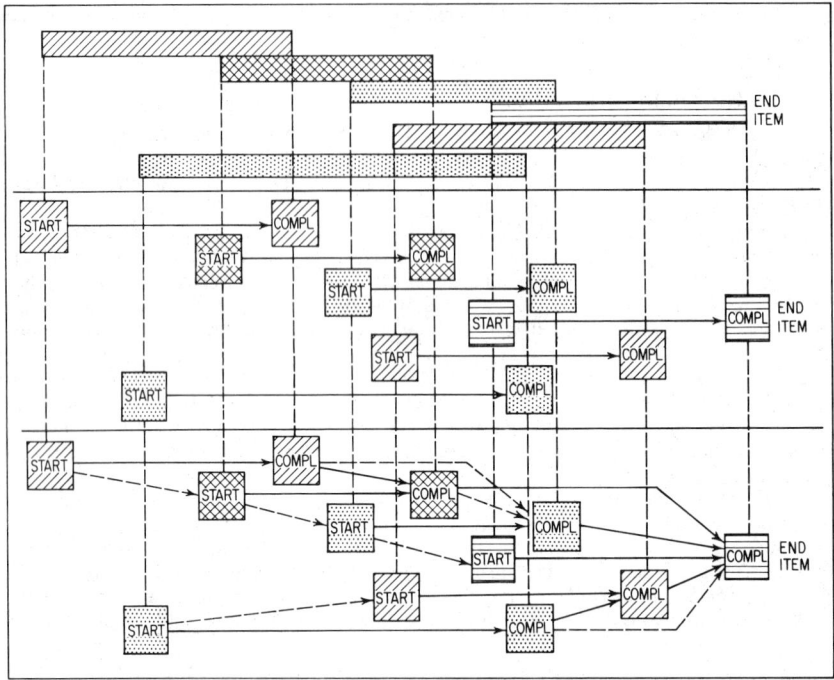

Fig. 3-2.   Relationship of networking to Gantt charting.

and scheduling tools for effective management planning and control. Milestone and network charts have been created to fill this need.

The relationship of the network approach developed and used in PERT and traditional Gantt charting is worth noting. Figure 3-2 illustrates this difference. The Gantt chart does not depict dependencies or interrelationships between activities shown; nor are coordinative functions and precedent relationships indicated. Such considerations are of major significance in research and development programs where many activities must be performed concurrently and coordinated properly. On the other hand, the Gantt chart is often easier to read quickly than a network chart. Consequently, network and bar charts are often used together in depicting a schedule of development activities.

An outgrowth of the simple bar technique is the "milestone" chart. A milestone may be described as an important event along the path to project completion. All milestones are not equally significant. The most important are termed "major milestones," usually representing the completion of an important group of activities. The method of collecting and organizing data for a milestone chart is similar to the Gantt technique. The primary difference is the graphic display. The milestone system offers no basic improvement over the Gantt chart except to provide focus on the events to be achieved.

PERT

The PERT technique was developed during 1958–1959 as a method of planning and controlling development progress on the complex Polaris Fleet Ballistic Missile Program for the Special Projects Office, U.S. Navy.[2]   PERT is designed for planning

[2] D. G. Malcolm, J. H. Roseboom, C. E. Clark, and W. Fazar, "Application of a Technique for R&D Program Evaluation," *Operations Research,* September-October, 1959.

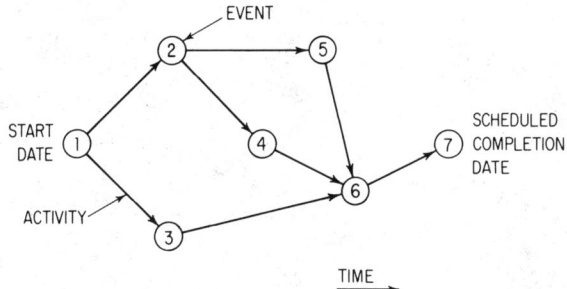

FIG. 3-3.   Network components.

and scheduling activities in the development phase of a new product and is not directly suitable for application to repetitive production operations.

PERT is a method of scheduling resources to accomplish a predetermined job within time constraints. The technique provides a means for minimizing delays, interruptions, and conflicts and expediting completion through coordination and synchronization of the various parts of the overall job. PERT was the first system which attempted to deal with scheduling uncertainties in a relatively sophisticated manner. As a communications device, it can report favorable and unfavorable progress to managers and supervisors.

Basic Elements of PERT. The basic and primary analytical tool in PERT is the network. The PERT network is a flow diagram which graphically depicts the sequences and interrelationships of selected activities and events to be achieved in realizing the stated objectives. An event is defined as a distinguishable, unambiguous point in time that coincides with the beginning or end of a specific task or activity. An event does not symbolize the performance of work, but represents the time at which an activity has been started or completed. Synonyms for event are "node" and "connector." An event may be depicted as a circle, although other geometric figures will serve the same purpose. Arrows connecting events are "activities" and represent performance of work or analyses necessary to accomplish an event. No event is considered accomplished until all work represented by arrows leading to it has been accomplished. In Figure 3-3, event 1 could represent the point in time work started, and event 7, work completed.

An activity always has a predecessor and a successor event. The work represented by the activity arrow cannot be started until the predecessor event has been completed. Activities may be related to such functional responsibilities as design, procurement, or production, or may even represent the time required for decision making processes. As such, activities may indicate the use of time, manpower, facilities, space, or other requirements. An activity may also indicate waiting time, and is shown as a dotted line in a network. A simplified network showing such dummy activities is set forth in Figure 3-4. In general, activities represent work and waiting time.

Typical network relationships are shown in Figure 3-5. In a, activity A immediately precedes activity B. In b, A precedes both B and C, while B and C may occur simultaneously. In c, A and B must be completed before C is started. In

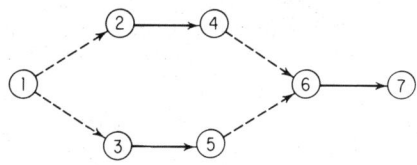

FIG. 3-4.   Simplified network waiting time.

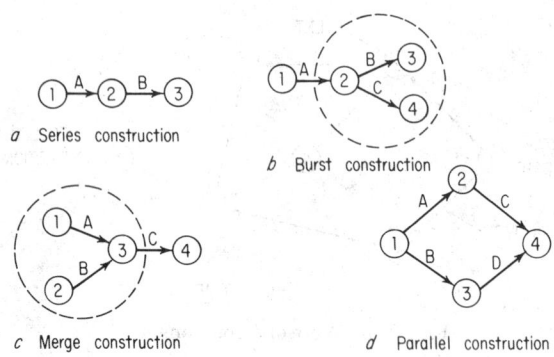

a Series construction

b Burst construction

c Merge construction

d Parallel construction

Fig. 3-5. Basic PERT patterns.

d, activities A, B, C, and D must be completed before an activity following event 4 can be started.

If an event represents the joint initiation of more than one activity, it is called a "burst" event, as shown in the circled portion of *b* in Figure 3-5. If an event represents the joint completion of more than one activity, it is called a "merge" event, as indicated in the circled portion of *c*.

Computation of Activity Elapsed Times. The time to complete a future activity in PERT is generally realistically stated in terms of a likelihood rather than a single best estimate. Three estimates, representing the range of time in which an activity may be accomplished, are obtained for each activity. The three time estimates—"optimistic," usually represented by the letter *a;* "most likely," by the letter *m;* and "pessimistic," by the letter *b*—are basic to the PERT methodology, although abbreviated implementations sometimes employ a single time estimate.[3]

Interpretation of the concepts of optimistic, most likely, and pessimistic time has varied, but generally accepted definitions are as follows:

1. The optimistic time, *a*, is an estimate of the minimum time an activity will take if unusually good luck is experienced (barring "acts of God"). The probability of this estimate being exceeded, that is, an activity completed more quickly, is on the order 0.01 to 0.05.

2. The most likely time, *m*, is an estimate of the normal time an activity will take—a result that would be expected to occur most often if the work were to be repeated under identical conditions.

3. The most pessimistic time, *b*, is an estimate of the maximum time an activity will take if unusually bad luck is experienced. This time should reflect the possibility of initial failure but not catastrophic events, unless specific hazards are inherent risks in the activity. The probability of this estimate being exceeded, that is, an activity requiring more time, is on the order of 0.01 to 0.05.

Other conditions usually specified when the activity duration times are secured are as follows:

1. Time estimates should be based on the assumption that all resources will be available on a normal basis.

2. The initial estimates should not be influenced by schedule or calendar dates.

Time estimates are based on a five-day workweek. Although other time units may be used, the prevalent unit is the week graduated in tenths. A time estimate of 0.1 week is equivalent to one-half day; an estimate of 0.2, to one day; and so on.

[3] When a single time estimate is used in PERT applications, experience has indicated that estimates frequently will be more optimistic, that is, shorter than the most likely value.

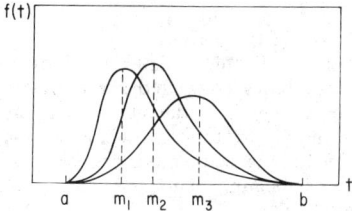

FIG. 3-6. Examples of beta distributions.

In PERT theory the three time estimates serve as three points on a nonsymmetrical distribution curve. The mode of this distribution is the most likely estimate, m, and the extremes are the optimistic and pessimistic values, a and b, respectively. Probability calculations in PERT are based on the assumption that the probable duration of an activity is beta distributed.[4] Three examples of the beta distribution are plotted in Figure 3-6, showing different m values which might be estimated for given a and b estimates.

A formula for approximating the mean of the beta distribution, t_e, referred to as the "expected time" or "expected elapsed time," is as follows:

$$t_e = \frac{a + 4m + b}{6}$$

Assume the three time estimates for an activity are

$$a = 2 \text{ weeks} \qquad m = 4 \text{ weeks} \qquad b = 8 \text{ weeks}$$

The expected time, t_e, is computed as follows:

$$t_e = \frac{2 + (4)(4) + 8}{6} = \frac{26}{6} = 4\frac{1}{3} \text{ weeks}$$

The statistical significance of t_e is that it represents an estimate of the particular elapsed time value for the activity being estimated, with a 0.5 probability of being exceeded.

Assuming that the beta distribution is a valid representation of the distribution of the estimates, the standard deviation, σ, of an activity may be approximated by the following formula:

$$\sigma = \frac{b - a}{6}$$

Assume again that the three time estimates for an activity are

$$a = 2 \text{ weeks} \qquad m = 4 \text{ weeks} \qquad b = 8 \text{ weeks}$$

The standard deviation is computed as follows:

$$\sigma = \frac{8 - 2}{6} = 1$$

The standard deviation is entirely a function of relative distance from the most optimistic estimate to the most pessimistic. Its use will be described subsequently. In the context of a PERT time estimate, the standard deviation may be described

---

[4] The probability density function of the beta distribution is $f(t) = K \cdot (t - a)^\alpha \cdot (b - t)^\beta$. The "tails" of the beta distribution do not approach infinity as in the case of the normal curve (see Figure 3-6). The mean of the beta distribution lies one-third of the distance from the mode (the most likely time, m) to the midpoint of the range. In practice, the skewness of activities tends to be toward the right.

as a measure of the probable range of uncertainty in the estimator's mind concerning his estimate of the elapsed time of the activity.

Network Time Calculations—Events. After the flow plan for a network has been developed and expected times computed for each activity, the calendar times for accomplishment of events may be calculated. The most time-consuming path of activities from the beginning to the end of a network is referred to as the critical path. After the network is completed, the critical path can be determined and monitoring and analysis begun. The following symbols are used for such purposes:

T_E = earliest possible date that an event can be attained or an activity completed
T_S = target completion date for the project
T_L = latest possible date that an event can be reached or an activity completed if the project completion date is to be met

Normally, the start of a project is associated with a specific calendar date, and then the elapsed time (activity duration) is added to that date to determine the calendar date of the succeeding event. The T_E value for a given event may be calculated as the sum of the expected elapsed times (t_e) for the activities on the longest path from the beginning of the project to a given event. This procedure is descriptively called the "forward pass" computation.

After the earliest possible completion date has been established for the end item in a project, the latest allowable occurrence date for each event can be determined. The T_L value for a given event may be calculated by subtracting the sum of the expected elapsed times (t_e) for the activities on the longest path from the given event to the end point of the project, from the latest date allowable for completing the project. This procedure is called the "backward pass" computation. T_L represents the latest date that an event can occur without jeopardizing the project completion date. The completion date may be a promised due date; or it may be the earliest possible completion date, T_E, in which case $T_S = T_E$ and the earliest and latest completion dates will be identical for events on the critical path.

Figure 3-7 illustrates the computation of T_E's for network events. Because no activities or events precede event 1, the T_E value is set at 0. The T_E value of event 2 is the value of event 1 plus the t_e of activity 1, which is 2; that is, $0 + 2 = 2$ weeks. The T_E of event 5 is equal to the value of the activity path 1-3-4-5; that is, $1 + 4 + 5 = 10$ weeks, because the T_E of an event is always the largest sum when multiple paths of activities lead to an event.

For purposes of illustration, the value of 10 weeks is assigned as the T_L of the network, as shown in Figure 3-8. Subtracting activity t_e's for $T_L = 10$ results in a value of 5 weeks at event 4, 1 week at event 3, and 6.4 weeks at event 2. When two activity paths merge into a single event, the smallest value of the two merging paths is chosen. Thus, in Figure 3-8, activity path 5-2-1 yields a T_L value of 4.4 weeks for event 1, while path 5-4-3 yields a T_L value of 0, which is assigned as the T_L value for event 1.

The slack for each event may now be obtained. Slack of an event is equal to

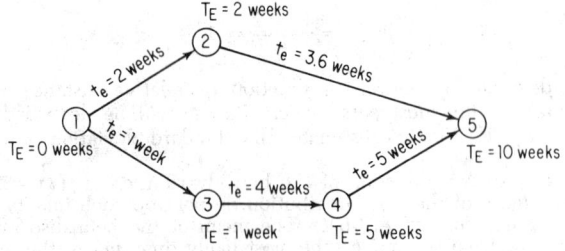

Fig. 3-7.   Computation of T_E's for network events.

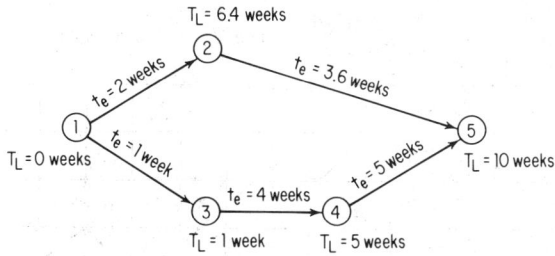

FIG. 3-8.   Computing T_L's for network events.

$T_L - T_E$.   By definition, the slack for any event along the critical path is equal to zero when T_L of the end event is set equal to T_E.   Under this condition, if T_L for a given event is later than T_E, positive slack exists and some leeway exists in scheduling the event.   If T_L for the end event is set equal to some T_S which in turn is earlier than T_E, negative slack exists and schedule slippage may be a distinct possibility.   Figure 3-9 presents the computation of slack using the T_E and T_L values of Figures 3-7 and 3-8.

Activity slack has a somewhat different meaning than slack concerning an event or path.   The slack value of an activity is equal to the T_L of its successor event minus the T_E of the activity.   The significance of activity slack can be seen from Figure 3-9.   The slack of activity 4-5 is the same as its successor event—zero—because this is the longest path leading to event 5.   The slack value of activity 2-5, however, is equal to 4.4 weeks, because its $T_E = 2 + 3.6 = 5.6$ weeks.   It may also be seen that event 2 could slip 4.4 weeks without affecting the completion of event 5 in 10 weeks.

Calendar dates are avoided in initial planning and estimating to prevent the biasing of time estimates by the estimator using those fixed dates to calibrate his estimate.   Also, negative slack should not exist in the planning phase.   Sometimes, however, management will agree to schedule a project in less time than is indicated by the preliminary estimates, for a variety of reasons.   A date imposed by an authority from outside the project may be called a directed date, T_D, to distinguish it from a scheduled date, T_S.   A major use of PERT under these conditions is to allocate resources to activities on the critical path in the attempt to reduce T_E to T_D.

One of the more frequently used reports in PERT applications is a Project Outlook Report, usually indicating the degree to which a project is ahead, behind, or on schedule.   An example of such a report prepared for one of the major events

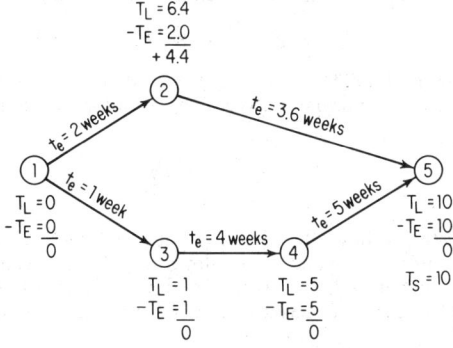

FIG. 3-9.   Computing slack where $T_L = T_E = T_S$.

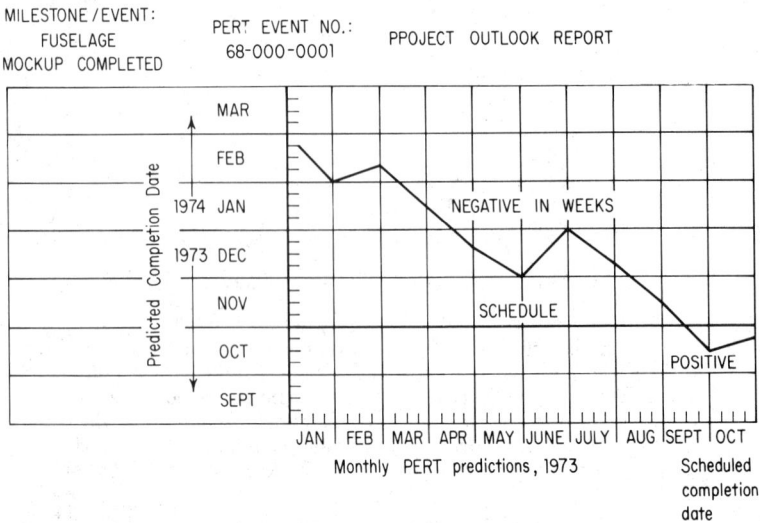

MILESTONE /EVENT:
   FUSELAGE
MOCKUP COMPLETED

PERT EVENT NO.:
  68-000-0001

PPOJECT OUTLOOK REPORT

FIG. 3-10.   PERT trend analysis—slack versus schedule.

in a network is presented in Figure 3-10.   The date scheduled for the given event is shown as a horizontal line.   Each reporting period, the amount of positive or negative slack associated with the event is plotted.   In the example, the projected time slippage has been reduced until, as the scheduled date approaches, the event is ahead of schedule.

Negative slack must be remedied if a project is to be completed on schedule.   If resources are transferable, they can be withdrawn from noncritical activities and allocated to more critical activities.   If feasible, additional labor may be procured, work subcontracted, overtime scheduled, and so on.

Probability Aspects of PERT.   One feature that distinguishes PERT from other management systems is the use of probability theory in forecasting the probable outcome of specific plans.

The variance associated with each activity in a network is as follows:

$$\sigma^2 = \left(\frac{b - a}{6}\right)^2$$

The standard deviation for any event in the network is the square root of the sum of the activity variance on the longest path to that event.[5]

$$\sigma_{T_E} = \sqrt{\Sigma\sigma_{t_e}{}^2}$$

A statement of the probability of meeting any date can be determined by using tables of areas under the normal curve.   The formula for this normalized statistic is

$$Z = \frac{T_S - T_E}{\sigma_{T_E}}$$

Figure 3-11 illustrates how the determination of meeting a scheduled date for an end event (event 5) is made.   The longest path to event 5 is 1-3-4-5.   The scheduled date for event 5 is 12 weeks, and the T_E is 10 weeks.   The variance of activity 1-3 is 0.0196; of activity 3-4, 0; and of activity 4-5, 2.8900.   The

[5] This assumption follows from the central limit theorem.

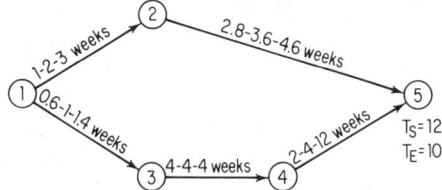

Fig. 3-11.   Determining the probability of meeting a scheduled date.

calculation to determine the Z value is

$$Z = \frac{12 - 10}{\sqrt{0.0196 + 2.8900}} = 1.134$$

From the tables of the area under the normal curve, a Z of 1.134 is equal to 0.87 of the total area under the curve. Based on the estimates, only 13 times out of 100 the event will not be completed on schedule.

The probability measure is not always employed in PERT applications. One feeling is that too much uncertainty exists in the entire estimating process for the statistical calculations to be meaningful. However, if properly used, the Z values are an index of relative uncertainty of meeting event schedules.

Network Processing. A decision to mechanize (computerize) a PERT application is dependent on the number of activities in a network. Experience has shown that the upper bound for manual processing is 200 events or less. Length and frequency of reports are also considerations. Nomographs and slide rules are available to facilitate manual computations. Calculations for T_E and T_L can be performed on a desk calculator.

Some computer programs are available from governmental agencies as well as business machine manufacturers; these must be adapted to the details of individual applications. Input data consist of the following:
1. Predecessor and successor event numbers
2. Activity descriptions
3. Time estimates
4. Scheduled or directed dates
5. Milestone event designation
6. Desired sorts (critical paths, departmental reports, and the like)

Most computer programs are designed to compute the expected elapsed time, earliest expected dates, latest allowable dates, and critical paths. Several general types of outputs are available. The event number report, for example, is a means of isolating any particular event and the activities leading to it. The slack time printout lists the series of activities composing each path through the network, starting with the critical path and ending with the path with the greatest amount of slack.

Most computer programs have built-in error messages, one of the most useful of which is the loop detector. This message indicates a circularity within a network or a returning from a successor event to an earlier predecessor. Other error messages may reveal incomplete networks and incompatible dates.

A close check must be made of input data against the network to ensure that event numbers are correct, that time estimates have been recorded correctly, and that all activities have been listed. Such checking will prevent many errors and improve the accuracy of the output.

Project Updating. PERT is a dynamic process involving change and constant replanning and scheduling. New networks must be formulated and existing networks modified to reflect revisions of planning and changes in schedules.

A basic step in project updating is the securing of time estimates for new activities and possible revision of those not yet accomplished. Updating also includes certain

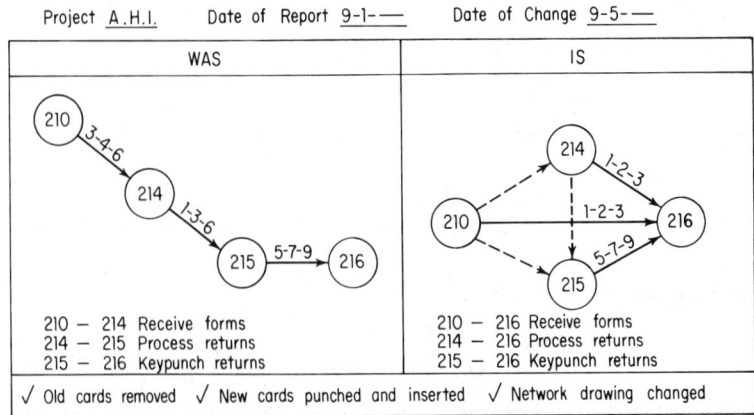

Fig. 3-12. Network change worksheet.

mechanical features attendant to revisions in the network. Major changes should be incorporated as soon as possible, while relatively minor changes need not result in a modification at the immediate reporting period. Such changes can be accumulated over a predetermined period of time and then incorporated as a group.

One practical way of keeping a record of both types of change is the use of a "Was-Is" chart, as illustrated in Figure 3-12. This chart provides necessary information concerning changes in the data deck for computer runs which can be transferred directly from the chart for keypunching purposes. A file of such charts forms an invaluable historical record of changes in the original plan and preserves documentation often important in the negotiation of contract changes.

Network Level of Detail. Individual networks for a total program can be grouped into a hierarchy of networks. The top level consists of one summary management network. Such a network is comprised of summarized key milestones, revealing the complete plan of the project or program. The next level includes the overall tasks necessary to complete the project. Generally, this level coordinates the key events or milestones set forth in the more detailed sublevel networks.

Following this grouping, the next tier of networks is usually defined as the lowest level cutting across departmental responsibilities. Individual task events may be described at the bottom level of the hierarchy. As a rule, lower level networks are activity oriented, while higher levels are event oriented.

General Output Display Criteria. Output displays are an important aspect of the PERT concept. Procedures must be established and equipment made available to add "quick response" for management reporting and decision making. An integrated, automatic facility may be developed, combining the conference room environment with a supporting display generation system. The ultimate PERT display system would utilize a Control Center Conference Room concept, permitting display of a variety of graphic media on a screen, including master and subnetwork charts, critical paths, summaries of major events, and the like. Facilities for producing hard copies of network charts and computer-produced tabular data and reports would also be provided. The Control Center Conference Room concept is portrayed in Figure 3-13.

The objectives of such an output display system, which would transmit PERT information dynamically to management, may be succinctly summarized as follows:

1. To provide a means for forecasting possible departures from a given plan
2. To direct managerial attention to the most critical areas
3. To provide a method for evaluation of the impact of changes upon a given plan
4. To help management in deciding among alternative possibilities

FIG. 3-13.   Automated management control center.

5. To aid in the achievement of timely and appropriate response to program needs

The Operating Phase.   The PERT technique can be applied to almost any project in which logical planning is required.   Applications can profitably be made, irrespective of the size of the task.   The PERT network planning cycle is summarized in Figure 3-14.   Figure 3-15 further illustrates the details of the decision making and program direction phase.

PERT COST

As originated in 1958, PERT may be viewed as a natural evolution in the systems approach to management planning and control needs.   It provided a technique for depicting the operation of a development-oriented organization as the operating entity which it is: an integrated assembly of interacting functions designed to achieve a predetermined goal at some specified time.

As first applied, however, PERT was oriented only toward planning and scheduling uses.   As the PERT Time system became established, the Department of Defense undertook the design and development of the cost aspect of PERT.   A document was issued in June, 1962, as a uniform approach to PERT Cost management, under the joint aegis of the Department of Defense and the National Aeronautics and Space Administration.   The manual, entitled *DOD and NASA Guide—PERT Cost System Design,*[6] was based on several limited-scale pilot tests on research and development projects.

PERT Cost is an extension of the PERT approach.   In the PERT Cost system, the overall program is divided into successively smaller pieces of prime hardware, support equipment, facilities, and services for costing purposes.   Estimates are made of manpower, material, and other resources necessary to perform groups of activities referred to as work packages.   These estimates are then converted to dollars, and the original estimates are compared with actual costs on a periodic basis.   In brief, PERT Cost seeks to integrate time and cost considerations on a common framework.

PERT Cost provides the means by which information is supplied in the varying levels of detail needed for evaluation of schedule and cost performance and for the prediction and control of time and cost variances.   Among the advantages claimed for PERT Cost is that it not only permits more accurate measurement

[6] *DOD and NASA Guide—PERT Cost System Design,* Department of Defense and National Aeronautics and Space Administration, June 1, 1962.

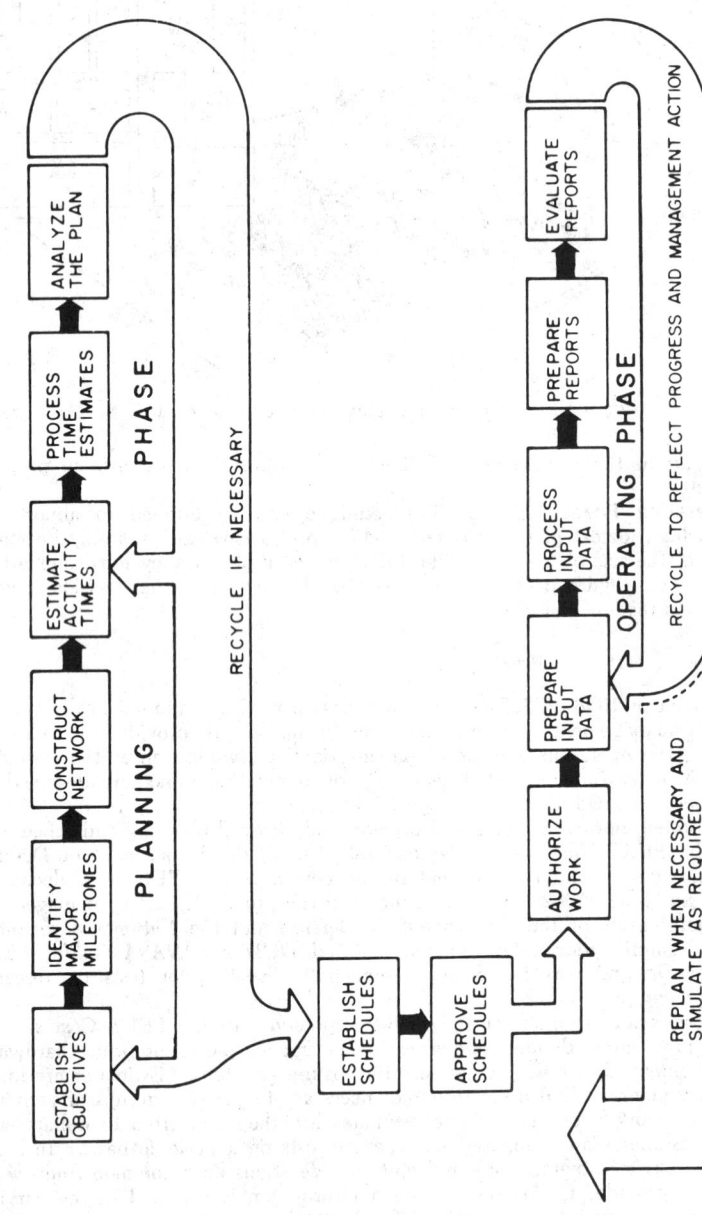

FIG. 3-14. PERT: planning and operating phases. (*Source*: PERT-Time System Description Manual, *vol.* 1, *U.S. Air Force*, 1963, p. v-3.)

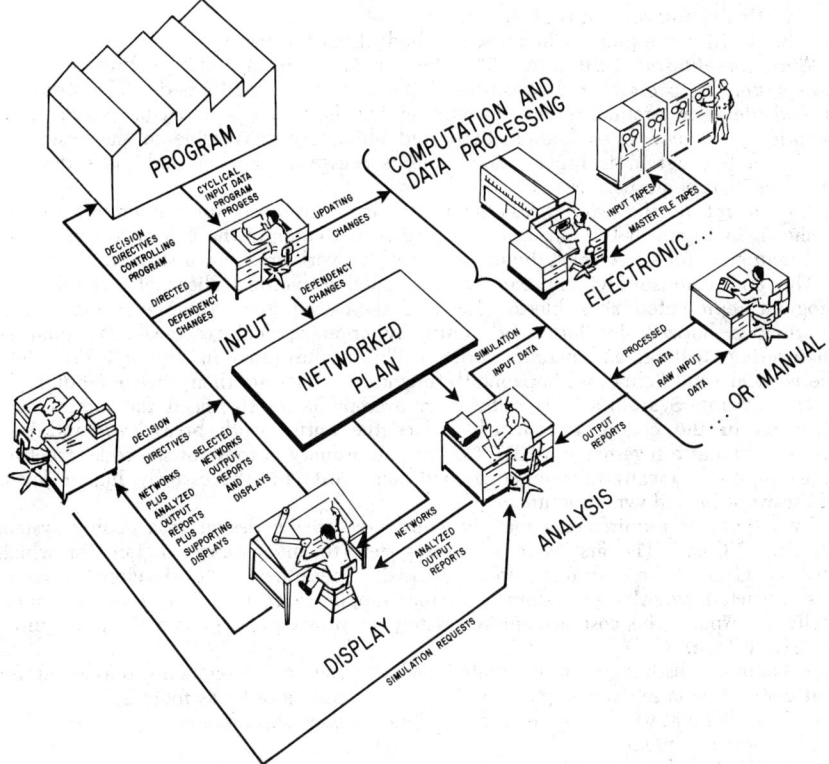

FIG. 3-15.   PERT decision making and program direction.

of progress, but also enables managers to appraise more realistically the relationships of accumulated and projected costs of the program. In addition, the expectation is that PERT Cost will provide time and cost data for decision makers weighing alternative courses of action.

The PERT Cost system as adopted by DOD and NASA had a planning cycle comprised of the following elements:

1. Establishing the work breakdown structure
2. Defining tasks to be accomplished
3. Preparing an account code structure
4. Constructing the PERT networks
5. Estimating the activity times
6. Preparing resources and cost estimates
7. Reviewing and revising the plan

These elements are considered to be the total requirement for planning and control of time, costs, and resources.

The corresponding management control cycle is comprised of the following activities:

1. Approving the program plan, schedule, and budget
2. Authorizing the work to be started
3. Accumulating actual time and cost information
4. Updating the plans as necessary
5. Preparing PERT Cost reports and information
6. Analyzing the PERT reports and information

7. Evaluating the status of the project
8. Revising the plans, schedules, and budgets as necessary

Work Breakdown Structure. The first major step in implementing a PERT Cost system is to establish a structure of the work to be performed. The structure is end item or product oriented, representing hardware subsystems, services, or facilities performed by a manufacturer and ultimately deliverable to the customer. The overall program is broken down into successively smaller and more detailed units for control purposes.

An excerpt of the work breakdown structure developed for an aircraft project is shown in Figure 3-16. Its primary purpose is to supply the means to summarize end item costs for successively higher levels of the work breakdown structure.

The end item subdivisions can be envisioned as vertical slices of a total work program represented as a block. Each of these end item rectangular blocks can be divided horizontally into work units by organizational responsibility, such as engineering, tooling, and manufacturing. This is illustrated in Figure 3-17. Both the vertical (end item) and horizontal (functional) slices are then further subdivided.

The Coding System. It is difficult to overemphasize the need for careful construction of the coding system which ties the entire work breakdown and cost element structure together in PERT Cost. The primary purpose of the code structure is to supply a means to summarize end item costs for successively higher levels of the work breakdown structure.

Two types of requirements may be considered in the design of a coding system for PERT Cost. The first is to adapt a system to an ongoing program for which another type of cost system already exists. The other is to develop a system less restricted by prior procedures. In most applications, it will be necessary minimally to expand the cost accounting system to provide the greater detail required for PERT Cost.

A summary listing of an integrated code of the major segments developed for collection of costs in one company are listed in ascending order as follows:

1. Individual worker (time card) or material issue slip number
2. Serial number
3. Work order number
4. PERT Cost code

The serial number is a basic unit of work performed at the lowest organizational subdivision with a median of perhaps 400 to 500 hours' duration. Serial numbers are assigned sequentially as individual units of work are released to engineering, shop, and other functions for accomplishment. These numbers are the basic building blocks of the PERT Cost system. At the time of release, serial numbers are assigned an appropriate work order number. In this manner, costs can then be identified to job and task.

For manufacturing operations, a traveler order, released against serial numbers, can be used for collection of costs. A computer "look-up" table may be employed to summarize labor and material charges to serial numbers by assembly, lots, and so on.

Because one of the objectives of the PERT Cost approach is to accomplish an integration of cost and scheduling considerations, the numbering system used to identify events in a PERT Time network is particularly pertinent.[7] Although the event code need not be entirely identical with the PERT Cost code, the two must be directly translatable by computer.

The work breakdown structure serves as a common framework for controlling both time and cost. It should be understood, however, that networks are not necessarily prepared supporting every item in the work breakdown structure, nor are all individual activities in a network necessarily costed.

---

[7] See L. S. Hill, "Some Cost Accounting Problems in PERT Cost," *The Journal of Industrial Engineering,* February, 1966, pp. 87–91, for an analysis of difficulties experienced in attempts to achieve this and other primary objectives in PERT Cost applications.

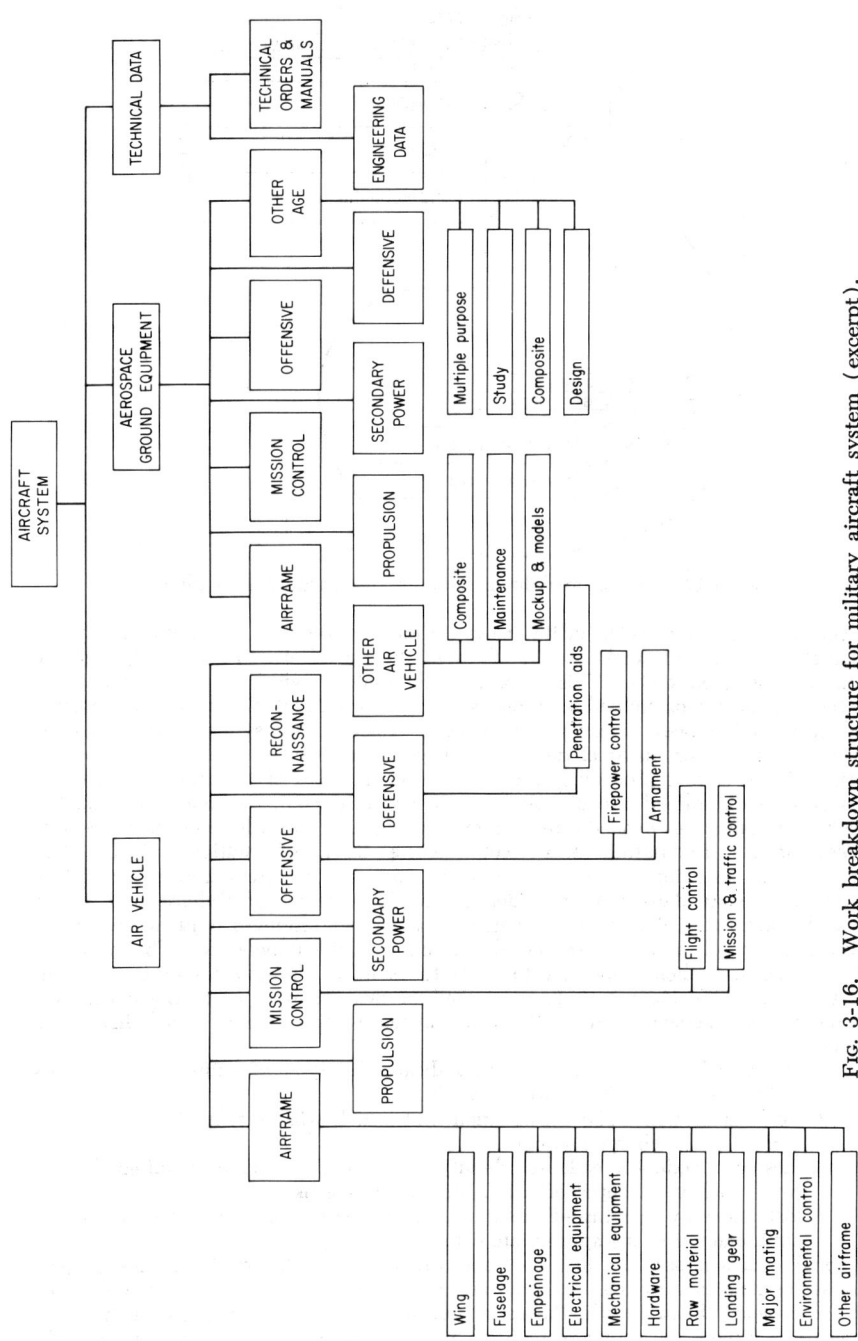

FIG. 3-16. Work breakdown structure for military aircraft system (excerpt).

Level of Detail. Frequently, it has been stated that it is not desirable to subdivide a cost accounting structure more finely than practical for collection of actual costs. However, this guideline is less of a limitation than in the past, because automation of accounting has made detailed cost collection feasible.

The level of detail which is desirable in PERT Cost is largely a matter of judgment and will vary from project to project. The depth of the work breakdown structure will differ on each program, depending on such factors as complexity, length and cost of program, a firm's organizational structure, and its traditional approach to management controls. Additionally, all work is not subdivided to a common level by all functional responsibilities in an organization.

All information accumulated during the operation of a PERT Cost application should be transmitted to higher management levels. On the other hand, considerable detail is required for analytical purposes. In the design of PERT Cost outputs, these differing requirements for information should be recognized.

Cost Reports. Cost control requires the periodic comparison of incurred costs with their original estimates. The work breakdown structure facilitates summarization of data for various management control needs. The basic information generated in the PERT Cost system may be summarized in several ways for reporting purposes. Although PERT Cost output reports may be interrelated, each can be designed to emphasize a different aspect of control. Table 3-1 illustrates the type of subsystem breakout which may be furnished (recall that only an excerpt from the total aircraft system is considered in this illustrative example). A breakout by cost element may also be provided. This cost element breakout is illustrated in Table 3-2. These output reports use the familiar variance analysis approach to identification of deviations from budgeted amounts. Current and projected schedules may also be included in columnar fashion in these reports. More detailed subsystem and cost element breakdowns may be available and provided management on an exception basis.

PERT Cost and Computers. PERT Cost computer programs have been originated by governmental agencies and industrial companies. Such programs are writ-

TABLE 3-1.   PERT Cost Management Summary Report Data

					Reporting period: ___	
					Report date: _____	
		(Thousands of dollars)				
Major end item	Current month	Cumulative amount	Total indicated cost	Total budget	Variance*	Change from previous month
Air vehicle						
Age						
Technical data						
Profit						
Total						
Air vehicle†						
Airframe						
Propulsion						
Secondary power						
Mission control						
Offensive						
Defensive						
Reconnaissance						
Other airframe						
Total air vehicle						

* The difference between total budget and total indicated cost.
† Similar breakouts are furnished for each major category.

TABLE 3-2. PERT Cost Operating Unit Status Report Data

| | | Reporting period: _____ | | |
| | | Report date: _____ | | |
Cost element (functional)	(Dollars) Current month	To date	Latest estimate	Variance
	Air vehicle*			
Material				
Outside production				
Subsystem				
Other				
Engineering				
Tooling, planning, and design				
Tool manufacturing				
Manufacturing				
Flight and shipping				
Quality assurance				
Plant engineering				
Logistics				
Total				

* Similar breakouts are furnished for each major category.

ten to summarize resource utilization by time period. Special computer routines have also been prepared—for example, programs that automatically convert time units to calendar dates. The calendar routines eliminate weekends and holidays to coincide with inputs of elapsed activity times which exclude similar data. Some computer programs have attempted to assign available resources against time to balance idle periods. Figure 3-19 is a brief diagrammatic summary of the PERT Cost approach.

CPM

The most widely applied network technique, next to PERT, is the critical path method (CPM). Whereas PERT has been utilized to a considerable degree on research and development programs, CPM has been applied most often to construction and maintenance projects. Where time and cost estimates can be obtained with a relatively high degree of certainty, as in construction, CPM may be preferred over PERT.

The CPM network is commonly referred to as an arrow diagram, events as nodes, and activities as jobs. Both normal and crash time and cost estimates are obtained for each activity in a CPM network.

Time-Cost Trade-off. The normal cost estimate for an activity is the lowest direct cost required to complete the activity. The corresponding activity duration is called the normal time. Crash time to complete an activity is the minimum time required to complete an activity, and the corresponding direct cost is called the crash cost. Thus, two pairs of time-cost estimates are required under the CPM system for each activity in the network. These estimates are illustrated in Figure 3-20. A cost penalty will be incurred if an attempt is made to complete an activity in a shorter elapsed time, as indicated in the figure.

In the CPM procedure, the time-cost trade-off points are assumed to be on a continuous linear decreasing curve, and the activities in the network are assumed to be independent in the sense that buying time on one activity does not in any manner affect the availability, cost, or need to buy time on some other activity. The slope of each activity cost line (using the straight-line approximation) is given

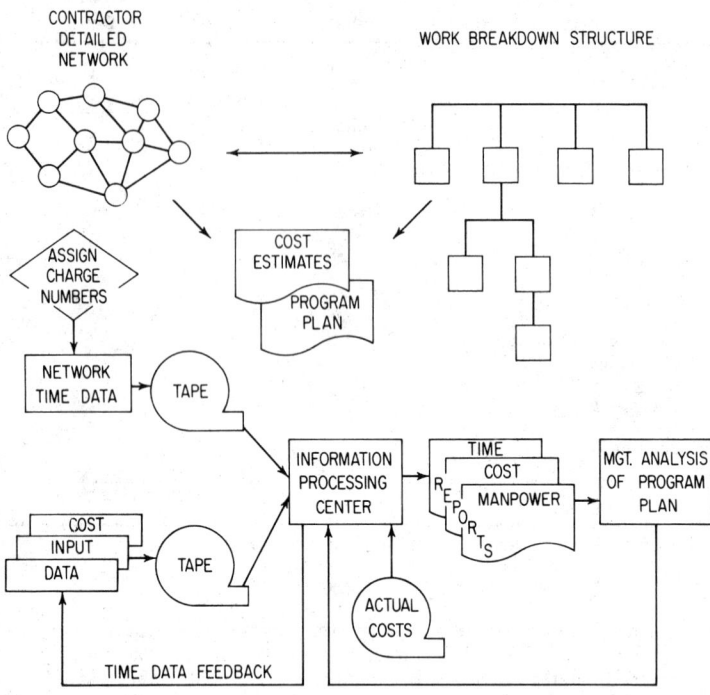

CONTRACTOR
DETAILED
NETWORK

WORK BREAKDOWN STRUCTURE

ASSIGN
CHARGE
NUMBERS

COST
ESTIMATES

PROGRAM
PLAN

NETWORK
TIME DATA

TAPE

COST
INPUT
DATA

TAPE

INFORMATION
PROCESSING
CENTER

TIME
COST
MANPOWER

REPORTS

MGT. ANALYSIS
OF PROGRAM
PLAN

ACTUAL
COSTS

TIME DATA FEEDBACK

FIG. 3-19. Brief summary of the PERT Cost system.

by the formula

$$\text{Slope of activity cost line} = \frac{C_2 - C_1}{t_2 - t_1}$$

Slope is expressed in cost per unit of time.   As in PERT, time is generally represented in weeks and tenths of weeks.   The slope, in effect, is a measure of cost increase associated with every unit of time decrease over the range $t_1 - t_2$.

To develop an optimal acceleration plan for a project, a study of each activity must be made to determine its slope.   The following steps may then be accomplished:

   1. Determine the normal critical path of the network, that is, the critical path if all activities are performed at the normal time, t_1.

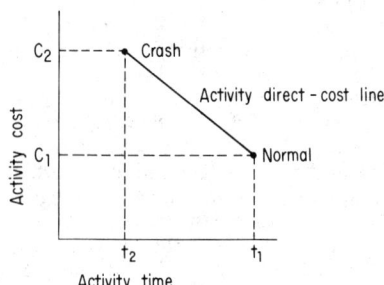

C_2 ----- Crash

Activity direct - cost line

C_1 ------------ Normal

Activity cost

t_2 t_1

Activity time

FIG. 3-20. CPM time-cost relationship.

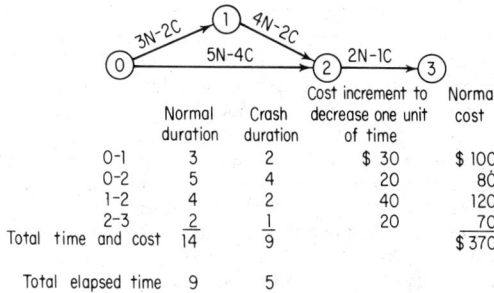

	Normal duration	Crash duration	Cost increment to decrease one unit of time	Normal cost
0-1	3	2	$ 30	$ 100
0-2	5	4	20	80
1-2	4	2	40	120
2-3	2	1	20	70
Total time and cost	14	9		$ 370
Total elapsed time	9	5		

FIG. 3-21.   Minimum cost schedule.

2. Examine the slope of each activity on the critical path and select the minimum value. Accelerate the chosen activity until either (a) another path becomes critical or (b) the minimum time has been attained for the accelerated activity.
3. Consider the new critical path(s), examine the cost coefficients for the critical activities, and choose the minimum slope.
4. Continue in the above manner until no further acceleration becomes possible. The task of calculating time-cost trade-offs in the manner briefly noted above or otherwise is formidable and difficult in practice. Procedures and computer programs[8] have been developed which will automatically schedule the project for the least-cost activities. Such programs generally incorporate the linear assumption. The resulting time-cost function for the entire project can provide useful trade-off information on the relative cost of reducing scheduled time in various activities.

Figure 3-21 illustrates a project with a normal duration of nine days and a cost of $370. The crash duration is five days with a cost of $520. The figure illustrates the ways in which the schedule may be decreased with attendant cost penalties. Only activities on the critical path should be changed. Activity 2-3 can be decreased at least cost, and activity 0-1, at the next lowest cost. The resultant schedule is now seven days' duration at a cost of $420.

The above procedure applied to additional days' crash produces a schedule, shown graphically in Figure 3-22. The total project cost curve is the summation of two

[8] See D. R. Fulkerson, "A Network Flow Computation for Project Cost Curves," *Management Science*, January, 1961, pp. 167–179.

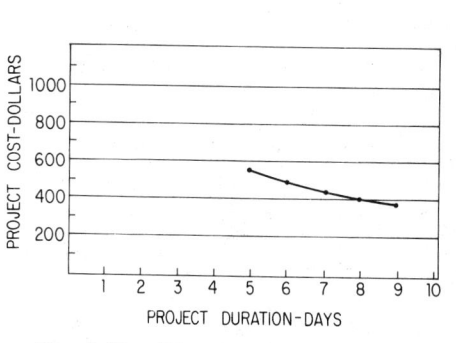

FIG. 3-22.   Direct project cost curve.

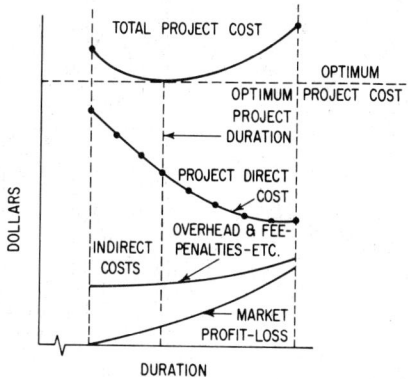

FIG. 3-23.   Total project cost curve.

and sometimes three curves: (1) the project direct cost curve, (2) the project indirect cost curve, and (3) perhaps a market loss and penalty cost curve as shown in Figure 3-23. Under proper conditions, management may choose to perform the project in the number of days which produces least total project costs.

Float. The leeway that exists in scheduling activities not on a critical path in a CPM network is called float. The technique for determining float is as follows. Starting at the beginning of the network, determine the earliest occurrence time for each event in the program. Because the first event must occur before any succeeding activities can begin, its earliest occurrence time (ES) is zero. Add to this time the duration of the activity leading to the next event, which is the ES for that event. If several activities lead to a given event, then its ES is the highest value obtained by adding the duration of each predecessor activity to the ES of the activity's beginning event. Thus, when an event is a part of two or more paths, the longest path to the event must be completed before any subsequent activities can be started. Continue the process until the final event has been reached; its ES becomes the earliest completion time for the project.

To determine the latest occurrence time (LC) for each event, begin with the time estimate for the completed project obtained from the ES procedure above, and assign this as the LC for the final event. Then subtract the time duration of the immediate predecessor activity from this to obtain the LC for the activity's beginning event. If an event has several succeeding activities, its LC is taken as the smallest value obtained by subtracting the duration of each of these activities from the LC of its ending event. In this manner, calculate the LC for each event, starting at the end of the network and working backward along activity paths until the beginning event is reached. The LC for this event will be equal to zero.

If both the earliest and the latest occurrence time for each event are available, the float or leeway in scheduling each can be readily calculated. Those events and activities with zero float are necessarily on the critical path.

An Example of CPM Computations. The actual procedure for computing float is as follows. Let i = an event signifying the origin of an activity; j = an event signifying the termination of the activity; and Y_{ij} = the activity time duration. Note that an activity's earliest start time (ES_{ij}) equals ES_i, the earliest occurrence time of event i; and the activity's latest completion time (LC_{ij}) equals LC_j, the latest occurrence time of event j.

Construct a matrix by entering the Y_{ij} for each activity in the proper cell. For example, using the arrow network shown in Figure 3-24, a matrix can be constructed as follows:

ES	i \ j	1	2	3	4	5
0	0	2	6	—	—	—
2	1	—	4	8	—	—
6	2	—	—	5	2	13
11	3	—	—	—	5	9
16	4	—	—	—	—	3
20						

	0	2	6	11	17	20	LC

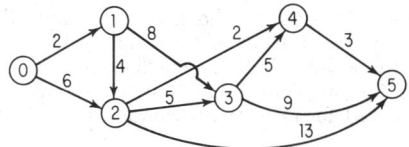

Fig. 3-24.   CPM arrow diagram.

The procedure for computing earliest occurrence time (ES) is as follows:
1. Enter a zero in the first cell of the ES column.   This represents the starting time of the project.
2. Add the corresponding values of Y_{ij} to the ES values column by column. In the example, $ES_0 = 0$ and $Y_{01} = 2$; $0 + 2 = 2$; and enter 2 in the ES column below the zero, indicating that 2 weeks are required before the activities immediately after event 1 can be started.
3. Continue this procedure for each column.
4. Where different times result from this summation process, select the longest time (path) and enter that number in the ES column.   For example, column 3 of the matrix has Y_{ij} values of 8 and 5; the corresponding ES values are 2 and 6.   By adding $8 + 2 = 10$ and $5 + 6 = 11$, the longest time path (11) can be determined and placed in the ES column.

The procedure for computing latest occurrence time (LC) is as follows:
1. Enter the longest time path in the project (20 weeks, taken from the last cell in the ES column) in the last cell of the LC row.
2. Subtract the corresponding values of Y_{ij} from the LC values row by row. In the example, $LC_5 = 20$, and $Y_{45} = 3$; $20 - 3 = 17$, and 17 is entered into the LC row to the left of the 20 weeks.   This means that event 4 must occur by the seventeenth week if the project is to be completed in 20 weeks. Continue this procedure for each row.
3. Where different times result from the subtraction process, select the shortest time (path) and enter that number in the LC row.   For example, row 3 of the matrix has Y_{ij} values of 5 and 9; the corresponding LC values are 17 and 20.   By subtracting $17 - 5 = 12$ and $20 - 9 = 11$, the shortest time path (11) can be determined and entered in the LC row.
4. The last entry in the LC row should be a zero, corresponding to the zero in the first cell of the ES column.

Every event that has an equal ES and LC time is on the critical path.   In the example, event 1 has an ES of 2 and an LC of 2; hence it is on the critical path.   Event 4 has an ES of 16 and an LC of 17; hence it is not on the critical path.   Accordingly, the critical path includes events 0, 1, 2, 3, and 5.

Total float for an activity is the amount of time available for an activity less the amount of estimated time required to complete the activity.   In the example, total float for an activity equals $(LC_j - ES_i) - Y_{ij}$.   Thus, for event 3, $LC_3 = 11$; $ES_1 = 2$; $Y_{13} = 8$; $(11 - 2) - 8 = 1$ week of float.

It may be necessary to determine how much a preceding activity may possibly be delayed without interfering with the earliest start of the succeeding activity.   This time is referred to as free float.   The concept of earliest completion time for an activity must now be introduced (EC_{ij}). EC_{ij} is derived by adding the estimated time required for an activity (Y_{ij}) to the activity's earliest start time (ES_{ij}).   Let ES_{12}, EC_{12}, LC_{12}, and Y_{12} apply to the preceding activity and let ES_{23}, EC_{23}, LC_{23}, and Y_{23} apply to the succeeding activity.   Then, $ES_{23} - (EC_{12} + Y_{12}) =$ free float for activity 1-2.

Interfering float is total float minus free float.   For example, any delay in activity 1-2 beyond the ES date of activity 2-3 will delay or interfere with activity 2-3. Hence, part of the total float for activity 1-2 is free float ($ES_{23} - EC_{12}$) and the remainder is interfering float ($LC_{12} - ES_{23}$).

Independent float is computed as $ES_{34} - LC_{12} - Y_{23}$.   For example, if all activities prior to activity 2-3 are completed by the LC_{12} date, and all activities succeeding activity 2-3 are started at the ES_{34} date, then $ES_{34} - LC_{12}$ is the amount of time available to perform activity 2-3. Independent float may be computed by subtracting the actual time required to perform the activity from the available time; that is, the activity can be displaced forward or backward within this time interval without interfering with any other event.

CPM and Uncertainty. The CPM technique provides only for the incorporation of single best elapsed time estimates in the network planning. If the time estimates are even slightly in error, the possibility exists that the critical path may be selected erroneously. Some hedging against this eventuality may be secured through determination of the second most critical path, third most critical, and so on, depending on the degree of uncertainty inherent in the program. The subsidiary critical paths may then be monitored along with the most critical. This generally would result in a more cumbersome and costly treatment of uncertainty than the method provided for in PERT.

LINE OF BALANCE

Line of balance (commonly referred to as LOB) is a production planning system which schedules key events necessary for completing an assembly with respect to the delivery dates for the completed system. Graphic displays are used to monitor progress achieved on a project and to indicate where an objective is not being met. In this manner, the LOB technique is based on the principle of management by exception, wherein management attention is directed to existing or potential problems. This method was developed by the U.S. Navy during World War II. The LOB technique is a useful complement to PERT and Gantt charts.

An LOB application is comprised of four primary elements:

1. Determination of the objective
2. Development of a program plan
3. Measurement of progress
4. Construction of the line of balance

The Objective Chart. The initial step in the application of the LOB method is to graph the cumulative delivery schedule of the end item. Such a schedule is often a straight line. As shown in Figure 3-25a, the ordinate represents cumulative deliveries and the abscissa represents time, usually expressed as a 22-day month. In any case, the number of working days per month should be held constant for the duration of a project. The schedule of actual deliveries is portrayed on the same chart as the schedule. The vertical distance between the two lines indicates the difference between scheduled and actual deliveries. The time lag between scheduled and actual deliveries is revealed by the horizontal distance between the two lines. Finally, the relative slopes of the two lines show whether they are diverging or converging.

The Program Plan. The second step in the implementation of an LOB system is to chart the program. The program, also called the production plan, depicts the stages in the manufacturing process and consists essentially of key manufacturing and assembly operations sequenced in the planned production scheme. The two basic sources of data for preparation of the program plan are the operation or routing sheets and bills of material, both of which should be analyzed on an aggregated basis. If available, detailed operation schedules may provide additional background.

As noted, the LOB technique is concerned only with principal and key events and limiting factors. The production plan consequently reflects only the critical assembly points and parts or subassemblies requiring significant lead times. These data are transferred to the chart, as shown in Figure 3-25b. Each point must be inserted in its correct temporal position and in its correct relationship to the preceding and succeeding points. The item requiring zero lead time is positioned directly above zero on the time scale. The other items are inserted in ascending order of lead time requirements. Symbols and color schemes can be used to depict

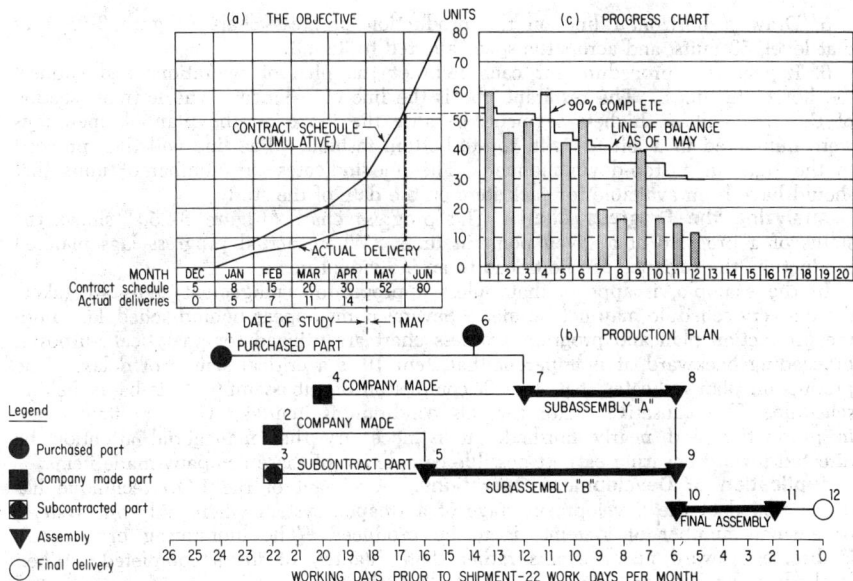

Fɪɢ. 3-25.   Line of balance charts.

different types of activities on the chart. The items are numbered beginning from left to right and top to bottom. The chart indicates, for example, that item 1 represents the receipt of a purchased part at a point which is 24 days prior to final delivery.

The Progress Chart and Construction of the Line of Balance. The third step in the LOB technique is to prepare a progress chart which shows the status of a program at a given point in time and a line of balance which represents the number of items that should pass through each control point at a given date if the delivery schedule is to be met.

Figure 3-25c illustrates a progress chart with the vertical axis representing units to the same scale as the corresponding axis of the objective chart (Figure 3-25a). The horizontal axis contains a position for a vertical bar for each item in the production plan, each bar representing the accumulated finished units of the item. The term "accumulated finished units" includes all units finished to date, those already delivered, and those currently in inventory. In instances in which units currently are in production but nearly finished, say, in excess of 90 percent completed, the convention is to indicate such status by a hollow bar. This technique is shown in item 4 of Figure 3-25c. Inventory records are utilized to determine the number of accumulated finished items.

The bars on the progress chart represent the status of items as they actually are, the line of balance indicates where they should be at a given date. The line of balance is constructed in the following manner:

1. Select a particular control point, say, item 3 of Figure 3-25.
2. From the production plan (Figure 3-25b), determine the number of days lead time required from completion of a unit to the end of the production plan, that is, 22 days.
3. Using this number, determine the date the units should be totally completed. This date is May 1, the date of the study, plus 22 working days, or June 1.
4. Find the point corresponding to the June 1 completion date on the contract schedule line of Figure 3-25a and determine the number of units that should be completed on that date if the delivery schedule is to be met, that is, 53 units.

5. Draw a horizontal line on the production progress chart (Figure 3-25c) at that level, 53 units, and across the space allotted to item 3.

6. Repeat this procedure for each item of the plan of operations and connect the horizontal lines. The resultant line is the line of balance. The leftmost section of the line will be highest segment, because the items on the plan of operations were numbered in a left-to-right, top-to-bottom fashion. The line will then proceed to the right in a step-down manner. The line indicates the number of units that should have been available for each item on the date of the study.

Analyzing the Progress Chart. The progress chart (Figure 3-25c) shows the status of a program at a given point in time. Where actual progress lags planned production, the variance can be traced to an individual item.

In the example, it appears that unless appropriate management steps are taken, the delivery schedule may not be met. Several items appear behind schedule. Both the production plan and program progress chart are utilized for analytical purposes. Proceeding backward, it is apparent that item 10 is a critical source of delay. The production plan indicates that item 8, completion of subassembly A, is badly behind schedule. The causative factor for this condition is item 4. Units of item 4 are in production and nearly finished. It is necessary that managerial attention be directed toward ensuring earliest possible completion of these company-made items.

Application to Development Operations. A variant of the LOB technique has been applied to the development stage of a weapon system where only one system, or a small number of systems, is to be produced. The monitoring of progress is directed toward major events rather than quantity of items completed. When applied in the development phase, the LOB technique must be altered, as indicated below.

For the development project, the objective chart reflects a delivery schedule based on the production of a single unit or a limited number of units. The chart will thus indicate the percent completion of individual activities and one plan for the overall project, as shown in Figure 3-26. Instead of lead time, the plan is laid out in terms of weeks of the project. Supporting data for Figure 3-26 is given in Table 3-3.

An overall development phase progress chart is developed from these data as shown in Figure 3-27. The vertical columns of the chart represent the activities of the project, with one column representing the total project. The vertical scale is also in terms of percentage of completion and should be equivalent to the scale of the objectives chart.

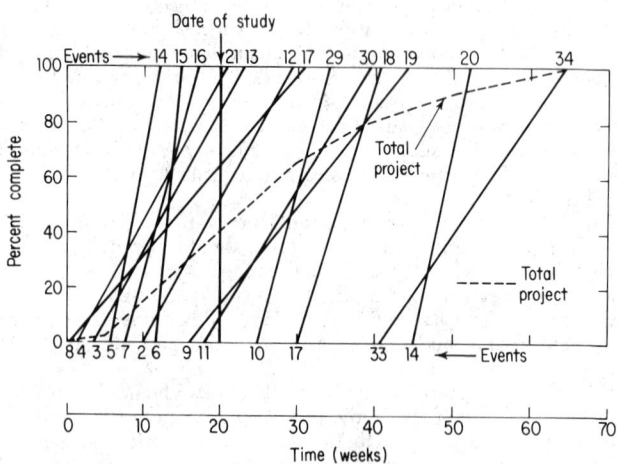

Fig. 3-26.   LOB development objectives chart.

TABLE 3-3.   Supporting Data for Figure 3-26

Activity no.	Activity	Estimated activity time (weeks)	Scheduled date Start	Scheduled date Complete
2-12	Fabricate maintenance equipment	19	10	29
3-13	Train operating personnel	19	4	23
4-21	Fabricate ground equipment	19	2	21
5-14	Fabricate installation and checkout equipment	6	6	12
6-15	Fabricate missile erection equipment	3	12	15
7-16	Fabricate missile transportation vehicle	9	8	17
8-17	Fabricate missile	30	0.2	30.2
9-19	Fabricate emplacement equipment	28	16	44
10-29	Train maintenance personnel	9	25	34
11-30	Construct launch site	21	18	39
14-20	Test installation and checkout equipment	7	45	52
17-18	Correct deficiencies in missile	10	30.2	40.2
33-34	Check out missile installation	24	40.6	64.6
Total project		204	0	64.6

The line of balance is derived from the objective chart as follows. A vertical line is constructed perpendicular to the abscissa at the date of the study. This line will intersect many, if not all, of the percent completion lines for the individual events at a point representing their currently scheduled completion status. Next, a horizontal line representing this scheduled percent completion point is drawn for each event on the progress chart (Figure 3-27). The heights of the vertical bars on the progress chart are derived from estimates made by project personnel. Height is calculated as follows:

$$\text{Percent completion} = 100 \left(1 - \frac{d}{A}\right)$$

where d = estimated weeks to complete an activity
      A = originally estimated weeks estimated for the activity

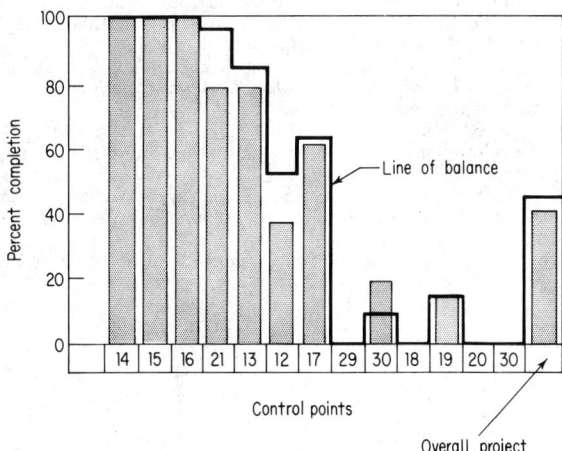

FIG. 3-27.   LOB development phase progress chart.

Both the scheduled status and the actual status of the events and of the overall study are shown for the dates of the study.

The LOB technique has several shortcomings when applied to development projects. Primary among these is a valid technique for determination of percent completion of components, because the method described above provides no means for dealing with uncertainty. In general, the line of balance technique affords no simulation capability when management desires to consider the effects of alternative approaches toward overcoming a problem area.

SUMMARY

Advancements in science and technology have resulted in research and development projects of unprecedented scope, size, and complexity. Network and graphical planning techniques have been conceived to improve communications, streamline organizations, link planning and operations more effectively, and provide a means of simulating the effect of proposed plans or changes in plans on cost, time, and product performance before taking action or enunciating policy. Progressive, competitive managements have been quick to realize and utilize the better decision making and managerial control made possible by the electronic computer, coupled with advanced management systems and used in conjunction with mathematical modeling.

BIBLIOGRAPHY

Cook, D. E., "Program Evaluation and Review Technique: Applications to Education," U.S. Department of Health, Education, and Welfare, Office of Education, OE-12024, 1966.

Fulkerson, D. R., "A Network Flow Computation for Project Cost Curves," *Management Science*, January, 1961.

Hein, Leonard W., *The Quantitative Approach to Management Decisions*, Prentice-Hall, Inc., Englewood Cliffs, N.J., 1967.

Hill, L. S., "Communications, Semantics, and Information Systems," *The Journal of Industrial Engineering*, March–April, 1965.

Hill, L. S., "Some Cost Accounting Problems in PERT Cost," *The Journal of Industrial Engineering*, February, 1966.

Hill, L. S., "Some Pitfalls in the Design and Use of PERT Networking," *Journal of the Academy of Management*, June, 1965.

Hill, L. S., "Toward an Improved Basis of Estimating and Controlling Research and Development Tasks," *The Journal of Industrial Engineering*, August, 1967.

Holtz, J. N., "An Analysis of Major Scheduling Techniques in the Defense Systems Environment," The RAND Corporation, RM-4697, October, 1966.

Macdonald, D., and D. G. Malcolm, "CPM Networking in Loan and Credit Operations," *Bulletin of the Robert Morris Associates*, September, 1964.

Malcolm, D. G., "New Tools in the Growing Management Technology," in G. N. Stilian et al (eds.), *PERT, A New Management Planning and Control Technique*, American Management Association, New York, 1962.

Malcolm, D. G., J. R. Hibbs, J. W. Taul, and M. J. Vaccaro, "GREMEX: A Research and Development Management Simulation Exercise," *Management Technology*, December, 1963.

Malcolm, D. G., J. H. Roseboom, C. E. Clark, and W. Fazar, "Application of a Technique for R&D Program Evaluation," *Operations Research*, September–October, 1959.

Malcolm, D. G., and A. J. Rowe, *Management Control Systems*, John Wiley & Sons, Inc., New York, 1960.

Chapter 4

Inventory Management and Control

ROY L. ALLEN

**Manager—Management Systems,
Columbus Division of North American Rockwell Corporation, Columbus, Ohio**

This chapter treats inventory management and control as an integrated activity. The systems approach to inventory management, computer systems, and return on investment objectives are all major forces working to broaden the necessity and the acceptance of improved inventory management techniques.

The chapter explains general principles of inventory management and discusses analysis of planning and forecasting techniques and data processing systems. Finally, ways are explored in which major objectives in inventory management can be achieved and improvements can be accomplished.

NATURE OF INVENTORY MANAGEMENT

An inventory control system may be very simple or it can be quite elaborate. Any system, however, can be thought of as essentially an integrated system of rules for deciding when and how much to order. The decision of when to order may be based on either a constant review time or a constant reorder time, the latter being favored by most companies. The order quantity decision is a problem of minimizing both the total cost of maintaining an inventory and the cost of placing an order. General management must establish operating objectives which can be translated into inventory objectives. The task of inventory management is one of controlling inventory through the selection of the time to order and the quantity to order, taking full account of future requirements and the uncertainties in their estimates, inventories, and ordering or setup costs.

Inventory is an aggregate or total mass of goods. It serves the function of making a company's internal operation relatively stable, while providing service to customers. It is possible to reduce inventory by purchasing more frequently, in smaller lots. However, the processing of many small orders to the distributor's vendors and the increased receiving load would be likely to result in serious disruption of operations. Further, a substantially smaller inventory might incur risk of unacceptable delays in filling customer orders.

On the other hand, if inventory were substantially larger, the operation would be much smoother but the capital investment might be intolerable. Management therefore attempts to strike some middle ground where an acceptable inventory investment buys an acceptable degree of smoothness in internal operations.

Although management may think of inventory as one large mass, the size of that mass is determined by the multitude of decisions relating to individual items during the course of a year. Policy directives, no matter how specific, must ultimately be reduced to the ordering strategy for a single item.

Inventory management is a critical function. It can either increase or decrease cash flow, can improve or destroy customer service, and can make or break company profits.

TYPES OF INVENTORIES

Classes of inventory found in the typical manufacturing company are of many types, and the reason for every type and variation is the final use. Inventories are classified and segregated according to the purpose to which they will finally be applied. Some of the more common types of inventories are as follows:

Raw Materials. Items acquired by the company in a form that needs further processing to make them a part of or to convert them into an end product. Examples are materials in a natural state, such as iron ore, crude oil, and wood fiber, or processed materials for general use, like steel rods or woolen garb.

Work-in-process. All product materials on which the company has performed some manufacturing, processing, or converting operations, but which are not yet in finished form, and are not ready for sale or for storage as component parts.

Finished Parts. Completed parts or components used in the manufacture of an end product. Strictly speaking, these are a part of the work-in-process. A separate inventory class of this type is usually set up, however, when detail parts or subassemblies are placed in storerooms, either for subsequent withdrawal and assembly into finished products or for sale as replacement parts.

Finished Goods. Completed products stored temporarily, awaiting sale or shipment.

Branch Office, Store, and Warehouse Inventories. Inventories maintained at a location away from central manufacturing plant or headquarters.

Consignment Stock. Merchandise belonging to the manufacturer but in the possession of the retailer or dealer on consignment.

Packaging Materials. Unit packages for finished products, nonreturnable containers, closures, labels, shipping cartons, and similar items. They are usually set up as a separate inventory class whenever they form a significant part of product costs.

Supplies. Items used to maintain operations either in the factory or in the office, but that do not become a part of the finished product. This category is known by a variety of names, including general stores, maintenance stores, and operating supplies. It includes the nonproduct items regularly stocked by the company and either consumed in operations of the plant or office or needed to maintain its buildings and equipment.

Service Parts. Parts used to maintain the equipment a company sells or services.

COST OF INVENTORY MAINTENANCE

There is always a maintenance cost in inventory. Inventory carrying costs are expressed as a percentage of the inventory value. Although this figure can vary greatly from one industry to another, the annual expense of inventory maintenance often goes as high as 25% of its value. That is, each $1 million in inventory costs $250,000 per year to maintain. There are four major costs of maintenance: obsolescence, interest, depreciation, and storage.

Obsolescence. Every business must face the grim possibility of obsolescence to some degree. Parts in stock suddenly become obsolete because of a model change or a new product. Needs cannot be estimated with perfect accuracy even in the most rigid inventory control systems. Well-managed companies ruthlessly weed out surplus inventory and dispose of it. A general rule is never to hold inventories for which there is no immediate need. Although some obsolescence is inevitable, it cannot be predicted. Therefore, a part of the cost of maintenance is an allowance to cover losses from obsolescence. The charge naturally varies widely, but few companies can hold it to less than 1% of the value of the inventory per year. Extreme conditions are illustrated in the garment and millinery industries. Manufacturing companies can experience obsolescence on the order of 10% of the value of the inventory per year.

Interest. Inventories tie up a company's most versatile asset, cash. Businesses have a limited amount of capital available to them from their owners and creditors, and each business tries to use it as efficiently as possible to earn bigger profits. Capital never is so readily available that it can be invested in inventory at no cost.

In a broad sense, the cost of interest as applied to inventory is the gain that would have been earned if the same amount of money had been invested in a different manner. Were it not in inventory, it could always earn a rate of return at least equal to the interest on government bonds. High-grade stocks can yield annual dividends in excess of 6%. In times of curtailed money supply, it could get a much higher return.

Depreciation. Normally, depreciation is considered as the reduction in value of a capital asset. In the case of inventories, however, depreciation relates to damage or deterioration due to storage, handling, weather, age, evaporation, or shrinkage. Depreciation can vary with type of inventory, inventory policies, and facilities for storage and handling. A normal amount is 5% of the value of the inventory per year.

Storage. Storage is the most obvious inventory carrying cost. It includes cost of storage and warehouse space, salaries of personnel and related storage expenses, insurance, and taxes. Insurance and taxes on inventory are a directly variable cost, because they are normally paid at a rate directly proportional to inventory value.

Storage costs can vary widely with the type of material stored and type of storage facilities used. Usually, the storage costs are equal to at least 4% of the value of material stored per year.

Cost of maintenance is expensive and can total to an amount by category as follows:

*Percent of value of inventory
per year*

Obsolescence	10
Interest	6
Depreciation	5
Storage	4
Inventory maintenance	25

These elements of risk must be carefully calculated and considered in inventory management.

INVENTORY CHARACTERISTICS

Inventories serve a variety of useful functions in an industrial economy. They can be used to get better prices, reduce costs, cover uncertainty, and reduce the need for organization.

Because investment in inventory can serve a useful purpose, there must be a proper level. Too much inventory is an unprofitable investment—a truism that leads many managers to try to force investment down. It is equally true, but not so well recognized, that too little stock is equally costly.

The manager wants a standard against which to measure actual performance. In the case of inventories, it is usually recognized that the level of sales should have some effect on the inventory required. Hence, the classical standard for judging inventory management is turnover—the volume of sales divided by the value of the inventory. In addition, a novel and more sensitive standard is known as the standard ratio.

Inventories characteristically include a large number of items. It is not unusual to find a wholesaler with from 10,000 to more than 100,000 items. Retailers can often count the number of stockkeeping units in the hundreds of thousands when considering different sizes and colors. Such a profusion of different items may appear to be a formidable obstacle to analysis. The cost of investigating thousands of items individually, many of which are of low value and contribute little to the company's revenue, may be prohibitive.

Fortunately, there is a simple approach to classifying items which makes the analysis job substantially easier. Equally important, it permits a measure of the change in the inventory value that management action will produce. The vehicle for this analysis is a listing known as distribution by value.

Distribution by Value. The listing desired as the first tool for inventory analysis is particularly easy to prepare if the company has its inventory records in machine-readable form. The items records are arranged and listed in descending sequence by annual dollar sales rates.

1. Calculate annual dollar sales for each item in inventory by multiplying the unit cost by the number of units sold each year.
2. Sort all items by annual dollar sales in descending sequence.
3. Print a list from these ranked items, including, as a minimum, the item number, the annual units sold, the unit cost, and the annual dollar sales.
4. Starting at the top of the list, compute a running total, item by item, of the item count, the dollar sales, and inventory value.
5. Compute and print for each item the cumulative percentages for the item count and cumulative dollar sales. These percentages are required only for a few selected items and may be easily computed by hand, if necessary.

Typically, it will be found that a small number of items provides a large proportion of the dollars taken as income. The top 1% of items might account for 15% of the dollar sales; the upper 5% might account for 40% of the sales; the upper 20% of the items might account for 70% of sales. The upper 60% of the items might account for 95% of the sales, or conversely, the lower 40% of the items for only 5% of the sales.

To many managers, these figures are astounding. However surprising these relationships may be on first exposure, it is a fact that they will invariably be found in any inventory. That is, a very few of the top items account for a majority of the sales, and the large majority of the items accounts for a small portion of the sales. Reducing the control effort devoted to many items produces large savings. If each item is unimportant, the loss from not devoting much effort to its control cannot total very much.

The ABC Method. The ABC method of inventory control is used to keep the amount of attention given an item somewhat in proportion to its importance. This method of inventory management was developed by H. Ford Dickie at the General Electric Company. The importance of an item is determined by the value of the use in a period of time, the time required to replenish depleted stocks of the item, and the costs caused by the occurrence of a stock-out. A small percent of the items in a stockroom makes up a large portion of the inventory investment. These are considered to be class A items. Special care is used to maintain the accuracy of the perpetual inventory records of these items. Both the suppliers and the manufacturing division of the company are required to use extra efforts to meet the scheduled dates of delivery. The engineering and manufacturing departments will seek designs and methods of processing that will hold the lengths of the delivery cycles of these items to a minimum.

The class B items have average importance and receive the normal amount of attention. It has been said that class B is made up of the items that are neither class A nor C.

The class C items are comparatively unimportant. Usually, well over half the items kept in stock are in this class. They make up only a small part of the inventory investment. Generally, the cost of having liberal quantities of these items in stock is less than the cost of maintaining close control over the ordering lot quantities, reserve stock, and in-stock balances. The inventory control system for these items may be a simple two-bin system. Many companies have realized worthwhile savings from the informed use of this method of inventory control.

Preparing the distribution by value may be justified for the sole purpose of developing the ABC classification. There is an additional profitable use of this tool. For example, it is possible to make estimates of the change in inventory value that might be the result of changes in total sales, number of items carried, different services to be provided, and the like.

The Standard Ratio. The standard ratio is a measure of how extreme the few items/many dollars relationship is. For example, in some inventories 2% of the items can account for 80 or 90% of the sales, while in others, the same 2% of the top items yields only 10% of the sales. The standard ratio for the first, most extreme case would be relatively high, perhaps 20 or 25. For the second case, the number would be relatively low, perhaps 2 or 3.

Inventories for various industries in the economy have typical standard ratios as follows:

Industry	Standard ratio
Technological.......	25
Industrial..........	10
Wholesale..........	4 to 7
Retail..............	2 or 3

The technological inventory is high because a few large components have a very high cost and appreciable volume. It is characteristic that the closer the inventory is to the consumer, the lower the standard ratio.

INVENTORY MANAGEMENT SYSTEM

The inventory management system is made up of three basic subsystems: forecasting, reviewing, and ordering, as shown in Figure 4-1. Balancing the costs of operating each subsystem against the sensitivity of the expected results leads to specifying a different frequency of use for each subsystem.

The forecasting subsystem has to do with the order point or when to order. To know when to order and to have some idea of how fast an item is to be used up, each item's usage rate must be forecast. To recognize changes in usage patterns, the forecast must be relatively frequent, a monthly or semimonthly interval being most common. These forecasts are used to set order points, taking the cost of lost sales and the cost of maintaining inventory into account.

In the reviewing subsystem, the order point set by the forecasting subsystem is compared with the available stock to determine whether stock is sufficiently low to order now. If it is not, no further action is indicated; if it is, the reviewing subsystem looks up the order quantity computed by the ordering subsystem. This

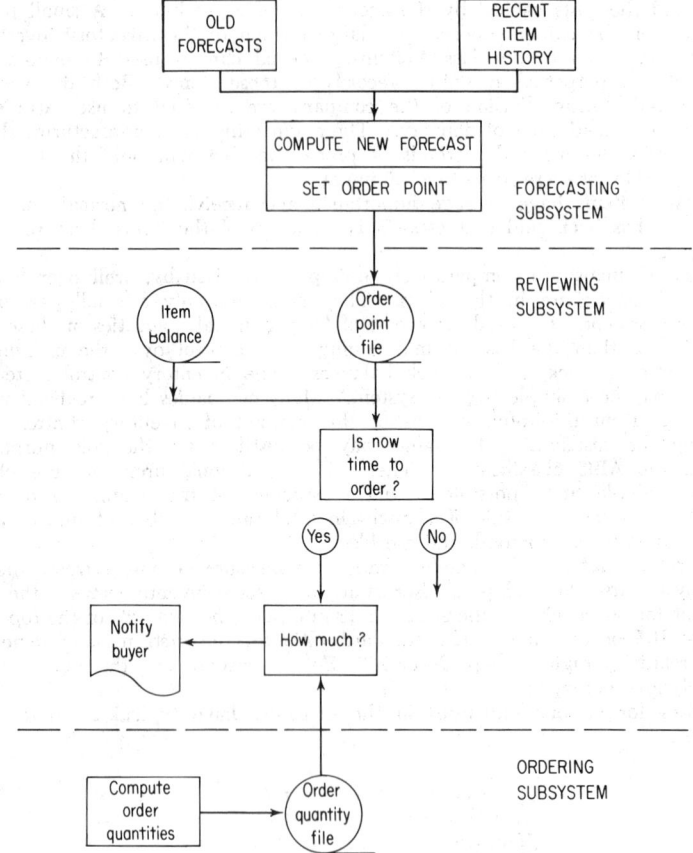

Fɪɢ. 4-1.   A representative inventory management system.

quantity is then sent to the buyer for his approval.   Reviewing should be done frequently because of the constant depletion of stock and the attendant possibility of reaching the order point.   It is not uncommon to check an item's status after every issue, although weekly and biweekly reviews are often encountered as well.

The ordering subsystem considers the order quantity or how much to order, balancing the cost factors relevant to ordering strategy to find the minimum cost strategy for each item.   It is usually not worthwhile to recalculate order quantities more than once or twice a year.   The pertinent cost factors are the cost of purchasing, the cost of maintaining inventory, the effective unit cost, and the sales rate.

Types of Control.   The process of controlling inventories breaks down into two types of control, which differ in both the time and the organization level at which they are performed.   These are:

   1. Short-term, front-line operating control.   This is exercised before inventory acquisitions on an individual item and unit transaction basis.
   2. Longer term, higher level management control.   This is exercised after inventory acquisition, usually through comparing actual results with budgeted inventory levels, turnover standards, delivery service policies, and the like.

This distinction emphasizes the fact that real control over inventories is exercised at the time purchase requisitions and manufacturing authorizations are issued.   Any control exercised after either of these steps has been taken can only be corrective rather than preventive.

Probably the fundamental task in managing inventories is to relate the short-term and the long-term controls to each other effectively. That is, the long-term controls need to include policies and mechanisms that indicate clearly how to make the day-by-day short-term inventory decisions for both manufactured and purchased items. These mechanisms are called order rules. Inventory policy defines the general types of order rules to be used for each class of inventory; the order rules for each item provide a guide for determining when to replenish and by what quantity.

Characteristics of Order Rules. An order rule is a statement of when to replenish inventories, in what quantity to replenish them, and how frequently to review the item to determine whether it should be ordered. An order rule might be stated as follows: "For item number 12345, order 100 units whenever the quantity on hand plus the quantity already on order falls below 80 units; review every time any units are withdrawn from stock."

There are several basic types of order rules and many variations of each. For any given situation and class of inventory, different types of order rules may cause significant differences in total costs and inventory levels. For this reason, order rules should be tailored to fit each class of inventory and each type of inventory situation. The objective in applying order rules is to secure the best balance among (1) costs of replenishing inventory, (2) costs of carrying inventory, and (3) costs of ordering and controlling inventory.

In securing this balance, it will often be desirable to break a given inventory into two or more groups and use a different type of order rule for each group. In the distribution by value discussion, it was shown that in many companies relatively few items account for the bulk of total inventory investment. Examples of a small percent of the items accounting for a large percent of the inventory investment were quoted. In such a case, it is usually appropriate to use a precise and elaborate order rule for the high investment items, and a relatively simple and easily applied order rule for the remainder.

Steps in Developing Order Rules. Several steps are involved in developing and applying order rules:

1. Determine a suitable method for forecasting future usage.
2. Determine the amount of time required to replenish each item.
3. Determine the basis for deciding how much to order by analyzing the cost factors for each class of inventory.
4. Establish the basis for deciding when to order by determining the appropriate level of protection against running out of stock.
5. Determine the appropriate form of order rule to apply by analyzing the characteristics of each class of inventory.

Any of the steps can be made the subject of a comprehensive and detailed analysis, and an extensive body of literature is available to provide guidance for this purpose. A partial listing is shown in the bibliography at the end of this chapter. In most situations, however, these steps can be handled quite adequately by a combination of judgment and some fairly simple analyses.

The aim in developing order rules should be to arrive at a practical compromise between the cost savings to be obtained by greater precision and the costs and time required to obtain such savings.

Forecasting. Reaching a sound decision on how much to order and when requires some kind of assumption about the future usage of the inventory item. This explicit or implicit forecast of future usage underlies any order rule. The following are typical, basic forms:

1. An explicit forecast
2. Projection of requirements based on orders already on hand
3. Direct projection of previous actual usage
4. Projection of previous usage adjusted for trends, seasonal variances, or other factors

The objective in selecting the type of forecast is to balance the value of forecast accuracy against the cost of making the forecast. The decision is usually an important one, because forecasts often must be made many times a year for thousands of inventory items, and the costs of elaborate forecasting can become substantial. And for efficiency, some order rules require relatively accurate forecasts, whereas others do not. Often, a direct projection of previous usage is sufficiently accurate for items having variable but sustained usage.

Replenishment Time. Having determined a basis for forecasting future usage, the next step is to determine accurately how long it takes to replenish an item. Replenishment time is defined as the total interval between the time at which some signal is given that an item needs to be ordered and the time the replenishment quantity actually arrives in stock and is available for use. Replenishment time calculations should therefore include all these elements:

1. Interval between time stock is physically moved out of inventory and time transaction is entered on inventory records
2. Interval between entry on records and review for reordering
3. Time required for processing and placing the order
4. Time from ordering until new stock physically enters inventory and is available for use

A significant part of total replenishment time is frequently taken up with reviewing inventory records and processing orders. If this is true, worthwhile reductions of replenishment times and inventories may be made by redesigning record processing procedures to reduce the time required for them or to make the time required more uniform.

Making the replenishment time more uniform does not reduce the expected or average replenishment time, but it does reduce the maximum time likely to be allotted to the inventory replenishment cycle. Because inventory coverage must be based in some way on the maximum probable demand and the maximum probable usage, reducing the maximum probable replenishment time causes a reduction in the inventory levels needed to provide the desired level of protection.

How Much to Order. The objective in deciding how much to order for replenishment is to achieve the lowest overall cost—not merely the lowest cost of carrying inventory. This requires establishing the most economical balance among the following conflicting factors:

1. The costs of carrying inventory
2. The clerical and other overhead costs of ordering an item
3. The costs of buying in smaller quantities (higher unit cost of purchased items or increased setup or learning costs for manufactured items)
4. Risks of loss through obsolescence, deterioration, or price decline
5. Costs of additional storage space if present warehouse capacity is exceeded

The problem of balance must be carefully considered. If smaller quantities are ordered, the average inventory level is lower and the carrying, space, and obsolescence costs decrease; yet, smaller quantities result in more frequent ordering, and the costs of the other factors increase. For items with highly erratic usage, the most economical balance will usually be achieved by ordering the quantity expected to be required in the near future. For items having sustained and regular usage, however, the most economical balance can usually be obtained by calculating the economic order quantity (EOQ) for the item. Many useful formulas have been developed for calculating EOQ's, and a number of techniques are available to simplify the actual computations. These include:

1. Hand computation, using the formula directly
2. EOQ slide rules
3. Nomographs
4. Charts
5. Tables

Three widely used methods are shown here. The first is used for items not subject to price differentials for different quantity purchases; it is based on a total cost formula.

Total annual cost = inventory carrying costs + ordering and setup costs

$$T = \frac{Q}{2} IC + \frac{S}{A} A$$

where T = total annual cost, dollars
Q = order quantity, units
I = inventory carrying cost as a decimal fraction
C = unit cost of the item, dollars
S = expected annual usage, units
A = out-of-pocket setup and ordering cost, dollars

The inventory carrying cost, I, includes only variable costs as a fraction of the inventory investment. These variable costs may include the cost of capital invested in the item, insurance, inventory taxes, and obsolescence or spoilage. Costs for space are not included because they do not normally vary short-term with changes in the level of inventories. If warehouse space is rented on a short-term basis and is variable, then a factor for the cost of space should be added to the formula.

The economic order quantity is determined by trying different values of order quantity, Q, in the formula until the lowest total cost, T, is obtained. The order quantity corresponding to the lowest total cost is the economic order quantity.

The second method is also used for items not subject to differential prices and is much easier to use. The formula for the second method is derived from the total cost formula above, but it determines the economic order quantity directly:

$$EOQ = \frac{2AS}{IC}$$

The third method is used when an item has differential prices for different order quantities. If out-of-pocket ordering costs are very small, the EOQ for an item can be calculated by using the following total cost formula:

$$T = \frac{Q}{2} IC + SC$$

The procedure consists of calculating the total cost separately for each price, using in each case the minimum order quantity associated with the price. The EOQ is then the order quantity corresponding to the lowest total cost.

Certain principles should be followed in applying EOQ formulas. First, the formula should be specifically designed for the class of inventory to which it is applied. This means that it should take into account all the cost factors that are significant, and to avoid unnecessary complexity, should exclude those which are insignificant. For example, if deterioration is a significant factor, it should be included because serious errors could result from its omission. Second, the cost figures used should be out-of-pocket or savable costs. Thus, clerical ordering costs, for instance, should include only those costs which will actually be eliminated if the order volume is reduced.

The Economic Lot Size Model. Evaluation of the economic lot size model proceeds from a very basic approach with broad assumptions to more sophisticated models with specific input data.

The objective in the determination of the economic lot size is to find the lowest annual inventory cost. The very basic model is the one more commonly used by industry, and with the statement of its broad assumptions, is the starting point leading to more elaborate approaches to optimize the lowest annual inventory cost. The assumptions for the basic approach are:

1. No consideration is taken of any emergency reserve—the order is placed so that it is received at the time the stock is depleted.
2. Inventory carrying cost is applied to the average inventory value.
3. Unit cost remains constant.

Total annual inventory cost is the sum of the cost of preparation for production, the cost of carrying the inventory, and the cost of the products in inventory.

When to Order. An order rule contains a signal or indicator of when to order, based on the replenishment time for the item. The indicator may be expressed in time—"Order five weeks before item is needed in stock"—or in units of inventory—"Order when stock on hand and already on order drops to 75 units." When the signal is in units of inventory, it is called an order point.

To assure that the item will be available when needed, the time must be the maximum probable replenishment time for the item, or the units must be the maximum probable number of units of usage during the maximum likely replenishment time. This number of units can be thought of as having two components, the average or expected number of units used during average expected replenishment time, and a safety stock to cover unexpected increases in usage or in replenishment time.

Typically, usage during replenishment time is variable and cannot be predicted exactly, and providing complete and total protection against running out of stock under all possible contingencies is not feasible. In general, as the level of protection is increased, the additional inventory required to provide still more protection increases disproportionately, and at some point becomes uneconomical. It is necessary, therefore, to choose a reasonable level of protection. Thus maximum probable usage during replenishment time becomes the value that will provide the appropriate level of protection required.

Establishing a precise measure for the appropriate level of protection is difficult, however. In most operating situations, a satisfactory approximation can be reached by using judgment in balancing the costs of running out of stock against the costs of carrying additional inventory. Different levels of protection are appropriate for different classes of inventory. For example, running out of an equipment spare part is likely to impose more severe cost and operating penalties than running out of a general supply item. For that reason, the reorder signal will often be set at a relatively higher level on equipment spare parts.

Order points can be established by a number of different methods ranging from rules of thumb to fairly complex statistical analyses. Although a complete coverage of these methods is beyond the scope of this chapter, one method is described to illustrate the general approach. This method can be used if the following conditions exist: records are available to show the past pattern of demand for the item; the pattern has not been distorted due to running out of stock on the item; the pattern of demand can be expected to remain the same in the future; and the replenishment time does not vary a great deal.

The method consists of these steps:

1. Determine the maximum expected replenishment time (using the time elements discussed earlier).
2. Using the previous year's sales history, calculate running totals of usage during the calculated replenishment time.
3. Select the highest running total as the order point.

If the maximum expected replenishment time for an item is eight weeks, the running totals are developed by successively calculating the usage during the first eight weeks of the previous year, the second through ninth weeks, the third through tenth weeks, and so on. The highest figure for the year can then be taken as the order point, with a high assurance of not running out of stock under the conditions specified above. This degree of assurance may be greater or less than necessary. If so, the method can be modified; for example, the second highest figure for the year can be taken as the order point.

SIX BASIC TYPES OF ORDER RULES

A system with a fixed order quantity can be specified by the lead time in weeks between placing an order and receiving it; the order size quantity; the safety stock quantity; and the expected demand rate. The expected inventory balance if demands are made uniformly at the expected rate is shown in Figure 4-2. The inventory balance, averaged over time, is the safety stock plus half the order size.

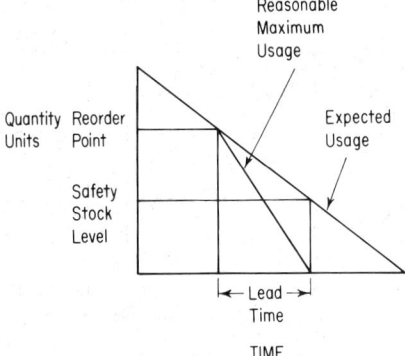

Fig. 4-2. Expected inventory balance if demands are made uniformly at the expected rate.

The reorder point is the inventory balance at which a new order is placed. The reorder point is reached when the inventory balance equals expected demand during the lead time plus the safety stock needed to protect against possible excess demand over that expected during the lead time.

However, when the lead time is long relative to the expected time between orders, care is needed in defining the rule governing reordering and the reorder point. If the lead time is three months, for example, and the amount purchased by each order is a one-month supply, this does not mean that it is necessary to place a new order when the amount on hand drops to a maximum three months' usage. Because an order will be placed once a month on the average, there will almost certainly be some orders outstanding all the time, which on being filled, will help replenish the inventory on hand. Thus it is characteristic of replenishment systems that the safety stocks, reorder points, and the like should be based on both the amount on hand and on order. Where the lead time is short compared with the usual time between orders, as is assumed in most factory two-bin systems, the amount on hand and the total on hand and on order are in fact equivalent at the time of reordering.

Six basic types of order rules are discussed here:

1. Fixed reorder point
2. Net requirements
3. Reservation
4. Fixed review time
5. Group ordering
6. Two-bin order control

The discussion of each rule shows how the rule is developed and when it should be used.

Order Diagrams. A simple and useful technique for understanding and visualizing order rules is the order diagram, Figure 4-3, which is a graph of the inventory levels against time for an item, assuming average usage and average replenishment time. Because neither the rate of usage nor the replenishment time is completely constant, a graph of actual inventory levels for an item is also shown. When an order rule is applied to a large number of items, an order diagram fairly represents the resulting inventory situation.

The safety stock level is based on the time it takes to obtain material in an emergency. A factory usually knows what delay can be expected in obtaining material in the ordinary course of events. It is also known that this time can be shaved considerably in an emergency by giving orders special rush priority, which may mean revising schedules, changing orders, shipping by air, or other

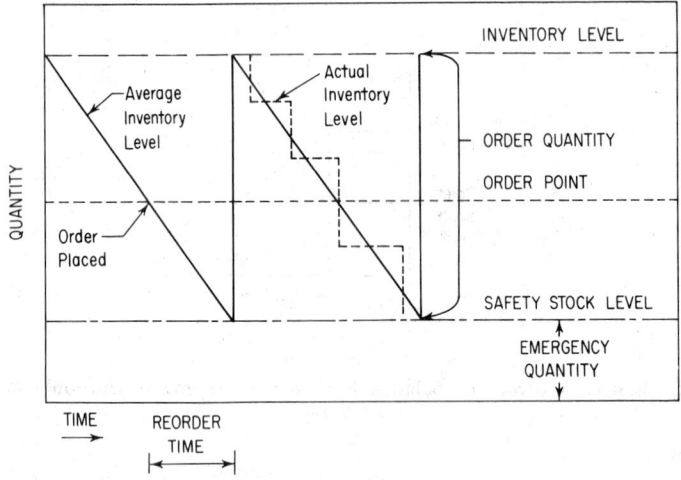

Fig. 4-3. Order diagram.

expedients.   Of course, this is very expensive and should not be permitted to occur too frequently.

If the order rule for an item is known, an order diagram can be constructed for it.   The diagram will show how the rule will operate and will also provide a convenient method for determining the average inventory level that will result from application of the rule.

Fixed Reorder Point.   In the fixed reorder point rule, Figure 4-4, when to order is established by the order point, which is an inventory level as large as the reasonable maximum usage of the item during the lead time required to replenish stock.   When the stock on hand plus stock already on order falls to the reorder point, the inventory planner places a new order.   How much to order is prescribed by a predetermined economic order quantity.   The economic order quantity is calculated as the quantity that will result in the lowest total costs of ordering, making, or procuring the item and carrying the resulting inventories.   Thus, the order quantity is fixed and the time interval between orders varies, depending on the rate of usage.

In the fixed reorder point rule, the quantity on hand is checked every time material is removed from the inventory.   When it reaches the fixed reorder point, a quantity of a fixed size is ordered.

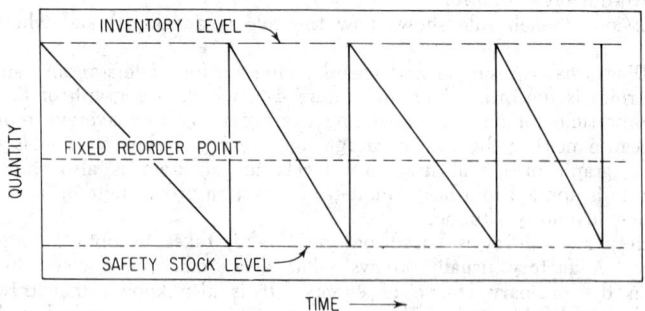

Fig. 4-4. Order diagram—fixed reorder point rule.

This type of order rule can be widely used. It is relatively simple to operate and to maintain, because order points and quantities for many items need be revised only infrequently. It is also flexible, because it automatically adjusts to changes in usage through more or less frequent ordering. Within the normal ranges of inventory carrying, ordering, and setup costs, the system provides economical ordering even when changes in usage are fairly significant. For severe changes in usage, the system can be modified. One method is to develop separate order points and order quantities for each season of the year in seasonal businesses.

Net Requirements. Under the net requirements rule, gross requirements in units are determined for each individual item—either from a forecast or from orders already on hand—for each period (such as each month) in the future; the gross requirements are then converted to net requirements by subtracting the number of unallocated units on hand and already on order. The inventory planner orders the net requirements, thereby adjusting for any errors in the forecast for the previous period. Thus inventory is ordered every time requirements are developed, usually at fixed intervals such as weekly or monthly, and the order quantity varies, depending on the net requirements, as shown in Figure 4-5.

The net requirements rule often demands more clerical or data processing work than other order rules, because requirements and ordering dates for individual parts and materials are usually obtained by exploding bills of material. For this reason, the system is generally best suited to situations where the potential savings offset the added clerical work—for example, where usage fluctuates radically or where it is important to keep inventory commitments at a minimum. Thus requirements ordering is often used for products or items that (1) are made to order, (2) are very expensive or bulky, (3) are used intermittently, or (4) offer no significant economies from larger lot orders. Requirements ordering is also used when the number of items—or at least the number to be exploded—is relatively small, for unusually large single orders, and for balancing out parts inventories for products that are being discontinued.

Under requirements ordering, deliveries can theoretically be scheduled for exactly when the item is needed, so that the resulting average inventory levels should be extremely low. In practice, however, some average inventory is carried when using this system. The inventory arises because stock is ordered for delivery earlier than it is needed or is used later than was anticipated. The inventory planner may order for earlier delivery for several reasons: to allow for greater than planned usage or longer than planned replenishment time; or to simplify the clerical process by not breaking the gross requirements into such fine increments of time; or by ordering short and long lead time items at the same time.

Reservation. The pure requirements system just described has the disadvantage of failing to provide the most economical ordering pattern for many items with sustained usage, because the order quantities are determined without regard for the costs affecting the individual items. This disadvantage can be overcome, however, by using what is often called reservation or mortgage ordering.

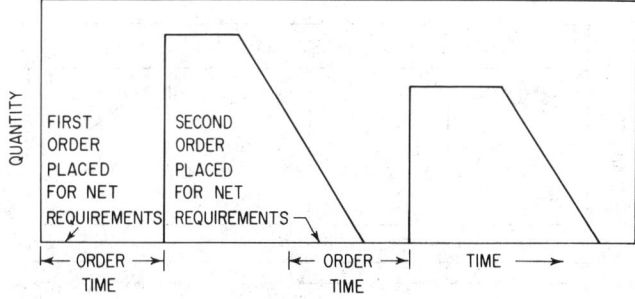

FIG. 4-5. Order diagram—net requirements rule.

In this order rule, how much to order is determined by the economic order quantity for each item. When to order is established by the requirements for the item. The gross requirement quantity for each item is reserved on the inventory record to indicate that that portion of the inventory is not available for other uses. When all the inventory quantity on hand plus that already on order has been reserved, it is time to reorder the item. This procedure is used when requirements are developed and reserved far enough in advance of actual use to allow time for replenishing items. In such a case, the item is ordered when the unreserved balance—on hand, plus already on order, less reservations—falls to zero. In some cases, the item must be ordered before the unreserved balance falls to zero, and an order point is established to signal the need to reorder. The order point is set at a level as high as the maximum probable demand during the replenishment time less the length of time in advance of actual use that reservations are made.

In effect, the reservation type of order rule requires that two sets of inventory records be kept for each item—one record of actual stock balances and another of net unreserved stock balances. In return for the extra record keeping, however, the reservation order rule allows the inventory planner to reduce or eliminate the safety stock portion of inventories. If firm reservations can be made reasonably far in advance, the reservation type of order rule should be considered for use.

Generally, a reservation type of order rule is most useful when potential economies from quantity ordering are significant and when the possible reductions in inventory levels for safety stocks are large in relation to the extra clerical effort. Thus it is most usually applied for high-cost items, items with moderate to high annual usage value, and items with variable but sustained usage.

Fixed Review Time. When the fixed review time rule is used, items are ordered at fixed, regular intervals such as every week. The specified interval indicates when to order. How much to order is determined by subtracting the stock on hand, plus that already on order, from a so-called "order guide figure" or "imprest level," Figure 4-6. Thus the order quantity varies, but the interval between orders is fixed.

Computers have stimulated an increasing interest in the fixed review time method because it is relatively easy to check a large number of inventory items periodically. Fixed review time can also be used in a manually operated system if the inventory consists of just a few items, or if the records can be displayed as bar charts which can be quickly reviewed.

The order guide is set large enough to cover the maximum probable demand for the item during the replenishment time. The replenishment time under this rule, however, is usually longer than under an order point system, because it includes the order interval in addition to the other elements of replenishment time. The fixed review time rule is most used in situations where (1) many items are ordered at the time, (2) quantity cost savings are based on the size of the total order for all items, and (3) there are no significant economies from ordering individual items in larger quantities.

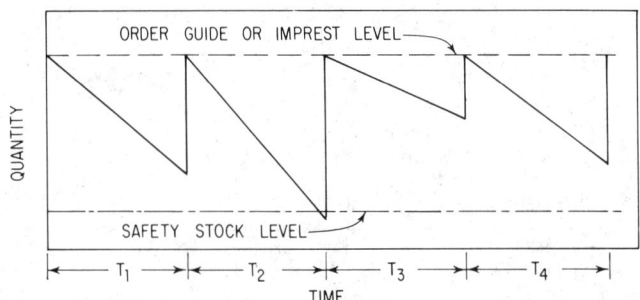

Fig. 4-6.   Order diagram—fixed review time rule.

In actual practice, there are problems which occur when the fixed review time method is used. As is apparent on the sawtooth chart, when the time is fixed, the reorder quantity will vary, which means that it will be difficult to take advantage of economic order quantities. It is true that the person in charge of the inventory may take the option of ordering or not ordering at the review period. When only a small quantity is needed, the decision may be not to reorder, but this brings up the chance of a human error, and the next review period might find the stock dangerously low.

The fixed review time rule is most useful in three applications. One is requisitioning low-value supply or expense items from a central stockroom. In this case, it is often less expensive to review stock levels of all items and place a consolidated requisition periodically rather than to order individual items at random intervals.

A second application of this rule is in replenishing regional or branch warehouses from a central factory or warehouse. Definite economies may be obtained here by making replenishment shipments large enough to gain favorable transportation rates. For instance, the order interval for a branch warehouse may be set to ensure that replenishment shipments from the factory can be made at carload or truckload freight rates. At the same time, the size of the order for each individual item may make little or no difference in costs.

The third application of the rule is in ordering several different purchased items from a vendor who gives quantity discounts based on the size of the total order without regard to the quantity of each item on the order.

Group Ordering. Group ordering order rules are used for ordering groups of items at the same time. Group ordering often but not always uses a variable interval between orders, a fixed total order quantity, and fixed order quantities for individual items.

Although group ordering rules have many variations, a typical arrangement is as follows:

1. All the items in a group are coded to designate the particular ordering group, and each item is assigned an individual order point and order quantity. In addition, a total order quantity is assigned for the group of items.

2. When the stock on hand and previously on order of any one of the items in the group falls to its order point, a new order is placed. The order is made up by ordering the established order quantity for each item in the group, starting with the item that has reached its order point and adding items that are near their order points until the size of the overall order reaches the assigned total order quantity. This type of order rule is used when significant economies are to be gained from ordering each item in an economic order quantity and from assuring that the total order is of a given size. It is used most for items whose combined annual usage value is large or for items that pose difficult delivery problems or scheduling problems within the plant.

Most variations of group order rules tend to be more difficult to develop and apply than other order rules. Some judgment is required both in determining the most economical form of rule and in assessing whether the potential savings are sufficient to justify the cost and complexity of group ordering.

Two-bin Order Control. The two-bin order control rule is designed to provide control at minimum cost. It relies on physical and visual control of inventory rather than on formal inventory records. The order diagram and principles of operation are similar to those used in fixed reorder point ordering.

The order rule operates as follows. The storekeeper maintains two adjacent bin locations for each item. One bin contains a quantity equal to the reorder point quantity, and this bin is sealed or taped shut so that all withdrawals of stock are made from the second bin. The signal to order comes when the second bin is emptied and the seal on the first bin is broken to continue further withdrawals. At that time, the storekeeper notes the order quantity (posted on the bin) and reorders. When the new order arrives in stock, he places it in the second bin and seals it, while continuing to make withdrawals from the first bin. When the first bin is emptied, it is again time to order more stock.

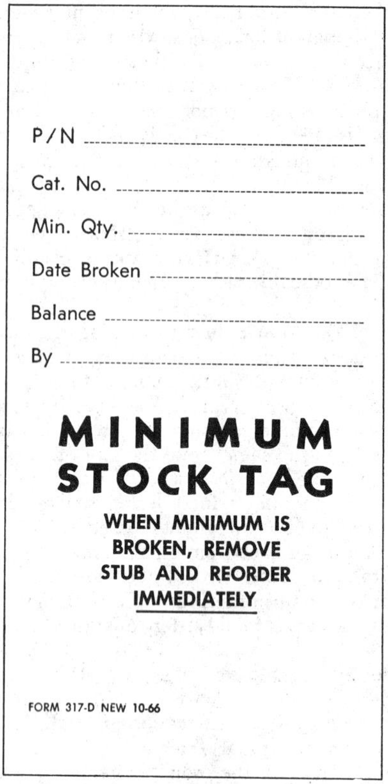

P / N _____

Cat. No. _____

Min. Qty. _____

Date Broken _____

Balance _____

By _____

MINIMUM
STOCK TAG

**WHEN MINIMUM IS
BROKEN, REMOVE
STUB AND REORDER
IMMEDIATELY**

FORM 317-D NEW 10-66

FIG. 4-7.   Variation of the two-bin order control rule using a minimum stock tag.

A variation of this order rule uses the minimum stock tag, Figure 4-7. The tag is placed in the bin, and when the minimum quantity is broken, the stub is removed and the parts are reordered. The tag includes the following information: part number, catalog number, minimum quantity, date broken, and balance.

This type of order rule is most often used for controlling stock of inexpensive items or expense-type items where the cost of regular inventory records would be excessive in relation to the value of the items. Because the costs of these items are small, the order points and order quantities may often be set arbitrarily at some convenient unit of measure such as a box, a package, or a bundle.

INVENTORY RECORDS

Many small companies operate efficiently without any written records of their inventories. The number of items is so few and the quantities so small that one man can keep track of them. This method depends upon an occasional physical count of the items, known as a physical inventory. For many small companies, this method is the least expensive and most efficient, and a complex and expensive system should not be imposed upon them.

Perpetual Inventory Record. The perpetual inventory record is a running account of incoming materials, outgoing materials, and the balance on hand. The most basic inventory record includes stock number, description, opening balance, receipts, issues, and quantity on hand. This basic record can take on many forms which assist in the rapid transfer of information, including simple card system, Kardex, facsimile posting, 80-column card, magnetic tape, and disc file. The perpetual inventory record with its three columns—receipts, issues, and balance on hand—while being sufficient for many smaller inventory systems, is too limited for control of larger inventories.

The basic inventory record is substantially expanded by the requirements for feedback at various points in the process, and it is further expanded by the inclusion of engineering change and lot control. Although the reporting and tracing of all the information shown below seems most complex, selection of reporting points can reduce the complexity. In almost every question cited later in the Data Collection Systems discussion, the area involved has a controlling desk, a supervisor, a dispatcher, or a clerk who is responsible for the checking in and out of material and people. The reporting can be done at these points.

The computer method by which inventory records are established provides a reliable basis for management decisions related to procurement, surplus disposition, and other action. Data accepted include information about requirements, procurements, receipts, issues, transfers, and adjustments. When processed, this information produces reports to indicate additional procurement action, quantity required, and specific dates by which material is needed. The receiving mechanism for the reporting function, in the form of a direct access file, can receive the information instantly and update the files immediately. This method is referred to as real-time or on-line processing.

Many concerns do not require such immediate updating; however, to minimize the chance of loss, strayed, or late reporting, they install the same transmission and collecting devices but accumulate the transactions on tape. At a predetermined time, the tape is sorted and passed against the basic inventory to post all the transactions. A detail transaction register is printed, and current balances are printed on a report or punched into a card for distribution.

The computer maintains perpetual inventory records or control of inventory balances as illustrated in Figure 4-8. These records provide an immediate reaction

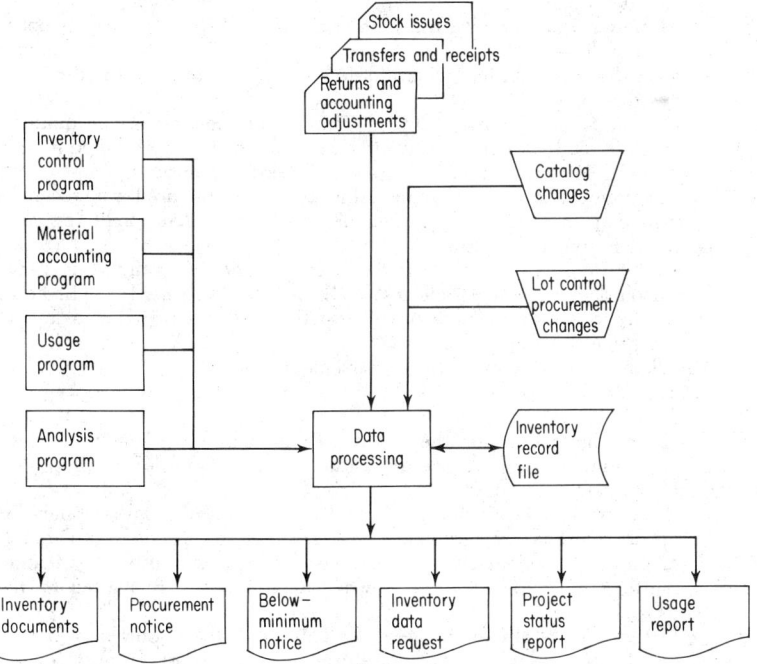

Fig. 4-8.   Maintenance of perpetual inventory records by computer.

system in which the varied effects of requirements, procurement, and usage activity are recorded, correlated, and analyzed to make available reports noting exception conditions. These reports of history, action, and operations are provided as the basis for future operations. They become the authority for exercising selective inventory control management. This management by exception concept, brought about by a combination of catalog data and a computer-programmed formula, ensures that action conditions are detected and immediately reported. It also ensures a consistent standard of performance regardless of volume.

The source documents illustrated in Figure 4-8 represent issues from stores, generated requirements, accountability transfers, interim physical inventory adjustments, returns of material to stores, and realignments of material and reworked parts. These transactions are fed into the computer and are processed to update the master records.

Information Recorded. Not all the information listed below will be found on any one continuous inventory record. Nevertheless, the list serves to illustrate the range of information that can be found among the records of different companies.

1. Identifying information.
   a. Name of material or part.
   b. Specifications or other descriptive information. This is the sort of information needed for purchase requisitioning and ordering.
   c. Identifying number. This may be a part number, drawing number, stock code number, or the like.
   d. Unit of measurement.
   e. Location in stockroom.
   f. Models or products on which used and the quantity used per unit. This sort of use or interchangeability information facilitates the estimating of future requirements.
2. Control information.
   a. Reorder point.
   b. Reorder quantity.
   c. Minimum quantity or danger point. Although this quantity is not needed for purchase requisition purposes, it is often shown on the record as the basis for determining when to follow up the purchasing department on overdue orders.
   d. Cumulative usage information. This usually consists of the quantity used or shipped by month, year, or other period. It is included to facilitate periodic reviews of reorder points and reorder quantities.
   e. Procurement cycle time, including the time required for receiving and incoming inspection. Inclusion of this information facilitates review of reorder points.
   f. Unit price. On parts manufactured by the company, this may be a standard or an actual unit cost. On purchased items, it might be the last invoice price; an average price; a first-in, first-out price; or a last-in, first-out price.
3. Detailed information on status or movement of stocks.
   a. Reorders.
      (1) Date.
      (2) Authorization for the entry. This might be a manufacturing order number or a purchase requisition or purchase order number.
      (3) Quantity ordered.
   b. Receipts. These include purchased items received from suppliers, finished goods returned by customers, parts or finished goods completed by the manufacturing department, and unused materials returned by manufacturing for credit. The information that may be posted to the record includes:
      (1) Date.
      (2) Authorization for the entry. This might be a purchase or manufacturing order number, a returned goods report number, or a credit slip number.

(3) Quantity received.

   c. Reservations. These are also referred to as inventory allocations, assignments, or mortgages. A reservation may but ordinarily does not involve a physical segregation of the quantity reserved; it indicates that that particular quantity has been earmarked for a specific job or manufacturing order and is therefore unavailable for other orders. This technique is often used where materials and parts of fairly high value are used on several products, where the usage varies substantially from week to week or month to month, and where the procurement cycle time is long. The information typically posted for this type of transaction includes:

(1) Date.

(2) Manufacturing order or lot number for which stock is being reserved.

(3) Quantity reserved.

   d. Withdrawals. These are also referred to as issues or disbursements. The information posted may include:

(1) Date.

(2) Authorization for the entry. This might include a stores requisition number or a manufacturing order or lot number.

(3) Quantity withdrawn.

   e. Balances. These could be any of the following four types:

   (1) Balance on hand. This is the actual physical quantity in the stockroom. Under a reservation or allocation system, part of this quantity ordinarily has been reserved.

   (2) Balance available. This type of balance is carried in inventory records only where a system of stock reservation is in effect. It shows the total quantity on hand and on order that has not yet been reserved. It is increased by the issuance of new replenishment orders and decreased by reservations.

   (3) Balance on order. This is the quantity still to be received against outstanding purchase or manufacturing orders that have not yet been completely filled. The balance is increased by the issuance of new orders and decreased by receipts.

   (4) Balance reserved. This is the total quantity that has been earmarked for future use but not yet withdrawn. It is increased by reservations and decreased by withdrawals from stock.

Methods of Posting and Housing Records. Continuous inventory records can be posted by hand or by bookkeeping machines or can be prepared in the form of a punched card, magnetic tape, disc, or through a data collection terminal.

Manually Posted Records. The manually posted records category breaks down into many types of records that differ from each other principally in the kind of equipment in which they are housed. These include:

1. Loose-leaf books in which the sheets covering individual inventory items are shingled so that the reference margin of each sheet remains visible.
2. Tub file of cards, usually indexed.
3. Wheel-attached cards, usually indexed.
4. Trays of visible cards housed in cabinets. This type of record is illustrated in Figure 4-9.

The choice among these various types of equipment depends on factors such as:

1. The number of inventory items and the volume of posting. These factors are a measure of the complexity and magnitude of the record keeping job. They have a bearing, therefore, on the size of the equipment investment that can be profitably made.
2. Comparative operating costs. These essentially are a measure of finding speed and the ease of posting.
3. The condition of the record after a period of normal use.

Records Posted by Bookkeeping Machine. The principal advantages of the bookkeeping machine method of posting are that the new balance is automatically

FORM-18-V-1 REV. 8-52

TOTAL CONTRACT REQUIREMENTS –   PAGE   OF

DATE	REFERENCE	ACCOUNTABILITY		A.F. REPAIR. BAL. OUT	INTERM CONTRACT TRANSFER IN		DISBURSEMENTS					STORES BALANCE	NOT AVAILABLE FOR DISBURSEMENT							AVAIL. FOR DISB.	BY
		DEBIT	CREDIT / NET RECD.		QUAN.	BAL.	PRODUCTION			MISC. CUM.			REJECTIONS		REJ. REP.		INT. CONT. TRF. OUT		MISC.		
							DEPT.	REL.	QUAN.	CUM.			IN	OUT	BAL.	QUAN.	BAL.	QUAN.	BAL.	QUAN.	

AER. NO.   PART NAME   TYPE   SPEC.   QUAN./SHIP.   STOCK NO.

882   0483

AF–
NA–

TYPIST PLEASE NOTE—THIS SCALE CORRESPONDS TO TYPEWRITER (PICA) SCALE—SET PAPER GUIDES SO THAT CARD SCALE WILL REGISTER WITH MACHINE SCALE WHEN CARD IS TURNED INTO WRITING POSITION. START INDEX THREE (3) POINTS FROM LEFT EDGE OF CARD. USE OTHER POINTS OF SCALE FOR OTHER DIVISIONS OF VISIBLE TITLE. SET TABULATORS TO INSURE PERFECT ALIGNMENT OF EACH DIVISION OF INFORMATION. FOLD BACK OR REMOVE STUB AFTER TYPING. USE NEW TYPEWRITER RIBBON.

Fig. 4-9. Visible inventory card.

computed and printed by the machine and a written record is provided of the entire posting run. This facilitates various balancing and checking operations that minimize clerical errors.

Computerized Inventory Records. The computerized inventory record and its method of preparation can be accomplished by card, tape, or disc files designed to cover all types of individual transactions affecting the inventory account on any one item. The basic information in source documents is entered into the data file. The information covering all types of transactions on each item can then be processed against a stock balance file covering the same item. From that point on, the data file can be automatically priced and extended and a new stock balance can be automatically created to incorporate the transactions covered by the data.

The computer method by which inventory records are established provides a reliable basis for management decisions related to procurement, surplus disposition, and other action. Data accepted include information about requirements, procurements, receipts, issues, transfers, and adjustments. When processed, this information produces reports to indicate additional procurement action, quantity required (adjusted by attrition allowances), and specific dates by which material is needed.

Data Collection Systems. The question basic to all inventory applications and one which involves the maintenance of a record is: "What is on hand?" Although at first this appears to be a simple question, it becomes more complicated with the introduction of details such as how much is at a supplier, in transit, on the receiving dock, in receiving, in inspection, rejected and can be reworked, rejected and can be accepted with deviation, en route to the stockrooms, in which stockroom, in transit to the manufacturing floor, in floor stock, hidden in personal stock, retrievable from higher assemblies, held for lot control, reserved for spares, earmarked for nonproductive activities, and so on. From this seemingly endless list, it appears that the answer to the basic question, amplified by the subsequent detailed questions, resolves itself into a need for a good feedback system through which the answers to all the questions can be transmitted rapidly into a data file. The file can be interrogated, and the latest information transmitted to the interrogator.

The data transmission system terminals can be either of two types:

1. Capable only of transmitting data. This type has the ability to accept information from cards, plastic identification badges, and manual entry.
2. Capable of receiving as well as transmitting data. This type is able to read and punch cards, to accept manual entry, and to create printed output on a printer.

The computer can accept data, process the information, and feed back instructions or other data to key points. In addition to immediate acceptance or rejection of the data, reports can be prepared or exceptions noted. By installing these data transmission devices at various locations, as the material passes any point, inventory information is transmitted via the terminal to the file to update the record.

As companies have progressed in their automation of inventory management systems, the methods and techniques have evolved from a manually posted record, to records posted by a bookkeeping machine, to the same records on IBM cards, to the same records on magnetic tape, to the same records on magnetic disc direct access files.

PHYSICAL INVENTORY VERIFICATION

To ensure the continued usefulness of perpetual inventory records as a control device, they must be verified periodically by a physical check of the inventories. Depending on the type of item involved, this check might be made by counting, weighing, measuring, or estimating. The purpose of such a verification is to ensure that the company's inventory assets are correctly stated by eliminating from both the individual item inventory records and the general ledger controlling accounts all cumulative clerical errors and heretofore undisclosed shrinkage or loss.

No matter how complete or how comprehensive the inventory book records may be, they should be verified against the inventory on hand by a physical count. To be of any value, the physical count should include every item of inventory. When the count has been compiled and verified, its value is computed and the book value adjusted if there is any difference.

Many plants suspend operations while the inventory is being taken. There are situations, however, where the inventory count on specific items must be taken too frequently to suspend operations; or the load may be so great that other measures must be taken not to disrupt operations.

The physical verification can be made by either one or a combination of the following two methods:

1. A continuous check made throughout the year and scheduled so that each stocked item is inventoried at least once during the year, and more frequently for items of greater value.
2. A one-time check covering one or more days during which the plant is shut down to facilitate the inventory work. This sort of periodic physical inventory is typically made over a weekend at the end of the company's fiscal year.

Both these methods have certain advantages, but when the methods are compared, the advantages weigh heavily in favor of the continuous form of inventory taking. The advantages of this method include the following:

1. The consequential costs of inventory taking are virtually eliminated. These are the costs that result from loss of production during a one-time physical inventory.
2. The direct costs of inventory taking are lower under the continuous method for the following reasons:
   a. The personnel engaged constantly in this work become more proficient at inventory taking than if they were recruited on a temporary basis from workers ordinarily performing other tasks.
   b. Through careful scheduling, each storeroom item can be counted when it is at or near its lowest point, that is, between the time the reorder point is reached and the quantity ordered is received.
3. The inventory results are likely to be more accurate because of the lack of pressure and the personnel's greater experience.
4. The inventory plan can be related to the value classification of the inventory, which requires the more valuable items to be counted more frequently. An example is shown as follows:

Dollar unit cost	Inventory interval
0–99	Annual
100–499	Semiannual
500–999	Three times per year
1,000–1,799	Quarterly
1,800 up	Bimonthly

The principal advantage of a one-time physical inventory is that the suspension of production operations makes possible a much more accurate check of work in process. In fact, in many situations, this is not only an advantage, but a necessity. In companies which have a long manufacturing cycle or which manufacture against specific customer orders instead of for stock, as much as half the total inventory investment might consist of items not recorded on the continuous inventory record. These might be located on the machining or assembly floors or other in-process locations, in inspection areas, in the receiving or shipping rooms, or in the plant of a subcontractor. Very often, the only practical way of verifying these in-process inventories is by a complete physical check during a year-end plant shutdown.

The Need for Thorough Planning. There are few business activities that require more planning and preparation in relation to the time spent on actual performance of the job than a work-in-process physical inventory. This is true because the inventory taking period must be short; because large numbers of employees, uninitiated and uninterested in inventory taking routines, must be relied upon to do the job; and because no compromise can be made with the requirement that the results obtained be complete and accurate.

The actual amount of planning and preparatory work required is influenced by a wide range of factors, including the amount of previous experience in inventory taking, the size and diversity of the inventories, the degree of orderliness that normally prevails in the company's shop, the number of different operations or stages of completion in the manufacturing cycle of each product, the stages of completion for which reliable cost data are available, and so on.

However, within broad limits of variation in the amount of planning and preparatory work required, some attention must generally be given to the following steps:

1. Determining the pricing methods to be used, that is, whether work in process will be priced on the basis of last operation completed, equivalent unit cost, estimated percentage of completion, or some other basis
2. Designing the forms to be used for recording the item identification and quantity
3. Planning the inventory taking procedures in detail and preparing a manual of instructions
4. Developing the inventory taking organization structure
5. Selecting and training the personnel
6. Arranging and identifying the materials to be inventoried so that the counting process will be facilitated

Figure 4-10 illustrates one of the types of cards used for inventory count. The normal procedure is to count the items and record the count on the card. The stub of the card is left with the items counted; the main card is processed as the inventory count. The stub is used to:

1. Allow the auditor to choose at random any item he desires to verify. The auditor selects the item, performs his count, and checks his count and identification on the stub. If the counts are identical, the auditor has proved the accuracy of the original count.
2. Allow any count supervisor, auditor, or manager to verify visually which items have been counted and which remain to be counted.
3. Prevent another clerk from counting the same item. If the items are tagged with a stub card, they have been counted; counting them again would distort the inventory figure.

Other card techniques include mark-sense cards; prepunched cards with part number and location; and prepunched cards used with data transmission terminals with the amount information keyed by the clerk.

The count cards from either the continuous count, the annual count, or both are processed through the computer as illustrated in Figure 4-11. The sum total from all the inventory locations of any particular part is compared with the inventory or record. For those items for which the count is outside the tolerance limits, a recount card is cut and a recount is performed. If the difference still exists, an adjustment is made. An additional exception report is made to allow an inventory clerk to attempt to reconcile the difference. The accepted cost of the item is on the inventory record. By extending the quantity obtained on the count, the value of the total inventory of that item is obtained.

The cycle inventory reconciliation sheet is illustrated in Figure 4-12. Information is recorded by material code for all parts by bin number; inventory tag number; unit; quantity counted, adjustment ledger, and over or under; unit price; and total value. The sheet also provides for listing those responsible for posting tags, posting ledger balances, reconciliation, checking, and approving.

INVENTORY MANAGEMENT AND CONTROL ORGANIZATION

The inventory control function can be in one of several places on the organization chart. Its exact location can depend upon:

1. The financial condition of the company

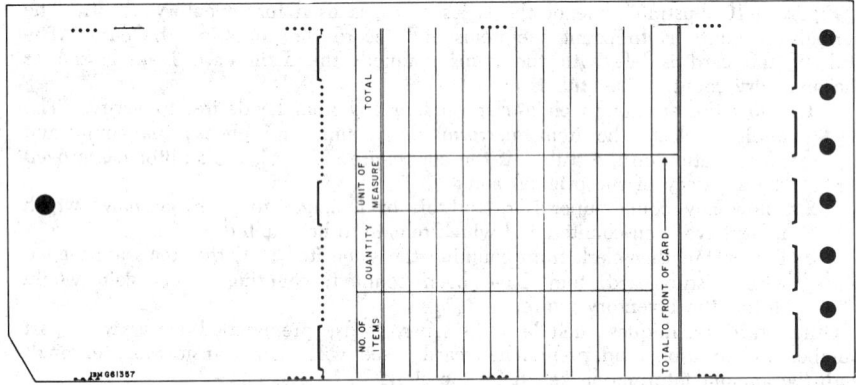

Fig. 4-10. IBM card for physical inventory input.

2. The product which is being produced
3. The people who are involved

From the manufacturing viewpoint, the best place for the inventory function is directly under manufacturing management. It is often under the supervision of accounting. In case a company is short of working capital or is buying in a speculative market, the treasurer should keep a close watch on the inventory. The materials management organization, closely connected with the vendor, is sensitive to new products coming out and can often buy to advantage if it knows the inventory.

The development of any kind of organization plan begins with a statement of the functions to be performed. In the field of inventory control, these functions consist of the following:

1. Determining what items are to be carried in inventory
2. Determining when and by what quantities to replenish inventories
3. Purchasing inventory replenishments
4. Receiving, storing, and issuing inventory items as needed
5. Maintaining records of inventory quantities and values
6. Verifying inventory quantities and condition by physical count and inspection

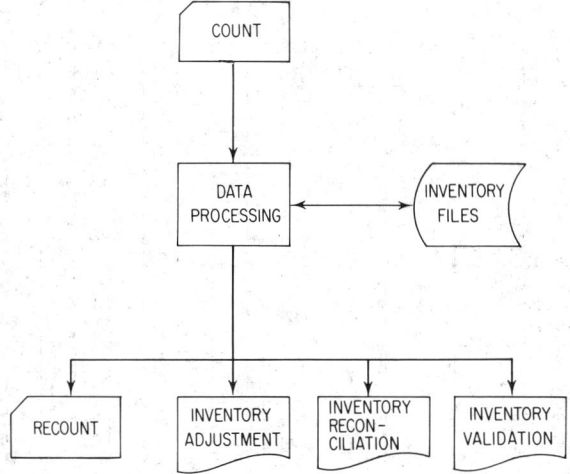

Fig. 4-11. Inventory card count processing.

7. Identifying and disposing of slow-moving, obsolete, or damaged inventories

8. Furnishing summary information on inventory position for control purposes

The position responsible for performing each of these functions and the location of that position in the organization structure will vary greatly from company to company. All that can profitably be done here, therefore, is to list some of the factors to be considered in assigning these responsibilities in any given situation.

Determination of the product materials to be carried in inventory is a more or less automatic result of product design or formulation. Determination of what supply items are to be carried in stores is a somewhat different matter. Those directly consumed in the manufacturing process are usually designated by methods or process engineering. The balance, representing the great bulk of supply items, is made up of repair parts and general supplies. Responsibility for determining the specific items to be carried in these categories may be assigned to the plant engineer, the maintenance or stores superintendent, or a material standardization engineer.

The responsibility for determining when and by what quantities to replenish inventories represents the heart of the organization problem in this whole area. Ordinarily, this responsibility will be assigned somewhat differently for different classes of inventory.

Because control of finished goods represents essentially a balancing of sales, financial, and production interests, the assignment of responsibility for this class of inventory must make provision for or at least reflect the necessary coordination among these groups. Where the demand for all products is fairly stable or where the product line consists of very few items, this coordination can be supplied through the medium of inventory policies. Under this circumstance, the responsibility for finished-goods inventories might be assigned wholly to one of the groups involved—either production or sales—because here the responsibility is merely one of administering policies which themselves are designed to protect and balance the interests of all groups.

In companies with a longer product line, however, and particularly where short-term demand fluctuates considerably among the product items, less and less reliance can be placed on broad policies as a means of achieving day-to-day coordination between sales and production. Here a more positive and more flexible coordination mechanism is needed. This frequently takes the form of a standing committee, known sometimes as the inventory or production planning committee, made up of representatives of all departments concerned. One of this group's primary working

FIG. 4-12. Computer printout sheet for cycle inventory reconciliation.

tools is a summary report of recent order bookings and current inventories or order backlog, all broken down by product class and item.

Occasionally, the problem of sales-production coordination is so acute that full responsibility for finished-goods inventory management is vested in a separate division, occupying a coordinate position with sales and manufacturing. This is the case in one manufacturing company which has a material control division, headed by a vice president, responsible for all inventory control and procurement activities. A possible danger in this arrangement is that it will create a wall between sales and manufacturing and therefore increase the risk that each will lose the feel of the other's problems, needs, and operating conditions.

With respect to raw materials and purchased parts, responsibility for determining when and how much to buy should ordinarily be assigned to the manufacturing division if that group is to be held accountable for meeting production schedules. Thus, on purchased materials, the most typical and probably the most desirable organization arrangement is to make manufacturing responsible for specifying what quantities are to be bought and when they are needed, and to make the purchasing department responsible for determining where each item will be bought and how much will be paid for it.

An occasional exception to this assignment of responsibility to manufacturing can be found in companies where one or a very few raw materials represent a large portion of manufacturing costs. This is often the case in the processing of basic commodities that are purchased on the open market. Here, to achieve the benefits of specialization, the determination of when and how much to buy may be withdrawn from manufacturing and assigned to a top-level individual or group having no other duties.

The only other inventory control function on which responsibility is frequently shared is that of maintaining records of inventory quantities and values. This happens because inventory records serve both operating and property protection objectives. For this reason, inventory records are of concern to both production and accounting. To minimize duplication of records between the two and yet serve the needs of both, the best arrangement is to:

1. Have the production department maintain whatever quantity records are kept on individual items
2. Have the accounting department maintain dollar controlling accounts on broad inventory groups
3. Reconcile the two records periodically

INVENTORY MANAGEMENT IMPROVEMENT

From a cost improvement point of view, inventory management offers both management and the industrial engineer a field where unusually large returns can be obtained. The pressure for operating capital has made business increasingly aware of inventory as a form of earnings-producing investment. In turn, this has stimulated extensive work in business, research, and academic circles to build an understanding of the function of inventory and to design techniques for improved management control.

Any comprehensive system of control covering all classes of inventory is aimed, directly or indirectly, at accomplishing a great variety of purposes. Most of them, however, fall readily into one of the following groups which represent the three basic objectives of inventory control:

1. Financial objectives. Here the goal is to keep investment in inventories within the limits of funds available so that the company's cash position is not jeopardized and the composition of its working capital is not thrown seriously out of balance.
2. Property protection objectives. These are twofold in terms of the end results to be achieved, namely:
   a. To safeguard an important tangible asset against theft, preventable waste, insurable damage, or unauthorized use.

   b. To make certain that, within reasonable tolerances, the value of this asset is correctly stated on the company's books.
3. Operating objectives. These cover a greater diversity of purposes than the other two objectives, but most of them can be classified under the following:
   a. To obtain the best overall balance between production and inventory carrying costs on the one hand and customer service on the other.
   b. To minimize losses resulting from inventory deterioration, obsolescence, or price declines.

Benefits. The three objectives listed above sound deceptively simple, but in actual practice they are usually difficult to achieve because each has so many ramifications. Nevertheless, their achievement is worth the expenditure of every possible effort because of the many worthwhile benefits that ordinarily result. The following list covers at least the major types of gains that can be achieved from a well-planned, competently administered system of inventory control:

1. Improvement of customer relations. This is achieved through faster, more dependable delivery service, which in turn results from:
   a. Maintenance of a better balance among the quantities of finished items on hand.
   b. Better geographic deployment of field warehouse inventories.
2. Improvement of labor and community relations. This results from a greater leveling of production peaks and valleys and the consequent increase in stability of employment.
3. Increase in the effectiveness of key personnel.
   a. Effective inventory control can save considerable executive time that otherwise would be spent on expediting critical raw materials, correcting repeated production interruptions, and straightening out serious back order problems with important customers.
   b. The effectiveness of field salesmen is also increased by a reduction in the amount of time they must spend in following up the home office and in pacifying customers.
4. Reduction of manufacturing costs. A well-balanced system of inventory control can reduce manufacturing costs in the following ways:
   a. By increasing the utilization of labor, supervision, and facilities through elimination of idle time caused by raw material shortages.
   b. By minimizing machine downtime caused by unavailability of critical spare parts.
   c. By making possible more economical manufacturing runs in place of the small lots, constant rescheduling, and expensive setup changes that are needed to compensate for unbalanced or hand-to-mouth inventories.
5. Reduction of purchased material costs. This results from elimination of much emergency purchasing and the consequent payment of overtime or special setup charges and similar premiums to suppliers.
6. Reduction of inventory costs and inventory losses. These are achieved by:
   a. Improving the inventory mix and thereby reducing the total quantity of finished goods needed to provide competitive service to customers.
   b. Maintaining the most economical balance between inventory carrying costs and inventory acquisition costs on purchased items.
   c. Preventing the build-up of work-in-process inventories that comes from unanticipated shortages of raw materials and purchased parts.
   d. Simplifying and standardizing the company's lines of component parts, raw materials, and supply items.
   e. Minimizing inventory losses that result from:
      (1) Declines in the market value of raw materials or finished goods.
      (2) Spoilage or deterioration.
      (3) Product obsolescence. This might be caused by shifts in market demand, competitive pressures, raw material shortages, or failure of the company to take existing inventory levels into consideration in planning the timing of design, style, or package changes.

(4) Failure to make physical verification of the quantity or condition of inventories or to provide adequate safeguards against pilferage or excessive waste.

7. Reduction of clerical and other office costs. This is achieved through a reduction in:

   a. The costs of purchasing follow-up and expediting activities.

   b. The costs of physical inventory verification.

   c. The time spent by office personnel in administering complex back order routines and in order tracing, rescheduling, and similar activities essential to answering order status inquiries and to expediting emergency orders of key customers.

8. Strengthening of financial position. This benefit of sound inventory control may result from:

   a. Preservation of a more liquid working capital position.

   b. Reduction of overall capital requirements for field warehouses and other storage space and for plants and equipment to meet peak production demand.

This list of potential benefits emphasizes two points that are fundamental in the development of a sound approach to inventory control problems. These are:

1. Inventory control does not necessarily mean keeping inventories at a minimum, for the lowest possible inventory is often not the best or least costly inventory. The real need is to achieve the best balance between too much inventory with all its financial hazards and too little inventory with all its unfavorable effects on customer relations, competitive position, and production stability. Even this is somewhat of an oversimplification, because both financial risks and competitive weaknesses are caused more by a lack of balance within the inventory than by the size of the inventory as a whole.

The same general observation applies to all the potential benefits already listed. Because many of them conflict with each other, they cannot all be achieved to the fullest. Thus, the design of a sound inventory control system is in large measure a balancing operation.

2. Although the relative importance of various inventory control objectives may change with shifts in business conditions, the need for effective control of inventories is constant. This need is most widely recognized during periods of falling prices or declining business activity. It is much less generally recognized, however, during upturns in the business cycle or during periods of threatened material shortages or rising prices. Under these conditions, few people would question the soundness of an increased investment in materials for which requirements are definite and continuing and on which price rises are probable. On the other hand, increased or even continued investment in materials for which there is not a carefully defined need is equally unsound. In addition, under these conditions, the systematic analysis of inventories not only serves to highlight critical items for advance buying, but also helps to hold the overall increase in inventory investment to a minimum by offsetting essential increases of some items with reductions of others. Finally, the existence of carefully developed mechanics of inventory control makes possible a prompt, intelligent, and orderly readjustment of inventory levels as soon as the need for forward buying or manufacturing diminishes. Thus, whether an increased or a decreased inventory investment is needed, the value of established procedures and fixed responsibility for determining actual inventory requirements remains unchanged.

Symptoms of Poor Inventory Management. Weaknesses in the planning and control of inventories are usually indicated by or expressed in complaints about specific symptoms or conditions rather than as an overall criticism of the inventory system. Some specific, frequently encountered symptoms are:

1. Periodic, severe back orders (inability to meet delivery promises)

2. Continuously growing inventory quantities while the order backlog is remaining constant or also increasing

3. High rate of customer turnover or order cancellations

4. Uneven production, with frequent layoffs and rehirings
5. Frequent need for uneconomical production runs to meet sales requirements
6. Excessive machine downtime because of material shortages
7. Periodic lack of adequate shortage space
8. Consistently large inventory write-downs because of price declines, distress sales, disposal of obsolete or slow-moving stocks, and the like
9. Widely varying rates of inventory loss or turnover among branch warehouses or widely varying rates of turnover among major inventory items
10. Consistently large write-downs at time of physical inventory taking

Typical Improvement Opportunities in Inventory Paper Work. Inventory control systems usually represent a very sizable clerical expense. Although the primary need, of course, is to maintain adequate financial and operating control of inventories, the ideal goal is to achieve this control at the lowest possible clerical expense. Reductions in this expense should never be made at the risk of losing much larger sums through weakened control. On the other hand, inventory control systems typically are more complicated and costly than they need to be to do a satisfactory job.

Following are some of the most common opportunities to cut these costs without impairing effective control.

Eliminate Duplication of Inventory Records. A number of companies maintain individual item inventory records which duplicate each other or at least overlap considerably in the information recorded. Sometimes, duplicate records are kept in accounting and in the material control section; in other instances, the storeroom duplicates, in the form of a bin card, the information kept by material control. Except in rare instances, this duplication arises because of jurisdictional disputes or lack of confidence in the records maintained by the other group. In any event, all essential financial and operating needs can usually be met by the maintenance of a single set of individual item records plus a relatively small number of dollar-controlling accounts.

Reduce Number of Items on Which Records Are Maintained. A number of companies have adopted the practice of eliminating all continuous inventory records on many inventory items and substituting a visual control in their place. This is done most frequently on fast-moving items used in fairly large quantities and generally having a low unit value. These may include both product materials and supply items; they typically consist of many hardware store articles as well as small parts such as bolts and nuts, pins, capscrews, rivets, plugs, washers, and some electrical parts.

One large company in the chemical industry reduced its inventory control clerical costs by 40 percent through discontinuing perpetual inventory records on all its stores items except equipment spare parts. The major steps of its revised procedure are as follows:

1. All stores items are divided into the following four classes:
   Class 1—pipe fittings, valves, and related items.
   Class 2—electrical supplies and related items.
   Class 3—hard items such as welding rods, bolts, nails, and workman's tools.
   Class 4—soft items such as safety clothing, paper and cloth products, and liquids.
2. For each of these four classes, a physical inventory is taken and priced every four months, and the total value of the class is used to adjust the controlling account record in step 3.
3. Individual inventory cards for all items in these four classes are no longer maintained. In their place, the company carries only one control card for each class, and all inventory debits and credits are made in summary form to this control card. For example, all stores withdrawal requisitions, after being priced and extended, are sorted by class, and the dollars added. The total amount is then posted to the control card as a single entry.
4. In the storeroom, the four major classes are segregated as much as possible to facilitate physical inventory taking.

5. A bin card for each individual item is attached to the bin shelf immediately adjacent to the item. The following information is entered on this form:
   a. Identifying number and description.
   b. Reorder point.
   c. Reorder quantity.
   d. Last invoice price.
   e. Quantities received.
   f. Physical inventory count.
   By reference to the two physical counts, four months apart, and the receipts during the intervening period, storekeepers can readily compute the usage information needed to adjust reorder points and reorder quantities.
6. At time of disbursing an item, stores countermen transcribe the unit price from the bin card to the withdrawal requisition.
7. As a basis for issuing stores replenishment requisitions, the quantity representing the reorder point on an item is identified or segregated in one of three ways:
   a. In some instances, a red line is painted on the bin to identify the reorder level.
   b. On other items, the reorder point may be identified by insertion of a divider either vertically or horizontally in the bin.
   c. On still other items, the quantity representing the reorder point is sealed in a container upon receipt of a replenishment order. When all other stock of the item is gone and this seal is broken, the item is reordered.

Charge Off Items at Time of Acquisition. Closely related to the simplification opportunity described above is the practice of charging off small-value materials, parts, and supplies at the time of receipt rather than on the basis of individual withdrawals from stores. Like the technique described above, this practice also eliminates the maintenance of a perpetual inventory record. In addition, however, it eliminates the need for pricing and extending withdrawal requisitions and charging them to another account.

One company charges to an overhead expense account at time of acquisition all supply items which have a unit-of-issue value of 35 cents or less and which also meet the following characteristics:

1. The items are used in a number of departments throughout the plant. This characteristic ensures that any method of prorating the accumulated charge-off of such items results in a reasonably equitable distribution.
2. The items are purchased at fairly regular intervals. This characteristic tends to ensure a fairly uniform purchase volume of the total of such items so that any one month is not charged with an unusually heavy expense.
3. The items are not typically withdrawn in large quantities; for example, withdrawals do not ordinarily exceed $5 in value. This limitation would exclude such items as lumber, brick, and I beams, all of which might have a unit of measure value of less than 35 cents but are ordinarily withdrawn in large quantities.

Simplify Inventory Record Posting. The clerical time spent in posting the inventory records that are maintained can often be reduced by the following three means:

1. Eliminate from the record all unnecessary or marginal-value information. This might include such information as money balances, total value of receipts and disbursements, and reservation postings on many items of moderate value and reasonable procurement cycle times.
2. Simplify inventory pricing operations by using last invoice price instead of computing a new average price with each receipt. Also, discontinue including small-value inbound freight charges in the unit price.
3. Lay out or design the inventory record form so that posting will be speeded and clerical errors reduced. This is largely a matter of providing adequate space for the entries and arranging them in a logical sequence.

Simplify Replenishment Requisitioning. On all items regularly carried in stock, a permanent or so-called "traveling" requisition card can be used to eliminate the handwriting or typing of single-use purchase requisitions.

These records should be kept by the material control unit, sent to the purchasing department when the item is to be reordered, and returned to the material control unit as soon as the order has been placed. The same type of form can be used to initiate manufacturing orders for replenishing company-made items controlled on a reorder point, reorder quantity basis. In the use of these forms, the amount of stock on hand and requisitioning and receiving information are entered by the material control unit; the ordering information is entered by the purchasing department.

In designing a traveling requisition form, the two basic purposes to be achieved should be kept in mind. These are:

1. To eliminate repetitive writing of fixed information by the purchase requisitioner.
2. To place in front of the buyer all the available internal information needed to take action on the requisition. This means that the traveling requisition should be complete enough so that separate commodity-purchase-history records need not be referred to. Also, enough space should be provided to record the address and telephone number of the supplier, provided this information is not a matter of memory. In this way, no separate look-up reference need be made by either the buyer or the purchase order typist.

Computerized Applications. A good computerized inventory control system is a management tool which provides accurate, useful, and timely data to all levels of an organization and supports the operational activities. Efficient data collection, ready access to comprehensive data, and elimination of redundant data handling and redundant data files are its principal features. Data files are organized for use by multiple system functions to achieve greater efficiency in processing and data organization. Through the data transmission terminal, data captured at original sources are entered into the files with reduced possibility of error. Coincident with the reduction in potential errors is the immediate updating of the master files. From the master files, reports are generated, parts are ordered, and decisions are made. This eliminates information file phasing problems. This objective of this approach indicates that:

1. Material planning cycles can be reduced, with a consequent reduction in cost and a substantial reduction in time through the use of computer processing.
2. Inventory balances and supporting records monitored by the computer can be constantly examined for new-order requirements, and any unusual inventory circumstance can be brought immediately to planning's attention.
3. Count requirements for constant verification of record balances are computer monitored to verify the record accuracy and to comply with auditor requirements.
4. The cost of maintaining inventory can be cut by closely matching the need dates to quantities of inventory on hand. This improves inventory turnover and aids in meeting production schedules.
5. The cost of operating the material release and control department can be reduced. Computer processing can handle varying production levels without wide fluctuation in personnel requirements.
6. An inventory control system can react to production problems, delays, and losses by triggering order adjustments as needed. Any major schedule change can be examined overnight. New releases are processed in a single cycle instead of requiring weeks of extra effort.

GLOSSARY OF INVENTORY TERMS

ABC Classification: Classification of the items in an inventory in decreasing order of annual dollar volume, split into three levels, called A, B, and C. Class A contains the items with the highest annual dollar volume and receives the most attention. The medium class, B, receives less attention; and class C, which contains the low dollar volume items, is controlled routinely.

Anticipated Stock: Stock built up to buffer seasonal fluctuations in sales or a planned intensive sales campaign, or to carry sales over a plant vacation or maintenance shutdown.

Available Stock: The algebraic sum of the stock on hand and the stock already on order, less any unfilled customer demand.   Thus, available stock may be greater than, equal to, or less than the physical inventory and may even be negative if there are back orders.

Backlog of Orders: The sum of all unfilled orders waiting to be filled or processed.

Basic Stock: The desired level of the average inventory, also known as normal or standard stock.

Commitments: Orders which have been accepted but not yet filled.   They include back orders and are part of the backlog.

Delivery Cycle: A general term for the normal interval required to manufacture or purchase an item.   It covers the interval between deciding to order an item and its delivery to the stockroom.

Delivery Ratio: The number of inbound orders out of 100 that are delivered on or before the day the material on the order is needed.

Delivery Variance: The difference in working days between the actual and scheduled dates of delivery.

Demand Variance: The difference in working days' supply between the actual and forecast amounts of use during the actual delivery cycle.

Economic Lot Quantity: The amount of an ordering lot quantity that will minimize the annual total of the inventory costs for the stated conditions, costs, and rates.

Excess Stock: The usable stock over and above the standard inventory that is in stock because of a change in the rate of use, an error in ordering, or a decision to acquire stock for special use.

Formula Rate: A fraction used in economic lot quantity and best length of reserve cycle increment formulas to estimate the change in the annual total of the inventory carrying costs that will come from a change in the amount of the inventory investment.

Held Stock: Stock that is being held for a decision or disposition.

In-process Stock: Stock that is in the processing departments of a company.

In-transit Cycle: The time required to prepare to move and to move stock between two locations.

In-transit Stock: Stock that is charged to and is being moved to a stockroom but is not available for use.

Inventory Carrying Costs: The costs that come from owning and possessing stocks of items.

Inventory Control: The technique of maintaining stockkeeping items at desired levels, whether they be raw materials, goods in process, or finished products.

Inventory Costs: The costs, expenses, and losses that arise from acquiring, owning, possessing, and running out of stocks of items.

Inventory Management: The branch of business management concerned with the development of policies to which the firm's inventory is meant to conform.

Inventory Valuation: The value of the inventory at either its cost or its market value.   Because inventory value can change with time, some recognition must be taken of the age distribution of inventory.   Therefore, the cost value of inventory, under accounting practice, is usually computed on a first-in, first-out (FIFO) or a last-in, first-out (LIFO) basis to establish cost of goods sold.

Lead Time: The length of the interval that is allowed to obtain a lot quantity of an item.

Line Balancing Stock: The inventory which is maintained between two operations in order to make it possible to control them independently.

Lot Quantity Cost: The total of the costs per order that have the characteristic of varying with the number of orders run rather than with the number of units of product made.

Manufacturing Cycle: The interval normally required to obtain a manufactured item.

Maximum Inventory: The sum of the reserve stock and ordering lot quantity.

Maximum Inventory Cycle: The interval covered by the maximum inventory.

Maximum Inventory Investment: The sum of the values of the reserve stock and ordering lot quantity in dollars.

Nonstandard Inventory: The sum of all classifications of stock excepting reserve, turnover, in-process, and in-transit.

Ordering Point: A low limit for the sum of the in-stock and on-order balances that is used to determine when to order another lot quantity of an item that is kept in stock.

Procurement Cycle: The interval normally required to obtain a purchased item.

Rate for Carrying Inventory: A fraction found by dividing the average inventory investment in a group of items into the annual total of the costs of carrying inventory of this group of items.

Required Reserve Cycle: The sum of the demand and delivery variances or the adjustment in the delivery cycle that, if known and made at the time of scheduling delivery, would have caused an order to be delivered on the day it was needed.

Reserve Cycle: The interval allowed between the scheduled date of delivery and expected date of use of the material on an order.

Reserve Stock: The in-stock balance at the time of receipt of the next order.

Salvage Stock: Stock being repaired or reworked.

Standard Inventory: The sum of the reserve, average turnover, and in-transit stocks.

Stockkeeping Unit (SKU): An item of stock that is completely specified as to style, size, color, and location.

Stockroom Cycle: The average length of time a lot quantity of an item is in a stockroom.

Surplus Stock: Stock for which there is no visible use.

Turnover Stock: Stock that is brought into the stockroom to meet expected requirements. It represents the difference between total stock on hand and reserve stock.

Turnover (of Stock): The ratio of total sales during a specified period, generally one year, to the average inventory on hand during that time period can be applied to an individual SKU or to the aggregate inventory.

BIBLIOGRAPHY

Ammer, D. S., *Materials Management,* rev. ed., Richard D. Irwin, Inc., Homewood, Ill., 1968.

Brown, Richard A., "The Systems Approach to Inventory Management," *Management Review,* March, 1969.

Carroll, Phil, *Practical Production and Inventory Control,* McGraw-Hill Book Company, New York, 1966.

Fabrycky, W. J., and Paul E. Jorgensen, *Operations Economy: Industrial Applications of Operations Research,* Prentice-Hall, Inc., Englewood Cliffs, N.J., 1966.

General Information Manual IMPACT: Inventory Management Program and Control Techniques, International Business Machines Corporation, Technical Publications Department, White Plains, N.Y.

Green, James H., *Operations Planning and Control,* Richard D. Irwin, Inc., Homewood, Ill., 1967.

Magee, John F., and David M. Boodman, *Production Planning and Inventory Control,* 2d ed., McGraw-Hill Book Company, New York, 1967.

Nath, Mahendra, and Sant R. Arora, "Computer Maker Upgrades Inventory System," *Journal of Industrial Engineering,* June, 1969.

Shriver, Richard H., and Russell C. White, *Distribution Planning and Control: Effective Use of Computer Systems and Models,* American Management Association, New York, 1969.

Wholesale IMPACT: Advanced Principles and Implementation Reference Material, International Business Machines Corporation, Technical Publications Department, White Plains, N.Y.

Chapter 5

Quality Control

DORIAN SHAININ

**Vice President and Director of Reliability and Quality Control,
Rath & Strong, Inc., Lexington, Massachusetts**

This chapter is designed to serve as an outline of the kind of thinking necessary to modify current quality programs to realize substantial improvements. Industry has entered an era of developing changes in quality requirements, reflected in the growing government, public, and industrial recognition of the need for increased:

Product safety
Product dependability or reliability
Protection against legal liability for inadequacies

Increasing costs and product complexities have added interest to methods for offsetting difficulties with such requirements. Quality control, practiced as a system, can contribute to surprisingly large savings.

For the industrial executive responsible for coping with these problems, this outline can be a catalyst for getting useful data and activity out of a mixture of fabricating products, fabricating equipment, kinds of customers, and people: from suppliers, through his own associates, on to customers—even those in the government regulating agencies.

Industrial engineers will be particularly interested in the cost reducing aspects of the procedures and techniques which will be discussed. The discussions cannot be exhaustive or even complete, but if the techniques which are briefly described appear to have application to a specific situation in which the industrial engineer is interested, he can find further information in the references which are given throughout the chapter.

EFFECTIVE QUALITY CONTROL

Effective quality control is the system of coordinated activities which prevents the manufacture of product from deviating from what is expected by the typical customer. Many current practices unfortunately violate this brief definition. Inspection is the act of comparing a product with accepted specifications or other recognized standards, but these quality limits are seldom realistic. They have generally been set with inadequate information and tend to be conservative. Sometimes they have been influenced by the complaining nontypical customer, and often they are set at values that have been "lived with" in the past on similar products. Dependence upon strong customer wishes or critical industry beliefs (aerospace, submarines) typically results in tight tolerances.

Quality control frequently tries to exercise compensating influences: allow acceptable quality levels (AQL's) in sampling; set the tolerance wide enough to accommodate the variability of the fabricating process—its "capability." These practices fit many suppliers who retain sufficient independence of customers. Hence, user needs are being rationalized extensively, almost always in error. Both the conservatively tight and the arbitrarily loose tolerances add to costs unnecessarily. Fortunately, realistic individual tolerances can be determined objectively, with economical internal system trade-offs to keep final product characteristics at values which give satisfaction to the typical customers.

Other clear indications of ineffective quality control are such practices as continuously detecting defective product, and with additional cost, going through procedures to accept much of it by special deviations. Charts or tabulations report the partial success or lack of success of attempted corrective actions. Even the sponsors of motivational Zero Defects programs, who set zero as an objective, admit that this level is unattainable across the board.

In contrast, the prevention of the manufacture of product which deviates from realistic limits can be approached more closely than ever before considered practical. The amount of necessary inspection in industry can be drastically reduced by the introduction of the system of coordinated activities mentioned in the brief definition of effective quality control.

To implement appropriate adaptation of the strategies which follow, and which supply the needed technical quality information currently not known, the line activities of many individual departments of the company must be involved. The staff function of quality management has to be successful in coordinating (1) the planning and gathering of some rather "unusual" data, (2) the analyses, and (3) the actions taken. These requirements should be kept in mind when choosing and training people for quality management positions, when stating the company's quality goals, and when assigning authority for taking action within the line departments, along with the responsibility for the effectiveness of those actions.

Some firms have cogently expressed the implications of the foregoing definition and discussion in the form of a word equation:

$$\text{Product value} = \frac{\text{quality} + \text{reliability}}{\text{cost}}$$

The departmental responsibilities for achieving better customer desire for the product have been tabulated in Figure 5-1 as primary and secondary. The primary method is the most efficient, least costly way of doing the job. The secondary method is less efficient and therefore more costly.

INSPECTION'S RESPONSIBILITIES

The inspection department of Figure 5-1 conducts its activities after a degree of quality has been built into the product which is submitted as lots, sublots, or in a continuous stream. Conformance requirements range from visual, through dimensional, to other characteristics measured with standard or special test equipment.

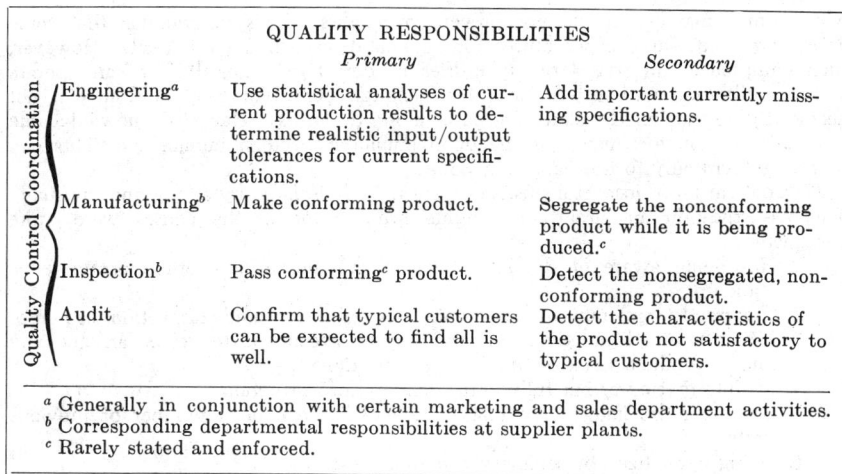

QUALITY RESPONSIBILITIES

		Primary	*Secondary*
Quality Control Coordination	Engineering[a]	Use statistical analyses of current production results to determine realistic input/output tolerances for current specifications.	Add important currently missing specifications.
	Manufacturing[b]	Make conforming product.	Segregate the nonconforming product while it is being produced.[c]
	Inspection[b]	Pass conforming[c] product.	Detect the nonsegregated, nonconforming product.
	Audit	Confirm that typical customers can be expected to find all is well.	Detect the characteristics of the product not satisfactory to typical customers.

[a] Generally in conjunction with certain marketing and sales department activities.
[b] Corresponding departmental responsibilities at supplier plants.
[c] Rarely stated and enforced.

FIG. 5-1.   Primary and secondary departmental responsibilities for quality control.

Because so few companies realize that inspection's primary responsibility is to pass conforming material, it is unfortunately common to use both 100 percent inspection and statistical sampling inspection ineffectively. Quality control should demonstrate progress, month by month, with its coordinating activities; less and less product should have to be rejected by inspection. Eventually, with realistic tolerances from engineering and control plans for manufacturing, on the few occasions inspection rejects an item there will be time for an investigation to determine why the secondary and primary quality responsibilities of manufacturing were not discharged. When submitted material lots often conform, 100 percent inspection unnecessarily adds cost, even for critical quality characteristics. A properly selected sampling plan (with the necessary discriminating power) will pass such submitted material and be ready to identify the rare lot or time when the material does not all conform.

Many organizations consider that even frequent rejections by sampling improve quality because those lots are screened 100 percent to remove defectives, thus reducing the average outgoing quality (AOQ) fraction defective figure. But the accepted lots are usually of corresponding quality, and customer shipments are also made from them! As far as any particular customer is concerned, the AOQ and its limiting value, average outgoing quality limit (AOQL) (see below), are paper numbers of not much help to him; he received a percent defective greater than the AOQL.

Moreover, it is well known that 100 percent inspection is seldom 100 percent effective, and so the costly screening itself falls short of its goal. For critical characteristics, commonly listed as requiring 100 percent inspection, one can then say that the requirement virtually guarantees that unless the lot is completely conforming, some defective material will be passed. In certain companies, several repeats of screening (200, 300, or even 500 percent) inspection are conducted for critical characteristics, all because they have not yet learned how to achieve the goal of manufacturing's quality responsibilities.

An organization must know how to prevent inspection oversights, both for 100 percent and for sampling inspection, and how to select the correct sampling plan—one with the necessary discriminating power. These two subjects follow.

Preventing Inspection Oversights and Carelessness. The usual prescription involves rest periods, pleasant environment but separation of inspectors prone to conversation, better lighting, and rotation of kind of product inspected—all of which unfortunately still do not prevent oversights. It is no wonder that some firms have made large expenditures for automatic inspection equipment. However, such equipment can pass large quantities of defective material if it can operate out of calibration undetected for a long enough period of time; and it will not necessarily repeat its decisions on parts close to limits. Happily, knowledgeable companies can have a new reliance upon human inspectors; human oversights can be brought virtually to the vanishing point.

The root causes of inspector mistakes were initially isolated by a long and unusually objective study in an aircraft jet engine firm. Some of the causes were quite unexpected:

1. Not being aware of the limitations of the measuring method or reference standard
2. Incorrectly remembering the point of an interrupted inspection sequence
3. Using unaided judgment in deciding to accept or to reject an anomaly on the product, not described by specifications
4. Misinterpreting or not fully understanding an instruction
5. Retaining the image of a feature repeatedly seen on a number of previous units
6. Being hypnotized by seeing very familiar features
7. Occasionally just handling rather than really inspecting the product

Some of the ways to minimize—and virtually prevent, if enforced with patient and continuing attention—each of these universal human troubles are, for causes 1 to 7 above, respectively:

1. Make each inspection supervisor responsible for:
   a. Studying each measuring device, standard sample, and inspector's work area and tabulating (for the review of his superior) the particular quality characteristics of the product which are not evaluated adequately if at all by the present method. There is really no such thing as a blank list.
   b. Advising the inspector of these limitations, discussing their relative importance, and setting up substitute ways of at least preventing gross cases of extensive lack of compliance whenever the quality characteristic is of more than trivial import.
2. Furnish each inspector with an appropriate marking device or its equivalent and require that it always be used to denote any stopping point of inspection on the product. One's memory always seems reliable to him, but it actually is notoriously deceptive.
3. Have each inspector get his supervisor's judgment on every unusual characteristic found on the product which has not been covered by the drawing, specification, or inspection methods sheet. Hardly a day goes by when something unusual and not defined could not be questioned. It is necessary to keep the inspection supervisor on the floor easily available to his inspectors, rather than to have him writing reports, attending meetings, and the like, away from continuous contact with his people. Otherwise, inspectors will have to use unaided judgment in deciding to accept or reject something unusual in order to get their expected day's work done.
4. Require the inspector to repeat back, in his own words, what he will do in following a supervisor's instruction just received. Frequently, some minor (and often major) points will have been misinterpreted or entirely misunderstood. Although sometimes it will be evident that the inspector was not completely attentive, it is often sobering to the supervisor to realize that at times he did not state what he wanted as clearly as he could have. A single instance of such lack of proper communication can result in a serious inspection error, as can the previous situations of using poor judgment, of incorrectly remembering where to resume inspection, and of naively expecting the available gage or conventional inspection method to detect all unusable product.
5. Because the human eye at times retains a previous image, use the sense of

touch to supplement the sense of sight. "If you see it, but you can't feel it, it isn't there!" Again this human characteristic is hard to believe, but by being forced to touch a snap ring, a soldered resistor lead, the back of a bored-through hole, or the like, it will not be long before the inspector will be convinced. He will encounter a missing snap ring which he saw but could not feel—a "good looking" unsoldered lead; he will have seen light in a hole that was not drilled through!

6, 7. Never inspect one unit at a time visually and dimensionally; always line up and gage two or more units simultaneously. This admonition is seldom heard, yet it can virtually prevent two of the most serious causes of inspector oversights:

  a. Hypnosis. A generally unused but useful characteristic of the human eye is its ability to react quickly, to detect even minor visual dissimilarities among supposedly similar sides of two or more objects being inspected. In dimensional inspection, the pair of needles on the dial indicators not rising together to a similar acceptance region will more positively emphasize the one which is out of tolerance. This form of comparison inspection overcomes hypnosis, the tendency to expect something before one actually sees it because that characteristic is so familiar.

  b. Part handling. The observed actions of the inspector in gaging the work all seem proper, but his mind wanders to any subject of momentary compelling interest, usually many times a day. The lack of agreement among the parts whenever a discrepant part is encountered breaks the spell, and the number of oversights is virtually brought to the vanishing point.

How to Select a Sampling Plan with the Necessary Discriminating Power. A quality characteristic is judged as being critical (safety considerations), major (functional but not safety considerations), or minor (neither). Then the extent of inspection may be specified as, perhaps:

Classification of defects	Inspection
Critical.............	100% insp.,
	or 0.65% AQL
Major...............	1.0 or 2.50% AQL
Minor...............	4.0 or 10.0% AQL

Depending upon product requirements, industry practice may be found at the left of the inspection column, or toward the right-hand AQL's, or in a few cases even beyond these popular ranges of control of sampling risk. Then reference to one of a number of published sampling tables[1]—for example, MIL-STD-105D—indicates, for the lot size being considered, the sample number of items to be selected and the acceptance number. Only if the sample is found to contain more than that number of defective units will that lot be rejected, meaning that it is to be destroyed, returned to the producer, or screened of all defective units, whichever may be appropriate.

Because, in general terms, this procedure describes rather universal practice, it will come as a surprise to most quality control managers to learn that it is fundamentally unsound. Consider first the varying risk of any sampling plan, depicted by its operating characteristic (OC) curve. It will clarify matters and avoid much of the seemingly involved mathematics of statistical sampling to derive the OC curve for a very small number situation, ineffective for industrial application, but completely rigorous in teaching some principles of effective (larger number) plans.

Lot size: 5 units
Sample size: 2 units
Acceptance number: zero defective units

[1] Eugene L. Grant, *Statistical Quality Control,* 3d ed., McGraw-Hill Book Company, New York, 1964, chaps. 14 to 16.

Operating procedure: From each lot of 5, select 2 at random and inspect them. If both are acceptable, accept the remaining 3 units without inspection. Otherwise, reject the lot of 5.

Rationale: Because the sample of 2 is taken at random, each one of these combinations of 2 is equally likely from a lot of 5:

1-2	2-3	3-4	4-5
1-3	2-4	3-5	
1-4	2-5		
1-5			

a total of 10 different possible samples.

Plot the true fraction defective of the lot submitted to inspection (horizontal scale) against the probability of acceptance of that lot (vertical scale). Determine the plotted points for the OC curve of Figure 5-2 by counting the number of samples which do not contain the single defective number ($p = 0.20$, or $\frac{1}{5}$). Say it is part no. 3. Six of the ten samples do not have part no. 3; hence, 0.60 ($\frac{6}{10}$) is the probability of acceptance P_A.

Mathematical statistics provide formulas which in effect do the counting for the much greater number of combinations, and form the appropriate ratios typical of practical industrial sampling situations.

Figure 5-3 depicts a typical OC curve for an acceptance sampling plan by attributes; that is, each item is recorded discretely, classified as either acceptable or not. In contrast, "by variables"[2] would mean each item would be recorded as a measurement on a continuous scale of the quality characteristic in question. Whenever the acceptance number of a plan is zero, its OC curve drops, as in Figure 5-2, from a P_A of 1.00 at a p of 0.00. The flat region at the top of the curve in Figure 5-3 shows that the plan has an acceptance number greater than zero—the bigger the number, the longer the flat portion. Finally, larger sample sizes steepen the slope of the OC curve.

On the vertical P_A scale of Figure 5-3, a pair of constant values are selected as reference points. General convention in the early days of statistical acceptance sampling established these values as:

Nomenclature	*Value*
α (producer's risk)	0.05 or (1.00 − 0.95)
β (consumer's risk)	0.10

The AQL and RQL (rejectable quality level, also called lot tolerance percent defective or LTPD) are the numbers which vary from plan to plan. Although

[2] *Ibid.*, chap. 17.

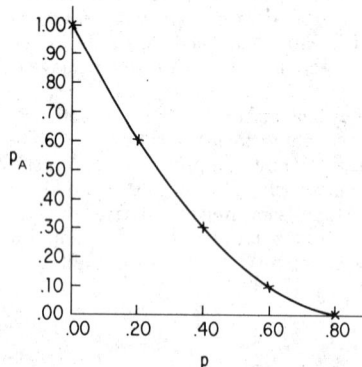

FIG. 5-2. The operating characteristic curve showing probability of acceptance versus fraction defective.

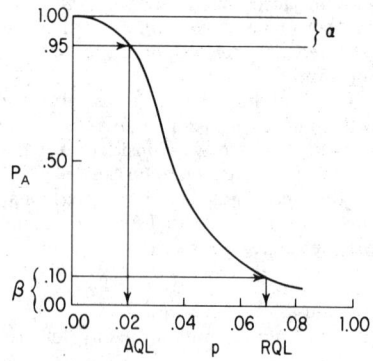

FIG. 5-3. Elements of the operating characteristic curve.

both these points describing a range of true fraction defective of material submitted for sampling inspection should be considered by anyone selecting a plan, the RQL value is the more important of the two. It describes a selected fraction defective which would be rejected 0.90 of the time. In Figure 5-3, that selected fraction defective is 0.07 (7 percent). Management selecting that sampling plan can state that "They are 90 percent confident that lots containing 7 or more percent defective units will be rejected." The RQL then is a number reflecting a point of control to a predictable degree.

The corresponding statement about the more popularly used AQL will be less informative: "They are 95 percent confident that lots containing 2 or less percent defective will be accepted."

Specifying 2 percent AQL alone gives no information about lots containing more than that percent defective. A glance at Figure 5-3 shows that 3½ percent defective lots will be passed half of the time! A typical booklet of sampling plans classified by AQL's will have a number of plans for the same AQL, with different sample sizes and acceptance numbers, hence having different slopes for their OC curves.

Moreover, it is seldom appreciated that the larger lot sizes do not require larger sample sizes. For all lot sizes greater than ten times the sample size, the OC curves will not be different with a fixed sample size and acceptance number. For reductions of lot sizes to less than ten times the sample size, the OC curve begins to steepen its slope slightly; so here the sample size can be reduced a little if it is desired to keep the curve from increasing its protection, that is, its discriminating power.

To select a sampling plan for a quality characteristic, maintain a prior record of lots with no defects and of the fraction defective found in others. In Figure 5-4 each X represents such a result.

The dashed OC curve fitted to these X's has the ability, with high confidence, to detect a future defective lot in that distribution of troubled lots. Lots which are trouble-free will continue to pass all the time. The AQL of the plan is meaningless. The RQL should be set at 0.04 for p, with an acceptance number of zero. The sample size calculated from the formula (the binomial expression[3] for the probability of no rejected units equaling a β of 0.10) is

$$0.10 = (1.00 - 0.04)^n$$
$$n = 56$$

For certain quality characteristics, the distribution of lots in trouble may extend closer to the origin, requiring still steeper OC curves to discriminate between good lots and troublesome ones. Notice how the sample size increases as the left tail

[3] Harry R. Larson, "A Nomograph of the Cumulative Binomial Distribution," *Industrial Quality Control*, December, 1966.

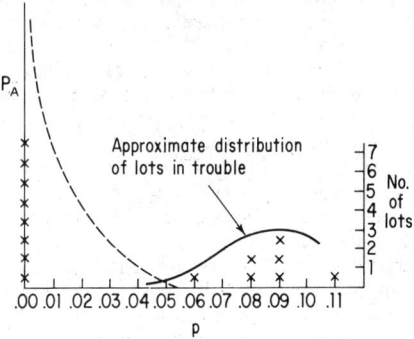

Fig. 5-4. Operating characteristic curve combined with lot fraction defective frequency distribution showing discrimination needed to separate good from bad lots.

of the approximate distribution gets closer to zero:

p	n
0.03	76
0.02	114
0.01	229

When troublesome lots can exist at very low p's, the formula is correctly announcing that the sample size will approach the lot size; do not use sampling when, instead, a quality improvement should be made to a chronically troublesome manufacturing process.

As the distribution moves away from the origin:

p	n
0.04	56
0.05	45
0.07	32
0.10	22
0.15	14
0.23	9

a smaller sample size with its shallower slope will still have adequate discriminating ability.

After the selected plan is in force, it remains effective as long as future lots are sampled and found to have no defects or more than the p fraction defective used to select the sample size. But a single case of a p greater than zero but less than the p used for the plan calls for action. The encountered fraction defective has not been characteristic of a troublesome lot from that process, and the plan will not adequately detect such a situation. Unless the cause can be rapidly isolated and its recurrence prevented, an appropriately larger sample size should now be used.

ENGINEERING'S RESPONSIBILITIES

Quality Standards. Three kinds of tolerances have to be realistic before any system of quality control can begin to operate with efficient economy: (1) tolerances for the characteristics of the final product (performance, appearance, and the like), (2) tolerances for the parts and manufacturing processes, raw material, and so on, which may have an effect upon the characteristics of the final product, and (3) tolerances which permit assembly and interchangeability of mechanical and electrical components. As discussed briefly earlier, past and current practices almost inevitably result in tolerances too tight or too loose or nominals at incorrect values. For manufactured and inspected quality to be meaningful, management needs ways to evaluate which tolerances, if any, are correct, and to know what changes to make to all the others.

Evaluating a Final Product Requirement. The characteristics of the final products, (1) above, may have been arbitrarily set by a customer or by engineering, based upon whatever information they had or could get. Although it may be desired to open up one of these tolerances, it is feared it might be unfavorable to the customer's use of the product. How many should be tested to accept or to deny the suggested increase, with everyone agreeing to the decision? Because, heretofore, there has been no clear answer to this question, most final tolerances probably remain unnecessarily tight.

A quality control activity within engineering, and with the customer, can answer the question with an appropriate level of statistical confidence. To work out the solution, an example will be considered. Say a spring manufacturer feels that a broader tolerance on finished valve springs will not adversely affect their life in his customer's product, an internal combustion engine. A problem, of course,

TABLE 5-1.   Probabilities in a Card Drawing
Simulation of Risk and Confidence

Risk: Probability of not agreeing	Confidence: Probability of agreeing	No. of cards
0.50	0.50	9
0.10	0.90	29
0.05	0.95	38
0.016	0.984	52
0.01	0.99	57

is that no two springs will exhibit the same life when run in an engine. Accordingly, a statistical method must be used to compare current springs (C) with new springs (N) manufactured to the broader tolerance.

Risk and Confidence.   As described previously for acceptance sampling, the consumer's risk has been conventionally set at 0.10, resulting in 0.90 confidence that the quality of accepted lots is better than the RQL.   But the tolerance decision could affect a manufacturing method, cost, and long-term customer satisfaction.   The risk of a wrong decision should perhaps be less than 0.10—perhaps 0.05, or even 0.01.

To be able to make that selection of risk from personal experience, the reader can make some trial runs with a deck of 52 shuffled cards.   With the deck face down, turn over the top card in sequence, calling out "ace, two, three, . . . , queen, king, ace," and so on.   Half of the time, the card called will be the one turned over for nine or less cards; nine times out of ten it will appear in position by the 29th card or less (Table 5-1).

For the entire deck of 52 cards, there is only 0.016 probability that one can go that far without having a card turn over as called.   Although this still represents a finite risk, the reader can spend evening after evening at this game and never seem to succeed in getting through the deck.   To demonstrate the 0.01 risk, the deck plus 5 cards are needed.

Conclusions: Risk and confidence always add to 1.00.   A confidence of 0.95 is always practical; 0.99 seems as close to certain as one could ever want.   When evaluating a final product requirement, such levels of confidence represent protection against poor decisions exceeding any afforded by one's judgment as to whether an increase in tolerance incurs an actual penalty in product performance.

A Test for a Significant Difference.   Select a measure of merit, usually an important property of the product—in the case of the valve springs, cycles to failure.   Design the test to show, with 0.95 confidence, that the tighter tolerance wanted by the customer (current C springs) does in fact lengthen spring life in comparison to those with the looser tolerance (new N springs).

The test will be planned around a hypothesis of no real difference in cycles to failure coming from the tolerance in question.   If that hypothesis were true and one N spring were compared with two C's, any one of the three possible ranked outcomes would be equally likely:

Outcome number	Longest life		Shortest life
1	C	C	N
2	C	N	C
3	N	C	C

TABLE 5-2. All Possible Ranking Combinations
of Three New and Three Current Spring
Manufacturing Methods

Outcome no.	Longest life					Shortest life
1	C	C	C	N	N	N
2	C	C	N	C	N	N
3	C	C	N	N	C	N
4	C	C	N	N	N	C
5	C	N	C	C	N	N
6	C	N	C	N	C	N
7	C	N	C	N	N	C
8	C	N	N	C	C	N
9	C	N	N	C	N	C
10	C	N	N	N	C	C
11	N	C	C	C	N	N
12	N	C	C	N	C	N
13	N	C	C	N	N	C
14	N	C	N	C	C	N
15	N	C	N	C	N	C
16	N	C	N	N	C	C
17	N	N	C	C	C	N
18	N	N	C	C	N	C
19	N	N	C	N	C	C
20	N	N	N	C	C	C

Outcome no. 1 would make it seem the customer was right; the two C springs were both better by test than the N spring. But that can happen by chance 0.33 of the time—the risk of making a wrong decision, one favoring C spring. The confidence, then, of a three-spring comparison test is only 0.67. That confidence can be raised to the desired 0.95 by planning for more than three tests. Try six tests, three C's and three N's.

A study of Table 5-2 reveals a methodical pattern of 20 different, equally likely outcomes. Now if outcome no. 1 does occur and all the C's place in the top three ranks, then it is either pure chance outcome ($\frac{1}{20} = 0.05$ risk) or it was caused by the tighter spring tolerance. By concluding the latter, one rejects the hypothesis of no favorable influence from the tighter tolerance with 0.95 confidence.

A complete statement of the test plan can now follow.

Fabricate, from randomly selected raw coils, random coiling machines, and so on, a number of springs so that three can be selected to represent the current tight tolerance, and three the new looser, proposed tolerance.

Code them with different color paint marks, say, red, white, and blue for the C springs and yellow, green, and black for the N springs. The code identification is held in a sealed envelope by an independent third party.

Go over Table 5-2 with the customer so that he is also convinced that outcomes no. 2 through 20 would mean that some variable other than the tolerance in question is having the controlling influence on spring life. Get him to agree to place these six springs in an engine and to record the cycles to failure for each color spring.

If desired, in advance of the testing, both parties can agree on a minimum amount of difference which must exist to justify keeping the tighter tolerance on the drawing in the event outcome no. 1 does occur. Say an improvement of at least 2 percent is so stated.

Call in the third party to reveal the code when the test results are in (see Table 5-3).

The C springs did come in the top three ranks. One is 0.95 sure that the

TABLE 5-3. Ranking of Spring Test Results

Millions of cycles to failure	Color	Tolerance	Rank
0.26	Yellow	N	6
4.30	Black	N	5
7.89	Green	N	4
8.53	White	C	3
9.22	Blue	C	2
12.75	Red	C	1

current tighter tolerance does improve life; but how about the requirement of at least 2 percent?

Multiply the rank 4 spring life, 7.89 million cycles by 1.02. If it remains in rank 4, then one is 0.95 sure of at least a 2 percent difference.

$$7.89 \times 1.02 = 8.05$$

which is not above rank 3's 8.53, and so in this case the customer's tighter requirement is justified; the tolerance cannot be broadened without paying a penalty in what the customer expects.

A Realistic System of Manufacturing Tolerances. Specifications for parts, components, manufacturing processes, and raw materials have been set by metallurgists, chemists, design engineers, and production engineers—usually based on quite meager information. These are the tolerances which seem to undergo most change as troubles of various sorts are encountered. Even the changed limits, at times, have to be changed again.

Each such proposed change could first be tested as so many N units against so many C units, by setting up a decision rule to change or not to change, based upon the relative number of equally likely favorable and unfavorable combinations, as illustrated in the previous paragraphs. One of two generally appropriate tests is frequently used for evaluating all potentially costly engineering and manufacturing changes.

A ratio of hypergeometric relations serves as the statistical and mathematical way to enlarge Table 5-4 readily for any special confidence, number of tests, or overlapping ranks situation needed.[4]

But a systems approach is always more effective than fire-fighting individual tolerances giving difficulties here and there. During the manufacture of a product, the materials, ingredients, parts, and the like go together under conditions (usually with some control to specified tolerances) and result in a finished product having properties y_1, y_2, y_3, etc., each of which is checked against specifications discussed

[4] Dorian Shainin, "How to Calculate the Risk of a Decision," *Quality Progress,* August, 1968.

TABLE 5-4. Required Tests and Rank
Criteria for Selected Confidence Levels

Confidence selected	No. of N tests	No. of C tests	Change if all the N results are in the top
0.95	3	6	4
0.99	4	8	5

above. But the material characteristics x_1, x_2, x_3, etc., part quality characteristics x_{10}, x_{11}, x_{12}, etc., and fabricating conditions (temperatures, pressures, flows, feeds) x_{20}, x_{21}, x_{22}, etc., also have tolerances of the type mentioned previously on page 8-126. They do not have equal influence upon a given y.

Are the nominal values for each x, and its limits, at the best values in combination with those for every other x which does have an influence upon that y? They have generally been set with inadequate information to answer that question. If they happen to be even in the vicinity of being best, it would be pure coincidence.

When these tolerances, the very reference points for quality actions, are thus incorrect, it becomes clear that management has to plan modifications of its current company quality program to set the "foundation for the house" right. This systems approach gives visibility to the relationship, if any actually exists, of each x to a given y. At the same time, it shows the separate effect upon that y of the combination of all the other or remaining x's, including that of neglected, unsuspected factors which are important enough to be discovered (see Variation Research, pages 8-138 and 8-139) and to be correctly toleranced.

The mathematical statistical tool to analyze the results of specially planned gathering of measurements goes under the name of regression analysis.[5] Two scatter diagrams, from a group of 56 plotted in such a realistic tolerance evaluation study, will be shown to illustrate the kind of thinking involved.

From randomly assigned, premeasured (and recorded) components, 30 consecutive assemblies were constructed. The result y (maximum allowed value for acceptable product equals 60 units) was measured on each assembly. Figures 5-5 and 5-6 show the 30 y results plotted against x_{10}, and against x_1. In Figure 5-5, imagine a central curved regression line, midway between the two parallel curved lines shown. This regression line estimates, on the average, what happens to the value of y as x_{10} alone changes. A similar regression line for the same 30 y points, but plotted against x_1 on Figure 5-6, would be substantially horizontal; x_1 is unimportant as far as y is concerned. Forty of the fifty-six charts were like x_1.

The present tolerance range for x_{10} runs from 0.874 to 0.884, and the points show it is often being violated, both below the low limit and above the high limit. Nevertheless, the tolerance of 0.010, if observed, will not assure the production of effective units.

The geometry of the two straight dash lines at the left corner of Figure 5-5 explains why. The product itself, from this controlled "traceability" study, is showing that the tolerance should be a "maximum of 0.852." Then virtually all the

[5] Wilfred J. Dixon and Frank J. Massey, Jr., *Introduction to Statistical Analysis*, 3d ed., McGraw-Hill Book Company, New York, 1969.

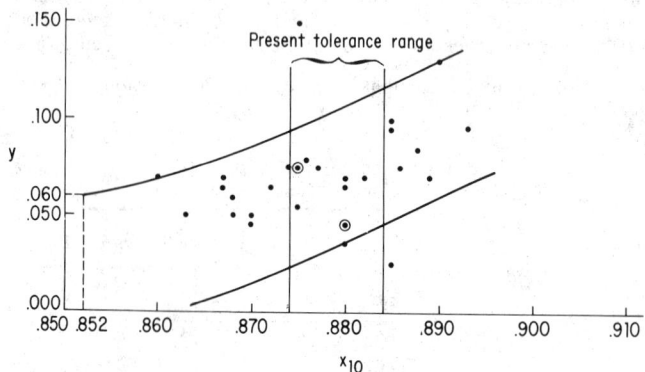

FIG. 5-5. Scatter diagram of y results for 30 assemblies, plotted against x_{10}.

product lying below the top curved line will meet the 60-unit maximum specification for y.

This sample of 30 assemblies had 19 rejected, measuring more than 60 units on the y scale. That ratio of 0.63 defective would be a plotted point on a fraction defective control chart (known as a p chart). In this process, a number of these points would all fall within statistically determined control limits, indicating "a stable operation—it would not be worthwhile to attempt to locate causes of the trouble." Modern statistical quality control has come a long way since those "textbook" days. Now one such p chart point, dissected by this tolerance study, has shown which 16 of the 56 x's can be modified, and by how much, to prevent future rejections of finished product.

Figure 5-5 illustrates one more important principle. The vertical distance between the two curved lines is not affected by x_{10}. The combined effect of all the other x's, known and unknown, causes it. There are 15 other charts in this study which have regression lines that are not horizontal. Hence, it is a matter of simple calculation from the 16 charts to determine the simultaneous realistic tolerances for all 16, which will result in a minimum total manufacturing cost with no product exceeding 60 y units. For example, as changes are made to one or more of the 15 other x tolerances, the vertical scatter of points on Figure 5-5 will decrease, the top curved line will be lower, and the current maximum value of 0.852 could be higher; this is desirable if it would save manufacturing cost. Electronic data processing facilities permit the rapid extension of the use of such data to rectify the great number of functional tolerances currently guiding quality incorrectly.

Larger Tolerances Which Retain Interchangeability. Tolerances for quality characteristics which combine with others have too long been set with calculations which permit components at their permissible extreme values to come together. Actually, each quality characteristic will vary somewhat, item by item, but it will not occur equally at all possible values. It will follow a distribution which almost always will be reasonably close to a normal or bell-shaped distribution[6] in the long run if the process generating that characteristic has been stable (see Pre-Control, below, for keeping it stable with time).

Consider the example of the fit of a shaft in a bore. If the normal distributions of each diameter used up their full tolerance and were properly centered, it would be a worst-case but entirely acceptable situation. In this case, the designer intended to have an assembled clearance range from 0.001 inch loose to 0.005 inch loose on the diameter. He elected to split the 0.004-inch difference equally, 0.002 inch for each tolerance. The diagram of Figure 5-7 depicts that situation. Such an engineering specification implies that manufacturing can fabricate its shafts anywhere

[6] Grant, *op. cit.*, chap. 3.

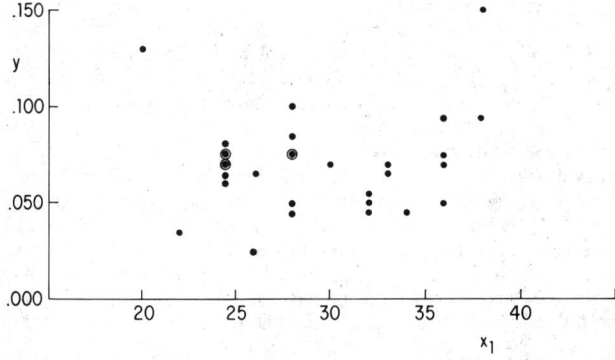

Fig. 5-6. Scatter diagram of y results for 30 assemblies, plotted against x_1.

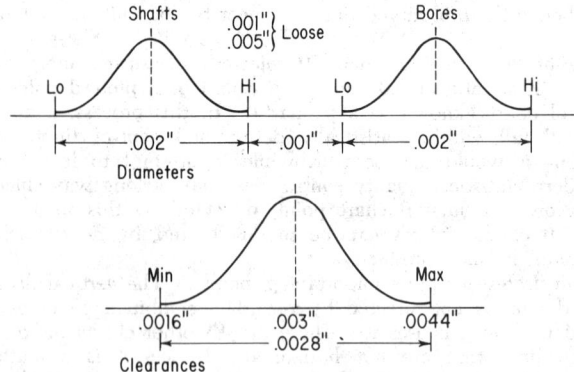

FIG. 5-7.   Shaft and bore diameters and fit.

within that 0.002-inch tolerance, all at the high limit if it pleases. But engineers now can be made responsible for eliminating this unnecessarily costly interpretation. They are being taught about distributions and about manufacturing control plans which, through simple statistical methods like Pre-Control, will tend to keep the distribution located close to the nominal dimension.

The engineer learns, from a sketch like Figure 5-7, that the probability of a randomly selected high-limit shaft coming together with a randomly selected low-limit bore would be the product of their frequencies of occurrence, about $1/1,000 \times 1/1,000 = 1/1,000,000$, or virtually nil. As has been illustrated in Table 5-1, even a probability of occurrence of the joint event of parts at opposite limits of 0.01 would be sufficiently rare to be quite acceptable. But here the statistical method sets the figure at 1/10 of that, at 0.001. When the two or more distributions are controlled, their base widths combine by chance by the square root of the sum of the squares[7] to result in a base width of the distribution of accumulated fit or clearance. Here

$$\sqrt{0.002^2 + 0.002^2} = 0.0028 \text{ inch}$$

and so a more liberal tolerance can be allowed for both parts, for the root answer to come out 0.004 inch. Notice that the 0.003-inch average clearance is the difference between the nominals, average bore and the average shaft sizes, and that the chance combination of several distributions tends to result in a symmetrical, normal distribution of clearances.

The nomograph of Figure 5-8 has been drawn to facilitate such square root computations involving trial and error steps. A straightedge quickly shows the answer on the center scale for the two part tolerances on the outer scales. Always use the same side of all three scales; the right-hand sides magnify the range from 0 to 5. All scales can be simultaneously multiplied, or divided, by 10, by 100, by 1,000, and so on. To accumulate more than two tolerances, the answer on the center scale for any two is moved to an outside scale, and a third tolerance can now be combined with that answer.

Try some examples on Figure 5-8. In the familiar 3, 4, 5 right triangle, check out that 3^2 plus 4^2 equals 5^2. Then show that 0.002^2 plus 0.002^2 gives 0.0028^2. To get an answer of 0.004 inch clearance variation, use a horizontal line to show that the tolerances for both the shaft and bore can be increased to 0.0028 inch, 40 percent more tolerance as a bonus for using the Pre-Control plan during the manufacture of parts. Suppose it is easy to hold the outside diameter of the shafts to a 0.002-inch tolerance, but difficult to hold the inside diameters of the bores. Connect 2 for the

[7] J. M. Juran (ed.), *Quality Control Handbook*, McGraw-Hill Book Company, New York, 1962, pp. 3-42 to 3-46.

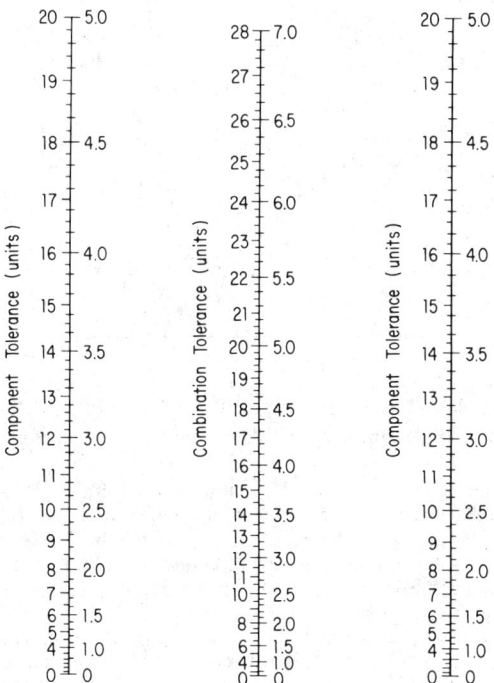

Fᴵɢ. 5-8. Design nomogram for computing rapidly how tolerances actually combine.

shafts with required clearance of 4 and reveal that as much as 0.0035 inch can be allowed for the bore tolerance, almost doubling the original 0.002-inch tolerance. The allowable increase can become surprisingly substantial if one of the other tolerances is relatively large. The nomogram permits rapid determination of combinations which economically allow the tolerances to accommodate the manufacturing process capabilities.

MANUFACTURING'S RESPONSIBILITIES

Figure 5-1 emphasized that the people who manufacture, fabricate, blend, and produce products have a primary quality responsibility. "Make it right the first time," originally a clever phrase for Zero Defects motivation programs, has worthwhile meaning. And a rising industrial productivity comes from product conforming to realistic specifications with no rework, no scrap, and practically no inspection by people other than those cutting chips, blending the ingredients, or building up assemblies. Manufacturing is gradually accepting a responsibility for managing its activities to make more initially acceptable work per dollar.

To manage anything well, one must have this complete information system. He must:

1. Know clearly and completely what is wanted
2. Know, in time, how he is doing in comparison with what is wanted
3. Know what action to take and when to take it to prevent deviations from what is wanted

Well-assigned realistic specifications ease the manufacturing task of meeting them, but too often one or more of the three requirements for managing is inadequately prepared; often the last one is completely missing.

Suppose a production operator does have the first two requirements as depicted

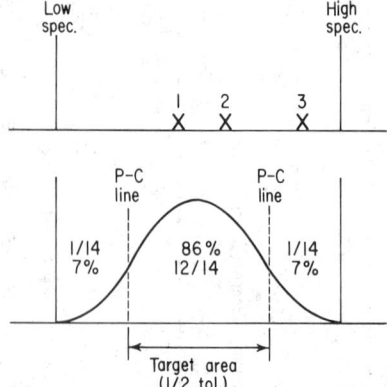

FIG. 5-9. Distribution of product with Pre-Control lines and specification limits.

in the top section of Figure 5-9. He knows the high and low limits, and he uses a gage and measures the first, second, and third unit produced. But the third requirement, a plan for effective control, is missing. He would be inclined to adjust the equipment setting toward the low-spec limit before processing the fourth unit. That action could be correct or wrong, and the operator has no idea what the odds might be for the adjustment to be correct.

The Pre-Control Plan. Incorrectly adjusting the process up and down (hunting) is virtually eliminated by the Pre-Control strategy. Divide the tolerance into three regions by placing P-C lines on the gage to mark off the central half of the full tolerance. Call it the target area and color it green for ready identification. Color the remaining outer quarters of the full tolerance yellow, for caution, and use red for the regions beyond the spec limits. The operating rule, simply, is to take appropriate action only when two consecutive units both measure in the yellow area, or when one red occurs (very rare when the entire strategy is properly applied).

The bottom section of Figure 5-9 has a borderline normal distribution superimposed upon the tolerance limits and P-C lines. It extends from one limit to the other—borderline, because any shift or widening of the distribution with time will cause some defective work to be produced. But right at the moment, virtually all results

TABLE 5-5. Probabilities of Events Beyond a Given Standard Deviation*

Proba-bility	0.00	0.0	0.1	0.2	0.3	0.4
0	. . . .	. . . .	1.28155	0.84162	0.52440	0.25335
1	3.09023	2.32635	1.22653	0.80642	0.49585	0.22754
2	2.87816	2.05375	1.17499	0.77219	0.46770	0.20189
3	2.74778	1.88079	1.12639	0.73885	0.43991	0.17637
4	2.65207	1.75069	1.08032	0.70630	0.41246	0.15097
5	2.57583	1.64485	1.03643	0.67449	0.38532	0.12566
6	2.51214	1.55477	0.99446	0.64335	0.35846	0.10043
7	2.45726	1.47579	0.95417	0.61281	0.33185	0.07527
8	2.40892	1.40507	0.91537	0.58284	0.30548	0.05015
9	2.36562	1.34076	0.87790	0.55338	0.27932	0.02507

* SOURCE: Columbia University, Statistical Research Group, *Techniques of Statistical Analysis*, McGraw-Hill Book Company, New York, 1947, p. 18.

will be acceptable within the tolerance; the process capability of plus and minus 3 standard deviations is equal to the tolerance. The entries in Table 5-5 are the number of standard deviations from the center of any normal distribution which will leave a tail area on that side equal to the probability that a result will occur that number of standard deviations or more away in that direction. Read the probability from the column heading, with the next digit being the number for the row on the left side. Notice that the probability for exceeding 3.09 standard deviations is 0.001, or virtually no probability of being beyond the 3 standard deviation (spec limit in this case) distance.

The P-C line, accordingly, is at 1.5 standard deviations from the center. Read 0.07, or 7 percent, for 1.48 standard deviations from Table 5-5. Seven percent means that 1 part in 14, in the long run, will occur beyond the green target area on the right, and 1 in 14 on the left.

At a random time, the operator takes two consecutive parts and measures the first one:

1. If it falls in the green target area, keep on running.
2. If it falls in the yellow area, measure the second one.
   a. If it is green, keep on running.
   b. If it is also yellow, STOP.

For the borderline distribution condition of Figure 5-9, all possible outcomes of a random consecutive pair of parts are listed in Figure 5-10, along with the products of the separate probabilities for each part, thus giving the probability of the joint event.

The action indicated by the two consecutive yellow results (top two lines of Figure 5-10) would be common only when the actual distribution is wider than

ACTION	YELLOW	GREEN	YELLOW	JOINT PROBABILITY

H
E
  L
    P
	1st
	2nd
			2nd
			1st
				$1/14 \times 1/14 = 1/196$
				$1/14 \times 1/14 = 1/196$

A
  D
   J
    U
     S
      T
	1st 2nd		1st 2nd	$1/14 \times 1/14 = 1/196$
				$1/14 \times 1/14 = 1/196$

K
 E
  E
   P
O
 N
R
 U
  N
   N
    I
    N
     G
	1st
	
	2nd
		1st 2nd
		2nd
		2nd
		1st
		1st
			
			1st
			2nd
				$12/14 \times 12/14 = 144/196$
				$1/14 \times 12/14 = 12/196$
				$12/14 \times 1/14 = 12/196$
				$12/14 \times 1/14 = 12/196$
				$1/14 \times 12/14 = 12/196$

PROBABILITY OF EACH PART	1/14	12/14	1/14	
				TOTAL 196/196

FIG. 5-10. Pre-Control sampling system showing all possible outcomes of a sample pair.

the tolerance—the operator needs help from a setup man, foreman, or manufacturing engineer. For the safe picture of Figure 5-9, that signal would be rare, occurring only 2 times in 196 (about 0.01 of the time). The signal shown on the next two lines, equally rare when all is well, would be common when a distribution shifts to one side, thus requiring adjustment of the process setting.

Altogether, Pre-Control works with a maximum risk of 0.02, a minimum confidence of 0.98.

The recommended frequency for taking the set of two consecutive parts depends upon the stability of the fabricating equipment. The operator should average 25 sets per process adjustment. That frequency will statistically keep the maximum long-term fraction defective output of the process at less than 1 percent.

In place of the inadequate procedure of first-piece inspection, Pre-Control requires that 5 consecutive pieces measure in the green target area to permit approval of the setup or of a process adjustment. The operator then knows, with high probability, that he has a nearly centered distribution narrower than the tolerance.

Pre-Control's performance in hundreds of companies has exhibited six important properties:

1. Easy understanding by operators, with no paper work or calculations needed.
2. The tails of the distribution are kept inboard and away from tolerance limits where the inability of a gage to repeat can be distressing.
3. Runs of several million parts with no defectives at all produced.
4. It can give a signal as early as the second part made, thereby also being useful for short-run jobs.
5. It shows cases where the variability of the process is too great for the tolerance allowed.
6. It improves productivity by greatly reducing the number of incorrect process adjustments and by producing a larger number of acceptable pieces per operating machine-hour.

QUALITY ENGINEERING'S RESPONSIBILITIES

The completely effective discharge of the primary quality responsibilities of Figure 5-1 should leave nothing further to be desired. Realistic tolerances have been objectively determined and economically allocated by engineering.

With practical specifications, manufacturing has a strong incentive to use Pre-Control to prevent the production of nonconforming product; this is not the case when everyone knows that the limits are arbitrary and approximate and tend to be conservative. A similar strong incentive can affect each inspector. He will practice the necessary precautions to avoid human oversights because the tolerances now mean something. And a group, acting independently of those operating departments, audits the degree to which the overall system of primary quality responsibilities produces the intended results for top management. But to the quality control function falls the responsibility of bringing about a smooth coordination of the many related activities which should be carried out well by independent departments. The knowledge of what needs to be done, and how it is to be done, comes from quality engineering's data gathering and analysis activities.

Recognition of the general lack of this knowledge is apparent from the activities listed in Figure 5-1 as secondary quality responsibilities. Rather than backup actions, these activities portray the continual presence of gaps somewhere in a primary task. Quality engineering must be able to show progress by closing old gaps faster than new ones appear. Its strategic tools are forged from some often neglected principles.

Repeatable and Nonrepeatable Tests. The sketches of Figure 5-11 display in graphical and tabular form the principle of repeatable and nonrepeatable results when searching for the root cause or causes of the variation of an effect y. Before one can conclude that he can predict a value of y with a practical degree of precision, he should certainly test each level of the suspected cause, x, more than once, and in random sequence of the levels, say, x', x'', and x''', on randomly assigned units

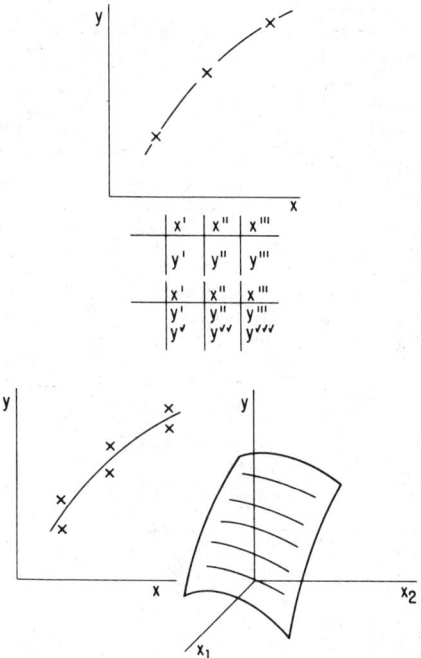

Fig. 5-11. Regression table, curve, and surface showing nonrepeatability and influence of additional unknown variable.

of test material.   If, with such an opportunity for some other cause of influence upon y to enter the picture, one nevertheless gets good agreement between such pairs of "replicated" results, y' and $y^\vee$, y'' and $y^{\vee\vee}$, y''' and $y^{\vee\vee\vee}$, then one can sense that x is having a controlling effect if something other than a horizontal line goes through the points.

The lower left sketch shows a lack of agreement (scatter) between the replicated results.   That could be caused by having neglected as little as only one other important cause, say x_2.   A confirmation of this fact would come from testing combinations of levels of x_1 and of x_2, and having those replications agree in the same three-dimensional response surface depicted at the lower right.   If they do not agree, it could be that only one other variable of importance, x_3, is being neglected.

The table for the three-dimensional x_1, x_2, y graph, called a matrix, shown at the top of Figure 5-12, again emphasizes that one test result, y, for each of the nine test combinations provides no information as to whether the search for additional important causes can be terminated.   Randomized replications serve as that check.

If the three-dimensional response surface at the bottom of Figure 5-11 consists of parallel elements (intersections of parallel planes cutting that surface), then it is a ruled surface; this means the main effect of x_1 is completely independent of the main effect of x_2.   But if that surface has a twist (see the central sketch of Figure 5-12), its elements all rotate one to another, out of parallel; the amount of the twist is designated as the "$x_1 \cdot x_2$ interaction"—an amount which is independent of the main effects of x_1 and of x_2.

This principle of repeatable and nonrepeatable results can, of course, continue; y values which do not repeat in a three-dimensional surface clearly indicate the existence of at least an x_3, which, as implied above, should be found.   Say it varies from a minus level (x_3^-) to a plus level (x_3^+).   But how can one graphically display four axes, each perpendicular to the other three?   The lower sketch of Figure 5-12

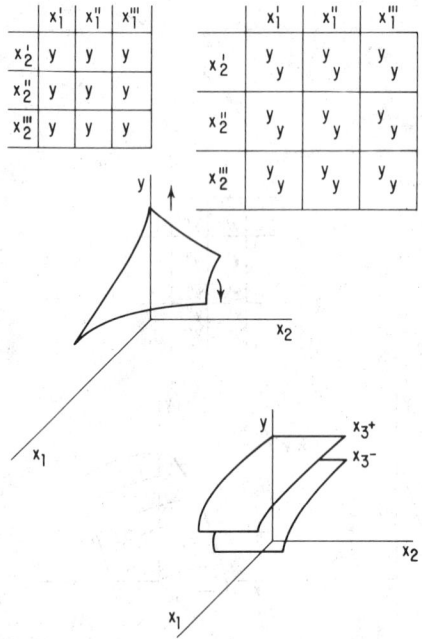

FIG. 5-12. Planned experiment matrices and surfaces showing $x_1 \cdot x_2$ interaction and third variable, ranging from x_3^- to x_3^+, in a four-dimensional representation.

illustrates how the four influences x_1, x_2, x_3, and y can be visualized (a four-dimensional picture)—three main effects: an $x_1 \cdot x_2$ interaction (the average twist of the two or more response surfaces), an $x_1 \cdot x_2 \cdot x_3$ three-factor interaction (the amount the response surfaces are not parallel to each other), and the remaining two-factor interactions $x_1 \cdot x_3$ and $x_2 \cdot x_3$ (best visualized by replotting the data so that the bottom plane is, respectively, x_1 and x_3, with two or more levels of x_2 as the twisted response surfaces, or x_2 and x_3, with x_1 response surfaces).

Isolating the Leading Cause. Actually, the variation of every y literally has a great number of causes, but these variables do not contribute equally. There is always a single greatest contributor, which may be an interaction or a main effect. But because such independent contributions from many combine by chance to give a total variation of y, they tend to compensate; some raise y, others simultaneously lower y. Accordingly, it can be shown statistically that the variation of y is equal to the square root of the sum of the squares of the variation from each contributor, as was the case described previously for tolerances contributing to fits and clearances. The few (or one) larger contributions thus control the y variation; the smaller ones relatively contribute so little (after being squared) that they do not cause any practical lack of agreement among replications. The task of the quality engineer is to isolate these few (or one) important interactions or main effects.

For most quality problem situations, when manufacturing cannot readily reduce the variation of y, it would be discouraging to attempt somehow to select the key cause without a special, strategic form of evidence.

Variation Research. By taking samples from the operating process in a planned manner, it becomes possible to obtain independent estimates of the variation coming from variables which change with time, from those which change cyclically (one unit to its neighbor), and from those which are a function of location on the unit (or of a lack of agreement of repeated measurements). Whichever of these three independent components of total variation exhibits the most variability must contain the single key cause being sought.

At times, this clue leads the quality engineer directly to a perhaps previously unsuspected variable, permitting more than adequate control of y. If several variables fitting the clue seem likely candidates, their main effects and interactions can be evaluated with an appropriate statistically designed experiment.[8]

CONCLUSION

The competent professional quality engineer potentially can become tomorrow's most effective industrial engineer. With these strategies, he develops objective, problem solving skills which accelerate his firm's progress in using men, machines, and materials with a coordinated degree of sophisticated control of unexpectedly important combined variables that exceeds the productivity and product value results of the fondest expectations of top management.

BIBLIOGRAPHY

Burk, E. C., and J. F. Chapman, *New Decision Making Tools for Managers*, Harvard University Press, Cambridge, Mass., 1963.

Columbia University, Statistical Research Group, *Techniques of Statistical Analysis*, McGraw-Hill Book Company, New York, 1947.

Dixon, Wilfred J., and Frank J. Massey, Jr., *Introduction to Statistical Analysis*, 3d ed., McGraw-Hill Book Company, New York, 1969.

Grant, Eugene L., *Statistical Quality Control*, 3d ed., McGraw-Hill Book Company, New York, 1964.

Juran, J. M. (ed.), *Quality Control Handbook*, 2d ed., McGraw-Hill Book Company, New York, 1962.

Larson, Harry R., "A Nomograph of the Cumulative Binomial Distribution," *Industrial Quality Control*, December, 1966.

Shainin, Dorian, "How to Calculate the Risk of a Decision," *Quality Progress*, August, 1968.

[8] E. C. Burk and J. F. Chapman, *New Decision Making Tools for Managers*, Harvard University Press, Cambridge, Mass., 1963, chap. 16.

Chapter 6

Reliability Engineering

JOSEPH J. NARESKY

Chief, Reliability and Compatibility Division, Rome Air Development Center, United States Air Force, Air Force Systems Command, Griffiss Air Force Base, Rome, New York

For all but the most recent years of human history, the performance expected from man's implements was quite low and the life realized was long, both because it just happened to be so in terms of man's lifetime and because he had no reason to expect otherwise. The great technological advances, beginning in the latter half of the twentieth century, have been inextricably tied to more and more complex implements or devices. In general, these have been synthesized from simpler devices having a satisfactory life. It is a well-known fact that any device which requires all its parts to function will always be less stable than any of its parts. Although significant improvements have been made in increasing the lives of basic components—for example, microelectronics—these have not usually been accompanied by corresponding increases in the lives of equipment and systems. In some cases, equipment and system complexity has progressed at so rapid a pace as to negate, in part, the increased life expected from use of the longer-lived basic components. In other cases, the basic components have been misapplied or overstressed so that their potentially long lives were cut short. In still other cases, management has been reluctant to devote the time and attention necessary to ensure that the potentially long lives of the basic components were achieved.

The military services, because they had the most acute problems, gave the impetus to the orderly development of the discipline of reliability engineering. It was they who were instrumental in developing mathematical models for reliability, as well as design techniques, to permit the quantitative specification, measurement, and prediction of reliability.

The commonly accepted definition of reliability is that "reliability is the probability that a device will satisfactorily perform its specified function for a specified period of time under a given set of operating conditions." Hence reliability differs from quality control in that quality control is a time-zero measurement of the quality of a product, whereas reliability is a time-dependent measurement of quality. Reliability can be considered as quality control plus time.

MATHEMATICS OF RELIABILITY

Basic Concepts. Because reliability is defined in terms of probability, probabilistic parameters such as random variables, density functions, and distribution functions are utilized in the development of reliability theory. Reliability studies are concerned with both discrete and continuous random variables. An example of a discrete variable is the number of failures in a given interval of time. Examples of continuous random variables are the time from part installation to failure, and the time between successive equipment failures.

The distribution function $U(t)$ is defined as the probability in a random trial that the random variable is not greater than t, or

$$U(t) = \int_{-\infty}^{t} u(t)\, dt \tag{1}$$

where $u(t)$ is the density function of the random variable, time to failure. This is termed the "unreliability function" when speaking of failure. It can be thought of as representing the probability of failure prior to some time t. If the random variable is discrete, the integral is replaced by a summation.

The reliability function, or the probability of a device not failing prior to some time t, is given by

$$R(t) = 1 - U(t) = \int_{t}^{\infty} u(t)\, dt \tag{2}$$

By differentiating equation (2) it can be shown that

$$\frac{dR(t)}{dt} = \frac{-dU(t)}{dt} = -u(t) \tag{3}$$

Failure and Hazard Rates. The probability of failure in a given time interval t_1 to t_2 can be expressed by the reliability function

$$\int_{t_1}^{\infty} u(t)\, dt - \int_{t_2}^{\infty} u(t)\, dt = R(t_1) - R(t_2) \tag{4}$$

The rate at which failures occur in the interval t_1 to t_2, the failure rate $\phi(t)$, is defined as the ratio of the probability that failure occurs in the interval, given that it has not occurred prior to t_1, the start of the interval, divided by the interval length. Thus,

$$\phi(t) = \frac{R(t_1) - R(t_2)}{(t_2 - t_1)R(t_1)} \tag{5}$$

or the alternative form

$$\phi(t) = \frac{R(t) - R(t + h)}{hR(t)} \tag{6}$$

where $t = t_1$ and $t_2 = h + t_1$. The hazard rate $z(t)$ or instantaneous failure rate is defined as the limit of the failure rate as the interval length approaches zero, or

$$z(t) = \lim_{h \to 0} \left[\frac{R(t) - R(t + h)}{h R(t)} \right] \tag{7}$$

$$= \frac{-1}{R(t)} \left[\frac{dR \ (t)}{dt} \right] = \frac{u(t)}{R(t)}$$

which can also be written as

$$z(t) = \frac{-d \ln R(t)}{dt} \tag{8}$$

The last differential equation tells us, then, that the hazard rate is nothing more than a measure of the change in survivor rate per unit change in time. By carrying this equation one step further, we can find a completely general expression for the reliability function:

$$R(t) = \exp \left[- \int_0^t z(t) \ dt \right]^* \tag{9}$$

Perhaps the simplest explanation of hazard and failure rate is made by analogy. Suppose a family takes an automobile trip of 200 miles and completes the trip in 4 hours. Their average rate was 50 mph, although they drove faster at some times and slower at other times. The rate at any given instant could have been determined by reading the speed indicated on the speedometer at that instant. The 50 mph is analogous to the failure rate and the speed at any point is analogous to the hazard rate.

Specific Density and Distribution Functions. It has been found that a relatively small number of functions satisfies most needs in reliability work. In this presentation, only the major aspects of each function will be discussed. Figures 6-1A and 6-1B give formulas and curves for some of the functions associated with the selected density functions.

Differences in failure rates and hazard rates are significant elements in the comparison of density functions. The exponential has a constant hazard rate, independent of time; this means that the probability of failure is independent of age. The Gaussian or normal density hazard rate increases with time; this means that the probability of failure increases with age. The hazard rate for the Weibull distribution depends upon the value of β, as can be seen in Figure 6-1A. For $\beta = 1$, the hazard rate and reliability function are identical to that of the exponential distribution; when β is approximately 3.25, the reliability function approaches that of the normal distribution.

The Weibull and exponential distributions have the most widespread applicability in reliability analysis. Survival curves for most systems and complex equipments are of the exponential form; survival curves for many component parts follow a Weibull distribution. Because of limited space, this chapter will treat the exponential and Weibull distributions. Although they have widespread applicability, the reader is cautioned not to apply them in every case. When in doubt, standard statistical methods should be used to verify the underlying distribution.

Computing Reliability. Consider an equipment or system made up of n parts in series in which the failure of any part causes equipment failure. Assume that the failure of any one part is independent of the failure of another. Denote the part reliability functions by $R_i(t)$, $i = 1, 2, \ldots, n$ and the equipment reliability

---

* Pure mathematicians object to the use of the same letter in the integral and also in the limits of the integral. This is done here in spite of the objection in order to simplify the reference to time as the variable in such functions as $U(t)$, $u(t)$, $R(t)$, and $z(t)$.

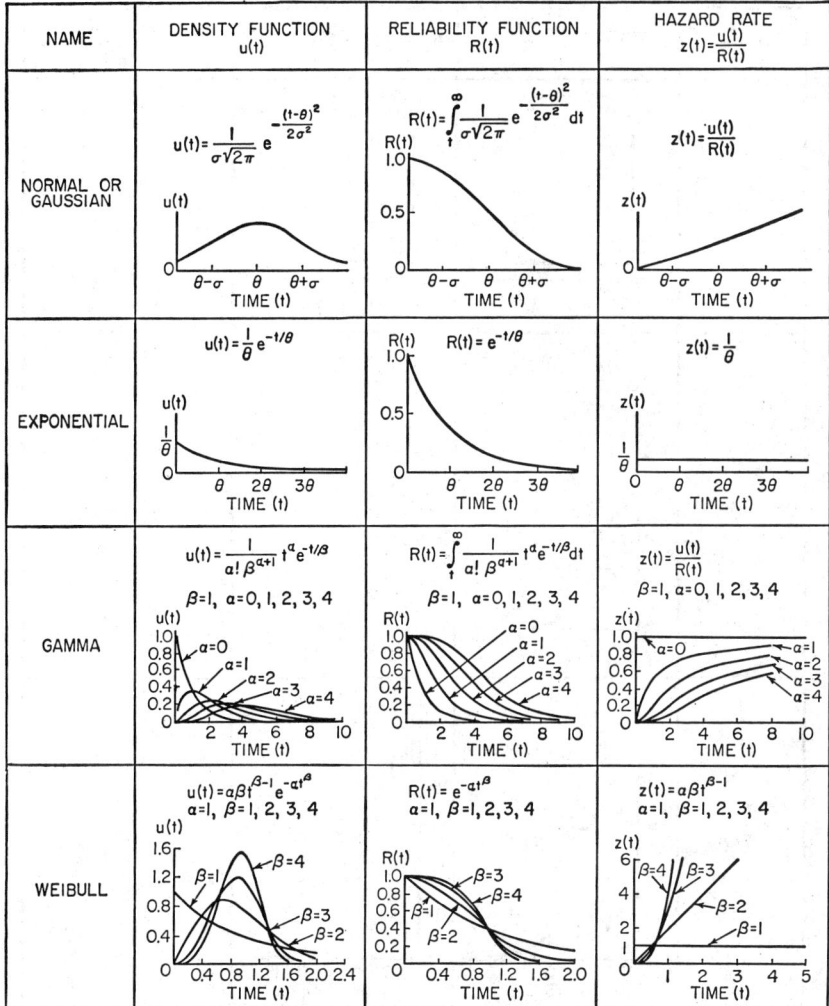

FIG. 6-1A. Density and reliability functions and hazard rates of the normal, exponential, gamma, and Weibull distributions.

function by $R(t)$. Thus the probability that the equipment will survive to time t without failure is given by

$$R(t) = R_1(t) \cdot R_2(t) \, \cdots \, R_n(t) \tag{10}$$

When each part has an exponential time-to-failure density, then

$$R(t) = e^{-\lambda_1 t} e^{-\lambda_2 t} \, \cdots \, e^{-\lambda_n t} = \exp\left(-\sum_{i=1}^{n} \lambda_i t\right) = e^{-\lambda t} \tag{11}$$

where $\lambda = \lambda_1 + \lambda_2 + \cdots + \lambda_n$, and $\lambda_i = 1/\theta_i$. Thus the system failure rate λ is the sum of the individual component failure rates and the system mean life $\theta = 1/\lambda$.

Consider a system composed of 400 component parts, each having an exponential

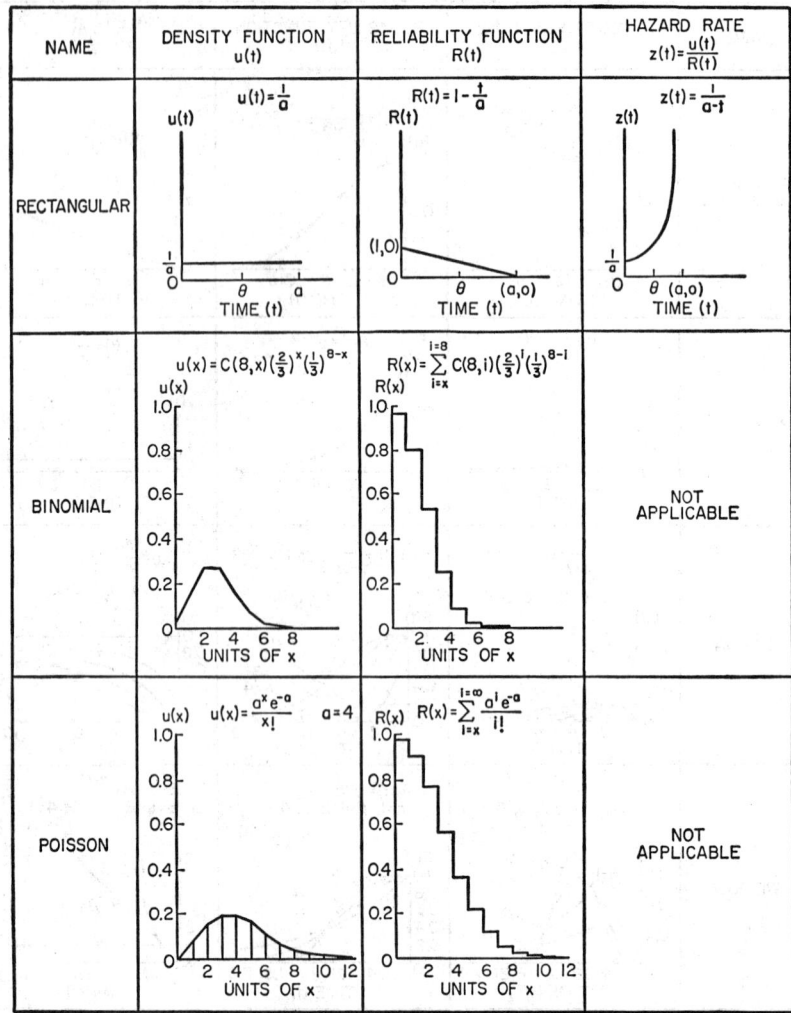

Fig. 6-1B. Density and reliability functions of the rectangular, binomial, and Poisson distributions; hazard rate of the rectangular distribution.

time-to-failure density. Further assume that each component part has a reliability of 99 percent for some time t. The system reliability for the same time t is

$$R(t) = 0.99^{400} = 0.018 = 1.8 \text{ percent}$$

Out of 1,000 such systems, 982 would fail to survive to time t.

For other types of distributions, equation (10) can be used except that $R(t)$ and $z(t)$ must be obtained from Figures 6-1A and 6-1B.

Mean Life and MTBF. The mean life θ is analogous to the life expectancy of an individual in a human population. It should not be confused with another term, mean time between failures (MTBF), which one regularly sees in the reliability literature. The term "mean life" is used to describe the case in which samples are not replaced upon failure, and it is merely the arithmetic mean of the time to failure of all the samples tested. Mean time between failures (MTBF), on the other hand, is used to describe the case in which components are replaced upon failure and is merely the ratio of total operating time to total number of failures. It can be shown that for the replacement case, MTBF represents exactly the same parameter as mean life. It is important to remember that only in a replacement model does MTBF have any meaning, and even more important is the fact that it can be appropriate only when the exponential law holds. Thus, ordinarily, for the nonreplacement case, either $R(t)$ or θ is given; whereas for the replacement case of an exponential distribution, MTBF can be used interchangeably with

$$R(t) = e^{-t/\theta} = e^{-t/MTBF} \qquad (12)$$

RELIABILITY PREDICTION AND ANALYSIS

General Discussion. Reliability prediction is the process whereby a numerical estimate is made of the ability, with respect to failure, of a design to perform its intended function. The measures used are $R(t)$, or the probability of survival without failure for a specified time; mean life θ or its reciprocal, failure rate λ, for the nonreplacement case; and mean time between failures (MTBF) for the case of replacement.

General Approach. The governing equation for reliability prediction was previously given as equation (10). It is

$$R(t) = R_1(t) \cdot R_2(t) \cdots R_n(t)$$

where $R(t)$ is the probability of survival of the device or system for time t, and $R_1(t)$, $R_2(t)$, $\ldots$, $R_n(t)$, is the individual element probability of survival for its required operating time. This assumes that the failure of any one element constitutes system failure, and that failures occur independently.

The simplicity of the approach utilizing the exponential distribution, as previously indicated, makes it extremely attractive. Fortunately, it is widely applicable for complex equipments and systems and is used almost exclusively for reliability prediction of electronic equipment. If complex equipments consist of many components, each having a different mean life and variance which are randomly distributed, then the system malfunction rate becomes essentially constant as failure parts are replaced. Thus, even though the failures might be wearout failures, the mixed population causes them to occur at random time intervals with a constant failure rate and exponential behavior. Figure 6-2 indicates this for a population of incandescent lamps in a factory. This has been verified for many equipments from electronic systems to bus motor overhaul rates.

Failure Modes of Parts. In prediction work, it is necessary to anticipate the frequency with which various modes or mechanisms of failure will occur. Catastrophic or chance part failures are generally defined as occurring when the parts become completely inoperative or exhibit a gross change in characteristics; they are characterized by a sudden breakdown, without preceding deterioration symptoms. This differs from wearout failures, which are indicated by slow deterioration with age.

Let us now plot a curve of the failure rate against the lifetime T of a very large sample of a homogeneous component population. The resulting failure rate graph is shown in Figure 6-3. At the time T_0, we place in operation a very large number of new components of one kind. This population will initially exhibit a high failure rate if it contains some proportion of substandard weak specimens. As these weak components fail one by one, the failure rate decreases comparatively rapidly during the so-called "burn-in" or debugging period, and stabilizes to an approximately constant value at time T_B, when the weak components have died out. After debugging, the component population reaches its lowest failure rate level, which is approximately constant. This is termed the useful life period, because it is in this period that the components can be utilized to the greatest advantage,

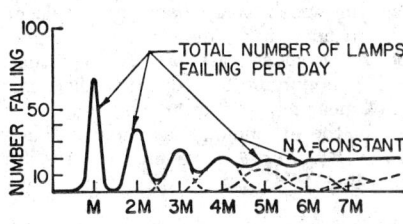

FIG. 6-2. Stabilization of failure frequency.

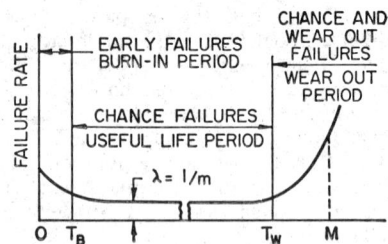

FIG. 6-3. Failure rate as a function of age.

and it is in this period that the exponential law is a good approximation. When the components reach the life T_w, wearout begins to make itself noticeable. The time at which the slope of the curve increases is of importance in prediction. Usually it occurs at a point far beyond the life expectancy of the equipment in which the part is used. Failure rate curves of complex equipments consisting of large numbers of heterogeneous components also have the same form. This assumes, of course, that preventive maintenance is used on short-lived parts exhibiting wearout failures, such as alternators and generators.

Another type of failure mode is incompatibility between system tolerance limits and individual or combined tolerance limits of parts within a system. In one case, failure may be considered due to not taking into account the initial (or time-zero) variability of part values in equipment design; in the other, it is due to not taking into account the variation of part values as a result of time and stress.

Prediction Procedure. The following steps outline a reliability prediction method for use as a guide, with variations dependent upon the specific application on hand.

1. Define the system and what constitutes a failure.
2. Draw a reliability block diagram showing, in sequence, the essential elements which must function for successful system operation, and also any redundant or alternate paths. Elements which are not essential to successful operation should not be included—for example, pilot lights.
3. List all parts within the block and, if possible, the stresses on each.
4. Select part reliability data. The required part data consist of information on catastrophic failures and on tolerance variations with respect to time under known operating and environmental conditions.
5. Determine appropriate failure rates for each part, part class, or module within the system.
6. Determine block and unit failure rates by adding the part failure rates.
7. Determine appropriate reliability index in terms of the reliability function $R(t)$, the mean time to failure θ, or the mean time between failures m.

Following are examples of various prediction techniques.

Example 1. "Ballpark" Technique, Electronic Equipment. This technique can be used to obtain a quick estimate of equipment reliability from a knowledge of the number of nonredundant active elements. Active elements are defined as vacuum tubes, transistors, relays, and rectifier diodes. The expected mean time between failures is plotted in Figure 6-4 as a function of the number of active elements. In Table 6-1, the various classes of equipment are subdivided into low, average, and high quality. The reliability function can be obtained from the mean time between failures by use of the formula

$$R(t) = e^{-t/m} \tag{13}$$

where t is the time period of interest and m is the mean time between failures.

Let us assume that one would like to estimate the reliability of an airborne

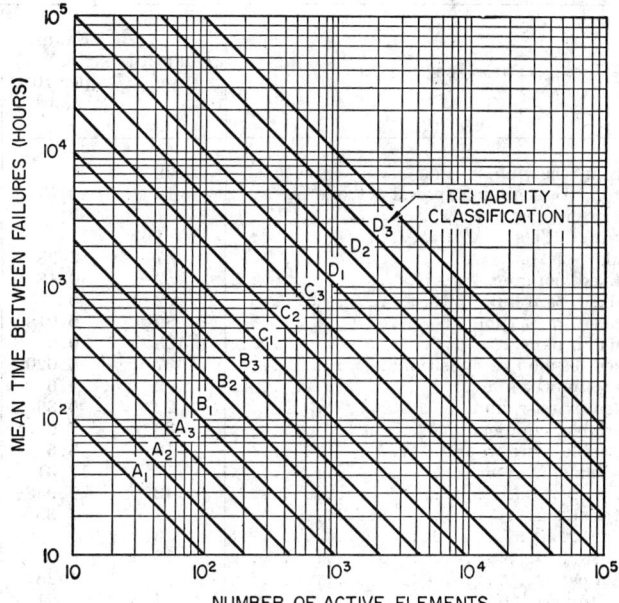

FIG. 6-4. Mean time between failures versus numbers of active elements for various reliability classes.

transistorized equipment of high quality, containing 500 active elements. Table 6-1 gives the classification B_3 for this case. Entering Figure 6-4 and finding the intersection of the diagonal B_3 line with the vertical line for 500 active elements, we read the estimated MTBF as 90 hours.

Example 2. Summation of Failure Rates. This technique is usually more accurate than that of example 1 and is based upon assumption of the exponential time-to-failure density function. This permits simple addition of average component failure

TABLE 6-1.   Electronic Equipment Reliability Classifications

Type of equipment	Reliability class		
	Low quality	Average quality	High quality
Airborne, vacuum tube..................	A_1	A_2	A_3
Airborne, transistorized................	B_1	B_2	B_3
Ground-based, vacuum tube.............	B_1	B_2	B_3
Ground-based, transistorized............	C_2	C_3	D_1
Mobile, vacuum tube...................	A_2	A_3	B_1
Mobile, transistorized..................	B_2	B_3	C_1
Missile, vacuum tube..................	A_1	A_2	A_3
Missile, transistorized..................	B_1	B_2	B_3
Ship-borne, vacuum tube...............	B_1	B_2	B_3
Ship-borne, transistorized..............	C_1	C_2	C_3
Space-borne, vacuum tube..............	B_1	B_2	B_3
Space-borne, transistorized.............	C_1	C_2	C_3

Part type	Quantity used	Failure rate per 10^6 hours	Total failures per 10^6 hours
Tubes, electron, receiving..........................	96	6	576.00
Tubes, electron, transmitting (power tetrode)........	12	40	480.00
Tubes, electron, magnetrons........................	1	200	200.00
Tubes, electron, CRT's............................	1	15	15.00
Crystals, diode....................................	7	2.98	20.86
Capacitors, fixed, ceramic, high K.................	59	0.18	10.62
Capacitors, fixed, tantalum foil....................	2	0.45	0.90
Capacitors, fixed, mica molded.....................	89	0.018	1.60
Capacitors, fixed, paper...........................	108	0.01	1.08
Resistors, fixed, carbon composition...............	467	0.0207	9.67
Resistors, fixed, power film.......................	2	1.6	3.20
Resistors, fixed, wire-wound.......................	22	0.39	8.58
Resistors, variable, composition....................	38	7.0	266.00
Resistors, variable, wire-wound....................	12	3.5	42.00
Connectors, coaxial................................	17	13.31	226.27
Inductors...	42	0.938	39.40
Meters, electrical.................................	1	1.36	1.36
Motors, blower....................................	3	630	1,890.00
Motors, synchro...................................	13	0.8	10.40
Relays, crystal can................................	4	21.28	85.12
Relays, contactor.................................	14	1.01	14.14
Switches, toggle...................................	24	0.57	13.68
Switches, rotary...................................	5	1.75	8.75
Transformers, power and filter.....................	31	0.0625	1.94
Summation..	...	....	3,926.57

$$\text{MTBF } (m) = \frac{10^6}{3,926.57} = 255 \text{ hr}$$

Probability of successful operation for 100 hours without failure:

$$R(100) = e^{-100/255} = e^{-0.392} = 0.676 = 67.6\%$$

FIG. 6-5.   Sample reliability calculation.

rates to arrive at the equipment or system failure rate, from which the MTBF or reliability function may be obtained. The mathematical basis for this approach is given in equation (11). In this technique, we merely count up the number of nonredundant components of each type, multiply this by the basic average failure rate for each type of component, and add these figures to obtain the equipment failure rate. The MTBF is then the reciprocal of the equipment failure rate, and the reliability function can be obtained from this by use of equation (12).

Figure 6-5 is an example of how this technique might be applied to a search radar using the average component failure rates from Tables 6-2 and 6-3. Tables 6-2 and 6-3 are supplied for use in reliability prediction; more extensive data and detailed procedures are contained in the source documents indicated.

Example 3. Prediction from Parts Rates with Stress Variations. Examples 1 and 2 have been based upon average part failure rates. It is a well-known fact that part failure rates vary significantly with applied stresses, sometimes by several orders of magnitude. For example, a 110-volt light bulb does not operate very long when subjected to 220 volts. It is this interaction between strength of the component and the stress level at which the component operates which determines the failure rate of a component in a given situation. Thus, at different stress

TABLE 6-2. Average Failure Rates for
Electronic Components by Part Category*

Part category	Failures per 10^6 hours†
Capacitors:	
Fixed:	
Aluminum, wet foil	1.42
Ceramic, high K	0.18
Ceramic, low K	0.322
Glass and porcelain enamel	0.032
Mica, button	0.63
Mica, dipped	0.0086
Mica, molded	0.018
Mylar—metalized	0.01
Mylar or Teflon	0.36
Paper	0.01
Plastic film	0.038
Polystyrene	0.41
Solid tantalum	0.024
Tantalum foil	0.45
Tantalum, wet slug	0.32
Variable:	
Air	0.13
Ceramic	5.68
Glass piston	0.39
Crystals (frequency control)	1.36
Diodes:	
(When used as mixers or detectors)	2.98
Logic switching	0.23
Power rectifier	1.10
Inductors (deflection, focus, r-f coils)	0.938
Magnetic amplifiers (<100 volts)	0.075
Microcircuits:	
Digital, average grade	0.84
Linear, average grade	2.52
Resistors:	
Fixed:	
Carbon composition	0.0207
Insulated fixed film	0.186
Power film	1.60
Precision film	0.015
Wire-wound	0.39
Variable:	
Composition	7.0
Wire-wound	3.5
Switches snap-action DPDT	4.50
Transformers:	
Audio	0.038
Power and filters	0.0625
Pulse (low level)	0.019
Transistors:	
Analog, silicon, npn	1.28
Digital, silicon, npn	0.39
Tubes:	
Special purpose:	
Backward wave oscillator	790
Cathode ray	15
Crossed field amplifier	600

* *RADC Reliability Notebook*, RADC-TR 67-108, vols. I and II,
AD-845-304 and AD-821-640, November, 1968.
† These failure rate figures represent average stress conditions.

TABLE 6-2. Average Failure Rates for Electronic
Components by Part Category* (Continued)

Part category	*Failures per 10^6 hours*†
Tubes (*continued*):	
Special purpose (*continued*):	
Kinescopes..................................	20
Klystron (<3,000 watts avg power).............	290
Klystron (>3,000 watts avg power).............	90
Magnetron.................................	200
Microwave switch tubes......................	100
Rectifiers, power...........................	15
Thyratrons................................	50
TR tube (dual).............................	3,700
Traveling wave tube.........................	120
Voltage regulator (glow discharge).............	5
Transmitting:	
Beam power...............................	100
Power tetrode.............................	40
Power triode..............................	15
Receiving:	
Miniature.................................	6
Nuvistors.................................	1.3
Subminiature.............................	4

* *RADC Reliability Notebook*, RADC-TR 67-108, vols. I and II,
AD-845-304 and AD-821-640, November, 1968.
† These failure rate figures represent average stress conditions.

levels, components necessarily assume different failure rates. Failure rate versus stress curves for the more significant stresses have been developed for a large number of components, particularly electronic components. Figure 6-6 illustrates some representative curves for capacitors, resistors, and tubes. Limited space does not permit the reproduction of all available failure rate versus stress curves in this chapter; the references, for example, *RADC Reliability Notebook*, contain rather extensive compilations of such curves. If one suspects that many components in an equipment are being overstressed, he should use the failure rate versus stress method to obtain a more accurate reliability prediction.

Application of this technique is the same as shown in example 2, except that the individual component failure rates are modified to reflect the anticipated or actual stress environment for each component. Thus it can be seen that the technique is much more laborious and time consuming, although it does provide the most accurate results.

Modification for Nonexponential Failure Densities (General Case). Although the exponential technique indicated in the previous section can be used in most applications with little error, it must be modified (1) if the system contains parts for which the density function of failure times cannot be approximated by an exponential over the time period of interest; or (2) if the parts which are the dominant factor in overall system unreliability do not follow an exponential density function of times to failure. Mechanical parts such as gears, motors, and bearings usually fall in this category.

In these cases, one cannot add the failure rates of all parts, because there are some parts whose failure rates vary significantly with time. The method used is to consider separately, within each block diagram, the portion of the block containing parts with constant failure rates, and the portion containing parts with varying failure rates. If the former portion contains x parts, then the reliability

TABLE 6-3. Average Failure Rates for
Nonelectronic Components by Part Category*

Part category	Failures per 10^6 hours†
Actuators, hydraulic	7.15
Alternators	775
Batteries (secondary)	1,429
Bearings	10–100
Connections:	
Crimped	0.0073
Soldered	0.0044
Welded	0.0022
Wire wrap	0.00000375
Connectors:	
Grade C (per mated pair)	0.324
Circular multipin	1.03
Coaxial	13.31
Counters, mechanical (tally register type)	4.54
Dynamotors	100
Fuses	8
Generators:	
AC	1,120
DC	480
Gyros:	
Integrating	410.26
Rate	163.15
Lamp (indicator)	4.5
Meters:	
Electrical	1.36
Mechanical	2.19
Motors:	
Blower and fan	630
Generator	38.793
Resolvers and synchros	0.80
Pumps:	
Fuel or booster	146.71
Hydraulic	1.68
Relays:	
Armature	12.54
Contactor	1.01
Crystal can	21.28
Reed	3.93
Thermal	13.07
Time delay	6.08
Switches:	
Push button	0.21
Rotary	1.75
Snap action	2.27
Toggle	0.57
Tanks (compressed gas)	506.33
Thermostats	4.08
Transducers (pressure)	860
Valves:	
Check	310
Control	1,740
Relief	714

* *Data Collection for Nonelectronic Reliability Handbook*, RADC-TR 68-114 (NEDCO I and NEDCO II), vols. I to III, AD-841-106,-AD 841-107, and AD-841-108, June, 1968.
† These failure rate figures represent average stress conditions.

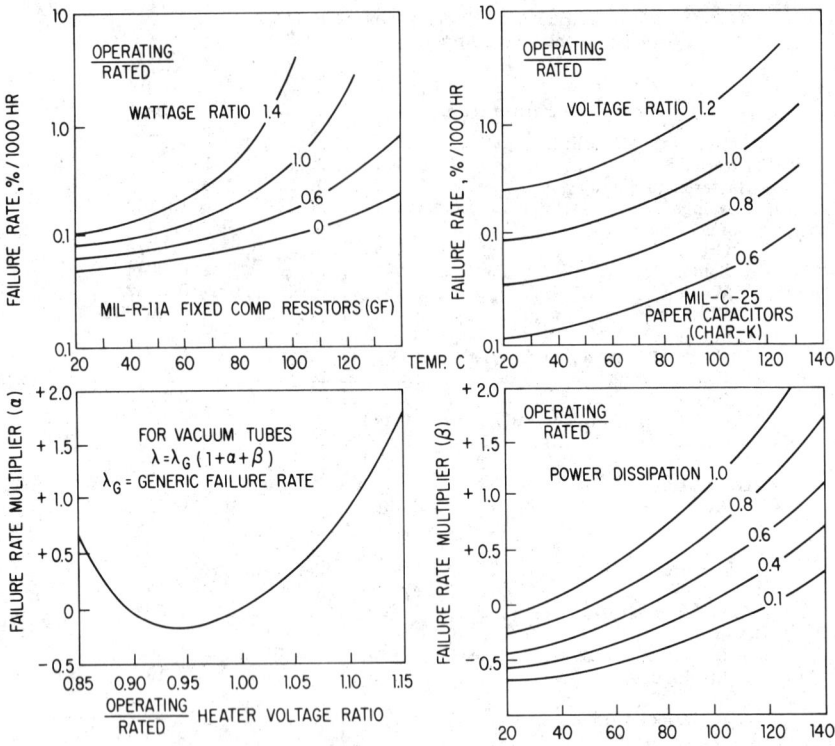

Fig. 6-6.   Failure rate versus stress curves.

of this portion is

$$R_1(t) = \exp\left[-\left(\sum_{i=1}^{x} \lambda_i\right) t\right] \tag{14}$$

The reliability of the second portion at time t is formed by using the appropriate failure density function for each part whose parameters have been determined through field experience or testing. If this portion contains B parts, then

$$R_2(t) = \prod_{i=1}^{B} R_i(t) \tag{15}$$

where

$$R_i(t) = \int_{t}^{\infty} u_i(t)\, dt \tag{16}$$

The reliability for the block diagram, under the assumption of independence between the two portions, is

$$R(t) = R_1(t) R_2(t) \tag{17}$$

By solving for various values of t, the block and system reliability function can be plotted. Often the shape of the curve will be very similar to that for an exponential, and the system mean life can be estimated graphically by finding the time interval over which the reliability is equal to 0.37.

For example, consider the failure rates of two elements X and Y that make up a system. Let X have a constant failure rate λ of $1{,}000 \times 10^{-6}$ failures per hour, and Y a hazard rate $z(t)$ that varies with time and is given by $(500 \times 10^{-6} + 0.01t)$. Thus,

$$R_y(t) = \exp\left[-\int_0^t (500 \times 10^{-6} + 0.01\tau)\, d\tau\right]$$

The reliability of a system composed of these two independent elements would be obtained through

$$R(t) = R_x(t)R_y(t)$$
$$= [\exp(-10^3 \times 10^{-6})t]\left\{\exp\left[-\int_0^t (500 \times 10^{-6} + 0.01\tau)\, d\tau\right]\right\}$$
$$= \exp\left[-\left(1{,}500 \times 10^{-6}t + \frac{0.01t^2}{2}\right)\right]$$

Evaluation of the above equation for several discrete points in time permits construction of the reliability function.

For those systems which have a reliability curve appreciably different from the exponential, the mean life is equal to

$$\theta_s = \int_0^\infty R(t)\, dt \tag{18}$$

RELIABILITY DESIGN

The preceding discussion dealt with the problem of predicting reliability when one has some knowledge of the numbers and types of basic components that make up the equipment being designed. Initially, this is not the case. One is usually presented with a quantitative reliability requirement for the desired equipment, such as a 200-hour MTBF or a 90 percent probability of failure-free operation for 20 hours.

Reliability Allocation. The first step in the design process is to allocate the equipment reliability requirement among the main elements that will constitute the equipment. From previous experience with similar types of equipment, one can usually estimate the complexity of each main element of the equipment. Once this has been done, it is possible to determine the average failure rate per part for each main element. These figures are then compared with available data on average failure rates to determine whether they are feasible. If not, then the designer must use one or any number of the following approaches (assuming that they are not mutually exclusive) to achieve the desired reliability:

1. Find more reliable component parts to use.
2. Simplify the design by using fewer component parts, if this is possible without degrading performance.
3. Apply component derating techniques to reduce the failure rates below the averages, as shown in Figure 6-6.
4. Use redundancy for those cases where 1, 2, and 3 do not provide acceptable failure rates.

For example, consider a subsystem consisting of three equipments: a power supply, a receiver, and a transmitter. An MTBF of 200 hours is desired, which means a maximum allowable failure rate of 0.005 failures per hour.

A failure in either one of the three equipments will cause a subsystem failure. The complexity of each equipment has been estimated as follows:

Power supply: 100 parts
Receiver: 255 parts
Transmitter: 560 parts

The weight assigned to each equipment is then

$$W_{\text{(power supply)}} = \frac{100}{100 + 255 + 560} = 0.11$$

$$W_{\text{(receiver)}} = \frac{255}{100 + 255 + 560} = 0.28$$

$$W_{\text{(transmitter)}} = \frac{560}{100 + 255 + 560} = 0.61$$

The failure rates per equipment can then be apportioned as follows:

Power supply: $0.11(0.005) = 0.00055$
Receiver: $0.28(0.005) = 0.00140$
Transmitter: $0.61(0.005) = 0.00305$
System: $= \overline{0.00500}$

Dividing by the number of estimated parts per equipment will then provide the average part failure rate required. For example, for the power supply

$$\lambda_{\text{avg}} = \frac{0.00055}{100} = 5.5 \times 10^{-6} \text{ failures/hr, or } 5.5 \text{ failures/}10^6 \text{ hr}$$

From Table 6-2, we note that power rectifiers have an average rate of 15 failures per 10^6 hours. Because the power supply will undoubtedly use one or more power rectifiers, it is obvious that the reliability allocated to the power supply cannot be met. Thus, the reliability must be reallocated or the alternatives previously mentioned must be explored.

Redundancy as a Design Technique. In reliability engineering, redundancy can be defined as the existence of more than one means for accomplishing a given task. In general, all means must fail before there is a system failure.

Thus, if we have a simple system of two parallel elements as shown in Figure 6-7, with A_1 having a probability of failure q_1 and A_2 having a probability of failure q_2, the probability of total system failure is

$$Q = q_1 q_2 \tag{19}$$

Hence the reliability or probability of no failure is

$$R = 1 - Q = 1 - q_1 q_2 \tag{20}$$

For example, assume that A_1 had a reliability r_1 of 0.9 and A_2 a reliability r_2 of 0.8. Then their unreliabilities q_1 and q_2 would be

$$q_1 = 1 - r_1 = 0.1$$
$$q_2 = 1 - r_2 = 0.2$$

and the probability of system failure would be

$$Q = (0.1)(0.2) = 0.02$$

Hence the system reliability would be

$$R = 1 - Q = 0.98$$

which is a higher reliability than either of the component parts acting singly. Parallel redundancy is therefore a design tool for increasing system reliability when all

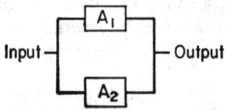

FIG. 6-7. Parallel redundancy.

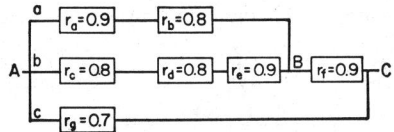

FIG. 6-8.   Series-parallel redundancy network.

other approaches have failed. In general, with m components in parallel, the overall probability of failure in time t is

$$Q(t) = q_1(t) \cdot q_2(t) \cdots q_m(t) \tag{21}$$

and the probability of operating without failure is

$$R(t) = 1 - Q(t) = 1 - q_1(t)q_2(t) \cdots q_m(t) \tag{22}$$

which can also be given as

$$R(t) = 1 - \{[1 - r_1(t)][1 - r_2(t)] \cdots [1 - r_m(t)]\} \tag{23}$$

because $q_i(t) = 1 - r_i(t)$ for each component. Where each of the component reliabilities is equal, the above equations reduce to

$$Q(t) = [q(t)]^m \tag{24}$$
$$R(t) = 1 - [q(t)]^m \tag{25}$$
$$= 1 - [1 - r(t)]^m \tag{26}$$

So far it has been assumed that parallel components do not interact and that they may be activated when required by ideal failure sensing and switching devices. Needless to say, the latter assumption in particular is difficult to meet in practice, and the potential benefits of redundancy are not realized fully. The reader is referred to the bibliography for detailed treatment of redundancy with sensing and switching devices which are not ideal.

Most cases of redundancy encountered will consist of various groupings of series and paralleled elements. Figure 6-8 typifies such a network. The basic formulas previously given can be used to solve for the overall network reliability R_{AC}. To gain familiarity with the application of the formulas given, the reader may wish to verify that $R_{AC} = 0.94$.

Redundancy in Time-dependent Situations. The previous discussion of reliability at a point in time did not consider the time-dependent reliability function. As a rule, the results given above can be extended to the time-dependent situation. For example, returning to Figure 6-7, assume that A_1 and A_2 have constant failure rates of λ_1 and λ_2 and exponential time-to-failure distributions. Then the overall reliability is given by

$$\begin{aligned} R(t) &= 1 - q_1(t)q_2(t) \\ &= 1 - (1 - e^{-\lambda_1 t})(1 - e^{-\lambda_2 t}) \\ &= e^{-\lambda_1 t} + e^{-\lambda_2 t} - e^{-(\lambda_1 + \lambda_2)t} \end{aligned} \tag{27}$$

because for each element $r(t) = e^{-\lambda t}$; hence $q(t) = 1 - e^{-\lambda t}$.

The basic redundancy formulas previously given can then be used to solve for the case of parallel components as well as any series-parallel combinations.

An important point to be remembered, however, is that the constant failure rates of the elements in a redundant configuration cannot be combined in the usual manner (addition) to obtain the system failure rate. This is so because the system failure rate is not constant but increases with time because the number of paths for successful operation decrease because of individual path failure. The system mean life, however, is found from equation (18), $\theta_s = \displaystyle\int_0^\infty R(t)\ dt$. For the example given in equation (27), the redundant system mean life would be

$$\theta_s = \int_0^\infty e^{-\lambda_1 t}\, dt + \int_0^\infty e^{-\lambda_2 t}\, dt - \int_0^\infty e^{-(\lambda_1 + \lambda_2) t}\, dt$$

$$= \frac{1}{\lambda_1} + \frac{1}{\lambda_2} - \frac{1}{\lambda_1 + \lambda_2} \quad \text{for } \lambda_1 \neq \lambda_2 \tag{28}$$

or $\qquad\qquad = \dfrac{3}{2\lambda} \quad \text{for } \lambda_1 = \lambda_2 = \lambda$

Thus it can be seen that the mean life of a redundant system containing two parallel elements of equal reliability is 1.5 times the mean life of a single element. For n equal components in parallel

$$\theta_p = \frac{1}{\lambda} + \frac{1}{2\lambda} + \cdots + \frac{1}{n\lambda} \tag{29}$$

$$R_p(t) = 1 - (1 - e^{-\lambda t})^n \tag{30}$$

Tolerances and Their Effects. No two parts made to the same specification are exactly alike. The variability in parts leads to a variability in systems composed of these parts. The designer can take this variability into account if armed with a knowledge of typical part variations, both initially and with time and stress. If they are not taken into account, unreliability due to parameter drift results. Even though none of the component parts may have failed catastrophically, the equipment or system performance is degraded below acceptable performance because the parts have drifted in value with time and stress. Electronic equipment is particularly susceptible to this mode of failure.

Many proved circuit analysis techniques suitable for degradation analysis are currently available for use. They are similar in that they involve a computer solution of a mathematical model describing circuit output variables in terms of several interrelated input parameters. Because of the large number and complexity of analysis techniques available, a detailed discussion of degradation analysis is not within the scope of this chapter. However, many of the available techniques can be identified as being developed from, or similar to, one of four general methods. These methods are (1) parameter variation, (2) worst case, (3) moment, and (4) Monte Carlo. The interested reader should consult the references, such as the *RADC Reliabilty Notebook,* for more details.

RELIABILITY TESTING

The ultimate purpose of a reliability test is to provide an estimate of the probability that the device in question will adequately perform its function for a specified period of time, when used in a specific environment. Hence reliability tests permit us to estimate statistically the reliability of a device.

To find the probability of an event, we must first compile statistically significant data of the event's occurrence. In the case of reliability measurements, statistical data on the failure-free performance of devices in the time domain are gathered and compiled. This is done by observing a number of the devices in operation, measuring the time of failure-free performance, and counting the number of failures, if any occur, during the observation period. When sufficient data on times to failure are taken, the mean time to failure or mean time between failures can be closely estimated.

The problem of reliability testing is further complicated when little or no knowledge is available of the distribution form taken by the times to failure of the components or device. In this case, the sample must be used to estimate the distribution form as well as the parameters of the distribution. For example, if a large enough sample exhibits a constant hazard rate, an exponential distribution may be safely assumed for times to failure, and the mean time to failure estimated. This chapter will not concern itself with the case of the unknown distribution but will treat the cases of known or assumed distribution forms. In fact, it will concentrate on the exponential distribution of times to failure (complex systems, systems with replacement) and the Weibull distribution (devices governed by wearout mechanisms) because of their widespread applicability. For treatment of other distributions, as well as methods for determining the distribution form, one should refer to the applicable publications in the bibliography.

Reliability testing, then, is concerned with the following:

1. Determining the distribution form of a statistical parameter (for example, mean time to failure) and estimating the value of that parameter from the samples tested. If the distribution form is known or assumed, then testing merely estimates the value of the parameters desired.
2. Determining what assurance there is that the actual value of the population parameter is within some specified value (such as ±20 percent of the measured parameter of the sample).
3. Determining whether a device or component has at least a certain mean life, and establishing how sure one can be that this is so.
4. Determining the sample size and test time necessary to do 1, 2, or 3.

Chance Failures and Complex Systems. For chance failures we are interested in a single parameter—mean time between failures m, mean life θ, or failure rate λ. When the parameter is known, the reliability for a given time can be calculated from equation (12). Because there are normally a limited number of samples to measure, or a limited time for measurement, the most that can be expected is a reasonably good estimate of the true value. These estimates will be indicated by $\hat{m}$, $\hat{\theta}$, and $\hat{\lambda}$.

When dealing with component parts, we usually try to get a large sample and test for a short time to be reasonably certain that no wearout failures will occur during the test period. Because components and equipment exhibit a high failure rate in early life (as shown in Figure 6-3), one must ensure that this portion of the life curve has been passed before attempting to test for reliability. For most equipments of any complexity, a maximum of 200 hours of debugging time is enough to ensure that one is on the constant failure rate portion of the life curve; the same time is also adequate for modern, good parts.

The sample size required (N) depends upon the available test time (t) and the confidence required in the failure rate measured. For example, a point estimate of failure rate ($\hat{\lambda}$) at about a 60 percent confidence level is given by

$$\hat{\lambda}_{60} = \frac{r}{Nt} \times 10^5 \qquad \text{percent/1,000 hr} \tag{31}$$

where $\hat{\lambda}$ = estimated failure rate in percent/1,000 hr
r = observed number of failures
N = sample size
t = total test time

(*Note:* For most practical tests, the value of r is so small compared with the sample size N that the reduction in sample size as the parts fail may be ignored.)

Equation (31) can be used to determine the test time required to estimate the failure rate to 60 percent confidence when testing N parts. For example, assume that we have 1,000 parts. How long should we test if the failure rate is no better than 0.1 percent per 1,000 hours? From equation (31)

$$t = \frac{r(10^5)}{N\hat{\lambda}_{60}}$$

If $r = 1$: $t = \dfrac{(1)(10^5)}{(10^3)(0.1)} = 1,000$ hr

If $r = 2$: $t = 2,000$ hr

If $r = 3$: $t = 3,000$ hr   and so on

TABLE 6-4. Failure Rate versus Test Item Hours Nt Data

Test item hours Nt ($\times 10^3$)	Failure rate $\hat{\lambda}$, percent/1,000 hr (greatest likelihood estimate), at approximately 60 % confidence for r failures				
	1	2	3	4	5
2,000	0.05	0.1	0.15	0.20	0.25
3,000	0.033	0.06	0.10	0.13	0.16
5,000	0.02	0.04	0.06	0.08	0.10
10,000	0.01	0.02	0.03	0.04	0.05
20,000	0.005	0.01	0.015	0.020	0.025
30,000	0.003	0.006	0.010	0.013	0.016
50,000	0.002	0.004	0.006	0.008	0.010
100,000	0.001	0.002	0.003	0.004	0.005

Table 6-4 is a handy reference to determine Nt quickly for a given failure rate and observed number of failures. For example, to demonstrate 0.01 percent per 1,000 hours failure rate with one observed failure, $Nt = 10 \times 10^6$. Thus, we would have to test 10,000 samples for 1,000 hours and experience no more than one failure.

In the case of equipment demonstration, the parameter of interest is usually mean time between failures, m. Because components are replaced as they fail,

$$\hat{m} = \frac{Nt}{r} \qquad (32)$$

For example, two radar sets which run for 1,000 hours each and experience a total of five failures would have an $\hat{m}$ of 400 hours. Keep in mind that we have not yet said anything about the accuracy of this estimate for this case; this comes later with the discussion of confidence intervals.

Confidence Limits—Chance Failures. We know that statistical estimates are more likely to be closer to the true value as the sample size increases. Only the impossible situation of having an infinitely large number of samples to test could give us 100 percent confidence or certainty that a measured value of m coincides with the true value. For any practical situation, therefore, we must establish confidence intervals or ranges of values between which we know, with a probability determined by the finite sample size, that the true value of m lies.

Confidence intervals around point estimates are defined in terms of a lower confidence limit L and an upper confidence limit U. If, for example, we calculate the confidence limits for a probability of, say, 95 percent, this means that in 95 percent of the cases we can be sure the true value of m will lie within the calculated limits, or in 5 percent of the cases it will lie outside these limits. If we want to be 99 percent sure that the true value lies within certain limits for a given sample size, we must widen the interval or test a larger number of samples if we wish to maintain the same interval. The problem, then, is reduced to one of either determining the interval within which m lies with a given probability for a given sample size, or determining the sample size required to assure us with a specified probability that m lies within a specified interval.

For the exponential distribution, we have only the single parameter, m, to measure in reliability testing. The task is to assign confidence limits to an estimate $\hat{m}$ of the true mean time between failures m when $\hat{m}$ was obtained from a test in which r failures were counted. One of the first things that must be established is the risk α that we shall accept that m is not within the specified confidence interval. In other words, let us say that we can accept m as being outside the interval in 5 percent of the cases, or there is a 5 percent probability that m will be outside the interval. Then, in this case, α is 0.05. Mathematically, this is

$$P(L \leq m \leq U) = 1 - \alpha \tag{33}$$

which states that we want to assure ourselves with a $(1 - \alpha)$ probability that the true mean time between failures lies between the specified upper and lower confidence limits.

It has been found that the ratio $2r(\hat{m}/m)$ has a chi-square distribution with $2r$ degrees of freedom when the test from which the estimate $\hat{m}$ was obtained was terminated as the rth failure occurred. The chi-square distribution is described in any standard statistical text. For the exponential case, equation (33) can be written

$$P\left(\frac{2r\hat{m}}{\chi^2_{\alpha/2;2r}} \leq m \leq \frac{2r\hat{m}}{\chi^2_{1-\alpha/2;2r}}\right) = 1 - \alpha \tag{34}$$

which establishes the limits, based upon the number of failures r and the estimated mean time between failures $\hat{m}$, between which there is a $(1 - \alpha)$ probability that the true value m lies. The terms $\chi^2_{\alpha/2;2r}$ and $\chi^2_{1-\alpha/2;2r}$ are the values of the chi-square distribution evaluated at these points; they can be obtained from standard chi-square tables.

"For example, a large electronic system showed $r = 20$ failures in 2,000 hours" of operation. The twentieth failure occurred at exactly $T = 2,000$ hours when the test was terminated. Between what limits can we be 95 percent sure that the true value of m lies?

First we estimate m for the data

$$\hat{m} = \frac{T}{r} = \frac{2,000}{20} = 100 \text{ hr}$$

Then we calculate the upper and lower confidence limits of equation (34) for

$$1 - \alpha = 0.95$$

that is, for the percentage points $\alpha/2 = 0.025$ and $1 - (\alpha/2) = 0.975$ and for $2r = 40$ degrees of freedom. From chi-square tables we find

$$\chi^2_{0.025;40} = 59.3 \qquad \chi^2_{0.975;40} = 24.4$$

Then the upper and lower confidence limits are

$$L = \frac{2r\hat{m}}{\chi^2_{0.025;40}} = \frac{(40)(100)}{59.3} = 67 \text{ hr}$$

$$U = \frac{2r\hat{m}}{\chi^2_{0.975;40}} = \frac{(40)(100)}{24.4} = 164 \text{ hr}$$

Thus we can state that there is a 95 percent probability that the true m of the system is between 67 and 164 hours.

For most cases, the graph of Figure 6-9 can be used instead of chi-square tables. It gives upper and lower percentage deviation from $\hat{m}$ for several confidence intervals of $100(1 - \alpha)$ percent for up to 1,000 failures observed.

In our example, the lower limit of 67 hours deviates from $\hat{m} = 100$ hours by -33 percent, whereas the upper limit of 164 hours deviates from m by 64 percent. These deviations can be obtained directly from the 95 percent confidence level curves for $r = 20$ failures.

In most reliability analyses, we are not so much concerned with the confidence interval as we are with the assurance, with some probability $(1 - \alpha)$, that m exceeds some specified minimum value. This is the case of the lower one-sided

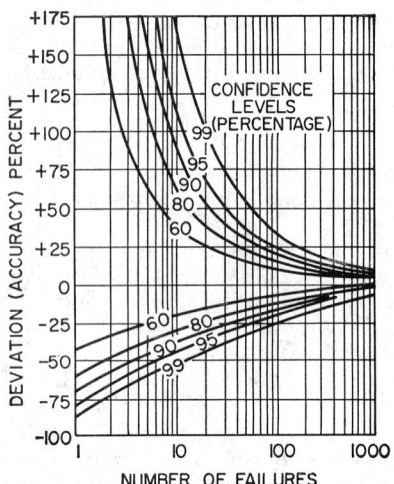

Fig. 6-9.   Confidence limits for measurement of mean time between failures.

confidence limit C_L which is given by

$$C_L = \frac{2r\hat{m}}{\chi^2_{\alpha;2r}} \tag{35}$$

We must then prove in a test that, with a specified probability $(1 - \alpha)$,

$$\hat{m} \geq C_L \frac{\chi^2_{\alpha;2r}}{2r} \tag{36}$$

For example, assume that we should like to assure ourselves with a 95 percent probability that the system given in the previous example had a mean time between failures of at least 70 hours.   Thus $\alpha = 1 - 0.95 = 0.05$ and

$$\chi^2_{\alpha;2r} = \chi^2_{0.05;40} = 55.8$$

From equation (35) we obtain the lower confidence limit

$$C_L = \frac{(40)(100)}{55.8} = 72 \text{ hr}$$

which satisfies our requirement.

Figure 6-9 can also be used to obtain one-sided confidence levels rapidly:

Two-sided confidence level, percent	One-sided confidence level, percent
60	80
80	90
90	95
95	97.5
99	99.5

For a lower one-sided 95 percent requirement in our example, we would use the 90 percent curve in the lower portion of the graph which, for $r = 20$, shows a deviation of -28 percent from $\hat{m}$ and therefore

$$C_L = \hat{m} - 0.28\hat{m} = 100 - 28 = 72 \text{ hr}$$

which is the same value as that obtained from chi-square tables.

Wearout Failures—Weibull Distribution. When failures are predominantly due to wearout rather than chance, the exponential distribution cannot be used for reliability testing and acceptance plans. One must gather enough failure data to ascertain the underlying distribution of time to failure and design the test using this distribution. Although the gamma and normal distributions have been used as models for reliability testing for these cases, the Weibull distribution, because of its flexibility, is the most popular model used in testing the reliability of families of component parts. Many mechanical and electromechanical components, such as pumps, relays, bearings, or switches, have been verified to have Weibull time to failure distributions.

The unreliability function for a Weibull distribution is given by

$$F(t) = 1 - \exp\left(-\frac{t^\beta}{\theta}\right) \tag{37}$$

Hence, the reliability function, or the probability of survival without failure to time t, is

$$R(t) = 1 - F(t) = \exp\left(-\frac{t^\beta}{\theta}\right) \tag{38}$$

and the hazard or instantaneous failure rate is

$$h(t) = \frac{\beta}{\theta} t^{\beta-1} \tag{39}$$

where β is the shape parameter and θ is the scale parameter.

As was shown in Figure 6-1A, the Weibull can be fitted to many failure distributions such as exponential, normal, lognormal, and the like. Hence the reason for its popularity. So as not to confuse the reader, the θ in the above equations is the reciprocal of the α, the Weibull scale parameter shown in Figure 6-1A.

Thus, what is done is to acquire failure data to estimate θ and β, from which the reliability, instantaneous failure rate, and other parameters of interest may be computed.

Although it is possible to solve mathematically for point estimates of β and θ, a much simpler graphical method using "Weibull probability paper" has been developed by Goode and Kao.[1] Equation (37) can be converted to the form

$$\ln \ln = \frac{1}{1 - F(t)} = \beta \ln t - \ln \theta \tag{40}$$

On ln versus ln-ln graph paper, equation (40) represents a straight line with slope β and intercept $-\ln \theta$. A sample example of such paper is shown in Figure 6-10. Note that failure age (time to failure) is plotted horizontally along the bottom of the figure and ln (failure age) is plotted horizontally along the top; similarly, percent failure is plotted vertically in the left and ln ln[100/(100 − percent failure)] is plotted vertically on the right.

Before being able to use the paper, one more bit of information is needed, that concerning the fitting technique known as median ranks. Briefly, median ranks are used to establish an estimate of the cumulative percent failed (the left ordinate of the Weibull chart). This value is ordinarily obtained from tables,[2] an example of which is shown as Table 6-5. Such tables are available for sample sizes up to forty.

[1] J. H. K. Kao, "A Summary of Some New Techniques on Failure Analysis," *Proceedings of the Sixth National Symposium on Reliability and Quality Control in Electronics*, 1960, pp. 190–201.
[2] L. G. Johnson, "The Median Ranks of Sample Values in Their Population with an Application to Certain Fatigue Studies," *Industrial Mathematics*, vol. 2, 1951, pp. 1–9.

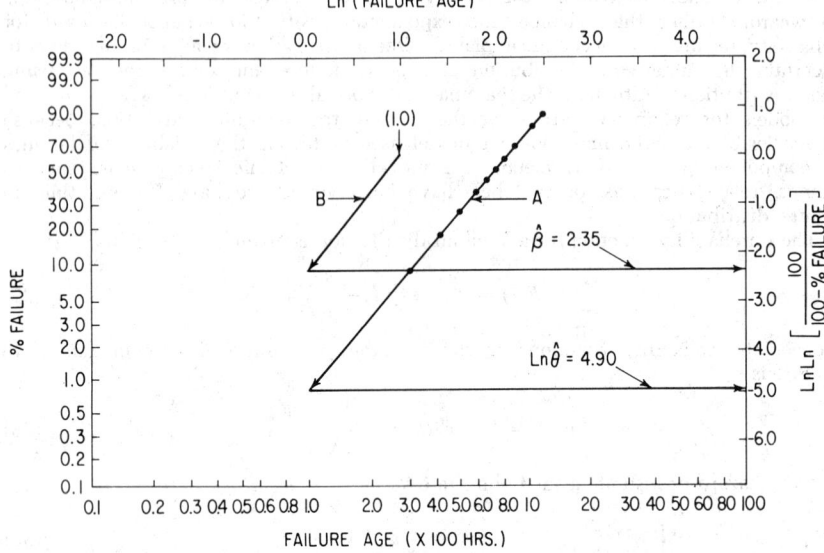

Fig. 6-10.   Graphical estimation of Weibull parameters.

For sample sizes outside of the tables, the percent failures may be approximated by

$$\text{Percent failures} = \frac{i}{N+1} \tag{41}$$

where the ith failure is that which occurs when they are tabulated in ascending order of time to failure, and N is the sample size.

As an example of the use of the Weibull chart, suppose that in a life test we have ten samples that fail at 300, 400, 500, 600, 675, 750, 825, 900, 1,050, and 1,200 hours. Assuming that they follow a Weibull distribution of time to failure, find $\hat{\beta}$, $\hat{\theta}$, $R(t)$, and $h(t)$. The steps are as follows:

1. Rank the times to failure in ascending order (as shown).
2. For $N = 10$, obtain from Table 6-5 the percent failure point for each of

TABLE 6-5.   Median Ranks

N	Percent failure points for sample size N						
	5	6	7	8	9	10	11
1	13	11	9.5	8.5	7.5	6.5	6
2	31	27	23	20	18	16.5	15
3	50	42	32	29	29	26	24
4	69	58	50	44	39	36	32
5	87	73	64	56	50	45	41
6	...	89	77	68	61	55	50
7	...	...	91	80	71	64	59
8	...	...	....	92	82	74	68
9	...	...	....	....	93	84	76
10	...	...	....	....	....	93	85
11	...	...	....	....	....	....	94

the samples. For example, the fourth failure time (600 hours) will be at the 36 percent failure point. See plot on Figure 6-10.

3. Plot failure age (lower abscissa) versus percent failures (left ordinate), for example, 600 hours versus 0.36.

4. Fit a straight line to the plotted points (line A, Figure 6-10).

5. Extend the straight line until it intersects the 0.0 line of the upper abscissa; read the intersection point on the right ordinate (−4.90). This is (−ln $\hat{\theta}$),

$$\hat{\theta} = e^{4.9} = 134.3$$

6. Draw a line from the point (1.0)—1 is the value of the upper abscissa, 0 the value of the right ordinate—parallel to the line through the sample points. This is line B. Extend this line until it intersects the 0.0 line of the upper abscissa.

7. Read the intersection point on the right ordinate. This is −$\hat{\beta}$, ∴ $\hat{\beta}$ = 2.35.

8. From equations (38) and (39) compute reliability $R(t)$ and instantaneous failure rate $h(t)$ for any desired time of interest.

$$R(t) = \exp\left(-\frac{t^{2.35}}{134.3}\right)$$

$$h(t) = \frac{2.35}{134.3}\, t^{1.35}$$

Actually, $R(t)$ can be read directly from Figure 6-10. For example, to find the probability of surviving 450 hours without failure, at the 450-hour point on the lower abscissa, project vertically to line A and horizontally to the left ordinate. The reliability is 100 minus the percent just read. For example,

$$R(450) = 100 - 20$$
$$= 80\%$$

Other parameters of interest, such as the population mean and standard deviation, initial failure rate, and characteristic life, can be obtained from the chart. For a discussion of these parameters and methods for determining the confidence intervals, the interested reader should consult the previously mentioned references.

Two pamphlets of acceptance sampling tables for reliability tests based on the Weibull distribution have been issued by the U.S. Department of Defense, Office of the Assistant Secretary of Defense.[3]

The acceptance sampling plans in these pamphlets assume that the shape parameter β is known. Both pamphlets contain tables to aid in the selection of reliability test plans for β = ⅓, ½, 1, 1⅔, 2, 2½, 3⅓, 4, and 5.

Sequential Reliability Tests. The purpose of a reliability test is to establish in the shortest possible time and at minimum cost whether or not the reliability of a type of component or of a system is equal to or better than a specified minimum. The sequential probability ratio tests have been devised for this. This method enables us to make one of three decisions as each failure occurs: (1) accept, (2) reject, (3) continue testing. Essentially, what we do is establish two values of mean life m_1 and m_2. m_1 is some minimum acceptable value, and m_2 (greater than m_1) is some chosen upper value. After r failures have occurred, we then compute the probability of r failures occurring for a mean life of m_1 versus the probability of r failures occurring for a mean life of m_2. For example, in the case of the Poisson distribution, the probability of r failures in time t for an equip-

[3] "Quality Control and Reliability Technical Report TR 3, Sampling Procedures and Tables for Life and Reliability Testing Based on the Weibull Distribution (Mean Life Criterion)"; "Quality Control and Reliability Technical Report TR 4, Sampling Procedures and Tables for Life and Reliability Testing Based on the Weibull Distribution (Hazard Rate Criterion)."

ment whose times to failure are exponentially distributed is

$$P_r = \frac{(t/m)^r e^{-t/m}}{r!} \tag{42}$$

where m is the chosen mean life. Having computed this for m_1 and m_2, we take the ratio Pm_1/Pm_2 and compare this against two selected positive constants A and B which are based on previously agreed upon risks—the consumer's risk β or probability of accepting equipment with $m = m_1$ and the producer's risk α or probability of rejecting equipment with $m = m_2$. These constants are given by

$$A = \frac{1 - \beta}{\alpha} \tag{43}$$

$$B = \frac{\beta}{1 - \alpha} \tag{44}$$

As each failure occurs, the ratio Pm_1/Pm_2 is computed, and the following decision rules are applied:

Accept if:
$$\frac{Pm_1}{Pm_2} \leqq B \tag{45}$$

Reject if:
$$\frac{Pm_1}{Pm_2} \geqq A \tag{46}$$

Continue testing if:
$$B < \frac{Pm_1}{Pm_2} < A \tag{47}$$

For a complex equipment in which we assume m to be exponentially distributed, the ratio Pm_1/Pm_2 is given by

$$p(r) = \frac{Pm_1}{Pm_2} = \left(\frac{m_2}{m_1}\right)^r \exp\left[-\left(\frac{1}{m_1} - \frac{1}{m_2}\right)t\right] \tag{48}$$

Thus, once m_1, m_2, α, and β have been specified or agreed upon, equations (45) through (48) can be used to arrive at a decision after r failures.

For example, let us assume that $\alpha = \beta = 10$ percent, so that $A = 9$ and $B = 0.111$, and that $m_1 = 100$ hours and we choose $m_2 = 200$ hours. Then

$$p(r) = 2^r e^{-t/200}$$

If no failure occurs up to 200 hours, $p(r) = 2^0 e^{-1} = 0.368$. The value still lies between A and B; so no decision can be made. If no failure occurred to 440 hours, however, then $p(r) = 0.111 = B$, and we would accept the equipment. Remember that t is the sum of the operating times of all the equipments under test and r is the total number of failures. If five equipments were under test, and no failures occurred, the accept decision could be made in 88 hours of test time.

To ease the problem of computation, a graphical technique has been developed which enables one to determine instantaneously whether to accept, reject, or continue testing. It can be shown for the exponential case that equation (47) is of the form

$$a + bt < r < c + bt \tag{49}$$

where the left and right sides are equations of two parallel straight lines with equal slopes b. When these two lines are plotted on paper with t as the abscissa and r as the ordinate, the constants a and c are the intercepts of the lines with

criteria based upon the number of failures and the test time in multiples of m_2. θ_0 on the chart is the same as m_2. MIL-STD-781A contains a number of sequential test plans for differing values of α, β, and discrimination ratio.

BIBLIOGRAPHY

Arinc Research Corporation, *Reliability Engineering*, Prentice-Hall, Inc., Englewood Cliffs, N.J., 1964.

Barlow, R. E., and F. Proschan, *Mathematical Theory of Reliability*, John Wiley & Sons, Inc., New York, 1965.

Bazovsky, I., *Reliability Theory and Practice*, Prentice-Hall, Inc., Englewood Cliffs, N.J., 1961.

Data Collection for Nonelectronic Reliability Handbook, NEDCO I and NEDCO II, vols. I to III, AD-841-106, AD-841-107, and AD-841-108, June, 1968.

Hahn, G. J., and S. S. Shapiro, *Statistical Methods in Engineering*, John Wiley & Sons, Inc., New York, 1967.

Ireson, W. G. (ed.), *Reliability Handbook*, McGraw-Hill Book Company, New York, 1966.

Lloyd, D. R., and M. Lipow, *Reliability: Management, Methods, and Mathematics*, Prentice-Hall, Inc., Englewood Cliffs, N.J., 1962.

MIL-STD-781A, *Reliability Tests, Exponential Distribution*, U.S. Department of Defense, Washington, D.C., 1965.

Pieruschka, E., *Principles of Reliability*, Prentice-Hall, Inc., Englewood Cliffs, N.J., 1963.

RADC Reliability Notebook, RADC TR-67-108, vols. I and II, AD-845-304 and AD-821-640, November, 1968.

Chapter **7**

Cost Control
and Profit Prediction

A. J. BERGFELD

President, Case and Company, Inc., New York, New York

D. W. SCHWEPPE

Principal, Case and Company, Inc., New York, New York

A. D. KIDD

Principal, Case and Company, Inc., New York, New York

Traditionally, the underlying principal of cost control has been that of management by exception. Standard costs, flexible budgets, and variance analysis were designed to provide management with information on how past actual performance departed from normal or standard. Exception reports were used to devise corrective measures to prevent recurrence of excessive costs.

An increasing number of managements have found that such after-the-fact information is no longer adequate. Change is the only thing that is recurrent; the exception is not going to happen again, and a lesson learned is an opportunity lost and wasted. These managements have further learned that it is practical to apply a system of control that corrects exceptions or variances before they occur. This system is called management by prediction or predictive management.

Predictive management considers that effective management is a learning process analogous to the learning process of scientific experimentation and making full use of management science techniques. The learning process begins with the recognition and conceptual definition of a management problem, drawing on past experience and available data. Here, the increasing ability of management scientists to describe business phenomena by mathematical models permits the form of deductive reasoning that has been so effective in science. Tools such as statistical decision theory, mathematical programming, and simulation models are useful in determining what to do about a problem. Finally, management learns from results how the decision should be modified and adds this knowledge to the store of experience on which future decisions may draw.

By looking at his actions as experiments, the predictive manager develops an increasing understanding of the interrelationships among the multiple variables in management problems. Only from knowledge of the causal relationships can he derive the ability to direct the enterprise toward its goals. It is purposeful direction that constitutes control.

Before control can exist, it is necessary to have a basis for sound measurement. Certainly it is not enough to say that the payroll is less this month than it was last month. If one can say that the payroll is less than it was last month, for the same number of units produced, a basis for comparison exists. If this information can be used to predict what the payroll will be next month for the same or different production, then the start of a system of predictive management exists.

The concepts presented are based on the economic principles of direct costs. Measurements used are those common to all cost accounting systems, including full absorption standard cost systems. Application of these concepts requires no change in current cost accounting and inventory valuation policies. The concepts are for management decision making and for predictions of the profit consequences of these decisions.

THE PRINCIPLES OF PREDICTIVE MANAGEMENT

Predictive management is a concept which the decision making unit of an enterprise employs to establish what is going to be the result of actions taken, how these results will be accomplished, what alternatives are available, and the conditions under which a decision to implement an alternative should be made.

The concept is a reasoned implementation of the objectives of the enterprise and the results expected. To support the predictive management concept, it is necessary to have an array of management measurements, techniques, methods, plans, and objectives. Without this support, it is difficult to have a reasoned prediction of results.

It is part of the underlying structure of predictive management that its predictions are based on reason. The structure is essential to the concept so that it may progress in logical sequence and contain within the structure the means for its correction. Such correction may be required because conditions and circumstances develop in a different pattern or with different effect on the enterprise than was originally expected. Thus, if there is a reasoned pattern to the prediction, the correction can follow logically and with minimum impact on the enterprise. If the prediction is not a reasoned one, then the results will be random and their correction will probably be through new methods, plans, and programs designed to fit only the requirements of the moment. These, in turn, will probably lead again to the need for further correction and different programs of instant availability.

In implementing any system for management control, it is useful to start from statements of principles which define what the system is trying to accomplish and which govern the details of the system. These principles follow.

A Business Enterprise Requires Objectives Stated in Terms of Profit and Time. A business enterprise is a collection of individuals working with finite resources, striving to deliver to the economy more in the form of goods and services than they take from the economy in the form of cost inputs. The difference between what is put into the economy and what is taken from it is profit. The objectives of the enterprise are most appropriately stated as profit objectives.

The normal segregation of cost producing (productive and administrative) functions and revenue producing (sales) functions within an enterprise may improperly allow cost control to become an end in itself. But costs are only one term of the profit equation, with volume being the other. Under a predictive management system, the interplay and interrelationship among the kinds and sources of revenue, cost, and volume are understood and integrated into the system to provide maximum effectiveness in meeting the profit objectives of the enterprise. The dimension of time is necessary to make the amount of profit meaningful and tangible.

The Management System Must Be Responsive to Both the Unique and the Repetitive Opportunity. A business enterprise constantly needs adjustment to adapt to changes in its environment. If this were not so, all decisions could be made once to meet the permanent, unvarying factors affecting its operations. There would be no need for management or a management system. Because this situation is not real, management must be prepared to deal with situations which are constantly changing in some degree.

Plans Are Necessary to Carry Out Objectives. An objective, without the means for accomplishing it, is only a wish. Plans are the means of assuring that objectives are realistic and attainable within the context of the current environment. The process of converting objectives into a plan forces critical examination of the underlying assumptions used in setting the objectives and forces revision of objectives to accommodate prevailing conditions. Plans also provide a means for communicating the overall objectives of the enterprise to individual functions in terms that are meaningful.

Effective Prediction and Control Involves All Levels of Management in the Preparation of Plans. The preparation of plans is not the exclusive prerogative of top management. Most behavioral scientists agree that when middle and lower level supervision is encouraged to participate directly in setting goals and formulating plans, the results are dramatic. A sense of pride in accomplishment is engendered, rather than one of imposed requirement. A line of communication is opened, because the plans are expressed in a language which is understood by all levels. There will be agreement on the practicability of the plans. There will be a high degree of commitment to the goals of the enterprise, which directly influences the quality of subsequent management performance.

Plans Are Made with Direct Relation to the Way Work Is Accomplished. Nothing is more frustrating to the operating levels of management than a plan which is at odds with the way the work must be done for practical reasons. The plan at once becomes only a piece of paper. It will be ignored.

What is wanted are plans that work, plans that make the job easier because they solve real problems. These plans have time measurements which can be met; they use methods that are well conceived and better than any other known available methods; the operations follow one another logically without backtracking or gaps; the tooling is a help, not a hindrance; and above all, the costs will be incurred in the manner, amount, and sequence which the plan calls for.

Management Decisions Carry Out Plans in the Most Effective Manner. If plans are necessary to carry out the profit objectives of the enterprise, their achievement requires a series of effective decisions by managers. Plans must be converted to action, and effective action requires decisions.

Decisions Are Based on Sound Measurements of the Profit Consequences of Alternatives. Plans are made to provide for different sets of conditions. Each plan presumes its own set of conditions; as many plans may be made as there are sets. New sets are constantly generated as time changes some assumptions into reality and others vanish. As new conditions arise, they must be considered. Their profit consequences must be evaluated, and a decision must be made which optimizes the overall effect on the enterprise.

The Responsibility Limits of Managers Are Explicit and Defined. The principle of responsibility limits for managers has been well understood since formal organization has been studied. These limits need to be defined and understood so that there can be no question about them and how they relate to the objectives and plans. When a manager clearly comprehends his responsibilities, he can make decisions and control people. The more these responsibilities are defined and the limits established, the less ambiguity and uncertainty will surround the manager and detract from his achievement of the stated objectives.

When applied to the subject of cost control, these principles place very specific requirements on management information. These requirements are that management information must be relevant, available before the fact, and based on measurements supported by facts or well-reasoned judgments. Management measurements reflect the true behavior of profits and costs as volume and other variables change. They should be based on standard conditions, materials, and methods and be available to those who manage the activity measured. The predictive system should be compatible with the cost reporting system so that the values of experience can be related to the predictions made.

THE RISK AND RECOVERY EQUATION

The risk and recovery equation is a working model of the economics of an enterprise. It displays the concept that there is no profit (pretax profit is always referred to in this chapter) until all the costs which are risked by the fact of being in business have been fully recovered. These are all time-oriented costs, commonly called the fixed, planned, and programmed costs. The fixed costs are oriented to the basic costs of being in business: depreciation and taxes, for example. Planned costs are decision and profit oriented: indirect labor and maintenance, for example. Programmed costs are policy and objective oriented: research and advertising, for example.

There is no automatic portion of an income producing unit of product which can be tagged "profit" when it is sold. Rather, a profit contribution is generated. This is the difference or margin between the income produced and the variable costs of producing this additional unit. It is the result of mathematical analysis of the costs of the enterprise. For this purpose, costs are separated into two categories:

1. The fixed, planned, and programmed costs, which tend to remain constant over the planning period
2. The variable costs, which are commonly thought of as being the direct labor used to produce a product (or service) and the material associated with it

Variable costs are treated as if they varied uniformly (a straight line). Actually, they do not, because of product mix and because they change at different rates at different volumes of the same product. Managers do not use the equation to forecast what cost will be, but to structure the risks and expectations of the enterprise. Over the range of variation normally anticipated, the linear assumption is adequate for planning purposes. Managers look at costs to manage them. They see costs in quantities determined by a planning period. Although the period is commonly a day, month, or year, it may also be the time to make a crop, complete a voyage, or cover the entire life cycle of a product.

The risk and recovery equation, illustrated by Figure 7-1, provides more information about an enterprise than its fixed, planned, and programmed costs; its profit; or its profit contribution rate. Once a sales level is assumed, the fixed, planned, and programmed costs calculated, and the profit (or loss) established, the equation of the enterprise is determined.

With the equation, it is now possible to calculate the margin (fixed, planned, and programmed costs plus profit), the PV ratio (margin divided by revenue), the break-even sales (fixed, planned, and programmed costs divided by PV ratio), and the margin of safety (total sales less break-even sales, divided by total sales).

Using the Risk and Recovery Equation. The suitability of the risk and recovery approach to the equation of the enterprise is apparent in its use as a tool of

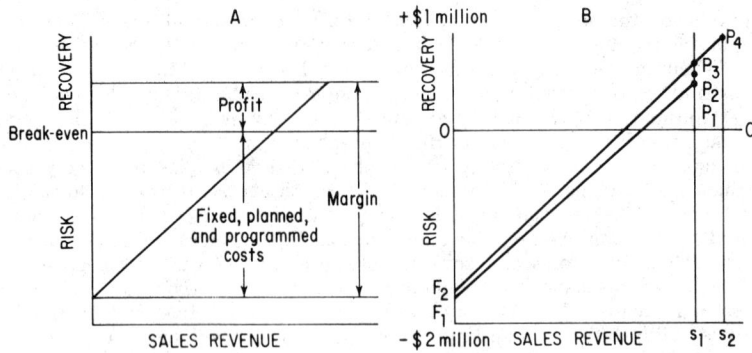

FIG. 7-1.   Graphical representation of the risk and recovery equation.

predictive management.   The objective is to establish the effect of changes in conditions.   On Figure 7-1B, it can be seen that a decrease in fixed, planned, and programmed costs (F_1 to F_2) appears directly as an increase in profit (P_1 to P_2), decreases the break-even point, and improves the margin of safety.   An increased PV ratio acts to increase the margin on all sales and increase profit (P_2 to P_3).   Finally, an increase in volume (S_1 to S_2) increases profit by an amount (P_3 to P_4) equal to the PV ratio times the increase in sales.

These relationships are fundamental to predictive management.   Assume that it is suggested that advertising (a programmed cost) be increased to obtain a higher sales volume.   With the equation, it is easy to calculate how much of a sales increase is necessary to recover the risk.   The same applies if a new facility is proposed which would increase fixed depreciation costs.

If a price increase is proposed and the change in volume for various price changes can be estimated, a pattern is determined which suggests the best price; there may be a combination of price and volume which, taken together, yield a higher total margin and therefore a greater profit.   A cost reduction similarly increases margin and also opens the opportunity for a price reduction, a volume increase, and a greater total margin and profit improvement than would be indicated by the cost reduction alone.

One of the anomalies of business is that management places greater emphasis on cost control when facilities are only partially utilized.   The risk and recovery equation shows that cost control is most important when capacity is fully utilized.   Productivity savings then not only appear directly as profit, but also increase the effective capacity of the facility.   For example, in Figure 7-1B, if the PV ratio (margin) increase results from an improvement in productivity, there will be an immediate profit increase from P_2 to P_3.   Less productive capacity will be required by the output S_1, and the freed capacity may be applied to producing the additional volume ($S_2 - S_1$).   This volume generates additional margin, increasing profit from P_3 to P_4.

All these relationships can be calculated with the mathematics of the equations.   The graphic representation aids in demonstrating what happens.

THE ROLE OF ACCOUNTING COST SYSTEMS IN PREDICTIVE MANAGEMENT

A cost system exists to serve management in three ways:
1. To supply information for management decisions.   These are the decisions which affect the risk and recovery equation of the enterprise.
2. To supply information for predictive cost control.
3. To provide the values of the work-in-process and the finished goods inventories.

All cost systems are not organized to provide information in these three areas. It is important to decide in advance what information management needs to have to do its job well, and then plan a cost system to provide it. The system should have an adequate provision for flexibility. The ability to manage change is as important as the ability to manage the present, and it is likely to be considerably more difficult.

Cost accounting systems are classified into three types: job order cost, process cost, and standard cost. Because this chapter deals with predictive cost control, it will not discuss the books of original entry, but rather the cost ledger, the data input and output, and the manner in which these data are treated.

It is generally possible to produce almost any information relating to costs outside the books of account. The problem is to determine which data are to be formalized, accumulated regularly, and presented to management for predictive control. It is important to recognize that historical data can be analyzed but not changed, but a results-oriented management seeks information that is futuristic.

So that the principles of predictive management may be better applied, the essential characteristics of the three types of cost systems are described briefly below. These descriptions are not intended to be full explanations. More detailed information may be found in references listed in the bibliography.

The Job Order Cost System. The job order cost system is designed to gather the actual costs which are incurred in processing an order. All costs directly applicable to the order are charged to it. Overhead costs are distributed on the basis of some characteristic common to all orders, frequently direct labor hours or dollars. Work-in-process values are the sum of all costs shown on the uncompleted job order cards.

The Process Cost System. The process cost system is designed to collect costs where all the products passing through a given department are identical. In this system, production costs are gathered by cost center, and other costs are allocated to these centers. The number of production units passing through each center is determined. Unit costs are determined by dividing the number of units into the production and overhead costs. The total product cost is the sum of the costs incurred in each department it passes through.

The Standard Cost System. The standard cost system is the beginning of predictive cost control. The system establishes what costs should be.

Of itself, a standard cost system cannot be distinguished from either a job order or a process cost system. The distinguishing feature is that there is a predicted unit cost which has been established in advance by using predetermined methods, material, and services (with their established cost rates) developed by analytical means.

Further, work-in-process and finished inventories are valued at the established standard cost rates. When an item is shipped, finished goods inventory is credited with the total standard cost of the item, and the cost of sales is debited. Because all variances which occur while producing this item are charged to profit and loss as they occur, only the standard cost of the item must be carried to inventory. This technique of charging inventory with total standard production by cost center and crediting inventory with total shipments at standard cost for each item shipped is the key to standard costs.

Direct versus Absorption Cost Accounting. Within the framework of the several cost systems, a differentiation is made between the concept that all manufacturing costs must be assigned to each unit of product (absorption costing) and the concept that only the variable manufacturing costs must be assigned (direct costing). Absorption costing is the traditional method, but for the principles of predictive management to work, direct costing techniques are essential.

Because one of the major problems of a direct cost system is the separation of fixed overhead from variable overhead, techniques have been evolved to do this with reasonable accuracy and ease. These range from considering all overhead as fixed or considering each category as either wholly fixed or wholly variable, to separating the portions which behave as either fixed or variable as determined by experience. The problem can be more readily solved by considering the nature of the cost: fixed, planned, or programmed. If the nature of the cost is understood, its classification is generally obvious.

Much of the confusion surrounding direct versus absorption costing systems stems from not understanding the purpose of the accounting system. Traditional accounting systems are concerned with inventory valuations, balance sheets, and income statements. Management, on the other hand, needs to understand the nature of the costs of the enterprise so that they may be managed and controlled and so that decisions are made on a more informed basis.

Traditional Control Techniques. The standard cost system is based on what costs ought to be. Control is exercised by examining deviations from standard on a post facto basis and then attempting to avoid deviations of this kind again. Many opportunities for control happen only once, and every deviation is for a different set of circumstances. This makes control so difficult that it approaches the impractical.

The first attempt to control costs and cost variances is generally the preparation of a budget. A budget is a plan for production which generates planned costs based on standards for all categories of costs: material, labor, and overhead. Because plans can never be carried out exactly, variances are incurred. These require explanation.

It is a logical next step, therefore, to develop a flexible budget which uses the same standards and the same plans, but the effects of volume differences on costs are segregated. Volume effects being obvious, no explanation is necessary. The variances which remain are true variances. They are the differences between costs as they should have been and actual costs. These are good controls to have, but they are post facto—the expense has been incurred, the variance generated, and the excess cannot be recovered this time. The opportunity has passed, perhaps never to return.

Realization of the post facto problem and its effect on management decisions has led to the development of predictive management principles and requirements. If management can predict with practical accuracy what costs, margins, and profit contribution will be under one assumed set of conditions, then it can alter the conditions if the "right" results are not forthcoming in the prediction. It can do so in advance and thus make its right prediction actually happen. This is predictive control.

For the responsible supervisor who takes action to incur cost, predictive control means that he will be evaluated on his performance, he will be visible, he needs alternatives, and he is aware of the consequences of his actions.

The Requirements of a Predictive Management System. The fifth principle of the predictive management concept states: "Plans are made with direct relation to the way work is accomplished." This implies time, sequence, methods, tooling, and cost incurrence. The principle can be paraphrased by saying that all costs must stand by themselves, separately and without encumbrances. It further implies that there should be management measurements to establish what the right and proper cost is under the conditions which are presumed to apply. Only the direct, standard cost system provides all these elements. The requirements of time, sequence, methods, and tooling can be provided by other cost systems, but not cost incurrence the way the work is accomplished. The requirements of the predictive management system are focused on variable costs without allocation of fixed costs.

MANAGEMENT MEASUREMENTS

Management measurements help those in supervisory or managerial positions manage their responsibilities by providing them with the means of testing and understanding the profit consequences of alternative courses of action. Management measurements are related to traditional cost control standards, but provide the ability to answer the question "How?" as well as "How much?"

For example, it is not enough for the supervisor responsible for the machine shop to know only the number of pieces to be made and the standard unit cost of each piece (or even the standard man-hours per piece). To plan and supervise his department properly, he should have available to him:

1. The breakdown between fixed (setup) and variable (handling and machining) time, so that he may determine whether or not it is economically sound to break large orders into segments to fit scheduling needs
2. The relative material and labor cost content, so that he may manage the balance between material waste and excessive labor cost
3. The materials and worker skills required and available, so that he can plan to avoid unnecessary delays and interferences
4. The elements of the standard cost and their relative importance, so that he may evaluate the opportunity or need for methods improvement
5. The value added to the product by his operation, so that he may evaluate important cost versus production volume trade-offs

Managers need, in advance of performing a job, the kind of information that traditionally becomes available only as a result of after-the-fact variance analysis. With such information, the manager can take action to avoid variances before they occur and optimize profits even on tasks that have never been done before and will probably occur only once.

After-the-fact variance analysis still has a place, and should be based on the same management measurements. Its purpose, however, is to improve the quality of the measurements and improve management's predictive abilities, rather than to fix the responsibility for errors when it is too late to correct them.

To facilitate feedback and reporting, management measurements should be compatible with accounting measurements and based upon the same standards of performance. This does not mean that they will be identical. Management measurements have as their purpose the management and control of costs and profits. Accounting measurements most often have as their purpose the proper valuation of inventory and the documentation of the performance of the enterprise for external reporting. It is too much to expect that the same measurements will be ideally suited to both purposes. It is reasonable, however, to insist that they have a common basis and be served by a common cost reporting system which duly recognizes the needs of each.

Management measurements should be tailored to the needs of the individual manager and the task at hand. To direct the worker in a more efficient method may require the most exacting detail of incremental times. Planning monthly production requires only an overall standard of what can be accomplished in an hour, a day, or a week. When measurements are of the wrong kind, either too general or too detailed, they fall into disuse. The needed information is then generated by some unknown, nonreasoned rule of thumb. Management measurements are uniquely fragile.

Management measurements should be accurate for the purpose for which they are intended. A measurement of the time required to machine an individual part should be accurately stated to several decimal places, because it will be multiplied by hundreds, thousands, or even millions of repetitions. For the measurement of a one-time programmed cost of a few hundred dollars, one significant figure will suffice. Above all, the statement of the measurement should not imply an accuracy that does not exist. The statement of amounts to the nearest cent, born of the need to account accurately for all moneys, is not appropriate to predictive management measurements. Predictive management measurements should focus manager attention on significant items.

Where possible, management measurements should be based on sound analysis supported by historical fact. This is not always possible and an estimate is preferable to no measurement at all. Estimates should be documented and the assumptions underlying them recorded. The well-reasoned judgment, even when incorrect, is superior to the guess, though correct, for only the former carries with it the mechanism for its own correction.

Management measurements will not of themselves provide the benefits of predictive management. Systems do not automatically solve problems. Solutions are derived from systems only after enthusiastic administration and intelligent interpretation. One of the most valuable resources a company possesses when developing a control system is the knowledge and intuition of its managerial personnel, developed through years of experience and known to be valid and correct. The development of management measurements is often the organization and formalization of what is already known.

Classification of Cost Measurements. Management measurements should recognize the fixed versus variable behavior of costs as volume varies. They should be separated into those that measure rates and those that measure amounts. The distinction is fundamental, and the two types of measurement must be treated separately. This is the principal difficulty encountered when using absorption cost systems for predictive management purposes. Absorption expresses an amount as a rate.

It would be convenient if it were possible to take a typical chart of accounts and classify each account as always variable or always fixed. Unfortunately, the behavior of costs depends on the basic nature of the enterprise and on the specific event under consideration.

Variable Costs—Rate Measurements. The variable costs of an enterprise are those that tend to vary in proportion to the volume of goods produced and sold or the volume of services rendered. For the typical business enterprise, materials, direct labor, and salesmen's commissions are normally variable costs. In a pure sense, the linear or direct proportion assumption is a simplification. Such things as volume discounts on purchased items, sliding-scale sales commissions, and differential wage rates based on seniority make the relationship curvilinear. Furthermore, the cost per unit is not the same for ten units of production as it is for one thousand or one million. It is possible with high-speed electronic data processing to reflect these nonlinear costs. Predictive management, however, deals with incremental changes in volume, and over the range of variation normally encountered, the linear assumption is sufficiently accurate. The critical test of a variable cost is the existence of a direct cause and effect relationship between volume and cost.

Fixed, Planned, and Programmed Costs—Amount Measurements. Fixed, planned, and programmed costs are costs associated with a period of time, and although they may change from time period to time period, they do not change within a period in response to changes in volume. "Period" is normally thought of as the normal accounting interval of one month, but for predictive management purposes, it should be the interval conforming to the cost involved. Temporary stenographic services are planned daily, while research and development programs may be planned over five years or more.

The fixed or constant costs of an enterprise are the basic costs of being in business. Depreciation, rent, taxes, and top management salaries are examples of such costs. They are passive in nature, that is, they are not responsive to management's need to reduce costs in periods of declining volume, and are not necessarily increased in periods of increasing volume.

The planned costs of an enterprise are the costs which are changeable from period to period, but once committed, become fixed for the period. Indirect labor, maintenance, and other service functions are examples of planned costs. They are the costs that management consciously decides upon in anticipation of the requirements of the coming period.

The programmed costs of an enterprise represent the costs that management incurs to further the objectives of the enterprise. Research, advertising, personnel training, and territory development are examples of such costs. Like planned costs, programmed costs result from conscious management decisions; unlike planned costs, they are not responsive to fluctuations in business conditions, and once committed, behave like fixed costs.

There is a further useful breakdown of fixed, planned, and programmed costs between those that are specific to a given activity or product and those that are common to many activities or products. Depreciation of the plant building is a typical common fixed cost, while depreciation of a new extrusion press is specific to the line of extruded products. Variable costs are by definition specific. Shared costs, such as the crane operator in a plant, are best viewed as a planned rather than a variable cost.

Cost classification is complicated by the fact that the same cost may have a different behavior in a different context. For example, from the viewpoint of the plant as a whole, maintenance labor is properly regarded as a planned or even a programmed cost. In the context of the maintenance department, as an enterprise in itself, labor is a variable cost depending on the volume of maintenance service rendered. Predictive management means the evaluation of alternatives, and it is necessary to consider only the limited number of costs that are affected by a decision. A large portion of the benefit of predictive management is the insight it gives managers into the true behavior of costs and profits. This insight comes from the process of systematically classifying costs according to their behavior.

THE TECHNIQUES OF MANAGEMENT MEASUREMENT

Industrial engineering and cost accounting have developed many techniques for measuring costs and establishing standards. In recent years, the management sciences, with the aid of the electronic computer, have expanded the utility of this body of knowledge.

The use of historical costs to develop standard rates is simple and easily understood. But it tells what a cost has been, not what it should or will be. Errors of the past are perpetuated, and opportunities for improvement are unrecognized. Historical data are best used, after careful analysis, to support standards developed by other methods.

Data based on sound work measurement and time study are preferable to historical data. Such standard data are recognized as a valid basis for wage incentives. Applications have been extended from manufacturing functions to other activities such as maintenance and clerical work. When obtained carefully and used according to the procedures discussed elsewhere in this Handbook, standards are invaluable as management measurements.

The application of sound standard data may be extended through the use of mathematical simulation models. Such a model shows how the cost or time for an operation behaves in response to changes in the variables (quantity, dimensions, materials, and the like) which affect it. When incorporated into computer programs, times for complex operations may be calculated in a matter of seconds. Development of management measurements for one-of-a-kind or complex operations then becomes practical.

Rate Measurements—Variable Costs

Direct Labor. The above comments are most specifically applicable to the measurement of direct labor rates. For predictive management purposes, the preferred management measurement is the standard hour rather than the piecework dollar. Direct labor measurements should include realistic allowances normally added to standards, but should exclude allocated fixed costs and the special allowances that creep in for take-home pay adjustment purposes. In measurement to be used by first-line supervision for planning and scheduling, unwarranted allowances mean double trouble. Not only is the worker paid more for the work done, but he is also asked to do less; the enterprise loses part of the capacity represented by its physical assets and fixed cost risks.

For workers operating several machines, interference allowances are appropriate to reflect the fact that two or more machines might require attention simultaneously. A special kind of interference arises when labor is specialized by trade or function. The productive worker is delayed if a serviceman is not readily available, and a serviceman is idle when he is not required. Too few service personnel cause excessive production delays; too many result in excessive service costs. There is an optimum excess of service personnel, which depends not on the individual job, but on the array of all jobs being worked on concurrently which require the service.

For nonrepetitive, job shop operations, service labor is not directly proportional to volume and is best treated as a planned cost.

In some industries, it may be desirable to establish direct labor measurements on the basis of a machine-hour rather than a labor-hour. This is particularly true in heavy industries where a crew of men operate a machine with a heavy depreciation rate. In other industries, notably chemical processing, the plant is kept running regardless of throughput, and direct labor as a whole becomes a planned rather than variable cost.

Materials. A bill of material should set forth the specific materials to be used in the exact quantities and at the standard purchase cost. Quantities should be stated in units meaningful to first-line supervision. For maximum usefulness, material standards should include the quantity of material per unit of product produced, the standard material cost per unit quantity of material, and the resulting cost per unit of product produced.

Expenses. Major items (such as power or fuel in hot-processing industries) should be separately analyzed and standards set for unit quantities, cost per unit, and cost per unit of product. Miscellaneous minor direct expense items may be grouped together in a historical cost based guideline.

Amount Measurement—Fixed, Planned, and Programmed Costs

Fixed Costs. Fixed costs tend to remain constant from control period to control period, and recent actual costs are valid for short-term predictive management purposes. Changes in fixed costs result from deliberate management actions.

Planned Costs. Every management responsibility has planned costs which represent the administrative and support functions necessary to keep an enterprise functioning smoothly. Engineered standards can be applied to measurements of how much work an individual can do, but the determination of how much work is required for a given enterprise and a given volume is often a matter of experience and judgment. Planned costs are best determined through careful analysis of costs by supervisors, assisted by cost control personnel, to determine the manageable cost elements and the extent to which they should be changed as volume levels are altered. For example, supervision costs change sharply when going from one- to two-shift operation; certain service personnel should sometimes be added at a rate greater than in proportion to volume increase to minimize interference delays to production personnel.

Planned costs which include items such as indirect labor are traditionally referred to as mixed costs. Mixed costs are partially fixed and partially variable with activity level.

Programmed Costs. There is no formal technique for deciding the proper amount of programmed costs. Items such as research and development, engineering, advertising and promotion, and the like must be based on management's judgment of what is required to meet enterprise objectives. Budgets are set on a project or program basis, the period being the duration of the project.

The Measurement of Margin.

The principles of predictive management require that operations be controlled and decisions made in terms of the profit consequences of alternative courses of action. The management measurement needed to accomplish this is margin, the difference between the value added to the economy by a product or service (its price) and the value taken from the economy in producing it (its variable costs).

To establish a standard margin for a product or service, it is necessary to establish a standard selling price, much as a standard purchase price is established for materials. Once this is established, standard margin is equal to standard selling price less the sum of all direct variable costs associated with the particular product or service. Standard margin represents the amount per unit of product or service that is available for the recovery of fixed, planned, and programmed costs and the accrual of profit. The ratio of standard margin to standard selling price is the profit-volume ratio of the product, or the slope of the line representing the product in the enterprise's risk and recovery equation.

In a predictive management control system, margin occurs twice. Margin is created when a product is made, and margin is realized when a product is sold. The standard margin per unit of product in each case is the same, but as production and sales for the period are not always the same, total margin created during a period may be different from total margin realized. Recognition of this dual nature of margin eliminates much of the confusion operating personnel feel when they encounter items such as "Profit Effects of Changes in Inventory" on period performance reports, or the wrong decisions that may be made when fixed costs are presumed to be recovered when the product is produced and profit realized when it is sold.

Variance Analysis. The emphasis on cost control as a predictive management tool does not eliminate the need for variance analysis. Variance analysis becomes all the more important. Applied as part of planning, it is used to identify the cause and responsibility for prevention of variances; applied after-the-fact, it is used to refine the measurement and the system, as well as to indicate managerial and supervisory performance. All variances may be divided into four classifications:

Margin variance is the margin gained or lost from variation in volumes from those planned or anticipated and is the direct responsibility of operating management. It is akin to volume variance in the full absorption cost system, except that it recognizes the profit consequences of volume changes and not just their effect on overhead absorption. It is mathematically the difference between the margin anticipated by the profit plan and the margin represented by actual volume at the same standard unit variable costs and selling prices.

Usage variance is the gain or loss resulting from requiring more or less than standard quantities of materials, labor hours, and the like. This also is the direct responsibility of operating management. It is mathematically the difference between actual and standard quantities of variable cost elements for the actual volume of product produced or sold.

Unit rate variance is the gain or loss resulting from changes in prices, purchased material costs, labor wage rates, and similar costs. It is frequently outside the responsibility and control of operating managers except in the case of sales, where departures from standard prices are the direct responsibility of sales personnel. Mathematically, it is equal to the difference between the actual and standard per unit costs of variable cost items.

Budget variances are the gains or losses resulting from the departure of fixed, planned, and programmed costs from those budgeted in the profit plan. These also are the responsibility of operating management. Mathematically, budget variances are the differences between actual fixed, planned, and programmed costs and those anticipated in the profit plan.

Allowances. A system of management measurement must provide for change. Methods, prices, and volume forecasts change frequently, and unless provision is made, changes will first be recognized as variances in after-the-fact period reports. To avoid the administrative problem of continually revising budgets and standards, it is useful to introduce allowances to adjust the period profit plan. These allowances represent variances, predicted before they occur.

Allowances may apply to margin, usage, and unit rates or fixed, planned, and programmed cost budgets. They are most useful for reflecting temporary or special situations such as the substitution of materials, the use of more or less efficient facilities, or design changes to meet an individual customer's requirements. When of a permanent nature, the change should be incorporated into revised standards, and the allowance eliminated as soon as practical.

A PREDICTIVE MANAGEMENT CONTROL SYSTEM

The Operating Profit Plan. In a predictive management control system, management measurements are combined with sales forecasts and production schedules to produce an operating profit plan. These plans are prepared for each department or management responsibility and are accumulated into the overall profit plan of the enterprise. Profit plans are based on management's reasoned judgment of how the risks of the enterprise should be structured to meet current objectives. They are prepared far enough in advance to permit review, identification of critical problems, and implementation of corrective measures.

The planning period depends on the nature of the events being planned. For compatibility, plans are broken down or built up into monthly plans which are refined and expanded into greater detail as the period approaches. Their preparation is the responsibility of operating management, financial, and control personnel, working under the overall guidance of the controller. A completed profit plan represents the concurrence of the responsible supervisor, the control executive, and higher management on what can be accomplished in the period ahead under prevailing conditions.

The concept of the operating profit plan is illustrated by the following pro forma system for a small foundry and machine shop fabricating subassemblies for several customers. To illustrate the use of the system as a predictive management tool, it is assumed that one of the customers requires accelerated shipments and is willing to pay a premium. The additional volume cannot be met entirely from inventory and requires increased production and overtime. The handling of the special allowances resulting from this change applies the concepts of variance analysis to the prediction of variances.

The Product Standard. The essential source document in the preparation of a profit plan is the product standard, which combines the many management measurements pertaining to a specific product. It shows the standard selling price, the elements of standard direct variable costs (expressed in both quantities and dollar cost per unit), the standard margin created by each unit produced, and the standard margin realized on each unit sold. To provide a basis for production scheduling and planned cost budgeting, it also includes information on services required and facilities used. For a job shop, the job work ticket replaces the product standard and should contain much the same information.

	Units	Quantity/piece		Cost/unit quantity		Dollars per piece			
		Std.	Allow.	Std.	Allow.	Std. cost	Allowances Usage	Allowances Rate	Curr. cost
PRODUCT A ASSEMBLY									
Selling price						$400.00			$400.00
Materials and parts									
Part #075324	2			$85 00	n.a.	170.00			170.00
Part #075325	1			40.00	n.a.	40.00			40.00
Purchased parts	4			12.50		50.00			50.00
Direct labor									
Assembly		3.50		4.00	+0.20	14.00		0.70	14.70
Shipping		0.25		2.80		0.70			0.70
Manufacturing supplies						1.50			1.50
Direct manufacturing cost						$276.20		0.70	$276.90
Other direct costs									
Commissions @ 5%						20.00			20.00
Direct total cost						$296.20		0.70	$296.90
Margin						$103.80		$(0.70)	$103.10
PART #075324									
Direct manufacturing costs									
Materials									
Aluminum	lb	20.00		$ 0.75		$ 15.00			$ 15.00
Bronze bushing		2		1.50	+0.50	3.00		1.00	4.00
Direct labor									
Foundry	hr	4.00		3.50		14.00			14.00
Machining	hr	12.00	−0.50	4.00	+0.20	48.00	(2.00)	2.30	48.30
Manufacturing supplies						5.00			5.00
Total						$ 85.00	$(2.00)	$3.30	$ 86.30
Planned services									
Material handling		0.1 hour per piece							

Fig. 7-2. Assembly product standard and cost standard for one of the parts of which it is comprised.

Figure 7-2 shows a product standard and a cost standard for one of the parts contained in the product. The overtime needed will increase the average hourly rate. The increase is partly offset by institution of a methods improvement. The effects of these changes, along with the effect of an increased cost for bronze bushings, are shown as allowances to the normal standard.

The Departmental Profit Plan. Product standards, production schedules, and sales forecasts are combined to form departmental profit plans. These plans should contain the following information:

1. The volume anticipated, measured in terms of some common unit such as man-hours. This provides the manager with a basis for scheduling activity and planning costs.
2. The direct variable cost inputs associated with these activities.
3. The margin output created or realized. This emphasizes the manager's responsibility to manage volume as well as control the costs of his department.
4. The fixed, planned, and programmed costs budgeted for the period. This represents how the risks of the department are to be structured in terms of the expected level of activity.

Not all departments will have all four types of information, and the exact form of each department's profit plan should be suited to its function.

Value-added departments with measurable output, such as production and sales, will normally contain all the above information. Where a product passes through several departments before it is completed, the margin is most appropriately included in the profit plan of the department completing the product. Margin allowances or variances, however, are the responsibility of the department limiting overall production.

Supporting functions with measurable output produce no margin. Their output is measured not in product or part volume, but in terms of some other activity. Maintenance is such a function, as are material handling and many functions traditionally included under indirect labor. Management's responsibility is to accomplish a variable amount of work at a total cost commensurate with the volume of work done. Such functions may be viewed as unit enterprises within an enterprise, having their own variable, fixed, planned, and programmed costs but zero revenue. The output or net result is a cost instead of a profit.

Supporting functions with no measurable output include the many administrative and supervisory functions necessary to the enterprise. They have only fixed, planned, and programmed costs. Here the responsibility of management is to accomplish the functions of the department within budgeted cost levels.

Figure 7-3 is the profit plan of a value-added department having measurable output. It shows the normal activity and that planned for the current period; the variable costs and margin associated with this activity; and the appropriate fixed, planned, and programmed costs. Allowances identify by cause the difference between the current plan and the normal month. The budget allowances reflect the supervisor's decision to divert normal training and miscellaneous housekeeping time to production during this period.

Study of Figure 7-3 will show that the principle involved is very much akin to that of flexible budgeting, with two significant exceptions: (1) it recognizes, in margin, the profit consequences of changes in volume and (2) it allows recognition of nonlinearities such as that introduced by overtime.

The Enterprise Profit Plan. The many departmental profit plans are combined to form the enterprise profit plan. For most companies, monthly production and sales are rarely identical. The problems raised by this are most easily handled by preparing separate production and sales profit plans based on different volumes.

Figure 7-4 illustrates such a difference. The margin summary shows the current manufacturing and sales volume by product as well as the standard revenue, variable costs, and margin associated with these volumes. These are the basis for the manufacturing and sales profit plans in which standard margins are combined with allowances and fixed, planned, and programmed costs to determine the period profit. The illustration is presented in profit and loss statement terms for ease of understanding. Margin allowances and departmental costs are the relevant information.

PROFIT PLAN   MILLING AND ASSEMBLY DEPARTMENT   June 19—

	Quantity		Allowed/unit		Total	
	Norm.	Curr.	Std.	Curr.	Std.	Curr.
Activity—man-hours						
Part #075324	180	200	12.00	11.50	2,160	2,300
Product A	180	210	3.50	3.50	630	735
					2,790	3,035

Normal activity: 2,790 man-hours @ $4.00/man-hour
Current activity: 3,035 man-hours @ $4.20/man-hour

$$\text{Activity ratio} = \frac{3,035}{2,790} = 1.088$$

Allowances

	Normal month	Margin	Usage	Rate	Budget	Current plan
Variable costs						
Materials						
Part #075324	$ 3,600			$ 200		$ 3,800
Product A	10,500					10,500
Direct labor						
Part #075324	9,600		(400)	460		9,660
Product A	2,940			147		3,087
Manufacturing supplies						
Part #075324	1,000					1,000
Product A	315					315
	$27,955		$(400)	$ 807		$28,362
Margin created						
Product A	$18,684	$ 3,114				$21,798
Manufacturing operations			$ 400	$(807)		(407)
	$18,684	$ 3,114	$ 400	$(807)		$21,391
Fixed, planned, and programmed costs						
Supervision	$ 1,100					$ 1,100
Methods improvement	64					64
Training	80				$ (80)	
Miscellaneous	80				(80)	
Depreciation	2,500					2,500
	$ 3,824				$(160)	$ 3,664
Net profit contribution	$14,860	$ 3,114	$ 400	$(807)	$ 160	$17,727

Fig. 7-3.   Departmental profit plan.

The use of allowances to represent predicted variances and their segregation by responsibility and cause are fundamental to predictive management. The same analysis concepts are applied to ex post facto variance analysis, except that in this case, the comparison would be between the current plan and actual results.

Profit plans should be prepared sufficiently in advance of a period to permit their use in planning and the prevention of variances. Otherwise, they become an academic exercise by control personnel and a burden to operating management.

Profit Consequence Analysis. Management's responsibility goes beyond meeting standards and profit plans. Managers should aggressively pursue opportunities through which profit plan performance can be bettered. The profit plan provides

STANDARD PRODUCT MARGIN SUMMARY   June 19—

	Per unit standards				Quant.	Period dollar totals			
	Rev.	Mfg.	Other	Margin		Rev.	Mfg.	Other	Margin
Normal monthly operation									
Product A	$400.00	$276.20	$20.00	$103.80	180	$ 72,000	$ 49,716	$ 3,600	$18,684
Product B	300.00	213.90	15.00	71.10	300	90,000	64,170	4,500	21,330
Product C	100.00	44.65	5.00	50.35	400	40,000	17,860	2,000	20,140
						$202,000	$131,746	$10,100	$60,154
Current–Manufacturing									
Product A					210	$ 84,000	$ 58,002	$ 4,200	$21,798
Product B					300	90,000	64,170	4,500	21,330
Product C					400	40,000	17,860	2,000	20,140
						$214,000	$140,032	$10,700	$63,268
Current–Sales									
Product A					220	$ 88,000	$ 60,764	$ 4,400	$22,836
Product B					300	90,000	64,170	4,500	21,330
Product C					400	40,000	17,860	2,000	20,140
						$218,000	$142,794	$10,900	$64,306

PROFIT PLANS	Normal month	Allowances				Current plan
		Margin	Usage	Rate	Budget	
MANUFACTURING						
Standard revenue	$202,000	$12,000				$214,000
Manufacturing direct costs	131,746	8,286	$(400)	$ 807		140,439
Other costs	10,100	600				10,700
Margin	$ 60,154	$ 3,114	$ 400	$ (807)		$ 62,861
Fixed, planned, programmed costs						
Milling and assembly department	3,824				$(160)	3,664
Foundry	9,250					9,250
Plant administration	7,300					7,300
General depreciation	3,250					3,250
	23,624				(160)	23,464
Gross manufacturing profit	$ 36,530	$ 3,114	$ 400	$ (807)	$ 160	$ 39,397
SALES						
Revenue	$202,000	$16,000		$1,100		$219,100
Manufacturing direct costs	131,746	11,048	$(400)	807		143,201
Other costs	10,100	800		55		10,955
Margin	$ 60,154	$ 4,152	$ 400	$ 238		$ 64,944
Fixed, planned, programmed costs						
Manufacturing	23,624				(160)	23,464
Sales department	4,000					4,000
Research and engineering	5,000					5,000
General administration	4,600					4,600
	37,224				(160)	37,064
Operating profit	$ 22,930	$ 4,152	$ 400	$ 238	$ 160	$ 27,880

FIG. 7-4. Enterprise profit plans for manufacturing and sales. Italicized numbers are redundant to the preparation of the plans or their use as a predictive management tool, but are included here to point up the relationship between the profit plan and the ultimate profit and loss statement.

a standard against which the profit consequences of alternatives can be tested, using the same standards and management measurements.

Alternatives can change variable costs (and therefore unit margins), planned costs, or volume. When an alternative affects only one of these, it is possible to make direct comparisons on the basis of the affected costs alone. When the alternative changes two or all three, there can be effects that are not immediately obvious, and simple comparisons can lead to the wrong conclusions. In these instances, the risk and recovery equation, graphically displayed, is especially useful.

Figure 7-5 illustrates a case of a methods change which increases variable costs, planned costs, and volume. In a chemical process, a new method is developed

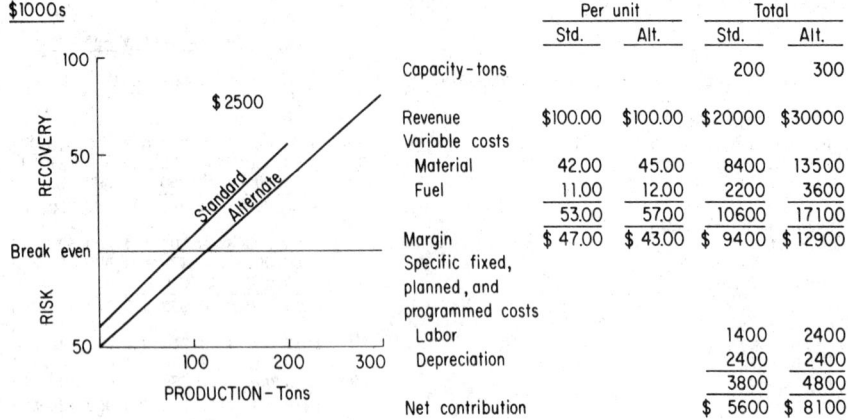

$1000s		Per unit		Total	
		Std.	Alt.	Std.	Alt.
Capacity – tons				200	300
Revenue		$100.00	$100.00	$20000	$30000
Variable costs					
Material		42.00	45.00	8400	13500
Fuel		11.00	12.00	2200	3600
		53.00	57.00	10600	17100
Margin		$ 47.00	$ 43.00	$ 9400	$12900
Specific fixed, planned, and programmed costs					
Labor				1400	2400
Depreciation				2400	2400
				3800	4800
Net contribution				$ 5600	$ 8100

FIG. 7-5.   Profit consequence analysis using the risk and recovery equation.

which consumes more material, requires greater energy input, adds an extra man to the operating staff (in chemical processing, labor is often a planned rather than variable cost), and increases production by 50 percent. The combination of effects provides the opportunity for a substantial increase in profit contribution. The importance of margin is seen in the fact that on a total cost basis, the standard method has a total cost of $14,400, or $72 per ton, while the more profitable alternative has a total cost of $21,900, or $73 per ton.

Multiple Plans.   The greatest danger in long-range plans is that they become rigid and unresponsive to change. Predictive management admits all types of changes to plans. These include not only changes to the economics of the enterprise and the management measurements applicable to it, but also changes in timing and sequencing. Plans that involve a series of steps, taken at specified times to attain an objective, should be reappraised if opportunities occur out of phase. A plan presents a constant challenge for improvement through change. Assumptions become facts; new data become available; better sequencing is made practical; better methods are developed.

A plan may be viewed as part of a set, a combination of interrelated elements. An enterprise profit plan; its component departmental plan, standards, production schedules, and sales forecasts; and assumptions on business conditions, all form a single set. At the same time, there are many other sets based upon different assumptions and conditions. For example, in the processing example above, the alternative would be attractive if there were a market for the increased production or if the capacity could be used for another product. If, however, the market was saturated and the plant was operating below capacity, the alternative would not be attractive.

Looking into the future, the number of sets becomes infinite. It is possible, however, with modern data processing to keep track of a reasonable number of sets associated with the immediate future, evaluate the profit consequences of each, and select the optimum plan in terms of an ever-changing reality.

THE ROLE OF COMPUTERS IN COST CONTROL

Although a computer is by no means essential to an effective system of cost control, its use can substantially increase the effectiveness of a control system. Through its ability to process large quantities of data quickly and inexpensively, the computer makes possible the use of methods and techniques that would otherwise be impractical. This is the real benefit to be derived from the application of computers to control systems. Managements which have applied computers to traditional ex post facto cost reporting and control systems have found that the more rapidly data are processed and the earlier reports are generated, the clearer becomes the need for information before the fact.

When a company plans to make extensive use of its computer facilities for dealing with operational and control problems, it should plan to construct a functional data base from which any system may draw. The data base concept is that all relevant data for operational, information, and control systems are stored in such a way that they are accessible. An item of data need be entered only once, even though it may be used in several ways.

Effective use of computers depends on the design of the system as a man-machine system. The computer handles the routine processing of large quantities of data and the application of well-defined decision rules. The man supplies the judgment required for the unique decision or the poorly defined situation. Installation of a computer will not by itself assure an effective predictive management control system; it is only a tool that makes a highly sophisticated and effective system practical.

CONCLUSION

The principles discussed in this chapter are contrary to some of the practices which still prevail. Many managements continue to control costs by comparing expenditures for one period with those of a period considered to be similar. Others attempt to measure performance by comparing current actual unit product costs with previous actual unit costs. Among those with standard cost systems, many have learned to think of profits as accruing automatically on a basis of so much per unit of product. None of these thought habits is practical in the dynamic environment of a competitive economy.

Sound predictive management control starts with measurements describing the true relationship of revenues and costs to volume and events. Careful study and classification of fixed and variable factors as well as timely preparation of profit plans is essential. Also important is the fact that the procedures involved should be fitted to each enterprise. Attempts to apply the details of a system developed for one company to another company are usually unsuccessful. In addition, there must be integration among the industrial engineering, planning and scheduling, and cost accounting departments. The combined efforts of all are required if the system is to yield its full benefits.

BIBLIOGRAPHY

Bergfeld, A. J., W. R. Knobloch, and J. S. Earley, *Pricing for Profit and Growth,* Prentice-Hall, Inc., Englewood Cliffs, N.J., 1962.
Bowman, E. H., and R. B. Fetter, *Analysis of Industrial Operations,* Richard D. Irwin, Inc., Homewood, Ill., 1959.
Bowman, E. H., and R. B. Fetter, *Analysis for Production Management,* Richard D. Irwin, Inc., Homewood, Ill., 1961.
Bunge, Walter R., *Managerial Budgeting for Profit Improvement,* McGraw-Hill Book Company, New York, 1968.
Crowningshield, Gerald R., *Cost Accounting Principles and Managerial Applications,* Houghton Mifflin Company, Boston, 1969.
Dearden, John, *Cost and Budget Analysis,* Prentice-Hall, Inc., Englewood Cliffs, N.J., 1962.
Drucker, Peter F., *Managing for Results,* Harper & Row, Publishers, Incorporated, New York, 1963.
Firestone, F. N., *Marginal Aspects of Management Practices,* Michigan State University, East Lansing, 1960.
Forrester, J. W., *Industrial Dynamics,* The M.I.T. Press, Cambridge, Mass., 1961.
Kelly, William F., *Management through Systems and Procedures,* John Wiley & Sons, Inc., New York, 1969.
Rautenstrauch, W., and R. Villers, *Budgetary Control,* Funk & Wagnalls Company, New York, 1950.

Chapter **8**

Budgetary Control

EDMUND J. McCORMICK

**Chief Executive Officer and Chairman,
McCormick & Company, Yonkers, New York**

The target of budgeting is to make the most effective use of resources and thereby to obtain the highest possible level of sustained profits. Budgeting is planning. It is profit planning.

In almost every industry, there are companies with relatively modest resources that are making profits equal to or greater than those of competitors who have more funds available. Nine times out of ten, the reason is that the more successful company is doing a better job of planning. It is using a budgeting system to chart the course that it means to follow.

Budgeting is often confused with standard costs—a mistake that obscures many of the greatest values that a budgeting system can produce. Budgeting is concerned with costs, naturally. But its real focus is on profits. Budgeting can be done without standard costs, though ideally, the two should be combined. With a good standard cost system as a basic control tool, budgeting will be more effective and the company will be more likely to achieve the profit it originally planned.

BUDGETING IS EMPHASIS ON PROFITS

A key tool for management planning and control is found in budgeting. By its very nature, budgeting forces management at all levels to plan for the year ahead. Collectively, the budget must be the organization's best estimate in terms of profits, sales, costs, and employment of resources. Budgeting forces emphasis on profits in all areas which can affect profits.

One of the commonest mistakes in budgeting is to concentrate on the figure representing profit as a percentage of sales. This figure is a poor tool for budgeting and it can lead to all sorts of false conclusions.

Take, for instance, two actual companies in a food processing industry. Their records for one year were as follows:

	Sales	Profits	Profit on sales, percent
Company A..............	$18,500,000	$4,700,000	25.4
Company B..............	20,300,000	750,000	3.7

On the face of it, Company A seems to be doing six times as well in profit planning as Company B.

But take a look at what happens when budgeting gets away from the percent profit on sales and gets back to its proper starting point—the comparison with the resources used in the business:

	Net worth	Profit	Profit on resources, percent
Company A..............	$10,000,000	$4,700,000	47
Company B..............	1,500,000	750,000	50

The picture is now reversed. Company B is actually doing slightly better than Company A when measured by the yardstick of resources. Figure 8-1 shows the comparison.

A sound budgeting system will tell Company B that its strategy is essentially right, that it is making better use of its resources than its competitor. More than that, a good budget system will help Company B to hold the advantageous spot it has gained.

Budgeting is to the planning of a company's operations what the architect's plan is to building a home. With a budget, there will be a minimum of surprises and hasty decisions. With a budget, management will be able to take action before the fact rather than after.

As an example of such planning before the fact, there is the case of a small textile converting mill. This example, incidentally, also demonstrates that budgeting is not a technique that is available only to big companies. The company in this case has less than $1,000,000 a year in sales volume.

When the company drew up its sales, direct labor, material, and overhead projections for a recent year, its management realized that there was trouble ahead. Adverse conditions confronted all wool and cotton converters in northeastern United States.

By the very fact that they had planned ahead, accepting the discipline of a budget, management asked the questions and did the research that gave them the advance picture of lean profits. Their picture was incomplete, of course, but it was available in time to do something about it. The company made drastic cuts in overhead, even to the extent of cutting officers' salaries. As a result, this small concern was able to soften the blow when the market for the year turned out far worse than previous years. It came out with a small profit—when much of the industry was showing losses.

The discipline of drawing up a profit plan can be useful all by itself—because it forces management to take a hard look at the conditions it will face in the coming budget period. But budgeting produces its greatest value only when

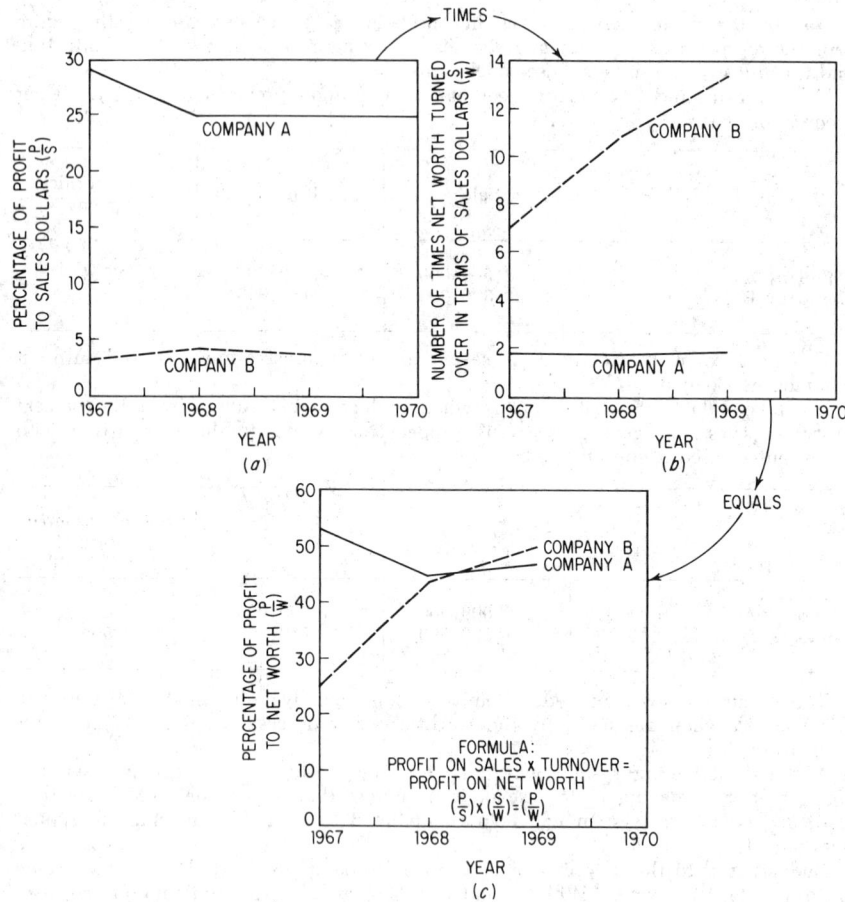

Fig. 8-1.   More accurate criteria for company comparisons than the profit-to-sales ratio (*a*) are the ratio of sales to net worth (*b*) and the ratio of profit to net worth (*c*).

it includes an adequate system of "budgetary controls" designed to keep actual performance in line with the plan.   The technique by which the planned destination is reached is as important as the planning itself.

It is in this area that the industrial engineer shines.   He has the training in the technique of putting plans into practice.   He can take the responsibility for showing how the budget plan can be converted into operating fact.

STEPS OF BUDGETARY CONTROL

Budgetary control, briefly, consists of these four simple steps:
1. Setting the targets (or preparing the budget)
2. Placing actual performance against these targets
3. Uncovering the causes of the deviations from target
4. Eliminating the cause of the variance or changing the target

In this area of control, the accountant and the industrial engineer make a fine team.   The strongest budgeting is found where accounting and industrial engineering work closely together. The accounting function is one of ascertaining and recording the facts about a business with consistency, timeliness, and accuracy. Thus, the accountant provides the facts about a company's operations. The engineer marshals these facts and draws them together so that they may be interpreted.

In budgetary control, accounting is clearly responsible for step 2, the vital comparison of actual performance with the targets that have been set.

The engineer is equipped to handle steps 3 and 4, the determination of the causes of the variances and the development of methods to correct them.

Step 1 is a joint responsibility. Both the accountant and the industrial engineer should collaborate in setting up the machinery through which management sets the targets for the year. The industrial engineer can make a major contribution to the budget planning through his unique understanding of cost behavior. Many industrial engineers may be surprised to learn that a well-organized plan of budgeting in actual operation is the exception rather than the rule.

SELLING EFFECTIVE BUDGETARY CONTROL

In selling a plan of budgeting to top management, the industrial engineer should be prepared to meet these objections:

1. It takes too much executive time to prepare a budget.
2. Our business is different; we cannot forecast sales.
3. Our company is too small to afford budgeting.
4. We are making good profits; we do not need a budget.

The remainder of this chapter will give the industrial engineer detailed replies to these objections. Briefly, here are the fallacies that lie behind each of the common objections to budgeting:

It Takes Too Much Executive Time to Prepare a Budget. If this is true, it is a valid complaint; but it is a symptom of faulty budget design, not an inherent drawback of budgeting. Top executives and supervisors should be asked to make only the key decisions. They should be aided in these decisions by detailed schedules prepared by the accountant and the industrial engineer.

Our Business Is Different; We Cannot Forecast Sales. This complaint is often heard from companies in a business where demand is based on acceptance of styling or on seasonal factors.

It is true that erratic demand makes budgeting difficult. But a well-prepared forecast is better than none. The action of making a sales forecast forces management to make the best use of all that it knows about the industry, the market, and the competition. A company can at least have a forecast that is as good as its toughest competitor's.

Our Company Is Too Small to Afford Budgeting. The wool converter mentioned earlier had sales of only $1,000,000. A system installed for a Connecticut machine tool builder has been operating for several years; this firm has sales of about $700,000 a year.

Budgeting is not a tool reserved for medium- and large-size firms. Good design of the system will keep it within bounds for any size firm.

We Are Making Good Profits; We Do not Need a Budget. This is one of the most difficult arguments the industrial engineer will have to deal with. It is often heard from companies enjoying the acceptance of new products, from new industries, or from firms in an old industry where there is a temporary shortage of capacity.

But being in a new industry does not guarantee a company against reverses. Television is an example. The booming period when almost any producer could make a handsome profit was followed by a general shake-out; profits dwindled under sharp competitive pricing.

One of the major television manufacturers made the mistake of thinking that the easy days would last indefinitely. It regarded budgeting as just a mechanical process. It saw no reason for profit planning when unit profits were enormous and the selling price could be almost anything the company wanted to ask. Instead of using the time to set up a system of planning, the company made only day-to-day decisions. And so, when the shake-out came, it took severe losses—which is almost unforgivable when a firm occupies a leading position in a new industry.

BASIC CONCEPTS TO REMEMBER

Budgeting is profit planning. The target is the most effective use of resources within the industry.

Planning is not enough; it must be followed by action to bring profit targets into actual being. To do this, control is necessary. Control consists of four steps:
1. Setting the target
2. Comparing actual versus target
3. Investigating the falldowns
4. Correction

In following through on these steps, the industrial engineer and the accountant make an effective team.

Budgeting must be sold. The best budget mechanism fails where management is not convinced.

Managements often fool themselves by thinking that budgeting is an extra burden. Actually, a clear-cut budgeting system will replace an expensive and unsatisfactory collection of uncoordinated controls.

THE INDUSTRIAL ENGINEER'S ROLE IN BUDGET DEVELOPMENT

An effective budget brings into focus the goals of all segments of the organization involving:
1. Operational planning
2. Financial planning
3. Preservation planning

Due to his function of project evaluation which becomes involved with all areas and divisions of the organization, the well-rounded industrial engineer is especially fitted to play an important role in budget development.

The budget is a complete organizational plan for the year ahead and it must be developed through the teamwork of all management personnel. Its development will have important psychological effects upon all who participate. This should be evident in terms of the budgeted profit results. It should be realized through the acceptance by individuals of responsibility for their portion of the actual performance.

A well-executed budgeting procedure involves basic management disciplines initiated or serviced by the industrial engineer as follows:
1. Organizational definition
2. Data processing techniques
3. Systems and procedures
4. Standards
5. Training and education

The industrial engineer is directly involved in development of the budget throughout the procedure. The initial step in budget development is the sales forecast. At this point, the industrial engineer determines feasibility of the forecast by interpreting the sales plan into the manufacturing plan.

This will involve functional tools for short-range planning, as they exist or as developed by the industrial engineer. These are:
1. Methods and materials determination
2. Production scheduling
3. Inventory movement and control
4. Capacity utilization analysis

Capacity is here defined not only as machine- or plant-hours, but also the involvement of the complete resources of space, equipment, people, materials, inventories, and funds. An effective plan will utilize all resources to the fullest practical point, will fit within the limits of these resources, and will provide the budgeted or desired return for their use.

It is important to consider capacity separately in terms of variable elements. Variable elements, such as inventories and payrolls (people), may be expanded within the limits of funds, borrowed or otherwise, and available skills. Fixed elements—such as equipment, and to a great extent, space—are much more rigid in dimension over the short term of a year.

During the development of the budget, the industrial engineer's role will involve many studies and much assistance to other management personnel in the accumulation, verification, and application of their budget data.

The industrial engineer contributes both to the budget and to the needs of other personnel for their role in the budget development and in budget attainment. Thus, it is necessary for the industrial engineer to communicate with other management personnel on a common plane.

USING THE BREAK-EVEN CHART IN BUDGETING

The break-even chart is a device to illustrate graphically the relationship between sales volume, costs, and profits. It performs this portrayal of economic behavior in picture form, providing the industrial engineer with an excellent tool for training as well as for operational use. The basic principles of these charts will be presented at this point because they make clear some of the terms and ideas in the remainder of this chapter.

To emphasize the profit relationship to volume, the use of positive and negative numbers on the vertical scale has come into use. The negative numbers represent constant or fixed costs that must be covered before profits accrue. The positive numbers represent profit. A single line then represents the amount by which losses are reduced or profits attained with each increase in the volume or sales figure (see Figure 8-2). Mathematically, this is represented by the formula

$$\text{Slope} = \frac{\text{fixed costs} + \text{profit (or loss)}}{\text{sales}}$$

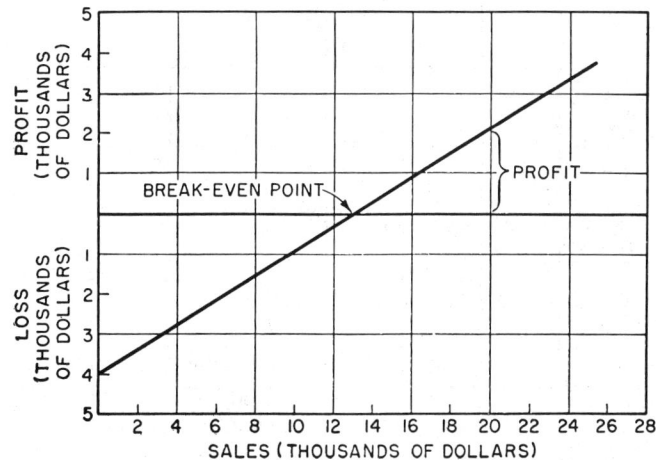

Fig. 8-2. A break-even chart using both positive and negative numbers emphasizes the relationship beween profit and sales volume.

Finding and Proving Fixed Expense. This type of break-even chart can be used in finding and proving fixed expenses from past history. By plotting the profit-sales volume relationship for several time periods, a trend line may be established. By extending this line back to the vertical scale, the fixed cost may be determined (see Figure 8-3). Mathematically, this line may be computed by the method of least squares. This information may be used to aid the setting of standards, determining past break-even points, or making a financial analysis.

The fixed expense found on the profit path should equal the total of the fixed expense found on the individual labor and expense graphs.

The Break-even Graph as a Management Tool. Fundamentally, management consists of planning operations and checking the results. The break-even chart can aid the performance of both these functions.

Planning. 1. The break-even chart can aid management because it pictures graphically the cost-sales volume-profit relationship. It can be readily seen that changes in costs and sales can have a great effect on profits. The planning of profits is then a matter of setting a profit target, calculating costs, and computing the necessary sales volume. Conversely, by plotting the fixed and variable costs, the break-even point and subsequent profits may be determined.

2. The break-even chart is an excellent means of predetermining the effects on profits of a pricing policy in relationship to other factors. The reduction or increase of profits and the changes that will result in costs and volume can be predicted. It must be emphasized that the break-even chart is not the whole story; but used in conjunction with other techniques, it can make decisions easier.

3. Expansion alters the cost figures of a business. The break-even chart can

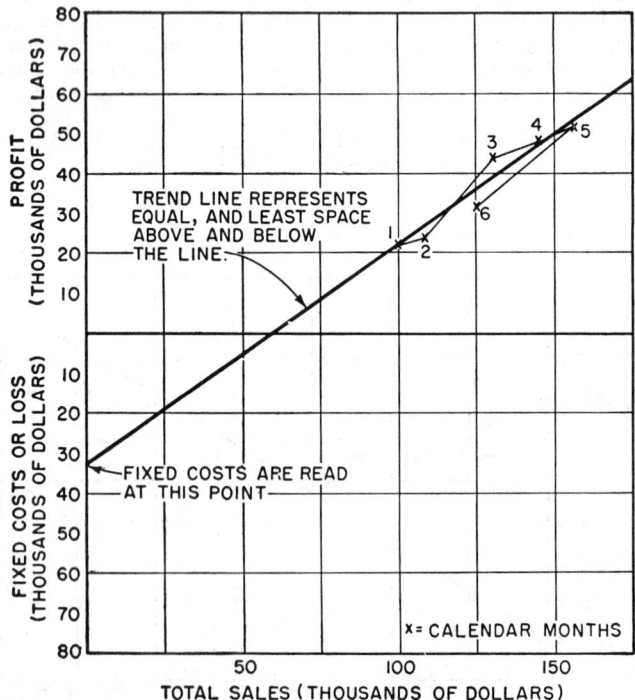

FIG. 8-3. A break-even chart used to determine fixed and variable costs from past records.

help evaluate the effect that capital expenditures will have on the financial structure before those expenditures are made.

4. The negotiation of labor agreements is one of the most important phases of business activity that management must contend with. Management is hampered in this endeavor by the misunderstanding on the part of both the public and labor as to the nature of profits. The break-even chart can serve as an educational tool to show that profits accrue only after the break-even point is exceeded and that proposed wage increases will raise that break-even point unless accompanying efficiencies are realized. The danger of being priced out of the market may be graphically revealed.

Checking. 1. The break-even chart is an important means of reporting performance to top management. It is an equally important means of reporting performance for lower levels of management. By adding details, it can be made to serve as a control mechanism for expenses, and as such, can guide operating personnel by providing specific information on performance.

2. Variable budgeting control is concerned with the expenses incurred at various levels of activity. The break-even chart helps show what expenses should be in relation to income and aids the differentiation of profit arising from volume and that arising from efficiency of operation.

Marginal Income. Marginal income (MI) is a concept closely associated with the break-even approach. It is the difference between the variable cost and the selling price. The fixed costs must still come out of it. Marginal income equals fixed costs and profits. Expressed on a per unit basis or as a percent of sales, it is a constant and the only correct measurement at all volumes. Obviously, a system using marginal income as a tool is dependent on separating fixed and variable costs. The terms are defined as follows:

1. Fixed costs, often called period costs—those accumulated by time and not by units produced
2. Variable costs—those which accumulate in direct proportion to number of units produced
3. Fixed and variable costs—costs which are a combination of fixed and variable

The following advantages are accrued by separating the fixed and variable costs:

1. Effective control of costs
2. Flexible budgets
3. Accurate cost projection at varying activity levels
4. Sharp, stable pricing at all levels
5. Defined product profitability
6. Scientific profit planning

By watching the MI percent closely, changes threatening the profit objectives are readily noted. A small variation in MI percent can have a great influence on profit. The following examples illustrate this point.

EXAMPLE 1

Annual sales......................	$4,000,000
Marginal income 20 percent.........	800,000
Less: fixed.......................	400,000
Profit before tax................	$ 400,000

EXAMPLE 2

Annual sales......................	$4,000,000
Marginal income 10 percent.........	400,000
Less: fixed.......................	400,000
Profit before tax................	$ 0

The marginal income in example 2 dropped 50 percent, but profit was wiped out 100 percent.

Notice how the MI percent signals trouble in the following examples while the conventional profit-to-sales ratio shows no danger.

EXAMPLE 3

Conventional Method		MI Method	
Total company sales........	$1,000,000		$1,000,000
Less material costs........	300,000		300,000
Less all labor and expense...	500,000	Less only variable labor and	
Profit.................	$ 200,000	expense.................	300,000
Profit-to-sales ratio.......	20 percent	Marginal income...........	$ 400,000
		Less all fixed expense.......	200,000
		Profit.................	$ 200,000
		MI ratio...............	40 percent

EXAMPLE 4

Conventional Method		MI Method	
Total company sales........	$2,000,000		$2,000,000
Less material costs.........	600,000		600,000
Less all labor and expense...	800,000	Less only variable labor and	
Profit.................	600,000	expense.................	600,000
Profit-to-sales ratio.......	30 percent	Marginal income...........	$ 800,000
		Less all fixed expense.......	200,000
		Profit.................	$ 600,000
		MI ratio...............	40 percent

EXAMPLE 5

Conventional Method		MI Method	
Total company sales........	$2,000,000		$2,000,000
Less material costs.........	600,000		600,000
Less all labor and expense...	1,000,000	Less only variable labor and	
Profit.................	400,000	expense.................	800,000
Profit-to-sales ratio.......	20 percent	Marginal income...........	$ 600,000
		Less all fixed expense.......	200,000
		Profit.................	$ 400,000
		MI ratio...............	30 percent

In examples 3 to 5, sales, material costs, and profit are recorded the same for both methods. In example 3, the conventional method records results as a profit-to-sales ratio of 20 percent and the MI method shows a marginal income-to-sales ratio of 40 percent. In example 4, sales and material costs double and profit triples from example 3. The profit-to-sales ratio increases to 30 percent but the marginal income ratio remains at 40 percent. Example 5 compared with example 4 shows sales and material costs the same and profit-to-sales ratio down to 20 percent, the same ratio as example 3. The marginal income, however, is only 30 percent, less than example 3. It quickly reveals variable labor and expense have not just doubled with twice the sales volume but have increased by 166 percent. This fact goes completely unnoticed when a profit-to-sales figure is used. The MI method enables management to be aware of changes in variable labor and expense and permits prompt remedial action to be taken on this controllable figure.

Graphic Presentation and Explanation of Terms. The marginal income method[1] is particularly adaptable to the use of the break-even graph or the profit path type of presentation. The following terms will be helpful in examining the profit path:

1. Marginal income is the amount of sales dollars left after subtraction of variable cost. Marginal income becomes profit after deducting fixed costs.
2. The break-even point is the point at which marginal income total equals fixed cost total. If the break-even point is not reached, the operation results in a loss equal to the difference between total fixed costs and total marginal income.

[1] The Marginal Income Control System is the name of the cost system embracing variable cost techniques developed by McCormick & Company.

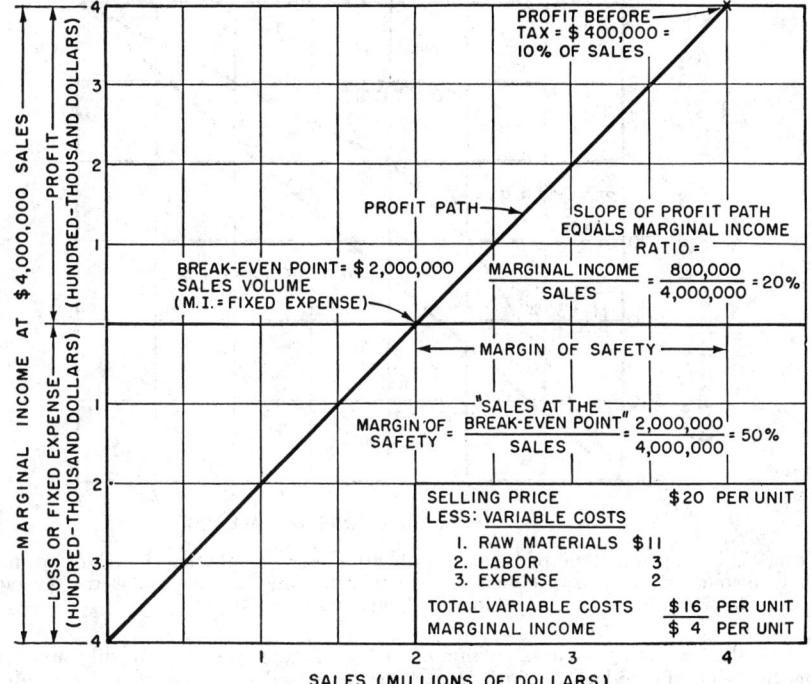

FIG. 8-4.   A break-even chart illustrating the concepts of marginal income.

3. The margin of safety is the percentage of sales beyond the break-even point or the percentage by which sales may drop before profit disappears (see Figure 8-4).

Hip Roof Chart.   The hip roof chart is an expansion of the basic concept of the profit path. It plots the behavior of several MI's and is extremely useful in evaluating the contribution of several products or operating divisions to a company. It consists of plotting individual MI's to obtain an overall MI for the company as a whole.

Figure 8-5 shows the result of plotting the following MI's in order of descending magnitude.

Product	Sales	MI, percent	MI, cumulative
A	$1,000,000	30	$300,000
B	500,000	20	400,000
C	2,000,000	10	600,000

This can lead to some very interesting manipulations of the product mix. Two approaches may be taken—manufacturing or sales.

Manufacturing.   The plotting of each product on the hip roof chart immediately tells management the items bringing the greatest profit. Armed with this knowledge, management can then determine the annual product mix most advantageous to the company. With a limited amount of capacity, the company can concentrate on the items that generate the greatest amount of marginal income and keep produc-

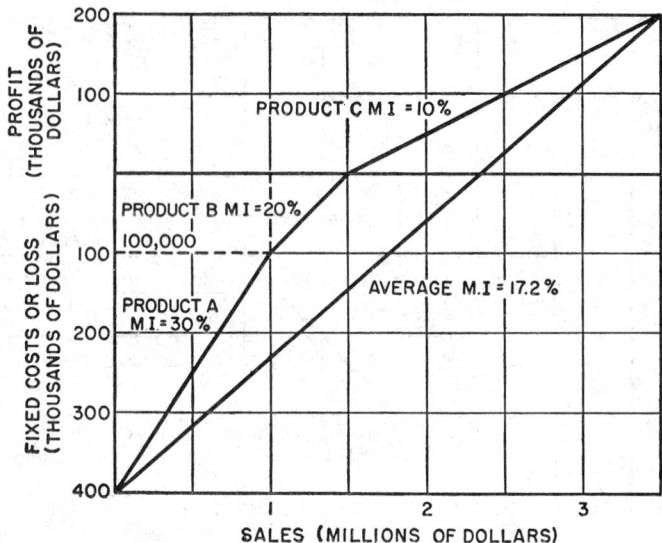

FIG. 8-5. A hip roof type of break-even chart used to emphasize the effect on marginal income of various products or departments and to determine the average marginal income.

tion of the remaining items at a minimum. In Figure 8-5, it is readily apparent that if sales of products A and B can be increased and the sales of product C limited, greater profits can be derived without increasing total sales.

Sales. Figure 8-6 shows the sales and marginal income attributable to three different products making up the total sales for a company. The solid white area for each product shows the minimum sales and marginal income for a specific

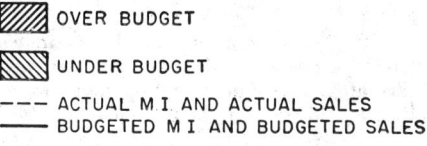

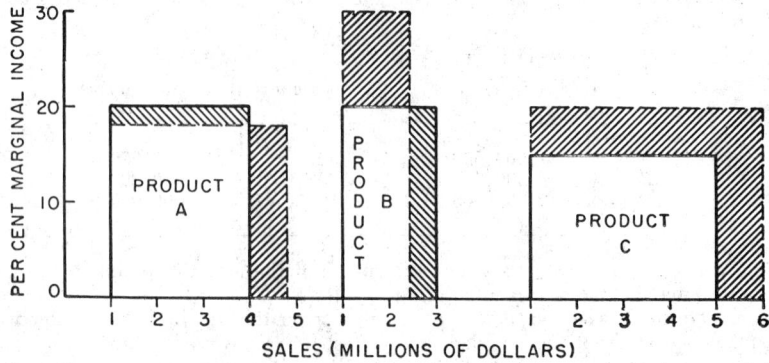

FIG. 8-6. Graphic representation of the sales and marginal income attributable to three products.

period by each product. In regard to each product, the sales manager has three courses of action:

1. Meet the sales quota for each product as represented by the areas bounded by the solid lines in Figure 8-6.
2. Raise the price of each product as is illustrated for products B and C in Figure 8-6 and meet the sales quota. This results in an increase of profits for each item but is not conducive to maintaining increasing sales volume. Product B actually suffered a reduction in total sales while the sales of product C increased.
3. Lower the prices as indicated for product A in Figure 8-6 and obtain a greater volume and subsequently either the same or more marginal income.

The sales manager's skill in combining actions 2 and 3 can improve the profit picture immensely. To do this, he must have information on the combinations of price and volume and the related amount of marginal income for each combination. This information should be prepared for him by the accounting department.

This same principle holds true when investigating the problem of planning more profit with limited capacity. The relationship between the time to manufacture and the marginal income generated by each product should be established by the industrial engineer. With this information, the most advantageous product mix can then be determined.

VARIABLE COST BUDGETING—A PRECISION TOOL

Some of the advantages of flexible budgeting in projecting the cost targets have already been mentioned. Flexible budgeting—or to give it a more descriptive name, variable cost budgeting—is worth looking at in a little more detail. It is far and away the best mechanism that is available, not only for budgeting (or profit planning) but also for the entire cost system of a company. It is a precision tool that enables management to separate profitable from unprofitable activities and open up neglected opportunities.

Most of the budgeting and standard cost systems set up in the past have used the device of under- and overabsorbed burden. To construct a system of this sort, first, some percentage of capacity should be settled on which is taken as "normal." Then the costs per unit at this level of operation can be calculated.

For example, say the capacity of the company is 12,000 tons, and operation at 80 percent of capacity is anticipated. Then 9,600 tons is the budgeted capacity. Costs per ton are obtained by dividing total costs by 9,600 tons.

But as any plant manager knows, costs per ton will actually come out to this figure only if production hits 9,600 tons on the nose. If it runs higher, per unit costs will be less; if it runs lower, they will be greater.

For this reason, the variable cost approach is rapidly being adopted for analyzing costs and projecting profits in a wide variety of industries. Among the advantages that it offers are:

1. Standards for control are good at all levels of production.
2. Costs are measured by actual behavior, hence are more accurate.
3. Because costs assigned to products are out-of-pocket costs, they represent the minimum price that will yield a profit.
4. Product cost ratios show accurately the contribution that each product makes to profit before fixed costs are distributed.
5. The necessity for tedious (and often inaccurate) distribution of overhead to departments is eliminated.

Why It Works. The theory behind variable cost control is simple. It is based on the sound assumption that fixed costs should not be translated to the product. By definition, fixed costs do not vary with production; they are there regardless of whether one unit or one hundred units are produced. Therefore they are a charge against the business as a whole and not against the individual product.

Variable costs, on the other hand, are plainly attached to the products that give rise to them. They represent the amount of money that the company puts out to manufacture that particular product.

The difference between the variable cost and the selling price is gross profit, more commonly called marginal income. This is the amount that each product contributes toward paying off the fixed expenses of the company and providing a profit on the operation as a whole. The more marginal income per unit a product contributes, the more profitable it is to sell.

This approach eliminates all argument and confusion over distributing fixed expenses. It avoids the arbitrary assessment of fixed costs to the various products.

How It Works. A look at two cases—one a prominent eastern railroad, the other a large steel company—gives an idea of how much difference the variable cost approach can make in company strategy.

The railroad, which uses variable cost analysis for profit planning, faced a problem in rate making. Historical analysis of freight train costs by its cost department showed that to haul a train 250 miles cost:

With 40 cars......... $2,400
With 50 cars......... $2,600

Dividing to obtain the cost per car, the cost department came up with $60 for cars in a 40-car train, $52 for cars in a 50-car train.

The question was whether or not the railroad could make a rate on hauling the trailers of a trucking company that was operating over the same route. The truck company figured that its costs were about $50 per truck for the trip.

As long as the railroad figured its costs at a minimum of $52 a car, it obviously could not quote a rate that would get the business. But the management decided to go into the problem further, using a variable cost analysis. Figure 8-7 shows the result.

The fixed cost of this train was $1,600; the variable cost was $20 per car. Therefore, instead of having to obtain $52 per car to break even in selling the 10 extra

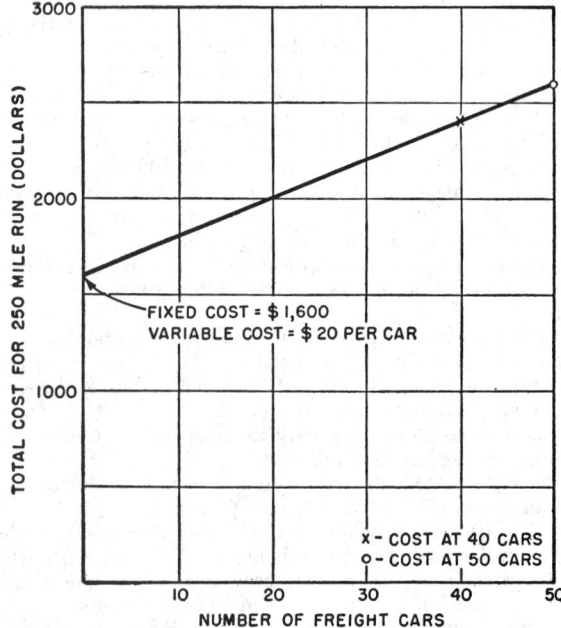

Fig. 8-7. Determining fixed and variable costs for a freight train haul of 250 miles.

cars to the trucking company, the railroad would be adding to profits if it got anything over $20.

The first 40 cars were sold to other shippers for $80 a car, or a gross income of $3,200. The cost was $2,400, leaving a profit of $800. By selling the next cars at $40 a car, the gross income would be increased by $400 to a total of $3,600. The cost was $2,600, giving a profit of $1,000. This was a gain of $200 in profit, or 25 percent—discovered by applying the principles of variable costs.

The marginal income of the first 40 cars was ($80 − $20), or $60 a car. The marginal income of the second 10 cars was ($40 − $20), or $20 a car. The profit path of their freight train would behave as shown in Figure 8-8.

The record of a midwestern steel company shows the same sort of thing. The company knew that its costs for making steel were:

	Total cost	Cost per ton
600,000 tons...	$60,000,000	$100
If they made 680,000 tons by adding 80,000 tons farm fence.	64,000,000	$94.10

Knowing that the best they could get for the added farm fence was $80 per ton, the company assumed that the sale of farm fence would be a losing proposition. From the above analysis, the evident loss to sell farm fence would be $14.10 per ton ($94.10 − $80.00) or the total loss due to manufacturing farm fence, 80,000 × $14.10, or $1,128,000.

But the executives put their heads together and used the principles of variable cost control. Plotting graphically the information given above, they found that the formula expressing the cost relationship was fixed, $30,000,000; variable, $50 per ton (Figure 8-9).

The graph shows that the out-of-pocket cost of producing a ton of steel was $50. The fixed costs were $30,000,000. Therefore, all tons sold over $50 would contribute marginal income to the company.

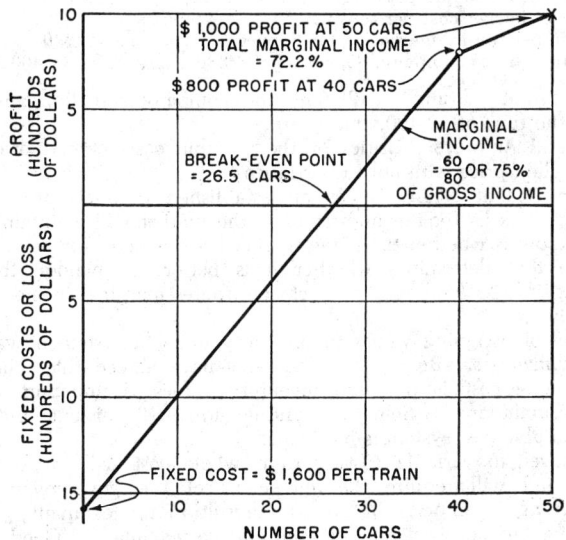

Fig. 8-8. A break-even chart showing the profit path for a freight train on a 250-mile haul.

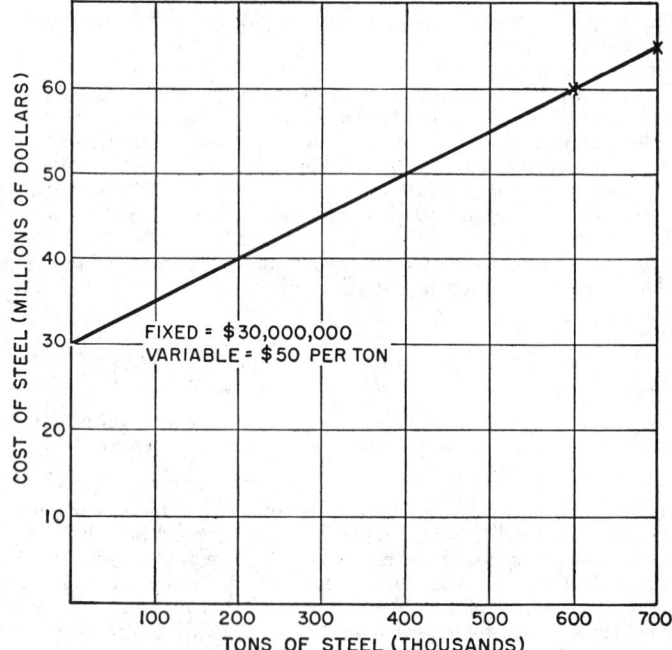

FIXED = $30,000,000
VARIABLE = $50 PER TON

FIG. 8-9.   A break-even chart of steel costs used to determine fixed and variable costs.

What actually occurred because of the sale of farm fence is shown as follows:

Farm fence selling price..........................	$80
Out-of-pocket cost per ton........................	$50
Gain per ton by marketing farm fence..............	$30
Total gain to company, 80,000 tons × $30...........	$2,400,000

Instead of apparently losing $1,128,000, the company actually gained by selling farm fence to the tune of $2,400,000.

Questions and Answers.   Critics of the variable cost method often ask, "How can we set a selling price with only a partial cost?"

The answer is that selling prices on established products are seldom set by costs.   A company sells for the highest price the market will maintain.   The important thing to know is whether the going market price is above or below out-of-pocket costs—because that determines whether it is better to abandon the product or keep on selling it.   As long as a product contributes marginal income to a business, it is worth selling.

The problem of assigning values to inventory under a system of variable costing may present difficulties.   But most accountants have agreed that whether a whole cost or a variable cost is put into inventory values is unimportant as long as consistency is maintained.   Some accounting firms still dissent, but the number who favor a variable cost system is increasing.

The changeover to variable costing from whole cost will involve a change in tax structure and will require the agreement of the government on the new method of valuing inventory.   But the net result to the company, profitwise, will always be the same except in the year the change was made.   Good control figures throughout the year are maintained by costing inventory at standard variable cost. At year-end, a fixed cost increment in inventory, held constant throughout the

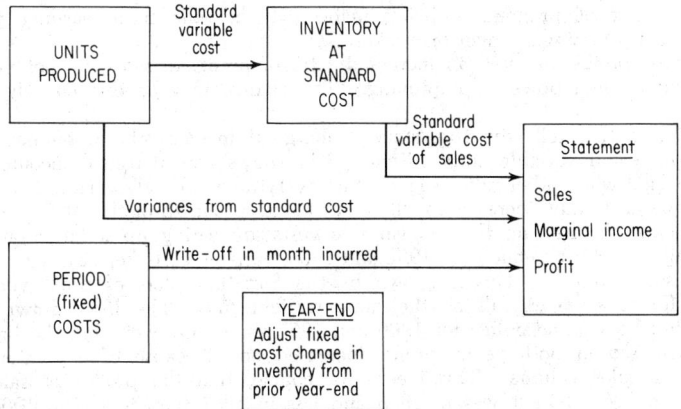

Fig. 8-10. Using variable costing during the year for sharp planning and control.

year, is adjusted upward or downward, based upon change in number of inventory units from the previous year-end. Thus, the performance at key review points during the year is not confused by under- or overabsorbed burden. (See Figure 8-10.)

Important Concepts. Flexible budgeting or variable cost budgeting is spreading rapidly among companies with advanced management methods because of the many advantages it offers.

In this system of budgeting, only variable costs are assigned to specific products. Fixed costs, which are properly a charge against the business as a whole, are not translated to the products.

Among the advantages that follow from this approach are that standards for control are good at all levels of production and there is no need for arbitrary and inaccurate assignment of overhead. From a planning standpoint, the great advantage of the variable cost approach is that it enables management to tell precisely which activities are profitable and which are not. By showing exactly how much a product contributes to income after out-of-pocket expenses are deducted, it reveals neglected profit opportunities and spotlights operations that are not paying their way.

Because of these advantages, the variable cost method is one of the most valuable tools of the profit planner.

SETTING THE TARGETS

A good budgeting system—one that will keep a company on the best and most profitable path—will set up five main targets. These targets and the company officers principally responsible for setting them are:

1. The profit target—the president
2. The sales target—the sales manager
3. The manufacturing target—the manufacturing manager
4. The financial target—the treasurer
5. The project target—the technical director

The Profit Target. There are two important yardsticks to use in setting up the profit target. One is what has been earned in the past; the other is what could be earned with the most effective possible use of the resources at the president's command. These two figures provide a minimum and maximum—a range of profit within which the target for the period will lie.

The earnings record of past years can best be understood if it is plotted on a break-even graph. This will show clearly which way the company is going and what sort of progress is being made—whether the basic earning power of the business is getting stronger or weaker.

Increased profits, or even an increased ratio of profits to sales, do not necessarily mean that earning power is improving. The change may be due entirely to rising volume.

The figure that tells the true story is marginal income—the difference between selling price and variable costs. Figure 8-11 shows how marginal income hoists a danger signal when other ratios suggest that everything is dandy. Here is a company that shows a steady increase in the amount of profit through 1967. But rising volume conceals the fact that its variable costs are eating up a larger and larger percentage of total revenue. In 1964 and 1965, variable costs kept the same relationship to sales volume. This is shown by the fact that plots of profit versus sales volume for these years fall on the same straight line. The lines drawn through the profit-sales volume points for 1966 and 1967 show, however, that the break-even point increased in both years; profits increased only because of a greater percent increase in sales volume. This means, of course, that the profit per sales dollar was less in 1966 than it was in 1965 and less in 1967 than it was in 1966. Such a situation is referred to as a dropping profit path. When volume fell off in 1968, the company made less profit than it did in 1966, though it was still ahead of 1966 in sales.

Earlier in this chapter under Using the Break-even Chart in Budgeting, several other important applications of the concept of marginal income were outlined. In setting a profit target, however, the great value of marginal income is that it helps a management to make a realistic estimate of the minimum profit the company can expect.

To get an idea of the maximum profit that might be made, a look should be taken at the competition. In most industries, there are enough companies whose earnings and balance sheets are a matter of public record to permit comparisons.

The best measure to use for such comparisons is return on net worth. For this purpose, net worth can be taken to consist of the stock and surplus accounts combined. Select the same year for comparison; if possible, use an average of two years.

The final profit target that a company sets up will fall somewhere between the theoretical maximum and the minimum demonstrated by experience. It should be realistic; there is no point in wishing for the moon when it cannot be reached.

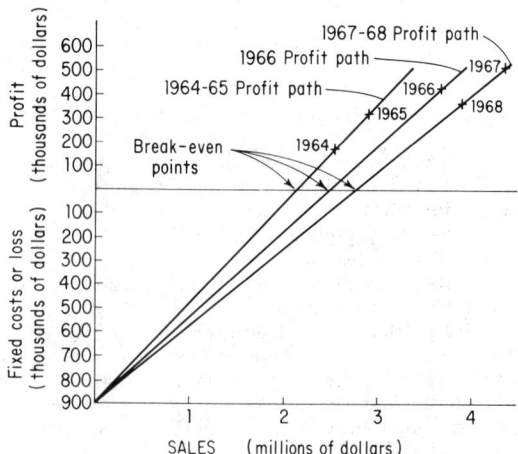

FIG. 8-11.   A break-even chart forcefully illustrates declining rates of profit (profit paths) and increasing break-even points.

But it should be high enough to keep the entire company on its toes, leaving no room for complacency and self-satisfaction.

An effective profit target used by many industrial companies is a percentage return on assets employed. For this purpose, the following assets are considered, being those most directly assignable to product:

1. Buildings and equipment
2. Inventories, materials, and goods
3. Accounts receivable

The industrial engineer can develop such a profit target for various product groups or lines within the overall company profit objective by separating the assets employed between those needed for specific product groups.

Although the marketplace is normally the final price level determinant, a study which indicates relative return on assets employed by various product groups can greatly affect marketing and product sales emphasis. The planned replacement of products with the weakest return must have a steady, though perhaps gradual, salutory effect upon company return on assets employed.

The employment of this technique serves the operational and financial goals mentioned previously, but in finality, is a serious part of preservation planning.

A bar chart, Figure 8-12, illustrating the measurement of several product lines against an overall company target return on assets employed, provides the industrial engineer with a visual tool with which to acquaint marketing and other management members with problems in this area.

Figure 8-12 readily illustrates that one product line, A, is at the company average and is holding its own. However, product line B is in trouble with only one-half the company average return. Serious study should be given this line in terms of cost, price, markets, volume, and mix. If nothing can be done to improve line B, it should become a phase-out candidate, or at best, a static situation with no increase in assets employed as long as opportunity exists to invest in more profitable lines such as line C or line A.

Line C, at a startling 50 percent higher than average company return, should be studied to determine possibilities for:

1. Expansion of the line
2. Expected life and ways to improve the life

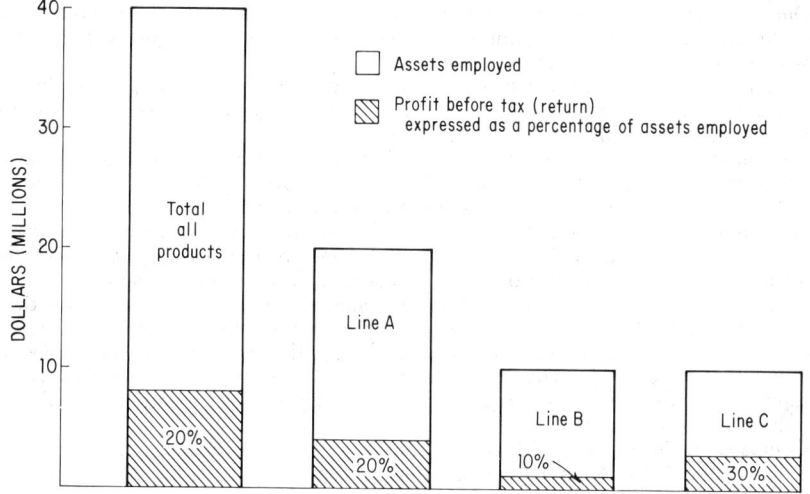

FIG. 8-12. A bar chart which matches total company return on assets employed with the return on specific product lines.

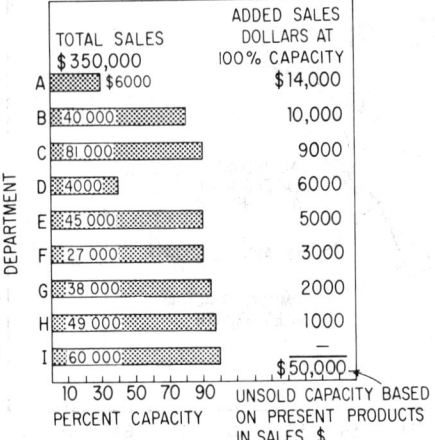

FIG. 8-14. A picture of available plant capacity, department by department, is useful information for planning a sales target.

simple ratio of profits to sales at a particular volume. The flexible ratio called the marginal income ratio (marginal income divided by sales volume) applies over a wide range of volume, whereas the ratio of profit to sales applies only at the particular volume for which it was calculated. The industrial engineer assures the effectiveness of the marginal income ratios by providing reliable product costs with which to compute marginal income.

When sales have finally been projected as accurately as possible, the final sales volume budget must be measured against the profit target. Often, it will be necessary to adjust the first draft to achieve the profit goal. These adjustments may call for expansion of the company itself or changes in its basic structure, as for example:

1. Additional capacity in certain departments
2. Larger sales force
3. Alterations or additions to sales territories
4. New products

When the sales budget has at last been keyed in with the profit budget, it then becomes the sales manager's target for the year. To allow for seasonal variations, it is usually best to set the budget up in terms of quarters. The sales manager will accept a quarterly breakdown; ordinarily he will regard a monthly quota as impractical. And so long as the combined sales for the quarter hit the target, it does not usually matter if individual months show fluctuations.

The Manufacturing Target. Once the sales target is set, the target for manufacturing costs can also be established.

Like the sales budget, the manufacturing budget should be stated briefly and simply. It should never attempt to do the detailed variance analysis job intended for the cost system. It should stick to the most significant targets and the broad planning necessary to achieve them. This is the purpose of a budget—to provide the big, basic planning for a company—and that purpose should never be buried in an avalanche of details.

Companies with a standard cost system, particularly if the system measures variances, have most of the tools they will need for the manufacturing budget. But, as contrasted with a cost system, the budget will stress such things as:

1. The allowance for staff departments, such as engineering, production control, purchasing
2. The target use of equipment centers

3. The price to be paid for production materials and supplies
4. The target allowance for rework and scrap
5. The rates to be paid for labor
6. The allowance for maintenance

Most of these items are picked up on a historic basis in the cost system. In budgeting, they are projected with the same care as other cost elements.

Flexible Budgeting. In projecting the cost targets for manufacturing, it is highly desirable, although not absolutely essential, to use a system of flexible budgeting.

Flexible budgeting is based on recognition of the fact that all manufacturing costs fall into one of three categories:

1. All fixed—no change with shifts in volume
2. All variable—increasing or decreasing with volume changes
3. Combination—part changing with volume, part not changing

Graphic analysis, using activity as the horizontal scale and cost as the vertical scale, will soon establish in which group a particular cost belongs. In the case illustrated by Figure 8-15, the best-fitting curve for the plotted points was a straight line. Its intersection with the "Overhead" axis indicates the portion of overhead that is a fixed cost, and its slope indicates how the variable cost portion changes with a change in direct labor costs. Overhead, therefore, is a combination cost.

The fixed costs developed from such analyses, plus the fixed portion of selling and administration costs, make up the total fixed costs used in break-even charts.

In flexible budgeting, the manufacturing cost targets are expressed in terms of fixed costs and variable costs. Thus, the targets in any period will vary as a function of volume, and they will always make a clear distinction between the costs that are under the control of the department manager and those that are not.

The older method of expressing manufacturing cost sets up a specific cost figure which is accurate only at the volume for which it is calculated. This method is perfectly clear in its handling of direct labor and materials, but it runs into trouble with overhead costs when volume does not coincide with the assumed figure. To overcome this difficulty, it is necessary to set up a special account for under-

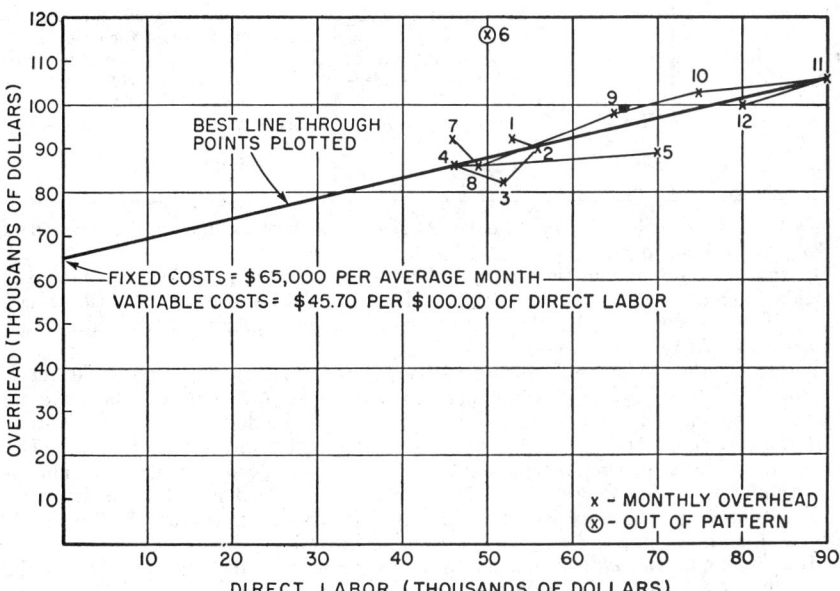

FIG. 8-15. By plotting overhead expense versus direct labor expense, fixed and variable costs can be determined.

and overabsorbed burden. If the sales volume is lower than the budget, not enough overhead has been allowed by the per unit cost assessments, and the balance has to be charged to the special account. If the budget volume is exceeded, too much overhead has been allowed, and it is said to be overabsorbed.

Either the flexible or the "over and under" system will permit effective profit planning. But the flexible system is far sounder and lends itself to wider use. It permits the profit planner to consider several alternative courses of action and select the most satisfactory one.

Under either system, the volume projection from sales can be applied to direct labor and material standards to arrive at a budgeted usage. This is a relatively simple operation.

Overhead targets are more difficult. They will require analysis, aided by graphs. The manufacturing manager will have to work out these figures in consultation with the department heads. He cannot expect the department heads to originate the estimates and send on their budgets without guidance. Because industrial engineering will later use these targets in analyzing variances, they should be dealt in at the time the targets are set up.

The final manufacturing targets should be expressed in terms of volume and in terms of units based on the categories used in the sales forecast. In addition, supplementary targets should be developed on projected material costs; this will be used by the purchasing department in preparing the schedule of material purchasing. Scrap and rework allowances should be approached with the idea of bettering the performance of the past.

When they are finally put together, the manufacturing targets will shape up like this:

Cost	Expressed in	Source
Direct labor.....	Per unit of product; total by quarter, by year	Cost system standards
Material........	Per unit of product; total by quarter, by year	Cost system standards for usage Purchasing forecast for cost
Indirect labor....	Variable portion per machine or direct labor hour or other production measure Fixed portion per day, week, month, or year	Graphic projection of standard crews at various volumes separating cost in fixed and variable portions
Expense.........	Same as indirect labor	Same as indirect labor

Each of these schedules can be summarized for most companies on one sheet of 8½- by 11-inch paper.

If time permits, a supplementary schedule should be prepared for projected machine center usage, based on budgeted volume. The back of this sheet can be used to present the sales department with a graphic picture of the sales value of unused capacity.

Before clearing the manufacturing budget, it is wise to make one simple check on the way the company's overall performance stacks up. This can be done by taking some unit of measure—pairs of shoes, yards of cloth, barrels of oil, tons of folding cartons—and dividing it into the total salaried and hourly workers. The resulting figure is an overall measure of productivity, a good common denominator for comparing a company's productivity with the productivity of competitors.

This simple check may flag the planner that something is out of line in his operation and lead him to set up higher goals for himself.

The Financial Target. The financial budget is clearly the responsibility of the company treasurer. It is the budget that provides the proper balance of moneys and thereby makes it possible for the company to achieve the profit target.

By watching inventories and receivables and by using bank credit, the treasurer makes an enormous contribution to efficient operation. He arranges for the fastest possible turnover of net worth. And, as pointed out earlier, it is not the percent of profit to sales alone that counts; it is the percent of profit in combination with turnover that makes a good profit performance.

Although the company treasurer performs many other services, including supervision of accounting, none is quite so important as providing the means by which the company can turn its resources over at the fastest possible rate. Behind companies rising with spectacular growth curves, there usually is an able treasurer who has learned the value of correctly placing inventory at the right point to service customers, who makes use of bank loans, and who keeps his receivables at a low level.

It is important to the treasurer that the sales forecast be presented by quarters, for the need for funds will fluctuate with the seasons. The treasurer will use the sales projection to prepare what is known as a cash budget. This will show by months how the varying requirements of income tax, vacation pay, seasonal shifts in inventory, and seasonal slumps in sales are projected in balance with the company's resources. Such a cash budget will show proposed timing and size of borrowings as well as payments on debt. It will also make provision for payments forecast by the project budget and try to provide for such contingencies as increases in labor rates.

The Project Target. The foresighted company that thinks of itself as a growing, dynamic organization is the exception. Far too often, a company will make no expenditures beyond daily operating costs—except on an emergency basis, with no thought or planning behind them.

These companies frequently complain about poor business or a bad industry condition. As a rule, there is nothing the matter with the industry. The trouble is with the company: it stopped growing organizationally years ago; lulled by good or fair profits, it has coasted along year after year making only day-to-day decisions. As a result, its profit position has gone to pieces bit by bit.

Dynamic budgeting, operating through a system of targets, automatically keeps a company from falling into this frame of mind. Planning the profit target and setting up the sales and manufacturing goals force the management to think about projects that go beyond the scope of day-to-day operations.

Such projects may include improvement of machinery, new equipment, better plant, project research, employee training—all things that the company could get along without on a temporary basis but things that are vital in the long run to a company that wants to stay alive and growing.

The scope of such projects among vigorous, ambitious companies can be enormous. In a single year, one company planned a reorganization study, researched simplification of a complex custom product and built a market for it, purchased a company to increase its sales territorial coverage, took on a new line of products, hired a division manager, broke ground for three new plants, opened a branch in Canada, and issued and sold $1,000,000 worth of stock.

The projects budgeted by many companies probably will be somewhat more modest than that. But they should be enough to make sure that by the end of the year the company will be on the upgrade in plant, organization, sales, and products.

Projects should be submitted by the department heads, supported by a computation to show the estimated necessity and probable payoff, as well as the amount to be spent and the expected plan of expenditures during the year. The industrial engineer can be of great help in devising a formula to compute the payout for the cost of those projects that are proposed with the idea of increasing profits.

The final project budget will be revised if necessary and approved by the president. The cash outlays that it calls for will, of course, be incorporated in the treasurer's cash budget.

Review of Budget Targets. A good budgeting system will set up five main targets:

1. The profit target

2. The sales target
3. The manufacturing target
4. The financial target
5. The project target

The profit target will fall somewhere between the minimum—as demonstrated by past performance—and the maximum—as shown by the efficiency with which competitors are using their resources.

The sales budget must be realistic, but it must also be ambitious enough to provide for achievement of the profit target. There are several devices that can be used to improve its accuracy.

The manufacturing target will be most effective if it is constructed on the principles of flexible budgeting, which make a distinction between fixed and variable costs.

The financial target is expressed in the cash budget prepared by the treasurer. This provides for the fastest possible turnover of the company's resources.

The project target provides for the company's growth and development.

FINALIZING THE PLANNED BUDGET

As indicated, the first step in budgeting is to establish the five main targets—for profits, sales, manufacturing costs, finances, and projects. The next step is to translate these major targets into specific departmental budgets. Each of the departmental budgets will be independent of the others in operation. But all of them will be coordinated, because they are derived from the same overall profit plan.

The final step in budgeting is to set up methods of reporting so that actual performance can be compared with the targets and variances corrected promptly. If the budget is properly drawn up and if the reporting system is well designed, each variance will stand out like a red flag. Then, at the regular variance meeting, the operating personnel—with the help of the industrial engineer—can analyze the cause of the trouble and start corrective action.

The Targets. In a good budgeting system, the five main targets are never set up independently and without regard for one another. They must be carefully coordinated and fitted together so that they form a completely consistent set of objectives. It is a sure sign of faulty budgeting if the project target, say, is established without regard to the sales projection.

The starting point is the profit target, which should be set to produce the highest possible return on investment. In setting up this target, it is important to take account of taxes as well as regular operating costs. Dividends are paid after taxes, and a comfortable income before taxes may be pitifully inadequate after the tax collector has taken his share.

Once the profit target has been fixed, the next step is to make the sales forecast. This must be cross-checked and tested to make sure that it is realistic.

When the sales forecast has taken shape, it should immediately be subjected to a three-way comparison:

1. Apply the product ratios from the cost system to translate sales into marginal income. This will show the profit that can be expected on this particular sales mix. It will also provide the basis for setting up the manufacturing cost target. Figure 8-16 shows the extension of a sales forecast into budgeting costs, marginal income, and profit, using variable cost standards provided by the industrial engineer. If more than one product was forecast, each product volume would be extended by its own standard variable costs and marginal income. The sum total of all products, then, in terms of variable costs and marginal income, would be the budget to this point. The subtraction of period costs from the total marginal income results in the profit content of the forecast.

2. Compare the profit expected on this sales mix with the profit target. Will sales, as forecast, yield the desired profit? If not, can something be done about it, either by putting on more sales pressure or by concentrating sales effort on the products that yield the greatest percent of marginal income?

3. Check the utilization of plant capacity that the sales forecast indicates. Will

STANDARDS	×	FORECAST VOLUME OF UNITS	=	BUDGET
Unit price	$10.00	1,000,000	Sales	$10,000,000
Unit costs				
Labor	2.00		Labor	2,000,000
Expense	1.00		Expense	1,000,000
Material	2.00		Material	2,000,000
Total variable cost	$ 5.00		Standard variable cost	$ 5,000,000
Unit marginal income	5.00		Standard marginal income	5,000,000
Marginal income ratio	50%		Period costs (fixed)	4,000,000
			Profit	1,000,000

Fig. 8-16. Using standard variable costs to determine the marginal income and the profit content of the sales forecast.

there be any significant amount of unused capacity? If so, can a use be found for it? Is capacity inadequate in any line? If so, how can enough be produced to make the sales target?

Obviously, these comparisons will force revisions and adjustments—sometimes in the sales target, sometimes in the profit target. After such revisions, both targets will be realistic, and both will be designed to yield the greatest possible profit.

Checking the sales forecast against plant capacity is a particularly important step because it will indicate whether or not there is a need for new capacity and will call attention to opportunities that may have been overlooked. With this information before it, management can then proceed to draw up the project budget for the period. It is important to remember, however, that the project budget may include many things besides additions to the physical plant. Reorganization of the company, development of new lines, and market research and development are all projects that may contribute to the long-run growth and prosperity of the company, even though they call for no expenditures on bricks and mortar.

Once the project target has been drawn up, the treasurer can make his cash forecast, balancing income and outgo by quarters, planning borrowings and repayments.

The Organization. Good budgeting is inseparably tied up with good organization and clear-cut assignment of functions. Each major area of the budget must be the responsibility of one of the top officers. Budgeting is a tool of management; it is not a substitute for management. In a company where the lines of authority and responsibility are unclear and confused, budgeting will be ineffectual.

The chief officers can very properly insist that by the time they have to make a decision, all the data should be in hand to show clearly what the choices are and what the consequences of each choice will be.

Top management cannot complain, however, if the budgeting process compels them to face up to some hard decisions. The purpose of budgeting is to see that such decisions are made—and made on the basis of the best possible information—instead of being allowed to go by default. Thus it may be uncomfortable for a president to face the fact that he must expand and modernize his plant if he is to maintain his profit position. And it may be hard for a marketing man to face the fact that his sales force is not producing the results that his competitor is getting. But facing facts like these and deciding what to do about them is the highest function of management.

Reporting and Measurement. A good budget demands a good cost system. It is impossible either to set up targets or to keep performance on target without adequate machinery for analyzing and controlling costs.

A discussion of cost systems is beyond the scope of this chapter. But it should be pointed out that, if a company has a satisfactory cost system, it can add a budgeting system with very little additional time and expense. The same breakdowns and divisions that it uses for costs can be used for the budget. Data on performance can be assembled and set up alongside the budget targets within forty-eight hours after a budget period ends.

In addition to the budget's setting targets for labor, expense, and material usage, there should be a purchasing budget to control:

1. Inventory turnover
2. Material costs

The material inventory or turnover budget is dependent on production usage requirements, cash position, lead time, storage space, and similar considerations. Material costs will, of course, depend on variations in the market. But setting up the budget should be the occasion for a careful review of buying practices and consideration of alternative sources of supply. The purchasing agent should then make the best possible forecast of what he will have to pay for materials in each quarter, and this should become the target.

Procedures will vary from company to company, but the process of reporting and review will be essentially the same for all companies. Reports from the individual departments will be used to make up summary reports on the main categories of costs—labor and expense, material usage, and general and administrative. These costs will then be combined with the sales report to make up the profit and loss statement for the period.

Selling and Administrative Costs. Selling and administrative costs contain a relatively low proportion of variable costs and therefore cannot be controlled by budgeting procedures as tightly as other costs. But they too will benefit from the application of budget methods. And in many cases, it is possible to achieve major economies by setting targets for these costs.

Administrative costs are those operating costs outside the manufacturing and sales department. For the sake of simplicity, it may be assumed that accounting functions fall in this category, although in larger organizations they may constitute another branch of the overhead budget.

The biggest part of the administrative budget consists of payments to people, most of whom are on a salary rather than a wage basis. The part of the budget that has to do with organization, mainly the salaries of department heads, is fixed and will vary little except to reflect the effects of salary changes from year to year. The part that represents the accounting function—bookkeeping costs, order handling, payroll, and billing—will vary partly with activity, partly with time.

Few companies realize that there is this element of variable cost in the administrative budget, principally because they do not know a simple method of uncovering it. The industrial engineer can fill in this gap in management's information by plotting activity (number of people on the payroll, number of invoices, or some similar measure) against hours of labor spent in performing these functions. For a quick start, the historical record of operations in recent years will be enough to uncover the relationship. Later, the standards derived in this way should be replaced by an exact study of the amount of paper work required and its relation to activity of the company.

A target should also be developed to set the number of supervisory personnel contained in the administrative budget. A relationship with production should be established and compared with that of competitors.

The expense portion of the administrative budget, consisting of taxes, insurance, depreciation, and the like, is largely fixed. It offers less opportunity for flexible budgeting.

The sales cost budget consists mainly of salaries, commissions, travel expense, and advertising. The determination of policy on most of these items lies outside the scope of budgeting.

The system of sales compensation, for example, is a subject apart from budgeting. The compensation plan that is in existence will simply be applied to the sales forecast, indicating how much to allow for compensation in the budget. If additions to the staff are planned to take care of additional volume beyond the capacity of the existing sales force, that should be included in the budget.

Travel expense rates for car allowance, hotels, and mileage per salesman can be budgeted. A great deal of information is available on these subjects, and there are tables that can be used in setting standards.

The amount of advertising and the advertising cost are beyond the scope of this chapter. They certainly should not be determined by a mechanical process of earmarking a fixed percentage of the sales dollar for advertising. They should be fixed only after careful study of markets, competition, and opportunities.

The profit and loss statement will show sales, the marginal income that the sales generate, the fixed costs that have to come out of that marginal income, and the profit that remains when the fixed costs are deducted. If actual performance is measuring up to the standards set in the budget, the profit figure will be the same in both columns. If something is wrong, there will be variances that pull the actual profit figure below the standard. If things have gone better than expected, the variances will put the actual above the standard.

One of the great virtues of a system organized along these lines is that it highlights the variances instead of letting them get lost in a mass of operating detail. At any level of management, from the department head on up, the warning sign will be visible whenever costs start to get out of line.

Budgeting Steps Reviewed. The first step in budgeting is to establish the overall targets for profits, sales, costs projects, and finances. These must be carefully cross-checked and coordinated. The next step is to translate the overall targets into departmental budgets. The final step is to set up procedures for reporting performance and comparing it with the budget.

With a good cost system, both the translation of sales goals into costs and the reporting of performance will take little additional time and work. The product ratios supplied by the cost system will do the job.

Good organization is essential to good budgeting. A good budget should not burden management with details. But the discipline of budgeting will make management face decisions that should be made.

MAKING THE PLANNED BUDGET PROGRAM WORK

The budget is a planned course of action to attain a profit goal. But a plan is useless if it merely stays on paper; there must be a definite procedure for seeing that operations follow the plan. Thus, the reporting of deviations from budget standards and their subsequent correction are the most important steps in the budget program. The industrial engineer plays a major role in the analysis of variances to determine cause and to find ways to eliminate variances of all types.

Tying in Responsibility for Deviations to the Organization. One of the prerequisites of a budget program is a clear-cut organization. When the responsibility for meeting a standard and correcting any deviations from it is vested in a specific individual, continuity of interest and action is assured. Every level of supervision should be charged with maintaining the standards that apply to its operations. This policy will hold good for both line and functional organizations. The maintenance of budget standards must be delegated just as duties and responsibilities are delegated. And just as the authority to carry out duties and responsibilities is delegated, authority to control the standards should also be delegated. In this manner, deviations will be discovered and dealt with by the people closest to the problem and by those with a direct interest in how things will look in the record. A system that ties standards and their deviations directly to the organization will produce specialization of effort and speed of action on deviations.

Some concerns use incentives as a means of assuring continued interest on the part of supervisory personnel in meeting or exceeding standards. The use of an incentive tied to budget performance is contingent on the following cautions:

1. Standards should be set so that they hold supervisors to the performance expected of them. The incentive should be used to reward supervisors exceeding the standards, not to pay for adequate but undistinguished performance.

2. The standards should apply only to factors over which the supervisors have full control.
3. The plan should be tailor-made to suit the conditions involved.
4. It should be weighted to give all the factors equal influence.

Perhaps the surest method of keeping the supervisors interested in reaching standards is to let them participate in the setting of the standards in the first place. When this method is used, it leaves little room for excuses or rationalizations to explain failures. Most people in a supervisory position take pride in achievement. If they think the standards are fair, they will have a powerful psychological incentive to meet or exceed them.

Deviations can be best analyzed and resolved when divorced from personalities. Supervisors should be trained to look at deviations as objectively as possible so that they can discover the real reason why a particular failure has occurred.

For example, a deviation might have occurred because a machine broke down. At first the supervisor might chalk the deviation up to mechanical failure. This might be absolutely true, but if the supervisor stepped back and looked at the situation objectively, he might consider the fact that one of his duties was to check the machine periodically to make sure that the proper preventive maintenance program was being carried out. Subsequent thought and investigation might reveal the fact that lack of such a check was the cause of the breakdown.

Reporting That Highlights the Variances. Successful control is dependent on the comparison of actual results with planned results. The presentation of results should be kept as simple and direct as possible. Actual and standard figures with the variance between them are the key points to be emphasized. Additional data may be interesting, but it is the variances that demand prompt action if the profit goal is to be reached.

Forms should be prepared by the accounting department to provide all the essential information and highlight the trouble spots. They should serve as the basis for discussion at the regular meeting on variances.

The Variance Meeting. The mere reporting of variances, however good the reporting system may be, is not enough. It is the beginning rather than the end of the process that makes good budgeting real profit planning. The object of budgeting is to see that the original plan is carried out and that the deviations are eliminated.

The mechanism for doing this is the variance meeting—the periodic meeting of the people charged with the immediate responsibility for putting the budget into practice.

The variance meeting should be attended by the budget director; the committee charged with reviewing actual results, including the industrial engineer; and the operating men who are responsible for meeting the standards that the budget sets. The makeup of the meeting reflects the fact that corrective action on variances must be taken at every level of supervision.

The proper timing of the variance meeting and the presentation of the results of a budget period is of the utmost importance. The sooner the results can be presented to the supervisors, the quicker corrective action can be taken and the easier the corrective action will be.

Thus, the lower level or departmental results should be presented to the people involved at the first moment the figures are available. While the consolidated results are being prepared for top management, remedial action on deviations may already be under way.

Any variance that shows up when actual results are compared with budget targets should be given thorough study. But not all variances reflect conditions that the company can do something about. The first thing that the budget committee must do is determine whether the cause of a deviation lies within the control of the business or outside of it.

Causes outside the control of the business cannot be remedied. They can only be recognized, recorded, and considered in making future budgets.

First Things First. Variances due to causes that can be controlled from within the business are something else again. All such deviations should be investigated, but deviations that have the greatest effect on profit should be considered first. An item such as material might deviate from the standard by a comparatively small amount, but in relation to the profit objective, the effect might be very large. Conversely, the deviation from the standard on a maintenance item might be very large in relation to that same standard but very small in its impact on profits.

This rule of first things first means that variations in direct labor should almost always receive immediate attention. Some budget programs call for checking actual labor cost versus standard labor cost a number of times within the regular budget period, so that deviations in this important item can be detected before they become too great.

The Glass House Approach. An effective mechanism for the control of labor cost is found in the "glass house" approach. This approach recognizes that labor supply and labor demand within any given department or plant will not always match ideally. Because it is impractical to lay off people in many areas and on a daily or an hourly basis, a medium is provided to permit a foreman to reduce his actual payroll cost by transfer of personnel. A control system cannot perform its function if explanations of variances include statements such as "I could not meet my standard because I had to hold onto my people."

An additional line on the labor control report entitled "Management Policy Account" is the place for labor costs to be transferred to whenever the foremen cannot find productive work for their people in their own or in any other department. The management policy account is a labor pool and should be the responsibility of the plant manager, who accepts the foremen's transfers into the account through a written or verbal contact.

In this manner, the foremen identify all excess labor and have no excuse for the nonattainment of their individual efficiency standards. The foremen also voluntarily identify for the plant manager the total excess people that are present in any given period of time within the plant. The plant manager now has several choices. One is to assign these people to varied "make work" projects such as painting, cleaning, or building maintenance, but still charged to him as a cost. Another is to make a permanent reduction in the payroll when sufficient management policy variance experience has been accumulated.

This variance is often called the "glass house," because it throws the spotlight on nonutilized labor. There is a large plant in England where the glass house physically exists. Foremen actually send unneeded people to a completely glass-enclosed room, centrally located in the plant. While there, the employee is permitted to rest while awaiting further assignment. It is the plant manager's responsibility to keep the glass house empty.

Sales Variance. The first things first approach also means that prompt action should be taken on any variation from the most important budget standard of all: the sales quota for the period. If sales fail to reach standard, the rest of the standards may be achieved but the profit target will be missed. The existence of a sales volume variance of any size should be a matter of concern for all top management but especially the sales executive. Often, extra effort on the part of the sales force will stave off a threatened slump in income and profits.

A carton plant once came up with a serious sales variance. Under prebudgeting procedure, it would have been chalked off as a bad month. Under a well-constructed budget system, however, the variance and its effect on overall objectives were immediately apparent. Top management applied heavy pressure on the sales division to increase sales for the next period so that quarterly standards could be met. This extra prod caused sales to contact a customer they had never been able to sell. Through the extra effort, they were able to land a substantial order from this customer, although ordinarily they would not even have called on him.

Beating Costs into Line. If sales are meeting their quotas, then the problem of achieving the budget targets resolves itself into a problem of keeping costs in line. Every company has certain costs, the control of which is the key to profits. These should be singled out early in the budget program and close watch should be kept on them.

It is in this area of cost control that the industrial engineer can make one of his greatest contributions. He is in touch with almost all phases of plant operations. His overall viewpoint, his training, and his experience help him determine the true reasons for cost variances and their consequences.

Often a particular variance will have several causes. And often the same variance will have multiple effects. Such cases require careful investigation and interpretation by someone whose knowledge is not confined to a single department. The industrial engineer is particularly well qualified to handle these cases.

In the course of a variance meeting, the industrial engineer who knows his company may find that he can make a wide variety of contributions:

   Interpretations of variances and the weight that each carries
   Suggestions to supervisors for remedying variances
   Analysis of major cost items and their components
   Suggestions for revisions of standards

In addition, the industrial engineer will probably find himself conducting studies and investigations on particular problems—ranging all the way from personnel practices to plant utilities. For a sample, here are some of the situations that often need investigation to keep costs in line with the budget targets:

1. Deviations are often attributable to the human element. Investigations and studies of employee attitude, morale, and labor relations in general can be highly revealing. Absenteeism, lateness, grievances, and other personnel problems can influence labor standards greatly.

2. Time studies, incentives, methods, and training procedures need periodic rechecking. The influence of time and new personnel can introduce subtle changes that go undetected but still influence established standards to a significant extent.

3. Methods employed in planning, scheduling, and controlling production need frequent restudy because they determine the utilization of men, materials, machines, plant, and time. A look at inventory and material controls is also important. It may reveal malfunctions affecting several different phases of operation.

4. Maintenance variations are becoming increasingly important as plants become more mechanized and automatically controlled. Negative and positive deviations from the standard have to be watched with equal alertness. In the long run, undermaintenance of facilities can be disastrous and overmaintenance very expensive. Deviations and maintenance requirements should be investigated by the industrial engineer.

5. In the average plant, a high percentage of direct costs is incurred by material handling operations. Handling does not enhance the value of an item. Deviations arising in this category can be curbed by studies of the layout, routing, and handling methods employed.

6. The industrial engineer can help the supervisor by ensuring that cost controls are being maintained and are doing the job they were set up to do.

7. Waste, scrap, and rework are key items. When they are traced to their sources and the reasons for their occurrence are determined, many other variations may be completely or partially explained. Industrial engineering will cooperate closely with quality control on such cases.

8. The industrial engineer can perform a great service to the supervisors and the plant as a whole by installing or maintaining a well-thought-out suggestion system. No one can see all the ways a job can be done. The person on the job should be encouraged to think about improvements and submit ideas for consideration. A well-run suggestion system can be a continuing source of new ideas.

Deviations Requiring New Equipment. By making a complete survey of the physical and economic aspects of the problem, the industrial engineer can be especially helpful in the handling of deviations requiring new equipment. Improved methods through the institution of new tools and jigs, improved equipment, and changes in material handling systems are all part of industrial engineering activities.

Take the case of a printing plant that had a large material handling variance. Investigation by the industrial engineer revealed that the addition of a forklift truck to the warehousing facilities would eliminate this variance. However, the lift truck would cost $8,000, and to justify such a capital expenditure, an increase of $40,000 in gross profits would have to be achieved. This was not possible at the time. But industrial engineering probed a little deeper into the problem and decided to recommend the rental of a forklift truck. The rental for each budget period was greater than the cost would have been if the company had purchased the truck, but the necessity of a large capital expenditure was avoided and immediate savings over the additional costs were realized.

One thing to remember in this connection is that equipment manufacturers are a valuable source of advice on the reduction of costs. Their representatives are usually available without cost and can make available a large fund of knowledge about similar situations. The advertisements and articles in pertinent trade publications are another source of real help.

Deviations in Materials and Supplies Cost. Frequently, in appraising the suitability of budgeting to its operations, management says, "The prices of our purchases jump around so much that any forecast would quickly lose its value." This objection can be met by providing for adjustment in the active budgeting period—from one year to six months and even down to three months if necessary. This will almost always do the trick. Few commodities have varied as much as wool, cocoa beans, and waste paper. Yet the erratic behavior of all three has been harnessed to the principles of budgeting.

One of the most effective correct steps in controlling purchase price variances is to free the buyer to buy. Surveys of purchasing practices reveal that few purchasing agents spend much time developing new and competitive sources of materials and supplies. Extending sources of supply and bringing into the spotlight items formerly given routine purchasing treatment pay fine profit dividends.

In any case, the purchasing procedure used to purchase each material should be studied with these questions in mind:

1. Can costs be reduced by buying in larger amounts?
2. Is it practical to lower quality?
3. Will long-term contracts be advantageous?
4. Can larger inventories be used to take advantage of off-season buying?
5. Are there substitutes or other sources available?

Variations in the cost of materials from the budget estimates may result from the way the materials are used rather than from changes in prices. For example, the real cause of a variance may be one of the following:

1. Excessive scrap and waste during processing
2. Introduction of new processes
3. Specification changes in products
4. Changes in materials used
5. Quality specifications above needs
6. Damages during handling and storage

Deviations of this sort require careful investigation. Each factor may involve secondary ramifications that range all the way from a breakdown in communications to poor supervision. Reasons must be established, recorded, and corrected. Here again, the industrial engineer is qualified to handle the analysis of the problem.

Meeting Pressure to Change Budget Targets. When variances occur, it is easy to say that the budget target was not set correctly and that it should be changed to match current results. The industrial engineer can best meet this pressure by being extra careful about setting standards in the first place.

It is a must that the person who will be responsible for meeting the target be consulted on the establishing of that target. In many cases, his opinion will completely contradict the facts. Every effort should be made to reconcile the two. This may take time and special studies, but when a man is convinced and approves the target at the start, little room is left for later changes in the standard.

Once the standards are set, they should not be changed except to meet changed conditions. Before making a change, it is well to consider the following basic questions:

1. Have conditions changed?
2. Have they changed to a significant extent?
3. Are there secondary effects on other standards?

The pressure on the industrial engineer may be very heavy and continue over a long period of time, but if the basic procedure is sound, the industrial engineer will be proved correct.

Take the case of a water standard in a paper mill. After careful consideration, industrial engineering set the water standard for the budget program. The first-period budget record showed a variance of several hundred percent. The standard was thoroughly restudied, but industrial engineering could find nothing to invalidate its initial conclusion. The variance continued for several years. Period after period it stuck out like a sore thumb. Management exerted heavy pressure on operating personnel to close the gap, and they in turn exerted pressure on industrial engineering to raise the standard. Industrial engineering stood firm. Gradually the gap lessened as more efficient methods were introduced and finally the standard was met. It took time and a great deal of effort by supervisory personnel, but in the end the standard was vindicated.

Major Points for Spotting and Correcting Deviations. To make a budget effective, there must be a system for reporting deviations from budget standards and taking action to correct them.

The comparison of actual results with planned results should be kept as simple as possible.

At the close of each budget period, there should be a variance meeting to discuss the deviations and start corrective action.

Investigation of variances should follow the rule of first things first, concentrating on the variances that have the greatest effect on profits.

The industrial engineer is particularly qualified to conduct such investigation and suggest methods of keeping costs in line.

Once the budget targets are set, they should not be lowered just because performance does not measure up to them.

SUMMARY

It cannot be repeated too often that a budget is a system of profit planning. And in this description, the emphasis should be on the word "planning" rather than on the word "system." The systems and procedures of budgeting should be designed to simplify and facilitate the task of setting profit goals, not to complicate it. A mass of uncoordinated forms and statistics is not a budget, and the mere accumulation of data is not planning.

It follows from this that budgeting should not be a burden upon top management. The system must be set up so that the chief officers of the company make only the broad and fundamental decisions. The attention of the top men should be focused on the vital questions of how the company can make the best possible use of the resources at its command, and how it can keep expanding and strengthening its position in the future.

In this chapter, the principles of such a budget system have been outlined, and the appropriate targets for budgeting and the best way to set up these targets have been discussed. It was then shown how these targets can be matched up with actual performance during the budget period so that prompt action can be taken to keep operations in line with the budget. The role of the industrial engineer, both in the development of the budget and in helping to attain the budget, has been stressed.

Although some of the tools used in budgeting have been discussed, a catalog of details has been deliberately avoided. Each separate budget system should be carefully tailored to the structure and needs of the individual company. Consequently, each will be different in details. The company that wants to install an effective budget system cannot expect to take a cut-and-dried formula and follow it step by step the way a cook follows a recipe. It will have to put the job in the hands of an expert and be guided by his advice.

All good budgeting systems, however, will have two things in common: they will set up targets, and they will provide machinery for making actual performance work toward these targets. Thus budgeting is not just a hopeful forecast of what the future will bring. It is a system of establishing goals and guiding the operations of a company toward those goals.

BIBLIOGRAPHY

Elliott, Norman J., "Training Management Personnel in Using Cost Reports," *Journal of Accountancy*, June, 1962, pp. 80–81.

Ferrara, William A., "What Managerial Functions Does Accounting Serve?" *Financial Executive*, July, 1964, pp. 27–30, 33.

McCormick, Edmund J., "Direct Costing," sec. 10, chap. 10, in H. B. Maynard (ed.), *Handbook of Business Administration*, McGraw-Hill Book Company, New York, 1967.

National Association of Accountants, Research Report 37, January, 1961.

Patrick, A. W., "Direct versus Absorption Costing," *The Controller* (now the *Financial Executive*), April, 1961, pp. 167–173.

Staehle, R. L., "The Human Side of Cost Control," paper presented before Financial Executives Conference, June, 1964.

Computers and the Industrial Engineer

Chapter 1

Computer Fundamentals

FRANK J. CARR

**Director, Information Systems Laboratory,
Westinghouse Electric Corporation, Pittsburgh, Pennsylvania**

The computer is a complex piece of machinery with a rich theoretical foundation. As a tool of management production, it has profoundly affected society during its relatively short modern history. Even to begin to talk about fundamentals, it is necessary to narrow the scope of the subject.

In the pages that follow, codes and machines are discussed to separate the computer itself from what makes it work conceptually. Subsequent comments relate codes to machine instructions and show what the computer does with the instructions it receives. Computer programming completes this initial set of computer fundamentals.

The discussion then moves into the economics of computation and its influence on the introduction of computers into industry. Typical relationships among various classes of computer expenditures also are reviewed.

Finally, the management decisions involved in the acquisition of a computer and the problems that have been faced by companies during this process are presented.

CODES AND MACHINES

There is a close relationship between communications equipment and the computer. This relationship offers a way of leading into the complex subject of computers by starting with a relatively simple device known as the telegraph.

Samuel Morse usually is credited with inventing the telegraph system and what

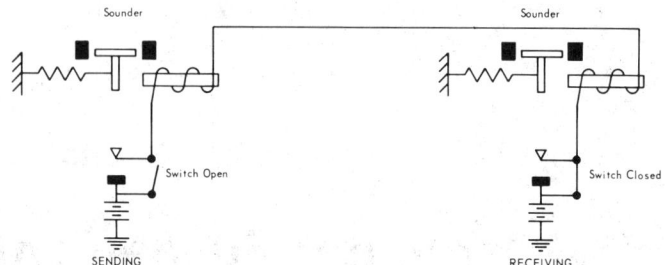

Fɪɢ. 1-1.   Telegraph system schematic.

we know today as the Morse code. Understanding the difference between the two provides a useful foundation in understanding how computers operate and how they can be programmed to do their work.

The clicking sound of the early telegraph system is familiar to most of us through exposure to western movies over the years. Figure 1-1 shows a typical telegraph system.

By depressing the key, the operator at the sending station causes a current to pass through the relays. As a relay closes, the armature strikes a stop, making a click. When the key is opened, the relay opens and another click is sounded. In this way, an operator can tell the position of the armature.

The telegraph system alone would be interesting, but not very useful. It was the code that Morse invented which added utility to the system.

A code is merely a system of symbols used for meaningful communication. The alphabet is a code which can be used with pen and ink and paper as a medium of communication. The alphabet uses a total of twenty-six symbols; Morse's code uses two symbols—the dot and the dash—in various combinations, which are familiar to most people (Table 1-1).

Of course, the symbols themselves are not transmitted; instead, they are represented physically by the position of the armature. A dot or dash is distinguished by the length of time the relay is in the closed position. Because the operator recognizes the time difference between the starting click and the ending click, he can distinguish between a dot and a dash by the relative difference in the length of time the sending operator allows for each symbol. The prescribed relationship is that the dash should be three times the length of the dot, with a period of time equal to the dot between each symbol and a period equal to the dash between each letter.

This code also is used for other means of transmission, for example, short and long flashes of light, or the more familiar short and long sounds used in radio transmission. A method used for training in the classroom is to substitute the spoken word "dit" for the dots and "dah" for the dashes. The letter V becomes "dit, dit, dit, dah," a very famous sound from World War II and Beethoven's Fifth Symphony.

TABLE 1-1.
Examples of
Morse Code

A   · —

B   — · · ·

C   — · — ·

D   — · ·

E

Computer Codes. Distinguishing between the code and the machine is not as difficult as it may appear at first. For example, the holes in a punched card or a punched paper tape are physical representations of a code. An electrical impulse to a print hammer is a physical representation of the code or symbol which will be printed on the paper when the hammer strikes. The arrangement of magnetized spots on a magnetic tape, drum, or disc and the arrangement of magnetized cores in an internal core storage are representations of codes.

The one common characteristic of all the above examples is that, by first defining the position or positions where the representation of the code is to appear in the physical equipment, the code itself can be represented on paper as a combination of 0s and 1s. Because 0 and 1 are used as digits in binary arithmetic (an arithmetic requiring only two symbols), they are usually referred to as bits (binary digits).

Any coding system in which the encoding is done through the use of bits (0 or 1) is called a binary code. The simplest is the straight binary, in which each decimal digit is represented by its equivalent binary representation, as shown in Table 1-2.

It is easy to see that counting in binary arithmetic involves only the following rules:

$$0 + 0 = 0$$

$$0 + 1 = 1$$

$$1 + 1 = 10 \text{ (or 0 and carry 1)}$$

For example:

3	11	$(1 + 1 = 0$, carry 1; the carried
+1	+ 1	$1 + 1 = 0$, carry 1; the carried $1 = 1)$
4	100	

5	101
+6	+110
11	1011

The straight binary code can be used to develop additional codes. For example, the binary coded decimal system uses a block of four binary digits to represent the decimal digits (0 through 9). Decimal numbers higher than nine are represented by combining appropriate blocks for that number. For example, the decimal number 573 is given by 0101 0111 0011. The decimal number 12 is simply 0001 0010. (Contrast this representation of 12 with that of the straight binary code as shown in Table 1-2.)

In Figure 1-2, a simple circuit consisting of a power source, lamps, switches, and diodes illustrates how the nine decimal digits can be transformed to a block of four binary digits. For example, the switch that turns on the lamp in the

TABLE 1-2. Straight Binary Codes

Decimal	Straight binary
0	0
1	1
2	10
3	11
4	100
5	101
6	110
7	111
8	1000
9	1001
10	1010
11	1011
12	1100

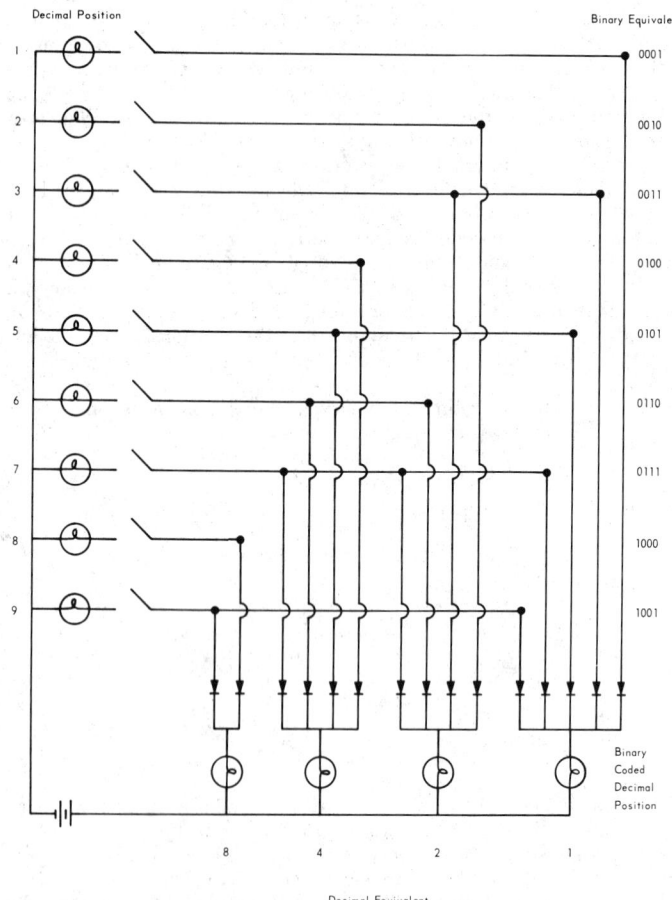

FIG. 1-2.   Circuit representation of decimal to binary conversion.

decimal position 6 will also turn on the two center lamps of the four binary coded decimal positions, indicating that 0110 is its binary coded decimal equivalent.

Note also in Figure 1-2 that each binary coded decimal position has a decimal equivalent associated with it.   It is therefore quite simple to determine the decimal value of the binary coded decimal.   For example, 0110 is equal to 4 + 2, or 6, while the value of 1001 is 8 + 1, or 9.

A more widely used version of the binary coded decimal system uses two zone bits, known as A and B bits, in addition to the four numeric bits described above. Adding a seventh, or C, position allows error checking.   A bit is placed in the C position only when the sum of zone and numeric bits in the first six positions is odd.   In this way, if an odd number of bits is detected in the seven positions, an error is indicated.

Because the A and B bits, in addition to the four numeric bits, permit up to 64 combinations of 0 and 1, the 54 positions remaining after 0 through 9 are coded can be used to represent the 26 letters of the alphabet and 28 other symbols such as punctuation marks, dollar and cent signs, or any other arbitrary mark.

One conclusion to be drawn from the preceding discussion is that the computer is a symbol manipulator. The manipulation may be simple, such as when a punched card is read into a computer and the symbols it contains are printed out on a piece of paper as letters, numbers, and punctuation marks, or it may be complex, such as when symbols are compared and rearranged.

It also should be obvious that the computer is not merely a giant calculator. In fact, computation may be thought of as one special case of symbol manipulation which the computer can do very well because of its speed and reliability.

Finally, it may occur to the reader that there is a strong relationship between Samuel Morse's telegraph system and the modern computer; in fact, there is. Computers are revolutionizing communications, and communications are revolutionizing the use of computers, as will be discussed subsequently.

COMPUTER INSTRUCTIONS

A further reinforcement of the relationship between communications and computation may be found in a few historical notes. As examples, the punched paper tape was developed by Jean Baudot in France during the 1870s as a means of transmitting telegraph messages between two unattended machines, and in 1944, the Mark I computer was completed by Dr. Howard H. Aiken of Harvard University, using paper tape teleprinters and other electromechanical equipment developed for communications systems.

Most of what has been said thus far about computer codes has been concerned with how data can be represented by codes that correspond to physical relationships in the computer. This permits the computer to manipulate symbols; however, a given computer can perform only certain types of manipulation, and it must be told which ones to perform on what pieces of data.

Although some instructions are received by the computer through the operator's console, the majority of instructions are read into the computer and are stored in computer memory in the same way as are data. The important consequence of this fact is that the instructions can be changed during the course of the operations being performed by the computer. This feature will be illustrated subsequently in the discussion of programming.

Data and instructions are read into the computer in the form of words, which are collections of bits of some specified length. Word lengths of 24, 36, or 48 bits are common. Each word represents a predetermined arrangement of binary numbers, binary coded decimal numbers, or alphanumeric characters. The computer differentiates between the two basic types of computation words (data words and instruction words) only as each word is being processed.

Instructions, therefore, also appear as codes made up of binary digits which the computer interprets as an instruction to read, write, add, subtract, or perform some other operation on specified data or to conduct some other function.

COMPUTER SYSTEMS

One of the difficulties in trying to explain computers is that one keeps getting ahead of the story. For example, despite the fact that we have concentrated on computer coding and instruction, we have had to refer to such things as punched paper cards and tapes; magnetic tapes, drums, and discs; and paper printers.

We also have introduced some of the functions the computer has to perform. For example, it must be able to read, store, and identify data and instruction words; perform mathematical and logical operations; and write out the results of its operations.

In the course of performing its functions, the computer must run error checks, translate from one code to another, find specific data, send signals to and receive instructions from the computer operator, change instructions already given to it, recognize the format of data and instruction words, and follow the correct sequence of instructions.

Although space does not permit going into detail about how this is done, it

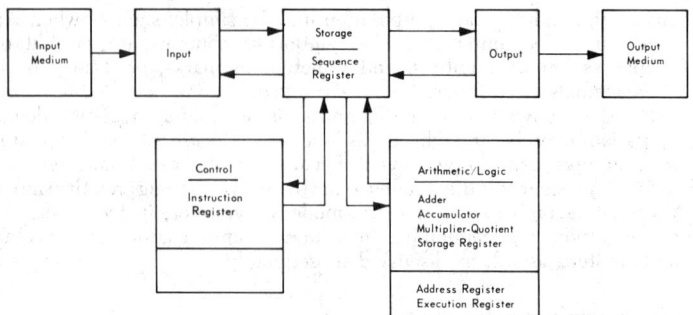

FIG. 1-3.   Main elements of a computer system.

is useful to consider how the main elements of a computer system are organized and how the computer carries out its instructions.

Figure 1-3 shows the organization of the main elements of a computer system. Contrary to all the cartoons, the computer is not one huge box with lights and switches on it; rather it is several separate pieces of equipment. The input devices vary, depending upon whether the input medium is punched cards, punched tape, or magnetic tapes. Similarly, the output device may be a printer, paper or card punch, or magnetic tape unit.

The input devices read both the data and instruction words, which then are placed in storage where they can be retrieved as needed. Their location in storage is identified by an address that is used by the control section for retrieval. Processed data also are stored until they are used for further processing or until they are delivered to one or more of the output devices. The storage device generally also contains a sequence register which initially is set to the address of the first instruction. As each instruction is executed, it steps up to the next instruction.

The control section, after receiving the address of the first instruction, retrieves the instruction and places it in an instruction register. The instruction word consists of an operand and an operation. The operand can be a storage address for data or instruction, or it can specify a control function. The operation indicates the function to be performed, such as read, write, add, or subtract.

COMPUTER PROGRAMS

A computer program is a list of instructions that are executed by the computer in the manner described earlier. There are many different types of programs whose names identify the purpose of the program or the form it takes.

A program written in the language of the computer is called an object program. The object program instructions appear in the form of computer words, which were described previously.

When a program is written using a set of instructions that, for the programmer's convenience, are in a mnemonic code (such as ADD for add and SUB for subtract), the program is called a source program. This program is read into the computer, and another program, known as an assembly program, treats it as data and translates it into a set of machine language instructions or an object program. An assembly program produces one machine instruction for each assembly language instruction.

The source program also can be written in codes in which a single statement covers several machine instructions. Such a code is referred to as a higher level language, because a program written in such a language, when converted to machine language, will have many more machine instructions in the object program than statements in the source program. Again, the source program is read into the computer as data, and another program (this time referred to as a compiler) converts the source program into the machine instructions of the object program.

The compiler differs from the assembly program in that it expands each instruction into a set of instructions, as well as performing the translation function.

In addition to assemblers and compilers which translate source programs into object programs, there are machine-oriented programs called executive programs which are designed to set up, monitor, and control the operation of other programs, including the assemblers and compilers. In controlling the execution of the program, an executive routine loads the program into storage, sets up counters and work areas, and controls the reading and writing of records.

Programming. There are a variety of problem-oriented languages, such as COBOL, FORTRAN, and BASIC, among the higher level computer languages. COBOL (Common Business Oriented Language) has become an international standard language for business applications where file and record processing are important. FORTRAN (Formula Translation) is designed for problems which can be expressed in algebraic notation. BASIC (Beginner's All-purpose Symbolic Instruction Code) has a simple vocabulary and grammar which resemble an ordinary mathematical notation.

To illustrate some of the features of a typical computer language and how a program is written, we will use a simple example where we want to calculate the mean value of a number of observations. We can assume our formula is

$$M = \frac{S}{N}$$

where M = mean
S = sum of values
N = number of observations

Let us also assume we have made nine observations whose values are 67.5, 82.0, 94.7, 86.2, 95.1, 35.0, 44.3, 72.0, and 66.7.

Figure 1-4 shows what the program looks like. Each statement was given a line number; for convenience, we numbered them in steps of 10, which will allow us to insert additional instructions between any two statements if we later discover we have overlooked some instructions which should have been given.

We start on line 10 by saying "LET N = 1." We are going to use N both as a counter to make sure we have added the values of all our observations and as in the formula M = S/N.

Then on line 20, we say "READ X." Note that on line 70 our data are listed. The first time the computer reaches this READ instruction, it will take the first value out of the list and set it aside for whatever computation is to be performed. The numeric value of 67.5 is effectively removed from the list; the next time any READ statement calls for a value, the following number, 82.0, is selected. This process is repeated until the list is exhausted.

On line 30, we say "LET S = S + X." On the first pass, this is equivalent to S = 0 + 67.5.

On line 40, we say "LET M = S/N." This is equivalent to M = 67.5/1 on the first pass. Line 50 advances our counter, N, by 1, and we are ready for a second pass through our loop. However, line 60 states "IF N < 10 THEN 20"; this means that if N is less than 10, go back to line 20 and repeat lines 20, 30, 40, and 50. On the ninth pass through this loop, M = S/N becomes M = 643.5/9 = 71.5.

```
10 LET N=1
20 READ X
30 LET S=S+X
40 LET M=S/N
50 LET N=N+1
60 IF N<10 THEN 20
70 DATA 67.5,82.0,94.7,86.2,95.1,35.0,44.3,72.0,66.7
80 PRINT "THE MEAN IS"; M
```

FIG. 1-4. Computer program for solving M = S/N.

TABLE 1-3. Explanation of Program Statements

LET	The LET statement instructs the computer to perform the indicated computation. Each LET statement is of the form LET [variable] = [formula].
READ and DATA	A READ statement is always accompanied by one or more DATA statements. The READ statement instructs the computer to assign to the variables listed in the READ statement the values listed in the DATA statements which have not already been assigned.
PRINT	The PRINT statement instructs the computer to print out verbatim the message included in the programs, the result of some computations, or a combination of both.
IF-THEN	The IF-THEN statement instructs the computer that, if a certain relationship holds, it should go to some other specified instruction. Each IF-THEN statement is of the form IF [formula] [relation] [formula] THEN [line number]. In effect, the IF-THEN statement permits the program, in the sequence of instructions, to change or be contingent upon the results of the computation being made.
END	The END statement instructs the computer to terminate the execution of the program.

Line 50 advances N to 10 and line 60 tells us not to return to 20, but to continue to the next instruction, which is line 80. (If there had been 17 observations instead of 9, the number used for comparison would have been $17 + 1$, or 18.) Line 80 tells the computer to print out the statement "THE MEAN IS 71.5." Table 1-3 lists the statements used in this program and gives a general explanation of what instruction is being given to the computer.

COMPUTER ECONOMICS

One characteristic of computers that has contributed to their expanding use has been the increased relative speed of the computer compared with cost over time. Generally speaking, the cost per unit of computation decreases as the size of the computer goes up. In most cases, it is less expensive to perform a given computation on a larger computer than on a smaller one. As larger and more powerful computers

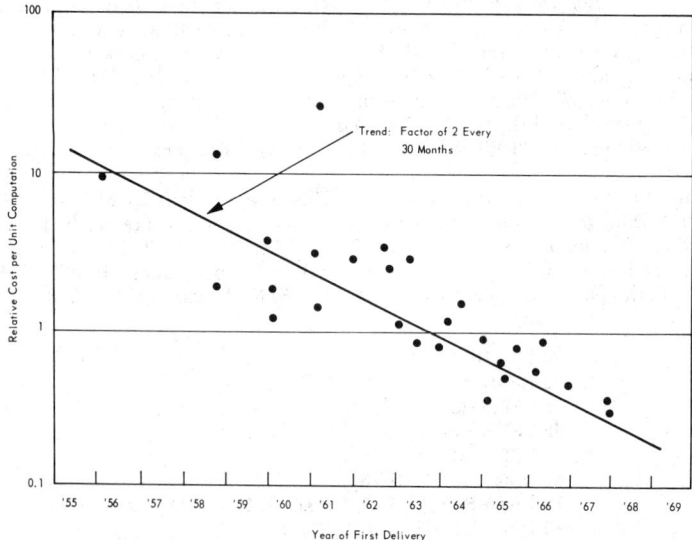

FIG. 1-5. Trend of cost per unit computation.

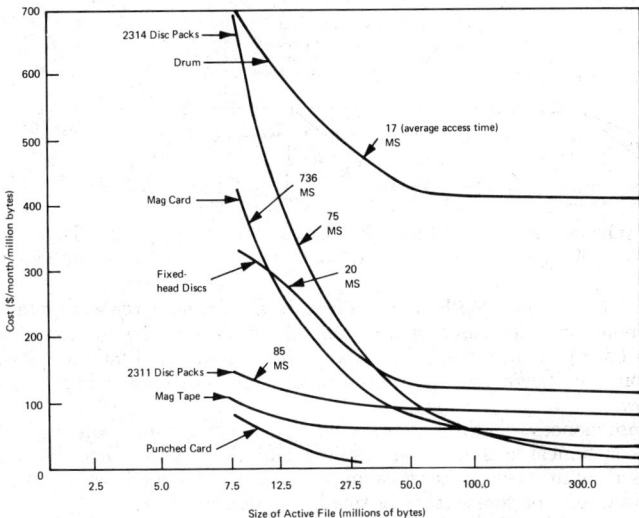

Fig. 1-6. Economics of computer storage media.

have been developed over the years, the increase in computational capability has resulted in a decreased cost per unit computation, despite the fact that the computer itself costs more. Figure 1-5 shows the relative cost per unit of computation for each new computer of greater capacity than its predecessors at the time it was introduced. It clearly shows the downward trend in cost per unit computation.

In addition, the cost of computer storage has improved. Figure 1-6 shows the cost of storage relative to the size of storage as a function of the medium used. For any given required file size, a variety of storage media are available, depending upon desired access time. (A byte is a term roughly equivalent to a single character.)

Figure 1-7 indicates how, for one large company having almost 100 computers of varying sizes, the total monthly rental and computer power have increased over

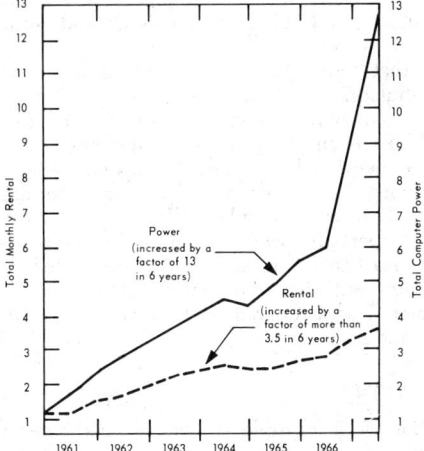

Fig. 1-7. Monthly rental versus computer power.

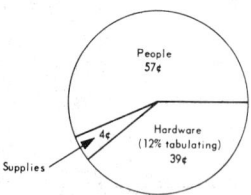

FIG. 1-8. Overall computer costs.

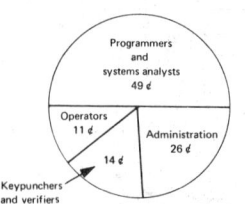

FIG. 1-9. Computer personnel costs.

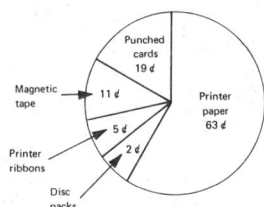

FIG. 1-10. Computer supplies costs.

a period of 6½ years. With a mix of both small and large computers, the total computer rental has increased more than 3.5 times, but the total computer power has increased by a factor of 13. Over this period of time, the cost per unit computation has decreased to approximately one-fourth of what it had been 6 years earlier.

The consequences of this situation are that applications which in past years were not economical can now be justified and those which had been economical can be used with even greater savings. A comparison of labor costs over this same period would, of course, show a steady per unit increase.

What is not evident from the chart is that the computers also can do more. The availability of random access memory may permit applications which would otherwise not be possible, and the availability of higher level languages will significantly reduce the cost of developing programs.

In addition, a company having five or more years of experience with a growing number of computers has two advantages that are difficult to evaluate but which are of obvious value. First, it has an applications base which permits it to move into higher payoff areas that are not penetrable with only a mechanized foundation. For example, a good production control and dispatching system is difficult to achieve without a mechanized bill of material system, but the bill of material system itself does not provide a substantial return compared with the production control system.

The second advantage is that such a company has learned how to plan systems, develop computer applications, and implement programs. It has experienced systems and programming people, and its computer people know what to expect, how to help make it work, and how to minimize the many problems involved in a transition from a manual system to a mechanized one.

It may be useful to look at a typical budget for a computer operation to get some idea of the various kinds of costs one finds and what their relative impact is on total costs.

Figure 1-8 shows that more than 10 percent of the hardware cost is required for punched card tabulating equipment, that the personnel costs far exceed the hardware costs, and that supplies constitute a significant part of the total.

Figure 1-9 shows how each computer personnel dollar is spent. Rather significantly, about half the personnel costs are for people other than the programmers and analysts who are normally associated with the computer activity.

Figure 1-10 shows how each supply dollar is spent and confirms what many of us have already suspected: when it comes to computers, paper is our most important product—at least that is what might be concluded.

It should be pointed out that the figures shown are for a typical case and should not be interpreted as standards. The actual requirements can differ widely from one situation to another.

COMPUTER ACQUISITION

Given an application, it is possible to select a computer which is best for that particular application. In acquiring a computer, however, the choice will have to be a compromise, because many different applications will have to be accommodated by it.

The choice also will be influenced by availability of software, training, delivery dates, price, and contractual terms. The latter includes such items as system performance guarantees; acceptance terms; definition of operational use time; programming assistance offered; maintenance provided; test, compile, and debug time available before installation and not charged as billable time afterward; liquidation damages; and purchase option credit.

Each of the major computer suppliers in the United States has a number of computers that vary in size, capability, and consequently, price. However, the rental range for a specific computer, even for the very large ones, can be two to five times the minimum monthly rental. What is popularly thought of as a supplier's computer is really a product line that may embrace several basic variations within each of several different models having a wide choice of features. The actual selection of the computer and the components of the computer system is a decision making process of considerable magnitude.

The decision to acquire a computer or to convert from one to another should be treated as a problem distinct from that of equipment selection, although they are obviously closely coupled. The first decision requires a realistic appraisal of what the firm hopes to achieve with automatic data processing systems and what is required to achieve those plans. This decision, usually documented along with the equipment selection, is what has come to be known as a feasibility study or computer acquisition proposal. The elements of a feasibility study are given in Table 1-4.

The most serious errors made in computer acquisition proposals are the failure

TABLE 1-4.   Elements to Be Considered During a
Feasibility Study

Systems plan
    Firm's overall objective
    Functional department objectives
    Planned information systems
    Specific application programs
Hardware and software requirements
    Computational and storage capacity
    Delivery
    Configuration
    Systems software
    Application software
    Equipment utilization
Equipment evaluation
    Alternatives considered
    Comparison of hardware performance characteristics
    Results with bench mark programs
    Comparison of computer throughput
    Cost comparison
    Equipment selection summary chart
Implementation schedule
    Facility requirements
    Personnel requirements
    Training schedules
    Programming and conversion
    Delivery schedule
    Application schedule
Economic analysis
    Lease versus rent
    Monthly operating costs
    Estimated savings
    Cost and savings versus financial guidelines
Organization
    Firm's organization
    Data processing organization
    Functional representatives
    Steering committee
Contractual terms

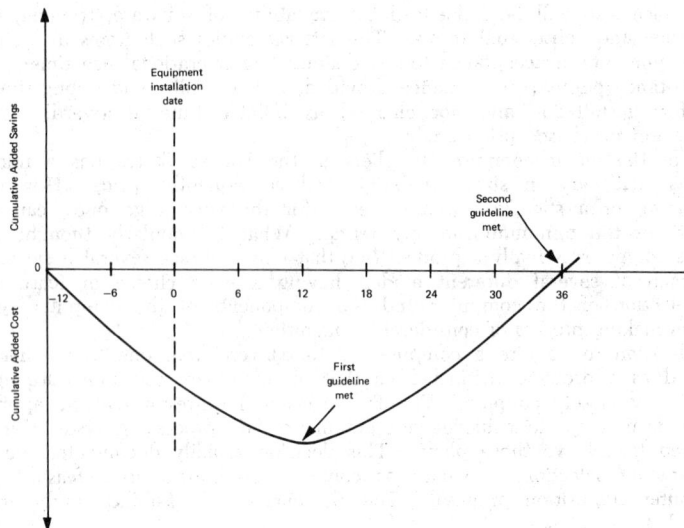

FIG. 1-11. Cumulative costs/savings versus time.

to make a realistic appraisal of the financial consequences of the proposed program and the failure to apply meaningful financial guidelines which will be followed during the implementation of the program.

One example of financial guidelines that have been used frequently is the following:

1. Within one year after installation, the monthly gross savings will equal monthly operating costs.

2. Within three years after installation, the cumulative gross savings will equal or exceed the cumulative gross costs.

Figure 1-11 shows how the cumulative monthly net cost or savings curve would appear for an installation satisfying these guidelines.

CONCLUSION

The most common error made in planning for new computer systems has been to underestimate the effort required to program and implement new applications. Much of this is caused by the combination of a lack of experience and a desire to justify the computer acquisition. When a new computer is a replacement for an older one, there usually is a serious underestimation of the conversion effort involved.

A contributing factor which is almost as serious has been the prevalent and foolish practice of making only a superficial analysis of competitive equipment and devoting most of the computer proposal effort toward justifying the choice which has already been made. The equipment evaluation section of such a report, interestingly enough, does not usually present a biased picture, except as it reflects the position taken in the hardware and software requirements section. In many cases, it is very clear that the hardware was chosen first and that the requirements which bias that choice were then developed. If this seems difficult to do, keep in mind that the computer man is a particularly talented individual.

Chapter **2**

Use of Computers in Manufacturing

R. H. NEWELL

Manager—Systems Research,
Goodyear Tire & Rubber Company, Akron, Ohio

The degree of utilization of computers in manufacturing depends on the state of (1) hardware technology, (2) relevant problem solving techniques, and (3) the risk taking proclivity of an organization.

Although hardware technology is reasonably uniform throughout most of the world, an organization's awareness of relevant problem solving techniques will vary from industry to industry as the problems vary, and from company to company within an industry, because individuals who have the required combination of technical competence, management orientation, and organizational influence are somewhat rare. The state of the risk taking proclivity of an organization varies from industry to industry, but competition enforces a good deal of uniformity within an industry. An announced level of achievement by one company will stir a rapid response from competitors within the industry.

Although some statements can be made as to the general level of use of computers in manufacturing in different industries, such statements will quickly be made obsolete by the rapid pace of progress in this field. Consequently, this chapter will take a stagewise approach to the subject. Four stages will be chosen to represent the increasing degree of complexity and scope in pursuing a computer-based technology by a company. These stages will be characterized as follows:

1. Basic computer applications
2. Intradivisional computer applications

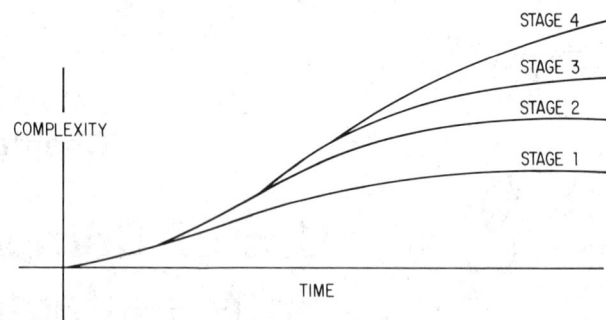

F<small>IG</small>. 2-1.   Four stages of computer application.

   3. Interdivisional computer applications
   4. Advanced computer applications
As shown in Figure 2-1, an overlap will occur, for earlier applications will persist in time.

Only representative applications will be highlighted in each stage, because any exhaustive presentation would require thousands of entries. Although some attempt will be made to give a fairly broad picture of computers in manufacturing, the major emphasis will be on the technological aspects.

BASIC COMPUTER APPLICATIONS—STAGE 1

The first areas of application in a manufacturing concern are usually those which require individual efforts, or at most, cooperation within small groups. As a result, the initial applications will be limited to projects of relatively small scope and complexity. In addition, computing hardware will be selected from among the less expensive alternatives. These factors are dictated by the normal tendency to limit financial risk in a new or unfamiliar technology.

Examples of these early applications can easily be grouped under traditional functional headings for they rarely cross functional lines.

Production.   As in many initial computer efforts, the emphasis on first production applications is to replace human labor in such activities as bookkeeping, sorting, report writing, and equation solving. These applications include:

   1. Automating the generation of work standards. Equations are derived which describe classes of work. By programming these equations and assembling the relevant parameters (dimensions, weight, and the like), the computer can generate work standards rapidly and accurately.
   2. Performance data reduction programs. These programs organize and tabulate performance data of the work force. One can organize the reports to reflect performance by crewing center, shift supervisor, and the like. Among the characteristics that might be reported are average crew size, average production by class of product, average scrap generation, run hours and setup hours, average quality of product, percent downtime, and percent performance against standard.
   3. Statistical computations of various kinds relating to quality control.
   4. Histograms and scatter diagrams for purposes of comparing various data of interest.
   5. Plant layout using simulation to analyze alternatives.

Sales.   Sales applications include various forms of simple data reduction leading to histograms of units and dollars of sales by line, outlet, region, time period, and so on.

Research and Development.   The research and development groups begin work in the computation and reduction of scientific data. The emphasis here is on replacing humans by a computer in the performance of well-defined computations in data reduction. The advantages are speed and accuracy, which permit the research and development efforts to proceed at accelerated rates. Few applications are likely to appear which represent a novel way of doing a task. However, the means of applying scientific approaches known in the past now become available. Various engineering analyses solve civil engineering problems of structural loads using "canned" routines like COGO and STRESS.

Development groups may help support sales personnel by generating tables of technical specifications from their equations.

Time Sharing. Another development affecting principally research and development or production groups in the early stages of the development of computerization is the availability of on-scene consoles which are linked to a large time sharing computer. This large computer can be owned and operated by the manufacturing company, but more commonly, it belongs to a company in the business of selling this service. The amount of investment in dollars and personnel and the need for a large market to offset the high fixed and low variable costs usually dictate that a company which specializes in computer work offer this service.

Both research and development and production find this type of service useful, because they usually have several problems which need a large amount of computation and rapid response time with limited input-output requirements.

Mathematical Techniques. Some of the computer-related mathematical techniques widely used during this stage are simulation, numerical analysis, optimum seeking methods, and curve fitting.

Simulation. Simulation has a popular appeal because its use does not necessarily require a high degree of mathematical proficiency. Using such computer languages as GPSS (General Purpose Systems Simulation), a system to be studied relative to some problem area can be modeled primarily through a knowledge of the process involved. The process can be physical, such as material handling, or abstract, where only information is involved. Other simulation languages, notably SIMSCRIPT, have a strong appeal to the mathematically oriented modeler.

The most common use for simulation is to solve specific problems where the analyst speculates that a change to a given process will result in a favorable response. The simulation is then used to test this speculation and to provide statistics on which to base new speculations.

For example, in an industrial process, one might test the plant's ability to manufacture using (1) a particular product mix, (2) a particular arrangement of machines, and (3) various policies of operation.

Simulation has an advantage in that it can be used to solve problems of a complex nature, where a general solution by mathematical analysis is very difficult. It has a disadvantage in that a limited number of speculations or alternative solutions are made available to the analyst.

Numerical Analysis. Perhaps the greatest initial impact of the computer on science and engineering in industry is that it provides a practical means for solving equations heretofore found difficult or almost impossible. Mathematical equations that represent physical phenomena for which explicit solutions are nonexistent are being solved by approximations. An entire field of numerical methods is available to solve problems for which only implicit solutions were formerly possible. Among these techniques are those utilizing convergent series, finite differences such as Runge-Kutta methods for numeric integration, Newton's methods of approximation by iteration, and solutions to problems by recursive methods.

Solutions of ordinary differential equations, with the desired accuracy, are possible if a sufficient number of points are used.

Data reduction from observations is automatic with the aid of computers. Large systems of simultaneous linear algebraic equations are being solved daily, from the solution of engineering problems in stress analysis to the determination of optimal mixes of animal feed products.

Optimization techniques for business management—such as linear and geometric programming, dynamic programming, networks for project scheduling such as PERT and CPM, and Monte Carlo methods using random numbers in computation—are new tools provided for more sophisticated utilization of computers in industry.

The whole area of statistics has taken on new dimensions as a result of the capability to deal with large volumes of data. Analysis of data with many variables, sampling methodology, testing for significance, least squares solutions to experimental data, and curve fitting are problems to which answers are obtained in a matter of seconds or minutes with the aid of computers.

Optimization. Optimization techniques are also in the first stages of computer use, because they generally require a rather limited amount of data and can be accomplished by a one-man effort. These techniques include the following: (1) zero derivative methods, (2) calculus of variations, (3) hill climbing techniques, (4) penalty methods, (5) linear programming, (6) dynamic programming, (7) cutting-plane methods, (8) branch and bound methods, and (9) geometric programming. Some of these techniques are described in some detail in Section 10.

The ability to apply linear and dynamic programming, even in their simpler forms, equips an analyst for solving problems of intermediate complexity. For problems of great complexity, an analyst may require a knowledge of some of the more advanced forms of mathematical programming. In this latter case, two or more of these techniques might be required in combination.

INTRADIVISIONAL COMPUTER APPLICATIONS—STAGE 2

A second discernible stage, usually involving personnel within a single division, is characterized by the maintenance and updating of files and a growing level of complexity. This usually occurs first in the financial arm of the organization. This leads to computer hardware with extensive input-output capability and serially accessible auxiliary storage, usually magnetic tape. Applications include payroll, accounts payable, accounts receivable, dividend issuance, stock registry, general ledgers, fixed asset accounting, interdivisional transfers, inventory records, and periodic closing of company books.

Production. Production applications involve extensive data analysis and record keeping. The previous performance analysis can be recorded from period to period, and year-to-date computations and trends can be reported and analyzed.

Scheduling. Although automated scheduling of the plant is generally not attempted at this stage, certain data processing aids to scheduling are produced.

One of these aids to manual scheduling is the so-called "parts explosion" where the manufactured product must be broken into various subassembly levels down to the smallest integral part. The subassemblies or components are then aggregated for manual scheduling at the various work centers. This is one of the most widely used programs in industry.

Another aid to scheduling is simulation. For certain applications, it is helpful to construct a simulation of the plant and use it to test the effectiveness of the manually desired schedule. A simulation can point out bottlenecks and work centers with poor utilization, and adjustments can be made before a schedule is implemented.

Numerical Control. Numerical control is a solidly based technology, with its use covering most of the metalworking industry. As originally conceived, the numerically controlled machine tool consisted of a machine tool, a controller for the tool which accepts instructions from a paper or magnetic tape, and a method of programming the controller and generating the tape. Although there have been some design changes required in metalworking tools to accept the numerical control technology, these changes have not been substantial. The machine tool controller has developed parallel to the digital computer, with the electronic advances of recent years affecting the size and the reliability of both in the same manner.

The method of programming the controller received its biggest boost from the development of the APT language in the 1950s. Various modifications and subsets of this language are in widespread use. Various other languages exist but are more limited in use and generality. Other developments include programming systems using cathode ray tube terminals attached to a computer. The widespread acceptance of these man-machine or interactive systems, as they are called, awaits reductions in computer and display costs.

Tool and controller companies as well as manufacturers of computers are developing systems which utilize a computer as a generalized controller to service a number of separate machines, providing both input to the tool and a path to receive feedback information. This concept has merit as an economic alternative to conventional and separate control and also permits consideration of adaptive control techniques. Further, information pertinent to the task can be entered and received by the computer from the production floor, giving the computer-based system a potential as a shop management aid. The implications in areas of production, inventory control, and scheduling are substantial.

At this second stage, inventory record keeping applications are implemented by production where financial inventory records are not suitable.

Sales. Sales personnel begin to keep historical records and identify marketing phenomena. The sales division is also interested in distribution patterns, lead times, and field inventory positions.

Research and Development. Research and development personnel tackle problems of increasing complexity. Certain classes of research and development work demand vast amounts of mathematical computation. Among these applications are those of crystallographers and other researchers in molecular structure. Computer capabilities required by research and development are acquired by these organizations. These computers vary substantially in overall capability but tend to have large memories and high computational speeds.

Remote Terminals. Terminals remote to a computing center within the company yet possessing adequate input-output facilities become a highly sought after tool by many departments. These terminals not only provide a convenience to the users but yield a sometimes necessary rapid turnaround time on the problem.

An example in production is solving for the minimum cost material additions which will bring a heat in an electric furnace up to the specifications required by the target specialty steel. This is a problem which must be solved at the furnace in a relatively short period of time. It is easily solved using linear programming, but computationally it demands a fair amount of main and auxiliary memory and rapid, on-site response.

INTERDIVISIONAL COMPUTER APPLICATIONS—STAGE 3

The third stage of computer application is characterized by an attempt at solving some of the more complex business problems requiring a limited amount of cooperation among divisions. There is a strong requirement to automate demand forecasts for the purpose of planning production and distribution. Order entry systems are built requiring various degrees of cooperation among sales, distribution, and finance to generate the maximum benefits.

There is a growing interest in the optimization of large, complex systems. The maximization of productivity in the petroleum industry is an example of this type of system. Other examples include maximizing the yield in the nuclear industries where the cost of processed materials is very high.

As a rule, the more capital-intensive industries (high ratio of fixed assets to employees) stand to gain the most through optimization of scheduling and use of equipment.

Computing Decentralization. As the work load builds up on a company's computing facilities, there is a tendency to favor the more urgent application, thus causing large queues to build up for others. One answer to this problem of insufficient response is the decentralization of computing responsibility. This is more the rule than the exception in large manufacturing companies. The prevalence of this mode of operation is due to the same causes that originally determined the organization to organize along functional lines.

Factory Scheduling. Optimal factory scheduling in even a moderately complex process can be a major effort. Such efforts are well supported by mathematical approaches reported in the management science journals; yet no generalized methodology has been developed, and each manufacturing situation must be treated uniquely. In addition to the development of an appropriate mathematical approach, a scheduling effort must deal with a vast amount of machine attributes, product specifications, and work standards.

Factory Monitoring. If the scheduling calculation is to be available on short notice, then in-process inventories must also be accurate and up to the minute. The latter requirement often prompts the installation of computerized factory monitoring systems which automatically sense the generation and the utilization of each component. Such inventories have been valid to 0.01 percent over a one-month period.

Production control centers based on a computer directly sensing actual factory machine production can give a complete synopsis of the factory status upon demand, but their primary usefulness lies in their exception reporting capability. Any deviation from schedule can be recognized along with the cause of such deviation, and remedial action can be taken promptly.

Factory Reporting. Statistical measures of efficiency can be maintained automatically and produced periodically or on demand. Such reports include downtime, quality control, and scrap. Others indicate such items as production summary, in-process inventory, and attendance.

Display Terminals. Some of the production monitoring systems make liberal use of the cathode ray tube display terminal. Such terminals are reliable and inexpensive and permit plant management a graphical or literal display of factory status in a quick and convenient fashion.

Machine Control. Associated with the above trend is the use of the very small and inexpensive computer (less than $10,000 to $15,000) to control individual or groups of factory machines directly or to reject products of insufficient quality. Such computers can report the status of machines under their control to the central monitoring computer. This network or arrangement of small computers reporting to a larger central computer is primarily an economic one, because a small computer without the extensive input-output hardware and the elaborate memory-occupying "software" of the central computer can have a cost effectiveness far greater than that achieved when employing the central computer for complex control calculations. The cost effectiveness is, of course, highly dependent on the state of the art in computing hardware and on individual applications.

Process Control. At the third stage, some of the more elementary process control applications are tried.

The term "computer process control" refers to the use of a computer to perform monitoring and regulating functions. The input to the process computer comes from on-line sensors that transmit information about the status of the process, material, product, or energy level.

Examples are chemical manufacturing, where the variables being controlled are reactant flows, temperature, pH, pressure, and material balance; food processing, where the controlled variables are dehydration, freezing, and warehousing; and energy production, where heat transfer and combustion are controlled.

Benefits. Process control has gained wide acceptance in industries where the products are produced either by a continuous flow or batch process or in discrete production steps, and where tight control of raw material together with close adherence to operating conditions can mean improved product quality and savings in raw materials and production costs.

Nonmanufacturing Examples. The traditional industries where process control had its start are the chemical, petroleum, steel, paper, cement, and utility industries. However, the application of process control has expanded into such nonorthodox applications as traffic control, biomedicine, stream pollution control, automatic warehousing, police dispatching, textile dyeing, and component testing.

Data Logging. The first applications of process control deal primarily with data logging and on-line process calculations. In such an installation, the computer receives signals from the various local process sensors, scans them, and prepares either a continuous or a periodical log. The computer also calculates and records the status of the process, the temperature profile, or the batch throughput, for example. The log will also alert the human operator if intolerable conditions or limit violations have occurred in the process. This function of process monitoring is fundamental to all process control applications.

Data Acquisition. On-line data acquisition and analysis systems are installed to benefit the research and development efforts.

On-line Experimentation. Notable among on-line experimentation applications is the gas chromatograph system, where the computer accepts signals from a gas chromatograph run, analyzes these signals, and prints a report of analyzed material. To such systems, other laboratory instruments, such as mass spectrometers and electron spin resonance devices, can be attached for similar on-line treatment. Other computing systems are dedicated to the digitizing of analog signals recorded on magnetic tape from remotely located test sites. Systems also exist that conduct complex experiments at the bench scale or pilot plant level.

The motivation behind most of these on-line data acquisition and experimentation systems is the same—the shortening of the total cycle from test through analysis. Usually attendant with this faster response are lower manpower requirements, greater accuracy, and greater permissible output.

Precedence Network Analysis. Some manufacturing companies find that precedence network analysis is useful to control time and costs. Examples are PERT and CPM. They are mainly of value when accomplishing a one-time job or when attempting a complex job for the first time. The aerospace industry finds these tools most useful. In fact, in many defense contracts, the use of these methods is mandatory. They are also used extensively in construction and other nonmanufacturing industries.

Purchasing Systems. Purchasing can maintain an automatic purchase order system if raw materials inventory records are automated and if order point and lot size determinations can be made. There will also be requirements for subsidiary reports made to departments responsible for company financial matters.

Profitability Analysis. The finance group utilizes the computer to evaluate investment alternatives, using various measures of profitability, such as return on investment, payout time, and cash flow. Such calculations, of course, require a high degree of interdivisional cooperation. For example, the sales division must estimate future demand and competitive influences; the development department becomes involved in process specifications; and the production engineering organization must consider plant layout, site locations, manning requirements, and the like.

Other finance applications include forecasting cash flows, which, as in conventional systems, demands input from the entire organization. However, there is a strong desire to capture this input in machine processable form.

ADVANCED COMPUTER APPLICATIONS—STAGE 4

At the end of stage 3, it becomes clear to most organizations that the piecemeal approach to problem solving and record keeping is not keeping pace with the organization's demand for information and answers. One response to this problem is to implement an integrated computerized information system.

Information Systems. In industries which have a relatively homogeneous product created with a relatively homogeneous technology, these systems can be devised without enormous difficulty. However, most manufacturing companies have a highly diversified technology of order entry, raw material acquisition, conversion processes, distribution and warehousing, and sales mechanisms. A simple measure of the heterogeneity of the information technology is the number of different paper forms used throughout the company.

Although these systems are often called management information systems, their common element is the need for an integrated data base. Constructing such a data base, where the meaning of each value is precisely understood by those personnel who wish to use it, is a major undertaking.

Fast Response. One of the main advantages of an integrated data base of any significant organizational scope is that it makes possible major fast response information and decision systems. The ethical drug industry, for example, has a very large number of products with different manufacturing lead times and differing shelf lives. The status of an inventory in this case is a dynamic and complex computation. However, much of the difficulty of obtaining fast, accurate answers to questions about inventory position can be handled by a computer.

Unanticipated Questions. Most information systems in use in manufacturing deal largely with responses to well-structured questions that solve specific problems. However, there is a requirement for top management to be in a position to ask "what if" questions of great detail. Furthermore, these questions can be answered in a relatively short time. An example of this kind of usage is an on-line response system which analyzes proposed changes in a union contract to evaluate their implications. This can be a powerful tool to aid negotiators in a bargaining situation.

Example of an Information System. In one application, an engineering information system was designed to conduct the business of constructing new plant capacity. The system is structured of forty "user" programs, each of which performs the function of a particular activity such as estimating, work order generation, project scheduling, cost reporting, and purchase order generation. Each activity is conducted by the engineers from cathode ray display terminals. In each of these activities, the engineer is guided through his work by a question and answer mode; the computer supplies alternative actions and calculations and the engineer makes the decisions. In the course of making his decision, the engineer has readily available historical information where similar decisions have been made in the past. All activities address a common data base by means of data acquisition routines (special programs written for this purpose). The management of information flow to and from the computer is conducted by a monitoring program, as illustrated by Figure 2-2.

One of the forty user programs is one with which the user can interrogate the data base with questions of a general or specific routine and do so in a simple "free format" language. Such information systems must be designed to encompass a wide area of use and should be capable of broad extension. Two factors which give the above example this capability are its modular design and very generalized data base structure.

Ready-made Information Systems. Information system packages are available

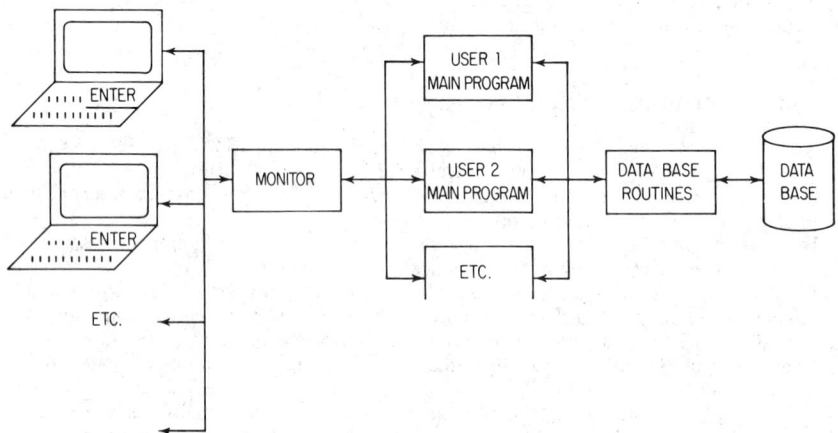

FIG. 2-2. Management of information flow by a monitoring program.

from major computer manufacturers; but in certain cases, these offerings may contain an excessive amount of computing overhead (time spent by the computer in managing its own operations). There is also a great amount of effort left to the user in adapting such generalized system design to his own operations.

An information system is almost by definition an interactive process of human judgment alternating with computer actions.

Computer Attributes. The computer provides such functions as:
1. Rapid calculation, routine processing, and fast editing
2. Fast reading of certain types of documents; high-speed printing, and therefore quick preparation of documents
3. Filing and maintenance of vast bulks of information and the retrieval of such information on displays and other terminals
4. Scanning, sorting, and searching for specific facts
5. Collecting information rapidly from many sources, by either local or remote communication links
6. Fast distribution of information to many locations
7. Giving immediate answers to routine inquiries
8. Surveillance for exception conditions needing human attention

Human Attributes. On the other hand, the following types of tasks are still best done by man:
1. Handling unforeseen events
2. Selecting goals and criteria
3. Selecting approaches to problems
4. Recognizing patterns in events and detecting and identifying relevance
5. Formulating questions and hypotheses
6. Producing new ideas; planning new approaches, products, and techniques

Automated Design and Production. One of the more advanced phases in the development of computer automation is the integration of design and production, particularly in areas where product homogeneity and standardization have reached a high level.

Design Consoles. Automated design received much of its inspiration from work done at the Massachusetts Institute of Technology. Much of the work is centered around the cathode ray tube display terminal. In this application, the designer sits at the display terminal, or design console as it is sometimes called, and constructs an image of his product on the display screen. He normally has at his disposal a typewriter keyboard for entering alphabetic or numeric information, a light pen (an electronic pencil-like device, which he can use by touching it to the screen in a manner analogous to an ordinary pen or pencil used on paper), and a function keyboard which he can use to actuate various preprogrammed computer functions.

The cathode ray tube display may look like Figure 2-3 to the designer. Here, in a typical application, there must be space allocated to a work area where the

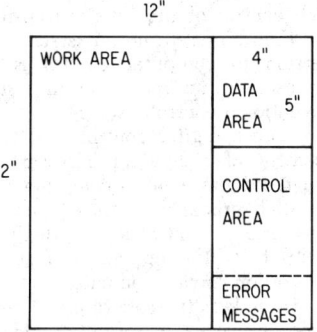

Fig. 2-3. Cathode ray tube display format.

design will be generated, a data area where data entered can be indicated, a control area where the various alternative functions can be selected with a light pen, and a space for error messages which indicate some infeasibility. The subject of computer-aided design is discussed more fully in Chapter 4 of this section.

Computer Graphics. The automated design process is one example of a growing tendency toward man-machine or interactive computer systems which are intended to solve very complex problems.

The value of interactive computer graphics lies in its capacity to augment human resources through:

1. Visual information displays—the form which is most readily comprehended by humans
2. The availability of a large store of information
3. The availability of otherwise impractical analysis techniques
4. The elimination of tedious and repetitive activity
5. The enforcement of constraints inherent in a particular process or activity

These capabilities are achieved in a time scale which permits a high level of concentration to be maintained by the user.

In one application, it was noted that only eleven pieces of data needed to be entered on a flat display to represent eleven elements in a complex design pattern. Symmetry and other known constraints then enabled the computer to generate the remaining 11,000 elements, a task which the human designer would normally be required to deal with. The burden of transforming each of these elements into a curvilinear coordinate system was also assigned to the computer. Thus the great economy in human labor was dramatically demonstrated.

Automated Fabrication. Because a complete design results in complete product specification, it is sometimes possible through numerical control techniques to command the computer to produce instructions for parts fabrication in certain metal products or to generate a mold in the case of molded goods. Sometimes it may be necessary to make certain additional machining decisions which cannot be explicitly derived from design specifications. In these cases, a separate computer graphics system can be developed for the entry of machining decisions.

These design manufacturing systems can be readily extended to other computerized subsystems such as production control and purchasing.

Advanced Process Control. Another noteworthy milestone in the growing sophistication in applying computer technology is the development of more advanced process control.

Supervisory Control. In supervisory or set-point control, the control action consists of a computer-initiated signal that changes the set-point of a current controller. The value of the set-point either can be calculated by the processor or can consist of a time variant value such as a predetermined temperature profile. The set-point change is accomplished by means of a signal output applied to and compatible with the process instrumentation. Special control loop set-point stations that will accept such signals are available. A manual override option is usually part of the control design in which control of the loop is transferred from the conventional controller; the set-point can then be reset and adjusted manually by the operator.

A more gradual conversion to computer control is one of the characteristics of this control scheme. However, no reduction in instrumentation cost is made, due to the retention of conventional controllers.

Direct Digital Control. Direct digital control (DDC) of one or several closed loops involves the replacement of individual hardware elements—mostly the controller—with components of the digital control computer.

Often, a computer is installed primarily to substitute for a number of conventional analog controllers in a conventional control loop. In the DDC mode, the computer will generate the pulse output for the operation of stepping motors, or an analog or voltage output for the operation of valve actuators.

The basic departure in this mode is the software and computational aspect whereby the calculation of the error between the set-point and the measured variable, the calculation of the valve position from the control equation, the limits, and the equations are stored in the computer rather than in the individual analog elements that have to be duplicated for each loop.

The control computer here not only obtains the input information directly from the process but also issues output commands to the final control elements.

DDC can decrease the instrumentation cost by the elimination of controllers; the greatest benefit, however, is the flexible computer decision logic that allows the use of various control strategies without costly replacement and modification of control elements.

Adaptive and Optimal Control. Finally, process computers have enabled the process industry to experiment, model, and develop a wide range of dynamic optimization and control algorithms, thereby improving process performance and control. An example of a more advanced control approach is the adaptive control concept.

Process equations are never completely accurate, because they are predicated on certain parameters which may change with gradual changes in the process. In view of these deviations, it is the aim of adaptive control to adjust its own parameters in response to the environment of the process to be controlled.

Report Generation. Management and report generating capabilities are basic and ever-present parts of each process control application.

Systems Engineering. The progress of a manufacturing company toward computerization can be considerably strengthened by the concept of systems engineering.

Characteristics. Here, the emphasis is placed on the goals of the total organization involved. Considerably greater changes in the operation are implied in this stage. Organizations must change in structure, traditional divisional goals must be altered, and new performance measures must be constructed.

The nature of this new technology is not yet settled. Is there a systems philosophy? Is systems engineering supported by a systems science? Or is systems engineering merely a collection of techniques? Many practitioners agree that each of these descriptors applies.

In general, systems engineering is the activity which attempts to arrive at the design of large-scale, complex systems. Recognition of this new field is evident in such undertakings as the Case Western Reserve Systems Research Center. Several journals are in existence, such as *Systems Science,* and the Institute of Electrical and Electronics Engineers publication, *Transactions on Systems Science and Cybernetics.*

Application Areas. Because many systems engineering concepts have evolved from control theory, many of the terms and techniques in the field stem from the latter activity. In a large manufacturing enterprise, at least three somewhat distinct systems engineering activities can arise. One is concerned with the design of complex electronic gear such as radar systems, computers, or aerospace missile guidance systems. Another activity is concerned with the direct control of manufacturing equipment such as chemical processes or machine tools. Still another general class of systems engineering effort is concerned with the optimal conduct of the business activity (management systems).

Sometimes all three classes of systems activity are merged, but more often they are separate efforts employing persons of different educational backgrounds. These activities vary considerably in the degree of uncertainty they deal with and in the degree of human interaction involved.

One of the greatest obstacles to the design of complex systems is the absence of mathematical optimization techniques of sufficient generality. The two most general and widely used methods, linear programming and dynamic programming, fall short when attempts are made to use them as a sole means of dealing with large-scale systems. Both, however, have broad areas where they can form the basis for efficient, problem solving algorithms.

These techniques and others can be effectively used in the design of complex systems if the system under consideration is appropriately decomposed into a hierarchy of subsystems, each having carefully defined goals and each level of subsystems being coordinated by a higher level.

Example. One computerized system, designed throughout by means of systems

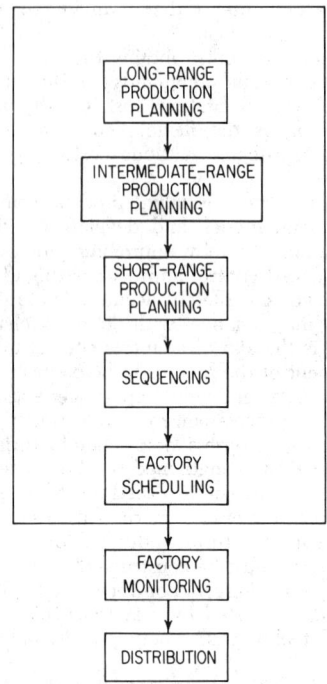

FIG. 2-4.   Part of the decision chain of a computerized planning and scheduling system.

engineering principles, conducts the production planning and distribution decisions of a large company almost without human intervention. Part of the main chain of decision is shown in Figure 2-4.

One could consider levels 1 through 5 in this hierarchy as one scheduling problem. The total problem, however, is so complex that unless one were able to redesign entire factories, their equipment, their warehousing policies, and so on, it would be nearly impossible to cast this problem into a monolithic form and consider its solution by some programming technique. By partitioning the problem as in Figure 2-4, it was possible to achieve a high degree of approach to optimality and yet use somewhat standard optimum seeking techniques.

Each level was assigned a distinct subgoal to achieve which provided guidance for a lower level. The objectives and mathematical techniques used for these levels are briefly described as follows:

1. Long-range production planning
   Objective: Provide the lowest cost plan over a nine-month horizon which will bring seasonal goods into production at the appropriate time to serve the market and yet avoid excessive lead time.
   Technique: Dynamic programming.
2. Intermediate-range production planning
   Objective: Accept mandate from the long-range plan, and provide lowest cost plan over a six-week horizon which considers plant capacity and many of the constraints inherent in the factory.
   Technique: Decomposition linear programming with dynamic programming subproblems.
3. Short-range production planning
   Objective: Accept direction from intermediate-range plan and provide lowest

cost day-by-day plan over a two-week horizon which balances the work load consistent with work force level.
Technique: Decomposition linear and dynamic programming subproblems.
4. Sequencing
Objective: Provide lowest cost day-by-day plan over a two-week horizon which avoids excessive setup time and considers constraints at component level.
Technique: Heuristic programming based on concepts of integer programming and transportation problems.
5. Factory scheduling
Objective: Provide lowest cost day-by-day plan which assigns to each factory machine the tasks specified in previous levels.
Technique: An assignment algorithm.

CONCLUSION

In a few decades, the acceptance of computers has gone from nonexistence to some degree of utilization in almost every facet of manufacturing. Although this unprecedented technological growth has occurred throughout industry in general, there are considerable differences in progress from industry to industry due to varying awareness and risk taking tendencies.

The growth in effective computer usage cannot take place in a short time span merely by the issuance of a directive to do so or by the allocation of large sums of money. It proceeds only with a complete reorientation within a total organization. Such an effort must be led by exceptionally inspired and dedicated management, willing to undergo, at times, a high degree of frustration. A successful effort which shows a strong return to a company cannot take place without a guiding philosophy, and the policy stemming from this philosophy cannot be formulated by an apathetic or uninformed management.

It has been demonstrated that with few exceptions each activity in a manufacturing concern can receive benefits from the use of computers. These activities can receive a synergistic bonus when the use of computers is widespread, because there is:

1. An economy of scale in computer hardware expense
2. An economy in programming costs due to standardization of techniques
3. An economy in training of large numbers of personnel
4. Commonality and data base requirements which create efficient information flow and storage
5. The avoidance of developing a methodology for each similar application, such as in the case of process control or numerical control, which may be applied in many separate plants

The compelling forces in pursuing a computer-based technology in manufacturing are:

1. The long lead time in achieving the advanced stage 4
2. The promise of richer returns from the use of computers, being brought about by constantly increasing computer hardware and software capabilities at declining costs

BIBLIOGRAPHY

Calica, A. B., "Long-range Planning Techniques for Discrete Production Operations," *Proceedings of IFIP Congress 65*, vol. 2, 1965, p. 390.
Duffy, G. F., and F. P. Gartner, "An On-line Information System for Management," *AFIPS Conference Proceedings*, SJCC, vol. 34, 1969, pp. 339–350.
Glinka, L. R., R. M. Brush, and A. J. Ungar, "Design, through Simulation of a Multiple-access Information System," *AFIPS Conference Proceedings*, FJCC, vol. 31, 1967, pp. 437-448.

Hamming, R. W., *Numerical Methods for Scientists and Engineers,* McGraw-Hill Book Company, New York, 1962.

Lee, T. H., G. E. Adams, and W. M. Gaines, *Computer Process Control, Modeling and Optimization,* John Wiley & Sons, Inc., New York, 1968.

Lind, Earl R., "Work Measurement in Data Processing," *Data Processing,* vol. 8, 1966, pp. 271–282.

McMillan, C., and R. F. Gonzalez, *Systems Analysis—A Computer Approach to Decision Models,"* Richard D. Irwin, Inc., Homewood, Ill., 1965.

Teichroew, D., "Examples of the Use of Matrices in Business Problems," *An Introduction to Management Science,* John Wiley & Sons, Inc., New York, 1964.

Chapter 3

The Role of the Computer in Work Measurement

GÖRAN HEDBERG
and
ULF SVENSÉN

AB Svenska MEC, Gothenburg, Sweden

The computer makes it possible to reduce drastically the need for human participation in routine clerical work. Many complicated work measurement systems would be impossible to develop and maintain without computer assistance. The computer also makes it possible to integrate work measurement activities with other closely related routines in the company, such as planning, wage systems, and bid proposals.

This chapter will describe briefly some of the ways in which computers may be used to assist the industrial engineer in his work measurement activities. No attempt will be made to present the computer applications in complete detail, for space does not permit. Enough will be said, however, to suggest to the industrial engineer how he may employ a computer to solve some of his own work measurement problems.

WORK MEASUREMENT SUBSYSTEM AS A PART OF AN INTEGRATED INFORMATION SYSTEM

Although many companies have computerized various subsystems and procedures, it is somewhat unusual to find all these subsystem applications linked together. Figure 3-1 shows an integrated information system designed for a medium-size

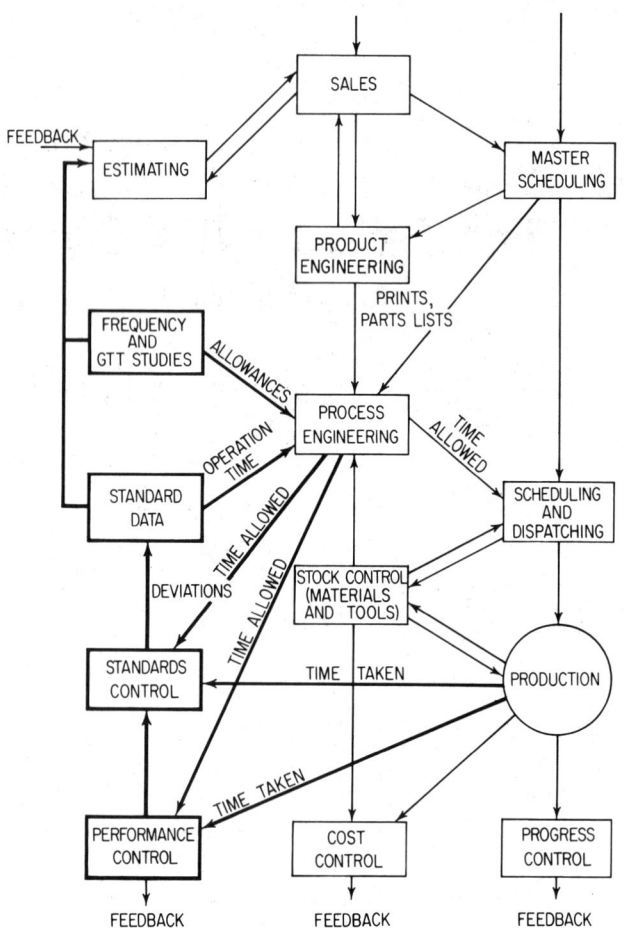

FIG. 3-1.   Integrated information system.

company.   The work measurement activities and their connections with other departments are shown by heavy lines.

The main data flow is as follows.   By means of master scheduling, the activities of sales, product engineering, process engineering, and scheduling and dispatching are linked together.   An order from a customer initiates collecting parts lists, materials, tools, and manufacturing time calculations.   These items are furnished by process engineering to scheduling and dispatching, which controls the production department's activities.

The flow from stock control and production represents data collection for calculation of costs and performance efficiency which become the feedback values for estimating, sales, and master scheduling.   As shown in the figure, frequency and GTT studies and standard data are the basis for the allowed time calculations and estimating.

Time-taken information obtained from production is the main control instrument for standards control.   Deviations from normal may initiate an updating of the standard data and also of the frequency and GTT studies.

CODING SYSTEMS SUITABLE FOR COMPUTER USE

During the planning stage of a new or revised work measurement system, the coding requirements of a computer should be considered. This should be done if the system is to be computerized either from the very beginning of the installation or at a later stage. The code should permit the classification of operations in a logical order and should make it possible to develop simulations and meaningful statistics. Special demands for each coding level must be considered. Consequently, a strictly logical way of building up the code between the different data levels cannot always be followed. Figures 3-2 and 3-3 show an example of how to code a computerized standard data bank. It was developed for MTM-based standard data for truck repair and service. The same procedure can also be applied to a stopwatch system. Figure 3-2 shows the five levels of data from which time standards are developed. Each higher level is developed from one or more lower levels. Figure 3-3 shows how the time for "Replace mounting rubber for muffler" is developed from lower levels of data. The time for the element "Tighten or loosen nut or bolt with medium-size wrench" (level 2) is developed from MTM data (level 1). This element becomes part of the operation element "Hand tools, fixed wrench" (level 3), which in turn becomes part of the operation "Remove mounting rubber for muffler" (level 4). This operation in turn becomes part of the final standard for "Replace mounting rubber for muffler" (level 5), which in this system is called "operation synthesis."

The codes assigned to levels 2 to 5 are shown in Figure 3-3. The rationale underlying them is explained in the following discussion.

Element, Level 2. On the element level, the basic MTM motions should be combined into as generally applicable sequences as possible. They should not be applicable to only a certain type of work and should preferably also be usable for more than one operation element within the same set of standard data. When building new time values from the standard data, it should be possible to visualize the code wanted and to search for the element in the element file to find out if it exists. If it does not exist, perhaps a closely related element can be used; therefore, related elements should be filed close to each other in the element file. A well-designed code should make it possible to get from the computer a list of related elements, such as all elements covering the act of hammering. Finally, it is desirable that the code should indicate the percentage of the total standard

Level	Element name	Synthesized of:	Approx. element size, TMU	No. of elements	Data type	Coding system
1	Basic movements		0–30	500	General	MTM
2	Element	Basic movements	30–100	1,000	General	Numerical
3	Operation elements	Elements and other operation elements	100–1,500	500	General, special	Alpha-numeric, mnemo-technical
4	Operations	Operation elements	4,000–100,000	10,000	Special	Numerical
5	Operation synthesis	Operations	10,000–	2,000	Special	

FIG. 3-2. Levels of standard data.

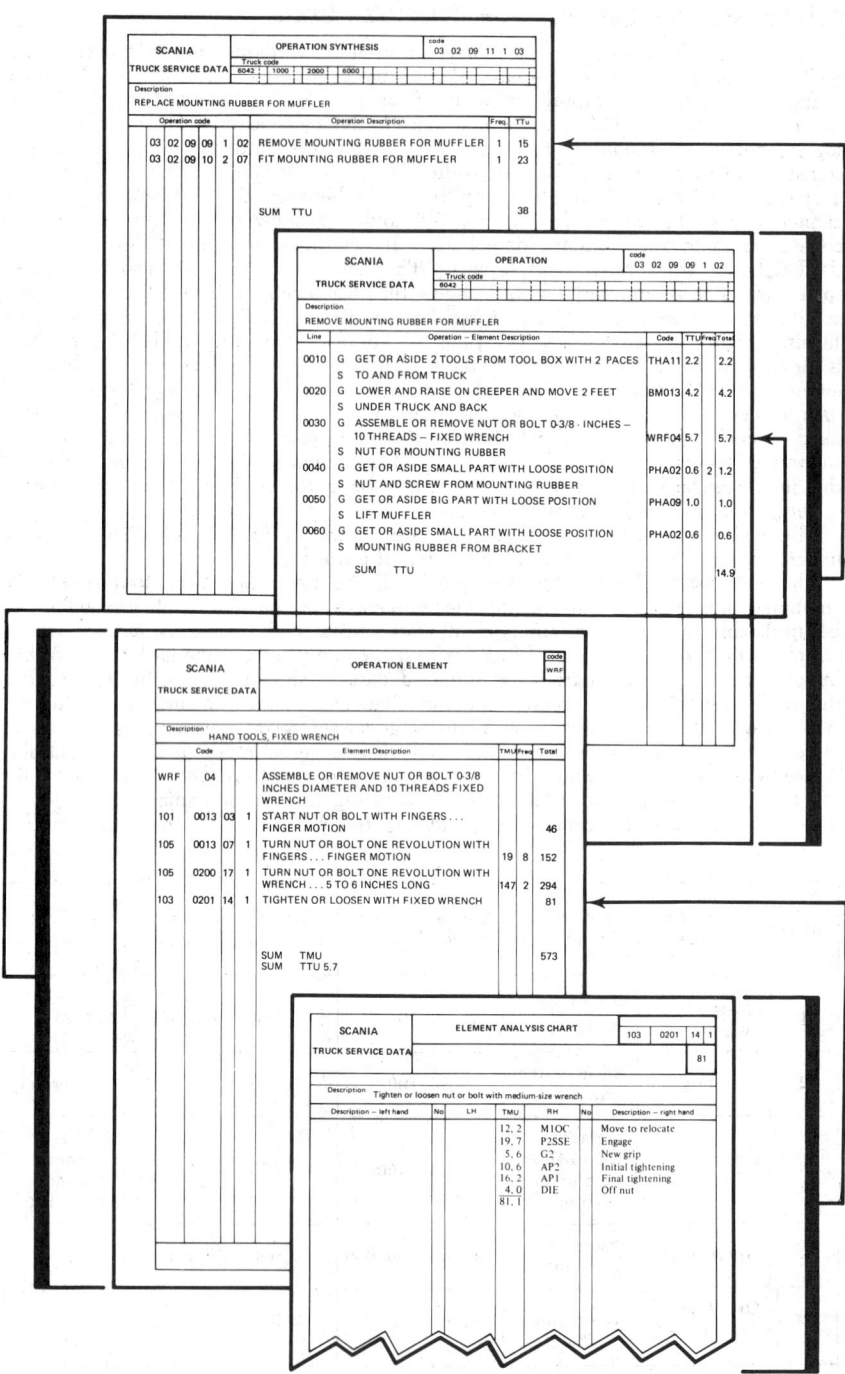

F<small>IG</small>. 3-3.   Example of standard data construction and coding.

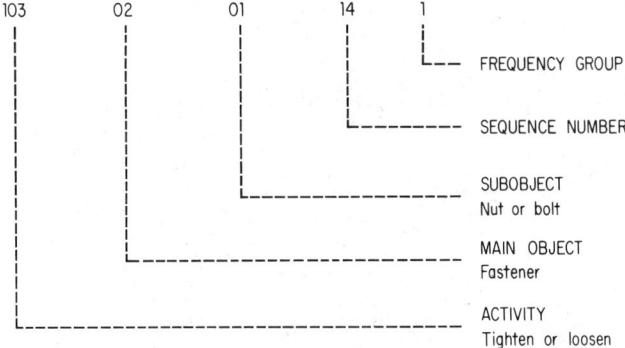

FIG. 3-4. The element code—level 2.

time per year that a certain element is used. This information, which affects the accuracy of the standards, may be obtained from the work order system and is discussed further later, under How to Maintain and Update a Standard Data System. Figure 3-4 is an example of the element code at level 2.

Operation Element, Level 3. For the operation element level, code objectives are dissimilar to those of the other levels. In priority order, the requirements for the third level are:

1. Easily remembered
2. Compact
3. Applicable to electronic work study methods
4. Designed to prevent errors
5. Element variables in the code
6. Logical

A mnemotechnical code best meets these requirements. Figure 3-5 is an example of code level 3.

This code is easy to remember, which means that the time required to learn the code will be reduced as compared with a nonmnemotechnical code. A three-letter code is sufficient to convey the necessary information and short enough to be easily memorized. The code should, of course, be as logical as possible. The code for wrench, for example, is WR. The third letter denotes the type of wrench as follows:

WRA   W̲rench, A̲djustable
WRF   W̲rench, F̲ixed

The code can also include the variables affecting the standard times if desired. MHA65 could be: material handling, move object up to six feet and up to five pounds. To prevent coding errors, only one of the letter pairs IJ and OQ should be used.

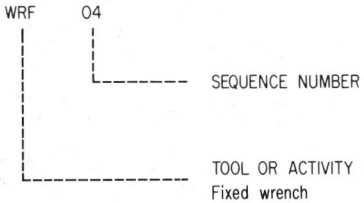

FIG. 3-5. The operation element code—level 3.

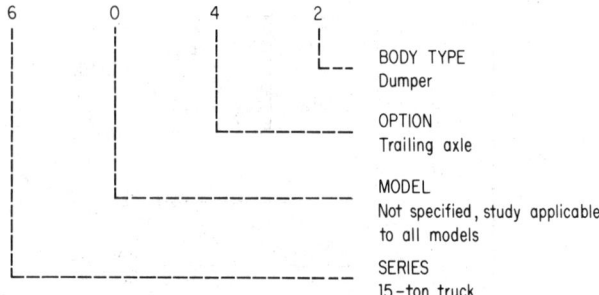

FIG. 3-6.   The truck code.

Operation and Operation Synthesis, Levels 4 and 5.   At the operation and operation synthesis levels, the code is expected to give answers to the following questions:

What truck model was studied?
What part of the truck was studied?
What job was performed?
What were the starting and the finishing positions for the job?
Is the study relevant for any other truck models?
Is the study relevant regardless of options and body types?

In addition to this, it is an advantage if the code conforms, in part at least, with other coding systems in the company, for instance, the spare parts list.

To meet these requirements, the code is divided into two parts. The first part, Figure 3-6, gives information about the truck studied and also the other models, options, and bodies where the study is applicable. This code is not involved in the computer updating program, which is discussed later. It merely gives information to the applicator.

By means of the second part of the code, Figure 3-7, every job done on the truck or any of its parts can be described. This code is part of the computer updating and simulation programs. Figure 3-8 shows an exploded view of the

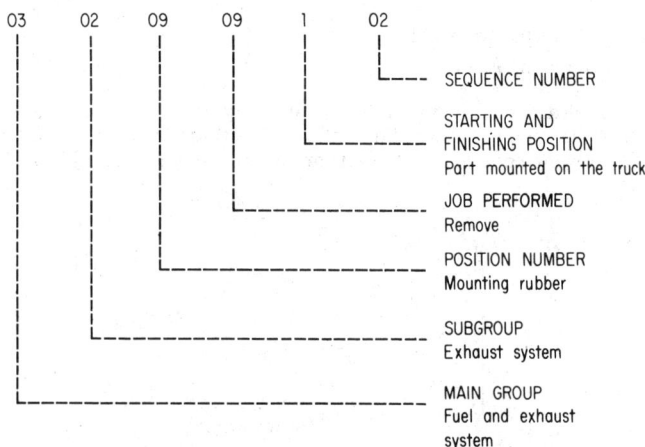

FIG. 3-7.   The operation and operation synthesis code—levels 4 and 5.

Main Group 03 Fuel and exhaust system
Subgroup 02 Exhaust system

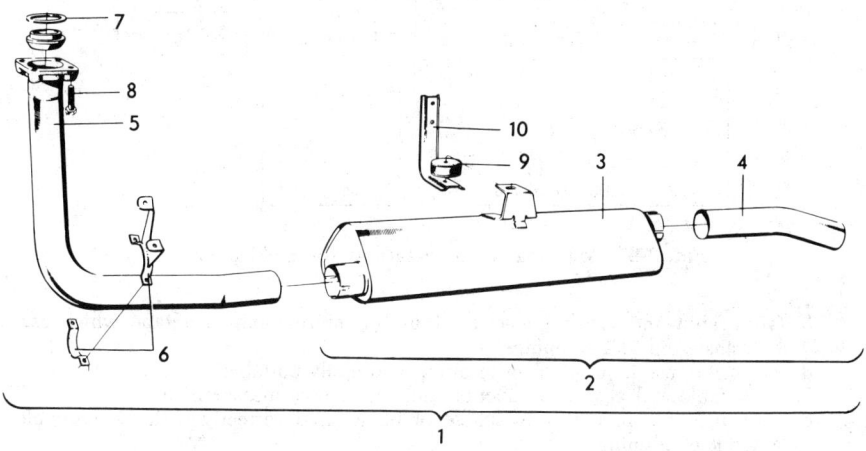

1	Exhaust system assembly	6	Attachment
2	Muffler with exhaust pipe, rear part	7	Gasket
3	Muffler	8	Flange bolt
4	Exhaust pipe, rear part	9	Mounting rubber
5	Exhaust pipe, front part	10	Bracket

FIG. 3-8.   Exploded view and position number list of the exhaust system.

truck fuel and exhaust system and illustrates graphically the meaning of the first part of the code, 03 02 09. In the following discussion, this coding system will be referred to further.

MAKING STUDIES THE ELECTRONIC WAY

When the industrial engineer is involved in establishing time standards, he spends most of his time analyzing and developing work standards in his office. The time he spends in the shop making studies is rather short. Not too many improvements can be made in this part of his work, except for planning the work to be studied so that it progresses without breaks or interruptions. But on the clerical side, electronic aids can do the job faster, more accurately, and more economically. Relations that are almost impossible to calculate by hand can also be investigated statistically quite easily with electronic aids. Making studies the electronic way requires a "work study machine," hereafter called the ETM (electronic time measurement) device, and a computer. To be universally useful in industry, the ETM device should:

1. Be applicable to stopwatch time study, predetermined time system-based standard data, work sampling, and the group timing technique
2. Eliminate the industrial engineer's calculating job

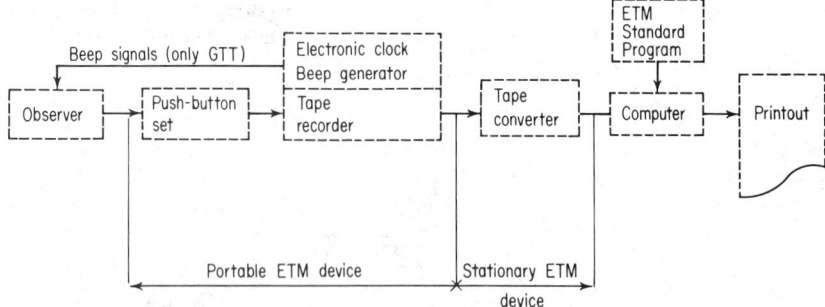

FIG. 3-9.   Main parts and data flow for ETM device.

3. Give rapid and reliable records of useful information on a tape which can be transferred to a computer
4. Be easy to use in all kinds of locations and easily portable
5. Be reliable with regard to function and accuracy of measurement
6. Be simple to operate and capable of being used correctly with a reasonable amount of training
7. Be acceptable to labor

A device meeting these demands is shown in Figures 3-9 and 3-10.

The portable ETM unit consists of two main parts, the push-button set with a built-in microphone and the cassette tape recorder. It is thus possible to register both spoken and digital information on the recorder tape. The push-button set for standard data studies is shown in Figure 3-10. As seen in the picture, there are two letters on each push button to keep the buttons at a convenient number. This means that pressing down buttons 1 and 2 can mean code AC, AD, BC, or BD. When constructing the code, only one of these combinations can be used. In a three-letter code, the conflicts encountered are very few.

For other types of studies, a different push-button set is used, containing the following abbreviations, together with the figures 0 to 9.

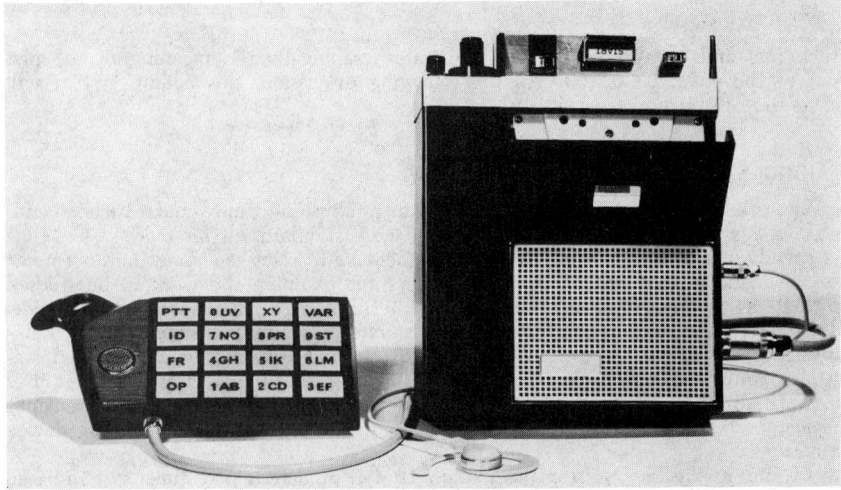

FIG. 3-10.   The ETM portable device with push-button set for standard data studies.

ID = identification
OP = operation
ACT = activity
VAR = variable
NU = number

In the following discussion, the use of the abbreviation implies the pressing of the button so marked.

The tape recorder is normally not running. When a button is pressed, the recorder starts, accelerates to normal speed, and records the button identification in digital form on the tape. If no other button is pressed, the recorder will stop again. Because of this method, each tape lasts about eight hours. When used for standard data studies, one tape cassette has a maximum capacity of 2,000 lines of work study and up to 20,000 readings for frequency or GTT studies. An electronic clock is included in the ETM device. It can be used to give signals to the observer when making the GTT studies described in Chapter 5 of Section 3. When the observer hears a beep, he takes a reading. The clock can also be used for conventional stopwatch time studies.

When the study is finished, the information on the tape is transformed in a tape converter—the ETM stationary part—into a computer tape. All spoken information is separated in the tape converter and recorded on another tape cassette, as will be discussed in more detail later. For calculating the results of the study, there are standard computer programs available. ETM standard programs are available for stopwatch time study, standard data study, work sampling, and group timing technique. An advantage of the ETM method of making studies is that all calculations are made by the computer and much more information can be extracted for no additional cost.

Work Sampling and GTT Studies. When making work sampling and GTT studies, the objective is generally one of the following:

1. To develop the allowances which should be added for fatigue and personal delays in incentive systems, or to measure the performance of a work group
2. To develop time standards as a base for an incentive payment system

Both these objectives can be accomplished with the ETM method.

The following example of a work group study comes from a plant manufacturing power cable, where power cable is manufactured on a processing line of four machines, each of them operated by one worker. The study time was 126 hours. The questions for which answers were desired were "How well does the operator utilize his machine?" and "What is the coordination between operator and machine?" The answers were obtained from a GTT study coded in the following way.

1. OPERATOR LIST
   01 Operator for plastic extruder
   02 Operator for cabling machine
   03 Operator for wire armoring machine
   04 Operator for cable jacketing extruder
2. OPERATOR'S ACTIVITY LIST
   All machines
   01 Check the cable being produced
   02 Operate controls
   03 Cut and join cables
   04 Change cable drums
   05 Change tools in machine
   06 Clean tools
   07 Use crane
   08 Paper work, clerical
   Plastic extruders
   10 Handling of plastics for cable jacket
   Cabling machine
   20 Join plastic tape

21   Change plastic tape pad
Armoring machine
30   Join copper wire, because of wire break
31   Join copper wire, because of changing of drums
32   Change copper wire drums
Allowances
50   Talk to supervisor
51   Talk to quality control personnel
52   Idle—unavoidable
53   Down—machine trouble
60   Personal time
61   Rest periods
70   Idle—avoidable
71   Start late, quit early

3. VARIABLE LIST
    01   Machine running at normal speed
    02   Machine running at reduced speed
    03   Machine not running

If, during the study, the observer saw the cabling machine operator changing cable drums and the cabling machine was not running, this observation would be recorded in three steps.

1. The operator is identified by pressing OP 02.
2. His activity is identified by pressing ACT 04.
3. Finally, the variable (what the machine is doing) is identified by pressing VAR 03.

The time required for this is on the average about five seconds, including the observer's searching time in the code list. Errors can be erased from the recorder tape by pressing the steering knob (OP, ACT, or VAR) three times. Figure 3-11 shows a portion of the printout when the study has been worked up by the com-

ORGANIZATION STUDY POWER CABLE DEPARTMENT.

OPERATOR NUMBER X02X – CABLING MACHINE.

VARIABLE	01		02		03		TOTAL	
	TH	P	TH	P	TH	P	TH	P
ACTIVITY								
01	72.5	57.5	6.7	5.3			79.2	62.8
02	0.3	0.2	0.8	0.6	0.4	0.3	1.5	1.1
03					7.9	6.3	7.9	6.3
04			3.9	3.1	6.8	5.4	10.7	8.5
05					2.3	1.8	2.3	1.8
06					0.3	0.2	0.3	0.2
SUM 01-06	72.8	57.7	11.4	9.0	17.7	14.0	101.9	80.7
07	1.5	1.2			1.1	0.9	2.6	2.1
08	0.1	0.1					0.1	0.1
10								
20					2.9	2.3	2.9	2.3
21					8.5	6.8	8.5	6.8
SUM 01-21	74.4	59.0	11.4	9.0	30.2	24.0	116.0	92.0
30								
31								
32								
SUM 01-32	74.4	59.0	11.4					
50	2.0	1.6						
51								

FIG. 3-11.   Example of GTT study result with ETM.

puter. It shows:

TH = time in hours on each activity

P = percent of time on each activity

Standard Data Studies. A major part of the cost when using a computer for maintaining and updating standard data banks, as will be discussed presently, is punching the input data. If the ETM device is used for establishing time standards from standard data, this cost can be considerably reduced because input will already be available in a form the computer can use. If the operation shown by Figure 3-3 had been studied using the ETM device, it would have been done as follows.

Making the Study. First, the observer presses the ID button, followed by the identification number of the study. The PTT (push to talk) button is pressed and the same number is talked into the microphone, together with the heading of the study, "Remove mounting rubber for muffler." The identification number must be recorded in both digital and spoken form to enable the computer to connect the spoken information with the right study. Numbers in digital and spoken form must also be recorded when putting spoken information into the study. If this is not done, the text cannot be attached to the right study line. The numbers need not necessarily be consecutive but must be given in ascending order. The mechanic now starts the job.

Mechanic:	Brings his tools to the truck.
Industrial engineer:	Pushes buttons: OP THA 11 ID 1.
	Talks: 1—to and from truck.
Mechanic:	Crawls under truck.
Industrial engineer:	Pushes buttons: OP BMO 13 ID 2.
	Talks: 2—under truck and back.
Mechanic:	Removes nut with fixed wrench.
Industrial engineer:	Pushes buttons: OP WRA.
	(This, however, is the wrong code; so he corrects by pressing OP OP OP.)
	Pushes buttons: OP WRF 04 ID 3.
	Talks: 3—nut for mounting rubber.
Mechanic:	Removes screw and nut from mounting rubber.
Industrial engineer:	Pushes buttons: OP PHA 02 FR 2 ID 4.
	Talks: 4—nut and screw from mounting rubber.
Mechanic:	Lifts muffler.
Industrial engineer:	No recording; element was not observed.
Mechanic:	Removes mounting rubber.
Industrial engineer:	Pushes buttons: OP PHA 00 ID 5.
	No figures; he has forgotten the numeric part of the code.
	Talks: 5—mounting rubber from bracket.

This concludes the study, and END is fed into the ETM by push button and microphone.

Working Up the Study. The work study recorded on the cassette is placed in the tape converter, which separates digital and spoken information, as shown by Figure 3-12. Spoken information is transferred to punched cards which, together with the data tape, become input to the computer. A list and a data tape which contain both digital and spoken information are produced by the computer. The list is checked, and missing codes and missing lines in the study are listed separately. Additions and corrections are transferred to punched cards which are fed into the computer at a convenient time. The final result is a complete and correct study on data tape as well as on a printed list for documentation. With this technique, the industrial engineer can devote his time to the higher level work for which he is trained and hand over the clerical work to the keypunch department and the computer.

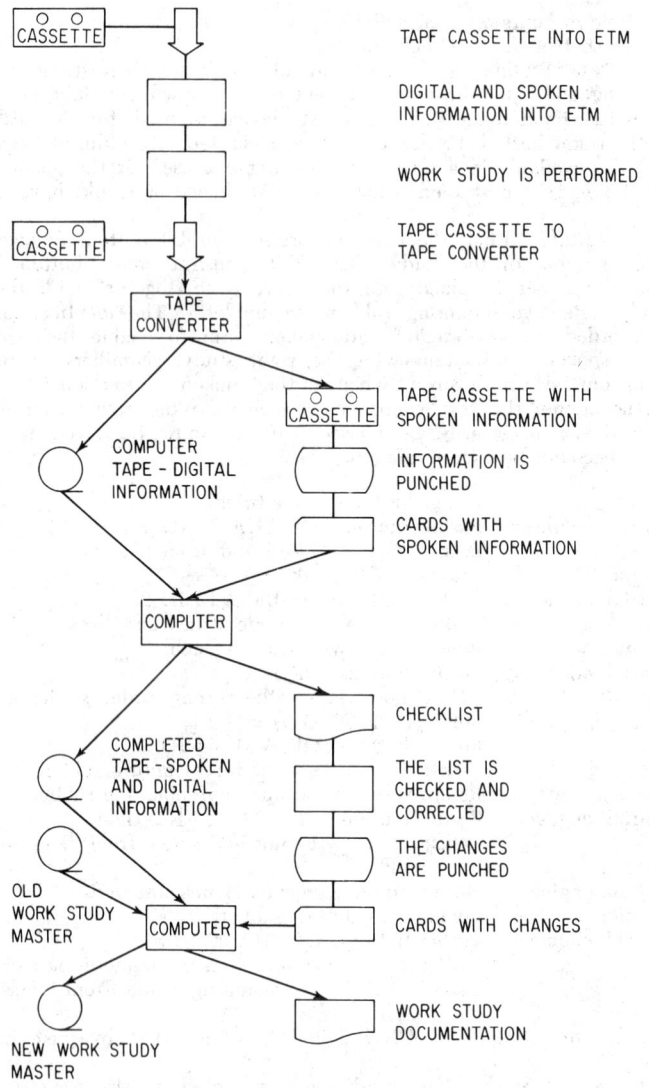

TAPF CASSETTE INTO ETM

DIGITAL AND SPOKEN
INFORMATION INTO ETM

WORK STUDY IS PERFORMED

TAPE CASSETTE TO
TAPE CONVERTER

TAPE CASSETTE WITH
SPOKEN INFORMATION

INFORMATION IS
PUNCHED

CARDS WITH
SPOKEN INFORMATION

CHECKLIST

THE LIST IS
CHECKED AND
CORRECTED

THE CHANGES
ARE PUNCHED

CARDS WITH CHANGES

WORK STUDY
DOCUMENTATION

Fig. 3-12.   Standard data studies with ETM.

HOW TO MAINTAIN AND UPDATE A STANDARD DATA SYSTEM

The structure of a standard data system has been illustrated by Figures 3-2 and 3-3. To give an idea of the size of the system, the number of individual elements on each level is shown in column 5 of Figure 3-2. Most of the time values are used many times. Not only is the size of the system significant but also the interaction between elements is important. Changing the value of one element on the second level may affect the time of more than fifty operation elements on the third level. These in turn may affect most operations on the fourth level.

Manual Handling of Standard Data. If the data system is handled manually, updating is a very tedious task. Assume that the time for an element must be changed because of a new method. The first step is to determine the new element time. Thereafter it is necessary to search all operation elements which are affected by the changed element. The sums of the changed operation elements are updated. The procedure is repeated, and operations which contain one or more of the changed operation elements are traced. The sums of these operations are then updated. When all updated elements are found and the whole recalculation is done, every page where there is a change must be rewritten.

Methods Improvements and Simulations. When a methods change is contemplated, it is important to know what the savings may be. To determine this requires the same procedure as updating the system, with the difference that the time saved must be multiplied by the number of times each operation is performed during a year. The total time saved must then be multiplied by the dollar value of every standard hour worked. Often this is too time consuming to justify a manual calculation. As a result, simulations are seldom performed in manual standard data systems.

Computer Solution. The problems encountered when revising standard data can be solved quite easily by using a computer. The principles of operation for the updating program are shown in Figure 3-13. The computer breaks down the standard data to contain only elements and no operation elements. This is

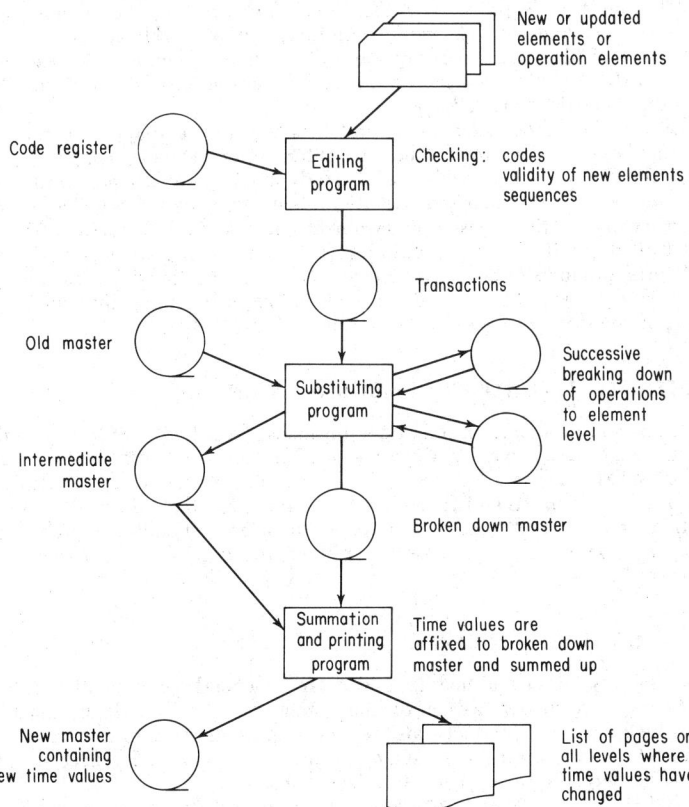

Fig. 3-13. Principles of operation for the updating program.

achieved by replacing the operation elements found within an operation with its elements. This goes on until the operations and operation syntheses consist of elements only. The end result is operation syntheses that are broken down to elements only. This breaking down procedure is performed only the first time when changing to computer handling of standard data or when new time values on any level are added to the system.

When an element time, a frequency, or any other change is made which affects time values anywhere in the system, all pages where there is a change are printed out by the computer. This means that the computer takes care of the tracing and the clerical work discussed earlier. The pages are printed in the same form as they were keypunched, not in the broken down version.

Simulations on Computer. To carry out a simulation, it is essential to know how many times per year each job is performed. The frequency can be taken directly from the work order system or it can be estimated. It is also necessary to know whether all jobs will be included in the simulation, or only a specified group such as jobs performed on a workbench. Assume that the question has been raised if the purchase of air impact wrenches is a sound investment in a truck repair shop. The simulation that will answer the question is carried out as follows:

1. Element times for fixed and adjustable wrenches are replaced by corresponding time values for air impact wrenches. A set of cards containing the wrench data code and the new time values is keypunched.
2. The type of work in which a change of method is possible is determined. This must be expressed by the code. In this case, the change is assumed to be possible for all work performed on a workbench but not on the truck. This condition is expressed by sorting on number 2 instead of number 1 in the "Starting and finishing position," in the job code in Figure 3-7. The others positions are left out.

The information under numbers 1 and 2 is punched and fed into the computer, and a simulation is ordered. The computer program will change all operation synthesis times which meet with the demands under point 2 above and which contain one or more of the changed operation elements or elements. The difference in operation synthesis time based on wrench and air impact wrench is calculated and multiplied by the number of times each specific operation is performed each year. These products are added and the sum is multiplied by savings per time unit. It is evident that the basis for decisions concerning investments in new tools and methods is improved by simulation.

COMPUTERIZED TIME FORMULA CONSTRUCTION

The operation elements in the time formulas described so far are built up by adding elements as shown by Figure 3-3. The elements may be stopwatch time values or MTM analyses. In some areas, the time values can be calculated from a mathematical formula which contains the variables that affect the time. These calculations are in most cases time consuming and should preferably be taken over by the computer. A discussion of this procedure may be found in Chapter 8 of Section 3.

CONTROL OF TIME STANDARDS

The reliability of established time standards depends among other factors upon the updating and modernization of the standards. New methods, machines, and tools as well as new products are regularly introduced into any company, and the time standards must be adapted to these changes. The necessity for good control is particularly great when using incentives.

The Reliability of the Organization. If the conditions and systems of the organization are in bad shape, it is very difficult to use a sophisticated work measurement system. Within a few years after such a system has been introduced, it is common to find that the installation has gradually degenerated from its initial level of effectiveness. Management must currently and frequently follow up and modernize the work order system, the planning and scheduling systems, and the wage control procedures. A computerized control system will give management guidance to a systematic follow-up of the standard data installation.

Control Model. The following example is taken from an incentive system for truck service and maintenance which is low repetitive work. The model controls the way the incentive system is handled by the applicators and the quality and reliability of the work standards. The coding system is the one described under Coding Systems Suitable for Computer Use.

Statistics Every Pay Period. The following information will help determine how well the installation is being administered.

1. The distribution of the working hours individually and by group, such as hours present, hours on standards, hours on other wage systems, and waiting hours.
2. Performance, individually and by group. Performance means the relationship between time allowed and time taken to do a job.
3. Standards coverage, individually and by group. Standards coverage means the percent of total working time covered by standards.

This information should be available in a report every pay period. When plotted on a graph, it will enable management to follow the trend from period to period.

Period Statistics. The above list gives only the average values. A more detailed analysis should be made once or twice a year. The performance distribution curve for each operator and group of operators serviced by one standards applicator is needed. This curve can be drawn by the computer, but often it is more convenient to get the values from the computer and plot a distribution curve of the most significant ones. If the quality of the time standards for low-quantity work is good and the applicator is doing his job well, the normal distribution curve peak will be between 105 and 115 percent performance on incentives. The curve will be even on both sides of the peak. In Figure 3-14, the jobs have been divided into performance groups to facilitate interpretation. The sum for "Total hours" gives an indication of how reliable the performance values are. If the number of hours used is small, the reliability is small and a good distribution curve cannot be drawn. If "Operator number" in Figure 3-14 is replaced by "Job code" or any wanted part of the job code, the printout will show performance deviations among different jobs or groups of jobs in the standard data bank.

Performance Distribution List—Period 12

Operator number	Applicator number	Performance group	Orders		Hours		Performance	
			Number	Percent	Total	Percent	Operator	Group
01	56	00–80	1	1.2	2.9	2.2		
		80–90	4	4.8	9.3	7.1		
		90–100	6	7.1	12.4	9.5		
		100–110	26	31.0	39.3	30.1		
		110–120	34	40.4	42.8	32.7		
		120–130	11	13.1	16.1	12.3		
		130–	2	2.4	7.9	6.1		
			84	100.0	130.7	100.0	114.8	109.6
02	56	00–80	3	3.5	6.9	4.3		

FIG. 3-14.   Performance distribution list.

Average Performance per Standard Work Group—Period 12

Standard work group	Time allowed	Perfor-mance	Order	
			Number	Percent of total
A	0.1	101.4	59	3.9
B	0.2	106.7	69	4.6
C	0.3	104.5	122	8.1
D	0.5	107.8	205	13.6

Fig. 3-15.   Average performance per standard work group.

If the range of time concept is used (see Chapter 8 of Section 3), the average performance in each standard work group, as shown by Figure 3-15, will indicate if the allowances and setup times are correct. A low performance in the groups with the smaller time ranges indicates too small allowances or setup times.

It should be remembered that in low-quantity work there are usually waiting times that do not occur in mass production, such as waiting for the foreman, a new assignment, or access to a workplace. A waiting time of about 2 to 4 percent should be expected on low-quantity work. It is often useful to report waiting times, as shown by Figure 3-16. The reasons for waiting times should be coded separately and recorded on the job card every pay period, together with time allowed, time taken, and so on. The waiting times of each job group should be shown separately for the group.

The control procedures mentioned briefly above are typical of those that can be used by management to determine how well a standards installation is being administered and maintained. A computer can generate any desired report rather easily. Care should be taken that only reports which are actually used for control purposes are regularly furnished.

CONCLUSION

This chapter has discussed briefly some of the more important areas where the computer and electronic aids for input of data to the computer are feasible. The

Waiting Time Report—Period 12

Job group	Waiting times									
	Parts		Assignment		Foreman		Access		Total	
	Hours	Per-cent	Hours	Per-cent	Hours	Per-cent	Hours	Per-cent	Hours	Per-cent
1	13.2	25.2	34.6	66.2	2.4	4.6	2.1	4.0	52.3	3.8
2	7.3	27.2	13.4	50.0	1.3	4.9	4.8	17.9	26.8	1.7
3	11.2	28.4	19.3	49.3	3.2	8.2	5.5	14.1	39.2	2.3
8	23.8	51.1	18.6	40.0	2.7	5.8	1.3	2.8	46.4	2.4
9	8.8	32.3	12.2	44.9	3.1	11.4	3.1	11.4	27.2	1.9
Total	176.4	38.1	201.3	43.5	46.2	10.0	38.7	8.4	462.6	2.9

Fig. 3-16.   Waiting time report.

savings in cost and time are obvious, but the biggest advantage is that the industrial engineering department which makes full use of the capabilities of the computer has a fair chance of keeping the work measurement system up to date. Over and above this, a systematic way of thinking is imposed upon the work measurement people. This will be of great value. With the aid of the computer, information can be stored and easily produced and updated, provided the coding system allows easy retrieval.

BIBLIOGRAPHY

IBM Data Processing Application, Methods and Standards Automation, International Business Machines Corporation Technical Publication E 20-0144-0, White Plains, N.Y., circa 1967.

IBM 1130 Work Measurements Aids—Application Description, International Business Machines Corporation Technical Publication H 20-0249-0, White Plains, N.Y., 1966.

Mattson, Rolf, *Arbetsmätning och ADB (Work Measurement and EDP),* SAF Technical Department, Stockholm, 1969.

Chapter **4**

Computer-aided Design

WALKER T. HOWELL

Corporate Manager—Industrial Engineering and Cost Control,
The Bendix Corporation, Southfield, Michigan

KERRY KILPATRICK

Research Assistant, Department of Industrial Engineering,
The University of Michigan, Ann Arbor, Michigan

JERRY KOVACH

Vice President—Management Operations,
Research for Management Science, Inc., Appleton, Wisconsin

A major automotive design group makes next year's body design changes by pointing at a TV tube. An electronics manufacturer obtains optimized printed circuit board layouts from an automatic plotting machine, and an industrial engineer saves time in the same way on workplace layouts. These are just a few examples of how computer-aided design has become a major technological advancement and a useful engineering tool.

Computer-aided design (CAD) is the application of computer sciences as an assistant in the design of products. The types of products or level of design may be quite broad. Automotive products, electronics, architecture, plant layout, and many other areas have all been affected by computer-aided design. In some cases,

9-46

the computer merely performs normal engineering calculations to save time. In others, it acts as a draftsman and information retrieval device, eliminating the drudgery of redundant activity and allowing more time for creativity. In still others, it goes beyond the normal capability of the man by performing complex calculations which yield optimized designs never before possible.

Computer-aided design has established itself not only as a significant engineering tool but also as a major industry. By 1970, over thirty companies were manufacturing a variety of equipment for this purpose. Examples include IBM, CALCOMP, BENDIX, and VARIAN. Furthermore, a multitude of software development firms emerged almost simultaneously with CAD capabilities.

It is obvious that CAD has become the third generation of applied industrial computer technology. First came the automated clerical systems which established the general field of data processing. Then came the MIS (management information systems) which recognized that the computer could also, with slight additional effort, sort, condense, and correlate data to provide previously unavailable information on a regular basis. Finally, it became recognized that certain routine thought processes, which are usually combined for unique situations, can be expressed as mathematical functions to optimize the general planning function. This latest step offers much to industrial engineering.

Because of the newness of this activity, some of the discussion that follows will cover areas which have not at the time of writing been fully researched and developed. Sufficient progress has been made, however, to indicate that all these procedures and more will become fully operational during the useful life of this Handbook.

There are several aspects of computer-aided design which should be recognized and given proper consideration by the potential user. The traditional equipment investment considerations are important, but just as important, if not more so, are the software considerations. In general, this is true for most computer-related implementations, but it is especially true for CAD. In the practice of CAD, a different kind of impact on organizational structure and planning procedures may well be experienced.

More than likely, the organization considering CAD will have substantial experience with other computer applications. For this reason, many of the details of basic computer technology have been omitted from this discussion. This chapter is intended primarily to review the basic fundamentals of computer-aided design only, and then to concentrate on the application aspects.

CAD HARDWARE

It may seem strange to the industrial engineer to put emphasis on the word "hardware"; the distinction is given, however, because hardware investment is just one of several important considerations which must be given to the potential CAD installation. Hardware is the equipment which is used to effect CAD. Moreover, it is equipment used primarily for input or output purposes. The computer is not featured in this case because it is assumed to be the essential ingredient of any CAD installation. A discussion of the computer itself would be like discussing the hydroelectric generators when considering the installation of a new light bulb socket. It is a utility. Furthermore, one would more than likely not consider using a computer just for CAD applications.

Modes of Operation. Before attention can be given to the specific types of hardware devices, their operational modes with the computer must be understood. In general, this situation can be explained as illustrated in Figure 4-1. The CAD installation may be on-site, that is, in the same general location as the computer; or it may be off-site with remote access by telephone lines. There is always a buffer between the CAD device and the computer which adds compatibility between the fast transfer rate of the computer and the relatively slow rate of most input-output devices. The manner of conversation with the computer may also be of two types: (1) off-line, where the operator must wait an indefinite

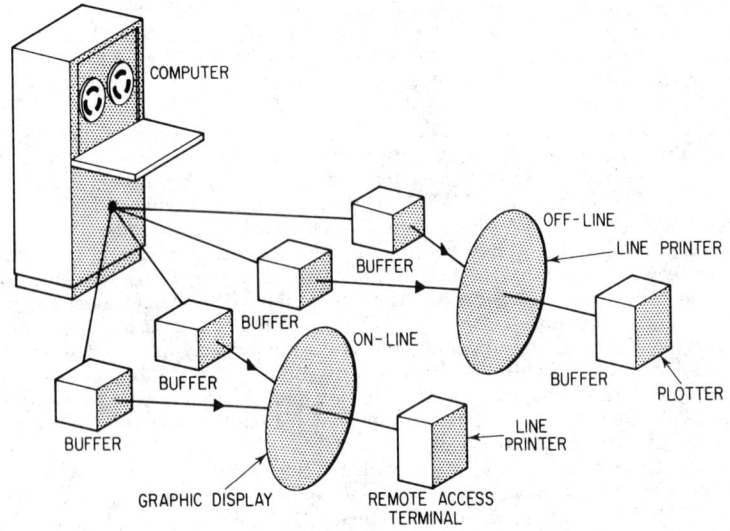

Fig. 4-1.   Modes of operation of CAD hardware.

time period for the computer to accept his input and send output and (2) on-line, interactive, where the operator can get instant access to and response from the computer.   The factors which determine the exact mode of operation are determined by the type of hardware used.

Types of Devices.   Although there are several manufacturers of CAD devices, the actual types are quite limited.   Their functions and other characteristics are discussed in the following paragraphs without detailed technical explanations of their operation.   One distinction which is important, however, is the d.fference between hard and soft copy output devices.   Hard copy devices provide permanent visual records, while soft copy devices only temporarily display CAD output.   A knowledge of these devices and their capabilities is vital to proper selection of a CAD system.

Line Printers.   It is only fitting that the most common form for all computer output, line printers, should also be used for computer-aided design.   The line printer which provides typewriter output is therefore the most available of all CAD devices.   It is limited in use, however, to applications which require only alphanumeric or simple coarse detail graphical output.   The line printer is a hard copy device providing printed output and is capable of providing multiple carbon copies.   The normal mode of operation for a line printer may be on-line or off-line.   Some line printers have built-in typewriter keyboards for on-line input.

Plotters.   To obtain complex graphical output in fine detail, it is necessary to use special-purpose plotting devices.   Plotters are essentially electromechanical pens which can move in two dimensions across a sheet of paper.   In some machines, one dimension of movement is obtained by moving the paper.   There is no special type of input device associated with plotters.   The normal mode of operation is off-line.

Cathode Ray Tube (CRT) Displays.   Although it provides only soft copy output, the CRT display, because of its versatility, offers much potential as a design tool. The CRT display is essentially an ordinary television with a digital analog converter which presents visual information to the user.   Since their introduction, several versions have been developed.   One type is the color CRT, which allows multicolor graphical presentations for special purposes. Another type allows temporary storage of a display on the cathode ray tube itself; updating with a new display, however, causes the one being stored to be destroyed.

Usually associated with CRT displays are various input devices which permit the on-line, interactive mode discussed above. One popular concept is the light pen concept, which executes a change in the display by touching the appropriate part of the display surface. For example, if an automobile body is being displayed, pointing to the taillight may cause it to be displayed in magnified form. Another type called Electrosketch, patented by IIT Research Institute, is a board with a movable cursor that sits in front of the display. Movement of the cursor causes corresponding movement of a dot on the display, and appropriate action can be executed. This type of device allows maps, curves, or other printed material to be placed under the cursor for tracing purposes. For precision computer input of large graphics, the proprietary Datagrid digitizer is available through the Advanced Products Division of the Bendix Corporation. It employs a free-moving cursor, or pointer, and a drafting board surface to support the graphic material. Dimensional data from graphics up to 60 by 60 inches can be transmitted to the computer with a resolution of 0.001 inch.

The major advantage of the cathode ray tube display is that it permits continuous monitoring of design changes until the desired results are obtained. At that point, something more permanent may be desired. One method of obtaining hard copy is taking photographs of the CRT display. This may be satisfactory if accuracy and clarity of detail are not important; if they are, it may be necessary to operate a plotter or some other hard copy device in parallel with the display.

SOFTWARE

Although hardware determines the form of computer-aided design output, software determines what problems will be solved and what the economic returns will be. Software refers to the computer programs and languages which make the hardware perform a prescribed function. Appropriate software is an essential ingredient to on-line versus off-line operation. Software generates the characters on a CRT display and controls the movement of a plotter. Software converts the formula and procedures of the engineers and managers to machine instructions. Possibly more than any other factor, the future of computer-aided design will depend on software developments.

Software exists at several levels. One level controls the mode of operation, assigns priority levels to various functions, and in general performs housekeeping functions which coordinate the computer and its peripheral equipment. This level is frequently referred to as the executive system. Another level is the language compiler, which is really a complex of predetermined machine instructions that can be recalled by the programmer in simplified terminology. Examples of programming languages familiar to most users are FORTRAN, COBOL, ALGOL, and PL/I. Although several of these can be used in CAD applications, some have been developed especially for conversational usage in graphic display systems. Examples of these include DIALOG and RUSH. These languages, however, provide only a base for CAD software. The software of primary concern to CAD are the special programs written in one of the preceding languages for solving specific problems.

During the 1960s, a variety of special-purpose programs for computer-aided design were developed. Unfortunately, many of them were the result of in-house projects by companies using CAD; therefore information about them is not available for this discussion. Several others have been developed by computer software suppliers. A brief review of typical CAD programs is presented here for the industrial engineer. The possibility of an available "canned" program should always be investigated before beginning a new program development.

PRINTED CIRCUIT BOARD DESIGN

Several large-scale computer circuit analysis programs have been developed, usually under the sponsorship of the federal government. These programs permit a designer to describe a proposed circuit in engineering terms, including the values of all components. The circuit then undergoes appropriate analysis—ac, dc, or transient—to evaluate operation of the circuit. This computer-designated performance is then compared with the required performance of the circuit, and engineering adjustments are made, either in the circuit design or in the selection of components, until analytical results correspond to the required performance.

All the design programs are written in terms of a few idealized components—resistors, capacitors, inductors, and voltage and current sources. Most of the products that are being designed, however, contain such nonlinear elements as transistors, diodes, and integrated circuits. To bridge this gap, these devices must be represented in the analysis programs by circuit models made up of components which the programs can handle. A growing inventory of such models exists, and through experimentation and measurement of actual devices, numerical values can be assigned to these models. The selection of an appropriate model, plus the determination of values for the model, is a critical and often difficult effort, perhaps requiring ancillary programs.

Each of the programs in use is special purpose and has inherent limitations which must be recognized. Of the more versatile, ECAP is intended for diode equations with capacitance effects included. Its normal version does not include a plotting capability, yet its output can be extremely bulky. Several users have added plotting capabilities. Other examples of circuit analysis programs include NASAP for small-signal dc transistors, SCEPTRE for hypoid-pi circuits, and NET-1.

PLANT LAYOUT

Plant layout is an industrial engineering function which since its beginning has relied on empirical techniques. With the computer, this activity has been automated and the results can be optimized. A number of computer programs have been developed for this purpose. One, called CRAFT, optimizes the location of work centers, given the building configuration. The objective function in this case is minimization of material handling costs.

Another program, called CORELAP, accounts for subjective conditions where material handling may not be the dominant factor. Other programs of a similar nature but of different capabilities have been developed for numerous applications.

WORKPLACE DESIGN

Current research indicates that computer-aided design methodology may also be applied to the design of manual workplaces, a task which consumes many industrial engineering man-hours annually. As in all design problems, workplace design requires the engineer to take certain available input information; satisfy a myriad of conflicting constraints; and by using whatever formal or heuristic techniques are available, produce an acceptable design. This general process appears schematically, essentially as in Figure 4-2.

Several features of the workplace design process as it is pursued at present should be noted. First, few engineers have the time to consider completely all the interdependent aspects of the process. Second, even if the time were available, the background data and catalog of techniques required to account for these interactions among worker, facilities, environment, product design, and process requirements are not in an easily usable form. Third, even given the required time and techniques, current practice makes it unlikely that the unaided engineer could synthesize more than a very few acceptable designs from which to choose the best design. Finally, the inability to determine the sensitivity of his conclusions to changes in inputs and constraints forces the designer either to recycle the entire design process each time a change does occur or to live with a suboptimal design.

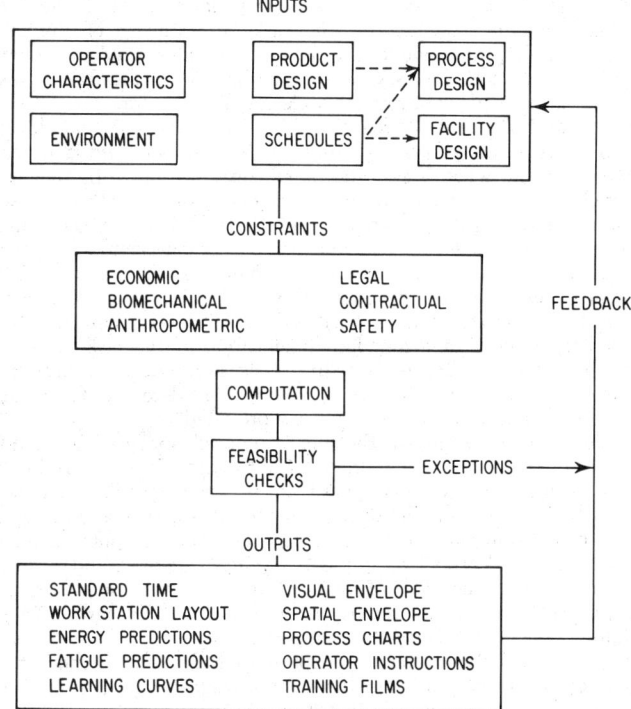

Fig. 4-2. Work station design process.

The effects of this undesirable situation are familiar to most industrial engineers. The workplace designer's inability to assess rapidly the economic consequences of alternative product or process designs results in a breakdown of interaction between those departments responsible for providing inputs to the workplace design process and the designer, who then must do the best he can with frozen product and process requirements. Another effect of the use of unaided design methodology is the widespread assumption that the environmental parameters such as light level, ventilation rate, or sound levels are set at normal values, because the designer does not have the ability to assess properly the effect of any deviations from normal. Perhaps the most frequently overlooked element in the design process is the operator who must perform the task. Problems of anthropometric and biomechanical feasibility of the task, effects of training on performance, energy expenditure rates and local muscle fatigue, and psychological factors such as information processing rates are rarely considered. The results of this neglect can range from local loss of production to employee disability. Although this is just a partial list of the problems with current practice, it is sufficient to indicate that the search for a better method is worthwhile.

Returning to Figure 4-2, the following features would be desirable if the workplace design process could be completely computerized. The system should be interactive with the user; it should allow simple English language inputs either through keyboard or light pen; it should produce on-line outputs in a form readily understood by the designer and have the option for hard copy if desired. The software package should ideally be able to take as input a parts list, an assembly sequence, a description of the tools and facilities, a set of environmental parameters, and a description of the operator or of the population from which an operator is to be selected;

and through its stored logic, produce, without violating any of the imposed constraints, a basic motion sequence for the task, an estimate of the cycle time, a line drawing of the optimal workplace layout, operator instructions and training aids, a learning curve for the task, energy consumption and local muscle fatigue predictions, and an optimal work-rest schedule for the operator.

Prior Efforts and Current Research. With this ideal in mind, a review of past and current research efforts will indicate how closely the goal has presently been achieved and in what areas additional research must be performed.

One generally accepted measure of effectiveness of an industrial workplace is the standard cycle time required to perform the task; so it is not surprising that considerable effort has been devoted to computerizing standard time prediction. Many companies have their own in-house computerized standard data systems, which generally resemble in structure the AUTORATE system[1] developed at IBM. Although these systems are a great aid to the development of time standards, they are not design tools, for a detailed specification of the task to be performed—usually in a highly coded form—is required as input and no provision is made for interactive use. Further, the workplace is considered a fixed input, and if constraints are violated, no error messages are provided.

A system closer in philosophy to the ideal proposed above is the ARMAN system developed at the University of Southern California.[2] It was originally designed as a tool to evaluate the time required to perform specified maintenance tasks on various types of electronic hardware, but it possesses many desirable features which permit broader application. In this system, the computer receives as input a description of the workplace, tools, and equipment required and a gross assembly-disassembly sequence for the task. From a file of stored micromotions, the computer selects the correct menu of micromotions to complete the task. These are then combined into an elemental sequence, using the knowledge of the workplace layout and heuristic rules of motion economy to produce a detailed methods time measurement analysis of the task. Although a major advance from the AUTORATE type of system, ARMAN is limited by the lack of interactive capability and by the absence of any considerations of the human element in the task.

At the time of writing, research to overcome the lack of interactive capability and representation of the operator is being pursued at the University of Michigan through a project sponsored by the MTM Association for Standards and Research. Also, models of the human operator in a working environment are being developed in at least two other locations: Boeing Aircraft in Seattle through the Boeing/JANAIR[3] cockpit geometry evaluation project and the SAMMIE project at the University of Nottingham, England.[4] Of these, the Boeing effort is currently the most ambitious, but unfortunately is not directly related to the workplace design problem.

A Computer-aided Workplace Design System. The structure of the University of Michigan system[5] in early 1970 was as shown in Figure 4-3. Inputs consisted of a description of a trial workplace, the associated tools and equipment, and an element sequence in English language (for example, GET BOLT). An example output from a remote teletype terminal is shown in Figure 4-4. Note that the feasibility of simultaneous motions is checked for each line of the analysis.

[1] M. F. Mobach, "AUTORATE—Computer Aid in Industrial Engineering," *Proceedings of the Sixteenth Industrial Engineering Institute,* University of California, Berkeley, 1963.

[2] A. K. Mason and D. M. Towne, "Toward Synthetic Methods Analysis," *Journal of Industrial Engineering,* January, 1967.

[3] L. F. Hickey, W. E. Springer, and F. L. Cundari, "A Development in Cockpit Geometry Evaluation," paper presented at the AGARD-NATO Symposium on Problems in Cockpit Environment, Amsterdam, November, 1968.

[4] M. C. Bonney, O. G. Evershed, and E. A. Roberts, "SAMMIE—A Computer Model of Man and His Environment," paper presented at the Ergonomics Research Society Annual Scientific Meeting, Bristol, England, March, 1969.

[5] K. E. Kilpatrick, "Computer-aided Workplace Design," *The Journal of Methods-Time Measurement,* vol. 14, no. 4, 1970.

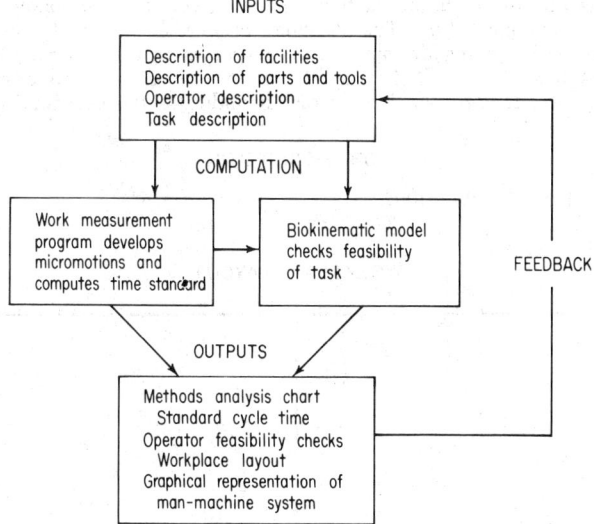

INPUTS

FIG. 4-3. Computer-aided design of workplace.

JOB DESCRIPTION
ASSEMBLE BOLT TO WASHERS
ANALYST A
MAY 20, 19—

METHODS ANALYSIS CHART

DESCRIPTION LH	NO.	L H		TMU	R H		NO.	DESCRIPTION RH
RCH RUBB AT LOC1	1	R 7.1C		# 10.9	R 7.1C		1	RCH RUBB AT LOC6
GRASP RUBB	1	G4B		* 18.2	G4B		1	GRASP RUBB
MVE RUBB TO ASS1	1	M10.3C	0.0	# 13.8	M10.3C	0.0	1	MVE RUBB TO ASS2
RELEASE RUBB	1	RL1		2.0	RL1		1	RELEASE RUBB
RCH WASH AT LOC2	1	R13.1C		# 15.1	R13.1C		1	RCH WASH AT LOC5
GRASP WASH	1	G4B		* 18.2	G4B		1	GRASP WASH
MVE WASH TO ASS1	1	M13.1C	0.0	# 16.2	M13.1C	0.0	1	MVE WASH TO ASS2
RELEASE WASH	1	RL1		2.0	RL1		1	RELEASE WASH
RCH BOLT AT LOC3	1	R13.0C		# 14.9	R13.0C		1	RCH BOLT AT LOC4
GRASP BOLT	1	G4B		* 18.2	G4B		1	GRASP BOLT
MVE BOLT TO ASS1	1	M13.0C	0.0	# 16.1	M13.0C	0.0	1	MVE BOLT TO ASS2
POS BOLT — ASS1	1	P3S D0.0		* 97.2	P3S D0.0		1	POS BOLT — ASS2
APPLY PRESSURE	1	AP1		16.2	AP1		1	APPLY PRESSURE
MVE BOLT TO ASS1	1	M 0.0C	0.0	# 2.0	M 0.0C	0.0	1	MVE BOLT TO ASS2
REGRASP ASSM	1	G2		5.6	G2		1	REGRASP ASSM
MVE ASSM TO HOL1	1	M 5.5B	0.0	8.5	M 5.5B	0.0	1	MVE ASSM TO HOL2
RELEASE ASSM	1	RL1		2.0	RL1		1	RELEASE ASSM
TOTAL				277.1				

* IN FRONT OF TMU MEANS LEFT AND RIGHT HAND MOTIONS CANNOT BE
PERFORMED SIMULTANEOUSLY.
IN FRONT OF TMU MEANS LEFT AND RIGHT HAND MOTIONS CAN BE
PERFORMED SIMULTANEOUSLY ONLY WITH PRACTICE.

FIG. 4-4. Computer-generated MTM analysis.

To illustrate the use of the interactive capability, consider the computer-generated layout shown in Figure 4-5. The designer may wish to modify some locations of this workplace by giving the appropriate responses to the computer as in Figure 4-6. This produces a new layout, as shown in Figure 4-7, and a new methods analysis, as in Figure 4-8, which immediately indicates the economic consequences

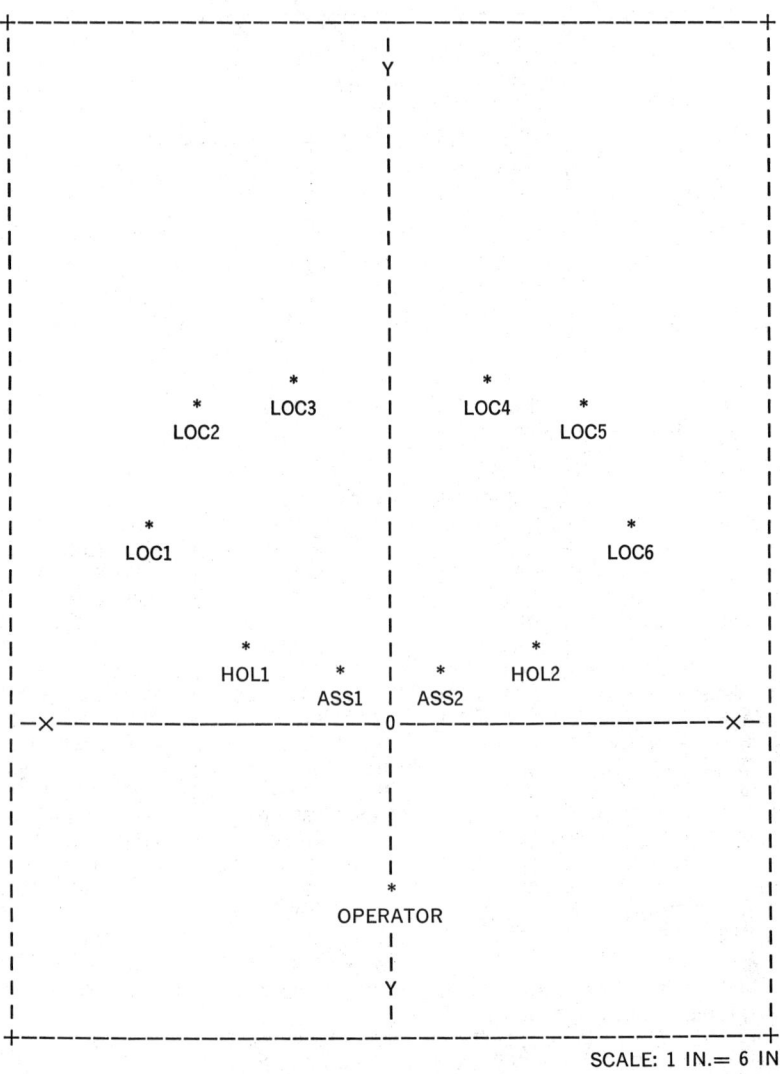

FIG. 4-5. Computer-generated workplace layout.

MTM FILE EDITOR
DO YOU WANT TO CHANGE A DATA LINE? 0=NO, 1=YES, 2=STOP
*1
ENTER LOCATION NAME (FORMAT A4).
*LOC3
ENTER CHANGES TO LOC3. EXAMPLE: X=5.1 0=NO MORE
*Y=25.
ENTER CHANGES TO LOC3. EXAMPLE: X=5.1 0=NO MORE
*0
DO YOU WANT TO CHANGE A DATA LINE? 0=NO, 1=YES, 2=STOP
*1
ENTER LOCATION NAME (FORMAT A4).
*LOC4
ENTER CHANGES TO LOC4. EXAMPLE: X=5.1 0=NO MORE
*Y=25.
ENTER CHANGES TO LOC4. EXAMPLE: X=5.1 0=NO MORE
*0
DO YOU WANT TO CHANGE A DATA LINE? 0=NO, 1=YES, 2=STOP
*2

* Indicates response by engineer.

FIG. 4-6. An interactive session with the computer.

of the design change. Once the designer is satisfied with the layout, he can request the computer to produce a more detailed layout of the type shown in Figure 4-9 for his permanent records.

The operator is represented by the stored logic of the computer as a biokinematic model of specified anthropometric characteristics. As the basic motion sequence of the task progresses, the biokinematic model assumes the required positions. If any portions of the task prove infeasible for the operator, appropriate comments are produced to alert the designer. In addition, a cathode ray tube display can be utilized to watch the simulated operator perform the task sequence. If hard copy is desired to study the effects of particular positions—for example, physical interference at the workplace—these can be produced as shown in Figure 4-10. If desired, one could produce a motion picture of the simulated operator performing the task, and this could be used for training purposes.

Extensions of Model and Future Research. Certain extensions of the current work can quite easily be accomplished. These include the incorporation of learning curve equations—described in Chapter 5 of Section 7—to produce a graphic representation of standard time expected as a function of the number of cycles completed, the computation of physical and visual envelopes for the operator, and the consideration of the interaction between information processing times and manual work elements.[6]

Areas where further research is required include the development of a general algorithmic solution of the basic motion sequence problem,[7] advances in our understanding of the relationship between the rate and level of effort and local muscle fatigue,[8] and the interaction between psychological and physiological factors

[6] T. L. Sadosky, "The Interaction of Sensory-decision and Motor Functions in the Performance of Manual Tasks," Ph.D. dissertation, The University of Michigan, Ann Arbor, 1969.

[7] J. W. Rigney and D. M. Towne, "Computer Techniques for Analyzing the Microstructure of Serial-action Work in Industry," *Human Factors*, vol. 11, no. 2, 1969.

[8] D. B. Chaffin, "Electromyography—A Method of Measuring Local Muscle Fatigue," *The Journal of Methods-Time Measurement*, vol. 14, no. 2, 1969.

at the workplace.[9] Regardless of these perceived shortcomings, one of the great values of the current work is that it holds forth the promise that the 1970s will see the emergence of many valuable design tools to help the industrial workplace designer do a better, more efficient job.

[9] L. L. Hoag, "Prediction of Physiological Strain and Performance under Conditions of High Psychological Stress," Ph.D. dissertation, The University of Michigan, Ann Arbor, 1969.

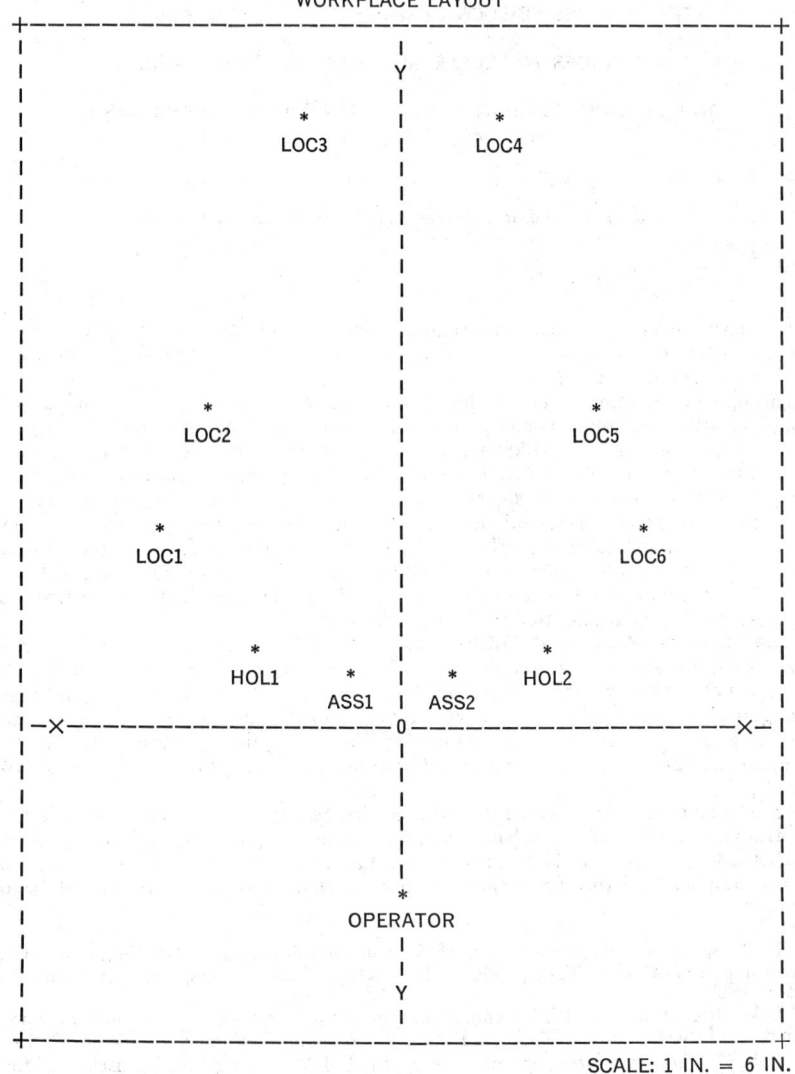

JOB DESCRIPTION

ASSEMBLE BOLT TO WASHERS
ANALYST A
MAY 20, 19__

WORKPLACE LAYOUT

SCALE: 1 IN. = 6 IN.

Fig. 4-7. Modified layout.

JOB DESCRIPTION

ASSEMBLE BOLT TO WASHERS
ANALYST A
MAY 20, 1969

METHODS ANALYSIS CHART

DESCRIPTION LH	NO.	L H		TMU	R H	NO.	DESCRIPTION RH
RCH RUBB AT LOC1	1	R 7.1C		# 10.9	R 7.1C	1	RCH RUBB AT LOC9
GRASP RUBB	1	G4B		* 18.2	G4B	1	GRASP RUBB
MVE RUBB TO ASS1	1	M10.3C	0.0	# 13.8	M10.3C 0.0	1	MVE RUBB TO ASS2
RELEASE RUBB	1	RL1		2.0	RL1	1	RELEASE RUBB
RCH WASH AT LOC2	1	R13.1C		# 15.1	R13.1C	1	RCH WASH AT LOC5
GRASP WASH	1	G4B		* 18.2	G4B	1	GRASP WASH
MVE WASH TO ASS1	1	M13.1C	0.0	# 16.2	M13.1C 0.0	1	MVE WASH TO ASS2
RELEASE WASH	1	RL1		2.0	RL1	1	RELEASE WASH
RCH BOLT AT LOC3	1	R24.3C		# 22.8	R24.3C	1	RCH BOLT AT LOC4
GRASP BOLT	1	G4B		* 18.2	G4B	1	GRASP BOLT
MVE BOLT TO ASS1	1	M24.3C	0.0	# 25.8	M24.3C 0.0	1	MVE BOLT TO ASS2
POS BOLT — ASS1	1	P3S D0.0		* 97.2	P3S D0.0	1	POS BOLT — ASS2
APPLY PRESSURE	1	AP1		16.2	AP1	1	APPLY PRESSURE
MVE BOLT TO ASS1	1	M 0.0C	0.0	# 2.0	M 0.0C 0.0	1	MVE BOLT TO ASS2
REGRASP ASSM	1	G2		5.6	G2	1	REGRASP ASSM
MVE ASSM TO HOL1	1	M 5.5B	0.0	8.5	M 5.5B 0.0	1	MVE ASSM TO HOL2
RELEASE ASSM	1	RL1		2.0	RL1	1	RELEASE ASSM
TOTAL				294.7			

* IN FRONT OF TMU MEANS LEFT AND RIGHT HAND MOTIONS CANNOT BE
PERFORMED SIMULTANEOUSLY.
IN FRONT OF TMU MEANS LEFT AND RIGHT HAND MOTIONS CAN BE
PERFORMED SIMULTANEOUS Y ONLY WITH PRACTICE.

FIG. 4-8.   Modified analysis.

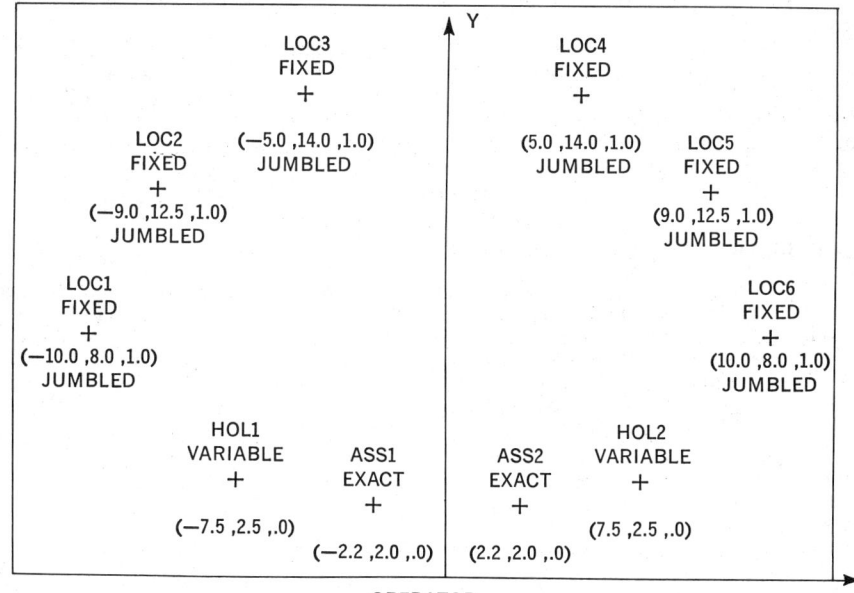

FIG. 4-9.   A line drawing layout.

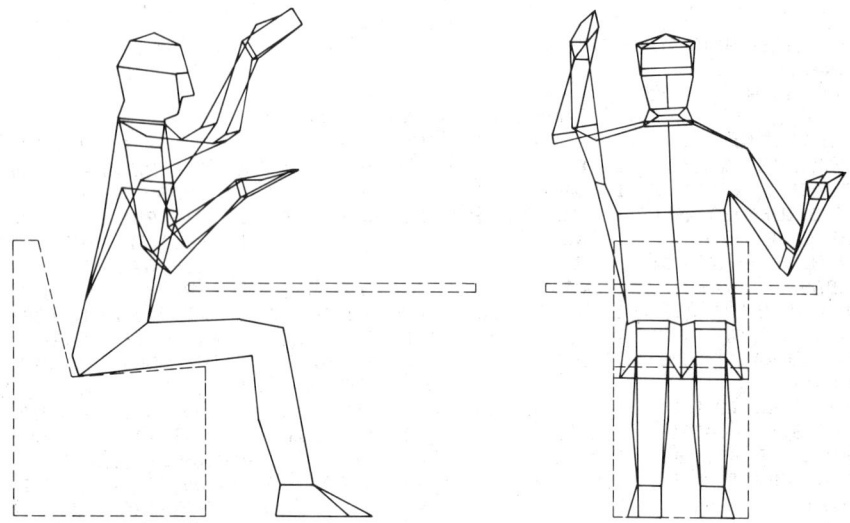

Fɪɢ. 4-10.   Computer graphics representation of seated operator.

IMPLEMENTATION PROBLEMS

Although many valuable CAD software packages have been developed, numerous difficulties can arise to hamper their effective use. Unfamiliarity with the programs often leads to frustration on the part of the designer. Some of these programs, promoted as near-panaceas for a broad range of design problems, simply do not meet the expectations of the men who casually pick them up without first learning their internal structure.

A considerably more significant problem may emerge because of the inherent lag of computer programs behind technological developments. Assume that an electronics firm supports the development of a computer code which it needs for the design of transistor circuits. This is not an inconsequential effort, financially or in technical skill. Then, quite suddenly, the company's product line changes, with integrated circuits replacing transistor circuits. Unless totally new models and measurement techniques can be devised, the code development will be wasted.

EVALUATING POTENTIAL APPLICATIONS

What constitutes good product design? If this question is asked of three men with different responsibilities in an industrial firm—a member of management, a product engineer, and a manufacturing engineer—there will probably be three different answers, reflecting the individual perspectives of their positions. The manager, while maintaining that dependability contributes mightily to his firm's product, might emphasize the aspect of cost effectiveness. Does the product contain needlessly expensive components that add cost but do not contribute to the function?

The product engineer might well respond to the question in terms of the intrinsic merits of the circuitry—how well it does the job for which it was intended. He is justifiably proud if the design incorporates the latest innovations of his craft, and his most splendid efforts can truly be called tour de force.

The design of a product is not entirely outside the manufacturing engineer's responsibility; rather than accept as a matter of course what is passed on to him, he often makes a positive contribution. For instance, the product engineer may have specified components with certain nominal values. The manufacturing engineer wants to know the permissible tolerance on those values: can they be 10, 5, or

2 percent? Cost is certainly a factor; components with a 2 percent tolerance are considerably more expensive than those with 10 percent. But the basic concern of the manufacturing engineer is the effect the choice of components will have on the production process. Perhaps a relatively simple change in design will eliminate severe problems in manufacture. To the manufacturing engineer, then, the effects of a good design can be evaluated by trading off manufacturing cost with performance.

One of the main benefits of computer-aided design is that the expectations of all these men can be met. The design of a circuit is more quickly and conveniently done, with choices possible among many alternative components and circuit elements. Production problems can be resolved more easily, for both designer and manufacturing engineer share a common tool—a computer program—with which they can make realistic appraisals of the product's performance. Cost effectiveness is also intrinsic to computer-aided design, eliminating, as it does, the unnecessary cost that results from overdesigning an element as a hedge against the fear of product failure.

CONCLUSION

In evaluating, a potential CAD installation, at least the following major points should be considered:

Does the design activity require substantial effort or could the design be significantly improved? Design effort of two types should be considered. If the design components are used repetitively, CAD may be justified even if the effort per time of use is low. In very complex designs, the calculations and so on may be both reduced in effort and improved if done on the computer.

What is the availability and cost of a CAD system? Both hardware and software must be considered. Proposal estimates for new developments in each area should be analyzed very carefully.

How flexible is the CAD system? Frequently, the systems must be updated to accommodate changing technology. The downfall of some installations has been the inflexibility associated with the updating activity.

Is the organization prepared to accept the full impact and benefits of the system? Product designers and manufacturing engineers must be prepared to interact earlier in the design process. The relative importance of these engineering functions may also fluctuate as a function of the particular product and process being considered. This, of course, results in a more fluid organizational structure. Without recognition of these potential problems and benefits, the full benefits of CAD may never be realized.

BIBLIOGRAPHY

Chaffin, D. B., "Electromyography—A Method of Measuring Local Muscle Fatigue," *The Journal of Methods-Time Measurement*, vol. 14, no. 2, 1969.
General Motors Corporation Research Laboratories, Computer Technology Department, *The GM DAC-1 System, Design Augmented by Computers*, GMR-430, Warren, Mich., 1964.
K. E. Kilpatrick, "Computer-aided Workplace Design," *The Journal of Methods-Time Measurement*, vol. 14, no. 4, 1970.
Mason, A. K., and D. M. Towne, "Toward Synthetic Methods Analysis," *Journal of Industrial Engineering*, January, 1967.
Mobach, M. F., "AUTORATE—Computer Aid in Industrial Engineering," *Proceedings of the Sixteenth Industrial Engineering Institute*, University of California, Berkeley, 1963.
Rigney, J. W., and D. M. Towne, "Computer Techniques for Analyzing the Microstructure of Serial-action Work in Industry," *Human Factors*, vol. 11, no. 2, 1969.
Sutherland, Ivan E., *SKETCHPAD, A Man-Machine Graphical Communication System*, Lincoln Laboratory Technical Report 296, Massachusetts Institute of Technology, Cambridge, Mass., 1963.

Chapter **5**

The Role of the Computer
in Numerical Control

JAMES J. CHILDS

James J. Childs Associates, Alexandria, Virginia

The function of the computer in numerical control is only slightly younger than numerical control itself. No sooner had the first numerical control machine successfully performed at the Massachusetts Institute of Technology in 1952 than plans were under way for the development of computer programs that would assist in the preparation of the coded tapes which are required to operate the numerical control machine tools automatically. It was deemed essential to use a computer to handle the voluminous amount of calculation required for describing the cutter paths of the NC machine tools. Without the aid of the computer, numerical control would surely never have progressed to its present state.

One of the problems facing the numerical control industry in its early sales efforts was the belief that a computer was essential for the operation of numerical control equipment. This is not the case. The misunderstanding undoubtedly arose from the fact that there are two types of numerical control: point-to-point and contouring. In point-to-point numerical control, a machine tool or other piece of numerical control equipment is directed to a specific point and then required to perform an operation. The operation may be drilling a hole, performing a spot weld, or inserting an electronic component into a predrilled circuit board. The part programming effort required in this instance is relatively simple, and the majority of tapes for point-to-point machines are prepared manually without the aid of a computer. Contouring numerical control requires that the path of the machine be prescribed to close tolerances. It generally requires a considerable amount of calculation which can best be performed by a computer. Applications include contour milling machines, lathes, drafting/plotting machines, flame cutters, and cutting material for clothing.

As the use of numerical control machines in the manufacturing process has increased, the use of the computer has increased also. More and more companies are finding that complex point-to-point requirements can be handled more economically with the assistance of a computer. The computer's role in the engineering design and manufacturing planning processes is being coupled to the numerical control programming requirements, the end result being a completely automated process beginning at the design intent and ending at the finished product.

This chapter describes the computer's role in the numerical control world, especially as applied to the part programming function, which is the process of preparing a coded tape for the automatic operation of the numerical control machine.

THE NUMERICAL CONTROL PROCESS

The numerical control process generally begins with the engineering drawing. Figure 5-1 illustrates the general flow pattern which begins at the blueprint or engineering drawing and proceeds to the finished part.

Methods Planning Function. Because there are very few shops that are comprised entirely of NC equipment, it is necessary to determine whether a given part is best suited for NC manufacturing or whether it can be more profitably produced on conventional equipment. The determination will depend primarily on the complexity and lot size of the part. The more complex the part and the smaller the lot size, generally the more suitable the job is for NC. The responsibility for this determination generally lies with the methods engineer. His responsibilities also include determining the other non-NC operations, such as drilling tooling holes and finishing operations, because there are few parts that do not require other operations in addition to the NC machining requirements. Additional responsibilities may include specifying any special tooling that may be required. Instructions coming from the methods engineer's desk are usually in the form of an operation sheet or routing.

Part Programmer's Function. After it has been decided that the part is to be manufactured by NC—and more specifically, what portions of the part are to be machined—the engineering drawings, together with the operation sheets and any other pertinent instructions, are passed on to the part programmer whose function is to list the detailed instructions necessary for machining the part on the NC machine. The instructions are transferred directly to the tape in manual programming. They are processed through a computer where computer-assisted part programming is employed. The instructions in the two cases are dissimilar because, in the latter instance, the computer performs the bulk of the calculations.

Qualifications of a Part Programmer. The part programming function is normally a full-time job in the larger shops. In smaller shops, or where the amount of NC work is relatively light, the part programmer may be expected to perform other duties such as methods engineering or tool design. Although generally not a graduate engineer, he should be expected to have a reasonably sound machine shop background; possess a high school diploma with preferably an associate degree in mechanical technology in addition; and have an aptitude for mathematics. The degree of mathematical training required will vary from trigonometry to complex analytic geometry and even calculus. The in-house formal training required may range from several days for the simpler point-to-point manual programming work to several weeks for the more complex contour programming which requires computer assistance. It may also require from one to two years of on-the-job training before a part programmer can become proficient in handling complex parts by computer-assisted programming. The part programmer is undoubtedly the key individual in the numerical control process, although many others such as maintenance personnel and operators contribute heavily. The part programmer's function is the most novel part of the operation and is usually an entirely new requirement in most companies.

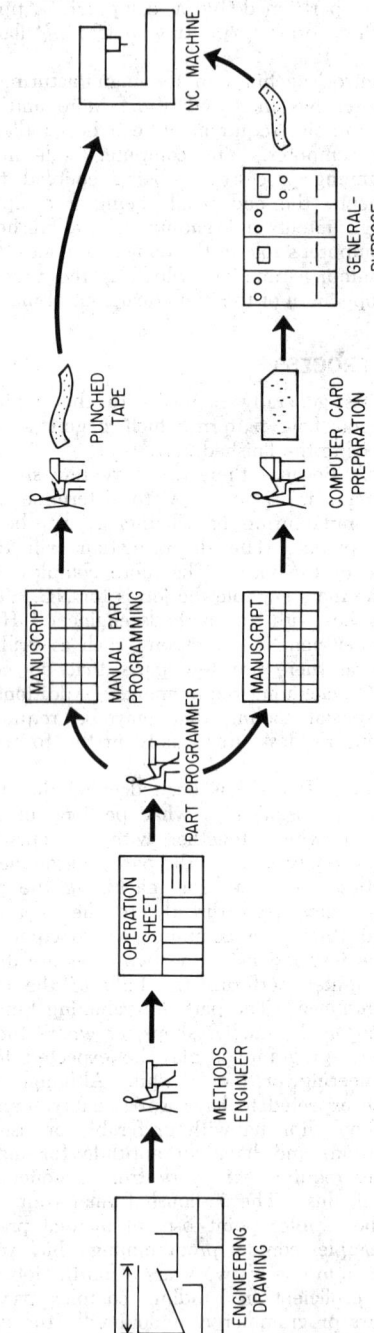

Fig. 5-1. Flow pattern, from engineering drawing to finished tape.

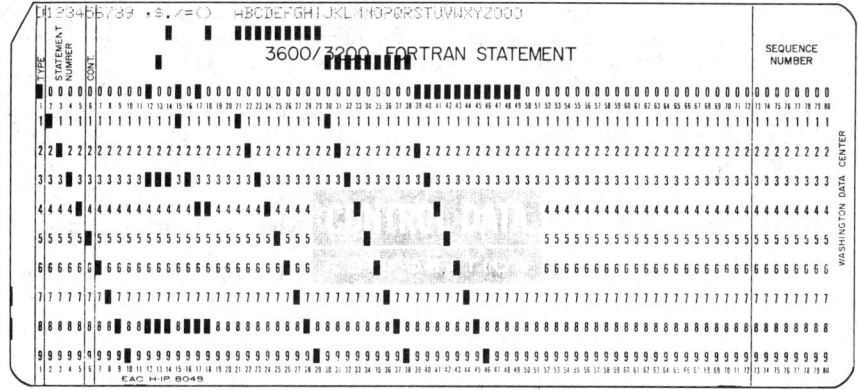

FIG. 5-2.   Computer input card showing the letters and numerals used in NC.

Tape Preparation.   Referring to Figure 5-1, it will be noted that there are two paths from the part programmer. The upper route involves the procedure for preparing a tape that has been planned by manual programming. The lower route describes the steps involved when a computer is utilized. Although there are fewer steps required when preparing a tape by the manual route, the effort of the part programmer is greater because he is required to perform all the mathematics required, in addition to noting the instructions as they are to appear on the tape. The part programmer's effort is considerably less when preparing a program for the computer, but the specialized training required for this work is greater. In either manual or computer programming, the instructions are noted on a form normally designated as a manuscript.

Manual Tape Preparation.   In manual tape preparation, instructions taken from the manuscript are typed directly onto the tape as a series of holes in coded form. A printout of the typed instructions is prepared simultaneously. This tape may then be inserted into the control system of the machine tool.

Computer Tape Preparation.   For computer tape preparation, the manuscript instructions, which are in a computer language, are typed onto IBM computer cards. The cards are then fed into the computer system, and a tape is prepared by the computer. An example of a computer card showing the punched alphanumeric notations used in NC computer statements is illustrated by Figure 5-2.

In either manual or computer programming, the tape is identical, for it is the electronic control system associated with the machine tool that dictates the format on the tape.

PUNCHED TAPE—THE CONTROLLING MEDIUM

In the early period of NC, there existed a sizable proliferation of input media. The assortment ranged from ½-inch magnetic analog tape, through 5-inch-wide punched plastic tape, to punched cards. Even motion picture tape was not overlooked.

To provide a reasonable consistency which would satisfy prospective users, it was imperative that standards be developed. This is not unlike the thread on an electric light bulb or the spacing on a wall socket for the insertion of a plug. Consider the situation if these two everyday items were not standardized. Then consider further the problems of a shop manager if he had to cope with five NC machines, each requiring different equipment to prepare the media.

The Electronics Industries Association.   The task of developing a standard for the input medium fell to an organization that had long had a reputation for preparing standards in the electronics industry, namely, the Electronics Industries Association (EIA).   Task groups, comprised of representatives from control system and machine tool manufacturing organizations, met regularly for several years to establish the standards that have now been almost universally adopted.

Tape standards published by the EIA may be procured from their offices at 2001 I Street NW, Washington, D.C., 20006.   A list of tape-related EIA standards and their prices is shown below.

> EIA Automation Bulletin 3B (Glossary of Terms). $.70
> RS 227   One-inch Perforated Paper Tape. $.50
> RS 244-A   Character Codes for Numerical Machine Tool Control Perforated Tape. $.50
> RS 273-A   Interchangeable Perforated Tape Variable Block Format for Posi-

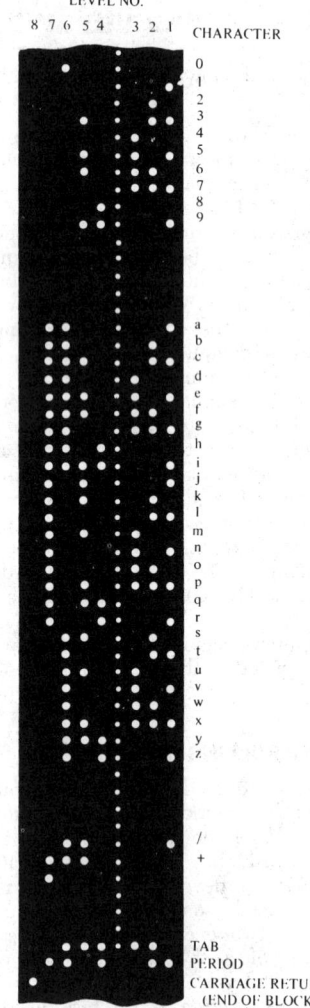

FIG. 5-3.   EIA binary-coded decimal arrangement.   (*Courtesy of The Industrial Press, New York.*)

tioning and Straight-cut Numerically Controlled Machines. $2.60

RS 274-B   Interchangeable Perforated Tape Variable Block Format for Contouring and Contouring/Positioning Numerically Controlled Machines. $2.80

RS 326   Interchangeable Perforated Tape Fixed Block Format for Positioning and Straight-cut Numerically Controlled Machines. $2.80

RS 358   Subset of USA Standards Code for Information Interchange for Numerical Machine Control Perforated Tape. $.60

Coding and Format.   The basic tape code is known as "binary coded decimal," which means that the hole pattern follows a combination of the binary code, based on 2 to a series of powers (2^2 equals 4, 2^4 equals 16, and so on) arranged in a decimal form. Figure 5-3 shows the standard 1-inch-wide tape, together with the standard hole patterns for numbers, letters, and required NC notations such as $+$ or $-$.   The tape material is 1 inch wide and may be paper, paper-coated Mylar plastic, or plastic-coated aluminum.   The strength, durability, and price also increase in this order.   There are eight rows of holes plus a row of smaller holes running down the tape.   When the tape is prepared by a typewriter-type unit, each hit of the selected key produces a combination of holes across the tape.   For example, by hitting the key letter "a," holes will be produced in columns 1, 6, and 7.   Computer-produced tape is handled automatically, usually by means of high-speed punches.   It should be noted that manual tape punching is a frequent source of error.   This problem is generally eliminated with computer-prepared tape.

Figure 5-4 shows a very simple block of data that contains sufficient instruction for the machine to perform at least a single operation.   The machine is instructed to move to a point that is 1.2345 inches in the x direction and 6.7890 inches in the y direction.   The EB designates the end of the block and stops the tape from moving through the reading head until the specified operation is completed. The reader then moves the tape sufficiently to read the next block, and the cycle repeats.

The block is subdivided into characters, which are assembled into a pattern

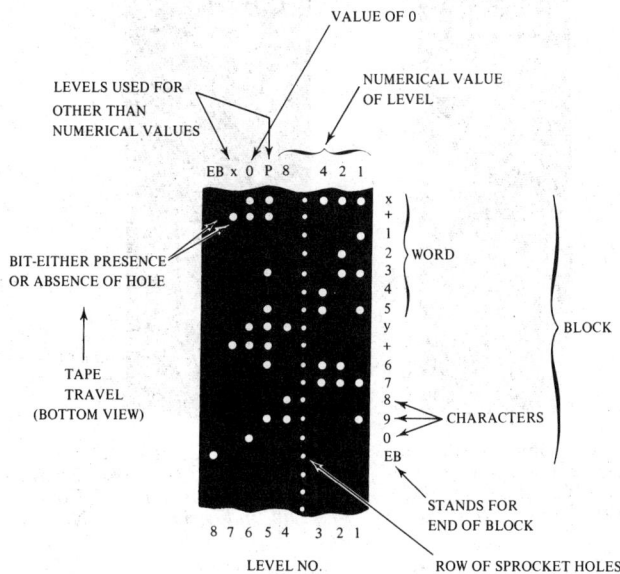

FIG. 5-4.   Single block of data.   (*Courtesy of The Industrial Press, New York.*)

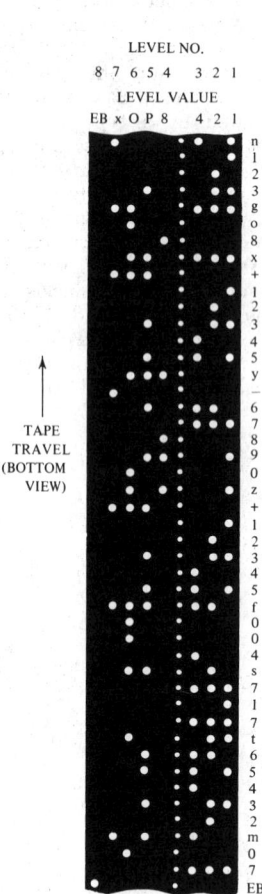

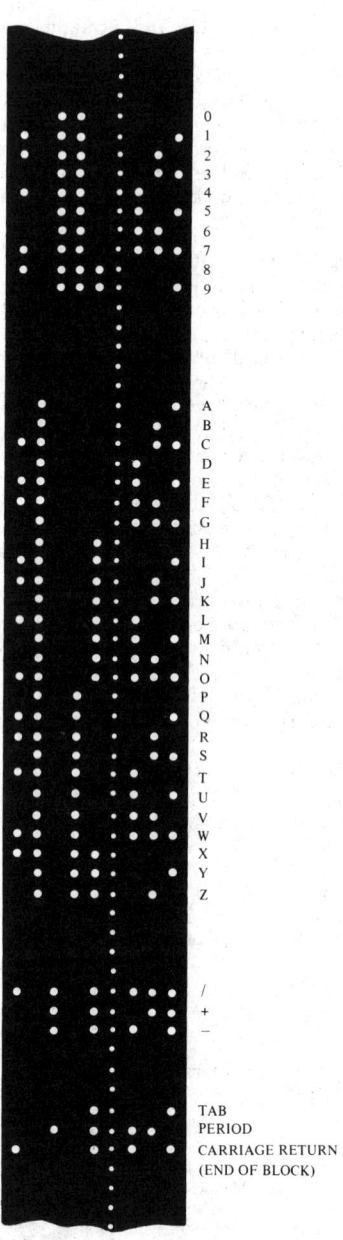

Fig. 5-5. Block format for a reasonably complex machine. (*Courtesy of The Industrial Press, New York.*)

Fig. 5-6. United States of America Standard Code for Information Interchange format. (*Courtesy of The Industrial Press, New York.*)

that forms words. A word has a distinct identity but cannot normally produce machine action unless combined with other words. The length of a block will vary, depending on the number of instructions or words required to operate the machine. As an example, the block shown in Figure 5-5 operates a machine having the capability for five simultaneous motions, changes tools, and specifies speeds and feeds. Words need not be shown in every block but only when action is required. In this instance, the "a" and "b" rotary axis words are absent because no motion is required of these axes.

Although practically all the machines operating in the United States abide by the EIA code, there is a strong influence being exerted to change the character coding to that shown in Figure 5-6. This code is being sponsored generally by the communications industries and is gaining acceptance in Europe through the International Standards Organization. The tape is still 1 inch wide, and the arrangement of the block information remains the same. The only change is the character coding.

PART PROGRAMMING

As pointed out above, the part programming function may be divided into two categories: point-to-point and contouring. Point-to-point programming, which is normally the simpler, will be reviewed first, following an explanation of the coordinate system.

Coordinate System. All numerical control programming has as its basis the orthogonal coordinate system. Positions in space are measured as distances from right angle lines noted as the X, Y, and Z axes in Figure 5-7. Actually, the distances are measured from planes formed by any two lines. Considering only the X and Y axes which form the XY plane, a point lying on this plane will be a distance x inches from the Y axis and y inches from the X axis. Figure 5-8 shows a point (PTA) whose coordinates are $x = 5$ (5 inches from the Y axis in the plus x direction) and $y = 3$ (3 inches from the X axis in the plus y direction). PTB is a second point whose coordinates are $x = -6$, $y = 4$. As a further example, the coordinates of PTC would be expressed as $x = -3$, $y = -4$.

Any point in space can be described by adding a z coordinate, which is the distance from the XY plane in Figure 5-7. The z coordinates of points PTA, PTB, and PTC are all zero in Figure 5-8 because the points lie on the XY plane. The z coordinate of any point lying on the side of the XY plane shown in the plus z direction is designated as positive, and any point lying on the far side shown as the minus z direction is noted as negative. The point where all three axes meet is called the origin, and its coordinates are $x = 0$, $y = 0$, $z = 0$.

Point-to-point Part Programming

The Manuscript. The detailed step-by-step machining instructions for point-to-

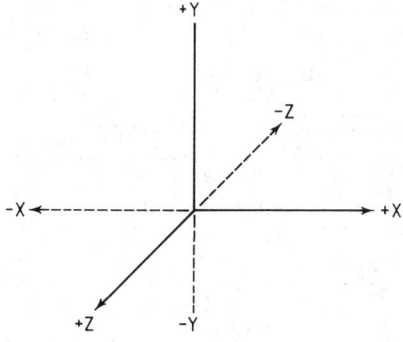

Fig. 5-7. The X, Y, Z coordinate system.

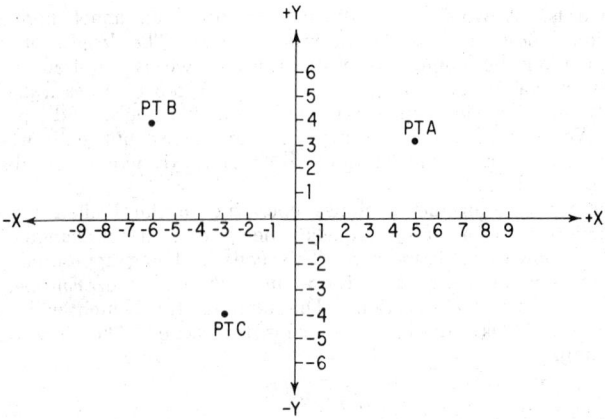

FIG. 5-8. Location of points *PTA*, *PTB*, and *PTC*.

point machines are listed on a form commonly known as a manuscript. Each line on the manuscript represents one block of instruction on the tape. In addition to the coordinates that are listed, other instructions in coded form are also noted. For example, the code word for starting the spindle in a clockwise direction is m03. The code word for stopping the machine is m30. This code will also rewind the tape automatically. As with the character coding, the code words have been standardized.

Figure 5-9 shows a relatively simple manuscript form for a point-to-point operation. The machine being controlled has two axes of control, *X* and *Y*, and the *Z* axis is set manually. This is the most common type of NC machine. The first column, designated as 1, describes the number of the block and is used by the operator as a reference if his machine is equipped with what is known as a sequence number readout. The readout is a lighted display panel that notes the particular sequence number of the block being worked on. Column 2 describes the *x* coordinates, and column 3 the *y* coordinates. Column 4 shows the auxiliary or m word codes when they are required. Most manuscript forms also have a "Comments" column wherein the programmer may note instructions for the operator, such as the changing of a tool or special setup requirements. Manuscript forms will differ, depending on the machine tool. They are normally furnished by the machine tool manufacturer.

If it is desired to move a machine to the points shown in Figure 5-8, and further, if these points lie on a part and holes are to be drilled at these points, the manuscript instructions would be as shown in Figure 5-9. Lines *A*, *B*, and

	①	②	③	④	⑤
	SEQUENCE NO.	x	y	AUXILIARY CODE	COMMENTS
Ⓐ	001	+5.000	+3.000	m 03	Use 1/2" drill
Ⓑ	002	−6.000	+4.000		
Ⓒ	003	−3.000	−4.000	m 30	Take out part.
					Place new
					material in
					holding fixture

FIG. 5-9. Example of simple manuscript for preparing tapes manually.

C refer to the three points *PTA*, *PTB*, and *PTC*. In this example, the machine and the part have been purposely kept simple.

After the manuscript is completed by the part programmer, it is generally turned over to a typist, who prepares the machine tool tape. Except for the comments column, the instructions on the tape are exactly as shown on the manuscript.

Contour Part Programming. The requirements for contour programming differ substantially from those of point-to-point part programming. With contouring, the entire path of the machine must be described because the machine performs an operation to close accuracy while in motion. In the point-to-point example described above, the path the tool takes in moving from one point to another is not too important because it is moving in air. Contouring applications are normally found with machines such as contour mills or lathes which perform cutting operations as they move along a path, or with drafting/plotting machines that guide a pen over a continuous and uninterrupted path.

Absolute versus Incremental Part Programming. Because most engineering drawings that call for point-to-point operations describe dimensions from a base line, it is generally more convenient for the part programmer if the points are specified as dimensions from the axes or coordinates. Most point-to-point control systems are therefore established in this way. The manuscript shown in Figure 5-9, which applies to Figure 5-8, follows this practice of specifying the coordinate points. This procedure is known as absolute programming.

Incremental programming describes the distances between the points. For example, the *x* distance, when moving from *PTA* to *PTB*, is 11 inches and the *y* distance is 1 inch. Also, because the motion is from right to left, the sign for the *x* incremental movement would be negative. And because the *y* incremental movement is up, or in the positive *y* direction, the *y* incremental movement sign would be positive. In moving from *PTB* to *PTC*, the *x* increment is plus 3 and the *y* increment is minus 8. Whereas absolute programming is the general practice with point-to-point equipment, incremental programming is popular with contouring applications. The reason for this is that contouring generally requires a much larger number of small movements. Very often these movements are in the order of 0.010 inch. Using an absolute system, if it were required to move from, say, 19.000 to 19.010 inches, all ten significant figures would have to be described on the tape. With the incremental system, only 0.010 inch need be noted. Therefore, the quantity of numbers involved with contouring has led most contour system builders to adopt an incremental programming approach. Also, because the majority of contour programming is handled by a computer, the part programmer is not required to perform the calculation of changing the common absolute form to incremental figures. This calculation is performed by the computer.

Linear Interpolation. Most contour systems move their associated machines in straight lines. A curved movement is actually a series of straight lines that approximate the curve within the tolerance requirements. The circular arc shown in Figure 5-10a has been divided into several straight-line segments, or chords. In this case, the chordal length (*L*) is calculated so that the tolerance (*t*) is not exceeded. Another method of approximating the circular arc is by means of straight-line segments calculated outside the arc, as in Figure 5-10b. In this instance, a cutter will leave an excess of material in moving around the arc, whereas in Figure 5-10a, the cutter will actually remove more than the requirement in order to develop the circular arc. Figure 5-10c is a compromise between undercutting the arc and leaving an excess amount and is the most widely adopted practice.

Theoretically, the length of the straight-line segments may be as short as the smallest programmable increment of the control system. In most cases, this is one ten-thousandth of an inch (0.0001 inch). This small an increment will be impractical in most cases because the tape reader could not keep up with the rapid movement of the blocks. Chord segments of 0.100 inch are more practical. The calculation of the chord length is normally handled by the computer routine and is dependent on the radius of the arc and the tolerance required. The computer also performs the function of calculating the path of the center line of the cutting

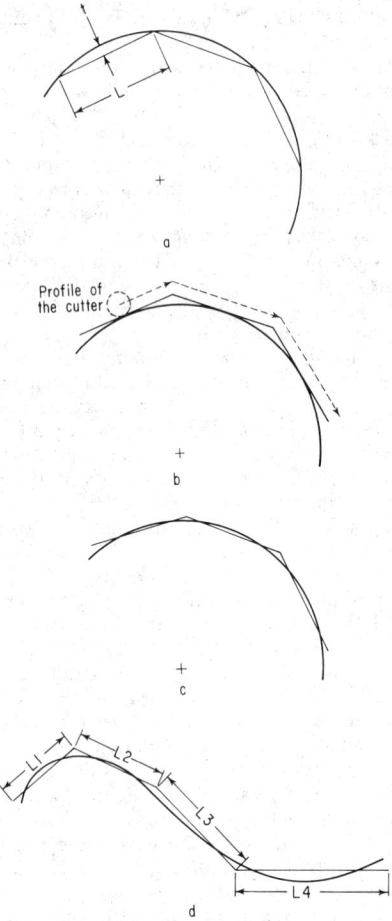

Profile of
the cutter

FIG. 5-10.   Approximating a curved movement by means of straight-line segments.

tool, as shown in Figure 5-10*b*. The path is offset by the amount of the radius of the cutter. With applications such as drafting/plotting machines, the radius of the cutter will be zero, and the offset calculation due to the cutter will not be required. With curves other than circular arcs, such as parabolas, the straight-line segments will vary in length to permit maintaining the calculated tolerance requirement. Curves that do not follow a formula pattern may also be calculated. Figure 5-10*d* illustrates such a situation. In this case, the straight-line segments $L1$, $L2$, $L3$, and $L4$ are of different lengths.

Circular Interpolation. Many control systems, especially those associated with numerical control lathes, incorporate a feature known as circular interpolation. This feature allows the part programmer to move the machine in a continuous circular arc without breaking the arc into the straight-line segments as would be necessary if the control system incorporated only linear interpolation. According to the standards and common practice, a block of information will move the machine in a circular arc not to exceed 90 degrees. The 90-degree movement will also have to be within any one of the four quadrants. Thus, if the circular arc were to cover the path shown in Figure 5-11, the programmer would have to program

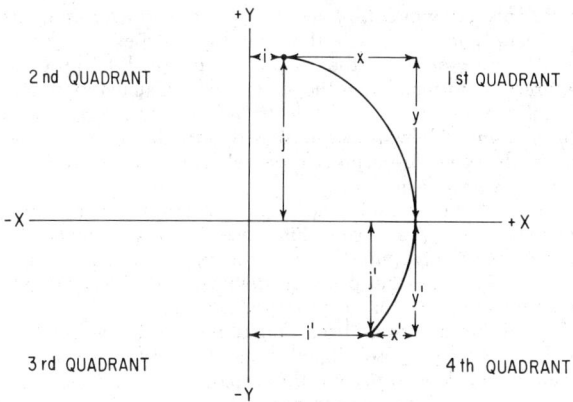

FIG. 5-11. Circular interpolation.

two blocks of information. Instead of noting the incremental straight-line motions, the programmer handling the arc shown in Figure 5-11 need only describe the distances i, j, x, and y in the first quadrant and i^1, j^1, x^1, and y^1 in the fourth quadrant. Each letter with its appropriate dimensions will represent a word in the block of information which is described on the tape. Circular interpolation is not suitable for a simultaneous motion involving all three axes, but is normally restricted to the XY plane. Certain control systems have allowed for switchable plane motions in which, for example, a circular motion can be performed in the XZ plane, or if desired, in the YZ plane. The reason for the popularity of circular interpolation with numerical control lathes is that most lathes operate in only two axes or one plane, and also because parts turned on lathes are comprised primarily of straight lines and circles. Although circular interpolation reduces the amount of part programming effort required if contour part programming were to be done manually, it is a moot point whether the saving is worth the extra cost of this feature when utilizing computers. It is generally felt that if the numerical control motion requirements call for a large number of circular arcs, circular interpolation will probably be a good investment. However, if most of the programming effort is confined to straight lines and curves other than circular arcs and there is a good deal of multiaxis programming wherein the machine moves in three or more axes simultaneously, it is doubtful if circular interpolation would be worth the added cost.

COMPUTERS SUITABLE FOR NUMERICAL CONTROL PART PROGRAMMING

Type of Computer. Computers may be divided into various categories, depending on their size and particularly their function. There are general-purpose computers that may be used to perform a wide range and variety of calculations. In contrast to this, there are special-purpose computers that are designed to perform restricted calculations. Examples are navigational or fire control computers used in the military services and special-purpose computers used in the process industry. The general-purpose computer can be used for various forms of calculations, provided the computer has been preprogrammed. In effect, the general-purpose computer can be converted to a special-purpose computer through the use of a specially stored program that accepts and operates on the particular problem.

Computers may also be divided into analog and digital. The analog computer normally produces answers that are approximate, yet within the requirements of the application. The numerical output in this instance is proportionate to some varying factor such as a voltage or power fluctuation. The analog computer is similar to a slide rule on which the accuracy of the answer is limited. This type of computer is generally less expensive than the digital type.

The digital computer, on the other hand, is designed to produce accuracies to within the required performance of the computer. Calculations are performed by the addition or subtraction of pulses, or electronic switching. The accuracy produced depends on the amount of electronics designed into the system. Generally, digital computers are of the general-purpose type whereas analog computers tend more toward special-purpose applications.

Computers may be further subdivided into business or accounting types and scientific or engineering types. The difference in this instance is not so much in the main section of the computer but more in the peripheral equipment associated with the computer, such as the printout devices. Also, scientific-type computers generally have larger memory storage capacity.

The type of computer that has been found most suitable for numerical control applications is the general-purpose, digital, scientific type. The size of the computer will vary considerably, depending on the program that is being utilized in the part programming effort. The storage capacity may range from 8,000 to 128,000 words or more, and the rental cost may range from approximately $20 an hour to over $1,000 an hour. It has generally been conceded that larger computers offer the best economy, but this will depend on the extent and complexity of the part program. The simpler programs, such as those used for point-to-point operations and simple two-axis contouring, may operate more efficiently on the smaller computer; whereas the three-, four-, and five-axis work will probably be more satisfactorily performed on the larger computer.

Computer Manufacturers. Although a number of computers are suitable for NC part programming, not all have the required computer programs that are necessary for acceptance of NC part programs.

Some of the companies that offer NC programs for their computers are:

IBM	360/30, 40, 50, 65, 75, 1130, 7090, 7094
UNIVAC	1107, 1108
GE	200, 400, 600
CDC	3600, 6600
RCA	Spectra 70
Philco	2000
SDS	900, 940
Burroughs	5500
Honeywell	200
PDS	1020

THE NUMERICAL CONTROL COMPUTER PROGRAM

Before a part program covering a particular part can be processed, it is necessary that a previously developed computer program be fed into the computer's storage system. This computer program is essentially a one-time development and may be used repeatedly to prepare tapes for machining or processing different parts. For the computer program to accept and process the part programming data, the input part programs must conform to the exacting requirements of the programming language of the computer. The general-purpose computer therefore must be primed or prearranged so as to be capable of handling the specific input program. In effect, the general-purpose computer is converted to a special-purpose computer by the insertion of the NC computer program.

The part program, when being processed by the computer, passes through three major sections or conversions. These are shown in Figure 5-12. The input translator converts the manuscript instructions into a binary-coded system that the computer can interpret. This is called machine language. Next, the machine language instructions are passed on to the arithmetic section, which performs the necessary mathematical and geometric computations to calculate the path of the

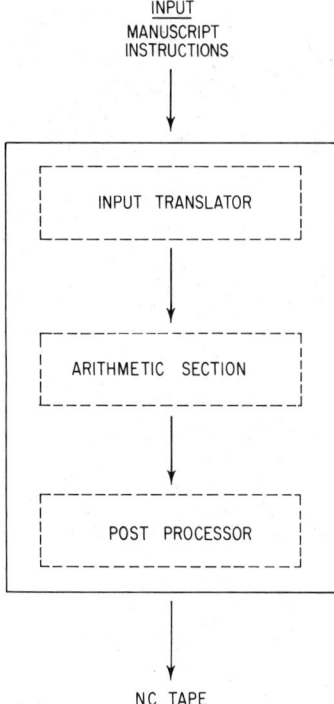

INPUT
MANUSCRIPT
INSTRUCTIONS

INPUT TRANSLATOR

ARITHMETIC SECTION

POST PROCESSOR

NC TAPE

FIG. 5-12.   Three main software sections through which a part program passes in a computer.   The computer program has been prepared previously and accepts and processes the particular part program.

NC machine.   It is generally possible, at this point, to obtain a computer printout of the coordinate points describing the calculated path of the machine, provided the input part program is correct.   It is essential that the programming rules be followed exactly for the program to operate.   The omission of a comma or a misspelled abbreviated word will nullify the entire program.   Should there be a definition error, it will usually be detected by the diagnostic portion of the input translator.   Should there be an error in the mathematical or geometric portion of the manuscript, this will normally be detected by the arithmetic section.   The post processor portion of the program will detect any unusual command such as attempting to run the machine at a speed above its limit or attempting to move the machine beyond its dimensional limits.

Should the part program satisfactorily pass through all three sections of the stored computer program, the output of the computer will be a punched tape suitable for insertion into the tape reader of the control system.

It should be noted that because of the normally high cost of operating the central processing unit of the computer, which is the main section of Figure 5-12, data are generally fed into and out of the main section by magnetic tape.   Therefore, the part program that is initially typed onto computer cards is automatically transferred to magnetic tape.   By the same token, the completed program emerges from the computer on magnetic tape and then is automatically transferred to punched tape.   Magnetic tape input and output are utilized because the data can be inserted and withdrawn from the central processing unit at far greater speeds than with cards or punched tape.

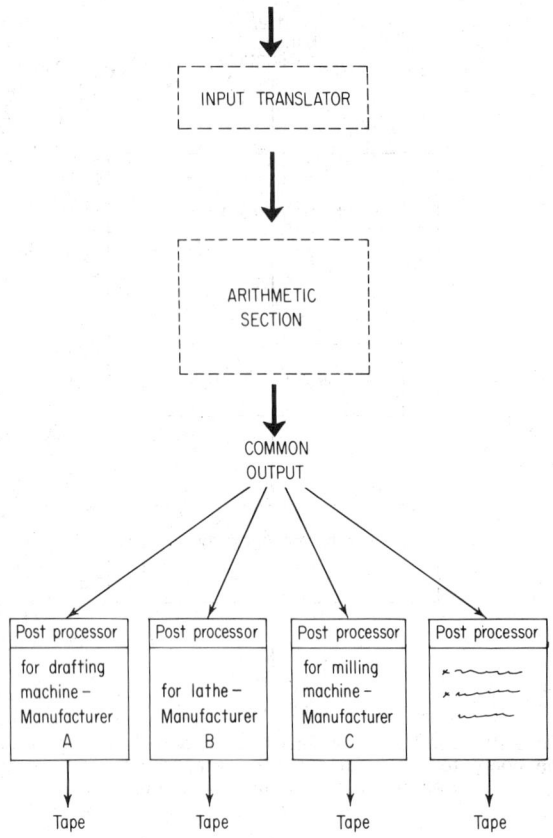

Fig. 5-13.   Several different post processors fed from the same output of the arithmetic section.

The Post Processor. If any phase of the NC computer operation has caused confusion among manufacturing personnel, it has been the post processor. Like the input translator and the arithmetic section, the post processor is software. Its purpose is to adapt the output from the arithmetic section, which describes the machine's path and is common to all machines, to suit the particular machine tool and associated control system in which it is to be used. The tape produced by the computer is identical with that produced manually by a tape-producing typewriter. Therefore, while one input translator and one arithmetic section may suffice for all machines, each machine tool and system combination must have its own individually designed post processor. This is illustrated in Figure 5-13.

The input translator and the arithmetic sections may generally be obtained from the computer manufacturer. The post processors are normally obtained from the machine tool builder. Although there is normally an extra charge for this software package, some machine tool builders offer it as part of the machine tool package.

PART PROGRAMMING LANGUAGES

There are numerous part programming languages available to enable the part programmer to describe the geometry of the part and the subsequent movements of the machine about the part in a far simpler manner than if he were required to program the part manually. The most useful language is one that allows the programmer to describe the characteristics of the part and machine motions in as near to ordinary English as possible. An English-like part programming language reduces the part programmer's memory requirements and thereby reduces a source of possible error. The most popular part programming language is the APT language. Other languages, many of which closely resemble the APT language, are shown in Table 5-1.

Development of the APT System. APT stands for Automatically Programmed Tool.

Shortly after the first numerical control machine was demonstrated at the Massachusetts Institute of Technology in 1952, it was realized that to handle the voluminous calculations and data required with contouring control machines, it would be necessary to enlist the aid of large, general-purpose computers. At that time, there were no commercial computers large enough; so the whirlwind vacuum tube computer which was developed at the Massachusetts Institute of Technology for research purposes was utilized. The development program was sponsored by the U.S. Air Force, which was also instrumental in financing the development of the first numerical control machine tool at the Massachusetts Institute of Technology. The objective sought by the Air Force was to reduce the

TABLE 5-1. Part Programming Languages

Part programming language	Comment	Percent popularity*
APT...............	Clearly the most popular program, both in the United States and abroad.	44
AUTOSPOT.........	An IBM proprietary program designed for point-to-point programming. It was one of the earlier developments in computer part programming languages.	20
ADAPT............	This is a junior version of APT and usable on smaller computers than are required for APT. As with the APT program, initial sponsorship was made by the U.S. government. The language is compatible with APT and almost identical. It does not have the full capabilities of APT and is restricted to two-dimensional and some modified three-dimensional moves.	15
FORTRAN.........	This usually involves special-type arrangements which are made up of FORTRAN language definitions. It has been found practical for specific requirements but lacks the geometric versatility and machine orientation of special numerical control languages.	12
SPLIT.............	This is a reasonably enduring language that was developed by the Sundstrand Corporation. It is essentially a point-to-point program but does have some contouring capability. It was designed originally for Sundstrand machine tools and has been further adopted and modified for more general use.	9
PROPRIETARY and IN-HOUSE	These are programs that have been developed by individual user companies and are normally confined within the company. A large number of companies utilize a combination of proprietary and nonproprietary programs, which explains why the percentages noted to the right exceed 100 percent.	46

* The figures listed in this column are the percentage of companies utilizing the language, as determined by a study made by the Numerical Control Society.

profile machining cost of airframe structures, at the same time increasing the strength-to-weight ratio of these structures. It was realized that as the speed of aircraft increased, the proportion of machining required increased in almost the same ratio. An airplane capable of traveling at Mach 2 required almost twice the machining time as one that traveled at Mach 1. To satisfy machining requirements, two choices were available. One was to double the number of machine tools required to produce supersonic aircraft; the other was to develop a more efficient machine tool system. The latter choice was adopted, and in time the original objective was achieved. Present numerical control machines produce three, four, and more times their conventional counterparts.

As MIT progressed in its research developments, it was joined by representatives of the aerospace industry under the sponsorship of the Aerospace Industries Association, which had set up a subcommittee devoted exclusively to numerical control. This industry group helped in establishing priorities for the development of the system most useful in practical shop operating environments. It was also necessary that the concept of the system be expanded to cover practical field applications. This was accomplished by the joint computer programming efforts of representatives of the numerous aerospace companies, who devoted their full time toward this program over a period of several years. Unquestionably, the APT program is the most versatile and universal system available. Its critics claim it is too encompassing and perhaps overpowered. To an extent, it is true that it does require a large computer, and for the simpler jobs, it may certainly be far too powerful. Its chief advantage, however, lies in its universal language and the fact that more programmers are familiar with this system than any other. It also has the largest number of post processors available and therefore is usable with the greatest number of machine tools. For those users who may not require the versatility and power of the APT system, there are other programs listed in Table 5-1 which may be utilized on smaller computers and which may satisfy particular requirements.

As APT progressed, it was recognized that industries outside the aerospace complex could readily benefit, especially the automotive people. The management of the program was therefore transferred to the Illinois Institute of Technology Research Institute (IITRI). IITRI's activities are supported by membership dues, paid on a yearly basis by those companies that wish to participate in the planning phases of the APT Long-range Program. It is not necessary that a company or individual be an APT member to use the APT program, because a number of computer companies and other users are members and are permitted to offer an APT data processing service to the general public.

APT System, Concept and Examples. The essence of the APT system and other APT-like programs lies in its close-to-English language. A point is described as PØINT; a line as LINE; a circle as CIRCLE. Other geometric definitions and instructions also follow their English counterparts as closely as possible. For example, the instruction "go right" is noted as GØRGT; "go left" as GØLFT; and "go forward" as GØFWD. The restriction is that a definition or command statement may not exceed six letters or a combination of six letters and numerals; abbreviated forms are thus required.

There are essentially two kinds of APT statements. One describes the geometry of the part in statements noting points, lines, circles, and other geometric forms; and the other describes the precise step-by-step motions required of a machine to move around the geometry that has been defined. The geometry statement is broken down into three parts which resemble an algebraic formula. Figure 5-14 explains the formula

$$P1 = PØINT/5,3$$

P1 is the symbol and notes the specific point. The symbol could be almost anything, provided there is at least one alphabetic letter included in the combination. CAT 1, MOUSE 2, RAT 3 would all be satisfactory. The purpose of the symbol is to identify the particular geometry configuration. Symbols are unique, and the

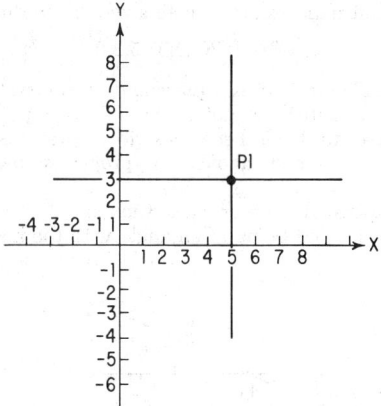

FIG. 5-14.   P1 is described by its coordinates in the APT language:
P1 = PØINT/5,3.

same symbol can be used only once in a program.   For example, if it was desired
to identify a second point, this could be designated P2.

Next, the symbol must be defined.   In the example, point readily becomes PØINT.

Next follows a slash line which separates the definition from the description.   The
description for a point, when the point lies on the XY plane, consists of the x
and y coordinates.   In this instance, $x = 5$ inches and $y = 3$ inches.   The coordinates may be listed as shown, the first number being the x coordinate and the second
being the y coordinate.   It is necessary that they be separated by a comma and
that no punctuation be placed after the y coordinate, except if there is a z coordinate,
which would be noted after a second comma in this way:

P1 = PØINT /5,3,2

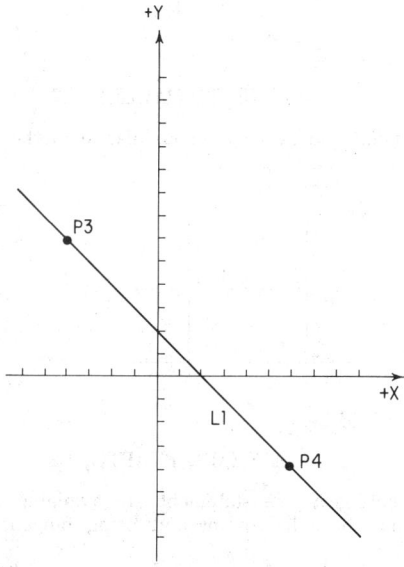

FIG. 5-15.   L1 is described as L1 = LINE/P3, P4 in the APT language.

It is also permissible but unnecessary to note a zero z coordinate:

$$P1 = PØINT/5,3,0$$

The line described in Figure 5-15 is represented by the symbol L1. The definition word is LINE and is described as going through two points, P3 and P4. Both these points would have to have been described previously in the program by a point statement as explained above. A proper statement would be L1 = LINE/P3, P4.

A line may also be described in other ways, such as:

1. Going through a point and being at an angle with the horizontal or X axis:

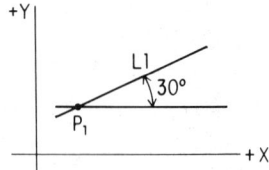

The APT statement would be

$$L1 = LINE/P1, ATANGL, 30$$

2. Going through a point and being parallel to another line:

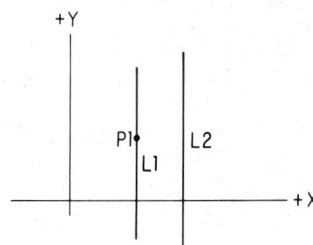

The APT statement would be

$$L1 = LINE/P1, PARLEL, L2$$

3. Going through a point and being perpendicular to another line:

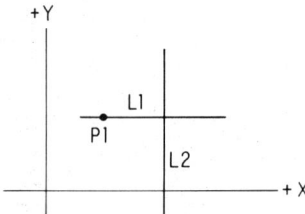

The APT statement would be

$$L1 = LINE/P1, PERPTØ, L2$$

Other methods of noting a line statement are available to the APT part programmer, such as going through a point and being tangent to a circle, or being tangent to two circles.

Although there are a number of ways of defining a circle, the most common is

$$C1 = CIRCLE/4,4,3$$

The first numeral 4 in the description is the value of the x coordinate, the second 4 is the y coordinate value, and the 3 is the value of the radius in inches.

There are numerous additional geometry statements covering configurations, such as parabolas and other definable curves, as well as those that are not algebraically defined. Three-dimensional surfaces may also be defined.

Motion statements differ somewhat from geometry statements in that there is no equals sign involved, the word before the slash mark is a direction command, and the symbol following the slash mark denotes where the machine is to go.

For example, the instruction GØ/TØ,L1 directs the machine to move to a line symbolized below:

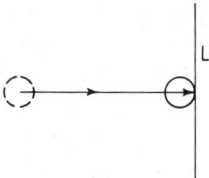

The instruction GØRGT/L1 directs the cutter to make a right turn along L1, as described below:

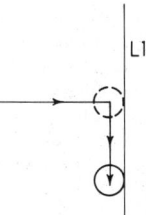

The part programmer is always looking in the direction of the travel of the cutter or machine, or in effect, imagines himself riding on the cutter.

The profile of a simple part and its complete program are shown in Figure 5-16.

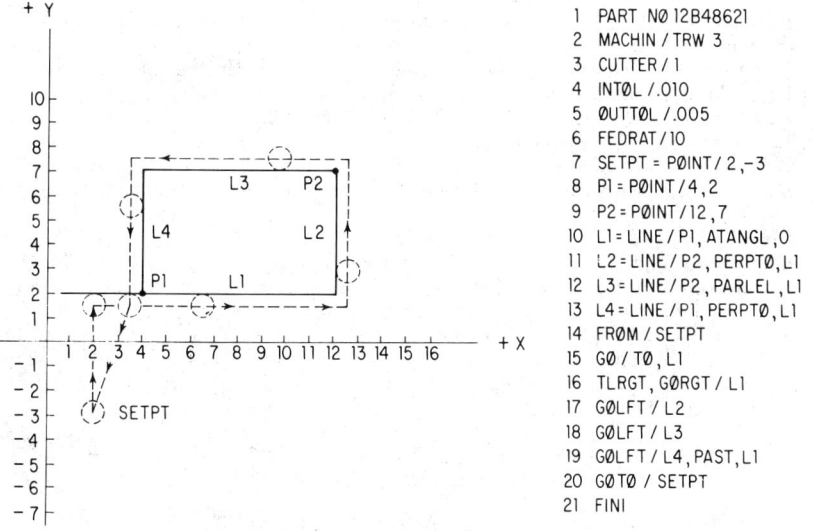

FIG. 5-16.   Example of an APT part program.

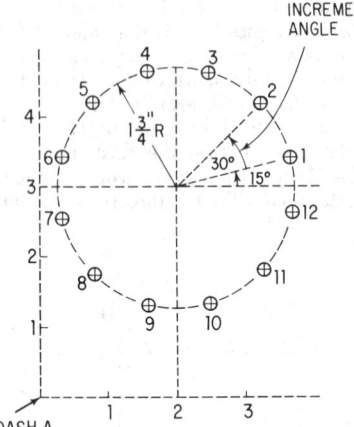

Fɪɢ. 5-17. Circular hole pattern to be programmed in AUTOSPOT language. Patterns are particularly well suited for computer-assisted part programming.

The cutter is planned to move counterclockwise around the rectangular part. It should be pointed out that a line extends indefinitely, and therefore, when the cutter moves to L1, it is moving to an extension of the portion of L1 that lies on the part. A description of each statement is as follows:

Statement 1. This is a required first statement for every part program. Any combination of letters and numerals may follow PART NØ.

Statement 2. Calls in the appropriate post processor.

Statement 3. The numeral 1 denotes the diameter of the cutter. This is required to describe the center line path of the cutter.

Statements 4 and 5. Although not required in this particular instance because all lines are straight, these statements denote the allowable tolerances of short straight lines approximating a circular arc (refer to Figure 5-10).

Statement 6. Describes the feed rate to be 10 inches per minute.

Statements 7 through 13. Describe the geometry of the part.

Statements 14 and 15. Instruct the machine (cutter) to move from the set-point to the line, which is designated by the symbol L1.

Statement 16. Translated, this statement means "With the tool on the right of L1, looking in the direction of travel, go right along L1."

Statement 17. Instructs the cutter to go left along L2.

Statement 18. This is similar to statement 17.

Statement 19. Translated, this statement means "Go left on L4 just past L1."

Statement 20. Instructs the machine to move back to the starting or set-point (SETPT).

Statement 21. FINI completes the program.

These statements are punched on computer cards, one line per card. The coded statements on the cards are then automatically transferred to magnetic tape for rapid input into the computer.

AUTOSPOT. AUTOSPOT, like other point-to-point computer programs, has been designed as an assist for complex or extensive patterns rather than for simple, indiscriminate-hole arrangements. The part program for the circular hole pattern shown in Figure 5-17 is as follows:

CIRC 1 SPDRL, 0209/DAA, R(1.75)
SA(15.0) 1A(30.0) NH(12)
AT(2.0, 3.0)
DRILL CN0130/CIRC 1/DP(0.25)$

CIRC 1	identifies the pattern and is similar to the symbol in the APT language.
SPDRL	means that the machine will make one pass and spot drill the points so that the drill which is to follow and drill the hole will not "walk."
0209	is a catalog number and refers to the cutting tool being used.
DAA	shifts the axes and allows the origin to be translated from the original. This translation, which is a simple x, y notation, would have to have been described earlier in the program.
R(1.75)	denotes the radius, which is $1\frac{3}{4}$ inches.
SA(15.0)	stands for starting angle, which is the angle between the horizontal and a line from the center of the circle to the center of hole 1, reading in a counterclockwise direction
IA(30.0)	stands for incremental angle, which in this case is 30 degrees.
NH(12)	describes the number of holes around the circumference of the circle, in this case 12.
AT(2.0, 3.0)	means that the coordinates of the center of the circle are $x = 2$ inches, $y = 3$ inches.
DRILL CN0130	denotes the operation as compared with SPDRL. 0130 designates the catalog number of the tool.
CIRC 1/DP(0.25)$	means that the circle which has been defined by the above statements is now to be drilled to a depth of $\frac{1}{4}$ inch, assuming the machine has Z axis or depth control.

REMOTE TIME SHARING

Numerical control data processing may be handled in one of two ways. One involves what is known as batch processing, where a single program is processed without interruption until completion. Programs are required to wait their turn while others are being run. The second method is known as time sharing, where a number of programs may be run on the computer during a single period. The computer generally operates on only one program at a time, but may alternate among various programs with such speed that it appears that the computer is operating on a number of programs simultaneously.

The input and output devices utilized with time sharing, called terminals, are generally arranged so that the programmer has direct on-line access to the computer and may operate in what is known as a conversational mode. The programmer may enter his program by a Teletype keyboard terminal or cathode ray tube arrangement by which he has the capability of writing on the face of the tube with an electronic pen; or he may prepare the program on tape or cards prior to transmission. The conversational capability allows the programmer to correct his program as it is being diagnosed by the computer, because errors will be noted almost immediately as they are made. The computer, in effect, will talk back to or converse with the operator.

Terminals may be located near the main computer or several thousand miles away. Because the communication is carried on over telephone lines, there is no technical restriction. The restraining factor is the cost of long distance telephone calls.

The number of terminals that may operate at any time is dependent on the storage capacity and speed of the computer and the complexity of the programs being run at the time. It is not uncommon for a computer to handle forty to fifty terminals in one period of time.

COMPUTER NUMERICAL CONTROL

Computer numerical control, sometimes referred to as direct numeric control, involves the direct operation of a number of machines simultaneously from a computer. The storage and logic units that are normally a part of the hardware of an individual control system beside the machine are, to a great extent, handled by the computer's software. The type and number of machines that may be operated by a single computer vary. Mills may be combined with drills, lathes, boring mills, or any other type of NC machines.

As with time sharing part programming, the computer shares its operation among several machine tools and produces the required commands for directing the combination of machines. Certain hardware controls, which enable the operator to maintain contact with the computer, are located beside the machine tool, for the computer itself may be at a remote site.

The Pros and Cons of Computer Numerical Control. There is little argument that the comparative hardware cost of a central computer operating a number of machines is less than individual control units operating the machines, because there is a good deal of redundancy involved with the individual control unit concept. Each unit, for example, requires a tape reader, storage units, and command hardware. A central computer, on the other hand, may share its storage and command hardware with a number of machines. The computer also eliminates the individual tape readers which contribute a sizable amount to the overall expense of a control system. The computer may require only one magnetic tape reader.

A graphic picture of this comparison is shown in Figure 5-18. At the top, the machines shown are each operated by their autonomous control system. The tapes may have been prepared manually or by a computer program, as described in this chapter. When the computer numerical control approach is used, a good share of the part programming calculations can be prepared by the same computer that is directing the machines. The input required of the central computer in this case will be less than required of the individual control systems.

One of the drawbacks is that, although the total cost of a complete arrangement as shown in the lower part of Figure 5-18 is less than if the machines and their control systems were bought individually, the initial outlay may be substantial.

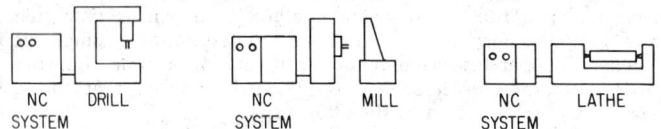

NC SYSTEM — DRILL   NC SYSTEM — MILL   NC SYSTEM — LATHE

a.   NC SYSTEMS OPERATING MACHINE TOOLS INDIVIDUALLY
AND INDEPENDENTLY

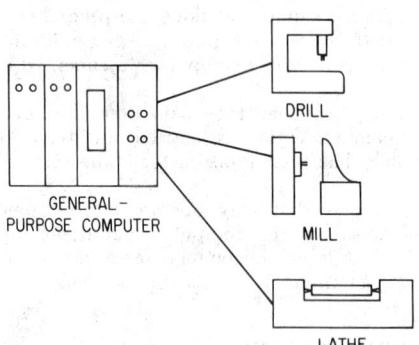

GENERAL-PURPOSE COMPUTER   DRILL   MILL   LATHE

b.   COMPUTER NUMERICAL CONTROL SYSTEM OPERATING A
NUMBER OF MACHINES

Fig. 5-18

Even though the computer may be purchased for one machine initially and additional machines are added later, the cost is still quite high.

Although computer numerical control improves management planning and control, it also requires a good deal of preplanning if the system is to operate efficiently.

CONCLUSION

The automated small lot factory is undoubtedly on its way. However, it will be some time before the individual hardware components, most of which now exist, can be assembled into a practical working system. What remains is the development of computer software, and of equal importance, management organization, planning, and foresight.

BIBLIOGRAPHY

Childs, James J., *Principles of Numerical Control,* 2d ed., The Industrial Press, New York, 1969.

Davis, Gordon B., *An Introduction to Electronic Computers,* McGraw-Hill Book Company, New York, 1966.

DeVries, Mary Ann (ed.), *Proceedings Sixth Annual Meeting and Technical Conference,* Numerical Control Society, Princeton, N.J., 1969.

McCarroll, John D., *Computer-aided Part Programming for Numerical Control,* The University of Michigan, Institute of Science and Technology, Ann Arbor, 1969.

"Time Sharing Service for NC Tape Preparation," *Automation,* May 1968.

Young, A. W., "Ideal Team: Numerical Control and Computers," *Iron Age,* Mar. 9, 1967, pp. 84–85.

Chapter 6

System Simulation

ARNOLD OCKENE

Vice President, Simulation Associates, Inc.,
White Plains, New York

Of the many possible definitions of system simulation, the one most germane to this discussion is "the construction of a dynamic model of a system and its operation on a computer." The objective is a thorough analysis of the behavior of a proposed or an existing system. A simulation model frequently can provide insights obtainable by no other means, and it is a powerful tool for systems analysis and design. The key words in the above definition are *system, model, dynamic,* and *computer.*

THE NATURE OF SYSTEM SIMULATION

The system under study might be a manufacturing plant, a retail organization, a vehicular traffic network, or a proposed capital investment. The objectives of these models might be the study of physical plant layout, the reduction of average inventory levels, optimum timing of traffic lights, and determination of return on investment, respectively. Virtually any complex system is suitable for analysis by simulation, and the key word here is *complex*. Simple systems can often be described in mathematical terms, for example, by means of a set of differential equations. The solution of these equations yields information on the behavior of the system, and this information is more exact and is obtained more rapidly and at lower cost than similar results from a simulation model. The problem with such analytic system descriptions is that they do not apply to a large class of problems. Most large-scale systems are too complex to be described analytically. There may be complex interaction between system elements, and some of the system variables may be stochastic (random), nonlinear, or time-varying. Many attempts at mathematical analysis break down under such conditions. The only alternative then left to the systems analyst is simulation—often called the technique of last resort. Because so many real-world systems fall into this "mathematically intractable" category, simulation has evolved into a powerful and extremely important tool for the industrial engineer.

A model is merely a representation of the essence of a system. By experimenting with the model, one obtains insights into the behavior of the real system. Models are useful in the study of both existing and proposed systems, and they may be classified as physical, mathematical, or procedural.

The physical model is the most easily understood and dates back many centuries to scale-model war games. By manipulating toy soldiers on a scale battlefield, military officers try out various strategies in an attempt to find an optimum. A more modern example of a physical model is the scale-model aircraft used in a wind tunnel to predict the aerodynamic performance of a proposed full-size airplane. Another familiar example is the solar system model found in every planetarium, which illustrates the relative speeds and positions of the planets as they revolve about the sun.

The mathematical or analytic model is frequently used to describe the behavior of a physical system. A very simple example is the differential equation describing the position of a weight at the end of a spring which is displaced and then permitted to oscillate. Solution of the equation permits determination of the position of the weight at any future time. Equations may also be written to describe the motion of the planets about the sun so that their future positions may be predicted. More recently developed analytic models such as linear programming permit optimization of system variables. The general rule to follow is that if an analytic model provides a satisfactory description of reality, it should be used. Only if such a model is unrealistic should a simulation model be constructed.

The third category of models is, for want of a better name, defined as procedural. Into this category fall models too complex to be defined analytically; they must be defined as a set of procedures—in a sense, a descriptive definition of the system and its operating rules. For example: "10 percent of all incoming orders are for product type 3; the sequence of machining operations for this product is . . . ; 2 percent of all items require rework;" and so on. It is for problems of this type, where neither physical models nor analytic models apply, that simulation models are of great value in determining improved system configuration and operating rules.

Simulation models are dynamic; that is, they operate over time and permit complex interactions to take place between elements of the system. This concept of moving the model through time does not exist in many analytic formulations. The term "dynamic" also implies that the system is studied under stress, where more than simple average behavior is of interest. Behavior under peak loads or under unusual conditions such as when portions of the system have failed may be studied by means of simulation.

Without the digital computer, large-scale systems simulation would not be practical. The procedural models defined above require much detailed bookkeeping to keep track of the many possible events, interactions, and status changes which may occur at any point in time. Such bookkeeping may, of course, be done manually, but this is not practical for any but trivial systems. Hence, the high-speed computer is a prerequisite for meaningful simulation studies.

Development of General-purpose Simulation Languages. Mathematical models are discussed in Section 10. This chapter will be concerned only with computer simulation. This is an appropriate point at which to draw the distinction between continuous and discrete event simulation. Continuous systems are those in which the variables change continuously with time; the state of the system changes from instant to instant. In modeling such a system, time is treated as flowing in a continuous fashion. Such systems are generally described by means of complex (nonlinear and time-dependent) differential equations, which are modeled on an analog computer or on a digital computer using languages such as the Continuous System Modeling Program (CSMP), PACTOLUS, and MIDAS. The representation of such systems is by means of differential equations, but the solutions are not analytic because the equations are mathematically intractable. Examples of cases where continuous models would be appropriate are studies of river pollution, chemical and biological systems, and flight trajectory problems.

Discrete event systems are those in which significant state changes occur only at discrete points in time rather than continuously. The system variables are discrete, individually distinguishable units rather than continuous variables. For example, events in an inventory control system include creation of a unit of demand, triggering of an order to the supplier when the inventory level falls below a critical point, and replenishment of inventory when a shipment arrives from the supplier. Between occurrences of these events, the system does not change state, so that the simulation model need not concern itself with points in time other than those at which an event occurs. Hence, the clock in such a model moves in unequal steps, each increment bringing it to a point at which the system is about to undergo a change in the status of one or more elements.

The discussion in this chapter will be confined to discrete event simulation. Early models of this type were written in machine language. When FORTRAN became available, it became the most popular language for model building. Both FORTRAN and lower level languages share certain disadvantages. Their use requires a high degree of programming proficiency. Because simulation models tend to be highly complex, one must be an expert programmer. Because the systems analyst or engineer performing the study is generally not an expert programmer, serious communication problems arise. Once the progammer understands what is required and develops a model, the need for constant interaction with the analyst remains during the debugging stage. Finally, when production runs are made as various alternatives are examined, the programmer must still be available to implement these changes.

FORTRAN has other shortcomings. Virtually all simulation models require certain facilities, such as the ability to account for the passage of time (a clock); the ability to sample from arbitrary probability distributions using random numbers; and most important, a sequencing algorithm, the internal mechanism responsible for maintaining the proper sequence of events in the model. Because these facilities are not provided in FORTRAN or any other general programming language, the user must program them.

These shortcomings provided the motivation for the development of general-purpose simulation languages. They are designed to be used by the analyst rather than by a professional programmer; they provide the facilities required by simulation programs; and they are designed to facilitate the accumulation and printing of output statistics required from a simulation model. The two principal languages in this field are the General Purpose Systems Simulation (GPSS) and SIMSCRIPT. These and several other languages are discussed below.

Reasons for Simulating. Simulation is a powerful tool for systems analysis and design. It permits the construction of models which are an accurate representation of the real world. By studying the behavior of these models—using a digital computer—detailed information on the behavior of the actual system may be obtained. Simulation is an analysis, as opposed to synthesis, technique—it will not design an optimum system. By means of an iterative procedure, using the results of one run to decide how the system should be modified for the next, one can "home in" on an "optimum" system design. However, it should be emphasized that simulation is not an optimization technique and consequently does not guarantee an optimal solution.

The reasons for simulating a proposed new system are obvious. It is less expensive and faster than physically constructing the real system, and it is far better to discover design errors in the model than in the actual system. Simulation of existing systems is also common, because it is a nondisruptive mechanism for experimentation. One may test new ideas for improving system performance without risking the disruption of existing operations. For example, an elevator manufacturer may have new ideas on how to improve the service provided in a building by better dispatching or scheduling rules. If he experiments with these rules by changing the actual system, he runs the risk of having a building full of irate tenants if the new rules turn out to be poor. A simulation model, on the other hand, permits experimentation with many possible dispatching rules at no greater cost than some additional computer time. Simulation may also be used to experiment with systems which cannot be experimented with in reality, if the process under study is destructive. To cite an extreme example, military strategists have studied the effects of general thermonuclear war by means of computer simulation.

As mentioned earlier, simulation is often the only technique available under conditions of uncertainty (due to stochastic variables), nonlinearity, or time variance of system elements. Mathematical treatment of such systems is frequently not possible. Computer simulation also gives the experimenter control over time, which may be compressed or expanded at will. For example, a simulation study of the reliability of a complex system such as an aircraft may gather data on many months of operation in a matter of minutes on a computer.

Aside from the benefits of having an experimental tool, the very act of building the model is of considerable value. It forces logical, well-organized system definition because every relevant aspect of the system must be specified quantitatively. It also forces collection of whatever data are necessary to design the system properly, such as traffic rates and timings in various portions of the system. "Traffic" may be shop orders, customers, vehicles, and the like, depending on the nature of the system. The discipline imposed by the modeling process is of value even if the model is never run on a computer.

Finally, simulation is useful as a training vehicle. It permits new managers to get the feel of a system by seeing how it responds to changes. Mistakes may be made with the model rather than in the actual system.

SIMULATION LANGUAGES

Computer simulation languages provide a "world view," or way of looking at systems in the real world, designed to facilitate the modeling process. As will be seen in the discussion to follow, each of these languages provides a basic system structure, diagnostics, and input and output facilities designed to satisfy the peculiar requirements of system simulation. Although the structure of different simulation languages may vary considerably, all have certain features in common. These are:

1. A timekeeping mechanism
2. The ability to represent uncertainty
3. The ability to account for resources of the system
4. A sequencing algorithm
5. Debugging facilities
6. Meaningful output

The Timekeeping Mechanism. A simulation program must account for the passage of time. This is done by means of a clock, one of the features provided in all simulation languages. A clock is merely a word in core storage; it is initially zero and is incremented as the simulation run progresses. The clock "unit" is problem-dependent; it may be microseconds, minutes, days, or other time interval, depending upon the nature of the system being modeled. One generally uses the largest clock unit possible, recognizing that it must be small enough to capture the finest level of detail deemed significant. For example, a clock measuring time in minutes would probably be adequate for a job shop simulation, but not for an urban traffic control model where traffic light timings are expressed in seconds. It would not be possible to specify a 30-second light if the smallest increment of time is 1 minute. Table 6-1 illustrates possible choices of clock units for a variety of systems.

In a discrete event simulation language, the clock is not incremented in fixed steps of one unit. The program looks for the "next event"—the earliest future time

TABLE 6-1. Examples of Choice of Clock Unit

Model type	Clock unit
Computer system	Microsecond or millisecond
Traffic control system	Second
Manufacturing operation	Minute
Inventory	Hour or day

at which something happens in the system. The clock is then set to that time. This avoids the wasted motion of advancing the clock (synchronous timing) in fixed time increments and searching for something to do each time. It is far more efficient to look ahead (asynchronous timing) to the most imminent future event. For example, the clock in a job shop model now reads 50, and two jobs are in process in the shop. One will not complete its current operation for another 20 clock units; the other for another 35 clock units. Because nothing in the shop changes until the clock reaches 70, it is incremented from 50 to 70 in a single step.

The Ability to Represent Uncertainty.[1] Much of what happens in the real world cannot be predicted in advance. The time for a particular operation in a job shop may average 12 minutes, but for any single sample, it might be anything between 10 and 15 minutes. The average number of vehicles entering at a particular highway ramp may be 4 per minute, but for any arbitrarily chosen minute, it might be anything between 0 and 15 vehicles. We may know that 80 percent of all customers entering a store will purchase something, but can we predict that the next customer to enter will make a purchase? These trivial examples illustrate the need for the ability to make decisions in the model based on the outcome of a chance event, that is, the random selection from a known probability distribution. All simulation languages provide pseudorandom number generators—pseudorandom because they are reproducible, although they should meet the various statistical tests for randomness.

A random number generator may be thought of as a roulette wheel, Figure 6-1, with a very large number of positions, say, ten thousand. If we wish to determine whether or not a particular customer makes a purchase (0.8 probability that he does), a "spin" of the wheel is made. A number in the 1 to 8,000 range is interpreted as a purchase; an outcome between 8,001 and 10,000 (0.2 probability) as no purchase. More complex distributions may be constructed by subdividing the wheel as finely as required. The percentages shown in Figure 6-1 might, for example, describe the job mix in a shop, and these percentages indicate the probability that any one of the five possible products manufactured in the shop will be the next order.

A simulation language provides both the pseudorandom number generator and the ability to select automatically a number from an arbitrary probability distribution. The distribution of Figure 6-1, for example, would be defined by means of a function such that, over a large number of trials, 20 percent of the values chosen will be one, 10 percent will be two, and so on. Reference to the function will, of course, result in a reference to the pseudorandom number generator, but this is done by the simulation system and the user need not concern himself with the details.

The Ability to Account for Resources of the System. A group of machine tools, a communication line, a bank teller, or a switching engine in a railroad yard are all examples of physical resources of different types of systems. In modeling physical systems, the user must maintain a record of the status of each resource (in use or available), its utilization (percentage of time in use), and the queue (or waiting line) statistics associated with the resource. A simulation language provides the internal bookkeeping required to maintain these and other relevant data.

---

[1] Uncertainty is interpreted here in the sense of risk, where the probabilities are assumed to be known. Some people interpret uncertainty as meaning that the probabilities of outcomes are completely unknown.

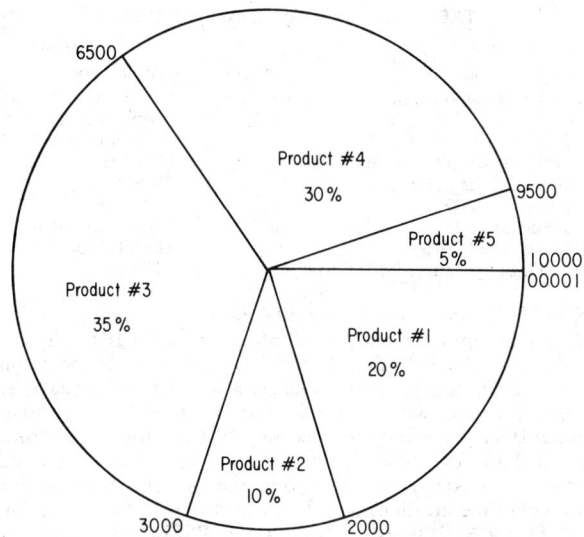

Fig. 6-1.   A roulette wheel as a random number generator.

Sequencing Algorithm. The mechanism for maintaining the proper sequence of occurrences or events in the system is the so-called sequencing algorithm. This is the single most important service provided by a simulation language. Events such as job arrival, job completion, shift change, and the like must be maintained in proper sequence by time and priority. Lists, such as queues, must be maintained in a FIFO, LIFO, or ranking order. Special conditions, such as high-priority interrupt of a resource or the synchronization of the movement of several jobs must be accounted for. The internal bookkeeping required to maintain such lists of activities and resources is considerable. It places a great burden upon the person who chooses to construct a simulation model in a nonsimulation language such as FORTRAN. If a simulation language is used, this burden is on the language designer. The sequencing algorithm built into the language maintains all internal lists of activities and resources and assures that the proper time and priority relationships are maintained.

Debugging Facilities. Facilities are provided in a simulation language for debugging. The program written by the user of the language must, of course, be debugged. This can be a tedious process if special diagnostic messages tailored to simulation applications are not provided. The use of a simulation language as opposed to a more general language such as FORTRAN assures that such diagnostics will be provided.

Meaningful Output. Output must be provided in a meaningful form. This means that statistics such as resource utilizations, queue lengths, and probability distributions must be accumulated; certain computations such as mean and standard deviation must be made to provide data for analysis; and results must be printed in a meaningful format. If FORTRAN is used, all statistical output must be computed and printed by the user. A simulation language, however, provides data accumulation, analysis, and output with a minimum of specification on the part of the user.

In summary, the use of a computer simulation language facilitates the model building process by providing a wide range of required services. The use of a language such as FORTRAN for simulation requires the user to program these services—a considerable expenditure of time and programming effort. There is only one significant disadvantage associated with simulation languages: they are generally less efficient in their use of the computer than FORTRAN. A given model coded

TABLE 6-2.   Existing Versions of GPSS

Machine	*Program name*
Burroughs 5500	GPSS/5500
Control Data 3600	GPSS III (available only through CDC data centers)
General Electric 600 Series	GESIM
Honeywell Series 200	GPS-K
IBM System/360	GPSS/360
RCA Spectra 70	Flow Simulator
UNIVAC 1100 Series	GPSS/1100
Xerox Data Systems Sigma Series	GPDS

in FORTRAN will, in most cases, execute faster and require less core storage than the same model programmed in GPSS or SIMSCRIPT. This should come as no surprise, because the FORTRAN model is custom-made for a specific application and contains no unused features, whereas a simulation language contains many features, not all of which will be used in every model. Performance, however, encompasses more than the computer time required for production runs. One must take into account both the time of the person doing the programming and the number of computer runs required to debug the model. Both model construction effort and debugging time are sharply reduced by the use of a simulation language.

The General Purpose Systems Simulation (GPSS). The most widely used of all simulation languages is the General Purpose Systems Simulation, or GPSS. The original version of this program was written for the IBM 7090 by Geoffrey Gordon in 1960. It has subsequently undergone three major modifications, and the current IBM version is known as GPSS/360. Many other manufacturers have produced similar programs, as listed in Table 6-2. The discussion to follow is valid for all versions of the program.

The "world view" of GPSS is that of units of traffic (transactions) flowing through a system (described by blocks) and interacting with other units of traffic by competing for resources (facilities and storages) of the system. Transactions represent physical units moving through the system; the exact nature of a transaction depends on the nature of the system being modeled. Table 6-3 illustrates this point for several system types. Blocks are instructions which perform operations analogous to operations in the actual system, such as joining a queue, competing for a resource, or making a decision based on the state of the system. Using a vocabulary of different block types, the GPSS user constructs a flow chart of the system, which describes the logical operation or decision rules of the system in addition to its physical structure. Resources are modeled as facilities capable of servicing one transaction at a time and storages capable of simultaneously servicing many transactions. Examples are given in Table 6-3.

TABLE 6-3.   Examples of GPSS Transactions, Facilities, and Storages

System type	Transaction	Resources*
Airport	Aircraft	Taxi runway [S]
		Takeoff runway [F]
	Passenger	Airplane seat [S]
Manufacturing system	Job order	Machines [F or S]
		People [F or S]
Computer system	Job	CPU [F]
	Real-time message	Disc drive [F]
		Core buffers [S]
Railroad yard	Train	Switching engine [F]
		Track sidings [S]

* F = facility, S = storage.

Note that even in a particular system model, there may be some ambiguity in the nature of the various GPSS entities. Depending on which portion of the model is under consideration, transactions in the airport model may represent either aircraft or passengers, and one group of storages might represent taxi runways while another group represents seating capacity on the various aircraft. Similarly, machines in a manufacturing model may be modeled as facilities if they are treated individually or as storages if several machines are identical and one does not care which specific machine in the group is actually used. The same is true of the workers who operate the machines—they may be modeled as individual facilities or grouped by skill level in a set of storages.

The structure described is unique to GPSS and similar programs listed in Table 6-2. The features present in all simulation languages are, of course, provided—clock, random number generators, sequencing algorithm, and so on. The predefined data structure of GPSS (blocks, transactions, facilities, and storages, among other entities) has both advantages and disadvantages. The language is easy to learn, use, and debug. The authors of GPSS were able to anticipate many types of errors and hence provide lucid diagnostic messages, many of which would not be possible with a user-defined data structure. However, the rigid structure may be a serious drawback in modeling very large, complex systems, because the user must force his problem to map into the predefined data structure of GPSS. This is sometimes quite difficult, and it is for problems of this type that a language such as SIMSCRIPT excels.

More detailed information on the GPSS language will be found in references 2, 6, and 7 in the bibliography at the end of the chapter.

SIMSCRIPT. SIMSCRIPT was originally written by the RAND Corporation for the IBM 7090, and is now available in a proprietary version known as SIMSCRIPT I.5 on Control Data, Philco, RCA, and UNIVAC equipment. A major new language, SIMSCRIPT II, was developed by RAND for the IBM System/360 in 1969 and has since been developed for the RCA Spectra 70 series. The discussion to follow applies to all versions of SIMSCRIPT.

SIMSCRIPT is a more conventional programming language when compared with GPSS. The user writes statements, as he would in a FORTRAN program, in which he describes both the static structure and the dynamic behavior of the system. Unlike GPSS, SIMSCRIPT does not have a rigid, predefined data structure; hence this structure must be defined by the user in a definition section of the program—the preamble section in SIMSCRIPT II. No such definition is required in GPSS, because the user has no control over the fundamental data structure of the system.

A SIMSCRIPT program is structured in terms of entities, attributes, and sets. Entities are "things" in the system—people, vehicles, machines, and the like. An entity may be either permanent or temporary, depending on whether or not it exists over the entire life of the simulation run. If one wishes to draw a parallel with GPSS, the GPSS transaction is a temporary entity while facilities and storages are examples of permanent entities. Each entity may have any number of attributes associated with it. For example, an entity named "vehicle" may have as its attributes speed, acceleration, capacity, and length, all defined by the user. Finally, entities may be grouped into sets. A GPSS queue would be defined as a SIMSCRIPT set.

After defining the static structure and performing whatever initialization is required, the dynamic behavior is defined by means of event subroutines. An event is a description of what happens at a point in time where the system changes state, such as arrival of an order, the start of a machining operation, or the completion of a shift. The event routines constitute the description of how the system changes over time, and all updating of entity attributes, creation and destruction of temporary entities, alteration of set memberships, and scheduling of future events are done in the event routines.

Output in SIMSCRIPT is not automatic as in GPSS. The user must define the statistics to be accumulated and the desired format of the output report. The report specification then becomes part of an event routine which may be called as desired. By defining several such output events, the user may call any one of several types of report.

An inevitable question in any discussion of simulation languages is "Which is better?" Obviously, there is no simple answer. If one language could be shown to be clearly superior to the other, the inferior language would rapidly fall into disuse. The only rational answer to the question is, "It depends on the problem." SIMSCRIPT is generally considered superior to GPSS for very large and complex systems for two reasons: (1) the fact that the data structure is user-defined permits more natural modeling of complex systems and (2) it is generally more efficient in its use of the computer, with respect to both execution time and core storage requirements. SIMSCRIPT is a compiler; GPSS is an interpretive program; the statements are not translated into machine code—they remain in core and act as calls to block subroutines which are written in machine language. Furthermore, the SIMSCRIPT mechanism for maintaining lists of scheduled events (sequencing algorithm) consumes less time in execution, although this efficiency is achieved at the expense of additional effort on the part of the user. SIMSCRIPT II also permits data packing to the bit level, which GPSS does not. It must be recognized that performance is a composite of design and coding time, program debugging time, and execution time for production runs on the computer. The relative efficiency of SIMSCRIPT in execution is paid for by increased time to code and debug the model.

GPSS, on the other hand, is distinctly easier to learn, use, and debug, and does not require the level of programming competence required of the SIMSCRIPT user. The author generally recommends the use of GPSS wherever appropriate. If it becomes too difficult to map the system into GPSS with its rigidly defined data structure, or if the model is large and will be run frequently so that performance considerations come to the fore, the user would be well advised to consider SIMSCRIPT. Additional information may be obtained in references 11 and 12.

Other Languages. In addition to GPSS and SIMSCRIPT, there are many other discrete event simulation languages in existence. Most have a rather small population of users. Three of the better-known languages in this group follow.

SIMULA. SIMULA is a language which views the world as a collection of processes which exist over time, rather than events occurring at points in time (the SIMSCRIPT view). In a sense, this is the same concept as that used in GPSS, although SIMULA is a statement-type language based on ALGOL. Developed at the Norwegian Computing Center, SIMULA enjoys considerably more popularity in Europe than in the United States. Details on this language may be found in reference 3.

CSL (Control and Simulation Language). CSL views the world as a collection of activities similar to the processes of SIMULA. Written as a cooperative venture by IBM United Kingdom and Esso Petroleum Limited, this language is also more popular in Europe than in the United States. Reference 1 in the bibliography is suggested for those desiring additional information.

GASP. GASP is a set of FORTRAN subprograms organized to assist in performing simulation studies. The various services required by the simulation programmer, such as a clock, random numbers, and a list sequencing mechanism, are provided. GASP provides discrete simulation capability on small machines, with GASP II available on the 8k IBM 1130 computer, the GE 225, and the CDC 3400. The complete language is documented in reference 10.

AN ELEVATOR PROBLEM

The application of both GPSS/360 and SIMSCRIPT II may be illustrated by an elevator problem. The complexity of this problem is such that the effort required to construct the model is approximately the same in either language. The intent of this illustration is to impart the flavor of GPSS and SIMSCRIPT II to the reader; a detailed discussion is beyond the scope of this chapter. The SIMSCRIPT

TABLE 6-4.   Data and Operating Rules—Elevator Model
Time unit = 0.001 hour
I. Cars

Number of cars = 4
Car capacity = 5
Move time = 1
Load or unload time = 2
Cars reverse direction at the first floor and the top floor.
Cars pick up only passengers wanting service in the direction the car is moving.
As many as half the cars are allowed to pause on the first floor or the top floor to wait for a
   signal for useful service.   In all other cases, cars keep moving until required to load
   or unload.
Half the cars start on the first floor and half on the top floor.
II. Building
Number of floors = 10
III. Passengers
Initial arrival of a group is at clock = 1
Group interarrival time = 25; interarrival time distribution = exponential
Group size: mean = 3 passengers; distribution = Poisson
Floor of origin = equiprobable
Floor of destination = equiprobable (excluding floor of origin)

II code was written by Philip J. Kiviat of Simulation Associates, Inc., in Los Angeles;
the GPSS code was written by the author.

Statement of the Problem.   A model is to be constructed to permit study
of the operation of a bank of elevators in a ten-story building.   Objectives of
the model include determination of the number of elevators required, optimum
elevator capacity, and optimum dispatching rules—the rules determining how the
system responds to a call, where elevators wait if the system is idle, and the like.
Table 6-4 is a summary of the data and operating rules used in the models to
be illustrated.

GPSS Solution.   To construct a GPSS model, the analyst must determine how
the actual system maps into GPSS, or how the various entities in the real system
are modeled in GPSS.   We shall represent both passengers and elevators as transactions.   Passenger transactions arrive and wait on one of two lists, up or down,
for an elevator transaction to pick them up.   Elevator transactions, which are always
active in the system, as opposed to passenger transactions which are temporary,
move from floor to floor examining the list of waiting passengers.   The elevators
are also represented by STORAGES for the purpose of accumulating occupancy
statistics.

Figure 6-2 is a listing of the entire GPSS model.   As an illustration of the
block diagram nature of GPSS, Figure 6-3 shows the eight blocks required to
simulate passenger arrivals.   A larger group of blocks is required to model the
elevator operations.   Once the block diagram is complete, the cards of Figure 6-2
are keypunched and the program is ready for the computer.

The program of Figure 6-2 comprises three major sections.   The first group of
cards—above the °RUN TIMER card—are definition cards, responsible for defining
certain arithmetic and logical features of the program.   These are followed by
the block cards, with blocks 1 and 2 serving as a run timer, blocks 3 to 10
modeling the passenger arrivals shown in Figure 6-3, and blocks 11 to 53 modeling
the elevator operations.   The third group of cards follows the REPORT card and
defines the structure of the output.   These cards are optional; if omitted, the standard
GPSS output will be produced.   Figure 6-4 is a portion of the output produced
by this program.

SIMSCRIPT II Solution.   Figure 6-5 is the SIMSCRIPT II solution to the elevator
problem.   Although the program is somewhat longer in terms of the number of
cards required, it has the great virtue of being somewhat readable to a person
with no prior knowledge of the language.   This is not true of the GPSS model.

A SIMSCRIPT II program has four parts: preamble, main routine, event routines
describing the system dynamics, and one or more event routines for producing

BLOCK NUMBER	*LOC	OPERATION	A,B,C,D,E,F,G	COMMENTS
	*	SIMULATE		
1		MATRIX	H,10,4	DEFINE 'PASSENGER OFF' MATRIX
		STORAGE	S1 — S4,5	ELEVATOR CAPACITY
1		FVARIABLE	25*FN1/10	USED TO OBTAIN NO. IN GROUP
2		VARIABLE	CH1 + CH2	TEST FOR ELEVATOR CALL
3		BVARIABLE	(P2'E'1) + (P2'E'10)	TEST FOR TOP OR BOTTOM
1		FUNCTION	RN2,C24	EXPONENTIAL
0	0	.1	.104 .2 .222 .3	.355 .4 .509 .5 .69
	.6	.915 .7	1.2 .75 1.38 .8	1.6 .84 1.83 .88 2.12
	.9	2.3 .92	2.52 .94 2.81 .95	2.99 .96 3.2 .97 3.5
	.98	3.9 .99	4.6 .995 5.3 .998	6.2 .999 7 .9997 8
2		FUNCTION	RN3,C2	FLOOR (1—10)
0	1 1.0		11	
	* RUN TIMER			
1		GENERATE	1000,,,,,0	RUN TIMER (ONE HOUR)
2		TERMINATE	1	
	*			
	* PASSENGER CREATION			
3		GENERATE	25,FN1,1,,1,3	PASSENGERS
4		ASSIGN	2,FN2	P2 = ORIGIN FLOOR
5	DEST	ASSIGN	3,FN2	P3 = DESTINATION FLOOR
6		TEST NE	P3,P2,DEST	
7		SPLIT	V1,DIREC	CREATE ONE TRANS. PER PERSON
8	DIREC	TEST G	P3,P2,DOWN	
9		LINK	1,P2	PERSONS WAITING FOR UP CAR
10	DOWN	LINK	2,P2	PERSONS WAITING FOR DOWN CAR
	*			
	* ELEVATORS			
11	ELEV	GENERATE	,,,4,,6	CARS
12		ASSIGN	3,N$ELEV	P3 = ELEVATOR NUMBER
13		TEST LE	P3,2,EDOWN	
14		ASSIGN	2,1	P2 = FLOOR (INITIALLY 1)
15	UPUP	ASSIGN	4,1	INDICATE GOING UP: P4 = CHAIN
16		ASSIGN	5,1	INDICATE GOING UP: INCR FLOOR
17	PIKUP	SAVEVALUE	5,P3	ELEVATOR NUMBER IN X5
18	MORE	UNLINK	*4,LOAD,1,2,,TIME	ONE PERSON ENTERS CAR
19		ASSIGN	1+,1	P1 = NUMBER OF PERSONS IN CAR
20		ASSIGN	6,1	INDICATE PASSENGER PICKUP
21		TEST E	P1,5,MORE	IS CAR FULL?
22	TIME	TEST E	P6,1,NEXT	SHOULD LOAD TIME BE SPENT?
23		ASSIGN	6,0	
24		ADVANCE	2	LOAD TIME
25	NEXT	ASSIGN	2+,P5	STEP FLOOR NUMBER
26		ADVANCE	1	MOVE TIME
27		TEST E	MH1(P2,P3),0,OFF	NO ONE OFF AT THIS FLOOR?
28		TEST E	BV3,0,TURN	TEST FOR TOP OR BOTTOM
29		TEST E	P1,5,PIKUP	TEST ELEVATOR FULL
30		TRANSFER	,NEXT	

Fig. 6-2.   GPSS elevator model.

31	EDOWN	ASSIGN	2,10	
32		ASSIGN	4,2	
33		ASSIGN	5—,1	
34		TRANSFER	,PIKUP	
35	OFF	ASSIGN	1—,MH1(P2,P3)	PASSENGERS LEAVE CAR
36		LEAVE	*3,MH1(P2,P3)	PASSENGERS LEAVE CAR
37		MSAVEVALUE	1,P2,P3,0,H	ZERO 'PASSENGER OFF' MATRIX CELL
38		TEST E	BV3,0,TURN	TEST FOR TOP OR BOTTOM
39		ADVANCE	2	UNLOAD TIME
40		TRANSFER	,PIKUP	
41	LOAD	MSAVEVALUE	1+,P3,X5,1,H	INCREMENT 'PASSENGER OFF' MATRIX
42		ENTER	X5	PASSENGER ENTERS ELEVATOR
43		TERMINATE		PASSENGER LEAVES SYSTEM
44	TURN	ADVANCE	2	CAR AT TOP OR BOTTOM
45		TEST E	X1,2,WAIT	TWO OTHER CARS WAITING?
46	PROC	TEST E	P2,10,UPUP	IS THIS TOP?
47		ASSIGN	4,2	INDICATE GOING DOWN: P4 = CHAIN
48		ASSIGN	5—,2	INDICATE GOING DOWN: DECR FLOOR
49		TRANSFER	,PIKUP	
50	WAIT	SAVEVALUE	1+,1	INCREMENT CARS WAITING
51		TEST NE	V2,0	WAIT FOR CALL
52		SAVEVALUE	1—,1	DECREMENT CARS WAITING
53		TRANSFER	,PROC	
		START	10,,5,1	
		REPORT		
		EJECT		
	11	TEXT	ELEVATOR NO.	AVG. NO. ON BOARD
		SPACE	1	
	15	FORMAT	1 — 4/S1,S3	
		SPACE	3	
		SPACE	3	
		TEXT	PERSONS WAITING FOR A CAR TO GO UP:	
	5	TEXT	AVERAGE NUMBER WAITING = #CH1,5/XXX.X#	
	5	TEXT	MAXIMUM NUMBER WAITING = #CH1,6/XXX.#	
	5	TEXT	AVERAGE WAITING TIME = #CH1,3/XXX.X# CLOCK UNITS	
		SPACE	3	
		TEXT	PERSONS WAITING FOR A CAR TO GO DOWN:	
	5	TEXT	AVERAGE NUMBER WAITING = #CH2,5/XXX.X#	
	5	TEXT	MAXIMUM NUMBER WAITING = #CH2,6/XXX.#	
	5	TEXT	AVERAGE WAITING TIME = #CH2,3/XXX.X# CLOCK UNITS	
		EJECT		
	BLO	TITLE	,	
		SPACE	3	
	STO	TITLE	,ELEVATOR STATISTICS	
		SPACE	3	
	CHA	TITLE	,STATISTICS FOR PERSONS WAITING FOR A CAR	
		SPACE	3	
	MHSA	TITLE	,"PASSENGER OFF" MATRIX	
		OUTPUT		
		END		

FIG. 6-2 (continued). GPSS elevator model.

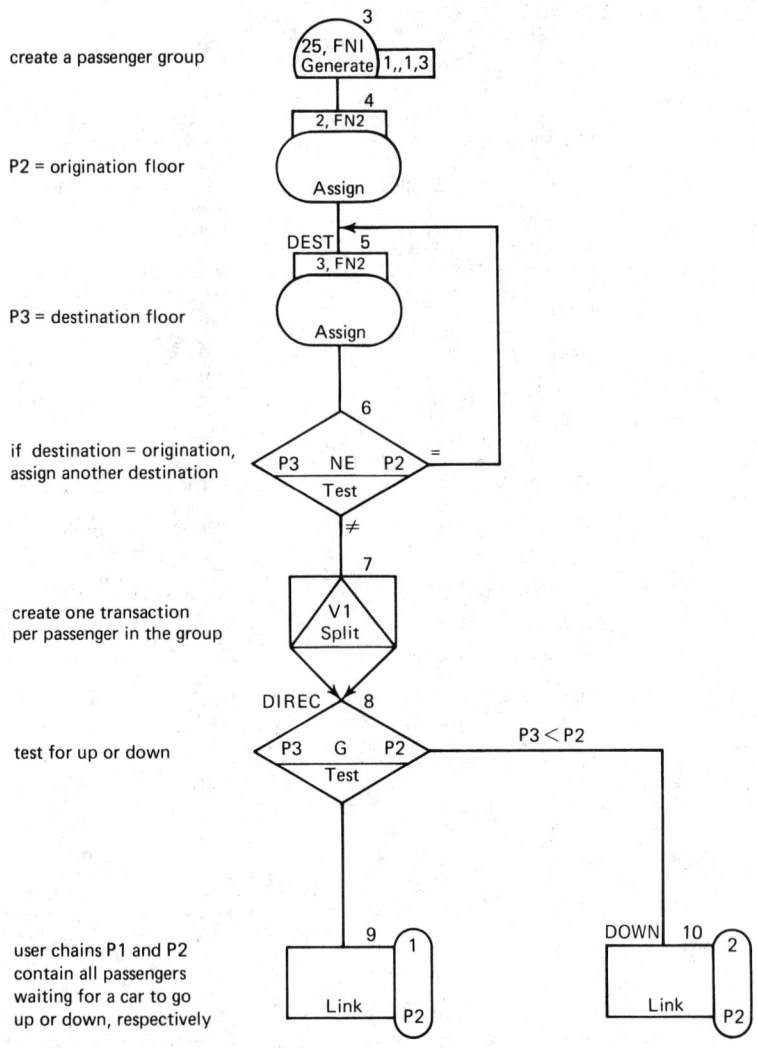

Fig. 6-3. Passenger portion of GPSS elevator model.

output. The preamble defines the static structure of the system to be modeled. The SIMSCRIPT II compiler uses this information to create a data structure for the system. All entities and their attributes are defined in the preamble, along with statistics to be accumulated, such as AVERAGE.OCCUPIED.

The main routine reads input data and performs initialization operations. It also starts the operation of the model by scheduling the first passenger arrival and the first elevator arrival. These are both coded as event routines, and subsequent passenger and elevator arrivals will be scheduled each time these routines are executed by the program. The main routine also schedules the STOP.SIMU-LATION event which produces the output report and ends the computer run.

The event PASS.ARRIVAL is analogous to the GPSS coding of Figure 6-3. The event ELEV.ARRIVAL describes the logic of elevator movement and passenger

ELEVATOR NO.   AVG. NO. ON BOARD

1	.157
2	.172
3	.188
4	.195

PERSONS WAITING FOR A CAR TO GO UP:
AVERAGE NUMBER WAITING =   .3
MAXIMUM NUMBER WAITING = 20.
AVERAGE WAITING TIME   = 5.4 CLOCK UNITS

PERSONS WAITING FOR A CAR TO GO DOWN:
AVERAGE NUMBER WAITING =   .3
MAXIMUM NUMBER WAITING = 14.
AVERAGE WAITING TIME   = 5.4 CLOCK UNITS

FIG. 6-4.   Portion of output of GPSS elevator model.

pickup and departure. As was the case in the GPSS model, this event comprises the bulk of the work required to describe the dynamics of the system.

The STOP.SIMULATION event produces all required output and ends the simulation run. This event was scheduled at the start of the simulation by the main routine. Accumulation of all statistics to be printed must be specified by the user in the preamble. All data are printed in the positions indicated by asterisks. The results obtained (Figure 6-6) compared very closely with those obtained from the GPSS run.

APPLICATIONS

The preceding portion of this chapter has been devoted to the nature of simulation and simulation languages. We now turn to an examination of the major application areas for this versatile technique. The discussion of each application is necessarily brief. Detailed descriptions of a large number of actual applications will be found in references 4, 8, 13, and 14.

Manufacturing. A wide range of manufacturing problems are amenable to analysis by simulation. Using a model of a manufacturing operation, whether it be in the nature of an assembly line or a job shop, the following benefits may be obtained.

Bottlenecks which may result from a change in existing operations can be anticipated. Any proposed change in operations, such as that produced by adding a new product or expanding the output of the present product line, will change the dynamics of the process. A simulation model incorporating the proposed changes will pinpoint potential trouble spots. By introducing appropriate changes—such as rerouting, altering the scheduling procedures, or adding facilities—and running the simulation again on an iterative basis, the necessary changes in structure and flow of the operation may be planned, resulting in minimum disruption of operations when these changes are implemented.

A totally new facility can be planned and costly mistakes can be avoided by observing the behavior of the plant before its construction by means of a simulation model. Scheduling procedures may be evaluated by comparing results obtained in several runs where the procedures are varied from run to run. Complex decision rules based on priority, time required for the various operations, due time, and the like can be formulated within the model.

Manpower allocation has been another fruitful area of study. For example, if production of an item is to be increased and additional manpower is required, should more people be hired, should the existing people go on overtime, or should second-shift operators be employed? By comparing the cost of each alternative with the increased output as indicated by the simulation, the most profitable decision can be made.

```
PREAMBLE
"       "THE ELEVATOR PROBLEM"
NORMALLY MODE IS INTEGER
THE SYSTEM HAS AN ORIGINATION
  DEFINE ORIGINATION AS A RANDOM STEP VARIABLE
PERMANENT ENTITIES...
  EVERY FLOOR HAS A TRAVEL
    DEFINE TRAVEL AS A RANDOM STEP VARIABLE
  EVERY FLOOR, UP.OR.DOWN OWNS SOME WAITING.GROUPS
  EVERY ELEVATOR HAS A CAPACITY, AN ON.BOARD AND A DIRECTION
  EVERY ELEVATOR, FLOOR HAS SOME DEPARTING "PASSENGERS
TEMPORARY ENTITIES.....
  EVERY GROUP HAS AN ORIGIN, A DESTINATION, A NUMBER AND BELONGS TO A
    WAITING. GROUPS
EVENT NOTICES INCLUDE PASS.ARRIVAL AND STOP.SIMULATION
  EVERY ELEV.ARRIVAL HAS AN IDENTIFIER AND A PLACE
DEFINE MEAN.GROUPSIZE, AVERAGE.SPACING, UNLOAD.TIME, LOAD.TIME AND
  MOVE.TIME AS REAL VARIABLES
DEFINE UP TO MEAN −1
DEFINE DOWN TO MEAN 1
ACCUMULATE AVERAGE.OCCUPIED AS THE AVERAGE OF ON.BOARD
ACCUMULATE AVERAGE.WAITING AS THE AVERAGE AND MAXIMUM.WAITING AS
  THE MAXIMUM OF N.WAITING.GROUPS
END

MAIN
DEFINE TSTOP AS A REAL VARIABLE
READ ORIGINATION
READ N.FLOOR
CREATE EVERY FLOOR
SCHEDULE A PASS.ARRIVAL NOW
READ N.ELEVATOR
CREATE EACH ELEVATOR
FOR EACH ELEVATOR, DO
  READ CAPACITY(ELEVATOR), ON.BOARD(ELEVATOR), DIRECTION(ELEVATOR),
    AND START.FLOOR
  SCHEDULE AN ELEV.ARRIVAL(ELEVATOR, START.FLOOR) NOW
LOOP
"ASSUME DEPARTING = 0
CREATE EACH UP.OR.DOWN(2)
READ MEAN.GROUPSIZE, AVERAGE.SPACING, UNLOAD.TIME, LOAD.TIME AND
  MOVE.TIME
READ TSTOP  SCHEDULE A STOP.SIMULATION IN TSTOP HOURS
START SIMULATION
END

EVENT PASS.ARRIVAL SAVING THE EVENT NOTICE
"CREATE A GROUP AND ESTABLISH ITS ATTRIBUTES
CREATE A GROUP
LET NUMBER = POISSON.F(MEAN.GROUPSIZE, 1) + 1
LET ORIGIN = ORIGINATION  LET FLOOR=ORIGIN
LET DESTINATION = TRAVEL(ORIGIN)
"DETERMINE THE DIRECTION OF TRAVEL
LET UP.OR.DOWN = 1
IF ORIGIN IS GREATER THAN DESTINATION, LET UP.OR.DOWN=2  REGARDLESS
FILE THIS GROUP IN WAITING.GROUPS
"GENERATE THE NEXT ARRIVAL
RESCHEDULE THIS PASS.ARRIVAL IN EXPONENTIAL.F(AVERAGE.SPACING,1) MINUTES
RETURN
END
```

FIG. 6-5.   SIMSCRIPT II elevator model.

```
EVENT ELEV.ARRIVAL(ELEVATOR, FLOOR) SAVING THE EVENT NOTICE
DEFINE TIME AS A REAL VARIABLE
LET TIME = MOVE.TIME
LET N = DEPARTING
IF N = 0, GO CHECK.DIRECTION  ELSE
"UNLOAD PASSENGERS BOUND FOR THIS FLOOR
ADD UNLOAD.TIME TO TIME
SUBTRACT N FROM ON.BOARD
LET DEPARTING = 0
'CHECK.DIRECTION'
IF FLOOR EQUALS 1,  GO UPWARD ELSE IF FLOOR EQUALS N. FLOOR, GO
    DOWNWARD ELSE
GO SAME
'UPWARD' LET DIRECTION = UP GO TO SAME
'DOWNWARD' LET DIRECTION = DOWN
'SAME' IF ON.BOARD = CAPACITY, GO FINISH ELSE
"ESTABLISH THE DIRECTION OF ELEVATOR TRAVEL
LET UP.OR.DOWN = 1 IF DIRECTION = DOWN, ADD 1 TO UP.OR.DOWN REGARDLESS
LET ENTER = 0
'LOAD.ELEVATOR'
IF WAITING.GROUPS IS EMPTY, GO FINISH ELSE
REMOVE THE FIRST GROUP FROM THE WAITING.GROUPS
LET ENTER = 1
LET N = MIN.F(NUMBER, CAPACITY − ON.BOARD)
ADD N TO ON.BOARD
ADD N TO DEPARTING(ELEVATOR, DESTINATION)
"IF ENTIRE GROUP HAS ENTERED TRY THE NEXT GROUP
IF NUMBER EQUALS N, DESTROY THIS GROUP  GO LOAD.ELEVATOR  ELSE
SUBTRACT N FROM NUMBER
FILE THIS GROUP IN THE WAITING.GROUPS
'FINISH' SUBTRACT DIRECTION FROM FLOOR
IF ENTER = 1, ADD LOAD.TIME TO TIME
REGARDLESS
RESCHEDULE THIS ELEV.ARRIVAL(ELEVATOR, FLOOR) IN TIME MINUTES
RETURN
END

EVENT STOP.SIMULATION
DEFINE T AS A REAL VARIABLE
START NEW PAGE
PRINT 3 LINES WITH MEAN.GROUPSIZE, AVERAGE.SPACING, UNLOAD.TIME,
    LOAD.TIME AND MOVE.TIME LIKE THIS
  RESULTS OF "AN ELEVATOR PROBLEM" EXPERIMENT
PASSENGER PARAMETERS:  AVG. GROUP SIZE = *.*  AVG.  TIME BETWEEN
    ARRIVALS= *.*
ELEVATOR PARAMETERS: UNLOAD TIME = *.** LOAD TIME = *.** MOVE
    TIME/FLOOR= *.**
SKIP 2 LINES
PRINT 1 LINE LIKE THIS
  ELEVATOR  AVERAGE NUMBER ON BOARD
FOR EACH ELEVATOR, PRINT 1 LINE WITH ELEVATOR AND AVERAGE.OCCUPIED THUS
      *                *.**
SKIP 3 LINES
PRINT 1 LINE THUS
  FLOOR  AVERAGE WAITING  MAXIMUM WAITING
FOR EACH FLOOR, FOR EACH UP.OR.DOWN, PRINT 1 LINE WITH FLOOR,
    AVERAGE.WAITING, AND MAXIMUM.WAITING THUS
      *        *.**          *
IF DATA IS ENDED, STOP  ELSE
READ T SCHEDULE A STOP.SIMULATION AT T
END
```

Fig. 6-5 (*continued*).   SIMSCRIPT II elevator model.

RESULTS OF "AN ELEVATOR PROBLEM" EXPERIMENT
PASSENGER PARAMETERS: AVG. GROUP SIZE = 3.0 AVG. TIME BETWEEN
   ARRIVALS = 1.5
ELEVATOR PARAMETERS: UNLOAD TIME = .12 LOAD TIME = .12 MOVE
   TIME/FLOOR = .06

ELEVATOR	AVERAGE NUMBER ON BOARD
1	.19
2	.18
3	.15
4	.17

FLOOR	AVERAGE WAITING	MAXIMUM WAITING
1	.01	1
1	0.	0
2	.02	1
2	.02	1
3	.01	1
3	.02	1
4	.01	1
4	.01	1
5	.02	1
5	.02	1
6	.02	1
6	.01	1
7	.02	1
7	.01	1
8	.04	2
8	.02	1
9	.02	1
9	.03	1
10	0.	0
10	.02	1

Fig. 6-6.   Portion of output of SIMSCRIPT II elevator model.

Inventory management considerations may be incorporated into a manufacturing simulation. The effect of changing reorder point and quantity for raw materials may be assessed. For finished goods, decisions may be made related to the size of a production run. For example, if a shop is set up to produce a particular item in response to an order, how much additional production of the item should be scheduled for inventory purposes? The answer is a function of both order pattern and the present status of the shop. The effect of changes in scheduling rules on in-process inventory may also be studied.

Manufacturing Data Requirements. The following data must be collected for a complete manufacturing model:

1. Product mix—the number of each item to be manufactured during the period under study—and information on the arrival distribution of orders
2. The sequence of operations for each product, including resources of machines, manpower, and material used and the time required for each operation
3. Facility scheduling and labor allocation rules
4. Job priorities
5. Inventory data on available space, lead times, and reorder criteria

Transportation and Material Handling. A wide range of problems involving physical movement has been studied by means of simulation. Some examples follow.

Railroad operations have been studied from the point of view of both an individual freight yard and an entire rail system. In the yard, problems of interest include number and location of sidings and allocation of switching engines, given a schedule of trains with cars to be added or removed, in addition to a schedule of through trains which must pass the area without delay. From a system-wide viewpoint, a model may answer questions such as:

1. How should trains be scheduled to provide the best service consistent with resource constraints? A secondary goal is the minimization of empty freight car movement.
2. What is the optimum train length for a given traffic mix?
3. How will a new yard affect operations elsewhere in the system?

Shipping models have been built to study both scheduling and harbor operations. The scheduling problem is concerned with the assignment of ships to a particular route. Factors to be considered are size of ship, volume and nature of cargo, present location of ship, location of pickup and destination ports, port capacity, and cargo priorities. Harbor models are concerned with the facilities available for docking ships and handling cargo, and are used to examine the behavior of proposed expansion plans.

Airline operations have been studied from a variety of viewpoints. The scheduling of aircraft, or rescheduling in the event that an aircraft goes out of service, is obviously an important area for study. The design of ground facilities for handling both passengers and cargo is another important application, especially when the introduction of very large aircraft forces the airlines to redesign their passenger terminals. From an airport operations viewpoint, the effect of an added runway or additional instrumentation to permit simultaneous use of several runways may be studied, along with alternative rules for maintaining "stacks" of aircraft waiting to land.

Vehicular traffic models may be used to study the timing of traffic lights, the effect of conversion from two-way to one-way streets, and the effect of imposing "no turn" rules at intersections. Highway models may be used as an aid in designing entrance and exit ramps.

Conveyor system models may be used as a design aid in the determination of speed, gate and switch configuration, and container size. As in all other simulation studies, there are two basic criteria: improving service provided and minimizing costs for a given level of service.

General Systems Reliability. The reliability, generally stated as an availability or percentage "up" requirement, of a complex system must often meet stringent requirements. This is generally true of military systems, aircraft, and on-line computer systems. Simulation is an excellent tool for obtaining a quantitative measure of system reliability if the characteristics of the individual components forming the system are known. Specifically, the mean time to failure and mean time to repair each component must be known, along with the probability distributions associated with these times, in addition to the time required to replace a component. A schedule for preventive maintenance should also be built into the simulation. Through trial and error iteration of the model, the required spare parts inventory may be determined, along with service crew and facility requirements. This may be done for several different availability requirements, thereby showing the trade-off between availability and total cost. The value of duplexing certain components of the system may also be assessed. Models of this nature have been used to study such diverse systems as data processing installations and a squadron of aircraft.

Enterprise Models. The area of enterprise or corporate modeling is one of the most interesting and challenging applications of simulation. Basically, an enterprise model is one which attempts to take into account the various factors that influence the performance of the corporation. The output of the model is in the form of financial statements, because earnings represent the ultimate criterion for the evaluation of performance. Using the model as an experimental tool, management can see the effect of various decisions upon profits.

Some of the factors which may be studied are:

1. The effect of delays in decisions, communications, and the like. An enterprise may be viewed as a feedback system, because its environment results in decisions which affect this environment, thereby resulting in new decisions. Delays in a

feedback system tend to produce instability; the longer the delays, the greater the instability. In a manufacturing operation, for example, production rates and inventory levels may experience wide swings if the delay in responding to market requirements is excessive. The value of taking measures to reduce this delay, such as installation of a real-time management information system, may be assessed by means of an enterprise model.

2. The effect of various manufacturing decisions such as production scheduling rules, hiring and manpower allocation rules, and inventory management. These will show up as factors affecting net profit. They will be built into a general enterprise model, although in less detail than if only the manufacturing operation itself were under study.

3. Advertising expenses. Expenditures for advertising will develop and broaden the existing market, resulting in increased demand for the company's products.

4. Capital investment, resulting in increased or more efficient production capacity and improved distribution capacity. This is another important ingredient of an enterprise model. Although the items discussed above are basically short-term factors, introduction of capital investment considerations provides a long-term planning tool for management.

An interesting aspect of capital investment simulation is risk analysis, which attempts to quantify the uncertainty inherent in forecasting the elements of revenue and expense which enter into the calculation of the rate of return on an investment. Rather than use an expected value estimate for each element, a probability distribution is used which indicates the range of possible values and the probability associated with each portion of that range. The model then produces as output the probability of meeting or exceeding any specified rate of return. Rather than a single expected rate of return, the simulation model presents a composite picture comprised of all possible outcomes.

Data Processing Systems. Because computer systems pervade so many aspects of industrial engineering, it is appropriate to spend a few moments on the simulation of such systems. In designing a data processing system, two criteria are of prime importance: throughput, which is defined as the volume of processing which can be completed in a unit time, such as messages per hour; and response time, which is usually given as a percentage of the total to be completed within a given time period, as for example, 90 percent within 10 minutes. Response time is most critical for inquiry-type, real-time systems—those where someone waits for a response after entering information at a terminal. Examples of such systems are on-line banking, airline reservations, and systems utilizing audio response. Response time is least critical for batch processing systems. It is the response time which generally determines the degree of satisfaction with system performance. Throughput can generally be calculated without a simulation model, but response time for all but trivial cases cannot, because complex systems quickly become intractable by analytic means.

Once the model has been constructed and tested, production runs will give information on queues and bottlenecks which develop in various portions of the system, and the results will indicate whether or not the throughput and response time criteria are satisfied. The output may also indicate areas of overdesign, where money can be saved without falling below specifications. Specifically, the following can be done.

1. The physical configuration of the system, including processor size, use of additional data channels, use of a high-speed drum for frequently referenced programs or data, and similar aspects, may be examined from two points of view: changes which improve performance without increasing cost beyond reason; and changes which eliminate overdesign beyond the required specifications, thereby reducing cost. Through the vehicle of the simulation model, the trade-offs between cost and performance become apparent.

2. The communication network in a real-time system can be studied to optimize the assignment of terminals to lines. The difference between a polling and a contention system can also be determined, and various polling sequences may be

tested. Buffered versus unbuffered terminals may be compared, along with the effect of holding a line during processing instead of releasing it to another terminal until output is ready.

3. Storage allocation can be planned. Buffer sizes, message segment size, program segmentation, and use of large core storage, either as an extension of main memory or as an input-output device, can be planned.

4. Organization of direct-access storage devices may be optimized to determine whether a given data set should reside on a single disc drive or be spread over several, thereby allowing more than one seek to proceed in parallel. The buffering scheme used may also be evaluated.

5. Alternative queuing disciplines may be studied. Should queues of output messages be maintained by terminal or by line? How much queuing should be in core and how much on disc or drum?

6. The effect of priorities on the performance of both the priority messages and the nonpriority messages may be measured. Alternatives such as priority for specific terminals or for specific output types may be evaluated.

7. The effect on system performance of the failure of a unit may be assessed, thereby giving a quantitative measure of the value of duplexing.

8. The operating system used can be changed, providing answers to such questions as "Is sequential processing adequate, or is multiprogramming required?" and "Is priority processing really necessary?"

PITFALLS AND PROBLEMS OF SIMULATION

Although simulation is a powerful and widely used technique, there are certain problems associated with its use.

Cost. Simulation studies are generally expensive. The talent required to construct and use a simulation model is relatively scarce and therefore high priced. Data collection is expensive. Because simulation is a statistical technique, computer runs must be long enough to assure that the output is statistically reliable, resulting in a significant expenditure of computer time. Furthermore, once a model is constructed, one becomes interested in studying the behavior of the system under a variety of conditions such as different system design, traffic rates, or operating rules. The net result is usually a significantly larger computer time cost than was originally anticipated. The value of such a model can be great, but the user must be prepared to pay the bill.

Data Collection. One of the major problems encountered by the simulation practitioner is the collection of data required for the model—traffic input rates, mix (percentage of each input type), operation times, and so on. A common tendency is to postpone data collection until after the model is constructed and debugged, using "dummy" or assumed data. This is generally unwise, because many an elegantly designed model has foundered on the problem of collecting meaningful data. Because the required information is rarely available in any well-organized form, if it is available at all, expensive and time-consuming work, often involving actual measurement and tabulation of field data, may be required. The cost of data collection often causes significant delay and occasionally cancellation of a simulation project; so it is wise to consider this aspect of the study as early as possible. If management is unwilling to provide the resources needed to obtain the required data, it is best to know this before going through the exercise of developing the model. Even if the data collection effort has full management support, it is wise to start early because of the time involved.

Complexity. The analyst should avoid the tendency to make his model overly complex. If in doubt about the inclusion of a particular variable, a sensitivity analysis should be performed to determine how sensitive the model's behavior is to a change in the variable. Assume extreme values for the variable—highest and lowest possible—and perform a computer run with each. Then observe the behavior of the output quantities of interest. If no significant differences are observed, one may conclude that the model is insensitive to the behavior of that variable, and the analyst need not be concerned with incorporating its effects. This permits a reduction in both complexity and data collection cost.

On the subject of complexity, there is a widespread tendency to attempt to model too much. Instead of starting with one aspect of a company's operations such as manufacturing or distribution, one is tempted to construct an all-encompassing corporate model. Most attempts of this nature are unsuccessful unless they are preceded by a series of smaller studies of various aspects of the system. The magnitude of the problem must be kept within reasonable bounds consistent with the experience of the people engaged in the simulation work, and the neophyte will be well advised not to attempt to take on the world in one sitting.

Statistical Considerations. Because simulation of the type discussed (stochastic) is a statistical technique, the analyst must be aware of some of the statistical problems inherent in the use of his model. Clearly, if the run duration is short (small sample size), it is quite possible to produce misleading results, because a relatively short run of numbers from a random number generator may produce a series which bears little relation to the longer term, steady state behavior. How long is long enough depends on the nature of the model. Unfortunately for the user, the standard statistical methods of confidence interval determination generally break down when attempts are made to apply them to the output of a simulation run, because the individual elements of such output are generally highly correlated rather than independent, as required. For example, it is obvious that the time spent by the tenth person on a queue is highly correlated with the time spent by the ninth.

The method generally used is quite empirical: the model is run with periodic "snaps" (output) until the statistics of interest no longer show any significant variation from snap to snap. A technique has been developed for statistical analysis during execution of the model, with run length determined by the analysis routine as a function of a specified confidence interval, but its intelligent use requires considerable knowledge of the technique. This work has been developed by Fishman and Kiviat (reference 5).

Assumptions. A final point to keep in mind is that every simulation model contains many assumptions about the structure of the system, the nature of the input, and the operating rules used. Such assumptions are necessary to reduce the model to the essence of the system under study. Without them, the model would become a hopeless tangle of irrelevant detail. The potential for trouble arises when gross assumptions are made without proper justification, often with the perfectly good intention of refining that portion of the model when additional information becomes available. If the assumption made is forgotten, all results obtained from the model may be, without the knowledge of the user, invalid. It is therefore imperative that the analyst maintain a list of all nontrivial assumptions made, so that the user of the results may judge whether they are reasonable.

CONCLUSION

Simulation, like the computer upon which it depends, is still evolving at a rapid pace. Use of the technique has reached the phase of explosive growth, and the new applications being developed will aid the entire engineering and management science community.

BIBLIOGRAPHY

1. Buxton, J. N., and J. G. Laski, "Control and Simulation Language," *The Computer Journal,* vol. 5, no. 3, 1962.
2. "Capital Investment Studies Using GPSS: Bulk Material Movement Problems," International Business Machines Corporation, Form E20-0313, White Plains, N.Y., 1968.
3. Dahl, O. J., and K. Nygaard, "SIMULA—An ALGOL-based Simulation Language," *Communications of the ACM,* vol. 9, September, 1966.

4. *Digest of the Second Conference on Applications of Simulation,* Association for Computing Machinery, New York, 1968.
5. Fishman, George S., and Philip J. Kiviat, *Digital Computer Simulation: Statistical Considerations,* The RAND Corporation, RM-5387-PR, Santa Monica, Calif., November, 1967.
6. "360 GPSS User's Manual," International Business Machines Corporation, Form H20-0326, White Plains, N.Y.
7. Gordon, Geoffrey, *System Simulation,* Prentice-Hall, Inc., Englewood Cliffs, N.J., 1969.
8. *IEEE Transactions on Systems Science and Cybernetics,* November, 1968.
9. Kiviat, Philip J., *Digital Computer Simulation: Computer Programming Languages,* The RAND Corporation, RM-5883-PR, Santa Monica, Calif., January, 1969.
10. Kiviat, Philip J., and Alan Pritsker, *Simulation with GASP II—A FORTRAN-based Simulation Language,* Prentice-Hall, Inc., Englewood Cliffs, N.J., 1969.
11. Kiviat, P. J., R. Villanueva, and H. M. Markowitz, *The SIMSCRIPT II Programming Language,* Prentice-Hall, Inc., Englewood Cliffs, N.J., 1969.
12. Markowitz, H. M., B. Hausner, and H. W. Karr, *SIMSCRIPT, A Simulation Programming Language,* Prentice-Hall, Inc., Englewood Cliffs, N.J., 1963.
13. *Proceedings of the Fourth Conference on Applications of Simulation,* Association for Computing Machinery, New York, 1970.
14. *Proceedings of the Third Conference on Applications of Simulation,* Association for Computing Machinery, New York, 1969.

Mathematical, Statistical, and Programming Procedures

Mathematics for the Industrial Engineer

BURTON V. DEAN

**Chairman, Department of Operations Research,
Case Western Reserve University, Cleveland, Ohio**

MARIA ALTSCHUL

**Department of Operations Research,
Case Western Reserve University, Cleveland, Ohio**

Many problems encountered by the industrial engineer are best solved by the application of mathematical, statistical, or programming procedures. There are a large number of such procedures available to the industrial engineer who has a sufficient grounding in mathematics and statistics to be able to apply them. This chapter and the other chapters of Section 10 assume that the industrial engineer has this grounding, for a handbook cannot serve as a text for teaching the fundamentals which are customarily taught in our educational institutions. Instead, the chapters of this section will describe, largely in mathematical terms, a number of the procedures which have been found useful by industrial engineers in solving the kinds of problems they may be expected to be confronted with in their never-ending search for improved ways of doing things. They will thus provide a useful source of information to the industrial engineer seeking the optimum procedure to apply to the specific problem with which he is dealing.

LOGIC

Proposition. A proposition is a sentence which we can assert is either true or false. In a proposition, a property of a certain object is mentioned.

Example:

1. The population of the United States is 200,000. This proposition is obviously false.

2. Neil Armstrong was the first man on the moon. As far as we know, this proposition is true.

Notation: The usual notation for a proposition is a lowercase letter such as p or q.

Operations between Propositions. There are a number of basic operations between propositions. By means of these and their combinations, we obtain complex propositional statements. These operations are the following:

1. And ($\wedge$): Given two propositions p and q, we define a new proposition $p \wedge q$ as the one which combines the statements given by the original propositions, and which will be true if and only if both are true. If at least one of them is false, then $p \wedge q$ is false. (See truth tables below.)

Example:

Let p be the proposition: Jones is a man.

Let q be the proposition: Jones is a college graduate.

Then $p \wedge q$ is the statement that Jones is a male college graduate.

2. Or (inclusive) ($\vee$): Given two propositions p and q, we define a new proposition p or q, which will be true if and only if at least one of the original propositions is true.

Example:

p: Jones went to a ball game.

q: Jones went to the movies.

$p \vee q$: Jones went to the movies or to a ball game.

3. Or (exclusive) ($\triangle$): Given p and q, we define the new proposition $p \triangle q$ as a combination of p and q, which will be true if and only if exactly one of the original propositions is true.

Example:

p: Jones is in Africa.

q: Jones is in Asia.

$p \triangle q$: Jones is in Africa or in Asia.

Note: In spoken language, we make no distinction between the two different or's. The difference in logic is clear from the examples given above. When we say Jones went to the movies or Jones went to a ball game, we are not specifying time. Jones could have been at the movies and at the ball game, but not at both places at the same time. The proposition Jones went to the movies or to the ball game will then be true if Jones went to the ball game, or if Jones went to the movies, or if Jones did both activities. When we say Jones is in Africa, we mean now. Jones is in Asia also means now. It is then clear that Jones is in Africa or in Asia, and the composite proposition will be true if Jones is in Africa or if Jones is in Asia; but the third possibility which was present before, the one saying Jones is both in Africa and in Asia, is absurd in this case.

4. Negation of a proposition ($\sim$): Given a proposition p, the negation of p is a proposition $\sim p$, which affirms the contrary to what p affirms. It is true if and only if p is false.

Example:

p: It is raining.

$\sim p$: It is not raining.

5. Implication ($\Rightarrow$): Given two propositions p and q, we say that $p \Rightarrow q$ if p implies q. Intuitively, it means that the truth of p implies the truth of q. For a clearer idea, see the truth tables below.

Truth Tables. Truth tables state the truth or falsehood of a proposition, given the truth or falsehood of its components. Furthermore, we can use these truth tables to define formally the operations listed above.

(T = true, F = false)

	p	q	$p \wedge q$	$p \vee q$	$p \triangle q$	$p \Rightarrow q$	$\sim p$	$\sim q$
1.	T	T	T	T	F	T	F	F
2.	T	F	F	T	T	F	F	T
3.	F	T	F	T	T	T	T	F
4.	F	F	F	F	F	T	T	T

All these are intuitively clear, except maybe the column for $p \Rightarrow q$. Cases 1 and 4 are straightforward. Case 2 is easily understood because a true statement cannot imply a false one. Case 3 is not intuitively obvious and should be considered as a logical definition.

In the same way as we construct a truth table for two simple propositions p and q, we can construct one for the more complex propositions that result from applying the operations defined above.

Equivalence ($\Leftrightarrow$). We say that two complex propositions are equivalent if their truth tables are the same. For example, consider the propositions

$$(\sim p) \vee q \qquad \text{and} \qquad p \Rightarrow q$$

p	q	$\sim p$	$(\sim p) \vee q$
T	T	F	T
T	F	F	F
F	T	T	T
F	F	T	T

This is the truth table for the proposition $(\sim p) \vee q$. If we check on the column $p \Rightarrow q$ of the previous table, we see that it coincides with the column for $(\sim p) \vee q$ in this one. Then $[(\sim p) \vee q] \Leftrightarrow [p \Rightarrow q]$.

SET THEORY

Notion of a Set. A set is a collection, conglomerate, or group of objects. Traditionally, in the literature, capital letters $(A,B,X, \ldots)$ are used to represent sets, and the objects or elements which form these sets are represented by lowercase letters $(a,b,x, \ldots)$.

Notion of Belonging. Given a certain element or object x and a set A, there are two possibilities. Either the element forms part of the set or the element does not form part of the set. To represent this idea, we use the following notation:

$$x \in A \qquad \text{if } x \text{ belongs to } A$$
$$x \notin A \qquad \text{if } x \text{ does not belong to } A$$

A set which has no elements is said to be an empty set, and is denoted by $\emptyset$.

Notation. A set is completely determined once we know all the elements that form it. Thus, we can characterize a set by enumerating or writing down all elements that determine it. That is, if X is formed by the elements x, y, and z, we write $X = \{x,y,z\}$.

Examples:

1. Let A be the set of all digits; then

$$A = \{0,1,2, \ldots ,9\}$$

2. Let B be the set of all positive numbers divisible by 2; then

$$B = \{2,4,6,8, \ldots \}$$

A more convenient way to characterize a set is by giving a property that the elements of the set, and only those elements, satisfy. That is, if A is the set of elements which satisfy property P, we write

$$A = \{x|x \text{ satisfies } P\}$$

Examples: Using the sets stated above,
1. $A = \{x|x \text{ integer and } 0 \leq x \leq 9\}$
2. $B = \{x|x \text{ divisible by 2 and } x \text{ positive}\}$

Notion of Inclusion. Given two sets X and Y, if all the elements of X are also elements of Y, we say that X is included (or contained) in Y. We also say that Y includes (or contains) X, or that X is a subset of Y. We write this as $X \subset Y$.

Example: Let

$$A = \{x|x \text{ integer}\}$$
$$B = \{x|x \text{ even}\}$$

then $B \subset A$.

If both $X \subset Y$ and $Y \subset X$ occur, we say the sets are identical, and we write

$$X = Y$$

Operations between Sets

1. *Union:* Given X and Y, we define the union as

$$X \cup Y = \{x|x \in X \text{ or } x \in Y\}$$

To illustrate this and the following operations, we will make use of the Venn diagrams. Consider the set U of all possible objects (called universe or universal set). We represent it by a rectangle as shown below. The union of X and Y is given by the shaded area.

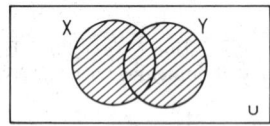

2. *Intersection:* The intersection of X and Y is defined as

$$X \cap Y = \{x|x \in X \text{ and } x \in Y\}$$

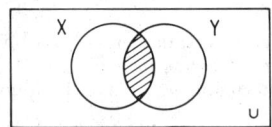

3. *Difference:* The difference between X and Y is defined as

$$X - Y = \{x|x \in X \text{ and } x \notin Y\}$$

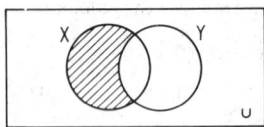

In particular, when X is the universal set, $\sim Y$ is denoted as $-Y$ and called the complement of Y; other common notations for the complement of Y are Y', Y^c, $\bar{Y}$.

Properties
1. $A \cap A = A \cup A = A$
2. $A \cap B = A$ if and only if $A \subset B$ if and only if $A \cup B = B$
3. $\left.\begin{array}{l} -(A \cup B) = (-A) \cap (-B) \\ -(A \cap B) = (-A) \cup (-B) \end{array}\right\}$ *De Morgan's laws*

4. $A \cap (B \cup C) = (A \cap B) \cup (A \cap C)$
   $A \cup (B \cap C) = (A \cup B) \cap (A \cup C)$
5. $A \cap (B \cap C) = (A \cap B) \cap C = A \cap B \cap C$
   $A \cup (B \cup C) = (A \cup B) \cup C = A \cup B \cup C$
6. $A \cap \varnothing = \varnothing$
   $A \cup \varnothing = A$
   $A \cap U = A$
   $A \cup U = U$
   $(-U) = \varnothing$ $(-\varnothing) = U$

CALCULUS

Functions. A function is a rule that assigns to each element of a set X an element of a set Y. X is called the domain; Y is called the range of the function.

Example: Let $f(x)$ be the function that to each real number assigns its square. Then $f(x) = x^2$ and

$$X = \{\text{real numbers}\}$$
$$Y = \{x | x \text{ real and } x \geq 0\}$$

Limits of Variables. Suppose we have a quantity x that varies, for instance, with time. If x approaches a constant value a in such a way that the difference $x - a$ gets eventually to be less than any preassigned number, we say that a is the limit of the variable x (in this case, as time approaches a certain value).

Note: If $x - a$ is a negative number, we consider $a - x$, which will then be positive. This is the definition of $|x - a|$ (absolute value or modulus of $x - a$):

$$|x - a| = \begin{cases} x - a & \text{if } x - a > 0 \\ a - x & \text{if } x - a < 0 \end{cases}$$

Properties:
1. The limit of a finite sum of variables is the sum of their limits.
2. The limit of the product of a finite number of variables is the product of their limits.
3. The limit of the product of a constant times a variable is the constant times the limit of the variable.
4. The limit of a quotient of two variables is the quotient of the limits, if this quotient is defined, that is, if the limit of the denominator is different from zero.

Definition: We say that a variable is "bounded from above" if there is a number b such that the variable is never greater than b.

The same definition holds for "bounded from below," replacing greater by smaller.

Property: If a variable is bounded from above and never decreasing, it approaches a limit that is never greater than the bound.

The same holds true for a variable bounded from below and never increasing, replacing greater by smaller.

Limits of Functions. We shall illustrate the idea of limits of functions by giving some examples.

Example 1: Consider the function $y = x^2$ whose graph is shown below.

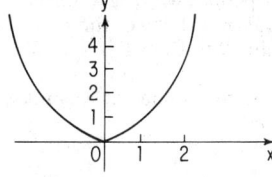

We can see that the lowest value that y takes is zero, when $x = 0$. As we move to the right, y increases indefinitely. We say that y tends to infinity as x tends to infinity and we write

$$\lim_{x \to \infty} y = \infty$$

Example 2:   Consider the function $y = 1 + 1/x^2$.

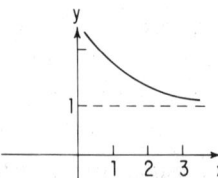

As x increases, y decreases and can be made arbitrarily close to 1.   We say that y tends to 1 when x tends to infinity, and write

$$\lim_{x \to \infty} y = 1$$

When x decreases to zero, y increases indefinitely, and we write that as

$$\lim_{x \to 0} y = \infty$$

Consider now a general function $y = f(x)$, and a fixed value a of the variable.   If the function $f(x)$ takes values very close to a value L when the values of x are close to a, then we say that $f(x)$ has limit L when x tends to a, and write it

$$\lim_{x \to a} f(x) = L$$

For the case of an infinite limit, if the function $f(x)$ takes values as large as we wish as the value of x gets closer to a, we say that $f(x)$ has limit infinity when x tends to a, and write

$$\lim_{x \to a} f(x) = \infty$$

Continuity.   The concepts of continuity and discontinuity of a function can be better understood by considering the following example.

Let $y = f(x)$ be the function whose graph is shown below.

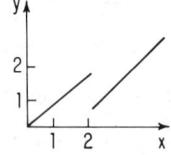

$$f(x) = \begin{cases} x & 0 \le x < 2 \\ x - 1 & 2 \le x \end{cases}$$

Intuitively, we see that at the point $x = 1$, the function is continuous, and at $x = 2$ it is discontinuous.   That is, at $x = 1$ it continues increasing smoothly, whereas at $x = 2$ it decreases from the value 2 to the value 1.   We are able to give the following definition.

Definition: A function $f(x)$ is continuous at a point $x = a$ when
  1. $f(x)$ is defined for $x = a$
  2. $\lim_{x \to a} f(x) = f(a)$

If this is not the case, we say that the function is discontinuous at $x = a$.

In the previous example, $f(x)$ is continuous at $x = 1$, and discontinuous at $x = 2$. Note that although $f(2) = 1$, $\lim_{x \to 2} f(x) = 2$ when we approach $x = 2$ from the left.

There are four possible types of discontinuity. We shall illustrate them with examples.

Case 1: In the case of the previous function, when

$$\lim_{x \to a^+} f(x) \neq \lim_{x \to a^-} f(x)$$

where $x \to a^+$ simply means that we approach x from the right and $x \to a^-$ that we approach it from the left.

Case 2: Let

$$f(x) = \begin{cases} x & 0 \leq x < 1 \\ -x + 2 & 1 < x \\ 2 & x = 1 \end{cases}$$

whose graph is shown below.

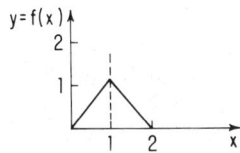

In this case, $\lim_{x \to 1^+} f(x) = \lim_{x \to 1^-} f(x) = 1$ but they are not equal to $f(1) = 2$. Formally, $\lim_{x \to a} f(x) = L$ exists, but $f(a) \neq L$. In this case, $f(a)$ is defined, but the same type of discontinuity exists when $\lim_{x \to a} f(x)$ exists but $f(a)$ is not defined.

Case 3: Let $y = f(x) = 1 + 1/x^2$, whose graph was shown on page 10-8, example 2. In this case, $\lim_{x \to 0} f(x) = \infty$, and the function is discontinuous at $x = 0$. Formally, $\lim_{x \to a} f(x) = \infty$.

Case 4:

$$\lim_{x \to a} f(x) = -\infty$$

As an example of this case, take $f(x) = -1 - 1/x^2$. So far, we have considered continuity only at a given point. We now give the following definition.

Definition: If $f(x)$ is continuous at every x such that $a < x < b$ for some given a, b, we say that $f(x)$ is continuous in the open interval (a,b). If, furthermore, $\lim_{x \to a^+} f(x) = f(a)$ and $\lim_{x \to b^-} f(x) = f(b)$, then the function $f(x)$ is continuous in the closed interval $[a,b]$. If $f(x)$ is continuous at x for all values of x, we say that it is continuous.

Derivative of a Function. When in industry we represent a relation between two variables in the form of a graph, we are often interested in the rate of change of the function represented, and especially in the range of values of the independent (or control) variable for which the rate of change may be positive or negative. The most important points are usually the extremes, where the rate of change is zero.

To illustrate this, consider the curve indicating total cost shown below.

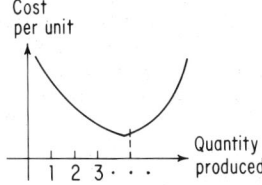

Production costs per unit decrease as we increase the production quantities, but stocking cost increases. The total unit costs are given in the figure. We are interested in that value of the production quantity that yields no change in the total costs. The value where the rate of change in costs with respect to production quantity is zero is the optimal amount to produce.

We shall now introduce the concept of a difference operator applied to a function so that the concept of a rate of change can be developed. Then, by use of the theory of limits of functions, we shall define the derivatives of operations. As was seen in the example above, the concept of derivative is used in problems of production to find an optimal decision rule.

Differences. Let u be a function of n ($u = f(n)$). It is common to use the sequence notation when the independent variable is a positive integer number ($n = 0,1,2,$ . . .). Instead of writing $u = f(n)$, we write u_n when we want to indicate that u varies with n. The change in the function u when n increases in value of one is called "first difference," and is written as

$$\Delta u_n = u_{n+1} - u_n$$

and is still a function of n. Note that Δu_n is not a product, it is a single entity; and the Δ in this case is not a number, it is called an "operator," because it stands for an operation or rule. Once we have the first difference Δu_n, we can compute the second difference.

$$\Delta^2(u_n) = \Delta(\Delta u_n)$$

In general, we can say

$$\Delta^k(u_n) = \Delta(\Delta^{k-1}u_n)$$

and the following properties hold:
1. $\Delta^r(\Delta^s u_n) = \Delta^r \Delta^s u_n = \Delta^{r+s} u_n$
2. $\Delta(u_n \pm v_n) = \Delta u_n \pm \Delta v_n$
3. $\Delta a u_n = a \Delta u_n$
4. $\Delta u_n v_n = u_{n+1} \Delta v_n + v_n \Delta u_n$
5. $\Delta \dfrac{u_n}{v_n} = \dfrac{v_n \Delta u_n - u_n \Delta v_n}{v_n v_{n+1}}$

Derivatives. We have defined the difference by finding $u_{n+1} - u_n$. We have assumed that our scale was such that it was convenient for us to count unit by unit. This, however, can be extended to a more general concept of differences, and so we define

$$\Delta u_n = u_{n+h} - u_n$$

where h is any arbitrary interval, positive or negative.

Definition: The rate of change of u_x over the interval x to $x + h$ (note that we consider x as the variable, instead of n, where x does not have to be integral) is defined as

$$\frac{\Delta u_x}{h} = \frac{u_{x+h} - u_x}{h}$$

This quotient gives us an approximation to how much our function is changing in a given interval x to $x + h$. Now we have to decide on the size we want our intervals to have. It seems only natural to think of intervals as small as we can possibly get them. For instance, when we talk about the speed of a vehicle, we are actually considering the change in distance divided by the change in time, and although we say the speed at a given instant, what we mean is the distance traveled by the vehicle in a very short interval of time, divided by the duration of the interval.

Even if we have limitations given by the measuring instruments as to how short an interval can be, theoretically we have already developed a weapon, the concept of a limit, which allows us to work with intervals as short as we want them. Consider the following example.

Let
$$u_x = x^2$$
$$\Delta u_x = u_{x+h} - u_x = (x + h)^2 - x^2 = 2hx + h^2$$

The rate of change is
$$\frac{\Delta u_x}{h} = \frac{2hx + h^2}{h} = 2x + h$$

The smaller h is, the closer the rate of change gets to be $2x$. We cannot say that for $h = 0$ the rate of change is $2x$, because for $h = 0$, $\Delta u_x = 0$, and the operation $0/0$ is not defined. But we can say that, as h tends to zero, the rate of change tends to $2x$. In symbols,

$$\lim_{h \to 0} \frac{\Delta x^2}{h} = \lim_{h \to 0} \frac{2xh + h^2}{h}$$

$$= \lim_{h \to 0} (2x + h) = 2x + \lim_{h \to 0} h$$

$$= 2x$$

Graphically, we can give an interpretation of the rate of change as shown. Let P be fixed. Let Q be a point that moves toward P; and $Q_1, Q_2, \ldots$ the positions Q takes.

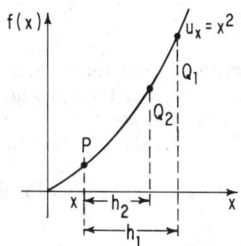

The slope of the chord PQ_1 is $[(x + h_1)^2 - x^2]/h_1$; for PQ_2 it is $[(x + h_2)^2 - x^2]/h_2$. As Q moves toward P, the interval h becomes smaller and smaller, and the chord PQ becomes the tangent line at P. We can now say that the rate of change at P, given by $\lim_{h \to 0} (\Delta x^2/h)$ is the slope of the tangent of the curve at P. At the minimum point of the curve, this slope is zero, and so is the rate of change. We give the following definition.

Definition: If $f(x)$ is a function, we say that

$$f'(x) = \lim_{h \to 0} \frac{f(x + h) - f(x)}{h}$$

is the derivative of $f(x)$. The limiting process is called differentiation. $f'(x)$ is sometimes denoted as df/dx, $(d/dx)(f(x))$, $Df(x)$, $\ldots$, where d/dx and D can be thought of as operations, as Δ was before. Just as with differences, we define the second derivative of a function as

$$f''(x) = \frac{d}{dx} (f'(x)) = \lim_{h \to 0} \frac{\Delta f'(x)}{h}$$

Operations. Using the properties of the Δ operator given before and using limiting operations, we have the following:

1. $\dfrac{d}{dx} [u(x) \pm v(x)] = \dfrac{d}{dx} u(x) \pm \dfrac{d}{dx} v(x)$

2. $\dfrac{d}{dx} [au(x)] = a \dfrac{d}{dx} [u(x)]$

3. $\dfrac{d}{dx} [u(x) \cdot v(x)] = u(x) \cdot \left[\dfrac{d}{dx} v(x)\right] + \left[\dfrac{d}{dx} u(x)\right] \cdot v(x)$

4. $\dfrac{d}{dx} \left[\dfrac{u(x)}{v(x)}\right] = \dfrac{\left[\dfrac{d}{dx} u(x)\right] \cdot v(x) - u(x) \cdot \left[\dfrac{d}{dx} v(x)\right]}{[v(x)]^2}$

Optimization. (Application of derivatives to find rules to minimize or maximize functions.) The problem of finding a maximum or a minimum of a function appears so often in industry that it is important to have a general procedure to find it. Note that finding a maximum of a function f is the same as finding a minimum for $-f$; so we need concern ourselves with solving only one of these problems.   We shall consider only the minimizing problem.   For example, if we are concerned with profits, our problem will be a maximization one; if we are concerned with costs, we try to solve a minimization problem.

Consider the case illustrated.

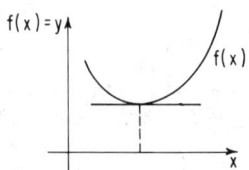

The function f starts decreasing, reaches a minimum, and then increases.   Before the minimum is reached, $f'(x) = dy/dx < 0$; after it, $dy/dx > 0$.   The conclusion then is that, for a minimum, it is necessary that $dy/dx = 0$.

Note: This, of course, holds if the derivative is a continuous function; otherwise, $f'(x)$ might not exist at the minimum.

Similarly, let us consider the necessary condition for determining the maximum of a function.

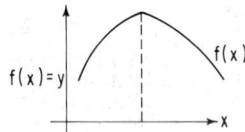

In the case illustrated, there is a maximum, and we have $dy/dx > 0$ before the maximum, and $dy/dx < 0$ after it.   We conclude that here also it is necessary to have $dy/dx = 0$ at the maximum.

Now let us consider the following function:

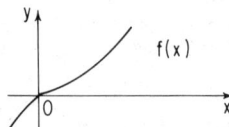

This function is $y = x^3$.   If we compute dy/dx at the point $x = 0$, we find that

$$\frac{dy}{dx}\bigg|_{x=0} = 3x^2\bigg|_{x=0} = 0$$

$\dfrac{dy}{dx}\bigg|_{x=0}$ reads $\dfrac{dy}{dx}$ at $x = 0$.   So in this case, we have $dy/dx = 0$ at a point which is clearly not a minimizing value.   This is called an inflection point.

Summarizing, we can state that a necessary condition for a minimum or a maximum of a function $f(x)$ at a point x_0 is that

$$\frac{d(f(x))}{dx}\bigg|_{x=x_0} = 0$$

It is not, as was seen for $y = x^3$ above, sufficient to have $dy/dx = 0$ in order to have a maximum or minimum.

Conditions for Maximum and Minimum.   Let $f(x)$ be a function and x_0 be a point at which $f'(x_0) = 0$. We wish to know if we are at an extreme point. If $f(x_0)$ is a minimum, then if h is small, we must have

$$f'(x_0 - h) < 0 \quad \text{and} \quad f'(x_0 + h) > 0$$

Then
$$\frac{f'(x_0 - h) - f'(x_0)}{-h} > 0$$

because $f'(x_0) = 0$, and

$$\frac{f'(x_0 + h) - f'(x_0)}{h} > 0$$

If we let $h \to 0$ in both inequalities, we get—provided it exists—$f''(x_0)$ as a second derivative, and analogous to the second difference Δ^2 defined before. So, if x_0 is a minimum, we conclude that $f''(x_0)$ has to be > 0. The same reasoning shows us that for x_0 to be a maximum, $f''(x_0)$ has to be < 0. If both $f'(x_0)$ and $f''(x_0)$ are zero, we cannot say anything. We might have a minimum as is the case with $f(x) = x^4$; we might have a maximum $f(x) = -x^4$; or we might have a shoulder or inflection point, as in $f(x) = x^3$. In these cases, then it is usually necessary to find higher order derivatives.

Before finishing our discussion about extreme points, we shall discuss functions with more than one extreme point, as the one whose graph is shown below.

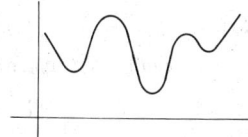

This function has three local minima and two local maxima. We usually want to find the global maximum or minimum. The way to do this is to examine the value of the function at each of these local extremes, and to select that one which gives us the maximum or minimum that is desired. However, in the discussion beginning on page 10-27, we see that if a function is convex (concave), it is guaranteed to have at most one minimum (maximum).

Integration. Suppose we have defined the operation sum $(+)$ between two quantities a and b, and we get as a result

$$a + b = c$$

This operation of addition is clearly applied to two factors. Consider now the operation square root, applied to a quantity whose square root is defined. Assume we have

$$\sqrt[2]{a} = b$$

and suppose we want to determine the value of the original quantity a. What we do is apply to the result b the inverse operation to square root; that is, we raise it to the exponent 2. We have

$$a = (\sqrt[2]{a})^2 = b^2$$

In a very similar fashion, we can think of the operation differentiation and define its inverse, which we shall call *integration*.

For example, if we have a relation between profit P and advertising expenditure x given by $P = f(x)$, we showed before that, to get the optimal value of x, we set

$$\frac{dP}{dx} = f'(x) = 0$$

where dP/dx was the rate of change. Suppose now that by some experimental means we can determine the rate of change. The question is, can we use this rate of change to determine the original relationship? The answer depends on having an operation which acts as the inverse of differentiation.   We could apply it to the rate of change and obtain as a result the original function.   We define the following.

Definition: If $f(x)$ is a given function, an integral of $f(x)$ is a function $y = F(x)$ whose derivative, $F'(x)$, is equal to $f(x)$, and we write

$$y = F(x) = \int f(x)\, dx \qquad y \text{ is the integral of } f(x)$$

where

$$\frac{dy}{dx} = f(x) = F'(x)$$

The symbol $\int$ indicates the operation being performed.   The symbol dx tells us to which variable it is applied where there is more than one variable involved.   Note that in the definition, we said an integral of $f(x)$, and not the integral for $f(x)$.   This can be made clear by noticing that if $F(x)$ is an integral of $f(x)$, then so is $F(x) + c$, where c is any constant.   This is because

$$\frac{d(y + c)}{dx} = \frac{dy}{dx} = f(x)$$

Integration, however, is not to be considered only as the inverse of differentiation.

Integration as the Limit of a Sequence of Sums.   In the previous discussion, we defined differences over an interval h as

$$\Delta f(x) = f(x + h) - f(x)$$

Then we said the rate of change was given by $\Delta f(x)/h$, and defined the derivative $f'(x)$ by letting h tend to zero.

$$f'(x) = \lim_{h \to 0} \frac{f(x + h) - f(x)}{h}$$

Now, suppose we want to find the results of the following sum:

$$S = \Delta f(x) + \Delta f(x + h) + \Delta f(x + 2h) + \cdots + \Delta f(x + nh)$$

Because

$$\Delta f(x) = f(x + h) - f(x)$$
$$\Delta f(x + h) = f(x + h + h) - f(x + h) = f(x + 2h) - f(x + h)$$
$$\cdots\cdots\cdots\cdots\cdots\cdots\cdots\cdots\cdots\cdots\cdots\cdots\cdots\cdots\cdots\cdots$$
$$\Delta f(x + nh) = f(x + (n + 1)h) - f(x + nh)$$

we get, when we cancel all the other terms,

$$S = f(x + (n + 1)h) - f(x)$$

That is, formally,

$$\sum_{r=0}^{n} \Delta f(x + rn) = f(x + (n + 1)h) - f(x)$$

Suppose now that we wish to find the sum of a function $g(x)$ between the limits $x = a$ and $x = b$.   Then

$$S = g(a) + g(a + h) + \cdots + g(b - h) + g(b)$$

In this case, $nh = b - a$.   If we know a function $G(x)$ such that $\Delta G(x) = G(x + h) - G(x) = g(x)$, then we have

$$S = G(b + h) - G(a)$$

That is, to find the sum of the function $g(x)$ between a and b, find a function $G(x)$ whose difference is $g(x)$.   Note that if we take limits, we get an interpretation very similar to the definition of integral stated above.   The function $G(x)$ is called an indefinite sum of $g(x)$—indefinite because if $\Delta G(x) = g(x)$, then $\Delta[G(x) + k] = g(x)$ also.   There is an infinite number of suitable functions $G(x)$.

Areas under Curves. Suppose we have the graph of the function $y = g(x)$ and we

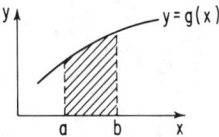

want to find out what is the area under this curve between the values $x = a$ and $x = b$.
To solve the problem graphically, we may divide the area in small rectangles, and add
up the areas of the rectangles. If the rectangles are narrow enough, the shaded area

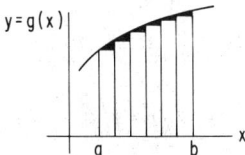

which is ignored when adding would be small enough to give a negligible error. If the
width of the rectangles is h, then the area is

$$S(b) = h[g(a) + g(a + h) + g(a + 2h) + \cdots + g(b - h)]$$
$$= h[G(b) - G(a)]$$

Now, we let h tend to zero, which will make the neglected areas in the above figure tend
to zero. To do this, let us increase the right-hand boundary by h; the new sum is
$S(b + h)$, and we get

1. $S(b + h) - S(b) = h[(G(b + h) - G(a)) - (G(b) - G(a))]$
    $$= h[G(b + h) - G(b)] = hg(b)$$

Similarly, we get

2. $S(b - h) - S(b) = -hg(b - h)$

Dividing equations 1 and 2 by h and $-h$, respectively, we get

1′. $\dfrac{S(b + h) - S(b)}{h} = g(b)$

2′. $\dfrac{S(b - h) - S(b)}{-h} = g(b - h)$

When we let $h \to 0$, we get

$$S'(b) = g(b)$$

But, when $h \to 0$, S tends to the area under the curve $g(x)$, between a and b. We thus
obtain the following results:

1. If $S(b)$ denotes the area under the curve $y = g(x)$ between $x = a$ and $x = b$,
    then $S'(b) = g(b)$.
2. To find the area, we must find a function $G(x)$ whose derivative is $g(x)$, and
    then compute $G(b) - G(a)$.

Such a function $G(x)$ is called an indefinite integral of $g(x)$ It is an indefinite integral,
again, because if $G(x)$ is such a function, then $G(x) + k$ also is an indefinite integral of
$g(x)$.

The process of finding $G(x)$ is called integration, and when we define the limits a and
b, we have a definite integral, which we write

$$\int_a^b g(x)\, dx$$

Properties:

1. $\displaystyle \int_a^b g(x)\, dx = -\int_b^a g(x)\, dx$

2. $\displaystyle \int_a^c g(x)\, dx + \int_c^b g(x)\, dx = \int_a^b g(x)\, dx$

LAPLACE TRANSFORM[1]

Let $f(t)$ be a given function which is defined for all positive values of t. We multiply $f(t)$ by e^{-st} and integrate with respect to t from zero to infinity. Then, if the resulting integral exists, it is a function of s, say $F(s)$:

$$F(s) = \int_0^\infty e^{-st} f(t)\, dt$$

The function $F(s)$ is called the Laplace transform of the original function $f(t)$, and will be denoted by $\mathcal{L}(f)$. Thus

$$F(s) = \mathcal{L}(f) = \int_0^\infty e^{-st} f(t)\, dt \qquad (1)$$

The described operation on $f(t)$ is called the Laplace transformation. Furthermore, the original function $f(t)$ in equation (1) is called the inverse transform or inverse of $F(s)$ and will be denoted by $\mathcal{L}^{-1}(F)$; that is, we shall write

$$f(t) = \mathcal{L}^{-1}(F)$$

Example: Let $f(t) = 1$ when $t > 0$. Then

$$\mathcal{L}(f) = \mathcal{L}(1) = \int_0^\infty e^{-st}\, dt = -\frac{1}{s} e^{-st} \Big|_0^\infty$$

Hence, when $s > 0$,

$$\mathcal{L}(1) = \frac{1}{s}$$

Properties of Laplace Transformation

Property 1.   The Laplace transformation is a linear operation; that is, for any functions $f(t)$ and $g(t)$ whose Laplace transforms exist and where a and b are constants, we have

$$\mathcal{L}[af(t) + bg(t)] = a\mathcal{L}(f) + b\mathcal{L}(g)$$

Property 2.   If $\mathcal{L}(f) = F(s)$ when $s > \alpha$, then

$$\mathcal{L}[e^{at}f(t)] = F(s - a) \qquad s > \alpha + a$$

That is, the substitution of $s - a$ for s in the transform corresponds to the multiplication of the original function by e^{at}.

We shall now see that differentiation and integration of $f(t)$ correspond to multiplication and division of the Laplace transform $F(s) = \mathcal{L}(f)$ by s. The significance of this property of the Laplace transformation is obvious, because in this way the operations of the calculus may be replaced by simple algebraic operations on the transforms.

Property 3.   Suppose that:
1. $f(t)$ is continuous for all $t \geq 0$.
2. $|f(t)| \leq M e^{at}$ for all $t \geq 0$ and for some constants α and M.
3. $f(t)$ has a derivative $f'(t)$ which is piecewise continuous on every finite interval in the range $t \geq 0$.

Then the Laplace transform of the derivative $f'(t)$ exists where $s > \alpha$, and

$$\mathcal{L}(f') = s\mathcal{L}(f) - f(0)$$

[1] Adapted partially from E. Kreyszig, *Advanced Engineering Mathematics*, John Wiley & Sons, Inc., New York, 1967.

Property 4. Let $f(t)$ and its derivatives—$f'(t)$, $f''(t)$, . . . , $f^{(n-1)}(t)$—be continuous functions for all $t \geq 0$, satisfying line 2 of property 3 for some α and M, and let the derivative $f^{(n)}(t)$ be piecewise continuous on every finite interval in the range $t \geq 0$. Then the Laplace transform of $f^{(n)}(t)$ exists and is given by the formula

$$\mathcal{L}(f^{(n)}) = s^n y(f) - s^{n-1} f(0) - s^{n-2} f'(0) - \cdots - f^{(n-1)}(0)$$

In this way, we obtain

$$\mathcal{L}(f'') = s^2 y(f) - sf(0) - f'(0) \tag{2}$$

Properties 3 and 4 may be used for determining transforms.

Example: Let $f(t) = t^2/2$. Find $\mathcal{L}(f)$. We have $f(0) = 0$, $f'(0) = 0$, $f''(t) = 1$. Because $\mathcal{L}(1) = 1/s$, we obtain from equation (2)

$$\mathcal{L}(f'') = \mathcal{L}(1) = \frac{1}{s} = s^2 \mathcal{L}(f)$$

or

$$\mathcal{L}\frac{t^2}{2} = \frac{1}{s^3}$$

Differentiation of $f(t)$ corresponds to multiplication of $\mathcal{L}(f)$ by s. Integration of $f(t)$ corresponds to division of $\mathcal{L}(f)$ by s.

Property 5. If $f(t)$ is piecewise continuous and satisfies an inequality of the form 2 of property 3, then

$$\mathcal{L}\left[\int_0^t f(t)\, dt \right] = \frac{1}{s} \mathcal{L}\,[f(t)] \qquad s > 0, s > \alpha$$

Transformation of Ordinary Differential Equations. Ordinary linear differential equations with constant coefficients can be reduced to algebraic equations of the transform. For example, consider the equation

$$y''(t) + w^2 y(t) = r(t) \tag{3}$$

where $r(t)$ and w are given. Applying the Laplace transformation and using equation (2), we obtain

$$s^2 y(s) - sy(0) - y'(0) + w^2 y(s) = R(s)$$

where $y(s)$ is the Laplace transform of the (unknown) function of $r(t)$. This algebraic equation is called the subsidiary equation of the given differential equation. Its solution is clearly

$$y(s) = \frac{sy(0) + y'(0)}{s^2 + w^2} + \frac{R(s)}{s^2 + w^2}$$

Note that the first term on the right is completely determined by means of given initial conditions, $y(0) = k$, $y'(0) = k^2$.

The last step of the procedure is to determine the inverse $\mathcal{L}^{-1}(y) = y(t)$, which is then the desired solution of equation (3).

Example: Find the solution of the differential equation

$$y'' + 9y = 0$$

satisfying the initial conditions $y(0) = 0$, $y'(0) = 2$. The subsidiary equation is

$$s^2 y(s) - 2 + 9y(s) = 0$$

Solving for $y(s)$, we obtain

$$y(s) = \frac{2}{s^2 + 9} = \frac{2}{3} \frac{3}{s^2 + 9}$$

From this and the table of some Laplace transforms (page 10–18), we find

$$y(t) = \mathcal{L}^{-1}(y) = \tfrac{2}{3} \sin 3t$$

$f(t)$	$F(s) = \mathfrak{L}[f(t)]$
1	$\dfrac{1}{s}$
t	$\dfrac{1}{s^2}$
$t^n \quad n = 1, 2, \ldots$	$\dfrac{n!}{s^{n+1}}$
$t^a \quad a \text{ positive}$	$\dfrac{\Gamma(a+1)}{s^{a+1}}$
$\dfrac{1}{\sqrt{\pi t}}$	$\dfrac{1}{\sqrt{s}}$
$2\sqrt{\dfrac{t}{\pi}}$	$\dfrac{1}{s^{3/2}}$
$\dfrac{t^{a-1}}{\Gamma(a)} \quad a > 0$	$\dfrac{1}{s^a}$
e^{at}	$\dfrac{1}{s-a}$
te^{at}	$\dfrac{1}{(s-a)^2}$
$\dfrac{1}{(n-1)!} t^{n-1} e^{at}$	$\dfrac{1}{(s-a)^n} \quad n = 1, 2, \ldots$
$\dfrac{1}{a-b}(e^{at} - b^{bt})$	$\dfrac{1}{(s-a)(s-b)} \quad a \neq b$
$\dfrac{1}{a-b}(ae^{at} - be^{bt})$	$\dfrac{s}{(s-a)(s-b)} \quad a \neq b$
$\dfrac{1}{w} \sin wt$	$\dfrac{1}{s^2 + w^2}$
$\cos wt$	$\dfrac{s}{s^2 + w^2}$
$\dfrac{1}{a} \sinh at$	$\dfrac{1}{s^2 - a^2}$
$\cos at$	$\dfrac{s}{s^2 - a^2}$
$\dfrac{1}{w} s^{at} \sin wt$	$\dfrac{1}{(s-a)^2 + w^2}$
$e^{at} \cos wt$	$\dfrac{s-a}{(s-a)^2 + w^2}$

DIFFERENCE EQUATIONS
The First Difference

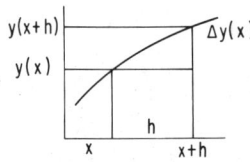

If a function $y(x)$ is given and h is a constant such that $x + h$ is in the domain of y, then Δy, the first difference of y, is a function whose value at x is given by

$$\Delta y(x) = y(x + h) - y(x)$$

Δ is called the difference operator and h the difference interval, which will be assumed to be constant (we are always using the same h) unless otherwise specified.

Example: If $y(x) = x + 1$ and $h = 1$, then

$$\Delta y(1) = y(2) - y(1) = 3 - 2 = 1$$
$$\Delta y(2.5) = y(3.5) - y(2.5) = 4.5 - 3.5 = 1$$

In general, we can get the value of Δy for every value of x, using the formula

$$\Delta y(x) = y(x + 1) - y(x) = (x + 2) - (x + 1) = 1$$

Second and Higher Differences. If a function y and its first difference are given, then the second difference of y, $\Delta^2 y$, is the difference of the first difference.

$$\Delta^2 y = \Delta(\Delta y)$$

or
$$\Delta^2 y(x) = \Delta y(x + h) - \Delta y(x)$$

Similarly, the third difference, $\Delta^3 y$, is the difference of the second difference:

$$\Delta^3 y = \Delta(\Delta^2 y) = \Delta(\Delta(\Delta y))$$

In general,
$$\Delta^n y = \Delta(\Delta^{n-1}(y))$$

The identity operator I is that operator which, applied to any function y, produces a new function Iy identical with y. That is, for any x,

$$Iy(x) = y(x)$$

The symbol Δ^0 is defined as the identity operator, that is,

$$\Delta^0 y = Iy = y$$

By using this last identity, we can state now that

$$\Delta^n y = \Delta(\Delta^{n-1} y)$$

is valid for every $n \geq 1$. (Consider $\Delta^1 = \Delta$.)

Example: Let $y(x) = x^2$.
$$\Delta y(x) = \Delta x^2 = (x + h)^2 - x^2$$
$$= x^2 + h^2 + 2xh - x^2$$
$$= h^2 + 2xh$$

Now
$$\Delta^2 y(x) = \Delta(\Delta y(x))$$
$$= \Delta(h^2 + 2xh)$$
$$= h^2 + 2(x + h)h - h^2 - 2xh$$
$$= 2h^2$$

and
$$\Delta^3 y(x) = \Delta(\Delta^2 y(x)) = 2h^2 - 2h^2 = 0$$

In general, if $y = x^2$, $\Delta^m y = 0$ for $m \geq 3$.

The Operator ε. If y is a given function and x a constant, then we define εy as the operator for which

$$\varepsilon y(x) = y(x + h)$$

So we can now write $\Delta y(x) = \varepsilon y(x) - y(x)$, and the following property is true:

$$\varepsilon^n y(x) = \varepsilon(\varepsilon^{n-1} y(x)) = y(x + nh)$$

where $\varepsilon^0 y(x) = y(x)$; $\varepsilon^1 = \varepsilon$.

Properties:
   1. $\Delta[cy(x)] = c\Delta y(x)$

2. $\Delta[y_1(x) + y_2(x)] = \Delta y_1(x) + \Delta y_2(x)$
3. If y is a polynomial of degree n, that is,

$$y(x) = a_0 + a_1 x + a_2 x^2 + \cdots + a_n x^n \qquad a_n \neq 0$$

then $\qquad \Delta^n y(x) = n! h^n a_n$

and $\qquad \Delta^p y(x) = 0 \qquad$ if $p > n$

4. If u and v are two functions, then

$$\Delta[u(x) \cdot v(x)] = \mathcal{E}u(x) \cdot \Delta v(x) + v(x) \cdot \Delta u(x)$$

Equivalence of Operations.　Two operations 0_1 and 0_2 are equivalent $(0_1 \equiv 0_2)$ if, when applied to a function, the functions $0_1 y$ and $0_2 y$ are equal.　We can say then that $\Delta \equiv \mathcal{E} - I$ because $\Delta y(x) = \mathcal{E} y(x) - I y(x)$.　Because the last identity can be written as $\mathcal{E} y(x) = \Delta y(x) + I y(x)$, we can say that

$$\mathcal{E} \equiv \Delta + I$$

We are assuming then that we can manipulate operators as we do algebraic quantities, which is indeed true.　With this new notation, we have

$$\Delta^0 \equiv I \qquad\qquad \mathcal{E}^0 \equiv I$$
$$\Delta^2 \equiv \mathcal{E}^2 - 2\mathcal{E} + I \qquad \Delta^3 \equiv \mathcal{E}^3 - 3\mathcal{E}^2 + 3\mathcal{E} - I$$

We define now $\mathcal{E}\Delta$ as the operator, so that

$$\mathcal{E}\Delta y(x) = \mathcal{E}[\Delta y(x)]$$

We can define $\Delta\mathcal{E}$ as $\Delta\mathcal{E} y(x) = \Delta[\mathcal{E} y(x)]$.　In general, the order in which we take two operators to form this product is not irrelevant, but for the case of the operators $\mathcal{E}$ and Δ, it is; so we have

$$\mathcal{E}\Delta y(x) = \Delta\mathcal{E} y(x) \qquad \text{or} \qquad \mathcal{E}\Delta \equiv \Delta\mathcal{E}$$

Another property of Δ and $\mathcal{E}$ is the following: If m and n are nonnegative numbers,

$$\Delta^m \Delta^n \equiv \Delta^n \Delta^m \equiv \Delta^{m+n}$$
$$\mathcal{E}^m \mathcal{E}^n \equiv \mathcal{E}^n \mathcal{E}^m \equiv \mathcal{E}^{m+n}$$

The Inverse Operator (Δ^{-1} and $\mathcal{E}^{-1}$).　If Y is a function whose first difference is y, then Y is called an indefinite sum of y and denoted by Δ^{-1}.　If

$$\Delta Y(x) = y(x)$$
then $\qquad\qquad \Delta^{-1} y(x) = Y(x)$

Property:　If Y_1 and Y_2 are indefinite sums of y_1 and y_2, respectively, and c_1 and c_2 are arbitrary constants, then

$$\Delta^{-1}(c_1 y_1 + c_2 y_2) = c_1 \, \Delta^{-1} y_1 + c_2 \, \Delta^{-1} y_2$$
$$= c_1 Y_1 + c_2 Y_2$$

In the same way as we defined Δ^{-1}, we can define $\mathcal{E}^{-1}$ as an operator such that if

$$\mathcal{E} Y(x) = y(x)$$
then $\qquad\qquad \mathcal{E}^{-1} y(x) = Y(x)$

We see that

$$\mathcal{E}^{-1} y(x) = y(x - h)$$

We can also define the operators $\Delta^{-2}, \Delta^{-3}, \ldots$ and $\mathcal{E}^{-2}, \mathcal{E}^{-3}, \ldots$.

Difference Equations.　An equation relating the value of a function y and one or more differences $\Delta y, \Delta^2 y, \ldots$ for each x value of some set of numbers S for which all these functions are defined is called a difference equation over the set S.

We are concerned, given a difference equation, with finding the function y for which the equation holds.

Example of a Difference Equation. $\Delta^2 y(x) + 2\Delta y(x) + y(x) = 0$, over the set of real numbers.

We are going to assume that our S set is either a finite or an infinite set of successive integers.   If this is the case, instead of using $y(x)$ for the values of the function y over the set S, we can use y_k, where k is a subindex which indicates which value of x we are using.   In any case, we must always specify the range of values of the x or of the index k.

Solutions of a Difference Equation.   Suppose we are given the following equation:

$$y_{k+1} - 2y_k = 0 \qquad k = 0, 1, 2, \ldots$$

What is meant by the solution of the difference equation?

The equation is a relation between the values of y at the points k and $k + 1$.   Is there a function y which makes this equation a true statement for every one of the k values over which the equation is defined?

The function $y_k = 2^k$, $k = 0, 1, 2, \ldots$ is such a function.   We can see that it satisfies the equation

$$y_{k+1} - 2y_k = 2^{k+1} - 2 \cdot 2^k = 0 \qquad k = 0, 1, 2, \ldots$$

The function y_k defined before is then said to be a solution to the difference equation: $y_{k+1} - 2y_k = 0$.   Note that it is a solution, but by no means the only one.   As a matter of fact, all functions of the form $y_k = c(2^k)$, where c is any constant, are solutions to the equation.   Any one of the solutions is called a particular solution, and the one containing the arbitrary constant c is called the general solution.

In general, a function y is a solution of a difference equation over a set S if the values of y reduce the difference equation to an identity over S.

Example: The function given by

$$y_k = 1 - \frac{2}{k} \qquad k = 1, 2, 3, \ldots$$

is a solution for the equation

$$(k + 1)y_{k+1} + ky_k = 2k - 3 \qquad k = 1, 2, \ldots$$

To prove it, we substitute in the previous equation.   We have

$$(k + 1)\left(1 - \frac{2}{k + 1}\right) + k\left(1 - \frac{2}{k}\right) = (k + 1) - 2 + k - 2$$

$$= k + 1 - 2 + k - 2 = 2k - 3$$

so the function y satisfies the equation.

Initial Conditions.   Given a difference equation with infinite solutions, we can determine which solution we want to choose by giving initial conditions that must be satisfied.   To have a uniquely determined solution, we need as many initial conditions as different constants appear in the general solution.

Example: Consider the following solution:

$$(*) \quad y_k = c(2^k) \qquad k = 0, 1, 2, \ldots$$

which satisfies the equation

$$y_{k+1} - 2y_k = 0 \qquad k = 0, 1, 2, \ldots$$

for every value of the constant c.

If we want to find a solution which satisfies the initial condition $y_0 = 3$, we find which one of all the possible $y_k = c2^k$ will satisfy our extra equation.

We see that $y_0 = c(2^0) = c$, replacing k by 0 in equation (*).   So if we choose $c = 3$, we have

$$y_k = 3(2^k)$$

which satisfies both the equation $y_{k+1} - 2y_k = 0$, $k = 0, 1, \ldots$, and the initial condition $y_0 = 3$.

Some Examples of Problems That May Be Solved Using Differences

Economic Dynamics.   One of the classical economic models gives national income Y as a function of consumption C and investment I.   We have

$$(*) \quad Y_t = C_t + I_t \qquad t = 0, 1, 2, \ldots$$

We assume consumption to vary linearly with Y, that is,

$$(**) \quad C_t = c + mY_t \qquad t = 0, 1, 2, \ldots$$

We have that $c \geq 0$, $0 < m < 1$.   We assume also that there exists a growth factor $r > 0$ such that

$$(***) \quad \Delta Y_t = Y_{t+1} - Y_t = rI_t$$

Using equations $(*)$, $(**)$, and $(***)$, we can state

$$Y_{t+1} - Y_t = rI_t = r(Y_t - C_t) = rY_t - r(c + mY_t)$$

or

$$Y_{t+1} = [1 + r(1 - m)]Y_t - rc \qquad t = 0, 1, 2, \ldots$$

which is a first order difference equation.   The solution to this equation is

$$Y_t = [1 + r(1 - m)]^t \left(Y_0 - \frac{c}{1 - m} \right) + \frac{c}{1 - m} \qquad t = 0, 1, 2, \ldots$$

where Y_0 is given as an initial condition.

Inventory.   Consumer goods are produced for sales and for maintaining inventory levels.   Assume, for simplicity, that they are produced only for sales.   Let u_t = number of units produced for sale in period t; let v_0 = net investment.   The total income y_0 produced in period t is equal to the total production of consumer goods plus net investment.

$$y_t = u_t + v_0 \qquad t = 0, 1, 2, \ldots$$

We assume that $u_t = \beta y_{t-1}$, $t = 1, 2, \ldots$, and β = marginal propensity to consumer.   We have the following difference equation:

$$y_t = \beta y_{t-1} + v_0 \qquad t = 2, 3, \ldots$$

which is a second order difference equation, whose solution is

$$y_t = \beta^t \left(y_0 - \frac{v_0}{1 - \beta} \right) + \frac{v_0}{1 - \beta}$$

where y_0 is given by the initial conditions.

LINEAR ALGEBRA AND CONVEXITY

Matrices.   A rectangular array of numbers is called a matrix.   The notation is as follows.   If

$$A = \|a_{ij}\| = \begin{bmatrix} a_{11} & \cdots & a_{1n} \\ \cdots & \cdots & \cdots \\ a_{m1} & \cdots & a_{mn} \end{bmatrix}$$

then A is an m by n matrix ($m \times n$).   The number a_{ij} is called an element of the matrix.

Operations and Properties

1. Two matrices A and B are equal if all corresponding elements are equal; that is, if $A = \|a_{ij}\|$ and $B = \|b_{ij}\|$, then $A = B$ if $a_{ij} = b_{ij}$ for all i, j.

2. Given matrices A and B, we define

$$C = A + B$$

as the matrix whose elements c_{ij} are given by

$$c_{ij} = a_{ij} + b_{ij}$$

Thus, two matrices cannot be added unless they have the same number of rows and the same number of columns.

3. $A + B = B + A$
4. $A + (B + C) = (A + B) + C = A + B + C$
5. Given a matrix A and a real number λ, then

$$\lambda A = \|\lambda a_{ij}\|$$

is the product of A and the real number λ (λ is often called a scalar).

6. $\lambda A = A\lambda$
7. The product AB of matrices A and B is defined only if the number of columns of A is equal to the number of rows of B. In that case, we define $C = AB$ as the matrix whose elements c_{ij} are

$$c_{ij} = \sum_{k=1}^{n} a_{ik} b_{kj} \qquad i = 1, \ldots, m \text{ and } j = 1, \ldots, r$$

where A is $m \times n$ and B is $n \times r$. C will be an $m \times r$ matrix. *Note:* BA is not necessarily defined when AB is defined.

8. $(AB)C = A(BC) = ABC$
$A(B + C) = AB + AC$
Note: In general, matrix multiplication is not commutative even when both AB and BA are defined.

We shall now define some special matrices.

Identity Matrix. A square matrix is of order n, having ones along the diagonal running from upper left to lower right, and zeros elsewhere, that is,

$$I = \begin{bmatrix} 1 & 0 & 0 & \cdots & 0 \\ 0 & 1 & 0 & \cdots & 0 \\ 0 & 0 & 1 & \cdots & 0 \\ \cdot & \cdot & \cdot & \cdots & \cdot \\ 0 & 0 & 0 & \cdots & 1 \end{bmatrix}$$

If we write it in element notation, we have

$$I = \|\delta_{ij}\|$$

where

$$\delta_{ij} = \begin{cases} 1 & i = j \\ 0 & i \neq j \end{cases}$$

This symbol δ_{ij} is called the Kronecker delta.

Properties:
1. $I^n = I$
2. If A is $m \times n$, then $I_m A = A I_n = A$.
3. $S = \|\lambda \delta_{ij}\| = \lambda I$ is a scalar matrix.
4. $D = \|\lambda_i \delta_{ij}\|$ is a diagonal matrix.

Null Matrix. A null matrix is a matrix whose elements are all zeros. It does not have to be a square matrix.

$$0 = \begin{bmatrix} 0 & \cdots & 0 \\ 0 & \cdots & 0 \\ \cdot & \cdots & \cdot \\ 0 & \cdots & 0 \end{bmatrix}$$

Properties: Whenever the operations are defined, we have
1. $A + 0 = A = 0 + A$
2. $A - A = 0$
3. $A0 = 0$
4. $0A = 0$

Note: The matrix equation $AB = 0$ does not imply that either A or B are equal to 0.

Example:

$$\begin{bmatrix} 1 & 2 \\ 0 & 0 \end{bmatrix} \begin{bmatrix} -2 & 0 \\ 1 & 0 \end{bmatrix} = \begin{bmatrix} 0 & 0 \\ 0 & 0 \end{bmatrix}$$

Transpose Matrix. The transpose of a matrix $A = \|a_{ij}\|$ is the matrix A', which has for columns the rows of A and for rows the columns of A. That is,

$$A' = \|a_{ij}'\| \qquad \text{where } a_{ij}' = a_{ji}$$

Properties:
1. $(A + B)' = A' + B'$
2. $(AB)' = B'A'$
3. $I' = I$
4. $(A')' = A$

Symmetric Matrix. A symmetric matrix is a matrix A such that $A' = A$. A symmetric matrix must be square, and $a_{ij} = a_{ji}$ for all i and j.

Determinants and Inverse Matrix. Given any square matrix A, we can find a number which is called its determinant $|A|$. To give a useful method for computing the determinant of a square $n \times n$ matrix, we shall first give a rule to find the determinant for a 2×2 matrix.

Rule 1:

$$\text{If } A = \begin{bmatrix} a_{11} & a_{12} \\ a_{21} & a_{22} \end{bmatrix} \text{ then } |A| = a_{11}a_{22} - a_{21}a_{12}$$

We define now the cofactor A_{ij} of an element a_{ij} of the matrix A as the determinant of the matrix formed by the rows and columns of A, except for row i and column j, multiplied by $(-1)^{i+j}$. We have the following rule.

Rule 2: To find the determinant of a square 3×3 matrix A,

$$A = \begin{bmatrix} a_{11} & a_{12} & a_{13} \\ a_{21} & a_{22} & a_{23} \\ a_{31} & a_{32} & a_{33} \end{bmatrix}$$

We compute the cofactors A_{ij} of the elements of any row (say i), multiply them by the corresponding a_{ij}, and add. For example, using row 1,

$$|A| = a_{11}A_{11} + a_{12}A_{12} + a_{13}A_{13}$$

$$= a_{11} \begin{bmatrix} a_{22} & a_{23} \\ a_{32} & a_{33} \end{bmatrix} (-1)^{1+1} + a_{12} \begin{bmatrix} a_{21} & a_{23} \\ a_{31} & a_{33} \end{bmatrix} (-1)^{1+2} + a_{13} \begin{bmatrix} a_{21} & a_{22} \\ a_{31} & a_{32} \end{bmatrix} (-1)^{1+3}$$

This is called an expansion in cofactors of row 1.

Note: It is possible to expand in cofactors of a column instead of a row.

Now we are able to give a rule to compute the determinant of any square matrix. Let A be an $n \times n$ matrix. Then, to find $|A|$, we expand in cofactors of any row, say row i.

$$|A| = a_{i1}A_{i1} + \cdots + a_{in}A_{in}$$

where $A_{i1}, \ldots, A_{in}$ are the determinants of order $n - 1$, which we compute by the same method. Because we are reducing the size of the matrices at each step, we shall arrive finally to a number of 2×2 matrices, whose determinants are easy to compute using Rule 1.

Definition: The adjoint A^+ of matrix A is the transpose of the matrix obtained by replacing each element a_{ij} by its cofactor A_{ij}.

Now we can define the inverse A^{-1} of a square matrix A as

$$A^{-1} = \frac{1}{|A|} A^+$$

A matrix A such that $|A| \neq 0$ is called nonsingular, and if $|A| = 0$, singular. Only nonsingular matrices have inverses, and every nonsingular matrix has an inverse.

Vectors. We can define an n vector either as a row matrix with n elements (or column matrix) or as a point in the Euclidean n space. Just as (a_1, a_2, a_3) can be considered a point in a three-dimensional space, $(a_1, a_2, \ldots, a_n)$ may be taken as a point in an n-dimensional space. We shall use row or column vectors according to notational

convenience. A vector $(a_1 \cdots a_n)$ is a row vector. A vector $\begin{bmatrix} a_1 \\ \cdot \\ \cdot \\ a_n \end{bmatrix}$ is a column

vector. We will use lowercase letters with a bar over them to indicate vectors.

Definition 1: A unit vector $\bar{e}_i$ is a vector with 1 as the value of its ith component and all other components equal to 0.

Definition 2: The null vector 0 is a vector whose components are zero.

Definition 3: The identity vector has all its components equal to 1.

We say that $\bar{a} \geq \bar{b}$, where $\bar{a}$ and $\bar{b}$ both have n components, if $a_i \geq b_i$ for all i. In the same way, $\bar{a} \leq \bar{b}$, $\bar{a} < \bar{b}$, $\bar{a} > \bar{b}$, $\bar{a} = \bar{b}$ are true if $a_i \leq b_i$, $a_i < b_i$, $a_i > b_i$, or $a_i = b_i$, for all i, respectively.

Definition 4: The scalar product of $\bar{a}$ and $\bar{b}$, both n component vectors, is $\displaystyle\sum_{i=1}^{n} a_i b_i$.

If $\bar{a}$ and $\bar{b}$ are both row vectors, the scalar product is denoted by $\bar{a}\bar{b}'$; if both are columns, by $\bar{a}'\bar{b}$; if $\bar{a}$ is row and $\bar{b}$ is column, by $\bar{a}\bar{b}$. Note that any vector $\bar{a}$ can be written in terms of the unit vectors $\bar{e}_i$ as

$$\bar{a} = a_1\bar{e}_1 + \cdots + a_n\bar{e}_n$$

We shall think, then, of the unit vectors as the unit coordinates of our n-dimensional space. In this coordinate system, a_i is the ith coordinate of the point $\bar{a}$.

Definition 5: n-dimensional Euclidean space (E^n) is a collection of vectors (points) $\bar{a} = (a_1 \cdots a_n)$, for which addition and multiplication by a scalar are defined by the rules of matrix operations. Associated with any two elements of the space is a non-negative number, called the distance between them, where, if

$$\bar{a} = (a_1 \cdots a_n) \qquad \text{and} \qquad \bar{b} = (b_1 \cdots b_n)$$

then

$$\text{Distance} = |\bar{a} - \bar{b}| = [(\bar{a} - \bar{b})'(\bar{a} - \bar{b})]^{1/2} = \left[\sum_{i=1}^{n} (a_i - b_i)^2 \right]^{1/2}$$

Linear Dependence

Definition 1: A vector $\bar{a}$ from E^n is a linear combination of vectors $\bar{a}_i, \ldots, \bar{a}_n$ from E^n if $\bar{a}$ can be written as

$$\bar{a} = \lambda_1\bar{a}_1 + \cdots + \lambda_n\bar{a}_n$$

for some set of scalars $\{\lambda_i\}$.

Definition 2: A set of vectors $\bar{a}_1, \ldots, \bar{a}_m$ of E^n is linearly dependent if we can find scalars λ_i not all zero such that

$$\lambda_1\bar{a}_1 + \cdots + \lambda_m\bar{a}_m = 0$$

If the only set of scalars for which this equality holds is the set $\lambda_1 = \lambda_2 = \cdots = \lambda_m = 0$, then the vectors are linearly independent.

Basis. A set of vectors $\bar{a}_1, \ldots, \bar{a}_n$ is a basis for E^n if (1) all the vectors are linearly independent and (2) every vector in E^n can be written as a linear combination of $\bar{a}_1, \ldots, \bar{a}_n$.

Note: This representation of a vector in terms of the basis is unique.

Vector Spaces, Subspaces, Rank

Definition 1: A vector space is a collection of vectors which is closed under the operations of addition and multiplication by a scalar. It is denoted by V_n.

Example: E^n is a vector space.

Definition 2: A subspace S_n of V_n is defined as a subset of V_n which is itself a vector space.

Example: E^3, the three-dimensional Euclidean space, is a vector space. E^2, the plane, is a subset of E^3, and a vector space; thus it is a subspace of E^3.

The expression $\lambda x_1 + (1 - \lambda)x_2$, $0 \le \lambda \le 1$ is referred to as a convex combination of the points x_1 and x_2.

Examples:
1. A circle in E^2 is a convex set.
2. Regular polygons in E^2 are convex sets.
3. The set drawn in the following figure is not a convex set.

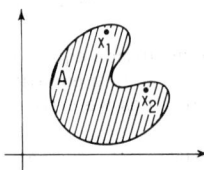

The line segment joining x_1 and x_2 is not included in A.

Definition 4:   A point x is an extreme point of a convex set if there do not exist any two points x_1, x_2, with $x_1 \ne x_2$ in the set, such that $x = \lambda x_1 + (1 - \lambda)x_2$ for some λ between 0 and 1, $\lambda \ne 0$ and $\lambda \ne 1$.   Intuitively, an extreme point cannot be on a line segment joining two other points that belong to the set.

An extreme point is a boundary point of a convex set, but not all boundary points need be extremes. In example 1 above, all boundary points are extremes. In example 2, the vertices of the polygons are extremes, but any point on the boundary between two vertices is not an extreme point.

Properties:
1. The intersection of two convex sets is convex.
2. Every vector space is convex. (Remember that a vector space is a set of all scalar combinations of vectors, and as such, it is convex.)

Definition 5:   The set of all convex combinations of a finite number of points is called the convex polyhedron spanned by these points.

Definition 6:   A cone C is a set of points with the following property: if x belongs to C, then μx belongs to C for all $\mu \ge 0$.

Definition 7:   A convex cone is a cone that is also a convex set.

Examples:
1.

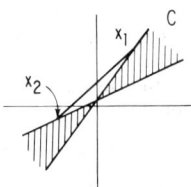

The shaded area is a cone, but is not a convex cone (we can see that the line segment between x_1 and x_2 is not in C).

2.

The shaded area is a cone, and is also a convex cone.

In the same way that the theory of basic solutions is the fundamental weapon used to develop the theory of the simplex method, the theory of convex sets, convex cones, and convex polyhedra is the basis for the geometrical interpretation of the same simplex methods.

It can be proved that there is a very well defined relation between systems of simultaneous linear equations (linear programming problems) and their basic solutions and convex polyhedral sets (the graphic representation of a system of simultaneous linear equations) and their extreme points.

The basic theory of the simplex algorithm consists in using the fact that all solutions to a system of simultaneous linear equations (linear program) are graphically the extreme points of a convex set that represents these equations, and thus only extreme points need be considered. But extreme points are the graphic equivalents of basic solutions, and then, of course, the problem of finding an optimal solution is simplified by the fact that only basic solutions need be considered as candidates.

Convex Functions

Definition 1: A function $f(x)$ is said to be convex over a convex set X in E^n if, for any two points x_1 and x_2 in X, and for all λ, $0 \leq \lambda \leq 1$:

$$f(\lambda x_1 + (1 - \lambda)x_2) \leq \lambda f(x_1) + (1 - \lambda)f(x_2)$$

Definition 2: A function $f(x)$ is concave over a convex set X in E^n if, given any two points x_1 and x_2 in X, and λ, $0 \leq \lambda \leq 1$:

$$f(\lambda x_1 + (1 - \lambda)x_2) \geq \lambda f(x_1) + (1 - \lambda)f(x_2)$$

Examples:
1.

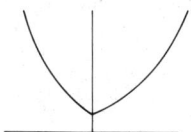

The function shown in the above figure is convex.

2.

The function whose graph is shown above is concave.

Graphically, a function is convex if the segment joining any two points on the curve lies entirely on or above the curve; it is concave if that same segment lies entirely on or below the curve.

Properties:
1. If $f(x)$ is convex then $-f(x)$ is concave.
2. A linear function is both convex and concave.

Definition 3: $f(x)$ is strictly convex over convex set X if, given any two points x_1 and x_2 in X, and λ, $0 < \lambda < 1$:

$$f(\lambda x_1 + (1 - \lambda)x_2) < \lambda f(x_1) + (1 - \lambda)f(x_2)$$

and is strictly concave if, with the same hypotheses,

$$f(\lambda x_1 + (1 - \lambda)x_2) > \lambda f(x_1) + (1 - \lambda)f(x_2)$$

Example:

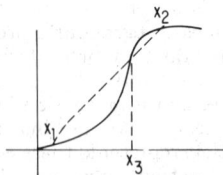

The function whose graph is shown in the above figure is neither convex nor concave. The segment joining x_1 and x_2 is above the curve to the left of x_3 and below the curve to the right of x_3.

This example shows that not all functions are convex or concave. In fact, most functions are neither convex nor concave. We ask, if the number of functions which have either of these properties is such a small number compared with the total number of functions we can find—both in theory and in real life—then why bother mentioning them? The answer to this question is given by the following.

Property: Let $f(x)$ be a convex (concave) function over a closed set X in E^n. Then, any local minimum (maximum) of $f(x)$ in X is also the absolute or global minimum (maximum).

On page 10-11, we talked about extremes, and we gave a method for finding minima (or maxima) of functions in E^2. We also mentioned the fact that if a function has more than one minimum (maximum), the method for finding the optimum global extremes is to compute the value of the function at each of these local extremes and select the desired value.

It is clear that if we know a function is convex (or concave), we can apply the methods given to get a minimum (or maximum) and then be absolutely sure that this min (or max) is the optimal solution to our problem.

As an example of the application of this, we can say that most of the theory of stochastic inventory is based on convexity, or properties associated with convexity, of the cost function.

GRAPHS

To illustrate a relation between certain situations, we can use a geometric representation, called a graph.

A graph consists of points, called nodes or vertices, and line segments, called arcs or edges.

Example:

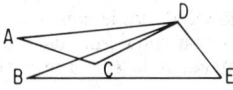

Many special graphs are used in graph theory:

1. A graph with no arcs is called a null graph.
2. A graph such that all the nodes are joined by one arc is called a complete graph.
3. If we have a graph G, with some arcs joining some nodes, then G', the complement of G, consists of the same nodes as G and all the arcs joining them that did not appear in G. That is, G and G' together would form a complete graph.
4. Two graphs G_1 and G_2 are isomorphic if, whenever A_1 and B_1 are joined by an arc in G_1, there are corresponding nodes A_2, B_2 in G_2, which are also joined by an arc in G_2.

Planar Graphs.   Planar graphs are graphs that can be drawn in a plane in such a way that the edges have no intersections other than the vertices.

Example: Maps of roads, if no bridges appear.

Number of Edges (or Arcs) in a Graph.   Where several arcs connect two vertices A and B, we say that the graph has multiple arcs.   In general, instead of drawing all the arcs, we can draw a single one and assign a number of multiplicity to it to indicate how many times the arc should be repeated.

At every nonisolated node A in a graph, there will be some arcs having A as an end point.   These arcs are incident to A.   The number of such arcs is denoted by $\rho(A)$ and called the local degree at A.

Example:

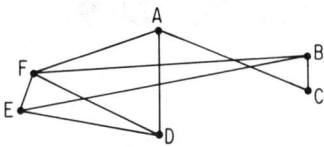

$$\rho(A) = \rho(B) = \rho(D) = \rho(E) = 3$$
$$\rho(F) = 4 \qquad \rho(C) = 2$$

The number of arcs in the graph is obtained as the sum of the number of arcs at each node, divided by 2 (because they are connected at both ends), that is, in this case,

$$\tfrac{1}{2}[\rho(A) + \rho(B) + \rho(C) + \rho(D) + \rho(E) + \rho(F)] = 9$$

In general, if a graph G has nodes $A_1, \ldots, A_n$ with local degrees $\rho(A_1), \ldots, \rho(A_n)$, then N = number of arcs in G is given by

$$N = \frac{1}{2}\left[\sum_{i=1}^{n} \rho(A_i)\right]$$

As a consequence, $\sum_{i=1}^{n} \rho(A_i)$ = even number.

Odd and Even Nodes.   The odd nodes A' are those for which $\rho(A')$ is an odd number. When $\rho(A'')$ is even, we say A'' is an even node.

Example: In the graph given above, A, B, D, and E are odd; F and C are even.

Property: Any graph has an even number of odd vertices (nodes).   It is easy to prove this property.   Because the sum of all the local degrees is even, then the sum of all the odd local degrees has to be even also.   Hence, there must be an even number of odd local degrees.

A graph where all local degrees are the same (say r) is called regular of degree r. The number of its arcs is then $N = \tfrac{1}{2}nr$, where n is the number of nodes.

Connected Graphs.   When each node of a graph is connected to every other node by a sequence of arcs, the graph is connected.

A circuit is a sequence of arcs such that no node is touched more than once, and the ending node is the same as the starting node.

A tree is a connected graph with no circuits.

A forest is a graph without circuits which has connected components.

Properties:

1. A tree with n nodes has $n - 1$ arcs.
2. A forest with k components and n nodes has $n - k$ arcs.

A graph is bipartite when the set of nodes is decomposed into two separate parts M and P such that there are arcs only between M and P.

Directed Graphs.   A graph G where a direction is indicated for every arc is called a directed graph.

Chapter **2**

Probability and Stochastic Processes

RALPH L. DISNEY

**Professor, Department of Industrial Engineering,
The University of Michigan, Ann Arbor, Michigan**

The purpose of this chapter is to present some concepts in the areas of probability and random processes which are of practical value to the industrial engineer.

From the earliest studies of industrial processes, it has been apparent that random behavior is more often than not the milieu within which the industrial engineer must work. The point is most clearly seen when one views the almost immediate, favorable response of industrial engineering to the Shewhart studies in statistical quality control. Hardly an issue of the journals in fields of direct concern to the industrial engineer appear without several articles on topics such as statistical quality control, queuing applications, Markov chain applications, probabilistic inventory and material handling applications, and time measurement applications. It has become clear that probability theory and random process theory are part of the foundation of industrial engineering.

PROBABILITY THEORY

Probability theory is concerned with a concept called a random experiment. It is supposed that the outcome of the experiment is not known with certainty prior to performing the experiment.

The set of all possible experimental outcomes is called the sample space S of the experiment. Any one possible outcome is a member of the sample space and is called a sample point s.

If the number of sample points in S is countable (finite or countable infinite), the sample space is said to be countable or discrete. If the number of points in S is noncountable, S is said to be continuous.

For most engineering studies, any subset of S is an event. Hence, for our purposes, an event is a subset A of S. It is supposed that S is an event and the null set $\emptyset$ consisting of no sample points is an event, respectively called the certain event and the impossible event. The set of all events is called an event space.

Algebra of Events

1. If A_1 and A_2 are two events, then $A_1 \cup A_2$ is an event, and in general, finite unions, $A_1 \cup A_2 \cup A_3 \cup \cdots \cup A_n$ are events. It is supposed that all countable unions are events. Hence

$$\bigcup_{i=1}^{\infty} A_i$$

is an event. The event $A_1 \cup A_2$ is the set of points contained in A_1 or A_2 or both.

2. If A_1 and A_2 are events, then $A_1 \cap A_2$ is an event and

$$\bigcap_{i=1}^{\infty} A_i$$

is an event. The event $A_1 \cap A_2$ is the set of points contained in both A_1 and A_2. If $A_1 \cap A_2 = \emptyset$, A_1 and A_2 are said to be mutually exclusive events.

3. Because $A \subset S$, $\bar{A} = S - A$ is an event, called the complementary event of A. $\bar{A}$ is the set of sample points not in A.

Probabilities.

To each event in the event space, one supposes that there can be attached a real number $\Pr[A]$ called the probability of the event A. These probabilities are determined by the axioms of probability:

1. $0 \leq \Pr[A] \leq 1$
2. $\Pr[S] = 1$
3. $\Pr[A \cup B] = \Pr[A] + \Pr[B]$ if $A \cap B = \emptyset$

Thus probabilities are nonnegative real numbers not greater than 1. The probability of the set S, the sample space, is 1. Hence the probabilities are normed. From axiom 3, one sees that the probability of A or B is the sum of the probabilities of the separate events A and B if A and B are mutually exclusive.

In general, one needs one more axiom which extends axiom 3:

4. $\Pr\left[\bigcup_{i=1}^{\infty} A_i\right] = \sum_{i=1}^{\infty} \Pr[A_i]; \quad A_i \cap A_j = \emptyset$ for $i \neq j$

Dependent and Independent Events.

If $A_i \cap A_j \neq \emptyset$ for $i \neq j$, one is interested in $\Pr[A_i \cap A_j]$, or for economy of notation, $\Pr[A_i A_j]$. This probability is called the joint probability of the events A_i and A_j. The events A_i and A_j are independent events if and only if

$$\Pr[A_i A_j] = \Pr[A_i]\,\Pr[A_j] \tag{1-1}$$

If A_i and A_j are not independent, then

$$\Pr[A_i A_j] = \Pr[A_i | A_j]\,\Pr[A_j] \tag{1-2}$$

where $\Pr[A_i | A_j]$ is the conditional probability of A_i, given A_j. If $A_i A_j \neq \emptyset$, then axiom 3 becomes

$$\Pr[A_i \cup A_j] = \Pr[A_i] + \Pr[A_j] - \Pr[A_i A_j]$$

Equation (1-1) generalizes immediately to many events. If

$$\Pr[A_1 A_2 \cdots A_n] = \Pr[A_1]\,\Pr[A_2] \cdots \Pr[A_n] \tag{1-3}$$

then the events $A_1, A_2, \ldots, A_n$ are said to be mutually independent.

A collection of events may be independent in the sense of equation (1-1) but not in the sense of equation (1-3), as the following example illustrates. If the events are independent in the sense of equation (1-3), they are independent in the sense of equation (1-1).

Example 1. Two coins are tossed, and one notes the occurrence of heads or tails on each toss. The sample space S for this simple experiment is

$$S = \{HH, HT, TH, TT\}$$

Let A_1 be the event "the 1st coin tossed is a head."
Let A_2 be the event "the 2nd coin tossed is a head."
Let A_3 be the event "the coins match."
One sees that

$$A_1 = \{HH, HT\}$$
$$A_2 = \{HH, TH\}$$

The event

$$A_1 \cup A_2 = \{HH, HT, TH\}$$

The event

$$A_1 \cap A_2 = \{HH\}$$

Hence, A_1 and A_2 are not mutually exclusive events. Let $E_1 = \{HH\}$, $E_2 = \{HT\}$, $E_3 = \{TH\}$, $E_4 = \{TT\}$. All pairs of events E_i are mutually exclusive. For obvious reasons, let $\Pr[E_j] = \frac{1}{4}$ for $j = 1, 2, 3, 4$. Notice that

1. $0 \leq \Pr[E_j] \leq 1$
2. $\Pr[E_i \cup E_j] = \frac{1}{2}$ for every pair E_i and E_j
3. $S = E_1 \cup E_2 \cup E_3 \cup E_4$

hence

$$\Pr[S] = \Pr[E_1 \cup E_2 \cup E_3 \cup E_4]$$

and because the events are mutually exclusive,

$$\Pr[S] = \sum_{i=1} \Pr[E_i] = 1$$

Hence, the assignment given for the $\Pr[E_j]$ satisfies the axioms of probabilities.

$$A_1 = E_1 \cup E_2$$
$$A_2 = E_1 \cup E_3$$
$$A_3 = E_1 \cup E_4$$

Also, because the E_j are mutually exclusive,

$$\Pr[A_1] = \Pr[E_1] + \Pr[E_2] = \frac{1}{2}$$
$$\Pr[A_2] = \Pr[E_1] + \Pr[E_3] = \frac{1}{2}$$
$$\Pr[A_3] = \Pr[E_1] + \Pr[E_4] = \frac{1}{2}$$

But

$$A_1 \cap A_2 = (E_1 \cup E_2) \cap (E_1 \cup E_3) = E_1$$
$$A_1 \cap A_3 = (E_1 \cup E_2) \cap (E_1 \cup E_4) = E_1$$
$$A_2 \cap A_3 = (E_1 \cup E_3) \cap (E_1 \cup E_4) = E_1$$
$$A_1 \cap A_2 \cap A_3 = (E_1 \cap E_2) \cup (E_1 \cap E_3) \cup (E_1 \cap E_4) = E_1$$

and

$$\Pr[A_1 \cap A_2] = \Pr[E_1] = \frac{1}{4} = \Pr[A_1] \Pr[A_2]$$
$$\Pr[A_1 \cap A_3] = \Pr[E_1] = \frac{1}{4} = \Pr[A_1] \Pr[A_3]$$
$$\Pr[A_2 \cap A_3] = \Pr[E_1] = \frac{1}{4} = \Pr[A_2] \Pr[A_3]$$

One sees immediately that A_1, A_2, A_3 are independent when taken in any pairs. The events are pairwise independent. But

$$\Pr[A_1 \cap A_2 \cap A_3] = \Pr[E_1] = \frac{1}{4} \neq \Pr[A_1] \Pr[A_2] \Pr[A_3] = \frac{1}{8}$$

Hence, even though the events are independent in pairs, they are not independent when considered three at a time.

Bayes' Theorem. Suppose one has events A_1, A_2, . . . , A_n that are mutually exclusive and

$$A_1 \cup A_2 \cup A_3 \cdots \cup A_n = S$$

(The events are "exhaustive.") Suppose one has another event, B. Bayes' theorem can then be stated as

$$\Pr[A_j|B] = \frac{\Pr[B|A_j]\,\Pr[A_j]}{\Sigma\,\Pr[B|A_j]\,\Pr[A_j]}$$

Example 2. Two machine centers are manufacturing identical parts. Center 1 has been running about 5 percent defective parts, and center 2 has been running about 1 percent defective parts. Seventy-five percent of the total production has been coming from center 2. (Perhaps center 1 is an old machine used only for standby operation.) At final inspection, a part is found to be defective. What is the probability that it came from center 1?

Let B be the event "the part is defective" and let A_j be the event "the part came from center j for $j = 1, 2$." Then $\Pr[A_1|B]$ is the probability sought. Using Bayes' theorem, one finds

$$
\begin{aligned}
\Pr[A_1|B] &= \frac{\Pr[B|A_1]\,\Pr[A_1]}{\Pr[B|A_1]\,\Pr[A_1] + \Pr[B|A_2]\,\Pr[A_2]} \\
&= \frac{0.05 \times 0.25}{0.05 \times 0.25 + 0.01 \times 0.75} \\
&= \frac{0.0125}{0.0125 + 0.0075} \\
&= \frac{0.0125}{0.0200} \\
&= 0.625
\end{aligned}
$$

Random Variables. In many applications, the outcomes of an experiment are nonnumeric. It is convenient to assign real numbers to sample points. One calls a function $X(s)$—a real valued function which maps the sample space into the real line—a random variable. Hence, a random variable is a function whose domain is the sample space and whose range is some subset of the real line.

If the range of X is continuous, we will say that X is a continuous random variable. If the range of X is a set of isolated points (usually the integers), we will say that X is a discrete random variable.

Distribution Functions. From the definition of a random variable, sample points are assigned real numbers. Collections of sample points are subsets of S and hence are events. Events have probabilities. Hence, one says that the random variable induces probabilities on the real line as follows. Let x be the set of points s such that $X(s) \leq x$ for a real number x. A is an event and has some probability assigned to it. We denote this probability by the following expression:

$$\Pr[A] = \Pr[X \leq x]$$

Notice that we have suppressed the variable s in this shorthand notation. $\Pr[X \leq x]$ is a function whose domain is the real line and whose range is the unit interval. We denote this function by $F(x)$ and call it the distribution function for the random variable X. Where no confusion is likely to arise, we use the abbreviated notation $F(x)$ for this function. The Poisson distribution (cumulative) function, shown in Table 2-1, has useful applications.

Some Properties of Distribution Functions
1. $F(x) = \Pr[X \leq x]$
2. $0 \leq F(x) \leq 1$
3. $F(-\infty) = 0$

TABLE 2-1.   Cumulative Poisson Probabilities (Probability of x or Less)

x \ λ	0.05	0.10	0.15	0.20	0.25	0.30	0.35	0.40	0.45	0.50	0.55	0.60	0.65	0.70	0.75	0.80	0.85	0.90	0.95	1.0	1.5	2.0	2.5	3.0	3.5	4.0	4.5	5.0	5.5	6.0	6.5	7.0	7.5	8.0	8.5	9.0	9.5	10.0
0	951	904	860	818	778	740	704	670	637	606	576	548	522	496	472	449	427	406	386	367	223	135	082	049	030	018	011	006	004	002	001	000	000	000	000	000	000	000
1	998	995	989	982	973	963	951	938	924	909	894	878	861	844	826	808	790	772	754	735	557	406	287	199	135	091	061	040	026	017	011	007	004	003	001	001	000	000
2	999	999	999	998	997	996	994	992	989	985	981	976	971	965	959	952	945	937	928	919	808	676	543	423	320	238	173	124	088	061	043	029	020	013	009	006	004	002
3				999	999	999	999	999	998	998	997	996	995	994	992	990	988	986	983	981	934	857	757	647	536	433	342	265	201	151	111	081	059	042	030	021	014	010
4									999	999	999	999	999	999	998	998	998	997	997	996	981	947	891	815	725	628	532	440	357	285	223	172	132	099	074	054	040	029
5															999	999	999	999	999	999	995	983	957	916	857	785	702	615	528	445	368	300	241	191	149	115	088	067
6																					999	995	985	966	934	889	831	762	686	606	526	449	378	313	256	206	164	130
7																						998	995	988	973	948	913	866	809	743	672	598	524	452	385	323	268	220
8																						999	998	996	990	978	959	931	894	847	791	729	661	592	523	455	391	332
9																							999	998	996	991	982	968	946	916	877	830	776	716	652	587	521	457
10																								999	998	997	993	986	974	957	932	901	862	815	763	705	645	583
11																									999	999	997	994	989	979	965	946	920	888	848	803	751	696
12																											999	997	995	991	983	972	957	936	909	875	836	791
13																												999	998	996	992	987	978	965	948	926	898	864
14																													999	998	996	994	989	982	972	958	939	916
15																														999	998	997	995	991	986	977	966	951
16																																999	999	998	996	993	988	982	972
17																																		999	998	997	994	991	985
18																																			999	998	997	995	992
19																																				999	998	998	996
20																																					999	999	998

4. $F(+\infty) = 1$
5. If $x_1 < x_2$ then $F(x_1) \leq F(x_2)$
6. $\lim_{x \to a^+} F(x) = F(a)$

For many applied problems, $F(x)$ is either a differentiable function of x for all x or $F(x)$ is a step function with jumps at the integers. If $F(x)$ is differentiable, then let

$$dF(x) = f(x)\, dx$$

$f(x)$ is called the probability density function of X. Some useful probability density functions are shown in Table 2-2. In terms of probabilities, it is taken as

$$f(x)\, dx = \Pr[x < X \leq x + dx]$$

Note that $f(x)$ itself is not a probability. If $F(x)$ is a step function with jumps at the integers, then

$$F(x + 1) - F(x) = p_{x+1}$$

is called a probability density function or a probability mass function. In terms of probabilities

$$p_x = \Pr[X = x]$$

for x, an integer. Because both $f(x)$ and p_x are defined in terms of probabilities, one has

1. $0 \leq f(x)$

2. $\displaystyle\int_{-\infty}^{\infty} f(x)\, dx = 1$

and similarly for p_x with the integral replaced by a sum over all x. In addition, $p_x \leq 1$.

TABLE 2-2.   A Short Table of Useful Probability Density Functions

Name	Density function	
Bernoulli	$p_j = \begin{cases} q & \text{if } j = 0 \\ p & \text{if } j = 1 \\ 0 & \text{otherwise} \end{cases}$ $p + q = 1$	
Binomial	$p_j = \begin{cases} \binom{n}{j} p^j q^{n-i} & \text{if } j = 0,1,2, \ldots, n \\ 0 & \text{otherwise} \end{cases}$ $p + q = 1$	
Geometric	$p_j = pq^i \qquad j = 1,2, \ldots$ $p + q = 1$	
Poisson	$p_j = \dfrac{\lambda^j e^{-\lambda}}{j!} \qquad j = 0,1,2, \ldots,$ and $\lambda > 0$	
Uniform	$p_j = \dfrac{1}{a + 1} \qquad j = 0,1,2, \ldots, a$	
Normal	$f(x) = \dfrac{1}{\sqrt{2\pi}\,\sigma} \exp\left[-\dfrac{(x - \mu)^2}{2\sigma^2} \right]$	$-\infty < x < \infty$
Gamma	$f(x) = \dfrac{a^n}{\Gamma(n)} x^{n-1} e^{-ax}$	$0 \leq x < \infty$
Exponential	$f(x) = ae^{-ax}$	$0 \leq x < \infty$
Beta	$f(x) = \dfrac{\Gamma(n + m)}{\Gamma(n)\Gamma(m)} x^{n-1}(1 - x)^{m-1}$	$0 \leq x \leq 1$
Weibull	$f(x) = \dfrac{b}{\theta - x_o} \left(\dfrac{x - x_o}{\theta - x_o} \right)^{b-1} \exp\left[-\dfrac{x - x_o}{\theta - x_o} \right]^b$	$x_0 \leq x < \infty$

TABLE 2-3.   Some Special Expectations*

$g(X)$	Name	Formula
X	Mean value or 1st moment about 0	$\mu = E[X] = \int x f(x)\, dx$
X^n	nth moment about 0	$\mu_n' = E[X^n] = \int x^n f(x)\, dx$
$[X - E(X)]^2$	Variance $V(X)$	$\mu_2 = E(X - E(X))^2$ $= \int (x - \mu)^2 f(x)\, dx$
$[X - E(X)]^n$	nth central moment	$\mu_n = E[(X - E(X))^n]$ $= \int (x - \mu)^n f(x)\, dx$
$aX + b$ a,b constants		$E(aX + b) = \int (ax + b) f(x)\, dy$ $= aE(X) + b$ $= b \quad \text{if } a = 0$

* In this table, we have presented only the formulas for the case of a continuous random variable. The corresponding formulas for a discrete random variable are apparent from formula (1-4).

Expectations.   If g is a function of a random variable X whose probability density function is $f(x)$ or p_x, then it can be shown that the expected value of the function is given by

$$E(g(X)) = \int_{-\infty}^{\infty} g(x) f(x)\, dx$$

if X is continuous, or

$$E(g(X)) = \sum_x g(x) p_x \qquad (1\text{-}4)$$

if X is discrete.   We suppose the function is absolutely integrable.

Table 2-4 gives the means and variances for some of the distributions given in Table 2-2.

TABLE 2-4.   Means and Variances for Some Distributions

Distribution	Mean	Variance
Bernoulli	q	pq
Binomial	np	npq
Geometric	q/p	q/p^2
Poisson	λ	λ
Uniform	$a/2$	$\dfrac{a(a+2)}{12}$
Normal	μ	σ^2
Gamma	n/a	n/a^2
Exponential	$1/a$	$1/a^2$

Two Other Useful Expectations

1. If X is a discrete random variable, one defines

$$E(z^X) = \sum_{i=0}^{\infty} z^i p_i$$

as the probability generating function.

2. A function used extensively in some work is the exponential form

$$E(e^{izX}) = \int e^{izx} f(x) \, dx$$

if X is continuous, or

$$E(e^{izX}) = \sum_{j=0}^{\infty} e^{izi} p_j$$

if X is discrete. Throughout, $i = \sqrt{-1}$. To abbreviate notation, let us agree to put

$$\phi(z) = E(e^{izX})$$

in either case. This expectation is called the characteristic function. Some properties of $\phi(x)$ that are useful in probability are:

1. For any purely discrete or purely continuous random variable, $\phi(z)$ exists.
2. $\phi(0) = 1$
3. $\dfrac{d^n \phi(z)}{i^n dz^n}\bigg|_{z=0} = E(X^n)$
4. If $\phi_X(z)$ and $\phi_Y(z)$ are the characteristic functions for independent random variables X and Y and if $Z = X + Y$, then

$$\phi_Z(z) = \phi_X(z)\phi_Y(z)$$

5. $f(x)$ or p_j can be retrieved from $\phi(z)$ using ordinary Fourier inversion methods.

An important theorem regarding characteristic functions is the characterization theorem. If X and Y are two random variables with distribution functions $F(x)$ and $G(y)$, respectively, and if

$$\phi_X(z) = \phi_Y(z)$$

then $F = G$.

Tchebyshev's Inequality. Knowing only the first and second control moments, one can obtain some limits on the probabilities for a random variable. There are

TABLE 2-5.   Probability Generating Functions
or Characteristic Functions for Some
Distributions*

Name	Probability generating function	Characteristic function
Bernoulli	$q + pz$	
Binomial	$(q + pz)^n$	
Geometric	$\dfrac{p}{1 - qz}$	
Poisson	$e^{-\lambda(1-z)}$	
Normal	....	$e^{-i\mu z + \frac{1}{2}(\sigma z)^2}$
Gamma	....	$\left(\dfrac{a}{a - iz}\right)^n$
Exponential	....	$\dfrac{a}{a - iz}$

* See Table 2-2.

several types of theorems of this kind.   We state only one here, called Tchebyshev's theorem:

$$\Pr[|X - \mu| > t] \leq \frac{\sigma^2}{t^2}$$

In many cases, Tchebyshev's theorem is quite rough and forms a very loose bound on the probability.   The theorem does lend some credence to the use of σ^2 as a measure of dispersion or spread of a density function, however.

Multivariate Distributions.   If X and Y are two (or a countable number in general) random variables defined on the same sample space, then one is interested in the event

$$[X(s) \leq x] \cap [Y(s) \leq y]$$

This event will have a probability associated with it as

$$\Pr[(X(s) \leq x) \cap (Y(s) \leq y)]$$

As before, we define a real function F, called the joint probability distribution function, by

$$F(x,y) = \Pr[(X(s) \leq x) \cap (Y(s) \leq y)]$$

$F(x,y)$ then determines the probability of the joint event $X(s) \leq x$ and $Y(s) \leq y$.   If

$$F(x,y) = F_1(x)F_2(y)$$

where
$$F_1(x) = \Pr[X(s) \leq x]$$
$$F_2(y) = \Pr[Y(s) \leq y]$$

then the random variables are independent random variables.   In general, if

$$F(x_1, x_2, \ldots, x_n) = F_1(x_1)F_2(x_2), \ldots, F_n(x_n)$$

the random variables $X_1(s)$, $X_2(s)$, $\ldots$, $X_n(s)$ are independent random variables.   If $F(x,y)$ is differentiable, then

$$\partial F(x,y) = f(x,y) \, \partial x \, \partial y$$

is the joint density function of the random variables.   In the discrete case, one defines the joint density function

$$p(x,y) = \Pr[X = x; \, Y = y]$$

The marginal distribution function of X is defined by

$$F_X(x) = F(x, \infty)$$

where the subscript X indicates a distribution function for the random variable X. Similarly, for Y,

$$F_Y(y) = F(\infty, y)$$

The corresponding marginal density functions can be defined as derivatives of F_X or F_Y or by

$$F_X(x) = \sum_y f(x,y)$$

$$F_Y(y) = \sum_x f(x,y)$$

Replacing the sums by integrals with respect to the indicated variables would yield corresponding formulas for continuous $f(x,y)$.   These definitions, of course, generalize to more than two random variables defined on the same sample space.

If X and Y are not independent, one defines the conditional probability density function by

$$\frac{f(x,y) \, dx \, dy}{f_X(x) \, dy} = f_{Y|X}(y|x) \, dy$$

and similarly,

$$\frac{f(x,y) \, dx \, dy}{f_Y(y) \, dy} = f_{X|Y}(x|y) \, dx$$

The probability meaning of $f_{X|Y}(x|y)$ is that

$$f_{X|Y}(x|y) \, dx \, dy = \Pr(x \leq X \leq x + dx | Y = y)$$

Often the subscript on the conditional densities is dropped and one writes $f(x|y)$ or $f(y|x)$. The conditional density functions are density functions in that

$$0 \leq f_{X|Y}(x|y) \, dx$$
$$\int f_{X|Y}(x|y) \, dx = 1$$

Similar definitions hold for $f_{X|Y}$ in the cases where X and Y are discrete random variables.

Moments of Multivariate Distributions. Just as in the case of a single random variable, one can define the expectation of a function of the random variables X, Y, say $g(X,Y)$, as

$$E[g(X,Y)] = \int\int g(x,y) \, f(x,y) \, dx \, dy$$

Of particular importance in many statistical studies is the covariance of X, Y. The value

$$E[(X - E(X))(Y - E(Y))] = \int\int_{yx} (x - E(X))(y - E(Y))f(x,y) \, dx \, dy$$

is called the covariance of X and Y. If X and Y are independent random variables, then the covariance of X and Y is 0. The converse is not true.

The value

$$\rho = \frac{\text{covariance of } X \text{ and } Y}{(\text{standard deviation of } X)(\text{standard deviation of } Y)}$$

is called the correlation coefficient and has the properties:
1. $-1 \leq \rho \leq +1$
2. If X and Y are independent, $\rho = 0$.
3. If $\rho = \pm 1$, then X and Y are two random variables related by $X = aX + b$.

The Bivariate Normal Density Function. Of considerable importance to the study of statistics is the bivariate normal density function.

The bivariate normal density is defined by

$$f(x,y) = (2\pi\sigma_x\sigma_y \sqrt{1 - \rho^2})^{-1} e^{-Q/2}$$

where $\quad Q = \dfrac{1}{1 - \rho^2} \left[\left(\dfrac{x - \mu_x}{\sigma_x} \right)^2 - 2\rho \left(\dfrac{x - \mu_x}{\sigma_x} \right) \left(\dfrac{y - \mu_y}{\sigma_x} \right) + \left(\dfrac{y - \mu_y}{\sigma_x} \right)^2 \right]$

and $\qquad -\infty < x < \infty, \, -\infty < y < \infty$

The means and variances are

$$E[X] = \mu_x \qquad V[X] = \sigma_x^2$$
$$E[Y] = \mu_y \qquad V[Y] = \sigma_y^2$$

The parameter ρ is the correlation coefficient. The probability density function is completely specified by μ_x, μ_y, σ_x, σ_y, ρ.

The Hazard Function. In many reliability studies, one is interested in the hazard function, defined as

$$\Pr[x < X \leq x + dx | X > x]$$

The hazard function is given in general by

$$\frac{\Pr[x < X \leq x + dx]}{\Pr[X > x]} = \frac{f(x) \, dx}{1 - F(x)}$$

For the Weibull density, for example, the hazard function is given by

$$\frac{b}{\theta - x_o} \left(\frac{x - x_o}{\theta - x_o}\right)^{b-1}$$

that is, the hazard function is a polynomial in x. In the case $b = 1$, the random variable is exponentially distributed, and the hazard function is independent of x. This property is unique to the exponential distribution (for continuous random variables) and is called the forgetfulness property of the exponential.

Sums of Two Random Variables. A common problem in probability is finding the distribution of a sum of independent random variables:

$$S_n = \sum_{j=1}^{n} X_j$$

One can show that

$$E(S_n) = \sum_{j=1}^{n} E(X_j)$$

"The expected value of a sum is the sum of the expected values." Furthermore, the variance of a sum—abbreviated $\mathrm{Var}(X)$—is given by

$$\mathrm{Var}(S) = \mathrm{Var}(X) + \mathrm{Var}(Y)$$

if X and Y are independent random variables. If X and Y are not independent variables, one has the more general formula

$$\mathrm{Var}(S) = \mathrm{Var}(X) + \mathrm{Var}(Y) + 2\, \mathrm{Covar}\,(X,Y)$$

The density function of S can be computed (though not always in closed form) using convolution arguments. For the case where $N = 2$ and X_j is discrete, one has

$$p_s = \sum_{j=0}^{s} p_j p_{s-j} \tag{1-5}$$

where
$$p_s = \Pr[S = s] \qquad s = 0, 1, 2, \ldots$$
$$p_j = \Pr[X_1 = j] \qquad j = 0, 1, 2, \ldots$$
$$p_k = \Pr[X_2 = k] \qquad k = 0, 1, 2, \ldots$$

Equation (1-5) is called the convolution of p_j and p_k. This can be generalized for any number of random variables in the sum S. If X_j is continuous,

$$f(s) = \int_0^s g(x_1)h(s - x_1)\, dx_1 \qquad 0 \le s < \infty \tag{1-6}$$

where
$$f(s)\, ds = \Pr[s < S \le s + ds]$$
$$g(x_1)\, dx_1 = \Pr[x_1 < X_1 \le x_1 + dx_1] \qquad 0 \le x_1 < \infty$$
$$h(x_2)\, dx_2 = \Pr[x_2 < X_2 \le x_2 + dx_2] \qquad 0 \le x_2 < \infty$$

The characteristic function gives a useful way to determine the characteristic function of the sum of two independent random variables. For if

$$Z = X + Y$$

then
$$\phi_Z(z) = \phi_X(z)\phi_Y(z) \tag{1-7}$$

This result generalizes to more than two independent random variables immediately.

Example 3. Let $X_1 =$ the time to set up one of two machines and $X_2 =$ the time to set up the other machine. Then $S = X_1 + X_2$ would be the time required for a work crew to set up the two machines. Suppose X_1 and X_2 are each negative exponentially distributed random variables with the same density functions

$$f_{X_1}(x) = f_{X_2}(x) = ae^{-ax} \qquad 0 \le x < \infty$$

Then, using equation (1-6),

$$f_S(s) = \int_0^s ae^{-ax} ae^{-a(s-x)} dx$$

$$= a^2xe^{-ax} \qquad 0 \le x < \infty$$

From Table 2-2, it is apparent that S is a gamma distributed random variable with $n = 2$.

Using formula (1-7) and Table 2-5, one has

$$\phi_{X_1}(z) = \frac{a}{a - iz} = \phi_{X_2}(z)$$

and
$$\phi_S(z) = \left(\frac{a}{a - iz}\right)^2$$

But if S is a gamma distributed random variable, one sees from Table 2-5 that

$$\phi_S(z) = \left(\frac{a}{a - iz}\right)^n$$

Hence, from the characterization theorem, one concludes that S has a gamma density function with $n = 2$.

A Central Limit Theorem. One of the remarkable properties of the normal density function is summarized in the central limit theorem, which can be stated as:

Let $\qquad\qquad\qquad\qquad X_1, X_2, \ldots , X_n$

be a sequence of independent random variables, each with the same density function (not necessarily normal) with finite mean μ and finite variance σ^2. Further, let

$$S_n = X_1 + X_2 + \cdots + X_n$$

be the nth partial sum of the X's. Then

$$\lim_{n \to \infty} \Pr\left[a \le \frac{S_n - n\mu}{\sigma \sqrt{n}} \le b \right] = \int_a^b N(z;0,1) \, dz$$

where $N(z;0,1)$ is the normal density function given in Table 2-2.

In many applied statistics problems, one "takes a random sample of size n" (that is, one observes n independent random variables) from a "population" (that is, each random variable has the same distribution). It is reasonable to assume that the population has a finite mean and variance. One then determines

$$\frac{\sum\limits_{i=1}^{n} X_i}{n}$$

called the "sample mean value." From this, one wants to make statements about whether a process which produces the observations is "in control"; or one wants to estimate a reasonable range of values that could be expected to contain the process true mean μ; or one wants to test a hypothesis regarding the behavior of the means of two processes or one process under two conditions. In such cases, one can make rather precise comments about the probabilities associated with the mean values by using the central limit theorem. Notice that the central limit theorem allows one to make rather precise statements about the probabilities associated with the mean values although knowing relatively little about the probabilities associated with the X's themselves. Hence, even though one might be able to say little about the probabilities of the X's, one can say a great deal about the probabilities of the mean of the X's by using the central limit theorem.

SPECIAL RANDOM PROCESSES

In many engineering studies, one is concerned with the time behavior of a process. Thus one is interested in $X(t)$ = the position of a particle at time t, or X_n = the time to produce a part in the nth day of production. In the study of random experiments, one is also interested in the time behavior of the process. Conceptually, one is concerned not with a single random variable X, but rather with an entire family of random variables, perhaps a sequence of random variables. Thus, one might define the sequence of daily demands for an item, $\{X_n:n = 1,2,3 \ldots\}$. n serves as a parameter which denotes which day we are concerned with, and X_n is the amount demanded on day n. The sequence $\{X_n\}$ is called a random sequence, or a discrete parameter (n) random process, or a stochastic process.

More generally one defines the family of random variables $\{X_t, t \in T\}$ as a stochastic process or a random process. The set T, the set of all possible parameter values, is called the parameter space. Often in application, the parameter set is taken to be a set of times. We shall be concerned only with the cases $T = \{0,1,2, \ldots\}$ or $T = (0, \infty)$. In the first case, $\{X_t\}$ is called a discrete parameter process, and in the second, it is called a continuous parameter process. Formally, $\{X_t\}$, or more properly $\{X_t(s)\}$, is a function X which maps points in the product space $S \times T$ into the real line, where S is the sample space for the experiment and T is the parameter space. One notes that if t is held fixed, then $X(s)$ is merely a random variable, as discussed under Probability Theory. On the other hand, if s is held fixed, $X_t(\cdot)$ is just a real valued function X whose domain is T and whose range is some subset of the real numbers.

The range of $\{X_t\}$ is called the state space of the random process. The state space may be continuous, as occurs in some reliability studies, or it may be discrete, for example, the integers. In the following discussion, we are concerned only with integer state random processes. If for some n (or t) $X_n = j$, $(X_t = j)$, we say that the "process is in state j at time n (or time t)."

Because $\{X_t\}$ is a collection of random variables, one must be able to determine the joint probability for all subcollections to make definite probability statements about the process. Hence, to completely define the probability structure of a random process, one must determine

$$\Pr[X_{t_1} = j_1; X_{t_2} = j_2; \ldots, X_{t_i} = j_i; \ldots ; X_{t_n} = j_n] \qquad (2\text{-}1)$$

for every value t_i, for all i, and all j_i in their respective spaces.

If the probability structure of the random process has the special property that

$$\Pr[X_1 = j_1, X_2 = j_2, X_3 = j_3, \cdots, X_n = j_n]$$
$$= \Pr[X_1 = j_1] \Pr[X_2 = j_2 | X_1 = j_1] \Pr[X_3 = j_3 | X_2 = j_2] \cdots$$
$$\Pr[X_n = j_n | X_{n-1} = j_{n-1}] \qquad (2\text{-}2)$$

then one says the random process is a discrete parameter Markov process. A similar property must hold for a continuous parameter Markov process, that is, if for every t and n

$$\Pr[X(t_1) = j_1, X(t_2) = j_2, X(t_3) = j_3, \cdots, X(t_n) = j_n]$$
$$= \Pr[X(t_1) = j_1] \Pr[X(t_2) = j_2 | X(t_1) = j_1] \Pr[X(t_3) = j_3 | X(t_2) = j_2]$$
$$\cdots \Pr[X(t_n) = j_n | X(t_{n-1}) = j_{n-1}] \qquad (2\text{-}3)$$

then $\{X(t)\}$ is a continuous parameter Markov process.

Discrete Parameter Markov Process. In the structure (2-2), one defines $\Pr[X_1 = j_1] = p_{j_1}$, the initial state probability for state j. The vector $\bar{p} = (p_{j_1}, p_{j_2}, \ldots)$ is called the vector of initial state probabilities. The probability

$$\Pr[X_n = j | X_{n-1} = i] = p_{ij}(n - 1, n)$$

is called a one-step transition probability of the process.

In most applications, the conditional probabilities in equation (2-2) do not depend on the particular values of n. Hence,

$$p_{ij} = \Pr[X_m = j_m | X_{m-1} = j_{m-1}] = \Pr[X_n = j_n | X_{n-1} = j_{n-1}]$$

for every $m - 1$, m, $n - 1$, n. In this case, the process is said to have a stationary transition mechanism.

By equation (2-2), the entire set of joint probabilities for the process can be obtained from the initial state probabilities and the set of transition probabilities, in the stationary transition mechanism case. We will discuss only stationary transition mechanism chains in what follows. Hence, the entire probability structure of the process is determined in terms of a vector $\bar{p}$ of the initial state probabilities and a matrix $\bar{P}$ of the transition probabilities in this case. The vector $\bar{p}$ is a probability vector. That is,

$$0 \le p_j \le 1 \qquad \text{for every element } p_j$$

and

$$\sum_j p_j = 1$$

The matrix $\bar{P}$ is a stochastic matrix. That is,

$$0 \le p_{ij} \le 1 \qquad \text{for every element } p_{ij}$$

and

$$\sum_j p_{ij} = 1$$

that is, each row of $\bar{P}$ sums to 1.

n-step Transition Probabilities and nth-step State Probabilities. The probability structure of a stationary transition mechanism, discrete parameter, Markov chain is completely determined by the vector of initial state probabilities and the matrix of one-step transition probabilities. One often needs to know, however,

$$p_{ij}{}^{(n)} = \Pr[X_{m+n} = j | X_m = i] \tag{2-4}$$

called the n-step transition probabilities. Because of stationarity of the transition probabilities, equation (2-4) is equivalent to

$$p_{ij}{}^{(n)} = \Pr[X_n = j | X_o = i]$$

The n-step transition probabilities satisfy the Chapman-Kolmogorov equations

$$p_{ij}{}^{(n+m)} = \sum_k p_{ik}{}^{(n)} p_{kj}{}^{(m)} \tag{2-5}$$

In matrix form, equation (2-5) is given by

$$\bar{P}^{(n+m)} = \bar{P}^{(n)} \bar{P}^{(m)} \tag{2-6}$$

It follows that the matrix of n-step transition probabilities can be obtained from $\bar{P}$ as

$$\bar{P}^{(n)} = \bar{P}^n \tag{2-7}$$

That is, the matrix of n-step transition probabilities is simply the nth power of the matrix of one-step transition probabilities. $\bar{P}^{(n)}$ is a stochastic matrix.

The probability

$$p_j{}^{(n)} = \Pr[X_n = j] \tag{2-8}$$

is called the nth-step state probability. These probabilities are obtained from

$$p_j{}^{(n)} = \sum_i p_i p_{ij}{}^{(n)} \tag{2-9}$$

using equation (2-7) and defining $\bar{p}^{(n)}$ as the vector whose elements are those probabilities given by equation (2-8). The vector $\bar{p}^{(n)}$ whose elements are the $p_j{}^{(n)}$ is obtained from

$$\bar{p}^{(n)} = \bar{p} \bar{P}^n$$

Passage Probabilities. One defines the probabilities

$$g_{ij}^{(n)} = \Pr[X_n = j,\ X_m \neq j \text{ for } m < n | X_o = i] \tag{2-10}$$

and

$$g_{ii}^{(n)} = \Pr[X_n = i,\ X_m \neq i \text{ for } m < n | X_o = i] \tag{2-11}$$

as the first passage probabilities (2-10) and first return probabilities (2-11). The $g_{ij}^{(n)}$ and $p_{ij}^{(n)}$ are connected by the relation

$$p_{ij}^{(n)} = g_{ij}^{(n)} + \sum_{m=1}^{n-1} p_{jj}^{(m)} g_{ij}^{(n-m)} \qquad n > 0 \tag{2-12}$$

The value

$$g_{jj} = \sum_{n=1}^{\infty} g_{jj}^{(n)} \tag{2-13}$$

is the probability of ever returning to j.
The value

$$\mu_j = \sum_{n=1}^{\infty} n g_{jj}^{(n)} \tag{2-14}$$

is called the mean return time to state j.

A Classification of State. States of the chain can be classified in several ways.
1. If, for some $n > 0$, $p_{ij}^{(n)} > 0$, then state j is reachable from state i.
2. If state j is reachable from state i and state i is reachable from state j, then states i and j communicate.
3. If all states communicate, the chain is said to be an irreducible chain.
4. If, for some set of states C and for every $i \in C$ and $k \notin C$, $p_{ik}^{(n)} = 0$ for every $n > 0$, then C is said to be a closed set of states.
5. If C is a closed set containing one state i, then i is called an absorbing state.
6. If return to state i has probability 0 except perhaps at times $n, 2n, 3n, \ldots$, then state i is called a periodic state with period n. Otherwise the state is aperiodic.
7. If $g_{jj} = 1$, then j is a recurrent state. Otherwise, j is a transient state.
8. If j is a recurrent state and $\mu_j < \infty$, then j is positive recurrent. Otherwise, j is a null recurrent state.
9. If state i is aperiodic, recurrent, and non-null, then it is called an ergodic state.

Three Important Theorems. The following are three important theorems for classifying states of a chain.
1. In an irreducible Markov chain, all states are of the same type. If one state is recurrent, they are all recurrent. If one is transient, they are all transient. If one is null recurrent, they are all null recurrent. If one is periodic with period n, they are all periodic with period n.
2. In a Markov chain with a finite number of states, not all states can be transient and no states can be null recurrent.
3. In a finite, irreducible Markov chain, all states are positive recurrent.

Limit Theorems. In a considerable number of engineering applications, one is concerned with the long-run behavior of a system. Quite often, systems are designed to optimize their behavior after such things as start-up effects have become negligible. In the study of Markov chains, the question of the limiting behavior of $p_{ij}^{(n)}$, or $p_j^{(n)}$, is of importance. We restrict attention here to aperiodic chains (every state is aperiodic).

A vector $\bar{x}$ is said to be a stationary probability vector if $\bar{x}$ is a probability vector satisfying

$$\bar{x} = \bar{x}\bar{P} \tag{2-15}$$

It can be shown that if the chain is irreducible and aperiodic, either of the two following conditions prevail:

1. All states are transient or null recurrent, in which case, $\lim\limits_{n \to \infty} p_{ij}^{(n)} = 0$ for every pair i,j, and there is no stationary probability vector $\bar{x}$.

2. All states are positive recurrent and $\lim\limits_{n \to \infty} p_{ij}^{(n)} = \pi_j$ exists. The π_j's are the unique stationary probability vector satisfying equation (2-15) above. In this case, $\pi_j = 1/\mu_j$. When π_j exists, one has $p_j^{(n)} \to \pi_j$ for $n \to \infty$.

Condition 2 above is often encountered in practice, and one is concerned with the elements π_j. These probabilities are called the steady state probabilities.

Continuous Parameter Markov Chains. In the structure of equation (2-3) for a continuous time Markov chain, one defines $\Pr[X_t = j] = p_j(t)$, the probability of being in state j at time t. In particular, $p_j(0)$ is the initial state probability for state j. The vector $\bar{p} = (p_o(0), \, p_1(0), \, \ldots \,)$ is called the vector of initial state probabilities. The probability

$$\Pr[X(\tau_2) = j | X(\tau_1) = i] = p_{ij}(\tau_1, \tau_2)$$

is called the transition probability. In many applications, one finds that the transition probability depends only on the difference $\tau_2 - \tau_1 = t$. In this case, one uses the abbreviated notation $p_{ij}(t)$ for the transition probabilities, and the resulting process is said to be time homogeneous or to have a stationary transition mechanism.

By equation (2-3), the entire set of joint probabilities for the process can be obtained from the vector of state probabilities and the matrix whose elements are transition probabilities, in the time homogeneous case. We will discuss only this case in what follows.

Chapman-Kolmogorov Equations. If one is concerned with an interval of length $t + s$, the transition probabilities $p_{ij}(t + s)$ are given by the Chapman-Kolmogorov equations.

$$p_{ij}(t + s) = \sum_k p_{ik}(t) p_{kj}(s) \tag{2-16}$$

The Instantaneous Transition Rates. One defines the instantaneous transition rates

$$\lambda_{ij} = \lim_{s \to 0} \frac{p_{ij}(s) - p_{ij}(0)}{s} = \frac{dp_{ij}(0)}{dt}$$

$$\lambda_{ii} = \lim_{s \to 0} \frac{p_{ii}(s) - 1}{s} = \frac{dp_{ii}(0)}{dt} \tag{2-17}$$

which we assume exist and are finite.

The λ_{ij}. The λ_{ij} are defined above. From their definition, it is clear that when they exist, the λ_{ij} are the derivatives of the $p_{ij}(t)$ functions evaluated at 0. There is another interpretation of this point that is extremely useful in building Markov models. From the definition of the λ_{ij}, one can argue that for some Δt near zero and sufficiently small,

$$p_{ij}(\Delta t) = \lambda_{ij} \, \Delta t + o(\Delta t)$$

That is, one can assume that, if the λ_{ij} exist in the sense of equation (2-17), then the transition probability function in the neighborhood of zero can be approximated by a straight line with slope λ_{ij}. The term $o(\Delta t)$ is the error of approximation and

$$\frac{o(\Delta t)}{\Delta t} \to 0 \qquad \text{for } \Delta t \to 0$$

It follows from the definition of λ_{ij} and λ_{ii} that

$$\lambda_{ij} \geq 0 \qquad \text{for } j \neq i$$
$$\lambda_{ii} \leq 0$$

For a conservative Markov chain,

$$\sum_j \lambda_{ij} = 0$$

and thus λ_{ii} can be obtained in any conservative process from

$$\lambda_{ii} = -\sum_{j \neq i} \lambda_{ij}$$

The Kolmogorov Equation. Using equation (2-17), it is shown that the transition probabilities satisfy the forward (2-18) and backward (2-19) Kolmogorov differential equations.

$$\frac{dp_{ij}(t)}{dt} = \sum_{k} p_{ik}(t)\lambda_{kj} \qquad (2\text{-}18)$$

$$\frac{dp_{ij}(t)}{dt} = \sum_{k} \lambda_{ik}p_{kj}(t) \qquad (2\text{-}19)$$

with initial conditions

$$p_{ij}(0) = \begin{cases} 1 & \text{if } i = j \\ 0 & \text{otherwise} \end{cases}$$

In matrix form, equations (2-18) and (2-19) become

$$\frac{d\bar{P}(t)}{dt} = \bar{P}(t)\bar{\Lambda} \qquad (2\text{-}18')$$

$$\frac{d\bar{P}(t)}{dt} = \bar{\Lambda}\bar{P}(t) \qquad (2\text{-}19')$$

with initial conditions

$$P(0) = \bar{I}$$

In this form, $\bar{I}$ is the identity matrix, $\bar{P}(t)$ is the matrix whose elements are the transition probabilities, $d\bar{P}(t)/dt$ is the matrix whose elements are the derivatives of those of $\bar{P}(t)$, and $\bar{\Lambda}$ is the matrix of the λ_{ij}.

It is clear from equations (2-18) and (2-19) and the initial conditions that the transition probabilities are known in terms of the instantaneous transition rates. Hence, the entire set of probability needed to define $\{X_t\}$ is known when $\bar{p}(0)$ and $\bar{\Lambda}$ are known.

The State Equations. In many cases, one is primarily concerned with the state probability $p_j(t)$. These state probabilities can be obtained from

$$p_j(t) = \sum_i p_i(0)p_{ij}(t)$$

If for some fixed i, $p_i(0) = 1$ and $p_k(0) = 0$, $k \neq i$, then $p_j(t) = p_{ij}(t)$, and the Kolmogorov equations (2-18) and (2-19) can be used to determine the state probabilities merely by suppressing the subscript i to obtain, for example,

$$\frac{dp_j(t)}{dt} = \sum_{k} p_k(t)\lambda_{kj} \qquad j = 0,1,2, \ldots \qquad (2\text{-}20)$$

This system of equations is often used in queuing theory to obtain the equation of state.

An Important Special Case—The Poisson Process. In the special case

$$\lambda_{kj} = \lambda \qquad \text{if } j = k + 1$$
$$\lambda_{jj} = -\lambda$$
$$\lambda_{kj} = 0 \qquad \text{otherwise}$$

one has

$$\bar{\Lambda} = \begin{pmatrix} -\lambda & \lambda & 0 & 0 & \cdots \\ 0 & -\lambda & \lambda & 0 & \cdots \\ 0 & 0 & -\lambda & \lambda & \cdots \end{pmatrix}$$

The state equations (2-20) yield for $p_o(0) = 1$, $p_k(0) = 0$, $k \neq 0$,

$$\frac{dp_o(t)}{dt} = -\lambda p_o(t)$$

$$\frac{dp_n(t)}{dt} = \lambda p_{n-1}(t) - \lambda p_n(t)$$

The solutions to the equations are

$$p_n(t) = \frac{(\lambda t)^n e^{-t}}{n!} \qquad n = 0,1,2, \ldots$$

and the Markov process is called a Poisson process.

Limiting Behavior. In an irreducible, time homogeneous Markov chain with a continuous parameter, the limits

$$\lim_{t \to \infty} p_{ij}(t)$$

always exist. If in addition the chain is positive recurrent, then

$$\lim_{t \to \infty} p_{ij}(t) = \pi_j > 0 \qquad \Sigma \pi_j = 1 \qquad (2\text{-}21)$$

That is, the limits exist; they are independent of the initial state and they form a probability density function.

The limiting probabilities π_j, when they exist, are the unique solution to the steady state equation

$$0 = \bar{\Pi}\bar{\Lambda} \qquad (2\text{-}22)$$

or in component form,

$$0 = \sum_j \pi_j \lambda_{jk} \qquad (2\text{-}23)$$

Equation (2-23) is a useful working tool. Under the stated conditions, these equations either have no solution that has the properties of a probability distribution or they have precisely one. The π_j are the steady state probabilities, and it can be shown that

$$\lim_{t \to \infty} p_j(t) = \pi_j$$

is satisfied, as well as equation (2-21).

Birth-Death Processes. Of considerable interest in queuing theory applications is the special Markov process whose instantaneous transition rates are given by

$$
\begin{aligned}
\lambda_{i,i+1} &= \lambda_i & i &= 0, 1, 2, \ldots \\
\lambda_{i,i-1} &= \mu_i & i &= 1, 2, \ldots \\
\lambda_{ij} &= -(\lambda_i + \mu_i) & i &= 1, 2, \ldots \\
\lambda_{ij} &= 0 & &\text{otherwise}
\end{aligned}
\qquad (2\text{-}24)
$$

A Markov process with this particular set of rates is called a birth-death process. The state equations are given by

$$\frac{dp_o(t)}{dt} = -\lambda p_o(t) + \mu_1 p_1(t)$$

$$\frac{dp_j(t)}{dt} = \lambda_{j-1} p_{j-1}(t) - (\lambda_j + \mu_j) p_j(t) + \mu_{j+1} p_{j+1}(t) \qquad j = 1, 2, \ldots$$

The steady state equations are given by

$$
\begin{aligned}
-\lambda_o \pi_o + \mu_1 \pi_1 &= 0 \\
\lambda_{j-1} \pi_{j-1} - (\lambda_j + \mu_j) \pi_j + \mu_{j+1} \pi_{j+1} &= 0
\end{aligned}
\qquad (2\text{-}25)
$$

The general solutions to the steady state equations are given by

$$\pi_j = \frac{\lambda_0 \lambda_1 \cdots \lambda_{j-1}}{\mu_1 \mu_2 \cdots \mu_j} \pi_0 \qquad j = 1, 2, \ldots \tag{2-26}$$

π_0 is determined from

$$\pi_0 = \left(1 + \frac{\lambda_0}{\mu_1} + \frac{\lambda_0 \lambda_1}{\mu_1 \mu_2} + \cdots + \frac{\lambda_0 \lambda_1 \cdots \lambda_{j-1}}{\mu_1 \mu_2 \cdots \mu_j} + \cdots \right)^{-1} \tag{2-27}$$

These steady state solutions exist if

$$\left(1 + \frac{\lambda_0}{\mu_1} + \frac{\lambda_0 \lambda_1}{\mu_1 \mu_2} + \cdots \right) < \infty$$

(a stronger condition than is necessary but one that is useful in application).

Four Special Cases of Birth-Death Processes. We note here four special cases of the birth-death process.

1. The $M/M/1$ queue. In the $M/M/1$ queue, one supposes that there is a "server" whose service times are negative exponentially distributed random variables with parameter μ and that successive service times are independent. "Arrivals" to the server form a Poisson process with parameter λ. Arrivals which occur when the server is busy form a single waiting line for their turn at service. It follows that the rate structure (2-24) is

$$\lambda_{i,i+1} = \lambda_i = \lambda \qquad \text{for } i = 0, 1, 2, \ldots$$
$$\lambda_{i,i-1} = \mu_i = \mu \qquad \text{for } i = 1, 2, \ldots$$
$$\lambda_{ii} = -(\lambda_i + \mu_i) = -(\lambda + \mu) \qquad \text{for } i = 1, 2, \ldots$$
$$\lambda_{ij} = 0 \qquad \text{otherwise}$$

The state equations are given by

$$\frac{dp_0(t)}{dt} = -\lambda p_0(t) + \mu p_1(t)$$

$$\frac{dp_i(t)}{dt} = \lambda p_{i-1}(t) - (\lambda + \mu) p_i(t) + \mu p_{i+1}(t) \qquad i = 1, 2, \ldots$$

The steady state probabilities are given by

$$\pi_j = \rho^j (1 - \rho) \qquad j = 0, 1, 2, \ldots$$

where $\rho = \lambda/\mu$ (the traffic intensity) upon using equations (2-26) and (2-27). These steady state solutions exist for $\rho < 1$.

2. If in case 1 above we retain the transition rate structure but require $\lambda_j = 0$ for $j \geq N$, and $\mu_j = 0$ for $j > N$, we are led for some $N > 0$ to the $M/M/1$ queue with finite size N waiting capacity. In this case, the steady state probabilities exist for every ρ and are given by

$$\pi_j = \frac{\rho^j (1 - \rho)}{1 - \rho^{N+1}} \qquad j = 0, 1, 2, \ldots, N$$

$$\pi_j = 0 \qquad \text{otherwise}$$

3. The multichannel queue. In the multichannel queue, one supposes that there are R servers sharing a common waiting line. Arrivals to the service system form a Poisson process with parameter λ. Each server performs service with service times negative, exponentially distributed with parameter μ. (The servers are "identical.") It follows that

$$\lambda_{i,i+1} = \lambda \qquad \text{for } i = 0, 1, 2, \ldots$$
$$\lambda_{i,i-1} = i\mu \qquad \text{for } i = 1, 2, \ldots, R - 1$$
$$\lambda_{i,i-1} = R\mu \qquad \text{for } i = R, R + 1, \ldots$$
$$\lambda_{ij} = 0 \qquad \text{for } i \neq j \text{ and } i \neq i + 1, i - 1$$

The steady state equations (2-25) are

$$-\lambda\pi_o + \mu\pi_1 = 0$$
$$-(\lambda + i\mu)\pi_i + \lambda\pi_{i-1} + (i + 1)\mu\pi_{i+1} = 0 \qquad \text{for } i < R$$
$$-(\lambda + R\mu)\pi_i + \lambda\pi_{i-1} + R\mu\pi_{i+1} = 0 \qquad \text{for } i \geq R$$

The steady state probabilities are

$$\pi_i = \frac{\pi_o\rho^i}{i!} \qquad \text{for } i < R$$

$$\pi_i = \frac{\rho^i}{R!R^{i-R}}\pi_o \qquad \text{for } i \geq R$$

and π_o is obtained from the condition

$$\Sigma\pi_i = 1$$

The π_i exist if $\lambda < R\mu$.

4. The machine repair problem. In a machine repair problem, we suppose that there are M identical machines, each subject to breakdown, that are repaired by one of R repair crews. Running times of each machine are random variables, exponentially distributed with parameter λ. The machines are identical. The repair structure is the same as in case 3. From this, it follows that if i machines are running,

$$\lambda_{i,i-1} = (M - i)\lambda \qquad i = 0, 1, 2, \ldots$$
$$\lambda_{i,i+1} = i\mu \qquad i = 1, 2, \ldots, R - 1$$
$$\lambda_{i,i+1} = R\mu \qquad i = R, R + 1, \ldots, M$$
$$\lambda_{ij} = 0 \qquad i \neq j, i - 1, i + 1$$

The steady state equations are

$$-M\lambda\pi_o + \mu\pi_1 = 0$$
$$(M - i + 1)\lambda\pi_{i-1} - [(M - i)\lambda + i\mu]\pi_i + (i + 1)\mu\pi_{i+1} = 0 \qquad i < R$$
$$(M - i + 1)\lambda\pi_{i-1} - [(M - i)\lambda + R\mu]\pi_i + R\mu\pi_{i+1} = 0 \qquad i \geq R$$
$$\lambda\pi_{M-1} + R\mu\pi_M = 0$$

Explicit formulas for π_i are quite complex but can be evaluated numerically. The π_i always exist in this case for finite M, R.

BIBLIOGRAPHY

Barlow, I., and F. Prochan, *Mathematical Theory of Reliability*, John Wiley & Sons, Inc., New York, 1965.

Bharucha-Reid, A., *Elements of the Theory of Markov Processes and Their Applications*, McGraw-Hill Book Company, New York, 1960.

Chung, K., *Markov Chains with Stationary Transition Probabilities*, Springer-Verlag New York Inc., New York, 1960.

Clarke, A. B., and R. L. Disney, *Probability and Random Processes for Engineers and Scientists*, John Wiley & Sons, Inc., New York, 1970.

Feller, W., *Introduction to Probability Theory and Its Applications*, 3d ed., vol. 1, John Wiley & Sons, Inc., New York, 1968.

Harris, B., *Theory of Probability*, Addison-Wesley Publishing Company, Inc., Reading, Mass., 1966.

Karlin, S., *A First Course in Stochastic Processes*, Academic Press Inc., New York, 1966.

Kemeny, J. G., and J. L. Snell, *Finite Markov Chains*, D. Van Nostrand Company, Inc., Princeton, N.J., 1959.

Meyer, P., *Introductory Probability and Statistical Applications*, Addison-Wesley Publishing Company, Inc., Reading, Mass., 1965.

Morse, P., *Queues, Inventories, and Maintenance*, John Wiley & Sons, Inc., New York, 1958.

Parzen, E., *Modern Probability Theory and Its Applications*, John Wiley & Sons, Inc., New York, 1960.

Parzen, E., *Stochastic Processes*, Holden-Day, Inc., San Francisco, 1962.

Pfeiffer, P. E., *Concepts of Probability Theory*, McGraw-Hill Book Company, New York, 1965.

Chapter **3**

Practical Statistics

RALPH L. DISNEY

**Professor, Department of Industrial Engineering,
The University of Michigan, Ann Arbor, Michigan**

The purpose of this chapter is to present some statistical tools of practical value to the industrial engineer.

The industrial engineer constantly finds himself in need of answering questions in the area where events are not constant from one observation to the next but appear to behave in a random manner—the time to produce a part is seldom the same from one part to the next; the number of defective parts is not the same from one lot to the next; the number of machines down for repairs varies from day to day. A large part of the industrial engineer's work is concerned with this kind of behavior.

In spite of the variability in his basic data, the industrial engineer must make decisions and help others make decisions. He must answer the questions placed in the context of varying information. He must, in short, make or help make decisions in the face of random behavior and uncertainty. Statistics offer a tool to assist in some of these situations.

This chapter discusses, in as brief a form as is practical, some of the basic tools of statistics that have been found useful in industrial engineering, with examples of their use.

Normal Density Function. The normal curve serves as one of the most important probability laws in all of statistics.

The general form of the normal curve is given by the density function formula

$$f(x) = \frac{1}{\sqrt{2\pi}\,\sigma'} e^{-\frac{(x-\bar{x}')^2}{2\sigma'^2}}$$

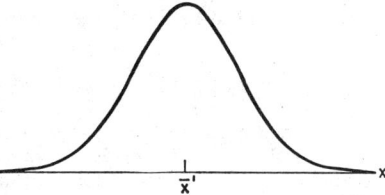

Fɪɢ. 3-1. Normal density function.

This curve has the familiar bell-shaped graph shown in Figure 3-1.

In the formula, $\bar{x}'$ is called the mean of the normal curve, σ' is called the standard deviation of the normal curve, and its square, σ'^2, is called the variance of the normal curve.

Properties of the Normal Density Function
    1. The normal density function or curve is symmetric about $\bar{x}'$.
    2. Changing $\bar{x}'$ slides the normal curve along the x axis without affecting its shape.
    3. Changing σ' spreads out or squeezes in the curve, without affecting $\bar{x}'$.

Figure 3-2a indicates these effects. In this figure, $\bar{x}_0' \leq \bar{x}_1'$. Thus, changing $\bar{x}'$ moves the curve along the x axis.

In Figure 3-2b, $\sigma_0' \leq \sigma_1'$. Thus, increasing σ' spreads out the curve and decreasing σ' squeezes in the curve.

In Figure 3-2c, $\bar{x}_0' \leq \bar{x}_1'$ and $\sigma_0' \leq \sigma_1'$. Thus, changing $\bar{x}'$ and σ' has the combined effect of changing each separately.

Normal Curve Tables.   The above results would indicate that a separate table of the normal curve would be required for each $\bar{x}'$ and σ'. Such is not the case, and Table 3-1 can serve as a table of normal curve areas for any $\bar{x}'$ and σ'. Through use of the formula

$$z = \frac{x - \bar{x}'}{\sigma'}$$

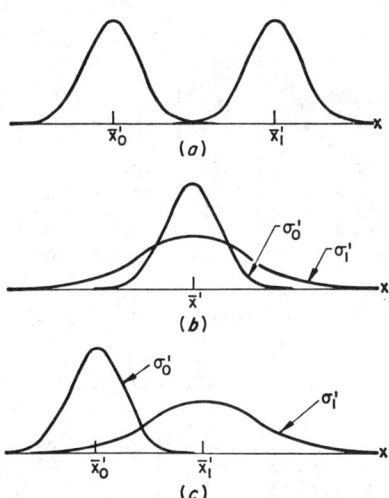

Fɪɢ. 3-2. Effect of varying $\bar{x}'$ and σ'
for the normal curve.

the general normal curve can be reduced to the standard normal curve

$$f(z) = \frac{1}{\sqrt{2\pi}} e^{-z^2/2}$$

It is the area of this standardized normal curve that is tabulated in Table 3-1.

Table 3-1 gives the area under the standard normal curve from 0 to some value of z of interest, say x. To use the table, z is calculated from the formula. If z is a positive number, the probability of z or less is found by taking the probability from the table and adding 0.5000 to the result. If z is a negative number, the probability of z or less is found by taking the probability from the table for positive z and subtracting it from 0.5000. If interest is in the probability of z or more, the above calculations are found for z or less and the result is subtracted from 1.

TABLE 3-1.   Normal Curve Areas (Area from 0.00 to z)

z	.00	.01	.02	.03	.04	.05	.06	.07	.08	.09
0.0	.0000	.0039	.0079	.0110	.0158	.0198	.0237	.0277	.0317	.0356
0.1	.0396	.0436	.0475	.0515	.0555	.0594	.0634	.0673	.0713	.0752
0.2	.0791	.0831	.0870	.0909	.0948	.0987	.1026	.1064	.1103	.1141
0.3	.1180	.1218	.1256	.1294	.1332	.1370	.1407	.1445	.1482	.1519
0.4	.1556	.1593	.1630	.1666	.1703	.1739	.1775	.1811	.1846	.1882
0.5	.1917	.1952	.1987	.2022	.2056	.2091	.2125	.2159	.2192	.2226
0.6	.2259	.2292	.2325	.2358	.2391	.2423	.2455	.2487	.2518	.2550
0.7	.2581	.2612	.2643	.2673	.2704	.2734	.2764	.2793	.2823	.2852
0.8	.2881	.2910	.2938	.2966	.2994	.3022	.3050	.3077	.3104	.3131
0.9	.3158	.3184	.3210	.3236	.3262	.3287	.3313	.3338	.3362	.3387
1.0	.3411	.3435	.3459	.3483	.3506	.3529	.3552	.3575	.3597	.3619
1.1	.3641	.3663	.3684	.3705	.3726	.3747	.3768	.3788	.3808	.3828
1.2	.3847	.3867	.3886	.3905	.3923	.3942	.3960	.3978	.3996	.4014
1.3	.4031	.4048	.4065	.4082	.4098	.4114	.4130	.4146	.4162	.4177
1.4	.4192	.4207	.4222	.4237	.4251	.4265	.4279	.4293	.4306	.4320
1.5	.4333	.4346	.4358	.4371	.4383	.4397	.4407	.4419	.4431	.4442
1.6	.4454	.4465	.4476	.4486	.4497	.4507	.4517	.4527	.4537	.4547
1.7	.4556	.4566	.4575	.4584	.4593	.4602	.4610	.4619	.4627	.4635
1.8	.4643	.4651	.4659	.4666	.4673	.4681	.4688	.4605	.4702	.4708
1.9	.4715	.4722	.4728	.4734	.4740	.4746	.4752	.4758	.4763	.4769
2.0	.4774	.4780	.4785	.4790	.4795	.4800	.4805	.4809	.4814	.4818
2.1	.4823	.4827	.4831	.4835	.4839	.4843	.4847	.4851	.4854	.4858
2.2	.4862	.4865	.4868	.4872	.4875	.4878	.4881	.4884	.4887	.4890
2.3	.4893	.4895	.4898	.4901	.4903	.4906	.4908	.4911	.4913	.4915
2.4	.4917	.4919	.4922	.4924	.4926	.4928	.4929	.4931	.4933	.4935
2.5	.4937	.4938	.4940	.4942	.4943	.4945	.4946	.4948	.4949	.4950
2.6	.4952	.4953	.4954	.4955	.4957	.4958	.4959	.4960	.4961	.4962
2.7	.4963	.4964	.4965	.4966	.4967	.4968	.4969	.4970	.4971	.4971
2.8	.4972	.4973	.4974	.4974	.4975	.4976	.4977	.4977	.4978	.4978
2.9	.4979	.4980	.4980	.4981	.4981	.4982	.4982	.4983	.4983	.4984
3.0	.4984	.4985	.4985	.4986	.4986	.4986	.4987	.4987	.4987	.4988
3.1	.4988	.4988	.4989	.4989	.4989	.4990	.4990	.4990	.4991	.4991
3.2	.4991	.4991	.4992	.4992	.4992	.4992	.4993	.4993	.4993	.4993
3.3	.4993	.4994	.4994	.4994	.4994	.4994	.4994	.4995	.4995	.4995
3.4	.4995	.4995	.4995	.4995	.4996	.4996	.4996	.4996	.4996	.4996
3.5	.4996									

Example 1.   If $z = +1.645$, the table gives the probability of $+1.645$ as 0.4500. Thus the probability of 1.645 or less is $0.4500 + 0.5000 = 0.9500$.

Example 2.   If $z = 1.96$, the table gives the probability of $+1.96$ as 0.4752.   Hence the probability of -1.96 or less is $0.5000 - 0.4752 = 0.0248$.

Example 3.   Historical data indicate that the diameter of a particular piston size is normally distributed with a mean of 4 inches and a variance of 9×10^{-6} inch.

Diameters greater than 4.006 or less than 3.994 are considered to be defects.   What proportion of finished pistons will be defects?   Here we use

$$\bar{x}' = 4.000 \qquad \sigma' = \sqrt{9 \times 10^{-6}} = 3 \times 10^{-3} = 0.003$$

Then
$$z = \frac{4.006 - 4.000}{0.003} = \frac{0.006}{0.003} = +2.0$$

From Table 3-1 the probability of 2.0 or less is $0.4774 + 0.5000$.   Hence the probability of a piston that is too large is $1 - 0.9774 = 0.0226$.   Now note that the lower limit is 0.006 on the opposite side of the mean from 4.006.   Because the normal is symmetric about its mean value, we can take the probability of falling below 3.994 to be the same as the probability of falling above 4.006.   Thus the probability of a defective part is 2×0.0226, or 0.0452.

Normal as an Approximation to the Binomial.   The binomial can be approximated by the normal density function under certain conditions.

For p' about 0.5 (0.3 to 0.7 might serve as a rough rule of thumb) and N large (greater than 30) the binomial probabilities can be approximated by the probabilities given in Table 3-1.

When using the normal to approximate the binomial, it can be shown that $\bar{x}' = Np'$ and $\sigma' = \sqrt{Np'q'}$.

Example 4.   What is the probability of getting fewer than 400 heads in 900 throws of a penny?   Here p' is 0.5 and N is large.   Thus the normal should serve as a good approximation to the binomial.   Take

$$\bar{x}' = (0.5)(900) = 450$$
$$\sigma' = \sqrt{(900)(\tfrac{1}{2})(\tfrac{1}{2})} = 30 \times \tfrac{1}{2} = 15$$

Then take
$$z = \frac{400 - 450}{15} = -\frac{50}{15} = -3.33$$

From the table, the probability that $z \leq 3.33$ is 0.4994.   Therefore, the probability of 400 or fewer heads in 900 tosses of a penny is $0.5000 - 0.4994 = 0.0006$.

Normal as an Approximation to the Poisson.   For sufficiently large values for λ, the normal density function can also serve as an approximation for the Poisson frequency function.   A rule of thumb is that for λ of 10 or more, the normal is a good approximation.

In using the normal as an approximation to the Poisson, $\bar{x}' = \lambda$ and $\sigma' = \sqrt{\lambda}$.

Example 5.   The daily demand for a product is known to be a Poisson variable with $\lambda = 25$.   What is the probability that a given day's demand will exceed 30 units? Here $\bar{x}' = \lambda = 25$; $\sigma' = \sqrt{\lambda} = 5$.   Then $z = (30 - 25)/5 = +1.00$.   From Table 3-1, the probability of 1.00 or less is $0.3411 + 0.5000$.   Therefore, the probability of exceeding 30 units is $1 - 0.8411 = 0.1589$.

Further Properties of the Normal Curve.   From Table 3-1, the following properties of the normal curve can be obtained.   Referring to Figure 3-3, note:

1. Approximately 68 percent of the area under the curve lies between $\bar{x}' - \sigma'$ and $\bar{x}' + \sigma'$.   About 5 percent of the area lies above $\bar{x}' + 1.645\sigma'$, and 5 percent lies below $\bar{x}' - 1.645\sigma'$.   2.5 percent of the area lies above $\bar{x}' + 1.96\sigma'$, and 2.5 percent lies below $\bar{x}' - 1.96\sigma'$.   Almost all the area lies within $\bar{x}' \pm 3\sigma'$.

2. The peak of the curve occurs at $\bar{x}'$.   The x value at which the peak occurs is called the mode.   For the normal curve, the mode and $\bar{x}'$ occur at the same point.

3. 50 percent of the area under the curve falls on either side of $\bar{x}'$.   The x value

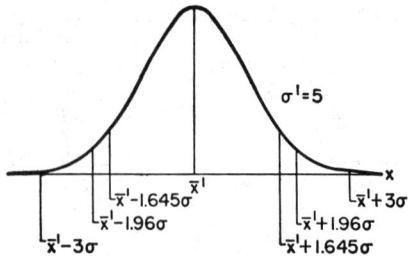

FIG. 3-3. Guidelines for a normal curve.

which divides the area under a density function in two is called the median. For the normal curve, the median and $\bar{x}'$ occur at the same point.

Two Theorems. The following theorems serve as a basis for much of statistics and involve the normal curve.

1. If N values are chosen at random from a normal curve and the number $\bar{x} = \Sigma x/N$ is computed, then $\bar{x}$ has its own density function and the density function is normal with mean value $\bar{x}'$ and standard deviation $\sigma_{\bar{x}} = \sigma'/\sqrt{N}$. This theorem serves for many tests of hypotheses.

2. If N values are chosen at random from a nonnormal curve and the number $\bar{x} = \Sigma x/N$ is computed, then $\bar{x}$ has its own density function and the density function is nearly normal for large values of N.

Other Density Functions. Three other density functions of importance are the t function, the χ^2 (chi-squared) function, and the F function. The use of these will be shown later in the chapter under the heading Statistics. We note here:

1. If N numbers are chosen at random from a normal universe and if $\bar{x}$ and s are calculated for these N numbers (how this is done will be discussed under Statistics), then the value $(\bar{x} - \bar{x}')/(s/\sqrt{N})$ has a density function called the t or Student t density function. It is tabulated in Table 3-2.

2. If N numbers are chosen at random from a normal universe and if s^2 is calculated for these N numbers (how this is done is discussed under Statistics), then the value $(N - 1)s^2/\sigma'^2$ has a density function called the χ^2 (chi-squared) density function. It is tabulated in Table 3-3.

3. If N_1 numbers are chosen at random from a normal universe and if another set of N_2 numbers is chosen from the same universe and if s_1^2/s_2^2 is computed (how this is done is discussed under Statistics), then s_1^2/s_2^2 has a density function called the F density function. It is tabulated in Table 3-4.

All these functions require that a parameter n be defined. The value n associated with the curve is called the degrees of freedom. In the following work, n is explicitly given when necessary.

SAMPLING METHODS

With few exceptions, the industrial engineer must gain knowledge of operations, processes, systems, elements, product quality, and quantity from viewing historical records, output of present activities, or estimates of future activities. Most often, data collection for later analysis must be done on some basis other than complete examination of all possible conditions and the direct measure of the variable of interest from these data. Complete examination is often impossible or undesirable because:

1. The entire process is not available within the time available for data collection.
2. Data collection is expensive.
3. Data collection may destroy the thing that gives the data.
4. Other methods may give better results.

For these reasons, much knowledge of activities is derived from sampling of processes. The most common sampling method is called simple random sampling.

Other sampling methods that are useful in some cases are called stratified sampling,

TABLE 3-2. Distribution of t*

n	Probability (includes both tails)												
	.9	.8	.7	.6	.5	.4	.3	.2	.1	.05	.02	.01	.001
1	.158	.325	.510	.727	1.000	1.376	1.963	3.078	6.314	12.706	31.821	63.657	636.619
2	.142	.289	.445	.617	.816	1.061	1.386	1.886	2.920	4.303	6.965	9.925	31.598
3	.137	.277	.424	.584	.765	.978	1.250	1.638	2.353	3.182	4.541	5.841	12.924
4	.134	.271	.414	.569	.741	.941	1.190	1.533	2.132	2.776	3.747	4.604	8.610
5	.132	.267	.408	.559	.727	.920	1.156	1.476	2.015	2.571	3.365	4.032	6.869
6	.131	.265	.404	.553	.718	.906	1.134	1.440	1.943	2.447	3.143	3.707	5.959
7	.130	.263	.402	.549	.711	.896	1.119	1.415	1.895	2.365	2.998	3.499	5.408
8	.130	.262	.399	.546	.706	.889	1.108	1.397	1.860	2.306	2.896	3.355	5.041
9	.129	.261	.398	.543	.703	.883	1.100	1.383	1.833	2.262	2.821	3.250	4.781
10	.129	.260	.397	.542	.700	.879	1.093	1.372	1.812	2.228	2.764	3.169	4.587
11	.129	.260	.396	.540	.697	.876	1.088	1.363	1.796	2.201	2.718	3.106	4.437
12	.128	.259	.395	.539	.695	.873	1.083	1.356	1.782	2.179	2.681	3.055	4.318
13	.128	.259	.394	.538	.694	.870	1.079	1.350	1.771	2.160	2.650	3.012	4.221
14	.128	.258	.393	.537	.692	.868	1.076	1.345	1.761	2.145	2.624	2.977	4.140
15	.128	.258	.393	.536	.691	.866	1.074	1.341	1.753	2.131	2.602	2.947	4.073
16	.128	.258	.392	.535	.690	.865	1.071	1.337	1.746	2.120	2.583	2.921	4.015
17	.128	.257	.392	.534	.689	.863	1.069	1.333	1.740	2.110	2.567	2.898	3.965
18	.127	.257	.392	.534	.688	.862	1.067	1.330	1.734	2.101	2.552	2.878	3.922
19	.127	.257	.391	.533	.688	.861	1.066	1.328	1.729	2.093	2.539	2.861	3.883
20	.127	.257	.391	.533	.687	.860	1.064	1.325	1.725	2.086	2.528	2.845	3.850
21	.127	.257	.391	.532	.686	.859	1.063	1.323	1.721	2.080	2.518	2.831	3.819
22	.127	.256	.390	.532	.686	.858	1.061	1.321	1.717	2.074	2.508	2.819	3.792
23	.127	.256	.390	.532	.685	.858	1.060	1.319	1.714	2.069	2.500	2.807	3.767
24	.127	.256	.390	.531	.685	.857	1.059	1.318	1.711	2.064	2.492	2.797	3.745
25	.127	.256	.390	.531	.684	.856	1.058	1.316	1.708	2.060	2.485	2.787	3.725
26	.127	.256	.390	.531	.684	.856	1.058	1.315	1.706	2.056	2.479	2.779	3.707
27	.127	.256	.389	.531	.684	.855	1.057	1.314	1.703	2.052	2.473	2.771	3.690
28	.127	.256	.389	.530	.683	.855	1.056	1.313	1.701	2.048	2.467	2.763	3.674
29	.127	.256	.389	.530	.683	.854	1.055	1.311	1.699	2.045	2.462	2.756	3.659
30	.127	.256	.389	.530	.683	.854	1.055	1.310	1.697	2.042	2.457	2.750	3.646
40	.126	.255	.388	.529	.681	.851	1.050	1.303	1.684	2.021	2.423	2.704	3.551
60	.126	.254	.387	.527	.679	.848	1.046	1.296	1.671	2.000	2.390	2.660	3.460
120	.126	.254	.386	.526	.677	.845	1.041	1.289	1.658	1.980	2.358	2.617	3.373
∞	.126	.253	.385	.524	.674	.842	1.036	1.282	1.645	1.960	2.326	2.576	3.291

* Table 3-2 is abridged from Table III of Fisher and Yates, *Statistical Tables for Biological, Agricultural, and Medical Research*, 4th ed., Oliver & Boyd, Ltd., Edinburgh, 1953, and used by permission of the authors and publishers.

systematic sampling, and cluster sampling. Each of these has properties that cannot be discussed here.

Simple Random Sampling. The basic concept of simple random sampling is that, if a sample of size N is drawn from a universe of size S, then each element in S should have the same chance of being in the sample. To ensure this equally probable criterion, simple random sampling often starts by numbering each element in the universe. Following this, it is an easy matter to consult a table of random numbers such as Table 3-5 and read N consecutive numbers. These N numbers then identify those elements in the universe that are to be placed in the sample. Because the random numbers are built on the concept of "equally probable," their use allows a sample to

TABLE 3-3. Distribution of χ²*

Probability

0.001	0.01	0.02	0.05	0.10	0.20	0.30	0.50	0.70	0.80	0.90	0.95	0.98	0.99	n
10.827	6.635	5.412	3.841	2.706	1.642	1.074	0.455	0.148	0.0642	0.0158	0.00393	0.0^3628	0.0^3157	1
13.815	9.210	7.824	5.991	4.605	3.219	2.408	1.386	0.173	0.446	0.211	0.103	0.0404	0.0201	2
16.266	11.345	9.837	7.815	6.251	4.642	3.665	2.366	1.424	1.005	0.584	0.352	0.185	0.115	3
18.467	13.277	11.668	9.488	7.779	5.989	4.878	3.357	2.195	1.649	1.064	0.711	0.429	0.297	4
20.515	15.086	13.388	11.070	9.236	7.289	6.064	4.351	3.000	2.343	1.610	1.145	0.752	0.554	5
22.457	16.812	15.033	12.592	10.645	8.558	7.231	5.348	3.828	3.070	2.204	1.635	1.134	0.872	6
24.322	18.475	16.622	14.067	12.017	9.803	8.383	6.346	4.671	3.822	2.833	2.167	1.564	1.239	7
26.125	20.090	18.168	15.507	13.362	11.030	9.524	7.344	5.527	4.594	3.490	2.733	2.032	1.646	8
27.877	21.666	19.679	16.919	14.684	12.242	10.656	8.343	6.393	5.380	4.168	3.325	2.532	2.088	9
29.588	23.209	21.161	18.307	15.987	13.442	11.781	9.342	7.267	6.179	4.865	3.940	3.059	2.558	10
31.264	24.725	22.618	19.675	17.275	14.631	12.899	10.341	8.148	6.989	5.578	4.575	3.609	3.053	11
32.909	26.217	24.054	21.026	18.549	15.812	14.011	11.340	9.034	7.807	6.304	5.226	4.178	3.571	12
34.528	27.688	25.472	22.362	19.812	16.985	15.119	12.340	9.926	8.634	7.042	5.892	4.765	4.107	13
36.123	29.141	26.873	23.685	21.064	18.151	16.222	13.339	10.821	9.467	7.790	6.571	5.368	4.660	14
37.697	30.578	28.259	24.996	22.307	19.311	17.322	14.339	11.721	10.307	8.547	7.261	5.985	5.229	15
39.252	32.000	29.633	26.296	23.542	20.465	18.418	15.338	12.624	11.152	9.312	7.962	6.614	5.812	16
40.790	33.409	30.995	27.587	24.769	21.615	19.511	16.338	13.531	12.002	10.085	8.672	7.255	6.408	17
42.312	34.805	32.346	28.869	25.989	22.760	20.601	17.338	14.440	12.857	10.865	9.390	7.906	7.015	18
43.820	36.191	33.687	30.144	27.204	23.900	21.689	18.338	15.352	13.716	11.651	10.117	8.567	7.633	19
45.315	37.566	35.020	31.410	28.412	25.038	22.775	19.337	16.266	14.578	12.443	10.851	9.237	8.260	20
46.797	38.932	36.343	32.671	29.615	26.171	23.858	20.337	17.182	15.445	13.240	11.591	9.915	8.897	21
48.268	40.289	37.659	33.924	30.813	27.301	24.939	21.337	18.101	16.314	14.041	12.338	10.600	9.542	22
49.728	41.638	38.968	35.172	32.007	28.429	26.018	22.337	19.021	17.187	14.848	13.091	11.293	10.196	23
51.179	42.980	40.270	36.415	33.196	29.553	27.096	23.337	19.943	18.062	15.659	13.848	11.992	10.856	24
52.620	44.314	41.566	37.652	34.382	30.675	28.172	24.337	20.867	18.940	16.473	14.611	12.697	11.524	25
54.052	45.642	42.856	38.885	35.563	31.795	29.246	25.336	21.792	19.820	17.292	15.379	13.409	12.198	26
55.476	46.963	44.140	40.113	36.741	32.912	30.319	26.336	22.719	20.703	18.114	16.151	14.125	12.879	27
56.893	48.278	45.419	41.337	37.916	34.027	31.391	27.336	23.647	21.588	18.939	16.928	14.847	13.565	28
58.302	49.588	46.693	42.557	39.087	35.139	32.461	28.336	24.577	22.475	19.768	17.708	15.574	14.256	29
59.703	50.892	47.962	43.773	40.256	36.250	33.530	29.336	25.508	23.364	20.599	18.493	16.306	14.953	30

* Table 3-3 is abridged from Table IV of Fisher and Yates, *Statistical Tables for Biological, Agricultural, and Medical Research*, 4th ed., Oliver & Boyd, Ltd., Edinburgh, 1953, and used by permission of the authors and publishers.

be drawn so that every element in the universe has the same chance of entering the sample and the criterion for random sampling is met.

Example. In a work sampling study, it is decided to take 10 readings over a period of one 8-hour day. Because of travel time and preparation time, the decision is made to have observations no less than 15 minutes apart. Because the observations are a small part of the 15-minute interval, it is decided to take a random sample of ten 15-minute intervals.

The universe size S (number of 15-minute intervals over an 8-hour day) is 32, from which a random sample of 10 intervals is to be selected. The question is which 10 intervals to include.

Numbering the intervals gives:

8:00–8:15	1	10:00–10:15	9	1:00–1:15	17	3:00–3:15	25
8:15–8:30	2	10:15–10:30	10	1:15–1:30	18	3:15–3:30	26
8:30–8:45	3	10:30–10:45	11	1:30–1:45	19	3:30–3:45	27
8:45–9:00	4	10:45–11:00	12	1:45–2:00	20	3:45–4:00	28
9:00–9:15	5	11:00–11:15	13	2:00–2:15	21	4:00–4:15	29
9:15–9:30	6	11:15–11:30	14	2:15–2:30	22	4:15–4:30	30
9:30–9:45	7	11:30–11:45	15	2:30–2:45	23	4:30–4:45	31
9:45–10:00	8	11:45–12:00	16	2:45–3:00	24	4:45–5:00	32

Having identified the periods, Table 3-5 gives a set of random numbers that can be used to choose the sample. Starting anywhere in the table (say, column 3 and row 2), ten 2-digit numbers are used. These are 17, 5, 30, 19, 13, 31, 15, 21, 28, and 10. Note that numbers not in the above numbering of intervals are passed over.

Then, for random sampling of these periods, the observations should be made at

9:00–9:15	11:00–11:15	1:00–1:15	2:00–2:15	4:15–4:30
10:15–10:30	11:30–11:45	1:30–1:45	3:45–4:00	4:30–4:45

STATISTICS

In many industrial engineering applications, one or often at most two values are chosen to represent the sample results. The most commonly used measures from the sample are (1) sample mean or (2) sample standard deviation or variance.

Sample Mean. 1. The formula for the sample mean for ungrouped data is

$$\bar{x} = \sum_{i=1}^{N} \frac{x_i}{N}$$

where the sub- and superscripts on Σ imply to add over all sample values. This value is also called the average.

Example 6. *a.* The mean of the following values is found as shown. Suppose the following represent the number of errors per page found in a sample of 25 pages for a typing pool. The problem is to determine the sample mean.

2	2	1	5	3
1	2	3	6	2
3	4	7	0	2
5	6	0	1	3
4	0	4	1	1

Then $\sum_{i=1}^{25} x_i = 2 + 1 + 3 + 5 + 4 + \cdots + 3 + 2 + 2 + 3 + 1 = 68$. (The dots here indicate that the sum is continued for all the remaining values but their precise values are not shown.) Thus

$$\bar{x} = \sum_{i=1}^{25} \frac{x_i}{N} = {}^{68}\!/_{25} = 2.72 \text{ errors per page}$$

TABLE 3-4. 5 and 1 Percent (Underlined) Points for the Distribution of F*

n_1 degrees of freedom (for greater mean square)

n_2	1	2	3	4	5	6	7	8	9	10	11	12	14	16	20	24	30	40	50	75	100	200	500	∞
1	161 / 4,052	200 / 4,999	216 / 5,403	225 / 5,625	230 / 5,764	234 / 5,859	237 / 5,928	239 / 5,981	241 / 6,022	242 / 6,056	243 / 6,082	244 / 6,106	245 / 6,142	246 / 6,169	248 / 6,208	249 / 6,234	250 / 6,258	251 / 6,286	252 / 6,302	253 / 6,323	253 / 6,334	254 / 6,352	254 / 6,361	254 / 6,366
2	18.51 / 98.49	19.00 / 99.00	19.16 / 99.17	19.25 / 99.25	19.30 / 99.30	19.33 / 99.33	19.36 / 99.34	19.37 / 99.36	19.38 / 99.38	19.39 / 99.40	19.40 / 99.41	19.41 / 99.42	19.42 / 99.43	19.43 / 99.44	19.44 / 99.45	19.45 / 99.46	19.46 / 99.47	19.47 / 99.48	19.47 / 99.48	19.48 / 99.49	19.49 / 99.49	19.49 / 99.49	19.50 / 99.50	19.50 / 99.50
3	10.13 / 34.12	9.55 / 30.82	9.28 / 29.46	9.12 / 28.71	9.01 / 28.24	8.94 / 27.91	8.88 / 27.67	8.84 / 27.49	8.81 / 27.34	8.78 / 27.23	8.76 / 27.13	8.74 / 27.05	8.71 / 26.92	8.69 / 26.83	8.66 / 26.69	8.64 / 26.60	8.62 / 26.50	8.60 / 26.41	8.58 / 26.35	8.57 / 26.27	8.56 / 26.23	8.54 / 26.18	8.54 / 26.14	8.53 / 26.12
4	7.71 / 21.20	6.94 / 18.00	6.59 / 16.69	6.39 / 15.98	6.26 / 15.52	6.16 / 15.21	6.09 / 14.98	6.04 / 14.80	6.00 / 14.66	5.96 / 14.54	5.93 / 14.45	5.91 / 14.37	5.87 / 14.24	5.84 / 14.15	5.80 / 14.02	5.77 / 13.93	5.74 / 13.83	5.71 / 13.74	5.70 / 13.69	5.68 / 13.61	5.66 / 13.57	5.65 / 13.52	5.64 / 13.48	5.63 / 13.46
5	6.61 / 16.26	5.79 / 13.27	5.41 / 12.06	5.19 / 11.39	5.05 / 10.97	4.95 / 10.67	4.88 / 10.45	4.82 / 10.27	4.78 / 10.15	4.74 / 10.05	4.70 / 9.96	4.68 / 9.89	4.64 / 9.77	4.60 / 9.68	4.56 / 9.55	4.53 / 9.47	4.50 / 9.38	4.46 / 9.29	4.44 / 9.24	4.42 / 9.17	4.40 / 9.13	4.38 / 9.07	4.37 / 9.04	4.36 / 9.02
6	5.99 / 13.74	5.14 / 10.92	4.76 / 9.78	4.53 / 9.15	4.39 / 8.75	4.28 / 8.47	4.21 / 8.26	4.15 / 8.10	4.10 / 7.98	4.06 / 7.87	4.03 / 7.79	4.00 / 7.72	3.96 / 7.60	3.92 / 7.52	3.87 / 7.39	3.84 / 7.31	3.81 / 7.23	3.77 / 7.14	3.75 / 7.09	3.72 / 7.02	3.71 / 6.99	3.69 / 6.94	3.68 / 6.90	3.67 / 6.88
7	5.59 / 12.25	4.74 / 9.55	4.35 / 8.45	4.12 / 7.85	3.97 / 7.46	3.87 / 7.19	3.79 / 7.00	3.73 / 6.84	3.68 / 6.71	3.63 / 6.62	3.60 / 6.54	3.57 / 6.47	3.52 / 6.35	3.49 / 6.27	3.44 / 6.15	3.41 / 6.07	3.38 / 5.98	3.34 / 5.90	3.32 / 5.85	3.29 / 5.78	3.28 / 5.75	3.25 / 5.70	3.24 / 5.67	3.23 / 5.65
8	5.32 / 11.26	4.46 / 8.65	4.07 / 7.59	3.84 / 7.01	3.69 / 6.63	3.58 / 6.37	3.50 / 6.19	3.44 / 6.03	3.39 / 5.91	3.34 / 5.82	3.31 / 5.74	3.28 / 5.67	3.23 / 5.56	3.20 / 5.48	3.15 / 5.36	3.12 / 5.28	3.08 / 5.20	3.05 / 5.11	3.03 / 5.06	3.00 / 5.00	2.98 / 4.96	2.96 / 4.91	2.94 / 4.88	2.93 / 4.86
9	5.12 / 10.56	4.26 / 8.02	3.86 / 6.99	3.63 / 6.42	3.48 / 6.06	3.37 / 5.80	3.29 / 5.62	3.23 / 5.47	3.18 / 5.35	3.13 / 5.26	3.10 / 5.18	3.07 / 5.11	3.02 / 5.00	2.98 / 4.92	2.93 / 4.80	2.90 / 4.73	2.86 / 4.64	2.82 / 4.56	2.80 / 4.51	2.77 / 4.45	2.76 / 4.41	2.73 / 4.36	2.72 / 4.33	2.71 / 4.31
10	4.96 / 10.04	4.10 / 7.56	3.71 / 6.55	3.48 / 5.99	3.33 / 5.64	3.22 / 5.39	3.14 / 5.21	3.07 / 5.06	3.02 / 4.95	2.97 / 4.85	2.94 / 4.78	2.91 / 4.71	2.86 / 4.60	2.82 / 4.52	2.77 / 4.41	2.74 / 4.33	2.70 / 4.25	2.67 / 4.17	2.64 / 4.12	2.61 / 4.05	2.59 / 4.01	2.56 / 3.96	2.55 / 3.93	2.54 / 3.91
11	4.84 / 9.65	3.98 / 7.20	3.59 / 6.22	3.36 / 5.67	3.20 / 5.32	3.09 / 5.07	3.01 / 4.88	2.95 / 4.74	2.90 / 4.63	2.86 / 4.54	2.82 / 4.46	2.79 / 4.40	2.74 / 4.29	2.70 / 4.21	2.65 / 4.10	2.61 / 4.02	2.57 / 3.94	2.53 / 3.86	2.50 / 3.80	2.47 / 3.74	2.45 / 3.70	2.42 / 3.66	2.41 / 3.62	2.40 / 3.60
12	4.75 / 9.33	3.88 / 6.93	3.49 / 5.95	3.26 / 5.41	3.11 / 5.06	3.00 / 4.82	2.92 / 4.65	2.85 / 4.50	2.80 / 4.39	2.76 / 4.30	2.72 / 4.22	2.69 / 4.16	2.64 / 4.05	2.60 / 3.98	2.54 / 3.86	2.50 / 3.78	2.46 / 3.70	2.42 / 3.61	2.40 / 3.56	2.36 / 3.49	2.35 / 3.46	2.32 / 3.41	2.31 / 3.38	2.30 / 3.36

13	2.21	2.22	2.24	2.26	2.28	2.32	2.34	2.38	2.42	2.46	2.51	2.55	2.60	2.63	2.67	2.72	2.77	2.84	2.92	3.02	3.18	3.41	3.80	4.67
	3.16	3.18	3.21	3.27	3.30	3.37	3.42	3.51	3.59	3.67	3.78	3.85	3.96	4.02	4.10	4.19	4.30	4.44	4.62	4.86	5.20	5.74	6.70	9.07
14	2.13	2.14	2.16	2.19	2.21	2.24	2.27	2.31	2.35	2.39	2.44	2.48	2.53	2.56	2.60	2.65	2.70	2.77	2.85	2.96	3.11	3.34	3.74	4.60
	3.00	3.02	3.06	3.11	3.14	3.21	3.26	3.34	3.43	3.51	3.62	3.70	3.80	3.86	3.94	4.03	4.14	4.28	4.46	4.69	5.03	5.56	6.51	8.86
15	2.07	2.08	2.10	2.12	2.15	2.18	2.21	2.25	2.29	2.33	2.39	2.43	2.48	2.51	2.55	2.59	2.64	2.70	2.79	2.90	3.06	3.29	3.68	4.54
	2.87	2.89	2.92	2.97	3.00	3.07	3.12	3.20	3.29	3.36	3.48	3.56	3.67	3.73	3.80	3.89	4.00	4.14	4.32	4.56	4.89	5.42	6.36	8.68
16	2.01	2.02	2.04	2.07	2.09	2.13	2.16	2.20	2.24	2.28	2.33	2.37	2.42	2.45	2.49	2.54	2.59	2.66	2.74	2.85	3.01	3.24	3.63	4.49
	2.75	2.77	2.80	2.86	2.89	2.96	3.01	3.10	3.18	3.25	3.37	3.45	3.55	3.61	3.69	3.78	3.89	4.03	4.20	4.44	4.77	5.29	6.23	8.53
17	1.96	1.97	1.99	2.02	2.04	2.08	2.11	2.15	2.19	2.23	2.29	2.33	2.38	2.41	2.45	2.50	2.55	2.62	2.70	2.81	2.96	3.20	3.59	4.45
	2.65	2.67	2.70	2.76	2.79	2.86	2.92	3.00	3.08	3.16	3.27	3.35	3.45	3.52	3.59	3.68	3.79	3.93	4.10	4.34	4.67	5.18	6.11	8.40
18	1.92	1.93	1.95	1.98	2.00	2.04	2.07	2.11	2.15	2.19	2.25	2.29	2.34	2.37	2.41	2.46	2.51	2.58	2.66	2.77	2.93	3.16	3.55	4.41
	2.57	2.59	2.62	2.68	2.71	2.78	2.83	2.91	3.00	3.07	3.19	3.27	3.37	3.44	3.51	3.60	3.71	3.85	4.01	4.25	4.58	5.09	6.01	8.28
19	1.88	1.90	1.91	1.94	1.96	2.00	2.02	2.07	2.11	2.15	2.21	2.26	2.31	2.34	2.38	2.43	2.48	2.55	2.63	2.74	2.90	3.13	3.52	4.38
	2.49	2.51	2.54	2.60	2.63	2.70	2.76	2.84	2.92	3.00	3.12	3.19	3.30	3.36	3.43	3.52	3.63	3.77	3.94	4.17	4.50	5.01	5.93	8.18
20	1.84	1.85	1.87	1.90	1.92	1.96	1.99	2.04	2.08	2.12	2.18	2.23	2.28	2.31	2.35	2.40	2.45	2.52	2.60	2.71	2.87	3.10	3.49	4.35
	2.42	2.44	2.47	2.53	2.56	2.63	2.69	2.77	2.86	2.94	3.05	3.13	3.23	3.30	3.37	3.45	3.56	3.71	3.87	4.10	4.43	4.94	5.85	8.10
21	1.81	1.82	1.84	1.87	1.89	1.93	1.96	2.00	2.05	2.09	2.15	2.20	2.25	2.28	2.32	2.37	2.42	2.49	2.57	2.68	2.84	3.07	3.47	4.32
	2.36	2.38	2.42	2.47	2.51	2.58	2.63	2.72	2.80	2.88	2.99	3.07	3.17	3.24	3.31	3.40	3.51	3.65	3.81	4.04	4.37	4.87	5.78	8.02
22	1.78	1.80	1.81	1.84	1.87	1.91	1.93	1.98	2.03	2.07	2.13	2.18	2.23	2.26	2.30	2.35	2.40	2.47	2.55	2.66	2.82	3.05	3.44	4.30
	2.31	2.33	2.37	2.42	2.46	2.53	2.58	2.67	2.75	2.83	2.94	3.02	3.12	3.18	3.26	3.35	3.45	3.59	3.76	3.99	4.31	4.82	5.72	7.94
23	1.76	1.77	1.79	1.82	1.84	1.88	1.91	1.96	2.00	2.04	2.10	2.14	2.20	2.24	2.28	2.32	2.38	2.45	2.53	2.64	2.80	3.03	3.42	4.28
	2.26	2.28	2.32	2.37	2.41	2.48	2.53	2.62	2.70	2.78	2.89	2.97	3.07	3.14	3.21	3.30	3.41	3.54	3.71	3.94	4.26	4.76	5.66	7.88
24	1.73	1.74	1.76	1.80	1.82	1.86	1.89	1.94	1.98	2.02	2.09	2.13	2.18	2.22	2.26	2.30	2.36	2.43	2.51	2.62	2.78	3.01	3.40	4.26
	2.21	2.23	2.27	2.33	2.36	2.44	2.49	2.58	2.66	2.74	2.85	2.93	3.03	3.09	3.17	3.25	3.36	3.50	3.67	3.90	4.22	4.72	5.61	7.82
25	1.71	1.72	1.74	1.77	1.80	1.84	1.87	1.92	1.96	2.00	2.06	2.11	2.16	2.20	2.24	2.28	2.34	2.41	2.49	2.60	2.76	2.99	3.38	4.24
	2.17	2.19	2.23	2.29	2.32	2.40	2.45	2.54	2.62	2.70	2.81	2.89	2.99	3.05	3.13	3.21	3.32	3.46	3.63	3.86	4.18	4.68	5.57	7.77
26	1.69	1.70	1.72	1.76	1.78	1.82	1.85	1.90	1.95	1.99	2.05	2.10	2.15	2.18	2.22	2.27	2.32	2.39	2.47	2.59	2.74	2.98	3.37	4.22
	2.13	2.15	2.19	2.25	2.28	2.36	2.41	2.50	2.58	2.66	2.77	2.86	2.96	3.02	3.09	3.17	3.29	3.42	3.59	3.82	4.14	4.64	5.53	7.72
27	1.67	1.68	1.71	1.74	1.76	1.80	1.84	1.88	1.93	1.97	2.03	2.08	2.13	2.16	2.20	2.25	2.30	2.37	2.46	2.57	2.73	2.96	3.35	4.21
	2.10	2.12	2.16	2.21	2.25	2.33	2.38	2.47	2.55	2.63	2.74	2.82	2.93	2.98	3.06	3.14	3.26	3.39	3.56	3.79	4.11	4.60	5.49	7.68

* Reprinted by permission from George W. Snedecor, *Statistical Methods*, 5th ed., The Iowa State University Press, Ames, 1956, pp. 246–249.

TABLE 3-4. (Continued)

n_1 degrees of freedom (for greater mean square)

n_2	1	2	3	4	5	6	7	8	9	10	11	12	14	16	20	24	30	40	50	75	100	200	500	∞
28	4.20 / 7.64	3.34 / 5.45	2.95 / 4.57	2.71 / 4.07	2.56 / 3.76	2.44 / 3.53	2.36 / 3.36	2.29 / 3.23	2.24 / 3.11	2.19 / 3.03	2.15 / 2.95	2.12 / 2.90	2.06 / 2.80	2.02 / 2.71	1.96 / 2.60	1.91 / 2.52	1.87 / 2.44	1.81 / 2.35	1.78 / 2.30	1.75 / 2.22	1.72 / 2.18	1.69 / 2.13	1.67 / 2.09	1.65 / 2.06
29	4.18 / 7.60	3.33 / 5.42	2.93 / 4.54	2.70 / 4.04	2.54 / 3.73	2.43 / 3.50	2.35 / 3.33	2.28 / 3.20	2.22 / 3.08	2.18 / 3.00	2.14 / 2.92	2.10 / 2.87	2.05 / 2.77	2.00 / 2.68	1.94 / 2.57	1.90 / 2.49	1.85 / 2.41	1.80 / 2.32	1.77 / 2.27	1.73 / 2.19	1.71 / 2.15	1.68 / 2.10	1.65 / 2.06	1.64 / 2.03
30	4.17 / 7.56	3.32 / 5.39	2.92 / 4.51	2.69 / 4.02	2.53 / 3.70	2.42 / 3.47	2.34 / 3.30	2.27 / 3.17	2.21 / 3.06	2.16 / 2.98	2.12 / 2.90	2.09 / 2.84	2.04 / 2.74	1.99 / 2.66	1.93 / 2.55	1.89 / 2.47	1.84 / 2.38	1.79 / 2.29	1.76 / 2.24	1.72 / 2.16	1.69 / 2.13	1.66 / 2.07	1.64 / 2.03	1.62 / 2.01
32	4.15 / 7.50	3.30 / 5.34	2.90 / 4.46	2.67 / 3.97	2.51 / 3.66	2.40 / 3.42	2.32 / 3.25	2.25 / 3.12	2.19 / 3.01	2.14 / 2.94	2.10 / 2.86	2.07 / 2.80	2.02 / 2.70	1.97 / 2.62	1.91 / 2.51	1.86 / 2.42	1.82 / 2.34	1.76 / 2.25	1.74 / 2.20	1.69 / 2.12	1.67 / 2.08	1.64 / 2.02	1.61 / 1.98	1.59 / 1.96
34	4.13 / 7.44	3.28 / 5.29	2.88 / 4.42	2.65 / 3.93	2.49 / 3.61	2.38 / 3.38	2.30 / 3.21	2.23 / 3.08	2.17 / 2.97	2.12 / 2.89	2.08 / 2.82	2.05 / 2.76	2.00 / 2.66	1.95 / 2.58	1.89 / 2.47	1.84 / 2.38	1.80 / 2.30	1.74 / 2.21	1.71 / 2.15	1.67 / 2.08	1.64 / 2.04	1.61 / 1.98	1.59 / 1.94	1.57 / 1.91
36	4.11 / 7.39	3.26 / 5.25	2.86 / 4.38	2.63 / 3.89	2.48 / 3.58	2.36 / 3.35	2.28 / 3.18	2.21 / 3.04	2.15 / 2.94	2.10 / 2.86	2.06 / 2.78	2.03 / 2.72	1.98 / 2.62	1.93 / 2.54	1.87 / 2.43	1.82 / 2.35	1.78 / 2.26	1.72 / 2.17	1.69 / 2.12	1.65 / 2.04	1.62 / 2.00	1.59 / 1.94	1.56 / 1.90	1.55 / 1.87
38	4.10 / 7.35	3.25 / 5.21	2.85 / 4.34	2.62 / 3.86	2.46 / 3.54	2.35 / 3.32	2.26 / 3.15	2.19 / 3.02	2.14 / 2.91	2.09 / 2.82	2.05 / 2.75	2.02 / 2.69	1.96 / 2.59	1.92 / 2.51	1.85 / 2.40	1.80 / 2.32	1.76 / 2.22	1.71 / 2.14	1.67 / 2.08	1.63 / 2.00	1.60 / 1.97	1.57 / 1.90	1.54 / 1.86	1.53 / 1.84
40	4.08 / 7.31	3.23 / 5.18	2.84 / 4.31	2.61 / 3.83	2.45 / 3.51	2.34 / 3.29	2.25 / 3.12	2.18 / 2.99	2.12 / 2.88	2.07 / 2.80	2.04 / 2.73	2.00 / 2.66	1.95 / 2.56	1.90 / 2.49	1.84 / 2.37	1.79 / 2.29	1.74 / 2.20	1.69 / 2.11	1.66 / 2.05	1.61 / 1.97	1.59 / 1.94	1.55 / 1.88	1.53 / 1.84	1.51 / 1.81
42	4.07 / 7.27	3.22 / 5.15	2.83 / 4.29	2.59 / 3.80	2.44 / 3.49	2.32 / 3.26	2.24 / 3.10	2.17 / 2.96	2.11 / 2.86	2.06 / 2.77	2.02 / 2.70	1.99 / 2.64	1.94 / 2.54	1.89 / 2.46	1.82 / 2.35	1.78 / 2.26	1.73 / 2.17	1.68 / 2.08	1.64 / 2.02	1.60 / 1.94	1.57 / 1.91	1.54 / 1.85	1.51 / 1.80	1.49 / 1.78
44	4.06 / 7.24	3.21 / 5.12	2.82 / 4.26	2.58 / 3.78	2.43 / 3.46	2.31 / 3.24	2.23 / 3.07	2.16 / 2.94	2.10 / 2.84	2.05 / 2.75	2.01 / 2.68	1.98 / 2.62	1.92 / 2.52	1.88 / 2.44	1.81 / 2.32	1.76 / 2.24	1.72 / 2.15	1.66 / 2.06	1.63 / 2.00	1.58 / 1.92	1.56 / 1.88	1.52 / 1.82	1.50 / 1.78	1.48 / 1.75
46	4.05 / 7.21	3.20 / 5.10	2.81 / 4.24	2.57 / 3.76	2.42 / 3.44	2.30 / 3.22	2.22 / 3.05	2.14 / 2.92	2.09 / 2.82	2.04 / 2.73	2.00 / 2.66	1.97 / 2.60	1.91 / 2.50	1.87 / 2.42	1.80 / 2.30	1.75 / 2.22	1.71 / 2.13	1.65 / 2.04	1.62 / 1.98	1.57 / 1.90	1.54 / 1.86	1.51 / 1.80	1.48 / 1.76	1.46 / 1.72

48	1.45 1.70	1.47 1.73	1.50 1.78	1.53 1.84	1.56 1.88	1.61 1.96	1.64 2.02	1.70 2.11	1.74 2.20	1.79 2.28	1.86 2.40	1.90 2.48	1.96 2.58	1.99 2.64	2.03 2.71	2.08 2.80	2.14 2.90	2.21 3.04	2.30 3.20	2.41 3.42	2.56 3.74	2.80 4.22	3.19 5.08	4.04 7.19
50	1.44 1.68	1.46 1.71	1.48 1.76	1.52 1.82	1.55 1.86	1.60 1.94	1.63 2.00	1.69 2.10	1.74 2.18	1.78 2.26	1.85 2.39	1.90 2.46	1.95 2.56	1.98 2.62	2.02 2.70	2.07 2.78	2.13 2.88	2.20 3.02	2.29 3.18	2.40 3.41	2.56 3.72	2.79 4.20	3.18 5.06	4.03 7.17
55	1.41 1.64	1.43 1.66	1.46 1.71	1.50 1.78	1.52 1.82	1.58 1.90	1.61 1.96	1.67 2.06	1.72 2.15	1.76 2.23	1.83 2.35	1.88 2.43	1.93 2.53	1.97 2.59	2.00 2.66	2.05 2.75	2.11 2.85	2.18 2.98	2.27 3.15	2.38 3.37	2.54 3.68	2.78 4.16	3.17 5.01	4.02 7.12
60	1.39 1.60	1.41 1.63	1.44 1.68	1.48 1.74	1.50 1.79	1.56 1.87	1.59 1.93	1.65 2.03	1.70 2.12	1.75 2.20	1.81 2.32	1.86 2.40	1.92 2.50	1.95 2.56	1.99 2.63	2.04 2.72	2.10 2.82	2.17 2.95	2.25 3.12	2.37 3.34	2.52 3.65	2.76 4.13	3.15 4.98	4.00 7.08
65	1.37 1.56	1.39 1.60	1.42 1.64	1.46 1.71	1.49 1.76	1.54 1.84	1.57 1.90	1.63 2.00	1.68 2.09	1.73 2.18	1.80 2.30	1.85 2.37	1.90 2.47	1.94 2.54	1.98 2.61	2.02 2.70	2.08 2.79	2.15 2.93	2.24 3.09	2.36 3.31	2.51 3.62	2.75 4.10	3.14 4.95	3.99 7.04
70	1.35 1.53	1.37 1.56	1.40 1.62	1.45 1.69	1.47 1.74	1.53 1.82	1.56 1.88	1.62 1.98	1.67 2.07	1.72 2.15	1.79 2.28	1.84 2.35	1.89 2.45	1.93 2.51	1.97 2.59	2.01 2.67	2.07 2.77	2.14 2.91	2.23 3.07	2.35 3.29	2.50 3.60	2.74 4.08	3.13 4.92	3.98 7.01
80	1.32 1.49	1.35 1.52	1.38 1.57	1.42 1.65	1.45 1.70	1.51 1.78	1.54 1.84	1.60 1.94	1.65 2.03	1.70 2.11	1.77 2.24	1.82 2.32	1.88 2.41	1.91 2.48	1.95 2.55	1.99 2.64	2.05 2.74	2.12 2.87	2.21 3.04	2.33 3.25	2.48 3.56	2.72 4.04	3.11 4.88	3.96 6.96
100	1.28 1.43	1.30 1.46	1.34 1.51	1.39 1.59	1.42 1.64	1.48 1.73	1.51 1.79	1.57 1.89	1.63 1.98	1.68 2.06	1.75 2.19	1.79 2.26	1.85 2.36	1.88 2.43	1.93 2.51	1.97 2.59	2.03 2.69	2.10 2.82	2.19 2.99	2.30 3.20	2.46 3.51	2.70 3.98	3.09 4.82	3.94 6.90
125	1.25 1.37	1.27 1.40	1.31 1.46	1.36 1.54	1.39 1.59	1.45 1.68	1.49 1.75	1.55 1.85	1.60 1.94	1.65 2.03	1.72 2.15	1.77 2.23	1.83 2.33	1.86 2.40	1.90 2.47	1.95 2.56	2.01 2.65	2.08 2.79	2.17 2.95	2.29 3.17	2.44 3.47	2.68 3.94	3.07 4.78	3.92 6.84
150	1.22 1.33	1.25 1.37	1.29 1.43	1.34 1.51	1.37 1.56	1.44 1.66	1.47 1.72	1.54 1.83	1.59 1.91	1.64 2.00	1.71 2.12	1.76 2.20	1.82 2.30	1.85 2.37	1.89 2.44	1.94 2.53	2.00 2.62	2.07 2.76	2.16 2.92	2.27 3.14	2.43 3.44	2.67 3.91	3.06 4.75	3.91 6.81
200	1.19 1.28	1.22 1.33	1.26 1.39	1.32 1.48	1.35 1.53	1.42 1.62	1.45 1.69	1.52 1.79	1.57 1.88	1.62 1.97	1.69 2.09	1.74 2.17	1.80 2.28	1.83 2.34	1.87 2.41	1.92 2.50	1.98 2.60	2.05 2.73	2.14 2.90	2.26 3.11	2.41 3.41	2.65 3.88	3.04 4.71	3.89 6.76
400	1.13 1.19	1.16 1.24	1.22 1.32	1.28 1.42	1.32 1.47	1.38 1.57	1.42 1.64	1.49 1.74	1.54 1.84	1.60 1.92	1.67 2.04	1.72 2.12	1.78 2.23	1.81 2.29	1.85 2.37	1.90 2.46	1.96 2.55	2.03 2.69	2.12 2.85	2.23 3.06	2.39 3.36	2.62 3.83	3.02 4.66	3.86 6.70
1,000	1.08 1.11	1.13 1.19	1.19 1.28	1.26 1.38	1.30 1.44	1.36 1.54	1.41 1.61	1.47 1.71	1.53 1.81	1.58 1.89	1.65 2.01	1.70 2.09	1.76 2.20	1.80 2.26	1.84 2.34	1.89 2.43	1.95 2.53	2.02 2.66	2.10 2.82	2.22 3.04	2.38 3.34	2.61 3.80	3.00 4.62	3.85 6.66
∞	1.00 1.00	1.11 1.15	1.17 1.25	1.24 1.36	1.28 1.41	1.35 1.52	1.40 1.59	1.46 1.69	1.52 1.79	1.57 1.87	1.64 1.99	1.69 2.07	1.75 2.18	1.79 2.24	1.83 2.32	1.88 2.41	1.94 2.51	2.01 2.64	2.09 2.80	2.21 3.02	2.37 3.32	2.60 3.78	2.99 4.60	3.84 6.64

* Reprinted by permission from George W. Snedecor, *Statistical Methods*, 5th ed., The Iowa State University Press, Ames, 1956, pp. 246–249.

TABLE 3-5.   Random Numbers

00101	13534	52050	10106	62164	52363	38603	34193	46146	67727
75907	71521	17604	41418	90770	05524	45262	30121	19543	34689
96398	92293	36287	75856	67701	13636	66903	31411	15429	99008
85709	93222	21884	51695	63138	87685	55818	82104	40942	28040
10698	86901	11278	92215	58348	92749	94661	18152	28188	93770
13757	75588	89781	22787	78603	39241	22627	72066	72534	42467
77203	37627	74981	22936	64291	21929	97054	46108	90487	82902
25846	71652	30039	97641	20382	27967	77213	34918	81453	40380
10019	98139	98552	25069	02313	33405	55957	80373	35190	10691
21274	47846	71501	19207	71689	01532	31001	12265	61637	77492
29949	02005	59845	62368	89197	86205	53247	79336	68361	14725
59135	57202	29612	30431	15741	21010	08814	45717	80221	16215
51098	89494	50119	91884	48429	99156	70800	01391	22863	39293
38726	66152	33216	67562	29055	54173	35477	76882	31141	18960
07668	94172	33827	77369	96792	31287	83791	18330	04761	21982
25843	34428	91056	73548	88145	55391	22413	33967	81533	41440
07065	63009	95527	75875	57770	08126	67396	75144	46919	00380
06381	21762	31977	84637	76228	85738	87746	71981	18634	46067
76793	36272	26474	51154	48147	75301	10193	36926	62101	18775
59537	81186	71423	37077	76520	06285	63001	19131	16278	95252
30954	47528	91125	53558	91407	71739	94055	62067	76471	21456
64507	79132	24837	80515	58837	78105	52902	23725	61703	38008
90354	43747	80735	53540	04334	43068	94161	15659	02793	39309
00487	82031	21064	51201	13095	62269	95672	27373	35444	50147
82495	65196	73160	08150	10412	21518	85685	56096	67230	09925
53373	36217	75548	85381	22510	06449	96035	58280	11337	80681
21483	45443	39557	74442	26425	52715	57203	35067	82548	84289
02360	08120	10432	26004	50198	91817	78816	64756	69919	92021
21422	24756	71181	16167	75857	75588	90592	31067	74909	93291
19945	57474	47861	16019	92775	59346	68838	88409	91724	46581
16514	50103	39188	90377	75492	29064	54328	82385	63778	87918
85324	44009	96578	91672	26757	76807	77526	64146	63990	06770
14165	61753	34835	54124	42114	45373	36986	67516	61919	92358
91092	34158	92128	89212	21960	05273	36403	33575	60551	21603
35178	93272	29277	77860	05907	76841	13742	31449	02195	64300
03047	74253	43255	58286	74999	05814	50193	45005	55245	60190
13758	84929	00370	11937	76955	62396	73174	45979	97372	27931
13783	37427	72102	28444	48992	32805	50421	19943	42007	78244
50607	76993	39371	20107	78743	34267	83755	62796	66912	27539
95797	76710	00814	49333	33854	47315	55211	13398	94673	44877
79931	16430	10735	59423	34853	38082	34064	46992	30280	14112
28284	51834	51535	57200	06101	10418	82686	66978	87080	09100
09133	36810	05719	93635	61770	12769	99275	56960	06940	11839
97141	22533	39708	88294	55648	89885	57670	08890	06462	33327
72417	78400	02178	86497	83279	97255	61264	51091	23056	71707
74865	55238	87478	92264	45613	33388	91588	87442	26657	79967
83888	88168	84668	88360	09362	30469	01908	87778	91392	31304
47349	02981	16989	98656	65910	08700	02563	37543	34900	04735
57648	92981	24345	62976	70205	51339	96383	38813	31412	21519
93792	32193	43684	50419	93057	76892	28819	00800	05027	76741

TABLE 3-5. (Continued)

21203	39809	99066	74097	82590	12192	33264	49187	75979	00039
93388	95908	83008	83801	10268	92336	68382	25522	24274	51698
95357	75619	00087	78775	63303	31857	81028	83992	31158	91052
25335	53856	70504	48449	00761	16105	52773	39916	66478	85318
87512	27279	05262	34063	35826	69362	28250	08321	16413	38846
68827	72471	14917	80603	34350	09270	06818	85591	18873	40013
31874	48458	85750	03711	16945	56934	50107	73866	64608	87710
03060	11483	38391	23239	00952	32903	35240	10854	48597	78194
49246	66022	26104	44773	36843	33502	28474	52418	81807	75231
13932	23925	56370	05877	77084	48077	76264	50614	42884	55574
47032	26940	10882	27699	05603	37994	51928	88969	00874	45697
80637	76229	97697	78173	39260	09163	34613	39778	88972	34564
52736	71481	17939	96143	40011	16037	82245	55312	27872	29996
75672	28127	77816	69676	70697	83933	35112	26655	53997	77578
87163	39911	18431	20556	70250	06561	14700	03139	95666	72417
78233	32651	18873	36734	47911	20059	02584	52852	25193	40326
64416	65800	03615	57990	13104	41791	24469	04241	17020	10547
80915	56233	36310	02195	60819	92014	48454	46577	83097	85221
11999	02815	51346	72197	85357	74058	86730	03611	11398	95375
56215	58605	52271	22407	72535	54169	04245	56139	01546	67974
50716	71123	39481	20230	03792	35588	90056	73040	04964	51087
83329	99498	95602	20217	79676	67597	81381	23146	67775	57698
94672	31094	52575	59139	98595	65310	02926	64220	10482	28673
41453	38646	68344	51975	58758	86669	97987	82505	55620	03405
57970	09595	61784	52678	86549	02151	20614	42245	54115	59020
09783	35780	06044	47227	79597	79157	78886	75825	52307	74207
74814	46894	51665	57897	84438	91740	11596	72204	48629	93117
74214	43031	17444	49023	32109	92733	42203	38013	31702	25815
59308	89959	00073	38550	09046	73005	57749	99092	26886	68551
13943	36232	31389	00009	92810	01870	12761	24077	81011	20642
32257	79058	91564	51688	90558	84874	47649	94698	94951	16210
01466	65716	63322	30034	44915	60939	01985	62226	67381	21100
07764	47759	03995	61063	39056	66274	49771	17417	72288	87082
29378	92113	35388	95346	64942	31013	34698	94068	84747	76201
13333	36331	15221	17027	78025	56134	45991	25638	91760	12055
63476	75156	69731	18792	36106	66463	35725	54509	00245	55056
69768	91740	09799	96732	30949	00081	24887	77756	66133	35418
89657	74577	81795	56722	28223	35061	16489	98560	07811	13378
91174	47556	65583	43831	14612	30345	54932	32161	14869	01458
93814	50108	87023	31983	37909	91179	01814	47727	74581	20279
98354	44045	60431	22074	52499	99252	24705	53945	57433	41064
48446	68181	20959	00922	28279	02590	09825	57018	87791	24770
12112	21477	78899	01915	54692	32330	09380	08201	13289	05223
32747	81828	85648	93088	95951	16965	58645	61342	27978	89264
53252	26919	92142	27005	51377	75209	90833	39920	08601	18123
32614	50781	16634	51430	04039	02076	73099	02306	68977	76173
42578	93638	91416	67795	66036	66814	50941	20895	62833	37452
27243	41471	22661	18894	49161	18996	71310	08119	98023	41209
94195	62204	47878	88039	02054	52776	71580	09518	82049	94098
95830	06932	25872	31729	94431	22096	71403	33788	95654	53483

b. If the data have been arranged in frequency function form before the mean is calculated, the following formula is useful:

$$\bar{x} = \sum_{i=1}^{M} \frac{f_i x_i}{N}$$

where f_i = number of cases in interval i
M = number of intervals

Example 7.   Using the above data, the following frequency function can be found:

Number of errors	*Frequency of occurrence*
0	3
1	5
2	5
3	4
4	3
5	2
6	2
7	1

N = total number 25

Then

$$\bar{x} = \sum_{i=1}^{8} \frac{f_i x_i}{N} = (3 \times 0 + 5 \times 1 + 5 \times 2 + 4 \times 3 + 3 \times 4 + 2 \times 5 + 2 \times 6$$

$$+ 1 \times 7)/25 = {}^{68}\!/_{25} = 2.72 \text{ errors per page}$$

c. A short-cut method of measuring $\bar{x}$ when the data are grouped in a frequency function is given by the formula

$$\bar{x} = \sum_{i=1}^{M} \frac{f_i y_i}{N} \Delta + A$$

Here A is any arbitrarily selected interval (most often chosen near the middle of the frequency function). $y_i = (x_i - A)/\Delta$. $\Delta = x_{i+1} - x_i$.

Example 8.   Using the same data as in example 7:

Number of errors x_i	y_i	Frequency of occurrence
0	−4	3
1	−3	5
2	−2	5
3	−1	4
4	0	3
5	1	2
6	2	2
7	3	1

Then $\displaystyle\sum_{i=1}^{8} \frac{f_i y_i}{N} \Delta = [-4 \times 3 + (-3) \times 5 + (-2) \times 5 + (-1) \times 4 + 0 \times 3$

$$+ 1 \times 2 + 2 \times 2 + 3 \times 1](1)/25 = -32/25 = -1.28$$

and $\displaystyle\sum_{i=1}^{8} \frac{f_i y_i}{N} \Delta + A = -1.28 + 4.00 = 2.72$

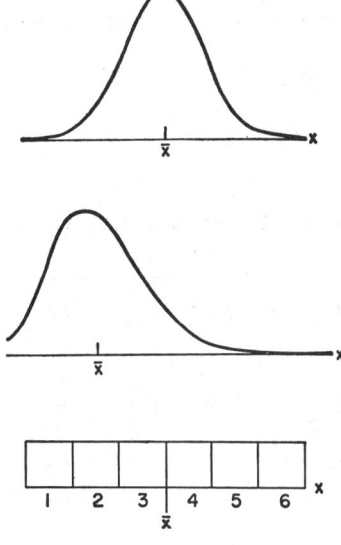

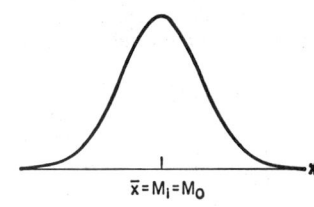

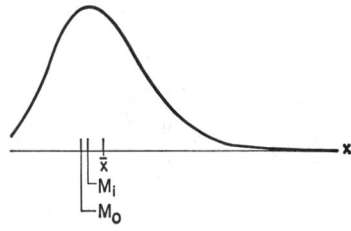

Fig. 3-4. Effect of distribution on $\bar{x}$.

Fig. 3-5. Relation of $\bar{x}$, M_i, M_0 for two curves.

The sample mean is spoken of as a measure of central value. In mechanical terms, the sample mean is similar to the center of gravity of a body. It is seriously affected by extremely large or small values of x_i and for this reason should not be considered as the only measure of central value for a curve. Other values of central measures discussed below may be more indicative of data for some uses. Figure 3-4 shows the behavior of $\bar{x}$ for several curves.

Other Central Measures. For specific purposes, other central measures may indicate behavior of the data not shown by the sample mean. Two of these other measures are median and mode.

Median (M_i). The median is defined as the value for which 50 percent of the cases are larger and 50 percent are smaller. When the frequency function has a long tail on one side, as for example, in income analysis, the median is useful as a central measure. The median does not weigh extreme values of x_i so heavily as does the mean.

Mode (M_0). The mode is defined as the value that is most frequently occurring. If data are arranged in a frequency function, the mode is the value of x which occurs most frequently. It is also used in income analysis and with frequency functions which have a long tail. The mode is not affected by large values of x_i.

Figure 3-5 gives the relative positions for $\bar{x}$, M_i, M_0 for two frequency functions.

Sample Proportions. In some applications, interest is in the proportion of favorable events that occur. A formula for calculating sample proportions is given by the ratio of the number of favorable events to the total number of observations. The resulting ratio is called p, the sample proportion.

Variability Measures. Another important measure for a random variable is its variability. The most commonly used measures of variability are the standard deviation and the range.

The Sample Standard Deviation. The formula for calculating the standard deviation σ for ungrouped data is

$$\sigma = \sqrt{\sum_{i=1}^{N} \frac{(x_i - \bar{x})^2}{N}}$$

The square of σ is called the variance.

An equivalent formula for σ that is often much easier to calculate in practice is

$$\sigma = \sqrt{\frac{\displaystyle\sum_{i=1}^{N} x_i^2 - \frac{\left(\sum_{i=1}^{N} x_i\right)^2}{N}}{N}}$$

Example 9.   Using the same data as in example 6, the standard deviation can be found as shown below.

x_i	x_i^2	x_i	x_i^2	x_i	x_i^2	x_i	x_i^2	x_i	x_i^2
2	4	2	4	1	1	5	25	3	9
1	1	2	4	3	9	6	36	2	4
3	9	4	16	7	49	0	0	2	4
5	25	6	36	0	0	1	1	3	9
4	16	0	0	4	16	1	1	1	1

Then, from example 6, $\displaystyle\sum_{i=1}^{25} x_i = 68$.   From summing the above x^2, the data show $\displaystyle\sum_{i=1}^{25} x_i^2 = 280$.   Then

$$\sigma = \sqrt{\frac{280 - (68)^2/25}{25}} = 1.96$$

The variance would be $(1.96)^2$ or 3.84.

The short formula when the data are in frequency function form gives

$$\sigma^2 = \frac{\displaystyle\sum_{i=1}^{M} f_i(y_i)^2}{N} \Delta^2 - \left(\frac{\displaystyle\sum_{i=1}^{M} f_i y_i}{N} \Delta\right)^2$$

The square root of this is then σ.

Example 10.   Using the results of example 8,

$$\left(\sum_{i=1}^{8} \frac{f_i y_i}{N} \Delta\right)^2 = (-1.28)^2 = 1.64$$

$$\sum_{i=1}^{8} \frac{f_i(y_i)^2}{N} \Delta^2 = (3 \times 16 + 5 \times 9 + 5 \times 4 + 4 \times 1 + 3 \times 0 + 2 \times 1 + 2 \times 4$$

$$+ 1 \times 9)(1)^2/25 = 136/25 = 5.44$$

Then $\sigma^2 = 5.44 - 1.64 = 3.80$ and $\sigma = 1.95$ (differences due to rounding).

The Range.   Another measure of variability is the sample range R.   The formula for calculating range is

$$R = x_h - x_l$$

where x_h is the largest value in the sample and x_l is the smallest value.   From the data of example 6, the range is 7 ($x_h = 7$ and $x_l = 0$).

The range and sample standard deviation can be used to estimate the universe standard deviation σ'. Each has some undesirable properties when used to estimate.

The standard deviation has a downward bias. This means that, if a large number of sample standard deviations were calculated from a universe whose standard deviation was σ', the average of the sample values would consistently underestimate σ'. The amount of bias can be measured, and it is found that the average of a large number of samples gives $\sqrt{(N-1)/N}\,\sigma'$. The bias is not serious if N is large.

The range is difficult to measure in large samples and is inefficient. However, for small samples the relative inefficiency is not great and the ease of calculating R makes it useful, as for example, in quality control where sample sizes may be small.

s^2, *an Unbiased Estimator.* Because of the bias in σ, a frequently used estimate of the universe standard deviation is s. The formula for s is

$$s = \sqrt{\frac{\sum\limits_{i=1}^{N}(x_i - \bar{x})^2}{N-1}} = \sqrt{\frac{\sum\limits_{i=1}^{N}x_i^2 - \dfrac{\left(\sum\limits_{i=1}^{N}x_i\right)^2}{N}}{N-1}}$$

Note that this is nearly the same formula as that for σ; therefore, all the short formulas used for calculating σ apply to s if the denominator N is changed to $N-1$.

Standard Deviation for Proportions. The standard deviation for sampling proportions is given directly from

$$\sigma_p = \sqrt{\frac{p(1-p)}{N}}$$

Confidence Limits. When sample values are used to estimate universe values, a measure of how much confidence can be placed in them is needed. This is answered through the use of confidence limits for the sample values.

Confidence limits are placed so that we are $(1-\alpha)$ percent sure that an interval based on the sample values includes the universe value being estimated. The choice of α can be made at will.

1. *For $\bar{x}'$.* When $\bar{x}$ is used to estimate $\bar{x}'$, the following formulas will give confidence limits:

   a. If σ' is known,

$$\bar{x} \pm \frac{z_{\alpha/2}\sigma'}{\sqrt{N}}$$

   Here z is a value chosen from the normal curve areas in Table 3-1 so that the probability of exceeding $+z$ is $\alpha/2$ and of falling below $-z$ is $\alpha/2$.

   b. If σ' is not known,

$$\bar{x} \pm \frac{t_{\alpha/2}s}{\sqrt{N}}$$

Here t is a value chosen from the t curve areas in Table 3-2 with $n = N-1$ so that the probability of exceeding $+t$ is $\alpha/2$ and of falling below $-t$ is $\alpha/2$.

Example 11. From example 6, $\bar{x}$ has been found to be 2.72. If σ' is known to be 2.0 then

$$2.72 \pm 1.96 \times \frac{2.0}{\sqrt{25}} = 1.94,\ 3.50$$

Here the value 1.96 was found from Table 3-1. It is the z value for which 2.5 percent of the curve lies above 1.96 and 2.5 percent lies below -1.96. Here α is 0.05.

It can be said then that we are 95 percent sure that the universe mean is somewhere in the interval 1.94 to 3.50.

Example 12.   Using the same data as in example 11, the following would be an appropriate analysis if σ' were not known but had to be estimated from the sample: Using s to estimate σ' let $s = 2$.   Then

$$2.72 \pm 2.064 \times \frac{2.0}{\sqrt{25}} = 1.89, 3.55$$

Here the value 2.064 was found from Table 3-2 using $n = N - 1$.   It is the t value for which 2.5 percent of the curve lies above 2.064 and 2.5 percent lies below -2.064.

It can be said that we are 95 percent sure that the interval 1.89 to 3.55 includes the universe mean.

Note that the lack of knowledge of σ' has caused a wider confidence interval for the same degree of confidence.

2. *For σ' (or σ'^2).*   When s is used to estimate σ', the following confidence limits can be used:

   a. If $N < 30$,

   $$\text{Upper limit} = \sqrt{\frac{(N-1)s^2}{\chi^2_{1-\alpha/2}}} \quad \text{for } \sigma' \quad \text{or} \quad \frac{(N-1)s^2}{\chi^2_{1-\alpha/2}} \quad \text{for } \sigma'^2$$

   $$\text{Lower limit} = \sqrt{\frac{(N-1)s^2}{\chi^2_{\alpha/2}}} \quad \text{for } \sigma' \quad \text{or} \quad \frac{(N-1)s^2}{\chi^2_{\alpha/2}} \quad \text{for } \sigma'^2$$

   Here $\chi^2_{1-\alpha/2}$ or $\chi^2_{\alpha/2}$ are values chosen from the χ^2 (chi-squared) curve in Table 3-3 so that the probability of exceeding $\chi^2_{\alpha/2}$ is $\alpha/2$ and the probability of falling below $\chi^2_{1-\alpha/2}$ is $\alpha/2$.   In Table 3-3, $n = N - 1$ for these limits.

   b. $N \geq 30$,

   $$\frac{s}{1 \pm z_{\alpha/2}/\sqrt{2N}}$$

   Here $z_{\alpha/2}$ is a value chosen from the normal curve areas in Table 3-1, so that the probability of exceeding $+z$ is $\alpha/2$ and of falling below $-z$ is $\alpha/2$.

Example 13.   In example 12, it was assumed that $s = 2$ for $N = 25$.   For confidence limits on σ',

$$\text{Upper limit} = \sqrt{\frac{(24)(4)}{12.4}} = 2.78$$

$$\text{Lower limit} = \sqrt{\frac{(24)(4)}{39.4}} = 1.56$$

Here the values 12.4 and 39.4 were found from Table 3-3 (table of χ^2) for

$$n = 25 - 1 = 24$$

12.4 is the table value $\chi^2_{0.975}$ and 39.4 is the table value $\chi^2_{0.025}$.

It can be said then that we are 95 percent sure that the interval 1.56 to 2.78 contains the universe standard deviation.   For the interval 2.44 [$= (1.56)^2$] to 7.74 [$= (2.78)^2$], we have the same confidence that the universe variance is included.

Example 14.   If in example 13 $N = 250$, confidence limits on σ' are

$$\frac{2}{1 \pm 1.96/\sqrt{500}} = 1.84, 2.19$$

Here the value 1.96 was found from the table of normal curve areas, Table 3-1.   It is the z value for which 2.5 percent of the area lies above $+1.96$ and 2.5 percent lies below -1.96.

It can be said that we are 95 percent sure that the interval 1.84 to 2.19 includes the universe standard deviation.

In these examples and in general, the confidence interval for the standard deviation and the variance are not symmetrically placed with respect to the sample values as is the case for the mean.

3. *For Proportions.* When p is used to estimate p', the following confidence limits can be set if N is large. More exact methods not given here are needed if N is small.

$$p \pm z_{\alpha/2} \sqrt{\frac{p(1-p)}{N}}$$

Example 15. If a work sampling study indicates that in a sample of 400 a machine was idle 10 percent of the time, then confidence limits for the proportion of idle time would be given as

$$0.10 \pm 1.96 \sqrt{\frac{(0.10)(0.90)}{400}} = 0.071, 0.129$$

Here the value 1.96 comes from Table 3-1 and is the z value associated with the normal curve for which 2.5 percent of the curve falls above $+1.96$ and 2.5 percent falls below -1.96.

It can be said that we are 95 percent sure that the universe proportion of idleness is between 0.071 and 0.129.

Sample Size. An important question in sampling is how large to make the sample. A general answer requires knowledge of:

1. The cost of sampling
2. The cost of imprecise estimates
3. The variability of the process
4. The amount of confidence required for the results of the sample

Neglecting cost factors, the confidence limits in the foregoing allow us to place the limits and determine the sample size required to get the desired confidence $(1 - \alpha)$ for these limits. For example, the formula

$$N = \left(\frac{z_{\alpha/2}\sigma'}{d}\right)^2$$

allows us to determine the sample size N for a confidence level $(1 - \alpha)$ if the interval can be $\pm d$ on either side of the mean when the universe variance is known.

Example 16. How large a sample must be taken to be 95 percent sure that the interval $\bar{x} \pm 0.1$ includes the universe mean?

Let $\sigma' = 0.2$. Then

$$N = \left[\frac{(1.96)(0.2)}{0.1}\right]^2 = 15.4 \text{ or } 16$$

Then to get the precision required of the estimate requires a sample of 16 units.

TESTING HYPOTHESES

One of the important uses of statistics is in the testing of hypotheses. We will discuss the concepts of testing hypotheses and give some of the more common tests with examples.

Theory of Testing. In testing hypotheses, decisions must be based on the results obtained from a sample from the process. Statistics—the quantity measured in the sample—then form the basis for drawing conclusions about the process or the universe. In the test, the hypothesis is generally in the form called a null hypothesis.

Examples of types of hypotheses that could be tested are (1) whether a methods change had significantly changed the average time to produce a part and (2) whether a new tool had significantly changed the variability of a dimension.

Here universe values which might be assumed to be known, perhaps from historical data or from targets set for the process, could be the universe mean or variance. The sample values used to test the hypothesis would be the sample mean or variance. In either event, the null hypothesis is: the sample results are from the universe whose corresponding values are known.

Other types of hypotheses that could be tested are (1) whether there is a significant difference in average production times between day and night shifts and (2) whether production method A is more variable than method B. In these events, the test is for the null hypothesis that the two sample values are from the same universe or process.

In these tests, the null hypothesis is either accepted or rejected on the basis of the test results. If the tests reject the hypothesis, then we say that the results are statistically significant. If the test does not reject the null hypothesis, then we say the results are not statistically significant, and hence there is not sufficient evidence to discard the null hypothesis.

If the tests indicate that the null hypothesis cannot be rejected, it must not be taken that it is true. The tests only indicate that "these results do not give enough evidence to discard the hypothesis." Thus the tests cannot be used to prove the null hypothesis. They can and are used, however, to guide action as if the hypothesis were true.

Implied in the reasoning of hypothesis testing is a problem of setting a measure of when to accept or reject a hypothesis. In testing a hypothesis, one of two errors can occur. The hypothesis might be correct, but sampling errors might cause the test to reject it as being false. This type of error, called an error of the first kind and specified as an α(alpha) error, can be controlled by the industrial engineer. In all the following tests, one of the first steps calls for setting α. In much industrial practice, α is set at 0.05. This means that, if the statistical tests were run many times (in practice this is not done) and if the hypotheses were true, then these tests would still reject the hypotheses 5 times out of 100. In some cases, it is important not to reject many true hypotheses, and hence α is set at a lower level. The following tests still apply to these cases although the comparisons made would be based on the α value of interest.

The other type of error arising in testing a hypothesis is called an error of the second kind and is specified as a β (beta) error. This type of error arises when a null hypothesis that is false is accepted by the test. In much practice, this type of error is difficult to control because the chance of accepting a false hypothesis depends in part on how false it is, and this may not be measurable. A tool to help disclose the probabilities of a β error is called the operating characteristic or OC curve. A full discussion of this is not possible here.

As a reasonable rule, errors of the second kind can be controlled by using large samples. The designer of the test of a hypothesis then must strike a balance between the protection afforded by the larger sample and the cost of sampling.

Testing a Sample Mean against a Known Universe Mean

Universe Standard Deviation Known—The Normal Curve Test. The test:

1. Hypothesis: The sample comes from a universe whose mean is $\bar{x}'$.
2. Set N.
3. Set α.
4. Calculate $z = \dfrac{\bar{x} - \bar{x}'}{\sigma'/\sqrt{N}}$.
5. Find from Table 3-1:
   a. z_α if interest is only in $\bar{x}$ less than $\bar{x}'$.
   b. $-z_\alpha$ if interest is only in $\bar{x}$ greater than $\bar{x}'$.
   Steps 5a and 5b are called one-tail tests.
   c. $\pm z_{\alpha/2}$ if interest is in $\bar{x}$ significantly less than or greater than $\bar{x}'$.
   Step 5c is called a two-tail test.
6. Compare z from step 4 against appropriate z_α in step 5.
7. a. In step 5a, reject the hypothesis if $z \leq z_\alpha$.
   b. In step 5b, reject the hypothesis if $z \geq -z_\alpha$.
   c. In step 5c, reject the hypothesis if $z \leq z_{\alpha/2}$ or if $z \geq -z_{\alpha/2}$.
   d. Otherwise accept the hypothesis.

Example 17. In the manufacture of certain electronic elements for missile use, it is important that the base material be free from dirt particles. Records for the past several years indicate that the dirt count per sample unit averaged 10.2 particles with a standard deviation of 3 particles. A methods study indicates that most of the dirt seems to come from an impregnating process. The process is changed to eliminate the dirt-producing step. A sample of 36 units is then taken and it is found that the dirt count has been reduced to 8.9 units. Because this is not a spectacular reduction, the question is raised as to whether this reduction might be due to sampling errors alone. The following analysis would be pertinent to test this question.

    1. Hypothesis: The sample comes from a universe whose mean is $\bar{x}' = 10.2$.

    2. $N = 36$

    3. Take $\alpha = 0.05$.

    4. $z = \dfrac{8.9 - 10.2}{3/\sqrt{36}} = \dfrac{-1.3}{0.5} = -2.6$

    5. From Table 3-1, $z_{0.05} = -1.64$ for a one-tail test because interest lies only in dirt reduction.

    6. Because $z = -2.6$ and $z_\alpha = -1.64$, $z \leq z_\alpha$, and the hypothesis is rejected.

From step 6 then, it can be argued that the process change has reduced the dirt count significantly.

Universe Standard Deviation Not Known and Sample Standard Deviation Must Be Used—The t Test. The test:

    1. Hypothesis: The sample comes from a universe whose mean is $\bar{x}'$.

    2. Set N.

    3. Set α.

    4. Compute:

       a. s, using the formula from page 10-69.

       b. $t = \dfrac{\bar{x} - \bar{x}'}{s/\sqrt{N}}$.

    5. Find from Table 3-2 using $n = N - 1$:

       a. $-t_\alpha$ or $+t_\alpha$ if interest is in a one-tail test.

       b. $\pm t_{\alpha/2}$ if interest is in a two-tail test.

    6. Compare t from step 4*b* against appropriate t_α from step 5.

    7. *a.* In step 5*a*, reject the hypothesis if $t \leq -t_\alpha$ or if $t \geq +t_\alpha$, whichever is of interest.

       b. In step 5*b*, reject the hypothesis if $t \leq -t_{\alpha/2}$ or if $t \geq +t_{\alpha/2}$.

       c. Otherwise accept the hypothesis.

Example 18. In the previous example, if the value given for σ' had not been known but had been calculated from the sample as s, the appropriate test would have been:

    1. Hypothesis: The sample comes from a universe whose mean is $\bar{x}'$.

    2. $N = 36$

    3. Take $\alpha = 0.05$.

    4. Compute $s = 3$,

$$t = \frac{8.9 - 10.2}{3/\sqrt{36}} = -2.6$$

    5. From Table 3-2, using $n = 35 = N - 1$,

$$t_{0.05} = -1.690$$

    6. Because $-2.6 \leq -1.690$, the hypothesis is rejected.

From step 6 then, it can be argued that the dirt count has been significantly reduced.

Testing a Sample Variance against a Known Universe Variance

Large Sample ($N \geq 30$)—The Normal Test. The test:

Hypothesis: The sample comes from a universe whose variance (or standard deviation) is σ'^2.

This test is the same as the normal curve test except $\bar{x}$ and $\bar{x}'$ are replaced by s and σ', respectively. Thus, for example, step 4 would calculate $z = \dfrac{s - \sigma'}{\sigma'/\sqrt{2N}}$. Note the denominator is $2N$ instead of N.

Small Sample ($N \leq 30$)—The χ^2 Test. The test:

    1. Hypothesis: The sample comes from a universe whose variance (or standard deviation) is σ'^2.

2. Set N.
3. Set α.
4. a. Calculate $s^2 = \dfrac{\Sigma(x - \bar{x})^2}{N - 1}$.

   b. Calculate $\dfrac{(N - 1)s^2}{\sigma'^2} = \chi^2$.

5. Find from Table 3-3, using $n = N - 1$:

   a. χ^2_α if interest is in s greater than σ'.

   b. $\chi^2_{1-\alpha}$ if interest is in s less than σ'.

   c. $\chi^2_{\alpha/2}$ and $\chi^2_{1-\alpha/2}$ if interest is in s both greater and less than σ'.

6. Compare step $4b$ against step 5:

   a. If $\chi^2 \geq \chi^2_\alpha$ for step $5a$, reject the hypothesis.

   b. If $\chi^2 \leq \chi^2_{1-\alpha}$ for step $5b$, reject the hypothesis.

   c. If $\chi^2 \leq \chi^2_{1-\alpha/2}$ or $\chi^2 \geq \chi^2_{\alpha/2}$ for step $5c$, reject the hypothesis.

   d. Otherwise accept the hypothesis.

Example 19.   In manufacturing a product, it is essential that its thickness be uniform to ensure that it does not jam the machinery using it.   0.05 inch standard deviation has been set as the standard for manufacturing the product.   A new supplier of base stock is under consideration.   To test his product, 10 units of final product are manufactured using his product as raw material.   The standard deviation of these 10 units is found to be 0.06, using the s formula on page 10-69.   The question arises whether this result is statistically significant.   If it is, then the product is too variable and hence unusable.   If it is not, then the company would consider changing suppliers.   The following test is pertinent.

1. Hypothesis: The sample comes from a universe whose variance is $(0.05)^2$, or alternatively, the sample comes from a universe whose standard deviation is 0.05.
2. $N = 10$
3. Take $\alpha = 0.05$.
4. $\chi^2 = \dfrac{(N - 1)s^2}{\sigma'^2} = \dfrac{9(0.06)^2}{(0.05)^2} = \dfrac{0.0324}{0.0025} = 12.96$
5. From Table 3-3, $\chi^2_{0.05} = 16.9$ for $n = 9$.
6. Because $\chi^2 \leq \chi^2_{0.05}$, accept the hypothesis.

From step 6, it can then be argued that there is not sufficient reason to reject the new product because it is too variable.

Testing Two Sample Mean Values

σ_1' and σ_2' *Known—The Normal Test.*   The test:

1. Hypothesis: $\bar{x}_1' = \bar{x}_2'$.
2. Set N_1 and N_2.
3. Set α.
4. Calculate $\sigma_{\bar{x}_1 - \bar{x}_2} = \sqrt{\dfrac{\sigma_1'^2}{N_1} + \dfrac{\sigma_2'^2}{N_2}}$.

5. Calculate $z = \dfrac{\bar{x}_1 - \bar{x}_2}{\sigma_{\bar{x}_1 - \bar{x}_2}}$.

6. Find from Table 3-1:

   a. $-z_\alpha$ or z_α if interest is only in a one-tail test.

   b. $\pm z_{\alpha/2}$ if interest is in a two-tail test.

7. a. Compare z from step 5 against appropriate z value from step 6 and reject the hypothesis if $z \geq -z_\alpha$ or $z \leq z_\alpha$, whichever is appropriate for the one-tail test, or if $z \geq -z_{\alpha/2}$ or $z \leq z_{\alpha/2}$ for a two-tail test.

   b. Otherwise accept the hypothesis.

Example 20.   In attempting to use the most economical manufacturing method, two plans were considered.   The following data were obtained during the manufacture of 16 pieces using plan 1.   The test was terminated after manufacturing 9 pieces using plan 2.   It was assumed that the plan requiring the least manufacturing time per piece would be the cheaper.   The question then is to find if there is a significant difference between the mean times to manufacture a piece.   The data shown in the table below are pertinent.

If it can be assumed that the universe standard deviations are known for these processes, the following analysis is pertinent.   (Let $\sigma_1' = 0.06$ and $\sigma_2' = 0.05$.)
1. Hypothesis: $\bar{x}_1' = \bar{x}_2'$.
2. $N_1 = 16$; $N_2 = 9$
3. Let $\alpha = 0.05$.
4. Calculate $\bar{x}_1 = \dfrac{36.66}{16} = 2.291$
5. Calculate $\bar{x}_2 = \dfrac{20.91}{9} = 2.323$.
6. Calculate $z = \dfrac{\bar{x}_1 - \bar{x}_2}{\sqrt{\dfrac{\sigma_1'^2}{N_1} + \dfrac{\sigma_2'^2}{N_2}}} = \dfrac{2.291 - 2.323}{\sqrt{\dfrac{(0.06)^2}{16} + \dfrac{(0.05)^2}{9}}} = -1.45.$

Manufacturing Time (in Minutes)

Plan 1	Plan 2
2.27	2.30
2.23	2.34
2.40	2.38
2.38	2.27
2.21	2.25
2.32	2.40
2.22	2.36
2.30	2.33
2.29	2.28
2.25	
2.35	
2.26	
2.34	
2.25	
2.30	
2.29	

7. From Table 3-1, $+z_{0.05/2} = -1.96$ for a two-tail test.
8. Because $z \geq z_{\alpha/2}$, accept the hypothesis.

From step 8 then, we can conclude the manufacturing methods are not significantly different.

σ_1' and σ_2' Not Known but Can Be Assumed Equal—The t Test.   The test:
1. Hypothesis: $\bar{x}_1' = \bar{x}_2'$.
2. Set N_1 and N_2.
3. Set α.
4. Calculate $s = \sqrt{\dfrac{(N_1 - 1)s_1{}^2 + (N_2 - 1)s_2{}^2}{N_1 + N_2 - 2}}.$

where $s_1{}^2$ and $s_2{}^2$ are found as on page 10-69.

5. Calculate $t = \dfrac{\bar{x}_1 - \bar{x}_2}{s\sqrt{\dfrac{1}{N_1} + \dfrac{1}{N_2}}}.$

6. Find from Table 3-2, using $n = N_1 + N_2 - 2$:
   a. $-z_\alpha$ or z_α if interest is only in a one-tail test.
   b. $\pm z_{\alpha/2}$ if interest is in a two-tail test.
7. a. Compare t from step 5 against appropriate t from step 6, and reject the hypothesis if $t \leq -t_\alpha$ or $t \geq t_\alpha$, whichever is appropriate for the one-tail test, or if $t \leq -t_{\alpha/2}$ or $t \geq t_{\alpha/2}$ for a two-tail test.
   b. Otherwise accept the hypothesis.

Example 21.   In example 20, it was assumed that σ_1' and σ_2' were known.   If this is not true, the following analysis would be pertinent to the problem if it is now assumed that $\sigma_1' = \sigma_2'$, but they must be estimated from the samples.

From example 20, it is known that $\bar{x}_1 = 2.291$ and $\bar{x}_2 = 2.323$. Using the short formula for s, we find that

$$(N_1 - 1)s_1{}^2 = 84.045 - \frac{(36.66)^2}{16} = 0.048$$

$$(N_2 - 1)s_2{}^2 = 48.603 - \frac{(20.91)^2}{9} = 0.022$$

The test then is:
1. Hypothesis: $\bar{x}_1' = \bar{x}_2'$.
2. $N_1 = 16$; $N_2 = 9$.
3. Let $\alpha = 0.05$.
4. Compute $t = \dfrac{\bar{x}_1 - \bar{x}_2}{s\sqrt{\dfrac{1}{N_1} + \dfrac{1}{N_2}}}$.

$$t = \frac{2.291 - 2.323}{\sqrt{\dfrac{0.048 + 0.022}{16 + 9 - 2}}\sqrt{\tfrac{1}{16} + \tfrac{1}{9}}}$$

$$= \frac{-0.032}{0.023} = -1.39$$

5. Find $t_{0.025}$ (for two-tail test) from Table 3-2. $-t_{0.025} = -2.069$ when

$$n = N_1 + N_2 - 2 = 23$$

6. Because $t \geq -t_{0.025}$, accept the hypothesis.

Because the hypothesis is accepted, the conclusion is that the two methods do give the same time of manufacturing.

σ_1' and σ_2' Not Known and Cannot Be Assumed Equal—The "Approximate t" Test (Aspin-Welsh Test).   The test:
1. Hypothesis: $\bar{x}_1' = \bar{x}_2'$.
2. Set N_1 and N_2.
3. Set α.
4. Calculate $s_1{}^2$ and $s_2{}^2$ for each sample, using the methods on page 10-69.
5. Calculate $t = \dfrac{\bar{x}_1 - \bar{x}_2}{\sqrt{s_1{}^2/N_1 + s_2{}^2/N_2}}$.
6. Calculate $c = \dfrac{s_1{}^2/N_1}{s_1{}^2/N_1 + s_2{}^2/N_2}$.
7. Find from Table 3-2, using $n = \dfrac{1}{c^2/(N_1 - 1) + (1 - c^2)/(N_2 - 1)}$:
   a. $-t_\alpha$ or t_α if interest is in a one-tail test.
   b. $\pm t_{\alpha/2}$ if interest is in a two-tail test.
8. a. Compare t found in step 5 against appropriate t_α from step 7 and reject the hypothesis if $t \leq -t_\alpha$ or $t \geq t_\alpha$, whichever is appropriate for a one-tail test, or if $t \leq -t_{\alpha/2}$ or $t \geq t_{\alpha/2}$ for a two-tail test.
   b. Otherwise accept the hypothesis.

Example 22.   In example 20, it was taken that the standard deviations of the universes, although not known, were equal.   If this assumption is not true, the following analysis would hold:
1. Hypothesis: $\bar{x}_1' = \bar{x}_2'$.
2. $N_1 = 16$; $N_2 = 9$.
3. Take $\alpha = 0.05$.
4. Compute $s_1{}^2 = \dfrac{0.048}{15}$ $s_2{}^2 = \dfrac{0.022}{8}$

$$s_1{}^2 = 0.00320 \qquad s_2{}^2 = 0.00275$$

5. Compute $t = \dfrac{\bar{x}_1 - \bar{x}_2}{\sqrt{\dfrac{s_1^2}{N_1} + \dfrac{s_2^2}{N_2}}}$.

$$t = -\frac{0.032}{\sqrt{\dfrac{0.0032}{16} + \dfrac{0.00275}{9}}} = -1.43$$

6. From Table 3-2, find $t_{\alpha/2}$ for

$$n = \frac{1}{\dfrac{c^2}{15} + \dfrac{(1-c)^2}{8}}$$

$$c = \frac{s_1^2/N_1}{\dfrac{s_1^2}{N_1} + \dfrac{s_2^2}{N_2}} = \frac{0.0032/16}{\dfrac{0.0032}{16} + \dfrac{0.00275}{9}} = 0.400$$

Thus

$$n = \frac{1}{\dfrac{(0.400)^2}{15} + \dfrac{(0.600)^2}{8}} = 17.9 \text{ (call this 18)}$$

Hence

$$-t_{0.025} = -2.101$$

7. Here $t \geq -t_{0.025}$. Hence accept the hypothesis.

From step 7 then, there is not sufficient evidence to reject the hypothesis that the methods are the same.

It is interesting to note here that, because of the amount of information that the sample must yield, the test is not so discerning as the test in examples 19 and 20. Because of this lack of discerning ability (precision), we may be making an error of the second kind here.

Because the value for α is wholly up to the industrial engineer to choose, it might have been desirable, in initiating the test, to settle for a larger α value. Such is not a statistical problem and is within the area of judgment for the engineer. Of course, α should be choosen before the test is run.

Testing Two Sample Variances—The F Test. The test:

1. Hypothesis: $\sigma_1'^2 = \sigma_2'^2$ or $\sigma_1' = \sigma_2'$.
2. Set N_1 and N_2.
3. Set α.
4. Calculate s_1^2 and s_2^2, using the methods on page 10-69.
5. Calculate $F = s_1^2/s_2^2$, using for s_1^2 the larger of the s values.
6. Using Table 3-4, find F_α. To find F_α, note that Table 3-4 uses two n values given as n_1 and n_2. Always choose n_1 as $N-1$ for the sample forming the numerator of the fraction in step 5.
7. If $F \geq F_\alpha$, reject the hypothesis.
8. Otherwise accept the hypothesis.

Example 23. In testing to determine the factors affected by fatigue, it is argued that fatigue has the effect of causing a highly variable performance. To test this assumption, a worker's output is measured closely in both the morning and the afternoon. It is expected that the afternoon performance will show a greater variability than the morning one. The data following are the result of 20 readings of production time on the same product. Set 1 readings were taken in the morning and set 2 in the afternoon. The question posed is whether these data are significantly different.

x_1		x_2	
8.37	8.35	8.33	8.38
8.32	8.34	8.39	8.40
8.32	8.36	8.30	8.37
8.35	8.35	8.31	8.38
8.33	8.30	8.32	8.35

The test appropriate here is:

1. Hypothesis: $\sigma_1'^2 = \sigma_2'^2$.
2. $N_1 = 10; N_2 = 10$.
3. Take $\alpha = 0.05$.
4. Calculate s_1^2 and s_2^2, using the formula for s given on page 10-69.

$$s_1^2 = 0.000455 \qquad s_2^2 = 0.00129$$

5. Compute $F = \dfrac{0.00129}{0.00046} = 2.8$.

6. From Table 3-4, compute F_α with $n_1 = 9; n_2 = 9$. $F_\alpha = 3.18$.
7. Because $F \leq F_\alpha$, accept the hypothesis that $\sigma_1'^2 = \sigma_2'^2$.

From step 7, we can argue that for this test there is not more variability in production times in the afternoon than in the morning.

Testing Sample Proportions.   If the sample size is large and if the proportions being tested are approximately 0.5, the following tests can be used to test sample proportions.   More exact methods, not given here, are required if the two conditions above do not hold.

Testing a Sample Proportion against a Universe Proportion.   The test:

1. Hypothesis: The sample comes from a universe whose proportion is p'.
2. Set N.
3. Set α.
4. *a.* Calculate sample proportion.

   b. Calculate $\sigma_{p'} = \sqrt{\dfrac{p'(1 - p')}{N}}$.

   c. Calculate $z = \dfrac{p - p'}{\sigma_{p'}}$.

5. Find from Table 3-1:
   a. z_α or $-z_\alpha$, if interest is in a one-tail test.
   b. $\pm z_{\alpha/2}$, if interest is in a two-tail test.
6. *a.* Compare z from step 4c against appropriate z_α from step 5 and reject the hypothesis if $z \geq -z_\alpha$ or if $z \leq z_\alpha$, whichever is appropriate for the one-tail test, or if $z \geq -z_{\alpha/2}$ or $z \leq z_{\alpha/2}$ for the two-tail test.
7. Otherwise accept the hypothesis.

Testing Two Sample Proportions.   The test:

1. Hypothesis: $p_1' = p_2'$.
2. Set N_1 and N_2.
3. Set α.
4. Calculate:

   a. $\bar{p} = \dfrac{N_1 p_1 + N_2 p_2}{N_1 + N_2}$

   b. $\sigma_{p_1 - p_2} = \sqrt{\bar{p}(1 - \bar{p})\left(\dfrac{1}{N_1} + \dfrac{1}{N_2}\right)}$

   c. $z = \dfrac{p_1 - p_2}{\sigma_{p_1 - p_2}}$

5. Find from Table 3-1:
   a. $-z_\alpha$ or z_α, if interest is in a one-tail test.
   b. $\pm z_{\alpha/2}$, if interest is in a two-tail test.
6. *a.* Compare z from step 4c against appropriate z_α from step 5 and reject the hypothesis if $z \geq -z_\alpha$ or if $z \leq z_\alpha$, whichever is appropriate for the one-tail test, or if $z \geq -z_{\alpha/2}$ or $z \leq z_{\alpha/2}$ for a two-tail test.
   b. Otherwise accept the hypothesis.

Example 24.   Management is disturbed by what appears to be a high level of absenteeism in the automatic screw machine department.   Over the last 200 man-days, absenteeism has been about 10 percent. The milling machine department, on the other hand, has reported only 5 percent absenteeism over a 500 man-day period. The question posed is whether these results are merely chance occurrences or whether something, perhaps supervision or working conditions, is causing these differences. The following analysis would be appropriate:

1. Hypothesis: There is no significant difference in the proportion of absenteeism from the two departments.
2. $N_1 = 200$; $N_2 = 500$.
3. Take $\alpha = 0.05$.
4. Calculate:

$$a.\ \bar{p} = \frac{20 + 25}{700} = 0.0643$$

$$b.\ \sigma_{p_1-p_2} = \sqrt{(0.0643)(0.9357)(\tfrac{1}{200} + \tfrac{1}{500})} = 0.02$$

$$c.\ z = \frac{0.10 - 0.05}{0.02} = 2.5$$

5. From Table 3-1, using $z_{0.05}$ because interest is in whether 1 is greater than 2, find $-z_\alpha = 1.645$.
6. Because $z \geq -z_\alpha$ or $2.5 \geq 1.645$, reject the hypothesis.

From step 6, it can be argued that the two departments have significantly different absentee rates.

Other Types of Hypotheses and Their Tests. The preceding sections gave some tests of hypotheses pertaining to the equality of means or variances. Other statistical hypotheses are frequently of interest. The test of dependence of factors affecting an outcome will be discussed.

Test of Independence—Contingency Tables. A two-way classification of factors affecting an outcome is discussed. The basic question posed is: Are the two factors used to classify the outcome independent of each other? The test relies on the fact that, if factors A and B are independent, then $P(AB) = P(A)P(B)$. From the data, $P(AB)$ is found. $P_i(A)$ is found by considering $\sum\limits_{j=1}^{c} x_{ij} \Big/ \sum\limits_{i=1}^{r} \sum\limits_{j=1}^{c} x_{ij}$ where x_{ij} is the outcome of the measurements when factor A is at level i and factor B is at level j. $\sum\limits_{j=1}^{c} x_{ij}$ then is the sum of row i. $\sum\limits_{i=1}^{r} \sum\limits_{j=1}^{c} x_{ij}$ is the sum of all cases examined. Next, $P_j(B) = \sum\limits_{i=1}^{r} x_{ij} \Big/ \sum\limits_{i=1}^{r} \sum\limits_{j=1}^{c} x_{ij}$. Then, if the events A and B are independent, $P_{ij}(AB) = P_i(A)P_j(B)$. If factors A and B are independent, it would be expected that $P(AB)$ found from the data and $P_{ij}(AB)$ as found above would be approximately equal. Because of sampling errors, these two values will not be identical. Therefore, the hypothesis of independence is tested by comparing $P(AB)$ and $P_i(A)P_j(B)$ (as defined above) using a χ^2 test of significance.

The test:

Given:

	Level of factor B				
	1	2	3	$\cdots$	c
1	x_{11}	x_{12}	x_{13}	$\cdots$	x_{1c}
2	x_{21}	x_{22}	x_{23}	$\cdots$	x_{2c}
3	x_{31}	x_{32}	x_{33}	$\cdots$	x_{3c}
Level of factor A	$\cdot$				
	$\cdot$				
	$\cdot$				
r	x_{r1}	x_{r2}	x_{r3}	$\cdots$	x_{rc}

1. Compute $P_1(A) = \sum_{j=1}^{c} x_{1j} \Big/ \sum_{i=1}^{r} \sum_{j=1}^{c} x_{ij}$.

2. Compute $P_2(A)$, $P_3(A)$, and so on, as in step 1, up to $P_r(A)$.

3. Compute $P_1(B) = \sum_{i=1}^{r} x_{i1} \Big/ \sum_{i=1}^{r} \sum_{j=1}^{c} x_{ij}$.

4. Compute $P_2(B)$, $P_3(B)$, and so on, as in step 3, up to $P_c(B)$.

5. Compute $P_{11}(AB) = P_1(A)P_1(B)$.

6. Compute $P_{ij}(AB)$ for each cell in the array.

7. Define $P_{ij}(AB)$ as T_{ij}.

8. Calculate $O_{ij} = x_{ij} \Big/ \sum_{i=1}^{r} \sum_{j=1}^{c} x_{ij}$ for each cell.   This is the cell entry divided by the grand sum.

9. Compute $(T_{ij} - O_{ij})^2 / T_{ij}$ for each cell.

10. Compute $\chi^2 = N \sum_{i=1}^{r} \sum_{j=1}^{c} (T_{ij} - O_{ij})^2 / T_{ij}$, which is the sum of step 9 for all cells multiplied by N, the grand sum.

11. Using Table 3-3, find χ^2_α for $n =$ (number of rows $-$ 1) $\times$ (number of columns $-$ 1).

12. If $\chi^2 \geq \chi^2_\alpha$, reject the hypothesis that factors A and B are independent.

13. If $\chi^2 \leq \chi^2_\alpha$, accept the hypothesis.

Because this test deals with ratios, apparent differences can occur because of small values of T_{ij}.   As a rule of thumb, if $NT_{ij} \leq 5$, it is best to combine cells in such a way as to make $NT_{ij} \geq 5$ for the smallest entry in the table.

Example 25.   In an attempt to increase productivity, a company instituted a non-financial incentive plan based on worker efficiency as measured by output/standard output.   After some time the question was raised as to whether the incentive had had any appreciable effect on the efficiency figures.   To answer the question, the company industrial engineer computed from historical records a "before-after" picture for three categories of performance.   "Less than 80 percent," "80 to 120 percent," and "Over 120 percent" served as the boundaries of the efficiency categories.   The following would be possible results:

	Number of men in each category			
	Less than 80 percent	80 to 120 percent	Over 120 percent	Total
Before incentive plan..........	30	164	36	230
After incentive plan..........	11	184	35	230
Total....................	41	348	71	460

If the incentive plan had no effect on efficiency, we would expect the following distribution of men to categories, as shown in the table on page 10-81.

The numbers in this table are derived on the assumption that the two factors—categories and before-after—are independent of each other.   On this assumption, probability theory tells us that $P_{ij}(AB) = P_i(A)P_j(B)$.   Taking i to be the category "Before incentive plan" and j to be "Less than 80 percent," it follows that

$$P_i(A) = {}^{41}\!/_{460}$$
$$P_j(B) = {}^{230}\!/_{460}$$

Therefore, $P_{ij}(AB) = \frac{41}{460} \times \frac{230}{460}$. To get the number in the category, these probabilities are considered to be proportions and $NP_{ij}(AB)$ is the expected number in each category.

Therefore, $\frac{41}{460} \times \frac{230}{460} \times 460$ gives 20.5 as the number of cases to expect in the ij category if the events are independent.

	Number of men expected in each category			
	Less than 80 percent	80 to 120 percent	Over 120 percent	Total
Before incentive plan..........	20.5	174	35.5	230
After incentive plan...........	20.5	174	35.5	230
Total.....................	41	348	71	460

The following test is appropriate to see if the differences shown in the tables are statistically significant.

1. Hypothesis: The factors are independent.
2. $r = 2; c = 3$.
3. Take $\alpha = 0.05$.
4. Compute $\chi^2 = \sum_{=1}^{2} \sum_{j=1}^{3} \frac{(T_{ij} - O_{ij})^2}{T_{ij}}$ over all cells.

$$\chi^2 = \frac{(20.5 - 30)^2}{20.5} + \frac{(20.5 - 11)^2}{20.5} + \frac{(174 - 164)^2}{174} + \frac{(174 - 184)^2}{174} + \frac{(35.5 - 36)^2}{35.5}$$
$$+ \frac{(35.5 - 35)^2}{35.5} = 9.97$$

5. From Table 3-3, $\chi^2_{0.05} = 5.99$ for $n = (r - 1)(c - 1) = 2$.
6. Because $\chi^2 \geq \chi^2_{0.05}$, reject the hypothesis of independence.

From step 6 then, we can argue that the categories are not independent. Hence there is a statistically significant relationship between efficiency and the presence or absence of the nonfinancial incentive system.

ANALYSIS OF VARIANCE

The discussion of testing sample mean values is applicable when there are no more than two means to be tested. The methods will not generalize simply to more than two sample means.

The industrial engineer is often faced with the problem of testing for significant differences between several mean values. Testing for differences among average cutting times for several raw materials; testing for differences among average times to complete a job using different methods; and testing for differences among average production times using different operators, different machines, and different raw materials are examples of situations that give rise to a need for analysis beyond testing differences between two means.

To test the differences among a group of mean values, one is tempted to use the previous tests pairwise for all or important pairs. Such practice introduces compound results that affect the α and β errors and hence leave the test of significance of the results at a level other than that for a single test. In general, the procedure is difficult to analyze; hence other methods, called analysis of variance, are used.

In addition to testing several means, the analysis allows a test of the effect of several different factors and their interactions, and a test of individual means against each other.

The purpose of the following discussion is to give the analysis of variance procedures for (1) one factor at several levels and (2) two factors, each at several levels. Interactions and partitioning effects are beyond the scope of this chapter.

One Factor

Given: c levels of the factor
r readings per factor

The test:
1. Hypothesis: $\bar{x}_1' = \bar{x}_2' = \cdots = \bar{x}_c'$.
2. Set α.
3. Compute the sum of the readings for each level and square this sum.
4. Compute the sum of the squares in step 3.
5. Divide step 4 by the number of readings for each factor r.
6. Compute the square of each reading and add all these together.
7. Compute the sum of all the readings and square this sum.
8. Compute step 7 divided by the total number of readings.
9. Compute the column sum of squares given by step 5 minus step 8.
10. Compute total sum of squares given by step 6 minus step 8.
11. Compute the residual sum of squares given by step 10 minus step 9.
12. Compute column mean square given by step 9 $\div$ $(c - 1)$.
13. Compute the residual mean square given by step 11 $\div$ $c(r - 1)$.
14. Compute $F = $ step 12 $\div$ step 13.
15. Find F_α from Table 3-4, using $n_1 = c - 1$, $n_2 = c(r - 1)$.
16. If $F \geq F_\alpha$, reject the hypothesis that the means are all equal.
17. If $F \leq F_\alpha$, accept the hypothesis of equality of means.

Example 26. Three manufacturing methods are available to produce a part. Because the order will be produced over a long period of time, it is desirable to use the most economical manufacturing method. It is assumed that the fastest method of production will give this most economical method. To see which method is fastest, four parts are manufactured using each method. The following data are the production times.

Times to Produce One Part

Method 1	Method 2	Method 3
8.6	8.3	8.9
8.4	8.2	8.8
8.9	8.4	8.8
8.8	8.4	8.6

The question then is whether these methods of manufacturing differ significantly. The following analysis of variance would be appropriate:
1. Hypothesis: $\bar{x}_1' = \bar{x}_2' = \bar{x}_3'$.
2. $r = 4; c = 3$.
3. Take $\alpha = 0.05$.
4. Compute sums of squares. To do this, the arithmetic is simplified if the data are coded. For this purpose, we subtract 8.5 from each reading. This will have no effect on the subsequent analysis. The new data are then:

Method 1	Method 2	Method 3
0.1	-0.2	0.4
-0.1	-0.3	0.3
0.4	-0.1	0.3
0.3	-0.1	0.1

To compute the sums of squares needed:

  a. Sum each column and square these sums. This results in:

Column 1		Column 2		Column 3	
Sum	Square of sum	Sum	Square of sum	Sum	Square of sum
0.7	0.49	−0.7	0.49	1.1	1.21

Sum of squares = 2.19

  b. Divide this sum of squares by the number of items in each column.

$$2.19 \div 4 = 0.5475$$

  c. Square all items and sum the squares. This gives

$$(0.1)^2 + (-0.1)^2 + (0.4)^2 + (0.3)^2 + (-0.2)^2 + \cdots + (0.1)^2 = 0.7700$$

  d. Sum all cases, square the sum, and divide the result by the total number of cases (12). Square of sum of all cases = $(1.1)^2 = 1.21$. Square of sum of all cases divided by number of cases = 0.1008.

Then the sums of squares pertinent to the analysis are:

  Column sum of squares = 0.5475 − 0.1008 = 0.4467.
  Total sum of squares = 0.7700 − 0.1008 = 0.6692.
  Residual sum of squares = total sum of squares − column sum of squares = 0.6692 − 0.4467 = 0.2225.

The following table simplifies further analysis:

	Sum of squares	d.f.	Mean square
Columns.........	0.4467	2	0.2234
Residual.........	0.2225	9	0.0247
	0.6692	11	

In both cases the mean square column is given by the sum of squares column divided by d.f. as given above.

  5. Compute $\dfrac{\text{column mean square}}{\text{residual mean square}} = \dfrac{0.2234}{0.0247} = 9.04.$

  6. Find F_α from Table 3-4, using $n_1 = c - 1 = 2$; $n_2 = c(r - 1) = 9$.

$$F_\alpha = 4.26$$

  7. Because $F \geq F_\alpha$, we reject the hypothesis.

From step 7 then, it can be argued that the manufacturing times are different. Inspection of the data indicates method 2 is the fastest.

Two Factors

Given: c levels of factor 1 (column factor)
  r levels of factor 2 (row factor)

The test:

  1. Hypothesis: (a) There is no effect of the row factor on the results. (b) There is no effect of the column factor on the results.

  2. Set α.

3. Compute the sum of readings for each column and square each of these sums.
4. Sum the squares in step 3 and divide the total by the number of readings in each column.
5. Compute the sum of each row and square these sums.
6. Sum the squares in step 5 and divide by the number of items in each row.
7. Square each element and sum the results.
8. Sum all the cases and square the sum.
9. Divide the sum of squares in step 8 by the total number of cases.
10. Compute the column sum of squares from step 4 minus step 9.
11. Compute the row sum of squares from step 6 minus step 9.
12. Compute the total sum of squares from step 7 minus step 9.
13. Compute the residual sum of squares from step 12 minus step 10 minus step 11.
14. Compute column mean square from step 10 $\div$ $(c - 1)$.
15. Compute row mean square from step 11 $\div$ $(r - 1)$.
16. Compute residual mean square from step 13 $\div$ $(r - 1)(c - 1)$.
17. Compute $F_{\text{cols}} =$ step 14 $\div$ step 16, $F_{\text{rows}} =$ step 15 $\div$ step 16.
18. To test for equality of column means, calculate F_α from Table 3-4 with $n_1 = (c - 1)$; $n_2 = (r - 1)(c - 1)$.

To test for equality of row means, calculate F_α for $n_1 = (r - 1)$ and $n_2 = (r - 1)(c - 1)$.

19. Compare F_{cols} and F_{rows} with F_α. If $F_{\text{cols}} \geq F_\alpha$, reject the hypothesis of equal-column means. If $F_{\text{cols}} \leq F_\alpha$, accept the hypothesis of equal-column means.
20. If $F_{\text{rows}} \geq F_\alpha$, reject the hypothesis of equal-row means. If $F_{\text{rows}} \leq F_\alpha$, accept the hypothesis of equal-row means.

	Time of first breakdown			
Driver	Truck no.			
	1	2	3	4
1	1.0	0.5	1.2	1.0
2	0.6	0.7	0.9	1.2
3	0.8	0.6	1.0	0.9
4	0.7	0.5	1.0	1.0

Example 27. In deciding on a replacement policy for material handling equipment, the following type of problem could arise:

Four types of forklift trucks are available, each of which can do all the work required of it. The question is raised as to the life of the machines. Some feel that one of the trucks has a shorter life than the others. Another view is that the apparent life of the one truck is shorter because of the rough treatment it gets from its driver. Yet another view is that both of these are contributing factors. To help answer the questions raised, each of the company's four drivers is given each of the four types of trucks to drive. Then, to measure the effects of drivers and trucks, each driver is told to drive as he would normally do until the first breakdown of his truck. The coded data shown in the table above were obtained for time to first breakdown for each driver on each truck.

Two questions are posed: Are the differences between trucks significant? Are the differences between drivers significant?

The following analysis of variance would be appropriate:

1. Hypothesis: There is no difference between mean times to first breakdown due to trucks. There is no difference between mean times to first breakdown due to drivers.
2. $r = 4; c = 4$.
3. Take $\alpha = 0.05$ for each hypothesis.
4. Calculate sums of squares:
   a. Sum each row, square the sums, add the squares, and divide by number of items in a row.

Row 1		Row 2		Row 3		Row 4	
Row sum	Square of row sum	Row sum	Square of row sum	Row sum	Square of row sum	Row sum	Square of row sum
3.7	13.6900	3.4	11.5600	3.3	10.8900	3.2	10.2400

Sum of squares of row sum $= 46.3800$
Sum of squares of row sum $\div$ number in each row

$$46.3800 \div 4 = 11.595$$

b. Sum each column, square the sums, add the squares, and divide by the number in each column.

Column 1		Column 2		Column 3		Column 4	
Sum	Square of sum	Sum	Square of sum	Sum	Square of sum	Sum	Square of sum
3.1	9.6100	2.3	5.2900	4.1	16.8100	4.1	16.8100

Sum of squares of column sum $= 48.5200$
Sum of squares of columns $\div$ number in each column

$$48.5200 \div 4 = 12.13$$

c. Square each item and sum the squares.

$$(1)^2 + (0.6)^2 + (0.8)^2 + (0.7)^2 + \cdots + (0.9)^2 + (1)^2 = 12.34$$

d. Sum all cases, square the sum, and divide by total number of cases.

$$\frac{(1 + 0.6 + 0.8 + 0.7 + \cdots + 0.9 + 1.0)^2}{16} = \frac{(13.6)^2}{16} = 11.56$$

e. Compute sums of squares.
   Row sum of squares: $11.595 - 11.560 = 0.035$
   Column sum of squares: $12.130 - 11.560 = 0.570$
   Total sum of squares: $12.340 - 11.560 = 0.780$
   Residual sum of squares: $0.780 - 0.570 - 0.035 = 0.175$

The following table summarizes the results:

	Sum of squares	d.f.	Mean square	F
Rows (men).............	0.035	3	0.0117	0.60
Columns (trucks).......	0.570	3	0.1900	9.79
Residual...............	0.175	9	0.0194	
Total..............	0.780	15		

f. From Table 3-4, $F_\alpha = 3.86$ for $n_1 = 3$; $n_2 = 9$.

g. $F_{\text{rows}} \leq F_\alpha$; $F_{\text{cols}} \geq F_\alpha$.

There is not sufficient evidence to reject the hypothesis regarding drivers.

There is reason to believe the truck types have an effect on mean time to first failure.

This result could be due to many factors. Because the F ratio is small, it could be argued that the effects of the men are not significant or that the residual mean square is too large to show the effect that is really present. Because the residual measures factors not accounted for directly in the test, it could be that some one or more factors of importance to truck life have not been taken into account in this test. Such a factor could be ton-miles to first breakdown or some measure of terrain covered to first breakdown. If the industrial engineer desires, he could continue the study by splitting out or controlling these factors in an attempt to reduce the residual mean square and hence make the test more discriminating.

REGRESSION AND CORRELATION

Regression. In some studies, the industrial engineer is required to discover relationships that exist between two or more quantities. The relationship may be needed so that measurements can be taken on the quantity that is easier to measure so as to yield information on a quantity more difficult to measure. In other cases, one quantity may be readily available and another inaccessible quantity may be desired for planning purposes. "How much indirect labor would be required to manufacture 100 units of product; 1,000 units?" "What kind of sales results can be expected if an advertising campaign is undertaken?" "What is the relation between tolerances and manufacturing times?" "What is the relation between estimated production times and actual production times?" In all such types of questions, there is implied a relationship between two or more variables. The following discussion covers one method that is useful in analyzing such questions. The method of analysis is called regression analysis and uses as a tool the method of least squares. An example of all the following calculations will be given on page 10-90 and following.

The Method of Least Squares (Two Variables). The method of least squares is a statistical method of fitting a curve to sample pairs of values. In the pair, one of the values is taken to be the independent variable x and the other value is taken to be the dependent variable y. It is often possible to reverse the roles of independent and dependent, and in general the results will be different. The nomenclature is a matter of convenience.

Given the independent and dependent variables, the industrial engineer must assume the general form of relationship between the variables. Most often it is assumed to be linear; that is, the dependent variable is given by

$$y = a' + b'x$$

where a' and b' are constants whose values will be determined by the analysis. a' is called the intercept and b' is called the slope of the line relating x and y.

If we let

$$y_r = a + bx$$

represent the straight line "fit" to the data points, the method of least squares then chooses a and b so that the sum of the squares of the differences between the value given by the equation above (called the regression line) and the actual values of the dependent variable y is as small as possible.

The method of least squares then is one of minimizing

$$\Sigma (y_i - y_r)^2$$

with respect to a and b, where the sum extends over all sample values.

In practice, the work of least squares calculation is reduced if the data are transformed using the relations $y_i = y_i - \bar{y}$ and $x_i = x_i - \bar{x}$ where $\bar{y}$ and $\bar{x}$ are the mean values of y and x. In effect, this transformation shifts the origin of the data from $(0,0)$ to $(\bar{x},\bar{y})$ and simplifies the calculations. Transformation back to $(0,0)$ is given in the following.

The simplified problem then is to find those values of a (intercept) and b (slope) for which

$$\Sigma (y_i - a - bx_i)^2$$

is minimum.

The following equations, called normal equations, give the values of a and b that minimize the sum.

$$Na = 0 \qquad a = 0$$

$$\sum x_i y_i - b \sum x_i^2 = 0 \qquad b = \frac{\Sigma x_i y_i}{\Sigma x_i^2}$$

Simplified calculations of these values are given on page 10-89. An example is given on page 10-91.

To transform the line back to the origin at $(0,0)$ requires a change in a but no change in b. An equation to get the origin to $(0,0)$ is

$$a = \bar{y} - b\bar{x}$$

a and b give the straight line that makes least the sum of squares of the deviations of the sample values of the dependent variable from the line that is used to estimate the relation between y and x.

Some Further Results. The foregoing has given only the "best" linear relation between x and y. Further questions regarding this relation have to do with placing confidence limits on the line or on the parameters a and b or on predictions of y for given x. There may also be questions as to whether a and b are significantly different from 0.

To answer such questions requires further knowledge of the behavior of the estimates a and b, and this in turn requires the following assumptions:

1. Deviations from the regression line are given by the normal density function.
2. The standard deviation of the normal curve giving the deviations is a constant for all x values.
3. Deviations due to sampling affect only y, not x.

In addition, the calculations necessary to answer questions about a and b are simplified if measurements are made using the point $(\bar{x},0)$ as the origin. This is accomplished by taking $a = \bar{y}$. The rest of this analysis will assume this.

a. Standard error of estimate. Because the variance of the normal curve of the deviations of y values is assumed to be constant for all x, the following variance $(s_{1.2}^2)$ is an estimate of this constant:

$$s_{1.2}^2 = \frac{\Sigma (y_i - y_r)^2}{N - 2}$$

The square root of $s_{1.2}^2$ is called the standard error of estimate.

A short formula to calculate $s_{1.2}^2$ is given by

$$\frac{\Sigma y_i^2 - b\Sigma x_i y_i}{N - 2}$$

If the original or untransformed data are used, this can be written as

$$\frac{\Sigma y_i^2 - a\Sigma y_i - b\Sigma x_i y_i}{N - 2}$$

b. *Standard deviations for a and b.*   Under the assumptions above, it follows that a is a sample value for the universe whose mean is a'. Further, the value a has a density function that is normal, with variance given by $s_a^2 = s_{1.2}^2/N$.

Under the same assumptions, d is a sample value from a universe whose mean is b' and whose variance is given by $s_b^2 = s_{1.2}^2/\Sigma x_i^2$.

From these facts, we can proceed to test hypotheses pertaining to the values of a' and b' and set confidence limits for a' and b' by the methods of pages 10-71 and 10-69.

c. *Confidence limits for a'.*   The following formula gives confidence limits for a':

$$a \pm t_{\alpha/2} \frac{s_{1.2}}{\sqrt{N}}$$

Recall a is taken equal to $\bar{y}$ for this analysis.

d. *Confidence limits on b'.*   The following formula gives confidence limits for b':

$$b \pm t_{\alpha/2} \frac{s_{1.2}}{\sqrt{\Sigma x_i^2}}$$

e. *Confidence limits for the regression line.*   The following gives confidence limits for the line of regression:

$$a + bx_i \pm t_{\alpha/2} \sqrt{s_{a2} + s_b^2 x_i^2}$$

s_a^2 and s_b^2 are defined above.

Note that these limits depend on x_i^2 as well as x_i (x_i being taken as $x_i - \bar{x}$). Hence these limits are not straight lines parallel to the regression line but are curvilinear and open as x moves away from $\bar{x}$ in either direction.

f. *Confidence limits for individual values.*   Another problem of importance is to estimate a value for y for a given value of x, using the regression line as an estimator. Confidence limits—called prediction limits—for this estimation are given by

$$a + bx_i \pm t_{\alpha/2} \sqrt{s_a^2 + s_b^2 x_i^2 + s_{1.2}^2}$$

g. *Tests of hypotheses for a'.*   The test:
 1. Hypothesis: a is a sample value from a universe whose mean value is a'.
 2. Set N.
 3. Set α.
 4. Calculate $t = \dfrac{a - a'}{s_{1.2}/\sqrt{N}}$.
 5. Find from Table 3-2, using $n = N - 2$:
   a. $-t_\alpha$ or t_α if interest is in a one-tail test (see page 10-73).
   b. $\pm t_{\alpha/2}$ if interest is in a two-tail test (see page 10-73).
 6. For conclusions, see page 10-73.

h. *Tests of hypotheses for b'.*   The test:
 1. Hypothesis: b is a sample value from a universe whose mean value is b'.
 2. Set N.
 3. Set α.
 4. Calculate $t = \dfrac{b - b'}{s_{1.2}/\sqrt{\Sigma x_i^2}}$.
 5. Find from Table 3-2, using $n = N - 2$:
   a. $-t_\alpha$ or t_α if interest is in a one-tail test (see page 10-73).
   b. $\pm t_{\alpha/2}$ if interest is in a two-tail test (see page 10-73).
 6. For conclusions, see page 10-73.

Practical Calculation.   The preceding analysis is simplified as noted if the variables

1	2	3	4	5	6	7	8
y_i	x_i	x_iy_i	y_i^2	x_i^2	y_i+x_i	$(y_i+x_i)y_i$	$(y_i+x_i)x_i$

FIG. 3-6. Computational form for regression calculations.

$y_i = y_i - \bar{y}$ and $x_i = x_i - \bar{x}$ for each y_i and x_i are used. Because the values of interest are $\Sigma x_i y_i$, Σx_i^2, and Σy_i^2, a shortcut method is presented here to find these directly from x and y without necessitating the subtraction implied above.

$$\bar{x} = \frac{\Sigma x_i}{N} \qquad \bar{y} = \frac{\Sigma y_i}{N}$$

$$\Sigma x_i y_i = \Sigma x_i y_i - N\bar{x}\bar{y}$$
$$\Sigma x_i^2 = \Sigma x_i^2 - N\bar{x}^2$$
$$\Sigma y_i^2 = \Sigma y_i^2 - N\bar{y}^2$$

These formulas then rely only on the original data and allow a computation form to be set up as shown in Figure 3-6.

After all columns have been summed, a check is:

1. Sum of column 6 = sum of column 1 + sum of column 2.
2. Sum of column 7 = sum of column 3 + sum of column 4.
3. Sum of column 8 = sum of column 5 + sum of column 3.

Analysis of Variance for Linear Regression and Correlation. Two questions remain after fitting the line of regression to the data. First, is there a significant relation between the two variables? Second, how much of the behavior of y is "explained" by x? Two types of tools are available to answer these questions: The first question can be answered using analysis of variance tools similar to those of page 10-81. The second type of question can be answered using correlation methods. Both tools are discussed below.

Analysis of Variance Related to Linear Regression. Using the nomenclature of page 10-82, the following analysis is pertinent to the question: Does the regression line express a statistically significant relation between the variables? This analysis can be used when at least one value of x has two or more y values associated with it. The test:

1. Compute the sum of squares for the variation explained by linear regression. This is given by

$$b(\Sigma x_i y_i - N\bar{x}\bar{y})$$

2. Compute the residual sum of squares. This is given by

$$(N - 2)s_{1.2}^2 \qquad \text{(see page 10-87)}$$

3. Check calculations by finding total sum of squares. This is given by

$$\Sigma y_i^2 - N\bar{y}^2$$

Step 1 + step 2 = step 3 is a check on the calculations.
4. Calculate mean square for the variation explained by linear regression. This is given by step 1 ÷ 1.
5. Calculate residual mean square. This is given by step 2 ÷ $(N - 2)$. (*Note:* This is $s_{1.2}^2$.)
6. Calculate F = mean square for the variation explained by linear regression ÷ residual mean square.
7. Find F_α from Table 3-4, using $n_1 = 1$; $n_2 = N - 2$.
8. If $F \leq F_\alpha$, accept the hypothesis that the linear relation is not statistically significant.
9. If $F \geq F_\alpha$, reject the hypothesis.

Correlation

a. Computation using regression results. In many studies, it is important to know how much association there is between the variables. A method to give some quantitative meaning to this is afforded by correlation and the coefficient of correlation r.

One method of computing the sample coefficient of correlation that is useful if a regression analysis has preceded the work is to take

$$r^2 = 1 - \frac{s_{1.2}{}^2}{\Sigma y_i{}^2/(N-1)}$$

The coefficient r is then the square root of this.

b. Properties of r^2

1. r^2 gives a measure of the amount of variation in the y data that is accounted for by the x data and hence can serve as a quantitative measure of how much relationship exists between the variables.
2. r is always between -1 and 1.
3. If $r \geq 0$, this implies y increases with increasing x. If $r \leq 0$, y decreases with increasing x. Figure 3-7 shows the relation of x and y for three values of r.
4. r', the universe correlation coefficient, is given by the relation

$$r' = b' \frac{\sigma_x{}'}{\sigma_y{}'}$$

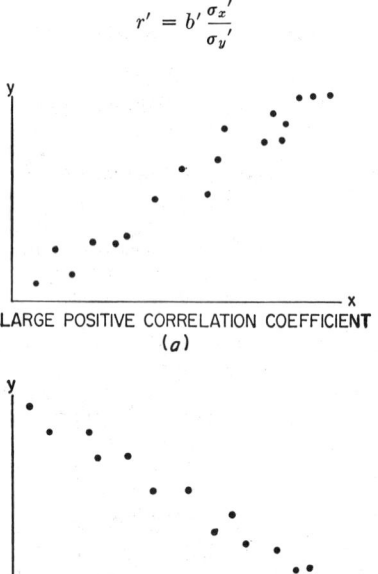

LARGE POSITIVE CORRELATION COEFFICIENT
(*a*)

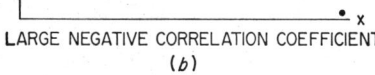

LARGE NEGATIVE CORRELATION COEFFICIENT
(*b*)

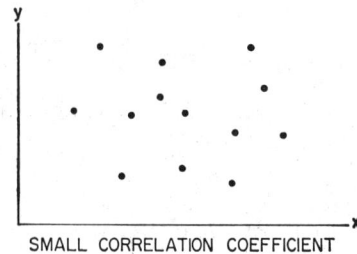

SMALL CORRELATION COEFFICIENT
(*c*)

FIG. 3-7. r, the correlation coefficient behavior.

where b' is the universe value of the slope of the regression line, σ_x' is the universe standard deviation of x, and σ_y' is the universe standard deviation of y.

Because of this relation, a test of hypothesis for $b' = 0$ can also be used to test the hypothesis $r' = 0$ (see page 10–88). This test can be used only to test for $r' = 0$.

c. *Rank correlation coefficient.* Another measure of correlation, called the Spearman rank correlation coefficient or the rank difference correlation coefficient, is useful with paired reading. In this measure, each set of data is ranked in some order. For example, sales records for all salesmen for two different years might form the basic data. The measure would rank the salesmen according to dollar volume of sales in ascending order for each year. Then, given the two sets of rankings, the correlation coefficient is given by

$$1 - \frac{6\Sigma D^2}{N(N^2 - 1)}$$

Here N is the number of pairs of observations; D is the difference between the rankings.

As in the case of r, the Spearman correlation coefficient is a number between $+1$ and -1.

Example 28. In determining an ordering policy for raw material, a company might consider the following:

Because of scrap, rework, variability of weight, and the like, the following analysis is used to determine the amount of material needed to manufacture a given number of finished parts. The data (hypothetical) from records would give number of parts manufactured x_i and the amount of raw material used (in feet of stock) y_i. Assume such a study has been conducted for two previous production runs.

x_i	y_i (run 1)	y_i (run 2)
100	80	100
200	200	230
300	350	380
400	420	370

The data are plotted in Figure 3-8.

The following table gives the short computational form for the following analysis.

Computational Form for Regression Analysis

1	2	3	4	5	6	7	8
y_i	x_i	$x_i y_i$	y_i^2	x_i^2	$y_i + x_i$	$(y_i + x_i)y_i$	$(y_i + x_i)x_i$
80	100	8,000	6,400	10,000	180	14,400	18,000
100	100	10,000	10,000	10,000	200	20,000	20,000
200	200	40,000	40,000	40,000	400	80,000	80,000
230	200	46,000	52,900	40,000	430	98,900	86,000
350	300	105,000	122,500	90,000	650	227,500	195,000
380	300	114,000	144,400	90,000	680	258,400	204,000
420	400	168,000	176,400	160,000	820	344,400	328,000
370	400	148,000	136,900	160,000	770	284,900	308,000
Sum 2,130	2,000	639,000	689,500	600,000	4,130	1,328,500	1,239,000

Check: Col. 6 sum = col. 1 sum + col. 2 sum
   4,130 = 2,130 + 2,000
 Col. 7 sum = col. 3 sum + col. 4 sum
  1,328,500 = 639,000 + 689,500
 Col. 8 sum = col. 3 sum + col. 5 sum
  1,239,000 = 639,000 + 600,000

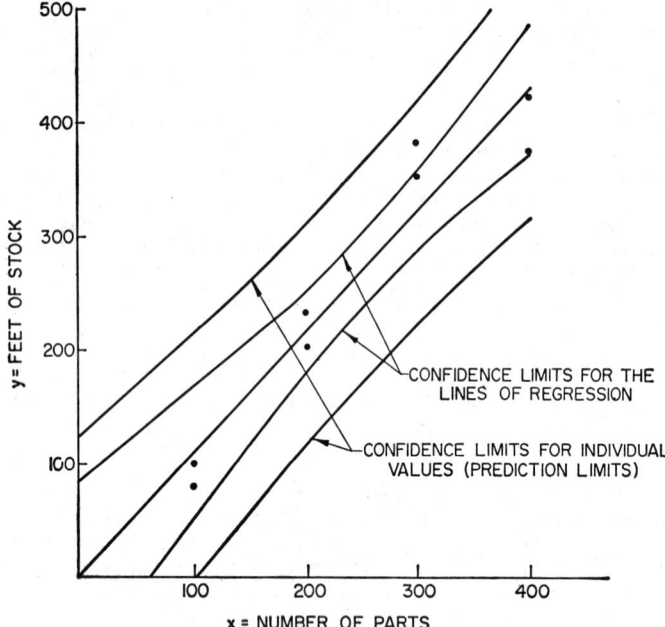

FIG. 3-8.   Regression study for number of parts manufactured and quantity of stock required.

Then, using the formulas on page 10-89,

$$\bar{x} = 2{,}000/8 = 250$$
$$\bar{y} = 2{,}130/8 = 266.25$$
$$\Sigma x_i y_i = 639{,}000 - (8)(250)(266.25)$$
$$= 106{,}500$$
$$\Sigma x_i^2 = 600{,}000 - (8)(250)^2$$
$$= 100{,}000$$
$$\Sigma y_i^2 = 689{,}500 - (8)(266.25)^2$$
$$= 122{,}387.5$$

Then, using the formulas on page 10-87,

1. $b = \sum x_i y_i \Big/ \sum x_i^2 = \dfrac{106{,}500}{100{,}000} = 1.065$

2. $a = \bar{y} - b\bar{x} = 266.25 - (1.065)(250) = 0$

The equation of the line is then

3. $y = a + bx = 1.065x$

The standard error of the estimate is

4. $s_{1.2}^2 = \dfrac{\Sigma y_i^2 - b\Sigma x_i y_i}{N - 2} = \dfrac{122{,}387.5 - 1.065(106{,}500)}{6} = 1{,}494.2$

and

$$s_{1.2} = \sqrt{1{,}494.2} = 38.06$$

5. $s_a^2 = \dfrac{s_{1.2}^2}{N} = \dfrac{1{,}494.2}{8} = 186.8$

6. $s_b^2 = \dfrac{s_{1.2}^2}{\Sigma x_i^2} = \dfrac{1{,}494.2}{100{,}000} = 0.015$

7. 0.95 confidence limits for a' are given by $266.25 \pm 2.447 \sqrt{186.8} = 298.23$, 234.27. Here the value 2.447 comes from Table 3-2 with $n = N - 2$.

8. 0.95 confidence limits for b' are given by $1.065 \pm 2.447 \sqrt{0.015} = 1.359, 0.771$.

9. To test the hypothesis that $b' = 1.0$,

$$t = \frac{0.065}{0.12} = 0.54$$

$t_{0.05}$ for $n = N - 2 = 6$ from Table 3-2 is 1.943. Hence the hypothesis that $b' = 1.0$ is accepted.

10. 0.95 confidence limits for the line of regression are given by $266.25 + 1.065x_i \pm 2.447 \sqrt{186.8 + 0.015x_i^2}$.
These limits are shown in Figure 3-8.

11. 0.95 confidence limits for individual values are given by $266.25 + 1.065x_i \pm 2.447 \sqrt{186.8 + 0.015x_i^2 + 1,494.2}$ or

$$266.25 + 1.065x_i \pm 2.447 \sqrt{1,681.0 + 0.015x_i^2}$$

These limits are shown in Figure 3-8.

12. To test if the regression fit is significant (see page 10-89):

a. Sum of squares for the variation explained by linear regression,

$$1.065[639,000 - 8(250)(266.25)] = 113,422.5$$

b. Residual sum of squares $6(1,494.2) = 8,965.2$

c. Total sum of squares $689,500 - 8(266.25) = 122,387.5$

Check: $113,422.5 + 8,965.2 = 122,387.7$ (error due to rounding). The following table simplifies further calculation:

Sum of squares	d.f.	Mean square
Linear regression = 113,422.5	1	113,422.5
Residual = 8,965.2	6	1,494.2
Total = 122,387.7	7	

Then
$$F = \frac{113,422.5}{1,494.2} = 75.9$$

$F_\alpha = 5.99$ for $\alpha = 0.05$ from Table 3-4, using $n_1 = 1$ and $n_2 = 6$. Because $F_\alpha = 5.99$, linear regression is significant.

13. Correlation coefficient for the data is given by

$$r^2 = 1 - \frac{1,494.2}{122,387.5/7} = 0.915$$

Because $r^2 = 0.915$, we can say that the finished number of pieces manufactured accounts for 91.5 percent of the variation in raw material used.

Extensions of Two-variable Linear Regressions

Multivariate Analysis. The method of least squares developed in this discussion is not restricted to two-variable problems. Extension to multivariable analysis proceeds with more extensive computational work but the same general concepts given here. A description of the necessary extension is beyond the scope of this chapter.

Nonlinear Regression. If the assumption of linearity cannot be made for a particular study (an example of a nonlinear relationship might be total costs and production over large ranges of production), it is often possible to transform the variables and proceed with the assumption of linearity on the transformed variables.

Taking natural logarithms of the variables converts a relation $y = ae^{bx}$ to

$$\log_e y = a + bx$$

so that the logarithm of the dependent variable y has a linear relation to the independent variable. Taking natural logarithms then would transform this nonlinear relation to a linear one.

Square roots, cube roots, inverses, squares, cubes, and so on, all furnish transformations that might allow data, that in their raw form do not meet the linearity assumption, to be transformed into new variables that do meet the linearity assumption. There is no guiding general theory in choosing a transformation. Often skill and luck plus intuition are the only guides available.

The method of least squares can be applied to the case of nonlinear forms if they are polynomials. In this case, for example, the nonlinear form $y = a + bx + cx^2$ becomes linear if x^2 is defined to be w. The problem then is reduced to a trivariate linear problem.

THE COMPUTER AS A TOOL FOR INSTRUCTION OF STATISTICAL TECHNIQUES AND FOR DATA ANALYSIS*

The utility of statistics to the industrial engineer in the analysis of real-life data has been recognized since World War II, at which time it became obvious that more objectivity should be included in the analysis of data for decision making purposes. Because many statistical computations are quite laborious and time consuming when done manually, greater use of statistical techniques was made as mechanical calculators became available. The significant impact of statistics as a problem solving tool for realistic problems, however, resulted from the introduction of high-speed digital computers in the early fifties. Another major thrust in the computer-dominated environment has been to use the computer to facilitate the learning process. This is commonly referred to as computer-assisted instruction (CAI).

Computer-assisted Instruction in Statistics. CAI is a more sophisticated approach to programmed learning, which has generated new ideas in instructional methods adapted to the individualized or tutorial type of instruction. The speed and storage capacity of the computer permit considerable flexibility in the structuring of learning modules. Two articles, one in *Automation* and the other in *Science*, give a good survey on CAI. At System Development Corporation, it was concluded that "CAI is most appropriately employed in the numerical demonstration of statistical concepts and for statistical laboratory exercise instruction."

The University of Pittsburgh has a computer laboratory for instruction of statistical methods in which a Monte Carlo approach is used by the student to determine the effects of parameters on various types of sampling distributions.

The programs and data files are in disc storage and are activated by simple commands from a remote terminal in the Pitt Time Sharing System. The statistical instruction consists of a sequence of computer experiments to provide learning experiences in the following statistical methodology:

1. Random number generation
2. Empirical and theoretical distributions
3. Sample statistics and population parameters
4. Study of sample variances
5. Symmetric and nonsymmetric binomial distributions
6. Central limit theorem and the normal distribution
7. Sampling distribution of the mean
8. The t distribution, power, type I and II errors
9. Sampling distribution of the correlation coefficient

These programs are helpful to the practicing industrial engineer in enabling him to obtain a feel for statistical techniques.

* This section was developed by Dr. A. G. Holzman of the University of Pittsburgh.

Data Analysis. A large number of statistical programs have been written since the inception of the computer era; in fact, thousands of statistical application programs are available from many different sources. One of the most prolific sources is SHARE; another is the biomedical programs. Some of these computer programs are designed for the more traditional batch processing mode, while others are developed for a real-time system. One example will be reviewed for each type of processing mode.

A batch processing system called STIL (statistical interpretive language) has been developed to minimize the need for the user to possess computer skills in communicating with the computer to solve a considerable number of statistical problems. In fact, the user must know only how to operate a key punch. Although many of the available subroutine packages are quite efficient in terms of computer time and computer storage but frustrating to use by one with little computer experience, STIL is practically format-free, and the commands can be started in any column of the punch card, with spaces ignored by the system interpreter. STIL commands enable the user to calculate the mean, standard deviation, variance, correlation, autocorrelation, regression, range, distribution, histogram, skewness, and kurtosis. Set commands are included which permit the creation of new arrays or the transformation of existing arrays.

Example 29. The following example illustrates the use of the time sharing mode for statistical data analysis for a single variable. This has been extended at the University of Pittsburgh to a two-term sequence in multivariate analysis. The ease of use of the time sharing program becomes obvious as one reviews this sample problem. The input data are entered in the computer at a remote teletypewriter or terminal. This remote station may be located hundreds of miles from the physical location of the computer itself.

The purpose of this program is to perform a statistical analysis on data for one variable. It computes 34 different measures for an array of weighted (as with frequencies) or unweighted values of the variable. It also gives a ten-class frequency distribution summary, and a recapitulation of the input data in terms of deviations from the mean and as an ordered array. For the problem solved, it is significant to point out that the computer time required for all computations was only 0.58 second.

To use this program, supply data in either of these two formats as determined by the problem:

  a. For unweighted values:
    *10 DATA 0, X(1), X(2), X(3), . . .
  b. For data with weights or frequencies:
    *10 DATA 1, X(1), F(1), X(2), F(2), X(3), F(3), . . .
The initial 0 or 1 signals the absence or the presence of weights. Lines 11 through 99 are available for additional input data.

Sample Problem. Suppose the following data points have been obtained as the result of an experiment:

261.4	252.1	255.5	258.3	253.2
270.8	268.3	249.6	256.3	266.4
265.4	250.3	280.9	259.3	
261.4	272.3	270.3	270.1	
258.1	262.8	263.2	259.3	

To use this program to get the statistical characteristics of the data, enter the following data:

*10 DATA 0, 261.4, 270.8, 265.4, 261.4, 258.1, 252.1, 268.3, 250.3
*11 DATA 272.3, 262.8, 255.5, 249.6, 280.9, 270.3, 263.2, 258.3, 256.3
*12 DATA 259.3, 270.1, 259.3, 253.2, 266.4

where the first data value (0) indicates unweighted data.

Sample Solution. These sample solutions are copies of the printout that will appear at the teletypewriter.

Note: All user-supplied information is underlined.

SYSTEM?..
  BAS STATAN*
*FILE-RETRIEVED

*1∅   DATA ∅,261.4,27∅.8,265.4,261.4,258.1,252.1,268.3,25∅.3
*11   DATA 272.3,262.8,255.5,249.6,28∅.9,27∅.3,263.2,258.3,256.3
*12   DATA 259.3,27∅.1,259.3,253.2,266.4
*RUN

STATAN 11-NOV-69 17:∅2

MODIFIED 05/29/68

COMPUTATIONS ON THE DATA ARRAY:

NUMBER OF VALUES = 22
NUMBER OF NONZERO WEIGHTS = 22
SUM OF WEIGHTS = 22
SUM OF UNWEIGHTED VALUES = 5765.3
WEIGHTED MEAN = 262.∅59
UNWEIGHTED MEAN = 262.∅59
STD DEVIATION OF SAMPLE = 7.7837
SAMPLE VARIANCE = 6∅.5861
SMALLEST VALUE = 249.6
LARGEST VALUE = 28∅.9
RANGE = 31.3
WEIGHTED SUM OF SQUARES = 1.51218E6

NOTE: THE FOLLOWING MEASURES PERTAIN TO THE MOST-LIKELY ESTIMATES
       FOR THE TOTAL POPULATION

VARIANCE = 63.4711
STANDARD DEVIATION = 7.96688
STANDARD ERROR OF MEAN = 1.69854
COEFFICIENT OF VARIATION = 3∅.4∅11E-2
STUDENT'S T = 154.285
MEAN SQUARE SUCCESSIVE DIFFERENCES = 145.4
(MEAN SQ SUCC DIFF)/(VARIANCE) = 2.29∅81
MEDIAN = 261.4
NUMBER OF RUNS UP AND DOWN = 11
EXPECTED NUMBER OF RUNS = 14.3333
STD ERROR OF NUMBER OF RUNS = 1.89444
(ACTUAL RUNS — EXP RUNS)/(STD ERR) = 1.75954

FREQUENCY DISTRIBUTION (TEN EQUAL CLASSES):
    3 2 3 4 2 3 3 1 ∅ 1

COMPUTATIONS ON DEVIATIONS FROM MEAN:

NUMBER OF + SIGNS IN DEVIATIONS = 1∅
NUMBER OF − SIGNS IN DEVIATIONS = 12
NUMBER OF RUNS (SIGN CHANGES + 1) = 12
EXPECTED NUMBER OF RUNS = 11.9∅91
STD ERROR OF NUMBER OF RUNS = 2.26883
(ACTUAL RUNS — EXP RUNS)/(STD ERR) = 4.∅∅687E-2
TREND VALUE = −5.7651E-2
STD ERROR OF TREND = 6.12764E-2
(TREND)/(STD ERROR) = −.94∅835

BETA ONE = .148616
BETA TWO = 2.67928
MEAN DEVIATION = 6.35537

RECAPITULATION OF INPUT:

VALUE	DEVIATIONS	WEIGHTS	ORDERED ARRAY
261.4	− .65902	1	249.6
270.8	8.74091	1	250.3
265.4	3.34091	1	252.1
261.4	− .65902	1	253.2
258.1	− 3.95909	1	255.5
252.1	− 9.95909	1	256.3
268.3	6.24091	1	258.1
250.3	−11.7591	1	258.3
272.3	10.2409	1	259.3
262.8	.74091	1	259.3
255.5	− 6.55909	1	261.4
249.6	−12.4591	1	261.4
280.9	18.8409	1	262.8
270.3	8.24091	1	263.2
263.2	1.14091	1	265.4
258.3	− 3.75909	1	266.4
256.3	− 5.75909	1	268.3
259.3	− 2.75909	1	270.1
270.1	8.04091	1	270.3
259.3	− 2.75909	1	270.8
253.2	− 8.85909	1	272.3
266.4	4.34091	1	280.9

BIBLIOGRAPHY

Alder, Henry L., and Edward B. Roessler, *Introduction to Probability and Statistics*, 4th ed., W. H. Freeman and Company, San Francisco, 1968.
Bowker, A. H., and G. J. Lieberman, *Engineering Statistics*, Prentice-Hall, Inc., Englewood Cliffs, N.J., 1959.
Cooley, William W., and P. R. Lohnes, *Multivariate Procedures for the Behavioral Scientist*, 2d ed., John Wiley & Sons, Inc., New York, 1970.
Dixon, W. J. (ed.), *BMD: Biomedical Computer Programs*, University of California, School of Medicine, Los Angeles, 1964.
Donaghey, C. E., and O. S. Ozkul, "STIL Users' Manual," Project THEMIS Information Processing Systems, ONR contract N00014-68-A-0151, University of Houston, Houston, Tex., August, 1969.
Duncan, A. J., *Quality Control and Industrial Statistics*, 3d ed., Richard D. Irwin, Inc., Homewood, Ill., 1965.
Freund, John E., *Mathematical Statistics*, Prentice-Hall, Inc., Englewood Cliffs, N.J., 1962.
Hadley, G., *Elementary Statistics*, Holden-Day, Inc., San Francisco, 1969.
Holzman, A. G., R. Glaser, and H. Schaefer, *Matrices and Mathematical Programming*, Encyclopaedia Britannica, Inc., Chicago, 1963.
Lohnes, Paul R., and W. W. Cooley, *Introduction to Statistical Procedures: With Computer Exercises*, John Wiley & Sons, Inc., New York, 1968.
Time-sharing Library Programs, On-Line Systems, Inc., Pittsburgh, Pa.

Chapter **4**

Bayesian Statistics and Modern Decision Theory

MORRIS H. DeGROOT

**Head, Department of Statistics,
Carnegie-Mellon University, Pittsburgh, Pennsylvania**

A large part of the work of an industrial engineer is concerned either with making decisions under the conditions of uncertainty and partial ignorance that necessarily exist in many of his activities or with collecting and processing information that will be helpful in making such decisions. For example, the quality of the items in a lot is typically not known with certainty, but the industrial engineer may have to decide whether or not the lot should be released for shipment. Or he may have to decide whether it would be worthwhile to inspect a random sample of items from the lot and, if so, how large a sample to inspect. Although the output of a plant varies from day to day, the industrial engineer may have to estimate the total output for a fifteen-day period three weeks in the future. He may have to decide whether a standard production process should be replaced by a new process for which the capabilities and costs are still not known with certainty.

The modern theory of Bayesian statistical inference and decisions deals with the development of techniques that are appropriate for making inferences and decisions in situations like these. The purpose of this chapter is to provide a survey of some of the basic techniques that can be helpful in formulating and solving decision problems in industrial engineering and related areas.

THE CHOICE OF A DECISION

Suppose that a malfunction that causes a certain system to break down and become inoperative can occur in either of two different parts of the system, part A or part B.

Suppose also that when the system does become inoperative, it is not immediately known whether the malfunction causing the breakdown has occurred in part A or in part B. We shall assume that the repair procedures are quite different for the two different parts, and that when a breakdown occurs in the system, one of the following three decisions must be chosen:

Decision d_1: The repair procedure for part A is activated immediately. If the malfunction causing the breakdown actually occurred in part B, then the cost of this decision in terms of unnecessary labor and lost time is $1,000. If the malfunction actually occurred in part A, then this decision leads to the repair of the malfunction in the most efficient manner and the cost is regarded as zero.

Decision d_2: The repair procedure for part B is activated immediately. If the malfunction actually occurred in part A, then the cost of this decision is $3,000. If the malfunction occurred in part B, then the cost is again regarded as zero.

Decision d_3: A test is applied to the system that will determine with certainty whether the malfunction has occurred in part A or part B. The cost of applying this test is $300.

Which decision should be chosen?

Specification of the Probabilities. The first step in the process for choosing among the three alternative decisions is a specification of the probability of each event whose occurrence is relevant to this choice. In this example, only the probability p that the malfunction occurred in part A needs to be determined. The probability q that the malfunction occurred in part B must then be equal to $1 - p$.

The specification of the value of the probability p must ultimately be made by the decision maker and must represent his belief of the likelihood that part A has malfunctioned. In this sense, the probability that is assigned will be a subjective probability. This probability will, however, incorporate the totality of information and data available to the decision maker at the time that he must specify the value of p. If that information is extensive and decisive, the subjective nature of the probability will be greatly reduced in the sense that most engineers would specify approximately the same value of p after being exposed to this same information. On the other hand, if few statistical data are available, the specification of the value of p must necessarily depend to a great extent on the knowledge and beliefs of the individual decision maker. In most cases, the value of p that is finally specified will be based on a combination of the following three types of reasoning:

1. *Relative frequency.* Suppose that a breakdown in the system has occurred several times in the past under conditions of uncertainty similar to those existing in the present problem. Then p is approximately equal to the proportion of those past breakdowns that were caused by a malfunction in part A.

2. *Equally likely possibilities.* Suppose that the breakdown in the system is caused by a defect in one of n similar components, all of which are equally likely to be defective. Suppose that m of these components are used in part A and the other $n - m$ components are used in part B. Then p is approximately equal to m/n.

3. *Subjective comparisons of relative likelihood.* The decision maker's overall subjective evaluation of p can be determined by his comparing the likelihood of a malfunction in part A to the likelihoods of various outcomes of an auxiliary experiment in which a number X is selected at random and with uniform probability density from the interval $0 \leq X \leq 1$. Let A denote the event that the malfunction is in part A; so $p = \Pr(A)$. Does the decision maker regard the event A as more likely to occur than the event that $0 \leq X \leq \frac{2}{3}$? If so, then $p > \frac{2}{3}$. If not, then $p \leq \frac{2}{3}$. Suppose finally that an interval $0 \leq X \leq x^*$ can be found such that the event A and the event that $0 \leq X \leq x^*$ are regarded as approximately equally likely to occur. Then p is approximately equal to x^*.

Expected Value. If a random variable can take the different values $x_1, x_2, \ldots , x_k$ with probabilities $p_1, p_2, \ldots , p_k$, respectively, then the expected value of that variable is defined to be the number $p_1 x_1 + p_2 x_2 + \cdots + p_k x_k$. If a random variable can take any value in some interval $a \leq x \leq b$ of the real line (this interval could possibly be the entire real line itself), and if the variable has a continuous distribution that can be represented by the density function $f(x)$ over the interval $a \leq x \leq b$, then the expected value is defined to be

$$\int_a^b xf(x)\ dx$$

The expected value of a variable is also called its mean, its mean value, or its expectation. All these terms may be used interchangeably. Furthermore, rather than speaking of the expected value of the variable itself, we may equivalently speak of the expected value of its probability distribution.

Thus, in the problem being considered here, the expected cost of choosing decision d_1 is

$$q(\$1{,}000) + p(\$0) = \$1{,}000q$$

Similarly, the expected cost of choosing d_2 is $\$3{,}000p$. The expected cost of choosing d_3 is just $\$300$, because d_3 will cost this amount with certainty.

Minimizing the Expected Cost. We shall now consider the decision rule that specifies that a decision should be chosen for which the expected cost is minimized. Suppose, for example, that $p = 0.8$ and $q = 0.2$. Therefore, the expected cost of d_1 is $\$200$, the expected cost of d_2 is $\$2{,}400$, and the expected cost of d_3 is $\$300$. Because d_1 has the smallest expected cost, d_1 should be chosen.

The expected cost for each decision is shown in Figure 4-1 as a function of p. It is seen from Figure 4-1 that if $0 \le p \le 0.1$, then d_2 has the smallest expected cost. If $0.1 \le p \le 0.7$, then d_3 has the smallest expected cost. If $0.7 \le p \le 1$, then d_1 has the smallest expected cost.

Advantages and Disadvantages of Minimizing the Expected Cost. The decision rule that specifies choosing a decision for which the expected cost is minimized is highly specialized. For each possible decision d, the only feature of the probability distribution of the cost of d that is considered by the decision maker is the mean of the distribution. If the decision which has the lowest expected cost also involves the largest probabilities of severe losses, the decision maker may prefer to choose a decision which has a larger expected cost but which is less risky. Thus, in our example, when $p = 0.8$, the decision maker may well prefer to choose d_3 and pay the certain cost of $\$300$ rather than to choose d_1, for which the expected cost is lower but for which there is probability 0.2 of losing $\$1{,}000$.

The use of the rule that specifies choosing a decision to minimize the expected cost is most easily justified and most clearly appropriate when the following two conditions are satisfied: (1) The same type or a similar type of decision problem will occur repeatedly in the future, and a decision will have to be chosen in each problem. (2) The possible cost that might be incurred in any individual problem is small relative to

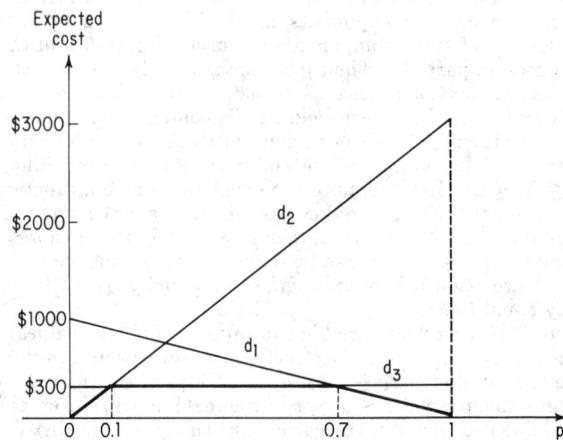

FIG. 4-1. Expected cost of each decision.

the resources of the decision maker or the organization he represents. Under these conditions, choosing a decision in each problem that minimizes the expected cost will result in the lowest average cost per problem over the entire sequence of future problems. Large costs in some individual problems will be compensated by small costs in other problems.

On the other hand, if the decision problem cannot be regarded as part of a repetitive sequence or if the possible outcomes of each decision cannot directly be assessed in terms of gains or losses, then another decision rule is needed. In such a problem, the appropriate decision procedure is determined by the theory of utility. For each possible outcome of each possible decision, the decision maker must specify a number called the "utility" of that outcome. This number represents the value of the outcome to his organization. He then chooses a decision for which the expected utility of the outcome is maximized. In other words, if the decision maker is not willing merely to compute the expected cost of each decision, then he must specify the value of each outcome in the appropriate units of utility for which he is willing to compute the expectation. We shall now consider an example.

A Betting Problem. Suppose that a gambler has a total fortune of $1,000 and that he may bet any part of this fortune on a certain gamble. We shall assume that if he bets an amount x and wins, then his fortune in dollars becomes $1,000 + x$; if he loses, then his fortune becomes $1,000 - x$. We shall assume that the probability of winning is $3/5$ and the probability of losing is $2/5$; so the bet is favorable to the gambler. How much of his total fortune should he bet?

If he bets the amount x, the expected value of his fortune is

$$\frac{3}{5}(1,000 + x) + \frac{2}{5}(1,000 - x) = 1,000 + \frac{1}{5}x$$

Therefore, to maximize the expected value of his fortune, he should choose x as large as possible. In other words, he should bet his entire fortune of $1,000. The gambler may not wish to apply this procedure, however, because there is probability $2/5$ that he will lose his entire fortune.

Now suppose that the gambler considers the utility $U(y)$ of having a fortune of y dollars to be $U(y) = \log y$. The use of the logarithmic function indicates that the gambler regards the utility of winning any fixed additional amount when he already has a large fortune to be less than the utility of winning the same amount when he has a smaller fortune. Also, the fact that $\log y \to -\infty$ as $y \to 0$ indicates that it is important for him to keep his fortune away from 0. If the gambler bets the amount x, then the expected utility of his fortune is

$$\frac{3}{5}\log (1,000 + x) + \frac{2}{5}\log (1,000 - x)$$

The value of x for which this expected utility is a maximum can be determined by setting the derivative with respect to x equal to 0. The optimum value of x is found to be

$$x = \left(\frac{3}{5} - \frac{2}{5}\right)(1,000) = 200$$

Hence, to maximize his expected utility, the gambler should bet $200.

CONDITIONAL PROBABILITY

We shall now consider the manner in which the probability of an event changes as information is gathered about the occurrence or nonoccurrence of related events. For any event A, we shall denote the probability of A by the symbol $\Pr(A)$. For two events A and B, $\Pr(A|B)$ denotes the conditional probability of A when B is known to have occurred. If $\Pr(B) > 0$, then

$$\Pr(A|B) = \frac{\Pr(AB)}{\Pr(B)}$$

where $\Pr(AB)$ is the probability that both A and B occur.

Example. Suppose that 72 percent of the plastic pieces produced on a certain machine have both the right shape and the right color, that 20 percent have the wrong shape but the right color, that 5 percent have the right shape but the wrong color, and that 3 percent have both the wrong shape and the wrong color. Thus all possible outcomes have been exhausted and the sum of the respective probabilities equals 1.00 (100 percent). If a piece selected at random has the right shape, what is the probability that it also has the right color?

Let S be the event that a piece has the right shape and C be the event that it has the right color. We wish to find $Pr(C|S)$. Because $P(CS) = 0.72$ and $Pr(S) = 0.72 + 0.05 = 0.77$, then

$$Pr(C|S) = \frac{0.72}{0.77} = 0.935$$

Bayes' Theorem. Suppose that $A_1, \ldots, A_k$ are k mutually exclusive and exhaustive events, exactly one of which must occur, and suppose that B is any other event. Suppose that $Pr(A_i) > 0$ for $i = 1, \ldots, k$ and $Pr(B) > 0$. The following relation is known as Bayes' theorem:

$$Pr(A_i|B) = \frac{Pr(B|A_i)Pr(A_i)}{\sum_{j=1}^{k} Pr(B|A_j)Pr(A_j)}$$

This relation provides a rule for computing the conditional probabilities $Pr(A_i|B)$ for $i = 1, \ldots, k$ from the conditional probabilities $Pr(B|A_i)$ and the probabilities $Pr(A_i)$ for $i = 1, \ldots, k$.

The values $Pr(A_i)$ for $i = 1, \ldots, k$ are often called the "prior probabilities" of the events A_i, because they are the probabilities before it is known whether the event B has occurred. The values $Pr(A_i|B)$ for $i = 1, \ldots, k$ are then called the "posterior probabilities," because they are the relevant values after it is learned that B has occurred.

Example. Suppose that three different machines were used for producing a large lot of similar tubes. Suppose that 20 percent of the tubes were produced by machine 1, that 30 percent were produced by machine 2, and that 50 percent were produced by machine 3. Suppose further that 4 percent of the tubes produced by machine 1 are defective, that 3 percent of the tubes produced by machine 2 are defective, and that 1 percent of the tubes produced by machine 3 are defective. If one tube selected at random from the entire lot is found to be defective, what is the probability that it was produced by machine 2?

Let A_i be the event that a tube selected at random is produced by machine i ($i = 1,2,3$), and let B be the event that the tube is defective. We wish to find $Pr(A_2|B)$. By Bayes' theorem,

$$\begin{aligned}
Pr(A_2|B) &= \frac{Pr(B|A_2)Pr(A_2)}{Pr(B|A_1)Pr(A_1) + Pr(B|A_2)Pr(A_2) + Pr(B|A_3)Pr(A_3)} \\
&= \frac{(0.03)(0.30)}{(0.04)(0.20) + (0.03)(0.30) + (0.01)(0.50)} \\
&= \frac{0.009}{0.022} = 0.4091
\end{aligned}$$

Computation of Posterior Probabilities in Successive Stages. Suppose that a box contains a fair coin (with a head and a tail) and a two-headed coin, and suppose that one of the coins is selected at random and flipped. If the coin shows heads, what is the probability that it is the fair coin?

Let A_1 be the event that the coin is fair, let A_2 be the event that the coin is two-headed, and let B be the event that the coin shows heads when it is flipped. We wish to find $\Pr(A_1|B)$. By Bayes' theorem,

$$\Pr(A_1|B) = \frac{(\tfrac{1}{2})(\tfrac{1}{2})}{(\tfrac{1}{2})(\tfrac{1}{2}) + (1)(\tfrac{1}{2})} = \frac{1}{3}$$

Hence, the posterior probability that the coin is fair, after one head has been observed, is $\tfrac{1}{3}$.

Suppose now that the same coin is flipped a second time and again shows heads. What is now the probability that it is the fair coin? There are two ways to solve this problem and both yield the same answer. First, we could return to the beginning and redefine B to be the event that the coin shows heads twice when it is flipped twice. Then $\Pr(B|A_1) = \tfrac{1}{4}$ and $\Pr(B|A_2) = 1$. Therefore, again by Bayes' theorem,

$$\Pr(A_1|B) = \frac{(\tfrac{1}{4})(\tfrac{1}{2})}{(\tfrac{1}{4})(\tfrac{1}{2}) + (1)(\tfrac{1}{2})} = \frac{1}{5}$$

Alternatively, we could proceed as follows. After the first flip of the coin, we have computed the posterior probability that the coin is fair to be $\tfrac{1}{3}$, and therefore the posterior probability that the coin is two-headed must be $\tfrac{2}{3}$. These posterior probabilities from the first flip must serve as the prior probabilities for the second flip. Thus, if B is now regarded as the event that the coin shows heads on the second flip, then by Bayes' theorem,

$$\Pr(A_1|B) = \frac{(\tfrac{1}{2})(\tfrac{1}{3})}{(\tfrac{1}{2})(\tfrac{1}{3}) + (1)(\tfrac{2}{3})} = \frac{1}{5}$$

Thus, from either point of view, the posterior probability that the coin is fair is $\tfrac{1}{5}$ after it has shown heads on two flips.

This example indicates that when observations are made in more than one stage, the posterior distribution can be computed in different stages by letting the posterior probabilities after each stage serve as the prior probabilities for the next stage. At each stage, the decision maker needs to keep in mind only his current probabilities of the events in which he is interested. Whenever he receives new relevant information, he revises these probabilities in accordance with Bayes' theorem. At any time that a decision must be chosen, he chooses one that is optimal with respect to his current probabilities.

PRIOR AND POSTERIOR DISTRIBUTIONS OF A PARAMETER

We shall now consider further problems involving a parameter whose exact numerical value is unknown. For example, the parameter might be the unknown proportion of defective items in a certain manufactured lot or the mean hardness of a certain alloy. We shall again assume that the decision maker's uncertainty about this value can be represented by a probability distribution—his prior distribution of the parameter. Furthermore, we shall assume that observations can be taken whose distribution depends on the parameter. In the examples just mentioned, it may be possible to observe the number of defective items in a random sample selected from the lot or to measure the hardness of each of several randomly selected specimens of the alloy. We shall then show how the values of the observations can be combined with the prior distribution of the parameter to obtain a new posterior distribution of the parameter.

Bayes' Theorem for the Posterior Probability Function of a Parameter. Consider a parameter W that can take only a finite number of different values $w_1, \ldots, w_k$, and suppose that the prior distribution of W is specified by the probability function $g(w_i) = \Pr(W = w_i)$. Suppose also that the value of a random variable X can be observed, where X may be either a discrete or a continuous random variable. If X is discrete, then for each possible value w_i of W, we shall let $f(x|w_i)$ denote the probability function of X when $W = w_i$. If X is continuous, then $f(x|w_i)$ will denote the density function of X when $W = w_i$. The random variable X may actually be an entire sample $X_1, \ldots, X_n$ of random variables. In this case, $f(x|w_i)$ represents either the joint probability function or the joint density function of $X_1, \ldots, X_n$.

Regardless of whether X is discrete or continuous, after the value $X = x$ has been observed, the posterior probability function $g(w_i|x) = \Pr(W = w_i|X = x)$ is specified by the following version of Bayes' theorem:

$$g(w_i|x) = \frac{f(x|w_i)g(w_i)}{\sum_{j=1}^{k} f(x|w_j)g(w_j)}$$

If the possible values of W form an infinite sequence $w_1, w_2, \ldots$, instead of being only finite in number, then this same formula can be applied, except that the summation in the denominator must be extended over all possible values w_i and hence becomes an infinite series.

Example 1. On any given day, a production system can operate at a low level w_1, at a medium level w_2, or at a high level w_3. The output of the system on any day is measured as a number X between 0 and 2. When the system is operating at a low level, the density function $f(x|w_1)$ of X is

$$f(x|w_1) = \frac{1}{2} \quad \text{for} \quad 0 \le x \le 2$$

When the system is operating at a medium level, the density function $f(x|w_2)$ is

$$f(x|w_2) = \frac{1}{2}x \quad \text{for} \quad 0 \le x \le 2$$

When the system is operating at a high level,

$$f(x|w_3) = \frac{3}{8}x^2 \quad \text{for} \quad 0 \le x \le 2$$

These three density functions are sketched in Figure 4-2.

Suppose that it is known that the system operates at a low level on 10 percent of the days, at a medium level on 70 percent of the days, and at a high level on 20 percent of the days. Then $g(w_1) = 0.1$, $g(w_2) = 0.7$, and $g(w_3) = 0.2$.

If the output on a certain day is observed to be $X = 1.5$, what is the probability that the system was operating at a high level on that day? The required probability is $g(w_2|1.5)$. By Bayes' theorem,

$$g(w_2|1.5) = \frac{(\tfrac{3}{8})(1.5)^2(0.2)}{(\tfrac{1}{2})(0.1) + (\tfrac{1}{2})(1.5)(0.7) + (\tfrac{3}{8})(1.5)^2(0.2)} = 0.227$$

The posterior probability that the system was operating at a low level or at a medium level can be calculated similarly.

Example 2. We shall now give an example that uses the normal density function. It is said that a random variable has a normal distribution with mean μ and variance σ^2 (that is, standard deviation σ) if its density function $f(x|\mu,\sigma^2)$ over the entire real

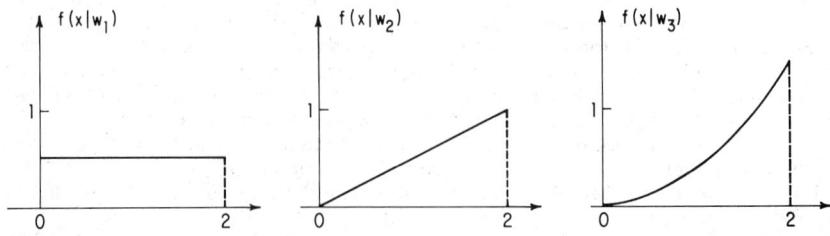

Fig. 4-2. Three possible density functions.

line is

$$f(x|\mu,\sigma^2) = \frac{1}{(2\pi)^{1/2}\sigma} \exp\left[-\frac{1}{2}\left(\frac{x-\mu}{\sigma}\right)^2 \right]$$

The symbol $\exp(u)$ is merely a convenient way of printing e^u.

Consider a process for manufacturing steel pins. Suppose that when the process is in control, the strength of the pins that are produced has a normal distribution with mean μ_1 and variance σ_1^2; but when the process is out of control, the strength has a normal distribution with mean μ_2 and variance σ_2^2. Suppose that the prior probability that the process is in control at any given time is p, and therefore the probability that it is out of control is $q = 1 - p$.

Suppose that the process is checked at a specified time by selecting a random sample of n pins and measuring their strengths $X_1, \ldots, X_n$. If the observed values are $x_1, \ldots, x_n$, what is the posterior probability that the process is out of control?

If each observation has a normal distribution with mean μ and variance σ^2, then the joint density function of the n observations is

$$\prod_{i=1}^{n} f(x_i|\mu,\sigma^2) = \frac{1}{(2\pi)^{n/2}\sigma^n} \exp\left[-\frac{1}{2}\sum_{i=1}^{n}\left(\frac{x_i-\mu}{\sigma}\right)^2 \right]$$

Therefore, by Bayes' theorem, the posterior probability that the process is out of control is

$$\frac{\dfrac{q}{\sigma_2^n} \exp\left[-\dfrac{1}{2}\sum_{i=1}^{n}\left(\dfrac{x_i-\mu_2}{\sigma_2}\right)^2 \right]}{\dfrac{p}{\sigma_1^n} \exp\left[-\dfrac{1}{2}\sum_{i=1}^{n}\left(\dfrac{x_i-\mu_1}{\sigma_1}\right)^2 \right] + \dfrac{q}{\sigma_2^n} \exp\left[-\dfrac{1}{2}\sum_{i=1}^{n}\left(\dfrac{x_i-\mu_2}{\sigma_2}\right)^2 \right]}$$

Bayes' Theorem for the Posterior Density Function of a Parameter. We shall now describe some problems in which a parameter W can take any value w in some interval $a < w < b$ of the real line and in which the prior distribution of W is specified in terms of a density function $g(w)$ for $a < w < b$. Either end point of the interval $a < w < b$ may be infinite, and in fact the interval may be the entire real line.

We shall again assume that a discrete or a continuous observation X can be taken. As before, for any given value w of W, $f(x|w)$ will denote either the probability function or the density function of X when $W = w$. In this case, the posterior density function $g(w|x)$ of W, after the value $X = x$ has been observed, is specified by the following equation:

$$g(w|x) = \frac{f(x|w)g(w)}{\int_a^b f(x|w')g(w')\,dw'} \qquad \text{for } a < w < b \qquad (1)$$

The denominator in equation (1) is merely the integral of the numerator over all values of w. It does not depend on the specific value of w at which the posterior density is being computed. Therefore, the denominator may be regarded as a constant with respect to w (although it depends on the observed value x), and we may write

$$g(w|x) = (\text{const.}) f(x|w)g(w) \qquad (2)$$

The constant in equation (2) is equal to the reciprocal of the denominator in equation (1). It is the constant which makes the posterior density function $g(w|x)$ into a legitimate density function by ensuring that $\int_a^b g(w|x)\,dw = 1$.

Another form for the posterior density $g(w|x)$ is

$$g(w|x) \propto f(x|w)g(w) \qquad (3)$$

In this relation, the proportionality symbol $\propto$ indicates that the right side of the relation differs from $g(w|x)$ only by a constant factor. The constant factor can always be determined from the requirement that $\int_a^b g(w|x)\, dw = 1$.

The Beta Distribution. A distribution on the interval $0 < w < 1$ is called a beta distribution if the density function $g(w)$ is of the following form:

$$g(w) = \frac{(\alpha + \beta + 1)!}{\alpha!\beta!}\, w^\alpha (1 - w)^\beta \qquad \text{for } 0 < w < 1$$

Here α and β can be any fixed nonnegative integers. It should be kept in mind that $0! = 1$. For $\alpha = \beta = 0$, the density function is merely the uniform density $g(w) = 1$ for $0 < w < 1$.

The mean of the beta distribution is

$$\frac{\alpha + 1}{\alpha + \beta + 2}$$

and the variance is

$$\frac{(\alpha + 1)(\beta + 1)}{(\alpha + \beta + 2)^2(\alpha + \beta + 3)}$$

The Beta Distribution as a Prior Distribution. Suppose that the proportion W of defective items in a large lot is unknown, but the prior distribution of W is a certain beta distribution with specified nonnegative integers α and β. Suppose that a random sample of n items is taken from the lot. For any fixed value of W, the proportion X of defective items in the sample has a binomial distribution. The probability $f(x|w)$ of obtaining exactly x defective items in the sample when $W = w$ is

$$f(x|w) = \binom{n}{x} w^x (1 - w)^{n-x}$$

Because the prior density $g(w)$ is a beta density, it now follows from equation (3) that the posterior density $g(w|x)$ satisfies the relation

$$g(w|x) \propto w^x (1 - w)^{n-x} w^\alpha (1 - w)^\beta$$

In this relation, all constant factors (factors that do not involve w) have been absorbed into the proportionality symbol. This relation can now be reduced to the form

$$g(w|x) \propto w^{\alpha+x}(1 - w)^{\beta+n-x}$$

Except for the appropriate constant factor, the right side of this relation is a beta density with α replaced by $\alpha + x$ and β replaced by $\beta + n - x$.

In other words, if the prior density of W is a beta density, then the posterior density will again be a beta density, with α increased by the number of defective items x in the sample and β increased by the number of nondefective items $n - x$ in the sample.

A similar result is valid if the observations are taken sequentially, one at a time. The posterior distribution is revised after each item is observed, but at each stage it will be a beta distribution; α is increased by one unit each time a defective item is observed and β is increased by one unit each time a nondefective item is observed.

The Normal Distribution as a Prior Distribution. Suppose that the quality of the items in a large lot, when measured in the appropriate units, has a normal distribution with unknown mean W and known variance v^2. Suppose also that the prior distribution of the unknown mean W is itself a normal distribution with mean μ_0 and variance σ_0^2. Suppose finally that a random sample of n items is taken from the lot and that the quality $X_1, \ldots, X_n$ of each of the n items is measured. Then it can be shown from the relations that have previously been developed that the posterior distributions of W, after the values $X_1 = x_1, \ldots, X_n = x_n$ have been observed, will again be a normal distribution. The mean μ_1 and the variance σ_1^2 of this posterior distribution are given by the following equations:

$$\mu_1 = \frac{\mu_0/\sigma_0^2 + n\bar{x}/v^2}{1/\sigma_0^2 + n/v^2}$$

and

$$\sigma_1^2 = \frac{1}{1/\sigma_0^2 + n/v^2}$$

where

$$\bar{x} = \frac{1}{n} \sum_{i=1}^{n} x_i$$

Features of the Posterior Distribution. It should be noted that the mean μ_1 of the posterior distribution is a weighted average of the mean μ_0 of the prior distribution and the sample mean $\bar{x}$. If the variance σ_0^2 of the prior distribution is small, then the prior distribution of W is tightly concentrated around μ_0. The posterior distribution will therefore be tightly concentrated around μ_1, and μ_0 will be weighted heavily in this posterior distribution. On the other hand, if σ_0^2 is not small but the variance v^2 of each observation is small or the number of observations n is large, then the posterior distribution will again be tightly concentrated around μ_1. Now, however, the sample mean $\bar{x}$ will be weighted heavily in the posterior distribution.

It should also be noted that in this example, although the mean of the posterior distribution depends on the observed value $\bar{x}$, the variance of the posterior distribution depends only on the number of observations n and not on their observed values. As each observation is taken, the variance of the posterior distribution decreases in a fixed, predetermined way. If it is desired to continue sampling until the variance of the posterior distribution is reduced to a specified small value, then the exact number of observations that will be required can be calculated before sampling begins.

Vague Prior Information. If only scant or vague prior information about the quality of the items in the lot is available in the example now being considered, then the prior distribution will be widely spread out over the real line. In other words, the variance σ_0^2 of the prior normal distribution will be very large. Rather than attempting to assign a specific very large value for σ_0^2 and a specific value for the mean μ_0 of the prior distribution on the basis of the vague prior information, the decision maker can approximate his prior distribution by assuming the density to be uniformly spread out over the entire line. In other words, the prior density is taken to be a uniform density over the entire real line.

This choice of prior density corresponds, in effect, to letting $\sigma_0^2 \to \infty$ in the prior normal density. Of course, the uniform density over the entire real line is not a legitimate probability density. The value of its integral over the whole line is not unity, and therefore it is not a density function for any probability distribution. For this reason, it is called an "improper" prior density. Nevertheless, this improper prior density does provide an approximate representation of vague prior information, and more importantly, it leads to proper and informative posterior distributions. Indeed, by letting $\sigma_0^2 \to \infty$ in the formulas we have given previously for μ_1 and σ_1^2, it is seen that the posterior distribution will be a normal distribution with mean $\mu_1 = \bar{x}$ and variance $\sigma_1^2 = v^2/n$. Thus, even though the prior distribution of W is represented by an improper uniform density over the entire line, after just a single observation x, the posterior distribution of W will be a proper normal distribution with mean x and variance v^2.

Prior Distributions of Several Parameters. If in a given problem there are k parameters $W_1, \ldots, W_k$ whose values are unknown, then it is necessary to specify their joint k-dimensional prior density $g(w_1, \ldots, w_k)$. Suppose that an observation X is obtained whose probability function or density function for any given values $w_1, \ldots, w_k$ of the parameters is $f(x|w_1, \ldots, w_k)$. Then the posterior joint density $g(w_1, \ldots, w_k|x)$ of $W_1, \ldots, W_k$ can again be obtained from Bayes' theorem by the relation

$$g(w_1, \ldots, w_k|x) \propto f(x|w_1, \ldots, w_k)g(w_1, \ldots, w_k)$$

In the example discussed in the preceding few paragraphs, if both the mean W and the variance V^2 of the quality of the items in the lot are unknown, then it is necessary to specify a joint prior density of W and V^2. Each time an item is selected from the lot and its quality x is measured, the joint density of W and V^2 is revised and a new posterior joint density is obtained.

STATISTICAL DECISION PROBLEMS

We shall now consider decision problems in which there is a parameter W whose k possible values are $w_1, \ldots, w_k$ and in which the decision maker must choose one of m available decisions $d_1, \ldots, d_m$. It has become traditional in statistical decision problems to specify the consequence of choosing any decision d when $W = w$ in terms of a loss function $L(w,d)$. In other words, $L(w,d)$ represents the loss or cost to the decision maker if he chooses d when $W = w$. If this choice actually results in a gain to the decision maker, then the value of $L(w,d)$ will be negative.

Suppose that the prior distribution of W is specified by the probabilities

$$g(w_i) = \Pr(W = w_i)$$

for $i = 1, \ldots, k$. Then the expected loss or risk $\rho(d)$ from any decision d will be

$$\rho(d) = \sum_{i=1}^{k} L(w_i,d)g(w_i)$$

A decision d_j should be chosen from among the available decisions $d_1, \ldots, d_m$ such that the risk $\rho(d_j)$ is minimized. A decision d^* that minimizes this risk is called a Bayes decision against the specified prior distribution of W. The risk $\rho(d^*)$ of a Bayes decision is called the Bayes risk.

As an example, we can consider again the problem illustrated in Figure 4-1. In terms of our present notation, the parameter W can take only two values, w_1 and w_2; there are three available decisions, d_1, d_2, and d_3; and the values of the loss function $L(w,d)$ are specified by Table 4-1. The prior probabilities are $p = \Pr(W = w_1)$ and $1 - p = \Pr(W = w_2)$.

The risk (or expected cost) $\rho(d_i)$ for each decision d_i is sketched in Figure 4-1 as a function of the prior probability p. It is seen from Figure 4-1 that d_2 is a Bayes decision if $0 \leq p \leq 0.1$, that d_3 is a Bayes decision if $0.1 \leq p \leq 0.7$, and that d_1 is a Bayes decision if $0.7 \leq p \leq 1$. For any given value of p, the Bayes risk will be the smallest value among the three straight lines in Figure 4-1 and is indicated by the heavy lines in that figure.

The Value of Observations. Consider again a standard decision problem in which the parameter W can take the values $w_1, \ldots, w_k$, one of the decisions $d_1, \ldots, d_m$ must be chosen, and the loss function is $L(w,d)$. In many problems, before the decision maker chooses a decision, he can take some observations whose values will give him information about the value of W. Hence, these observations will help him in choosing an effective decision.

Suppose that a random variable X can be observed whose probability function or density function for any given value w of W is $f(x|w)$. If the value $X = x$ is observed, then the posterior probability $g(w_i|x)$ that $W = w_i$ can be computed from the equation

TABLE 4-1

	d_1	d_2	d_3
w_1	0	3,000	300
w_2	1,000	0	300

given earlier. For any decision d, the risk with respect to this posterior distribution will be

$$\sum_{i=1}^{k} L(w_i,d)g(w_i|x)$$

Hence, after the observation $X = x$ has been obtained, a Bayes decision will be a decision that minimizes this risk.

The Bayes decision will depend on the particular value x of X that is observed. For each possible value x, let $\delta(x)$ denote a Bayes decision. Thus, $\delta(x)$ is a function that specifies an optimal decision for each possible outcome x of the observation. Such a function is called a Bayes decision function.

The Risk of a Bayes Decision Function. In some problems, the decision maker may have a choice among (1) taking the observation X at some specified cost, (2) taking some other observation at some other cost, or (3) taking no observation at all and merely choosing an immediate decision. In such a problem, the value of taking the observation X must be determined beforehand by computing the overall risk of the Bayes decision function $\delta(x)$. This risk can be computed as follows.

First, for each fixed value w_i of W, compute the expected value of $L[w_i,\delta(x)]$ with respect to the distribution $f(x|w_i)$ of X. In other words, if X has a discrete distribution, compute

$$\rho_i = \sum_{x} L[w_i,\delta(x)]f(x|w_i)$$

If X has a continuous distribution, compute

$$\rho_i = \int_{x} L[w_i,\delta(x)]f(x|w_i)\,dx$$

The risk ρ^* of the Bayes decision function $\delta(x)$ is then given by the equation

$$\rho^* = \sum_{i=1}^{k} \rho_i g(w_i)$$

Example. Suppose that the proportion W of defective items in a large lot is known to be either $\frac{3}{4}$ or $\frac{1}{3}$, and suppose that the prior probabilities are $\Pr(W = \frac{3}{4}) = \frac{2}{3}$ and $\Pr(W = \frac{1}{3}) = \frac{1}{3}$. A decision must be made either to reject or to accept the lot, and the loss function is specified by Table 4-2. Suppose that a single item can be selected from the lot and inspected. We shall let $X = 1$ if the item is defective and $X = 0$ if the item is nondefective, and we shall compute the Bayes decision function.

Suppose first that $X = 1$. Then by Bayes' theorem, the posterior probabilities will be

$$\Pr(W = \tfrac{3}{4}|X = 1) = \frac{(\tfrac{3}{4})(\tfrac{2}{3})}{(\tfrac{3}{4})(\tfrac{2}{3}) + (\tfrac{1}{3})(\tfrac{1}{3})} = \frac{9}{11}$$

and

$$\Pr(W = \tfrac{1}{3}|X = 1) = \tfrac{2}{11}$$

TABLE 4-2

	Reject	Accept
$W = \frac{3}{4}$	0	5
$W = \frac{1}{3}$	10	0

By Table 4-2, if the lot is rejected, the risk will be

$$0 \cdot \tfrac{9}{11} + 10 \cdot \tfrac{2}{11} = \tfrac{20}{11}$$

If the lot is accepted, the risk will be

$$5 \cdot \tfrac{9}{11} + 0 \cdot \tfrac{2}{11} = \tfrac{45}{11}$$

Therefore, the lot should be rejected.

Suppose next that $X = 0$. Then by Bayes' theorem, the posterior probabilities will be

$$\Pr(W = \tfrac{3}{4} | X = 0) = \frac{(\tfrac{1}{4})(\tfrac{2}{3})}{(\tfrac{1}{4})(\tfrac{2}{3}) + (\tfrac{2}{3})(\tfrac{1}{3})} = \frac{3}{7}$$

and

$$\Pr(W = \tfrac{1}{3} | X = 0) = \tfrac{4}{7}$$

By Table 4-2, if the lot is rejected, the risk will be

$$0 \cdot \tfrac{3}{7} + 10 \cdot \tfrac{4}{7} = \tfrac{40}{7}$$

If the lot is accepted, the risk will be

$$5 \cdot \tfrac{3}{7} + 0 \cdot \tfrac{4}{7} = \tfrac{15}{7}$$

Therefore, the lot should be accepted.

In summary, the Bayes decision function is to reject the lot if $X = 1$ and to accept the lot if $X = 0$. We shall now determine the risk of this decision function. If $W = \tfrac{3}{4}$, there is probability $\tfrac{3}{4}$ that the lot will be rejected and probability $\tfrac{1}{4}$ that the lot will be accepted. Therefore, from Table 4-2, the risk ρ_1 when $W = \tfrac{3}{4}$ is found to be

$$\rho_1 = 0 \cdot \tfrac{3}{4} + 5 \cdot \tfrac{1}{4} = \tfrac{5}{4}$$

If $W = \tfrac{1}{3}$, there is probability $\tfrac{1}{3}$ that the lot will be rejected and probability $\tfrac{2}{3}$ that it will be accepted. Therefore, the risk ρ_2 when $W = \tfrac{1}{3}$ is again found from Table 4-2 to be

$$\rho_2 = 10 \cdot \tfrac{1}{3} + 0 \cdot \tfrac{2}{3} = \tfrac{10}{3}$$

When the values ρ_1 and ρ_2 are weighted by the prior probabilities, the overall risk ρ^* of the Bayes decision function becomes

$$\rho^* = \tfrac{2}{3}\rho_1 + \tfrac{1}{3}\rho_2 = \tfrac{35}{18}$$

Suppose now that no items were available for inspection and the decision to reject or to accept the lot had to be made on the basis of no information other than that expressed by the prior probabilities. Then the risk from rejecting the lot would be

$$0 \cdot \tfrac{2}{3} + 10 \cdot \tfrac{1}{3} = \tfrac{10}{3}$$

The risk from accepting the lot would be

$$5 \cdot \tfrac{2}{3} + 0 \cdot \tfrac{1}{3} = \tfrac{10}{3}$$

Hence, either decision is a Bayes decision and the risk from an immediate decision without any observations is $\tfrac{10}{3}$.

By being able to inspect just a single item, the decision maker can reduce the risk from $\tfrac{10}{3}$ to $\tfrac{35}{18}$. Therefore, he should be willing to pay any cost less than $\tfrac{10}{3} - \tfrac{35}{18} = \tfrac{25}{18}$ for the opportunity of inspecting a single item selected at random from the lot.

Choosing the Number of Observations to Be Inspected. Let us now suppose that in this example, instead of merely deciding whether or not to inspect a single item, the decision maker can actually specify the number of items n that will be selected at random from the lot and inspected. We shall assume that the sampling and inspection cost is 0.01 unit per item.

As mentioned earlier, if the proportion of defective items is w, then the probability $(y|w)$ that exactly y items will be defective in a random sample of n items is

$$f(y|w) = \binom{n}{y} w^y (1 - w)^{n-y}$$

Suppose that the observed number of defective items in the sample is y. The derivation will not be given here, but it can be shown from this distribution, from the values of the loss function, and from the prior probabilities that the Bayes decision function is to accept the lot if

$$y < 0.5474n$$

and to reject the lot if

$$y > 0.5474n$$

The risk ρ^* of this decision function will be

$$\rho^* = 5 \left(\frac{2}{3}\right) \Pr(Y < 0.5474n|W = \tfrac{3}{4}) + 10 \left(\frac{1}{3}\right) \Pr(Y > 0.5474n|W = \tfrac{1}{3})$$

Values of ρ^* for various values of n are shown in Table 4-3. These values were com puted with the aid of a table of binomial probabilities.

It is seen from Table 4-3 that the risk ρ^* decreases as the sample size n increases. However, because the cost of taking a sample of n items is $0.01n$, the total risk or expected cost of taking a sample of size n and using the Bayes decision function will be $\rho_T = \rho^* + 0.01n$.

Values of the total risk ρ_T are also given in Table 4-3. It is seen from this table that ρ_T is minimized when $n = 25$. For a sample of size $n = 25$, $\rho^* = 0.0903$, the sampling cost is 0.25, and the total risk is therefore $\rho_T = 0.3403$.

ESTIMATION PROBLEMS

A special but important class of statistical decision problems is known as estimation problems. In an estimation problem, there is a parameter W that can take any value on the real line or in some interval of the real line, and the value of W must be estimated from the value of an observation or a sample of observations. The loss function $L(w,d)$ will typically be a function that depends on the magnitude of the difference between the actual value w of W and the estimated value d. A large value of $|w - d|$ will result in a large loss $L(w,d)$.

TABLE 4-3

n	ρ^*	ρ_T
21	0.1394	0.3494
22	0.1370	0.3570
23	0.1116	0.3416
24	0.1056	0.3456
25	0.0903	0.3403
26	0.0822	0.3422
27	0.0738	0.3438
28	0.0646	0.3446
29	0.0609	0.3509
30	0.0513	0.3513
31	0.0506	0.3606
32	0.0411	0.3611
33	0.0402	0.3702
34	0.0334	0.3734
35	0.0313	0.3813

Squared-error Loss. The most widely used loss function in estimation problems is the squared-error loss function $L(w,d) = (w - d)^2$. Suppose that the parameter W has a prior density $g(w)$ over the interval $a \leq W \leq b$. If an estimate d must be chosen without any further observations, then a Bayes decision or, in this case, a Bayes estimate will be a number d that minimizes the risk

$$\int_a^b (w - d)^2 g(w) \, dw$$

By differentiating this integral with respect to d, it can be shown that the minimum risk is attained when

$$d = \int_a^b wg(w) \, dw$$

In other words, the Bayes estimate d is equal to the mean of W.

Suppose now that before estimating W, the decision maker can take an observation X whose probability function or density function for any given value w of W is $f(x|w)$. After the value of X has been observed, the Bayes estimate will be the mean of the posterior distribution W. Hence, a Bayes decision function $\delta(x)$ specifies that for any observed value x, the parameter W should be estimated by the mean of its posterior distribution.

Example 1. Suppose that the prior distribution of the unknown proportion W of defective items in a large lot is a beta distribution with specified nonnegative integers α and β. If in a random sample of n items there are exactly x defectives, then as shown earlier, the posterior distribution of W will again be a beta distribution with α replaced by $\alpha + x$ and β replaced by $\beta + n - x$. The mean of this posterior distribution is

$$\frac{\alpha + x + 1}{\alpha + \beta + n + 2}$$

Therefore, when the squared-error loss function is used, this value is the Bayes estimate of W. In particular, if the prior density of W is the uniform density ($\alpha = \beta = 0$), then the Bayes estimate of W is $(x + 1)/(n + 2)$.

Example 2. Suppose that the breaking strength of fibers produced by a certain machine has a normal distribution with unknown mean W and known variance 0.25, and suppose that the prior distribution of W is itself a normal distribution with mean 6 and variance 0.4. Suppose that the breaking strengths of a random sample of n fibers produced by the machine are measured, and that the average of the n values is found to be $\bar{x}$. Then, as discussed earlier on pages 10-106 and 10-107, the posterior distribution of W will again be a normal distribution with mean

$$\mu_1 = \frac{15 + 4n\bar{x}}{2.5 + 4n}$$

and variance

$$\sigma_1^2 = \frac{1}{2.5 + 4n}$$

Therefore, when the squared-error loss function is used, the Bayes estimate of W is μ_1.

If the sampling and measurement cost is 0.0001 unit per fiber, how large a sample of fibers should be selected for measurement? To answer this question, we must first compute the risk of the Bayes decision function for each sample size n, then find the total risk for each value of n by adding in the sampling cost $0.0001n$, and finally determine the value of n for which this total risk is minimized. In this example, it can be shown that the risk of the Bayes estimate will be the variance σ_1^2 of the posterior distribution of W. Therefore, when the sampling cost is added, the total risk will be

$$\frac{1}{2.5 + 4n} + 0.0001n$$

By differentiating this total risk with respect to n, it can be shown that the minimum is attained when $n = 49.375$. Because the actual sample size must be an integer, the optimal sample size is either $n = 49$ or $n = 50$. Both values yield approximately the same total risk, 0.01. At the optimal sample size, this total risk will be equally divided between the risk from estimation error and the sampling cost.

Absolute Value of the Error. In certain estimation problems, it may be more realistic to assume that the loss is specified by the absolute value of the error $L(w,d) = |w - d|$ rather than to use the squared-error loss $L(w,d) = (w - d)^2$. The use of this loss function will lead in general to a different Bayes estimate from the one obtained with squared-error loss.

The Median. If a parameter W has density $g(w)$ over the interval $a \leq w \leq b$, then a number m in this interval is said to be a median of the distribution of W if $\Pr(W \leq m) = \Pr(W \geq m) = \frac{1}{2}$. If W has a discrete distribution, then a slightly different definition of the median is needed, but we shall not consider that possibility here.

If $L(w,d) = |w - d|$, then it can be shown that for any given distribution of W, the median of that distribution will be the Bayes estimate of W. Thus, if an observation X is available, then a Bayes decision function specifies that for any observed value of X, the parameter W should be estimated by the median of its posterior distribution.

Example 2 Reconsidered with a Different Loss Function. Let us again consider example 2 in which the breaking strength of the fibers produced by a certain machine has a normal distribution with unknown mean W, and the prior distribution of W is also a normal distribution. Let us now suppose, however, that the loss function is $L(w,d) = |w - d|$ rather than $L(w,d) = (w - d)^2$.

The posterior distribution of W for any observed sample values will be a normal distribution. Because the mean and the median of any normal distribution coincide, it follows that regardless of the observed sample values, the Bayes estimates for the new loss function and the old loss function will coincide.

However, if it is again assumed that the sampling cost is 0.0001 unit per fiber, then the new loss function will lead to a different value for the optimal sample size n. When the loss function is the absolute value of the error, then for any given sample size n, it can be shown that the risk from the Bayes decision function will be

$$\left[\frac{2}{\pi(2.5 + 4n)} \right]^{\frac{1}{2}}$$

The derivation of this value is omitted here. After the sampling cost $0.0001n$ is added to this value to obtain the total risk, the value of n for which the total risk is minimized can then be found by differentiation. The solution is $n = 157.84$. Because the actual sample size must be an integer, the optimal value of n will be either $n = 157$ or $n = 158$. Both values yield approximately the same total risk 0.0475. At the optimal sample size, two-thirds of the total risk will be the risk from estimation error and one-third of the total risk will be sampling cost.

In this example, the optimal sample size when the loss function is taken to be the absolute value of the error is larger than the optimal sample size when the loss function is taken to be squared-error. This relation occurs in this example because the product $c\sigma^2$ of the sampling cost c per observation and the variance σ^2 of each observation is small. Specifically, the relation obtained in this example will be obtained whenever $c\sigma^2 < 1/(2\pi)^2$. If $c\sigma^2 > 1/(2\pi)^2$, then the relative magnitudes of the optimal sample sizes required by the two different loss functions will be reversed.

BIBLIOGRAPHY

Blackwell, David, *Basic Statistics*, McGraw-Hill Book Company, New York, 1969.
Chernoff, Herman, and Lincoln E. Moses, *Elementary Decision Theory*, John Wiley & Sons, Inc., New York, 1959.
DeGroot, Morris H., *Optimal Statistical Decisions*, McGraw-Hill Book Company, New York, 1970.

Hadley, G., *Introduction to Probability and Statistical Decision Theory*, Holden-Day, Inc., San Francisco, 1967.

Lindley, D. V., *Introduction to Probability and Statistics from a Bayesian Viewpoint: Part 1, Probability; Part 2, Inference*, Cambridge University Press, Cambridge, 1965.

Pratt, John W., Howard Raiffa, and Robert Schlaifer, *Introduction to Statistical Decision Theory* (preliminary edition), McGraw-Hill Book Company, New York, 1965.

Raiffa, Howard, *Decision Analysis*, Addison-Wesley Publishing Company, Inc., Reading, Mass., 1968.

Raiffa, Howard, and Robert Schlaifer, *Applied Statistical Decision Theory*, Harvard University, Graduate School of Business Administration, Division of Research, Boston, Mass., 1961.

Schlaifer, Robert, *Introduction to Statistics for Business Decisions*, McGraw-Hill Book Company, New York, 1961.

Schmitt, Samuel A., *Measuring Uncertainty: An Elementary Introduction to Bayesian Statistics*, Addison-Wesley Publishing Company, Inc., Reading, Mass.,1969.

Chapter **5**

Operations Research

SALAH E. ELMAGHRABY

**Professor and Chairman, Operations Research Committee,
North Carolina State University, Raleigh, North Carolina**

The formal inception of operations research can be traced back only to World War II Then, the Allied Military Command, faced with acute shortages of resources of men and materiel and working against the clock, sought the assistance of the scientific and engineering community in the resolution of some knotty operational problems. The teams that formed to study and resolve these problems were composed of experts from a variety of fields such as physics, mathematics, statistics, and mechanics, as well as psychology, economics, and sociology. They studied these operational problems; hence the original (English) name operational research, which is still the dominant name on the Continent. It was later modified, with typical American unconcern for syntax, to operations research (OR).

HISTORICAL DEVELOPMENT OF OPERATIONS RESEARCH

Perhaps one of the earliest examples of the military turning to science to increase the effectiveness of existing weapons during World War II was concerned with how to set the time fuse of a bomb to be dropped from an aircraft onto a submarine. The Coastal Command of the RAF had decided that the fuse should be set to explode at a depth of 100 feet below the water surface, on the plausible expectation that the submarine would sight the approaching plane about two minutes before the instant of the attack and thus dive that far below the water surface by the time the encounter with the bomb could take place. But as the actual results of the offensive were disappointing, the problem was handed over to a team of scientists, since nicknamed Blackett's Circus.

After exhaustive field observations, Blackett's team found that in actual combat the bomber had only a small chance of aiming the bomb right if the submarine dived as promptly as the theory underlying the existing practice assumed. It had a good chance of hitting the target only if the submarine remained close to the surface at the time of the encounter. It therefore followed that when the aim could be true, the bomb exploded too far below the surface to affect the submarine. On the other hand, when the aim was necessarily poor, it exploded at the right depth but at the wrong location. It was a case of heads I lose, tails you win. It is obvious that in such a situation, it would be better to set the fuse to detonate almost on impact instead of 100 feet below water, because it is precisely on such occasions, when the submarine has not yet dived fully, that there is any chance of seeing the submarine and aiming true. No wonder the result of the subsequent change doubled the number of submarine kills.

There were plenty of other examples. They were concerned with the optimal formation of bombers as a function of the target shape; the best bomber-fighter combination to achieve maximum security and still accomplish the mission; the measurement of the effectiveness of arming merchant ships against enemy aircraft; the optimal location of radar stations; and similar problems. They constitute a delightful and penetrating reading in the application of the scientific approach to warfare. See the pamphlet by Mehta, Thiagarajan, and Jaiswal referenced at the end of the chapter.

The involvement of science in industrial problems is due to the advent of the Second Industrial Revolution. In the late 1940s, the new revolution began when electronic computers became commercially available. The potentialities of these electronic brains as a new tool for management were broadcast far and wide, and nontechnically trained executives began to look for help in the selection and utilization of computers. The emerging search for assistance was accelerated by the outbreak of the Korean conflict, which placed increased demands for greater productivity on a large part of American industry. Therefore, in the early 1950s, industry began to employ OR men. A few consulting firms, universities, research institutes, and governmental agencies did likewise. Thus OR began to spread and expand in the United States.

THE NATURE AND STRUCTURE OF OR

It is not difficult to see that OR is characterized by being scientific in its approach. Furthermore, because it deals or attempts to deal with the total system, the study is usually conducted by a team of experts in different specialized areas who share a common interest in pooling their resources to attack and resolve a specific problem. It is basically an interdisciplinary approach to the scientific study and resolution of operational man-machine systems.

Almost by its definition as being the "scientific" study of systems with the objective of optimizing performance, to the extent it does succeed in discovering ways of improvement, it naturally implies a tacit if friendly criticism of the existing state of affairs. A true OR man is always dissatisfied with the status quo, even if it is of his own doing. This is, after all, the basic motif of the scientific mind: always questioning, probing, changing, experimenting, and striving to achieve a better model. Consequently, an OR worker has to be extremely tactful and cautious in making such criticism when presenting his discovery.

It is true that scientific neutrality and executive involvement do not go together easily. Nevertheless, it is important for an operations research worker to blend the two. He is pragmatic and is interested in having his way and his ideas implemented here and now; if they are not, he has labored in vain. If Blackett, for example, had not succeeded in persuading the Coastal Command to accept the recommended change in the setting of the time fuse, no one would have known the value of his undoubtedly great discovery. Nor could posterity have evaluated it later, as with the concepts Babbage and Leonardo presented ahead of their time. Operations research that is not acted upon at once is like an unborn idea that is never missed.

However, before an OR expert has anything to communicate, he must first discover it. How does one make an OR discovery worth communicating? Through the careful study of the operational system from all its relevant aspects, constructing a model of the system, and then solving the model to arrive at an answer. Such a procedure usually involves rather esoteric mathematical analysis. However, it is important to realize that, in the final analysis, the quality of an OR project is judged, not by the power of mathematics it employs, but by the practical dividends, including fringe benefits, its adoption yields. Nearly all the classical OR work done during World War II that has given OR its present prestige was the application of disciplined common sense plus expert training in some scientific field to wartime problems, often without recourse to recondite mathematics.

However, common sense, although necessary, is not sufficient for serious OR work. It must be tempered by training in some specialized areas.

THE METHODOLOGY AND ART OF MODELING

In the discussions that follow, we shall be concerned with some of the better-known OR disciplines and techniques that have been developed over the past two or three decades. But the reader must not confuse the tool with the process, although there is little harm in associating one with the other. It is a common phenomenon of human thinking to try to summarize, abstract, condense, or simplify. Usually, this is conducted through association. We tend to associate the stethoscope with the physician, the slide rule with the engineer, the gavel with the judge, and so forth. It is just our way of summarization. But woe to the man who identifies the stethoscope as the physician, or the slide rule as the engineer. Similarly, OR is associated in many minds with, say, optimization through linear programming. But make no mistake, linear programming is not OR, nor is queuing, dynamic programming, geometric programming, theory of games, industrial dynamics, activity networks, or any of the specific approaches and tools to be discussed later in this chapter.

The central idea in OR is the same idea in scientific study in general, namely, that of constructing a model. The main difference is that OR is concerned with operational systems, systems that can be manipulated, altered, or controlled by us. This is a vast umbrella that covers many sins, but it certainly excludes nonoperational systems such as stellar systems (the domain of astronomers), historical systems (the domain of historians), physiological and biological systems (the domain of the psychologist, psychiatrist, biologist, and physician), and the like.

A model is a representation of a real thing (that is, a physical reality, where "real" is understood in the sense of being capable of construction rather than in the sense of past existence) or an idea. There is a vast array of models which can be classified in several ways, such as:

1. Iconic (look like the real thing) versus analog (in which we substitute one property for another) versus symbolic (mathematical)
2. Structural (define the relations among components) versus functional (give the output as a function of the input)
3. Descriptive (describe without prescribing) versus normative (imply a measure of value)
4. Algorithmic (well-defined computational schemes usually involving iteration) versus simulation (imitate the essence of, without the substance) versus heuristic (incorporate rules that modify the behavior of the systems in a commonsense manner)

We shall confine our discussion to mathematical models of the normative type. For a more detailed discussion of models and modeling, see the publication by Elmaghraby listed in the bibliography at the end of the chapter. Suffice it here to say that modeling, which is basically an art disciplined by scientific training in methodology, plays a central and dominant role in OR. The reason is that in all but very few instances, it is either impossible to experiment with the real-life system or it is economically infeasible to do so.

Uses of Models. A model is helpful in at least the following seven functions: (1) as an aid to thought, (2) as an aid to communication, (3) as a tool for prediction, (4) as a tool for control purposes, (5) as a tool for training and instruction, (6) as a vehicle for sensitivity analysis, and (7) as an aid to decision making.

Consider the seventh function—the model as an aid to decision making. One recognizes that there are, in general, three levels of decisions: strategic, tactical, and technical. Since its inception, OR has been applied to a wide variety of problems, helping in making decisions in each. Most of these, however, have been tactical rather than strategic in nature. Rarely has OR been used to reach a technical decision. The distinction between the strategic and tactical decisions is based on at least three characteristics of the problems, each of which involves a matter of degree. A strategic decision is ends-oriented, that is, related to an objective; of farther reach in impact; and of longer range. A tactical decision, in contrast, is ordinarily confined to optimization to achieve the previously stated objectives, and is characterized by a narrower scope of impact and a shorter range. Only recently has OR been applied to strategic decisions as well in government and industry alike.

Classification of OR Problems.   Every operational problem has form as well as content. Form refers to the way in which the parameters and variables are related to each other, for example, linearly. Content refers to the nature of these entities— x represents distance and c represents cost.

In one sense, no two problems are ever exactly the same; in fact, even the same problem is not really the same at two different points in time. In another sense, problems tend to cluster into a few well-defined types. The sense in which no two problems are exactly the same refers to their content. The sense in which they tend to fall into clusters is one that refers to their form.

We have already encountered such clustering when we referred to groupings such as queuing problems, linear programming problems, dynamic programming problems, and the like. This is a classification according to methodology.

There is another possible classification and this time it is by field of application. To name a few, there are problems in allocation, capital budgeting and investment, distribution, facility planning, inventory, production control, reliability, replacement, sequencing and scheduling, and so on.

Any area of application may claim several methodologies. For instance, one may solve an inventory problem by linear programming, or by dynamic programming, or by queuing. Alternatively, any methodology may claim several areas of application. For instance, dynamic programming is applied to problems in inventory, production, and replacement, to name but a few.

Approach to the Study of Operations Research.   In approaching an expository statement on the subject of OR, which is the main thrust of this chapter, one is tempted to follow the line of least resistance and present a methodological classification of OR. This is the standard approach, as a glance at any book on the techniques of OR will show. But such classification tends to emphasize the tool rather than the problem. It reinforces the misconception that OR is the union of a set of disparate techniques. And what is worse still, the reader—as well as the page allotment to the subject— is exhausted at the end of such a treatise on methodology, leaving a bitter taste of disappointment because the question of "What do I do with all this theory?" has never been answered.

Consequently, we shall couch our discussion in terms of the operational system and the problem it poses. This helps motivate the development of the model or models that can be used in its study. The reader then knows, at all times, what necessitated such development of tools and techniques, and hence their use.

OPERATIONS RESEARCH AND SYSTEMS THEORY

The terms "operations research" and "systems theory" have often been used interchangeably, although many people suspect that there is a distinction between the two. In fact, there is. Sometimes the difference is a matter of degree, and sometimes it is in basic outlook. As always, the content of either is highly conditioned by its historical development.

Traditionally, systems theory has been closely related to hardware systems (the first systems engineering, as such, was in telephony), while OR is identified with the software. Operations research tends to be concerned with problems which can be represented by mathematical models, which in turn can be analytically optimized. Systems theory, on the other hand, although formal in nature, is concerned with problems of greater complexity, and is more global and abstract in its approach. Its components may consist of mathematical models but may also incorporate social and biological factors which have not been successfully quantified. Operations research tends to be mostly concerned with problems in the small, while systems theory usually connotes a larger, total organizational activity.

A system is any cohesive collection of items that are dynamically related. Every system does something, and what it does can be regarded as its purpose. Systems theory is primarily concerned with the discovery of the mechanisms by which such purpose is achieved and by which equilibrium or self-regulation is maintained. Operations research is one of the most important inputs to such a theory, providing the quantitative models of the mechanisms that are amenable to such a treatment.

OPERATIONS RESEARCH AND INDUSTRIAL ENGINEERS

It is not difficult to see that of all engineering disciplines, industrial engineering is most dependent on and can profit the most from the development and application of OR. The reason is simple: industrial engineers are concerned with the design of man-machine systems, whether in the small, as in the design of an individual workplace, or in the large, as in the design of computer-based management control systems. OR is the applied science of the study and design of such systems, and industrial engineering is the engineering of such systems. In other words, OR bears the same relationship to the industrial engineer as thermodynamics to the mechanical engineer or aerodynamics to the aeronautical engineer.

The expanding horizons ahead of industrial engineers can be directly related, in the years to come, to their mastery of the philosophy and methods of OR.

A PRODUCTION PLANNING PROBLEM:
INTRODUCING THE THEORY OF DYNAMIC PROGRAMMING

The Problem. A manufacturer of a homogeneous product, such as chocolate, ice cream, cereal, or copper tubing, is faced with a known demand, denoted by r_n, for $n = 1, 2, \ldots, N$ weeks. His productive facility is expensive to set up, and his product is expensive to store—chocolate, ice cream, and cereal require climate control. Because the selling price is assumed more or less fixed, he is interested in minimizing his total production cost. What is his optimal production plan?

Assume that the cost of producing the total demand over the planning horizon of N time periods (weeks, say) is fixed and therefore is not subject to decision making. Consequently, the minimization of the total cost is crucially dependent on the pattern of production. This determines the number and cost of setups and the quantities held in inventory at the end of each week. Let I_n denote the on-hand inventory at the end of week n, $n = 1, 2, \ldots, N$. Let q_n denote the quantity of the commodity produced in week n, assumed available before the end of the week to satisfy demand in week n and all subsequent weeks. The material balance equation yields that

$$I_n = I_{n-1} + q_n - r_n \qquad n = 1, 2, \ldots, N$$

Without loss of generality, we can assume that the initial inventory I_0 is equal to zero. If this were not the case, a simple process of "netting out" the on-hand inventory against future demands terminates with zero initial inventory in some future period j, where $0 \leq j \leq N$, in which case the manufacturer's problem extends over the sub-horizon $j, j + 1, \ldots, N$. And, also without loss of generality, we may assume that the final inventory $I_N = 0$. If this were not the case, add the desired end-of-horizon inventory I_N to the demand in the last week to obtain a total final demand of $r_N + I_N$, which is our new r_N; the cost of stocking the quantity I_N at the end of the planning horizon is again a fixed cost and hence not subject to decision. Let the setup cost be a per setup, and the inventory cost be h per unit of the product carried over from one week to the next. Let $\delta(q_n)$ be Kroenecker delta: equal to 1 if $q_t > 0$ and equal to 0 otherwise. Clearly, then, in any week n, the cost of setup is $a\delta(q_n)$; and the cost of inventory (carried over to the following week) is $hI_n = h(I_{n-1} + q_n - r_n)$. Consequently, the total cost over the complete planning horizon is given by

$$C = \sum_{n=1}^{N} [a\delta(q_n) + hI_n] \qquad \text{to be minimized} \tag{1}$$

where
$$I_n = I_{n-1} + q_n - r_n \geq 0 \qquad n = 1, 2, \ldots, N \tag{2}$$

and
$$I_0 = I_N = 0 \qquad\qquad q_n \geq 0 \tag{3}$$

The model of equations (1) to (3) is a complete mathematical model: equation (1) is the criterion function; equations (2) and (3) are the relations and constraints of the operational system. The inequality in (2) reflects the desire to have no back orders (no out-of-stock condition). The minimization of (1) subject to constraints (2) and (3) can be achieved through application of dynamic programming.

The Elements of Dynamic Programming. Dynamic programming (DP) is the name given to a particular kind of sequential decision process in which it is desired to optimize (maximize or minimize) a criterion function subject to constraints. The form of both criterion and constraints is, in general, nonlinear. The programming problem of equations (1) to (3) is an excellent example of such processes. The criterion function is certainly nonlinear because of the 0,1 character of Kroenecker delta. In fact, it was this integer type of nonlinearity that prevented us from using the standard techniques of calculus in this problem. The constraints of equations (2) and (3) are linear.

The DP problem is defined in terms of five entities: the state, the stage, the decision space, the transformation function, and the criterion function. The state is specified by the parameter or set of parameters which contain all the information necessary to make the current and all future decisions, hopefully in an optimal fashion. In other words, the state separates the future from the past and acts as a regeneration point for the process. We shall denote the state by s. The stage or epoch n is defined in terms of the process of decision making: a stage exists whenever a decision is called for. The decision space D is the space of all possible decision variables, which may vary from stage to stage and also may be a function of the state of the system at any stage; that is, $D = D(n,s_n)$. The transformation function ξ relates the new state of the system to the old state, naturally as a function of the decision made, $s_{n+1} = \xi(s_n,d_n)$, where $d_n \in D(n,s_n)$. Finally, the criterion function is the measure of the performance of the system and is a function of all the decisions made and the initial state of the system, because all subsequent states are themselves functions of the initial stage and the decisions made. We denote the criterion function by C and note that $C = C(s_0,\{d_n\})$. Of course, we are always interested in the optimum of C, which we shall denote by f; f is the minimum (or maximum) value of $C(s_0,\{d_n\})$, and to emphasize the fact that the initial state is s_0 and the optimization is over a finite planning horizon of length N, we write $f_N(s_0)$.

Applying these concepts to our production planning problem, we find that:

The state is the on-hand inventory I_{n-1} at the beginning of the week n.

The stage is the week n. Each week calls for a decision, and there are, in all, N weeks (stages).

The decision space is the space of all feasible production quantities $\{q_n\}$. If the system is in week n with on-hand inventory I_{n-1}, it is absurd to produce in week n more than the total net demand in periods $n, n + 1, \ldots, N$. Consequently,

$$0 \leq q_n \leq \sum_{j=n}^{N} r_j - I_{n-1};$$ that is, the space of decisions $D(n,s_n)$ is the set of

integer points between 0 and $\sum_{j=n}^{N} r_j - s_{n-1}$. In this example, the decision space

is dependent on the state and the stage.

The transformation function is precisely the material balance equation,

$$s_n = s_{n-1} + q_n - r_n$$

because the state s_{n-1} is equated to the on-hand inventory I_{n-1}.
The criterion function is to minimize the total cost, that is, determine

$$f_N(I_0) = \min_{q_n \in D(n,I_{n-1})} \sum_{n=1}^{N} \{ a\delta(q_n) + h(I_{n-1} + q_n - r_n) \} \qquad (4)$$

$$n = 1, 2, \ldots, N$$

In equation (4), we made the expression explicit in the decision variables $\{q_n\}$.

The Functional Equation of Dynamic Programming. The solution of the above DP problem, which is typical of the solution of all finite horizon discrete DP problems, is achieved through the following reasoning. If $f_N(I_0)$ is the minimum cost for an N-week planning horizon with initial inventory s_0, we can similarly define $f_{N-n+1}(I_{n-1})$ to be the minimum cost for weeks n, $n+1$, $\ldots$, N, assuming the starting state (the on-hand inventory at the beginning of week n) is exactly $I_{n-1} \geq 0$. Let the cost incurred in week n be denoted by $g_n(I_{n-1};q_n) = a\delta(q_n) + h(I_{n-1} + q_n - r_n)$. Equation (4) can be rewritten as

$$f_N(I_0) = \min_{q_n \in D(n,I_{n-1})} [g_1(I_0,q_1) + g_2(I_1,q_2) + \cdots + g_N(I_{N-1},q_N)] \qquad (4')$$

$$n = 1, 2, \ldots, N$$

Because the functions $g_2, g_3, \ldots, g_N$ are independent of the decision q_1, and because the cost is an additive function of the g_n's, we can write equation (4') as follows:[1]

$$f_N(I_0) = \min_{q \in D(1,I_0)} \left\{ g_1(I_0,q_1) + \min_{q_n \in D(n,I_{n-1})} \sum_{n=2}^{N} g_n(I_{n-1},q_n) \right\} \qquad n = 2, 3, \ldots, N$$

But the second term between braces is $f_{N-1}(I_1)$, by definition. Hence, we have the functional equation of DP (so called because the same function appears on both sides of the equality sign),

$$f_N(I_0) = \min_{q_1 \in D(1,I_0)} \{ g_1(I_0,q_1) + f_{N-1}(I_1) \}$$

where $\qquad g_1 = a\,\delta(q_1) + hI_1 \qquad I_1 = I_0 + q_1 - r_1 \qquad I_0 = I_N = 0$

Of course, to evaluate $f_N(I_0)$, we need to have the values of $f_{N-1}(I_1)$ for all possible values of I_1 that can result from the decision q_1 and the demand r_1. But, by similar reasoning to the above, we can write

$$f_{N-1}(I_1) = \min_{q_2 \in D(2,I_1)} \{ g_2(I_1,q_2) + f_{N-2}(I_2) \}$$

from which we conclude that, to know $f_{N-1}(I_1)$, we must have the values of $f_{N-2}(I_2)$ for all possible values of I_2 that can result from any given value of I_1 and the decision q_2. Recursive reasoning leads to the inevitable conclusion that we need $f_0(I_N)$, which is identically equal to zero because there is no cost incurred beyond the planning horizon. Hence, we are led to conclude that we must first evaluate $f_1(I_{N-1})$ for all values of I_{N-1}—the so-called "one-stage optimization problem"—which is used in determining $f_2(I_{N-2})$ for all values of I_{N-2}, and so forth, until $f_N(I_0)$ is determined, which is the desired minimum cost and which yields the optimal production schedule $(q_1^*, q_2^*, \ldots, q_N^*)$.

Dynamic programming reasoning is forward reasoning: it transformed a problem of search for the optimum schedule over an N-dimensional space (the space of $q_1, q_2,$

---

[1] The "pushing through" with the "min" operation to separate the cost of the first stage from the cost of all other stages would still be legitimate as long as the total cost is composed of individual stage costs by a binary operator that is transitive and maintains inequalities on the real line.

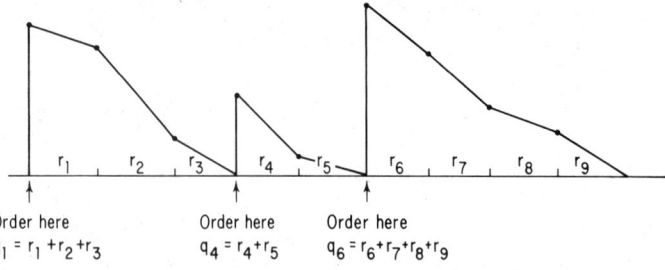

FIG. 5-1.   Typical inventory fluctuation curve.

. . . ,q_N) into N consecutive searches, each over a one-dimensional space (search over q_n at stage n).   That this affords a fantastic reduction in the computing burden should not be doubted.   For suppose $N = 26$ weeks (a planning horizon of six months), and each q_n can assume any of 100 values.   Then the optimization problem posed in equation (1) requires the search over $(100)^{26} = 10^{52}$ different possible values of the vector $(q_1, q_2, . . . ,q_N)$.   This is a number larger than the total number of atoms on earth and its atmosphere.   The functional equation of DP, equation (4), requires searching over 100 values at each stage and repeating that search 26 times (for the 26 stages), a total of 2,600 evaluations for each value of I_n.   If one allows 100 different values of I_n, the result is still a meager total of 25×10^4 comparisons.

Figure 5-1 shows a typical result of such optimization—the quantities ordered, the period ordered, and the resulting inventory fluctuations.[2]

Dynamic Programming: An Approach, not an Algorithm.   The functional equation of DP is the embodiment of the so-called principle of optimality, advanced by Bellman in 1952, and subsequently extensively studied and revised.   See the papers by Mitten and by Mitten and Denardo.   The principle can be stated as follows:

> *The principle of optimality:* An optimal sequence of decisions in a multistage decision process problem has the property that, whatever the initial stage, state, and decision are, the remaining decisions must constitute an optimal sequence of decisions for the remaining problem, with the stage and state resulting from the first decision considered as initial conditions.

However, the reader must be warned against the impression that developing the mathematical model reaches its climax at the statement of the functional equation. This is a false impression.   Although it is true that a great deal of ingenuity and artfulness goes into such a development, the analysis rarely stops there.   Unlike linear programming, the specification of the functional equation gives little clue about how to proceed with the calculations in the most efficient manner.   This is because DP is an approach to viewing the problem, a way of looking at it, not an algorithm which embodies a step-by-step recipe for solution.   There are many problems for which the functional equations have been written but no solution reached in the sense of achieving a numerical answer, because no computationally efficient procedure for their solution has been devised yet.   A great deal of analysis usually follows the formulation of the functional equation of DP.

To illustrate, the functional equation of the production planning problem, equation (4), has been studied by Wagner and Whitin, and they narrowed down the domain of search to no more than $N(N - 1)/2$ calculations.   For the numbers quoted above, the minimal cost, as well as the optimal schedule, can be obtained in no more than $50 \times 49/2 = 1,225$ calculations.   The interested reader can refer either to the original article by Wagner and Whitin or to Elmaghraby, page 15.

More Complicated Production Planning Problems.   For ease of exposition, the production planning problem described above was simplified to its bare essentials.

---

[2] For a numerical example, see Salah E. Elmaghraby, *The Design of Production Systems*, Reinhold Publishing Corporation, New York, 1966.

After all, if one cannot resolve the simpler problem, one cannot hope to resolve the more difficult ones. But we assumed:

1. The cost of setup a, and also the cost of carrying inventory h, are constant over the planning horizon. This need not be true; for example, we may have both parameters variable with the week: $a(n)$ and $h(n)$.

2. The cost of production to be fixed and hence outside the decision process. But this assumption can be dropped; we may have the cost of production dependent on the quantity produced, such as in economies of scale. Let the cost of producing q_n be given by $p(n,q_n)$, which is dependent on the stage n as well as on the quantity produced during that stage.

3. A nonstockout policy was adopted, which gave rise to the condition $I_n \geq 0$ for all n. This condition can also be dropped, in which case we permit stockout at a penalty, say, $c_n(I_n)$, $I_n \leq 0$. Now positive I_n denotes on-hand stock and negative I_n denotes shortage.

Although this is a seemingly formidable complication of the problem, the DP approach goes through intact. The resultant functional equation is indeed complicated, composed as it is of the sum of the cost of setup, $a_n\delta(q_n)$; the cost of production, $p_n(q_n)$; the cost of holding stock, $h_n(I_n)$, $I_n \geq 0$; and the penalty for shortage, $c_n(I_n)$, $I_n \leq 0$.

A basic assumption made in all the above is that:

4. Future demand is known. In many instances, especially where the planning horizon is of short duration and accurate forecasts are available, this is a very good working assumption. In other cases, demand is known only in a probabilistic sense. So we are not given a demand r_n for week n, but a probability distribution function $\Psi_n(r)$ for a demand of r in week n.

There is no question that the presence of indeterminate future demand changes the picture in a radical fashion, for now we must change the criterion itself from the minimization of total cost to the minimization of total expected (average) cost. However, the DP approach again goes through intact. We exhibit the functional equation for the case of "lost sales":

$$f_N(I_0) = \min_{q_1 \in D(1,I_0)} \left\{ a_1\delta(q_1) + p_1(q_1) + \int_{r=0}^{I_0+q_1} h_1(I_0 + q_1 - r)\, d\Psi_1(r) \right.$$

$$+ \int_{r=I_0+q_1}^{\infty} \pi_1(r - I_0 - q_1)\, d\Psi_1(r) \left. \right\} + \int_{r=0}^{I_0+q_1} f_{N-1}(I_0 + q_1 - r)\, d\Psi_1(r)$$

$$+ f_{N-1}(0) \int_{r=I_0+q_1}^{\infty} d\Psi_1(r) \quad (5)$$

The first term between braces is the setup cost; the second the production cost; the third is the average carrying charges (assuming $I_0 + q_1 - r > 0$ is carried over to week 2); and the fourth is the average shortage penalty [assuming shortage of $r - (I_0 + q_1) > 0$]. These four terms represent the immediate consequences of the decision to produce q_1 in week 1. The next two terms evaluate the average optimal cost for the planning subhorizon starting from week 2 onward to the end of the horizon. For either there is some stock left over at the end of the first period equal to $(I_0 + q_1 - r)$, in which case the optimal is $f_{N-1}(I_0 + q_1 - r)$, which is averaged over all r in the range 0 to $I_0 + q_1$; or the demand was in excess of the available $I_0 + q_1$, in which case week 2 starts with zero on-hand inventory. The probability of such an event is $\int_{r=I_0+q_1}^{\infty} d\Psi_1(r)$, and the optimal cost over the subhorizon is $f_{N-1}(0)$.

Finally, although all engineering and business applications deal with finite albeit sometimes long horizons, the question may still linger: can one carry out the analysis assuming an infinitely long horizon? This is an important question from at least two points of view. First, there is always the uneasy feeling that restriction to a finite horizon somehow implies that "we are going out of business" at the end of the horizon. This rarely fails to leave a bad taste in the mouths of practical people, even though they are made aware that N can be prolonged to, say, a thousand weeks or a thousand years. Second, and most important, the analysis of infinitely long horizons is often simpler than the analysis of finite N, except where N is very small.

The reason is the elimination of the distinction between stages. Appealing to well-developed concepts of stationarity, simple and sometimes elegant solutions are derived, in spite of the presence of rather delicate mathematical questions of convergence, existence of optimal policies, relevance of criterion functions, and the like.

AN INVENTORY PROBLEM:
INTRODUCING THE THEORY OF MARKOV PROGRAMMING

The Problem.   The scenario in this discussion concerns inventory and a particular problem that arises in its management.

There are many instances in which it is simply not correct to assume that the demand in future periods is known with certainty. The demand is uncertain, and it is felt that such uncertainty (risk) must be reflected in the model constructed for the system. In the overwhelming majority of cases, we have a fairly good idea about the probability of the realization of any future demand. Even in the cases of new products in completely new fields, it is possible, through the expenditure of some effort and money, to gain some idea about the probability of demand realization. This is what market research is all about.

Now, there is an important difference between what we must know about the probability distribution function of demand in the dynamic case (the case of a planning horizon of $N > 1$ time periods) and what we have to know about it in the static case (the one-period case). In the latter, we need only knowledge of one demand probability distribution. This is not so in the case of dynamic problems. Here we need to know what the probability distribution of demand will be for varying-length time intervals. It is not sufficient, for example, to know the probability distribution function of demand for one-week period, because we may, in our analysis, need to know it for three-week or for five-month periods. It is obvious that there must exist some relationship between the probability distribution function of demand for each week and that, for example, for each month, which is merely the sum of four weekly demands. It is always possible to describe this relationship mathematically, which gets more and more complicated as we permit varying degrees as well as varying types of dependence between the demand in one week and the demand in another.

Before letting these considerations carry the discussion far afield from its main course, we hasten to introduce a major assumption which is customarily made in this context: we assume that the demand probability distribution function is stationary and that the demand in following periods is independent of the demand level in preceding periods. Stationarity refers to invariance over time, and independence decouples the demand in one period from the demand in preceding or succeeding periods. This assumption is admittedly a ruthless one, but it is hoped that the statements in the previous paragraph convince the reader that it is possible to analyze inventory problems for which this assumption cannot be made, though the analysis in that case is certainly more difficult.

Granted this assumption, it is evident that although the demand in one period is decoupled from the demand in other periods, the performance of the system, as reflected in the criterion function, is certainly dependent on the demand realized in all periods, as well as on the decisions made concerning replenishment in all periods. Put differently, although the demand (as the exogenous input to the inventory system from the outside world) is independent from week to week, the decision made in any week is conditioned by previously made decisions and realized demand (through the residual on-hand inventory) and colors all future decisions. In other words, independence of demand does not imply independence of decision, and the system cannot be merely decomposed into infinitely many independent weeks. The question that is raised is: "What is the manager's optimal policy for replenishing the inventory?"

The Cost Structure.   To be able to answer the foregoing question intelligently, we must first know the structure of the manager's costs. To this end, assume that he incurs a cost $\$a$ per order every time he orders; that a unit left in inventory at the end of the week and carried over to the following week costs $\$h$ in carrying charges; and that a unit sold nets $\$b$ over and above its purchase price. For simplicity, assume the demand to be discrete and let p_r denote the probability that r units will be demanded in one period (week); $r = 0,1,2, \ldots$. Finally, assume that the lead time for supply or the time between placing an order and receiving it is exactly one week. We shall see below that this assumption can be generalized to any τ weeks, $\tau \geq 0$.

Clearly, the manager's policy will be of the following form: order a quantity $Q(i)$ if the on-hand inventory at the end of a week (which coincides with the beginning of the following week) is exactly i units. Naturally, $Q(i)$ depends on the demand probability distribution function.

To illustrate, let the weekly demand and its probability be given by the following table:

$r =$	0	1	2	3
$p_r =$	0.1	0.4	0.3	0.2

Then a possible policy is $Q(0) = Q(1) = 3$, and $Q(i) = 0$ for $i > 1$. This policy effectively states that as soon as the end-of-period inventory is 1 or 0, order 3 units; otherwise do not order. Another possible policy is $Q(0) = 4$, $Q(1) = 2$, $Q(2) = 1$, and $Q(i) = 0$ for $i > 2$. Of course there are infinitely many policies to choose from. The question is "Which is the optimal policy that will result in maximal average profit?"

The answer to this question lies in applying dynamic programming theory to Markov chains with rewards—the category in which our problem lies. We have introduced DP in the previous discussion; now we introduce the concepts of Markov chains. Then, if we combine the two concepts, we arrive at the desired solution.

Markov Chains.   For any given policy $Q \triangleq \{Q(i), i = 0,1,2, \ldots\}$, and any given state of the on-hand inventory at the beginning of a week, we know exactly the probability of transition from any initial state, s_i, to some other state, s_j, at the end of the period. In particular, because we assumed the lead time to be one week, no shipment will arrive except at the end of the current week, which is identical with the beginning of the next week. Then, clearly,

$$s_j = \begin{cases} s_i + Q(i) - r & \text{if } > 0, \text{ with probability } p_r \\ Q(i) & \text{otherwise, with probability } \Sigma\, p_r \text{ for } r = s_i, s_i + 1, \ldots \end{cases} \quad (6)$$

Notice that we implicitly assumed that the order $Q(i)$ cannot be used to satisfy the demand in the same week, hence the assumed lead time of one week. To emphasize that p_r is the probability of moving from state s_i to state s_j, we call it the "transition probability" from s_i to s_j and denote it by p_{ij}. The evaluation of the transition probabilities from all feasible states to all other states results in the matrix of transition probabilities $[p_{ij}]$, or the stochastic matrix, as it is sometimes called. For example, with the given demand distribution function and the first policy $Q = (3,3,0,0, \ldots)$, the matrix of transition probabilities is given by

Subsequent state
s_j

		0	1	2	3	4
	0	0	0	0	1.0	0
Current	1	0	0	0	0.9	0.1
state	2	0.5	0.4	0.1	0	0
s_i	3	0.2	0.3	0.4	0.1	0
	4	0	0.2	0.3	0.4	0.1
	$\Pi =$	(.17641	.20289	.18925	.40891	.02254)

The numbers along the row labeled Π will be explained presently. But first we remark that the level of the on-hand inventory at the beginning of each week varies in a stochastic fashion, depending only on the state of the inventory at the beginning of the previous week and the demand in that period, and is independent of all previous history of inventory fluctuations. This defines a Markov process, so called after the Russian probabilist who first investigated this elementary form of dependence. The matrix of transition probabilities, an example of which is given above, is often also referred to as the Markov matrix:

Now, it can be proved that for a finite state Markov process of the type (ergodic states) we are concerned with here, there exists a steady state probability distribution function $\Pi = (\pi_i)$, where π_i denotes the probability of finding the system in state s_i (the probability that on-hand inventory at the beginning of any period is equal to i units). The probabilities (π_i) are obtained as the (unique) solution to the set of simultaneous equations

$$\pi_i = \sum_j \pi_j p_{ji} \qquad \text{for all states } s_i$$

and

$$1 = \sum_i \pi_i$$

In our numerical example, the solution to the set of equations is given by the vector II shown under the stochastic matrix.

The Optimal Policy. The above discussion serves as necessary background for our main concern, the determination of the optimal reorder policy. One thing is immediately obvious: given any policy Q, it is an easy matter to determine the expected value of net profit after operating costs, for there is little doubt that knowing the on-hand inventory (the state of the system) at the beginning of any period enables us to determine the expected gain (or loss) in that period. Because the steady state probabilities (π_i) give the probability of finding the system in state s_i, a simple averaging calculation yields the desired result. For example, under the assumed probability, distribution function of demand and assuming the other cost parameters to be $a = 5$, $b = 10$, $h = 1$, and for $Q = (3,3,0,0, \ldots)$, if the initial state is $i = 2$, then the expected gain is equal to

$$\underbrace{[(2 \times 10) \times 0.5]}_{\substack{\text{Sale of} \\ \text{2 units}}} + \underbrace{[(1 \times 10 - 1 \times 1) \times 0.4]}_{\text{Sale of 1 unit}} + \underbrace{[(-2 \times 1) \times 0.1]}_{\text{No sale}} = 13.4$$

while if the initial state is $i = 1$, the expected gain is given by

$$\underbrace{-5}_{\substack{\text{Order} \\ \text{cost}}} + \underbrace{[(1 \times 10) \times 0.9]}_{\substack{\text{Sale of} \\ \text{1 unit}}} + \underbrace{[(-1 \times 1) \times 0.1]}_{\text{No sale}} = 3.9$$

Continuing in this fashion for the remaining four states, we determine the average one-period cost in each case. Multiplying by the steady state probabilities already determined for this particular Q, we obtain the following results:

State	0	1	2	3	4
Steady state problem π_i	.17641	.20289	.18925	.40891	.02254
Average gain per week	-5	3.9	13.4	14.6	13.6

$\therefore$ Weekly expected profit $= 8.7218$

Naturally, if we change Q, we change the result. The problem is to determine the optimal Q; and it is at this juncture that we utilize the theory of dynamic programming.

Let $h(i,Q)$ denote the one-period average profit (a function of the state s_i and of the policy Q). Because we are dealing with an infinite planning horizon, the total expected profit is always infinite, unless it is identically equal to zero in each period no matter what state the system is in—a highly undesirable situation indeed. To circumvent such difficulty, we truncate the planning horizon to some suitably large N, carry out the analysis under such assumption, and then observe the behavior of the solution as we let N escape to infinity. This is a standard technique in mathematical analysis. Let $f_n(s_i)$ denote the maximum expected gain when starting in state s_i and adopting an optimal policy for the remaining n stages of the horizon. Then applying the principle of optimality of DP, we have (see above)

$$f_n(s_i) = \max_Q \{ h(i,Q) + \sum_j f_{n-1}(s_j) p_{ij}(Q) \} \qquad f_0(s_i) \equiv 0 \qquad (7)$$

in which the new state s_j is given by equation (6) and the matrix of transition probabilities is dependent on the policy Q. Iteration is started with $f_1(s_i)$ and proceeds to the general $f_n(s_i)$ through recursive substitution.

Without cluttering this discussion with detailed calculations, application of this procedure results in the same optimal policy Q* repeated in stages 3 and 4. Consequently, it will repeat for all $n \geq 4$, independent of the length of the planning horizon. In particular, we can now speak of the infinitely long horizon and assert that the optimal policy is given by

$$Q^* = \{2,3,2,2,1,0, \ . \ . \ .\}$$

The optimal return for state s_i is, of course, $f_\infty(s_i)$, the optimal return $f_n(s_i)$ which repeated itself under policy Q* now extended to the infinite horizon. But for this particular policy Q*, the steady state probabilities are given by II* = $(0,0,4/14,4.6/14,4.4/14,1/14)$. The following table summarizes the $f_\infty(s_i)$, their probabilities, and the average weekly gain following the optimal policy Q*.

State s_i	0	1	2	3	4	5
Steady state probabilities π_i^*	0	0	.2857	.3285	.3143	.0714
$f_\infty(s_i)$	−5	3.9	8.4	9.6	8.6	12.6

∴ Weekly expected profit = 9.65

THE MACHINE REPAIRMAN'S PROBLEM:
INTRODUCING THE THEORY OF QUEUES

A perennial problem that has plagued industrial engineers since the inception of industrial engineering itself—especially industrial engineers who are concerned with the determination of "a fair day's pay for a fair day's work"—is that of "allowance for unavoidable delays." Here, the term "unavoidable" is descriptive although somewhat misleading. What is really meant is that, for a given design of the system, the delay is unavoidable; but certainly, if one changes the design of the system, the delay is changed. Consequently, the central issue is to determine the optimal design of the system; optimal in the sense of some overall criterion that considers the workman's delays as well as the multitude of other factors that enter into the design of such a system. For it must be intuitively clear that it would be foolhardy to consider the objective to be merely the minimization of the workman's delay. Under such a single-minded and very shortsighted criterion, the answer would be simply to overload him to death, and then we would be sure that he will not have a minute of idle time.

A moment's reflection will reveal that the essential factor, which is at play behind the scenes and gives rise to idleness on the part of the workman and delay in production, is the unpredictability in the performance of the workman as well as the unpredictability in the flow of work to him and in the demands of each job on his time. If his work is arranged so that new tasks arrive exactly at predetermined instants of time, and if each new job requires a fixed and predetermined time to complete, and if the workman is always available, never distracted, and maintains a steady rate of output that never varies, then we can design a system in which he is never idle.

But these if's are not always realized, even in highly automated systems, and they are certainly not realized in systems which involve human beings. Consequently, unpredictability and uncertainty enter the picture in a very forceful manner, and cannot be ignored.

Clearly, then, the question of unavoidable delays is intimately related to other questions concerning the desired level of service, the desired rate of output of both workman and machinery, and the very manner in which jobs are selected for processing.

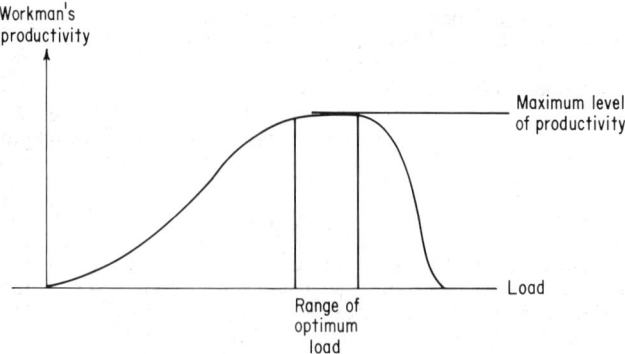

Fig. 5-2.   Probable relationship between productivity and load.

In all probability, the relationship between productivity and load is an inverted U, such as shown in Figure 5-2, in which the workman's productivity increases to a maximum and then deteriorates rapidly to zero under the stress and strain of excessive overload.   At the optimal loading level, the workman is idle some of the time and the work is not progressing at its theoretical ultimate rate; or put differently, the work encounters some delay, but an optimal balance or equilibrium between the availability of service and the flow of products is present.

The study of the conditions of such equilibrium and the manner in which equilibrium is achieved, as well as of the control of the load and number of workmen to guarantee the continued existence of such equilibrium, is the proper though not exclusive domain of the theory of queues or waiting line theory.

The Machine Repairman's Problem.   To illustrate the factors at play in the situations of concern to us here, consider the case of M automatic screw machines which are attended by one expert mechanic.   As is well known, once an automatic screw machine is set up, it continues to work on its own for a long period of time thereafter.   In fact, apart from regular and preventive maintenance performed at regular and well-known intervals of time, a machine requires service only (1) when the feed raw material is exhausted, or (2) when the job is completed and the machine needs resetting for a new job, or (3) if it fails, thus requiring repair.   This latter condition is a highly unpredictable situation, because the failure may be the result of any one of several causes.   What is even more confounding is that the time it takes the repairman to return a machine from the inoperative to the operative condition is equally highly unpredictable, depending, as it does, on the cause of failure.   The interaction between the randomness in the input to the repairman and the randomness in his own output gives rise to a peculiar situation whose treatment requires a queuing approach.

Notice that a machine which breaks down is serviced immediately unless the repairman is servicing another machine, in which case a waiting line is formed.   Hence, over a long period of time, the system will experience both idle time for the repairman and idle time for the machines (waiting) because of interference among the machines and the subsequent formation of queues.

A number of questions are usually asked of the designer of such a system whose answer would establish a quantitative base for decision.   One asks, for example, (1) how much idleness one expects on the part of the repairman and how it varies with the number of machines assigned to him; (2) in what manner additional repairmen would alleviate the load; (3) how much waiting is experienced by the machines due to the repairman or repairmen being occupied with another machine.

The Elements of a Queuing System.   In general, a queuing system is defined by three characteristics:

  1. The pattern of arrival
  2. The service mechanism
  3. The queue discipline

Customers (the failed machines in our case) can arrive equally spaced in time. But perhaps the most interesting and certainly the most widely studied pattern of arrival is the completely random pattern, which leads to the conclusion that the probability of r arrivals in a finite period of time of length h when the mean rate of arrival is λ is given by

$$p_r = \frac{(\lambda h)^r}{r!} e^{-\lambda h} \qquad r = 0, 1, 2, \ldots$$

the well-known Poisson distribution with mean λh. We can also conclude that the interarrival times are exponentially distributed with mean $1/\lambda$; that is, if Υ is the elapsed time between two arrivals, then

$$\Pr(\Upsilon \leq \tau) = 1 - e^{-\lambda \tau}$$

Another pattern of arrival is the distributed arrivals, in which the length of the interval between arrivals follows a known probability distribution function other than the exponential. Furthermore, there are batch arrivals, dependent arrivals, and many other patterns.

In discussing the service mechanism (the repairman or repairmen in our case), one wishes to distinguish among three of its facets: its availability, its capacity, and the duration of service. These are three independent variables; any one of them may be deterministically or probabilistically known. All three must be specified to achieve a complete definition of the service mechanism.

The service mechanism may be available at certain times but not available at others. Its capacity may be fixed or may be varying over time. Finally, a discussion similar to that presented above concerning the arrival distribution can be made concerning the service time. The service time can be a constant for all customers, or possess the exponential distribution function with mean $1/\mu$ (which implies complete randomness with parameter μ), or possess any other distribution function. We may also add the complication of different types of customers, each type possessing its own distribution function, or we may stipulate that the time of service is a function of the queue length: the longer the queue, the faster the service, and so on.

The final element in a queuing system is the discipline of the queue, by which is meant the manner in which customers are to be selected for service or given priority once they are in the queue.

Here one must distinguish between two cases: the single-server system and the multiple-server system.

In the single-server system (unit capacity), the customers are assigned priorities and they are serviced in the order of their priorities. There are several priority rules whose variety, needless to say, is bounded only by the imagination and ingenuity of the analyst, such as the random rule, or priority established as a function of time spent in the queue, or as a function of the economic worth of the customer. A peculiar type of priority assignment gives rise to the so-called preemptive priorities, in which the service mechanism would cease working on the customer it had been servicing and service the newly arrived customer if the latter is of a higher priority than the former.

A multiserver system (multiunit capacity) gives rise to somewhat different problems. Some units of the facility may specialize in certain services, and customers requiring that service are forced to go there. If more than one server is capable of performing the required service, jockeying (moving from one server to another) becomes a factor to be considered. A second phenomenon is that of balking or not joining a queue longer than a certain length. Reneging is leaving the queue after having joined it and having waited for a certain length of time.

Analysis of the Single Repairman's Case. With this terminology and structure in mind, we return to our repairman's problem. Assume that the machines arrive for service in a completely random fashion at the rate of λm machines if m machines are in operation. Furthermore, assume that the repairman's service time is characterized by a negative exponential with parameter μ. Let $p_m(t)$ denote the probability of m machines failed at time t; $m = 0, 1, \ldots, M$. Then, if there are M machines and one repairman, it can be shown that the probabilities $\{p_m(t)\}$ are related by the difference-differential equations

$$\frac{d}{dt}\,p_0(t) = -M\lambda p_0(t) + \mu p_1(t)$$

$$\frac{d}{dt}\,p_m(t) = -\{(M-m)\,\lambda + \mu\}\,p_m(t) + \{(M-m+1)\lambda\}p_{m-1}(t)$$
$$+ \mu p_{m+1}(t) \qquad 1 \leq m \leq n - 1$$

$$\frac{d}{dt}\,p_M(t) = -\mu p_M(t) + \lambda p_{M-1}(t)$$

This is a finite system of differential equations and can be solved by ordinary methods. The limits

$$\lim_{t \to \infty} p_m(t) = p_m$$

exist and satisfy the same set of equations with the left-hand side equal to zero. The solution is given by

$$p_0 = \frac{1}{S(\rho,M)} \qquad p_m = \frac{M!}{(M-m)!}\,\rho^m p_0 \qquad \rho = \frac{\lambda}{\mu} = \text{utilization factor}$$

and p_0 is obtained from the condition that the p_m's add to unity; hence the factor $S(\rho,M)$ is given by

$$S(\rho,M) = \{1 + Mp + M(M-1)\rho^2 + \cdots + M!\rho^M\}$$

Typical numerical values are exhibited in Table 5-1.

The probability p_0 may be interpreted as the probability of the repairman being idle (in the example of Table 5-1, he would be idle about 41 percent of the time). This is the figure of unavoidable idle time which was asked for at the beginning of this discussion. The average number of machines in the waiting queue is

$$\bar{Q} = \sum_{j=1}^{M} (j-1)p_j = M - \frac{1+\rho}{\rho}\,(1 - p_0)$$

In the example of Table 5-1, we have $\bar{Q} = 0.499 = 7 \times 0.071$. Thus, 0.071 is the average contribution of the machine to the waiting line. Of course, the average length of the line, including the machine being serviced, is $L = \bar{Q} + 1$.

The average utilization of the operator is the proportion of time that at least one

TABLE 5-1. Steady State
Probabilities p_m for the Case
$\rho = \lambda/\mu = 0.1$ $M = 7$

m	Machines in waiting line	p_m
0	0	0.4090
1	0	0.2863
2	1	0.1718
3	2	0.0859
4	3	0.0343
5	4	0.0103
6	5	0.0022
7	6	0.0002

TABLE 5-2.   Operator Utilization with M Machines per Operator
and Traffic Intensity ρ

ρ / M	0.1	0.2	0.4	0.6	0.8
1	0.092	0.167	0.286	0.375	0.444
2	0.180	0.324	0.528	0.658	0.742
4	0.353	0.602	0.850	0.938	0.903
6	0.516	0.808	0.972	0.994	0.998
8	0.662	0.930	0.997	$\simeq 1.000$	$\simeq 1.000$
10	0.785	0.982	$\simeq 1.000$	.	.
12	0.880	0.997	$\simeq 1.000$	.	.
14	0.943	$\simeq 1.000$	$\simeq 1.000$	$\simeq 1.000$	$\simeq 1.000$

machine is stopped, and is therefore $1 - p_0 \simeq 0.59$.   The rate of production of the operator is

$$\text{Operator average utilization} \times \frac{1}{\rho} = 0.59 \times 10 = 5.9$$

and the rate of production per machine (machine efficiency),

$$\text{Operator average utilization} \times \frac{1}{M\rho} = 0.59 \times \frac{10}{7} = 0.84$$

The most convenient way of summarizing this solution is by giving the repairman's utilization as a function of M (the number of machines under his supervision) and ρ. Table 5-2 is abstracted from the tables of Benson and Cox (see the bibliography at the end of this chapter).

The Case of Many Repairmen.   Would more than one server do any better?   Consider the case of s servers, $1 < s \leq M$; then we have the steady state probabilities

$$
p_m = \begin{cases}
\dfrac{M!}{(M-m)!m!}\,\rho^m p_0 & \text{if } 0 \leq m \leq s \\[2ex]
\dfrac{M!}{s!(M-m)!s^{m-s}}\,\rho^m p_0 & \text{if } s \leq m \leq M \\[2ex]
0 & \text{if } m > M
\end{cases}
$$

where

$$p_0 = \left[\sum_{m=0}^{s-1} \frac{M!}{m!(M-m)!}\rho^m + \sum_{m=s}^{M} \frac{M!}{s!(M-m)!s^{m-s}}\rho^m \right]^{-1}$$

The average length of the line (including the customer being served) is obtained from $\sum_{m=s}^{M} (m-s)p_m$.   In this case, the average operator utilization is given by

$$\sum_{m=1}^{s} \frac{mp_m}{s} + \sum_{m=s+1}^{M} p_m$$

where p_m is the probability that m machines are stopped.   Furthermore, the rate of production per machine (the average machine efficiency) is

$$\text{Average rate of production per machine} = \text{average operator utilization} \times \frac{s}{M\rho}$$

TABLE 5-3. Steady State Probabilities p_m for the Case
$\rho = \lambda/\mu = 0.1$ $M = 24$ $s = 3$

m	Machines in waiting line	p_m	m	Machines in waiting line	p_m
0	0	0.0839	13	10	0.0003
1	0	0.2014	14	11	0.0001
2	0	0.2317	15	12	0.00004
3	0	0.1699	16	13	0.00001
4	1	0.1189	17	14	.
5	2	0.0793	18	15	.
6	3	0.0502	19	16	.
7	4	0.0301	20	17	.
8	5	0.0170	21	18	.
9	6	0.0091	22	19	.
10	7	0.0045	23	20	.
11	8	0.0021	24	21	.
12	9	0.0009			

For instance, suppose there are 3 repairmen and 24 machines, so that, on the average, the load is one more machine per operator than in the previous example. Then typical numerical values are exhibited in Table 5-3. The expected total idle time is $(3 \times 0.0839) + (2 \times 0.2014) + (1 \times 0.2317) = 0.8862$, or 0.2954 per operator. Notice that, on the average, an operator is less idle when there are three of them than when there was only one. On the other hand, the average number of machines waiting is $\bar{w} = 2.865$, signifying a contribution of $2.865/24 = 0.1193$ per machine to the waiting line, which is smaller than in the case when only one operator was present. In other words, by pooling the resources of three repairmen, they can handle a larger load more efficiently.

Other Queuing Problems. Thus far our discussion has been motivated by the case of machine maintenance. This is a special case in many respects, the most important of which is the restriction to a finite maximum queue size because the population from which arrivals came was itself finite.

The important generalization of the above treatment to the case of infinite population, which is the more common case, leads to the following results:

Single server $(s = 1)$

$$p_n = \rho^n p_0 \qquad n = 1, 2, \ldots$$

$$p_0 = 1 - \rho \qquad \rho = \frac{\lambda}{\mu} < 1$$

Average length of the queue, $\bar{Q} = \dfrac{\lambda^2}{\mu(\mu - \lambda)}$ (excluding the customer being served)

Average waiting time per customer (excluding service time), $\bar{W}_q = \dfrac{\lambda}{\mu(\mu - \lambda)}$

Multiple server $(s > 1)$

$$p_n = \begin{cases} \dfrac{\rho^n}{n!} p_0 & 0 \leq n \leq s \\[2ex] \dfrac{\rho^n}{s! s^{n-s}} p_0 & n \geq s \end{cases}$$

$$\bar{Q} = \frac{\rho^{s+1}}{s! s (1 - \rho/s)^2} p_0$$

$$\bar{W}_q = \frac{\bar{Q}}{\lambda}$$

The alert reader will notice that throughout our previous discussions we have consistently assumed the completely random patterns of arrival and departure, with means equal to λ and μ, respectively. This assumption simplifies the analysis to a considerable degree. However, other probability distribution functions have been assumed and analytical results obtained in some special cases, despite the formidable magnitude of work involved.

As an example, consider a single-server system with Poisson input (i.e., completely random pattern of arrival) with mean λ and Erlang service time. The Erlang distribution function is a gamma distribution function, so called in honor of the pioneering work of the Danish engineer A. K. Erlang, of the Copenhagen Telephone Company, from 1909 to 1920. The probability density function for the Erlang distribution is

$$f(t) = \frac{(\mu k)^k}{(k - 1)!}\, t^{k-1} e^{-k\mu t} \qquad \text{for } t \geq 0$$

where μ and k are positive parameters of the distribution. The parameter k is usually integer. The mean and variance are $1/\mu$ and $1/k\mu^2$, respectively. Under these conditions, the average length of the queue is given by

$$\bar{Q} = \frac{1 + k}{2k}\,\frac{\lambda^2}{\mu(\mu - \lambda)} = \lambda \bar{W}_q$$

where $\bar{W}_q$ is the average waiting time in the queue.

RIVALRY, COMPETITION, AND CONFLICT: INTRODUCING THE THEORY OF GAMES

Game theory is the name given to a particular model of competitive systems and represents a model of such situations. It is perhaps one of the most elegant and useful branches of modern mathematics and was anticipated in the early 1920s by the French mathematician Émile Borel. But it was not until 1926 that John von Neumann gave his proof of the minmax theorem, the fundamental theorem of game theory (to be described below), and thus laid the foundation for the great structure of the theory. This came in the form of his classic 1944 work, *Theory of Games and Economic Behavior*, written jointly with the economist Oskar Morgenstern. This monumental work created a tremendous stir in economic and mathematical circles. Since then, the theory has developed into a fantastic amalgam of algebra, geometry, set theory, and topology, with applications to competitive situations in business, warfare, and politics, as well as economics.

Game theory did not succeed in making the inroads into industrial engineering thinking that it really deserves. This may be attributed to at least two factors: first, the theory is usually presented in the context of economic competition, and it is sometimes difficult for industrial engineers to visualize themselves in such a situation in their professional career. Second, interestingly enough, game theory introduces a concept of "optimal" or "equilibrium," which is radically different from those concepts familiar to engineers over the ages. The radical departure from the common path is usually a cultural shock that is often resented and consequently ignored by the industrial engineer.

Engineers in general are accustomed to optimization as the process of reaching the top of a mountain (maximizing) or the bottom of a valley (minimizing). And that also represents the notion of equilibrium. But that is not applicable to the case in which two or more forces are acting simultaneously with different objectives. And yet, it is precisely in the development of a new concept of equilibrium under such new conditions that the genius of von Neumann lies.

The Elements of Games. There are three elements of a game: (1) the number of players, (2) the strategies available to each player, and (3) the payoff function. A player P is a decision maker whose interests are at variance with each other player in at least one instance. A strategy is a complete specification of one player's decisions and actions under a particular set of decisions of the other players. An important qualification of strategies is that, for each player, the set of his strategies completely specifies his actions under all conceivable eventualities of the game (that is, of other players' decisions). The payoff is the gain, monetary or otherwise, to each player and is a function of the set of decisions made by each player.

Two-person Zero-sum Games.   Consider first the case of the so-called two-person zero-sum game, meaning that there are only two players and that the gain of one is the loss of the other, so that we need write only the payoff function to one of them. Because there are only two players, denote the strategies of player P_1 by $i = 1$, $2, \ldots, m$ and the strategies of player P_2 by $j = 1, 2, \ldots, n$, and let the payoff function be given by a_{ij}. The system can be represented by a matrix in which the strategies for P_1 are enumerated along the rows and the strategies for P_2 are enumerated along the columns. The matrix entry a_{ij} is, by convention, the payment to player P_1 if he chooses strategy i and P_2 chooses strategy j. It is always understood that the payoff to P_2 is given by $-a_{ij}$ under the same set of conditions: $i = 1, 2, \ldots, m$; $j = 1, 2, \ldots, n$.

Because the matrix $A = [a_{ij}]$ represents the payoff to P_1, it is logical to expect that P_1 wishes to maximize his reward. But his reward is conditioned upon P_2's actions, and P_2 is certainly interested in minimizing P_1's reward. Hence, the criterion adopted by P_1 is to maximize the minimum (maxmin) expected reward in anticipation of P_2's actions. Naturally, P_2's criterion is to minimize the maximum (minmax) of P_1's expected reward.

In general, each player will randomize among his strategies (which henceforth will be referred to as pure strategies to distinguish them from the randomized case). Let $X = (x_1, x_2, \ldots, x_m)$ be a point on the simplex in the m-dimensional space of P_1; that is, x_i denotes the probability that pure strategy i will be played by P_1, $x_i \geq 0$, $\sum_{i=1}^{m} x_i = 1$. Similarly, let $Y = (y_1, y_2, \ldots, y_n)$ be a point on the simplex in the n-dimensional space of P_2; that is, y_j denotes the probability that pure strategy j will be played by P_2, $y_j \geq 0$, $\sum_{j=1}^{n} y_j = 1$. We shall refer to X as the policy of P_1 and to Y as the policy of P_2. It is easy to see that a policy of P_1 which calls for playing only one (pure) strategy all the time corresponds to the degenerate case of a randomized policy X in which all the x_i's are equal to 0 except one, which is equal to 1; similarly for P_2.

Let $E(X,Y)$ denote the expected reward to P_1 when he plays policy X and P_2 plays policy Y. Hence

$$E(X,Y) = \sum_{ij} a_{ij} x_i y_j$$

Then the fundamental theorem of matrix games asserts that every two-person zero-sum game with a finite payoff matrix A possesses a solution in the following sense: there exists a pair of policies X^* for P_1 and Y^* for P_2 such that

$$E(X,Y^*) \leq E(X^*,Y^*) \leq E(X^*,Y) \tag{8}$$

for all policies X and Y. The real number given by $E(X^*,Y^*)$ is called the value of the game and is usually denoted by v^*. It represents the maximum expected gain that P_1 can ever hope to achieve if his opponent is rational. By symmetry, v^* also represents the minimum loss that P_2 can ever hope to achieve if his opponent is rational. If either P_1 deviates from his optimal policy X^* or P_2 deviates from his optimal policy Y^*, he stands to lose if his opponent plays his optimal policy.

Note that v^* may be positive—signifying a net average gain to P_1 per play of the game—or negative—signifying a net average loss to P_1 per play of the game—or, of course, zero—in which case, the game is fair. Moreover, it is interesting to remark that the game is absolutely stable even under complete information. In other words, the outcome of the game is unaltered by either player declaring his policy.

The Solution of Two-person Zero-sum Games. The fundamental theorem of matrix games is an existence theorem—it leaves open the question of how to arrive at the optimal policies X^* for P_1 and Y^* for P_2. Fortunately, this is easily answered through a linear programming formulation. In particular, $X^* = (x_1^*, x_2^*, \ldots, x_m^*)$ is the solution, not necessarily unique, to the following linear programming problem:

Maximize $\qquad\qquad\qquad\qquad \hat{v} \geq 0$

such that $\qquad\qquad\qquad \displaystyle\sum_{i=1}^{m} \hat{a}_{ij} x_i \geq \hat{v} \qquad j = 1, 2, \ldots, n$

$$\sum_{i=1}^{m} x_i = 1 \qquad x_i \geq 0, \hat{v} \geq 0$$

Here $\hat{a}_{ij} = a_{ij} + M$, where M is some large positive number, $M \geq |\min_{i,j} a_{ij}|$, which is introduced to guarantee that the value of the game evaluated from this LP problem is nonnegative. Of course, the value of the original game is simply $v^* = v^* - M$.

The optimal policy of player P_2 is similarly obtained as the solution to the dual LP problem:

Minimize $\qquad\qquad\qquad\qquad \hat{v} \geq 0$

such that $\qquad\qquad\qquad \displaystyle\sum_{j=1}^{n} \hat{a}_{ij} y_j \leq \hat{v}$

$$\sum_{j=1}^{n} y_j = 1 \qquad y_j \geq 0, \hat{v} \geq 0$$

Completely Determined Matrix Games. A special case of the general two-person zero-sum game deserves attention. It is the case in which the optimal policy of player P_1 is to play a pure strategy, say i^*, all the time, and the optimal policy of player P_2 is to play a pure strategy, say j^*, all the time. Such games are called "completely determined matrix games" because no randomization among the strategies is involved. The payoff $a_{i^*j^*}$ is called the saddle point of the game, and it has the peculiar property of being simultaneously the maximum value of its column and the minimum value of its row; that is, it satisfies the inequality

$$a_{ij^*} \leq a_{i^*j^*} \leq a_{i^*j} \qquad\qquad\qquad (9)$$

If the reader will compare these inequalities with those of equation (8), he will immediately recognize the similarity of the two, with equation (9) being more special than equation (8).

Example. Consider the game whose payoff matrix is

$$A = \begin{bmatrix} 2 & 4 & 0 \\ 1 & 0 & 4 \end{bmatrix}$$

Clearly, P_1 has two pure strategies, P_2 has three pure strategies, and the game has no saddle point. The optimal policy for P_1 is the solution of the LP problem:

Maximize $\qquad\qquad\qquad\qquad v$
such that $\qquad\qquad\qquad 2x_1 + x_2 \geq v$
$$4x_1 \qquad\quad \geq v$$
$$4x_2 \geq v$$
$$x_1, x_2, v \geq 0$$

The optimal is given by $x_1^* = \frac{3}{5}$, $x_2^* = \frac{2}{5}$, and $v^* = \frac{8}{5}$. This LP problem also yields the optimal policy for P_2 and is given by $y_1^* = \frac{4}{5}$, $y_2^* = 0$, $y_3^* = \frac{1}{5}$.

Suppose P_1 plays his optimal strategy $x_1^* = \frac{3}{5}$, $x_2^* = \frac{2}{5}$, and P_2 always plays his pure strategy 1. The value of the game is $(2)(\tfrac{3}{5}) + (1)(\tfrac{2}{5}) = \tfrac{8}{5}$, which equals v^*. If P_2 plays his pure strategy 3 at all times, the value of the game is $(0)(\tfrac{3}{5}) + (4)(\tfrac{2}{5}) = \tfrac{8}{5}$, which again equals v^*. However, if P_2 plays his pure strategy 2, which is an inactive strategy denoted by $y_2^* = 0$, then the value of the game is $(4)(\tfrac{3}{5}) + (0)\tfrac{2}{5} = 1\tfrac{2}{5}$, which is greater than v^*.

Thus far we have talked about only two-person zero-sum games. Other forms of games have been defined and studied with varying degrees of success. For example, there are two-person general-sum games, n-person games, infinite games, multistage games, and differential games. We will discuss the first of these generalizations and use it as a vehicle to demonstrate the complexities that enter the analysis.

Two-person General-sum Games. The most natural extension of the basic two-person zero-sum game is the two-person nonzero-sum, or general-sum, game. Such conflict situations are, in fact, the more common because it is not necessarily true that one player's gain is the other player's loss. Economic situations abound in which two competing firms can adopt strategies such that both increase their income, but not necessarily by the same degrees.

Here, three new considerations must be taken into account. First, we are no longer dealing with one payoff matrix, but with two such matrices: the payoff matrix to player P_1, say the matrix A; and the payoff matrix to player P_2, say the matrix B. An entry a_{ij} in A denotes the payoff to P_1 if he plays pure strategy i and P_2 plays his pure strategy j. Under the same set of strategies, the payoff to player P_2 is b_{ij}, an element of B. This is why such games are usually referred to as "bimatrix" games.

Second, we must recognize the possibility of cooperation between the two players. In other words, the game now has two equilibrium points instead of one: the non-cooperative equilibrium, in which each player is out to secure the maximum for himself without the prospect of reaching an understanding with the other player; and the cooperative equilibrium, in which such an understanding is sought and implemented. In the case of noncooperative equilibrium, we are in essence seeking a pair of mixed strategies (X^*, Y^*) for the bimatrix game (A,B) such that for any other mixed strategies X and Y the following is true:

$$XAY^{*t} \leq X^*AY^{*t}$$
$$X^*BY^t \leq X^*BY^{*t} \tag{10}$$

The case of cooperative equilibrium is somewhat more involved because now we must hypothesize on the mode of behavior of the two players, assuming that they will act in consonance. Such hypotheses are usually referred to as "axioms" of behavior. These are, in the final analysis, reasonable conditions to impose on any such behavior. For instance, although it is impossible to determine how a person will act, we can nevertheless set a minimum to the amount that a player will accept for himself. This is the amount he can obtain by unilateral action, whatever the other player does. This is, of course, the maxmin value of the game for that player. If we call these two values u^* and v^*, we have

$$u^* = \max_{X} \min_{Y} XAY^t$$
$$v^* = \min_{Y} \max_{X} XBY^t$$

We then demand that if $(\bar{u}, \bar{v})$ are the pair of values in the cooperative solution, then we should have

$$(\bar{u}, \bar{v}) \geq (u^*, v^*)$$

This is the so-called "axiom of individual rationality."

The reader must be warned that, although in the statement of these conditions they are called axioms, thus implying that they are "self-evident truths that need no justification," they are, in fact, not so self-evident in some cases and have actually been challenged.

Third, we must also consider the possibility of threats. A plausible scenario would be as follows:

1. Player P_1 announces a threat strategy X.

2. Player P_2, in ignorance of X, announces a threat strategy Y.
3. Players P_1 and P_2 bargain. If they can come to an agreement, then that agreement takes effect. If they do not come to an agreement, then they must use their threat strategies, X and Y. The payoffs to the two players will be determined in this manner.

This means that the maxmin values u^* and v^* defined in equation (10) above are replaced by XAY^t and XBY^t.

Games against Nature and Decision Theory. It is possible to interpret several decision situations as "games against nature," in which the decision maker is player P_1, say, and Nature (or Fate or the World) is the other player P_2. Then the decision maker's strategies are precisely the avenues open to his choice. Nature, on the other hand, confronts him with any of several possibilities, or metaphorically speaking, Nature may choose any one of its pure strategies. The outcome of this game is given by the payoff matrix $A = [a_{ij}]$, which now represents the benefit P_1 can extract from Nature.

Many industrial engineering problems in investment analysis, equipment replacement, quality control, and the like can be cast in this format. Such a format is both instructive and easy to study. Its major drawback stems from a purely philosophical point of view: many people cannot conceive of a supreme being who is malicious to the point that no matter what decision P_1 makes he is always handed down the worst possible outcome. This is too pessimistic a view of the world to be palatable to many analysts.

THEORY OF NETWORKS

Among the advances in the field of operations research, it can be truly said that the theories of networks are among the simplest, most elegant subjects which possess a wide variety of applications. Consequently, it is being recognized as one of the most important unifying concepts that underlie several seemingly separate and independent applications. For example, a traffic network may represent in physical reality a radio communication, a railway transportation, a highway, or an airline network. All these different physical realities possess one feature in common: their elements can be represented by a graph with stations and links between stations. The topological structure of such networks can be represented by a graph with vertices and branches corresponding to the stations and links, respectively.

At the outset, we make an explicit distinction between graph theory and network theory. A graph defines the purely structural relationships between the nodes, while a network bears also the quantitative characteristics of the nodes and arcs. Consequently, we take a graph-theoretic problem to be a problem related to pure structure. On the other hand, a network problem is related to the quantitative characteristics defined on the graph.

For example, the following is a graph-theoretic question: In a connected graph G of n nodes and A arcs, what is the minimum number of nodes that must be removed to eliminate all the arcs in the graph? This is the well-known "minimal cover" problem. Notice that the answer to the problem is evidently dependent only on the structure of the graph. On the other hand, the following is a network problem: Given a graph G of n nodes and A arcs and given that arc (i,j) is of length d_{ij}, what is the shortest path between two given nodes s and t? This is the well-known "shortest path" problem. Here, it is equally obvious that the answer depends not only on the structure of the graph, but also on the very specific values (the lengths) defined on the arcs of the graph.

We shall discuss three kinds of network models. They are:
1. Shortest path problems
2. Maximal flow models
3. Activity networks

In each case, a function is defined on the arcs of the network, but the algebra for the manipulation of these quantitative measures is different from model to model. A key concept in network models is: although the structure of several networks may be identical, the analysis of the functional relations defined on the network may be different for different models; hence the results of the analysis are different.

The Shortest Path Problem.   Generally speaking, one can classify problems of shortest path into four main categories:

Problem 1. Find the shortest path between two specified nodes s and t.

Problem 2. Find the m-shortest paths between two specified nodes s and t.

Problem 3. Find the shortest paths from an origin s to all other nodes of the network, or alternatively, the shortest paths from all nodes to a terminal t.

Problem 4. Find the shortest paths between all pairs of nodes.

Problem 2 is a generalization of problem 1 and requires some explanation.   Suppose we enumerate all the paths from s to t and compute their lengths; let Π_j denote the jth path and $L(\Pi_j)$ denote its length.   Suppose further that we rank the paths in ascending magnitude of their lengths so that we have $L(\Pi_1) \leq L(\Pi_2) \leq \cdots \leq L(\Pi_m)$. Path Π_1 is called the shortest path between s and t, path Π_2 the second shortest, and so on.   In problem 2, our task is to determine the first m-shortest loopless paths from s to t.   This is not a trivial problem, for we wish to avoid complete enumeration of paths from s to t.

As for problem 3, it turns out that almost all algorithms that solve problem 1 also solve problem 3.   It seems inherent in combinatorial approaches to the solution of the shortest path between s and t to evaluate simultaneously the shortest paths between s and all other nodes of the network.

Problem 4, that of determining the shortest paths between all pairs of nodes, is evidently a generalization of problem 1, when one lets the pair (s,t) range over all such possible pairs.   This problem is intriguing from a computational point of view, for it is clear that repeated n applications of the algorithm for problem 1 would provide the answer to problem 4.   A pertinent question is: can we do better?   Fortunately, the answer is yes, and an approach developed by R. W. Floyd solves the problem in one sweep and requires only $N(N - 1)(N - 2)$ elementary operations. We illustrate this class of problems by giving a solution to problem 1.

The Shortest Path between s and t.   This algorithm is due to E. W. Dijkstra, and capitalizes on the fact that if j is a node on the minimal path from s to t, knowledge of the latter implies knowledge of the minimal path from s to j.   In the algorithm, the minimal paths from s to other nodes are constructed in order of increasing length until t is reached.

In the course of the solution, the nodes are subdivided into three sets:

A—the nodes for which the path of minimum length from s is known.   Nodes will be added to this set in order of increasing minimum path length from node s.

B—the nodes from which the next node to be added to set A will be selected. This set comprises all those nodes that are connected to at least one node of set A but do not yet belong to A themselves.

C—the remaining nodes.

The arcs are also subdivided into three sets:

I—the arcs occurring in the minimal paths from node s to the nodes in set A.

II—the arcs from which the next arc to be placed in set I will be selected.   One and only one arc of this set will lead to each node in set B.

III—the remaining arcs (rejected or not yet considered).

To start with, all nodes are in set C and all arcs are in set III.   We now transfer node s to set A and then repeatedly perform the following steps:

Step 1.   Consider all arcs a connecting the node just transferred to set A with nodes j in sets B or C.   If node j belongs to set B, we investigate whether the use of arc a gives rise to a shorter path from s to j than the known path that uses the corresponding arc in set II.   If this is not so, arc a is rejected; if, however, the use of arc a results in a shorter connection between s and j than hitherto obtained, it replaces the corresponding arc in set II and the latter is rejected.   If the node j belongs to set C, it is added to set B and arc a is added to set II.

Step 2.   Every node in set B can be connected to node s in only one way if we restrict ourselves to arcs from set I and one from set II. In this sense, each node in set B has a distance from node s: the node with minimum distance from s is transferred from set B to set A, and the corresponding arc is transferred from set II to set I. We then return to step 1 and repeat the process until node t is transferred to set A. Then the solution has been found.

Related Topics. The shortest path problem bears close relationship to other problems which, as originally formulated, do not show any shortest path characteristics.

Relationships usually come about in one of two forms. Either the problem is recast or reformulated so that it yields a shortest path formulation or the methodologies of the two problems are in fact the same except that one problem (say, the shortest path problem) is a special case (or a general case) of the other.

Four such related problems will be discussed. By necessity, the discussion is brief. For a more detailed study of these problems and of related topics to them, the reader is advised to consult the references listed at the end of this chapter.

The Most Reliable Route. Suppose the network $G \equiv (N,A)$ is a representation of a large-scale system, where the arcs represent components of the system, and nodes represent junction points among the components. Suppose further that the origin s now designates the input junction point to the system, and the terminal t designates the output junction point. The input must follow one and only one path from s to t. The problem is to determine the most reliable path from s to t, where reliability is defined as the probability of nonfailure, and the reliability $R(\Pi)$ of a path Π composed of arcs a, b, . . . , w is the product of the reliabilities of the individual arcs.

This problem can be reformulated to correspond exactly to the longest route problem, because we are interested in the maximal reliability. Let $d_{ij} = \log p_{ij}$, where p_{ij} is the reliability of arc (i,j). Obviously, d_{ij} is ≤ 0. We have

$$\log R = \sum_{(i,j)\in\Pi} d_{ij}$$

and the most reliable route in the network is the longest loopless path(s) from s to t (the most negative path).

If the network is directed and acyclic, this problem can be solved very easily by the method outlined above. If, on the other hand, the network is of the general type, the solution of this problem still awaits an efficient algorithm for solving the longest route problem.

The Maximum Capacity Route. In this problem, every arc (i,j) of the network has associated with it a capacity $c_{ij} < \infty$. The arc capacity indicates the maximum amount of flow that can pass from i to j. The problem is to find a route from s to t such that

$$\min \{c_{s1}, c_{12}, \ . \ . \ . \ , c_{nt}\}$$

is maximum.

If the network is undirected, it can be easily shown that the maximum capacity routes between all pairs of nodes is a maximal spanning tree: it is a tree such that, for any arc (i,j) not in the tree, we have

$$c_{ij} \leq \min \{c_{i1}, c_{12}, \ . \ . \ . \ , c_{nj}\}$$

In other words, any arc not in the tree must be of capacity not larger than the minimal capacity of the (unique) path in the tree joining the terminal nodes of the arc.

The manner in which such a maximal spanning tree is constructed is very simple, and was developed by J. B. Kruskal, Jr.; see also R. C. Prim. Begin by selecting the arc with the largest capacity; at each successive stage, select from all arcs not previously selected the largest capacity arc that completes no cycle with previously selected arcs. After $N - 1$ arcs have been selected, a maximal spanning tree has been constructed.

For the general type of network—directed networks with loops and nonsymmetric capacity matrices—solution is obtained by application of the revised cascade method developed by J. D. Murchland.

The Transshipment Problem. The transshipment problem is a generalization of the well-known transportation problem of linear programming. In the transshipment problem, it is desired to ship a commodity which is available in quantities $a_1, \ldots, a_m$ at the m sources $s_1, \ldots, s_m$, respectively, to a number of terminals $t_1, \ldots, t_n$, where it is demanded in quantities $d_1, \ldots, d_n$, respectively: and minimize the total cost of transportation, where the cost of transporting a unit of the commodity in arc (i,j) is c_{ij}. The term "transshipment" stems from the fact that one need not ship directly from a source s_i to a terminal t_j, but may ship via another node or nodes.

In the special case of one source s and one terminal t, the solution of the transshipment problem is really given by the shortest path from s to t, and conversely.

The Traveling Salesman Problem. The traveling salesman problem is the problem of a salesman who starts at one city s and visits each of $N - 1$ cities exactly once and finally returns to his starting position with minimum distance traveled. Mathematically, given a finite set $\{1,2, \ldots, i, \ldots, N\}$ of nodes and the distance matrix $D = [d_{ij}]$ between every ordered pair (i,j), find the sequence of nodes $(i_1, i_2, \ldots, i_N)$ such that:

1. Every node appears in the sequence exactly once.
2. The total length $d_{1,i_2} + \sum\limits_{k=2}^{N-1} d_{i_k,i_{k+1}} + d_{i_N,1}$ of the sequence is minimal.

These two requirements imply that the sequence is a cycle. In other words, it is desired to find the shortest cycle from 1 and back to 1 which passes by all other nodes exactly once.

The simplicity of the statement of the problem is certainly misleading. Until very recently, there existed no computationally feasible algorithm for its solution. In fact, the practical resolution of the problem was achieved through the use of reliable heuristics of the branch and bound variety.

What relationship, if any, is there between the traveling salesman problem and the shortest path problem? On the surface, it seems absurd to ask such a question, because the answer seems to be self-evident. In fact, there is no relationship at all between the two.

There is, however, a strong relationship between the traveling salesman problem and the longest path problem. The following has been established:

1. It is possible to reformulate a minimum cycle problem of a network G into a maximum distance problem of another network G' constructed from G in the following fashion: (a) the nodes of G' are the nodes of G, plus one additional node denoted by s'; (b) the arcs of G' are the arcs of G except that all arrows going into 1 are replaced by arrows going into s'; (c) the distances d_{ij}' of G' are defined by

$$d_{ij}' = \begin{cases} 0 & \text{if } i = j \\ K - d_{ij} & \text{if } i \neq j \text{ and } j \neq s' \\ K - d_{is} & \text{if } i \neq j \text{ and } j = s' \end{cases}$$

and K is a constant strictly greater than σ, σ being the sum of the N longest d_{ij}'s in G.

2. It is possible to reformulate a minimum tour[3] problem of a network H into a minimum cycle problem of a network G which is constructed from H in the following fashion: (a) the nodes of G are the nodes of H; (b) in G, nodes i and j are joined by a directed arc (i,j) if and only if there is a directed path in H from i to j; (c) the lengths d_{ij} of G are the lengths of the shortest path in H between i and j.

Consequently, we can go from a tour formulation—which is a meaningful problem in its own right and has received little, if any, attention—to a cycle formulation to a longest path formulation. A computationally good solution to the latter problem provides a solution to the traveling salesman problem.

Maximal Flow Models. A natural though by no means elementary question that comes to mind when studying networks is that of the maximum possible flow between two specified nodes of the network. The physical problem arises in almost every instance in which commodities—physical or otherwise—flow from a source s to a destination t. Thus, in a traffic network, we may be interested in the maximum rate

---

[3] In the definition of a tour, condition 1 is relaxed to "Every node appears in the sequence at least once." Hence, the salesman may visit the same city more than once before returning to his starting position.

of traffic flow between two cities; in an electric power distribution network, we may be interested in the maximum power transmitted from a generating station to a particular location; in a natural gas distribution network, we may be interested in the maximum rate at which gas can be supplied to a particular consumer.

This is the most basic problem in network flows, whose solution is readily available through a labeling procedure developed by L. K. Ford, Jr., and D. R. Fulkerson.

There are two important extensions to the one-source-to-one-terminal flow problem.

The first is the multiterminal flow problem, in which several nodes are designated as (s,t) pairs, with the same commodity flowing through the network. For example, this is the case of a telephone network in which any unordered pair of the n cities covered by the network may indeed serve as the (s,t) pair.

One's first impulse when confronted with the problem of determining the maximum flow between all ordered pairs of nodes in a multiterminal flow network is to repeat the procedure for the single (s,t) pair as many times as is needed. A little reflection will indicate that, in an undirected network of n nodes, one will perform $n(n-1)/2$ such optimum evaluations. If n is large, which is usually the case, the amount of work involved can certainly be staggering. Fortunately, this is not necessary because a most elegant procedure developed by R. E. Gomory and T. C. Hu requires only the solution of $n-1$ maximum flow problems, and what is more pleasing, these problems usually get successively smaller in size as calculation proceeds.

The second extension is the multicommodity flow problem, in which several commodities flow simultaneously from designated sources to designated terminals. The sources and the terminals may be different for different commodities. Usually, two tasks are to be accomplished: (1) to maximize the total sum of the different flows—this is called the maximum multicommodity flow problem; (2) to prescribe lower bounds on each of the flow values and ask if it is feasible—this is called the feasibility problem.[4]

The solution of the multicommodity flow problem involves some delicate arguments of duality which, once understood and mastered, lead immediately to a simple procedure requiring no more than the determination of the shortest path in the network at each iteration. The length of the arcs at each iteration turns out to be the dual variables of the corresponding linear program.

Generally speaking, there are two kinds of flow problems which present a classification different from that discussed above. First, one may be interested in maximizing the flow. The problem of maximum flow (whether in single terminal or multiterminal networks; in single commodity or multicommodity flows) is meaningful only if the arcs of the network, or a key subset thereof, possess upper limits, called the "capacities" of the arcs and designated by $c(i,j)$ or c_{ij}, which no flow can exceed in that arc in the direction $i \to j$. Such networks are called capacitated networks.

Second, one may be interested in maximizing the value of the flow. In other words, if the network has A arcs and we let the vector $X = (x_{ij})$, $(i,j) \in A$, denote the vector of flow, then there is a function $g(X)$ which maps such flow into the real line and gives the value of the flow. Examples of such functions g are the cost of flow $g(X) = pX$, a linear function; or $g(X) = a\delta(X) + pX$, where $\delta(X) = 0$ if $X = 0$ and $= 1$ if $X > 0$, a concave function; and the like. In such a case, the arcs of the network may or may not be capacitated, depending on the context of the problem.

Once the concept of flow networks is visualized, the questions related to flow that can be legitimately asked seem to multiply in an abundant number. Even more interesting, many physical nonflow problems can be interpreted in flow terms to great advantage with respect to both insight into the behavior of the system and the ease of computation. The following problems, among many others, have been given flow network representation.

Examples of Flow Theory Applications. 1. The personnel assignment problem: n graduates have been hired to fill n vacant jobs. Aptitude tests, college grades, and recommendations of professors help assign a proficiency index a_{ij} to candidate i

---

[4] A feasibility problem has also been formulated for the single commodity flow network, of which this is a generalization.

on job j. What is the assignment of all n candidates that maximizes the total proficiency score over all jobs?

2. A plant location-allocation problem: A producer wishes to manufacture a new product in a number of plants to be built specifically for the purpose. There shall be no more than m such plants, and m locations have already been selected as the most suitable sites. There are n markets or demand centers; the demand r_j for the jth market is known over the planning horizon of T years. How many plants should be built, and what is the allocation of markets to plants which maximizes the total discounted net revenue over the production and distribution costs?

3. A production-inventory problem: The market requirements $\{r_t\}$ for a particular product over a finite planning horizon are known, $t = 1, 2, \ldots, T$. If x_t units are produced in period t, the cost of production is $p_t(x_t)$, and if I_t units are held in inventory at the end of period t, a holding cost equal to $h_t(I_t)$ is incurred, where

$$I_t = I_0 + \sum_{i=1}^{t} (x_i - r_i).$$ All demand must be satisfied in the period of its occurrence.

What is the production schedule (the x_i's) which maximizes the overall gain?

4. The warehousing problem: A wholesaler purchases, stores, and sells, in each of T successive periods, some commodity that is subject to known fluctuations in purchasing costs and selling prices. The wholesaler has a warehouse of fixed capacity in which new purchases and holdovers from the previous period are stored before selling. What is his strategy of buying, storing, and selling which maximizes his profit?

A Fundamental Theorem. The following theorem states a property that is of fundamental importance in the theory of flow:

> For any network, the maximum flow from s to t is equal to the value of the capacity of the minimal cut-set, where a cut-set is defined as any set of arcs that separate s from t; and the capacity of a cut-set is the sum of the capacities of its arcs in the direction s to t.

This theorem was known to graph theorists as early as 1924. Unfortunately, it is an existence theorem which tells nothing about the construction of the minimal cut-set or the determination of the maximum flow. Such construction was achieved by Ford and Fulkerson through the so-called labeling procedure.

Example. Consider the network shown in Figure 5-3, in which the capacities of the arcs are indicated on the arcs. Assume node 1 to be the source s with infinite availability, and node 7 to be the terminal t. The maximum flow from s to t is 18 units; the flow itself is as shown in Figure 5-4. The minimal cut-set is the set of arcs $\{(1,2), (2,3), (3,6), (4,6), (4,7)\}$. The reader can easily verify that the capacity of this cut-set in the direction s to t is indeed equal to 18 units.

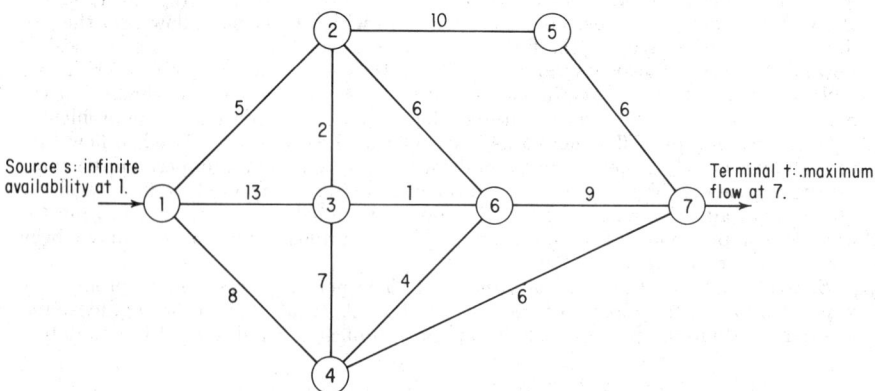

Fig. 5-3.   The network.

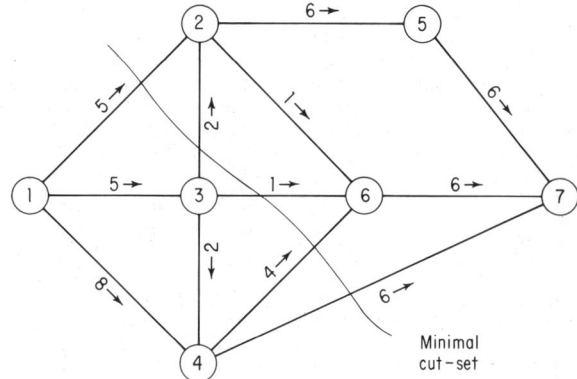

FIG. 5-4.   Maximal flow and minimal cut-set.

Activity Networks and Their Generalizations.   Traditionally, an activity network is a representation of two particular aspects of a project: the precedence relationship among the activities and the duration of each activity.   Precedence comes about from technological and other considerations; for example, one cannot mount the roof of a house without first having built the walls, which in turn must be preceded by the foundation.   In the network representation, the nodes usually depict events—which are considered as well-defined occurrences in time, such as "shipment received" or "test completed"—and the arcs depict activities—such as "build foundation" or "test for quality"—although the roles of these two components can certainly be interchanged.   An activity usually consumes something, such as materials, energy, skills, money, or the like, and it usually takes time to accomplish.

Because activity networks receive special attention in this Handbook in Chapter 3 of Section 8, we content ourselves here with this brief mention and refer the reader to that chapter as well as to the references cited at the end of this chapter.

THE EXPANDING HORIZONS OF OPERATIONS RESEARCH

Thus far, we have presented a few of the branches of study in OR, together with a discussion of the basic philosophy of OR.   Limitations on space forbid the detailed exposition of these topics.   Even more, it prevents more than the mere mentioning of the other branches of OR and their applications.   The following discussion is devoted to doing just that.   It is not intended to be comprehensive, but the discussion will underline some of the more important branches and point out their relationship to other, more traditional, fields of study.

Linear Programming (LP) and Its Offshoots.   Linear programming is the theory of maximizing or minimizing a linear function $C = \Sigma_j c_j x_j$ subject to a set of m constraints, a typical member of which is $\Sigma_j a_{ij} x_j = b_i; i = 1, 2, \ldots, m$.   The a_{ij}'s, b_i's, and c_j's are given constants.   The optimization process is further restricted by the requirement that all the unknown variables be nonnegative, that is, $x_j \geq 0$ for all $j = 1, 2, \ldots, n$.

Linear programming is more fully discussed in the next chapter.

Integer Linear Programming.   A class of problems has come to be known as integer LP problems in spite of the fact that, by insisting on integer variables, the problems are no longer linear.   This is because linearity implies convexity, which is not satisfied if the programming problem requires integer variables.

The mathematical statement of the integer LP model is identical with that of the ordinary LP statement, with the addition of the constraint that x_j is integer for all or some $j, j = 1, \ldots, n$.

Although the mathematical statement seems to be identical with ordinary LP, the solution of integer LP problems is radically different.   The standard simplex algorithm is not sufficient because it does not guarantee integer answers.   A different theory, based on cutting planes, was developed by Gomory for the solution of such programs.

Of special interest are integer LP problems in which the integer variables can take on only the values 0 or 1.   These are called 0, 1 integer LP problems.   Such problems emanate naturally in many practical applications, such as all problems which require yes-no, open-closed, or true-false answers.   But more interestingly, any LP problem which requires the variables to be integers can be transformed into a 0, 1 integer LP. For example, let x_j be an integer variable.   Suppose that the upper bound on the value of x_j is M.   Then x_j can be represented by the binary expansion $x_j = \sum_{k=0}^{K} 2^k y_{kj}$, $K \leq (\log_2 M)$, in which the new y variables are 0, 1 variables.   Substituting for x_j in the objective function and constraints, we complete the translation for the general integer LP into a 0, 1 integer LP.

Nonlinear Programming.   Leaving behind the strict adherence to the linear form, we encounter a great variety of mathematical programming models.   Of special interest are (1) quadratic programming, a quadratic objective function but still linear constraints; (2) convex programming, a convex—not necessarily quadratic—objective function subject to convex constraints; (3) stochastic LP, LP with indeterminacy in the parameters a_{ij}, b_i, or c_j; (4) chance constrained LP, where the constraints are themselves in the form of probability statements; (5) geometric programming, where both objective function and constraints are sums of polynomials; and (6) dynamic programming, which we have discussed before.   A more complete discussion of nonlinear programming is given in Chapter 7 of this section.

Optimum Search Theory.   Although the theory of mathematical programming is concerned with the optimization of a known function subject to constraints, there are many practical situations in which the form of the function to be optimized is not fully known.   Sometimes, all that is known about the function is that it does possess an optimum—a maximum or a minimum—in the range of interest.

The search for the optimum of a function more or less unknown to the observer necessarily involves experimentation, for the only way to gain information about such a function is by direct measurement.   The central problems in the theory of optimum search are thus seen to be how many experiments to conduct, where to conduct them, and how to interpret the results obtained from such experiments, in the most economical manner.

Control Theory and Cybernetics.   At the outset, we assert that there is no control without feedback.   But control theory involves much more than classical feedback control, important as that may be.   Unfortunately, engineers usually interpret control to be synonymous with exercising authority over, directing, or commanding, with all the direction and command coming from outside the system.   In fact, society in general normally accepts as control a kind of regulation which is no more than coercion.

We wish to suggest that there are more subtle notions of control available, in particular, the concept of homeomorphic control or self-regulating systems.   The system to be controlled emerges from this concept as a special kind of system; it is a tightly knit network of information which can hardly be discussed with any meaning at all except as a whole organism because the very functioning, the very existence, of each component is dependent on the functioning of the other components.

Any discussion of control leads directly to a discussion of cybernetics because: "Cybernetics is the science of communication and control.   The applied aspects of this science relate to whatever field of study one cares to name: engineering, or biology, or physics, or sociology, or the like.   The formal aspects of the science seek a general theory of control, abstracted from the applied fields and appropriate to them all."

In this general, all-encompassing area, operations research contributes significantly by providing the models and the tools of analysis that develop many of the building blocks needed for the general theory.

BIBLIOGRAPHY

Operations Research

Elmaghraby, Salah E., "The Role of Modeling in I.E. Design," paper presented at the 1967 Annual Meeting, American Society of Electrical Engineers, East Lansing, Mich.; also appeared in *Journal of Industrial Engineering*, vol. 19, no. 6, June, 1968.
Flagle, Charles D., William H. Huggins, and Robert H. Roy, *Operations Research and Systems Engineering*, The Johns Hopkins Press, Baltimore, Md., 1960.
Hillier, Frederick, and Gerald J. Lieberman, *Introduction to Operations Research*, Holden-Day, Inc., San Francisco, February, 1968.
Kaufmann, A., and R. Faure, *Introduction to Operations Research*, Academic Press Inc., New York, 1968.
Mehta, A., T. R. Thiagarajan, and N. K. Jaiswal, "A Collection of Some Operational Research Problems from World War II," Government of India, Defense Research and Development Organization of the Ministry of Defense, New Delhi.
Mesarovic, Mihajlo D. (ed.), "Views On General Systems Theory," *Proceedings of the Second Systems Symposium*, April, 1963, John Wiley & Sons, Inc., New York, 1964.
Singh, Jagjit, *Great Ideas of Operations Research*, Dover Publications, Inc., New York, 1968.
Wagner, H. M., *Principles of Operations Research*, Prentice-Hall, Inc., Englewood Cliffs, N.J., 1969.

Dynamic Programming

Bellman, Richard, *Dynamic Programming*, Princeton University Press, Princeton, N.J., 1957.
Bellman, Richard, and Stuart E. Dreyfus, *Applied Dynamic Programming*, Princeton University Press, Princeton, N.J., 1962.
Elmaghraby, Salah E., *The Design of Production Systems*, Reinhold Publishing Corporation, New York, 1966.
Hadley, G., and T. M. Whitin, *Analysis of Inventory Systems*, Prentice-Hall, Inc., Englewood Cliffs, N.J., 1963.
Kaufmann, A., and R. Cruon, *Dynamic Programming, Sequential Scientific Management*, Academic Press Inc., New York, 1967.
Mitten, Loring G., "Composition Principles for Synthesis of Optimal Multistage Processes," *Operations Research*, July–August, 1964.
Mitten, Loring G., and E. V. Denardo, "Elements of Sequential Decision Processes," *Journal of Industrial Engineering*, January, 1967.
Nemhauser, George L., *Introduction to Dynamic Programming*, John Wiley & Sons, Inc., New York, 1966.

Markov Programming

Feller, William, *An Introduction to Probability Theory and Its Applications*, John Wiley & Sons, Inc., New York, 1957.
Howard, Ronald A., *Dynamic Programming and Markov Processes*, John Wiley & Sons, Inc., New York, 1960.
Kemeny, John G., and J. Laurie Snell, *Finite Markov Chains*, D. Van Nostrand Company, Inc., Princeton, N.J., 1959.
Starr, Martin K., and David W. Miller, *Inventory Control: Theory and Practice*, Prentice-Hall, Inc., Englewood Cliffs, N.J., 1962.

Theory of Queues

Benson F., and D. R. Cox, "Productivity of Machines Requiring Attention at Random Intervals," *Journal of Royal Statistical Society*, series B, vol. 13, 1951.
Cox, D. R., and Walter L. Smith, *Queues*, John Wiley & Sons, Inc., New York, 1961.
Morse, Philip M., *Queues, Inventories, and Maintenance: The Analysis of Operational Systems with Variable Demand and Supply*, John Wiley & Sons, Inc., New York, 1958.
Prabhu, N. W., *Queues and Inventories: A Study of Their Basic Stochastic Processes*, John Wiley & Sons, Inc., New York, 1965.
Saaty, Thomas L., *Elements of Queueing Theory*, McGraw-Hill Book Company, New York, 1961.

Theory of Games

Isaacs, Rufus, *Differential Games*, John Wiley & Sons, Inc., New York, 1965.
McKinsey, J. C. C., *Introduction to the Theory of Games*, McGraw-Hill Book Company, New York, 1952.
Owen, Guillermo, *Game Theory*, W. B. Saunders Company, Philadelphia, 1968.
Rapoport, Anatol, "The Use and Misuse of Game Theory," *Science American*, December, 1962.
von Neumann, John, and Oskar Morgenstern, *Theory of Games and Economic Behavior*, Princeton University Press, Princeton, N.J., 1953.

Theory of Networks

Dijkstra, E. W., "A Note on Two Problems on Connexion with Graphs," *Numerische Mathematik*, vol. 1, 1959, pp. 269–271.
Elmaghraby, Salah E., "Some Network Models in Management Science," vol. 29 in *Lecture Notes in Operations Research*, Springer-Verlag, New York, 1970.
Floyd, R. W., "Algorithm 97, Shortest Path," *Communications of the ACM*, vol. 5, 1962, p. 345.
Ford, L. K., Jr., and D. R. Fulkerson, *Flows in Networks*, Princeton University Press, Princeton, N.J., 1962.
Kaufmann, A., *Graphs, Dynamic Programming, and Finite Games*, Academic Press Inc., New York, 1967.
Kruskal, J. B., Jr., "On the Shortest Spanning Subtree of a Graph and the Traveling Salesman Problem," *Proceedings, American Mathematical Society*, vol. 7, 1956, pp. 48–50.
Prim, R. C., "Shortest Connection Networks and Some Generalizations," *Bell System Technical Journal*, vol. 37, 1957, pp. 1389–1401.

Chapter **6** *

Linear Programming

PETER V. NORDEN
Program Administrator, Management Sciences,
Data Systems Division, International Business
Machines Corporation, White Plains, New York; and
Professor of Industrial and Management Engineering,
Columbia University, New York, New York

WILLIAM W. WHITE
New York Scientific Center, Data Processing Division,
International Business Machines Corporation, New York, New York;
and Assistant Professor of Mathematical Methods and Operations Research,
Columbia University, New York, New York

Of all the quantitative techniques which have enhanced the practice of management and industrial engineering in the last decade, linear programming has proved to be one of the most widely accepted, applicable, and profitable. Problems as diverse in nature and scale as models of the entire United States economy; personnel assignments; scheduling of milk delivery trucks; and optimal blending of ingredients in sausages, aviation gasoline, and chicken feed have been analyzed by means of linear programming. The common denominator of these and many kindred problems derives from their underlying structure: within broad limits, the relationships describing the flows or combinations of ingredients are linear; likewise, the constraints or restrictions of capacity or availability to which the ingredients are subject are linear.

* The contribution of James E. Kokie, coauthor of the original version of this chapter in the second edition, is gratefully acknowledged.

In practice, this means that the various production systems such as warehouses, machines in a machine shop, and the like are independent, and that the ingredients or elements of the throughput obey laws of strict proportionality. For example, if it takes one pound of pigment and one pound of vehicle to mix one gallon of a certain type of paint, then it will take two pounds of pigment and two pounds of vehicle to mix two gallons of this paint, and so on. The costs of these ingredients must also be describable by linear functions; that is, we must be able to say that two gallons of this paint will cost us exactly twice as much to produce as one gallon. In actuality, in large quantities of purchase, we may be able to avail ourselves of rebates and quantity discounts. However, to use linear programming computations, we must represent the actual (nonlinear) condition by a suitable linear approximation.

Linear programming is a part of the larger area of mathematical programming which includes such topics as integer programming, nonlinear programming, stochastic programming, and others. These techniques are all concerned with the optimal allocation of scarce resources under constraints imposed by technological, economic, or other practical considerations. In practice, a model of the situation of interest is developed, and the interrelationships of all pertinent factors are set down with great precision. Manipulation of this model then allows us to examine the operating and financial implications of the stated conditions and gain insights which are not intuitively obvious if the situation is—as it usually is—at all complex. Computer codes (ready-made programs) have been developed to handle many of the above-mentioned techniques and problems and are commercially available from computer manufacturers, software service firms, and other sources. This chapter discusses the history, fundamental principles, key topics, representative computer codes, and usage of linear programming.

History. V. Riley and S. Gass, in discussing the history of linear programming,[1] have noted that:

> Much of the mathematical theory of linear programming is contained in the theory of linear inequalities and convex sets formulated over the past century. In recent years, the search for techniques to aid administrators in determining optimal programs and feasible alternatives has given impetus to further research . . . .
>
> Foremost among parallel lines of research contributing to the method of linear programming were concepts formulated by von Neumann in 1928 in the application of the minmax theorem to games of strategy, and in the transportation problem posed by Hitchcock in 1941, which also was independently developed by Koopmans in 1947, and in the diet problem stated by Stigler in 1945.
>
> George Stigler, in 1945, studied the problem of determining adequate diet at minimum cost with 77 different foods in terms of nine nutrients contained in them. He found that, for a minimum of $39.93 for the year 1939, optimum diet consisted of wheat flour, cabbage, and dried navy beans. With 1944 prices, navy beans were eliminated in the solution and pork liver added, at a cost of $59.88 for the year.[2]
>
> It was recognized that elements of these problems—forerunners of the linear programming model—dealt with optimization of a linear function subject to linear constraints.
>
> Historically, however, the formulation of the general linear programming problem was primarily the work of George B. Dantzig, Marshall Wood, and their associates of the U.S. Department of the Air Force. In 1947 this group was called upon to investigate the feasibility of applying mathematical and related techniques to military programming and planning problems. In the course of a consideration of possible approaches, Dantzig was led to propose "that interrelations between activities of a large organization be viewed as a linear programming type model and the optimizing program determined by minimizing a linear objective function."
>
> In order to develop and extend this idea further, the Air Force organized a research group under the title of Project SCOOP (Scientific Computation of Optimal Programs).[3]

[1] V. Riley and S. Gass, *Linear Programming and Associated Techniques*, The Johns Hopkins Press, Baltimore, 1958, p. 3.
[2] Thomas L. Saaty, *Mathematical Methods of Operations Research*, McGraw-Hill Book Company, New York, 1959.
[3] Riley and Gass, *op. cit.*, pp. 3–4.

The outstanding contribution of this product was the simplex method developed by George Dantzig and its application to the solution of large linear programming problems on the UNIVAC and SEAC. Sufficient interest was stimulated in the field so that activity in research and applications has become widespread. The useful simplicity of these ideas has undoubtedly contributed to the outcome.[4]

Applications. The types of problems to which linear programming computations are being applied are very numerous and growing every day. However, the following listing of typical applications, given by Robert W. Metzger,[5] is representative:

1. *Product Allocation:* with a number of jobs that can run on a number of different machines, it is possible to determine how to best allocate the work to the machines so as to minimize either the total time or total cost to produce the entire work load.

2. *Distribution and Shipping:* with a product demand at various locations and a supply of products at several warehouses, it is possible to determine which warehouse should ship how much product to which customer so that the total distribution costs are a minimum.

3. *Market Research* (specifically locating factories, warehouses, and outlets): it is possible to determine the best of several possible warehouses, or factory or outlet locations, from various facts about each location.

4. *Job and Salary Evaluation:* here mathematical programming is used in place of multiple correlation analysis to determine the relative weights of the factors considered. This applies to salary and executive-type jobs. A similar analysis can be applied to any testing situation to give a better overall evaluation.

5. *Blending:* applied to blending oils, gasolines, alloy elements, etc. It is possible to determine either how to blend available ingredients or what ingredients to obtain to meet at minimum cost a specific end-product demand.

6. *Material Handling:* one of the newer areas of application. This presents an approach that can increase hand or nonautomated material handling utilization upward to 80%.

7. *Production Planning:* it is possible to develop the lowest cost producing plan, starting with a sales forecast, available plant capacity, and the tangible cost factors.

Methods of Computation. General linear programming problems can be solved by either manual or computer means. The algorithms (schemes for computation) are the same, but computers are indicated when the size of the problem to be solved becomes large. This is discussed in greater detail in the latter part of this chapter. Metzger illustrates the following techniques:[6]

1. Distribution methods
    a. Stepping-stone method
    b. Modified distribution method
    c. Vogel's approximation method
2. Simplex method
3. Approximation methods

These are techniques that lend themselves to hand computation. The next two parts of this chapter treat the concepts involved in solving problems of this type.

A tendency currently is to treat problems of greater and greater complexity. Multiple-objective functions, partitioning and decomposition of general linear programming problems, nonlinear programming, and dynamic programming are examples of this. In practical applications, more and more use is made of computers for both the complex and simpler problem solutions. The structure of the algorithms involved is of primary interest to the mathematician but beyond the scope of this material. Limitations of space prevent "walking" through a typical problem, but for interested readers, this type of example can be found in most of the references listed at the end of the chapter.

---

[4] Saaty, *op. cit.*
[5] Robert W. Metzger, *Elementary Mathematical Programming*, John Wiley & Sons, Inc., New York, 1958, p. 3.
[6] *Ibid.*, pp. 3–4.

HEURISTIC APPROACH TO LINEAR PROGRAMMING

To gain an intuitive grasp of linear programming, let us study a simple example and follow it through to completion. Forgoing mathematical rigor, our approach will be to state three postulates and from these develop a problem employing one of the basic theorems of linear programming. The problem will be two-dimensional so that it can be illustrated graphically.

Constraints. Linear programming deals with a basic economic problem: the allocation of limited resources to a given end or objective. It is customary to formulate the objective in economic terms: maximize profits or revenue or minimize costs. More recent developments involve more complex end objectives such as minimizing risk or partitioning large problems into subproblems.

Depending on the objective which is to be optimized (in the sense of a maximum or minimum), the problem has certain conditions or constraints placed upon it. These fall into three broad categories:

1. Restrictions limiting the usage of the resources singly or in groups
2. Conditions which the totality of resources must fulfill
3. Limitations defining interrelationships among the resources or groups of resources

These constraints are stated as equalities or inequalities in the formulation of the problem.

Equalities and Inequalities. Suppose we wish to produce a meat mixture which contains two nutrients: vitamin A and protein. The statement, "The final meat mixture must contain exactly 200 grams of vitamin A," states a strict equality: vitamin A = 200 grams.

In Figure 6-1, this relationship is shown by the line A. This line consists of all points which contain exactly 200 grams of vitamin A. However, the statement, "The final meat mixture must contain at least 200 grams of vitamin A," is an inequality: vitamin A $\geq$ 200 grams.

In Figure 6-2, this relationship is represented by the line A and the shading to the right of this line. This entire area consists of all points which have at least 200 grams of vitamin A.

The statement, "The final meat mixture must contain at most 200 grams of vitamin A," is also an inequality. In this case, the shaded portion would fall between the line A and the edge of the graph at 0 grams of vitamin A. Similarly, the restriction, "The final meat mixture must contain exactly 1.5 pounds of protein," is a strict equality and is represented by the line P in Figure 6-1. The inequality, "The final meat mixture must contain at least 1.5 pounds of protein," is shown by the striped area in Figure 6-2.

Permissible Region. An examination of Figure 6-2 will show that there is an area (doubly shaded) which fulfills the two separate conditions simultaneously; that is, this region has all those points which have at least 200 grams of vitamin A and at least

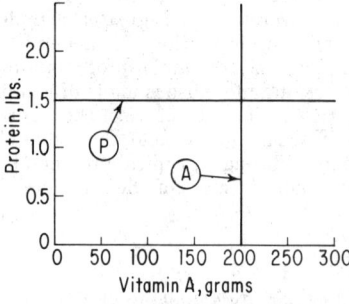

FIG. 6-1. Linear constraints that are strict equalities.

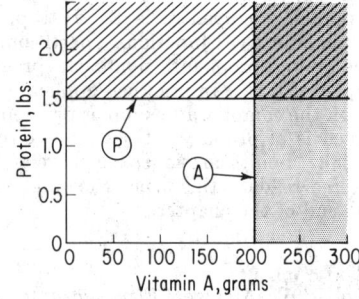

FIG. 6-2. Linear constraints that are inequalities.

1.5 pounds of protein. When considering both the restrictions simultaneously, this is the valid region of interest. This area is called the permissible region. On the other hand, in Figure 6-1 the permissible region is the point of intersection of the two lines. Only at this one point are both equalities satisfied.

Linear programming deals with constraints which can be represented by straight lines or enclosed spaces (the permissible regions) with straight-line boundaries; that is, the constraints are linear. A properly formulated problem with linear constraints will form a permissible region which has the following characteristics:

1. The boundaries are straight lines.
2. There are corners at the intersection of the straight-line boundaries.
3. The region is convex; that is, it is enclosed in such a way that, once leaving the region on a straight-line path, one will never reenter it.

Sausage-blending Problem. Let us now look at a very simple example of linear programming, in which only two restrictions are placed on the problem, and there are only three variables to consider. The problem to be examined is a simplified blending problem in which a sausage mixture is to be made from three ingredients: beef, pork, and mutton. It is usual in blending problems to have an input/output restriction on the total weight of the finished blend. To keep the example simple, this restriction is not imposed on the problem.

Each of these three ingredients can be analyzed into its protein content and its vitamin A content. In the case of beef, the protein content is 50 percent or 0.50 pound of protein in 1 pound of beef. Similarly, 1 pound of beef also contains 40 grams of vitamin A. This same type of analysis is carried out for the other two ingredients and is presented in tabular form in Figure 6-3. Each of the meat ingredients costs 10 cents per pound.

The restrictions on the problem are that the final mixture is to contain at least 1.5 pounds of protein and at least 200 grams of vitamin A. The final mixture is to be the least-cost formulation that meets these restrictions.

Three Postulates. We can accept the following three postulates as proved statements and use them in the solution of our problem:

1. The greatest number of ingredients we will use will be equal to the number of restrictions in the problem. In our case, with two restrictions, two meats will be used at most, despite the fact that we may be able to select from a very large number.
2. If at any time we think we have an optimum solution to the problem, all other meat ingredients will be examined to see whether we would realize a net gain by their use. If a meat ingredient exists which offers a net gain, the present solution is not optimum, and this ingredient will be incorporated into the blend.
3. Negative amounts of meat cannot be used. This, in effect, states that the permissible region contains only positive amounts of the meat ingredients.

Geometric Solution to the Problem. To solve this problem by graphic means, let us begin by ignoring the third ingredient, mutton. We can do this because we know that at most only two ingredients will be used. We can then set up a graph (Figure 6-4) which has as one axis the number of pounds of beef and as the other axis the number of pounds of pork. Now we fill in the graph by drawing the vitamin A restriction and the protein restriction.

Vitamin A Restriction. The vitamin A restriction tells us that there are to be at least 200 grams of vitamin A in the final mixture. If our final formula were to use no beef, just pork, we would have to use at least 4 pounds of pork to meet the vitamin A

	Beef	Pork	Mutton
Protein	.50 lbs.	.30 lbs.	.35 lbs.
Vitamin A	40 g.	50 g.	45 g.
Cost	10 ¢	10 ¢	10 ¢

FIG. 6-3. Ingredients in the sausage-blending problem.

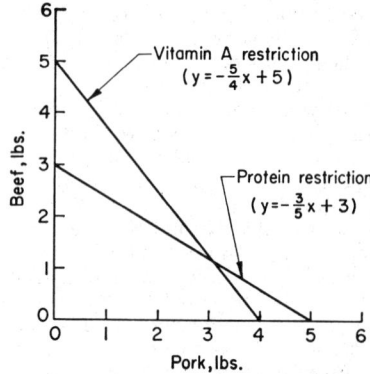

Fig. 6-4.   Vitamin and protein restrictions.

restriction.   Similarly, if no pork is used, at least 5 pounds of beef must be used,
These two facts give the two points (0,5) and (4,0) which we can use to draw a straight
line.   Notice that this line really represents the strict equality, namely, that we use
exactly 200 grams of vitamin A in the final blend.   Because we are dealing with
inequalities, the straight line we have drawn and the entire region to the right of the
line satisfy the inequality of the vitamin A restriction.

Protein Restriction.   We can find the straight line corresponding to the restriction
on protein in a similar way.   If we use no beef in our final blend, we must use at least
5 pounds of pork.   On the other hand, if we use no pork, we can make the blend with a
minimum of 3 pounds of beef.

If the two points (0,3) and (5,0) are plotted, we get the second straight line shown on
the graph.   Again, not only the line but also the entire region to the right of the line is
the permissible region, because it will satisfy the requirement of at least 1.5 pounds of
protein in the final meat mixture.

Now let us consider these restrictions simultaneously, as in Figure 6-5.   For
the problem as a whole (for the final mixture), both these restrictions have to be
satisfied at the same time.   This means that the answer can:

1. Lie to the right of both lines
2. Lie on one line and be to the right of the other
3. Lie at the intersection of the two lines

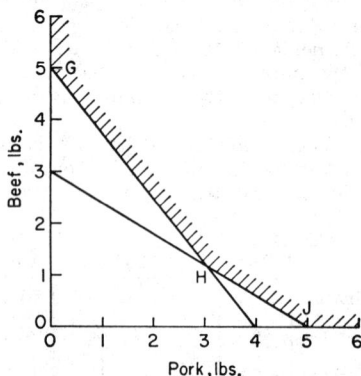

Fig. 6-5.   Permissible region—the only area which (1) has no negative values for
ingredients and (2) fulfills the restrictions.

All this area is the permissible region and is the shaded portion in the graph.   In our two-dimensional case, this shaded section is a convex polygon even though some of its sides are at infinity.   In the general case, the permissible region is a convex polyhedron.

A Basic Theorem of Linear Programming.   One of the basic theorems of linear programming tells us that in maximization- or minimization-type problems, the solution takes on its optimum value at one of the corners of the convex polyhedron.   For our problem, this means that the optimum solution to the problem (the lowest cost blend) will be located at corners *G*, *H*, or *J*.   If it should turn out that the optimum solution occurs at two adjacent corners, we shall assume that the optimum solution occurs at all points along the line joining the two adjacent corners.   In practice, though, this situation is extremely unlikely to occur in blending problems.   The important point is that the optimum solution will not occur inside the shaded area but will instead be at a corner, or in rare circumstances, along an edge.   This means that to solve the problem, we need not look at the infinite number of combinations of amounts of beef and amounts of pork that satisfy the restrictions; instead, we need look only at the three corners and pick that corner which yields the lowest cost.   In choosing the corners to evaluate, it might happen that, after we evaluate one corner, the next corner we pick may have an even higher cost.   If we had some systematic procedure for finding corners which have progressively lower costs, we could eliminate a large amount of unnecessary work.   The simplex method of linear programming is a strict mathematical method which allows us to go from one corner to another adjacent corner in such a way that we never backtrack but instead attempt to decrease the cost, or at least not increase the cost, at each step.

Let us evaluate the three corners in our problem:

Corner G.   At corner *G*, we use no pork and 5 pounds of beef.   Beef cost is 10 cents per pound.   Hence, total cost at *G* is 50 cents.

Corner J.   At corner *J*, we use no beef and 5 pounds of pork.   Pork cost is 10 cents per pound.   Hence, total cost at *J* is 50 cents.

Corner H.   At corner *H*, we use a combination of beef and pork.   To find out how much beef and pork to use, we must find what the point *H* is in terms of pounds of beef and pounds of pork.   We can do this by solving the two equations for the point of intersection.   Each equation is of the form $y = ax + b$ or

$$y = -\frac{y \text{ axis intercept}}{x \text{ axis intercept}} x + y \text{ axis intercept}$$

This gives the following equations:

$$y = -\tfrac{5}{4}x + 5 \qquad \text{vitamin A restriction}$$
$$y = -\tfrac{3}{5}x + 3 \qquad \text{protein restriction}$$

Solving these two simultaneous equations, we get the point $y = 1\tfrac{5}{13}$, and $x = 4\tfrac{0}{13}$. Translating this into pounds of meat, we get

$$
\begin{aligned}
&3.0769 \text{ pounds of pork}\\
&\underline{1.1539 \text{ pounds of beef}}\\
&4.2308 \text{ pounds of meat, at 10 cents per pound}
\end{aligned}
$$

and total cost is 42.308 cents.   Because this cost is lower than the total cost at the other two corners, the point *H* is the optimum solution to the problem.

Economic Interpretations of the Problem.   We limited ourselves to two ingredients in our graphical treatment of the problem and found that both were used.   If these had been the only two meat ingredients from which to choose, the restrictions on the problem and the market prices were such that the use of both ingredients was indicated.

If one or another of the prices had been too high, only one ingredient would have been used.   Similarly, if the price of one or another of the meat ingredients were to drop enough, the solution would again use only one ingredient.   Thus, when the final optimum solution is obtained, there exists an economic cost range for each meat ingredient used. If the open market cost goes out of this range for this ingredient, all other things remaining unchanged, the problem will have to be recomputed. A price increase will cause less of the ingredient to be used. A price decrease will allow more of that ingredient in the sausage blend.

What we have now found is that, within the restrictions of this problem, beef and pork were economical items to buy on the open market at the prices shown. However, because we did not include mutton in our graphical solution, we do not know if mutton would be more economical to use. Now we shall attempt to derive this information.

Imputed Costs. The aforementioned solution indicates that both beef and pork were economical at their current prices. Because both ingredients were used, the price of beef and the price of pork were considered "fair." Because what we are really buying is an amount of protein and an amount of vitamin A, our solution tells us that the prices are fair for the amounts of protein and vitamin A that the meat ingredients contain. From this fact, we can work backward and impute a cost to the protein and a cost to the vitamin A. The following two relationships hold:

$$0.5 \text{ pound protein} + 40 \text{ grams vitamin A} = 10 \text{ cents}$$
$$0.3 \text{ pound protein} + 50 \text{ grams vitamin A} = 10 \text{ cents}$$

We can then solve these two equations to get an idea of what protein costs and what vitamin A costs. Solving this set of simultaneous equations, we get

Imputed costs: 1 gram of vitamin A = 0.15385 cent
                   1 pound protein   = 7.6923 cents

Because our final mixture contains 200 grams of vitamin A and 1.5 pounds of protein, we can calculate the final cost of the mixture on this basis and find that we get the same total cost as before, 42.308 cents. Furthermore, we see that, based on the imputed costs of the nutrients, our two meat ingredients should cost 10 cents per pound. That is, based on the imputed costs of the nutrients that we have found, the prices of beef and pork are fair prices. We could now go back and consider any number of ingredients to determine if they should be used instead of beef and pork, and our rule will be very simple.

Based on the imputed prices of the nutrients in our present solution, find what should be the cost of these new ingredients. If their cost on the open market is lower, they represent a bargain and should be used instead. If their price on the open market is higher, they are uneconomical to use. Let us examine mutton in this light.

Shadow Prices. Based on the imputed costs when beef and pork were used, mutton should cost 0.35 (imputed cost of protein) + 45 (imputed cost of vitamin A). Solving this, we find that mutton should cost only 9.6155 cents per pound on the open market. Hence, because it costs 10 cents a pound, mutton is uneconomical and should not be used in the final formula. The difference between what it should cost based on the imputed prices and what it does cost represents the net penalty we would have to pay if we were to use 1 pound of this meat ingredient in our final blend.

$$9.6155 - 10.0000 = -0.3845 \text{ cents per pound}$$
$$= \text{cost penalty per pound of mutton used in final mixture}$$

This penalty is often referred to as the "shadow price" of the ingredient. If the shadow price is positive (indicates a net gain per unit used), it will be used in the blend.

The simplex method of linear programming finds a workable solution to the problem in terms of m ingredients if there are m restrictions. When we have a feasible solution, the method is an iterative method which works with this feasible solution (not necessarily optimum) and proceeds systematically to another feasible solution with at least no higher cost. We are then at a corner of the convex polyhedron (feasible region) and want to know two things:
   1. Is this the optimum solution?
   2. To which adjacent corner shall we proceed to decrease or at least not increase the cost?

The simplex procedure provides answers to both questions. We are not at an optimum solution if there exists some ingredient not presently being used which has a shadow price greater than zero. Its use would contribute a net gain. If we are not at an optimum solution, the adjacent corner to which we shall go next involves that unused ingredient which has the largest positive shadow price.

The Dual Problem of Linear Programming. In many problems in linear programming, the determination of the imputed costs of items such as the nutrients is a very important question. When the problem is reformulated in such a way that our final answers are to be not how much to use of the ingredients, but rather what costs to impute to the nutrients, then we are dealing with another aspect of the same problem. This aspect is called the dual problem. Every problem in linear programming has its dual problem. Furthermore, whenever we solve the one problem (the primal problem), we automatically solve the other problem (the dual problem). From a study of the answers of the one problem, we can work backward to the answers of the other problem.

The following symbols are often used in the description of linear programming problems in the literature:

Nomenclature

x variables	Solution to the primal problem (the amounts of each of the ingredients to use)
w variables	Solution to the dual problem (the imputed costs to be assigned to the nutrients)
c	Cost of an ingredient as determined on the open market
z	Cost of an ingredient as determined by the imputed costs of its nutrients
$(z - c)$	Shadow price or cost penalty per unit associated with an ingredient not used in the solution (The shadow price of an ingredient used in the solution is always zero.)

From our study of the shadow price of mutton, we can say two things:

1. We will pay a penalty of 0.3845 cent per pound for each pound of mutton we use in the final mixture in place of some other ingredient.
2. Because the shadow price of mutton is negative, it is uneconomical to use. However, if its price on the open market drops by 0.3845 cent per pound, we will begin to consider it if none of the other prices changes.

The shadow price has two interpretations:

1. The penalty for using one unit of this ingredient in the final answer (the shadow price is always per unit)
2. The change in cost that must occur for this ingredient (other costs remaining unchanged) before the ingredient will become economical

MATHEMATICAL STATEMENT OF LINEAR PROGRAMMING

For reasons of accuracy, completeness, uniformity, and convenience, linear programming problems are commonly formulated in mathematical notation. In this discussion, such notation is illustrated, and the example just discussed is cast in this form.

The treatment will present the problem in two ways—the horizontal approach which stresses the algebraic relationships of the row equations and the vertical approach which is in terms of columns and matrix relationships.

The Horizontal Approach. The general linear programming problem is to find a solution to the set of equations

$$a_{11}x_1 + a_{12}x_2 + a_{13}x_3 + \cdots + a_{1n}x_n = b_1$$
$$a_{21}x_1 + a_{22}x_2 + a_{23}x_3 + \cdots + a_{2n}x_n = b_2$$
$$\cdots\cdots\cdots\cdots\cdots\cdots\cdots\cdots\cdots\cdots$$
$$a_{1m}x_1 + a_{m2}x_2 + a_{m3}x_3 + \cdots + a_{mn}x_n = b_m$$

where all x's must be nonnegative and each of the m equations represents a linear constraint placed upon the problem. We are interested in the optimum (maximum or minimum) solution which is given as

$$c_1x_1 + c_2x_2 + c_3x_3 + \cdots + c_nx_n = \text{optimum}$$

Note that all the rows are expressed as equalities. This means that if our original problem contained inequalities, these must be changed to equalities by the insertion of dummy or slack variables. The values of the x variables are the solution, and we know that at most only m of the x variables will be greater than zero. Our example will illustrate these points.

Meat-blending Example. Our problem can be stated as follows:

$$(0.50)(\text{lb beef}) + (0.30)(\text{lb pork}) + (0.35)(\text{lb mutton}) \geq 1.5 \text{ lb protein}$$
$$(40)(\text{lb beef}) + (50)(\text{lb pork}) + (45)(\text{lb mutton}) \geq 200 \text{ g vitamin A}$$

and is subject to the fact that we want to minimize the cost c

$$(10 \text{ cents})(\text{lb beef}) + (10 \text{ cents})(\text{lb pork}) + (10 \text{ cents})(\text{lb mutton})$$

To simplify matters call

$$\text{lb beef} = x_1$$
$$\text{lb pork} = x_2$$
$$\text{lb mutton} = x_3$$

Then we have
$$0.50x_1 + \quad 0.30x_2 + \quad 0.35x_3 \geq 1.5 \text{ lb protein}$$
$$40x_1 + \quad 50x_2 + \quad 45x_3 \geq 200 \text{ g vitamin A}$$
and
$$10 \text{ cents } x_1 + 10 \text{ cents } x_2 + 10 \text{ cents } x_3 = \min$$

Slack Variables. Because the linear programming approach works with equalities, we have to change the inequalities to equalities. An inequality means we have leeway or freedom; that is, the amount of vitamin A can be 200 grams or more. The phrase "or more" is the freedom explicit in an inequality. To remove this freedom, we must introduce a new variable which takes up the slack in the inequality. For each inequality in our example, we can introduce a slack variable x_4 and x_5 (or s_1 and s_2, if you prefer). The inequalities can now be written as equations as follows:

$$0.50x_1 + 0.30x_2 + 0.35x_3 - 1x_4 = 1.5$$
$$40x_1 + \quad 50x_2 + \quad 45x_3 - 1x_5 = 200$$

Take the vitamin A equation for an example. When x_1, x_2, and x_3 take on values such that the mixture has exactly 200 grams of vitamin A, then x_5 will be zero, and there will be no slack or freedom in the equations. However, what if the total amount of vitamin A is greater than 200 grams (say 300 grams)? Then x_5 will no longer be zero but will be just large enough (100 grams) so that when subtracted from the amount of vitamin A used (300 grams − 100 grams) the result will be exactly 200 grams—and the equation holds.

The next question to consider is what cost to assign to the slack. In our case, the slack contributes nothing but simply indicates an oversupply of protein or vitamin A. Because its contributory effect is nothing, it is common to assign a zero cost to slack. The objective we wish to minimize then becomes

$$10 \text{ cents } x_1 + 10 \text{ cents } x_2 + 10 \text{ cents } x_3 + 0x_4 + 0x_5 = \min$$

(*Note:* Slack can be additive slack or subtractive slack. When the constraints are in the form of "at least" statements, the slack is subtractive; when the constraints are in the form of "at most" statements, the slack is additive.)

Artificial Variables. There is one more demand that linear programming makes: we must give it a good start—the mathematics will then carry on from there. The common way that this is done is to introduce an artificial variable[7] for each row in the problem.

For our example, we would have the following array:

$$0.50x_1 + 0.30x_2 + 0.35x_3 - 1x_4 \qquad + 1x_6 \qquad = 1.5$$
$$40x_1 + \ 50x_2 + \ 45x_3 \qquad - 1x_5 \qquad + 1x_7 = 200$$

We know that at any step in a linear programming problem, at most two of the variables are nonzero. Considering the slack variables and the artificial variables as ingredients also, we can give the problem a good start by acting as though the two artificial variables are the only two variables being used. The variable x_6 will be used in an amount of 1.5 pounds and the variable x_7 will be used in an amount of 200 grams.

But these variables are artificial constructs and we do not want them to appear in our final optimum solution. To ensure this, the cost assigned to each of the artificial variables is made prohibitively high—symbolically shown as costing an infinite amount. The linear programming approach starts with these two ingredients costing an infinite amount. The objective of minimizing the cost has not yet been achieved. However, these two ingredients represent a starting basis from which we can work to find a new basis. All variables—including the slack variables—are considered as candidates in the formation of the new basis. The shadow prices of all nonbasis variables are calculated, and the variable with the largest net gain per unit used $(z - c)$ is selected as becoming part of the new basis. If a new one is selected, one of the old ones is deleted; the process of determining which of the new is to enter and which of the old is to leave the basis is often called "pivot selection."

Pivot selection has an interesting geometric interpretation. When we have established any basis, we are located at one corner of our permissible region. We know that there are a number of corners to which we can proceed next. To choose from among these corners, we need a method of pivoting on our corner to get to a definite one of the new corners. The pivot selection stratagem chooses one corner from the many corners to which one can proceed. This corner will always be an adjacent extreme point. Once there, the entire sequence of calculations is repeated. This procedure is flow-charted later in the chapter.

The Vertical Approach. Referring to Figure 6-3, we can imagine each of the three columns—beef, pork, and mutton—being multiplied by a number which tells the number of pounds of each of the meat ingredients to use. This can be arranged as follows:

$$x_1 \begin{bmatrix} 0.50 \\ 40 \\ 10\cancel{c} \end{bmatrix} + x_2 \begin{bmatrix} 0.30 \\ 50 \\ 10\cancel{c} \end{bmatrix} + x_3 \begin{bmatrix} 0.35 \\ 35 \\ 10\cancel{c} \end{bmatrix} \begin{matrix} \geq \\ \geq \\ = \end{matrix} \begin{bmatrix} 1.5 \\ 200 \\ min \end{bmatrix}$$

As before, we introduce slack variables to make the inequalities into equalities and artificial variables to give the solution a starting basis as shown below:

$$x_1 \begin{bmatrix} 0.50 \\ 40 \\ 10\cancel{c} \end{bmatrix} + x_2 \begin{bmatrix} 0.30 \\ 50 \\ 10\cancel{c} \end{bmatrix} + x_3 \begin{bmatrix} 0.35 \\ 45 \\ 10\cancel{c} \end{bmatrix} + x_4 \begin{bmatrix} -1 \\ 0 \\ 0 \end{bmatrix} + x_5 \begin{bmatrix} 0 \\ -1 \\ 0 \end{bmatrix}$$

↖ ↗
Slack
variables

$$+ x_6 \begin{bmatrix} 1 \\ 0 \\ \infty \end{bmatrix} + x_7 \begin{bmatrix} 0 \\ 1 \\ \infty \end{bmatrix} = \begin{bmatrix} 1.5 \\ 200 \\ min \end{bmatrix}$$

↖ ↗
Artificial
variables

[7] If the slack variable for a particular row is positive, an artificial variable is not added to that row.

Matrix Notation. The columns can be compressed into a very compact notation:

$$\begin{bmatrix} 10 & 10 & 10 & 0 & 0 & \infty & \infty \\ 0.50 & 0.30 & 0.35 & -1 & 0 & +1 & 0 \\ 40 & 50 & 45 & 0 & -1 & 0 & +1 \end{bmatrix} \begin{bmatrix} x_1 \\ x_2 \\ x_3 \\ x_4 \\ x_5 \\ x_6 \\ x_7 \end{bmatrix} = \begin{bmatrix} \text{min} \\ 1.5 \\ 200 \end{bmatrix}$$

Note that the costs have been moved to the top and made a separate row by themselves. In matrix notation, the problem is stated tersely as follows:

$$cx = \text{min}$$
$$Ax = b$$

where c = cost row

x = column of x variables

b = column of restraint values

A = tabular array consisting of the data elements (except costs)

A is a matrix consisting of m rows (in our case, two) and n columns (in our case, seven). At any given step in the calculations, our linear programming problem always has a basis consisting of two of the seven columns. In our example, the starting basis consists of the last two columns of the matrix A. These last two columns have an interesting property—they form a submatrix (2 × 2) which has +1's on the main diagonal and zeros everywhere else. This is called a unit matrix and is denoted by I.

Initially, the starting basis is formed from such a unit matrix to begin the problem.

Dual Problems. Recall now that with every primal problem, there is associated the dual problem. The solution to one yields the solution to the other. In the example just considered, columns 4 and 5 also form a diagonal matrix except that now −1's appear on the diagonal. This type of matrix is called a negative unit matrix and is symbolically called −I.

If a linear programming problem contains a negative unit matrix in its initial formulation, each of whose columns is at a zero cost, the solution of the primal problem gives the values of the x variables, and the shadow prices of the columns of the negative unit matrix are the values of the dual variables (the w variables).

It sometimes happens that by recasting a problem in terms of its dual problem:

1. It may have fewer rows.
2. It may be easier to solve.
3. It may be faster to solve.

One such case is the transportation problem, which is faster to solve and more compact when recast in a special tabular form.

THE TRANSPORTATION PROBLEM

Consider the case of a shipper who has a number of plants and warehouses distributed throughout the country. For a given month, each warehouse has a demand for a given number of units of the shipper's product. The plants have the product available. The problem is for the shipper to fulfill the demand by shipping from plants to warehouses in such a way as to minimize the freight costs. By reference to freight rate tables, a shipping charge can be associated with each route from a plant to a warehouse.

The transportation problem makes two basic assumptions:

1. The freight costs are linear (it costs twenty times as much to ship twenty times the number of units—there are no freight breaks).
2. The total supply equals the total demand.

Although both the restrictions may seem severe, it is possible to relax them somewhat by slight changes in the formulation of the problem. To arrive at a solution, the information for the problem is arranged in a tabular fashion. Figure 6-6 shows this matrix layout for a simple case of three plants and four warehouses. The plants are represented by the columns and the warehouses are represented by the rows.

	P₁	P₂	P₃	
A / D	4	5	6	
W₁	3	8	4	1000
W₂	3	12	9	18
W₃	4	15	4	6
W₄	5	17	4	3

Fig. 6-6. Cost matrix.

The top row contains the availabilities at each of the plants P_1, P_2, and P_3; the demands at each of the warehouses W_1, W_2, W_3, and W_4 are represented by the first column in the matrix. Note that the sum of the availabilities equals the sum of the demands—15 units. The body of the matrix is filled in with the shipping costs per unit; that is, to ship from plant 1 to warehouse 1 costs $8 per unit shipped. In those cases where a route is impossible or not permitted, a prohibitively high cost is assigned to this route. In our example, shipping from plant 3 to warehouse 1 is not allowed; the extremely high cost of $1,000 has been assigned to this route. This forces the mathematical technique to avoid use of this route.

If this problem were formulated as a linear programming problem, it would be a fairly large matrix. If m is the number of origins (the plants) and n is the number of destinations (the warehouses), the linear programming matrix will have $(m + n)$ rows and $(m \times n)$ columns; for our example, the linear programming matrix is 7 rows and 12 columns, in addition to the artificial columns needed to start the problem. An examination of the equivalent linear programming matrix shown in Figure 6-7 reveals some interesting facts.

1. In any column, there are only two nonzero numbers, both of which are $+1$'s. One is for the origin and one for the destination of a given route.
2. One row equation is redundant and may be removed entirely.
3. The greatest number of routes used will be equal to the number of row equations after the redundant one is removed. This means that only $(m + n - 1)$ columns at most will be in the basis. For our example, only six routes will be used.
4. The structure of the matrix is such that, once the redundant equation is removed, the problem can be solved without using division. Technically, the matrix is unimodular. This means that if none of the numbers is a fraction, no fractions will result. The answers will always be integers.

Routes

Costs	8	12	15	17	4	9	4	4	1000	18	6	3	min
P₁	1	1	1	1									4
P₂					1	1	1	1					5
P₃									1	1	1	1	6
W₁	1				1				1				3
W₂		1				1				1			3
W₃			1				1				1		4
W₄				1				1				1	5

Fig. 6-7. Linear programming formulation of the transportation problem.

In this linear programming formulation of the problem, the amount of each column to use represents the number of units to ship by the route represented by the column. However, the linear programming approach is overly general in this case because:

1. There are many zeros in the matrix.
2. The only nonzero numbers appearing are $+1$'s.

The tabular array shown in Figure 6-6 is much more compact and eliminates the zeros. In fact, the cost is used as a tag to show which origin ships to which destination. The notation used in the transportation problem is reversed from the notation of linear programming arrays: m is the number of columns, and n is the number of rows. Both methods need a starting point. In the linear programming treatment, the starting basis is the set of artificial variables, all priced at an extremely high cost. Because the artificial variables will not be used in the final answer, it is wasteful to start with a solution that represents impossible routes. The transportation problem begins with an initial shipping assignment that may be close to the final solution.

The Initial Assignment.   The initial assignment can be obtained in two ways:

1. An initial assignment may be given which may represent a close approximation by a skilled analyst or the traffic department. In other cases, the given assignment may be a previous optimum solution obtained under a different set of freight rates.
2. Some rule may be used to obtain an initial assignment. One commonly used rule is the "northwest corner" rule—a simple, easy-to-apply rule.

In all calculations of the transportation problem, one set of relations must always hold: the sum of the assignment values of any row must equal the number in the stub of the tableau—the number in the leftmost column; the sum of the assignment values in any column must equal the number in the head of the tableau—the number at the top of the column.

Solution by the Stepping-stone Procedure.   Let us now ask if the transportation problem has an optimum solution as shown by the distribution in Figure 6-8. This assignment has a total cost of $87:

$$[(3 \times \$8) + (1 \times \$12) + (2 \times \$9) + (3 \times \$4) + (1 \times \$6) + (5 \times \$3)]$$

Now let us examine the unused routes to see if they would be cheaper to use. The route from plant 2 to warehouse 1 is not used. If one unit were to be shipped this way, the column and row would balance only if the P_1W_1 route and the P_2W_2 route were reduced by one unit each. Then, to restore the balance to the problems, the P_1W_2 route could be increased by one unit. The net result is adding:

Add $4.00 of cost
Subtract   9.00 of cost
Add   12.00 of cost
and   Subtract   8.00 of cost
    $-\$1.00$ of cost

	P_1	P_2	P_3
A / D	4	5	6
W_1	3	3	
W_2	3	1	2
W_3	4		3	1
W_4	5			5

Fig. 6-8.   Initial assignment matrix.

	P_1	P_2	P_3
A / D	4	5	6
W_1	3	3	-1	+
W_2	3	1	2	+
W_3	4	+	3	1
W_4	5	+	+	5

Fig. 6-9.   Matrix showing gain or loss by considering the unused routes.

	P_1	P_2	P_3
A / D	4	5	6
W_1 3	1	2	+
W_2 3	3	+	+
W_3 4	+	3	1
W_4 5	+	+	5

Fig. 6-10. Final assignment matrix.

Using route P_2W_1, we could decrease the cost by \$1 for each unit shipped. In Figure 6-9, we show not only the routes used, but also the gain $(+)$ or loss $(-)$ by considering the unused routes.

Similarly, route P_2W_1 is not used:

$$P_3W_2 \text{ is positive (loss)}$$
$$P_1W_3 \text{ is positive (loss)}$$
$$P_2W_4 \text{ is positive (loss)}$$
and
$$P_1W_4 \text{ is positive (loss)}$$

This last unused route (P_1W_4) requires considerable juggling to ensure that the rows and columns preserve their equality relationships. This involves:

	Add	\$17.00 of cost
	Subtract	3.00
	Add	6.00
	Subtract	4.00
	Add	9.00
and	Subtract	12.00
		+\$13.00

Now we pick the unused route with the largest gain (in our case, there is only one) and use as much as possible of this route. We can ship two units by this route because route P_2W_2 will become zero units if we subtract from it. We end up with the final assignment shown in Figure 6-10.

Again, to see if this is an optimum solution, we evaluate all the unused routes. This time, the unused routes will all be positive, indicating that we have reached an optimum solution. Because we have shipped two units by a route with a gain of \$1 per unit shipped, our total gain has been \$2 and the total cost of shipping is now \$85.

Note that we have solved the problem in one iteration, and this was a relatively simple set of calculations to make. If we had used the regular linear programming approach, we would have needed at least six iterations to remove the artificial columns from the basis. Hence, with less labor and in shorter time, the transportation problem has given us an answer. In addition, this answer is integral because no divisions were performed.

Relaxing the Equality Restriction. Recall that in a transportation problem, total demand must equal total supply. If the real problem does not meet this condition, it is possible to effect this equality by introducing dummy sources of supply (plants or factories) or dummy demand centers (warehouses). The amount of goods at these plants or warehouses, respectively, must, of course, be made equal to the net difference between the real total supply and total demand.

We want now to assure ourselves that the excess supply (or demand) is taken from that source (or assigned to that destination) which will make the greatest contribution to total profit by absorbing the excess production or shortage. This is done by assigning a zero cost to all routes emanating from a dummy plant, or respectively, to all routes leading to a dummy warehouse.

Suppose, for example, that supply exceeds demand. This would call for a dummy warehouse with all zero cost routes leading to it. In the course of computation, these zero cost routes will attract the excess supply from precisely that plant or plants which would incur the greatest incremental cost by shipping to any other (real) destination. The dummy warehouse thus forces the uneconomic plants to be "stuck" with the overage. By analogous reasoning, the most costly real warehouses to ship to must absorb the residual shortage if demand exceeds supply.

NETWORK FLOW

Because of the very special structure of the transportation problem, many special-purpose algorithms have been developed for solving it. One particular method deserves further discussion, for it can be extended to more general problems. This method utilizes a network flow technique as proposed by Ford and Fulkerson. This technique views the transportation problem as a problem of finding the flow of goods over a network. The network consists of points from which flow is to be sent (plants), points to which flow is to be sent (warehouses), and routes over which flow can be sent which connect plants and warehouses. The points are called nodes and the routes are called arcs. The network problem then becomes a problem of sending flow between nodes by way of arcs, subject to the restriction that the flow which leaves each plant node must equal the amount supplied by the plant, and that the flow which arrives at each warehouse node must equal the amount demanded by the warehouse.

Network Flow Problems. As we have seen, transportation problems can be viewed as network problems. However, more general network flow problems can occur. Consider the problem, for instance, of not only having to supply warehouses from plants, but also of having to supply customers from the warehouses. If more than one warehouse can supply the same customer, this sets up an intermediate stage which does not fit in directly with our transportation problem. Or consider a more general physical network structure, such as a railroad network, which includes many points at which the goods may be transshipped, merged with other goods going in the same direction, or split off from other goods coming from the same direction. These situations have the effect of introducing intermediate nodes and connecting arcs into the problem, thus creating a more general network flow problem.

The basic relationship in a network problem derives from the principle of conservation of flow. We can associate with each node a number which is to be the net amount of the goods supplied to the network by that node. An intermediate node has a net supply of zero, and a location with a demand can be represented by a node with a negative supply. The conservation of flow principle then states that, for each node in the network, the total amount sent away from the node minus the total amount sent to the node equals the net amount of supply at the node.

This principle is a direct generalization of the restriction in the transportation problem that the amount of flow which leaves each plant must equal the total amount available to be supplied by the plant, and that the total amount sent to a warehouse must equal the total amount required by the warehouse. In addition, for intermediate points, the total amount arriving must equal the total amount shipped on.

It can be readily seen that the total amount supplied to the network must equal the total amount demanded within the network. Where the real problem restrictions do not meet this requirement, a dummy source of supply or a dummy demand location can be added to the network to achieve this equality, as was done in the transportation problem.

This network flow problem can be formulated as a linear program by associating a variable with each arc, representing the amount of the goods sent over that arc. The cost of that variable would be the unit cost of sending the goods over that arc. There would be a row for each node of the network, representing the conservation of flow restriction at that node. As with the transportation problem, however, it is much more efficient to treat the problem directly. A procedure for doing this has been developed by Ford and Fulkerson, and is called the "out of kilter" method. This algorithm proceeds by rerouting flow in such a fashion that, at each step, flow is sent which either will help in satisfying some node's requirements that are currently unsatisfied or else will maintain the flow arriving at the node but will do so at less cost.

LARGE OR STRUCTURED LINEAR PROGRAMS

Many linear programming problems which originate from real-world problems have a special structure which reflects the original problem. In some cases, this structure results from combining several smaller problems into one interrelated problem. In other cases, some of the problem constraints themselves have special meaning. Techniques have been developed for incorporating many of these properties within the solution procedure itself, thus leading to more efficient means of solving these problems or toward handling large problems that would be intractable if treated directly.

Lower- and Upper-bound Techniques. One of the basic postulates of linear programming has been that none of the variables, such as sausage ingredients, can occur in negative amounts. This implies that the range of values that a variable can take on is between zero and some extremely high value. It happens frequently, however, that the allowed range which a variable may take on in reality is much narrower; that is, it is defined by both a lower bound and an upper bound. This is a common type of problem, and techniques have been devised, therefore, for handling it. For the lower bounds, a very simple reformulation of the problem usually permits them to be restated in such a way that they are considered zero.

Let us consider the case where a variable has a lower-bound constraint which is greater than zero, such as $x_j \geq l_j$, where $l_j > 0$. An example is $x_1 \geq 150$. For each lower-bound constraint where $l_j > 0$, we introduce a new variable y_j, so that $x_j = l_j + y_j$; now every place that x_j appears in the other constraints of the problem, we substitute $(l_j + y_j)$ for x_j, with $y_j \geq 0$. This reduces the computational effort because the lower-bound constraints are eliminated.

The situation is not so simple in the case of upper bounds, because a mere reformulation of the problem will not obviate the need for explicitly writing out the constraint equations. Furthermore, the inclusion of upper bounds means that it may be possible for some of the variables to achieve nonpermitted values, and these restraints must be stated within the matrix.

From a practical point of view, this is a drawback, because explicit inclusion of these upper-bound constraints within the matrix can enlarge the problem considerably, especially if many of the variables are confined by upper bounds. Whereas the addition of columns is a relatively simple matter for the computer and does not increase the running time of the program appreciably, the inclusion of extra row equations for each of the upper-bound conditions significantly lengthens the running time of a problem.

The structure of these upper-bound constraints has made it possible to incorporate the upper-bound restraining equations in a special computational procedure which exploits this structure and yet at the same time obviates the need for enlarging the matrix by the addition of extra row equations.

The principal method devised for solving problems of this sort deals with the upper bounds in an implicit fashion without stating the equations. In this technique, the upper-bound constraints are handled within the original system by merely changing some of the decision rules employed by the standard simplex algorithm.

The transportation problem is often presented as a problem which has many upper bounds placed upon it. In particular, in the case of a plant shipping goods to warehouses, there may be definite capacity restrictions upon the number of units that may be shipped by a certain route. The upper-bound constraints thus arise quite naturally in such problems.

The flow method of solution of the transportation problem is readily adaptable to the upper-bound technique. In particular, once the flow network is constructed, we

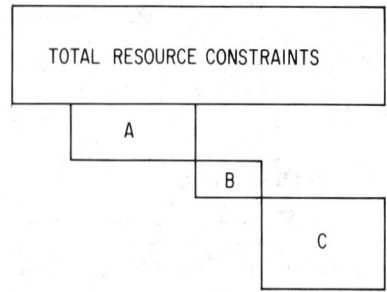

FIG. 6-11.   Diagram of a multisector model.

merely impose capacity limitations upon each of the individual links within the network.   The treatment of such transportation problems is quite simple, and special routines have been devised for handling the transportation problem with capacity limitations.   This problem is generally referred to as the capacitated transportation problem.

Generalized Upper Bounding.   There is a wide class of problems which have restrictions that generalize the concept of upper bounding.   These are problems which constrain the sums of variables instead of the individual variables themselves. Many scheduling and distribution problems are of this form.

Consider, for instance, a production distribution problem in which each variable represents the amount of a particular product produced at a particular plant destined for a particular customer.   Production capacity at the plant may restrict the total amount of production of an individual product.   There would then be a constraint for each plant-product pair requiring that the sum of the amount of the product produced at that plant for all customers should not exceed the capacity for producing that product at that plant.   The number of constraints of this type can easily be larger than the number of other constraints in this problem.

The technique devised for solving this problem treats these constraints implicitly. The decision rules of the simplex method are altered to allow such a treatment in a rather complicated way.

Structured Problems.   Linear programming problems formed by interrelating smaller linear programs usually have a special structure which can be taken advantage of in attempting to solve such problems.   Problems of this type often arise from consideration of multisector or multitime models.   Multisector models are problems in which there is a separate linear program for each sector of the problem, and then there are overall restrictions which constrain the interactions of the sectors.   This occurs in many economic models.   A typical constraint matrix may be diagramed as shown in Figure 6-11.   The three submatrices A, B, and C might describe the technological constraints of, say, production at plants A, B, and C, while the top set of constraints would then describe the total resource constraints on raw materials and finished products for the total enterprise.

Multitime models arise from the consideration of the production or distribution of goods over time.   For such purposes, time is considered in segments.   A common example is the problem of quarterly production plans over a year's time.   A typical constraint matrix may be diagramed as shown in Figure 6-12.   Here, the four sub-

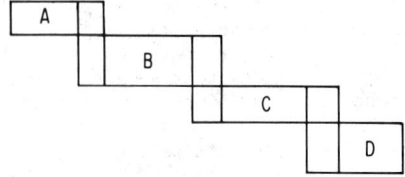

FIG. 6-12.   Diagram of a multitime model.

matrices A, B, C, and D might describe the technological constraints of a plant during quarters A, B, C, and D, while the remaining submatrices would then describe the inventories of raw materials and finished goods held over at the end of one quarter's production for input to the next quarter's production.

Other special structures can occur, although these are the two most prevalent. Mathematically, each of these forms may be converted by defining new variables and relations into the other forms.

The motivation in devising special techniques for solving these problems rather than solving them as a large linear program is threefold:

1. The problem as a whole might be too large to be solved on existing computers.
2. Even though the problem might fit in a computer, it might be solved more efficiently with less possibility of numerical error by taking advantage of the structure.
3. There is something appealing in solving the whole problem by trying to link together solutions obtained by examining the subproblems.

As a result, two classes of algorithms—partitioning and decomposition—have been developed for treating these structured problems.

Partitioning Algorithms. Partitioning algorithms treat structured problems by further specializing the decision rules of the simplex method. They essentially perform the simplex method on the whole problem, but they do so in such a fashion as to consider the subproblems individually, periodically taking into account the linkage between them. How often they consider this linkage depends both upon the strategy of the particular algorithm and upon the nature of the algorithm itself (for instance, primal algorithms must look at this linkage in a different way than dual algorithms).

Decomposition Algorithms. Decomposition algorithms treat structured problems by generating and solving a new linear program. This new linear program is created by solving the subproblems, each solution to a subproblem giving rise to a column of the new problem. The column itself consists of the total contribution of that subproblem solution to the linkage between the subproblems, if that solution were to be used 100 percent. The new problem consists of trying to determine what percentage of the various subproblem solutions should be used to achieve the best overall solution while still satisfying the linkage requirements. The algorithm depends highly on the property that, if one has two valid solutions to a subproblem, then, by utilizing p percent of one solution and 2 percent of the other (where $p + 2 = 100$ percent), the resulting solution is also a valid solution.

The new problem created by decomposition algorithms generally has a great many columns. However, it is not necessary to obtain all columns at the beginning, because the algorithm allows one to generate the columns only as they are needed, by successively solving the subproblems to obtain different solutions at different stages. In fact, some linear programs can be formulated this way directly, by specifying how to generate only the next column needed in the course of the computation. Such programs are known as generalized linear programs, and the procedures for automatically creating these columns are known as column generation procedures. An excellent example of such a problem is the cutting stock problem of Gilmore and Gomory, in which each column corresponds to a pattern of cutting a length of stock into smaller pieces. In this case, the subproblem which one solves to generate a column is not even a linear program. This is one of the main reasons for the formulation of many problems within a decomposition framework: the algorithm will work for any problem for which the subproblems have the property that a solution obtained by utilizing various percentages of valid subproblem solutions is itself a valid subproblem solution when those percentages add to 100 percent.

LINEAR PROGRAMMING COMPUTER CODES

Few linear programming problems encountered in real life are small enough to be solved by manual computation. Most linear programming problems of practical interest require a large number of complex calculations and generally are solved by means of electronic computers because of the speed, accuracy, and memory of computers.

The use of computers provides solutions to linear programming problems in shorter times than could possibly be realized by manual means. The arithmetic computations in a typical linear programming problem would require many man-hours of manual effort even for a small problem. Digital computers, with their high internal computing speeds, often solve these problems in a matter of minutes or even seconds. Furthermore, digital computers are reliable and usually contain self-checking devices which ensure the accuracy of the results at each step of the way. Anyone attempting to undertake a linear programming solution by hand will soon find that it is easy to omit a calculation accidentally at one step or another. A computer will not do this. It is important to remember that a linear programming problem involves a great mass of information which must be retained somewhere. With manual methods, this means keeping all the information before the analyst on paper. As the problems get larger and larger, the amount of information necessary to be presented to the analyst at each step is extremely large. Digital computers, with their vast memory capabilities, permit the storage of either the entire problem or at least major segments of the problem.

Development of Linear Programming Codes

Simplex Method. The first generally available computer programs utilized the simplex method of linear programming. In the simplex method, the entire problem is contained within the memory of a computer. This means that all the data, the constraints, and the costs are held by the computer in a tabular array or matrix form.

At any given step in the simplex method, however, only a portion of the information is actually being used in the solution of the problem. The information pertaining to the solution of the problem is called the basis. In the meat-blending problem, although there might be 500 meat ingredients to be considered, if there are only two restraints placed upon the problem at any time, only two of the ingredients will be used in the basis. All other ingredients will be nonbasis elements and are not of interest at that particular time. Figure 6-13 is a flow diagram of the basic simplex method.

The simplex method, although relatively fast, requires a great deal of memory, because basis and nonbasis information is contained within the computer at the same time. Consequently, other computer codes have been developed which allow the retention of only the essential information necessary to carry on the computation.

Revised Simplex Method. The simplex method has the disadvantage that it has to compute and record many numbers which either are not used at all or are used only in an indirect way. For example, if our particular meat problem contained only two restrictions but had 500 meat ingredients which could go into the final blend, we know that out of the 500, at most only two meat ingredients (the basis) will be used for any one tableau. It is superfluous, therefore, to carry the other information continually.

The revised simplex method allows retention in the computer of only the basis information, or more technically, the inverse of the basis. The computer begins with an artificial basis which is in effect a trial solution to the problem. Next, a new meat ingredient is presented to the computer. The computer calculates its shadow price, and if this meat ingredient is found to have a positive shadow price, the meat ingredient will be incorporated into the basis. At this point, the calculations are performed on only a small number of data elements within the computer, and the new meat ingredient becomes part of the basis. The computer then considers another meat ingredient and calculates its shadow price. If it finds that the shadow price of this ingredient is also positive, it is incorporated into the basis. On the other hand, if the shadow price is negative, this meat ingredient is rejected, and immediately the computer goes to the next meat ingredient.

The revised simplex method considers each of the individual columns or meat ingredients separately, one after the other, calculating the shadow price of each of the columns in turn. When the revised simplex method finally comes to the point where there are no more columns with positive shadow prices, this is a signal that the optimum value has been reached. Figure 6-14 is a flow diagram of the revised simplex method.

Regarding the memory size of the computer, the simplex method will have within the computer all the rows and all the columns of the matrix. The number of rows and columns in the linear programming matrix therefore limits the size of the problem that can be handled by the simplex method. With the revised simplex method, however, because the columns can be stored external to the machine and brought in by reading the external storage medium on which the individual columns are stored, there is virtually no limit to the number of columns that can be handled. The basis information (inverse of the basis) is a square matrix $m \times m$ in size. Thus the limitation in the revised simplex method is the number of rows which are in the formulated linear programming problem.

Product Form of the Inverse. We have seen from the above discussions that the memory capacity of the computer is the prime limitation on the size of the problem that can be handled by a given linear programming routine. To circumvent this problem, an extension of the revised simplex method has been used successfully. This form is known as the product form of the inverse.

In the revised simplex method, we are able to consider the columns as being external to the machine and only the inverse of the basis remaining inside the computer. However, a method exists for decomposing the inverse matrix of the basis into a set of columns. When these columns are multiplied together, their product forms the inverse of the basis. With this method of decomposing the inverse matrix into a series

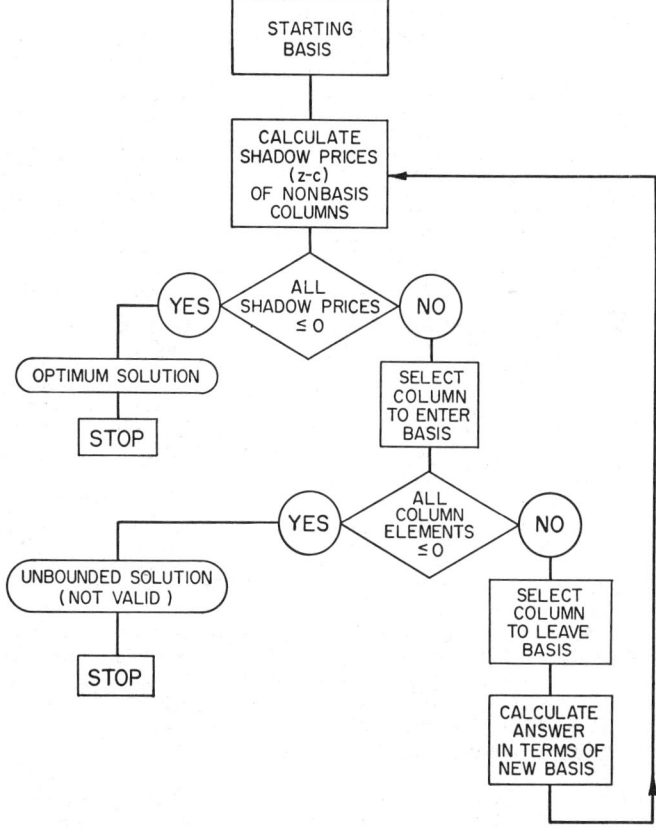

Fig. 6-13. Simplex method flow chart.

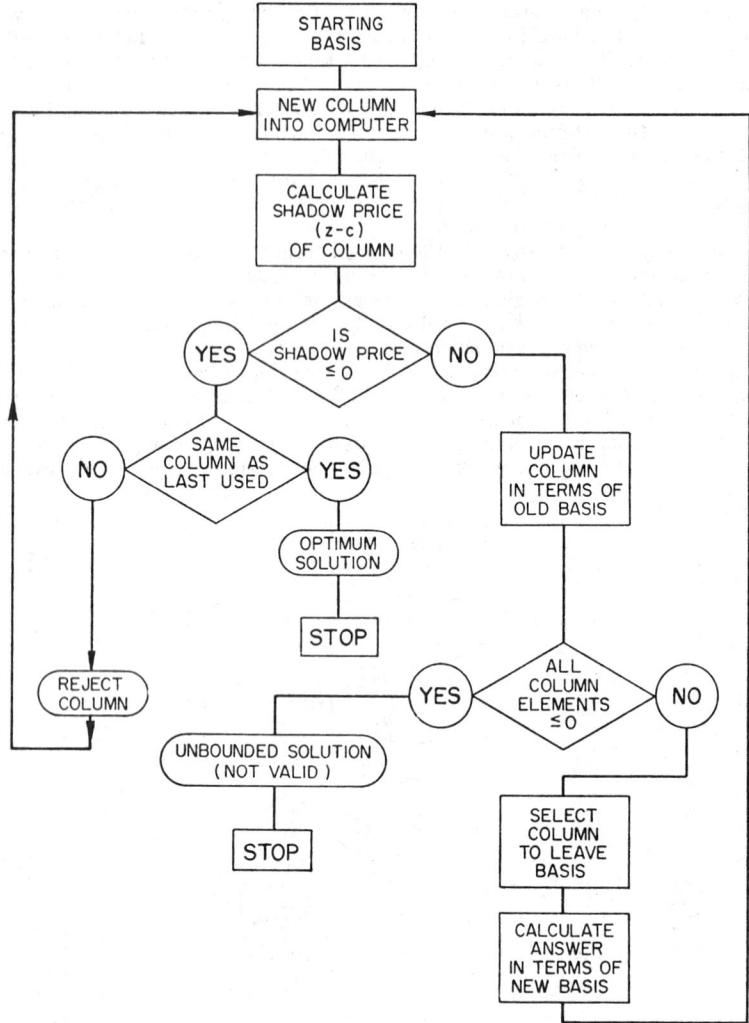

Fig. 6-14.   Revised simplex method flow chart.

of columns, these individual columns can then also be considered external to the machine.

The product form of the inverse needs, within the memory of the computer, only enough working storage to be able to read in the various columns of the inverse in a serial fashion. By means of matrix manipulations, the revised simplex algorithm can be applied to the columns of the entire matrix individually. The method is highly dependent on good input–output devices. The product form of the inverse lends itself very well to medium- or large-scale systems with fast input–output devices.

The Phase I–Phase II Process.   For the simplex method to work, a feasible basis must be supplied. It was seen earlier that this can always be done by introducing artificial variables as necessary. This is automatically done within most computer codes. To obtain a feasible solution in terms of the original problem variables, the artificial variables must be driven to zero.   The way most modern linear programming codes approach this problem is, whenever an artificial variable is at a positive level, to minimize the sum of the artificial variables rather than optimize the original objective function.   As soon as this sum is driven to zero, one proceeds to optimize, using the original objective function (if the sum cannot be driven to zero, the original problem is infeasible).   This process of first getting feasible and then getting optimal is known as the phase I–phase II procedure.

Naturally, if one has an initially feasible basis, one can bypass the phase I portion. This is often not the case, however, and sometimes a considerable amount of work must be done just to get feasible.   This work can easily exceed the amount of time to get optimal, although it often happens that after first getting feasible, it is still a long way to the optimal solution.

Linear Programming Extensions.   Most linear programming codes contain a simplex algorithm for solving linear programs.   However, a number of codes also contain algorithms for analyzing the linear programming solution in more detail.

Ranging.   Ranging is a technique for analyzing the sensitivity of the optimal linear programming solution to changes.   Linear programming codes in which ranging is implemented allow cost ranging, and some allow ranging for the values assigned to the problem requirements or restrictions.   Ranging as applied to individual matrix coefficients is not usually implemented.

Cost ranging gives, for each variable in the problem, the range that its cost may be in and still have the same optimal basic variables.   Should the cost change to another number outside the range, the optimal basis will change.   Thus cost ranging allows one to examine the sensitivity of the cost coefficient.   If the range is small, then only a small change in cost could cause a different solution to result; while if the range is large, the optimal solution is reasonably stable for small fluctuations in costs.

Parametric Linear Programming.   The problem in parametric linear programming is obtaining the accurate and reliable values for the data elements in the body of the matrix, establishing correct costs for the cost rows, and, finally, imposing restraints having definite numerical values.

It is often of interest to examine the behavior of solutions to linear programming problems when any of these values are allowed to vary.   In such cases, we have a problem in parametric linear programming.   There are three areas in which we may permit varying parameters:

1. Information in the body of the matrix
2. Values assigned to the cost of the individual columns
3. Values assigned to the requirements or restrictions imposed upon the problem

Parametric linear programming differs from ranging in that parametric linear programming actually traces out the changes in the problem solution as the parameters are changed outside the range given by ranging.

THE USE OF COMPUTERS FOR LINEAR PROGRAMMING SOLUTIONS

Linear programming has become widely used as a tool in selecting the most profitable or the least costly mode of industrial operations.   However, the efficiency of reaching an optimum is affected by four major factors:

1. Proper formulation of the problem.
2. Length of arithmetic calculations involved.   This is in part a function of the computer routine used.
3. Analysis of the output, especially a postoptimum analysis of the solution.
4. Optimization of the computer routine itself.

The following discussion deals with these four areas.   The proper formulation of the problem and its solution on a computer frequently benefit from the aid of a consultant who has the requisite applications or computer knowledge.   Some computer techniques have become of prominent importance in improving the running of the problem or reducing the running time of the problem, thereby obtaining answers more quickly and at less cost.   Once the solution is obtained, postoptimum analysis can take place to yield additional information without having to solve separate problems.   This can be done by introducing small changes into the problem or allowing the parameters of the problem to vary.

Problem Modeling and Formulation.   Modeling the real-world problem and then formulating it as a linear program is sometimes a tricky business.   There are fine points which can materially affect the efficiency, accuracy, and realism of the model and its solution.

When a nonexpert realizes that his optimization problem may be solvable by linear programming techniques, his most profitable next step could be to contact a computing service organization or an industrial consultant.   In some cases, this type of service may very well be found within his own company.   It is becoming more apparent that a knowledge of mathematics alone is not enough to solve linear programming problems.   It is also necessary to be familiar with the nature of the application involved and to be aware of various shortcut techniques which can be used to obtain the most efficient solution to the problem.

Applications Consultant.   The applications consultant has the knowledge and practical experience of the industry and application and consequently can often provide a better formulation of the problem.   The formulation of a linear programming model is not as simple as it may appear.   An expert will usually devise a model which more closely approximates reality.   At the same time, the consultant can take into account the many interrelationships and constraining equations and reformulate them so that they are more readily handled by the computer routine.   The applications consultant is also aware of the fact that the major restriction placed on the solution of linear programming problems is the number of rows in a matrix, and he may be able to restate the problem so that there will be fewer rows.   Consequently, the problem may run faster or be able to run on a smaller computer.   A good guide rule is that the running time always increases with the cube of the number of rows.

The application consultant may also be familiar with what type of answer to expect. It may, for instance, be possible to reduce the running time of a problem by providing an intuitive guess as to which variables should appear in the final solution and actually force these variables to appear in the final solution.   This technique, known as providing an advanced starting solution, relies on the ability to predetermine a good solution from which one can proceed toward the optimal solution.   His experience may be such that he will be able to guess up to 90 percent of the optimum solution to a particular problem.   In doing so, he may cut down the running time of a problem by as much as 50 percent.

The Computer Consultant.   The computer consultant is interested in the mathematical techniques to be used on this particular problem and the computer routine used.   He will look at ways of combining various row equations or combining various columns to come up with a smaller matrix that can be handled more efficiently by the computer.

If the computer consultant is told the type of information wanted in the final output, he can suggest a routine that will be best suited for this particular problem. A knowledge of the type of computer upon which the problem will be run may lead to specific advice regarding the routine.   Such consideration may come from a knowledge of the type of input and output devices available on the computer.   The computer consultant will also examine the data and the structure of the problem itself.   He can advise on the appropriateness of various algorithms and codes.   For instance, solving a small linear program many times in a semireal time environment calls for a different code than that for solving a large linear program once a month for planning purposes. Furthermore, he may know of some experimental codes which may be of value for a particular application; this is particularly true if the problem is not a standard linear program, but lies in the area of integer or nonlinear programming.

Problem Processing.   In this discussion, we shall look at various strategy considerations in the use of codes for solving linear programming problems.   The basic problem of running a linear programming problem is that of the balance between the size of the matrix and the speed and accuracy of the results one wishes to obtain.

As the size of the matrix grows larger, the running time of the problem increases significantly, and in some cases, the accuracy of the results begins to deteriorate.

Consequently, every effort is made to reduce the size of the matrix. These three factors—size of the matrix, running time of the program, and the accuracy of the answers—form a delicate balance which must be assayed by both the applications consultant and the computer consultant.

Crashing. It has been seen that the simplex method requires a feasible basis to work with. Unless otherwise specified, such a basis is automatically generated by the computer code by the creation of artificial variables as necessary, and the phase I process then attempts to achieve a feasible solution in terms of the real problem variables. Because this can be a lengthy process, the technique of crashing has been devised. This technique attempts to force the artificial variables out of the basis as fast as possible, usually with little regard as to how the incoming variables affect the objective function. By providing the simplex method with a starting basis which more accurately reflects the real problem, crashing can significantly reduce running time.

Scaling. For real problems, it is quite likely that the coefficients of the matrix take on a wide range of values. This can cause numerical difficulties, because the computer must round off fractions. To help ameliorate this problem, the coefficient matrix can often be scaled before starting to solve the problem. This has the effect of modifying the matrix entries so that they are more nearly equal in magnitude. At the end, of course, the answers are converted back to correspond to the original problem data. Scaling is useful for large problems, which sometimes cannot be solved in their original unscaled form.

Matrix Reinversion. To circumvent some of the problems involved in the round-off errors and accuracy problems in linear programming, special routines are often incorporated into the linear programming codes which allow the program to "clean itself up," as it were, by the use of a matrix reinversion program.

This can be accomplished in two ways:

1. After the optimum solution is attained, the computer considers only the basis matrix and reinverts it to obtain some of the original values that were in the matrix. At this point, a cleanup routine takes place which minimizes the round-off error.
2. After a certain number of iterations (for example, every 25 iterations), the computer routine automatically takes whatever basis it has at that time, reinverts it, and adjusts it to eliminate some of the round-off error. This scheme also has the effect of reducing the number of columns stored when the code uses the product form of the inverse.

Matrix reinversion is so useful in decreasing running time and increasing accuracy that most commercially available codes automatically use it. The user himself can sometimes set the number of iterations at which reinversion must occur.

Cycling. Most large, commercially available codes use the revised simplex method with the product form of the inverse. The revised simplex method considers the columns in the order they appear, one after another, in calculating the shadow prices, and stops when it finds the first positive shadow price, for if that variable is to be brought into the basis, it will improve the objective function. The process, then, is one of cycling through the columns, finding shadow prices as one goes. However, it is not necessary to stop at the first column which gives a positive shadow price. One may wish to look for other columns with a positive shadow price and then choose the best column from this set of selected columns to be the incoming column. Most commercially available codes have this provision, which has been shown to be particularly effective when used in conjunction with multiple pricing.

Multiple Pricing. Multiple pricing allows one to concentrate upon trying to put as many good columns into the basis as possible. Instead of selecting a single variable to be the incoming variable, select a small set of variables, each one of which has a positive shadow price. Then successively try to pivot in these variables, as long as they continue to lead to objective function improvement (the shadow prices, of course, change each time a pivot is made). When there is no more possible improvement in the objective function (the shadow prices are nonpositive), the subset of variables has been optimized, and one can proceed with cycling. This technique has been highly effective, particularly on large problems in the early stages of computation.

Solution Output and Analysis.   The computer codes are not only capable of producing the answer to the problem itself, but can also provide at the same time some auxiliary information which is of great interest and value.   Some of this is automatically provided in the output report, and some is provided in supplemental reports.

The Output Report.   The minimum output of most linear programming codes consists of the following:

1. Identification of the variables—either numeric or alphanumeric descriptions.
2. The status of each variable in the final solution—basic or nonbasic; feasible or infeasible; if bounded, whether at upper or lower bound.
3. Certain input information—cost; bounds, if any; constraint term for each row; and the like.
4. Final activity level or the amount of each variable to be used.
5. Shadow cost—marginal cost, dual variable—for each row and column in the problem.   For columns, the shadow cost indicates the marginal amount of change in the objective function if the corresponding variable changes its level by one unit; major changes in level could lead to completely different solutions.   For rows, the shadow price can be interpreted in terms of the penalty which we are paying for a given restriction.
6. Final cost of optimum solution.

The Range Report.   The range report gives the output of ranging, and for most linear programming codes, gives the cost range of each of the variables in the final solution.   This is essentially the upper and lower limits on the cost of each of the ingredients in the final solution.   If an ingredient used in the final solution passes outside the cost range, all other things being equal, a change in the solution will occur.   If the cost of an ingredient exceeds the upper limit cost, this variable will not be used any longer in the same amount as it is being used in the present solution.   If the cost drops below the lower limit, the problem will be reformulated and more of this variable will be used.

This report can be considered as an analysis of the sensitivity of the optimal solution to fluctuations in the costs, and hence as an indication of the stability of the solution.

Some codes also furnish reports on ranging for the constraint values, providing a similar analysis of sensitivity of the solution to fluctuations in the constant terms of the problem restrictions.

Parametric Reports.   Parametric linear programming is a general procedure for changing either the values in the cost row, the constraint values on the right-hand side, or data elements in the body of the matrix.   So far, workable parametric linear programming codes have been devised for only the first two cases, that is, incremental changes to cost and to the right-hand side values.

Parametric linear programming allows one to vary the parameters of the problem over their entire range; for example, in our particular problem of meat blending, we have specified that the final mixture is to contain at least 200 grams of vitamin A.   Parametric linear programming methods automatically scrutinize the entire range from zero grams of vitamin A up to an unlimited number of grams of vitamin A.   The parametric linear programming method gives a different solution at each of the breaking points where a change in the value of the cost data or the right-hand side causes a new optimum solution to be generated.   Because this technique is automatic, it does not require the manual intervention of an operator to introduce the necessary changes into the cost elements or the constraint equation values on the right-hand side.

From the mathematical point of view, there are still a number of problems to be solved in the area of parametric linear programming.   However, from the user's point of view, some important aspects of interest of parametric linear programming have already been incorporated into some linear programming codes for the large-scale systems.

Problem Modification.   In many cases, we are not interested in solving a single linear programming problem, but a whole set of similar linear programs.   Some of these may be straightforward modifications of the original linear program, and some may have major differences.   The basic property used in solving such a series of problems is that the optimal solution to one problem can often be used as a good starting solution for the next problem in the series.

Advanced Starting Solutions. A good starting solution can help considerably in progressing toward an optimal solution. This solution might be chosen either by educated guess or by the solution of a similar problem. In either case, it avoids having to start with artificial variables. This solution is usually inserted into the problem, at which point the simplex algorithm performs a matrix reinversion and then proceeds. In some cases, where the starting solution was a solution to a similar problem and the changes to the new problem are minor, the new problem can be solved starting directly from the inverse of the old problem.

Merging. One can often obtain a good starting solution for a large problem by breaking the large problem up into a set of small problems and finding the best solution to each of the small problems. These small problems can then be merged back into the large problem, and the merged solutions for the small problems can be used as a starting solution for the large problem.

Solution of a Subproblem of the Original Problem. There are a number of ways by which one can break up a large problem into subproblems. One can always formulate the smaller problems directly. However, there are ways in which one can have the whole problem but examine only a portion of it. Two such ways are by using selection lists and curtaining.

1. Selection lists. One method of shortcutting a linear programming solution is to assign a "prohibited" tag to selected columns which, from previous experience, will not appear in the final solution. Because these columns are prohibited, the computer program as a whole can eliminate these from the starting matrix. Such selection lists have long been used to speed up linear programming solutions.

This is an example of intelligent use of the man-machine combination: superior insight of the analyst can so structure a problem that the computer does not blindly examine and manipulate variables which are known ahead of time to have little or no likelihood of appearing in the final solution.

2. Curtaining. In this method, a number of columns are thought to be logical choices for the final basis; other columns are thought to be possible but not likely candidates. The nonlikely candidates are then set behind a "curtain" inside the computer, and the linear programming routine at first is able to consider only those columns which are in front of the curtain. After the problem has obtained an optimum solution in terms of the columns in front of the curtain, the linear programming routine then can "peek" behind the curtain and scrutinize the remaining columns to see whether or not they would improve the optimum solution. The curtaining technique is widely used on large-scale problems, especially those problems in which a set of priorities or a priority ordering scheme can be assigned to a set of columns.

We can now investigate some of the changes in the system which can be made. They can be divided into the following five categories:
1. Cost changes
2. Changes in the constraint equation values
3. Changes in the matrix elements
4. Addition of new column variables
5. Addition of constraint equations

In all cases (with most computer codes), if the changes can be made after an optimum solution has been determined, the reworking of the problem does not require going back to the initial artificial basis. The changes can be considered as simple extensions of the problem itself. In many cases, making an incremental change to the system under study requires only one or two more iterations after the optimum solution has been reached.

Cost Changes. Cost changes are some of the most interesting changes which can be performed. If the costs represent market prices, it is not usual that only one market price will change at a time. Instead, it is usual that all market prices will fluctuate rapidly; the period of time may be measured in terms of hours or weeks. Once an optimum solution is attained, the cost can be varied, changes introduced, and the linear programming routine—by means of a few more iterations—will come up with either the same solution or a different solution in terms of the new set of prices introduced.

Some computer codes allow for changes in the cost data by allowing multiple cost rows to be used in the computer at once. An optimum solution is obtained in terms of one of these cost rows, and then another is obtained in terms of the second cost row, the third in terms of the third cost row, and so on. The provision for multiple cost rows is an important feature of many linear programming codes.

Changes in the Constraint Equation Values. Many linear programming codes have the ability to allow changes to occur to the constraint equation values of the optimum solution. Because the constraint values are usually tabulated on the right-hand side of the linear programming matrix, such changes are many times referred to as changes in the right-hand side.

Some computer codes also have the ability to incorporate a number of right-hand sides. This gives the ability to solve a problem in terms of one set of constra:nts, then in terms of another set of constraints, and so on. The ability to have multiple right-hand sides is an extremely important factor in experimental studies when some developmental work is still taking place on the linear programming model.

Changes in the Matrix Elements. Some linear programming codes also have the ability to change the various elements in the matrix itself. Again, a sensitivity analysis can be performed in terms of the eigenvalues of the final solution matrix, to determine whether or not the answer obtained is reliable and valid.

Addition of New Column Variables. The addition of new columns to the linear programming problem once an optimum has been achieved is simple. In fact, the procedure is very similar to that followed when a revised simplex method is used. After a column has been added, its shadow price is calculated, and if it is negative, this column is not considered to be important. If the shadow price is positive, another iteration takes place in which this column is added into the basis of the final solution.

Addition of Constraint Equations. Some codes have the ability to add new constraint equations very readily to the system. The procedure or algorithm can in some cases be cumbersome, but this difficulty is outweighed by many of its advantages.

Automation of Input and Output. Linear programming is being used more and more as a management tool for the analysis and solution of problems dealing with complex systems. In such an environment, gathering the data and reporting the solutions themselves become important and highly involved tasks.

Information Systems. Many corporations are in the process of setting up or have already set up information systems of various sorts. These systems include a data base of pertinent information and access methods by which the data can be examined and reports made. It is clear that many of the data necessary for creating a linear programming problem can exist within such a data base. Intelligent use of sorting and maintaining appropriate data in this base and of providing access to these data can relieve the user of much of the spadework in having to determine the actual problem coefficients and their statistical properties.

Matrix Generators and Report Writers. Given the existence of a mass of data, a great deal of work must often be done in order to assemble these data into a form meaningful for input into a linear programming code. If these data originally are organized in some fashion, as they would be in the data base of an information system, it then becomes desirable to automate the generation of the input coefficients, particularly for large problems. Such matrix generators can be implemented as computer codes to link a data base to a linear programming code. Some general-purpose matrix generators are available.

Similarly, the output of a linear programming problem is often not in a form which can be read by a layman unfamiliar with linear programming terminology. Particularly when the linear programming solutions are used by top management as a base for decision making, this drawback can lead, on the one hand, to extensive reworking of the output report, and on the other hand, to a lack of confidence in being able to realize the import of the solution. This interpretation process has been automated in many locations by implementing report writers to translate the linear programming output into any of a variety of reports such as summaries or emphasis on critical products. The kind of reports available from such a program depends upon the user's knowledge of how to use and interpret linear programming solutions.

Typical Computer Codes for Mathematical Programming

Comparison of Three Linear Programming Codes. A large number of linear programming codes are available for existing digital computers of most manufacturers, and new codes embodying mathematical and computational refinements and innovations are generally developed for each new line of computers appearing on the market.

Computer efficiency is increasing markedly. Linear programming problems which once required hours of time for solution on the older machines can be solved in minutes on the newer machines. Thus the cost per solution is constantly decreasing, and the area of economic application of linear programming techniques is expanding. The arsenal of computer routines for solving linear programming problems is formidable, and selection of the most appropriate one for a given application depends on a number of considerations. Many programs are extremely flexible and generate auxiliary information which can be very useful. Some of these features, such as shadow prices and economic cost ranges, have been discussed in previous sections.

The three codes shown in Table 6-1 were written at approximately the same time in the development of computer routines. The table compares features to be found on codes in three different computer categories: small-, medium-, and large-scale systems. No running times are given because the linear programming algorithms used are different in each case. This table is intended for comparison purposes of typical sizes of programs only, and is not an exhaustive selection of codes for machines of all manufactures. However, the features are representative.

CONCLUSION

In summary, the attractiveness of mathematical programming techniques rests on the penetrating insights the formulation of a comprehensive resource allocation model

TABLE 6-1. Typical Computer Codes for Mathematical Programming

	Small	Medium	Large
Machine......................	1130	360	360
Operating system...............	Monitor II	DOS	OS
Name of code*..................	LPS/1130	LPS/360	MPS/360
Features:			
Maximum rows...............	500/8,000	200/32,000	1,800/100,000
Core availability..............	1,500/16,000–32,000	1,500/64,000	4,400/200,000 8,191/400,000
(Units of core)................	(Total words)	(Total bytes)	(Partition bytes)
Postoptimal capabilities:			
Range report..................	Yes	Yes	Yes
Parametrics...................	On any parts of problem	On any parts of problem	Five types
Matrix generation aids:			
Scrambled input acceptable......	Yes	Yes	No
Row may = function (other rows)	Yes	Yes	No
Revise problem file.............	Yes	Yes	Yes
Combine existing problems.......	MERGE	MERGE	No
Special languages..............	None	None	MARVEL
Report writing aids: special languages	None	None	(1) FORTRAN interface (2) Report generator (3) MARVEL

* IBM Corporation.
Source: H. V. Smith.

can give a businessman, and the power of large-scale electronic computers to handle problems of impressive size and complexity. This chapter has dealt with the description of problems requiring an optimal numerical solution, the computational characteristics of linear programming and kindred algorithms, and the role of the computer in solving this type of problem. Further information can be found in the bibliography or from computer manufacturers and other sources of computing hardware and software.

BIBLIOGRAPHY

Beale, E. M. L., *Mathematical Programming in Practice*, Sir Isaac Pitmann & Sons, Ltd., London, 1968.

Berge, C., and A. Ghouila-Houri, *Programming Games and Transportation Networks*, John Wiley & Sons, Inc., New York, 1965.

Charnes, A., and W. W. Cooper, *Management Models and Industrial Applications of Linear Programming*, vols. 1 and 2, John Wiley & Sons, Inc., New York, 1961.

Dorfman, R., P. A. Samuelson, and R. M. Solow, *Linear Programming and Economic Analysis*, McGraw-Hill Book Company, New York, 1958.

Garvin, W. W., *Introduction to Linear Programming*, McGraw-Hill Book Company, New York, 1960.

Gass, S. I., *Linear Programming*, 3d ed., McGraw-Hill Book Company, New York, 1969.

Hadley, G., *Linear Programming*, Addison-Wesley Publishing Company, Inc., Reading, Mass., 1962.

Kunzi, A. P., H. G. Tzschach, and C. A. Zehnder, *Numerical Methods of Mathematical Optimization with ALGOL and FORTRAN Programs*, Academic Press Inc., New York, 1968.

Meisels, K., *Primer of Linear Programming*, New York University Press, New York, 1962.

Metzger, R. W., *Elementary Mathematical Programming*, John Wiley & Sons, Inc., New York, 1958.

Orchard-Hays, W., *Advanced Linear-programming Computing Techniques*, McGraw-Hill Book Company, New York, 1968.

Simonnard, M., *Linear Programming*, Prentice-Hall, Inc., Englewood Cliffs, N.J., 1966.

Chapter **7**

Nonlinear Programming

A. G. HOLZMAN

Professor and Chairman, Department of Industrial Engineering, Systems Management Engineering, and Operations Research, University of Pittsburgh, Pittsburgh, Pennsylvania

Due to the development by G. B. Dantzig in 1947 of a very effective solution technique called the simplex method, most mathematical programming problems have been formulated as a linear programming model and solved by the simplex procedure. Although it is generally recognized that few real-life problems do satisfy the linearity assumption, nevertheless, due to the mathematical complexity and concomitant computational difficulties encountered in solving nonlinear programming problems, most of these problems have been attacked by linear programming.

The underlying theory for much of the research done in nonlinear programming was presented in a paper by Kuhn and Tucker[1] in 1951. However, the major thrust in solving real-life nonlinear programming problems was made after 1960, and even though no efficient procedure such as the simplex method for linear programming is on the horizon for nonlinear programming, significant progress has been made.

This chapter will address itself to the following areas:
1. Definition of the nonlinear programming model and its contrast with linear programming to indicate the essential differences which project a problem with a considerable higher level of difficulty
2. The fundamental concepts of nonlinear programming
3. The algebraic structure of basic algorithms designed for nonlinear programming problems

---

[1] H. W. Kuhn and A. W. Tucker, "Nonlinear Programming" in J. Neyman (ed.), *Proceedings of the Second Berkeley Symposium on Mathematical Statistics and Probability*, University of California Press, Berkeley, 1951, pp. 481–492.

4. Nonlinear programming under risk
5. Computer codes for solving nonlinear problems

LINEAR PROGRAMMING VERSUS NONLINEAR PROGRAMMING

Linear Programming Model.   The most familiar mathematical programming model is based on the assumption that the objective function and the constraints are linear. This problem is defined as follows:
Maximize (or minimize) the function

$$f(x_1, \ldots, x_n) = \sum_{j=1}^{n} c_j x_j,$$

subject to the constraints

$$g_i(x_1, \ldots, x_n) = \sum_{j=1}^{n} a_{ij} x_j \leq b_i \qquad \text{for } i = 1, \ldots, m$$

and subject to the nonnegativity restrictions

$$x_j \geq 0 \qquad \text{for } j = 1, \ldots, n$$

where the x_j's are the unknowns or decision variables, and the c_j's, a_{ij}'s, and b_i's are known constants.
Nonlinear Programming Model.   The general mathematical programming problem has the same format as the linear programming problem, with the exception that the objective function and the constraints can be any function of the decision variables. Thus, it is expressed as maximize (or minimize) the criterion

$$f(x_1, \ldots, x_n)$$

subject to the constraints

$$g_i(x_1, \ldots, x_n) \leq b_i \qquad \text{for } i = 1, \ldots, m$$

and subject to the nonnegativity restrictions

$$x_j \geq 0 \qquad \text{for } j = 1, \ldots, n$$

It is obvious that the linear programming problem is just a special case of this general model which permits nonlinearities.
Examples of Nonlinear Programming.   To cite an example of nonlinearity in the objective function, suppose that a company manufactures three products denoted by x_1, x_2, and x_3. The unit profit for x_1 is fixed at $c_1 = 2$, but x_2 and x_3 are sold at profits which decrease linearly as x_2 and x_3 increase. The unit profits are $c_2 = 100 - 2x_2$ and $c_3 = 150 - 3x_3$. The criterion function is then:

Maximize

$$2x_1 + (100 - 2x_2)x_2 + (150 - 3x_3)x_3$$

which gives the quadratic function:

Maximize

$$2x_1 + 100x_2 - 2x_2{}^2 + 150x_3 - 3x_3{}^2$$

Linear programming cannot be used to solve a problem having this objective function.
In many engineering design problems, the constraints as well as the objective function are nonlinear. For example, the design of a pressurized-water nuclear reactor system has nonlinearities in the cost components of the objective function and in the technological requirements of the constraints. Certainly, it is not appropriate here to develop the mathematical model for this complex system, but it is desirable to indicate the high degree of nonlinearity for some of the model components.

The cost component for the main coolant piping had been determined to be

$$x_{11}(150,000 + 37.0x_6)\left(\frac{x_{10}}{2.29}\right)^2$$

where x_6 = maximum operating pressure of primary system
x_{10} = diameter of reactor coolant piping
x_{11} = number of reactor coolant loops
For the reactor vessel, the component cost is

$$\frac{(447,000/x_7x_{14}) + 0.667x_{14}{}^3}{(2.767)[2\sqrt{(0.00041)(26,500)} + 2.5]^3} \times 460x_6 + 30,000x_{11}$$

where x_6 and x_{11} are defined above and
x_7 = number of fuel rods in reactor core
x_8 = width of thermal shield flow passage x_2
$x_{14} = 2\sqrt{0.00041x_7} + 1.5 + x_8$
An acceptable feasible solution has been obtained for this highly nonlinear problem consisting of 12 variables and 8 restrictions by use of the minimal code developed by Weisman.[2]

The Difficulties of Nonlinear Programming. The presence of nonlinearities in a mathematical programming problem results in the elimination of the "niceties" of the linear programming approach to obtain the optimum solution. The major difficulties can be traced to the absence of one or more features of the linear programming formulation.

Nonlinear Constraints. In linear programming, a convex set is formed by the intersection of a finite number of linear constraints, and only a finite number of extreme points exists. However, when at least one of the constraints is nonlinear, then it is possible that an infinite number of extreme points exists or the feasible solution area is nonconvex.

Figure 7-1 graphically shows a set of mathematical programming problems which have a single constraint and where the nonnegativity restrictions apply. The feasible solution areas are shaded. In Figure 7-1a, if the objective function is linear, then we have a linear programming problem and the optimum solution will be located at one of the three extreme points O, A, B. However, while the feasible solution area for Figure 7-1b is a convex set, an infinite number of extreme points exists. In fact, any point on the boundary of the curve AB is an extreme point.

Because the simplex method is an iterative procedure which moves from one extreme point to an adjacent one, it is not possible to use this method when an infinite number of extreme points exists. It is obvious that the feasible solution area denoted by Figure 7-1c is nonconvex because it is not a collection of points such that, for each pair

[2] Joel Weisman, "Engineering Design Optimization under Conditions of Risk," Ph.D. dissertation, University of Pittsburgh, Pittsburgh, Pa., 1968, pp. 162–179.

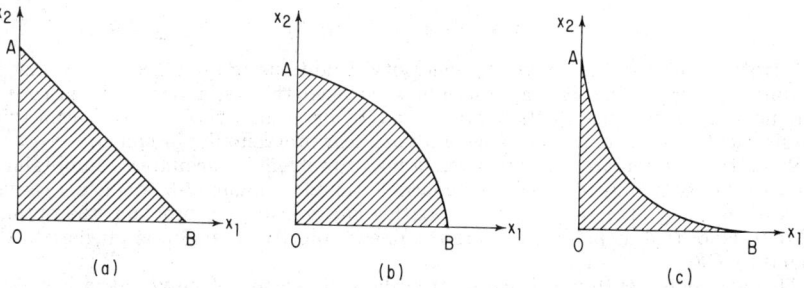

Fig. 7-1. Graphical presentation of a set of mathematical programming problems which have a single constraint and where the nonnegativity restrictions apply.

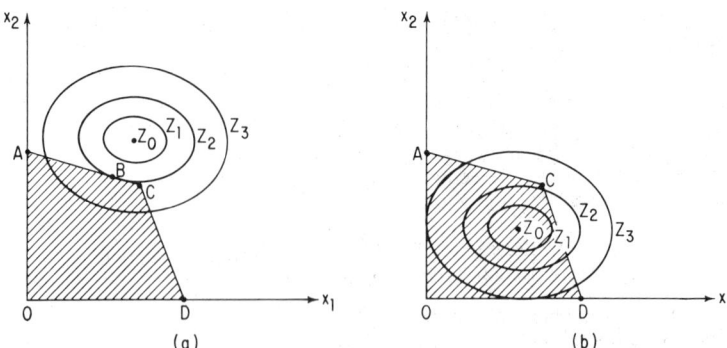

Fig. 7-2. Optimum solutions for nonlinear objective functions.

of points in the collection, the entire line segment joining these two points is also in this set. Again, the simplex method is not applicable for nonconvex sets.

Nonlinear Objective Function. Now consider the nonlinear programming problem which has linear constraints and a nonlinear objective function.

In Figure 7-2, the curves represent isoprofit lines, Z_0 being the highest profit for this function, with $Z_0 > Z_1 > Z_2 > Z_3$. However, in Figure 7-2a, the highest profit point Z_0 is located at a point outside the feasible solution area, but the optimum solution for this constrained problem is located at point B, where the isoprofit contour is tangent to the constraint AC. It is seen that the optimum solution is not located at an extreme point of the convex set. Suppose that the nonlinear objective function is shifted as shown in Figure 7-2b, again with Z_0 denoting the highest profit and $Z_0 > Z_1 > Z_2 > Z_3$. Now the optimum solution for this problem is located at an interior point of the feasible solution area, and the solution variables x_1 and x_2 are those which correspond to the solution value Z_0. Again, the feature of the linear programming problem that all solutions are at extreme points of the convex set is violated by both problems exhibited in Figure 7-2. It should be pointed out here, however, that modifications in the simplex method have been made to utilize this technique for mathematical programming problems which have a nonlinear objective function but linear constraints.

Convexity-Concavity Conditions. In the linear programming problem, the convexity-concavity conditions are automatically satisfied: the feasible solution area is a convex set, and the linear objective function can be considered a concave function (when maximizing) and a convex function (when minimizing). This is because the equality condition is met for both the concave function

$$f[\lambda x' + (1 - \lambda)x''] \geq \lambda f(x') + (1 - \lambda)f(x'') \qquad \text{with } 0 \leq \lambda \leq 1$$

and the convex function

$$f[\lambda x' + (1 - \lambda)x''] \leq \lambda f(x') + (1 - \lambda)f(x'') \qquad \text{with } 0 \leq \lambda \leq 1$$

However, because the convexity-concavity conditions are not necessarily met in nonlinear programming, it is appropriate to consider the distinction between convex and nonconvex feasible solution areas, and to determine the convexity-concavity requirements for optimum solutions of a nonlinear programming problem.

When the solution area is nonconvex, a major problem is encountered in obtaining the global optimum as opposed to the local optimum. Figure 7-3a depicts a convex set with a linear objective function. If this is a maximizing problem, the optimal solution is located at point D, where the linear objective function is tangent to the boundary CE.

This is a global optimum because, regardless of the size of move taken from this point, the value for the objective function can never be greater. Now contrast this to Figure 7-3b, where the solution set is nonconvex. At point B, an optimum solution

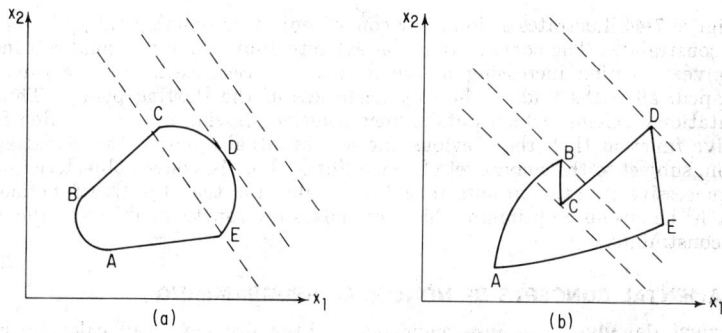

FIG. 7-3. Convex and nonconvex sets with a linear objective function.

exists, because as we move toward A or C, the value of the objective function decreases; however, at vertex D, another optimum solution exists and the value of the objective function is larger than at the previous optimum at B. Because no other solution exists which gives a larger value for the objective function, the solution at D is called the global (absolute) optimum. The solution at point B is called a local (relative) optimum. Most of the solutions for nonlinear programming problems which do not have a convex solution area give only local optima. However, by changing the starting point for the initial solution, we may be able to determine other local optima and then consider the local optimal solution that gives the largest value for the objective function to be the global optimum. This is not necessarily true, but such solutions are frequently acceptable in actual practice. Myopic computational methods, such as the simplex method, which look only at nearby points in testing for optimality cannot be relied upon to give global optima for nonconvex sets.

A similar problem exists if the nonlinear objective function does not satisfy certain conditions. Assuming that the feasible solution area is convex, then a global optimum is obtained for a maximizing problem if the objective function is concave (hill-shaped) and for a minimizing problem if the criterion is convex (valley-shaped). Figure 7-4a shows a nonlinear concave objective function subject to a single linear constraint,

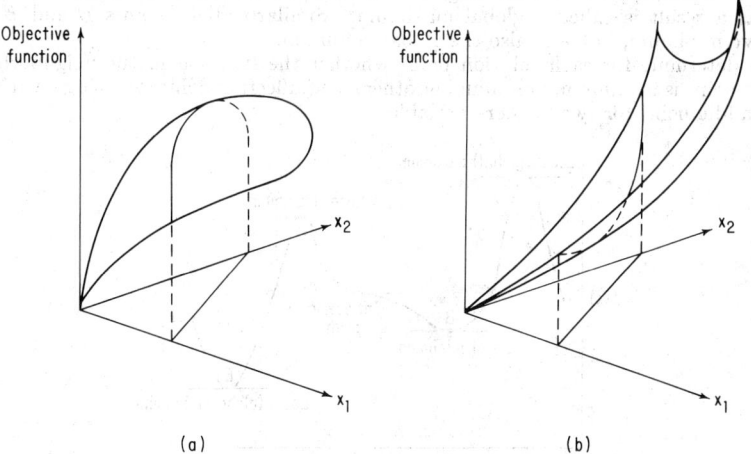

FIG. 7-4. Concave and convex objective functions subject to a single linear constraint.

and Figure 7-4*b* illustrates a nonlinear convex objective function subject to a single linear constraint. The concave function exhibits diminishing marginal returns, and the convex function increasing marginal returns. Because a concave function is hill-shaped, all paths lead to the top, regardless of the starting point. Thus, any computational scheme which obtains new solutions having a greater value for the objective function than the previous one will eventually obtain the global optimal solution, subject to the convex set of constraints. For the convex objective function, each successive iteration results in a lesser value for the objective function until we reach the absolute optimum which minimizes the function subject to the convex set of constraints.

FUNDAMENTAL CONCEPTS OF NONLINEAR PROGRAMMING

Classical Calculus. Because engineers traditionally have had calculus in their academic preparation, the question raised is "Why not use the classical optimization methods of calculus to solve nonlinear problems?" Certainly, it is true that aspects of calculus are used in some methods to solve nonlinear programming problems, but these are usually included as a subset of more efficient procedures. Classical optimization theory may be considered a point of departure for developing some of the newer techniques, thus forming a basis in the attack on various types of nonlinear programming problems. Consequently, a brief review of the classical calculus method to solve these problems will be presented with the concomitant computational difficulties encountered.

1. Determine all stationary points of the objective function. This is accomplished by taking the partial derivatives of $f(x_1, \ldots, x_n)$ with respect to each of the unknown x_j's, setting each derivative function equal to zero, and solving for the set of stationary points that satisfy these equations. Thus,

$$\frac{\partial f(x_1, \ldots, x_n)}{x_j} = 0 \qquad \text{for } j = 1, \ldots, n$$

It is to be emphasized that we must have a real valued differentiable function. Also, the task to solve a system of highly nonlinear derivatives may be quite difficult.

2. Once the solution is obtained, the next step is to ascertain whether the stationary points are relative (local) maximum, relative (local) minimum, or inflection or saddle points. For a single variable function, the possible results are shown in Figure 7-5.

Although points A and D are both maximum relative to the neighborhood of these respective points, point A is the largest value for $f(x)$ over the range of the variable x, and as a result is called a global maximum. Similarly, both points B and E are relative minimum, but E is also the global minimum.

We determine for each solution point whether the function in the neighborhood of this point is maximum, minimum, or otherwise (inflection point for a single variable and saddle point for two or more variables).

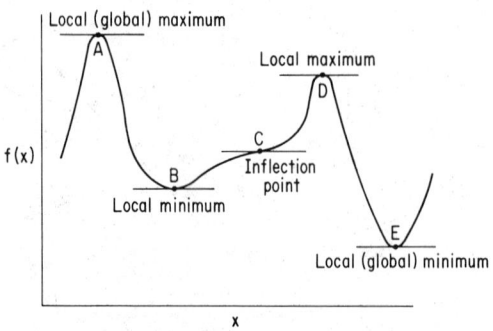

FIG. 7.5. Maximums and minimums for a given function.

By definition, let

$$
F_1 = \frac{\partial^2 f}{\partial x_1^2} \qquad
F_2 = \begin{vmatrix} \dfrac{\partial^2 f}{\partial x_1^2} & \dfrac{\partial^2 f}{\partial x_1 \, \partial x_2} \\[2ex] \dfrac{\partial^2 f}{\partial x_2 \, \partial x_1} & \dfrac{\partial^2 f}{\partial x_2^2} \end{vmatrix} \qquad
F_3 = \begin{vmatrix} \dfrac{\partial^2 f}{\partial x_1^2} & \dfrac{\partial^2 f}{\partial x_1 \, \partial x_2} & \dfrac{\partial^2 f}{\partial x_1 \, \partial x_3} \\[2ex] \dfrac{\partial^2 f}{\partial x_2 \, \partial x_1} & \dfrac{\partial^2 f}{\partial x_2^2} & \dfrac{\partial^2 f}{\partial x_2 \, x_3} \\[2ex] \dfrac{\partial^2 f}{\partial x_3 \, \partial x_1} & \dfrac{\partial^2 f}{\partial x_3 \, \partial x_2} & \dfrac{\partial^2 f}{\partial x_3^2} \end{vmatrix}
$$

The determinant is structured in a like manner for F_n, where n is the number of variables. A stationary point $(x_1, \ldots ,x_n)$ is a strictly concave function if $F_1 < 0$, $F_2 > 0, F_3 < 0, \ldots$, and is a strictly convex function if $F_1 > 0, F_2 > 0, F_3 > 0, \ldots$.

Sufficient conditions for the point $X = (x_1, \ldots ,x_n)$ to be a local maximum are that all first partial derivatives be zero and that the function be strictly concave in the neighborhood of that point. Similarly, a point is a local minimum if all first partial derivatives are zero and if the function is strictly convex in the neighborhood of that point.

3. When all the local or relative maximums (minimums) are obtained, the global optimum is obtained by selecting the solution which is the absolute maximum (minimum), provided all restrictions are satisfied.

4. Unfortunately, in a constrained problem, this optimum solution may be outside the feasible solution region. If this is true, the absolute optimum may be located on one of the boundaries of the feasible solution area. The procedure to determine the global maximum (minimum) follows: (a) by the method discussed in step 2 find all local solutions lying in the feasible solution area; (b) determine the maximum (minimum) value of $f(x_1, \ldots ,x_n)$ along each boundary. This greatly complicates the problem, especially if there are many boundaries (constraints) and variables.

For example, suppose the feasible solution is restricted to the nonnegative orthant. It is possible that the global maximum (minimum) may be located on the boundary of this feasible region, which means that at least one $x_j = 0$. We define n functions f_j of $n - 1$ variables, where $f_j = (x_1, \ldots , x_j = 0, \ldots ,x_n)$. Then, for each f_j, the solutions to the system of partial derivatives $\partial f_j / \partial x_i = 0$ for all $i \neq j$ are determined. This same procedure must be continued for $n - 2$ variables set equal to zero, then $n - 3$ variables set equal to zero, and on up until all variables are equal to zero. Consider the magnitude of this problem: in the case where there was a single variable set to zero, we had n functions for f_j; in the case where there are three variables set equal to zero, the number of functions increases combinatorially, as we now have $[(n)(n - 1)(n - 2)]/6$ functions denoted by f_{ijk}, for $x_i = x_j = x_k = 0$.

Thus, as is evident from the above discussion, the computational requirements to solve a nonlinear programming problem by classical calculus are indeed quite formidable.

Lagrange Multipliers. The classical calculus method has been extended to a method called Lagrange multipliers to solve a nonlinear programming problem in which the objective function is to be maximized (minimized) subject to a set of equality constraints. It should be pointed out here that by the application of the Kuhn-Tucker conditions to be discussed later, inequality constraints are also permitted. The Lagrange multiplier technique is a forerunner to new attacks on solving nonlinear problems by changing a constrained optimization problem to an unconstrained maximizing (minimizing) problem.

Consider the nonlinear programming problem:

Max

$$ Z = f(x_1, \ldots ,x_n) $$

subject to

$$ g_1(x_1, \ldots ,x_n) - b_1 = 0 $$
$$ \cdots\cdots\cdots\cdots\cdots $$
$$ g_m(x_1, \ldots ,x_n) - b_m = 0 $$

with all functions being differentiable. This problem is cast as an unconstrained problem by adding a new set of variables $\lambda_1, \ldots, \lambda_m$, as follows:

$$L(X,\lambda) = f(x_1, \ldots, x_n) - \lambda_1(g_1(x_1, \ldots, x_n) - b_1) -,$$
$$\ldots, -\lambda_m(g_m(x_1, \ldots, x_n) - b_m)$$

$$= f(X) - \sum_{i=1}^{m} \lambda_i(g_i(X) - b_i)$$

An example of a constrained problem expressed as an unconstrained problem by the Lagrangian expression is given now.

Max
$$2x_1{}^2 + 3x_1x_2 + 10x_3{}^6$$

subject to

$$x_1{}^3 - 2x_1x_3 + x_2{}^2 = 42$$
$$2x_1 + x_2{}^3 - x_3x_4 = 63$$

$$L(X,\lambda) = f(x_1,x_2,x_3) - \lambda_1(g_1(x_1,x_2,x_3) - 42) - \lambda_2(g_2(x_1,x_2,x_3) - 63)$$
$$= 2x_1{}^2 + 3x_1x_2 + 10x_3{}^6 - \lambda_1(x_1{}^3 - 2x_1x_3 + x_2{}^2 - 42)$$
$$- \lambda_2(2x_1 + x_2{}^3 - x_3x_4 - 63)$$

This is called the Lagrangian function. The problem is solved by (1) taking the partial derivatives of L with respect to the $n + m$ variables, $x_1, \ldots, x_n, \lambda_1, \ldots, \lambda_m$, (2) setting these expressions equal to zero, and (3) solving this system of equations. A stationary point obtained by the Lagrangian multiplier method may also be a local maximum, a local minimum, or neither. As stated before, the global optimum is the solution which gives the largest value (when maximizing) for the objective function or the least value (when minimizing) from among all local optima. If the convexity-concavity conditions as discussed under classical calculus hold, then the solution by Lagrange multipliers will give the global optimum. The unknown variable λ_i can be interpreted as a dual variable, a shadow price, or an imputed value. The Lagrange multiplier is the partial derivative of the global optimum with respect to b_i, evaluated at $b = (b_1, \ldots, b_m)$.

Kuhn-Tucker Conditions. The generalizing of the Lagrange multiplier technique to inequality constraints and nonnegative variables was made possible by the formulation of a set of necessary and sufficient conditions by Kuhn and Tucker. In fact, these conditions are used as a basis for further development of many new techniques. An important feature of the Kuhn-Tucker theorem is that they can be considered as existence theorems. These theorems enable us to state that, for the class of problems for which they hold, the problem has a solution if and only if the corresponding conditions are satisfied.

Saddle Point. A function $\phi(x_1, \ldots, x_n, \lambda_1, \ldots, \lambda_m)$ is said to have a saddle point at (X_0,λ_0) if $\phi(X,\lambda_0) \leq \phi(X_0,\lambda_0) \leq \phi(X_0\lambda)$. If this inequality holds for all X and λ, then we have a global saddle point at (X_0,λ_0); otherwise, it is a local saddle point. It is worthwhile to point out here that the saddle point also provides a solution to the von Neumann two-person zero-sum game.

The necessary and sufficient conditions for a saddle point are now stated. Let $\phi(X,\lambda)$ be a differentiable function with $x_j \geq 0$ and $\lambda_i \geq 0$, and let

$$\hat{\phi}_X = \frac{\partial \phi}{\partial x_j} \quad j = 1, \ldots, n \qquad \hat{\phi}_\lambda = \frac{\partial \phi}{\partial \lambda_i} \quad i = 1, \ldots, m$$

be partial derivatives evaluated at a particular point (X_0,λ_0). $\hat{\phi}_X$ and $\hat{\phi}_\lambda$ are gradient vectors, because $\hat{\phi}_X$ is the gradient of $\phi(X,\lambda)$ with respect to X evaluated at (X_0,λ_0), and $\hat{\phi}_\lambda$ is the gradient of $\phi(X,\lambda)$ with respect to λ at (X_0,λ_0).

The necessary conditions are:

1. $\hat{\phi}_X \leq 0, \qquad \hat{\phi}_X{}' X_0 = 0, \qquad X_0 \geq 0$
2. $\hat{\phi}_\lambda \geq 0, \qquad \hat{\phi}_\lambda{}'\lambda_0 = 0, \qquad \lambda_0 \geq 0$

and the sufficient conditions are:

3. $\phi(X,\lambda_0) \leq \phi(X_0,\lambda_0) + \hat{\phi}_X{}'(X - X_0)$
4. $\phi(X_0,\lambda) \geq \phi(X_0,\lambda_0) + \hat{\phi}_X{}'(\lambda - \lambda_0)$

Conditions 3 and 4 are always satisfied if $\phi(X,\lambda_0)$ is a concave function of X and $\phi(X_0,\lambda)$ is a convex function of λ. Thus it follows that if these convexity-concavity conditions are satisfied at $\phi(X,\lambda)$, then conditions 1 and 2 are both necessary and sufficient.

Kuhn-Tucker Conditions Applied to Lagrange Multipliers. If we let $\phi(X,\lambda)$ equal the Lagrangian expression, then

$$\phi(X,\lambda) = L(X,\lambda) = f(X) - \sum_{i=1}^{m} \lambda_i(g_i(X) - b_i)$$

Assuming $f(X)$, $g_1(X)$, . . . , $g_m(X)$ are differentiable functions, then the necessary conditions for a Lagrangian saddle point can now be translated to the Kuhn-Tucker conditions.

The vector $X_0 = (x_1^0, . . . ,x_m^0)$ is an optimal solution to the problem only if there exists a vector $\lambda = (\lambda_1, . . . ,\lambda_m)$ such that the succeeding conditions hold:

1. If $x_j^0 = 0$, then $\dfrac{\partial f(X)}{\partial x_j} - \displaystyle\sum_{i=1}^{m} \lambda_i \dfrac{\partial g_i(X)}{\partial x_j} \leq 0$ at $x_j = x_j^0, j = 1, . . . , n$

2. If $x_j^0 > 0$, then $\dfrac{\partial f(X)}{\partial x_j} - \displaystyle\sum_{i=1}^{m} \lambda_i \dfrac{\partial g_i(X)}{\partial x_j} = 0$

3. If $\lambda_i = 0$, then $g_i(X_0) - b_i \leq 0, i = 1, . . . , m$
4. If $\lambda_i > 0$, then $g_i(X_0) - b_i = 0, i = 1, . . . , m$

and the nonnegativity restrictions,
5. $x_j^0 \geq 0, \lambda_i \geq 0$

If the convexity-concavity conditions are satisfied, these five conditions are both necessary and sufficient and the solution is global optimal.

Consider the following nonlinear programming problem:
Max

$$x_1^2 + x_2^2$$

subject to

$$2x_1^3 - x_2 \leq 60$$
$$x_1 \geq 0$$
$$x_2 \geq 0$$

The Kuhn-Tucker conditions state that for an optimal solution to exist for this problem, the following conditions must be satisfied:
1. If $x_1 = 0$, then $2x_1 - \lambda_1(6x_1) \leq 0$
   If $x_2 = 0$, then $2x_2 - \lambda_1(-1) \leq 0$
2. If $x_1 > 0$, then $2x_1 - \lambda_1(6x_1) = 0$
   If $x_2 > 0$, then $2x_2 - \lambda_1(-1) = 0$
3. If $\lambda_1 = 0$, then $2x_1^3 - x_2 - 60 \leq 0$
4. If $\lambda_1 > 0$, then $2x_1^3 - x_2 - 60 = 0$
5. $x_1 \geq 0, x_2 \geq 0$

As stated previously, the Kuhn-Tucker conditions are not directly useful in solving nonlinear programming problems, but they do provide insights on dual prices (Lagrange multipliers) of the nonlinear programming problem, and also, quite importantly, they often enable us to develop alternative computational methods for a nonlinear programming problem. This is especially true for gradient methods.

ALGEBRAIC STRUCTURE OF BASIC ALGORITHMS
FOR NONLINEAR PROGRAMMING

In the following discussion, the focus is on a skeletal presentation of the algebraic structure of basic algorithms designed for nonlinear programming problems. Because there is no standard nonlinear programming algorithm, such as the simplex method for linear programming, an unusually large number of algorithms have been developed.

As a result, considerable selectivity has been exercised by the author in determining which methods should be reviewed. Although the simplex method was developed primarily to solve linear programming problems, various modified forms of this technique nevertheless have been developed to solve nonlinear problems.

Quadratic Programming. The quadratic programming problem is defined as follows:

Max

$$Z = cX + X'DX$$

subject to

$$AX \leq b$$
$$X \geq 0$$

It can be seen that the only difference between the linear programming problem and the quadratic problem is the addition of the term $X'DX$ in the objective function, where $X'DX$ is the strictly quadratic part of the criterion. Without loss of generality, the matrix D can be considered to be a symmetric matrix. In the literature one will frequently see the objective function expressed as $cX + \frac{1}{2}X'DX$. The reason for this is evident from the following example:

Max

$$4x_1{}^2 + 2x_1x_2 + 4x_2x_1 + x_2{}^2$$

In the form $X'DX$, we have

$$[x_1,x_2] \begin{bmatrix} d_{11} = 3 & d_{21} = 4 \\ d_{12} = 2 & d_{22} = 1 \end{bmatrix} \begin{bmatrix} x_1 \\ x_2 \end{bmatrix}$$

Because $x_ix_j = x_jx_i$, the terms $2x_1x_2 + 4x_2x_1$ can be written as $6x_1x_2$. When placed in symmetrical form $\frac{1}{2}X'DX$, the above function would be shown as

$$\frac{1}{2}[x_1x_2] \begin{bmatrix} 8 & 6 \\ 6 & 2 \end{bmatrix} \begin{bmatrix} x_1 \\ x_2 \end{bmatrix} = \frac{1}{2}(8x_1{}^2 + 12x_1x_2 + 2x_2{}^2)$$
$$= 4x_1{}^2 + 6x_1x_2 + x_2{}^2$$

Solution procedures have been developed for the case where the objective function is concave when maximizing or convex when minimizing. This is equivalent to saying that $X'DX$ is negative semidefinite (concave) or positive semidefinite (convex).

Wolfe's Method. The solution procedure developed by Wolfe, which is based on the work of Barankin and Dorfman, is one of the more efficient ways of solving the quadratic programming problem because it reduces the problem to an "almost linear" programming problem, thus permitting the use of a modified simplex procedure.

The Lagrangian function for the quadratic programming problem is

$$L(X,\lambda) = CX + X'DX + \lambda'(AX - b)$$

Applying the Kuhn-Tucker necessary conditions gives the following expressions:
1. $C + DX - A'\lambda + U = 0$, where U is a slack vector
2. $b - AX - V = 0$, where V is a slack vector
3. $U'X = 0$
4. $V'\lambda = 0$

Equations 3 and 4 are commonly called the complementary slackness conditions.

The Wolfe computational procedure uses the linear programming trick of introducing artificial variables to generate an initial solution, and then forces these variables to zero. The simplex method is modified to account for the two nonlinear restrictions $U'X = 0$ and $V'\lambda = 0$. Refer to Wolfe[3] for a complete description of the computational procedure.

Separable Programming. We define the separable programming problem as

---

[3] Philip Wolfe, "The Simplex Method for Quadratic Programming," *Econometrica*, vol. 27, no. 3, July, 1959, pp. 382–398.

Max

$$Z = \sum_{j=1}^{n} f_j(x_j)$$

subject to

$$\sum_{j=1}^{n} g_{ij}(x_j) \leq b_i \qquad \text{for } i = 1, \ldots, m$$

$$x_j \geq 0 \qquad \text{for } j = 1, \ldots, m$$

We assume that all functions are separable. The technique is an approximation method, because all functions $f_j(x_j)$ and $g_{ij}(x_j)$ are replaced by polygonal approximations. The problem then reduces to a linear programming problem, but usually at the expense of adding a great number of variables to the initial problem. Again, the global optimum solution is ensured only when the objective function is concave (maximizing) or convex (minimizing) and the constraints are convex.

The interval over which each variable x_j is permitted to range is subdivided by k mesh points x_{kj}, and then piecewise linear approximations are made by connecting the functional values for every two adjacent mesh points with a straight line. This approximation method for solving a separable programming problem can be expressed in λ form, with the variables x_j of the original problem replaced by the variables λ_{kj}.

The problem then is

Max

$$Z = \sum_{j=1}^{n} \sum_{k=0}^{r_j} f_{kj}\lambda_{kj}$$

subject to

$$\sum_{j=1}^{n} \sum_{k=0}^{r_j} g_{kij}\lambda_{kj} \leq b_i \qquad \text{for } i = 1, \ldots, m$$

$$\sum_{k=0}^{r_j} \lambda_{kj} = 1 \qquad \text{for } j = 1, \ldots, n$$

$$\lambda_{kj} \geq 0 \qquad \text{for all } k, j$$

where

$$f_{kj} = f_j(x_k) \qquad g_{kij} = g_{ij}(x_k)$$

This is a linear programming problem which can be solved by the simplex method with restricted basis entry. To be sure that all solution points will lie on the polygonal approximation, it is essential that no more than two λ_{kj}'s be positive and that they correspond to adjacent mesh points.

Hadley outlines this method in detail, and also proposes an alternative formulation called the δ method.

SUMT—Sequential Unconstrained Minimization Technique. This method presents an approach to the solution of nonlinear programming problems which is different from the modified simplex methods discussed for quadratic programming and separable programming. The method, much like the Lagrange multiplier concept, transforms a constrained optimization problem into an unconstrained one. An immediate advantage of this transformation is that gradient methods can be used. One of the most promising methods using this approach is SUMT, which is primarily the work of Fiacco and McCormick.[4] The basic ideas for this penalty function technique come from the theoretical work of Courant,[5] and more recently from Kelley[6] and Carroll.[7]

[4] A. V. Fiacco and G. P. McCormick, "The Sequential Unconstrained Minimization Technique for Nonlinear Programming: A Primal-Dual Method," *Management Science*, vol. 10, no. 2, 1964, p. 360.

[5] R. Courant, "Variational Methods for the Solution of Problems of Equilibrium and Vibration," *Bulletin of the American Mathematical Society*, vol. 49, 1943, pp. 1–23.

[6] H. J. Kelley, "Method of Gradients," in G. Leitmann (ed.), *Optimization Techniques*, Academic Press Inc., New York, 1962, p. 213.

[7] C. W. Carroll, "The Created Response Surface Technique for Optimizing Nonlinear Restrained Systems," *Operations Research*, vol. 9, no. 2, 1961, pp. 169–184.

SUMT has been applied successfully to solve some problems with both nonlinear objective function and nonlinear constraints.   If the usual convexity-concavity conditions hold, this method converges to a global optimum.   Computational efficiencies have been made in this algorithm to provide solutions to larger problems.   A certain type of problem with 100 variables has been solved.

Consider the general nonlinear programming problem:

Min[8]

$$Z = f(X)$$

subject to

$$g_i(X) \geq b_i \qquad \text{for } i = 1, \ldots, l$$
$$g_i(X) = b_i \qquad \text{for } i = l + 1, \ldots, m$$

with at least one point existing so that $g_i(X) > b_i$, for $i = 1, \ldots, l$.

This constrained problem is transformed to an unconstrained problem with penalty functions by defining the P function

$$P(X, r_k) = f(X) + r_k \sum_{i=1}^{l} \frac{1}{g_i(X) - b_i} + r_k^{1/2} \sum_{i=l+1}^{m} g_i^2((X) - b_i)$$

where r_k is a positive number.

Briefly, the procedure is:

1. Select a starting point X_0 so that $g_i(X_0) - b_i > 0$.   Repeated applications of the method itself can be used to obtain it if such a starting point is not available.
2. Select initial value of r_k, call it r_1.   Two criteria have been established for determining r_1.[9]
3. Minimize P function for this particular r_k.   Cauchy's first order gradient method and a second order gradient method have been used, with the latter method being more efficient.
4. If the convergence criterion is not satisfied, select r_2 so that $0 < r_2 < r_1$.
5. Determine minimum of P function using r_2.
6. Continue by selecting strictly monotonically decreasing r_k until convergence criteria are satisfied.

The rationale for this P function follows.

The second term $r_k \sum_{i=1}^{l} \dfrac{1}{g_i(X) - b_i}$ is a penalty factor which guarantees that the P function minimum is in the interior of the inequality-constrained region.   In the third term $r_k^{1/2} \sum_{i=1+l}^{m} g_i^2((X) - b_i)$, minimizing the P function forces $g_i(X) - b_i$ to approach zero.   This is essential; otherwise, as $r_k \to 0$, the third term would approach infinity.

For a theoretical discussion of SUMT, refer to Fiacco and McCormick, and for a discussion on applications, see Bracken and McCormick, as listed in the bibliography. Zangwill has also extended the theoretical discussion of penalty functions.[10]

Geometric Programming.   Geometric programming is one of the mathematical programming techniques developed most recently.   The original work in this area was stimulated by the need to solve engineering design problems at the Westinghouse Electric Corporation.   Frequently, the total cost of an engineering design (and other

---

[8] Because this is called a minimization technique, we will consider the minimizing function, although the technique can be applied to a maximizing problem with the appropriate changes in the inequalities.

[9] A. V. Fiacco and G. P. McCormick, "Programming under Nonlinear Constraints by Unconstrained Minimization: A Primal-Dual Method," Research Analysis Corporation Technical Paper RAC-TP-96, September, 1963, pp. 46–47.

[10] W. I. Zangwill, "Nonlinear Programming via Penalty Functions," *Management Science*, vol. 13, no. 5, 1967, pp. 344–358.

problems, too) can be expressed as

$$g = \mu_1 + \cdots + \mu_i + \cdots + \mu_n$$

where μ_i is considered to be a component cost which can be stated as a power function

$$\mu_i = c_i t_1{}^{a_{i1}} \cdots t_m{}^{a_{im}}$$

The coefficient constant c_i and the design variables $t_1, \ldots, t_m$ must be positive, but the exponents $a_{i1}, \ldots, a_{im}$ are arbitrary real constants. (In our past discussions, the variables were denoted by $x_1, \ldots, x_n$.)

The power function μ_i with the stated restrictions is called a posynomial (positive polynomial). The approach taken to solve a geometric programming problem is different from the usual mathematical programming attack in that the solution is addressed first to determining the relative contributions of each μ_i, and only then to obtaining values for the variables $t_1, \ldots, t_m$. One of the very attractive and unique features of geometric programming is that if both the objective function and the constraints can be expressed in posynomial form, we are guaranteed a global optimum solution, and we obtain this solution by solving a set of linear equations. It can be proved that the posynomials of geometric programming are convex functions of the variables on a log-log scale.

Thus, by geometric programming, the engineer can assess the contribution to the total cost (profit) made by each component in the system design.

The basis for this new method is the general geometric inequality

$$\delta_1 U_1 + \cdots + \delta_n U_n \geq U_1{}^{\delta_1} \cdots U_n{}^{\delta_n}$$

where $U_1, \ldots, U_n$ are arbitrary nonnegative numbers and $\delta_1, \ldots, \delta_n$ are arbitrary positive weights which satisfy the normality condition

$$\delta_1 + \cdots + \delta_n = 1$$

If we let the cost component $\mu_i = \delta_i U_i$, then we have, by substituting in the above expression for the general geometric inequality,

$$\mu_1 + \cdots + \mu_n \geq \left(\frac{\mu_1}{\delta_1}\right)^{\delta_1} \cdots \left(\frac{\mu_n}{\delta_n}\right)^{\delta_n}$$

with the left side of this inequality called the primal function, and the right side the predual function.

By substituting in the right side $\mu_i = c_i t_1{}^{a_{i1}} \cdots t_m{}^{a_{im}}$, the predual function then is

$$\left(\frac{c_1}{\delta_1}\right)^{\delta_1} \cdots \left(\frac{c_n}{\delta_n}\right)^{\delta_n} t_1{}^{D_1} \cdots t_m{}^{D_m}$$

with the exponents $D_j = \sum\limits_{i=1}^{n} \delta_i a_{ij}$, for $j = 1, \ldots, m$. If we select weights δ_i so that all the exponents $D_j = 0$, the variable t is eliminated, and we have the following dual function with $\delta_1, \ldots, \delta_n$ the unknowns:

$$\left(\frac{c_1}{\delta_1}\right)^{\delta_1} \cdots \left(\frac{c_n}{\delta_n}\right)^{\delta_n}$$

It is important to note that we determine the solution $\delta_1, \ldots, \delta_n$ by solving the system of linear equations

$$D_1 = 0 = \sum_{i=1}^{n} \delta_i a_{i1}$$

$$\cdots \cdots \cdots \cdots \cdots$$

$$D_m = 0 = \sum_{i=1}^{n} \delta_i a_{im}$$

$$1 = \delta_1 + \cdots + \delta_n$$

The first m equations are called the orthogonality condition, and the last equation the normality condition.

With a knowledge of $\delta_1, \ldots, \delta_n$, it is possible to calculate the respective component costs $\mu_1, \ldots, \mu_n$, and then the solution can be determined with respect to the original unknowns $t_1, \ldots, t_m$.

Constraints are expressed in the form

$$g_k(t) \leq 1 \qquad k = 1, \ldots, p$$

where $g_k(t)$ is a posynomial and p is the number of constraints.

Further research has resulted in the elimination of some restrictions imposed on geometric programming when first conceived. For a rigorous description of both theory and application, refer to Duffin, Peterson, and Zener. A good single chapter presentation is given by Wilde and Beightler. These sources are identified in the bibliography.

Integer Linear Programming. Integer linear programming is frequently classified as a nonlinear programming problem. Basically, it is a linear programming problem with the additional restriction that one or more of the variables must take on integral values.

This problem is stated as

Max

$$Z = cX$$

subject to

$$AX \leq b$$
$$X \geq 0$$
$$x_j \text{ an integer, } j \in J_1$$

If the set J_1 is empty, then it reduces to a linear programming problem.

In many practical problems, resources are indivisible, taking on only integer values. Examples are the number of airplanes assigned to a squadron in the military, the number of specific types of vehicles for an urban transportation system, or the number of workers by job classification assigned to various maintenance functions. Satisfactory answers may be obtained by merely rounding off noninteger solutions if the round-off is small relative to the values of the variables. Gomory[11] and Balinski[12] have made important contributions in this field, and the detailed procedure to solve an integer programming problem is included in their work.

A most important application of integer programming is where the integral values can take on only the values 0 or 1. Fixed-charge problems and problems with dichotomous constraints which define nonconvex feasible regions are examples where 0, 1 integer programming has been used successfully. Balas has made significant contributions in this field.[13]

Branch and Bound Method. Because a bounded integer programming problem has only a finite number of solutions, an intelligent partial enumeration technique can be utilized. One of the best known and most effective is the branch and bound method. The fundamental idea of this technique is to partition the set of all feasible solutions into several subsets, and (assuming the function is to be maximized) an upper bound is determined for each of the subsets. Then subsets which have upper bounds greater than the current upper bound for the objective function are eliminated from further consideration. Lawler and Wood[14] have developed an excellent survey on branch and bound methods.

[11] R. E. Gomory, "An Algorithm for Integer Solutions to Linear Programs," in R. L. Graves and P. Wolfe (eds.), *Recent Advances in Mathematical Programming*, 2d ed., McGraw-Hill Book Company, New York, 1963, pp. 269–302.

[12] M. L. Balinski, "Integer Programming: Methods, Uses, Computation," *Management Science*, vol. 12, no. 3, 1965, pp. 253–313.

[13] E. Balas, "An Additive Algorithm for Solving Linear Programs with Zero-One Variables," *Operations Research*, vol. 13, no. 4, 1965, pp. 517–546.

[14] E. L. Lawler and D. E. Wood, "Branch and Bound Methods: A Survey," *Operations Research*, vol. 14, no. 4, 1966, pp. 699–719.

Fundamental Problem. Dantzig, pioneer of mathematical programming, has stated that one of the most important new ideas for solving nonlinear programming problems is the use of complementary pivot theory applied to the fundamental problem. Given a real p vector q and a real $p \times p$ matrix M, the objective is to find vectors w and z such that

$$w = q + Mz$$
$$zw = 0$$
$$w \geq 0 \qquad z \geq 0$$

It can be shown that linear programming, quadratic programming, and bimatrix games can be structured in this format. The solution technique is based on the work of Lemke, using the complementary pivot theory approach. Excellent chapters on the work of Dantzig and Lemke in this area are contained in the book edited by Dantzig and Veinott which is listed in the bibliography.

NONLINEAR PROGRAMMING UNDER RISK

The nonlinear programming methods previously discussed all considered the problems to be deterministic, thus assuming that the values of the parameters are known precisely; in fact, most of the effort in solving nonlinear programming problems has been directed to the problem where risk is ignored. The inclusion of risk adds another dimension of difficulty to the already complex nonlinear problem. A concise review of a selected group of optimization studies under risk is presented here.

The general mathematical programming problem when deterministic conditions are assumed is defined to be

Max
$$Z = f(X)$$
subject to
$$g_i(X) \leq b_i \qquad \text{for } i = 1, \ldots, m$$
$$X \geq 0$$

When some or all of the quantities appearing in the problem are random variables with known distributions, we have a programming problem under risk. The simplest condition is where only the coefficients in the objective function are subject to random variation. This problem is formulated as

Max
$$Z = E(f(X))$$
subject to
$$g_i(X) \leq b_i \qquad \text{for } i = 1, \ldots, m$$
$$X \geq 0$$

where $E(f(X))$ is the mathematical expectation of $f(X)$. Dantzig considered this problem under the conditions of linear constraints and an objective function that would have been linear had all parameters been deterministic.[15]

Chance Constrained Programming. The problem becomes more difficult when all the problem quantities are permitted to be random variables, which is frequently the case for engineering design. The technique frequently applied to this problem is called chance constrained programming, in which the deterministic constraints are replaced by their probabilistic counterparts. For example, $g_i(X) \leq b_i$ is replaced by $P(g_i(X) \geq b_i) \leq \epsilon$, where P is the probability of the occurrence and ϵ is some arbitrarily selected small value. Charnes solved this problem by replacing the probability constraints with certainty equivalents.[16] The previously stated probability

[15] G. B. Dantzig, "Linear Programming under Uncertainty," *Management Science*, vol. 1, nos. 3 and 4, 1955, p. 197.

[16] A. Charnes, W. W. Cooper, and G. H. Symonds, "Cost Horizons and Certainty Equivalents: An Approach to Statistical Programming of Heating Oil," *Management Science*, vol. 4, no. 3, 1958, p. 235.

constraint is now written as

$$g_i(X) + K\partial(g_i) \leq b_i$$

where K is a constant assigned to give the desired confidence interval, and the standard deviation $\partial(g_i)$ is evaluated on the basis of past statistical experience. This problem has been most extensively examined where all relationships would be linear but for the presence of random variables.

Naslund[17] has made an extensive investigation of the application of chance constrained programming to problems in economics.

Van de Panne and Popp[18] solved a problem with probabilistic linear constraints and a linear objective function, using the method of feasible directions. Zoutendijk[19] has done considerable work in this area.

Hillier's approach to solving a chance constrained programming problem is unique in that he generates constraints which are uniformly tighter and looser than the actual constraints, but by alternate use of the tight and loose linear constraints, he can solve problems where the decision variables are constrained to be zero or one.[20]

Programming under Uncertainty. Ferguson and Dantzig[21] developed an alternative approach to chance constrained programming called programming under uncertainty. Probability constraints are replaced by penalty costs. For example, if the maximum available supply of resource i is b_i and penalty costs are proportional to the demand in excess of b_i, with other costs linear functions of the decision variables, then the objective function is

$$\sum_j c_j x_j + \sum_i \pi_i \int_{b_i}^{\infty} (D_i - b_i) g(D_i) d(D_i)$$

where $\sum_j c_j x_j$ are the operating costs, D_i is the demand for resource i, $g(D_i)$ is the density function for D_i, and π_i is the penalty cost per unit. Normality is assumed for the cost distribution.

Weisman's Algorithm. In summarizing the work on mathematical programming under risk, it is quickly obvious that most of the problems solved would be typical linear programming problems if it were not for the risk condition. However, Weisman[22] has developed an algorithm and computational procedure to solve risk problems which have both nonlinear objective function and nonlinear constraints expressed in chance constrained form. The algorithm contains options which will handle deterministic nonlinear programming problems, as well as the stochastic type, and will also optimize nonlinear risk-type problems where selected expected utility functions are to be optimized.

NONLINEAR PROGRAMMING CODES

A listing and source identification of computer codes which are being used to obtain solutions for mathematical programming problems is presented for the industrial engineer interested in obtaining answers. These problems are deterministic in nature but may have nonlinearity in both the objective function and the constraints. It must be emphasized that none of these codes guarantees a global optimum solution.

[17] B. Naslund, "Decisions under Risk," ONR Research Memorandum 134, Carnegie-Mellon University, Pittsburgh, Pa., 1964.
[18] C. Van de Panne and W. Popp, "Minimum Cost Cattle Feed under Probabilistic Protein Constraints," *Management Science*, vol. 9, no. 3, 1963, p. 405.
[19] G. Zoutendijk, *Methods of Feasible Directions*, Elsevier Publishing Company, Amsterdam, 1960.
[20] F. S. Hillier, "Chance Constrained Programming with 0, 1 or Bounded Continuous Decision Variables," *Management Science*, vol. 14, no. 1, 1967, p. 34.
[21] A. R. Ferguson and G. B. Dantzig, "Allocation of Aircraft to Routes: Example of Linear Programming of Uncertain Demand," *Management Science*, vol. 3, no. 1, 1956, p. 45.
[22] Joel Weisman, "Engineering Design Optimization under Conditions of Risk," Ph.D. dissertation, University of Pittsburgh, Pittsburgh, Pa., 1968, p. 137.

However, satisfactory solutions for problems with as many as 100 variables have been obtained for some of the codes. Richmond has made a comparative analysis of these codes relative to their capability to solve fifty-nine test problems.[23] Most of the codes use some form of search or gradient methods.

The following list of codes is not all-inclusive but does represent a good size sample of the more useful programs to solve practical nonlinear problems.

  1. Hooke and Jeeves' PATTRN
     Source: R. Hooke and T. A. Jeeves, "Direct Search Solution of Numerical and Statistical Problems," *Journal of the Association of Computing Machinery,* vol. 8, no. 2, 1961, p. 212.
  2. Rosenbrock's ORTHOG
     Source: H. H. Rosenbrock, "An Automatic Method for Finding the Greatest or Least Value of a Function," *Computer Journal,* vol. 3, no. 3, 1960, p. 175.
  3. Fletcher and Reeves' GRAD
     Source: R. Fletcher and C. M. Reeves, "Function Minimization by Conjugate Gradient," *Computer Journal,* vol. 7, no. 2, 1964, p. 149.
  4. Fletcher and Powell's VARMET
     Source: R. Fletcher and M. J. D. Powell, "A Rapidly Convergent Descent Method for Minimization," *Computer Journal,* vol. 6, no. 2, 1963, p. 163.
  5. Shah, Buehler, and Kempthorne's PARTAN
     Source: D. J. Wilde, *Optimum Seeking Methods,* Prentice-Hall, Inc., Englewood Cliffs, N.J., 1964.
  6. Shanno and Smith's REEP
     Source: R. A. Cox, "Comparison of the Performance of Seven Optimization Algorithms on Twelve Unconstrained Optimization Problems," Gulf Research and Development Company Ref. 1335CN04, Pittsburgh, Pa., January, 1969.
  7. Gottfried's COMPUTE
     Source: R. A. Cox, "Comparison of the Performance of Three Optimization Techniques on Ten Constrained Optimization Problems," Gulf Research and Development Company Ref. 133T9215, Pittsburgh, Pa., March, 1969.
  8. Fiacco and McCormick's SUMT
     Source: A. V. Fiacco and G. P. McCormick, "The Slacked Unconstrained Minimization Technique for Convex Programming," *SIAM Journal of Applied Mathematics,* vol. 15, no. 3, 1967, p. 505.
  9. Smith's POP-II
     Source: R. A. Cox, "Comparison of the Performance of Three Optimization Techniques on Ten Constrained Optimization Problems," Gulf Research and Development Company Ref. 133T9215, Pittsburgh, Pa., March, 1969.
  10. Glass and Cooper's Sequential Search
     Source: H. Glass and L. Cooper, "Sequential Search: A Method for Solving Constrained Optimization Problems," *Journal of the Association of Computing Machinery,* vol. 12, no. 1, 1965, p. 71.
  11. Himmelblau's MGST
     Source: W. R. Klingman and D. M. Himmelblau, "Nonlinear Programming with the Aid of a Multiple-gradient Summation Technique," *Journal of the Association of Computing Machinery,* vol. 11, no. 4, 1964, p. 400.
  12. Davies' Davidon-CRST
     Sources: A. R. Colville, "A Comparative Study on Nonlinear Programming Codes," IBM New York Scientific Center Report 320-2949, New York, June, 1968; and C. W. Carroll, "The Created Response Surface Technique for Optimizing Nonlinear Restrained Systems," *Operations Research,* vol. 9, no. 2, 1961, p. 169.
  13. Mugele's PROBE
     Source: R. A. Mugele, "A Program for Optimal Control of Nonlinear Processes," *IBM System Journal,* vol. 1, no. 1, 1962, p. 2.

[22] E. R. Richmond, "A Comparative Analysis among Selected Nonlinear Programming Methods and Weisman's 'Minimal' Code," M.S. thesis, University of Pittsburgh, Pittsburgh, Pa., 1969.

14. Weisman's MINIMAL
Source: J. Weisman, "Engineering Design Optimization under Conditions of Risk," Ph.D. dissertation, University of Pittsburgh, Pittsburgh, Pa., 1968.

BIBLIOGRAPHY

Boot, John C. G., *Quadratic Programming, Algorithms—Anomalies—Applications*, Studies in Mathematical and Managerial Economics, vol. 2, Rand McNally & Company, Chicago, 1964.

Bracken, J., and G. P. McCormick, *Selected Applications of Nonlinear Programming*, John Wiley & Sons, Inc., New York, 1968.

Carr, C. R., and C. W. Howe, *Introduction to Quantitative Decision Procedures in Management and Economics*, McGraw-Hill Book Company, New York, 1964.

Dantzig, G. B., and A. F. Veinott, Jr. (eds.), *Mathematics of the Decision Sciences*, part 1, American Mathematical Society, Providence, R.I., 1968.

Dorfman, R., P. A. Samuelson, and R. M. Solow, *Linear Programming and Economic Analysis*, McGraw-Hill Book Company, New York, 1958.

Duffin, R. J., E. L. Peterson, and C. Zener, *Geometric Programming: Theory and Application*, John Wiley & Sons, Inc., New York, 1967.

Fiacco, A. V., and G. P. McCormick, *Nonlinear Programming: Sequential Unconstrained Minimization Techniques*, John Wiley & Sons, Inc., New York, 1968.

Graves, R. L., and P. Wolfe, *Recent Advances in Mathematical Programming*, McGraw-Hill Book Company, New York, 1963.

Gue, R. L., and M. E. Thomas, *Mathematical Methods in Operations Research*, The Macmillan Co. of Canada, Limited, Toronto, 1968.

Hadley, G., *Nonlinear and Dynamic Programming*, Addison-Wesley Publishing Company, Inc., Reading, Mass., 1964.

Hillier, F. S., and G. J. Lieberman, *Introduction to Operations Research*, Holden-Day, Inc., San Francisco, 1968.

Hu, T. C., *Integer Programming and Network Flows*, Addison-Wesley Publishing Company, Inc., Reading, Mass., 1969.

Kunzi, H. P., and W. Krelle, *Nonlinear Programming*, Blaisdell Publishing Company, Waltham, Mass., 1966.

Teichroew, D., *An Introduction to Management Science: Deterministic Models*, John Wiley & Sons, Inc., New York, 1964.

Wilde, D. J., and C. S. Beightler, *Foundations of Optimization*, Prentice-Hall, Inc., Englewood Cliffs, N.J., 1967.

Zangwill, Willard I., *Nonlinear Programming: A Unified Approach*, Prentice-Hall, Inc., Englewood Cliffs, N.J., 1969.

Equipment and Facilities

Site Selection

LEONARD C. YASEEN

Chairman, The Fantus Company, New York, New York

Increased sales, variance in labor costs and productivity, higher freight costs, shifting markets, and the need for more efficient production facilities have been the principal causes for an unprecedented expansion and decentralization of American industry. Unfortunately, most plant location research has been conducted in an atmosphere of guesswork, and most decisions are reached without the application of scientific principles.

The object of this chapter, therefore, will be to outline for the industrial engineer a series of suggested steps that are considered accepted practice in determining the best location for a new plant.

RAW MATERIALS AND MARKETS

All manufacturing operations require inbound movement of raw materials and outbound movement of the finished product to the market. Accordingly, it is vital for the engineer contemplating new plant location to understand the intricacies of the complex transportation system at his disposal.

In plant location study, distance is measured in terms of freight rates. Because the relative costs of assembling raw materials and distributing finished products will vary from industry to industry, the location pull of raw materials and markets will differ in each problem. No simplified formula can be provided. Generally, all other things being equal, an industry will tend to locate at that point where it will possess the lowest aggregate transportation cost.

In studying the transportation costs of raw materials, it is suggested that the following data be gathered:

1. Location of source of each material
2. Availability
3. Price
4. Terms of sale
5. Freight rates to site

In modern industrial terminology, raw materials include commodities which are purchased in a partially processed state or as completed components. The machinery manufacturer considers his raw material to be iron and steel mill shapes rather than the ore, fuel, and flux used in their production. The final inputs of the computer manufacturer include main frames, peripheral equipment, and subassemblies. At intermediate stages of production, these require integrated circuits, modules, memory devices, and other components. Hence, in the locational sense, raw materials include all purchased materials and supplies necessary to manufacture the product.

It may become advisable in some studies (particularly those involving divisions of national corporations) to divide raw materials into static and dynamic supplies. The former group represents captive producers or others with long-term commitments which cannot be readily shifted. Dynamic sources include those which are entirely dependent upon location of the plant and can be adjusted without difficulty.

In the selection of raw material sources, there is a tendency to overemphasize the competitive position of the local producer. Actually the more distant source may be able to compete effectively, owing to a favorable freight rate relationship. Most raw materials have been accorded low rate levels under various commodity rate structures. Designed to reflect the delicate balance between competitive sources, these rates place less emphasis upon distance than upon the ability to reach the market. Commodity rates are dynamic, and new rates can be readily established to meet shipper and carrier needs at new manufacturing locations.

Basing-point pricing was eliminated in some industries following a post-World War II Supreme Court decision. However, the practice of equalization, freight allowance, and zone pricing has not been disturbed. On raw materials involving such adjustments, the competitive relationship among producers must be examined to determine whether the practice will be continued, curtailed, or expanded at the new location.

In determining the area which offers the greatest marketing advantages, the rate levels of competitors should be analyzed to determine their natural market. New facilities should be established at that point where a freight advantage can ensure control of a substantial market and where important adjoining markets can be reached competitively.

In determining the best market location, the plant location investigator should attempt to neutralize his competitor's freight advantage. As an example, consider a manufacturer of electrical apparatus who is considering central Ohio as a location for his new plant. The majority of his distribution is in Ohio and neighboring states, but he has an important carload movement to a distributor in Los Angeles. His major competitor is located in Cleveland, Ohio.

If he selects Marysville, Ohio, as his location, he and his competitor will have a rate of 440 cents per 100 pounds to Los Angeles. Owing to rate blanketing, his location 100 miles nearer the West Coast will have no effect on his competitive position. However, if he shifts only 7 miles farther northwest of Marysville to Peoria, Ohio, group C rates will replace group B, and his rate will be 421 cents. On every carload to Los Angeles, he will enjoy a differential of $76 over his competitor.

Freight rates are not determined by means of mathematical formulas but are constructed to reflect the revenue needs of the carriers and the competitive relationship among products, producers, and regions. Their thousands of entirely distinct substructures and delicate interterritorial relationships are beyond the scope of this chapter. However, the engineer considering plant location should be familiar with the general aspects of rate making and traffic geography.

The railroads divide the nation into three major classification territories with general boundaries as follows:

1. *Official Classification Territory* consists of the northeastern portion of the United States lying north of the Ohio and Potomac Rivers and east of the Mississippi River. Omitted from the territory are the northern peninsula of Michigan, the states of Wisconsin and Minnesota, and the northwestern portion of the state of Illinois.
2. *Southern Classification Territory* consists of that portion of the country lying south of the Ohio and Potomac Rivers and east of the Mississippi.
3. *Western Classification* is the balance of the United States west of the Mississippi River and including those states east of the Mississippi which are not part of Official Classification Territory.

Motor carriers have renamed the Official, Southern, and Western Classification Territories as East, South, and West, respectively. However, their boundaries remain virtually unchanged.

Within these regional boundaries, the carriers have organized themselves into rate making associations (specifically exempted from antitrust laws). They deal with common problems of carriers serving economic subdivisions of the major classification territories and usually publish rate books called tariffs.

Freight rates are usually related to distance, but they do not increase in direct ratio with increasing mileage. Each rate involves a terminal cost at origin, a line-haul cost, and a terminal cost at destination. As hauls increase in length, these terminal costs assume less and less importance. Hence, the rate of increase per mileage block is lower on longer hauls than on shorter hauls. It is this paradox that requires careful consideration of actual freight rates rather than the scale of miles in appraising prospective plant locations.

Three basic types of rates have emerged:

1. *Class rates.* Every article produced in the country is classified into one of a limited number of class ratings, and a list of the ratings assigned is published in a classification. The ratings are percentages of the first-class rate (100 percent) or multiples of it. In turn, the first-class rates are published between virtually all origins and destinations in the country in class rate tariffs. The applicable rate is determined by finding the first-class rate from origin to destination and then applying the assigned rating or percentage. Class rates can be superseded by exception ratings.
2. *Exception ratings.* Because of commercial and operating conditions within rate making territories, the carriers may wish to alter the original rating assigned to an article. This is accomplished by publication of an exception rating. When applied as a percentage of the first-class rate between the stations involved, a lower (or higher) rate results than from similar use of the classification rating. Both class rates and exception ratings can be superseded by commodity rates.
3. *Commodity rates.* To meet special needs of the carriers or shipping public, commodity rates are published. Usually they apply on a specific commodity moving from one named station to another.

LABOR

The traffic study is a prerequisite for the selection of a broad geographic area. Once such an area has been determined, the engineer is then ready for exploration of specific communities within that area. Perhaps the most important single phase of the search will revolve around labor and all its ramifications—cost, availability, stability, and productivity.

It is important that a community have a true and permanent labor surplus if it is to receive serious consideration. The following is a series of steps that are recommended to determine labor availability:

1. Contact the local office of the state employment service. Personally inspect

applicants' cards, including age, former position, previous wages, skills, and residence.

2. Determine qualified applicants by weeding out applicants who are interested only in domestic, professional, or commercial work; those overage; those eligible for military service; and laborers who may be drawing temporary unemployment compensation owing to off-season layoffs.

3. The caliber of available help can often be determined by the number of unfilled jobs. Hence, compare the number of job openings currently available.

4. Interview present manufacturers to determine what their experiences have been in the community.

5. Examine newspaper want ads over a long period of time to ascertain the skills most in demand.

6. In marginal cases where field investigation and statistical analyses fail to produce conclusive measurement of available resources, it may be necessary to resort to a labor registration. The purpose of such a registration is actually to solicit applicants for a proposed new plant.

Following is a questionnaire recapitulation of an actual labor registration conducted by the Fantus Company:

	Female	
Total applicants	1,373	
Age 18–35	841	
Percent of total returns	.....	61.3
Over 35	532	
Percent of total returns	.....	38.7
Age 18–35, presently employed	472	
Percent of total in age group	.....	56.1
Over 35, presently employed	274	
Percent of total in age group	.....	51.5
Age 18–35, willing to work shifts	530	
Percent of total in age group	.....	63.0
Over 35, willing to work shifts	374	
Percent of total in age group	.....	70.3
Age 18–35, married	477	
Percent of total in age group	.....	56.7
Over 35, married	387	
Percent of total in age group	.....	72.7
Age 18–35, not employed and willing to work shifts	224	
Percent of total in age group	.....	26.6

Residence	Age distribution						Total
	18–25	26–30	31–35	36–40	41–45	46–over	
Local	310	162	140	131	95	173	1,011
Suburban	56	25	26	25	13	31	176
County	63	28	31	31	21	12	186
Total	429	215	197	187	129	216	1,373

In evaluating the foregoing registration, it is evident that despite the impressive total of 1,373 female applicants, only 224 who do not object to shift work, are not presently employed, and are in the preferred age bracket are qualified for consideration. An aptitude test for the specific type of industrial process involved might eliminate at least half of these remaining applicants.

LABOR COSTS

Wide variations in wage rates exist among geographic areas, between large cities and small communities, and often between neighboring communities. Wage scales for unskilled labor and many skilled occupations are traditionally lowest throughout the southeastern area of the United States. Highest wages appear most often on the Pacific Coast.

Estimates made in 1967 reveal that median income for all United States families was $9,019 as compared with $9,873 in the West and $7,881 in the South. In mid-1969, average hourly earnings and average weekly hours for production workers in manufacturing plants approximated the following:

State	Average weekly hours	Average hourly earnings
California................	40.6	$3.62
Oregon....................	39.2	$3.54
Illinois...................	41.3	$3.43
New York.................	39.6	$3.21
Pennsylvania..............	40.5	$3.17
Massachusetts.............	40.0	$3.02
Tennessee................	40.5	$2.56
Mississippi...............	40.5	$2.32
North Carolina............	41.1	$2.30

Some economists have advanced the theory that when rural areas become industrialized, wage rate differentials cease to exist. However, North Carolina's low average hourly rate of $2.30 has been maintained despite the fact that over 275,000 are now employed in the manufacture of textile mill products alone in that state.

Wage differentials among cities are just as pronounced. In mid-1969, the following rates were in effect:

City	Average weekly hours	Average hourly earnings
Detroit, Mich.............	42.6	$4.16
San Francisco, Calif.........	39.5	$4.03
Portland, Ore.............	39.4	$3.58
Rochester, N.Y............	41.0	$3.58
Boston, Mass.............	39.5	$3.26
Newark, N.J..............	41.1	$3.25
Atlanta, Ga..............	41.4	$3.17
Chattanooga, Tenn..........	41.9	$2.73
Providence, R.I............	40.2	$2.70
Little Rock, Ark...........	40.6	$2.50
Charlotte, N.C.............	41.6	$2.44

In spite of unionization, vast differences exist in wage rates for identical work performed in separated locations. A carpenter receives $6.67 in Detroit, $6.71 in Newark, $4.98 in Baltimore, $5.13 in Dallas, and $4.65 in Memphis.

Wage rates in large metropolitan areas are usually higher than in less developed regions. Union wage scales in the building trades as of July 1, 1967, averaged $5.34 in the highly urbanized mid-Atlantic states. In the predominantly rural Southeast, identical jobs averaged $4.02, a $1.32 differential.

It is obvious, therefore, that if labor cost is a vital percentage factor in total delivered-to-customer cost of a given product, the community finally chosen will exert tremendous influence on the competitive position of the new plant.

LIVING COSTS

A very important influence on wage rates is the spread in expenditure standards from one community to another. BLS statistics developed in 1967 define three standards of living for a wife and two children supported by an employed husband.

On the national average, the income required was $5,915 for a low standard of living, $9,076 for a moderate standard, and $13,050 for a good standard. Nationally, these costs are increasing at an annual rate of 7.5 percent a year.

Nationally, the highest and lowest cost cities for a low standard of living are Honolulu, $7,246, and Austin, Texas, $5,237. For a good standard of living, the spread is from Honolulu's $16,076 to Austin's $11,299.

Contributing to variations in expenditure standards and buying tempo are such factors as differences in heating costs, need for winter apparel in colder climes, rentals, community utility and service costs, and state and local taxes.

To compare the relationship of earnings and selected living cost items more specifically, in Rochester, New York, manufacturing production workers in 1967 had average earnings of $3.17 per hour. In contrast, factory production workers in Nashville, Tennessee, received $2.46. The following table reveals two basic living cost differentials which allow the Nashville worker with lower earnings to live as comfortably as the Rochester employee with higher wages.

City	Median monthly rent (1960 census)	500 kwhr residential electric bill	Number of degree-days*
Rochester, N.Y.........	$80	$11.62	6,748
Nashville, Tenn........	59	6.90	3,578

* Determines heating costs.

LABOR STABILITY

Thorough precautions to assure low production costs are of no avail unless the proposed new plant can operate with continuity and with tranquil labor-management relations. More than one company has been forced out of business because of unreasonable or prohibitive labor demands fostered, in extreme cases, by community-wide antagonisms.

As unions become increasingly stronger, relatively fewer work stoppages are due to the fundamental question of union organization. Wage increase demands and jurisdictional disputes continue to be important points of conflict, and many stoppages are caused by fringe benefit demands such as pension and insurance plans.

The question of labor stability must be approached from a positive standpoint. There are certain strong indices of community attitude that should influence its selection. Perhaps the most crucial question that can be asked about a community is "What is its past history?"

POWER

The cost of power as a percentage of total delivered-to-customer costs in most industries is not significant. Hence, the average industrialist will not relocate his plant or establish a branch unit solely because of a power differential. Nevertheless, power costs do constitute a sizable, constantly recurring expense for many industries and should be carefully compared along with adequacy, reliability, and type of service available in the area under consideration.

Most companies considering plant location are prepared to purchase rather than manufacture their own power. The size of the average utility, the diversified load, and particularly the increased efficiency of generating plants have maintained the cost of purchased power at comparatively low levels.

If the plant is in operation 24 hours a day and a sizable amount of exhaust steam can be used for processing, refrigeration, cooling, or heating, it may be advisable to generate rather than purchase power. Gas companies have successfully marketed the "total energy concept" in many parts of the country to customers who satisfy all their energy needs with natural gas. Substantial utility cost savings are possible where the customer's requirements for electric power and steam can be engineered to match the performance output of a natural gas-fed turbine and exhaust heat recovery system unit. Some gas companies offer financing arrangements to reduce the customer's capital outlay. In many parts of the country, commercial operations—hotels, resort complexes, apartment house complexes, and shopping centers—have had notable cost savings employing the total energy concept. Even so, industrial firms must give careful consideration to the capital outlay, cost of failures and repair, and cost of standby services, assessing all these factors against the total kilowatt-hour annual bill.

Some of the factors to be considered when examining the power situation in a given area may be found in the following suggested checklist:

1. Type of service
   a. Hydroelectric
   b. Steam
   c. Other
2. Reliability of service, history of stoppages
3. Adequacy of supply, seasonal restrictions
4. Kind
   a. Phase
   b. Cycle
   c. Voltage
5. Rates
6. Availability of off-peak contracts
7. Fuel adjustment
8. Lighting allowances
9. Discounts and penalties

The type of generation will have an effect on power bills. Hydroelectric power is usually associated with cheap rates, although the original installation cost of the generating plant is considerably more than for steam plants with similar capacity.

At most hydroelectric sites, stream flow fluctuates widely. Unless steam-developed power is available to carry part of the load during deficient flow periods, industry may be subject to interruptions in service. Interconnection of transmission lines with systems utilizing steam generation can ensure continuous supply. These tie-in arrangements should be carefully investigated in hydroelectric districts. The remoteness of the hydroelectric installation from consuming areas and difficult terrain conditions may increase the frequency of service interruptions due to transmission line failures.

It is important to ascertain the reserve power of a community before definite steps are taken to establish a manufacturing plant. Despite an installed capacity of electric utility generating plants in 1968 of 290 million kilowatts and an estimated annual production of 1,400 billion kilowatt-hours, some areas of the United States had insufficient power to cope with mushrooming industrial expansion. Even certain portions of the huge TVA project could not immediately provide additional power to potential new customers. As a stopgap measure, TVA purchased power from privately owned utility companies on its periphery.

The phase, cycle, and voltage of the supply must be carefully checked. Motor wiring in plants in long-established areas of the country may not conform to the power furnished in decentralized areas. Accordingly, the manufacturer faces either an expensive conversion process or a sizable transformer loss. In rural areas, the distribution lines are normally of insufficient capacity to serve large manufacturing plants. Although power companies are willing to make reasonable improvements to accommodate industrial plants, responsible utility executives must be consulted to determine the extent of cooperation which will be extended.

FUEL

Fuel as a locational factor varies in importance from industry to industry. In those processes utilizing fuel as a basic raw material, the plant may be entirely oriented to low-cost supply of the commodity. Substitution of other fuels may be difficult or impossible. Typical examples are the manufacture of coke and coking by-products from bituminous coal and carbon black from natural gas.

Considered along with assembly cost of raw materials and freight rates to markets, fuel has influenced the location of an important segment of heavy industry. For example, the production of 1 ton of pig iron requires the assembly of 1.73 tons of iron ore, 0.41 ton of limestone, and 0.93 ton of coke. Derived from certain grades of bituminous coal, coke is used in the smelting process both as a reducing agent and as a fuel. Orientation to coke sources has accounted for the development of iron- and steelmaking in Pittsburgh and the Mahoning Valley. However, even in this basic industry, changes in pricing policy and shifting markets are reducing the influence of fuel supply on the location of blast furnaces.

Vast differences exist in the quality of coal mined in various areas. Coals found in the central part of the country, for example, are usually of inferior quality to coal found in the Appalachian Highlands, having less fixed carbon and more volatile matter and moisture. Coal in southern Illinois has a much higher carbon content than coal mined just a few hundred miles north. Many manufacturers in comparing the cost of delivered coal fail to assess comparative heating value in terms of Btu's.

Over 800,000 miles of mains deliver natural gas to every important metropolitan area of the United States. Reserves in the Gulf Coast and Southwest areas are soon to be augmented by overseas supplies which will be transported and stored at cryogenic temperatures as liquid natural gas (LNG). In 1966, 17.2 trillion cubic feet of gas were consumed. An average of 850,000 new natural gas customers a year had been added in the previous six years.

Most companies considering plant locations select natural gas, when available, for their space heating and process fuel requirements. Even in regions where coal is available at lower delivered unit cost on a heating value basis, natural gas provides lower total investment and operating costs for most cases. Where economic evaluation balances options among coal, fuel oil, and natural gas, careful consideration must be given to air pollution regulations, disposal of solid waste ashes, controllability and turndown ratio requirements, and provisions for expansion.

When considering natural gas as a source of fuel for the proposed plant, the following suggestions may be of value:

1. The presence of a trunk line is no indication of availability.
2. Despite the fact that the community is presently served by natural gas, it may not have sufficiently large allocations to supply a new industry.
3. Users must be exceptionally large to allow economical tapping of a main line in the absence of local distribution facilities.
4. Only points near the gas fields can guarantee a firm supply. In other areas, service is on an interruptible basis.
5. If a firm supply is offered, the rates quoted should be checked, as interruptible rate schedules may be lower.
6. Where interruptible supplies are to be utilized, the cost of standby or storage facilities must be considered part of the fuel cost.
7. The source of the supply must be carefully checked. There is danger in depending on a single field unless tie-in arrangements are assured.
8. The pumping capacity at the site can be as important as the size of the line.
9. In comparing costs, the average Btu content must be ascertained.

WATER

As the population of the nation expands, the demand for water increases. The average United States city requires about 113 gallons per day per capita. Four acres of land are required to support each person in our economy, and of necessity, over 22 million acres of land in the nation must be irrigated.

New industrial techniques, including the harnessing of atomic power and the production of synthetic fuels, require more and more water. In many industries, the problem of securing usable water at reasonable rates is a pressing one. In fact, water supply is a prerequisite in site selection in steel, paper pulp, paperboard, wool scouring, food, and chemical processes. Typical water requirements for various industries are shown below:

	Gallons
Produce a ton of aluminum	80,000
Make a ton of steel plate from ore	40,000
Produce a ton of paper from cordwood	30,000
Produce a ton of sugar from sugar beets	25,000
Dehydrate 1,000 pounds of potatoes	12,000
Produce a motor vehicle	9,000
Pack 1,000 pounds of meat	3,000
Brew a barrel of beer	500
Produce a case of canned tomatoes	100
Generate 1 kilowatt-hour from coal	40

In the majority of manufacturing operations, public water supplies will prove satisfactory directly from the tap or with minor treatment. The quality of the supply will depend upon the source from which it is derived. Shown below is a comparative study of public water supplies in one region of a southeastern state. The table illustrates the wide variations in characteristics to be found within a limited area.

Comparative Typical Analysis of Public Water Supplies
(Piedmont province, southeastern state)

	Source		
	Surface	Springs	Wells
Total solids (180°C)	58.8	83.0	484.4
Volatile residue	12.0	28.5	188.4
Mineral residue	46.8	54.5	296.0
Dissolved silica (SiO_2)	11.6	24.2	30.6
Iron and aluminum oxides (Fe_2O_3 and Al_2O_3)	1.6	2.2	2.4
Calcium (Ca)	3.6	7.3	55.9
Magnesium (Mg)	4.3	1.4	16.8
Chlorides (Cl)	4.8	3.12	125.8
Nitrates (NO_3)	0.06	0.12	3.6
Sulfates (SO_4)	4.8	14.9	24.7
Bicarbonates (HCO_3)	35.0	45.0	107.4
Soap hardness	11.1	78.6	145.8
Calculated hardness	28.0	74.4	208.6
Iron (Fe) colorimetric	None	....	Trace
Free carbon dioxide (CO_2)	2.6	....	174.3
Normal carbonates (CO_3)	None	None	None
Sodium (Na), calculated	6.4	15.0	39.3
pH	6.8	6.8	6.4

Basically, water is available from three sources:

Surface (water from lakes, streams, and the like)
Ground (springs and wells)
Rain

Impounded supplies (lakes and reservoirs) are usually clear, soft, and high in oxygen, as impurities tend to settle out during storage. Of the various possible sources, supplies taken from impounded waters show the most consistency in composition. During summer months, however, microscopic organisms and vegetation may add taste and color harmful to certain manufacturing processes. Decomposing organic matter easily combines with the free oxygen to form carbon dioxide, which is injurious to pipelines and industrial equipment.

Rivers and streams vary greatly in their analyses, even during relatively short periods of time. The character of the drainage area will dictate relative hardness, turbidity, and other characteristics. Surface waters are subject to contamination from animal wastes, sewage, and seepage from coal mines.

The prudent investigator will determine maximum and minimum flow, seasonal variations, tidal influence, temperature, and the like. Several analyses should be studied from samples taken at various times of the year. On larger streams, lengthy flow records may be available from numerous gaging stations, but data on smaller tributaries may be completely lacking.

Experienced engineers can estimate flow data by examining topographic conditions, rainfall in the drainage area, and like factors. Similar study will reveal the most advantageous location for the erection of private dams to impound the supply.

The dispersal of industry into semirural areas beyond city water mains has revived keen interest in the study of groundwater supplies. The geology of an area will dictate the quality, quantity, and location of underground sources. If wells are contemplated, a basic understanding of groundwater principles is necessary.

Precipitation is the principal source of groundwater supply. Rain and melted snow percolate down through soil particles and through cracks, joints, and bedding planes of rocks. At the zone of saturation, all cracks and pores become filled with water. The upper limit of this zone is commonly called the water table. Depending upon geographic location, the water table may lie a few inches under the land surface or hundreds of feet below it and will be constantly fluctuating.

The underlying water-bearing formations are called aquifers. The quantity of water which they contain will depend upon the characteristics of the rock, which are known as porosity, specific yield, and permeability. The most important property of the underlying aquifer to the industrialist is its permeability or ability to transmit water. Clays, for example, transmit water very slowly. Clean, medium-grained sands give up water very rapidly. In rocks, the size and number of openings will determine the yield. The tilt of the bedding planes also affects the yield, as the number of planes intersected by a well is greater in areas of moderate tilt than where the planes are steeper.

Removal of water from a well causes a decrease in pressure, and the water table in the vicinity will have a shape similar to an inverted cone, known as the cone of depression. If two or more wells are drilled in close proximity, these cones may overlap, interfering with each other and lowering the yield of both wells. This limits the number of wells which may be sunk on a given site.

In some densely populated areas, this mutual interference of wells has been of grave concern to industry. Some coastal cities report a drastic lowering of the water table to the point where all supplies have been contaminated by seawater drawn into the low-pressure areas.

The necessity of finding a water supply that does not require extensive conditioning is important. High-pressure and low-pressure steam are important throughout industry in general, particularly in chemical and food industries and paper and textile finishing. Large amounts of cooling water are required in the manufacture of iron and steel, in metal processes, and in modern plants using air conditioning and other refrigeration devices.

Even rainwater is contaminated, and most raw water requires some conditioning before it is fit for industrial or domestic use. As rain falls, it absorbs the gases of the atmosphere. When this water comes in contact with soils and rocks, it dissolves considerable mineral matter owing to the solvent activity of the atmospheric gases in solution. As water passes over the ground or percolates through it, it will therefore undergo many changes in its chemical composition.

Set out below are the minerals usually found in water and their effect upon industrial processes and equipment.

Mineral	Effect on industrial equipment and products
Bicarbonate (HCO_3)	Affects taste
Calcium (Ca)	Forms an insoluble curd in pipes and boiler tubes, soap consuming
Chloride (Cl)	Affects taste, increases corrosiveness
Fluoride (F)	Amounts larger than 1.5 parts per million mottle enamelware
Iron (Fe)	About 0.3 part per million stains cloth, porcelain fixtures, and other materials
Magnesium (Mg)	Forms an insoluble curd in pipes and boiler tubes, soap consuming
Manganese (Mn)	About 0.3 part per million stains cloth, porcelain fixtures, and other materials
Nitrate (NO_3)	Large amounts indicate pollution
Potassium (K)	Large amounts cause foaming in boilers
Silica (SiO_2)	Results in boiler scale and destructive hard deposits on equipment
Sulfate (SO_4)	Can form permanent hardness and scale

Hard water can eventually create havoc in steam boilers, hot water pipes, pumps and circulating systems, diesel engines and other water-jacketed equipment, bleach tanks, and many industrial processes. The corrosion and scale resulting from its use mean increased costs for replacements and disruptions in operations. A helpful chart on relative hardness is shown below:

Hardness, parts per million	Classification
Less than 15	Very soft water
15–50	Soft water
50–100	Medium-hard water
100–200	Hard water
Over 200	Very hard water

The results of water analysis are usually expressed in parts per million (ppm). An exception is the indication of the pH factor. This is a measure of hydrogen-ion concentration of water and is an expression of its acidity or alkalinity. Neutral water has a pH of 7.0. Values below 7.0 and approaching zero denote increasing acidity, and values above 7.0 and 14 indicate increasing alkalinity.

Manufacturers who propose to utilize water for cooling purposes will do well to examine carefully climatic data for the region under consideration. The temperature of surface water follows change in area temperature. However, groundwater temperature generally is about the same as the mean annual air temperature.

TAXATION

Few industries have relocated their plants solely because of unfavorable state taxes. It is rather the cumulative effect of this and other high costs that prompts a manufacturer to consider relocation.

Because of huge budgetary increases, the various states have sought new forms of taxation and have increased existing sales, property, alcoholic beverage, and other taxes. All states have property taxes, but the principal variation between states is most apparent in the corporate income tax. As of July, 1969, eight states assessed no corporate income tax, and 42 states, plus the District of Columbia, imposed a tax ranging from 1 to 9 percent. Because of differences in application of the income tax and the wide variation in regulations, it is difficult, if not completely misleading, to compare simple tax rates. Some states employ a formula for the apportionment of income, and some do not. Some states permit deduction of Federal income tax, and some do not.

On a statewide basis, the principal source of revenue is the ad valorem or property tax, based on the use or ownership of property and measured by the value of the property. Many states impose some type of real property tax, and some states

1969 Special Features Concerning State Corporate Income Taxes

State	Net operating loss carry-over	Federal income tax deductible	Date due for report of change in Federal Reserve	Federal income used as state tax base	Follow the Uniform Division of Income for Tax Purposes Act
Alabama	No	Yes	None	No	Yes
Alaska	Yes	No	20 days	Yes	Yes
Arizona	Yes	Yes	90 days	No	No
Arkansas	Yes	No	30 days	No	Yes
California	No	No	90 days	No	Yes
Colorado	Yes	No	30 days	Yes	No
Connecticut	No	No	90 days	Yes	No
Delaware	Yes	No	90 days	Yes	No
District of Columbia	No	No	30 days	No	Yes
Georgia	Yes	No	2 years	Yes	No
Hawaii	Yes	No	90 days	Yes	Yes
Idaho	Yes	No	Immediately	Yes	Yes
Illinois	No	No	20 days	Yes	Yes
Indiana	Yes	No	None	Yes	Yes
Iowa	Yes	Yes	[a]	Yes	No
Kansas	Yes	Yes	90 days	Yes	Yes
Kentucky	No[b]	Yes[c]	30 days	Yes	Yes
Louisiana	No	Yes	60 days	No	No
Maine	No	No	90 days	Yes	Yes
Maryland	No	No	30 days	Yes	No
Massachusetts	No	No	1 year	Yes	No
Michigan	Yes	No	20 days	Yes	Yes
Minnesota	Yes	Yes	90 days	No	No
Mississippi	Yes	No	60 days	No	No
Missouri	No	Yes	None	No	No
Montana	No	No	90 days	Yes	Yes
Nebraska	Yes	No	90 days	Yes	Yes
New Jersey	No	No	90 days	Yes	No
New Mexico	Yes	No	3 days	Yes	Yes
New York	Yes	No	90 days	Yes	No
North Carolina	Yes	No	2 years	Yes	Yes
North Dakota	Yes	Yes	90 days	Yes	Yes
Oklahoma	No	Yes	1 year	No	Yes[d]
Oregon	Yes	No	90 days	No	Yes
Pennsylvania	No	No	30 days	Yes	No
Rhode Island	No	No	60 days	Yes	No
South Carolina	Yes	No	None	No	Yes
Tennessee	Yes	No	None	No	No
Utah	No	Yes	90 days	No	Yes
Vermont	Yes	No	30 days	Yes	No
Virginia	Yes[e]	No	None	No	Yes
West Virginia	Yes	No	90 days	Yes	Yes[d]
Wisconsin	Yes	Yes[c]	None	No	No

[a] Amended state return must be filed whenever an amended Federal return is filed.
[b] Allowed for new business in first year of operation.
[c] Deductions limited.
[d] Applicable to allocation of nonbusiness income.
[e] Carry-over allowed manufacturing business only.

impose a tax on tangible personal property. Still others assess intangible personal property as well, such as money, investments, and the like.

The general property tax rate is the total of all school, county, township, and state levies assessed against the property. Because most property valuations for assessment purposes are made on a local level, municipal or county practices are very important when estimating the total property tax to which a new plant might be subject.

LABOR LAWS

Virtually all industry is subject to state labor controls, legislation for which may be found in the statutes, codes, and session laws of each state. Because state laws are frequently more stringent and specific, they frequently supersede Federal laws.

The regulation of labor unions, strikes, picketing and boycotting, collective bargaining agreements, unfair employment practices, anti-injunction laws, and wage and hour laws differs radically from state to state.

Interested manufacturers can secure complete digests of state labor legislation from the Department of Labor (or Industrial Commission) of the state capitals. States having labor relations acts include:

Colorado	New York
Connecticut	Oregon
District of Columbia	Pennsylvania
Kansas	Rhode Island
Massachusetts	Utah
Michigan	Wisconsin
Minnesota	

The states of Alaska, Arizona, California, Colorado, Connecticut, Delaware, District of Columbia, Hawaii, Idaho, Illinois, Indiana, Iowa, Kansas, Kentucky, Maine, Maryland, Massachusetts, Michigan, Minnesota, Missouri, Montana, Nebraska, Nevada, New Hampshire, New Jersey, New Mexico, New York, North Dakota, Ohio, Oklahoma, Oregon, Pennsylvania, Rhode Island, Utah, Vermont, Washington, West Virginia, Wisconsin, and Wyoming have Fair Employment Practice Acts.

Those states having minimum wage orders, affecting either all or some phases of industry, include Alaska, Arkansas, California, Connecticut, Delaware, District of Columbia, Hawaii, Idaho, Illinois, Indiana, Kentucky, Maine, Maryland, Massachusetts, Michigan, Minnesota, Nebraska, Nevada, New Jersey, New Mexico, New York, North Carolina, North Dakota, Oregon, Pennsylvania, Rhode Island, South Dakota, Texas, Vermont, Washington, Wisconsin, and Wyoming.

WORKMEN'S COMPENSATION INSURANCE LAWS

Workmen's compensation insurance rates are applied against every single dollar the manufacturer expends in payroll. Geographically, there can be a sizable differential to the manufacturer in his annual workmen's compensation insurance bill due to varying state laws. The maximum period for temporary total disability in California is 240 weeks, whereas the worker in the state of New Mexico is entitled to a maximum of 500 weeks. Maximum percentage of wages range from 90 percent in Illinois down to 50 percent in Montana and Oregon.

Rates payable on each dollar of payroll can produce sizable differentials, depending upon the state under consideration. Iron foundries (classification code no. 3081) are subject to the following rates for each $100 of payroll expended (base rates without giving effect to the individual company's experience benefits):

New Jersey	$9.23
Oregon	8.11
Rhode Island	3.81
Texas	3.33
Illinois	3.06
Alabama	2.25
Indiana	2.07
Pennsylvania	1.95
Virginia	1.71

UNEMPLOYMENT TAX

States vary in their minimums and the employers' actual rate of contribution. State averages range from 0.4 percent to 4.0 percent.

CLIMATE

It is a proved fact that weather exerts great influence on human efficiency and behavior. More crimes are committed during the hot, humid period in July and August than at any other time of the year. The desire to work and the capacity to produce are affected by weather. Air pressure, humidity, and climatic conditions play a large role in a thousand different industrial processes. Snow, rain, haze, and sunshine all have an effect on both employer and employee, on total costs of doing business, and even on the type of structure necessary to house an industrial operation.

One of the nation's principal airplane manufacturers, when a choice had to be made between a location in either of two fine communities, one in Pennsylvania and the other in Tennessee, finally selected the Tennessee city because of its higher percentage of days with sunshine. He realized that grounding of his planes would cost him money.

Each area of the United States has a characteristic climate. If certain industrial processes require minimal seasonal differences, the manufacturer would do well to examine southern California. No other section of the country offers so slight a difference between winter and summer temperatures. On the other hand, the Pacific Coast region experiences the greatest degree of fog in the United States.

If a manufacturer requires an area in which he can recruit common labor at competitively low rates, he would consider the cotton belt, whose climate makes it possible to subsist at lower cost than perhaps anywhere else in the country. Heavy winter clothing is unnecessary, heating bills are insignificant, and food crops can be raised at low cost.

Manufacturers utilizing large amounts of floor space in proportion to total number of employees, where maintenance is a factor, will prefer areas where there is no frost-line problem and no freeze and thaw cycle—thus eliminating continuous pointing up of brick work, removal of ice and snow, enclosed loading platforms, and the winterizing of trucks and other plant equipment.

In areas where temperatures seldom fall to the freezing level, concerns requiring open storage of drums and other materials can operate with no time lost owing to weather conditions.

The velocity and direction of prevailing winds may affect the ventilating problems of large plants, and the orientation of a new plant on a proposed site will vary greatly, depending upon the characteristics of the area. This is especially true if any phase of the operation or neighboring plants produce noxious odors, fumes or heat. In the city of Chicago, average wind velocity in the month of June is 15 miles per hour and wind direction is southwest. In Kansas City, average velocity is 9.1 miles per hour and wind direction is north. In Albany, New York, velocity is 6.7 miles per hour and wind direction is south. The weatherman is not at all satisfied with the above inconsistencies—in many areas, wind direction reverses or shifts during periods throughout the year. Prevailing winds in Seattle, for example, are south in June and north in July.

Manufacturers interested in ideal working conditions should seek a climate with frequent but moderate weather changes and gradual seasonal changes. Periods of continuous wet spells or constant sunshine lead to weather monotony and may reduce initiative and accomplishment. Approximately 40° during the winter season and 64° in the summer are considered ideal temperatures.

From a personnel standpoint, geographic extremes should be avoided. Exceptional dryness—anything below 25 percent relative humidity—promotes susceptibility to the common cold and other respiratory diseases. It is a well-known fact that pneumonia is more prevalent in large cities than in rural areas, owing perhaps to dust.

The type of construction necessary, the amount and kind of insulation required, and the heating costs of a proposed plant in any area can be determined quite accurately through the utilization of data on degree-days. Engineers have discovered that the fuel required to heat an enclosed area is contingent upon the number of degrees by which the average temperature falls below 65°F.

The degree-day is officially defined as "a departure of one degree per day in the mean daily temperature from an adopted standard reference temperature, usually 65°F." For instance, if the mean outside temperature were 45, 65 − 45, or 20 degree-days, would be recorded.

Today, heating engineers depend upon degree-day computation to determine heating capacity and size of boilers, radiators, and all other heating devices. In areas with mild weather, space or wall heaters are used instead of costlier central heating plants. As a result, less costly above-grade utility rooms are usually installed in such homes instead of basements.

FINAL ANALYSIS OF COMMUNITIES

Up to this point in his plant location analysis, the investigator has analyzed his markets; the geographic pull of his necessary raw materials; his labor requirements; the power, fuel, and water costs of various areas; and the effect of various state taxes and laws in his specific operation. He has, through a process of careful elimination, selected a general area within which he must indicate the one outstanding community for his specific manufacturing requirements.

The gathering of objective information is fundamental in any important step of this nature. Promotional conversation on the part of civic representatives, no matter how well intended, cannot be substituted for fact. Many questions should be answered before a final decision is reached: What is the true nature of the people of the community? Do they own their own homes? Do they have deeply embedded roots in the city, or is the working force transient and disinterested in the general community good? Have there been many strikes? What were the issues, and what was the reaction of local law enforcing agencies? How have the tax assessors treated present industry? Have zoning boards been sympathetic to the unique problems of industry? Are there hidden but powerful antagonisms and resentment to industry that do not appear on the surface? Is there a traffic problem in the community that has been met with forceful action? Or will employees meet with congestion and delay in getting to and from the proposed plant?

An investigation of the civic administration of the community is important. The form of government, police and fire personnel and equipment, streets and highways, sewers, and of course, the taxes and budget set up to activate community services all have a bearing on the ultimate safety and protection of a new plant. Cities with a high incidence of crime, abnormal losses from theft and attack against property, or a high ratio of juvenile delinquency should be avoided. Police personnel may be inadequate or civic authorities may be lethargic in combating these abortive influences.

As a yardstick for comparing the civic awareness of communities, the following factors are suggested:

Administration
  1. Political party
  2. Form of government
     a. Police
        (1) Personnel
        (2) Equipment
        (3) Do industrial properties receive patrol service?
        (4) Incidence of crime
     b. Fire
        (1) Personnel
        (2) Equipment
        (3) Annual losses

  (4) Insurance class
  (5) Do existing plants have sprinkler tanks?
  (6) Water pressure
  c. Streets and highways
   (1) How cleaned
   (2) Miles paved and unpaved
   (3) Contemplated building program
  d. Sewers
  e. Garbage disposal
  f. Hospital facilities
   (1) Number of doctors
   (2) Number of beds
  g. Judiciary
3. Taxes
  a. Rate
   (1) Real estate
   (2) Personal property
  b. Assessments, percentage of value
  c. Rates for
   (1) Township taxes
   (2) Municipal taxes
   (3) County taxes
   (4) Park board taxes
   (5) School taxes
   (6) Other taxes
  d. Poll tax
  e. Business license fees
  f. Exemptions
  g. Contemplated expansion of city facilities affecting tax rate
Budget
 1. Income and expenditures
 2. Indebtedness
When preliminary territorial analysis has revealed that the general area in which the community is located has a superficially satisfactory labor background, a more detailed study is then indicated. Total supply of labor, local skills available, wage rates, and characteristics of the labor supply should be ascertained. Seasonal fluctuations, labor turnover, degree of labor stability, worker productivity, and training facilities are also elements that should be appraised.

Suggested labor worksheets for a comparison of communities within a specific area follow:

Labor
 1. Total employment
 2. Chart supply of suitable labor available
 3. Chart supply of unskilled labor
 4. Elements of labor unrest
  a. Labor organizations
  b. Radical groups
  c. Unemployed councils
 5. Past history of labor disturbances
  a. History of
   (1) Strikes
   (2) Threatened strikes
   (3) Wage disputes
   (4) Walkouts
   (5) Lockouts
  b. Determine number of employees directly and indirectly affected in each instance
 6. Prevailing wage scale for all classes and types of employment

7. Minimum, average, and maximum hour shifts
8. Labor turnover
9. Characteristics of labor
   a. Rural
   b. Urban
   c. Agricultural-lumber-mining-industrial
   d. Percent illiteracy
   e. Percent foreign-born, by nationalities
10. Sex and type
    a. Percent male gainfully employed
    b. Percent female gainfully employed
11. Efficiency of labor
12. Bonus systems
    a. Describe piecework and bonus systems now in effect
13. Seasonal variations
    a. If seasonal variations occur, inquire as to cause and effect
14. Training facilities
    a. Apprenticeship courses
    b. Trade schools
    c. Foremen's courses
15. Housing
    a. Number of units available for lease and rentals
    b. Number of units available for sale and prices

TRANSPORTATION FACILITIES

The manufacturer considering decentralization from urban centers must reorient his thinking. He has become accustomed to picking up his telephone and summoning any and all types of transportation service to his plant. In heavily industrialized areas, there is seldom a problem in receiving service. Rather, the only problem is choosing one of the many competitive carriers he wishes to use.

In less industrialized communities, transportation facilities are usually less numerous. It is preferable to locate in a community that is served by more than one major railroad to obtain the element of competition, which ensures good service.

For those industries seeking to utilize transit or stop-off privileges for partial loading or unloading en route, it is imperative that responsible carrier representatives be consulted to determine whether or not such privileges will be extended in the community.

For those industries using LCL services, the existence of scheduled merchandise trains is of utmost importance. In this manner, full cars for important key cities will be loaded directly from the local railroad station. Time in transit will be competitive with the most efficient trucking services between the community and these major cities.

The existence of arrangements for store-door pickup and delivery service is sometimes overlooked but is a vital factor for those companies utilizing LCL service. The railroads have begun to assess charges for such service. In most instances, the cost will be less than if the industry were forced to use common carrier truck service between the railroad station and its plant. It is assumed that the freight rate relationship of the community has been analyzed in advance of field inspection. If this has not been done, it is imperative that the project be completed prior to site selection.

The existence of motor truck facilities in the community will tend to reduce time in transit and inbound and outbound movements. Accordingly, those locations should be selected which are trucking gateways, if there is much dependence on this type of service. Trucking gateways can be determined readily from various public routing guides. The normal definition is as follows: A point at which carriers have pickup facilities including telephones listed in their company names.

One major caution should be respected. There may be restrictions on the service offered by the carriers in terms of the minimum weight for which they will place a vehicle into the plant. Some carriers insist upon a truckload. Many others have restrictions, such as 10,000 or 5,000 pounds. Those areas served by waterways have distinct competitive advantages. If the industry is of sufficient magnitude so that it can utilize barge loads of inbound commodities or even ship outbound products in this manner, the total annual freight bill can be substantially reduced. The presence of the waterway is another competitive factor in rate making. The industry should be able to negotiate lower rates via all competing types of transportation service because of this situation.

Those companies presently located near Atlantic port and Gulf port cities have become accustomed to shipping via the intercoastal and coastwise steamships with corresponding savings compared with overland routes. Removal of the plant facilities to inland communities requires a complete revaluation of the cost of serving markets which were normally reached by use of the lower cost water service. For example, the differential in rail rates between Ohio points and eastern seaboard cities is frequently insufficient to offset shipping costs from present tidewater locations to West Coast points served via the intercoastal waterways.

Manufacturers are frequently disturbed to find an absence of freight forwarder service in less industrialized areas. This is a natural consequence of the insufficient volume of freight business in and out of the communities to sustain a carloading operation. In other instances, rates from major cities in the manufacturing belt to the smaller communities may be too low to provide a sufficient margin between carload and less-than-carload rates to pay for the costs in assembling and distributing miscellaneous merchandise.

The absence of freight forwarder service is one of those disadvantages of removal to areas with less industrial population. It can be partially overcome by careful review of the company's distribution policy. Pool cars can be consigned to points normally served by forwarder service, and the savings may be of such magnitude that the company will wonder why it has not resorted to such practices in the past. Similarly, local nonprofit pooling associations can be joined which will reduce the total costs on distributing LCL freight.

The investigator should carefully check facilities for parcel post and railway express shipments, as they will undoubtedly differ from the present location.

The existence of public warehouses can be of utmost importance. There are those inevitable periods in every manufacturing operation when inventories of raw materials or finished products exceed space availability in the plant. During such periods, the manufacturer seeks elbow room in either public space or some short-term lease arrangement. In small communities, the availability of loft space and unused, antiquated industrial structures is limited compared with their availability for storage purposes in urban areas. Unless the manufacturer has local warehouses offering these facilities, he may find it necessary to build such emergency space into his plant at additional cost.

EDUCATIONAL, RECREATIONAL, AND CIVIC DATA

Enrollment in schools, quality of newspapers, hotel facilities, and cultural and recreational advantages all play a part in attracting and holding personnel in a new location, especially when the demand for certain skills far outstrips the supply. The following list will serve as an outline for community comparisons:

Educational, Recreational, and Civic Data
  1. Schools (with enrollment)
      a. Primary
      b. High schools
      c. Junior college
      d. Universities
      e. Parochial
      f. Facilities for occupational training
  2. Churches

    3. Fraternal organizations
    4. Libraries
    5. Parks
    6. Playgrounds
    7. Motion picture theaters
    8. TV stations
    9. Facilities for
       a. Golf
       d. Fishing
       c. Tennis
       d. Fishing
       e. Hunting
       f. Boating
   10. Newspapers
       a. Weekly
       b. Daily
       c. Circulation
   11. Special agencies
   12. Hotels
   13. Hospitals
   14. Public buildings

POPULATION STATISTICS, COST OF LIVING, AND CLIMATE

Population statistics and expenditure standards are excellent indices of growth and stability of the area under consideration. Data on climate are easily secured and may be important in certain processes.

The following schedule lists the most important factors to consider:

Population Statistics
    1. Growth
    2. Corporate limits
    3. Suburban
    4. Labor-drawing area
    5. Brief historical sketch

Climate
    1. General weather conditions, elevation of community
    2. Temperature
       a. Annual average
       b. Seasonal average
       c. Maximum temperature
       d. Minimum temperature
    3. Precipitation
       a. Average (annual) rainfall
       b. Average (annual) snowfall
    4. Humidity
       a. Average relative humidity
    5. Days with sunshine, rain, or fog

Cost of Living (per Capita)
    1. Rent
    2. Food
    3. Clothing
    4. Other necessities
    5. Luxuries
    6. Department and merchandise store expenditure
    7. Residential rates
       a. Power
       b. Gas
       c. Water

8. Transportation
    a. Streetcar fares
    b. Bus fares
    c. Bridge tolls

FINAL SELECTION OF THE COMMUNITY

A proper summation of all the cost factors will provide a dollar-and-cents analysis of the present industrial location and calculated costs at proposed other points of operation. The comparison chart, Figure 1-1, will serve as a guide in comparing the recurring annual totals existing in each community.

COMPARISON CHART

BASIC FACTORS	Present Location	City A	City B	City C	City D	City E	City F	City G	City H	City I	City J
Total Transportation Costs											
Inbound materials.....											
Outbound products....											
Total	$										
Labor											
Direct production.....											
Nonproductive........											
Total	$										
Plant Overhead											
Rent or carrying costs, excluding taxes											
Additional costs due to inefficient layout, lack of siding, etc.											
Real estate taxes......											
Personal property taxes, etc..........											
Fuel for heating purposes only..........											
Total	$										
Utilities											
Power..............											
Gas................											
Water..............											
Total	$										
State Factors											
State taxes..........											
Workmen's compensation...........											
Insurance...........											
Total	$										
Miscellaneous											
Other cost factors inherent or peculiar to your present location(s).............											
Total	$										
GRAND TOTAL	$										

FIG. 1-1. Comparison chart useful for analyzing the theoretical cost factors of several possible plant locations.

This chart does not take into account the nonrecurring costs of site acquisition, building costs, and the like. Neither does it consider the many intangible factors that deserve great weight and attention, such as community attitudes, labor productivity, background of labor stability, and general surroundings for executives.

Occasionally, superior intangibles, together with outstanding community or state cooperation, in the form of bond issuance for new plant construction, tax consideration and financial aid in the building of access roads, and installation of water mains, sewers, and railroad siding as well as free sites, may actually motivate the selection of a location having a less favorable dollar-and-cents advantage. Because every case is different, the engineer must correctly evaluate all interrelated factors and reach a sound conclusion.

SELECTING THE SITE

Probably more mistakes are made because of the temptation offered by a fine site or an attractive building than any other single phase of plant location engineering. An investigation of the specific site is recommended only after a community has been chosen that combines the most favorable economic features. Rarely is it necessary to reject a community because of the lack of sites—especially with the definite trend toward peripheral rather than central city operations.

Contrary to general procedure, the first and most important consideration in orienting the proposed plant in the community is the labor pool that is to be employed. Whether the new production facilities are to be near the central district of the city or in one direction or another from the city will depend upon a number of labor-related factors.

Female employees and unskilled male workers usually depend upon local transit systems to reach their place of employment. Where the majority of workers fall into these classifications, the site should be oriented to local bus routes. The amount of the fare, relationship of the site to fare zone boundaries, and the availability of free transfers will affect recruitment. Although transit authorities are often willing to extend routes to serve a plant offering new riders, careful investigation of the system may reveal blind spots requiring excess transit time and extra fares between the majority of worker residences and the plant site. Female labor will be easier to obtain if the plant location offers opportunities for noon-hour shopping.

A factor commonly overlooked is the influence of neighboring industries. Wage rates, radical tendencies, working hours, shift schedules, and even fringe benefit patterns of manufacturing plants in the immediate neighborhood may have as much effect on the future of the proposed plant as one's own carefully worked out personnel policies.

When the most strategic area has been determined based upon studies of local labor conditions, examination of specific plant sites may be begun.

A list of general specifications should be prepared as follows:
1. Description of building to be constructed (including sketch)
2. Size of plot
3. Necessary railroad, highway, and waterway facilities
4. Minimum size of water mains, gas line, and power line
5. Volume of groundwater to be utilized
6. Sewage and effluent disposal requirements
7. Safety area for offensive odors, noise, smoke, and other obnoxious conditions
8. Provisions for sprinkler pressure (gravity tank or local water mains)

Generally, a site of not less than five times the actual size of the plant is considered the minimum to allow for sidings, loading platforms, truck ingress and egress, parking facilities, storage area, and future expansion. If possible, open land should be available on two or more sides to permit future site expansion.

From a topographical map (showing configuration of the land surface), it is possible to determine not only the relationship of highways and railroads but also the terrain conditions in the area. One can almost select possible site areas from his desk. At least the amount of time spent in the field can be substantially reduced by eliminating from consideration those sectors which lie in swampland or rough terrain.

Topographic maps will reveal important information on drainage conditions in the vicinity of sites. From a study of land gradient and drainage basins of creeks, the experienced engineer can predict the absence or presence of flood danger at specified points.

Determination of the character of the ground for building and equipment foundations during initial inspections can save considerable expense, for the occurrence of rock and soft underlying formations may necessitate a change of location in the interest of building economy. A helpful hint can be obtained from the local agricultural agent or from the state geologist's office.

Safe bearing values of foundation soils are set forth in the following table:

Material	Tons per square feet
Granite rock	30
Limestone, compact beds	25
Sandstone, compact beds	20
Shale formation or soft friable rock	8–10
Gravel and sand, compact	6–10
Gravel, dry and coarse, packed and confined	6
Gravel and sand mixed with dry clay	4–6
Sand, compact, well cemented and confined	4
Earth, solid, dry, and in natural beds	4
Sand, clean and dry, in natural beds and confined	2
Clay, soft	1–1½
Quicksand, alluvial soils	1

The adequacy, reliability, and cost of power, water, and gas play an important role in the final selection of the site. If large quantities of water are required for process, orientation to the supply will be a paramount consideration. The underlying aquifers will dictate the amount of groundwater which may be available and the spacing of wells. If surface water is to be utilized, it will be important to calculate the pumping head of the site above the river or stream. The possibility of impounding supplies must be investigated from both a topographical and legal viewpoint.

Barring unusually heavy loads, the local power company will normally extend or improve lines to the site without cost to the industry. City water mains, sewerage lines, and gas mains are less flexible, and their extension may require considerable contribution on the part of the new industry.

It is important for those industries releasing extensive effluent to investigate the local sewage disposal plant to determine whether or not the type of installation is adequate to neutralize the wastes which the plant will release. If streams are to be used for disposal, it is imperative that the industry clear its proposed waste releases with state authorities, perhaps even interstate commissions. Any waste which increases the biological oxygen demand in the stream is certain to meet opposition from state wildlife and game authorities. Discoloration will anger downstream residents.

In an era of uncertainty, it is advisable to minimize the danger of atomic bomb attack by choosing a site at least 3 miles from a vital target, especially if the community has a population of over 50,000 and harbors vital war industries. Bridges, airports, railroad marshaling yards, or military installations of any kind are to be avoided.

The following checklist should be used in final evaluation of the site.

1. Will local building code and zoning regulations allow the construction of the type of building proposed?
2. Are there area restrictive covenants, easements, or other legal entanglements that will interfere with property and maximum use of the property?
3. Who will remove snow, where necessary, and maintain access roads?
4. Can definite long-term commitments be received on the valuation of a proposed plant for tax purposes?

5. Even though it may be outside city limits, will the plant receive full police and fire protection?

BIBLIOGRAPHY

Hoch, L. C., "Intergovernmental Fiscal Competition," Tax Institute of America, New York, October, 1965.

Hornbruch, Frederick W., Jr., "Plant Location," sec. 5, chap. 1, in H. B. Maynard (ed.), *Handbook of Business Administration*, McGraw-Hill Book Company, New York, 1967.

Hoyt, Park R., "Relocation of Manufacturing Facilities," sec. 5, chap. 12, in H. B. Maynard (ed.), *Handbook of Business Administration*, McGraw-Hill Book Company, New York, 1967.

"Old Factors—New Facets," *Plant Location*, Simmons-Boardman Publishing Corporation, New York, 1965.

Oppel, Edwin I., *A Plant Location Road Map*, American Institute of Banking, New York, May, 1963.

Yaseen, Leonard C., *Plant Location*, American Research Council, Larchmont, N.Y., 1960.

Yaseen, Leonard C., "Techniques of Plant Location," in Carl Heyel (ed.), *Encyclopedia of Management*, Reinhold Publishing Corporation, New York, 1963.

Chapter 2

Plant Layout

RICHARD MUTHER

Richard Muther & Associates, Inc.,
Kansas City, Missouri

Plant layout embraces the physical arrangement of industrial facilities. This arrangement, either installed or in plan, includes the spaces needed for material movement, storage, indirect laborers, and all other supporting activities or services, as well as for operating equipment and personnel.

The term "plant layout" sometimes means the existing arrangement, sometimes the proposed new layout plan, and often the area of study or the work of making a plant layout. Hence, plant layout may be an actual installation, a plan, or a job.

OBJECTIVES

Making a layout plan is not the end result, even of those responsible for the planning. Rather, improved operations, increased output, reduced costs, better service to customers, and convenience and satisfaction for company personnel are likely to be the chief objectives. It is important to target on these real aims; otherwise it becomes easy to drift into the viewpoint that the plan—rather than what the plan can accomplish when properly installed—is the only accomplishment required.

Each planning or rearrangement project will have its own individual objectives, and these will vary with different management viewpoints or operating policies, as well as with the current considerations surrounding each project. For efficiency in planning layouts and plant designs, it is important that the real objectives be clearly stated early in the planning.

Essentially, a plant facility is a combination of objectives and considerations, and its planning rests on a compromise of various isolated benefits and limitations, which in turn are modified by time, degree of relative importance, and management attitude or policy.

Still, a planner should aim at certain general objectives in his layout. These include:

1. Integration—an integration of all pertinent factors affecting the layout
2. Utilization—an effective utilization of machinery, people, and plant space
3. Expansion—easy to expand
4. Flexibility—easy to rearrange
5. Versatility—readily adaptable to changes in product design, sales requirements, and process improvements
6. Regularity—a regular or straight division of areas and relatively even sizes of areas, especially when separated by building walls, floors, main aisles, and the like
7. Closeness—a practical minimum distance for moving materials, supporting services, and people
8. Orderliness—a sequence of logical work flow and clean work areas with suitable equipment for scrap, trash, and wastes
9. Convenience—for all employees, in both day-to-day and periodic operations
10. Satisfaction and safety for all employees

TYPES OF ARRANGEMENTS

The classic types of layout are three in number.

First, there is layout by fixed position or by fixed material location. This is a layout where the material or major component remains in a fixed place. It does not move. All tools, machinery, men, and other pieces of material are brought to it. The complete job is done or the product is made with the major component staying in the one location. One man or crew makes the complete assembly, bringing all parts to each assembly point. The workmen may or may not move from one assembly location to the others. Advantages are:

1. Handling of major assembly unit is reduced (through increased parts handling to assembly point).
2. Highly skilled operators are allowed to complete their work at one point, and responsibility for quality is fixed on one person or assembly crew.
3. Frequent changes in products or product design and in sequence of operations are possible.
4. The arrangement is adapted to variety of product and intermittent demand.
5. It is more flexible in that it does not require highly organized or expensive layout engineering, production planning, or provisions against breaks in work continuity.

Second, there is layout by process or layout by function. Here all operations of the same process or type of process are grouped together. All welding is in one area, all drilling in another, all stitching in the stitching room, and all painting in a paint shop. This layout has these advantages:

1. Better machine utilization allows lower machine investment.
2. It is adapted to a variety of products and to frequent changes in sequence of operations.
3. It is adapted to intermittent demand (varying production schedules).
4. The incentive for individual workers to raise the level of their performance is greater.
5. It is easier to maintain continuity of production in event of:
   a. Machine or equipment breakdown
   b. Shortages of material
   c. Absent workers

Third, there is line production or layout by product. Here one product or one type of product is produced in one area. But unlike layout by fixed position, the material moves. This layout places one operation immediately adjacent to the next. It means that any equipment used to make the product, regardless of the process it performs, is arranged according to the sequence of operations. Advantages of this layout generally include:

1. Reduced handling of material
2. Reduced amounts of material in process, allowing reduced production time (time in process) and lower investment in materials
3. More effective use of labor:
   a. Through greater job specialization
   b. Through ease of training
   c. Through wider labor supply (semi- and unskilled)
4. Easier control:
   a. Of production allows less paper work
   b. Over workers and fewer interdepartmental problems; allows easier supervision
5. Reduced congestion and floor space otherwise allotted to aisles and storage

Which Type to Use. Use layout by fixed position or fixed material location when:

1. Material forming or treating operations require only hand tools or simple machines.
2. Making only one or a few pieces of an item.
3. The cost of moving the major piece of material is high.
4. The skill of workmanship lies in the abilities of the workers or it is desired to fix responsibility for product quality on one workman.

Use layout by process or function when:

1. Machinery is highly expensive and not easily moved.
2. Making a variety of products.
3. There are wide variations in times required for different operations.
4. There is a small or intermittent demand for the product.

Use line production or layout by product when:

1. There is a large quantity of pieces or products to make.
2. The design of the product is more or less standardized.
3. The demand for it is fairly steady.
4. Balanced operations and continuity of material flow can be maintained without difficulty.

In actual practice, most layouts are a combination of these classic types. They are made to utilize the advantages of all three types.

GUIDING FUNDAMENTALS TO SUCCESSFUL LAYOUT

Basically, every layout involves three fundamentals:

1. Relationships—closeness desired between various activities or functional areas
2. Space—in amount, kind, and shape for each activity or functional area
3. Adjustment—of the activity areas into a layout plan (see Figure 2-1)

The solution to any layout problem will of necessity be a compromise of the various considerations and of the various objectives of good plant layout. The relation of machinery to handling, of service to building, of change to man are all tied together. One feature or consideration influences many others. And constantly in layout work, engineers end up "robbing Peter to pay Paul."

Weaknesses in layouts come because the compromise that is worked out overlooks some feature that should be provided or fails to recognize some consideration that has an important effect. Careful following of the fundamentals can help avoid weaknesses in layout plans.

1. *Plan the whole and then the details.* Begin with the layout of the site or plant as a whole, and then work toward the details. First, determine the general requirements in relation to the volume of production anticipated. From this, develop a general overall layout. After approval of the overall layout, proceed with the detailed arrangement within each area. This is the actual positioning of the men,

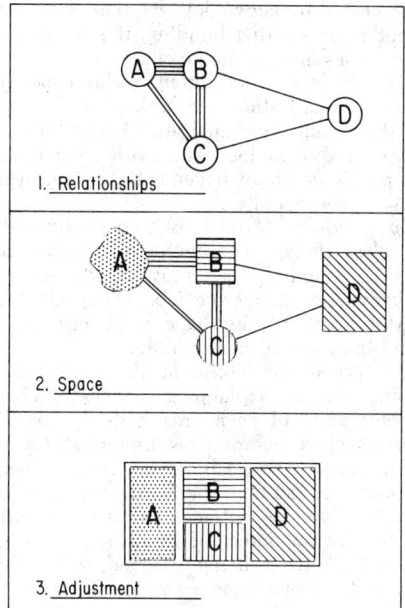

I. Relationships

2. Space

3. Adjustment

Fig. 2-1. Fundamentals of layout planning.

materials, machines, and supporting activities which becomes the detailed layout plan.

2. *Plan the ideal and from it the practical.* The initial concept of the layout should represent a theoretically ideal plan, without regard for existing conditions and irrespective of cost. Later, make adjustments to incorporate the practical limitations of buildings and other factors. By this means, the possibility of a good layout is not lost through an early misbelief that certain features are necessary.

3. *Follow the cycles of layout development and make the phases overlap.* The cycles of layout development follow a sequence of four phases. The first phase is to determine where the layout shall be—where the facilities to be laid out are to be located. This may involve plant location (see Section 11, Chapter 1) or merely location within the existing site or plant. The second phase is to plan an overall layout for the new production area. Then comes the detailed layout plan, and finally the installation. In practice, there should be an overlapping from one phase to the next.

4. *Plan the process and machinery around the material requirements.* The material factor is basic. The product design and manufacturing specifications largely determine what processes to use. And the quantities or rates of production of the various products or parts must be known to calculate what processes are needed. The process and machinery must be built around the material requirements.

5. *Plan the layout around the process and machinery.* After the proper production processes are selected, the layout planning begins. The demands of the equipment itself—weight, size, shape, movement to and away, and the like—must be considered. The space and location of the production processes or machinery (including tools and equipment) are the heart of the layout plan.

6. *Plan the building around the layout.* Where machinery, service equipment, and layout are to be more permanent than the building, the building should be set around the most efficient layout. Where the layout is less permanent than the building, still follow this guiding fundamental but alter it to read "layouts." Because there will inevitably be changes, design a general-purpose plant, planning

the building around several prospective layouts that may successively occupy the building. For the layout in an existing building, this guiding fundamental becomes: Plan the building modifications around the layout.

7. *Plan with the aid of clear visualization.* The experienced layout specialist knows that aids to clear visualization are a key to his work. They help him gather his facts, and they help him analyze. In addition, clear visualization is essential when he wants to discuss his plans with foremen and service personnel, when he presents his proposals to management for approval, or when he shows the workers how the layout will operate.

8. *Plan with the help of others.* Layout is a cooperative affair. The best layout will not be obtained unless the cooperation of all persons concerned is built into it. Their ideas should be solicited; they must be drawn into the project. They already have detailed knowledge about the job. They are the ones who will make the layout operate. Moreover, if they have a chance to take part in planning the layout, they will tend to accept it more readily.

9. *Check the layout.* When one phase of the project is developed, it should be approved before going too far in planning the next. This will save later headaches and ensure an integration of each area with the overall plans. Each phase of the layout must be checked before it is presented for approval. This check assures that the layout is sound or reveals further improvements that can be made.

10. *Sell the layout plan.* Sometimes the hardest part of layout work is getting others to buy it. It may be good. But it is still a compromise; it means changing people around; it will require an outlay of funds. Therefore, keep enthusiastically plugging the benefits of the layout being planned; take additional time to bring the operating people into the project; get everyone to participate; take time making ready to present the layout to the prospects who will in the end put up the money for it.

METHODS OF APPROACHING PLANT LAYOUT PROJECTS

Instinct and Intuition. Layouts can be planned by instinct and intuition. This is often fast, direct, and timesaving; but it is limited generally to simple or emergency situations or where there is deep experience and a record of sound decisions in the past.

Find One Ready-made. Magazine articles, visits to other plants, discussions with planners from other companies, social events, trade shows, or professional society meetings may lead to finding a layout—one that is spoken of enthusiastically and could be "just the thing." New ideas and methods are essential in this day of rapid change and certainly should be sought out; but remember that what is good for someone else is not necessarily suitable for a different situation, and without at least some modifications, is likely not to be.

Full Participation or "Keep Everyone Happy" Approach. This approach involves the democratic process: get all ideas from everyone, discuss them, and translate them into a visual presentation; then call the group together for comment; make changes; and again solicit agreement of the group. This gives everyone involved a chance to participate and therefore to support the ultimate plan. But this approach draws only on past experience, it is usually time consuming, and it does not take advantage of the analytical techniques so important to moving the company forward at the very time it has the opportunity to do something progressive and constructive. Additionally, it tends to put emphasis on discussion and visualization rather than on problem analysis.

Flow of Materials. Centuries ago, engineers discovered that moving material directly from each operation to the next afforded a logical sequence for control and reduced the cost of handling materials. By analyzing the sequence of necessary moves and arranging the layout accordingly, these benefits are gained. This is the approach most frequently taught. It is ideal for process-type industries such as oil refineries or flour mills. But this approach is limited to those situations where there are dominant patterns of material flow, for it does not fully recognize that relationships other than flow of materials may be equally or more important.

Organized Systematic Methodology. Systematic Layout Planning (SLP) is a universally applicable approach. It incorporates the benefits of the other approaches and organizes the whole planning process into a rational system. It is generally recognized as the most realistically analytical of any approach yet developed. As a result, it develops plans more soundly and gets approvals faster. Learning the approach initially takes time and some training, and once learned, there is a tendency to become intrigued with the methodology itself and to substitute mechanics of problem solving for the intelligent analysis and creative synthesis that should accompany the procedures.

SYSTEMATIC LAYOUT PLANNING*

Systematic Layout Planning (SLP) is equally applicable to office, laboratory, service, warehouse, or manufacturing operations. It is also equally applicable to minor and major rearrangements, to existing or new building, or to new plant site planning.

SLP consists of a framework of phases, a pattern of SLP procedures, and a set of conventions. See Figure 2-2.

The Four Phases of Layout Planning. As each layout project runs its course from initially stated objective to its installed physical reality, it passes through the four phases of layout planning.

Phase I is that of location. Here must decided where the area to be laid out will be. This is not necessarily a new site problem. More often it is one of determining whether the new layout of layout will be in the same place it is now, in a present storage area which be made free for the purpose, in a newly acquired building, or in a similar type potentially available area.

Phase II is that of planning the general overall layout. This establishes the basic flow pattern(s) for the area being laid out also indicates the size, relationship, and configuration of each major activity, department, or area.

Phase III is the preparation of detailed layouts and includes planning where each piece of machinery and equipment will be placed.

Phase IV is the installation. This involves planning the installation and physically making the necessary moves.

These phases come in sequence, and for best results, they should overlap each other.

Phases I and IV are frequently not part of the planning engineer's specific project—even though his project must in every pass through these first and last phases. Therefore, the layout planner concentrates his attention on the layout planning phases: II, general overall layout, III, detailed layout.

Basic Input Data for Layout Planning. Before looking at phases II will more closely, the basic input data or factors on facts and be needed should be recognized. These are remember the "alphabet of the facilities planning engineer"—PQ

P—product or material, including variations and machinery
Q—quantity or volume of each variety or item operations
R—routing or process: the operations, their and how these
S—services or supporting activities which elements
T—time or timing as it relates PQRS The analytical often study of the input becomes an analysis many supporting layouts,

Practically every layout plan starts e many supporting as a basis for its planning. fact, many layouts,

The Pattern of Procedures—Phase Layout Planning, part of planning the general ove ass., 1961. data. Refer to the SLP capsul of the flow of materials. But service areas must be integr

* This portion of this chap. by Richard Muther, Industrial

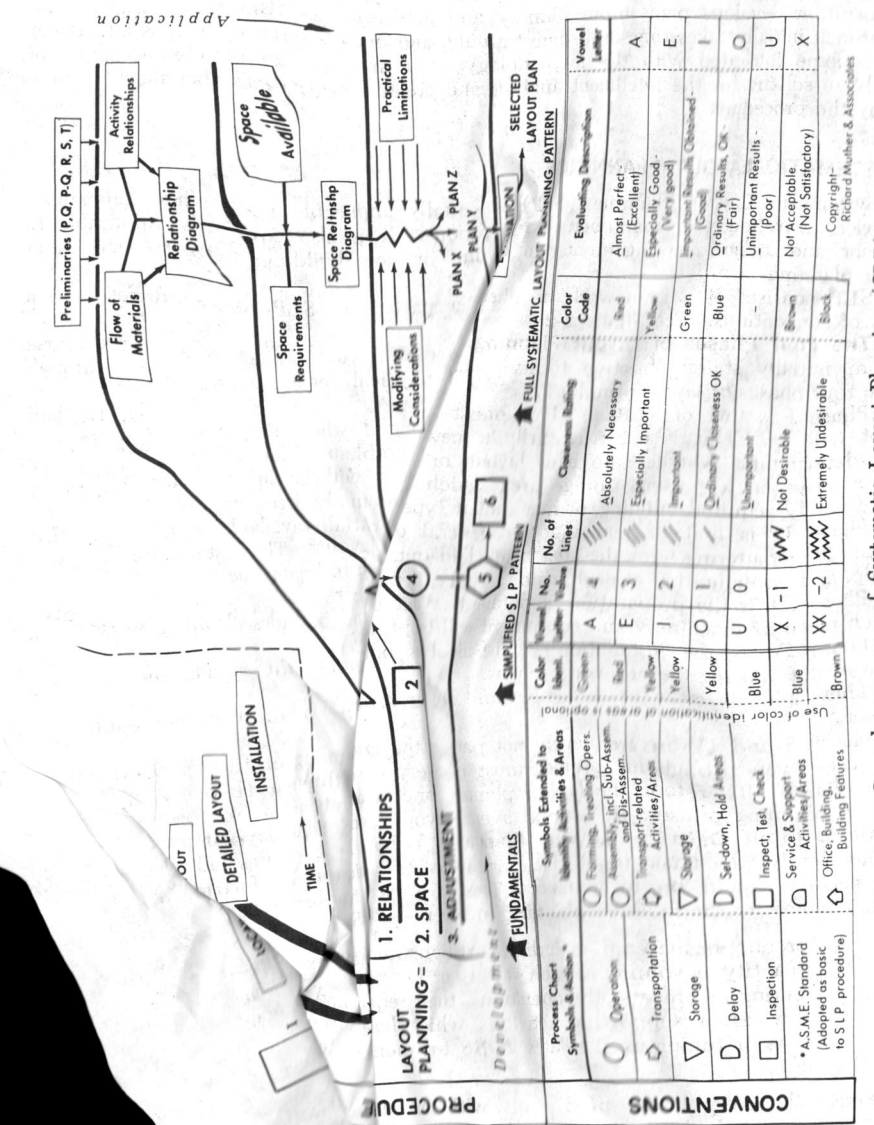

FIG. 2-2. Capsule summary of Systematic Layout Planning (SLP).

such as offices or laboratories, and many plants producing small articles, such as miniature components, do not have a traditional flow of materials from which a meaningful flow analysis can be made. As a result, developing or charting activity relationships for service or reasons other than flow of materials is frequently of equal importance.

These two investigations then are combined into a flow or activity relationship diagram. In this process, the various activity areas or departments are geographically diagramed without regard to the actual floor space each requires. To arrive at the space requirements, analysis must be made of process machinery and equipment necessary and of the service facilities involved. These area requirements must be balanced against the space available. Then the area allowed for each activity will be "hung on" the activity relationship diagram to form a space relationship diagram.

Relationships and space are essentially "married" at this point. The space relationship diagram almost becomes a layout. But it is not an effective layout until it is adjusted and manipulated to integrate with its arrangement the modifying considerations that also affect it. These include such basic considerations as handling methods, operating practices, safety considerations, and the like. And as each potentially good idea concerning these features is thought up, it must face the challenge of practicality—represented by the practical limitations at the right in the pattern of procedures in Figure 2-2.

As the integrating and adjusting of the modifying considerations and the practical limitations are worked out, one idea after another is probed and examined. The ideas that have practical value are retained, and those that do not stand the test are discarded. Finally, after abandoning those plans which do not seem to stand up, two, three, four, or maybe five alternative layout proposals may remain. Each of these will work; each has value. The problem lies in deciding which of these alternative layout plans should be selected.

These alternative plans may be termed plan X, plan Y, and plan Z. At this point, a cost analysis of some kind should be made, together with an evaluation of intangible factors. As a result of this evaluation, a choice is made in favor of one alternative or the other, although in many cases the evaluation process itself suggests that a new, even better layout could be a combination of two or more of the alternative layouts being evaluated.

The Pattern of Procedures—Phase III, Detailed Layout. The next phase, detailed layout, involves the spotting of each specific piece of machinery and equipment, each aisle, and each storage rack—and doing this for each of the activity areas or departments which have been blocked out in the previous general overall plan. As mentioned before, phase III overlaps phase II. This means that before actually finalizing the general overall layout, certain details will have had to be looked into. For example, the actual orientation of a conveyor may have to be analyzed before the general overall layout can be determined. This is the kind of overlapping investigation that takes the layout planning engineer into detailed layout planning in certain areas before his phase II is completed.

Note that a detailed layout plan must be made for each departmental area involved. This means that some adjustment may have to be made between departmental blocks as the detailed areas are being planned; that is, some readjustment of the general layout may be called for. Indeed, it is important not to be governed by too rigid an application of the general overall layout worked out in phase II. It can be adjusted and changed within limits, as the details within each area are worked out.

In planning each detailed layout, the same pattern of procedures used in phase II is repeated. However, the flow of materials now becomes the movement of materials within the department. The department relationships now become relationships of the equipment within the department. Similarly, the space requirement now becomes the space required for each specific piece of machinery and equipment and its immediate supporting area. And the space relationship diagram now becomes a rough arrangement of templates or other replicas of machinery and equipment, men, and materials or products. As in phase II, several alternative layouts may

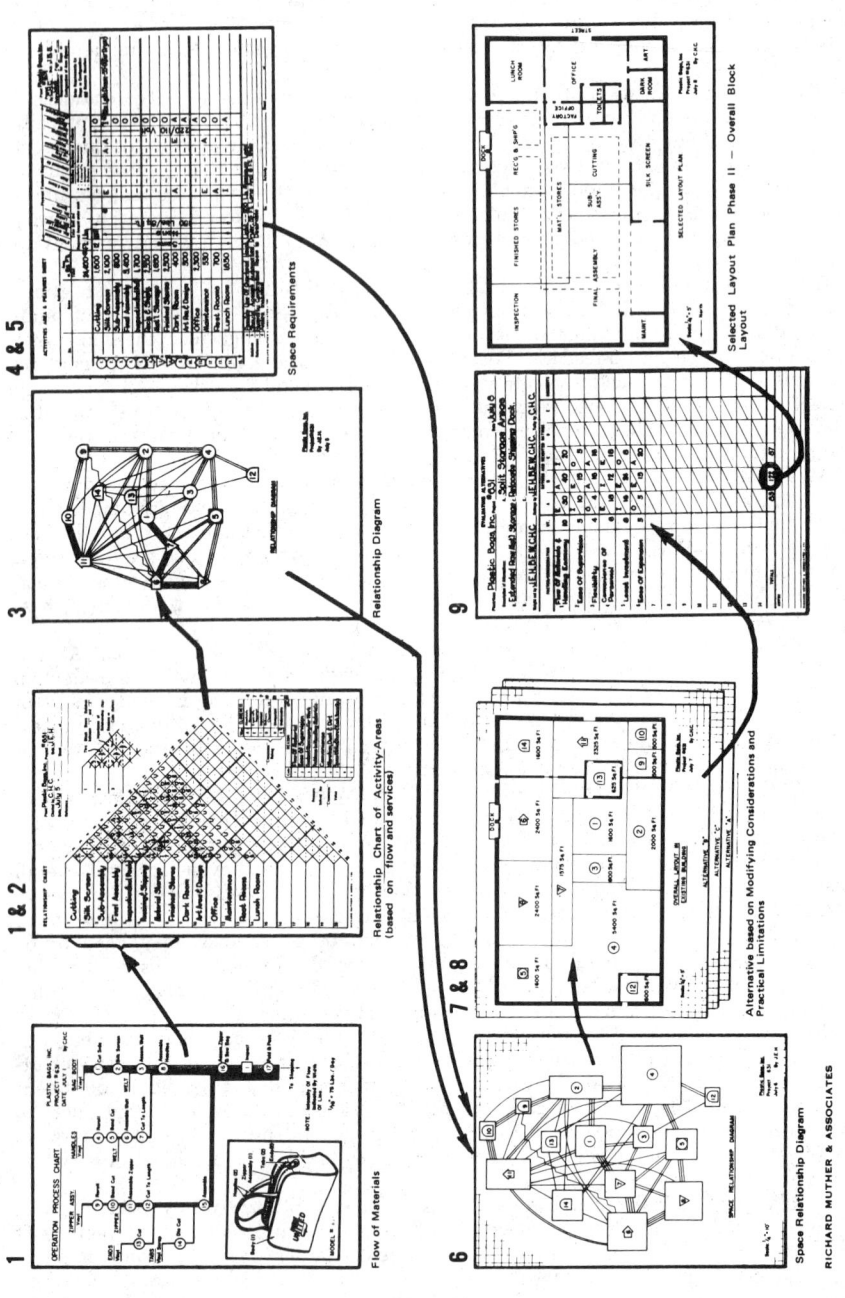

FIG. 2-3. Example of SLP, overall block layout, phase II.

result. This leads to an evaluation to select the most satisfactory departmental layout.

This SLP pattern of procedures provides a basic planning discipline while at the same time allowing for logically different content of the PQRST input data. And just as the flow of materials analysis will become less important and the activity relationship study will become more important in office or laboratory areas, so the entire pattern has the flexibility to be modified for the needs of any layout project. It becomes a matter of adjusting the importance of each box rather than changing the sequence or arrangement of boxes.

Set of Conventions. A set of conventions is utilized to aid in planning, understanding, and communicating. The conventions are used throughout each step of the previously described pattern of procedures for diagraming, rating, visualizing, and evaluating. The conventions are shown in Figure 2-2. They consist of seven symbols, seven letters, seven line ratings, and five colors plus black and white. These are cross-integrated for multiple use in any application employing SLP.

Examples of SLP. Figure 2-3 shows an example of the several steps (or boxes) in the pattern of procedures. Here is a company making plastic bags of various kinds. The planner followed these steps in developing his overall (block) layout. He then followed the same sequence—with different emphasis and different data, of course—to develop the layout for each departmental area.

Figure 2-4 shows a conceptual example of an SLP project. It shows, in simplified form, first the phase I problem of location, then the phase II overall layout, followed by the phase III detailed layouts of each department, and finally, the phase IV installation.

LAYOUT PLANNING TECHNIQUES

Determine the Flow. The sequence of operations as the basis for the flow of materials is the heart of many layout plans. After gathering data and facts, the layout analysis should begin here. As a result, the process chart in its various forms is the most useful of all layout planning devices.

1. *One-product Analysis of Flow.* Operation process charts are of particular value for assembly or disassembly of products. Figure 2-5 shows how the operation process chart practically leads to the layout plan. The right-hand side of the chart is practically the production line layout; the horizontal material feeding lines on the chart become delivery racks or conveyors in the layout; and the operations charted to the left become subassembly benches or equipment. When a process involves forming or treating only, the operations and the information can be listed in columnar worksheets without the necessity of chart symbols. As a general rule, however, start with an operation process chart in any layout work. Even if making a half dozen different products, begin with charts for each one. A separate area and a separate layout may be needed for each product. Or a combined layout may be needed for all of them. This cannot be determined until the data are assembled in a form convenient for analysis.

2. *Multiple-product Analysis of Flow.* When there are a number of products or parts, use a combined, multiproduct, or gang process chart. The problem here is to combine products or classes or groups of products so that together they will give volume enough to justify an effective flow of material. Figure 2-6 shows an example of such a chart for the five different products of a small manufacturing plant. It was made to analyze the layout of a factory producing small metal sign plates and shields. The first and last operations are shaded for quick spotting. Operations out of normal sequence are indicated by a slant line joining them with the preceding operation. Volume of business was determined from past records and sales forecasts.

From this chart, the layout can be seen already taking shape. Using the operation numbers at the very left for identification, arrange the operations in a list for the best flow of material.

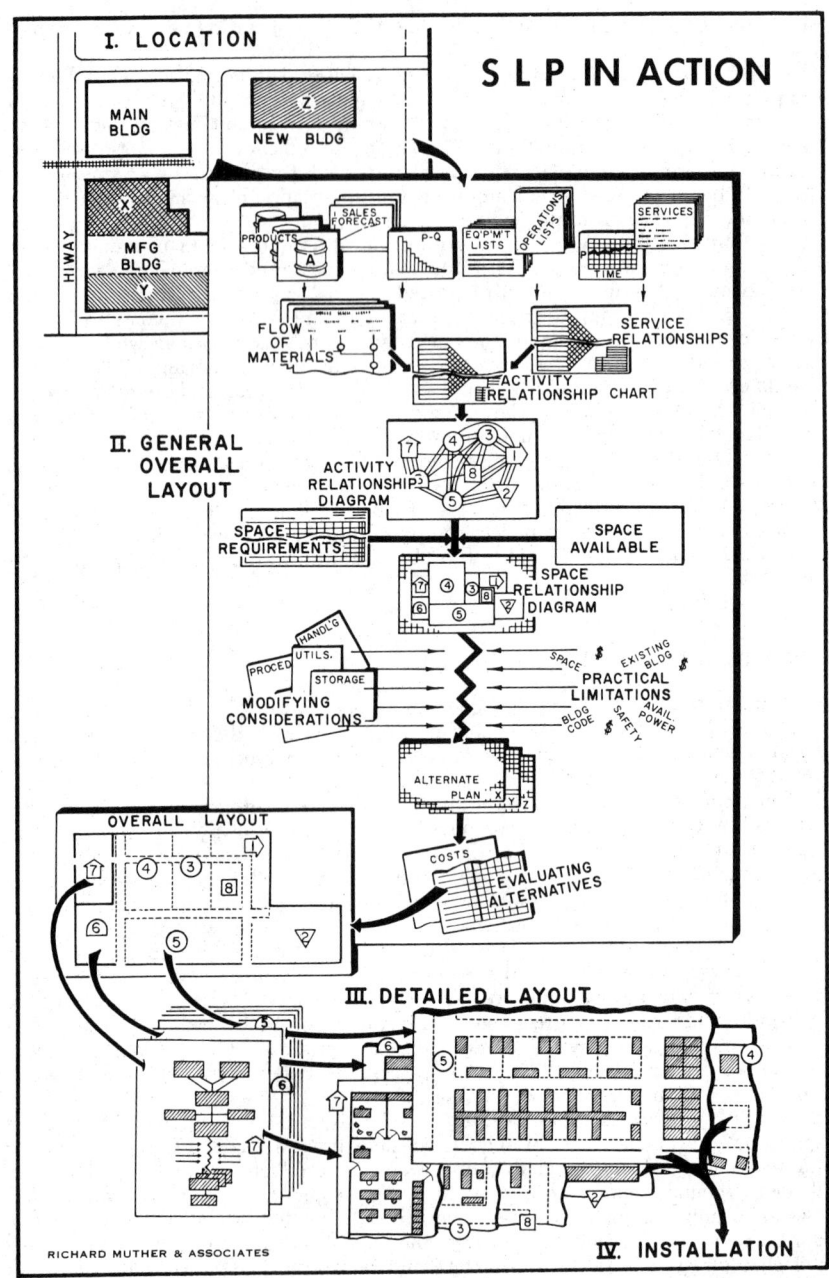

Fig. 2-4.   Conceptual diagram of all four phases for a complete SLP project.

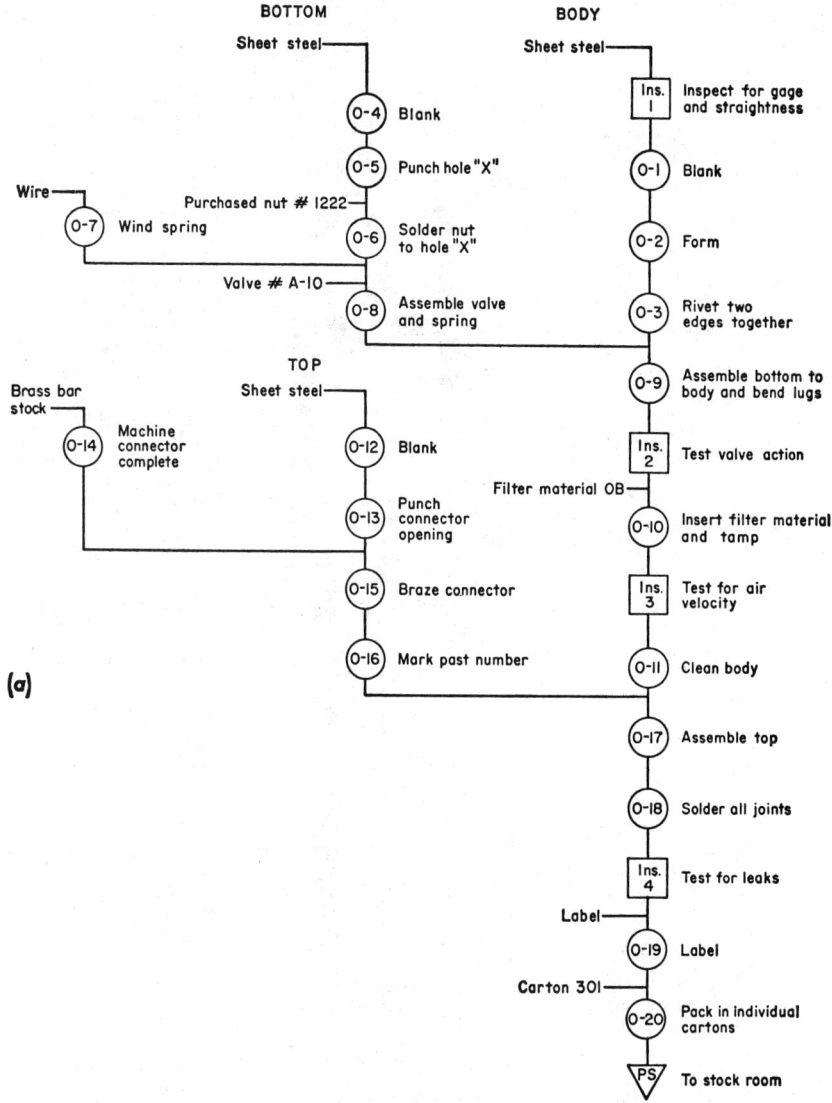

FIG. 2-5.   The operation process chart as a basis for layout.

If the products do not fall into natural groups, as in this example, juggle many different combinations to get a right grouping.

In classing various products for flow possibilities, look for the following:

1. Products requiring similar machinery
2. Products requiring similar operations
3. Products requiring similar sequence of operations
4. Products requiring similar operation times
5. Products of similar shape, size, or purpose
6. Products requiring similar degree of quality
7. Products of same material

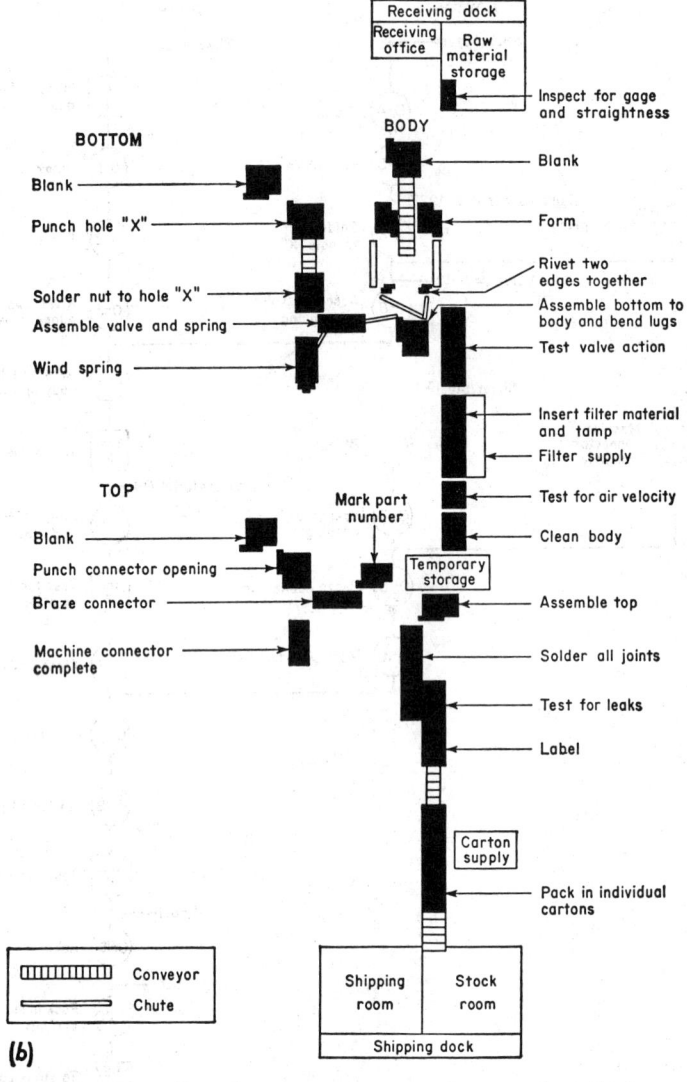

Fig. 2-5 (*continued*).   The operation process chart as a basis for layout.

Figures 2-7 and 2-8 show how in this way one plant combined into a flow pattern all its production of shafts. The engineers on this project grouped certain products together to take advantage of flow and minimum distance. First, they gathered together operation sheets showing operations, sequence, machine type, and machining time. Then they classified each one of the parts using the following considerations:

1. Parts completed in one machine, such as automatically turned parts
2. Parts almost completed in one machine but requiring simple additional machining, such as parts turned on automatic or turret lathes with subsequent simple drilling or milling

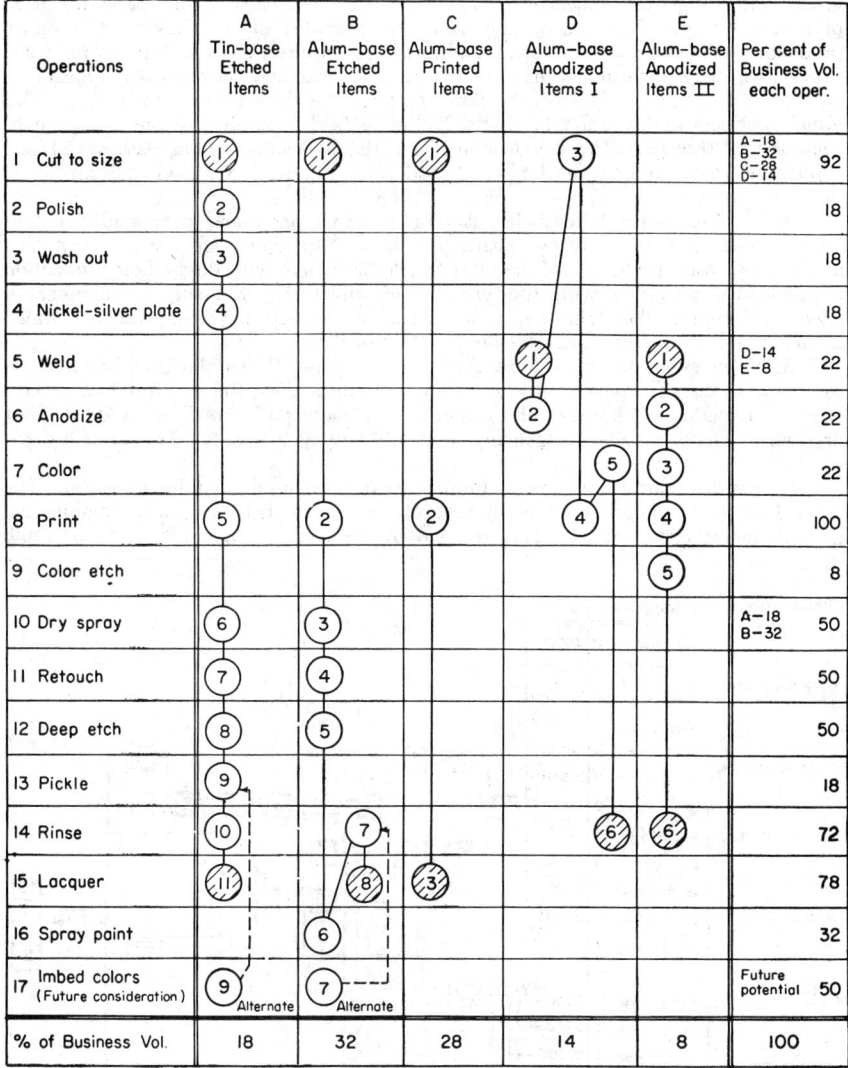

Fig. 2-6. Multiproduct chart showing combined operations for five different types of products.

3. Parts of similar nature, especially main components, such as shafts, gears, flanges, levers, and the like

4. Other parts requiring similar operation sequences, such as parts turned, milled, and drilled; parts milled and drilled; and the like

In establishing subdivisions of these four classes, the engineers brought together parts which:

1. Required special-purpose machines for some of the operations

2. Were of similar size

3. Had to be machined to similar accuracy

Their chief aims were to obtain full utilization of the key machines in each group and to get a sequence of operations that was nearly the same for each part in a group. To do this, they compiled planning sheets showing the yearly production demands and the corresponding hours required on each type of machine. From this, they determined how many of each machine they should allocate to each group. Figure 2-7 shows the flow chart for the shaft production group. This group produces eighty different shafts, including fairly complicated ones with gears, splines, and threads. The key machines are the hydraulic contour lathes (2) and spline and gear milling machines (4 and 9). Figure 2-8 shows the layout of the shaft group.

A similar approach is to pick out representative parts and draw individual or multiproduct operation process charts for them. In doing this, select, say, five of the most costly parts, five of the most fragile parts, five with the highest production requirements, and five with the greatest manufacturing difficulty or number of rejects. Compare the operation process charts of each by laying them parallel to one another, and from this, develop a pattern of flow.

3. *Many-parts or Many-products Analysis of Flow.* When the products become too many to classify conveniently into some pattern of flow, the multiproduct process chart is inadequate. Because of the number of route paths involved, a cross chart or from-to chart is better—especially in plants having a variety of nonstandardized products.

The from-to chart takes several forms and can be used in different ways. The main idea is to determine the amount of movement between each combination of two operations or areas. This is done by referring to operations lists or route

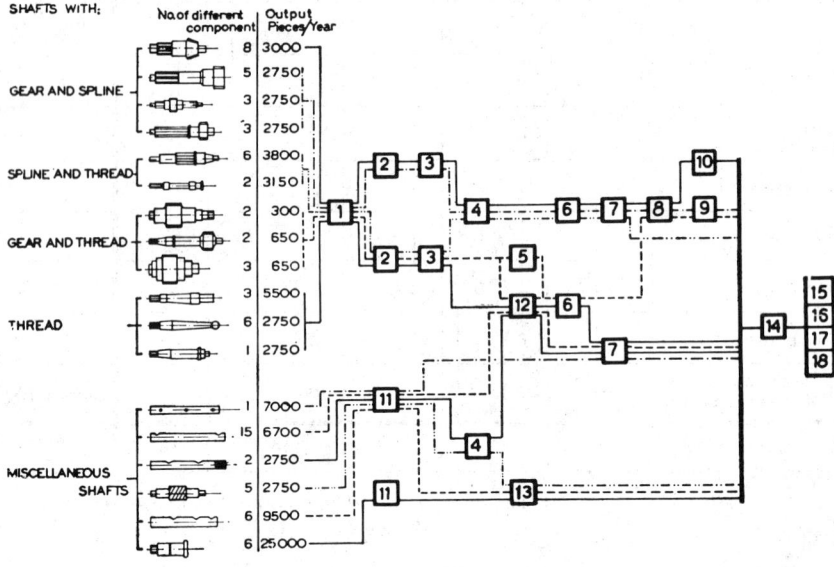

1	CENTERING LATHE.	7 3 sp.DRILLING PRESS.	13 HYDR. PLAIN MILLING MACHINE.
2	HYDR. CONTOUR LATHE.	8 CYL.GRINDER.	14 INSPECTION.
3	FINISH ENGINE LATHE.	9 GEAR MILLING MACHINE.	15 HARDENING DEPARTMENT.
4	SPLINE MILLING MACHINE.	10 GEAR CUTTING MACHINE.	16 GRINDING DEPARTMENT.
5	KEY MILLING MACHINE.	11 TURRET LATHE.	17 FINAL INSPECTION.
6	THREAD MILLING MACHINE.	12 UNIV. MILLING MACHINE.	18 ASSEMBLY STOCK.

FIG. 2-7.   Shaft production group flow chart.   (*Courtesy of A. B. Scania-Vabis.*)

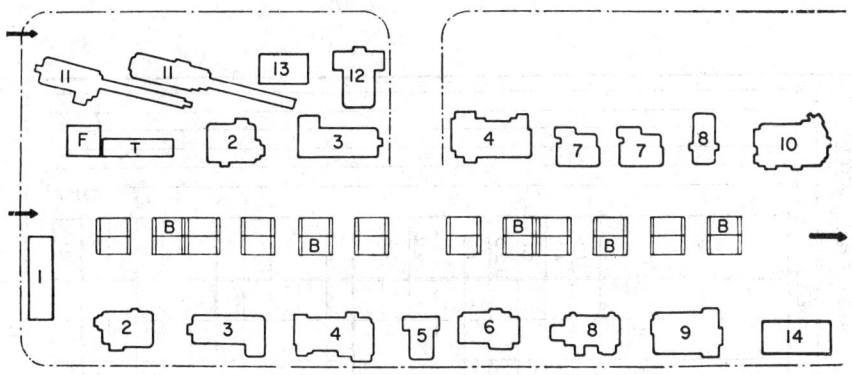

T – TOOLS AND FIXTURE • F – FOREMAN'S STAND • B – TRANSPORT BOXES

FIG. 2-8.   Shaft production group layout.   (*Courtesy of A. B. Scania-Vabis.*)

sheets. The chart can be built for all parts that are involved or a representative selection of parts. Each move is recorded opposite the appropriate "from" and "to" columns, and the moves from each activity to each other activity are then tallied and totaled. Figure 2-9 shows a typical from-to chart in which unit loads per day is the charted value.

Flow Alone Not Best Basis for Layout. There are several reasons why the flow of material—as determined predominantly by the routing—cannot be the sole basis for layout arrangements.

1. The supporting services must integrate with the flow in an organized way. The maintenance crib, the superintendent's office, the locker and rest rooms, and the transformer bank all have a relatively preferred closeness to each of the producing areas. They are all part of the layout; they must be planned into it, yet they are not part of the flow of materials.

2. Flow of materials is often relatively unimportant. In some electronic and jewelry plants, only a few pounds of material will be transported during an entire day. In other industries, materials are piped, or one skid load lasts a worker all week.

3. In completely service industries, office areas, or maintenance and repair shops, there is often no real or definitive flow of materials, even if paper work, equipment, or people are regarded as the "materials" that will flow.

4. Additionally, in heavy material movement plants, where the influence of material flow will dominate the layout, flow will not be the sole basis for arranging the process operations and equipment. Flow of materials is only one reason for the closeness of certain operations to each other. There are many others. And these may conflict with, or at least cause adjustments in, the closeness as based on the analysis of flow. For example, the routing may call for this sequence: form, trim, treat, subassembly, assembly, and pack. For best flow of materials, treating should lie between trimming and subassembly. But treating is both a very dirty and a dangerous operation. Therefore, it should be kept away from the delicate subassembly area and its high concentration of workers. The effect of factors such as these—or the distribution of utilities, the cost of controlling quality, the contamination of the product, and the like—must be compared with the importance of material flow, and adjustments made as practical.

In any case, some systematic way is needed of relating service activities to each other and of integrating supporting services with the flow of materials. The relationship chart is the best method of meeting this need.

The Relationship Chart. The relationship chart is a cross-section form where the relationship between each activity (or function or area or machine) and all other activities can be recorded. The basis of the form is shown in Figure 2-10.

FROM–TO CHART

Item(s) Charted ALL ITEMS   Basis of Values UNIT LOADS/DAY

Plant ABC MECH. WORKS   Project NEW LAYOUT
By RMH   With DL
Date JAN. 9   Page 1   of 1

FROM \ TO (Activity or Operation)	1 PRESS	2 WELD	3 MACHINE	4 ASSEMBLY	5 PAINT	6 STEEL STOR.	7 PARTS & SUPPLY	8 FINISHED STOR	9 DRIVEWAY	TOTALS
1 PRESS	=	15	12	—	—	—	—	—	—	27
2 WELD	—	=	—	—	11	—	—	2	—	13
3 MACHINE	—	3	=	15	—	—	2	12	—	32
4 ASSEMBLY	—	—	—	=	3	—	—	18	—	21
5 PAINT	—	—	7	13	=	—	—	—	—	20
6 STEEL STOR.	12	5	1	—	—	=	—	—	—	18
7 PARTS & SUPPLY	—	1	4	3	1	—	=	—	—	9
8 FINISHED STOR	—	—	2	—	—	—	—	=	20	22
9 DRIVEWAY	—	—	—	—	—	6	11	—	=	19
TOTALS	12	24	26	31	15	6	13	32	20	181

NOTES:

RICHARD MUTHER & ASSOCIATES

FIG. 2-9. From-to chart showing the flow of materials between each origin point and each destination.

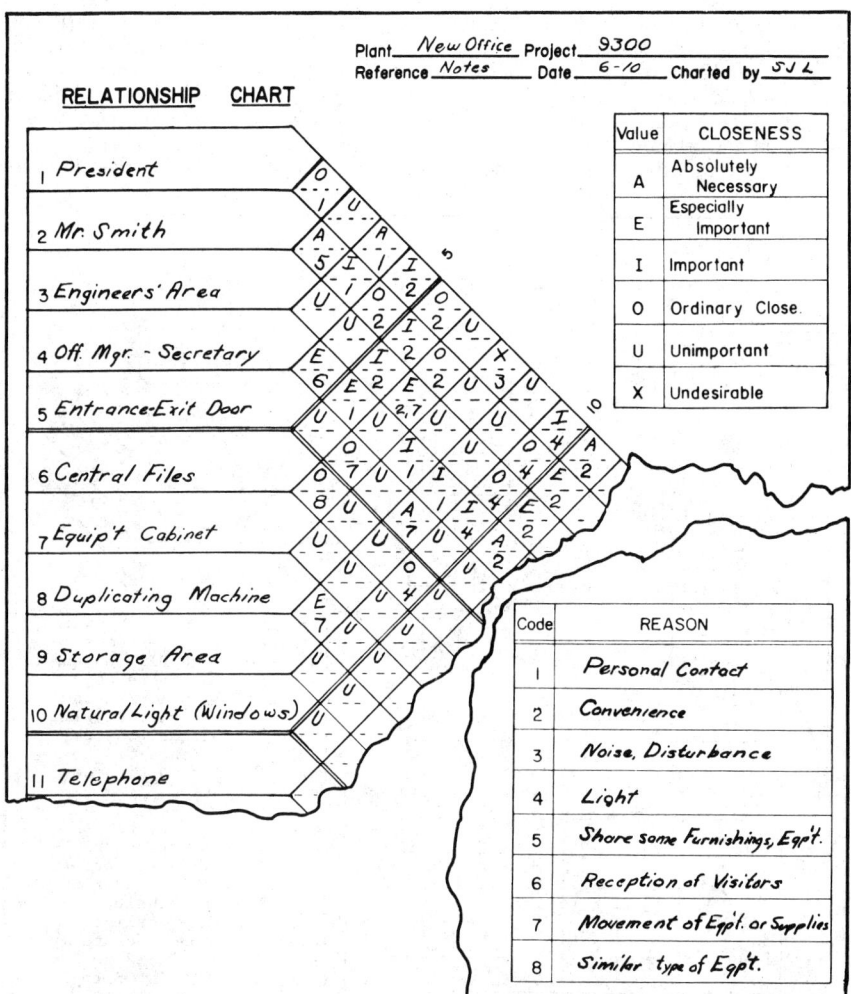

FIG. 2-10. The relationship chart is extremely effective for planning all activities not tied together with a significant flow pattern. This chart was prepared for an office of consulting soil test engineers. It indicates that Mr. Smith must be near the Engineers' Area, to the telephone to a lesser extent, still less to the office manager–secretary and central files, and not at all near the duplicating machine or storage area. Reasons are filled in and recorded in lower half of the appropriate boxes. (*Source:* Systematic Layout Planning, *Richard Muther, Industrial Education Institute,* Boston, 1961.)

The chart itself is almost self-explanatory. Where the activity on downsloping line 1 intersects the activity represented by upsloping line 3, the relationship between activity 1 and activity 3 is recorded. In this way, there is an intersecting box for each pair of activities involved. The basic idea is to show which activities should be located close together and which far apart, with all in-between relationships also rated and recorded.

Note that each box is divided horizontally. The upper part is for the closeness

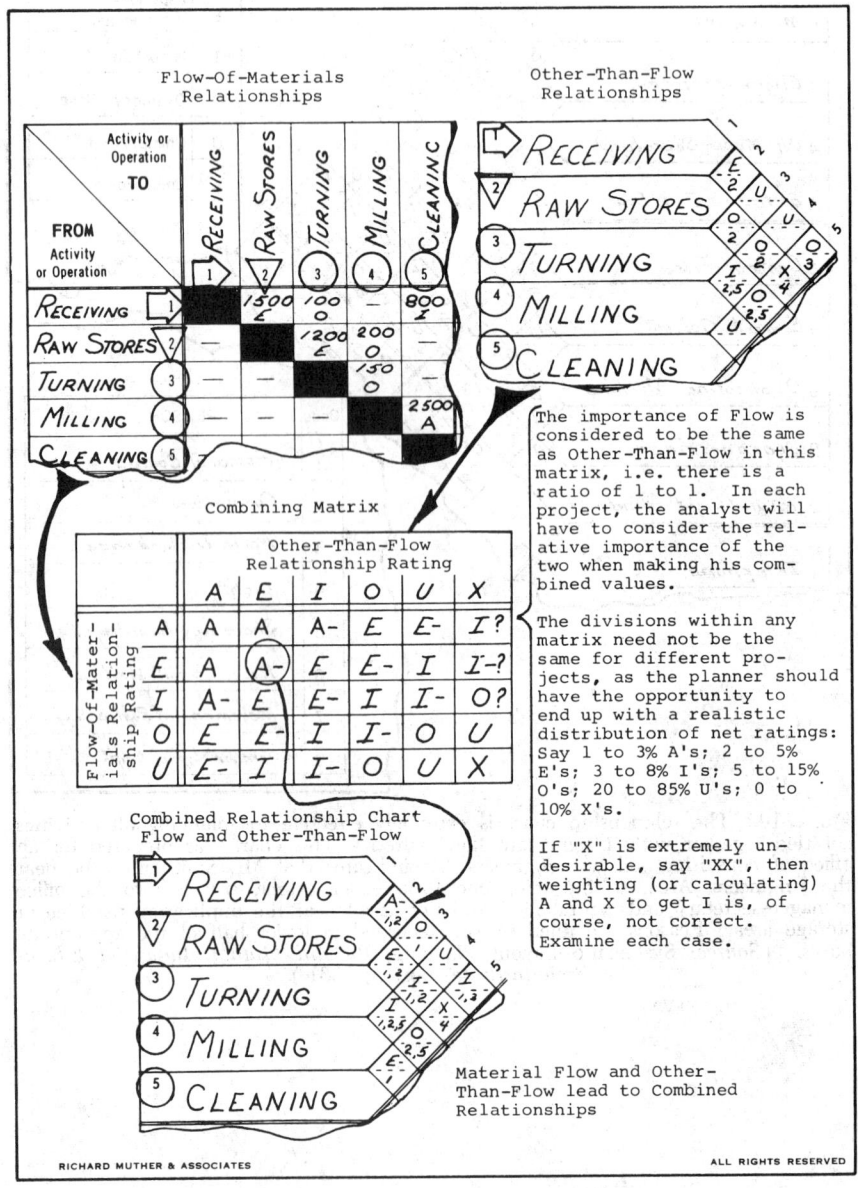

FIG. 2-11. Combining flow and other-than-flow relationships.

rating value. The lower half is for recording the reason causing the designated closeness value. This gives a rate and reason for each relationship.

Typical reasons supporting relationship ratings include the following, although many terms may be used and many other reasons are possible:

1. Flow of materials
2. Degree of personal contact
3. Degree of communicative or paperwork contact
4. Use of same equipment or facilities
5. Use of common records
6. Sharing of same personnel
7. Specific management desires or personal convenience
8. Supervision or control
9. Noise, dust, dirt, fumes, hazards
10. Distractions or interruptions

Combining Flow and Other-than-flow Relationships. One reason the relationship chart is so effective is that it records relationships based on flow of materials and other-than-flow reasons. Calibrating these two and getting on a common denominator rating for the combination demands rational analysis. Figure 2-11 shows a worksheet example of how this can be kept track of during the sequence of calibrating flow to vowel letters, developing ratings for other-than-flow reasons, and combining the two into a net resultant relationship.

Diagraming the Relationships. A diagram is a visual drawing of charted data. It can be made on a floor plan of an existing layout or on a blank sheet. In the former case, the flow will be tracked on a scale drawing of the area involved. In the latter, conceptual diagrams can be developed and analyzed. Figure 2-12 shows

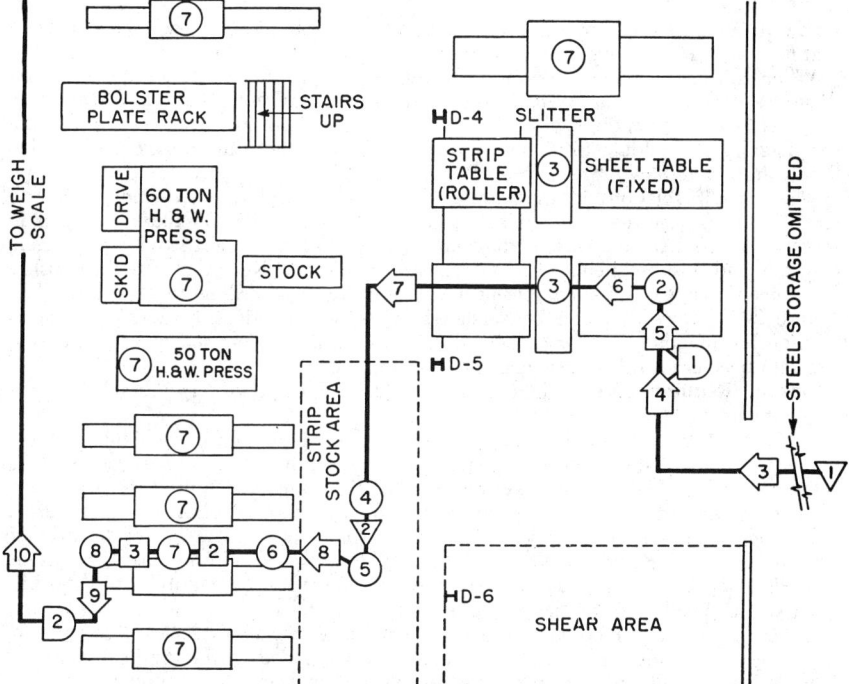

FIG. 2-12. Portion of a detailed flow diagram showing what happens and where it happens to the sheet steel, strip stock, and laminations produced in a plant making small electrical transformers.

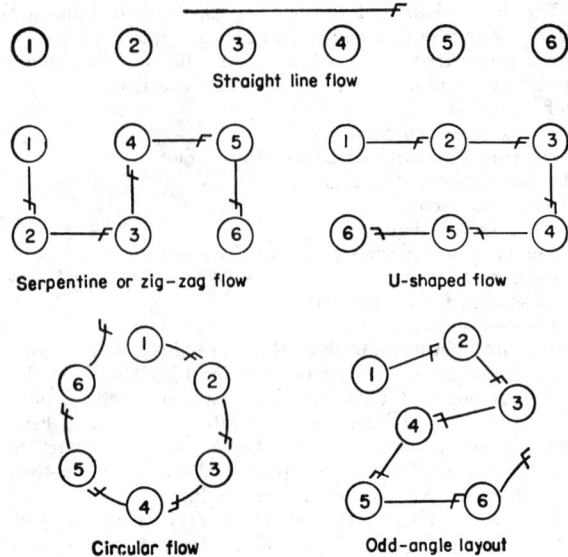

Fig. 2-13.   Conceptual flow diagrams for the five basic flow patterns.

a flow diagram of the former type for a detailed layout.   Figure 2-13 shows conceptual flow diagrams for five basic flow patterns.

When flow and other-than-flow are combined in the same relationship data, the diagram of these is usually made first as a best-fit concept.   This is usually done progressively by working with the high-rated relationships first, then redrawing the diagram and adding to it lower rated relationships until all are included.   Then the diagram is redrawn for final best fit.

The activity relationship diagram is drawn with symbols only—independent of space.   After space requirements are established, they can be added and the diagram redrawn as a space relationship diagram.   Figure 2-3 shows an activity relationship diagram and a subsequent space relationship diagram.   Note that the relationships have been drawn in the number-of-lines convention, according to SLP procedure. Figure 2-14 shows a similar conversion from activity relationship diagram to space relationship diagram, but for a detailed layout situation.   Here the space relationship diagram becomes a rough arrangement of templates.

Space Requirements.   There   are   at   least   five   ways   to   establish   space requirements.

1. *Calculation.*   Determine the amount of space required for each piece of machinery or equipment, including areas for workers, maintenance service, material setdown, and access to aisle; extend this by the number required of each piece of machinery; and add in space allowances for aisles and general or support areas.

2. *Conversion.*   Determine the amount of space now used for each machine, machine group, or activity area; adjust this to what should be used to do the job efficiently at present; then convert this by some factor or multiplier to determine what will be needed for the new requirements.

3. *Rough layout.*   Prepare a rough detailed layout plan to scale of a proposed or at least a possible arrangement.   It will in all likelihood not be the final, approved layout; but it will indicate approximate spacing between the equipment involved and will allow measuring the rough plan for total area requirements.

4. *Space standards.*   In cases where certain types of areas are subject to repetitive layout planning, it is practical to develop standard amounts of space.   This is

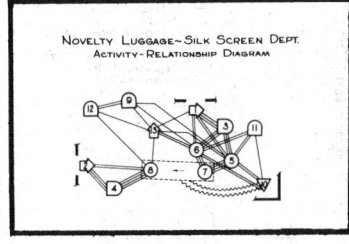

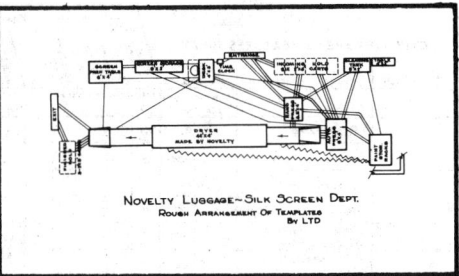

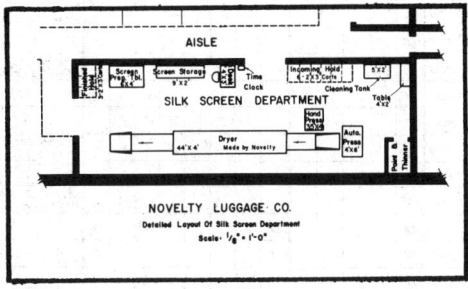

FIG. 2-14. Conversion from activity relationship diagram to space relationship diagram for a detailed layout situation.

particularly applicable for office areas or standard assembly bench layouts. But there is danger in using any standard if it is not understood.

5. *Ratio trend and projection.* There are a number of ratios that can be of value. From a plot of each ratio against time, a trend of that ratio is noted. This in turn can be projected into the future. Then, by knowing the projected ratio, the square feet required for any projected denominator can be calculated. For example, if 135 gross square feet per office employee is projected, 135 times a five-year-plan figure of 100 office employees means that 13,500 square feet of office will be required to meet the five-year plan.

In practice, space requirements are not established quite this simply. In fact, several of the five methods may be employed on the same project. Moreover, space requirements must be balanced against space availability. Here, it may be most helpful to rate each of the activity areas as to the relative importance of maintaining its space requirement. The same SLP letter rating can be used effectively: A—absolutely necessary to honor the requirements, and so on. The areas rated O and U are squeezed the most when reducing area requirements to a smaller area available. In industrial plants, these easily squeezed areas usually end up being storage, office, and flexible service areas, as compared with production areas or fixed equipment services.

In any case, it is important to summarize the total plan-for space figures. Space comes in three basic forms: amount, kind or nature, and shape or configuration. Most experienced planners want to know early in their projects all three of these aspects of the space with which they are working. As a result, the activities area and features sheet is recommended. Figure 2-15 shows how these data can be recorded.

Space Relationship Diagram. Working from the activity relationship diagram, each activity symbol is converted into its specific area. This may be done on cross-section paper at a convenient scale. Each activity will continue to be identified by symbol, number, and possibly name; but in addition, it will be diagramed

Activity No.	Activity Name	Area in Sq. Ft. Total (Aisle incl.)	Ft.	O'Head Clearance	Max. Overhead Supported Load	Max. Floor Loading	Min. Column Spacing	Water & Drains	Steam	Compressed Air	Foundations – Pits	Fire or Explosion Hazard	Special Ventilation	Special Electrification	Requirements for Shape or Configuration of Area (Space) and Reasons therefore
1	Receiving	450	15					–	–	–	–	–	–	–	–	Min. of 2 Dock Doors Width: 10 Ft. Min.
2	Material Std.	11 500	18					–	–	–	–	–	–	–	–	
3	Machining	3 000	15					–	–	E	–	–	–	–	–	
4	Wire Stringing	400	"					–	–	–	–	–	–	–	–	
5	Small Parts Subassembly	500	"					–	–	E	–	–	–	–	–	
6	Fluorescent Assembly	1 800	"					–	–	E	–	–	–	–	–	Long, Narrow Areas Suitable for Assembly Lines
7	Mercury Vapor Assembly	1 500	"					–	–	E	–	–	–	–	–	
8	Facade Light Assembly	2 100	"					–	–	E	–	–	–	–	–	Long, Narrow Areas Suitable for Assembly Lines
9	Finished Fixture Storage	3 500	18					–	–	–	–	–	–	–	–	
10	Pipe Receiving & Storage	2 800	"	3 Ton				–	–	–	–	–	–	–	–	Crane Way not Less than 30' Wide. 1 Truck Space-Underoof.
11	Pipe Bending	1 000	"					–	–	–	–	–	–	–	–	
12	Welding	5 500	"					–	–	I	–	E	E	A	–	
13	Paint	800	"					I	–	–	–	E	I	–	–	Must Accept Paint Dip Tank 30' Long.
14	Outside Pole Storage	–	–					–	–	–	–	–	–	–	–	
15	Shipping	450	15					–	–	–	–	–	–	–	–	Min. 2 Dock-High Doors – Must be conv. for loading poles ft. yd.
16	Tooling, Maint. & Test Lab.	700	12					O	–	A	–	–	–	–	O	
17	Employee Services	900	10					A	–	–	–	–	–	–	I	
18	General Offices	3 200	10					I	–	–	–	–	–	–	I	

ACTIVITIES AREA & FEATURES SHEET — Physical Features Required

Plant White Lighting
Project
By M.D.M.   With J.W.M.
Date 7-26

Enter Unit and Amount under each Ft.
Enter Relative Importance: A-ABSOL. NECESS. E-ESPEC. IMPORT. O-ORDIN. IMPORT. I-IMPORTANT —NOT REQUIRED

Total 40,100

RMA – 192-P   COPYRIGHT 1963 – RICHARD MUTHER & ASSOC. K.C. MO.

Fig. 2-15.   Activities area and features sheet.

to scale and the actual square feet will be shown.   This way, the space is recorded both in actual numbers and visually in relative size.

Many refinements can be made in space relationship diagrams to show specific information pertinent to the project at hand—refinements involving existing buildings versus new construction, number of employees involved, need for space relief, or cost to relocate.   These can all be coded into the diagram by use of colors, symbols, letters, and the like.   In complex layout problems, it may be better to prepare the diagram on a reproducible print and, using several copies, show different significant information on each copy.

To get the full potential of planning the layout to the theoretically ideal conditions without the limitations of existing columns, walls, rail sidings, and the like, it is best to hang or superimpose the space as directly as possible right on the final redrawn activity relationship diagram.   Later, when adjusting the space diagram to the modifying considerations and their practical limitations, there will be ample opportunity to bring the theoretical diagram back into the constraints of existing buildings or other fixed features.

On the other hand, if it is known that certain fixed building features such as walls, columns, or floor loads definitely cannot be changed as part of the layout planning project, it may seem an unnecessary step to go all the way toward the fully ideal space diagram.   So long as it is certain that there is no possibility of missing a real improvement, the fully ideal space diagram can be bypassed, but it should be remembered that many potential savings have been lost at this same shortcut.

Adjusting the Diagram. When the space relationship diagram is available, it can be adjusted and manipulated to create various possible arrangements.   At this point, the operating and service managers should be brought back into the project, for there is now something for them to visualize.   Furthermore, much of the adjustment must come as a result of the desires or practices of these people.

The space relationship diagram is almost a layout plan. It is not a very good one, in all likelihood, because the modifying considerations and the practical limitations which hold within bounds the ideas which come to mind as possible modifications have not yet been incorporated into it.

There are many modifying considerations for different projects. Typical ones include:

1. Method of handling
2. Storage facilities
3. Personnel requirements
4. Building features
5. Utilities and auxiliaries
6. Procedures and controls

For each modifying consideration, there will be a set of practical limitations which must be weighed against it. This is a process of making many compromises. The objective is to get an arrangement of activities which will give the most practical overall combination of all considerations and limitations.

By integrating the modifying considerations with the space relationship diagram and discarding all impractical ideas, the planner usually arrives at two to five alternative plans. Any of these can be made to work. The next problem is to decide which layout alternative to adopt.

Visualize the Layout. Experienced layout men know that a clear understanding of the plan they are making is the only way to get a sound layout. They have to visualize how the layout will look and how it is going to work. They also have to have some clear picture or reproduction of their layout to discuss with other people. They must have something others can see clearly.

The three common ways of visualization involve:

1. Drawings and diagrams
2. Templates and layout board
3. Three-dimensional models

Of these three, drawings and diagrams are the most basic. They are readily made, easily altered, and inexpensive. In actually putting together a reproduction of a detailed layout, templates are the most valuable. There are many kinds, and they can be used in many ways. Essentially, they permit reproduction of as many different layout proposals as desired merely by rearranging the templates on the board.

To prove and check the layout plan or to help others visualize what is planned, the three-dimensional model comes into play. For these latter purposes, the model is supreme. But the great misunderstanding is that models themselves are a substitute for layout planning. They are not. They glamorize the layout job; they help sell the layouts proposed; they develop interest; they help in training workers, foremen, and staff personnel; and they act as a check on the planning of the layout engineer. But, for his own visualization or planning, he does not need them and in fact is usually better off without them, except in complex three-dimensional situations.

The most common pitfall into which the inexperienced layout man falls is jumping into the use of templates and models too early in the project before he has the necessary facts to evaluate the various layout proposals his visualization may create.

The purpose of visualization is to assist in developing a sound layout. After getting information about the various input data involved, determining and diagraming the relationships, and conceiving the various ways to arrange these physical features, check them by seeing what they actually look like. Reproduce a likeness of each arrangement to see if it is as good as it first appeared. Adjust and change the arrangements; this is the time to move equipment without cost. These adjustments—still in the paper stage—lead to the arrangement that gives the best layout compromise. The planner uses a physical means of visualizing each logical improvement to check his thinking; he does not start right in with sketching or templates and merely cut and try until by trial and error he comes up with a layout that looks as if it will work.

The stages in developing a detailed layout with templates and models include the following:

1. Engineer works up his proposed arrangement by analyses, diagrams, and templates on layout board.
2. He then sets up his scale models on a layout plan according to his planned arrangement or places the models on the layout board. He checks his plan. He then explains his ideas and invites suggestions from all concerned.
3. Suggested improvements are tried out by moving models or otherwise evaluating the idea.
4. When the final arrangement is decided upon, the layout board is rearranged to agree or the models are removed and templates fastened securely to undersheet or floor plan and the board is made ready for photographing or reproduction.

Templates and models should be made up early enough to be ready when needed. Moreover, if a few of them are set out on a layout board in plain sight, it is surprising how much more interested and helpful staff and operating personnel become in the layout project.

Although the scale model is clearest to see, it is not easy to carry around. A photographic reproduction of the model, even when taken directly overhead, will hide many of the details. Also, it is difficult to put dimensions and other information on models. Some plants overcome these difficulties by adding identification tags to models and by photographing from more than one angle.

Because of these and other reasons, a combination of the two-dimensional template and three-dimensional model is frequently the answer, but this is not always so. Drawings and sketches, diagrams, and block templates all serve their purpose. Simple unit area templates are just as helpful in the early stages of overall layout planning as models later on. The best solution is to be able to use all of them and select the proper device for the proper purpose.

The device itself should be readily movable so it can record quickly every idea and proposed plan. In addition, the layout engineer should make provision to keep track of ideas as they develop. Otherwise, the templates or models may be moved and drawings marked up without capturing the suggestions. Some layout engineers periodically supply each interested department head and management person with copies of their layout proposals as these are worked out. A few dollars in reproducing costs is a good investment, they feel, for the cooperation and valuable suggestions they get in return.

Actually, once drawings, templates, and models have been made up, they should be kept on hand if the plant anticipates any layout changes in the future. They should be properly identified and filed away, ready for use again. This is all part of good plant layout practice.

When planning is finished and the detailed layout is installed, be sure to bring the permanent layout record up to date. Keep the existing layout record untouched while planning a new layout. If no other record is available and there is only one layout board, to start changing the existing plan around will destroy the record. This is why most experienced layout engineers have a layout board or model that acts as the current record. They do all new layout planning on smaller boards which they can easily move about and change as necessary.

A scale model has to be kept horizontal. Some template layout boards are also safer if kept flat. Others, that can be supported on the wall, take less floor space and can be reached at any point with a stool or ladder. But whether flat or on a wall, if the unit is large, it should be made in sections. Each section should be small enough to remove and work on conveniently and transport to the photography room or the president's office, if necessary. Boards that are enclosed or covered have the very practical advantage of keeping out dust and dirt, which can often soil a board so much that photographic reproductions are not clear.

Evaluate Alternative Layouts. The best layout is a compromise—a compromise of the various factors, considerations, layout objectives, and types of layout. To select the best compromise, plan alternative layout proposals and eliminate those or the portions of them that compare unfavorably. Evaluating alternative plans should determine which proposal offers the best layout.

Various techniques of evaluation have been used successfully. Here are several:

Ranking based on selected considerations
Tally of gains and losses expected
Value rating of pros and cons
Rating of alternatives versus objectives
Audit of alternatives against established check questions

Perhaps the most frequently used methods, however, are the following:

1. List the advantages and disadvantages. This is the simplest way to evaluate alternatives: merely write down the advantages and disadvantages of each layout being evaluated. It is surprising how often such a listing quickly clarifies which alternative should be selected.

2. Factor analysis. This method selects the factors or considerations on which the decision will be made. Each factor is given a weight value according to its importance (10, 9, 8, . . .). All the alternatives are then rated on one factor at a time. The rating (in SLP the vowel letter ratings are used) is converted to a number and multiplied by the weight value. The weighted ratings are totaled for each alternative and a numerical comparison is made. This increases the objectiveness of what can be a very subjective decision making process. Moreover, it offers an excellent way of involving management in the selection and weighting of the factors, and the operating and support supervisors in rating the alternatives on each factor.

3. Cost comparison. In important projects, costs will nearly always become a basis for selecting the best alternative. This means everything that goes into the cost of installation and operation. In establishing costs, the layout engineer should consider the following list and charge against his installation every one that should be included. Moreover, he should make his comparison or justification in accordance with methods of cost analysis approved by his company's accounting or financial officials. Cost to be considered include:

   a. Investment:
     (1) Initial cost of new facilities of all kinds
     (2) Accessory costs
     (3) Installation costs
     (4) Depreciation and obsolescence costs
   b. Operating costs:
     (1) Material
     (2) Labor
     (3) General and burden

No matter how many layouts are investigated, none of them will have everything. There must be a compromise somewhere to get a practical solution. As a result, develop from the theoretical layout two or three practical solutions. Evaluate these, pick one that looks good, then develop it. Otherwise, too much time may be spent debating which solution is best, and insufficient time will be left for developing the details.

Engineers are notorious for wanting to weigh meticulously every scrap of fact or influencing detail. Be accurate and sound, but do not deliberate so long that the next phase of the layout is held up.

INSTALLING THE LAYOUT

Installing the layout is the fourth phase of layout work. It follows location of area to be laid out, general overall layout, and detailed layout plan. The person making the layout is sometimes responsible for seeing that the layout is properly installed. More often, he is an advisor or coordinator, and the installation work rests with the plant engineer or maintenance department. As a very minimum, the layout engineer is called upon to supply the details of what the new installation should be like.

Information needed for layout installation usually includes:

A list of all new machinery and equipment to be installed or existing equipment to be moved or changed in location

A layout print, drawing, or photograph explaining details of new locations

A schedule of moves

A specifications sheet to show how each machine should be disconnected, moved, and hooked up

Figure 2-16 illustrates the practice used by one automobile engine plant where almost every weekend is moving day. Because interference with production schedules cannot be tolerated, all moves must be carefully planned and scheduled. The procedure shown here makes sure that floor spaces are empty before trying to put machines into them.

1. Proposed new layout is shown in black on diagram delivered to plant engineer. He marks existing location (cross sectioned) of machines to be relocated and shows route to follow and any intermediate locations (broken cross section).
2. List of moves is issued by plant superintendent. It tells what is to be done, but not how.
3. Moving order tells how and in what sequence moves shown on proposed layout are to be made.

Other significant points in installation are:

1. *Condition Employees for the Change.* The layout installation is a disrupting time for employees. If a good job has been done so far, they are generally familiar with the proposed new layout. When the installation plans are nearing completion, invite employees to come in and see the layout model or move a copy of the template layout out into the plant. Identify each department or code mark each work station so each worker can find where he is going to be. Have supervisors review the details with their workers. Talk enthusiastically and ask for comments. Give employees the details about the new layout in writing with diagrams and photographs.

2. *Basic Relayout Problem.* Making a relayout is like a game of checkers—one move is made into the spot presently occupied by some other piece of equipment. Here the sequence of all moves must be planned so the movers do not try to move something into space already occupied. This problem, together with that of keeping up production during the change, is frequently a major obstacle limiting the design of the layout itself.

3. *When to Install.* Although the time to install the layout is important, it is generally a matter of selecting the least inconvenient time rather than finding one that is fully acceptable to everyone. Frequently preferred times include:

a. Time of annual changes in product design
b. Time of plant shutdown for vacation
c. Slack season
d. Weekends or extended weekend

4. *Planning the Installation.* R. H. McCarthy of Western Electric Company recommends seven stages in layout installation: plan, provide, prepare, move, install, start up, and clean up. These seven might include:

a. Plan:
(1) Start planning early. Sound planning ahead saves time during the fast-action stages.
(2) Determine the sequence of moves, weighing practical operating problems.
(3) Make up an inventory of everything to be relocated. Get disposition of the balance.
(4) Schedule the moves in detail. Set up a timetable with specific dates and times.

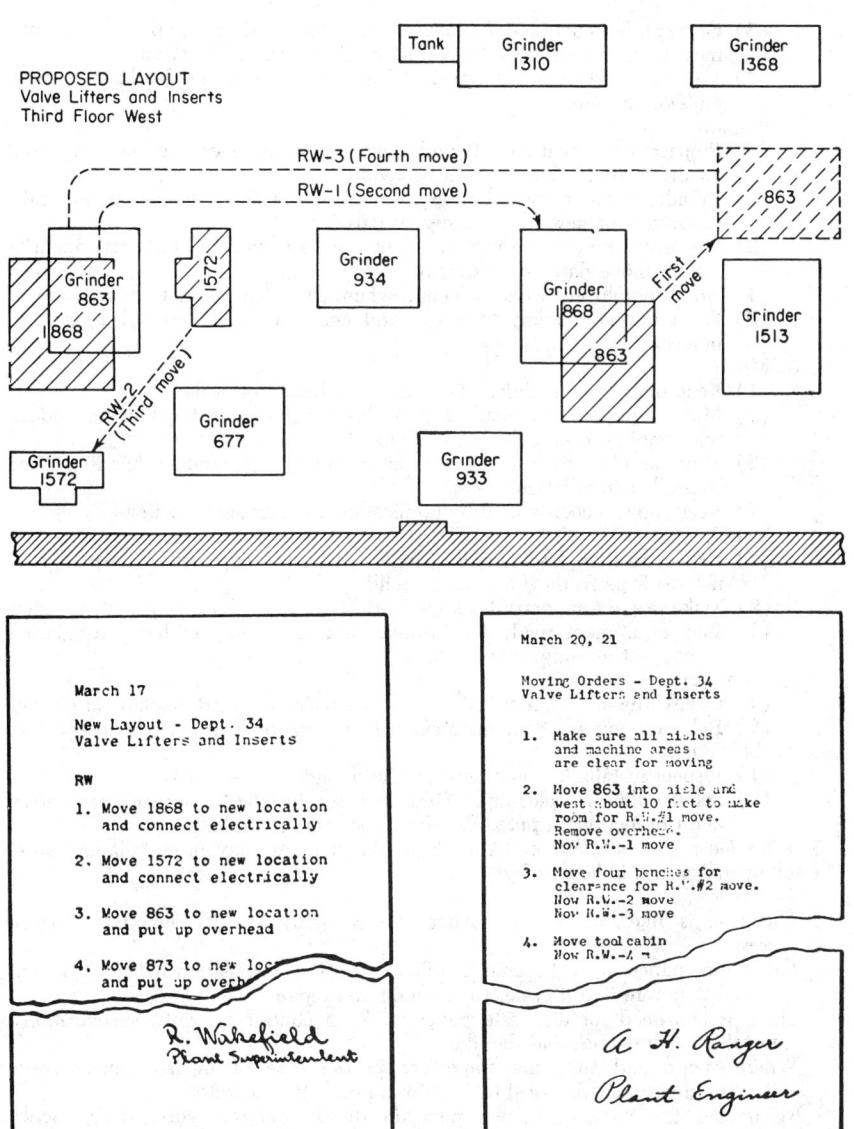

FIG. 2-16.   Method of planning small layout changes used by an automobile engine plant.

(5) Assign a move number to each item, mark it on the inventory sheet, and check it against the machine tag.
b. Provide:
(1) Consider the use of outside moving and installation contractors and get bids from several.
(2) Have adequate help. Set up key men in each department or area involved.

(3) Get ample moving equipment. Consider renting equipment to assist moving or to keep operations running during the installation.

(4) Provide good communications. Have telephones and leadmen at both ends of the move.

c. Prepare:

(1) Prepare new location. Foundations, partitions, cleaning, painting, and auxiliary service lines should be ready.

(2) Broadcast the plans. Let everyone know what is going on and take advantage of new ideas or suggestions.

(3) Tag every item to be moved. Use color and coding, and mark identification, move date, and destination.

(4) Notify employees what to do and when, where, and how to do it.

(5) Get equipment ready to move, and check it out before releasing it to movers.

d. Move:

(1) Keep move on schedule. Post accomplishment each day.

(2) Move equipment intact. Try to keep equipment together to reduce reassembling time before it can operate.

(3) Move as close to installation point as practical to reduce handling time by skilled installation crews.

(4) Keep movers coordinated by notification and frequent briefing.

e. Install:

(1) Expect to have last-minute changes. Do not get upset if the plan does not work perfectly, for it never will.

(2) Make use of temporary hookups, with permanent service connections later.

(3) Flag equipment ready for installation inspection, and have installation crews post accomplishment daily.

f. Start Up:

(1) Check the installation. Be sure the placement and hookup are right.

(2) Release equipment for maintenance tryout and for foreman's acceptance.

g. Clean Up:

(1) Inspect installation, and note any loose ends.

(2) Set deadline for cleanup. Otherwise the installation remains temporary and production output suffers from the same attitude.

5. *Who Does the Installation.* Most firms do their own layout installation work. The following reasons indicate why:

The cost is likely to be less when the company has its own maintenance crew.

The maintenance men become familiar with the installation when it is put in and therefore find it easier to maintain and repair.

There is less need for elaborate paper work on contracts, prints, specifications, installation drawings, and the like.

Where speed and time are important in the case of hurried changeovers, it is frequently not practical to wait for an outside contractor.

By having the company's own men do the installation work, the presence of maintenance men is subsequently assured in case of emergency.

There are, however, many advantages to employing outside contractors to make the installation:

Contractors are often highly skilled and familiar with layout installation work and techniques; they will have the proper equipment available and can do a safe, efficient job.

Frequently, the company does not have a large enough plant engineering staff to handle the rather infrequent task of layout installation. For new layouts, there may be no company staff at all.

A company's own construction and maintenance crews will have a lot of other details to attend to during a relayout, and they do not have time to handle the installation itself.

Where outside contractors are employed, it is practical to have one or more company men work closely with them. Also be sure to specify for both groups what details of the installation each group will do. Thus if there is an omission, one group will not assume that the other is handling it.

6. *Identify Locating Points.* Before beginning to move anything, experienced layout engineers get their major aisleways marked in. Otherwise, installation men will set down equipment temporarily in the aisles, and by clogging the aisle, make it necessary to move through areas where equipment is being installed. This leads to much shuffling. Also, columns should be identified before the installation if not already marked. Most equipment is going to be located from these aisle lines and columns. Whenever floor space is clear, the exact location of each piece of equipment in the new layout should be marked on the floor.

7. *Coordinating the Installation.* When a major move is involved and plans for its execution are completed, conduct a conference of the heads of each function concerned. Following this meeting, advance written notification of the move schedule should be posted, again allowing sufficient time for possible conflicts to appear. When all are ready, the actual moves are usually initiated by a work order or an equipment move notice together with an accompanying list of equipment.

Perhaps the easiest way to schedule and control the installation of a new layout is the Gantt chart. This shows on the same sheet both the plan and the accomplishment to date. Checks of action completed should be made daily.

8. *Layout Engineer Should Be on Hand.* Layout men should be ready to make any necessary changes during the installation. No matter how good the layout engineering has been, there will be adjustments at installation time. The engineer should expect such problems and anticipate any necessary changes.

9. *The Follow-up Check.* There will always be bugs, regardless of how thorough the planning has been. The layout engineer should check the actual layout—as installed and operating—against his approved plan. He should recognize any differences and either accept them as satisfactory or make arrangements with the installation crews to reset equipment as called for.

10. *The Layout Record.* Even if for no other reason, a check of the actual installed layout is needed to bring the layout board or prints up to date. Only in this way will a plan of the existing layout be ready and available for future reference.

OFFICE LAYOUTS

Systematic Layout Planning (SLP) has a simplified version that is particularly adaptable to office and laboratory layouts. Figure 2-17 shows an example of applying Simplified SLP to a small office layout. Although the simplified version is especially suited to small areas not having a dominant flow of materials, many analysts have found it valuable for large office layouts. They can "subcontract" smaller, less complicated areas to the area supervisors by training them with the booklet, *Simplified Systematic Layout Planning.*[1] The supervisors follow the simple, six-step procedure to lay out their own areas. Tremendous time savings can be achieved on involved projects by using this technique for the less vital areas—and without default on the part of the total project leader.

The six-step pattern is shown in the center of the capsule summary in Figure 2-2. The six steps include:

1. Chart the relationships.
2. Establish space requirements.
3. Diagram activity relationships.
4. Draw space relationship layouts.

[1] R. Muther and J. D. Wheeler, *Simplified Systematic Layout Planning,* Management and Industrial Research Publications, Kansas City, Mo., 1962.

SIMPLIFIED SYSTEMATIC LAYOUT PLANNING . . .

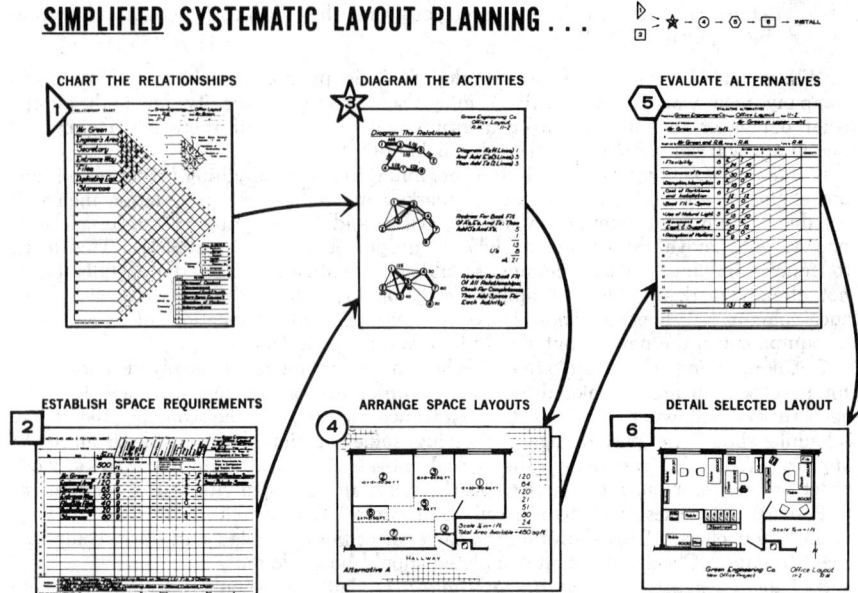

FIG. 2-17. The six steps of Simplified Systematic Layout Planning. (*Source: Simplified Systematic Layout Planning, R. Muther and J. D. Wheeler, Management and Industrial Research Publications, Kansas City, Mo., 1962.*)

5. Evaluate alternative arrangements.
6. Detail the selected layout plan.

Although Simplified SLP rests on the basic fundamentals of relationships, space, and adjustment, it is applicable to only a limited number of activities. As a result, it is used frequently in phase III detailed layouts—or where the area involved is generally not more than 5,000 to 8,000 square feet.

COMPUTER-AIDED LAYOUT PLANNING

Basically, there are four computer-aided layout planning programs generally available in the United States. They are CRAFT (Computerized Relative Allocation of Facilities Technique), CORELAP (Computerized Relationship Layout Planning), ALDEP (Automated Layout Design Program), and RMA Comp. I (Richard Muther & Associates contribution to the field). There are other programs, both in the United States and abroad, but they are not readily available to industry.

CRAFT was developed by Messrs. Armour, Buffa, and Vollmann and is available through the SHARE program library. A fairly complete description of CRAFT was published in the *Harvard Business Review*, March–April, 1964. CORELAP was developed at the Industrial Engineering Department of Northeastern University, headed up by Professor James Moore, and is available from Engineering Management Associates at Northeastern University, Boston. A description was published in the *Journal of Industrial Engineering*, March, 1967. ALDEP was developed by IBM, Rochester, Minnesota, and is available through IBM sales offices. The *Proceedings* of the 1967 annual meeting of AIIE in Toronto give a more detailed account of ALDEP. RMA Comp. I was developed by the staff of Richard Muther & Associates, Kansas City, Missouri, and information about application of the program is available through them.

Computer Input Data. In general, the input data for all four programs is similar. All four programs require the fundamental inputs of relationships and space. See Figure 2-18.

How the Layout Programs Work. Basically, CRAFT computes the product of the flow data, multiplied by the handling cost, multiplied by the distance between centers of activities given in the initial layout. In this way, a total, initial cost is computed. Then the program considers exchanges of activity locations; it examines two-way and three-way interchanges. The interchange involving the greatest cost reduction is made, and a new total cost is calculated. This process is repeated through successive iterations (selected interchanges) until no significant cost reduc-

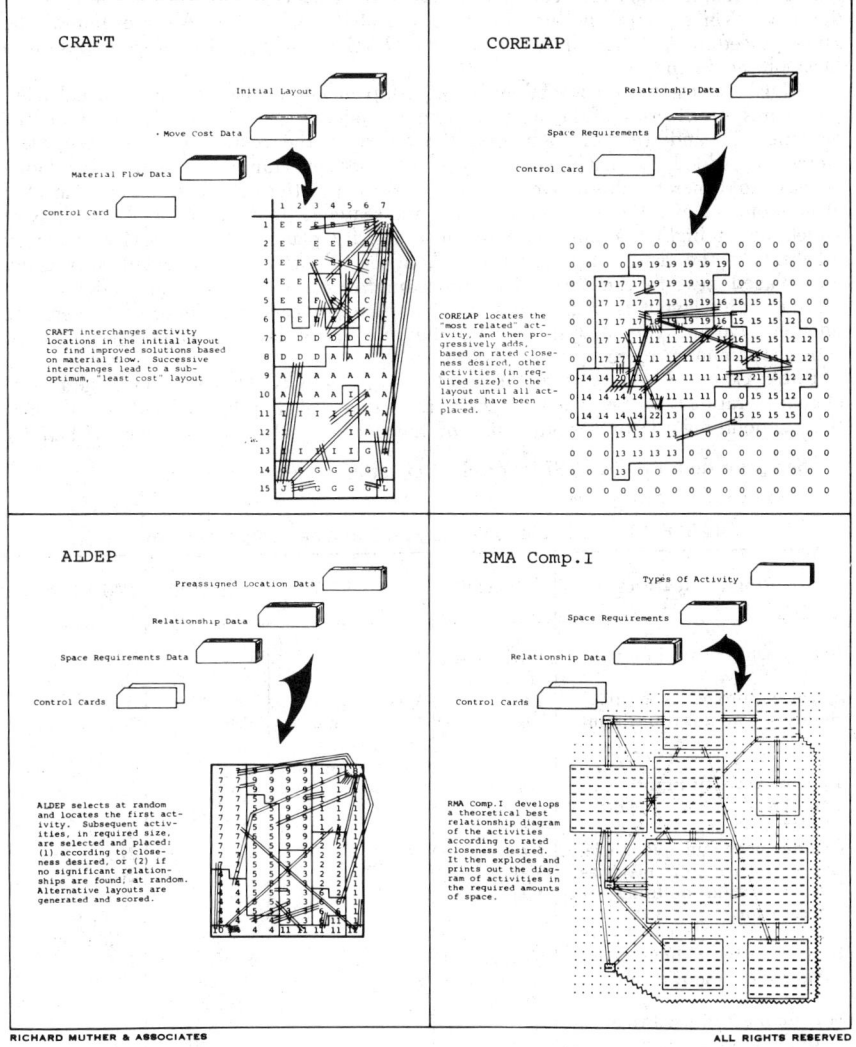

Fig. 2-18. Computer-aided layout planning—input-output for available programs. Area outlines and relationship lines have been added—they are not part of the original printouts.

tion can be found. The program is path oriented, so all possible interchanges are not examined.

CORELAP begins by calculating which of the activities to be placed in the layout are the busiest or most related. The sums of each activity's closeness relationships with all other activities are compared, and the activity with the highest total closeness relationship (TCR) count is selected and located first in the layout matrix. This activity is named "Winner." Next, an activity which has an A relationship[2] to the Winner is selected and placed as close as possible to the Winner. This activity is named "Victor." A search is then made of Winner's remaining relationships for more A-related Victors, and these are placed, again, as nearly adjacent as possible. If no more A's can be found, the Victors become potential Winners, and their relationships are searched for A's. If an A is found, the Victor becomes the new Winner, and the procedure is repeated. When no A's are found, the same procedure is followed for E, I, and O relationships until all activities have been placed in the layout.

ALDEP uses a preference table (relationship values in matrix form) to calculate the scores of a series of layouts which are randomly generated. If, for example, activities 11 and 19 were adjacent, the value of the relationship between them would be added to that layout's score. A modified random selection technique is used to generate alternative layouts. The first activity is selected and located at random. Next, the relationship data are searched to find an available activity which has a high relationship to the first activity placed. This activity is placed adjacent to the first. If none is found, a second activity is selected at random and placed next to the first. This procedure is continued until all activities are placed. This layout is then scored, and the entire procedure is repeated to generate another layout. The analyst specifies the number of layouts wanted which must satisfy a minimum score.

RMA Comp. I also selects the most related activity (having the largest total closeness rating) to be placed first in the center of the layout matrix. It does this without regard for the area size of the activity. As each subsequent activity

[2] See prior descriptions of SLP vowel-letter rating scales.

TABLE 2-1.   Computer-aided Layout Planning Comparison Table

Program feature	CRAFT	CORELAP	ALDEP	RMA Comp. I
Relationship chart input	With modification	Yes	Yes	Yes
Space requirements input	Yes	Yes	Yes	Yes
Building configuration input	Yes	No; optional with time-shared version	Yes	No
Type of activity input	No	No	No	Yes
Can consider X (negative) relationships	No	No	No	Yes
Can consider multiple stories	No	No	Yes	No
Can fix location of activities	Yes	No	Yes	No
Output within building configuration	Yes	No	Yes	No; space relationship diagram
Honors activity shape or configuration requirements	No; yes, if location is fixed	No	No; yes, if location is fixed	No
Scoring method used to evaluate layouts	Yes	No	Yes	No
Honors all relationships for practical problems	No	No	No	No

is placed, all its relationships are considered before it is placed. In other words, before a given activity can be located, its relationships to activities not yet placed as well as to those activities already placed are considered. When placing each activity, room is left for related activities which will be placed later. At the same time, a check is made to see that X relationships will be satisfied. In this manner, A-related activities are selected and placed, followed by E-, I-, and O-related activities. A relationship diagram is thus formed—still exclusive of area requirements. The diagram is then exploded and the required amount and type of space are assigned to each activity in the diagram.

Evaluation. Layout planners who have used computer programs find that they aid planning, as shown by Table 2-1. None of them actually prints out an acceptable layout, however; they all require further adjustment. Additionally, it is common to find computer layouts where some of the relationships have not been honored. Still, run in parallel with or as a check on manual planning, they can bring to light an independent alternative that might not have been considered.

WHO LAYS OUT THE PLANT

To find out who does industry's layout planning and to identify the problems that plague the planners, the American Material Handling Society (now IMMS) conducted a three-year survey (1961–1964) of American and European firms.

Perhaps the most notable fact uncovered in this survey is that the plant layout engineering is indeed a staff function. More than half of the people who do this work are staff specialists.

Organizationally speaking, the layout function is evenly split between industrial and plant engineering. However, as might be expected, there are many other groups or individuals who have on occasion been given the responsibility for this function.

And as for the problem areas, management criticizes layout planners more for their abuse of time than for matters of money. Layout planners in turn criticize their management most often for their poor planning and poor communications.

Largely, the people surveyed were from manufacturing firms, with a few process industry firms and public or commercial warehouses. They were located in the United States, Canada, England, Sweden, Norway, The Netherlands, and Switzerland. Roughly half the questionnaires were filled out by Americans, and half by Europeans, 463 in all.

The results are summarized in Figure 2-19. Further information on the results is reported in *Modern Materials Handling*, April, 1964.

MANAGEMENT INTEREST IN PLANT LAYOUT

Most managers realize that processes and methods are constantly changing and that product improvement is essential to a plant's existence. These conditions lead to continuous projects in relayout and in adjustments to existing layouts. Thus, a layout planned and installed is not necessarily completed and can be considered complete only when it is entirely replaced by a new layout.

When Is a New Layout Needed? Because plant layout is a compromise of many factors, there will always be something about each layout that is imperfect. For this reason, one can criticize something in every layout. The most logical time for a layout change is when making improvements in methods or machinery. Layout changes and improvements in process machinery or equipment go hand in hand.

Because changes in other activities can be or will be made at the time of a relayout, management should note three further points:

1. A request for layout change may be made to permit a change in another activity.
2. A layout change gives management the opportunity to make other changes it may may have been holding back.
3. A suggested change in layout may not always solve the real problem; a

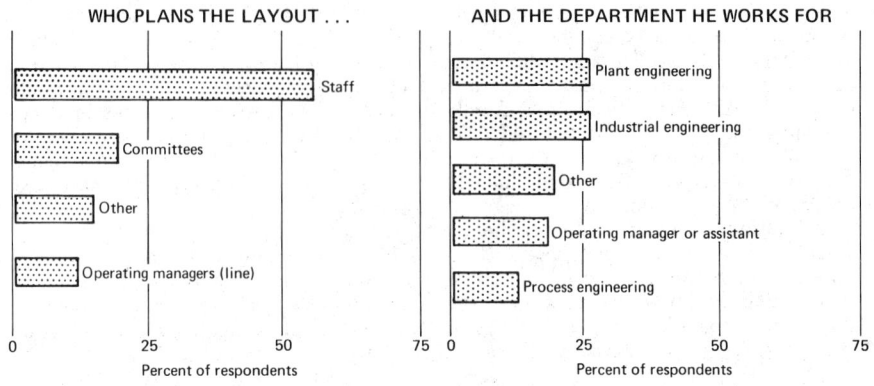

WHO PLANS THE LAYOUT . . .

Staff
Committees
Other
Operating managers (line)

0   25   50   75

Percent of respondents

STAFF SPECIALISTS handle more than half the layout work. In American firms, even more are staff (64 percent).

AND THE DEPARTMENT HE WORKS FOR

Plant engineering
Industrial engineering
Other
Operating manager or assistant
Process engineering

0   25   50   75

Percent of respondents

ENGINEERING, in some form, most often responsible for layout. Europeans lean more heavily on committees.

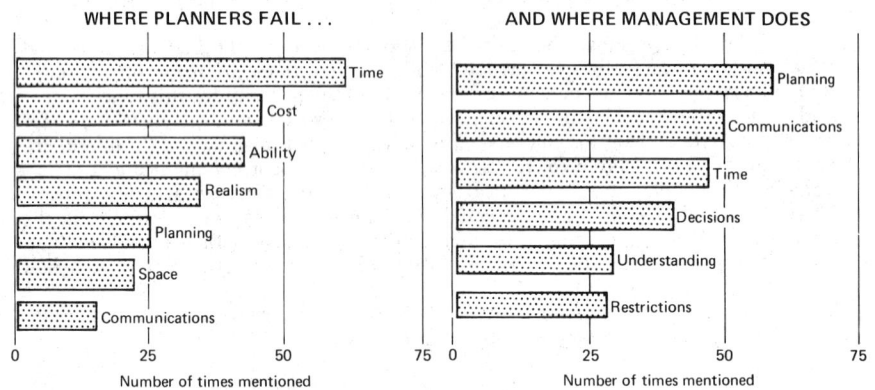

WHERE PLANNERS FAIL . . .

Time
Cost
Ability
Realism
Planning
Space
Communications

0   25   50   75

Number of times mentioned

USE OF TIME is layout planners' chief fault in the eyes of management, with cost problems, surprisingly, second.

AND WHERE MANAGEMENT DOES

Planning
Communications
Time
Decisions
Understanding
Restrictions

0   25   50   75

Number of times mentioned

POOR PLANNING, on management's part, causes most problems for layout planners, with lack of information next.

*DATA FROM SURVEY BY A.M.H.S., TECHNICAL SERVICES DIVN., AS PUBLISHED IN MODERN MATERIAL HANDLING, APRIL 1964.

FIG. 2-19. Who lays out the plant?

bad situation may be the result of several factors of which layout is but a part.

MANAGEMENT SUPERVISION OF LAYOUT PROJECTS

Management should conduct its assignment, supervision, and execution of the layout projects in ways that will tend to ensure success.

1. *Appoint a Group or Individual to Head up the Layout Project.* Fix the responsibility in someone who knows layout problems and techniques.   Other alternatives include:

   a. Assign the project to a committee of various people closest to the layout.

   b. Employ an outside specialist.

c. Develop layout engineers from within the firm.

d. Plan visits to other companies.

2. *Supervise the Layout Work as Necessary.*

a. Follow the guiding fundamentals of good layout.

b. Make a clear statement of the project.

c. Schedule the project.

d. Make a check of progress as each phase of the layout work nears completion. Management should review progress in these four areas:

Review phase I (location): When the area to be laid out has been selected.

Review phase II (overall layout): When approval is requested for the general overall layout plan.

Review phase III (detailed layout plan): When approval is requested for the detailed layout plans.

Review phase IV (installation): When the layout is installed and ready to release to production personnel.

3. *Balance Forcing Accomplishment with Getting Good Results.* A manager will lose out in the long run if he presses his layout engineers for some visible plan too quickly. At the same time, endless refinement is not wanted. All too often, meticulous perfection of details is the chief cause of delay.

4. *Plan for the Future.* Management must consider the long-range plans of the business and question whether the layout suggested will be as good at the halfway point in its life as it looks at the beginning. In developing layout plans, the wise manager will call for a planned analysis of anticipated requirements. He will project his sales plans for many years ahead. In this way, a basic plan can be drawn up for developing the entire area according to a sound overall layout. The initial building and all subsequent additions can then be integrated into the planned pattern. Managers will be interested to know that one of the most difficult problems faced by a layout man is to get management to look far enough ahead when planning their facilities.

5. *Approve the Layout.* Executives or department heads who are asked to approve layout plans normally consider three questions:

a. What will be gained from this layout?

b. What are the risks in this layout?

c. How does this affect me personally and my group?

The manager will certainly want to see analyses of the costs required and of the savings expected. He should watch for hidden costs in terms of increased maintenance, greater demands on service equipment, and the like. Three other factors should be checked: safety, convenience for personnel, and changes in organization structure or procedures.

There are two ways for management to be sure that all groups agree on the layout:

Hold a meeting of all personnel involved to go over the layout plans together. Make sure all department heads or a given list of them have initialed the layout plan.

Besides a good layout being more or less guaranteed by these methods, there is the important factor of group participation. In the eyes of a department head or supervisor, the layout engineer represents change. And with every change comes fear—fear that the new will mean less importance in the organization, more difficult operation of the department, greater dependence on others, or more pressure and rush service. Much of this can be avoided by building from the start the idea that the layout is a company project.

6. *Build Morale with a New Layout.* By encouraging and developing cooperative activity in the layout project, a manager is building up his team. Be sure operating and service heads are included, and let employees in on the plans. Besides these internal activities to build employee enthusiasm, favorable publicity outside the plant can help swell employees' pride in "their" new layout.

BIBLIOGRAPHY

Apple, James M., *Plant Layout and Materials Handling*, 2d ed., The Ronald Press Company, New York, 1963.

Moore, James M., *Plant Layout and Design*, The Macmillan Company, New York, 1962.

Muther, Richard, *Practical Plant Layout*, McGraw-Hill Book Company, New York, 1955.

Muther, Richard, *Systematic Layout Planning*, Industrial Education Institute, Boston, 1961.

Muther, Richard, and John D. Wheeler, *Simplified Systematic Layout Planning*, Management and Industrial Research Publications, Kansas City, Mo., 1962.

Reed, Ruddell, Jr., *Plant Layout: Factors, Principles, and Techniques*, Richard D. Irwin, Inc., Homewood, Ill., 1961.

Chapter 3

Economics of Equipment Selection*

RALPH O. SWALM

**Professor, Department of Industrial Engineering,
Syracuse University, Syracuse, New York**

Most companies are in business to make profits, although they may have other motives as well. Also, most companies are in business on a long-range basis. That is, they plan to operate indefinitely into the future and perhaps to expand their operations. In spite of these objectives, many companies direct too little of their attention to the basic means by which they can reach these goals. These means are the buildings, machinery, and similar capital assets which they use in the production of their product and which make it possible for them to have an income.

In the United States, industrial firms may have from 10 to 90 percent of their total assets tied up in plant and equipment. From the standpoint of both stockholder and manager, either extreme warrants careful attention. To the stockholder, protection of his original investment, as well as continuance of his current income, is of extreme importance. To the manager, maintenance of his job is at stake, in addition to his pride in a competitive operation.

This problem is of particular interest to American industry. There is a continuing trend toward broadening competition from foreign producers. Many of these foreign producers are building new plants and equipping them with the very latest automatic and high-speed machinery and tools. With both an equipment advantage and

---

* The discussion on pages 11-63 to 11-64 is adapted from Harry T. Schwan, "Replacement of Machinery and Equipment," chap. 3 of sec. 8 in the second edition of the *Industrial Engineering Handbook*.

lower labor costs, competition is extremely keen. It is true that American manufacturers have purchased better equipment heavily. It is also true that analyses of machinery and equipment will usually show a vast amount of production capacity that is over fifteen years old.

Only a small percentage of manufacturers have a planned and systematic way of recognizing and doing something about this problem. There is a definite trend, however, toward facing the problem realistically and taking systematic action toward a solution. Companies are checking their present position, stating policies for the guidance of operating management, and substituting a planned approach for the old hunch and spur-of-the-moment decisions.

THE BASIC PROBLEMS OF EQUIPMENT STUDIES

Management's problem in equipment studies is essentially the selection of one plan of action from among two or more alternatives. This seems simple until the factors that are involved are considered. Here are some thoughts that may run through a manager's mind as he tries to decide what to do:

1. I am about to spend some of the company's money. If I make a decision which turns out poorly, will I be censured more than if I let things go on as they are?
2. Am I considering all the good alternatives, or is there a better one which has not been presented?
3. Many of the figures presented to me are based on estimates of the future. Will things actually happen that way, or will something come up to change the situation?
4. Will the new machine or procedure actually do what is claimed for it, and are the claimed savings a real possibility?
5. Will a new alternative present itself in a year or so and cause my decision to turn out wrong, even though it appears to be good now?

Many other disturbing thoughts can come up, but the foregoing are sufficient to point out that as many human problems are involved as technical ones. Fear of being wrong and fear of the unknown—the future—are the basic stumbling blocks.

The fact that these thoughts do exist—in addition to the fact that many factors in an equipment study are complicated, based on estimates, and not subject to precise, quantitative expression—has caused many management men to shy away from systematic studies. They have said, "Because there are so many variables and so much estimating, why waste time on careful computations? I'll just use my long experience and intuition."

That is where their logic falls apart. The very fact that these decisions are complicated calls for systematic study and the use of all the tools available. Even the most brilliant intuition and longest experience are insufficient to handle all the factors in their proper perspective. This in no way minimizes the use and value of astute judgment. All the facts available must be obtained, tempered with estimates and judgment where needed, and then used.

There is evidence of increasing use of careful computations rather than intuition in this area. But the most careful computations may fail to lead to a good decision if the computations are used to address the wrong question or if the criteria chosen are inadequate.

ASKING THE RIGHT QUESTION

An engineer once proposed that an automatic method costing $100,000 be used to replace a hand method. His careful computations showed that this could save $70,000 per year for at least five years. Now almost any criterion applied to the question "Should automatic equipment which costs $100,000 and offers a savings of $70,000 per year for five years be installed?" would yield an affirmative signal. But when it was recognized as more meaningful to rephrase the question to "How can we best improve the present method?" the following alternatives were examined.

Alternative	Investment	Savings
Methods improvement......	$ 5,000	$50,000 per year
Semiautomation...........	40,000	60,000 per year
Automation..............	100,000	70,000 per year

Looking at all this information, should automation be recommended?

Compared with the methods improvement option, it costs an additional $95,000 to obtain an additional yearly return of only $20,000. In five years, the company would just about get its money back.

The automatic equipment would undoubtedly be recommended if the engineer's predictions were accepted and used to fill out a typical authorization for expenditure form, making a comparison of the present with the proposed method. It is true that an improvement would be made by automating, but it is also clear that this would not be the best course of action.

Nor would the decision to automate be shown to be suboptimal by a follow-up study. Such a study might confirm that automating did cost $100,000 and that the predicted savings of $70,000 per year were realized—but it could never point out that savings almost as great could have been made for a much smaller investment. It is impossible to postaudit a decision—at best, it is only possible to postaudit predictions.

To sum up: Before invoking any criterion, one should first ask, "Have I considered all the potentially optimal ways of solving my problem?"

And, of course, one should be sure that he is trying to solve the right problem. The chapters on methods study and value engineering, Chapters 1 and 7 of Section 2, have much to say about this.

CHOOSING A CRITERION

Ideally, a criterion should lead to a methodology that:
1. Is simple to use and easy to understand
2. Addresses the proper question
3. Takes all relevant factors into consideration
4. Can be used to select among multiple alternatives
5. Accepts probabilistic inputs
6. Leads to an optimal choice

It is obvious that trade-offs must be made among these desirable attributes. The art of analysis consists in judging when increasing sophistication ceases to pay its way.

Many businessmen will wisely prefer a simple, reasonably logical criterion to one that is more logical but much more complex. They feel that it is better to have a reasonably good decision rule that is used than an elegant one that none but the analyst can accept or understand.

On the other hand, one should refuse to buy simplicity at too great a price. With proper effort, that which is not acceptable today may become so tomorrow.

A case in point is the use of probabilistic inputs in decision analysis. Many find it difficult enough to estimate the life of a project, and when asked to estimate the probability that the project will last four years, then that it will last five years, and so on, they throw up their hands in despair. Yet on further thought, they realize that the second or probabilistic type of prediction conveys considerably more information, and with experience, they find it more comfortable to live with.

METHODOLOGIES FOR CAPITAL EXPENDITURE ANALYSIS

Short Payback Approach. According to all published surveys, more people use some form of the short payout criterion than any other method of judging the wisdom of a proposed capital expenditure. The question most commonly asked is "Will it pay for itself in two years?" but there is less than complete agreement on the two-year period or on how to evaluate the amount of the payback.

Forgetting these matters for the moment, the two-year payback criterion would signal the acceptance of a project costing $50,000 and paying $30,000 per year, but would reject a similar investment paying only $20,000 per year.

How does this criterion measure up to the suggested list of desirable properties? It is certainly both simple and easy to understand. But does it answer the proper question? Can it be used to select among multiple alternatives? Does it consider all relevant factors? And does it accept probabilistic inputs?

These questions can be examined through the use of simple examples. Suppose it is desired to choose among the following investments which would result in certain cash flows of the amount and at the time indicated.

Cash flow	A	B	C	D	E
Time zero	-100	-100	-100	-100	-100
End of year					
1	$+ 50$	$+ 48$	$+100$	$+ 30$	$+100$
2	$+ 50$	$+ 48$	$+ 10$	$+ 30$	$+100$
3	0	$+ 48$	$+ 10$	$+ 30$	$+100$
4 to infinity	0	0	0	$+ 30$	0

Defining the payback period as the first cost divided by the annual cash inflow, project A shows a payback of 2 years, and project B a payback of about 2.1 years, yet most would prefer B to A. Project C gives more difficulty—should the incomes for the three years be averaged, or should that of the first year be used in the denominator? The first choice leads to a 2½-year payback, the second to a 1-year payback.

Using a slightly different definition of payback and asking how long a time it is until the investment is paid back, C offers a 1-year payback, as does E. Yet C is clearly less desirable than E, which pays $90 a year more after the first year, and most would consider it inferior to B, which has only a 2.1-year payback.

Project D shows the longest payback of all—3.3 years—yet many would find it more desirable than any alternative but E.

Ranking these alternatives in order of increasing payback:

E (and possibly C) . . . .   1 year
A . . . .   2 years
B . . . . 2.1 years
C (possibly) . . . . 2.5 years
D . . . . 3.3 years

If all projects with 2-year or better payback were accepted, E, A, and possibly C would make it; B, D, and possibly C would not. Yet most analysts would prefer B, C, and D to A. Note that A is equivalent to lending $100 for 2 years at zero interest rate.

What can be learned from these examples?

1. As seen from the problem of defining the payout for C, the short payback criterion is not as simple as it first appears.
2. The payback criterion fails to recognize that B is better than A, E better than C, and C better than B, because it does not take into consideration all the relevant factors—here, the pattern and deterioration of the benefits produced.

Furthermore, this approach obviously does not permit probabilistic inputs. It attempts to take risk into account by demanding a short payback period; but this does not lead to a wise selection among projects having varying degrees of risk. It is felt to be conservative, yet it accepts A, in which the money invested is paid back over a 2-year period at zero interest rate, and rejects D, which pays a 30% return on invested capital.

Returning to the question of how the payback period should be calculated, the following variants, in different combinations, are in current use:

1. Divide the cost of a proposed asset by the average annual predepreciation incomes it will produce.
2. Divide the first cost by the annual income after depreciation.
3. Determine the length of time for the anticipated income to equal the first cost.
4. Determine the length of time for the incomes, discounted at a nominal interest rate, to equal the first cost.

Each variant, of course, yields a different payout period. Because none is to be recommended except as a rough screening device, they will not be further discussed.

Return on Investment Approach. Perhaps the next most commonly used approach is some variant of the so-called return on investment method, sometimes called the accounting method. In this approach, the annual return is divided by the amount invested to yield the return on investment.

Again, variations of this general method are legion. Some users calculate the return before depreciation, and others after; some use the initial investment in the denominator, and others the average investment.

Note that in its simplest form—return before depreciation divided by first cost—the return on investment will be merely the reciprocal of the payback period. Because the reciprocal of a number conveys exactly the same information as the number itself, this method has all the advantages and limitations of the payback approach.

One of these limitations is that neither the payback nor the simple rate of return method is sensitive to the fact that for a given investment and annual return, the longer the life of the asset the greater is its desirability. Thus an asset costing $100,000 and producing $20,000 savings per year for five years has the same payback and return on investment as another costing the same amount but producing savings of $20,000 per year for twenty years.

The subtraction of the straight-line depreciation amount from the numerator of the return on investment calculation (or the denominator of the payback) tends to take the greater value of long-lived assets into consideration, because the amount subtracted is less as the lives increase.

Thus, defining the return on investment (ROI) as the ratio of income after depreciation to the first cost, the ROI in the two cases above would be:

If life is 5 years: $$\frac{20,000 - 100,000/5}{100,000} = 0\% \text{ ROI}$$

If life is 20 years: $$\frac{20,000 - 100,000/20}{100,000} = 15\% \text{ ROI}$$

This is a step in the right direction—a 20-year asset producing the same annual income before depreciation as a 5-year asset costing the same amount is surely a better bargain. However, proponents of this method get into difficulty in a realistic prediction when incomes are not projected as uniform throughout the project life, when depreciation is based on other than straight line, and when project life and the depreciation period permitted by the internal revenue service differ. As in the payback approach, this method is not well suited to probabilistic inputs or to examining tax effects in situations of any complication.

Discounted Cash Flow and Present Worth Approaches. The next most widely used method, and one whose use is rapidly increasing, is the so-called discounted cash flow method. This method is also known by such other names as the return on investment method, internal rate of return method, profitability index method, and the investors method.

In essence, the users of this approach determine the interest rate that discounts the sum of the cash inflows resulting from a project to the cash outflow required to obtain the project. This is done in recognition of the fact that a dollar in the future is worth less than a dollar today.

A close relation to the discounted cash flow (DCF) method is known as the present worth method. It answers the question, "At a given interest rate, will the present worth of the cash inflows resulting from a project meet or exceed the cash outflow required to obtain that project?"

Both these methods use the fundamental equation of compound interest:

$$F = P(1 + i)^N$$

where P = an amount of money at the present time

   i = an interest rate per period, generally stated as a percent, but used as a decimal in the formula

   N = number of interest periods

   F = an amount of money in the future, N interest periods away

In actual practice, this equation is more often used in its reciprocal form:

$$P = F\frac{1}{(1 + i)^N}$$

To derive this equation, note that if P dollars are placed in an account at interest rate i per period, it grows to $P(1 + i)$ at the end of the first period. At the end of the second, $P(1 + i)$ is increased by a second interest payment; it becomes $P(1 + i)$ $(1 + i)$, or $P(1 + i)^2$. Similarly, at the end of the third period, it grows to $P(1 + i)^3$, and at the end of the Nth period, to $P(1 + i)^N$. Calling this future amount F yields the equation

$$F = P(1 + i)^N$$

Solving this for P yields

$$P = F\frac{1}{(1 + i)^N}$$

To illustrate the use of the present worth method in a very simple case, suppose it were proposed to buy a calculator for $1,000 and that it would pay a net saving, compared with the present method, of $500 per year for 3 years. If a 25% return (before taxes) is required, should the calculator be purchased? In asking this question, it is assumed that the present method cannot be improved on and that no other logical alternatives exist.

Assuming the incomes to occur at the end of each year, purchase of the calculator would lead to the following cash flow pattern:

Time zero..........	−1,000
End of year 1.......	+ 500
End of year 2.......	+ 500
End of year 3.......	+ 500

The user of the present worth method would discount each of the three $500 incomes to their value at time zero, using the formula shown above. He would examine whether or not

$$500\frac{1}{1 + i} + 500\frac{1}{(1 + i)^2} + 500\frac{1}{(1 + i)^3} \geq 1,000 \qquad \text{when } i = 25\%$$

Because tables of $1/(1 + i)^N$ (which is known as the single payment present worth factor) are widely available, it would not be necessary to calculate the parenthetic factors.   In practice, the value on the left of the expression above would be found as follows:

Time of cash inflow	Amount	25 % PW factor*	PW at 25 %
1	$500	0.800	$400
2	500	0.640	320
3	500	0.512	256
Total..........	....	.....	$976

* These factors are found in Table 3-1.

Because the present worth (PW) of the cash inflow ($976) is less than the cash outflow ($1,000), the purchase would not be indicated.

A discounted cash flow solution would approach this same problem by asking, "What interest rate would make the present worth of the cash inflows just equal to the outflow at time zero?"   For all but the simplest problems, this question must be answered by using a trial and error method, which can be performed easily on a computer.   A simplified method based on a graphical solution that has been developed by Raymond I. Reul will be described in a moment.

To obtain a straightforward, hand solution to the calculator problem, note that at 25% the present worth of the inflows is a bit less than the $1,000 outflow.   Therefore, a 20% discounting rate is tried:

Time of cash inflow	Amount	20 % PW factor	PW at 20 %
End of year 1........	$500	0.833	$ 416.50
End of year 2........	500	0.694	347.00
End of year 3........	500	0.579	289.50
Total.............	....	.....	$1,053.00

At 20%, the PW of the inflows is greater than $1,000.   Thus, the solution rate is seen to be between 20 and 25%.   Interpolating, it is

$$20\% + \left[\frac{53}{53 + 24} \times 5\% \right], \text{ or about } 23.4\%$$

That the money in such a project would actually earn 23.4% while invested in it is seen by the following sequential calculations:

$1,000 invested at 23.4% for 1 year becomes $1,234.
If $500 is withdrawn, there is left $734.
$734 invested for 1 year at 23.4% becomes $905.
If $500 is withdrawn, there is left $405.
$405 invested for 1 year at 23.4% becomes $500.
If $500 is withdrawn, there is left nothing.

Note that in this problem the payout period is 2 years; the return on investment method that is the reciprocal of the payout period, of course, yields 50%; that in which straight-line depreciation is subtracted from the cash inflows yields a result of $(500 - 1,000/3)/1,000$, or 16.7%; and that in which the return is calculated on the average investment yields 33.3%.

To be realistic, an evaluation system should take into account all cash flows produced by an investment; taxes are such a cash flow and should be explicitly considered. This is accomplished in both the discounted cash flow and its close relative, the present worth method, by using after-tax cash flows in the calculations. Furthermore, because cash incomes are often more nearly distributed uniformly throughout the year than at the end of the year, this fact should also be taken into consideration. This is best accomplished by assuming that cash flows are distributed uniformly throughout the year and that interest is compounded continuously.

The three interest tables—Tables 3-1, 3-2, and 3-3—are intended to be as condensed as possible while permitting solutions to almost all practical problems to the degree of accuracy justified by available data. These objectives result in the following features:

1. The tables are limited to three significant figures in the major portion of the table and a corresponding number of figures after the decimal point in the balance of the tables.
2. The annual compounding portion of the tables is limited to present worth factors (single payment), capital recovery factors, and series gradient factors, because all other factors can easily be obtained from these factors.
3. They are limited to 20 interest periods because:
   a. Predictions beyond 20 years are almost meaningless.
   b. Particularly for the more meaningful higher interest rates, as the number of interest periods reaches 20, present worth factors tend to converge to zero, capital recovery factors to i, and series gradient factors to $1/i$, where i is in decimal form.
   c. Results for problems in which compounding is more frequent than twice a year are quite well approximated by assuming continuous compounding and continuous cash flow.

TABLE 3-1.   Present Worth Factors (Single Payment) or $(P/F, i\%, N)$

N \ i	6%	8%	10%	12%	15%	20%	25%	30%	35%	40%	45%	50%
1	0.943	0.926	0.909	0.893	0.870	0.833	0.800	0.769	0.741	0.714	0.690	0.667
2	0.890	0.857	0.826	0.797	0.756	0.694	0.640	0.592	0.549	0.510	0.476	0.444
3	0.840	0.794	0.751	0.712	0.658	0.579	0.512	0.455	0.406	0.364	0.328	0.296
4	0.792	0.735	0.683	0.636	0.572	0.482	0.410	0.350	0.301	0.260	0.226	0.198
5	0.747	0.681	0.621	0.568	0.497	0.402	0.328	0.269	0.223	0.186	0.156	0.132
6	0.705	0.630	0.564	0.507	0.432	0.335	0.262	0.207	0.165	0.133	0.108	0.088
7	0.665	0.583	0.513	0.452	0.376	0.279	0.210	0.159	0.122	0.095	0.074	0.058
8	0.627	0.540	0.466	0.404	0.327	0.323	0.168	0.123	0.091	0.068	0.051	0.039
9	0.592	0.500	0.424	0.361	0.284	0.194	0.134	0.094	0.067	0.048	0.035	0.026
10	0.558	0.463	0.386	0.322	0.247	0.162	0.107	0.072	0.050	0.035	0.024	0.017
11	0.527	0.429	0.350	0.288	0.215	0.135	0.086	0.056	0.037	0.025	0.017	0.012
12	0.497	0.397	0.319	0.257	0.187	0.112	0.069	0.043	0.027	0.018	0.012	0.008
13	0.469	0.368	0.290	0.229	0.162	0.094	0.055	0.033	0.020	0.013	0.008	0.005
14	0.442	0.340	0.263	0.205	0.141	0.078	0.044	0.025	0.015	0.009	0.006	0.003
15	0.417	0.315	0.239	0.183	0.123	0.065	0.035	0.020	0.011	0.006	0.004	0.002
16	0.394	0.292	0.218	0.163	0.107	0.054	0.028	0.015	0.008	0.005	0.003	0.002
17	0.371	0.270	0.198	0.146	0.093	0.045	0.022	0.012	0.006	0.003	0.002	0.001
18	0.350	0.250	0.180	0.130	0.081	0.038	0.018	0.009	0.004	0.002	0.001	0.001
19	0.330	0.232	0.164	0.116	0.070	0.031	0.014	0.007	0.003	0.002	0.001	0.000
20	0.312	0.214	0.149	0.104	0.061	0.026	0.012	0.005	0.002	0.001	0.001	0.000
$(F/A, i\%, N)$	1.030	1.039	1.049	1.059	1.073	1.097	1.120	1.143	1.166	1.189	1.211	1.233

4. Interest rates are shown from 6 to 50%. It is felt that a project paying less than 6% is obviously uninteresting; one paying more than 50% is obviously attractive, if the data can be believed.
5. The use of a single additional factor for each interest rate permits the use of annual compounding tables in solving continuous flow problems. This factor is called the continuous compounding compound amount factor (continuous, uniform payments) and it is designated as $(F/\bar{A}, i\%, N)$. It is needed only for the case when N is one year.

Thus, to find the present worth at time zero, at an effective annual interest rate i, of a series of cash flows of X_N flowing continuously throughout year N, one would calculate

$$PW = (F/\bar{A}, i\%, 1) \sum_{N=1}^{N} X_N(P/F, i\%, N)$$

Continuous compounding tables exist in which interest rates are tabulated under both nominal and effective rates. The latter is preferred because:
1. The results are more readily compared with annual rates.
2. Conventional annual compounding tables can be used for cash flows that occur instantaneously at the end of a year.
3. Tabulating by nominal rates requires an extensive, relatively complex set of tables. As will be shown later, tabulating by effective rates permits the use of annual compounding tables plus a single additional factor for each interest rate to cover all cases where flows are assumed to be uniform throughout the year. This concept is discussed in detail on page 11-80.

More complex cases, involving nonuniform or exponentially increasing or decreasing flows, can be handled mathematically, but the accuracy of the predictions on which the results are to be based seldom justifies such refinements.

TABLE 3-2. Capital Recovery Factors $(A/P, i\%, N)$

N \ i	6%	8%	10%	12%	15%	20%	25%	30%	35%	40%	45%	50%
1	1.060	1.080	1.100	1.120	1.150	1.200	1.250	1.300	1.350	1.400	1.450	1.500
2	0.545	0.561	0.577	0.592	0.615	0.655	0.694	0.735	0.776	0.817	0.858	0.900
3	0.374	0.388	0.402	0.416	0.438	0.475	0.512	0.550	0.590	0.629	0.669	0.710
4	0.289	0.302	0.315	0.329	0.350	0.386	0.423	0.462	0.501	0.541	0.582	0.623
5	0.237	0.250	0.264	0.277	0.298	0.334	0.372	0.411	0.450	0.491	0.533	0.576
6	0.203	0.216	0.230	0.243	0.264	0.301	0.338	0.378	0.419	0.461	0.504	0.548
7	0.179	0.192	0.205	0.219	0.240	0.277	0.316	0.357	0.399	0.442	0.486	0.531
8	0.161	0.174	0.187	0.201	0.223	0.261	0.300	0.342	0.385	0.429	0.474	0.520
9	0.147	0.160	0.174	0.188	0.210	0.248	0.289	0.331	0.375	0.420	0.466	0.513
10	0.136	0.149	0.163	0.177	0.199	0.238	0.280	0.323	0.368	0.414	0.461	0.509
11	0.127	0.140	0.154	0.168	0.191	0.231	0.273	0.318	0.363	0.410	0.458	0.506
12	0.119	0.133	0.147	0.161	0.184	0.225	0.268	0.313	0.360	0.407	0.455	0.504
13	0.113	0.127	0.141	0.156	0.179	0.221	0.265	0.310	0.357	0.405	0.454	0.503
14	0.108	0.121	0.136	0.151	0.175	0.217	0.261	0.308	0.355	0.404	0.453	0.502
15	0.103	0.117	0.131	0.147	0.171	0.214	0.259	0.306	0.354	0.403	0.452	0.501
16	0.099	0.113	0.128	0.143	0.168	0.211	0.257	0.305	0.353	0.402	0.451	0.501
17	0.095	0.110	0.125	0.140	0.165	0.209	0.256	0.304	0.352	0.401	0.451	0.501
18	0.092	0.107	0.122	0.138	0.163	0.208	0.255	0.303	0.352	0.401	0.451	0.500
19	0.090	0.104	0.120	0.136	0.161	0.206	0.254	0.302	0.351	0.401	0.450	0.500
20	0.087	0.102	0.118	0.134	0.160	0.205	0.253	0.301	0.351	0.400	0.450	0.500

TABLE 3-3. Series Gradient Factors or $(A/G, i\%, N)$

N	6%	8%	10%	12%	15%	20%	25%	30%	35%	40%	45%	50%
1	0	0	0	0	0	0	0	0	0	0	0	0
2	0.49	0.48	0.48	0.47	0.47	0.45	0.44	0.43	0.43	0.42	0.41	0.40
3	0.96	0.95	0.94	0.92	0.91	0.88	0.85	0.83	0.80	0.78	0.76	0.74
4	1.43	1.40	1.38	1.36	1.33	1.27	1.22	1.18	1.13	1.09	1.05	1.02
5	1.88	1.85	1.81	1.77	1.72	1.64	1.56	1.49	1.42	1.36	1.30	1.24
6	2.33	2.28	2.22	2.17	2.10	1.98	1.87	1.77	1.67	1.58	1.50	1.42
7	2.77	2.69	2.62	2.55	2.45	2.29	2.14	2.01	1.88	1.77	1.66	1.56
8	3.20	3.10	3.00	2.91	2.78	2.58	2.39	2.22	2.06	1.92	1.79	1.68
9	3.61	3.49	3.37	3.26	3.09	2.84	2.60	2.40	2.21	2.04	1.89	1.76
10	4.02	3.87	3.73	3.58	3.38	3.07	2.80	2.55	2.33	2.14	1.97	1.82
11	4.42	4.24	4.06	3.90	3.65	3.29	2.97	2.68	2.44	2.22	2.03	1.87
12	4.81	4.60	4.39	4.19	3.91	3.48	3.11	2.80	2.52	2.28	2.08	1.91
13	5.19	4.94	4.70	4.47	4.14	3.66	3.24	2.89	2.59	2.33	2.12	1.93
14	5.56	5.27	5.00	4.73	4.36	3.82	3.36	2.97	2.64	2.37	2.14	1.95
15	5.93	5.59	5.28	4.98	4.56	3.96	3.45	3.03	2.69	2.40	2.17	1.97
16	6.28	5.90	5.55	5.21	4.75	4.09	3.54	3.09	2.72	2.43	2.18	1.98
17	6.62	6.20	5.81	5.44	4.93	4.20	3.61	3.13	2.75	2.44	2.19	1.98
18	6.96	6.49	6.05	5.64	5.08	4.30	3.67	3.17	2.78	2.46	2.20	1.99
19	7.29	6.77	6.29	5.84	5.23	4.39	3.72	3.20	2.79	2.47	2.21	1.99
20	7.61	7.04	6.51	6.02	5.37	4.46	3.77	3.23	2.81	2.48	2.21	1.99

Sources of compound interest tables using varying conventions can be found as follows:

1. Annual compounding—any text on engineering economy, such as references 1, 2, 3, and 12 in the bibliography at the end of the chapter
2. Continuous compounding, tabulated by nominal rates—references 1 and 12
3. Continuous compounding, tabulated by effective rates—references 2 and 3

None of the conventions can be claimed to offer a completely valid model of the real world, and it is almost never true that the choice of convention can affect the decision that is indicated. Thus it is not surprising that most analysts feel that the choice of convention is more a matter of individual taste than of seeking the most valid representation of the real world.

One problem common to both the discounted cash flow and the present worth approaches is that both require a common life for all alternatives under consideration. In an absolute sense, this is seldom realistic. In comparing a new method with the present method, it is seldom true that the present method can continue without change for the life of the new method; in choosing between a gasoline or an electric industrial truck, it is not logical to assume identical lives for each.

Another limitation of these methods is that, as commonly used, they fail to take into consideration that the economic life (defined as that period for which the equivalent uniform annual costs are lowest) of an asset is a function of its cost pattern and not an independent variable to which one is free to assign a value.

Fortunately, it can be shown that neither of these limitations poses serious problems in most practical cases, provided that:

1. The economic lives of the alternatives under consideration are reasonably similar.
2. The study period is reasonably close to these lives.

3. The terminal value of each asset is explicitly included in the calculation.
4. It can be assumed that the need for service will be some common multiple of the lives (including infinity) and that cost and technological history will repeat.

Note that a replacement study, particularly of like with like, typically violates point 1 above; in such studies, the economic life of the existing equipment is generally much less than that of the challenger.

In this case, it can be shown that the hypothesis that the existing machine is less expensive can be tested by recycling it, that is, by assuming it is sold at the end of its economic life (generally one more year) and at that time buying a used machine just like today's machine, at a cost equal to the terminal value of today's machine. This methodology does not assume that such a purchase is possible; it is merely a method of testing a hypothesis.

It should be emphasized here that the discounted cash flow approach is often misused in selecting among multiple alternatives. It is not correct to calculate the return on each of several mutually exclusive alternatives and to select that alternative having the highest return. Instead, pair by pair comparisons on an incremental basis must be made.

The logic behind this statement is as follows. Suppose one wishes to choose among three increasingly expensive methods (A, B, and C) to reduce the labor content of an operation.

One should first ask if the least costly method, A, is better than the present method. If it is, it becomes the new standard of comparison and an additional investment in B should be made only if the additional return received justifies this. If, on the other hand, A is not better than the present method, B should be compared with the present method.

Suppose that B proves acceptable in either case. This means that it has been decided not to use the present method or A; therefore the only logical alternative to C is B. C should therefore be compared with B on an incremental basis. But merely calculating the return on investment (ROI) on C would be comparing it with the present method—an illegitimate alternative.

The following numerical example illustrates these ideas. Suppose A costs $10,000 and pays $5,000 a year forever; B costs $20,000 and pays $8,000 a year forever; and C costs $25,000 and pays $8,500 forever. If the minimum rate of return is 20% before taxes, which, if any, should be installed?

The rate of return on A, compared with doing nothing, is 50%,[1] that on B compared with doing nothing is 40%, and that on C is 34%. But the 50% figure indicates that A or something better should be chosen. It is therefore misleading to compare B or C with doing nothing. The choice is not between B or nothing, but between B and A. Such a comparison shows that an additional investment of $10,000 pays an additional $30,000 a year forever—a 30% rate of return. Because the return on the incremental investment is better than the minimum attractive rate of return, B is preferred to A and becomes the new standard. Comparing C with B shows that an additional investment of $5,000 leads to an additional return of only $500, which represents an incremental rate of return of only 10%. Thus B is the optimal choice. Had the minimum ROI been 35%, A would have been optimal.

THE PROFITABILITY INDEX SIMPLIFICATION OF DISCOUNTED CASH FLOW CALCULATIONS

The following excerpt from Raymond I. Reul's chapter in the *Handbook of Business Administration*[2] offers a succinct description of the methodology he originated. Note that Reul uses interest rates tabulated by the nominal annual rate for continuous compounding.

---

[1] For perpetual lives, the rate of return is the income divided by first cost.
[2] Raymond I. Reul, "Techniques for Evaluating Profitable Investments," sec. 3, chap. 8, in H. B. Maynard (ed.), *Handbook of Business Administration*, McGraw-Hill Book Company, New York, 1967.

TABLE 3-4

| Timing | Actual cash flows | |
	Disbursements	Receipts
Start 1965.	$1,000	
Start 1966.	2,000	
End 1966.	. . . . . .	$ 800
End 1969.	. . . . . .	2,000
End 1970.	. . . . . .	3,904
Total.	$3,000	$6,704

Simplified Computation of the True Rate of Return. The profitability index approach to the calculation of the true rate of return was first published in *Factory Management and Maintenance* in October, 1955. It introduced the idea of using pretabulated worksheets and graphical interpolation as a means of reducing the time-consuming and often confusing trial and error computation of rate of return to a simple, routine clerical operation. The use of this technique to solve the demonstration problem in Table 3-4[3] is illustrated in Figure 3-1.

Description of the Worksheet. The worksheet provides separate schedules for disbursements and receipts. Both, however, are oriented to the same zero point, which is identified by a heavy line with a diamond at the left. The form is designed to facilitate the assumption of the beginning of the first year in which receipts occur as the zero point. (With most problems this will result in minimum computational effort for specified accuracy of answer.) Single payment compounding and discounting present worth factors for three different rates are provided. (On this demonstration form, the pretabulated present worth factors are based upon annual compounding and instantaneous cash flows. Conversion to the more frequently used convention of continuous compounding and during-the-year receipt is merely a matter of substituting different present worth factors. A blank form with such factors is illustrated in Figure 3-2.)

Directions for the Use of the Pretabulated Worksheet. The first step in the use of this worksheet is to enter each of the anticipated disbursements and receipts at its scheduled time in the column headed "Trial 1 @ 0% Interest Rate." Note that by making this comparison of the totals of actual disbursements and receipts, we are literally making a trial at 0 percent interest rate. That is, because the sum of the actual receipts in this demonstration problem is greater than the sum of the actual disbursements, merely setting down the figures in this form demonstrates that the earning rate or rate of return of this project is greater than zero.

The next trial, 2 at 10% interest rate, is made by multiplying each of the actual amounts of disbursements and receipts each year by the adjacent applicable present worth factor. The figures for trial 3 at 25% and 4 at 40% are calculated and entered in the same fashion. Then at each trial rate, the sum of the present worths of disbursements is divided by the sum of the present worths of receipts. The ratios thus obtained are entered in spaces labeled "A/B."

The solution is found by determining the interest rate at which this A/B ratio is equal to unity. This is done, using the chart provided, by plotting each of these ratios against the interest rate at which it was calculated (including the ratio at zero interest rate) and drawing a smooth curve through these points. A horizontal line drawn through the point at which this curve intersects the unity line indicates the answer on the vertical scale. This is the true earning rate or rate of return of the project.

[3] For the convenience of the reader, the numbers of the illustrations in the excerpt have been changed to conform with the illustration numbers in this chapter.

TIMING		TRIAL 1 0% INTEREST RATE	TRIAL 2 10% INTEREST RATE		TRIAL 3 25% INTEREST RATE		TRIAL 4 40% INTEREST RATE	
CAL. YEAR	PERIOD	ACTUAL AMOUNT OF DISBURSEMENTS	FACTOR	PRESENT WORTH	FACTOR	PRESENT WORTH	FACTOR	PRESENT WORTH
BEFORE	1ST YR. AT ST.	1,000	1.10	1,100	1.25	1,250	1.40	1,400
	MIDDLE		1.05		1.13		1.20	
AFTER	1ST YR. AT ST.	2,000	1.00	2,000	1.00	2,000	1.00	2,000
	MIDDLE		0.95		0.91		0.83	
	TOTALS (A)	3,000		3,100		3,250		3,400
CAL. YEAR	PERIOD	ACTUAL AMOUNT OF RECEIPTS	FACTOR	PRESENT WORTH	FACTOR	PRESENT WORTH	FACTOR	PRESENT WORTH
BEFORE ZERO POINT	1ST YEAR END	800	0.91	728	0.80	640	0.71	568
	2ND YEAR END		0.83		0.54		0.51	
	3RD YEAR END		0.75		0.51		0.36	
	4TH YEAR END	2,000	0.68	1,360	0.41	820	0.26	520
	5TH YEAR END	3,904	0.62	2,420	0.33	1,288	0.19	742
	6TH YEAR END		0.56		0.25		0.13	
AFTER ZERO POINT	7TH YEAR END		0.51		0.21		0.09	
	8TH YEAR END		0.47		0.17		0.07	
	9TH YEAR END		0.42		0.13		0.05	
	10TH YEAR END		0.39		0.11		0.03	
	11TH YEAR END		0.35		0.09		0.02	
	12TH YEAR END		0.32		0.07		0.02	
	13TH YEAR END		0.29		0.06		0.01	
	14TH YEAR END		0.26		0.04		0.01	
	15TH YEAR END		0.24		0.04		0.01	
	16TH YEAR END		0.22		0.03			
	17TH YEAR END		0.20		0.02			
	18TH YEAR END		'0.18		0.02			
	19TH YEAR END		0.16		0.01			
	20TH YEAR END		0.15		0.01			
	TOTALS (B)	6,704		4,508		2,748		1,830
	RATIO A/B	0.47		0.69		1.18		1.86

INTERPOLATION CHART

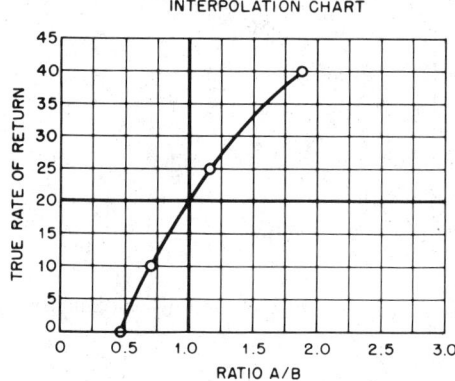

TRUE RATE OF RETURN, 20 %

Fig. 3-1. Worksheet for computing true rate of return—annual compounding.

The answer thus obtained by graphical interpolation is not approximately correct, but is as exact as the scale of the plotting chart will permit us to read. It is, to a certain extent, self-checking. If a smooth curve cannot be drawn through the four points plotted, it is positive evidence that a mistake has been made in the computations or plotting. Only three points are required to determine a curve. The fourth serves as a check on accuracy. In addition

TIMING		TRIAL 1 0% INTEREST RATE	TRIAL 2 10% INTEREST RATE		TRIAL 3 15% INTEREST RATE		TRIAL 4 25% INTEREST RATE		TRIAL 5 40% INTEREST RATE	
	PERIOD	ACTUAL AMOUNT OF DISBURSEMENTS	FACTOR	PRESENT WORTH	FACTOR	PRESENT WORTH	FACTOR	PRESENT WORTH	FACTOR	PRESENT WORTH
SECTION A OUTFLOW — BEFORE ZERO	2 AT ST.		1.221		1.350		1.649		2.225	
	2 DURING		1.162		1.253		1.459		1.834	
	1 AT ST.		1.105		1.162		1.284		1.492	
	1 DURING		1.052		1.079		1.136		1.230	
SECTION A OUTFLOW — AFTER ZERO	1 AT ST.		1.000		1.000		1.000		1.000	
	1 DURING		0.952		0.929		0.885		0.824	
	2. DURING		0.861		0.799		0.689		0.553	
	3. DURING		0.779		0.688		0.537		0.370	
	4. DURING		0.705		0.592		0.418		0.248	
	5. DURING		0.638		0.510		0.326		0.166	
TOTALS (A)										
		ACTUAL AMOUNT OF RECEIPTS	FACTOR	PRESENT WORTH	FACTOR	PRESENT WORTH	FACTOR	PRESENT WORTH	FACTOR	PRESENT WORTH
SECTION B CASH INFLOW — AFTER ZERO	1. DURING		0.952		0.929		0.885		0.824	
	2. DURING		0.861		0.799		0.689		0.553	
	3. DURING		0.779		0.688		0.537		0.370	
	4. DURING		0.705		0.592		0.418		0.248	
	5. DURING		0.638		0.510		0.326		0.166	
	6. DURING		0.577		0.439		0.254		0.112	
	7. DURING		0.522		0.378		0.197		0.075	
	8. DURING		0.473		0.325		0.154		0.050	
	9. DURING		0.428		0.280		0.119		0.034	
	10. DURING		0.387		0.241		0.093		0.023	
	11. DURING		0.350		0.207		0.073		0.015	
	12. DURING		0.317		0.178		0.057		0.010	
TOTALS (B)										
RATIO A/B		0%	10%		15%		25%		40%	

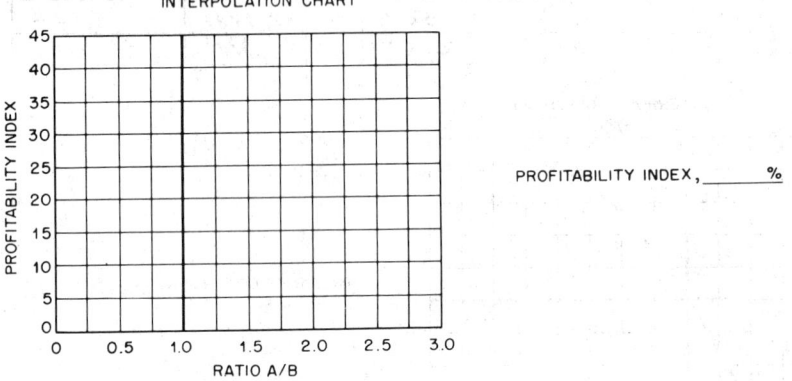

INTERPOLATION CHART

PROFITABILITY INDEX, _____ %

FIG. 3-2. Alternative worksheet for computing true rate of return—continuous compounding.

to simplicity, speed, and minimum possibility for errors, this method offers the maximum accuracy of answers with a given number of decimal places in the present worth factors. This is because the ratio at zero interest rate has maximum sensitivity to variations in cash flows regardless of their timing.

EQUIVALENT UNIFORM ANNUAL COST APPROACH

Some analysts use what is known as the equivalent uniform annual cost approach. In this approach, one uses compound interest theory to calculate the equivalent uniform annual cost of all cash flows for each alternative at the minimum attractive rate of return; that having the lowest annual cost or the highest annual income is recommended.

This approach involves the use of additional compound interest formulas and their associated tables; it is obviously best suited to situations in which most costs can be assumed to be uniform throughout the life of the proposal. This is seldom the case in realistic after-tax studies; perhaps for that reason the equivalent uniform annual cost approach is not very widely used in industry.

COMPOUND INTEREST THEORY

A discussion of compound interest theory in sufficient depth to understand the equivalent uniform annual cost approach requires fairly extensive terminology and symbology. A set of standards for this terminology and symbology has recently been presented by the Engineering Economy Division, American Society for Engineering Education. Because that report is useful both to define the symbols and as an excellent, concise summary of the formulas used, it is here presented. Following this report is a discussion of the formulas and their use.

INTRODUCTION FOR AUTHORS AND PUBLISHERS

This manual is the product of a long-felt need and considerable effort by numerous persons to effect a reasonable standardization of notation in the literature of engineering economy.

The standards suggested herein were, after considerable deliberation, endorsed by the Engineering Economy Division of the American Society for Engineering Education.

STANDARDS

Table 3-5[4] lists suggested standard definitions and symbols for common parameters.

Table 3-6[4] lists two alternative suggested standard notational formats for twenty common interest factors. The two alternative formats are commonly

[4] *Ibid.*

TABLE 3-5. Definitions and Symbols Used for Parameters (Suggested Standards)

No.	Definition of parameter	Symbol
1	Effective interest rate per interest period.	i
2	Nominal interest rate per year.	r
3	Number of compounding periods.	N
4	Number of compounding periods per year.	M
5	Present sum of money. The letter P implies present (or equivalent present value).	P
6	Future sum of money. The letter F implies future (or equivalent future value).	F
7	End-of-period cash flows (or equivalent end-of-period values) in a uniform series continuing for a specified number of periods. The letter A implies annual or annuity.	A
8	Uniform period-by-period increase or decrease in cash flows (or equivalent values); the arithmetic gradient.	G
9	Amount of money (or equivalent value) flowing continuously and uniformly during a given period.	$\bar{P}$ or $\bar{F}$
10	Amount of money (or equivalent value) flowing continuously and uniformly during each and every period continuing for a specific number of periods.	$\bar{A}$

TABLE 3-6. Mnemonic/Functional Forms of Compound
Interest Factors (Suggested Standards)

Ref. no.	Name of factor	Mnemonic format	Functional format
	Group I. All cash flows discrete: end-of-period compounding		
1	Compound amount factor (single payment)	(CA-$i\%$-N)	(F/P, $i\%$, N)
2	Present worth factor (single payment)	(PW-$i\%$-N)	(P/F, $i\%$, N)
3	Sinking fund factor	(SF-$i\%$-N)	(A/F, $i\%$, N)
4	Capital recovery factor	(CR-$i\%$-N)	(A/P, $i\%$, N)
5	Compound amount factor (uniform series)	(SCA-$i\%$-N)	(F/A, $i\%$, N)
6	Present worth factor (uniform series)	(SPW-$i\%$-N)	(P/A, $i\%$, N)
7	Arithmetic gradient conversion factor (to uniform series)	(GUS-$i\%$-N)	(A/G, $i\%$, N)
8	Arithmetic gradient conversion factor (to present value)	(GPW-$i\%$-N)	(P/G, $i\%$, N)

called *mnemonic* and *functional*. It is suggested that each author choose one or the other format based on his tastes and use it consistently. The names of each of the factors are also given as suggested standards.

The fundamental equation of compound interest has already been presented in two reciprocal forms, namely,

$$F = P(1 + i)^N \tag{1}$$

$$P = F \frac{1}{(1 + i)^N} \tag{1a}$$

The expression $(1 + i)^N$ in equation (1) is often termed the compound amount factor (single payment); that in equation (1a), $1/(1 + i)^N$, the present worth factor (single payment). These and the names of other commonly used factors are clearly too bulky to use in writing equations; they are therefore designated by abbreviation. As suggested above, these abbreviations can be based on either a mnemonic or a functional format. Each of these has its advantages. The mnemonic focuses attention on the traditional names given these factors; once these names are known, it becomes easy to write the mnemonic abbreviation. Conversely, the use of this format aids in the memorization of these names.

The functional format is far easier to learn. One need only note that $P(F/P)$ equals F to see the logic on which this system is based. In using this format, one never need learn the traditional names of the factors; if (F/P, $i\%$, N) is read "F is calculated given P, i, and N," it is obvious that this is used when a future amount is wanted and a present amount is given.

Although it has become customary to offer tables of both present worth and compound amount single payment factors, their reciprocal nature obviously makes this unnecessary. In Tables 3-1 and 3-2, therefore, only the more commonly used present worth factor (single payment)—(P/F, $i\%$, N)—is shown.

To make annual cost comparisons, it is frequently necessary to convert from a single payment at the beginning or at the end of the period under consideration to its uniform, end-of-period equivalent. This can be accomplished by repeated use of the fundamental equation of compound interest plus some algebra. Because this operation is repeated so often, it is simpler to develop the algebra in generalized form and to use the results to define new factors. Thus, to find the uniform, end-of-year amount (for N years) of P dollars invested now, when interest rate is $i\%$ per year, the following formula is used:

$$A = P \left[\frac{i(1 + i)^N}{(1 + i)^N - 1} \right] \tag{2}$$

The bracketed factor is called the capital recovery factor. Table 3-2 is a tabulation of capital recovery factors for various i's and N's. To simplify, this equation is written either as $A = P(CR\text{-}i\%\text{-}N)$ or as $A = P(A/P, i\%, N)$. To solve for amount A in a practical case, tabulated values of the factors are used.

It is sometimes desired to convert from a uniform series to its equivalent present worth; to do so, equation (2) could be written in its reciprocal form:

$$P = A \left[\frac{(1 + i)^N - 1}{i(1 + i)^N} \right] \tag{2a}$$

The bracketed factor is termed the present worth factor (uniform series); and the equation is abbreviated, using the mnemonic format, as $P = A(SPW\text{-}i\%\text{-}N)$, or using the functional format, as $P = A(P/A, i\%, N)$.

Again, although it is customary to tabulate both the reciprocal forms relating a uniform, end-of-period series to its equivalent present worth, this is not really necessary because $(P/A, i\%, N) = 1/(A/P, i\%, N)$. As it is more common to seek A when given P, only the capital recovery factor $(A/P, i\%, N)$ is usually tabulated.

Another commonly recurring problem is to convert from a single amount occurring in the future to a uniform, end-of-period series of payments, starting at the end of the first period and continuing until the time of the single sum.

To do this, the formula used is

$$A = F \left[\frac{i}{(1 + i)^N - 1} \right] \tag{3}$$

The bracketed factor is called the sinking fund factor; depending on whether mnemonic or functional notation is used, this is simplified by writing $A = F(SF\text{-}i\%\text{-}N)$ or $A = F(A/F, i\%, N)$. Again, this equation is sometimes used in its reciprocal form to find F (a future, end-of-series sum) when A (a uniform, end-of-period amount continuing for N periods) is given. The factor used is termed the compound amount factor (uniform series) and abbreviated as $(SCA\text{-}i\%\text{-}N)$ or $(F/A, i\%, N)$.

Simple algebra shows that the sinking fund or A/F factor is equal to the capital recovery or A/P factor, minus the interest rate. Thus all series factors can readily be found from the capital recovery or A/P factor:

$$(P/A, i\%, N) = \frac{1}{(A/P, i\%, N)}$$

$$(F/P, i\%, N) = (A/P, i\%, N) - i$$

$$(P/F, i\%, N) = \frac{1}{(A/P, i\%, N) - i}$$

For this reason, only the capital recovery or A/P factor among the series factors is here tabulated.

The other annual compounding factors mentioned in the standardization report are the two arithmetic gradient conversion factors, one to a uniform series and one to a present value. These convert from a series of payments of $0, g, 2g, 3g,$ $\ldots, (n - 1)g$ at the end of the 1st, 2nd, 3rd, $\ldots, n$th period to either a uniform end of period series or to a present worth. As shown in the report, these are abbreviated as $(GUS\text{-}i\%\text{-}N)$ and $(GPW\text{-}i\%\text{-}N)$ or as $(A/G, i\%, N)$ and $(P/G, i\%, N)$, respectively, in the mnemonic and functional formats.

Also listed in the report is a series of factors used in continuous compounding. If the convention of tabulating is by effective rates rather than by nominal rates, most of these are unnecessary and will not be discussed here.

The assumption of continuous compounding leads to the following relations:

$$F = Pe^{rN} \tag{4}$$
$$P = Fe^{-rN} \tag{4a}$$
$$i = e^r - 1 \tag{5}$$
$$r = \log_e (1 + i) \tag{5a}$$

where F, P, and r are defined in the ASEE standard and N is the number of years. Note that equations (4) and (4a) are the instantaneous compounding equivalents of equations (1) and (1a).

To find the present worth of a series of cash flows flowing uniformly throughout each of N years, one could convert the uniform flow during each year to its end-of-year equivalent; this would entail the use of the continuous compounding compound amount factor (continuous uniform payments). If this is done by using the appropriate effective rate, these end-of-year equivalents can be discounted to time zero, using annual compounding present worth (single payment) tables.

Algebraically, this leads to

$$\text{Present worth} = \sum_{N=1}^{N} X_N(F/\bar{A}, i\%, N)(P/F, i\%, N) \tag{6}$$

where X_N is the uniform, continuous flow in year N and $(F/\bar{A}, i\%, N)$ is the continuous compounding compound amount factor (continuous uniform payments). This expression can be rewritten as

$$\text{Present worth} = (F/\bar{A}, i\%, N) \sum_{N=1}^{N} X_N(P/F, i\%, N) \tag{7}$$

Equation (7) leads to a simple method for calculating the present worth of a series of uniform, continuous cash flows: treat the cash flows as if they were end-of-period payments but multiply the present worth thus obtained by $(F/\bar{A}, i\%, N)$.

Note that this result infers that if, instead of an instantaneous investment at time zero followed by continuous cash flows, the original investment were made uniformly throughout the year ending at time zero, the results obtained by using the end-of-year convention would be identical with those using the seemingly more appropriate continuous convention. (The author wishes to express his debt to Mr. Laurance Bell of the General Electric Company for first calling these relationships to his attention.)

To evaluate $F/\bar{A}$, note that the compound amount factor, continuous flow, for one year is

$$\frac{e^r - 1}{r} \tag{8}$$

Substituting for r the value obtained from equations (5) and (5a) yields an expression in terms of the effective interest rate:

$$S = \bar{A}\left[\frac{i}{\log_e(1+i)}\right] \tag{9}$$

Several problems will serve to illustrate the use of compound interest factors in engineering situations. (In all problems, the end-of-year convention for cash flows will be used.) The summary on page 11-81, using functional notation, will aid in setting up solutions.

Example 1. An investment offers a return of $10,000 ten years from now. If alternative investments of like risk yield a before-tax rate of return of 15% per year, how much can you afford to pay for this investment?

First note that $10,000 is a single, future sum; it therefore corresponds to F. Other factors given are

$$N = 10 \text{ years}$$
$$i = 15\% \text{ per year}$$
$$P = \text{amount to be found}$$

Summary of Compound Interest Equations (Annual Compounding)

To obtain	When given	Use equation	Or
P	F	$P = F(P/F, i\%, N)$	
F	P	$F = P(F/P, i\%, N)$	$F = \dfrac{P}{(P/F, i\%, N)}$
A	P	$A = P(A/P, i\%, N)$	
P	A	$P = A(P/A, i\%, N)$	$P = \dfrac{A}{(A/P, i\%, N)}$
A	F	$A = F(A/F, i\%, N)$	$A = F(A/P, i\%, N) - i$
F	A	$F = A(F/A, i\%, N)$	$F = \dfrac{A}{(A/P, i\%, N) - i}$
A	G	$A = G(A/G, i\%, N)$	
P	G	$P = G(P/G, i\%, N)$	

Referring to the above summary or remembering the general pattern, the equation to use to obtain P when given F is

$$P = F(P/F, i\%, N)$$

Substituting values,

$$P = \$10,000 \; (P/F, 15\%, 10)$$

From Table 3-1, the value of the parenthetic amount is seen to be 0.247; therefore the result is $P = \$2,470$.

Example 2. \$1,000 invested in a bank paying 6% interest, compounded annually, will amount to how much if left in for 12 years?

$$P = \$1,000$$
$$N = 12$$
$$i = 6\%$$
$$F = \text{unknown}$$

From the formula,

$$F = P(F/P, i\%, N) = \frac{P}{(P/F, i\%, N)}$$

Because only the second factor is tabulated, it is used to write

$$F = \frac{\$1,000}{0.497}$$

$$F = \$2,140$$

Example 3. What is the uniform annual cost of ownership (ACO) of a truck costing \$10,000 and having a life of 5 years, with a terminal value at the end of that time of \$2,000, if interest is 20%?

This problem is actually solved in several stages.   First, the annual equivalent of the initial investment is found; next, the annual equivalent of the income from the terminal value is found.   The answer is the difference between these amounts.

Calling the terminal value T,

$$\text{ACO} = P(A/P, i\%, N) - T(A/F, i\%, N)$$

Because tables of $(A/F, i\%, N)$ are not provided, the alternative form of $(A/P, i\%, N) - i$ can be used to evaluate the second term.

$$\text{ACO} = \$10,000\ (0.334) - \$2,000\ (0.334 - 0.20)$$
$$= \$3,072$$

In much of the literature, the formula for the annual cost of ownership of an asset having a terminal value is given as

$$\text{ACO} = (P - T)(A/P,\ i\%,\ N) + Ti$$

Note that the solution shown above does, in effect, use this same formula because $\$10,000(0.334) - \$2,000(0.334 - 0.20)$ can be rewritten as $(\$10,000 - \$2,000)(0.334) + \$2,000(0.20)$.

Example 4. A full service lease costing $2,400 per year at the beginning of each year is offered as an alternative to buying a fork truck. Purchase will cost $6,000 now, plus maintenance estimated at $500 in the first year and increasing by $200 in each successive year. The truck will be kept in service for 6 years; its estimated terminal value at the end of that period is $1,000.

a. To aid in the lease versus buy decision, it is proposed to find the equivalent uniform annual pretax cost of the purchase option. If interest is taken at 10%, what is this uniform annual cost?

Again, the equivalent uniform annual cost (EUAC) sought is made up of several parts; an equation yielding the total would be

$$\text{EUAC} = \$6,000\ (A/P,\ 10\%,\ 6) - \$1,000\ [(A/P,\ 10\%,\ 6) - i]$$
$$+ \$500 + \$200\ (A/G,\ 10\%,\ 6)$$

Looking up the appropriate factors and performing the multiplication gives

$$\text{EUAC} = \$6,000\ (0.230) - \$1,000\ (0.130) + \$500 + \$200\ (2.22)$$
$$= \$2,194$$

b. Can this figure legitimately be compared with the $2,400 per year lease cost? If not, how should the comparison be made?

No, it cannot be compared, even on a pretax basis. One is a beginning-of-year payment, the other an end-of-year payment. But the end-of-year payment equivalent to $2,400 at the beginning of the year can be found by dividing $2,400 by the present worth factor (single payment), or P/F factor, for one year and 10%. This yields $2,400/0.909 = $2,640/year.

Example 5. A temporary building will cost $50,000, last for 10 years, and have maintenance costs of $1,000 per year and zero terminal value. A permanent building will cost $200,000, last 50 years, and have maintenance costs of $800 per year. Compare the pretax annual costs of the two alternatives at an interest rate of 12%. Note that $(A/P,\ i\%,\ N) \rightarrow i$ when $N \rightarrow \infty$. For most practical purposes, it can be assumed the $(A/P,\ i\%,\ N) \rightarrow i$ when $N > 20$ and $i \geq 10\%$.

Solution:

	Temporary	Permanent
P	50,000	200,000
N	10	50
i	12 %	12 %
T	0	0
A	?	?

$$\text{EUAC for temporary} = \$60,000\ (A/P,\ 12\%,\ 10) + \$10,000$$
$$= \$60,000\ (0.177) + \$10,000$$
$$= \$20,620$$
$$\text{EUAC for permanent} = \$200,000\ (A/P,\ 12\%,\ 50) + \$8,000$$
$$= \$200,000\ (0.120) + \$8,000$$
$$= \$32,000$$

The temporary building has much lower annual costs.

Note that each of the above problems was solved using the equivalent uniform annual cost approach; all could equally well have been solved using the present worth approach. Indeed, once the EUAC is found, the present worth can be obtained by PW = EUAC $(P/A, i\%, N)$.

Thus it is seen that the present worth and equivalent uniform annual cost approaches must lead to the same decision. In a similar way, the equivalent uniform annual cost approach could be used to solve for an unknown interest rate; it would then lead to the same result as a discounted cash flow study. But in general, a discounted cash flow study is simpler to use.

The Problem of Taxes. Income taxes are clearly an important factor in many economic decisions. Therefore, most analysts prefer to make an after-tax study. The methodologies discussed above all readily permit this; it is only necessary to determine after-tax cash flows before using them.

In determining after-tax cash flows, it should be kept in mind that depreciation is not a cash flow; it is a bookkeeping entry. But it does affect taxable income and therefore income taxes—and taxes represent a very real cash flow.

To illustrate the effect of taxes, consider the following examples.

An automatic machine costing $60,000, and having a useful life of 3 years with zero terminal value at the end of that time, will produce a saving of $30,000 per year. What after-tax cash flow would this yield if straight-line depreciation is used? Assume a 50% income tax rate.

Solution: Set up a table showing
1. Pretax cash flow
2. Depreciation }
3. Taxable income } *not* cash flows!
4. Tax
5. After-tax cash flow

Time	Pretax cash flow	Not cash flows!		Tax	After-tax cash flow
		Deprecia-tion	Taxable income		
0	−60,000	......	......	....	−60,000
1	+30,000	20,000	10,000	−5,000	+25,000
2	+30,000	20,000	10,000	−5,000	+25,000
3	+30,000	20,000	10,000	−5,000	+25,000

Had sum-of-year digits depreciation been used, the depreciation figures would have been 30,000, 20,000 and 10,000, respectively, leading to taxable incomes of 0, 10,000, and 20,000 in years 1, 2, and 3. This in turn would produce after-tax cash flows of 30,000, 25,000, and 20,000. The total income over 3 years would remain unchanged, but the apparent timing would change in a way favorable to the taxpayer.

Indeed, if the entire $60,000 could be expensed (for example, if it were a research expenditure), the incremental tax effect would be to reduce the taxes paid at time zero by $15,000; in return, half of each of the yearly pretax cash flows would be taxed. After-tax cash flows would be −30,000, 15,000, 15,000, and 15,000 at time zero and the ends of years 1, 2, and 3, respectively; and this would produce the same after-tax return as pretax, because each item is just halved. For further discussion of this important topic, see reference 3, Chap. 16.

THE MAPI APPROACH

Another approach, widely used in the machine tool industry but relatively unknown outside it,[5] is the MAPI (Machinery and Allied Products Institute) approach.

[5] An adaptation of this approach has been introduced into the military services in AMC pamphlet AMCP 700-2, published by Headquarters, U.S. Army Materiel Command, July, 1966.

Developed for that institute by George Terborgh, this formula approach explicitly considers the additional complication of continuous obsolescence. It has gone through repeated revisions since its introduction in 1949; the latest version is presented in *Business Investment Management.*[6] The revisions have not altered the basic concepts, but the format and degree of flexibility available have varied considerably.

The current version appears to consider only next year's costs and incomes for both the present situation and the proposed alternative; actually, the underlying

[6] George Terborgh, *Business Investment Management,* Machinery and Allied Products Institute, Washington, D.C., 1967.

PROJECT NO._____   SHEET I

MAPI SUMMARY FORM
(AVERAGING SHORTCUT)

PROJECT_____

ALTERNATIVE_____

COMPARISON PERIOD (YEARS)   (P)_____

ASSUMED OPERATING RATE OF PROJECT (HOURS PER YEAR)   _____

I. OPERATING ADVANTAGE
(NEXT YEAR FOR A 1-YEAR COMPARISON PERIOD,* ANNUAL AVERAGES FOR LONGER PERIODS)

A. EFFECT OF PROJECT ON REVENUE

		INCREASE	DECREASE	
1	FROM CHANGE IN QUALITY OF PRODUCTS	$	$	1
2	FROM CHANGE IN VOLUME OF OUTPUT			2
3	TOTAL	$ X	$ Y	3

B. EFFECT ON OPERATING COSTS

		INCREASE	DECREASE	
4	DIRECT LABOR	$	$	4
5	INDIRECT LABOR			5
6	FRINGE BENEFITS			6
7	MAINTENANCE			7
8	TOOLING			8
9	MATERIALS AND SUPPLIES			9
10	INSPECTION			10
11	ASSEMBLY			11
12	SCRAP AND REWORK			12
13	DOWN TIME			13
14	POWER			14
15	FLOOR SPACE			15
16	PROPERTY TAXES AND INSURANCE			16
17	SUBCONTRACTING			17
18	INVENTORY			18
19	SAFETY			19
20	FLEXIBILITY			20
21	OTHER			21
22	TOTAL	$ Y	$ X	22

C. COMBINED EFFECT

23	NET INCREASE IN REVENUE (3X−3Y)	$	23
24	NET DECREASE IN OPERATING COSTS (22X−22Y)	$	24
25	ANNUAL OPERATING ADVANTAGE (23+24)	$	25

* Next year means the first year of project operation. For projects with a significant break-in period, use performance after break-in.

FIG. 3-3. MAPI Summary Form, sheet 1.

formula projects future benefits based on this information and the service life that is assumed. Built into the formula is the assumption that an improved alternative will become available each year. It is further assumed that each year's alternative will be superior to the previous year's alternative by the same fixed amount in each year of their respective lives. In earlier versions, this amount was termed the obsolescence gradient; in all versions except the first, the numerical value for the total of this gradient and that due to deterioration produced by aging is inferred from the next year's operating advantage of the proposed equipment, its assumed service life, and disposal value at the end of that life. Unfortunately, the method does not permit the ready determination of the gradient that is inferred.

A recommended format is shown as Figures 3-3 and 3-4. Figure 3-3 contains

SHEET 2

II. INVESTMENT AND RETURN

A. INITIAL INVESTMENT

26 INSTALLED COST OF PROJECT $ _____
    MINUS INITIAL TAX BENEFIT OF $ _____   (Net Cost) $ _____   26
27 INVESTMENT IN ALTERNATIVE
    CAPITAL ADDITIONS MINUS INITIAL TAX BENEFIT $ _____
    PLUS: DISPOSAL VALUE OF ASSETS RETIRED
    BY PROJECT * $ _____ $ _____   27
28 INITIAL NET INVESTMENT (26−27) $ _____   28

B. TERMINAL INVESTMENT

29 RETENTION VALUE OF PROJECT AT END OF COMPARISON PERIOD
    (ESTIMATE FOR ASSETS, IF ANY, THAT CANNOT BE DEPRECIATED OR EXPENSED. FOR OTHERS, ESTIMATE OR USE MAPI CHARTS.)

Item or Group	Installed Cost, Minus Initial Tax Benefit (Net Cost) A	Service Life (Years) B	Disposal Value, End of Life (Percent of Net Cost) C	MAPI Chart Number D	Chart Percent- age E	Retention Value $\left(\dfrac{A \times E}{100}\right)$ F
	$					$

ESTIMATED FROM CHARTS (TOTAL OF COL. F) $ _____
    PLUS: OTHERWISE ESTIMATED $ _____ $ _____   29
30 DISPOSAL VALUE OF ALTERNATIVE AT END OF PERIOD * $ _____   30
31 TERMINAL NET INVESTMENT (29−30) $ _____   31

C. RETURN

32 AVERAGE NET CAPITAL CONSUMPTION $\left(\dfrac{28-31}{P}\right)$ $ _____   32

33 AVERAGE NET INVESTMENT $\left(\dfrac{28+31}{2}\right)$ $ _____   33

34 BEFORE-TAX RETURN $\left(\dfrac{25-32}{33} \times 100\right)$ % _____   34

35 INCREASE IN DEPRECIATION AND INTEREST DEDUCTIONS $ _____   35
36 TAXABLE OPERATING ADVANTAGE (25−35) $ _____   36
37 INCREASE IN INCOME TAX (36×TAX RATE) $ _____   37
38 AFTER-TAX OPERATING ADVANTAGE (25−37) $ _____   38
39 AVAILABLE FOR RETURN ON INVESTMENT (38−32) $ _____   39

40 AFTER-TAX RETURN $\left(\dfrac{39}{33} \times 100\right)$ % _____   40

* After terminal tax adjustments.

Copyright 1967, Machinery and Allied Products Institute

FIG. 3-4. MAPI Summary Form, sheet 2.

an excellent checklist for establishing the operating benefits, regardless of the analytical procedure used. Figure 3-4 contains the heart of the MAPI system; except for the MAPI charts referred to, it is relatively self-explanatory. There are six of these charts; to use them intelligently, a careful reading of *Business Investment Management* in which they appear is strongly recommended.

These charts, in effect, project the capital consumption in the first year based on the standard MAPI assumption. This capital consumption is divided by the average net investment to get the before-tax rate of return (if desired); a somewhat more complex series of calculations is required to yield the after-tax rate of return. These results are comparable with those obtainable from a discounted cash flow or equivalent uniform annual cost approach if these approaches are modified to include the concept of obsolescence.

Although the forms appear somewhat formidable to one who has never used them, this method is actually quite simple to use once the amounts on which it is based are determined. Unlike the discounted cash flow approach, it requires, in general, only a first-year estimate for the cost difference between the proposed alternatives. It is based on the assumption that the appropriate comparison period for the present method is one more year; it thus avoids the common mistake in discounted cash flow studies of assuming that, if the current asset is retained, it will be kept for that period of years assumed as the life of its proposed replacement. But although it is easy to use the MAPI method, it is quite difficult to understand it fully. This seems to be one of the major deterrents to its wider use. It is also relatively difficult to use when the standard assumptions appear to be inappropriate.

SELECTING AN INTEREST RATE

The selection of an appropriate cutoff rate, or minimum attractive rate of return (MARR), is important but most difficult. Indeed, no generally acceptable method has yet appeared in the literature.

There is general agreement that a lower limit is offered by a firm's cost of capital, but there is no agreement as to how this cost of capital should be determined, and little agreement on how far above this limit the minimum attractive rate of return should be.

One widely discussed approach to the establishment of the minimum attractive rate of return follows a suggestion first popularized by Joel Dean. Briefly, the reasoning is as follows: A firm should rank all the capital expenditure proposals in the coming year in decreasing order of the rate of return they promise. One should then proceed down this list, cumulating the investment costs, until these reach the capital expenditure budget for the year. The rate of return on the final project included then becomes the minimum attractive rate of return.

This methodology has both theoretical and practical drawbacks. For example, it does not consider the fact that a high-risk investment promising a high rate of return may be less attractive than a low-risk investment promising a lower rate of return. And, as Terborgh has pointed out, it might be unwise for an executive officer to settle for a minimum attractive rate of return of, say, 50%, established in this fashion. He would then have to reject a proposal promising a 45% return. It might be wiser, notes Terborgh, to increase the capital budget.

From a practical point of view, it is seldom possible to determine all the potential capital expenditure opportunities a year in advance. Faced with such formidable difficulties in establishing an appropriate minimum attractive rate of return, most firms seem to use an empirical approach. They ask, in effect, "If these difficulties could be overcome, what do I think the minimum attractive rate of return would be?" They then try this value in practice and adjust it upward or downward, depending on whether too much or too little of the budget is being spent. Generally, however, the minimum attractive rate of return is not permitted to go below what they consider to be their cost of capital.

RISK AND UNCERTAINTY

The criteria discussed thus far lead to mathematical models that are properly described as models based on certainty. That is, the models treat each input as if it were certain to occur. However, in interpreting the results obtained when using these models, it is generally recognized that risks are involved—the data that the model treated as certain to occur may not, in fact, eventuate. Thus a two-year payback is sought because of the risk of obsolescence; a relatively high-risk investment is approved only if it pays a relatively high rate of return.

A more explicit way of taking risk into consideration is to build into the model itself considerations of the probabilistic nature of almost all predictions. There is a philosophical problem here; such models demand the acceptance of the concept of subjective probability. This important idea is discussed in reference 8.

Sensitivity Studies. Sensitivity studies are useful either as a first step for those who accept subjective probability notions or as a final step for those who hold to more traditional objective probability concepts. A sensitivity study consists in examining the variation in results obtained when key parameters take on their lowest expected value, their most likely value, and their highest expected value.

Thus, in a discounted cash flow study, it may be felt that the most critical prediction is that of the period of use of a proposed asset. If, for example, the best estimate of this period is 10 years but it is felt that it might drop to as low as 7 or rise to as much as 15, a rate of return based on each of these values should be calculated. Properly done, the study should recognize such factors as the increasing maintenance costs and lower salvage values as the asset ages.

Sensitivity studies are a step in the direction of the explicit consideration of risk, but they suffer from the following shortcomings:

1. They offer high, low, and most likely estimates, but as generally used, give no information regarding the probability of each.
2. If, as is usually the case, possible variation in more than one parameter is of concern, there are two unattractive options:
   a. Make a separate study of the variation in results produced while varying each parameter in turn, holding all others constant (usually at their most likely value). This seems unrealistic, is time consuming, and is difficult to interpret.
   b. Make a study in which all parameters simultaneously are assigned successively their worst, most likely, and best values. This, too, is unrealistic. If, for example, worst is defined as that value which has less than a 1 percent chance of occurring and there are five independent parameters, then the probability that all will be in their lowest 1 percent is only 0.0000000001. To report a value having that low a probability is meaningless; and if, as usually happens, it is merely reported as lowest, it is grossly misleading.
      A minor problem in using this option is that it is sometimes difficult to see the relationship between direction of change in a parameter and in the final result.

Risk Analysis. For those willing to use subjective probabilities, the way to surmount these difficulties is obvious: devise a model that accepts explicit probability statements about each parameter. In general, present worth or discounted cash flow methods are better suited to accepting such inputs than is the equivalent uniform annual cost approach.

Probabilistic inputs lead to models that cannot be solved analytically in most interesting problems of capital expenditure analysis. Therefore, simulation or gaming techniques are employed, usually with the help of a computer. Essentially, this is what is done: Each factor in the analysis—such as first cost, cash flow in each year (either in total or by component, life, and corresponding terminal value)—is stated as a probability distribution. The computer selects at random a single value from each distribution (which can be continuous or discrete) and performs an analysis based on these values. It then repeats this procedure many times, each time selecting at random a new value from each of the appropriate probability distributions. The results obtained from each of these successive passes are arranged into a probability or probability density distribution as output.

Thus, in a present worth study, the output might give the probability of each of a number of ranges of present worths, or alternatively, yield a table or graph showing the probability of a present worth of X dollars or less.

Detailed description of such a methodology, which is rapidly gaining in acceptance, is found in references 5 and 6.

Decision Analysis. A still further refinement in the use of probabilistic inputs—in which it is recognized that a decision often results in a series of sequential, probabilistically related events—is known as decision analysis. In this approach, the sequence of anticipated potential events and decisions is generally portrayed by use of a decision tree. Practitioners of this approach frequently use Bayesian statistics to revise probability estimates in the light of additional information and to determine the value of such additional information. An example is the determination of whether to buy market research information or to proceed directly to one of the alternative courses of action, such as build a large plant, build a small plant, or do nothing.

Another hallmark of this technique is the use of utility theory in recognition of the fact that, in nonrepetitive decisions involving large amounts, few are willing to take that course of action which optimizes the expected value of the outcome. The expected value can be loosely defined as what the average would be if the decision were to be repeated many times. To give a concrete illustration of the problem involved, a large number of people have been asked to choose between getting a certain $1 million, tax free, or taking a 50–50 chance on receiving either nothing or $5 million, tax free. Almost all preferred the sure million, despite the fact that the expected value of the gamble is clearly 2½ times as great.

An excellent introductory article on decision analysis is offered in reference 4, and an introductory discussion of utility theory and its industrial implications in reference 10. A more detailed discussion of both is found in reference 8.

CONCLUSION

The real art of analysis lies in the determination of the degree of sophistication that it is wise to employ. It is easy to fall into the trap of figuratively using an elephant gun to shoot a mouse or a shotgun on an elephant. And the decision on the degree of sophistication justified is complicated by rapidly changing technology. Thus the computer opens up vistas for risk analysis in situations where only a certainty model would previously have been feasible.

As these new vistas open, it must be remembered that the best analytical technique can offer results no better than the predictions on which the results are based, and none can offer a meaningful result if applied to the wrong question. Too often, it is asked, "Should this machine be purchased?" rather than "How can this function be best accomplished?"

Decision making is both an art and a science. It is an art to know when a decision is needed and the questions that must be asked. Art and science meet in the selection of a criterion and a methodology to apply. No methodology can hope to capture all the richness of the real world; so the analyst must decide when additional complexity in his methodology pays dividends and when it buys only additional confusion.

The application of the methodology chosen is a science, but the interpretation of the results remains an art, for no analytical technique can offer more than a guide to an intelligent decision.

BIBLIOGRAPHY

1. Barish, Norman N., *Economic Analysis for Engineering and Managerial Decision Making*, McGraw-Hill Book Company, New York, 1962.

2. Fleischer, Gerald A., *Capital Allocation Theory*, Appleton-Century-Crofts, Inc., New York, 1969.
3. Grant, Eugene L., and W. Grant Ireson, *Principles of Engineering Economy*, 5th ed., The Ronald Press Company, New York, 1970.
4. Hammond, John S., III, "Better Decisions with Preference Theory," *Harvard Business Review*, November–December, 1967.
5. Hertz, David B., "Investment Policies That Pay Off," *Harvard Business Review*, January–February, 1968.
6. Hertz, David B., "Risk Analysis in Capital Investment," *Harvard Business Review*, January–February, 1964.
7. Merrit, A. J., and Allen Sykes, *The Finance and Allocation of Capital Projects*, Longmans, Green, & Co., Ltd., London, 1962.
8. Raiffa, Howard, *Decision Analysis: Introductory Lectures on Choice under Uncertainty*, Addison-Wesley Publishing Company, Inc., Reading, Mass., 1968.
9. Reul, Raymond I., "Profitability Index for Machine Justification," *Automation*, March, 1965.
10. Swalm, R. O., "Utility Theory: Insights into Risk Taking," *Harvard Business Review*, November–December, 1966.
11. Terborgh, George, *Business Investment Management*, Machinery and Allied Products Institute, Washington, D.C., 1967.
12. Thuesen, H. G., and W. J. Fabrycky, *Engineerng Economy*, 3d ed., Prentice-Hall, Inc., Englewood Cliffs, N.J., 1964.

Chapter **4**

Production Line Techniques*

RICHARD MUTHER

**Richard Muther & Associates, Inc.,
Kansas City, Missouri**

The production line is recognized as the chief way to produce large quantities of standardized items at low costs. Chapter 2 of Section 11 indicates that line production is basically an arrangement of work areas.

In its most refined state, line production is an arrangement of work areas where subsequent operations are located immediately adjacent to each other, where the material moves continuously and at a uniform rate through a series of balanced operations which permit simultaneous performance throughout, the work moving toward completion along a reasonably direct path. These complete refinements, however, are not necessarily required.

Underlying Principles. Line production takes advantage of the following principles:

1. *The Principle of Minimum Distance Moved.* With work areas immediately adjacent, one operation begins where the preceding one finishes; one worker picks up where the previous worker sets aside.

2. *The Principle of Flow of Work.* Flow involves continuous movement, at a uniform rate, which the production line provides. Flow is measured by the rate of production rather than a specific quantity.

* Several illustrations in this chapter are from *Production Line Technique* by Richard Muther, McGraw-Hill Book Company, New York, 1944.

3. *The Principle of Division of Labor.* The most efficient use of labor is to give specific small portions of a job to each of the workers, to divide the work and assign one operation or skill to one worker.

4. *The Principle of Simultation or Simultaneous Operation.* Up and down the line, workers are continuously performing their operations; the first, last, and middle operations are being performed simultaneously without the completion of one operation at a time on all pieces.

5. *The Principle of Unit Operation.* The line is considered as a single producing unit—one series of operations or one group of workers solely assigned to one product. The entire line performs as one producing unit.

6. *The Principle of Fixed Routing.* The routine is preestablished when the line is set up, and the opportunity for diversions or lost work is minimized. The idea here is to set up the fixed route and then keep material moving over it.

7. *The Principle of Minimum Time or Material in Process.* Line production achieves a thin, swiftly flowing stream of material—a fixed operation sequence—with simultaneous operations. This allows minimum time and minimum material in the producing unit at one time.

8. *The Principle of Interchangeability.* Interchangeable parts and components are a must—line production takes advantage of this similarity, at the same time being largely dependent on it.

There are various types of line production:

Flow within one machine or within several integrated machines (automation)
Sporadic lines within process or functional departments set up for occasional runs
Line assembly with forming and treating done in process departments
Line production throughout—forming, treating, and assembly

Prerequisites. Certain conditions must exist for line production to be practical. These prerequisites involve:

Quantity. The quantity or volume of production must be sufficient to cover the cost of setting up the line. This is dependent, then, on the rate of production and the length of time the job will last. The quantity must be of a single standardized part or product or group of basically standardized products.

Balance. The required times for each operation on the line must be approximately the same. Times must be available for each operation and the equipment and manpower synchronized to a common balancing factor—usually expressed as time per work station.

Continuity. Once started, the line must continue to flow, for a stop at one point starves the rest of the operations. This means that precautions must be taken to be assured of a continuous supply of material, parts, and subassemblies and of freedom from breakdown of equipment.

How to Plan a Line. The steps the industrial engineer or others must take to get a line into operation are illustrated in Figure 4-1. Although no two companies operate exactly the same and the nature of the product, requirements, and facilities vary with every line, these steps cover the basic functions. These will be covered in the following discussion.

The Product or Material. Every plant layout—which includes the arrangement of production lines—begins with analysis of the product or material. The product itself should be checked to be sure it is designed for ease of production, not merely for functioning. Can the parts be made and assembled according to the drawings and specifications? Can they be changed in order to be made and assembled more easily?

In addition, the specific rate of production and the total quantity scheduled must be obtained or developed. Not until he has these facts can the industrial engineer begin to plan the operations, their sequence, and the equipment required.

In dividing the product into components, generally, try to keep as many operations as possible off the final assembly line. It is easier to do the operation as part

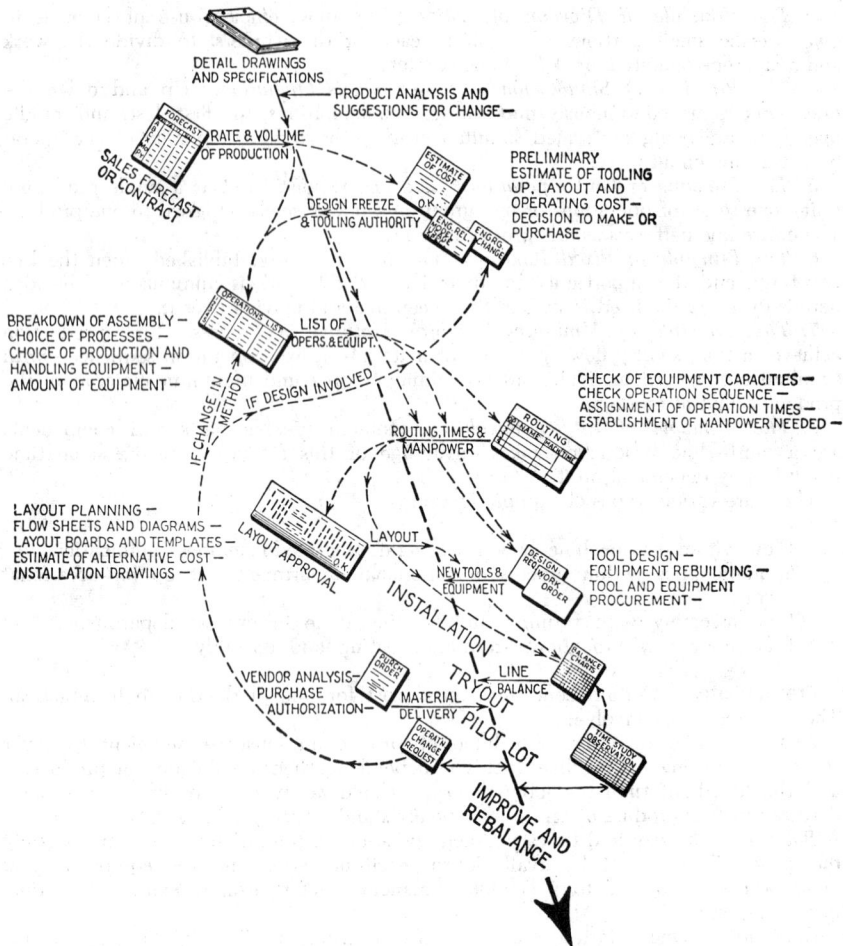

DETAIL DRAWINGS
AND SPECIFICATIONS
PRODUCT ANALYSIS AND
SUGGESTIONS FOR CHANGE —

SALES FORECAST OR CONTRACT

FORECAST

RATE & VOLUME
OF PRODUCTION

ESTIMATE OF COST

PRELIMINARY
ESTIMATE OF TOOLING
UP, LAYOUT AND
OPERATING COST—
DECISION TO MAKE OR
PURCHASE

DESIGN FREEZE
& TOOLING AUTHORITY

O.K. ENGRG. REL.
ENGRG CHANGE

BREAKDOWN OF ASSEMBLY —
CHOICE OF PROCESSES —
CHOICE OF PRODUCTION AND
HANDLING EQUIPMENT —
AMOUNT OF EQUIPMENT —

OPERATIONS LIST

LIST OF
OPERS. & EQUIPT.

IF CHANGE IN METHOD

IF DESIGN INVOLVED

ROUTING, TIMES & MANPOWER

ROUTING

CHECK OF EQUIPMENT CAPACITIES —
CHECK OPERATION SEQUENCE —
ASSIGNMENT OF OPERATION TIMES —
ESTABLISHMENT OF MANPOWER NEEDED —

LAYOUT PLANNING —
FLOW SHEETS AND DIAGRAMS —
LAYOUT BOARDS AND TEMPLATES —
ESTIMATE OF ALTERNATIVE COST —
INSTALLATION DRAWINGS —

LAYOUT APPROVAL

LAYOUT

NEW TOOLS & EQUIPMENT

DESIGN REWORK ORDER

TOOL DESIGN —
EQUIPMENT REBUILDING —
TOOL AND EQUIPMENT
PROCUREMENT —

INSTALLATION

VENDOR ANALYSIS —
PURCHASE
AUTHORIZATION —

PURCHASE ORDER

MATERIAL
DELIVERY

TRYOUT

LINE
BALANCE

BALANCE CHART

PILOT LOT

OPER'N
CHANGE
REQUEST

OPERATION OBSERVATION

IMPROVE AND
REBALANCE

FIG. 4-1.   Production engineering steps to get a production line into operation.

of the forming or subassembly operations where the material is more accessible. The fewer the operations, the shorter the line and the less the investment in expensive assembly fixtures. Where synchronization of subassemblies is relatively difficult compared with the cost of setting up the line, this does not form a rule.

Methods and Equipment. Lining up the operations required, the capacity or number of pieces of each type of equipment selected, and the sequence of operations is the next step in planning the line. Right from the beginning, tie this back to the rate of production required.

Units per hour = (quantity per month + scrap)

÷ (number of working days per month

× number of hours per day the line will operate)

This is the rate which all work stations and equipment must meet if the line is to function as a unit.

Many times it is better to supply a line with equipment that is slower than the most efficient machine might be; there is little sense in planning over capacity for one operation. Many plants slow down machines to the capacity of the line rather than have an operation that is so out of phase that it results in rehandling.

It is especially important that adequate operation analysis and methods study go into the job before it starts into production. With individual assembly in a fixed location—layout by fixed position—subsequent methods correction work is always helpful. With a line, methods work done after the line is in operation should be limited to bottleneck operations. If reliance is placed on methods work done after production begins, trouble will result, for every such change in one portion of a synchronized producing unit causes disruptions up and down the line.

Production lines frequently call for special equipment. This does not mean that specially designed machinery is essential. It is usually better to work from universal, general-purpose equipment and modify it or tool it for the special job.

One major point: in tooling for line production, it is fundamental that the product design be fixed. To get all tooled up and then have design changes come along is far more serious than in layout-by-process plants. With line production, just one change may upset an entire sequence of highly synchronized operations.

Movement of Material. The movement of material is a fundamental part of planning the production line. There will be short moves, to be sure, and relatively less time and money tied up in handling. But material movement is the thing that ties all operations together. It is relied on to maintain the continuity of the line, and it must deliver parts and subassemblies reliably as needed. The handling device must therefore be planned right into the line from the beginning. Some lines, in fact, are built entirely around a basic handling device. Figure 4-2 is an example of house assembly on a conveyor. Four- and six-room houses are assembled at the rate of one a day on a 600-foot-long conveyor running throughout the full assembly area.

The sketch shows this more clearly than a picture could because of the length of the line and because the conveyor is inside a building up to the station where the roof is put on.

Conveyors are not the special gadget that brings all the economies. Conveyors do form a useful part of many lines, but simple hand-passing, chutes, or portable, wheeled fixtures are just as often more economical and usually more flexible. In either event, the handling device is part of tooling up the line just as a die or jig is part of tooling up a single operation.

Handling devices serve several purposes that should be recognized:

Transporting—movement to, from, and along the line
Pacing—maintaining a steady, uniform output
Holding the work—convenience and reduction of nonproductive hand handling
Storing—especially for temporary reserves or cushions between operations

The selection of the device will depend on the purpose it should serve. Where the handling device is a portable rack or fixture, do not overlook planning for its return to the head of the line.

1. *Moving a Single Piece or a Lot.* Move individually if the part or product is:
   a. Large, bulky, or heavy
   b. Held in a working fixture
   c. Uniform and the operating times are consistently the same
Transport in a lot or group when the parts are:
   a. Small, compact, and light
   b. Irregular and the time per piece varies
   c. Worked on more efficiently when working several together
Generally, there is a tendency to use a lot handling when the part or product:
   a. Is easily misplaced or stolen
   b. Is odd-shaped or easily damaged
   c. Needs drying or curing time
Where both individual and lot are used on the same line, try to keep the two

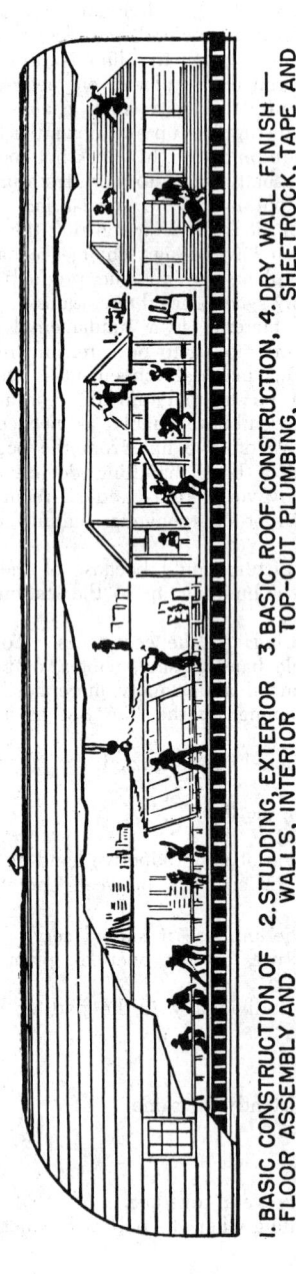

1. BASIC CONSTRUCTION OF FLOOR ASSEMBLY AND ROUGH PLUMBING

2. STUDDING, EXTERIOR WALLS, INTERIOR PARTITIONS

3. BASIC ROOF CONSTRUCTION, TOP-OUT PLUMBING, WIRING, SHINGLING

4. DRY WALL FINISH — SHEETROCK, TAPE AND TEXTURE WALLS

5. FINISH CARPENTRY, TILING, LINOLEUM, PLUMBING FIXTURES

6. PAINTING FINISHING, HARDWARE, ELECTRIC FIXTURES, FLOOR FINISH

7. DELIVERY BY TRUCK AND SPECIALLY DESIGNED TRAILER

8. CREW INSTALLS THE HOME ON PRE-BUILT FOUNDATION

TO SITE COMPLETE

Fig. 4-2. Conveyor line assembly of houses.

types of handling separate. Where different-sized lots are used, keep them in standard sizes, preferably in multiples of each other.

2. *Removal or Nonremoval of Work from the Conveyor.* Normally, work is not removed from the conveyor when:

    a. It passes directly into and through the operation's work area

    b. Parts are large, bulky, or heavy

    c. Parts are easily lost, damaged, or stolen

Parts are frequently removed when:

    a. Hand handling is a small part of a relatively long operation

    b. They cannot be worked on conveniently on the conveyor

    c. Operations must be done at a stationary point but the conveyor moves continuously

3. *Continuous or Intermittent Movement.* Continuous movement offers the major advantage of constant pacing and a more or less guaranteed output. It also offers automatic control with easier supervision. Conveyor costs are less.

Factors favoring intermittent movement include the following cases:

    a. Where a powered conveyor is not justified

    b. Where indexing several pieces at a time allows operations to be done more easily when a piece is stationary

    c. Where attachments, access ladders, and the like have to be set up or hooked to (and removed from) the product

    d. Where the workers cannot do a satisfactory job when they must move during the operation

4. *Recognition of Work.* On many conveyors, the identification of the proper parts is a problem that leads to errors and diverts the attention of the operator from his work. Where possible, automatic discharging devices can help here. Limit switches or light-beam stopping and starting mechanisms also can help. Where neither of these is warranted, use a color code, tag, mark, or number to identify the work or the conveyor space or hook involved. Still another means of identification is by positioning the work placed on the conveyor—up or down, across or lengthwise, and the like.

In distributing the work along a line from one group of workers to another, automatic sweep-offs or diverters can be used to feed the parts to an accumulation area close to the point of use for the following operation. The ingenuity of industrial engineers here is important. Figure 4-3 illustrates an assembly line on which balanced distribution to the proper workers is automatically accomplished. The electrical inspection operation is four times as fast as the soldering. After testing four units, the inspector releases them all at one time by means of the sliding device. Parts are picked off at each of the cover-soldering stations. When the units are soldered, operators return them to the far side of the conveyor belt.

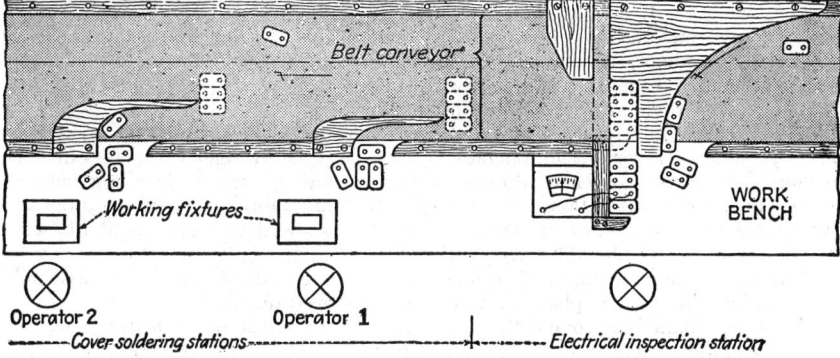

Fig. 4-3. Sketch of conveyor incorporating automatic line balancing devices.

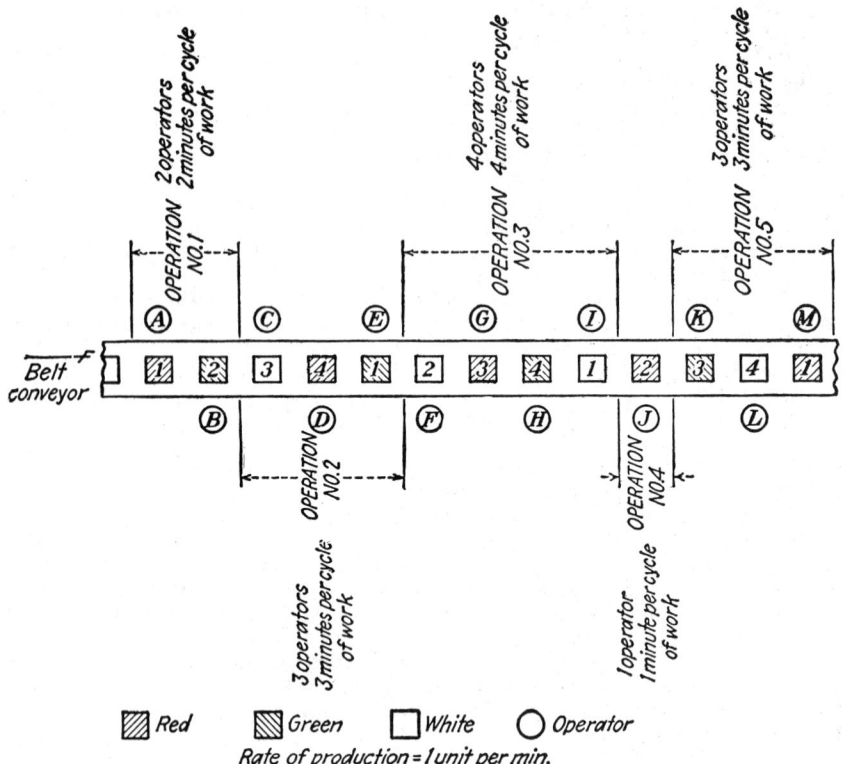

Red ▨ Green ▩ White ☐ Operator ◯

Rate of production = 1 unit per min.

FIG. 4-4. Line balancing can be facilitated by identifying the pieces on which each operator is to work.

Marking the conveyor or moving fixtures allows easy identification of each operator's proper work and permits balancing of workers as shown in Figure 4-4. Operator A works on pieces marked 1 and 3, operator C on all white markings, and operator F handles every piece marked 2.

Line Balance. The balancing of operations in terms of equal times and in terms of the time required to meet the desired rate of production is the problem of line balance. It is part and parcel of line production. The desired rate of production is converted to a time per work station; this is called by various engineers the balancing factor, balancing time, cycle time, or station time. This balancing factor or station time is equal to the reciprocal of the rate of production; that is, it equals one divided by the rate of production.

Perfect balance is rarely achieved; there is always some extra time in at least one operation. However, an operator with idle time to balance may often be assigned additional work not necessary to his operation, such as handling material to the line; extra inspections of his work; applying lubricant, label, or tag; and even more lengthy operations when the work is allowed to bank up at his work station. Because of the difficulty of dividing machine operations, it is far harder to balance forming or fabrication lines than assembly lines, where the assembly time can be split at many places and the workers moved readily.

The term "bank" is commonly used for the accumulation of material waiting for an operation. The use of banks is common where parts are put over a line and with irregular amounts of work to be done on each part. They are also used where parts are handled as a lot or batch on the line and again where materials are delivered to or removed from the line at irregular intervals or where a given machine or work station is also used for some other part or product periodically.

Banks may be located directly in the line. Examples of this are hoppers between punch presses into which parts from the previous operation are elevated, and additional lengths of chute or roller conveyor on which parts can accumulate but index themselves forward by sliding. Or the bank may be beside the line on a shunt, over-the-line shelf, or spare conveyor. These arrangements are far better than removing the bank to a separate storage point.

Where banks are held for protection rather than balancing, they must be adequate to protect the operations subsequent to the point of interruption of flow. This means that the anticipated interruption or delay time divided by the station time or balancing time per piece will determine the number of pieces in the bank.

$$\text{Bank size (no. of pieces)} = \text{rate of production (pieces/hr)} \times \text{time of interruption or delay (hr)}$$

$$\text{Bank size (no. of pieces)} = \frac{\text{time of interruption or delay (hr)}}{\text{balancing factor or station time (hr/pieces)}}$$

Line Speed and Length. Speed of flow bears a direct relation to rate of production and space per work station.

$$\text{Speed of line (ft/hr)} = \text{rate of production (pieces/hr)} \times \text{station length or space per piece (ft/pieces)}$$

$$= \frac{\text{station length or space per piece (ft/piece)}}{\text{balancing factor or station time (hr/piece)}}$$

The station length or space is dependent on the size of the part or unit, the room required by workers and equipment, and the amount of work to be performed there.

Length of line = station length or space × no. of stations
= station length or space (ft) × overall time for completing piece or unit (total line hr/piece) ÷ balancing factor or station time (hr/piece)

Methods of Getting Balance for Material Forming Operations

Improving the Operation. This is the best way to balance material forming lines. Concentrate on methods improvements for the bottleneck operations.

Change Machine Speeds. When a slow operation can be brought up to the line rate or a machine is slowed down to the same rate as other operations, the problem of balance is easily solved. Usually, it is easier to let the fast machine stand idle part-time.

Bank Material and Operate the Slower Machines Overtime or on Extra Shift. This increases both floor space requirements and material in process at the bottleneck operations, but it is relatively simple when there are only one or two machines out of balance on the slow side.

Divert Excess Pieces to Other Machines Not in the Line. Accumulate surplus pieces from the long operation, move them to another machine, complete them on a job lot basis, and bring them back to the line when finished.

Multiple Items or Combination Lines. Sometimes it is possible to group similar items and produce them on a combination line. The theory here is to spread idle machine time of one product against that of another.

Methods of Getting Balance for Assembly Operations

Divide Operations and Apportion the Elements. This is the most common way of balancing assembly operations. Because assembly operations usually can be easily divided, a high degree of balance with little idle time is attained.

Combine Operations and Balance Groups. Often it is not practical to divide an operation further. Here a group of five workers can be spread over, say, three operations, alternating on the operations they cover. The idle time of one operation is balanced against the idle time of other operations without subdividing any of the operations. A variation of this is shown in Figure 4-5. In this example of loose balance, several stations on operation 1 (gas welding) are balanced against several stations on operations 2 and 3 (hammering and welding). When one operator gets so far ahead of the work station he is feeding that the bank is filled, he moves over to the next position beside him and starts working on the chuteful of parts that have banked up ahead of that work area.

Have Operators Move. When operations take less time than the station or balancing time, the worker may (1) move with the work and do several operations, (2) work on two different lines, (3) assist at tight work stations, or (4) do material handling or other work outside his immediate workplace.

Improve Operations. Where an operation is longer than the station or balancing time, balance it by improving the method. Many industrial engineers will spend costs far beyond the saving possible on that one operation if they can reduce a bottleneck in a line of several operations.

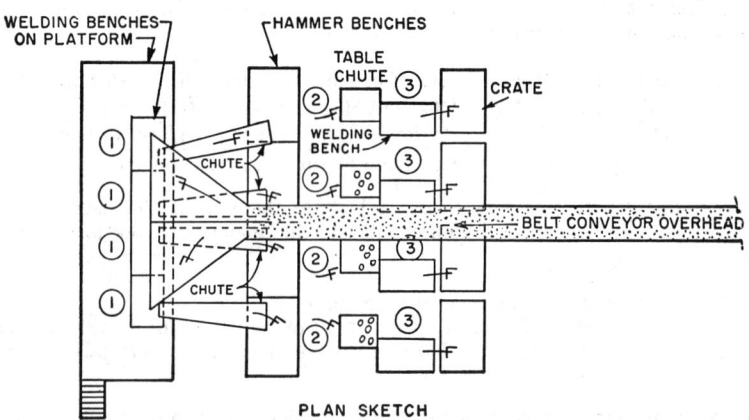

PLAN SKETCH

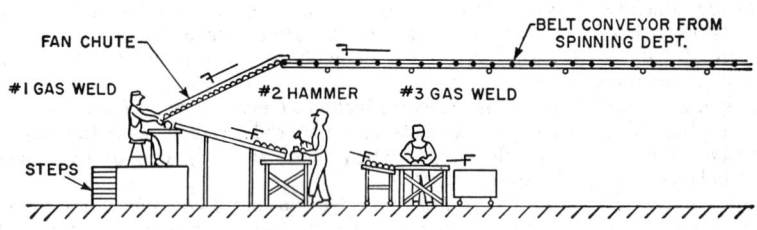

ELEVATION SKETCH

LEGEND
①, ② & ③ REPRESENT A WORKING POSITION NOT NECESSARILY A WORKER

Fɪɢ. 4-5.   A loosely balanced production line.

Bank Material and Do Slower Operations on Extra Time. This sacrifices floor space and increases materials in process at the bottleneck operations.

Improve Operator Performance. Sometimes a bottleneck operation can be balanced by (1) picking the fastest or most able workman for it, (2) giving an added bonus for anyone on the bottleneck operation who keeps pace with the other operations, or (3) assigning extra help there, such as a utility operator or a material handler to position work exactly.

Mechanics of Balancing. Balancing a line involves establishing a relationship among:

1. The rate of production
2. The operations necessary and their required sequence
3. The time necessary to perform each operation and preferably each element of operation

These become prerequisites. For the line to be set up, it must be designed for a given rate of production; for it to operate as a unit, the operation times must be such as to let the material flow evenly. Time study, predetermined motion times such as methods time measurement, standard data, and machine capacity come into their own here. They are fundamental to establishing a balanced line, except in those cases where an overall time is estimated and the workers are left to balance out the operations themselves. This latter can be done on simple, man-paced operations when there is room for banking material between operations and when there is some form of group incentive. Although it works in some situations, it usually results in overly large protective banks and cannot be considered a precise method of organizing the line.

The steps in balancing involve the following:

1. List the required operations or elements thereof.
2. List them in sequence, noting beside each the limiting operation or element it must precede or follow.
3. Show the operation or elemental time for each, indicating which is the controlling or bottleneck operation.
4. Adjust operations by combining or dividing operations or changing the methods indicated above so as to get a total time for each operation that is equal to the balancing factor or station time.
5. Draw, diagram, or otherwise put down on paper the operations, times, and manpower of the balanced line, and be sure it is understood by operating foremen or leadmen.

This is called "paper balance" and should be worked out before the line is laid out. After the line is installed, rebalancing or readjusting the balance is common, the extent of this being directly related to how accurately the original balancing was done. For examples of balancing, see Figure 4-6. The top example illustrates the balancing sheet used to work up and record the balancing of a line set up to meet a specific rate of production—22.4 pieces per hour. This is the second of two sheets used to balance the production of one part.

The lower example shows a balancing graph. It indicates the number of workers for each operation, the time required for each operation, the operations performed in each station, and the amount of idle balancing time.

In this wing line, the change in output called for meant that the speed of the conveyor and therefore the time in each station had to be changed. A line representing the station time was marked on the chart, and each operation (including allowances) was composed of elements that would approach this maximum as closely as possible.

Here the rate of production established a balancing factor of 32 minutes. Against this, the time required per operation is entered with the personal and contingency allowance shown in white above it. The rest of the time up to the 32 minutes is idle time due to balancing. In new operations 50 and 80, it would seem that one less operator is needed to do the work, thus reducing the balancing time. However, the mechanics of these operations make it necessary for the operators to work in pairs. Therefore, crews of an odd number of operators are not possible.

(a)

PART NO. 7768	PART NAME Instrument Case	DEPT. 92	LINE STANDARD 22.4 - 1/2 M & 70	DATE 3-27	SHEET 2 OF 2

OPER. NO.	OPERATION NAME	CLASS RATE	SELECT TIME (mins)	STANDARD TIME (mins)	ACTUAL STANDARD (pcs/hr)	LINE STANDARD (pcs/hr)	WORK	COMBINATION OF OPERATION	% OF TIME WORKING	OVER	UNDER	CAPACITY OF BANK	NUMBER OF CHANGES PER HOUR	TIME ALLOWED TO CHANGE (PER HOUR)	MONEY PER OPERATION PER HOUR	PRICE PER 100
50	Centralize part with tapered plug in .2187 diameter hole - clamp and bore .2187 diameter hole. Hand tap 1/4-32 thread.	1.19	1.1607	1.3345	22.4* 1/2 M	22.4 1/2 M	30.00		100%	-	-	60	None	None	.5958	2.66

(b)

OUTER WING ASSEMBLY COMPLETE

WING SPACE ON CONVEYOR = 26 FEET
STATION LENGTH = 52 FEET
CONVEYOR SPEED = 9.8 IN. PER MIN.
STATION TIME = 32 MIN.(WITH ALLOW.)
NEW RATE OF PRODUCTION = 15 WINGS PER 8 HRS.

MALE OPERATORS
FEMALE OPERATORS
ALLOWANCE-5% PERSONAL - 10% CONTINGENCY
TIME TO BALANCE
POSSIBLE STATION TIME (COLORS ARE USED IN ACTUAL WORK FOR ALL THESE DESIGNATIONS)

STATION #	1	1	1	1	2	2	2	2	3	3	3
NEW OPER.#	10	20	30	40	50	60	70	80	90	100	110
PREVIOUS OPER.#	10	10/20	10/20	20	30	30	40	50	30/60	70	70

STATION MIN. INCLUDING ALLOWANCES=32.00
POSSIBLE STATION MINUTES = 33 1/3

NUMBER OF OPERATORS IN CREW	4	4	4	3	4	6	8	4

Fig. 4-6. (a) A balancing sheet used to work up and record the balancing of a line; (b) a balancing graph.

The line marked "possible station minutes" indicates how well the line is balanced. It represents the average station time if all idle balancing time could be eliminated and is calculated as the ratio of the total man-minutes to produce a complete unit to the actual man-minutes expected to be taken by this arrangement of workers. A chart like this helps in attaining the balance, and it also shows the foreman what operations are assigned to each work station and how he should assign his workers.

In balancing with a paced conveyor, care should be taken to arrange the line so an operator who occasionally cannot finish in time does not starve the following operator. To overcome this, some lines hold in each work station a small bank of pieces that have been previously completed. When there is a delay, one of these pieces is placed on the conveyor. Where large assemblies are involved, utility men may step in and help on operations that are temporarily taking more time than the station time allowed. Still other plants have warning signals, time-indicating lights, bells, clocks, markings on the conveyor, and the like to keep operators aware of the time they have left to complete their cycle.

Further points in balancing include:

1. Save light operations—those with idle time to balance—for breaking in or training stations for new operators.

2. Keep the first operation light to be sure the head of the line always feeds in the work on time.

3. Balance into the line the inspection, material handling, and other supporting or service operations so they are synchronized with the line.

Computer Application. There is also a ranked positional weight technique for assembly line balancing that can be programmed for complete application.[1] The approach is relatively easy to understand and can also be done by hand. In general, there are two types of situations:

1. Minimize the number of work stations for a given cycle time.

2. Minimize the cycle time for a given number of work stations.

The second situation is described in the following example. See Figure 4-7. The sequence of work units (operations or groups of suboperations) is shown on a precedence graph (top). For computer solution, the precedence graph is put into matrix form, as shown lower left. In the matrix, a "1" denotes that the work unit in the row must precede the work unit in the column; a "−1" denotes that the row must not precede the column; and "0" denotes a no difference relationship.

The next step is to calculate positional weights for all work units. This is done by adding together all the time values for the specific work unit and all other work units that must follow it as defined in the precedence matrix. The work units are then sorted and listed in descending order of positional weights.

A simple method for finding the minimum cycle time for a given number of work stations is as follows: determine a realistic cycle time; balance the line according to the specific assignment rules in the next paragraph; determine the limiting work station (station with the maximum assigned time); rebalance the line using a cycle time one time increment smaller than the limiting station time; and continue through successive iterations of reduced cycle times, holding the number of work stations constant, until the minimum acceptable cycle time is reached, at which condition the total idle-time-to-balance (leftover unassigned times) is a minimum that meets the precedence requirement and number-of-stations requirement.

The specific assignment rules to follow are:

1. Select a cycle time.

2. Select the work unit with the highest positional weight and assign that work unit to the first work station.

3. Calculate the unassigned time (portion of the cycle time left over) for the first work station.

---

[1] This discussion is based on W. B. Helgeson and D. P. Birnie, "Assembly Line Balancing Using the Ranked Positional Weight Technique," *Journal of Industrial Engineering*, November–December, 1961.

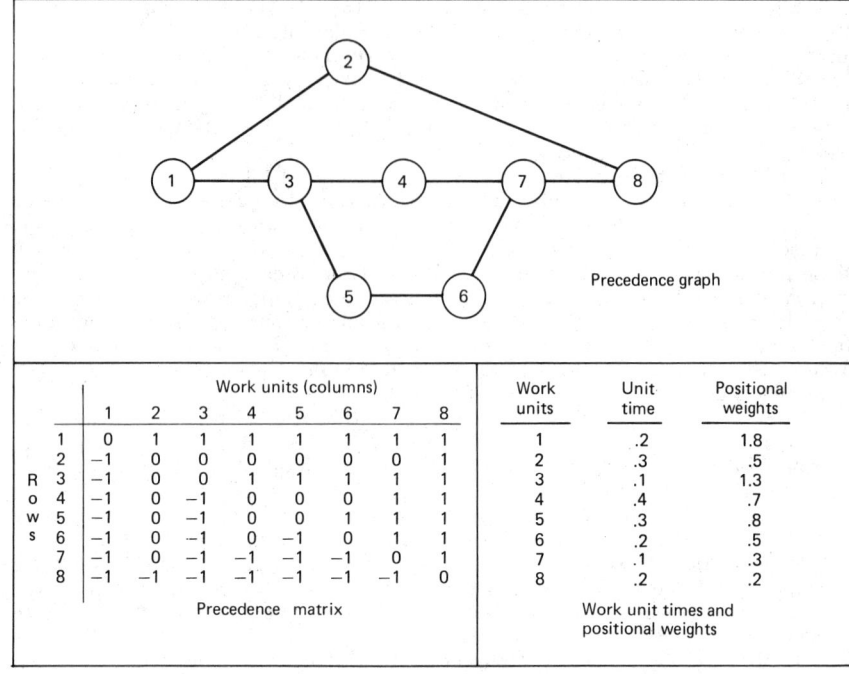

FIG. 4-7.   Primary inputs for computer application to line balancing.

4. Select the work unit with the next highest positional weight and attempt to assign it to the first work station after making the following checks:
   a. Check to ensure the proper precedence is being followed.
   b. Check to ensure the work unit time is less than the unassigned portion of the cycle time.
   If the work unit is rejected for either reason, hold it for later assignment to the second work station. If the work unit is acceptable in precedence requirements and is within the unassigned time, assign it to the first work station and proceed to step 5.
5. Continue to select work units, check, and assign if possible, until one of two conditions has been met:
   a. All work units have been assigned.
   b. No unassigned work unit remains that can satisfy both the precedence requirement and the "less than the unassigned time" requirement.
6. Assign the unassigned work unit with the highest positional weight (including any rejected work unit from the first work station) to the second work station, and proceed through the preceding steps in the same manner.
7. Continue assigning work units to the work stations until all work units have been assigned.

Layout.   Although there is more to line production than layout, basically the arrangement of equipment and work stations is the factor that distinguishes line production as a method of manufacture. Layout, being fundamental and involving so many factors, demands more coordination than any other activity in establishing the line. It affects all persons responsible for setting up and operating the line.

Besides the actual equipment and work space required for the operations, several other factors are vital yet frequently overlooked.

1. Allow adequate access space for maintenance or repair.
2. Provide for handling material properly to various work stations on assembly lines.
3. Let subassembly lines deliver work close to the point of assembly on the final line.
4. Allow adequate storage space and access to it for holding any banks along the line.
5. Provide for inspection points and makeup or repair stations in the line.

The factory building and the nature of the part or product itself influence the shape that the line will take. Tied in with this is the nature of the equipment required and the means of handling. Lines may be laid out horizontally or vertically; and they may be straight, circular or oval, serpentine or zigzag, U-shaped or odd-angle. As a general rule the straight line is preferred. It is simple, systematic, easy to install, and easy to service; it reduces costs of the conveyor system and raises no problems of transfers at corners. However, when the following situations prevail, one of the other shapes should be used.

1. When line is so long that supervision is difficult, use of floor space is wasteful, feeding in of parts or subassemblies is awkward, use U-shaped or circular line.
2. When an expensive machine is needed for two different operations on the part at widely separated steps in the process, use U-shaped line.
3. When one operator tends several machines, use odd-angle, serpentine, or U-shaped line.
4. When return of empty fixtures is involved, use U-shaped or circular line.
5. When overhead crane covers a bay wide enough for more than one row of machines, use U-shaped or serpentine line.
6. Whenever available space requires more compact arrangement, use one of the other shapes rather than the straight line.
7. When electric, air, or other connections have to be attached to the work for more than one or two stations, use circular, U-shaped, or serpentine line.
8. When a forming or fabricating line has several parts or products going over it but does not use all machines, use a combination.

In arranging the lines within the plant, generally work backward from the end point of the line. That is, determine where the line should end and plan the layout from this point. When tying in the feeding lines, their end point or point of use should be the key location for that line (see Figure 4-8). Where process departments are used for forming and lines for assembly, it is better to plan the assembly flow in general before planning the location of process departments.

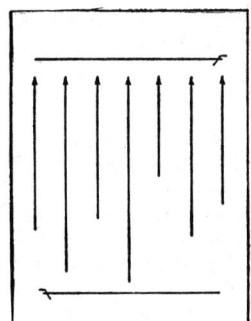 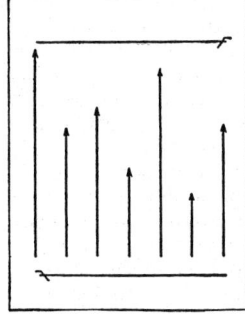

Fig. 4-8.   The right and wrong of feeding or subassembly flow.

Because line production involves larger quantities and a fixed routine, the work of layout engineering is far more important than in plants having layout by process. With line production, layout is a recurring problem, it involves more specialized equipment, and it must be planned so the entire line will function as a unit.

Manning the Line. With line production, an operator generally does only one operation. Usually this is on but one product. This means (1) that a less skilled operator can be used, (2) that operators can be trained in a short time, (3) that the labor market is widened, (4) that production can be gotten under way quicker, and (5) that waste motions can be reduced by keeping the work moving to the operator and taking it away.

The number of operators required for any line equals the total time (in standard hours allowed) required to perform all operations on one unit multiplied by the rate of production scheduled (in units per hour) divided by the length of the shift or period worked (in hours per man). This assumes no idle time to balance. All calculations of this type, therefore, are a minimum or target rather than actual.

Some operators prefer not to work on a line. Their reasons include the following:

1. Their output is limited to the speed of the slowest operator on the line.
2. Their opportunity for variety of work and therefore for self-improvement is restricted.
3. Their skill can easily be replaced, so they feel less secure.
4. There is pressure of compulsion, for they must keep pace with the speed of the line.
5. They have less personal freedom.

Other workers feel that line production offers certain advantages:

1. The work is easy to learn, and they can pay their way earlier or earn a bonus sooner.
2. The job becomes more or less automatic, rhythmic, and free from problems or mental concerns. The work is brought in and taken away with little planning, print reading, or judging of tolerances on the part of the worker.
3. Because less attention is required and operators are usually located close to each other, there is more sociability among them.
4. There is greater uniformity in the work, in the effort and skill required of the operator, in his pay check, and because there is less pressure to keep a whole line overtime, in the length of his working day.
5. There are fewer nonproductive delays and irritating interruptions.

In freeing workers from boredom or fatigue, fixed rest periods are generally recommended. Healthy competition or a wage incentive gives workers greater interest. Rhythmic patterns of work and the rotation or shifting of operators also can be used.

Assuring Continuity of Manpower. The absent worker raises a problem of how to keep the line going. Here are the principal techniques used to overcome absences, either unforeseen or anticipated.

Transfers of Workers. Borrowing of personnel from other departments usually leads to lending the poorest workers. However, where a pool of loanable labor is maintained primarily for this purpose, they will be more skilled and be paid a higher wage. Usually such men can work out of a small-order, customer repair, or miscellaneous machining department.

Shifting Workers. It is possible simply to shift the workers into a new balancing whenever a man is absent. Where operators are familiar with all jobs and there is a group incentive, this works fairly well. In other cases, it is possible to shift workers from subassembly operations when enough subassemblies are banked ahead.

Predetermined Speeds and Manpower Assignment. Where there are many workers, the line can be operated at one of several preplanned speeds. Each speed will have its predetermined manpower assignment, and if some men are off one day, the line drops back to the next lower speed and the workers are reassigned according to plan. This is especially useful where the quantity of product scheduled varies from week to week.

Buffer Banks. Where there are banks between operations or an extra large float of in-process material, operators can leave the line more or less at will. Others may spurt and build up a temporary bank so they can take a brief time off.

Rest Periods. Rest periods are a means of preventing irregular absence from the line. Although everyone stops at the same time, the loss is less than it appears, for the men would otherwise be away at varying times. Rest periods synchronize all personnel so the line can flow more uniformly. They are definitely recommended.

Relief Operators. Workers who specifically substitute for regular operators are called relief operators. They relieve each operator on the line at specified times for a specified period. This avoids congestion in rest rooms often caused by rest periods. Where processes must continue to run when once started, as in conveyorized paint booths, the use of relief operators is preferred.

Utility Operators. Utility operators on the line regularly are used to fill in for absences, although normally they are there to pick up any mistakes or omissions, to train new workers, to check stock of material, to help out on a troublesome operation, or to act as assistant to the group leaders or leadmen. In other cases, spare operators, irregulars, or substitutes may be called on to fill in for the day or for the relief schedule period.

Others. Other methods of overcoming the operator-away-from-the-line problem include: attendance bonus or tardiness penalty, adequate training for extra operators, a daily early check of manpower attendance, social pressure from others in the pay group, and duplicate lines.

Incentives. Many plants feel that a wage incentive is unnecessary on production lines. The pace of a conveyor or the inherent pull of the line, they say, holds output at the normal amount. Besides, the increased effort or skill of one individual worker cannot alter the output.

However, some paced production conveyors can have a wage incentive. The speed of the line can be set as fast as the operators wish it to be set, but any worker can stop the line momentarily by merely pulling a cord above the line in the event of minor delays in his operation. The line may restart automatically after several seconds or may be set to begin again only after one of the operators again pulls the cord.

An incentive can be used to advantage, even with paced conveyors. However, it should generally be a group incentive. Because the line performs as a unit, the output of the unit measured against its standard is the basis for paying all members of the unit. In a group, workers have an interest in helping each other in minor variations of balancing, in breaking in new workers, and keeping leadmen, foremen, or stockmen notified of impending delays. Group incentive also has these further advantages: (1) timekeeping and payroll costs are less than individual incentive and (2) workers will exert pressure on the slow operator.

The size of the group should be kept small, and its members should not be scattered. Including the leadmen and stock suppliers in the group is a further aid to increased efficiency.

When the balancing of the line requires an operator to use several skills or to work in two or more different job classifications, the pay rate of the highest class job will normally have to be used, although where rates are not fixed in step increments, a compromise rate is logical and may be used. Utility operators, relief operators, and workers transferred from the pool of loanable labor will usually carry the rate of the highest class of work they must do or will be given a slight premium for their versatility.

Piece Counts. Counting of production is generally the responsibility of the production control group or the inspection department. However, automatic counters are commonplace on production lines. Some plants use tickets that are punched or from which stubs are torn at certain points on the line.

Lengthy lines should be divided into sections ending at a count point or break point. This allows the smaller group just above the count point to get credit for what it has produced. Count points are set at points where an easy count can be made, where the work is inspected, where the supervision over the group changes, or where the type of work performed changes materially.

Potential credits may have to be counted where there are two shifts, even to the extent of counting each piece in each station of the line and assigning a percentage completion factor to each. Some long, slow lines carry a temporary hourly pay rate during the initial starting period. Later the group is put on incentive. During the closing down period, the entire line should participate in the bonus paid—not merely those at the end of the line.

Assuring Continuity of Material. To keep a line flowing, the material must be right—in quantity, in quality, and in location. The functions of planning, purchasing, material control, tool maintenance, product engineering, and inspection assume a great importance in line production plants.

In other words, planning must synchronize and schedule the work so everything flows together as needed. Purchasing must be as cognizant of quality and service of suppliers as of price. Material control and production control must be sure the material is at the right place at the right time. Inspection must be sure material that reaches the line is correct and will not cause delays. In addition, tool maintenance must be sure the tools produce correctly, and product engineering must be sure it has supplied a design that will remain constant during the life of the run.

Planning. Timing is perhaps the most significant feature of line production. It is timing rather than speed that makes for efficiency. This applies to both the preproduction planning of getting the line into production and the planning and scheduling of the actual operations.

In the first case, the many technical steps of establishing the line must be integrated to meet a time schedule. Any delay will multiply the difficulties later on, so each step of the schedule has to be met. In the second case, planning for line operation must tie back to an overall master plan and schedule of requirements. The master plan gives the operating times and component quantity requirements; the schedule is tied to the forecast or contract.

It is best to make the planning department responsible for synchronizing the entire flow of material and thus keeping the plant tuned to the rate of production required. Its plans must be laid out well in advance. This should be in general terms. As each specific operating period approaches, the plans become more detailed and more definite. For example, one automobile assembly plant bases its operations on five schedules: (1) broad plans for the entire year, (2) tentative 90-day schedule, (3) definite program for each 30 days, (4) shipping schedules for 10-day intervals, and (5) daily schedules with complete specifications.

Purchasing. In selecting suppliers, the ability of the vendor to supply the correct parts when they are needed is as important as price. If suppliers are not reliable, because of either quality or delivery, it is better to change sources than to disrupt the plant's continuity of operation.

To ensure adequate material supply, the following practices are frequently followed:

1. Investigate each supplier's plant facilities and reliability before orders are placed.
2. Check carefully on all deliveries from any new supplier.
3. Split the source of supply, and place orders with two or more suppliers for the same items.
4. Place blanket orders for a guaranteed purchase quantity, yet call for the material to be delivered according to schedules set up when the time comes.
5. Respect delivery promises, and do not schedule production based on wishful thinking.
6. Wait until all parts are received before scheduling production.
7. Maintain an adequate reserve bank of parts for protection.
8. Maintain a careful purchase follow-up with reports of impending shortages issued well ahead of time. Figure 4-9 is a shortage report intended for daily issue, based on an actual check by the stock checker on the line. Parts are listed on the shortage report before they are actually completely out, but rather when the supply gets below a predetermined quantity.

October 27
Page #2

LIBERTY PRODUCTS CORPORATION
DAILY SHORTAGE REPORT

FOLLOW UP FILE	PART NO.	PART NAME	USAGE AMT.	USAGE MODEL	AMOUNT RECEIVED	AMOUNT ON HAND	VENDOR	REMARKS & PROMISES	DATE 1st REPORTED
ASSEMBLY STATION 3 CONT.									
3	BAC12ER	Bolt 3/4x7x2½	2	all		0	Buffalo Bolt	Picking up 2000 10/26	10/24
6	C-126472	Cross Shaft	1	all	640	1390	Highland	More tonight	10/20
ASSEMBLY STATION 4									
4	* 616792	Mounting Bracket	2	T,W,M	470	470	Grand River		10/27
8	A-294628	Light Ass'y	1	T,A	60	58	Long Bros.	Will ship 120 today	10/22
ASSEMBLY STATION 5									
1	* 164821	Gas Line	1	all	500	1800	Thompson Tube	More on 10/29	10/27

FIG. 4-9. A daily shortage report.

The important relationships in scheduling any material requirements are:

$$\frac{\text{Monthly production}}{\text{Working days per month}} = \text{daily rate of production}$$

Usage per unit $\times$ daily rate of production = daily requirements
Daily requirements $\times$ days between deliveries = quantity per delivery

Material Control. Within the plant, the material control group must plan its deliveries of material to the line with equal care. And it must also take precautions to be sure material is on hand when needed. Some techniques for this include the following:

1. Store material at the point where it is to be used rather than in centralized storerooms.
2. Predetermine and identify all storage places along the line, the normal and minimum quantity of each item that should be held there, and the normal delivery quantity and frequency.
3. Have stock checkers who inventory the stock along the line and see that it is provided before or certainly when the minimum bank is reached.
4. Replenish material along the line each night, being sure that enough material is issued for a full day's run.
5. Have material handlers report organizationally to the planning or material control supervisor.
6. Synchronize delivery conveyors to the speed of the final line, but hold a protective bank near the point of use.
7. Hold oversize banks at the line or adjacent to it.
8. Have an adequate communication system available for use of the stockmen.
9. Issue the stock to the line in sets or lists so that all parts for one unit or lot are checked out at the stockroom.
10. Make the material handlers and stockmen lean over backward to take proper care of material; otherwise workers will become wasteful of material and careless with their work.

Production Control. Scheduling, dispatching, and control of production become very much easier as a plant moves toward complete line production. The routing of the material is fixed, operations are adjacent to one another, and the entire line is scheduled and controlled as a unit. Much of the clerical work of dispatching and follow-up, move orders, forwarding tags, and individual shop orders is eliminated.

On assembly lines, the synchronizing of feeding or subassembly lines is the only intricacy. This is largely a matter of adjusting the speeds of all lines to the rate required. Once done, the float or system needs only to be filled with components according to schedule and the sequence of products thereafter maintained. In this way, different colors, styles, sizes, and the like of a basic product can be controlled so as to be assembled on the same line at one time. Communicating systems like telautographs or teletypewriters that report the sequence desired to several points throughout the plant are usually necessary for this type of control.

Where the variety of product is too great to allow economic balance by scheduling several varieties over the line at the same time, blocks of units or items are run off together. For example, make twelve of A, then twelve of B, then twelve of C, then twelve of A again, or make a day's run of A, then a day's run of B, and so on. The concept here is holding the product design constant for a given period, then converting to some other product or design and concentrating on it. This means runs or lots covering the blocks scheduled.

The control of forming or fabricating lines requires only that material be at the starting point when needed. Where process departments feed parts to the line and the operations, done in lots, are faster than line speed, the technique of cycling is used. A group of machines in a process department will be scheduled to produce lots large enough to keep the line supplied for a certain period of time. This period is called the cycle time. The machines will operate on several parts, one after another, returning at the end of the cycle to the first part. This procedure is illustrated by Figure 4-10, which shows the method of cycling one machine which can operate faster than the required rate of production. It represents a fairly loose cycle with an ample safety bank of eight days' protection. The shaded section of the figure shows finished inventory of part 2. The bars indicate machine time allotted to each part. In this way, fast operations are synchronized with the rate of production on the line.

One thing is vital: the production planning and control function must have the authority to tell operating departments what, when, and how much to produce.

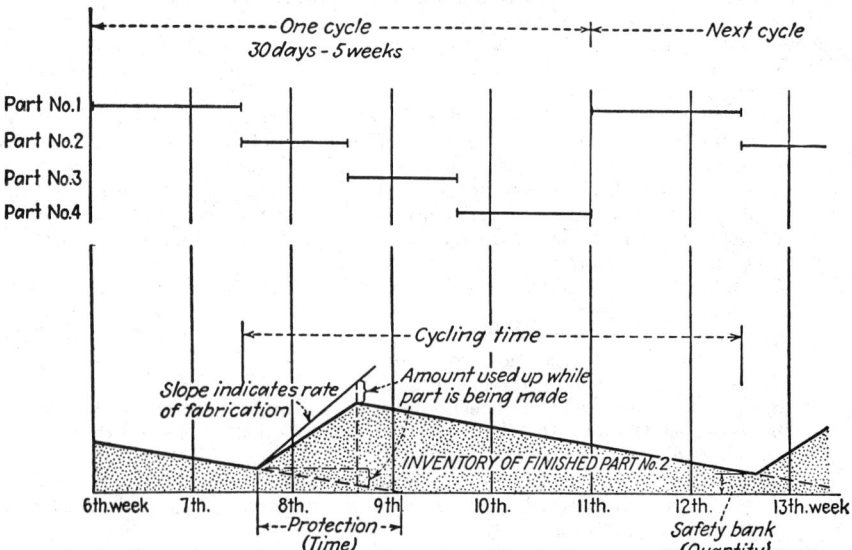

FIG. 4-10.   Machine cycling to synchronize a fast machine to the rate of the production line.

Quality. On an assembly line, there is usually little time for fitting and adjusting. The parts must be interchangeable. Moreover, if the parts reaching a production line are not correct and must be scrapped, the entire line will be shut down. This places greater responsibility on the inspection department.

With line production, there will be no centralized inspection point; inspectors must be spread throughout the plant. Centralized inspection increases handling, the very thing line production aims to reduce.

Line production may aid quality in the following ways:

1. Large quantities of bad work or material cannot be made before the error is caught.
2. The same operator continually doing the same operation makes for uniformity and regularity of work.
3. Machine setups and tools remain unchanged for long periods, again making for uniformity and reduced errors.
4. Inspectors, too, remain on the same job and know the problems, trouble spots, and idiosyncrasies of individual operators.

On the other hand, line production may allow quality to suffer for:

1. The work is spread among all persons in the line so workers do not feel individually responsible, and thus it is hard to place individual blame.
2. Operators are normally unskilled or semiskilled.

In operating production lines, the following techniques should be followed:

1. Make sure that the product is designed correctly by testing sample parts.
2. Make sure that all new tools, dies, and fixtures will produce correct parts.
3. Put confidence in tooling, inspection gages, and fixtures rather than individuals. Tools and gages assume a far greater importance with line production than is generally realized. Make them simple and sturdy.
4. Plan for the inspection group to be busier than usual during the pilot lot or trial run period.
5. Assign inspectors along the line:
   a. So they are balanced in
   b. So they cover the same stations as are assigned to specific leadmen or foremen
   c. So they are at a point where defects can be pulled off, where a bank of corrects can be held, or where repair stations can be maintained
   d. So they can identify defective work to specific workers
   e. So they can catch errors before the piece becomes covered, painted, or otherwise made inaccessible to later inspection
6. Carefully identify any matched sets or selective assemblies.
7. Charge cost of repair station or repair operator against the group on the line.
8. Place emphasis on prohibiting any incorrect parts from getting to production lines.
9. Have an inspection tag or log sheet on each assembly, to stay with it as in Figure 4-11. The inspector in each station marks down the defect. When the correction has been made, the repair operator writes in his badge number. If the repair is satisfactory, the inspector OK's the work by initialing the last column. Obviously this type of inspection log can be used on slow-moving lines only.
10. Maintain cleanup or touch-up stations at the end of the line to care for minor defects that can occur anywhere along the line.
11. Use utility operators to pick up occasional defects along the line.
12. Stop production and move line operators to the end of the line to correct units when an epidemic of difficulties occurs, or move work back over the line. This is largely psychological and seldom economical.

All this sounds as though everything must be in apple-pie order before a line should start. Quite the contrary; force the starting of the line ahead of time, and pick up the lagging points later on. Otherwise, it will never get started.

Maintenance. Proper maintenance of tools and equipment is essential to achieve

INSPECTOR'S NAME _J Byrne_ , SCHEDULE NO. OF THIS UNIT _51_

BADGE NO. _17-68_ DATE _11/19_ SHEET NO _1_ OF_____

DEFECT NO.	NATURE OF DEFECT	REPAIR BY NO.	INSPECTOR'S O.K.
1	Grind front support part flush	8-103	J.B.
2	Grind 63" dimension to size.	8-47	
3	Clean slag off welds	8-103	
4	Gas line bracket in motor compartment shy	8-63	J.B.

Fig. 4-11.   Inspection log sheet traveling with each assembly unit on the line.

continuity of operation. If any machine breaks down, the flow is interrupted and the line can no longer function as a unit. This problem is most serious on the powered conveyor line where there is little if any float between operations. As a result, maintenance is more important, is more costly, is more decentralized, and is more dependent on preventive measures. It is the effort to prevent breakdown that really makes line maintenance.

Several methods of minimizing the number, duration, and cost of breakdowns may be used:

1. A sound preventive program involving periodic inspection, repair, or replacement of tools, equipment, and handling devices.
2. Protective banks held at critical points.
3. Rapid communication system or automatic signaling devices to indicate a breakdown and where it has occurred.
4. Fuses, overload switches, shear pins on conveyor drives, electric eye or limit-switch controls to stop overrunning or self-indexing parts, and the like.
5. Pay a group incentive based on output or on freedom from downtime.
6. Standby equipment, inventory of spare parts for equipment, and extra tools of all kinds.
7. Bypass the operation that is down, and make up the postponed operation off the line.
8. Duplicate lines so that work and operators may be switched to the second line or so that at least part of the production is continued. For long interruptions, the down-line workers may be switched to a second shift on the other line.
9. Guards and protective devices to avoid machine or conveyor jamming.

Engineering Changes. Changes in product design and in methods or tooling must be kept to a minimum with line production. Indiscriminate engineering changes can upset the whole line; a methods change at one point may throw the balance out of phase. Take care, therefore, to be sure that the design is right before tooling up. Then be sure that the tooling and methods are right before releasing the line to production.

Precautions and techniques to employ here include:

1. Thorough laboratory and field test of the product before releasing it for tooling.
2. Complete tool tryout and assembly of parts produced from these tools.
3. Pilot lot or trial run under the direction of industrial engineers, with adequate attention to debugging.
4. Station-by-station checkout by industrial engineers and leadmen, with sign-off by both, before releasing the line to production.
5. Freezing design for a full run, block, or period and then incorporating all engineering changes at one time.
6. Accept the fact that it is often more costly to make a change than to live with a design that is not perfect or a line that is not ideally balanced.
7. Maintain a small-order department where new products are tried out and leadmen trained, the product later being moved to a regular production line.

Line Flexibility. The great limitation to the use of line production is variety of product. When there are many products, it is difficult to make them on the same line. This can be overcome in several ways:

1. Build a basically standardized product part of the way down the line. Then divide the line into spurs or distributaries to pick up the special features, send the product to a small-order department or other modification point, or enclose the special attachments with the main unit to be assembled in the field.
2. Build a more or less standardized product, but offer variety through selection by customers of standardized optional parts, colors, trim, or other features.
3. Combine all similar parts into groups or families of parts, and create enough quantity to justify a line.
4. Make the line serve several products by changing over from one to the other but holding one design constant at any one period. Such changeovers may be only minutes apart and may involve complete rearrangement of the plant.
5. Freeze the basic design fundamentals, but periodically change the outward appearance.
6. Schedule special workers into the line to pick up special features only.

The ingenuity with which engineers can work the above ideas into reality is one of the chief problems of mass and medium-volume producers today.

Changes in volume or output may be taken care of by:

1. Altering the period of time the line operates by overtime, second shifting, or reduction of working hours.
2. Changing the number of workers combined with changing the speed of the conveyor and the spacing of the units on the line or altering the station assignment.
3. Leaving room for expansion.
4. Having parallel lines making different products but which can be converted to either product.

Line changeovers usually come at a low point in the production season or time of month. Weekends are also used. Simple lines that are changed over frequently may be so handled at any time during the working day, the line operators themselves serving as stock handlers and tool adjusters during the change. When the change involves an entire plant and several days, the operators are usually given their vacations at that time and the changeover is made by maintenance and machine repair groups, often supplemented by outside contractors.

In restarting a line, the group leaders, leadmen, and utility operators are usually brought into the picture at the beginning. Aim for a high rate of production right from the start, or there may be difficulty exceeding what workers figure is a satisfactory rate.

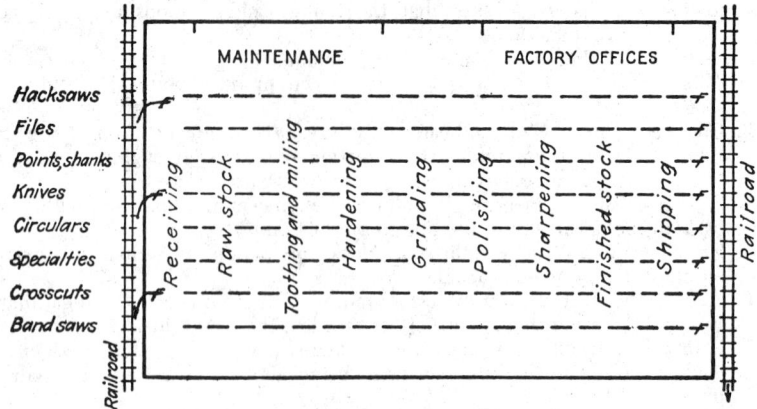

FIG. 4-12.   Layout of an entire plant according to progressive machine groupings.

Variations and Modifications. Line production is applied to operations other than production. Disassembly, inspection lines, repair and overhaul work, and salvage and teardown operations all can use lines. The cafeteria, processing a soldier through a physical examination, and even the family dishwashing project may utilize these principles.

There are several other types of line variations:

Progressive Machine Groupings. Here a few machines of each type are set up together, and the work progresses from one group of machines to the next. Figure 4-12 shows this carried to the point where the entire plant is laid out on this idea. The layout of this plant making a variety of saws is really flow through process departments. Vertically the departments are by type of operation; horizontally each type of product flows in its more or less assigned route. It offers considerable flexibility and good machine utilization by keeping functional or process departments, yet it affords a high degree of material flow.

Broken or Extended Line. This exists whenever a line is broken by movement of material to a storeroom or process-controlled layout (like heat-treating or plating). As long as the flow is broken and there must be a special handling operation, it makes little difference where the line begins again. Broken or extended lines are desirable where the building space is limited, where flexibility in the sequence of the float is needed, or where the product must be segregated or a dangerous operation performed. Short lines, sometimes referred to as unit production layouts, avoid the refinements of the highly balanced line and offer more flexibility in product design and scheduling.

The Common Machine and Multiproduct Line. Where there is an expensive operation that can serve two or more lines, the layout should feed the product to this operation so it can be a part of each line. When this idea is carried further, it results in several products moving through the sequence of machines or work stations. Figure 4-13 shows a portion of a press room for the forming of sheet metal. The products made are side, top, and bottom panels; inner panels; and doors for kitchen cabinets of various types and sizes. The line is set up for small and medium doors and can be readily changed over to produce large doors. The press operators themselves can make the changeover.

The first line is established as in (*a*). Three lengths of standard gravity wheel conveyor are used to make the second line (*b*). The Toledo drawing press is not used on the large doors; one length of conveyor carries parts directly through this press, it being set so the dies are kept apart and there is ample room for the parts to roll through on the conveyor. The other two lengths of conveyor carry parts to the Milwaukee brake for the fourth operation.

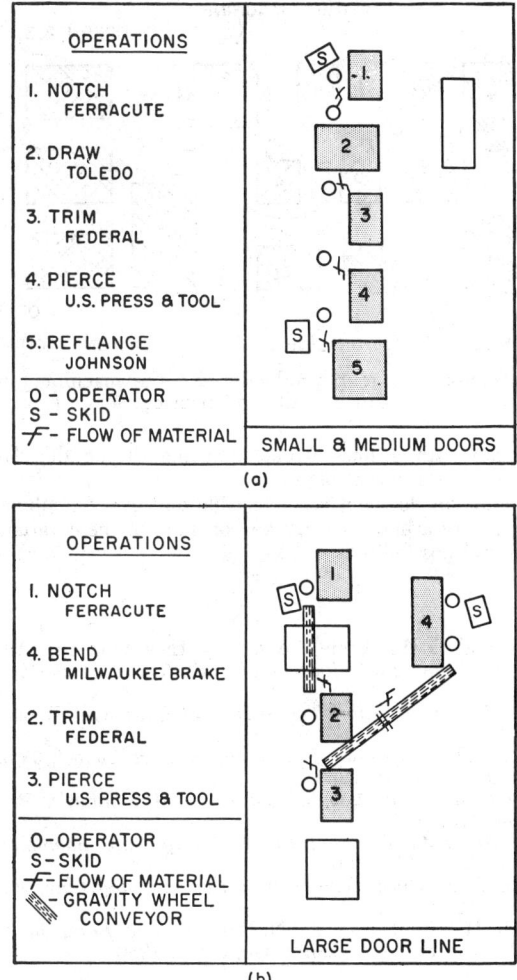

Fig. 4-13.   Layout of a press room by use of a common machine in a multiproduct line.

The Milwaukee brake is used for other work when not a part of the large-door line.   Similarly, the Johnson press, used for reflanging the small and medium doors, is used for other parts while the rest of the machines make large doors.

Common Machine Grouping or Group Production.   When a grouping of machines common to several products is set up as part of one or more lines, the layout is termed group production.   It is usually a combination of line production and process layout.   See Figure 4-14.

Moving Workers through Fixed-position Assembly.   Instead of moving the product to the men, the men move progressively from one fixed-position assembly station to the next.   Each man performs his same assigned job on the several products, and although he must carry his tools with him and have his material distributed progressively to different points, this saves the necessity of setting up a costly device to move the major component.   This moving of men can sometimes be

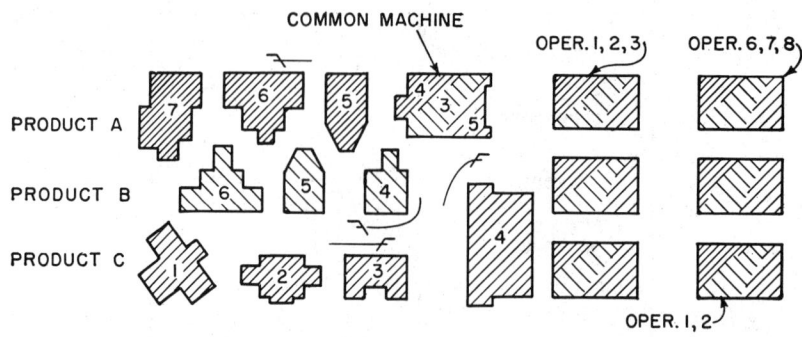

FIG. 4-14. An example of group production for the machining of a small rotor, flange shaft, and housing.

used in combination with a line—workers moving during the day and the units being moved up several stations at night.

Where line production does not seem feasible and every combination of products has been tried to justify a line, then try one or more of these variations or modifications. They offer real possibilities.

BIBLIOGRAPHY

Helgeson, W. B., and D. P. Birnie, "Assembly Line Balancing Using the Ranked Positional Weight Technique," *Journal of Industrial Engineering,* November–December, 1961.

Muther, Richard, *Production Line Technique,* McGraw-Hill Book Company, New York, 1944.

Salverson, M. E., "The Assembly Line Balancing Problem," *Transactions of the ASME,* vol. 77, no. 6, August, 1955.

Starr, M. K., *Production Management Systems and Synthesis,* Prentice-Hall, Inc., Englewood Cliffs, N.J., 1964.

Tonge, F. M., *A Heuristic Program for Assembly Line Balancing,* Prentice-Hall, Inc., Englewood Cliffs, N.J., 1961.

Walker, C. R., and R. H. Guest, *The Man on the Assembly Line,* Harvard University Press, Cambridge, Mass., 1952.

Walker, C. R., R. H. Guest, and A. N. Turner, *The Foreman on the Assembly Line,* Harvard University Press, Cambridge, Mass., 1956.

Chapter **5**

Material Handling

JAMES M. APPLE

**Professor of Industrial and Systems Engineering,
Georgia Institute of Technology, Atlanta, Georgia**

There are several basic operations in any production activity, including processing, assembling, inspecting, and material handling. What is not generally recognized is that the last of these—material handling—more often than not accounts for the largest portion of the total production cost. This chapter is devoted to a discussion of material handling, the design of handling systems, and a general introduction to material handling equipment.

DEFINITION OF MATERIAL HANDLING

A search of the literature produces a good many definitions of material handling, most of which are filled with qualifications as to what it is or is not. For the purposes of this chapter, it is sufficient to suggest that "material handling is the handling of materials."

The extent of the material handling activity in any company depends on such things as the type of company, its product, the size of the company, the value of the product or the activity being performed, the relative importance of handling to the enterprise, the personalities of the individuals involved in handling, and the organization of the enterprise.

A limited interpretation of material handling deals primarily with the movement of items from one point to another within the confines of a plant. Here, the problem is to move something from point A to point B. The concern of the material handling engineer is most commonly with individual, isolated, independent

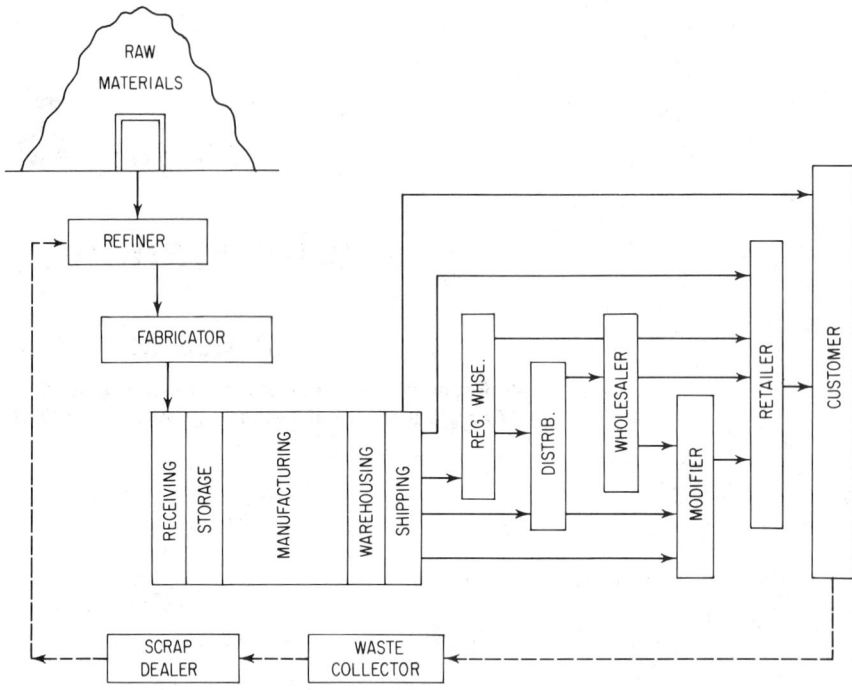

Fig. 5-1.   Material flow cycle.

material handling problem situations.   A more desirable point of view involves a total, plant-wide concern for the flow of materials.   The analyst is concerned with interrelating all plant handling problems, establishing a general material handling plan, and tying each problem solution into all others.   A still broader point of view, called the systems concept, visualizes handling and distribution as one all-encompassing system.   This viewpoint gives consideration to all handling activities involved in (1) the movement of materials from all sources of supply, (2) all intraplant handling, and (3) the distribution of finished goods to all customers.   The goal is to conceptualize a solution to the total handling problem and to design and implement those portions of the systems feasible at the moment, while continuing to work on other phases of the system and implementing them as it becomes practicable.

The extent to which the systems concept should be carried out depends upon the importance of handling in the enterprise and the practical economics of attempting to extend the handling system back to the sources of materials and forward to the customers' locations.   These interrelationships are shown in Figure 5-1.   The concern of material handling is for the entire flow as shown, from the sources of raw materials to the delivery of goods to the customer, and in some cases, the flow of scrap, waste, or returned goods back through appropriate channels and into the system again.

Material Handling Activities.   The areas of activity and responsibility with which the material handling function is generally concerned include:

1. Transportation from vendor
2. Unloading
3. Receiving
4. Stores (indoors and out)
5. Issuing to production
6. In-process handling
7. In-process storage
8. Workplace handling
9. Intradepartmental handling
10. Interdepartmental handling

11. Intraplant handling
12. Handling related to auxiliary functions
13. Packaging (consumer)
14. Packing (protective)
15. Finished goods warehousing
16. Loading
17. Shipping
18. Transportation
19. Record keeping

In dealing with these activities, the engineer will find himself involved in studying some or all of the following functions:

1. Material handling
2. Storage
3. Loading and unloading
4. Testing
5. Specifications and standards
6. Equipment feasibility
7. Handling and storage equipment
8. Auxiliary equipment
9. Selection of containers (shop, packing, shipping)
10. Packaging (consumer)
11. Packing (protective)
12. Equipment repair and maintenance
13. Damage prevention (materials and product)
14. Safety
15. Training
16. Surveys
17. Costs and cost control
18. Keeping up to date on equipment, methods, procedures, and the like
19. Related paperwork, control, and communication systems

The number and extent of the activities performed depend on the individual plant and the importance of handling.

ORGANIZING FOR MATERIAL HANDLING

With material handling sometimes taking up to 90 percent of the labor dollar, more and more companies are placing emphasis on material handling as a separate function. Some companies have given it the staff responsibility accorded to production or quality control, and others have preferred to distribute the responsibility for efficient material handling among manufacturing supervisors and material handling consultants. Quite often, it is the responsibility of the industrial engineering organization.

The Industrial Engineer and Material Handling. If the industrial engineer is held responsible for efficient material handling, he will be largely involved in the analysis of material handling operations. These analyses will not necessarily cover only small, individual sections of the plant's operations; one analysis may cover the entire manufacturing operation. In addition to the material handling activities listed above, the industrial engineer will be involved in inventory control, safety programs, labor relations, purchasing, plant layout, and transportation. In general, his duties will consist of establishing and maintaining the proper flow of materials through the plant in the most efficient and economical manner.

Frequently, industrial engineers are selected for material handling assignments because of their broad background and plant-wide experience. They can, however, be specifically trained for the job. A number of courses are offered in material handling, and many companies have added them to their training programs.

The Material Handling Department. When material handling is given a staff position in a company, it is necessary to organize a new department. Although the responsibilities of this department are not hard to define, it is sometimes difficult to set up an organization which does not conflict unduly with responsibilities of other activities. The organization of the department will depend somewhat on the organization of the company itself, but a few general rules will apply. First, the material handling department must be given the authority to make material handling changes wherever they are needed. Second, it should be permitted to act as a consulting service to anyone having a material handling problem or a problem related to material handling. Third, it should be set up to integrate material handling in all plant operations.

Cost of Material Handling. Figures from various industries throughout the United States indicate that the cost of material handling in industry varies between 10 and 90 percent of the total labor cost. Probably most of these figures are "guessti-mates," and few if any are the result of exhaustive, accurate, or thorough studies. This is primarily because of a general feeling that material handling costs are difficult to determine and that the time and effort necessary to document them accurately are not justified. However, even casual observation of any production activity will convince the observer that only a very small percentage of the total activity is actually productive time devoted to making products. It stands to reason, then, that the balance of the activity is a combination of handling, or handling-related activities, such as in-process storage, and delays of one sort or another. It should also be evident that this portion of the activity occupies a major portion of the total time involved in the production process, and for that reason constitutes a major cost which, in many companies, has never been either properly identified or accurately determined. If this were done, it should easily persuade any manager that material handling deserves serious attention, because of the tremendous improve-ment and cost saving opportunities inherent in it.

PRINCIPLES OF MATERIAL HANDLING

As in many other fields of endeavor, certain fundamentals exist in the areas of material handling which might be referred to as principles. They have been developed over a period of years and therefore represent the accumulated experience of many handling experts. The following list of principles is based on that proposed by the College Industry Committee on Material Handling Education:

1. Related to planning
   a. Planning principle. All handling activities should be planned.
   b. Systems principle. Plan a system integrating as many handling activities as is practical and coordinating the full scope of operations (receiving, storage, production, inspection, packing, warehousing, shipping, and transportation).
   c. Material flow principle. Plan an operation sequence and equipment arrangement optimizing material flow.
   d. Simplification principle. Reduce or eliminate unnecessary movements and equipment.
   e. Gravity principle. Whenever practicable, utilize gravity to move material.
   f. Space utilization principle. Make optimum utilization of building cube.
   g. Unit size principle. Increase quantity, size, weight of load handled.
   h. Safety principle. Provide for safe handling methods and equipment.
2. Related to equipment
   a. Mechanization/automation principle. Use mechanized or automated han-dling equipment when practicable.
   b. Equipment selection principle. In selecting handling equipment, consider all aspects of the material to be handled, the move to be made, and the methods to be utilized, in terms of the lowest overall cost.
   c. Standardization principle. Standardize methods as well as types and sizes of handling equipment.
   d. Flexibility principle. Use methods and equipment that can perform a variety of tasks and applications.
   e. Dead-weight principle. Reduce the ratio of equipment dead weight to payload.
   f. Motion principle. Keep in motion equipment designed to transport materials.
   g. Idle time principle. Reduce idle or unproductive time of both handling equipment and manpower.
   h. Maintenance principle. Plan for preventive maintenance and scheduled repair of all handling equipment.
   i. Obsolescence principle. Replace obsolete handling methods and equip-ment when newer methods or equipment will pay off in a reasonable time.

3. Related to operations
   a. Control principle. Use material handling equipment to improve production control, inventory control, and other handling.
   b. Capacity principle. Use handling equipment to help achieve full production capacity.
   c. Performance efficiency principle. Determine efficiency of handling performance in terms of expense per unit handled.

HOW FACILITY LAYOUT AFFECTS MATERIAL HANDLING

If the engineer determines in great detail all the facilities required for a given operation but fails to arrange them properly, he will not have a good layout. Perhaps the greatest justification for any layout analysis stems from its ability to eliminate or reduce material handling operations. By the same reasoning, it is only through effective plant layout analysis that many material handling operations can be eliminated or improved.

It is important, therefore, to keep in mind the fact that the arrangement of equipment determines the amount of material handling that is necessary. Coincident with equipment layout is the location of service areas such as storerooms, toolrooms, lavatories, offices, test floors, shipping, wrapping, packing, and the like.

A layout determines to a large extent how efficiently operations are performed, because the layout generally influences the pace, motions, effort, and safety with which the employees work.

Material Handling Considerations in Facilities Planning. To incorporate effective material handling methods into a layout, the engineer responsible should realize the following:

1. It is impossible to process materials without handling operations. However, surveys reveal that a large portion of handling is entirely unnecessary. The number of pickups, setdowns, and transports which many materials undergo is far out of proportion to the actual processing time. These are the movements that can be eliminated, simplified, mechanized, or automated.

2. For many years, layouts were designed around the process or activity requirements with little or no thought given to material logistics. Layouts are often designed with templates of the processing equipment amply supplied but without corresponding templates of either materials in process or handling equipment and facilities.

3. In reviewing layouts, supervision is usually concerned with the adequacy of processing capacity and arrangement. Activities such as receiving, temporary storage, movement to processing, movement through processing, storerooms, inspection, dispatching, testing, packing, warehousing, and shipping do not receive the same layout scrutiny. This results in the so-called hidden costs of handling.

4. The facilities planning engineer should at all times keep in mind that the building materials, building designs, and modern material handling equipment make it unnecessary and even undesirable to follow precedent and tradition in layout designs. Patterns dictated by line-shaft-driven machines, buildings with close column spacing, and natural lighting belong to bygone days. Each column requires 1 square foot of actual space but invalidates about 10 to 15 square feet of effective space in the layout.

5. In planning a building addition or new building, it should be remembered that available materials make it practical to have clear spans of 80 to 100 feet without complicating the building design or increasing costs. Bay sizes under 30 by 50 feet should be questioned.

6. Modern lighting and ventilating and air-conditioning equipment seldom make it necessary to position equipment in other than the most effective location.

7. Improved power distribution and protective devices permit arranging equipment and controls in the most effective manner even though heavy power loads are encountered or the controls are complicated.

8. In making equipment layouts, the engineer should consider the following list of handling axioms:

a. Provide aisles wide enough to accommodate safely the latest types of mobile material handling equipment. This should take into consideration both the largest loads anticipated and pedestrian traffic.

b. Provide space to set down unit loads of work in process in such a pre-positioned manner as to eliminate the need to rehandle them.

c. Keep work at a convenient working level.

d. Never set materials directly on the floor unless absolutely necessary. It requires manual labor, as a rule, to unload and reload each time.

e. Eliminate isolated, fenced-in storerooms wherever possible, unless one of the following conditions which make them mandatory prevails.

(1) Materials must be kept in strict inventory.

(2) Materials are easily lost, damaged, or pilfered.

(3) Materials are not readily obtained except on long delivery dates.

(4) Safety problems.

Usually, storerooms of this kind require additional handling to receive and dispense materials, and often additional paper work.

f. Plan first operations as near to the point of receiving as possible. If possible, move material directly to the first point of receiving inspection.

g. Determine whether or not the material can be received in containers or in a manner which does not require its removal or rehandling between receipt and first use.

h. In processing materials with high scrap ratios, determine whether or not the largest amount of scrap can be removed early in the processing cycle to eliminate much of the handling of the material which will eventually become scrap.

i. Provide an adequate scrap removal system. Waste materials must be removed constantly and without delay, or production cannot be carried on.

j. Wherever possible, keep materials flowing from one work station to the next without intermediate setdown. Where this can be done, mechanized handling is warranted.

k. To conserve floor space, utilize overhead means of conveying or storage.

l. Handle materials in bulk or unit loads when they must be intermittently handled and stored.

m. Plan inspection stations in the flow of work if possible, to avoid lateral movements of materials.

n. Plan packing operations as an integral part of the process. Avoid repacking and rehandling at another location.

o. Ship direct from the packing floor if loads are repetitive to one destination and quantities warrant. It is less expensive to handle paper than materials.

p. On intraplant movements of materials, provide for unit mass handling in standardized containers and avoid packaging wherever possible. This saves the cost of packing materials and labor. If material requires subsequent operations, determine if it can be shipped in a manner that permits its use without rehandling at the receiving point.

q. In planning receiving and shipping areas, provide for recessed docks to permit entry into trucks and railroad cars with material handling equipment. Receive and ship in unit loads wherever possible.

r. Use yard storage when materials do not require protection from weather.

s. Materials which require protection and which must be stored for long periods of time can often be stored outdoors with temporary protection.

t. Plan operations in such a manner that undue paper work is not necessary. Automate the clerical function where practical. This often saves much messenger service and production delay.

u. Consider mechanical aids when workmen:

(1) Must lift more than 75 pounds, or female operators more than 35 pounds.

(2) Must handle the same kind of materials more than ½ hour each day.

(3) Must move materials more than 50 feet.

(4) Are exposed to unusual safety hazards.

v. In planning the layout, consideration should be given to the utilization of the following types of mobile handling equipment where skids or pallets can be utilized.

(1) Movement within a group—hand-powered lift jack.

(2) Movement within a section or department (not over 150 feet)—motorized, hand-steer lift truck.

(3) Movement and stacking within a section or department (not over 150 feet)—motorized, hand-steer, elevating fork truck.

(4) Movement between departments and buildings (about 300 feet)—power-operated, rider-type, fork truck or elevating platform truck.

(5) Movement between departments and buildings of large loads for distances over 300 feet—tractor-trailer trains. An operator with a hand truck costs considerably less per hour than an operator with a fork truck, but performs about 10 percent as much work.

w. Wherever possible, power-operated fork trucks should be available to unload trailer trains and move and stack loads at work stations.

x. Where materials follow a fixed route repetitively, even for appreciable distances (up to 1,000 feet), conveyors should be considered. In some instances, longer distances can be effectively conveyorized.

y. Where large quantities of material are used constantly but purchased, handled, and stored in small containers, study the possibility of mass purchase and storage.

z. In planning the layout, provide for maintenance and service.

A Basis for Material Handling Analysis. The material handling equation, Figure 5-2, may be helpful in interrelating the many factors inherent in the analysis of a handling problem. It will be seen that the basic elements of material handling are materials, move, and methods. Each of these implies many questions, as indicated by the why, what, where, when, how, and who shown directly above the three basic elements. Beneath the three elements will be found ten major categories of factors which should be considered when analyzing material handling problems. One approach to these factors is shown in Table 5-1, where a few of the questions that should be asked and answered are inferred in the body of the table.

Actually, the first problem is the proper classification of the material to be handled, because its characteristics, along with those of the move to be made, will determine the method to be used. Table 5-2 is a checklist for determining the properties of the material to be handled, which is useful when determining and recording material characteristics.

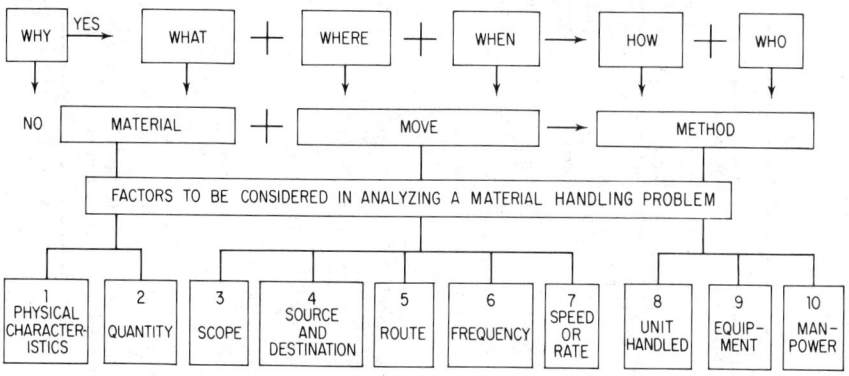

Fig. 5-2. Material handling equation.

TABLE 5-1. Factors of Material Handling*

Move what? Materials	In what? Containers	Where? Routing	How often? Frequency	How far? Distance	How fast? Speed	Upon what? Environs	By whom? Labor	With what? Equipment
11. High-pressure gas	21. High-pressure vessels	31. To storage	41. Uniform regularity	51. Horizontal	61. Uniform	71. Roadways—highways, trails, outside pavements	81. Without tools	91. Conveyors
12. Low-pressure gas	22. Low-pressure vessels	32. To preprocess	42. Intermittent or irregular	52. Vertical	62. Variable	72. Inside pavements—floors and aisles	82. With hand tools	92. Cranes, elevators, hoists, and winches
13. Unstable liquids	23. Tight containers	33. To process	43. Combination of regular and intermittent	53. Combination of horizontal and vertical	63. Synchronized	73. Levels—ramps, platforms, ceilings, etc.	83. Manually operated vehicles	93. Positioning and transferring equipment
14. Stable liquids	24. Entirely enclosed containers	34. Between production units	44.	54. Incline or slope	64.	74. Openings—doors, passages, column spacing	84. Manually operated unpowered equipment	94. Industrial vehicles
15. Semiliquids	25. Open-topped containers	35. To packing	45.	55.	65.	75. Off-highway terrain—hilly, flat, marshy, sandy	85. With walk-along powered equipment	95. Motor vehicles
16. Bulk solids	26. Platform supports	36. To warehouse	46.	56.	66.	76. Railroads	86. Riding operation of powered equipment	96. Railroad cars
17. Formed solids	27. Coil supports	37. To shipping	47.	57.	67.	77. Waterways—ocean, lake, river, canal	87. Remote control of powered equipment	97. Marine carriers
18. Living things	28. Securements	38. Ship	48.	58.	68.	78. Airways	88. Crew operation	98. Air transport
19. Other miscellaneous material	29. Other containers	39. Other routing	49.	59.	69.	79. Other environs	89. Other labor	99. Other equipment

* Courtesy of International Material Management Society.

TABLE 5-2.   A Checklist for Determining the Properties of Material to Be Handled*
Indication of degree: H = high, M = medium, L = low

A. Chemical properties
  1. Acidity..........................
  2. Alkalinity.......................
  3. Corrosiveness...................
  4. Solubility.......................
  5. Explosiveness...................
  6. Noxiousness or poisonousness.......
  7. Odoriferousness.................
  8. Light sensitivity...............
  9. Perishability...................
B. Physical properties
  1. Hardness—Rockwell scale........
  2. Density—pounds per cubic foot.....
  3. Specific gravity.................
  4. Compressibility.................
  5. Elasticity......................
  6. Ductility.......................
  7. Volatility......................
  8. Porosity.......................
  9. Permeability...................
C. Mechanical properties
  1. Abrasiveness...................
  2. Slipperiness....................
  3. Stickiness or tackiness............
  4. Ruggedness or toughness..........
  5. Fragility or brittleness............
  6. Viscosity—Baumé...............
  7. Free flowing....................
  8. Pressure—psi or inches mg........
  9. Moisture content—relative humidity
D. Electrical properties
  1. Conductivity...................
  2. Resistance.....................

  3. Magnetism......................
  4. Capacity.......................
  5. Radioactivity...................
  6. Affected by static...............
E. Thermal properties
  1. Heat conductivity...............
  2. Expansion—coefficient...........
  3. Specific heat...................
  4. Boiling point...................
  5. Melting point..................
  6. Latent heat (boiling).............
  7. Latent heat (melting)............
  8. Deterioration from exposure to heat
F. Size and shape
  1. Dust..........................
  2. Powder........................
  3. Granular.......................
  4. Lumpy.........................
  5. Mixed sizes....................
  6. Uniformity of particle size
  7. Nonuniformity of particle shape....
  8. Simplicity of shape or form........
  9. Complexity of manufactured form..
G. Dimensions and weight
  1. Length........................
  2. Width.........................
  3. Height.........................
  4. Diameter.......................
  5. Cross-sectioned area............
  6. Surface area...................
  7. Volume........................
  8. Weight........................
  9. Screen size....................

* Courtesy of International Material Management Society.

MATERIAL HANDLING PROBLEM ANALYSIS PROCEDURES

There are two different kinds of handling situations. One is the typical material handling problem in which the scope is relatively limited. The solution procedure is less rigorous than if the problem were of the second problem type—the so-called systems type. In the latter case, the problem is not so easily identified nor is the answer easily obtained. In fact, it may be a multiple problem, involving several interrelated handling activities. The first solution procedure can be referred to as the "problem analysis procedure," and the second and more complex as the "systems design procedure."

Problem Analysis Procedure. The problem analysis procedure consists of the following steps:

  1. Identify the problem.
  2. Determine the scope.
  3. Establish objectives.
  4. Define the problem.
  5. Determine what data are needed.
  6. Establish work plan and schedule.
  7. Collect data.
  8. Develop, weigh, and analyze data.
  9. Develop improvements.
  10. Prepare justification report and presentation.
  11. Obtain approvals.

12. Revise as necessary.
13. Work out procedures for installation.
14. Supervise the installation.
15. Follow up.

It is not always necessary to follow this procedure in a one, two, three fashion. Some problems will not require consideration of all the steps, while in other cases, it may be found necessary to perform certain steps out of sequence because of problem requirements or limitations. In any case, it is wise to check each step, to be sure that no important factor has been overlooked.

Identify the Problem. The most important step in analyzing a handling problem is to identify the problem. If the problem is not immediately apparent, it may be helpful to use a systematic approach such as the preliminary survey check sheet shown in Figure 5-3. In using the check sheet, each indicator observed in the problem area should be checked in the "yes" column. Each check mark should then be investigated until the situation observed can be properly identified as a handling problem. Each problem area so identified then becomes a prospect for a material handling improvement project.

Determine the Scope. The next step is to determine the complete scope of the problem to be sure that the problem is not solved out of context, but will be analyzed and solved within its total framework and in proper perspective to related activities.

Establish Objectives. Objectives should be clearly stated in such terms that it is possible to check on the degree to which they are achieved, when auditing the problem solution to see if the solution is solving the problem.

Define the Problem. Only after the problem has been properly identified and its complete scope established can the total problem be accurately defined. Also, redefining the problem at this point serves as a review of previous thinking and aids in establishing parameters within which the investigation should proceed.

Determine What Data Are Needed. The basic data form, shown in Figure 5-4, should be useful at this point, as it carries out the theme implied in the material handling equation. This form will not solve a problem, but it will serve as a guide to organizing data collection. It will also assure that no significant item of information is overlooked.

Establish Work Plan and Schedule. Depending on the scope and complexity of the problem, establishing the work plan and schedule may include the following:

1. Meeting with persons who will be concerned with the problem
2. Grouping similar or related materials, problems, or activity areas
3. Breaking down the overall problem into smaller segments
4. Preparing plans to study the problem
5. Establishing a detailed work schedule
6. Approving the work schedule
7. Assigning responsibilities

Collect Data. The actual collection of the necessary data may be accomplished as outlined below:

1. Review sources of data. This may involve checking with related departments or functions; talking with vendors, sales representatives, and others; or investigating books, periodicals, or commercial literature.
2. Establish project relationships with other functions and activities.
3. Carefully observe the activities being analyzed.
4. Carefully plan the problem situation.
5. Obtain complete data on the material, move, and method. Record it in a convenient form such as the basic data form.
6. Secure supplementary data such as schedules, layouts, flow patterns, building drawings, equipment details, and the like.
7. Obtain a layout of the area under study.
8. Obtain or plan data on the material flow, covering the move phase of the material handling equation. The following sources of information or analytical techniques may be helpful at this point.

PRELIMINARY SURVEY CHECK SHEET

Company_____ Plant_____

Building_____ Area_____

Compiled by_____ Date_____

Indicators of ineffective material handling (if checked in YES column, INVESTIGATE!)	Check		Comments, suggestions
	Yes	No	
GENERAL			
1. Crowded conditions..........................			
2. Unexpected delays..........................			
3. Empty floor space..........................			
4. Poor housekeeping..........................			
5. Excessive temporary storage..................			
6. Materials piled directly on floor..............			
7. Wasted cube..............................			
8. "Vanishing" aisles..........................			
MATERIAL			
1. Characteristics of materials cause handling problems			
2. Quantity justifies mechanical handling...........			
3. Too much/too little on hand..................			
4. Damaged materials..........................			
5. Excessive scrap............................			
MOVE			
1. Scope of move beyond area under investigation....			
2. Building characteristics restrict move...........			
3. Carrier characteristics restrict move............			
4. Move appears too long.......................			
5. Move not in direct path......................			
6. Flow pattern complicates material handling.......			
7. Crowded conditions..........................			
8. Backtracking..............................			
9. Lack of alternate paths......................			
10. Cross-traffic impedes flow....................			
11. Too much distance between operations...........			
12. Material delivered to wrong place on first move....			
13. Obstacles in material flow....................			
14. Poor location of service areas.................			
15. Nonuniform rate of flow......................			
16. Related work scattered.......................			
17. Traffic jams..............................			
18. Slow material movement......................			
METHOD OF HANDLING			
General			
1. Moving one item at a time....................			
2. Not using gravity...........................			
3. Excess manual handling......................			
4. Manual feeding to operations..................			
5. Manual removal of finished items..............			
6. Inadequate scrap removal.....................			
7. Excess storage at workplace...................			
8. Insufficient storage at workplace...............			
9. Poor flow between work areas.................			
10. Material piled on floor......................			
11. Stock control difficulties.....................			
12. Bottlenecks in production....................			
13. Scheduling difficulties.......................			
14. Low production output/sq ft..................			
15. Use of process layout when product layout possible.			
16. Improper location of feeder or subassembly lines...			

FIG. 5-3. Preliminary survey check sheet for a material handling analysis.

Indicators of ineffective material handling (if checked in YES column, INVESTIGATE!)	Check		Comments, suggestions
	Yes	No	
17. Inspection not integrated with production.........			
18. Hard, hazardous work done by hand.............			
19. Safety hazards...............................			
20. Unnecessary handling.........................			
21. Two-man lifting job..........................			
22. Rehandling.................................			
23. Item not delivered right to workplace............			
24. Unplanned material handling methods...........			
25. Frequent, short, repetitive moves by hand........			
26. Makeshift methods............................			
27. High load/unload time........................			
28. Difficult handling............................			
29. No alternative method........................			
30. Overmechanized handling......................			
Unit Handled			
1. Items not moved in unit loads..................			
2. Unit received not utilized in subsequent moves.....			
3. Unit received inefficient for handling.............			
4. Unit received too large or too small..............			
Equipment Used			
1. Idle equipment.............................			
2. Excessive equipment repairs....................			
3. Nonstandard equipment.......................			
4. Equipment not compatible with system..........			
5. Overloaded equipment........................			
6. Underloaded equipment.......................			
7. Equipment obsolete..........................			
8. Unsafe equipment...........................			
9. Fixed speed equipment.......................			
10. Shortage of equipment—no spares...............			
11. Inflexible equipment..........................			
12. Mobile equipment not moving..................			
13. Operating over or under rated speed.............			
14. Inadequate maintenance or repairs..............			
Containers			
15. No container used in move.....................			
16. Frequent change of containers..................			
17. Noncollapsible containers......................			
18. Heavy container.............................			
19. Nonstandardized containers....................			
20. Excess number of containers...................			
21. Shortage of containers........................			
22. Costly containers............................			
23. Container not suitable for mechanized handling....			
Utilization of Manpower			
1. Excessive injuries...........................			
2. Frequent complaints..........................			
3. Large number of men doing handling............			
4. Handling done by direct labor.................			
5. Men walking for material.....................			
6. Operators waiting for material.................			
7. Heavy physical exertion......................			
Costs			
1. High overhead costs..........................			
2. High indirect labor cost......................			
3. Unexplainable cost increases...................			
4. High unit handling costs......................			

Fɪɢ. 5-3 (*continued*). Preliminary survey check sheet for a material handling analysis.

BASIC DATA REQUIRED

IDENTIFICATION

Company_____Plant_____Compiled by_____

Building_____Location in building_____Date_____

Statement of problem_____

_____

(Should be accompanied by process chart or flow diagram)

(Check and fill in as applicable)	Remarks, explanation, etc.
MATERIAL—Part no(s)._____	

Description_____
1. Physical characteristics
   Form: solid_____liquid_____gas_____other_____
   Type of material, load:
      individual item_____unit load_____
      packaged_____bulk_____
   How received_____
                    (carton, pallet, drum, bag, or the like)
   Nature: fragile_____sturdy_____bulky_____
   Properties: dimensions_____
      shape_____ wt./item_____
      density_____
2. Quantity
   Annual quant._____
   Quant./deliver_____
   Lot size_____
   Max. inventory_____

MOVE
3. Scope (check each phase applicable)
   Packaging by vendor_____Interdept. handl._____
   Packing by vendor_____Service & auxil. oper.____
   Loading by vendor_____Qual. contr. activ._____
   Common carrier_____Packaging_____
   External transp._____Packing_____
   Unloading _____Fin. gds. whsg._____
   Receiving_____Stock picking_____
   Materials storage_____Order assembly_____
   Materials issue_____Loading oper._____
   Production activ._____Shipping oper._____
   Intra-dept. handl._____Carrier-from plant_____
   Workplace handl._____Intra-plt. hdlg._____
   In-process storage_____
4. Source and Destination
   Source   Destination
      Vendor_____   Receiving_____
      Carrier_____   Storage_____
      Storage area_____   Point-of-use_____
      Work station_____   Work station_____
      Fixed_____   Fixed_____
      Variable_____   Variable_____
      Other ( )_____   Other ( )_____
   Building characteristics
   Source   Destination

   _____ Aisle width   _____
   _____ Column height   _____
   _____ Truss height   _____
   _____ Truss capacity   _____
   _____ Floor construc.   _____
   _____ Floor load capacity _____
   _____ Floor condition   _____

FIG. 5-4. Basic data form for a material handling analysis.

11-127

	Power avail./req'd.	
_____	Openings (no., size)	_____
_____	Dock	_____
_____	Duct work	_____
_____	Sprinklers	_____
_____	Elevators	_____
_____	Other	_____

Common carrier characteristics
   Type_____Volume_____
   Dimensions_____Load capacity_____
   Door size_____
Storage Area and Volume
   _____L by_____W by_____H

5. Route
   Distance_____
   Area covered_____
      Size_____Fixed_____Variable_____
   Path
      Straight_____Curved_____
      Fixed_____Obstacles_____
      Variable_____Combination_____
   Course
      Fixed pt./fix. pt._____Var. pt./fix. pt._____
      Fixed pt./var. pt._____Two-way_____
      Var. pt./var. pt._____Other_____
   Direction/plane
      Horizontal_____Multilevel_____
      Vertical_____Combination_____
      Incline_____Ramp____L by_____W
      Single level_____Ramp_____% grade
   Level
      On floor_____Overhead_____
      Working ht._____Level/level_____
   Location
      Inside_____Between bldgs._____
      Outside_____Beyond bldgs._____
   Operations in Transit_____
   Cross Traffic_____

6. Frequency
   Regular_____Continuous_____
   Irregular_____Intermittent_____
   Unpredictable_____Reciprocating_____

7. Speed or Rate
   Speed
      Uniform_____Synchronized_____
      Variable_____Other_____
   Rate:_____items/_____(time period)
      Fixed_____Pounds/hr_____
      Variable_____Ft/min_____

8. Load handled: uniform_____, variable_____

Type	Alternative 1	Alternative 2	Alternative 3
Size			
Construction			
Tare weight			
Items/load (contr.)			
Weight of load			
Loads/tot. quant.			
How carried			
Disposal			
Cost			
Prorated cost/load			
	$_____	$_____	$_____

Fig. 5-4 (*continued*). Basic data form for a material handling analysis.

BASIC DATA REQUIRED (*continued*)

9. Equipment
   Desired characteristics
   Powered_____   Self-loading_____
   Operator req'd._____   Tiering/stacking_____
   Mobile_____   Elevate/lower_____
   Control_____   Positioning_____
     Manual_____   Transferring_____
     Automatic_____   _____
     Remote_____   _____
   Equipment indicated
   None_____   Combination_____
   Manual handl._____   Common carrier_____
   Conveyor_____   Rack_____
   Crane/hoist_____   Other_____
   Truck_____   (see no. 8 for containers)
   Cost/hour $_____
   Time/move_____equip._____man
   % Capacity req'd. for this move:
   Wt. capacity_____%   Time capacity_____%
10. Manpower (load, move, unload)
    Time/load_____Hourly rate $_____
    Cost/load $_____

Fig. 5-4 (*continued*). Basic data form for a material handling analysis.

    a. Bills of material
    b. Assembly chart
    c. Production routings
    d. Operation process chart
    e. Multiproduct process chart
    f. Process chart
    g. Flow process chart
    h. Flow diagram
    i. From-to chart
    j. Activity relationship chart
    k. Activity relationship diagram
    l. String diagram
    m. Memomotion pictures
    Detailed information on the above will be found in plant layout and motion study textbooks or in Sections 2, 3, and 11 of this Handbook.

   9. Obtain information on each item of existing material handling equipment.
  10. Procure data on manpower requirements.
  11. Tabulate all information pertaining to storage aspects of the problem.
  12. Investigate and analyze communication and control aspects of the problem.

Develop, Weigh, and Analyze Data. The next step is to study, weigh, and analyze the facts, figures, drawings, and other information on hand. Some helpful suggestions are:

   1. Sort and classify all information into major aspects of the problem—material, move, and method.
   2. Check all information for accuracy.
   3. Summarize data on each phase.
   4. Determine practical ranges of data.
   5. Average, weigh, or otherwise treat data to put them in the most usable form.
   6. Develop charts, graphs, and the like.
   7. Check for inconsistencies, omissions, errors, and irrelevant data.
   8. Summarize in appropriate form.

Develop Improvements. The solution to many material handling problems involves a comparison of the facts and data accumulated with the material handling background and knowledge of the analyst. The problem becomes one of carefully comparing the summarized data which represent material and move characteristics with the characteristics or capabilities of possible alternative solutions. A detailed discussion of the improvement process is beyond the scope of this chapter, but the following list of suggestions may be helpful:

1. Redefine the problem.
2. Investigate, evaluate, and summarize the effect of anticipated changes in sales volume, capacity, product design, processing, tooling, and the like.
3. Select possible solutions.
4. Reexamine related activities.
5. Make detailed plans for the methods to be followed.
6. Select equipment.
7. Check the proposed methods against the principles of material handling and the preliminary survey check sheet.
8. Determine what construction work will be required by the proposed changes.
9. Design required communication and control procedures and techniques.
10. Review the entire project, checking it against original goals and objectives.
11. Develop operating procedures.

Prepare Justification Report and Presentation. Next, it is necessary to prepare a justification of the plan for management approval as described below:

1. Compare the costs of the present and the proposed methods.
2. Determine the rate of return on the investment.
3. Review any intangible gains.
4. Review the proposed solution in terms of the original problem definition and objectives.
5. Present the justification.

The above completes the formal work involved in solving the problem. Company policy will usually determine whether the problem solution should be presented in the form of a report, an oral presentation, or a combination of the two. The content of the report will vary with the individual problem as well as with company practice. Plans should be made for oral presentation of the report, using appropriate visual aids.

Obtain Approvals. Approval is not only necessary but is usually advisable to create an awareness of the problem among higher levels of management. In fact, acceptance and successful operation of the proposed plan may hinge on the fact that the management personnel affected by the proposed plans were given an opportunity to be a part of the decision making group.

Revise as Necessary. Immediate approval of the proposal may be withheld pending a further investigation. If changes become necessary, they too should be approved and then incorporated into the solution.

Work Out Procedures for Installation. Plans for the installation of the problem solution may involve:

1. Obtaining firm quotations and delivery dates on items to be purchased
2. Tabulating information on quotations
3. Selecting vendors and issuing purchase orders (or initiating fabrication of components)
4. Establishing a time schedule for the installation, usually in the form of a CPM network, a Gantt chart, or both

Other activities may be necessary, depending on the nature and complexity of the project.

Supervise the Installation. Probably no other step in the entire procedure is more important than effective supervision of the installation of the problem solution. The material handling engineer, having spent many hours working out exactly how the new method will benefit the company, should be sure that the installation is made in accordance with his plans.

Follow Up. The material handling engineer should carefully observe the new installation to see that everything works as planned and everyone does what he is supposed to do. If not, he should take the necessary corrective action.

Although the above procedure may appear to be a rather long and drawn out way to solve a material handling problem, it will ensure that a complete analysis will be made.

MATERIAL HANDLING SYSTEMS DESIGN PROCEDURE

The relatively straightforward procedure just outlined will not adequately cover the complex interrelationships of the many overlapping aspects of a multiple problem situation. In this type of complex problem, the systems approach can frequently lead to a more thorough, integrated solution to the problem.

The Systems Concept. In comparison with the material handling problem analysis procedure, the systems approach is a larger-scale, results-oriented point of view which is broader in scope, greater in depth, and interdisciplinary in approach.

One of the distinguishing characteristics of the systems concept[1] is that it consciously exploits gaps between traditional categories of knowledge and cuts across the commonly accepted boundaries of functions and departments. This is done in an attempt to discover and sort out the factors involved in a complex situation and find relevant and reliable interrelationships among them. This more detailed investigation and analysis is necessary if it is desired to conceptualize, design, and evaluate complex combinations of sophisticated hardware, organizational interrelationships, and information flow.

A material handling system, then, is a carefully and thoroughly researched solution to a material handling problem situation. It usually results in an integrated composite of facilities, activities, and information flow encompassing as much of the total problem scope and environment as is feasible and economical.

The overall objective of the systems approach is to conceptualize a total solution to the overall problem and then implement those portions which are technologically possible and economically feasible.

Because the systems approach involves a more sophisticated analysis of a situation and a more thorough and rigorous investigation, the solution should be of a higher caliber than if the problem was approached from the more limited point of view.

SYSTEMS ANALYSIS PROCEDURE

The systems analysis procedure follows the same general pattern as the problem analysis procedure, with the first eight steps just as applicable to the more complex situation, but with more care required in carrying them out. However, in step 9—develop improvements—the systems procedure calls for a considerably more detailed treatment. This more extended treatment is developed in three major areas: systems synthesis, systems design, and systems implementation. Figure 5-5 will aid in visualizing some of the differences between a limited material handling problem and a systems-type problem situation.

Systems Synthesis. In synthesizing a system, the analyst should follow the thinking pattern outlined below:

Conceptualize System Possibilities. One of the distinguishing features of the systems approach is that it requires the analyst to stretch his imagination far beyond the boundaries within which he is used to working. It also implies an obligation to conceptualize a larger number of alternative possibilities for the solution. This is frequently done in the form of a flow chart, as shown in Figure 5-6.

The next step may involve conceptualizing subsystems for implementing individual phases of the project, which are then integrated into the total system.

Structure Alternative Systems. At this point, the systems approach becomes more pronounced as it may involve the simulation of the proposed alternatives with

[1] Partially adapted from Van Court Hare, Jr., *Systems Analysis: A Diagnostic Approach,* Harcourt, Brace & World, Inc., New York, 1967.

PROBLEM DEFINITION	1. Identify problem 2. Determine scope 3. Define problem 4. Establish objectives
PROBLEM INVESTIGATION	5. Determine what data are needed 6. Establish work plan and schedule 7. Collect data 8. Develop, weigh, and analyze data

LIMITED PROBLEM (Individual)		COMPLEX PROBLEM (System)
PROBLEM SOLUTION		
9. Develop improvements	*SYSTEMS SYNTHESIS*	9. Conceptualize system possibilities 10. Structure alternative systems 11. Select feasible system 12. Prepare preliminary justification
	SYSTEMS DESIGN	13. Define proposed system 14. Establish functional specifications 15. Develop preliminary budget for implementation 16. Develop and design system components 17. Evaluate development progress
10. Prepare justification report and presentation		18. Prepare justification report and presentation
11. Obtain approvals		19. Obtain approvals
12. Revise as necessary		20. Revise as necessary
INSTALLATION		
13. Work out procedure for implementation	*SYSTEMS IMPLEMENTATION*	21. Organize for procurement 22. Procure equipment 23. Plan for training of operators and managers 24. Prepare for human and public relations considerations
14. Supervise installation		25. Supervise installation of equipment
15. Follow up		26. Start up and debug 27. Audit and follow up

FIG. 5-5. Approaches to limited and complex material handling problems.

the aid of a computer. Developing the type of flow chart shown in Figure 5-6 is the most common preliminary step in structuring a problem solution for computer simulation. If simulation is found desirable, the procedure is somewhat as follows:

1. Sketch a preliminary flow chart of the entire system.
2. For each step shown on the flow chart, determine parameters, characteristics, factors, and the like required to reach system objectives.
3. Establish constraints to be complied with in reaching the objectives.
4. Develop a block diagram of the logic, such as that shown in Figure 5-7.
5. Convert model to computer language. The block diagram shown in Figure 5-7, frequently referred to as a model, must be converted to a computer language. Figure 5-8 represents the conversion of several steps from the preceding block diagram into the GPSS language.
6. Punch computer cards to represent each of the steps in computer language.

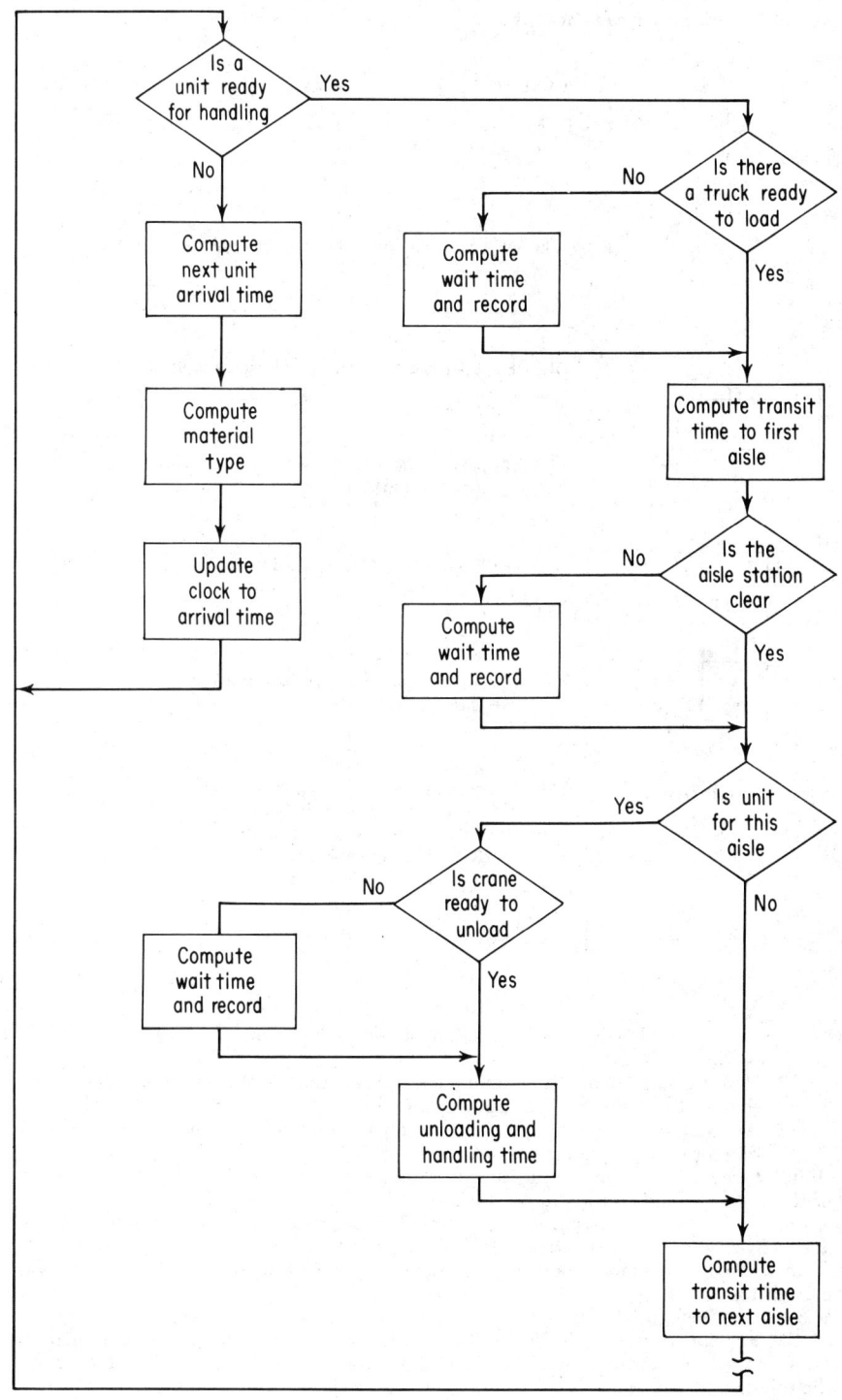

FIG. 5-6. Flow chart of alternative possibilities.

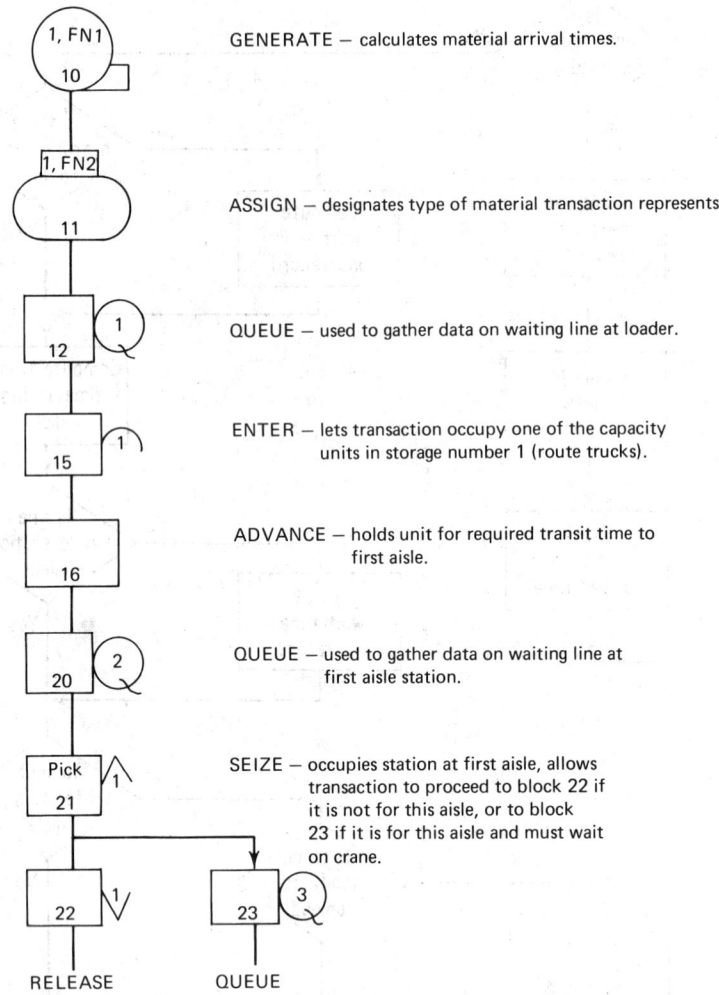

Fig. 5-7.   Portion of block diagram.

7. Run the model on the computer to check the parameters, constraints, and the results of the data manipulation and to test out the proposed system.

8. Analyze the computer runs and compare simulations using several sets of data or of alternative systems or subsystems.

Select Feasible System.   Next, it will be necessary to make a preliminary selection of the system which appears to be feasible at this point in the analysis.

Prepare Preliminary Justification.   Work out a preliminary cost estimate for the alternatives under consideration, remembering that an evaluation of the intangible factors should be made, for some of them may be of more importance than the cost factors.

Systems Design.   Reference to Figure 5-5 will show that it is the design segment of the systems procedure which differs most from that required for the analysis of a limited problem. Therefore, step 9 of the limited problem is replaced by the following five steps:

Define Proposed System.   Although the problem definition phase of the analysis

PROGRAMMER: M. P. O.

PROGRAM TITLE: M. H. SIMULATION

PROJECT NO: 1 2 3 4

VERIFY ____

DATE: ____

PAGE ____

LOCATION	NAME (LEFT JUSTIFIED)	X	Y	Z	SELECTION MODE	NEXT BLOCK A	NEXT BLOCK B	MEAN TIME	MODIFIER
1	JOB								
1	FUNCTION	RN1							
.20		1	.60						
10	GENERATE	2	90	3	1.0	4			FN1
11	ASSIGN	1			1.0				
2	FUNCTION	RN1	FN2						
.03		1	.17	5	.19	4		5	6
.44		8	.64	9	.82	10	.26		.35
2	FUNCTION	RN1	D3			15	1.0		
12	QUEUE	1				16			
15	ENTER					20	23		
16	ADVANCE	2	0	100		22			
20	QUEUE				PICK	30			
21	QTABLE					31			
24	SEIZE	3							
22	RELEASE								
23	QUEUE								

FIG. 5-8. GPSS worksheet.

deals with the definition of the problem, it is now necessary to define the system. This definition may consist of a written description, a detailed step-by-step outline, or a flow chart or block diagram similar to those used in structuring the alternative systems during the synthesis phase of the problem analysis.

Establish Functional Specifications. Next, it is necessary to detail the functional specifications to be met by the system as a whole, as well as for each individual segment of the system. Specifications may be in such terms as physical dimensions, overall configuration, output per hour, and costs per unit.

Develop Preliminary Budget for Implementation. Quotations should be requested from selected suppliers, with their full knowledge that figures are tentative and are to be used only as a basis for preliminary approval. Quotations should be evaluated and the conclusions used as a basis for preparing a preliminary budget to be submitted to appropriate persons for review and necessary approvals prior to the actual design process.

Develop and Design System Components. In this step, the actual design process takes place. The hardware aspects of the system are developed in the form of blueprints; the control system is designed and specified in an appropriate manner; and the information system is developed and reduced to a form capable of implementation.

Evaluate Development Progress. Accomplishments to date should be checked or measured against the original system objectives as well as against the development schedule, if one was established.

Prepare Justification Report and Presentation.

Obtain Approvals.

Revise as Necessary.

Note: The activities necessary for carrying out these three steps are not significantly different from those in the procedure for analyzing the limited problem, as described on page 11-130.

Systems Implementation. The complex problem analysis procedure requires a more detailed consideration of the implementation phase than does the limited problem situation. The following seven steps are required:

Organize for Procurement. Organizing for procurement involves making detailed plans for the investigation of each major segment of the system, arranging contacts with those involved in the procurement process—both inside and outside the buying organization—and properly scheduling all procurement efforts to assure coordination.

An important aspect of the procurement problem is that of establishing the responsibilities of the user, vendor, consultant, and architect. They should be spelled out in detail so that they can be adequately followed up during the implementation process.

Procure Equipment. Procuring equipment involves all the activities necessary in step 13 of the problem analysis procedure, Work Out Procedures for Installation, on page 11-130. It may be wise in some cases to include, as part of the purchase order, a note regarding vendor follow-up at predetermined intervals, or to require proper operation for a specified period of time—without breakdowns—before payment is made.

Plan for Training of Operators and Managers. The proper operation of a complex system requires orientation and training in its use. It will be necessary to establish training programs for all potential operators of the equipment as well as those who will be managing the installation.

Prepare for Human and Public Relations Considerations. In addition to the training activities, it is frequently advisable to prepare for human relations problems which might arise due to the introduction of the new equipment, as well as the possible public relations aspects of the installation.

Supervise Installation of Equipment.[2] The first step often required in planning

---

[2] Adapted from the outline of a presentation, "The Systems Approach" by Allan Harvey at the Materials Handling Management Course, Lake Placid, N.Y., June, 1967.

for the installation of the equipment is the construction of a CPM network to ensure that equipment is installed in the right sequence and on a predetermined schedule. If the building is new, it will be necessary to plan for the site to be ready to receive the new equipment. On the other hand, if the equipment is to be installed in an existing building, it will be necessary to develop plans so that current operations can continue during the installation of the new equipment as much as is practicable.

Start Up and Debug. No system is immune to the unpredictable difficulties which are bound to develop during the start-up process. For this reason, it is necessary to establish a procedure for handling them. It will be helpful to log the types of problems encountered and record the downtime caused by each. After the installation has been operating for a sufficient period and management is satisfied with the performance, it is appropriate to accept the completed installation officially.

Audit and Follow Up. The final step involves an audit of the entire operation to ensure that the system is operating properly. In preparation for audit, it will be found desirable to pay close attention to:

1. Establishing responsibilities for specific segments of the system
2. Assigning personnel to each
3. Equipment warranties
4. Unforeseeable events which could not have been known until the system was in operation
5. Operating or other conditions which may change after the installation
6. The responsibilities of all individuals to assure that they adhere to the responsibilities assigned when drawing up the purchase order, where responsibilities should have been carefully defined
7. Evaluation of accomplishment

It should be obvious that there is a considerable difference between the analysis of an individual problem of relatively limited scope and a complex handling situation requiring the development of a system. The above discussion has pointed out those phases of the analysis and design which are different for the complex problem.

PACKAGING*

Packaging serves the purpose of protecting goods, making them attractive to the consumer, and making them readily movable. Packaging is used in a more restricted sense than packing, which covers all the materials and procedures used in preparing goods for handling, shipping, storing, and marketing.

Decorative Packaging. When a package is required to furnish sales appeal, it can be classed as decorative. The decorative package does not overly concern the industrial engineer making material handling studies because the difference involved is often only in the provision made for that sales appeal. This is usually the sales promotion or advertising man's responsibility.

Protective Packing. The principal purpose of a protective package or container is to provide protection, control the quantity or size of the unit, and provide a suitable means of containing the material during movement or storage. The protective package may be expendable or it may be a reusable item.

PACKAGING FUNCTIONS

Several years ago, a well-known packaging engineer, W. B. Lincoln, Jr., mentioned fifteen functions of packaging for shipped goods by containers or other packages. These make an excellent checklist for a designer. They are listed below:

1. Enclosure. The degree to which this principle should be applied, as in the case of all the others, varies considerably. The enclosure may be required

* Pages 11-137 to 11-166 have been adapted from the chapter by Randolph W. Mallick and Walter R. Turkes, "Material Handling," in the second edition of the *Industrial Engineering Handbook.*

as protection against dust, dirt, moisture, water, heat, cold, insects or rodents, pilferage, or foreign bodies. Sometimes the enclosure material can be used also to strengthen the basic framework of the container.

2. Compatibility. Container and contents can be of such natures that one has a deleterious effect on the other. If foodstuffs or drugs are being packaged, for example, the container must be sterile. There must be no odor in packaging material used on butter. Certain soaps must be enclosed in moldproof boxes. Generally, this problem does not arise, but when it does, it is important that the principle be applied.

3. Retention. Obviously, the container must not permit the contents to fall out. It would be unusual to have a new container which did not retain its contents, but it is not unusual to find containers broken, cut, or torn after they have been subjected to the hazards of handling, storing, and shipping. Unless containers in actual use or in laboratory tests simulating real hazards have been observed, it is difficult to appreciate the disruptive forces which occur in transportation. Failure to retain is a frequent fault in containers.

4. Restraint. Movement within the container should be kept to a minimum. A small amount of shifting grows to be too much during shipment.

5. Separation. When more than one article is packed within a container, the units should be separated if they are fragile or have a damageable finish. The possible condition of the separator at the end of the journey should be kept in mind when it is being selected.

6. Cushion. All fragile articles must be cushioned against shock and vibration. Sometimes the breakable part is inside the object shipped, as in a television set or an electronic testing instrument. The cushion serves to bring the moving article to rest over a sufficient distance to reduce the forces of impact below its breaking strength. It is important, therefore, to have the right amount of "give" in the cushion. Here is an area where real engineering is required for precise results. Without such knowledge, experience and testing must be relied upon.

7. Clearance. Articles inside a crate or other container will move in relation to the inside of the wall of the container when the package is distorted by the stresses of handling, storing, and shipping. To prevent contact, sufficient space between the article and the container must be allowed. Generally, this principle applies to large items such as furniture, refrigerators, and machines.

8. Support. The container must be strong enough to prevent normal outside loads from affecting the contents. Sometimes parts of the article shipped must be supported so that they will not bend or break off during shipment.

9. Position. Usually, an article will ride best when standing on its base or bottom, but not all items have bases, and in such cases, the best riding position must be found.

10. Nonabrasion. Highly finished surfaces must be protected against damage by abrasion, imprinting, indentation, and so forth. Smooth materials, cushions, and air spaces are used for this purpose.

11. Distribution. The weight of the contents must be distributed over a bearing area large enough to prevent damage to the contents and to the container.

12. Suspension. Many irregularly shaped articles or those with highly finished surfaces can be packaged best by securely fastening one surface to the container or a suitable frame within the container and keeping the other container surfaces away from the article.

13. Visible Container. Occasionally, an extremely fragile item can be packed most effectively in an open crate so that the handlers can observe its fragile nature.

14. Closure. The closure of the package must be made so that it will be effective until it reaches the end of the journey. In certain cases, reinforcement, such as provided by tape or woven or metal strapping, is desirable for increased strength at relatively low cost.

15. Instructions. Packages must be marked at least to indicate destination, but other instructions often should be used. For example, acids and explosives must be labeled. Indications of proper positioning of slings on an export container could be essential. "Fragile" and "This Side Up" markings often are ignored, but there is no doubt that they are better than no markings.

Receiving Containers. The engineer responsible for material handling should check the following points to determine if there are trouble spots in receiving containers:

1. Are the materials received in good condition with a minimum of defects or losses?
2. Are the materials received in such a quantity and type of container that they can be handled, stored, and used with minimum rehandling or breaking down of units?
3. Are materials overpacked; that is, can less packing material be used, at a cost saving, with no sacrifice to quality?
4. Can packing materials be salvaged and reused or returned for credit?

If possible, shipments being received should be packaged in such a way that they can be used in the first manufacturing operation without requiring extra operations. In the case of long-term contracts, it is sometimes possible to use returnable containers. This, of course, is by agreement with the supplier and can usually be effected at considerable cost savings.

Containers and Supports for Storage or Movement between Workplaces. When material is not unitized, that is, contained as a movable unit, it becomes relatively immobilized and usually requires manual handling to move it. The use of skids and pallets has paid gratifying returns in industry and will continue to play an important role with the growth of automation.

Skids should be used when loads must be transported by a lift process and when they cannot or should not be tiered. Pallets are for use under loads that can be transported by fork truck and tiered, although the tiering condition is not absolutely necessary for palletizing to be effective. Palletized loads can be placed on trucks, trailers, or conveyors to reduce manual handling.

Pallets are versatile supports. They can be designed with corner posts so they can be stacked without straining the palletized material. They can be made with portable partitions so materials, for instance tote pans, can be added to and removed from the pallet without disturbing the entire load. Or they can be built with detachable side frames to convert them to pallet boxes.

In addition to skids and pallets, a small tote pan, say 6 by 12 by 18 inches, is widely used for handling small parts that cannot be easily damaged. When parts are fragile or of such design that entanglement might result from random piling in tote pans, specially designed trays or fixtures are used. Such special handling devices—in addition to protecting the part or material—also often increase the efficiency of the subsequent operation by presenting the part or material properly pre-positioned.

Shipping or End-use Containers. Material going to some distant point often requires protection against rust, corrosion, breakage, moisture, and so forth. As a rule of thumb, minimum amounts of packaging should be used to provide this protection. Packaging, a close relative to material handling, is a science in itself, and some study may be needed to determine just what a minimum of packaging means. There are, however, some general points that apply to most questions:

1. Use containers of standard size and construction where possible.
2. To reduce the number of container sizes needed to accommodate a wide range of products, use pads, fillers, and inserts in standard boxes.
3. Before establishing packing specifications, determine all the conditions to be met. Often a less costly package will do a satisfactory job. The packaging industry is constantly developing new containers and new packaging materials, and the engineer should determine the best one for each specific application.
4. Utilize the largest size of container or unit load possible. It is usually less costly to put a larger number of units in one container than to package each one separately.
5. Investigate the applicability of other types of packing material.

6. Remember that packing and shipping costs alone average from 5 to 10 percent of product costs. This is a cost which must be added to a product already completed in the form that the customer desires, be it a feeder part or the final apparatus.

RECEIVING AND SHIPPING AREAS

Receiving has an effect on the subsequent operations, and as the first operation, frequently sets the pace and continuity of flow. Shipping, on the other hand, has less direct effect on other operations but has a decided effect on profits and customer relations. Both receiving and shipping involve material handling to a high degree.

Effect of Layout. As with other material handling operations, layout is extremely important in a receiving and shipping department. Receiving and shipping departments should be built around the layout and not conversely. When a receiving or shipping department is being planned, the engineer responsible should insist on these points:

1. The packing area should be located adjacent to the shipping department so that a minimum of material movement is involved. In repetitive operations, mechanical or automatic equipment can often be used to eliminate handling altogether.
2. Dock heights should be built to accommodate the average railroad car and the average road truck. Where a choice is involved, railroad car heights should be specified and leveling docks or other raising mechanisms provided for road trucks. Hydraulically operated ramps can often be employed to advantage.
3. If the material being handled justifies it, mechanical handling or automated means should be provided.
4. If materials are received in more than one type of transportation vehicle, such as in both rail cars and road trucks, the receiving dock should be laid out to avoid conflict between the different types and yet provide flexibility. This aspect of the layout is often vital, because a variety of vehicles will be involved.
5. The receiving area should be located so that the movement of materials to the first operation can be made most efficiently. If necessary, several receiving points can be set up in a plant to satisfy this condition.

In an attempt to obtain an efficient shipping or receiving layout, the engineer responsible should be prepared to give information on how much and what kind of material will be received daily; where it will come from and how it will arrive; how the material will be disbursed to the plant; at what hours traffic is heaviest; what rail connections will be required; how the product will be shipped and in what quantity; the packing facilities required; what temporary and permanent storage areas will be required around the docks; what crating materials will be required; and other similar data.

Integrating Receiving and Shipping. A shipping and receiving department should not be considered an added operation. Rather, it should be integrated into the processing function. In general, receiving operations should be integrated with suppliers' methods, raw material inspection, carriers, packaging, and processing. To do this, receiving should be approached as the first stage of the processing function, and shipping as the last stage.

STORAGE

Although not strictly material handling, storage is such a closely allied function that it should be considered here.

In general, storage can be classed as one of two types:

1. Temporary—considered to be the temporary holding of material in any location due to inability to utilize it immediately. The duration is usually short.

2. Permanent—considered to be a planned storage of the material, usually in a predetermined location, for the purpose of controlling the supply or to hold materials not required for immediate use. The time interval of storage is usually long.

Temporary Storage. In-process storage is the best and most prevalent example of temporary storage; characteristically, it occurs during the flow of the material through the processes. Temporary storage is the more expensive of the two types of storage and is a large contributor to hidden production cost. The extent of temporary storage can easily be determined by examining a flow process chart.

Because reduction of temporary storage depends very largely on the process involved, there is no procedure that will apply in every case. However, there are a few rules which are generally applicable.

1. *Production Scheduling.* Production schedules should match material flow to avoid temporary storages. Material must reach the right place at the right time to eliminate in-process storage.

2. *Facilities Layout.* Layout should be such that smooth flow to all sections is possible. Work stations should be so arranged that material can be brought in and removed without disturbing other operations.

3. *Inventory Control.* Chapter 4 of Section 8 of this Handbook gives the details of inventory control. Essentially, it involves having the correct amount of materials on hand to permit productive loading.

4. *Work Planning.* Daily work should be planned so that every employee and work station can function without undue interruptions from either too few or too many materials in and around the workplace.

5. *Housekeeping.* Good housekeeping is dependent on proper attitude. This attitude is best established by management through urging by safety supervisors. Housekeeping is sometimes a key element in keeping temporary storage at a minimum.

6. *Supervision and Training.* Indifference on the part of the supervisor toward in-process or temporary storage will result in general indifference in employee attitude. Where necessary, key people should be trained to recognize and take steps to eliminate unnecessary temporary storage.

If it becomes necessary to store materials temporarily, the cost of doing so can be minimized if one or more of the following practices can be employed:

1. Keep materials off the floor and ready for mobilization. Store them on pallets, skids, skid boxes, trays, or pans, where possible. Store them in such a location as to require a minimum of movement to the next work station or process. Stack materials where possible to conserve space.

2. Contain or unitize materials wherever possible.

3. Record the location of all temporarily stored material to make a minimum of expediting and searching. See that materials are adequately protected to prevent deterioration, corrosion, loss due to pilferage, or breakage.

4. Store material in such a location as to be easily accessible to mechanized equipment.

Permanent Storage. Permanent storage is a necessity because of the need for reserve supplies. This does not mean, however, that permanent storage should be a burden. It should be treated as a technical problem and consequently deserves a technical, planned approach. Although there may be other avenues of approach, a form is a sure method for coordinating all the information, and in certain instances, can suggest immediate solutions. Such a form is shown by Figure 5-9.

If all the information required for the completion of this form is filled in, it will undoubtedly help solve the storage problem. In specific cases, it may be that additional information will be required, but this can be easily added to the usual form.

Storage Practices. In providing for the storage of materials, several conditions should be explored and considered.

1. Can the storage take place either indoors or outdoors?

2. Is the material in bulk or packaged form, and can it be stored suitably in its present form?

MATERIAL STORAGE SURVEY

Description of material _____

Condition: wet__, hot__, fragile__, sticky__, fluid__, explosive__, packed__, loose__,
flammable__, rough__, finished__, perishable__, other _____

Affected by: heat__, cold__, dampness__, dust__, dirt__, fumes__, other _____

Size of piece: min. _____ max. _____   Wt. of piece: min. _____ max. _____

Drawing no. _____ Item no. _____ Style no. _____

Handled and stored, individually: tote pan ___ size _____; carton ___ size _____;
skid ___ size _____; pallet ___ size _____; box ___ size _____;
can or drum ___ size _____; other _____ size _____

Quantity in container _____ Avg. wt. of load _____

Material received from _____ via _____ stored in (place) _____
by (method) _____

Quantity normally on hand _____ Frequency of receipts _____
_____ Frequency of issues _____

Method of disbursing _____

Disbursed to (location) _____

Can material be stacked? _____   How high? _____

Floor type: concrete __, wood __, asphalt __, woodblock __, brick __, dirt __, other __

Floor condition: good __, bad __, smooth __, rough __, wet __, dry __, inclined __

Are cranes available? _____ Kind _____ Capacity _____

Elevators: Kind _____ Capacity _____ Platform size _____

Can trucks be unloaded at floor level? _____

Allowable load on floor _____ on trusses _____

Can racks, bins, shelving be used? _____

Is there any unusual hazard involved? _____
Is there any unusual quality control problem? _____
Is there any unusual production or accounting control involved? _____

Remarks: _____
_____

Compiled by _____ Date _____

Fig. 5-9. A material storage survey form.

3. What degree of protection is necessary?
4. What quantities will be handled into and out of storage?
5. What is the frequency of turnover?
6. What storage and handling facilities are necessary for economical utilization of space and labor?

If floor space indoors is limited, it is often advisable to consider outdoor storage. Leased storage space at a removed location is not only expensive but also often inconvenient and an impediment to good inventory control. Although the nature and the quantity of the material will determine if the material can be stored outdoors, far more material can stand outdoor storage than is normally realized. Often, the only added protective measure that must be taken is against damage by weather.

If the material is in bulk form, a study should be made to determine the most advantageous method of storing it. In some cases, bulk materials can be made up into unit loads and stored outdoors, thus providing ready means for rehandling.

The type of storage method selected will also be dependent upon the turnover of the material in storage and the frequency of receipt and issue. If movements are frequent, the location and type of facilities are important. If quantities are large, mechanical handling and unit loads must be considered. Also, turnover will partially determine the type of material handling equipment required.

When all the information about the material to be stored has been collected, the final step is the determination of the storage location—indoor or outdoor in geographic relation to the point of receipt and disbursement of the material. The selection of a storage method and the facilities to be provided can also be determined after all the information has been collected. The guide to follow is to select the nearest site that provides suitable conditions. Naturally, locations far removed from the point of activity give rise to problems of movement, supervision, and control.

If the volume of traffic is known or can be calculated from the daily transactions of receipts, issues, and quantities, the type of handling and storage equipment required will become more evident. This information, augmented by the knowledge of the material characteristics, will finally determine the requirements for facilities.

Types of Storage Facilities. The principal types of storage facilities can be classified as follows:

For large quantities of loose materials	For small quantities of loose materials	For unit or packaged materials
Bulk bins	Shelves	Bins
Silos	Drawers	Racks
Hoppers	Pans	Pallets
Bunkers	Trays	Skids
Open bins	Boxes	Platforms
Open pits	Rotary bins	Lockers
	Small racks	Cabinets
	Cans	Stands
	Drums	
	Counters	
	Portable racks	

Because the above facilities are constructed in such a variety of designs and types, only a basic classification is practicable. The above classification is intended as a general guide only.

Of importance in the selection of equipment of this kind is the specification of standard models wherever possible, to improve both price and delivery. This should not, of course, be done at a sacrifice of utility; where standard design and construction are suitable, economic benefits result from their use. Progressively more items are made in standard increments of size.

Developments in Storage and Warehousing Techniques. The trend in storeroom practice has been toward complete mobility of both equipment and materials. The old type of storeroom with fixed binnage and solid racks is fast disappearing where the nature of the material stored does not preclude such equipment.

Layout is having a more pronounced effect on warehouse operations. Considerable thought is also being given to the advantages of single-story warehouses.

Storage equipment is becoming more versatile and generally wider to fit the various types of materials that have to be stored. Adjustable shelf racks, in which the spacing of shelves can be varied to suit the height of the load, are being used extensively. Where racks or bins can be dispensed with and tierable containers used, no fixed structures are installed. The containers form their own bins or racks. Stacking racks which can be built up to any reasonable height and width by means of interlocking features provide the same tiering facility for tote pans as do bins, yet they are completely portable. Rotating circular bins are replacing rectangular fixed bins and saving many steps and much handling. Stacking cranes now perform the operation of placing palletized loads into racks or of removing them from the rack much the same as a fork truck operation and with a considerable saving in space.

SAFETY AND MATERIAL HANDLING

According to the records of the National Safety Council, a very large percentage of all plant accidents involves material handling operations. There is always considerable danger involved in the handling of material, and when poor layout or faulty material handling techniques are involved, the danger is greatly increased. Accordingly, the industrial engineer in charge of material handling is frequently called upon to help initiate safety programs. He should, in any event, try to integrate safety into every material handling improvement he makes. The engineer should attempt to mechanize or automate every material handling operation involving safety hazards to the extent economically justified.

Selection of Equipment. The characteristics of the buildings involved should be considered before any mobile material handling equipment is selected. Part of these considerations may come under the analysis of the material handling project, but all should be reconsidered with an eye toward providing better safety. Some of the characteristics that will have to be considered are ceiling heights, aisle widths, overhead obstructions, floor load capacity, door clearances, column spacing, structural strength, and others.

When purchasing fixed equipment, the engineer can often add small accessories that greatly enhance safety. In purchasing conveyors, counterweighted conveyor gates can often be had to assure that the gate will stay either up or down. Sheet metal guards can be added to the outer edge of a conveyor curve to prevent material from falling from the conveyor and injuring workers. Nonskid treads between rollers on conveyor lines can be provided for conveyor crossings or for the necessary inspection operations. The height of a conveyor should be fixed so that the material being handled can be put on the conveyor without undue reaching.

High-lift fork trucks or other powered trucks that will be required to negotiate in close quarters should be protected with guards to prevent the driver from being crushed. Naturally, only fork trucks or powered trucks which can be used with the particular building characteristics should be selected.

Layout. Layout is extremely vital in providing safe working conditions. Not only should the building itself be constructed so that it will provide structural safety, but the actual layout of equipment should be such that easy and safe flow of material is permitted to all work stations and process areas. Pedestrian traffic should be allowed for, as well as vehicular traffic. Aisles should be sufficiently wide to permit trucks to pass one another safely and to permit employees to move about freely while handling materials in and out of racks, bins, or piles. Machines should be located to allow ample room for repair and maintenance men.

Shipping areas should be planned so that plenty of dock space is allowed for highway trailers and freight cars. Dock levels should be built to average truck bed and railroad car levels if hydraulic ramps are not employed.

AIR CARGO HANDLING

The impetus of World War II gave a rapid rise to air cargo handling. With the advent of four-engine aircraft—and later, the jet—having enormous weight-carrying capabilities, the movement of commodities by air rather than by surface vehicles has become economical and expeditious.

The anticipated growth of air cargo handling is shown by Figure 5-10. The industrial engineer must remain cognizant of this trend if he is to plan facilities

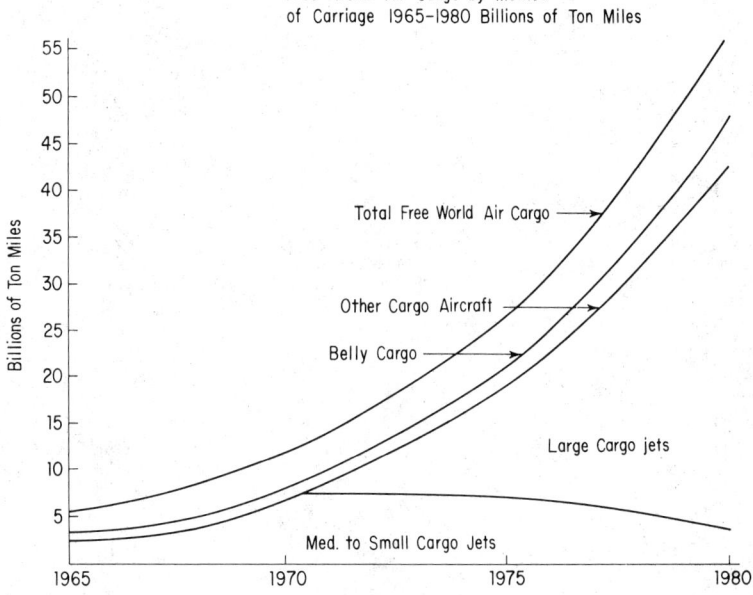

FIG. 5-10.   Projected air cargo growth.   (*Courtesy of Lockheed-Georgia Company.*)

for the receiving or shipping of goods at his plant.   Many new industrial plants are being located near airports in anticipation of this trend, and some companies are installing their own private airstrips.

The airlines have developed improved air cargo handling methods and equipment. Nearly all air cargo is unitized before loading into the airplane.   Most major airlines have all-cargo aircraft, and some airlines handle cargo exclusively.   Figure 5-11 illustrates the typical procedures followed in preparing shipments for unit handling at terminals prior to loading onto aircraft.

Although all airlines carry cargo on passenger airplanes, this practice is giving way to all-cargo flights.   Both the airlines and airplane builders have recognized the need for aircraft that are specifically designed for single-purpose operation. Where the volume warrants, air freight can be moved at a lower cost per ton-mile in all-cargo aircraft.   With the advent of such planes as the Boeing 747 and the Lockheed L-500, a new era in air freight transportation may be expected.   For example, the Lockheed plane, with a cargo space 150 feet long, 19 feet wide, and 13½ feet high, has a payload capacity of 300,000 pounds and a usable volume of 52,256 cubic feet.   Its main cargo deck accommodates twenty-eight containers 8 by 8 by 10 feet, with twenty-two containers 4 by 8 by 10 feet on overhead rails and an additional sixteen pallets 86 by 88 by 125 inches on the upper cargo deck.   Figures 5-12 and 5-13 indicate the enormous cargo capacity of this airplane.   By way of comparison, it will accommodate six Greyhound busses (two abreast) on its main deck.

Nearly all airlines and air terminals have mechanized loading and unloading of air cargo with which an airfreighter can be loaded or unloaded in a matter of minutes.   For example, the L-500 can be loaded as shown in Figures 5-14 and 5-15.

MATERIAL HANDLING TRAINING

In the discussion of in-process handling, it was pointed out that very often supervisors need material handling training.   They are in the best position to recog-

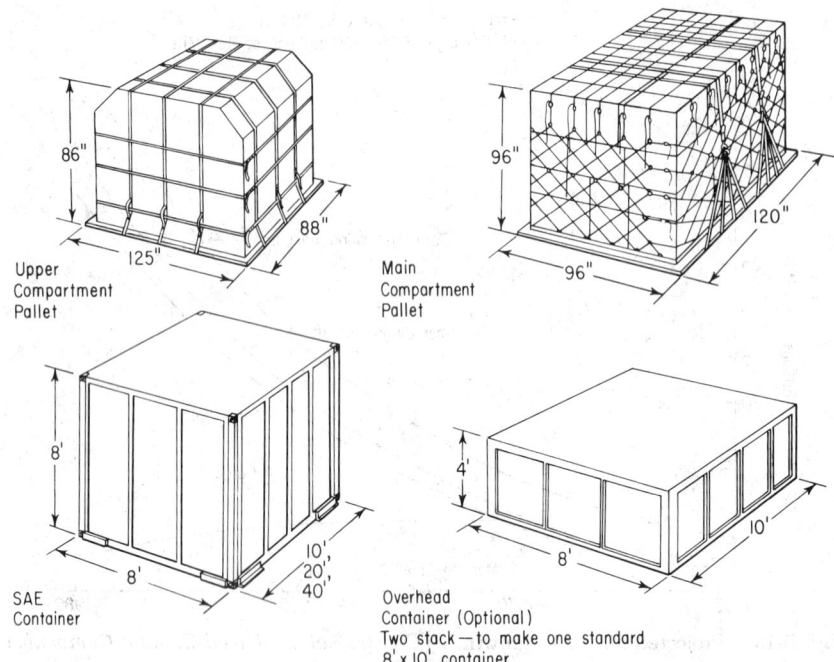

Upper Compartment Pallet

Main Compartment Pallet

SAE Container

Overhead Container (Optional)
Two stack — to make one standard
8' x 10' container

FIG. 5-11. Methods of unitizing air cargo. (*Courtesy of Lockheed-Georgia Company.*)

nize the need for and institute material handling improvements, because they are closest to the problems and are probably most interested in seeing them corrected. Courses can also be made profitable to a company by making the stipulation that every participant must undertake a material handling project and complete it.

Management personnel should be given the rudiments of material handling—not with the thought of making them experts in material handling, but rather to give them enough knowledge to recognize material handling problems in their area.

A material handling course should be conducted with the cooperation of management. If this is done, periodic reports on the performance of each participant can be routed through his supervisor for his approval or suggestions. The participant then feels that his supervisor is taking a personal interest in his progress.

If the organization is small and there are no engineers specializing in material handling available to teach the course, the services of a professor from a local college or university or of a professional material handling consultant or a local society can usually be obtained. A thorough course, economically planned, can be given for relatively little money compared with the returns in savings from the projects undertaken.

Savings continue year after year as the same supervisors detect more and more faulty techniques in the in-process handling in their areas. As a conservative estimate, a material handling course for supervisors should return in savings approximately ten times its cost.

MATERIAL HANDLING EQUIPMENT

In general, there are three major classifications of material handling equipment: conveyors, cranes and hoists, and industrial trucks.

Conveyors—gravity or power devices commonly used for moving uniform loads

Notes:

1. Container volumes are corrected for walls and structure, approximating a 3.5-inch wall thickness. Pallet volumes are based on 1.5-inch setback, 96-inch load height on main deck, 2-inch pallet + roller height and 2-inch clearance from aircraft contour on upper deck.

2. Main compartment pallet load is rectangular, upper compartment pallet load contoured for 707 or DC-8 all-cargo aircraft.

3. The Optional Overhead containers are stackable and can be rolled on their bottoms. Two mounted together are handled as one standard intermodal 8' x 8' x 10' container in the L-500 or any other vehicle.

	Unit	Height	Width	Length	Volume cu ft	Weight lb
1	Container	8'	8'	10'	504	980
2	Container	8'	8'	40'	2,270	4,150
3	Overhead Container	4'	10'	8'	250	700
4	Pallet (Main Compt)	96"	96"	120"	605	375
5	Pallet (Upper Compt)	86"	88"	125"	465	300

Upper Deck— Side loaded (two separate compartments). Door opening in each compartment is 92 in. high x 120 in. wide.

Main Deck— Raised nose exposes approximately 19.5 ft x 13.5 ft cross-section for straight-in loading.

Bulk Cargo Compartment — Side-loaded. Door Opening is 7' x 5' wide.

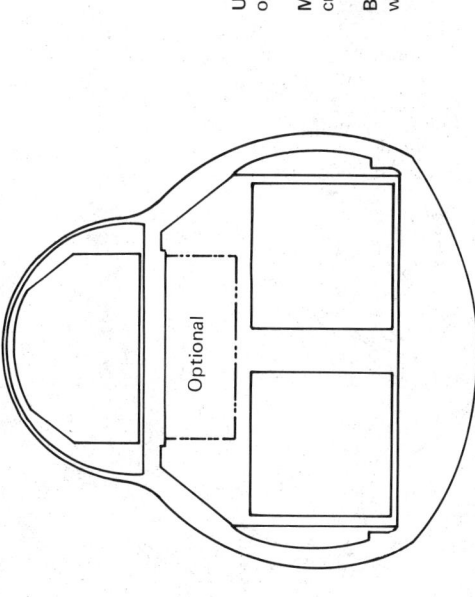

Fig. 5-12. Container and pallet sizes for the L-500 airplane. (*Courtesy of Lockheed-Georgia Company.*)

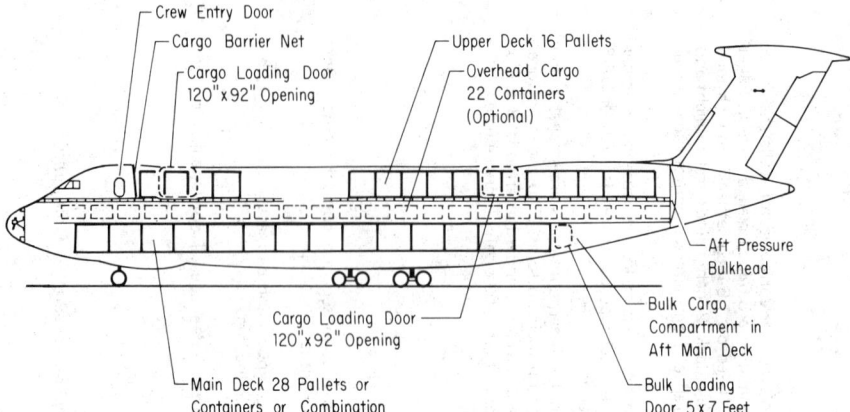

Fɪɢ. 5-13.   Inboard profile of the Lockheed L-500 airplane. (*Courtesy of Lockheed-Georgia Company.*)

continuously from point to point over fixed paths where the primary function is movement.

    1. Common examples:
        a. Roller conveyor
        b. Belt conveyor
        c. Chute
        d. Monorail
        e. Trolley conveyor
    2. Conveyors are generally useful when:
        a. Loads are uniform.
        b. Materials move continuously.
        c. Routes do not vary.

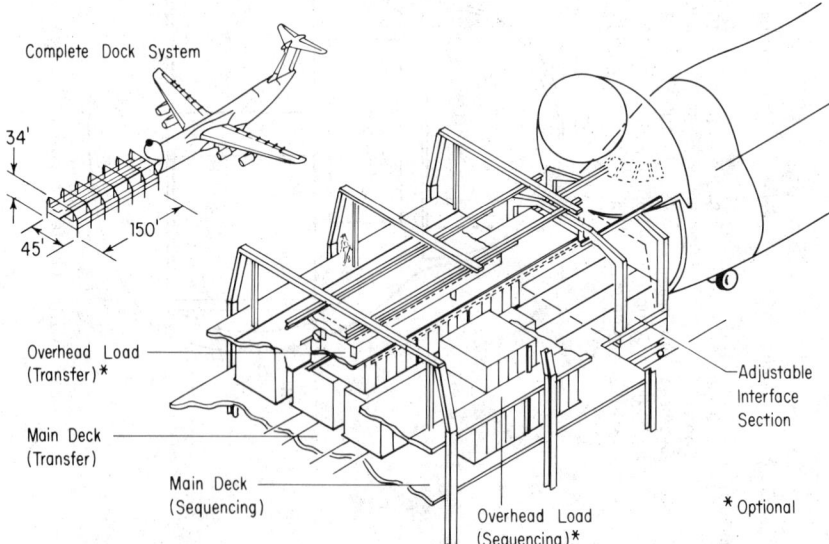

Fɪɢ. 5-14.   L-500 cargo dock. (*Courtesy of Lockheed-Georgia Company.*)

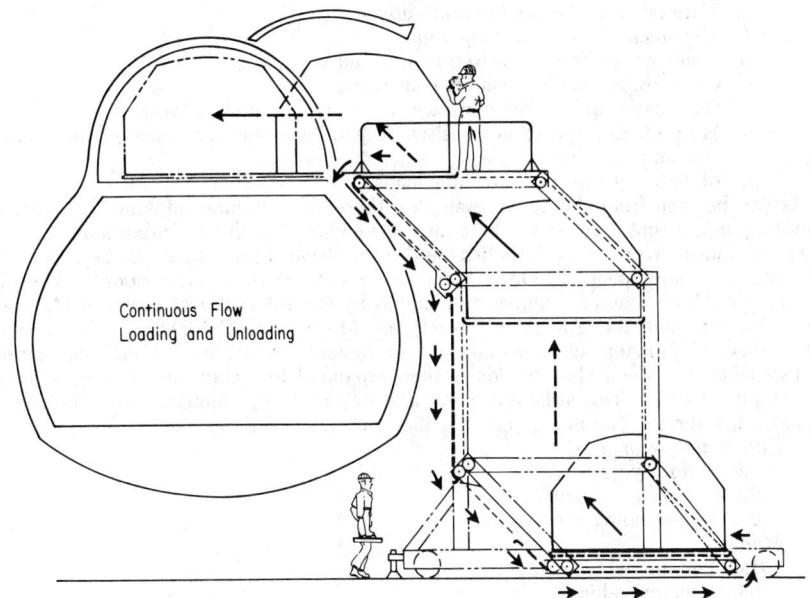

Continuous Flow
Loading and Unloading

FIG. 5-15.   L-500 upper lobe loader.   (*Courtesy of Lockheed-Georgia Company.*)

    d. Loads are constant.
    e. Movement rate is relatively fixed.
    f. Conveyors can bypass cross traffic.
    g. Path to be followed is fixed.
    h. Movement is from one point to another point.
  Cranes and hoists—overhead devices usually utilized to move varying loads intermittently between points within an area fixed by the supporting and guiding rails, where the primary function is transferring.
    1. Common examples:
      a. Overhead traveling crane
      b. Gantry crane
      c. Jib crane
      d. Hoist
      e. Stacker crane
    2. Cranes and hoists are most commonly used when:
      a. Movement is within a fixed area.
      b. Moves are intermittent.
      c. Loads vary in size and weight.
      d. Cross traffic will interfere with conveyors.
      e. Loads handled are not uniform.
  Industrial trucks—hand or powered vehicles (nonhighway) used for movement of mixed or uniform loads intermittently over varying paths which have suitable running surfaces and clearances and where the primary function is maneuvering or transporting.
    1. Common examples:
      a. Forklift truck
      b. Platform truck
      c. Two-wheel hand truck
      d. Tractor-trailer train
      e. Hand stacker
    2. Industrial trucks are generally used when:

  a. Materials are moved intermittently.
  b. Movement is over varying routes.
  c. Loads are uniform or mixed in size and weight.
  d. Cross traffic would prohibit conveyors.
  e. Clearances and running surfaces are adequate and suitable.
  f. Most of the operation consists of handling—lifting, maneuvering, stacking, and the like.
  g. Material can be put into unit loads.

As can be seen from the above examples, there are a number of kinds of material handling equipment. In fact, there are somewhere in the neighborhood of 400 different kinds, types, and varieties, and they have been classified in a variety of ways by many people. One of the more exhaustive classifications was made by Herbert H. Hall and subsequently adopted by the International Material Management Society and the American Society of Mechanical Engineers. This system established nine major classifications of equipment under three broad categories, as listed below. Each classification is then expanded to include nearly 1,000 pieces of equipment or devices, although many are only remotely connected with industrial material handling. The nine major classifications are as follows:

Either mobile or fixed
  1.000 conveyors
  2.000 cranes
  3.000 positioning equipment
Mobile
  4.000 industrial vehicles
  5.000 motor vehicles
  6.000 railroad cars
  7.000 marine vehicles
  8.000 air transports
Fixed
  9.000 containers and supports

Each of the nine major classifications of material handling equipment will be defined and described here, but no attempt will be made to list all the items that are contained within them.

Conveyors—Classification 1.000. There are nine general types of conveyors: belt, elevating, carrier-chain and cable, haulage, roller, screw, pipeline, reciprocating, and vibrating. In addition, there are special conveyors and auxiliary equipment.

Belt conveyors consist essentially of endless moving belts which carry materials within a supporting framework. The belt itself is the transporting medium. It may be a fabric, rubber, leather, plastic, or metal belt, and it may or may not be equipped with cleats or other grabbing devices. The belt may be supported by either rollers, roller idlers, or solid beds of wood or metal; it may be either a flat belt or troughed in a U shape. All belt conveyors, and there are many kinds, have a power-driven pulley at one end which moves the belt. A take-up pulley is often located at the other end to adjust for wear and belt stretch. In some designs, the take-up device is superimposed between the end pulleys.

Belt conveyors are most frequently used to move quantities of either packaged or bulk materials over fixed routes at desired speeds. The capacity of a belt conveyor will depend on the width of the belt, its speed, and the density of the material. Troughed belt conveyors are universally used for handling bulk materials such as coal, cement, glass, ore, and sand in either a moist or dry state. The trough is usually formed by three rollers, with the outer rollers usually set at an angle of 20 degrees to provide maximum carrying volume. Portable belt conveyors are generally used for handling bulk material, the unloading of cars, the loading of trucks, and other carriers. Portable belt conveyors, as well as most other types of belt conveyors, can be used to elevate material, but the maximum elevation angle without cleats is between 20 and 25 degrees. Portable conveyors can also be connected together to form a very flexible transportation system.

Elevating conveyors are lifting conveyors which transport materials vertically, although they may move up an inclined plane. The actual transporting medium may be buckets, flights, trays, arms, or fabric pockets. These transporters are powered by lifting chains or belts. Elevating conveyors may be either enclosed or open and may be either portable or fixed.

Among the most commonly used kinds of equipment in the elevating conveyor classification are the bucket elevators. These conveyors are mostly vertical, and consist essentially of one or two endless chains or a belt to which buckets are attached. They are most often applied to the elevation of fairly dry, granular or semigranular materials.

Carrier-chain and cable conveyors is the class designation given to all horizontal conveyors which are driven by chains or cables or in which the load is actually carried by chains or cables. This flexible type of construction makes it adaptable to a multitude of material handling problems. Because this classification is broad, it is well to specify the subtypes of carrier-chain and cable conveyors. They are apron conveyors, slat conveyors, crossbar conveyors, carrier-chain conveyors, pallet conveyors, car conveyors, trolley conveyors, and aerial tramways.

Among the most common of these are the apron conveyor and the slat conveyor. The apron conveyor consists essentially of overlapping plates or aprons powered by a chain to form a continuous moving bed to carry material in a sort of continuous moving box. Apron conveyors are often installed flush with the floor or just slightly above it; this is especially true of the shallow apron-pan conveyors and slat conveyors. These kinds of conveyors are used for moving barrels and boxes of merchandise and are particularly effective when installed flush with the floor. Some automobile assembly line conveyors are of the slat type. They have been used as moving floors in foundries and the like. In smaller single-chain form, they are used extensively in the food processing industry. The slat conveyor can be made either of wood or of steel slats driven by chains. Slat conveyors are extensively used through storage areas as mobile floors. They may be set flush with the floor or at any desired height. Those of wood are particularly adapted to handling packages and bulky articles.

The pallet conveyor is nothing more than a series of flat plates which are powered by an endless chain; they are used for unit handling. The car conveyors are a series of four-wheel cars propelled around an endless track by a chain. The trolley conveyor is a series of trolleys supported from an overhead track and connected by a chain. They are particularly advantageous when a floor space shortage is involved. Aerial tramways are overhead conveyor systems where the trolleys are supported by cables and so driven; ski lifts belong to this type of conveyor.

Haulage conveyor is the generic term applied to any conveyor whose prime duty it is to move large quantities of bulk materials. They are used extensively in the mining and refining industries, though not exclusively. Again, this classification is broad; so the various types within the classification will be enumerated. They are drag conveyors, flight conveyors, cable tramways, and car hauls. Drag conveyors are used in literally hundreds of different applications. They consist essentially of power-operated chains designed to move materials by dragging action through an open or enclosed trough. Pusher-bar conveyors—parallel chains connected at intervals by bars—are drag conveyors, and the rope-and-button conveyor is a drag conveyor. It is often used for moving abrasive material and has been installed for moving grain, pulpwood, and lumber mill waste.

Flight conveyors are very much like drag conveyors, with the exception that all use flights, that is, plain or shaped steel plates. These conveyors are universally used for moving medium-fine bulk materials. Tow conveyors, as the name implies, consist of either overhead or underfloor tracks, with powered chains used to tow trailers along the floor. Cable tramways are overhead conveyor systems supported or driven by cables for more than one span. A car haul is very much like the tow conveyor, except that the moving mechanism is always overhead.

The *roller conveyor* is undoubtedly the best known and most widely used type of conveyor. It consists of rollers supported in a frame over which materials are moved by either power or gravity. The rollers need not be straight or symmetrical.

Roller conveyors have been adapted to handle various shapes of objects in different manners by changing the cross section of the rollers or their alignment in the framework. Because the gravity roller conveyor requires no power and very little maintenance, it is considered the most economical type of conveyor. Gravity roller conveyor systems are often combined with belt conveyor boosters to lift objects to another elevation. The so-called V-trough conveyor is a gravity roller conveyor used for heavy-duty jobs and is especially made to handle cylindrical objects of up to 20 tons. Spiral roller conveyors are conveyors spiraled around a vertical center line and are used in moving materials from one floor to a lower floor at moderate speeds.

The live roller conveyor is one which is power driven by underside belts, gears on the roller shafts, or chains and sprockets. It is generally used for moving materials horizontally. Chains and gears are usually used on heavy-duty, live roller conveyors. Belts are used to power moderate- or light-duty live roller conveyors.

Roller conveyors, as a class, are used in almost every industry. Their versatility and flexibility make them a foremost material handling tool. Accessories such as Y's, switches, and turntables facilitate the construction of a conveyor system.

Screw conveyors are conveyors which utilize a large screw or spiral. Often, the screw moves bulk material by rotating in a semicircular trough or tube, thus advancing the material. (There are, however, screw conveyors in which the envelope or tube rotates.) The helix or screw is rotated by power supplied by belt or chain or by a motor directly connected to it. Screw conveyors can operate on a horizontal plane or an incline up to 30 degrees with standard pitch screws. However, special applications requiring steeper angles are not uncommon. Screw conveyors are made in sections of 10 to 12 feet, and several sections can be connected together to attain the desired length.

Pipeline conveyors carry material through pipes or tubes either by gravity, vacuum, air pressure, or a mechanical means. The use of an air stream for moving materials dates back to about 1890, when the pneumatic tube and carrier were developed to convey sales transactions in department stores. This type of system is applied to the handling of dry bulk materials. Pneumatic systems, which move materials on an air stream provided by a vacuum pump or suction fan, find ideal application for unloading grain. They can handle over 1,000 tons per hour. They are widely used for transporting dry pulverized and granular material and are particularly well suited for handling dusty or irritating chemical powders. They are virtually silent in operation, are fast and clean, and require very little maintenance and space. In industry, unit-handling pneumatic systems handle mail, paper work, small tools, and parts. Among other pipeline conveyors are the hydraulic or liquid-carrying type. These pipelines carry petroleum products for literally thousands of miles, and although not so simple in operation as the pneumatic system, they must be regarded as a comparatively thrifty material handling device. More of these pipelines are being applied by the petroleum industry, but the future may see application by other industries whose products are principally in the liquid form, such as the dairy, chemical, and irrigation industries. Pipeline conveyors are also used to handle semiliquids such as slurries, waste materials, clays, and the like.

Reciprocating conveyors usually consist of one or more reciprocating beams which move material by a reciprocating motion, with some grabbing device moving the material in the desired direction only. A reciprocating beam conveyor has tilting dogs between parallel reciprocating beams to advance objects in the proper direction. A reciprocating flight conveyor has hinged flights properly arranged to advance material in the desired direction. Reciprocating conveyors can be portable or fixed and are often used as bulk feeders to other conveyors.

Vibrating conveyors move material by vibrating it in one direction. The material is usually placed in a tube or channel, and the tube itself vibrated. The vibrations may be induced electrically or mechanically, and they may be of either high frequency or low frequency, depending on the material being moved. The units may be either portable or fixed. They are suitable for moving bulk material.

Special conveyors and auxiliary equipment form the last classification for conveyors. This classification includes special feeders; screens; hopper car shakeout unloaders; chutes, hoppers, troughs, and spouts; and flumes and sluices.

Cranes, Elevators, and Hoists—Classification 2.000. The crane is one of the oldest devices used for handling material. Although it has been improved a great deal and some have been made portable, its uses are basically the same as they were years ago. Many types of cranes are in use today. Some are of standard design, while many others are special-purpose machines.

The prime duty of any crane is to lift a heavy weight vertically. Some cranes can swing a load, and others can travel with the load in several directions. Transporting material by crane is not the most economical method of movement, although in some instances, great weight and mass may prevent the use of any other device. Most cranes are of the traveling type, either on a special runway or with the aid of powered mountings. The traveling bridge crane, so widely used in manufacturing plants, is limited in the service it can render by the fixed path it must traverse. It must be supplemented by other handling equipment to give aisle-to-aisle service. This stop-and-go operation results in making crane handling of material expensive when compared with some other applicable means.

Some low-capacity suspended crane installations can be made which can provide complete coverage of a manufacturing floor with the aid of interlocking sections. Sound planning should precede the installation of overhead cranes because of their cost and the service limitations involved.

There are nine headings in the cranes, elevators, and hoists classification: fixed cranes and derricks; traveling cranes; portable cranes; elevator and lifts; cableways, cable scrapers, and shifters; hoists; winches; crane, elevator, and hoist auxiliary equipment; and special cranes, elevators, and hoists.

Fixed cranes and derricks are the nonmobile class of cranes such as pedestal cranes and stationary jibs. Fixed cranes must be considered necessities rather than the most efficient choices in many cases, because they have limited flexibility and their prime duty is to lift. The derricks—stiff leg, A frame, and guyed—come under this classification, and because of their wide use in construction industries and yard activities, deserve mention here. An A-frame derrick boom can be raised and lowered and rotated through an arc of 180 degrees. The boom of the stiff-leg derrick can likewise be raised and lowered but can be rotated through 270 degrees, thus giving it slightly more flexibility. The guyed derrick boom usually cannot be rotated as far as either the A-frame boom or the stiff leg boom because of interference from guy wires which support the main mast. Jib cranes belong to this classification of cranes and are widely used in industry to supplement overhead cranes. The simplest form of the jib is the wall-bracket jib, which has one end of its beam hinged at a wall or a column. It is economical in first cost and floor space and can usually swing through an arc of slightly less than 180 degrees.

Traveling cranes are perhaps the most prominent type in the crane family. Any crane that travels a fixed path must be considered a traveling crane. The most widely used of these is the overhead bridge, which consists of one or more horizontal bridging girders supported by trolleys on both sides of an aisle or by other vertical supports. The hook may move laterally and vertically and the carriage can move longitudinally, providing maximum flexibility. Usually, however, it is limited in that it cannot reach both walls, due to its construction. The traveling gantry group of cranes also belongs to this classification. There are several kinds of gantry cranes—portal, semiportal, cantilever, wharf—and, as gantry implies, they all have a traveling framework from which a crane is suspended or attached. The cantilever gantry, instead of having a rectangularlike structure, consists of traveling legs, on one or both sides of which cantilever beams extend. It is used to reach over inaccessible areas. The semigantry has one of its legs at ground level, although both legs run on tracks to present a traveling triangle structure. It is often used as an auxiliary to large overhead traveling cranes and often also along the sides of buildings. Overhead monorail cranes, and the many types of overhead traveling cranes so often found in light- or moderate-duty applications in industry, also belong to this group.

Portable cranes, as distinguished from overhead traveling cranes, are those which are mounted on a fully mobile carriage and do not travel fixed paths. The railroad cranes, although somewhat fixed in travel, must be classified here rather than as traveling cranes because of their greater mobility. Crawler cranes are boom cranes mounted on crawlers or endless treads and are used wherever terrain is rough. They are seldom if ever used for indoor work. Another type of portable crane that finds less application than either crawler cranes or railroad cranes is the truck-mounted crane. The portable floor cranes used in industry are versatile pieces of equipment with either fixed or rotating boom, and are often used for positioning work. The swinging-boom portable floor crane is perhaps the most versatile of these and is used as a portable jib crane. The portable gooseneck crane is a crane with a curved supporting boom for offsetting the load. It is extensively used in industry and often found in garages also.

Elevators and lifts are platform-type devices used to lift materials. Freight elevators and passenger elevators are classified here. The freight elevator is an extremely vital integer of a material handling system and comes in various capacities from 2,000 to over 10,000 pounds and in various platform sizes and speeds. Special control systems which provide automatic or semiautomatic operation have also been developed. Skip hoists, which are motor-operated hoists consisting of bucket cars driven up sharp inclines by a powered chain or cable, are included in this classification. Bulk material is loaded into the buckets from a feeder, hoisted to the head end, and automatically dumped, the bucket returning by an adjacent rail. These hoists are used extensively for elevating bulk materials into bins and for charging cupolas.

Cableways, cable scrapers, and shifters have been lumped into one crane type because of their common construction; that is, they are all cable powered or supported. A cableway system is a combination of cables to permit lifting or dragging a grab, bucket, sling, or cab. The drag-line bucket used extensively in the mining industry is a cableways system consisting of a drag-line bucket attached to an endless cable suspended between two towers and powered from one of the towers. Cableways can be either taut-line cableways, that is, with the cable tight, or slack-line cableways, with the cable loose between the supporting media. Cable-drag scrapers are cableway systems set up to scrape the surface of the material being worked. The monorail-trolley drag scraper is a scraper attached to a cable, which is in turn attached to a trolley that rides on a monorail. It is used where a more rigid drive is required and usually in more permanent installations than the totally cable-supported and cable-driven scraper.

Hoists are lifting mediums—usually for light-duty work. They are most frequently used to lift loads for short vertical distances as opposed to elevators. Hoists are among the most widely used pieces of material handling equipment and can be either powered or manual, floor or cab operated. The cylinder hoist consists of a motor or manually driven cylindrical drum on which rope, chain, or cable is wound and paid out. The screw hoist, on the other hand, is a hoist in which the load drum is driven by a power drum through a worm and gear. More often than not, the screw hoist is manual and the power drum is driven by an endless chain pulled by hand. It is best used for lifting of a light-duty nature, such as the occasional lifting of loads in an industrial plant. Hoists also use chains, ropes, or cables for their lifting medium. Chain hoists are not usually multiple-strand hoists, due to the difficulties involved in winding chain.

Although usually powered by electric motors or manual power, there are some hoists which operate on compressed air. These air hoists consist of cylinders in which compressed air is injected to work pistons to lift the load. Because of their definitely explosionproof, sparkproof nature, they are often used in hazardous locations in the petroleum and chemical processing industries. Some hoists are operated by air motors, that is, motors rotated by pressure of compressed air on a series of vanes or blades. They are usually built to rotate in either direction.

Winches are power- or hand-operated drums on which are wound ropes, chains, or cables for lifting or towing. The drums are usually operated through a set of gears to give a better mechanical advantage. The capstan, originally made for shipping and wharf operations, is commonly used for moving railroad cars at sidings and terminals.

Crane, elevator, and hoist auxiliary equipment forms a classification itself. These auxiliary items are manufactured and used because they add to the versatility, effectiveness, and efficiency of crane equipment. They include grabs, grips, tongs, slings, bridles, spreaders, nets, clamshell buckets, orange-peel buckets, drag-line buckets, grapple buckets, cabs, cars, gages, magnets, steel balls, pile drivers, drills, clocks, kevels, and many other items.

Positioning Equipment—Classification 3.000. Positioning equipment has been acquiring increasing recognition. The equipment itself is as varied as the work done and applies to every industry. In this classification also must be placed special steel mill equipment—the almost human, giant manipulators and furnace chargers used in open-hearth, blast-furnace, and forging and rolling mill operations. By far the majority of the equipment, however, is the small positioning equipment used in industry, such as elevating tables, table positioners, welding positioners, and the like.

Remembering that a single pickup and laydown operation costs $0.005 at 1969 labor rates, anyone responsible for material handling operations should give positioning equipment a very careful look. Besides increasing the efficiency of any work station or the effectiveness of any material handling system, positioning equipment can be relied on to enhance safety and housekeeping and directly affect quality.

Manipulators and positioners form the first classification in the positioning equipment group. Manipulators are devices that grab, rotate, turn over, slide, or otherwise similarly handle a solid object. They are extensively used in the steel industry and sometimes are of giant proportions. Manipulators are always remotely operated and may be either fixed or portable. Positioners are devices that hold objects in one position or put objects through a prescribed set of motions. The universal positioner is used widely in the welding industry; it will hold the piece being welded firmly in any one position at the proper working level. Although the universal positioner probably finds its greatest application in welding processes, it is not strictly confined to these processes alone. Bench-type and floor models, relatively small universal positioners, are available for holding work in a positive position over a prolonged period of time. They can be used for inspection, maintenance, and assembly operations and provide maximum flexibility through their ability to rotate about three axes. Positioning devices are also made in the form of rotating tables which can be locked in place. These are likewise used for assembly operations, the work being placed on the table and rotated to the desired position.

Upenders and dumpers form the second classification in the positioning equipment group. Upenders, as the name denotes, are devices for setting on end items of material. Upenders are both hand and motor powered. Although they find their greatest application for turning over coils, rolls, or drums, they can be used on odd-shaped pieces. Some upenders are designed with elevating features for floor-to-floor transfer. Some industrial vehicles are equipped with upending attachments or attachments to handle rolls and cylinders. The usual fork truck upending attachment is able to pick rolls or drums from either vertical or horizontal positions and rotate the load 90 degrees.

Dumpers used for dumping loads come in a wide range of sizes. They vary all the way from comparatively small barrel-and-bag dumpers up to the open-car turnover dumpers which turn over an entire railroad car.

The *furnace-charging equipment* classification is limited almost exclusively to the iron and steel industry. These huge pieces of equipment are usually mounted on tracks and have a wide assortment of controls that allow them to perform a variety of operations.

Fixed positioning tables and platforms are positioning devices that are fixed in place. They are widely used throughout industry. Positioning tables are often hydraulically operated. Fixed positioning platforms can handle 50 tons or more. They are widely used at large permanent machines such as punch presses, shears, or rollers. They are also used at loading docks. The hydraulic service-rack lift found in almost every filling station also belongs to this classification.

Portable positioning tables and platforms are perhaps more widely used than their fixed counterparts but do not have the wide capacity range. Positioning tables come in a variety of sizes, capacities, and shapes. They are often hydraulically operated. Portable positioning tables and platforms are especially helpful in die handling, installing and removing dies from presses and storage racks, sheet feeding, welding, and work positioning.

Also classified as portable positioning platforms are portable telescoping scaffolds and portable jacks, both hydraulic and mechanical.

Fixed positioning bridges and ramps form the sixth positioning equipment classification. To industry, the most important of these are movable crossover bridges and dock-leveling bridge ramps. Movable crossover bridges are used to provide passenger access over continuous conveyor lines, temporary industrial vehicle aisles, or other places where pedestrian traffic is either unsafe or undesirable. Dock-leveling bridge ramps are adjustable ramps that make it possible to load and unload directly from all types of trucks and trailers by providing a gradual runway. They can be manually set in place or they can be hydraulically operated, permanent parts of the dock. These adjustable ramps are a considerable asset to any loading dock because truck and trailer beds vary in height as much as 1 foot. The hydraulically operated ramps become a part of the dock when not in use and do not in any way interfere with traffic.

Positioning transfers include pieces of equipment, either fixed or mobile, that transfer material or other objects from one location or system to another or change the direction of movement of material.

Turntables of all sizes belong in the transfer group. They range in size from small sections of conveyor systems to heavy floor-type tables with capacities that will handle locomotives.

Ball transfers are used on conveyor systems to transfer flat-bottomed objects. They consist of steel balls set at close spacings in flat metal plates. They are frequently used at conveyor intersections. Wheel or caster transfers are similar pieces of equipment, with the exception that the transfer surface is made of small wheels or casters held on rods in a steel framework. There are numerous other items of transferring equipment.

Industrial Vehicles—Classification 4.000. The industrial power truck is one of the great contributors to lower material handling costs. Its development dates back to about 1906, when a battery-operated nonlift platform truck was built and widely used at freight terminals for hauling merchandise and baggage.

With the advent of mass production and the necessity of moving large quantities of material quickly, the skid platform and self-loading, low-lift platform truck were developed. This led directly to the stacking of loads, using a high-lift platform truck and a telescoping fork truck.

Shallow load-bearing platforms, that is, pallets, then entered the material handling picture, and their use in the unit load system with a high-tiering fork truck has provided for better utilization of storage areas and floor capacity. Large unit loads can be moved more efficiently by this method.

The industrial tractor is a result of a development of the nonlift, load-bearing platform truck. The trailers used with tractors can be ordinary four-wheel floor trucks or they can be specially designed for high load carrying capacity. The tractor-trailer system of handling materials was developed for transporting large quantities of material over long distances within a plant and between plants.

Many manufacturing plants produce a variety of parts in large quantities. Some parts must be moved considerable distances to complete a process, and the continual handling of such material can be troublesome and time consuming. The use of the tractor-trailer system in such a plant permits efficient pickup and delivery service on a scheduled basis. An industrial fork truck can be an important supplement to a tractor-trailer system for loading and unloading trailers. One of the requisites for tractor-trailer operation is a smooth, hard-surface roadway.

The nine subclassifications in the industrial vehicles classification are powered industrial trucks, hand trucks, industrial tractors and locomotives, industrial trailers, industrial cars, bulk material handling vehicles, and agricultural vehicles and attachments.

Powered industrial trucks carry a very large percentage of the material handled in industry. Some of the important types are fixed platform, lift platform, fork lift, side self-loading, straddle, and powered industrial scooters. Each of these types has various modifications.

Fixed or nonlift powered platform trucks can be of either high- or low-platform variety. The high-platform type, sometimes called a burden carrier, is capable of carrying heavy loads and requires some external means of loading and unloading. The operator rides along with the truck. The platform sizes vary and can be secured as large as 42 inches wide by 87 inches long. The trucks are usually powered by storage batteries but gas-electric power units can be obtained. Four-wheel drives have been developed to obtain maximum traction. This type of truck is best adapted to heavy hauls and is used mostly by industry in conjunction with cranes for loading and unloading. As the platform is fixed, this truck cannot be used for lifting pallets or skids from the floor.

The low-platform type is very much the same as the high-platform type, except that it can be had in the four- or six-wheel variety and the platforms are usually smaller and of less capacity. The platforms are usually 11 to 17 inches from the floor, with capacities of from 2,000 to 10,000 pounds. These low-platform trucks are powered by either a gasoline motor, storage battery, or gas-electric power unit. The low-platform type, as contrasted to the high-platform type, is best adapted to transporting light articles, particularly if they can be easily loaded and unloaded by hand. This type of equipment should be used for short- or medium-distance hauls. For long hauls, the tractor-trailer type is less costly to operate.

Trucks which have platforms that can lift as well as carry loads also belong to this first type—powered industrial trucks. There are low-lift and high-lift platform trucks. Both are self-loading, power-propelled vehicles with extended front platforms that can be run under a skid or platform which is subsequently lifted and moved to another location. The low-lift platform type of truck is constructed with a main frame on small wheels under the front of the lift platform and on large drive wheels at the rear. A second frame carrying the platform is incorporated with linkage and lifting mechanisms. When it is positioned under the skid and the power is applied, it raises the load for transport. Lowering is done mechanically or by a hydraulic check system. Lift trucks are equipped with gasoline engine, storage battery, or gas-electric power units. Their platforms vary in size and in elevation from 7 to 17 inches above the floor in a lowered position, with a lift of 3½ to 6 inches for the low-lift platform trucks. Capacities range from 2,000 to 30,000 pounds. The trucks operate at speeds up to 6 miles per hour. The low-lift platform truck should always be used in conjunction with skid platforms and can be operated interchangeably with hand-lift trucks. Their main function is to carry unit loads and aid in keeping material off the floor to reduce handling.

The high-lift platform truck is similar to the low-lift, with the exception that it has a different lifting element consisting of two upright channel masts rigidly attached to a main frame. The elevating platform unit is guided in the mast with wide-face rollers and is raised and lowered by chain or hydraulic mechanisms. The lifting height for a high-lift platform truck is about 60 inches. High-lift platform trucks can be used to advantage over the low-lift platform type for load tiering. Telescopic high-lift trucks are also available for extra-high stacking of skids.

The fork truck also comes under the powered truck classification. It has many variations, some of which will be discussed here.

The most simple type of fork truck is the nontelescoping variety. This is a truck chassis carrying a channel mast on which is mounted an adjustable cantilevered fork. The fork is designed to support and carry loads and to stack one load on top of another. Fork trucks are available in capacities up to 30,000 pounds. Ram trucks, which are an adaptation of the fork truck, have been built with capacities of 100,000 pounds for handling coils in steel mills. The mast carrying the fork can be tilted approximately 5 degrees forward and 12 degrees backward for easy loading and unloading and for greater stability during travel. Speed of travel under full load varies from 4 to 7 miles per hour. Hoist speed varies from 12 to 50 feet per minute. Maximum elevation of the fork is approximately 6 feet in the nontelescoping hoist types. These units are available in gasoline, electric, and gas-electric powered models. They are best adapted for moving materials that are placed on pallets and for stacking such pallet loads in storage.

There are also telescoping fork trucks which stack to greater heights than the ordinary nontelescoping type. This truck in the rider type has four-point ground contact. Standard forks are 36 inches long, with an adjustable spread of up to 38 inches and a lift of up to 144 inches. The carrying capacity ranges from 2,000 to 20,000 pounds, and the upright tilt is 5 degrees forward and 12 degrees backward. The power unit is mounted over the rear steering wheels. The lift mechanism is usually mounted over the front wheels. Separate control units are provided for driving, tilting, and lifting operations. They can be powered by gasoline engine, diesel engine, or battery. The telescoping-type fork truck is mostly used for handling unit loads into storage or for moving loads between operations. Fork trucks can often be used in conjunction with a tractor-trailer train for loading and unloading. These trucks should not be used in place of tiering machines or stackers where movement is not the prime purpose. They can be used in handling boxes or containers equipped with skids or for handling boxes, bales, reels, and so forth without pallets.

There is also a low-lift pallet fork truck of the walk-along or operator-riding type. This type weighs about 900 pounds and has a lifting capacity of about 3,000 pounds, an unloaded-pallet fork height of about 3 inches, and an elevated height of about 7 inches. They can be had in models which steer by either their front or rear wheels. Fork elevation is hydraulically operated by a hand pump. These small trucks are extremely easy to operate and handle in restricted spaces and consequently find their greatest application there.

There are also other cantilever or outdoor-type lift trucks. One of these is the telescoping ram type of high-lift truck with the fork equipment replaced by a ram attachment as the load-lifting and load-carrying member. Ram trucks are built on the cantilever principle, with the load being carried and supported in front of the telescoping uprights. The uprights tilt forward 5 degrees and backward 12 degrees. Capacities range from 3,000 to 100,000 pounds. The load is carried directly on the ram, which is cylindrical in form and is mounted on the elevating backplate. The load is hoisted by means of a chain or a hydraulic and chain-lift combination. The load is counterbalanced by the weight of the truck. General dimensions approximate those for a fork truck. Telescoping ram trucks are most efficiently used in the handling of cylindrical material with an open center, such as tires, wire coils, heavy spools, and so forth. They can be used in production operations, for storing materials in warehouses and storage yards, for loading and unloading cars, and so forth. When a long-distance movement is required, this type of truck should be supplemented by a tractor-trailer train, industrial cars, or some other means of transportation.

A wide variety of attachments and other devices is available for fork trucks to make them more versatile. Some of these are rams, cranes, fork extensions for open-face pallets and skids, hydraulically operated scoops, rotating mechanisms that will allow the forks to be rotated 360 degrees, rotating roll clamps, side shifters that move the forks laterally, clamps for hard-to-handle loads, snowplows, unloaders, load-stabilizing devices, and many other special devices.

Another type of self-loading power truck is the side loader. These are four-wheel fork trucks which load and unload from the side. The forks move outward from the side to load, lift the load until it is clear of the side-bed of the truck, then move it in and lower it onto the bed of the truck. The lift tower tilts slightly forward or backward. It is equipped with jacks for stability while loading or unloading and is available in capacities up to 30,000 pounds. It is particularly suited for the handling of long, narrow materials.

Still another type of powered industrial truck is the straddle type. This type of truck has four-wheel steering with a frame constructed to permit the wheels to straddle the load and then lift and carry it in suspension on hoisting shoes supported by the framework. Loads are arranged on bolsters so that hoisting shoes which are supported by the straddle framework will come up under the bolsters from the side and pick up the load for transportation. The truck has four-wheel steering, one in each corner of the framework, so it is possible to turn in a very small space. Capacities range from 10,000 to 40,000 pounds, and speeds are geared to 40 miles per hour for highway service. The units are mounted on pneumatic tires and are gasoline powered. The length of the load it can handle is a matter of turning radius and balance. Although straddle trucks were primarily designed for the handling of lumber, they are now used in industrial plants for handling rods; bars; beams; long containers for bulk material such as chemicals, coal, rock, steel chips, and refuse; and similar commodities. Another important area of use is in handling the "jumbo" shipping containers on and off highway trailers and railroad cars, and in and around air and marine terminals. The straddle truck has a great advantage in that it can be loaded and unloaded by the operator without leaving his control position.

Hand trucks form the second classification in this industrial vehicles group. Although some of them are powered, they differ from the previous classification in that none are rider-type vehicles. Platform trucks of the hand type form part of this classification.

A power-driven, low-lift platform truck is a motorized version of the efficient foot-operated hydraulic lift platform. The height of the truck platform from the floor can vary from 6 to 11 inches, and the platform has a lift of about 3½ inches. It comes in platform widths of from 20 to 26½ inches, and the length varies from 36 to 72 inches. Overall lengths range up to about 104 inches. These trucks can handle a capacity load of from 4,000 to 6,000 pounds, and the maximum speed is set at a comfortable walking speed for the operator. They are extremely useful where elevator and floor capacities are insufficient for heavier trucks, and also for moving work-in-process and handling materials from points of storage to work stations.

Another type of hand truck belonging to this classification is the hand-pull pallet lift truck, which is a completely hand-operated lift truck. The load is raised by a hand lever which also steers the truck. The truck frame is arranged so that it can enter between the top and bottom boards of the pallet. This type of hand pallet truck can be used only in conjunction with pallets constructed with bottom openings. It works well in conjunction with power-elevating fork trucks in that it can be used where power trucks cannot be operated, in packing rooms, at freight terminals, or for the movement of pallet loads in boxcars and onto trailers.

Another type of hand truck that should be mentioned here is the power-driven pallet truck. This is very similar to the hand pallet truck in construction, except that the front wheels are driven by an electric motor placed directly in back of the steering handle. Push-button control in the handle enables operators to move the truck forward or backward at will. After the truck has been placed into the pallet, the lifting mechanism is operated by either a foot pedal or a push button connected to a hydraulic system. The small rear wheels are lowered through the pallet opening sufficiently to raise the pallet for transportation. Lowering of the pallet is done by gravity. This type of truck can be used only with pallets constructed with bottom openings, unless the legs are long enough to extend all the way through the pallets. They are used for short horizontal movements, but are also frequently used in conjunction with fork lift trucks in packing rooms and freight terminals. They are also often used in conjunction with freight elevator loading, as they are relatively light in comparison with the loads handled.

There are many types of two-wheel hand-powered trucks. The western-type warehouse hand truck has its wheels on the inside of the frame for greater maneuverability, and the eastern type has wheels on the outside of the frame for greater wheel width.

There are special trucks for handling appliances, cotton bales, cylinders, barrels, stacks of cases, bags, and many other items. One type of two-wheel hand truck used extensively in industry is the tote box type, which has two prongs that grip the tote box beneath its handles and lift it when leverage is applied to the truck handle.

The dolly must be classified as a hand truck and is in essence a small, single-roller truck platform which consists basically of a low, heavy rectangular frame, on the underside of which are secured bearings for carrying the shaft or axle of a wide-faced wheel or roller. Due to the design of the dolly, it can easily be shoved under the object to be moved, so that the center of the object is located over the roller. The object can then be moved forward or backward or pivoted. It is generally used for transporting heavy objects for short distances in locations such as freight terminals, piers, wharves, and other places where it can take the place of the old roller and pinch-bar method of laboriously and slowly moving heavy objects. It can be used only on a smooth surface.

One of the oldest handling devices, the wheelbarrow, is also classified as a hand truck.

Industrial tractors and locomotives form the next classification in the industrial vehicles group and include crawler tractors, wheeled tractors, industrial locomotives, underground mining locomotives, and rack and pinion tractors.

The applications for industrial tractors can be divided into two types: pulling and pushing. The pulling-type applications for industrial tractors are usually assemblages of wheeled vehicles which can be moved over level or slightly ramped paved surfaces. One popular use is the scheduled movement of large amounts of materials from or to loading centers. The number of trailers served by one tractor is governed by the working conditions; normally, one tractor will serve three fleets of four trailers each for loading, transit, and unloading. Maximum hauling distance depends on physical conditions and the number of jobs or stops involved.

Pushing applications are more often involved with necessary plant services rather than with the actual moving of materials. Because a variety of attachments is available for industrial tractors, they can be used for sweeping operations, scoop and shoveling operations, snow removal, scrap iron pickup with an electromagnet, and other operations. They are sometimes used for positioning heavy loads and for moving long objects carried by special trailers.

Industrial trailers come in a wide variety of types and sizes. They are used in conjunction with industrial tractors and fork trucks or are sometimes drawn by hand.

The caster steer trailer is very widely used with industrial tractors. It has rear wheels on rigid axles and front wheels on caster forks. The drawbar is on the platform, and the steering is done through the caster wheels. The rear wheels are larger than the front wheels. This type of trailer is built in capacities of up to 3 tons and generally has a platform 3 feet wide and 6 feet long. The rigid axle is located one-third of the way forward under the platform and carries two-thirds of the load. Caster wheels are located near the front of the trailer. Location of the wheels in this manner throws less weight on the casters, making them easier to swivel under the load. The drawbar and rear hitch can be of either the pintle-and-eye type, the hook-and-eye type, or the automatic coupler type. The platform is equipped with stick pockets uniformly spaced so that racks can be easily shifted from one point to another to support the load. Most platforms are 14 inches above the floor. Because there is a tendency for this type of trailer to creep slightly on right-angle turns, care must be taken to see that they are operated in trains of the proper length to prevent damage. Loading should be done from front to rear to prevent tipping of the truck.

The so-called fifth-wheel steer industrial trailer is another popular type. It is a four-wheel trailer with the rear wheels on rigid axles, the front wheels mounted on an axle with a drawbar attachment, and a fifth-wheel steering unit. These trailers are built in capacities of from 1 to 5 tons for industrial use. Although smaller-capacity trailers have only four wheels, large-tonnage trailers can have multiple-wheel mountings. Trailers can be equipped with various types of racks, dump bodies, removable box sidings, and other attachments. The coupler is generally of the pintle, self-locking type, although eye-and-bolt couplers are sometimes used. These trailers are used within industrial plants and warehouses and on wharves for making up tractor-trailer trains. They should be used on hard-surface roads or good floors and not on soft ground.

Bulk material handling vehicles form an important part of the industrial vehicles classification. They include the big earth-moving equipment used in the construction industry and in road work.

Motor Vehicles—Classification 5.000. The subclassifications listed under motor vehicles are highway semitrailers, highway full trailers, special-purpose motor vehicles, motor truck auxiliary equipment, automobiles and motorcycles, buses, motor trucks, and highway tractors. Three of these classifications—automobiles and motorcycles, buses, and motor trucks—will not be discussed because of their almost universally known characteristics and capabilities and because of their lesser influence in industry.

Highway tractors are the pulling devices used in semitrailer road trucks. They can be gasoline engine, diesel engine, or gas turbine powered. They are built for long periods of heavy duty and contain a complex gear mechanism to provide the wide power range necessary in this type of hauling. They are equipped with a universal table which serves as the hitch to the semitrailer. Sometimes, two sets of dual wheels are used in the rear of the truck for added power and support. Though diesel engines always are in front of the cab, some gasoline engine models have the cab over the engine.

The *highway semitrailers* that the above-mentioned tractors haul are of wide variety and of varying capacities. They include van and refrigerator, flatbed and stake, livestock, dump, tank, auto transport, and heavy-machinery trailers, and semitrailers with pintle hitches.

The semitrailer is the detachable two-wheel trailing unit of an over-the-road motor transportation system. The trailer consists of two main elements: the underconstruction and the body. The underconstruction is designed with a fifth-wheel attachment at its front, so that it and the power unit can be quickly coupled or detached. This feature makes a hinged-in-the-middle vehicle that is easy to maneuver. The underconstruction at the rear is supported upon an axle having one or more pairs of wheels and supporting about 60 percent of the load, the other 40 percent being carried on the rear of the hauling or power-driven unit. When detached, the trailer is still capable of supporting its load by means of permanently attached retractable supports or wheel jacks. The devices for uncoupling and retracting the vertical supports are automatically controlled from the cab of the towing vehicle. The upper structure or body is made in designs ranging from a flat platform 8 feet wide and up to 40 feet long, with bed heights from 30 to 48 inches, to covered, with side or rear doors. Payloads on these trailers range from 10 to 15 tons. Because they are separable, trailer units may be left standing for loading and unloading while the power unit is busy hauling other trailers. Their most economical use is for long distance transportation, but they may be used between buildings of large industrial plants. Some bodies have been equipped with roller conveyors. For use with fork trucks, a body with side doors as well as rear doors is preferred.

Highway full trailers are load-sustaining, over-the-road vehicles designed to be pulled or towed by another power-driven unit. Four-wheel trailers are either reversible or nonreversible types. The reversible type is equipped for steering at either end and has detachable drawbars which permit the trailer to be attached to the pulling unit from either end and without the necessity of turning around. Nonreversible trailers have the steering mechanism and drawbar at one end only, and the rear axle is permanently fixed in position. The body should be designed to suit particular cases; that is, if the material must be enclosed, a panel-type body should be used. Because four-wheel trailers are used for over-the-road travel as well as within large industrial plants, it is necessary to give thought to the body design so it can work in conjunction with other material handling equipment. The four-wheel trailer, because of its self-supporting arrangement, can be moved with any towing vehicle of sufficient drawbar pull. Therefore its use in industrial plants is flexible. It is usually disconnected during loading and unloading, thus releasing the towing vehicle for other uses.

Motor truck auxiliary equipment is growing in importance with the increasing application of motor trucks. Available are elevating tail gates and side platforms that are a permanent part of the truck and are hydraulically operated. They are of great value in loading, particularly in areas where loading docks, elevating platforms, or other loading devices are either difficult or impossible to obtain. They are particularly suitable in yard handling of selected objects. There are also trolley-loading attachments for motor trucks which may be either permanent or portable. Jib-loading devices are sometimes employed by mounting the crane directly on the truck bed and using it for light-duty loading.

Railroad Cars—Classification 6.000. The railroad car classification includes passenger cars, baggage cars, house cars, flat cars, gondola cars, hopper cars, tank cars, locomotives, and special railroad cars.

A *gondola car* is a railroad car with sides and ends and an open top. It is used for carrying freight in bulk. Gondola cars are constructed with steel or wood sides and floor. These are carried on steel frames mounted on two-, four-, or six-wheel trucks. Cars are available with inside dimensions ranging from approximately 40 to 65 feet long, approximately 8 to 10 feet wide, with side walls from 2½ to about 8 feet high. Carrying capacities range from 50 to 120 tons. Standard gondola cars have no doors; however, cars known as general-service cars are available with either drop ends and drop doors in the floor or both. So-called stake pockets are built into some cars for the insertion of vertical posts which are used for fastening high loads securely. Gondolas are used for railroad movement of bulk materials in large quantity and for transporting material which does not require protection from the weather. These cars are usually loaded by overhead cranes or crane trucks, bulk materials being loaded by chutes or conveyors. Cars with drop ends can be loaded by vehicles such as fork or platform trucks. They can also be unloaded by car dumpers if desired. Coal, sand, bricks, and lumber are some of the materials that can be shipped in these cars.

The most important item in the *house cars* classification is the boxcar—a completely enclosed railroad car provided with entrances for loading and unloading freight. Boxcars are constructed of a steel box frame built on two four-wheel railroad trucks. The car body is a steel frame covered with wood or steel sheeting. The interior of the car is wood floored and has unfinished walls or wood walls. Normal boxcar inside dimensions range from 40 to 50 feet long, 8 to 9 feet wide, and 9 to 10 feet high. These are approximate figures. Carrying capacities range from 40 to 50 tons. Standard door equipment consists of two sliding or rolling side doors, centrally located, about 6 feet wide. Special boxcars, known as automobile cars, have double doors on each side. These door openings vary in width from 10 to 15 feet. Boxcars of this type are sometimes provided with double swinging doors in one end of the car.

A more recent development in railroad material handling is the supercapacity car—which is just that. It is 86 feet, 6 inches long; 12 feet, 9 inches high; 9 feet, 2 inches wide; and has a door 20 feet wide for large, bulky freight. Piggybacking is another fairly new practice, which involves transporting truck-trailers on flatbed cars, with delivery to the railroad on both ends by highway tractor. Also, large shipping containers, currently "semistandardized" at 8 by 8 by 10 feet, 20 feet, 30 feet, or 40 feet, are widely used for less than carload (LCL) lots. They have the primary advantage of traveling as unit loads from origin to destination, with no intermediate handling of individual items inside. The economy of loading, handling, and unloading makes possible sharply reduced freight rates.

Boxcars are used for the railroad shipment of materials which require protection from the weather or pilferage. Examples are packaged or loose bulk material. To reduce the packing and loading of expensive material that is shipped from one plant to another, boxcars can be provided with special loading racks designed to handle parts or assemblies as they come off the production line. Because these racks are designed for continuous use, they can be built into the car and protect the product mounted in them from damage.

A *flatcar* is a railroad car with a platform or deck surface for transporting large, bulky freight. It usually consists of a flat platform or deck, a steel supporting frame, and a pair of four-wheel railroad truck units. The platform or deck is usually constructed of wood and is securely fastened to the supporting frame of the car. The supporting frame is mounted on the four-wheel trucks, which vary in number with the load-carrying capacity of the car. Some of the largest flatcars are built with two four-wheel trucks at each end and have a load capacity of 200 tons. Flatcar sizes vary in length from 44 to about 65 feet and in width from 9 to 10½ feet. Industrial trucks can be used for loading or unloading material from these cars if a loading platform or a ramp is available, or if a suitable clear area can be provided on each side of the car. Side stakes of various lengths are used to help secure the load. Weather protection can be obtained by the use of roofing paper or canvas or plastic tarpaulins.

Hopper cars are special railroad cars with hopper bottoms. With the exception of the hopper bottom, they are built very much like gondola cars and have about the same dimensions. They are used exclusively for hauling loose bulk materials such as coal, grain, and iron ore. They can be either covered or open and can also be obtained with a drop end. The hopper bottom does away with the need for railroad car dumpers, as it is easily opened and closed.

Tank cars are cylindrical tanks mounted on railroad car frames and used for transporting liquids and semisolid materials. The frame has two four-wheel railroad trucks. The tank can be built of steel or wood and can be glass or rubber lined. It can also be coated with various materials to prevent corrosion of the tank and the contamination of its contents. The frame can be constructed to absorb damaging shocks and vibration. The dome on the top of a tank car is provided with a loading port and an arrangement of valves for the release of pressure generated by some liquids while in transit. Compartmented tank cars are available for less than carload shipments; each compartment has its own loading port and safety devices. Tank cars are sometimes unloaded from the bottom by gravity through a connection provided for this purpose, or they may be unloaded by air pressure. Several vertical wood tanks can be secured to a suitable car frame to provide another form of tank car. When semisolids are transported, they are unloaded by preheating the material with steam or by extruding it with air pressure.

Marine Vehicles—Classification 7.000. In very early America, waterways and marine transportation were so important that, for many years, the only locations settled were those which had convenient access to marine transportation. At first, the sea, the rivers, and the lakes were used to get the necessities of life to the people.

Marine transportation is a vital link in the import-export business and has a decided effect on the American market. So marine transportation affects industry, not in its plant operation but in its marketing operations. These in turn affect plant operation. Some marine carriers also have a more direct effect on industry. They are a main transportation medium for carrying very large bulk shipments. For instance, they carry millions of tons of iron ore through the Great Lakes from the iron ore ranges in Minnesota to the eastern steel mills. They are also used within the country to carry very large items of finished products on the navigable rivers.

Some of these marine carriers will be discussed here. This classification includes ocean vessels, lake vessels, river and canal boats, barges and lighters, tow boats, small craft, mobile floating equipment, and dredges.

Of the *ocean vessels*, the one most affecting industry is the freighter. It is exclusively a cargo-carrying ship and has a very large capacity. Freighters carry most of the finished products that are exported and imported.

The bulk carrier is another type of ocean vessel used extensively by industry. It is a heavy-duty, large cargo ship that carries bulk materials, usually in raw form and unpackaged. These ships are used to import and export sugar, hemp, bananas, grain, and so forth.

The tanker is still another type of ocean vessel used extensively by industry. Tanks fill virtually the entire ship. Tankers are used extensively by the petroleum industry and are largely responsible for the rapid growth of oil fields all over the world. They are equipped with valve mechanisms for easy loading, unloading, and release of gas pressure in transit.

Lake vessels are very much like ocean vessels, although often much smaller. Freight boats, bulk carriers, and tankers are also used on lakes. Naturally, the Great Lakes see a large percentage of the lake traffic in the United States, and a ship that constitutes a considerable percentage of the Great Lakes traffic is the bulk carrier for iron ore.

Barges and lighters are widely used around seaports, on rivers, and on lakes. Barges come in many forms. The hopper barge, with the hopper-shaped bottom, is used for carrying bulk materials. The deck barge, as the name implies, carries its cargo above the waterline. It is used for carrying either unit load or bulk materials, although more often the former. There are also tank barges; self-unloading bulk barges; covered, dry cargo barges; and other types.

Air Transports—Classification 8.000. This classification was discussed earlier in the chapter, under Air Cargo Handling.

Containers and Supports—Classification 9.000. The storage of material is an ever-present problem in an industrial plant. Several factors can influence the method of storage which may be most suitable to a particular activity. Some of these are the size, weight, characteristics of the material, and the method of handling employed.

The efficient storage of material involves the utilization of floor space, vertical space, and some sort of rack or container. Many kinds of storage containers and racks are available and can be adapted to most any kind of storage problem. Portable storage facilities can be used if the condition is of a temporary nature, or a combination of fixed and portable equipment can be utilized to produce the desired results.

Under the final main classification of containers and supports, the main subclassifications are pressure vessels, tight containers, entirely enclosed containers, open-top containers, platform supports, coil supports, securements, weighing and packaging equipment, and special containers and supports.

Pressure vessels are those used for storing gas or liquid under pressure—acetylene, inert gases, oxygen, nitrogen, and other gases used in many industries. They are thick-walled vessels, the thickness of the wall being a function of the pressure contained. This group also includes comparatively small, compressed gas cartridges and bottles often used on portable tools.

Tight containers are units that are either hermetically sealed or sealed with a breather. They may be made of metal, glass, plastic, or rubber. The group includes metal drums and tight barrels, hogsheads, casks, kegs, tubes, bottles, flasks, carboys, vials, ampoules, and other items. This group also includes the so-called tin can, which has been such an invaluable asset not only to the food processing industry but to every American home as well. In general, tight containers are used in the packaging of perishables, volatiles, foods, and other materials that for one reason or another cannot be exposed to air.

Entirely enclosed containers differ from tight containers in that, although they are entirely enclosed, they are not hermetically sealed. This classification includes barrels, kegs, drums, boxes, chests, trunks, cases, crates, cartons, cans with loose lids, bags, wrappings, and many others. Much attention has been given to the design of the corrugated fiberboard boxes used so extensively in industry. They are available in many types and are enclosed by means of glue, gum tape, or staples. Wood boxes are used for heavier items, and many shipping departments have nailing machines for the rapid building of these boxes. In some cases, containers made of either fiberboard or corrugated paper are reinforced with wood cleats to increase their strength. This method is less expensive than all-wood crating, and they can be made with excellent eye appeal and still have minimum weight. Most entirely enclosed containers have been made with some provision for weatherproofing them. This can be done either by outside or interior coatings or by enclosing the actual material in a special weatherproof inside container.

Open-top containers are, as the name connotes, containers which are not totally enclosed but have an open top. This group includes shop boxes and containers, tote boxes, pans and baskets, bins, silos, bunkers, hoppers, and many others. Very often, it is possible to ship partially processed parts or materials in these open-top containers. They are particularly advantageous in the transfer of material between widely separated processes. Wire mesh boxes or bins are sometimes made so they can be set directly on a pallet or made with a pallet bottom. Often, bins and boxes are made portable, with an open side or end designed for easy movement by handling equipment, and sometimes are designed for vertical stacking. They are available in several styles and sizes according to the material being handled and the method of handling. Portable bins and boxes provide a flexible storage facility which can be changed in size and location to meet production needs while still taking advantage of vertical storage space. The extremely useful shop box is also a member of this group. These range from the open-top, all-steel tote box to the wire, canvas, and perforated metal models. The type of box used will again depend upon the product being handled and the method of handling.

Platform supports are structures used for the support of another load. This group includes the pallet, the skid platform, separators, racks, skids, and others. The pallet is undoubtedly the most prominent member of this group. Pallets come in all sizes, are made of many materials, and are of several styles and for many uses. There are assembly pallets, transportation pallets, expendable pallets, and others. They can be made to permit either two- or four-way fork truck approach. They can be made with collapsible boxes on the usual supporting structure. These boxes can be made of steel, paper, wood, or wire. The pallets themselves may be made of aluminum, wood, corrugated paper, fiberboard, plywood, steel, or other materials. Steel skeleton pallets have been made of welded steel construction with semiopen framework for use in palletizing and depalletizing with a lift truck. Other important members of this family are the skids and skid platforms. Skid platforms are low, flat tables equipped with legs 6 to 11 inches high. They can be made of metal or reinforced wood. When they are equipped with wheels or casters under their legs, they are referred to as live skids or platforms. Live skids are valuable because they can be handled manually as well as by lift trucks. Skids can also be made with either collapsible or permanent boxes on the main supporting structures. They have literally hundreds of applications in general industry.

There are a wide variety of *securements* for strapping, bracing, and fastening. They range in size from small wire staples to large chains and cables, and their application depends upon the product and the method of handling. Also in the classification of securements must come the portable bulkheads used in trucks, railroad cars, and airplanes for securing equipment in transit. Some of these bulkheads are permanently attached, some are temporary. Also in the securement classification comes dunnage for padding material in transit. Dunnage is an important problem to shipping operations.

Although more than 50 percent of the containers and supports on the market are universally used, it is possible to obtain a special container for the particular product either by adapting one of these more or less universally used containers or by having a container designed around the product. A well-designed container can often mean the saving of a great deal of money in reduced breakage, handling, and loss.

BIBLIOGRAPHY

Books

Apple, James M., *Plant Layout and Material Handling*, 2d ed., The Ronald Press Company, New York, 1963.

Bolz, H. A., and G. E. Hagemann, *Materials Handling Handbook*, The Ronald Press Company, New York, 1958.

Briggs, Andrew J., *Warehouse Operations, Planning and Management*, John Wiley & Sons, Inc., New York, 1960.

Conveyor Terms and Definitions, Conveyor Equipment Manufacturers Association, Washington, D.C., 1966.

Haynes, D. O., *Material Handling Applications*, Chilton Company, Philadelphia, Pa., 1958.

Haynes, D. O., *Material Handling Equipment*, Chilton Company, Philadelphia, Pa., 1957.

An Introduction to Material Handling, Material Handling Institute, Inc., Pittsburgh, Pa., 1967.

Material Handling Engineering Directory and Handbook, Industrial Publishing Company, Cleveland, Ohio, biennial.

Warehouse Operations, U.S. Government Printing Office, Washington, D.C., 1969.

Periodicals

Automation
Distribution Worldwide
Handling & Shipping
Material Handling Engineering
Mechanical Handling
Modern Manufacturing
Modern Materials Handling
Plant Engineering
Transportation & Distribution Management

Industrial Engineering Tools

Chapter 1

Industrial Engineering Terminology

RITA M. CARLSON

**Executive Assistant, Maynard Research
Council Incorporated, Pittsburgh, Pennsylvania**

The key to cooperation among people, among segments of an organization, or among organizations is complete understanding. This understanding can be reached only through effective communication, which in turn is achieved through the use of a clearly defined, standardized terminology. To be able to communicate effectively, people must talk the same language.

As industrial engineering has developed into an interdisciplinary activity, its terminology has expanded rapidly. This rapid growth has caused some ambiguity and lack of clarity in the use and interpretation of terms.

Recognizing the need to define the language of the industrial engineer and to establish terminology standards, the Work Standardization Committee of the Management Division of the American Society of Mechanical Engineers pioneered in the effort to standardize industrial engineering terminology. After twelve years of research, development, and review, the final definitions—although still considered tentative—were approved as a partial ASME standard and were published in 1955. The American Institute of Industrial Engineers and the American Society of Mechanical Engineers jointly sponsor, as a continuing project, the development of new definitions.

The following definitions will be useful in interpreting the meanings of many terms used throughout this Handbook. Many of the definitions are based on the ASME publication,[1] but to keep pace with the new techniques which have been added to the field of engineering since 1955, several other publications were used as sources for expansion of the terminology list. Among them were:

> *Handbook of Business Administration,* H. B. Maynard (ed.), McGraw-Hill Book Company, New York, 1967.
> *Handbook of Modern Manufacturing Management,* H. B. Maynard (ed.), McGraw-Hill Book Company, New York, 1970.
> "Industrial Engineering Terminology Manual," *The Journal of Industrial Engineering,* November–December, 1965.

Abnormal Reading: See Abnormal Time.

Abnormal Time: The elapsed time for any element recorded during a time study which, being excessively longer or shorter than the majority or median of the elapsed times, is judged at the time of the study to be not representative for the element, and which may be excluded in determining the most typical elapsed time (or the average time) for the element.

Absorption Costing: A cost method that charges to inventory, in addition to direct labor and direct material, all manufacturing overhead.   These overhead costs are considered to be inventoriable until the unit is sold.

Acceptable Quality Level (AQL):   Usually, a stated fraction defective which has been determined tolerable with respect to functional and economic requirements of the item. In some cases, it may also refer to the acceptable universe average of a variable property of the items.

Acceptance Sampling: The extraction of a portion of a lot of material to be inspected for the purpose of determining whether the entire lot will be accepted or rejected.

Accident Rate: A measure of the disabling accidents occurring in any specified exposure of workers to employment hazards.   1. Frequency rate—the number of disabling or lost-time accidents in an exposure of one million man-hours worked.   2. Severity rate—the total number of lost man-days charged to disabling accidents during an exposure of one million man-hours worked.

Accumulative Timing: A time study technique utilizing two stopwatches connected so that when one is stopped, the other is simultaneously started.   Each watch is thus read alternatively while its hand is stationary.

Accuracy: Degree of correctness, exactness, or precision.   The relationship between the mean value of a large number of measurements and the objective true value of the quality measured.

Action-Demand Chart: Relates performance variances from projected standards to the degree of action demanded from management.

Activity (time study usage):   The number of times a given operation or occurrence is repeated during a given period, usually a year.

Activity Network: A representation of two particular aspects of a project: (1) the precedence relationship among the activities and (2) the duration of each activity.

Activity Sampling: See Work Sampling Study.

Actual Cost: An acceptable approximation of the true cost of producing a part, product, or group of parts or products, including all labor and material costs and a reasonable allocation of overhead charges.

Actual Hours: See Actual Time.

Actual Time: The time taken by a workman to complete a task or an element of a task.

Addition Theorem: A law of chance that states that if an event may occur in several different ways, then the probability of the event occurring should be stated as the sum of the probabilities of each of the different ways.

Administration: The function which is concerned with the determination of the general objectives, major policies, and organizational structure of an enterprise.

Algorithm: A method for deriving a solution to a problem by using the results from each cycle to refine the following cycle.   The process is repeated until a satisfactory answer is obtained.

Allowance: A time increment included in the standard time for an operation to compensate the workman for production lost due to fatigue and normally expected interruptions, such as for personal and unavoidable delays.   It is usually applied as a percentage of the normal or leveled time.

---

[1] *ASME Standard Industrial Engineering Terminology,* The American Society of Mechanical Engineers, New York, 1955.

Allowed Hours: See Allowed Time.

Allowed Time: The leveled time plus allowances for fatigue and delays. See Standard Time.

Alternate Time Standard: A standard allowed time developed for use with a method of performing a task other than the established standard method.

Aptitude: The physical and psychological potentiality of an individual to perform a specific type of work.

Arrival Rate: The mean number of units arriving at a service facility during a given interval of time.

Assemble (n.): The basic element employed when one or more objects are put on or into another object so that they fit or contact each other in a predetermined relation to form a unit.

Assembly: Two or more parts or subassemblies joined together to form a complete machine, structure, or other article.

Assembly Line: The arrangement of machines, equipment, material, and workers which permits the work in process to progress sequentially from operation to operation until the product (or product component) has been assembled.

Assignable Cause: A significant cause of variation in a process which is either much larger than or of a different origin from the random causes. "Assignable" means that it may be possible to identify the nature of this cause.

Assignment Problem: A special case of the mathematical programming problem concerned with the optimum assignment of resources (facilities or personnel) to needs (jobs), where each resource must be assigned to one and only one need.

Attitude Survey: A study of the opinions and attitudes of employees concerning established policies, practices, working conditions, or some other facet of employee relations; a morale survey.

Automaticity: The ability to perform hand, arm, leg, or body motions or motion patterns, without apparent mental direction, as a result of practice.

Automation: A substitution of machine labor for human labor, either manual or intellectual.

Auxiliary Department: A department that provides services such as maintenance, material handling, warehousing, and the like to the production departments. It rarely performs manufacturing operations on the product to be marketed.

Average Earned Rate: The total earnings of an individual or group of individuals for a period divided by the number of man-hours worked during the period. Total earnings include all the components which are a function of pay per hour, such as base rate earnings, shift differentials, incentive earnings, overtime premiums, and the like, but not profit sharing bonuses, Christmas bonuses, or other bonuses that are not a function of pay per hour.

Average Earnings: The total earnings of an individual or group of individuals during a specified period divided by the number of man-hours, man-days, man-weeks, man-pay periods, or any similar measure of the time elapsed during the specified period.

Average Elemental Time: 1. The sum of all the unleveled, individual actual times recorded for an element divided by the number of unleveled, individual actual times. 2. The sum of all the consistent unleveled, individual actual times recorded for an element divided by the number of consistent unleveled, individual actual times.

Average Sample Number: The expected number of pieces which must be inspected to determine the acceptability of a lot. This is a function of the sampling plan used and the incoming quality.

Average Selected Time: See Average Elemental Time and Average Time.

Average Time: The arithmetical average of all the actual times, or of all except the abnormal times, taken by a workman to complete a task or an element of a task.

Avoidable Delay: Any time during an assigned work period which is within the control of the workman and which he uses for idling or for doing things unnecessary to the performance of the operation. Such time does not include allowance for personal requirements, fatigue, and unavoidable delays.

Balance: 1. That quality of a motion sequence which promotes the development of rhythm and automaticity. 2. As applied to progressive related operations, it is the condition in which the standard times required for each successive operation are approximately equal and the work flows steadily or at a desired rate from one operation to the next.

Balanced Line: A series of progressive related operations with approximately equal standard times for each, arranged so that work flows at a desired steady rate from one operation to the next.

Balanced Motion Pattern: A motion sequence that promotes the development of rhythm and automaticity.

Balancing Delay: The delay which occurs when one body member performs its work faster than another body member because of different motions due to the requirements of the layout or the required sequence of motions, and therefore must wait for the slower member or must work more slowly so as to finish its work simultaneously with the slower body member.

Base Pay: 1. See Base Wage Rate. 2. The product of a workman's base wage rate and the time he worked during a pay period, when expressed in the proper measurement units.

Base Points: 1. The minimum point values assigned to the factors of a job evaluation plan. 2. The minimum points assigned to any job by an evaluation system.

Base Time: 1. See Standard Time. 2. In the piecework system of wage payment, that time value which, when multiplied by the applicable base wage rate, gives the piece rate.

Base Wage Rate: The amount of pay per hour, or other unit of time, established to compensate the workman for the requirements and conditions associated with a job. It is generally determined through job evaluation.

Basic Division of Accomplishment (Work): The smallest elements of human activity or inactivity used in any particular system of motion analysis. See Therblig.

Basic Element: See Basic Division of Accomplishment.

Basic Motion: See Basic Division of Accomplishment.

Basic Motion Timestudy (BMT): A system of predetermined motion time standards. The essence of the system lies in the arbitrary definition of a basic motion as one that commences from rest and ends at rest. The system's purpose is to establish time standards for procedures that are composed of human motions controlled only by the individual performing them, and to do so without resorting to time study. An allied purpose is to facilitate the analysis of methods.

Batch Processing: Collection of data over a period of time to be sorted and processed as a group during a particular machine run.

Behavioral Sciences: Sciences dealing with human behavior, such as psychology, sociology, or anthropology.

Benefit-Cost Ratio: The dollar estimate of benefits, advantages, or gains from a project divided by the dollar costs of the project.

Binomial Distribution: A frequency distribution which is applicable to events or phenomena which can be classified dichotomously.

Bit: A binary digit; a quantum of information; a single pulse in a group of pulses.

Bonus: The portion of wages, in excess of base wages and overtime earnings, derived from incentive plan payments.

Bonus Plan: See Financial Incentive Plan.

Break-even Chart: A graphic representation of the relation between total income and total costs for various levels of production and sales indicating areas of profit and loss.

Break-even Point: The level of sales where profits start.

Break Point: 1. The end of an element in a work cycle and the point at which a time study reading is made. 2. A specified place in a computer routine where the routine may be interrupted by external intervention or by a monitor routine.

Budget: An organized statement of expected income and expenditures for a definite period, usually a month or a year, made to assist in controlling expenditures and to provide a criterion for judging performance during that period.

Burden: See Overhead.

Byte: A sequence of adjacent binary digits operated upon as a unit and usually shorter than a word.

Camera Study: See Micromotion Study, definition 2.

Chance Cause: A cause of variation in a process which comes and goes in an unpredictable manner and cannot be identified as being present or absent at any given time.

Chance Variation: Variation due to the combination of chance causes.

Change Direction: The basic element employed to change the line or plane along which a Reach or Move (or a Transport Empty or Transport Loaded) is made.

Check Study: See Restudy.

Chronocyclegraph Technique: A special type of cyclegraph technique in which an interrupter is placed in the electric circuit with a light bulb, and the light is flashed on and off. The slow cooling of the filament causes the path of the bulb when photographed to appear to be pear-shaped dots which indicate the direction and the path of the motion. Because the spots of light are spaced according to the speed of the movement—being widely separated when the workman moves fast and close together when the workman's movement is slow—it is possible, from the photograph, to obtain an approximation of time, speed, acceleration, and retardation and to show direction and the path of motion.

Compiler: A computer program that will translate a set of statements written in a specific language into a program in the machine language of a computer.

Consistency: The degree of uniformity or agreement which exists among the actual times recorded for two or more repetitions of the same element during a time study.

Constant Element: 1. An element for which the leveled or normal time is always the same, regardless of the characteristics of the parts being worked upon, as long as the method and the working conditions are unchanged.   2. An element for which, under a specified set of conditions, the standard time allowance should always be the same.   Example: Raise spindle a definite distance on a drill press of a certain size and make.

Continuous Cost System: A cost procedure used in continuous manufacturing environments where costs are added as the manufacturing process progresses from beginning to end.

Continuous Method: The procedure of timing, used in making time studies, whereby the watch is permitted to run continuously throughout the period of study while the observer notes and records the reading of the watch at the end of each element, delay, or any other occurrence happening in the study, regardless of whether or not it has a direct bearing on the job.   The elapsed times are secured by subtracting the successive readings after the timing has been completed.

Continuous Timing: See Continuous Method.

Control Charts: A chart in which some observed or computed property of a product or process is plotted, usually in the order of production, for the purpose of ascertaining the nature of the variation in the process and the possible need for corrective action.

Control System: An administrative system that has as its primary function the collection and analysis of feedback from a given set of functions for the purpose of controlling those functions.   Control may be implemented by monitoring and systematically modifying parameters or policies used in those functions, or by preparing control reports that initiate useful action with respect to significant deviations and exceptions.

Cost Center: Any subdivision of an organization comprised of workmen, equipment, areas, activities, or combination of these that is established for the purpose of assigning or allocating costs.

Cost Reduction Report: A form designed to allow easy comparison of two or more methods, plans, designs, and the like on the basis of known or anticipated costs and savings.

Critical Path: The sequence of jobs or activities in a network analysis project such that the total duration of the project is equal to the sum of the durations of individual jobs in the sequence.   There is no time leeway or slack (float) in any activity along the critical path.   That is, if the time to complete one or more jobs in the critical path increases, the total project time increases.   See Network Analysis.

Critical Path Method (CPM):   A network technique for scheduling resources to accomplish a certain job within time constraints, preferably where time and cost estimates can be obtained with a relatively high degree of certainty.

Cutting Speed: The relative velocity, usually expressed in feet per minute, between a cutting tool and the surface of the material from which it is removing stock.

Cyclegraph Technique: The method devised by the Gilbreths and used in motion study by which a three-dimensional pattern of a motion path may be recorded by attaching a small electric light bulb to the finger, hand, or other part of the workman's body and photographing, with a stereoscopic camera, the path of the light as it moves through space.

Cycle Timing: 1. Observance of the total time required to complete a cycle.   2. A time study technique used to time work elements that are too short to time in the usual manner.   It consists of timing a cycle or periodically recurring series of elements, first including and then excluding the element for which the time is needed. The needed time for this element is then obtained by subtraction.

Data Processing: The performance of certain clerical functions upon a selected body of information. Four basic clerical functions are involved, whether they are performed manually or mechanically: *input,* the introduction of raw data into the system; *manipulation,* the arrangement of data into a desired pattern; *computation,* the process by which arithmetic operations are performed on the data; and *output,* the presentation of the results of manipulation and computation in the required form. These functions, together with the controls exercised over them, constitute a data processing system.

Day Rate: Rate of compensation for daywork as differentiated from incentive work. Usually expressed in terms of money paid per period of time.

Daywork: Work for which the hourly or daily compensation is not directly dependent upon the quantity of production, as is the case in incentive work.

Debug: To isolate and remove all malfunctions from machines or equipment or all mistakes from a routine.

Decimal Hour Stopwatch: A two-handed timing device whose movement may be started or stopped manually and whose large outer dial is divided into 100 spaces, each of which represents 0.0001 hour.   The position of the large hand on the large dial indicates time in hours to four decimal places, and the position of the small hand on a smaller, inner dial indicates time in hours to two decimal places.

Decimal Minute Stopwatch: A two-handed timing device whose movement may be started or stopped manually and whose large outer dial is divided into 100 spaces, each of which represents 0.01 minute. The position of the large hand on the large dial indicates time in minutes to two decimal places, and the position of the small hand on a smaller, inner dial indicates time in whole minutes.

Delay: A period during which conditions (except those which intentionally change the physical or chemical characteristics of an object) do not permit or require immediate performance of the next planned action.

Delay Allowance: 1. A time increment included in a time standard to allow for contingencies and minor delays beyond the control of the workman.   2. A separate credit (in time or money) to compensate the workman on incentive for a specific instance of delay not covered by the piece rate or standard.

Depletion: A lessening of the value of an asset due to a decrease in the quantity available. It is similar to depreciation except that it refers to such natural resources as coal, oil, and timber in forests.

Depreciable Value: 1. The difference between the first cost of an asset to the current owner and the net recoverable value at the time of its disposal. 2. The recorded or book value of an asset at any time.

Depreciation: The actual decline in the value of an asset due to exhaustion, wear and tear, and obsolescence.

Depreciation Base: The actual or adjusted initial cost of an asset to which a depreciation rate is applied in computing depreciation cost or expense.

Depreciation Expense: A periodic accounting charge or operating cost arising from the systematic writing off of assets in the accounting records.

Differential Piecework: A wage incentive plan that employs two or more piece rates. One piece rate is paid if the expected output is not attained. A higher piece rate is paid if the expected output is attained or exceeded.   While originally devised by F. W. Taylor to provide only two piece rates, modifications of the plan provide for more than two.

Differential Time Plan: See Multiple Time Plan.

Differential Timing: See Cycle Timing.

Direct Costing: A cost method which charges to inventory, in addition to direct labor and direct material, those manufacturing costs which vary directly as a result of manufacturing the product.

Direct Labor: Work which alters the composition, condition, conformation, or construction of the product, the cost of which can be identified with and assessed against a particular part, product, or group of parts or products accurately and without undue effort and expense.

Direct Labor Standard: A specified output or a time allowance established for a direct labor operation.

Direct Material: All material that enters into and becomes part of the finished product (including waste), the cost of which can be identified with and assessed against a particular part, product, or group of parts or products accurately and without undue effort and expense.

Disassemble (n.):   The basic element denoting the removal of a part of a unit or assembly.

Disengage (n.):   The basic element employed to break the contact between one object and another. It is characterized by an involuntary recoil caused by the sudden ending of resistance.

Dispose (n.):   An element of a total operation that involves the laying aside or releasing or otherwise getting rid of a part, assembly, tool, or other object during or at the end of the operation.

Do (n.):   The basic element that accomplishes in full or in part the purpose of the operation. It includes the basic elements Use and Assemble and may sometimes be expressed in terms of other basic elements.

Downtime: A period of time that is usually equal to or greater than a specified minimum during which an operation is halted due to a lack of materials, a machinery breakdown, or the like.

Drop Delivery: 1. The method of introducing an object to the workplace by gravity. 2. *a.* A method whereby a chute or container is so placed that, when work on a part in question is finished, it will fall or drop into a chute or container or onto a conveyor with little or no "transport" by the workman.   *b.* The laying aside of a part by releasing it so that it falls or moves away from the work area either through the force of gravity or by mechanical or other means.

Dynamic Programming: A mathematical programming technique involving sequential multistage problem situations. At each stage, a decision (solution) must be made among alternatives.   However, as each decision is made, the parameters of the remaining stages of the problem change.   Dynamic programming is thus the most sophisticated of the mathematical programming techniques.

Earned Hours: The time in standard hours credited to a workman or group of workmen as a result of their completion of a given task or group of tasks.

Earned Rate: See Average Earned Rate.

Economic Lot Size: That number of units of material or a manufactured item that can be purchased or produced within the lowest unit cost range.   Its determination involves reconciling the decreasing trend in preparation unit costs and the increasing trend in unit costs of storage, interest, insurance, depreciation, and other costs incident to ownership as the size of the lot is increased.

EDP (Electronic Data Processing):   The use of electronic equipment to perform integrated data processing.

Efficiency: 1. The ratio of standard performance time to actual performance time, usually expressed as a percentage. 2. The ratio of actual performance numbers (for example, number of pieces) to standard performance numbers, usually expressed as a percentage.

Effort: 1. The evidence of the will to work as manifested by a workman performing an operation. 2. The sum total of the mental absorption and physical participation which may be required by a workman on a given operation.

Effort Rating: 1. That part of any performance rating technique concerned with evaluating the extent or degree to which the will to work is exhibited by a workman. 2. See Speed Rating.

Elapsed Time: 1. The actual time taken by a workman to complete a task, an operation, or an element of an operation. 2. The total time interval from the beginning to the end of a time study.

Element: A subdivision of the work cycle, composed of a sequence of one or several fundamental motions and/or machine or process activities, which is distinct, describable, and measurable.

Elemental Motion: See Basic Division of Accomplishment.

Elemental Time Value: See Element Time.

Element Breakdown: The subdivisions of an operation, each of which is composed of a distinct, describable, and measurable sequence of one or several fundamental motions and/or machine or process activities.

Element Time: The term used to indicate either the actual, observed, selected, normal, or standard time to perform an element of an operation.

Emerson Efficiency Plan: A wage incentive plan providing a guaranteed base wage rate and an increasing rate of incentive payment beginning when performance reaches 67 percent of standard.   Under the original plan, incentive payment became constant at 120 percent of the base wage rate when standard performance was reached. Modifications of this plan provide increasing incentive payment for performance above standard.

Evaluated Wage Rate: A wage rate for a job or position, determined through a job evaluation plan.

Examine (n.): The basic element employed when comparing the quality of an object with a definite standard by means of any of the sense organs. (Examine was formerly called Inspect, but the name has been changed to avoid confusion with factory inspection operations.)

External Element: An element of a processing operation performed by the workman outside of the machine cycle. It usually begins with "stop machine" and ends with "start machine."

Factor Comparison: A type of job evaluation plan in which relative values for each of a specified number of factors of a job are established by direct comparison with the values established for these same factors on selected key jobs.

Failure Analysis Activity: A program of systematic collection and analysis of data, the detection and selection of significant deviations or variations from established limits, isolation of the cause, analysis of the defect, and finally, the recommendations for corrective action with time follow-up.

Fair Day's Work: The amount of work that can be produced during a working day by a qualified individual with average skill who follows a prescribed method, works under specified conditions, and exerts average effort.

Fall Down (v.): To fail to meet the standard or expected performance level.

Falldown (n.): A performance that is less than standard.

Fatigue: A physical or mental weariness, real or imaginary, adversely affecting a person's ability to perform work.

Fatigue Allowance: Time included in the production standard to allow for decreases or losses in production which might be attributed to fatigue. Usually applied as a percentage of the leveled, normal, or adjusted time.

FIFO (First In, First Out): A method of determining the cost of inventory used in a product. In this method, the costs of material are transferred to the product in chronological order. Also used to describe the movement of materials.

Film Analysis: A frame-by-frame study of a motion picture of an operation to determine the motions used, their sequence, and the time taken for each. See Micromotion Study, definition 2.

Film Analysis Chart: A graphic representation of the activities of the various body members as determined by film analysis. Often referred to as right-and-left-hand chart or simo chart.

Film Analysis Record: A tabular record of the data obtained from a film analysis.

Financial Incentive: A monetary inducement, other than base or overtime wages, bearing some predetermined relationship to performance, such as quantity or quality of work, reduction of spoilage, or some other desired result, paid to the workman as a reward for meeting or exceeding an established standard of performance.

Financial Incentive Plan: A method of systematic financial remuneration in relation to specified standards of performance. Performance may relate to quantity, quality, control of waste, costs, or other factors. Commonly, the systems provide extra remuneration for achievement in excess of a prescribed base.

Find (n.): The basic element following the search element which denotes the mental recognition of the desired part or object for which one is searching.

First-piece Time: The time required to produce the first of a number of identical units, including all necessary setup and makeready time.

Fixed Expense: Expenditures that remain constant with respect to time, regardless of the volume of production, such as taxes (on land and buildings), insurance, certain administrative salaries, and the like.

Fixture: A device which (1) holds material in a desired position but does not guide the machine or tools performing the necessary operations or (2) holds two or more parts in prescribed positions relative to each other while the parts are being assembled or joined together.

Flow Diagram: A graphic representation on a floor plan of the work area involved of the locations of work stations and the paths of movement of men and materials.

Flow Process Chart: A graphic representation of the sequence of all operations, transportations, inspections, delays, and storages occurring during a process or procedure. It includes information considered desirable for analysis, such as time required and distance moved. *a.* The material type presents the process in terms of the events which occur to the material. *b.* The man type presents the process in terms of the activities of the man.

Foreign Element: An interruption which is not a regular occurrence in the work cycle or operation, and one for which no provision was made in the normal sequence of elements of a time study.

Frame: 1. An individual picture or space for a single picture on a motion picture film.   2. In motion study work, the time which elapses between two successive exposures of a motion picture film.   This time depends upon the speed with which the picture is taken, usually 16 frames per second.

Frame Counter: A mechanical counter which can be used to determine the number of frames that have passed a predetermined point in a motion picture.   The frame counter may be attached to any device for showing or viewing motion pictures.

Frequency: 1. The number of times an element occurs during an operation cycle. 2. The number of times a specific value occurs within a sample of several measurements of the same dimension or characteristic of several similar items.

Frequency Study: A study made to determine the number of occurrences of elements during a given period.

Functional Layout: See Process Layout.

Gain Sharing: A feature of some wage incentive plans which divides the bonus as computed by the wage incentive plan formula in some predetermined proportion between management and the workman.

Game Theory: A method of determining the best strategy in competitive situations where the outcome of various strategies can be precisely predicted.   It generally has limited application to real-world problems, although it has been used extensively as the basis for executive development and training programs involving simulation of competitive business situations.

Gantt Chart: A graphic representation on a time scale of the current relationship between actual and planned performance.

Gantt Task and Bonus Plan: A wage incentive plan that originally provided a low guaranteed rate but offered a strong inducement to meet or better standard performance by paying a step bonus when standard performance was reached and the equivalent of a piece rate thereafter.   Modifications of the plan tend to provide higher guaranteed rates while maintaining its strong incentive feature.

Gilbreth Basic Element: The name given by the American Society of Mechanical Engineers to each of the basic divisions of accomplishment defined and used by Frank B. and Lillian M. Gilbreth to classify physical motions and associated mental processes.   Synonym: Therblig.

Going Rate: The base rate or wage commonly paid for a given job or class of work in a community, area, or industry.

Graphic Rating Scale: A method of rating or appraising characteristics of individuals, jobs, organizations, and the like.   For each characteristic to be rated, several descriptive statements are arranged in either ascending or descending order of quality.   A check mark placed opposite the statement best indicating the degree to which each characteristic is possessed permits a visual comparison with the extremes to which the characteristic could be possessed.

Grasp (n.):   The basic element employed when the predominant purpose is to secure sufficient control of one or more objects with the fingers or the hand to permit the performance of the next required basic element.

Gravity Feed: A method of supplying materials into a machine or to a work station by the force of gravity. Generally, a hopper and/or a chute is used to store and to guide the materials to the point of use.

Group Incentive: Any financial incentive plan under which the output of workmen performing the same, related, or interdependent operations is pooled and their earnings resulting from production above the established standard are distributed to the members of the group according to some predetermined plan.

Group Leader: A member of a production unit or crew who is responsible for the coordination of the unit's efforts and who may assist other members in the performance of their assigned tasks.

Group Payment: See Group Incentive.

Group Timing Technique: A work measurement procedure for multiple activities that enables one observer using a stopwatch to make a detailed elemental time study on from two to fifteen men or machines at the same time.

Guaranteed Annual Wage: A minimum amount of money which an employee is assured he will receive during a given year.

Guaranteed Hourly Rate: See Guaranteed Wage Rate.
Guaranteed Time Standard: An established expected performance level which management assures will not be changed, regardless of workmen's earnings, unless there is a significant change in quality requirements, method, materials, tools, layout, equipment, teeds, speeds, design, or working conditions.
Guaranteed Wage Rate: The assured minimum amount of compensation per hour or other unit of time paid under a financial incentive plan even though the workman fails to reach the established standard or specified level of performance.

Halo Effect: That tendency on the part of a rater to rate an individual the same on all traits because of the general impression he has or forms immediately, or because of the effect of one particularly dominating characteristic of the person being rated. It is present in all ratings to a greater or lesser degree and is caused by the rater's inability to isolate and evaluate independently all the various traits or characteristics which an individual may possess.
Handling Time: 1. The time required to perform the manual portion of an operation. 2. The time required to move materials or parts to or from a work station.
Hardware (computer usage):   The physical equipment, such as mechanical, magnetic, electrical, and electronic devices, associated with a computer.
Hold (n.):   The basic element employed when the hand maintains static control of an object while work is being performed on it.
Human Factors Engineering: A merging of those branches of engineering and the behavioral sciences which concern themselves principally with the human component in the design and operation of man-machine systems.   Based on a fundamental knowledge and study of man's physical and mental abilities and emotional characteristics.
Hundred Percent Incentive: A feature of some wage incentive plans which gives the workman the entire monetary value of the time which he saves by exceeding a specified level of performance.

Idle Time: 1. A time interval during which either the workman, the equipment, or both do not perform useful work.   2. In motion study, the interval during which a body member does not perform useful work.
Incentive: Any factor which motivates a workman to maintain or exceed an established standard of performance.   May be financial or nonfinancial in nature.
Incentive Earnings: The amount of money paid to a workman in excess of the guaranteed hourly rate for performance at or above the established standard.
Incentive Operation: Work compensated for in a manner that motivates those executing it to maintain or exceed an established standard performance level.
Incentive Opportunity: The possibility for an individual to earn more than his guaranteed base wage rate by maintaining or exceeding the established standard performance level.
Incentive Pace: The performance level at which a qualified workman works when earning incentive pay.
Incentive Performance: The execution of work by a qualified individual following a specified method in such a way that his average output during a specified period of time equals or exceeds the established standard level of output.
Incentive Rate: The hourly wage rate used for incentive calculations.
Income Statement: A report of the revenue and expenses of an accounting entity for a specified period of time.
Indirect Expense: Costs necessary in manufacturing which cannot be readily identified with or charged to a particular part, product, or group of parts or products.
Indirect Labor: 1. Work which is performed rendering services necessary to production, the cost of which cannot be assessed against any part, product, or group of parts or products accurately or without undue effort and expense.   2. Necessary work which does not alter the composition, condition, conformation, or construction of the product.
Indirect Manufacturing Expense: See Overhead.
Indirect Material: Material consumed in the process of production or manufacture that does not become a part of the finished product or cannot be readily identified with or charged to a particular part, product, or group of parts or products.
Industrial Engineer: One who has the necessary education, training, experience, and personal attributes to perform the work included in the field of industrial engineering.
Industrial Engineering: The design, improvement, and installation of integrated systems of men, materials, and equipment.   It draws upon specialized knowledge and skill in the mathematical, physical, and social sciences, together with the principles

and methods of engineering analysis and design, to specify, predict, and evaluate the results to be obtained from such systems.

Industrial Relations: The management function that deals with all phases of employee-management relationships.   Its objective is to devise and administer plans and procedures that engender and stimulate employee productive effort, cooperation, and job satisfaction.

Inspect: See Examine.

Inspection: Examining an object for identification or checking it for verification of quality or quantity in any of its characteristics.

Installation: 1. The execution of the steps or measures necessary to introduce a technique, procedure, or proposed course of action into an organization and to get it functioning properly.   2. A technique, procedure, or equipment arrangement that is being set up or used by a company.

Instruction Card: Written information supplied to a workman that specifies method, machines—and when appropriate, their speeds, feeds, and depth of cut—tools, fixtures, specification limits, and the like to be used in performing a task.

Instruction Sheet: See Instruction Card.

Integrated Data Processing: A system designed as a whole, which permits the data, once recorded, to be used for whatever purpose is required, as often as necessary, without manual copying.   An integrated system may be manual, mechanized, or a combination of both.

Interference Time: A period of time during which one or more machines are not operating because the workman or workmen assigned to operate them are busy operating other machines in their assignment or are performing necessary duties related to operating such other machines, for example, making repairs, cleaning the machines, or inspecting completed work.

Intermittent Element: An element essential to an operation, which occurs at less regular intervals than those of the regular or basic cycle of elements.

Internal Element: 1. Any element performed by a workman while the machine he controls is operating automatically.   2. A short-duration element performed by one hand while the other hand is performing a more time-consuming element.

Inventory: 1. (n.) All the materials, parts, supplies, expense tools, and in-process or finished products recorded on the books by an organization and kept in its storerooms, warehouses, or plants.   2. (n.) A list of the names, quantities, or monetary values of all or any group or classification of the items specified in definition 1.   3. (v.) To identify; count, weigh, or measure; and evaluate the worth of all or any group or classification of the items specified in definition 1 and to record this information.

Inventory Control: The technique of maintaining stockkeeping items at desired levels, whether they are raw materials, goods in process, or finished products.   A glossary of inventory control terms is given on pp. 8-116 to 8-118 of Section 8.

Irregular Element: See Intermittent Element.

Jig: A device which holds a piece of work in a desired position and guides the tool or tools which perform the necessary operations.

Job: 1. A position or post of employment.   2. A group of tasks assigned to an employee or group of employees.

Job Analysis: A detailed examination of a job to determine the duties, responsibilities, and specialized requirements necessary for its performance.

Job Breakdown: A description of a task in terms of its elements.

Job Characteristic: See Job Factor.

Job Class: 1. A group of jobs or positions having approximately the same relative worth as determined by a job evaluation plan.   2. A group of jobs or positions with duties and responsibilities so similar that individuals with approximately equivalent education, experience, skills, and the like are required for their satisfactory performance.

Job Classification: A systematic arrangement of jobs into groups, categories, or classes according to the relative requirements necessary for their performance.

Job Comparison Scale: A listing of job factors and the points or money assigned to key jobs under each factor.

Job Content: The duties, functions, and responsibilities constituting a given job.

Job Description: A written statement covering the essential features of a job, including its purpose, duties, skill requirements, effort and responsibility, working conditions, and relation to other jobs.

Job Enlargement: The expansion of job content to include a wider variety of tasks and to increase the worker's freedom of pace, responsibility for quality, and discretion of method.

Job Evaluation: 1. A systematic procedure following job analysis for comparing, for wage and salary determination, the relative worth to the employer of two or more jobs or positions. 2. A weighting of all the factors in the various jobs or positions in an enterprise so that, directly or indirectly, an objective scale may be established whereby commensurate pay or pay rates may be assigned on the basis of job content.

Job Factor: Any characteristic of a job which influences its relative worth or value and provides a basis for the selection, training, placement, and compensation of workmen. Major job characteristics or factors are skill required, responsibility exercised, physical and mental effort involved, working conditions, and experience and education required.

Job Lot: A relatively small number of a specific type of part or product that is produced at one time. The part or product may be a standard item that has been and will again be produced, or it may be a special item destined for a specific customer who has not ordered it before and may not order it again.

Job Lot Layout: An arrangement of machines, equipment, and facilities specially set up or arranged to handle job lot production.

Job Lot Production: The manufacturing of parts or products to customer or stock orders in small quantities.

Job Rating: See Job Evaluation.

Job Shop: A manufacturing enterprise devoted to producing special or custom-made parts or products, usually in small quantities for specific customers.

Job Specification: A detailed written statement of the physical and mental attributes required of a person to perform a specific job competently.

Job Standardization: The establishment of a prescribed method for performing an operation or procedure and the specifying of its minimum requirements.

Joint Time Study: 1. A time study technique that utilizes more than one observer and is often used to study large, complex operations performed by more than one workman. 2. A time study made by both company and union representatives to prevent or to resolve disagreements over time standards.

Key Job: A job used in job evaluation systems as a guide or bench mark when evaluating other jobs. It is selected because its duties and responsibilities are known and are relatively stable, and because both management and labor agree that the evaluation of and the wage paid for the job are fair and equitable. The wage paid for key jobs serves as a basis for setting wage rates for other jobs.

Kumograph: An electrical recording device developed by Dr. Ralph M. Barnes, used chiefly in laboratory micromotion studies to measure extremely short (0.0001 minute) time intervals. Solenoid-operated pencils, when actuated by push buttons, photocells, or other contacting devices, make jog marks on a motor-driven tape which moves at a constant velocity. The distance between the jogs indicates the elapsed time interval.

Labor: 1. (n.) The mental and physical effort and energy expended by humans to produce and distribute materials, goods, and services. 2. (n.) Employees with little or no supervisory responsibility whose sole or main task is to aid in the production of materials, goods, or services. 3. (v.) To work or toil.

Labor Class: See Job Class.

Labor Cost: That part of the cost of goods, services, and the like attributable to wages. It commonly refers only to direct workmen, but may include indirect workmen as well.

Labor Grade: See Job Class.

Labor Productivity: The rate of output of a workman or group of workmen per unit of time, usually compared with an established standard or expected rate of output.

Laborsaving Ratio: The relation of the unit labor cost of an improved method to the unit labor cost of another method.

Lay Out (v.): To draw lines on material as guides for subsequent operations.

Layout (n.): The arrangement of items within an area. The items may include roads, railroads, buildings, offices, departments, warehouses, equipment, machinery, furniture, facilities, parts, aisles, and so on. See Plant Layout and Work Station Layout.

Layout Template (Templet): See Template, definition 2.

Learner's Wage Rate: The wage rate paid to an inexperienced employee while he is gaining the skill and experience in his job that is needed to meet the standard or expected performance level. This wage rate is usually lower than the one paid an experienced employee performing the same job. It is paid either for a specified period of time or until the employee attains the desired performance level or ability.

Learning Curve: A plot of productive output or unit work times of an individual or group as a function of time or output per unit time. A typical output versus time plot curves up and to the right from the origin and gradually levels off.

Leveled Elemental Time: See Rated Average Elemental Time.

Leveled Time: The average time, adjusted to account for differences in skill, effort, conditions, and consistency between workmen and the factors surrounding an operation. See Normal Time.

Leveling: A method of performance rating in which the causes for the observed performance, considered to be skill, effort, conditions, and consistency, are evaluated. The algebraic sum of the point values assigned to each factor is used in adjusting the time taken by the workman being time studied to the time required by a workman working at the average performance level under the usually prevailing conditions.

LIFO (Last In, First Out): A method of determining the cost of inventory used in a product. In this method, the costs of material are transferred to the product in reverse chronological order. Also used to describe the movement of goods.

Linear Programming: A mathematical programming technique involving those problem situations in which all relationships are directly proportional (linear). This means that they can be plotted as a straight line. For example, one unit of output costs $2 and fifty units of output cost $100. When all variables behave in this manner, a linear programming problem results.

Line Balance: See Balanced Line.

Line Layout: See Line Production.

Line Production: A method of plant layout in which the machines and other equipment required, regardless of the operations they perform, are arranged in the order in which they are used in the process (layout by product).

Loose Standard: (Colloq.) An allowed or standard time greater than that required by a qualified workman performing his job with normal skill and effort and following the prescribed method.

Machine Attention Time: That portion of a machining operation during which the workman performs no physical work yet must watch the progress of the work and be available to make necessary adjustments, initiate subsequent steps or stages of the operation at the proper time, and the like.

Machine Center Template (Templet): A machine template to which additional floor area has been added to provide space necessary for operating, servicing, and maintaining the machine as well as for material. Often, space for aisles is also included.

Machine-controlled Time: That part of a work cycle that is entirely controlled by a machine and therefore is not influenced by the skill or effort of the workman.

Machine Downtime: Any time during a regular working period that a machine cannot be operated.

Machine Element: A work cycle subdivision that is distinct, describable, and measurable, the time for which is entirely controlled by a machine and therefore is not influenced by the skill or effort of the workman.

Machine Flow Diagram: See Flow Diagram.

Machine-hour: A unit for measuring the availability or utilization of machines. It is equivalent to one machine working for 60 minutes, two machines working for 30 minutes, or an equivalent combination of machines and working time.

Machine Idle Time: That portion of a regular working period during which a machine that is capable of operating is not being used.

Machine Interference: The time that a machine is idle because the operator is servicing another machine in the group.

Machine Layout: See Plant Layout.

Machine Template (Templet): A two-dimensional scale representation of the maximum floor area occupied by a machine.

Machine Time: See Machine-controlled Time.

Magnetic Disc: A flat, circular plate with a magnetic surface on which information may be placed in the form of magnetically polarized spots.

Magnetic Tape: A tape or ribbon with a magnetic surface on which information may be placed in the form of magnetically polarized spots.

Makeup Pay: In an incentive wage payment plan, the difference between the guaranteed base wage rate earnings of an employee who fails to attain the established standard output and the earnings he would have received had there been no guaranteed base wage rate.

Makeup Time: The difference between the standard time earned by an employee failing to attain the established standard output and the actual time taken.

Management: 1. The art and science of directing and controlling human effort so that the established objectives of an enterprise may be attained in accordance with accepted policies. 2. The group of people who direct and control human effort toward the attainment of the objectives of an enterprise.

Management Audit: A systematic appraisal or evaluation of the worth or quality of the various management functions or activities or procedures in an organization, made by comparing them to established normals for good management or theoretical measures of management perfection.

Management Consultant: One who counsels or advises on a professional basis in organization, administration, and operational activities such as production, marketing, research, industrial relations, and accounting.

Management Counselor: See Management Consultant.

Management Engineer: One who has the necessary education, training, and experience to perform the functions of management engineering. See Management Engineering.

Management Engineering: The application of engineering principles to all phases of planning, organizing, and controlling a project or enterprise.

Management Information System: A system which meets the information needs of executives and managers at all levels of an enterprise.

Man and Machine Chart: See Multiple Activity Process Chart.

Man-hour: A unit for measuring work. It is equivalent to one man working at normal pace for 60 minutes, two men working at normal pace for 30 minutes, or some similar combination of men working at normal pace for a period of time.

Manit: A contraction for man-minute. Originally used in the Haynes incentive plan.

Man-minute: A unit used for measuring work. It is equivalent to one man working at normal pace for 1 minute, two men working at normal pace for 30 seconds, or an equivalent combination of men working at normal pace for a period of time.

Manual Element: A distinct, describable, and measurable subdivision of a work cycle or operation performed by one or more human motions that are not controlled by process or machine.

Marginal Analysis: An economic concept concerned with those elements of costs and revenue which are associated directly with a specific course of action, normally using available current costs and revenue as a base and usually independent of traditional accounting allocation procedures.

Marstochron: An instrument used in time study work to measure and record short (0.01 minute) time intervals. Two or more finger-controlled type bars make instantaneous marks on a motor-driven tape which moves at a constant velocity. The distance between the marks indicates the elapsed time interval.

Mass Production: A method of quantity production in which a high degree of planning, specialization of equipment and labor, and integrated utilization of all productive factors are the outstanding characteristics.

Master Table of Detail Time Studies: A master record of time study data arranged so that times for the same elements can be compared. It is used to collect, analyze, and develop standard time data.

Material Flow: The progressive movement of material, parts, or products toward the completion of a production process between work stations, storage areas, machines, departments, and the like. See Flow Diagram, Flow Process Chart, and Routing.

Material Handling: The movement of materials, parts, subassemblies, or assemblies either manually or through the use of powered equipment.

Mathematical Programming: Certain techniques which are useful for solving allocation problems where limited resources must be allocated between competing demands. Mathematical programming can be linear, nonlinear, or dynamic.

Maximum Working Area: 1. (Horizontal plane.) The area at the workplace which is bounded by the imaginary arc drawn by the workman's fingertips moving in the horizontal plane with the arm fully extended and moving about the shoulder as a pivot. The section where the maximum areas of the right and left hands overlap constitutes the maximum working area for the two hands. 2. (Vertical plane.) The space on the surface of the imaginary sphere which would be generated by rotating, about the workman's body as an axis, the arc traced by the workman's fingertips of the right or the left hand when the arm is fully extended and is moved vertically about the shoulder as a pivot. 3. (Three-dimensional.) The space within reach of a workman's fingertips as they develop arcs of revolution when the workman's hands are extended and moving about the shoulder as a pivot.

Measured Daywork: 1. The establishment of standard or allowed times for operations without providing the opportunity for incentive earnings. 2. A type of wage incentive plan in which each employee's performance record is reviewed periodically and his base wage rate is adjusted upward or downward from his base wage rate for the previous period as warranted by his average performance over the previous period but never below his guaranteed base wage rate.

Measure of Effectiveness: The criterion used in evaluating alternative solutions to a problem to select the optimum one.

Memomotion Study: A motion study technique that utilizes a motion picture camera operating at slower than normal speeds, such as one frame per second or one frame per one hundredth minute.

Memory (computer usage): The capacity of a machine to store information subject to recall; the component of the computer system in which such information is stored.

Merit Rating: An organized and systematic evaluation of an employee's ability and job performance in terms of such factors as quality and quantity of work, knowledge, initiative, and dependability. The rating is made periodically to determine if the employee's services are better or poorer than the accepted norm.

Method: 1. The procedure or sequence of motions used by one or more individuals to accomplish a given operation or work task. 2. The sequence of operations or processes used to produce a given product or accomplish a given job. 3. A specific combination of layout and working conditions; materials, equipment, and tools; and motion pattern involved in accomplishing a given operation or task.

Methods Engineer: The title given a member of that subclassification of industrial engineering comprised of individuals qualified by training, education, and experience to establish methods and the means by which they can be made most effective.

Methods Engineering: The technique that subjects each operation of a given piece of work to close analysis to eliminate every unnecessary element or operation and to approach the quickest and best method of performing each necessary element or operation. It includes the improvement and standardization of methods, equipment, and working conditions; operator training; the determination of standard times; and occasionally devising and administering various incentive plans.

Methods Study: The analysis of the sequence of motions used or proposed for use in performing an operation and of the tools, equipment, and work station layout used or proposed for use.

Methods Time Measurement (MTM): A system of predetermined motion time standards. It is a procedure which analyzes any operation into certain classifications of human motions required to perform it and assigns to each motion controlled only by the individual performing it a predetermined time standard determined by the nature of the motion and the conditions under which it is made.

Methods Training: 1. Detailed instruction and guided practice given workmen to ensure that they use the proper methods to perform their jobs. 2. Courses or programs of instruction given in the techniques of scientific management as related to methods engineering.

Microchronometer: 1. A two-handed clock whose dial is divided into 100 divisions, the large hand of the clock usually geared to make 20 revolutions in one minute and the small hand to make 2 revolutions each minute. 2. A clock devised by Frank B. Gilbreth which is used in micromotion study by placing it in the foreground when photographing an operation so that the time is recorded on the film. 2. An accurate timepiece which measures time in units of 1/2,000 part of a minute and fractions thereof.

Micromotion Study: 1. That phase of motion study which divides manual work into fundamental elements—often called therbligs or Gilbreth Basic Elements—analyzes these elements separately and relatively, and from this analysis, establishes more efficient methods. 2. The analysis of elements of motions too short or rapid for the eye to distinguish, by the use of motion pictures, sometimes in combination with an adequate time-indicating device. Because the motion picture camera itself can indicate time intervals, an additional timing device is often dispensed with in micromotion study.

Minimum Time: The shortest elapsed time recorded for a particular element of a time study excepting those known to be incorrect.

Modal Selection: The elapsed time that appears most frequently for an element in a time study.

Model: A representation of reality to facilitate analysis, experimentation, and comprehension.

Monte Carlo Method: A combination of probability mathematics, and the statistics of sampling used to solve problems that are too complex to be solved by pure mathematics.

Motion: A movement of the human body or any of the body members.

Motion Analysis: See Motion Study.

Motion Cycle: A complete series of motion elements involved in performing an operation, beginning with a motion connected with the production of the unit and ending when the same motion is about to be repeated with the next unit.

Motion Study: The analysis of the manual and the eye movements occurring in an operation or work cycle, for the purpose of eliminating wasted movements and establishing a better sequence and coordination of movements.

Motion Time Analysis (MTA): A system of predetermined motion time standards used for describing and recording an operation in terms of its motions. The value of each motion is predetermined both as to utility and time allowance.

Move (n.): The basic element employed when the predominant purpose is to transport an object to a destination. See Transport Loaded.

Multiple Activity Process Chart: A synchronized graphic representation of operations performed simultaneously by two or more men, two or more machines, or a combination of men and machines.

Multiple Piece Rate Plan: See Differential Piecework.

Multiple Time Plan: A type of wage incentive plan that provides for the payment of higher base wage rates at progressively higher levels of production. For example, a workman may be paid $3.00 per hour at an output of 60 to 65 pieces per hour, $3.30 per hour at an output of 66 to 71 pieces per hour, $3.65 per hour at an output of 72 to 77 pieces per hour, and so on. Synonym: Differential Time Plan.

Multiple Watch Timing: See Accumulative Timing.

Network Analysis: A technique of analysis, useful in planning a project, that consists of showing the sequence of activities (tasks or jobs) and their interrelationship within a network of activities making up a project. By computing the cumulative time for each path (chain of activities) through the network from the starting event to the terminal event, the extent or cost of the critical path is determined. See Critical Path.

Nomograph: A graphical set of straight lines and/or curves representing equations and arranged so that a solution can be obtained by successively connecting points representing the independent variables. Used to save calculation time when a series of solutions is desired.

Nonfinancial Incentive: All influences, other than financial, which tend to stimulate workmen to greater exertion, for example, promotion, training, competition with others, recognition, or praise.

Nonlinear Programming: A mathematical programming technique involving those problem situations in which the relationships of the variables are exponential. Graphically, they would plot as a curve. For example, one unit of output costs $2 while fifty units of output cost $75.

Nonproductive Labor: See Indirect Labor.

Nonrepetitive: A descriptive term applied to a type of work, operation, part, or the like that does not recur frequently or in any reasonable regular sequence.

Normal Elemental Time: The selected or average elemental time, adjusted by leveling or other methods of adjustment to obtain the time required by a qualified workman to perform a single element of an operation.

Normal Pace: The work rate ordinarily used by workmen performing under capable supervision but without the stimulus of an incentive wage payment plan. This pace can easily be maintained day in and day out without undue physical or mental fatigue and is characterized by the fairly steady exertion of reasonable effort.

Normal Time: 1. The time required by a qualified workman, working at a pace which is ordinarily used by workmen when capably supervised, to complete an element, cycle, or operation when following the prescribed method. 2. The sum of all the normal elemental times which constitute a cycle or operation.

Normal Working Area: 1. (Horizontal plane.) The area at the workplace which is bounded by the imaginary arc drawn by the workman's fingertips moving in the horizontal plane with the elbow as a pivot when the workman is standing or is seated in the normal working position and when the upper arm is hanging from the shoulder in a relaxed position. The section where the normal areas of the right and left hands overlap in front of the workman constitutes the optimum normal working area for the two hands. 2. (Vertical plane.) The space on the surface of the imaginary sphere which would be generated by rotating, about the workman's body as an axis, the arc traced by the workman's fingertips of the right or the left hand when the forearm is moved vertically about the elbow as a pivot. 3. (Three-dimensional.) The space within reach of a workman's fingertips as they develop arcs of revolution, the elbow acting as a pivot when the workman is standing or is seated in the normal working position and when the upper arm is hanging from the shoulder in a relaxed position. *Numerical Control:* Any system of control plus controlled equipment which accepts commands—data and instructions—in symbolic form as an input and converts this information into a physical output—into physical values such as dimensions or quantities.

Observation: 1. In time study, the act of noting and recording the time taken by a workman performing an operation or an element of an operation. 2. In motion study, the act of noting and recording the motions used by a workman to perform an operation or an element of an operation. 3. In work sampling, the act of noting and recording what a workman is doing at a specific instant.
Observation Board: A portable, flat, rigid backing designed to be held by one hand and to support observation forms while information is written on them. Usually, a spring clip attached to the board holds the forms firmly in place. A device for holding a stopwatch may or may not be attached.
Observation Form: A sheet of paper used to record data taken during time studies, methods studies, or work sampling studies, specifically ruled into titled lines, columns, and spaces to suit the specific requirements of the study.
Observation Period: See Elapsed Time.
Observer: An individual who makes an observation or collects data.
Obsolescence: A decrease in the value of an asset brought about by the development of new and more economical methods, processes, or machinery.
Occupational Rate: The average hourly or daywork rate of pay received by workmen in a specific trade or type of employment within an area, industry, or plant, exclusive of incentive bonuses, overtime premiums, or shift differentials.
One Best Way: 1. A term originally used by Frank and Lillian Gilbreth to describe a method of performing an operation that cannot be economically improved at the moment by those attempting to do so. 2. The ideal method of performing an operation that is the goal of all methods engineering work. 3. A figure of speech which denotes the optimum method of performing an operation under present conditions, at the present time, and in the opinion of this particular analyst.
Operation: The intentional changing of an object in any of its physical or chemical characteristics; the assembly or disassembly of parts or objects; the preparation of an object for another operation, transportation, inspection or storage; planning, calculating, or the giving or receiving of information.
Operation Analysis: 1. A study of the factors which affect the performance of an operation, such as purpose of the operation, other operations on the part, inspection requirements, materials used, manner of handling material, setup and tool equipment, existing working conditions, and methods employed. 2. A procedure employed in studying the major factors which affect the general method of performing a given operation.
Operation Analysis Chart: A form that lists all the important factors affecting the effectiveness of an operation and is used to guide the progress and ensure the completeness of an operation analysis.
Operation Breakdown: See Job Breakdown.
Operation Element: See Element.
Operation Instruction Card: See Instruction Card.
Operation Instruction Sheet: See Instruction Card.
Operation Process Chart: A graphic representation of the points at which materials are introduced into the process and of the sequence of inspections and all operations except those involved in material handling. It may include other information considered desirable for analysis, such as time required and location.
Operations Research (OR): The application of scientific methods, tools, and techniques to the analysis of the relationships and functions of a system for the purpose of determining quantitatively the conditions under which optimum results should be achieved.
Operator Process Chart: A motion study aid used to chart the time relationship of

the movements made by the body members of a workman performing an operation. See Right- and Left-hand Chart; Simo Chart, Full; and Simo Chart, Simple.

Organization: 1. The process of determining the necessary activities and positions within an enterprise, department, or group; arranging them into the best functional relationships; clearly defining the authority, responsibilities, and duties of each; and assigning them to individuals so that the available effort can be effectively and systematically applied and coordinated. 2. The group of people who have been brought together to conduct a business or enterprise.

Organization Chart: A graphic representation of the formal organizational structure of an enterprise, showing lines of authority, responsibility, and coordination.

Output: 1. See Production, definition 3. 2. (Computer usage.) *a.* Data that have been processed. *b.* Information transferred from the internal storage of a computer to secondary or external storage. *c.* Information transferred to any device exterior to the computer.

Output Standard: See Performance Standard.

Overall Study: A time study made of one or more operation cycles without an element breakdown.

Overhead: Costs or expenses which are not directly identifiable with or chargeable to the manufacture of a particular part or product. For example, items such as taxes, insurance, supplies, supervisory and clerical charges, and the like. Synonyms: Burden, Indirect Manufacturing Expense.

Overtime: Time worked in excess of regular working time as established by agreement or law, usually paid for by a premium in addition to the base wage rate.

Pace Rating: See Speed Rating.

Pareto's Law: Sometimes called the "law of the trivial many and the critical few." A law which states that in most business activities, a small fraction (commonly estimated at 20 percent) of the total item count produces the major portion (commonly estimated at 80 percent) of the work, cost, profit, or other measure of importance. Often shown as a plot of total output against cumulative frequency.

Parkinson's Law: Work expands to fit the organization that is developed to perform it, and there is a tendency for each unit within an organization to try to build up its importance by expanding the number of its personnel.

Performance: The degree with which a workman applies his skill and effort to an operation under the conditions prevailing. This degree is expressed in terms of a performance efficiency or defined bench marks such as good, average, and poor.

Performance Efficiency: A ratio—usually expressed as a percentage—of actual output to a bench mark or standard output when both are measured on the same basis.

Performance Rating: The act of comparing an actual performance by a workman against a defined concept of a normal performance. Various methods of performance rating are in use, differing primarily as to the bases on which comparison is made.

Performance Rating Factor: 1. A numerical index that relates an observed performance to a defined normal performance. 2. Any of the terms or elements used for the comparison of performance.

Performance Rating Scale: 1. A series of descriptive graduated statements or numbers serving as a guide by which an observed performance may be estimated and its worth more objectively determined. 2. A series of graduated statements or numbers that describe a specific characteristic and serve as a guide for making more objective estimates of the degree to which that characteristic is exhibited in any specific instance.

Performance Standard: A criterion or bench mark with which actual performance is compared.

Personal Allowance: Time included in the production standard to permit the workman to attend to personal necessities such as obtaining drinks of water, making trips to the rest room, and the like. Usually applied as a percentage of the leveled, normal, or adjusted time.

Personal Time: See Personal Allowance.

PERT (Program Evaluation and Review Technique): A network technique designed for planning and scheduling activities to accomplish a predetermined job within time constraints. It is most widely used in the development of a new product and is not directly suitable for application to repetitive production operations.

Piece Rate: The amount of money paid for a unit of production. It serves as the basis for determining the total remuneration paid an employee working under a piecework incentive plan.

Piece Rate Wage Plan: See Piecework.

Piecework: A wage incentive plan which pays a definite sum of money for each unit of production.

Pilot Lot: An experimental or preliminary order for a product. Such an order is relatively small and is produced to correlate the product design with the development of an efficient manufacturing process. It may also be used to establish time standards for the operations that make up the developed process. Synonym: Pilot Order.

Pilot Order: See Pilot Lot.

Pilot Plant: A plant devoted to the production of pilot lots or to the continuous production of small quantities of a product for the purpose of experimenting with its design and production methods.

Plan: The basic element which denotes the mental act, previous to the physical movement, of determining a method of proceeding with the work.

Planning: The procedure for determining a course of action intended to accomplish a desired result.

Plant Layout: The physical arrangement, either existing or in plans, of industrial facilities.

Point of Make-out: The performance level where calculated earnings are exactly equal to the guaranteed base wage rate earnings.

Point System: A method of job evaluation in which a range of point values is assigned to each of several job factors. The wage rates for specific jobs are then determined by comparing the total points each receives with the point values and wage rates of key jobs.

Position: 1. (Time study usage.) The element which consists of aligning, orienting, or locating one object in relation to another. 2. (Motion study usage.) The basic element which consists of aligning, orienting, and engaging one object with another where the motions used are so minor that they do not justify classification as other basic elements.

Position Class: See Job Class.

Predetermined Motion Time System: A procedure in which (1) all manual motions are analytically subdivided into the basic elements required for their performance and (2) predetermined time values are assigned to the basic elements.

Premium: See Bonus.

Pre-position (n.):   The basic element employed when the transporting device or the object transported is prepared for the next basic element, which is usually position. (Note the difference between the basic element, pre-position, and the pre-positioning of tools and materials in laying out the work station. The latter term, denoting a function of general planning, involves a number of basic elements.)

Principles of Motion Economy: The rules and their corollaries applying to human motions, which guide toward development of the optimum way of accomplishing a given job.

Probability Theory: A set of mathematical theorems and practices dealing with the expected long-run relative frequency of events. Often used to describe the short-run likelihood of an event and to help determine whether a short run typifies the long run, as in quality control and other sampling matters.

Process: 1. A planned series of actions or operations which advances a material or procedure from one stage of completion to another. 2. A planned and controlled treatment that subjects materials to the influence of one or more types of energy for the time required to bring about the desired reactions or results. Examples include the curing of rubber, mixing of compounds, heat-treating of metals, machining of metals, and the like.

Process Chart: A graphic presentation of events occurring during a series of actions or operations and of information pertaining to those events.

Process Chart Symbols: Graphic symbols or signs used on process charts to depict the types of events that occur during a process. Five such symbols have been defined and approved by the American Society of Mechanical Engineers. Their names and symbols are:

Operation	◯	Delay	D
Inspection	▢	Storage	▽
Transportation	⇨		

Process Engineer: An individual qualified by education, training, and experience to prescribe efficient production processes to produce a product as designed and who specializes in this work. This work includes specifying all the equipment, tools, fixtures, and the like that are to be used, and often, the estimated cost of producing the product by the prescribed process.

Processing: 1. The act of prescribing the production process to produce a product as designed. This may include specifying the equipment, tools, fixtures, machines, and the like required; the methods to be used; the workmen necessary; and the estimated or allowed time. 2. The carrying out of a production process.

Process Layout: A method of plant layout in which the machines, equipment, and areas for performing the same or similar operations are grouped together. For example, all welding is done in one area or department, all painting in another, and so on. Layout by function.

Process Time: The time required to complete a specified series of progressive actions or operations on one unit of production. 2. That portion of a work cycle during which the material or part is being machined or treated according to a specification or recipe designed to produce the desired reaction or result. The time required is controlled by the machine, specification, or recipe and not by the workman.

Production: 1. The manufacturing of goods. 2. The act of changing the shape, composition, or combination of materials, parts, or subassemblies to increase their value. 3. The quantity of goods manufactured.

Production Center: 1. A group of productive facilities (machines, auxiliary tools, and the like) which, for administrative and accounting purposes, are considered a unit. 2. The area containing the machine or machines operated by a workman or workmen as well as the space required for the storage of materials at the machine and for loading and unloading it; auxiliary tools, benches, jigs, and the like; and the free and safe movement of the workman while working.

Production Control: The procedure of planning, routing, scheduling, dispatching, and expediting the flow of materials, parts, subassemblies, and assemblies within the plant from the raw state to the finished product in an orderly and efficient manner.

Production Department: 1. That part of a manufacturing organization responsible for the actual processing of materials or parts. 2. That subdivision of management responsible for planning how, where, and at what cost to manufacture or assemble a product. See Production Engineering.

Production Engineer: An individual qualified by education, training, and experience to perform production engineering functions and who specializes in this work.

Production Engineering: 1. The function of planning where and when to perform work necessary to produce a product and of coordinating internal and external orders, delivery dates, workmen, machines, and the like, thereby promoting efficient operation. 2. A term used as a synonym for industrial engineering, methods engineering, and manufacturing engineering. 3. Designing products to be manufactured, utilizing materials, equipment, methods, processes, and skills that are available.

Production Load: The demand for output established by scheduling, based on consumer orders or sales forecasts. Usually, it is stated in terms of the time required to produce the demanded output or as a percent of capacity output, normal output, available machine-hours, or the like.

Production Planning: 1. The systematic scheduling of men, materials, and machines by using lead times, time standards, delivery dates, work loads, and similar data for the purpose of producing products efficiently and economically and meeting desired delivery dates. 2. Routing and scheduling.

Production Report: A formal written statement giving information on the output of an organization or one or more of its subdivisions for a specified period. The information normally includes the type and quantity of output; workmen's efficiencies; departmental efficiencies; costs of direct labor, direct material, and the like; overtime worked; and machine downtime.

Production Standard: See Performance Standard.

Production Study: A detailed record, often in the form of a time study or work sampling study, kept of an activity, operation, or group of activities or operations, for a period of time to obtain reliable data concerning working time, idle time, downtime, personal time, machine breakdowns, amount produced, and so on.

Production Unit: 1. The workmen, equipment, and areas involved in performing a given task. 2. A measure of a product expressed in terms of weight, volume, quantity, dollar value, or the like.

Productive Labor: See Direct Labor.

Productive Time: 1. Elapsed time during which useful work is performed in a manufacturing process. 2. That portion of an operation cycle during which the workman's time is utilized effectively. The balance of his time is considered idle or unproductive.

Productivity: The actual rate of output or production per unit of time worked.

Product Layout: See Line Production.

Profit and Loss Statement: See Income Statement.

Profitgraph: See Break-even Chart.

Profit Sharing: A plan whereby employees receive periodic payments in cash, stock, or future credits over and above their regular wages and overtime and incentive earnings, the earnings being related in some way to the past or present earnings of the business.

Program (computer usage): A sequence of instructions or commands that will cause a computer to perform a prescribed set of data processing functions.

Progress Chart: A graphic representation of the status or extent of completion of work in process. See Gantt Chart.

Qualified Operator: A person who has the mental and physical characteristics, the job knowledge, and the experience required by the work he is to perform and who should be able to meet or exceed the performance level expected on that work without undue mental or physical fatigue.

Quality Control: The procedure of establishing acceptable limits of variation in size, weight, finish, and so forth for products or services and of maintaining the resulting goods or services within these limits.

Queuing Theory: The theory involving the use of mathematical models, theorems, and algorithms typically used in the analysis of systems in which some service is to be performed under conditions of randomly varying demand, and where waiting lines or queues may form due to lack of control over either the demand for service, the amount of service required, or both. Utilization of the theory extends to process, operation, and work studies.

Random Access (computer usage): Access to storage under conditions in which the next position from which information is to be obtained is in no way dependent on the previous one.

Random Element: See Foreign Element.

Ranking Method: A job evaluation system wherein each job as a whole is ranked with respect to all the other jobs and no attempt is made to establish a measure of value.

Rate: 1. (n.) See Base Wage Rate. 2. (n.) See Allowed Time and Standard Time. 3. (v.) To estimate the worth or value of anything by comparing it with a standard or scale, as, for instance, in performance rating.

Rate Change: 1. An upward or downward adjustment of a production standard generally made because of a revision in product design, quality requirements, production methods, materials, or conditions. 2. An upward or downward adjustment in wages paid per unit of time or unit of output.

Rate Cutting: A term used when referring to the act of reducing the production standard in terms of standard hours or dollars for an operation, particularly when the amount of the reduction is thought to be unjustified by changes in product design, quality requirements, production methods, materials, or conditions.

Rated Average Elemental Time: The result of adjusting by a performance rating factor the mathematical average of the times obtained for one element of a time-studied operation. Usually, any abnormal time values are excluded in calculating the mathematical average.

Rated Selected Elemental Time: See Normal Elemental Time.

Rate Setting: The establishing of production standards by time study, predetermined motion times, standard data, time formulas, or some other means.

Rating: See Merit Rating and Performance Rating.

Rating Scale: A graduated series of criteria or reference points used as a standard for measuring, evaluating, or classifying, as those often used, for example, in performance rating, merit rating, or job evaluation.

Ratio Delay Study: See Work Sampling Study.

Reach (n.): The basic element employed when the predominant purpose is to move the hand to a destination or general location. See Transport Empty.

Real Time: The performance of a computation during the actual time that the related physical process transpires, so that results of the computations are useful in guiding the physical process.

Regular Element: An element of an operation or process that occurs either every cycle of that operation or process or occurs frequently and in a fixed pattern with the cycles of that operation or process, as for example, once every third cycle.

Release (n.):   (Abbreviated term for "release load.")   The basic element employed when the hand or body member relinquishes control of an object.

Reliability (quality usage):   The probability that a device will perform satisfactorily in its specified function for a specific period of time under a given set of operating conditions.

Re-operation: See Rework.

Repetitive: The general term used when referring to processes, operations, elements of operations, or the products resulting therefrom that occur or are produced over and over again with negligible variation.   The term must be qualified or explained when it is used, in order to have a concrete meaning.

Repetitiveness of an Operation: See Activity.

Replacement Theory: Methods of analysis which deal with two types of situations involving the life pattern of the equipment involved. One pertains to equipment that gradually deteriorates in value or efficiency; the other involves equipment that breaks down permanently.

Required Idle Time: See Unavoidable Delay.

Reserve for Depreciation: That part of the first or base cost of an asset which has been periodically written off as an expense and accumulated in the accounting records because of wear, tear, obsolescence, or inadequacy of the asset.

Restricted Element: An operation element for which the performance time is not completely under the control of the workman but is instead governed or paced by a machine, process, or other element.   See Machine-controlled Time and Process Time.

Restricted Job: A job for which the performance time is not completely under the control of the workman but is instead governed or paced by a machine, process, other job, or the nature of the job itself.   See Machine-controlled Time and Process Time.

Rest to Overcome Fatigue: An allowance or delay allowed workmen for the purpose of recovering from the effects of exertion or sustained mental or visual attention. It is usually included in the general allowance, but on work of a particularly exhausting nature, it may be included in the job time standard as a separate allowance or element.

Restudy: Any study of the performance time for an operation that is made to verify or disprove a previous study.   The technique used may be the same or different from that used in the previous study.

Retime: To make a time study of an operation to check the validity or application of a previous time study.

Rework: 1. The term applied to an operation when it is performed on the same material or item more than once to correct the result of performing it the first time. 2. Any work done on material or an item to correct work done improperly or to comply with revisions in design or specifications.   Synonym: Re-operation.

Right- and Left-hand Chart: A form of operator process chart on which the motions made by one hand in relation to those made by the other hand are recorded, using standard process chart symbols or basic therblig abbreviations or symbols.

Route: 1. (n.) The path followed by a man, material, part, or the like in a particular production process.   2. (v.) To prescribe the above path.

Routing: 1. (n.) A form listing for the manufacture of a particular item the sequence of operations, transportations, storages, and inspections to be used, and usually also the standard times applicable, and the machines, equipment, tools, work stations, number of workmen, materials, parts, and the like that are to be used.   2. (v.) Establishing the sequence of processes, operations, transportations, and storages to be followed, and the machines, tools, work stations, and miscellaneous equipment that will be used in producing a particular part, product, or job lot.

Salary: Fixed compensation paid weekly, monthly, or yearly for services rendered, usually based on a certain minimum number of hours per day or week.

Sampling: The practice of selecting a small portion (usually determined statistically) of the total group under consideration, for the purpose of inferring the value of one or several characteristics of the group.

Scheduling: 1. The prescribing of when and where each operation necessary to the manufacture of a product is to be performed. 2. The establishing of times at which to begin or complete each event or operation constituting a procedure.

Search (n.): The basic element employed to locate an object with the eyes or fingers.

Select (n.): The element of choosing between several items which may be alike or different. It appears as a hesitation in the motion sequence.

Selected Average Time: See Average Time and Average Elemental Time.

Selected Elemental Time: The raw or unadjusted time which is chosen as being representative of the actual times taken to perform a single element of an operation.

Selected Time: That time value chosen by the time study observer from those obtained for an element of a time-studied operation as being representative of the time used by the workman when he performed the element correctly.

Service Department: See Auxiliary Department.

Setup: Making ready or preparing for the performance of a job or operation. Machine setup involves equipping a machine with the appropriate accessories, tools, and fixtures, setting the proper feed, speed, and depth of cut, and so forth. In manual work, setup is the arrangement, prior to commencing the work, of the tools, accessories, component parts, and details involved. It also includes the teardown to return the machine or work area to its original or normal condition.

Simo Chart (Simultaneous Motion Cycle Chart): A graphic representation of an operation, generally, although not necessarily, made from a motion picture film in which the basic motions, such as therbligs, used to perform the operation by the right- and left-hand members of the body are separately plotted in columns scaled to time, using standard symbols for the elements.

Simo Chart, Full: An extremely detailed simo chart containing columns to represent every body member directly or indirectly entering the operation.

Simo Chart, Simple: A simplified form of simo chart, usually sufficient for most practical purposes, on which are recorded only the basic motions of the body members immediately or directly engaged in the operations, as the actions of the two hands, or of the hands and feet, or of the wrists, lower arms, and upper arms.

Simplex Method: A systematic procedure for solving linear programming problems.

Simulation: A technique which employs mathematical models as analogs of real-world systems. Inputs to the models are varied to establish the effect of decisions on conditions described by the model. Manipulation of the models may take various forms.

Skill: 1. The ability to use one's knowledge, technical proficiency, and developed or acquired ability to devise an efficient method of accomplishing a given objective. 2. (Leveling method of performance rating.) Proficiency at following a given method, good or bad, developed as the result of aptitude and practice.

Snapback Method: The procedure of timing, used in making time studies, whereby the stopwatch is read and the watch hand returned to zero at the termination of each element or work cycle.

Snapback Reading: See Snapback Method.

Software: The collection of programs and routines, such as compilers and assemblers, associated with a computer program.

Solid State Devices: Devices which utilize the electric, magnetic, and photic properties of solid materials.

Special Time Allowance: A temporary time value applying to an operation in addition to or in place of a standard allowance to compensate for a specified, temporary, nonstandard production condition.

Speed Rating: A method of performance rating that compares the speed or tempo with which a workman performs the motions necessary to execute an operation against the observer's concept of standard or normal tempo.

Standard: 1. See Performance Standard, Standard Time, Standard Hour, Direct Labor Standard, Guaranteed Time Standard, and so on. 2. Any established or accepted rule, model, or criterion against which comparisons are made.

Standard Allowance: The established or accepted amount by which the normal time for an operation is increased within an area, plant, or industry to compensate for the usual amount of fatigue or personal or unavoidable delays.

Standard Cost: The normal expected cost of an operation, process, or product—including labor, material, and overhead charges—computed on the basis of past performance costs, estimates, or work measurement.

Standard Data: See Standard Time Data.

Standard Elemental Time: The normal elemental time plus allowances for fatigue and delays.

Standard Hour: An hour of time during which a specified amount of work of acceptable quality is or can be performed by a qualified workman following the prescribed method, working at normal pace, and experiencing normal fatigue and delays.
Standard Hour Plan: A wage incentive plan having standard times expressed as standard hours. The hourly base wage rate is paid for standard hours earned rather than actual hours worked. Usually, the plan provides for a guaranteed minimum wage based on the hours worked if they exceed the standard hours earned.
Standardization: A management-sponsored program to establish criteria or policies that will promote uniform practices and conditions within the company and permit their control through comparisons. It deals with such areas as work quality and quantity, working conditions, wage rates, and production methods.
Standard Performance: The performance which must be given by a workman to accomplish his work in the standard time allowed.
Standard Practice: The established or accepted procedure used within an area, plant, or industry for carrying out a specified task or assignment.
Standard Rate: See Going Rate.
Standard Time: 1. The time which is determined to be necessary for a qualified workman, working at a pace ordinarily used under capable supervision and experiencing normal fatigue and delays, to do a defined amount of work of specified quality when following the prescribed method. 2. The normal or leveled time plus allowances for fatigue and delays.
Standard Time Data: A compilation of all the elements that are used for performing a given class of work with normal elemental time values for each element. Without making actual time studies, the data are used as a basis for determining time standards on work similar to that from which the data were determined.
Starting Rate: The rate of pay per unit of time that a workman is guaranteed when first hired to perform a specified job or class of work. See Going Rate.
Start-up Curve: A learning curve applied to a job or process to adjust for work times longer than standard or average, as a result of the introduction of a new job or new worker(s).
Static Work: Work performed by the hands or arms where no significant motion occurs, such as holding.
Statistical Quality Control: A means of controlling the quality of a product or process by the application of the laws of probability and statistical techniques to the observed characteristics of such product or process.
Statistics: A field of scientific endeavor that is central to the scientific method in the proper design of experiments and the drawing of valid inferences about events in sampling from a given population or so-called universe.
Step Bonus: A feature of some wage incentive plans whereby a substantial increase is made in incentive payment when the quantity or quality of output reaches a specified level. Such increases are ordinarily expressed as percentages of either the base wage rate or the incentive rate.
Stock-out: The lack of materials or products which are normally expected to be on hand in stores or stock.
Storage: 1. Keeping and protecting an object against unauthorized removal. 2. (Computer usage.) The components which enable the computer to retain data internally by electromechanical, magnetic, or electronic devices until needed.
Straight Time Rate: See Day Rate.
Stretch Out: 1. The act of extending the time taken to do a job by malingering. 2. Term referring to giving a workman additional work to do without a compensating increase in earnings or change in conditions, methods, or the like.
Subassembly: Two or more parts joined together to form a unit which is only a part of a complete machine, structure, or other article.
Subsidized Time: See Makeup Time.
Subtracted Time: On a time study conducted using the continuous timing method, the elapsed time obtained for an element of an operation by subtracting the watch reading recorded at the beginning of an element from the watch reading recorded at the end of that element during the same cycle.
Suggestion System: A procedure designed to encourage employees to submit their ideas for improving working conditions, production methods, or other aspects of the company's activities, and usually to reward them in some manner for the ideas which are adopted.
Supersession: The replacement of one piece of equipment by another that performs the same work but with greater efficiency.
Supervision: 1. Guidance and direction given to one or more individuals performing

assigned tasks or operations. 2. The group of individuals within an area, plant, or industry who are responsible for giving guidance and direction to one or more individuals performing assigned tasks.

Survey: A general check or brief investigation of an activity or organization, made to evaluate the existing situation and usually the opportunities for improvement that are present.

Synthesis of Elemental Times: 1. The act of selecting and combining the proper elemental times obtained from time studies or predetermined elemental motion time studies of actual operations to obtain the normal or standard time for an operation without making a time study of it. 2. In time formula development, the combining and simplifying of the mathematical or graphical expressions for determining individual elemental times.

Synthetic Data: 1. Work measurement time values not obtained from direct measurement of the work to which they are applied. They generally represent values for task elements sufficiently basic as to occur in several jobs. Obtained from measuring task elements in similar jobs or from predetermined motion time systems. 2. Any production data not measured directly from but applicable to a given situation.

Synthetic Time Standard: A time standard developed for an operation by utilizing predetermined elemental time data or standard data rather than by making a time study.

Take-home Pay: 1. The total amount of wages earned by an individual during a pay period, less all deductions from this amount. 2. The total amount of wages earned by an individual during a pay period prior to any deductions for taxes, insurance, savings, union dues, and so on.

Task: 1. The amount of work established as standard in any particular instance. 2. A specifically assigned amount of work.

Task and Bonus Plan: Any wage incentive plan that pays a specific percent of the base wage rate in addition to the base wage rate when a specified level of output is maintained or exceeded for a specified period of time.

Template (Templet): 1. A gage or pattern used as a guide to the size and shape of a part to be made. 2. A two-dimensional cutout—representing the area required by a workman; by a machine when operating; or by a bench, desk, or temporary storage bin at the work station and other equipment—used in planning and layout of plant and office facilities.

Temporary Rate: 1. See Temporary Standard. 2. A special wage rate that is paid for a limited period of time to compensate for a nonstandard condition or an unusual type of work such as that of an experimental or developmental nature.

Temporary Standard: A time standard applied to an operation for a limited period pending the development of a more accurate time standard, the development of a new method, or the correction of abnormal conditions affecting the operation.

Therblig: 1. The name of the basic work elements which are used in varying sequences and combinations to perform all manual and mental work. 2. The term coined by Frank B. Gilbreth to designate subdivisions of work in his classification of physical motions and associated mental processes. Synonym: Gilbreth Basic Element.

Tight Rate: See Tight Standard.

Tight Standard: (Colloq.) A time standard that provides a qualified workman with insufficient time to do a defined amount of work of specified quality when following the prescribed method, working at normal pace, and experiencing normal fatigue and delays.

Time Allowance: See Allowance, Special Time Allowance, and Standard Time.

Time Formula: A collection of standard time data arranged in the form of an algebraic expression for determining the standard time for an operation.

Timekeeping: A phase of wage administration and cost accounting which involves the recording of total time worked to compute wages earned and charges to various jobs. Also, these records are used to prove compliance with wage and hour laws.

Time Standard: See Standard Time.

Time Study: The procedure by which the actual elapsed time for performing an operation or subdivisions or elements thereof is determined by the use of a suitable timing device and recorded. The procedure usually but not always includes the adjustment of the actual time as the result of performance rating to derive the time which should be required to perform the task by a workman working at a standard pace and following a standard method under standard conditions.

Time Study Comparison Sheet: See Master Table of Detail Time Studies.

Time Study Observation Sheet: A form for the systematic, detailed recording of elemental time values, pace and effort rating estimates, delays, and irregular occurrences during a time study. Generally, space is also provided for entering other pertinent information and for computation of standard times from the data.

Time Study Recap Sheet: See Master Table of Detail Time Studies.

Time Study Summary Sheet: See Master Table of Detail Time Studies.

Time Taken: See Actual Time.

Time Ticket: Any form on which a workman's name or identifying number is recorded that serves as the original or source document for information on the operations or types of work he did, when he did them, and the time he used doing them. Other information such as applicable job or order numbers, base wage rates, standard times, and the like is also often recorded.

Time Used: See Actual Time.

Tolerance: A permissible variation in a characteristic of a product or process, usually shown on a drawing or specification.

Tool Design: That division of mechanical design which specializes in the design of tools. The complete specification of materials, dimensions and tolerances, finishes, heat treatment, and so on for a given tool. The use of the tool from a methods design viewpoint may also be a consideration.

Total Information System: A reasonably large, computer-based information handling system which supplies the information needs of the entire corporation.

Transportation: Moving an object from one place to another, except when such movements are a part of the operation or are caused by the workman at the work station during an operation or an inspection.

Transport Empty: The basic element employed when the hand, or a transporting device held in the hand, is moved from one point or object to another unresisted and without load. See Reach.

Transport Loaded: The basic element employed to move a part or object with the hand or other transporting device to a desired location. See Move.

Travel Time: Time used by or granted a workman for walking or riding between any two designated locations, as between two work stations, between the factory entrance and a work station, between two plants, or the like.

Trunk Movement: Any motion made by that portion of the human body located above the hips and below the neck but excepting the arms, hands, and fingers.

Two-handed Chart: See Right- and Left-hand Chart.

Unavoidable Delay: An occurrence which is essentially outside the workman's control or responsibility that prevents him from doing productive work.

Unavoidable Delay Allowance: Time included in the production standard to allow for time lost which is essentially outside the workman's control, for example, interruption by supervision for instruction, waits for crane, or minor adjustments to machines or tools. Usually applied as a percentage of the leveled, normal, or adjusted time.

Uniform Flow Convention: In economy studies, the practice of assuming that receipts and disbursements are made at a constant rate throughout the time period of concern so as to facilitate calculation or analysis.

Unrestricted Element: An operation element that is completely under the control of the workman. Synonym: Manual Element.

Unrestricted Job: A job that is completely under the control of the workman.

Upgrading: Promoting a workman to a higher paid and usually more demanding job.

Use (n.): The basic element employed to perform the activity which is the purpose of the operation other than assembly. The time required for use is usually but not necessarily markedly controlled by the requirements of the activity rather than by the workman.

Value Analysis: An arrangement of techniques which makes clear precisely the functions that the customer wants; establishes the appropriate cost for each function by comparison; and causes required knowledge, creativity, and initiative to be used to accomplish each function for that cost.

Variable Element: An element for which the leveled or normal time, under the same methods and working conditions, will change because of the varying characteristics of the parts being worked upon, as size, weight, shape, density, hardness, viscosity, tolerance requirements, finish, and so on.

Variable Expense: Expenditures that vary in proportion to the volume of production.

Variable Time Element: See Variable Element.

Variance: The difference between any standard or expected value and an actual value. For example, the difference between the established standard cost and the cost actually incurred in performing a job or operation.

Wage and Salary Administration: The management function concerned with the systematic determination and equitable application, adjustment, and control of the compensation structure of an organization.

Wage and Salary Structure: The established or existing compensation or ranges of compensation within an area, industry, or plant related to the levels of job values to which they apply.   Such compensation or ranges of compensation are usually established in relationship to the predominant wage rates for equal-valued jobs in the surrounding areas, the same industry or profession, or plants with comparable work.

Wage and Salary Survey: The gathering and comparing of base wage or salary rates or ranges for comparable jobs within an area, industry, plant, or profession.

Wage Arbitration: The hearing and settlement of wage disagreements between management and labor, by one or more individuals whose judgment or decision all parties have agreed to accept.

Wage Curve: A plot or graph showing the relationship between jobs, job classes, job evaluation point ratings, or the like and their corresponding wage rates or ranges.

Wage Differential: An increment in the pay received by an individual or group of individuals due to such factors as seniority, shift worked, geographic location, or other considerations.

Wage Incentive: See Financial Incentive.

Wage Incentive Plan: A method of payment which directly relates earnings to production.   A system which enables workmen to increase their earnings by maintaining or exceeding an established standard of performance.

Wage Rate: See Base Wage Rate.

Wages (Monetary):   The monetary compensation paid to an employee for services rendered, usually based on the hourly or daily wage rate in effect and the number of hours worked or the amount of work accomplished.

Wages (Real):   The value or purchasing power of the income received as measured by the goods and services that it will purchase, usually related to the amount which it would purchase during some previous period or in some other country.

Waiting Line: A line formed by units waiting for service; a queue.

Waiting Line Theory: See Queuing Theory.

Waiting Time: See Downtime.

Wink: A unit of time equal to 1/2,000 minute which the Gilbreths developed and used in motion and time study.

Wink Counter: An electrically or mechanically driven timing device indicating time in winks.

Work Content: See Job Content.

Work Cycle: 1. A pattern of motions or processes that is repeated with negligible variation each time an operation is performed.   2. A succession of operations or processes that is repeated with negligible variation each time a unit of production is completed.

Work-Factor: A system of predetermined motion time standards employing the Work-Factor as an index of motion difficulty (that is, demonstrating that time is proportional to specific factors in work, such as body member, distance, direction, weight, control, and the like, and that the relationship is consistent and interchangeable).   The system is used for determining efficient methods and setting performance time standards.

Working Area: See Maximum Working Area, Normal Working Area, and Production Center.

Working Conditions: Factors such as light, temperature, smoke, safety, hazards, noise, dust, and the like that affect the performance of a job or the general well-being of the employee.

Workplace: See Work Station.

Workplace Layout: See Work Station Layout.

Work Sampling Study: A statistical sampling technique employed to determine the proportion of delays or other classifications of activity present in the total work cycle.

Work Simplification: See Methods Engineering.

Work Station: That section of a production center where the workman performs his assigned tasks, including the space required for his auxiliary equipment such as tools; a workbench; or a machine with any stands, containers, conveyors, and the like for the material being worked on.

Work Station Layout: The arrangement of the tools, fixtures, bins, chutes, and other equipment at a specific work station.

Work Task: See Task.

Written Standard Practice: A standard practice that has been recorded and approved by the proper authority or authorities.   See Standard Practice.

Chapter **2**

Slide Rule Operation

PAUL E. MACHOVINA

Late Professor of Engineering Drawing,
The Ohio State University, Columbus, Ohio

The general-purpose slide rule enables the user to perform numerical calculations with considerable speed through mechanical manipulation of logarithmic scales. On the usual 10-inch rule, answers can be read to three or four significant digits, resulting in an average accuracy of 1 part in 1,000.

Slide Rule Components and Models. Slide rules of the straight-line style have three main components: the *stock*, which constitutes the frame or body; the *slide*, a single strip fitting and sliding in a groove of the stock; and the *indicator*, a transparent plate which slides along the face of the rule. Scales appear on both stock and slide, while the indicator carries a hairline used for setting and reading values.

The various slide rule models commonly encountered differ mainly in the number and variety of scales provided. In general, models with more scales permit solutions with fewer settings. Most manufacturers produce comparable models, and scales are fairly well standardized. In this discussion, the scales described are those appearing on standard rules. Scales are identified by letters printed on the rule near the end of the scale.

Reading Slide Rule Scales—C and D Scales. With the exception of the L scale, general-purpose slide rule scales are graduated to represent logarithmic functions, hence are nonuniformly divided. The pattern of subdividing varies somewhat with the available space; otherwise the setting and reading of numbers on these scales are the same as with ordinary uniformly divided decimal scales. The C and D scales are fairly typical of subdividing patterns.

Figure 2-1 shows the C or D scale (in segments) as it usually appears on a 10-inch

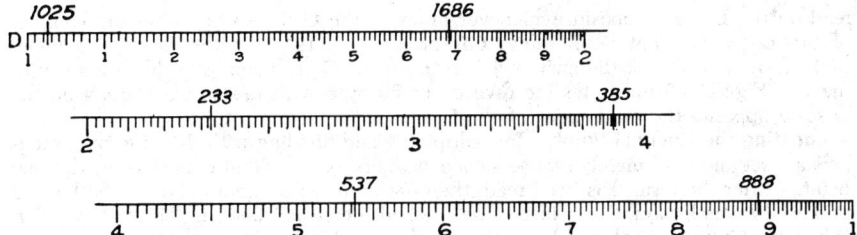

FIG. 2-1. The D scale (in segments) as it usually appears on a 10-inch slide rule. Note the logarithmic divisions.

rule.   Fewer graduations are employed with shorter rules, and more with longer ones. The graduations may be classified as *primary, secondary,* and *tertiary.*   The scale is divided in 9 main spaces with 10 primary graduations calibrated from 1 to 10 inclusive.   The extreme left- and right-hand primary graduations are called the *left index* and *right index,* respectively.   Secondary graduations subdivide each of the spaces between primary graduations into 10 spaces which are, in turn, subdivided by tertiary graduations.   The secondary spaces between primaries 1 and 2 are divided into 10 spaces each; between primaries 2 to 4 into 5 spaces each, with the graduations being read as 2, 4, 6, and 8; and between primaries 4 to 10 into 2 spaces each, with the single tertiary graduation read as 5.

Several numbers have been indicated in Figure 2-1 as they would be set or read. At the left end of these scales, three digits can be read and a fourth interpolated, while at the right end, two digits are read and a third interpolated.   In using these and similar scales, numbers are set and read as a sequence of digits *without regard to the decimal point.*   For example, the setting 1025 may represent the value 0.001025. 0.1025, 1.025, or 1,025,000.

Multiplying and Dividing—C and D Scales.   The C and D scales are fundamental with the slide rule.   They are used together when multiplying and dividing and also enable other calculations to be made when supplemented with additional scales.   The C scale is on the slide, while the D scale is on the stock.   They are identical in graduation.

Figure 2-2 illustrates the procedure when multiplying.   (For simplicity, scales employing only primary graduations have been used in this and future illustrations.) In Figure 2-2, the C scale has been set with its left index opposite 4 on the D scale. This permits the multiplication of 4 by a second factor to be set on the C scale.   If the second factor were 2, the product 8 is read on the D scale opposite 2 on the C scale.   With this slide setting and a multiplier larger than 2.5, the answer falls outside the D scale.   To remedy this, the *right* index of the C scale is set opposite 4 on the D scale.   Figure 2-3 illustrates this procedure where, for example, the product of 4 × 5 is read as 20 on the D scale opposite 5 on the C scale.   The placement of the decimal point, which is involved here, will be considered later.

In dividing with the C and D scales, the procedure for multiplying is reversed. The divisor is set on the C scale opposite the dividend on the D scale; the answer is

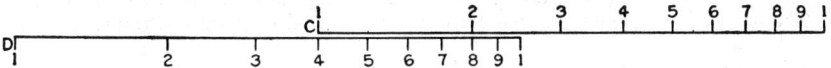

FIG. 2-2.   The C and D scales as they would be used for multiplying 4 by 2.

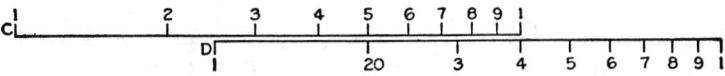

FIG. 2-3.   When dividing 20 by 5, the C and D scales are used in this manner.

read on the D scale opposite whichever index of the C scale lies within the confines of the D scale. Figure 2-2 shows this process. Here, 8 on the D scale is being divided by 2 on the C; the answer 4 is read on the D scale opposite the left index of the C. Figure 2-3 illustrates the division of 20 by 5 with the answer 4 read on the D scale opposite the right index of the C.

Locating the Decimal Point. In multiplying and dividing with the slide rule, numbers are set and read merely as a sequence of digits, paying no attention to the decimal point. After the result has been read, the position of its decimal point can be located in one of several ways. The inspection method, described here, is the most popular. When working practical problems, the number of decimal places in the answer frequently is already known.

With the inspection method, the problem is usually solved mentally, using rounded-off values to obtain an approximate answer indicating the decimal place in the result. As an example, multiply 43.5 by 79.5. These numbers are set as sequences of digits on the C and D scales, and the answer 346 is read as a sequence of digits on the D scale by following the multiplying procedure described above. By rounding the factor 43.5 to 40 and the factor 79.5 to 100, mental multiplication of the rounded values indicates the result to be approximately 4,000. Consequently, the answer is 3,460.

It is important to be aware that exactly the same settings and readings would be made if the preceding example were changed by moving the decimal points in the factors. Thus, 43,500 times 0.000795 would be set, and the result, 346, read as before. In this case, the factors may be rounded to 40,000 and 0.001 where it will be noted that the answer is approximately one one-thousandth of 40,000, or 40. Hence, the answer to this problem is 34.6.

With division, the approximate cancellation of values is often helpful.

The Folded Scales, CF and DF. The folded scales, CF and DF, are useful when the answer cannot be read on either the C or D scale without moving the slide. These scales are identical in graduation. The CF scale is on the slide, and the DF scale is on the stock of the rule. These scales can be produced by severing C or D scales at π (3.1416) and interchanging the segments; the terminal graduations have the value of π.

Since π on the DF scale is in a fixed position opposite the indexes of the D scale, a number set on D will be read opposite (use indicator hairline) on the DF scale as multiplied by π. Conversely, a number set on the DF scale will appear opposite on the D scale as divided by π. These relationships also exist between the C and CF scales of the slide.

The principal function of the folded scales is to minimize settings when multiplying with the C and D scales. Observe in Figure 2-4 that numbers opposite each other on the C and D scales, respectively, are also opposite on the CF and DF scales, respectively. Since the folded scales are out of phase with the C and D scales, it is frequently possible to read an answer on the DF scale which falls outside the limits of the D scale. For example, in Figure 2-4, 2 on the D scale is to be multiplied by 7. Since the left index of the C scale has been used, the answer cannot be read on the D scale. To avoid changing the slide, the answer 14 may be read on the DF scale opposite 7 on the CF scale.

The Inverted Scales, CI and CIF. These are simply C and CF scales, respectively, that are graduated in the reverse direction so they read from *right to left*. Because of this inversion, care must be exercised when setting numbers on these scales.

The inverted scales may be used with the noninverted scales to determine reciprocal

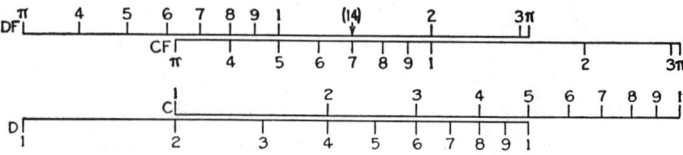

Fig. 2-4. Folded scales, CF and DF.

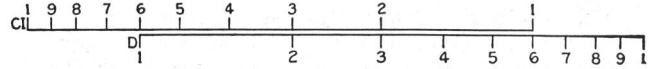

Fig. 2-5.   Inverted scales, CI and CIF

values.   Numbers opposite each other on the C and CI scales are reciprocals if the decimal points are correctly located.   The same is true for the CF and the CIF scales.

The inverted scales are useful in simple multiplication and in combined multiplication and division problems by reducing the number of slide settings.   The CI scale is used with the D scale in place of the C, although the C scale may also be employed in combined-type problems.   In simple multiplication, using CI and D, the two factors are set opposite each other and the answer read on D under the index of the CI.   For example, in Figure 2-5, 3 on the CI scale has been set opposite 2 on the D scale.   The product 6 is found on the D scale opposite the index of the CI.

Division using the CI and D scales is also illustrated in Figure 2-5.   For example, 6 is to be divided by 3.   The index of the CI scale is set to 6 on the D scale, and the answer 2 read on the D scale opposite 3 on the CI.   Had the problem been to divide 6 by 8, the answer could not be read on the D scale without shifting the slide in order to use the other index of the CI scale.   To avoid this, 8 may be set on the *folded* CI scale, i.e., the CIF, and the answer 0.75 read on the DF scale.

Squares and Square Roots, A and B Scales.   The A scale, on the stock, and the B scale, on the slide, are identical in graduation.   These scales compare with two C or D scales placed end to end and compressed to the length of one C or D scale.   Thus each has two sections, a left and a right.   These scales may be used to multiply and divide the same as the C and D scales; however, less accuracy is to be expected.

The A and B scales are intended primarily to assist in the squaring and the extraction of the square roots of numbers.   In this connection, the A scale functions with the D scale and the B scale operates in exactly the same manner with the C scale.   To square a number, it is set on the D scale and the square read opposite on the A.   For example, 4 will be found on the A scale opposite 2 on the D scale, and 25 on the A scale opposite 5 on the D.   The decimal point in the square can be placed by inspection.

To extract a square root, the process of squaring is reversed.   That is, the number is set on the A scale and the square root read opposite on the D.   It is possible to set the number in either section of the A scale.   *However, only one will yield the correct root.*   To determine the proper section, point the number off into groups of *two* digits, starting at the decimal point, as in longhand extraction.   If the first (left) group containing significant digits has *one* such digit, the number is set in the *left* section of the A scale; if there are *two* significant digits in the group, the number is set in the *right* section.   For example, these numbers would be set in the left section: 67,200; 450; 4.00; 0.037; 0.00086; and these would be set in the right section: 947,000; 25; 0.726; 0.0000187.

The decimal point in a square root can be placed by inspection or through observing that there will be one digit in the root for each of the above-mentioned groups of digits in the number.   Thus the square root of a whole or mixed number will have the same number of digits to the left of its decimal point as there are groups of digits to the left of the decimal point in the number.   If the given number is wholly decimal, the square root is also wholly decimal and will have one cipher between its decimal point and first significant digit for each group of two ciphers immediately following the decimal point in the number.

Cubes and Cube Roots, the K Scale.   The K scale is ordinarily located on the stock.   In this case, it functions with the D scale to cube and extract cube roots of numbers.   If it is on the slide, the C scale is used in place of the D.   The K scale is comparable to three D scales placed end to end and compressed to the length of one D scale.   Consequently three sections, a left, a middle, and a right, will be found.

To cube a number, it is set on the D scale and the cube read opposite on the K.   For example, 8 on the K scale is opposite 2 on the D, and 125 on the K is opposite 5 on the D. The decimal point in the cube can be located by inspection. To extract a cube root, the number is set on the K scale and the root read opposite on the D. However, as with square root extraction, the number must be set within the proper section of the K scale to obtain the correct root. To determine the proper section, point the number off into groups of *three* digits each, starting at the decimal point, as in long-hand extraction. If the first (left) group containing significant digits has *one* such digit, the number is set in the *left* section of the K scale; if there are *two* significant digits in the group, set the number in the *middle* section; if there are *three* significant digits, set in the *right* section. For example, these numbers would be set in the left section: 3,020,000; 6,210; 7.38; 0.00936; and 0.00000521; these in the middle section: 22,600,000; 84,700; 22.6; 0.0757; and 0.0000965; and these in the right section: 842,000,000; 123,000; 593; 0.376; 0.000000624.

The decimal point in a cube root can be placed by inspection or through observing that there will be one digit in the root for each of the above-mentioned groups of digits in the number. Thus the cube root of a whole or mixed number will have the same number of digits to the left of its decimal point as there are groups of digits to the left of the decimal point in the number. If the given number is wholly decimal, the cube root is wholly decimal and will have one cipher between its decimal point and first significant digit for each group of three ciphers immediately following the decimal point in the number.

The Trigonometric Scales, S, ST, and T. These scales are used when trigonometric functions of angles occur as factors of problems; consequently, the scales are graduated in angles. Two types of graduating are used: (1) angles in degrees and minutes and (2) angles in degrees and decimal parts thereof. The regular form S, ST, and T scales, which will be considered here, operate with the D scale. The S and ST scales are actually one continuous scale presented in two segments. Rules lacking a scale marked ST combine this scale with the S into a single scale (marked S) which operates with the A scale in the same manner as the S and ST scales operate with the D. The trigonometric scales are ordinarily on the slide and will be so considered here.

Sines and cosines may be read on the C scale opposite the proper angle on S or ST. Both the S and ST scales ordinarily have one set of calibration values reading from *left to right* serving as a sine scale and another set reading from *right to left* serving as a cosine scale. *For sines* (left to right), the ST scale ranges from 0.57 to 5.74 degrees (sines from 0.010 to 0.100) and the S scale ranges from 5.74 to 90 degrees (sines from 0.100 to 1.000). *For cosines* (right to left), the S scale ranges from 0 to 84.26 degrees (cosines from 1.000 to 0.100) and the ST scale ranges from 84.26 to 89.43 degrees (cosines from 0.100 to 0.010). For example, set 30 degrees (left to right range) on the S scale and read 0.500 on the C as the value of sine 30 degrees. Observe that this setting corresponds to 60 degrees in the right-to-left (cosine) range of the S scale; hence, cosine 60 degrees also equals 0.500.

If possible, problems involving trigonometric functions are manipulated by setting the angle on the proper trigonometric scale to the other factor on the D scale, thereby substituting the trigonometric scale for the C. For example, solve 540 × sine 23 degrees. Set the index of the S scale opposite 540 on the D; opposite 23 degrees on the S read 211 (answer) on the D. Note that the value of sine 23 degrees is on the C scale opposite 23 degrees on the S. Consequently the operation is a simple C and D scale multiplication; no attention is paid to the actual value of the sine. If a division by the sine is involved, the angle on the S scale is set opposite the dividend on the D and the answer is read on the D scale opposite the index of the S. The ST scale functions in a similar manner. Decimal points can be fixed through inspection by using the values given above for the functions of the angles covered by the scale used.

Tangents of angles appear on either the C or the CI scale opposite the angle on the T scale. The range of the T scale from *left to right* is 5.71 to 45 degrees (tangents from 0.100 to 1.000), and the tangents for these angles are on the C *scale.* The range of the T scale from *right to left* is 45 to 84.29 degrees (tangents from 1.000 to 10.00) and the tangents for these angles are on the CI *scale.* When using the T scale in multiplication or division, it must be borne in mind whether the value of the tangent appears on the C or the CI scale so that the proper setting will be made between the T and D scales. Use of the T scale follows procedures similar to the use of the S. Decimal points may be placed by inspection. The ST scale is used directly for tangents of small angles, because the sines and tangents of these angles are equivalent to several decimal places.

Cotangents are obtained by setting the angle on the T scale as for the tangent; the value of the cotangent appears on the reciprocal scale to the one on which the tangent appears, thus following the relationship cot $\theta = 1/\tan \theta$. Explicitly, tangents of angles between 5.71 and 45 degrees are on the C scale, while the cotangents of these angles are on the CI; tangents of angles between 45 and 84.29 degrees are on the CI scale, while the cotangents are on the C. A cotangent appearing on the CD will lie between 10.00 and 1.000, while one appearing on the C scale will be between 1.000 and 0.100.

Combined Operations. Problems of three and sometimes more factors can ordinarily be solved with one slide setting by combining several scales in the operation. Problems with many factors can be handled with greatest efficiency by repeating the procedures.

Combining operations is based on these principles:
1. Set the first factor, and read the answer on a scale of the stock (D or DF). All other factors are set on a scale of the slide (C, CF, CI, CIF, S, T, and so on).
2. Perform the first operation so that the answer to this step (on the D or DF scale) is under an index of a scale on the slide. *This is important* because the intermediate answer (no need to read it) is then automatically set to be *multiplied* by a number on the C or CF or to be *divided* by a number on the CI or CIF. No change in slide setting is necessary. Note that either a division using the C and D scales or a multiplication using the CI and D scales results in the desired condition.
3. Use the folded scales to avoid resetting the slide and observe that problems can be set initially with the folded scales. Advantage should be taken of the relationship between the D and DF scales when π is a factor. Principles for utilizing the A, B, K, and trigonometric scales in combined operations are illustrated in the following examples.

Example. Solve

$$\frac{35 \times 110 \times \pi}{4.2 \times 68 \times 8.7}$$

Opposite 35 on the D scale, set 4.2 on the C (division), move the indicator to 110 on the CF scale (multiplication), set 68 on the CF to the indicator (division), move the indicator to 8.7 on the CI (division), read the result 487 on the DF scale under the hairline, and place the decimal point by inspection. The answer is 4.87.

Example. Solve

$$\left(\frac{\sqrt[3]{8,500}}{\sqrt[2]{24 \times 1.5}}\right)^3$$

Opposite 8,500 on the K scale (left section), set 24 on the B (right section); opposite 1.5 on the CI, read the answer 21.5 on the K scale.

Example. Solve

$$\frac{250 \times \tan 3°}{\cos 65° \times 2.1 \times 1.4}$$

Opposite 250 on the D scale, set 65 degrees (for cosine) on the S; move the indicator to 3 degrees on the ST, set 2.1 on the C to the indicator; opposite 1.4 on the CIF, read the answer 10.0 on the CF.

Logarithms and the L Scale. The L scale is a uniformly divided scale occurring on the stock of some rules and on the slide of others. If on the stock, as will be assumed in this discussion, it is used in conjunction with the D scale. If on the slide, it is ordinarily used with the C.

The L scale serves with the D as a graphical table of common logarithms. To obtain the logarithm of a number, the number is set on the D scale and the mantissa of the logarithm read opposite on the L scale. To complete the logarithm, the characteristic is supplied in the usual manner. As an example, find the logarithm of 300. Opposite 300 on the D scale, read 0.477 on the L. Since the characteristic of 300 is 2, the logarithm is 2.477.

To find an antilogarithm, the procedure is reversed by reading the number on the D scale opposite the mantissa on the L; the decimal point in the number is located from the characteristic of the logarithm.

The Log Log Scales. The log log scales are very useful when solving problems involving roots, powers, and exponents. The five to eight LL scales provided are segments of one continuous scale ranging from 0.0000454 to 22,026 with a gap in the vicinity of 1.000. Numbers are set on these scales with regard to the decimal point. The scales marked LL1, LL2, and LL3 are used when setting numbers greater than 1, while the other LL scales are used with numbers less than 1, i.e., decimal fractions. On most modern rules, all LL scales operate with the C or D scales. With older slide rules, the scales for fractions operate with the A or B scales and are usually marked LL0 and LL00. The three basic operations performed with the assistance of the LL scales are (1) raising numbers to powers, (2) extracting roots, and (3) finding exponents. The operations are complemental.

To raise a number to a power, the index of the C scale is set to the number on the LL scale and the power is read on an LL scale opposite the exponent on the C scale. Figure 2-6 shows the solution of $4^{2.5}$. The index of the C scale is set to 4, which is found on the LL3; the answer 32 is read on the LL3 opposite the exponent 2.5 on the C. If the exponent is changed to 0.25 in this example, the same setting is employed but the power is read on the LL2 scale as 1.414. Similarly, if the exponent is 0.025, the power is read on the LL1 as 1.0353. It may be necessary to use the right index of C in order to bring certain exponents into play. Thus to raise $4^{0.8}$, the right index of the C is set to 4 on the LL3 and the answer 3.03 read on the LL3 scale. Observe that raising 4^8 yields an answer beyond the limits of the LL scales. In problems involving powers and roots where the numbers are too large to be set or read on the LL scales, the problems can be handled by factoring, thus:

$$a^x = (b \times c)^x = b^x \times c^x \quad \text{and} \quad \sqrt[x]{a} = \sqrt[x]{b \times c} = \sqrt[x]{b} \times \sqrt[x]{c}$$

To extract a root, the procedure for finding a power is reversed. As an example, solve $\sqrt[2.5]{32}$ (see Figure 2-6). Set 2.5 on the C scale opposite 32 on the LL3. The root 4 is read on the LL3 opposite the index of the C. Note that with this setting, the twenty-fifth root of 32 is read on the LL2 scale as 1.149 and the two-hundred-fiftieth root read on the LL1 as 1.014. When finding powers and roots, the LL scale on which the answer appears can be determined by a rough mental calculation.

To find the exponent in $4^x = 32$ (Figure 2-6), set the index of the C scale to 4 on the LL3; the exponent 2.5 is read on the C scale opposite 32 on the LL3. Decimal points in problems where exponents are sought may be located by inspection.

When working with powers of e (2.718) and natural logarithms, observe that e serves as a terminal graduation for the LL3 scale; consequently, powers of e may be

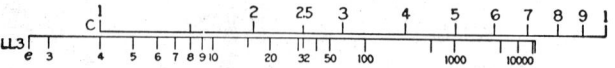

Fig. 2-6. Log-log scales.

read on the LL scales opposite the exponent on the D.   Similarly, natural logarithms may be read on the D opposite the number on an LL scale.

BIBLIOGRAPHY

Arnold, Joseph N., *Complete Slide Rule Handbook*, Prentice-Hall, Inc., Englewood Cliffs, N.J., 1962.
"Logarithms and the Slide Rule" in Abraham Sperling and Monroe Stuart, *Mathematics Made Simple*, Cadillac Publishing Company, New York, 1961.
Saxon, James A., *The Slide Rule*, Prentice-Hall, Inc., Englewood Cliffs, N.J., 1966.

Chapter **3**

Methods of
Measuring Time

JOSEPH E. SCOTT

**Vice President, Meylan Stopwatch
Corporation, New York, New York**

The greatest timing device, the universe, began with creation. As the universe measures off the millennia, man has produced the calendar and the chronograph to divide his life and history into useful and understandable segments of time.

Industrial engineers have applied the measurement of time and motion to the solution of the production and cost problems of industry. The success of time and motion study techniques is historical. However, they require constant attention, application, and refinement. As the more complex, high-speed, short-cycle jobs are time studied, the demand for more accurate and versatile timers is heard.

The employment of technicians to replace industrial engineers in the basic work measurement positions has created the need for simple, accurate, easy-to-read timing devices. The situation does not eliminate the responsibility of the engineers for the accuracy and the installation of the resulting time standards. The producers of timers and timing devices have kept pace with the professional needs of the industrial engineer. A brief description of innovations in timers and timing techniques is presented to assist in selecting the best available instrument for any special requirement.

STOPWATCHES

The decimal minute and decimal hour stopwatches were adopted for time study because the whole-digit readings can readily be converted into pieces per hour or hours per piece. Fractional seconds invite arithmetical errors.

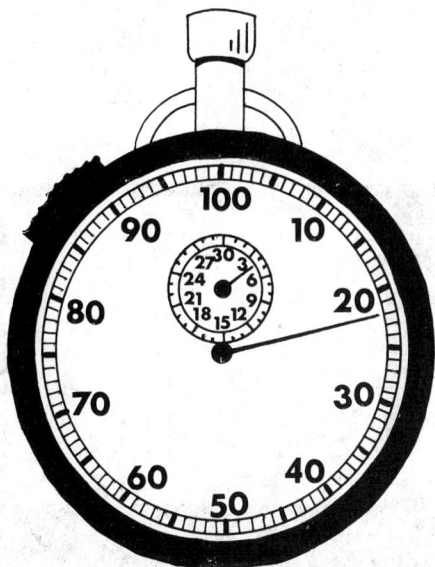

FIG. 3-1. Conventional decimal minute stopwatch with small 30-minute totalizer.

The decimal minute stopwatch, Figure 3-1, is the most commonly used. The large hand makes one revolution in 1 minute, and each graduation represents $\frac{1}{100}$ of a minute. The small hand totalizes to 30 minutes. Figure 3-2 shows the identical stopwatch with two concentric hands. Again, the larger hand makes one revolution in 1 minute, but the totalizing hand simplifies reading the elapsed minutes because of the larger scale. This is a comparison of plain stopwatches, past and present.

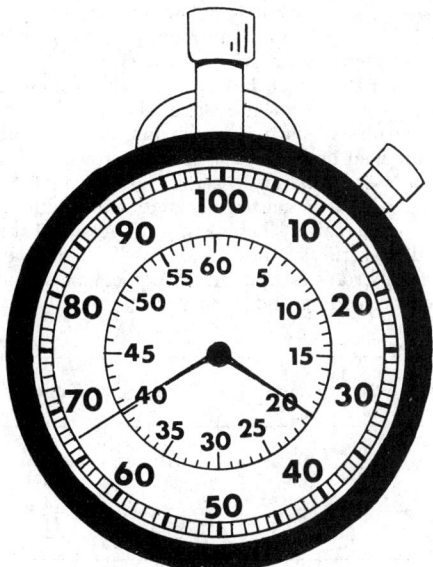

FIG. 3-2. Decimal minute stopwatch with concentric totalizer reading to 60 minutes.

FIG. 3-3. Time study board with stopwatch holder and paper clamp.

The obvious advantage of the larger minute register is to increase the ease of reading elements exceeding 1 minute.

In subsequent paragraphs, the terms "element" and "cycle" will be used frequently. For clarity in this chapter, an element is defined as one component of an operation. For example, performing an operation may require pick up piece, position, machine, and remove. These four elements constitute a cycle.

A time study board with stopwatch holder and paper clamp to hold the time study form is a common accessory. A typical board is shown in Figure 3-3.

The Figure 3-1 stopwatch is started by moving the side-slide toward the crown. When the side-slide is moved down, away from the crown, the hands stop. Depressing the crown will then reset the hands to zero. One time study practice is to operate this stopwatch "snapback." It is started conventionally with the slide at the beginning of the time study. Thereafter, the crown is depressed at the end of each element while simultaneously the position of the hand is read. This is known as the snapback technique. The most important advantage of the decimal minute stopwatch (Figure 3-2) is that an element exceeding one minute is more easily read. Incidentally, this

Element	Time	Time	Time	Time
Pick up...........	.07	.07	.06	(.17)
Position...........	.03	.05	.04	.04
Machine.........	.14	.15	.14	.15
Remove..........	.06	.05	(.11)	.05

FIG. 3-4. Watch readings obtained from the snapback method of timing.

Element	Time calc.	Time calc.	Time calc.	Time calc.
Pick up..........	.07	.37	.68	1.14
Position..........	.10	.42	.72	1.18
Machine.........	.24	.57	.86	1.33
Remove..........	.30	.62	.97	1.38

FIG. 3-5.   Watch readings obtained from the continuous method of timing.

stopwatch has a side-button rather than a side-slide.   It starts, stops, and resumes by successive depressions of the crown.   The side-button is the zero reset.   Depressing the side-button while the hands are in motion provides snapback operation.

Snapback versus Continuous Timing.   Using the snapback method, the observer times each element as an entirety.   The total cycle time is unknown until the elemental times are added later for each cycle.   Actually, this is not a disadvantage, because the time study observer is not concerned with cycle time at this point.   By recording elemental times, any unusual variation in the time of an element is apparent and the observer is alerted to determine the reason.

Figure 3-4 illustrates a snapback time study with four elements in each cycle.   The time study observer has circled two elements that reflect inconsistencies which will be explained on the time study form.   The snapback method exposes such inconsistencies as they occur.   The continuous method of reading a stopwatch does not, because these inconsistencies are revealed only when subtractions and calculations are made, after the study is completed.   Figure 3-5 shows the same four cycles using continuous timing.   The abnormal elements are generally not apparent until the elemental times are determined by subtraction.   The first element subtracted from the second, the second from the third, and so on provide individual elemental times, but the observer must depend upon his alertness to detect departures from the established method. Continuous timing, however, has a number of important advantages:

1. All time is accounted for.   The stopwatch is started, the elements read, and the study completed while the watch runs continuously.   Nothing timewise can be omitted.   A two-hour study must account for two hours.   Using the snapback method, the observer can lead or lag (read ahead or behind the graduation of the dial) or arbitrarily stop the stopwatch when the worker pauses, and resume timing only when the worker resumes his normal duties and pace.   Accordingly, 2 hours of time study may add up to only 1½ hours of recorded elemental times.

2. Labor unions generally accept elemental times only when overall study time is proved.   Leveling and allowances, discussed in other chapters in this Handbook, are less discretionary because the observer rates and allows on a noncomparative basis; that is, any time inconsistency has no influence because there is no immediate awareness that the .09 element just recorded was .06 on the previous cycle.

In a continuous study, two types of stopwatches are generally used.   Figure 3-6 shows a conventional decimal minute stopwatch with a sweep hand and auxiliary 30-minute, small totalizing hand.   It is started by depressing the crown.   The next depression of the crown stops the hands, and the third depression resets both hands to zero.   In continuous time study, these sequences are not used successively.   Instead, the stopwatch is started at the beginning of the study and it is not stopped until the last element to be timed terminates.   Accordingly, successive readings are taken, such as .07, .10, .24, .30, and so on.   The hand is read "on the fly" (while in motion). Subtractions provide the elemental readings; that is, .07 is the time for the first element, .10 minus .07 is .03 for the second element, .24 minus .10 is .14 for the third element, and so on.   Although this ensures accountability for cycle time, the elemental times can be misread because a moving hand is being read.

Timing continuously is more accurate when using the split-hand stopwatch shown by Figure 3-7.   This stopwatch has two concentric sweep hands.   Both start simul-

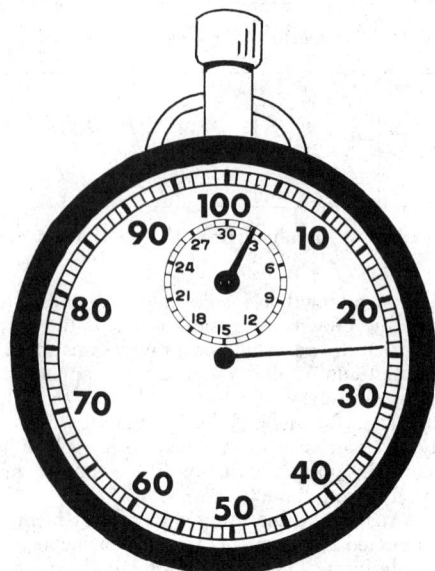

FIG. 3-6.   Decimal minute stopwatch with single sweep hand.

taneously from zero by depressing the crown.   Thereafter, the crown is not used until the final element terminates.   At the end of the first element, the side-button is depressed.   This stops the lower concentric hand which is read.   Then this same side-button is again depressed.   This causes the stopped hand to rejoin (catch up to) the hand in motion.   This procedure is followed for each element.   Here the advantage

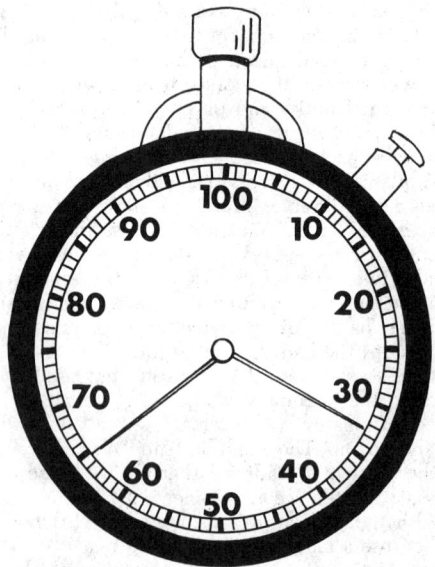

FIG. 3-7.   Decimal minute stopwatch with split hands.

is that elemental times are locked in place. Obviously, by reading a stopped hand, the reading is more accurate. However, in either case, the overall time is the same, but the accuracy of the elemental readings is less subject to error. Finally, the overall study time is registered by depressing the crown to stop all hands when the last element terminates.

There is another feature available on many split-hand stopwatches. The small minute-register hand on most stopwatches generally moves progressively; that is, it moves in proportion to the progress of the sweep hand. On split-hand stopwatches, it can remain stationary until the sweep hand completes one revolution and then it "jumps" to the next graduation. Thus, elapsed minutes are easier to read because the register hand is always on an increment, never in between.

Special Techniques. Direct elemental reading has been enhanced by three innovations. Snapback, as explained earlier, is reading the elapsed time and depressing the reset control simultaneously. An experienced time study man can misread the stopwatch, due to diversions, lighting, and the like. Figure 3-8 shows a stopwatch which has concentric sweep hands—the same as the instrument in Figure 3-7—but the operation differs. This stopwatch is started by moving the side-slide toward the crown. At the end of the first element, the crown is depressed and released in the usual snapback manner. One hand stops and stays in place; the other returns to zero and instantaneously begins timing the next element. The stopped hand is read, and the crown is then depressed again. This causes the stopped hand to rejoin the hand in motion (timing the next element). This is repeated for each succeeding element, and the elemental times are read directly without ever reading a hand in motion.

A time study board which operates three stopwatches controlled by one lever is illustrated in Figure 3-9. The first stopwatch is set at zero, the second is stopped at any point other than zero, the third watch is put in motion. When the lever is depressed, the first watch starts timing the element, the second resets to zero, and the third stops. The lever is again depressed to stop the first watch, start the second, and reset the third to zero. The first watch is now read to record the elemental time. The sequence is repeated for each element in the operation by reading the watch stopped after depressing the lever at the end of each element. This system provides direct elemental times without reading a hand in motion.

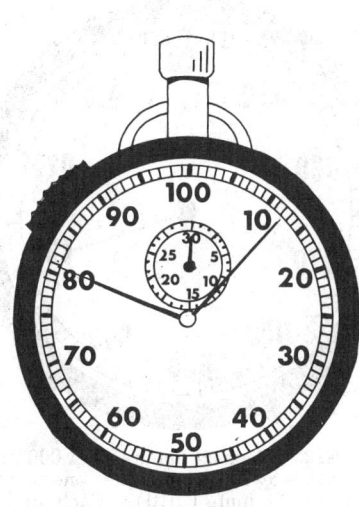

Fig. 3-8. Another decimal minute stopwatch with split hands.

Fig. 3-9. Time study board with three stopwatches.

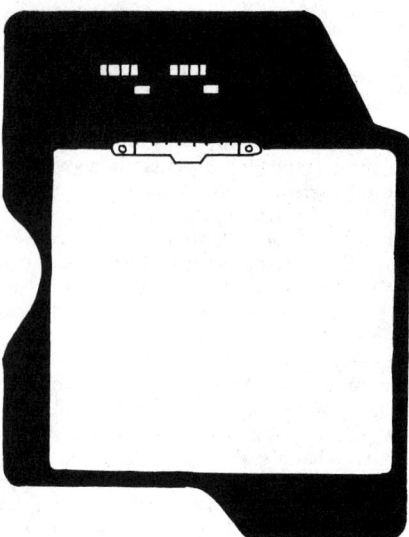

Fig. 3-10.   Battery-powered time study board with two digital counters reading in decimal minutes.

A battery-powered board with two digital counters reading in decimal minutes is shown in Figure 3-10. The counter on the left is a totalizer for overall time. Depressing the lever starts both counters. At the elemental break point, the lever is depressed partially, to stop the right-side counter. It is read. Then, depressing the lever the rest of the way causes the electronic memory in this device to reset to the time consumed in reading and holding the observed elemental time (generally, .02 or .03 decimal minute). The process is then repeated to time the next element. This

Fig. 3-11.   Decimal minute (1/100 minute) dial. Sweep hand makes one revolution in one minute. Each graduation represents 1/100 minute. Reading: .03 minute.

Fig. 3-12.   Decimal minute (1/1,000 minute) dial. Sweep hand makes one revolution in 1/10 minute (.010). Each graduation represents 1/1,000 minute. Reading: .030 minute.

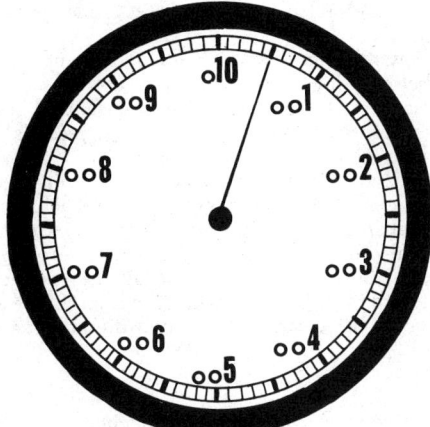

FIG. 3-13. Decimal hour (1/100 hour) dial. Sweep hand makes one revolution in 1/100 hour (36 seconds). Smallest graduation represents .0001 hour. Reading: .0005 hour.

procedure is continued indefinitely, with the left counter, unaffected by the lever during the study, accumulating the overall time. The battery-powered board with two digital counters provides direct elemental readings and easy-to-read digits.

Other Dials and Readings. Stopwatches for time study also read in decimal hours, 1/1,000 minute, and TMU. These dials are illustrated in Figures 3-11 to 3-14. On each, the reading is identical in terms of elapsed time: .03 minute :: .030 minute :: .0005 hour :: 50 TMU. These times are equivalent to 1.8 seconds.

It is evident that no stopwatch should be used whose calibration is not understood.

Stopwatch Maintenance. The following practices will help keep a stopwatch in good operating condition.

1. Treat a stopwatch as a precision instrument. Protect it from moisture, dust, and sudden temperature changes. Condensation can be ruinous.

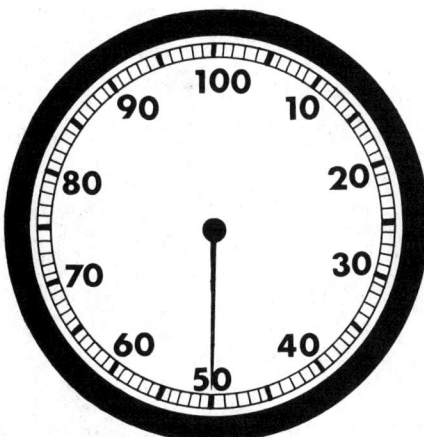

FIG. 3-14. Special dial for MTM. Sweep hand makes one revolution in 3.6 seconds (100 TMU), which is .001 hour, or .06 minute. Smallest graduation represents .00001 hour. Reading: 50 TMU.

2. Have it cleaned and oiled periodically. A one-year guarantee is normal protection.
3. Wind it and allow it to run down once a month if not used regularly.
4. Wind it fully before using.
5. When not in use, keep it in an appropriately cushioned box.

Accuracy. An error not exceeding plus or minus 1.5 seconds (.025 minute) over 1 hour is stipulated in many U.S. government specifications. Therefore, using the continuous method of time study, the possible mechanical error is considerably less than the probable human reading error. However, in snapback timing, the cumulative errors (human and mechanical) increase this tolerance. Government specifications recognize the mechanical error by allowing an average deviation of .3 second (.005 minute) per operation of 30 seconds. Thus, using the continuous method of time study, the mechanical (stopwatch) error will be within 1.5 seconds (.025 minute). The mechanical error could be considerably greater using the snapback method.

RECORDING TIME

Chart Recorders. When using a stopwatch or other manually controlled timer, it is necessary for the user to read and record. Devices (chart recorders) are available to furnish automatically a printed record of elapsed time and its usage. These instruments generally use time-graduated circular or strip charts. The circular chart is easier to analyze but does not provide as close definition. An interval recorded on an 8-hour chart usually cannot be read accurately to less than 1 minute; and on a 24-hour chart, a definition to 5 minutes should be expected. Charts usually revolve over 8, 12, or 24 hours. Therefore, on a 24-hour chart, 1 degree represents 25 minutes.

Representative circular chart and strip chart recorders are shown in Figures 3-15 and 3-16. Strip chart recorders print on long rolls of time-graduated paper about the size of adding machine tape. Various speeds are available, depending upon the definition required. At 12 inches per hour, 1 inch equals 5 minutes. At 60 inches per hour, 1 inch equals 1 minute. At 120 inches per hour, 1 inch equals 30 seconds and therefore can be accurately interpreted to 2 seconds. Thus, a strip chart recorder can, if necessary, define machine on-off time and operator performance in terms of usage to a remarkable degree of accuracy.

Chart recorders can be used to reveal the work pattern before a study is initiated,

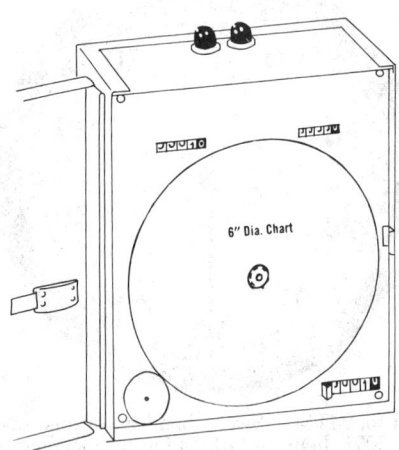

Fɪɢ. 3-15. Circular chart recorder.

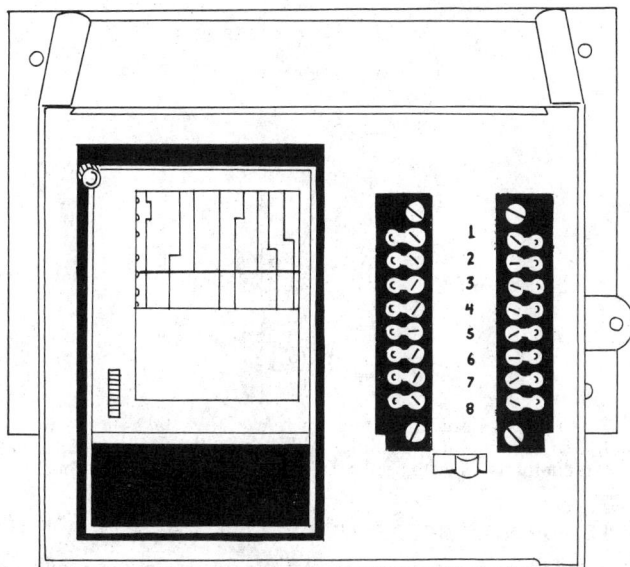

Fig. 3-16.   Strip chart recorder.

but generally are used to reveal discrepancies after the performance standard has been established.   Thus, they are fundamentally work time versus downtime analysts— silent sentinels or "watchdogs" to record performance.   It must be remembered that a day or work shift expires before the chart can be analyzed.   The chart is, in actuality, a postmortem.   It cannot correct what already happened but indicates the areas that need remedial action or where established performance standards are not being achieved.

Chart recorders both supplement and implement work standards established through conventional stopwatch studies or predetermined time systems.

Because of the accuracy of synchronous motors and the availability of electric current, most chart recorders are electrically driven.   To differentiate between on-time and downtime, electric sensors are used to control the styli, microswitches, limit switches, photocells, and the like.

Incidentally, the motion picture camera is also a form of chart recorder.

TIME IN TERMS OF SPEED

Tachometers.   A tachometer is as essential to work measurement as the speedometer is to the automobile.   The automobile speedometer reads in miles per hour (mph) and is a constant reminder of safety and optimum speed.   Tachometers generally read in revolutions per minute (rpm) or feet per minute (fpm).   A time study might indicate that optimum performance is achieved at a certain speed.   Using a tachometer, both the operator and the supervisor can verify the speed.

For continuous speed indication, a stationary (permanently mounted) tachometer, shown by Figure 3-17, is used.   A less expensive hand tachometer, Figure 3-18, will provide intermittent readings.   The decision which to use depends on two considerations: is it critical to output or quality that the speed be indicated continuously, or are occasional speed checks sufficient?

Tachometers operate on various principles: mechanical-centrifugal, chronometric, electrical, reed-relay, photocell, and the like.

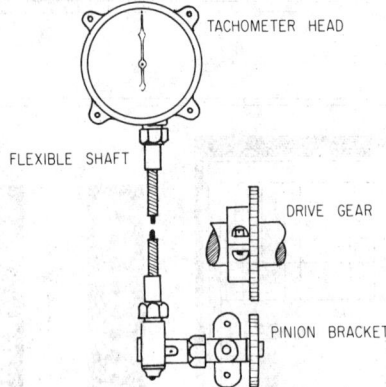

FIG. 3-17.   Continuously indicating tachometer.   Drive take-off gear is attached to machine shaft.   In turn, pinion, flexible cable, and other components feed speed to tachometer head precalibrated for a specific installation.

ALTERNATIVES FOR TIME ACCOUNTING

Frequently, a rough rather than a precise measurement of time will provide sufficient definition for the task at hand.

Elapsed-time Meters.   Using the automobile as the analogy, the automobile odometer can register miles traveled only when the car is in motion.   The elapsed-time meter, according to how it is connected, registers downtime or productive time, based on the activity or lack of activity of the machine.   These meters are available for both alternating current and direct current in most voltages and read in minutes

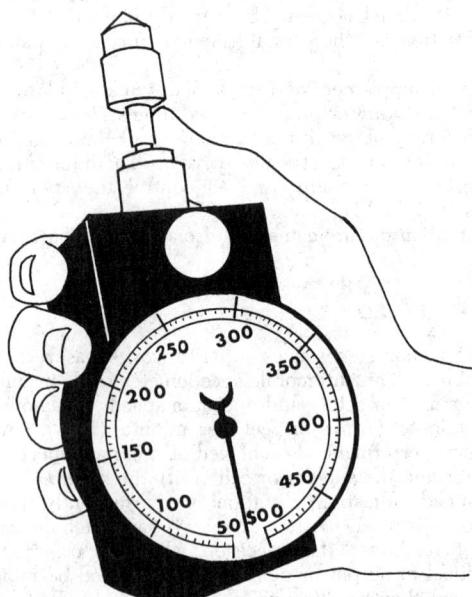

FIG. 3-18.   Hand tachometer which comes with an assortment of tips to engage shaft properly to measure rpm, fpm, or other variable.

FIG. 3-19. Elapsed-time meter which runs to maximum reading and begins anew from 0.

and tenths, hours and tenths, or other units as desired.   Figure 3-19 shows a commonly used elapsed-time meter.

RANDOM-TIME ANALYSIS

Work Sampling.   The work sampling technique of measuring performance through the percentage of time spent performing operations and activities is discussed in detail in Chapter 4 of Section 3.   Essentially, work sampling is a method of determining the frequency and duration of activities by random observations.   The number of random observations required is determined statistically.   The exact times for these observations are then selected from a table of random numbers.   The observer must then devise some means of being alerted to make his tour of observations at the random times selected.   Clock watching or repeated setting of an interval timer can be extremely time consuming. The instrument in Figure 3-20 is a random-time signal

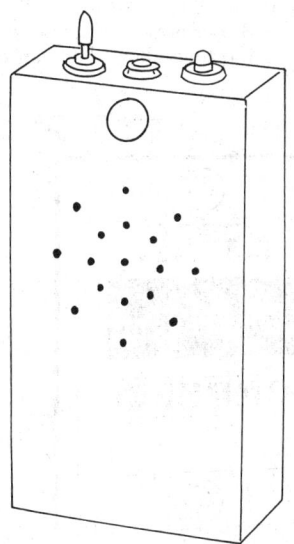

FIG. 3-20.   Random-time signal generator.

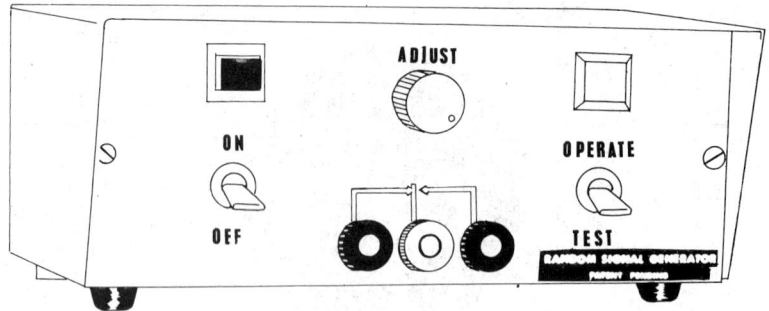

Fig. 3-21.   Random-pulse generator.

generator. It automatically generates random times (eliminating the table of random numbers) and audibly signals the observer. Thus, the observer can concentrate on other tasks, knowing that the device has generated and will signal the next random time. A device of this nature can usually be adjusted to deliver 2 to 30 random signals per hour.

The random-pulse generator, Figure 3-21, instead of emitting an audible signal, operates a self-contained relay to open or close an external circuit momentarily. Camera actuation at random times is one application of this device. Ejecting a product from an assembly or production line at random times is another. Memomotion is enhanced by random pulses.

TIME IN REVERSE

Often, allowed time is as important in work measurement as elapsed time. A predetermined counter, Figure 3-22, can be set to show the number of pieces or cycles wanted and to record numerically the number of pieces produced or remaining. When the preset figure is reached, the counter actuates a signal or relay to alert the operator or automatically stop the machine.

An interval timer, Figure 3-23, performs the same counting function in terms of segments of time. It is set for the time allowed for an operation or process, and at

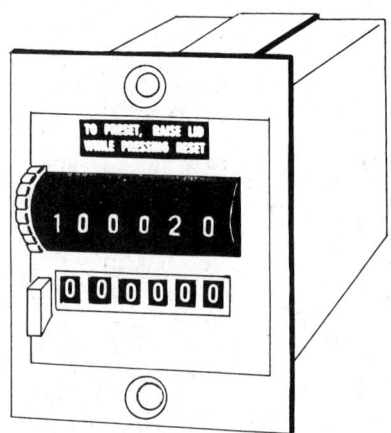

Fig. 3-22.   Predetermined counter.

PORTABLE
7 INCH DIAL,
10 AMP. CAPACITY

Fig. 3-23. Interval timer.

the expiration of the preset time, emits a signal, and, if desired, discontinues power to the equipment.

SUMMARY

In this chapter, a description has been presented of some basic timing devices for work measurement, analysis, and control. Other instruments are available to record clerical, administrative, and executive time utilization. A careful search will usually locate the timing device needed to solve a particular measurement problem. Suppliers of measuring devices can give instrument recommendations for special or new timer applications. Usually, a standard unit can be found or adapted to perform the job.

Chapter **4**

Making and Using
Industrial Motion Pictures

GERALD J. SKERRETT

**Technical Editor, Eastman Kodak
Company, Rochester, New York**

The most important advantage of motion pictures to industrial engineers is the ability to record or manipulate time. A motion picture consists of a series of still pictures called frames taken at regular intervals and viewed at the same or at different regular intervals. To state it simply, time is made to appear normal by taking the pictures at the same rate that it is intended to project them. To slow down time, pictures are taken at a rate faster than the projection rate will be. To compress time, pictures are taken at a slower rate than the projection rate. Thus, actions or movements that occur too quickly for an observer to follow can be slowed down so that all aspects of the action can be studied. In this case, it may be said that time has been expanded or magnified. The following equation can be used to determine time magnification for any combination of camera speed and projection speed:

$$\text{Time magnification} = \frac{\text{camera speed}}{\text{projector speed}}$$

Events that occur over an extended period of time, such as days or weeks, can be speeded up so the observer can appreciate the entire motion continuum of the event. Motion pictures taken and projected at the same rate are, perhaps, most familiar to everyone. Such pictures not only provide an accurate record of events, but the motion picture can be projected over and over in order to study each portion of the event.

TABLE 4-1 Exposure Time

Formula for determining exposure times for motion picture cameras with rotating shutters:

$$\text{Exposure time (in seconds)} = \frac{\text{shutter opening (in degrees)}}{\text{no. of frames per second} \times 360°}$$

Example: Exposure times for various shutter openings at 24 frames per second.

Shutter opening (in degrees)	Exposure time (in seconds)	Shutter opening (in degrees)	Exposure time (in seconds)
235	$\frac{1}{37}$	120	$\frac{1}{72}$
200	$\frac{1}{43}$	100	$\frac{1}{86}$
180	$\frac{1}{48}$	90	$\frac{1}{96}$
175	$\frac{1}{49}$	80	$\frac{1}{108}$
170	$\frac{1}{51}$	60	$\frac{1}{144}$
160	$\frac{1}{54}$	45	$\frac{1}{192}$
150	$\frac{1}{58}$	22½	$\frac{1}{384}$
135	$\frac{1}{64}$	10	$\frac{1}{864}$

The motion picture can also be used as a training film to teach principles or practices in an interesting, informative, and controlled manner.

Time Resolution. One of the most important factors in almost any engineering study is time—when did an event occur, how long did it take, and so on. An important concept to the engineer considering the use of motion pictures is that of the time resolution of the data provided. Because motion pictures are really a series of still pictures, each frame or picture is capable of recording information for only the duration of the exposure time. For most motion picture cameras operating at 24 frames per second, the actual exposure time of each frame is about $\frac{1}{50}$ second. During the interval between frames, no data can be recorded. At 24 frames per second, this interval is also about $\frac{1}{50}$ second.

If a rapid event is being photographed at 24 frames per second, it may be that $\frac{1}{50}$ second is so long that the image is recorded as a blur and also that the interval between frames is so long that much of the event is not photographed at all. In other words, the time resolution is not good enough to record the data desired.

For those cameras that have 170- to 180-degree shutters, exposure time in seconds is approximately equal to the reciprocal of twice the frame rate. The formula and table in Table 4-1 can be used for these, as well as for other cameras with rotating disc shutters.

Table 4-2 can be used to convert camera speed settings to number of frames per minute and to determine the number of frames per foot of film.

A little time spent in considering the time resolution and frame rate of a particular photographic job before starting the actual photography will prevent many disappointments.

INDUSTRIAL MOTION PICTURE TECHNIQUES

Normal Speed. Normal-speed motion pictures are usually characterized as those which are projected at the same rate as they were exposed. However, there is considerable range for this rate. Generally speaking, the lower frame-frequency limit for motion picture photography is 8 frames per second. The frequency standard is 16 frames per second for silent pictures and 24 frames per second for sound pictures. A frame-frequency limit of 250 frames per second has been suggested by some as an upper limit for normal-speed motion pictures. At 250 frames per second, the exposure time approaches 1/1,000 second.

Normal-speed motion pictures are the ones which are most familiar. They are used

TABLE 4-2. Camera Speed Setting and Film
Size Conversion Table

Camera speed setting (frames per second)	Number of frames per minute
8	480
16	960
18	1,080
24	1,440
32	1,920
48	2,880
64	3,840

Film size	Number of frames per foot of film
16mm	40
8mm	80
Super 8	72
35mm	16

extensively in industry to make reports to management and presentations to the public. As an audiovisual tool, they are an excellent medium for the communication of ideas.

Slow Motion. In slow-motion photography, the camera speed is greater than that of the projector and gives the effect of magnifying time. Slow-motion pictures are usually made at speeds not greater than 200 frames per second. The slowdown effect is obtained when the exposing speed exceeds the projection speed by ratios of 2:1 to 4:1.

This technique allows the engineer to observe motions and relationships which occur too quickly for the eye to follow.

High-speed Motion Pictures. Increasing the frame rate of the camera above that used for slow-motion photography results in the technique known as high-speed motion picture photography. In general, this area consists of frame rates from about 275 frames per second and up. Frame rates of over 10 million per second have been accomplished with some special cameras. Obviously, high-speed photography is used when the motion of the subject is so rapid that neither the eye nor normal-speed photography is of any benefit. Actions of spray nozzles, high-speed machine operations, ballistic projectiles, and many others are suitable subjects for high-speed photography.

Time-lapse Photography. At the other end of the scale from high-speed photography is the technique known as time-lapse photography. In this technique, the camera speed is less than the projection speed. It is possible, therefore, to compress several hours, days, or even weeks into a few minutes' projection time. This time compression reveals gross movements, patterns, and events that happen too slowly or too infrequently to be observed by the human eye.

The technique requires a motion picture camera capable of exposing a single frame at a time. In addition, a device to actuate the single-frame release on the camera plus a time device called an intervalometer to control the time interval between successive frames are required. A special camera, called a "pulse camera," exposes one frame at a time in response to a signal applied to the camera's electric motor. These cameras do not require devices to attach to the shutter release, but they do require an intervalometer.

A variation of the time-lapse technique consists of exposing short bursts of normal-speed motion pictures at random or fixed intervals. These motion pictures give all the information of a time-lapse movie, with the added advantage of motion of the subject. A burst of exposures can be programmed for any duration and frequency desired. Although the bursts are usually shot at normal speed, short bursts of slow-motion pictures can be used as well.

Application of this technique was used by industrial engineers at Eastman Kodak Company to investigate a jamming problem on a machine in which material was moved between tracks through a narrow tunnel with close tolerance. Jamming was occurring randomly—once per 10,000 production units or about once a week. Direct observation was ineffective because of this infrequency. Continuous motion picture filming was unacceptable because of the quantity of film which would be required to film the entire operation—about 144,000 feet of film per week. The advance cycle of able because the jamming could be missed during the camera "off" cycle.

the machine was ¼ second; thus normal time-lapse photography was also unaccept-

A unique burst technique was used to solve the problem. Because jamming occurred only during the advance cycle, the camera was focused on the tunnel and was set up to be pulsed from this portion of the machine operation. As the advance cycle began, the camera and the lights were turned on. When the advance was completed, the camera and lights were turned off. Because only 5 percent of the total machine operation involved advancing the material, this technique reduced the film usage by 95 percent.

Analysis was made of those rolls of film which were being exposed at the time of the malfunction. The result of this analysis showed the problem to be a slight misalignment of the material just prior to entering the tunnel.

Micromotion Photography. Micromotion pictures are pictures taken at governed speeds for the purpose of making actual counts of the film frames which, in this case, represent units of time. Or a timing device can be included in the scene to relate the action directly with elapsed time.

Micromotion analysis can be done in three ways. In one method, a special camera is used. The camera speed is governed at a constant rate, such as 1,000 frames per minute. Each frame then represents 0.001 minute. A motion picture projector, called an analyzer, is used to view the finished film. This type of projector has a frame counter which is used to count the number of frames during a particular action.

The second method is to photograph a clock or microchronometer in the scene with the operation being studied, as illustrated by Figure 4-1. Another method of including a clock in the picture is shown in Figure 4-2. This is a double-exposure technique which superimposes the face of a clock or stopwatch over the picture. To

Fig. 4-1. Industrial operation photographed with microchronometer included in the camera's field of view.

Fig. 4-2.   Industrial operation photographed with stopwatch superimposed.

use this technique, photograph the clock against a black velvet background and under-expose by one stop.   The clock image can be placed in one corner of the frame or in the center as shown.   Expose the entire roll of film.   In the dark, rewind the film onto the camera spool in such a way that the head end of the film is out.   The event can then be filmed normally.

When filming fast-cycle operations or operations where slow-motion photography is necessary, it is best to use a timepiece with a fast movement, for example, 1 revolution per 6 seconds.   A timepiece with a black case will minimize unwanted reflections.   Care should also be taken in placing lights so that reflections from the face of the timepiece do not occur.

The third method involves a special camera that has provision for putting a timing signal on the edge of the film.   A signal generator triggers a small timing light inside the camera.   Each time the light blinks, a small spot is exposed on the film.   If the light blinks at a known frequency, the average camera speed between any two spots can be determined by using the following equation:

$$\text{Camera speed} = \frac{\text{no. of frames between signals}}{\text{time interval between signals (seconds)}}$$

A typical example of micromotion photography and analysis was done at Eastman Kodak Company on a production operation in which an operator controlled a cycle involving six elemental functions, accomplished in ⅔ second at 100 percent normal effort. Because of the variability of the operation, it was impossible to automate fully. The cycle was repeated normally between 50 to 75 times per minute.

This operation was ideal for micromotion analysis because it had all the classic characteristics; it was:

    1. Short-cycle
    2. Highly repetitive
    3. High-volume

The operation was filmed and analyzed, taking into consideration the following points:

    1. Because the operation was rapid, faster-than-normal camera speed ($4\times$) was used to slow the action down.
    2. A watch having a fast, sweep-second hand was placed in the camera's field of view to check camera speed.
    3. The operation had to be carried out in subdued light because the operator had to view film which was backlighted. No supplemental light could be used. A fast black and white film (KODAK TRI-X Film) was used so that the operation could be filmed with existing light.
    4. A 16mm analyzer-projector with a frame counter was used for analysis and micromotion charting.

The results of the analysis enabled engineers to suggest modifications to the equipment design. This led to an easier cycle which involved only three elemental functions and which was accomplished in ½ second at normal 100 percent effort. Labor savings amounting to thousands of dollars were realized.

Infrared Motion Pictures. Time and motion studies are sometimes required in areas of subdued lighting or where special lighting conditions are necessary and supplemental lighting for photography cannot be used. In these cases, motion pictures can still be made by using infrared-sensitive films. The area can be "lighted" using supplementary infrared radiation.

Color versus Black and White. For most studies, black and white film is the logical choice. It is economical to use and duplicate, and special lighting conditions are not necessary.

Color film is the logical choice if color is an important characteristic of the subject or if the film is to be used for public presentation or other uses where good appearance and quality are important. It is sometimes useful to paint various parts of complex mechanisms different colors to permit easier identification in the finished film. Remember, however, that black and white duplicate prints can be made from color originals, but of course the opposite cannot be done.

If pictures are to be made in an area where the light level is low or where supplementary lighting is impossible or inconvenient, the answer is to use high-speed or very high speed black and white film. The ability to make pictures using the available work area lighting also contributes to the comfort of the operators and minimizes distraction.

EQUIPMENT

Film Types Available. There are only a few film types available in 8mm or super 8 sizes as a stock item in most photographic stores or from most film manufacturers. Color films balanced for daylight or tungsten illumination are readily available in these sizes.

The number of types of 16mm and 35mm films is almost impossible to count. In these sizes also are many special-purpose films such as infrared-sensitive, high-contrast, and extremely high speed films.

Picture Quality and Detail. The larger size films have better picture quality and will record more detail. However, super 8 films will reproduce detail in this situation better than 8mm films. 16mm and 35mm films are designed for large-screen projection.

The larger the film size, the larger the image possible and the easier the film is to analyze in a viewer, film editor, or by projection. If great detail in the film is not required, then 8mm or super 8 will probably be satisfactory. Otherwise, 16mm or 35mm film should be used.

Cameras. Within the scope of 8mm, super 8, 16mm, and 35mm sizes, there is a wide range of motion picture cameras to choose from. Most high-speed motion picture cameras are designed for the smaller film sizes because it is simpler to achieve the high speeds at which the film must be transported through the camera and easier to achieve the longer film running times often desired. Most high-speed cameras require an external power supply to drive them.

In general, the smaller the film size, the smaller the camera and the easier it is to operate. 8mm and super 8 cameras are designed for simple operation. This means that it is not necessary to have a trained photographer do the actual filming. 16mm and 35mm cameras tend to be more versatile and thus more adaptable to a wide variety of picture-taking applications.

The length of the film load is often another important criterion in the task to be studied. The largest film loads are available on 16mm and 35mm cameras. There are cameras in these formats that will accept 1,200-foot rolls. 8mm cameras usually accept 25-foot (double width, which results in 50 feet of processed film) or 100-foot rolls. Super 8 cameras take 50-foot magazines, which eliminates the need to thread the film through the camera. Many 8mm and super 8 cameras have battery-powered drive motors for continuous filming.

In engineering work, the 16mm size is the most popular. The 8mm and super 8 sizes, primarily designed for amateur use, are gaining in popularity, and the ease of operation and low cost associated with this format should be given serious consideration.

There are firms which rent camera and lighting equipment. Regular motion picture cameras and specialized equipment such as high-speed or pulse cameras can be rented from these organizations. The cost of infrequent film studies, or studies which require special cameras or lights, can be kept to a minimum by renting the equipment rather than purchasing it. Rental services can also make it possible to try out a particular photographic technique and fully evaluate it before committing funds for the purchase of equipment.

Lights. Adequate lighting of the subject or scene is extremely important in photography. For daylight photography, the time of day is an important consideration because the amount of light and the color quality of daylight change throughout the day. Photography at midday with strong sunlight can result in dark shadows. If the subject or area to be photographed is not too large, flat or umbrella-type folding reflectors made of plastic or fabric can be used to reflect light and fill in the shadow areas.

Photography indoors usually requires supplemental lighting, depending on the amount of available light and the speed of the film. Photoflood lamps (3200 or 3400°K) are useful for most situations. For situations which require more light, halogen-cycle lamps (quartz iodide, tungsten iodide, or bromine cycle) are also available, with color temperatures of 3200 and 3400°K. These lamps provide more lumens per kilowatt and are light in weight and small in size. Lighting large areas or rooms will require multiple units. In these cases, careful consideration must be given to the total electric power consumption of the units, and an adequate power source must be provided.

Table 4-3 shows the average lighting levels found in industry for certain types of seeing tasks.

Illumination for high-speed photography poses an additional problem. The high framing rates of high-speed cameras require a great deal of light to illuminate the subject. Three approaches are possible. One is the use of a continuous gaseous-discharge source. Mercury-vapor lamps are of this type and are widely used. The second approach is the use of a pulsed, gaseous-discharge source. Using this approach, the exposure is controlled by the length of the light pulse. The third is the use of high-intensity, long-duration flash bulbs such as the Sylvania FF-33 lamp.

Lenses. Lenses used on motion picture cameras can generally be placed in one of three categories: normal, wide-angle, or telephoto. Each has its own characteris-

TABLE 4-3.   Average Lighting Level for Various Industrial Operations

Foot-candles	Types of seeing tasks	Typical examples
30	Extra rough (easy or casual seeing)	Loading (inside freight cars or truck bodies), labeling, cartoning, plating, warehousing (large, bulky stock)
50	Rough (rough material, difficult seeing)	Auto frame assembly, reading gages or meter panels, most rough foundry tasks, rough inspection, rough bench or machine work, sheet-metal punches, presses, shears
100	Medium	Most general inspection, medium benchwork, machine tasks, drilling, riveting, screw fastening, auto chassis assembly, body parts manufacturing, electrical equipment testing, foundry grinding and chipping, glass grinding, polishing, beveling
200–500	Fine	Auto final assembly and inspection; embossing; cloth cutting, sewing, and pressing; glass etching, decorating, and inspection; very difficult inspection tasks
1,000 or more	Extra fine (intricate detail)	Cloth inspection, the most difficult or critical inspection, extra-fine bench and machine work, precision manual arc welding, hospital surgery, autopsy

tics and uses, but the greatest versatility is achieved by having all three available for use.   Most work can be done with a lens of normal focal length.   A wide-angle lens (focal length shorter than normal) can be used to photograph an area of greater scope than can be photographed by a normal lens.   This is often the case in situations where the working distance between lens and subject is limited.   Telephoto lenses (focal length larger than normal) are used to narrow the field of view without moving the camera closer to the subject.   Thus, a person can be photographed from a distance without being distracted.   Also, the telephoto lens permits placement of the camera in a safe location where machine chips or chemical sprays will not damage it.

Most 16mm and 35mm cameras have a provision for interchanging lenses.   This is accomplished either by removing one and inserting another or by including all three in a movable turret on the front of the camera.

Most 8mm or super 8 cameras do not have removable lenses.   However, an 8mm camera with a zoom lens—the focal length of a zoom lens is variable and therefore can be considered adjustable—can do approximately the same job.

Exposure Control.   Almost all manufacturers of 8mm and super 8 cameras supply some models with built-in, automatic exposure control.   Again, this type of camera is designed for ease of operation to produce good results on the first take.   Automatic exposure control is particularly useful in situations where the amount of illumination changes during the shooting period.

Most 16mm and 35mm cameras require manual setting of the lens for the proper exposure.   Film manufacturers publish exposure-index values for their films.   These are intended to serve only as guides in obtaining correct exposure.   Because of differences in cameras, meters, and methods of use, exact exposure settings should be determined by tests whenever possible.

Table 4-4 shows the intensity of illumination in foot-candles required for a given film exposure index at various lens openings.

There are two general types of exposure meters.   The incident meter measures the illumination falling on the subject.   An incident-light measurement is taken with the meter at the subject, pointed toward the camera.   Reflected-light meters are used at the camera, pointed at the subject.   Because readings taken from the camera position

TABLE 4-4. Incident-light Exposure Table for a Shutter Speed of 16 fps

Normal 175° opening, approximately equivalent to $\frac{1}{28}$ second

Film exposure index	Foot-candles at lens opening of:						
	$f/1.4$	$f/2$	$f/2.8$	$f/4$	$f/5.6$	$f/8$	$f/11$
25	60	125	250	500	1000	2000	4000
40	40	80	160	320	640	1250	2500
64	25	50	100	200	400	800	1600
125	12	25	50	100	200	400	800
160	10	20	40	80	160	320	640
320	5	10	20	40	80	160	320
1000	1.6	3.2	6.4	12.5	25	50	100
1250	1.3	2.5	5	10	20	40	80
4000	0.4	0.8	1.5	3	6	12.5	25
8000	0.2	0.4	0.8	1.5	3	6	13

cover an area greater than the area of the subject, the usual technique is to measure the light from a gray test card at the subject position. Such a test card reflects 18 percent of the light. This amount is equivalent in reflecting power of the average subject.

APPLICATIONS AND TECHNIQUES

Methods Measurement and Analysis. Most motion picture photography done by industrial engineering departments is people-job oriented. There usually exists a need to evaluate a job, set standards, raise operator efficiency, or relieve a bottleneck in production. Motion pictures can be the means to fulfill that need.

Micromotion pictures of an operator at a workplace can be used for time and motion studies and for methods analysis. They are most effective on highly repetitive, short-cycle operations. When the operator's actions are rapid and intricate, close-up motion pictures will reveal details unnoticed by the eye.

Time-lapse pictures can be used for studies such as elevator usage, movement patterns of loading docks, conveyor operations, and use of storage areas. They can also be used in determining a machine's effectiveness, the operator's idle time, the cause of delays, the arrival of raw materials, and the removal of finished goods.

Design of new machines can be facilitated by photographing an operator at a machine mock-up. This allows rapid evaluation of the placement of machine controls.

Interplant Transmission of Information. New techniques or important modifications or process changes can be photographed and distributed to associated plants. Such films can serve two purposes: first, they inform others of the changes in a clear, graphic manner; and second, they serve as a training film for those responsible for instituting the new procedures. These films can also result in reducing costs of time and travel.

Employee Training. Motion pictures for training new employees or training old employees in new methods are often well worth the cost of making them. This is particularly true when large groups of employees have to be trained. Some firms place a motion picture projector with a rear-projection, daylight-viewing screen in front of an operator at the workplace. The new operator learns his job by following instructions and examples from the movie in front of him.

Informing Management. In the development of new products or methods, there comes the inevitable time when the facts have to be presented to management for a decision. Selected portions of the motion picture footage used in the development or analysis can serve to illustrate both the problem and the solution.

An effective technique for the purpose of comparing production methods is that of split-frame or before-and-after photography. In this method, an operator is first trained to an equal level of proficiency in each of the two methods to be compared. He then processes the same number of units while being photographed with a camera fitted with a mask covering one half of the film. After the first method has been photographed, the film is rewound back to a starting frame by hand or on rewind equipment in a darkroom. The mask is repositioned to cover the previously exposed half of the film and to uncover the unexposed portion. The second method is then filmed on this half of the film. When completed, the film will show the two methods side by side on the screen, and the faster method will finish before the slower method.

Traffic Studies. Time-lapse motion pictures are one of the best ways of recording vehicular and pedestrian traffic studies. When the compressed-time motion pictures are projected, traffic patterns are clearly evident; when examined frame by frame, counts and types of traffic can be recorded.

Generally, a high viewpoint is necessary to get a good view of the area to be studied. If the study is being made outdoors where lighting conditions can vary considerably in the course of a day, a camera with automatic exposure control is desirable.

For those instances when a continuous record is not required—for example, to count only trucks using a particular gate at a plant—a pulse camera can be controlled by a photocell placed about 7 feet off the ground.

Time-lapse techniques can be used to record river traffic, the action of tides, the water level variation in a creek, the usage of a parking lot, or almost any motion or pattern that takes place too slowly to be seen.

IMPORTANT CONSIDERATIONS BEFORE FILMING

It was stated earlier that a few moments of careful consideration of all aspects of the filming situation can ensure that all the details have been taken care of that are necessary for an effective film. Sometimes the most obvious thing does not become apparent until the finished film is viewed. The following considerations should be checked before photographing the study.

1. Select a place in the camera's field of view for a timepiece. It is a good idea to include a timepiece whenever possible. Some future study of the film may require it.
2. Make sure there is an adequate supply of film. Preliminary study of the entire operation will give a fairly accurate idea of the footage required.
3. Check the lens opening and focus to make sure they are correct.
4. Check the viewfinder carefully for good coverage of the event. Usually, a high camera angle is best if the person is standing or walking; an eye-level angle if he is sitting.
5. Select a camera viewpoint that will allow the most important motions to be perpendicular to the camera axis.
6. At the start of each roll of film, shoot a few frames of a card showing the date, plant number, machine number, and the like.
7. Always use a tripod.
8. Consider using a high camera speed, especially for close-ups of intricate and rapid finger motions.
9. Arrange with production people to have material in process, operators, and so on available.
10. If using supplementary lighting, check the electrical supply and make sure that sufficient power is available.
11. Do not forget the shop steward. Explain to him what is being done and why. Invite him to be present at the shooting.
12. Sell the operators on the idea. If possible, show a movie from a previous study. Offer to let them see the movies of themselves.
13. Tell operators to dress normally.
14. Get a release from the operators being photographed.
15. Make as few changes as possible in normal working conditions. Do not

move people or things around or change the lighting more than is absolutely necessary. Motion pictures of operations performed under normal conditions provide more reliable data than those made under changed conditions.

HUMAN FACTORS

Photographing the operation of a machine is a relatively simple matter. However, if it is a situation where a machine operator is a necessary part of the action, or if a number of people are necessary to show the complete story, a whole new dimension is added to the industrial engineer's preparations for filming.

Supervisors and shop foremen should be given a full explanation of the purpose and the goal of the film beyond just obtaining the proper clearances to make the film. The machine operator deserves a good deal of consideration in preparing to film the job. He can cause the filming session to be a success or a failure. Occasionally, the industrial engineer encounters an operator who objects to being photographed. This can be due to the fact that he does not understand why he is being photographed. Tell the operator what type of photograph is to be taken and why. Reassure him that the objective is not to try to catch him in some error. If possible, show him some examples of the same type of motion pictures which are to be taken. Agree to let him see the film in which he appears.

Because indoor filming will probably require supplemental lighting, let the machine operator run through the action a few times under the lights. In this way, he can get used to the filming environment and will be less distracted during the actual filming.

ANALYSIS OF THE FILM

There are two types of analysis that can be made of a motion picture. One type is a quantitative analysis, which is made by measuring quantities such as image density, blur, displacement, and image size to determine actual values of velocity, position, time, temperature, and the like. The other type of analysis is qualitative, and it can be made by viewing the film by projecting it either at normal rate or frame by frame.

Industrial engineers most often need to know: "What is the order of the events?" "When does each event occur?" "How long does the event last?" These questions can be answered by a qualitative analysis, and there are several techniques available.

Standard Motion Picture Projector. The standard motion picture projector, operating at 18 or 24 frames per second, provides a convenient way for the engineer to study the film for the first time. General patterns of behavior and the general order of events can be noted at this time.

Stop-motion Projector. The stop-motion projector is perhaps the best way to study motion pictures. Most of these are 16mm projectors, but equipment has been developed for projecting super 8 film. Such a projector is capable of projecting film, both forward and backward, at speeds from about 2 to 6 frames per second up to 24 frames per second. The projector can also advance the film a single frame at a time if desired. Special provisions for protecting the film from heat, such as extra-heat-absorbing glass and heavy-duty blowers, allow stopping on a particular frame for prolonged study. Some models include a frame counter, and others are available with an extra-wide aperture for viewing timing marks along the edge of the film. With these projectors, the image can be projected on a conventional screen, but it is usually more convenient to use one of the daylight-viewing, rear-projection viewers. Some stop-motion projectors include such viewers as part of their carrying cases.

Motion Analyzer. A motion analyzer is an instrument used to obtain quantitative data from a film. The motion analyzer consists of two parts, a projection head, which provides the film transport and projection system, and a projection box, which provides the viewing screen and cross-hair measuring system. By using the measuring system of a motion analyzer, very accurate incremental changes can be measured from frame to frame on the motion picture film.

Film Loops. If it is desirable to study a particular piece of film many times, the film can be cut out and spliced to form a loop. By draping the film over an accessory roller near the projector, the film can be projected over and over without rewinding and rethreading.

Microfilm Readers. Microfilm readers can also be used to study motion pictures frame by frame. Although these readers do not have a provision for making quantitative measurements, they do provide an enlarged view of each frame of the motion picture.

TRAINING FILMS

The motion picture is an effective medium for instructing people in techniques or skills. The motion picture can be shown to an individual or to a group of people. Proper consideration of the action beforehand ensures that all the important points of the topic will be brought before the audience. Each successive group that views the film will see the same story presented in the same manner.

Using Existing Films. It is often the case that motion study films have been taken previously of the action in question. By editing such a film, appropriate footage showing the technique can be used and made into a training film for other operators. Care should be taken that good quality footage is used. Footage that is incorrectly exposed or poorly lighted should be avoided. Such defects are sure to disrupt the viewer's concentration.

The training film need not be a polished production. However, it is useful if simple, appropriate titles are included to identify and explain the important parts of the action. Simple titles can be made with a camera mounted on a tripod and the artwork tacked or taped to a wall. Some suggestions to aid in producing simple titles are:

1. Use 10- by 12-inch cardboard covered with colored paper to produce a background color. The colored paper should have a slightly matte surface.
2. The subject matter should be composed to fit a 6¾- by 9-inch area.
3. The minimum letter height should be approximately one-twentieth of the short dimension.
4. The number of lines should not exceed eight.
5. The total number of words in a single card title should not exceed thirty.
6. A title, chart, or diagram should be on the screen for twice the average time it takes to read the words. Shoot more footage than will probably be required. Excess footage can be edited out of the final film.

Commercial Motion Picture Laboratory Services. Motion picture film processing and other services for producing high-quality industrial films for training or corporate use are available from commercial laboratories. Very few industrial plants have the specialized equipment and personnel that are required and therefore must rely on the services provided by these laboratories.

The following is a list of these services. All laboratories may not offer all the services listed, but most will provide a major portion of them.

Processing. Developing negative, positive, color, or black and white camera films shot by in-plant photographers.

Printing and Duplicating. Producing prints or duplicates from original camera films for use as work prints or prints for release.

Black and White Duplicates from Color Originals. A black and white work print saves wear and tear on the original and is less costly than a color duplicate.

Editing. Cutting, splicing, and assembling the film to provide smooth continuity.

Producing Opticals. Producing special effects and scene transitions, such as fades, dissolves, freezes, and wipes.

Sound Production. Recording, narrating, editing, and sound effects.

Titling. Preparing all lettering and artwork.

Animation. Supplying original artwork and photography to add animation to the film.

Special Photography. Providing equipment or facilities not available to the plant photographer.

Script Writing. Preparing an original script from a story line or editing from a rough script.

BIBLIOGRAPHY

Hyzer, W. G., *Engineering and Scientific High-speed Photography,* The Macmillan Company, New York, 1962.

Industrial Motion Pictures, Kodak Publication P-18, Eastman Kodak Company, Rochester, N.Y., 1969.

Moon, Irwin A., and F. Alton Everest, "Time-lapse Cinematography," *Journal of the Society of Motion Picture and Television Engineers,* February, 1967.

Motion Pictures for Methods Measurements, Kodak Publication P-100-10, Eastman Kodak Company, Rochester, N.Y. 1968.

Shaften, Kenneth, and Dean Hawley, *Photographic Instrumentation,* Society of Photographic Instrumentation Engineers, Redondo Beach, Calif., 1962.

Chapter 5

Curves and Nomographs*

R. P. HOELSCHER

**Professor Emeritus, General Engineering Department,
University of Illinois, Urbana, Illinois**

In engineering offices, it is frequently desirable to have equations which express the results of test data. When it is impossible or impractical to derive rational equations mathematically, empirical equations may frequently be developed from curves drawn through the plotted data. The purpose of the first part of this chapter is to explain a few methods for obtaining such equations. It is assumed that the reader is familiar with the methods of plotting data on rectangular, semilogarithmic, and logarithmic coordinate paper.

CURVE PLOTTING AND EQUATION DERIVATION

The method is to plot the data on a type of coordinate paper which will permit the drawing of a straight line through the average of the points. Ideally, this means that the sum of the distances from the line to the plotted points above it will equal the sum of the distances from the line to the points below it. Practically, the procedure is as follows:

   1. Plot the data on rectangular coordinate paper and draw a smooth curve
      through the average of the points.

* Most of the illustrations of this chapter and much of the discussion have been abstracted by permission of the copyright owners from *Graphic Aids in Engineering Computation* by R. P. Hoelscher, J. N. Arnold, and S. H. Pierce, Balt Publishing Co., Lafayette, Ind., 1952. For a more complete discussion, consult this text and others listed in the bibliography at the end of this chapter.

2. If the curve is not a straight line, make an intelligent guess as to what it may be, as for example, a parabola or other exponential form.
3. If the curve did not come out as a straight line on rectangular coordinate paper, replot the data on semilogarithmic or logarithmic coordinate paper until a straight line is secured.

If, in replotting, the line is still slightly curved, it may be possible to rectify it by adding or subtracting a constant. This expedient is illustrated by the discussion of Case IV later in the chapter.

TYPE FORMS OF EQUATIONS

Although many mathematical equations are possible, it is fortunate that much of the test data obtained in engineering work and the sciences will fall into some of the simpler forms enumerated in the list below.

Case I. $Y = mX + b$. If the data plotted on rectangular coordinate paper give a straight line, the equation is of this form.

Case II. $Y = A(10)^{mX}$. If this equation is rewritten in logarithmic form, it appears as log $Y = $ log $A + mX(\log 10)$. Because log $10 = 1$ and A is a constant, it will be noted that log Y is linear with X . Hence, if the plotted data give a straight line on semilog paper, the equation will have this form.

Case III. $Y = AX^m$. If this equation is written in logarithmic form, it appears as log $Y = $ log $A + m$ log X , which indicates that log Y is linear with log X . Hence, if the data, plotted on logarithmic paper, give a straight line, the equation will have this form.

Case IV. $Y = AX^m + c$. Proceeding as before, the equation may be written

$$\log (Y - c) = \log A + m \log X$$

which indicates that log $(Y - c)$ is linear with log X . Hence, if the data after subtracting a constant give a straight line on logarithmic paper, this form of equation results. For the method of determining the constant c , refer to the later discussion of Case IV.

Case V. $Y = A(10)^{mX} + c$. Again, by rewriting the equation,

$$\log (Y - c) = \log A + mX \log 10$$

is obtained, which shows that log $(Y - c)$ is linear with X . A straight line on semilogarithmic paper may give this form. For the method of determining c , refer to the later discussion of Case V.

METHODS OF DETERMINING THE EQUATION

Having rectified the curve on one of the types of coordinate paper, the equation representing the line may be obtained by any one of the following methods. These have been listed in the order of increasing difficulty.

1. Slope and intercept method
2. Method of selected points
3. Method of averages
4. Method of moments
5. Method of least squares

Because the accuracy of test data will in many cases not exceed the exactness achieved by one of the first three methods, only these three are presented. All three will be used in the following discussion of Case I. Thereafter, only one of the methods will be used, although all are applicable.

Case I. $Y = mX + b$. Straight line on rectangular coordinates.

Slope and Intercept Method. In this equation, m is the slope of the line and b is the intercept on the Y axis at $X = 0$. It is necessary only to determine the value of these quantities to complete the equation.

As an illustration, the data from Table 5-1 showing the horsepower required

TABLE 5-1.   Horsepower Required for ½-inch Drill in Cast Iron*

Horsepower required...............	1.82	2.20	2.35	2.75	3.26	4.20	4.50	5.50
Rate of feed, thousandths of an inch per revolution..................	7.0	9.7	13.3	18.0	24.9	34.6	42.0	47.4

*From F. A. Halsey, *Handbook for Machine Designers and Draftsmen*, McGraw-Hill Book Company, New York, 1916, p. 277.

for various rates of feed of a ½-inch drill in cast iron have been plotted as shown in Figure 5-1.

From the figure, the intercept b on the Y axis can be observed to be 1.32. Observations as close as this can be made only on large-scale charts, but Figure 5-1 illustrates the method.

The slope can be determined by the usual mathematical method. Thus, two points such as P_1 at $X_1 = 20$, $Y_1 = 3$ and P_2 at $X_2 = 44$, $Y_2 = 5$ can be used in the equation for slope, $m = (Y_2 - Y_1)/(X_2 - X_1)$, giving

$$m = \frac{5 - 3}{44 - 20} = 0.0833$$

The same result may be obtained by first taking the trigonometric slope

$$\frac{1}{2.4} = 0.4166$$

as shown in Figure 5-1 and then dividing it by the ratio of the scale moduli used on the two axes. Thus, if $Y = \frac{1}{2}$ inch per unit and $X = \frac{1}{10}$ inch per unit, then $\frac{1}{2} \div \frac{1}{10} = 5$. Hence, the slope is

$$m = 0.4166 \div 5 = 0.0833$$

These values may now be substituted in the equation to give

$$Y = 0.0833X + 1.32$$

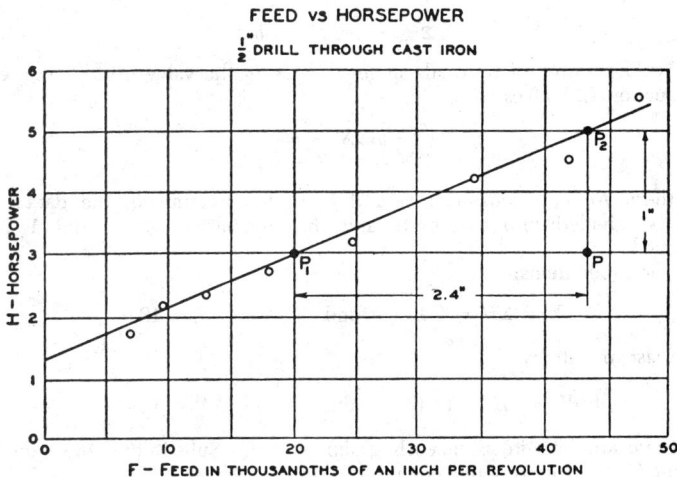

FEED vs HORSEPOWER

½" DRILL THROUGH CAST IRON

F - FEED IN THOUSANDTHS OF AN INCH PER REVOLUTION

Fig. 5-1.   Graph of horsepower required at various rates of feed for a ½-inch drill in cast iron.

or using the symbols of the chart,

$$H = 0.0833F + 1.32$$

Method of Selected Points.   Because there are two quantities m and b to be determined in the equation, the usual method of analytical geometry may be used to determine these values.   If the known values of X and Y for any two points are inserted in the equation $Y = mX + b$, two equations may be obtained.

Using the same points as before, P_1 where $X = 20$, $Y = 3$ and P_2 where $X = 44$ and $Y = 5$ gives

$$3 = 20m + b \qquad \text{and} \qquad 5 = 44m + b$$

Solving these equations simultaneously, $m = 0.0833$ and $b = 1.336$.

In terms of the symbols used in the chart, this gives the equation

$$H = 0.0833F + 1.336$$

It should be noted here that the two points chosen should be on the line and that they may or may not be a part of the original data.   The points should also be as far apart as is conveniently possible.

In both of the foregoing methods, the accuracy of the results depends upon two factors:

1. How well the straight line was drawn to go through the average of the points.   This depends upon the judgment of the draftsman.
2. How accurately the coordinates of the points and the Y intercept are read from the chart.

Method of Averages.   The third method, though still requiring the plotting of the data on a type of coordinate paper which will produce a straight line, does not depend upon the accuracy of that line for the determination of the equation.

If the distances (hereinafter called residuals) from the line to the points above it are designated plus and distances from the line to points below it are called minus, the ideal line would be so drawn that the algebraic sum of all these distances would be zero.   If Y represents the plotted ordinate and Y_c the corresponding ordinate on the line, then the foregoing sentence can be expressed mathematically as

$$\Sigma Y - \Sigma Y_c = 0 \tag{1}$$

Hence, in Case I, where $Y = mX + b$, the equation may be written

$$\Sigma Y_c = m\Sigma X + nb \tag{2}$$

where n is the number of residuals summed.   Using the value of ΣY_c from equation (2) in equation (1) gives

$$\Sigma Y - m\Sigma X - nb = 0$$

or
$$\Sigma Y = m\Sigma X + nb \tag{3}$$

Because there are two constants, m and b, to be determined, the data of Table 5-1 may be divided into two parts and the summations of X and Y for each part obtained.

For the first four items:

$$Y = \Sigma H = 9.12 \qquad \text{and} \qquad X = \Sigma F = 48.0$$

For the last four items,

$$Y = \Sigma H = 17.46 \qquad \text{and} \qquad X = \Sigma F = 148.9$$

Because there are four items in each group, $n = 4$.   Substituting the values above in equation (3) and writing one equation for each set gives

$$9.12 = 48.0m + 4b \qquad \text{and} \qquad 17.46 = 148.9m + 4b$$

The coefficient of b is 4 because these are four items summed up. Solving these two equations simultaneously,

$$m = 0.0827 \quad \text{and} \quad b = 1.29$$

Hence, again using the symbols of the chart in Figure 5-1, the equation may be written

$$H = 0.0827F + 1.29$$

An equation has now been determined for the data of Table 5-1 by three different methods and three slightly different equations have been obtained. To determine whether any of these equations is satisfactory and if so which of the three is the best in this case, it is necessary to:

1. Compute the values of H from the equation and find the residuals for each plotted point.
2. Find the summation of the residuals to see how far it departs from zero and whether the plus and minus residuals are about equally distributed. This has been done and the values listed in Table 5-2.

It will be noted that the method of averages gives the smallest summation of residuals and the best distribution of plus and minus values. Hence it is the best of the three equations.

Case II. $Y = A(10)^{mX}$. Straight line on semilog paper. The data in Table 5-3 were plotted first on rectangular coordinate paper with the resulting curve parabolic or exponential in form. The next plotting, on semilog paper with values of G on the logarithmic scale and values of R on the uniform scale, gave a somewhat flatter but still very definite curve. The third plotting, with values of R on the logarithmic scale and values of G on the uniform scale, gave the result shown in Figure 5-2.

Method of Averages. This case will be solved by the method of averages only. Because $\log 10 = 1$, the equation above may be written

$$\log Y = \log A + mX \tag{4}$$

For the sum of residuals to equal zero,

$$\Sigma \log Y - \Sigma \log Y_c = 0 \tag{5}$$

where
$$\Sigma \log Y_c = n \log A + m\Sigma X \tag{6}$$

Hence
$$\Sigma \log Y = n \log A + m\Sigma X \tag{7}$$

TABLE 5-2. Data for Selecting the Most Satisfactory Equation for the Curve in Figure 5-1

Observed data		Computations for residuals					
		Slope and intercept method		Method of selected points		Method of averages	
F	H	H_c	Residual	H_c	Residual	H_c	Residual
7.0	1.82	1.90	−0.08	1.92	−0.10	1.87	−0.05
9.7	2.20	2.13	+0.07	2.15	+0.05	2.09	+0.11
13.3	2.35	2.43	−0.08	2.45	−0.10	2.39	−0.04
18.0	2.75	2.82	−0.07	2.84	−0.09	2.78	−0.03
24.9	3.26	3.39	−0.13	3.41	−0.15	3.35	−0.09
34.6	4.20	4.20	0.00	4.22	−0.02	4.15	+0.05
42.0	4.50	4.82	−0.32	4.84	−0.34	4.76	−0.26
47.4	5.50	5.26	+0.24	5.28	+0.22	5.21	+0.29
		Sum =	−0.37	Sum =	−0.53	Sum =	−0.02

TABLE 5-3. Data on Circulating Pump Speed
and Capacity Used in Plotting Figure 5-2

R = revolutions per minute		G = gallons per minute	
R	$\log R$	G	$\log G$
500	2.699	1,000	3.000
600	2.778	1,500	3.176
700	2.845	1,950	3.195
800	2.903	2,300	3.362
900	2.954	2,600	3.415
1,000	3.000	2,900	3.462
1,100	3.041	3,200	3.505
1,200	3.079	3,500	3.542
1,300	3.114	3,750	3.574
1,400	3.146	3,900	3.591

Because there are ten items in the table, the data may be divided into two groups of five each, thus making n in the equation above equal to 5. For the first five items,

$$\Sigma \log Y = \Sigma \log R = 14.179 \quad \text{and} \quad \Sigma X = \Sigma G = 9{,}350$$

For the second five items,

$$\Sigma \log Y = \Sigma \log R = 15.380 \quad \text{and} \quad \Sigma X = \Sigma G = 17{,}250$$

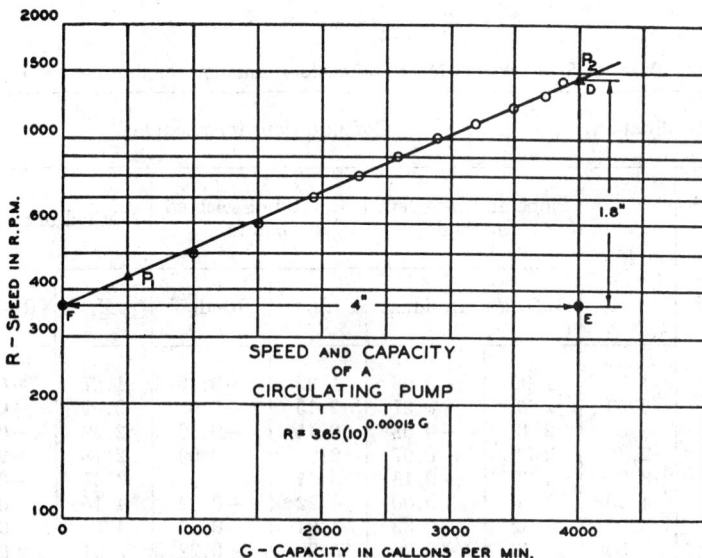

Fig. 5-2. Chart of capacity of a circulating pump at various speeds.

TABLE 5-4. Relationship between Values of R
Calculated from the Equation for the Curve in
Figure 5-2 and the Values of R Used in Plotting
the Curve

Observed data		Method of averages	
R	G	R_c	Residual
500	1,000	505	− 5
600	1,500	601	− 1
700	1,950	704	− 4
800	2,300	797	+ 3
900	2,600	885	+15
1,000	2,900	990	+10
1,100	3,200	1,090	+10
1,200	3,500	1,210	−10
1,300	3,750	1,320	−20
1,400	3,900	1,394	+ 6
		Sum =	+4

Substituting these values in equation (7) gives

$$14.179 = 5 \log A + 9,350m \quad \text{and} \quad 15.380 = 5 \log A + 17,250m$$

Solving these equations simultaneously,

$$m = 0.000152 \quad \text{and} \quad A = 356$$

The equation for the data, therefore, is

$$R = 356(10)^{0.000152G}$$

Computing the values of R from this equation and determining the residuals gives
the results shown in Table 5-4. The summation of residuals is quite small, and
the distribution of plus and minus residuals is good. The departure of the largest
residual from the plotted value is less than 1.6 percent.

Case III. $Y = AX^m$. Straight line on logarithmic coordinate paper. After
several plottings, it was found that the data of Table 5-5 plotted as a straight line
on logarithmic paper as shown in Figure 5-3.

To give another illustration of the method of selected points, this method will

TABLE 5-5. Data on Friction Losses in an Air Duct
Used in Plotting Figure 5-3*

V	F
1.30	0.195
1.45	0.245
1.60	0.290
1.70	0.320
1.80	0.380
1.95	0.430
2.20	0.550
2.60	0.710

* From R. H. Heilman, *Power Plant Engineering*, March, 1939, p. 215.

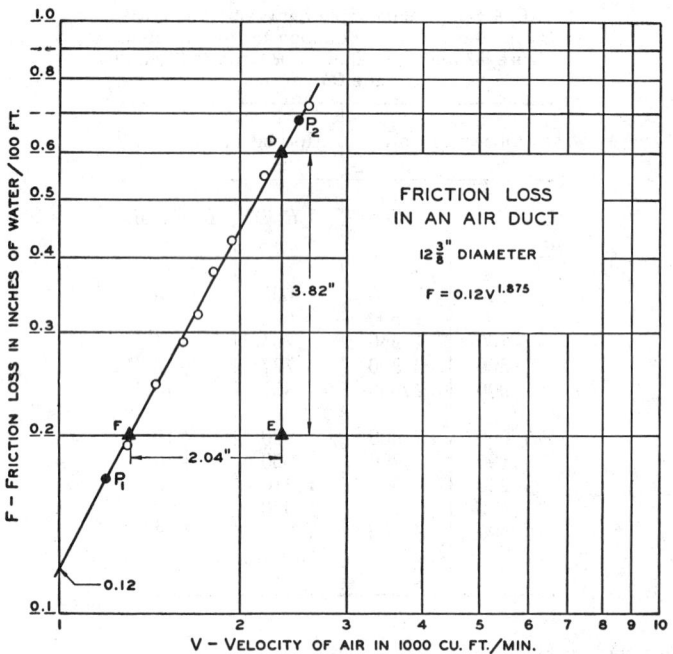

FIG. 5-3.   Friction loss for given velocity in an air duct.

be applied to the following problem.   The points P_1 at $V = 1.2$, $F = 0.175$ and P_2 at $V = 2.5$, $F = 0.68$ in Figure 5-3 will be used.

Taking logarithms to place equation $Y = AX^m$ in linear form gives

$$\log Y = \log A + m \log X \qquad (8)$$

Substituting the values of coordinates for P_1 and P_2 in equation (8) gives

$$\log 0.175 = \log A + m \log 1.2 \qquad \text{and} \qquad \log 0.68 = \log A + m \log 2.5$$

Solving these equations simultaneously gives

$$m = 1.85 \qquad \text{and} \qquad A = 0.122$$

Using the symbols of the chart,

$$Y = F \qquad \text{and} \qquad X = V$$

the equation for the data may be written

$$F = 0.122 V^{1.85}$$

Again, to check the accuracy of the equation, solve for F using the corresponding values of V, and then determine the residuals and their summation as shown in Table 5-6.

Case IV. $Y = AX^m + c$.   Modified curve on logarithmic coordinates.   Various plottings of the data in Table 5-7 failed to give a straight line.   The nearest approach to a straight line was obtained on logarithmic paper as shown by curve (a) in Figure 5-4.   Because the equation above may be written $Y - c = AX^m$, it appeared that the subtraction of a constant from Y might rectify the curve.

TABLE 5-6. Relationship between Values of
F Calculated from the Equation for the Curve
in Figure 5-3 and the Values of F Used in
Plotting the Curve

Observed data		Method of selected points	
V	F	Friction loss from equation F_c	Residual
1.30	0.195	0.198	−0.003
1.45	0.245	0.242	+0.003
1.60	0.290	0.291	−0.001
1.70	0.320	0.326	−0.006
1.80	0.380	0.361	+0.019
1.95	0.430	0.419	+0.011
2.20	0.550	0.525	+0.025
2.60	0.710	0.713	−0.003
		Sum =	+0.045

One of the simplest methods for determining the proper value of c is given by Dale S. Davis.[1] This method follows:

Plot Y against X on ordinary rectangular coordinate paper and draw a smooth curve through the points as shown in Figure 5-5. Select two points X_1 and X_2 near the ends of the curve and a third point X_3 such that $X_3 = \sqrt{X_1 X_2}$. Then

$$c = \frac{Y_1 Y_2 - Y_3^2}{Y_1 + Y_2 - 2Y_3} \tag{9}$$

where Y_1, Y_2, and Y_3 are the ordinates corresponding to X_1, X_2, and X_3.

The points X_1 and X_2, must, of course, be on the curve. They may or may not be a part of the original data. From the first plotting on rectangular coordinates, $X_1 = 1.5$, $Y_1 = 30$ and $X_2 = 15$, $Y_2 = 810$ were chosen. These points happen to be a part of the original data, but the only requisite is that they lie on the curve. From these values X_1 and X_2, the value of $X_3 = \sqrt{1.5 \times 15} = 4.74$. Hence, from equation (9)

$$c = \frac{30 \times 810 - 136^2}{30 + 810 - 2(136)} = \frac{5,804}{568} = 10.22$$

[1] Dale S. Davis, *Empirical Equations and Nomography,* McGraw-Hill Book Company, New York, 1943.

TABLE 5-7. Data for Plotting Figure 5-5

X	1.5	3.0	4.0	6	8	10	12	15
Y	30	71	107	196	305	430	570	810

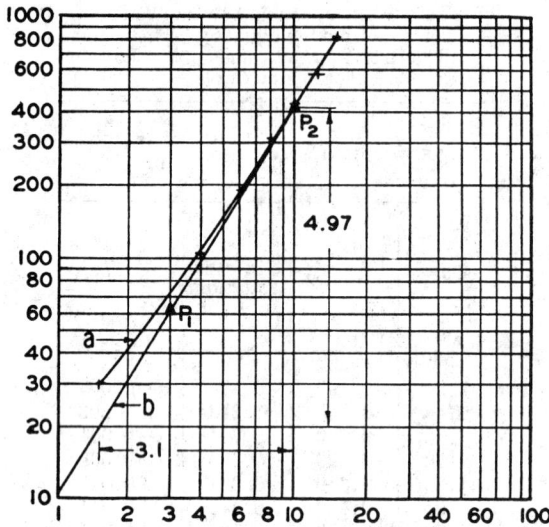

FIG. 5-4.   Rectification of curve from Figure 5-5 on logarithmic coordinates.

Using this value of c, the curve (b) of Figure 5-4 was plotted with $Y - c$ on the vertical axis.   A satisfactory straight line resulted.

Having the line rectified, the method of selected points may now be used to determine the equation.   The two points P_1 $(X = 3.0,\ Y - c = 60.35)$ and P_2 $(X = 10,\ Y - c = 419.35)$ shown in Figure 5-4 were used.

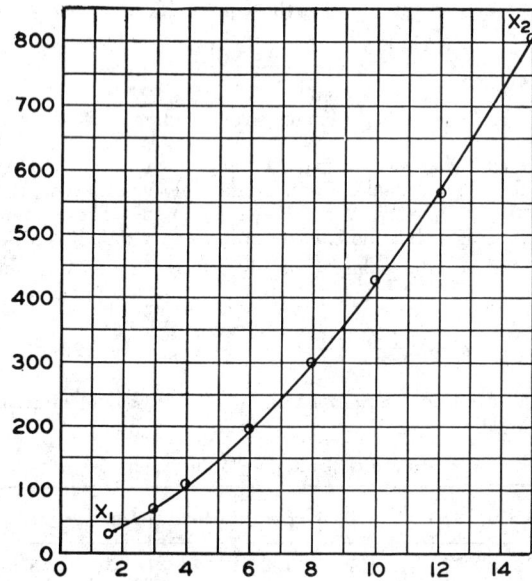

FIG. 5-5.   Data of Table 5-7 plotted on rectangular coordinates.

TABLE 5-8.   Relationship between Values of Y Calculated from the Equation for the Curve in Figure 5-5 and the Values of Y Used in Plotting the Curve

Observed data		$Y - c$ $c = 10.2$	log $(Y - c)$	log X	Computations for residuals	
					Method of selected points	
X	Y				Y_c	Residual
1.5	30.5	20.3	1.308	0.176	30.0	−0.5
3.0	70.4	60.4	1.781	0.477	70.6	+0.2
4.0	108	97.8	1.991	0.602	106.3	−1.7
6.0	194	183.8	2.265	0.778	191.2	−2.8
8.0	307	296.8	2.473	0.903	302.2	−4.8
10.0	428	417.8	2.622	1.000	432.2	+4.2
12.0	575	564.8	2.752	1.079	572.5	−2.5
15.0	805	794.8	2.900	1.176	811.2	+6.2
					Sum = −1.7	

Substituting these values in the equation $\log (Y - c) = \log A + m \log X$ gives

$$\log 60.35 = \log A + m \log 3.0$$
and
$$\log 419.35 = \log A + m \log 10$$
or
$$1.781 = \log A + 0.477m$$
and
$$2.622 = \log A + 1.000m$$
from which
$$m = 1.61 \qquad \log A = 1.012$$
and
$$A = 10.28$$

Placing these values in the equation gives

$$Y = 10.28 \times 1.61 + 10.22$$

Computation of the values of Y_c from the equation and the determination of residuals are shown in Table 5-8. The values of $\log (Y - c)$ and $\log X$ have been included in the table so that the reader may try the method of averages for himself.

Case V. $Y = A(10)^{mX} + c$.   Modified curve on semilogarithmic paper. Because $\log 10 = 1$, this equation may be rewritten in logarithmic form as

$$\log (Y - c) = \log A + mX$$

This indicates that $\log (Y - c)$ is linear with X. Therefore, if $(Y - c)$ is plotted on the logarithmic scale and X on the linear scale, a straight line will result.

Before the quantity $Y - c$ can be plotted, the value of the constant c must be determined. As in Case IV, it can be shown[2] that if a point X_3 on the curve is so chosen that $X_3 = (X_1 + X_2)/2$, where X_1 and X_2 are points on the curve near its opposite extremities, then

$$c = \frac{Y_1 Y_2 - Y_3{}^2}{Y_1 + Y_2 - 2Y_3} \tag{10}$$

Where Y_1, Y_2, and Y_3 are the ordinates corresponding to X_1, X_2, and X_3. These values are taken from the original curve of Y plotted against X on rectangular coordinate paper, as shown in Figure 5-6.

[2] *Ibid.*, p. 5.

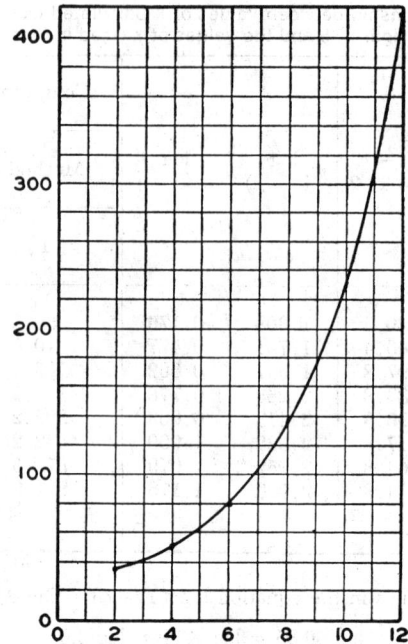

FIG. 5-6.   Data of Table 5-9 plotted on rectangular coordinates.

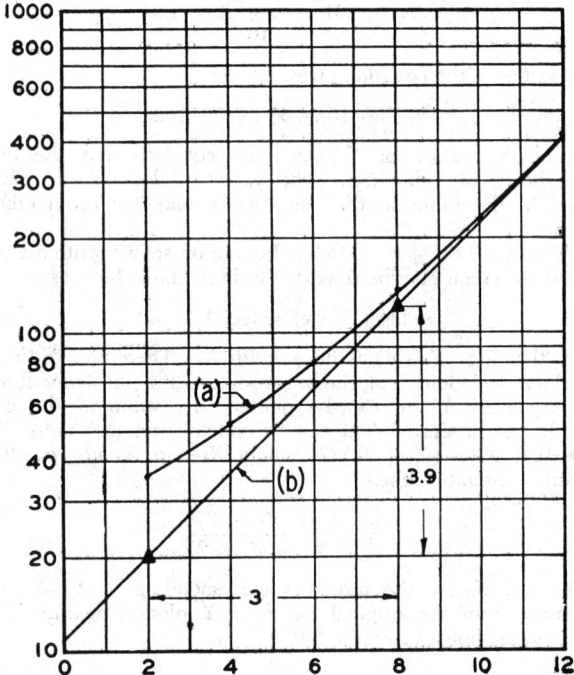

FIG. 5-7.   Rectification of curve of Figure 5-6 on semilog paper.

TABLE 5-9. Data Used for Plotting the Curves in Figures 5-6 and 5-7

X	2	4	6	8	10	12
Y	35	52	80	137.5	230	415

As an illustration, the equation for the curve to fit the data of Table 5-9 will be determined by the method of averages. From the plot of the curve shown in Figure 5-6, it was assumed that this curve might be rectified on semilogarithmic paper. In the replotting, the curve (a) of Figure 5-7 resulted. Because this line was still slightly curved, the addition of a constant was suggested.

From the original curve in Figure 5-6, two points $X_1 = 2$, $Y_1 = 35$ and $X_2 = 12$, $Y_2 = 415$ were selected. Then $X_3 = (2 + 12)/2 = 7$, and from the curve, $Y_3 = 106$. Therefore, from equation (10)

$$c = \frac{35 \times 415 - 106^2}{35 + 415 - 2(106)} = \frac{3,289}{238} = 13.82$$

Using this value, the second curve (b) in Figure 5-7 was plotted, resulting in a straight line. Because the form of the equation is $\log (Y - c) = \log A + mX$, it is necessary to have the values of $Y - c$ as shown in Table 5-10. With six items in the table, the first three are summed in one group and the last three in the second. For the first three, the summation gives

$$4.729 = 3 \log A + 12m$$

The summation of the last three gives

$$7.031 = 3 \log A + 30m$$

Solving these equations for m,

$$m = \frac{2.305}{18} = 0.128$$

and $\log A = 3.192/3 = 1.064$, from which

$$A = 11.605$$

TABLE 5-10. Relationship between Values of Y Calculated from the Equation for the Curve in Figure 5-7 and the Values of Y Used in Plotting the Curve

Observed data		$Y - c$	$\log (Y - c)$	Method of averages	
X	Y			Y_c	Residual
2	35	21.18	1.326	34.7	+0.3
4	52	38.18	1.582	51.5	+0.5
6	80	66.18	1.821	81.7	−1.7
8	137.5	123.68	2.092	129.9	+7.6
10	230	216.18	2.336	234.4	−4.4
12	415	401.18	2.603	411.2	+3.8
				Sum =	+6.1

Therefore, the equation becomes

$$Y_c = 11.605(10)^{0.128X} + 13.82$$

The results of computations for Y_c and the residuals are also shown in Table 5-10.

For further information concerning methods of determining empirical equations, consult the texts listed in the bibliography at the end of this chapter.

NOMOGRAPH CONSTRUCTION

The use of nomographs can save a great deal of time in an engineering office where the repeated solution of an equation is often required. Anyone can be quickly trained to use a nomograph, but its construction is another matter.

In this discussion, it is assumed that the reader understands the mathematical meaning of the terms "constant" and "variable." The equations necessary for the construction of nomographs are given without proof. For an understanding of these proofs, the reader is referred to the texts in the bibliography of this chapter.

Definition of Terms

Function of a Variable. If a mathematical expression contains a variable in such a way that the value of the expression is determined when any value is assigned to the variable, such an expression is called a function of the variable. The function will be designated as $f(r)$ where the function may be r^2, sin r, log r, $(2r^2 + 1)$, or any other quantity involving r as noted above.

Functional Modulus. In any alignment chart, the length of the scales is usually limited to the range of values for the various functions which have practical value. If the upper limit is designated by $f(r_2)$ and the lower limit by $f(r_1)$, then the difference between these two values must have a definite length, L, which can be chosen to suit the designer. If L is divided by the difference $f(r_2) - f(r_1)$ the exact value of the proportionality factor m can be determined. The length of the scale can then be written

$$L = m[f(r_2) - f(r_1)] \tag{11}$$

This proportionality factor m is called the functional modulus.

The Functional Scale. All the scales on a standard slide rule are functional scales. The meaning of a functional scale and the method of its construction can best be illustrated by an example. Let it be required to construct a scale for $f(r) = r^2$ for values of r from 0 to 5.

Step 1.   List the values of the variable.

Step 2.   Compute the values of the function.

Step 3.   Knowing the approximate length which the scale may have, assume a value for the functional modulus m and multiply the function by this factor.

Step 4.   Plot the distances determined in step 3 along a line.

Step 5.   Mark the values of the variable (not the function) at the plotted points.

These steps are illustrated below.

Step 1.   Values of r, 0, 1, 2, 3, 4, 5.

Step 2.   Values of r^2, 0, 1, 4, 9, 16, 25.

Step 3.   If the scale is to be 5 inches long, then $m = 0.2$ inch is satisfactory and $mr^2 = 0$, 0.2 inch, 0.8 inch, 1.8 inches, 3.2 inches, 5.0 inches.

Step 4.   Lay off these lengths as shown in Figure 5-8.

Step 5.   Mark the values of r (not r^2) at the plotted points.

The Scale Modulus. If a function has a constant coefficient, it is frequently simpler to multiply the functional modulus by this constant and use this value, M, which is called a scale modulus.

For example, if $f(r) = 2$ log r with r varying between the limits 0.2 and 2.0, then from equation (11)

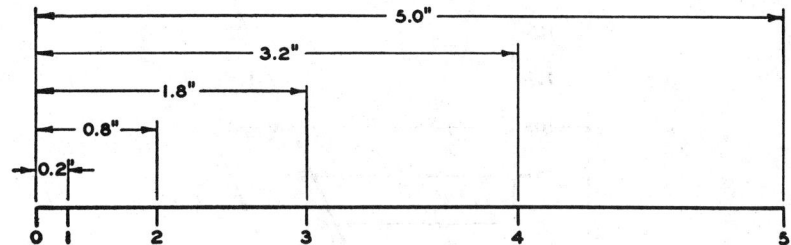

FIG. 5-8. Functional scale for r^2.

$$L = m(2 \log 2.0 - 2 \log 0.2)$$

or
$$L = m[0.602 - (-1.398)] = 2m \tag{11a}$$

If L is chosen as 10 inches, then

$$M = 10 = 2m \quad \text{and} \quad m = 5$$

It will be noted that in the above equation (11a), it was necessary to multiply each logarithm by 2. The labor could have been reduced by placing the 2 outside the bracket and writing

$$L = 2m(\log 2.0 - \log 0.2)$$

Using a scale modulus $M = 2m$, the logarithms of the numbers between 0.2 and 2.0 could be plotted directly without the multiplication operation. If there is no constant coefficient of the function, then $M = m$.

Selection of the Modulus. The work of plotting scales can be reduced if some thought is given to the selection of the functional modulus so that the engineer's scale units 1, ½, ⅓, ¼, ⅕, or ⅙, or powers of 10 times these factors such as ¹⁄₂₀ and ¹⁄₃₀, may be used. Thus, in the illustration of $f(r) = r^2$, the functional modulus was chosen as 0.2 inch, or ⅕. The engineer's 50 scale could have been used to plot the function by simply marking off the values 1, 4, 9, 16, and 25 from the scale, thus saving the computation of the third step. Obviously this happy choice cannot always be made, but it is worth considerable effort to get moduli of this kind whenever possible.

When an odd value for the scale modulus M must be used as is sometimes the case, for example, 4.67 = 10 units, the construction can be made as shown in Figure 5-9.

Technique of Graduating Scales. A careful examination of the illustrations used in this discussion will indicate the proper technique in laying out and marking scales. The title, the equation solved by the chart, and a key to the use of the chart should always be included.

General Practical Principles. Before undertaking the construction of a chart for a specific equation, certain principles which apply to all parallel scale charts should be thoroughly understood. These are illustrated and explained in the six parts of Figure 5-10. Beginning at (A), which represents the chart for the original equation, and following through the sequence, the significance of reversing the direction of scales, shifting scales, and changing scale spacing may be noted.

The application of these principles and of equations (11), (12), and (13) in the construction of a nomograph is illustrated in the following paragraphs.

Parallel Scale Charts with Natural Scales. A chart of this kind is used for equations of the type $f(r) + f(s) = f(t)$. For an equation of this type, the term on one side of the equation by itself is placed on the center scale. The other two terms are placed on the outside scales and may be on either side.

If m_r is chosen to give the desired length for the scale of $f(r)$ and m_s is chosen in a similar manner for the scale $f(s)$, then it can be shown that the scale spacing is

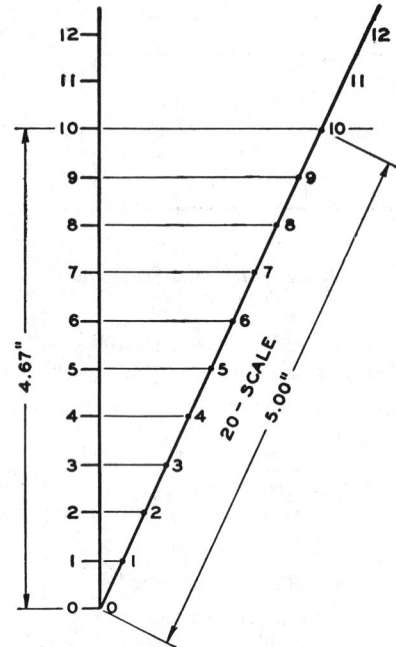

FIG. 5-9. Construction of a scale with an odd-scale modulus.

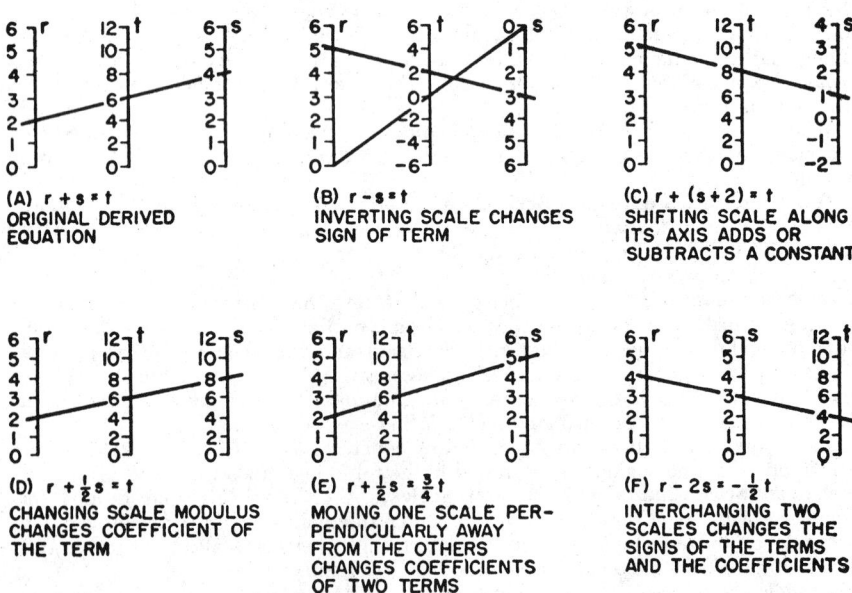

FIG. 5-10. Effect of changing scale positions and moduli.

proportional to the moduli selected or

$$\frac{m_r}{m_s} = \frac{a}{b} \tag{12}$$

where a is the distance from the scale for $f(r)$ to the center scale and b is the distance from the center scale $f(t)$ to the scale for $f(s)$. Although the moduli for the outside scales may be chosen at will, the modulus for the center scale is determined by the equation

$$m_t = \frac{m_r m_s}{m_r + m_s} \tag{13}$$

As an illustration of these principles, a chart for the equation $4r = \sqrt{D^2 + d^2}$, which gives the radius of gyration of a hollow circular column, will be designed. Assume that the range of the variable is to be 4 to 10 inches for D, the outside diameter, and 3 to 8 inches for d, the inside diameter.

Step 1. Get the equation into proper form by eliminating the radical sign.

$$16r^2 = D^2 + d^2$$

This makes r the middle scale.

Step 2. Choose values of m and L to give a chart of the desired size, say for use on 8½- by 11-inch paper. The length, L, could be chosen and the value of m determined from equation (11), but it is better to write the equation and then select a convenient value for m which will give approximately the length desired. Using the letters for the variables in the equation as subscripts for values of L and m, the result from equation (11) is

$$L_D = m_D(100 - 16) = 84m_D \quad \text{if } m_D = \tfrac{1}{10}, \text{ then } L_D = 8.4 \text{ in.}$$
and
$$L_d = m_d(64 - 9) = 55m_d \quad \text{if } m_d = \tfrac{1}{6}, \text{ then } L_d = 9.17 \text{ in.}$$

The modulus for the center scale, r, is obtained from equation (13).

$$m_r = \frac{m_D \times m_d}{m_D + m_d} = \frac{\tfrac{1}{10} \times \tfrac{1}{6}}{\tfrac{1}{10} + \tfrac{1}{6}} = \frac{1}{16}$$

The length of the middle scale need not be computed except as a check, because its length must lie somewhere between the lengths of the outside scales.

Step 3. Compute scale spacing. Using the values of m_D and m_d, the scale spacing can be computed from equation (12).

$$\frac{m_D}{m_d} = \frac{a}{b} = \frac{\tfrac{1}{10}}{\tfrac{1}{6}} = \frac{6}{10}$$

The scale spacings must therefore have this ratio. Choosing 3 and 5 inches would make the chart too wide. Using 2.4 and 4 inches, which are in proper ratio, gives a total chart width of 6.4 inches, which is suitable.

Step 4. Determine the scale modulus. Because the coefficients of the variables D^2 and d^2 are 1, their scale moduli are equal to their functional moduli. For r^2, the coefficient is 16; hence the scale modulus is $16 \times \tfrac{1}{16} = 1$.

Before going to the labor of laying out the scales, it is well to place all data in tabular form as a check upon the accuracy of the work. This step is shown in Table 5-11.

Step 5. Construct the chart. a. Draw three vertical lines spaced 2.4 and 4.0 inches apart as shown in Figure 5-11. Because all variables are positive, the scales will all be upward.

b. Draw a horizontal line across the chart as a base line for the outside scales, and on the left-hand line lay out the scale for D. Use the 10 scale on an engineer's scale, and with 16 on the base line, plot successively 25, 36, 49, 64, 81, and 100, which are the values of D^2.

TABLE 5-11. Data for Constructing Nomograph in Figure 5-11
Scale spacing $a/b = 2.4$ to 4.0 inches

Variable	Function	m	M	Length, inches	Direction
D	D^2	$\frac{1}{10}$	$\frac{1}{10}$	8.4	↑
d	d^2	$\frac{1}{6}$	$\frac{1}{6}$	9.17	↑
r	$16r^2$	$\frac{1}{16}$	1	8.69	↑

c. Similarly, on the right-hand line, using the engineer's 60 scale and beginning with 9 on the base line, plot 16, 25, 36, 49, and 64, which are the values of d^2.

d. To locate a starting point for the center scale, substitute the values of 3 and 4 in the original equation and solve for r. This gives 1.25 as the value of r. Place the square of this value (1.56) of the engineer's 10 scale on the line joining 3 and 4 of the outside scales, and plot 4 and 9 on the 10 scale. Four and 9 are the squares of 2 and 3 marked on the center scale.

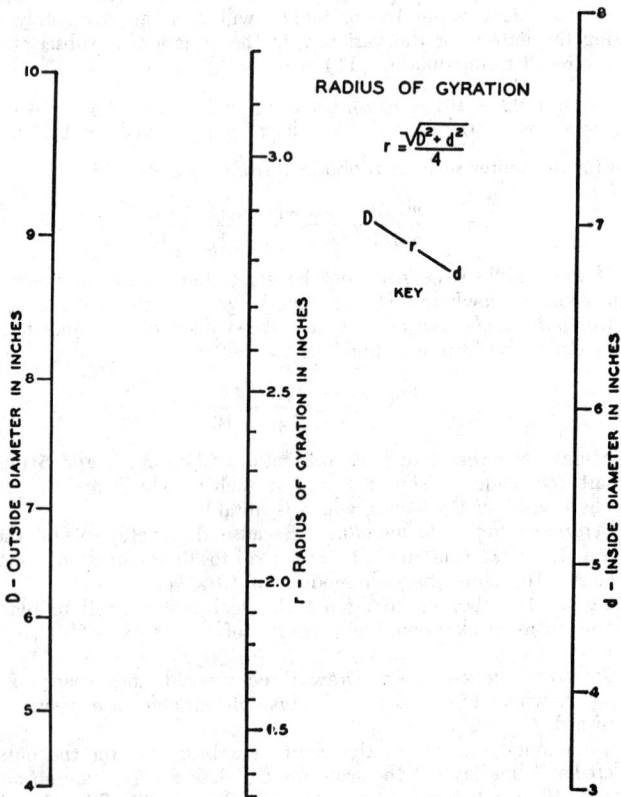

FIG. 5-11. Nomograph for equation $r = \sqrt{D^2 + d^2}/4$.

The finished chart is shown in Figure 5-11. The outside scales could be further subdivided by the use of the fan chart shown in Figure 5-12, but the spaces on the center scale are too large. It is therefore necessary to compute a number of intermediate values and plot them, after which the fan chart could be used. For greater accuracy on all scales, plotted values should be computed.

Parallel Scale Charts with Logarithmic Scales.   Nomographs of this type are used for equations of the type $f(r)f(s) = f(t)$. This equation may be placed in proper form for a parallel scale chart by taking logarithms of both sides.

$$\log f(r) + \log f(s) = \log f(t)$$

As an illustration of a problem of this type, the equation for wind pressure on a flat surface, $F = 0.004AV^2$, will be used. Let the limits for A be 1 to 10 square feet and for V from 20 to 80 miles per hour.

Step 1.   Change the equation to proper form.

$$\log F = \log 0.004 + \log A + 2 \log V$$

The constant, $\log 0.004$, can be neglected in the first computation and then accounted for by shifting the last scale to be plotted. (See General Practical Principles on page 12-79.)

Step 2.   Determine m and L.

$$L_A = m_A (\log 10 - \log 1) = m_A \times 1$$

If a log scale 8 inches long for one cycle is chosen, then $L_A = 8$

$$L_v = m_v (2 \log 80 - 2 \log 20) = m_v \times 1.204$$

If the same scale is chosen for m_v, then $8 \times 1.204 = 9.63$ inches, which is satisfactory for $8\frac{1}{2}$- by 11-inch paper. It is convenient to have the moduli on the outside scales

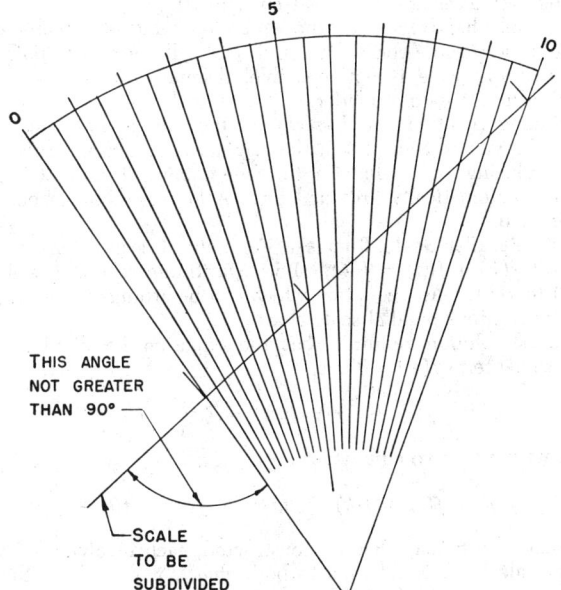

FIG. 5-12.   Fan chart for subdividing nonuniform spaces.

TABLE 5-12.   Data for Constructing Nomograph in Figure 5-13

Variable	Function	m	M	Length, inches	Direction
A	$\log A$	8	8	8	↑
V	$2 \log V$	8	16	9.63	↑
F	$\log F$	4	4	8.82	↑

the same, because this makes the modulus for the center scale just one-half that of the outside scales.

$$m_F = \frac{m_A \times m_v}{m_A + m_v} = \frac{8 \times 8}{8 + 8} = \frac{64}{16} = 4$$

Step 3.   Determine scale moduli.

$$M_A = m_A \times 1 = 8 \text{ in.} \qquad M_v = m_v \times 2 = 16 \text{ in.}$$
$$M_F = m_F \times 1 = 4 \text{ in.}$$

This requires that three logarithmic scales be used.

Step 4.   Determine scale spacing. $m_A/m_v = 8/8$. A ratio of 3 to 3 inches. All computations are now summarized in Table 5-12.

Step 5.   Construct the chart.   a. Draw three vertical lines spaced 3 inches apart and then draw a horizontal base line across the bottom as a starting point for the outside scales as shown in Figure 5-13. It may be noted that the base line could be sloping as well as horizontal if this will give a more pleasing shape to the chart.

b. Placing the A scale on the left, bring the 1 on an 8-inch log scale to the base line and plot all necessary divisions from 1 to 10.

c. Using the right-hand line for the V scale, bring the point 2 (for 20) of a 16-inch log scale to the base line and plot all necessary points from 20 to 80. If a 16-inch log scale is not available, distances taken from an 8-inch scale can be stepped off twice with a divider.

d. The beginning point of the F scale on the middle line must be determined by substituting the values 1 and 20 in the original equation and solving for F. This gives 1.6. By placing this value on the line joining 1 and 20 (base line), the constant 0.004 is automatically provided for. A log scale having one cycle 4 inches long should be used.

Four Variable Parallel Scale Charts.   For natural scales, the equation will have the form $f(q) + f(r) + f(s) = f(t)$. For logarithmic scales, it will have the form $f(q) f(r) f(s) = f(t)$. Any equation which can be arranged in either of these forms can be solved with a four parallel scale chart.

The method of solution requires that the equation be divided into two parts, each equal to a new term $f(k)$. Thus,

$$f(q) + f(r) = f(k) = f(t) - f(s)$$

which may be written as two equations

$$f(q) + f(r) = f(k) \qquad \text{and} \qquad f(k) = f(t) - f(s)$$

Two three-scale charts may then be constructed, each involving a scale for $f(k)$.

Because the scale for k is common to both charts, it must have the same modulus and the same direction in both charts. This restriction must be kept in mind in designing the two parts of the chart.

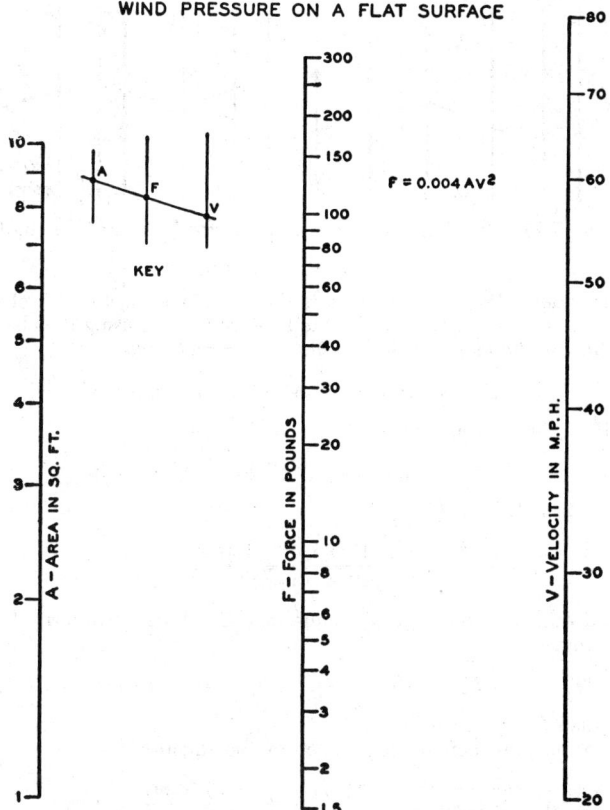

WIND PRESSURE ON A FLAT SURFACE

$F = 0.004 AV^2$

KEY

A – AREA IN SQ. FT.

F – FORCE IN POUNDS

V – VELOCITY IN M.P.H.

FIG. 5-13.   Chart for wind pressure on a flat surface.   Logarithmic scales.

As an illustration for this type of chart, the equation for the cooling surface of a condenser will be used.

$$C = 43.6 \frac{dL}{p^2}$$

Assume the range of variables to be as follows: C from 200 to 500 square feet, d from 0.5- to 1.0-inch diameter, L from 5 to 16.0 feet between two end plates, and p from 0.8- to 1.5-inch pitch of tubes.

Step 1.   Place the equation in proper form.   The original equation may be divided as follows:

$$Cp^2 = k = 43.6dL$$

and then written in logarithmic form

$$\log C + 2 \log p = \log k \qquad \text{and} \qquad \log k = \log 43.6 + \log d + \log L$$

Step 2.   Arrange the scales.   In this situation it is necessary to add an additional step, namely, the arrangement of the scales of the two charts. While several arrangements could be made, the one in Figure 5-14c seems best. With the equations written as above, all scales are positive upward and the k scale is in the center in both cases. It is necessary therefore only to arrange the scale spacing

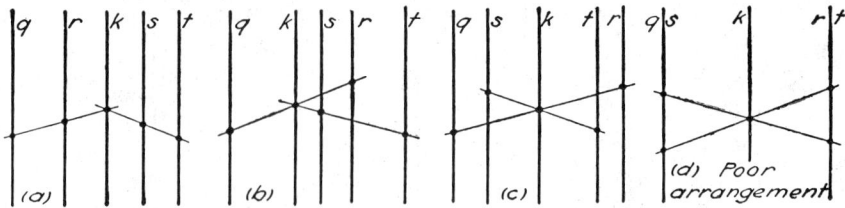

FIG. 5-14.   Possible scale arrangements for four variable charts.

so that the outside scales of the two charts do not coincide with each other.   Coincident outside scales as in Figure 5-14d make for confusion in using the chart.

Step 3.   Choose the moduli and determine L for each scale.

$$L_c = m_c(\log 500 - \log 200) = m_c \times 0.398$$

If $m_c = 15$, $L_c = 5.96$ in.

$$L_p = 2m_p(\log 1.5 - \log 0.8) = m_p \times 0.546$$

If $m_p = 10$, then $L_p = 5.46$ in.

Then
$$m_k = \frac{10 \times 15}{10 + 15} = \frac{150}{25} = 6$$

In the second equation, only one modulus may be chosen arbitrarily because m_k is already determined.

$$L_d = m_d(\log 1 - \log 0.5) = 0.301m_d$$

If $m_d = 15$, then $L_d = 4.52$ in.

The value of m_L must now be determined by the equation

$$m_k = \frac{m_d \times m_L}{m_d + m_L} = 6 = \frac{15 \times m_L}{15 + m_L}$$

or
$$90 + 6m_L = 15m_L \qquad m_L = 10$$

Then $\qquad L_L = 10(\log 16 - \log 5) = 10(1.204 - 0.699) = 5.05$ in.

Step 4.   Determine scale spacing.   From equation (12) the scale spacing is

$$\frac{m_c}{m_p} = \frac{a}{b} = \frac{15}{10} \quad \text{or} \quad \frac{3 \text{ in.}}{2 \text{ in.}}$$

Because the moduli for both parts of the chart are the same, the spacing for both parts is also the same.

As a check upon these computations, all data are now assembled in Table 5-13.

Step 5.   Construct the chart.   For the first chart, draw three vertical lines respectively 3 and 2 inches apart and draw a horizontal base line, as shown in Figure 5-15.

On the left-hand line, plot a log scale having a length of 15 inches for one cycle, with 2 on the base line representing 200 square feet of cooling surface.   On the right-hand line, plot a log scale having a length of 20 (or double the distances on a 10-inch scale) beginning with 0.8 on the base line.

The k scale need not be calibrated, because it is not necessary to read values on this scale.   The second chart may now be constructed using the same k scale.   Although the scale spacing is the same as for the first chart, the outside scales for the second are reversed so that the spacing from left to right is 2 and 3 inches from the k scale.   This places the d scale having the 15-inch modulus

TABLE 5-13.   Data for Constructing Nomograph in Figure 5-15

First chart $a/b = {}^{15}\!/_{10} = 3$ inches/2 inches

Second chart $a/b = {}^{15}\!/_{10} = 3$ inches/2 inches

Variable	Function	m	M	Length, inches	Direction
C	$\log C$	15	15	5.95	↑
p	$2 \log p$	10	20	5.46	↑
k	$\log k$	6	6	....	↑
d	$\log d$	15	15	4.52	↑
L	$\log L$	10	10	5.06	↑

on the extreme right. Using a log scale 15 inches long, plot the d scale beginning with 0.5 on the base line.

Before the last scale for L can be plotted, the values of the variables on the base line must be substituted in the equation and the value of L determined.

$$L = \frac{Cp^2}{43.6d} = \frac{200 \times 0.8^2}{43.6 \times 0.5} = 5.872$$

By placing this value of L on the base line connecting the corresponding values of the other variable and using a 10-inch log scale, the scale for L may be completed. The location of the L scale from the solution of the equation automatically provides for the constant 43.6 which has been ignored up to this point. The finished chart is shown in Figure 5-15.

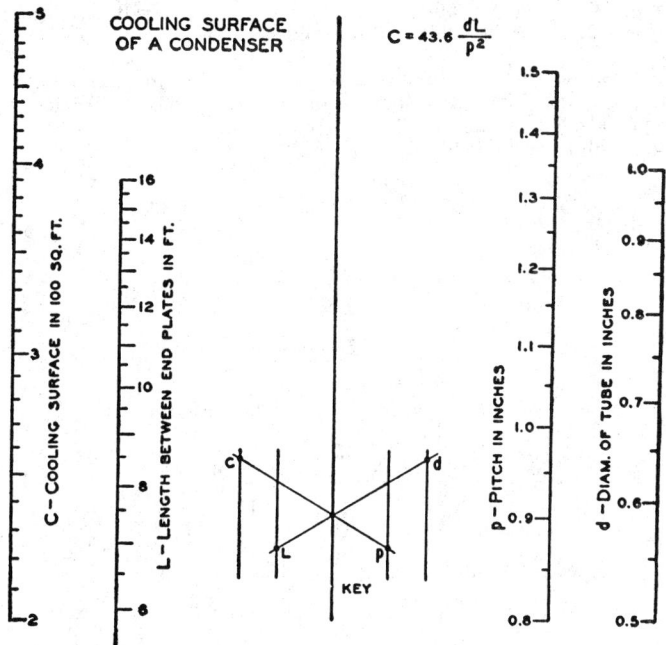

Fig. 5-15.   Nomograph for equation $C = 43.6(dL/p^2)$.

N **Chart with Natural Scales.** For equations of the type $f(t) = f(r)/f(s)$, the N chart with natural scales is often used. By placing this equation in logarithmic form, it can be solved by a three parallel scale chart as previously illustrated. There are situations, however, in which it is desirable to have natural scales.

The following facts about the N chart, or Z chart as it is sometimes called, should be noted:

1. The moduli for the two outside scales may be arbitrarily chosen at a value which keeps them within the limits of the paper to be used.
2. The scales begin at the zero values of the function and run in opposite directions.
3. The term on one side of the equation by itself is always on the diagonal scale which connects the zero values of the other functions on the parallel scales.
4. The length of the diagonal between zero points of the other scales may be arbitrarily chosen.

For some equations, the length of the diagonal may be such that the zero points of the two vertical scales are inaccessible on the drawing board. A graphical method for plotting the position of the diagonal and the scale on it is illustrated in the following example.

The equation for the air density correction factor is $D = 0.392B/(273 + T)$. If the limits for B, air pressure, are 710 to 770 millimeters of mercury and for T, temperature, from -30 to $+30°C$, the corresponding upper and lower limits for D are 0.92 and 1.24.

Step 1. Place the equation in proper form.

$$\frac{D}{0.392} = \frac{B}{273 + T}$$

Step 2. Choose values of m, L, and the length of the diagonal H.

$L_B = m_B(770 - 710) = 60m_B$ if $m_B = 0.1$, $L_B = 6$ in.
$L_T = m_T[(273 + 30) - (273 - 30)] = 60m_T$ if $m_T = 0.1$, $L_T = 6$ in.

Although the zero values of the scales are not included in the computations above, it is necessary to determine the distance from the zero point to the lowest plotted point on each scale as illustrated in Figure 5-16.

Thus, the length from 0 to 710 for the B scale is

$$L = 0.1(710 - 0) = 71 \text{ in.}$$

and for the T scale,

$$L = 0.1(273 - 30) - 0 = 24.3 \text{ in.}$$

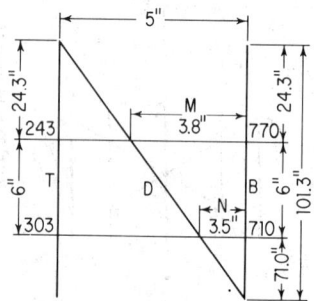

Fig. 5-16. Method of locating diagonal scale.

Therefore, the overall height of the chart from zero to zero values is

$$71 + 6 + 24.3 = 101.3$$

as shown in Figure 5-16. Because the length of the diagonal scale from zero to zero may be arbitrarily chosen, the width of the chart may be selected to be 5 inches and the diagonal let be what it will.

The position of the diagonal scale may be determined by similar triangles as illustrated in Figure 5-16. Thus,

$$\frac{101.3}{5} = \frac{77.0}{M} \qquad M = 3.8 \text{ in.}$$

and

$$\frac{101.3}{5} = \frac{71.0}{N} \qquad N = 3.5 \text{ in.}$$

Step 3. Construct the chart. Draw two vertical lines 5 inches apart and 6 inches long. Lay out the temperature and pressure scales in the usual manner, making the lower values at the same horizontal level.

Plot the position of the diagonal from the data above, measuring horizontally from 710 a distance of 3.5 inches and from 770 a distance of 3.8 inches. Connecting these points gives the position of the diagonal. The scale on the diagonal may be determined graphically as follows: Choose the values of D which are to be plotted, and with any convenient value of B (750 was used in this case), solve the original equation for values of T. The results are shown in Table 5-14.

To locate any point on the D scale, draw a line from 750 on the B scale to the corresponding value on the T scale as shown in Table 5-14. Thus, to locate 1.05 on the D scale, connect 750 with 280 on the T scale. The crossing point of this line with the diagonal locates point 1.05. Other points are located in the same manner as illustrated in Figure 5-17.

The method described above also may be used for charts which have the zero values of the vertical scales within the limits of the chart.

Proportional Charts. Charts for equations of the type $f(q)/f(r) = f(s)/f(t)$ may be made in a variety of forms as illustrated in Figure 5-18. In forms (a) and (b), the isopleths pass through a common point whereas in forms (c) and (d) the isopleths are parallel. An isopleth is the straight line or lines drawn across the chart to solve the equation.

It can be shown that the functional moduli for the scales must have the following relationship

$$\frac{m_q}{m_r} = \frac{m_s}{m_t}$$

Obviously, any three moduli may be chosen arbitrarily and the fourth computed from these three. The diagonal in forms (a) and (b) is merely a pivot scale.

Example. Make a chart for the equation $S = Mc/I$ with the following limits for the variables: M from 0 to 16,000,000 inch-pounds, I from 0 to 20,000 inch⁴, S from 0 to 16,000 pounds per square inch, and c from 0 to 18 inches.

Step 1. Place the equation in proper form.

$$\frac{M}{S} = \frac{I}{c}$$

TABLE 5-14. Calculated Data for Determining the Scale for the Diagonal of the Nomograph in Figure 5-17

$$f(T) = 273 + T = 0.392 \times 750/D$$

D	1.25	1.20	1.15	1.10	1.05	1.00	0.95	0.90
T	235	245	256	267	280	294	310	327

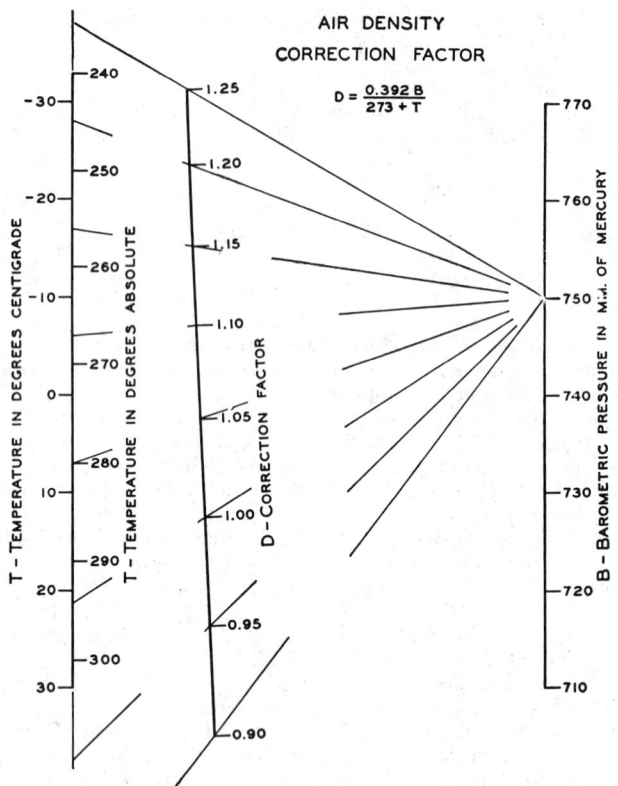

FIG. 5-17.   Construction of chart for equation $D = 0.392B/(273 + T)$.

Step 2. Choose values of m and L. If either form (a) or (b) is used, the diagonal is uncalibrated and may have any length. The zero values of all scales are at the ends of the diagonals.

$$L_M = m_M(16,000,000 - 0) \qquad \text{if } m_M = 0.0000005, \text{ then } L_M = 8 \text{ in.}$$
$$L_S = m_S(16,000 - 0) \qquad \text{if } m_S = 0.0005, \text{ then } L_S = 8 \text{ in.}$$
$$L_I = m_I(20,000 - 0) \qquad \text{if } m_I = 0.0005, \text{ then } L_I = 10 \text{ in.}$$

$$m_c = \frac{m_S m_I}{m_M} = \frac{0.0005 \times 0.0005}{0.0000005} = 0.5$$

$$L_c = m_c(18 - 0) = 18 \times 0.5 = 9 \text{ in.}$$

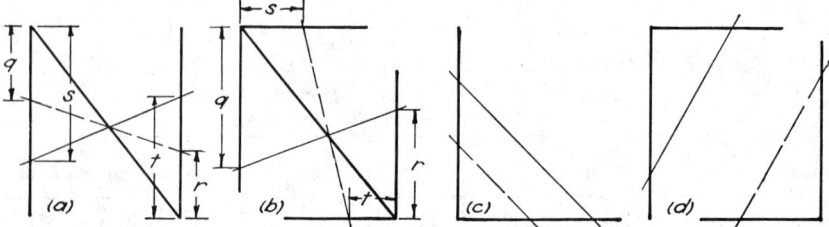

FIG. 5-18.   Various forms of proportional charts.

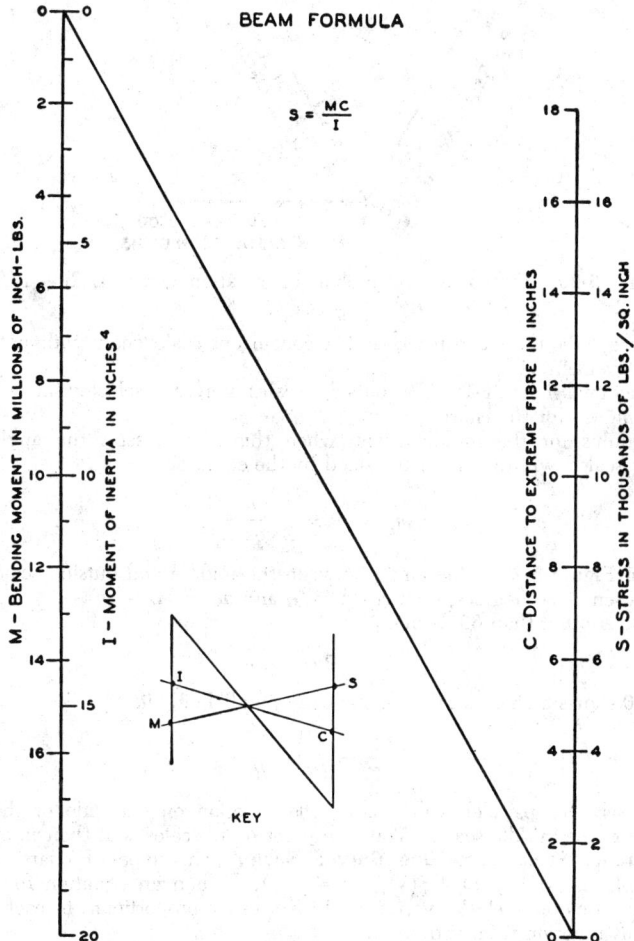

FIG. 5-19. Nomograph for equation $S = Mc/I$.

Step 3. *Construct the chart.* Because all scales are natural scales and the zero values are at the ends of the diagonal, the construction of these scales offers no new problem. The width of the chart or the length of diagonal may have any value.

It should be noted that the functions in the numerators of the two fractions are on one axis while the denominators are on the other, as shown in Figure 5-19.

Thus, in the equation

$$\frac{M}{S} = \frac{I}{c}$$

M and I are on one axis and S and c on the other.

Concurrent Scale Charts. An equation of the type $[1/f(r)] + [1/f(s)] = [1/f(t)]$ is most often charted by use of concurrent scale charts. By using reciprocal scales, this equation can be solved by a parallel scale chart. Reciprocal scales,

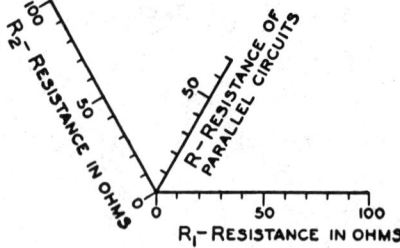

F‍ɪɢ. 5-20.   Concurrent scale chart of equation $1/R = 1/R_1 + 1/R_2$.

however, are tedious to construct, and a concurrent scale chart with natural scales is more convenient.

The functional moduli for the outside scales may be selected at will to give a convenient size for the chart.

The modulus for the middle scale, when this scale bisects the angle between the outside scales, will then be determined by the equation

$$m_r = m_s = \frac{m_t}{2 \cos A}$$

where A in Figure 5-20 is the angle between the middle and outside scales.   If this angle is chosen as 60 degrees, then $\cos A = \frac{1}{2}$ and $m_r = m_s = m_t$.

For angle A other than 60 degrees,

$$m_t = 2 \cos A m_r$$

Figure 5-20 shows a chart for the resistance in parallel circuits

$$\frac{1}{R} = \frac{1}{R_1} + \frac{1}{R_2}$$

Note that, as with parallel scale charts, the function on one side of the equation by itself is on the middle scale.   The zero point of all scales is at their intersection.

Two Parallel Scales and One Curved Scale.   This type of chart is used for equations of the form $f(r) + f(s)f_1(t) = f_2(t)$.   When an equation for a variable contains two functions of the variable which are not proportional to each other, the scale for this function is curved.

The moduli for the parallel scales and their spacing may be chosen arbitrarily to give a convenient size for the chart.   The curved scale may be plotted by ordinates from the line joining the zero values of the functions on the parallel scales and the other parallel to these scales.

A simpler method, largely graphical and involving less computation, is illustrated in the following example.

Example.   The volume of a segment of a sphere is given by the equation

$$V = 0.5236(3R^2 + D^2)D$$

where R is the radius of the horizontal surface of the liquid and D is the depth of the liquid.   For this illustration, assume both R and D vary from 0 to 10 feet. Then V will vary from 0 to 2,094 cubic feet.

Step 1.   Place the equation in proper form.

$$V - 1.5708R^2D = 0.5236D^3$$

Step 2.   Choose values of m and L for the parallel scales.   Because D is on the curved scale, V and R will be on the parallel scales.

$$L_V = m_V(2,094 - 0) \quad \text{if } m_V = 0.005, \text{ then } L_V = 10.47 \text{ in.}$$

TABLE 5-15. Data Required for the Plotting of the Curve
for Values of D in Figure 5-21

$$V - 1.5708R^2D = 0.5236D^3$$

Values of D to be plotted	Pole R	V	Pole R	V
1	5	39.8	10	157.6
2	5	82.8	10	318.2
3	5	132.0	10	485
4	5	190.5	10	662
5	5	261.8	10	851
6	10	1,059	12	1,474
7	10	1,279	12	1,761
8	10	1,525.5	12	2,080
9	10	1,796	12	2,399
10	10	2,095	12	2,787

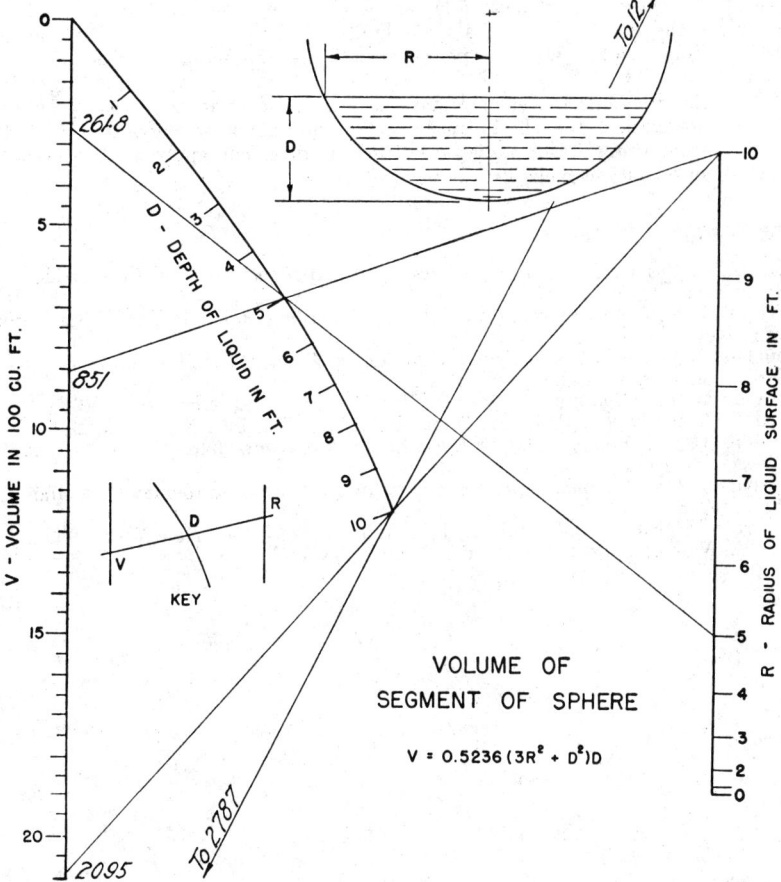

Fɪɢ. 5-21. Construction of chart for the volume of a segment of a sphere.

The function of R is $1.5708R^2$: therefore,

$$L_R = m_R(157.08 - 0) \qquad \text{if } m_R = 0.05, \text{ then } L_R = 7.85 \text{ in.}$$

Step 3. Construct the chart. The method of construction involves the repeated solution of the equation, using the values of D for the major divisions of the scale which are to be plotted, and two convenient values of R for each value of D as shown in Table 5-15.

It is now a simple matter to draw two lines connecting the values of R with the corresponding values of V. The intersection of these lines in pairs locates the point for the value of D on the curve as shown in Figure 5-21.

Thus, if a line is drawn from 5 on the R scale to 262 (261.8) on the V scale and a second line is drawn from 10 on the R scale to 851 on the V scale, the intersection of these lines locates the point 5 on the curved scale. Other points on the curved scale are determined in the same manner and a smooth curve drawn through them.

Subdivisions on the curve scale can be made with a fan chart because the curve between any three consecutive points is so nearly straight.

If the curvature is sharp, other points must be computed and plotted as before.

The Choice of Limits for Variables. The choice of moduli in the foregoing illustrations, which seemed to work out so conveniently, is not entirely a matter of chance.

The limits chosen for the variables are a governing consideration, as is the size chosen for the chart. In general, the limits of the variables will be chosen to give the range within which experience indicates the most frequent use of the equation lies.

When the fundamental assumptions used in developing an equation limit its range of application, the chart can be made for this range only, thus preventing a novice from using it for ranges to which it does not apply or beyond which practical considerations make its use inexpedient.

BIBLIOGRAPHY

Adams, Douglas P., *Nomography: Theory and Applications,* The Shoe String Press, Inc., Hamden, Conn., 1964.

Davis, Dale S., *Nomography and Empirical Equations,* Reinhold Publishing Corporation, New York, 1955.

Douglass, Raymond D., and Douglas P. Adams, *Elements of Nomography,* McGraw-Hill Book Company, New York, 1947.

Epstein, L. I., *Nomography,* John Wiley & Sons, Inc., New York, 1958.

Fasal, J., *Nomography,* Frederick Ungar Publishing Co., Inc., New York, 1968.

Johnson, L. H., *Nomography and Empirical Equations,* John Wiley & Sons, Inc., New York, 1952.

Kharbanda, O. P., *Nomograms for Chemical Engineers,* Academic Press Inc., New York, 1958.

Levens, A. S., *Nomography,* 2d ed., John Wiley & Sons, Inc., New York, 1959.

Otto, E., *Nomography,* Pergamon Press, Inc., New York, 1964.

Chapter 6

Estimating Data Including Speed and Feed Tables

C. F. STEPHENSON

President, The John B. Adt Co., York, Pennsylvania

The industrial engineer is often called upon to make estimates of labor and material costs. If he has any appreciable volume of estimating to do, he will find it quite useful to have tables of various kinds of data to assist him in his computations and estimates.

The data and tables included in this chapter are consolidations of generally available information which experience has shown are needed repeatedly by any estimator. The data are presented in a form which has been found to be convenient for quick application.

The industrial engineer who wishes to make use of these tables to help him with his estimating work should first study them carefully to become familiar with their arrangement and the information which they contain. Then when he begins to make his estimates, he will be able to find, without unnecessary search, any information he may need that is provided by the tables.

TABLE 6-1. Weights and Measures

United States System
Length

12	inches	= 1 foot
3	feet	= 1 yard
5½	yards or 16½ feet	= 1 rod
40	rods	= 1 furlong
8	furlongs	= 1 mile (statute)

Dry Measure

2 pints = 1 quart
8 quarts = 1 peck
4 pecks = 1 bushel (2,150.42 cubic inches)
36 bushels = 1 chaldron

Area   *Liquid Measure*

144	square inches = 1 square foot	4	gills	= 1 pint
9	square feet = 1 square yard	2	pints	= 1 quart
30¼	square yards = 1 square rod	4	quarts	= 1 gallon (231 cubic inches)
40	square rods = 1 rood	31½	gallons	= 1 barrel
4	roods = 1 acre	2	barrels	= 1 hogshead
640	acres = 1 square mile			

Volume of Capacity

1,728	cubic inches	= 1 cubic foot
27	cubic feet	= 1 cubic yard
128	cubic feet	= 1 cord of wood
24¾	cubic feet	= 1 perch of stone

Weight Avoirdupois

16 ounces   = 1 pound (7,000 grains)
100 pounds   = 1 hundredweight
20 hundredweight = 1 ton

Metric System
Length

10 millimeters (mm)	= 1 centimeter
10 centimeters	= 1 decimeter
10 decimeters	= 1 meter
10 meters	= 1 decameter
10 decameters	= 1 hectometer
10 hectometers	= 1 kilometer

Volume

1,000 cubic millimeters	= 1 cubic centimeter
1,000 cubic centimeters	= 1 cubic decimeter or 1 liter
1,000 cubic decimeters	= 1 cubic meter
1,000 cubic meters	= 1 cubic decameter
1,000 cubic decameters	= 1 cubic hectometer
1,000 cubic hectometers	= 1 cubic kilometer

TABLE 6-1 (Continued)

Metric System
Area

100 square millimeters (mm²)	= 1 square centimeter
100 square centimeters	= 1 square decimeter
100 square decimeters	= 1 square meter or centare
100 square meters or centares	= 1 are
100 ares	= 1 hectare
100 hectares	= 1 square kilometer

Weight

10 milligrams (mg)	= 1 centigram
10 centigrams	= 1 decigram
10 decigrams	= 1 gram
10 grams	= 1 decagram
10 decagrams	= 1 hectogram
10 hectograms	= 1 kilogram
1,000 kilograms	= 1 metric ton

Miscellaneous

Inches		*Articles*	
3 inches = 1 palm		2 articles = 1 pair	
4 inches = 1 hand		12 articles = 1 dozen	
9 inches = 1 span		12 dozen = 1 gross	
18 inches = 1 cubit		12 gross = 1 quire	
21.8 inches = 1 Bible cubit		20 quires = 1 ream	
30 inches = 1 military pace			

TABLE 6-2. Equivalents of Measure—Lengths*

1 meter	= 10 decimeters, 100 centimeters, 1,000 millimeters
1 meter	= 0.1 decameter, 0.01 hectometer, 0.001 kilometer
1 meter	= 39.37 inches, U.S. Standard; 39.370113 inches, British Standard
1 millimeter	= 1,000 microns, 0.03937 inch, 39.37 mils

Meters	Inches	Feet	Yards	Rods	Chains	Miles, U.S. Statute	Miles, U.S. Nautical	Kilometers
1	39.37	3.28083	1.09361	0.19884	0.04971	$0._0^3 6214$	$0._0^3 5396$	0.001
0.02540	1	0.08333	0.02778	$0._0^2 5051$	$0._0^2 1263$	$0._0^4 1578$	$0._0^4 1371$	$0._0^4 2540$
0.30480	12	1	0.33333	0.06061	0.01515	$0._0^3 1894$	$0._0^3 1645$	$0._0^3 3048$
0.91440	36	3	1	0.18182	0.04545	$0._0^3 5682$	$0._0^3 4934$	$0._0^3 9144$
5.02921	198	16.5	5.5	1	0.25	$0._0^2 3125$	$0._0^2 2714$	$0._0^2 5029$
20.1168	792	66	22	4	1	0.01250	0.01085	0.02012
1,609.35	63,360	5,280	1,760	320	80	1	0.86839	1.60935
1,853.25	72,962.5	6,080.20	2,026.73	368.497	92.1243	1.15155	1	1.85325
1,000	39,370	3,280.83	1,093.61	198.838	49.7096	0.62137	0.53959	1

1 cable length, U.S. = 120 fathoms, 960 spans, 720 feet, 219.457 meters
1 league, U.S. = 3 statute miles, 24 furlongs
1 international geographical mile = $\frac{1}{15}$ degree at equator, 7,422 meters
= 4.611808 U.S. statute miles
1 international nautical mile = $\frac{1}{60}$ degree at meridian, 1,852 meters
= 0.999326 U S. nautical miles

*Notations $\frac{2}{0}, \frac{3}{0}, \frac{4}{0}$, etc., indicate that the $\frac{2}{0}, \frac{3}{0}, \frac{4}{0}$, etc., are to be replaced by 2, 3, 4, etc., ciphers.

TABLE 6-3. Equivalents of Measure—Surfaces and Areas*

1 square meter   = 100 square decimeters, 10,000 square centimeters
1 square meter   = 0.01 are, 0.0001 hectare
1 square millimeter = 0.01 square centimeter, 0.00155 square inch, 1973.5 circular mils
1 are   = 1 square decameter, 0.0247104 acre

Square meters	Square inches	Square feet	Square yards	Square rods	Acres	Hectares	Square miles	Square kilometers
1	1,550.00	10.7639	1.19599	0.03954	$0.{}^{3}_{0}2471$	0.0001	$0.{}^{6}_{0}3861$	$0.{}^{5}_{0}1$
$0.{}^{3}_{0}6452$	1	$0.{}^{2}_{0}6944$	$0.{}^{3}_{0}7716$	$0.{}^{4}_{0}2551$	$0.{}^{6}_{0}1594$	$0.{}^{7}_{0}6452$	$0.{}^{9}_{0}2491$	$0.{}^{9}_{0}6452$
0.09290	144	1	0.11111	$0.{}^{2}_{0}3673$	$0.{}^{4}_{0}2296$	$0.{}^{5}_{0}9290$	$0.{}^{7}_{0}3587$	$0.{}^{7}_{0}9290$
0.83613	1,296	9	1	0.03306	$0.{}^{3}_{0}2066$	$0.{}^{4}_{0}8361$	$0.{}^{6}_{0}3228$	$0.{}^{6}_{0}8361$
25.2930	39,204	272.25	30.25	1	0.00625	$0.{}^{2}_{0}2529$	$0.{}^{5}_{0}9766$	$0.{}^{4}_{0}2529$
4,046.87	6,272,640	43,560	4,840	160	1	0.40469	$0.{}^{2}_{0}1563$	$0.{}^{2}_{0}4047$
10,000	15,499,969	107,639	11,959.9	395.366	2.47104	1	$0.{}^{2}_{0}3861$	0.01
2,589,999		27,878,400	3,097,600	102,400	640	259.000	1	2.59000
1,000,000		10,763,867	1,195,985	39,536.6	247.104	100	0.38610	1

* Notations $\frac{2}{0'}\,\frac{3}{0'}\,\frac{4}{0'}$, etc., indicate that the $\frac{2}{0'}\,\frac{3}{0'}\,\frac{4}{0'}$ etc., are to be replaced by 2, 3, 4, etc., ciphers.

TABLE 6-4. Equivalents of Measure—Volume and Capacity*

1 cubic meter = 1,000 cubic decimeters, 1,000,000 cubic centimeters
1 liter   = 10 deciliters, 100 centiliters, 1,000 milliliters, 1,000 cubic centimeters
1 liter   = 0.1 decaliter, 0.01 hectoliter, 1 cubic decimeter

Cubic decimeter	Cubic inches	Cubic feet	Cubic yards	U.S. quarts		U.S. gallons		U.S. bushels
				Liquid	Dry	Liquid	Dry	
1	61.0234	0.03531	$0.{}^{2}_{0}1308$	1.05668	0.90808	0.26417	0.22702	0.02838
0.01639	1	$0.{}^{3}_{0}5787$	$0.{}^{4}_{0}2143$	0.01732	0.01488	$0.{}^{2}_{0}4329$	$0.{}^{3}_{0}3720$	0.4650
28.3170	1,728	1	0.03704	29.9221	25.7140	7.48055	6.42851	0.80356
764.559	46,656	27	1	807.896	694.279	201.974	173.570	21.6962
0.94636	57. 5	0.03342	$0.{}^{2}_{0}1238$	1	0.85937	0.25	0.21484	0.02686
1.10123	67.2006	0.03889	$0.{}^{2}_{0}1440$	1.16365	1	0.29091	0.25	0.03125
3.78543	231	0.13368	$0.{}^{2}_{0}4951$	4	3.43747	1	0.85927	0.10742
4.40492	268.803	0.15556	$0.{}^{2}_{0}5761$	4.65460	4	1.16365	1	0.125
35.2393	2,150.42	1.24446	0.04609	37.2368	32	9.30920	8	1

U.S. dry measure: 1 bushel = 4 pecks, 8 gallons, 32 quarts, 64 pints
U.S. liquid measure: 1 gallon = 4 quarts, 8 pints, 32 gills, 128 fluid ounces
British Imperial gallon dry and liquid measure = 1.03202 U.S. dry gallon, 1.20091 U.S. liquid gallon
1 gallon, U.S. liquid = 8.34545 pounds, 3.78543 kilograms
1 cubic foot = 62.4283 pounds avoirdupois, 28.3170 kilograms
1 cubic inch = 0.57804 ounce avoirdupois, 16.3872 grams

* Notations $\frac{2}{0'}\,\frac{3}{0'}\,\frac{4}{0'}$ etc., indicate that the $\frac{2}{0'}\,\frac{3}{0'}\,\frac{4}{0'}$ etc., are to be replaced by 2, 3, 4, etc., ciphers.

TABLE 6-5. Equivalents of Measure—Weights*

1 gram = 10 decigrams, 100 centigrams, 1000 milligrams
1 gram = 0.1 decagram, 0.01 hectogram, 0.001 kilogram
1 kilogram = 1 cubic decimeter of water or liter 4°C, 45 degrees Latitude and sea level, 15,432.35639 U.S. and British Standard.

| Kilograms | Grains | Ounces | | Pounds | | Tons | | |
		Troy	Avoirdupois	Troy	Avoirdupois	Net short 2,000 pounds	Gross long 2,240 pounds	Metric
1	15,432.4	32.1507	35.2740	2.67923	2.20462	0.$^{2}_{0}$1102	0.$^{3}_{0}$9842	0.001
0.$^{4}_{0}$6480	1	0.$^{2}_{0}$2083	0.$^{2}_{0}$2286	0.$^{3}_{0}$1736	0.$^{3}_{0}$1429	0.$^{7}_{0}$7143	0.$^{7}_{0}$6378	0.$^{7}_{0}$6480
0.03110	480	1	1.09714	0.08333	0.06857	0.$^{4}_{0}$3429	0.$^{4}_{0}$3061	0.$^{4}_{0}$3110
0.02835	437.5	0.91146	1	0.07595	0.06250	0.$^{4}_{0}$3125	0.$^{4}_{0}$2790	0.$^{4}_{0}$2835
0.37324	5,760	12	13.1657	1	0.82286	0.$^{3}_{0}$4114	0.$^{3}_{0}$3674	0.$^{3}_{0}$3732
0.45359	7,000	14.5833	16	1.21528	1	0.00050	0.$^{3}_{0}$4464	0.$^{3}_{0}$4536
907.185	14,000,000	29,166.7	32,000	2,430.56	2,000	1	0.89286	0.90719
1,016.05	15,680,000	32,666.7	35,840	2,722.22	2,240	1.12	1	1.01605
1,000	15,432,356	32,150.7	35,274.0	2,679.23	2,204.62	1.10231	0.98421	1

* Notations $^{2}_{0}$, $^{3}_{0}$, $^{4}_{0}$, etc., indicate that the $^{2}_{0}$, $^{3}_{0}$, $^{4}_{0}$, etc., are to be replaced by 2, 3, 4, etc., ciphers.

TABLE 6-6. Temperature Conversion—Centigrade to Fahrenheit

| Degrees centigrade →
 grade ↓ | 0 | 10 | 20 | 30 | 40 | 50 | 60 | 70 | 80 | 90 |
	Degrees Fahrenheit									
0	32	50	68	86	104	122	140	158	176	194
100	212	230	248	266	284	302	320	338	356	374
200	392	410	428	446	464	482	500	518	536	554
300	572	590	608	626	644	662	680	698	716	734
400	752	770	788	806	824	842	860	878	896	914
500	932	950	968	986	1004	1022	1040	1058	1076	1094
600	1112	1130	1148	1166	1184	1202	1220	1238	1256	1274
700	1292	1310	1328	1346	1364	1382	1400	1418	1436	1454
800	1472	1490	1508	1526	1544	1562	1580	1598	1616	1634
900	1652	1670	1688	1706	1724	1742	1760	1778	1796	1814
1000	1832	1850	1868	1886	1904	1922	1940	1958	1876	1994
1100	2012	2030	2048	2066	2084	2102	2120	2138	2156	2174
1200	2192	2210	2228	2246	2264	2282	2300	2318	2336	2354
1300	2372	2390	2408	2426	2444	2462	2480	2498	2516	2534
1400	2552	2570	2588	2606	2624	2642	2660	2678	2696	2714
1500	2732	2750	2768	2786	2804	2822	2840	2858	2876	2894
1600	2912	2930	2948	2966	2984	3002	3020	3038	3056	3074
1700	3092	3110	3128	3146	3164	3182	3200	3218	3236	3254
1800	3272	3290	3308	3326	3344	3362	3380	3398	3416	3434
1900	3452	3470	3488	3506	3524	3542	3560	3578	3596	3614
2000	3632	3650	3668	3686	3704	3722	3740	3758	3776	3794

Conversion formula:

Fahrenheit = (centigrade × ⅑) + 32
Centigrade = (Fahrenheit − 32) × ⅝

TABLE 6-7. Hardness Conversion Table

Values vary depending on grades and conditions of material involved. Rockwell B scale should not be used over B-100. The C scale should not be used under C-20.

Brinell hardness No.	Shore scleroscope hardness No.	Rockwell		Brinell hardness No.	Shore scleroscope hardness No.	Rockwell		Brinell hardness No.	Shore scleroscope hardness No.	Rockwell	
		B scale	C scale			B scale	C scale			B scale	C scale
782	107	.....	72	277	39	.....	29	137	....	75	.....
744	100	.....	69	269	38	.....	28	134	....	74	.....
713	96	.....	67	262	37	.....	27	131	....	72	.....
683	92	.....	65	255	36	.....	26	128	....	71	.....
652	88	.....	63	248	36	.....	25	126	....	70	.....
627	85	.....	61	241	35	100	24	124	....	69	.....
600	81	.....	59	235	34	99	23	121	....	67	.....
578	78	.....	58	228	33	98	22	118	....	66	.....
555	75	.....	56	223	33	97	21	116	....	65	.....
532	72	.....	54	217	32	96	20	114	....	64	.....
512	70	.....	52	212	31	95	.....	112	....	62	.....
495	68	.....	51	207	30	94	.....	109	....	61	.....
477	66	.....	49	202	30	93	.....	107	....	59	.....
460	64	.....	48	196	29	92	.....	105	....	58	.....
444	61	.....	47	192	29	91	.....	103	....	57	.....
430	59	.....	45	187	28	90	.....	101	....	56	.....
418	57	.....	44	183	28	89	.....	99	....	54	.....
402	55	.....	43	179	27	88	.....	97	....	53	.....
387	53	.....	41	174	27	87	.....	96	....	52	.....
375	52	.....	40	170	26	86	.....	95	....	51	.....
364	50	.....	39	166	26	85	.....	93	....	50	.....
351	49	.....	38	163	25	84	.....	92	....	49	.....
340	47	.....	37	159	25	83	.....	90	....	48	.....
332	46	.....	36	156	24	82	.....	88	....	47	.....
321	45	.....	35	153	24	81	.....	87	....	46	.....
311	44	.....	34	149	23	80	.....	86	....	45	.....
302	42	.....	33	146	23	78	.....	85	....	44	.....
293	41	.....	31	143	22	77	.....	83	....	43	.....
286	40	.....	30	140	....	76	.....	82	....	42	.....

TABLE 6-8. Decimal Equivalents—Inches to Millimeters

Fraction	Decimal	Millimeter	Fraction	Decimal	Millimeter	Fraction	Decimal	Millimeter
1/64	0.015625	0.397		0.35433	9.0	11/16	0.6875	17.462
1/32	0.03125	0.794	23/64	0.359375	9.128	45/64	0.703125	17.859
	0.03937	1.0	3/8	0.375	9.525		0.70866	18.0
3/64	0.046875	1.191	25/64	0.390625	9.922	23/32	0.71875	18.256
1/16	0.0625	1.587		0.39370	10.0	47/64	0.734375	18.653
5/64	0.078125	1.984	13/32	0.40625	10.319		0.74803	19.0
	0.07874	2.0	27/64	0.421875	10.716	3/4	0.75	19.05
3/32	0.09375	2.381		0.43307	11.0	49/64	0.765625	19.447
7/64	0.109375	2.778	7/16	0.4375	11.112	25/32	0.78125	19.844
	0.11811	3.0	29/64	0.453125	11.509		0.7874	20.0
1/8	0.125	3.175	15/32	0.46875	11.906	51/64	0.796875	20.241
9/64	0.140625	3.572		0.47244	12.0	13/16	0.8125	20.637
5/32	0.15625	3.969	31/64	0.484375	12.303		0.82677	21.0
	0.15748	4.0	1/2	0.5	12.7	53/64	0.828125	21.034
11/64	0.171875	4.366		0.51181	13.0	27/32	0.84375	21.431
3/16	0.1875	4.762	33/64	0.515625	13.097	55/64	0.859375	21.828
	0.19685	5.0	17/32	0.53125	13.494		0.86614	22.0
13/64	0.203125	5.159	35/64	0.546875	13.891	7/8	0.875	22.225
7/32	0.21875	5.556		0.55118	14.0	57/64	0.890625	22.622
15/64	0.234375	5.953	9/16	0.5625	14.287		0.90551	23.0
	0.23622	6.0	37/64	0.578125	14.684	29/32	0.90625	23.019
1/4	0.25	6.35		0.59055	15.0	59/64	0.921875	23.416
17/64	0.265625	6.747	19/32	0.59375	15.081	15/16	0.9375	23.812
	0.27559	7.0	39/64	0.609375	15.478		0.94488	24.0
9/32	0.28125	7.144	5/8	0.625	15.875	61/64	0.953125	24.209
19/64	0.296875	7.541		0.62992	16.0	31/32	0.96875	24.606
5/16	0.3125	7.937	41/64	0.640625	16.272		0.98425	25.0
	0.31496	8.0	21/32	0.65625	16.669	63/64	0.984375	25.003
21/64	0.328125	8.334		0.66929	17.0	1	1.000	25.400
11/32	0.34375	8.731	43/64	0.671875	17.066			

TABLE 6-9. Conversion of Millimeters to Decimals of Inches

Milli-meters	Inches	Milli-meters	Inches	Milli-meters	Inches	Milli-meters	Inches
1	0.03937	26	1.02362	51	2.00787	76	2.99213
2	0.07874	27	1.06299	52	2.04724	77	3.03150
3	0.11811	28	1.10236	53	2.08661	78	3.07087
4	0.15748	29	1.14173	54	2.12598	79	3.11024
5	0.19685	30	1.18110	55	2.16535	80	3.14961
6	0.23622	31	1.22047	56	2.20472	81	3.18898
7	0.27559	32	1.25984	57	2.24409	82	3.22835
8	0.31496	33	1.29921	58	2.28346	83	3.26772
9	0.35433	34	1.33858	59	2.32283	84	3.30709
10	0.39370	35	1.37795	60	2.36220	85	3.34646
11	0.43307	36	1.41732	61	2.40157	86	3.38583
12	0.47244	37	1.45669	62	2.44094	87	3.42520
13	0.51181	38	1.49606	63	2.48031	88	3.46457
14	0.55118	39	1.53543	64	2.51968	89	3.50394
15	0.59055	40	1.57480	65	2.55906	90	3.54331
16	0.62992	41	1.61417	66	2.59843	91	3.58268
17	0.66929	42	1.65354	67	2.63780	92	3.62205
18	0.70866	43	1.69291	68	2.67717	93	3.66142
19	0.74803	44	1.73228	69	2.71654	94	3.70079
20	0.78740	45	1.77165	70	2.75591	95	3.74016
21	0.82677	46	1.81102	71	2.79528	96	3.77953
22	0.86614	47	1.85039	72	2.83465	97	3.81890
23	0.90551	48	1.88976	73	2.87402	98	3.85827
24	0.94488	49	1.92913	74	2.91339	99	3.89764
25	0.98425	50	1.96850	75	2.95276	100	3.93701

TABLE 6-10. Functions of Numbers

Number	Square	Cube	Square root	Cube root	Logarithm	1,000 × reciprocal	Number = diameter	
							Circumference	Area
1	1	1	1.0000	1.0000	0.00000	1,000.000	3.142	0.7854
2	4	8	1.4142	1.2599	0.30103	500.000	6.283	3.1416
3	9	27	1.7321	1.4422	0.47712	333.333	9.425	7.0686
4	16	64	2.0000	1.5874	0.60206	250.000	12.566	12.5664
5	25	125	2.2361	1.7100	0.69897	200.000	15.708	19.6350
6	36	216	2.4495	1.8171	0.77815	166.667	18.850	28.2744
7	49	343	2.6458	1.9129	0.84510	142.857	21.991	38.4845
8	64	512	2.8284	2.0000	0.90309	125.000	25.133	50.2655
9	81	729	3.0000	2.0801	0.95424	111.111	28.274	63.6173
10	100	1,000	3.1623	2.1544	1.00000	100.000	31.416	78.5398
11	121	1,331	3.3166	2.2240	1.04139	90.9091	34.558	95.0332
12	144	1,728	3.4641	2.2894	1.07918	83.3333	37.699	113.097
13	169	2,197	3.6056	2.3513	1.11394	76.9231	40.841	132.732
14	196	2,744	3.7417	2.4101	1.14613	71.4286	43.982	153.938
15	225	3,375	3.8730	2.4662	1.17609	66.6667	47.124	176.715
16	256	4,096	4.0000	2.5198	1.20412	62.5000	50.265	201.062
17	289	4,913	4.1231	2.5713	1.23045	58.8235	53.407	226.980
18	324	5,832	4.2426	2.6207	1.25527	55.5556	56.549	254.469
19	361	6,859	4.3589	2.6684	1.27875	52.6316	59.690	283.529
20	400	8,000	4.4721	2.7144	1.30103	50.0000	62.832	314.159
21	441	9,261	4.5826	2.7589	1.32222	47.6190	65.973	346.361
22	484	10,648	4.6904	2.8020	1.34242	45.4545	69.115	380.133
23	529	12,167	4.7958	2.8439	1.36173	43.4783	72.257	415.476
24	576	13,824	4.8990	2.8845	1.38021	41.6667	75.398	452.389
25	625	15,625	5.0000	2.9240	1.39794	40.0000	78.540	490.874
26	676	17,576	5.0990	2.9625	1.41497	38.4615	81.681	530.929
27	729	19,683	5.1962	3.0000	1.43136	37.0370	84.823	572.555
28	784	21,952	5.2915	3.0366	1.44716	35.7143	87.965	615.752
29	841	24,389	5.3852	3.0723	1.46240	34.4828	91.106	660.520
30	900	27,000	5.4772	3.1072	1.47712	33.3333	94.248	706.858
31	961	29,791	5.5678	3.1414	1.49136	32.2581	97.389	754.768
32	1,024	32,768	5.6569	3.1748	1.50515	31.2500	100.53	804.248
33	1,089	35,937	5.7446	3.2075	1.51851	30.3030	103.67	855.299
34	1,156	39,304	5.8310	3.2396	1.53148	29.4118	106.81	907.920
35	1,225	42,875	5.9161	3.2711	1.54407	28.5714	109.96	962.113
36	1,296	46,656	6.0000	3.3019	1.55630	27.7778	113.10	1,017.88
37	1,369	50,653	6.0828	3.3322	1.56820	27.0270	116.24	1,075.21
38	1,444	54,872	6.1644	3.3620	1.57978	26.3158	119.38	1,134.11
39	1,521	59,319	6.2450	3.3912	1.59106	25.6410	122.52	1,194.59
40	1,600	64,000	6.3246	3.4200	1.60206	25.0000	125.66	1,256.64
41	1,681	68,921	6.4031	3.4482	1.61278	24.3902	128.81	1,320.25
42	1,764	74,088	6.4807	3.4760	1.62325	23.8095	131.95	1,385.44
43	1,849	79,507	6.5574	3.5034	1.63347	23.2558	135.09	1,452.20
44	1,936	85,184	6.6332	3.5303	1.64345	22.7273	138.23	1,520.53
45	2,025	91,125	6.7082	3.5569	1.65321	22.2222	141.37	1,590.43
46	2,116	97,336	6.7823	3.5830	1.66276	21.7391	144.51	1,661.90
47	2,209	103,823	6.8557	3.6088	1.67210	21.2766	147.65	1,734.94
48	2,304	110,592	6.9282	3.6342	1.68124	20.8333	150.80	1,809.56
49	2,401	117,649	7.0000	3.6593	1.69020	20.4082	153.94	1,885.74
50	2,500	125,000	7.0711	3.6840	1.69897	20.0000	157.08	1,963.50
51	2,601	132,651	7.1414	3.7084	1.70757	19.6078	160.22	2,042.82
52	2,704	140,608	7.2111	3.7325	1.71600	19.2308	163.36	2,123.72
53	2,809	148,877	7.2801	3.7563	1.72428	18.8679	166.50	2,206.18
54	2,916	157,464	7.3485	3.7798	1.73239	18.5185	169.65	2,290.22
55	3,025	166,375	7.4162	3.8030	1.74036	18.1818	172.79	2,375.83
56	3,136	175,616	7.4833	3.8259	1.74819	17.8571	175.93	2,463.01
57	3,249	185,193	7.5498	3.8485	1.75587	17.5439	179.07	2,551.76
58	3,364	195,112	7.6158	3.8709	1.76343	17.2414	182.21	2,642.08
59	3,481	205,379	7.6811	3.8930	1.77085	16.9492	185.35	2,733.97
60	3,600	216,000	7.7460	3.9149	1.77815	16.6667	188.50	2,827.43
61	3,721	226,981	7.8102	3.9365	1.78533	16.3934	191.64	2,922.47
62	3,844	238,328	7.8740	3.9579	1.79239	16.1290	194.78	3,019.07
63	3,969	250,047	7.9373	3.9791	1.79934	15.8730	197.92	3,117.25
64	4,096	262,144	8.0000	4.0000	1.80618	15.6250	201.06	3,216.99
65	4,225	274,625	8.0623	4.0207	1.81291	15.3846	204.20	3,318.31

TABLE 6-10 (Continued)

Number	Square	Cube	Square root	Cube root	Logarithm	1,000 × reciprocal	Number = diameter	
							Circumference	Area
66	4,356	287,496	8.1240	4.0412	1.81954	15.1515	207.35	3,421.19
67	4,489	300,763	8.1854	4.0615	1.82607	14.9254	210.49	3,525.65
68	4,624	314,432	8.2462	4.0817	1.83251	14.7059	213.63	3,631.68
69	4,761	328,509	8.3066	4.1016	1.83885	14.4928	216.77	3,739.28
70	4,900	343,000	8.3666	4.1213	1.84510	14.2857	219.91	3,848.45
71	5,041	357,911	8.4261	4.1408	1.85126	14.0845	223.05	3,959.19
72	5,184	373,248	8.4853	4.1602	1.85733	13.8889	226.19	4,071.50
73	5,329	389,017	8.5440	4.1793	1.86332	13.6986	229.34	4,185.39
74	5,476	405,224	8.6023	4.1983	1.86923	13.5135	232.48	4,300.84
75	5,625	421,875	8.6603	4.2172	1.87506	13.3333	235.62	4,417.86
76	5,776	438,976	8.7178	4.2358	1.88081	13.1579	238.76	4,536.46
77	5,929	456,533	8.7750	4.2543	1.88649	12.9870	241.90	4,656.63
78	6,084	474,552	8.8318	4.2727	1.89209	12.8205	245.04	4,778.36
79	6,241	493,039	8.8882	4.2908	1.89763	12.6582	248.19	4,901.67
80	6,400	512,000	8.9443	4.3089	1.90309	12.5000	251.33	5,026.55
81	6,561	531,441	9.0000	4.3267	1.90849	12.3457	254.47	5,153.00
82	6,724	551,368	9.0554	4.3445	1.91381	12.1951	257.61	5,281.02
83	6,889	571,787	9.1104	4.3621	1.91908	12.0482	260.75	5,410.61
84	7,056	592,704	9.1652	4.3795	1.92428	11.9048	263.89	5,541.77
85	7,225	614,125	9.2195	4.3968	1.92942	11.7647	267.04	5,674.50
86	7,396	636,056	9.2736	4.4140	1.93450	11.6279	270.18	5,808.80
87	7,569	658,503	9.3274	4.4310	1.93952	11.4943	273.32	5,944.68
88	7,744	681,472	9.3808	4.4480	1.94448	11.3636	276.46	6,082.12
89	7,921	704,969	9.4340	4.4647	1.94939	11.2360	279.60	6,221.14
90	8,100	729,000	9.4868	4.4814	1.95424	11.1111	282.74	6,361.73
91	8,281	753,571	9.5394	4.4979	1.95904	10.9890	285.88	6,503.88
92	8,464	778,688	9.5917	4.5144	1.96379	10.8696	289.03	6,647.61
93	8,649	804,357	9.6437	4.5307	1.96848	10.7527	292.17	6,792.91
94	8,836	830,584	9.6954	4.5468	1.97313	10.6383	295.31	6,939.78
95	9,025	857,375	9.7468	4.5629	1.97772	10.5263	298.45	7,088.22
96	9,216	884,736	9.7980	4.5789	1.98227	10.4167	301.59	7,238.23
97	9,409	912,673	9.8489	4.5947	1.98677	10.3093	304.73	7,389.81
98	9,604	941,192	9.8995	4.6104	1.99123	10.2041	307.88	7,542.96
99	9,801	970,299	9.9499	4.6261	1.99564	10.1010	311.02	7,697.69
100	10,000	1,000,000	10.0000	4.6416	2.00000	10.0000	314.16	7,853.98
101	10,201	1,030,301	10.0499	4.6570	2.00432	9.90099	317.30	8,011.85
102	10,404	1,061,208	10.0995	4.6723	2.00860	9.80392	320.44	8,171.28
103	10,609	1,092,727	10.1489	4.6875	2.01284	9.70874	323.58	8,332.29
104	10,816	1,124,864	10.1980	4.7027	2.01703	9.61538	326.73	8,494.87
105	11,025	1,157,625	10.2470	4.7177	2.02119	9.52381	329.87	8,659.01
106	11,236	1,191,016	10.2956	4.7326	2.02531	9.43396	333.01	8,824.73
107	11,449	1,225,043	10.3441	4.7475	2.02938	9.34579	336.15	8,992.02
108	11,664	1,259,712	10.3923	4.7622	2.03342	9.25926	339.29	9,160.88
109	11,881	1,295,029	10.4403	4.7769	2.03743	9.17431	342.43	9,331.32
110	12,100	1,331,000	10.4881	4.7914	2.04139	9.09091	345.58	9,503.32
111	12,321	1,367,631	10.5357	4.8059	2.04532	9.00901	348.72	9,676.89
112	12,544	1,404,928	10.5830	4.8203	2.04922	8.92857	351.86	9,852.03
113	12,769	1,442,897	10.6301	4.8346	2.05308	8.84956	355.00	10,028.7
114	12,996	1,481,544	10.6771	4.8488	2.05690	8.77193	358.14	10,207.0
115	13,225	1,520,875	10.7238	4.8629	2.06070	8.69565	361.28	10,386.9
116	13,456	1,560,896	10.7703	4.8770	2.06446	8.62069	364.42	10,568.3
117	13,689	1,601,613	10.8167	4.8910	2.06819	8.54701	367.57	10,751.3
118	13,924	1,643,032	10.8628	4.9049	2.07188	8.47458	370.71	10,935.9
119	14,161	1,685,159	10.9087	4.9187	2.07555	8.40336	373.85	11,122.0
120	14,400	1,728,000	10.9545	4.9324	2.07918	8.33333	376.99	11,309.7
121	14,641	1,771,561	11.0000	4.9461	2.08279	8.26446	380.13	11,499.0
122	14,884	1,815,848	11.0454	4.9597	2.08636	8.19672	383.27	11,689.9
123	15,129	1,860,867	11.0905	4.9732	2.08991	8.13008	386.42	11,882.3
124	15,376	1,906,624	11.1355	4.9866	2.09342	8.06452	389.56	12,076.3
125	15,625	1,953,125	11.1803	5.0000	2.09691	8.00000	392.70	12,271.8
126	15,876	2,000,376	11.2250	5.0133	2.10037	7.93651	395.84	12,469.0
127	16,129	2,048,383	11.2694	5.0265	2.10380	7.87402	398.98	12,667.7
128	16,384	2,097,152	11.3137	5.0397	2.10721	7.81250	402.12	12,868.0
129	16,641	2,146,689	11.3578	5.0528	2.11059	7.75194	405.27	13,069.8
130	16,900	2,197,000	11.4018	5.0658	2.11394	7.69231	408.41	13,273.2

TABLE 6-11. Specific Gravities and Weights*

Substance	Specific gravity	Weight, lb per cu ft	Substance	Specific gravity	Weight, lb per cu ft
Metals, alloys, ores:			Timber, U.S. seasoned:		
Aluminum, cast hammered........	2.55– 2.75	165	Moisture content by weight:		
Aluminum, bronze...............	7.7	481	Seasoned timber 15–20%		
Antimony......................	6.62	416	Green timber up to 50%		
Arsenic.......................	5.73	358	Ash, white, red.................	0.60	40
Bismuth.......................	9.79	608	Cedar, white, red...............	0.32–0.38	22
Brass, cast-rolled..............	8.4 – 8.7	534	Chestnut.......................	0.39	41
Bronze, 7.9 to 14% Sn...........	7.4 – 8.9	509	Cypress........................	0.46	30
Chromium.....................	6.93	428	Fir, Douglas spruce.............	0.48	32
Cobalt........................	8.72– 8.95	552	Fir, eastern....................	0.40	25
Copper, cast-rolled.............	8.8 – 9.0	556	Elm, white.....................	0.50	45
Copper, ore, pyrites.............	4.1 – 4.3	262	Hemlock.......................	0.42–0.52	29
Gold, cast-hammered............	19.25–19.35	1,205	Hickory.......................	0.72	49
Iron, cast, pig.................	7.2	450	Locust.........................	0.73	46
Iron, wrought.................	7.6 – 7.9	485	Maple, hard....................	0.63	43
Iron, steel....................	7.8 – 7.9	490	Maple, white...................	0.45	33
Iron, spiegeleisen..............	7.5	468	Oak, chestnut..................	0.86	54
Iron, ferrosilicon..............	6.7 – 7.3	437	Oak, live......................	0.95	59
Iron, ore, hematite.............	5.2	325	Oak, red, black................	0.63	41
Iron, ore, hematite in bank.......	....	160–180	Oak, white.....................	0.68	46
Iron, ore, hematite loose	....	130–160	Pine, Oregon...................	0.51	32
Iron, ore, limonite.............	3.6 – 4.0	237	Pine, red......................	0.48	30
Iron, ore, magnetite............	4.9 – 5.2	315	Pine, white....................	0.41	26
Iron, slag.....................	2.5 – 3.0	172	Pine, yellow, long-leaf..........	0.58	44
Lead.........................	11.28–11.35	706	Pine, yellow, short-leaf.........	0.51	38
Lead ore, galena...............	7.3 – 7.6	465	Poplar.........................	0.42	30
Magnesium....................	1.74	109	Redwood, California............	0.40	26
Manganese....................	7.20– 7.42	456	Spruce, white, black............	0.40–0.46	27
Manganese ore, pyrolusite........	3.7 – 4.6	259	Walnut, black..................	0.55	38
Mercury......................	13.55	848	Walnut, white..................	0.41	26
Molybdenum..................	10.2	562	Various liquids:		
Nickel........................	8.57– 8.90	545	Alcohol, 100%.................	0.79	49
Nickel monel metal.............	8.8 – 9.0	556	Acids, muriatic, 40%...........	1.20	75
Platinum, cast-hammered........	21.1 –21.5	1,330	Acids, nitric, 91%..............	1.50	94
Silver, cast-hammered..........	10.4 –10.6	656	Acids, sulfuric, 87%............	1.80	112
Tin, cast-hammered............	7.2 – 7.5	459	Lye, soda, 66%.................	1.70	106
Tin, babbitt metal.............	7.1	443	Oils, vegetable.................	0.91–0.94	58
Tin, ore, cassiterite............	6.4 – 7.0	418	Oils, mineral, lubricants.........	0.90–0.93	57
Tungsten......................	18.7 –19.1	1,180	Petroleum.....................	0.88	55
Vanadium....................	5.5 – 5.7	350	Gasoline.......................	0.66–0.69	42
Zinc, cast-rolled...............	6.9 – 7.2	440	Water, 4°C, maximum density....	1.0	62.428
Zinc, ore, blende...............	3.9 – 4.2	253	Water, 100°C...................	0.9584	59.830
Minerals:			Water, ice.....................	0.88–0.92	56
Asbestos......................	2.1 – 2.8	153	Water, snow, fresh fallen........	0.125	8
Barytes.......................	4.50	281	Water, sea water...............	1.02–1.03	64
Basalt........................	2.7 – 3.2	184	Gases, air = 1:		
Bauxite.......................	2.55	159	Air, 0°C, 760 mm..............	1.0	0.08071
Borax........................	1.7 – 1.8	109	Ammonia......................	0.5920	0.0478
Chalk.........................	1.8 – 2.6	137	Carbon dioxide.................	1.5291	0.1234
Clay, marl....................	1.8 – 2.6	137	Carbon monoxide...............	0.9673	0.0781
Dolomite......................	2.9	181	Gas, illuminating...............	0.35–0.45	0.028–0.036
Feldspar, orthoclase............	2.5 – 2.6	159	Gas, natural...................	0.47–0.48	0.038–0.039
Gneiss, serpentine..............	2.4 – 2.7	159	Hydrogen......................	0.0693	0.00559
Granite, syenite................	2.5 – 3.1	175	Nitrogen.......................	0.9714	0.0784
Greenstone, trap...............	2.8 – 3.2	187	Oxygen........................	1.1056	0.0892
Gypsum, alabaster..............	2.3 – 2.8	159	Bituminous substances:		
Hornblende....................	3.0	187	Asphaltum.....................	1.1 –1.5	81
Limestone, marble..............	2.5 – 2.8	165	Coal, anthracite................	1.4 –1.7	97
Magnesite.....................	3.0	187	Coal, bituminous...............	1.2 –1.5	84
Phosphate rock, apatite..........	3.2	200	Coal, lignite...................	1.1 –1.4	78
Porphyry......................	2.6 – 2.9	172	Coal, peat, turf, dry............	0.65–0.85	47
Pumice, natural................	0.37– 0.90	40	Coal, charcoal, pine............	0.28–0.44	23
Quartz, flint..................	2.5 – 2.8	165	Coal, charcoal, oak.............	0.47–0.57	33
Sandstone, bluestone............	2.2 – 2.5	147	Coal, coke.....................	1.0 –1.4	75
Shale, slate...................	2.7 – 2.9	175	Graphite.......................	1.9 –2.3	131
Soapstone, talc................	2.6 – 2.8	169	Paraffin.......................	0.87–0.91	56
Coal and coke, piled:			Petroleum.....................	0.87	54
Coal, anthracite................	....	47–58	Petroleum, refined..............	0.79–0.82	50
Coal, bituminous, lignite.........	....	40–54	Petroleum, benzine..............	0.73–0.75	46
Coal, peat, turf................	....	20–26	Petroleum, gasoline.............	0.66–0.69	42
Coal, charcoal.................	....	10–14	Pitch..........................	1.07–1.15	69
Coal, coke....................	....	23–32	Tar, bituminous................	1.20	75

*The specific gravities of solids and liquids refer to water at 4°C; those of gases to air at 0°C and 760 millimeter pressure. The weights per cubic foot are derived from average specific gravities, except where stated that weights are for bulk, heaped or loose material, etc.

TABLE 6-12. Tapers and Angles

Taper per foot	Taper per inch	Taper per inch from center line	Included angle	With center line
1/8	0.010416	0.005203	0°35'48"	0°17'54"
3/16	0.015625	0.007812	0°53'42"	0°26'51"
1/4	0.020833	0.010416	1°11'38"	0°35'49 '
5/16	0.026042	0.013021	1°29'32"	0°44'46"
3/8	0.031250	0.015625	1°47'22"	0°53'41"
7/16	0.036458	0.018229	2° 5'14"	1° 2'37"
1/2	0.041667	0.020833	2°23'10"	1°11'35"
9/16	0.046875	0.023438	2°41' 4"	1°20'32"
5/8	0.052084	0.026042	2°59' 2"	1°29'31"
11/16	0.057292	0.028646	3°16'54"	1°38'27"
3/4	0.062500	0.031250	3°34'48"	1°47'24"
13/16	0.067708	0.033854	3°52'40"	1°56'20"
7/8	0.072917	0.036456	4°10'32"	2° 5'16"
15/16	0.078125	0.039063	4°28'25"	2°14'13"
1	0.083330	0.041667	4°46'16"	2°23' 8"
1 1/4	0.104666	0.052084	5°57'48"	2°58'54"
1 1/2	0.125000	0.062500	7° 9'10"	3°34'35"
1 3/4	0.145833	0.072917	8°20'28"	4°10'14"
2	0.166666	0.083332	9°31'36"	4°45'48"
2 1/2	0.208333	0.104166	11°53'36"	5°56'48"
3	0.250000	0.125000	14°15' 0"	7° 7'30"
3 1/2	0.291666	0.145833	16°35'40"	8°17'50"
4	0.333333	0.166666	18°55'30"	9°27'45"
4 1/2	0.375000	0.187500	21°14' 2"	10°37' 1"
5	0.416666	0.208333	23°32'12"	11°46' 6"
6	0.500000	0.250000	28° 4' 2"	14° 2' 1"

TABLE 6-13. Area and Dimensional Relationships for Various Regular Polygons

No. of sides	Name	Area when diameter of inscribed circle = 1	Area when sides = 1	Length of side when perpendicular from center of side to center of circumscribed circle = 1	Perpendicular when sides = 1	Radius of circumscribed circle when side = 1	Length of side when radius of circumscribed circle = 1
3	Triangle	1.299	0.433	3.464	0.289	0.577	1.732
4	Square	1.000	1.000	2.000	0.500	0.707	1.414
5	Pentagon	0.908	1.720	1.453	0.688	0.851	1.176
6	Hexagon	0.866	2.598	1.155	0.866	1.000	1.000
7	Heptagon	0.843	3.634	0.963	1.039	1.152	0.868
8	Octagon	0.828	4.828	0.828	1.207	1.307	0.765
9	Nonagon	0.819	6.182	0.728	1.374	1.462	0.684
10	Decagon	0.812	7.694	0.650	1.539	1.618	0.618
11	Nudecagon	0.807	9.366	0.587	1.703	1.775	0.563
12	Dodecagon	0.804	11.196	0.536	1.866	1.932	0.518

TABLE 6-14. Formulas for Right- and Oblique-angle Triangles

Key to Symbols

$\text{Sin} = \text{sine} = \dfrac{\text{side opposite}}{\text{hypotenuse}}$

$\text{Cos} = \text{cosine} = \dfrac{\text{side adjacent}}{\text{hypotenuse}}$

$\text{Tan} = \text{tangent} = \dfrac{\text{side opposite}}{\text{side adjacent}}$

$\text{Cot or cotan} = \text{cotangent} = \dfrac{\text{side adjacent}}{\text{side opposite}}$

$\text{Sec} = \text{secant} = \dfrac{\text{hypotenuse}}{\text{side adjacent}}$

$\text{Cosec} = \text{cosecant} = \dfrac{\text{hypotenuse}}{\text{side opposite}}$

$S = \frac{1}{2}(a + b + c)$

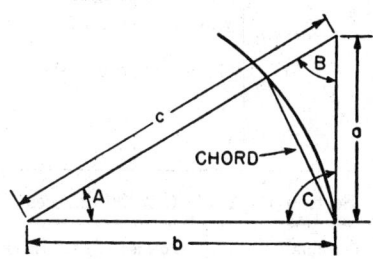

Section I. Angle $C = 90$ degrees

To find	Given	Formula	To find	Given	Formula
sin A	Sides ac	$\dfrac{a}{c}$	cosec A	Sides a, c	$\dfrac{c}{a}$
	cosec A	$\dfrac{1}{\text{cosec } A}$		sin A	$\dfrac{1}{\sin A}$
	tan A, cos A	tan $A \times$ cos A		sec A, cot A	sec $A \times$ cot A
	tan A, sec A	$\dfrac{\tan A}{\sec A}$		sec A, tan A	$\dfrac{\sec A}{\tan A}$
	cos A, cot A	$\dfrac{\cos A}{\cot A}$		cot A, cos A	$\dfrac{\cot A}{\cos A}$
	cos A	$\sqrt{1 - \cos A^2}$		cot A	$\sqrt{\cot A^2 + 1}$
cosin A	Sides b, c	$\dfrac{b}{c}$	Versed sine A	Sides b, c	$\dfrac{c - b}{c}$
	sec A	$\dfrac{1}{\sec A}$		cos A	$1 - \cos A$
	cot A, sin A	cot $A \times$ sin A	Coversed sine A	Sides c, a	$\dfrac{c - a}{c}$
	sin A, tan A	$\dfrac{\sin A}{\tan A}$		sin A	$1 - \sin A$
	cot A, cosec A	$\dfrac{\cot A}{\text{cosec } A}$	Side a	Sides b, c	$\sqrt{c^2 - b^2}$
				Side c, sin A	c sin A
				Side b, tan A	b tan A
	sin A	$\sqrt{1 + \sin A^2}$		Side b, cot A	$\dfrac{b}{\cot A}$
tan A	Sides a, b	$\dfrac{a}{b}$		Side c, cosec A	$\dfrac{c}{\text{cosec } A}$
	cot A	$\dfrac{1}{\cot A}$	Side b	Sides c, a	$\sqrt{c^2 - a^2}$
	sec A, sin A	sec $A \times$ sin		Side c, cos A	c cos A
				Side a, cot A	a cot A
	sin A, cos A	$\dfrac{\sin A}{\cos A}$		Side a, tan A	$\dfrac{a}{\tan A}$
	sec A, cosec A	$\dfrac{\sec A}{\text{cosec } A}$		Side c, sec A	$\dfrac{c}{\sec A}$
	sec A	$\sqrt{\sec A^2 + 1}$	Side c	Sides a, b	$\sqrt{a^2 + b^2}$
cotan A	Sides a, b	$\dfrac{b}{a}$		Side b, sec A	b sec A
				Side a, cosec A	a cosec A
	tan A	$\dfrac{1}{\tan A}$		Side a, sin A	$\dfrac{a}{\sin A}$
	cosec A, cos A	cosec $A \times$ cos A		Side b, cos A	$\dfrac{b}{\cos A}$
	cos A, sin A	$\dfrac{\cos A}{\sin A}$		Sides b, c Versed sin A	$\dfrac{c - b}{\text{Vers. } A}$
	cosec A, sec A	$\dfrac{\text{cosec } A}{\sec A}$		Sides a, c Coversed sin A	$\dfrac{c - a}{\text{Covers. } A}$
	cosec A	$\sqrt{\text{cosec } A^2 - 1}$			

TABLE 6-14 (Continued)
Section I. Angle $C = 90$ degrees

To find	Given	Formula	To find	Given	Formula
sec A	Sides b, c	$\dfrac{c}{b}$	Angles A B C	Angles B, C Angles A, C Angles A, B	$C - B$ $C - A$ $A + B$
cos A		$\dfrac{1}{\cos A}$	Area	Sides a, b	$\dfrac{ab}{2}$
cosec A, tan A		cosec $A \times \tan A$	Chord	Side b, sin A	$2b \sin \dfrac{A}{2}$
tan A, sin A		$\dfrac{\tan A}{\sin A}$			
cosec A, cot A		$\dfrac{\operatorname{cosec} A}{\cot A}$			
tan A		$\sqrt{\tan A^2 + 1}$			

Section II. Angle $C > 90$ degrees

To find	Given	Formula	To find	Given	Formula
A	BC	$180° - (B + C)$	Tangent C	bcA	$\dfrac{c \sin A}{b - (c \cos A)}$
sin A	abB	$\dfrac{a \sin B}{b}$	a	bAB	$\dfrac{b \sin A}{\sin B}$
sin A	acC	$\dfrac{a \sin C}{c}$	a	cAC	$\dfrac{c \sin A}{\sin C}$
cos A	abc	$\dfrac{b^2 + c^2 - a^2}{2bc}$	a	bcA	$\sqrt{b^2 + c^2 - (2bc \cos A)}$
tan A	abC	$\dfrac{a \sin C}{b - a \cos C}$	b	aAB	$\dfrac{a \sin B}{\sin A}$
tan A	acB	$\dfrac{a \sin B}{c - (a \cos B)}$	b	cBC	$\dfrac{c \sin B}{\sin C}$
B	AC	$180° - (A + C)$	b	acB	$\sqrt{a^2 + c^2 - (2ac \cos B)}$
sin B	abA	$\dfrac{b \sin A}{a}$	c	aAC	$\dfrac{a \sin C}{\sin A}$
sin B	bcC	$\dfrac{b \sin C}{c}$	c	bBC	$\dfrac{b \sin C}{\sin B}$
cos B	abc	$\dfrac{c^2 + a^2 - b^2}{2ac}$	c	abC	$\sqrt{a^2 + b^2 - (2ab - \cos C)}$
tan B	abC	$\dfrac{b \sin C}{a - b \cos C}$	Area	abC	$\dfrac{ab \sin C}{2}$
tan B	bcA	$\dfrac{b \sin A}{c - (b \cos A)}$	Area $S = \frac{1}{2}$ $(a + b + c)$	abc	$\sqrt{s(s - a)(s - b)(s - c)}$
C	AB	$180° - (A + B)$			
sin C	acA	$\dfrac{c \sin A}{a}$			
sin C	bcB	$\dfrac{c \sin B}{b}$			
cos C	abc	$\dfrac{a^2 + b^2 - c^2}{2ab}$			
tan C	acB	$\dfrac{c \sin B}{a - (c \cos B)}$			

TABLE 6-15.   Comparison of Standard Gages

Name of gage	United States Standard Gage U.S. Std	Birmingham (or stubs iron) wire gage BWG	American or Browne & Sharpe wire gage B & S	American Steel & Wire Co., formerly Washburn & Moen	British Imperial or English legal standard wire gage SWG	New Birmingham standard sheet and hoop gage BG	Music wire
Principal use	Uncoated steel sheet and light plates	Strips, bands, hoops and wire	Non-ferrous sheets and wire	Steel wire except music wire	Wire	Iron and steel sheets and hoops	
Gage No.	Thickness, inches	Thickness or diameter, inches					
7-0s	0.4902	.....	....	0.4900	0.500	0.6666	.....
6-0s	0.4596	.....	0.580000	0.4615	0.464	0.6250	.....
5-0s	0.4289	0.500	0.516500	0.4305	0.432	0.5883	.....
4-0s	0.3983	0.454	0.460000	0.3938	0.400	0.5416	.....
3-0s	0.3676	0.425	0.409642	0.3625	0.372	0.5000	0.007
2-0s	0.3370	0.380	0.364796	0.3310	0.348	0.4452	0.008
1-0	0.3064	0.340	0.324861	0.3065	0.324	0.3964	0.009
1	0.2757	0.300	0.289297	0.2830	0.300	0.3532	0.010
2	0.2604	0.284	0.257627	0.2625	0.276	0.3147	0.011
3	0.2451	0.259	0.229423	0.2437	0.252	0.2804	0.012
4	0.2298	0.238	0.204307	0.2253	0.232	0.2500	0.013
5	0.2145	0.220	0.181940	0.2070	0.212	0.2225	0.014
6	0.1991	0.203	0.162023	0.1920	0.192	0.1981	0.016
7	0.1838	0.180	0.144285	0.1770	0.176	0.1764	0.018
8	0.1685	0.165	0.128490	0.1620	0.160	0.1570	0.020
9	0.1532	0.148	0.114423	0.1483	0.144	0.1398	0.022
10	0.1379	0.134	0.101897	0.1350	0.128	0.1250	0.024
11	0.1225	0.120	0.090742	0.1205	0.116	0.1113	0.026
12	0.1072	0.109	0.080808	0.1055	0.104	0.0991	0.029
13	0.0919	0.095	0.071962	0.0915	0.092	0.0882	0.031
14	0.0766	0.083	0.064084	0.0800	0.080	0.0785	0.033
15	0.0689	0.072	0.057068	0.0720	0.072	0.0699	0.035
16	0.0613	0.065	0.050821	0.0625	0.064	0.0625	0.037
17	0.0551	0.058	0.045257	0.0540	0.056	0.0556	0.039
18	0.0490	0.049	0.040303	0.0475	0.048	0.0495	0.041
19	0.0429	0.042	0.035890	0.0410	0.040	0.0440	0.043
20	0.0368	0.035	0.031961	0.0348	0.036	0.0392	0.045
21	0.0337	0.032	0.028462	0.03175	0.032	0.0349	0.047
22	0.0306	0.028	0.025346	0.0286	0.028	0.03125	0.049
23	0.0276	0.025	0.022572	0.0258	0.024	0.02782	0.051
24	0.0245	0.022	0.020101	0.0230	0.022	0.02476	0.055
25	0.0214	0.020	0.017900	0.0204	0.020	0.02204	0.059
26	0.0184	0.018	0.015941	0.0181	0.018	0.01961	0.063
27	0.0169	0.016	0.014195	0.0173	0.0164	0.01745	0.067
28	0.0153	0.014	0.012641	0.0162	0.0148	0.015625	0.071
29	0.0138	0.013	0.011257	0.0150	0.0136	0.0139	0.075
30	0.0123	0.012	0.010025	0.0140	0.0124	0.0123	0.080
31	0.0107	0.010	0.008928	0.0132	0.0116	0.0110	0.085
32	0.0100	0.009	0.007950	0.0128	0.0108	0.0098	0.090
33	0.0092	0.008	0.007080	0.0118	0.0100	0.0087	0.095
34	0.0084	0.007	0.006305	0.0104	0.0092	0.0077	0.100
35	0.0077	0.005	0.005615	0.0095	0.0084	0.0069	0.106
36	0.0069	0.004	0.005000	0.0090	0.0076	0.0061	0.112
37	0.0065	.....	0.004453	0.0085	0.0068	0.0054	0.118
38	0.0061	.....	0.003965	0.0080	0.0060	0.0048	0.124
39	0.0057	.....	0.003531	0.0075	0.0052	0.0043	0.130
40	0.0054	.....	0.003144	0.0070	0.0048	0.00386	0.138

TABLE 6-16. Color Code for Marking Steel Bars

	Aluminum	Black	Blue	Bronze	Brown	Green	Orange	Pink	Red	White	Yellow
Aluminum	1095		3240 3245	9255 9260	7260	4340 4345	3435		2350	6125	
Black		5120	3115 3120	3450	52100	4615 4620	3312 3325		2515	5140 5150	X1335 X1340
Blue	3240 3245	3115 3120	1030 1035	3250	3150	3130 3135	3335 3340	3415	2315 2320	3140 X3140 3145	X1314
Bronze	9255 9260	3450	3250	1050	71660	X4130			2115	6140	
Brown	7260	52100	3150	71660	X1020 1020	4140 4150	71360		2015	6115 6120	1120
Green	4340 4345	4615 4620	3130 3135	X4130	4140 4150	1040 X1040	T1330 T1340 T1335	4640	2340 2345	4130	4135
Orange	3435	3312 3325	3335 3340		71360	T1340 T1335 T1330	1045 X1045		T1345 T1350	6145 6150	
Pink			3415			4640		3125			
Purple			3215 3220 3230			4315 4320				6195	
Red	2350	2515	2315 2320	2115	2015	2340 2345	T1345 T1350		1025 X1025	2330 2335	X1315
White	6125	5140 5150	3140 X3140 3145	6140	6115 6120	4130	6145 6150		2330 2335	1010 1015 X1015	6130 6135
Yellow		X1335 X1340	X1314		1120	4135			X1315	6130 6135	1112 X1112

TABLE 6-17. Distance across Corners

Square = distance across flats $\times$ 1.4142
Hexagon = distance across flats $\times$ 1.1547
Octagon = distance across flats $\times$ 1.0824
Decagon = distance across flats $\times$ 1.0515
Dodecagon = distance across flats $\times$ 1.0353

TABLE 6-18. Distance across Flats

Square = distance across corners × 0.70711
Hexagon = distance across corners × 0.86603
Octagon = distance across corners × 0.92388
Decagon = distance across corners × 0.95106
Dodecagon = distance across corners × 0.96593

TABLE 6-19. Proper Machine to Use

Timewise

$$\frac{\text{Setup of more efficient machine} - \text{setup of less efficient machine}}{\text{Per-piece time on less efficient machine} - \text{per-piece time on more efficient machine}}$$
= No. pieces that can be done at the same time on either machine

Example

Less efficient machine	More efficient machine
Setup time.................. 60.0 min	Setup time.................. 90.0 min
Per-piece time.............. 20.0 min	Per-piece time.............. 10.0 min

$$\frac{90 - 60}{20 - 10} = 3 \text{ pieces same time each machine}$$

Costwise

$$\frac{\text{Cost more efficient machine} \times \text{setup time more efficient machine}}{\text{Cost} \times \text{time per piece on less efficient machine}} - \frac{\text{cost less efficient machine} \times \text{setup time less efficient machine}}{\text{cost} \times \text{time per piece on more efficient machine}}$$
= No. pieces that can be done for the same cost on either machine

Example

Less efficient machine	More efficient machine
Hourly rate............... $1.80	Hourly rate................ $1.60
Setup time............... 60.0 min	Setup time................ 90.0 min
Per-piece time............ 20.0 min	Per-piece time............ 10.0 min

$$\frac{(160 \times 90) - (180 \times 60)}{(180 \times 20) - (160 \times 10)} = \frac{14{,}400 - 10{,}800}{3{,}600 - 1{,}600} = \frac{3{,}600}{2{,}000}$$
= 1.8 pieces done at the same cost on either machine

TABLE 6-20. Approximate Hourly Production*

Seconds to make one piece	Gross production per hour	Gross hours per 1,000 pieces	Seconds to make one piece	Gross production per hour	Gross hours per 1,000 pieces	Seconds to make one piece	Gross production per hour	Gross hours per 1,000 pieces
½	7,200	0.14	12½	288	3.47	50	72	13.9
⅝	5,760	0.17	13	276	3.62	52	69	14.5
¾	4,800	0.21	13½	267	3.76	54	66	15.0
⅞	4,114	0.24	14	257	3.89	56	64	15.6
1	3,600	0.28	14½	248	4.03	58	62	16.1
1¼	2,880	0.35	15	240	4.17	60	60	16.7
1½	2,400	0.42	15½	232	4.31	62	58	17.2
1¾	2,057	0.49	16	225	4.44	64	56	17.8
2	1,800	0.55	16½	218	4.58	66	54	18.4
2¼	1,600	0.62	17	212	4.72	68	53	18.9
2½	1,440	0.69	17½	206	4.86	70	51	19.5
2¾	1,309	0.76	18	200	5.00	72	50	20.0
3	1,200	0.83	18½	195	5.14	74	49	20.6
3¼	1,107	0.90	19	189	5.28	76	47	21.1
3½	1,028	0.97	19½	185	5.42	78	46	21.7
3¾	960	1.04	20	180	5.56	80	45	22.2
4	900	1.11	21	171	5.83	82	44	22.8
4¼	847	1.18	22	164	6.12	84	43	23.3
4½	800	1.25	23	156	6.40	86	42	23.9
4¾	757	1.32	24	150	6,67	88	41	24.5
5	720	1.39	25	144	6.95	90	40	25.0
5¼	686	1.46	26	138	7.22	92	39	25.5
5½	654	1.53	27	133	7.50	94	38	26.1
5¾	626	1.60	28	128	7.78	96	37	26.7
6	600	1.67	29	124	8.06	98	37	27.2
6¼	576	1.73	30	120	8.33	100	36	27.8
6½	553	1.81	31	116	8.62	105	34	29.2
6¾	533	1.87	32	112	8.90	110	33	30.6
7	514	1.94	33	109	9.17	115	31	32.0
7¼	497	2.01	34	106	9.45	120	30	33.3
7½	480	2.08	35	103	9.73	125	29	34.7
7¾	465	2.15	36	100	10.00	130	28	36.1
8	450	2.22	37	97	10.30	135	27	37.5
8¼	436	2.29	38	95	10.56	140	26	38.9
8½	423	2.36	39	92	10.83	145	25	40.3
8¾	411	2.43	40	90	11.11	150	24	41.6
9	400	2.50	41	88	11.39	155	23	43.1
9¼	389	2.57	42	86	11.67	160	22	44.4
9½	379	2.64	43	84	11.94	165	22	45.8
9¾	369	2.71	44	82	12.22	170	21	47.2
10	360	2.78	45	80	12.50	175	21	48.6
10½	342	2.92	46	78	12.78	180	20	50.0
11	327	3.05	47	77	13.05	185	20	51.4
11½	313	3.19	48	75	13.34	190	19	52.8
12	300	3.33	49	73	13.61	195	18	54.2

*Data through the courtesy of *Steel Handbook*, No. 42, The Union Drawn Steel Co.

TABLE 6-21. Diameters of Number and Letter Drills

Number Drills						Letter Drills	
Drill No.	Diam, in.	Drill No.	Diam, in.	Drill No.	Diam, in.	Drill letter	Diam, in.
80	0.0135	52	0.0635	24	0.1520	A	0.2340
79	0.0145	51	0.0670	23	0.1540	B	0.2380
78	0.0160	50	0.0700	22	0.1570	C	0.2420
77	0.0180	49	0.0730	21	0.1590	D	0.2460
76	0.0200	48	0.0760	20	0.1610	E	0.2500
75	0.0210	47	0.0785	19	0.1660	F	0.2570
74	0.0225	46	0.0810	18	0.1695	G	0.2610
73	0.0240	45	0.0820	17	0.1730	H	0.2660
72	0.0250	44	0.0860	16	0.1770	I	0.2720
71	0.0260	43	0.0890	15	0.1800	J	0.2770
70	0.0280	42	0.0935	14	0.1820	K	0.2810
69	0.0292	41	0.0960	13	0.1850	L	0.2900
68	0.0310	40	0.0980	12	0.1890	M	0.2950
67	0.0320	39	0.0995	11	0.1910	N	0.3020
66	0.0330	38	0.1015	10	0.1935	O	0.3160
65	0.0350	37	0.1040	9	0.1960	P	0.3230
64	0.0360	36	0.1065	8	0.1990	Q	0.3320
63	0.0370	35	0.1100	7	0.2010	R	0.3390
62	0.0380	34	0.1110	6	0.2040	S	0.3480
61	0.0390	33	0.1130	5	0.2055	T	0.3580
60	0.0400	32	0.1160	4	0.2090	U	0.3680
59	0.0410	31	0.1200	3	0.2130	V	0.3770
58	0.0420	30	0.1285	2	0.2210	W	0.3860
57	0.0430	29	0.1360	1	0.2280	X	0.3970
56	0.0465	28	0.1405			Y	0.4040
55	0.0520	27	0.1440			Z	0.4130
54	0.0550	26	0.1470				
53	0.0595	25	0.1495				

TABLE 6-22. Surface Feet per Minute for Drilling Using High-speed Drills

Material	Hardness		Surface feet
	Brinell	Rockwell	
Allegheny metal........................	146–149	B78–80	50
Aluminum..............................	99–101	B54–56	200–250
Brass.................................			200
Bronze, common........................	166–183	B85–89	200–250
Bronze, phosphor, half hard.............	187–202	B90–94	175–180
Cast iron, soft.........................	126	B70	140–150
Cast iron, medium......................	196	B93	80–110
Cast iron, hard........................	293–302	C32–33	45–50
Cast steel.............................	286–302	C30–33	40–50
Copper................................	80–85	B40–44	70
Duralumin.............................	90–104	B48–58	200
Everdure..............................	179–207	B88–95	60
Machine steel..........................	170–196	B86–93	110
Manganese copper 30%..................	134	B74	15
Malleable iron.........................	112–126	B66–70	85–90
Mild steel 0.2–0.3 carbon...............	170–202	B86–93	110–120
Molybdenum steel......................	196–235	B92–99	55
Monel metal...........................	149–170	B80–86	50
Nickel steel 3½%.......................	196–241	B93–100	60
Permalloy 77% nickel...................	131–163	B72–84	50
Spring steel............................	402	C43	20
Stainless steel.........................	146–149	B78–80	50
Steel 0.4–0.5 carbon....................	170–196	B86–93	80
Tool steel.............................	149	B80	75

TABLE 6-23. Feeds for Drilling

Drill No.	Feed, inches per revolution	Drill diameter	Feed, inches per revolution
60	0.0005	$\frac{1}{16}$	0.0015
55	0.0010	$\frac{1}{8}$	0.0025
50	0.0015	$\frac{3}{16}$	0.004
45	0.0020	$\frac{1}{4}$	0.005
40	0.0020	$\frac{5}{16}$	0.005
35	0.0025	$\frac{3}{8}$	0.006
30	0.0030	$\frac{7}{16}$	0.007
25	0.0030	$\frac{1}{2}$	0.008
20	0.0035	$\frac{9}{16}$	0.008
15	0.0035	$\frac{5}{8}$	0.009
10	0.0040	$\frac{11}{16}$	0.009
5	0.0040	$\frac{3}{4}$	0.010
1	0.0040	$\frac{13}{16}$	0.010
		$\frac{7}{8}$	0.011
		$\frac{15}{16}$	0.012
		1	0.013
		$1\frac{1}{16}$	0.013
		$1\frac{1}{8}$	0.014
		$1\frac{3}{16}$	0.014
		$1\frac{1}{4}$	0.015
		$1\frac{5}{16}$	0.015
		$1\frac{3}{8}$	0.015
		$1\frac{7}{16}$	0.015
		$1\frac{1}{2}$	0.015
		$1\frac{9}{16}$	0.016
		$1\frac{5}{8}$	0.016
		$1\frac{11}{16}$	0.016
		$1\frac{3}{4}$	0.016
		$1\frac{13}{16}$	0.016
		$1\frac{7}{8}$	0.018
		$1\frac{15}{16}$	0.018
		2	0.018

TABLE 6-24. Reduction in Drill Feed and Speed for Successive Increases in Hole Depth

Depth of hole, in multiples of drill diameter	Reduction of speed, percent	Reduction of feed, percent
3	10	10
4	20	10
5	30	20
6	35–40	20
7	35–40	20
8	35–40	20

TABLE 6-25. Conservative Reaming Speeds

Material	Surface feet per minute	Lubricant
Aluminum	225–250	Lard oil, or $\frac{1}{2}$ lard oil and $\frac{1}{2}$ kerosene
Brass	150–160	Soluble oil
Cast iron	65–75	Dry*
Mild steels	70–75	Lard oil
Machinery steels	70–75	Lard oil
Steel, 40–50 carbon	55–60	Soluble oil
Malleable iron	55–60	Soluble oil
Tool steel, 120 carbon	30–40	Light sulfur-base oil
Nickel steel	30–40	Light sulfur-base oil
Molybdenum	30–40	Light sulfur-base oil
Drop forgings	30–40	Light sulfur-base oil
Stainless steel	30–35	Sulfur-base oil
Monel metal	30–35	Sulfur-base oil
Nickel-chromium alloys	30–35	Sulfur-base oil

* When taper reamers are used in cast iron, lard or soluble oils are sometimes used.

TABLE 6-26. Tap Drill Sizes for U.S. Standard Threads*

Thread diameter	No. of threads	Tap-drill diameter	Thread diameter	No. of threads	Tap-drill diameter	Thread diameter	No. of threads	Tap-drill diameter
$\frac{1}{4}$	20	$\frac{13}{64}$	$\frac{13}{16}$	10	$\frac{45}{64}$	$1\frac{3}{8}$	6	$1\frac{7}{32}$
$\frac{5}{16}$	18	$\frac{1}{4}$	$\frac{7}{8}$	9	$\frac{49}{64}$	$1\frac{1}{2}$	6	$1\frac{11}{32}$
$\frac{3}{8}$	16	$\frac{5}{16}$	$\frac{15}{16}$	9	$\frac{53}{64}$	$1\frac{5}{8}$	$5\frac{1}{2}$	$1\frac{7}{16}$
$\frac{7}{16}$	14	$\frac{23}{64}$	1	8	$\frac{7}{8}$	$1\frac{3}{4}$	5	$1\frac{35}{64}$
$\frac{1}{2}$	13	$\frac{27}{64}$	$1\frac{1}{16}$	8	$\frac{15}{16}$	$1\frac{7}{8}$	5	$1\frac{43}{64}$
$\frac{9}{16}$	12	$\frac{15}{32}$	$1\frac{1}{8}$	7	$\frac{63}{64}$	2	$4\frac{1}{2}$	$1\frac{25}{32}$
$\frac{5}{8}$	11	$\frac{17}{32}$	$1\frac{3}{16}$	7	$1\frac{3}{64}$	$2\frac{1}{4}$	$4\frac{1}{2}$	$2\frac{1}{32}$
$1\frac{1}{16}$	11	$1\frac{9}{32}$	$1\frac{1}{4}$	7	$1\frac{7}{64}$	$2\frac{1}{2}$	4	$2\frac{1}{4}$
$\frac{3}{4}$	10	$\frac{41}{64}$	$1\frac{5}{16}$	7	$1\frac{11}{64}$	$2\frac{3}{4}$	4	$2\frac{1}{2}$

* These tap drill diameters allow approximately 75 percent of a full thread.

TABLE 6-27. Tap Drill Sizes for SAE Standard Threads

Diameter of tap	No. of threads	Diameter of tap drill	Diameter of tap	No. of threads	Diameter of tap drill	Diameter of tap	No. of threads	Diameter of tap drill
$\frac{1}{4}$	28	0.213	$\frac{9}{16}$	18	0.500	1	14	0.921
$\frac{5}{16}$	24	0.272	$\frac{5}{8}$	18	0.562	$1\frac{1}{8}$	12	1.031
$\frac{3}{8}$	24	0.332	$\frac{11}{16}$	16	0.625	$1\frac{1}{4}$	12	1.156
$\frac{7}{16}$	20	0.386	$\frac{3}{4}$	16	0.687	$1\frac{3}{8}$	12	1.281
$\frac{1}{2}$	20	0.437	$\frac{7}{8}$	14	0.796	$1\frac{1}{2}$	12	1.406

TABLE 6-28. Tap Drills for Pipe Taps

Size of tap	Drills for Briggs pipe taps	Drills for Whitworth pipe taps	Size of tap	Drills for Briggs pipe taps	Drills for Whitworth pipe taps	Size of tap	Drills for Briggs pipe taps	Drills for Whitworth pipe taps
$\frac{1}{8}$	$\frac{11}{32}$	$\frac{5}{16}$	$1\frac{1}{4}$	$1\frac{1}{2}$	$1\frac{15}{32}$	$3\frac{1}{4}$	....	$3\frac{1}{2}$
$\frac{1}{4}$	$\frac{7}{16}$	$\frac{27}{64}$	$1\frac{1}{2}$	$1\frac{23}{32}$	$1\frac{25}{32}$	$3\frac{1}{2}$	$3\frac{13}{16}$	$3\frac{3}{4}$
$\frac{3}{8}$	$1\frac{9}{32}$	$\frac{9}{16}$	$1\frac{3}{4}$	......	$1\frac{15}{16}$	$3\frac{3}{4}$	......	4
$\frac{1}{2}$	$2\frac{3}{32}$	$1\frac{1}{16}$	2	$2\frac{3}{16}$	$2\frac{5}{32}$	4	$4\frac{1}{4}$	$4\frac{1}{4}$
$\frac{5}{8}$	......	$2\frac{5}{32}$	$2\frac{1}{4}$	......	$2\frac{13}{32}$	$4\frac{1}{2}$	$4\frac{3}{4}$	$4\frac{3}{4}$
$\frac{3}{4}$	$1\frac{5}{16}$	$2\frac{9}{32}$	$2\frac{1}{2}$	$2\frac{11}{16}$	$2\frac{25}{32}$	5	$5\frac{5}{16}$	$5\frac{1}{4}$
$\frac{7}{8}$	......	$1\frac{1}{16}$	$2\frac{3}{4}$	......	$3\frac{1}{32}$	$5\frac{1}{2}$	......	$5\frac{3}{4}$
1	$1\frac{5}{32}$	$1\frac{1}{8}$	3	$3\frac{5}{16}$	$3\frac{9}{32}$	6	$6\frac{3}{8}$	$6\frac{1}{4}$

TABLE 6-29. Tapping Speeds and Lubricants

Material	Suggested tapping speed	Suggested lubricant
Aluminum.................	95–105	Kerosene and lard oil
Brass....................	95–105	Soluble or light-base oil
Bronze, soft.............	55– 65	Soluble or light-base oil
Bronze, hard.............	40– 50	Light-base oil
Copper..................	50– 60	Light-base oil
Die castings.............	65– 75	Kerosene and lard oil or soluble
Duralumin...............	95–105	Kerosene and lard oil or soluble
Iron, cast................	75– 85	Dry or soluble oil
Iron, malleable...........	50– 65	Sulfur-base oil or soluble
Magnesium...............	95–105	Dry
Monel metal..............	25– 30	Sulfur-base oil or kerosene and lard oil
Plastics..................	70– 80	Dry
Steel, cold-rolled.........	50– 65	Soluble or sulfur-base oil
Steel, alloy grades........	25– 35	Sulfur-base oil
Steel, cast...............	25– 35	Sulfur-base oil
Steel, tool...............	25– 35	Sulfur-base or kerosene and lard oil
Stainless steel............	20– 30	Sulfur-base oil

TABLE 6-30. Cutting Speeds in Feet per Minute—for Planing

Material	For high-speed steel tools				For cast-alloy tools				For carbide tools			
	Depth of cut, inches											
	$\frac{1}{8}$	$\frac{1}{4}$	$\frac{1}{2}$	1	$\frac{1}{8}$	$\frac{1}{4}$	$\frac{1}{2}$	1	$\frac{1}{16}$	$\frac{3}{16}$	$\frac{3}{8}$	$\frac{3}{4}$
	Feed, inches											
	$\frac{1}{32}$	$\frac{1}{16}$	$\frac{3}{32}$	$\frac{1}{8}$	$\frac{1}{32}$	$\frac{1}{16}$	$\frac{3}{32}$	$\frac{1}{8}$	$\frac{1}{32}$	$\frac{1}{32}$	$\frac{1}{16}$	$\frac{1}{16}$
Cast iron, soft...........	95	75	60	50	160	135	110	95	300	240	195	165
Cast iron, medium.......	70	55	45	35	125	105	90	75	240	195	160	130
Cast iron, hard...........	45	35	25	..	95	80	65	..	165	130	105	
Steel, free-cutting........	90	70	55	40	140	105	85	65	350	270	210	155
Steel, average...........	70	55	40	30	105	80	60	45	300	225	175	130
Steel, low machinability..	40	30	25	..	65	50	40	..	215	160	125	
Bronze...............	150	150	125	..	*	*	*	*	*	*	*	*
Aluminum.............	200	200	150	..	*	*	*	*	*	*	*	*

* Maximum table speed.

TABLE 6-31. Planer Formulas

$$\frac{\text{Planer ratio}}{\text{Cutting feet}} = \text{minutes per foot}$$

$$\frac{\text{Minutes per foot}}{12} = \text{minutes per inch}$$

$$2\!:\!1 \text{ ratio} = \tfrac{1}{2} = 0.5 + 1 = 1.5$$

Example: $\dfrac{1.5}{20} = 0.075 \div 12 = 0.00625$ minutes per inch

$$3\!:\!1 \text{ ratio} = \tfrac{1}{3} = 0.333 + 1 = 1.333$$

Example: $\dfrac{1.333}{20} = 0.0666 \div 12 = 0.00555$ minutes per inch

$$4\!:\!1 \text{ ratio} = \tfrac{1}{4} = 0.250 + 1 = 1.250$$

Example: $\dfrac{1.250}{20} = 0.0625 \div 12 = 0.00521$ minutes per inch

TABLE 6-32. Suggested Feed per Tooth for High-speed Steel Milling Cutters*

Material	Face mills	Helical mills	Slotting and side mills	End mills	Form-relieved cutters	Circular saws
Plastics	0.013	0.010	0.008	0.007	0.004	0.003
Magnesium and alloys	0.022	0.018	0.013	0.011	0.007	0.005
Aluminum and alloys	0.022	0.018	0.013	0.011	0.007	0.005
Free-cutting brasses and bronzes	0.022	0.018	0.013	0.011	0.007	0.005
Medium brasses and bronzes	0.014	0.011	0.008	0.007	0.004	0.003
Hard brasses and bronzes	0.009	0.007	0.006	0.005	0.003	0.002
Copper	0.012	0.010	0.007	0.006	0.004	0.003
Cast iron, soft (150–180 Brinell hardness No.)	0.016	0.013	0.009	0.008	0.005	0.004
Cast iron, medium (180–200 Brinell hardness No.)	0.013	0.010	0.007	0.007	0.004	0.003
Cast iron, hard (220–300 Brinell hardness No.)	0.011	0.008	0.006	0.006	0.003	0.003
Malleable iron	0.012	0.010	0.007	0.006	0.004	0.003
Cast steel	0.012	0.010	0.007	0.006	0.004	0.003
Low-carbon steel, free-machining	0.012	0.010	0.007	0.006	0.004	0.003
Low-carbon steel	0.010	0.008	0.006	0.005	0.003	0.003
Medium-carbon steel	0.010	0.008	0.006	0.005	0.003	0.003
Alloy steel, annealed (180–200 Brinell hardness No.)	0.008	0.007	0.005	0.004	0.003	0.002
Alloy steel, tough (220–300 Brinell hardness No.)	0.006	0.005	0.004	0.003	0.002	0.002
Alloy steel, hard (300–400 Brinell hardness No.)	0.004	0.003	0.003	0.002	0.002	0.001
Stainless steels, free-machining	0.010	0.008	0.006	0.005	0.003	0.002
Stainless steels	0.006	0.005	0.004	0.003	0.002	0.002
Monel metals	0.008	0.007	0.005	0.004	0.003	0.002

* Data through the courtesy of The Cincinnati Milling Machine Co.

TABLE 6-33. Milling Feed Factor

Milling feeds can be determined by multiplying the factor as given in this table by surface feet per minute times the feed per tooth. As an example, assume a 10-inch-diameter cutter with 30 teeth. The tooth load is 0.003 inch, and the speed is 80 surface feet per minute.

Sample calculation: Factor from table (10-inch-diameter cutter with 30 teeth) = 11.45

Feed (inches per minutes) = 11.45 × 0.003 × 80 = 2.75

$$\text{Minutes per inch} = \frac{1}{2.75} = 0.36$$

In cases where cutter diameter or number of teeth is beyond the range of the table, the factor can be calculated as follows:

$$\text{Factor} = 3.817 \times \frac{\text{number of teeth}}{\text{cutter diameter}}$$

Cutter diameter	Number of teeth per cutter																										
	2	3	4	5	6	8	10	11	12	13	14	15	16	17	18	19	20	21	22	23	24	25	26	27	28	29	30
	Factor (revolutions per minute at 1 surface foot per minute × number of teeth)																										
¼	30.53	45.80	61.07	76.34	91.60																						
⅜	20.36	30.53	40.71	50.89	61.07	81.42																					
½	15.27	22.90	30.53	38.17	45.80	61.07	76.34	83.97	91.60	99.24																	
⅝	12.21	18.32	24.43	30.53	36.64	48.85	61.07	67.18	73.28	79.39	85.50																
¾	10.18	15.27	20.36	25.45	30.53	40.71	50.89	55.98	61.07	66.16	71.25	76.34															
⅞	8.72	13.09	17.45	21.81	26.17	34.90	43.62	47.98	52.34	56.71	61.07	65.43															
1	7.63	11.45	15.27	19.08	22.90	30.53	38.17	41.98	45.80	49.62	53.44	57.25	61.07	64.89	68.70	72.52	76.34	80.15	83.97	87.79	91.60	95.42	99.24	103.05	106.87	110.69	114.50
1¼	6.11	9.16	12.21	15.27	18.32	24.43	30.53	33.59	36.64	39.69	42.75	45.80	48.85	51.91	54.96	58.02	61.07	64.12	67.18	70.23	73.28	76.34	79.39	82.44	85.50	88.55	91.60
1½	5.09	7.63	10.18	12.72	15.27	20.36	25.45	27.99	30.53	33.08	35.62	38.17	40.71	43.26	45.80	48.35	50.89	53.44	55.98	58.52	61.07	63.61	66.16	68.70	71.25	73.79	76.34
1¾	4.36	6.54	8.72	10.91	13.09	17.45	21.81	23.99	26.17	28.35	30.53	32.72	34.90	37.08	39.26	41.44	43.62	45.80	47.98	50.16	52.34	54.53	56.71	58.89	61.07	63.25	65.43
2	3.82	5.73	7.63	9.54	11.45	15.27	19.08	20.99	22.90	24.81	26.72	28.63	30.53	32.44	34.35	36.26	38.17	40.08	41.98	43.89	45.80	47.71	49.62	51.53	53.44	55.34	57.25
2½	3.05	4.58	6.11	7.63	9.16	12.21	15.27	16.79	18.32	19.85	21.37	22.90	24.43	25.95	27.48	29.01	30.53	32.06	33.59	35.11	36.64	38.17	39.69	41.22	42.75	44.27	45.80
3	2.54	3.82	5.09	6.36	7.63	10.18	12.72	13.99	15.27	16.54	17.81	19.08	20.36	21.63	22.90	24.17	25.45	26.72	27.99	29.26	30.53	31.81	33.08	34.35	35.62	36.90	38.17
3½	2.18	3.27	4.36	5.45	6.54	8.72	10.91	12.00	13.09	14.18	15.27	16.36	17.45	18.54	19.63	20.72	21.81	22.90	23.99	25.08	26.17	27.26	28.35	29.44	30.53	31.62	32.72
4	1.91	2.86	3.82	4.77	5.73	7.63	9.54	10.50	11.45	12.40	13.36	14.31	15.27	16.22	17.18	18.13	19.08	20.04	20.99	21.95	22.90	23.85	24.81	25.76	26.72	27.67	28.63
4½	1.70	2.54	3.39	4.24	5.09	6.79	8.48	9.33	10.18	11.02	11.87	12.72	13.57	14.42	15.27	16.12	16.96	17.81	18.66	19.51	20.36	21.20	22.05	22.90	23.75	24.60	25.45
5	1.53	2.29	3.05	3.82	4.58	6.11	7.63	8.40	9.16	9.92	10.69	11.45	12.21	12.98	13.74	14.50	15.27	16.03	16.79	17.56	18.32	19.08	19.85	20.61	21.37	22.14	22.90
5½	1.39	2.08	2.78	3.47	4.16	5.55	6.94	7.63	8.33	9.02	9.72	10.41	11.10	11.80	12.49	13.19	13.88	14.57	15.27	15.96	16.66	17.35	18.04	18.74	19.43	20.12	20.82
6	1.27	1.91	2.54	3.18	3.82	5.09	6.36	7.00	7.63	8.27	8.91	9.54	10.18	10.81	11.45	12.09	12.72	13.36	13.99	14.63	15.27	15.90	16.54	17.18	17.81	18.45	19.08

TABLE 6-33 (Continued)

Number of teeth per cutter

Factor (revolutions per minute at 1 surface foot per minute × number of teeth)

Cutter diameter	2	3	4	5	6	8	10	11	12	13	14	15	16	17	18	19	20	21	22	23	24	25	26	27	28	29	30
6½	1.17	1.76	2.35	2.94	3.52	4.70	5.87	6.46	7.05	7.63	8.22	8.81	9.40	9.98	10.57	11.16	11.74	12.33	12.92	13.51	14.09	14.68	15.27	15.85	16.44	17.03	17.62
7	1.09	1.64	2.18	2.73	3.27	4.36	5.45	6.00	6.54	7.09	7.63	8.18	8.72	9.27	9.81	10.36	10.91	11.45	12.00	12.54	13.09	13.63	14.18	14.72	15.27	15.81	16.36
7½	1.02	1.53	2.04	2.54	3.05	4.07	5.09	5.60	6.11	6.62	7.12	7.63	8.14	8.65	9.16	9.67	10.18	10.69	11.20	11.70	12.21	12.72	13.23	13.74	14.25	14.76	15.27
8	0.95	1.43	1.91	2.39	2.86	3.82	4.77	5.25	5.73	6.20	6.68	7.16	7.63	8.11	8.59	9.06	9.54	10.02	10.50	10.97	11.45	11.93	12.40	12.88	13.36	13.84	14.31
8½	0.90	1.35	1.80	2.25	2.69	3.59	4.49	4.94	5.38	5.84	6.29	6.74	7.18	7.63	8.08	8.53	8.98	9.43	9.88	10.33	10.78	11.23	11.67	12.12	12.57	13.03	13.47
9	0.85	1.27	1.70	2.12	2.54	3.39	4.24	4.66	5.09	5.51	5.94	6.36	6.79	7.21	7.63	8.06	8.48	8.91	9.32	9.78	10.18	10.60	11.03	11.45	11.87	12.30	12.72
10	0.76	1.15	1.53	1.91	2.29	3.05	3.82	4.20	4.58	4.96	5.34	5.73	6.11	6.49	6.87	7.25	7.63	8.02	8.40	8.78	9.16	9.54	9.92	10.31	10.69	11.07	11.45
11	0.69	1.04	1.39	1.73	2.08	2.78	3.47	3.82	4.16	4.51	4.86	5.20	5.55	5.90	6.25	6.59	6.94	7.29	7.63	7.98	8.33	8.67	9.02	9.37	9.72	10.06	10.41
12	0.64	0.95	1.27	1.59	1.91	2.54	3.18	3.50	3.82	4.13	4.45	4.77	5.09	5.41	5.73	6.04	6.36	6.68	7.00	7.32	7.63	7.95	8.27	8.59	8.91	9.22	9.54
13	0.59	0.88	1.17	1.47	1.76	2.35	2.94	3.23	3.52	3.82	4.11	4.40	4.70	4.99	5.28	5.58	5.87	6.17	6.46	6.75	7.05	7.34	7.63	7.93	8.22	8.51	8.81
14	0.55	0.82	1.09	1.36	1.64	2.18	2.73	3.00	3.27	3.54	3.82	4.09	4.36	4.63	4.91	5.18	5.45	5.73	6.00	6.27	6.54	6.82	7.09	7.36	7.63	7.91	8.18
15	0.51	0.76	1.02	1.27	1.53	2.04	2.54	2.80	3.05	3.31	3.56	3.82	4.07	4.33	4.58	4.83	5.09	5.34	5.60	5.85	6.11	6.36	6.62	6.87	7.12	7.38	7.63
16	0.48	0.72	0.95	1.19	1.43	1.91	2.39	2.62	2.86	3.10	3.34	3.58	3.82	4.06	4.29	4.53	4.77	5.01	5.25	5.49	5.73	5.96	6.20	6.44	6.68	6.92	7.16
17	0.45	0.67	0.90	1.12	1.35	1.80	2.25	2.47	2.69	2.92	3.14	3.37	3.59	3.82	4.04	4.26	4.49	4.71	4.94	5.16	5.39	5.61	5.84	6.06	6.29	6.51	6.74
18	0.42	0.64	0.85	1.06	1.27	1.70	2.12	2.33	2.54	2.76	2.97	3.18	3.39	3.60	3.82	4.03	4.24	4.45	4.66	4.88	5.09	5.30	5.51	5.73	5.94	6.15	6.36
20	0.38	0.57	0.76	0.95	1.15	1.53	1.91	2.10	2.29	2.48	2.67	2.86	3.05	3.24	3.44	3.63	3.82	4.01	4.20	4.39	4.58	4.77	4.96	5.15	5.34	5.53	5.73
22	0.35	0.52	0.69	0.87	1.04	1.39	1.73	1.91	2.08	2.26	2.43	2.60	2.78	2.95	3.12	3.30	3.47	3.64	3.82	3.99	4.16	4.34	4.51	4.68	4.86	5.03	5.20
24	0.32	0.48	0.64	0.80	0.95	1.27	1.59	1.75	1.91	2.07	2.23	2.39	2.54	2.70	2.86	3.02	3.18	3.34	3.50	3.66	3.82	3.98	4.13	4.29	4.45	4.61	4.77
26	0.29	0.44	0.59	0.73	0.88	1.17	1.47	1.61	1.76	1.91	2.06	2.20	2.35	2.50	2.64	2.79	2.94	3.08	3.23	3.38	3.52	3.67	3.82	3.96	4.11	4.26	4.40
28	0.27	0.41	0.55	0.68	0.82	1.09	1.36	1.50	1.64	1.77	1.91	2.04	2.18	2.32	2.45	2.59	2.73	2.86	3.00	3.14	3.27	3.41	3.54	3.68	3.82	3.95	4.09
30	0.25	0.38	0.51	0.64	0.76	1.02	1.27	1.40	1.53	1.65	1.78	1.91	2.03	2.16	2.29	2.42	2.54	2.67	2.80	2.93	3.05	3.18	3.31	3.44	3.56	3.69	3.82

TABLE 6-34. Conversion Table for Surface Feet and Revolutions per Minute for Given Diameters

The range of this table is from 1 to 900 surface feet per minute accurate to 1 revolution per minute. The table is direct reading in the range from 10 to 90 surface feet per minute. 1 to 9 surface feet and 100 to 900 surface feet can be read by moving the decimal as required in the appropriate column.

Example: For a 1-inch diameter—When desired surface feet per minute = 2, revolutions per minute = 7.63
When desired surface feet per minute = 20, revolutions per minute = 76.3
When desired surface feet per minute = 200, revolutions per minute = 763.0

Diameter, inches	\[Surface feet per minute\] 10	11	12	13	14	15	16	17	18	19	20	21	22	23	24	25	26	28	30	35	40	45	50	60	70	80	90
											\[Revolutions per minute\]																
¼	152.7	167.9	183.2	198.5	213.7	229.0	244.3	259.5	274.8	290.1	305.3	320.8	335.9	351.1	366.4	381.7	396.9	427.5	458.0	534.4	610.7	687.0	763.4	916.0	1068.7	1221.4	1374.0
⅜	101.8	112.0	122.2	132.3	142.5	152.7	162.9	173.0	183.2	193.4	203.6	213.8	223.9	234.1	244.3	254.5	264.7	285.0	305.3	356.3	407.2	458.1	509.0	610.8	712.6	814.4	916.1
½	76.3	84.0	91.6	99.2	106.9	114.5	122.1	129.8	137.4	145.0	152.7	160.3	167.9	175.6	183.2	190.8	198.5	213.7	229.0	267.0	305.3	343.5	381.7	458.0	534.4	610.7	687.0
⅝	61.1	67.1	73.3	79.4	85.5	91.6	97.7	103.8	109.9	116.0	122.1	128.2	134.4	140.5	146.6	152.7	158.8	171.0	183.2	213.7	244.3	274.8	305.3	366.4	427.5	488.5	549.6
¾	50.9	56.0	61.1	66.1	71.2	76.3	81.4	86.5	91.6	96.7	101.8	106.8	112.0	117.0	122.1	127.2	132.3	142.5	152.6	178.1	203.5	229.0	254.4	305.3	356.1	407.0	457.9
⅞	43.6	48.0	52.4	56.7	61.1	65.4	69.8	74.2	78.5	82.9	87.3	91.6	96.0	100.3	104.7	109.1	113.4	122.2	130.9	152.7	174.1	196.3	218.1	261.8	305.4	349.0	392.6
1	38.2	42.0	45.8	49.6	53.4	57.3	61.1	64.9	68.7	72.5	76.3	80.2	84.0	87.8	91.6	95.4	99.2	106.9	114.5	133.6	152.7	171.8	190.8	229.0	267.1	305.3	343.5
1¼	30.5	33.6	36.6	39.7	42.7	45.8	48.9	51.9	55.0	58.0	61.1	64.1	67.2	70.2	73.3	76.3	79.4	85.5	91.6	106.9	122.1	137.4	152.7	183.2	213.7	244.3	274.8
1½	25.4	28.0	30.5	33.1	35.6	38.2	40.7	43.3	45.8	48.4	50.9	53.5	56.0	58.6	61.1	63.6	66.2	71.3	76.4	89.1	101.8	114.6	127.3	152.7	178.2	203.7	229.1
1¾	21.8	24.0	26.2	28.3	30.5	32.7	34.9	37.0	39.2	41.4	43.6	45.8	47.9	50.1	52.3	54.5	56.7	61.0	65.5	76.3	87.2	98.1	109.0	130.8	152.6	174.4	196.1
2	19.1	21.0	22.9	24.8	26.7	28.6	30.5	32.4	34.4	36.3	38.2	40.1	41.9	43.9	45.8	47.7	49.6	53.4	57.3	66.8	76.3	85.9	95.4	114.5	133.6	152.7	171.8
2¼	16.9	18.6	20.3	22.0	23.7	25.4	27.1	28.8	30.5	32.2	33.9	35.6	37.3	39.0	40.7	42.4	44.1	47.5	50.8	59.3	67.8	76.3	84.7	101.7	118.6	135.6	152.5
2½	15.3	16.8	18.3	19.8	21.4	22.9	24.4	26.0	27.5	29.0	30.5	32.1	33.6	35.1	36.6	38.2	39.7	42.7	45.8	53.4	61.1	68.7	76.3	91.6	106.9	122.1	137.4
2¾	13.9	15.3	16.7	18.0	19.4	20.8	22.2	23.6	25.0	26.4	27.8	29.1	30.5	31.9	33.3	34.7	36.1	38.9	41.6	48.6	55.5	62.5	69.4	83.3	97.1	111.1	124.9
3	12.7	14.0	15.3	16.5	17.8	19.1	20.4	21.6	22.9	24.2	25.4	26.7	28.0	29.3	30.5	31.8	33.0	35.6	38.2	44.5	50.9	57.3	63.6	76.3	89.0	101.9	114.5
3½	10.9	12.0	13.1	14.2	15.3	16.4	17.5	18.6	19.6	20.7	21.8	22.9	24.0	25.1	26.2	27.3	28.4	30.6	32.7	38.2	43.7	49.1	54.6	65.5	76.4	87.3	98.2
4	9.5	10.5	11.5	12.4	13.4	14.3	15.3	16.2	17.2	18.1	19.1	20.0	21.0	21.9	22.9	23.9	24.8	26.7	28.6	33.4	38.2	42.9	47.7	57.3	66.8	76.3	85.9
4½	8.5	9.3	10.2	11.0	11.9	12.7	13.6	14.4	15.3	16.1	17.0	17.8	18.6	19.5	20.3	21.2	22.1	23.7	25.4	29.7	34.0	38.2	42.4	50.9	59.3	67.9	76.3
5	7.6	8.4	9.2	9.9	10.7	11.5	12.2	13.0	13.7	14.5	15.3	16.0	16.8	17.6	18.3	19.1	19.8	21.4	22.9	26.7	30.5	34.4	38.2	45.8	53.4	61.1	68.7
5½	6.9	7.6	8.3	9.0	9.7	10.4	11.1	11.8	12.5	13.2	13.9	14.6	15.3	16.0	16.7	17.4	18.1	19.5	20.8	24.3	27.8	31.3	34.7	41.7	48.6	55.6	62.5
6	6.4	7.0	7.6	8.3	8.9	9.5	10.2	10.8	11.5	12.1	12.7	13.4	14.0	14.6	15.3	15.9	16.5	17.8	19.1	22.3	25.4	28.6	31.8	38.2	44.5	50.9	57.3

TABLE 6-34 (Continued)

Surface feet per minute

Revolutions per minute

Diameter, inches	10	11	12	13	14	15	16	17	18	19	20	21	22	23	24	25	26	28	30	35	40	45	50	60	70	80	90
6½	5.9	6.5	7.1	7.6	8.2	8.8	9.4	10.0	10.6	11.2	11.8	12.3	12.9	13.5	14.1	14.7	15.3	16.5	17.6	20.6	23.5	26.5	29.4	35.3	41.1	47.0	52.9
7	5.5	6.0	6.5	7.1	7.6	8.2	8.7	9.3	9.8	10.4	10.9	11.5	12.0	12.5	13.1	13.6	14.2	15.3	16.4	19.1	21.8	24.5	27.3	32.7	38.2	43.6	49.1
7½	5.1	5.6	6.1	6.6	7.1	7.6	8.1	8.6	9.1	9.6	10.2	10.7	11.2	11.7	12.2	12.7	13.2	14.2	15.2	17.8	20.3	22.8	25.4	30.5	35.5	40.6	45.7
8	4.8	5.2	5.7	6.2	6.7	7.2	7.6	8.1	8.6	9.1	9.5	10.0	10.5	11.0	11.5	11.9	12.4	13.4	14.3	16.7	19.1	21.5	23.9	28.6	33.4	38.2	42.9
8½	4.5	5.0	5.4	5.9	6.3	6.8	7.2	7.7	8.1	8.6	9.0	9.5	9.9	10.4	10.8	11.3	11.7	12.6	13.5	15.8	18.0	20.3	22.5	27.0	31.5	36.0	40.5
9	4.2	4.7	5.1	5.5	5.9	6.4	6.8	7.2	7.6	8.1	8.5	8.9	9.3	9.8	10.2	10.6	11.0	11.9	12.7	14.8	17.0	19.1	21.2	25.4	29.7	33.9	38.2
9½	4.0	4.4	4.8	5.2	5.6	6.0	6.4	6.8	7.2	7.6	8.0	8.4	8.8	9.2	9.6	10.0	10.4	11.2	12.0	14.0	16.0	18.0	20.0	24.0	28.0	32.1	36.1
10	3.8	4.2	4.6	5.0	5.3	5.7	6.1	6.5	6.9	7.3	7.6	8.0	8.4	8.8	9.2	9.5	9.9	10.7	11.5	13.4	15.3	17.2	19.1	22.9	26.7	30.5	34.4
12	3.2	3.5	3.8	4.1	4.5	4.8	5.1	5.4	5.7	6.0	6.4	6.7	7.0	7.3	7.6	8.0	8.3	8.9	9.5	11.1	12.7	14.3	15.9	19.1	22.3	25.4	28.6
14	2.7	3.0	3.3	3.5	3.8	4.1	4.4	4.6	4.9	5.2	5.5	5.7	6.0	6.3	6.5	6.8	7.1	7.6	8.2	9.5	10.9	12.3	13.6	16.4	19.1	21.8	24.5
16	2.4	2.6	2.9	3.1	3.3	3.6	3.8	4.1	4.3	4.5	4.8	5.0	5.2	5.5	5.7	6.0	6.2	6.7	7.2	8.3	9.5	10.7	11.9	14.3	16.7	19.1	21.5
18	2.1	2.3	2.5	2.8	3.0	3.2	3.4	3.6	3.8	4.0	4.2	4.5	4.7	4.9	5.1	5.3	5.5	5.9	6.4	7.4	8.5	9.5	10.6	12.7	14.8	17.0	19.1
20	1.9	2.1	2.3	2.5	2.7	2.9	3.1	3.2	3.4	3.6	3.8	4.0	4.2	4.4	4.6	4.8	5.0	5.3	5.7	6.7	7.6	8.6	9.5	11.5	13.4	15.3	17.2
22	1.7	1.9	2.1	2.3	2.4	2.6	2.8	3.0	3.1	3.3	3.5	3.6	3.8	4.0	4.2	4.3	4.5	4.9	5.2	6.1	6.9	7.8	8.7	10.4	12.1	13.9	15.6
24	1.6	1.8	1.9	2.1	2.2	2.4	2.5	2.7	2.9	3.0	3.2	3.3	3.5	3.7	3.8	4.0	4.1	4.5	4.8	5.6	6.4	7.2	8.0	9.5	11.1	12.7	14.3
26	1.5	1.6	1.8	1.9	2.1	2.2	2.4	2.5	2.6	2.8	2.9	3.1	3.2	3.4	3.5	3.7	3.8	4.1	4.4	5.1	5.9	6.6	7.3	8.8	10.3	11.7	13.2
28	1.4	1.5	1.6	1.8	1.9	2.0	2.2	2.3	2.5	2.6	2.7	2.9	3.0	3.1	3.3	3.4	3.5	3.8	4.1	4.8	5.5	6.1	6.8	8.2	9.5	10.9	12.3
30	1.3	1.4	1.5	1.7	1.8	1.9	2.0	2.2	2.3	2.4	2.5	2.7	2.8	2.9	3.1	3.2	3.3	3.6	3.8	4.5	5.1	5.7	6.4	7.6	8.9	10.2	11.4
32	1.2	1.3	1.4	1.6	1.7	1.8	1.9	2.0	2.2	2.3	2.4	2.5	2.6	2.7	2.9	3.0	3.1	3.3	3.6	4.2	4.8	5.4	6.0	7.2	8.4	9.5	10.7
34	1.1	1.2	1.4	1.5	1.6	1.7	1.8	1.9	2.0	2.1	2.3	2.4	2.5	2.6	2.7	2.8	2.9	3.1	3.4	3.9	4.5	5.1	5.6	6.7	7.9	9.0	10.1
36	1.1	1.2	1.3	1.4	1.5	1.6	1.7	1.8	1.9	2.0	2.1	2.2	2.3	2.4	2.5	2.7	2.8	3.0	3.2	3.7	4.2	4.8	5.3	6.4	7.4	8.5	9.5

TABLE 6-35. Cutting Time for Turning, Facing, and Boring

The range of this table is from 1 to 900 surface feet per minute. The minutes per inch, at 1-inch diameter, are read directly from the chart from 10 to 90 surface feet per minute at the respective feeds. The minutes per inch, at 1-inch diameter, for 1 to 9 surface feet and 100 to 900 surface feet, can be read by moving the decimal as required in the appropriate column.

Therefore:
At 0.015 feed and 2 surface feet per minute, time per inch at 1-inch diameter = 8.7 minutes
At 0.015 feed and 20 surface feet per minute, time per inch at 1-inch diameter = 0.87 minute
At 0.015 feed and 200 surface feet per minute, time per inch at 1-inch diameter = 0.087 minute

Formula: Diameter of work in inches × length of cut in inches × minutes per inch at 1-inch diameter of selected surface feet and feed
Example: 8-inch diameter, 10 inch length, 80 surface feet and 0.015 feed or 8 inch × 10 inch × 0.22 = 17.60 minutes per cut

Surface feet per minute

Feed	10	11	12	13	14	15	16	17	18	19	20	21	22	23	24	25	26	28	30	35	40	45	50	60	70	80	90
	38.2	42.0	45.8	49.6	53.4	57.3	61.1	64.9	68.7	72.5	76.3	80.2	84.0	87.8	91.6	95.4	99.2	106.9	114.5	133.6	152.7	171.8	190.8	229.0	267.1	305.3	343.5

Revolutions of 1-inch diameter (row above). Minutes per inch at 1-inch diameter:

Feed	10	11	12	13	14	15	16	17	18	19	20	21	22	23	24	25	26	28	30	35	40	45	50	60	70	80	90
0.002	13.10	11.91	10.92	10.08	9.36	8.73	8.19	7.71	7.28	6.89	6.55	6.24	5.95	5.70	5.46	5.24	5.04	4.68	4.37	3.74	3.27	2.91	2.62	2.18	1.87	1.64	1.46
0.003	8.73	7.94	7.28	6.72	6.24	5.82	5.46	5.14	4.85	4.60	4.37	4.16	3.97	3.80	3.64	3.49	3.36	3.12	2.91	2.50	2.18	1.94	1.75	1.46	1.25	1.09	0.97
0.004	6.55	5.95	5.46	5.04	4.68	4.37	4.09	3.85	3.64	3.45	3.28	3.12	2.98	2.85	2.73	2.62	2.52	2.34	2.18	1.87	1.64	1.46	1.31	1.09	0.93	0.82	0.73
0.005	5.24	4.76	4.37	4.03	3.74	3.49	3.28	3.08	2.91	2.76	2.62	2.50	2.38	2.28	2.18	2.10	2.02	1.87	1.75	1.50	1.31	1.16	1.05	0.87	0.75	0.66	0.58
0.006	4.37	3.97	3.64	3.36	3.12	2.91	2.73	2.57	2.43	2.30	2.18	2.08	1.98	1.90	1.82	1.75	1.68	1.56	1.46	1.25	1.09	0.97	0.87	0.73	0.62	0.55	0.48
0.007	3.74	3.40	3.12	2.88	2.67	2.50	2.34	2.20	2.08	1.97	1.87	1.78	1.70	1.63	1.56	1.50	1.44	1.34	1.25	1.07	0.94	0.83	0.75	0.62	0.53	0.47	0.42
0.008	3.28	2.98	2.73	2.52	2.34	2.18	2.05	1.93	1.82	1.72	1.64	1.56	1.49	1.42	1.36	1.31	1.26	1.17	1.09	0.94	0.82	0.73	0.66	0.55	0.47	0.41	0.36
0.009	2.91	2.65	2.43	2.24	2.08	1.94	1.82	1.71	1.62	1.53	1.46	1.39	1.32	1.27	1.21	1.16	1.12	1.04	0.97	0.83	0.73	0.65	0.58	0.49	0.42	0.36	0.33
0.010	2.62	2.38	2.18	2.01	1.87	1.75	1.63	1.54	1.46	1.38	1.31	1.25	1.19	1.14	1.09	1.05	1.01	0.94	0.87	0.75	0.66	0.58	0.52	0.44	0.37	0.33	0.29
0.011	2.38	2.17	1.98	1.83	1.70	1.59	1.49	1.40	1.32	1.25	1.19	1.13	1.08	1.04	0.99	0.95	0.92	0.85	0.79	0.68	0.60	0.53	0.45	0.40	0.34	0.30	0.26
0.012	2.18	1.98	1.82	1.68	1.56	1.46	1.36	1.28	1.21	1.15	1.09	1.04	0.99	0.95	0.91	0.87	0.84	0.78	0.73	0.62	0.55	0.49	0.44	0.36	0.31	0.27	0.24
0.013	2.02	1.83	1.68	1.55	1.44	1.34	1.26	1.19	1.12	1.06	1.01	0.96	0.92	0.88	0.84	0.81	0.78	0.72	0.67	0.58	0.50	0.45	0.40	0.34	0.29	0.25	0.22
0.014	1.87	1.70	1.56	1.44	1.34	1.25	1.17	1.10	1.04	0.98	0.94	0.89	0.85	0.81	0.78	0.75	0.72	0.67	0.62	0.53	0.47	0.42	0.37	0.31	0.27	0.23	0.21
0.015	1.75	1.59	1.46	1.34	1.25	1.16	1.09	1.03	0.97	0.92	0.87	0.83	0.79	0.76	0.73	0.70	0.67	0.62	0.58	0.50	0.44	0.39	0.35	0.29	0.25	0.22	0.19

TABLE 6-35 (Continued)

	Surface feet per minute																										
	10	11	12	13	14	15	16	17	18	19	20	21	22	23	24	25	26	28	30	35	40	45	50	60	70	80	90
	Revolutions of 1-inch diameter																										
	38.2	42.0	45.8	49.5	53.4	57.3	61.1	64.9	68.7	72.5	76.3	80.2	84.0	87.8	91.6	95.4	99.2	106.9	114.5	133.6	152.7	171.8	190.8	229.0	267.1	305.3	343.5
Feed	Minutes per inch at 1-inch diameter																										
0.016	1.64	1.49	1.36	1.26	1.17	1.09	1.02	0.96	0.91	0.86	0.82	0.78	0.74	0.71	0.68	0.66	0.63	0.58	0.55	0.47	0.41	0.36	0.33	0.27	0.23	0.20	0.18
0.017	1.54	1.40	1.28	1.19	1.10	1.03	0.96	0.91	0.86	0.81	0.77	0.73	0.70	0.67	0.64	0.62	0.59	0.55	0.51	0.44	0.39	0.34	0.31	0.26	0.22	0.19	0.17
0.018	1.46	1.32	1.21	1.12	1.04	0.97	0.91	0.86	0.81	0.77	0.73	0.69	0.66	0.63	0.61	0.58	0.56	0.52	0.49	0.42	0.36	0.32	0.29	0.24	0.21	0.18	0.16
0.019	1.38	1.25	1.15	1.06	0.98	0.92	0.86	0.81	0.77	0.73	0.69	0.66	0.63	0.60	0.57	0.55	0.53	0.49	0.46	0.39	0.34	0.31	0.28	0.23	0.20	0.172	0.153
0.020	1.31	1.19	1.09	1.01	0.94	0.87	0.82	0.77	0.73	0.69	0.66	0.62	0.60	0.57	0.55	0.52	0.50	0.47	0.44	0.37	0.33	0.29	0.26	0.22	0.187	0.164	0.145
0.022	1.19	1.08	0.99	0.92	0.85	0.79	0.74	0.70	0.66	0.63	0.60	0.57	0.54	0.52	0.50	0.48	0.46	0.43	0.40	0.34	0.30	0.26	0.24	0.20	0.170	0.149	0.132
0.024	1.09	0.99	0.91	0.84	0.78	0.73	0.68	0.64	0.61	0.57	0.55	0.52	0.50	0.47	0.45	0.44	0.42	0.39	0.36	0.31	0.27	0.24	0.22	0.18	0.156	0.138	0.121
0.026	1.01	0.92	0.84	0.78	0.72	0.67	0.63	0.59	0.56	0.53	0.50	0.48	0.46	0.44	0.42	0.40	0.39	0.36	0.34	0.29	0.25	0.22	0.20	0.168	0.144	0.126	0.112
0.028	0.94	0.85	0.78	0.72	0.67	0.63	0.58	0.55	0.52	0.49	0.47	0.45	0.43	0.41	0.39	0.37	0.36	0.33	0.31	0.27	0.23	0.21	0.187	0.156	0.134	0.117	0.104
0.030	0.87	0.79	0.73	0.67	0.62	0.58	0.55	0.51	0.49	0.46	0.44	0.42	0.40	0.38	0.36	0.35	0.34	0.31	0.29	0.25	0.22	0.194	0.175	0.146	0.125	0.109	0.097
0.035	0.75	0.68	0.62	0.58	0.53	0.50	0.47	0.44	0.42	0.39	0.37	0.36	0.34	0.33	0.31	0.30	0.29	0.27	0.25	0.21	0.187	0.166	0.150	0.125	0.107	0.094	0.083
0.040	0.66	0.60	0.55	0.50	0.47	0.44	0.41	0.39	0.36	0.34	0.33	0.31	0.30	0.28	0.27	0.26	0.25	0.23	0.22	0.187	0.164	0.146	0.131	0.109	0.094	0.082	0.073
0.045	0.58	0.53	0.49	0.45	0.42	0.39	0.36	0.34	0.32	0.31	0.29	0.28	0.26	0.25	0.24	0.23	0.22	0.21	0.194	0.166	0.146	0.129	0.116	0.097	0.083	0.073	0.065
0.050	0.52	0.48	0.44	0.40	0.37	0.35	0.33	0.31	0.29	0.28	0.26	0.25	0.24	0.23	0.22	0.21	0.20	0.187	0.175	0.150	0.131	0.116	0.105	0.087	0.075	0.065	0.058
0.055	0.48	0.43	0.40	0.37	0.34	0.32	0.30	0.28	0.26	0.25	0.24	0.23	0.22	0.21	0.198	0.191	0.183	0.170	0.159	0.136	0.119	0.106	0.095	0.079	0.068	0.060	0.053
0.060	0.44	0.40	0.36	0.34	0.31	0.29	0.27	0.26	0.24	0.23	0.22	0.21	0.198	0.190	0.182	0.175	0.168	0.156	0.146	0.125	0.109	0.097	0.087	0.073	0.062	0.055	0.049
0.065	0.40	0.37	0.34	0.31	0.29	0.27	0.25	0.24	0.22	0.21	0.20	0.192	0.183	0.175	0.168	0.161	0.155	0.144	0.134	0.115	0.101	0.090	0.081	0.067	0.058	0.050	0.045
0.070	0.37	0.34	0.31	0.29	0.27	0.25	0.23	0.22	0.21	0.197	0.187	0.178	0.170	0.163	0.156	0.150	0.144	0.134	0.125	0.107	0.094	0.083	0.075	0.062	0.053	0.047	0.042
0.075	0.35	0.32	0.29	0.27	0.25	0.23	0.22	0.21	0.194	0.184	0.175	0.166	0.159	0.152	0.146	0.140	0.134	0.125	0.116	0.100	0.087	0.078	0.070	0.058	0.050	0.044	0.039

TABLE 6-36. Cutting Speeds for Milling*
(Surface feet per minute)

Work material	High-speed steel tools		Carbide-tipped tools		Coolant
	Rough mill	Finish mill	Rough mill	Finish mill	
Cast iron............	50–60	80–110	180–200	350–400	Dry
Semisteel............	40–50	65–90	140–160	250–300	Dry
Malleable iron......	80–100	110–130	250–300	400–500	Soluble, sulfurized or mineral oil
Cast steel.........	45–60	70–90	150–180	200–250	Soluble, sulfurized, mineral or mineral lard oil
Copper.............	100–150	150–200	600	1,000	Soluble, sulfurized or mineral lard oil
Brass..............	200–300	200–300	600–1,000	600–1,000	Dry
Bronze.............	100–150	150–180	600	1,000	Soluble, sulfurized or mineral lard oil
Aluminum..........	400	700	800	1,000	Soluble or sulfurized oil, mineral oil and kerosene
Magnesium........	600–800	1,000–1,500	1,000–1,500	1,000–1,500	Dry, kerosene, mineral lard oil
SAE steels:					
1020 (coarse feed).	60–80	60–80	300	300	⎫
1020 (fine feed)...	100–200	100–200	450	450	⎪
1035..............	75–90	90–120	250	250	⎪
X-1315............	175–200	175–200	400–500	400–500	⎬ Soluble, sulfurized, mineral or mineral lard oil
1050..............	60–80	100	200	200	⎪
2315..............	90–110	90–110	300	300	⎪
3150..............	50–60	70–90	200	200	⎭
4150..............	40–50	70–90	200	200	
4340..............	40–50	60–70	200	200	⎫ Sulfurized and mineral oils
Stainless steel......	60–80	100–120	240–300	240–300	⎭

* Data through the courtesy of *American Machinist*.

TABLE 6-37. Recommended Surface Feet for Turning Various Metals with Approximately $3/32$-inch Depth of Cut and $1/32$-inch Feed Using High-speed Steel Tools

Metal	*Surface Feet*
Aluminum and alloys....................	400–1,000
Bakelite, plastics, hard rubber............	100
Brass................................	300
Bronze silicon, everdure.................	70
Bronze..............................	200–400
Cast iron, medium.....................	125
Copper, rolled........................	80–150
Copper, cast..........................	200
Magnesium and alloys...................	200–600
Malleable cast iron.....................	90
Monel, cast...........................	60
Monel, rolled.........................	70
Semisteel............................	80
Steel, free cutting.....................	150
Steel, low carbon......................	90
Steel, medium carbon, annealed..........	70
Steel, high carbon, annealed.............	50
Steel, very hard.......................	25
Steel, 12% Mn, 1.2% carbon.............	12
Steel, stainless 18 Cr, 8 Ni..............	30
Zinc, base die castings..................	180

TABLE 6-38. Recommended Surface Feet for Turning Various Metals with Approximately $\frac{3}{32}$-inch Depth of Cut and $\frac{1}{32}$-inch Feed Using Carbide Tools

Metal	Surface Feet
Aluminum and alloys.....................	1,000–3,000
Bakelite, plastics and hard rubber..........	300
Brass....................................	700
Bronze silicon, everdure...................	219
Cast iron, medium........................	240–350
Copper, rolled...........................	400
Malleable cast iron.......................	200
Monel cast...............................	180
Semisteel................................	200
Steel, low carbon.........................	260
Steel, medium carbon, annealed...........	200
Steel, high carbon, annealed...............	150
Steel, very hard..........................	75
Steel, 12% Mn, 1.2% carbon...............	40
Steel, stainless 18 Cr, 8 Ni.................	150

TABLE 6-39. Surface Feet for Turret Lathes Using High-speed Steel Tools

Material	Surface speed, feet per minute	Material	Surface speed, feet per minute
Soft cast iron, roughing.......	50–60	Soft machine steel...........	80–100
Soft cast iron, finishing.......	60–80	Medium-hard machine steel...	60–80
Hard cast iron, roughing......	35–50	Hard machine steel..........	40–60
Hard cast iron, finishing......	60–80	Tool steel, annealed..........	60–80
Malleable cast iron...........	80–90	Tool steel, unannealed........	25–35
Steel casting.................	50–60	Alloy steel, annealed.........	50–60
Brass.......................	150–250	Alloy steel, treated..........	30–40
Bronze......	100–150	Cutting threads on brass.....	60–150
Hard bronze................	80–100	Cutting threads on steel and cast iron.................	25–40
Copper......................	150–200		
Aluminum...................	250–400		

TABLE 6-40. Machinability Rating Chart

SAE	Brinell	Percent
1112 (0.08–0.16% carbon)	178–228	100 (leaded 130%)
X1112 (0.08–0.16% carbon)*	178–228	135 (leaded 155%)
1120 (0.15–0.25% carbon)	143–179	80 (leaded 110%)
X1315 (0.10–0.20% carbon)	143–179	85 (leaded 115%)
X1335 (0.30–0.40% carbon)	187–229	70 (leaded 95%)
X1020 (0.15–0.25% carbon)	159–192	75 (leaded 100%)
4120 (0.15–0.25% carbon)	179–223	70
6115 (0.10–0.20% carbon, 0.80–1.10% chromium, 0.15% vanadium)	179–223	70
Malleable iron	....	75–150
1020 (0.15–0.25% carbon)	134–166	65 (leaded 90%)
1030 (0.25–0.35% carbon)	170–212	70 (leaded 95%)
1040 (0.35–0.45% carbon)	179–228	60 (leaded 80%)
X1040 (0.35–0.45% carbon)	179–228	70
2315 (0.10–0.20% carbon, 3.25–3.75% nickel)	174–217	55
3130 (0.25–0.35% carbon, 1.00–1.50% nickel, 0.45–0.75% chromium)‡	179–217	55
3140 (0.35–0.45% carbon, 1.00–1.50% nickel, 0.45–0.75% chromium)‡	187–228	55
4130 (0.25–0.35% carbon, 0.50–0.80% chromium, 0.15–0.25% molybdenum)‡	187–228	65
4615 (0.10–0.20% carbon, 1.65–2.00% nickel, 0.20–0.30% molybdenum)	174–217	60
4640 (0.35–0.45% carbon, 1.65–2.00% nickel, 0.20–0.30% molybdenum)‡	187–235	55
4815 (0.10–0.20% carbon, 3.25–3.75% nickel, 0.20–0.30% molybdenum)	187–228	65
5120 (0.15–0.25% carbon, 0.60–0.90% chromium)	170–212	60
Stainless iron F.C.	160–217	60
1010 (0.05–0.15% carbon)	131–170	50
1050 (0.45–0.55% carbon)‡	179–228	50
2330 (0.25–0.35% carbon, 3.25–3.76% nickel)‡	179–217	50
2340 (0.35–0.45% carbon, 3.25–3.75% nickel)‡	184–235	50
3115 (0.10–0.20% carbon, 1.00–1.50% nickel, 0.45–0.75% chromium)	174–217	50
3220 (0.15–0.25% carbon, 1.50–2.00% nickel, 0.90–1.25% chromium)	179–228	50
3230 (0.25–0.35% carbon, 1.50–2.00% nickel, 0.90–1.25% chromium)‡	184–235	45
4140 (0.35–0.45% carbon, 0.80–1.10% chromium, 0.15–0.25% molybdenum)‡	184–235	50
4150 (0.45–0.55% carbon, 0.80–1.10% chromium, 0.15–0.25% molybdenum)‡	196–235	45
5140 (0.35–0.45% carbon, 0.80–1.10% chromium)‡	187–228	45
6130 (0.25–0.35% carbon, 0.80–1.10% chromium, 0.15% min vanadium)‡	179–217	50
6140 (0.35–0.45% carbon, 0.80–1.10% chromium, 0.15% min vanadium)‡	187–228	40
T1330 (0.25–0.35% carbon, 1.60–1.90% manganese)†.‡	179–217	50
T1340 (0.35–0.45% carbon, 1.60–1.90% manganese)‡	187–228	40

TABLE 6-40 (Continued)

SAE	Brinell	Percent
9260 (0.55–0.65% carbon, 0.60–0.90% manganese, 1.80– 2.20% silicon)‡	196–241	45
Ingot iron	101–131	50
Wrought iron	101–131	50
Stainless 18/8F.M	179–212	45
Cast iron		
2350 (0.45–0.55% carbon, 3.25–3.75% nickel)‡	196–235	35
2515 (0.10–0.20% carbon, 4.75–5.25% nickel)	179–228	30
3240 (0.35–0.45% carbon, 1.50–2.00% nickel, 0.90– 1.25% chromium)‡	184–235	40
52100 (0.95–1.10% carbon, 1.20–1.50% chromium)¶	184–217	30
Ni-Resist		30
Stainless 18/8 austenitic‡		30
Steel, manganese, oil-hardening‡		35
Tool steel, low tungsten-chromium‡		25
Carbon tool steels‡		30
High-speed steel‡		25
Steel, high-carbon, high-chromium‡		20
Nonferrous:		
Dowmetal		100
Magnesium and alloys		200–500
Aluminum 11-S		200–500
Aluminum 2-S		100–300
Aluminum 17-S		100–300
Brass-leaded F.C. C.D		150
Brass-yellow		100
Brass-red		100
Bronze, leaded bearing		100
Zinc		200
Gun metal		60
Bronze manganese		40
Copper-cast		70
Copper-rolled		60
Nickel		20
Monel-cast		35
Monel-rolled		45
Monel—"K"		50
Inconel		45
Everdur		60

* The prefix X is used in numerous instances to denote variations in the range of elements, e.g., in manganese, sulfur, etc.

† The prefix T is used with the manganese steels (1300 series) to avoid confusion with steels of somewhat different manganese range which have been identified by the same numerals but without the prefix.

‡ Annealed.

¶ Spheroidized annealed.

TABLE 6-41. Useful Information

To find circumference of circle, multiply diameter by 3.1416.
To find diameter of circle, multiply circumference by 0.31831.
To find area of circle, multiply square of diameter by 0.7854.
Doubling the diameter of a circle increases its area four times.
To find the surface of a ball, multiply square of diameter by 3.1416.
To find side of an inscribed square, multiply diameter by 0.7071 or multiply circumference by 0.2551 or divide circumference by 4.4428.

To find side of an equal area square, multiply diameter by 0.8862.
Area of square multiplied by 1.273 equals squared diameter of an equal area circle.
To find cubic inches in a ball, multiply cube of diameter by 0.5236.
To find cubic contents of a cone, multiply area of base by one-third the altitude.
A gallon of water (U.S. Standard) weighs 8⅓ pounds and contains 231 cubic inches.
A cubic foot of water contains 7½ gallons and weighs 62½ pounds.
To find the pressure in pounds per square inch of a column of water, multiply the height of the column in feet by 0.434.
To find the capacity in U.S. gallons of cylindrical tanks: diameter squared in inches multiplied by length in inches multiplied by 0.0034.
The cubical content of any container multiplied by 0.004329 equals its capacity in U.S. gallons.
To find weight per foot of round steel bars, square the diameter and multiply by 2.67.
To find weight per foot of square or flat iron, multiply the width in inches by the thickness in inches by 3.4.
To find width of leather belt required for a given horsepower:

$$\frac{\text{Hp} \times 2{,}750}{\text{Diam of driven pulley in inches} \times \text{rpm}}$$

To find diameter of driving pulley:

$$\frac{\text{Diam of driven pulley in inches} \times \text{rpm required}}{\text{rpm of driving pulley}}$$

To find diameter of driven pulley:

$$\frac{\text{Diam of driving pulley in inches} \times \text{rpm of driving pulley}}{\text{Rpm of driven pulley}}$$

To find revolutions per minute of driven pulley:

$$\frac{\text{Diam of driving pulley in inches} \times \text{rpm of driving pulley}}{\text{Diam of driven pulley in inches}}$$

$$\text{Time} = \frac{\text{travel}}{\text{rpm} \times \text{feed}} \qquad \text{Surface feet} = 0.262 \times \text{diam} \times \text{rpm}$$

$$\text{Feed} = \frac{\text{travel}}{\text{rpm} \times \text{time}} \qquad \text{Rpm} = \frac{\text{surface feet}}{\text{diam} \times 0.262}$$

$$\text{Rpm} = \frac{\text{travel}}{\text{feed} \times \text{time}} \qquad \text{Diameter} = \frac{\text{surface feet}}{\text{rpm} \times 0.262}$$

Chapter 7

Effective Communication

JACKSON E. MORRIS

Long Beach, California

All of us would like to write engineering text that is clear, brief, and full of interest. When we seek easy methods of doing this, however, we are often told, in effect, simply to write "clearly, briefly, and naturally." We are usually told this in a manner which arouses such great enthusiasm for the power and the beauty of good prose that we begin with great energy. But enthusiasm wanes, and so do our efforts as we confront the tough task every day. To persist fruitfully, we need some guidelines more discriminating than enthusiasm; we need a critical apparatus that will inform us when we are doing wrong and show us how others do well. And we need all the theoretical information we can get. We cannot spurn the contributions of grammar, particularly of syntax (sentence structural theory), because the structure of our sentences should parallel the logical structure of the ideas we are expressing.

In what follows, we shall develop structural rules to apply to sentences, keeping our rules as free of bristling grammatical nomenclature as we can. We shall apply the rules to examples of faulty engineering writing to demonstrate usage. Finally, we shall introduce examples of good communication in the spoken, the informal, and the formal modes to show how certain requirements shape the style for each mode. It will appear that the inventiveness, the flexibility, and the directness of good speech can be translated into formal engineering writing with profit. If something of this sort is not done, formal engineering writing can end as "engineering Latin," similar to the cramped, inaccessible scholar's Latin, seldom spoken, in which Newton and other early scientists were forced to write.

CONTRIBUTIONS OF GRAMMAR

Grammar is a double study dealing with the classification and proper form of either words (inflection) or word groups (syntax). Inflection, dealing with word variation, treats topics we all probably encountered in junior high school along with the school yells and pig Latin, that is, with the declension of nouns and pronouns, the conjugation of verbs, and the comparison of adjectives. Inflection is a topic we shall not say much about because (1) the proper inflections are understood instinctively by most native speakers, (2) improper inflections are not an important problem in engineering writing, and (3) no one wants to hear about inflection anyway, it being the arithmetic of language, the set of tedious small manipulations we should have perfected as children and then relegated to our unconscious minds along with nicknames.

Syntax, the structural study of word groups, cannot be lightly disposed of. Sentences and paragraphs model the world of our common experience, somewhat as mathematical equations do, and syntax itself is a form of functional analysis. We shall try to develop a critical apparatus here to examine the methods by which English sentences model our logical thoughts about the world. The phrase, the clause, and the sentence are all formal building blocks of language structure, and we must examine them in particular.

Sentence Forms. The sentence is the mind's basic unit of information transmittal; no information is transmitted at all until the sentence is complete and the framework of words is defined. There is something magic about how a sentence accomplishes this, some intimate connection between our brain and sentence structure. Take the elementary sentence "Spacecraft visits moon" and observe what a profoundly complex, dynamic operation has been modeled by three words in a given, simple relationship. Or take the sentences "Test verifies design" and "Transistor fails" and notice how the few words have suggested very subtle ideas by their structural relationship. Notice that all these sentences change the world, in effect. They do this by stating the occurrence of some action. Because they report a change in the configuration of the world, so to speak, we shall call them dynamic sentences.

Dynamic sentences exist in two forms, the transitive and the intransitive. In the transitive form ("Spacecraft visits moon"), one object, idea, or process (something we shall generally call a substantive) interacts with another to produce the change. In the intransitive sentence, some substantive interacts in effect with itself to produce the change ("Boat rocks" or "Transistor fails").

Sentences which leave the world unchanged by their report are static sentences (or linking sentences, in grammatical terminology). Examples are "Capacitors are circuit elements" and "Radars are accurate." Notice that the static sentence always employs some form of the verb *to be*, for instance, *is, are, was, will have been.* Static sentences are of two forms: classificatory and attributive. "Capacitors are circuit elements" is classificatory because it assigns capacitors as one of a class of similar objects in the world of our experience. The other sentence, "Radars are accurate," does not assign radars to a more extended class but names an interior or self-aspect of radars. Attributive sentences thus are used to develop the interior rather than the exterior world of the radar. They are a tool of analysis or "taking apart," rather than of synthesis or "putting together" which the classificatory sentence does. Static sentences thus can be related either to analyzing or to synthesizing the substantives they deal with. Dynamic sentences relate either to the interactions of several substantives or to the self-actions of one substantive.

We now state generally that the entire spectrum of English sentences is either static or dynamic. This means that there are only four basic information frames supplied by language for the transmission of any possible idea about the world of our common experience: the transitive, intransitive, classificatory, and attributive. These are the modeling molds which language forces on our ideas when we try to express them. The good writer will select the information frame which best matches the idea he is trying to transmit; the poor writer usually exercises no discrimination whatsoever, arbitrarily forcing any idea into the first information frame which happens to occur to him, something it is possible to do only at a great expense in words and in sentence complexity.

Some sentences are exceptions to a strict interpretation of the above forms, but they are not really critical exceptions. The passive sentence is the worst anomaly, and some remarks must be made about it. A sentence with a passive structure, such as "Moon is visited by spacecraft," can always be considered as an inversion of the basic transitive form; it involves two interacting bodies, but the action flows in a reverse sense from the predicate into the subject. The basic perversion of the form becomes clearer when we note that we can actually suppress one of the interacting bodies and still transmit information, as "Moon is visited." Now a second body is not mentioned although we can infer that one exists. Still, our original conclusion remains if we grant the passive inversion. Four basic sentence frames exist in English to model the infinite variety of sentences we can transmit. These frames deal with the dynamic interactions and self-actions of substantives and with the static synthesis and analysis of substantives.

Levels of Structure. Static and dynamic sentences can all be represented at the principal structural level by one formula, which we shall set down here arbitrarily as substantive-verb-complement. The substantive (subject noun or pronoun) and the verb are probably familiar already to most engineers. The complement is the key word which follows the verb in most cases, as "moon," "elements," and "accurate," respectively, in "Spacecraft visits moon," "Capacitors are circuit elements," and "Radars are accurate." Intransitive and passive sentences have no complement, and we shall simply consider the position as a blank in these instances. The substantive-verb-complement structure of a simple sentence we shall describe hereafter as the principal structure, and everything else we shall describe as the modifying or subordinate structure of the sentence. "Spacecraft visits moon" happens to have no modifying structure, but a more probable form of the sentence, as "An Apollo spacecraft will visit the moon in July," has both principal and modifying structure. The modifiers in any sentence consist of those words and word groups attaching to the substantive and verbs which sharpen or focus their application. The substantive "dog," for instance, is a class word which includes all dogs in the world, but in "little brown dog down the street," the class of dogs is restricted to one member. This restriction is accomplished by use of the modifying adjectives "little" and "brown" and the modifying prepositional phrase "down the street," which we understand as applying to "dog" by their proximity and order.

Single-word modifiers include, of course, the adjectives and adverbs; adjectives modify substantives and adverbs modify verbs as well as adjectives and other adverbs. Modifying word groups can be used in either the adjectival or adverbial sense. These groups include phrases of various kinds and dependent or modifying clauses which have a complete principal sentence structure but cannot stand alone as sentences because of the subordinating conjunction which introduces them. Examples are "because there are more positive than negative ions," "although the dielectric breaks down at 400 volts," and "if we consider all the alternatives." It is the subordinating conjunction alone which makes all these clauses modifiers. Without this connective word, the clause could stand alone as an independent clause or simple sentence.

Coordination. Other conjunctions must be discussed before we possess a complete apparatus for the study of sentence form. These are the coordinating conjunctions—usually "and," "but," and "or"—and they have the property of connecting equal grammatical structures at any level of use, rather than connecting a principal to a subordinate structure as the subordinating conjunctions do. "And," for instance, is additive in effect when placed between two grammatical structures, as "around the corner and under the tree," where we have added or compounded two prepositional phrases. "Or" gives us our choice of the two grammatical structures between which it is placed, as "the circuit on page 3 or the circuit on the blackboard."

We now have specified the four basic sentence forms in English, discussed the principal and modifying levels of construction, and examined the processes of coordination and subordination of structure. With so much of syntax on call in our personal file, we are in a position to begin to relate sentence structure to sentence content or meaning, all with the aim of getting clarity and brevity into the sentences we ourselves write as engineers.

STRUCTURE AND MEANING

We now shall state four rules which can guide us in matching the structure of the sentences we write to the ideas we are trying to express in these sentences—rules which relate the syntax of the sentence to the logical structure of the idea we are trying to model.   Here are the rules.

   Rule 1. Choose the sentence form (transitive, intransitive, classifying, attributive) which best models the natural structure of the idea being expressed (interaction, self-action, synthesis, analysis).

   Rule 2. Construct all sentences so that the principal information elements occupy the principal structural positions (subject-verb-complement), and the subordinate information elements appear in the subordinate or modifying structure.

   Rule 3. Construct all sentences so that equal or like ideas at any level are expressed by like grammatical forms.

A fourth rule of composition will be added now for ready reference though its meaning must be clarified on later pages.   The fourth rule stems from the fact that good sentence modification and reference (to be discussed later) depend on nearness or proximity of the coupled elements, and in long sentences, this proximity or close coupling cannot always be obtained.   This occurs because in most sentences there is only one principal structure, but modifying and reference word groups can be added indefinitely.   Thus the associations depending on proximity tend to break down as sentence length increases.

   Rule 4. Construct all sentences so that the association of coupled word groups depending on proximity is unambiguous.

   Application of Rules.   With these four rules, we have developed a critical apparatus which we can apply in assessing and improving our own sentences.   Some of the rules may seem obvious or platitudinous, but they can be applied in very subtle ways.   The reader is warned that they can, in fact, cover an astonishing variety of situations requiring long study and experience with writing problems to appreciate. Indeed, most problems of modern engineering writing—problems of sentence complexity and ambiguity—stem directly from violations of these four rules.   Here is a simple application of Rule 1.

   Faulty: There is another communications satellite which is sent into orbit from Cape Kennedy every month.

In our criticism of this sentence, we will confine our remarks to an examination of its structure.   The sentence attempts to model a dynamic (transitive) situation with a static structure ("There is . . .").   It is better to use a dynamic structure for this idea.

   Correct: Another communications satellite rises into orbit from Cape Kennedy every month.

Here is an application of Rule 1 in the inverse sense.

   Faulty: The design of a radar usually provides high accuracy.

We have an essentially static idea expressed here with a transitive sentence structure.   An attributive statement would be more apt.

   Correct: A radar is usually highly accurate.

Writers who like to make careful distinctions in meaning might maintain that any structure with which we express an idea can lend a subtly individual emphasis or weighting to the different sentence elements, and that this different emphasis can be a distinct part of the meaning.   This conclusion may be true of some of the sentences we examine; we do not know what was in the writer's mind. But we are developing a critical apparatus here to apply to our own sentences, and we can conclude that when we ourselves write indirect, diffuse, wordy, and stale sentences, we had better be ready with more convincing justification of what is essentially a bad practice. The writer of any poor sentence will usually have only poor ideas to defend it, subtleties of emphasis notwithstanding.

Here are applications of Rule 2 relating to principal and subordinate structure.

> *Faulty:* Flight to the planets and beyond certainly can be accomplished in this decade with human pilots aboard.

Because flight to the planets has already been accomplished with unmanned spacecraft, the writer must be predicting something about the flight of manned spacecraft. Yet we do not discover this anywhere in the principal elements of this passive sentence structure ("Flight . . . can be accomplished . . ."). We discover this important idea only in the adjective "human" occurring in a modifying prepositional phrase at the end of the sentence—a case of an important logical element buried in a modifier buried in another modifier. We shall adopt a transitive structure.

> *Correct:* Men certainly can pilot spacecraft to the planets and beyond during this decade.

Rule 3 is an extension of the rule of parallel phrasing to be found in most books of composition. It can be applied in an astonishing variety of ways, either at the principal or at the subordinate level of structure. It is a coordinate rule, in effect, applying to any structure signaled by the presence of a coordinating conjunction (and, or, but).

> *Faulty:* Early pins were made of wood, horn, bone, and later of metal.

The "and" in this sentence signals the occurrence of a coordinate construction. On first examining it, perhaps we conclude that we are dealing with four equal ideas: wood, horn, bone, and metal pins. But more thought shows us that this conclusion is in error. We are dealing logically with two ideas: "early pins," of which we are given three examples, and "later pins," of which we are given one example. Thus we have a branching or parallel structure within a parallel structure. Here is the sentence rewritten to show this more complex logic.

> *Correct:* Early pins were made of wood, horn, and bone, but later pins were made of metal.

Notice we have two coordinate structures here, the major one signaled by "but" and the minor one by the "and" of "wood, horn, and bone." Now the structure of the sentence parallels the structure of the ideas; that is, equivalent ideas are expressed by equivalent grammatical structures.

Parallel Structures. The parallel structures of sentences can be diagrammed with profit, somewhat like the parallel structures of circuit theory, but a few preliminary remarks should be made. Sentences as they appear in print are word sequences, with relative word order usually being important. But in parallel structures, the order of the repeated structures may be unimportant. That is, we can freely interchange the order of "wood," "horn," and "bone" without damage to the meaning of the given sentence. The first part of the corrected sentence is diagrammed in Figure 7-1 to show its parallel elements explicitly.

A fundamental principle of parallel structures will now be stated. A proper parallel structure can be read independently along any one of its branches alone, as "Early pins

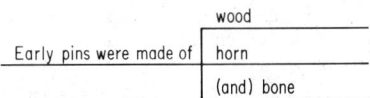

Fig. 7-1. Simple parallel structure.

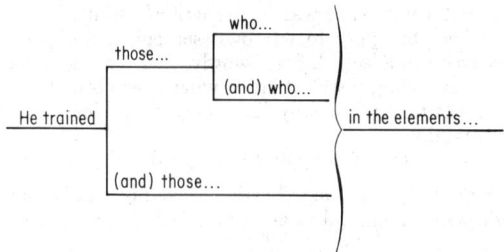

FIG. 7-2. Complex parallel structure.

were made of wood." This result will also apply to the second and third branches if we adopt the convention that "and" in the third branch is merely a structural signal indicating that the parallelism is being concluded. Notice that, if we apply this principle of parallelism to the original faulty sentence, we find we have four parallel branches, with the last branch reading something like this (leaving off the "and"): "Early pins were made of later of metal." Clearly, our test of branch independence shows that we have a faulty parallelism here.

Diagraming is a powerful method for revealing the series-parallel structure of any sentence, but we shall pursue it in only one more example: "He trained those launch engineers who had come in from Vandenburg Air Base and who had come back from Cape Kennedy, and those interested local design engineers, in the elements of mission control."

Notice in Figure 7-2 the role of "and" in signaling the end of each parallel structure. Also notice that logical sentences can be read simply by proceeding serially through the structure along any given path, this being the rule of independence. The reader is referred to books on composition for more extended examples of sentence diagraming. He is also advised that powerful new methods of treating syntax by network and tree structures have been developed. These are the methods of transformational grammar, devised to make sentence structure amenable for computer processing. The elements of transformational grammar are being taught in junior high and high school classes throughout the country. This modern grammar is also being applied with some success to the problem of automatic translation between modern languages. Reference books appear in the bibliography at the end of the chapter.

Normal Order. Another important syntactical concept is normal sentence order. Normal order is the familiar serial order in which we are used to writing the words of normal sentences. Usually, we place the substantive before the verb (except in interrogative sentences) and let the complement follow the verb; this constitutes normal order at the principal sentence level of organization.

The proper order of the modifying elements of a sentence (should modifiers normally precede or follow the words they modify?) will give us more trouble in statement. Normally, an adverb or adjective precedes the word it modifies and a phrase and clause follow in that order, as "little boy with black hair who cries all the time." The relative positions of the modifiers and the modified word provide us with a sort of proximity coding to enable us to make the proper modifying associations. However, this sort of coding is not always adequate to sort out the modifying connections in English, as we will see.

Faulty: The technician wired a resistor in the chassis which was substandard.

Here we have a modifying phrase, "in the chassis," and a modifying clause, "which was substandard," following the modified word, "resistor," all being in normal order. Notice, however, that we cannot tell whether the resistor or the chassis is substandard, and it may be important to know.

Correct: The technician wired a substandard resistor in the chassis.

Reducing the modifying clause to a single word, "substandard," allows us to change its position with respect to "resistor" and make the relationships clear. It is easy to write sentences, however, in which normal sentence order cannot be manipulated to sort out all the linked pairs.

> *Faulty:* The technician removed the nozzle of the rocket engine from the spacecraft which was damaged in the test chamber.

There are many different ways in which this sentence can be interpreted. The confusion occurs because some of the phrase and clause modifiers ("from the spacecraft," "which was damaged," "in the test chamber") can logically link to the verb "removed," to the complement "nozzle," or to the substantive "spacecraft," which is here part of a modifying phrase. Normal order concepts simply cannot sort out all these associations uniquely because there is too much strain at the sentence position following the verb, where the complement and the phrase and clause modifiers all must jockey for position.

Inverted Order. English syntax employs a clever dodge for avoiding the traffic jams which can occur after the verb in many sentences. The writer in these cases is at liberty to insert the verb modifiers as introductory phrases if he finds it useful. The faulty sentence will be rewritten here to take advantage of this flexibility in the positioning of verb modifiers.

> *Correct:* From the spacecraft which was damaged, the technician removed the nozzle of the rocket engine in the test chamber.

There may be other ways to resolve the ambiguity, but we have provided a possible solution. It would also have been easy to write two sentences and thus absorb the modifiers by means of two principal structures, as "The spacecraft was damaged. In the test chamber, the technician then removed the nozzle of the rocket engine."

The curious flexibility of positioning of verb modifiers is made possible by the fact that there is usually only one verb to be modified in a normal sentence. There are usually many substantives, on the other hand; so adjective placement must be rigidly controlled to be clear.

In discussing structure, we have tried to relate sentence syntax and meaning. The concepts of sentence form, coordination, subordination, parallel structure, normal order, and inverted verb modification are all basic to this relationship. Most good writers understand these concepts instinctively, as they do the concepts of inflection. But typical sentences of engineering writing have become so long and complex that an intuitive understanding no longer suffices. To construct long sentences correctly, we should know explicitly what we are about.

Reference. There are other instances besides modification where words must be linked or associated by proximity in our sentences. These are the associations of reference, where a pronoun or pronominal adjective such as "he," "it," "his," "theirs," "mine," "they," "we," "this," and "that" must be linked to some substantive (the antecedent) which it replaces. Usually, these reference words refer to the subject of the sentence or to the substantive immediately preceding them, but whatever the antecedent, the association between the two words must be carefully controlled to be made clear by the writer. Otherwise, ambiguity can quickly result. Here is an example:

> *Faulty:* The divisions of our corporation halted research on dish antennas when it appeared the Air Force no longer favored them.

We may correctly inquire what went out of favor, dish antennas or the divisions of our corporation? We shall have to avoid use of the ambiguous reference word "them."

> *Correct:* The divisions of our corporation halted research on dish antennas when it appeared the Air Force no longer favored dish antennas.

Problems of Long Sentences. It should be clear to the reader now that, in addition to having a formal structure, every English sentence has arbitrary interior linkages which the writer must control chiefly by means of word order and word proximity. These order and proximity linkages can be severely strained, or even dissolved entirely, in long sentences because there may be too many modifying word groups and reference pairs for the principal structure to accommodate. In other words, modification and reference tend to break down in long sentences. The obvious solution for the engineering writer—write short sentences—is not always as easy as advertised. Using two principal structures where one might do is often distasteful because it seems to increase the total word count. This increase might occur, for instance, when we try to simplify a sentence with a compound predicate by rewriting it as two sentences with a repeated subject. And minimizing the total word count is often such a strong requirement in writing papers for journals, whose publishing costs depend on word count, that the writer may be driven into much compounding; to save space, he may even be tempted into slipping important ideas conveniently into his modifying structure where they do not belong.

Engineering Writing Requirements. Besides the requirement for minimizing total word count, certain others exist to force the writer into producing long sentences. One of these is the genuine complexity of modern engineering systems, requiring genuine complexity of explanation. Another is the "rule of impersonality" in engineering writing. This rule prohibits the use of names and personal pronouns and promotes much passive sentence structure as a consequence, the passive being well known as a wordy, diffuse type of sentence structure. The final modern requirement we might describe as the "rule of caution," a rule which seems to bar the engineering writer from making qualitative statements or using metaphors and comparisons, even in explanatory passages where a proof is not involved. Explanation is an art and should rely a great deal on suggestion and example, but proof should rely straightforwardly on inductive and deductive logic. Confusion between the proper requirements of explanation and of proof leads to much excessive qualification and caution in explanatory writing. Engineers appear to write explanations as if they were conducting a continual proof, and this is what we have called the "rule of caution."

The possibility of relieving some of the cramping requirements of these two so-called rules will be examined on following pages. Some solutions which have been adopted in engineering speech and in informal writing will be applied to formal engineering writing. Before we do this, it may be opportune to recall Rule 4. Perhaps its importance can be better appreciated after the discussion of modification and reference we have been through. Rule 4 stated in essence that we must manage word order in our sentences so that the proximity associations of modification and reference are unambiguous.

MODES OF COMMUNICATION

We shall examine engineering communications in three modes now: spoken, informal, and formal. Our fundamental problem is the excessively long, ambiguous sentence of formal engineering writing. We shall examine some of the fruitful practices of speech and informal communication to determine if these can be applied to ease the problems of formal engineering writing.

Spoken Communication. It is the speakers of any language who keep it flexible in the long run, constantly adapting it to new requirements, and it is probably true to say that engineering writing suffers a great deal from disdaining the inventiveness and flexibility of the spoken idiom. Poetry, the drama, and the novel are "revolutionized" perhaps every twenty or thirty years by lively infusions of words and useful practices from the spoken idiom. Engineering writing has not been so receptive to invention. At its worst, like Pentagonese and gobbledygook, it is an example of a language, never spoken, which has become so rigid and unresponsive to change that it is in danger of complete stultification, perhaps to be replaced by some form of communication between computers. On the other hand, engineering speech and informal writing are often very inventive modes of communication, as will be shown. The first example is from the conversation of the astronauts of Apollo 8—Frank Borman, Bill Anders, and James Lovell—during the famous Christmas mission which circled the moon. James Lovell, in particular, is quoted as he communicated with CapComm at Mission Control in Houston to set up the second television broadcast from 200,000 miles out. Notice the direct, clear, vigorous presentation of a complex technical situation.

> Lovell: "We are maneuvering for the TV. Bill has got it set up in Frank's left rendezvous window, and I'm over in Bill's spot looking out the right rendezvous window. The earth is now passing through my window. It's about as big as the end of my thumb. Waters are all sort of a royal blue; clouds, of course, are bright white. The reflection off the earth is much greater than the moon. The land areas are generally sort of dark brownish to light brown. What I keep imagining is, if I were a traveler from another planet, what would I think about the earth at this altitude? Whether I think it would be inhabited."
>
> CapComm: "Don't see anyone waving. Is that what you are saying?"
>
> Lovell: "I was just kind of curious if I would land on the blue or the brownish part."
>
> Borman: "You better hope we land on the blue part."

This is excellent communication under the circumstances, being clear, brief, and full of interest. Notice that the speaker introduced personal pronouns into his report, greatly simplifying his syntax because he did not have to use indirect, passive sentence forms. He also gave qualitative impressions and used questions and comparisons (". . . about as big as the end of my thumb."). There are even two incomplete sentences ("Whether I think it would be inhabited," "Don't see anyone waving"), this truncation of syntax being a useful device to obtain brevity provided there is no loss of clarity.

The directness and vigor of Lovell's speech have been gained at only small loss in exactness of statement, as ". . . land areas are generally sort of dark brownish." We accept this lack of precision because we understand that an exact statement of color cannot be quickly made, and the qualitative impression is valuable in itself. We conclude that the clarity and brevity of good oral communication are gained perhaps at some loss in exactness of statement and precision of syntax, but that this loss is small cost for the advantages gained. It is only in statements of proof—argumentative writing—and in the observations and descriptions on which argumentative writing is based that we should require absolute exactness of statement. Explanations, which make up most of engineering writing, do not need to be written with such extreme caution; they may with profit employ some of Lovell's techniques: an idiomatic vocabulary, personal pronoun structure, qualitative statements, comparisons and analogies, and within reason, a truncated syntax. This is not to say that explanatory writing does not need to be correct, only that it can be informal and suggestive.

Informal Writing. The great English geologist T. H. Huxley was one of the masters of engineering writing. In his time, through his essays, he was credited with stimulating immense popular interest in science, particularly in evolution. We shall present an informal passage of Huxley's in which he explains the logical principles of induction and deduction for a wide audience. Notice that Huxley, writing in another century, uses many of the techniques of good speech: personal pronouns and a transitive sentence structure, familiar comparisons and analogies, an idiomatic vocabulary, and qualitative statements. In this way he is able to explain a difficult pair of ideas with grace and dispatch.

> Suppose you go into a fruiterer's shop, wanting an apple—you take one up, and, on biting, you find it is sour; you look at it, and see that it is hard, and green. You take up another one and that too is hard, green, and sour. The shop man offers you a third; but before biting it, you examine it, and find that it is hard and green, and you immediately say that you will not have it, as it must be sour, like those you have already tried.
>
> Nothing can be more simple than that, you think; but if you will take the trouble to analyze and trace out into its logical elements what has been done by the mind, you will be greatly surprised. In the first place, you have performed the operation of induction. You find that, in two experiences, hardness and greenness in apples went together with sourness. It was so in the first case, and it was confirmed by the second. True, it is a very small basis, but still it is enough to make an induction from; you generalize the fact, and you expect to find sourness in apples where you get hardness and greenness. You found upon that a

general law, that all hard and green apples are sour; and that, so far as it goes, is a perfect induction. Well, having got your natural law in this way, when you are offered another apple which you find is hard and green, you say, "All hard and green apples are sour." That train of reasoning is what logicians call a syllogism, and has all of its various parts and terms—its major premise, its minor premise, and its conclusion. And, by the help of further reasoning, which if drawn out, would have to be exhibited in two or three other syllogisms, you arrive at your final determination. "I will not have that apple." So that, you see, you have, in the first place, established a law by induction, and upon that you have founded a deduction, and reasoned out the special conclusion of the particular case.

Huxley's essay is a good example of informal writing suitable for explanation and for the type of reasoning we employ in letters, speeches, and essays. Though some of the vocabulary may be old-fashioned—"fruiterer" perhaps for "supermarket"—it is clearly good explanatory writing, achieving directness and intimacy. This quality is obtained because Huxley does not feel called upon to browbeat the reader, that is, support his explanation at every point with the qualifications and weighty references which a lesser man might feel were necessary to defend his position as a heavy thinker. This is not a proof Huxley offers, not argumentative writing. But proof is not necessary in explanations. Any technique which engages the reader's attention and brings the explanation home to him in terms he is already familiar with should be suitable.

Formal Writing. In treating formal engineering report writing, we may be forced willy-nilly to accept the conventions of the typical corporation style guide. These style guides may not permit the use of personal pronouns such as the editorial "we." But if "we" is permitted, it should be used because it gives the writer flexibility in using dynamic rather than passive sentence forms. The imperative sentence form ("Rotate the control knob to obtain the best focus") is a special form of the dynamic sentence with the implied subject "you" which can always be used in giving directions and specifying processes. Use it in formal writing whenever the occasion arises because it offers the simplest form of English sentence structure. Use of the implied personal subject "you" also solves many problems of ambiguous modification.

An example of poor formal engineering writing will be given; then an amended version will be developed, using some of the techniques we have mentioned.

It has come to the attention of this office that many problems of vehicle test have developed recently and that severe schedule slippage must now be faced as a potential. A constant progress report of test status requirements shall be maintained by the engineer to ensure that supporting activities will complete their assignments in due time. Matters which cannot be satisfactorily accomplished in a timely manner must be reported through internal supervision accompanied by a statement of actions that have been applied, as well as recommendations for management action of a type to facilitate coordinated efforts in alleviating the lack of timeliness.

Reading this involved, turgid passage, letting it wash through our brains like a flood of dirty water, makes us wonder what happened to the language since such crystalline streams flowed from the great English scientists of Huxley's period. Let us rewrite the passage as an editor might, applying some of our rules of syntax.

Reports have come to this office recently of many vehicle test problems and of severe potential schedule slippage. The individual engineer must frequently report his status so that the scheduling of support tasks can be realistic. He should report any possible slippage through his supervisor; the report must describe actions he has taken to alleviate the slippage, and it must recommend actions the manager can take to coordinate an overall solution of the problem.

With our "before" and "after" passages now on display like the faces of two small boys in a soap ad—both perhaps looking a bit unnatural—let us assess what we have accomplished in terms of sentence structure. Close reading will show that the second passage employs dynamic sentence forms throughout and that the burden of modification and compounding in each sentence has thereby been reduced. In particular, unneeded passive forms of the original have been changed; "must be faced," "shall be maintained," "cannot be . . . accomplished," "must be reported," and "have been applied" were all switched into dynamic sentence forms (Rule 1). A particularly flagrant violation of Rule 2 occurred in the second sentence: "Matters which cannot be satisfactorily accomplished in a timely manner. . . ." Such a weak choice of subject as "matters" always requires a great deal of modification to limit it to something definite; in addition, "matters" is compounded by the tag connective "as well as" inserted at the end of the sentence. This makes the entire original subject "Matters which cannot be accomplished in a timely manner as well as recommendations for management action of a type to facilitate coordinated efforts in alleviating the lack of timeliness must be reported . . ."—a true verbal monstrosity. Rule 2 required us to put principal ideas in the principal structural slots, and lesser ideas in the modifiers; fortunately, this advice has been followed in the cleaned-up version.

Rule 3 is the rule of parallel structure, requiring us to express equivalent ideas by equivalent grammatical structures. No parallelism of this sort appeared at all in the original passage. In the rewritten passage, an example of parallel structure appears in "The report must describe . . . must recommend. . . ." Rule 4, requiring the close management of modifiers and reference words so that they do not become ambiguous, is not clearly violated in the original passage, which is about as much as can be said for it.

Word Choice. Something must be said about word choice—diction—at this point. Errors of faulty diction are commonplace in the original. They are not necessarily all cleared up in the amended version because we have had to guess at meaning. What is meant by "constant progress report," for instance? Does "constant" imply twenty-four hours a day, or shall we substitute "daily," "hourly," "periodic," "frequent," "written," or "oral," or what? We do not know. What are "status requirements"? Is either "status" or "requirements" alone sufficient here? Does "due time" mean "on time"? Why does anyone say, ". . . facilitate coordinated efforts in alleviating the lack of timeliness"? This sounds like baroque poetry in its worst stage of decadence, surely unlike honest, rough-and-ready American engineering speech. We are tempted to conclude that the man who wrote this has other degraded habits of even worse sort. But no, upon seeking out the author, we probably find some poor fellow with unmatching tie and socks who is forthright about sports and politics and trying earnestly to pay off a mortgage. Answering our questions about how he managed to produce such degraded text, he may tell us he "writes like that" because the boss wants him to. We conclude his boss is the worst sort of windbag and hypocritical, pompous, vain, affected, posing, secretive, and a cousin of the director, to boot. But no, we are told, the boss really is okay; he's just funny about some things. So it goes. Who is responsible for all this hypocrisy of statement? All of us are, and we must all try to do better. We must all write as clearly and briefly as we can, employing our four clever rules of syntax and taking as many hints from good speech patterns as possible.

Writing Models. In conclusion, we recommend that the practicing writer read and observe the practices of skilled engineering writers. We have already mentioned T. H. Huxley; another great English science writer was J. B. S. Haldane. The modern Americans, Linus Pauling, Nobel prize winner in chemistry, and John R. Pierce, credited with systems engineering design of the first communications satellite, are others. The magazines *Scientific American* and *Science* present consistently well-written and well-edited scientific and engineering articles, and *The New Yorker* presents many skillfully written pieces covering topics of interest from plastics technology to spacecraft design. On the other hand, the engineer who reads only the newspapers, trade magazines, and government and corporation technical reports may find he is using a set of faulty writing models. It is all like trying to improve your tennis game; it may be wiser to bat the ball around with a Davis Cup champion than with your neighbor with the heavy, limp rackets if you want to improve your game.

SUMMARY

We considered the problem of writing clear and brief engineering text and concluded that the fundamental problem is one of poor syntax. Many sentences of engineering text are too long, too complex, and too confusing to be readily understood.   We examined English syntax and developed four rules of good usage which engineering writers can apply to their own text.   These rules relate to (1) the proper sentence form, (2) principal and subordinate structure, (3) parallel structure, and (4) order and proximity requirements.   In developing these rules, we examined many subsidiary concepts, including dynamic and static structure, passive structure, principal and modifying structure, coordination and subordination, normal order, modification, and reference.   We next studied certain artificial requirements on formal engineering text, particularly the rule of impersonality and the rule of caution, and concluded that these requirements are not necessary to serve our purposes and they possibly defeat it.

BIBLIOGRAPHY

Bach, Emmon, *An Introduction to Transformational Grammar,* Holt, Rinehart & Winston, Inc., New York, 1964.

Baugh, Albert C., *A History of the English Language,* 2d ed., Appleton-Century-Crofts, Inc., New York, 1957.

Bloomfield, Leonard, *Language,* Holt, Rinehart & Winston, Inc., New York, 1933.

Haldane, J. B. S., *Possible Worlds,* Harper & Row, Publishers, Incorporated, New York, 1927.

Huxley, T. H., in Loren Eiseley (ed.), *On a Piece of Chalk,* Charles Scribner's Sons, New York, 1965.

Morris, Jackson E., *Principles of Scientific and Technical Writing,* McGraw-Hill Book Company, New York, 1966.

Pauling, Linus C., *Chemical Bond: A Brief Introduction to Modern Structural Chemistry,* Cornell University Press, Ithaca, N.Y., 1967.

Pierce, John R., *Science, Art and Communication,* Clarkson N. Potter, Inc. (Crown Publishers, Inc.), New York, 1968.

Chapter **8**

Practical Training Methods

JOHN W. HANNON

**Executive Vice President,
Maynard Research Council Incorporated, Pittsburgh, Pennsylvania**

Training is an important section of the whole problem of communications. Proper training methods and attitudes affect vital areas of production, quality, and labor relations. This is a how-to-do-it discussion of training needs, industrial applications, and methods. Because of his natural fact finding ability and information gathering techniques, the industrial engineer represents a key figure in proper training activities. Discussed are training needs and philosophy, master training concept methods, various job training ideas and skill presentation methods, and intrinsic programming or machine teaching applications.

KEY ROLE OF THE INDUSTRIAL ENGINEER IN TRAINING

Although there are many facets in the development and administration of proper and productive training, much of the success in training depends upon skills known to industrial engineers.

Fundamental to all training is fact finding, that is, discovering exactly what is involved in each job, what tasks are required, what methods are used, and what results are expected. Obvious fact finding techniques are process charting, operation analysis, motion study, time study, and MTM. Expected results are determined by establishing production and quality standards, and efficiency is measured by operator performance records. It is no wonder that companies turn to industrial engineering departments to select training engineers. Here they have men trained and experienced in "getting the facts."

It is very important that the trainers themselves be thoroughly trained. Just because a man is an expert in his field, it does not follow that he can successfully impart his knowledge to others. Companies must always make sure that the men doing the training are themselves qualified in the techniques of training.

NEED FOR OPERATOR TRAINING

Training is or should be a continuing process. The reasons are fundamental. Consider the following normal possible changes or situations in the typical firm over even the relatively short term of one year:

1. Number of new employees
2. Number of employees retired
3. Turnover rate
4. Job transfers
5. Job promotions
6. Introduction of new products or processes increasing job skills required
7. New machines added
8. Number of methods changes
9. Increases in scrap or rejection rates
10. Increases in customer service costs
11. "Makeup" costs; determined by the number of employees working at below-standard performance
12. Seniority provisions causing continuous job changes because of a "bumping" procedure

These and other situations are common. A closer look at just the last two should emphasize training needs.

One purpose of training, and certainly a management objective, is to get all employees working at a performance level of 100 percent (standard) or better if incentives are used.

Consider the situation where 50 people in a department are, for perhaps a number of reasons, averaging an 80 percent performance level. If their average wage guaranteed is $3.00 an hour, they are actually earning $2.40 an hour. To state it another way, they are being overpaid $0.60 an hour. In one week, "makeup" costs (wages paid for work not done) would amount to $1,200 ($0.60 per employee × 50 employees × 40 hours). This situation indicates a potential training need.

Take another plant situation. Certain seniority provisions had been agreed to over a long period of time. These rules evolved to the point where any man with more seniority could "bump" a man with less seniority out of his job. This resulted in constant job changes (weekly and sometimes daily) as production levels went up or down and an almost continual training problem on nearly all jobs. When a carefully planned training plan was proposed, the operating vice president saw in it not only a means of solving the current training problem, but also a way of bringing to the attention of the union the almost ruinous cost of constant training.

Training needs are continuous, but well-organized training will meet the needs and lead to cost reduction and control.

TRAINING PEOPLE—NOT LABOR

Operators are thinking human beings who have pride, dignity, and a well-developed ego. They respond in accordance with the way they are treated. Years ago, all too often a man was considered only as a supply of energy to be used at management's discretion to get a job done. Present-day industry has found that it no longer is practical to ignore the man as a unique individual when hiring him. If a person is properly prepared for his responsibilities, the results are measurable in dollars and cents by reduced turnover and increased output.

In an article published in *Management News*, Lawrence A. Appley, president of the American Management Association, wrote:

Individuals usually produce more when they know:

What they are supposed to do
What authority they have
What their relationships are with other people
What constitutes a job well done in terms of specific results
What they are doing exceptionally well
Where they are falling short
That there are just rewards for jobs well done and exceptionally well done
That what they are doing and thinking is of value
That the boss has a deep interest in and concern for them
That the boss is anxious for them to succeed and progress.

These statements are amplified and explained by the master training concept or accelerated training methods. This concept, developed originally in Holland by the management consulting firm, Raadgevend Bureau Ir. B. W. Berenschot, contains all the elements of sound training, and it will therefore be discussed in some detail.

MASTER TRAINING CONCEPT (MTC)

The master training concept is a way of solving problems involving people. And what problem does not have people at the root of it? They are problems caused by the fact that someone did the wrong thing—or did not do what should have been done. MTC recognizes that most such actions happen because the person did not really understand what was the right thing to do. "Few people want to do a bad job." An operator may produce poor quality because he or she does not realize how it affects the customer or later operations. An operator may agitate against the company because he or she does not know why the company has a certain policy. An operator may turn out less than standard production because he or she does not realize that other operators suffer when they do not have sufficient work—or perhaps because it was left up to the operator to figure out how to do important details of the job. Of course, there are many reasons why the right thing was not done when it should have been.

It is management's job to teach employees exactly what to do—in a way that makes the correct things to do ingrained in them—so they just cannot bring themselves to do the wrong things.

This requires training—training of a kind that is different and more comprehensive. The somewhat unusual word "concept" is used because MTC is much more than setting up another training course. Indeed, MTC is not a course at all, although courses designed to solve specific problems are part of the concept. MTC conceives that most problems can be eliminated or at least reduced in severity by getting together with the people involved and helping them overcome their difficulties. Of necessity, MTC must be broad in its approach. Problems affecting training results must be considered and dealt with. Training should not be an end in itself. MTC is designed to solve a management problem.

The formal definition of MTC is: MTC consists of the selection of applicable training and other related procedures and their integration into a training program designed for a specific situation to the end that the trainee will become in the shortest possible time a well-adjusted, competent part of the industrial civilization in which he or she lives.

"Applicable training and other related procedures" refers to the use of various techniques to train or to deal with problems affecting the training results. Included are such things as detailed analysis of job method, attitude surveys, review of employee selection procedures, stress on trainee participation, use of visual aids, use of guest instructors, extra practice on the "difficult to learn" parts of the job, teaching until the trainee has learned, teaching the trainee to want to do the job, careful follow-up procedures, and supervisory training. There are many others.

". . . designed for a specific situation" means that each problem involving training is usually different and must be dealt with individually. Differences in people, jobs, materials, places, and the like means that each program must be custom-built. You cannot get the solution "off the shelf."

". . . in the shortest possible time" means to concentrate the training in as efficient and economical a program as possible. Results show that reducing conventional training time by one-third or one-half is not unusual.

". . . a well-adjusted, competent part of the industrial civilization in which he or she lives" means it is important to know how but it is just as important to want to do the job and to be a teammate.

MTC then is an approach—a broad approach, involving training, designed to solve a management problem. Specific solutions are developed by the people involved—by and with the people who are going to have to follow through and live with the solutions.

Uses of MTC. The most common uses of MTC are:
1. To train new operators
2. To retrain experienced operators in new methods
3. To give present satisfactory operators a refresher or tune-up course
4. To improve low performance and to eliminate makeups
5. To upgrade present operators to jobs of greater importance

New Operators. The training of new people is perhaps the major purpose of MTC. The object is to train the new worker "to become a well-adjusted, competent part" of the company.

When a new man is hired, he knows little or nothing about the company, its purposes and problems, and the people in it. He may know nothing at all about the work or he may know something about how the work is done elsewhere. Training the new worker involves a complex and comprehensive training program. This is often the easiest to conduct, especially when the trainee earnestly desires to learn and has little or nothing to unlearn.

Retraining Experienced Operators. It is often difficult to retrain experienced operators when methods or conditions have changed. The difficulties usually stem from attitudes rather than the physical problem of replacing an old skill with a new one. A man may not want to change his methods because he does not see what he can gain by doing so. Often he fears for his own security. He may recognize that the new method is better, safer, less tiring, and so on. But if he feels that he or his friends may be less secure in their jobs or that his familiar environment or work relationship will be changed, he may resent the change emotionally and strongly resist it.

MTC corrects the effects by straightening out the causes. It trains attitudes as well as skills, making it exceptionally effective for retraining.

Refresher or Tune-up Training. Some highly complex jobs, such as certain types of maintenance work, are never truly learned under conventional training methods. Although an experienced operator can get the work done, each new job requires a great deal of planning, trial and error work, and discussions with others before it can be properly executed. The operator has no clearly established thought patterns or work patterns, and each new job is tackled almost as if it were a completely new problem. This frequently leads to a good deal of lost time and false motion.

Using MTC, the industrial engineer establishes the best thought patterns and work patterns for doing the job. He sees that they are taught so clearly and thoroughly that they actually become a part of the operator's makeup. Thereafter, the operator will use these patterns automatically and correctly and will be able to do his work in substantially less time.

Improving Low Performances. The problem of improving low performances and eliminating the cost of makeups is similar to the training and retraining problems. Low performance may be caused by unwillingness to use the correct methods or by lack of knowledge of how to do the work properly. It may also be caused by wrong attitudes toward the job.

Regardless of the reasons, other than complete unfitness for the job, the MTC approach can be used to solve the problem. The industrial engineer must determine the causes for the low performances and then design a proper training program which will correct these causes.

Upgrading Present Operators. MTC can be used to upgrade people to positions of greater responsibility. This may involve training an operator to do more complex jobs in the same class of work, or advancing a man to a higher classification of work. In both cases, the training will result in a promotion and a corresponding increase in pay for the man.

MTC may also be used to increase the versatility of a group of operators. Industrial engineers frequently encounter a situation where all the men in a group of workers are doing the same general class of work but each member is capable of doing only a limited kind of work. Regardless of what causes this situation, it results in an inflexibility which causes scheduling difficulties. MTC can be used to correct this situation by developing versatility in the workers. What appears to be a production control and scheduling problem can thus often be solved by MTC.

Benefits of MTC. The benefits resulting from MTC can be summarized as follows:
1. Respect for and loyalty toward the company is greatly increased.
2. Labor turnover is reduced.
3. Training time and cost are sharply reduced.
4. Greater production is obtained from a given set of facilities.
5. The quality of the work is greatly improved.
6. Scrap and waste are reduced both during and after training.
7. Greater profits are obtained because more units are available for sale.
8. Operators reach their normal earning power more quickly.
9. A more versatile work force is developed.

Specific Elements of MTC. MTC is a complex of training procedures woven into a training program designed to (1) orient the employee to his work environment, (2) develop in him positive attitudes, and (3) develop within him the skills necessary to do the work. None of the procedures used is particularly new, but it is the integration of all necessary procedures into a well-designed program and the correct presentation of the program that get results.

A successful program requires infinite attention to detail. If the learner is to be helped to (1) become oriented to his environment, (2) develop proper attitudes toward the company, his work, and his associates, and (3) learn his job in a few short days, weeks, or months, there are any number of details that must be:
1. Identified as being necessary if the objectives of training are to be met
2. Woven properly into the training design
3. Taught, with all that this implies

Figure 8-1 will help give a quick if incomplete impression of the component parts of an MTC program. Some of the steps shown are discussed briefly in the paragraphs which follow.

Analysis. Before a training program can be designed, the training engineer must determine what it is necessary that the program accomplish. This involves surveying the work which is to be learned, defining the training problem, analyzing the work in considerable detail, and identifying or developing the best methods for doing the work. This involves many of the techniques of methods engineering and may require the use of process charts, operation analysis, and motion study or MTM. A solid methods engineering background is necessary to perform this step properly. Present performance records are obtained in the areas of production, quality control, and personnel. Personnel records include such factors as turnover rate, grievances filed, and the like. These same records will indicate training results after programs are conducted.

At the same time, the analysis made by the training engineer must often go further than the analysis conventionally made by the methods engineer. The training engineer must be able to distinguish between elements that are difficult and elements that are not difficult. He is primarily interested more in what must be taught rather than in how long it takes to perform at least certain of the elements. The running time of the machine, for example, is not of much interest to him, because once the machine is set right, there is no training problem involved.

On the other hand, he is interested in what it is that causes the operator to do certain things. For example, he will ask first, "Why did he go there?" and only then "What is the quickest way to do it?" It may appear that the experienced operator

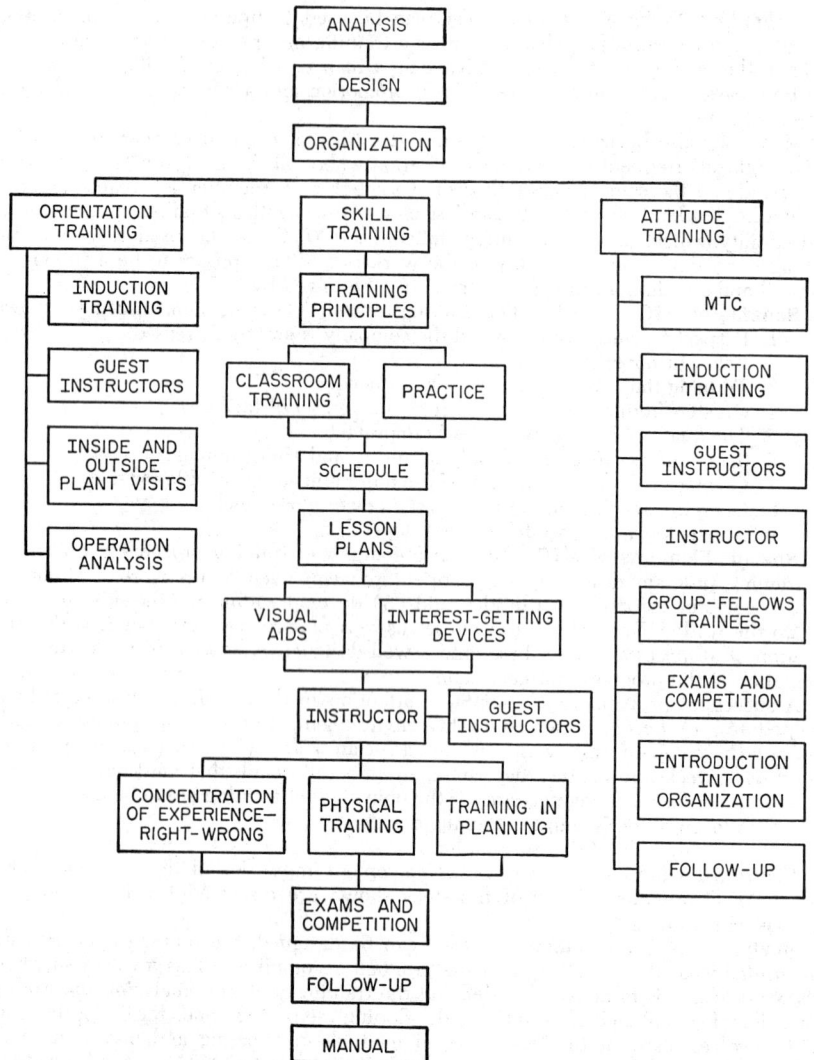

Fig. 8-1. Chart illustrating the component parts of an MTC program.

apparently has "eyes in the back of his head" and is able to see machine trouble without looking at it. Further analysis will show that the operator consciously or unconsciously is guided by sounds which would not be noticed by the untrained. Thus the problem becomes one of training the new operator to hear sounds correctly.

It is doubtful if the methods engineer would get into such matters in the course of his regular studies. Perhaps he should, but in all probability his attention will be fixed on what the operator is doing and how long it is taking him to do it. It is for this reason that the training engineer must go further in his analysis work than the methods engineer. Therefore, in addition to having a sound methods engineering background, the training engineer must possess the initiative, imaginative, and deductive traits that are necessary for success in making the kind of analyses he is called upon to handle.

Design. Based upon a knowledge of the training problem and the methods by which the work is done, the training engineer is in a position to begin the design of the master training program. This is a more or less continuing process from this point until the very end of the program. Experience has shown that it is not necessary to have the program worked out in every detail when the training begins. Indeed, a certain amount of flexibility must be maintained if the training is to be accomplished fully and in the shortest possible time. Variations in the human element as represented by the instructor or the trainees or both will cause the optimum length of the training program to vary from group to group. By the same token, the design of the program may have to be varied to fit the capabilities, past experience, and knowledge of the people involved.

Design, therefore, must be regarded as a continuing process. It requires a thorough knowledge of the psychology of learning, training methods, and human and group relations. In addition, the training engineer must be imaginative and able to make an original approach to new problems if he is to be able to do a satisfactory job of training program design.

Organization. The typical organizational unit for an MTC program is shown by Figure 8-2.

The coordinator is usually a line department head or assistant department head—someone with line authority. It is his function to verify that the training course is being developed and presented properly in light of the department needs. It is also his function to work with the training engineer to see that proper training aids such as mock-ups and sample materials are made available during training. He should indicate or approve who is to be trained and when. He should approve on-the-job training arrangements and arrange for supervisory cooperation wherever necessary.

The training engineer designs the specific program. In preparation for this, he studies methods and improves them whenever possible. He trains the instructors in training methods and continues in an advisory capacity throughout the program. The training engineer can work with several instructors after the programs have once been designed.

The instructor is usually a skilled operator or foreman who has had special instruction from the training engineer on how to train. The instructor is a key person in the program and must be carefully selected and adequately trained. He must be 100 percent qualified from a job knowledge and experience standpoint, for it is this person who must transmit all technical knowledge about the job to the trainees.

The guest instructors are other people from the company's organization, called in to present special information to the trainees. It may be technical information or information which the management or in some cases the union wishes to present. For example, the chief industrial engineer may be called upon to handle a session on standards setting or wage incentives, the chief inspector a session on quality, and so on. It is usually desirable to have the material to be presented by the guest instructors well spelled out, particularly when matters of company policy are involved.

The trainees are the people to be trained. One instructor can handle up to ten trainees at a time. The trainees are made to feel very much a part of the program.

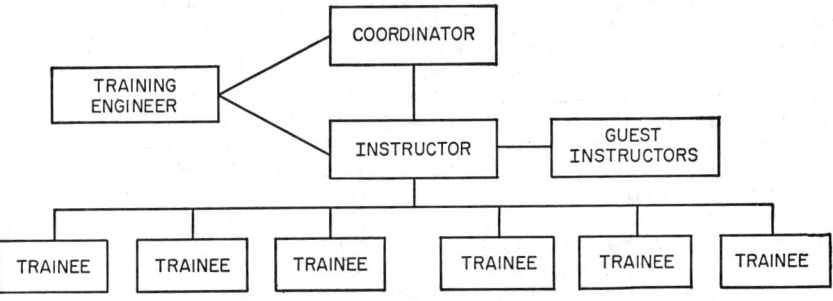

Fig. 8-2. Chart of a typical organizational unit for an MTC program.

They recognize that the company is interested in them as individuals, and they in turn become interested in the company and its problems.

Orientation Training. With this organizational picture in mind, we are now ready to turn our attention to the first of the three subareas of training orientation. To orient means to find one's proper place or relationship relative to other objects or persons. Orientation is accomplished by induction training where the trainee is introduced to the company and his surroundings, by guest instructors who present both management information and technical information on matters which will affect the trainee, and by plant visits both within the company and to outside plants which supply materials or use the company's products. Company benefits, general rules and regulations, and union-management policies are also covered during induction training.

Attitude Training. Attitude training is in part accomplished during orientation training and skill training. The procedure itself which deals in a constructive manner with the problems of each trainee results in a form of attitude training. So does the induction training and the training given by guest instructors.

The instructor can do a great deal to mold the attitude of the trainees. Hence he should be given coaching by the training engineer in this respect. Attitude development will also come from fellow trainees. During the program, an effort is made to build up a positive group spirit. If a person knows, likes, and respects his fellow workers, he is more likely to find the company a good place to work.

The quizzes, examinations, and competitive situations which are introduced during training are likely to have an effect on attitudes, particularly if the trainee handles them successfully. When he meets the challenges that they impose, he knows, and he knows that his boss knows, that he has demonstrated his ability to do the work successfully.

Finally, the procedures which are followed in introducing the new person into the organization and in following up afterward to make sure that he is getting along all right have a positive effect on attitude. The new person feels that he is important to the organization and that others care about how he is getting along.

Skill Training. Training to develop job knowledge and skills is, of course, the heart of the MTC program. That is why it occupies the central position in Figure 8-1. Certain psychologically sound training principles are woven into all phases of skill training.

Part of the training is done in a classroom or training center. This is set up with tables, chairs, blackboards, and the like in typical classroom fashion. Here are taught matters of theory which can only be covered properly in the classroom.

The major part of the training, however, is done in a location where the trainee may learn by doing. The location may be one corner of the classroom, a special area set aside for training purposes in the shop, or in the regular shop department, depending upon the nature of the work to be learned and the equipment involved.

The equipment may be the actual equipment as it is used in the shop, the actual equipment modified for training purposes, or a simulated workplace set up for training purposes.

Usually, the actual shop equipment is better to use than specially designed gadgets, but it may be desirable to modify it for training purposes. In the operation of spinning, for example, a new girl is likely to be frightened by the big, noisy spinning machine. Furthermore, it would be impractical to move it into a training area; so if it were used for training purposes, it would have to be done amid the distractions of a busy department. In addition, production would be interfered with as the new girl fumbled about with a regular piece of equipment.

This whole situation can be easily resolved by moving one spindle of a spinning machine to the training room. There the new girl can learn in peace and tranquility. Everything she needs to know about spinning can be learned on the single spindle without fright or distraction.

A general schedule is made up before the program begins, but no fixed overall timetable is prepared. The schedule is used to keep track of the time spent on various activities, but it serves as a historical record rather than a future plan.

It is a fundamental principle to take as much or as little time as is required to cover any given subject satisfactorily. The amount of time spent on any lesson is determined on a day-to-day basis by the instructor, the coordinator, and the training engineer. Of course, after the same program has been given several times, the training engineer will have a pretty good idea of how long the various segments of the course will take. He can then set general time budgets which the instructor can use as a guide. The principle of teaching until the trainee has learned and no longer, however, still applies.

Instruction guides are prepared by the training engineer in outline form with the help of the instructor and other experienced operators. They may be quite detailed, but it is more common to find them rather sketchy, with the details to be filled in by the instructor from his own experience. Supervisors are questioned on technical points as instruction guides are developed. Their answers, however, are critically evaluated by the training engineer. Experience has shown that supervisors are sometimes inclined to be unrealistic in their requirements. They are inclined to confuse what a person must know to do the job successfully with what the supervisor thinks it would be good for the trainee to know on general principles.

Extensive use of visual aids is made during MTC, but only if visual aids will get across points which cannot be taught directly from the workplace itself. Many interest-getting devices are used, particularly for teaching some of the duller subjects. A card game used to teach an employee the names of the parts of a machine is a typical example.

Much of the actual teaching is done by the conference technique. The classroom training periods are conferences rather than lectures. The instructor lets the group members tell him the answers rather than giving them himself. He must, of course, pass on information which is completely new to everyone in the group. But if at least one member of the group knows something about the subject under discussion, the instructor will encourage him to explain it to the others. This ensures constant participation in the discussions on the part of the trainees and keeps interest high.

Everything possible is done to keep up the morale of the group. Recognizing that physical fatigue is likely to get a person down, special physical training exercises can be introduced periodically either to develop the muscles that a person will use on the job or to help to relax muscles that may have become too tense during a prolonged period of activity.

Some jobs require some rather unusual skills. In shipbuilding work, certain jobs require walking on planks at rather dizzy heights. To condition trainees for this, special equilibrium exercises can be developed.

The program is designed to give experience in concentrated doses, which is one reason trainees learn so quickly. For example, in operating newspaper printing presses, a reel operator during the course of a week may change fifty rolls of paper. On the other hand, the same operator may get the opportunity to change an impression roller only once a month. If each of these jobs must be performed fifty times before an operator is considered proficient, then it is obvious that it will take fifty months to train an operator, if the training is done on the job. If, however, the training program is so arranged that the operator is given the opportunity to practice each element fifty times in a period of two weeks, then the training time can be reduced from fifty months to two weeks.

Besides providing for concentration of experience, MTC stresses the errors which are commonly made on the work and how to avoid them. During the practice sessions, the instructor watches the trainees carefully. If he sees a person about to make a mistake, he forces himself "to keep his hands in his pockets" and lets the trainee go ahead unless danger or damage is involved. Then as soon as trouble develops, he stops that trainee and asks what happened. He tries to let the trainee figure out the remedy for himself, giving him as little guidance as possible. In this way, the trainee learns and remembers.

Training in planning is another extremely important part of the program. There are two kinds of situations for which a person must be given training in planning or, perhaps better, thinking: reaction situations and analysis situations. Reaction situations are those in which a person should react rather than think. They occur on the repetitive part of any task. Analysis situations require thinking or knowing how to think in logical thought patterns. They occur on nonrepetitive work. Training in planning then becomes a part of the training.

During the progress of the program and at the end, things are enlivened by the introduction of quizzes, examinations, and competitions. These are used as a means of checking on the effectiveness of the instructor and the progress of the trainees. The weekly quizzes are usually prepared by the instructor for the first two or three weeks. Thereafter, the trainees themselves suggest the questions which should be asked, giving them a further sense of responsibility and participation. The examinations at the end of the program may be prepared by the supervisor under whom the trainee is to work. Small prizes are sometimes offered to quicken the competitive spirit of the group.

After the program has concluded, the instructor follows up periodically to see that the trainees are getting along all right. The training engineer also keeps in touch with their progress and at the same time makes up a manual which captures on paper the experiences gained from the program. The manual includes the schedule, instruction guides, quizzes, and exams, and anything else that may be helpful the next time the same program is given.

Equipment for Practice Work. It will be noted from the foregoing that a good deal of stress is placed on practice in doing the parts of the work that require the development of skill. It is important, therefore, that adequate equipment be provided to give each trainee plenty of practice opportunity.

When the equipment involved is expensive, it is often felt that trainees can double up. Experience has shown that it is better not to conduct a program than to attempt it with too little equipment.

On some classes of work, it is desirable to have actual jobs to do to have effective training. To learn to lay a floor, you should preferably have an actual floor to lay. In other cases, however, it is possible to simulate a workplace which will give all the practice opportunity necessary.

Summary. From this discussion, it may be seen that, although MTC is a complex thing to describe, it is complex because there are so many features to it and not because it is complicated. There are many things that a new person must learn before he can do any job properly. If we wish the trainee to learn them correctly and quickly, we must be prepared to do a thorough training job.

Some of the basic elements we have talked about are:

1. Detailed analysis of the work methods on the job to be learned
2. Full use of the knowledge and experience of present employees
3. Use of a skilled operator as the instructor
4. Use of guest instructors for special subjects
5. The building of a positive group spirit
6. Liberal use of quizzes, competitive situations, and prizes
7. Teaching both on and off the job—theory for part of the day, learning by doing for the remainder
8. Teaching until the trainee has learned and no longer
9. Visual aids and interest-getting devices whenever applicable
10. The use of the conference technique for much of the teaching
11. Weekly critical review of the program by the trainees
12. Physical training exercises to help develop the muscles needed on the job
13. Concentration of experience by giving emphasis to the difficult-to-learn elements of a job
14. Stress on errors commonly made and how to avoid them
15. Development of logical thought patterns for handling situations requiring analysis
16. Maximum participation by trainees in decision making relating to content and conduct of the training
17. Flexible pattern for the training schedule
18. Continuing development of training manual
19. Insistence upon adequate equipment for practice work

PROGRAM DEVELOPMENT PROCEDURE
FOR THE TRAINING ENGINEER

The industrial engineer may because of his specialized knowledge and abilities be called upon to act as the training engineer in developing training programs in accordance with the master training concept. The procedure which he should use is as follows:

Survey

1. Find out from affected management:
    a. Why training is needed
    b. Where and when training is needed
    c. What type of training is believed to be needed

2. Find out what everyone's job is in the department where training is to be done as well as the department's relationship to the whole plant.

3. Become familiar with the function of the department, including the purpose of each operation. Find out how these operations fit in with operations of other departments.

4. Get applicable figures on output, material and labor costs, scrap, waste, turnover, number of employees, accidents, number of grievances, and the like.

5. See that a coordinator is assigned and work with him.

6. Find out selection and induction procedures for persons to be trained. Determine standards of selection, including muscular, sensory, and mental skills required.

7. If applicable, become familiar with union contract. Talk with union officials to gain their cooperation, interest, and feeling of participation.

8. Become familiar with the wage incentive system.

9. Learn present training methods.

10. Meet with all department personnel. Explain what is to be done, how they will benefit, and how they can help. Meet individually with all supervisors.

11. Visit related line and staff departments to determine their relationships and their ideas about possible problems and solutions.

12. Using data gathered so far, design applicable survey forms to get more information from the people in the department. Conduct a survey and tally, analyze, and report results.

13. If necessary, redefine the problem that the program is to solve, prepare a statement of objectives, and get approval from all concerned.

Analysis

1. Prepare subject outline for training course under the headings of orientation, theory, and practice. Get necessary approval.

2. Select instructors on the basis of psychological testing, length of service, knowledge of the job, and opinions of supervisors.

3. Schedule and conduct training of instructors.

4. Make detailed study of the operations in which training is to be conducted. Get detailed information about knowledge and skills required.

5. Determine standardized job method. Get participation of instructors, supervisors, and operators.

6. Make any obvious methods improvements.

7. Determine production standards for qualified operators.

Design and Organization

1. Prepare tentative day-by-day schedule. Determine tentative starting and ending dates of the course.

2. Order special training materials needed such as mock-ups, photographs, and the like—anything which takes time to prepare.

3. Determine follow-up training to be done until operators are qualified.

Introduction

1. Arrange for suitable classroom facilities.

2. Make arrangements for selecting trainees. If new employees, use applicable tests to assist in selection. If present employees, select trainees on the basis of similar experience.

3. Develop subject outlines for beginning the course. Explain what is needed to those who will be guest instructors.

4. Advise all affected persons as to progress and their future responsibilities with the trainees—especially among the supervisors.

5. Complete instructor training prior to beginning course.

6. Initiate desirable public relations activities.

Coaching

1. Participate in first-day ceremonies.

2. Sit in on course as necessary to evaluate progress, coach instructors, and modify schedule.

3. See that outlines and quizzes are prepared properly and in time for use.

4. Prepare training manual.

5. Advise all concerned about progress. Take necessary steps to provide for acceptance of the trainees among present employees.

Evaluation

1. Follow up trainee progress at least until they are qualified.

2. Modify course contents and schedules based on experience and future needs.

3. Record evaluation factors for measuring training results—immediate and long-range. Get necessary approval.

4. Report training results to all concerned.

SUMMARY OF DUTIES OF THE TRAINING ENGINEER

1. Identify the training problem.

2. See that the coordinator is selected, get his assistance where necessary, and advise him about training progress from time to time.

3. Collect information from every likely source.

4. Design the training program, including skill, attitude, and orientation training.

5. Organize the training program.

6. Help select the instructors, train them initially, and coach and guide them thereafter.

7. Specify the guest instructors needed and coach them.

8. Prepare the training schedule.

9. Supervise the preparation of the training facilities.

10. Obtain materials and jobs to work on.

11. Initiate public relations activities in connection with the program.

12. Schedule plant visits.

13. Help select trainees upon request.

14. Develop the detailed instruction guides and quizzes before and during the presentation of the training program.

15. Participate in first-day ceremonies.

16. Modify course and subject material as needed.

17. Train supervisors as necessary in any job method changes, course content, new training procedures, and trainee follow-up.

18. Participate in follow-up.

19. Finalize the training manual.

TRAINING METHODS

There are a number of specific training methods which may be incorporated into an MTC program, or they may be used by themselves to accomplish a specific training objective. Among the more effective training methods are:

1. TWI methods

2. Motion picture methods

3. Multimedia instruction methods

TWI Methods. The main features of the TWI principles of job training are:

1. The preparation of the instruction

2. The instruction process

The "preparation" is broken down into:
Step 1. Analyze the job.
Step 2. Prepare a training time schedule.
Step 3. Set up the workplace properly.
The "instruction" process is broken down into the following four steps:
Step 1. Prepare the operator.

Put him at ease.
State the job and find out what he already knows about it.
Get him interested in learning the job.
Place in correct position.

Step 2. Present the operation.

Tell, show, and illustrate one important step at a time.
Stress each key point.
Instruct clearly, completely, and patiently, but no more than he can master.

Step 3. Try out performance.

Have him do the job. Correct errors.
Have him explain each key point to you as he does the job again.
Make sure he understands.
Continue until you know he knows.

Step 4. Follow up.

Put him on his own. Designate to whom he goes for help.
Check frequently. Encourage questions.
Taper off extra coaching and close follow-up.

The JIT method has broad application in job training, and the analysis is performed by tools well known to the industrial engineer. There are a number of practical ways the instruction material can be presented. One is the job instruction breakdown. Important steps (elements) of the operation are listed in one column. Across from each job step are key points. Operation analysis, motion study, and MTM can be used to prepare the job breakdown.

Figure 8-3 shows a sample job instruction breakdown for the operation Lay Out Parts for Glove. A more detailed job breakdown arrangement is shown in Figure 8-4. Because of the intricate nature of the operation, the action of each finger is described.

Another type of job breakdown is the job instruction card. Figure 8-5 shows an arrangement in right- and left-hand form. This job instruction card was prepared from a detailed MTM analysis and is in the general form of the methods analysis sheet used in making MTM studies.

Whatever the form, the basic principle of showing job steps and key points applies, and industrial engineering techniques are used to prepare the job analysis.

Motion Picture Methods. The use of motion pictures is valuable in training people on jobs where a high degree of manual skill and dexterity is required, the work is repetitive, and the cycle time is relatively short. This method is also desirable in situations where sudden increases in the work force are occurring, producing a heavy training need, and when a series of new products and methods are being introduced. Although not overly expensive, film production takes time, and standard methods need to be established before jobs are filmed for training purposes.

These statements point out one of the advantages of the use of films. When good methods have been established and a number of people are to be trained, they must be taught exactly the desired method. When the desired method is determined, it is then photographed. A single cycle can be cut out of the film, and by joining the ends of the film together, a loop may be made which can be projected continuously. This gives the effect of the operation being performed over and over again. In training new operators, the instructor usually first shows the film, explaining any special points that need emphasizing. The film is shown a number of times at varying

LAY OUT PARTS FOR GLOVES

Steps	Key points
1. Get box.............	1. Using both hands, hang box on rack. Second and fourth fingers should be closest to machine. Outside edge of box should be even with edge of table.
2. Examine box ticket...	2. Look at lot number on box ticket to see that you have the correct parts. Replace ticket in box until box is emptied.
3. Lay out cut of parts..	3. There are three stacks of parts—parts for twenty-four pairs of gloves in a stack.
	a. Pick up one stack at a time—need not be exactly twenty-four gloves if there are strays. Your stack should be paired though—nap on top and bottom of stack. Use both hands to grab parts, to sort out strays, and to even up sides of parts.
	b. If there are loose or stray parts in the box, leave them until you have removed the last stack. Then place them on the proper stacks at the machine.
	c. Lay out parts in the following order:
	(1) First finger
	(2) Second finger
	(3) Third finger
	(4) Fourth finger
	(5) Thumb
	(6) Palm
	d. Place parts on table as shown in drawing of work layout.

FIG. 8-3.   Sample job instruction breakdown sheet for the operation Lay Out Parts for Glove.

speeds until the operators are familiar with it. They are then permitted to try the operation under guidance. In many cases, the film is continuously projected so that the trainees have it as a guide while learning. It is also desirable to project the film at increasing speeds. At first, film projection is slow while trainees learn the method. When this is accomplished, film speed is increased to bring the trainees as close as possible up to standard performance which will be expected of them.

Another form of motion picture training is the use of a special-purpose projector called the Perceptoscope.

The Perceptoscope is a projector that incorporates the features of film strip, motion picture, and tachistoscopic projection, all in a single continuous film. This 16mm electronic device is equipped to read cues on the film sound track to stop motion automatically for still projection. The operator can show motion at any of 16 speeds from 1 to 24 frames per second with a remote control unit. The projector is also equipped, by means of a second back-film loop, to (1) project patterns over major image and (2) carry program cues to advance the front film automatically in a variable time pattern.

These features give the Perceptoscope several distinct advantages as a training device. It enables an instructor to present live action in short sequences and still pictures, animated pictures, or diagrams. The instructor can inject his own comment or invite group participation at any point in the presentation. Thus a whole session can be programmed around a single short length of film.

The Perceptoscope combines the features of motion pictures, slide projectors, flannel boards, and flip charts, because all their features can be encompassed on a single film. The back-film programming and variable-speed functions also make the projector especially suitable for drill or for pacing training for clerical work. To be able to stop live action automatically makes it very versatile for training in machine opera-

WHAT FINGERS OF BOTH HANDS SHOULD BE DOING AS YOU SEW THUMB IN

R.H.	First Starting Position	Sew 1st Crotch	Sew Tip	Second Starting Position	Sew 2nd Crotch and Finish Thumb
Thumb	Center of palm in front of needle	Stays out of way			
1st	Just below crotch	Push pleat away	Out of way to left of presser foot	Just below crotch	Push pleat away
2nd	Just above crotch on edge			Just above crotch on edge	
3rd and 4th	Halfway down thumb	Pull thumb steadily to left of gage	Located center of tip. Guide tip along gage and keep pleat away	Halfway down last part sewn	Pull material steadily to left of gage
L.H.					
Thumb	Underneath thumb and *your* 1st finger	Follows underneath edge of material under *your* 1st finger. Also helps break off thread with your finger			
1st	1 in. below top of thumb and presser foot	Guides material along gage: Slides along edge until crotch is reached. Guides crotch closely along gage (next to presser foot). Then slides along edge again	Guides material along gage	At point that forms crotch	Guides material to end of seam. Then follows material to cut off thread
2nd	At top of thumb next to presser foot	Out of way	At center of tip next to presser foot. Keeps tip from pleating. Swings glove around tip	Out of way next to your 1st finger	Out of way
3rd and 4th	Out of way	Out of way	Two-thirds way up tip, then grab outer edge of seam and pull material toward you so it feeds along gage properly	Out of way	Out of way

FIG. 8-4. Sample of a more detailed job instruction breakdown sheet.

tions for factory or office. Machine operations can be demonstrated on the screen in short, serially presented sequences while trainees work at real machines.

Multimedia Instruction Methods. Many companies have achieved the most effective training results by combining various training techniques. Trainees tend to become bored very easily. By varying the techniques, the trainee's interest is held for long periods of time, and he is more likely to retain what he has learned.

Multimedia techniques cut the time required to teach the material and are more practical and effective than any single technique can be.

Department: Machine Shop _____ Date: _____

Operation: Drill, counterbore, and burr two holes in brass contact

Tools: 3-spindle drill press, one F drill, one ½-in. drill, one 1.005-in. counterbore

Left Hand		Right Hand	
Key Points	Description	Description	Key Points
1. Insert two fingers in bottom slots, push pcs. upward.	Raise both contacts so they protrude from jig.	Grasp both contacts with fingers.	Release L.H. pc. as soon as L.H. grasps.
2.	Remove L.H. contact from jig.	Remove R.H. contact from jig.	
3. Take 2nd pc. from R.H. during move. Drop pcs. no more than 6 inches.	Put 2 finished pcs. in tote pan.	Get 2 pcs. to be drilled.	Transfer R.H. pc. to L.H. during Reach. Pick up 1st new pc., palm, pick up 2nd.
4. Take pc. from R.H., Position with short side of angle up and away from body.	Place 1st pc. in jig.	Place 2nd pc. in jig.	Transfer one pc. to L.H., Position other pc. same as L.H.
5.	Close jig cover		
6. Slide jig on table.	Move jig to spindle No. 1.	Move jig to spindle No. 1.	Turn and walk to spindle No. 1.
7. Reposition for each hole drilled.	Hold jig for drilling of holes.	Drill 4 holes.	
8. Slide jig on table then rotate over on side so cover is at the rear.	Move to spindle No. 2.	Move to spindle No. 2.	Sidestep to 2nd spindle, then rotate jig 90° so holes for counterbore are up.
9. Position jig for each hole, then move toward 3rd spindle.	Hold jig.	Counterbore two holes.	
10. Slide jig on table, then rotate another 90° so that bottom side of 4 holes is up.	Move to spindle No. 3.	Move to spindle No. 3.	Sidestep to 3rd spindle, aid L.H. turn jig, reach for drill handle.

Fig. 8-5. A job instruction card utilizing the right- and left-hand form arrangement.

Left Hand		Right Hand	
Key Points	Description	Description	Key Points
11. Reposition as necessary for burring.	Hold jig.	Burr 4 holes.	Light touch only.
12. Rotate jig 180° so that cover side is up.	Turn and hold jig.	Open jig cover.	Aid L.H. rotate jig.
13. Position each hole under spindle.	Hold jig.	Burr 4 holes.	Light touch only.
14.		Brush chips off table under spindles No. 1 and No. 2.	
15.		Brush oil on drill in spindle No. 1 and on counterbore.	

FIG. 8-5 (*continued*). A job instruction card utilizing the right- and left-hand form arrangement.

TRAINING TECHNIQUES

Some of the training techniques which can be successfully combined are:
1. Sound slide
2. Programmed instruction
3. Computers
4. Learn-by-doing sessions
5. Video tape
6. Motion pictures
7. On-site training
8. Texts
9. Lectures
10. Closed-circuit television
11. Role playing

Sound Slide Presentation. The various sound slide techniques utilize a combination of tapes and slides. The sound can be on the slide itself or the slide can change on a signal from the tape. The sound slide presentation can include worksheets which help the trainee remember what he is seeing and hearing. An instruction on the tape can tell the trainee to write down the key points of the presentation.

Programmed Instruction. An effective way of training large numbers of people in a short amount of time is programmed instruction. This method works best where the material to be taught remains stable over a period of time.

Programmed instruction requires that the training objectives be defined beforehand

and that performance before and after the training be compared. Learning is controlled so that the trainee is not given more material than he can handle. He advances only when he is ready to do so.

Teaching machine devices are frequently used to present programmed instruction. These devices provide a study-practice combination which simulates the functions of a private tutor in recitation and practice. The program immediately corrects errors, shows successes (correct answers), and feeds new information to the trainee in carefully planned steps.

In practice, the teaching machine presents the trainee with certain information, questions, answers, problems, and practice exercises. There is always some type of feedback or correction so that the trainee is immediately informed of his progress.

Thus, programmed learning differs markedly from TV presentation, films, and other audiovisual media.

In basic form, the elements of programmed learning are as follows:

1. A small amount of new information is presented.
2. The trainee is directed to respond immediately by filling in an answer or choosing among several suggested answers.
3. The trainee responds by writing or selecting an answer. This may be done mechanically, electronically, or by handwriting.
4. The trainee is informed as to whether the answer is correct or incorrect.
5. New information is again presented.

Textbooks, manuals, slides, motion pictures, and sound material have of course been available for years. The actual value of programmed materials, however, lies in the provision for individual trainee feedback in the learning program, and the difference achieved is quite striking. Of course, the machine itself does not teach. It merely brings the trainee into constant contact with the person who composed the material the machine presents. It is a laborsaving device and can bring one programmer into contact with an infinite number of trainees.

The industrial engineer may decide that this method suggests mass production. This is true and important, but the effect on an individual trainee is surprisingly like having his own private tutor.

This comparison holds in several ways.

1. There is constant interchange between programmed information and the trainee.
2. The machine insists that a given point be completely understood before the trainee is permitted to go on to something else.
3. The machine presents just the material for which the trainee is ready.
4. The machine helps the trainee decide upon the right answer.
5. The machine reinforces the trainee for each correct response; that is, it informs him he is right, inferentially praises him, and holds his interest. This is accomplished by utilizing immediate feedback.

Basic to programmed learning and to all effective methods of instruction is the broad principle that the trainee learns by doing and really only by doing, mentally or physically. In the words of the psychologist, training must be organized in such a way that the trainee undergoes a "learning experience." The training process must therefore be interpreted not in the light of what a person reads, sees, or hears, but rather in the light of his response and subsequent action.

The use of programmed learning methods suggests the need for additional skills required of the training engineer. Indeed, know-how and skill in programming is a whole skill-knowledge area in itself. The method also suggests enlargement of the idea of team development of training material. This is true with other methods as well; that is, the training engineers get program information from technical and line personnel responsible for operating an area or department. They then design the training methods to be used to impart information to employees requiring training.

When programmed methods are selected, it will be necessary to separate the line or technical person completely from his regular activity for a period of time during which he works regularly with the training engineer (programmer) in the development of training material.

Teaching machines are actually only one way of presenting a program. Programmed instruction can be presented by almost any device that allows a student to advance a step at a time and to get immediate confirmation of his response by receiving the correct answer. Any teaching machine used should be educationally sound. It should not be used to add glamor to the course or because it is popular to use teaching machines.

Computers. The use of computers as a training device is growing. Computers are effective in aiding programmed instruction, where they can be used for complicated branching techniques. They are also an aid to visuals and rote learning and can be used to set up entire training programs.

Learn by Doing. An important aid to operator training is learn-by-doing or laboratory sessions. The trainee gets the chance to do what he is being taught. This is very practical and should be included whenever possible. Learn-by-doing sessions simulate the actual job conditions to minimize the transition the trainee has to make when he actually starts to work.

Video Tape. A quick and accurate method of showing trainees the best method of performing operations is video tape. The operator can see exactly what he is required to do to perform on the job. He does not have to depend upon observation, which may not give him the exact method of performing his duties. Video tape has an advantage over motion pictures in that a tape can be made and played back immediately. It is a versatile technique and can be used to allow the trainee to see himself as others see him.

Motion Pictures. Motion pictures are similar to video tape as an aid in showing the trainee exactly how to perform a given operation. With both techniques, it is important to give the trainee practice immediately following the film. This enables him to use what he has just learned while it is still fresh in his mind.

On-site Training. Training the operator at the site where his job is to be performed gives him the advantage of being in the work environment. He can practice on equipment that he will actually use on the job. In this case, it is important to make sure the trainee is shown the best way of performing the job.

Texts. The traditional textbook is still an important training aid. By itself, it serves as a reference. In multimedia training, texts supplement material taught by other methods.

Lectures. Lectures can be used either by themselves or to supplement other training techniques. They are a popular way of teaching a number of people at the same time. The effectiveness of this type of training depends on the effectiveness of the instructor. Not everyone who is technically competent is a good instructor.

Closed-circuit Television. Closed-circuit television is a good technique for getting information to a number of different groups in scattered locations. It can be used effectively for interviews and case studies. To get the full effect of television, it is good to have more than one camera.

Role Playing. Role playing is used most frequently in training supervisors. It gives the trainee a chance to become a part of his job and make decisions as he will have to make them on the job. Role playing can be used most effectively with video tape. The trainee can immediately see how he reacts to certain situations and can attempt to adjust his behavior accordingly. Care should be taken with this technique, however, because it also carries a risk of potential damage to the confidence of the trainee.

No matter which method is used, it is very important to get feedback on what the employee has learned. Whenever possible, it is best to make the employee actually perform what he has been taught to do. This is the best way of making sure that he has learned the material.

The aim of multimedia instruction is to keep the trainee so busy that he has no time to get bored. The trainee reads, writes, listens, inserts tapes, changes reels, tries for himself what he has just seen on the screen, and trades ideas with others.

Thus, the variety of training techniques makes the trainee more than just a passive observer. It helps him become personally involved in what he is learning.

GUIDELINES FOR DEVELOPING A TRAINING PROGRAM

With a variety of training methods available, the person developing a training program must become skilled at picking the most effective method for teaching each subject. In other words, some subjects can be taught best by a sound slide presentation while others can be taught most effectively by video tape or programmed instruction. Within the limits of the budget and facilities, the instruction should be varied as much as possible.

Selecting Objectives. During planning and throughout the training program, the main purpose of the training—the final behavior of the trainee—should always be kept in mind. No matter how elaborate and well conceived the training program is, it will be a failure if it does not accomplish its objective of teaching the trainee to perform his job in the desired manner.

When testing a training program, it is necessary to establish criteria for measuring the trainee's progress toward the desired behavior. One criterion, for example, might be how many parts per hour the trainee is expected to produce on the job. The training must be designed to enable him to meet this standard.

Much effective training is aimed toward teaching concepts to the employee. As industry becomes more sophisticated, more is demanded of each worker. Many workers have to make decisions that they were not previously required to make. It takes effective training to teach the trainee the concepts behind what he is doing.

It is far better to teach the trainee why he must do something than it is merely to tell him what to do. Once he knows why he is doing a certain thing and how it relates to his job, he is more likely to be able to make a decision when something different comes up.

Choosing a Training Program. When choosing a training program, the following questions should be considered:

1. How much will it cost?
2. Where will the money come from?
3. What is the best time to do the training?
4. What is the best way of allocating the training funds?

Choosing a Medium. When choosing which medium to use for a particular subject or training program, consider the following:

1. Who is to be trained—what type of trainee?
2. What is the objective of the training?
3. What is the subject to be taught?
4. Where is the best place to teach the material?
5. What is the best time to teach the material?

Source of Programs. There are three principal sources of training programs. The first is off-the-shelf courses, which are marketed by numerous companies. Because they are already prepared, such programs will not usually be very costly. They are particularly useful with subjects which are common to many companies.

The second source of programs is from companies specializing in the preparation of customized training. These specialists prepare programs to meet companies' specific training needs.

The third general source of training programs is from within a company's own organization. Many companies have training departments which write programs to fill the company's needs. The advantage of this source is that the company's own employees are closer to its needs.

Training Centers. Many companies which do not have facilities for in-plant training are turning to training companies which have set up regional training centers. These training centers may present either classroom training or individualized instruction. They also have counselors available to assist the trainee. The advantage of the training center is that companies do not have to invest their time and personnel in training.

Timing. It is very important that training be given at the time it is needed, not necessarily when it is most convenient for the company. The trainee is likely to forget what he has learned if he cannot put his training to immediate use. Thus, the closer the training is to the time when the employee will use what he has learned, the more effective the training will be.

Individualized Instruction. As industry recognizes the importance of effective training, it is putting more and more emphasis on making the instruction fit the student, rather than vice versa.

Individualized instruction enables each trainee to work at his own best pace. Because the student competes only with himself, the fast learner is not held back and the slow learner is not pushed too hard. This self-pacing quality is a special advantage for older workers or men who have not been in school for some period of time.

Individually Prescribed Instruction. This method provides training for each employee when he needs it and has the time to take it. This is a change from the old days when classes were set up only at the convenience of the schedule rather than when they were needed.

Another advantage of individually prescribed instruction is that each trainee learns exactly the same material. The basic material is down in black and white, and does not depend upon instructors, who can inadvertently alter the original intent with the passage of time.

Field of Application. The variety of programs that can be prepared in programmed teaching form is limited only by the ingenuity of the programmer or training engineer. The value of the program depends upon the skill of the programmer to teach the desired behavior to the trainee. It does not depend upon the cost or complexity of the training machine.

Some of the industrial applications of programmed instruction include:

1. Electronic troubleshooting
2. Electrical technology
3. Electrical engineering
4. Mechanical assembly
5. Wiring diagrams
6. Inspection methods
7. Quality control techniques
8. Engineering drawings
9. Sheet metal layout
10. Construction skills
11. Time and motion study
12. Shop mathematics
13. Inventory control
14. Transistors
15. Fundamentals of computer operation
16. Fundamentals of management science
17. Operations research techniques
18. Computer arithmetic
19. Mathematics for technicians
20. First aid
21. Algebra
22. Statistics
23. Physics
24. Chemistry
25. PERT
26. Slide rule use
27. Bookkeeping
28. Insurance sales
29. Methods time measurement
30. Metric analysis
31. Foreign languages

In addition, it is possible to develop programmed learning techniques for training trade apprentices, machine operators, clerical personnel, supervisors, sales personnel, and industrial engineers.

BIBLIOGRAPHY

Barber, J. W. (ed.), *Industrial Training Handbook,* A. S. Barnet Company, Cranbury, N.J., 1969.

Brethower, D. M., D. G. Markle, G. A. Rummler, A. W. Schrader, and D. E. P. Smith, *Programmed Learning, A Practicum,* Ann Arbor Publishers, Ann Arbor, Mich., 1964.

Craig, Robert L., and Lester R. Bittel, *Training and Development Handbook,* McGraw-Hill Book Company, New York, 1967.

Mesics, Emil A., *Education and Training for Effective Manpower Utilization (An Annotated Bibliography),* New York State School of Industrial and Labor Relations, Ithaca, N.Y., 1969.

Murphy, John R., and Irving A. Goldberg, "Strategies for Using Programmed Instruction," *Harvard Business Review,* May–June, 1964.

Northeastern University, *Programmed Instruction Guide,* Entelek Incorporated, Newburyport, Mass., compiled yearly.

Stokes, Paul M., *Total Job Training,* American Management Association, New York, 1967.

Chapter **9**

Suggestion Systems

STANLEY J. SEIMER

Professor of Organization and Management, College of Business Administration, Syracuse University, Syracuse, New York

Among the tools that are available to the industrial engineer to increase the effectiveness of operations is the employee suggestion system or plan. The history of this tool began in 1880 when the Yale and Towne Manufacturing Company introduced the first recorded suggestion plan. Interest in suggestion systems has increased since 1880 until in 1968 it was estimated that there were in excess of 6,000 plans in operation in more than 10,000 plants in the United States.[1]

The impact which suggestion systems can have upon industry is indicated by the report that employees of General Motors (United States and Canada) earned awards on a total of 247,109 suggestions in 1968 and that a record amount of $14,295,387 was paid out for these awards. The total number of suggestions submitted, from which the above were selected, was in excess of 948,000. In addition to the direct financial benefit to the employees and to the company, there is, of course, considerable indirect benefit which results from the involvement of employees in the concerns and interests of the company.

Although these remarks refer to suggestion systems in industry in the United States and Canada, it should be noted that their use is much more widely spread. For many years, plans have been operated by various government agencies at city, county, state, and Federal levels. Also, plans have been introduced and are expanding in various business and government activities overseas; data about suggestion plans over-

[1] *Annual Statistical Report,* National Association of Suggestion Systems, Chicago, 1968.

seas are solicited, analyzed, and reported by the Deutsches Institut fur Betriebswiert-scheft E. V. (German Institute for Business Administration).

The following discussion is organized around and focuses upon the assumed desire of a company to organize, introduce, and operate an effective suggestion system.

OBJECTIVES

The objectives underlying the decision to introduce an employee suggestion system tend to be identified as having an industrial engineering (cost reduction) orientation or a personnel (employee relations) orientation. As might be expected, in the vast majority of instances, a combination of both of these objectives is desirable. The following excerpts from company manuals reflect this composite set of objectives.

We have a suggestion plan . . . for two big reasons: first, we need the fresh ideas that will help us keep improving our products, our methods, our cost position. Second, it's a simple fact that interested, intelligent employees *do* think about improvement and when they come up with good ideas, they deserve a courteous, thorough hearing and a fair reward for their contribution. *IBM Manager's Guide*

and

The . . . suggestion plan is an employee activity which accomplishes objectives of great benefit to (company) men and women as well as to the plants and divisions of (the company).

For employees, the plan is designed to help them develop and demonstrate their initiative and ingenuity by providing them with an opportunity to offer their constructive ideas for consideration by management.

To give them personal recognition for constructive thinking and the very real satisfaction of seeing their own ideas in use.

To give them a tangible share in the benefits which result from the adoption of their ideas.

For plants and divisions of (the company), the plan is designed . . .

To serve as a means of good two-way communications between employees and management and as a tool which supervision can use to improve relations with employees.

To improve employee attitudes by directing their attention to the positive and progressive aspects of their jobs.

To make use of the constructive ideas of employees which contribute to technological progress, improved tools, methods, processes, and equipment and thus help produce more with the same amount of human effort. *The General Motors Suggestion Plan Operating Manual*

A statement of the purpose of the incentive awards program from the *Federal Personnel Manual,* April 17, 1969, reads:

The Government employees' incentive awards program is established for the purpose of improving government operations and recognizing employees through the medium of incentive awards. The awards under this program are designed:

(1) To encourage employees to participate in improving the efficiency and economy of Government operations;

(2) To recognize and reward employees, individually or in groups, for their suggestions, inventions, superior accomplishments, or other personal efforts which contribute to the efficiency, economy, or other improvements in Government operations;

(3) To recognize and reward employees, individually or in groups, who perform special acts or services in the public interest in connection with or related to their employment.

ORGANIZATION

For the most part, the suggestion system organization is relatively simple. The number of personnel involved, of course, will vary with the quantity of suggestions which need to be processed.

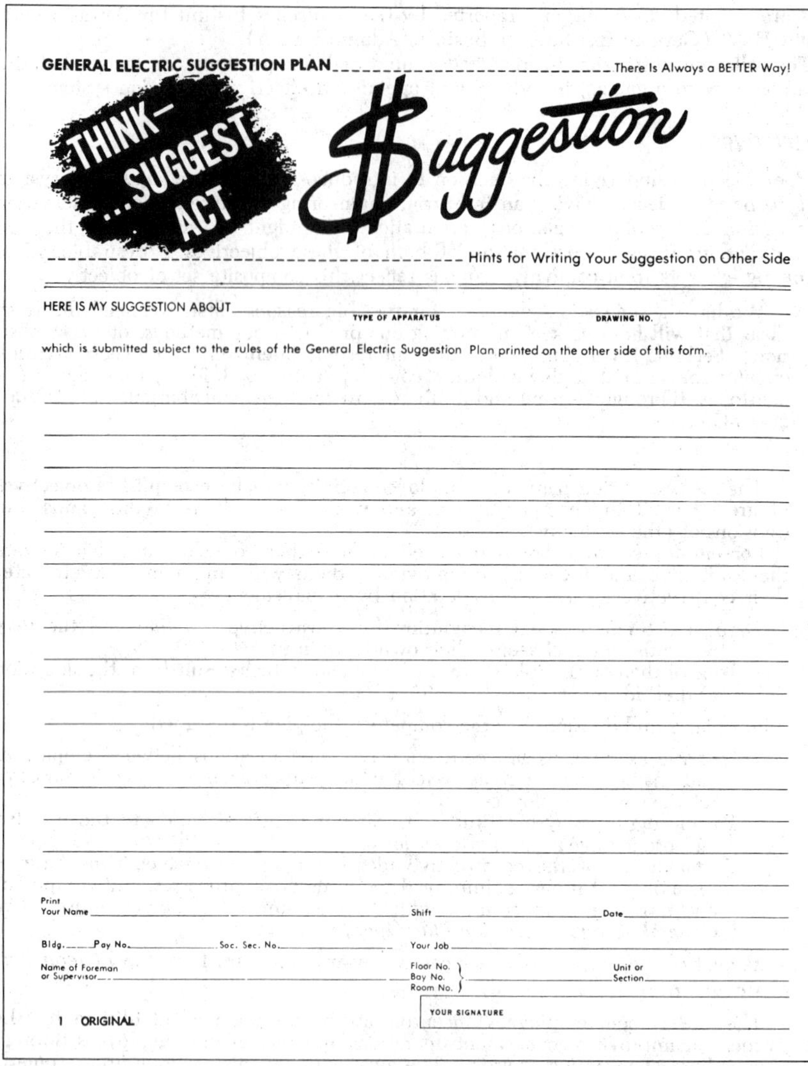

GENERAL ELECTRIC SUGGESTION PLAN_____ There Is Always a BETTER Way!

THINK—
...SUGGEST
ACT

$uggestion

_____ Hints for Writing Your Suggestion on Other Side

HERE IS MY SUGGESTION ABOUT_____

TYPE OF APPARATUS   DRAWING NO.

which is submitted subject to the rules of the General Electric Suggestion Plan printed on the other side of this form.

Print
Your Name _____   Shift _____ Date_____

Bldg.____Pay No._____ Soc. Sec. No._____   Your Job _____

Name of Foreman
or Supervisor___ ._____   Floor No.)
Bay No. }_____   Unit or
Room No.)   Section_____

1   ORIGINAL   YOUR SIGNATURE

FIG. 9-1. Suggestion form. (*Used by permission of General Electric Company.*)

A key decision prior to the formation of the actual organization is the determination of the activity or function which will have overall administrative responsibility for the operation of the plan. In the majority of instances, the suggestion system is attached to industrial relations or personnel; second in popularity is industrial engineering or manufacturing. In some instances, administration is assigned to corporate staff, administration, or top management. The importance of this decision is obvious when it is realized that the activity having primary responsibility for the suggestion system will strongly influence the focus of those directly responsible for its promotion and administration. If attached to the industrial engineering function, the primary emphasis is likely to be upon efficiency and cost reduction; if it is attached to the

RULES OF THE SUGGESTION PLAN
The following rules apply to all suggestions submitted:

ADMINISTRATION

Each Department manager has responsibility consistent with these rules for the administration of the Suggestion Plan in his Department. The Department manager may designate an individual or a group to process suggestions received under this Plan. For the purpose of the following rules of this Plan, any person or persons so designated shall be considered to be included in the term "manager."

ELIGIBILITY

All employees of the Company other than supervisory, professional, and methods and planning personnel are eligible to submit suggestions under the Plan and to receive awards for those beyond the scope of their assigned duties which are adopted.

AWARDS

In no event shall an award be made, nor shall the Company be obligated, for an amount of more than $25,000 for or in connection with any suggestion, including all applications of ideas embodied in the suggestion

RETENTION OF SUGGESTIONS

A suggestion as to which no award has been made will be kept in an active file for a period of two years from the date it is received by the Company. An award will be made only if the suggestion is adopted during that period. A submitter of the same or of a similar suggestion after the expiration of the two-year period may be treated as an original submitter.

JOINT OR SIMILAR SUGGESTIONS

When two or more employees submit a suggestion jointly, or submit at approximately the same time suggestions which, in the judgment of the appropriate manager, are substantially similar, a single award will be granted upon the adoption of any such suggestion or suggestions, which award may be divided among the submitters in such proportions as the manager may determine.

CLAIMS

The Company shall not be liable with respect to any claim made in connection with any suggestion unless such claim shall have been made in writing to the Company not later than three years from the date on which the suggestion was submitted to the Company.

OWNERSHIP OF SUGGESTIONS

A suggestion and all ideas embodied therein become the absolute and exclusive property of the Company upon submission.

USE OF FORMS

All suggestions should be submitted on printed Suggestion Forms.

DECISIONS

All decisions of the Company with respect to eligibility, the adoption of suggestions, the making of awards or the interpretation of the Suggestion Plan shall be final and binding on all parties.

CHANGE OF RULES—AMENDMENT OR TERMINATION OF PLAN

No change in, nor commitment inconsistent with, these rules shall be binding upon the Company unless made in writing and signed by the President of the Company. The Company reserves the right to amend or terminate the Plan at any time, except that no amendment or termination shall adversely affect any eligible person with respect to any suggestion submitted by him prior to the effective date of such amendment or termination.

WANTED

SUGGESTIONS WHICH WILL -
Improve methods, products, working conditions, etc.
Reduce costs, losses, errors, etc.
Eliminate safety hazards, waste, etc.
Benefit the Company or employees

TO MAKE SURE YOUR SUGGESTION IS UNDERSTOOD - - -
IDENTIFY THE SUBJECT. What is it—where is it—identification numbers, etc.?
WHAT IS DONE NOW? Describe present method. Tell what is wrong with it, and what you are trying to correct.
WHAT DO YOU SUGGEST? Describe your proposal and if possible tell how it can be put into effect.
WHAT ARE THE ADVANTAGES? Name the benefits of your suggestion.

REMEMBER!
An idea is worthless until it is put to work.
Write only one suggestion on each sheet. Use extra sheets if necessary.

Make sketches on a separate sheet of paper.
If you want help in preparing your suggestion, see your foreman or supervisor, or call your Suggestion Plan Administrator.
Place your suggestion in a suggestion box or mail it to your local Suggestion Plan Administrator.

SUGGESTION PLAN

Fig. 9-1 (*continued*).   Suggestion form.   (*Used by permission of General Electric Company.*)

industrial relations function, the primary concern is likely to be the stimulation of good communications and high employee morale.

The organization of the suggestion activity itself will be composed, essentially, of the director or coordinator, the award committee, the suggestion investigator, and the clerical staff.

Director or Coordinator.   The director or coordinator is the key individual in the operation of a suggestion system.   He is responsible for developing and maintaining the overall effectiveness of the plan and for recommending changes that need to be made from time to time.   It is crucial that the director or coordinator have the complete and visible support of top management.   He must also have a keen appreciation of the important contribution that the suggestion system can make to operating

SUGGESTION PLAN

REQUEST TO INVESTIGATE

SUGGESTION NO._____

Date_____

Mr._____

Please investigate the following Suggestion, submitting your report promptly in order that an early decision may be forwarded to the Suggester. To assure the most favorable results to the Suggester, this suggestion should be reviewed with all components of the Company in mind. If to your knowledge this idea has application in other areas, please advise Suggestion Office concerning such facts.

SUGGESTION BOARD_____

NAME   LOCATION

HERE IS MY SUGGESTION ABOUT_____

TYPE OF APPARATUS   DRAWING NO.

which is submitted subject to the rules of the General Electric Suggestion Plan.

Print
Your Name_____   Shift _____   Date_____

B!dg._____ Pay No._____ Soc. Sec. No. _____   Your Job _____

Name of Foreman
or Supervisor_____   Floor No.
                                             Bay No.
                                             Room No.   Unit or Section_____

2   INVESTIGATOR   YOUR SIGNATURE

FIG. 9-2.   Investigation form.   (*Used by permission of General Electric Company.*)

effectiveness and to the improvement of employee understanding of and concern for company activities.   The suggestion system director or coordinator is the hub around which the entire suggestion system activities revolve.

Award Committee.   A vital part of any suggestion system is the award committee. Its decisions have a major impact on the suggestion plan itself, the enthusiasm of the employees for the plan, the extent to which ideas for improvements in operations will be obtained, and the promptness with which improvements will be put into effect. Although the size of the committee can vary substantially—as many as forty in at least one case—it probably should include from three to seven individuals.   The exact number depends upon the functions or activities which can most effectively make a contribution to the review and evaluation of employees' suggestions.   Among

INSTRUCTIONS TO INVESTIGATORS

Your reply should be courteous and complete giving detailed reasons for your recommendations.
If previous consideration has been given to this idea, include reference to specific correspondence, drawings, shop orders, etc

PLEASE REPLY TO ALL QUESTIONS AS EXPLICITLY AS POSSIBLE.

1. Is the idea new in our practice? _____

2. Is it already under consideration from another source? _____

3. If so, who has the matter in hand? _____

4. Adoption? (Answer "Yes", "In part" or "No") _____

5. Improvement in quality of product? _____ Estimated value? _____

6. Reduction of material? _____ Yearly saving? _____

7. Saving of labor? _____ Yearly saving? _____

8. Other savings? _____

9. Total yearly saving? _____

10. Show in moderate detail how savings are figured _____

_____

_____

_____

11. If suggestion is adopted, when will it be put into effect? _____

12. Who will be responsible for putting it into effect? _____

13. Have instructions been issued to put it into effect? _____

14. Estimated cost of putting it into effect? _____

15. Other information of interest _____

_____

_____

_____

_____

_____

DATE _____ SIGNED _____

FIG. 9-2 (*continued*). Investigation form. (*Used by permission of General Electric Company.*)

the functions generally represented are personnel, manufacturing, industrial engineering, plant engineering, and finance or accounting. Sometimes, but not often, employee or union representatives are included.

Suggestion Investigator. The suggestion investigator may be a permanent member of the suggestion system organization or he may be appointed on a temporary basis when the need for his particular ability arises. Under either situation, it is common that he carries out his responsibilities in conjunction with the line supervisor responsible for the area affected by the suggestion. Whether full-time or part-time, it is essential that the investigator be competent to determine the data and information necessary to evaluate the suggestion fairly or that he have the knowledge and experience necessary to identify the activity or individual that can properly carry out this responsibility. Incidentally, because a suggestion investigator must acquaint himself with all facets of the company's operations which relate to the suggestion under con-

SUGGESTION PLAN

ACKNOWLEDGMENT TO SUGGESTER

SUGGESTION NO._____

Date_____

TO THE SUGGESTER

We thank you for your suggestion submitted below and wish to assure you that it is being considered in accordance with the rules of the suggestion plan printed on the reverse side of this sheet.

Please retain this acknowledgment, and use the above number whenever referring to this suggestion. You will be notified of the Board's decision as soon as our investigation has been completed.

SUGGESTION BOARD_____

SECRETARY   LOCATION

HERE IS MY SUGGESTION ABOUT_____

TYPE OF APPARATUS   DRAWING NO.

which is submitted subject to the rules of the General Electric Suggestion Plan printed on the other side of this form.

Print
Your Name_____   Shift_____   Date_____

Bldg.____Pay No._____ Soc. Sec. No._____   Your Job_____

Name of Foreman
or Supervisor_____   Floor No. ⎫
                             Bay No. ⎬_____   Unit or Section_____
                             Room No. ⎭

3 ACKNOWLEDGMENT   YOUR SIGNATURE

FIG. 9-3. Acknowledgement form. (*Used by permission of General Electric Company.*)

sideration, an assignment to this role is frequently considered an excellent opportunity for training junior management personnel.

Clerical Staff. Clerical staff refers to a secretary who is responsible for carrying out the day-to-day activities related to the operation of the suggestion plan. In addition to the competence which is characteristically expected, the suggestion system secretary should be sensitive to the feelings—concerns, fears, hopes, and so on—of those who become involved in or are related to the suggestion program. The requirement for tact which this implies should be augmented by the ability to distinguish those matters which can be dealt with directly and those which must be forwarded to the director, the investigator, or the award committee.

RULES OF THE SUGGESTION PLAN
The following rules apply to all suggestions submitted:

ADMINISTRATION
Each Department manager has responsibility consistent with these rules for the administration of the Suggestion Plan in his Department. The Department manager may designate an individual or a group to process suggestions received under this Plan. For the purpose of the following rules of this Plan, any person or persons so designated shall be considered to be included in the term "manager."

ELIGIBILITY
All employees of the Company other than supervisory, professional, and methods and planning personnel are eligible to submit suggestions under the Plan and to receive awards for those beyond the scope of their assigned duties which are adopted.

AWARDS
In no event shall an award be made, nor shall the Company be obligated, for an amount of more than $25,000 for or in connection with any suggestion, including all applications of ideas embodied in the suggestion.

RETENTION OF SUGGESTIONS
A suggestion as to which no award has been made will be kept in an active file for a period of two years from the date it is received by the Company. An award will be made only if the suggestion is adopted during that period. A submitter of the same or of a similar suggestion after the expiration of the two-year period may be treated as an original submitter.

JOINT OR SIMILAR SUGGESTIONS
When two or more employees submit a suggestion jointly, or submit at approximately the same time suggestions which, in the judgment of the appropriate manager, are substantially similar, a single award will be granted upon the adoption of any such suggestion or suggestions, which award may be divided among the submitters in such proportions as the manager may determine.

CLAIMS
The Company shall not be liable with respect to any claim made in connection with any suggestion unless such claim shall have been made in writing to the Company not later than three years from the date on which the suggestion was submitted to the Company.

OWNERSHIP OF SUGGESTIONS
A suggestion and all ideas embodied therein become the absolute and exclusive property of the Company upon submission.

USE OF FORMS
All suggestions should be submitted on printed Suggestion Forms.

DECISIONS
All decisions of the Company with respect to eligibility, the adoption of suggestions, the making of awards or the interpretation of the Suggestion Plan shall be final and binding on all parties.

CHANGE OF RULES—AMENDMENT OR TERMINATION OF PLAN
No change in, nor commitment inconsistent with, these rules shall be binding upon the Company unless made in writing and signed by the President of the Company. The Company reserves the right to amend or terminate the Plan at any time, except that no amendment or termination shall adversely affect any eligible person with respect to any suggestion submitted by him prior to the effective date of such amendment or termination.

FIG. 9-3 (*continued*). Acknowledgement form. (*Used by permission of General Electric Company.*)

PROMOTION

Promotion is designed to accomplish two things: stimulate participation in the suggestion program and stimulate the submission of high-quality suggestions—those which will result in tangible or intangible benefits to the company. Prominently displayed posters, which are available from commercial suppliers, and news items or feature articles in company house organs are the most common methods used to draw attention to the suggestion system. A third approach which is popular is the scheduling of a ceremonial presentation of awards for accepted suggestions. These ceremonies range from a handshake from an executive to a dinner or banquet for special accomplishments such as the largest number of suggestions submitted during a month or year, the most suggestions accepted for an award, the most unusual suggestion, or the highest award. In addition to these three commonly used publicity techniques, there are numerous others, such as displays, banners, and handouts, which are often used for special promotions. Continuous and imaginative promotion is a vital element in the operation of a successful suggestion system, and the various techniques for promotion should be planned carefully to maximize their effectiveness.

INTRODUCTION OF A SUGGESTION SYSTEM

Prior to the actual operation of a suggestion system, there are a number of decisions which should be made and activities that should be completed. These relate to:

1. Establishment of policies
2. Preparation of an employee booklet
3. Selection or preparation of necessary forms
4. Installation of boxes into which employees can place their suggestions
5. Education of company personnel who will be responsible for the administration of the program

Policies. Careful determination of the policies which will guide the suggestion system director or coordinator, the award committee, investigators, clerical staff, line supervision, and others is of crucial importance if confusion, uncertainty, and conflict are to be minimized when the suggestion system becomes operative. Illustrative of what might be included in a suggestion plan policy manual are the topics noted below which are stated in the manual of a large United States company:

SUGGESTION PLAN POLICY MANUAL

I. The Objectives of the Suggestion Plan
II. The Basic Responsibilities for the Administration of the Suggestion Plan
III. The Rules of the Suggestion Plan
IV. The Procedures That Are Recommended
V. The Reports or Statistics That Are Needed to Evaluate Effectiveness
VI. The Promotional Media and Opportunities Available

General Electric Policy Manual, 1965

In the elaboration and explanation of these policies, the manual discusses in considerable detail various subtopics which are judged to be of importance. Among these are:

1. Eligibility of employees
2. Awards
3. Retention of suggestions
4. Joint or similar suggestions
5. Ownership of suggestions
6. Use of forms
7. Finality of decisions, change of rules, amendment or termination of plan

Items such as the above should be considered for inclusion in the suggestion system manual.

Employee Booklet.   An employee booklet is needed to facilitate the promotion and understanding of the suggestion system and to develop enthusiasm for its benefits and use. It should be written in a style which is interesting and informative. Generous use of symbols, cartoons, and examples of the forms which are used is desirable. An illustration of the items which are likely to be of concern to employees and the tone of the booklet are indicated by the following headings found in the employee booklet distributed by a company which operates a very successful program:

What is a good suggestion?
Who may submit suggestions?
Who may receive awards?
When is a suggestion adopted?
What suggestions may be awarded?

IBM, *Ideas for Improvement*

Each of these questions is followed by a carefully written and comprehensive answer. In addition, this particular booklet contains a discussion of how ideas are to be submitted, the use of the suggestion form, ineligible suggestions, the procedure followed in investigation, the determination of the amount of the award for an accepted suggestion, and the procedure to follow in resubmitting a suggestion. Of course, additional items such as those mentioned above under policies might also be included.

Forms.   The forms typically needed in the operation of a suggestion system are:

Suggestion form on which the employee submits his ideas (Figure 9-1)
Investigation form on which the factors (criteria) to be considered by the investigator as he accumulates information and data for submission to the award committee are noted (Figure 9-2)
Acknowledgement form, which is sent to the suggester upon receipt of his idea on a suggestion form (Figure 9-3)
Award committee worksheet, which facilitates an orderly evaluation of a suggestion (Figure 9-4)
Award notice, which is sent to the suggester when the amount of the award to be granted has been determined (Figure 9-5)
Rejection notice, which is sent to the suggester to advise him that his suggestion has not been accepted (Figure 9-6)
Suggestion status notice, which advises the suggester of delays which have occurred in processing his suggestion (Figure 9-7)

```
                              AWARD GUIDE          Suggestion No._____
SUGGESTION:


ACTION:          "When suggestions cause action, an award based upon the full value of
                 the resulting action will be made."   Policy statement.


ADVANTAGES:


AWARD:           "The Committee balances any differences between management and the
                 suggester, but in case of doubt, acts for the suggester in matters of eligibil-
                 ity, investigation and awards."   Policy statement.


CLASS:           Type                                                      Minimum
   1.            Product Improvement                                          $50
                 New Product—3 % estimated net sales for 1 year
   2.            Cost Reduction—15 % net savings for 2 years                  $20
                 Calculate savings for 1 year as follows:

                 a. Materials saved                             _____
                 b. Direct labor saved: $____hourly rate × ____hours_____
                 c. Employees' benefits—33 % of direct labor saved  _____
                 d. Less: Cost of implementation $_____
                          Divide by estimated life_____years
                          Annual pro-rata share  =             _____
                 Net savings for 1 year—total         $_____ × 15% =
                                                 AWARD  _____1st year
                 Pay for first year only and review, unless
                 award for 2 years is less than $250.   AWARD $_____2nd year
                                                 TOTAL  $_____
   3.            Method Improvement                                           $20
   4.            Safety                                                       $15
   5.            Miscellaneous                                                $10
```

FOR EVALUATING SUGGESTIONS OTHER THAN CLASS 2:

Extent of Use

	Minor	Fair	Average	Good	Outstanding
Small	$10	$15	$20	$25	$30
Near and up to ½ of Dept., Plant or Co.	$15	$20	$25	$30–40	$45–65
Over ½ of Dept., Plant or Co.	$20	$25	$30	$35–55	$60—

Extra Credits, if Justified
Notable completeness of idea in presentation—$10–25
Exceptional initiative by suggester   —$10–50
Exceptional ingenuity of device or idea   —$10–75

1/2/_____

FIG. 9-4. Award committee worksheet.

Master file record, which is maintained in the suggestion system office for purposes of control (Figure 9-8)
Suggestion statistical and follow-up record (Figure 9-9)
Approved for installation form, which, when signed by the proper authority,
authorizes the installation of the change that has been investigated and approved (Figure 9-10)

As participation in the suggestion program increases, the burden of work is such
that the development of standardized forms is practically a necessity. Regardless of
this practical aspect, however, it should be noted that careful preparation of standard

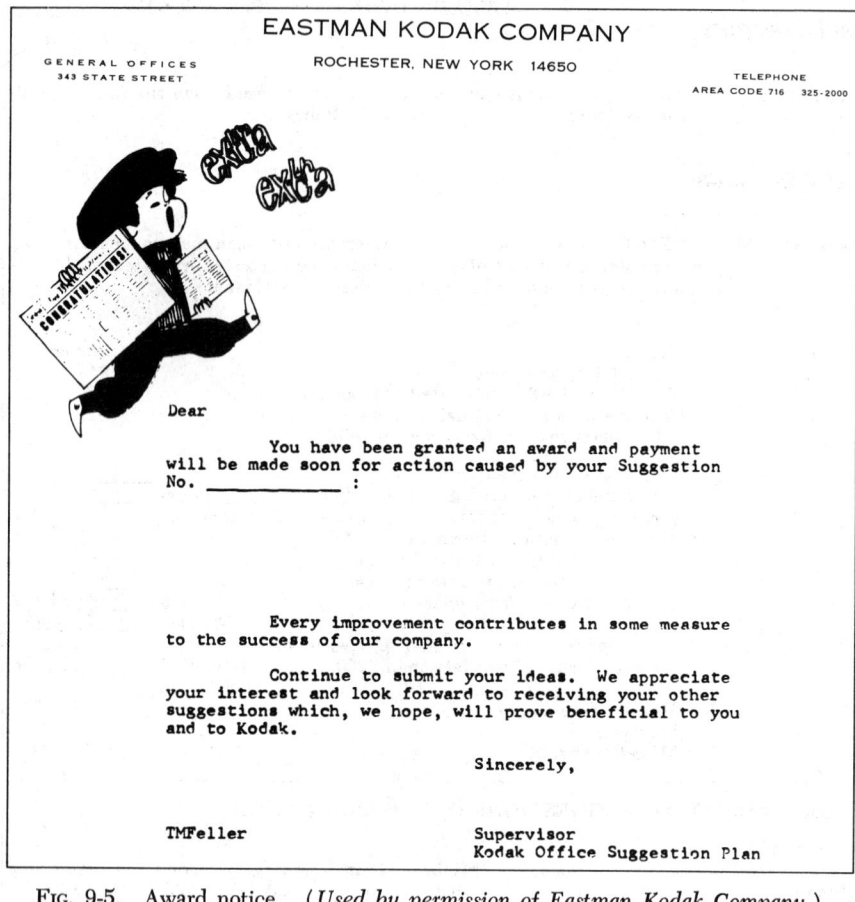

FIG. 9-5.   Award notice.   (*Used by permission of Eastman Kodak Company.*)

forms minimizes the possibility of error, oversight, or misunderstanding.   Some, if not all, of these forms are available from commercial supply firms.

Suggestion Boxes.   Suggestion boxes are an obvious necessity in the operation of a suggestion plan.   They may be simple wooden boxes with lids, or metal boxes which have been obtained from a commercial supplier and which incorporate spaces for promotional posters and for a supply of suggestion forms.   Regardless of the design, the suggestion box should have several important features.   First, it should have a lid which is easily opened but which can be locked to ensure that once a suggestion has been placed in the slot at the top, there will be no likelihood of loss of the suggestion or disclosure of the identity of the suggester.   Second, the box should accommodate a supply of suggestion forms so that a suggester can obtain new blanks when he submits a completed form.   Finally, the feature of availability should not be overlooked.   The boxes should be located in various high-density traffic areas so that the suggester can submit his idea with a minimum of effort and inconvenience.

Education of Personnel Involved.   Education of personnel involved is important, because it will strongly influence the environment within which a suggestion system will operate.   All persons involved in the operation of the plan must be made thoroughly aware of the impact they will have upon the operating effectiveness of the

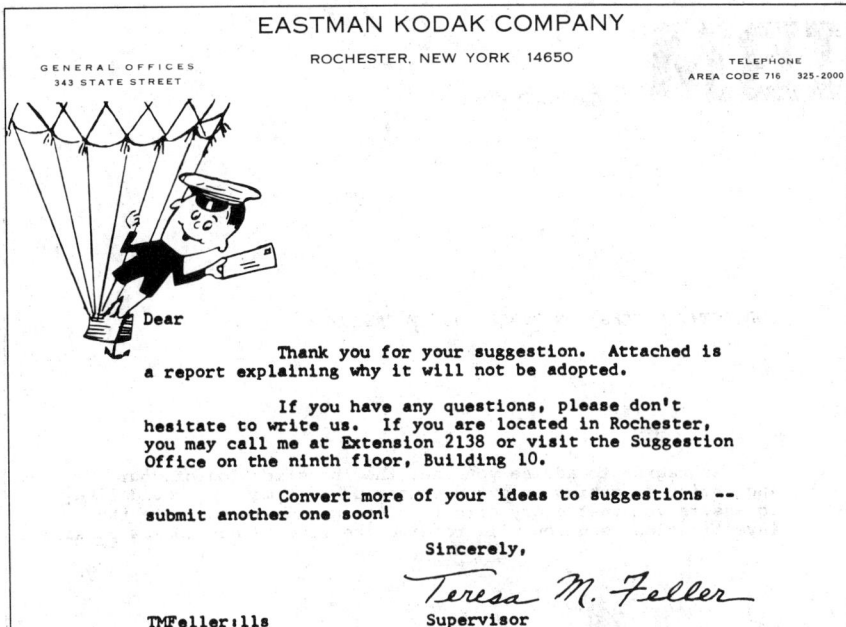

FIG. 9-6. Rejection notice. (*Used by permission of Eastman Kodak Company.*)

plan. The objectives, policies, and administrative procedures and practices must be completely understood and endorsed. Continuous and uniformly high support for the plan must be assured.

OPERATING PROCEDURES

To ensure fair, equitable, and consistent administration of the suggestion system, the establishment and pursuit of a prescribed procedure is desirable. The following procedure indicates the decisions which have to be made and the actions which must be taken.

Collect Suggestions. Once a suggestion has been placed in the box, the employee—and the company—want to keep it moving. Regular collection of suggestions from the boxes should be made at least once a week, and more often if the situation warrants.

Review Suggestions for Eligibility. In the policy statement, eligibility of suggesters and suggestions should be clearly spelled out. In general, all company employees are eligible except those in managerial, engineering, and methods improvement jobs. A suggestion is generally eligible for consideration unless it is included in the suggester's defined job responsibility, is classified as routine maintenance, is already a documented project currently under investigation by the company, or is a duplication of a suggestion previously submitted.

Identify Suggestion. This involves the assignment of an identifying number to the suggestion which will provide a basis for control during further processing. This number and other pertinent information are recorded on the master file card.

Prepare and Send Acknowledgement. Once the employee suggestion has been accepted for processing, an acknowledgement of this action should be sent to the suggester immediately. When a multisheet ("snap" type) suggestion form is used, this merely involves removing the appropriate copy and forwarding it. This acknowl-

Suggestion Plan

SUBJECT: SUGGESTION STATUS NOTIFICATION

        We regret to advise you that the investigation of your
suggestion is taking an extended period of time. We would like
to assure you that every effort is being made to complete the
investigation, and you will receive the results as soon as possible.

R. D. McLaughlin
R. D. McLaughlin
Administration Manager
Field Suggestions
Endicott

st

FIG. 9-7.   Suggestion status notice.   (*Used by permission of International Business Machines Corporation.*)

edgement advises the suggester that the idea which he presented is being processed.

Conduct an Investigation.   Prompt investigation of the suggester's idea is important, and delays—a common weakness in the administration of suggestion systems—should be strongly resisted.   The investigator is usually appointed by the suggestion program director or a line supervisor.   It is his responsibility to determine accurately all the cost savings and other information or data which will be needed by the award committee when it conducts its deliberations to determine the merits of the idea. The time required for investigation is dependent upon the nature of the data, information needed, and its availability.   If considerable refinement of the idea is required and engineering design time is involved or if other time-consuming activities are necessary, the time for investigation can be quite long.   Experience indicates that processing time ranges from 10 to 360 days, the average being 62 days.

Evaluate Suggestion and Determine the Award.   Suggestion evaluation is typically the responsibility of an award committee.   Both tangible savings and intangible benefits are considered, and an appropriate award must be determined.

When tangible savings can be calculated, the amount of the award is generally based on a percentage of the estimated first year's savings.   This percentage ranges from 10 to 20 percent, although in some instances it is much higher.   Different time periods for calculating savings can also be used, such as six months or the life of the suggestion.   Typically, minimum and maximum amounts of an award are established by policy statements, such as from as low as $5 to as high as $75,000.

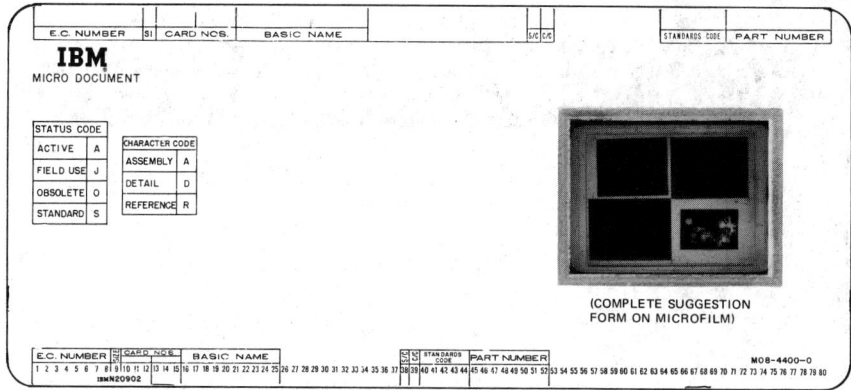

FIG. 9-8. Master file record. (*Used by permission of International Business Machines Corporation.*)

When the benefits from a suggestion are intangible, the suggestion committee determines an amount which in its judgment is appropriate. Intangible suggestions have to do with such things as health and safety, improved housekeeping or maintenance, employee relations, and security.

Although an award is usually made in cash, various other forms of payment, such as bonds, merchandise, or trading stamps, may be used. When the award is monetary and the amount can be estimated accurately, it is likely to be paid in full; when accurate determination is dependent upon verification of the savings from actual operations, a portion—typically a half—is paid immediately and the remainder is paid upon completion of the study.

Notify Suggester of Award Committee Action. The suggester should be notified promptly of the final action taken by the award committee. Furthermore, if there is likely to be a substantial lag between the suggester's receipt of an acknowledgement and the notification of final action, an interim report should be forwarded to him explaining the cause for delay and the estimated time of completion.

Present the Award. Although there is little likelihood that the award presentation step of the procedure will be overlooked, it is possible that full advantage will not be taken of this opportunity for publicity. A high-level line officer should participate

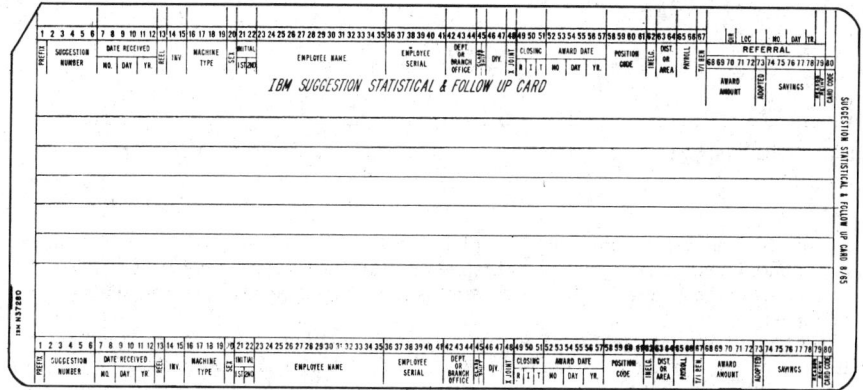

FIG. 9-9. Suggestion statistical and follow-up record. (*Used by permission of International Business Machines Corporation.*)

ACCEPTED SUGGESTION - HELD FOR ADOPTION

TERNSTEDT
SUGGESTION PLAN

TO _____   DATE _____

DEPARTMENT _____

The Suggestion Committee has reviewed and accepted for adoption Suggestion Number _____ relative to _____

_____

_____

Before an award is paid, this suggestion must be put into effect by your department. Please notify us when this is done so that the committee may make the award to the suggester.

_____
Chairman Suggestion Committee

NOTE: If a change has occurred in the original estimate of the award calculation, please indicate the change and attach to this letter.

NOTIFICATION TO COMMITTEE

The above suggestion has been put into effect _____
(Date)

Signed _____

Date _____

70A695 4-56

FIG. 9-10. Approved for installation form. (*Used by permission of Ternstedt Division, General Motors Corporation.*)

in making the award, and photographers and reporters from the company house organ should be present. Frequently, there is also a release to the local press. Of course, when an award is of unusual size or importance, special arrangements should be made.

Close Out the Suggestion Record. Upon completion of the award presentation, the record concerning the suggestion should be reviewed and completed. Notations (cross-indexing) should be made to facilitate the finding or the identification of the suggestion should a need arise, such as the receipt of a similar suggestion or a question from an employee.

Reopening of Suggestions.   The suggestion plan should be designed to provide as much flexibility as possible and to permit a suggester to pursue further the evaluation of an idea which has been declined for processing.   A suggestion is commonly classified as active for three years.   However, typically, at the end of this time, the suggester is authorized to reactivate his suggestion if he so wishes.   This opportunity should be made known to him, and he should be encouraged to take advantage of it if he still thinks his idea has merit.

SUMMARY

Suggestion systems are of substantial benefit to both employees and the company. According to recent statistics, an average of about forty suggestions per hundred eligible employees is received as the result of the operation of an employee suggestion plan.   Of those which are accepted for investigation, 24 percent receive awards. The average award paid in 1968 was about $50, and the maximum award was $75,000.

Appreciation by industry of the benefits which can be obtained from an effective suggestion system is reflected in the increase in the number of plans in operation. Although many plans fail—due largely to lack of attention, weak or spotty promotion, or inadequate top management support and participation—an indication of the durability of a well-run plan is the continuous operation of plans which were established many years ago.   As a tool available to the industrial engineer for improving operations, the suggestion system has a potential that should not be overlooked.

BIBLIOGRAPHY

Egbert, W. H., "Selling the Suggestion System," *Supervision,* February, 1963.
"Installing and Maintaining an Employee Suggestion Program," Dartnell Corporation Report 589, Chicago.
Montana, J. A., "Managing an Effective Suggestion System," *Administrative Management,* October, 1966.
National Association of Suggestion Systems, Chicago, pamphlets:
    "Administration of a Suggestion System"
    "Objectives of a Suggestion System"
    "Principles and Practices of a Suggestion System"
    "Starting a Suggestion System"
    "Statistical Report" (members only)
    "What's the Big Idea" (16mm color film)
Northrup, H. R., *Suggestion Systems,* National Industrial Conference Board, New York, 1953.
Seimer, S. J., *Suggestion Plans in American Industry,* Syracuse University Press, Syracuse, N.Y., 1959.
Seinwerth, H. W., *Getting Results from Suggestion Plans,* McGraw-Hill Book Company, New York, 1948.

Industrial Engineering Applications

Chapter 1

Design of Production and Distribution Systems

RALPH A. MAGGIO

Associate Professor, Department of Industrial Engineering, Systems Management Engineering, and Operations Research, University of Pittsburgh, Pittsburgh, Pennsylvania

Historically, industrial firms have adopted a functional organization structure (production, marketing, and finance) supported by technical and coordinative staff groups. The objective of this chapter is to examine the interface of what might normally be considered marketing and production functions. For present purposes, this interface may be termed logistics. It serves as "the functional bridge that provides for the physical movement (and coordination) of goods. . . ."[1]

This examination is begun by defining the elements and subsystems of a logistics system, their interrelationships, and the constraints within which they operate. Next, the methodology for analyzing such systems is detailed. The chapter concludes with a review of the organization implications of the analysis, and a brief summary.

ELEMENTS AND SUBSYSTEMS OF A LOGISTICS SYSTEM

The central elements of a logistics system are schematically shown in Figure 1-1. These are, as noted, the number, location, and size of raw material sources, plants,

[1] Alan H. Gepfert, "Business Logistics for Better Profit Performance," *Harvard Business Review*, November–December, 1969, p. 75.

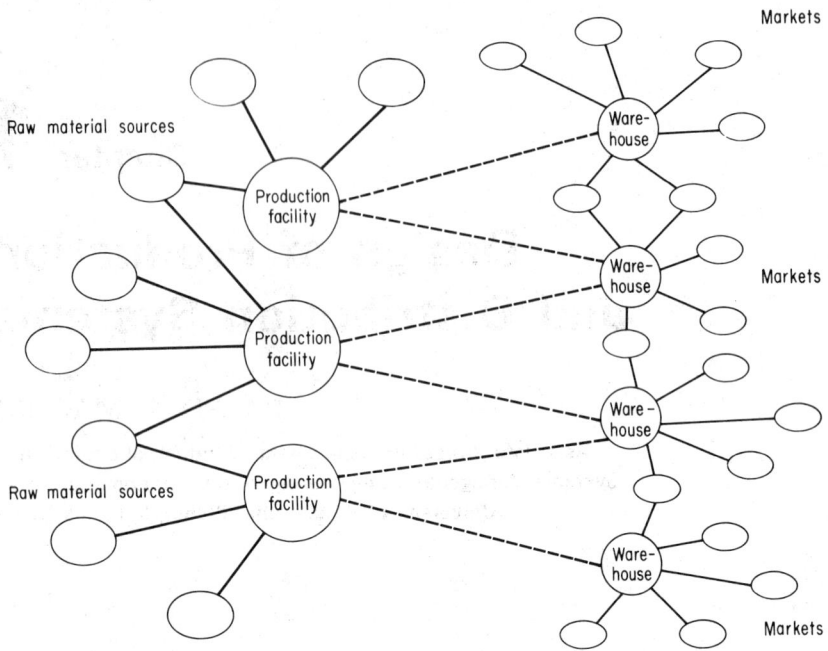

Fɪɢ. 1-1.   Central elements of a logistics system.

and warehouses. In practice, these elements are structurally connected by several subsystems, including:

> *Transportation subsystem*—consists of the selection of transportation mode (rail, truck, air freight, ship, or barge) and the routing for movement of goods between two or more of the central elements as well as interplant and interwarehouse shipments.
>
> *Inventory subsystem*—pertains to all inventory points at and within the several nodes in the logistics network.
>
> *Material handling subsystem*—consists of loading and unloading equipment, dock facilities, packaging and palletization, containerization, and the like.
>
> *Scheduling subsystem*—is concerned primarily with actual loading of intraplant operations, but also includes coordination of movement between system elements such as the purchasing of raw materials.
>
> *Information subsystem*—consists of the flow of information associated with the various subsystems, and particularly processing of customer orders.
>
> All the above subsystems are illustrated in Figure 1-2.

INTERRELATIONSHIPS BETWEEN ELEMENTS AND SUBSYSTEMS

The principal advantage in considering materials management as a distinct function called logistics rests in the ability to recognize and resolve conflicts which normally arise within this element subsystem complex. Dependent upon the particular firm, an acceptable resolution of these conflicts, in terms of the firm as a whole, may be quite difficult without such an integrated approach. As a means of clarification, several examples are enumerated below.

Example 1: Inventory versus Transportation. There is strong motivation for main-

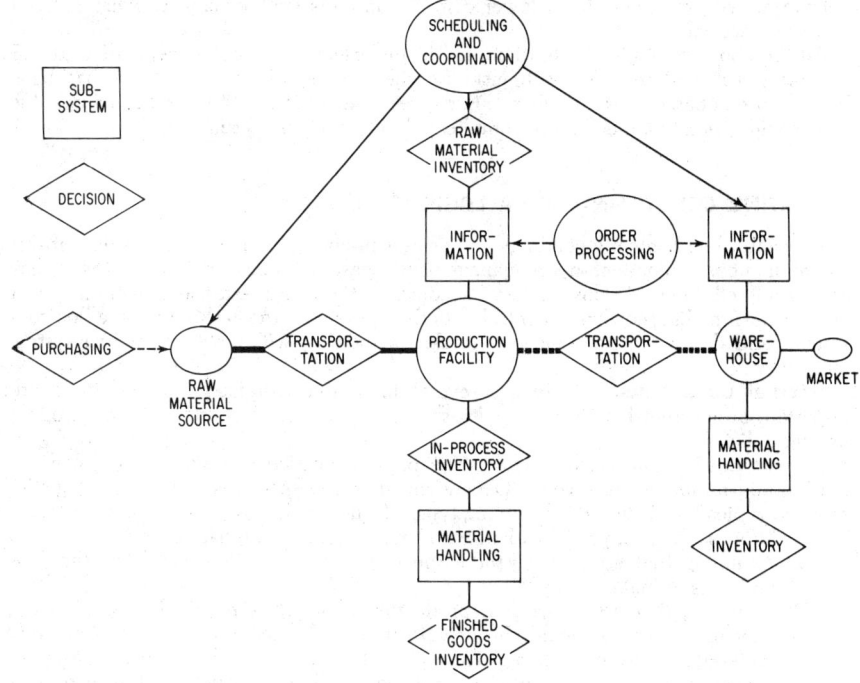

Fig. 1-2.   Subsystems.

taining a low inventory of finished goods.   Selection of a transportation mode with short transit time will permit such lower levels.   However, shorter transit time usually corresponds to higher transportation costs.   A trade-off must be established between inventory and transportation costs, with simultaneous recognition of the level of customer service desired.

Example 2: Material Handling Subsystem versus Scheduling Subsystem.   There is little to be gained by rapid, frequent scheduling of trucks to a company's warehouse if there is not an equivalent unloading dock capacity.   Such a situation requires a balancing of the cost of idle service facilities as opposed to the cost of waiting trucks.   In addition, the impact of any adopted policy must likewise consider, at a minimum, the impact on inventory levels, plant storage capacity, production scheduling, and again, level of customer service.

Example 3: Packaging versus Transportation Subsystem.   Product packaging must be consistent with the transportation mode selected.   The various transportation modes have different requirements relative to packaging design.   Package design must be compatible with the appropriate characteristics of the mode, and vice versa.

Example 4: Information Subsystem versus Other Subsystems.   The initializing information in a logistics system is a customer order.   It activates all elements and subsystems, and these must be capable of adequate time response.   There is little achieved if the documentation phase of the order cycle is expedited by utilization of sophisticated and expensive electronic data processing equipment if this advantage is more than offset by excessive time requirements elsewhere in the system.

Example 5: Increase in Plant Capacity versus Addition of a New Plant.   In a growth company, the question of how to increase production capacity ultimately arises.   May such an increase be best achieved by expanding present plant and thus achieving economies of scale; or should a new plant be constructed closer to the dispersing market, with associated lower distribution costs and perhaps a higher level of customer service?

All the above examples have attempted to underline the basic concepts of trade-off and total cost analyses. In each instance, the number of possible effects has been limited to emphasize these notions. For a real-life situation, the number of possible effects will generate a much more complicated although similar analysis.

OPERATING CONSTRAINTS OF A LOGISTICS SYSTEM

An established company conducts its business in an environment. A portion of this environment is dependent upon management's past decisions and goals; these may be considered internal constraints. In contrast, there are external constraints over which the firm has no direct control. Both types limit the flexibility of a logistics system, and determine, to some extent, the boundaries within which such a system must operate.

Internal Constraints. There are several internal constraints. Among the more important are current location of fixed assets, customer service standards, and product pricing policies.

Current Location of Fixed Assets. The present location and size of plants, warehouses, and terminals are given. Dependent upon the specific situation, the logistics problem typically involves the following types of questions:

1. What is the best product-plant-warehouse-market combination?
2. Assuming that additional production capacity is required, what are the pros and cons of make or buy?
3. Assuming that an increase in production capacity is required by building a new plant, where should the new facility be located relative to existing raw material sources, plants, and markets?
4. Assuming a system having decentralized warehouses, what are the incentives for centralization?
5. Assuming geographical growth of a market, when and where should a new warehouse be added to the system, and what should its capacity be?

Although not exhaustive, these questions underline the effects of an existing system on current and future logistics planning. On the one hand, this constraint decreases the magnitude of such planning and thus permits examination of fewer alternatives. On the other hand, it also decreases the degrees of freedom available to the planner.

Customer Service Standards. The second internal constraint reflects the firm's ability to provide a level of service consistent with customer needs. These needs are usually expressed along several lines, including:

1. Length of order cycle. The length of time from order placement to actual physical delivery, or the lead time from the customer's viewpoint, is paramount in his internal planning. The logistics system, which is spanned by this cycle, must be minimally capable of this level of flow time.
2. Multiproduct line. Customers often desire a broad line of products to increase the amount of product differentiation which they in turn may offer their customers. Such situations compound requirements on the flexibility of production processes, transportation facilities, and warehousing capabilities.
3. Maintenance of inventory by supplier. This is a means of decreasing the need for internal fixed assets such as warehouse facilities, and improving cash flow by a customer. This constraint not only may increase inventory levels at various internal points carried by the supplier, but may also affect decisions as to the type of production process necessary.
4. Condition of landed goods. This refers to damage that may be incurred in transit from supplier to customer. In the event of serious damage, it is clear that the order cycle is at least doubled, and may result in lost sales. This places a lower boundary on the adequacy of package design as well as on the method of transportation.
5. Selection of transportation mode. In some cases, the customer maintains

control over the mode of transportation to be used by a supplier. The rationale for such control is based partly on the customer's unloading facilities and partly on the customer's desire to negotiate with the transportation mode directly. The supplier consequently is left with less opportunity to consolidate shipments to several customers in the same market area, as well as less ability to take advantage of other cost reducing possibilities..

Product Pricing Policies. In general, those aspects of pricing policies which affect the logistics system involve conditions of sale and quantity discounts.

The point of transfer of legal responsibility for goods is normally stipulated under the condition of sale, together with statement of price, method of payment, and the like, and affects the length of span of a logistics system. Quantity discounts, in contrast, affect inventory policies, production scheduling, and transportation decisions.

All the above constraints are internal in nature, and essentially reflect a firm's orientation, that is, whether it is process or marketing oriented. In total, once again, they also form a basis within which the logistics system must operate.

External Constraints. External constraints are not under a firm's direct control, and indeed, the firm may not have any influence on them. From a logistics perspective, the more critical include:

Strategy of Competition. Strategies of competition usually relate to level of customer service, pricing, and technological change. Although, under conditions of competition, it is impossible for a firm to avoid exposure to these actions by competition, it is extremely important that its counteraction be timely and appropriate. Such counteraction depends heavily upon the ability of the logistics system to adjust accordingly.

Government Regulation. Regulation at the Federal level affects the cost of a logistics system, assuming interstate commerce to be present. This effect is expressed essentially along two lines:

1. Robinson-Patman Act. Prohibits discrimination in price among different purchasers for goods in like grade and quality and in like quantity. For example, quantity discounts must be justified by cost savings which are not less than the discount. These savings may be derived from an analysis of corresponding transportation and material handling cost reductions.

2. Regulation of transportation industry. Common and contract carriers, again assuming interstate movement of goods, are subject to economic regulation by the Federal government. Although such regulation acts as a constraint on the logistics system, it is mandatory that system personnel be well versed in this subject.[2] Evaluation of a firm establishing its own private transportation fleet may be accomplished only after a sound analysis of the costs of applicable rates and support services afforded by regulated carriers has been made.

Capabilities of Transportation Modes. The various transportation modes have different capabilities relative to speed, frequency of service, dependability of service, capability for movement of goods, and availability. These capabilities may vary by geographical location and to this extent limit the transportation subsystem. Equally important is the degree of technical innovation within and between the various modes. For example, the development of coordinated systems, such as piggyback[3] or trailer on flat car, allows the shipper to utilize the advantages of two or more different modes. Although the existing transportation subsystem is in fact a constraint, the logistician should maintain awareness of these developments as a possible basis for system modification and improvement.

Both internal and external constraints are summarized in Figure 1-3.

---

[2] Suggested references are D. Philip Locklin, *Economics of Transportation*, Richard D. Irwin, Inc., Homewood, Ill., 1966; and Marvin L. Fair and Ernest W. Williams, Jr., *Economics of Transportation*, Harper & Row, Publishers, Incorporated, New York, 1959.

[3] For the extent of piggybacking, see *76th Annual Report of the Interstate Commerce Commission*, U.S. Government Printing Office, Washington, D.C., 1962.

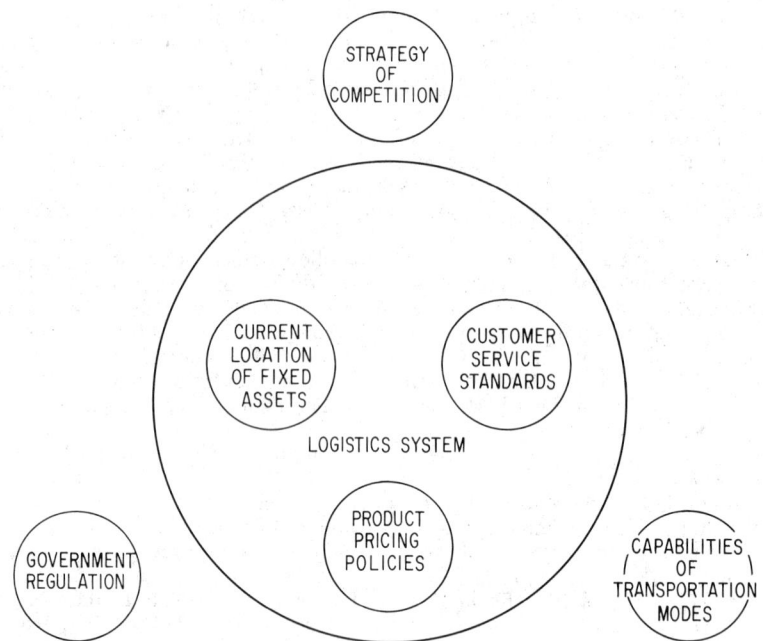

Fig. 1-3. Constraints on a logistics system.

ANALYSIS OF LOGISTICS SYSTEMS

At this stage, the elements and subsystems of a logistics system, their interrelationships, and typical internal and external constraints have been defined. The system described has been simplified by (1) reducing the number of interrelationships and constraints considered and (2) ignoring the position of the firm with respect to the total economy. The latter is treated in depth by several of the references at the conclusion of this chapter, as well as in the work by Isard.[4]

Even with the simplifications noted, it is clear that a logistics system may still be quite complex and have many aspects. At the moment, there is no cure-all method or technique which has the capability of analyzing all such aspects. However, in terms of methodology, three approaches appear to be quite useful. These are operations research, simulation, and heuristics.

Operations Research. The subject of operations research (OR) has been explored in Chapter 5 of Section 10 and will not be repeated here. Instead, an application of linear programming to a specific problem will be considered.

Problem Scenario. A company has m autonomous plants producing the same product at different geographical points. Due to different local costs, the total production cost per hundredweight of product, C_i, is different for each plant.

Any plant may serve any of n distinct market areas, some of which correspond to the plant locations. The transportation cost for each plant-market combination is c_{ij} per hundredweight. It will be observed that although a plant may be located in a market, a transportation cost is still incurred due to local delivery expenses.

Each plant has a limited capacity (in hundredweight) of P_i, and each market a demand of D_i (in hundredweight).

[4] Walter Isard, *Methods of Regional Analysis: An Introduction to Regional Science,* The M.I.T. Press, Cambridge, Mass., 1963.

Problem Statement. Due to a forecasted substantial increase in one of the markets not having a plant, as well as a lesser increase at all other markets, the company is considering the construction of a plant at the former location. Should this plant be built?

Problem Solution. Based upon the above description, we have the ingredients of a transportation problem in which we are attempting to find the least cost trade-off between production and transportation cost elements. Using the notation given, we may develop a matrix as shown in Figure 1-4 for the existing plant-market situation, that is, without a new plant and at current demand levels. Solution of this matrix will permit evaluation of present performance without the aid of linear programming.

In similar fashion, Figure 1-4 may be appropriately modified to evaluate the alternatives of (1) current demand with the new plant, (2) forecasted demand without the new plant, and (3) forecasted demand with the new plant. Anticipated savings with the new plant may be derived from these results, and a return on investment analysis performed.

The linear programming output in this problem will also indicate which plants are marginal and perhaps merit a closer examination by management.

Additional Commentary. This problem has been intentionally limited for purposes of discussion. For example, the alternative of expanding one or more of the existing plants has not been considered. In addition, the analysis rests on several assumptions:

1. There is no change in unit production cost at various operating levels.
2. There is no change in transportation costs reflecting loads of varying sizes; in practice, there are different rates for truckload versus less than truckload.
3. Plant capacity has been limited to 100 percent.
4. Per unit costs at the time at which the forecasted demand will be experienced are the same as current costs.

These assumptions have been enumerated to stress the danger of blindly applying an operations research model. The extent to which each of these assumptions has compromised the final results must be considered.

Simulation. Malcolm[5] has stated:

> Systems simulation is useful in the study of a class of problems wherein the operating rules, policies, procedures, and other elements that control production,

[5] D. G. Malcolm, "The Use of Simulation in Management Analysis: A Survey," *Report of the Second Simulation Symposium,* American Institute of Industrial Engineers, New York, 1958, p. 18.

MARKET

PLANT	1	2	3	$\cdots$	j	$\cdots$	n	Capacity
1	$C_1 + c_{11}$	$C_1 + c_{12}$	$C_1 + c_{13}$	$\cdots$	$C_1 + c_{1j}$	$\cdots$	$C_1 + c_{1n}$	P_1
2	$C_2 + c_{21}$	$C_2 + c_{22}$	$C_2 + c_{23}$	$\cdots$	$C_2 + c_{2j}$	$\cdots$	$C_2 + c_{2n}$	P_2
3	$C_3 + c_{31}$	$C_3 + c_{32}$	$C_3 + c_{33}$	$\cdots$	$C_3 + c_{3j}$	$\cdots$	$C_3 + c_{3n}$	P_3
$\vdots$	$\vdots$	$\vdots$	$\vdots$	$\vdots$	$\vdots$	$\vdots$	$\vdots$	$\vdots$
i	$C_i + c_{i1}$	$C_i + c_{i2}$	$C_i + c_{i3}$	$\cdots$	$C_i + c_{ij}$	$\cdots$	$C_i + c_{in}$	P_i
$\vdots$	$\vdots$	$\vdots$	$\vdots$	$\vdots$	$\vdots$	$\vdots$	$\vdots$	$\vdots$
m	$C_m + c_{m1}$	$C_m + c_{m2}$	$C_m + c_{m3}$	$\cdots$	$C_m + c_{mj}$	$\cdots$	$C_m + c_{mn}$	P_m
Demand	D_1	D_2	D_3	$\cdots$	D_j	$\cdots$	D_n	

Fig. 1-4. Transportation problem matrix.

inventory, etc., are under question . . . and in which the number of variables involved, the uncertain nature of inputs, among other things, makes these problems, which are referred to generally as a system, difficult to analyze.

In contrast to operations research models, a simulation model is run, not solved. The simulation technique in itself is not an optimizing procedure, even though the results of a simulation (just like real-life data) can be subjected to an optimizing procedure such as linear programming. Input data to the model are usually generated by some random number generator, and the mathematical structure of the model may or may not be specified. Accordingly, simulation is most helpful where a more formal method of mathematical analysis is not possible or convenient. Simulation places (1) less emphasis on the mathematical background of the problem solver and (2) more stress on his ability to reason in a logical manner.

One of the most powerful types of simulation is known as Monte Carlo simulation and involves variables having probability patterns. Monte Carlo may be described in a step-by-step procedure as follows:

1. Determine which system variables or components are significant.
2. Determine a measure of effectiveness which incorporates these variables for the system under study.
3. Plot the cumulative probability distribution of each of these variables.
4. Establish ranges of random numbers which are in direct correspondence with the cumulative probability distribution of each variable.
5. Based upon examination of the data, establish possible solutions for the problem.
6. Generate a set of random numbers, using random number tables or some other suitable device.
7. Utilizing each random number and the corresponding cumulative probability distribution of each variable, determine variable values.
8. Substitute the result of step 7 in step 2 and compute the value of the measure of effectiveness.
9. Perform several trials of steps 6 to 8 for each possible solution as stated in step 5.
10. Based upon the results of step 9, make a decision relative to optimum solution.

It is clear that we are executing a sample of trials using a model. The mean value of the measure of effectiveness may be determined from sample results for each potential solution stipulated in step 5. Because these means are sample means, confidence intervals for the true means of the measure of effectiveness and required sample sizes may be determined by statistical methods dealing with estimation and confidence limits.

Problem Scenario. As a means of clarifying these ideas, the following problem suggested by Bowman and Fetter[6] is presented.

The Cleveland Steel Company must store iron ore from Minnesota for use during the winter while the lakes are frozen and lake traffic is not possible. The amount of iron ore necessary to supply the winter's needs is uncertain for two reasons: the season of ice is a variable and the amount of iron ore consumed per unit time is also variable.

If too much iron ore is stored during the winter, unnecessary storage costs will be incurred. If too little iron ore is stored during the winter, it is necessary to bring the needed ore in by train, a more expensive operation. Applicable costs are:

Transport a ton of ore by ship = $3
Transport a ton of ore by train = $5
Hold a ton of ore for entire winter = $1

[6] E. H. Bowman and R. B. Fetter, *Analysis for Production and Operations Management,* Richard D. Irwin, Inc., Homewood, Ill., 1967, p. 444.

The probability distribution for the number of weeks in the ice season is:

Probability, %.............	0	8	47	31	11	2	1	0
Weeks....................	11	12	13	14	15	16	17	18

The probability distribution for average ore demand (in tons) per week during the winter season is:

Probability, %.	0	4	16	42	21	9	4	2	1	1
Tons..........	7,800	8,000	8,200	8,400	8,600	8,800	9,000	9,200	9,400	9,600

Problem Statement. Determine the optimum number of tons of ore to store.

Problem Solution. Using the step-by-step procedure outlined, we have:

Step 1: Determination of system variables. Examination of the problem indicates two significant variables: weeks in ice season and average ore demand (in tons) per week. These variables, together with appropriate cost information, may be defined symbolically as:

s = cost to transport a ton of ore by ship = \$3
h = cost to hold a ton of ore for entire winter = \$1
t = cost to transport a ton of ore by train = \$5
D = demand for entire season = Nd = (no. weeks in season) × (avg. demand/week)

Step 2: Determination of measure of effectiveness. Letting I represent the inventory stocked at the beginning of the season, we may develop two total cost expressions relating the variables of step 1. These cost expressions are the measure of effectiveness for the possible outcome of any amount stocked and may be defined as:

a. Amount in inventory less than demand, $I \leq D$:

Total cost = cost/ton by ship × tons in inventory + cost/ton by train
                                                         × additional tons needed

or in symbols,

$$TC = sI + t(D - I) = \text{measure of effectiveness } A$$

b. Amount in inventory greater than demand, $I \geq D$:

Total cost = cost/ton by ship × tons in inventory + holding cost/ton
                                                      × excess tons in inventory

or in symbols,

$$TC = sI + h(I - D) = \text{measure of effectiveness } B$$

Step 3: Cumulative probability distribution of significant variables. The cumulative probability distribution for weeks in the ice season and average ore demand per week may be developed from the original data. These are shown in Figures 1-5 and 1-6, respectively.

Step 4: Assignment of random numbers. At this stage, random numbers are assigned for the various values of each significant variable as shown in Tables 1-1 and 1-2. In each of the assignments in Tables 1-1 and 1-2, the range of random numbers corresponds to the probability of a given value of the variable as specified in the original data. For example, the random number range 08 to 54 includes 47 random numbers, and because the entire range of random numbers is 00 to 99, this corresponds to the 47 percent probability of having 13 weeks in the ice season as stated in the original data. Similar reasoning may be used in obtaining the remaining random number ranges.

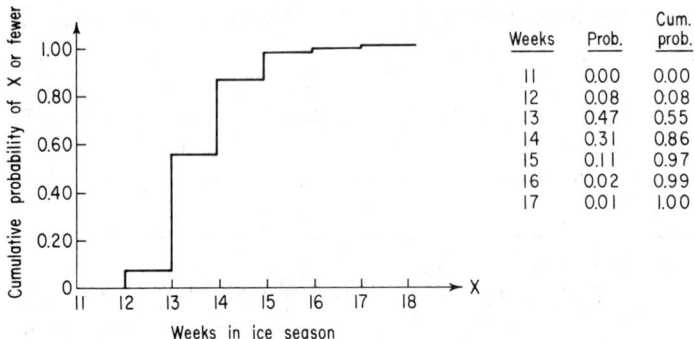

Weeks	Prob.	Cum. prob.
11	0.00	0.00
12	0.08	0.08
13	0.47	0.55
14	0.31	0.86
15	0.11	0.97
16	0.02	0.99
17	0.01	1.00

FIG. 1-5. Cumulative probability distribution of weeks in ice season.

Step 5: Establishing possible solutions. The expected number of weeks in the ice season may be derived by multiplying the probability of a given number of weeks in the season by the number of weeks, and summing this product for all possible season lengths:

$$(0\%)(11) + (8\%)(12) + (47\%)(13) + \cdots + (0\%)(18) = 13.55 \text{ weeks}$$

In similar fashion, the expected demand per week is found to be 8,592 tons. The expected demand for the entire ice season may be calculated to be

$$(13.55)(8,592) = 116,421 \text{ tons}$$

On this basis, reasonable inventory stock levels to be simulated might be $I_1 = 110,000$, $I_2 = 112,000$, $I_3 = 114,000$, $I_4 = 116,000$, $I_5 = 118,000$, and $I_6 = 120,000$.

Steps 6 through 9: Execution of simulation. The results from simulating each of the inventory levels are shown in Table 1-3. Calculations for two levels for trial 1 are presented below.

Trial 1: A random number, 66, is drawn from a random number table; and from Table 1-1, this corresponds to an ice season of 14 weeks. A second random number, 83, is also drawn from a random number table; and from Table 1-2, this is associated with an average demand per week of 8,800 tons.

These results indicate a demand for the entire season of

$$(14)(8,800) = 123,200 \text{ or } 1.23 \times 10^5 \qquad \text{(to nearest thousand)}$$

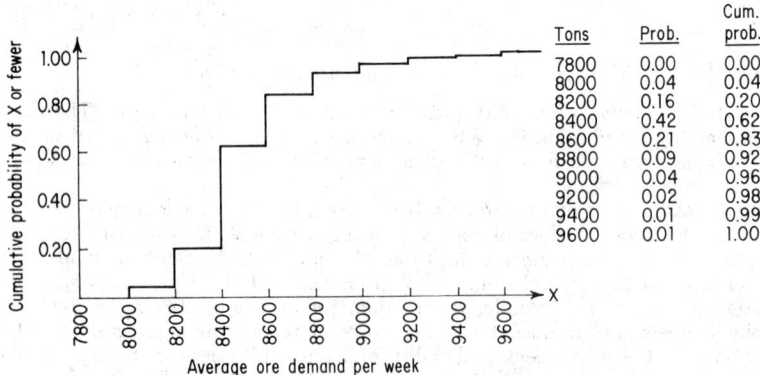

Tons	Prob.	Cum. prob.
7800	0.00	0.00
8000	0.04	0.04
8200	0.16	0.20
8400	0.42	0.62
8600	0.21	0.83
8800	0.09	0.92
9000	0.04	0.96
9200	0.02	0.98
9400	0.01	0.99
9600	0.01	1.00

FIG. 1-6. Cumulative probability distribution of average ore demand per week.

TABLE 1-1. Assignment of Random Numbers for
Weeks in Ice Season

Random number	Weeks in ice season
00–07	12
08–54	13
55–85	14
86–96	15
97–98	16
99	17

Because this demand is greater than each of the given stock levels, measure of effectiveness A is used for determining the total cost in each case. These are:

For $I_2 = 112,000$,
$$TC = sI + t(D - I)$$
$$= (3)(1.12 \times 10^5) + (5)(1.23 \times 10^5 - 1.12 \times 10^5)$$
$$= \$3.91 \times 10^5$$

For $I_6 = 120,000$,
$$TC = sI + t(D - I)$$
$$= (3)(1.20 \times 10^5) + (5)(1.23 \times 10^5 - 1.20 \times 10^5)$$
$$= \$3.75 \times 10^5$$

It will be noted from Table 1-3 that, dependent upon the relative magnitude of demand versus inventory stock level, measure of effectiveness A is required in some instances, and in others—such as trial 3 for I_4, I_5, and I_6—measure of effectiveness B must be used.

Step 10: Decision relative to optimum solution. At the base of each cost analysis column in Table 1-3 will be found the sum of the total cost for all trials per inventory stock level. The lowest such sum is that associated with a stock level of 112,000 tons. The expected average seasonal cost for this optimum level is 9,033,000/25, or approximately \$361,000.

Because there is only a slight difference between the total cost for an inventory level of 1.10×10^5 and that indicated for the 1.12×10^5 level, it would be well to continue the simulation beyond 25 trials. This may readily be done using a digital computer.

Additional Commentary. In addition to this solution, the problem has several other notable points:

1. The search for an optimum solution involves a balancing of costs. A solution is being sought that will minimize total cost. Total cost in this case depends upon the level of inventory. The level of inventory determines both excess holding cost and excess transportation cost. These are contracosts because as inventory increases, excess storage cost increases and excess transportation cost decreases (see Table 1-4). The optimum solution is therefore the inventory level that minimizes the sum of these two costs.

TABLE 1-2. Assignment of Random Numbers for
Average Demand per Week

Random number	Average demand per week
00–03	8,000
04–19	8,200
20–61	8,400
62–82	8,600
83–91	8,800
92–95	9,000
96–97	9,200
98	9,400
99	9,600

TABLE 1-3. Trials to Solve Simulation Problem

Trial no.	Weeks in ice season Random number	No. weeks = N	Demand/week Random number	Ton = d	Demand for season = D ($\times 10^5$)	$I_1 = 1.10$ ($\times 10^5$)	$I_2 = 1.12$ ($\times 10^5$)	$I_3 = 1.14$ ($\times 10^5$)	$I_4 = 1.16$ ($\times 10^5$)	$I_5 = 1.18$ ($\times 10^5$)	$I_6 = 1.20$ ($\times 10^5$)
1	66	14	83	8,800	1.23	\$ 3.95	\$ 3.91	\$ 3.87	\$ 3.83	\$ 3.79	\$ 3.75
2	89	15	97	9,200	1.38	4.70	4.66	4.62	4.58	4.54	4.50
3	32	13	88	8,800	1.14	3.50	3.46	3.42	3.50	3.58	3.66
4	37	13	12	8,200	1.06	3.34	3.42	3.50	3.58	3.66	3.74
5	39	13	22	8,400	1.09	3.31	3.39	3.47	3.55	3.63	3.71
6	74	14	16	8,200	1.15	3.55	3.51	3.47	3.49	3.57	3.65
7	76	14	24	8,400	1.18	3.70	3.66	3.62	3.58	3.54	3.62
8	26	13	00	8,000	1.04	3.36	3.44	3.52	3.60	3.68	3.76
9	18	13	64	8,600	1.12	3.40	3.36	3.44	3.51	3.60	3.68
10	28	13	37	8,400	1.09	3.31	3.39	3.47	3.55	3.63	3.71
11	29	13	62	8,600	1.12	3.40	3.36	3.44	3.52	3.60	3.68
12	06	13	58	8,400	1.01	3.39	3.47	3.55	3.63	3.71	3.79
13	20	13	09	8,200	1.06	3.34	3.42	3.50	3.58	3.66	3.74
14	09	13	64	8,600	1.12	3.40	3.36	3.44	3.52	3.60	3.68
15	56	14	74	8,600	1.20	3.80	3.76	3.72	3.68	3.64	3.60
16	66	14	15	8,200	1.15	3.55	3.51	3.47	3.49	3.57	3.65
17	87	15	47	8,400	1.26	4.10	4.06	4.02	3.98	3.94	3.90
18	94	15	86	8,800	1.32	4.40	4.36	4.32	4.28	4.24	4.20
19	11	13	79	8,600	1.12	3.40	3.36	3.44	3.52	3.60	3.68
20	22	13	43	8,400	1.09	3.31	3.39	3.47	3.55	3.63	3.71
21	50	13	35	8,400	1.09	3.31	3.39	3.47	3.55	3.63	3.71
22	59	14	12	8,200	1.15	3.55	3.51	3.47	3.49	3.57	3.65
23	81	14	25	8,400	1.18	3.70	3.66	3.62	3.58	3.54	3.62
24	76	14	64	8,600	1.20	3.80	3.76	3.72	3.68	3.64	3.60
25	59	14	65	8,600	1.20	3.80	3.76	3.72	3.68	3.64	3.60
					Total cost....	\$90.37	\$90.33	\$90.77	\$91.50	\$92.43	\$93.59

Cost analysis per inventory level

TABLE 1-4. Minimum Total Cost

	$I_1 = 1.10$ ($\times 10^5$)	$I_2 = 1.12$ ($\times 10^5$)	$I_3 = 1.14$ ($\times 10^5$)	$I_4 = 1.16$ ($\times 10^5$)	$I_5 = 1.18$ ($\times 10^5$)	$I_6 = 1.20$ ($\times 10^5$)
Total excess transportation cost..........	$63.70	$63.02	$49.06	$34.87	$34.51	$27.15
Total excess inventory cost.	26.67	27.31	41.71	56.63	57.92	66.44
Total cost.....	$90.37	$90.33	$90.77	$91.50	$92.43	$93.59

2. The optimum solution is not at the expected demand level. In fact, the expected average seasonal cost for an inventory level of 116,000 tons is $366,000, an increase of about $5,000.

3. Possible inventory levels slightly below and above 112,000 tons could and should be simulated. This will more sharply pinpoint the optimum solution.

4. The solution shown is based on a limited number of trials. The sample of trials could and should be increased to gain reliability in results. This reliability improves only as the square root of the number of trials.

5. As in all models, a sensitivity analysis of inputs could be performed. This will determine the range of each variable within which the optimum solution of 112,000 tons will still be applicable. For example, one could increase the cost of transporting ore by train in increments until a new optimum solution results.

6. Sensitivity analyses could be extended to the original probability distributions. This might be accomplished by purposely building shifts into these distributions and observing the magnitude of change in the optimum solution and its associated cost.

In concluding this example, we note an implication to a noncaptive ore supplier. Ore suppliers typically ship in large tonnage quantities. This permits lower transportation cost. On the buyer's side, this means a higher average inventory and consequent storage costs. How should this be reflected in ore price?

Heuristics. A heuristic is a rule of thumb or principle which reduces the amount of search in problem solving. This approach, as compared with an algorithm, does not guarantee an optimal solution. Nevertheless, it is still preferable to a random or unstructured search. The central idea is to progress toward an acceptable, as opposed to an optimum, solution. This methodology has been found especially useful in problems where either no algorithm has been developed or an algorithm exists but the magnitude of the problem is too large for present digital computer systems with respect to size or time. It differs from simulation in that the latter usually is furnished with fixed alternatives whereas these are built into heuristic models.

Rather than using an example, a typical application of heuristics from the literature will be discussed.

Location of Warehouses. Kuehn and Hamburger[7] have developed a subject model which includes multiple products, the capability of handling nonlinear transportation costs, warehousing costs, and delivery time to customer.

The model's solution rests on two key heuristics:

1. Locations with the greatest promise are those at or near concentrations of demand.
2. Near-optimum warehousing systems can be developed by locating warehouses one at a time and adding at each stage of the analysis that specific warehouse which produces the greatest cost saving for the entire system.

Procedurally, m possible warehouse sites are selected from n possible locations. Each of these m sites is evaluated, and a warehouse is located at that site yielding

[7] Alfred A. Kuehn and Michael J. Hamburger, "A Heuristic Program for Locating Warehouses," *Management Science,* July, 1963, pp. 643–666.

the largest reduction in distribution costs. The remaining $(m - 1)$ sites are then reassessed, and a warehouse is again located if there is an indicated reduction in costs. Noneconomical warehouses are discarded after this sequential procedure is completed. Feldman, Lehrer, and Ray[8] have extended this work to handle concave warehousing cost functions.

Summary of Methodology. The preceding methodologies—operations research, simulation, and heuristics—appear to offer much in the analysis of logistical systems, and an attempt has been made to indicate briefly the possibilities of each. Clearly, there are advantages and disadvantages with any given methodology, and it is left to the analyst to determine which is most appropriate in a given application. An excellent comparative summary of these methods will be found in the work by Stasch.[9]

ORGANIZATIONAL IMPLICATIONS

As indicated earlier, the logistics function serves as a bridge between the traditional functions of marketing and production. Unfortunately, in organizations in which materials management is not a distinct function, there tends to be a conflict of interests that inhibits proper functioning of the logistics system. As Magee[10] has pointed out, this is due to:

1. Failure to achieve the proper degree of coordination for logistics planning
2. Failure to designate proper planning and control responsibilities
3. Failure to establish proper measures of effectiveness, that is, measures of performance that are consistent with designated responsibilities as well as the basic economics of the firm

In overview, these failures may be defined by the term "suboptimization," namely, optimization of the marketing and production functions. Clearly, such suboptimization is not necessarily best for the firm.

Dependent upon the degree to which they are present, factors favoring the proposed reorganization include:

1. Need for a systems approach. Based upon the contents of this chapter, it is clear that the methodology required involves systems analyses which cut across organizational lines.
2. Adoption of EDP equipment. The use of electronic data processing for materials control has shown much growth. The expansion of this application from raw materials through the entire logistics network to finished product distribution is inevitable.
3. Influence of expanding markets. As firms become more nationally oriented in market, with an increase in number of regional warehouses, there is a corresponding concern with logistical costs. These costs include not only the fixed assets such as warehouses, but also a large investment in finished inventories and additional transportation costs.
4. Increasing competition. Many firms, as well as entire industries, have come to realize that the buyer, within limitations, must be adequately served. In many situations, this has resulted in a highly competitive level of service and has necessitated a reevaluation of the internal organization structure.

All these factors, together with others, exert pressures in the direction of a unified function for the physical movement and coordination of materials. Logically, this is resulting in the formation of a cost or profit center in this area.

Implications for Industrial Engineering. Due to its emphasis on methods, systems analysis, and knowledge of internal operations, industrial engineering has a role in the logistics function. Its contribution is especially important with respect to the

[8] E. Feldman, F. A. Lehrer, and T. L. Ray, "Warehouse Location under Continuous Economies of Scale," *Management Science,* May, 1966, pp. 670–684.

[9] Stanley F. Stasch, "Distribution Systems Analysis: Methods and Problems," *The Logistics Review,* March–April, 1968, pp. 7–34.

[10] John F. Magee, "The Logistics of Distribution," *Harvard Business Review,* July–August, 1960.

systems analyses themselves and also in the development of input information for these analyses.

As an example of the latter, it is well known that cost data for financial statement purposes are not necessarily useful for the types of analysis under discussion. In particular, variable costs applicable to different load levels on the logistical system are critical for a sound analysis. Experience reveals that this type of information is too often lacking and that it is typically the industrial engineer who must provide such data.

Finally, there will be a need for industrial engineers to broaden their knowledge of other aspects of the firm in which they have not normally been engaged.

CONCLUSION

In summary, this chapter has been concerned with presenting the logistics concept, its components and methodology, and the organizational implications of its adoption. The influence of this approach on financial management has not been directly discussed, but rather has been handled implicitly through the total cost basis underlying logistical analyses.

For further material on this subject, the reader is referred to the bibliography below. In addition, it is suggested that chapters of the Handbook included in Sections 8 to 11 should also be examined.

BIBLIOGRAPHY

Brewer, Stanley H., *Rhochrematics,* University of Washington, Bureau of Business Research, Seattle, June, 1960.

Busacker, Robert G., and Thomas L. Saaty, *Finite Graphs and Networks,* McGraw-Hill Book Company, New York, 1965.

Clevett, Richard M., *Marketing Channels,* Richard D. Irwin, Inc., Homewood, Ill., 1954.

Conway, Richard W., William L. Maxwell, and Louis W. Miller, *Theory of Scheduling,* Addison-Wesley Publishing Company, Inc., Reading, Mass., 1967.

Forrester, Jay W., *Industrial Dynamics,* John Wiley & Sons, Inc., New York, 1961.

Hertz, David B., and Roger T. Eddison, *Progress in Operations Research,* John Wiley & Sons, Inc., New York, 1964, vol. II, chap. 4.

Heskett, J. L., Robert M. Ivie, and Nicholas A. Glaskowsky, *Business Logistics: Management of Physical Supply and Distribution,* The Ronald Press Company, New York, 1964.

Hillier, Frederick S., and Gerald J. Liebermann, *Introduction to Operations Research,* Holden-Day, Inc., San Francisco, 1968.

Johnson, Richard A., Fremont E. Kast, and James E. Rosenzweig, *The Theory and Management of Systems,* McGraw-Hill Book Company, New York, 1963, chap. 8.

Mossman, Frank H., and Newton Morton, *Logistics of Distribution Systems,* Allyn and Bacon, Inc., Boston, 1965.

Plowman, Grosvenor, *Elements of Business Logistics,* Stanford University, Graduate School of Business, Stanford, 1962.

Smykay, Edward W., Donald J. Bowersox, and Frank H. Mossman, *Physical Distribution Management,* The Macmillan Company, New York, 1964.

Chapter **2**

Work Simplification—A Program
of Continuous Improvement

ALLAN H. MOGENSEN

Founder and Director,
Work Simplification Conferences, Lake Placid, New York

For many years, work simplification has been regarded as a training program—by some a supervisory training program only, by others a program to be given eventually to all employees. During World War II, this was accentuated by the incorporation of the basic work simplification pattern into a 10-hour course designed to help supervisors in plants engaged in defense production to improve methods. Disillusioned by the absence of a magic formula, many executives assumed that the only way to improve methods was to revert to the "expert" approach. It is apparent now that the failure of both these attempts to secure continuing and substantial improvement in methods has brought into sharp focus the success of some firms that have seen in work simplification a philosophy of management.

And it is significant that most of the chief executives of companies that have found this to be the case say that, despite the millions of dollars they have saved each year through a program of work simplification, they feel that the greatest benefit has come through the change in attitude of all their people.

The late Professor Erwin H. Schell, when asked what characteristic of the industrial future will have the most profound effect on work simplification in particular and plant management in general, said:

> One of industry's most amazing oversights has been the emergence of the
> hourly employee as a manager! While he doesn't supervise people, he does

manage his machine, his tools, his materials, his output, his time, his workmanship, his health. Work simplification has recognized this trend for years, and is in a better position to capitalize on this "managerial" skill than ever before. How profoundly Karl Marx erred when he said men would be slaves of machines!

Therefore, it should be realized that work simplification, to be successful, must consist of three parts—all equally important:

1. The philosophy of work simplification
2. The work simplification pattern—the tools and techniques
3. The plan of action—the actual program

THE WORK SIMPLIFICATION PHILOSOPHY

Perhaps the best way to emphasize the importance of the proper understanding of the importance of this part of a successful program of improvement is to quote from Dr. Rensis Likert, in his book, *New Patterns of Management,* published by McGraw-Hill. In it he says:

> The facts seem clear. Work simplification has been conspicuously successful, but its underlying philosophy has not been generalized and used as a basic principle of management. There are few, if any, companies which are using a managerial system that applies participation to all management procedures to the extent to which Mogensen feels is necessary for the successful operation of work simplification. The forces which have prevented an integration of the two management systems of job organization and cooperative motivation apparently are so strong that even the impressive success of work simplification has failed to overcome them.
>
> At present, all of the costs which a company incurs because of its particular management system are rarely considered in choosing between alternate systems of management. The costs of building and maintaining an effective human organization are usually ignored in the accounting methods of most companies. Similarly, spurious earnings achieved by liquidating some of the company's investment in the human organization are not charged against the operation and used in evaluating which system of management works best. There are other large costs which are also neglected. For example, the long and bitter strikes over such questions as local working agreements and the extension of work standards have rarely been charged against the job organization system of management by companies using that system. Nor has it been charged with the costs of slowdowns, excessive grievances, and similar developments. Yet the underlying cause of these slowdowns and strikes, which has cost companies, labor, and the public hundreds of millions of dollars and much suffering, appears to be the hostility, resentment, and distrust produced by aspects of the job organization system of management.
>
> When the accounting procedures fully charge each system of management with all of the costs for which it is responsible, the evidence is likely to indicate that companies should consider, seriously, the newer systems of management.
>
> Integration of the two systems has actually occurred in the case of one managerial process, namely, "work simplification." The sensitive insight of Allan Mogensen led him many years ago to change motion study and related industrial engineering procedures into "work simplification." Mogensen's essential idea is that the power of industrial engineering methods in simplifying tasks and eliminating wasteful activity should be used not by the staff of the industrial engineering department alone, but by all the members of the organization. His view is that the industrial engineering department should train supervisors and workers in work simplification methods and give them all the technical assistance and consultation they may require and seek, but the actual application of the methods are to be done by the workers and supervisors themselves.
>
> As Mogensen says: "Work simplification *always* introduces the human element; it's always designed for foreman and employee participation." Erwin H. Schell, who worked closely with Mogensen and who coined the name "Work Simplification," felt as Mogensen does: "He (Mogensen) used the principles of motion study originally developed by the Gilbreths for structuring a program in which every member of the organization might participate."
>
> The success of work simplification, on the one hand, the slow spread of partici-

pation to other managerial processes, on the other, raises an important and perplexing question: Why has not the basic philosophy and the general principle of participation upon which work simplification is based spread more rapidly to other processes of management? Mogensen feels that it actually has spread to other procedures of management and to management processes generally in the form of "consultative" management. But consultative management falls far short of the amount of participation which Mogensen insists is necessary in work simplification. In using work simplification, the industrial engineering department does not discuss a problem with employees and then, itself, make the decision. Where work simplification is functioning well, the employees and the foremen are fully involved in making the decision and often make it entirely on their own. In consultative management, the higher echelon may discuss a problem with one or all persons on the lower echelon(s), but the decision is often made without any real participation by the lower echelon(s).

It may be well to stress the fact that Dr. Likert's viewpoint is not that of an active practitioner in any of the conventional means of increasing productivity, such as industrial engineering, systems and procedures, methods improvement, and the like, but that of a social scientist who has for many years concerned himself with some of the reasons why the usual attempts of management to improve its performance have not been too successful. Since his first book was written, further research and field tests have enabled him to add to the theory and describe more fully a workable management system which can be used by any enterprise to achieve high productivity, above average financial success, and improved labor relations. Working with accounting experts, Dr. Likert has evolved a plan for developing accounting procedures to enable dollar estimates to be attached to the value of the human organization.

Since Professor Schell spoke of industry's amazing oversight, quite a few companies have successfully changed their whole management system to one where every employee is really a manager. Dr. M. Scott Myers[1] says:

> The informed manager no longer needs to be convinced of the merits of job enrichment. Experiments at Texas Instruments and elsewhere have shown tangible improvements in terms of such diverse criteria as reduced costs, higher yields, less scrap, accelerated learning time, fewer complaints and trips to the health center, reduced anxiety, improved attitudes and team efforts, and increased profits. Hence, the desirability of job enrichment is no longer in question, but, rather, the quest now is for definitions and implementation procedures.
>
> A self-managed job is one which provides a realistic opportunity for the incumbent to be responsible for the total plan-do-control phases of his job. Though many jobs in their present forms cannot be fully enriched, most can be improved and some can be eliminated. Whether the mission be to enrich, improve, or eliminate the job, it is achieved best by utilizing the talents of the incumbents themselves, provided, of course, this involvement will lead to equivalent or better opportunity. Job enrichment is an iterative process. Though it finds most dramatic expression at the lower levels, it depends on supportive climate and action at the top. When achieved at the lower levels, its impact in terms of both organizational and human criteria reinforce its support from the top.
>
> The application of meaningful work offers substantial short-range incentive for managers to support it. Judged as they are, year-to-year, in terms of profit, cost reduction, cash flow, and return-on-assets criteria, job enrichment is seen as a significant resource for achieving success. But it offers even greater rewards on a long-term basis, particularly if criteria of success are broadened to include aspects of human effectiveness, such as self-actualization of employees, responsible civic and home relationships, and the profitable and self-renewing growth of the organization. The role of business and industry in an entrepreneurial society such as the United States has a profound influence on the health of that society. Approximately eighty percent of people at work are in traditional conformity-oriented, nonexempt job categories. Hence, the implementation of job enrichment principles in industry has great potential for developing a pattern of respon-

[1] M. Scott Myers, "Every Employee a Manager," *California Management Review,* Spring, 1968, pp. 9–20.

sible behavior learned through a way of life at work which can influence people's behavior in their multiple roles in the community and family.

In actual practice, at Texas Instruments, greatly increased responsibility has been given employees for managing their own work. It is part of a long-term effort directed toward having people at every level in the company more involved in planning, doing, and controlling their own and the work of their natural work group to solve problems and match and achieve personal and company goals. It is in contrast to the tradition that most work in industry requires only the special skills of workers and that it is the responsibility of management supervision alone to organize and direct the work of others.

TI is taking a team approach which calls for working groups to analyze their own jobs and suggest improvements, set individual and group performance goals, measure their own achievement against these goals, and learn from their own mistakes.

The results are striking where team improvement efforts have been established. Over an eighteen-month period, operators on a germanium transistor manufacturing line exceeded by 20 percent their own performance goals for reducing production costs and increasing output. These goals were higher than standards previously set by supervisors. Absenteeism during the period averaged less than 1 percent of scheduled work hours. Team members on another mechanized assembly line introduced equipment improvements which cut labor costs at their point in the process by 40 percent. Assemblers working on magnetic and rotary components for radar equipment cut production time from 2.5 to 1.2 hours per unit over a fifteen-month period. Members of these groups reported that interest in their work increased sharply.

This team approach to work, however, must be developed. Therefore, they established training programs to educate managers and supervisors for the broader leadership role the approach requires, and they began to encourage employees at all levels to participate as team members in individual and group goal setting and problem solving activities.

It is our feeling that, unless such a change in attitude can be secured from the top down, all the "programs" of work simplification, work design, value analysis, operations research, systems and procedures, and so on, will have little lasting benefit.

THE WORK SIMPLIFICATION PATTERN

The work simplification pattern started out as a four-step pattern in JMT (job methods training), moved to five steps through many years of work simplification, and has been amplified by some to a six- or seven-step pattern. However, it is believed that it is still the basic approach of the engineer to the solution of any problem. Recently, much emphasis has been placed on problem solving, problem analysis, and problem prevention. The basic patterns in use are almost identical with the five-step work simplification pattern. This is:

1. Select a job to improve
2. Get all the facts—make a process chart
3. Challenge every detail

      What—WHY?
      Where—WHY?
      When—WHY?
      Who—WHY?
      How—WHY?

List possibilities—can we:

      Eliminate
      Combine
      Change $\begin{cases} \text{Sequence} \\ \text{Place} \\ \text{Person} \end{cases}$
      Improve necessary details

4. Develop the preferred method
5. Install it—check results

Once the "student" has learned the basic pattern, it is relatively simple to give him the tools and techniques that he or she will be most likely to use in improving methods, whether a top-flight research scientist, an operator in a factory, or a clerk in an office.

Such tools may be listed here in part only:

Flow process chart	Memomotion study
Flow diagram	Motion pictures
Multiple activity chart	Video tape
Operator chart	Micromotion study
Procedure flow chart	Value analysis or engineering
Multicolumn chart	Work design
Gang chart	Operations research
Work distribution chart	Mathematical models
Statistical work sampling	Simulation

Certainly, on the new equipment being developed as a result of the intensive activity in the field of automation, in both the factory and office, new and more accurate as well as valuable tools will be added to this list. In addition, some of these tools will be adopted by others than the technicians—witness the use of the memomotion camera by hourly workers on one railroad to secure the data they needed for a work sampling study in a classification yard, resulting in their request to management for two-way radio. An attempt by management to introduce the same equipment a few years prior was met by the insistence of the union that a full day's pay be given each time anyone made a call.

Years ago, most of the delegates to work simplification conferences had never used a process chart—today most supervisors in business and industry are familiar with this basic tool.

Use of the five-step pattern by people at all levels of management replaces the too-frequent snap judgment or flash improvement with the same type of project analysis used by the industrial engineer. Such an analysis sheet, used at the apparatus division of Texas Instruments and shown in Figure 2-1, reduces the number of judgments or decisions that have to be made by top management. Once a worker or supervisor has been trained in the use of such an analysis, the old "I suggest . . ." (which could be and frequently is a completely unworkable brainstorm) is replaced by "I see two possibilities, one that can be installed at once at little or no expense, with such and such savings, or a more radical improvement, which might take two years to perfect and install, but saves so much more. I propose we adopt the first, and for such reasons."

This explains why work simplification not only results in amazing savings but develops people as well. True, if the industrial engineer develops the improvement, the savings are sometimes realized; but if it is the product of participation, with the help and advice of the expert, when requested, the person making the improvement has grown in stature, and management can then spot those who show outstanding ability, with a view to future advancement in the organization.

THE PLAN OF ACTION

Too often, training programs end with the graduation ceremonies of each group trained. In effect, management has said, "You are now trained in work simplification, go out on the job and use what you have been taught!" As someone once said, training unfortunately is often the process where information gets from the instructor's manual into the notebook of the student without going through the heads of either one. Too often, there is no carry-over from the artificial atmosphere of the conference room to the day-to-day job situation.

TI FORM 1K3735 10–61

WORK SIMPLIFICATION METHODS IMPROVEMENT REPORT
TEXAS INSTRUMENTS INCORPORATED
APPARATUS DIVISION

SUBMITTED BY.	COST CENTER	SUBJECT OF REPORT		(USE PENCIL OR BLACK INK)

		BASIS OF SAVINGS ANNUAL / CONTRACT	IF CONTRACT – LIST NAME	DATE OF REPORT	DATE INSTALLED

PART NUMBER_____   DESCRIPTION OF IMPROVEMENT: _____
CONTRACT NO._____
W.O. NUMBER _____
OTHER _____
AREA–LOCATION OF
IMPROVEMENT

SUMMARY OF SAVINGS

	PRESENT METHOD	PROPOSED METHOD

MATERIAL OR OTHER COSTS: (M)

1. MATERIAL, SUPPLIES, ETC. PER _____ (DAY OR UNIT) $_____ $_____
2. TOTAL PRODUCTION OF _____ DAYS, OR _____ UNITS.
3. TOTAL MATERIAL COSTS (LINE 1 X LINE 2) $_____ $_____

LABOR: (L)

4. MAN HOURS (MH) PER_____ (DAY OR UNIT)   _____ MH   _____ MH
5. TOTAL PRODUCTION OF _____ DAYS, OR _____ UNITS.
6. TOTAL MAN HOURS (LINE 4 X LINE 5)   _____ MH   _____ MH
7. LABOR COST PER MAN HOUR $_____ (RATE PER HOUR)
8. TOTAL LABOR COST (LINE 6 X LINE 7) $_____ $_____

OVERHEAD: (O)

9. _____ % OF TOTAL LABOR COST (THIS % X LINE 8) $_____ $_____
10. TOTAL MLO COSTS (LINE 3 + LINE 8 + LINE 9) $_____ $_____
11. TOTAL MLO GROSS SAVINGS (DIFFERENCE OF PRESENT AND PROPOSED METHOD) $_____
12. COST OF INSTALLATION: (JIGS, FIXTURES, TOOLS, ETC.)

    MATERIAL $_____
    LABOR $_____
    _____%OH $_____
    (THIS % X LABOR)
13. TOTAL COST OF INSTALLATION (SUM OF ITEM 12) $_____
14. NET SAVINGS (LINE 11–LINE 13) $_____

IMMEDIATE SUPERVISOR

IS THIS METHOD WORKABLE AND SOUND?_____

WHAT IMPROVEMENTS, IF ANY, CAN YOU ADD?_____

IMPROVEMENT HAS BEEN ☐ OR WILL BE ☐ INSTALLED

DATE_____   SIGNATURE_____

FIG. 2-1. Work simplification methods improvement report. (*Courtesy of Texas Instruments Incorporated.*)

To correct this, many years ago we began to stress what we call the three phases of work simplification. To understand this, look at Figure 2-2. This is from *Work Simplification: An Effective Program of Improvement,* by Professor Herbert F. Goodwin, published originally by the Society for the Advancement of Management in January, 1957, and amplified in a *Factory Magazine* article on work simplification, which appeared in 1958. It has been our experience that the failure properly to carry out each of these three important phases is like trying to sit on a three-legged stool without all three solid legs. It is possible, to be sure, to balance oneself on a

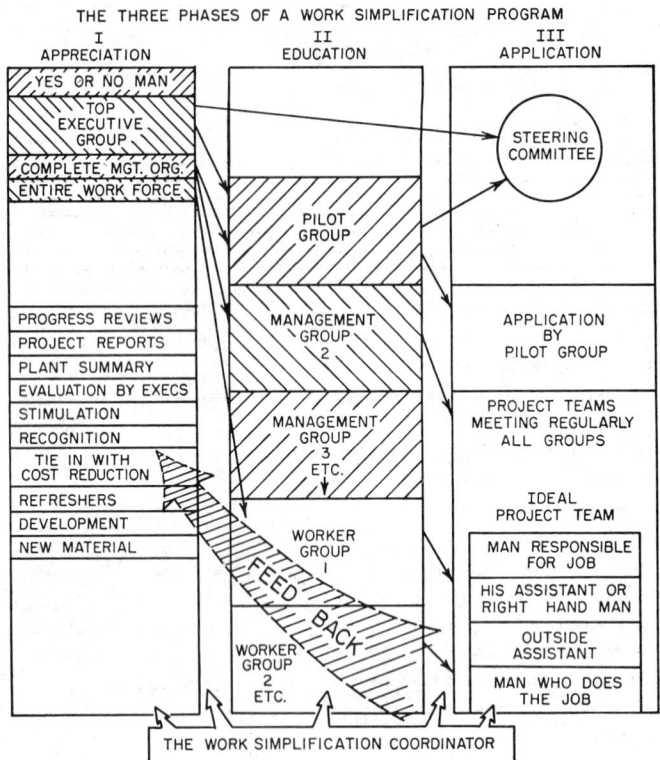

Fig. 2-2.   The three phases of a work simplification program.

stool with only one leg, easier to do it with two, but it should be unnecessary to go further with the analogy.   Yet, far too many firms believe they have a work simplification program, without such a solid plan of action to ensure perpetual continuation of improvement.   As a result, they are always complaining, "our foremen will not make process charts" or are asking for new gimmicks to revive a program that has become, in the eyes of all concerned, "just another training program."

In the previous editions of this handbook, we cited the success of work simplification at H. P. Hood & Sons, with headquarters in Boston.   Nearly thirty years have passed since their activity was begun, and they are more convinced than ever that this is the way to continuous improvement at all levels.   While they do continue with the educational aspect of their program, refreshing previous graduates with new tools and techniques as well as giving basic training to new employees, the bulk of the effort has been for many years devoted to what they call "developmental conferences" that are held on a monthly basis with people from all over their widespread system.   Hourly workers join with the supervisors and managers in the solution of not only production problems, but those in the fields of clerical, engineering, and sales.

It is our feeling that only by providing the two routes of communication of ideas for improvement, both for the individual and for those working as a project team, with the proper incentives to encourage and stimulate all concerned, can a company be certain that it can remain continually competitive.   Often those firms which have had a suggestion system with cash awards in the past will integrate it into a good work simplification program, but the realization that it is not the cash award that is really the important motivating force seems to be gaining in acceptance.   We find

that people give their ideas for improvement for these reasons:
1. They want to see the idea in and working.
2. They like the recognition they receive.
3. If an equitable cash reward is forthcoming, they like this too, but it is a mistake to assume that this is the most important motivating force.

As a result, the most successful companies see to it that any idea that should be installed is done as soon as possible, regardless of its magnitude.   They see that the idea is dramatized, if possible, and that recognition is adequately given by a handshake, a letter from the boss, or a picture on the bulletin board or in the company paper or magazine.

Naturally, no one means of recognition will work for all, or for too long in any one company.   New means of stimulating interest must constantly be developed, and what may work well in one company or plant may fail in another.

At Procter & Gamble, their Deliberate Methods Change Program started in 1946, with formal work simplification training being offered to management representatives of each manufacturing plant.   The basic objective of this program is simple: to increase profits through the philosophy of deliberate change.   The key points, based on Procter & Gamble's experience, are these:
1. Form methods teams.
2. Establish dollar goals.
3. Provide positive recognition.

They say:

> At Procter & Gamble, since 1946, the rate of return—using first-year savings only—has been around 500 to 1,000 percent.   In other words, $5 to $10 of profit is returned for every $1 spent.   No other portion of management can match this rate of return.

And despite the seeming emphasis on dollar goals, Procter & Gamble stresses the importance of positive recognition by saying:

> Taking good methods for granted will not produce outstanding results.   Sincerely done, it is almost impossible to overdo proper recognition.   People like to be recognized in the presence of their associates—be part of a winning team— feel important by being important.

Finally, anyone going into work simplification as a top management philosophy will have to take another look at the whole matter of traditional engineered work standards.   Few issues have aroused as much controversy in manufacturing management.   Dr. David Sirota[2] says:

> The debate takes place not only between management and employees, but also within the ranks of management itself.   It is difficult to find anyone in industry taking a neutral position on work standards.   At best, people are ambivalent. They have often witnessed large productivity improvements accompanying the introduction of standards in plants, but they are uneasy about the costs of the system.   The latter include the obvious expenses of maintaining large industrial engineering staffs and the less obvious, but no less real, costs of the time consumed managing the conflicts between employees and management so frequently engendered by standards.   The most prominent of these hidden costs is the time spent dealing with the large number of worker grievances about standards.

CONCLUSION

Progress reviews, project reports, written or on film or video tape, and continual evaluation by top management will ensure that goals that have been set by the people themselves will usually not only be met but exceeded.   Only by having an active three-phase program can work simplification be a substantial factor in continuing improvement in any organization.

[2] David Sirota, "Productivity Management," *Harvard Business Review*, September–October, 1966, pp. 111–116.

BIBLIOGRAPHY

Gellerman, Saul W., *Motivation and Productivity,* American Management Association, New York, 1963.

Goodwin, H. F., and A. H. Mogensen, "Work Simplification," *Factory,* July, 1958.

Goodwin, H. F., and Leo Moore, *Management Thought and Action, in the Words of Erwin H. Schell,* The M.I.T. Press, Cambridge, Mass., 1967.

Likert, Rensis, *The Human Organization: Its Management and Value,* McGraw-Hill Book Company, New York, 1967.

Myers, M. Scott, *Every Employee a Manager,* McGraw-Hill Book Company, New York, 1970.

Uris, Auren, "Mogy's Work Simplification Is Working Miracles," *Factory,* September, 1965.

Chapter **3**

Office Cost Control

WILLIAM M. AIKEN

Senior Vice President, H. B. Maynard and Company, Incorporated,
Pittsburgh, Pennsylvania

KIPLING ADAMS

Senior Consultant, H. B. Maynard and Company, Incorporated,
Pittsburgh, Pennsylvania

The value of cost controls for direct labor operations has been recognized for many years, probably because at one time, direct labor costs represented the major expense of most companies. Factory management has learned to use effectively cost controls based on work measurement and to extend them to other cost areas such as indirect labor. Administrative management, while recognizing the value of cost controls for factory work, did not naturally think of its own work in terms of measurement and control. Unlike factory managers, most office managers have had no exposure to this type of cost control and are unaware of the benefits that can result from a well-organized cost control program. A similar attitude existed in factory managers' minds when controls were first introduced to them. Through experience, they have learned the real value of controls. Office managers have not had this same experience to influence them. However, as administrative costs have approached or exceeded direct labor costs, more and more managements have turned their attention to the control of office costs.

From a measurement viewpoint, there is considerable difference between factory and office work. Office people rarely produce a tangible physical product. There is less tendency to associate a given output of clerical work with a corresponding cost. In addition, office work requires a variety of measurement techniques and approaches because of its variable nature.   Rarely are two tasks identical in every respect.   For example, if a company produces an order of 5,000 light-metal assemblies, there will be 5,000 nearly identical sets of parts to be processed.   The time required to perform each step in the process will be practically identical for each of the 5,000 assemblies.   If the same company sends out 5,000 invoices, however, the chances are that each invoice will be a distinct document requiring different address information and different items and quantities than any other invoice.

The objective of this chapter is to explain how to reduce and control office costs with a carefully planned approach and the application of some basic industrial engineering techniques in spite of the technical and psychological obstacles normally encountered.

THE THREE COMPONENTS OF OFFICE COST CONTROL

Effective office cost control is more than just work measurement.   Corrective action must follow the application of measurement, and the result of this action or lack of it must be reported.   This is how control is obtained.

There are three equally important components of successful cost control.

1. Complete, accurate measurement
2. Accurate, uncumbersome reporting
3. Corrective action

A program that is weak in any one of the three components can never achieve its full cost control potential.

Complete, Accurate Measurement.   The basis of cost control is knowing how long it should take to do what must be done.   If a system reports how long it should take to do all the jobs done by only half the people, then the control information is limited. The same is true if the work of all the people is only half covered.   Complete coverage is a necessity, and this requires the application of various industrial engineering techniques.   It must always be remembered that effective office cost control depends upon complete coverage obtained by the practical application of measurement.

Obviously, control can be no better than the accuracy of the measurement.   Competent analysts, well trained in the specific techniques to be applied, are essential. Because of the basic differences from factory situations, guidance should be provided by someone specifically experienced in office cost control procedures.

Accurate, Uncumbersome Reporting.   Operating data from the application of measurement must be expressed in a carefully designed reporting system that gathers work load information to determine utilization of the staff.   This system should express how well the staff is supervised and how effectively they handle work load variations.

To promote acceptance by employees and supervision, the reporting system should be simple.   It must give timely, pertinent information to key people.   It must be easily understood and designed to be a tool of management.   Supervisors should have standards available for determining individual performances if necessary, but overall manpower requirements for managerial action must be provided to determine actual and required costs as part of the complete picture.

The reporting procedure must be carefully and thoroughly installed, fully informing supervisors and managers of the mechanics of reporting, of the meaning of the reports, and of their responsibilities as the key people in controlling clerical costs.

Corrective Action.   Gaining control by motivating managers to take positive steps is the most difficult of the three components of clerical cost control.   After performance standards have been determined and the reporting system installed, the analyst must follow up constantly to assist managers in responding to reported information. This involves orienting managers and supervisors in the techniques of clerical cost control and preparing specific recommendations.   Reporting the cost of not taking corrective action is one of the most effective methods of getting their attention.

BENEFITS OF OFFICE COST CONTROL

There are direct and indirect benefits that accrue, gradually at first, when conducting an office cost control program. The indirect benefits are difficult to evaluate, involving such things as improved morale of the work force, a better method of evaluating the abilities of supervisors and managers, and managerial experience for the program analysts. These are long-term benefits. Direct benefits can be condensed into five items which can be realized in a relatively short time.

1. Cost reduction
2. Methods improvement
3. Documented procedures
4. Evaluation of equipment needs
5. Improved supervision

Cost Reduction. The major benefits of an office cost control program are usually a direct result of improved utilization from the application of work measurement. When a department of clerical workers is informally supervised by a person who has many responsibilities in addition to supervising (as most supervisors do), utilization of the group will generally average between 50 and 70 percent. With strong supervision, some type of historical operating data, and up-to-date work counts, utilization will average 70 to 85 percent. Properly engineered standards can help strong supervisors who are trained in the use of controls to achieve 100 percent utilization.

Cost reductions are achieved in two ways:

1. Doing the same work with fewer hours. This is the most common way to achieve savings. It is based on the principle of proper staffing for work load variations.
2. Doing more work with the same hours. This assumes that there is a heavy backlog or that necessary work is not being done.

It is safe to estimate the potential return from an office cost control study at 20 percent of the labor cost in the department studied. Cost improvements of 30 percent are more likely, but the lower figure is conservative and may be used to indicate the value of making the study.

Methods Improvement. In a detailed examination of clerical tasks, it is impossible not to question why some tasks are being done the way they are or even why they are being done at all. By documenting present methods carefully, measuring the time to do the job, and comparing the results to the measured or estimated time to do the task in a revised way, the benefit of a method improvement can be accurately determined. When the cost (based on measurement) of performing a job is known, management can decide if that job is satisfactory as it is presently being done or if it can or should be mechanized.

Simplifying or eliminating a task will only create leisure time and save nothing unless there is an appropriate reduction of hours paid. Measurement of the remaining tasks and a good utilization report will provide control of this. Savings due to methods and systems improvements are difficult to achieve without this information, because they usually involve only portions of jobs rather than complete functions or easily identified whole positions.

Documented Procedures. The value of a function is determined partly by the careful measurement of its components. In doing this, the analyst carefully documents the methods used. The result is a manual of all clerical procedures used by the department, logically arranged for easy reference and maintenance. Instead of hand-me-down methods from old clerk to new clerk, this manual should be used as a training device to assure consistent operating procedures regardless of personnel turnover.

Evaluation of Equipment Needs. Decisions to mechanize can be made on a factual basis when the real cost of doing an operation manually is known. It is also important, in order to evaluate alternative methods, to know the cost of doing the operation mechanically. If labor requirements for manual and automated systems (including peripheral duties) are measured first, there will be many cases where new equipment cannot be economically justified. In other cases, work measurement will reveal the true cost of an operation, clearly establishing the desirability of a mechanized procedure.

Improved Supervision. Properly applied, measurement evaluates the effectiveness of supervision. In addition to being technically competent, a supervisor should be an able leader. He must make assignments to his people to utilize them fully. He must be aware of what his labor requirements are for varying work loads and what to do in peak and valley situations. He must know quantitatively what to do.

Without measurement information, a supervisor can hardly be expected to know what his costs should be. When he does have this information, he can begin to supervise effectively. This is more than a by-product of clerical cost control; it is the heart of it.

PLANNING AN OFFICE COST CONTROL PROGRAM

There is little doubt that a measurement and control program can benefit an organization, but it must be well organized and properly introduced to be successful.

Support from Top Management. In undertaking a clerical cost control program, management should realize that the program may not always be met with enthusiasm by everyone in the organization. Office supervisors and managers are naturally in favor of economies and desire to hold down expenses wherever they can. Almost everyone will give vocal support to the program in the beginning. Later, however, the results of the program may indicate the need for changes or reductions in their own groups. At this point, if not before, many supervisors can find a multitude of reasons to prove that the program does not apply to their work or that it is unsound for some other reason.

There is no certain way to prevent this from occurring, but one thing that can help—and without it no program can be completely successful—is strong support from top management.

First, the program must be sold to top management. To attract its attention, the cost of the studies (considering the time it will take) and the anticipated return will be useful.

The next step is to work closely with the top management representative of the areas to be studied, usually the treasurer or controller, to identify areas for the initial studies. He should be informed of the importance of his role and the objectives of the studies. When he is convinced of the benefits of the program, he should provide the introductions to his organization to get the program started.

The need for active support from the top cannot be overemphasized. This support must continue through the period of implementing changes. Without this support, the program can flounder from human inertia and resistance to change, and eventually be discarded as impractical or unsuited to the conditions of the organization.

Planning a Complete Program. The principles of office cost control are easily understood by an experienced industrial engineer. Technically, this work is not complicated; but from an administrative viewpoint, there are many pitfalls, difficult to foresee, that can slowly render useless even a vigorously conducted and technically sound program. The best safeguard against this is careful, thorough planning of the complete study with someone experienced in the specific application of measurement in clerical situations.

In selling a program, too, it is beneficial to present to top management a timetable of events with anticipated key steps and the time required to take them. If the events of conducting a program (outlined below), including development of reporting procedures and training of supervisors, are carefully planned in advance, the chances of selling and conducting a successful program are much greater.

Selecting and Training a Team of Analysts. Although industrial engineers may not be familiar with office procedures, they are generally the best qualified to be trained as analysts because of their familiarity with work measurement. Systems men and accountants have become excellent analysts after training in measurement techniques. The basic qualifications for a good analyst are generally as follows:

1. Above average intelligence or ability

2. Imagination
3. Ability to get a job done, thoroughly and on schedule
4. Salesmanship or tact
5. Ability to follow directions
6. Willingness to learn to work with people at all levels in the organization
7. A friendly approach

Regardless of background, training in the application of measurement in clerical areas is essential. An analyst must know the details of the study, the mileposts to measure his progress, and the objectives of his work. This training requires about two weeks plus experienced guidance through at least one complete pilot installation.

Identifying Areas for Study. As part of the planning process, a survey of all clerical groups eligible for study should be made. This is to identify the groups that will provide the greatest return on investment. It is important to select initial areas for study carefully to maximize chances of a successful installation. This will pave the way for selling the program in other areas.

The survey should include the following points:

1. List all departments. From this list, some groups can be eliminated immediately as being too small or too complex or having inadequate or indifferent supervision. Counsel on these matters should be sought from top management.
2. Survey the departments considered for study as follows:
   a. Function. What does the group do, regardless of its title?
   b. Number of people. Beware of estimates or payroll figures—a person charged to one place may work in another.
   c. Place in the organization, organization of the department. Who is in charge, to whom does he report, how many "assistants" does he have, what is the ratio of chiefs to Indians?
   d. Estimate of number of major tasks. This is the figure most indicative of the time required to make the study. It also indicates the relative complexity of the groups being surveyed.
   e. Determine current production of the group. If group members know how much they accomplish or what their work loads are from day to day, they are probably fairly well supervised and will be more receptive to learning how to deal with work load variations.
   f. Review forms used and reports prepared. A review of the forms and reports with a knowledgeable person will help the analyst understand specifically what work is being done.
   g. Evaluate quality of supervision. Probably the most important consideration in determining where to start an office cost control program lies with the supervisors and their attitudes. The success or failure of installing a program of office controls depends largely upon the immediate supervisor. Initial studies should be made in areas where supervision is strong. When good results are shown in areas considered to be well supervised (and they are more likely to be shown for that reason), then management will be more confident about moving into areas with weaker supervision.

CONDUCTING THE PROGRAM

After the initial areas to be studied have been identified and the analyst training is complete, the major events of each study should be scheduled. This schedule may be modified as the study proceeds, but a formal plan must be prepared to guide the analysts.

Program Procedure. Figure 3-1 is an activity chart showing the major events of a study in roughly chronological order.

Analyst training—the first step in a program; this usually requires about two weeks.

Work load surveys—the development of task lists (described below) and installation of counts to survey volume of paper work or frequency of occurrence.

Operation analysis—the review and documentation of all procedures.

Standard times—the development of the time it takes to do each job.

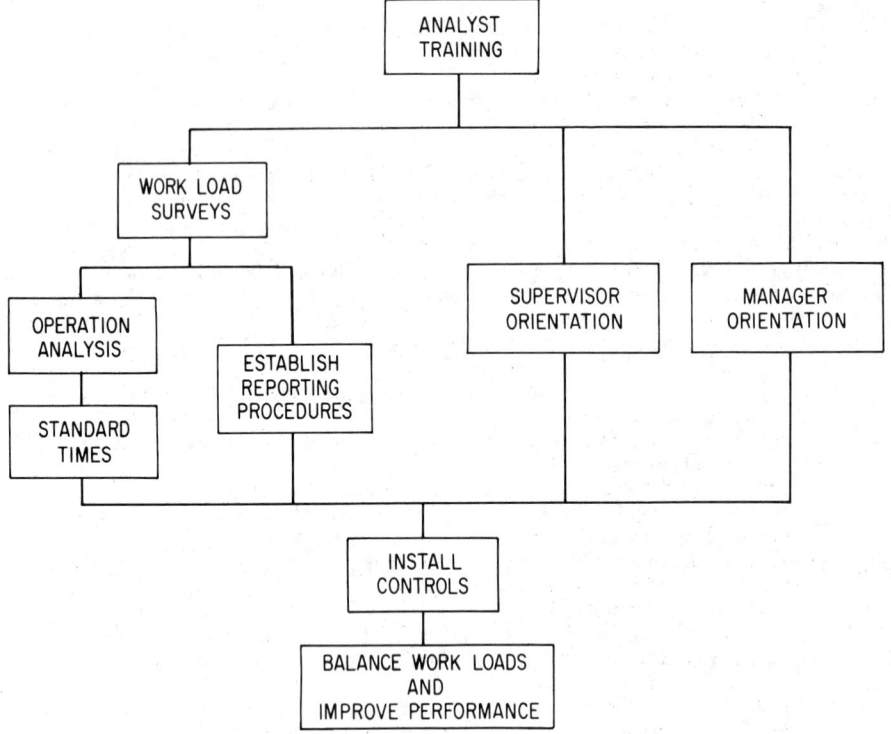

Fig. 3-1.   Office cost control program procedure.

Establish reporting procedures—arranging the standard time and work counts in a useful manner for reporting utilization and operating cost.

Supervisor orientation and *manager orientation*—during the development of standard times, the project leader should meet with supervision and management to inform them of the mechanics of measurement, the purpose of the program, and their responsibilities as part of clerical cost control.

Install controls—applying the standard times to the groups measured, obtaining operating data, and distributing performance reports.

Balance work loads and improve performance—the corrective action which supervision and management must be motivated to take as a result of the performance reports.

Introduction of the Program.   Before the study can begin, the supervisors involved must be informed of the program. If their cooperation is to be expected, they must have a basic understanding of what is to be done. They must also be in a position to explain and sell the program to their people. The emphasis at this stage should be to help the supervisors understand that the objective of the program is to provide information to enable them to be more effective supervisors.

Next, the program must be explained by the supervisor to the clerical people involved. It is important not to try to explain too much but to emphasize the obvious benefits they can understand. For example:

    1. Measurement will help management recognize good performance.

    2. The program is not a "speed-up."

    3. A result of the program will be more equitable distribution of work.

    4. Jobs and not people are being measured.

If it is the policy—and it should be—that no one will be discharged as a result of the program, assurance on this point should be given to the clerical people. Attrition may reduce the work force, and reassignments may be necessary, and it is best to clear the air on this subject. Honesty with everyone is the best policy.

Obtaining the Facts. Three techniques are required to obtain sufficient information for formulating specific recommendations:

1. Task lists and work distribution
2. Work measurement
3. Checklist of operating data

Task Lists and Work Distribution. The first information the analyst must have is what work is presently being done by each person and his estimate of how much time he spends doing it. To obtain this information, each employee is interviewed to determine how he spends his time. This information is compiled on a task list for each employee.

The initial interviews should be brief, usually less than half an hour, and must include everyone. The person being interviewed is asked to identify the major tasks he performs and to estimate how much time, usually on a weekly basis, he spends on each task. Estimating time is difficult for some people, but a skillful analyst will use various approaches—such as percentages, time of the day or week the task is done, or time per item or occurrence—to help the person decide on a roughly correct figure. This must be done tactfully, explaining that only a ball-park figure is sought to help separate large jobs from small ones. It is helpful to remind the person that jobs and not people are being studied.

Figure 3-2 is a sample task list. It shows that the clerk does four major tasks and spends half her time on one of them. It ignores break time, discussions, and other details that are not helpful for an overall survey of major tasks.

After task lists have been prepared for the group, they should be summarized on a work distribution sheet. This lists the people across the top and the jobs down the side. The estimated time spent on each job is posted in the proper place, as shown on Figure 3-3.

Preparation of task lists and work distribution sheets should be completed in one to two days. In effect, they photograph a department at the beginning of a study. It is with this information that the analyst will compare his findings.

After task lists have been prepared, the analyst will have a good idea of what the group does and who does it. He will know everyone in the group, and they will know him.

Again it should be remembered that the success of office cost control depends upon the practical application of measurement. It is very easy to overanalyze clerical work. This results in lost confidence and unnecessary emphasis on details, and it slowly erodes the vitality and effectiveness of a program. The last column of Figure

| Department | Accounting | Name | Sally Forth |
| Section | Order Service | Date | April 23, 19__ |

Task no.	Task	Hours per week
1	Edit orders	20
2	Check typed orders against original	10
3	Post releases	5
4	Type correspondence to customers	5
	Total hours	40

Fig. 3-2. Task list.

DEPARTMENT   Accounting
SECTION   Order Service

Task	Forth	Smith	Johnson	Jones	Bentley	Fitzpatrick	TOTAL	%
Edit orders	20	10		15	15	20	80	33
Check typed orders	10	10					20	8
Post releases	5		5	10	5	5	30	13
Type orders		15	10	15	20	5	65	27
Process inquiries		5	5				10	4
Prepare backlog report						10	10	4
Type correspondence	5		20				25	11
Total	40	40	40	40	40	40	240	100%

FIG. 3-3.   Work distribution sheet.

3-3 is the percent of the total hours of work that each task represents. This is to guide the analyst in selecting the measurement techniques to use and in budgeting the time he spends studying a task. The cumulative percentages of measured tasks indicate his progress toward complete coverage.

Work Measurement. Determining the time it should take to perform the tasks is the most technically demanding step in a clerical cost control program. The analyst should be familiar with the application of the various techniques which can be used to establish these times. The work distribution sheet is his guide in selecting the technique to use. There are four basic techniques for measuring tasks:

1. Time study
2. Predetermined times
3. Work sampling
4. Allowances

Time study is a valuable technique that has been used successfully in measuring direct labor work for many years. It has not been used as widely in measuring office work, largely because of the feeling of management that office workers are completely unlike factory workers and not susceptible to labor controls based on what is considered to be a factory-oriented technique. There is also a question about the economics of using time study to measure the highly variable work characteristics of clerical operations.

The most common approach is to make a time study of individual tasks for the development of specific standards. The procedure is identical with the time study procedure used in the shop. An operator is selected, the task is observed, elements are identified, and the elemental times the analyst observes are recorded. The number of cycles observed depends upon the length of the cycle and the amount of variation in the observed times. After rating the skill and effort of the operator, the necessary allowances are applied, and the standard for the task is computed. This procedure is repeated for each task in the office.

The drawbacks to time study in the office include the predictable adverse reaction by clerical workers to being timed. In addition, performance rating clerical tasks is far more difficult and less reliable than rating factory operations.

Predetermined times are time values for all basic manual motions. To apply them, a task must be divided into its basic component motions and the time values applied to each. The development of predetermined time systems—MTM being the most widely used—represented a breakthrough in the development of objective time standards. Performance rating is eliminated, and the time values are universally applicable.

But basic predetermined time values are tedious to apply to clerical tasks that are characteristically long-cycle and variable. Clerical tasks do, however, have basic motion patterns that constantly recur in varying sequences. It is certainly uneconomical to analyze the same motions over and over again. The versatility of predetermined times and the fact that different clerical tasks are composed of the same motions have led to the development of standard data for clerical operations.

Standard data are predetermined times prearranged in common motion patterns. They cover virtually all clerical work such as get, put away, file, read, staple, walk, calculate, operate common equipment such as calculators and copiers, and many more. The tables of standard data are arranged with a number of time values to cover elemental variables such as distance, number of digits, number of items in a file drawer, and the like. Every value should be supported by a detailed analysis, preferably by MTM or an equally valid technique. Figure 3-4 is a typical table of standard data for sorting sheets into piles. The variables for this operation are the number of piles (the more piles, the longer it takes per sheet) and whether the sort can be made quickly or if careful identification of sorting information is required.

To apply standard data, the analyst:
1. Selects a major task such as checking invoices or filing correspondence
2. Determines frequencies—how often certain variables occur in the operation
3. Documents the elements
4. Applies the time values to each element
5. Adds the time values to get the standard time

Figure 3-5 shows a typical analysis of a clerical operation, using standard data. The time values in this example were taken from the standard data developed by H. B. Maynard and Company and originally published in 1960 by the Management Publishing Company in the book, *Practical Control of Office Costs*.[1] The frequency data were derived by the analyst. He determined, for example, that the average number of items processed in a batch was ten. The time to get and lay aside a pencil with which to post items was distributed to each item by dividing the time for that element by 10 and recording the result in the "TMU per item posted" column.

By the same reasoning, there are three items per order number on the average, and the time for copying the order number must be divided by 3 to put it on a per item posted basis. However, there are three colors represented by an item; so the time to copy the items per box by color, to subtract from open balance, and to write the new balance must be multiplied by 3 to get the allowed time per item.

[1] H. B. Maynard, William M. Aiken, and J. F. Lewis, *Practical Control of Office Costs*, available through Maynard Research Council Incorporated, Pittsburgh, Pa., 1960.

SORT SHEETS INTO PILES
Start of Element: Sheets held in hand
End of Element: Release sheet in pile

Number of piles	Time per sheet (TMU)	
	Quick decision	Hesitation
2	33	40
3, 4	40	47
5, 6	45	52
7–10	52	59

1 TMU = 0.00001 hour

FIG. 3-4. Standard data example.

Operation *POST ITEMS TO CONTRACT*

Element			Frequency	TMU per item posted
No.	Description	TMU		
1	Get and lay aside pencil	54	$\frac{1}{10}$	5
2	Copy order number	144	$\frac{1}{6}$	24
3	Write date entered	68	$\frac{1}{3}$	23
4	Copy items per box by color	51	3	153
5	Subtract to get new balance, write balance	95	3	285
6	Check off item as complete	17	1	17
7	Lay aside item sheet	38	$\frac{1}{10}$	4

1 TMU = 0.00001 hour

TOTAL   511 TMU

FIG. 3-5.   Analysis of clerical operation using standard data.

This example shows the variability in clerical work and demonstrates why it is necessary to determine frequencies of occurrence. The need for thoroughness and accuracy when using standard data is obvious, but with practice, actual and proposed methods can be analyzed quickly by visualization.

Work sampling in the office permits great coverage in a short time. It can be used in two ways: to set standards and to measure overall productivity.

For setting standards, the volume of work completed is compared with productivity for the period during which observations are made. This approach can be used, but it presents problems in isolating the time worked on various tasks, determining if what appears to be productive effort really is effective, and rating performance.

In measuring overall productivity, no attempt is made to relate what is done to the productive effort observed. Unproductive time alone, usually expressed in euphemistic terms, is determined by random observations.

Work sampling is a rather special-purpose technique in clerical cost control. A group to be studied may be encountered that is characterized by:

1. Day- or week-long projects
2. Lack of easily obtained key work counts
3. Much effort spent troubleshooting
4. Many small tasks
5. Lack of definite job assignments

Under these circumstances, it is impractical to measure by using standard data or time study. Here work sampling can be helpful. It will not tell supervision what to do. There will be no methods descriptions or improvements as a result of sampling, but repeated sampling and a good supervisor can improve productivity.

Figure 3-6 compares characteristics of measurement techniques. Experience has shown that the use of standard data based on predetermined times is far superior in terms of continuing return on investment.

Allowances must be made for interruptions. No one works without interruption, and office workers usually have more interruptions than factory workers. In addition to personal time, there are delays such as phone calls for information, exceptions to check, special tasks, and other unpredictable demands.

For miscellaneous related duties—and by this is meant necessary tasks that are impractical to measure in detail but are routine and constitute a relatively small portion of a job—allowances can be established to cover:

1. Official discussions with supervisors, clerks, and outside contacts
2. Related duties or small housekeeping chores not practical to measure

Factor	Work sampling	Time study	Standard data based on predetermined times
Skill of analysts	Low	Average	High
Time to install	Low-average	High	Average
Administrative cost	Average	Average	Average
Savings potential	Low	Average	High
Methods improvement potential	Low	Average	High
Accuracy of standards	Low	Average	High
Applicability to:			
High volume	High	High	High
Low volume	Average	Average	Average
Employee acceptance	Average	Low-average	Average
Supervisory acceptance	Average-high	Low-average	Average-high
Flexibility (as changes occur)	Low	Average	High

FIG. 3-6.   Comparison of measurement techniques.

3. Variable elements that can occur randomly which also are not practical to measure

The values for these allowances can be measured, and this has been done many times with time studies and work sampling. Experience shows that it is far more economical to use table values for standard allowances under various conditions, as shown, for example, in Part II, "Universal Office Controls," in *Practical Control of Office Costs.*[2]

To the resulting standard time, including the miscellaneous related duties allowances, must be added the allowance for personal time. This is almost always 10 percent. It must be added to the other allowances for miscellaneous related duties, too, for they represent a significant amount of necessary work.

Checklist of Operating Data. A package of recommendations for improving the effectiveness of an operation is incomplete until the facts of the present situation and operating policies are known. Knowledge of these facts and policies will modify impulsive reactions to the results of measurement. Recommendations can be made. carefully considering all influencing factors, improving acceptance by management and increasing chances of a successful program.

An outline of basic information that an analyst should know about the group he is studying is shown in Figure 3-7.

Reporting Effectiveness. When the measurement is complete and full coverage has been obtained, a method for applying the standards must be devised. This is determined from the quantity of key items completed, the standard time for each item, and the number of man-hours required; in other words, how heavy was the workload in standard hours. Before this can be done, it must be decided how the standards are to be applied.

Some reporting systems collect counts and hours worked for each employee so that individual performances can be calculated. This is fine for special, highly repetitive jobs that consume 100 percent of a day with few separate duties. It can be helpful for training clerks. For incentive payment, this type of reporting is necessary but is effective only under the above circumstances. In typical individual performance applications, however, there is likely to be complex tallying and record keeping for the clerk, unreliable results, and low coverage because of all the "special projects" that are reported.

There is no reason for a complicated reporting system just to report individual performances. The effectiveness of the supervisor is more important than the effective-

[2] *Ibid.*

1. *Personnel*
   a. Trend of staff size for past 12 months.
   b. Present pay rates.
   c. Average length of service of employees, present ages.
   d. Number of part-time employees used—when and why.
   e. Training provided for new employees.
   f. Present method of evaluating employee performance.
   g. Method of granting pay increases.
   h. Personnel problems, if any.
2. *Hours of work*
   a. Number of shifts, time and hours worked.
   b. Authorized breaks, how controlled.
   c. Holidays.
   d. Policies for paid time off for sickness and vacation.
   e. Policies for other absences.
   f. Attendance records—how kept and reviewed.
3. *Hours worked, payroll costs*
   a. Straight-time hours and payroll costs per week.
   b. Average overtime costs per week.
   c. Average hours-paid-not-worked costs per week.
4. *Work performed*
   a. What is done by the group?
   b. Are there counts to indicate volumes—peaks and valleys?
   c. Upon what or whom does the department depend for its work and who depends upon it?
   d. What are the time requirements for processing the work?
   e. What are the quality standards?
   f. Are there any outstanding problems in the department?
   g. How did the present organization evolve?
5. *Effort toward improvement*
   a. What action or programs have recently been initiated to improve the operation?
   b. What results have been achieved compared with the goals?
   c. What schedule is there for current programs?

Fig. 3-7.   Checklist of operating data.

ness of individual clerks. The supervisor must have a list of individual task times to apply as he wishes, but clerical cost control is far more effective as a tool of management, reporting labor requirements based on standards, budgeting allowances for projects and special tasks, and considering variations in the work load.

To do this, standard times for individual tasks should be combined according to frequency of occurrence to give overall standards for key items. Ideally, for example, an invoice department might report only the number of invoices processed. The standard time would include allowances and standards for all required peripheral duties to produce an invoice. Similarly, although there are many tasks in producing checks in a payroll department, only the number of checks produced might be reported. Easily counted, time-consuming ancillary duties that might vary significantly from week to week should be reported as separate items. Tasks that require a constant amount of time, such as periodic report preparation, should be reported as a single value allowed for that week.

A system properly designed to report overall effectiveness will compare all hours worked with standard hours produced. Many programs have failed over this point; they have attempted to report too much or to overinterpret incomplete facts, which resulted in misdirected emphasis and limited results. It is the supervisor who must be more effective before individuals can be. This is why coverage—and not all of it can be detailed measurement—must be at least 90 percent and preferably 100 percent.

Figure 3-8 is a simplified weekly summary of counts (volume) and standard hours produced. The number of items counted has been reduced by combining as many tasks as possible. Notice that the preparation of the weekly shipment report, a major task that requires the same work regardless of invoice volume, has been given an allowance of 12.5 standard hours, probably based on estimates or observation.

INVOICING DEPARTMENT
Weekly Summary

Major task	Standard hours per 100	Volume	Standard hours
Prepare invoices	7.45	3,780	281.6
Issue credits	10.20	410	41.8
Process errors	3.41	234	8.0
Prepare weekly shipment report	—	—	12.5

TOTAL   343.9

Fig. 3-8.   Weekly summary of counts and standard hours produced.

The remaining step in reporting results is gathering and displaying pertinent data to evaluate operating effectiveness.   The most successful method of doing this is to:
1. Avoid reporting the number of people because it ignores work load conditions
2. Avoid percent "performance" because this always causes misunderstandings about who is working how hard and whether the standards are fair
3. Concentrate everyone's attention on a subject of common concern and understanding—the cost, in dollars, of doing the job

Figure 3-9 is a typical cost per standard hour report.   Utilization percent is a comparison of hours worked to work load (which is the standard hours produced). The total work pay is based on an average wage rate and the hours worked plus overtime premium.   (In some applications, it would be useful to include paid time off and call this "total cost.")   Cost per standard hour is total work pay divided by the standard hours produced.

The base period is an average of the initial weeks of an installation, usually about four.   It is with this average week for the base that subsequent weeks are compared. In the example, last week's cost per standard hour was improved by $.75.   This represents a cost improvement for that week of $.75 times the number of standard hours, 330.7, or $248.   To explain the improvement, it is observed that the work load increased 11.7 standard hours or 3.7 percent, and the hours worked decreased 75.5, or 17 percent.

The performance reports should be prepared by the supervisor.   He can more readily identify with the program if he is a participant in providing data.   By doing

INVOICING DEPARTMENT

	Base period	Last week	Current week
Utilization	72 %	90 %	94 %
Total hours worked	443.0	367.5	366.0
Overtime hours	41.0	7.5	10.0
Total work pay	$1043	$834	$835
Standard hours produced	319.0	330.7	343.9
Base cost per standard hour	$3.27	$3.27	$3.27
Current cost per standard hour	—	$2.52	$2.43
Saving (loss)	—	$0.75	$0.84
Saving (loss) for the week	—	$248	$289

Fig. 3-9.   Cost per standard hour performance report.

so, he is aware of the factors that affect his effectiveness, and as a result of this, understands what he must do to improve.

Installation. Installation follows the development of performance standards. The objective of the installation period is to apply the information learned during the study and to assist supervision and management in taking corrective action.

Elements of Installation

1. Prepare and distribute performance reports in final form. If the analyst has planned carefully, he will be obtaining key counts long before the study is complete and installation begins. When installation begins, all work counting is done by the clerks, using the permanent form, which will include only the necessary items. Weekly summaries and cost per standard hour reports are prepared for the previous weeks of history to obtain a base period. The information should be reported and displayed as it will be after installation is complete.

2. Instruct the supervisor on the preparation of the reports and what they mean. Patience and practice for several weeks will be required for this.

3. Review the operation analyses and standard times with the supervisor. Give him the opportunity to check over the methods and to question the standards. The purpose in doing this is to give him confidence in the study and to use his experience to locate oversights or errors.

4. Check any operations, methods, or standards the supervisor questions. The check studies, which need not be long or complicated, will give the analyst and the supervisor greater confidence in the standards. The supervisor must feel that the analyst has been fair in every case, and it is worth the analyst's time to prove that he has been.

5. Explain, in a general way, the results of the study to the department that has been studied. After the supervisor has an understanding of the standards, the reports, and his role, he should have a short meeting with his people. This is to explain that:

   a. The study is complete.

   b. The jobs—not the people—have been studied carefully.

   c. Standard times have been developed for each task, and they will be used to evaluate his ability as a supervisor and only occasionally to measure individual performance.

   d. There will be changes in procedure and job assignments.

The last point about job assignments is very important   Honesty is the best policy.

6. Develop a scheduling system. To improve performance systematically, a supervisor may need a device for anticipating his manpower needs. This can be:

   a. A backlog reporting system

   b. An item received report

   c. A forecasting device

   d. A combination of the above

Of course, some provision must be made for calculating required hours from the standards. These systems must be custom-made for each application. A good forecasting system, used regularly, can be a most useful tool for improving utilization.

7. Review preliminary results with the managers and explain the reports. The analyst must review the development and applications of standards and the meaning of the cost per standard hour report with the key people at the levels above the supervisor. In doing this, he can develop their confidence in the program, explain the mechanics of reporting results, and indicate their role as an important part of the program.

8. Pursue methods improvement opportunities noted during the study. During the study of a major function in an organization, it is impossible not to question the way something is being done or even why it is being done at all. It is similarly almost impossible to avoid the temptation to become completely involved in the pursuit of each improvement idea. There are good reasons why this temptation must be avoided.

There is no way of knowing accurately what the value of an improvement is until the real cost of the present method is known. In other words, it must first be measured.

Experience has shown, generally, that most of the benefits from a clerical cost control program are derived from the application of measurement, control reports, and improved supervision, and not from savings made through methods improvements or simplification of work.

When a methods improvement is developed, there is no way to anticipate how long it will take to install. This is especially true for suggestions involving systems changes. If it will take more than one day, as a rule, it will take many, because of inevitable complications with other departments, resistance to change, and other unforeseeable factors. To protect the return on program investment by avoiding misdirected effort, it is wise to rule that if an analyst can recommend a change, have it accepted, and install it in a day, then he should do so. If not, the recommendation should be documented in his final report with estimated benefits and left for future action by management. If this rule is firmly stated and understood at the start of a program, it will keep the program moving steadily toward its objective.

9. Suggest specific reassignments. A supervisor's familiarity with his own department inhibits his ability to see objectively where changes can be made. By piecing together tasks to form new job assignments using standard times and average work loads, the analyst can recommend specific reassignments. Tasks are like stones— not bricks. They are of all different sizes, and it requires careful examination to arrange them in groups of forty hours each. These documented recommendations are to act as guidelines for the supervisor.

10. Obtain commitments for reassignments. It is common for a program to lose momentum during installation, due to lack of corrective action by management. From a personnel relations viewpoint, achieving manpower reductions is best accomplished by attrition and reassignment. This can take time. But first, the personnel department should be informed as soon as it is known that there is available manpower in an area, so that hiring may be suspended. It is very important also that the key people specifically commit themselves in writing to the action they will take, including a projected target date. Otherwise, it is possible that no action at all will be taken.

Supervisory Training. The success or failure of installing a clerical cost control program depends more upon the immediate supervisor than anyone else. The alert supervisor who is willing to control or reduce his staff size according to the results of the study will succeed. However, if the supervisor is defensive or is put on the defensive by an unobjective analyst, the program will have difficulty. In this case, the supervisor will feel that his staff is already overworked, that management does not appreciate how hard he and his people work, and that the analyst is only trying to please management at his expense. He will look everywhere for excuses for low effectiveness. Standards will be too tight; claims will be made that many tasks were not completely covered in the analysis; complaints may be registered about the competence of the work force or the wage scale. This is all a form of buck-passing, an attempt to focus attention elsewhere.

How can these common attitudes be combated? The most effective method is for the leader of the team of analysts to conduct training sessions for the supervisors currently involved in a study. These sessions are more than just training or explanations of what the studies are about. They are an opportunity for the supervisors to exchange ideas, feelings, problems, and experiences and to see the attitudes of other supervisors. The team leader can get a reading of the attitudes and relative abilities of each supervisor during these sessions. If he is a good leader, he will promote exchanges between participants and recognize every opportunity to point out how measurement and the program in general can help them supervise more effectively.

These sessions should be designed to get across the following key points:
1. The objectives of the program—in honest terms
2. The responsibilities of being a supervisor
3. The fact that the program is designed to help them supervise more effectively and is not a criticism of the way they are presently operating
4. The mechanics of measurement and reporting
5. The corrective action that must be taken

6. The fact that they are the key people in the cost control effort and that they will be fully informed of the work done in their areas and of proposed changes

The program must have the full endorsement of management before supervisory sessions are begun. It is helpful to have a higher level management representative introduce the sessions with comments about management's attitude toward the program.

These sessions are the formal or scheduled portion of supervisory indoctrination. The informal portion is the daily contact of the analyst with each supervisor to review operations, explain the program procedure, and bring the supervisor into the program gradually. Overall, the program probably benefits most from these informal contacts between supervisor and analyst. The formal sessions, however, make the supervisors aware of what is expected of them and provide prepared instruction in the responsibilities of supervision.

Program Management. In a manner similar to the use of measurement for office supervisors, there are planning and measuring techniques that the program manager can use to assure good utilization of the analysts.

The first thing to guard against is a tendency, manifested particularly by industrial engineers, to overanalyze clerical tasks. This can result in spending too much time studying each motion or small task in detail, sometimes unnecessarily employing basic motion times. The "cost reduction by simplifying or eliminating" instinct of some analysts can cause misdirected effort on projects with less return than expected, or can postpone nearly immediate results that can be accomplished through measurement. It is the team leader's responsibility to see that overall objectives are kept in perspective and to guide the program toward its goal.

The work distribution summary, which was used to identify the people and the relatively significant tasks, is useful in controlling the progress of the study. Notice on the sample work distribution sheet, Figure 3-3, that the last column indicates the approximate portion of the department's time spent on each task. As each task is measured, the percentages of all measured tasks can be accumulated to show the progress being made toward complete coverage. Experience in typical clerical operations indicates that on the average, for the study phase (excluding installation), an analyst should be able to measure a task a day. Of course, some important tasks such as edit orders—33 percent in Figure 3-3—will receive more attention than one day. Others will receive less, but with this approximation, a schedule can be set. For example, a department with twenty tasks should be scheduled as a four-week study.

Other considerations can modify this, but usually not much. A larger department may take longer for the same number of tasks than a small one, but it is tasks and

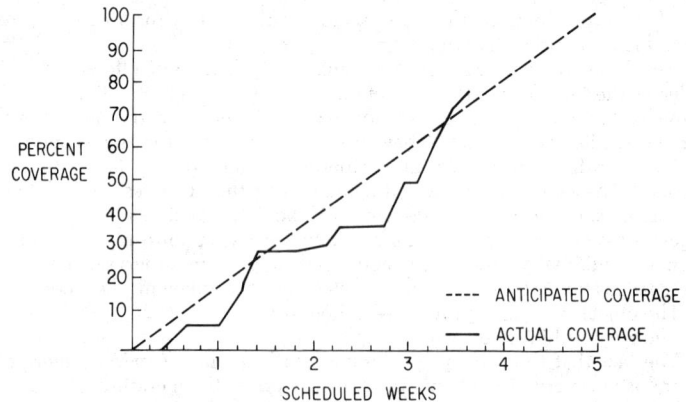

Fig. 3-10. Plot of coverage progress.

not people that take time to study. More complex tasks take longer to study than simple ones, but the rule of thumb of a day per task is still a good one to follow.

Progress toward 100 percent coverage can be shown graphically if desired. On a graph divided into scheduled dates on the horizontal axis, and with percent (0 to 100) on the vertical axis, a line may be drawn from day 0, 0 percent, to the final day at 100 percent. This is the straight dashed line shown on Figure 3-10. Each day, the analyst adds the percent coverage from the work distribution and plots it on the graph. Almost automatically, he can judge how much of his time a task is worth. This helps him decide what techniques to use in studying the task, and he knows on a daily basis how well he is doing according to his schedule.

Figure 3-10 is a typical plot of coverage progress. A slow start is to be expected due to getting acquainted in a new area, making task lists, and studying large tasks carefully. Good coverage comes quickly after that and then tapers off as many less important tasks are studied. It is during this period that strong guidance will see the study through to completion on schedule.

It is important to be able to anticipate the length of the study and to finish on schedule. This creates an atmosphere of confidence in the team and avoids giving the impression of lingering too long in an area.

ACHIEVING AND MAINTAINING RESULTS

A good study and installation are characterized by harmony among analysts, supervision, management, and clerks. People outside the study, and even the clerks, may be unaware that any changes have been made, except possibly that some different job assignments have resulted.

The reports will tell the story. Cost per standard hour reports can be summarized for the man in charge of each group or division on a single report showing utilization percent, standard hours, and savings by department for that week. Another provocative figure is the excess cost for representing unnecessary expenditures. The totals for the division or group will reflect overall operating effectiveness.

It remains for the person responsible for the cost control program to report to top management what it is getting for its investment. There is a subtle and more important purpose than this alone for reporting costs and savings. The report should indicate potential savings revealed by the studies, actual savings achieved, and the difference or excess cost. Potential savings indicate not necessarily how well the analysts have done their part, but that they have done it, and that this figure is what can be achieved. The actual savings relative to potential savings is how well supervision and management have responded to the cost information now available to them. Excess cost shows what they have not done. This figure should be of interest to someone at the top.

Figure 3-11 is an example of a cost and savings report for a four-week period. The figures represent the cost of four analysts and a supervisor. Their coverage to date is approximately one hundred clerical workers. The report shows that supervision and management in the areas covered have achieved less than half the potential cost improvements that are possible. These are the results of completed studies. As more departments presently being studied are added to the report and performances improve, the value of the program and its costs will be very clear. If, for some reason, the program is unable to justify itself, then it is a poor investment indeed. Either assistance in making this effort profitable is required or it should be discontinued.

SYMPTOMS OF THE NEED FOR OFFICE COST CONTROL

It has been stated above that in the typical informally supervised office situation, staff utilization will be between 50 and 70 percent. Study after study has verified this rule of thumb. But a program cannot be sold to management on this basis alone; more tangible indicators are needed. Possibly the office under consideration

<u>   6th   </u>   FOUR-WEEK PERIOD ENDING   <u>   </u>

PROGRAM COSTS	ACTUAL SAVINGS	POTENTIAL SAVINGS
Four-week period $ _5,500_	Savings for period $ _4,100_	Potential savings for period $ _9,300_
	Annualized $ _53,300_	Annualized $ _120,900_
Program to date $ _33,000_	Cumulative savings to date $ _11,100_	Excess cost for period $ _5,200_
		Annualized $ _67,600_

FIG. 3-11.   Cost and savings report for a four-week period.

is a beehive of activity with large backlogs of work and a supervisor that always seems to be out of breath. Under these circumstances, few people familiar with the group would accept the thought that they are 50 to 70 percent utilized.

Appearances can be completely deceptive even to an experienced industrial engineer. For example, key departments in two different companies were estimated to be at full effectiveness. After study, their base period performances supported the rule of thumb indicating excellent improvement potential. These offices suffered from misdirected clerical effort, uncontrolled personal time and breaks, and lack of organized duty assignments. A few methods changes and some reassignments based on measurement helped both companies improve paper work deliveries, reduce overtime, improve staff utilization, and control clerical costs at any work load volume.

There are six questions that can help identify symptoms of the need for clerical cost control:

1. What is the trend of administrative costs over the past three years? It is useful to compare the trend of administrative costs with direct costs or sales volume. If administrative costs have grown faster than sales or other costs for a certain recent period, further specific explanations should be sought. The answer to bringing administrative costs into line again may be an office cost control program.

2. Is overtime being paid constantly, regardless of sales or work load fluctuations? Excessive overtime every week is a sure symptom of ineffectiveness. There are, of course, times when peak loads require overtime that may be preferable to adding full-time people. Constant overtime, however, becomes a privilege and then a right, and it can be an insidious form of cost increase. Office cost control will indicate proper staffing for all work load conditions, permitting the regulated use of overtime.

3. Are reports and other paper work often late? If an office is consistently late in processing paper and backlogs are always high, it is an indication that control is lacking. In this situation, there probably is a constant struggle between supervision who wants more people and management who does not. Supervision's appeals are meant honestly enough but without the facts of what is required to do the job. Remember,

the rule of thumb indicates that their staff utilization is probably between 50 and 70 percent.

4. Are there written procedures?   Written procedures help assure consistent methods and are an indication that some analysis has been made of what is being done.   This is not a substitute for measurement and control reports, but where written procedures exist, it can be expected that the office will be better organized and more effective than one without them.

5. Are there productivity measures for the staff?   If there are definite job reassignments made according to work load volumes determined by a system of counts, then the office is likely to be much more productive than one where no counts exist. Counts alone are not a good indicator of effectiveness, but evidence of flexibility in job assignments made according to work load information is a good indicator.

6. What is the supervisor's work pace?   Very often an extremely busy supervisor is mistakenly assumed to be a good supervisor.   Frequently, supervisors are appointed on the basis of detailed knowledge of a function and personal diligence.   Attention to details and diligence may then increase.   Instead, supervisors should be delegating routine tasks and planning the work of the department so that as a team it will be effective.   If the supervisor is well organized, relaxed, and has time to devote to his real job of supervising, it is likely that his group will have above average effectiveness.

The answers to the above six questions are, especially from a manager's viewpoint, good indicators of what can be accomplished with an office cost control program.

CONCLUSION

Office cost control involves the specialized application of industrial engineering techniques.   It requires many technical skills plus imagination, aggressive leadership, and ability to plan a complete program and sell the concepts at all levels. Measurement and reporting must be done in a manner significantly different than in the factory, and the analyst must learn to think in much larger terms to develop controls for economic and effective management.   In addition, failure to recognize the difficulty of selling and conducting an office cost control program in an atmosphere of reluctant acceptance has caused some attempts to have short-lived or limited results.

For a program to produce a good return, the effort must, from the beginning, have the appearance of being well organized, objective, aggressive, thorough, and capable of producing results.   If the program is viewed as a weak, plodding effort with poorly defined objectives, results will always be just out of reach.   There is no substitute for experienced guidance in office cost control.

When the office cost control program has proved its value, management should quickly recognize the benefit of coordinating this with the budgeting function, cost accounting, systems development, and profit planning.

Another significant benefit of office cost control is that it provides excellent training for future managers.   It is an opportunity for them to work with people at all levels and to become intimately acquainted with the function, details, problems, and management of many groups within the organization.

If management is willing to have the patience and foresight to carry out a program properly, the investment will be repaid many times over by the resulting savings and intangible benefits.

BIBLIOGRAPHY

Aiken, W. M., and J. F. Lewis, "Office Work Measurement," sec. 14, chap. 6, in H. B. Maynard (ed.), *Handbook of Business Administration*, McGraw-Hill Book Company, New York, 1967.
Eisenberg, J., *Cost Controls for the Office*, Prentice-Hall, Inc., Englewood Cliffs, N.J., 1968.
Johnson, H. W., and W. G. Savage, *Administrative Office Management*, Addison-Wesley Publishing Company, Inc., Reading, Mass., 1968.

Karger, D. W., and F. H. Bayha, *Engineered Work Measurement,* The Industrial Press, New York, 1957.

Maynard, H. B., W. M. Aiken, and J. F. Lewis, *Practical Control of Office Costs,* Maynard Research Council Incorporated, Pittsburgh, Pa., 1960.

Mills, G., and O. Standingford, *Office Organization and Method,* 4th ed., Pitman Publishing Corporation, New York, 1968.

Pemberton, L. A., and E. D. Gibson, *Administrative Systems Management,* Wadsworth Publishing Company, Inc., Belmont, Calif., 1968.

Walley, B. H., *Manual of Office Administration,* Business Publications, London, 1968.

Chapter **4**

Cost Estimating

KENNETH J. SLEYMAN

**Director of Management Systems Development,
H. B. Maynard and Company, Incorporated, Pittsburgh, Pa.**

Estimating combines the science of skillful analysis with the art of experienced, mature judgment. If it has been done properly within a required time period, it produces a reasonably accurate prediction of the eventual manufacturing cost of a product. The method is by detailed study of component parts. The result is a statement of cost that will serve as a realistic basis for making business decisions.

ESSENTIAL ELEMENTS

Timing, historical evidence, and judgment are indispensable elements in cost estimating. The purpose for which the estimate will be used determines which of these is the most essential in any given case. For example, timing is the most critical factor in a competitive sales situation where the cost estimate is needed for setting a price. Historical evidence or judgment might be more significant if the estimate is required for a feasibility study or a process change or a product modification decision. However, all three are absolutely essential to every cost estimate.

Timing. The expression, "Time is money and success," is particularly appropriate to cost estimating. Obviously, the function adds no value to the product. Therefore, it is an expense that increases in direct ratio to the time taken. The more time required, the greater the cost. Similarly, there can be no argument that time becomes a marketing imperative when cost estimates must be available on a given date. An estimate completed after that deadline is of no use to anyone. This means that good cost estimating requires a sense of proportion and discipline regarding time. It follows that this balancing of effort and resources against practical business needs is a vital element in the process of cost estimating.

Historical Evidence. Valuable increments of estimating data are gained almost every time a new cost estimate is prepared or an actual cost report is received. If this information is accumulated and organized in a systematic manner, it frequently will eliminate the need to "reinvent the wheel." In other words, if the project, product, or component part can be identified with historical data in the files, a considerable amount of time and effort in preparing the required cost estimate can be saved. However, the most important function of these data is to supply values which can be used to test and check the validity of new cost estimates. The greater the amount of historical detail available, the smaller the margin of error between predicted manufacturing costs and the actual costs. Without this factual basis of historical evidence, cost estimating often loses its objectivity and degenerates into a guessing game.

Judgment. There is no substitute for experience or skill in cost estimating. The work requires both systematic analysis and study of job specifications and close attention to the terms and conditions of sale. It also requires a managerial attitude and viewpoint. High intelligence and a good background in engineering, manufacturing, and accounting may equip a man for the analytical work, but experience is the only teacher when it becomes a matter of applying the basic cost elements to a specific situation. Labor and material escalations, unusual transportation charges, contract penalties, weather conditions, and legal constraints, along with manpower and facility requirements, are only a few of the factors that may have to be evaluated and quantified in the final estimate. Therefore, only those in the organization who have the requisite experience are capable of exercising good judgment during the cost estimating process.

Paradoxically, the need for practical business judgment to avoid serious errors in making cost determinations increases if the process is automated. The effort required to define a system, code and collect data, and interpret results adds a new dimension to the estimator's work. Utilizing a device like a computer offers the potential of significant savings in clerical costs, timelier data, and new and better information not previously available, but it also means much more decision making for the estimator to discipline the details of the systems information flow.

SOURCE DATA

The information required to prepare a cost estimate must be obtained from the functional departments responsible for such things as product and tool design, production planning, vendor selection, and process and methods engineering. One method of preparing an estimate, and quite possibly the only one in the case of a new company or division, is to route the request to each concerned department for their inputs. In practice, many companies have found that this departmentalized approach lacks responsiveness and frequently overstates manufacturing costs. To correct these shortcomings they have established separate, self-sustained estimating departments that make the estimate from information which is already in their files or which they obtain from the functional departments.

A fully integrated estimating department will normally require two or three years to develop. Even then, it will never be completely independent of its need for assistance from engineering, manufacturing, and purchasing. However, it should be adequately staffed and geared to the task of preparing estimates when they are wanted or needed. The estimators become the final arbiters of the quality and completeness of the data used and the amount of information required to produce a good estimate. This automatically reduces the possibility that estimates will be inflated with protective cushions or delayed by misguided attempts to achieve a greater degree of accuracy. Estimates are either good or bad, and they fully serve their purpose if they maintain a high confidence level within the organization that actual costs work out reasonably close to the estimates.

TABLE 4-1.   Types and Sources of Cost Estimating Data

Description of data	Sources
General design specifications.............	Product engineering or sales department
Quantity and rate of production...........	Request for estimate or sales department
Assembly or layout drawings.............	Product engineering or sales department or customer's contact man
General tooling plans and list of proposed subassemblies of product	Product engineering or manufacturing engineering
Detail drawings and bill of material.......	Product engineering or sales department
Test and inspection procedures and equipment	Quality control or product engineering or sales department
Machine tool and equipment requirements..	Manufacturing engineering or vendors of materials
Packaging and transportation requirements.	Sales department or shipping department or product engineering (government specifications)
Manufacturing routings and operation sheets	Manufacturing engineering or methods engineering
Detail tool, gage, machine, and equipment requirements	Manufacturing engineering or material vendors
Operation analysis and workplace studies...	Methods engineering
Standard time data.....................	Special charts, tables, time studies, and technical books and magazines
Material release data...................	Manufacturing engineering or purchasing department or materials vendors
Subcontractor cost and delivery data......	Manufacturing engineering or purchasing department or customer
Area and building requirements...........	Manufacturing engineering or plant layout or plant engineer
Historical records of previous cost estimates (for comparison purposes)	Manufacturing engineering or cost department or sales department
Current costs of items presently in production................................	Cost department or treasurer or comptroller

SOURCE: Frank Wilson, *Manufacturing Planning and Estimating Handbook*, McGraw-Hill Book Company, New York, 1963.

Types of Estimating Data.   Information requirements for a good cost estimate vary considerably, depending on the nature of the business.   A company producing massive, one-of-a-kind engineered products will have different information needs from a simple job shop, a converter, or a processing plant.   Manufacturers of mass-produced industrial or consumer products will have information requirements quite similar to the engineered products company, but the difference in emphasis will be substantial. Table 4-1 covers the full range of conventional information needs.   How many of these items are needed in a given case will be determined by the nature of the business and the purpose for which the estimate will be used.

Processing.   Regardless of whether the cost estimate is based on departmental inputs or a data file maintained by a special department, information must be collected and arranged in a manner that facilitates manipulation and analysis.   This can be done manually, mechanically, or electronically.   Reporting requirements and the quantity of information to be processed will determine which method is best for a particular application.   Figure 4-1 shows the characteristics and advantages of each method.

PREPARING COST ESTIMATES

Essentially, there are only two ways in which an estimate can be prepared.   If it is an entirely new product or part with no historical precedents or data, it will be necessary to use basic estimating techniques.   On the other hand, if it is a product or part which is a near duplicate of a previous project or is something which can be defined

	Manual	Mechanical	Punched card	Computer
File medium (the material on which the file data are recorded)	Paper or card documents on which data are hand-posted or typed. In a mechanical addressing system, the file medium may be fiber or metal plates on which names and addresses are either typed or embossed.	Ledger cards on which data are machine-posted.	Punched cards on which data are recorded in the form of holes in the cards.	Reels of magnetic tape on which data are recorded in the form of magnetized spots. Also computer storage units such as discs, drums, or cores.
Method for updating file data	Records are manually filed; records to be updated are retrieved from the file by manual search; new data are hand-posted or typed on the document.	Combination of manual and mechanical methods in which records are manually filed, records to be updated are found by manual search of file, and new data are posted on the ledger card by accounting machine.	Mechanical methods, after the input data have been keypunched into the cards. The insertion of new data into the file and the deletion of old data is done mechanically.	Electronic methods, after input data have been entered into the system. Input data may be on punched cards, paper tape, magnetic tape, or in a form readable by a scanner. The insertion of new data into the file and the deletion of old data is done electronically.
Method for processing file data to produce output data (These operations are performed to meet requirements for output from the system. Output data are usually prepared in hard-copy form, but they may also be recorded in punched card or tape form.)	Data are retrieved from the file manually, processed manually, and reported only by hand-posting or by typing.	Data are retrieved from the file manually and processed mechanically. There are limitations to the type and complexity of processing capabilities of mechanical equipment. Data are printed by accounting machine. Recording of data in punched card or paper tape form is also possible.	Data are selected mechanically, processed mechanically, and printed mechanically. Recording of data in punched card form is also possible.	Data are selected electronically, processed electronically, and printed mechanically or electronically. Recording of data in punched card, paper tape, or magnetic tape form is also possible.

FIG. 4-1. Classification of estimating systems. (*Source: Beryl Robichaud,* Understanding Modern Business Data Processing, *McGraw-Hill Book Company, New York, 1966.*)

in terms of familiar components, the estimate can be prepared by comparison. In practice, because of cost and time factors, detail estimating should be used only when there is no alternative or when it serves a special purpose such as a training exercise for fledgling estimators.

Guidelines. There are four important rules or guidelines that should be followed in preparing cost estimates. If the organizational patterns or conventions within a company are in conflict with any of these general policies, the system will not produce good cost estimates consistently.

1. *Secure Sufficient Information to Define the Product.* Drawing specifications, bills of material, weights, and shop hours are the most critical facts. This information must be supplied by the originator of the inquiry and should be checked by him for ambiguities and errors before it is released for cost estimating. When the originator is outside the company and beyond its discipline, sales or engineering will have to perform this review. An important and often neglected point here is the need to set acceptable data standards so that customers routinely supply sufficient information, and just as important, provide it in a way in which it can be used without costly conversion or rework.

2. *Match Product Requirements against Available Resources.* Determining what will be needed in space, tools, machines, and manpower is the next important phase of estimating. Process engineering and production planning and control are responsible for making this assessment. Estimating, sales, and top management all have a right and need to know what kind of an impact the proposed product or part is likely to have on the current and near future status of the shop. In many companies, this need extends to the effect the new item will have on the engineering load.

3. *Always Make a Comparative Review and Secure Additional Data Only Where Necessary.* While estimates are still in a preliminary stage, they should be checked against an accumulation of past experience and data. Injecting the judgment of senior estimators or department managers at this point has two advantages. First, gross errors, omissions, or inconsistencies will be detected early and resolved before a great deal of expense is incurred pursuing false starts. Second, the preliminary estimate may be entirely adequate for the purpose at hand, making additional work unnecessary. This is particularly true when all that sales needs is a rough figure which a customer will use in an appropriation request, or where engineering is content with an approximation that will help them decide between two alternatives. Almost every inquiry tests the skill of the estimator in this respect, as he strives to decide just how much time and money will be required to produce a good cost estimate.

4. *Review Final Cost Estimates with Accounting before They Are Released.* Cash requirements for the manufacture of the proposed item and a careful audit of such significant items as the overhead rates or inventory values used are aspects best left to accounting. By inclination and practice, accounting personnel can be more objective than anyone else in determining additional capital required for tools, facilities, or materials. Another factor that should not be overlooked is the advantage of having a disinterested second party involved with sales, contract administration, and manufacturing in the calculation of anticipated costs. In other words, accounting can add a stabilizing influence as well as checking or testing the figures.

Procedures. The size and nature of the company dictate the operational procedures required to achieve effective cost estimating. Procedures in a small shop with fifty to one hundred employees where only one or two men do the estimating can be much less formal than in a plant employing thousands of people that has a dozen or more estimators. Conditions and needs in a job shop are very different from those in a process industry or a converter. Listed below are the critical areas and types of formal procedures generally required to control waste and inefficiency in estimating. Specific conditions will decide which should be structured into any new system or integrated into an existing system to improve its efficiency.

Requesting Cost Estimates. Most inquiries for cost estimates will originate with sales, but they may also come from engineering or manufacturing. Whatever the circumstances, it is mandatory to have a written inquiry form, with all concerned parties disciplined to its use. Figure 4-2 is a typical example of such a form. The

REQUEST FOR COST ESTIMATE

DESCRIPTION		NO.	PART NAME	WEIGHT		TOTAL MAN-HOURS	TOTAL COST		
				UNIT	TOTAL		LABOR	MATERIAL	TOTAL

SHEET NO. ____ OF ____

SUBJECT	CUSTOMER	ESTIMATE NO.
	LOCATION	
	REQUESTED BY	ESTIMATOR
	APPROVED BY	CHECKED BY
	DATA FURNISHED	DATE IN / DATE DUE

FIG. 4-2. Request for cost estimate form.

principal benefits are that it (1) establishes valid channels of inquiry, (2) facilitates getting an adequate definition of what is required, (3) communicates information needed to set priorities and schedules, and (4) makes all these facts a matter of record. Refinements should be added as necessary to reinforce the basic form. For example, if there is a high incidence of change notices due to vague or incomplete specifications, a checklist of required inquiry data should be prepared and used to get a better definition of requirements. Or if scheduling conflicts develop, a priority subsystem based on sensible criteria of value and need will be required.

Screening. Somewhere in the organization, there must be a responsible authority who approves or rejects requests for cost estimates. Normally, this will be done by sales. The purpose is to avoid wasting time and money on projects that are either not feasible or not attractive economically. Therefore, this screening must be done before estimating or anyone else in the organization becomes involved. If there is some concern that this is too arbitrary a use of authority or that the responsible individual may not have sufficient technical knowledge to make this judgment, two alternatives are available. One is that sales can be required to prepare and distribute periodic listings of all rejected inquiries. This type of automatic review will act as a deterrent to shortsighted or irresponsible behavior. The other way is to have a committee-type approval procedure in which all parties participate in the decision.

Recording Data. A cost folder should be prepared for each estimate. Blueprints, specifications, correspondence, and notes of any kind should be kept in it. Special forms that standardize and summarize the data will make it easier to find and use pertinent data. If several people must have access to the files, an information retrieval system should be organized so that anyone can locate folders not in the files. A set of summary cards that break down product costs into functional increments is also a particularly valuable data record. Catalogs of commonly used purchased materials are other useful and important records to keep.

Pricing. The most common and likely purpose of preparing a cost estimate is to establish a selling price. Although estimators ordinarily should not participate in pricing, they must design the format of the estimate to simplify the process. The most basic and essential item in the estimate required for pricing is the direct cost of the product or order. This is the value of materials, labor, and shop expense that vary directly with production. Competitive pricing is difficult to achieve if the cost accounting system does not keep these figures separated from fixed expenses. Therefore, they should usually also be kept separate in estimating instead of being combined with prorated fixed expenses into a unit cost. Profit responsibility rests with management, and it—not the estimators—is responsible for pricing. Actually, the prime concern of the estimators is the effect of volume on production costs; management on the other hand, is analyzing the effect of all costs on volume when it sets the selling price. Most companies recognize this distinction, even if in practice they involve the estimators in calculating suggested selling prices by applying general administrative and profit factors.

Revisions. Considerable care must be taken in handling change notices, to be sure charges are properly applied and to prevent spending an unreasonable amount of time and money in their preparation. All information and correspondence regarding the change should be kept in the cost folder or in a supplement folder indexed to the original. Some type of standard form is desirable, so that there is a written record of the change for sales, contract administration, the customer, and anyone else who has a need to know the details. Implementing change notices can be quite costly. Therefore sales and estimating should cooperate to minimize the expense. Left uncontrolled, the cost of preparing change notices can match and even exceed the time and effort put into the original estimate.

ESTIMATING

Cost estimators are expected to have a general knowledge of manufacturing methods and plant facilities, but it is not expected that they can perform their function without consulting with other departments. Also, a great deal of valuable and helpful information is available from sources outside the company, such as vendors, technical societies, trade associations, and even professional acquaintances. The amount of information kept on file by the estimator and the methods used to update these files for changes in methods or technology determine the degree of dependency on these outside sources of information. Therefore, the process of accumulating the information needed for a specific cost estimate can require much effort. On the other hand, it can be as simple a process as making a comparison with a similar product where the dependency is only on the estimator's own knowledge, or where he merely feeds variable data into a computer program that creates the final cost estimate.

Detailed or Basic Estimating. When the estimator deals with a new product of which he has limited knowledge, or even when it is familiar but he has a meager amount of data with which to work, he must use basic estimating techniques. The steps in this process are roughly as follows:

1. If all he has to work with are drawings, the estimator must make a manual take-off of materials and calculate what must be fabricated and what can be purchased. The same procedure must be followed in the case of parts and assemblies designed internally, if they are not accompanied by a bill of material or a detailed listing of components. Figure 4-3 is typical of the type of information that is extremely helpful in this kind of estimating.

2. Next, material costs must be developed. Prices for raw materials such as steel, brass, and plastics can be taken from price lists provided by suppliers or may be obtained from purchasing if there is sufficient time. Price breaks and handling or administrative costs are additional factors the estimator must take into consideration. Make-or-buy decisions will be required on standard purchased parts, and subcontracting must be arranged for special items that cannot be purchased or manufactured internally.

3. A manufacturing or processing plan must be prepared that details the major operations required to manufacture, assemble, and if necessary, install the product.

To find	Known factors	Formula
Surface speed, in feet per minute. Round bars	RPM and diameter	$\dfrac{\text{Diameter} \times 3.1416 \times \text{RPM}}{12}$
Surface speed, in feet per minute. Hexagon bars. (Based on distance across corners)	RPM and size or RPM and distance across corners	$\dfrac{\text{Size} \times 3.1416 \times \text{RPM} \times 1.155}{12}$ $\dfrac{\text{Distance} \times 3.1416 \times \text{RPM}}{12}$
Surface speed, in feet per minute. Square bars. (Based on distance across corners)	RPM and size or RPM and distance across corners	$\dfrac{\text{Size} \times 3.1416 \times \text{RPM} \times 1.414}{12}$ $\dfrac{\text{Distance} \times 3.1416 \times \text{RPM}}{12}$
Number of revolutions per minute. Round bars	SFM and diameter	$\dfrac{\text{SFM} \times 12}{\text{Diameter} \times 3.1416}$
Number of revolutions per minute. Hexagon bars	SFM and size or SFM and distance across corners	$\dfrac{\text{SFM} \times 12}{\text{Size} \times 3.1416 \times 1.155}$ $\dfrac{\text{SFM} \times 12}{\text{Distance} \times 3.1416}$
Number of revolutions per minute. Square bars	SFM and size or SFM and distance across corners	$\dfrac{\text{SFM} \times 12}{\text{Size} \times 3.1416 \times 1.414}$ $\dfrac{\text{SFM} \times 12}{\text{Distance} \times 3.1416}$
Feed, in inches per revolution	RPM and feed in inches per minute	$\dfrac{\text{Feed inches per minute}}{\text{RPM}}$
Feed, in inches per revolution	Diameter, SFM, and feed (inches per minute)	$\dfrac{\text{Diameter} \times 3.1416 \times \text{feed}}{\text{SFM} \times 12}$
Feed per tooth	Feed (inches per revolution) and number of teeth in tool	$\dfrac{\text{Feed}}{\text{No. of teeth}}$
Length of stock required to make one part (inches)	Length of part (inches), width of cutoff (inches), and facing allowance (inches)	Length + cutoff + facing
Footage of stock required to make 1,000 parts (feet)	Length of stock for 1 part (inches), bar end loss (inches), scrap loss (inches)	$\dfrac{\text{Stock for 1 part} \times 1,000 + \text{end loss} + \text{scrap loss}}{12}$
Weight of stock required to make 1,000 parts (pounds)	No. feet required for 1,000 parts (feet), wt. per foot of bar used	No. feet × wt. per foot
Weight (pounds) of one foot of stainless steel. Round bar	Diameter (inches)	$\text{Diameter}^2 \times 2.67$
Weight (pounds) of one foot of stainless steel. Hexagon bar	Bar size (inches)	$\text{Size}^2 \times 2.94$
Weight (pounds) of one foot of stainless steel. Square bar	Bar size (inches)	$\text{Size}^2 \times 3.4$
Weight (pounds) of one foot of stainless steel. Flat bars	Bar width (inches), bar thickness (inches)	Width × thickness × 3.4
Time (in seconds) for actual machining	Number of revolutions, RPM	$\dfrac{\text{Revolutions required} \times 60 \text{ sec.}}{\text{RPM}}$
Machine time	Time for machining, idle time	Time for machining + idle time
Tapping or threading time (in seconds)	No. of threads, threading speed in RPM	$\dfrac{\text{No. of threads} \times 60 \text{ sec.}}{\text{Actual threading speed in RPM}}$
Feed rate (inches/minute)	Feed (inches per revolution), RPM	Feed (ipr) × RPM
Removal rate (cubic inches/minute)	Feed (inches per revolution), depth of cut, SFPM	Feed (ipr) × depth (inches) × SFPM × 12

FIG. 4-3. Formulas useful for detailed estimating. (*Source: Norman McClymonds, Machining Estimating Manual for Stainless Steel, Joslyn Stainless Steel, Ft. Wayne, Ind., 1968.*)

This plan should encompass tooling, equipment, and all other needed facilities, and the effect of volume should be given special attention.   If additional equipment or space must be obtained, its full cost should be incorporated into the estimate, including the cost of services required for building or installation.   Tooling includes cutters, jigs, fixtures, gages, and machine tools.   Facilities include plant layout, flow patterns, equipment relocations, material handling equipment, and building services.

4. Direct labor costs are estimated on the basis of the manufacturing or processing plan.   Setup time must be included, and the effect of learning curves must be applied to different volumes.   The cost of each operation is then the time it takes, multiplied by the workman's wages.   Burden factors are applied either by operation or cost center to make a final extension of labor costs.

5. The final cost estimate combines the material, labor, and burden cost of manufacturing.   A set of estimate sheets should be prepared that provides the complete cost details of every feature of the product.   Normally, a summary or final cost sheet, Figure 4-4, is provided on which general administrative costs have been added into the totals.   If the estimator is also responsible for establishing the selling price, a profit factor is added at this time.

Comparative Estimating.   The completeness of the information on file for the estimator determines if the estimate can be prepared by comparing the inquiry with a product in process or previously completed.   In cases where this is practical or necessitated by a time constraint, judgment must be used to determine the percentage of deviation.   This judgment may be made by the individual estimator, or where desirable, by a group or committee assembled for the task.   Some special techniques that apply are:

1. File estimates in a manner that keeps the estimator aware of parts previously estimated, so they can be related to similar current requests.
2. Correlate certain design requirements with the total weight of the article to be produced.   Very rapid estimating can be accomplished in this manner if good historical data are available and the estimator has detailed experience and confidence in this system.
3. Standard data can be established for the elements that are most used in the production of each given facility.   A historical record of standards is the best source of manufacturing times for estimating.
4. Mathematical curves can be developed for machines, tools, equipment, size, or capacity by plotting actual costs to a matrix.   Estimators can then use the equation of the curves as a simple means of estimating costs.
5. Forms should be designed to provide a sufficient description of a part so that a process engineer or production supervisor gets a quick, clear picture of the part.
6. Procedures should be developed so that costs can be obtained on a rush basis from the responsible departments when data must come from supporting departments.

Automated Estimating.   The computer can improve cost estimating because (1) more data can be considered, (2) greater speed is possible, (3) more alternatives can be explored, and (4) a more comprehensive grasp of a decision's impact is possible. It can read and write, compare, make yes-or-no choices, and transfer information from one location to another, in addition to storing vast quantities of data and performing a bewildering variety of mathematical analyses.   In other words, it can be programmed to perform all the activities noted above under basic and comparative estimating.

In addition, the following applications can be made:

1. Engineering drawings can be translated into computer programs that geometrically define the part to be manufactured by tool movements and operations.
2. Programs can be written that use the type of materials required and the machine tools available to sequence machine operations; to select the cutter size, speeds, feeds, and coolants; and to specify necessary operator instructions.
3. Process sheets and standards can also be built into the system, and price exten-

SUMMARY SHEET

XYZ CORPORATION

Customer   ABC COMPANY   Estimate No. 6820-4/   Date 12/7/—

Product   FABRICATED STRUCTURE   Estimated by EAM   Checked by KJS

Drawings   Approved by   Quoted

DESCRIPTION	Finished Weight	RAW MATERIAL Weight	RAW MATERIAL Rate	RAW MATERIAL Amount	PROCESSED MATL Amount	LABOR ANALYSIS Hours	LABOR ANALYSIS Rate	LABOR ANALYSIS Amount	BURDEN %	BURDEN Amount	Component Cost Incl. B & S, Conty.
Structural Parts	1755900	1883080		152340	51190	38380					617300
Boom Hoist	29630	21410		3200	5950	1090					21170
Reeving System	62280	48180		9340	8710	1650					36600
Boom Conveyor Drive	20180	17340		1780	6410	1230					21600
Trucks & Equalizers	317100	272500		23030	49720	10900					192100
Hyd. Legs	29850	48230		7500	10320	1540					35200
Wheel, Buckets, & Drive	90340	96390		11590	11480	3210					58200
Wire Rope Items	3000	-		-	5250	-					5670
Rail Clamps	17000	13000		1080	4070	840					14300
Slew Ring, Gear & Drive	68470	64950		18560	49940	5010					126000
Conveyor Parts & Misc.	183870	44060		4610	119220	1470					159900
Misc. Items	25060	14980		1310	16530	760					27200
Motors, Brakes, & Controls ⎫	50700	-		-	239230	-			Dwgs. &		258400
Wiring Matls. ⎭	-	-		-	-	-			Temp.		-
Lube Oil Filters	-	-		-	500	-	L	318000		30130	540
Matl. Increase — Steel — 5%	-	-		11400	8200	-	OH	323000		23600	21170
Jigs 300/2	-	-		500	-	150	SS	57300		5430	2100
Clean & Paint	-	-		-	16350	5070		695300		59160	70200
Inspection	-	-		-	-	810	B&S	45400		3870	-
Supervision	-	-		-	-	700		743700		63030	-
	-	-		-	-	72810					
S.S. Handle 28 1/2%	-	-		-	-	8290					
O.S. Handle 13 1/2%	-	-		-	-	3460					
Boat Yd. Handle 17%	-	-		-	-	3080					
Load Out	-	-		-	1490	1700					
Template Matérials	-	-		-	280	1/2	89340				
Weld Wire	-	37260		5950	-	1/2	86700				
Overrun Allowance	120	620		-	-	-					
TOTALS	2653400	2562000		252190	604840						

DRAWINGS AND TEMPLATES				BURDEN		GENERAL EXPENSE ITEMS	Amount
Shop & Class of Work	Hours	Rate	Amount	%	Amount	Freight & Trucking	1800
Design	-		-		-	Premium on Bonds	-
Shop Drawings	5265	5³⁷	28300	65	18400	Insurance	-
Drawings Premium			-		3200	Delivery	34000
Templates	450	4⁰⁷	1830	110	2000	Photos, Tracings, Manuals	800
			-		-	Computer	600
TOTALS			30130		23600	Erection Superv. & Expenses	11000
LABOR ANALYSIS				BURDEN		Erection Back Charges	15000
Productive Units	Hours	Rate	Amount	%	Amount	Fees	-
Structural-Fabricate	13950		49800		89700	Misc. Insp. & Tests	2500
— Assembly & Weld	15130		54350		65220	Salaries	-
— Handling	8290		27610		-		-
			-		-	TOTAL	65700
Machine — Hand	5060		17450		21850	RECAPITULATION	NET COST O.H. Incl.
— Machine	13540		54300		76020		
— Handling	2510		8590		-	Raw Materials	252190
			-		-	Processed Materials	604840
Boat Yard	18090		68400		54710	Shop & Yard Labor	318000
B.Y. — Superv. & Handling	3080		11670		-	Shop & Yard Burden	323000	323000
			-		-	Drawing & Templ. Labor	30130
Mechanical	-		-		-	Drawing & Templ. Burden	23600	23600
Labor					-	General Expense Items	65700
Paint	5070		17600		12300	Employees Security	62730
Shipping ⎫	-		-		-	Bids and Sales, G & A	109220	109220
Electric ⎪	-		-		-	Allowance for Contingencies	26890
Marine Ways ⎬	1160		4400		4400		1816300
Carpenter ⎪	-		-		-	Erection	275200	16600
Pipe ⎭	-		-		-	Material Savings	— 12000	— 730
Other Shops Handling	950		3230		-	Escalation To Make Firm	63900	21000
Inspection	810		4550		-	Applevage Fee	52200	3130
Load Out	1700		6050		9100		-
TOTALS 1 - Basis - 1	89340		328000		333300	Total Estimated Cost	2195600	495820
1 - Basis - 2	86700	3⁶⁸	318000	101⁶	323000	19 Burden Rates	20600 ⎫	
						Mark-up	116800 ⎭ 6³% —
Erection:		Escal. To Make Firm					
Struct. & Mech. Firm	177560					1 - Basis - 2 BIDS	2333000
Electrical & Escal	77000	Misc. Matls.	2%	7200			× 2
B & S	16600	Elec. MBC	4 1/2%	9400			$4666000
Conty	4040	Elec. Wiring	7 1/2%	3100	Conditions of Bid:—	
	275200	1.25 / 2.100	Shop Labor	5⁶%	32000		
		Eng.		2500		
				54200		
		Applevage Fee		1500		
		Elec. Erect.		6000		
		Freight		2200		
				63900		

FIG. 4-4.   Estimate summary sheet.

```
                                                                                              338279
KAYDON              091669    10 8  HP D    4705   SER            MAT SP 8662   BHN  285   MAT 2   MACH 2

START OD  57.375   ID 51.625   W 12.875   FINISH OD   56.313   ID 52.688   W  3.625   FINISH CODE F  500  T  5008   500
                                           TOL. OD    0.031    ID  0.031   W  0.031                C    2 H    0

 TOOL       CUT          DIAMETER     STOCK  DEPTH    SPEED      FEED  HOURS LENGTH  HOURS  NO.  TOTAL       HEADS        MAX.
MAT NO. DESCRIPTION  START   FINISH          OF CUT SFM   RPM         /INCH OF CUT   /CUT  CUTS  HOURS      I       II   STOCK
CA   1   RF    OD   57.375  51.625  2.375  0.250  175.   11.7 0.042 0.034   2.875  0.098  2    0.196  0.000 -0.196 13.12
CA   2   RT    OD   57.375  56.563  0.406  0.250  175.   11.7 0.042 0.034   9.875  0.336  2    0.673  0.000 -0.673 57.56
CA   3   RB    OD   51.625  52.438  0.427  2.250  157.   11.7 0.042 0.034  12.875  0.439  2    0.877  0.877  0.000 51.44
CA   4   FF    OD   56.563  52.438  0.125  0.125  225.   15.2 0.031 0.035   2.062  0.073  1    0.073  0.073  0.000
CA   5   FT    OD   56.563  56.313  0.125  0.125  225.   15.2 0.031 0.035   9.500  0.336  1   -0.336  0.000 -0.336
CA   6   FB    OD   52.438  52.688  0.125  0.125  209.   15.1 0.031 0.036  12.500  0.444  1    0.444  0.444  0.000
CA   2  SLIC PC 1
                    57.375  52.688  0.000  1.812   90.    0.0 0.012 0.868   1.812  1.573  0    0.787 -0.787
CA   7   FF    OD   56.563  52.688  0.125  0.125  225.   15.2 0.031 0.035   1.938  0.069  1    0.069 -0.069  0.000
CA   8   FT    OD   56.563  56.313  0.125  0.125  225.   15.2 0.031 0.035   9.375  0.332  1    0.332  0.000  0.332

CA   9   RF    OD   57.375  52.688  2.375  0.250  175.   11.7 0.042 0.034   2.343  0.080  2    0.160  0.000 -0.160  0.00
CA  10   RT    OD   57.375  56.563  0.406  0.250  175.   11.7 0.042 0.034   3.000  0.102  2    0.204  0.204  0.000  0.00
CA  11   FF    OD   56.563  52.688  0.125  0.125  225.   15.2 0.031 0.035   1.938  0.069  1    0.069  0.069  0.000
CA  12   FT    OD   56.563  56.313  0.125  2.125  225.   15.2 0.031 0.035   2.562  0.091  2    0.181  0.000  0.181
CA   4  SLIC PC 3
                    57.375  52.688  0.000  1.812   90.    0.0 0.012 0.868   1.812  1.573  0    0.787 -0.787
      PIECE 2
      PIECE 1
      PIECE 3

                                                                                 TOTAL HOURS   2.112

                                                                                              338280
SHOP SET-UP - FIRST CHUCKING                   K  2   0.156    1  0.156

GET AND ASIDE PART
OD TO 72 IN   HT. TO 18 IN                                             K  7   0.061     2  0.122
PUT ASIDE FIN. MACH. PART                                              K144   0.031     2  0.062

CLAMP AND TRUE

ROUGH MACHINE-NO SPEC. FINISH
RATIO OD/WALL TO 30/1 FIRST CHUCKING           K 55   0.435    1  0.435  K 83   0.134     1  0.134

TOOLING

SET UP TURNING TOOL                                                    K267   0.017  12-2  0.170
SET UP FORMING OR SLICING TOOL                                         K271   0.026     4  0.104
SET TOOL TO FIRST CUT                                                  K 69   0.013    16  0.208
SET TOOL TO ADDITIONAL CUT                                             K 72   0.009     6  0.054
TRIAL CUT-CALIPER - WIDTH                                              K 35   0.022     1  0.022
TRIAL CUT  TOL. OVER 1/64                                              K 31   0.023     7  0.161
GAUGING

TRAMMEL OR ROD                                 K 98   0.025    2  0.050  K 70   0.015     3  0.045
TEMPLATE OR SQUARE                                                     K138   0.012     4  0.048
CALIPERS - WIDTH                               K187   0.010    0  0.000  K134   0.015     6  0.090

MISCELLANEOUS
CHANGE SPEED                                                           K 64   0.002     6  0.012
CHANGE FEED                                                            K 62   0.001     4  0.004

TOTAL LEVEL MANUAL TIME                                   0.641                            1.236

TOTAL MACHINE TIME                                                                         2.112



TOTAL LEVEL TIME                                          0.641                            3.348

PER CENT ALLOWANCE                                        0.064                            0.335

PER CENT ALLOWANCE                                        0.070                            0.422

TOTAL ALLOWED TIME                                        0.775                            4.105
```

FIG. 4-5. Estimate of allowed time produced by a computer.

sions made so that cost estimates are developed concurrently with establishing the production and machining steps.

In other words, if there is a way to feed such information as machine type, material group, start and finish dimensions, tolerances, and finishes to the computer on three or four data cards, the preparation of process sheets, drawings, standards, run times, and the like can be automated. Figure 4-5 shows an example of how allowed time can be determined by a computer. Cost estimating can be a by-product of this system, or it can work as a less complex, stand-alone computer routine.

COST ESTIMATING MANAGEMENT

Estimating is a valuable business tool, but it must be managed wisely to get the most return from the amount of time expended. This means recruiting competent people who have been exposed to all facets of engineering and manufacturing and seeing that they select the best estimating method for the task at hand. Some type of base figure or cost of estimating index should be kept to control the function. Records should be kept of the deviation between estimates and actual costs. A final point to be considered is the value of the estimator as a source of cost reductions. Although others should do the actual value analysis, suggestions from estimators frequently make significant contributions to corporate profits.

BIBLIOGRAPHY

McClymonds, Norman, *Machining Estimating Manual for Stainless Steel,* Joslyn Stainless Steel, Ft. Wayne, Ind., 1968.

Robichaud, Beryl, *Understanding Modern Business Data Processing,* McGraw-Hill Book Company, New York, 1966.

Turner, Spencer, *Cost Estimating and Pricing with Machine-hour Rates,* Prentice-Hall, Inc., Englewood Cliffs, N.J., 1962.

Wilson, Frank, *Manufacturing Planning and Estimating Handbook,* McGraw-Hill Book Company, New York, 1963.

Chapter **5**

Profit Improvement Procedures

SIR WALTER SCOTT

**Governing Director, W. D. Scott & Co. Pty. Ltd.,
North Sydney, Australia**

Cost reduction is a vitally important activity and one which can have a marked influence on the future of any industrial enterprise.   Although few will deny its importance, it is surprising how seldom it is consistently tackled to the extent of forming an integral part of a company's policy.   Few doubt its general logic, but many overlook the strength of its claims, what it can contribute, and its value to an organization.

THE LOGIC OF COST REDUCTION

In many industries, break-even points are high.   For any organization wishing to increase its profits, the usual methods are to:

1. Increase prices without offsetting reductions in dollar volume of sales.   For example,

	Present	*10 % sales price increase*
Sales—Product A	$ 80,000,000	$ 88,000,000
Product B	20,000,000	22,000,000
	$100,000,000	$110,000,000
Less		
Costs to make and sell	92,000,000	92,000,000
Profit before tax	$ 8,000,000	$ 18,000,000

13-59

Increases in selling prices can result in a reduction in the volume of units sold, thus enabling competitors to increase their share of the market. This is essentially a marketing problem.

2. Increase dollar volume of sales without offsetting reduction in prices. For example,

	Present volume	*10 % increase in volume of units sold*
Sales	$100,000,000	$110,000,000
Less variable cost	69,000,000	75,900,000
Profit contribution	$ 31,000,000	$ 34,100,000
Less fixed costs	23,000,000	23,000,000
Profit before tax	$ 8,000,000	$ 11,100,000

Profit is increased due to the fixed nature of costs such as depreciation, plant insurance, and administration, and not because of management action on the part of the industrial engineer. Again, this is essentially a marketing problem.

3. Alter the product mix by concentration on the more profitable lines. For example,

	Present mix	*Proposed mix*
Sales—Product A	$ 80,000,000	$ 60,000,000
Product B	20,000,000	40,000,000
	$100,000,000	$100,000,000
Cost—Product A		
Variable	$ 57,000,000	$ 38,000,000
Fixed	19,000,000	19,000,000
Product B		
Variable	12,000,000	24,000,000
Fixed	4,000,000	4,000,000
Total	$ 92,000,000	$ 85,000,000
Profit—Product A	$ 4,000,000	$ 3,000,000
Product B	4,000,000	12,000,000
Total before tax	$ 8,000,000	$ 15,000,000

A complete analysis of the product mix by models and product lines can reveal avenues for profit improvement. The analysis will cover market share, plant capacity, costs, and profitability. The industrial engineer will be mainly concerned with the probable effect of the change on plant capacity. Is there sufficient plant to produce the product volumes? Will plant become redundant? Can plant be adapted to produce the required product? He can also be called upon to improve the profitability of a product through better methods, alternative materials, and the like. Finally, it will depend on marketing ability to sell the more profitable product.

4. Reduce cost. For example,

	Present	*10 % cost reduction*
Sales	$100,000,000	$100,000,000
Cost		
Manufacturing	60,000,000	54,000,000
Distributing and selling	22,000,000	19,800,000
Administration	10,000,000	9,000,000
Total	$ 92,000,000	$ 82,800,000
Profit before tax	$ 8,000,000	$ 17,200,000

The industrial engineer can play an important part in this area of profit improvement, particularly as regards manufacturing and distribution costs.

In a competitive economy, the only practical solution may often be found in cost reduction, particularly as this is often more under the control of the organization than any of the other listed factors.

Indeed the case for cost reduction is unanswerable, because:

1. New machines, equipment, processes, materials, and methods, and often new personnel, demand a continuing approach to cost reduction.
2. Business is a complex affair, and cost extravagances are difficult to find and avoid.
3. Many concerns when insulated against competition—usually a temporary condition—lose the competitive urge to keep costs down.
4. Many successful organizations tend to drift into extravagances.
5. In a time of inflation, all kinds of extra costs creep in.

With the individual organization, the urgency of cost reduction may arise from the preparedness necessary to meet a coming competitive position or consumer resistance to higher prices.

From a national angle, cost reduction is a vital factor in raising the standard of living. Lower costs make for lower prices, can facilitate improved quality, and therefore raise and stimulate consumer demand.

These reasons support a view that cost reduction should be practiced much more strongly and on a much wider scale than it actually is. The reason it is often neglected is that cost reduction is a complex matter. Its complexity matches the increasing complexity of business. Savings may often be made in every section of an organization, but to do a thorough job requires careful planning and programming.

Cost reduction can also be a psychologically sound personnel practice. Many employees think management wastes too much money through extravagance, poor planning, remote judgment, and lack of adequate controls. Properly applied, a cost reduction procedure represents an opportunity for management to seek and obtain the support of employees in something which can so often be shown to be of mutual benefit.

THE OBSTACLES TO COST REDUCTION

Although cost reduction can be complex and difficult, its advantages are such as to appear acceptable without reservation by management and labor alike. This is far from the case. Indeed, obstacles are found that are both serious and effective.

On the part of labor:

1. Cost reduction programs are often interpreted as chiefly designed to reduce the number of jobs available.
2. There is the usual psychologically explainable resistance to change.
3. Fear is engendered through lack of education and training.
4. There is often apathy and lack of interest.
5. As a defensive measure, there is sometimes a deliberate restriction of output.
6. Unions often distrust management's motives.
7. Unions and employees are skeptical of management's earnestness.
8. Past failures result in a derisive attitude.

On the part of management:

1. Cost reduction programs have to compete with many other demands on executive time and attention.
2. Suitable personnel are not always available—indeed are rarely so—as the persons most qualified are wanted in other spheres.
3. Jealousies and sectional interests are aroused.
4. There is reluctance on the part of some executives to support cost reduction because they feel their own work may be subject to criticism.
5. Some executives are unable to see the widespread opportunities for cost reduction and to point out where and why savings can be made.
6. Some executives desire to control cost reduction programs in their own domains, with consequent lack of action.
7. Middle management apathy—their own operations are efficient, it is the other fellow who is at fault.

Both on the part of labor and of management, these attitudes often end in passive resistance.

On account of physical and operating conditions:

1. Financial obstacles impede progress, for example, shortage of working capital.
2. Physical conditions, such as lack of working space which limits relocation and relayout.

When limiting factors such as these are recognized, they may be difficult to overcome, because there is often an unfortunate tendency to assign to all cost reduction possibilities the difficulties experienced in some.

OVERCOMING THE OBSTACLES TO COST REDUCTION

The answers to obstacles to cost reduction are along three lines:

Understand What Cost Reduction Is Not. Morse and Wyatt[1] have emphasized this:

1. Planned cost reduction is not a series of letters from the company president directing subordinates to save. The authors have seen this casual approach used with a sincere expectancy of positive results. Some presidents have felt quite satisfied that word from the front office is all that is needed for a company Cost Reduction program. Lacking the substance of any further organized planning, the letters unhappily served only as a warning and a passing rebuke. Real accomplishment: nothing.
2. Planned cost reduction is not strictly an accounting function. Reports from accounting form the basis for planning cost reduction efforts, it is true. They form the basis for judging results, also. They do not, however, constitute a Cost Reduction program. There is a common fallacy that cost reduction begins and ends in company ledgers.

   We submit earnestly, that to take a strictly accounting approach to cost reduction is to confine cost reduction efforts to only one group of experts and thereby to submerge the efforts of other talented members of the company, and depress their initiative. While better than president's letters, accounting reports are still only useful tools, not a complete program, and the distinction between tool and program should be kept clearly in mind.
3. Planned cost reduction is not the executive directive which arbitrarily dictates—cut personnel 20 percent. Such a shotgun blast reduces costs, yes. It may even be necessary. If it is regrettably necessary, it is stark evidence that no planned Cost Reduction program had been previously effective.
4. Planned cost reduction is not a program based on personal whims, idiosyncrasies of one or two individuals powerfully enough situated within a company to make their directives law. Such an approach indicates unhealthy development of managerial authority at all levels, and even if practiced, would hardly result in a continuing program.

Adopt a Careful Psychological Approach

1. Cost reduction should be a continuing function; therefore, promote it as a state of mind.
2. Make it the province of the whole organization by endeavoring to bring everyone into it. Be generous in explanations. People often oppose the things they do not understand. Ask for cooperation and suggestions; promote suggestion systems. Often the men most likely to think of possible solutions are those closest to the jobs. A man will never oppose a change he himself suggests, but he will often resent something forced upon him.
3. Explain it to the unions and solicit their cooperation.
4. Approach with benefits by making the project a salable package. Show what cost reduction can mean to the worker, to his security, to his job, to the competitive position of the organization, to stability and continuity of employment, toward wage increases. Acquire the worker's confidence by being

[1] H. C. Morse and E. E. Wyatt, *Cost Reduction Guide for Manufacturing Management,* Wyatt & Morse, Inc., Chicago, 1961.

frank about objectives and avoiding the appearance of knowing all the answers. Use supervisors to the greatest degree possible to ensure cooperation and to facilitate their explanations to the workers.

5. Give it the active, continuous, and unqualified support of top management. Make sure that this active support figures in the minds of the workers.
6. Foster a competitive spirit. Properly conceived, competition can overcome obstacles and promote enthusiasm to an astonishing degree.
7. Give full credit where credit is due. Praise usually gets much better results than criticism. A certain amount of periodical acknowledgment of results by management with exhortations toward the next step is psychologically sound.
8. Make certain that all executives understand that the program is not instituted to engender criticism and that support and cooperation are essential.

Adopt a Practical Approach
1. Senior management should give its obvious support to the program continuously.
2. Define the program and its scope most carefully.
3. Give to its planning the most thoughtful consideration warranted.
4. Give it authority, organization, and direction.
5. Most carefully choose the personnel to conduct the program.
6. Give it all desirable publicity.
7. Set up necessary targets.
8. Control the performance.
9. Measure the savings.
10. Publicize the results.
11. Use successes as the starting point for the next program.
12. Follow up to ensure performance of savings.

The *Conservation Handbook* of the Aerospace Industries Association of America, Inc., has some excellent advice to offer on the practical approach to cost reduction.

The function of a conservation staff involves getting a lot of people to change their accustomed ways of doing things. It's just human nature to resist change—especially when the change suggested implies criticism of what we have been doing. Waste is like sin—everybody's "agin it" providing it's the other fellow's waste. So, of course, a good deal of tact and diplomacy are needed.

There are only 40 hours in a workweek. Usually, there are more problems involving conservation opportunities than there are hours to work them out. Conservation of time, then, demands our selection of the projects having the most fruitful possibilities, leaving the lesser, and sometimes easier, projects for a time when they become the most worthwhile.

A word of caution might prove helpful: avoid change for the sake of change.

Many of the existing methods have evolved through trial and error over a period of years. Consequently, due to the circumstances involved, they may represent the best method of handling the problem even though it may not appear so on first examination.

It is almost a foregone conclusion that people, resisting change, will sometimes accuse you of wanting to spend more to save something than it is worth. Therefore, it is important to have all the pertinent facts in hand to back up your proposed action.

Here is another way of putting this problem of keeping the demand for change in balance. Too much or too rapid change tends to make an organization unstable. The most successful changes seem to evolve naturally out of the past experience of the organization. For that reason, it is seldom practical to transplant in toto the successful system of one plant over to another, if radical changes are involved. The change should be rooted in collective experience.

NECESSARY TOP MANAGEMENT DECISIONS

When instituting a cost reduction program, there are certain decisions which must be made by top management if the work is to progress smoothly. These are:
1. The scope of the cost reduction program

2. The objectives and goals of the program
3. The amount of expenditure to be allocated for the program
4. The type of cost reduction organization to be set up
5. The scope of authority for action involved in the program
6. The controls necessary to cover the program
7. The extent of participation by top management
8. The time factors involved in the program

TYPES OF COST REDUCTION PROGRAMS

Different organizations impart different meanings to cost reduction. A cost reduction program can mean one to lessen cost in connection with:

1. A single item of expense (cost of power)
2. An element of cost (indirect materials)
3. A group of operations (assembly line)
4. A single product (manufacturing 3-horsepower motors)
5. A department or section (maintenance)
6. A particular situation (abnormal waste increase)
7. An improved technique (provision of incentive payment)
8. A list of specific projects
9. All factory operations
10. All administrative and clerical costs
11. Modern developments (use of air freight to eliminate warehouses and miniaturization)

COST REDUCTION PERSONNEL

The foregoing may be accomplished by an individual, a committee, a cost reduction section or organization. The size of the organization and the size of the problem are the usual factors that determine which one is best. The work can be covered by a person or personnel within the organization, outside consultants, or a combination of both.

A decision as to who should carry out the work depends upon a number of factors, two of the most important being: are suitable men available from within the organization, and what are the type and extent of the work to be carried out? Personnel from within the organization have the following advantages as against outside consultants:

1. They probably have a much more thorough knowledge of all factors relating to and including personnel in the business.
2. They are usually less costly.
3. More control over them can be exercised.
4. Sometimes more people can be allocated to the various tasks in hand.
5. In some cases, they may be able to engender more teamwork.

Outside consultants have the following advantages:

1. They probably have more specific experience and are more expert in the way to conduct the program, to expedite its completion, and to carry out specific assignments.
2. They usually have within their organization people specifically trained for the kind of work required and with a highly developed aptitude for digging into problems, unearthing facts, and analyzing and interpreting them.
3. They are not beset by organizational or personal loyalties which may interfere with or unduly influence their actions or findings and can therefore adopt an interested but impartial attitude. They can sometimes say from the outside what it is difficult to say from the inside.
4. They can devote all their time and attention to the task involved and are not beset by the host of everyday problems.

5. They frequently have a knowledge and experience of what has happened in other places in comparable situations.
6. They often have the ability to work quickly and effectively within the client's organization, thus obtaining earlier results and consequently earlier savings. This may ultimately make use of the consultant less expensive.

Outside consultants are generally used for specific assignments such as a sectional methods improvement program, the installation of an incentive plan, a training program to improve productivity, or the overhaul of production planning procedures. For wider programs involving committees and extensive operational work, the personnel are usually chosen solely from within the organization or the work is divided between such personnel and outside consultants.

SHORT- AND LONG-RANGE COST REDUCTION PROGRAMS

The program can be tackled in different ways but may subscribe to some overall consideration, such as short-range procedures, long-range procedures, or both. In the last case, the purpose may be to cut costs immediately wherever possible but at the same time to spend time and money on programs which will result in lower unit costs in the future.

A short-range program may arise from such considerations as:
1. A fall in profits which it is desired to correct as quickly as possible
2. An impending unfavorable competitive situation
3. Some items which come under scrutiny because competitors are underselling on the particular product
4. Some operations or functions which appear to be costing too much
5. Adverse cost variances which invite speedy attention
6. Some operations which look to be inefficient and seem to offer possibilities for quick and substantial savings
7. The suggested program being the current assignment of a studied approach to cost reduction

When some or all of these and similar conditions exist, a number of cost reduction projects may result and may be undertaken concurrently.

Short-range programs often have an element of expediency and usually concentrate upon those problems which have a degree of obvious urgency.

Long-range cost reduction programs often arise similarly to long-range budgets. They often involve capital expenditure designed toward a continual reduction in costs, dependent upon better layout, relocation of plant, modernization programs, material handling improvements, and personnel and supervisory training. They sometimes operate as a kind of x-year plan during which time it is hoped to cover the whole of the operations and every aspect of the business. This approach is necessary because it is so often virtually impossible with the resources and personnel at hand to cover the program in a lesser period.

In addition to the long-term program, most organizations also arrange for a short-term effort to make improvements of a more general, continuous, or psychological nature. The result will be a composite short- and long-range program.

THE COMPANY PLAN

A total company plan for cost reduction should be determined, or at least carefully approved, at either directorial or top management level. The management services department will have prepared the details of the plan for such consideration and approval. An example is given in Figure 5-1.

ORGANIZING FOR COST REDUCTION

The organization required for cost reduction depends upon the type, scope, and extent both of the program and of the organization. For a single item (reduction

Area	Action to be taken, techniques which might be used
1. Company policy	Profit optimization—correct product mix (linear programming), pricing policy, factory and warehouse locations, company strategy.
2. Organization (a) Structure (b) Personnel (c) Training	Challenge effectiveness of current company organization. Organization planning, management audit, job study, management by objectives, training, merit rating.
3. Overheads (a) Factory (b) Administration	Work measurement, work simplification, office layout, data processing, variable factor programming.
4. Factory labor (a) Direct (b) Indirect	Work measurement, incentives, fatigue study, motion study, economics, queuing theory, linear programming, work simplification. Supervisor training.
5. Factory methods	Work study/method study, plant layout, methods engineering, tool design, planned plant maintenance.
6. Materials management (a) Raw material (b) Finished goods (c) Purchasing	Different approach for high value and low value items. Purchasing control. Value analysis, materials and stock control, set up a reclamation and reprocessing section, classification and coding, material utilization and scrap control (material yield), standardization, and variety reduction.
7. Control systems	Budgetary control and standard costing, labor control and efficiency reporting, quality control, production planning and control. Material utilization control, inventory control.
8. Selling (a) Marketing (b) Warehousing (c) Transport	Appropriate sales organization—territories, factors, agent representatives, appropriate remuneration. Material handling, transport control, vehicle routing schedules. Packaging policy.
9. New products New projects	Project and product research, use of break-even analysis, appropriate payback method for new item (D.C.F.), profit planning, replacement theory, design theory, market analysis (market research), production engineering, standardization, and variety reduction. PERT.

FIG. 5-1.   Cost reduction plan.   (*Source:* Work Study, *October,* 1967.)

in cost of power), a single engineer may be assigned. For an element of cost (indirect materials), a committee may be necessary, comprising representatives of (1) purchasing, (2) factory, and (3) costs. With a group of operations (assembly line), there may be representatives of (1) factory, (2) design, (3) tooling, and (4) costs.

A simple chart showing the organization of several cost reduction committees and the areas in which each would function is given in Figure 5-2.

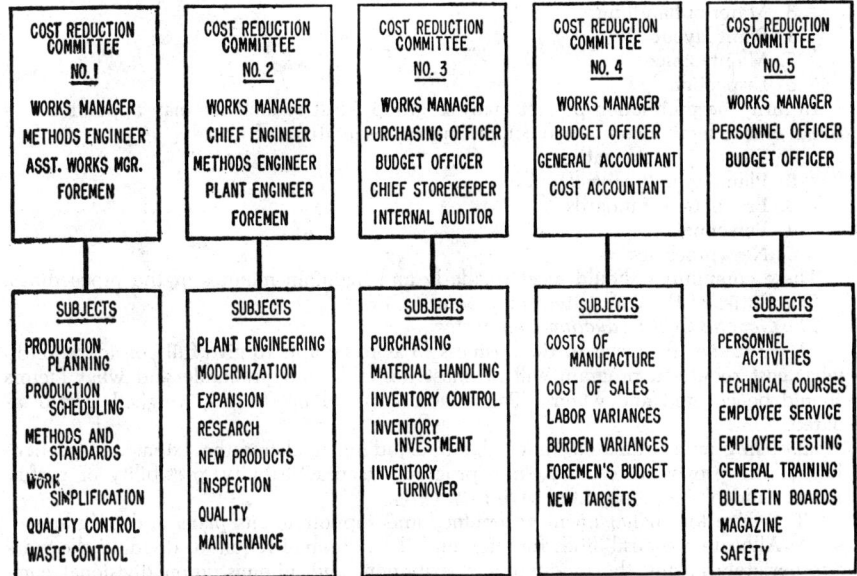

Fig. 5-2.   Typical cost reduction committee organizations and the areas in which each would function.

COST REDUCTION COMMITTEES

If all the operations of the business are to be covered (and too much attention is often paid to only one aspect, section, department, or branch), it may require a special organization for cost reduction.   As an example, this may comprise:

General Cost Reduction Committee.   The general cost reduction committee would report to the president or vice president and could comprise:

1. Works manager
2. Design engineering manager
3. Sales manager
4. Works accountant
5. Purchasing agent
6. Office manager

Often a vice president may head this committee as chairman.

It is also advisable to have a cost reduction control officer whose liaison duties are similar to those of the budget officer in the budgetary control committee.   The orders of this committee are transmitted to the divisional committee (cost reduction).

Divisional Committee (Cost Reduction).   The divisional committee (cost reduction) would report to the general cost reduction committee and would comprise representatives of:

1. Production
2. Distribution
3. Administration

Sometimes engineering and design would also be represented.

Project Analysis Group.   Each of the major sections of the divisional committee (cost reduction) would appoint its project analysis group (cost reduction).   In the case of production, the project analysis group (cost reduction) would consist of representatives from:

1. Production
2. Methods engineering

3. Material handling
4. Plant layout
5. Maintenance
6. Inspection

In turn, the production project analysis group (cost reduction) may appoint a sectional cost reduction committee comprising representatives of:

1. Process reorganization
2. Planning and control
3. Production standards
4. Personnel
5. New processes

These committees should meet regularly and maintain minutes of the proceedings. The duties of the committees may be as follows:

The General Cost Reduction Committee

1. Examine all aspects of the business so as to be able to give full consideration to what cost reduction program will be most desirable and profitable and what factors should be covered and when. These factors should be studied, weighed, and evaluated.

2. Using its coordinated knowledge and judgment, define the extent and particulars of the proposed cost reduction program, its practicability, possibility of performance, and probable potential savings.

3. Secure top management acceptance and support for its plans and programs.

4. Allocate responsibilities for such and the personnel required, dividing the work appropriately among the production, distribution, and administration divisional committees.

5. Set a timetable for progressive completion of the program.

6. In due course, receive from the divisional committee (cost reduction) its report upon the feasibility, practicability, savings potential, and timetable of the cost reduction program.

7. Examine, reject, adopt, amend, or otherwise deal with the comments of the divisional committee (cost reduction), ultimately forwarding to them the final agreed programs for their attention and action.

8. Arrange for all necessary advisory assistance to be available where required.

9. Receive periodic reports from each of the divisional committees upon progress in relation to achievements, savings, and timetable.

10. Pass judgment on the results achieved in terms of target and actual cost savings and timetable.

11. Advise top management on results achieved.

The Production Divisional Committee (Cost Reduction)

1. Through the divisional committee, receive from the general cost reduction committee those sections of the cost reduction program which it must carry out.

2. Study the program, its extent, scope, and particulars, together with the established timetable.

3. Set up a project analysis group (cost reduction).

4. Decide how the work is to be allocated among the members of the project analysis group.

5. Receive from the project analysis group an assessment of the proposed program, including its practicability, possibility of performance, a suggested timetable, and estimated savings.

6. Set up cost reduction targets.

7. Through the divisional committee, advise the general cost reduction committee of its judgment, comments, and acceptance of the program either as received or as amended.

8. Set up timetables for progressive completion of each portion of the work.

9. Arrange for all necessary assistance to each group, with liaison and coordination among them.

10. Exercise close and continuous overall control over the progress of the work in relation to both savings and timetable.

11. Receive periodic reports from each project analysis group, study them, and evaluate performance.

12. Prepare reports on all projects for submission to the general cost reduction committee.

The Project Analysis Group (Cost Reduction)

1. Receive from the production divisional committee (cost reduction) those sections of the cost reduction program for which it is to be responsible.

2. Study the program, its extent, scope, and particulars, together with the timetable allowed.

3. Set up a sectional cost reduction committee.

4. Decide how the work is to be allocated among the members of the sectional cost reduction committee.

5. Set up the cost reduction target.

6. Pass judgment upon that part of the cost reduction program submitted to the production divisional committee.

7. Set up timetables for progressive completion of each portion of the work.

8. Arrange for all necessary assistance to all groups, with close liaison and coordination among them.

9. Commence to put the program into effect.

10. Exercise overall control over the progress of the work in relation to both savings and the time schedule.

11. Receive periodic reports from each sectional committee, carefully study them, and evaluate performance.

12. Prepare reports on all projects for submission to the production divisional committee (cost reduction).

The duties of the project analysis group (cost reduction) are very similar to those of the divisional committee (cost reduction) except that it must study in much greater detail the practical working requirements of each project; decide how it is to be done, when, and by whom; and obtain the complete support and cooperation of supervisors and workmen. It must, in turn, complete its progress reports, which should be prepared much more often than with higher committees. It should ask for all the assistance it requires.

Finally, the process reorganization section (cost reduction) is charged with the performance of its task, which may be to examine the processes involved in the manufacture of product A. It could, for example, lay down a program:

1. To obtain all possible facts and factors regarding the manufacturing processes of product A

2. To make a step-by-step factual record of the existing production process

3. To make a critical examination of all operations

4. To decide which of these processes and operations could be
   a. Eliminated
   b. Reduced
   c. Combined with others
   d. Simplified
   e. Altered to increase automation or mechanization and to reduce supervision or inspection—seeking fullest supervisory and workmen cooperation in doing so

5. To decide what form reorganization should take and lay down the processes for it such as the impact of the change on the operation, the extent of training courses necessary, whether implementation is immediate or deferred, in one operation or in a series, or whether the new and the old systems should be carried simultaneously until the new system reaches required standards

6. To estimate the savings in costs resulting therefrom

7. To submit its findings to the project analysis group (cost reduction)

8. To receive from the project analysis group its approval to put the proposed plan into operation

9. In conjunction with factory management, to set up the procedures to put the plan into effect

10. To police the plan and see that it is properly established

11. To calculate the actual savings and compare these with the estimates

12. To prepare a complete report on the whole project for submission to the project analysis group

13. To receive the next assignment

To achieve the best results, three factors are essential in the planning of the cost reduction program and in the personnel to carry it through:

1. Capable men having the necessary time for the purpose should be appointed. It is of little use utilizing people already so overburdened that their everyday problems make cost reduction seem of secondary importance. These people are too busy to be efficient cost reducers. Only those able to spend sufficient time on the activity should be chosen. Furthermore, they should be capable of being welded into a team. It is teamwork that gets results in cost reduction.

2. Close liaison with and education by advisory groups. These latter can be executive, technical, and supervisory. Executive advice is essential where policies are affected and where it is necessary to go outside the scope and responsibility given or where other sections of the organization are vitally affected. Technical advice may be required from:

a. Manufacturing, in connection with all manufacturing processes

b. Time study, in relation to time study data and analyses

c. Design, in relation to suggested design changes

d. Tooling, where changes are suggested

e. Purchasing, in relation to qualities, changes in materials, and transport

f. Inspection, in relation to quality control standards, tolerances, and defective workmanship

g. Cost accounting, in relation to cost analysis, estimates, and results

h. Supervision, in relation to detailed operations and personnel

3. Close coordination with all other groups. Unless this is forthcoming, overlapping will develop, as will also the promotion of sectional instead of organizational interests, duplication of effort, and competition for technical resources. Business is really an integral whole, and one function can rarely carry out its work in the best way without close cooperation with others. Furthermore, interchange of information and assistance is promoted, thus avoiding needless effort and wasted time.

SKILLS REQUIRED FOR COST REDUCTION WORK

A cost reduction program should be initiated with as clear a conception as possible of what results should be aimed at, not only through a lessening of cost by the reduction in wasted time, motions, expenditure, and other cost making factors, but also through:

1. Increasing productivity

2. Eliminating unprofitable or expensive projects

3. Improving techniques and training

4. Replacing products or programs

5. Changing overall policies

Partly because of these factors, but also to obtain the best results, members of the various cost reduction committees should possess or have access to certain skills and tools. These former have been listed by *Factory Management and Maintenance* [2] as:

1. General technical and engineering knowledge

2. Properly cataloged manufacturing knowledge

3. Intimate knowledge of the products, their design, application, functions, manufacturing details

4. Knowledge of available kinds, types, and shapes of materials

5. Knowledge of systematic analysis techniques

6. Process charts: operation, flow, man-and-machine, and operator

7. Skill in analyzing manufacturing operations

[2] *Factory Management and Maintenance,* vol. 107, no. 5, May, 1949, p. 71.

8. The right mental attitude toward cost reduction analysis work
9. Professional honesty
10. Ability to evaluate productive output
11. Ability to estimate manufacturing time and costs
12. Ability to classify and catalog information
13. Ability to search out facts from data obtained
14. Ability to summarize facts and form sound conclusions
15. Ability to do creative thinking
16. Ability to disassemble apparatus and visualize all operations performed for producing similar parts or doing similar operations
17. Ability to organize groups for analysis work
18. Ability to stimulate people to study, analyze, and think, also to get action when work slows down and interest lags
19. Ability to delegate work
20. Personal initiative for starting work on new investigations and finding new approaches for studying difficult problems
21. Fortitude that gives a continuing confidence in the end result
22. Determination that will see a project through to completion

THE TOOLS OF COST REDUCTION

In addition to the foregoing skills, it is fundamental that all personnel engaged upon cost reduction procedures should either know or have access to certain management tools of great value in suggesting cost reduction possibilities. Some of the following tools may overlap somewhat but each may have its purpose in a comprehensive program.

1. Cost Accounting and Analyses. At the outset, it should be noted that cost reduction is not the same as cost control. The latter is more concerned with the maintenance of costs in accordance with established standards; the former with pushing costs downward.

Cost control takes the established cost standards and endeavors to keep operation costs in line with those standards. Cost reduction challenges all cost standards and endeavors to reduce these continuously. Standards are targets in cost control procedures but are suspect in the case of cost reduction. In cost control, there is more of an element of the present and the past, while with cost reduction, the emphasis is much more upon the present and the future. Cost control is often limited to those items for which standards have been set, and this all too frequently means productive operations. Cost reduction should be applied to every area of the business. It is not dependent upon standards, though target amounts may be set. Cost control seeks to get the best results at the lowest possible cost under existing conditions. Cost reduction recognizes no condition as permanent where a change will secure a lower cost figure.

Professor Juran[3] has argued that the approach to better management should be based in large part on the belief that survival and growth of an organization make it virtually mandatory to break through to higher levels of performance. He defines breakthrough as "the creation of good, or at least necessary, change" and "a dynamic, decisive movement to new higher levels of performance." This he maintains should be counterbalanced by control, which is defined as the prevention of bad or detrimental change.

Certainly, these two concepts and their meanings should be remembered in profit improvement and in cost reduction and can be profitably studied in industrial engineering projects.

Costs must not be confused with expenditures. Cost reduction may increase expenditure, as it is sometimes necessary to spend more money to reduce costs. Greater sums may be spent on research or tooling, but the final cost of the product may be reduced. It is therefore important to note that it is not total cost but unit cost which really counts.

Although there is a difference between cost control and cost reduction, the analyses

[3] Joseph M. Juran, *Managerial Breakthrough: A New Concept of the Manager's Job,* McGraw-Hill Book Company, New York, 1964.

from cost accounting are among the most valuable information available to cost reduction committees because:

  a. They give an analysis of cost which may lead a cost reduction committee directly to answers to questions such as by how much, where, and why costs are high. They thus enable concentration upon quick savings.

  b. They divide costs into the elements of material, labor, and expense. This is important because if in $1 of production cost, materials amount to 80 cents, labor to 12 cents, and expenses to 8 cents, there is obviously a clear call for a great deal of attention to be given in the first place to what can be saved in material costs. In another organization, if materials were 15 cents, labor 30 cents, and overhead 55 cents, there would be a preliminary invitation to examine overhead.

  c. If standard costs are in operation, losses exhibited through variances may also show where examination and effort are invited. In such a case, a cost reduction committee may be called upon partly to reduce costs to the standards set and partly to reduce those standards.

  d. Some standard cost analyses are prepared in such a way as to show up reasons and causes. This type of approach makes for speedy results. The cost of idle time may, for example, invite immediate attention and may be capable of speedy reduction.

  e. In addition to the types of information already mentioned, cost analyses give valuable information for the purpose of arriving at alternatives and their cost, as well as calculating savings made through cost reduction procedures. Differential or marginal costs are also valuable in this regard.

Cost analyses are one of the most valuable tools available for analyzing potentialities and possibilities of reducing the cost of the product, the process and operating procedure, or the general factory expenses. Cost accounting reports are necessary to give an appraisal of the effectiveness of cost reduction and control as initiated by cost reduction personnel.

2. Budgets and Budgetary Control. Many organizations make their cost reduction program part of their budgetary control system. It is argued that a budget should not be prepared unless and until a very thorough investigation is made of every area of the business affected by budgetary figures and all possible economies and efficiencies are introduced into those figures. It is therefore easy to understand the arguments advanced that cost reduction is best integrated into budgetary control. Many other organizations, however, consider that this is not the best procedure, and there are valid reasons for their objections. These are:

  a. Although all care should be given to the preparation of budget figures, the time allowed for budget preparation does not usually permit the same approach to cost reduction as is possible when a planned cost reduction procedure is set up.

  b. A cost reduction program focuses attention directly and specifically upon cost reduction and not on preparation of budgets.

  c. The people who prepare the budgets may not be the best people to plan, set in motion, control, and effect cost reduction.

  d. A cost reduction program will cover expenses which can be reduced and efficiencies which can be increased which may never appear as part of the budgets or budgetary control system.

Despite the fact that it is wise to keep in mind all possible economies while the budget figures are being prepared, the cost reduction program should be a separate activity. Sometimes an advantage is gained by using the budget control officer as the focal point for the cost reduction program where he has time to attend to both.

Nevertheless, the budget plan and budgetary control figures do provide a storehouse of cost reduction possibilities. The budget committee will have information gained during the preparation of the budgets which will be invaluable to the cost reduction committee in seeking matters which should be examined. The budget committee cannot have prepared budgets without finding some of the weaknesses and some of the costly phases within the organization. The urgency of budget preparation may have precluded it from giving necessary attention to these phases, but the figures are ready-made for the cost reduction committee.

Furthermore, the operation of budgetary control will show where operations are not going according to plan. In many cases, the attention of the cost reduction committee may be invited to these aspects so as to bring the budget back to its targets. The budgetary control plan will also (if it is complete) show up the various activity areas of the business, namely, the production, engineering, distribution, and administration. Actually, budgets are often the best source of information in relation to distribution and administration costs. The cost reduction committee will itself use budgetary control by setting up targets for savings.

3. Break-even Charts. Break-even charts are closely allied to cost accounting and budgetary control. Although they may show up phases of the business with which the cost reduction committee is not concerned, break-even charts do indicate many factors in the profit structure which deserve closest scrutiny. Very often, too, these charts give the clearest picture of what is contributing to high break-even points. They give a perspective in relation to shrinking profits. The technique is valuable in graphing alternatives, and a break-even chart can often help to avoid misconceptions. Furthermore, it is used in connection with every aspect of the business, including distribution and administration, and in these latter particularly, it may be very valuable. For example, a perspective of the major functional cost of the business is often best obtained from a break-even chart. In a bakery business, distribution costs may be 40 percent of the total. In shipbuilding, they may be 1 percent. An acceptance and a knowledge of these factors will enable the time of the cost reduction committee to be conserved and used to best advantage.

Break-even charts should also be used for analyses of the fixed and variable expense ratios. Sometimes cost reduction can be effected by turning fixed costs into variable costs when business volume is shrinking, and turning variable costs into fixed costs when business volume is mounting. Furthermore, the division between fixed and variable costs is important because the former are very often the outcome of policy decisions. An attack upon them may therefore well require the assistance of top management.

4. Methods Engineering, Work Simplification, and Standardization. Although the tools already mentioned are designed to provide knowledge and to be used for the purpose of calculating savings, these tools and those immediately following are designed more as agencies by which cost reduction is actually effected. Many of them, however, overlap. In many organizations, the tool of methods engineering, more than any other, will be valuable for:

   a. Analyzing present methods
   b. Suggesting alternatives
   c. Effecting savings through the reduction or elimination of motions
   d. Prompting ideas for quicker and easier methods
   e. Effecting possible savings

The most fruitful sources for cost reduction will vary in different organizations, but it is probable that over all, the biggest source of savings may come from methods engineering. This term, however, frequently includes some aspects often considered separately, as for example, plant layout and material handling. No cost reduction activity seems to be complete without very full and efficient use being made of methods engineering.

Work simplification and standardization are very closely related to methods engineering and often form part of it. Both these factors have, in certain cases, offered startling cost reductions and should be used by every cost reduction committee.

5. Time and Motion Studies. Time study gives much factual information which is of value in cost reduction. Most attractive results in lowered costs have been secured through the use of studies obtained either by predetermined elemental time procedures or stopwatches. The value of such techniques as tools for cost reduction is self-evident. Motion studies are closely akin to and used for methods engineering, which is in itself a necessary preliminary to the preparation and installation of an incentive plan.

6. Statistical Analyses. This is a tool which no cost reduction committee can afford to ignore, if only because of its great value in connection with quality control. Cost reductions often follow a thorough statistical analysis of trends. Furthermore, correlation studies and research into cost factors rely upon statistical analyses. In fact, such analyses enter into various aspects of cost reduction procedures.

7. Flow Process Charts. Flow process charts are often used in connection with methods engineering and are most valuable in connection with reduction of material handling costs. They are also valuable, however, in the case of office costs, warehousing, and transportation and are useful in effecting reductions in these fields, which often do not come under so careful scrutiny as does production.

8. Quality Control Analyses. Quality control is discussed in Chapter 5 of Section 8. Inferior-quality goods attract many costs which quality control seeks to eliminate. The costs of materials scrapped including labor and burden attaching to the scrapped work, excess inspection, costs of reworking, discounts on seconds and allowances necessary, all offer a rich field for control and improvement. Cost reductions often follow the institution of a better balance in the quality control factor as between inspection and defective work. If this balance can be obtained and preserved, the net ratio between inspection and defective work results in lowest cost.

9. Value Engineering. Although value engineering can overlap other techniques, it may well approach the subject of cost reduction in a different way and present some overall aspects not otherwise covered, for example: Does it contribute value? Is its cost proportionate to its usefulness? Does it need all its features? Is there anything better for the intended use? Can a standard product be found which will be usable? Will another dependable supplier provide it for less? Is anyone buying it for less? Former Defense Secretary Robert S. McNamara stated, "By eliminating 'gold-plated' specifications, the Defense Department was saving about $1 million per week and expected this reduction in costs to triple in two years."[4]

10. Organization Charts, Manuals, and Surveys. Very often, inefficiency and therefore wasteful expenditure occur because of faults in organization. Duplication of control, overlapping responsibilities, and wasteful efforts may all result in excessive costs. Organization manuals and charts will disclose some of these reasons for cost excesses and may indicate ways of avoiding them.

11. Management Audits. No cost reduction committee can afford to ignore the very fruitful source of savings where a management audit has been carried out and the report is available to the committee. Usually, a management audit comprises a thorough and comprehensive survey of every aspect of the organization. It is generally carried out by outside consultants, and valuable sources for investigation and subsequent cost savings are often found.

Care must be exercised in the use of these tools (*a*) to see that they are thoroughly understood, (*b*) to be sure that action prompted by them has not already been taken, and (*c*) to make certain that they are used in liaison with other personnel who may also use or act upon them. Properly understood, correctly applied, and carefully used, they are a fruitful source of cost reduction and time saving.

Two other tools in a somewhat different capacity may prove of value.

12. Industrial Mathematics and Operations Research. A description of mathematical and statistical procedures appears in Section 10. They become of interest and importance in cost reduction because making better decisions and increasing profits can often result in lowering costs. The reduction of inventories through the use of linear programming, for example, may result in lower inventory costs, including storage and handling, waste, and pilfering costs, as well as a saving in interest on bank loans. Industrial mathematics will be found to be particularly useful when used to reduce marketing and warehousing costs.

13. Radioisotopes. With the splitting of the atom and the harnessing of radioisotopes (often used in conjunction with the advanced development of electronic technology), new improvements in such important areas as quality control, plant

---

[4] *Administration of Cost Reduction Programs,* Studies in Business Policy, No. 117, National Industrial Conference Board, New York, 1965.

utilization, product development, and operating efficiency have become possible through their application. A survey of 523 companies revealed that they had achieved annual net savings of $39 million for an outlay of $3.74 million annual investment in equipment and facilities.[5]

14. Computers. Computers assist management decision making in areas such as:
  a. Profit control
  b. Inventory control
  c. Production planning and control
  d. Product sales analyses
  e. Distribution analyses

The industrial engineer is particularly interested in the use of the computer in production planning and inventory control and also in its use to control complex manufacturing operations where quality is paramount. (Refer to Section 9, Computers and the Industrial Engineer.)

COST REDUCTION REPORTS

The failure to record results achieved is like commencing a race and refusing to complete the course. The accurate, prompt, and continuous recording of savings is necessary for:

1. Measuring savings against the cost of making the savings
2. Determining just how successful the cost reduction program has been and using this information to improve subsequent procedures
3. Publicizing the results to secure even greater enthusiasm and cooperation for future programs
4. Ascertaining which areas of the business have been most fruitful in yielding savings

The National Industrial Conference Board's Studies in Business Policy, No. 117,[6] shows a form (Figure 5-3) which may be used to control each project.

The *NAA Bulletin*[7] cites three forms (Figures 5-4 to 5-6) for recording savings, each form illustrating a report prepared from a different basis.

In addition to the requirement that each committee should advise its parent committee of savings effected, provision must also be made for monthly reports of progress achieved on all current uncompleted projects and for a list of abandoned projects and the reasons for abandonment.

RESULTS ACHIEVED FROM COST REDUCTION PROGRAMS

All kinds of examples could be given to illustrate savings made in a wide variety of industries and in many countries of the world. The following examples are widespread, both geographically and in type of industry, and indicate the different tools and agencies which have been used to reduce costs.

Methods engineering is one of the most profitable tools to use for cost reduction. Thus a cost reduction program of an agricultural machinery organization in Australia embraced methods engineering, plant relocation, analysis of all excess costs disclosed through variations from standards, closer inventory control, and an attack on every cost item. In three months, this resulted in the following:

Man-hours saved per week	900
Man-hours saved per 50-week year	45,000
Cost saved per year at $1.50 per man-hour	$ 67,500
Increased weekly production	13.2 units
Increased weekly dollar volume	$ 5,227
Increased yearly dollar volume (50-week year)	$261,350

[5] *Radioisotopes in Industry,* Studies in Business Policy, No. 93, National Industrial Conference Board, New York, 1959.

[6] *Op. cit.*

[7] "Cost Improvement for Profit Improvement," Accounting Practice Report 8, *NAA Bulletin,* October, 1959.

REPORT OF FINAL SAVINGS ON COST REDUCTION CASE

TO _____ DATE _____

COST REDUCTION CASE NO. _____ ☐ PARTIALLY EFFECTIVE ☐ 100% EFFECTIVE

COST EST. NO. _____ APP. OR EQPT. _____

SUBJECT _____

PIECE PART, APPARATUS OR R.M. NOS.	ANNUAL REQUIREMENTS			RATE OF SAVINGS PER 100			ANNUAL SAVINGS						EFFECTIVE	
	USED FOR PREDICTION 5 YR. AVERAGE	USED FOR FINAL SAVINGS		PREDICTED	FINAL	PREDICTED 5 YR. AVERAGE	FINAL				CURRENT TOTAL		DATE	PER-CENT
		5 YR. AVERAGE	CURRENT				5 YR. AVERAGE							
							LABOR	EXPENSE	MATERIAL	TOTAL				

THE DIFFERENCE BETWEEN THE PREDICTED AND FINAL SAVINGS IS DUE TO:

☐ CHANGE IN REQUIREMENTS.
☐ LABOR FOR NEW METHOD (MORE) (LESS) THAN PREDICTED.
☐ (HIGHER) (LOWER) LABOR RATES.
☐ AMOUNT OF MATERIAL FOR NEW METHOD (MORE) (LESS) THAN PREDICTED.
☐ (HIGHER) (LOWER) MATERIAL PRICES.
☐

☐ VARIATION HAS BEEN REPORTED.
☐ VARIATION HAS NOT BEEN REPORTED BECAUSE:
 ☐ SAVINGS ARE REFLECTED IN CURRENT BULLETIN.
 ☐ SAVINGS ARE REFLECTED IN NEW CODES.
 ☐ ACCOUNTING IS ON DEFINITE ORDER BASIS.
 ☐ AMOUNT INVOLVED IS TOO SMALL.
 ☐ NONOPERATING VARIATION IS AFFECTED.

ACTUAL COST OF NEW PLANT REQUIRED AS A RESULT OF THIS CASE (NOW INCLUDING EXPERIMENTAL PLANT):
(A) LAND, BLDGS, SERVICE EQUIP. & MACHINERY (INCL. DESIGN) _____ (B) SMALL TOOLS _____ (C) FURN. & FIXT. _____

COMMENTS: _____

COPY TO: _____ ACCOUNTANT-ENGINEERING COSTS. _____

FIG. 5-3. Cost reduction report prepared upon completion of the project.

PROJECT	Estimated Annual Rate				Realizable in Current Year			
	Jan. 1	April 1	July 1	Oct. 1	Jan. 1	April 1	July 1	Oct. 1
MATERIAL								
Project #1	$16,800	$16,800	$16,800	$16,800	$ 5,600	$ 5,600	$ 5,600	$ 2,800
Project #2		51,000	51,000	51,000		4,000	4,000	2,000
Project #3			5,000	5,000				
Project #4			2,000	2,000				
Project #5				2,000				
TOTAL MATERIAL	$16,800	$67,800	$74,800	$76,800	$ 5,600	$ 9,600	$ 9,600	$ 4,800
LABOR								
Project #1	$ 2,000	$ 2,000	$ 2,000	$ 2,000	$ 2,000	$ 2,000	$ 2,000	$ 2,000
Project #2	4,812	4,812	4,812	4,812	4,812	4,812	4,812	4,812
TOTAL LABOR	$ 6,812	$ 6,812	$ 6,812	$ 6,812	$ 6,812	$ 6,812	$ 6,812	$ 6,812
BURDEN								
Project #1	$ 6,500	$ 6,500	$ 6,500	$ 6,500	$ 2,378	$ 2,378	$ 2,378	$ —
Project #6	1,600	1,600	1,600	1,600	1,000	1,000	1,000	
Project #7	6,000	6,000	6,000	6,000	3,000	3,000	3,000	
Project #8			1,000	1,000				500
Project #9				500				
TOTAL BURDEN	$14,100	$14,100	$15,100	$15,600	$ 6,378	$ 6,378	$ 6,378	$ 500
TOTAL DEPARTMENT	$37,712	$88,712	$96,712	$99,212	$18,790	$22,790	$22,790	$12,112

FIG. 5-4. Progress report of cost improvement projects grouped by cost element. Prepared quarterly by departments.

Inventory Reduction. The following list indicates some of the specific ways in which inventories have been reduced as the result of cost reduction studies:

1. By returning surplus stock through supplier channels as well as utilizing surpluses between company divisions
2. By reviewing monthly the turnover of major raw material commodities, so that if purchase commitments are in excess of planned requirements, they are promptly discovered and excess stocks avoided
3. By consolidating purchase orders to permit purchasing in larger quantities
4. By letting jobbers carry stock, especially on standard items
5. By increased use of market forecasts

	Current Month					Year to Date				
	Direct Material	Direct Labor	Expense	Total	R %	Direct Material	Direct Labor	Expense	Total	R %
1. Appropriation savings	$ —	$ 4,485	$ —	$ 4,485		$ —	$ 63,319	$ 49,186	$112,505	
2. Elimination of waste	2,880	—	—	2,880		35,536	6,454	589	42,579	
3. Elimination of spoilage	—	—	—	—		—	—	—	—	
4. More accurate standards	—	—	—	—		—	—	—	—	
5. Application of new cutting metals	—	—	—	—		—	—	—	—	
6. Improved methods	—	990	—	990		163,174	326,550	35,830	525,554	
7. Change in design	47,977	275	—	48,252		72,238	9,693	3,942	85,873	
8. Miscellaneous	—	175	4,281	4,456		821	25,448	103,657	129,926	
Total	$50,857	$ 5,925	$4,281	$61,063		$271,769	$431,464	$193,204	$896,437	
Budget	$ 9,812	$39,412	$9,576	$58,800	103.8%	$110,385	$443,385	$107,730	$661,500	135.5%

FIG. 5-5. Cost improvement report by type of saving. Prepared monthly.

Chairman	Committee	Goal for 1958	Goal to Date	Realization	% of Goal to Date	Amount at Audit
XXX	1	$175,000	$157,800	$290,649	184.2	$10,894
XXX	2	50,000	45,000	51,200	113.8	1,600
XXX	8	150,000	135,000	204,465	151.5	51,482
XXX	9	60,000	54,000	43,546	80.6	4,063
Total		$995,000	$898,800	$1184,076	131.7	$79,304

FIG. 5-6. Summary of cost improvement achieved by committees. Prepared monthly.

6. By centralizing all purchasing operations

7. By standardization of items to be stocked

An organization specializing in canned fruits and vegetables and working somewhat on a seasonal pattern undertook an inventory reduction program. Linear programming techniques were used and the net results meant that average inventory was reduced from $3,000,000 to $1,800,000, with savings of $130,000 per annum. Possibly the most striking result was the fact that for the first time in sixteen years, the organization found it unnecessary to call on the bank for any funds, its accommodation in the past having averaged over a million dollars. Incidentally, by scientific checking, it was disclosed that its service to customers was greatly improved and its out-of-stock items materially reduced.

Standardization and Simplification. Standardization and simplification should not be overlooked. Recently, a British investigation into orders received revealed:

1. 81 percent (in quantity) of orders received were for 15.5 percent of the designs in the firm's range.

2. 19 percent were scattered over the remaining 84.5 percent of the designs.

3. The costs of the 81 percent left a profit of 2.532 cents per square yard; the costs of the 19 percent left a loss of 2.165 per square yard.

An examination of the foregoing figures would indicate drastic action in connection with the 19 percent on which losses were made. It is true that under some circumstances, more profit could result by retaining the 19 percent upon which losses were being made. On the other hand, it would probably be found that some of this 19 percent could be turned into profitable channels with acceptable cost savings resulting.

Although single techniques are often used, particularly where no coordinated program is in operation and also in smaller businesses, cost savings often result from a combination of what can be offered through two or three techniques. In the case of an electrical organization in Australia, a product was selling in very large quantities at 33 cents each. The organization had no accurate costs and felt it was producing this product at a profit. The installation of a standard cost system soon showed it to be in error. Actually it was losing 11 cents on every product sold, amounting to more than $44,000. This obviously called for immediate action. The costs were examined, but it was found that they could not be reduced. The organization was under long-term contract to supply large quantities. It was decided that the only way to avoid losses was (1) to redesign the whole product with a view to cutting the cost considerably and (2) to give to their contractors a longer guarantee for their new product as an incentive to allow the change. The design section, knowing what was at stake, made every effort. As a result, a totally new product was produced, and as most careful attention had been given to the required production processes, the cost dropped from 44 to 21 cents. The organization was then able to go to the contractors and not only sell the new product at 33 cents but point out that it was an improved product, and instead of the 5-year guarantee, to guarantee the product for 15 years. Actually none of this would have happened unless the loss had been displayed through the cost accounting system. Then a combination of design and production produced an article which meant a difference of nearly $110,000 per year in profit.

Distribution. In the distribution field, more and more emphasis should be placed on cost reduction, and more and more opportunities for savings will result. Cost analysis will often disclose conditions ripe for distribution cost reduction. Warehousing and delivery costs are largely untouched fields in many organizations. Some remarkable results have been achieved from the analysis of profitable and unprofitable accounts, sizes of orders, territories, and products.

For example, a manufacturer found that 68 percent of his accounts, which brought in only 10 percent of his volume, were unprofitable. His selling effort was therefore shifted from unprofitable to profitable customers. Over a period of five years, sales increased 75 percent as a result of more effective use of selling effort. Marketing expenses were cut nearly in half, from 22.8 to 11.5 percent of sales, and a net loss of 2.9 percent was turned into a profit of 15 percent of sales.

There is a very fruitful field for investigation, and often considerable savings can be made by the institution of a campaign to increase the size of orders. The high cost of the small order is often overlooked, but it can have a very serious effect upon marketing expenses. This is shown particularly by an investigation of the relative expenses of two orders—one five times as large as the other. The following results were established:

1. Office costs
   a. The cost of recording the small order was 88 percent of that of the large order.
   b. The cost of invoicing the small order was 82 percent of that of the large order.
   c. The cost of granting credit for the small order was 97 percent of that of the large order.
   d. The cost of collecting the account of the small order was 101 percent of that of the large order.
2. Handling costs
   a. The cost of packing the small order was 54 percent of that of the large order.
   b. The cost of trucking the small order was 52 percent of that of the large order.
   c. The cost of shipping the small order was 56 percent of that of the large order.

Office Costs. Office costs are also a field offering great cost reduction possibilities. P. Mulligan, in the 1952 *NACA Yearbook,* points out that in the years 1900 to 1940, factory costs in the United States went up by 75 percent, but office costs increased by 352 percent in the same period. Since World War II, the trend has if anything been accelerated. Although there may be reasons for a greater increase than in the case of factory costs, no serious attack on costs in the office has been made that is comparable with the attacks made on manufacturing costs. A study of an office in Australia disclosed the following:

> The present procedure for preparing invoices reveals that the time required under existing conditions is 1.23 minutes per invoice set consisting of an average of six invoices and one envelope.
> The present processing time involved is excessive, owing mainly to the poor workplace layout, and with some attention to this aspect, the processing time can be reduced by 38.5% as described below.
> In addition to unsatisfactory workplace arrangement, it would appear that time is lost due to the number of interruptions to the operator for such causes as requests for the processing of "priority" invoices urgently needed and the need to price other invoices before completely processing a batch of invoices.

With the present layout, the processing time is excessive mainly because of the following reasons.

a. Reaches up to 42" are involved.

b. Insufficient space is made available, so that when a "Master" has been processed, it is necessary to place it at the *bottom* of a pile of masters.

c. During the 1¾ seconds when the machine is processing a copy, the operator could be procuring a blank invoice form.

Further savings in addition to the 38.5% mentioned above can be effected if these excesses are rectified.

Measurement of clerical activities has greatly increased the opportunities for the reduction of office costs. Work simplification and methods improvement programs generally provided the first attack on office costs, but the knowledge of what staff should be used to cover a given volume of work has resulted in significant cost reductions in offices in various parts of the world. A cotton textile company in one country has reported a saving of 30 percent in office staff and 16 percent in office space; a life insurance company in a different continent has indicated savings of $675,000 spread over three years, and there are additional recorded examples of savings running into some millions of dollars.

The use of radioisotopes (and electronic technology) should not be overlooked for the part they can play in cost reduction.

The technical literature of most countries, particularly that of the United States and Great Britain, is full of examples of what is being achieved through cost reduction techniques in a wide range of industries varying greatly in size, complexity, and geographical location. The benefits thus expressed are too striking to be ignored, and the variety of tools used shows how valuable a part each one can play. Indeed, it is not infrequently found that a systematic, scientific, and studied plan to effect maximum cost reduction can in some cases result in savings in excess of the dividend requirements of the organization.

COST REDUCTIONS IN SMALL ORGANIZATIONS

In every country in the world, in point of numbers, small businesses far outweigh large ones. Cost reduction is not limited to the large organization. It can be practiced as effectively in small as in large businesses and with the same techniques, although on the modified scale warranted by smaller operations. The committee structure mentioned previously would not be suitable for the small organization. It would be top-heavy and clumsy. The smaller the organization, the more simple the cost reduction procedure should be. The personnel devoted to cost reduction should be much more versatile and perhaps have an all-around knowledge of cost reduction techniques. The procedures, too, should be much simplified. A suitable organization is shown in Figure 5-7.

The following are the steps taken in methods engineering for cost reduction purposes in one small business:

1. Briefing of the senior methods engineer by the works manager
2. Selection and briefing of a methods engineer by the senior methods engineer
3. Discussion of the project with the plant superintendent by the senior methods engineer and methods engineer
4. Discussion of the project with the foreman concerned
5. Familiarization investigation by the methods engineer
6. Preparation of a draft plan, assessment of overall time to be allowed for the department, and timetable of actual operations by the methods engineer in conjunction with the senior methods engineer
7. Approval of plan by works manager
8. Discussion of plan with plant superintendent and foremen
9. Detailed methods investigations with regular reference to the foremen on the lines of investigations being followed
10. Explanation of the advantages of the method to the foremen and the obtaining of their opinions and suggestions thereon

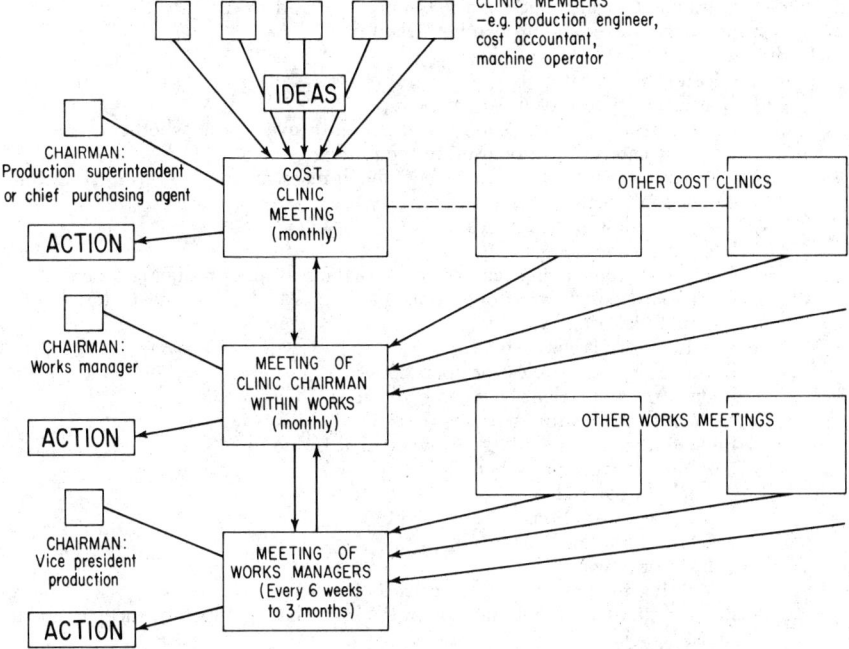

Fig. 5-7. Cost clinic organization for cost reduction in small companies. (*Source:* "*Cost Clinics,*" Business Management, *October,* 1967.)

11. Obtaining the approval of the works manager, and plant superintendent where necessary, for the recommended methods change
12. Assistance in the training of the operator in the new method and explanation of the advantages to the operator
13. Confirmation to the design office of the changes finally agreed upon (design office will have been consulted at appropriate stages during development of the method)
14. Preparation of a time formula as required
15. Notification to the cost office of changes involved in time standard and routing
16. Notification to the pay office of the incentive basis decided upon
17. Notification to production control of any changes made to operation times

Summarizing the procedure, the methods engineer should:

Discuss the project with appropriate personnel
Analyze the operation
Eliminate the unnecessary
Improve the necessary
Standardize the conditions
Train the operator
Establish a time standard
Notify all concerned

Very often in the small business, the cost reduction section may consist of two or three people taken from production, with an indication being given to them that they must "see what money can be saved." Here is a cost reduction checklist given by the top executive of a comparatively small factory to three people specially chosen to see what savings could be made. It is couched in general terms, but as can be expected, it is designed to introduce a note of urgency and to indicate the more usual places where quick cost reductions can be picked up.

1. In the factory:
    a. In connection with materials:
        (1) Check the amount of waste and scrap.
        (2) Be ruthless about inventory losses, and investigate whether better or cheaper materials can be used in the product.
        (3) Cut down inventories. Goods cost much more money to keep in store than most people realize.
        (4) Get rid of slow-moving lines.
        (5) Look at the storekeeping costs.
        (6) Decide whether savings can be made by increasing or reducing orders.
        (7) If standard costs are in operation, have a good look at the material price and material usage variations.
    b. In connection with labor costs:
        (1) Improve the time cycles for operations.
        (2) Reduce the time spent between operations.
        (3) Reduce the setup times.
        (4) Reduce idleness by finding out whether it is due to:
            Available work not scheduled
            Lack of materials
            Lack of instructions
            Lack of tooling
            Lack of power
            Waiting for repairs or other causes of delays
        (5) Reduce times by combining operations or change the sequence of operations.
        (6) Examine indirect labor—can it be changed to direct labor?
        (7) Note the efficiency of operators.
        (8) Decide if further operator training will help.
        (9) Decide if it would pay to use subcontractors.
        (10) Examine extra mechanization and the possibilities of the development of better tools and better tool and machine design or, perhaps, special-performance machines.
        (11) Try redesigning the product to make it easier and less expensive to produce.
2. In the matter of expense:
    a. Institute a critical audit of every item of expense that appears in the classification of accounts, and when this is being done, be sanely ruthless about every account.
    b. See if fixed costs can be changed into variables.
    c. Improve the quality of supervision as that will reduce its cost.
    Within this category also come items upon which nearly everyone can save money: wasted time and materials in repairs and maintenance, power and steam wastes, and costly material handling. In many cases, more savings can be made in these items than anywhere else in the factory.
3. In the case of delivery costs:
    a. Ascertain if the delivery services can be streamlined so as to avoid waste running and unnecessary deliveries.
    b. Give thought to better utilization of delivery services by better balanced and larger loads.
    c. Go over every item of the delivery expenses and critically analyze each.
    d. Have a look at which trucks are costing money through excessive repairs or fuel costs.
    e. Examine packing and crating costs.
    f. Examine afresh the costs by road, by rail, and by ship. Perhaps things have changed since they were last looked at.
    g. Develop comparative costs per ton mile or hour of operation.

4. Examine sales costs:
Because sales are hard to get, it does not necessarily mean that sales expenses cannot be reduced.
  a. Critically examine the cost of selling to see if every dollar is pulling its weight.
  b. Compare the costs of various sales areas, and ascertain if some areas are justifiably costing more than others.
  c. Compare the costs of various representatives, and decide if differences are justified.
  d. Investigate advertising costs in the light of sales pulling power.
  e. If sales servicing operates, make sure the expenditure is wisely incurred.
  f. Have a look at the channels of distribution. Perhaps the method in operation is more expensive than another type which will do the job equally well.
  g. Examine territory costs. Is it worthwhile holding on to the heavy-cost territory?
  h. Have a careful look at classes of customers. Many an organization has considerably reduced costs by eliminating unprofitable customers.
  i. Critically examine the cost of small orders, as this is a most fruitful method of reducing selling expenses.
5. Do not forget the office expense:
  a. Start off by having an audit of every form prepared in the office, and decide to scrap every one which cannot be fully justified.
  b. Prepare process charts to find out whether there is wasted effort.
  c. Have a careful look at more or better mechanization.
  d. Examine ordering procedures.
  e. Critically examine every item of expense in the office.
It should not be thought that attractive savings are confined to larger organizations. Indeed, it is frequently easier to obtain greater proportionate cost reductions in smaller businesses than in large ones.

142 Causes of Waste on Which Foremen Should Take Action NOW [8]

The Waste of Time

1. Lack of proper planning; keeping workers waiting between jobs or waiting for material.
2. Failure on the foreman's part to thoroughly understand orders and instructions received.
3. Lack of knowledge of what constitutes a full day's work.
4. Failure to make orders and instructions clear to workers.
5. Failure to insist that tools, supplies, and portable equipment be kept in proper places.
6. Ordering overtime work that could have been avoided.
7. Not seeing that men are supplied with proper tools and equipment for every job.
8. Allowing workers intentionally to do less work than they can.
9. Failure to notify employment department when more men are needed.
10. Keeping too many men at work.
11. Failure to write records and requisitions intelligibly.
12. Failure to question and correct workers who lay off.
13. Allowing workers to get habit of talking, visiting, killing time.
14. Failure to get workers started on time; slack supervision.
15. Delay in making decisions.
16. Unnecessary absenteeism or tardiness on the foreman's part.
17. Being late with reports.
18. Not investigating immediately when repairs are needed.
19. Unnecessary visiting and conversation on the job.
20. Failure on the foreman's part to organize his own time and work.

[8] "Waste Which Foremen Can Prevent." (*Copyright Elliott Service Company, Inc., Mount Vernon, N.Y.*)

The Waste of Ideas

1. Failure to listen and comment when workers offer suggestions.
2. Failure to encourage workers to offer suggestions.
3. Not asking worker's advice on problems.
4. Failure to read and study about the work and about business methods generally.
5. Failure to get from new men helpful ideas which they may bring from previous employment.
6. Not consulting enough with other departments, as engineering, etc.
7. Failure to consider or refer to the proper person all usable suggestions no matter where they come from.
8. Failure to take proper interest in foremen's meetings.

The Waste of Materials and Supplies

1. Inadequate supervision resulting in spoilage of material.
   (a) New men not thoroughly instructed.
   (b) Men not instructed on new work.
   (c) Blueprints or sketches torn or illegible.
   (d) Machines out of order or not adjusted.
   (e) Failure to follow each job through.
2. Failure to explain money value of materials and supplies to workers.
3. Failure to give orders and instructions clearly.
4. Permitting improper or rough handling of materials and supplies.
5. Not paying attention to workers' eyesight and health as possible causes of spoiled work.
6. Lack of discipline among workers, thereby encouraging carelessness and off-quality work.
7. Allowing men to use supplies unsuited for the work: too good or not good enough.
8. Inability to trace defective work to the man who did it so that it can be corrected.
9. Taking men's ability for granted; not making sure that workers are qualified for the work they are to do; especially new men.
10. Not knowing right kind of supplies to order.
11. Ordering more materials and supplies than necessary and not returning excess to stock.
12. Failure to see that materials are piled and stored properly.
13. Failure to investigate all bare wires, leaky valves, pipes, fittings on steam, water, gas, electric, and compressed air lines, etc.
14. Allowing workers to use oil, compressed air, small tools, chemicals, etc., for personal use.
15. Letting defective material go through as standard.
16. Lack of system in controlling outgo of supplies to prevent loss and theft.
17. Scrapping materials that could be salvaged.
18. Permitting the waste or abuse of such supplies as brooms, stationery, oilers, light globes, shovels, rubber hose, etc.

The Waste of Machinery and Equipment

1. Failure to plan work so that full and proper use may be made of all available machinery.
2. Failure to inspect machinery to keep it in good condition and to prevent breakdowns.
3. Foreman's lack of knowledge on possible use and capacity of various machines.
4. Failure to make regular examinations of wire ropes, belts, chain drives, gear drives, conveyors, lubrication systems, valves, etc.
5. Using unnecessarily large and powerful machines for small work.
6. Lack of cooperation with maintenance department; upkeep, repairs, painting, etc.
7. Not protecting idle machinery from weather, dust, dirt, rust, fumes, etc.
8. Allowing machinery to stay dirty by lack of periodical cleaning.
9. Failure to inspect for proper lubrication of all moving parts.

10. Failure to make needed repairs promptly.
11. Lack of instruction to men on the proper operation of machinery.
12. Lack of proper discipline to prevent abuse of machinery or equipment.
13. Allowing men to make "shoestring" repairs.
14. Failure by the foreman to keep informed on latest types of machinery and equipment.
15. Failure to pay attention to workers' opinions on value and condition of machinery.
16. Abusing small machines on large work.
17. Repairing machinery that should be scrapped; may cost more than new machinery.
18. Scrapping machinery that should be repaired.

The Waste of Manpower

1. Failure to control turnover of capable workers because of the following:
   (a) Not appreciating the direct and indirect costs of labor turnover.
   (b) Too much "bossing" and not enough intelligent direction.
   (c) Too strict or too lax enforcement of discipline.
   (d) Not keeping promises which could be fulfilled.
   (e) Making promises which can't be fulfilled in regard to wages, promotion, etc.
   (f) Discharging men without sufficient cause; improper use of the discharge slip as a penalty.
   (g) Keeping a worker on a job for which he has a violent dislike.
   (h) Treating one man better or worse than others; favoritism.
   (i) Taking sides in workers' arguments.
   (j) Criticizing one worker to another.
   (k) Failure to question men who leave of their own accord.
   (l) Failure to interpret correctly management's real aims and policies to workers.
   (m) Failure of the foreman to do all he can to fairly adjust wages and working conditions.
2. Failure to get full production as quickly as possible from new workers.
   (a) Not receiving new workers in kindly, helpful manner.
   (b) Incomplete job instruction of new workers.
   (c) Failure to impress on new worker the necessity of a full day's work, and what it consists of.
   (d) Failure to select new men with proper qualifications for work to be done.
   (e) Impatience with new men who learn slowly.
   (f) Failure to get other workers to show a friendly, helpful attitude to new men.
   (g) Failure to contact new worker as often as may be required.
   (h) Not informing new worker as to plant living conditions and regulations, as safety, paydays, lavatories, drinking water, lockers, washrooms, etc.
   (i) Lack of information to new worker about unpleasant or dangerous parts of his work.
3. Failure to get the best efforts of which workers are capable.
   (a) Failure to commend men for doing good work.
   (b) Failure to explain as much about the work as possible in order to make it interesting.
   (c) Lack of interest in worker's progress and personal affairs.
   (d) Failure on the foreman's part to admit a mistake to worker.
   (e) Lack of attention to worker's ability and temperament in assigning work to him.
   (f) Failure to study men as individuals in order to get their best efforts.
   (g) Countenancing the formation of cliques or groups among workers.
   (h) Rating men on any grounds but competence; racial, religious, fraternal, etc.
   (i) Keeping a man in a job for which he is physically or mentally unsuited.
   (j) Permitting a man to remain at work when he is sick.
   (k) Not giving men all the help they need.
   (l) Failure to promote workers when it is possible and advisable.

(*m*) Lack of due consideration of problems affecting wages and working conditions.

(*n*) Failure to train an understudy.

The Waste of Accidents

1. Failure to recognize accident prevention as part of production.
2. Failure to give all men thorough instruction in safe practices.
3. Failure to install mechanical safeguards and to keep them in repair.
4. Allowing men to work with guards out of place.
5. Failure to display danger signs at proper places and to see that they are clean and legible.
6. Failure to thoroughly understand indirect accident costs.
7. Poor housekeeping.
8. Lack of understanding of what constitutes an accident hazard.
9. Failure to keep records of accidents, to analyze them, and to use the information gained.
10. Not setting a good example in the matter of safe practices.
11. Lack of regular and conscientious safety inspection.
12. Failure to enforce consistently all safety rules and regulations.
13. Allowing men to work without necessary protective devices such as goggles, welding helmets, safety shoes, safety belts, etc.
14. Failure by the foreman to recognize his responsibility for accidents in his department.
15. Failure to stimulate and maintain interest of employees in accident prevention.
16. Lack of cooperation with state and insurance inspectors.

The Waste of Noncooperation

1. Failure to cooperate:
   (*a*) With other foremen and departments.
   (*b*) With clerical, engineering, sales, employment departments, etc.
2. Lack of thorough understanding of company policies and failure to explain them to workers.
3. Failure to deal sensibly with gossip and tale bearing.
4. Passing the buck; to other foremen, to workers, or to management.
5. Not adequately representing the workers to the management.
6. Permitting disgruntled employees to agitate against the company.
7. Failure by foremen to give full support to unpopular company regulations.
8. Failure to promote friendliness and cooperation among workers.
9. Thoughtless criticism by the foreman himself of any company policy or of any individual in the organization.
10. Not cooperating wholeheartedly with management in its educational activities such as apprentice training, bulletin boards, employee's magazines, suggestion systems, safety meetings, etc.

The Waste of Space

1. Improper piling or storage of materials.
2. Not enough attention paid to routing of materials through plant.
3. Wrong placement of machines and other permanent equipment.
4. Allowing men to leave portable tools, ladders, wheelbarrows, etc., in way of other workers; failure to keep passageways clear.
5. Keeping material which should be scrapped.
6. Lockers, oil tanks, stock supplies, etc., in inconvenient places.
7. Letting unused machinery and equipment take up valuable space.
8. Leaving needed space unused for want of needed repairs to roof, floor, etc.
9. Allowing "dark spots" in plant; inefficient lighting.
10. Failure to maintain order and good housekeeping in department.
11. Keeping unnecessary materials at workplaces.

Industrial waste costs millions of dollars yearly. Most of it can be stopped. Management is doing its part. Workers are helping. But positive, active cooperation by foremen is the surest method of producing immediate results. Sometimes, particularly with (though not confined to) small business, splendid results in cost reduction can be obtained by concentrating the effort on one section of the personnel, and often the most logical is the foremen.   By careful planning, assisting, and policing, the 142 causes of waste which foremen can prevent can form the basis of an effective and vital campaign.

CONCLUSION

In any approach to cost reduction, there are a number of fundamental aspects which should be borne in mind.   These are:

1. No organization is too small or too large to rule out the attractive possibilities of savings through cost reduction.   Opportunities exist in every organization irrespective of size.

2. Cost reduction is a state of mind and should be promoted as such.

3. Involve as many employees as possible in cost reduction.   Psychologically it is sound to spread the degree of participation so that as many as possible feel that, to some extent, the success of the program rests with them.

4. It should be a continuing program, for it is never finished.   A product line that is competitive today needs analysis tomorrow.   Cost reduction should therefore be a permanent, alive, and vital part of a company's organization.

5. Concentration should be made upon controlling the controllables.   Unless care is exercised, much time and money can be wasted in endeavoring to control the uncontrollables.   Much of the success in cost reduction comes from judgment exercised in relation to categories in which various expenses lie and the respective opportunities offered.   Although the day-to-day approach is that every cost reduction tool should be used wherever practicable, over the years, cost reduction has largely come through better methods and better equipment.

6. Costs are not reduced by raising voices against people.   Often there is too much talk about cost reduction and too little planning and concentrated effort.

7. Cost reduction is the real key to continued national prosperity under a rising standard of living.

8. No concern can afford to ignore cost reduction.   Its competitor will not do so.

9. Money must often be spent in one place to save more in another.

10. Cost reduction requires resourcefulness, imagination, and enthusiasm.   Success in effecting savings today merely presages extra effort tomorrow.

11. Cost reduction leadership flows down from the top.   It will rarely seep up from the bottom.

BIBLIOGRAPHY

Administration of Cost Reduction Programs, Studies in Business Policy, No. 117, National Industrial Conference Board, New York, 1965.
Control of Nonmanufacturing Costs, Special Report 26, American Management Association, New York, 1957.
"Cost Clinics," *Business Management,* October, 1967.
"Cost Improvement for Profit Improvement," Accounting Practice Report 8, *NAA Bulletin,* October, 1959.
Cost Reduction, Institute of Cost & Works Accountants, Ltd., United Kingdom, 1956.
Cost Reduction at Work, Management Report 28, American Management Association, New York, 1959.
Cutting Production Costs, Special Report 4, American Management Association, New York, 1955.
"An Integrated Approach to Cost Reduction," *Work Study,* October, 1967.
Morse, H. C., and E. E. Wyatt, *Cost Reduction Guide for Manufacturing Management,* Wyatt & Morse, Inc., Chicago, 1961.
Radioisotopes in Industry, Studies in Business Policy, No. 93, National Industrial Conference Board, New York, 1959.
"Vigilant Cost Control Sustains Profitability," *NAA Bulletin,* December, 1964.

Chapter **6**

The Control Group

WARREN D. KNAPP

Manager, Armaflex & Expanded Products, Millroom & Service Operations, Armstrong Cork Company, Braintree, Massachusetts

The control group approach to the more effective use of engineering staff techniques, tools, and talents has been developed in a continuing effort to challenge the engineering group to greater involvement in the management of business. The concept is relatively simple. The control group consists basically of key engineering personnel such as the supervisor of industrial engineering, the supervisor of quality control, and the supervisor of chemical engineering. This group may be expanded as other talents are required. Management establishes the basic group, but the inclusion of additional talent is the prerogative of the basic group.

RESPONSIBILITY OF THE CONTROL GROUP

The control group is held responsible for the evaluation, standardization, control, and improvement of all phases of a process, including materials, equipment, and people.

The members of the control group literally live together. Any processing problem is their problem. The line group, consisting of leaders, foremen, general foremen, and the like, is held responsible only for making sure the process is being run according to the established process procedure. The foreman is not expected to deviate from procedure, but rather he has the directive to shut down his operation and not try to improvise his way out of his difficulty, regardless of the situation. A process that fails is then set up to run under the direction of the control group. Experience has shown that the need for deviation from an approved procedure is usually caused by some other unauthorized deviation which may never be found if subsequent process deviations are allowed.

Under this method of operation, a meeting is held each morning at 9 A.M. of the entire plant operating group, including the control group and the general foreman, product controller, scheduler, mechanical engineer, and others as appropriate. This operating group meets with the product manager to discuss the events of the previous twenty-four hours. Each member of the group is given the opportunity to discuss any specific problem he might be concerned about and also such subjects as results of tests, new product development, plans for test runs, missed schedules, and the like.

Any problem discussed at the meeting which involves processing difficulties immediately becomes the concern of the control group. The operating group discusses what the problem is; the control group decides how the problem will be handled. This morning meeting, which normally lasts thirty minutes, is one of the most important features of the control group technique. It provides the basic source of information about day-to-day problems and plans for the entire operation.

MULTIDISCIPLINE CAPABILITY OF THE CONTROL GROUP

The following is a list of some of the arsenal of tools, techniques, and approaches that are available to management for the solution of production problems. Some are better known than others; some are highly technical, involved, and costly to use; others involve little more than a casual look at an operation or problem.

Mathematical Tools
Sampling theory
Hypothesis testing
Analysis of variance
Regression and correlation analysis
Distribution analysis
Design of experiments
Analysis of time series
Bayesian decision theory
Statistical forecasting
Decision tree methodology
Predicting systems
Linear programming
Dynamic programming
PERT
CPM
Game theory
Direct digital control
Reliability theory
Queuing theory
Input-output analysis
Simulation
Network theory
Inventory theory
Exponential smoothing
Modeling
Techniques
Methods analysis
Time study techniques
Work sampling
Methods time measurement (MTM)
Work factor analysis
Engineering economy
Value engineering and value analysis

Cost estimating
Plant layout and design
Information retrieval
Industrial dynamics
Automation or mechanization
Operational analysis
Resource utilization
Computer applications
    Order entry
    Perpetual inventories
    Warehouse inventories
    Scrap systems
    Scheduling
    Forecasting
    Machine loading
    Standard cost building
    Incentive calculations
    Payroll and labor controls
    Operating results file (ORF)
    Management information systems and control reports
    Quality control applications
Management Approaches
Management by results
Management by exception
Management by objectives
Zeugma cost analysis
Control group
Historical standard cost analysis
Inverted spiral theory
Theory of a planned approach to product cost reduction
Computer utility
Manager evaluation formula
Cost reduction
Theory of organization
Daily staff and line coordinating meeting

All these tools, techniques, and approaches are available, but someone must determine which approach is going to be used, when it is to be used, and the degree or depth of involvement. Also, someone must decide the schedule for the utilization of all the various engineering techniques and functions, and must coordinate and evaluate the results as related to the interaction of the various engineering groups.

With the heavy demand on engineering talent which always exists, it is essential that engineering services be utilized to the fullest extent and not allowed to be dissipated through the lack of coordination or duplication. When an engineering study of a process or operation is proposed, it should be designed to accumulate the greatest amount of information that can be used to fill all engineering needs, and it should not be confined to a particular engineering service.

As the problem of who coordinates the various engineering services in any plant is examined, it becomes obvious that coordination is usually one of the weakest functions in the typical plant.

In the control group approach, coordination is accomplished through self-discipline. An examination of the workings of the group reveals that leadership varies within the group, depending primarily on the approach that will be used to evaluate or solve a problem. If, for example, an industrial engineering problem is encountered, then the leadership will fall to the supervisor of industrial engineering. The value of having the opportunity to draw on the many varied engineering techniques that each of the individual members can bring to the group cannot be overemphasized, for this synergistic effect is one of the real strengths of the control group technique.

THE CONTROL GROUP APPROACH TO ACTION

Evaluation, standardization, control, and improvement are the four basic steps involved in the control group's method of operation. This is true whether the problem involves the introduction of a new product, the solving of an unexpected scrap problem, or a concern for increasing the capacity of a specific piece of equipment or an entire process line.

At the outset, a set of conditions exists that must be evaluated and, hopefully, improved. It becomes the responsibility of the control group to determine the tools and techniques that are to be used to solve the problem, together with the timing of the study and the depth of penetration.

When the evaluation study has been completed and the findings are available for inclusion in the process, the control group proceeds to the second step, which is standardization of the process.

The standardization procedure may include written specifications, operator process instructions, or elaborate training programs, but in any event, the tools that will be used, including the degree of sophistication, are the complete responsibility of the control group.

The third step involves control—control of equipment, material, and people. Again, the necessity, types, and degree of control—whether it is of a mechanical or personnel type—and the degree of involvement are the responsibility of the control group.

The fourth and final step is improvement. Here again, the procedures, tools, and techniques to be used for improving an operation or process are the sole responsibility of the control group.

An examination of the nature of the four steps of evaluation, standardization, control, and improvement shows that the use of all the tools and techniques available to management falls into one or more of the four steps. It further shows that, as the tools and techniques are assigned to one or more of the categories, a logical sequence for an orderly use of these tools and techniques emerges.

An even more interesting observation reveals that, when a basic improvement for an operation or process is established, it is subsequently necessary again to go through the same basic steps of evaluation, standardization, control, and improvement. This is a never-ending tightening cycle, gradually closing toward the center which represents the elimination of the operation or process involved. Because of the structure of the cycle, it has been called the "inverted spiral theory."

THE PLANNED APPROACH TO COST REDUCTION

Cost reduction, under the control group theory, is not set up as a separate program with specific meeting times and a planned agenda. Cost reduction objectives and ideas stem from a constant examination of all costs that combine to make up the total cost of a product.

The inverted spiral approach provides the mechanism by which all members of management are involved in a continual reduction of all costs concerned with the process. Any ideas for change in a process, regardless of where they might originate, are turned over to the control group for test and evaluation. If tests are favorable, the process is changed in accordance with the four steps of the inverted spiral concept.

A basic responsibility of the control group is to evaluate continually the capability of a process for increased capacity of equipment and improved material utilization with an ever-improved scrap control. All proposals by vendors for use of substitute materials are cleared through the control group. All purchase specifications incorporate the basic control information prepared by the control group, and no deviation from specification, regardless of economic benefits, is allowed unless evaluated by the control group.

Under the control group technique, the scope of responsibility for the members of the control group may appear to outweigh the usual responsibilities of the normal plant engineering staff group, and yet, if each responsibility is examined, it becomes evident that these responsibilities already exist. The control group technique is merely a method of handling them in an orderly, systematic manner designed to maximize the efficiency of each of the engineers assigned to the group.

THE ROLE OF THE INDUSTRIAL ENGINEER IN THE CONTROL GROUP

The control group theory does not recognize industrial engineering as a tool or technique. It does not identify an industrial engineer as a methods man or a layout, time study, or value engineer.

A mechanical engineer, an electrical engineer, or an aeronautical engineer is known for his responsibility in a given area and not for any particular tool or technique he can bring to that area. Under the control group theory, the industrial engineer is also known for his own area of concern. This area is costs. Under this method of operation, any cost assigned to a product becomes the concern of the industrial engineer. His objectives are twofold: (1) to establish methods of control for all cost areas, plus adequate dissemination of cost variance information to management, and (2) to establish a continuing program of cost reduction for all areas of cost. Put another way, the first responsibility of the industrial engineer is to identify and establish controls for all costs involved in the manufacture of a product and then to determine the type of action that should be taken toward a continued reduction of every individual cost.

The industrial engineer is assigned, as an objective under this technique, the responsibility for establishing each year lower cost standards than were in effect for the previous year. To accomplish this objective, he must be given the freedom and responsibility to investigate any area of manufacturing, even those that may be closed to him under the conventional method of organization. For example, under the control group concept, it is not unusual to find the industrial engineer traveling to the supplier's plant to establish product controls.

Equally important is the establishment of a closer relationship with the sales department, aimed at resulting in a better understanding of the final use of a product. Experience has shown that, with this opening up of new channels of information, the opportunities for lowering product costs open up correspondingly.

This does not mean that, under the control group concept, the industrial engineer is freed from the conventional responsibilities of time study, methods studies, layouts, and the like, but because these are merely tools of industrial engineering, they are used at the discretion of the engineer in his ever-widening search for improved performance and lower costs.

ADVANTAGES RESULTING FROM THE USE OF THE CONTROL GROUP

With new tools and techniques being introduced almost daily, it is virtually impossible for any single engineer, industrial or otherwise, to know what is the best technical approach available for solving a given problem. The pooling of technical information in a never-ending search for the truth will expose more and more plant personnel to what is possible in each engineering field and the degree that each engineering group can contribute as a problem is explored in depth.

Engineers must be challenged. The objectives of management must stimulate a manufacturing organization to greater in-depth evaluation of the true potential of any manufacturing process. If an engineering group does not seek and use the most advanced engineering techniques, then it is reasonable to believe that they have not been sufficiently challenged and the real potential of the process has not been adequately explored.

Management, under the control group theory, establishes for the control group the objective of lower costs each year. The control group, under this management challenge, must explore the use of the latest tools and techniques as the obvious opportunities for lowering costs become less apparent.

With management objectives clearly established and the control group organized to meet the objectives, the economic advantages that will accrue are exciting to contemplate.

Chapter **7**

Industrial Engineering in the Service Industries

HAROLD E. SMALLEY

Regents' Professor of Industrial Engineering;
Director, Health Systems Research Center,
Georgia Institute of Technology, Atlanta, Georgia

JOHN R. FREEMAN

Director, Systems Development,
The Medicus Corporation,
Dallas, Texas

MARVIN E. MUNDEL

M. E. Mundel and Associates, Silver Spring, Maryland

LOWRIE W. McINTOSH

Weston, Connecticut (formerly Vice President,
The Northern Trust Company, Chicago, Illinois)

J. N. SALAPATAS

Florida Power and Light Company,
Miami, Florida

JOHN H. HILDENBIDDLE, JR.

Director, Industrial Engineering,
Penn Central Company, Philadelphia, Pennsylvania

EARL K. BOWMAN

U.S. Department of Agriculture,
University of Florida, Gainesville, Florida

FREDERICK H. YOUNG

Manager, Corporate Industrial Engineering,
Stop and Shop, Inc., Boston, Massachusetts

The genesis of the field of industrial engineering lies with the creative ideas and work of Taylor in the 1880s and other pioneers who applied scientific management to problems within that segment of industry known as "manufacturing." And even though the Gilbreths, in the early years of the twentieth century, broadened the application of these principles and practices to managerial problems in other industries—the construction trades, hospital and medical areas, and work with the physically handicapped—the mainstream of activity continued to be in manufacturing. Accordingly, as industrial engineering began to be recognized as a profession in the 1930s, this branch of engineering, in the minds of most people, had come to be associated exclusively with manufacturing industry. Since World War II, however, industrial engineering has been practiced at an accelerating rate in other segments of the product-oriented industries (agriculture, transportation, retailing), and in recent years, in the vastly expanding service industries. The present chapter is concerned with such nonmanufacturing applications of industrial engineering, particularly in the service industries.

MODERN INDUSTRIAL ENGINEERING*

Modern industrial engineering, as the term is used here, is a broad discipline encompassing the analysis, design, and improvement of any and all productive elements (systems) of any enterprise, indeed of any organized human endeavor. Given this broad interpretation, modern industrial engineering includes the use of the philosophies, approaches, and techniques of operations research, systems engineering, and management science, and in concert with other relevant disciplines, aims toward improvements in the productivity (ratio of results to costs) of those production and management systems that provide society with needed products and services.

> The appropriateness of the term, industrial engineering, to describe this broad function may be seen by examining the true meanings of the words, *industrial* and *engineering*. *Industrial* is the adjective form of the noun, industry, which means skill, cleverness, ingenuity, diligence in any employment or pursuit, and human exertion for the creation of value. Contrary to popular usage, the term is not limited to manufacturing or to the factory; industry includes all aspects of the instrumentalities relating to the creation of value satisfactions. *Engineering* is that branch of applied science which attempts to utilize the resources of nature and of human nature for the benefit of mankind, with due regard for the relative scarcity of such resources. Accordingly, industrial engineering is the use of engineering principles and practices to facilitate the creation of value satisfactions.[1]

TRENDS TOWARD THE SERVICE INDUSTRIES*

Three major trends account for an increasing emphasis upon improvements in the production and management systems of service industries.

Changes in Manufacturing. The first of these trends is the high level of mechanization and automation being reached within manufacturing industry, a development which reduces the labor content of factory systems and induces a movement from blue-collar to white-collar occupations. Not only is there relatively less human work to be studied and improved in factories, but modern manufacturing firms are larger, are more sophisticated in their management practices, and tend to assign industrial engineers to the broad, high-level, complex problems of top management. At this level, administrative and managerial problems tend to be corporate in scope,

---

* Written by Harold E. Smalley.
[1] Harold E. Smalley and John R. Freeman, *Hospital Industrial Engineering*, Reinhold Publishing Corporation, New York, 1966, p. 11.

are not uniquely associated with the manufacturing division, and share a certain commonality with the complex problems of any organization, manufacturing or otherwise.

Modern Social Problems. The second major trend inducing more attention to the service industries is the seriousness and complexity of problems within modern society. These include the population explosion and concentration in urban areas, the coexistence of affluence and poverty, rising expectations for better health and education, concern about pollution of the physical environment, and a growing hostility toward the *modus operandi* of traditional institutions. Increasingly, there seems to be a belief that the technology that defeated the enemy in World War II, created the atom bomb, put a man on the moon, and gave most Americans an astonishingly high level of materialistic living, can be put to work to solve current social problems. And even though it is no panacea, industrial engineering is a part of that technology.

Governmental Influences. The third major trend is the increasing involvement of government, essentially a service industry, in all facets of modern life. The Federal government is perhaps the largest and most complex organization within American industry, and its political and economic influences permeate the entire economy. Federal, state, and local governments constitute a sizable segment of industry in terms of their own activities. In addition, they have varying degrees of control over a multitude of quasi-governmental agencies and institutions and significantly affect the private sector of the economy through regulatory legislation, grants and subsidies, and contracts with a vast number of privately owned corporations, institutions, and other organizations. As the magnitude of governmental intervention in American life has increased, more attention or concern has been directed toward the effectiveness and efficiency of governmental programs, and hence more use has been made or advocated of industrial engineering as a means of wisely allocating public resources of all kinds.

NEW HORIZONS FOR INDUSTRIAL ENGINEERS*

As a part of the general trend advocating improvements in the service industries, large numbers of industrial engineers have been demanded and attracted to careers in exciting, challenging, and rewarding new fields of professional practice. Fields such as health, education, urbanization, and environmental control invite creative talents, call for high degrees of professional and technical competence, offer competitive salaries, and provide social relevance.

Perhaps the most advanced of these new career fields is in the health services, one of the top five industries as measured by assets, annual expenditures, or number of employees. Industrial engineers practice their profession in hundreds of individual hospitals throughout the United States and abroad; in multihospital systems programs sponsored by various state hospital associations; as commissioned officers in the U.S. Public Health Service; for health-related programs of the Veterans Administration, the armed forces, and other government agencies; in professional consulting firms with hospital clients; and as faculty members in various university programs of health systems research and education. Applications of industrial engineering in the health services are described more fully later in this chapter.

Another service industry in which industrial engineering has gained acceptance and wide use is government, a career field also described later in this chapter. In considering industrial engineering activity within government, one should keep in mind the broad scope of modern governmental influences, as pointed out previously, as well as the magnitude and diversity of major governmental operations such as those of the military; NASA; the Post Office; the Bureau of the Budget; the Department of Health, Education, and Welfare; and the vast array of office and clerical tasks.

In addition to well-established governmental programs involving industrial engineering, there are numerous government-related problem areas attracting the atten-

* Written by Harold E. Smalley.

tion of some industrial engineers. These include public educational institutions, libraries, urban problems, housing, traffic congestion, welfare systems, community planning, conservation of natural resources, taxation, voting and representation, court procedures, police methods, penal systems, and even strategies for political campaigns.[2]

Normally regarded as a part of the private sector of the economy but with significant public control are three other service industries actively utilizing industrial engineering resources. These are banking, public utilities, and transportation. And even though product oriented, agriculture and retailing are included because of the rapidly expanding role of industrial engineers in these nonmanufacturing industries. Each of these five career fields is described later in the chapter.

As industrial engineers continue to be attracted to new areas of application, many other interesting problems will be recognized and attacked. Opportunities include insurance, communications, architecture and building construction, hotels and food service, homemaking,[3] recreation, community and civic affairs, cultural pursuits, and perhaps the most challenging of all, organized religion.[4]

The remainder of this chapter is devoted to descriptions of industrial engineering applications in seven nonmanufacturing (or service) industries: health services, government, banking, public utilities, transportation, agriculture, and retailing.

INDUSTRIAL ENGINEERING IN THE HEALTH SERVICES*

Significant changes have taken place within the health industry, and corresponding differences are being witnessed within that segment of the industrial engineering profession devoting its energies and talents to health systems problems. In 1963, there were fewer than 100 industrial engineers seriously devoting their time to the health industry, and the general employment situation for industrial engineers in that industry continued to be characterized by reluctance and skepticism on the part of health leaders.

Approximately six years later, there were widespread use and acceptance of industrial engineers within the health industry, not only in hospitals but also in a broad spectrum of other health institutions and services. In fact, the employment situation became such that the demand for competent engineers exceeded the available supply by many times.

Increased Understanding and Wider Acceptance. Perhaps the most significant development contributing to wider acceptance and use of industrial engineers within the health field has been an increased understanding by health leaders of the industrial engineering function. This has come slowly and has been accomplished only after years of devoted effort by societies and individuals committed to this cause.

Society Activities. Educational activities by societies representing industrial engineers in the health field have contributed substantially to increased acceptance. Such activities have been undertaken primarily by the Hospital Management Systems Society[5] (now an affiliated society of the American Hospital Association) and by the Hospital and Health Services Division of the American Institute of Industrial Engi-

[2] For example, see William C. Smith, "Politics and the Industrial Engineer," *Technical Papers of the 20th Annual Conference,* American Institute of Industrial Engineers, New York, May, 1969, pp. 109–117.

[3] For example, see Lillian M. Gilbreth, Orpha M. Thomas, and Eleanor Clymer, *Management in the Home,* Dodd, Mead, & Company, Inc., New York, 1955.

[4] For example, see James P. Topp, "Industrial Engineering and the Institutional Church," *Technical Papers of the 20th Annual Conference,* American Institute of Industrial Engineers, New York, May, 1969, pp. 119–127.

* Written by John R. Freeman.

[5] For more information, contact the Secretary, Hospital Management Systems Society, American Hospital Association, 840 North Lake Shore Drive, Chicago, Ill. 60611.

neers.[6] Conferences, workshops, and institutes sponsored by these two societies have done much to bring the benefits of industrial engineering to the attention of the health field.

Education of Hospital Administration Students. Increased acceptance of industrial engineering has also been enhanced by the introduction of industrial engineering course material into the curricula of various graduate programs in hospital administration. In this manner, the graduates gain an understanding and appreciation of the industrial engineering function and are equipped to be discerning consumers of industrial engineering studies and results. It is axiomatic that increased understanding removes unwarranted fears and can lead only to greater interest and participation by future hospital administrators in the activities of the industrial engineer. Without this interest and participation, even the best of industrial engineering studies may never be implemented.

Information Explosion. A relative explosion of information describing industrial engineering results within the health field has obviously had a profound influence upon increased use and acceptance. Prominent among this literature are several books referenced in the bibliography on page 13-99, virtually hundreds of articles published in the trade periodicals of the health industry, and the quarterly publication of *Abstracts of Hospital Management Studies* by the Cooperative Information Center for Hospital Management Studies, University of Michigan.

The latter publication deserves special comment, not only for its usefulness as a source document for literature research, but also because it represents an attribute which is rather unique to the health industry. This publication annually includes abstracts of well over 1,000 published and unpublished studies, with a sizable percentage of these by industrial engineers; and arrangements are made whereby copies of studies of interest may be purchased in either hard-copy or microfilm form.[7] In what other industry is information so readily exchanged among constituent organizations? The need to withhold or classify certain information is nonexistent, and the result has had an obvious and highly desirable impact upon industrial engineering work in the health field.

Among these abstracts, one can find a set of time standards for virtually any method of performing almost every type of laboratory test. This set of standard data is readily obtained from its developer. It does not require much of an imagination to conclude that this type of information exchange can easily lead to considerable savings of what would otherwise be duplicated industrial engineering effort, especially when one considers that such standards might be redeveloped in thousands of hospital laboratories.

Widespread Use. As a result of efforts such as these, the use of industrial engineers within the health industry has significantly increased. The number of industrial engineers seriously involved in health systems work in 1969 was probably in the neighborhood of 500. And as was mentioned previously, the current demand is going largely unsatisfied.

Multihospital Industrial Engineering Programs. A development in the health industry has been the joining together of two or more hospitals to share the services of industrial engineers and allied professionals. In at least one respect, this means that the hospital industrial engineering movement has come full circle—the first industrial engineer to accept a full-time hospital position was shared by two hospitals in Cleveland.[8]

Most of the multihospital industrial engineering programs are under the auspices of state or local hospital associations or councils. In some cases, the association has assisted in the creation of a separate nonprofit organization to provide industrial engineering services on a shared basis. Alternatively, multihospital programs can

---

[6] For more information, contact the Executive Director, American Institute of Industrial Engineers, 345 East 47th Street, New York, N.Y. 10017.

[7] Available from University Microfilms, 300 North Zeeb Road, Ann Arbor, Mich. 48106.

[8] Smalley and Freeman, *op. cit.*, p. 70.

be established through professional consulting firms, university-based systems groups, and other arrangements.

A 1968 survey by the Hospital Management Systems Society showed that there were active, recognized multihospital programs on a regional or state-wide level in sixteen states. The same survey indicated that, in twenty-seven other states, various health care organizations were seriously considering the establishment of similar programs. Thus, at that time, only seven states were not known to be considering a multihospital industrial engineering program of some type.

Most of the shared industrial engineering programs emphasize one or more of three major activities: (1) customized engineering analyses and studies, (2) development and implementation of group methodologies, and (3) training and education of hospital employees.

This approach generally provides concentrated amounts of staff time to individual hospitals for short periods of time for the conduct of specific projects tailored to the particular needs of that institution. Services are usually rendered through the creation of one or more continuous engineers-in-residence with the individual hospital on a contract basis, or with a daily consulting rate. In some cases, the project may deal with a problem of immediate interest to more than one of the cooperating hospitals; and in almost all cases, the project provides certain insights and experiences of value in serving other member institutions later.

The shared, centralized service develops study materials, conducts courses in basic industrial engineering techniques for hospital management personnel, and provides follow-up assistance to participants as they attempt to institute improvements in their respective institutions. Some programs have emphasized the training of management analysts who are placed in the participating hospitals to carry out studies under the supervision of professional engineers from the shared service.

Multihospital industrial engineering programs offer to the industrial engineers in their employ several advantages in terms of career development which are sometimes unavailable to the engineer working alone. The centralized, shared organization provides professional training and supervision to new employees and serves as a vehicle for ready exchange of information and experiences among engineers. Further, in a growing engineering organization, the opportunities for advancement in position and salary seem clearer. Also, the shared arrangement offers an advantage to the institutions it serves, providing a broader base of talents which are available as unique problems arise. It is generally agreed, however, that the most successful utilization of a shared program is made when the participating hospital has its own staff industrial engineer to relate to the engineers at the centralized service. Furthermore, the most suitable approach would seem to be one in which the shared service offers all three of the activities described above.

Broadened Vistas for Industrial Engineers. Most of the industrial engineers employed in the health field are associated with an individual hospital or with a central industrial engineering service established to serve a group of hospitals. However, recent years have seen increasing numbers of industrial engineers employed by other health institutions and services or by commercial firms serving primarily the health industry. Table 7-1 shows a number of jobs held by industrial engineers in the health industry. The list is by no means complete; nor does it emphasize the various jobs in proportion to the numbers of persons engaged in that type of activity. Rather, the jobs were selected for their diversity to illustrate the broad range of opportunities for industrial engineers in the health industry.

It should be evident from Table 7-1 that the opportunities for industrial engineers in the health field are expanding rapidly, and that sufficient variety is available to satisfy almost any taste.

Conclusion. A number of developments in the health industry have resulted in a much wider acceptance and use of industrial engineers in hospitals, with a movement toward cooperative programs in which groups of hospitals share industrial engineering services. At the same time, other organizations, associations, services, agencies, and academic institutions within the health industry have become aware of the benefits of industrial engineering, thus expanding available opportunities.

TABLE 7-1.   Representative Positions Held by Industrial Engineers in the Health Industry

Systems engineer for a large community hospital
President of a management consulting firm specializing in services to health institutions
Staff member of the management engineering division of the American Hospital Association
Director of data processing for a group of hospitals constituting a large medical center
Marketing consultant for a firm supplying equipment to health institutions
Systems specialist for a state department of mental hygiene
Director of management engineering for a large Veterans Administration hospital
Industrial engineer employed by a large manufacturing firm, but on a one-year loan to a
   community hospital
Systems consultant for a large architectural firm specializing in health facilities
Principal for a major management consulting firm with some hospital clients
Administrator of a hospital
Director of planning for a hospital council of a metropolitan area
Management analyst for one of the Canadian provincial hospital insurance services
Director of a laundry facility which is shared by a number of hospitals
Chief industrial engineer for a division of the Department of Health, Education, and
   Welfare
Graduate student in industrial engineering with master's thesis on a health systems
   problem
Director of management engineering for a state hospital association
Systems engineer for a large medical center hospital
Public Health Service commissioned officer assigned to systems work with the Indian
   Health Service
Regional representative with the accounting service of the American Hospital Association
Project engineer for a firm providing industrial engineering services for a group of hospitals
Director of health planning for a major state university
Senior systems engineer for a new school of dentistry
Management analyst in the budget analysis department of a populous county
Professor of industrial engineering serving as principal investigator on a health systems
   research project
Director of organization providing supervisory training for hospitals in a large metropolitan
   area
Systems analyst for a state Blue Cross Affiliate
Project engineer for a group involved in planning major expansions in health education
   facilities
Director of organization implementing group staffing methodologies in various hospitals
Staff member in the Operations Research and Systems Analysis Branch of the Division
   of Regional Medical Programs, Public Health Service
Undergraduate student in industrial engineering collecting data in a physician's office
   for government-sponsored research project

Bibliography

Bennett, Addison C., *Methods Improvement in Hospitals,* J. P. Lippincott Company,
   Philadelphia, 1964.
"Cooperative Multihospital Management Engineering Programs," Hospital Manage-
   ment Systems Society, Chicago, 1968.
Flagle, Charles D., and John P. Young, "Application of Operations Research and
   Industrial Engineering to Problems of Health Services, Hospitals, and Public
   Health," *The Journal of Industrial Engineering,* vol. 17, no. 11, November, 1966,
   pp. 609–614.
Frederick, Earl J., "Adapting the Industrial Engineer's Many-sided Skills to Hospital
   Management," *Hospitals,* vol. 39, no. 20, Oct. 16, 1965, pp. 73–75.
Smalley, Harold E., "Hospital Industrial Engineering," *The Journal of Industrial
   Engineering,* vol. 17, no. 10, October, 1966, pp. 511–518.
Smalley, Harold E., "Professional Methods Improvement Is Ultimate Path to More
   Efficient Hospitals," *The Modern Hospital,* vol. 105, no. 3, September, 1965, pp.
   107–110.
Smalley, Harold E., and John R. Freeman, *Hospital Industrial Engineering,* Reinhold
   Publishing Corporation, New York, 1966.

INDUSTRIAL ENGINEERING IN GOVERNMENT*

In any activity or occupation, industrial engineering can be usefully employed to help select the most appropriate inputs or raw material, design the outputs, find a preferred way of doing the work, and assist in effectively managing or controlling the activity. The approach of industrial engineering fits equally well when applied to heavy or light factory, office, production, maintenance, staff, or supervisory work; farm work; housework; surgery; cafeteria work; department store or hotel work; the whole range of government activities; battle activities; or any other human activity.

What is accomplished may vary from organization to organization. The nature of the raw material will vary widely: in one case, it may be information; in another, it may be some simple or complex substantive material. The outputs may be services, responses to another group's actions, or any one of the almost infinite variety of products found in a modern society. The varieties of process are almost limitless, and the varieties of things worked with and places worked at are enormous. However, the requisite human efforts will in all cases be composed of the same basic acts, and the information relating to the effective use of human effort will be universally applicable.

The problem of determining a feasible and preferable output to achieve the objective and a feasible and preferable method of accomplishing the work will always be present. The problem of determining the required amount of human work time will be a normal concomitant, and the problems of constraining the organization are inevitable. Regardless of the variation in accomplishment or field of knowledge, some procedure must be selected from among the broad variety of available industrial engineering procedures to assist in applying available knowledge to the solution of the numerous managerial design and measurement problems.

Government's Contribution to Industrial Engineering. The range of activity encompassed by a modern government is probably as extensive as the range of activity within society; so the range of industrial engineering techniques employed in government is as extensive as that found in organizations external to government. Further, the U.S. government has long been in the forefront of much industrial engineering work. Early studies of metal cutting were government sponsored at Watervliet Arsenal about 1910; early motion studies were government sponsored—Gilbreth's work with the World War I handicapped; mathematical modeling was brought to a useful state by government activity—World War II operations research; and network diagrams were developed in connection with the Navy Polaris missile. Data processing activities of government have spurred data processing equipment designers and systems technologists.

Personnel activities, plant or office location problems, and distribution problems of government are so complex, because of the integrated nature of government, that their solution requires not only the application of normal industrial engineering techniques to an extremely broad range of work, but also the application of these techniques to problems of such enormous size as to require the development of much new industrial engineering technology. Government is not only an area of activity in which much use is made of industrial engineering, but an area where much industrial engineering has been developed.

Service Activities in Government. Government activity can be divided into two broad sectors: one producing substantive outputs and one producing services. The sector producing substantive outputs—that is, arsenals, shipyards, and so forth—uses industrial engineering technology, as does its external counterpart in private business, with one important differentiating characteristic: the criterion of success on a government industrial engineering project is almost always something other than "money profit." However, the techniques described elsewhere in this Handbook with respect to manufacturing may be applied readily to government activity and have been.

---

* Adapted by Marvin E. Mundel from his book, *Motion and Time Study: Principles and Practice*, 4th ed., Prentice-Hall, Inc., Englewood Cliffs, N.J., 1970, chap. 1.

Service activities of government present a distinctly different problem and hence are the main concern here.

Differences with Respect to Manufacturing. Almost any service-type output has an important difference as compared with a substantive output. A substantive output can be represented by an engineering drawing; a service-type output cannot. Because of this, substantive outputs are easily described in quantitative terms related to both objectives and internal production problems; service-type outputs do not lend themselves readily to quantification. With service-type outputs, totally different dimensions or characteristics of the outputs relate to objectives as compared with the method of quantification which relates to hardware-type outputs.

For instance, an organization making home appliances has no difficulty in recognizing that there are different profits, materials, work loads, and so forth, connected with the production of each of their different appliances such as washing machines, dryers, and dishwashers. In the process of quantifying either profit or work load, each type of appliance must be counted separately. Further, each type of appliance, such as a washing machine, can be represented by a unique engineering drawing of the total assembled unit. The washing machine assembly drawing may be supported by a set of part drawings, one for each part. Each part drawing and the information concerning how many of the part are needed for each assembly may be supported by a material requirement statement and by a process chart or other presentation of the production sequence for the part.

It is important to note the relatively fixed relationship (for a given method and process) between the number of units of final product (washing machines) and the tasks on the parts. Of course, if a large number of different models of different kinds of appliances are made, relevant calculations may involve much detail. Make-or-buy decisions must also be made, value analysis should be applied, and so forth; but difficulties will not be encountered with respect to what to count; a convenient procedure is obvious and direct.

With the many service activities of government, however, there is no substantive (hardware-type) output. Consequently, there is no engineering drawing to assist in determining what to count, although the manpower resources are usually a larger portion of the required resources than with substantive outputs and hence are an even more important problem for managerial control. Further, there is seldom as direct a relationship between what appears to be the final outputs and the manpower or tasks required. Some additional analytical aids are needed to assist in deciding what and how to count and how to summarize.

Quantifying Service Outputs. Central to the problem of applying industrial engineering to the production of services is the problem of quantifying outputs. What seems to be needed is a concept involving a hierarchy of work units. A work unit is any amount of work or the results of such work which are convenient to use as an integer when quantifying work.[9] Such a concept of work units should have three useful properties or criteria as follows:

1. They should provide a clear hierarchy of countable, convertible units of quantification from the objective to the work load; the resources required for the larger or higher orders of work units should be divided into that needed for the smaller or lower orders of work units; the resources required for the smaller ones should readily be aggregated into the resources needed for the larger work units.

2. They should permit a meaningful forecast of the work load to be made in terms related to the required resources, or convertible to such terms.

3. At one or more levels, a firm relationship of the work unit to required resources should be established; at these levels, a work count—the number of times a work unit is to be performed or produced—should be a meaningful number with respect to required resources.

---

[9] See M. E. Mundel, *A Conceptual Framework for the Management Sciences,* McGraw-Hill Book Company, New York, 1967, pp. 166–167.

Applying Industrial Engineering. If a work unit hierarchy is delineated for a government activity, the whole range of industrial engineering technology can be usefully applied. Of course, as in any industrial engineering problem, cognizance must be taken of (1) the nature of the outputs and appropriate criteria of success, (2) the underlying production–subject-matter technology, and (3) the human problems internal to the organization.

Bibliography

Dunn, D. W., "From Idea to Reality: The Apollo Program Documentation System," *Proceedings of the 16th Annual Conference,* American Institute of Industrial Engineers, New York, May, 1965.

Hannon, E., "Administering Federal Property and Records," *Proceedings of the 17th Annual Conference,* American Institute of Industrial Engineers, New York, May, 1966.

"Measuring Productivity of Federal Government Organizations," Executive Office of the President, Bureau of the Budget, Government Printing Office, Washington, D.C., 1964.

Moundalexis, J., and W. Lichtenberg, "Input/Output Analysis of Organizations Having Intangible Outputs: AO 10 Technique," *Proceedings of the 17th Annual Conference,* American Institute of Industrial Engineers, New York, May, 1966.

Mundel, M. E., "Objective Oriented Applications of Work Measurement in Government," *Proceedings of the 20th Annual Conference,* American Institute of Industrial Engineers, New York, May, 1969.

"Work Load Analysis for the Office of the Solicitor," U.S. Department of the Interior, Washington, D.C., 1967.

INDUSTRIAL ENGINEERING IN BANKING*

Because banking is an industry, it is natural that industrial engineering should have become an important resource for bank management. After all, banking is large-scale business, dealing with the production and delivery of vital services throughout the economy.

The computer, as a tool of industrial engineering, is making the major impact on banking. It provides the capability and incentive to do things not possible in the past. Production problems reminiscent of job shop manufacturing have existed in the past, but with the passage of time, these problems have been mutated and have become more complex, particularly in the larger installations, because of the simultaneous and multiprogramming capabilities of present computers. Initially, the computer was used as a device to accelerate clerical operations. Little rethinking of applications to the computer was undertaken as banks struggled to cope with an alarming growth in check volumes during the late 1950s and early 1960s. However, a trend is under way for a completely new look at the older systems and a larger role for the industrial engineer.

Training Needs. It seems obvious that the normal sources of technically trained bank personnel necessary to carry on the increasingly specialized trend in bank operations are not sufficient to satisfy the demand. Also, the supply of industrial engineers is not likely to satisfy the growing demand for such professionals in banking. For these reasons, it is almost certain that banks will have to place greater emphasis on training their own people in an attempt to satisfy their needs. Although banks have long been in the business of training clerical personnel to the special conditions of banking, the necessity to create banking-oriented technicians and, at the same time, technically oriented bankers is bound to be a challenging and difficult task.

The industrial engineer with an interest in training should be able to make a significant contribution either in his traditional staff role or as a full-fledged member of the training staff. Because of the need to prepare personnel for change, a major adjunct to the skills of the industrial engineer should be training capabilities. In any

* Written by Lowrie W. McIntosh.

event, the pressure for broad, in-depth training is only beginning to develop in many banks. This type of training will become more vital as the technician/banker combination becomes a necessity for operating a bank. Increased hiring in the industry should mean increased opportunity for the technically trained.

Operations Research. For those industrial engineers who plan to go into operations research, the opportunities are very promising. It is clear that although other industry groups were moving ahead in the science of management, the motivation to make changes in banking practices and operations was weak and changes were late in arriving. It is expected that although the pressure for change has only just begun, the rate of acceleration for change will be dramatic.

One of the primary reasons for the slow growth of operations research in banking is that banks have rarely employed engineers in a technical capacity on their staffs. Where manufacturing firms were able to relate previous engineering work directly to top management operations, banks were slow to accept the newly developing approaches. As a result, the vast body of mathematical and statistical methodologies used in many industries to provide improved answers to management problems is largely unknown to most bank managers.

Cost Systems. Cost systems are rapidly evolving in the larger banks—"evolving" because, as of now, none of the systems being installed has been developed sufficiently for full operation. Because cost input historically has had very little part in management decisions, the use of the data developed from these new cost systems requires a period of adaptation and adjustment. Computers make possible economic handling of the high volume of data required to support such a cost system. Also, standard cost approaches involve a project effort of some magnitude, including:

1. Developing work standards suitable for cost input
2. A reporting system designed to show key variances
3. A forecasting and budgeting system to feed back improvements and monitor the needs of management
4. A maintenance system designed to signal problem areas during the routine functioning of the system

A serious need exists for engineers with a conceptional awareness of interfacing among the various bank management functions—engineers with the ability to bring together the related functions of the bank, either in an actual or simulated mode, and to present to management an objective appraisal of the individual contribution of each unit. A well-designed cost system provides the basis for such an evaluation. Coupled with a profit planning and forecasting system, this becomes the backbone of a management information system. Responsibility-performance evaluation, service or product profitability, and account profitability require a soundly engineered cost system.

It is imperative, therefore, that the banking industrial engineer have an intimate knowledge of cost systems; a mechanical knowledge of costs is not adequate. He must be able to interpret management's needs before they become needs, and provide for those needs objectively.

Information Systems. There are pressing needs for objectively generated information in banks. Because both the art and science of good management are required for sound decision making, however, a management information system, even with all the proper checks and balances, will not function properly unless the participants in the system fully understand the logic of its design and what it is intended to accomplish. This, in part, requires that users of the system have some understanding of how data are generated within the system. Orientation and training, therefore, must be given important consideration in the implementation schedule. This training will accelerate management's use of the system, but more importantly, will assure a more consistent response by members of management to the signals generated by the system.

Reports produced by the information system require skilled and thoughtful design. The format and presentation of data should be simple and should motivate the reader to action. If written interpretation is required beyond graphic and columnar portrayal, time should be available in the report production schedule to allow for this.

The following is a checklist which should help in testing the design logic of a reporting system.

1. The design should protect employees, supervisors, and managers from the incentive to manipulate results for individual gain (cheating).
2. The system should encourage unlimited opportunity for increasing productivity.
3. The reports in the system should lend themselves to positive and constructive, not punitive, administration.
4. The system should show indices of performance in units that truly reflect the job being done, such as actual dollars and hours, budget dollars and hours, or variances in corresponding units.
5. The reports should be designed to present results objectively rather than subjectively.
6. The presentation of results should be action oriented, not primarily for information purposes.
7. Reports to an individual manager should be primarily oriented to items for which he has responsibility and upon which he can take action.
8. General rules of management intent should be in writing to explain the purpose and use of each item in the control report.
9. Continuity of reported data should exist vertically throughout each organizational unit and should increase in scope with responsibility.
10. The system should provide a valid basis for comparing the performance of the various organizational groups although their operations may be completely dissimilar.

Pricing Bank Services.   It is becoming increasingly important for the industrial engineer to participate in the process which results in decisions on the pricing of bank services. This calls for a systematic, analytical, and objective approach. Better methods of assisting with the marketing of bank services in a competitive environment should be a major goal of the industrial engineering function.

There are at least three incentives that are motivating banks to develop systems which aim to provide better marketing criteria:

1. The intensified *competition* to which banking is subject
2. The increasing *complexity* of performing banking services
3. The rapid pace of *change* in the economy and marketplace

All these factors—competition, complexity, and change—have a direct bearing on the problems of this industry and form the groundwork for which industrial engineering technology can be applied.

A number of tools are used to determine product or service costs; of these, profit planning, cost accounting, and work measurement are basic. In some cases, even improved data on profit or loss, historically applied, are not sufficient to motivate changes in the price structure. Specific examples must be developed and amplified,

TABLE 7-2.   Stock Transfer—Corporate Services

Company A		Company B	
Fee (based on number of shareholders and new accounts)	$4,250	*Fee* (based on number of shareholders and new accounts)	$4,050
High activity		Low activity	
10,000 shareholders		10,000 shareholders	
1,500 new accounts annually		500 new accounts annually	
Cost		*Cost*	
8,030 bookkeeping entries at 70¢ each =	$5,621	3,876 bookkeeping entries at at 70¢ each =	$2,713
Loss on account	($1,371)	Profit on account	$1,337

such as the stock transfer case, Table 7-2, which shows that fee schedules for services are often keyed to the wrong activity.

Once improved cost data are available and the pricing schedules analyzed for fee-generating elements, the next consideration should be estimating the value of the service to the user. Although the cost should have some input to pricing decisions, price is not just a summation of production costs, overhead, and markup for profit. Effective pricing requires (1) a clear determination of specific pricing objectives; (2) a sound, coordinated approach; and (3) continued application of refined statistical information and techniques to the marketing situation involved. Following this, it is possible to select the proper pricing objective to use in determining the best price for products or services.

Before prices can be established and used, there is the necessity of defining a standard service. Often this can prevent developing a problem such as that illustrated by this lock box example:

Lock Box Service
(Assume the standard price to all customers for photocopy is 6¢.)

Company A—actual cost	*Cost per photostat*
1. Open envelope and remove check	$.010
2. Add checks	.010
3. Photostat checks	.020
4. Type credit ticket	.005
5. Bundle photostats and send to company; also included is postage expense	.005
Total expense per photostat	$.050

Company A's processing was standard. Now look at Company B, which asked for and received extra processing services:

Company B—actual cost	*Cost per photostat*
1. Open envelope and remove check	$.010
2. Sort checks into four divisional groups	.010
3. Check list of delinquent customers and notify company if check was received	.050
4. Add checks by divisional group	.020
5. Photostat checks, leaving a 2-inch stub	.030
6. Type credit ticket for each divisional group	.020
7. Bundle photostats and send to company; also included is postage expense	.010
Total expense per photostat	$.150

Result: If Company B's business is obtained on a fee basis of $.06, the bank will incur a loss per item of $.09. If costs had been considered in making the bid, no loss would have been incurred and nonstandard high cost business would have been discouraged.

Bibliography

Cohen, Kalman J., and Frederick S. Hammer, "Operations Research: A New Approach to Bank Decision Making," *Bankers Magazine,* Summer, 1965.

Dooley, Richard E., and Robert E. Moll, "Information System: Planning for Success," *Bankers Monthly,* June 15, 1967.

Hilburn, Earl D., "Management Must Lead Information Revolution," *American Banker,* Jan. 31, 1967.

McIntosh, Lowrie W., "Financial Control Systems," *Bankers Monthly,* June 15, 1968.

Perrett, Joseph A., "EDP Customer Services: Producing at a Profit," *Bankers Monthly,* July 15, 1967.
Randall, K. A., "Comparing Bank Efficiency for Services Rendered," *Commercial and Financial Chronicle,* Jan. 19, 1967.

INDUSTRIAL ENGINEERING IN PUBLIC UTILITIES*

"Public utility" is a term commonly applied to a company that furnishes a community with an essential commodity such as water, gas, electricity, or telephone service. Public utility companies, unlike other corporations which engage in manufacturing, mercantile, or financial matters, operate with government approval as monopolies and supply a service designated by our courts and laws as "business affected with a public interest."

For the most part, public utilities are unique because they are (1) considered absolute monopolies, (2) free from normal business competition, (3) obligated to provide service on demand, (4) required to charge reasonable rates, and (5) allowed to earn, but not guaranteed, a reasonable profit. Because public utilities do not operate in a normal competitive marketplace which influences other businesses in setting terms of trade, government regulation has been adopted as the market substitute. These regulations are designed to achieve the objectives of competition which are (1) quality service, (2) reasonable prices or rates, and (3) reasonable profits. Optimizing these objectives in a public utility is the challenge which offers significant opportunities for the industrial engineer.

Public Utility Functions. The major functions of a public utility—water, gas, electricity, or telephone service—can be described as follows:

1. Customer sales and service
2. Planning and construction of facilities
3. Acquisition or generation of energy or resource
4. Transmission and distribution of energy or resource

In public utilities, as in any other industry, the business activity normally follows an orderly sequence of events. It usually begins with an engineering design which is needed to meet the public demand for service. Providing service to an individual customer or even a manufacturing complex does not constitute the entire scope of public demand. Before this can be accomplished, a public utility must be prepared by having the capacity in reserve long before a customer asks for service. This preparation begins with extensive research in population growth and industrial development for the geographic area which will be served. Forecasting these capacity requirements in terms of engineering design and alternative costs is a fundamental task of the management decision process in public utilities.

Public utilities are characterized by requiring a proportionately heavier investment in fixed assets than any other form of business enterprise. These fixed assets comprise the facilities which generate energy or store resources and distribute them to the customer for use at his residence or place of business. The engineering and design of these facilities require the specialized skills of many engineering disciplines, and the planning and construction of the facilities involve the coordination of men, materials, and equipment—a function which summons the industrial engineer to the forefront with his specialized knowledge and skill in work measurement, production planning, and systems engineering.

Furthermore, operation of these facilities is accompanied by substantial fixed costs which are a function of plant size. The cost per unit of service to the customer decreases only as production approaches ultimate capacity for a given plant size and the scope of customer sales goes beyond the normal service connection. Customer service reliability, economic analysis studies, and industrial development studies offer opportunities for using the techniques of operations research and engineering economy.

Although public utilities have been somewhat slow in establishing industrial engineering departments, many industrial engineers can be found throughout the utility

---

* Written by J. N. Salapatas.

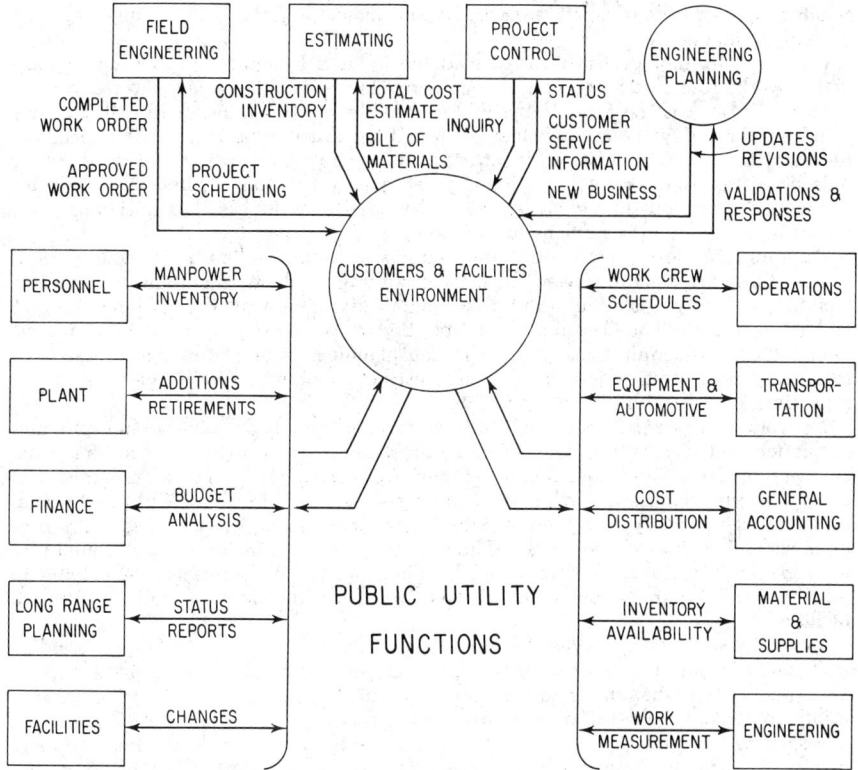

FIG. 7-1.   Integrated systems of men, material, and equipment.

industry, combining their specialized talents in a cooperative effort with those of other employees in each of the public utility functions.   (See Figure 7-1.)

Industrial Engineering Activity.   The activities of industrial engineering in public utilities center primarily around the construction and management of facilities as they relate to the cost of providing service to the customer.   In the discussion that follows, the focus will be on those areas which rely heavily on industrial engineering principles, enhanced through the use of computers, for improving public utility operations.

Management Information.   Investment analysis and economic studies using operations research have served in developing data to aid management in decision making. Industrial engineers in the gas industry have developed budgetary performance controls through improved management reporting systems.   Some of these studies feature alternative expansion proposals for optimizing gas transmission operations. Once the engineering design has been determined, it must be reduced to writing and transmitted to all other departments involved in doing the work.   Beginning with the cost estimate which management needs to authorize the work, the information must be translated into manpower requirements, material and equipment needs, and finally, detailed job instructions for construction.

Systems Engineering.   The first step toward effective management and control of facilities construction is a system for handling the technical information from the conception of engineering design to the final act of providing service to the customer. Because of their size and the obligation to provide service wherever needed, public utilities carry on construction and maintenance over a wide geographic area.   This results in a multitude of small work orders and major projects, each having a specific beginning and ending.

Computer technology has greatly aided the industrial engineer in developing integrated work order systems to handle the enormous work load. A significant advancement in this area is the MECA[10] system used in the electric utility field. It provides engineers with a symbolic language for describing distribution engineering specifications and uses the symbolism as direct input for computer processing of cost estimates, bills of material, work load projections, and job instructions. Similar work order systems using computers are also employed by other electric, gas, and telephone companies to aid in performing construction work.

Planning. Major projects constitute a large volume of public utilities construction. Among these are electric power plants, water filtration plants, main-line transmission facilities, compressor stations, central communications exchanges, switching stations, and many more. The design and construction of a major project is an enormous undertaking and requires many years in the planning stage. Moreover, because of their size, most utilities are engaged in construction of many major projects at the same time, usually with different priorities.

Network analyses by means of the critical path method (CPM) and program evaluation and review technique (PERT) are being used to plan and coordinate the engineering design and construction of major projects. Allocation of resources has become a very important part of this planning activity. The use of CPM and PERT in the utility industry has evolved into multiproject networks combined with programmed allocation of resources. Through the use of third-generation computers, networks in excess of 20,000 activities, handling up to 1,000 projects, are helping to schedule and control millions of dollars worth of transmission and distribution facilities.

Work Measurement. Perhaps the most significant contribution made by the industrial engineer in the utility industry has been in the field of work measurement. Everyone knows what the industrial engineer can do in a factory. Considering that utility construction consists of many operations that add up to millions of dollars, it is easy to find those which are repeated thousands of times, only in different arrangements. It is in this area, in the so-called "factory," that industrial engineering continues to produce effective results in work measurement and methods engineering.

Techniques ranging from stopwatch time studies to time-lapse photography to linear programming are used by these industrial engineers to measure work and establish operating standards. Methods studies of construction work have resulted in developing optimum crew sizes to perform on-site work. Off-site, factory-type production lines have also been designed to handle repetitive, high-volume subassemblies.

Inventory Control. The control of inventory has commonly been associated with the production and sale of finished goods. Traditional concepts in formulating inventory policies have held to the idea that production and sales must be balanced to minimize inventory. Increasing production over sales results in overstocking and increased expenses. Increasing sales over production results in stock-out and loss of business. Whatever the situation, a tangible cost value can be applied to each to develop an optimal inventory control policy for the business enterprise.

In public utilities, the primary inventory consists of component parts and subassemblies to be used for the construction and maintenance of facilities. A public utility has the same opportunity to balance its construction work against customer sales to minimize inventory. However, this is not the major consideration in developing inventory policy. Because of its geographic size and obligation to provide reliable service to the customer, public utilities are unable to fix a tangible value on the cost of stock-out. This presents a formidable problem in developing models to optimize inventory calculations. Significant opportunities for industrial engineering exist in this

[10] Mechanized Engineering Construction Assemblies (MECA), copyrighted 1966 by Florida Power and Light Company, Miami, Fla. Rights granted to utilities in the United States, Canada, England, Scotland, Japan, the Philippines, and Australia.

area for the use of simulation techniques and the cardinal utility theory to quantify the inventory process for public utilities by these modeling techniques.

Conclusion. Several important conclusions can be drawn concerning industrial engineering in public utilities. The applications of systems engineering, work measurement, and operations research are recognized as potent techniques in public utilities to help management obtain results, despite differences between the characteristics of public utilities and other forms of business enterprise. Establishing company objectives and striving to meet them are really the same in every industrial organization. Significant opportunities exist for the industrial engineer in public utilities once he can identify these differences and apply industrial engineering principles to the "business affected with a public interest."

Bibliography

Baxter, Samuel S., "Industrial Engineering in a Large Water Utility," *Proceedings of the 15th Annual Conference,* American Institute of Industrial Engineers, New York, May, 1964.

Garfield, P. J., and W. F. Lovejoy, *Public Utility Economics,* Prentice-Hall, Inc., Englewood Cliffs, N.J., August, 1964.

Henessy, Robert L., "Industrial Engineering at Southern California Gas Company," *The Journal of Industrial Engineering,* January–February, 1965.

Justin, Frederic D., "The Industrial Engineer in a Utility," *Proceedings of the 15th Annual Conference,* American Institute of Industrial Engineers, New York, May, 1964.

Lindahl, A. W., "Electronic Data Processing Speeds Distribution Line Construction," *Transmission and Distribution,* June, 1967.

Raney, L. E., K. A. Rist, and W. D. Wyatt, Jr., "Industrial Engineering in an Integrated Natural Gas Company," *The Journal of Industrial Engineering,* March–April, 1965.

Salapatas, J. N., "Reddy Information System," *Proceedings of the 19th Annual Conference,* American Institute of Industrial Engineers, New York, May, 1968.

Salapatas, J. N., and J. S. Elwell, "Field Coding System Cuts Engineering Time," *Electrical World,* October, 1967.

INDUSTRIAL ENGINEERING IN TRANSPORTATION*

Industrial engineering has experienced a rapid growth within the transportation industry. United Airlines was one of the pioneers when it organized a group in 1941. Most industrial engineering functions relating to the transportation industry have been formulated since about 1955.

The relative use of industrial engineering by companies in different modes of transportation has not been identified in any known survey, but industrial engineering does seem to exist in some companies in the following modes:

Airline	Pipeline
Forwarding	Railroad
Ocean shipping	Trucking

Other organizations known to utilize industrial engineering in transportation activities are the U.S. Post Office, Federal and state departments of transportation, and manufacturing companies in dealing with distribution problems.

Industrial Engineering Practices. The literature indicates that most of the known industrial engineering practices are used in the transportation industry.

Considerable effort has been directed toward providing management controls for labor expenditures in nonincentive situations. Labor control systems have been designed and installed in railroad switching operations and in railroad and airline equip-

---

* Written by John H. Hildenbiddle, Jr.

ment maintenance. The railroads that are subsidiaries of the United States Steel Corporation have developed standards as part of their fully integrated standard cost system.

Efforts are being directed toward improvement in the utilization of equipment in which transportation companies have tremendous investments. The operations of most transportation companies are complex and are made up of many interacting variables. As a result, the design of improvements in many cases requires the use of operations research techniques. Simulation of operations is being used to determine economic additions to railroad plants; and linear programming is being used to optimize the distribution of empty freight cars and to determine the optimal number of aircraft pilot domiciles and the optimal number of pilots assigned to each.

Publications indicate that applications of many other techniques are being made in transportation. These include methods engineering, plant layout, material handling, value analysis, systems design, and economic feasibility studies. Condensed examples in the railroad, airline, ocean shipping, and traffic management sectors follow.

Labor Standards for Railroad Switching Operations. Much of the impetus to industrial engineering within the transportation industry has resulted from management's recognized need for standards in planning, scheduling, and budgeting operations and as a basis for measuring performance. One application of work measurement is a study of switching operations of Penn Central.

After a train reaches a yard, the cars get many handlings. The individual handling of each car depends on many things, such as whether the car is destined for delivery to a local plant, is in a nonserviceable condition, is empty or loaded, or is destined for sorting and regrouping with other cars to a destination beyond the yard. However, each operation consists of work elements which are highly repetitive and susceptible to standard data application.

Switching operations were analyzed in considerable depth, and work elements were selected and defined with beginning and ending points. Through time study, thousands of elemental times were accumulated for analysis and development of standard data tables, such as Table 7-3. The resultant standards are used in various ways.

TABLE 7-3. Example of Standard Data, Switching Freight Cars

Work element	Normal minutes
1. Couple to car..................	0.20 per occurrence
2. Uncouple from car:	
Air in cars..................	0.35 per occurrence
No air in cars..............	0.20 per occurrence
3. Work on ground beside cars.....	0.25 per car
4. Throw switch................	0.20 per occurrence
5. Locomotive movement—yard:	

Feet		Minutes		
Over	To	Without cars	Shoving cars	Pulling cars
100	150	0.35	0.45	0.45
150	200	0.45	0.60	0.60
200	250	0.55	0.80	0.75
250	300	0.65	0.95	0.90
300	350	0.70	1.10	1.00
350	400	0.80	1.25	1.10

Work loads are scheduled for switching crews; crew performance and management performance are measured; and capability is provided for accurate cost determination. At the initial installation, the company saved 9.5 percent in crew payroll. The system is presently being expanded throughout the company.

Simulation of Railroad Operations. One of the first efforts to use simulation in the transportation industry was the work of the Battelle Memorial Institute, under contract with seven major railroads.[11] This project included the development of four computer-based simulation models. These models simulated terminal operations, train operation on main tracks, locomotive processing through an inspection and servicing area, and the distribution assignment and utilization of locomotives over an entire railroad network.

Other organizations are now using or developing models with the newer simulation-oriented programming languages such as SIMSCRIPT, TRANSIM, and GPSS. Examples are as follows:

Single-track capacity analyzer—Canadian National
Car flow through network diagram—St. Louis–San Francisco
Train operation, main tracks—Department of Transportation and Penn Central
Locomotive terminal—Penn Central
Car flow through railroad network in United States—Association of American
   Railroads

The industrial engineering function involves the evaluation of alternatives in the design of methods or systems of doing work. This evaluation has been extremely cumbersome in the area of complex railroad operations. It is apparent that computer-based simulation techniques are quite useful in dealing with the complexities of such operations.

Control of Aircraft Maintenance. The airlines have expended large amounts of industrial engineering effort on aircraft maintenance. One of the major reasons for this is the need for reducing downtime and thus improving the utilization factor of the increasingly expensive aircraft. As just one example, American Airlines owns 900 turbine engines at $250,000 each and has estimated the cost of future engines at about $750,000 each.

In recognition of the need to improve utilization, American Airlines embarked on the design and implementation of a computer-based system to plan, monitor, direct, and report all maintenance activities on aircraft and their components. This effort was undertaken by a task group of representatives from all divisions within the company, including not only industrial and systems engineers, but also personnel from purchasing, other branches of engineering, planning, finance, field maintenance, and many others. The objective was to produce a totally integrated system requiring participation from all affected groups.

The activity studied included the maintenance of three separate but related operational environments, namely, aircraft, engines, and components. Each environment had significant elements within the maintenance function, namely, long-range forecast, short-range forecast, scheduling, operating control, action, and administration. The system requirements were to serve these elements of activity.

The design provided subsystems to accomplish this. All subsystems use common files, some of which are:

Master data record—relates each work item to all necessary information
Personnel locator file—assigned shop, shift
Master material file—service limits for components and periodic maintenance
Work-in-process file
Master inventory file
History file

[11] Canadian National, Chesapeake and Ohio, Great Northern, New York Central, Norfolk and Western, St. Louis–San Francisco, and Southern.

In total, seventeen subsystems were designed to interact with the maintenance operations to perform the elements of activity using common files. The subsystems will be installed over a period of seven years, and at the conclusion of the implementation, American Airlines expects to have a system completely responsive to the needs of the maintenance planners and operators.

United Air Lines used industrial engineering to develop network diagrams for determining the critical paths during aircraft overhaul. These were analyzed on a computer to identify those activities which had the potential for delaying the completion of the overhaul.

United also used industrial engineering to perform long-range planning for turbine engine maintenance. This study determined the adequacy of the overhaul capability for a period of ten years. The findings resulted in an expenditure of $8 million for building additional capacity in machinery and 130,000 square feet of floor space.

Integrated Freight Payment System. Not all industrial engineering applications to transportation have been made by transportation companies. E. I. Du Pont de Nemours, Inc., annually uses in excess of $100 million worth of freight service. Although their traffic department is not directly responsible for the purchase of service, they do conduct distribution studies recommending changes that will lower freight costs and improve service.

One of their major projects was the design of a computer system to provide freight statistics which could be used for control and also for physical distribution studies. The individual transactions in the system comprised 9,000 carrier billing locations to over 1,000 shipping points and the annual handling of more than 1.2 million freight bills.

The key to this system was the design of a freight detail slip which contains the essential shipper, consignee, and commodity information and the freight charges. The latter are calculated by the person filling out the slip from available rate/route sheets and the bill of lading at the time of shipping. The slip is mailed immediately to a data center at Wilmington, where it is converted to punched cards. Because the freight charges are available, all statistics can be processed without waiting for the carrier's freight bill. In addition, an audit is available when the carrier presents its bill. This system did not create any savings in clerical costs, but it did reduce freight costs.

Design of Container Freight Station. Matson Navigation Company has been converting a fleet of conventional cargo ships into container-type ships. Parallel with this conversion, the shoreside facilities had to be expanded to accommodate the new fleet. These facilities include a container yard for handling full-lot containers and a freight station for handling less-than-container-load shipments.

The problem presented to industrial engineering was the determination of the size of the freight house. This required consideration of many factors such as the increasing volume of business, turnaround times, increases in number and size of ships, limited business hours for customers versus around-the-clock ship handling, and varying freight house storage times of different products. The problem was solved by use of manual simulation techniques. The model included a fleet simulator which recognized differing ship transit times, stevedoring times, capacities, and priorities. Also included was a container flow simulator for handling containers on shore and into the freight house.

As a result of the study, it was determined that 74 percent of the time less than the full capacity of a 50,000-square-foot building would be occupied by cargo. Original estimates of required floor space had been twice this amount. The 50,000-square-foot building was adequate two years later when actual volume reached the projected volume.

Summary. The preceding case studies present an exposure to problems susceptible to solution using industrial engineering approaches and techniques. The problems are real and in many cases complex. The public has recognized these problems, and as a result, departments of transportation at the Federal and state levels have been created. The industrial engineer can readily satisfy his need for challenge by working with transportation problems in either the private or public sectors.

Bibliography

Clausing, E. L., "Use of Computer to Analyze Train Delay Problems," *Annual Proceedings*, vol. 69, American Railway Engineering Association, Chicago.

Evans, Truman, "A Control System for Airline Maintenance," *Proceedings of the 19th Annual Conference*, American Institute of Industrial Engineers, New York, May, 1968.

Griffin, Walter C., "Potential Usefulness for Mass Transportation Facilities," *The Journal of Industrial Engineering*, November, 1966.

Macomber, F. S., "Use of Industrial Engineering Techniques in Transportation," *Proceedings of the 16th Annual Conference*, American Institute of Industrial Engineers, New York, May, 1965.

Rosenshine, Matthew, "Operations Research in the Solution of Air Traffic Control Problems," *Proceedings of the 18th Annual Conference*, American Institute of Industrial Engineers, New York, May, 1967.

Shelton, John R., "Industrial Engineering at the Port of New York Authority," *The Journal of Industrial Engineering*, September–October, 1964.

Thoolen, S. I., "Modern Industrial Engineering Techniques in a Marine Terminal," *Proceedings of the 17th Annual Conference*, American Institute of Industrial Engineers, New York, May, 1966.

INDUSTRIAL ENGINEERING IN AGRICULTURE*

Industrial engineering in relation to agriculture may be approached by considering some of the background and trends of agricultural operations. In early days, the growing, harvesting, and marketing of things produced on the land was considered a unique kind of effort, mainly guided by experience and intuition in dealing with Nature and the items produced by harnessing her capabilities for the purposes of mankind. In the U.S. Department of Agriculture Report for 1862, the first issued by the Department after it was created by an Act of Congress, Commissioner of Agriculture Isaac Newton wrote: "Agriculture in its first inception could scarcely be considered as an art or even occupation." Thoughts of efficient methods and management procedures seemed remote and out of place for agriculture because of the great variability in the products of Nature, the wide range of unexpected and uncontrollable conditions associated with most agricultural operations, and the availability of only crude equipment and facilities.

The need to eliminate heavy physical labor and the search for a better way encouraged progress. "Mechanization in the United States followed a logical course, beginning with the operations—mainly plowing, cultivating, and harvesting—that require the most physical effort or involve the most time in harvesting a product."[12] The steel plow was first made in 1837 by John Deere, and as early as 1838, plows were being manufactured in factories. In addition to making better equipment available, the development of farm implement manufacturing progressively freed farmers from the time-consuming effort of making their own equipment, thus permitting them to grow crops more effectively.

Industrial engineering as a specific field of work, emerging to meet needs for effective management in manufacturing enterprises as they increased in size and sophistication, was taking shape in America about the turn of the century when the early farm management efforts were under way. In agriculture, excluding manufacturing situations such as plants producing related equipment and supplies, the attention focused on management and methods was primarily, until the 1940s, by people based in agricultural disciplines. The U.S. Department of Agriculture first employed industrial engineers about 1946, and efforts in this field of work have been expanded since that time. The Agricultural Marketing Act of 1946 called for such work.

Many opportunities for industrial engineering in agriculture are evident in the developments described herein, not to mention the vast operations in which consumer

---

* Written by Earl K. Bowman.
[12] E. G. McKibben and W. M. Carleton, "Engineering in Agriculture," *1964 Yearbook of Agriculture*, U.S. Department of Agriculture, Washington, D.C., 1964, p. 86.

items such as breads, cereals, and cookies are made; the input commodities have already moved through separate facilities, such as those for making flour from grain and butter from milk, which further extend the field of opportunity.

Clearly, the full range of possibilities for industrial engineering involvement in agriculture is great, and there is a challenge for workers in the field representing widely diverse specialties or technique areas.

Work Measurement. The application of industrial engineering procedures in the realm of agriculture began mainly with the measurement of the time required for various activities. On the farm proper, measurement of time for given activities involving different methods and equipment is needed to provide input requirements which are important in managing the farm. The manufacturing of farm equipment, supplies, and chemicals may be considered essentially the same as other manufacturing as regards work measurement and other phases of industrial engineering. Generally, work measurement is regularly used in these firms.

Essentially, the conventional time study principles and procedures are followed, while adapting to the particular conditions which may be involved in agriculture, to obtain information for the desired end—good management. Detailed information concerning techniques and procedures should be obtained from recognized reference sources. Work measurement in agriculture may involve stopwatch time study; work sampling studies, where there are groups of workers and either it is not practical to study individuals or the desired information can be obtained more economically by sampling; and motion picture camera techniques.

Micromotion studies have been used extensively by industrial engineers in poultry (broiler) processing plants to make studies of workers who perform work on each bird as it moves on a conveyor by their work station. Memomotion studies have been used where their slower speed can serve the purpose, while saving both film and analysis labor. Predetermined elemental time data can be used in cases where the motions and activity involved in the work are within the scope of the data.

The involvement of work measurement procedures in agriculture may be better understood in the light of efforts which have gone before, such as the amount of accomplishment which represents a normal day's work for given farm operations.

Methods Improvement. Methods analysis and work simplification in agriculture will in most instances be limited only by the ingenuity and imagination of the person concerned with methods improvement. Although work measurement was mentioned first, this does not mean that final values for time requirements are obtained before improved methods are put into use. Data are usually needed on the existing as well as on the proposed methods. Recognized techniques for methods analysis and improvement, as covered in available references, should be used, augmented by specialized references in the field of agriculture.

An actual case to show the large saving possible through methods improvement is a research study on celery handling and packing, which brought savings of about 40 percent through new methods, including mechanical instead of manual cutting of stalks in the field, bulk handling to a central packing-precooling plant instead of field packing followed by transport to the precooling plant, and unit load precooling and in-plant handling of the packed crates instead of manual handling of individual crates.

Layout. The procedure normally referred to by industrial engineers as plant layout needs attention in agriculture also. Established techniques and procedures for layout can be applied advantageously.

The largest farm unit for consideration is the farmstead—the base of operations—which includes all buildings, lots, equipment, and service areas. Special attention has been given to farmstead engineering through agricultural engineering channels, and all recommendations from this should be taken into consideration. Individual facilities in the farmstead offer opportunity for the basic plant layout approach.

Fruit and vegetable packinghouses and processing plants, bakeries, farm equipment factories and repair facilities, and many other facilities, all need efficient layouts. For these, established procedures may be used, taking into account any needs which are unique because of the agricultural character of the business activity. A good body of reference material is available covering established layout principles and procedures.

Engineering Economy. The major emphasis of engineering economy is on procedures for evaluating the cost effect of changes in equipment, that is, the relationship of the total cost to own and operate given equipment as compared with the returns which are expected to result from the use of the equipment.

Managers of all kinds of enterprises in the realm of agriculture must make decisions involving engineering economy. There is an ever-recurring need to answer the question "Will it pay?" Steps must be taken on the basis either that returns will be sufficient to justify a given equipment change or that they will not. Various formulas are available with which to explore investment-return relationships and to develop answers which are vital in decision making. Adequate reference material is available to provide information on mathematical relationships and procedures.

If the farm manager considers producing a new crop, he must take into account the investment which will be involved in necessary equipment and facilities relative to anticipated revenue from the crop as well as direct costs such as labor, seed, fertilizer, spray materials, and irrigation. Processing operations—poultry, fruit, or vegetables, for example—need to follow engineering economy procedures when considering changes in equipment or facilities. The same applies to packinghouses, bakeries, dairies, equipment plants, terminal markets, and in fact, the entire gamut of agricultural operations.

There are many choices related to engineering economy facing management in the field of agriculture, for example, more expensive equipment to obtain more accurately filled containers versus savings from better yield of packed product; precooling of product and additional costs for the treatment versus possible better response in the market, bringing increased revenue; the cost effect of shipping products to market on expendable pallets versus handling of individual containers; installation of a sampling procedure to permit differentiated payment for incoming product according to quality classification versus the cost of sampling.

Systems Engineering. The systems engineering approach is aimed at more effective evaluation of different combinations of methods in a sequence of operations and responses to the interactions involved. Systems engineering procedures may be brought into play in agriculture for combined activities such as the production, handling, preparation for market, and shipping out of virtually any commodity produced on a commercial scale. Although this example is taken from the viewpoint of the production area, opportunities may be presented for similar work in other situations—terminal market activities as one instance. Again, the manufacturing, processing, and all agriculture-related operations besides those actually involved with fresh food items can be included.

Operations Research. Operations research emphasizes the mathematical approach to problems which concern managers and can supply results which point the way in certain decision making situations. It involves the use of various techniques such as linear programming, mathematical statistics, queuing theory, and systems simulation. A mathematical model may be structured to represent the factors in a complex problem requiring decisions in virtually any enterprise.

Noteworthy work has been done in the area of operations research in connection with agriculture. Examples include activity network techniques in a farm machinery selection problem, simulation of truck arrivals at country grain elevators, and development of bases for selecting the best capacity for truck receiving facilities. Actually, there have been applications of linear programming in agriculture since 1956.

Reference material is available concerning the use of operations research, but successful application will depend heavily upon the availability of suitably trained personnel.

Material Handling. Material handling is the "thread" which is commonly interwoven into operations and into systems. In general, belt and roller conveyors and forklift trucks are very adaptable to the handling needs of fruit and vegetable packinghouses, processing and equipment-making plants in agriculture, and terminal market operations.

Discussion in detail, even for certain categories, of the many variations available in material handling equipment will not be attempted. References are available which cover this field of knowledge thoroughly and may be adapted readily to the peculiarities of agricultural operations.

Bibliography

Crossan, R. M., and H. W. Nance, *Master Standard Data,* McGraw-Hill Book Company, New York, 1962.

French, C. E., M. M. Snodgrass, and J. C. Snider, "Application of Operations Research in Farm Operation in Agricultural Marketing," *Operations Research,* vol. 6, no. 5, pp. 766–775.

Link, D. A., "Activity Network Techniques Applied to a Farm Machinery Selection Problem," *ASAE Transactions,* vol. 10, no. 3, ref. 1967, pp. 310–317.

Preston, T. A., "Systems and Industrial Engineering Techniques," *Agricultural Materials Handling Manual,* National Committee on Agricultural Engineering, The Queen's Printer, Ottawa, Canada.

Sammett, L. L., "Systems Engineering in Agriculture," *1960 Yearbook of Agriculture,* USDA, pp. 427–434.

Vaughn, L. M., and L. S. Hardin, *Farm Work Simplification,* John Wiley & Sons, Inc., New York, 1949.

Vincent, S. E., "An Application of Linear Programming to Agricultural Economics," *Applied Statistics,* vol. 9, no. 1, March, 1960, pp. 28–36.

Young, E. C., and L. S. Hardin, "Simplifying Farm Work," *1943–1947 Yearbook of Agriculture,* USDA, pp. 817–823.

INDUSTRIAL ENGINEERING IN RETAILING*

Industrial engineering is firmly established in manufacturing industry and is becoming increasingly valuable to organizations primarily concerned with the sale of goods or services. More industrial engineers will enter these service-oriented organizations, and many will move into upper management as they have in manufacturing. The future for the profession is bright in this segment of the economy; it is limited only by the industrial engineer's ability to produce meaningful results. One such opportunity is in the retailing industry.

A Definition of Retailing. Retailing is essentially the sale of goods and services to the consuming public. Because other parts of this chapter are devoted to industries emphasizing the sale of services, attention here will be directed toward that portion of the retailing industry concerned with the sale of goods. Primary attention will be given to those activities that take place within the sales unit or store. Thus, retailing, for the purposes of this discussion, resolves itself into those activities that must be performed within a store to consummate the sale of goods to the general public.

Merchandising and Operations. One of the major subgroups of the retailing activity is that required to induce the customer to purchase particular items in preference to others. This is known as merchandising. The objective is to sell those items which return the greatest profit to the store. Industrial engineering techniques can play an important part in planning the most effective utilization of the man-hours available for those tasks and also in determining specifically what items should be emphasized.

Another major subgroup consists of those activities required to process the goods from the receiving door of the store to the display fixture on the sales floor. This is referred to as operations. Included in this group are all those activities related to the processing of information and money within the store. Some of the typical operating tasks common to most retail stores are receiving, marking, stocking, customer transactions, waste handling, and ordering. Depending on the level of detail required, an industrial engineer can identify from twenty-five up to several hundred operations-oriented tasks necessary to the processing of goods for sale.

Other activities may be related to both merchandising and operations. There are,

* Written by Frederick H. Young.

for example, significant hours spent each week on telephone conversations, walking, training, and attendance at meetings, which are necessary for the orderly and efficient conduct of the business. Should the industrial engineer uncover activities that are neither merchandising nor operations oriented, he should give serious consideration to their elimination. They are probably not serving a useful or profitable purpose.

The Need for Measurement. In most retail organizations, the industrial engineer will be faced with the immediate need to measure the activities presently being performed. Unlike manufacturers, retailers have not developed a long history of labor standards and performance levels. Therefore there is no readily available basis for evaluating the effectiveness or the productivity of retailing systems.

An effective way to initiate the measurement program is to design and conduct a detailed work sampling study. The benefits of this approach lie primarily in its ability to cover a broad spectrum of tasks on a continuing basis and to relate them both to each other and to the total weekly work load. A secondary benefit is the ability to utilize nontechnical personnel as observers. In practice, it is more important that the observers be familiar with the operation itself than that they be technically trained. However, the design of the study can be quite involved, particularly if activity ratios or variances are to be forecast or sample sizes are to be adjusted as actual study data are obtained. This design work, along with study supervision, requires a professional industrial engineer.

In most retail stores, it is not practical to reduce accountability to an operator level. In practice, accountability at some group level, such as the receiving department or the men's clothing department, is sufficient for effective management. Accordingly, work sampling designs can utilize rather gross work categories, and accuracy tolerances and confidence levels can be relatively loose.

Stopwatch time study and predetermined time systems have to date been of somewhat limited usefulness in retail store environments. Their principal applications have been in the verification of work sampling results and in the development of cost justifications for specific projects such as the purchase of a particular piece of equipment. As retail store management becomes more adept as consumers of industrial engineering services, more sophisticated and useful forms of work measurement information can be expected to be generated and used.

Methods Improvement. The use of flow process charts to analyze the operations of a retail store will reveal many opportunities for methods improvement. It will be apparent that the classical possibilities of "eliminate, combine, change sequence, and simplify" will result in significant profit contributions.

Invariably, retail operations have developed in a patchwork manner. Any activity may be in itself efficient; but of even greater importance, the total operations procedure may be found to be unnecessarily costly to the business. Multiple rehandling of goods is a prevalent condition found in retail stores. By merely smoothing the flow from receiving to display area, the industrial engineer should be able to reduce the cost of operations significantly. Some utilization of new material handling equipment may be called for, but this should not be considered until all the possibilities for improvement without capital investment are exhausted. The retailer's present inability to cope with sophisticated equipment and the inadvisability of imposing such equipment on an otherwise poorly engineered system make this a good rule.

Scheduling. Perhaps the most profitable assignment an industrial engineer can undertake in his initial encounter with retailing is the development of effective methods of scheduling store personnel. The key to success in retailing is having the right number of people assigned to the right task at the right time, a situation which has been characterized by a low degree of predictability. The work load at specific points varies considerably through the week. Customer traffic flow, for example, is a function of weather, competition, available disposable income, and many other factors. The saving grace is that cumulative demand is fairly easy to predict, and total work force requirements for a week or month can be forecast with reasonable accuracy.

Under these conditions, the essentials of effective scheduling become the development of a cross-trained work force—one in which individual operators can perform multiple tasks within the store; careful tracking of the cumulative level of demand at any point through the week; and a system of signals that indicate when operators should be moved from one task to another. An example of such a signal would be a rule that when a fifteenth customer is waiting for service at a bank of checkstands, an operator would be removed from the receiving operations and assigned to handle customer orders. Much industrial engineering effort has been spent on this area with excellent results.

Fiscal Controls. One of the most critical needs of retailers is the need for effective in-store fiscal planning and control. Traditionally, retailers performed these two tasks on the basis of percentages. Wage percentages, gross profit percentage mark-ups, and sales percentages of various department sales to total store sales are typical examples. The influence of industrial engineering has induced a movement away from these concepts toward ones more closely related to profits. Many retailers now measure labor utilization on a sales per man-hour basis. This still leaves a great deal to be desired but does represent a distinct improvement.

Better still is the movement toward planning and control on a production unit basis. Retailers are beginning to recognize that the production units a clerk can process per hour are the real measure of efficiency. The units can be pounds, garments, customers, or any other applicable entity. They are also beginning to recognize the advantages of evaluating unit profit dollars rather than gross percentages. Given an elastic demand and other things equal, it is better to sell a unit yielding 20 percent markup and $1.00 profit than one yielding 30 percent markup and only $.75 profit. Industrial engineers in retailing can serve their organizations well by developing planning and control systems on a unit rather than percentage basis.

Summary. Industrial engineering is just starting to influence the thinking of retail operators. The techniques and approaches most needed are those that would be considered fundamental or even out of date by more advanced practitioners in manufacturing environments. The reason that they are most needed is that they will yield significant profit contributions at relatively little cost to the retailer. He will be able to understand them and digest them without extending his knowledge of basic business practices. Once the role and contribution of industrial engineering are firmly established, the more advanced techniques may prove useful. Until then, it seems wise to stay with the basic fundamentals of the profession and apply them in all areas of the retailing industry.

Bibliography

Buzzell, Robert D., Walter J. Salmon, and Richard F. Vancil, *Product Profitability Measurement and Merchandising Decisions,* Harvard Business School, Division of Research, Boston, 1965.
Harwell, Edward M., *Checkout Management,* Chain Store Publishing Corporation, New York, 1963.

Index

Prepared by Rita M. Carlson, Executive Assistant,
Maynard Research Council Incorporated,
Pittsburgh, Pennsylvania

1